BASIC ALGEBRA FORMULAS

Arithmetic Operations

$$a(b + c) = ab + ac, \qquad \frac{a}{b} \cdot \frac{c}{d} = \frac{ac}{bd}$$

$$\frac{a}{b} + \frac{c}{d} = \frac{ad + bc}{bd}, \qquad \frac{a/b}{c/d} = \frac{a}{b} \cdot \frac{d}{c}$$

Laws of Signs

$$-(-a) = a, \qquad \frac{-a}{b} = -\frac{a}{b} = \frac{a}{-b}$$

Zero Division by zero is not defined.

$$\text{If } a \neq 0: \frac{0}{a} = 0, \quad a^0 = 1, \quad 0^a = 0$$

For any number a: $a \cdot 0 = 0 \cdot a = 0$

Laws of Exponents

$$a^m a^n = a^{m+n}, \qquad (ab)^m = a^m b^m, \qquad (a^m)^n = a^{mn}, \qquad a^{m/n} = \sqrt[n]{a^m} = \left(\sqrt[n]{a}\right)^m$$

If $a \neq 0$,

$$\frac{a^m}{a^n} = a^{m-n}, \qquad a^0 = 1, \qquad a^{-m} = \frac{1}{a^m}.$$

The Binomial Theorem For any positive integer n,

$$(a + b)^n = a^n + na^{n-1}b + \frac{n(n-1)}{1 \cdot 2} a^{n-2}b^2 + \frac{n(n-1)(n-2)}{1 \cdot 2 \cdot 3} a^{n-3}b^3 + \cdots + nab^{n-1} + b^n.$$

For instance,

$$(a + b)^2 = a^2 + 2ab + b^2, \qquad (a - b)^2 = a^2 - 2ab + b^2$$

$$(a + b)^3 = a^3 + 3a^2b + 3ab^2 + b^3, \qquad (a - b)^3 = a^2 - 3a^2b + 3ab^2 - b^3.$$

Factoring the Difference of Like Integer Powers, $n > 1$

$$a^n - b^n = (a - b)(a^{n-1} + a^{n-2}b + a^{n-3}b^2 + \cdots + ab^{n-2} + b^{n-1})$$

For instance,

$$a^2 - b^2 = (a - b)(a + b),$$
$$a^3 - b^3 = (a - b)(a^2 + ab + b^2),$$
$$a^4 - b^4 = (a - b)(a^3 + a^2b + ab^2 + b^3).$$

Completing the Square If $a \neq 0$,

$$ax^2 + bx + c = au^2 + C \qquad \left(u = x + (b/2a), C = c - \frac{b^2}{4a}\right)$$

The Quadratic Formula If $a \neq 0$ and $ax^2 + bx + c = 0$, then

$$x = \frac{-b \pm \sqrt{b^2 - 4ac}}{2a}.$$

GEOMETRY FORMULAS

A = area, B = area of base, C = circumference, S = lateral area or surface area, V = volume

Triangle

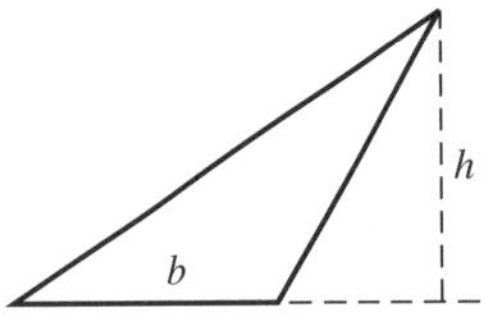

$A = \frac{1}{2}bh$

Similar Triangles

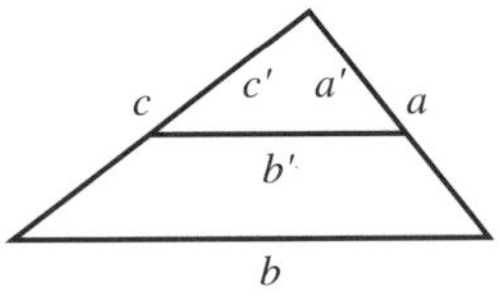

$\frac{a'}{a} = \frac{b'}{b} = \frac{c'}{c}$

Pythagorean Theorem

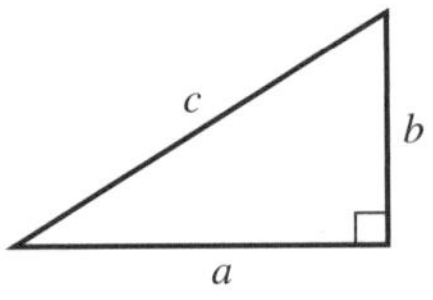

$a^2 + b^2 = c^2$

Parallelogram

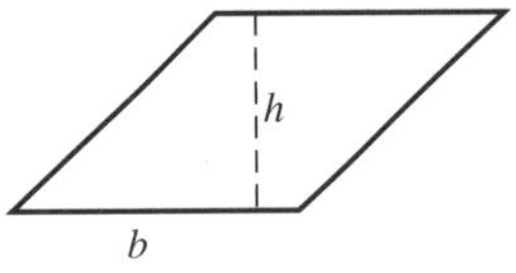

$A = bh$

Trapezoid

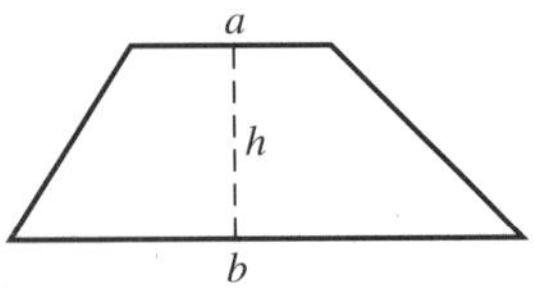

$A = \frac{1}{2}(a + b)h$

Circle

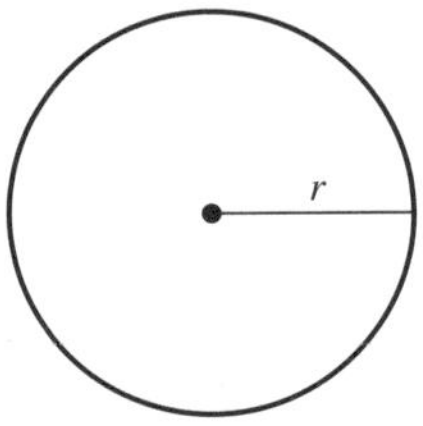

$A = \pi r^2$,
$C = 2\pi r$

Any Cylinder or Prism with Parallel Bases

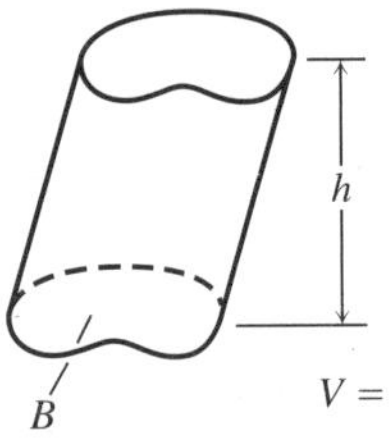

$V = Bh$

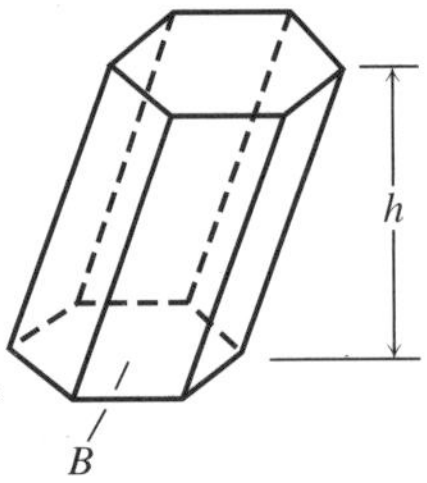

Right Circular Cylinder

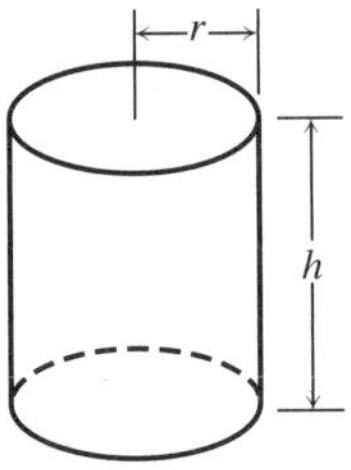

$V = \pi r^2 h$
$S = 2\pi rh$ = Area of side

Any Cone or Pyramid

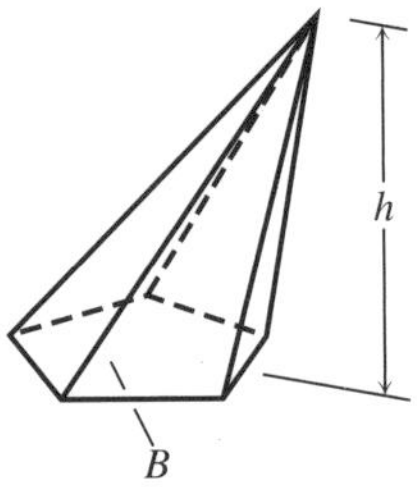

$V = \frac{1}{3}Bh$

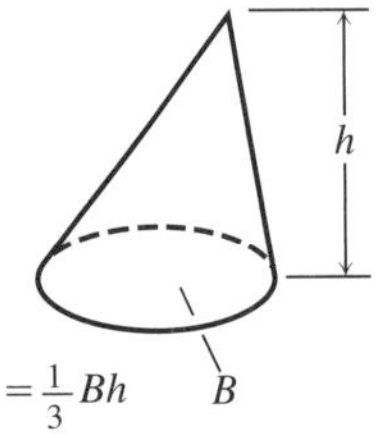

Right Circular Cone

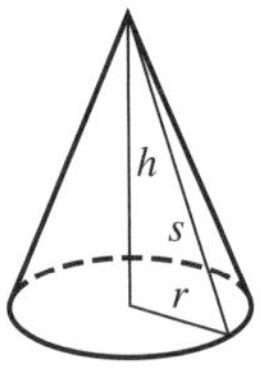

$V = \frac{1}{3}\pi r^2 h$
$S = \pi rs$ = Area of side

Sphere

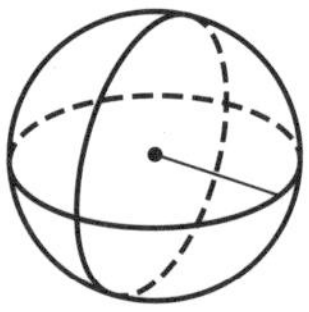

$V = \frac{4}{3}\pi r^3$, $S = 4\pi r^2$

UNIVERSITY CALCULUS

ELEMENTS WITH EARLY TRANSCENDENTALS

UNIVERSITY CALCULUS

ELEMENTS WITH EARLY TRANSCENDENTALS

Joel Hass
University of California, Davis

Maurice D. Weir
Naval Postgraduate School

George B. Thomas, Jr.
Massachusetts Institute of Technology

PEARSON
Addison Wesley

Boston San Francisco New York
London Toronto Sydney Tokyo Singapore Madrid
Mexico City Munich Paris Cape Town Hong Kong Montreal

Editor in Chief: Deirdre Lynch
Senior Acquisitions Editor: William Hoffman
Associate Editor: Caroline Celano
Senior Project Editor: Rachel S. Reeve
Senior Managing Editor: Karen Wernholm
Senior Production Supervisor: Sheila Spinney
Senior Designer: Barbara T. Atkinson
Digital Assets Manager: Marianne Groth
Media Producer: Christine Stavrou
Software Development: Mary Durnwald and Bob Carroll
Marketing Assistant: Jon Connelly
Senior Author Support/ Technology Specialist: Joe Vetere
Senior Prepress Supervisor: Caroline Fell
Manufacturing Manager: Evelyn Beaton
Production Coordinator: Mary Sanger
Composition: Nesbitt Graphics, Inc.
Illustrations: Techsetters, Inc.

Cover Image: River of Silk, 1938
Three outboards in the Eastern Outboard Championships in Philadelphia.
© Mystic Seaport, Rosenfeld Collection, Mystic, Conn.

Library of Congress Cataloging-in-Publication Data

Hass, Joel.
University calculus elements with early transcendentals / Joel Hass, Maurice D. Weir, George B. Thomas, Jr.
p. cm.
Includes index.
ISBN-10: 0-321-53348-8
ISBN-13: 978-0-321-53348-7
1. Calculus—Textbooks. I. Weir, Maurice D. II. Thomas, George B. (George Brinton), 1914–2006. III. Title.

QA303.2.H373 2007
515—dc22 2007060057

1 2 3 4 5 6 7 8 9 10—CRK—11 10 09 08

Contents

PREFACE

OVERVIEW *University Calculus: Elements with Early Transcendentals* offers a streamlined text to allow for great flexibility in teaching an early transcendentals course. It still maintains the high standards, careful development, and focused applications that have been the hallmark of the *Thomas' Calculus* series, but this text presents the essential elements of calculus in a highly pointed way.

An increasing number of today's students have become familiar with the terminology and operational methods of calculus in high school. However, many of these students have a limited conceptual understanding of calculus when they enter college. Recognizing these dualities, we review functions quickly and then proceed to limits immediately in the first chapter. Differentiation and integration are presented next, with the transcendental functions fully incorporated into their developments and the exercises. As in our previous texts, we continue to focus on concepts and their applications throughout.

In *University Calculus: Elements with Early Transcendentals*, we treat the various techniques of integration immediately after the chapter defining the integral. This approach allows us to combine various techniques within a particular application section, and gives the student more flexibility and breadth in solving problems that require use of the definite integral.

To achieve the goals of *University Calculus: Elements with Early Transcendentals*, we consulted with many colleagues and reviewers across the country. This resulted in the changes already discussed and a reduction in the overall number of exercises. We continued to refine and streamline the presentation wherever possible, but the pace and rigor remain unchanged, and the figure art is still outstanding. We omitted the geometric treatment of the conics, but all of the standard early transcendentals topics are still covered. For those instructors interested in a later treatment of the transcendental functions (with the logarithm defined as an integral), but still wanting a streamlined text, we recommend looking at our book titled *University Calculus: Alternate Edition.*

Pedagogical Features

EXERCISES Exercises and examples play a crucial role in learning calculus. Within each section, we have organized and grouped the exercises by topic, progressing from computational problems to applied and theoretical problems. This arrangement gives students the opportunity to develop skills in using the methods of calculus and to deepen their understanding of its applications within a wide variety of abundant exercises.

RIGOR The level of rigor is consistent throughout. We give both formal and informal discussions, making clear the distinction between the two, and we include precise definitions

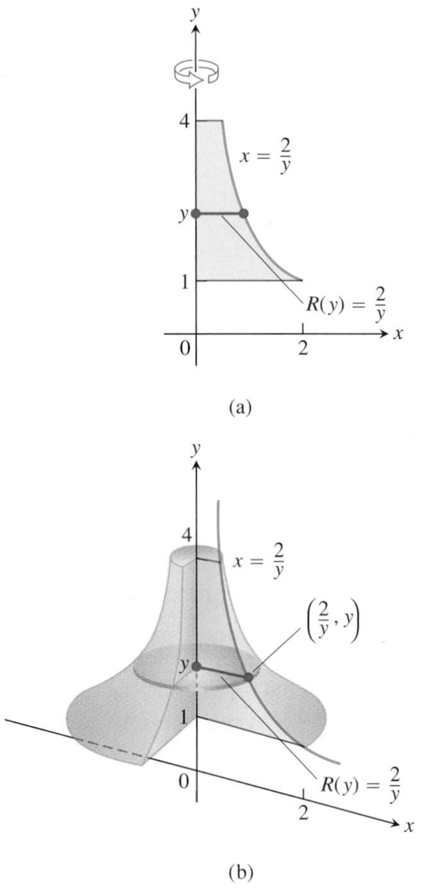

FIGURE 6.11, page 362
Finding the volume of the solid generated by revolving the region (a) about the y-axis.

and many accessible proofs for the students. The text is organized so the material can be covered informally, giving the instructor a degree of flexibility. For example, while we do not prove that a continuous function on a closed and bounded interval has a maximum there, we do state this theorem carefully and use it to prove several subsequent results.

ART We realize that figures and illustrations are a critical component to learning calculus. We paid special attention to the clarity with which the figures illustrate their associated concepts. This is especially evident with the three-dimensional graphics, where we were able to better indicate depth, layering, and rotation (see figures on this page).

END-OF-CHAPTER REVIEWS AND PROJECTS In addition to problems appearing after each section, each chapter culminates with review questions and practice exercises. Student projects are available over the Internet at **www.aw-bc.com/thomas**.

WRITING EXERCISES Writing exercises placed throughout the text ask students to explore a variety of calculus concepts and applications. In addition, each chapter contains questions asking students to summarize what they have learned. Many of these questions ask for written descriptions that test conceptual understanding.

ANSWERS Answers are provided for all odd-numbered exercises when appropriate, and these have been carefully checked for correctness.

MATHEMATICAL CORRECTNESS We have been careful to say only what is true and mathematically sound. Every definition, theorem, corollary, and proof has been reviewed for clarity and mathematical correctness.

WRITING AND APPLICATIONS This text continues to be easy to read, conversational, and mathematically rich. Each new topic is motivated by clear, easy-to-understand examples and applications.

TECHNOLOGY Technology can be incorporated according to the taste of the instructor. Each section contains exercises requiring the use of technology; these are marked with a **T** if suitable for calculator or computer usage or are labeled **Computer Explorations** if a computer algebra system (CAS, such as *Maple*® or *Mathematica*®) is required.

FIGURE 6.13, page 363
The cross-sections of the solid of revolution generated here are washers, not disks.

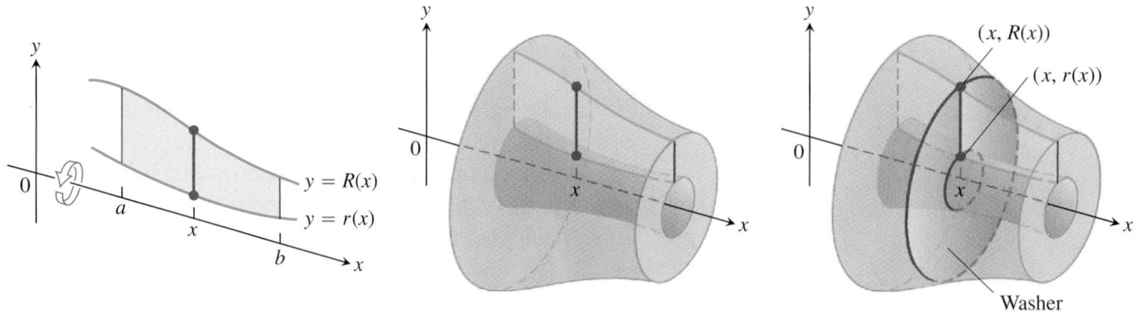

Print Supplements

STUDENT EDITION OF UNIVERSITY CALCULUS: ELEMENTS WITH EARLY TRANSCENDENTALS
ISBN 13: 978-0-321-53348-7; ISBN 10: 0-321-53348-8

INSTRUCTOR'S SOLUTIONS MANUAL
Part One (Chapters 1–8), ISBN 13: 978-0-321-53627-3; ISBN 10: 0-321-53627-4
Part Two (Chapters 9–13), ISBN 13: 978-0-321-53623-5; ISBN 10: 0-321-53623-1
The *Instructor's Solutions Manual* by William Ardis, et al., contains complete worked-out solutions to all of the exercises in the text.

STUDENT'S SOLUTIONS MANUAL
Part One (Chapters 1–8), ISBN 13: 978-0-321-53608-2; ISBN 10: 0-321-53608-8
Part Two (Chapters 9–13), ISBN 13: 978-0-321-55917-3; ISBN 10: 0-321-55917-7
The *Student's Solutions Manual* by William Ardis, et al., is designed for the student and contains carefully worked-out solutions to all the odd-numbered exercises in the text.

JUST-IN-TIME ALGEBRA AND TRIGONOMETRY FOR EARLY TRANSCENDENTALS CALCULUS, Third Edition
ISBN 13: 978-0-321-32050-6; ISBN 10: 0-321-32050-6
Sharp algebra and trigonometry skills are critical to mastering calculus, and *Just-in-Time Algebra and Trigonometry for Early Transcendentals Calculus,* Third Edition, by Guntram Mueller and Ronald I. Brent is designed to bolster these skills while students study calculus. As students make their way through calculus, this text is with them every step of the way, showing them the necessary algebra or trigonometry topics and pointing out potential problem spots. The easy-to-use table of contents has algebra and trigonometry topics arranged in the order in which students will need them as they study calculus.

PEARSON'S CALCULUS REVIEW CARD
The Calculus Review Card is a resource for students containing important formulas, functions, definitions, and theorems that correspond precisely to *University Calculus: Elements with Early Transcendentals.* This card can work as a reference for completing homework assignments or as an aid in studying and is available bundled with a new text. Contact your Pearson sales representative for more information.

Media and Online Supplements

TECHNOLOGY RESOURCE MANUALS
Maple Manual by Donald Hartig, California Polytechnic State University

Mathematica Manual by Marie Vanisko, California State University Stanislaus, and Lyle Cochran, Whitworth College

TI-Graphing Calculator Manual by Luz DeAlba, Drake University

These manuals cover *Maple* 10, *Mathematica* 5, and the TI-83 Plus/TI-84 Plus, TI-85/TI-86, and TI-89/TI-92 Plus, respectively. Each manual provides detailed guidance for integrating a specific software package or graphing calculator throughout the course, including syntax and commands. These manuals are available to qualified instructors through **http://www.aw-bc.com/irc/**.

MYMATHLAB®

MyMathLab is a series of text-specific, easily customizable online courses for Pearson Education's textbooks in mathematics and statistics. Powered by CourseCompass™ (our online teaching and learning environment) and MathXL® (our online homework, tutorial, and assessment system), MyMathLab gives you the tools you need to deliver all or a portion of your course online, whether your students are in a lab setting or working from home. MyMathLab provides a rich and flexible set of course materials, featuring free-response exercises that are algorithmically generated for unlimited practice and mastery. Students can also use online tools, such as video lectures, animations, a multimedia textbook, and Maple/Mathematica projects, to independently improve their understanding and performance. Instructors can use MyMathLab's homework and test managers to select and assign online exercises correlated directly to the textbook, and they can also create and assign their own online exercises and import TestGen® tests for added flexibility. MyMathLab's online gradebook—designed specifically for mathematics and statistics—automatically tracks students' homework and test results and gives the instructor control over how to calculate final grades. Instructors can also add offline (paper-and-pencil) grades to the gradebook. MyMathLab also includes access to the **Pearson Tutor Center** (**www.pearsontutorservices.com**). The Tutor Center is staffed by qualified mathematics instructors who provide textbook-specific tutoring for students via toll-free phone, fax, email, and interactive Web sessions. MyMathLab is available to qualified adopters. For more information, visit our Web site at **www.mymathlab.com** or contact your Pearson sales representative.

MATHXL

MathXL is a powerful online homework, tutorial, and assessment system that accompanies Pearson Education's textbooks in mathematics or statistics. With MathXL, instructors can create, edit, and assign online homework and tests using algorithmically generated exercises correlated at the objective level to the textbook. They can also create and assign their own online exercises and import TestGen tests for added flexibility. All student work is tracked in MathXL's online gradebook. Students can take chapter tests in MathXL and receive personalized study plans based on their test results. The study plan diagnoses weaknesses and links students directly to tutorial exercises for the objectives they need to study and retest. Students can also access supplemental animations and video clips directly from selected exercises. MathXL is available to qualified adopters. For more information, visit our Web site at **www.mathxl.com**, or contact your Pearson sales representative.

TESTGEN

TestGen enables instructors to build, edit, print, and administer tests using a computerized bank of questions developed to cover all the objectives of the text. TestGen is algorithmically based, allowing instructors to create multiple but equivalent versions of the same question or test with the click of a button. Instructors can also modify test bank questions or add new questions. Tests can be printed or administered online. The software and testbank are available for download from Pearson Education's online catalog.

WEBASSIGN®

WebAssign is an online homework, quizzing, and testing management system. As a hosted application service, WebAssign gives professors the tools they need to assign their students algorithmically generated questions. WebAssign also includes a grade book that is automatically populated with student answers. WebAssign works with personal computers using any recent operating system and browser.

WEBWORK®

WeBWorK is an Internet-based method for delivering homework problems to students online. It automatically grades homework and provides immediate feedback, allowing students to correct mistakes while they are still working on the problem. WeBWorK works with personal computers using any recent operating system and browser.

WEB SITE **www.aw-bc.com/thomas**

The *University Calculus: Elements with Early Transcendentals* Web site provides the expanded historical biographies and essays referenced in the text. Also available is a collection of *Maple* and *Mathematica* modules that can be used as projects by individual students or groups of students.

Acknowledgments

For his careful checking of this text for accuracy, we extend our gratitude to Thomas Wegleitner. We also extend our sincere thanks for the suggestions and contributions made by the following reviewers:

Harry Allen, *Ohio State University*
Mohammad Amin, *National University*
Edoh Amiran, *Western Washington University*
Anthony Bedenikovic, *Bradley University*
Joseph E. Borzellino, *California Polytechnic State University*
Deborah Brandon, *Carnegie Mellon University*
Jen-Mei Chang, *Colorado State University*
Dennis DeTurk, *University of Pennsylvania*
Said Fariabi, *San Antonio College*
Christopher Johnston, *University of Missouri—Columbia*
Robert Hesse, *Saint John's University*
Krystyna Kuperberg, *Auburn University*
Jack Mealy, *Austin College*
Kasso A. Okoudjou, *Cornell University*
Paul Sacks, *Iowa State University*
Stephen Summers, *University of Florida*
Blake Thornton, *Washington University in St. Louis*
Ilie Ugarcovici, *Rice University*
Gary L. Walls, *West Texas A&M University*
Clifford Weil, *Michigan State University*

Finally, we are grateful for the advice, suggestions, and encouragment provided by our development editor, David Chelton.

UNIVERSITY CALCULUS

ELEMENTS WITH EARLY TRANSCENDENTALS

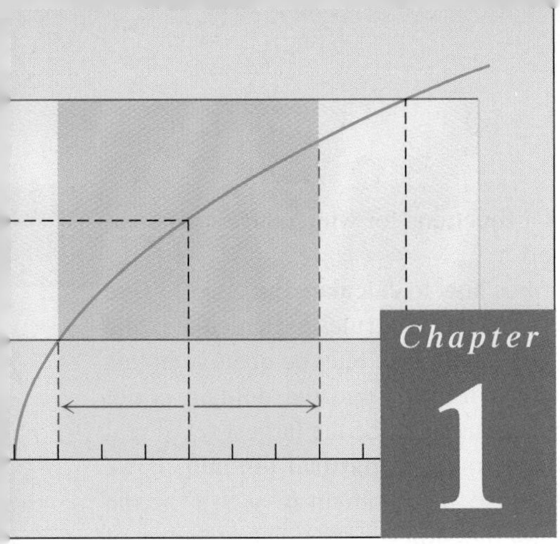

Chapter 1 FUNCTIONS AND LIMITS

OVERVIEW Functions are fundamental to the study of calculus. Limits are the key to finding the velocity of a moving object and the tangent to a curve. In this chapter we review the functions frequently encountered in calculus and the ways they are combined and transformed. We then develop the limit concept, which gives a precise way to distinguish between functions that vary *continuously*, have values that jump, vary erratically, or tend to increase or decrease without bound.

1.1 Functions and Their Graphs

Functions are a tool for describing the real world in mathematical terms. A function can be represented by an equation, a numerical table, a graph, or by verbal description; we will use all four representations throughout this book. This section reviews these function ideas.

Functions

The temperature at which water boils depends on the elevation above sea level (the boiling point drops as you ascend). The interest paid on a cash investment depends on the length of time the investment is held. The area of a circle depends on the radius of the circle. The distance an object travels at constant speed along a straight line path depends on the elapsed time.

In each case, the value of one variable quantity, say y, depends on the value of another variable quantity, which we might call x. We say that "y is a function of x" and write this symbolically as

$$y = f(x) \qquad (\text{“}y \text{ equals } f \text{ of } x\text{”}).$$

In this notation, the symbol f represents the function, the letter x is the **independent variable** representing the input value of f, and y is the **dependent variable** or output value of f at x.

> **DEFINITION** A **function** f from a set D to a set Y is a rule that assigns a *unique* (single) element $f(x) \in Y$ to each element $x \in D$.

The set D of all possible input values is called the **domain** of the function. The set of all values of $f(x)$ as x varies throughout D is called the **range** of the function. The range may not include every element in the set Y. The domain and range of a function can be any sets of objects, but often in calculus they are sets of real numbers interpreted as points of a coordinate line. (In Chapters 10–13, we will encounter functions for which the elements of the sets are points in the coordinate plane or in space.)

Often a function is given by a formula that describes how to calculate the output value from the input variable. For instance, the equation $A = \pi r^2$ is a rule that calculates the area A of a circle from its radius r (so r, interpreted as a length, can only be positive in this formula). When we define a function $y = f(x)$ with a formula and the domain is not stated explicitly or restricted by context, the domain is assumed to be the largest set of real x-values for which the formula gives real y-values, the so-called **natural domain**. If we want to restrict the domain in some way, we must say so. The domain of $y = x^2$ is the entire set of real numbers. To restrict the domain of the function to, say, positive values of x, we would write "$y = x^2, x > 0$."

FIGURE 1.1 A diagram showing a function as a kind of machine.

Changing the domain to which we apply a formula usually changes the range as well. The range of $y = x^2$ is $[0, \infty)$. The range of $y = x^2, x \geq 2$, is the set of all numbers obtained by squaring numbers greater than or equal to 2. In set notation (see Appendix 1), the range is $\{x^2 | x \geq 2\}$ or $\{y | y \geq 4\}$ or $[4, \infty)$.

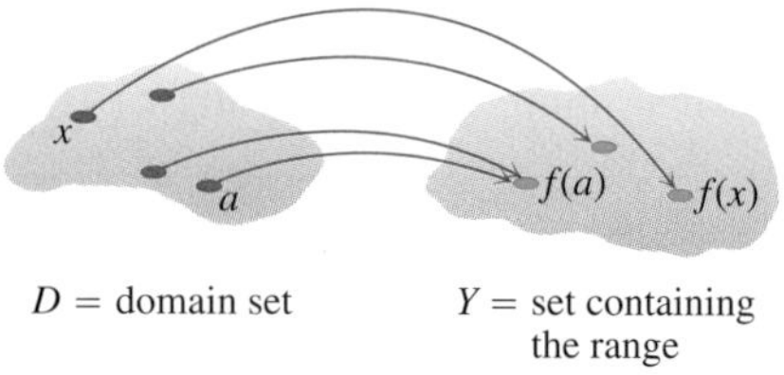

FIGURE 1.2 A function from a set D to a set Y assigns a unique element of Y to each element in D.

When the range of a function is a set of real numbers, the function is said to be **real-valued**. The domains and ranges of many real-valued functions of a real variable are intervals or combinations of intervals. The intervals may be open, closed, or half open, and may be finite or infinite. The range of a function is not always easy to find.

A function f is like a machine that produces an output value $f(x)$ in its range whenever we feed it an input value x from its domain (Figure 1.1). The function keys on a calculator give an example of a function as a machine. For instance, the $\sqrt{x}$ key on a calculator gives an output value (the square root) whenever you enter a nonnegative number x and press the $\sqrt{x}$ key.

A function can also be pictured as an **arrow diagram** (Figure 1.2). Each arrow associates an element of the domain D to a unique or single element in the set Y. In Figure 1.2, the arrows indicate that $f(a)$ is associated with a, $f(x)$ is associated with x, and so on. Notice that a function can have the same *value* at two different input elements in the domain (as occurs with $f(a)$ in Figure 1.2), but each input element x is assigned a *single* output value $f(x)$.

Another way to visualize a function is by its graph. If f is a function with domain D, its **graph** consists of the points in the Cartesian plane whose coordinates are the input-output pairs for f. In set notation, the graph is

$$\{(x, f(x)) \mid x \in D\}.$$

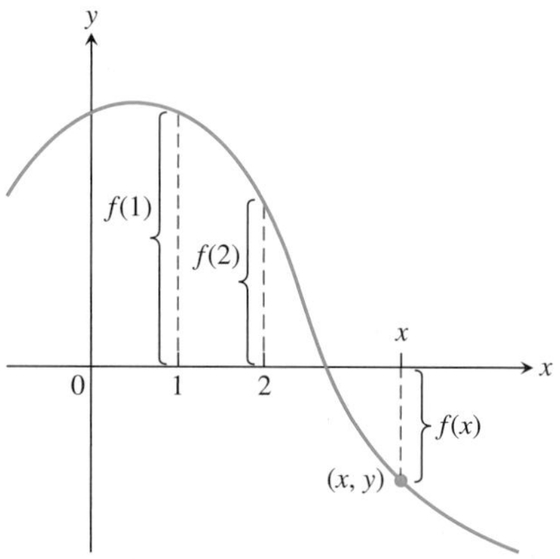

FIGURE 1.3 If (x, y) lies on the graph of f, then the value $y = f(x)$ is the height of the graph above the point x (or below x if $f(x)$ is negative).

If (x, y) is a point on the graph, then $y = f(x)$ is the height of the graph above the point x. The height may be positive or negative, depending on the sign of $f(x)$ (Figure 1.3).

EXAMPLE 1 Graph the function $y = x^2$ over the interval $[-2, 2]$.

Solution Make a table of xy-pairs that satisfy the equation $y = x^2$. Plot the points (x, y) whose coordinates appear in the table, and draw a *smooth* curve (labeled with its equation) through the plotted points (see Figure 1.4).

x	-2	-1	0	1	3/2	2
$y = x^2$	4	1	0	1	9/4	4

■

How do we know that the graph of $y = x^2$ doesn't look like one of these curves?

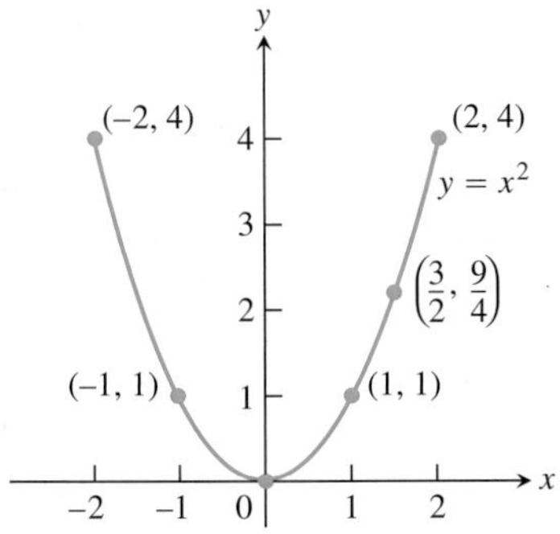

FIGURE 1.4 Graph of the function in Example 1.

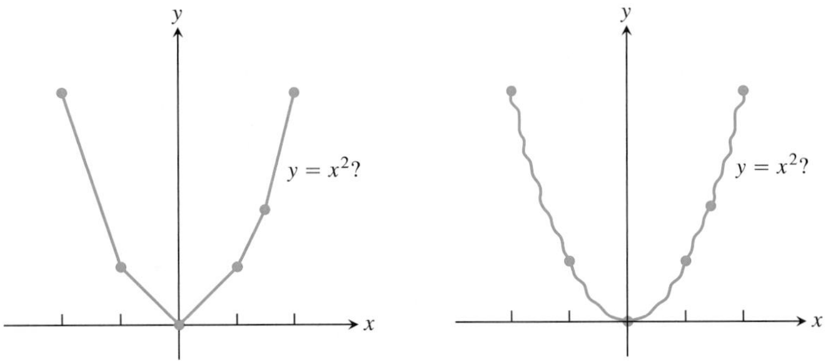

To find out, we could plot more points. But how would we then connect *them*? The basic question still remains: How do we know for sure what the graph looks like between the points we plot? The answer lies in calculus, as we will see in Chapter 3. Meanwhile we will have to settle for plotting points and connecting them as best we can.

Representing a Function Numerically

We have seen how a function may be represented algebraically by a formula (the area function) and visually by a graph (Example 1). Another way to represent a function is **numerically**, through a table of values. Numerical representations are often used by engineers and scientists. From an appropriate table of values, a graph of the function can be obtained using the method illustrated in Example 1, possibly with the aid of a computer. The graph of only the tabled points is called a **scatterplot**.

EXAMPLE 2 Musical notes are pressure waves in the air that can be recorded. The data in Table 1.1 give recorded pressure displacement versus time in seconds of a musical note produced by a tuning fork. The table provides a representation of the pressure function over time. If we first make a scatterplot and then connect the data points (t, p) from the table, we obtain the graph shown in Figure 1.5.

TABLE 1.1 Tuning fork data

Time	Pressure	Time	Pressure
0.00091	−0.080	0.00362	0.217
0.00108	0.200	0.00379	0.480
0.00125	0.480	0.00398	0.681
0.00144	0.693	0.00416	0.810
0.00162	0.816	0.00435	0.827
0.00180	0.844	0.00453	0.749
0.00198	0.771	0.00471	0.581
0.00216	0.603	0.00489	0.346
0.00234	0.368	0.00507	0.077
0.00253	0.099	0.00525	−0.164
0.00271	−0.141	0.00543	−0.320
0.00289	−0.309	0.00562	−0.354
0.00307	−0.348	0.00579	−0.248
0.00325	−0.248	0.00598	−0.035
0.00344	−0.041		

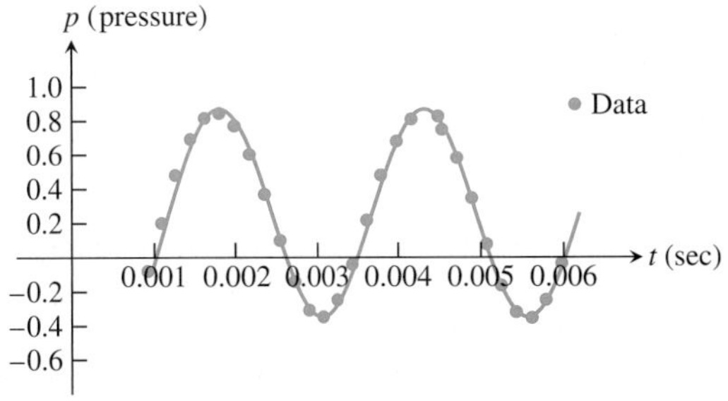

FIGURE 1.5 A smooth curve through the plotted points gives a graph of the pressure function represented by Table 1.1 (Example 2).

■

The Vertical Line Test

Not every curve in the coordinate plane can be the graph of a function. A function f can have only one value $f(x)$ for each x in its domain, so *no vertical* line can intersect the graph of a function more than once. If a is in the domain of the function f, then the vertical line $x = a$ will intersect the graph of f at the single point $(a, f(a))$.

A circle cannot be the graph of a function since some vertical lines intersect the circle twice. The circle in Figure 1.6a, however, does contain the graphs of *two* functions of x; the upper semicircle defined by the function $f(x) = \sqrt{1 - x^2}$ and the lower semicircle defined by the function $g(x) = -\sqrt{1 - x^2}$ (Figures 1.6b and 1.6c).

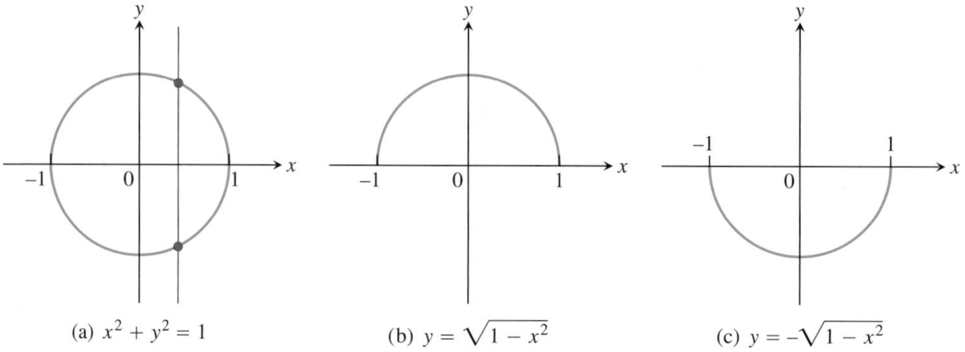

FIGURE 1.6 (a) The circle is not the graph of a function; it fails the vertical line test. (b) The upper semicircle is the graph of a function $f(x) = \sqrt{1 - x^2}$. (c) The lower semicircle is the graph of a function $g(x) = -\sqrt{1 - x^2}$.

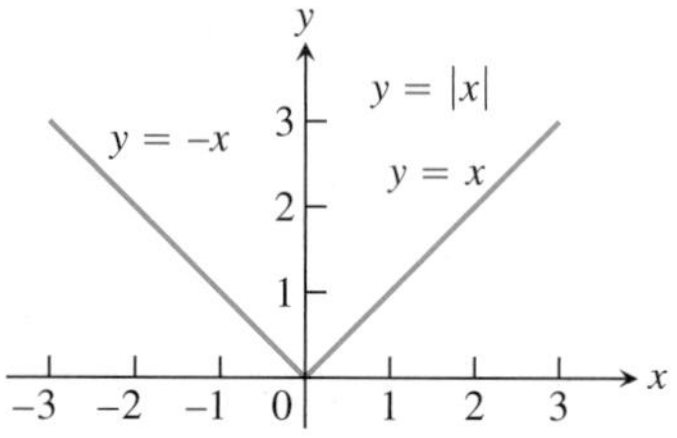

FIGURE 1.7 The absolute value function has domain $(-\infty, \infty)$ and range $[0, \infty)$.

Piecewise-Defined Functions

Sometimes a function is described by using different formulas on different parts of its domain. One example is the **absolute value function**

$$|x| = \begin{cases} x, & x \geq 0 \\ -x, & x < 0, \end{cases}$$

whose graph is given in Figure 1.7. The right-hand side of the equation means that the function equals x if $x \geq 0$, and equals $-x$ if $x < 0$. Here are some other examples.

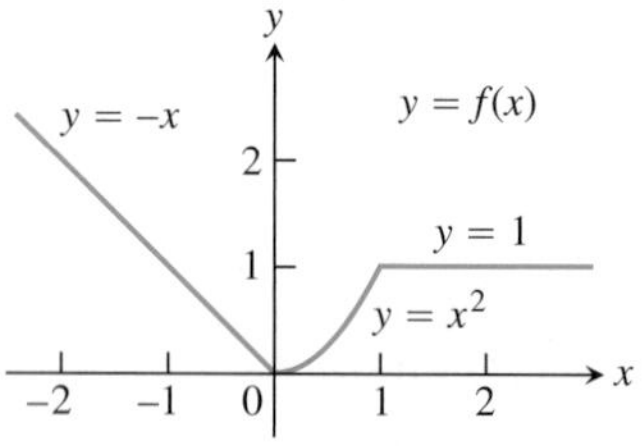

FIGURE 1.8 To graph the function $y = f(x)$ shown here, we apply different formulas to different parts of its domain (Example 3).

EXAMPLE 3 The function

$$f(x) = \begin{cases} -x, & x < 0 \\ x^2, & 0 \leq x \leq 1 \\ 1, & x > 1 \end{cases}$$

is defined on the entire real line but has values given by different formulas depending on the position of x. The values of f are given by: $y = -x$ when $x < 0$, $y = x^2$ when $0 \leq x \leq 1$, and $y = 1$ when $x > 1$. The function, however, is *just one function* whose domain is the entire set of real numbers (Figure 1.8). ■

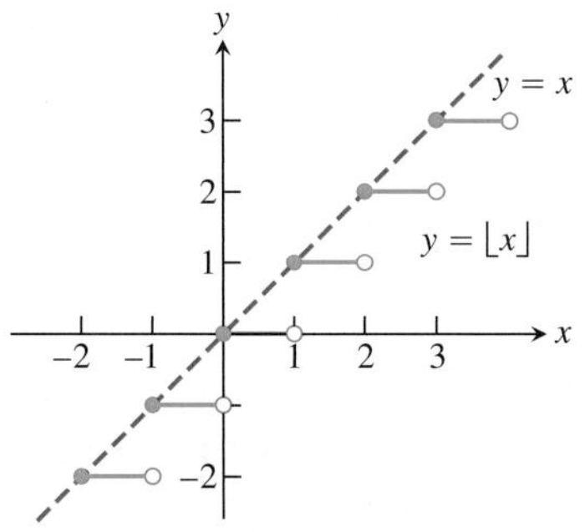

FIGURE 1.9 The graph of the greatest integer function $y = \lfloor x \rfloor$ lies on or below the line $y = x$, so it provides an integer floor for x (Example 4).

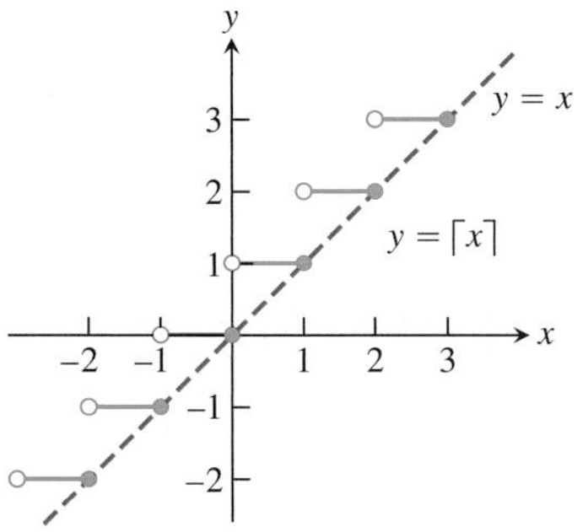

FIGURE 1.10 The graph of the least integer function $y = \lceil x \rceil$ lies on or above the line $y = x$, so it provides an integer ceiling for x (Example 5).

EXAMPLE 4 The function whose value at any number x is the *greatest integer less than or equal to* x is called the **greatest integer function** or the **integer floor function**. It is denoted $\lfloor x \rfloor$. Figure 1.9 shows the graph. Observe that

$$\begin{array}{llll} \lfloor 2.4 \rfloor = 2, & \lfloor 1.9 \rfloor = 1, & \lfloor 0 \rfloor = 0, & \lfloor -1.2 \rfloor = -2, \\ \lfloor 2 \rfloor = 2, & \lfloor 0.2 \rfloor = 0, & \lfloor -0.3 \rfloor = -1 & \lfloor -2 \rfloor = -2. \end{array}$$ ■

EXAMPLE 5 The function whose value at any number x is the *smallest integer greater than or equal to* x is called the **least integer function** or the **integer ceiling function**. It is denoted $\lceil x \rceil$. Figure 1.10 shows the graph. For positive values of x, this function might represent, for example, the cost of parking x hours in a parking lot which charges \$1 for each hour or part of an hour. ■

Increasing and Decreasing Functions

If the graph of a function *climbs* or *rises* as you move from left to right, we say that the function is *increasing*. If the graph *descends* or *falls* as you move from left to right, the function is *decreasing*.

> **DEFINITIONS** Let f be a function defined on an interval I and let x_1 and x_2 be any two points in I.
>
> **1.** If $f(x_2) > f(x_1)$ whenever $x_1 < x_2$, then f is said to be **increasing** on I.
> **2.** If $f(x_2) < f(x_1)$ whenever $x_1 < x_2$, then f is said to be **decreasing** on I.

It is important to realize that the definitions of increasing and decreasing functions must be satisfied for *every* pair of points x_1 and x_2 in I with $x_1 < x_2$. Because of the inequality $<$ comparing the function values, and not $\leq$, it is sometimes said that f is *strictly* increasing or decreasing on I. The interval I may be finite or infinite.

EXAMPLE 6 The function graphed in Figure 1.8 is decreasing on $(-\infty, 0]$ and increasing on $[0, 1]$. The function is neither increasing nor decreasing on the interval $[1, \infty)$ because of the strict inequalities used to compare the function values in the definition. ■

Even Functions and Odd Functions: Symmetry

The graphs of *even* and *odd* functions have characteristic symmetry properties.

> **DEFINITIONS** A function $y = f(x)$ is an
>
> **even function of** x if $f(-x) = f(x)$,
> **odd function of** x if $f(-x) = -f(x)$,
>
> for every x in the function's domain.

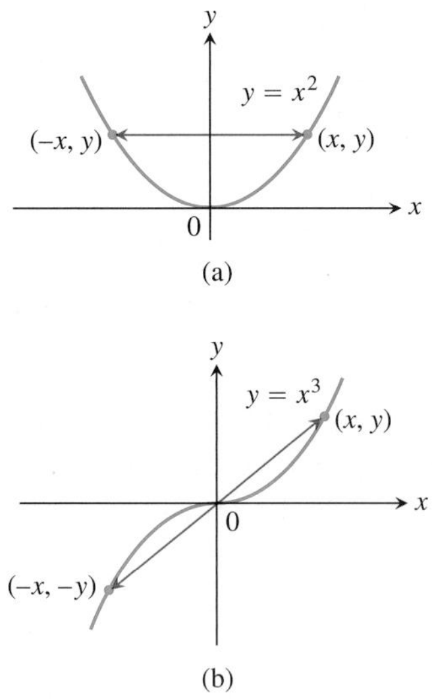

FIGURE 1.11 (a) The graph of $y = x^2$ (an even function) is symmetric about the y-axis. (b) The graph of $y = x^3$ (an odd function) is symmetric about the origin.

The names even and odd come from powers of x. If y is an even power of x, as in $y = x^2$ or $y = x^4$, it is an even function of x because $(-x)^2 = x^2$ and $(-x)^4 = x^4$. If y is an odd power of x, as in $y = x$ or $y = x^3$, it is an odd function of x because $(-x)^1 = -x$ and $(-x)^3 = -x^3$.

The graph of an even function is **symmetric about the y-axis**. Since $f(-x) = f(x)$, a point (x, y) lies on the graph if and only if the point $(-x, y)$ lies on the graph (Figure 1.11a). A reflection across the y-axis leaves the graph unchanged.

The graph of an odd function is **symmetric about the origin**. Since $f(-x) = -f(x)$, a point (x, y) lies on the graph if and only if the point $(-x, -y)$ lies on the graph (Figure 1.11b). Equivalently, a graph is symmetric about the origin if a rotation of 180° about the origin leaves the graph unchanged. Notice that the definitions imply both x and $-x$ must be in the domain of f.

EXAMPLE 7

$f(x) = x^2$	Even function: $(-x)^2 = x^2$ for all x; symmetry about y-axis.
$f(x) = x^2 + 1$	Even function: $(-x)^2 + 1 = x^2 + 1$ for all x; symmetry about y-axis (Figure 1.12a).
$f(x) = x$	Odd function: $(-x) = -x$ for all x; symmetry about the origin.
$f(x) = x + 1$	Not odd: $f(-x) = -x + 1$, but $-f(x) = -x - 1$. The two are not equal. Not even: $(-x) + 1 \neq x + 1$ for all $x \neq 0$ (Figure 1.12b).

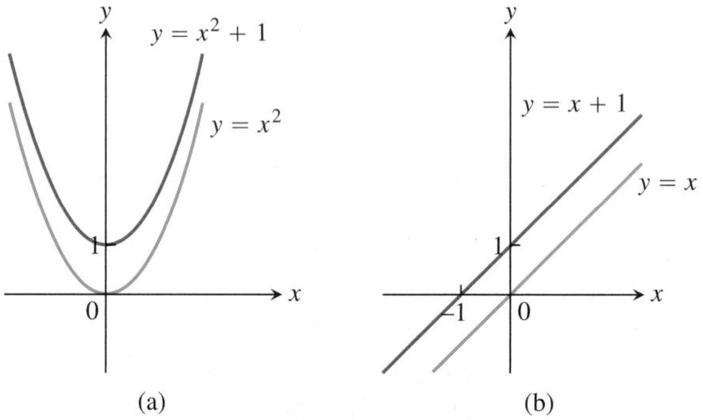

FIGURE 1.12 (a) When we add the constant term 1 to the function $y = x^2$, the resulting function $y = x^2 + 1$ is still even and its graph is still symmetric about the y-axis. (b) When we add the constant term 1 to the function $y = x$, the resulting function $y = x + 1$ is no longer odd. The symmetry about the origin is lost (Example 7). ■

Common Functions

There are a number of important types of functions frequently encountered in calculus. We identify and briefly describe them here.

Linear Functions A function of the form $f(x) = mx + b$, for constants m and b, is called a **linear function**. Figure 1.13a shows an array of lines $f(x) = mx$ where $b = 0$,

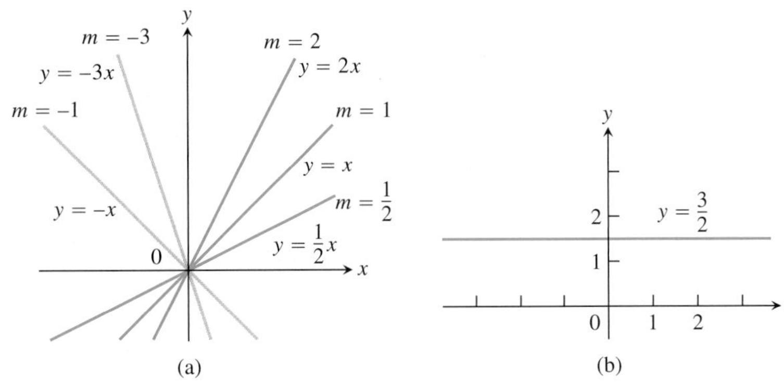

FIGURE 1.13 (a) Lines through the origin with slope m. (b) A constant function with slope $m = 0$.

so these lines pass through the origin. The function $f(x) = x$ where $m = 1$ and $b = 0$ is called the **identity function**. Constant functions result when the slope $m = 0$ (Figure 1.13b). A linear function with positive slope whose graph passes through the origin is called a *proportionality* relationship.

> **DEFINITION** Two variables y and x are **proportional** (to one another) if one is always a constant multiple of the other; that is, if $y = kx$ for some nonzero constant k.

Power Functions A function $f(x) = x^a$, where a is a constant, is called a **power function**. There are several important cases to consider.

(a) $a = n$, a positive integer.

The graphs of $f(x) = x^n$, for $n = 1, 2, 3, 4, 5$, are displayed in Figure 1.14. These functions are defined for all real values of x. Notice that as the power n gets larger, the curves tend to flatten toward the x-axis on the interval $(-1, 1)$, and also rise more steeply for $|x| > 1$. Each curve passes through the point (1, 1) and through the origin. The graphs of functions with even powers are symmetric about the y-axis; those with odd powers are symmetric about the origin. The even-powered functions are decreasing on the interval $(-\infty, 0]$ and increasing on $[0, \infty)$; the odd-powered functions are increasing over the entire real line $(-\infty, \infty)$.

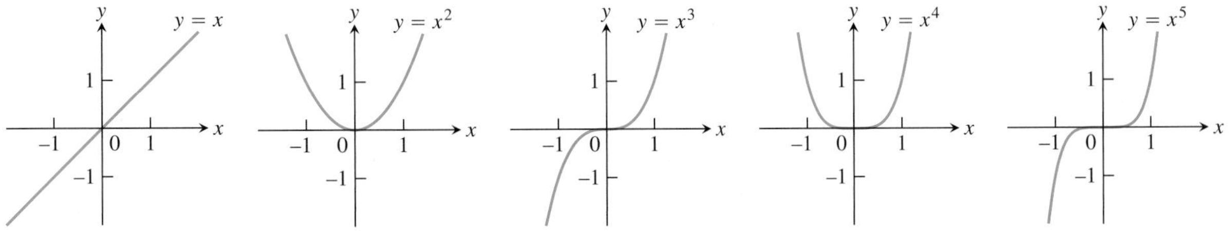

FIGURE 1.14 Graphs of $f(x) = x^n$, $n = 1, 2, 3, 4, 5$ defined for $-\infty < x < \infty$.

(b) $a = -1$ or $a = -2$.

The graphs of the functions $f(x) = x^{-1} = 1/x$ and $g(x) = x^{-2} = 1/x^2$ are shown in Figure 1.15. Both functions are defined for all $x \neq 0$ (you can never divide by zero). The graph of $y = 1/x$ is the hyperbola $xy = 1$, which approaches the coordinate axes far from the origin. The graph of $y = 1/x^2$ also approaches the coordinate axes. The graph of the function f is symmetric about the origin; f is decreasing on the intervals $(-\infty, 0)$ and $(0, \infty)$. The graph of the function g is symmetric about the y-axis; g is increasing on $(-\infty, 0)$ and decreasing on $(0, \infty)$.

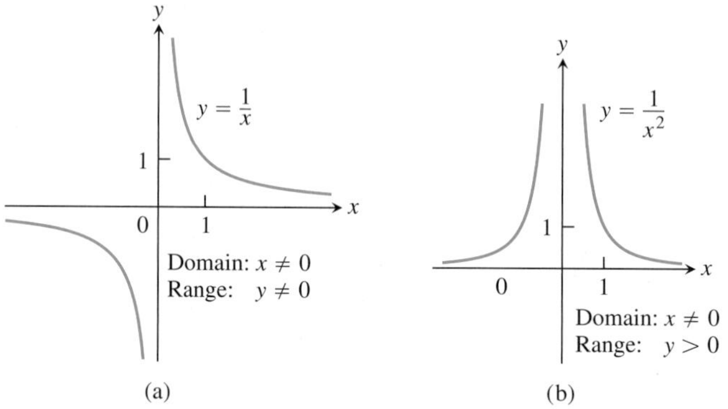

FIGURE 1.15 Graphs of the power functions $f(x) = x^a$ for part (a) $a = -1$ and for part (b) $a = -2$.

(c) $a = \frac{1}{2}, \frac{1}{3}, \frac{3}{2}$, and $\frac{2}{3}$.

The functions $f(x) = x^{1/2} = \sqrt{x}$ and $g(x) = x^{1/3} = \sqrt[3]{x}$ are the **square root** and **cube root** functions, respectively. The domain of the square root function is $[0, \infty)$, but the cube root function is defined for all real x. Their graphs are displayed in Figure 1.16 along with the graphs of $y = x^{3/2}$ and $y = x^{2/3}$. (Recall that $x^{3/2} = (x^{1/2})^3$ and $x^{2/3} = (x^{1/3})^2$.)

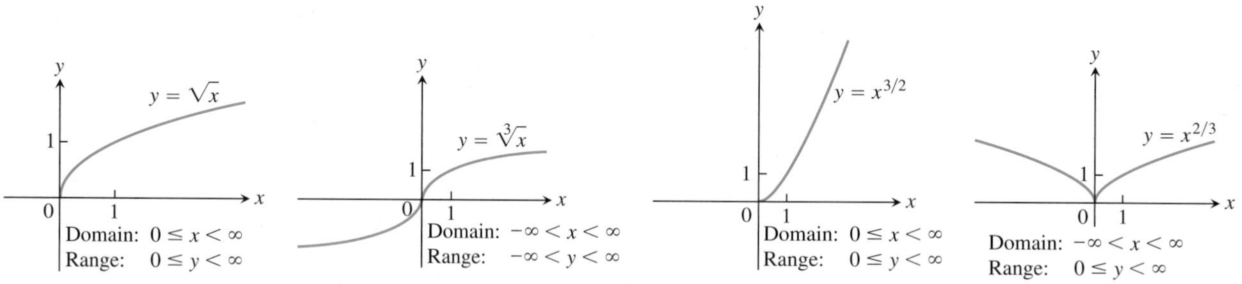

FIGURE 1.16 Graphs of the power functions $f(x) = x^a$ for $a = \frac{1}{2}, \frac{1}{3}, \frac{3}{2}$, and $\frac{2}{3}$.

Polynomials A function p is a **polynomial** if

$$p(x) = a_n x^n + a_{n-1}x^{n-1} + \cdots + a_1 x + a_0$$

where n is a nonnegative integer and the numbers $a_0, a_1, a_2, \ldots, a_n$ are real constants (called the **coefficients** of the polynomial). All polynomials have domain $(-\infty, \infty)$. If the leading coefficient $a_n \neq 0$ and $n > 0$, then n is called the **degree** of the polynomial. Linear functions with $m \neq 0$ are polynomials of degree 1. Polynomials of degree 2, usually written as $p(x) = ax^2 + bx + c$, are called **quadratic functions**. Likewise, **cubic functions** are polynomials $p(x) = ax^3 + bx^2 + cx + d$ of degree 3. Figure 1.17 shows the graphs of three polynomials. Techniques to graph polynomials are studied in Chapter 3.

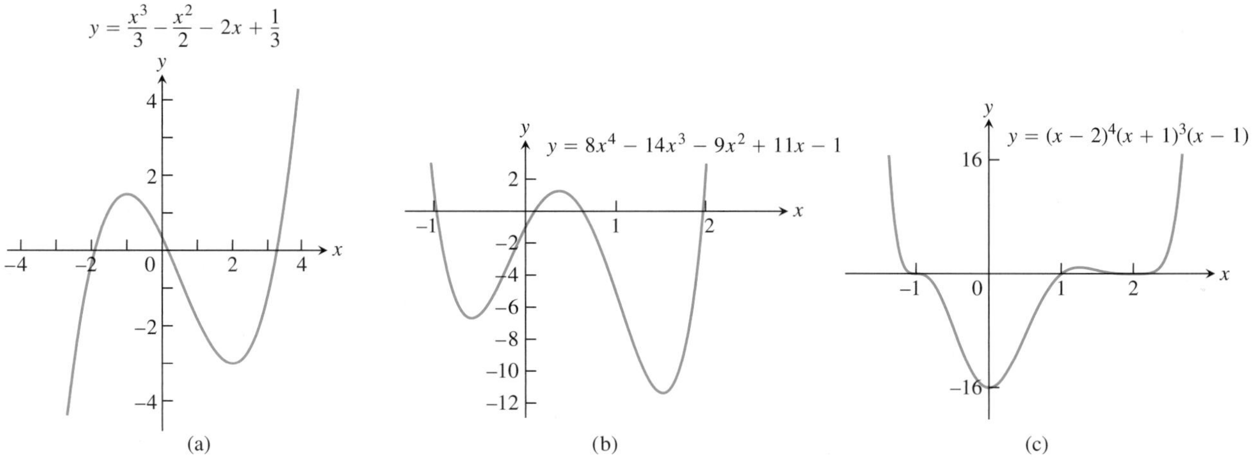

FIGURE 1.17 Graphs of three polynomial functions.

Rational Functions A **rational function** is a quotient or ratio $f(x) = p(x)/q(x)$, where p and q are polynomials. The domain of a rational function is the set of all real x for which $q(x) \neq 0$. The graphs of several rational functions are show in Figure 1.18.

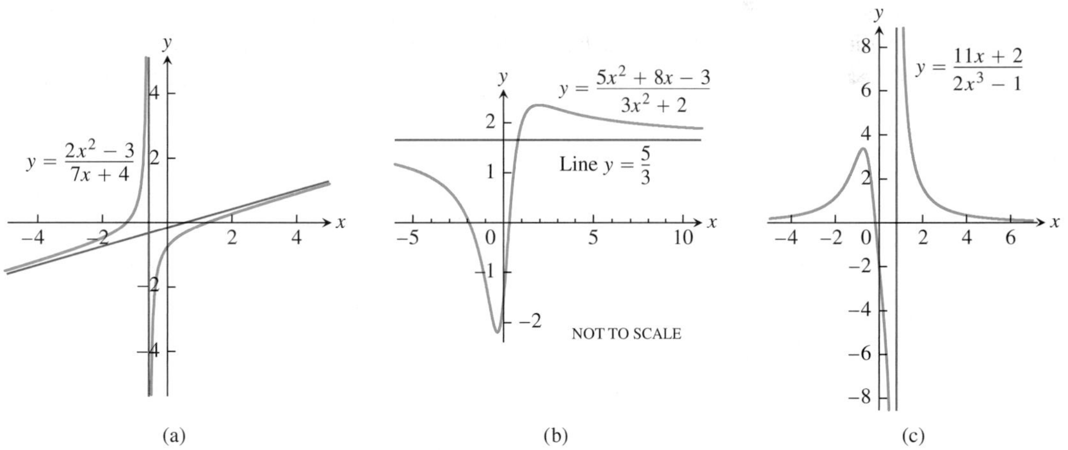

FIGURE 1.18 Graphs of three rational functions.

Algebraic Functions Any function constructed from polynomials using algebraic operations (addition, subtraction, multiplication, division, and taking roots) lies within the class of **algebraic functions**. All rational functions are algebraic, but also included are more complicated functions (such as those satisfying an equation like $y^3 - 9xy + x^3 = 0$, studied in Section 2.8). Figure 1.19 displays the graphs of three algebraic functions.

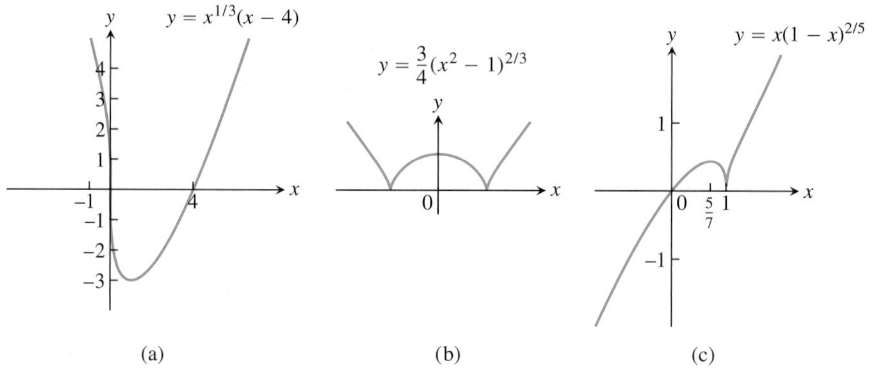

FIGURE 1.19 Graphs of three algebraic functions.

Trigonometric Functions The six basic trigonometric functions are reviewed in Appendix 4. The graphs of the two special sine and cosine functions are shown in Figure 1.20.

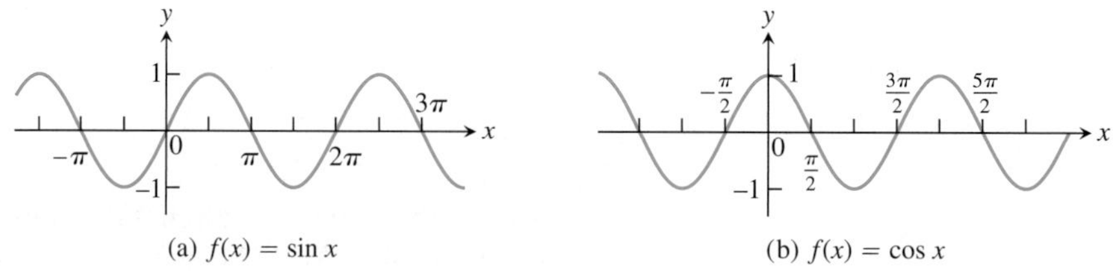

FIGURE 1.20 Graphs of the sine and cosine functions.

Exponential Functions Functions of the form $f(x) = a^x$, where the base $a > 0$ is a positive constant and $a \neq 1$, are called **exponential functions**. All exponential functions have domain $(-\infty, \infty)$ and range $(0, \infty)$, so an exponential function never assumes the value 0. We introduce exponential functions in Section 2.6. The graphs of some exponential functions are shown in Figure 1.21.

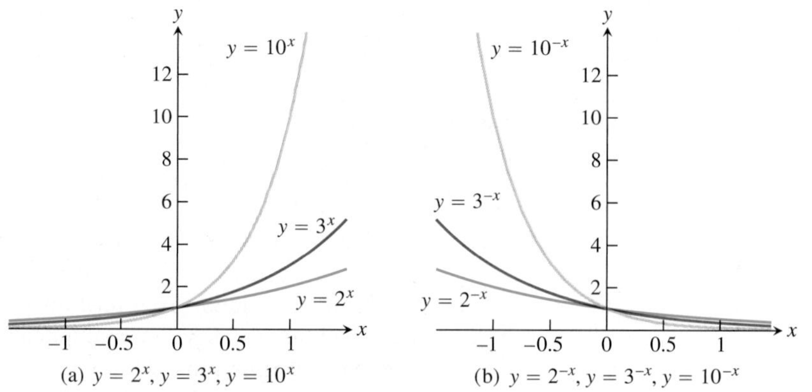

FIGURE 1.21 Graphs of exponential functions.

Logarithmic Functions These are the functions $f(x) = \log_a x$, where the base $a \neq 1$ is a positive constant. They are the *inverse functions* of the exponential functions, and the calculus of these functions is studied in Chapter 2. Figure 1.22 shows the graphs of four logarithmic functions with various bases. In each case the domain is $(0, \infty)$ and the range is $(-\infty, \infty)$.

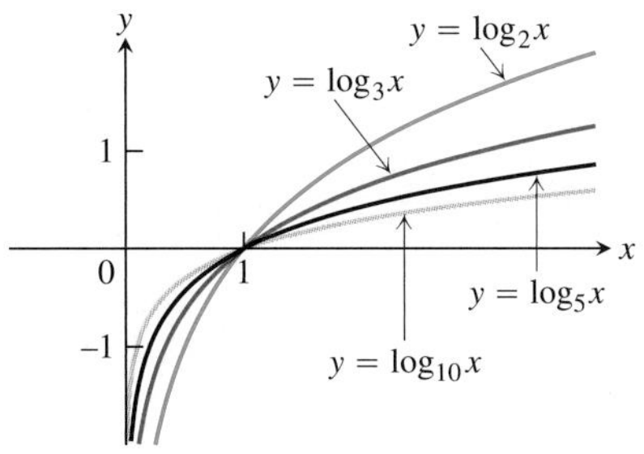

FIGURE 1.22 Graphs of four logarithmic functions.

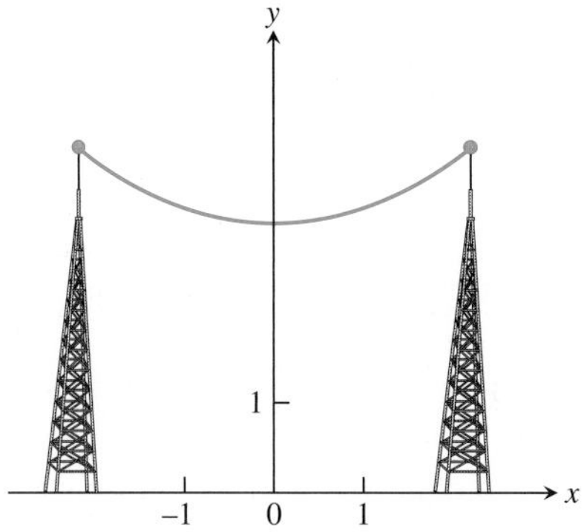

FIGURE 1.23 Graph of a catenary or hanging cable. (The Latin word *catena* means "chain.")

Transcendental Functions These are functions that are not algebraic. They include the trigonometric, inverse trigonometric, exponential, and logarithmic functions, and many other functions as well. A particular example of a transcendental function is a **catenary**. Its graph takes the shape of a cable, like a telephone line or TV cable, strung from one support to another and hanging freely under its own weight (Figure 1.23). The function defining the graph is discussed in Section 3.9.

EXERCISES 1.1

In Exercises 1–4, find the domain and range of each function.

1. $f(x) = 1 + x^2$

2. $f(x) = 1 - \sqrt{x}$

3. $F(t) = \dfrac{1}{\sqrt{t}}$

4. $g(z) = \dfrac{1}{\sqrt{4 - z^2}}$

In Exercises 5 and 6, which of the graphs are graphs of functions of x, and which are not? Give reasons for your answers.

5. a.

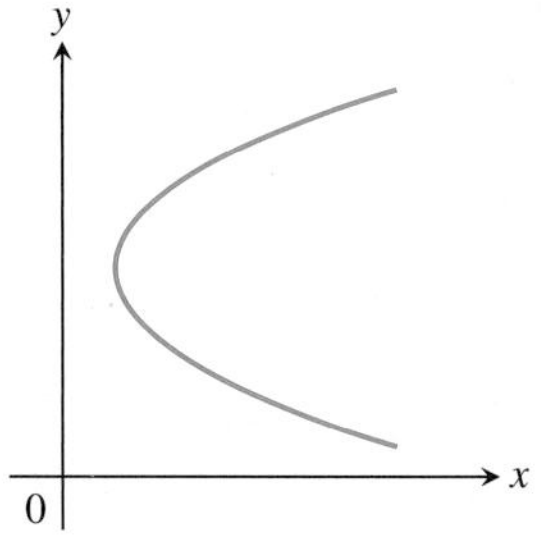

b.

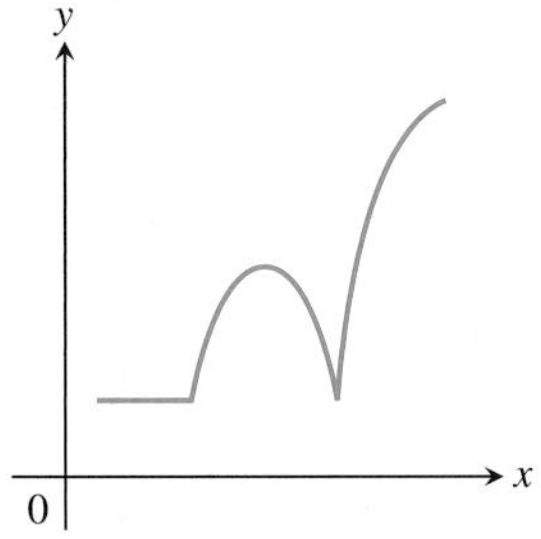

6. a.

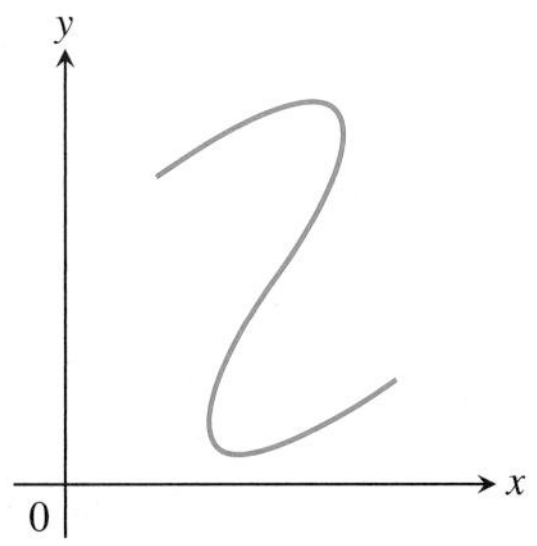

b.

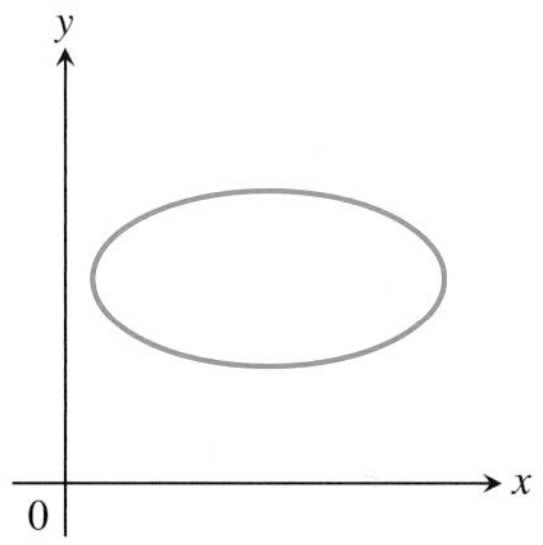

7. Express the area and perimeter of an equilateral triangle as a function of the triangle's side length x.

8. Express the side length of a square as a function of the length d of the square's diagonal. Then express the area as a function of the diagonal length.

9. Express the edge length of a cube as a function of the cube's diagonal length d. Then express the surface area and volume of the cube as a function of the diagonal length.

10. A point P in the first quadrant lies on the graph of the function $f(x) = \sqrt{x}$. Express the coordinates of P as functions of the slope of the line joining P to the origin.

Find the domain and graph the functions in Exercises 11–16.

11. $f(x) = 5 - 2x$

12. $f(x) = 1 - 2x - x^2$

13. $g(x) = \sqrt{|x|}$

14. $g(x) = \sqrt{-x}$

15. $F(t) = t/|t|$

16. $G(t) = 1/|t|$

17. Graph the following equations and explain why they are not graphs of functions of x.

a. $|y| = x$ **b.** $y^2 = x^2$

18. Graph the following equations and explain why they are not graphs of functions of x.

a. $|x| + |y| = 1$ **b.** $|x + y| = 1$

Graph the functions in Exercises 19–22.

19. $f(x) = \begin{cases} x, & 0 \le x \le 1 \\ 2 - x, & 1 < x \le 2 \end{cases}$

20. $g(x) = \begin{cases} 1 - x, & 0 \le x \le 1 \\ 2 - x, & 1 < x \le 2 \end{cases}$

21. $F(x) = \begin{cases} 3 - x, & x \le 1 \\ 2x, & x > 1 \end{cases}$

22. $G(x) = \begin{cases} 1/x, & x < 0 \\ x, & 0 \le x \end{cases}$

23. Find a formula for each function graphed.

a.

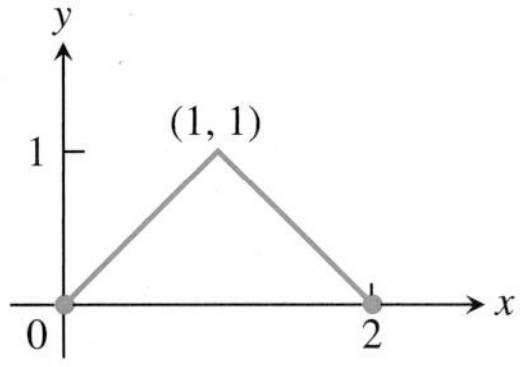

b.

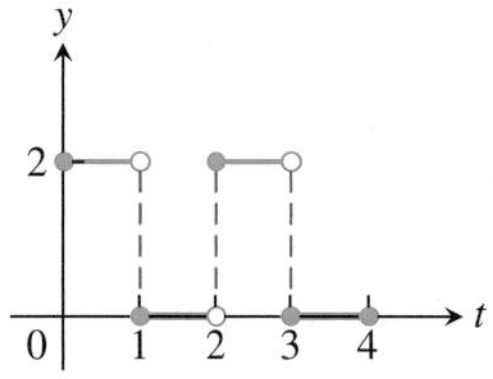

24. a.

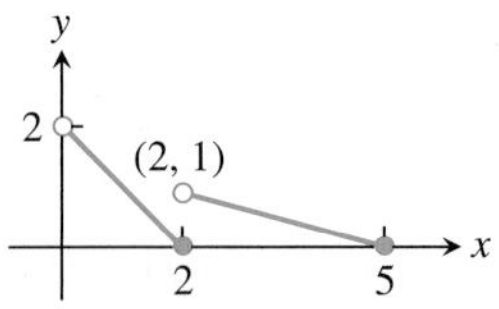

b.

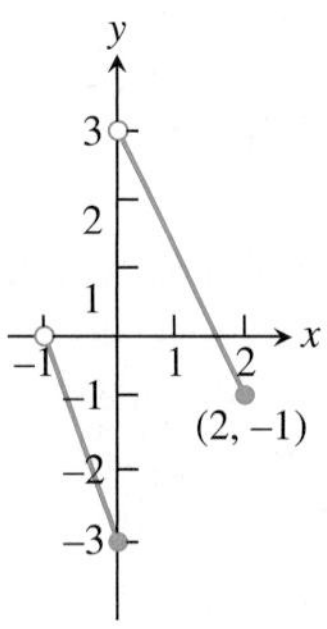

25. a.

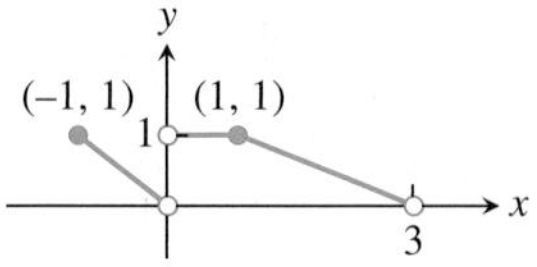

b.

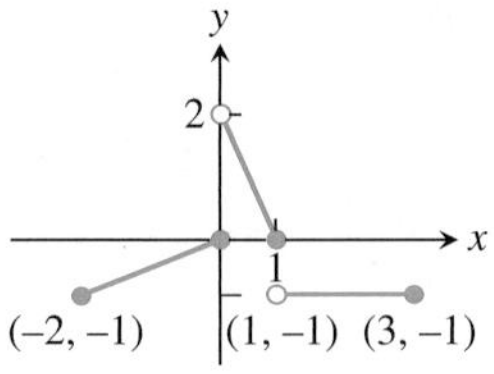

26. a.

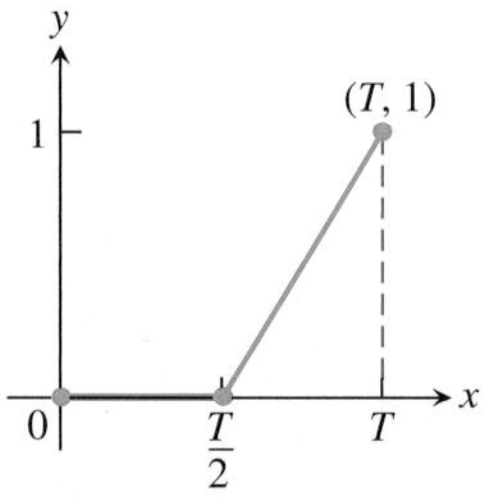

b.

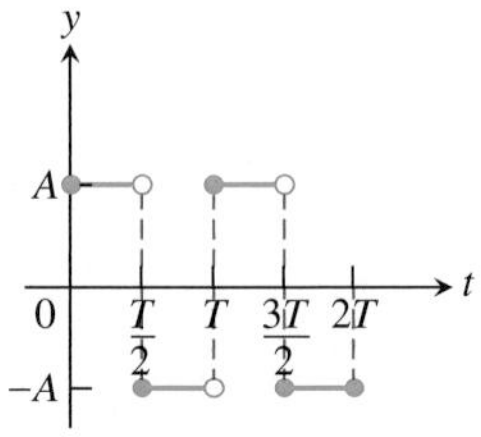

27. For what values of x is

a. $\lfloor x \rfloor = 0$? **b.** $\lceil x \rceil = 0$?

28. What real numbers x satisfy the equation $\lfloor x \rfloor = \lceil x \rceil$?

29. Does $\lceil -x \rceil = -\lfloor x \rfloor$ for all real x? Give reasons for your answer.

30. Graph the function

$$f(x) = \begin{cases} \lfloor x \rfloor, & x \ge 0 \\ \lceil x \rceil, & x < 0 \end{cases}$$

Why is $f(x)$ called the *integer part* of x?

Graph the functions in Exercises 31–42. What symmetries, if any, do the graphs have? Specify the intervals over which the function is increasing and the intervals where it is decreasing.

31. $y = -x^3$

32. $y = -\dfrac{1}{x^2}$

33. $y = -\dfrac{1}{x}$

34. $y = \dfrac{1}{|x|}$

35. $y = \sqrt{|x|}$

36. $y = \sqrt{-x}$

37. $y = x^3/8$

38. $y = -4\sqrt{x}$

39. $y = -x^{3/2}$

40. $y = (-x)^{3/2}$

41. $y = (-x)^{2/3}$

42. $y = -x^{2/3}$

In Exercises 43–58, say whether the function is even, odd, or neither. Give reasons for your answer.

43. $f(x) = 3$

44. $f(x) = x^{-5}$

45. $f(x) = x^2 + 1$

46. $f(x) = x^2 + x$

47. $g(x) = x^3 + x$

48. $g(x) = x^4 + 3x^2 - 1$

49. $g(x) = \dfrac{1}{x^2 - 1}$

50. $g(x) = \dfrac{x}{x^2 - 1}$

51. $h(t) = \dfrac{1}{t - 1}$

52. $h(t) = |t^3|$

53. $h(t) = 2t + 1$

54. $h(t) = 2|t| + 1$

55. $f(t) = \sin 2t$

56. $f(t) = 3 \cos t$

57. $g(t) = t\cos t$

58. $g(t) = \sin t - t$

59. **a.** Graph the functions $f(x) = x/2$ and $g(x) = 1 + (4/x)$ together to identify the values of x for which

$$\frac{x}{2} > 1 + \frac{4}{x}.$$

b. Confirm your findings in part (a) algebraically.

60. For a curve to be *symmetric about the x-axis*, the point (x, y) must lie on the curve if and only if the point $(x, -y)$ lies on the curve. Explain why a curve that is symmetric about the x-axis is not the graph of a function, unless the function is $y = 0$.

1.2 Combining Functions; Shifting and Scaling Graphs

In this section we look at the main ways functions are combined or transformed to form new functions.

Sums, Differences, Products, and Quotients

Like numbers, functions can be added, subtracted, multiplied, and divided (except where the denominator is zero) to produce new functions. If f and g are functions, then for every x that belongs to the domains of both f and g (that is, for $x \in D(f) \cap D(g)$), we define functions $f + g$, $f - g$, and fg by the formulas

$$(f + g)(x) = f(x) + g(x).$$
$$(f - g)(x) = f(x) - g(x).$$
$$(fg)(x) = f(x)g(x).$$

Notice that the + sign on the left-hand side of the first equation represents the operation of addition of *functions*, whereas the + on the right-hand side of the equation means addition of the real numbers $f(x)$ and $g(x)$.

At any point of $D(f) \cap D(g)$ at which $g(x) \neq 0$, we can also define the function f/g by the formula

$$\left(\frac{f}{g}\right)(x) = \frac{f(x)}{g(x)} \qquad (\text{where } g(x) \neq 0).$$

Functions can also be multiplied by constants: If c is a real number, then the function cf is defined for all x in the domain of f by

$$(cf)(x) = cf(x).$$

EXAMPLE 1 The functions defined by the formulas

$$f(x) = \sqrt{x} \qquad \text{and} \qquad g(x) = \sqrt{1 - x},$$

have domains $D(f) = [0, \infty)$ and $D(g) = (-\infty, 1]$. The points common to these domains are the points

$$[0, \infty) \cap (-\infty, 1] = [0, 1].$$

The following table summarizes the formulas and domains for the various algebraic combinations of the two functions. We also write $f \cdot g$ for the product function fg.

Function	Formula	Domain
$f + g$	$(f + g)(x) = \sqrt{x} + \sqrt{1 - x}$	$[0, 1] = D(f) \cap D(g)$
$f - g$	$(f - g)(x) = \sqrt{x} - \sqrt{1 - x}$	$[0, 1]$
$g - f$	$(g - f)(x) = \sqrt{1 - x} - \sqrt{x}$	$[0, 1]$
$f \cdot g$	$(f \cdot g)(x) = f(x)g(x) = \sqrt{x(1 - x)}$	$[0, 1]$
f/g	$\frac{f}{g}(x) = \frac{f(x)}{g(x)} = \sqrt{\frac{x}{1 - x}}$	$[0, 1)$ ($x = 1$ excluded)
g/f	$\frac{g}{f}(x) = \frac{g(x)}{f(x)} = \sqrt{\frac{1 - x}{x}}$	$(0, 1]$ ($x = 0$ excluded)

■

Composite Functions

Composition is another method for combining functions.

DEFINITION If f and g are functions, the **composite** function $f \circ g$ ("f composed with g") is defined by

$$(f \circ g)(x) = f(g(x)).$$

The domain of $f \circ g$ consists of the numbers x in the domain of g for which $g(x)$ lies in the domain of f.

The definition says that $f \circ g$ can be formed when the range of g lies in the domain of f. To find $(f \circ g)(x)$, *first* find $g(x)$ and *second* find $f(g(x))$. Figure 1.24 pictures $f \circ g$ as a machine diagram and Figure 1.25 shows the composite as an arrow diagram.

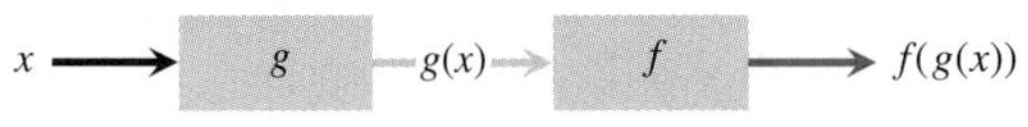

FIGURE 1.24 Two functions can be composed at x whenever the value of one function at x lies in the domain of the other. The composite is denoted by $f \circ g$.

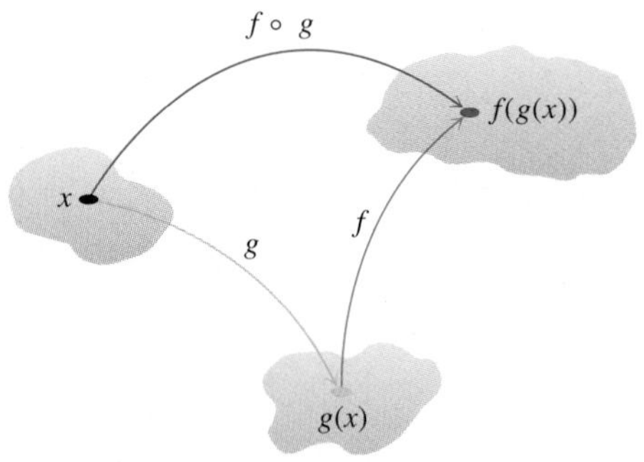

FIGURE 1.25 Arrow diagram for $f \circ g$.

To evaluate the composite function $g \circ f$ (when defined), we find $f(x)$ first and then $g(f(x))$. The domain of $g \circ f$ is the set of numbers x in the domain of f such that $f(x)$ lies in the domain of g.

The functions $f \circ g$ and $g \circ f$ are usually quite different.

EXAMPLE 2 If $f(x) = \sqrt{x}$ and $g(x) = x + 1$, find

(a) $(f \circ g)(x)$ **(b)** $(g \circ f)(x)$ **(c)** $(f \circ f)(x)$ **(d)** $(g \circ g)(x)$.

Solution

Composite	**Domain**
(a) $(f \circ g)(x) = f(g(x)) = \sqrt{g(x)} = \sqrt{x+1}$	$[-1, \infty)$
(b) $(g \circ f)(x) = g(f(x)) = f(x) + 1 = \sqrt{x} + 1$	$[0, \infty)$
(c) $(f \circ f)(x) = f(f(x)) = \sqrt{f(x)} = \sqrt{\sqrt{x}} = x^{1/4}$	$[0, \infty)$
(d) $(g \circ g)(x) = g(g(x)) = g(x) + 1 = (x+1) + 1 = x + 2$	$(-\infty, \infty)$

To see why the domain of $f \circ g$ is $[-1, \infty)$, notice that $g(x) = x + 1$ is defined for all real x but belongs to the domain of f only if $x + 1 \geq 0$, that is to say, when $x \geq -1$. ■

Notice that if $f(x) = x^2$ and $g(x) = \sqrt{x}$, then $(f \circ g)(x) = (\sqrt{x})^2 = x$. However, the domain of $f \circ g$ is $[0, \infty)$, not $(-\infty, \infty)$ since $\sqrt{x}$ requires $x \geq 0$.

Transformations of Functions

A common way to obtain a new function from an existing one is by adding a constant to each output of the existing function, or to its input variable. The graph of the new function is the graph of the original function shifted vertically or horizontally, as follows.

Shift Formulas

Vertical Shifts

$y = f(x) + k$ — Shifts the graph of f *up* k units if $k > 0$
Shifts it *down* $|k|$ units if $k < 0$

Horizontal Shifts

$y = f(x + h)$ — Shifts the graph of f *left* h units if $h > 0$
Shifts it *right* $|h|$ units if $h < 0$

Figures 1.26 and 1.27 display the results of vertical and horizontal shifts for the function $y = x^2$.

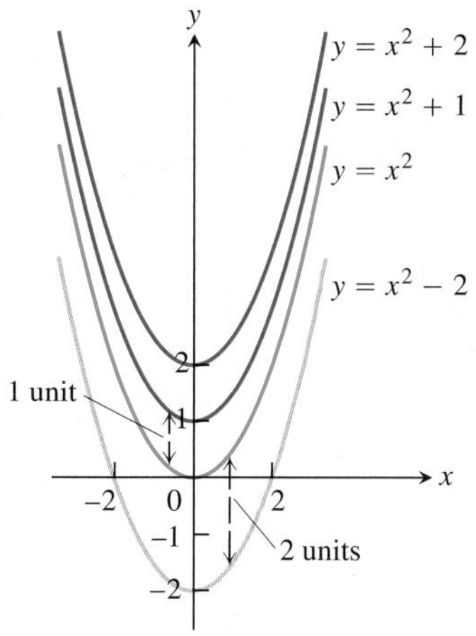

FIGURE 1.26 To shift the graph of $f(x) = x^2$ up (or down), we add positive (or negative) constants to the formula for f.

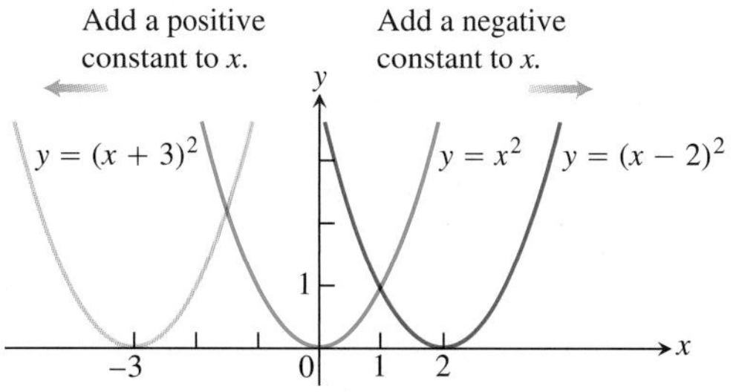

FIGURE 1.27 To shift the graph of $y = x^2$ to the left, we add a positive constant to x. To shift the graph to the right, we add a negative constant to x.

To *scale* the graph of a function $y = f(x)$ is to stretch or compress it, vertically or horizontally. This is accomplished by multiplying the function f, or the independent variable x, by an appropriate constant c. Reflections across the coordinate axes are special cases where $c = -1$.

Vertical and Horizontal Scaling and Reflecting Formulas

For $c > 1$,

$y = cf(x)$	Stretches the graph of f vertically by a factor of c.
$y = \frac{1}{c}f(x)$	Compresses the graph of f vertically by a factor of c.
$y = f(cx)$	Compresses the graph of f horizontally by a factor of c.
$y = f(x/c)$	Stretches the graph of f horizontally by a factor of c.

For $c = -1$,

$y = -f(x)$	Reflects the graph of f across the x-axis.
$y = f(-x)$	Reflects the graph of f across the y-axis.

Figures 1.28, 1.29, and 1.30 display the results of applying the scaling and reflecting formulas to the function $y = \sqrt{x}$.

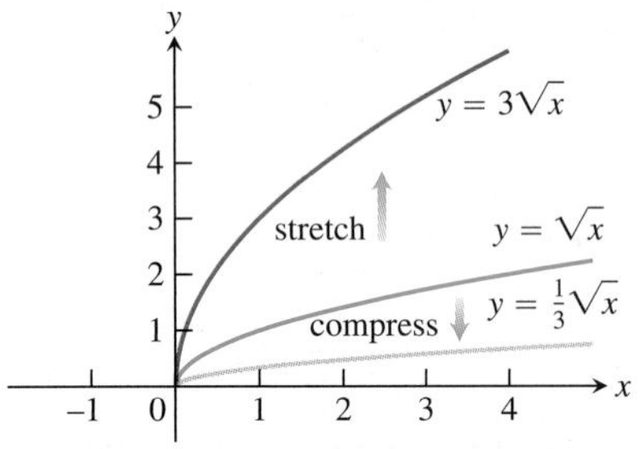

FIGURE 1.28 Vertically stretching and compressing the graph $y = \sqrt{x}$ by a factor of 3.

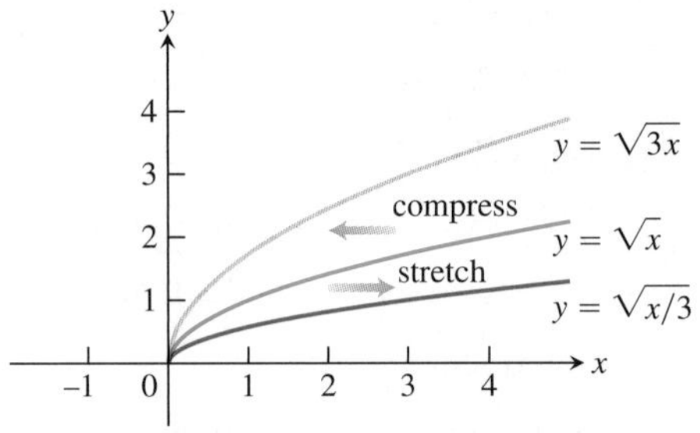

FIGURE 1.29 Horizontally stretching and compressing the graph $y = \sqrt{x}$ by a factor of 3.

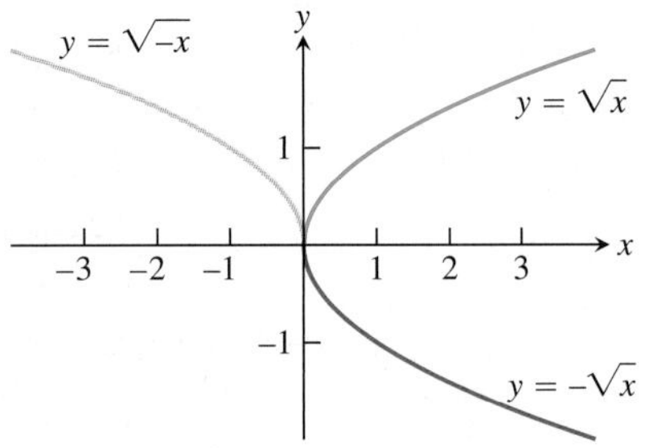

FIGURE 1.30 Reflections of the graph $y = \sqrt{x}$ across the coordinate axes.

EXERCISES 1.2

In Exercises 1 and 2, find the domains and ranges of $f, g, f + g$, and $f \cdot g$.

1. $f(x) = x, \quad g(x) = \sqrt{x - 1}$
2. $f(x) = \sqrt{x + 1}, \quad g(x) = \sqrt{x - 1}$

In Exercises 3 and 4, find the domains and ranges of $f, g, f/g$, and g/f.

3. $f(x) = 2, \quad g(x) = x^2 + 1$
4. $f(x) = 1, \quad g(x) = 1 + \sqrt{x}$
5. If $f(x) = x + 5$ and $g(x) = x^2 - 3$, find the following.
   a. $f(g(0))$ **b.** $g(f(0))$
   c. $f(g(x))$ **d.** $g(f(x))$
   e. $f(f(-5))$ **f.** $g(g(2))$
   g. $f(f(x))$ **h.** $g(g(x))$

6. If $f(x) = x - 1$ and $g(x) = 1/(x + 1)$, find the following.

a. $f(g(1/2))$ **b.** $g(f(1/2))$
c. $f(g(x))$ **d.** $g(f(x))$
e. $f(f(2))$ **f.** $g(g(2))$
g. $f(f(x))$ **h.** $g(g(x))$

7. If $u(x) = 4x - 5$, $v(x) = x^2$, and $f(x) = 1/x$, find formulas for the following.

a. $u(v(f(x)))$ **b.** $u(f(v(x)))$
c. $v(u(f(x)))$ **d.** $v(f(u(x)))$
e. $f(u(v(x)))$ **f.** $f(v(u(x)))$

8. If $f(x) = \sqrt{x}$, $g(x) = x/4$, and $h(x) = 4x - 8$, find formulas for the following.

a. $h(g(f(x)))$ **b.** $h(f(g(x)))$
c. $g(h(f(x)))$ **d.** $g(f(h(x)))$
e. $f(g(h(x)))$ **f.** $f(h(g(x)))$

Let $f(x) = x - 3$, $g(x) = \sqrt{x}$, $h(x) = x^3$, and $j(x) = 2x$. Express each of the functions in Exercises 9 and 10 as a composite involving one or more of f, g, h, and j.

9. a. $y = \sqrt{x} - 3$ **b.** $y = 2\sqrt{x}$
c. $y = x^{1/4}$ **d.** $y = 4x$
e. $y = \sqrt{(x - 3)^3}$ **f.** $y = (2x - 6)^3$

10. a. $y = 2x - 3$ **b.** $y = x^{3/2}$
c. $y = x^9$ **d.** $y = x - 6$
e. $y = 2\sqrt{x - 3}$ **f.** $y = \sqrt{x^3 - 3}$

11. Copy and complete the following table.

	$g(x)$	$f(x)$	$(f \circ g)(x)$
a.	$x - 7$	$\sqrt{x}$	?
b.	$x + 2$	$3x$	?
c.	?	$\sqrt{x - 5}$	$\sqrt{x^2 - 5}$
d.	$\dfrac{x}{x - 1}$	$\dfrac{x}{x - 1}$	?
e.	?	$1 + \dfrac{1}{x}$	x
f.	$\dfrac{1}{x}$	?	x

12. **(a)** Write a formula for $f \circ g$ and $g \circ f$ and find the **(b)** domain and **(c)** range of each, where $f(x) = \sqrt{x + 1}$ and $g(x) = \dfrac{1}{x}$.

13. Match the equations listed in parts (a)–(d) to the graphs in the accompanying figure.

a. $y = (x - 1)^2 - 4$ **b.** $y = (x - 2)^2 + 2$
c. $y = (x + 2)^2 + 2$ **d.** $y = (x + 3)^2 - 2$

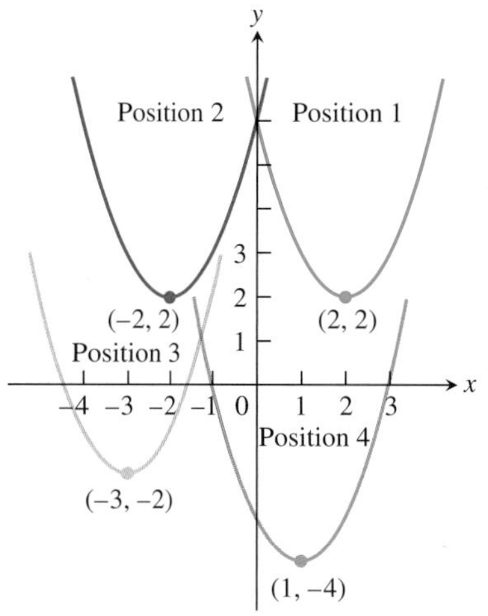

14. The accompanying figure shows the graph of $y = -x^2$ shifted to four new positions. Write an equation for each new graph.

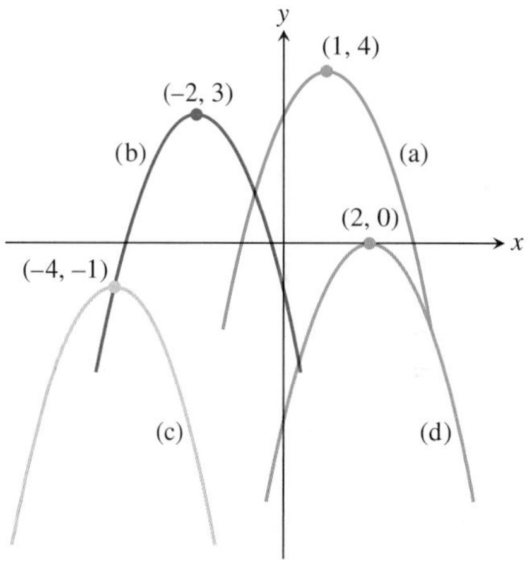

Exercises 15–20 tell how many units and in what directions the graphs of the given functions or equations are to be shifted. Give an equation for the shifted graph. Then sketch the original and shifted graphs together, labeling each graph with its equation.

15. $x^2 + y^2 = 49$ Down 3, left 2

16. $y = x^3$ Left 1, down 1

17. $y = \sqrt{x}$ Left 0.81

18. $y = -\sqrt{x}$ Right 3

19. $y = 1/x$ Up 1, right 1

20. $y = 1/x^2$ Left 2, down 1

Graph the functions in Exercises 21–34.

21. $y = \sqrt{x + 4}$ **22.** $y = \sqrt{9 - x}$

23. $y = |x - 2|$ **24.** $y = |1 - x| - 1$

25. $y = 1 + \sqrt{x - 1}$ **26.** $y = 1 - \sqrt{x}$

27. $y = (x + 1)^{2/3}$ **28.** $y = 1 - x^{2/3}$

29. $y = \sqrt[3]{x - 1} - 1$ **30.** $y = (x + 2)^{3/2} + 1$

31. $y = \frac{1}{x} + 2$ **32.** $y = \frac{1}{x + 2}$

33. $y = \frac{1}{(x - 1)^2}$ **34.** $y = \frac{1}{x^2} - 1$

35. The accompanying figure shows the graph of a function $f(x)$ with domain $[0, 2]$ and range $[0, 1]$. Find the domains and ranges of the following functions, and sketch their graphs.

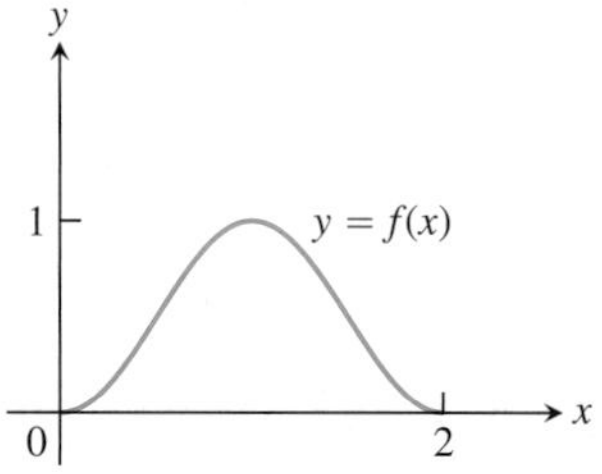

a. $f(x) + 2$ **b.** $f(x) - 1$
c. $2f(x)$ **d.** $-f(x)$
e. $f(x + 2)$ **f.** $f(x - 1)$
g. $f(-x)$ **h.** $-f(x + 1) + 1$

36. The accompanying figure shows the graph of a function $g(t)$ with domain $[-4, 0]$ and range $[-3, 0]$. Find the domains and ranges of the following functions, and sketch their graphs.

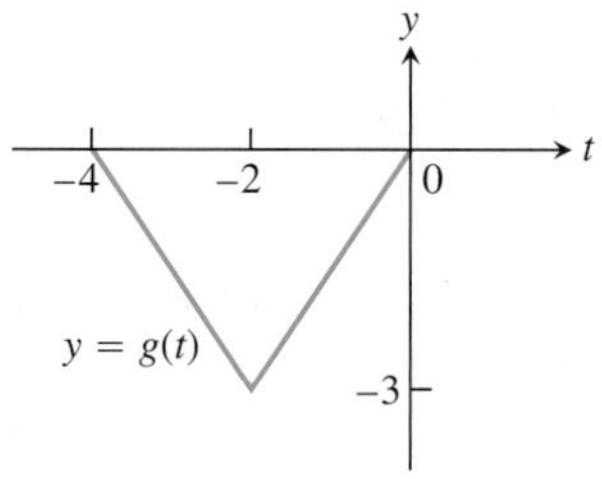

a. $g(-t)$ **b.** $-g(t)$
c. $g(t) + 3$ **d.** $1 - g(t)$
e. $g(-t + 2)$ **f.** $g(t - 2)$
g. $g(1 - t)$ **h.** $-g(t - 4)$

Exercises 37–46 tell by what factor and direction the graphs of the given functions are to be stretched or compressed. Give an equation for the stretched or compressed graph.

37. $y = x^2 - 1$, stretched vertically by a factor of 3

38. $y = x^2 - 1$, compressed horizontally by a factor of 2

39. $y = 1 + \frac{1}{x^2}$, compressed vertically by a factor of 2

40. $y = 1 + \frac{1}{x^2}$, stretched horizontally by a factor of 3

41. $y = \sqrt{x + 1}$, compressed horizontally by a factor of 4

42. $y = \sqrt{x + 1}$, stretched vertically by a factor of 3

43. $y = \sqrt{4 - x^2}$, stretched horizontally by a factor of 2

44. $y = \sqrt{4 - x^2}$, compressed vertically by a factor of 3

45. $y = 1 - x^3$, compressed horizontally by a factor of 3

46. $y = 1 - x^3$, stretched horizontally by a factor of 2

In Exercises 47–54, graph each function, not by plotting points, but by starting with the graph of one of the standard functions presented in Figures 1.14–1.16, and applying an appropriate transformation.

47. $y = -\sqrt{2x + 1}$ **48.** $y = \sqrt{1 - \frac{x}{2}}$

49. $y = (x - 1)^3 + 2$ **50.** $y = (1 - x)^3 + 2$

51. $y = \frac{1}{2x} - 1$ **52.** $y = \frac{2}{x^2} + 1$

53. $y = -\sqrt[3]{x}$ **54.** $y = (-2x)^{2/3}$

55. Graph the function $y = |x^2 - 1|$.

56. Graph the function $y = \sqrt{|x|}$.

57. Assume that f is an even function, g is an odd function, and both f and g are defined on the entire real line $\mathbb{R}$. Which of the following (where defined) are even? odd?

a. fg **b.** f/g **c.** g/f
d. $f^2 = ff$ **e.** $g^2 = gg$ **f.** $f \circ g$
g. $g \circ f$ **h.** $f \circ f$ **i.** $g \circ g$

58. Can a function be both even and odd? Give reasons for your answer.

T **59.** (*Continuation of Example* 1.) Graph the functions $f(x) = \sqrt{x}$ and $g(x) = \sqrt{1 - x}$ together with their **(a)** sum, **(b)** product, **(c)** two differences, **(d)** two quotients.

T **60.** Let $f(x) = x - 7$ and $g(x) = x^2$. Graph f and g together with $f \circ g$ and $g \circ f$.

1.3 Rates of Change and Tangents to Curves

In this section we look at average and instantaneous rates of change, ideas that are closely related to the slope of a curve at a point P on the curve. These concepts lead naturally to the concept of the *limit*, which will be investigated in the next several sections.

Average and Instantaneous Speed

HISTORICAL BIOGRAPHY*

Galileo Galilei
(1564–1642)

In the late sixteenth century, Galileo discovered that a solid object dropped from rest (not moving), near the surface of the earth and allowed to fall freely, will fall a distance proportional to the square of the time it has been falling. This type of motion is called **free fall**. It assumes negligible air resistance to slow the object down, and that gravity is the only force acting on the falling body. If y denotes the distance fallen in feet after t seconds, then Galileo's law is

$$y = 16t^2,$$

where 16 approximates the constant of proportionality. (If y is measured in meters, the constant is about 4.9.) We use these constants throughout the text as the values for gravitational effects near the earth's surface.

A moving body's **average speed** during an interval of time is found by dividing the distance covered by the time elapsed. The unit of measure is length per unit time: kilometers per hour, feet (or meters) per second, or whatever is appropriate to the problem at hand.

EXAMPLE 1 A rock breaks loose from the top of a tall cliff. What is its average speed

(a) during the first 2 sec of fall?

(b) during the 1-sec interval between second 1 and second 2?

Solution The average speed of the rock during a given time interval is the change in distance, Δy, divided by the length of the time interval, Δt. Measuring distance in feet and time in seconds, we have the following calculations:

(a) For the first 2 sec: $$\frac{\Delta y}{\Delta t} = \frac{16(2)^2 - 16(0)^2}{2 - 0} = 32\,\frac{\text{ft}}{\text{sec}}$$

(b) From sec 1 to sec 2: $$\frac{\Delta y}{\Delta t} = \frac{16(2)^2 - 16(1)^2}{2 - 1} = 48\,\frac{\text{ft}}{\text{sec}}$$ ■

The next example examines what happens when we look at the average speed of a falling object over shorter and shorter time intervals. Our discussion is informal here but will be made precise in Chapter 2.

EXAMPLE 2 Find the speed of the falling rock at $t = 1$ and $t = 2$ sec.

Solution We can calculate the average speed of the rock over a time interval $[t_0, t_0 + h]$, having length $\Delta t = h$, as

$$\frac{\Delta y}{\Delta t} = \frac{16(t_0 + h)^2 - 16{t_0}^2}{h}. \tag{1}$$

*To learn more about the historical figures and the development of the major elements and topics of calculus, visit **www.aw-bc.com/thomas**.

We cannot use this formula to calculate the "instantaneous" speed at t_0 by substituting $h = 0$, because we cannot divide by zero. But we *can* use it to calculate average speeds over increasingly short time intervals starting at $t_0 = 1$ and $t_0 = 2$. When we do so, we see a pattern (Table 1.2).

TABLE 1.2 Average speeds over short time intervals in ft/sec

$$\text{Average speed: } \frac{\Delta y}{\Delta t} = \frac{16(t_0 + h)^2 - 16t_0^2}{h}$$

Length of time interval h	Average speed over interval of length h starting at $t_0 = 1$	Average speed over interval of length h starting at $t_0 = 2$
1	48	80
0.1	33.6	65.6
0.01	32.16	64.16
0.001	32.016	64.016
0.0001	32.0016	64.0016

The average speed on intervals starting at $t_0 = 1$ seems to approach a limiting value of 32 as the length of the interval decreases. This suggests that the rock is falling at a speed of 32 ft/sec at $t_0 = 1$ sec. Let's confirm this algebraically.

If we set $t_0 = 1$ and then expand the numerator in Equation (1) and simplify, we find that

$$\frac{\Delta y}{\Delta t} = \frac{16(1 + h)^2 - 16(1)^2}{h} = \frac{16(1 + 2h + h^2) - 16}{h}$$

$$= \frac{32h + 16h^2}{h} = 32 + 16h.$$

For values of h different from 0, the expressions on the right and left are equivalent and the average speed is $32 + 16h$ ft/sec. We can now see why the average speed has the limiting value $32 + 16(0) = 32$ ft/sec as h approaches 0.

Similarly, setting $t_0 = 2$ in Equation (1), the procedure yields

$$\frac{\Delta y}{\Delta t} = 64 + 16h$$

for values of h different from 0. As h gets closer and closer to 0, the average speed has the limiting value 64 ft/sec when $t_0 = 2$ sec. ■

Average Rates of Change and Secant Lines

Given an arbitrary function $y = f(x)$, we calculate the average rate of change of y with respect to x over the interval $[x_1, x_2]$ by dividing the change in the value of y, $\Delta y = f(x_2) - f(x_1)$, by the length $\Delta x = x_2 - x_1 = h$ of the interval over which the change occurs.

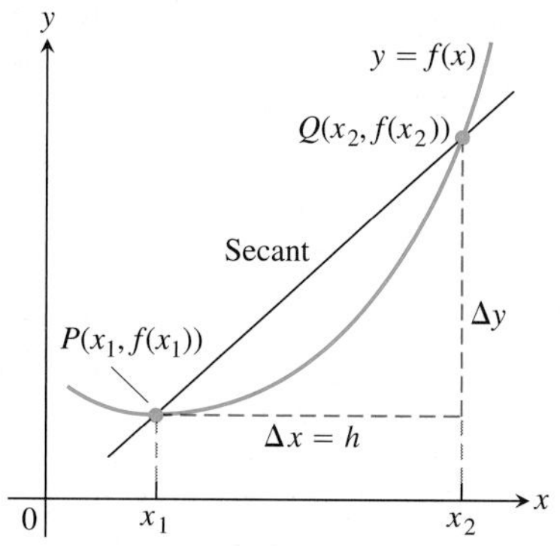

FIGURE 1.31 A secant to the graph $y = f(x)$. Its slope is $\Delta y/\Delta x$, the average rate of change of f over the interval $[x_1, x_2]$.

DEFINITION The **average rate of change** of $y = f(x)$ with respect to x over the interval $[x_1, x_2]$ is

$$\frac{\Delta y}{\Delta x} = \frac{f(x_2) - f(x_1)}{x_2 - x_1} = \frac{f(x_1 + h) - f(x_1)}{h}, \qquad h \neq 0.$$

Geometrically, the rate of change of f over $[x_1, x_2]$ is the slope of the line through the points $P(x_1, f(x_1))$ and $Q(x_2, f(x_2))$ (Figure 1.31). In geometry, a line joining two points of a curve is a **secant** to the curve. Thus, the average rate of change of f from x_1 to x_2 is identical with the slope of secant PQ. Let's consider what happens as the point Q approaches the point P along the curve, so the length h of the interval over which the change occurs approaches zero.

Slope of a Curve

We know what is meant by the slope of a straight line, which tells us the rate at which it rises or falls—its rate of change as the graph of a linear function. But what is meant by the *slope of a curve* at a point P on the curve? If there is a *tangent* line to the curve at P—a line that just touches the curve like the tangent to a circle—it would be reasonable to identify *the slope of the tangent* as the slope of the curve at P. So we need a precise meaning for the tangent at a point on a curve.

To define tangency for general curves, we need an approach that takes into account the behavior of the secants through P and nearby points Q as Q moves toward P along the curve (Figure 1.32). It goes like this:

1. We start with what we *can* calculate, namely the slope of the secant PQ.
2. Investigate the limiting value of the secant slope as Q approaches P along the curve. (We clarify the *limit* idea in the next section.)
3. If the *limit* exists, take it to be the slope of the curve at P and define the tangent to the curve at P to be the line through P with this slope.

This procedure is what we were doing in the falling-rock problem discussed above. The next example illustrates the geometric idea.

HISTORICAL BIOGRAPHY

Pierre de Fermat
(1601–1665)

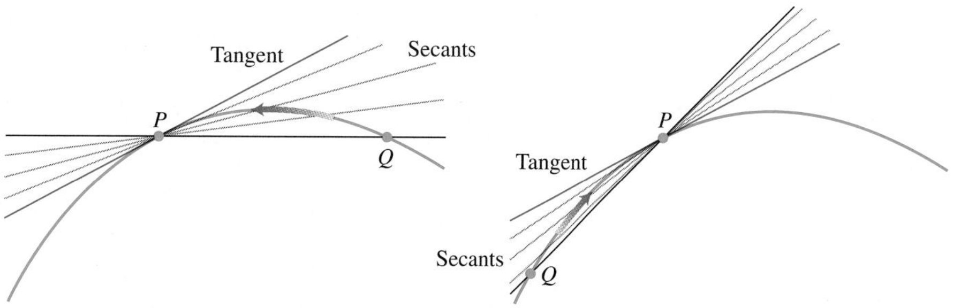

FIGURE 1.32 The tangent to the curve at P is the line through P whose slope is the limit of the secant slopes as $Q \to P$ from either side.

EXAMPLE 3 Find the slope of the parabola $y = x^2$ at the point $P(2, 4)$. Write an equation for the tangent to the parabola at this point.

Solution We begin with a secant line through $P(2, 4)$ and $Q(2 + h, (2 + h)^2)$ nearby. We then write an expression for the slope of the secant PQ and investigate what happens to the slope as Q approaches P along the curve:

$$\begin{aligned} \text{Secant slope} = \frac{\Delta y}{\Delta x} = \frac{(2 + h)^2 - 2^2}{h} &= \frac{h^2 + 4h + 4 - 4}{h} \\ &= \frac{h^2 + 4h}{h} = h + 4. \end{aligned}$$

If $h > 0$, then Q lies above and to the right of P, as in Figure 1.33. If $h < 0$, then Q lies to the left of P (not shown). In either case, as Q approaches P along the curve, h approaches zero and the secant slope approaches 4:

As h approaches 0, the sum $(h + 4)$ approaches the limiting value 4.

We take 4 to be the parabola's slope at P.

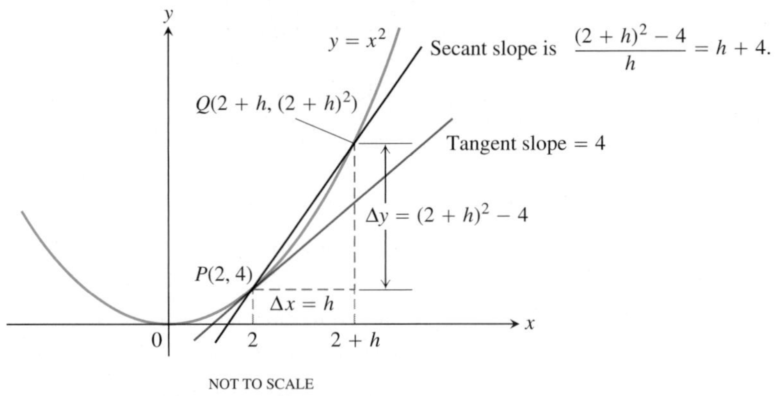

FIGURE 1.33 Finding the slope of the parabola $y = x^2$ at the point $P(2, 4)$ as the limit of secant slopes (Example 3).

The tangent to the parabola at P is the line through P with slope 4:

$$y = 4 + 4(x - 2) \qquad \text{Point-slope equation}$$

$$y = 4x - 4.$$

■

Instantaneous Rates of Change and Tangent Lines

The rates at which the rock in Example 2 was falling at the instants $t = 1$ and $t = 2$ are called **instantaneous rates of change.** The instantaneous rates were found to be the values of the average speeds, or average rates of change, as the time interval of length h approached 0.

That is, the instantaneous rate is just the value of the average rate as the length h of the interval over which the change occurs approaches zero. The average rate of change corresponds to the slope of a secant line; the instantaneous rate corresponds to the slope of the tangent line as the independent variable approaches a fixed value. In Example 1, the independent variable t approached the values $t = 1$ and $t = 2$. In Example 3, the independent variable x approached the value $x = 2$. So we see that instantaneous rates and slopes of tangent lines are intimately connected. We investigate this connection thoroughly in the next chapter, but to do so we need the concept of a *limit.*

EXERCISES 1.3

In Exercises 1–6, find the average rate of change of the function over the given interval or intervals.

1. $f(x) = x^3 + 1$;

a. $[2, 3]$ **b.** $[-1, 1]$

2. $g(x) = x^2$;

a. $[-1, 1]$ **b.** $[-2, 0]$

3. $h(t) = \cot t$;

a. $[\pi/4, 3\pi/4]$ **b.** $[\pi/6, \pi/2]$

4. $g(t) = 2 + \cos t$;

a. $[0, \pi]$ **b.** $[-\pi, \pi]$

5. $R(\theta) = \sqrt{4\theta + 1}$; $[0, 2]$

6. $P(\theta) = \theta^3 - 4\theta^2 + 5\theta$; $[1, 2]$

7. Speed of a car The accompanying figure shows the time-to-distance graph for a sports car accelerating from a standstill.

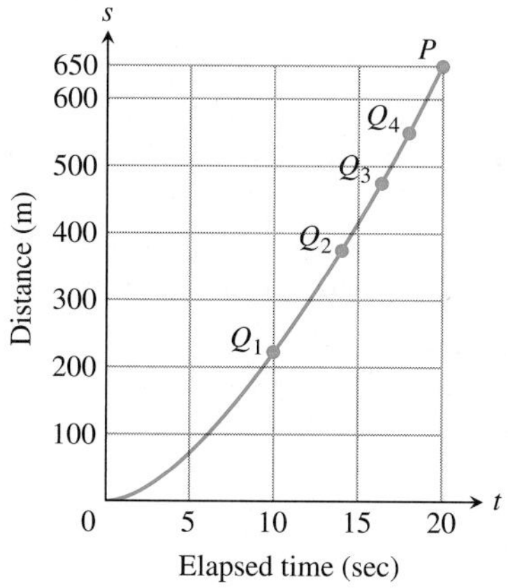

a. Estimate the slopes of secants PQ_1, PQ_2, PQ_3, and PQ_4. What are the appropriate units for these slopes?

b. Then estimate the car's speed at time $t = 20$ sec.

8. The accompanying figure shows the plot of distance fallen versus time for an object that fell from the lunar landing module a distance 80 m to the surface of the moon.

a. Estimate the slopes of the secants PQ_1, PQ_2, PQ_3, and PQ_4.

b. About how fast was the object going when it hit the surface?

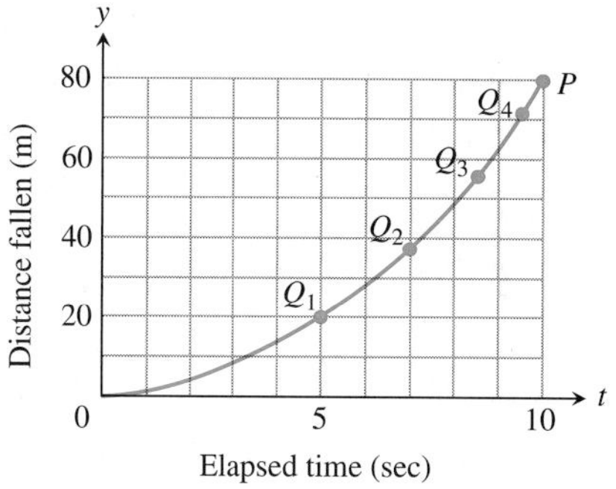

In Exercises 9–16, use the method in Example 3 to find **(a)** the slope of the curve at the given point P, and **(b)** an equation of the tangent line at P.

9. $y = x^2 - 3$, $P(2, 1)$

10. $y = 5 - x^2$, $P(1, 4)$

11. $y = x^2 - 2x - 3$, $P(2, -3)$

12. $y = x^2 - 4x$, $P(1, -3)$

13. $y = x^3$, $P(2, 8)$

14. $y = 2 - x^3$, $P(1, 1)$

15. $y = x^3 - 12x$, $P(1, -11)$

16. $y = x^3 - 3x^2 + 4$, $P(2, 0)$

T **17.** The profits of a small company for each of the first five years of its operation are given in the following table:

Year	Profit in $1000s
2000	6
2001	27
2002	62
2003	111
2004	174

a. Plot points representing the profit as a function of year, and join them by as smooth a curve as you can.

b. What is the average rate of increase of the profits between 2002 and 2004?

c. Use your graph to estimate the rate at which the profits were changing in 2002.

T **18.** Make a table of values for the function $F(x) = (x + 2)/(x - 2)$ at the points $x = 1.2$, $x = 11/10$, $x = 101/100$, $x = 1001/1000$, $x = 10001/10000$, and $x = 1$.

a. Find the average rate of change of $F(x)$ over the intervals $[1, x]$ for each $x \neq 1$ in your table.

b. Extending the table if necessary, try to determine the rate of change of $F(x)$ at $x = 1$.

1.4 Limit of a Function and Limit Laws

In Section 1.3 we saw that limits arise when finding the instantaneous rate of change of a function or the tangent to a curve. Here we begin with an informal definition of *limit* and show how we can calculate the values of limits. A precise definition is presented in the next section.

HISTORICAL ESSAY

Limits

Limits of Function Values

Let $f(x)$ be defined on an open interval about x_0, *except possibly at x_0 itself.* If $f(x)$ is arbitrarily close to L (as close to L as we like) for all x sufficiently close to x_0, we say that f approaches the **limit** L as x approaches x_0, and we write

$$\lim_{x \to x_0} f(x) = L,$$

which is read "the limit of $f(x)$ as x approaches x_0 is L." Essentially, the definition says that the values of $f(x)$ are close to the number L whenever x is close to x_0 (on either side of x_0). This definition is "informal" because phrases like *arbitrarily close* and *sufficiently close* are imprecise; their meaning depends on the context. To a machinist manufacturing a piston, *close* may mean *within a few thousandths of an inch*. To an astronomer studying distant galaxies, *close* may mean *within a few thousand light-years*. Nevertheless the definition is clear enough to enable us to recognize and evaluate limits of specific functions. We will need the precise definition of Section 1.5, however, when we set out to prove theorems about limits.

EXAMPLE 1 How does the function

$$f(x) = \frac{x^2 - 1}{x - 1}$$

behave near $x = 1$?

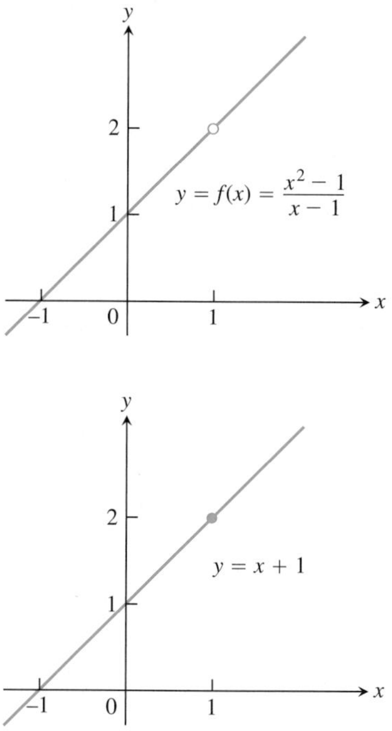

FIGURE 1.34 The graph of f is identical with the line $y = x + 1$ except at $x = 1$, where f is not defined (Example 1).

Solution The given formula defines f for all real numbers x except $x = 1$ (we cannot divide by zero). For any $x \neq 1$, we can simplify the formula by factoring the numerator and canceling common factors:

$$f(x) = \frac{(x - 1)(x + 1)}{x - 1} = x + 1 \quad \text{for} \quad x \neq 1.$$

The graph of f is thus the line $y = x + 1$ with the point (1, 2) *removed*. This removed point is shown as a "hole" in Figure 1.34. Even though $f(1)$ is not defined, it is clear that we can make the value of $f(x)$ *as close as we want* to 2 by choosing x *close enough* to 1 (Table 1.3).

TABLE 1.3 The closer x gets to 1, the closer $f(x) = (x^2 - 1)/(x - 1)$ seems to get to 2

Values of x below and above 1	$f(x) = \frac{x^2 - 1}{x - 1} = x + 1, \quad x \neq 1$
0.9	1.9
1.1	2.1
0.99	1.99
1.01	2.01
0.999	1.999
1.001	2.001
0.999999	1.999999
1.000001	2.000001

We say that $f(x)$ approaches the *limit* 2 as x approaches 1, and write

$$\lim_{x \to 1} f(x) = 2, \qquad \text{or} \qquad \lim_{x \to 1} \frac{x^2 - 1}{x - 1} = 2. \qquad \blacksquare$$

EXAMPLE 2 The function f in Figure 1.35 has limit 2 as $x \to 1$ even though f is not defined at $x = 1$. The function g has limit 2 as $x \to 1$ even though $2 \neq g(1)$. The function h

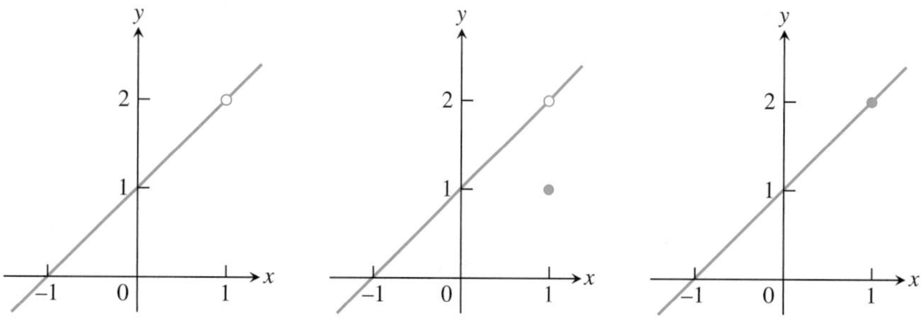

(a) $f(x) = \frac{x^2 - 1}{x - 1}$ (b) $g(x) = \begin{cases} \frac{x^2 - 1}{x - 1}, & x \neq 1 \\ 1, & x = 1 \end{cases}$ (c) $h(x) = x + 1$

FIGURE 1.35 The limits of $f(x)$, $g(x)$, and $h(x)$ all equal 2 as x approaches 1. However, only $h(x)$ has the same function value as its limit at $x = 1$ (Example 2).

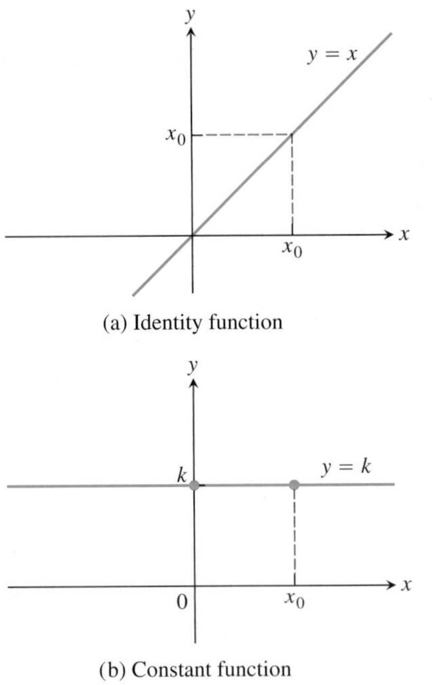

FIGURE 1.36 The functions in Example 3 have limits at all points x_0.

is the only one of the three whose limit as $x \to 1$ equals its value at $x = 1$. For h, we have $\lim_{x\to1} h(x) = h(1)$. This equality of limit and function value has to do with *continuity* of the function, and we return to it in Section 1.7. ■

EXAMPLE 3

(a) If f is the **identity function** $f(x) = x$, then for any value of x_0 (Figure 1.36),

$$\lim_{x\to x_0} f(x) = \lim_{x\to x_0} x = x_0.$$

(b) If f is the **constant function** $f(x) = k$ (function with the constant value k), then for any value of x_0 (Figure 1.36b),

$$\lim_{x\to x_0} f(x) = \lim_{x\to x_0} k = k.$$

For instance,

$$\lim_{x\to 3} x = 3 \quad \text{and} \quad \lim_{x\to -7} (4) = \lim_{x\to 2} (4) = 4.$$

We prove these results in Example 3 in Section 1.5. ■

Some ways that limits can fail to exist are illustrated in Figure 1.37 and described in the next example.

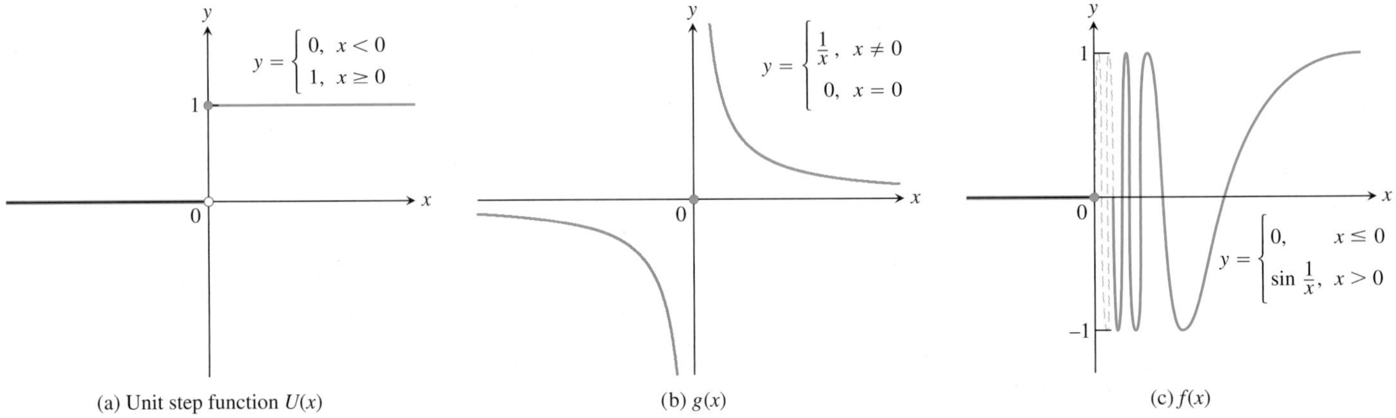

FIGURE 1.37 None of these functions has a limit as x approaches 0 (Example 4).

EXAMPLE 4 Discuss the behavior of the following functions as $x \to 0$.

(a) $U(x) = \begin{cases} 0, & x < 0 \\ 1, & x \ge 0 \end{cases}$

(b) $g(x) = \begin{cases} \dfrac{1}{x}, & x \ne 0 \\ 0, & x = 0 \end{cases}$

(c) $f(x) = \begin{cases} 0, & x \le 0 \\ \sin\dfrac{1}{x}, & x > 0 \end{cases}$

Solution

(a) It *jumps:* **The unit step function** $U(x)$ has no limit as $x \to 0$ because its values jump at $x = 0$. For negative values of x arbitrarily close to zero, $U(x) = 0$. For positive values of x arbitrarily close to zero, $U(x) = 1$. There is no *single* value L approached by $U(x)$ as $x \to 0$ (Figure 1.37a).

(b) It *grows too large to have a limit*: $g(x)$ has no limit as $x \to 0$ because the values of g grow arbitrarily large in absolute value as $x \to 0$ and do not stay close to *any* fixed real number (Figure 1.37b).

(c) It *oscillates too much to have a limit*: $f(x)$ has no limit as $x \to 0$ because the function's values oscillate between $+1$ and -1 in every open interval containing 0. The values do not stay close to any one number as $x \to 0$ (Figure 1.37c). ■

The Limit Laws

The following theorem tells how to calculate limits of functions that are arithmetic combinations of functions whose limits we already know.

THEOREM 1—Limit Laws If L, M, c and k are real numbers and

$$\lim_{x \to c} f(x) = L \quad \text{and} \quad \lim_{x \to c} g(x) = M, \quad \text{then}$$

1. *Sum Rule*: $\lim_{x \to c} (f(x) + g(x)) = L + M$
2. *Difference Rule*: $\lim_{x \to c} (f(x) - g(x)) = L - M$
3. *Constant Multiple Rule:* $\lim_{x \to c} (k \cdot f(x)) = k \cdot L$
4. *Product Rule*: $\lim_{x \to c} (f(x) \cdot g(x)) = L \cdot M$
5. *Quotient Rule*: $\lim_{x \to c} \dfrac{f(x)}{g(x)} = \dfrac{L}{M}, \quad M \neq 0$
6. *Power Rule:* $\lim_{x \to c} [f(x)]^n = L^n$, n a positive integer
7. *Root Rule:* $\lim_{x \to c} \sqrt[n]{f(x)} = \sqrt[n]{L} = L^{1/n}$, n a positive integer

(If n is even, we assume that $\lim_{x \to c} f(x) = L > 0$.)

In words, the Sum Rule says that the limit of a sum is the sum of the limits. Similarly, the next rules say that the limit of a difference is the difference of the limits; the limit of a constant times a function is the constant times the limit of the function; the limit of a product is the product of the limits; the limit of a quotient is the quotient of the limits (provided that the limit of the denominator is not 0); the limit of a positive integer power (or root) of a function is the integer power (or root) of the limit (provided that the root of the limit is a real number).

We can convince ourselves that the properties in Theorem 1 are true (although these intuitive arguments do not constitute proofs). If x is sufficiently close to c, then $f(x)$ is close to L and $g(x)$ is close to M, from our informal definition of a limit. It is then reasonable that $f(x) + g(x)$ is close to $L + M$; $f(x) - g(x)$ is close to $L - M$; $kf(x)$ is close to kL; $f(x)g(x)$ is close to LM; and that $f(x)/g(x)$ is close to L/M if M is not zero. We prove the Sum Rule in Section 1.5, based on a precise definition of limit. Rules 2–5 are proved in Appendix 5. Rule 6 is obtained by applying Rule 4 repeatedly. Rule 7 is proved in more advanced texts.

EXAMPLE 5 Use the observations $\lim_{x\to c} k = k$ and $\lim_{x\to c} x = c$ (Example 3) and the properties of limits to find the following limits.

(a) $\lim\limits_{x\to c}(x^3 + 4x^2 - 3)$ **(b)** $\lim\limits_{x\to c}\dfrac{x^4 + x^2 - 1}{x^2 + 5}$ **(c)** $\lim\limits_{x\to -2}\sqrt{4x^2 - 3}$

Solution

(a)
$$\begin{aligned}\lim_{x\to c}(x^3 + 4x^2 - 3) &= \lim_{x\to c} x^3 + \lim_{x\to c} 4x^2 - \lim_{x\to c} 3 && \text{Sum and Difference Rules}\\ &= c^3 + 4c^2 - 3 && \text{Product and Multiple Rules}\end{aligned}$$

(b)
$$\begin{aligned}\lim_{x\to c}\frac{x^4 + x^2 - 1}{x^2 + 5} &= \frac{\lim\limits_{x\to c}(x^4 + x^2 - 1)}{\lim\limits_{x\to c}(x^2 + 5)} && \text{Quotient Rule}\\ &= \frac{\lim\limits_{x\to c} x^4 + \lim\limits_{x\to c} x^2 - \lim\limits_{x\to c} 1}{\lim\limits_{x\to c} x^2 + \lim\limits_{x\to c} 5} && \text{Sum and Difference Rules}\\ &= \frac{c^4 + c^2 - 1}{c^2 + 5} && \text{Power or Product Rule}\end{aligned}$$

(c)
$$\begin{aligned}\lim_{x\to -2}\sqrt{4x^2 - 3} &= \sqrt{\lim_{x\to -2}(4x^2 - 3)} && \text{Root Rule with } n = 2\\ &= \sqrt{\lim_{x\to -2} 4x^2 - \lim_{x\to -2} 3} && \text{Difference Rule}\\ &= \sqrt{4(-2)^2 - 3} && \text{Product and Multiple Rules}\\ &= \sqrt{16 - 3}\\ &= \sqrt{13}\end{aligned}$$

■

Two consequences of Theorem 1 further simplify the task of calculating limits of polynomials and rational functions. To evaluate the limit of a polynomial function as x approaches c, merely substitute c for x in the formula for the function. To evaluate the limit of a rational function as x approaches a point c *at which the denominator is not zero*, substitute c for x in the formula for the function. (See Examples 5a and 5b.)

THEOREM 2—Limits of Polynomials
If $P(x) = a_n x^n + a_{n-1}x^{n-1} + \cdots + a_0$, then
$$\lim_{x\to c} P(x) = P(c) = a_n c^n + a_{n-1}c^{n-1} + \cdots + a_0.$$

THEOREM 3—Limits of Rational Functions
If $P(x)$ and $Q(x)$ are polynomials and $Q(c) \neq 0$, then
$$\lim_{x\to c}\frac{P(x)}{Q(x)} = \frac{P(c)}{Q(c)}.$$

EXAMPLE 6

$$\lim_{x \to -1} \frac{x^3 + 4x^2 - 3}{x^2 + 5} = \frac{(-1)^3 + 4(-1)^2 - 3}{(-1)^2 + 5} = \frac{0}{6} = 0$$ ■

Eliminating Zero Denominators Algebraically

Identifying Common Factors
It can be shown that if $Q(x)$ is a polynomial and $Q(c) = 0$, then $(x - c)$ is a factor of $Q(x)$. Thus, if the numerator and denominator of a rational function of x are both zero at $x = c$, they have $(x - c)$ as a common factor.

Theorem 3 applies only if the denominator of the rational function is not zero at the limit point c. If the denominator is zero, canceling common factors in the numerator and denominator may reduce the fraction to one whose denominator is no longer zero at c. If this happens, we can find the limit by substitution in the simplified fraction.

EXAMPLE 7 Evaluate

$$\lim_{x \to 1} \frac{x^2 + x - 2}{x^2 - x}.$$

Solution We cannot substitute $x = 1$ because it makes the denominator zero. We test the numerator to see if it, too, is zero at $x = 1$. It is, so it has a factor of $(x - 1)$ in common with the denominator. Canceling the $(x - 1)$'s gives a simpler fraction with the same values as the original for $x \neq 1$:

$$\frac{x^2 + x - 2}{x^2 - x} = \frac{(x - 1)(x + 2)}{x(x - 1)} = \frac{x + 2}{x}, \qquad \text{if } x \neq 1.$$

Using the simpler fraction, we find the limit of these values as $x \to 1$ by substitution:

$$\lim_{x \to 1} \frac{x^2 + x - 2}{x^2 - x} = \lim_{x \to 1} \frac{x + 2}{x} = \frac{1 + 2}{1} = 3.$$

See Figure 1.38. ■

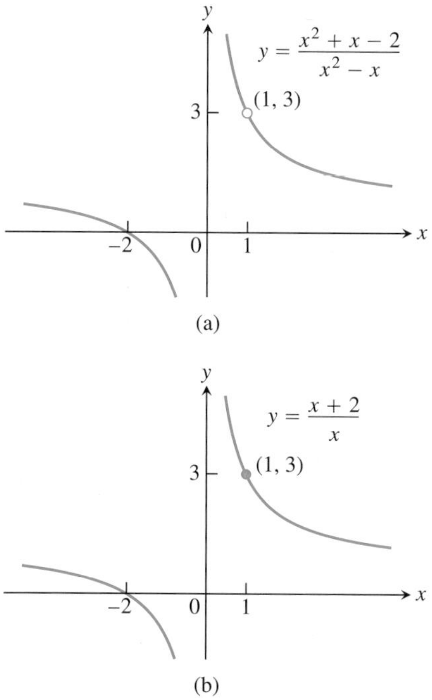

FIGURE 1.38 The graph of $f(x) = (x^2 + x - 2)/(x^2 - x)$ in part (a) is the same as the graph of $g(x) = (x + 2)/x$ in part (b) except at $x = 1$, where f is undefined. The functions have the same limit as $x \to 1$ (Example 7).

EXAMPLE 8 Evaluate

$$\lim_{x \to 0} \frac{\sqrt{x^2 + 100} - 10}{x^2}.$$

Solution We can create a common factor by multiplying both numerator and denominator by the expression $\sqrt{x^2 + 100} + 10$ (obtained by changing the sign after the square root). The preliminary algebra rationalizes the numerator:

$$\begin{aligned}
\frac{\sqrt{x^2 + 100} - 10}{x^2} &= \frac{\sqrt{x^2 + 100} - 10}{x^2} \cdot \frac{\sqrt{x^2 + 100} + 10}{\sqrt{x^2 + 100} + 10} \\
&= \frac{x^2 + 100 - 100}{x^2\left(\sqrt{x^2 + 100} + 10\right)} \\
&= \frac{x^2}{x^2\left(\sqrt{x^2 + 100} + 10\right)} && \text{Common factor } x^2 \\
&= \frac{1}{\sqrt{x^2 + 100} + 10}. && \text{Cancel } x^2 \text{ for } x \neq 0
\end{aligned}$$

Therefore,

$$\lim_{x \to 0} \frac{\sqrt{x^2 + 100} - 10}{x^2} = \lim_{x \to 0} \frac{1}{\sqrt{x^2 + 100} + 10}$$

$$= \frac{1}{\sqrt{0^2 + 100} + 10}$$ Denominator not 0 at $x = 0$; substitute

$$= \frac{1}{20} = 0.05.$$ ■

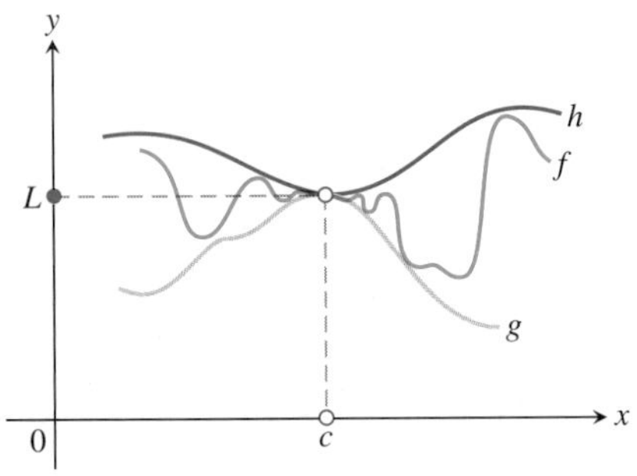

FIGURE 1.39 The graph of f is sandwiched between the graphs of g and h.

The Sandwich Theorem

The following theorem enables us to calculate a variety of limits. It is called the Sandwich Theorem because it refers to a function f whose values are sandwiched between the values of two other functions g and h that have the same limit L at a point c. Being trapped between the values of two functions that approach L, the values of f must also approach L (Figure 1.39). You will find a proof in Appendix 5.

THEOREM 4—The Sandwich Theorem Suppose that $g(x) \le f(x) \le h(x)$ for all x in some open interval containing c, except possibly at $x = c$ itself. Suppose also that

$$\lim_{x \to c} g(x) = \lim_{x \to c} h(x) = L.$$

Then $\lim_{x \to c} f(x) = L$.

The Sandwich Theorem is also called the Squeeze Theorem or the Pinching Theorem.

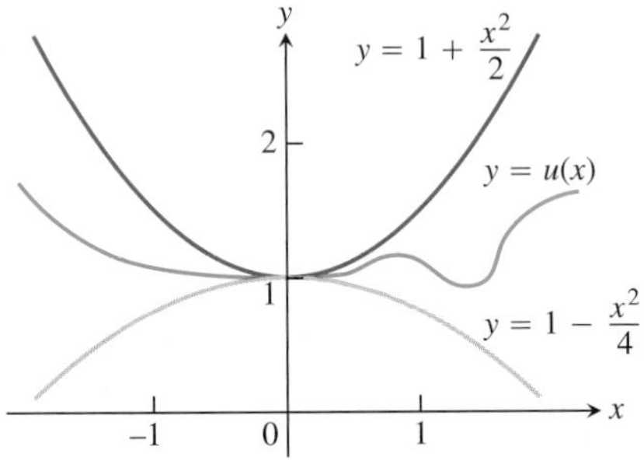

FIGURE 1.40 Any function $u(x)$ whose graph lies in the region between $y = 1 + (x^2/2)$ and $y = 1 - (x^2/4)$ has limit 1 as $x \to 0$ (Example 9).

EXAMPLE 9 Given that

$$1 - \frac{x^2}{4} \le u(x) \le 1 + \frac{x^2}{2} \quad \text{for all } x \ne 0,$$

find $\lim_{x \to 0} u(x)$, no matter how complicated u is.

Solution Since

$$\lim_{x \to 0} (1 - (x^2/4)) = 1 \quad \text{and} \quad \lim_{x \to 0} (1 + (x^2/2)) = 1,$$

the Sandwich Theorem implies that $\lim_{x \to 0} u(x) = 1$ (Figure 1.40). ■

EXAMPLE 10

(a) (See Figure 1.41a.) In Appendix 4 we establish that $-|\theta| \le \sin\theta \le |\theta|$ for all θ. Since $\lim_{\theta \to 0} (-|\theta|) = \lim_{\theta \to 0} |\theta| = 0$, we have

$$\lim_{\theta \to 0} \sin\theta = 0.$$

(b) (See Figure 1.41b.) From Appendix 4, $0 \le 1 - \cos\theta \le |\theta|$ for all θ, and we have $\lim_{\theta \to 0} (1 - \cos\theta) = 0$ or

$$\lim_{\theta \to 0} \cos\theta = 1.$$

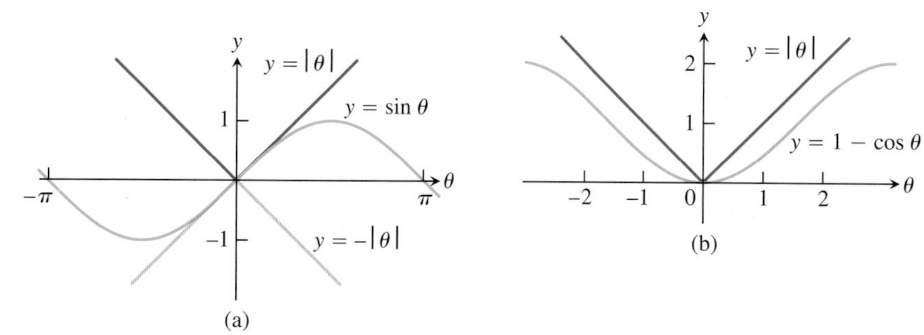

FIGURE 1.41 The Sandwich Theorem shows that (a) $\lim_{\theta \to 0} \sin\theta = 0$ and (b) $\lim_{\theta \to 0}(1 - \cos\theta) = 0$ (Example 10).

(c) For any function $f(x)$, if $\lim_{x \to c} |f(x)| = 0$, then $\lim_{x \to c} f(x) = 0$. The argument: $-|f(x)| \le f(x) \le |f(x)|$ and $-|f(x)|$ and $|f(x)|$ have limit 0 as $x \to c$. ■

Another important property of limits is given by the next theorem. A proof is given in the next section.

THEOREM 5 If $f(x) \le g(x)$ for all x in some open interval containing c, except possibly at $x = c$ itself, and the limits of f and g both exist as x approaches c, then

$$\lim_{x \to c} f(x) \le \lim_{x \to c} g(x).$$

The assertion resulting from replacing the less than or equal to ($\le$) inequality by the strict less than ($<$) inequality in Theorem 5 is false. Figure 1.41a shows that for $\theta \ne 0$, $-|\theta| < \sin\theta < |\theta|$, but in the limit as $\theta \to 0$, equality holds.

EXERCISES 1.4

1. For the function $g(x)$ graphed here, find the following limits or explain why they do not exist.

a. $\lim_{x \to 1} g(x)$ **b.** $\lim_{x \to 2} g(x)$ **c.** $\lim_{x \to 3} g(x)$

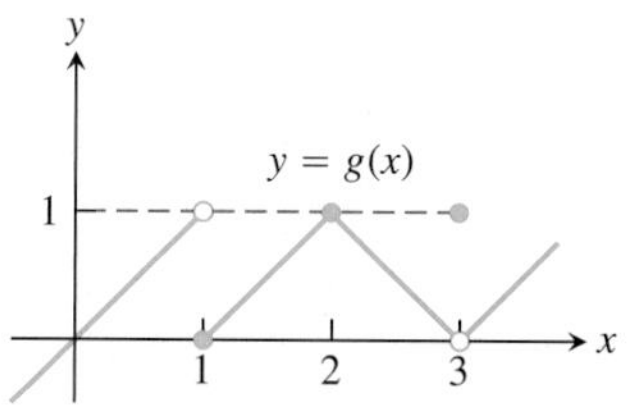

2. For the function $f(t)$ graphed here, find the following limits or explain why they do not exist.

a. $\lim_{t \to -2} f(t)$ **b.** $\lim_{t \to -1} f(t)$ **c.** $\lim_{t \to 0} f(t)$

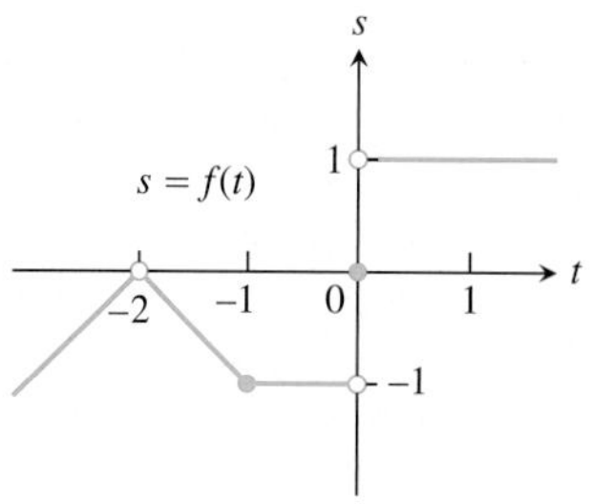

3. Which of the following statements about the function $y = f(x)$ graphed here are true, and which are false?

a. $\lim_{x\to 0} f(x)$ exists.

b. $\lim_{x\to 0} f(x) = 0$.

c. $\lim_{x\to 0} f(x) = 1$.

d. $\lim_{x\to 1} f(x) = 1$.

e. $\lim_{x\to 1} f(x) = 0$.

f. $\lim_{x\to x_0} f(x)$ exists at every point x_0 in $(-1, 1)$.

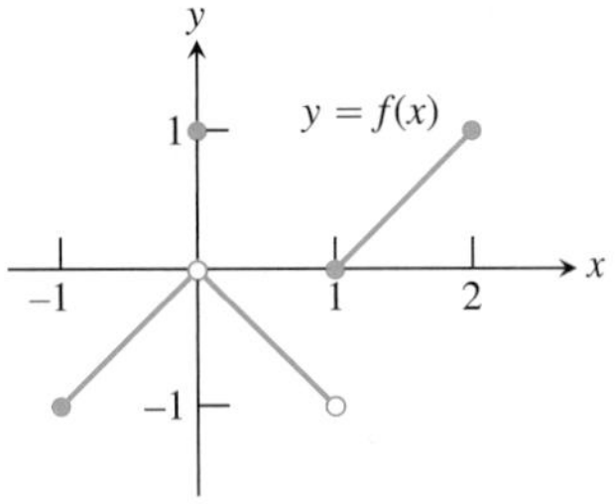

4. Which of the following statements about the function $y = f(x)$ graphed here are true, and which are false?

a. $\lim_{x\to 2} f(x)$ does not exist.

b. $\lim_{x\to 2} f(x) = 2$.

c. $\lim_{x\to 1} f(x)$ does not exist.

d. $\lim_{x\to x_0} f(x)$ exists at every point x_0 in $(-1, 1)$.

e. $\lim_{x\to x_0} f(x)$ exists at every point x_0 in $(1, 3)$.

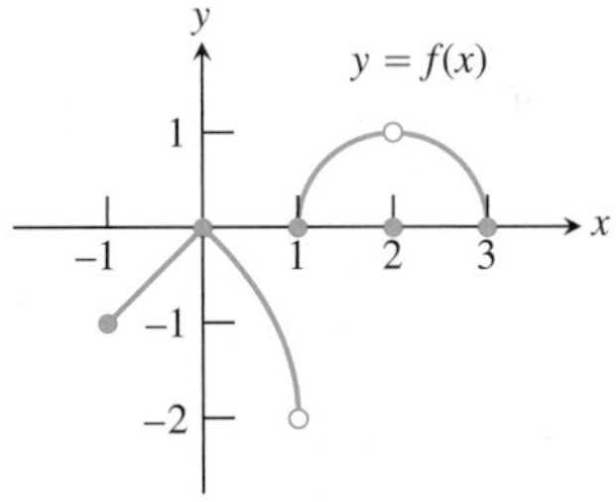

In Exercises 5 and 6, explain why the limits do not exist.

5. $\lim_{x\to 0} \dfrac{x}{|x|}$

6. $\lim_{x\to 1} \dfrac{1}{x-1}$

7. Suppose that a function $f(x)$ is defined for all real values of x except $x = x_0$. Can anything be said about the existence of $\lim_{x\to x_0} f(x)$? Give reasons for your answer.

8. Suppose that a function $f(x)$ is defined for all x in $[-1, 1]$. Can anything be said about the existence of $\lim_{x\to 0} f(x)$? Give reasons for your answer.

9. If $\lim_{x\to 1} f(x) = 5$, must f be defined at $x = 1$? If it is, must $f(1) = 5$? Can we conclude *anything* about the values of f at $x = 1$? Explain.

10. If $f(1) = 5$, must $\lim_{x\to 1} f(x)$ exist? If it does, then must $\lim_{x\to 1} f(x) = 5$? Can we conclude *anything* about $\lim_{x\to 1} f(x)$? Explain.

Find the limits in Exercises 11–28.

11. $\lim_{x\to -7} (2x + 5)$

12. $\lim_{x\to 12} (10 - 3x)$

13. $\lim_{x\to 2} (-x^2 + 5x - 2)$

14. $\lim_{x\to -2} (x^3 - 2x^2 + 4x + 8)$

15. $\lim_{t\to 6} 8(t - 5)(t - 7)$

16. $\lim_{s\to 2/3} 3s(2s - 1)$

17. $\lim_{x\to 2} \dfrac{x+3}{x+6}$

18. $\lim_{x\to 5} \dfrac{4}{x-7}$

19. $\lim_{y\to -5} \dfrac{y^2}{5-y}$

20. $\lim_{y\to 2} \dfrac{y+2}{y^2+5y+6}$

21. $\lim_{x\to -1} 3(2x - 1)^2$

22. $\lim_{x\to -4} (x + 3)^{1984}$

23. $\lim_{y\to -3} (5 - y)^{4/3}$

24. $\lim_{z\to 0} (2z - 8)^{1/3}$

25. $\lim_{h\to 0} \dfrac{3}{\sqrt{3h+1}+1}$

26. $\lim_{h\to 0} \dfrac{5}{\sqrt{5h+4}+2}$

27. $\lim_{h\to 0} \dfrac{\sqrt{3h+1}-1}{h}$

28. $\lim_{h\to 0} \dfrac{\sqrt{5h+4}-2}{h}$

Find the limits in Exercises 29–46.

29. $\lim_{x\to 5} \dfrac{x-5}{x^2-25}$

30. $\lim_{x\to -3} \dfrac{x+3}{x^2+4x+3}$

31. $\lim_{x\to -5} \dfrac{x^2+3x-10}{x+5}$

32. $\lim_{x\to 2} \dfrac{x^2-7x+10}{x-2}$

33. $\lim_{t\to 1} \dfrac{t^2+t-2}{t^2-1}$

34. $\lim_{t\to -1} \dfrac{t^2+3t+2}{t^2-t-2}$

35. $\lim_{x\to -2} \dfrac{-2x-4}{x^3+2x^2}$

36. $\lim_{y\to 0} \dfrac{5y^3+8y^2}{3y^4-16y^2}$

37. $\lim_{u\to 1} \dfrac{u^4-1}{u^3-1}$

38. $\lim_{v\to 2} \dfrac{v^3-8}{v^4-16}$

39. $\lim_{x\to 9} \dfrac{\sqrt{x}-3}{x-9}$

40. $\lim_{x\to 4} \dfrac{4x-x^2}{2-\sqrt{x}}$

41. $\lim_{x\to 1} \dfrac{x-1}{\sqrt{x+3}-2}$

42. $\lim_{x\to -1} \dfrac{\sqrt{x^2+8}-3}{x+1}$

43. $\lim_{x\to 2} \dfrac{\sqrt{x^2+12}-4}{x-2}$

44. $\lim_{x\to -2} \dfrac{x+2}{\sqrt{x^2+5}-3}$

45. $\lim_{x\to -3} \dfrac{2-\sqrt{x^2-5}}{x+3}$

46. $\lim_{x\to 4} \dfrac{4-x}{5-\sqrt{x^2+9}}$

Find the limits in Exercises 47–54.

47. $\lim_{x\to 0} (2\sin x - 1)$

48. $\lim_{x\to 0} \sin^2 x$

49. $\lim_{x\to 0} \sec x$

50. $\lim_{x\to 0} \tan x$

51. $\lim_{x\to 0} \dfrac{1+x+\sin x}{3\cos x}$

52. $\lim_{x\to 0} (x^2 - 1)(2 - \cos x)$

53. $\lim_{x\to 0} \sqrt{x+1}\,\cos^{1/3} x$

54. $\lim_{x\to 0} \sqrt{1+\cos^2 x}$

55. Suppose $\lim_{x\to c} f(x) = 5$ and $\lim_{x\to c} g(x) = -2$. Find

a. $\lim_{x\to c} f(x)g(x)$

b. $\lim_{x\to c} 2f(x)g(x)$

c. $\lim_{x\to c}(f(x) + 3g(x))$ d. $\lim_{x\to c}\dfrac{f(x)}{f(x) - g(x)}$

56. Suppose $\lim_{x\to 4} f(x) = 0$ and $\lim_{x\to 4} g(x) = -3$. Find

a. $\lim_{x\to 4}(g(x) + 3)$ b. $\lim_{x\to 4} xf(x)$

c. $\lim_{x\to 4}(g(x))^2$ d. $\lim_{x\to 4}\dfrac{g(x)}{f(x) - 1}$

Because of their connection with secant lines, tangents, and instantaneous rates, limits of the form

$$\lim_{h\to 0}\frac{f(x + h) - f(x)}{h}$$

occur frequently in calculus. In Exercises 57–62, evaluate this limit for the given value of x and function f.

57. $f(x) = x^2,\quad x = 1$ **58.** $f(x) = x^2,\quad x = -2$

59. $f(x) = 3x - 4,\quad x = 2$ **60.** $f(x) = 1/x,\quad x = -2$

61. $f(x) = \sqrt{x},\quad x = 7$ **62.** $f(x) = \sqrt{3x + 1},\quad x = 0$

63. If $\sqrt{5 - 2x^2} \le f(x) \le \sqrt{5 - x^2}$ for $-1 \le x \le 1$, find $\lim_{x\to 0} f(x)$.

64. If $2 - x^2 \le g(x) \le 2\cos x$ for all x, find $\lim_{x\to 0} g(x)$.

65. **a.** It can be shown that the inequalities

$$1 - \frac{x^2}{6} < \frac{x\sin x}{2 - 2\cos x} < 1$$

hold for all values of x close to zero. What, if anything, does this tell you about

$$\lim_{x\to 0}\frac{x\sin x}{2 - 2\cos x}?$$

Give reasons for your answer.

T **b.** Graph $y = 1 - (x^2/6)$, $y = (x\sin x)/(2 - 2\cos x)$, and $y = 1$ together for $-2 \le x \le 2$. Comment on the behavior of the graphs as $x \to 0$.

66. **a.** Suppose that the inequalities

$$\frac{1}{2} - \frac{x^2}{24} < \frac{1 - \cos x}{x^2} < \frac{1}{2}$$

hold for values of x close to zero. (They do, as you will see in Section 7.9.) What, if anything, does this tell you about

$$\lim_{x\to 0}\frac{1 - \cos x}{x^2}?$$

Give reasons for your answer.

T **b.** Graph the equations $y = (1/2) - (x^2/24)$, $y = (1 - \cos x)/x^2$, and $y = 1/2$ together for $-2 \le x \le 2$. Comment on the behavior of the graphs as $x \to 0$.

T You will find a graphing calculator useful for Exercises 67–72.

67. Let $f(x) = (x^2 - 9)/(x + 3)$.

a. Make a table of the values of f at the points $x = -3.1$, -3.01, -3.001, and so on as far as your calculator can go. Then estimate $\lim_{x\to -3} f(x)$. What estimate do you arrive at if you evaluate f at $x = -2.9, -2.99, -2.999, \ldots$ instead?

b. Support your conclusions in part (a) by graphing f near $x_0 = -3$ and using Zoom and Trace to estimate y-values on the graph as $x \to -3$.

c. Find $\lim_{x\to -3} f(x)$ algebraically, as in Example 7.

68. Let $G(x) = (x + 6)/(x^2 + 4x - 12)$.

a. Make a table of the values of G at $x = -5.9, -5.99, -5.999$, and so on. Then estimate $\lim_{x\to -6} G(x)$. What estimate do you arrive at if you evaluate G at $x = -6.1, -6.01, -6.001, \ldots$ instead?

b. Support your conclusions in part (a) by graphing G and using Zoom and Trace to estimate y-values on the graph as $x \to -6$.

c. Find $\lim_{x\to -6} G(x)$ algebraically.

69. Let $f(x) = (x^2 - 1)/(|x| - 1)$.

a. Make tables of the values of f at values of x that approach $x_0 = -1$ from above and below. Then estimate $\lim_{x\to -1} f(x)$.

b. Support your conclusion in part (a) by graphing f near $x_0 = -1$ and using Zoom and Trace to estimate y-values on the graph as $x \to -1$.

c. Find $\lim_{x\to -1} f(x)$ algebraically.

70. Let $F(x) = (x^2 + 3x + 2)/(2 - |x|)$.

a. Make tables of values of F at values of x that approach $x_0 = -2$ from above and below. Then estimate $\lim_{x\to -2} F(x)$.

b. Support your conclusion in part (a) by graphing F near $x_0 = -2$ and using Zoom and Trace to estimate y-values on the graph as $x \to -2$.

c. Find $\lim_{x\to -2} F(x)$ algebraically.

71. Let $g(\theta) = (\sin\theta)/\theta$.

a. Make a table of the values of g at values of θ that approach $\theta_0 = 0$ from above and below. Then estimate $\lim_{\theta\to 0} g(\theta)$.

b. Support your conclusion in part (a) by graphing g near $\theta_0 = 0$.

72. Let $G(t) = (1 - \cos t)/t^2$.

a. Make tables of values of G at values of t that approach $t_0 = 0$ from above and below. Then estimate $\lim_{t\to 0} G(t)$.

b. Support your conclusion in part (a) by graphing G near $t_0 = 0$.

73. If $x^4 \le f(x) \le x^2$ for x in $[-1, 1]$ and $x^2 \le f(x) \le x^4$ for $x < -1$ and $x > 1$, at what points c do you automatically know $\lim_{x\to c} f(x)$? What can you say about the value of the limit at these points?

74. Suppose that $g(x) \le f(x) \le h(x)$ for all $x \ne 2$ and suppose that

$$\lim_{x\to 2} g(x) = \lim_{x\to 2} h(x) = -5.$$

Can we conclude anything about the values of f, g, and h at $x = 2$? Could $f(2) = 0$? Could $\lim_{x\to 2} f(x) = 0$? Give reasons for your answers.

75. **a.** If $\lim_{x\to 2} \frac{f(x) - 5}{x - 2} = 3$, find $\lim_{x\to 2} f(x)$.

b. If $\lim_{x\to 2} \frac{f(x) - 5}{x - 2} = 4$, find $\lim_{x\to 2} f(x)$.

76. If $\lim_{x\to -2} \frac{f(x)}{x^2} = 1$, find

a. $\lim_{x\to -2} f(x)$ **b.** $\lim_{x\to -2} \frac{f(x)}{x}$

T **77.** **a.** Graph $g(x) = x \sin(1/x)$ to estimate $\lim_{x\to 0} g(x)$, zooming in on the origin as necessary.

b. Confirm your estimate in part (a) with a proof.

T **78.** **a.** Graph $h(x) = x^2 \cos(1/x^3)$ to estimate $\lim_{x\to 0} h(x)$, zooming in on the origin as necessary.

b. Confirm your estimate in part (a) with a proof.

COMPUTER EXPLORATIONS

In Exercises 79–84, use a CAS to perform the following steps:

a. Plot the function near the point x_0 being approached.

b. From your plot guess the value of the limit.

79. $\lim_{x\to 2} \frac{x^4 - 16}{x - 2}$

80. $\lim_{x\to -1} \frac{x^3 - x^2 - 5x - 3}{(x + 1)^2}$

81. $\lim_{x\to 0} \frac{\sqrt[3]{1 + x} - 1}{x}$

82. $\lim_{x\to 3} \frac{x^2 - 9}{\sqrt{x^2 + 7} - 4}$

83. $\lim_{x\to 0} \frac{1 - \cos x}{x \sin x}$

84. $\lim_{x\to 0} \frac{2x^2}{3 - 3\cos x}$

1.5 The Precise Definition of a Limit

We now turn our attention to the precise definition of a limit. We replace vague phrases like "gets arbitrarily close to" in the informal definition with specific conditions that can be applied to any particular example. With a precise definition, we can prove the limit properties given in the preceding section and establish many important limits.

To show that the limit of $f(x)$ as $x \to x_0$ equals the number L, we need to show that the gap between $f(x)$ and L can be made "as small as we choose" if x is kept "close enough" to x_0. Let us see what this would require if we specified the size of the gap between $f(x)$ and L.

EXAMPLE 1 Consider the function $y = 2x - 1$ near $x_0 = 4$. Intuitively it appears that y is close to 7 when x is close to 4, so $\lim_{x\to 4}(2x - 1) = 7$. However, how close to $x_0 = 4$ does x have to be so that $y = 2x - 1$ differs from 7 by, say, less than 2 units?

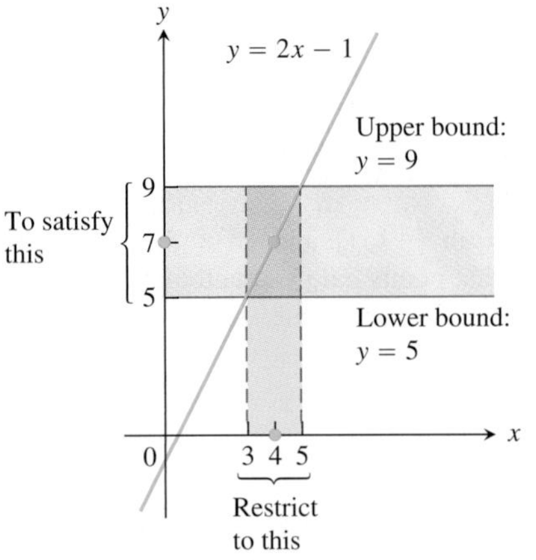

FIGURE 1.42 Keeping x within 1 unit of $x_0 = 4$ will keep y within 2 units of $y_0 = 7$ (Example 1).

Solution We are asked: For what values of x is $|y - 7| < 2$? To find the answer we first express $|y - 7|$ in terms of x:

$$|y - 7| = |(2x - 1) - 7| = |2x - 8|.$$

The question then becomes: what values of x satisfy the inequality $|2x - 8| < 2$? To find out, we solve the inequality:

$$\begin{aligned} |2x - 8| &< 2 \\ -2 < 2x - 8 &< 2 \\ 6 < 2x &< 10 \\ 3 < x &< 5 \\ -1 < x - 4 &< 1. \end{aligned}$$

Keeping x within 1 unit of $x_0 = 4$ will keep y within 2 units of $y_0 = 7$ (Figure 1.42). ■

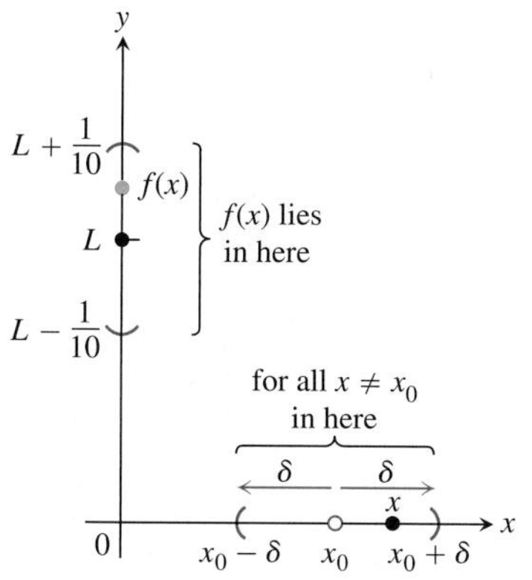

FIGURE 1.43 How should we define $\delta > 0$ so that keeping x within the interval $(x_0 - \delta, x_0 + \delta)$ will keep $f(x)$ within the interval $\left(L - \frac{1}{10}, L + \frac{1}{10}\right)$?

In the previous example we determined how close x must be to a particular value x_0 to ensure that the outputs $f(x)$ of some function lie within a prescribed interval about a limit value L. To show that the limit of $f(x)$ as $x \to x_0$ actually equals L, we must be able to show that the gap between $f(x)$ and L can be made less than *any prescribed error*, no matter how small, by holding x close enough to x_0.

Definition of Limit

Suppose we are watching the values of a function $f(x)$ as x approaches x_0 (without taking on the value of x_0 itself). Certainly we want to be able to say that $f(x)$ stays within one-tenth of a unit from L as soon as x stays within some distance δ of x_0 (Figure 1.43). But that in itself is not enough, because as x continues on its course toward x_0, what is to prevent $f(x)$ from jittering about within the interval from $L - (1/10)$ to $L + (1/10)$ without tending toward L?

We can be told that the error can be no more than $1/100$ or $1/1000$ or $1/100{,}000$. Each time, we find a new δ-interval about x_0 so that keeping x within that interval satisfies the new error tolerance. And each time the possibility exists that $f(x)$ jitters away from L at some stage.

The figures on the next page illustrate the problem. You can think of this as a quarrel between a skeptic and a scholar. The skeptic presents ϵ-challenges to prove that the limit does not exist or, more precisely, that there is room for doubt. The scholar answers every challenge with a δ-interval around x_0 that keeps the function values within ϵ of L.

How do we stop this seemingly endless series of challenges and responses? By proving that for every error tolerance ϵ that the challenger can produce, we can find, calculate, or conjure a matching distance δ that keeps x "close enough" to x_0 to keep $f(x)$ within that tolerance of L (Figure 1.44). This leads us to the precise definition of a limit.

DEFINITION Let $f(x)$ be defined on an open interval about x_0, except possibly at x_0 itself. We say that the **limit of $f(x)$ as x approaches x_0 is the number L**, and write

$$\lim_{x \to x_0} f(x) = L,$$

if, for every number $\epsilon > 0$, there exists a corresponding number $\delta > 0$ such that for all x,

$$0 < |x - x_0| < \delta \quad \Rightarrow \quad |f(x) - L| < \epsilon.$$

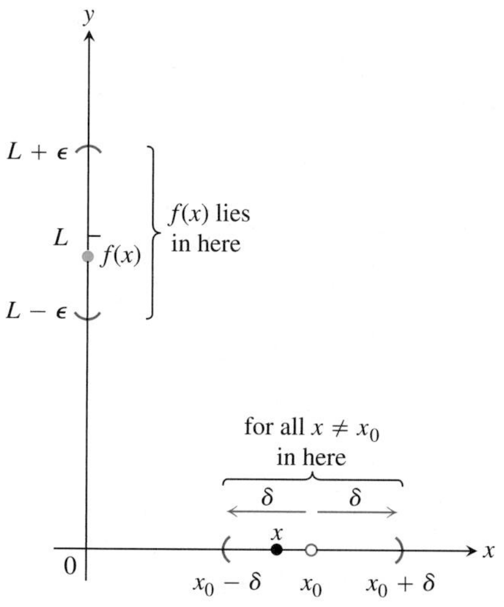

FIGURE 1.44 The relation of δ and ϵ in the definition of limit.

One way to think about the definition is to suppose we are machining a generator shaft to a close tolerance. We may try for diameter L, but since nothing is perfect, we must be satisfied with a diameter $f(x)$ somewhere between $L - \epsilon$ and $L + \epsilon$. The δ is the measure of how accurate our control setting for x must be to guarantee this degree of accuracy in the diameter of the shaft. Notice that as the tolerance for error becomes stricter, we may have to adjust δ. That is, the value of δ, how tight our control setting must be, depends on the value of ϵ, the error tolerance.

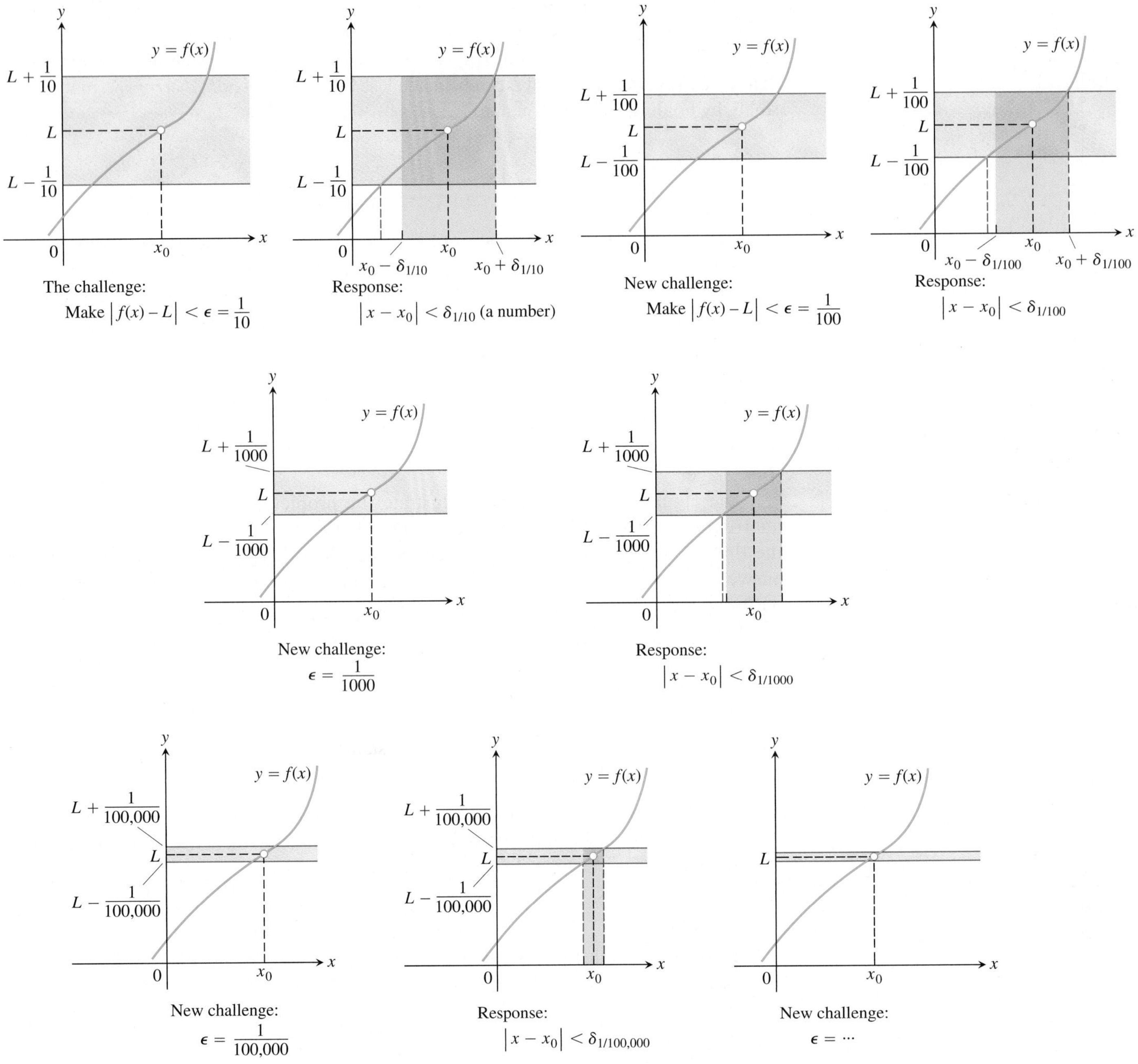

Examples: Testing the Definition

The formal definition of limit does not tell how to find the limit of a function, but it enables us to verify that a suspected limit is correct. The following examples show how the definition can be used to verify limit statements for specific functions. However, the real purpose of the definition is not to do calculations like this, but rather to prove general theorems so that the calculation of specific limits can be simplified.

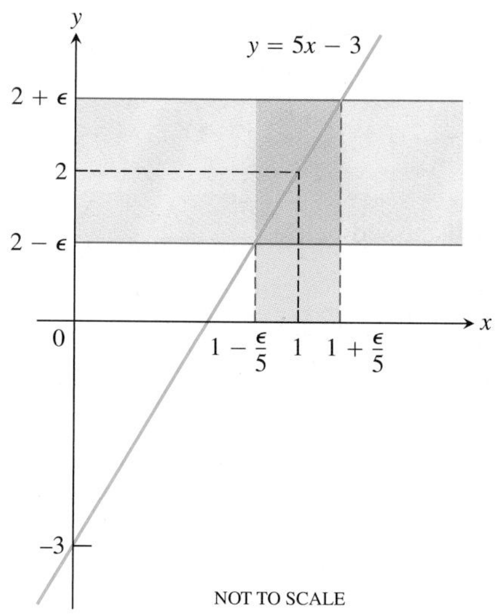

FIGURE 1.45 If $f(x) = 5x - 3$, then $0 < |x - 1| < \epsilon/5$ guarantees that $|f(x) - 2| < \epsilon$ (Example 2).

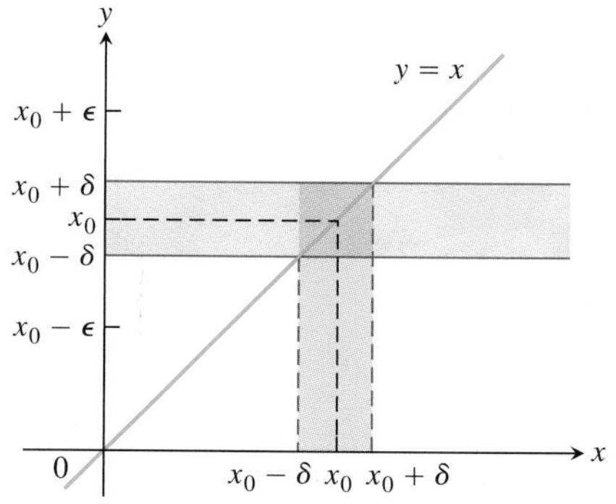

FIGURE 1.46 For the function $f(x) = x$, we find that $0 < |x - x_0| < \delta$ will guarantee $|f(x) - x_0| < \epsilon$ whenever $\delta \le \epsilon$ (Example 3a).

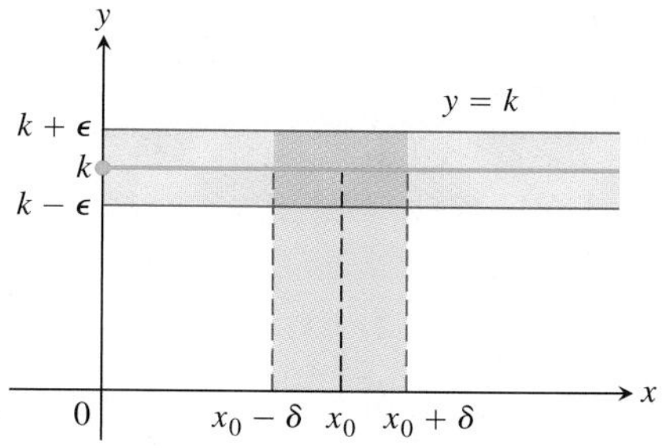

FIGURE 1.47 For the function $f(x) = k$, we find that $|f(x) - k| < \epsilon$ for any positive δ (Example 3b).

EXAMPLE 2 Show that

$$\lim_{x \to 1} (5x - 3) = 2.$$

Solution Set $x_0 = 1$, $f(x) = 5x - 3$, and $L = 2$ in the definition of limit. For any given $\epsilon > 0$, we have to find a suitable $\delta > 0$ so that if $x \neq 1$ and x is within distance δ of $x_0 = 1$, that is, whenever

$$0 < |x - 1| < \delta,$$

it is true that $f(x)$ is within distance ϵ of $L = 2$, so

$$|f(x) - 2| < \epsilon.$$

We find δ by working backward from the ϵ-inequality:

$$\begin{aligned} |(5x - 3) - 2| = |5x - 5| &< \epsilon \\ 5|x - 1| &< \epsilon \\ |x - 1| &< \epsilon/5. \end{aligned}$$

Thus, we can take $\delta = \epsilon/5$ (Figure 1.45). If $0 < |x - 1| < \delta = \epsilon/5$, then

$$|(5x - 3) - 2| = |5x - 5| = 5|x - 1| < 5(\epsilon/5) = \epsilon,$$

which proves that $\lim_{x \to 1}(5x - 3) = 2$.

The value of $\delta = \epsilon/5$ is not the only value that will make $0 < |x - 1| < \delta$ imply $|5x - 5| < \epsilon$. Any smaller positive δ will do as well. The definition does not ask for a "best" positive δ, just one that will work. ■

EXAMPLE 3 Prove:

(a) $\lim_{x \to x_0} x = x_0$ **(b)** $\lim_{x \to x_0} k = k$ (k constant).

Solution

(a) Let $\epsilon > 0$ be given. We must find $\delta > 0$ such that for all x

$$0 < |x - x_0| < \delta \quad \text{implies} \quad |x - x_0| < \epsilon.$$

The implication will hold if δ equals ϵ or any smaller positive number (Figure 1.46). This proves that $\lim_{x \to x_0} x = x_0$.

(b) Let $\epsilon > 0$ be given. We must find $\delta > 0$ such that for all x

$$0 < |x - x_0| < \delta \quad \text{implies} \quad |k - k| < \epsilon.$$

Since $k - k = 0$, we can use any positive number for δ and the implication will hold (Figure 1.47). This proves that $\lim_{x \to x_0} k = k$. ■

Finding Deltas Algebraically for Given Epsilons

In Examples 2 and 3, the interval of values about x_0 for which $|f(x) - L|$ was less than ϵ was symmetric about x_0 and we could take δ to be half the length of that interval. When such symmetry is absent, as it usually is, we can take δ to be the distance from x_0 to the interval's *nearer* endpoint.

EXAMPLE 4 For the limit $\lim_{x\to 5}\sqrt{x-1} = 2$, find a $\delta > 0$ that works for $\epsilon = 1$. That is, find a $\delta > 0$ such that for all x

$$0 < |x - 5| < \delta \quad \Rightarrow \quad |\sqrt{x-1} - 2| < 1.$$

Solution We organize the search into two steps, as discussed below.

1. *Solve the inequality* $|\sqrt{x-1} - 2| < 1$ *to find an interval containing* $x_0 = 5$ *on which the inequality holds for all* $x \neq x_0$.

$$\begin{aligned} |\sqrt{x-1} - 2| &< 1 \\ -1 < \sqrt{x-1} - 2 &< 1 \\ 1 < \sqrt{x-1} &< 3 \\ 1 < x - 1 &< 9 \\ 2 < x &< 10 \end{aligned}$$

The inequality holds for all x in the open interval (2, 10), so it holds for all $x \neq 5$ in this interval as well.

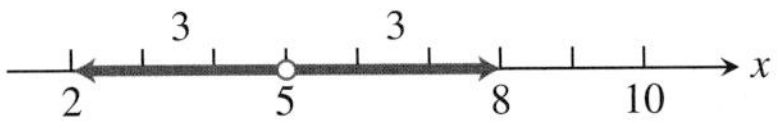

FIGURE 1.48 An open interval of radius 3 about $x_0 = 5$ will lie inside the open interval (2, 10).

2. *Find a value of* $\delta > 0$ *to place the centered interval* $5 - \delta < x < 5 + \delta$ (centered at $x_0 = 5$) *inside the interval* (2, 10). The distance from 5 to the nearer endpoint of (2, 10) is 3 (Figure 1.48). If we take $\delta = 3$ or any smaller positive number, then the inequality $0 < |x - 5| < \delta$ will automatically place x between 2 and 10 to make $|\sqrt{x-1} - 2| < 1$ (Figure 1.49), so

$$0 < |x - 5| < 3 \quad \Rightarrow \quad |\sqrt{x-1} - 2| < 1.$$ ■

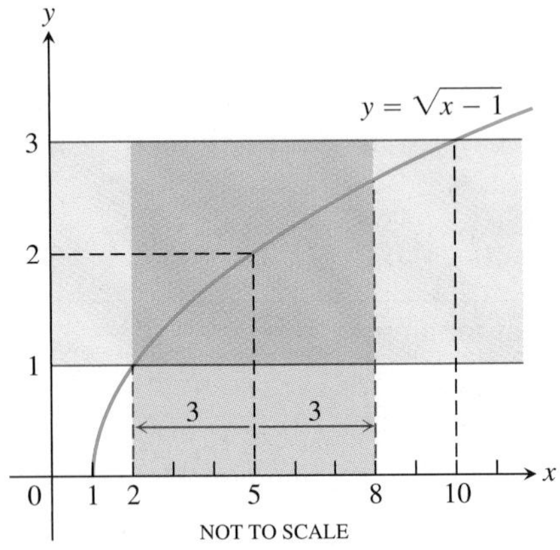

FIGURE 1.49 The function and intervals in Example 4.

Using the Definition to Prove Theorems

We do not usually rely on the formal definition of limit to verify specific limits such as those in the preceding examples. Rather we appeal to general theorems about limits, in particular the theorems of Section 1.4. The definition is used to prove these theorems (Appendix 5). As an example, we prove part 1 of Theorem 1, the Sum Rule.

EXAMPLE 5 Given that $\lim_{x\to c} f(x) = L$ and $\lim_{x\to c} g(x) = M$, prove that

$$\lim_{x\to c}(f(x) + g(x)) = L + M.$$

Solution Let $\epsilon > 0$ be given. We want to find a positive number δ such that for all x

$$0 < |x - c| < \delta \quad \Rightarrow \quad |f(x) + g(x) - (L + M)| < \epsilon.$$

Regrouping terms, we get

$$\begin{aligned} |f(x) + g(x) - (L + M)| &= |(f(x) - L) + (g(x) - M)| \\ &\leq |f(x) - L| + |g(x) - M|. \end{aligned}$$

Triangle Inequality: $|a + b| \leq |a| + |b|$

Since $\lim_{x\to c} f(x) = L$, there exists a number $\delta_1 > 0$ such that for all x

$$0 < |x - c| < \delta_1 \quad \Rightarrow \quad |f(x) - L| < \epsilon/2.$$

Similarly, since $\lim_{x \to c} g(x) = M$, there exists a number $\delta_2 > 0$ such that for all x

$$0 < |x - c| < \delta_2 \qquad \Rightarrow \qquad |g(x) - M| < \epsilon/2.$$

Let $\delta = \min\{\delta_1, \delta_2\}$, the smaller of δ_1 and δ_2. If $0 < |x - c| < \delta$ then $|x - c| < \delta_1$, so $|f(x) - L| < \epsilon/2$, and $|x - c| < \delta_2$, so $|g(x) - M| < \epsilon/2$. Therefore

$$|f(x) + g(x) - (L + M)| < \frac{\epsilon}{2} + \frac{\epsilon}{2} = \epsilon.$$

This shows that $\lim_{x \to c}(f(x) + g(x)) = L + M$. ■

Let's also prove Theorem 5 of Section 1.4.

EXAMPLE 6 Given that $\lim_{x \to c} f(x) = L$ and $\lim_{x \to c} g(x) = M$, and that $f(x) \le g(x)$ for all x in an open interval containing c (except possibly c itself), prove that $L \le M$.

Solution We use the method of proof by contradiction. Suppose, on the contrary, that $L > M$. Then by the limit of a difference property in Theorem 1,

$$\lim_{x \to c} (g(x) - f(x)) = M - L.$$

Therefore, for any $\epsilon > 0$, there exists $\delta > 0$ such that

$$|(g(x) - f(x)) - (M - L)| < \epsilon \qquad \text{whenever} \quad 0 < |x - c| < \delta.$$

Since $L - M > 0$ by hypothesis, we take $\epsilon = L - M$ in particular and we have a number $\delta > 0$ such that

$$|(g(x) - f(x)) - (M - L)| < L - M \qquad \text{whenever} \quad 0 < |x - c| < \delta.$$

Since $a \le |a|$ for any number a, we have

$$(g(x) - f(x)) - (M - L) < L - M \qquad \text{whenever} \quad 0 < |x - c| < \delta$$

which simplifies to

$$g(x) < f(x) \qquad \text{whenever} \quad 0 < |x - c| < \delta.$$

But this contradicts $f(x) \le g(x)$. Thus the inequality $L > M$ must be false. Therefore $L \le M$. ■

EXERCISES 1.5

In Exercises 1–6, sketch the interval (a, b) on the x-axis with the point x_0 inside. Then find a value of $\delta > 0$ such that for all x, $0 < |x - x_0| < \delta \Rightarrow a < x < b$.

1. $a = 1, \quad b = 7, \quad x_0 = 5$

2. $a = 1, \quad b = 7, \quad x_0 = 2$

3. $a = -7/2, \quad b = -1/2, \quad x_0 = -3$

4. $a = -7/2, \quad b = -1/2, \quad x_0 = -3/2$

5. $a = 4/9, \quad b = 4/7, \quad x_0 = 1/2$

6. $a = 2.7591, \quad b = 3.2391, \quad x_0 = 3$

In Exercises 7–10, use the graphs to find a $\delta > 0$ such that for all x

$$0 < |x - x_0| < \delta \quad \Rightarrow \quad |f(x) - L| < \epsilon.$$

7.

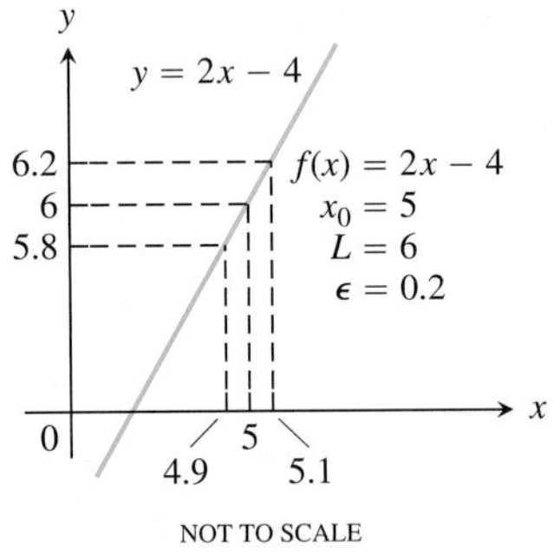

8.

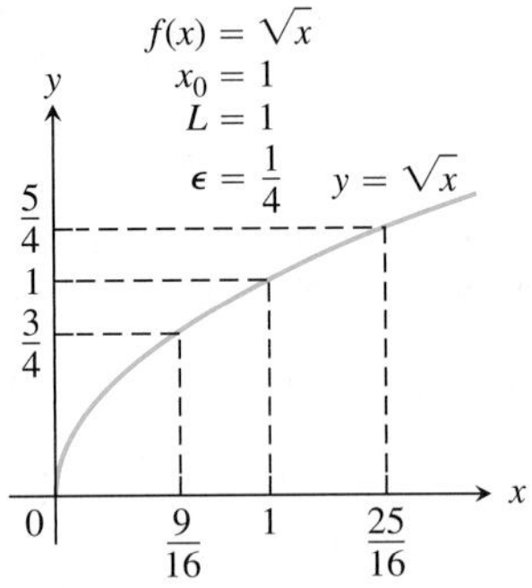

9.

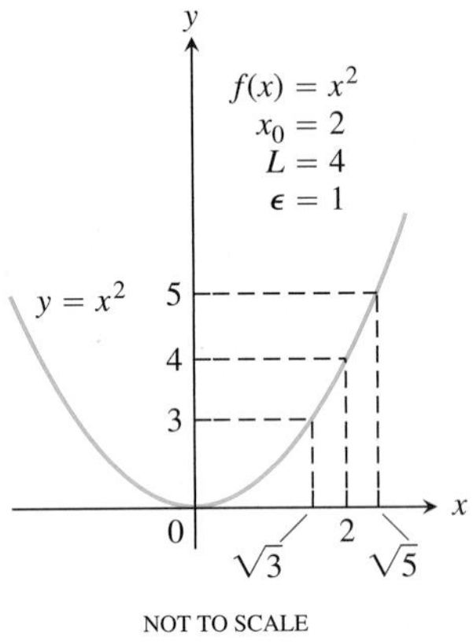

10.

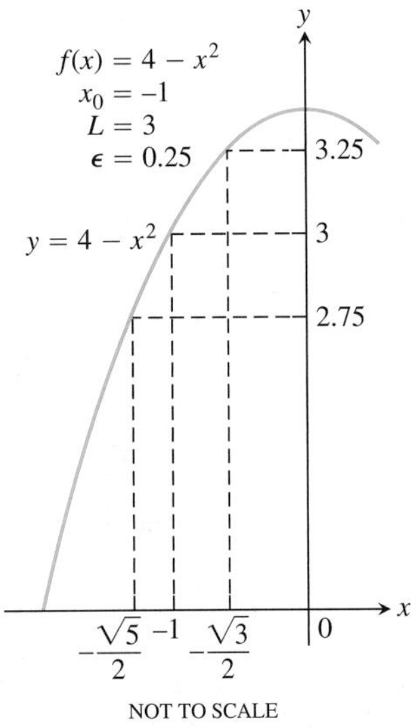

Each of Exercises 11–22 gives a function $f(x)$ and numbers L, x_0, and $\epsilon > 0$. In each case, find an open interval about x_0 on which the inequality $|f(x) - L| < \epsilon$ holds. Then give a value for $\delta > 0$ such that for all x satisfying $0 < |x - x_0| < \delta$ the inequality $|f(x) - L| < \epsilon$ holds.

11. $f(x) = x + 1, \quad L = 5, \quad x_0 = 4, \quad \epsilon = 0.01$

12. $f(x) = 2x - 2, \quad L = -6, \quad x_0 = -2, \quad \epsilon = 0.02$

13. $f(x) = \sqrt{x + 1}, \quad L = 1, \quad x_0 = 0, \quad \epsilon = 0.1$

14. $f(x) = \sqrt{x}, \quad L = 1/2, \quad x_0 = 1/4, \quad \epsilon = 0.1$

15. $f(x) = \sqrt{19 - x}, \quad L = 3, \quad x_0 = 10, \quad \epsilon = 1$

16. $f(x) = \sqrt{x - 7}, \quad L = 4, \quad x_0 = 23, \quad \epsilon = 1$

17. $f(x) = 1/x, \quad L = 1/4, \quad x_0 = 4, \quad \epsilon = 0.05$

18. $f(x) = x^2, \quad L = 3, \quad x_0 = \sqrt{3}, \quad \epsilon = 0.1$

19. $f(x) = x^2, \quad L = 4, \quad x_0 = -2, \quad \epsilon = 0.5$

20. $f(x) = 1/x, \quad L = -1, \quad x_0 = -1, \quad \epsilon = 0.1$

21. $f(x) = x^2 - 5, \quad L = 11, \quad x_0 = 4, \quad \epsilon = 1$

22. $f(x) = mx + b, \quad m > 0, \quad L = m + b, \quad x_0 = 1, \quad \epsilon = 0.05$

Each of Exercises 23–26 gives a function $f(x)$, a point x_0, and a positive number ϵ. Find $L = \lim_{x \to x_0} f(x)$. Then find a number $\delta > 0$ such that for all x

$$0 < |x - x_0| < \delta \quad \Rightarrow \quad |f(x) - L| < \epsilon.$$

23. $f(x) = 3 - 2x, \quad x_0 = 3, \quad \epsilon = 0.02$

24. $f(x) = -3x - 2, \quad x_0 = -1, \quad \epsilon = 0.03$

25. $f(x) = \dfrac{x^2 - 4}{x - 2}, \quad x_0 = 2, \quad \epsilon = 0.05$

26. $f(x) = \sqrt{1 - 5x}, \quad x_0 = -3, \quad \epsilon = 0.5$

Prove the limit statements in Exercises 27–38.

27. $\lim_{x \to 4} (9 - x) = 5$

28. $\lim_{x \to 3} (3x - 7) = 2$

29. $\lim_{x \to 9} \sqrt{x - 5} = 2$

30. $\lim_{x \to 0} \sqrt{4 - x} = 2$

31. $\lim_{x \to 1} f(x) = 1 \quad \text{if} \quad f(x) = \begin{cases} x^2, & x \neq 1 \\ 2, & x = 1 \end{cases}$

32. $\lim_{x \to -2} f(x) = 4 \quad \text{if} \quad f(x) = \begin{cases} x^2, & x \neq -2 \\ 1, & x = -2 \end{cases}$

33. $\lim_{x \to 1} \dfrac{1}{x} = 1$

34. $\lim_{x \to \sqrt{3}} \dfrac{1}{x^2} = \dfrac{1}{3}$

35. $\lim_{x \to -3} \dfrac{x^2 - 9}{x + 3} = -6$

36. $\lim_{x \to 1} \dfrac{x^2 - 1}{x - 1} = 2$

37. $\lim_{x \to 0} x \sin \dfrac{1}{x} = 0$

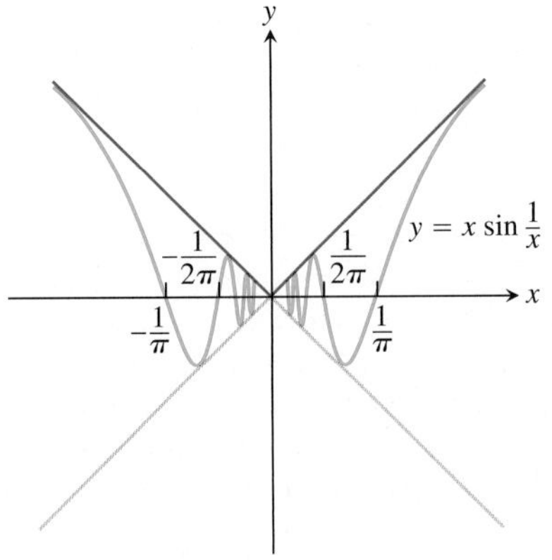

38. $\lim_{x \to 0} x^2 \sin \frac{1}{x} = 0$

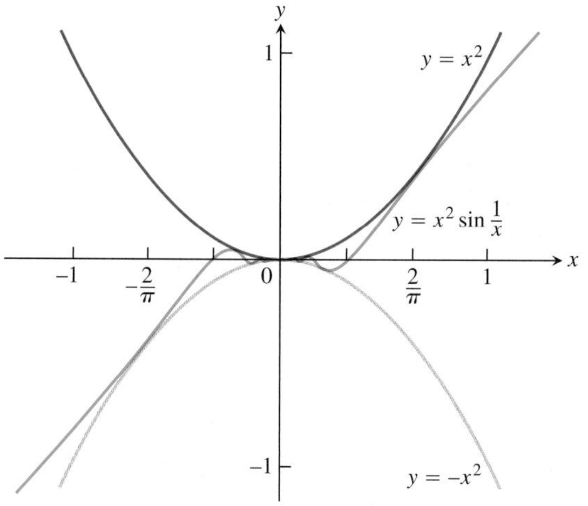

39. Define what it means to say that $\lim_{x \to 0} g(x) = k$.

40. Prove that $\lim_{x \to c} f(x) = L$ if and only if $\lim_{h \to 0} f(h + c) = L$.

41. A wrong statement about limits Show by example that the following statement is wrong.

> The number L is the limit of $f(x)$ as x approaches x_0 if $f(x)$ gets closer to L as x approaches x_0.

Explain why the function in your example does not have the given value of L as a limit as $x \to x_0$.

42. Another wrong statement about limits Show by example that the following statement is wrong.

> The number L is the limit of $f(x)$ as x approaches x_0 if, given any $\epsilon > 0$, there exists a value of x for which $|f(x) - L| < \epsilon$.

Explain why the function in your example does not have the given value of L as a limit as $x \to x_0$.

T **43. Grinding engine cylinders** Before contracting to grind engine cylinders to a cross-sectional area of 9 in^2, you need to know how much deviation from the ideal cylinder diameter of $x_0 = 3.385$ in. you can allow and still have the area come within 0.01 in^2 of the required 9 in^2. To find out, you let $A = \pi(x/2)^2$ and look for the interval in which you must hold x to make $|A - 9| \le 0.01$. What interval do you find?

44. Manufacturing electrical resistors Ohm's law for electrical circuits like the one shown in the accompanying figure states that $V = RI$. In this equation, V is a constant voltage, I is the current in amperes, and R is the resistance in ohms. Your firm has been asked to supply the resistors for a circuit in which V will be 120 volts and I is to be 5 ± 0.1 amp. In what interval does R have to lie for I to be within 0.1 amp of the value $I_0 = 5$?

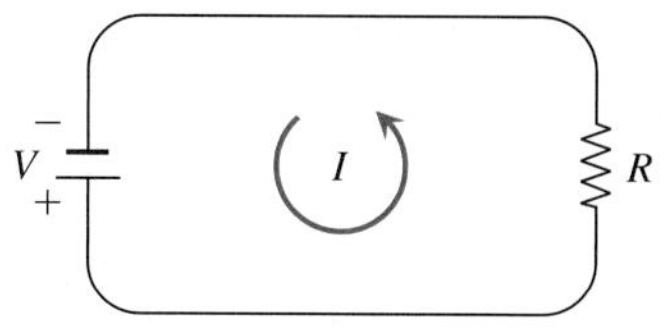

Showing L is not a limit We can prove that $\lim_{x \to x_0} f(x) \ne L$ by providing an $\epsilon > 0$ such that no possible $\delta > 0$ satisfies the condition

$$\text{For all } x, \quad 0 < |x - x_0| < \delta \quad \Rightarrow \quad |f(x) - L| < \epsilon.$$

We accomplish this for our candidate ϵ by showing that for each $\delta > 0$ there exists a value of x such that

$$0 < |x - x_0| < \delta \quad \text{and} \quad |f(x) - L| \ge \epsilon.$$

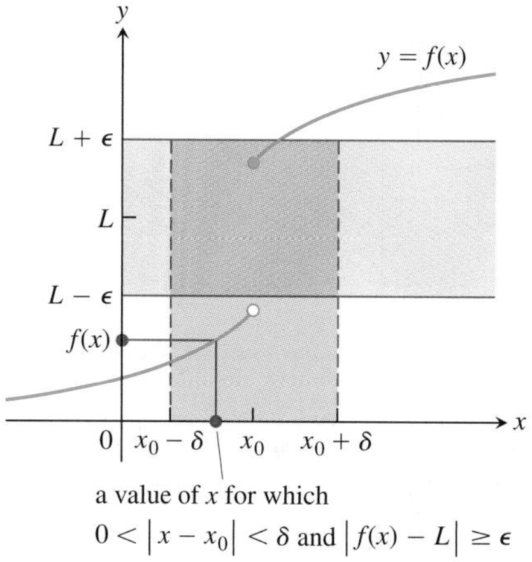

45. Let $f(x) = \begin{cases} x, & x < 1 \\ x + 1, & x > 1. \end{cases}$

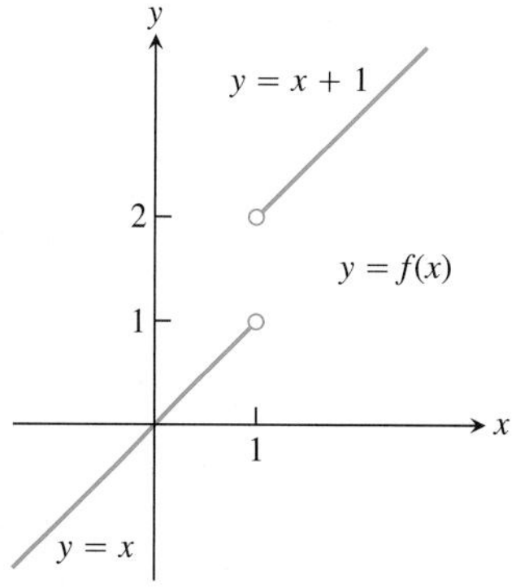

a. Let $\epsilon = 1/2$. Show that no possible $\delta > 0$ satisfies the following condition:

For all x, $0 < |x - 1| < \delta \quad \Rightarrow \quad |f(x) - 2| < 1/2$.

That is, for each $\delta > 0$ show that there is a value of x such that

$$0 < |x - 1| < \delta \quad \text{and} \quad |f(x) - 2| \geq 1/2.$$

This will show that $\lim_{x\to 1} f(x) \neq 2$.

b. Show that $\lim_{x\to 1} f(x) \neq 1$.

c. Show that $\lim_{x\to 1} f(x) \neq 1.5$.

46. Let $h(x) = \begin{cases} x^2, & x < 2 \\ 3, & x = 2 \\ 2, & x > 2. \end{cases}$

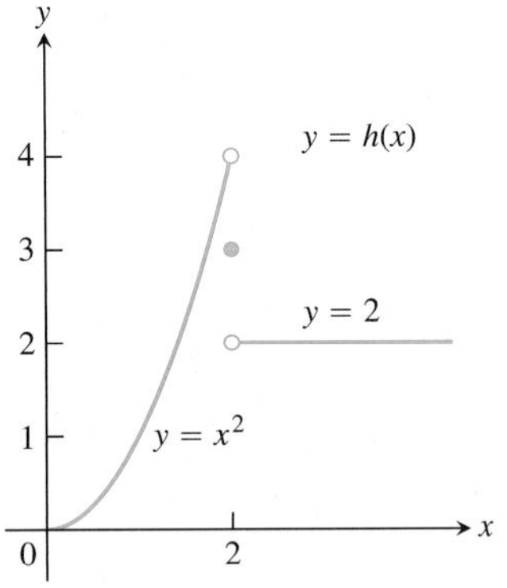

Show that

a. $\lim_{x\to 2} h(x) \neq 4$

b. $\lim_{x\to 2} h(x) \neq 3$

c. $\lim_{x\to 2} h(x) \neq 2$

COMPUTER EXPLORATIONS

In Exercises 47–52, you will further explore finding deltas graphically. Use a CAS to perform the following steps:

a. Plot the function $y = f(x)$ near the point x_0 being approached.

b. Guess the value of the limit L and then evaluate the limit symbolically to see if you guessed correctly.

c. Using the value $\epsilon = 0.2$, graph the banding lines $y_1 = L - \epsilon$ and $y_2 = L + \epsilon$ together with the function f near x_0.

d. From your graph in part (c), estimate a $\delta > 0$ such that for all x

$$0 < |x - x_0| < \delta \quad \Rightarrow \quad |f(x) - L| < \epsilon.$$

Test your estimate by plotting f, y_1, and y_2 over the interval $0 < |x - x_0| < \delta$. For your viewing window use $x_0 - 2\delta \leq x \leq x_0 + 2\delta$ and $L - 2\epsilon \leq y \leq L + 2\epsilon$. If any function values lie outside the interval $[L - \epsilon, L + \epsilon]$, your choice of δ was too large. Try again with a smaller estimate.

e. Repeat parts (c) and (d) successively for $\epsilon = 0.1, 0.05$, and 0.001.

47. $f(x) = \dfrac{x^4 - 81}{x - 3}, \quad x_0 = 3$

48. $f(x) = \dfrac{5x^3 + 9x^2}{2x^5 + 3x^2}, \quad x_0 = 0$

49. $f(x) = \dfrac{\sin 2x}{3x}, \quad x_0 = 0$

50. $f(x) = \dfrac{x(1 - \cos x)}{x - \sin x}, \quad x_0 = 0$

51. $f(x) = \dfrac{\sqrt[3]{x} - 1}{x - 1}, \quad x_0 = 1$

52. $f(x) = \dfrac{3x^2 - (7x + 1)\sqrt{x} + 5}{x - 1}, \quad x_0 = 1$

1.6 One-Sided Limits

In this section we extend the limit concept to *one-sided limits*, which are limits as x approaches the number x_0 from the left-hand side (where $x < x_0$) or the right-hand side ($x > x_0$) only.

One-Sided Limits

To have a limit L as x approaches c, a function f must be defined on *both sides* of c and its values $f(x)$ must approach L as x approaches c from either side. Because of this, ordinary limits are called **two-sided**.

If f fails to have a two-sided limit at c, it may still have a one-sided limit, that is, a limit if the approach is only from one side. If the approach is from the right, the limit is a **right-hand limit**. From the left, it is a **left-hand limit**.

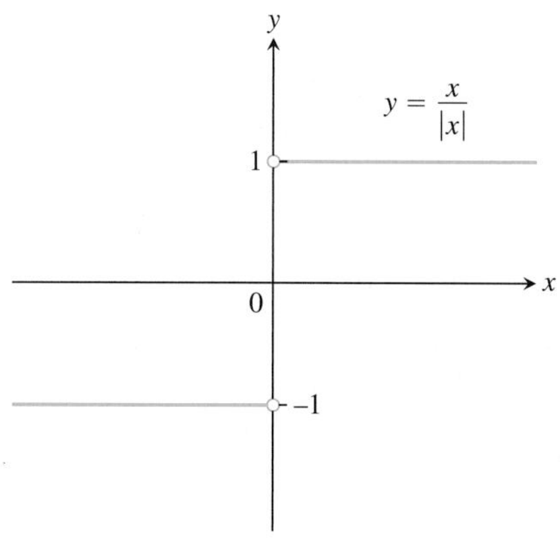

FIGURE 1.50 Different right-hand and left-hand limits at the origin.

The function $f(x) = x/|x|$ (Figure 1.50) has limit 1 as x approaches 0 from the right, and limit -1 as x approaches 0 from the left. Since these one-sided limit values are not the same, there is no single number that $f(x)$ approaches as x approaches 0. So $f(x)$ does not have a (two-sided) limit at 0.

Intuitively, if $f(x)$ is defined on an interval (c, b), where $c < b$, and approaches arbitrarily close to L as x approaches c from within that interval, then f has **right-hand limit** L at c. We write

$$\lim_{x \to c^+} f(x) = L.$$

The symbol "$x \to c^+$" means that we consider only values of x greater than c.

Similarly, if $f(x)$ is defined on an interval (a, c), where $a < c$ and approaches arbitrarily close to M as x approaches c from within that interval, then f has **left-hand limit** M at c. We write

$$\lim_{x \to c^-} f(x) = M.$$

The symbol "$x \to c^-$" means that we consider only x values less than c.

These informal definitions are illustrated in Figure 1.51. For the function $f(x) = x/|x|$ in Figure 1.50 we have

$$\lim_{x \to 0^+} f(x) = 1 \qquad \text{and} \qquad \lim_{x \to 0^-} f(x) = -1.$$

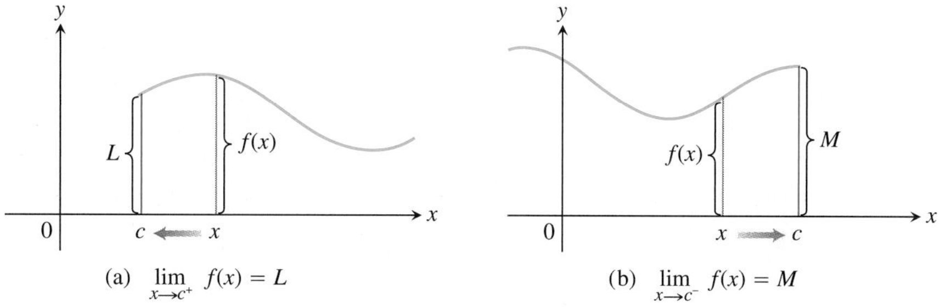

FIGURE 1.51 (a) Right-hand limit as x approaches c. (b) Left-hand limit as x approaches c.

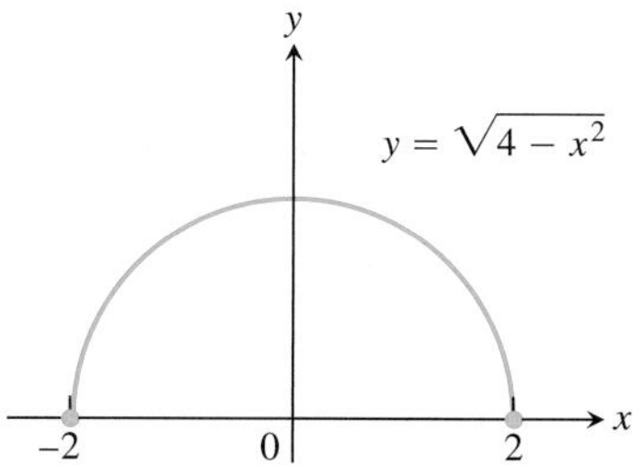

FIGURE 1.52 $\lim_{x \to 2^-} \sqrt{4 - x^2} = 0$ and $\lim_{x \to -2^+} \sqrt{4 - x^2} = 0$ (Example 1).

EXAMPLE 1 The domain of $f(x) = \sqrt{4 - x^2}$ is $[-2, 2]$; its graph is the semicircle in Figure 1.52. We have

$$\lim_{x \to -2^+} \sqrt{4 - x^2} = 0 \qquad \text{and} \qquad \lim_{x \to 2^-} \sqrt{4 - x^2} = 0.$$

The function does not have a left-hand limit at $x = -2$ or a right-hand limit at $x = 2$. It does not have ordinary two-sided limits at either -2 or 2. ■

One-sided limits have all the properties listed in Theorem 1 in Section 1.4. The right-hand limit of the sum of two functions is the sum of their right-hand limits, and so on. The theorems for limits of polynomials and rational functions hold with one-sided limits, as does the Sandwich Theorem and Theorem 5. One-sided limits are related to limits in the following way.

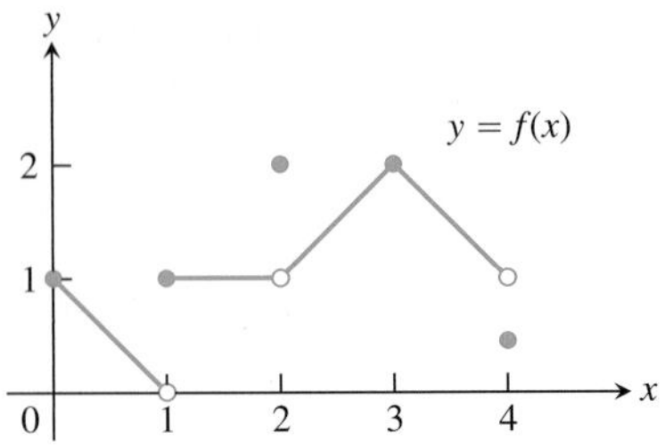

FIGURE 1.53 Graph of the function in Example 2.

THEOREM 6 A function $f(x)$ has a limit as x approaches c if and only if it has left-hand and right-hand limits there and these one-sided limits are equal:

$$\lim_{x \to c} f(x) = L \quad \Leftrightarrow \quad \lim_{x \to c^-} f(x) = L \quad \text{and} \quad \lim_{x \to c^+} f(x) = L.$$

EXAMPLE 2 For the function graphed in Figure 1.53,

At $x = 0$: $\lim_{x \to 0^+} f(x) = 1$,
$\lim_{x \to 0^-} f(x)$ and $\lim_{x \to 0} f(x)$ do not exist. The function is not defined to the left of $x = 0$.

At $x = 1$: $\lim_{x \to 1^-} f(x) = 0$ even though $f(1) = 1$,
$\lim_{x \to 1^+} f(x) = 1$,
$\lim_{x \to 1} f(x)$ does not exist. The right- and left-hand limits are not equal.

At $x = 2$: $\lim_{x \to 2^-} f(x) = 1$,
$\lim_{x \to 2^+} f(x) = 1$,
$\lim_{x \to 2} f(x) = 1$ even though $f(2) = 2$.

At $x = 3$: $\lim_{x \to 3^-} f(x) = \lim_{x \to 3^+} f(x) = \lim_{x \to 3} f(x) = f(3) = 2$.

At $x = 4$: $\lim_{x \to 4^-} f(x) = 1$ even though $f(4) \neq 1$,
$\lim_{x \to 4^+} f(x)$ and $\lim_{x \to 4} f(x)$ do not exist. The function is not defined to the right of $x = 4$.

At every other point c in [0, 4], $f(x)$ has limit $f(c)$. ■

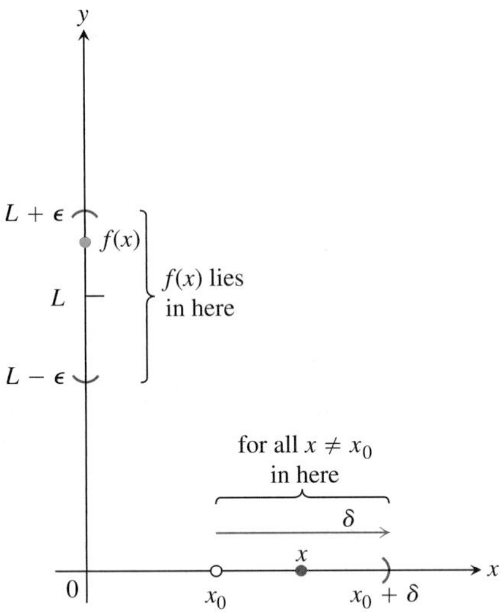

FIGURE 1.54 Intervals associated with the definition of right-hand limit.

Precise Definitions of One-Sided Limits

The formal definition of the limit in Section 1.5 is readily modified for one-sided limits.

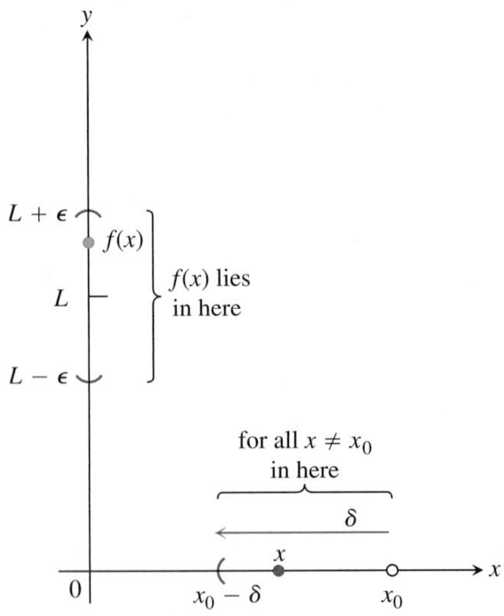

FIGURE 1.55 Intervals associated with the definition of left-hand limit.

DEFINITIONS We say that $f(x)$ has **right-hand limit L at x_0**, and write

$$\lim_{x \to x_0^+} f(x) = L \qquad \text{(See Figure 1.54)}$$

if for every number $\epsilon > 0$ there exists a corresponding number $\delta > 0$ such that for all x

$$x_0 < x < x_0 + \delta \quad \Rightarrow \quad |f(x) - L| < \epsilon.$$

We say that f has **left-hand limit L at x_0**, and write

$$\lim_{x \to x_0^-} f(x) = L \qquad \text{(See Figure 1.55)}$$

if for every number $\epsilon > 0$ there exists a corresponding number $\delta > 0$ such that for all x

$$x_0 - \delta < x < x_0 \quad \Rightarrow \quad |f(x) - L| < \epsilon.$$

EXAMPLE 3 Show that $y = \sin(1/x)$ has no limit as x approaches zero from either side (Figure 1.56).

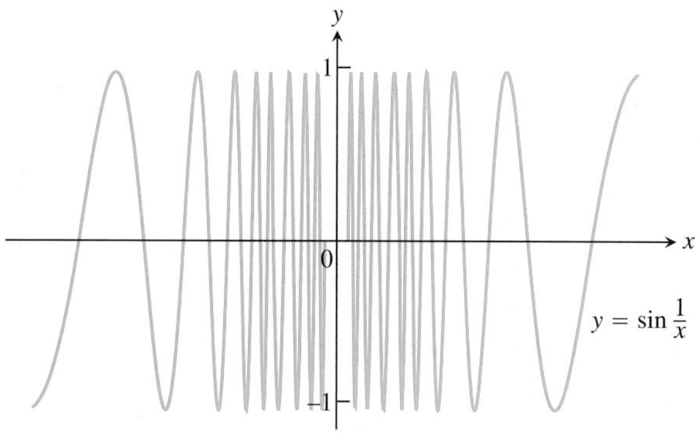

FIGURE 1.56 The function $y = \sin(1/x)$ has neither a right-hand nor a left-hand limit as x approaches zero (Example 3).

Solution As x approaches zero, its reciprocal, $1/x$, grows without bound and the values of $\sin(1/x)$ cycle repeatedly from -1 to 1. There is no single number L that the function's values stay increasingly close to as x approaches zero. This is true even if we restrict x to positive values or to negative values. The function has neither a right-hand limit nor a left-hand limit at $x = 0$. ■

Limits Involving $(\sin \theta)/\theta$

A central fact about $(\sin \theta)/\theta$ is that in radian measure its limit as $\theta \to 0$ is 1. We can see this in Figure 1.57 and confirm it algebraically using the Sandwich Theorem. You will see the importance of this limit in Section 2.5, where instantaneous rates of change of the trigonometric functions are studied.

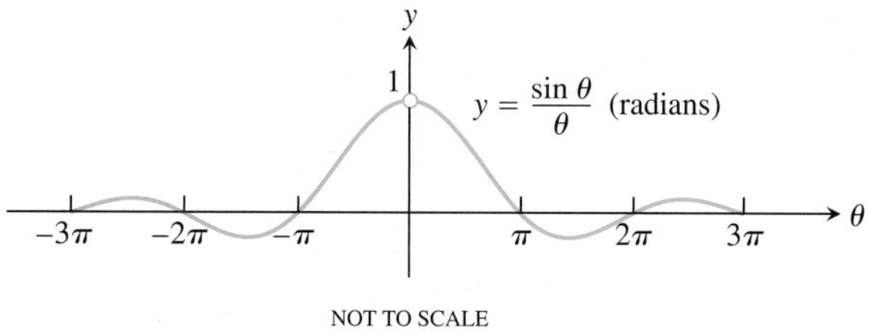

FIGURE 1.57 The graph of $f(\theta) = (\sin \theta)/\theta$ suggests that the right- and left-hand limits as θ approaches 0 are both 1.

THEOREM 7

$$\lim_{\theta \to 0} \frac{\sin \theta}{\theta} = 1 \qquad (\theta \text{ in radians}) \tag{1}$$

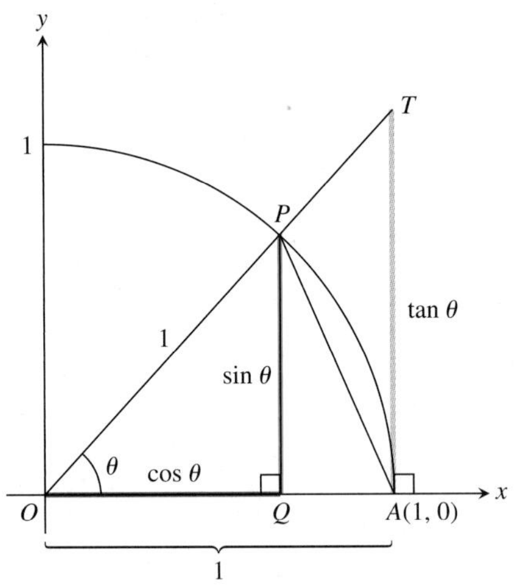

FIGURE 1.58 The figure for the proof of Theorem 7. $TA/OA = \tan\theta$, but $OA = 1$, so $TA = \tan\theta$.

Equation (2) is where radian measure comes in: The area of sector OAP is $\theta/2$ only if θ is measured in radians.

Proof The plan is to show that the right-hand and left-hand limits are both 1. Then we will know that the two-sided limit is 1 as well.

To show that the right-hand limit is 1, we begin with positive values of θ less than $\pi/2$ (Figure 1.58). Notice that

$$\text{Area } \Delta OAP < \text{ area sector } OAP < \text{ area } \Delta OAT.$$

We can express these areas in terms of θ as follows:

$$\text{Area } \Delta OAP = \frac{1}{2}\text{base} \times \text{height} = \frac{1}{2}(1)(\sin\theta) = \frac{1}{2}\sin\theta$$

$$\text{Area sector } OAP = \frac{1}{2}r^2\theta = \frac{1}{2}(1)^2\theta = \frac{\theta}{2} \tag{2}$$

$$\text{Area } \Delta OAT = \frac{1}{2}\text{base} \times \text{height} = \frac{1}{2}(1)(\tan\theta) = \frac{1}{2}\tan\theta.$$

Thus,

$$\frac{1}{2}\sin\theta < \frac{1}{2}\theta < \frac{1}{2}\tan\theta.$$

This last inequality goes the same way if we divide all three terms by the number $(1/2)\sin\theta$, which is positive since $0 < \theta < \pi/2$:

$$1 < \frac{\theta}{\sin\theta} < \frac{1}{\cos\theta}.$$

Taking reciprocals reverses the inequalities:

$$1 > \frac{\sin\theta}{\theta} > \cos\theta.$$

Since $\lim_{\theta\to 0^+}\cos\theta = 1$ (Example 10b, Section 1.4), the Sandwich Theorem gives

$$\lim_{\theta\to 0^+}\frac{\sin\theta}{\theta} = 1.$$

Recall that $\sin\theta$ and θ are both *odd functions* (Section 1.1). Therefore, $f(\theta) = (\sin\theta)/\theta$ is an *even function*, with a graph symmetric about the y-axis (see Figure 1.57). This symmetry implies that the left-hand limit at 0 exists and has the same value as the right-hand limit:

$$\lim_{\theta\to 0^-}\frac{\sin\theta}{\theta} = 1 = \lim_{\theta\to 0^+}\frac{\sin\theta}{\theta},$$

so $\lim_{\theta\to 0}(\sin\theta)/\theta = 1$ by Theorem 6. ■

EXAMPLE 4 Show that **(a)** $\lim_{h\to 0}\frac{\cos h - 1}{h} = 0$ and **(b)** $\lim_{x\to 0}\frac{\sin 2x}{5x} = \frac{2}{5}$.

Solution

(a) Using the half-angle formula $\cos h = 1 - 2\sin^2(h/2)$, we calculate

$$\begin{aligned}\lim_{h\to 0}\frac{\cos h - 1}{h} &= \lim_{h\to 0} -\frac{2\sin^2(h/2)}{h}\\ &= -\lim_{\theta\to 0}\frac{\sin\theta}{\theta}\sin\theta \qquad \text{Let } \theta = h/2.\\ &= -(1)(0) = 0.\end{aligned}$$

(b) Equation (1) does not apply to the original fraction. We need a $2x$ in the denominator, not a $5x$. We produce it by multiplying numerator and denominator by $2/5$:

$$\begin{aligned}\lim_{x\to 0}\frac{\sin 2x}{5x} &= \lim_{x\to 0}\frac{(2/5)\cdot \sin 2x}{(2/5)\cdot 5x}\\ &= \frac{2}{5}\lim_{x\to 0}\frac{\sin 2x}{2x}\\ &= \frac{2}{5}(1) = \frac{2}{5}\end{aligned}$$

Now, Eq. (1) applies with $\theta = 2x$.

■

EXERCISES 1.6

1. Which of the following statements about the function $y = f(x)$ graphed here are true, and which are false?

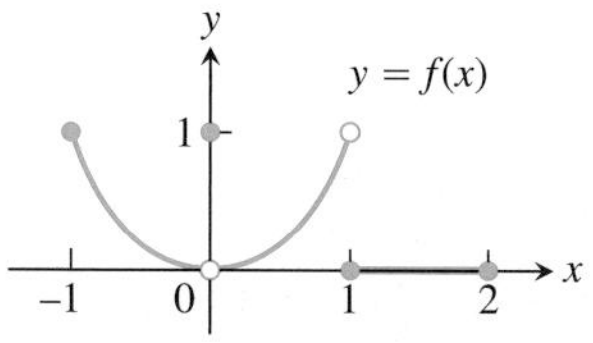

a. $\lim_{x\to -1^+} f(x) = 1$ **b.** $\lim_{x\to 0^-} f(x) = 0$

c. $\lim_{x\to 0^-} f(x) = 1$ **d.** $\lim_{x\to 0^-} f(x) = \lim_{x\to 0^+} f(x)$

e. $\lim_{x\to 0} f(x)$ exists **f.** $\lim_{x\to 0} f(x) = 0$

g. $\lim_{x\to 0} f(x) = 1$ **h.** $\lim_{x\to 1} f(x) = 1$

i. $\lim_{x\to 1} f(x) = 0$ **j.** $\lim_{x\to 2^-} f(x) = 2$

k. $\lim_{x\to -1^-} f(x)$ does not exist. **l.** $\lim_{x\to 2^+} f(x) = 0$

2. Which of the following statements about the function $y = f(x)$ graphed here are true, and which are false?

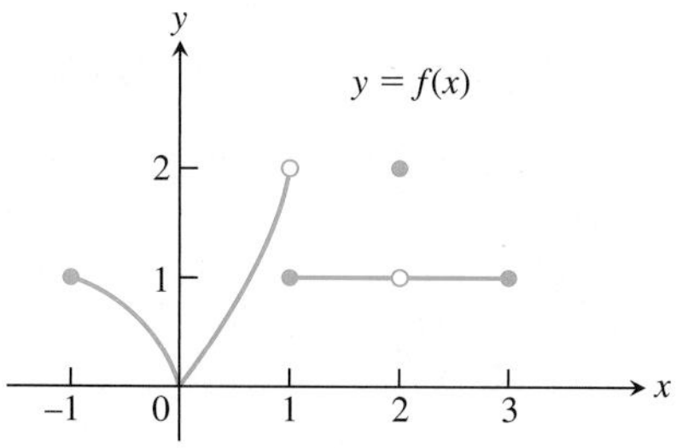

a. $\lim_{x\to -1^+} f(x) = 1$ **b.** $\lim_{x\to 2} f(x)$ does not exist.

c. $\lim_{x\to 2} f(x) = 2$ **d.** $\lim_{x\to 1^-} f(x) = 2$

e. $\lim_{x\to 1^+} f(x) = 1$ **f.** $\lim_{x\to 1} f(x)$ does not exist.

g. $\lim_{x\to 0^+} f(x) = \lim_{x\to 0^-} f(x)$

h. $\lim_{x\to c} f(x)$ exists at every c in the open interval $(-1, 1)$.

i. $\lim_{x\to c} f(x)$ exists at every c in the open interval $(1, 3)$.

j. $\lim_{x\to -1^-} f(x) = 0$ **k.** $\lim_{x\to 3^+} f(x)$ does not exist.

3. Let $f(x) = \begin{cases} 3 - x, & x < 2 \\ \dfrac{x}{2} + 1, & x > 2. \end{cases}$

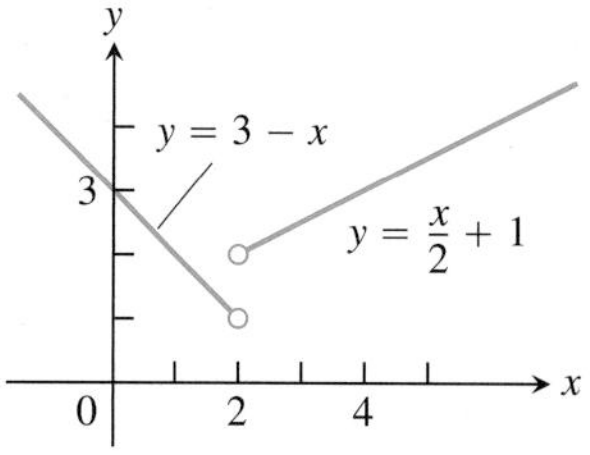

a. Find $\lim_{x\to 2^+} f(x)$ and $\lim_{x\to 2^-} f(x)$.

b. Does $\lim_{x\to 2} f(x)$ exist? If so, what is it? If not, why not?

c. Find $\lim_{x\to 4^-} f(x)$ and $\lim_{x\to 4^+} f(x)$.

d. Does $\lim_{x\to 4} f(x)$ exist? If so, what is it? If not, why not?

4. Let $f(x) = \begin{cases} 0, & x \le 0 \\ \sin \frac{1}{x}, & x > 0. \end{cases}$

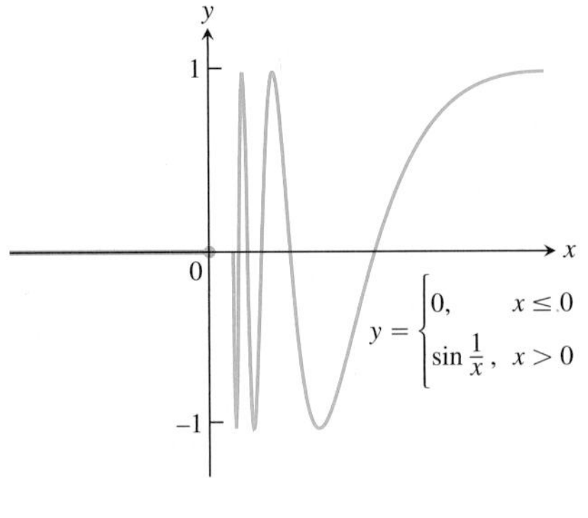

a. Does $\lim_{x\to 0^+} f(x)$ exist? If so, what is it? If not, why not?

b. Does $\lim_{x\to 0^-} f(x)$ exist? If so, what is it? If not, why not?

c. Does $\lim_{x\to 0} f(x)$ exist? If so, what is it? If not, why not?

5. a. Graph $f(x) = \begin{cases} x^3, & x \ne 1 \\ 0, & x = 1. \end{cases}$

b. Find $\lim_{x\to 1^-} f(x)$ and $\lim_{x\to 1^+} f(x)$.

c. Does $\lim_{x\to 1} f(x)$ exist? If so, what is it? If not, why not?

6. a. Graph $f(x) = \begin{cases} 1 - x^2, & x \ne 1 \\ 2, & x = 1. \end{cases}$

b. Find $\lim_{x\to 1^+} f(x)$ and $\lim_{x\to 1^-} f(x)$.

c. Does $\lim_{x\to 1} f(x)$ exist? If so, what is it? If not, why not?

Graph the functions in Exercises 7 and 8. Then answer these questions.

a. What are the domain and range of f?

b. At what points c, if any, does $\lim_{x\to c} f(x)$ exist?

c. At what points does only the left-hand limit exist?

d. At what points does only the right-hand limit exist?

7. $f(x) = \begin{cases} \sqrt{1 - x^2}, & 0 \le x < 1 \\ 1, & 1 \le x < 2 \\ 2, & x = 2 \end{cases}$

8. $f(x) = \begin{cases} x, & -1 \le x < 0, \text{ or } 0 < x \le 1 \\ 1, & x = 0 \\ 0, & x < -1 \text{ or } x > 1 \end{cases}$

Find the limits in Exercises 9–16.

9. $\lim_{x\to -0.5^-} \sqrt{\frac{x + 2}{x + 1}}$

10. $\lim_{x\to 1^+} \sqrt{\frac{x - 1}{x + 2}}$

11. $\lim_{x\to -2^+} \left(\frac{x}{x + 1}\right)\left(\frac{2x + 5}{x^2 + x}\right)$

12. $\lim_{x\to 1^-} \left(\frac{1}{x + 1}\right)\left(\frac{x + 6}{x}\right)\left(\frac{3 - x}{7}\right)$

13. $\lim_{h\to 0^+} \frac{\sqrt{h^2 + 4h + 5} - \sqrt{5}}{h}$

14. $\lim_{h\to 0^-} \frac{\sqrt{6} - \sqrt{5h^2 + 11h + 6}}{h}$

15. a. $\lim_{x\to -2^+} (x + 3)\frac{|x + 2|}{x + 2}$ **b.** $\lim_{x\to -2^-} (x + 3)\frac{|x + 2|}{x + 2}$

16. a. $\lim_{x\to 1^+} \frac{\sqrt{2x}\,(x - 1)}{|x - 1|}$ **b.** $\lim_{x\to 1^-} \frac{\sqrt{2x}\,(x - 1)}{|x - 1|}$

Use the graph of the greatest integer function $y = \lfloor x \rfloor$, Figure 1.9 in Section 1.1, to help you find the limits in Exercises 17 and 18.

17. a. $\lim_{\theta\to 3^+} \frac{\lfloor \theta \rfloor}{\theta}$ **b.** $\lim_{\theta\to 3^-} \frac{\lfloor \theta \rfloor}{\theta}$

18. a. $\lim_{t\to 4^+} (t - \lfloor t \rfloor)$ **b.** $\lim_{t\to 4^-} (t - \lfloor t \rfloor)$

Find the limits in Exercises 19–34.

19. $\lim_{\theta\to 0} \frac{\sin\sqrt{2}\theta}{\sqrt{2}\theta}$

20. $\lim_{t\to 0} \frac{\sin kt}{t}$ (k constant)

21. $\lim_{y\to 0} \frac{\sin 3y}{4y}$

22. $\lim_{h\to 0^-} \frac{h}{\sin 3h}$

23. $\lim_{x\to 0} \frac{\tan 2x}{x}$

24. $\lim_{t\to 0} \frac{2t}{\tan t}$

25. $\lim_{x\to 0} \frac{x \csc 2x}{\cos 5x}$

26. $\lim_{x\to 0} 6x^2(\cot x)(\csc 2x)$

27. $\lim_{x\to 0} \frac{x + x\cos x}{\sin x \cos x}$

28. $\lim_{x\to 0} \frac{x^2 - x + \sin x}{2x}$

29. $\lim_{t\to 0} \frac{\sin(1 - \cos t)}{1 - \cos t}$

30. $\lim_{h\to 0} \frac{\sin(\sin h)}{\sin h}$

31. $\lim_{\theta\to 0} \frac{\sin\theta}{\sin 2\theta}$

32. $\lim_{x\to 0} \frac{\sin 5x}{\sin 4x}$

33. $\lim_{x\to 0} \frac{\tan 3x}{\sin 8x}$

34. $\lim_{y\to 0} \frac{\sin 3y \cot 5y}{y \cot 4y}$

35. Once you know $\lim_{x\to a^+} f(x)$ and $\lim_{x\to a^-} f(x)$ at an interior point of the domain of f, do you then know $\lim_{x\to a} f(x)$? Give reasons for your answer.

36. If you know that $\lim_{x\to c} f(x)$ exists, can you find its value by calculating $\lim_{x\to c^+} f(x)$? Give reasons for your answer.

37. Suppose that f is an odd function of x. Does knowing that $\lim_{x\to 0^+} f(x) = 3$ tell you anything about $\lim_{x\to 0^-} f(x)$? Give reasons for your answer.

38. Suppose that f is an even function of x. Does knowing that $\lim_{x\to 2^-} f(x) = 7$ tell you anything about either $\lim_{x\to -2^-} f(x)$ or $\lim_{x\to -2^+} f(x)$? Give reasons for your answer.

39. Given $\epsilon > 0$, find an interval $I = (5, 5 + \delta)$, $\delta > 0$, such that if x lies in I, then $\sqrt{x - 5} < \epsilon$. What limit is being verified and what is its value?

40. Given $\epsilon > 0$, find an interval $I = (4 - \delta, 4)$, $\delta > 0$, such that if x lies in I, then $\sqrt{4 - x} < \epsilon$. What limit is being verified and what is its value?

Use the definitions of right-hand and left-hand limits to prove the limit statements in Exercises 41 and 42.

41. $\lim_{x\to 0^-} \dfrac{x}{|x|} = -1$

42. $\lim_{x\to 2^+} \dfrac{x-2}{|x-2|} = 1$

43. Greatest integer function Find **(a)** $\lim_{x\to 400^+} \lfloor x \rfloor$ and **(b)** $\lim_{x\to 400^-} \lfloor x \rfloor$; then use limit definitions to verify your findings. **(c)** Based on your conclusions in parts (a) and (b), can anything be said about $\lim_{x\to 400} \lfloor x \rfloor$? Give reasons for your answers.

44. One-sided limits Let $f(x) = \begin{cases} x^2 \sin(1/x), & x < 0 \\ \sqrt{x}, & x > 0. \end{cases}$

Find **(a)** $\lim_{x\to 0^+} f(x)$ and **(b)** $\lim_{x\to 0^-} f(x)$; then use limit definitions to verify your findings. **(c)** Based on your conclusions in parts (a) and (b), can anything be said about $\lim_{x\to 0} f(x)$? Give reasons for your answer.

1.7 Continuity

When we plot function values generated in a laboratory or collected in the field, we often connect the plotted points with an unbroken curve to show what the function's values are likely to have been at the times we did not measure (Figure 1.59). In doing so, we are assuming that we are working with a *continuous function*, so its outputs vary continuously with the inputs and do not jump from one value to another without taking on the values in between. The limit of a continuous function as x approaches c can be found simply by calculating the value of the function at c. (We found this to be true for polynomials in Section 1.4.)

Intuitively, any function $y = f(x)$ whose graph can be sketched over its domain in one continuous motion without lifting the pencil is an example of a continuous function. In this section we investigate more precisely what it means for a function to be continuous. We also study the properties of continuous functions, and see that many of the function types presented in Section 1.1 are continuous.

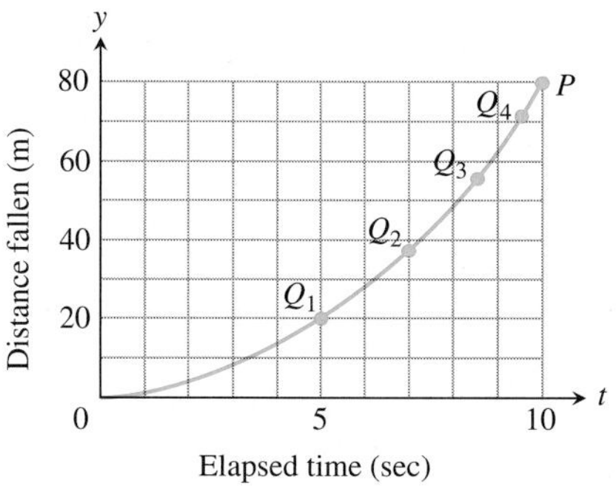

FIGURE 1.59 Connecting plotted points by an unbroken curve from experimental data $Q_1, Q_2, Q_3, \ldots$ for a falling object.

Continuity at a Point

To understand continuity, it helps to consider a function like the one in Figure 1.60 whose limits we investigated in Example 2 in the last section. (The figure is repeated here.)

EXAMPLE 1 Find the points at which the function f in Figure 1.60 is continuous and the points at which f is not continuous.

Solution The function f is continuous at every point in its domain [0,4] except at $x = 1, x = 2$, and $x = 4$. At these points, there are breaks in the graph. Note the relationship between the limit of f and the value of f at each point of the function's domain.

Points at which f is continuous:

At $x = 0$, $\quad \lim_{x\to 0^+} f(x) = f(0)$.

At $x = 3$, $\quad \lim_{x\to 3} f(x) = f(3)$.

At $0 < c < 4, c \neq 1, 2$, $\quad \lim_{x\to c} f(x) = f(c)$.

Points at which f is not continuous:

At $x = 1$, $\quad \lim_{x\to 1} f(x)$ does not exist.

At $x = 2$, $\quad \lim_{x\to 2} f(x) = 1$, but $1 \neq f(2)$.

At $x = 4$, $\quad \lim_{x\to 4^-} f(x) = 1$, but $1 \neq f(4)$.

At $c < 0, c > 4$, $\quad$ these points are not in the domain of f. ■

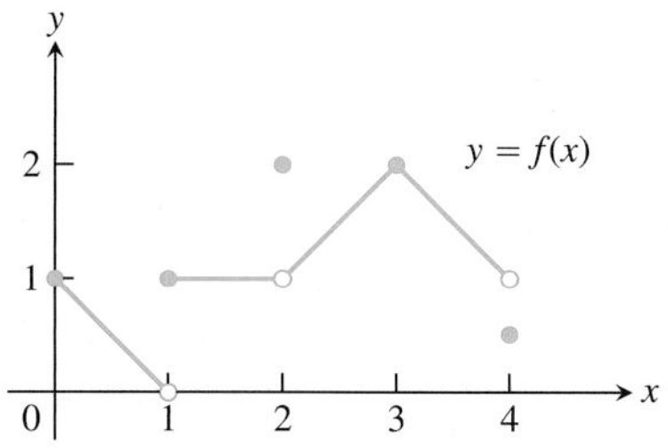

FIGURE 1.60 The function is continuous on [0, 4] except at $x = 1, x = 2$, and $x = 4$ (Example 1).

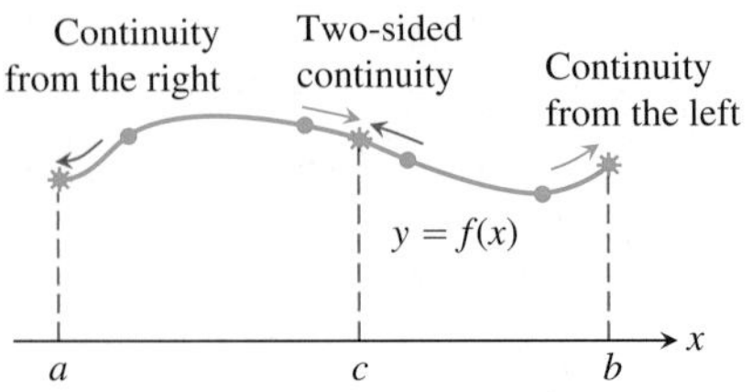

FIGURE 1.61 Continuity at points a, b, and c.

To define continuity at a point in a function's domain, we need to define continuity at an interior point (which involves a two-sided limit) and continuity at an endpoint (which involves a one-sided limit) (Figure 1.61).

> **DEFINITION**
> *Interior point*: A function $y = f(x)$ is **continuous at an interior point c** of its domain if
> $$\lim_{x\to c} f(x) = f(c).$$
> *Endpoint*: A function $y = f(x)$ is **continuous at a left endpoint a** or is **continuous at a right endpoint b** of its domain if
> $$\lim_{x\to a^+} f(x) = f(a) \quad \text{or} \quad \lim_{x\to b^-} f(x) = f(b), \quad \text{respectively.}$$

If a function f is defined on an open interval containing c, except possibly at c itself, and f is not continuous at c, then we say that f is **discontinuous at c** (or c is a **point of discontinuity** of f). Note that c need not be in the domain of f.

A function f is **right-continuous (continuous from the right)** at a point $x = c$ in its domain if $\lim_{x\to c^+} f(x) = f(c)$. It is **left-continuous (continuous from the left)** at c if $\lim_{x\to c^-} f(x) = f(c)$. Thus, a function is continuous at a left endpoint a of its domain if it is right-continuous at a and continuous at a right endpoint b of its domain if it is left-continuous at b. A function is continuous at an interior point c of its domain if and only if it is both right-continuous and left-continuous at c (Figure 1.61).

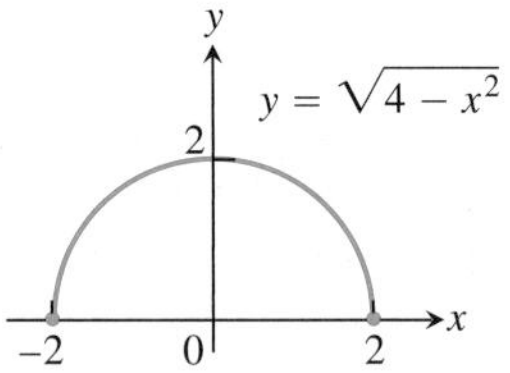

FIGURE 1.62 A function that is continuous at every domain point (Example 2).

EXAMPLE 2 The function $f(x) = \sqrt{4 - x^2}$ is continuous at every point of its domain $[-2, 2]$ (Figure 1.62), including $x = -2$, where f is right-continuous, and $x = 2$, where f is left-continuous. ■

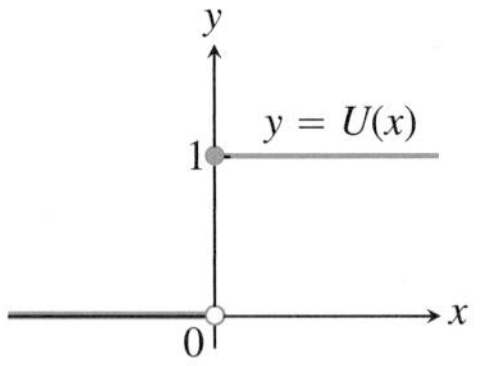

FIGURE 1.63 A function that is right-continuous, but not left-continuous, at the origin. It has a jump discontinuity there (Example 3).

EXAMPLE 3 The unit step function $U(x)$, graphed in Figure 1.63, is right-continuous at $x = 0$, but is neither left-continuous nor continuous there. It has a jump discontinuity at $x = 0$. ■

We summarize continuity at a point in the form of a test.

> **Continuity Test**
> A function $f(x)$ is continuous at an interior point $x = c$ of its domain if and only if it meets the following three conditions.
>
> 1. $f(c)$ exists (c lies in the domain of f)
> 2. $\lim_{x\to c} f(x)$ exists (f has a limit as $x \to c$)
> 3. $\lim_{x\to c} f(x) = f(c)$ (the limit equals the function value)

For one-sided continuity and continuity at an endpoint, the limits in parts 2 and 3 of the test should be replaced by the appropriate one-sided limits.

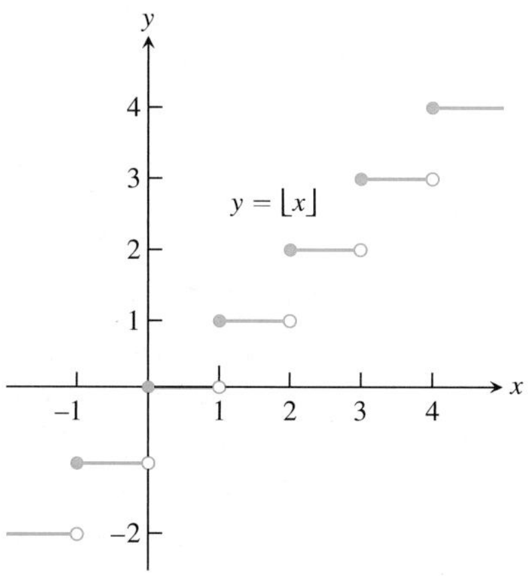

FIGURE 1.64 The greatest integer function is continuous at every noninteger point. It is right-continuous, but not left-continuous, at every integer point (Example 4).

EXAMPLE 4 The function $y = \lfloor x \rfloor$ introduced in Section 1.1 is graphed in Figure 1.64. It is discontinuous at every integer because the left-hand and right-hand limits are not equal as $x \to n$:

$$\lim_{x \to n^-} \lfloor x \rfloor = n - 1 \quad \text{and} \quad \lim_{x \to n^+} \lfloor x \rfloor = n.$$

Since $\lfloor n \rfloor = n$, the greatest integer function is right-continuous at every integer n (but not left-continuous).

The greatest integer function is continuous at every real number other than the integers. For example,

$$\lim_{x \to 1.5} \lfloor x \rfloor = 1 = \lfloor 1.5 \rfloor.$$

In general, if $n - 1 < c < n$, n an integer, then

$$\lim_{x \to c} \lfloor x \rfloor = n - 1 = \lfloor c \rfloor.$$

■

Figure 1.65 displays several common types of discontinuities. The function in Figure 1.65a is continuous at $x = 0$. The function in Figure 1.65b would be continuous if it had $f(0) = 1$. The function in Figure 1.65c would be continuous if $f(0)$ were 1 instead of 2. The discontinuities in Figure 1.65b and c are **removable**. Each function has a limit as $x \to 0$, and we can remove the discontinuity by setting $f(0)$ equal to this limit.

The discontinuities in Figure 1.65d through f are more serious: $\lim_{x \to 0} f(x)$ does not exist, and there is no way to improve the situation by changing f at 0. The step function in Figure 1.65d has a **jump discontinuity**: The one-sided limits exist but have different values. The function $f(x) = 1/x^2$ in Figure 1.65e has an **infinite discontinuity**. The function in Figure 1.65f has an **oscillating discontinuity**: It oscillates too much to have a limit as $x \to 0$.

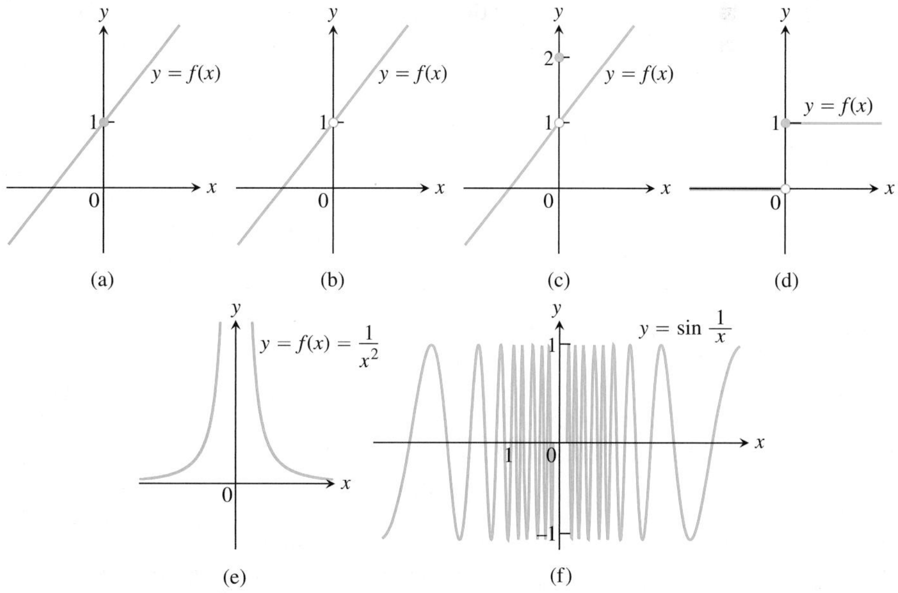

FIGURE 1.65 The function in (a) is continuous at $x = 0$; the functions in (b) through (f) are not.

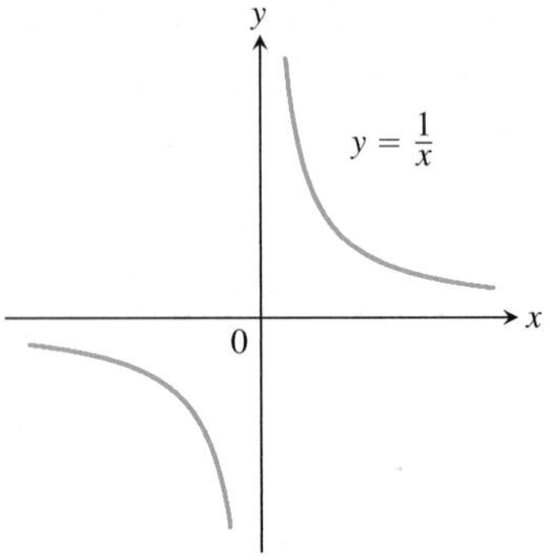

FIGURE 1.66 The function $y = 1/x$ is continuous at every value of x except $x = 0$. It has a point of discontinuity at $x = 0$ (Example 5).

Continuous Functions

A function is **continuous on an interval** if and only if it is continuous at every point of the interval. For example, the semicircle function graphed in Figure 1.62 is continuous on the interval $[-2, 2]$, which is its domain. A **continuous function** is one that is continuous at every point of its domain. A continuous function need not be continuous on every interval.

EXAMPLE 5

(a) The function $y = 1/x$ (Figure 1.66) is a continuous function because it is continuous at every point of its domain. It has a point of discontinuity at $x = 0$, however, because it is not defined there; that is, it is discontinuous on any interval containing $x = 0$.

(b) The identity function $f(x) = x$ and constant functions are continuous everywhere by Example 3, Section 1.5. ■

Algebraic combinations of continuous functions are continuous wherever they are defined.

> **THEOREM 8—Properties of Continuous Functions** If the functions f and g are continuous at $x = c$, then the following combinations are continuous at $x = c$.
>
> 1. *Sums:* $f + g$
> 2. *Differences:* $f - g$
> 3. *Constant multiples:* $k \cdot f$, for any number k
> 4. *Products:* $f \cdot g$
> 5. *Quotients:* f/g, provided $g(c) \neq 0$
> 6. *Powers:* f^n, n a positive integer
> 7. *Roots:* $\sqrt[n]{f}$, provided it is defined on an open interval containing c, where n is a positive integer

Most of the results in Theorem 8 follow from the limit rules in Theorem 1, Section 1.4. For instance, to prove the sum property we have

$$\begin{aligned}
\lim_{x\to c}(f+g)(x) &= \lim_{x\to c}(f(x)+g(x)) \\
&= \lim_{x\to c} f(x) + \lim_{x\to c} g(x), && \text{Sum Rule, Theorem 1} \\
&= f(c) + g(c) && \text{Continuity of } f, g \text{ at } c \\
&= (f+g)(c).
\end{aligned}$$

This shows that $f + g$ is continuous.

EXAMPLE 6

(a) Every polynomial $P(x) = a_n x^n + a_{n-1}x^{n-1} + \cdots + a_0$ is continuous because $\lim_{x\to c} P(x) = P(c)$ by Theorem 2, Section 1.4.

(b) If $P(x)$ and $Q(x)$ are polynomials, then the rational function $P(x)/Q(x)$ is continuous wherever it is defined ($Q(c) \neq 0$) by the Quotient Rule in Theorem 8. ■

EXAMPLE 7 The function $f(x) = |x|$ is continuous at every value of x. If $x > 0$, we have $f(x) = x$, a polynomial. If $x < 0$, we have $f(x) = -x$, another polynomial. Finally, at the origin, $\lim_{x\to 0}|x| = 0 = |0|$. ■

The functions $y = \sin x$ and $y = \cos x$ are continuous at $x = 0$ by Example 10 of Section 1.4. Both functions are, in fact, continuous everywhere (see Exercise 48). It follows from Theorem 8 that all six trigonometric functions are then continuous wherever they are defined. For example, $y = \tan x$ is continuous on $\cdots \cup (-\pi/2, \pi/2) \cup (\pi/2, 3\pi/2) \cup \cdots$.

Composites

All composites of continuous functions are continuous, as stated in the next theorem. Figure 1.67 gives an arrow diagram of this result. In this case, the limit of $g(f(x))$ as $x \to c$ is $g(f(c))$.

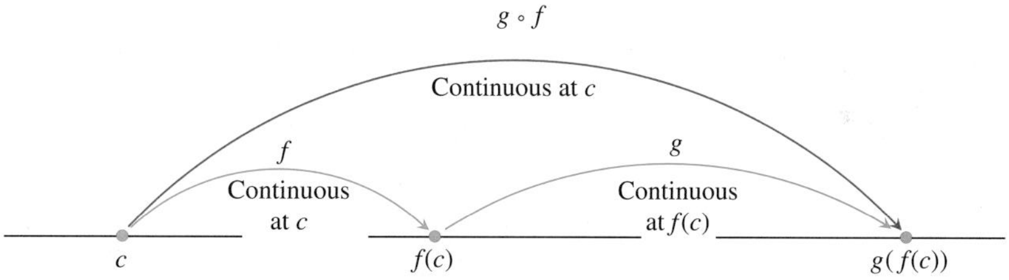

FIGURE 1.67 Composites of continuous functions are continuous.

> **THEOREM 9—Composite of Continuous Functions** If f is continuous at c and g is continuous at $f(c)$, then the composite $g \circ f$ is continuous at c.

Intuitively, Theorem 9 is reasonable because if x is close to c, then $f(x)$ is close to $f(c)$, and since g is continuous at $f(c)$, it follows that $g(f(x))$ is close to $g(f(c))$.

The continuity of composites holds for any finite number of functions. The only requirement is that each function be continuous where it is applied.

EXAMPLE 8 Show that the following functions are continuous everywhere on their respective domains.

(a) $y = \sqrt{x^2 - 2x - 5}$ **(b)** $y = \dfrac{x^{2/3}}{1 + x^4}$ **(c)** $y = \left|\dfrac{x-2}{x^2-2}\right|$ **(d)** $y = \left|\dfrac{x\sin x}{x^2+2}\right|$

Solution

(a) The square root function is continuous on $[0, \infty)$ because it is a root of the continuous identity function $f(x) = x$ (Part 7, Theorem 8). The given function is then the composite of the polynomial $f(x) = x^2 - 2x - 5$ with the square root function $g(t) = \sqrt{t}$, and is continuous on its domain, the points where $f(x) \geq 0$.

(b) The numerator is the cube root of the identity function squared; the denominator is an everywhere-positive polynomial. Therefore, the quotient is continuous.

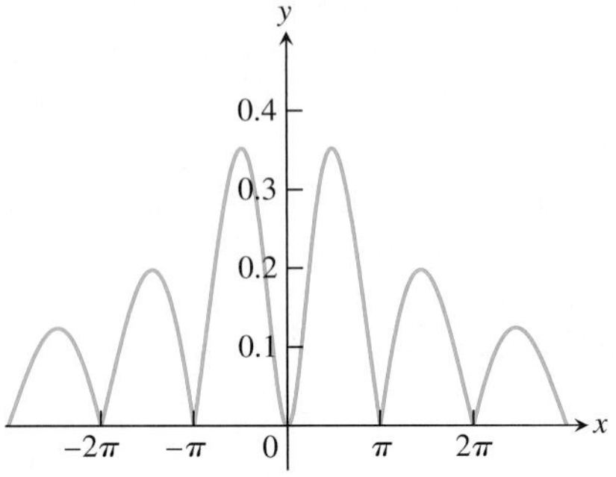

FIGURE 1.68 The graph suggests that $y = |(x \sin x)/(x^2 + 2)|$ is continuous (Example 8d).

(c) The quotient $(x - 2)/(x^2 - 2)$ is continuous for all $x \neq \pm\sqrt{2}$, and the function is the composition of this quotient with the continuous absolute value function (Example 7).

(d) Because the sine function is everywhere-continuous (Exercise 48), the numerator term $x \sin x$ is the product of continuous functions, and the denominator term $x^2 + 2$ is an everywhere-positive polynomial. The given function is the composite of a quotient of continuous functions with the continuous absolute value function (Figure 1.68). ■

Theorem 9 is actually a consequence of the following more general result. A proof is given in Appendix 5.

THEOREM 10—Limits of Continuous Functions If g is continuous at the point b and $\lim_{x\to c} f(x) = b$, then

$$\lim_{x\to c} g(f(x)) = g(b) = g(\lim_{x\to c} f(x)).$$

EXAMPLE 9 As an application of Theorem 10, we have

$$\begin{aligned}\lim_{x\to\pi/2} \cos\left(2x + \sin\left(\frac{3\pi}{2} + x\right)\right) &= \cos\left(\lim_{x\to\pi/2} 2x + \lim_{x\to\pi/2} \sin\left(\frac{3\pi}{2} + x\right)\right)\\ &= \cos(\pi + \sin 2\pi) = \cos \pi = -1.\end{aligned}$$ ■

Intermediate Value Theorem for Continuous Functions

Functions that are continuous on intervals have properties that make them particularly useful in mathematics and its applications. One of these is the *Intermediate Value Property*. A function is said to have the **Intermediate Value Property** if whenever it takes on two values, it also takes on all the values in between.

THEOREM 11—The Intermediate Value Theorem for Continuous Functions If f is a continuous function on a closed interval $[a, b]$, and if y_0 is any value between $f(a)$ and $f(b)$, then $y_0 = f(c)$ for some c in $[a, b]$.

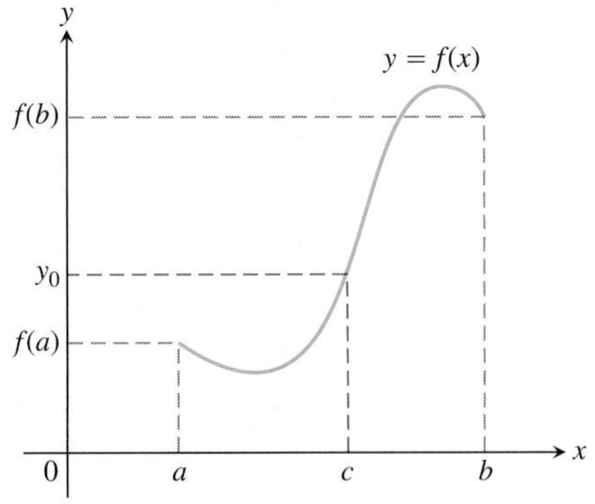

Geometrically, the Intermediate Value Theorem says that any horizontal line $y = y_0$ crossing the y-axis between the numbers $f(a)$ and $f(b)$ will cross the curve $y = f(x)$ at least once over the interval $[a, b]$.

The proof of the Intermediate Value Theorem depends on the completeness property of the real number system (Appendix 7) and can be found in more advanced texts.

The continuity of f on the interval is essential to Theorem 11. If f is discontinuous at even one point of the interval, the theorem's conclusion may fail, as it does for the function graphed in Figure 1.69.

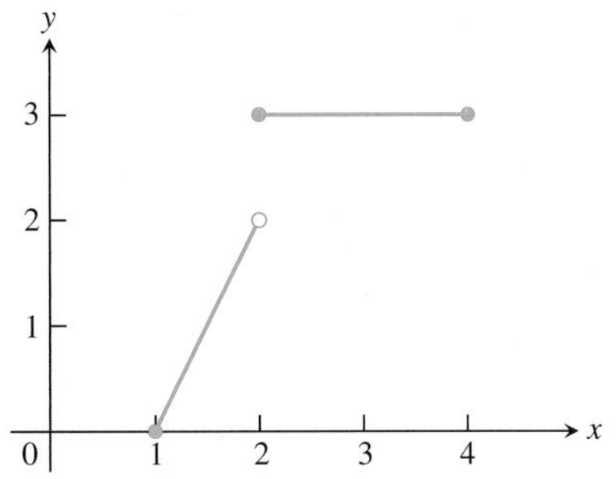

FIGURE 1.69 The function
$$f(x) = \begin{cases} 2x - 2, & 1 \le x < 2 \\ 3, & 2 \le x \le 4 \end{cases}$$
does not take on all values between $f(1) = 0$ and $f(4) = 3$; it misses all the values between 2 and 3.

A Consequence for Graphing: Connectedness Theorem 11 implies that the graph of a function continuous on an interval cannot have any breaks over the interval. It will be **connected**—a single, unbroken curve. It will not have jumps like the graph of the greatest integer function (Figure 1.64), or separate branches like the graph of $1/x$ (Figure 1.66).

A Consequence for Root Finding We call a solution of the equation $f(x) = 0$ a **root** of the equation or **zero** of the function f. The Intermediate Value Theorem tells us that if f is continuous, then any interval on which f changes sign contains a zero of the function.

EXAMPLE 10 Show that there is a root of the equation $x^3 - x - 1 = 0$ between 1 and 2.

Solution Let $f(x) = x^3 - x - 1$. Since $f(1) = 1 - 1 - 1 = -1 < 0$ and $f(2) = 2^3 - 2 - 1 = 5 > 0$, we see that $y_0 = 0$ is a value between $f(1)$ and $f(2)$. The Intermediate Value Theorem says there is a zero of f between 1 and 2. Figure 1.70 shows the result of zooming in to locate the root near $x = 1.32$.

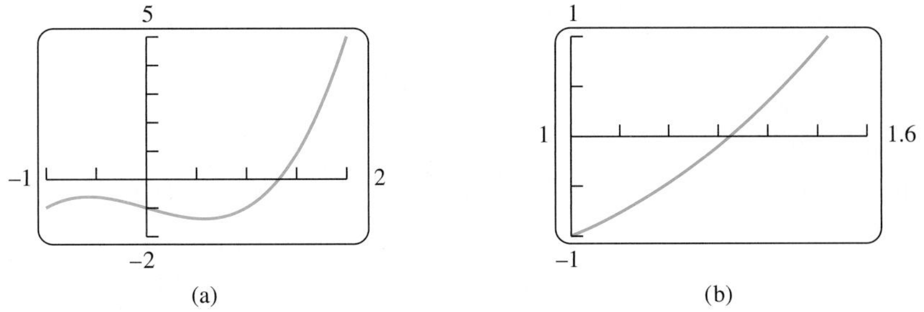

FIGURE 1.70 Zooming in on a zero of the function $f(x) = x^3 - x - 1$. The zero is near $x = 1.32$. ■

EXERCISES 1.7

In Exercises 1–4, say whether the function graphed is continuous on $[-1, 3]$. If not, where does it fail to be continuous and why?

1.

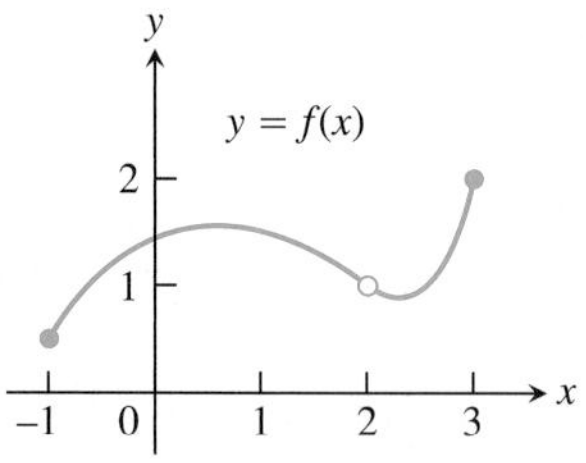

2.

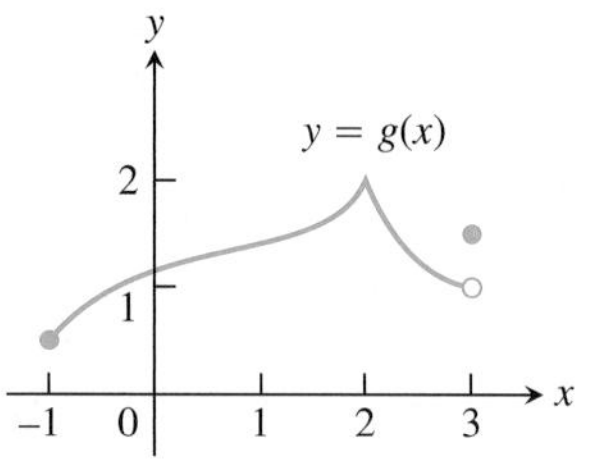

3.

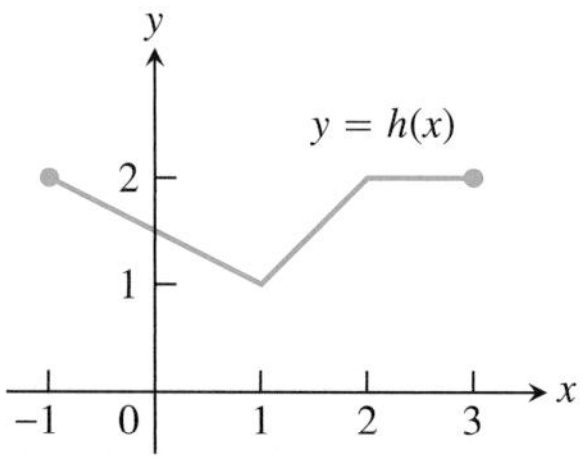

4.

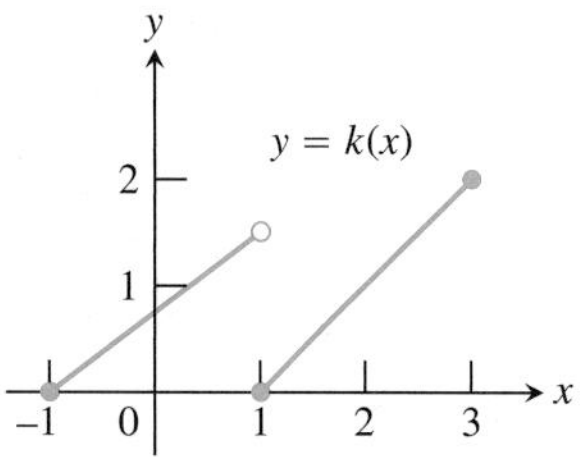

Exercises 5–8 are about the function

$$f(x) = \begin{cases} x^2 - 1, & -1 \le x < 0 \\ 2x, & 0 < x < 1 \\ 1, & x = 1 \\ -2x + 4, & 1 < x < 2 \\ 0, & 2 < x < 3 \end{cases}$$

graphed in the accompanying figure.

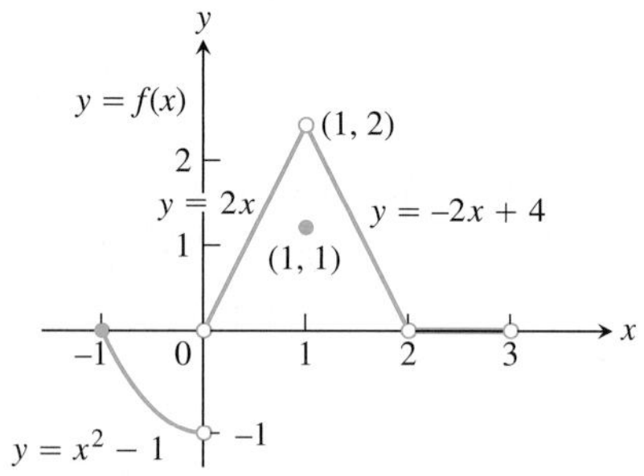

The graph for Exercises 5–8.

5. a. Does $f(-1)$ exist?

b. Does $\lim_{x\to -1^+} f(x)$ exist?

c. Does $\lim_{x\to -1^+} f(x) = f(-1)$?

d. Is f continuous at $x = -1$?

6. a. Does $f(1)$ exist?

b. Does $\lim_{x\to 1} f(x)$ exist?

c. Does $\lim_{x\to 1} f(x) = f(1)$?

d. Is f continuous at $x = 1$?

7. a. Is f defined at $x = 2$? (Look at the definition of f.)

b. Is f continuous at $x = 2$?

8. To what new value should $f(1)$ be changed to remove the discontinuity?

At which points do the functions in Exercises 9 and 10 fail to be continuous? At which points, if any, are the discontinuities removable? Not removable? Give reasons for your answers.

9. Exercise 1, Section 1.6

10. Exercise 2, Section 1.6

At what points are the functions in Exercises 11–26 continuous?

11. $y = \dfrac{1}{x - 2} - 3x$

12. $y = \dfrac{1}{(x + 2)^2} + 4$

13. $y = \dfrac{x + 1}{x^2 - 4x + 3}$

14. $y = \dfrac{x + 3}{x^2 - 3x - 10}$

15. $y = |x - 1| + \sin x$

16. $y = \dfrac{1}{|x| + 1} - \dfrac{x^2}{2}$

17. $y = \dfrac{\cos x}{x}$

18. $y = \dfrac{x + 2}{\cos x}$

19. $y = \csc 2x$

20. $y = \tan \dfrac{\pi x}{2}$

21. $y = \dfrac{x \tan x}{x^2 + 1}$

22. $y = \dfrac{\sqrt{x^4 + 1}}{1 + \sin^2 x}$

23. $y = \sqrt{2x + 3}$

24. $y = \sqrt[4]{3x - 1}$

25. $y = (2x - 1)^{1/3}$

26. $y = (2 - x)^{1/5}$

Find the limits in Exercises 27–30. Are the functions continuous at the point being approached?

27. $\lim_{x\to\pi} \sin (x - \sin x)$

28. $\lim_{t\to 0} \sin\left(\dfrac{\pi}{2} \cos (\tan t)\right)$

29. $\lim_{y\to 1} \sec (y \sec^2 y - \tan^2 y - 1)$

30. $\lim_{x\to 0} \tan\left(\dfrac{\pi}{4} \cos (\sin x^{1/3})\right)$

31. For what value of a is

$$f(x) = \begin{cases} x^2 - 1, & x < 3 \\ 2ax, & x \ge 3 \end{cases}$$

continuous at every x?

32. For what value of b is

$$g(x) = \begin{cases} x, & x < -2 \\ bx^2, & x \ge -2 \end{cases}$$

continuous at every x?

33. Roots of a cubic Show that the equation $x^3 - 15x + 1 = 0$ has three solutions in the interval $[-4, 4]$.

34. A function value Show that the function $F(x) = (x - a)^2 \cdot (x - b)^2 + x$ takes on the value $(a + b)/2$ for some value of x.

35. **Solving an equation** If $f(x) = x^3 - 8x + 10$, show that there are values c for which $f(c)$ equals **(a)** π; **(b)** $-\sqrt{3}$; **(c)** 5,000,000.

36. Explain why the following five statements ask for the same information.

a. Find the roots of $f(x) = x^3 - 3x - 1$.

b. Find the x-coordinates of the points where the curve $y = x^3$ crosses the line $y = 3x + 1$.

c. Find all the values of x for which $x^3 - 3x = 1$.

d. Find the x-coordinates of the points where the cubic curve $y = x^3 - 3x$ crosses the line $y = 1$.

e. Solve the equation $x^3 - 3x - 1 = 0$.

37. **Removable discontinuity** Give an example of a function $f(x)$ that is continuous for all values of x except $x = 2$, where it has a removable discontinuity. Explain how you know that f is discontinuous at $x = 2$, and how you know the discontinuity is removable.

38. **Nonremovable discontinuity** Give an example of a function $g(x)$ that is continuous for all values of x except $x = -1$, where it has a nonremovable discontinuity. Explain how you know that g is discontinuous there and why the discontinuity is not removable.

39. **A function discontinuous at every point**

a. Use the fact that every nonempty interval of real numbers contains both rational and irrational numbers to show that the function

$$f(x) = \begin{cases} 1, & \text{if } x \text{ is rational} \\ 0, & \text{if } x \text{ is irrational} \end{cases}$$

is discontinuous at every point.

b. Is f right-continuous or left-continuous at any point?

40. If functions $f(x)$ and $g(x)$ are continuous for $0 \le x \le 1$, could $f(x)/g(x)$ possibly be discontinuous at a point of [0, 1]? Give reasons for your answer.

41. If the product function $h(x) = f(x) \cdot g(x)$ is continuous at $x = 0$, must $f(x)$ and $g(x)$ be continuous at $x = 0$? Give reasons for your answer.

42. **Discontinuous composite of continuous functions** Give an example of functions f and g, both continuous at $x = 0$, for which the composite $f \circ g$ is discontinuous at $x = 0$. Does this contradict Theorem 9? Give reasons for your answer.

43. **Never-zero continuous functions** Is it true that a continuous function that is never zero on an interval never changes sign on that interval? Give reasons for your answer.

44. **Stretching a rubber band** Is it true that if you stretch a rubber band by moving one end to the right and the other to the left, some point of the band will end up in its original position? Give reasons for your answer.

45. **A fixed point theorem** Suppose that a function f is continuous on the closed interval [0, 1] and that $0 \le f(x) \le 1$ for every x in [0, 1]. Show that there must exist a number c in [0, 1] such that $f(c) = c$ (c is called a **fixed point** of f).

46. **The sign-preserving property of continuous functions** Let f be defined on an interval (a, b) and suppose that $f(c) \neq 0$ at some c where f is continuous. Show that there is an interval $(c - \delta, c + \delta)$ about c where f has the same sign as $f(c)$.

47. Prove that f is continuous at c if and only if

$$\lim_{h \to 0} f(c + h) = f(c).$$

48. Use Exercise 47 together with the identities

$$\sin(h + c) = \sin h \cos c + \cos h \sin c,$$
$$\cos(h + c) = \cos h \cos c - \sin h \sin c$$

to prove that $f(x) = \sin x$ and $g(x) = \cos x$ are continuous at every point $x = c$.

T Use a graphing calculator or computer grapher to solve the equations in Exercises 49–56.

49. $x^3 - 3x - 1 = 0$

50. $2x^3 - 2x^2 - 2x + 1 = 0$

51. $x(x - 1)^2 = 1$ (one root)

52. $x^x = 2$

53. $\sqrt{x} + \sqrt{1 + x} = 4$

54. $x^3 - 15x + 1 = 0$ (three roots)

55. $\cos x = x$ (one root). Make sure you are using radian mode.

56. $2 \sin x = x$ (three roots). Make sure you are using radian mode.

1.8 Limits Involving Infinity

In this section we investigate the behavior of a function when the magnitude of the independent variable x becomes increasingly large, or $x \to \pm\infty$. We further extend the concept of limit to *infinite limits*, which are not limits as before, but rather a new use of the term limit. Infinite limits provide useful symbols and language for describing the behavior of functions whose values become arbitrarily large in magnitude. We use these limit ideas to analyze the graphs of functions having *horizontal* or *vertical asymptotes*.

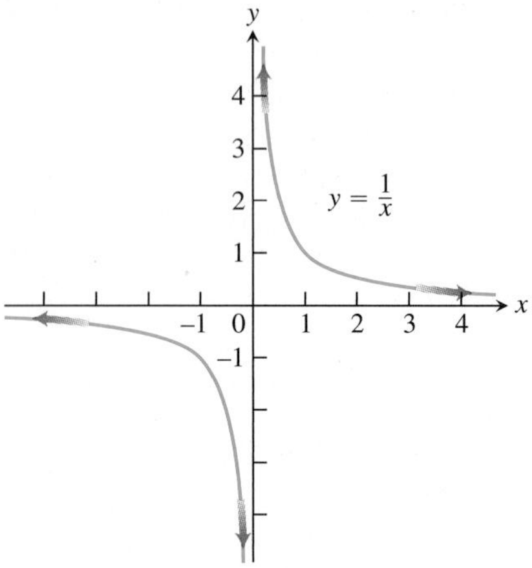

FIGURE 1.71 The graph of $y = 1/x$ approaches 0 as $x \to \infty$ or $x \to -\infty$.

Finite Limits as $x \to \pm\infty$

The symbol for infinity (∞) does not represent a real number. We use ∞ to describe the behavior of a function when the values in its domain or range outgrow all finite bounds. For example, the function $f(x) = 1/x$ is defined for all $x \neq 0$ (Figure 1.71). When x is positive and becomes increasingly large, $1/x$ becomes increasingly small. When x is negative and its magnitude becomes increasingly large, $1/x$ again becomes small. We summarize these observations by saying that $f(x) = 1/x$ has limit 0 as $x \to \infty$ or $x \to -\infty$, or that 0 is a *limit of* $f(x) = 1/x$ *at infinity and negative infinity*. Here is a precise definition.

DEFINITIONS

1. We say that $f(x)$ has the **limit L as x approaches infinity** and write
$$\lim_{x \to \infty} f(x) = L$$
if, for every number $\epsilon > 0$, there exists a corresponding number M such that for all x
$$x > M \quad \Rightarrow \quad |f(x) - L| < \epsilon.$$
2. We say that $f(x)$ has the **limit L as x approaches minus infinity** and write
$$\lim_{x \to -\infty} f(x) = L$$
if, for every number $\epsilon > 0$, there exists a corresponding number N such that for all x
$$x < N \quad \Rightarrow \quad |f(x) - L| < \epsilon.$$

Intuitively, $\lim_{x\to\infty} f(x) = L$ if, as x moves increasingly far from the origin in the positive direction, $f(x)$ gets arbitrarily close to L. Similarly, $\lim_{x\to-\infty} f(x) = L$ if, as x moves increasingly far from the origin in the negative direction, $f(x)$ gets arbitrarily close to L.

The strategy for calculating limits of functions as $x \to \pm\infty$ is similar to the one for finite limits in Section 1.4. There we first found the limits of the constant and identity functions $y = k$ and $y = x$. We then extended these results to other functions by applying a theorem about limits of algebraic combinations. Here we do the same thing, except that the starting functions are $y = k$ and $y = 1/x$ instead of $y = k$ and $y = x$.

The basic facts to be verified by applying the formal definition are

$$\lim_{x \to \pm\infty} k = k \quad \text{and} \quad \lim_{x \to \pm\infty} \frac{1}{x} = 0. \tag{1}$$

We prove the latter and leave the former to Exercises 71 and 72.

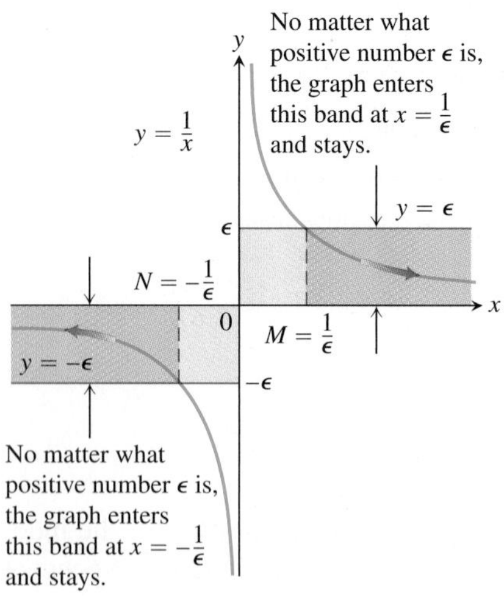

FIGURE 1.72 The geometry behind the argument in Example 1.

EXAMPLE 1 Show that

(a) $\lim_{x \to \infty} \frac{1}{x} = 0$ **(b)** $\lim_{x \to -\infty} \frac{1}{x} = 0.$

Solution

(a) Let $\epsilon > 0$ be given. We must find a number M such that for all x
$$x > M \quad \Rightarrow \quad \left|\frac{1}{x} - 0\right| = \left|\frac{1}{x}\right| < \epsilon.$$

The implication will hold if $M = 1/\epsilon$ or any larger positive number (Figure 1.72). This proves $\lim_{x\to\infty} (1/x) = 0$.

(b) Let $\epsilon > 0$ be given. We must find a number N such that for all x

$$x < N \quad \Rightarrow \quad \left|\frac{1}{x} - 0\right| = \left|\frac{1}{x}\right| < \epsilon.$$

The implication will hold if $N = -1/\epsilon$ or any number less than $-1/\epsilon$ (Figure 1.72). This proves $\lim_{x\to-\infty}(1/x) = 0$. ■

Limits at infinity have properties similar to those of finite limits.

THEOREM 12 All the limit laws in Theorem 1 are true when we replace $\lim_{x\to c}$ by $\lim_{x\to\infty}$ or $\lim_{x\to-\infty}$. That is, the variable x may approach a finite number c or $\pm\infty$.

The properties in Theorem 12 are used to calculate limits in the same way as when x approaches a finite number.

EXAMPLE 2

(a) $$\lim_{x\to\infty}\left(5 + \frac{1}{x}\right) = \lim_{x\to\infty} 5 + \lim_{x\to\infty}\frac{1}{x} \qquad \text{Sum Rule}$$

$$= 5 + 0 = 5 \qquad \text{Known limits}$$

(b) $$\lim_{x\to-\infty}\frac{\pi\sqrt{3}}{x^2} = \lim_{x\to-\infty}\pi\sqrt{3}\cdot\frac{1}{x}\cdot\frac{1}{x}$$

$$= \lim_{x\to-\infty}\pi\sqrt{3}\cdot\lim_{x\to-\infty}\frac{1}{x}\cdot\lim_{x\to-\infty}\frac{1}{x} \qquad \text{Product Rule}$$

$$= \pi\sqrt{3}\cdot 0\cdot 0 = 0 \qquad \text{Known limits}$$ ■

Limits at Infinity of Rational Functions

To determine the limit of a rational function as $x \to \pm\infty$, we can divide the numerator and denominator by the highest power of x in the denominator. What happens then depends on the degrees of the polynomials involved.

EXAMPLE 3

(a) $$\lim_{x\to\infty}\frac{5x^2 + 8x - 3}{3x^2 + 2} = \lim_{x\to\infty}\frac{5 + (8/x) - (3/x^2)}{3 + (2/x^2)} \qquad \text{Divide numerator and denominator by } x^2.$$

$$= \frac{5 + 0 - 0}{3 + 0} = \frac{5}{3} \qquad \text{See Fig. 1.73}$$

(b) $$\lim_{x\to-\infty}\frac{11x + 2}{2x^3 - 1} = \lim_{x\to-\infty}\frac{(11/x^2) + (2/x^3)}{2 - (1/x^3)} \qquad \text{Divide numerator and denominator by } x^3.$$

$$= \frac{0 + 0}{2 - 0} = 0 \qquad \text{See Fig. 1.74.}$$ ■

Examples of the case for which the degree of the numerator is greater than the degree of the denominator are given in Exercises 83–88.

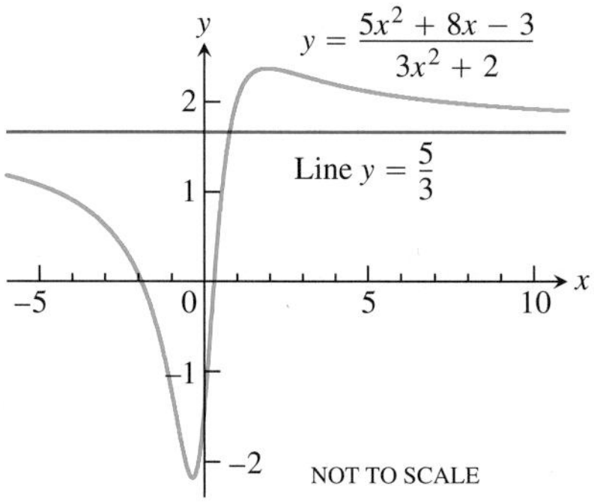

FIGURE 1.73 The graph of the function in Example 3a. The graph approaches the line $y = 5/3$ as $|x|$ increases.

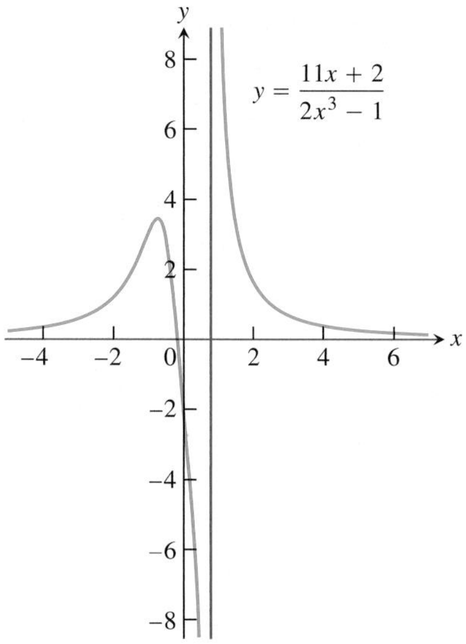

FIGURE 1.74 The graph of the function in Example 3b. The graph approaches the x-axis as $|x|$ increases.

Horizontal Asymptotes

If the distance between the graph of a function and some fixed line approaches zero as a point on the graph moves increasingly far from the origin, we say that the graph approaches the line asymptotically and that the line is an *asymptote* of the graph.

Looking at $f(x) = 1/x$ (see Figure 1.71), we observe that the x-axis is an asymptote of the curve on the right because

$$\lim_{x\to\infty} \frac{1}{x} = 0$$

and on the left because

$$\lim_{x\to-\infty} \frac{1}{x} = 0.$$

We say that the x-axis is a *horizontal asymptote* of the graph of $f(x) = 1/x$.

DEFINITION A line $y = b$ is a **horizontal asymptote** of the graph of a function $y = f(x)$ if either

$$\lim_{x\to\infty} f(x) = b \quad \text{or} \quad \lim_{x\to-\infty} f(x) = b.$$

The curve

$$f(x) = \frac{5x^2 + 8x - 3}{3x^2 + 2}$$

sketched in Figure 1.73 (Example 3a) has the line $y = 5/3$ as a horizontal asymptote on both the right and the left because

$$\lim_{x\to\infty} f(x) = \frac{5}{3} \quad \text{and} \quad \lim_{x\to-\infty} f(x) = \frac{5}{3}.$$

EXAMPLE 4 Find the horizontal asymptotes of the graph of

$$y = \frac{x^3 - 2}{|x|^3 + 1}.$$

Solution We calculate the limits as $x \to \pm\infty$.

For $x \geq 0$: $$\lim_{x\to\infty} \frac{x^3 - 2}{|x|^3 + 1} = \lim_{x\to\infty} \frac{x^3 - 2}{x^3 + 1} = \lim_{x\to\infty} \frac{1 - (2/x^3)}{1 + (1/x^3)} = 1.$$

For $x < 0$: $$\lim_{x\to-\infty} \frac{x^3 - 2}{|x|^3 + 1} = \lim_{x\to-\infty} \frac{x^3 - 2}{(-x)^3 + 1} = \lim_{x\to-\infty} \frac{1 - (2/x^3)}{-1 + (1/x^3)} = -1.$$

The horizontal asymptotes are $y = -1$ and $y = 1$. The graph is displayed in Figure 1.75. Notice that the graph crosses the horizontal asymptote $y = -1$ for a positive value of x. ■

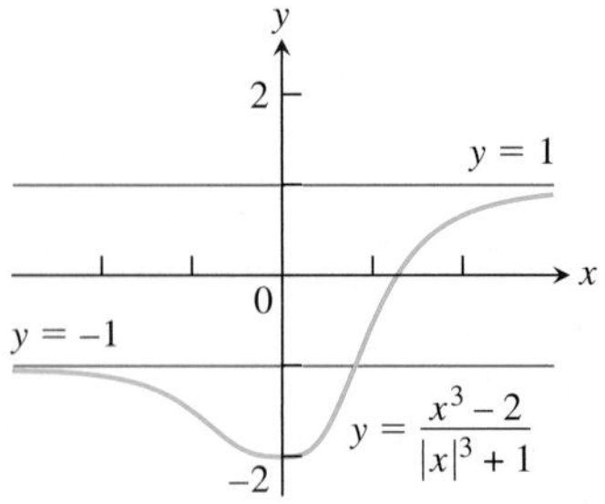

FIGURE 1.75 The graph of the function in Example 4 has two horizontal asymptotes.

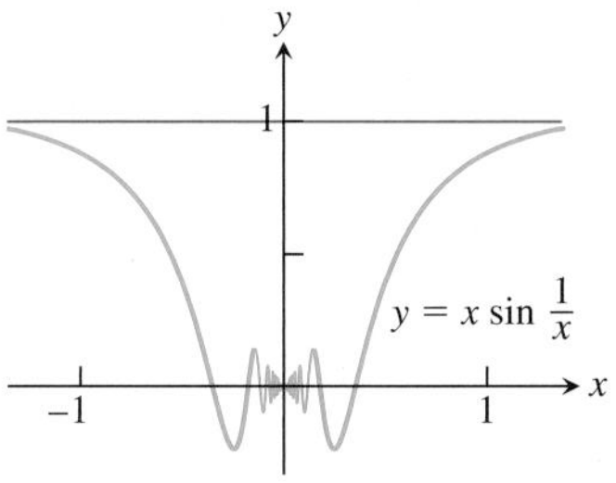

FIGURE 1.76 The line $y = 1$ is a horizontal asymptote of the function graphed here (Example 5b).

EXAMPLE 5 Find **(a)** $\lim_{x \to \infty} \sin(1/x)$ and **(b)** $\lim_{x \to \pm\infty} x \sin(1/x)$.

Solution

(a) We introduce the new variable $t = 1/x$. From Example 1, we know that $t \to 0^+$ as $x \to \infty$ (see Figure 1.71). Therefore,

$$\lim_{x \to \infty} \sin \frac{1}{x} = \lim_{t \to 0^+} \sin t = 0.$$

Likewise, we can investigate the behavior of $y = f(1/x)$ as $x \to 0$ by investigating $y = f(t)$ as $t \to \pm\infty$, where $t = 1/x$.

(b) We calculate the limts as $x \to \infty$ and $x \to -\infty$:

$$\lim_{x \to \infty} x \sin \frac{1}{x} = \lim_{t \to 0^+} \frac{\sin t}{t} = 1 \quad \text{and} \quad \lim_{x \to -\infty} x \sin \frac{1}{x} = \lim_{t \to 0^-} \frac{\sin t}{t} = 1.$$

The graph is shown in Figure 1.76, and we see that the line $y = 1$ is a horizontal asymptote. ■

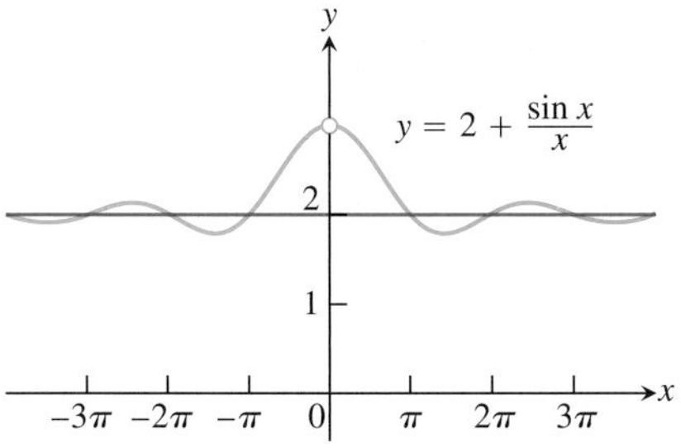

FIGURE 1.77 A curve may cross one of its asymptotes infinitely often (Example 6).

The Sandwich Theorem Revisited

The Sandwich Theorem also holds for limits as $x \to \pm\infty$.

EXAMPLE 6 Using the Sandwich Theorem, find the horizontal asymptote of the curve

$$y = 2 + \frac{\sin x}{x}.$$

Solution We are interested in the behavior as $x \to \pm\infty$. Since

$$0 \le \left| \frac{\sin x}{x} \right| \le \left| \frac{1}{x} \right|$$

and $\lim_{x \to \pm\infty} |1/x| = 0$, we have $\lim_{x \to \pm\infty} (\sin x)/x = 0$ by the Sandwich Theorem. Hence,

$$\lim_{x \to \pm\infty} \left(2 + \frac{\sin x}{x} \right) = 2 + 0 = 2,$$

and the line $y = 2$ is a horizontal asymptote of the curve on both left and right (Figure 1.77).

This example illustrates that a curve may cross one of its horizontal asymptotes many times. ■

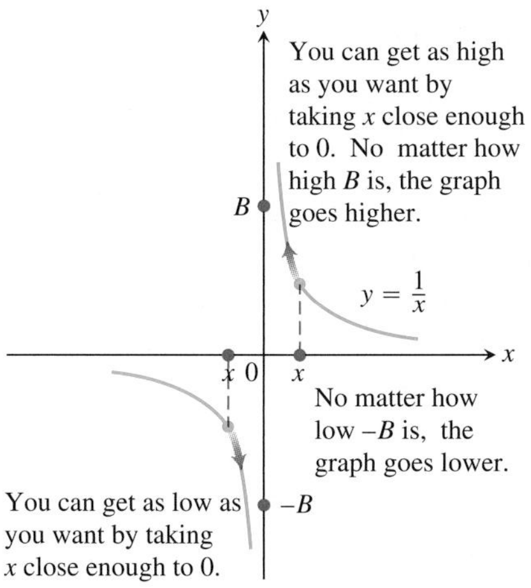

FIGURE 1.78 One-sided infinite limits: $\lim_{x \to 0^+} \frac{1}{x} = \infty$ and $\lim_{x \to 0^-} \frac{1}{x} = -\infty$.

Infinite Limits

Let us look again at the function $f(x) = 1/x$. As $x \to 0^+$, the values of f grow without bound, eventually reaching and surpassing every positive real number. That is, given any positive real number B, however large, the values of f become larger still (Figure 1.78). Thus, f has no limit as $x \to 0^+$. It is nevertheless convenient to describe the behavior of f by saying that $f(x)$ approaches ∞ as $x \to 0^+$. We write

$$\lim_{x \to 0^+} f(x) = \lim_{x \to 0^+} \frac{1}{x} = \infty.$$

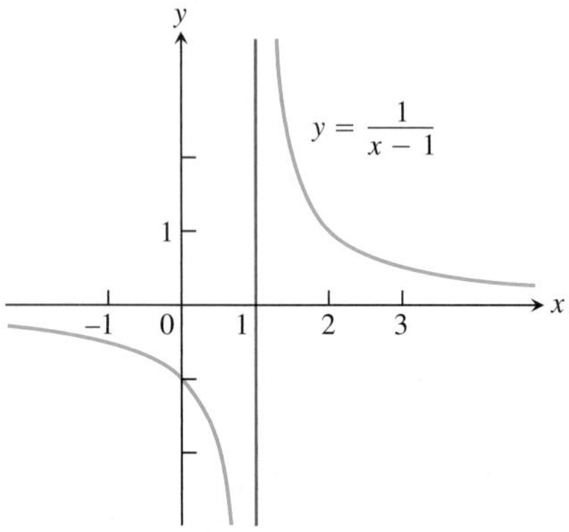

FIGURE 1.79 Near $x = 1$, the function $y = 1/(x - 1)$ behaves the way the function $y = 1/x$ behaves near $x = 0$. Its graph is the graph of $y = 1/x$ shifted 1 unit to the right (Example 7).

In writing this, we are *not* saying that the limit exists. Nor are we saying that there is a real number ∞, for there is no such number. Rather, we are saying that $\lim_{x\to 0^+}(1/x)$ *does not exist because* $1/x$ *becomes arbitrarily large and positive as* $x \to 0^+$.

As $x \to 0^-$, the values of $f(x) = 1/x$ become arbitrarily large and negative. Given any negative real number $-B$, the values of f eventually lie below $-B$. (See Figure 1.78.) We write

$$\lim_{x\to 0^-} f(x) = \lim_{x\to 0^-} \frac{1}{x} = -\infty.$$

Again, we are not saying that the limit exists and equals the number $-\infty$. There *is* no real number $-\infty$. We are describing the behavior of a function whose limit as $x \to 0^-$ *does not exist because its values become arbitrarily large and negative.*

EXAMPLE 7 Find $\lim_{x\to 1^+} \frac{1}{x-1}$ and $\lim_{x\to 1^-} \frac{1}{x-1}$.

Geometric Solution The graph of $y = 1/(x - 1)$ is the graph of $y = 1/x$ shifted 1 unit to the right (Figure 1.79). Therefore, $y = 1/(x - 1)$ behaves near 1 exactly the way $y = 1/x$ behaves near 0:

$$\lim_{x\to 1^+} \frac{1}{x-1} = \infty \qquad \text{and} \qquad \lim_{x\to 1^-} \frac{1}{x-1} = -\infty.$$

Analytic Solution Think about the number $x - 1$ and its reciprocal. As $x \to 1^+$, we have $(x - 1) \to 0^+$ and $1/(x - 1) \to \infty$. As $x \to 1^-$, we have $(x - 1) \to 0^-$ and $1/(x - 1) \to -\infty$. ■

EXAMPLE 8 Discuss the behavior of

$$f(x) = \frac{1}{x^2} \qquad \text{near} \qquad x = 0.$$

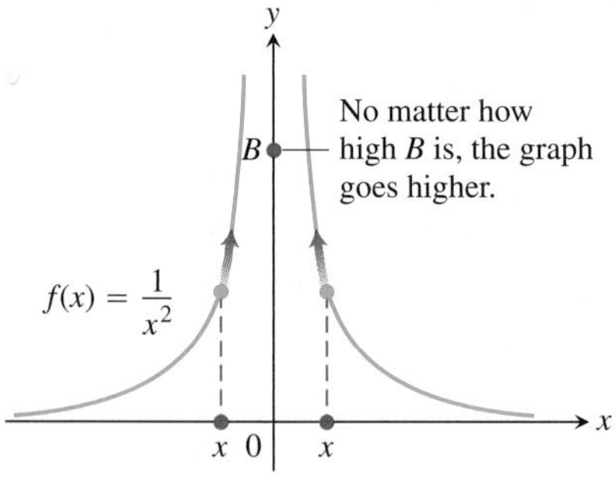

FIGURE 1.80 The graph of $f(x)$ in Example 8 approaches infinity as $x \to 0$.

Solution As x approaches zero from either side, the values of $1/x^2$ are positive and become arbitrarily large (Figure 1.80):

$$\lim_{x\to 0} f(x) = \lim_{x\to 0} \frac{1}{x^2} = \infty.$$ ■

The function $y = 1/x$ shows no consistent behavior as $x \to 0$. We have $1/x \to \infty$ if $x \to 0^+$, but $1/x \to -\infty$ if $x \to 0^-$. All we can say about $\lim_{x\to 0}(1/x)$ is that it does not exist. The function $y = 1/x^2$ is different. Its values approach infinity as x approaches zero from either side, so we can say that $\lim_{x\to 0}(1/x^2) = \infty$.

EXAMPLE 9

(a) $$\lim_{x\to 2} \frac{(x-2)^2}{x^2-4} = \lim_{x\to 2} \frac{(x-2)^2}{(x-2)(x+2)} = \lim_{x\to 2} \frac{x-2}{x+2} = 0$$

(b) $$\lim_{x\to 2} \frac{x-2}{x^2-4} = \lim_{x\to 2} \frac{x-2}{(x-2)(x+2)} = \lim_{x\to 2} \frac{1}{x+2} = \frac{1}{4}$$

(c) $\lim_{x\to 2^+} \dfrac{x-3}{x^2-4} = \lim_{x\to 2^+} \dfrac{x-3}{(x-2)(x+2)} = -\infty$ The values are negative for $x > 2$, x near 2.

(d) $\lim_{x\to 2^-} \dfrac{x-3}{x^2-4} = \lim_{x\to 2^-} \dfrac{x-3}{(x-2)(x+2)} = \infty$ The values are positive for $x < 2$, x near 2.

(e) $\lim_{x\to 2} \dfrac{x-3}{x^2-4} = \lim_{x\to 2} \dfrac{x-3}{(x-2)(x+2)}$ does not exist. See parts (c) and (d).

(f) $\lim_{x\to 2} \dfrac{2-x}{(x-2)^3} = \lim_{x\to 2} \dfrac{-(x-2)}{(x-2)^3} = \lim_{x\to 2} \dfrac{-1}{(x-2)^2} = -\infty$

In parts (a) and (b) the effect of the zero in the denominator at $x = 2$ is canceled because the numerator is zero there also. Thus a finite limit exists. This is not true in part (f), where cancellation still leaves a zero factor in the denominator. ■

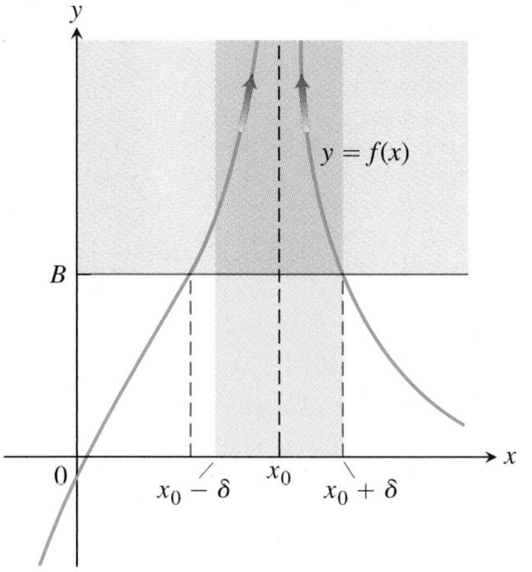

FIGURE 1.81 For $x_0 - \delta < x < x_0 + \delta$, the graph of $f(x)$ lies above the line $y = B$.

Precise Definitions of Infinite Limits

Instead of requiring $f(x)$ to lie arbitrarily close to a finite number L for all x sufficiently close to x_0, the definitions of infinite limits require $f(x)$ to lie arbitrarily far from zero. Except for this change, the language is identical with what we have seen before. Figures 1.81 and 1.82 accompany these definitions.

DEFINITIONS

1. We say that ***f*(*x*) approaches infinity as *x* approaches x_0**, and write

$$\lim_{x\to x_0} f(x) = \infty,$$

if for every positive real number B there exists a corresponding $\delta > 0$ such that for all x

$$0 < |x - x_0| < \delta \quad \Rightarrow \quad f(x) > B.$$

2. We say that ***f*(*x*) approaches negative infinity as *x* approaches x_0**, and write

$$\lim_{x\to x_0} f(x) = -\infty,$$

if for every negative real number $-B$ there exists a corresponding $\delta > 0$ such that for all x

$$0 < |x - x_0| < \delta \quad \Rightarrow \quad f(x) < -B.$$

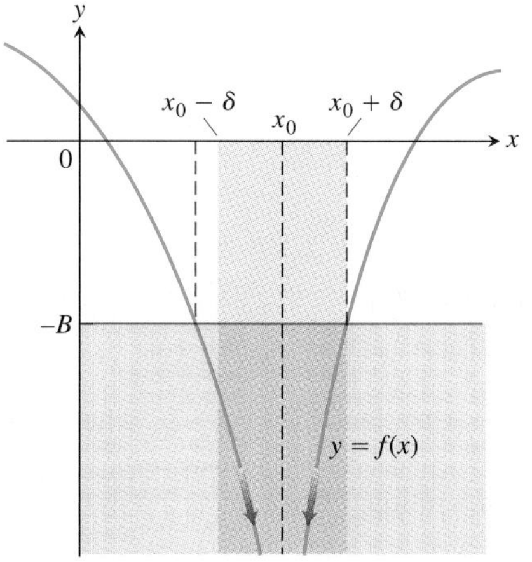

FIGURE 1.82 For $x_0 - \delta < x < x_0 + \delta$, the graph of $f(x)$ lies below the line $y = -B$.

The precise definitions of one-sided infinite limits at x_0 are similar and are stated in the exercises.

EXAMPLE 10 Prove that $\lim_{x\to 0} \dfrac{1}{x^2} = \infty$.

Solution Given $B > 0$, we want to find $\delta > 0$ such that

$$0 < |x - 0| < \delta \quad \text{implies} \quad \frac{1}{x^2} > B.$$

Now,

$$\frac{1}{x^2} > B \qquad \text{if and only if } x^2 < \frac{1}{B}$$

or, equivalently,

$$|x| < \frac{1}{\sqrt{B}}.$$

Thus, choosing $\delta = 1/\sqrt{B}$ (or any smaller positive number), we see that

$$|x| < \delta \quad \text{implies} \quad \frac{1}{x^2} > \frac{1}{\delta^2} \geq B.$$

Therefore, by definition,

$$\lim_{x \to 0} \frac{1}{x^2} = \infty.$$

■

Vertical Asymptotes

Notice that the distance between a point on the graph of $y = 1/x$ and the y-axis approaches zero as the point moves vertically along the graph and away from the origin (Figure 1.83). This behavior occurs because

$$\lim_{x \to 0^+} \frac{1}{x} = \infty \qquad \text{and} \qquad \lim_{x \to 0^-} \frac{1}{x} = -\infty.$$

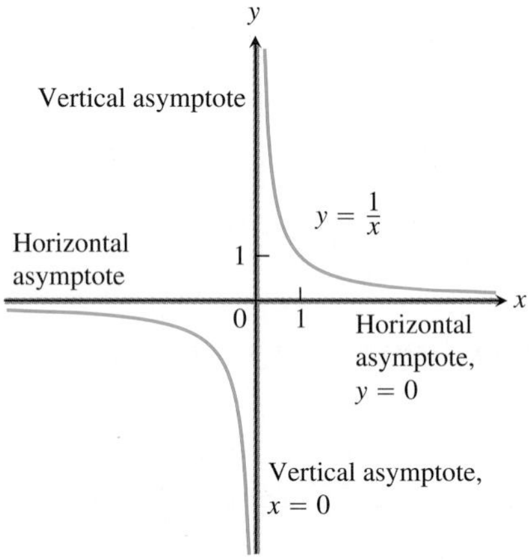

FIGURE 1.83 The coordinate axes are asymptotes of both branches of the hyperbola $y = 1/x$.

We say that the line $x = 0$ (the y-axis) is a *vertical asymptote* of the graph of $y = 1/x$. Observe that the denominator is zero at $x = 0$ and the function is undefined there.

DEFINITION A line $x = a$ is a **vertical asymptote** of the graph of a function $y = f(x)$ if either

$$\lim_{x \to a^+} f(x) = \pm\infty \qquad \text{or} \qquad \lim_{x \to a^-} f(x) = \pm\infty.$$

EXAMPLE 11 Find the horizontal and vertical asymptotes of the curve

$$y = \frac{x+3}{x+2}.$$

Solution We are interested in the behavior as $x \to \pm\infty$ and as $x \to -2$, where the denominator is zero.

The asymptotes are quickly revealed if we recast the rational function as a polynomial with a remainder, by dividing $(x + 2)$ into $(x + 3)$.

$$\begin{array}{r} 1 \\ x+2 \overline{)\, x+3} \\ \underline{x+2} \\ 1 \end{array}$$

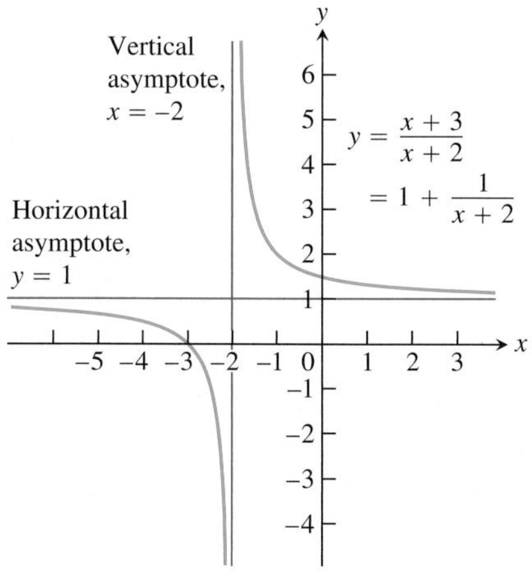

FIGURE 1.84 The lines $y = 1$ and $x = -2$ are asymptotes of the curve $y = (x + 3)/(x + 2)$ (Example 11).

This result enables us to rewrite y:

$$y = 1 + \frac{1}{x + 2}.$$

We now see that the curve in question is the graph of $y = 1/x$ shifted 1 unit up and 2 units left (Figure 1.84). The asymptotes, instead of being the coordinate axes, are now the lines $y = 1$ and $x = -2$. ■

EXAMPLE 12 Find the horizontal and vertical asymptotes of the graph of

$$f(x) = -\frac{8}{x^2 - 4}.$$

Solution We are interested in the behavior as $x \to \pm\infty$ and as $x \to \pm 2$, where the denominator is zero. Notice that f is an even function of x, so its graph is symmetric with respect to the y-axis.

(a) *The behavior as* $x \to \pm\infty$. Since $\lim_{x\to\infty} f(x) = 0$, the line $y = 0$ is a horizontal asymptote of the graph to the right. By symmetry it is an asymptote to the left as well (Figure 1.85). Notice that the curve approaches the x-axis from only the negative side (or from below).

(b) *The behavior as* $x \to \pm 2$. Since

$$\lim_{x\to 2^+} f(x) = -\infty \quad \text{and} \quad \lim_{x\to 2^-} f(x) = \infty,$$

the line $x = 2$ is a vertical asymptote both from the right and from the left. By symmetry, the same holds for the line $x = -2$.

There are no other asymptotes because f has a finite limit at every other point. ■

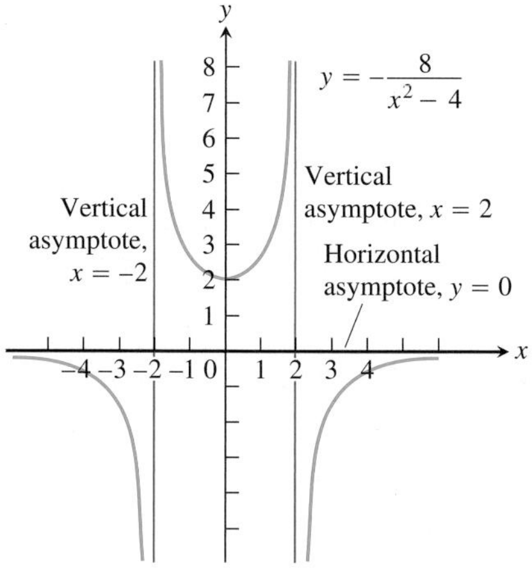

FIGURE 1.85 Graph of $y = -8/(x^2 - 4)$. Notice that the curve approaches the x-axis from only one side. Asymptotes do not have to be two-sided (Example 12).

EXAMPLE 13 The curves

$$y = \sec x = \frac{1}{\cos x} \quad \text{and} \quad y = \tan x = \frac{\sin x}{\cos x}$$

both have vertical asymptotes at odd-integer multiples of $\pi/2$, where $\cos x = 0$ (Figure 1.86).

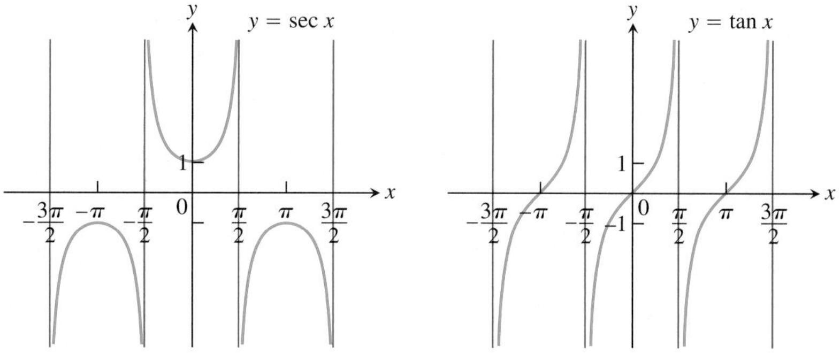

FIGURE 1.86 The graphs of sec x and tan x have infinitely many vertical asymptotes (Example 13). ■

EXERCISES 1.8

In Exercises 1–6, find the limit of each function **(a)** as $x \to \infty$ and **(b)** as $x \to -\infty$. (You may wish to visualize your answer with a graphing calculator or computer.)

1. $f(x) = \frac{2}{x} - 3$

2. $f(x) = \pi - \frac{2}{x^2}$

3. $g(x) = \frac{1}{2 + (1/x)}$

4. $g(x) = \frac{1}{8 - (5/x^2)}$

5. $h(x) = \frac{-5 + (7/x)}{3 - (1/x^2)}$

6. $h(x) = \frac{3 - (2/x)}{4 + (\sqrt{2}/x^2)}$

Find the limits in Exercises 7–10.

7. $\lim_{x \to \infty} \frac{\sin 2x}{x}$

8. $\lim_{\theta \to -\infty} \frac{\cos \theta}{3\theta}$

9. $\lim_{t \to -\infty} \frac{2 - t + \sin t}{t + \cos t}$

10. $\lim_{r \to \infty} \frac{r + \sin r}{2r + 7 - 5 \sin r}$

In Exercises 11–20, find the limit of each rational function **(a)** as $x \to \infty$ and **(b)** as $x \to -\infty$.

11. $f(x) = \frac{2x + 3}{5x + 7}$

12. $f(x) = \frac{2x^3 + 7}{x^3 - x^2 + x + 7}$

13. $f(x) = \frac{x + 1}{x^2 + 3}$

14. $f(x) = \frac{3x + 7}{x^2 - 2}$

15. $h(x) = \frac{7x^3}{x^3 - 3x^2 + 6x}$

16. $g(x) = \frac{1}{x^3 - 4x + 1}$

17. $g(x) = \frac{10x^5 + x^4 + 31}{x^6}$

18. $h(x) = \frac{9x^4 + x}{2x^4 + 5x^2 - x + 6}$

19. $h(x) = \frac{-2x^3 - 2x + 3}{3x^3 + 3x^2 - 5x}$

20. $h(x) = \frac{-x^4}{x^4 - 7x^3 + 7x^2 + 9}$

The process by which we determine limits of rational functions applies equally well to ratios containing noninteger or negative powers of x: divide numerator and denominator by the highest power of x in the denominator and proceed from there. Find the limits in Exercises 21–26.

21. $\lim_{x \to \infty} \frac{2\sqrt{x} + x^{-1}}{3x - 7}$

22. $\lim_{x \to \infty} \frac{2 + \sqrt{x}}{2 - \sqrt{x}}$

23. $\lim_{x \to -\infty} \frac{\sqrt[3]{x} - \sqrt[5]{x}}{\sqrt[3]{x} + \sqrt[5]{x}}$

24. $\lim_{x \to \infty} \frac{x^{-1} + x^{-4}}{x^{-2} - x^{-3}}$

25. $\lim_{x \to \infty} \frac{2x^{5/3} - x^{1/3} + 7}{x^{8/5} + 3x + \sqrt{x}}$

26. $\lim_{x \to -\infty} \frac{\sqrt[3]{x} - 5x + 3}{2x + x^{2/3} - 4}$

Find the limits in Exercises 27–38.

27. $\lim_{x \to 0^+} \frac{1}{3x}$

28. $\lim_{x \to 0^-} \frac{5}{2x}$

29. $\lim_{x \to 2^-} \frac{3}{x - 2}$

30. $\lim_{x \to 3^+} \frac{1}{x - 3}$

31. $\lim_{x \to -8^+} \frac{2x}{x + 8}$

32. $\lim_{x \to -5^-} \frac{3x}{2x + 10}$

33. $\lim_{x \to 7} \frac{4}{(x - 7)^2}$

34. $\lim_{x \to 0} \frac{-1}{x^2(x + 1)}$

35. **a.** $\lim_{x \to 0^+} \frac{2}{3x^{1/3}}$ **b.** $\lim_{x \to 0^-} \frac{2}{3x^{1/3}}$

36. **a.** $\lim_{x \to 0^+} \frac{2}{x^{1/5}}$ **b.** $\lim_{x \to 0^-} \frac{2}{x^{1/5}}$

37. $\lim_{x \to 0} \frac{4}{x^{2/5}}$

38. $\lim_{x \to 0} \frac{1}{x^{2/3}}$

Find the limits in Exercises 39–42.

39. $\lim_{x \to (\pi/2)^-} \tan x$

40. $\lim_{x \to (-\pi/2)^+} \sec x$

41. $\lim_{\theta \to 0^-} (1 + \csc \theta)$

42. $\lim_{\theta \to 0} (2 - \cot \theta)$

Find the limits in Exercises 43–48.

43. $\lim \frac{1}{x^2 - 4}$ as

a. $x \to 2^+$ **b.** $x \to 2^-$

c. $x \to -2^+$ **d.** $x \to -2^-$

44. $\lim \frac{x}{x^2 - 1}$ as

a. $x \to 1^+$ **b.** $x \to 1^-$

c. $x \to -1^+$ **d.** $x \to -1^-$

45. $\lim \left(\frac{x^2}{2} - \frac{1}{x}\right)$ as

a. $x \to 0^+$ **b.** $x \to 0^-$

c. $x \to \sqrt[3]{2}$ **d.** $x \to -1$

46. $\lim \frac{x^2 - 1}{2x + 4}$ as

a. $x \to -2^+$ **b.** $x \to -2^-$

c. $x \to 1^+$ **d.** $x \to 0^-$

47. $\lim \frac{x^2 - 3x + 2}{x^3 - 2x^2}$ as

a. $x \to 0^+$ **b.** $x \to 2^+$

c. $x \to 2^-$ **d.** $x \to 2$

e. What, if anything, can be said about the limit as $x \to 0$?

48. $\lim \frac{x^2 - 3x + 2}{x^3 - 4x}$ as

a. $x \to 2^+$ **b.** $x \to -2^+$

c. $x \to 0^-$ **d.** $x \to 1^+$

e. What, if anything, can be said about the limit as $x \to 0$?

Find the limits in Exercises 49–52.

49. $\lim \left(2 - \frac{3}{t^{1/3}}\right)$ as

a. $t \to 0^+$ **b.** $t \to 0^-$

50. $\lim \left(\dfrac{1}{t^{3/5}} + 7\right)$ as

a. $t \to 0^+$ **b.** $t \to 0^-$

51. $\lim \left(\dfrac{1}{x^{2/3}} + \dfrac{2}{(x-1)^{2/3}}\right)$ as

a. $x \to 0^+$ **b.** $x \to 0^-$

c. $x \to 1^+$ **d.** $x \to 1^-$

52. $\lim \left(\dfrac{1}{x^{1/3}} - \dfrac{1}{(x-1)^{4/3}}\right)$ as

a. $x \to 0^+$ **b.** $x \to 0^-$

c. $x \to 1^+$ **d.** $x \to 1^-$

Graph the rational functions in Exercises 53–58. Include the graphs and equations of the asymptotes.

53. $y = \dfrac{1}{x-1}$ **54.** $y = \dfrac{1}{x+1}$

55. $y = \dfrac{1}{2x+4}$ **56.** $y = \dfrac{-3}{x-3}$

57. $y = \dfrac{x+3}{x+2}$ **58.** $y = \dfrac{2x}{x+1}$

In Exercises 59–62, sketch the graph of a function $y = f(x)$ that satisfies the given conditions. No formulas are required—just label the coordinate axes and sketch an appropriate graph. (The answers are not unique, so your graphs may not be exactly like those in the answer section.)

59. $f(0) = 0$, $f(1) = 2$, $f(-1) = -2$, $\lim_{x\to-\infty} f(x) = -1$, and $\lim_{x\to\infty} f(x) = 1$

60. $f(0) = 0$, $\lim_{x\to\pm\infty} f(x) = 0$, $\lim_{x\to0^+} f(x) = 2$, and $\lim_{x\to0^-} f(x) = -2$

61. $f(0) = 0$, $\lim_{x\to\pm\infty} f(x) = 0$, $\lim_{x\to1^-} f(x) = \lim_{x\to-1^+} f(x) = \infty$, $\lim_{x\to1^+} f(x) = -\infty$, and $\lim_{x\to-1^-} f(x) = -\infty$

62. $f(2) = 1$, $f(-1) = 0$, $\lim_{x\to\infty} f(x) = 0$, $\lim_{x\to0^+} f(x) = \infty$, $\lim_{x\to0^-} f(x) = -\infty$, and $\lim_{x\to-\infty} f(x) = 1$

In Exercises 63–66, find a function that satisfies the given conditions and sketch its graph. (The answers here are not unique. Any function that satisfies the conditions is acceptable. Feel free to use formulas defined in pieces if that will help.)

63. $\lim_{x\to\pm\infty} f(x) = 0$, $\lim_{x\to2^-} f(x) = \infty$, and $\lim_{x\to2^+} f(x) = \infty$

64. $\lim_{x\to\pm\infty} g(x) = 0$, $\lim_{x\to3^-} g(x) = -\infty$, and $\lim_{x\to3^+} g(x) = \infty$

65. $\lim_{x\to-\infty} h(x) = -1$, $\lim_{x\to\infty} h(x) = 1$, $\lim_{x\to0^-} h(x) = -1$, and $\lim_{x\to0^+} h(x) = 1$

66. $\lim_{x\to\pm\infty} k(x) = 1$, $\lim_{x\to1^-} k(x) = \infty$, and $\lim_{x\to1^+} k(x) = -\infty$

67. Suppose that $f(x)$ and $g(x)$ are polynomials in x and that $\lim_{x\to\infty} (f(x)/g(x)) = 2$. Can you conclude anything about $\lim_{x\to-\infty} (f(x)/g(x))$? Give reasons for your answer.

68. Suppose that $f(x)$ and $g(x)$ are polynomials in x. Can the graph of $f(x)/g(x)$ have an asymptote if $g(x)$ is never zero? Give reasons for your answer.

69. How many horizontal asymptotes can the graph of a given rational function have? Give reasons for your answer.

70. Find $\lim_{x\to\infty} \left(\sqrt{x^2+x} - \sqrt{x^2-x}\right)$.

Use the formal definitions of limits as $x \to \pm\infty$ to establish the limits in Exercises 71 and 72.

71. If f has the constant value $f(x) = k$, then $\lim_{x\to\infty} f(x) = k$.

72. If f has the constant value $f(x) = k$, then $\lim_{x\to-\infty} f(x) = k$.

Use formal definitions to prove the limit statements in Exercises 73–76.

73. $\lim_{x\to0} \dfrac{-1}{x^2} = -\infty$ **74.** $\lim_{x\to0} \dfrac{1}{|x|} = \infty$

75. $\lim_{x\to3} \dfrac{-2}{(x-3)^2} = -\infty$ **76.** $\lim_{x\to-5} \dfrac{1}{(x+5)^2} = \infty$

77. Here is the definition of **infinite right-hand limit**.

> We say that $f(x)$ approaches infinity as x approaches x_0 from the right, and write
>
> $$\lim_{x\to x_0^+} f(x) = \infty,$$
>
> if, for every positive real number B, there exists a corresponding number $\delta > 0$ such that for all x
>
> $$x_0 < x < x_0 + \delta \quad \Rightarrow \quad f(x) > B.$$

Modify the definition to cover the following cases.

a. $\lim_{x\to x_0^-} f(x) = \infty$

b. $\lim_{x\to x_0^+} f(x) = -\infty$

c. $\lim_{x\to x_0^-} f(x) = -\infty$

Use the formal definitions from Exercise 77 to prove the limit statements in Exercises 78–82.

78. $\lim_{x\to0^+} \dfrac{1}{x} = \infty$ **79.** $\lim_{x\to0^-} \dfrac{1}{x} = -\infty$

80. $\lim_{x\to2^-} \dfrac{1}{x-2} = -\infty$ **81.** $\lim_{x\to2^+} \dfrac{1}{x-2} = \infty$

82. $\lim_{x\to1^-} \dfrac{1}{1-x^2} = \infty$

If the degree of the numerator of a rational function is one greater than the degree of the denominator, the graph has an **oblique** or **slant asymptote**. To find it, use long division to divide the numerator by the denominator to express the rational function as a linear function plus a remainder that goes to zero as x goes to plus or minus infinity. Graph the rational functions in Exercises 83–88. Include the graphs and equations of the asymptotes.

83. $y = \dfrac{x^2}{x-1}$ **84.** $y = \dfrac{x^2+1}{x-1}$

85. $y = \dfrac{x^2 - 4}{x - 1}$

86. $y = \dfrac{x^2 - 1}{2x + 4}$

87. $y = \dfrac{x^2 - 1}{x}$

88. $y = \dfrac{x^3 + 1}{x^2}$

T Graph the curves in Exercises 89–92. Explain the relation between the curve's formula and what you see.

89. $y = \dfrac{x}{\sqrt{4 - x^2}}$

90. $y = \dfrac{-1}{\sqrt{4 - x^2}}$

91. $y = x^{2/3} + \dfrac{1}{x^{1/3}}$

92. $y = \sin\left(\dfrac{\pi}{x^2 + 1}\right)$

T Graph the functions in Exercises 93 and 94. Then answer the following questions.

a. How does the graph behave as $x \rightarrow 0^+$?

b. How does the graph behave as $x \rightarrow \pm\infty$?

c. How does the graph behave at $x = 1$ and $x = -1$?

Give reasons for your answers.

93. $y = \dfrac{3}{2}\left(x - \dfrac{1}{x}\right)^{2/3}$

94. $y = \dfrac{3}{2}\left(\dfrac{x}{x - 1}\right)^{2/3}$

Chapter 1 Questions to Guide Your Review

1. What is a function? What is its domain? Its range? What is an arrow diagram for a function? Give examples.
2. What are the important types of functions frequently encountered in calculus? Give an example of each type.
3. What is meant by an increasing function? A decreasing function? Give an example of each.
4. What is an even function? An odd function? What symmetry properties do the graphs of such functions have? What advantage can we take of this? Give an example of a function that is neither even nor odd.
5. How do you change the equation $y = f(x)$ to shift its graph vertically up or down by a factor $k > 0$? Horizontally to the left or right? Give examples.
6. How do you change the equation $y = f(x)$ to compress or stretch the graph by $c > 1$? Reflect the graph across a coordinate axis? Give examples.
7. What is the average rate of change of the function $g(t)$ over the interval from $t = a$ to $t = b$? How is it related to a secant line?
8. What limit must be calculated to find the rate of change or slope of a function $g(t)$ at $t = t_0$?
9. What is an informal or intuitive definition of the limit
$$\lim_{x \rightarrow x_0} f(x) = L?$$
Why is the definition "informal"? Give examples.
10. Does the existence and value of the limit of a function $f(x)$ as x approaches x_0 ever depend on what happens at $x = x_0$? Explain and give examples.
11. What function behaviors might occur for which the limit may fail to exist? Give examples.
12. What theorems are available for calculating limits? Give examples of how the theorems are used.
13. How are one-sided limits related to limits? How can this relationship sometimes be used to calculate a limit or prove it does not exist? Give examples.
14. What is the value of $\lim_{\theta \rightarrow 0}((\sin\theta)/\theta)$? Does it matter whether θ is measured in degrees or radians? Explain.
15. What exactly does $\lim_{x \rightarrow x_0} f(x) = L$ mean? Give an example in which you find a $\delta > 0$ for a given f, L, x_0, and $\epsilon > 0$ in the precise definition of limit.
16. What conditions must be satisfied by a function if it is to be continuous at an interior point of its domain? At an endpoint?
17. How can looking at the graph of a function help you tell where the function is continuous?
18. What does it mean for a function to be right-continuous at a point? Left-continuous? How are continuity and one-sided continuity related?
19. What does it mean for a function to be continuous on an interval? Give examples to illustrate the fact that a function that is not continuous on its entire domain may still be continuous on selected intervals within the domain.
20. What are the basic types of discontinuity? Give an example of each. What is a removable discontinuity? Give an example.
21. What does it mean for a function to have the Intermediate Value Property? What conditions guarantee that a function has this property over an interval? What are the consequences for graphing and solving the equation $f(x) = 0$?
22. What exactly do $\lim_{x \rightarrow \infty} f(x) = L$ and $\lim_{x \rightarrow -\infty} f(x) = L$ mean? Give examples.
23. What are $\lim_{x \rightarrow \pm\infty} k$ (k a constant) and $\lim_{x \rightarrow \pm\infty}(1/x)$? How do you extend these results to other functions? Give examples.
24. How do you find the limit of a rational function as $x \rightarrow \pm\infty$? Give examples.
25. What are horizontal and vertical asymptotes? Give examples.

Chapter 1 Practice and Additional Exercises

In Exercises 1–6, find the **(a)** domain and **(b)** range.

1. $y = |x| - 2$

2. $y = -2 + \sqrt{1 - x}$

3. $y = \sqrt{16 - x^2}$

4. $y = \tan(2x - \pi)$

5. $y = 2\sin(3x + \pi) - 1$

6. $y = -1 + \sqrt[3]{2 - x}$

In Exercises 7 and 8, find the **(a)** domain and **(b)** range.

7. $y = \begin{cases} \sqrt{-x}, & -4 \le x \le 0 \\ \sqrt{x}, & 0 < x \le 4 \end{cases}$

8. $y = \begin{cases} -x - 2, & -2 \le x \le -1 \\ x, & -1 < x \le 1 \\ -x + 2, & 1 < x \le 2 \end{cases}$

In Exercises 9 and 10, write a piecewise formula for the function.

9.

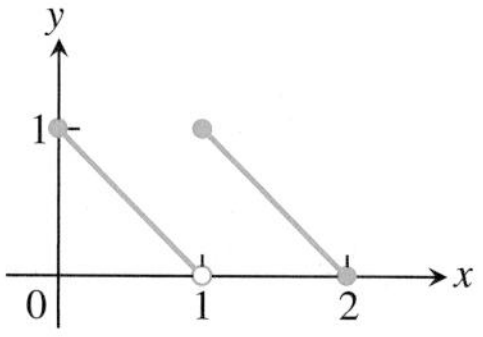

10.

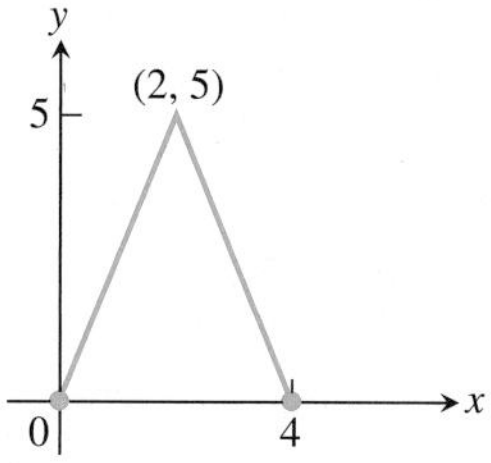

In Exercises 11 and 12, find

a. $(f \circ g)(-1)$.

b. $(g \circ f)(2)$.

c. $(f \circ f)(x)$.

d. $(g \circ g)(x)$.

11. $f(x) = \dfrac{1}{x}, \quad g(x) = \dfrac{1}{\sqrt{x + 2}}$

12. $f(x) = 2 - x, \quad g(x) = \sqrt[3]{x + 1}$

In Exercises 13 and 14, **(a)** write a formula for $f \circ g$ and $g \circ f$ and find the **(b)** domain and **(c)** range of each.

13. $f(x) = 2 - x^2, \quad g(x) = \sqrt{x + 2}$

14. $f(x) = \sqrt{x}, \quad g(x) = \sqrt{1 - x}$

In Exercises 15–20 determine whether the function is even, odd, or neither.

15. $y = x^2 + 1$

16. $y = x^5 - x^3 - x$

17. $y = \dfrac{x^4 + 1}{x^3 - 2x}$

18. $y = 1 - \sin x$

19. $y = x + \cos x$

20. $y = \sqrt{x^4 - 1}$

21. Graph the function

$$f(x) = \begin{cases} 1, & x \le -1 \\ -x, & -1 < x < 0 \\ 1, & x = 0 \\ -x, & 0 < x < 1 \\ 1, & x \ge 1. \end{cases}$$

Then discuss, in detail, limits, one-sided limits, continuity, and one-sided continuity of f at $x = -1, 0$, and 1. Are any of the discontinuities removable? Explain.

22. Repeat the instructions of Exercise 21 for

$$f(x) = \begin{cases} 0, & x \le -1 \\ 1/x, & 0 < |x| < 1 \\ 0, & x = 1 \\ 1, & x > 1. \end{cases}$$

23. Suppose that $f(t)$ and $g(t)$ are defined for all t and that $\lim_{t \to t_0} f(t) = -7$ and $\lim_{t \to t_0} g(t) = 0$. Find the limit as $t \to t_0$ of the following functions.

a. $3f(t)$

b. $(f(t))^2$

c. $f(t) \cdot g(t)$

d. $\dfrac{f(t)}{g(t) - 7}$

e. $\cos(g(t))$

f. $|f(t)|$

g. $f(t) + g(t)$

h. $1/f(t)$

24. Suppose that $f(x)$ and $g(x)$ are defined for all x and that $\lim_{x \to 0} f(x) = 1/2$ and $\lim_{x \to 0} g(x) = \sqrt{2}$. Find the limits as $x \to 0$ of the following functions.

a. $-g(x)$

b. $g(x) \cdot f(x)$

c. $f(x) + g(x)$

d. $1/f(x)$

e. $x + f(x)$

f. $\dfrac{f(x) \cdot \cos x}{x - 1}$

In Exercises 25 and 26, find the value that $\lim_{x \to 0} g(x)$ must have if the given limit statements hold.

25. $\lim_{x \to 0} \left(\dfrac{4 - g(x)}{x} \right) = 1$

26. $\lim_{x \to -4} \left(x \lim_{x \to 0} g(x) \right) = 2$

27. On what intervals are the following functions continuous?

a. $f(x) = x^{1/3}$

b. $g(x) = x^{3/4}$

c. $h(x) = x^{-2/3}$

d. $k(x) = x^{-1/6}$

28. On what intervals are the following functions continuous?

a. $f(x) = \tan x$

b. $g(x) = \csc x$

c. $h(x) = \dfrac{\cos x}{x - \pi}$

d. $k(x) = \dfrac{\sin x}{x}$

In Exercises 29–40, find the limit or explain why it does not exist.

29. $\lim \dfrac{x^2 - 4x + 4}{x^3 + 5x^2 - 14x}$

a. as $x \to 0$

b. as $x \to 2$

30. $\lim \dfrac{x^2 + x}{x^5 + 2x^4 + x^3}$

a. as $x \to 0$

b. as $x \to -1$

31. $\lim\limits_{x\to 1} \dfrac{1-\sqrt{x}}{1-x}$

32. $\lim\limits_{x\to a} \dfrac{x^2-a^2}{x^4-a^4}$

33. $\lim\limits_{h\to 0} \dfrac{(x+h)^2-x^2}{h}$

34. $\lim\limits_{x\to 0} \dfrac{(x+h)^2-x^2}{h}$

35. $\lim\limits_{x\to 0} \dfrac{\dfrac{1}{2+x}-\dfrac{1}{2}}{x}$

36. $\lim\limits_{x\to 0} \dfrac{(2+x)^3-8}{x}$

37. $\lim\limits_{x\to 0} \dfrac{\tan(2x)}{\tan(\pi x)}$

38. $\lim\limits_{x\to \pi^-} \csc x$

39. $\lim\limits_{x\to \pi} \sin\left(\dfrac{x}{2}+\sin x\right)$

40. $\lim\limits_{x\to \pi} \cos^2(x-\tan x)$

In Exercises 41–44, find the limit of $g(x)$ as x approaches the indicated value.

41. $\lim\limits_{x\to 0^+} (4g(x))^{1/3} = 2$

42. $\lim\limits_{x\to \sqrt{5}} \dfrac{1}{x+g(x)} = 2$

43. $\lim\limits_{x\to 1} \dfrac{3x^2+1}{g(x)} = \infty$

44. $\lim\limits_{x\to -2} \dfrac{5-x^2}{\sqrt{g(x)}} = 0$

Find the limits in Exercises 45–54.

45. $\lim\limits_{x\to \infty} \dfrac{2x+3}{5x+7}$

46. $\lim\limits_{x\to -\infty} \dfrac{2x^2+3}{5x^2+7}$

47. $\lim\limits_{x\to -\infty} \dfrac{x^2-4x+8}{3x^3}$

48. $\lim\limits_{x\to \infty} \dfrac{1}{x^2-7x+1}$

49. $\lim\limits_{x\to -\infty} \dfrac{x^2-7x}{x+1}$

50. $\lim\limits_{x\to \infty} \dfrac{x^4+x^3}{12x^3+128}$

51. $\lim\limits_{x\to \infty} \dfrac{\sin x}{\lfloor x \rfloor}$ (If you have a grapher, try graphing the function for $-5 \le x \le 5$.)

52. $\lim\limits_{\theta\to \infty} \dfrac{\cos\theta - 1}{\theta}$ (If you have a grapher, try graphing $f(x) = x(\cos(1/x) - 1)$ near the origin to "see" the limit at infinity.)

53. $\lim\limits_{x\to \infty} \dfrac{x+\sin x+2\sqrt{x}}{x+\sin x}$

54. $\lim\limits_{x\to \infty} \dfrac{x^{2/3}+x^{-1}}{x^{2/3}+\cos^2 x}$

T **55.** Let $f(x) = x^3 - x - 1$.

a. Show that f has a zero between -1 and 2.

b. Solve the equation $f(x) = 0$ graphically with an error of magnitude at most 10^{-8}.

c. It can be shown that the exact value of the solution in part (b) is

$$\left(\frac{1}{2}+\frac{\sqrt{69}}{18}\right)^{1/3} + \left(\frac{1}{2}-\frac{\sqrt{69}}{18}\right)^{1/3}.$$

Evaluate this exact answer and compare it with the value you found in part (b).

T **56.** Let $f(\theta) = \theta^3 - 2\theta + 2$.

a. Show that f has a zero between -2 and 0.

b. Solve the equation $f(\theta) = 0$ graphically with an error of magnitude at most 10^{-4}.

c. It can be shown that the exact value of the solution in part (b) is

$$\left(\sqrt{\frac{19}{27}}-1\right)^{1/3} - \left(\sqrt{\frac{19}{27}}+1\right)^{1/3}.$$

Evaluate this exact answer and compare it with the value you found in part (b).

In Exercises 57–60, use the formal definition of limit to prove that the function is continuous at x_0.

57. $f(x) = x^2 - 7,\quad x_0 = 1$

58. $g(x) = 1/(2x),\quad x_0 = 1/4$

59. $h(x) = \sqrt{2x-3},\quad x_0 = 2$

60. $F(x) = \sqrt{9-x},\quad x_0 = 5$

61. A function continuous at only one point Let

$$f(x) = \begin{cases} x, & \text{if } x \text{ is rational} \\ 0, & \text{if } x \text{ is irrational.} \end{cases}$$

a. Show that f is continuous at $x = 0$.

b. Use the fact that every nonempty open interval of real numbers contains both rational and irrational numbers to show that f is not continuous at any nonzero value of x.

62. Uniqueness of limits Show that a function cannot have two different limits at the same point. That is, if $\lim_{x\to x_0} f(x) = L_1$ and $\lim_{x\to x_0} f(x) = L_2$, then $L_1 = L_2$.

63. Bounded functions A real-valued function f is **bounded from above** on a set D if there exists a number N such that $f(x) \le N$ for all x in D. We call N, when it exists, an **upper bound** for f on D and say that f is bounded from above by N. In a similar manner, we say that f is **bounded from below** on D if there exists a number M such that $f(x) \ge M$ for all x in D. We call M, when it exists, a **lower bound** for f on D and say that f is bounded from below by M. We say that f is **bounded** on D if it is bounded from both above and below.

a. Show that f is bounded on D if and only if there exists a number B such that $|f(x)| \le B$ for all x in D.

b. Suppose that f is bounded from above by N. Show that if $\lim_{x\to x_0} f(x) = L$, then $L \le N$.

c. Suppose that f is bounded from below by M. Show that if $\lim_{x\to x_0} f(x) = L$, then $L \ge M$.

64. Limits and continuity Which of the following statements are true, and which are false? If true, say why; if false, give a counterexample (that is, an example confirming the falsehood).

a. If $\lim_{x\to a} f(x)$ exists but $\lim_{x\to a} g(x)$ does not exist, then $\lim_{x\to a}(f(x)+g(x))$ does not exist.

b. If neither $\lim_{x\to a} f(x)$ nor $\lim_{x\to a} g(x)$ exists, then $\lim_{x\to a}(f(x)+g(x))$ does not exist.

c. If f is continuous at x, then so is $|f|$.

d. If $|f|$ is continuous at a, then so is f.

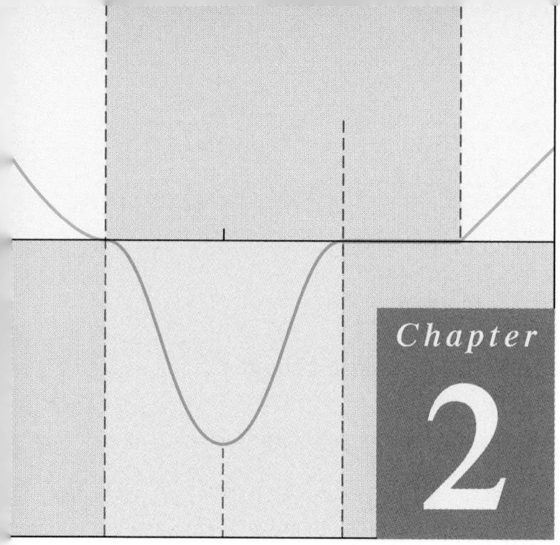

Chapter 2

DIFFERENTIATION

OVERVIEW In this chapter we introduce the derivative of a function at a point, making precise our discussion of slopes and tangent lines to a curve in Section 1.3. The derivative measures the rate at which a function changes and gives the slope of its graph. We then extend this concept from a single point to the *derivative function*, and we develop rules for finding this derivative function easily, without having to calculate any limits directly. These rules are used in problems involving rates of change and also to approximate complicated functions. Along the way we introduce exponential, logarithmic, and inverse trigonometric functions.

2.1 Tangents and Derivatives at a Point

This section makes precise our discussion in Section 1.3 of slopes and tangent lines to a curve. We saw how these notions are related to the instantaneous rate of change of a function, which is an interpretation of the *derivative* of the function. We define the derivative here.

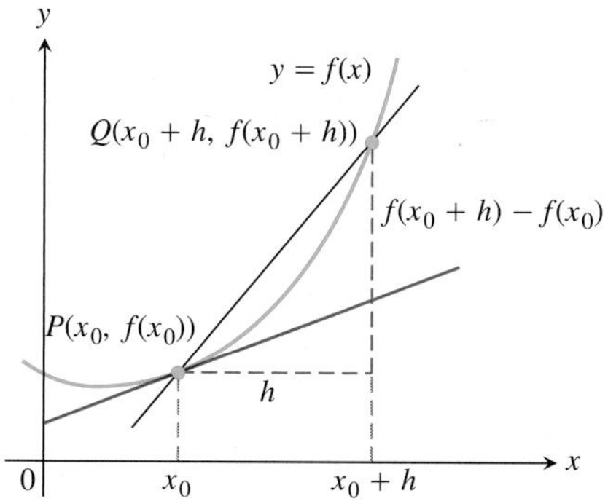

FIGURE 2.1 The slope of the tangent line at P is $\lim_{h\to 0} \dfrac{f(x_0 + h) - f(x_0)}{h}$.

Finding a Tangent to the Graph of a Function

To find a tangent to an arbitrary curve $y = f(x)$ at a point $P(x_0, f(x_0))$, we use the procedure introduced in Section 1.3. We calculate the slope of the secant through P and a nearby point $Q(x_0 + h, f(x_0 + h))$. We then investigate the limit of the slope as $h \to 0$ (Figure 2.1). If the limit exists, we call it the slope of the curve at P and define the tangent at P to be the line through P having this slope.

DEFINITIONS The **slope of the curve** $y = f(x)$ at the point $P(x_0, f(x_0))$ is the number

$$m = \lim_{h\to 0} \frac{f(x_0 + h) - f(x_0)}{h} \quad \text{(provided the limit exists).}$$

The **tangent line** to the curve at P is the line through P with this slope.

In Section 1.3, Example 3, we applied these definitions to find the slope of the parabola $f(x) = x^2$ at the point $P(2, 4)$ and the tangent line to the parabola at P. Let's look at another example.

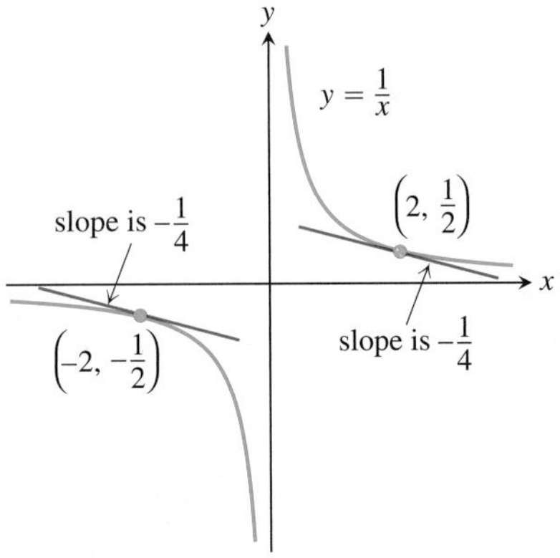

FIGURE 2.2 The two tangent lines to $y = 1/x$ having slope $-1/4$ (Example 1).

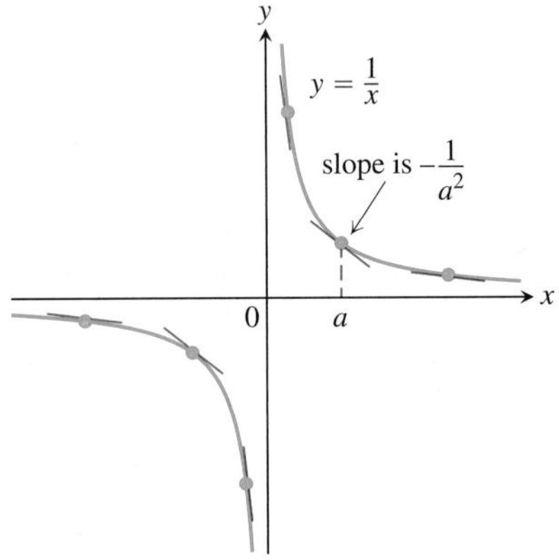

FIGURE 2.3 The tangent slopes, steep near the origin, become more gradual as the point tangency moves away.

EXAMPLE 1

(a) Find the slope of the curve $y = 1/x$ at $x = a \neq 0$.

(b) Where does the slope equal $-1/4$?

(c) What happens to the tangent to the curve at the point $(a, 1/a)$ as a changes?

Solution

(a) Here $f(x) = 1/x$. The slope at $(a, 1/a)$ is

$$\lim_{h\to 0} \frac{f(a+h) - f(a)}{h} = \lim_{h\to 0} \frac{\frac{1}{a+h} - \frac{1}{a}}{h} = \lim_{h\to 0} \frac{1}{h}\frac{a-(a+h)}{a(a+h)}$$

$$= \lim_{h\to 0} \frac{-h}{ha(a+h)} = \lim_{h\to 0} \frac{-1}{a(a+h)} = -\frac{1}{a^2}.$$

Notice how we had to keep writing "$\lim_{h\to 0}$" before each fraction until the stage where we could evaluate the limit by substituting $h = 0$. The number a may be positive or negative, but not 0.

(b) The slope of $y = 1/x$ at the point where $x = a$ is $-1/a^2$. It will be $-1/4$ provided that

$$-\frac{1}{a^2} = -\frac{1}{4}.$$

This equation is equivalent to $a^2 = 4$, so $a = 2$ or $a = -2$. The curve has slope $-1/4$ at the two points $(2, 1/2)$ and $(-2, -1/2)$ (Figure 2.2).

(c) The slope $-1/a^2$ is always negative if $a \neq 0$. As $a \to 0^+$, the slope approaches $-\infty$ and the tangent becomes increasingly steep (Figure 2.3). We see this situation again as $a \to 0^-$. As a moves away from the origin in either direction, the slope approaches 0 and the tangent levels off to become horizontal. ■

Rates of Change: Derivative at a Point

The expression

$$\frac{f(x_0 + h) - f(x_0)}{h}$$

is called the **difference quotient of f at x_0 with increment h**. If the difference quotient has a limit as h approaches zero, that limit is given a special name and notation.

DEFINITION The **derivative of a function f at a point x_0**, denoted $f'(x_0)$, is

$$f'(x_0) = \lim_{h\to 0} \frac{f(x_0 + h) - f(x_0)}{h}$$

provided this limit exists.

If we interpret the difference quotient as a secant slope, the derivative gives the slope of the curve $y = f(x)$ at the point $P(x_0, f(x_0))$. Exercise 31 shows that the derivative of the linear function $f(x) = mx + b$ at any point x_0 is simply the slope of the line,

$$f'(x_0) = m,$$

as it should be.

If we interpret the difference quotient as an average rate of change (Section 1.3), the derivative gives the function's instantaneous rate of change with respect to x at the point $x = x_0$.

EXAMPLE 2 In Examples 1 and 2 in Section 1.3, we studied the speed of a rock falling freely from rest near the surface of the earth. We knew that the rock fell $y = 16t^2$ feet during the first t sec, and we used a sequence of average rates over increasingly short intervals to estimate the rock's speed at the instant $t = 1$. Exactly what *was* the rock's speed at this time?

Solution We let $f(t) = 16t^2$. The average speed of the rock over the interval between $t = 1$ and $t = 1 + h$ seconds was

$$\frac{f(1 + h) - f(1)}{h} = \frac{16(1 + h)^2 - 16(1)^2}{h} = \frac{16(h^2 + 2h)}{h} = 16(h + 2).$$

The rock's speed at the instant $t = 1$ was

$$\lim_{h \to 0} 16(h + 2) = 16(0 + 2) = 32 \text{ ft/sec}.$$

Our original estimate of 32 ft/sec was right. ■

Summary

We have been discussing slopes of curves, lines tangent to a curve, the rate of change of a function, the limit of the difference quotient, and the derivative of a function at a point. All of these ideas refer to the same limit:

> The following are all interpretations for the limit of the difference quotient,
>
> $$\lim_{h \to 0} \frac{f(x_0 + h) - f(x_0)}{h}$$
>
> 1. The slope of $y = f(x)$ at $x = x_0$
> 2. The slope of the tangent to the curve $y = f(x)$ at $x = x_0$
> 3. The rate of change of $f(x)$ with respect to x at $x = x_0$
> 4. The derivative $f'(x_0)$ at a point

The derivative is one of the two most important mathematical concepts in calculus. We begin a thorough study of it in the next section. The other important concept is the *integral*, and we initiate its study in Chapter 4.

EXERCISES 2.1

In Exercises 1–4, use the grid and a straight edge to make a rough estimate of the slope of the curve (in y-units per x-unit) at the points P_1 and P_2. Graphs can shift during a press run, so your estimates may be somewhat different from those in the back of the book.

1.

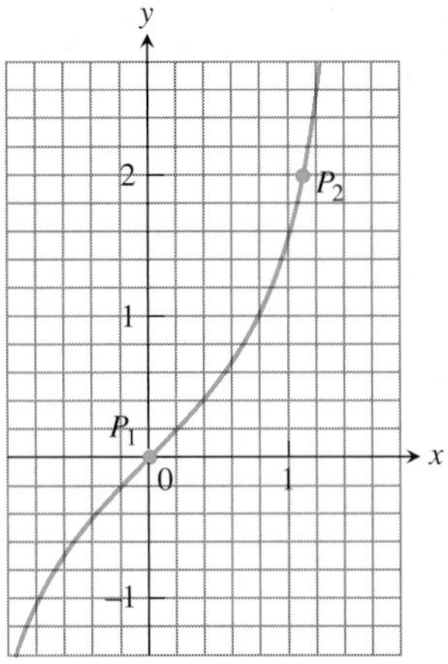

2.

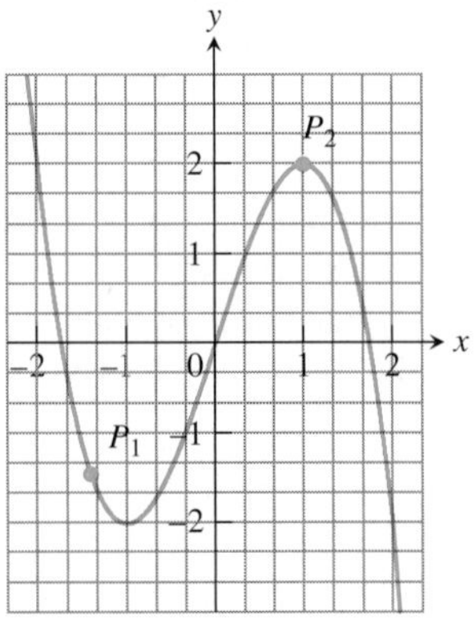

3.

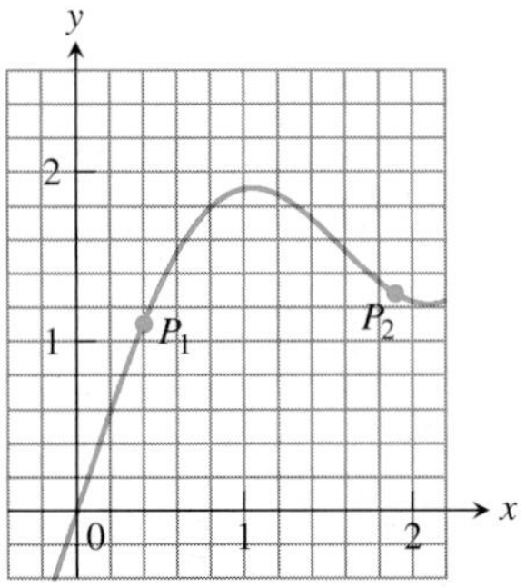

4.

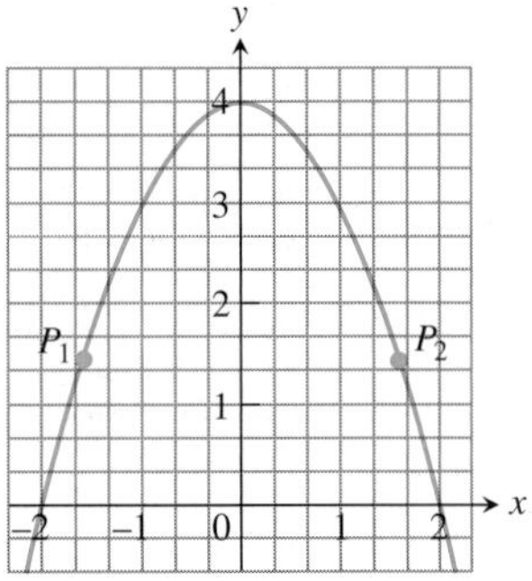

In Exercises 5–10, find an equation for the tangent to the curve at the given point. Then sketch the curve and tangent together.

5. $y = 3 - x^2, \quad (-1, 2)$

6. $y = (x - 1)^2 + 1, \quad (1, 1)$

7. $y = 2\sqrt{x}, \quad (1, 2)$

8. $y = \dfrac{1}{x^2}, \quad (-1, 1)$

9. $y = x^3 + 1, \quad (-2, -7)$

10. $y = \dfrac{1}{x^3}, \quad \left(-2, -\dfrac{1}{8}\right)$

In Exercises 11–18, find the slope of the function's graph at the given point. Then find an equation for the line tangent to the graph there.

11. $f(x) = x^2 + 1, \quad (2, 5)$

12. $f(x) = x - 2x^2, \quad (1, -1)$

13. $g(x) = \dfrac{x}{x - 2}, \quad (3, 3)$

14. $g(x) = \dfrac{8}{x^2}, \quad (2, 2)$

15. $h(t) = t^3 - t, \quad (2, 6)$

16. $h(t) = t^3 + 3t, \quad (1, 4)$

17. $f(x) = \sqrt{x}, \quad (4, 2)$

18. $f(x) = \sqrt{x + 1}, \quad (8, 3)$

In Exercises 19–22, find the slope of the curve at the point indicated.

19. $y = 5x^2, \quad x = -1$

20. $y = 1 - x^2, \quad x = 2$

21. $y = \dfrac{1}{x - 1}, \quad x = 3$

22. $y = \dfrac{x - 1}{x + 1}, \quad x = 0$

At what points do the graphs of the functions in Exercises 23 and 24 have horizontal tangents?

23. $f(x) = x^2 + 4x - 1$

24. $g(x) = x^3 - 3x$

25. Find equations of all lines having slope -1 that are tangent to the curve $y = 1/(x - 1)$.

26. Find an equation of the straight line having slope $1/4$ that is tangent to the curve $y = \sqrt{x}$.

27. Object dropped from a tower An object is dropped from the top of a 100-m-high tower. Its height above ground after t sec is $100 - 4.9t^2$ m. How fast is it falling 2 sec after it is dropped?

28. Speed of a rocket At t sec after liftoff, the height of a rocket is $3t^2$ ft. How fast is the rocket climbing 10 sec after liftoff?

29. Circle's changing area What is the rate of change of the area of a circle ($A = \pi r^2$) with respect to the radius when the radius is $r = 3$?

30. Ball's changing volume What is the rate of change of the volume of a ball ($V = (4/3)\pi r^3$) with respect to the radius when the radius is $r = 2$?

31. Show that the line $y = mx + b$ is its own tangent at any point $(x_0, mx_0 + b)$.

32. Find the slope of the tangent to the curve $y = 1/\sqrt{x}$ at the point where $x = 4$.

33. Does the graph of

$$f(x) = \begin{cases} x^2 \sin(1/x), & x \neq 0 \\ 0, & x = 0 \end{cases}$$

have a tangent at the origin? Give reasons for your answer.

34. Does the graph of

$$g(x) = \begin{cases} x \sin(1/x), & x \neq 0 \\ 0, & x = 0 \end{cases}$$

have a tangent at the origin? Give reasons for your answer.

Vertical Tangents We say that the curve $y = f(x)$ has a **vertical tangent** at the point where $x = x_0$ if $\lim_{h \to 0} (f(x_0 + h) - f(x_0))/h = \infty$ or $-\infty$. For example, $y = x^{1/3}$ has a vertical tangent at $x = 0$ (see accompanying figure):

$$\lim_{h \to 0} \frac{f(0 + h) - f(0)}{h} = \lim_{h \to 0} \frac{h^{1/3} - 0}{h} = \lim_{h \to 0} \frac{1}{h^{2/3}} = \infty$$

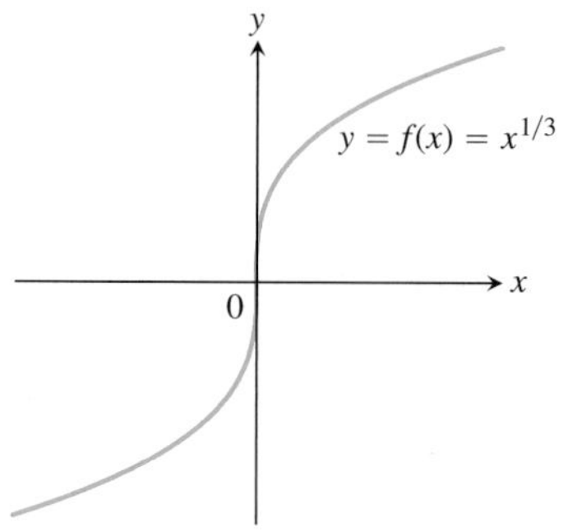

VERTICAL TANGENT AT ORIGIN

However, $y = x^{2/3}$ has *no* vertical tangent at $x = 0$ (see next figure):

$$\lim_{h\to 0}\frac{g(0+h)-g(0)}{h} = \lim_{h\to 0}\frac{h^{2/3}-0}{h} = \lim_{h\to 0}\frac{1}{h^{1/3}}$$

does not exist, because the limit is ∞ from the right and $-\infty$ from the left.

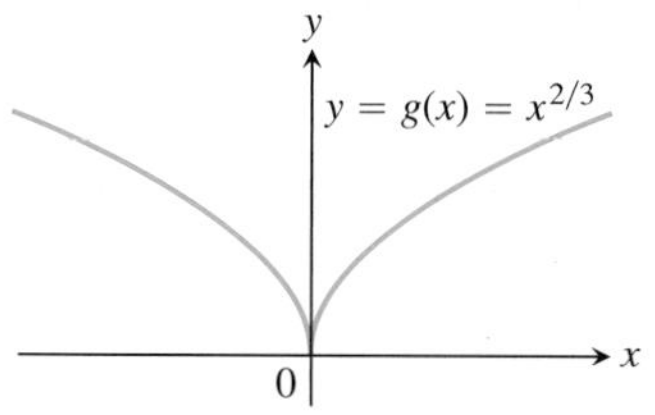

NO VERTICAL TANGENT AT ORIGIN

35. Does the graph of

$$f(x) = \begin{cases} -1, & x < 0 \\ 0, & x = 0 \\ 1, & x > 0 \end{cases}$$

have a vertical tangent at the origin? Give reasons for your answer.

36. Does the graph of

$$U(x) = \begin{cases} 0, & x < 0 \\ 1, & x \ge 0 \end{cases}$$

have a vertical tangent at the point (0, 1)? Give reasons for your answer.

T **a.** Graph the curves in Exercises 37–46. Where do the graphs appear to have vertical tangents?

b. Confirm your findings in part (a) with limit calculations. But before you do, read the introduction to Exercises 35 and 36.

37. $y = x^{2/5}$

38. $y = x^{4/5}$

39. $y = x^{1/5}$

40. $y = x^{3/5}$

41. $y = 4x^{2/5} - 2x$

42. $y = x^{5/3} - 5x^{2/3}$

43. $y = x^{2/3} - (x - 1)^{1/3}$

44. $y = x^{1/3} + (x - 1)^{1/3}$

45. $y = \begin{cases} -\sqrt{|x|}, & x \le 0 \\ \sqrt{x}, & x > 0 \end{cases}$

46. $y = \sqrt{|4 - x|}$

COMPUTER EXPLORATIONS

Use a CAS to perform the following steps for the functions in Exercises 47–50.

a. Plot $y = f(x)$ over the interval $(x_0 - 1/2) \le x \le (x_0 + 3)$.

b. Holding x_0 fixed, the difference quotient

$$q(h) = \frac{f(x_0 + h) - f(x_0)}{h}$$

at x_0 becomes a function of the step size h. Enter this function into your CAS workspace.

c. Find the limit of q as $h \to 0$.

d. Define the secant lines $y = f(x_0) + q \cdot (x - x_0)$ for $h = 3, 2,$ and 1. Graph them together with f and the tangent line over the interval in part (a).

47. $f(x) = x^3 + 2x, \quad x_0 = 0$

48. $f(x) = x + \dfrac{5}{x}, \quad x_0 = 1$

49. $f(x) = x + \sin(2x), \quad x_0 = \pi/2$

50. $f(x) = \cos x + 4\sin(2x), \quad x_0 = \pi$

2.2 The Derivative as a Function

HISTORICAL ESSAY

The Derivative

In the last section we defined the derivative of $y = f(x)$ at the point $x = x_0$ to be the limit

$$f'(x_0) = \lim_{h\to 0}\frac{f(x_0 + h) - f(x_0)}{h}.$$

We now investigate the derivative as a *function* derived from f by considering the limit at each point x in the domain of f.

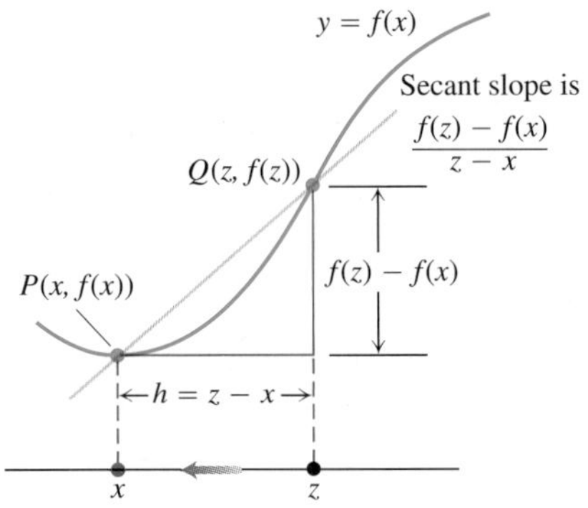

Derivative of f at x is

$$f'(x) = \lim_{h \to 0} \frac{f(x+h) - f(x)}{h}$$

$$= \lim_{z \to x} \frac{f(z) - f(x)}{z - x}$$

FIGURE 2.4 Two forms for the difference quotient.

> **DEFINITION** The **derivative** of the function $f(x)$ with respect to the variable x is the function f' whose value at x is
>
> $$f'(x) = \lim_{h \to 0} \frac{f(x+h) - f(x)}{h},$$
>
> provided the limit exists.

We use the notation $f(x)$ in the definition to emphasize the independent variable x with respect to which the derivative function $f'(x)$ is being defined. The domain of f' is the set of points in the domain of f for which the limit exists, and the domain may be the same or smaller than the domain of f. If f' exists at a particular x, we say that f is **differentiable (has a derivative) at x**. If f' exists at every point in the domain of f, we call f **differentiable**.

If we write $z = x + h$, then $h = z - x$ and h approaches 0 if and only if z approaches x. Therefore, an equivalent definition of the derivative is as follows (see Figure 2.4).

> **Alternative Formula for the Derivative**
>
> $$f'(x) = \lim_{z \to x} \frac{f(z) - f(x)}{z - x}.$$

Calculating Derivatives from the Definition

The process of calculating a derivative is called **differentiation**. To emphasize the idea that differentiation is an operation performed on a function $y = f(x)$, we use the notation

$$\frac{d}{dx} f(x)$$

as another way to denote the derivative $f'(x)$. Example 1 of Section 2.1 illustrated the differentiation process for the function $y = 1/x$ when $x = a$. For x representing any point in the domain, we get the formula

$$\frac{d}{dx}\left(\frac{1}{x}\right) = -\frac{1}{x^2}.$$

Derivative of the Reciprocal Function

$$\frac{d}{dx}\left(\frac{1}{x}\right) = -\frac{1}{x^2}, \; x \neq 0$$

Here are two more examples in which we allow x to be any point in the domain of f.

EXAMPLE 1 Differentiate $f(x) = \dfrac{x}{x-1}$.

Solution Here we have $f(x) = \dfrac{x}{x-1}$ and $f(x+h) = \dfrac{(x+h)}{(x+h)-1}$, so

$$f'(x) = \lim_{h \to 0} \frac{f(x+h) - f(x)}{h} \qquad \text{Definition}$$

$$= \lim_{h \to 0} \frac{\dfrac{x + h}{x + h - 1} - \dfrac{x}{x - 1}}{h}$$

$$= \lim_{h \to 0} \frac{1}{h} \cdot \frac{(x + h)(x - 1) - x(x + h - 1)}{(x + h - 1)(x - 1)} \qquad \frac{a}{b} - \frac{c}{d} = \frac{ad - cb}{bd}$$

$$= \lim_{h \to 0} \frac{1}{h} \cdot \frac{-h}{(x + h - 1)(x - 1)} \qquad \text{Simplify}$$

$$= \lim_{h \to 0} \frac{-1}{(x + h - 1)(x - 1)} = \frac{-1}{(x - 1)^2}. \qquad \text{Cancel } h \neq 0$$

■

EXAMPLE 2 Find the derivative of $f(x) = \sqrt{x}$ for $x > 0$.

Solution We use the alternative form to calculate f':

$$f'(x) = \lim_{z \to x} \frac{f(z) - f(x)}{z - x}$$

$$= \lim_{z \to x} \frac{\sqrt{z} - \sqrt{x}}{z - x}$$

$$= \lim_{z \to x} \frac{\sqrt{z} - \sqrt{x}}{\left(\sqrt{z} - \sqrt{x}\right)\left(\sqrt{z} + \sqrt{x}\right)}$$

$$= \lim_{z \to x} \frac{1}{\sqrt{z} + \sqrt{x}} = \frac{1}{2\sqrt{x}}.$$

■

Derivative of the Square Root Function

$$\frac{d}{dx}\sqrt{x} = \frac{1}{2\sqrt{x}}, \; x > 0$$

Notations

There are many ways to denote the derivative of a function $y = f(x)$, where the independent variable is x and the dependent variable is y. Some common alternative notations for the derivative are

$$f'(x) = y' = \frac{dy}{dx} = \frac{df}{dx} = \frac{d}{dx}f(x) = D(f)(x) = D_x f(x).$$

The symbols d/dx and D indicate the operation of differentiation. We read dy/dx as "the derivative of y with respect to x," and df/dx and $(d/dx)f(x)$ as "the derivative of f with respect to x." The "prime" notations y' and f' come from notations that Newton used for derivatives. The d/dx notations are similar to those used by Leibniz. The symbol dy/dx should not be regarded as a ratio (until we introduce the idea of "differentials" in Section 2.13).

To indicate the value of a derivative at a specified number $x = a$, we use the notation

$$f'(a) = \left.\frac{dy}{dx}\right|_{x=a} = \left.\frac{df}{dx}\right|_{x=a} = \left.\frac{d}{dx}f(x)\right|_{x=a}.$$

For instance, in Example 2

$$f'(4) = \left.\frac{d}{dx}\sqrt{x}\right|_{x=4} = \left.\frac{1}{2\sqrt{x}}\right|_{x=4} = \frac{1}{2\sqrt{4}} = \frac{1}{4}.$$

Graphing the Derivative

We can often make a reasonable plot of the derivative of $y = f(x)$ by estimating the slopes on the graph of f. That is, we plot the points $(x, f'(x))$ in the xy-plane and connect them with a smooth curve, which represents $y = f'(x)$.

EXAMPLE 3 Graph the derivative of the function $y = f(x)$ in Figure 2.5a.

Solution We sketch the tangents to the graph of f at frequent intervals and use their slopes to estimate the values of $f'(x)$ at these points. We plot the corresponding $(x, f'(x))$ pairs and connect them with a smooth curve as sketched in Figure 2.5b. ■

What can we learn from the graph of $y = f'(x)$? At a glance we can see

1. where the rate of change of f is positive, negative, or zero;
2. the rough size of the growth rate at any x and its size in relation to the size of $f(x)$;
3. where the rate of change itself is increasing or decreasing.

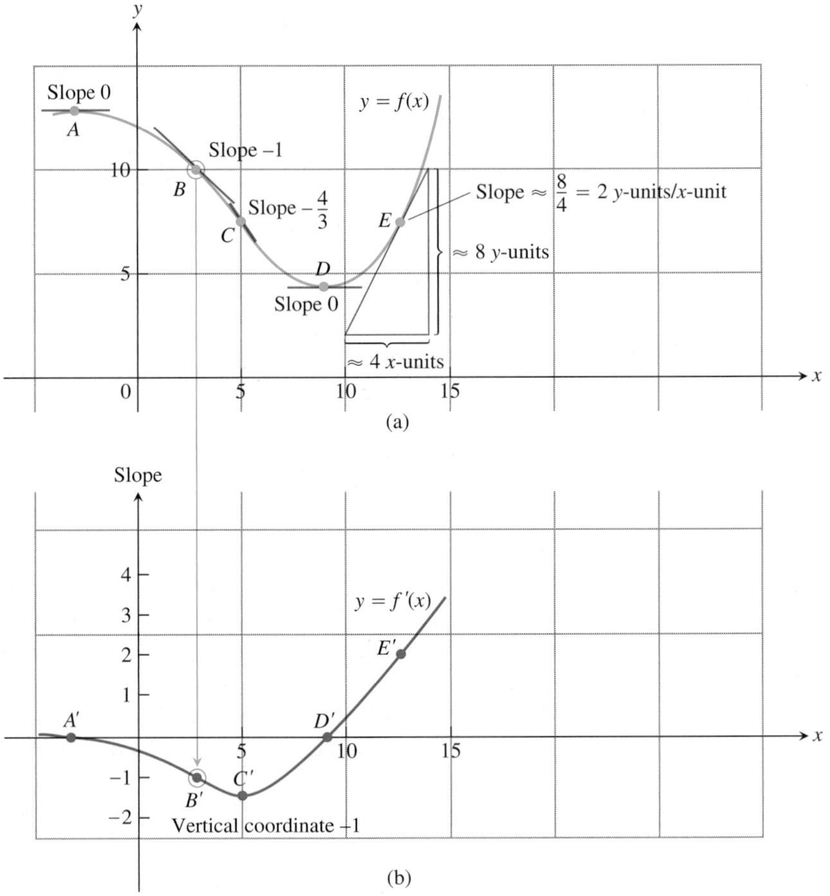

FIGURE 2.5 We made the graph of $y = f'(x)$ in (b) by plotting slopes from the graph of $y = f(x)$ in (a). The vertical coordinate of B' is the slope at B and so on. The graph of f' is a visual record of how the slope of f changes with x.

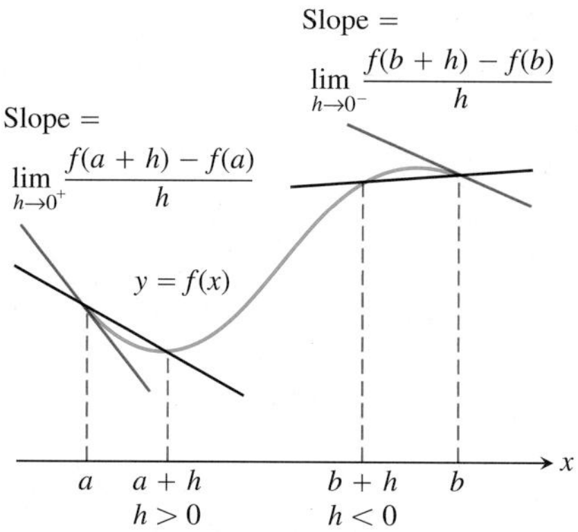

FIGURE 2.6 Derivatives at endpoints are one-sided limits.

Differentiable on an Interval; One-Sided Derivatives

A function $y = f(x)$ is **differentiable on an open interval** (finite or infinite) if it has a derivative at each point of the interval. It is differentiable on a closed interval $[a, b]$ if it is differentiable on the interior (a, b) and if the limits

$$\lim_{h\to 0^+} \frac{f(a+h)-f(a)}{h} \qquad \textbf{Right-hand derivative at } a$$

$$\lim_{h\to 0^-} \frac{f(b+h)-f(b)}{h} \qquad \textbf{Left-hand derivative at } b$$

exist at the endpoints (Figure 2.6).

Right-hand and left-hand derivatives may be defined at any point of a function's domain. The usual relation between one-sided and two-sided limits holds for these derivatives. Because of Theorem 6, Section 1.6, a function has a derivative at a point if and only if it has left-hand and right-hand derivatives there, and these one-sided derivatives are equal.

EXAMPLE 4 Show that the function $y = |x|$ is differentiable on $(-\infty, 0)$ and $(0, \infty)$ but has no derivative at $x = 0$.

Solution From Section 2.1, the derivative of $y = mx + b$ is the slope m. Thus, to the right of the origin,

$$\frac{d}{dx}(|x|) = \frac{d}{dx}(x) = \frac{d}{dx}(1\cdot x) = 1. \qquad \frac{d}{dx}(mx+b) = m,\ |x| = x$$

To the left,

$$\frac{d}{dx}(|x|) = \frac{d}{dx}(-x) = \frac{d}{dx}(-1\cdot x) = -1 \qquad |x| = -x$$

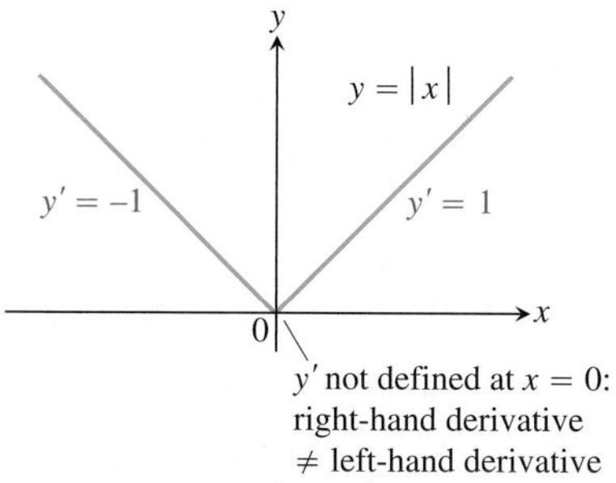

FIGURE 2.7 The function $y = |x|$ is not differentiable at the origin where the graph has a "corner."

(Figure 2.7). There can be no derivative at the origin because the one-sided derivatives differ there:

$$\begin{aligned}
\text{Right-hand derivative of } |x| \text{ at zero} &= \lim_{h\to 0^+} \frac{|0+h|-|0|}{h} = \lim_{h\to 0^+} \frac{|h|}{h} \\
&= \lim_{h\to 0^+} \frac{h}{h} \qquad |h| = h \text{ when } h > 0 \\
&= \lim_{h\to 0^+} 1 = 1 \\
\text{Left-hand derivative of } |x| \text{ at zero} &= \lim_{h\to 0^-} \frac{|0+h|-|0|}{h} = \lim_{h\to 0^-} \frac{|h|}{h} \\
&= \lim_{h\to 0^-} \frac{-h}{h} \qquad |h| = -h \text{ when } h < 0 \\
&= \lim_{h\to 0^-} -1 = -1.
\end{aligned}$$

■

EXAMPLE 5 In Example 2 we found that for $x > 0$,

$$\frac{d}{dx}\sqrt{x} = \frac{1}{2\sqrt{x}}.$$

We apply the definition to examine if the derivative exists at $x = 0$:

$$\lim_{h \to 0^+} \frac{\sqrt{0 + h} - \sqrt{0}}{h} = \lim_{h \to 0^+} \frac{1}{\sqrt{h}} = \infty.$$

Since the (right-hand) limit is not finite, there is no derivative at $x = 0$. Since the slopes of the secant lines joining the origin to the points $(h, \sqrt{h})$ on a graph of $y = \sqrt{x}$ approach ∞, the graph has a *vertical tangent* at the origin. (See Exercises 35–46, Section 2.1). ■

When Does a Function *Not* Have a Derivative at a Point?

A function has a derivative at a point x_0 if the slopes of the secant lines through $P(x_0, f(x_0))$ and a nearby point Q on the graph approach a finite limit as Q approaches P. Whenever the secants fail to take up a limiting position or become vertical as Q approaches P, the derivative does not exist. Thus differentiability is a "smoothness" condition on the graph of f. A function can fail to have a derivative at a point for several reasons, such as at points where the graph has

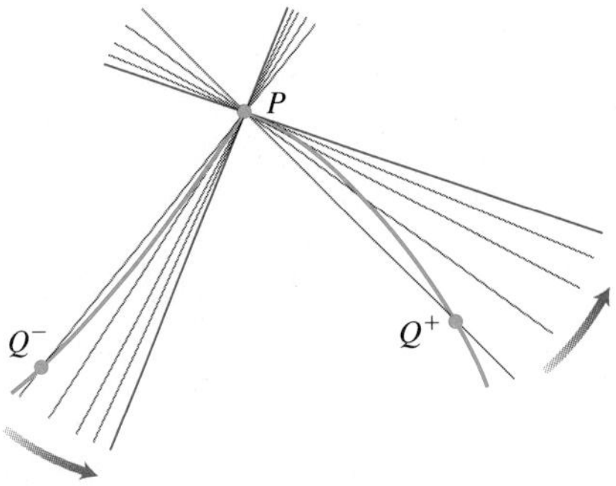

1. a *corner*, where the one-sided derivatives differ.

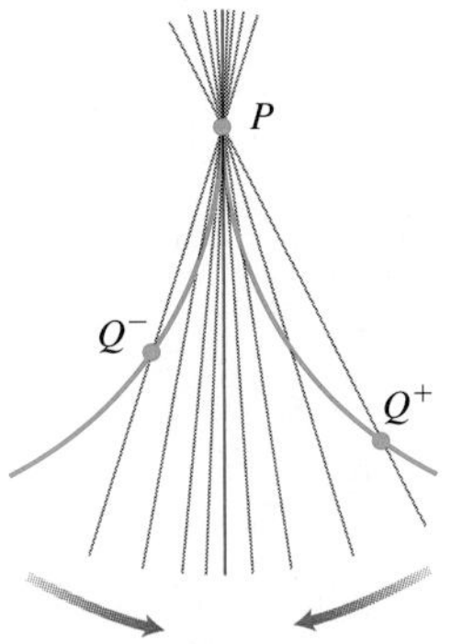

2. a *cusp*, where the slope of PQ approaches ∞ from one side and $-\infty$ from the other.

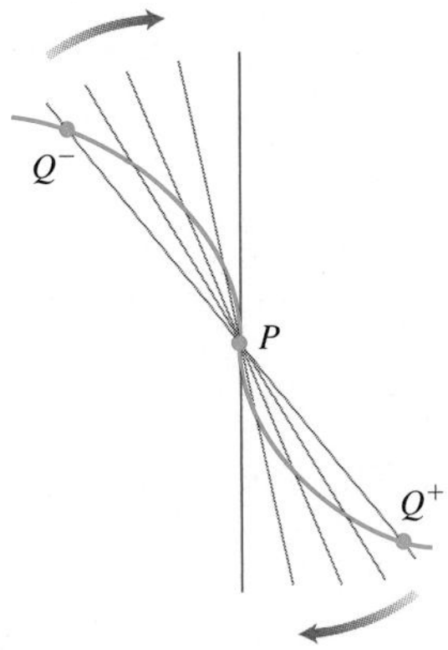

3. a *vertical tangent*, where the slope of PQ approaches ∞ from both sides or approaches $-\infty$ from both sides (here, $-\infty$).

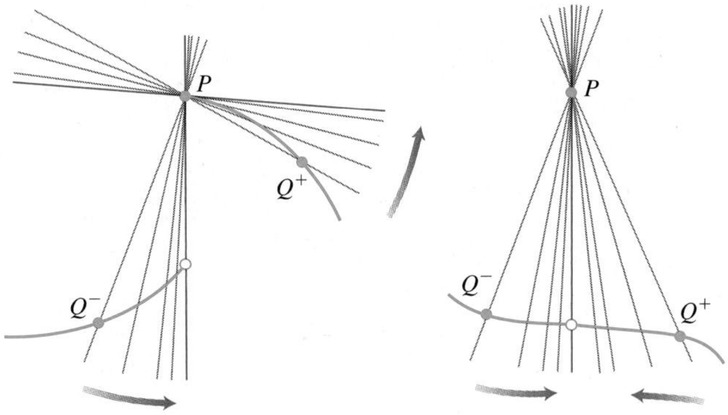

4. a *discontinuity* (two examples shown).

Differentiable Functions Are Continuous

A function is continuous at every point where it has a derivative.

> **THEOREM 1—Differentiability Implies Continuity** If f has a derivative at $x = c$, then f is continuous at $x = c$.

Proof Given that $f'(c)$ exists, we must show that $\lim_{x\to c} f(x) = f(c)$, or equivalently, that $\lim_{h\to 0} f(c + h) = f(c)$. If $h \neq 0$, then

$$\begin{aligned} f(c + h) &= f(c) + (f(c + h) - f(c)) \\ &= f(c) + \frac{f(c + h) - f(c)}{h} \cdot h. \end{aligned}$$

Now take limits as $h \to 0$. By Theorem 1 of Section 1.4,

$$\begin{aligned} \lim_{h\to 0} f(c + h) &= \lim_{h\to 0} f(c) + \lim_{h\to 0} \frac{f(c + h) - f(c)}{h} \cdot \lim_{h\to 0} h \\ &= f(c) + f'(c) \cdot 0 \\ &= f(c) + 0 \\ &= f(c). \end{aligned}$$

■

Similar arguments with one-sided limits show that if f has a derivative from one side (right or left) at $x = c$ then f is continuous from that side at $x = c$.

Theorem 1 says that if a function has a discontinuity at a point (for instance, a jump discontinuity), then it cannot be differentiable there. The greatest integer function $y = \lfloor x \rfloor$ fails to be differentiable at every integer $x = n$ (Example 4, Section 1.7).

Caution The converse of Theorem 1 is false. A function need not have a derivative at a point where it is continuous, as we saw in Example 4.

The Intermediate Value Property of Derivatives (Darboux's Theorem)

Not every function can be some function's derivative, as we see from the following theorem, first proved in 1875 by the French mathematician Jean Gaston Darboux (1842–1917).

> **THEOREM 2—Darboux's Theorem** If a and b are any two points in an interval on which f is differentiable, then f' takes on every value between $f'(a)$ and $f'(b)$.

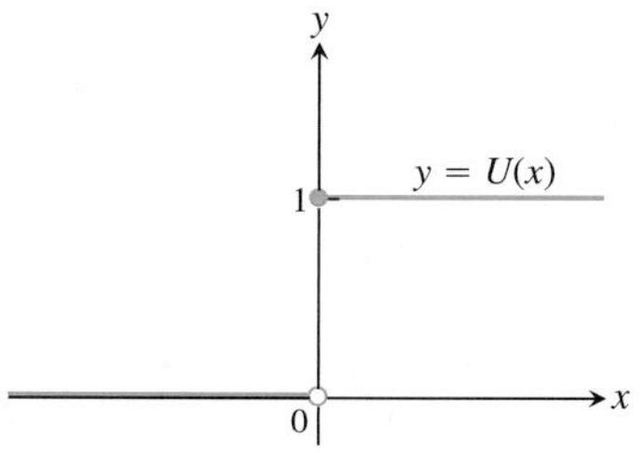

FIGURE 2.8 The unit step function does not have the Intermediate Value Property and cannot be the derivative of a function on the real line.

Theorem 2 (which we will not prove) says that a function cannot *be* a derivative on an interval unless it has the Intermediate Value Property there. For example, the unit step function in Figure 2.8 cannot be the derivative of any real-valued function on the real line. In Chapter 4 we will see that every continuous function is a derivative of some function.

In Section 3.4, we invoke Theorem 2 to analyze what happens at a point on the graph of a twice-differentiable function where it changes its "bending" behavior.

EXERCISES 2.2

Using the definition, calculate the derivatives of the functions in Exercises 1–6. Then find the values of the derivatives as specified.

1. $f(x) = 4 - x^2;\quad f'(-3), f'(0), f'(1)$
2. $F(x) = (x-1)^2 + 1;\quad F'(-1), F'(0), F'(2)$
3. $g(t) = \dfrac{1}{t^2};\quad g'(-1), g'(2), g'(\sqrt{3})$
4. $k(z) = \dfrac{1-z}{2z};\quad k'(-1), k'(1), k'(\sqrt{2})$
5. $p(\theta) = \sqrt{3\theta};\quad p'(1), p'(3), p'(2/3)$
6. $r(s) = \sqrt{2s+1};\quad r'(0), r'(1), r'(1/2)$

In Exercises 7–12, find the indicated derivatives.

7. $\dfrac{dy}{dx}$ if $y = 2x^3$
8. $\dfrac{dr}{ds}$ if $r = \dfrac{s^3}{2} + 1$
9. $\dfrac{ds}{dt}$ if $s = \dfrac{t}{2t+1}$
10. $\dfrac{dv}{dt}$ if $v = t - \dfrac{1}{t}$
11. $\dfrac{dp}{dq}$ if $p = \dfrac{1}{\sqrt{q+1}}$
12. $\dfrac{dz}{dw}$ if $z = \dfrac{1}{\sqrt{3w-2}}$

In Exercises 13–16, differentiate the functions and find the slope of the tangent line at the given value of the independent variable.

13. $f(x) = x + \dfrac{9}{x},\quad x = -3$
14. $k(x) = \dfrac{1}{2+x},\quad x = 2$
15. $s = t^3 - t^2,\quad t = -1$
16. $y = (x+1)^3,\quad x = -2$

In Exercises 17–18, differentiate the functions. Then find an equation of the tangent line at the indicated point on the graph of the function.

17. $y = f(x) = \dfrac{8}{\sqrt{x-2}},\quad (x, y) = (6, 4)$
18. $w = g(z) = 1 + \sqrt{4-z},\quad (z, w) = (3, 2)$

In Exercises 19–22, find the values of the derivatives.

19. $\left.\dfrac{ds}{dt}\right|_{t=-1}$ if $s = 1 - 3t^2$
20. $\left.\dfrac{dy}{dx}\right|_{x=\sqrt{3}}$ if $y = 1 - \dfrac{1}{x}$
21. $\left.\dfrac{dr}{d\theta}\right|_{\theta=0}$ if $r = \dfrac{2}{\sqrt{4-\theta}}$
22. $\left.\dfrac{dw}{dz}\right|_{z=4}$ if $w = z + \sqrt{z}$

Use the formula

$$f'(x) = \lim_{z \to x} \frac{f(z) - f(x)}{z - x}$$

to find the derivative of the functions in Exercises 23–26.

23. $f(x) = \dfrac{1}{x+2}$
24. $f(x) = \dfrac{1}{(x-1)^2}$
25. $g(x) = \dfrac{x}{x-1}$
26. $g(x) = 1 + \sqrt{x}$

Match the functions graphed in Exercises 27–30 with the derivatives graphed in the accompanying figures (a)–(d).

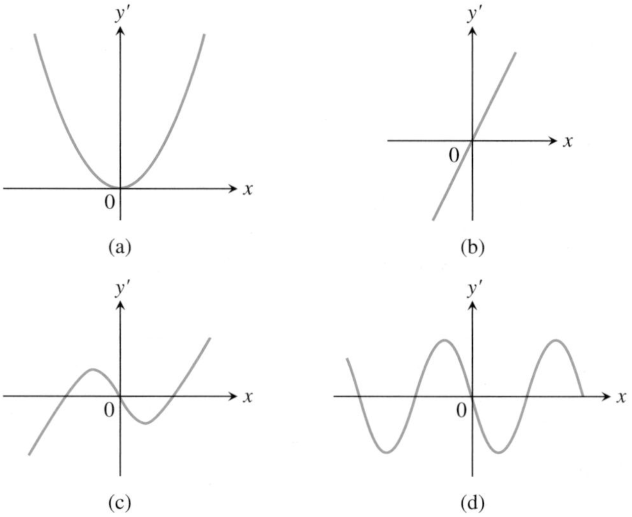

27.

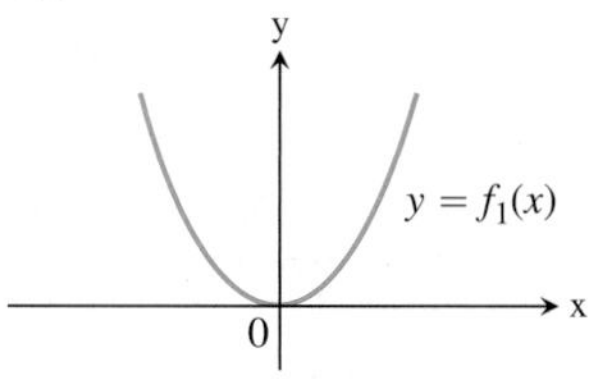

28.

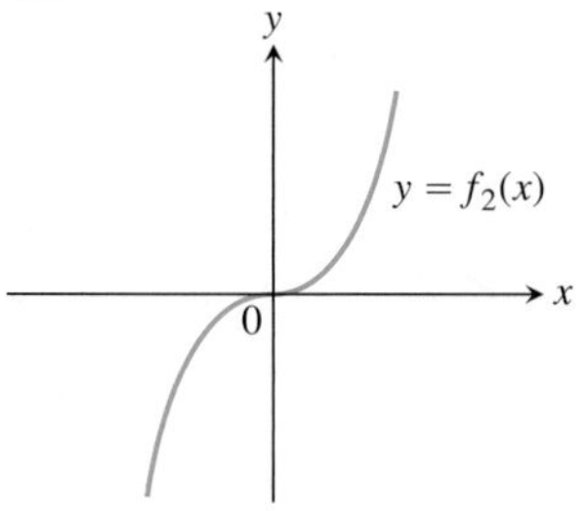

29.

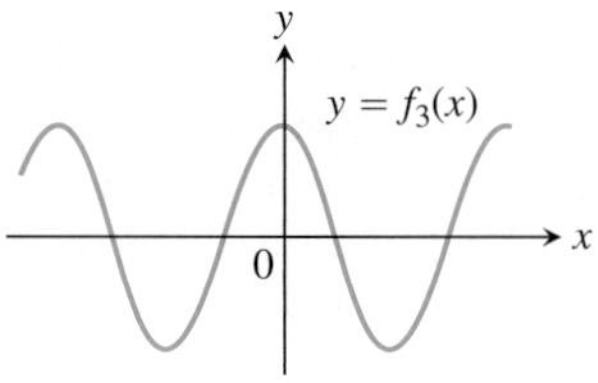

30.

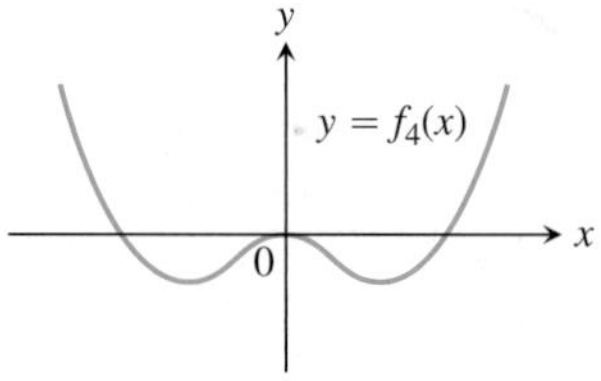

31. **a.** The graph in the accompanying figure is made of line segments joined end to end. At which points of the interval $[-4, 6]$ is f' not defined? Give reasons for your answer.

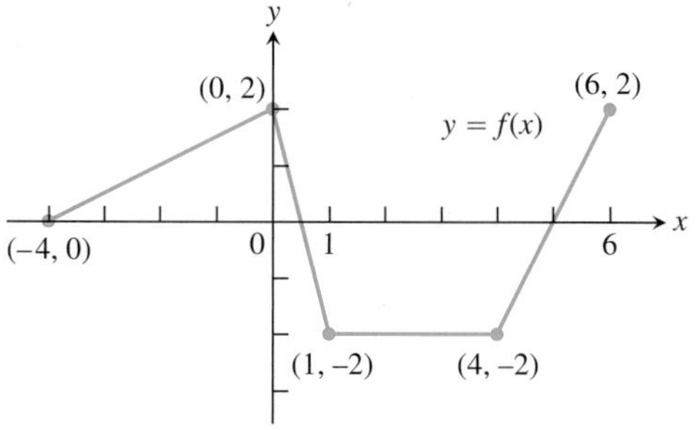

b. Graph the derivative of f.
The graph should show a step function.

32. Growth in the economy The graph in the accompanying figure shows the average annual percentage change $y = f(t)$ in the U.S. gross national product (GNP) for the years 1983–1988. Graph dy/dt (where defined). (*Source*: *Statistical Abstracts of the United States*, 110th Edition, U.S. Department of Commerce, p. 427.)

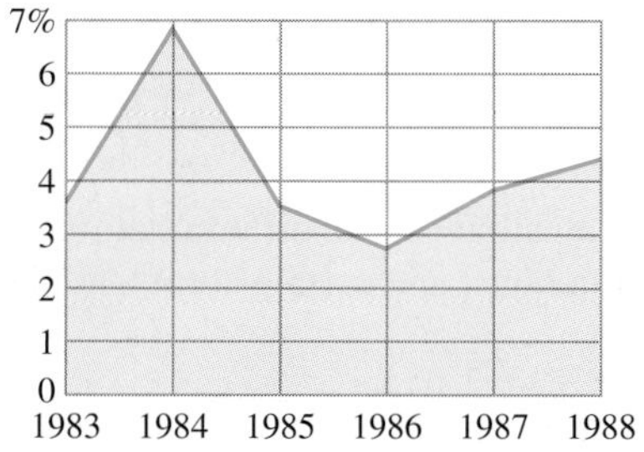

Compare the right-hand and left-hand derivatives to show that the functions in Exercises 33–36 are not differentiable at the point P.

33.

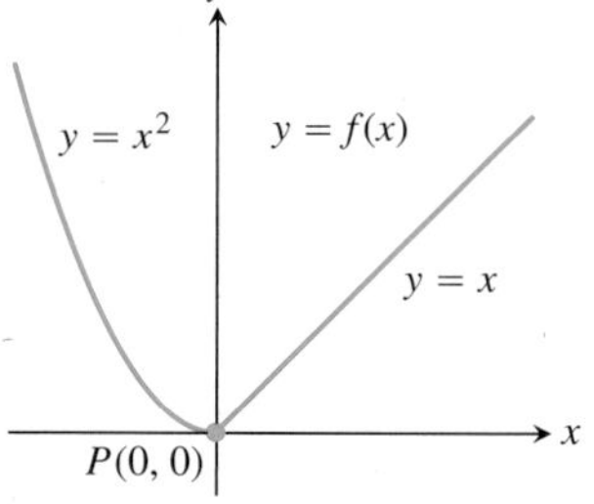

34.

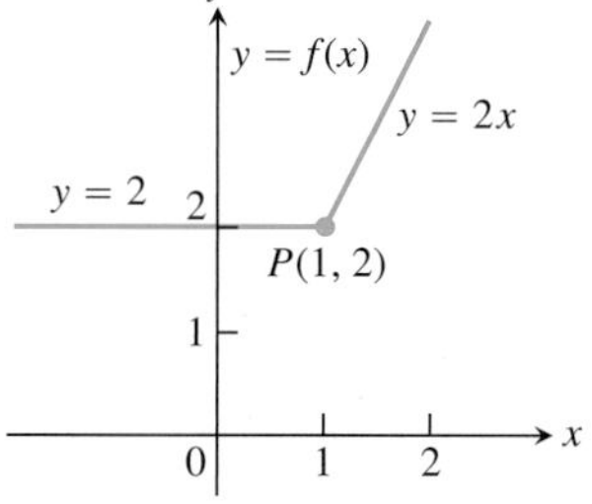

35.

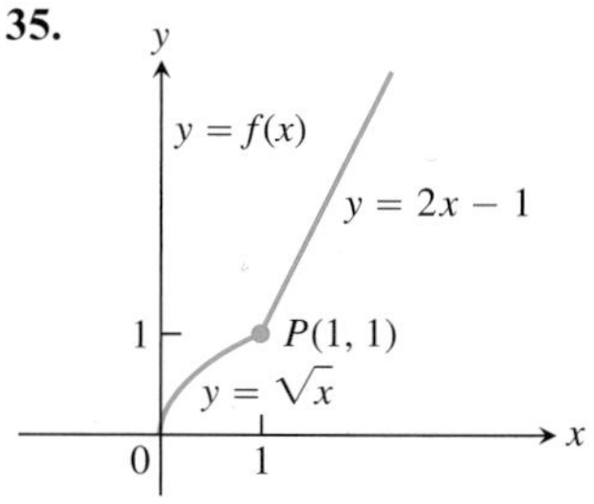

36.

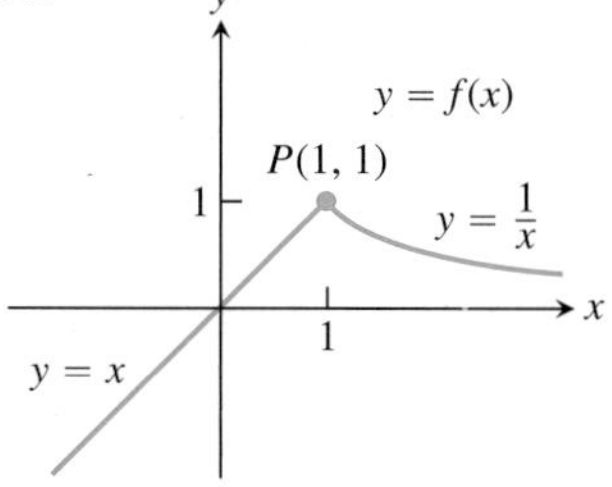

Each figure in Exercises 37–42 shows the graph of a function over a closed interval D. At what domain points does the function appear to be

a. differentiable?

b. continuous but not differentiable?

c. neither continuous nor differentiable?

Give reasons for your answers.

37.

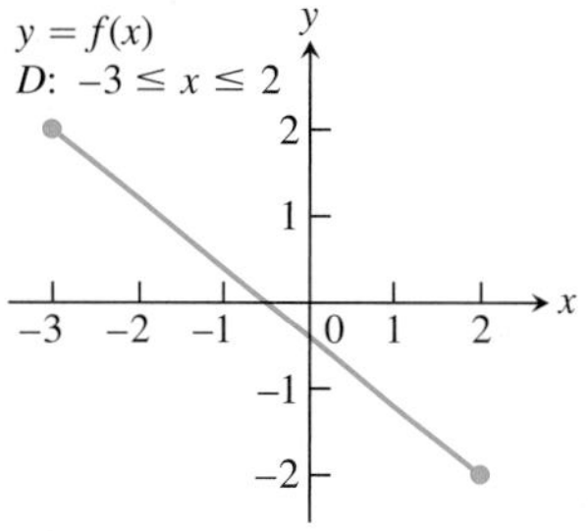

38.

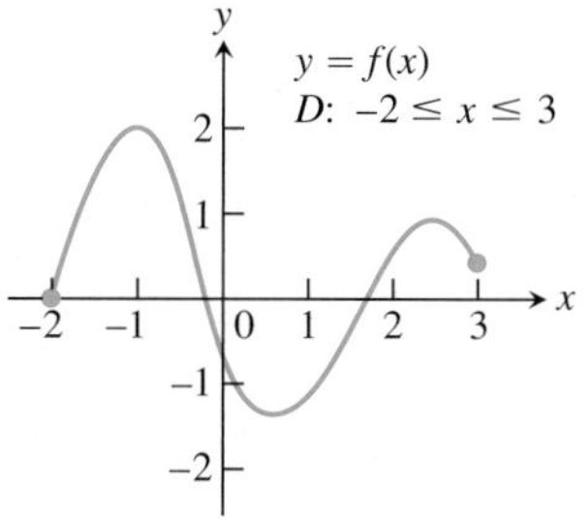

39.

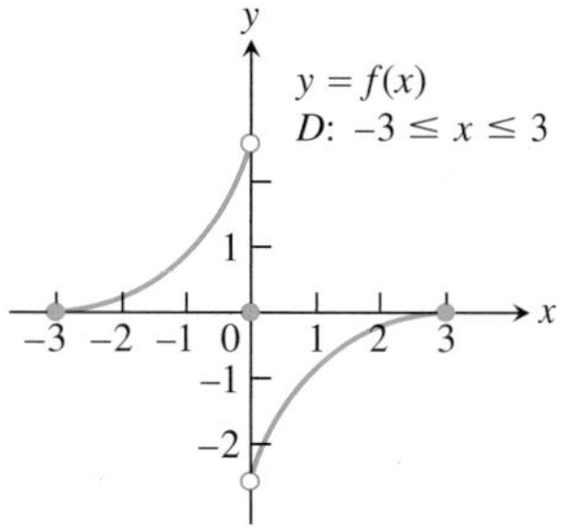

40.

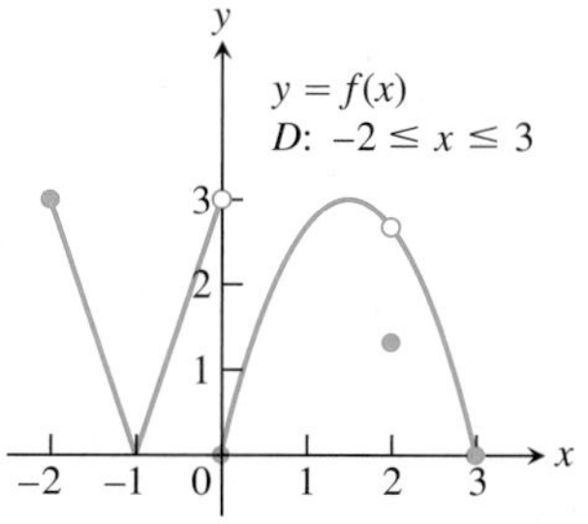

41.

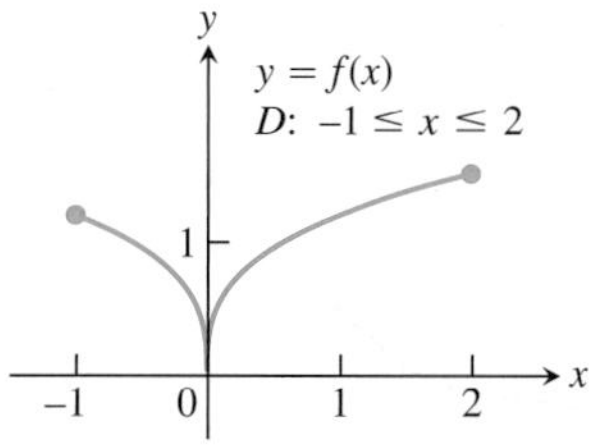

42.

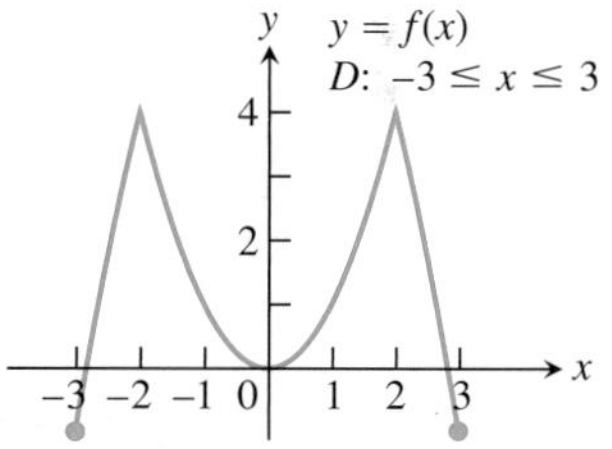

43. Tangent to a parabola Does the parabola $y = 2x^2 - 13x + 5$ have a tangent whose slope is -1? If so, find an equation for the line and the point of tangency. If not, why not?

44. Tangent to $y = \sqrt{x}$ Does any tangent to the curve $y = \sqrt{x}$ cross the x-axis at $x = -1$? If so, find an equation for the line and the point of tangency. If not, why not?

45. **a.** Let $f(x)$ be a function satisfying $|f(x)| \le x^2$ for $-1 \le x \le 1$. Show that f is differentiable at $x = 0$ and find $f'(0)$.

b. Show that

$$f(x) = \begin{cases} x^2 \sin \dfrac{1}{x}, & x \ne 0 \\ 0, & x = 0 \end{cases}$$

is differentiable at $x = 0$ and find $f'(0)$.

T **46. Weierstrass's nowhere differentiable continuous function** The sum of the first eight terms of the Weierstrass function $f(x) = \sum_{n=0}^{\infty} (2/3)^n \cos(9^n \pi x)$ is

$$g(x) = \cos(\pi x) + (2/3)^1 \cos(9\pi x) + (2/3)^2 \cos(9^2\pi x) + (2/3)^3 \cos(9^3\pi x) + \cdots + (2/3)^7 \cos(9^7\pi x).$$

Graph this sum. Zoom in several times. How wiggly and bumpy is this graph? Specify a viewing window in which the displayed portion of the graph is smooth.

COMPUTER EXPLORATIONS

Use a CAS to perform the following steps for the functions in Exercises 47–52.

a. Plot $y = f(x)$ to see that function's global behavior.

b. Define the difference quotient q at a general point x, with general step size h.

c. Take the limit as $h \to 0$. What formula does this give?

d. Substitute the value $x = x_0$ and plot the function $y = f(x)$ together with its tangent line at that point.

e. Substitute various values for x larger and smaller than x_0 into the formula obtained in part (c). Do the numbers make sense with your picture?

f. Graph the formula obtained in part (c). What does it mean when its values are negative? Zero? Positive? Does this make sense with your plot from part (a)? Give reasons for your answer.

47. $f(x) = x^3 + x^2 - x, \quad x_0 = 1$

48. $f(x) = x^{1/3} + x^{2/3}, \quad x_0 = 1$

49. $f(x) = \dfrac{4x}{x^2 + 1}, \quad x_0 = 2$

50. $f(x) = \dfrac{x - 1}{3x^2 + 1}, \quad x_0 = -1$

51. $f(x) = \sin 2x, \quad x_0 = \pi/2$

52. $f(x) = x^2 \cos x, \quad x_0 = \pi/4$

2.3 Differentiation Rules

This section introduces several rules that allow us to differentiate constant functions, power functions, polynomials, rational functions, and certain combinations of them.

Powers, Multiples, Sums, and Differences

The first rule of differentiation is that the derivative of every constant function is zero.

> **RULE 1 Derivative of a Constant Function**
> If f has the constant value $f(x) = c$, then
>
> $$\frac{df}{dx} = \frac{d}{dx}(c) = 0.$$

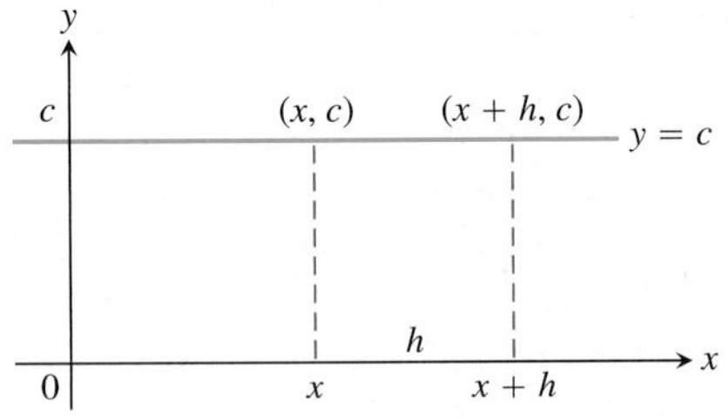

FIGURE 2.9 The rule $(d/dx)(c) = 0$ is another way to say that the values of constant functions never change and that the slope of a horizontal line is zero at every point.

Proof of Rule 1 We apply the definition of derivative to $f(x) = c$, the function whose outputs have the constant value c (Figure 2.9). At every value of x, we find that

$$f'(x) = \lim_{h \to 0} \frac{f(x + h) - f(x)}{h} = \lim_{h \to 0} \frac{c - c}{h} = \lim_{h \to 0} 0 = 0. \qquad \blacksquare$$

From Section 2.1, we know that

$$\frac{d}{dx}(x) = \frac{d}{dx}(1 \cdot x) = 1 \qquad \frac{d}{dx}(mx + b) = m = 1$$

and

$$\frac{d}{dx}\left(\frac{1}{x}\right) = -\frac{1}{x^2}, \quad \text{or} \quad \frac{d}{dx}\left(x^{-1}\right) = -x^{-2}$$

From Example 2, Section 2.2, we also know that

$$\frac{d}{dx}\left(\sqrt{x}\right) = \frac{1}{2\sqrt{x}}, \quad \text{or} \quad \frac{d}{dx}\left(x^{1/2}\right) = \frac{1}{2}x^{-1/2}.$$

These three examples illustrate a general rule for differentiating a power x^n. We first prove the rule when n is a positive integer.

Power Rule for Positive Integers:
If n is a positive integer, then

$$\frac{d}{dx}x^n = nx^{n-1}.$$

HISTORICAL BIOGRAPHY

Richard Courant
(1888–1972)

Proof of the Positive Integer Power Rule The formula

$$z^n - x^n = (z - x)(z^{n-1} + z^{n-2}x + \cdots + zx^{n-2} + x^{n-1})$$

can be verified by multiplying out the right-hand side. Then from the alternative form for the definition of the derivative,

$$\begin{aligned} f'(x) &= \lim_{z \to x} \frac{f(z) - f(x)}{z - x} = \lim_{z \to x} \frac{z^n - x^n}{z - x} \\ &= \lim_{z \to x} (z^{n-1} + z^{n-2}x + \cdots + zx^{n-2} + x^{n-1}) \qquad n \text{ terms} \\ &= nx^{n-1}. \end{aligned}$$

■

The Power Rule is actually valid for all real numbers n. We have seen examples for a negative integer and fractional power, but n could be an irrational number as well. To apply the Power Rule, we subtract 1 from the original exponent n and multiply the result by n. Here we state the general version of the rule, but postpone its proof until Section 2.10.

RULE 2 Power Rule (General Version)
If n is any real number, then

$$\frac{d}{dx}x^n = nx^{n-1},$$

for all x where the powers x^n and x^{n-1} are defined.

EXAMPLE 1 Differentiate the following powers of x.

(a) x^3 **(b)** $x^{2/3}$ **(c)** $x^{\sqrt{2}}$ **(d)** $\frac{1}{x^4}$ **(e)** $x^{-4/3}$ **(f)** $\sqrt{x^{2+\pi}}$

Solution

(a) $\dfrac{d}{dx}(x^3) = 3x^{3-1} = 3x^2$ **(b)** $\dfrac{d}{dx}(x^{2/3}) = \dfrac{2}{3}x^{(2/3)-1} = \dfrac{2}{3}x^{-1/3}$

(c) $\dfrac{d}{dx}\left(x^{\sqrt{2}}\right) = \sqrt{2}x^{\sqrt{2}-1}$ **(d)** $\dfrac{d}{dx}\left(\dfrac{1}{x^4}\right) = \dfrac{d}{dx}(x^{-4}) = -4x^{-4-1} = -4x^{-5} = -\dfrac{4}{x^5}$

(e) $\dfrac{d}{dx}(x^{-4/3}) = -\dfrac{4}{3}x^{-(4/3)-1} = -\dfrac{4}{3}x^{-7/3}$

(f) $\dfrac{d}{dx}\left(\sqrt{x^{2+\pi}}\right) = \dfrac{d}{dx}\left(x^{1+(\pi/2)}\right) = \left(1 + \dfrac{\pi}{2}\right)x^{1+(\pi/2)-1} = \dfrac{1}{2}(2 + \pi)\sqrt{x^{\pi}}$ ■

The third rule says that when a differentiable function is multiplied by a constant, its derivative is multiplied by the same constant.

RULE 3 Constant Multiple Rule
If u is a differentiable function of x, and c is a constant, then

$$\frac{d}{dx}(cu) = c\frac{du}{dx}.$$

In particular, if n is any real number, then

$$\frac{d}{dx}(cx^n) = cnx^{n-1}.$$

Proof of Rule 3

$$\begin{aligned}\frac{d}{dx}cu &= \lim_{h\to 0}\frac{cu(x+h) - cu(x)}{h} && \text{Derivative definition with } f(x) = cu(x)\\ &= c\lim_{h\to 0}\frac{u(x+h) - u(x)}{h} && \text{Limit property}\\ &= c\frac{du}{dx} && u \text{ is differentiable.}\end{aligned}$$

■

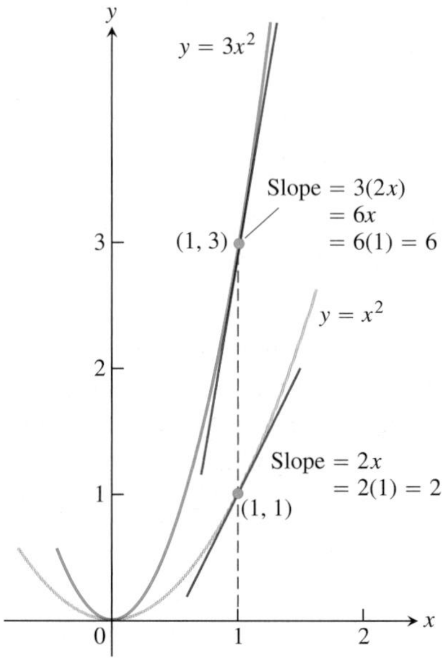

FIGURE 2.10 The graphs of $y = x^2$ and $y = 3x^2$. Tripling the y-coordinates triples the slope (Example 2).

EXAMPLE 2

(a) The derivative formula

$$\frac{d}{dx}(3x^2) = 3\cdot 2x = 6x$$

says that if we rescale the graph of $y = x^2$ by multiplying each y-coordinate by 3, then we multiply the slope at each point by 3 (Figure 2.10).

(b) Negative of a function

The derivative of the negative of a differentiable function u is the negative of the function's derivative. Rule 3 with $c = -1$ gives

$$\frac{d}{dx}(-u) = \frac{d}{dx}(-1\cdot u) = -1\cdot\frac{d}{dx}(u) = -\frac{du}{dx}.$$

■

Denoting Functions by u and v

The functions we are working with when we need a differentiation formula are likely to be denoted by letters like f and g. When we apply the formula, we do not want to have it using these same letters in some other way. To guard against this problem, we denote the functions in differentiation rules by letters like u and v that are not likely to be already in use.

The next rule says that the derivative of the sum of two differentiable functions is the sum of their derivatives.

RULE 4 Derivative Sum Rule

If u and v are differentiable functions of x, then their sum $u + v$ is differentiable at every point where u and v are both differentiable. At such points,

$$\frac{d}{dx}(u + v) = \frac{du}{dx} + \frac{dv}{dx}.$$

For example, if $y = x^4 + 12x$, then y is the sum of $u(x) = x^4$ and $v(x) = 12x$. We then have

$$\frac{dy}{dx} = \frac{d}{dx}(x^4) + \frac{d}{dx}(12x) = 4x^3 + 12.$$

Proof of Rule 4 We apply the definition of derivative to $f(x) = u(x) + v(x)$:

$$\begin{aligned}\frac{d}{dx}[u(x) + v(x)] &= \lim_{h\to 0} \frac{[u(x+h) + v(x+h)] - [u(x) + v(x)]}{h} \\ &= \lim_{h\to 0} \left[\frac{u(x+h) - u(x)}{h} + \frac{v(x+h) - v(x)}{h}\right] \\ &= \lim_{h\to 0} \frac{u(x+h) - u(x)}{h} + \lim_{h\to 0} \frac{v(x+h) - v(x)}{h} = \frac{du}{dx} + \frac{dv}{dx}. \quad \blacksquare\end{aligned}$$

Combining the Sum Rule with the Constant Multiple Rule gives the **Difference Rule**, which says that the derivative of a *difference* of differentiable functions is the difference of their derivatives.

$$\frac{d}{dx}(u - v) = \frac{d}{dx}[u + (-1)v] = \frac{du}{dx} + (-1)\frac{dv}{dx} = \frac{du}{dx} - \frac{dv}{dx}$$

The Sum Rule also extends to finite sums of more than two functions. If $u_1, u_2, \ldots, u_n$ are differentiable at x, then so is $u_1 + u_2 + \cdots + u_n$, and

$$\frac{d}{dx}(u_1 + u_2 + \cdots + u_n) = \frac{du_1}{dx} + \frac{du_2}{dx} + \cdots + \frac{du_n}{dx}.$$

EXAMPLE 3

$$\begin{aligned} y &= x^3 + \frac{4}{3}x^2 - 5x + 1 \\ \frac{dy}{dx} &= \frac{d}{dx}x^3 + \frac{d}{dx}\left(\frac{4}{3}x^2\right) - \frac{d}{dx}(5x) + \frac{d}{dx}(1) \\ &= 3x^2 + \frac{4}{3}\cdot 2x - 5 + 0 = 3x^2 + \frac{8}{3}x - 5 \quad \blacksquare\end{aligned}$$

We can differentiate any polynomial term by term, the way we differentiated the polynomial in Example 3. All polynomials are differentiable everywhere.

EXAMPLE 4 Does the curve $y = x^4 - 2x^2 + 2$ have any horizontal tangents? If so, where?

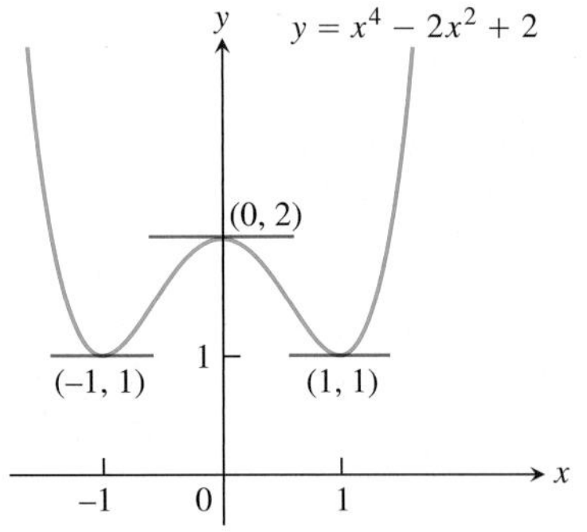

FIGURE 2.11 The curve $y = x^4 - 2x^2 + 2$ and its horizontal tangents (Example 4).

Solution The horizontal tangents, if any, occur where the slope dy/dx is zero. We have

$$\frac{dy}{dx} = \frac{d}{dx}(x^4 - 2x^2 + 2) = 4x^3 - 4x.$$

Now solve the equation $\dfrac{dy}{dx} = 0$ for x:

$$\begin{aligned} 4x^3 - 4x &= 0 \\ 4x(x^2 - 1) &= 0 \\ x &= 0, 1, -1. \end{aligned}$$

The curve $y = x^4 - 2x^2 + 2$ has horizontal tangents at $x = 0, 1$, and -1. The corresponding points on the curve are $(0, 2)$, $(1, 1)$, and $(-1, 1)$. See Figure 2.11. ■

Products and Quotients

While the derivative of the sum of two functions is the sum of their derivatives, the derivative of the product of two functions is *not* the product of their derivatives. For instance,

$$\frac{d}{dx}(x \cdot x) = \frac{d}{dx}(x^2) = 2x, \qquad \text{while} \qquad \frac{d}{dx}(x) \cdot \frac{d}{dx}(x) = 1 \cdot 1 = 1.$$

The derivative of a product of two functions is the sum of *two* products, as we now explain.

> **RULE 5 Derivative Product Rule**
> If u and v are differentiable at x, then so is their product uv, and
>
> $$\frac{d}{dx}(uv) = u\frac{dv}{dx} + v\frac{du}{dx}.$$

The derivative of the product uv is u times the derivative of v plus v times the derivative of u. In *prime notation*, $(uv)' = uv' + vu'$. In function notation,

$$\frac{d}{dx}[f(x)g(x)] = f(x)g'(x) + g(x)f'(x).$$

EXAMPLE 5 Find the derivative of $y = (x^2 + 1)(x^3 + 3)$.

Solution

(a) From the Product Rule with $u = x^2 + 1$ and $v = x^3 + 3$, we find

$$\begin{aligned} \frac{d}{dx}\left[(x^2 + 1)(x^3 + 3)\right] &= (x^2 + 1)(3x^2) + (x^3 + 3)(2x) \qquad \frac{d}{dx}(uv) = u\frac{dv}{dx} + v\frac{du}{dx} \\ &= 3x^4 + 3x^2 + 2x^4 + 6x \\ &= 5x^4 + 3x^2 + 6x. \end{aligned}$$

(b) This particular product can be differentiated as well (perhaps better) by multiplying out the original expression for y and differentiating the resulting polynomial:

$$y = (x^2 + 1)(x^3 + 3) = x^5 + x^3 + 3x^2 + 3$$

$$\frac{dy}{dx} = 5x^4 + 3x^2 + 6x.$$

This is in agreement with our first calculation. ■

Proof of the Product Rule

$$\frac{d}{dx}(uv) = \lim_{h \to 0} \frac{u(x+h)v(x+h) - u(x)v(x)}{h}$$

To change this fraction into an equivalent one that contains difference quotients for the derivatives of u and v, we subtract and add $u(x+h)v(x)$ in the numerator:

$$\begin{aligned}
\frac{d}{dx}(uv) &= \lim_{h \to 0} \frac{u(x+h)v(x+h) - u(x+h)v(x) + u(x+h)v(x) - u(x)v(x)}{h} \\
&= \lim_{h \to 0} \left[u(x+h)\frac{v(x+h) - v(x)}{h} + v(x)\frac{u(x+h) - u(x)}{h} \right] \\
&= \lim_{h \to 0} u(x+h) \cdot \lim_{h \to 0} \frac{v(x+h) - v(x)}{h} + v(x) \cdot \lim_{h \to 0} \frac{u(x+h) - u(x)}{h}.
\end{aligned}$$

As h approaches zero, $u(x+h)$ approaches $u(x)$ because u, being differentiable at x, is continuous at x. The two fractions approach the values of dv/dx at x and du/dx at x. In short,

$$\frac{d}{dx}(uv) = u\frac{dv}{dx} + v\frac{du}{dx}.$$ ■

Picturing the Product Rule

Suppose $u(x)$ and $v(x)$ are positive and increase when x increases, and $h > 0$.

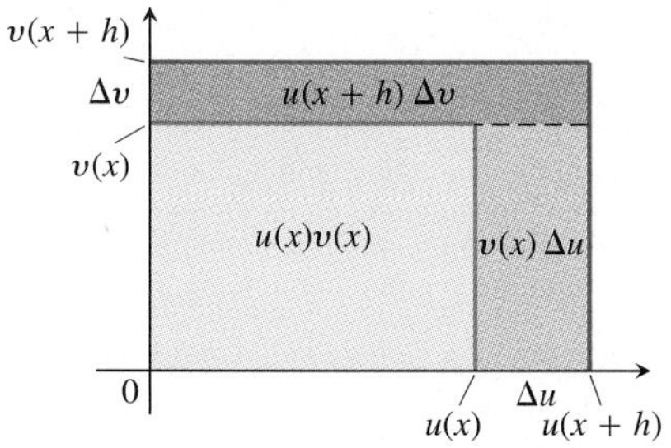

Then the change in the product uv is the difference in areas of the larger and smaller "squares," which is the sum of the upper and right-hand reddish-shaded rectangles. That is,

$$\begin{aligned}
\Delta(uv) &= u(x+h)v(x+h) - u(x)v(x) \\
&= u(x+h)\Delta v + v(x)\Delta u.
\end{aligned}$$

Division by h gives

$$\frac{\Delta(uv)}{h} = u(x+h)\frac{\Delta v}{h} + v(x)\frac{\Delta u}{h}.$$

The limit as $h \to 0^+$ gives the Product Rule.

Just as the derivative of the product of two differentiable functions is not the product of their derivatives, the derivative of the quotient of two functions is not the quotient of their derivatives. What happens instead is the Quotient Rule.

RULE 6 Derivative Quotient Rule

If u and v are differentiable at x and if $v(x) \neq 0$, then the quotient u/v is differentiable at x, and

$$\frac{d}{dx}\left(\frac{u}{v}\right) = \frac{v\dfrac{du}{dx} - u\dfrac{dv}{dx}}{v^2}.$$

In function notation,

$$\frac{d}{dx}\left[\frac{f(x)}{g(x)}\right] = \frac{g(x)f'(x) - f(x)g'(x)}{g^2(x)}.$$

EXAMPLE 6 Find the derivative of $y = \dfrac{t^2 - 1}{t^2 + 1}$.

Solution We apply the Quotient Rule with $u = t^2 - 1$ and $v = t^2 + 1$:

$$\begin{aligned}\frac{dy}{dt} &= \frac{(t^2 + 1)\cdot 2t - (t^2 - 1)\cdot 2t}{(t^2 + 1)^2} \\ &= \frac{2t^3 + 2t - 2t^3 + 2t}{(t^2 + 1)^2} \\ &= \frac{4t}{(t^2 + 1)^2}.\end{aligned}$$

$$\frac{d}{dt}\left(\frac{u}{v}\right) = \frac{v(du/dt) - u(dv/dt)}{v^2}$$

■

Proof of the Quotient Rule

$$\begin{aligned}\frac{d}{dx}\left(\frac{u}{v}\right) &= \lim_{h\to 0}\frac{\dfrac{u(x + h)}{v(x + h)} - \dfrac{u(x)}{v(x)}}{h} \\ &= \lim_{h\to 0}\frac{v(x)u(x + h) - u(x)v(x + h)}{hv(x + h)v(x)}\end{aligned}$$

To change the last fraction into an equivalent one that contains the difference quotients for the derivatives of u and v, we subtract and add $v(x)u(x)$ in the numerator. We then get

$$\begin{aligned}\frac{d}{dx}\left(\frac{u}{v}\right) &= \lim_{h\to 0}\frac{v(x)u(x + h) - v(x)u(x) + v(x)u(x) - u(x)v(x + h)}{hv(x + h)v(x)} \\ &= \lim_{h\to 0}\frac{v(x)\dfrac{u(x + h) - u(x)}{h} - u(x)\dfrac{v(x + h) - v(x)}{h}}{v(x + h)v(x)}.\end{aligned}$$

Taking the limits in the numerator and denominator now gives the Quotient Rule. ■

The choice of which rules to use in solving a differentiation problem can make a difference in how much work you have to do. Here is an example.

EXAMPLE 7 Rather than using the Quotient Rule to find the derivative of

$$y = \frac{(x - 1)(x^2 - 2x)}{x^4},$$

expand the numerator and divide by x^4:

$$y = \frac{(x-1)(x^2-2x)}{x^4} = \frac{x^3-3x^2+2x}{x^4} = x^{-1} - 3x^{-2} + 2x^{-3}.$$

Then use the Sum and Power Rules:

$$\begin{aligned}\frac{dy}{dx} &= -x^{-2} - 3(-2)x^{-3} + 2(-3)x^{-4} \\ &= -\frac{1}{x^2} + \frac{6}{x^3} - \frac{6}{x^4}.\end{aligned}$$

■

Second- and Higher-Order Derivatives

If $y = f(x)$ is a differentiable function, then its derivative $f'(x)$ is also a function. If f' is also differentiable, then we can differentiate f' to get a new function of x denoted by f''. So $f'' = (f')'$. The function f'' is called the **second derivative** of f because it is the derivative of the first derivative. It is written in several ways:

$$f''(x) = \frac{d^2y}{dx^2} = \frac{d}{dx}\left(\frac{dy}{dx}\right) = \frac{dy'}{dx} = y'' = D^2(f)(x) = D_x^{\,2} f(x).$$

The symbol D^2 means the operation of differentiation is performed twice.

If $y = x^6$, then $y' = 6x^5$ and we have

$$y'' = \frac{dy'}{dx} = \frac{d}{dx}(6x^5) = 30x^4.$$

Thus $D^2(x^6) = 30x^4$.

If y'' is differentiable, its derivative, $y''' = dy''/dx = d^3y/dx^3$ is the **third derivative** of y with respect to x. The names continue as you imagine, with

$$y^{(n)} = \frac{d}{dx}y^{(n-1)} = \frac{d^ny}{dx^n} = D^ny$$

denoting the ***n*th derivative** of y with respect to x for any positive integer n.

We can interpret the second derivative as the rate of change of the slope of the tangent to the graph of $y = f(x)$ at each point. You will see in the next chapter that the second derivative reveals whether the graph bends upward or downward from the tangent line as we move off the point of tangency. In the next section, we interpret both the second and third derivatives in terms of motion along a straight line.

How to Read the Symbols for Derivatives

y'	"y prime"
y''	"y double prime"
$\frac{d^2y}{dx^2}$	"d squared y dx squared"
y'''	"y triple prime"
$y^{(n)}$	"y super n"
$\frac{d^ny}{dx^n}$	"d to the n of y by dx to the n"
D^n	"D to the n"

EXAMPLE 8 The first four derivatives of $y = x^3 - 3x^2 + 2$ are

First derivative:	$y' = 3x^2 - 6x$
Second derivative:	$y'' = 6x - 6$
Third derivative:	$y''' = 6$
Fourth derivative:	$y^{(4)} = 0.$

The function has derivatives of all orders, the fifth and later derivatives all being zero. ■

EXERCISES 2.3

In Exercises 1–12, find the first and second derivatives.

1. $y = -x^2 + 3$
2. $y = x^2 + x + 8$
3. $s = 5t^3 - 3t^5$
4. $w = 3z^7 - 7z^3 + 21z^2$
5. $y = \frac{4x^3}{3} - x$
6. $y = \frac{x^3}{3} + \frac{x^2}{2} + \frac{x}{4}$
7. $w = 3z^{-2} - \frac{1}{z}$
8. $s = -2t^{-1} + \frac{4}{t^2}$
9. $y = 6x^2 - 10x - 5x^{-2}$
10. $y = 4 - 2x - x^{-3}$
11. $r = \frac{1}{3s^2} - \frac{5}{2s}$
12. $r = \frac{12}{\theta} - \frac{4}{\theta^3} + \frac{1}{\theta^4}$

In Exercises 13–16, find y' **(a)** by applying the Product Rule and **(b)** by multiplying the factors to produce a sum of simpler terms to differentiate.

13. $y = (3 - x^2)(x^3 - x + 1)$
14. $y = (x - 1)(x^2 + x + 1)$
15. $y = (x^2 + 1)\left(x + 5 + \frac{1}{x}\right)$
16. $y = \left(x + \frac{1}{x}\right)\left(x - \frac{1}{x} + 1\right)$

Find the derivatives of the functions in Exercises 17–28.

17. $y = \frac{2x + 5}{3x - 2}$
18. $z = \frac{2x + 1}{x^2 - 1}$
19. $g(x) = \frac{x^2 - 4}{x + 0.5}$
20. $f(t) = \frac{t^2 - 1}{t^2 + t - 2}$
21. $v = (1 - t)(1 + t^2)^{-1}$
22. $w = (2x - 7)^{-1}(x + 5)$
23. $f(s) = \frac{\sqrt{s} - 1}{\sqrt{s} + 1}$
24. $u = \frac{5x + 1}{2\sqrt{x}}$
25. $v = \frac{1 + x - 4\sqrt{x}}{x}$
26. $r = 2\left(\frac{1}{\sqrt{\theta}} + \sqrt{\theta}\right)$
27. $y = \frac{1}{(x^2 - 1)(x^2 + x + 1)}$
28. $y = \frac{(x + 1)(x + 2)}{(x - 1)(x - 2)}$

Find the derivatives of all orders of the functions in Exercises 29 and 30.

29. $y = \frac{x^4}{2} - \frac{3}{2}x^2 - x$
30. $y = \frac{x^5}{120}$

Find the first and second derivatives of the functions in Exercises 31–38.

31. $y = \frac{x^3 + 7}{x}$
32. $s = \frac{t^2 + 5t - 1}{t^2}$
33. $r = \frac{(\theta - 1)(\theta^2 + \theta + 1)}{\theta^3}$
34. $u = \frac{(x^2 + x)(x^2 - x + 1)}{x^4}$
35. $w = \left(\frac{1 + 3z}{3z}\right)(3 - z)$
36. $w = (z + 1)(z - 1)(z^2 + 1)$
37. $p = \left(\frac{q^2 + 3}{12q}\right)\left(\frac{q^4 - 1}{q^3}\right)$
38. $p = \frac{q^2 + 3}{(q - 1)^3 + (q + 1)^3}$

39. Suppose u and v are functions of x that are differentiable at $x = 0$ and that

$$u(0) = 5, \quad u'(0) = -3, \quad v(0) = -1, \quad v'(0) = 2.$$

Find the values of the following derivatives at $x = 0$.

a. $\frac{d}{dx}(uv)$ **b.** $\frac{d}{dx}\left(\frac{u}{v}\right)$ **c.** $\frac{d}{dx}\left(\frac{v}{u}\right)$ **d.** $\frac{d}{dx}(7v - 2u)$

40. Suppose u and v are differentiable functions of x and that

$$u(1) = 2, \quad u'(1) = 0, \quad v(1) = 5, \quad v'(1) = -1.$$

Find the values of the following derivatives at $x = 1$.

a. $\frac{d}{dx}(uv)$ **b.** $\frac{d}{dx}\left(\frac{u}{v}\right)$ **c.** $\frac{d}{dx}\left(\frac{v}{u}\right)$ **d.** $\frac{d}{dx}(7v - 2u)$

41. a. Normal to a curve Find an equation for the line perpendicular to the tangent to the curve $y = x^3 - 4x + 1$ at the point $(2, 1)$.

b. Smallest slope What is the smallest slope on the curve? At what point on the curve does the curve have this slope?

c. Tangents having specified slope Find equations for the tangents to the curve at the points where the slope of the curve is 8.

42. a. Horizontal tangents Find equations for the horizontal tangents to the curve $y = x^3 - 3x - 2$. Also find equations for the lines that are perpendicular to these tangents at the points of tangency.

b. Smallest slope What is the smallest slope on the curve? At what point on the curve does the curve have this slope? Find an equation for the line that is perpendicular to the curve's tangent at this point.

43. Find the tangents to *Newton's serpentine* (graphed here) at the origin and the point $(1, 2)$.

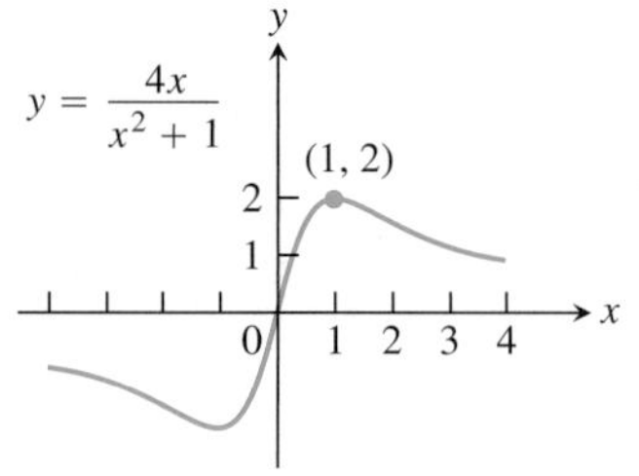

44. Find the tangent to the *Witch of Agnesi* (graphed here) at the point (2, 1).

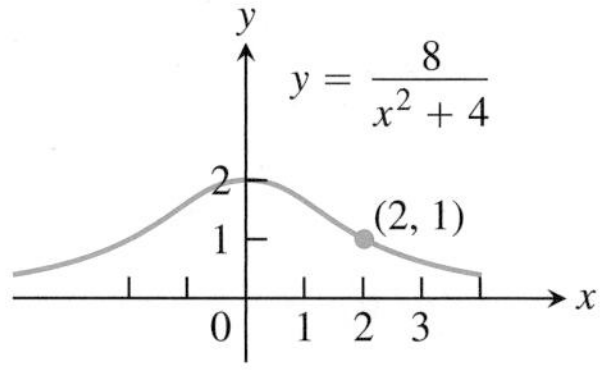

45. Quadratic tangent to identity function The curve $y = ax^2 + bx + c$ passes through the point (1, 2) and is tangent to the line $y = x$ at the origin. Find a, b, and c.

46. Quadratics having a common tangent The curves $y = x^2 + ax + b$ and $y = cx - x^2$ have a common tangent line at the point (1, 0). Find a, b, and c.

47. a. Find an equation for the line that is tangent to the curve $y = x^3 - x$ at the point $(-1, 0)$.

T **b.** Graph the curve and tangent line together. The tangent intersects the curve at another point. Use Zoom and Trace to estimate the point's coordinates.

T **c.** Confirm your estimates of the coordinates of the second intersection point by solving the equations for the curve and tangent simultaneously (Solver key).

48. a. Find an equation for the line that is tangent to the curve $y = x^3 - 6x^2 + 5x$ at the origin.

T **b.** Graph the curve and tangent together. The tangent intersects the curve at another point. Use Zoom and Trace to estimate the point's coordinates.

T **c.** Confirm your estimates of the coordinates of the second intersection point by solving the equations for the curve and tangent simultaneously (Solver key).

49. The general polynomial of degree n has the form

$$P(x) = a_n x^n + a_{n-1}x^{n-1} + \cdots + a_2 x^2 + a_1 x + a_0$$

where $a_n \neq 0$. Find $P'(x)$.

50. The body's reaction to medicine The reaction of the body to a dose of medicine can sometimes be represented by an equation of the form

$$R = M^2\left(\frac{C}{2} - \frac{M}{3}\right),$$

where C is a positive constant and M is the amount of medicine absorbed in the blood. If the reaction is a change in blood pressure, R is measured in millimeters of mercury. If the reaction is a change in temperature, R is measured in degrees, and so on.

Find dR/dM. This derivative, as a function of M, is called the *sensitivity* of the body to the medicine.

51. Suppose that the function v in the Product Rule has a constant value c. What does the Product Rule then say? What does this say about the Constant Multiple Rule?

52. The Reciprocal Rule

a. The *Reciprocal Rule* says that at any point where the function $v(x)$ is differentiable and different from zero,

$$\frac{d}{dx}\left(\frac{1}{v}\right) = -\frac{1}{v^2}\frac{dv}{dx}.$$

Show that the Reciprocal Rule is a special case of the Quotient Rule.

b. Show that the Reciprocal Rule and the Product Rule together imply the Quotient Rule.

53. Cylinder pressure If gas in a cylinder is maintained at a constant temperature T, the pressure P is related to the volume V by a formula of the form

$$P = \frac{nRT}{V - nb} - \frac{an^2}{V^2},$$

in which a, b, n, and R are constants. Find dP/dV.

54. Power Rule for negative integers Use the Quotient Rule to prove the Power Rule for negative integers, that is,

$$\frac{d}{dx}(x^{-m}) = -mx^{-m-1}$$

where m is a positive integer.

2.4 The Derivative as a Rate of Change

In Section 1.3 we introduced average and instantaneous rates of change. In this section we study further applications in which derivatives model the rates at which things change. It is natural to think of a quantity changing with respect to time, but other variables can be treated in the same way. For example, an economist may want to study how the cost of producing steel varies with the number of tons produced.

Instantaneous Rates of Change

If we interpret the difference quotient $(f(x + h) - f(x))/h$ as the average rate of change in f over the interval from x to $x + h$, we can interpret its limit as $h \to 0$ as the rate at which f is changing at the point x.

> **DEFINITION** The **instantaneous rate of change** of f with respect to x at x_0 is the derivative
>
> $$f'(x_0) = \lim_{h \to 0} \frac{f(x_0 + h) - f(x_0)}{h},$$
>
> provided the limit exists.

Thus, instantaneous rates are limits of average rates.

It is conventional to use the word *instantaneous* even when x does not represent time. The word is, however, frequently omitted. When we say *rate of change*, we mean *instantaneous rate of change*.

EXAMPLE 1 The area A of a circle is related to its diameter by the equation

$$A = \frac{\pi}{4} D^2.$$

How fast does the area change with respect to the diameter when the diameter is 10 m?

Solution The rate of change of the area with respect to the diameter is

$$\frac{dA}{dD} = \frac{\pi}{4} \cdot 2D = \frac{\pi D}{2}.$$

When $D = 10$ m, the area is changing with respect to diameter at the rate $(\pi/2)10 = 5\pi\ \mathrm{m}^2/\mathrm{m}$. ■

Motion Along a Line: Displacement, Velocity, Speed, Acceleration, and Jerk

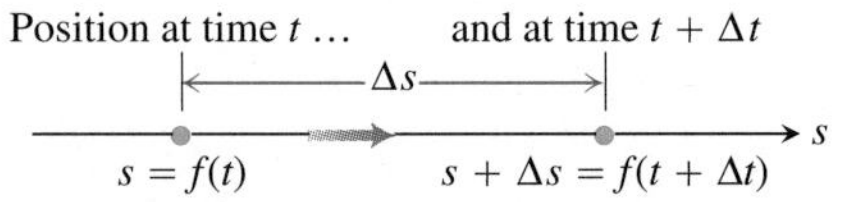

FIGURE 2.12 The positions of a body moving along a coordinate line at time t and shortly later at time $t + \Delta t$.

Suppose that an object is moving along a coordinate line (say an s-axis) so that we know its position s on that line as a function of time t:

$$s = f(t).$$

The **displacement** of the object over the time interval from t to $t + \Delta t$ (Figure 2.12) is

$$\Delta s = f(t + \Delta t) - f(t),$$

and the **average velocity** of the object over that time interval is

$$v_{av} = \frac{\text{displacement}}{\text{travel time}} = \frac{\Delta s}{\Delta t} = \frac{f(t + \Delta t) - f(t)}{\Delta t}.$$

To find the body's velocity at the exact instant t, we take the limit of the average velocity over the interval from t to $t + \Delta t$ as Δt shrinks to zero. This limit is the derivative of f with respect to t.

DEFINITION **Velocity (instantaneous velocity)** is the derivative of position with respect to time. If a body's position at time t is $s = f(t)$, then the body's velocity at time t is

$$v(t) = \frac{ds}{dt} = \lim_{\Delta t \to 0} \frac{f(t + \Delta t) - f(t)}{\Delta t}.$$

Besides telling how fast an object is moving, its velocity tells the direction of motion. When the object is moving forward (s increasing), the velocity is positive; when the body is moving backward (s decreasing), the velocity is negative (Figure 2.13).

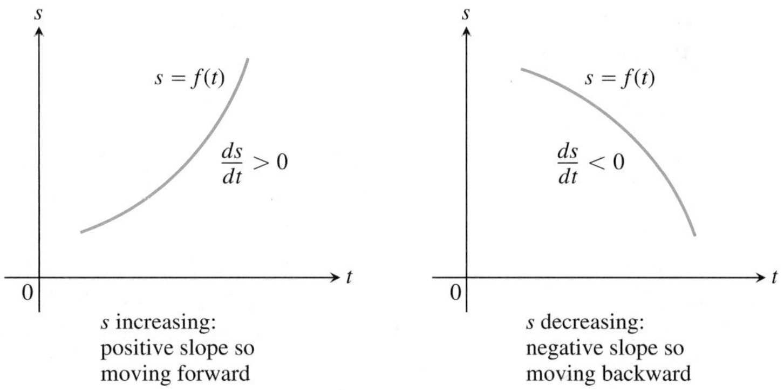

FIGURE 2.13 For motion $s = f(t)$ along a straight line, $v = ds/dt$ is positive when s increases and negative when s decreases.

If we drive to a friend's house and back at 30 mph, say, the speedometer will show 30 on the way over but it will not show -30 on the way back, even though our distance from home is decreasing. The speedometer always shows *speed*, which is the absolute value of velocity. Speed measures the rate of progress regardless of direction.

DEFINITION **Speed** is the absolute value of velocity.

$$\text{Speed} = |v(t)| = \left|\frac{ds}{dt}\right|$$

EXAMPLE 2 Figure 2.14. shows the velocity $v = f'(t)$ of a particle moving on a coordinate line. The particle moves forward for the first 3 sec, moves backward for the next 2 sec, stands still for a second, and moves forward again. The particle achieves its greatest speed at time $t = 4$, while moving backward. ■

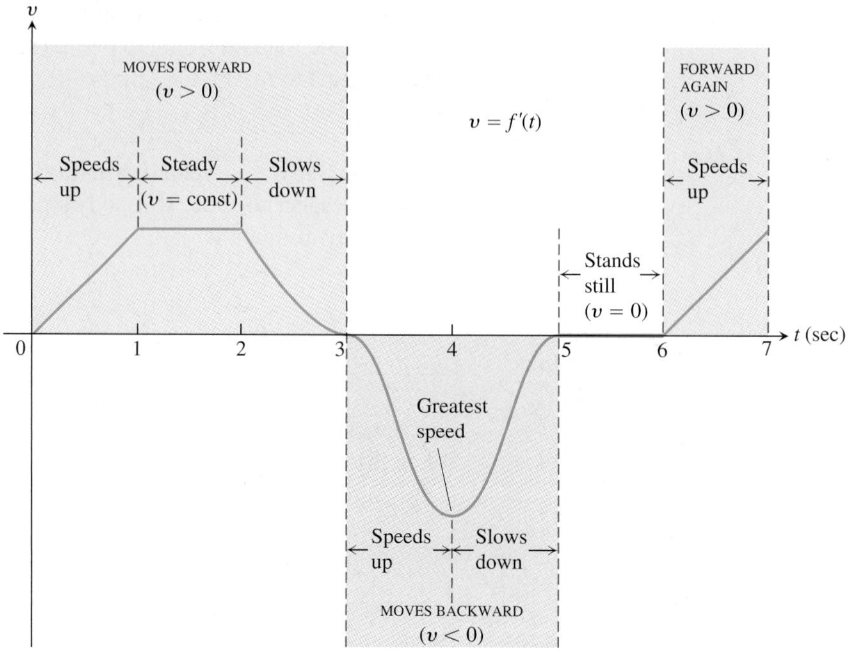

FIGURE 2.14 The velocity graph for Example 2.

HISTORICAL BIOGRAPHY

Bernard Bolzano
(1781–1848)

The rate at which a body's velocity changes is the body's *acceleration*. The acceleration measures how quickly the body picks up or loses speed.

A sudden change in acceleration is called a *jerk*. When a ride in a car or a bus is jerky, it is not that the accelerations involved are necessarily large but that the changes in acceleration are abrupt.

DEFINITIONS **Acceleration** is the derivative of velocity with respect to time. If a body's position at time t is $s = f(t)$, then the body's acceleration at time t is

$$a(t) = \frac{dv}{dt} = \frac{d^2s}{dt^2}.$$

Jerk is the derivative of acceleration with respect to time:

$$j(t) = \frac{da}{dt} = \frac{d^3s}{dt^3}.$$

Near the surface of the Earth all bodies fall with the same constant acceleration. Galileo's experiments with free fall (Example 1, Section 1.3) lead to the equation

$$s = \frac{1}{2}gt^2,$$

where s is the distance fallen and g is the acceleration due to Earth's gravity. This equation holds in a vacuum, where there is no air resistance, and closely models the fall of dense, heavy objects, such as rocks or steel tools, for the first few seconds of their fall, before the effects of air resistance are significant.

The value of g in the equation $s = (1/2)gt^2$ depends on the units used to measure t and s. With t in seconds (the usual unit), the value of g determined by measurement at sea level is approximately 32 ft/sec^2 (feet per second squared) in English units, and $g = 9.8$ m/sec^2 (meters per second squared) in metric units. (These gravitational constants depend on the distance from Earth's center of mass, and are slightly lower on top of Mt. Everest, for example.)

The jerk of the constant acceleration of gravity ($g = 32$ ft/sec^2) is zero:

$$j = \frac{d}{dt}(g) = 0.$$

An object does not exhibit jerkiness during free fall.

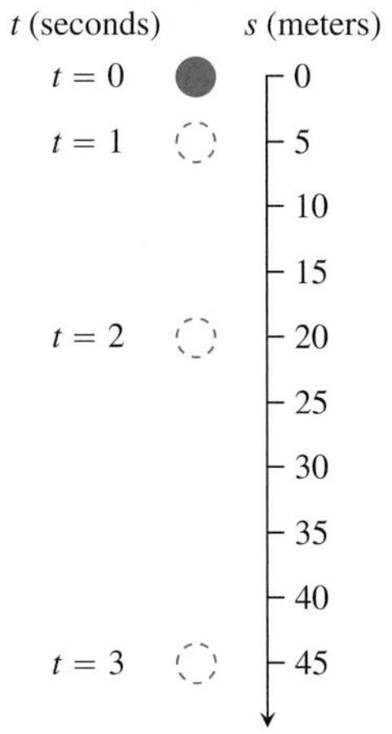

FIGURE 2.15 A ball bearing falling from rest (Example 3).

EXAMPLE 3 Figure 2.15 shows the free fall of a heavy ball bearing released from rest at time $t = 0$ sec.

(a) How many meters does the ball fall in the first 2 sec?

(b) What is its velocity, speed, and acceleration when $t = 2$?

Solution

(a) The metric free-fall equation is $s = 4.9t^2$. During the first 2 sec, the ball falls

$$s(2) = 4.9(2)^2 = 19.6 \text{ m}.$$

(b) At any time t, *velocity* is the derivative of position:

$$v(t) = s'(t) = \frac{d}{dt}(4.9t^2) = 9.8t.$$

At $t = 2$, the velocity is

$$v(2) = 19.6 \text{ m/sec}$$

in the downward (increasing s) direction. The *speed* at $t = 2$ is

$$\text{Speed} = |v(2)| = 19.6 \text{ m/sec}.$$

The *acceleration* at any time t is

$$a(t) = v'(t) = s''(t) = 9.8 \text{ m/sec}^2.$$

At $t = 2$, the acceleration is 9.8 m/sec^2. ■

EXAMPLE 4 A dynamite blast blows a heavy rock straight up with a launch velocity of 160 ft/sec (about 109 mph) (Figure 2.16a). It reaches a height of $s = 160t - 16t^2$ ft after t sec.

(a) How high does the rock go?

(b) What are the velocity and speed of the rock when it is 256 ft above the ground on the way up? On the way down?

(c) What is the acceleration of the rock at any time t during its flight (after the blast)?

(d) When does the rock hit the ground again?

Solution

(a) In the coordinate system we have chosen, s measures height from the ground up, so the velocity is positive on the way up and negative on the way down. The instant the rock is

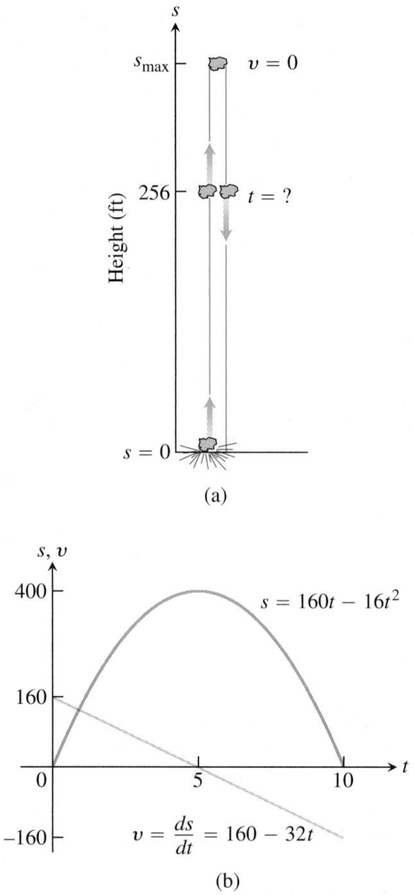

FIGURE 2.16 (a) The rock in Example 4. (b) The graphs of s and v as functions of time; s is largest when $v = ds/dt = 0$. The graph of s is *not* the path of the rock: It is a plot of height versus time. The slope of the plot is the rock's velocity, graphed here as a straight line.

at its highest point is the one instant during the flight when the velocity is 0. To find the maximum height, all we need to do is to find when $v = 0$ and evaluate s at this time.

At any time t, the velocity is

$$v = \frac{ds}{dt} = \frac{d}{dt}(160t - 16t^2) = 160 - 32t \text{ ft/sec}.$$

The velocity is zero when

$$160 - 32t = 0 \quad \text{or} \quad t = 5 \text{ sec}.$$

The rock's height at $t = 5$ sec is

$$s_{max} = s(5) = 160(5) - 16(5)^2 = 800 - 400 = 400 \text{ ft}.$$

See Figure 2.16b.

(b) To find the rock's velocity at 256 ft on the way up and again on the way down, we first find the two values of t for which

$$s(t) = 160t - 16t^2 = 256.$$

To solve this equation, we write

$$\begin{aligned} 16t^2 - 160t + 256 &= 0 \\ 16(t^2 - 10t + 16) &= 0 \\ (t - 2)(t - 8) &= 0 \\ t = 2 \text{ sec}, t &= 8 \text{ sec}. \end{aligned}$$

The rock is 256 ft above the ground 2 sec after the explosion and again 8 sec after the explosion. The rock's velocities at these times are

$$v(2) = 160 - 32(2) = 160 - 64 = 96 \text{ ft/sec}.$$
$$v(8) = 160 - 32(8) = 160 - 256 = -96 \text{ ft/sec}.$$

At both instants, the rock's speed is 96 ft/sec. Since $v(2) > 0$, the rock is moving upward (s is increasing) at $t = 2$ sec; it is moving downward (s is decreasing) at $t = 8$ because $v(8) < 0$.

(c) At any time during its flight following the explosion, the rock's acceleration is a constant

$$a = \frac{dv}{dt} = \frac{d}{dt}(160 - 32t) = -32 \text{ ft/sec}^2.$$

The acceleration is always downward. As the rock rises, it slows down; as it falls, it speeds up.

(d) The rock hits the ground at the positive time t for which $s = 0$. The equation $160t - 16t^2 = 0$ factors to give $16t(10 - t) = 0$, so it has solutions $t = 0$ and $t = 10$. At $t = 0$, the blast occurred and the rock was thrown upward. It returned to the ground 10 sec later. ■

Derivatives in Economics

In a manufacturing operation, the *cost of production* $c(x)$ is a function of x, the number of units produced. The **marginal cost of production** is the rate of change of cost with respect

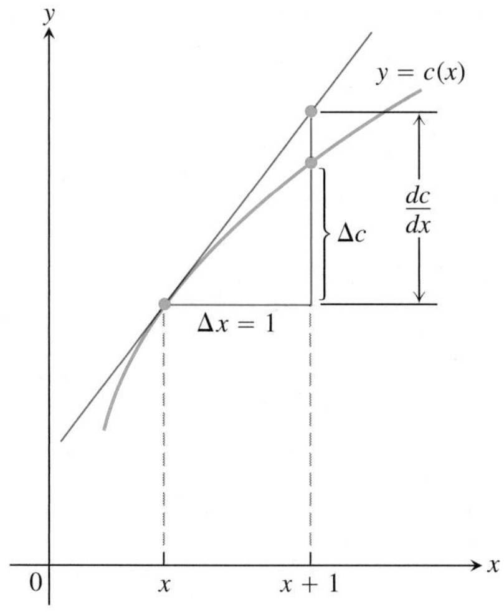

FIGURE 2.17 The marginal cost dc/dx is approximately the extra cost Δc of producing $\Delta x = 1$ more unit.

to level of production, so it is dc/dx. Sometimes the marginal cost of production is loosely defined to be the extra cost of producing one additional unit:

$$\frac{\Delta c}{\Delta x} = \frac{c(x + 1) - c(x)}{1},$$

which is approximated by the value of dc/dx at x. This approximation is acceptable if the slope of the graph of c does not change quickly near x. Then the difference quotient will be close to its limit dc/dx, which is the rise in the tangent line if $\Delta x = 1$ (Figure 2.17). The approximation works best for large values of x.

EXAMPLE 5 Suppose that it costs

$$c(x) = x^3 - 6x^2 + 15x$$

dollars to produce x radiators when 8 to 30 radiators are produced and that

$$r(x) = x^3 - 3x^2 + 12x$$

gives the dollar revenue from selling x radiators. Your shop currently produces 10 radiators a day. About how much extra will it cost to produce one more radiator a day, and what is your estimated increase in revenue for selling 11 radiators a day?

Solution The cost of producing one more radiator a day when 10 are produced is about $c'(10)$:

$$c'(x) = \frac{d}{dx}\left(x^3 - 6x^2 + 15x\right) = 3x^2 - 12x + 15$$

$$c'(10) = 3(100) - 12(10) + 15 = 195.$$

The additional cost will be about $195. The marginal revenue is

$$r'(x) = \frac{d}{dx}(x^3 - 3x^2 + 12x) = 3x^2 - 6x + 12.$$

The marginal revenue function estimates the increase in revenue that will result from selling one additional unit. If you currently sell 10 radiators a day, you can expect your revenue to increase by about

$$r'(10) = 3(100) - 6(10) + 12 = \$252$$

if you increase sales to 11 radiators a day. ■

EXERCISES 2.4

Exercises 1–6 give the positions $s = f(t)$ of a body moving on a coordinate line, with s in meters and t in seconds.

a. Find the body's displacement and average velocity for the given time interval.

b. Find the body's speed and acceleration at the endpoints of the interval.

c. When, if ever, during the interval does the body change direction?

1. $s = t^2 - 3t + 2, \quad 0 \le t \le 2$

2. $s = 6t - t^2, \quad 0 \le t \le 6$

3. $s = -t^3 + 3t^2 - 3t, \quad 0 \le t \le 3$

4. $s = (t^4/4) - t^3 + t^2, \quad 0 \le t \le 3$

5. $s = \dfrac{25}{t^2} - \dfrac{5}{t}, \quad 1 \le t \le 5$

6. $s = \dfrac{25}{t + 5}, \quad -4 \le t \le 0$

7. Particle motion At time t, the position of a body moving along the s-axis is $s = t^3 - 6t^2 + 9t$ m.

a. Find the body's acceleration each time the velocity is zero.

b. Find the body's speed each time the acceleration is zero.

c. Find the total distance traveled by the body from $t = 0$ to $t = 2$.

8. Particle motion At time $t \geq 0$, the velocity of a body moving along the s-axis is $v = t^2 - 4t + 3$.

a. Find the body's acceleration each time the velocity is zero.

b. When is the body moving forward? Backward?

c. When is the body's velocity increasing? Decreasing?

9. Free fall on Mars and Jupiter The equations for free fall at the surfaces of Mars and Jupiter (s in meters, t in seconds) are $s = 1.86t^2$ on Mars and $s = 11.44t^2$ on Jupiter. How long does it take a rock falling from rest to reach a velocity of 27.8 m/sec (about 100 km/h) on each planet?

10. Lunar projectile motion A rock thrown vertically upward from the surface of the moon at a velocity of 24 m/sec (about 86 km/h) reaches a height of $s = 24t - 0.8t^2$ meters in t sec.

a. Find the rock's velocity and acceleration at time t. (The acceleration in this case is the acceleration of gravity on the moon.)

b. How long does it take the rock to reach its highest point?

c. How high does the rock go?

d. How long does it take the rock to reach half its maximum height?

e. How long is the rock aloft?

11. Free fall from the Tower of Pisa Had Galileo dropped a cannonball from the Tower of Pisa, 179 ft above the ground, the ball's height above the ground t sec into the fall would have been $s = 179 - 16t^2$.

a. What would have been the ball's velocity, speed, and acceleration at time t?

b. About how long would it have taken the ball to hit the ground?

c. What would have been the ball's velocity at the moment of impact?

12. Speeding bullet A 45-caliber bullet fired straight up from the surface of the moon would reach a height of $s = 832t - 2.6t^2$ feet after t sec. On Earth, in the absence of air, its height would be $s = 832t - 16t^2$ ft after t sec. How long will the bullet be aloft in each case? How high will the bullet go?

13. The accompanying figure shows the velocity $v = ds/dt = f(t)$ (m/sec) of a body moving along a coordinate line.

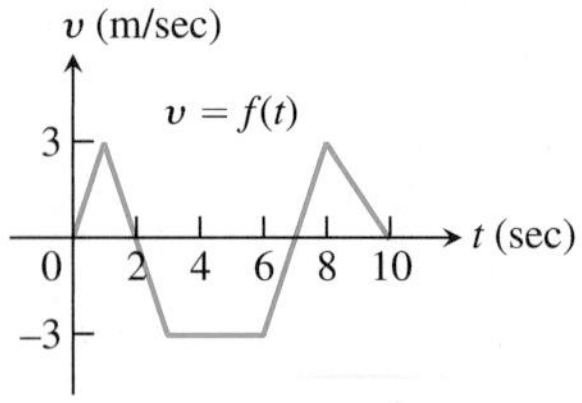

a. When does the body reverse direction?

b. When (approximately) is the body moving at a constant speed?

c. Graph the body's speed for $0 \leq t \leq 10$.

d. Graph the acceleration, where defined.

14. A particle P moves on the number line shown in part (a) of the accompanying figure. Part (b) shows the position of P as a function of time t.

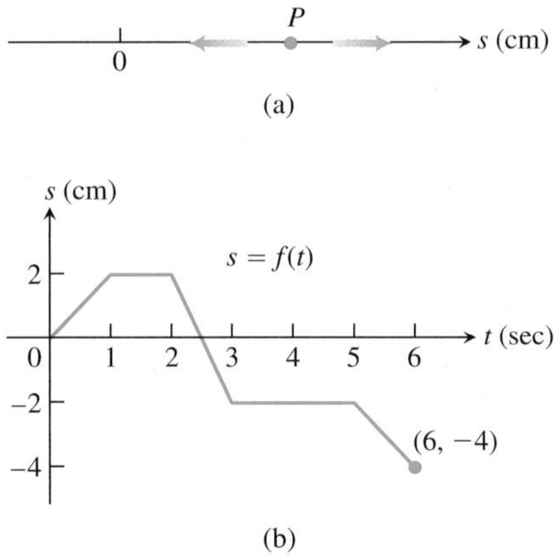

a. When is P moving to the left? Moving to the right? Standing still?

b. Graph the particle's velocity and speed (where defined).

15. Launching a rocket When a model rocket is launched, the propellant burns for a few seconds, accelerating the rocket upward. After burnout, the rocket coasts upward for a while and then begins to fall. A small explosive charge pops out a parachute shortly after the rocket starts down. The parachute slows the rocket to keep it from breaking when it lands.

The figure here shows velocity data from the flight of the model rocket. Use the data to answer the following.

a. How fast was the rocket climbing when the engine stopped?

b. For how many seconds did the engine burn?

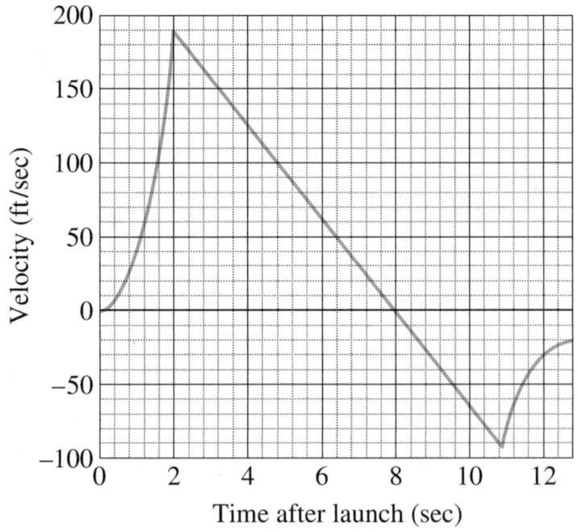

c. When did the rocket reach its highest point? What was its velocity then?

d. When did the parachute pop out? How fast was the rocket falling then?

e. How long did the rocket fall before the parachute opened?

f. When was the rocket's acceleration greatest?

g. When was the acceleration constant? What was its value then (to the nearest integer)?

16. The accompanying figure shows the velocity $v = f(t)$ of a particle moving on a coordinate line.

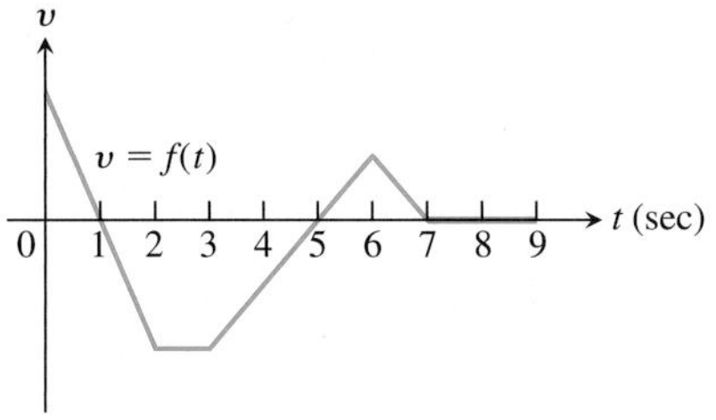

a. When does the particle move forward? Move backward? Speed up? Slow down?

b. When is the particle's acceleration positive? Negative? Zero?

c. When does the particle move at its greatest speed?

d. When does the particle stand still for more than an instant?

17. Two falling balls The multiflash photograph in the accompanying figure shows two balls falling from rest. The vertical rulers are marked in centimeters. Use the equation $s = 490t^2$ (the free-fall equation for s in centimeters and t in seconds) to answer the following questions.

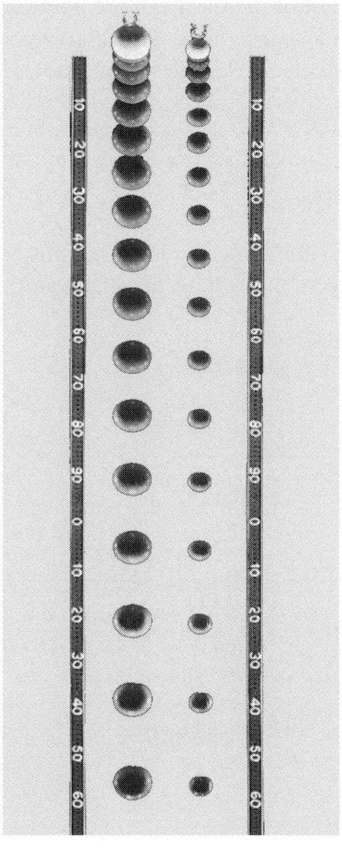

a. How long did it take the balls to fall the first 160 cm? What was their average velocity for the period?

b. How fast were the balls falling when they reached the 160-cm mark? What was their acceleration then?

c. About how fast was the light flashing (flashes per second)?

18. Volcanic lava fountains Although the November 1959 Kilauea Iki eruption on the island of Hawaii began with a line of fountains along the wall of the crater, activity was later confined to a single vent in the crater's floor, which at one point shot lava 1900 ft straight into the air (a world record). What was the lava's exit velocity in feet per second? In miles per hour? (*Hint*: If v_0 is the exit velocity of a particle of lava, its height t sec later will be $s = v_0 t - 16t^2$ ft. Begin by finding the time at which $ds/dt = 0$. Neglect air resistance.)

19. Marginal cost Suppose that the dollar cost of producing x washing machines is $c(x) = 2000 + 100x - 0.1x^2$.

a. Find the average cost per machine of producing the first 100 washing machines.

b. Find the marginal cost when 100 washing machines are produced.

c. Show that the marginal cost when 100 washing machines are produced is approximately the cost of producing one more washing machine after the first 100 have been made, by calculating the latter cost directly.

20. Marginal revenue Suppose that the revenue from selling x washing machines is

$$r(x) = 20{,}000\left(1 - \frac{1}{x}\right)$$

dollars.

a. Find the marginal revenue when 100 machines are produced.

b. Use the function $r'(x)$ to estimate the increase in revenue that will result from increasing production from 100 machines a week to 101 machines a week.

c. Find the limit of $r'(x)$ as $x \to \infty$. How would you interpret this number?

21. Bacterium population When a bactericide was added to a nutrient broth in which bacteria were growing, the bacterium population continued to grow for a while, but then stopped growing and began to decline. The size of the population at time t (hours) was $b = 10^6 + 10^4 t - 10^3 t^2$. Find the growth rates at

a. $t = 0$ hours.

b. $t = 5$ hours.

c. $t = 10$ hours.

22. Inflating a balloon The volume $V = (4/3)\pi r^3$ of a spherical balloon changes with the radius.

a. At what rate (ft^3/ft) does the volume change with respect to the radius when $r = 2$ ft?

b. By approximately how much does the volume increase when the radius changes from 2 to 2.2 ft?

23. Airplane takeoff Suppose that the distance an aircraft travels along a runway before takeoff is given by $D = (10/9)t^2$, where D is measured in meters from the starting point and t is measured in seconds from the time the brakes are released. The aircraft will become airborne when its speed reaches 200 km/h. How long will it take to become airborne, and what distance will it travel in that time?

24. Draining a tank The number of gallons of water in a tank t minutes after the tank has started to drain is $Q(t) = 200(30 - t)^2$. How fast is the water running out at the end of 10 min? What is the average rate at which the water flows out during the first 10 min?

T Exercises 25–28 give the position function $s = f(t)$ of a body moving along the s-axis as a function of time t. Graph f together with the velocity function $v(t) = ds/dt = f'(t)$ and the acceleration function $a(t) = d^2s/dt^2 = f''(t)$. Comment on the body's behavior in relation to the signs and values of v and a. Include in your commentary such topics as the following:

a. When is the body momentarily at rest?

b. When does it move to the left (down) or to the right (up)?

c. When does it change direction?

d. When does it speed up and slow down?

e. When is it moving fastest (highest speed)? Slowest?

f. When is it farthest from the axis origin?

25. $s = 200t - 16t^2, \quad 0 \le t \le 12.5$ (a heavy object fired straight up from Earth's surface at 200 ft/sec)

26. $s = t^2 - 3t + 2, \quad 0 \le t \le 5$

27. $s = t^3 - 6t^2 + 7t, \quad 0 \le t \le 4$

28. $s = 4 - 7t + 6t^2 - t^3, \quad 0 \le t \le 4$

2.5 Derivatives of Trigonometric Functions

Many of the phenomena of nature we want information about are approximately periodic (electromagnetic fields, heart rhythms, tides, weather). The derivatives of sines and cosines play a key role in describing periodic changes. This section shows how to differentiate the six basic trigonometric functions.

Derivative of the Sine and Cosine Functions

To calculate the derivative of $f(x) = \sin x$, for x measured in radians, we combine the limits in Example 4a and Theorem 7 in Section 1.6 with the angle sum identity for the sine:

$$\sin(x + h) = \sin x \cos h + \cos x \sin h.$$

If $f(x) = \sin x$, then

$$f'(x) = \lim_{h \to 0} \frac{f(x+h) - f(x)}{h} = \lim_{h \to 0} \frac{\sin(x+h) - \sin x}{h} \qquad \text{Derivative definition}$$

$$= \lim_{h \to 0} \frac{(\sin x \cos h + \cos x \sin h) - \sin x}{h} = \lim_{h \to 0} \frac{\sin x(\cos h - 1) + \cos x \sin h}{h}$$

$$= \lim_{h \to 0} \left(\sin x \cdot \frac{\cos h - 1}{h}\right) + \lim_{h \to 0} \left(\cos x \cdot \frac{\sin h}{h}\right)$$

$$= \sin x \cdot \lim_{h \to 0} \frac{\cos h - 1}{h} + \cos x \cdot \lim_{h \to 0} \frac{\sin h}{h} = \sin x \cdot 0 + \cos x \cdot 1 = \cos x.$$

Example 4a and Theorem 7, Section 1.6

The derivative of the sine function is the cosine function:

$$\frac{d}{dx}(\sin x) = \cos x.$$

With the help of the angle sum formula for the cosine,

$$\cos(x + h) = \cos x \cos h - \sin x \sin h,$$

we have

$$\begin{aligned}
\frac{d}{dx}(\cos x) &= \lim_{h \to 0} \frac{\cos(x + h) - \cos x}{h} && \text{Derivative definition} \\
&= \lim_{h \to 0} \frac{(\cos x \cos h - \sin x \sin h) - \cos x}{h} && \text{Cosine angle sum identity} \\
&= \lim_{h \to 0} \frac{\cos x(\cos h - 1) - \sin x \sin h}{h} \\
&= \lim_{h \to 0} \cos x \cdot \frac{\cos h - 1}{h} - \lim_{h \to 0} \sin x \cdot \frac{\sin h}{h} \\
&= \cos x \cdot \lim_{h \to 0} \frac{\cos h - 1}{h} - \sin x \cdot \lim_{h \to 0} \frac{\sin h}{h} \\
&= \cos x \cdot 0 - \sin x \cdot 1 && \text{Example 4a and Theorem 7, Section 1.6} \\
&= -\sin x.
\end{aligned}$$

The derivative of the cosine function is the negative of the sine function:

$$\frac{d}{dx}(\cos x) = -\sin x.$$

EXAMPLE 1

(a) $y = x^2 - \sin x$:

$$\begin{aligned}
\frac{dy}{dx} &= 2x - \frac{d}{dx}(\sin x) && \text{Difference Rule} \\
&= 2x - \cos x.
\end{aligned}$$

(b) $y = \sin x \cos x$:

$$\begin{aligned}
\frac{dy}{dx} &= \sin x \frac{d}{dx}(\cos x) + \cos x \frac{d}{dx}(\sin x) && \text{Product Rule} \\
&= \sin x(-\sin x) + \cos x(\cos x) \\
&= \cos^2 x - \sin^2 x.
\end{aligned}$$

(c) $y = \dfrac{\cos x}{1 - \sin x}$:

$$\begin{aligned}\frac{dy}{dx} &= \frac{(1 - \sin x)\frac{d}{dx}(\cos x) - \cos x \frac{d}{dx}(1 - \sin x)}{(1 - \sin x)^2} &&\text{Quotient Rule}\\ &= \frac{(1 - \sin x)(-\sin x) - \cos x(0 - \cos x)}{(1 - \sin x)^2}\\ &= \frac{1 - \sin x}{(1 - \sin x)^2} &&\sin^2 x + \cos^2 x = 1\\ &= \frac{1}{1 - \sin x}.\end{aligned}$$

■

Simple Harmonic Motion

The motion of a body bobbing freely up and down with no resistance on the end of a spring or bungee cord is an example of *simple harmonic motion*. The next example describes a case in which there are no opposing forces such as friction or buoyancy to slow the motion.

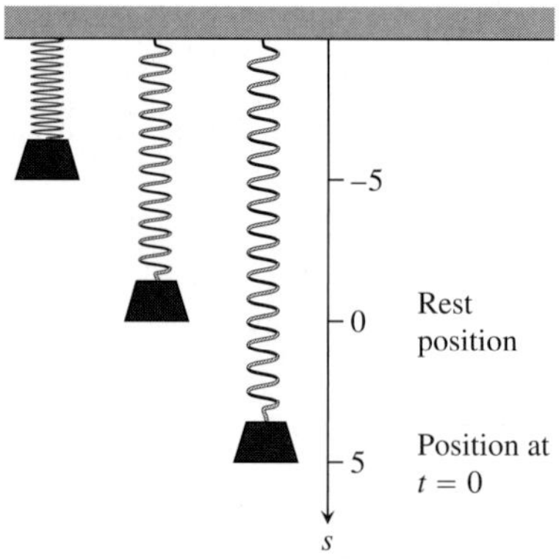

FIGURE 2.18 A body hanging from a vertical spring and then displaced oscillates above and below its rest position. Its motion is described by trigonometric functions (Example 2).

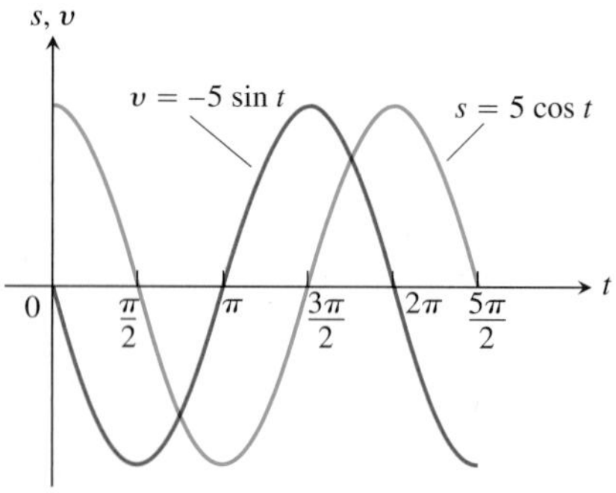

FIGURE 2.19 The graphs of the position and velocity of the body in Example 2.

EXAMPLE 2 A body hanging from a spring (Figure 2.18) is stretched down 5 units beyond its rest position and released at time $t = 0$ to bob up and down. Its position at any later time t is

$$s = 5 \cos t.$$

What are its velocity and acceleration at time t?

Solution We have

Position: $s = 5 \cos t$

Velocity: $v = \dfrac{ds}{dt} = \dfrac{d}{dt}(5 \cos t) = -5 \sin t$

Acceleration: $a = \dfrac{dv}{dt} = \dfrac{d}{dt}(-5 \sin t) = -5 \cos t.$

Notice how much we can learn from these equations:

1. As time passes, the weight moves down and up between $s = -5$ and $s = 5$ on the s-axis. The amplitude of the motion is 5. The period of the motion is 2π.
2. The velocity $v = -5 \sin t$ attains its greatest magnitude, 5, when $\cos t = 0$, as the graphs show in Figure 2.19. Hence, the speed of the weight, $|v| = 5|\sin t|$, is greatest when $\cos t = 0$, that is, when $s = 0$ (the rest position). The speed of the weight is zero when $\sin t = 0$. This occurs when $s = 5 \cos t = \pm 5$, at the endpoints of the interval of motion.
3. The acceleration value is always the exact opposite of the position value. When the weight is above the rest position, gravity is pulling it back down; when the weight is below the rest position, the spring is pulling it back up.
4. The acceleration, $a = -5 \cos t$, is zero only at the rest position, where $\cos t = 0$ and the force of gravity and the force from the spring balance each other. When the weight is anywhere else, the two forces are unequal and acceleration is nonzero. The acceleration is greatest in magnitude at the points farthest from the rest position, where $\cos t = \pm 1$. ■

Derivatives of the Other Basic Trigonometric Functions

Because $\sin x$ and $\cos x$ are differentiable functions of x, the related functions

$$\tan x = \frac{\sin x}{\cos x}, \qquad \cot x = \frac{\cos x}{\sin x}, \qquad \sec x = \frac{1}{\cos x}, \qquad \text{and} \qquad \csc x = \frac{1}{\sin x}$$

are differentiable at every value of x at which they are defined. Their derivatives, calculated from the Quotient Rule, are given by the following formulas. Notice the negative signs in the derivative formulas for the cofunctions.

The derivatives of the other trigonometric functions:

$$\frac{d}{dx}(\tan x) = \sec^2 x \qquad \frac{d}{dx}(\cot x) = -\csc^2 x$$

$$\frac{d}{dx}(\sec x) = \sec x \tan x \qquad \frac{d}{dx}(\csc x) = -\csc x \cot x$$

To show a typical calculation, we find the derivative of the tangent function. The other derivations are left to Exercise 48.

EXAMPLE 3 Find $d(\tan x)/dx$.

Solution

$$\begin{aligned}\frac{d}{dx}\left(\tan x\right) = \frac{d}{dx}\left(\frac{\sin x}{\cos x}\right) &= \frac{\cos x \frac{d}{dx}\left(\sin x\right) - \sin x \frac{d}{dx}\left(\cos x\right)}{\cos^2 x} \qquad \text{Quotient Rule}\\ &= \frac{\cos x \cos x - \sin x\,(-\sin x)}{\cos^2 x}\\ &= \frac{\cos^2 x + \sin^2 x}{\cos^2 x}\\ &= \frac{1}{\cos^2 x} = \sec^2 x\end{aligned}$$

■

EXAMPLE 4 Find y'' if $y = \sec x$.

Solution

$$\begin{aligned} y &= \sec x\\ y' &= \sec x \tan x\\ y'' &= \frac{d}{dx}(\sec x \tan x)\\ &= \sec x \frac{d}{dx}(\tan x) + \tan x \frac{d}{dx}(\sec x) \qquad \text{Product Rule}\\ &= \sec x(\sec^2 x) + \tan x(\sec x \tan x)\\ &= \sec^3 x + \sec x \tan^2 x\end{aligned}$$

■

The differentiability of the trigonometric functions throughout their domains gives another proof of their continuity at every point in their domains (Theorem 1, Section 2.2).

EXERCISES 2.5

In Exercises 1–12, find dy/dx.

1. $y = -10x + 3\cos x$
2. $y = \frac{3}{x} + 5\sin x$
3. $y = \csc x - 4\sqrt{x} + 7$
4. $y = x^2 \cot x - \frac{1}{x^2}$
5. $y = (\sec x + \tan x)(\sec x - \tan x)$
6. $y = (\sin x + \cos x)\sec x$
7. $y = \frac{\cot x}{1 + \cot x}$
8. $y = \frac{\cos x}{1 + \sin x}$
9. $y = \frac{4}{\cos x} + \frac{1}{\tan x}$
10. $y = \frac{\cos x}{x} + \frac{x}{\cos x}$
11. $y = x^2 \sin x + 2x\cos x - 2\sin x$
12. $y = x^2 \cos x - 2x\sin x - 2\cos x$

In Exercises 13–16, find ds/dt.

13. $s = \tan t - t$
14. $s = t^2 - \sec t + 1$
15. $s = \frac{1 + \csc t}{1 - \csc t}$
16. $s = \frac{\sin t}{1 - \cos t}$

In Exercises 17–20, find $dr/d\theta$.

17. $r = 4 - \theta^2 \sin\theta$
18. $r = \theta\sin\theta + \cos\theta$
19. $r = \sec\theta\csc\theta$
20. $r = (1 + \sec\theta)\sin\theta$

In Exercises 21–24, find dp/dq.

21. $p = 5 + \frac{1}{\cot q}$
22. $p = (1 + \csc q)\cos q$
23. $p = \frac{\sin q + \cos q}{\cos q}$
24. $p = \frac{\tan q}{1 + \tan q}$

25. Find y'' if
    a. $y = \csc x$.
    b. $y = \sec x$.
26. Find $y^{(4)} = d^4y/dx^4$ if
    a. $y = -2\sin x$.
    b. $y = 9\cos x$.

In Exercises 27–30, graph the curves over the given intervals, together with their tangents at the given values of x. Label each curve and tangent with its equation.

27. $y = \sin x, \quad -3\pi/2 \le x \le 2\pi$
    $x = -\pi, 0, 3\pi/2$
28. $y = \tan x, \quad -\pi/2 < x < \pi/2$
    $x = -\pi/3, 0, \pi/3$
29. $y = \sec x, \quad -\pi/2 < x < \pi/2$
    $x = -\pi/3, \pi/4$
30. $y = 1 + \cos x, \quad -3\pi/2 \le x \le 2\pi$
    $x = -\pi/3, 3\pi/2$

T Do the graphs of the functions in Exercises 31–34 have any horizontal tangents in the interval $0 \le x \le 2\pi$? If so, where? If not, why not? Visualize your findings by graphing the functions with a grapher.

31. $y = x + \sin x$
32. $y = 2x + \sin x$
33. $y = x - \cot x$
34. $y = x + 2\cos x$
35. Find all points on the curve $y = \tan x$, $-\pi/2 < x < \pi/2$, where the tangent line is parallel to the line $y = 2x$. Sketch the curve and tangent(s) together, labeling each with its equation.
36. In the accompanying figure, find an equation for **(a)** the tangent to the curve at P and **(b)** the horizontal tangent to the curve at Q.

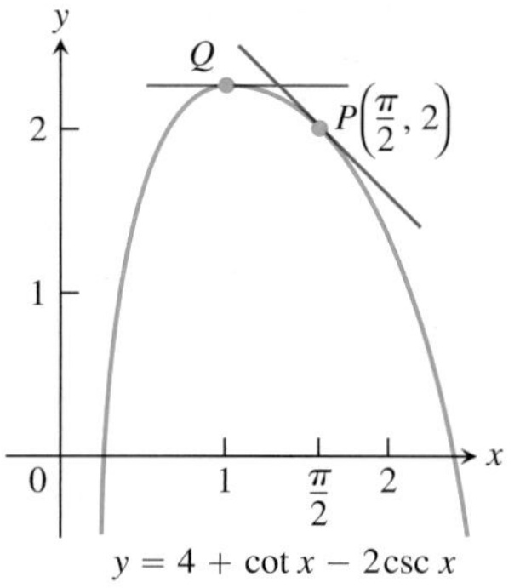

Find the limits in Exercises 37–42.

37. $\lim_{x\to 2} \sin\left(\frac{1}{x} - \frac{1}{2}\right)$
38. $\lim_{x\to -\pi/6} \sqrt{1 + \cos(\pi\csc x)}$
39. $\lim_{x\to 0} \sec\left[\cos x + \pi\tan\left(\frac{\pi}{4\sec x}\right) - 1\right]$
40. $\lim_{x\to 0} \sin\left(\frac{\pi + \tan x}{\tan x - 2\sec x}\right)$
41. $\lim_{t\to 0} \tan\left(1 - \frac{\sin t}{t}\right)$
42. $\lim_{\theta\to 0} \cos\left(\frac{\pi\theta}{\sin\theta}\right)$

The equations in Exercises 43 and 44 give the position $s = f(t)$ of a body moving on a coordinate line (s in meters, t in seconds). Find the body's velocity, speed, acceleration, and jerk at time $t = \pi/4$ sec.

43. $s = 2 - 2\sin t$
44. $s = \sin t + \cos t$
45. Is there a value of c that will make

$$f(x) = \begin{cases} \frac{\sin^2 3x}{x^2}, & x \ne 0 \\ c, & x = 0 \end{cases}$$

continuous at $x = 0$? Give reasons for your answer.

46. Is there a value of b that will make

$$g(x) = \begin{cases} x + b, & x < 0 \\ \cos x, & x \geq 0 \end{cases}$$

continuous at $x = 0$? Differentiable at $x = 0$? Give reasons for your answers.

47. Find $d^{999}/dx^{999}(\cos x)$.

48. Derive the formula for the derivative with respect to x of

a. $\sec x$. **b.** $\csc x$. **c.** $\cot x$.

T **49. Centered difference quotients** The *centered difference quotient*

$$\frac{f(x + h) - f(x - h)}{2h}$$

is used to approximate $f'(x)$ in numerical work because (1) its limit as $h \to 0$ equals $f'(x)$ when $f'(x)$ exists, and (2) it usually gives a better approximation of $f'(x)$ for a given value of h than the difference quotient

$$\frac{f(x + h) - f(x)}{h}.$$

See the accompanying figure.

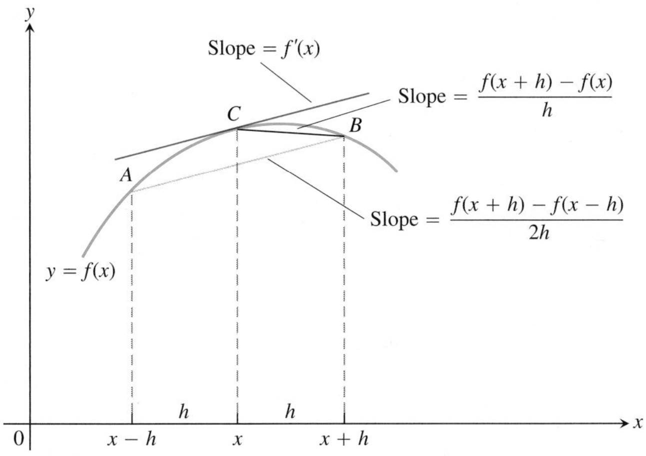

a. To see how rapidly the centered difference quotient for $f(x) = \sin x$ converges to $f'(x) = \cos x$, graph $y = \cos x$ together with

$$y = \frac{\sin(x + h) - \sin(x - h)}{2h}$$

over the interval $[-\pi, 2\pi]$ for $h = 1, 0.5$, and 0.3.

b. To see how rapidly the centered difference quotient for $f(x) = \cos x$ converges to $f'(x) = -\sin x$, graph $y = -\sin x$ together with

$$y = \frac{\cos(x + h) - \cos(x - h)}{2h}$$

over the interval $[-\pi, 2\pi]$ for $h = 1, 0.5$, and 0.3.

50. A caution about centered difference quotients *(Continuation of Exercise 49.)* The quotient

$$\frac{f(x + h) - f(x - h)}{2h}$$

may have a limit as $h \to 0$ when f has no derivative at x. As a case in point, take $f(x) = |x|$ and calculate

$$\lim_{h \to 0} \frac{|0 + h| - |0 - h|}{2h}.$$

As you will see, the limit exists even though $f(x) = |x|$ has no derivative at $x = 0$. *Moral:* Before using a centered difference quotient, be sure the derivative exists.

T **51. Slopes on the graph of the tangent function** Graph $y = \tan x$ and its derivative together on $(-\pi/2, \pi/2)$. Does the graph of the tangent function appear to have a smallest slope? A largest slope? Is the slope ever negative? Give reasons for your answers.

T **52. Slopes on the graph of the cotangent function** Graph $y = \cot x$ and its derivative together for $0 < x < \pi$. Does the graph of the cotangent function appear to have a smallest slope? A largest slope? Is the slope ever positive? Give reasons for your answers.

2.6 Exponential Functions

In this section we take an intuitive view of exponential functions to introduce their basic properties and derivatives.

Exponential Behavior

When a positive quantity P doubles, it increases by a factor of 2 and the quantity becomes $2P$. If it doubles again, it becomes $2(2P) = 2^2P$, and a third doubling gives $2(2^2P) = 2^3P$. Continuing to double in this fashion leads us to the consideration of the function $f(x) = 2^x$. We call this an *exponential* function because the variable x appears in the

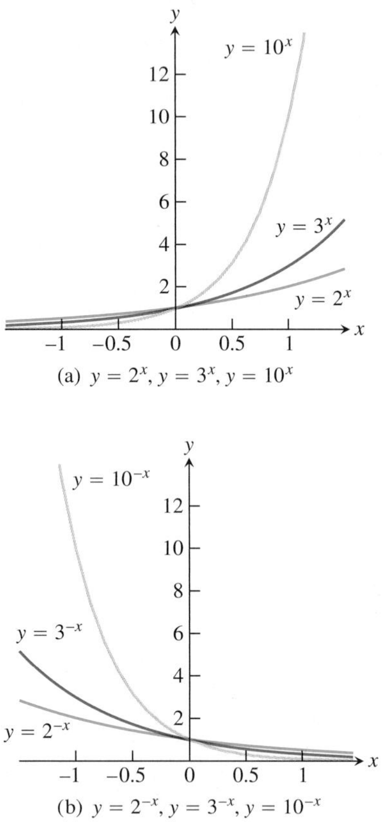

FIGURE 2.20 Graphs of exponential functions.

exponent of 2^x. Functions such as $g(x) = 10^x$ and $h(x) = (1/2)^x$ are other examples of exponential functions. In general, if $a \neq 1$ is a positive constant, the function

$$f(x) = a^x$$

is the **exponential function with base *a*.**

For integer and rational exponents, the value of an exponential function $f(x) = a^x$ is obtained arithmetically as follows. If $x = n$ is a positive integer, the number a^n is given by multiplying a by itself n times:

$$a^n = \underbrace{a \cdot a \cdot \cdots \cdot a}_{n \text{ factors}}.$$

If $x = 0$, then $a^0 = 1$, and if $x = -n$ for some positive integer n, then

$$a^{-n} = \frac{1}{a^n} = \left(\frac{1}{a}\right)^n.$$

If $x = 1/n$ for some positive integer n, then

$$a^{1/n} = \sqrt[n]{a},$$

which is the positive number that when multiplied by itself n times gives a. If $x = p/q$ is any rational number, then

$$a^{p/q} = \sqrt[q]{a^p} = \left(\sqrt[q]{a}\right)^p.$$

If x is *irrational*, the meaning of a^x is not so clear, but its value can be defined by considering values for rational numbers that get closer and closer to x. This informal approach is based on the graph of the exponential function.

We display the graphs of several exponential functions in Figure 2.20. These graphs describe the values of the exponential functions for all real inputs x. The value at an irrational number x is chosen so that the function $f(x) = a^x$ is continuous. This choice ensures the graph retains its increasing behavior when $a > 1$, or decreasing behavior when $0 < a < 1$ (see Figure 2.20). Also note that the x-axis is a horizontal asymptote of the graph of each exponential function.

Arithmetically, the graphical idea can be described in the following way, using the exponential $f(x) = 2^x$ as an illustration. Any particular irrational number, say $x = \sqrt{3}$, has a decimal expansion

$$\sqrt{3} = 1.732050808\ldots.$$

We then consider the list of numbers, given as follows in the order of taking more and more digits in the decimal expansion,

$$2^1, 2^{1.7}, 2^{1.73}, 2^{1.732}, 2^{1.7320}, 2^{1.73205}, \ldots. \tag{1}$$

We know the meaning of each number in list (1) because the successive decimal approximations to $\sqrt{3}$ given by 1, 1.7, 1.73, 1.732, and so on, are all *rational* numbers. As these decimal approximations get closer and closer to $\sqrt{3}$, it seems reasonable that the list of numbers in (1) gets closer and closer to some fixed number, which we specify to be $2^{\sqrt{3}}$. This is in fact true because of the *completeness property* of the real numbers discussed in Appendix 7 (although it is beyond the scope of this text to give a proof here). In a similar way, we can identify 2^x (or a^x, $a > 0$) for any irrational number x. That is, in general

$$a^x = \lim_{r \to x} a^r, \qquad r \text{ rational.} \tag{2}$$

Equation (2) makes the function $f(x) = a^x$, $a > 0$, continuous. In Appendix 8 we establish the continuity of exponential functions which guarantees that the limit in Equation (2) actually exists.

Exponential functions obey the familiar rules of exponents listed below. It is easy to check these rules using algebra when the exponents are integers or rational numbers. We prove them for all real numbers in Appendix 8.

Rules for Exponents

If $a > 0$ and $b > 0$, the following rules hold true for all real numbers x and y.

1. $a^x \cdot a^y = a^{x+y}$
2. $\dfrac{a^x}{a^y} = a^{x-y}$
3. $(a^x)^y = (a^y)^x = a^{xy}$
4. $a^x \cdot b^x = (ab)^x$
5. $\dfrac{a^x}{b^x} = \left(\dfrac{a}{b}\right)^x$

EXAMPLE 1

1. $3^{1.1} \cdot 3^{0.7} = 3^{1.1+0.7} = 3^{1.8}$
2. $\dfrac{\left(\sqrt{10}\right)^3}{\sqrt{10}} = \left(\sqrt{10}\right)^{3-1} = \left(\sqrt{10}\right)^2 = 10$
3. $\left(5^{\sqrt{2}}\right)^{\sqrt{2}} = 5^{\sqrt{2}\cdot\sqrt{2}} = 5^2 = 25$
4. $7^{\pi} \cdot 8^{\pi} = (56)^{\pi}$
5. $\left(\dfrac{4}{9}\right)^{1/2} = \dfrac{4^{1/2}}{9^{1/2}} = \dfrac{2}{3}$ ■

The Natural Exponential Function e^x

The most important exponential function used for modeling natural, physical, and economic phenomena is the **natural exponential function**, whose base is the special number e. The number e is irrational, and its value is 2.718281828 to nine decimal places. It might seem strange that we would use this number for a base rather than a simple number like 2 or 10. The advantage in using e as a base is that it simplifies many of the calculations in calculus.

If you look at Figure 2.20a you can see that the graphs of the exponential functions $y = a^x$ get steeper as the base a gets larger. Figure 2.21 shows the tangent line to the graph of $y = a^x$, together with its estimated slope, for several values of a, as the graph crosses the y-axis. The particular number e is chosen as the base for which the slope of the graph of a^x is exactly 1 when it crosses the y-axis. The slope is smaller than 1 if $a < e$, and larger than 1 if $a > e$. This is the property that makes the number e so useful in calculus: **The graph of $y = e^x$ has slope 1 when it crosses the y-axis.**

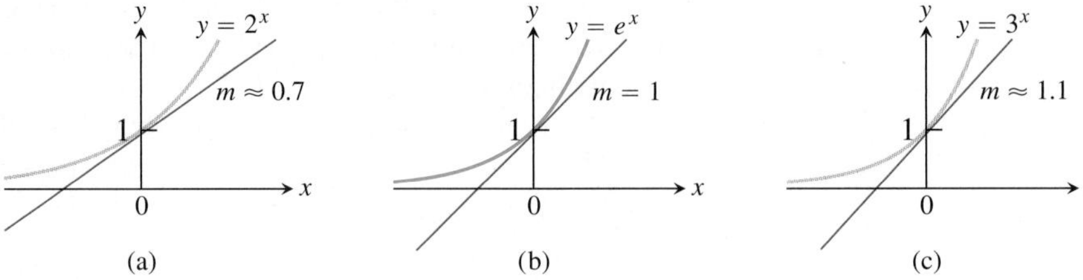

FIGURE 2.21 Among the exponential functions, the graph of $y = e^x$ has the property that the slope m of the tangent line to the graph is exactly 1 when it crosses the y-axis. The slope is smaller for a base less than e, such as 2^x, and larger for a base greater than e, such as 3^x.

In Section 2.10 we use that slope property to prove e is the limit of $(1 + 1/x)^x$ as x tends to ∞. This limit provides one way to compute the value of e, at least approximately. The graph and table in Figure 2.22 show the behavior of this expression and how it approaches the line $y = e \approx 2.718281828$ asymptotically as x tends to ∞.

x	$(1 + 1/x)^x$
1000	2.7169
2000	2.7176
3000	2.7178
4000	2.7179
5000	2.7180
6000	2.7181
7000	2.7181

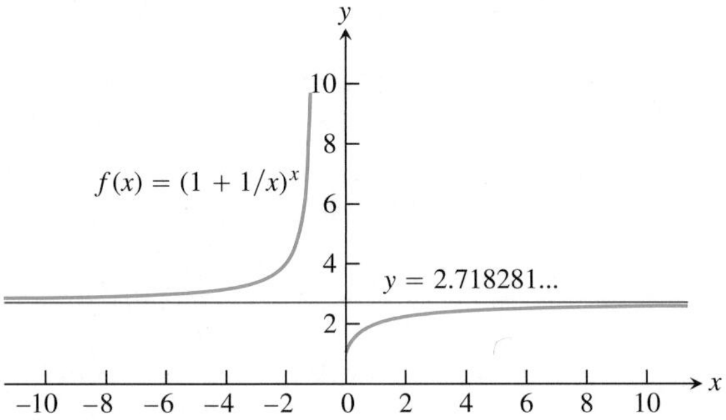

FIGURE 2.22 A graph and table of values for $f(x) = (1 + 1/x)^x$ both suggest that as x tends to ∞, $f(x)$ approaches $e \approx 2.7182818\ldots$ asymptotically.

Derivatives of Exponential Functions

When we apply the definition of the derivative to $f(x) = a^x$, we get

$$\frac{d}{dx}(a^x) = \lim_{h \to 0} \frac{a^{x+h} - a^x}{h} \qquad \text{Derivative definition}$$

$$= \lim_{h \to 0} \frac{a^x \cdot a^h - a^x}{h} \qquad a^{x+h} = a^x \cdot a^h$$

$$= \lim_{h \to 0} a^x \cdot \frac{a^h - 1}{h} \qquad \text{Factoring out } a^x$$

$$= a^x \cdot \lim_{h \to 0} \frac{a^h - 1}{h} \qquad a^x \text{ is constant as } h \to 0.$$

$$= \underbrace{\left(\lim_{h \to 0} \frac{a^h - 1}{h} \right)}_{\text{a fixed number } L} \cdot a^x. \qquad (3)$$

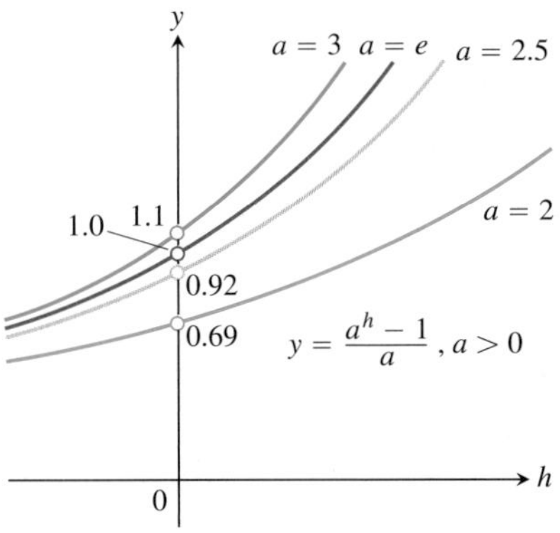

FIGURE 2.23 The position of the curve $y = (a^h - 1)/h$, $a > 0$, varies continuously with a.

Thus we see that the derivative of a^x is a constant multiple L of a^x. The constant L is a limit unlike any we have encountered before. (In Appendix 8 we establish the existence of L.) Note that the number L equals the value of the derivative of $f(x) = a^x$ at $x = 0$:

$$f'(0) = \lim_{h \to 0} \frac{a^h - a^0}{h} = \lim_{h \to 0} \frac{a^h - 1}{h} = L.$$

The limit L is therefore the slope of the graph of the continuous function $f(x) = a^x$ where it crosses the y-axis. We investigate the values of L by graphing the function $y = (a^h - 1)/h$ and studying its behavior as h approaches 0.

Figure 2.23 shows the graphs of $y = (a^h - 1)/h$ for four different values of a. The limit L is approximately 0.69 if $a = 2$, about 0.92 if $a = 2.5$, and about 1.1 if $a = 3$. It appears that the value of L is 1 at some number a chosen between 2.5 and 3. That number is given by $a = e \approx 2.718281828$. With this choice of base we obtain the natural exponential function $f(x) = e^x$ and see that it satisfies the property

$$f'(0) = \lim_{h \to 0} \frac{e^h - 1}{h} = 1. \tag{4}$$

That the limit is 1 implies an important relationship between the natural exponential function e^x and its derivative:

$$\frac{d}{dx}(e^x) = \lim_{h \to 0} \left(\frac{e^h - 1}{h} \right) \cdot e^x \qquad \text{Eq. (3) with } a = e$$

$$= 1 \cdot e^x = e^x. \qquad \text{Eq. (4)}$$

Therefore the natural exponential function is its own derivative.

Derivative of the Natural Exponential Function

$$\frac{d}{dx}(e^x) = e^x$$

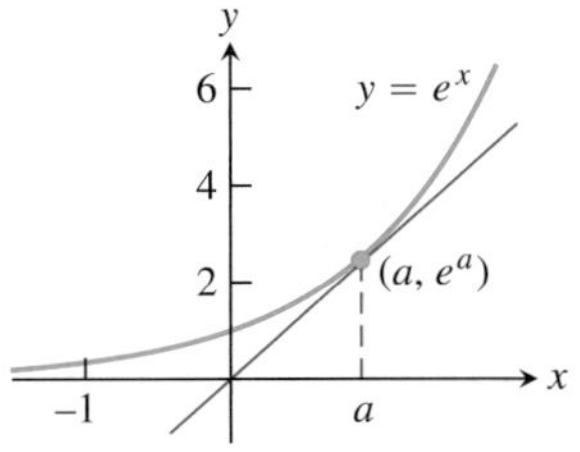

FIGURE 2.24 The line through the origin is tangent to the graph of $y = e^x$ when $a = 1$ (Example 2).

EXAMPLE 2 Find an equation for a line that is tangent to the graph of $y = e^x$ and goes through the origin.

Solution Since the line passes through the origin, its equation is of the form $y = mx$, where m is the slope. If it is tangent to the graph at the point (a, e^a), the slope is $m = (e^a - 0)/(a - 0)$. The slope of the natural exponential at $x = a$ is e^a. Because these slopes are the same, we then have that $e^a = e^a/a$. It follows that $a = 1$ and $m = e$, so the equation of the tangent line is $y = ex$. See Figure 2.24. ■

We might ask if there are functions *other* than the natural exponential function that are their own derivatives. The answer is that the only functions that satisfy the property that $f'(x) = f(x)$ are functions that are constant multiples of the natural exponential function, $f(x) = c \cdot e^x$, c any constant. We prove this fact in Section 6.4. Note from the Constant Multiple Rule that indeed

$$\frac{d}{dx}(c \cdot e^x) = c \cdot \frac{d}{dx}(e^x) = c \cdot e^x.$$

EXAMPLE 3 Find the derivative of $y = e^{-x}$.

Solution We apply the Quotient Rule to $e^{-x} = 1/e^x$:

$$\frac{d}{dx}(e^{-x}) = \frac{d}{dx}\left(\frac{1}{e^x}\right) = \frac{e^x \cdot 0 - 1 \cdot e^x}{(e^x)^2} = \frac{-1}{e^x} = -e^{-x}$$ ■

The following example illustrates a limit calculation involving the natural exponential function.

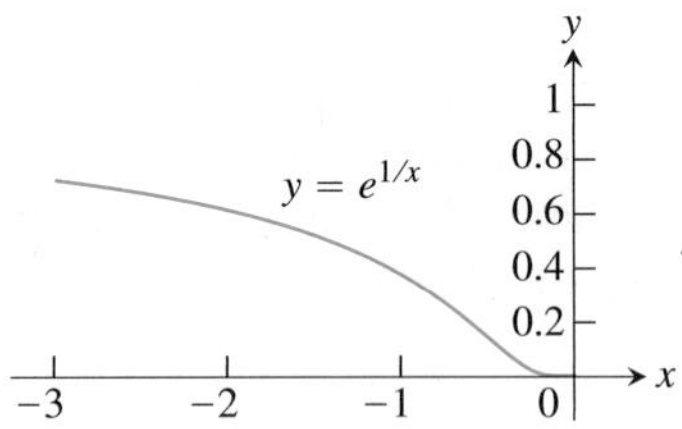

FIGURE 2.25 The graph of $y = e^{1/x}$ for $x < 0$ shows $\lim_{x\to 0^-} e^{1/x} = 0$ (Example 4).

EXAMPLE 4 Find $\lim_{x\to 0^-} e^{1/x}$ (Figure 2.25).

Solution We let $t = 1/x$. We know that $t \to -\infty$ as $x \to 0^-$. Therefore,

$$\lim_{x\to 0^-} e^{1/x} = \lim_{t\to -\infty} e^t = 0$$

because $y = 0$ is a horizontal asymptote of $y = e^t$. ■

The following example illustrates the importance of the exponential function in modeling the decay of radioactive substances. Carbon-14 decay is used to date the remains of dead organisms such as shells, seeds, and wooden artifacts.

EXAMPLE 5 Laboratory experiments indicate that some atoms emit a part of their mass as radiation, with the remainder of the atom re-forming to make an atom of some new element. For example, radioactive carbon-14 decays into nitrogen; radium eventually decays into lead. If y_0 is the number of radioactive nuclei present at time zero, the number still present at any later time t will be

$$y = y_0 e^{-rt}, \qquad r > 0.$$

The number r is called the **decay rate** of the radioactive substance. (We will see how this formula is obtained in Section 6.4.) For carbon-14, the decay rate has been determined experimentally to be about $r = 1.2 \times 10^{-4}$ when t is measured in years. Predict the percent of carbon-14 present after 866 years have elapsed.

Solution If we start with an amount y_0 of carbon-14 nuclei, after 866 years we are left with the amount

$$\begin{aligned} y(866) &= y_0 e^{(-1.2\times 10^{-4})(866)} \\ &\approx (0.901)y_0. \qquad \text{Calculator evaluation} \end{aligned}$$

That is, after 866 years, we are left with about 90% of the original amount of carbon-14, so about 10% of the original nuclei have decayed. In Example 2 in Section 2.10, you will see how to find the number of years required for half of the radioactive nuclei present in a sample to decay (called the *half-life* of the substance). ■

EXERCISES 2.6

In Exercises 1–6, sketch the given curves together in the appropriate coordinate plane and label each curve with its equation.

1. $y = 2^x, y = 4^x, y = 3^{-x}, y = (1/5)^x$

2. $y = 3^x, y = 8^x, y = 2^{-x}, y = (1/4)^x$

3. $y = 2^{-t}$ and $y = -2^t$

4. $y = 3^{-t}$ and $y = -3^t$

5. $y = e^x$ and $y = 1/e^x$

6. $y = -e^x$ and $y = -e^{-x}$

In each of Exercises 7–10, sketch the shifted exponential curves.

7. $y = 2^x - 1$ and $y = 2^{-x} - 1$

8. $y = 3^x + 2$ and $y = 3^{-x} + 2$

9. $y = 1 - e^x$ and $y = 1 - e^{-x}$

10. $y = -1 - e^x$ and $y = -1 - e^{-x}$

Use the laws of exponents to simplify the expressions in Exercises 11–20.

11. $16^2 \cdot 16^{-1.75}$

12. $9^{1/3} \cdot 9^{1/6}$

13. $\dfrac{4^{4.2}}{4^{3.7}}$

14. $\dfrac{3^{5/3}}{3^{2/3}}$

15. $\left(25^{1/8}\right)^4$

16. $\left(13^{\sqrt{2}}\right)^{\sqrt{2}/2}$

17. $2^{\sqrt{3}} \cdot 7^{\sqrt{3}}$

18. $\left(\sqrt{3}\right)^{1/2} \cdot \left(\sqrt{12}\right)^{1/2}$

19. $\left(\dfrac{2}{\sqrt{2}}\right)^4$

20. $\left(\dfrac{\sqrt{6}}{3}\right)^2$

Find the domain and range for each of the functions in Exercises 21–24.

21. $f(x) = \dfrac{1}{2 + e^x}$

22. $g(t) = \cos(e^{-t})$

23. $g(t) = \sqrt{1 + 3^{-t}}$

24. $f(x) = \dfrac{3}{1 - e^{2x}}$

Find the derivatives of the functions in Exercises 25–34.

25. $y = x^3 e^x$

26. $w = re^{-r}$

27. $y = x^{9/4}$

28. $y = x^{-3/5} + \pi^{3/2}$

29. $s = 2t^{3/2} + 3e^2$

30. $w = \dfrac{1}{z^{1.4}} + \dfrac{\pi}{\sqrt{z}}$

31. $y = \sqrt[7]{x^2} - x^e$

32. $y = \sqrt[3]{x^{9.6}} + 2e^{1.3}$

33. $r = \dfrac{e^s}{s}$

34. $r = e^\theta\left(\dfrac{1}{\theta^2} + \theta^{-\pi/2}\right)$

Find the derivatives of all orders of the functions in Exercises 35 and 36.

35. $y = \dfrac{x^4}{2} - \dfrac{3}{2}x^2 - x$

36. $y = \dfrac{x^5}{120}$

Find the first and second derivatives of the functions in Exercises 37–46.

37. $y = \dfrac{x^3 + 7}{x}$

38. $s = \dfrac{t^2 + 5t - 1}{t^2}$

39. $r = \dfrac{(\theta - 1)(\theta^2 + \theta + 1)}{\theta^3}$

40. $u = \dfrac{(x^2 + x)(x^2 - x + 1)}{x^4}$

41. $w = 3z^2 e^z$

42. $w = e^z(z - 1)(z^2 + 1)$

43. $p = \left(\dfrac{q^2 + 3}{12q}\right)\left(\dfrac{q^4 - 1}{q^3}\right)$

44. $p = \dfrac{q^2 + 3}{(q - 1)^3 + (q + 1)^3}$

45. $y = e^{-x}(x^2 - 3)$

46. $y = \dfrac{1 - e^{-x}}{x^2 + 1}$

T In Exercises 47–52, use a graphing calculator to estimate the answer.

47. $2^x = 5$

48. $e^x = 4$

49. Population growth The population of Knoxville is 500,000 and is increasing at the rate of 3.75% each year. Approximately when will the population reach 1 million?

50. Radioactive decay The half-life of phosphorus-32 is about 14 days. There are 6.6 grams present initially.

a. Express the amount of phosphorus-32 remaining as a function of time t.

b. When will there be 1 gram remaining?

51. Doubling your money Determine how much time is required for an investment to double in value if interest is earned at the rate of 6.25% compounded annually.

52. Cholera bacteria Suppose that a colony of bacteria starts with 1 bacterium and doubles in number every half hour. How many bacteria will the colony contain at the end of 24 hr?

Find the limits in Exercises 53–56.

53. $\lim_{x\to\infty} e^{-x} \sin x$

54. $\lim_{x\to-\infty} e^x \cos\left(\dfrac{1}{x}\right)$

55. $\lim_{x\to-\infty} \dfrac{e^x - e^{-x}}{e^x + e^{-x}}$

56. $\lim_{x\to\infty} \dfrac{3x^2 + e^{-x}}{\sin(1/x) - 2x^2}$

T **57.** Compare the graphs of the exponential function $y = 2^x$ and the power function $y = x^2$. For what value of x does it appear that the exponential starts to grow far more rapidly than the parabola $y = x^2$?

T **58.** The equation $x^2 = 2^x$ has three solutions: $x = 2, x = 4$, and one other. Estimate the third solution as accurately as you can by graphing.

2.7 The Chain Rule

How do we differentiate $F(x) = \sin(x^2 - 4)$, which is the composite $f \circ g$ of the functions $y = f(u) = \sin u$ and $u = g(x) = x^2 - 4$? The answer, given by the *Chain Rule*, says that the derivative is the product of the derivatives of f and g. We will learn the rule in this section.

FIGURE 2.26 When gear A makes x turns, gear B makes u turns and gear C makes y turns. By comparing circumferences or counting teeth, we see that $y = u/2$ (C turns one-half turn for each B turn) and $u = 3x$ (B turns three times for A's one), so $y = 3x/2$. Thus, $dy/dx = 3/2 = (1/2)(3) = (dy/du)(du/dx)$.

Derivative of a Composite Function

The function $y = \frac{3}{2}x = \frac{1}{2}(3x)$ is the composite of the functions $y = \frac{1}{2}u$ and $u = 3x$. We have

$$\frac{dy}{dx} = \frac{3}{2}, \qquad \frac{dy}{du} = \frac{1}{2}, \qquad \text{and} \qquad \frac{du}{dx} = 3.$$

Since $\frac{3}{2} = \frac{1}{2} \cdot 3$, we see in this case that

$$\frac{dy}{dx} = \frac{dy}{du} \cdot \frac{du}{dx}.$$

If we think of the derivative as a rate of change, our intuition allows us to see that this relationship is reasonable. If $y = f(u)$ changes half as fast as u and $u = g(x)$ changes three times as fast as x, then we expect y to change 3/2 times as fast as x. This effect is much like that of a multiple gear train (Figure 2.26). Let's look at another example.

EXAMPLE 1 The function

$$y = (3x^2 + 1)^2$$

is the composite of $y = f(u) = u^2$ and $u = g(x) = 3x^2 + 1$. Calculating derivatives, we see that

$$\begin{aligned}\frac{dy}{du} \cdot \frac{du}{dx} &= 2u \cdot 6x \\ &= 2(3x^2 + 1) \cdot 6x \\ &= 36x^3 + 12x.\end{aligned}$$

Calculating the derivative from the expanded formula $(3x^2 + 1)^2 = 9x^4 + 6x^2 + 1$ gives the same result:

$$\begin{aligned}\frac{dy}{dx} &= \frac{d}{dx}(9x^4 + 6x^2 + 1) \\ &= 36x^3 + 12x.\end{aligned}$$

■

The derivative of the composite function $f(g(x))$ at x is the derivative of f at $g(x)$ times the derivative of g at x. This is known as the Chain Rule (Figure 2.27).

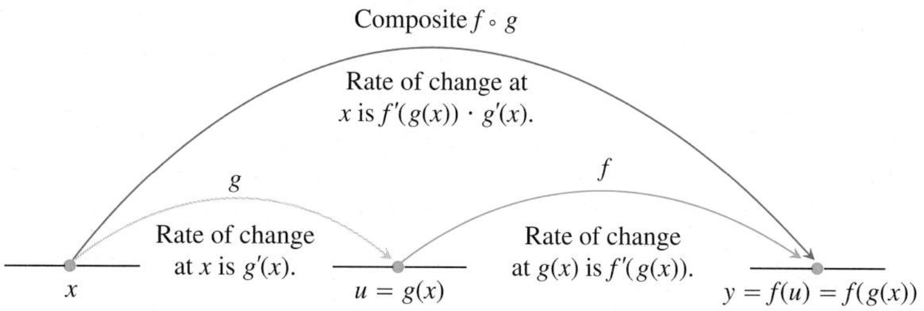

FIGURE 2.27 Rates of change multiply: The derivative of $f \circ g$ at x is the derivative of f at $g(x)$ times the derivative of g at x.

THEOREM 3—The Chain Rule If $f(u)$ is differentiable at the point $u = g(x)$ and $g(x)$ is differentiable at x, then the composite function $(f \circ g)(x) = f(g(x))$ is differentiable at x, and

$$(f \circ g)'(x) = f'(g(x)) \cdot g'(x).$$

In Leibniz's notation, if $y = f(u)$ and $u = g(x)$, then

$$\frac{dy}{dx} = \frac{dy}{du} \cdot \frac{du}{dx},$$

where dy/du is evaluated at $u = g(x)$.

Intuitive "Proof" of the Chain Rule:

Let Δu be the change in u corresponding to a change of Δx in x, that is

$$\Delta u = g(x + \Delta x) - g(x).$$

Then the corresponding change in y is

$$\Delta y = f(u + \Delta u) - f(u).$$

It would be tempting to write

$$\frac{\Delta y}{\Delta x} = \frac{\Delta y}{\Delta u} \cdot \frac{\Delta u}{\Delta x} \tag{1}$$

and take the limit as $\Delta x \to 0$:

$$\begin{aligned}
\frac{dy}{dx} &= \lim_{\Delta x \to 0} \frac{\Delta y}{\Delta x} \\
&= \lim_{\Delta x \to 0} \frac{\Delta y}{\Delta u} \cdot \frac{\Delta u}{\Delta x} \\
&= \lim_{\Delta x \to 0} \frac{\Delta y}{\Delta u} \cdot \lim_{\Delta x \to 0} \frac{\Delta u}{\Delta x} \\
&= \lim_{\Delta u \to 0} \frac{\Delta y}{\Delta u} \cdot \lim_{\Delta x \to 0} \frac{\Delta u}{\Delta x} \qquad \text{(Note that } \Delta u \to 0 \text{ as } \Delta x \to 0 \text{ since } g \text{ is continuous.)} \\
&= \frac{dy}{du}\frac{du}{dx}.
\end{aligned}$$

The only flaw in this reasoning is that in Equation (1) it might happen that $\Delta u = 0$ (even when $\Delta x \neq 0$) and, of course, we can't divide by 0. The proof requires a different approach to overcome this flaw, and we give a precise proof in Section 2.13. ■

EXAMPLE 2 An object moves along the x-axis so that its position at any time $t \geq 0$ is given by $x(t) = \cos(t^2 + 1)$. Find the velocity of the object as a function of t.

Solution We know that the velocity is dx/dt. In this instance, x is a composite function: $x = \cos(u)$ and $u = t^2 + 1$. We have

$$\frac{dx}{du} = -\sin(u) \qquad x = \cos(u)$$

$$\frac{du}{dt} = 2t. \qquad u = t^2 + 1$$

By the Chain Rule,

$$\begin{aligned}\frac{dx}{dt} &= \frac{dx}{du} \cdot \frac{du}{dt} \\ &= -\sin(u) \cdot 2t \qquad \frac{dx}{du} \text{ evaluated at } u \\ &= -\sin(t^2 + 1) \cdot 2t \\ &= -2t \sin(t^2 + 1).\end{aligned}$$

■

"Outside-Inside" Rule

A difficulty with the Leibniz notation is that it doesn't state specifically where the derivatives in the Chain Rule are supposed to be evaluated. So it sometimes helps to think about the Chain Rule using functional notation. If $y = f(g(x))$, then

$$\frac{dy}{dx} = f'(g(x)) \cdot g'(x).$$

In words, differentiate the "outside" function f and evaluate it at the "inside" function $g(x)$ left alone; then multiply by the derivative of the "inside function."

EXAMPLE 3 Differentiate $\sin(x^2 + e^x)$ with respect to x.

Solution

$$\frac{d}{dx}\sin\underbrace{(x^2 + e^x)}_{\text{inside}} = \cos\underbrace{(x^2 + e^x)}_{\substack{\text{inside}\\\text{left alone}}} \cdot \underbrace{(2x + e^x)}_{\substack{\text{derivative of}\\\text{the inside}}}$$

■

EXAMPLE 4 Differentiate $y = e^{\cos x}$.

Solution Here the inside function is $u = g(x) = \cos x$ and the outside function is the exponential function $f(x) = e^x$. Applying the Chain Rule, we get

$$\frac{dy}{dx} = \frac{d}{dx}(e^{\cos x}) = e^{\cos x}\frac{d}{dx}(\cos x) = e^{\cos x}(-\sin x) = -e^{\cos x}\sin x.$$

■

Generalizing Example 4, we see that the Chain Rule gives the formula

$$\frac{d}{dx}e^u = e^u\frac{du}{dx}.$$

Thus, for example,

$$\frac{d}{dx}(e^{kx}) = e^{kx} \cdot \frac{d}{dx}(kx) = ke^{kx}, \quad \text{for any constant } k$$

and

$$\frac{d}{dx}\left(e^{x^2}\right) = e^{x^2} \cdot \frac{d}{dx}(x^2) = 2xe^{x^2}.$$

Repeated Use of the Chain Rule

We sometimes have to use the Chain Rule two or more times to find a derivative.

EXAMPLE 5 Find the derivative of $g(t) = \tan(5 - \sin 2t)$.

HISTORICAL BIOGRAPHY

Johann Bernoulli
(1667–1748)

Solution Notice here that the tangent is a function of $5 - \sin 2t$, whereas the sine is a function of $2t$, which is itself a function of t. Therefore, by the Chain Rule,

$$\begin{aligned} g'(t) &= \frac{d}{dt}(\tan(5 - \sin 2t)) && \text{Derivative of } \tan u \text{ with } u = 5 - \sin 2t \\ &= \sec^2(5 - \sin 2t) \cdot \frac{d}{dt}(5 - \sin 2t) && \text{Derivative of } 5 - \sin u \text{ with } u = 2t \\ &= \sec^2(5 - \sin 2t) \cdot \left(0 - \cos 2t \cdot \frac{d}{dt}(2t)\right) \\ &= \sec^2(5 - \sin 2t) \cdot (-\cos 2t) \cdot 2 \\ &= -2(\cos 2t)\sec^2(5 - \sin 2t). \end{aligned}$$

■

The Chain Rule with Powers of a Function

If f is a differentiable function of u and if u is a differentiable function of x, then substituting $y = f(u)$ into the Chain Rule formula

$$\frac{dy}{dx} = \frac{dy}{du} \cdot \frac{du}{dx}$$

leads to the formula

$$\frac{d}{dx} f(u) = f'(u)\frac{du}{dx}.$$

Here's an example of how it works: If n is any real number and $f(u) = u^n$, the Power Rule (Rule 2) tells us that $f'(u) = nu^{n-1}$. If u is a differentiable function of x, then we can use the Chain Rule to extend this to the **Power Chain Rule**:

$$\frac{d}{dx} u^n = nu^{n-1}\frac{du}{dx}. \qquad \frac{d}{du}(u^n) = nu^{n-1}$$

EXAMPLE 6

(a)
$$\begin{aligned} \frac{d}{dx}(5x^3 - x^4)^7 &= 7(5x^3 - x^4)^6 \frac{d}{dx}(5x^3 - x^4) && \text{Power Chain Rule with } u = 5x^3 - x^4, n = 7 \\ &= 7(5x^3 - x^4)^6(5 \cdot 3x^2 - 4x^3) \\ &= 7(5x^3 - x^4)^6(15x^2 - 4x^3) \end{aligned}$$

(b) $$\frac{d}{dx}\left(\frac{1}{3x-2}\right) = \frac{d}{dx}(3x-2)^{-1}$$
$$= -1(3x-2)^{-2}\frac{d}{dx}(3x-2)$$ Power Chain Rule with $u = 3x - 2, n = -1$
$$= -1(3x-2)^{-2}(3)$$
$$= -\frac{3}{(3x-2)^2}$$

In part (b) we could also find the derivative with the Quotient Rule.

(c) $$\frac{d}{dx}(\sin^5 x) = 5\sin^4 x \cdot \frac{d}{dx}\sin x$$ Power Chain Rule with $u = \sin x$, $n = 5$, because $\sin^n x$ means $(\sin x)^n$, $n \neq -1$.
$$= 5\sin^4 x \cos x$$

(d) $$\frac{d}{dx}\left(e^{\sqrt{3x+1}}\right) = e^{\sqrt{3x+1}} \cdot \frac{d}{dx}\left(\sqrt{3x+1}\right)$$
$$= e^{\sqrt{3x+1}} \cdot \frac{1}{2}(3x+1)^{-1/2} \cdot 3$$
$$= \frac{3}{2\sqrt{3x+1}}e^{\sqrt{3x+1}}$$ ■

Derivative of the Absolute Value Function

$$\frac{d}{dx}(|x|) = \frac{x}{|x|},\ x \neq 0$$

EXAMPLE 7 In Section 2.2, we saw that the absolute value function $y = |x|$ is not differentiable at $x = 0$. However, the function *is* differentiable at all other real numbers as we now show. Since $|x| = \sqrt{x^2}$, we can derive the following useful formula:

$$\frac{d}{dx}(|x|) = \frac{d}{dx}\sqrt{x^2}$$
$$= \frac{1}{2\sqrt{x^2}} \cdot \frac{d}{dx}(x^2)$$ Power Chain Rule with $u = x^2, n = 1/2, x \neq 0$
$$= \frac{1}{2|x|} \cdot 2x$$ $\sqrt{x^2} = |x|$
$$= \frac{x}{|x|}, \quad x \neq 0.$$ ■

EXAMPLE 8 The formulas for the derivatives of both $\sin x$ and $\cos x$ were obtained under the assumption that x is measured in radians, *not* degrees. The Chain Rule gives us new insight into the difference between the two. Since $180° = \pi$ radians, $x° = \pi x/180$ radians where $x°$ means the angle x measured in degrees.

By the Chain Rule,

$$\frac{d}{dx}\sin(x°) = \frac{d}{dx}\sin\left(\frac{\pi x}{180}\right) = \frac{\pi}{180}\cos\left(\frac{\pi x}{180}\right) = \frac{\pi}{180}\cos(x°).$$

See Figure 2.28. Similarly, the derivative of $\cos(x°)$ is $-(\pi/180)\sin(x°)$.

The factor $\pi/180$, annoying in the first derivative, would compound with repeated differentiation. We see at a glance a compelling reason for the use of radian measure. ■

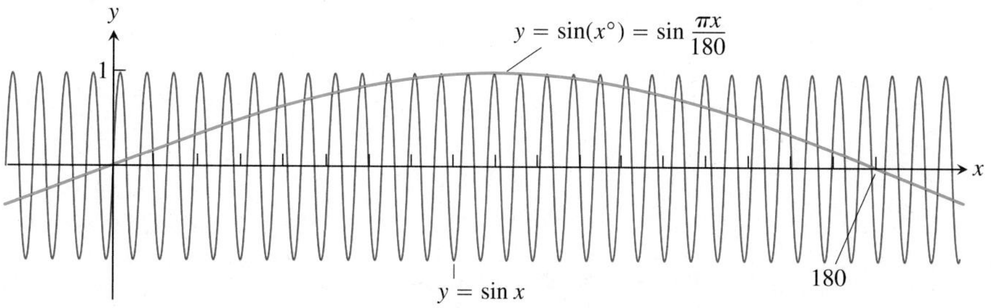

FIGURE 2.28 Sin$(x°)$ oscillates only $\pi/180$ times as often as $\sin x$ oscillates. Its maximum slope is $\pi/180$ at $x = 0$ (Example 8).

EXERCISES 2.7

In Exercises 1–8, given $y = f(u)$ and $u = g(x)$, find $dy/dx = f'(g(x))g'(x)$.

1. $y = 6u - 9, \quad u = (1/2)x^4$
2. $y = 2u^3, \quad u = 8x - 1$
3. $y = \sin u, \quad u = 3x + 1$
4. $y = \cos u, \quad u = -x/3$
5. $y = \cos u, \quad u = \sin x$
6. $y = \sin u, \quad u = x - \cos x$
7. $y = \tan u, \quad u = 10x - 5$
8. $y = -\sec u, \quad u = x^2 + 7x$

In Exercises 9–22, write the function in the form $y = f(u)$ and $u = g(x)$. Then find dy/dx as a function of x.

9. $y = (2x + 1)^5$
10. $y = \sqrt[3]{2x}$
11. $y = \left(1 - \frac{x}{7}\right)^{-7}$
12. $y = \left(\frac{x}{2} - 1\right)^{-10}$
13. $y = (1 - 6x)^{2/3}$
14. $y = \left(\frac{x}{5} + \frac{1}{5x}\right)^5$
15. $y = \sec(\tan x)$
16. $y = \cot\left(\pi - \frac{1}{x}\right)$
17. $y = \sin^3 x$
18. $y = 5\cos^{-4} x$
19. $y = e^{-5x}$
20. $y = e^{2x/3}$
21. $y = e^{5-7x}$
22. $y = e^{\left(4\sqrt{x}+x^2\right)}$

Find the derivatives of the functions in Exercises 23–48.

23. $p = \sqrt{3 - t}$
24. $y = x(x^2 + 1)^{1/2}$
25. $s = \frac{4}{3\pi}\sin 3t + \frac{4}{5\pi}\cos 5t$
26. $y = \sin[(2t + 5)^{-2/3}]$
27. $r = (\csc\theta + \cot\theta)^{-1}$
28. $r = -(\sec\theta + \tan\theta)^{-1}$
29. $y = x^2\sin^4 x + x\cos^{-2} x$
30. $z = \cos[(1 - 6t)^{2/3}]$
31. $y = \frac{1}{21}(3x - 2)^7 + \left(4 - \frac{1}{2x^2}\right)^{-1}$
32. $g(x) = 2(2x^{-1/2} + 1)^{-1/3}$
33. $y = (4x + 3)^4(x + 1)^{-3}$
34. $y = x(x^2 + 1)^{-1/2}$
35. $y = xe^{-x} + e^{3x}$
36. $y = (1 + 2x)e^{-2x}$
37. $y = (x^2 - 2x + 2)e^{5x/2}$
38. $y = (9x^2 - 6x + 2)e^{x^3}$
39. $h(x) = x\tan\left(2\sqrt{x}\right) + 7$
40. $k(x) = x^2\sec\left(\frac{1}{x}\right)$
41. $f(\theta) = \left(\frac{\sin\theta}{1 + \cos\theta}\right)^2$
42. $g(t) = \left(\frac{1 + \cos t}{\sin t}\right)^{-1}$
43. $h(\theta) = \sqrt[3]{1 + \cos 2\theta}$
44. $r = \sec\sqrt{\theta}\tan\left(\frac{1}{\theta}\right)$
45. $q = \sin\left(\frac{t}{\sqrt{t + 1}}\right)$
46. $q = \cot\left(\frac{\sin t}{t}\right)$
47. $y = \cos\left(e^{-\theta^2}\right)$
48. $y = \theta^3 e^{-2\theta}\cos 5\theta$

In Exercises 49–60, find dy/dt.

49. $y = \sin^2(\pi t - 2)$
50. $y = \sec^2 \pi t$
51. $y = (1 + \cos 2t)^{-4}$
52. $y = (1 + \cot(t/2))^{-2}$
53. $y = e^{\cos^2(\pi t - 1)}$
54. $y = \left(e^{\sin(t/2)}\right)^3$
55. $y = \sin(\cos(2t - 5))$
56. $y = \cos\left(5\sin\left(\frac{t}{3}\right)\right)$
57. $y = \left(1 + \tan^4\left(\frac{t}{12}\right)\right)^3$
58. $y = \frac{1}{6}(1 + \cos^2(7t))^3$
59. $y = \sqrt{1 + \cos(t^2)}$
60. $y = 4\sin\sqrt{1 + \sqrt{t}}$

Find y'' in Exercises 61–66.

61. $y = \left(1 + \frac{1}{x}\right)^3$

62. $y = \left(1 - \sqrt{x}\right)^{-1}$

63. $y = \frac{1}{9}\cot(3x - 1)$

64. $y = 9\tan\left(\frac{x}{3}\right)$

65. $y = e^{x^2} + 5x$

66. $y = \sin(x^2 e^x)$

In Exercises 67–72, find the value of $(f \circ g)'$ at the given value of x.

67. $f(u) = u^5 + 1, \quad u = g(x) = \sqrt{x}, \quad x = 1$

68. $f(u) = 1 - \frac{1}{u}, \quad u = g(x) = \frac{1}{1 - x}, \quad x = -1$

69. $f(u) = \cot\frac{\pi u}{10}, \quad u = g(x) = 5\sqrt{x}, \quad x = 1$

70. $f(u) = u + \frac{1}{\cos^2 u}, \quad u = g(x) = \pi x, \quad x = 1/4$

71. $f(u) = \frac{2u}{u^2 + 1}, \quad u = g(x) = 10x^2 + x + 1, \quad x = 0$

72. $f(u) = \left(\frac{u - 1}{u + 1}\right)^2, \quad u = g(x) = \frac{1}{x^2} - 1, \quad x = -1$

73. Suppose that functions f and g and their derivatives with respect to x have the following values at $x = 2$ and $x = 3$.

x	$f(x)$	$g(x)$	$f'(x)$	$g'(x)$
2	8	2	$1/3$	-3
3	3	-4	2π	5

Find the derivatives with respect to x of the following combinations at the given value of x.

a. $2f(x), \quad x = 2$

b. $f(x) + g(x), \quad x = 3$

c. $f(x) \cdot g(x), \quad x = 3$

d. $f(x)/g(x), \quad x = 2$

e. $f(g(x)), \quad x = 2$

f. $\sqrt{f(x)}, \quad x = 2$

g. $1/g^2(x), \quad x = 3$

h. $\sqrt{f^2(x) + g^2(x)}, \quad x = 2$

74. Suppose that the functions f and g and their derivatives with respect to x have the following values at $x = 0$ and $x = 1$.

x	$f(x)$	$g(x)$	$f'(x)$	$g'(x)$
0	1	1	5	$1/3$
1	3	-4	$-1/3$	$-8/3$

Find the derivatives with respect to x of the following combinations at the given value of x,

a. $5f(x) - g(x), \quad x = 1$

b. $f(x)g^3(x), \quad x = 0$

c. $\frac{f(x)}{g(x) + 1}, \quad x = 1$

d. $f(g(x)), \quad x = 0$

e. $g(f(x)), \quad x = 0$

f. $(x^{11} + f(x))^{-2}, \quad x = 1$

g. $f(x + g(x)), \quad x = 0$

75. Find ds/dt when $\theta = 3\pi/2$ if $s = \cos\theta$ and $d\theta/dt = 5$.

76. Find dy/dt when $x = 1$ if $y = x^2 + 7x - 5$ and $dx/dt = 1/3$.

What happens if you can write a function as a composite in different ways? Do you get the same derivative each time? The Chain Rule says you should. Try it with the functions in Exercises 77 and 78.

77. Find dy/dx if $y = x$ by using the Chain Rule with y as a composite of

a. $y = (u/5) + 7 \quad \text{and} \quad u = 5x - 35$

b. $y = 1 + (1/u) \quad \text{and} \quad u = 1/(x - 1)$.

78. Find dy/dx if $y = x^{3/2}$ by using the Chain Rule with y as a composite of

a. $y = u^3 \quad \text{and} \quad u = \sqrt{x}$

b. $y = \sqrt{u} \quad \text{and} \quad u = x^3$.

79. a. Find the tangent to the curve $y = 2\tan(\pi x/4)$ at $x = 1$.

b. Slopes on a tangent curve What is the smallest value the slope of the curve can ever have on the interval $-2 < x < 2$? Give reasons for your answer.

80. Slopes on sine curves

a. Find equations for the tangents to the curves $y = \sin 2x$ and $y = -\sin(x/2)$ at the origin. Is there anything special about how the tangents are related? Give reasons for your answer.

b. Can anything be said about the tangents to the curves $y = \sin mx$ and $y = -\sin(x/m)$ at the origin (m a constant $\neq 0$)? Give reasons for your answer.

c. For a given m, what are the largest values the slopes of the curves $y = \sin mx$ and $y = -\sin(x/m)$ can ever have? Give reasons for your answer.

d. The function $y = \sin x$ completes one period on the interval $[0, 2\pi]$, the function $y = \sin 2x$ completes two periods, the function $y = \sin(x/2)$ completes half a period, and so on. Is there any relation between the number of periods $y = \sin mx$ completes on $[0, 2\pi]$ and the slope of the curve $y = \sin mx$ at the origin? Give reasons for your answer.

81. Particle motion The position of a particle moving along a coordinate line is $s = \sqrt{1 + 4t}$, with s in meters and t in seconds. Find the particle's velocity and acceleration at $t = 6$ sec.

82. Constant acceleration Suppose that the velocity of a falling body is $v = k\sqrt{s}$ m/sec (k a constant) at the instant the body has fallen s m from its starting point. Show that the body's acceleration is constant.

83. Falling meteorite The velocity of a heavy meteorite entering Earth's atmosphere is inversely proportional to $\sqrt{s}$ when it is s km from Earth's center. Show that the meteorite's acceleration is inversely proportional to s^2.

84. Particle acceleration A particle moves along the x-axis with velocity $dx/dt = f(x)$. Show that the particle's acceleration is $f(x)f'(x)$.

85. Chain Rule Suppose that $f(x) = x^2$ and $g(x) = |x|$. Then the composites

$$(f \circ g)(x) = |x|^2 = x^2 \quad \text{and} \quad (g \circ f)(x) = |x^2| = x^2$$

are both differentiable at $x = 0$ even though g itself is not differentiable at $x = 0$. Does this contradict the Chain Rule? Explain.

86. Tangents Suppose that $u = g(x)$ is differentiable at $x = 1$ and that $y = f(u)$ is differentiable at $u = g(1)$. If the graph of $y = f(g(x))$ has a horizontal tangent at $x = 1$, can we conclude anything about the tangent to the graph of g at $x = 1$ or the tangent to the graph of f at $u = g(1)$? Give reasons for your answer.

Using the Chain Rule, show that the Power Rule $(d/dx)x^n = nx^{n-1}$ holds for the functions x^n in Exercises 87 and 88.

87. $x^{1/4} = \sqrt{\sqrt{x}}$

88. $x^{3/4} = \sqrt{x\sqrt{x}}$

2.8 Implicit Differentiation

Most of the functions we have dealt with so far have been described by an equation of the form $y = f(x)$ that expresses y explicitly in terms of the variable x. We have learned rules for differentiating functions defined in this way. Another situation occurs when we encounter equations like

$$x^3 + y^3 - 9xy = 0, \qquad y^2 - x = 0, \qquad \text{or} \qquad x^2 + y^2 - 25 = 0.$$

(See Figures 2.29, 2.30, and 2.31.) These equations define an *implicit* relation between the variables x and y. In some cases we may be able to solve such an equation for y as an explicit function (or even several functions) of x. When we cannot put an equation $F(x, y) = 0$ in the form $y = f(x)$ to differentiate it in the usual way, we may still be able to find dy/dx by *implicit differentiation*. This section describes the technique.

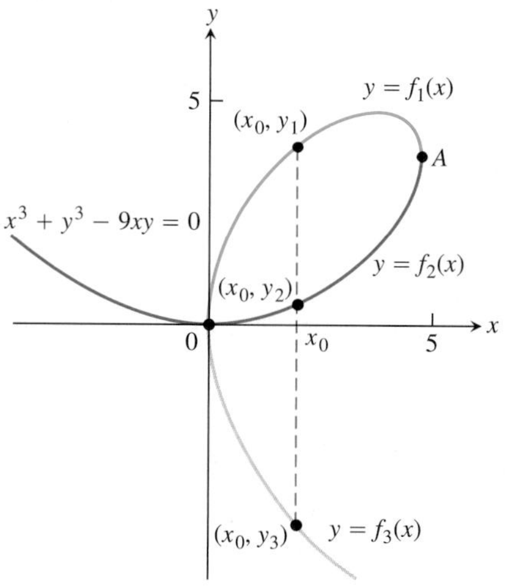

FIGURE 2.29 The curve $x^3 + y^3 - 9xy = 0$ is not the graph of any one function of x. The curve can, however, be divided into separate arcs that *are* the graphs of functions of x. This particular curve, called a *folium*, dates to Descartes in 1638.

Implicitly Defined Functions

We begin with examples involving familiar equations which we can solve for y as a function of x to calculate dy/dx in the usual way. Then we differentiate the equations implicitly, and find the derivative to provide a comparison of the two methods. Following the examples, we summarize the steps involved in the new method. In the examples and exercises, it is always assumed that the given equation determines y implicitly as a differentiable function of x so that dy/dx exists.

EXAMPLE 1 Find dy/dx if $y^2 = x$.

Solution The equation $y^2 = x$ defines two differentiable functions of x that we can actually find, namely $y_1 = \sqrt{x}$ and $y_2 = -\sqrt{x}$ (Figure 2.30). We know how to calculate the derivative of each of these for $x > 0$:

$$\frac{dy_1}{dx} = \frac{1}{2\sqrt{x}} \quad \text{and} \quad \frac{dy_2}{dx} = -\frac{1}{2\sqrt{x}}.$$

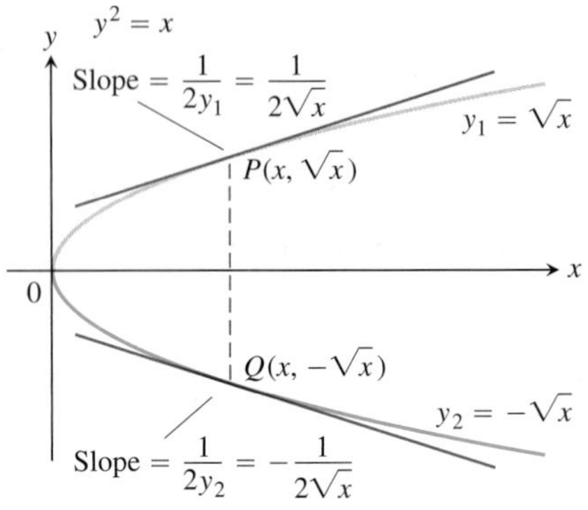

FIGURE 2.30 The equation $y^2 - x = 0$, or $y^2 = x$ as it is usually written, defines two differentiable functions of x on the interval $x \geq 0$. Example 1 shows how to find the derivatives of these functions without solving the equation $y^2 = x$ for y.

But suppose that we knew only that the equation $y^2 = x$ defined y as one or more differentiable functions of x for $x > 0$ without knowing exactly what these functions were. Could we still find dy/dx?

The answer is yes. To find dy/dx, we simply differentiate both sides of the equation $y^2 = x$ with respect to x, treating $y = f(x)$ as a differentiable function of x:

$$y^2 = x$$

$$2y\frac{dy}{dx} = 1 \qquad \text{The Chain Rule gives } \frac{d}{dx}(y^2) = \frac{d}{dx}[f(x)]^2 = 2f(x)f'(x) = 2y\frac{dy}{dx}.$$

$$\frac{dy}{dx} = \frac{1}{2y}.$$

This one formula gives the derivatives we calculated for *both* explicit solutions $y_1 = \sqrt{x}$ and $y_2 = -\sqrt{x}$:

$$\frac{dy_1}{dx} = \frac{1}{2y_1} = \frac{1}{2\sqrt{x}} \quad \text{and} \quad \frac{dy_2}{dx} = \frac{1}{2y_2} = \frac{1}{2\left(-\sqrt{x}\right)} = -\frac{1}{2\sqrt{x}}.$$ ■

EXAMPLE 2 Find the slope of the circle $x^2 + y^2 = 25$ at the point $(3, -4)$.

Solution The circle is not the graph of a single function of x. Rather it is the combined graphs of two differentiable functions, $y_1 = \sqrt{25 - x^2}$ and $y_2 = -\sqrt{25 - x^2}$ (Figure 2.31). The point $(3, -4)$ lies on the graph of y_2, so we can find the slope by calculating explicitly:

$$\left.\frac{dy_2}{dx}\right|_{x=3} = -\left.\frac{-2x}{2\sqrt{25 - x^2}}\right|_{x=3} = -\frac{-6}{2\sqrt{25 - 9}} = \frac{3}{4}.$$

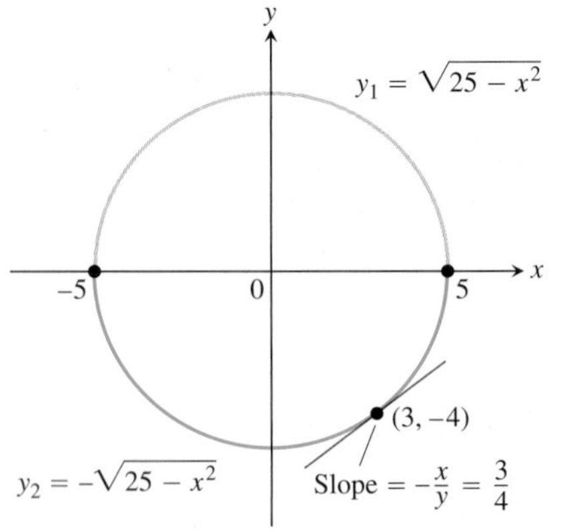

FIGURE 2.31 The circle combines the graphs of two functions. The graph of y_2 is the lower semicircle and passes through $(3, -4)$.

We can solve this problem more easily by differentiating the given equation of the circle implicitly with respect to x:

$$\frac{d}{dx}(x^2) + \frac{d}{dx}(y^2) = \frac{d}{dx}(25)$$

$$2x + 2y\frac{dy}{dx} = 0$$

$$\frac{dy}{dx} = -\frac{x}{y}.$$

The slope at $(3, -4)$ is $\left.-\frac{x}{y}\right|_{(3,-4)} = -\frac{3}{-4} = \frac{3}{4}.$

Notice that unlike the slope formula for dy_2/dx, which applies only to points below the x-axis, the formula $dy/dx = -x/y$ applies everywhere the circle has a slope. Notice also that the derivative involves *both* variables x and y, not just the independent variable x. ■

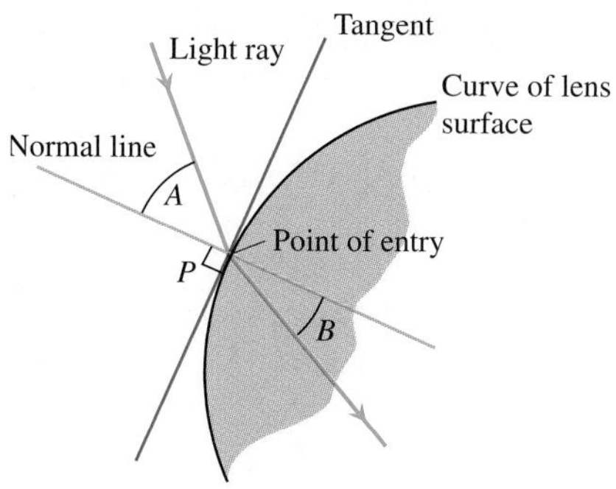

FIGURE 2.32 The profile of a lens, showing the bending (refraction) of a ray of light as it passes through the lens surface.

Implicit Differentiation

1. Differentiate both sides of the equation with respect to x, treating y as a differentiable function of x.
2. Collect the terms with dy/dx on one side of the equation.
3. Solve for dy/dx.

Lenses, Tangents, and Normal Lines

In the law that describes how light changes direction as it enters a lens, the important angles are the angles the light makes with the line perpendicular to the surface of the lens at the point of entry (angles A and B in Figure 2.32). This line is called the *normal* to the surface at the point of entry. In a profile view of a lens like the one in Figure 2.32, the **normal** is the line perpendicular to the tangent of the profile curve at the point of entry.

EXAMPLE 3 Show that the point (2, 4) lies on the curve $x^3 + y^3 - 9xy = 0$. Then find the tangent and normal to the curve there (Figure 2.33).

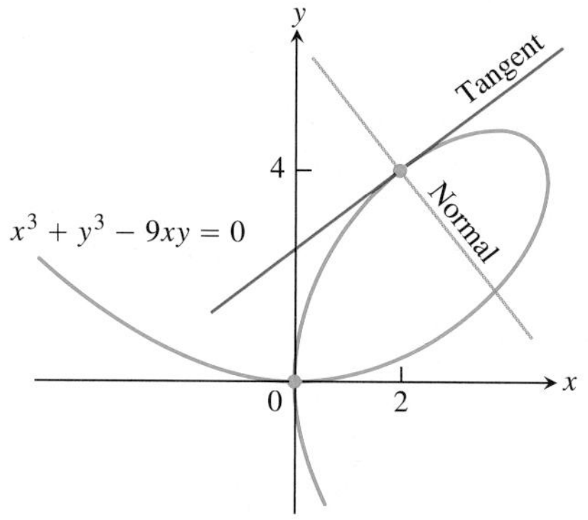

FIGURE 2.33 Example 3 shows how to find equations for the tangent and normal to the folium of Descartes at (2, 4).

Solution The point (2, 4) lies on the curve because its coordinates satisfy the equation given for the curve: $2^3 + 4^3 - 9(2)(4) = 8 + 64 - 72 = 0$.

To find the slope of the curve at (2, 4), we first use implicit differentiation to find a formula for dy/dx:

$$x^3 + y^3 - 9xy = 0$$

$$\frac{d}{dx}(x^3) + \frac{d}{dx}(y^3) - \frac{d}{dx}(9xy) = \frac{d}{dx}(0)$$

$$3x^2 + 3y^2\frac{dy}{dx} - 9\left(x\frac{dy}{dx} + y\frac{dx}{dx}\right) = 0$$

Differentiate both sides with respect to x.

$$(3y^2 - 9x)\frac{dy}{dx} + 3x^2 - 9y = 0$$

Treat xy as a product and y as a function of x.

$$3(y^2 - 3x)\frac{dy}{dx} = 9y - 3x^2$$

$$\frac{dy}{dx} = \frac{3y - x^2}{y^2 - 3x}.$$

Solve for dy/dx.

We then evaluate the derivative at $(x, y) = (2, 4)$:

$$\left.\frac{dy}{dx}\right|_{(2,4)} = \left.\frac{3y - x^2}{y^2 - 3x}\right|_{(2,4)} = \frac{3(4) - 2^2}{4^2 - 3(2)} = \frac{8}{10} = \frac{4}{5}.$$

The tangent at (2, 4) is the line through (2, 4) with slope 4/5:

$$y = 4 + \frac{4}{5}(x - 2)$$

$$y = \frac{4}{5}x + \frac{12}{5}.$$

The normal to the curve at (2, 4) is the line perpendicular to the tangent there, the line through (2, 4) with slope $-5/4$:

$$y = 4 - \frac{5}{4}(x - 2)$$

$$y = -\frac{5}{4}x + \frac{13}{2}. \qquad \blacksquare$$

Derivatives of Higher Order

Implicit differentiation can also be used to find higher derivatives. Here is an example.

EXAMPLE 4 Find d^2y/dx^2 if $2x^3 - 3y^2 = 8$.

Solution To start, we differentiate both sides of the equation with respect to x in order to find $y' = dy/dx$.

$$\frac{d}{dx}(2x^3 - 3y^2) = \frac{d}{dx}(8)$$

$$6x^2 - 6yy' = 0 \qquad \text{Treat } y \text{ as a function of } x.$$

$$y' = \frac{x^2}{y}, \quad \text{when } y \neq 0 \qquad \text{Solve for } y'.$$

We now apply the Quotient Rule to find y''.

$$y'' = \frac{d}{dx}\left(\frac{x^2}{y}\right) = \frac{2xy - x^2y'}{y^2} = \frac{2x}{y} - \frac{x^2}{y^2}\cdot y'$$

Finally, we substitute $y' = x^2/y$ to express y'' in terms of x and y.

$$y'' = \frac{2x}{y} - \frac{x^2}{y^2}\left(\frac{x^2}{y}\right) = \frac{2x}{y} - \frac{x^4}{y^3}, \quad \text{when } y \neq 0 \qquad \blacksquare$$

EXERCISE 2.8

Use implicit differentiation to find dy/dx in Exercises 1–14.

1. $x^2y + xy^2 = 6$
2. $x^3 + y^3 = 18xy$
3. $2xy + y^2 = x + y$
4. $x^3 - xy + y^3 = 1$
5. $x^2(x - y)^2 = x^2 - y^2$
6. $(3xy + 7)^2 = 6y$
7. $y^2 = \dfrac{x - 1}{x + 1}$
8. $x^2 = \dfrac{x - y}{x + y}$
9. $x = \tan y$
10. $xy = \cot(xy)$
11. $e^{2x} = \sin(x + 3y)$
12. $x + \sin y = xy$
13. $y \sin\left(\dfrac{1}{y}\right) = 1 - xy$
14. $e^{x^2y} = 2x + 2y$

Find $dr/d\theta$ in Exercises 15–18.

15. $\theta^{1/2} + r^{1/2} = 1$
16. $r - 2\sqrt{\theta} = \dfrac{3}{2}\theta^{2/3} + \dfrac{4}{3}\theta^{3/4}$
17. $\sin(r\theta) = \dfrac{1}{2}$
18. $\cos r + \cot\theta = e^{r\theta}$

In Exercises 19–26, use implicit differentiation to find dy/dx and then d^2y/dx^2.

19. $x^2 + y^2 = 1$
20. $x^{2/3} + y^{2/3} = 1$
21. $y^2 = e^{x^2} + 2x$
22. $y^2 - 2x = 1 - 2y$
23. $2\sqrt{y} = x - y$
24. $xy + y^2 = 1$

25. If $x^3 + y^3 = 16$, find the value of d^2y/dx^2 at the point (2, 2).

26. If $xy + y^2 = 1$, find the value of d^2y/dx^2 at the point $(0, -1)$.

In Exercises 27 and 28, find the slope of the curve at the given points.

27. $y^2 + x^2 = y^4 - 2x$ at $(-2, 1)$ and $(-2, -1)$

28. $(x^2 + y^2)^2 = (x - y)^2$ at $(1, 0)$ and $(1, -1)$

In Exercises 29–38, verify that the given point is on the curve and find the lines that are **(a)** tangent and **(b)** normal to the curve at the given point.

29. $x^2 + xy - y^2 = 1$, $(2, 3)$

30. $x^2 + y^2 = 25$, $(3, -4)$

31. $x^2y^2 = 9$, $(-1, 3)$

32. $y^2 - 2x - 4y - 1 = 0$, $(-2, 1)$

33. $6x^2 + 3xy + 2y^2 + 17y - 6 = 0$, $(-1, 0)$

34. $x^2 - \sqrt{3}xy + 2y^2 = 5$, $(\sqrt{3}, 2)$

35. $2xy + \pi \sin y = 2\pi$, $(1, \pi/2)$

36. $x \sin 2y = y \cos 2x$, $(\pi/4, \pi/2)$

37. $y = 2 \sin (\pi x - y)$, $(1, 0)$

38. $x^2 \cos^2 y - \sin y = 0$, $(0, \pi)$

39. Parallel tangents Find the two points where the curve $x^2 + xy + y^2 = 7$ crosses the x-axis, and show that the tangents to the curve at these points are parallel. What is the common slope of these tangents?

40. Power rule for rational exponents Let p and q be integers with $q > 0$. If $y = x^{p/q}$, differentiate the equivalent equation $y^q = x^p$ implicitly and show that, for $y \neq 0$,

$$\frac{d}{dx}x^{p/q} = \frac{p}{q}x^{(p/q)-1}.$$

41. The eight curve Find the slopes of the curve $y^4 = y^2 - x^2$ at the two points shown here.

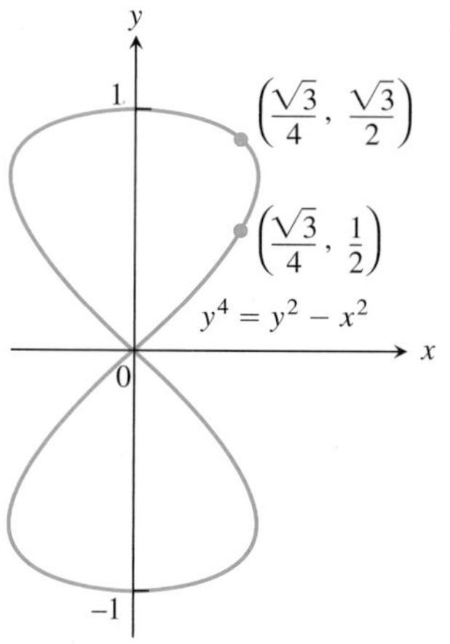

42. The folium of Descartes (See Figure 2.29.)

a. Find the slope of the folium of Descartes $x^3 + y^3 - 9xy = 0$ at the points (4, 2) and (2, 4).

b. At what point other than the origin does the folium have a horizontal tangent?

c. Find the coordinates of the point A in Figure 2.29, where the folium has a vertical tangent.

43. Intersecting normal The line that is normal to the curve $x^2 + 2xy - 3y^2 = 0$ at (1, 1) intersects the curve at what other point?

44. Normals parallel to a line Find the normals to the curve $xy + 2x - y = 0$ that are parallel to the line $2x + y = 0$.

T In Exercises 45 and 46, find both dy/dx (treating y as a differentiable function of x) and dx/dy (treating x as a differentiable function of y). How do dy/dx and dx/dy seem to be related? Explain the relationship geometrically in terms of the graphs.

45. $xy^3 + x^2y = 6$

46. $x^3 + y^2 = \sin^2 y$

COMPUTER EXPLORATIONS

Use a CAS to perform the following steps in Exercises 47–54.

a. Plot the equation with the implicit plotter of a CAS. Check to see that the given point P satisfies the equation.

b. Using implicit differentiation, find a formula for the derivative dy/dx and evaluate it at the given point P.

c. Use the slope found in part (b) to find an equation for the tangent line to the curve at P. Then plot the implicit curve and tangent line together on a single graph.

47. $x^3 - xy + y^3 = 7$, $P(2, 1)$

48. $x^5 + y^3x + yx^2 + y^4 = 4$, $P(1, 1)$

49. $y^2 + y = \dfrac{2 + x}{1 - x}$, $P(0, 1)$

50. $y^3 + \cos xy = x^2$, $P(1, 0)$

51. $x + \tan\left(\dfrac{y}{x}\right) = 2$, $P\left(1, \dfrac{\pi}{4}\right)$

52. $xy^3 + \tan(x + y) = 1$, $P\left(\dfrac{\pi}{4}, 0\right)$

53. $2y^2 + (xy)^{1/3} = x^2 + 2$, $P(1, 1)$

54. $x\sqrt{1 + 2y} + y = x^2$, $P(1, 0)$

2.9 Inverse Functions and Their Derivatives

A function that undoes, or inverts, the effect of a function f is called the *inverse* of f. Many common functions, though not all, are paired with an inverse. Important inverse functions often show up in applications. Inverse functions also play a key role in the development and properties of the logarithmic functions studied in the next section.

One-to-One Functions

A function is a rule that assigns a value from its range to each element in its domain. Some functions assign the same range value to more than one element in the domain. The function $f(x) = x^2$ assigns the same value, 1, to both of the numbers -1 and $+1$; the sines of $\pi/3$ and $2\pi/3$ are both $\sqrt{3}/2$. Other functions assume each value in their range no more than once. The square roots and cubes of different numbers are always different. A function that has distinct values at distinct elements in its domain is called one-to-one. These functions take on any one value in their range exactly once.

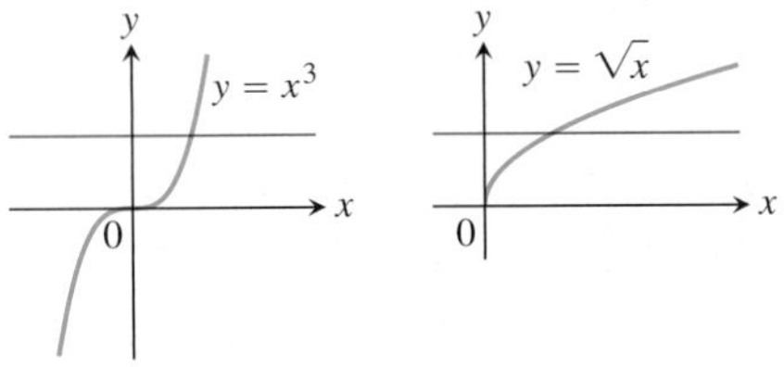

(a) One-to-one: Graph meets each horizontal line at most once.

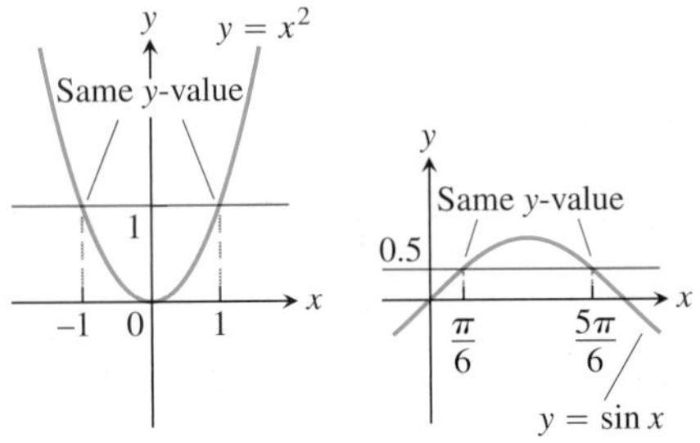

(b) Not one-to-one: Graph meets one or more horizontal lines more than once.

FIGURE 2.34 (a) $y = x^3$ and $y = \sqrt{x}$ are one-to-one on their domains $(-\infty, \infty)$ and $[0, \infty)$. (b) $y = x^2$ and $y = \sin x$ are not one-to-one on their domains $(-\infty, \infty)$.

DEFINITION A function $f(x)$ is **one-to-one** on a domain D if $f(x_1) \neq f(x_2)$ whenever $x_1 \neq x_2$ in D.

EXAMPLE 1

(a) $f(x) = \sqrt{x}$ is one-to-one on any domain of nonnegative numbers because $\sqrt{x_1} \neq \sqrt{x_2}$ whenever $x_1 \neq x_2$.

(b) $g(x) = \sin x$ is *not* one-to-one on the interval $[0, \pi]$ because $\sin(\pi/6) = \sin(5\pi/6)$. The sine function *is* one-to-one on $[0, \pi/2]$ however, because it is an increasing function on $[0, \pi/2]$ giving distinct outputs for distinct inputs. ■

The graph of a one-to-one function $y = f(x)$ can intersect a given horizontal line at most once. If it intersects the line more than once, it assumes the same y-value for at least two different x-values and is therefore not one-to-one (Figure 2.34).

The Horizontal Line Test for One-to-One Functions

A function $y = f(x)$ is one-to-one if and only if its graph intersects each horizontal line at most once.

Inverse Functions

Since each output of a one-to-one function comes from just one input, the effect of the function can be inverted to send an output back to the input from which it came.

DEFINITION Suppose that f is a one-to-one function on a domain D with range R. The **inverse function** f^{-1} is defined by

$$f^{-1}(a) = b \text{ if } f(b) = a.$$

The domain of f^{-1} is R and the range of f^{-1} is D.

The symbol f^{-1} for the inverse of f is read "f inverse." The "-1" in f^{-1} is *not* an exponent; $f^{-1}(x)$ does not mean $1/f(x)$. Notice that the domains and ranges of f and f^{-1} are interchanged.

If we apply f to send an input x to the output $f(x)$ and follow by applying f^{-1} to $f(x)$ we get right back to x, just where we started. Similarly, if we take some number y in the range of f, apply f^{-1} to it, and then apply f to the resulting value $f^{-1}(y)$, we get back the value y with which we began. Composing a function and its inverse has the same effect as doing nothing.

$$(f^{-1} \circ f)(x) = x, \qquad \text{for all } x \text{ in the domain of } f$$

$$(f \circ f^{-1})(y) = y, \qquad \text{for all } y \text{ in the domain of } f^{-1} \text{ (or range of } f)$$

Only a one-to-one function can have an inverse. The reason is that if $f(x_1) = y$ and $f(x_2) = y$ for two distinct inputs x_1 and x_2, then there is no way to assign a value to $f^{-1}(y)$ that satisfies both $f^{-1}(f(x_1)) = x_1$ and $f^{-1}(f(x_2)) = x_2$.

A function that is increasing on an interval so it satisfies the inequality $f(x_2) > f(x_1)$ when $x_2 > x_1$, is one-to-one and has an inverse. Decreasing functions also have an inverse. Functions that are neither increasing nor decreasing may still be one-to-one and have an inverse, as with the function $f(x) = 1/x$ for $x \neq 0$ and $f(0) = 0$, defined on $(-\infty, \infty)$ and passing the horizontal line test.

Finding Inverses

The graphs of a function and its inverse are closely related. To read the value of a function from its graph, we start at a point x on the x-axis, go vertically to the graph, and then move horizontally to the y-axis to read the value of y. The inverse function can be read from the graph by reversing this process. Start with a point y on the y-axis, go horizontally to the graph, and then move vertically to the x-axis to read the value of $x = f^{-1}(y)$ (Figure 2.35).

We want to set up the graph of f^{-1} so that its input values lie along the x-axis, as is usually done for functions, rather than on the y-axis. To achieve this we interchange the x and y axes by reflecting across the 45° line $y = x$. After this reflection we have a new

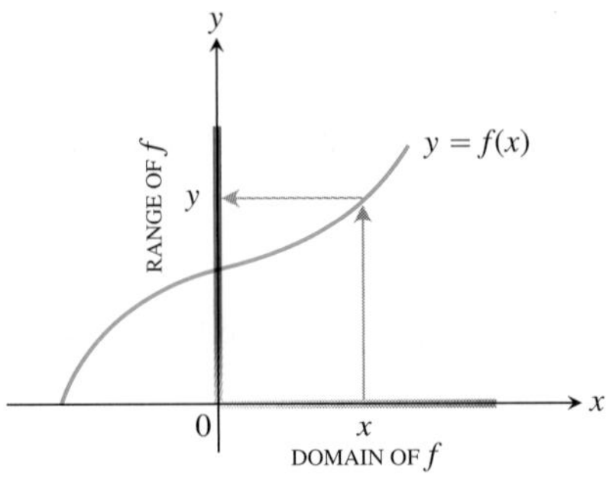

(a) To find the value of f at x, we start at x, go up to the curve, and then over to the y-axis.

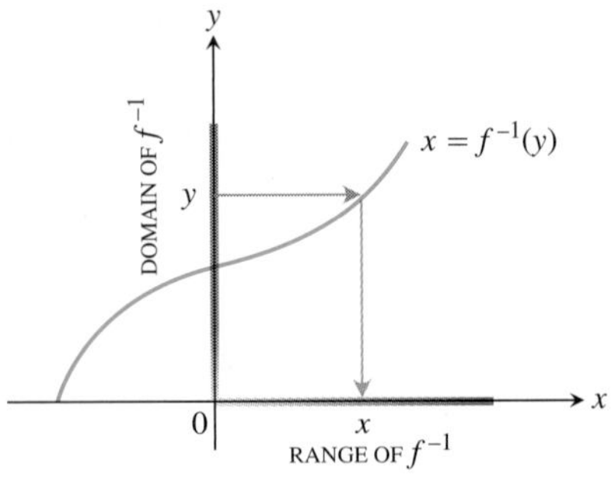

(b) The graph of f is already the graph of f^{-1}, but with x and y interchanged. To find the x that gave y, we start at y and go over to the curve and down to the x-axis. The domain of f^{-1} is the range of f. The range of f^{-1} is the domain of f.

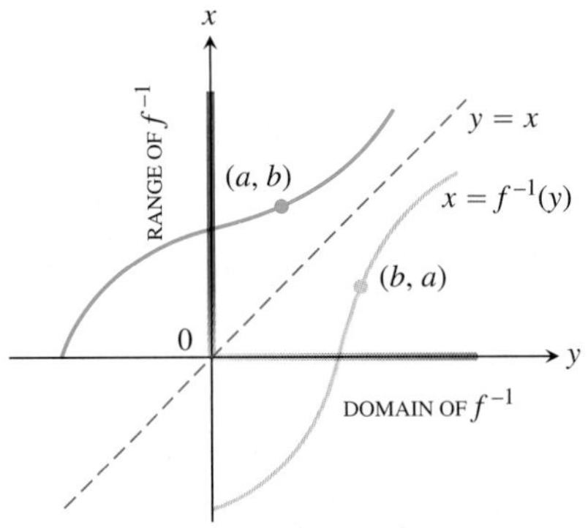

(c) To draw the graph of f^{-1} in the more usual way, we reflect the system across the line $y = x$.

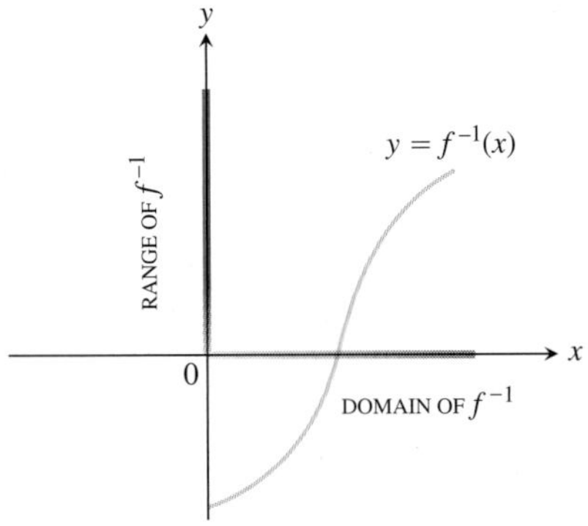

(d) Then we interchange the letters x and y. We now have a normal-looking graph of f^{-1} as a function of x.

FIGURE 2.35 Determining the graph of $y = f^{-1}(x)$ from the graph of $y = f(x)$. The graph of f^{-1} is obtained by reflecting the graph of f about the line $y = x$.

graph that represents f^{-1}. The value of $f^{-1}(x)$ can now be read from the graph in the usual way, by starting with a point x on the x-axis, going vertically to the graph, and then horizontally to the y-axis to get the value of $f^{-1}(x)$. Figure 2.35 indicates the relation between the graphs of f and f^{-1}. The graphs are interchanged by reflection through the line $y = x$. Although this geometric treatment is not a proof, it does make reasonable the fact that the inverse of a continuous function is also continuous.

The process of passing from f to f^{-1} can be summarized as a two-step process.

1. Solve the equation $y = f(x)$ for x. This gives a formula $x = f^{-1}(y)$ where x is expressed as a function of y.
2. Interchange x and y, obtaining a formula $y = f^{-1}(x)$ where f^{-1} is expressed in the conventional format with x as the independent variable and y as the dependent variable.

EXAMPLE 2 Find the inverse of $y = \frac{1}{2}x + 1$, expressed as a function of x.

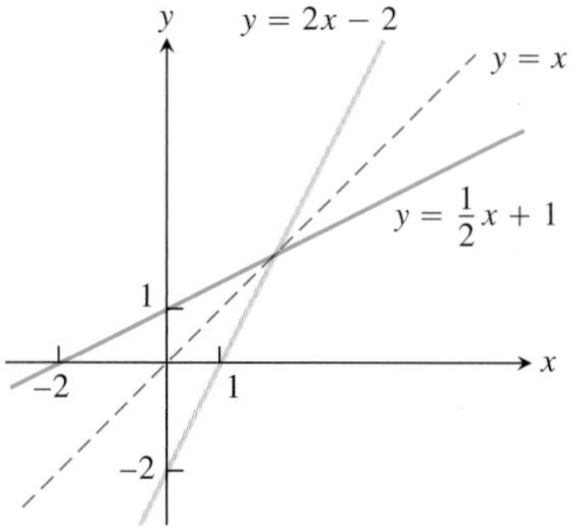

FIGURE 2.36 Graphing $f(x) = (1/2)x + 1$ and $f^{-1}(x) = 2x - 2$ together shows the graphs' symmetry with respect to the line $y = x$ (Example 2).

Solution

1. *Solve for x in terms of y*:
$$y = \frac{1}{2}x + 1$$
$$2y = x + 2$$
$$x = 2y - 2.$$

2. *Interchange x and y*: $y = 2x - 2.$

The inverse of the function $f(x) = (1/2)x + 1$ is the function $f^{-1}(x) = 2x - 2$. To check, we verify that both composites give the identity function:

$$f^{-1}(f(x)) = 2\left(\frac{1}{2}x + 1\right) - 2 = x + 2 - 2 = x$$

$$f(f^{-1}(x)) = \frac{1}{2}(2x - 2) + 1 = x - 1 + 1 = x.$$

See Figure 2.36. ■

EXAMPLE 3 Find the inverse of the function $y = x^2, x \geq 0$, expressed as a function of x.

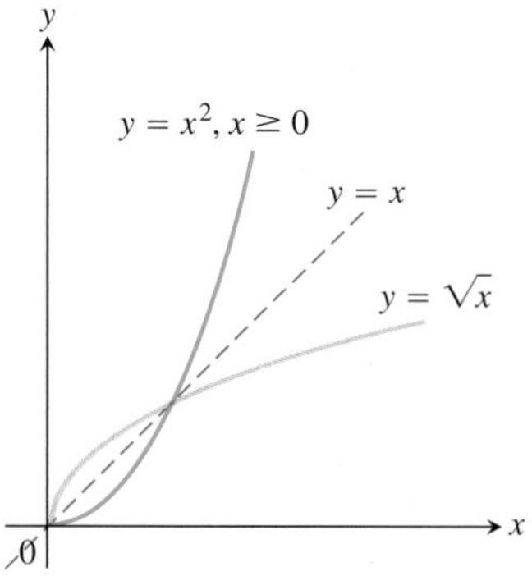

FIGURE 2.37 The functions $y = \sqrt{x}$ and $y = x^2, x \geq 0$, are inverses of one another (Example 3).

Solution We first solve for x in terms of y:

$$y = x^2$$
$$\sqrt{y} = \sqrt{x^2} = |x| = x \qquad |x| = x \text{ because } x \geq 0$$

We then interchange x and y, obtaining

$$y = \sqrt{x}.$$

The inverse of the function $y = x^2, x \geq 0$, is the function $y = \sqrt{x}$ (Figure 2.37).

Notice that the function $y = x^2, x \geq 0$, with domain *restricted* to the nonnegative real numbers, *is* one-to-one (Figure 2.37) and has an inverse. On the other hand, the function $y = x^2$, with no domain restrictions, *is not* one-to-one (Figure 2.34) and therefore has no inverse. ■

Derivatives of Inverses of Differentiable Functions

If we calculate the derivatives of $f(x) = (1/2)x + 1$ and its inverse $f^{-1}(x) = 2x - 2$ from Example 2, we see that

$$\frac{d}{dx}f(x) = \frac{d}{dx}\left(\frac{1}{2}x + 1\right) = \frac{1}{2}$$

$$\frac{d}{dx}f^{-1}(x) = \frac{d}{dx}(2x - 2) = 2.$$

The derivatives are reciprocals of one another, so the slope of one line is the reciprocal of the slope of its inverse line. (See Figure 2.36.)

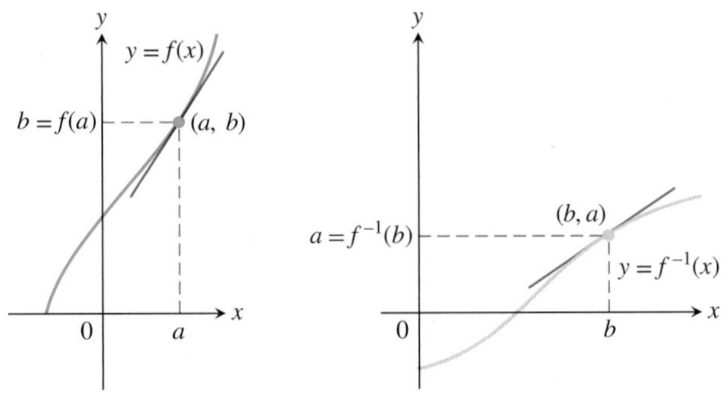

The slopes are reciprocal: $(f^{-1})'(b) = \dfrac{1}{f'(a)}$ or $(f^{-1})'(b) = \dfrac{1}{f'(f^{-1}(b))}$

FIGURE 2.38 The graphs of inverse functions have reciprocal slopes at corresponding points.

This is not a special case. Reflecting any nonhorizontal or nonvertical line across the line $y = x$ always inverts the line's slope. If the original line has slope $m \neq 0$, the reflected line has slope $1/m$.

The reciprocal relationship between the slopes of f and f^{-1} holds for other functions as well, but we must be careful to compare slopes at corresponding points. If the slope of $y = f(x)$ at the point $(a, f(a))$ is $f'(a)$ and $f'(a) \neq 0$, then the slope of $y = f^{-1}(x)$ at the point $(f(a), a)$ is the reciprocal $1/f'(a)$ (Figure 2.38). If we set $b = f(a)$, then

$$(f^{-1})'(b) = \frac{1}{f'(a)} = \frac{1}{f'(f^{-1}(b))}.$$

If $y = f(x)$ has a horizontal tangent line at $(a, f(a))$ then the inverse function f^{-1} has a vertical tangent line at $(f(a), a)$, and this infinite slope implies that f^{-1} is not differentiable at $f(a)$. Theorem 4 gives the conditions under which f^{-1} is differentiable in its domain (which is the same as the range of f).

THEOREM 4—The Derivative Rule for Inverses If f has an interval I as domain and $f'(x)$ exists and is never zero on I, then f^{-1} is differentiable at every point in its domain (the range of f). The value of $(f^{-1})'$ at a point b in the domain of f^{-1} is the reciprocal of the value of f' at the point $a = f^{-1}(b)$:

$$(f^{-1})'(b) = \frac{1}{f'(f^{-1}(b))} \tag{1}$$

or

$$\left.\frac{df^{-1}}{dx}\right|_{x=b} = \frac{1}{\left.\dfrac{df}{dx}\right|_{x=f^{-1}(b)}}$$

The main importance of Theorem 4 is its asssertion of the existence of the derivative of the inverse function. A formula for the derivative can then be obtained directly using Equation (1) or by the process of implicit differentiation.

The proof of Theorem 4 is omitted. Evidence for the result can be seen in the following way. When $y = f(x)$ is differentiable at $x = a$ and we change x by a small amount Δx, the corresponding change in y is approximately

$$\Delta y \approx f'(a)\,\Delta x. \qquad \frac{\Delta y}{\Delta x} \approx f'(a)$$

This means that y changes about $f'(a)$ times as fast as x when $x = a$ and that x changes about $1/f'(a)$ times as fast as y when $y = b$. Since $a = f^{-1}(b)$, we are led to Equation (1): The derivative of f^{-1} at b is the reciprocal of the derivative of f at a.

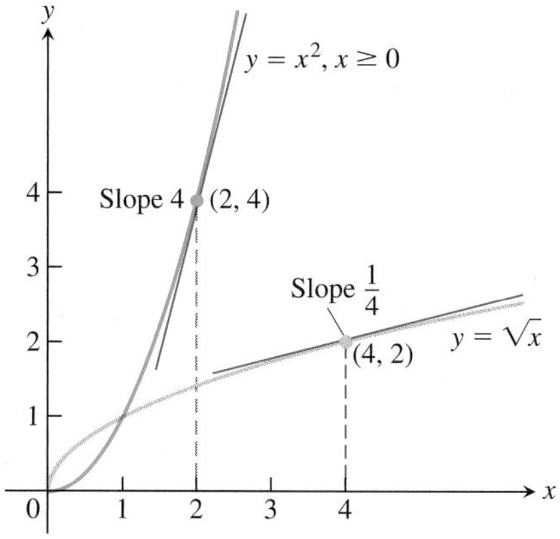

FIGURE 2.39 The derivative of $f^{-1}(x) = \sqrt{x}$ at the point (4, 2) is the reciprocal of the derivative of $f(x) = x^2$ at (2, 4) (Example 4).

EXAMPLE 4 The function $f(x) = x^2, x \geq 0$ and its inverse $f^{-1}(x) = \sqrt{x}$ have derivatives $f'(x) = 2x$ and $(f^{-1})'(x) = 1/(2\sqrt{x})$.

Theorem 4 predicts that the derivative of $f^{-1}(x)$ is

$$\begin{aligned}(f^{-1})'(x) &= \frac{1}{f'(f^{-1}(x))}\\ &= \frac{1}{2(f^{-1}(x))} && f'(x) = 2x \text{ with } x \text{ replaced by } f^{-1}(x)\\ &= \frac{1}{2(\sqrt{x})}.\end{aligned}$$

Theorem 4 gives a derivative that agrees with our calculation using the Power Rule for the derivative of the square root function.

Let's examine Theorem 4 at a specific point. We pick $x = 2$ (the number a) and $f(2) = 4$ (the value b). Theorem 4 says that the derivative of f at 2, $f'(2) = 4$, and the derivative of f^{-1} at $f(2)$, $(f^{-1})'(4)$, are reciprocals. It states that

$$(f^{-1})'(4) = \frac{1}{f'(f^{-1}(4))} = \frac{1}{f'(2)} = \left.\frac{1}{2x}\right|_{x=2} = \frac{1}{4}.$$

See Figure 2.39. ■

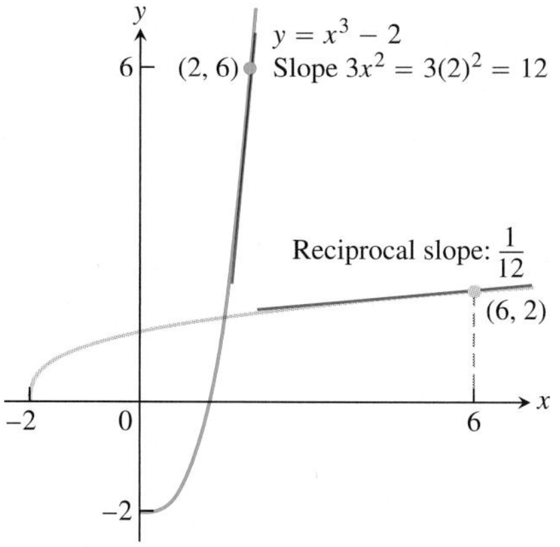

FIGURE 2.40 The derivative of $f(x) = x^3 - 2$ at $x = 2$ tells us the derivative of f^{-1} at $x = 6$ (Example 5).

Equation (1) sometimes enables us to find specific values of df^{-1}/dx without knowing a formula for f^{-1}.

EXAMPLE 5 Let $f(x) = x^3 - 2$. Find the value of df^{-1}/dx at $x = 6 = f(2)$ without finding a formula for $f^{-1}(x)$.

Solution

$$\left.\frac{df}{dx}\right|_{x=2} = \left.3x^2\right|_{x=2} = 12$$

$$\left.\frac{df^{-1}}{dx}\right|_{x=f(2)} = \frac{1}{\left.\dfrac{df}{dx}\right|_{x=2}} = \frac{1}{12} \qquad \text{Eq. (1)}$$

See Figure 2.40. ■

EXERCISES 2.9

Which of the functions graphed in Exercises 1–4 are one-to-one, and which are not?

1.

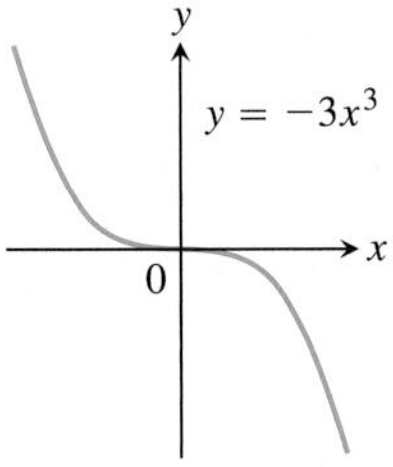

2.

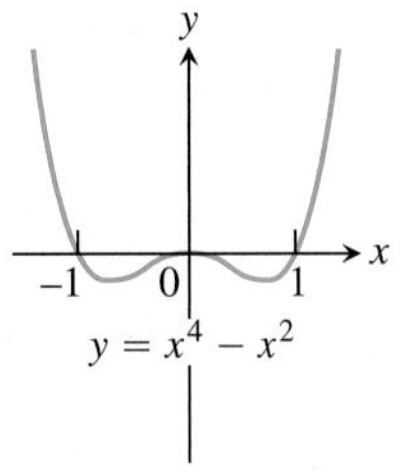

3.

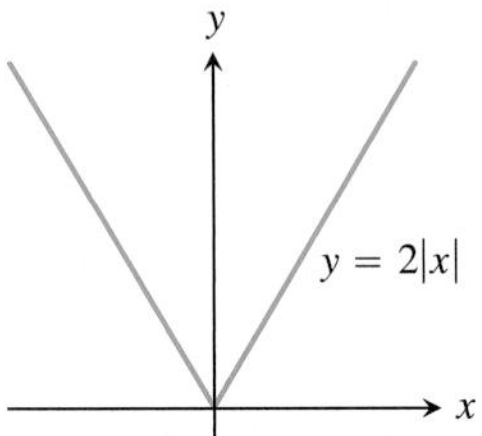

4.

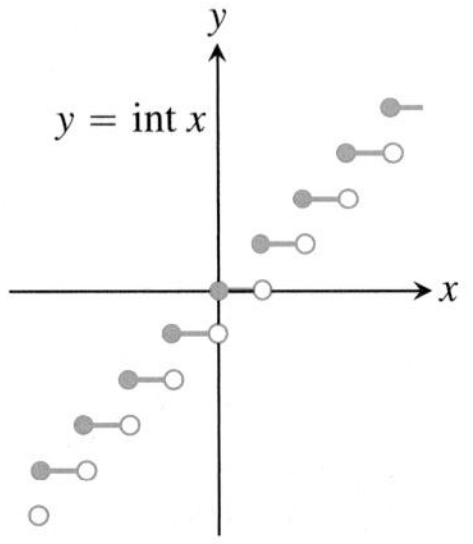

Each of Exercises 5 and 6 shows the graph of a function $y = f(x)$. Copy the graph and draw in the line $y = x$. Then use symmetry with respect to the line $y = x$ to add the graph of f^{-1} to your sketch. (It is not necessary to find a formula for f^{-1}.) Identify the domain and range of f^{-1}.

5.

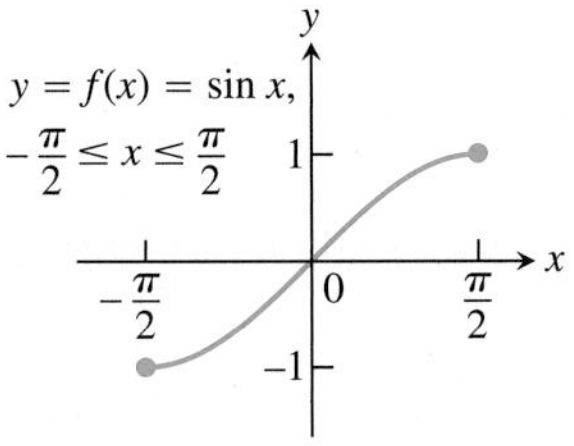

6.

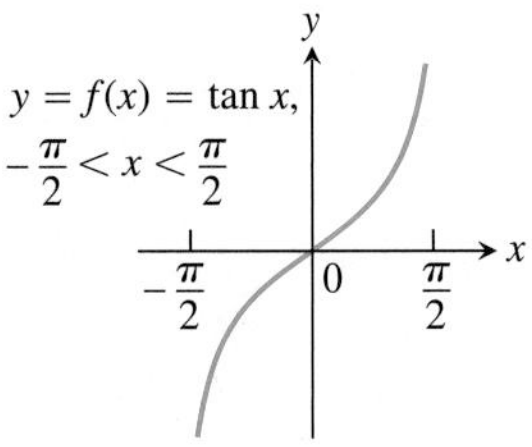

Each of Exercises 7–12 gives a formula for a function $y = f(x)$ and shows the graphs of f and f^{-1}. Find a formula for f^{-1} in each case.

7. $f(x) = x^2 + 1, \quad x \geq 0$

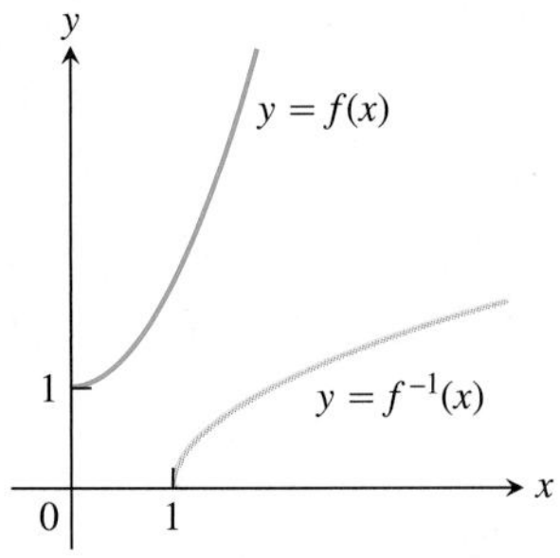

8. $f(x) = x^2, \quad x \leq 0$

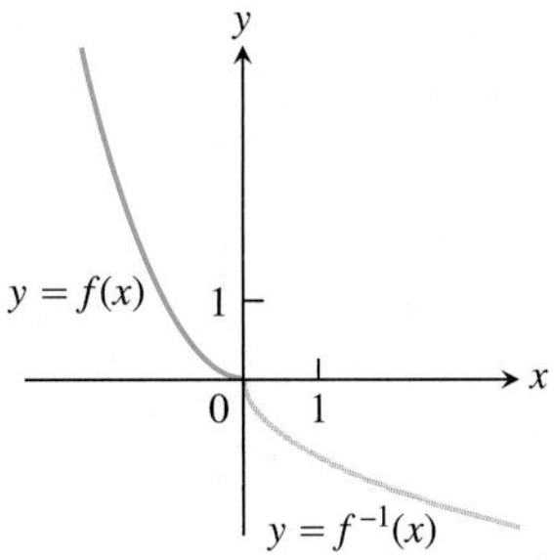

9. $f(x) = x^3 - 1$

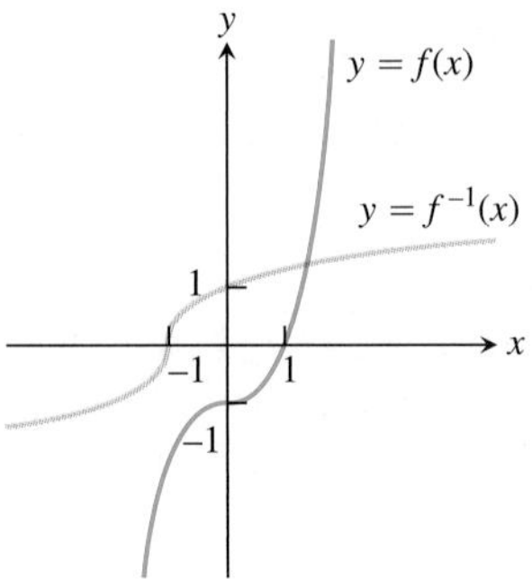

10. $f(x) = x^2 - 2x + 1, \quad x \geq 1$

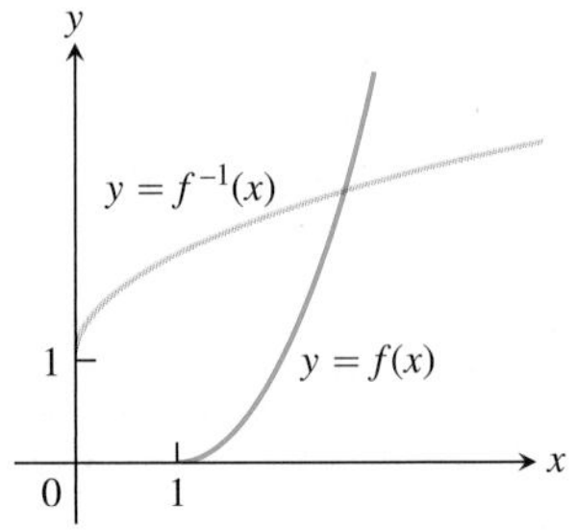

11. $f(x) = (x + 1)^2, \quad x \geq -1$

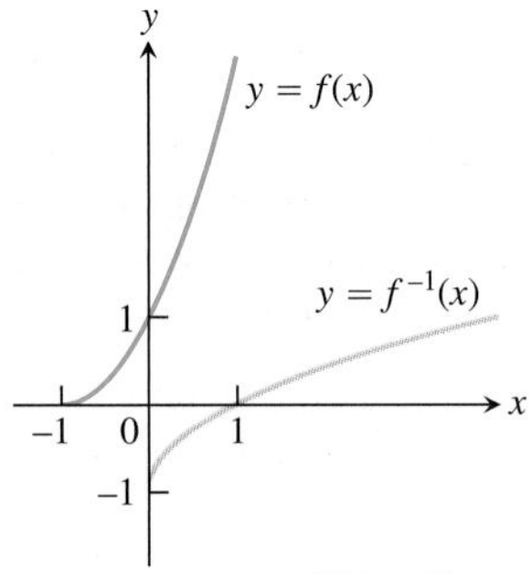

12. $f(x) = x^{2/3}, \quad x \geq 0$

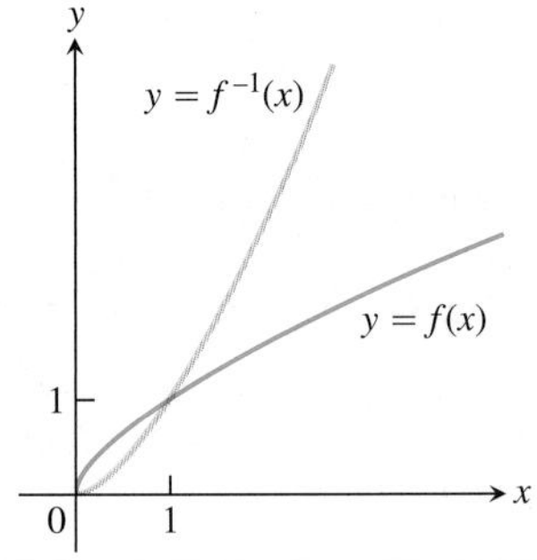

Each of Exercises 13–16 gives a formula for a function $y = f(x)$. In each case, find $f^{-1}(x)$ and identify the domain and range of f^{-1}. As a check, show that $f(f^{-1}(x)) = f^{-1}(f(x)) = x$.

13. $f(x) = x^5$

14. $f(x) = x^3 + 1$

15. $f(x) = 1/x^2, \quad x > 0$

16. $f(x) = 1/x^3, \quad x \neq 0$

In Exercises 17–20:

a. Find $f^{-1}(x)$.

b. Graph f and f^{-1} together.

c. Evaluate df/dx at $x = a$ and df^{-1}/dx at $x = f(a)$ to show that at these points $df^{-1}/dx = 1/(df/dx)$.

17. $f(x) = 2x + 3, \quad a = -1$

18. $f(x) = (1/5)x + 7, \quad a = -1$

19. $f(x) = 5 - 4x, \quad a = 1/2$

20. $f(x) = 2x^2, \quad x \geq 0, \quad a = 5$

21. **a.** Show that $f(x) = x^3$ and $g(x) = \sqrt[3]{x}$ are inverses of one another.

b. Graph f and g over an x-interval large enough to show the graphs intersecting at $(1, 1)$ and $(-1, -1)$. Be sure the picture shows the required symmetry about the line $y = x$.

c. Find the slopes of the tangents to the graphs of f and g at $(1, 1)$ and $(-1, -1)$ (four tangents in all).

d. What lines are tangent to the curves at the origin?

22. a. Show that $h(x) = x^3/4$ and $k(x) = (4x)^{1/3}$ are inverses of one another.

b. Graph h and k over an x-interval large enough to show the graphs intersecting at $(2, 2)$ and $(-2, -2)$. Be sure the picture shows the required symmetry about the line $y = x$.

c. Find the slopes of the tangents to the graphs of h and k at $(2, 2)$ and $(-2, -2)$.

d. What lines are tangent to the curves at the origin?

23. Let $f(x) = x^3 - 3x^2 - 1, x \geq 2$. Find the value of df^{-1}/dx at the point $x = -1 = f(3)$.

24. Let $f(x) = x^2 - 4x - 5, x > 2$. Find the value of df^{-1}/dx at the point $x = 0 = f(5)$.

25. a. Find the inverse of the function $f(x) = mx$, where m is a constant different from zero.

b. What can you conclude about the inverse of a function $y = f(x)$ whose graph is a line through the origin with a nonzero slope m?

26. Show that the graph of the inverse of $f(x) = mx + b$, where m and b are constants and $m \neq 0$, is a line with slope $1/m$ and y-intercept $-b/m$.

27. Show that increasing functions and decreasing functions are one-to-one. That is, show that for any x_1 and x_2 in I, $x_2 \neq x_1$ implies $f(x_2) \neq f(x_1)$.

Use the results of Exercise 27 to show that the functions in Exercises 28–32 have inverses over their domains. Find a formula for df^{-1}/dx using Theorem 4.

28. $f(x) = (1/3)x + (5/6)$

29. $f(x) = 27x^3$

30. $f(x) = 1 - 8x^3$

31. $f(x) = (1 - x)^3$

32. $f(x) = x^{5/3}$

33. If $f(x)$ is one-to-one, can anything be said about $g(x) = -f(x)$? Is it also one-to-one? Give reasons for your answer.

34. If $f(x)$ is one-to-one and $f(x)$ is never zero, can anything be said about $h(x) = 1/f(x)$? Is it also one-to-one? Give reasons for your answer.

35. Suppose that the range of g lies in the domain of f so that the composite $f \circ g$ is defined. If f and g are one-to-one, can anything be said about $f \circ g$? Give reasons for your answer.

36. If a composite $f \circ g$ is one-to-one, must g be one-to-one? Give reasons for your answer.

37. Assume that f and g are differentiable functions that are inverses of one another so that $(g \circ f)(x) = x$. Differentiate both sides of this equation with respect to x using the Chain Rule to express $(g \circ f)'(x)$ as a product of derivatives of g and f. What do you find? (This is not a proof of Theorem 4 because we assume here the theorem's conclusion that $g = f^{-1}$ is differentiable.)

38. Let $g(x) = x^{1/n}$ and $f(x) = x^n$. Calculate a formula for $g'(x)$ using Theorem 4. Does your formula agree with the Power Rule?

COMPUTER EXPLORATIONS

In Exercises 39–46, you will explore some functions and their inverses together with their derivatives and linear approximating functions at specified points. Perform the following steps using your CAS:

a. Plot the function $y = f(x)$ together with its derivative over the given interval. Explain why you know that f is one-to-one over the interval.

b. Solve the equation $y = f(x)$ for x as a function of y, and name the resulting inverse function g.

c. Find the equation for the tangent line to f at the specified point $(x_0, f(x_0))$.

d. Find the equation for the tangent line to g at the point $(f(x_0), x_0)$ located symmetrically across the 45° line $y = x$ (which is the graph of the identity function). Use Theorem 4 to find the slope of this tangent line.

e. Plot the functions f and g, the identity, the two tangent lines, and the line segment joining the points $(x_0, f(x_0))$ and $(f(x_0), x_0)$. Discuss the symmetries you see across the main diagonal.

39. $y = \sqrt{3x - 2}, \quad \frac{2}{3} \leq x \leq 4, \quad x_0 = 3$

40. $y = \frac{3x + 2}{2x - 11}, \quad -2 \leq x \leq 2, \quad x_0 = 1/2$

41. $y = \frac{4x}{x^2 + 1}, \quad -1 \leq x \leq 1, \quad x_0 = 1/2$

42. $y = \frac{x^3}{x^2 + 1}, \quad -1 \leq x \leq 1, \quad x_0 = 1/2$

43. $y = x^3 - 3x^2 - 1, \quad 2 \leq x \leq 5, \quad x_0 = \frac{27}{10}$

44. $y = 2 - x - x^3, \quad -2 \leq x \leq 2, \quad x_0 = \frac{3}{2}$

45. $y = e^x, \quad -3 \leq x \leq 5, \quad x_0 = 1$

46. $y = \sin x, \quad -\frac{\pi}{2} \leq x \leq \frac{\pi}{2}, \quad x_0 = 1$

In Exercises 47 and 48, repeat the steps above to solve for the functions $y = f(x)$ and $x = f^{-1}(y)$ defined implicitly by the given equations over the interval.

47. $y^{1/3} - 1 = (x + 2)^3, \quad -5 \leq x \leq 5, \quad x_0 = -3/2$

48. $\cos y = x^{1/5}, \quad 0 \leq x \leq 1, \quad x_0 = 1/2$

2.10 Logarithmic Functions

In this section we present the natural logarithm function $y = \ln x$ as the inverse of the exponential function $y = e^x$. This is one of the most important function-inverse pairs in mathematics and science. We learned how to differentiate the exponential function in Section 2.6, so we can apply Theorem 4 to find the derivative of the natural logarithm function.

Logarithms

If a is any positive real number other than 1, the base a exponential function $f(x) = a^x$ is one-to-one. It therefore has an inverse. Its inverse is called the *logarithm function with base a*.

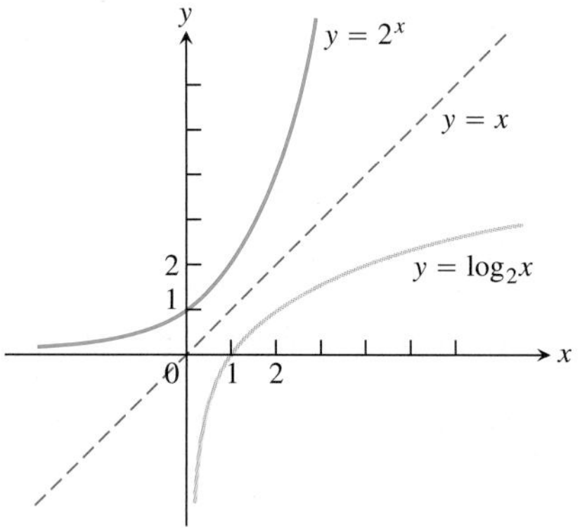

FIGURE 2.41 The graph of 2^x and its inverse, $\log_2 x$.

> DEFINITION The **logarithm function with base *a***, $y = \log_a x$, is the inverse of the base a exponential function $y = a^x$ $(a > 0, a \neq 1)$.

The domain of $\log_a x$ is $(0, \infty)$, the range of a^x. The range of $\log_a x$ is $(-\infty, \infty)$, the domain of a^x.

Figure 1.22 in Section 1.1 shows the graphs of four logarithmic functions with $a > 1$. Figure 2.41 shows the graph of $y = \log_2 x$. The graph of $y = a^x$, $a > 1$, increases rapidly for $x > 0$, so its inverse, $y = \log_a x$, increases slowly for $x > 1$.

Because we have no technique yet for solving the equation $y = a^x$ for x in terms of y, we do not have an explicit formula for computing the logarithm at a given value of x. Nevertheless, we can obtain the graph of $y = \log_a x$ by reflecting the graph of the exponential $y = a^x$ across the line $y = x$. Figure 2.41 shows the graph for $a = 2$.

Logarithms with base 2 are commonly used in computer science. Logarithms with base e and base 10 are so important in applications that calculators have special keys for them. They also have their own special notation and names:

$$\log_e x \quad \text{is written as} \quad \ln x.$$
$$\log_{10} x \quad \text{is written as} \quad \log x.$$

The function $y = \ln x$ is called the **natural logarithm function**, and $y = \log x$ is often called the **common logarithm function**. For the natural logarithm,

$$\ln x = y \iff e^y = x.$$

In particular, if we set $x = e$, we obtain

$$\ln e = 1$$

because $e^1 = e$.

Properties of Logarithms

HISTORICAL BIOGRAPHY

John Napier
(1550–1617)

Logarithms, invented by John Napier, were the single most important improvement in arithmetic calculation before the modern electronic computer. What made them so useful is that the properties of logarithms reduce multiplication of positive numbers to addition of their logarithms, division of positive numbers to subtraction of their logarithms, and exponentiation of a number to multiplying its logarithm by the exponent.

We summarize these properties as a series of rules that we prove in Appendix 8.

Algebraic Properties of Logarithms

For any numbers $x > 0$ and $y > 0$,

1. *Product Rule:* $\log_a xy = \log_a x + \log_a y$
2. *Quotient Rule:* $\log_a \frac{x}{y} = \log_a x - \log_a y$
3. *Reciprocal Rule:* $\log_a \frac{1}{y} = -\log_a y$ — Rule 2 with $x = 1$
4. *Power Rule:* $\log_a x^y = y \log_a x$

In particular, these algebraic properties are valid for the natural logarithmic function. We now illustrate how the rules apply.

EXAMPLE 1

(a) $\log_2 4 + \log_2 \sin x = \log_2 (4 \sin x)$ — Product

(b) $\ln \frac{x+1}{2x-3} = \ln (x+1) - \ln (2x-3)$ — Quotient

(c) $\ln \frac{1}{8} = -\ln 8$ — Reciprocal

$= -\ln 2^3 = -3 \ln 2$ — Power ■

Because a^x and $\log_a x$ are inverses, composing them in either order gives the identity function.

Inverse Properties for a^x and $\log_a x$

1. Base a: $a^{\log_a x} = x$, $\quad \log_a a^x = x$, $\quad a > 0, a \neq 1, x > 0$

2. Base e: $e^{\ln x} = x$, $\quad \ln e^x = x$, $\quad x > 0$

Substituting a^x for x in the equation $x = e^{\ln x}$ enables us to rewrite a^x as a power of e:

$$a^x = e^{\ln (a^x)} \qquad \text{Substitute } a^x \text{ for } x \text{ in } x = e^{\ln x}.$$
$$= e^{x \ln a} \qquad \text{Power Rule for logs}$$
$$= e^{(\ln a)x}. \qquad \text{Exponent rearranged}$$

Thus, the exponential function a^x is the same as e^{kx} for $k = \ln a$.

> Every exponential function is a power of the natural exponential function.
>
> $$a^x = e^{x \ln a}$$
>
> That is, a^x is the same as e^x raised to the power $\ln a$: $a^x = e^{kx}$ for $k = \ln a$.

For example,

$$2^x = e^{(\ln 2)x} = e^{x \ln 2}, \qquad \text{and} \qquad 5^{-3x} = e^{(\ln 5)(-3x)} = e^{-3x \ln 5}.$$

Returning once more to the properties of a^x and $\log_a x$, we have

$$\begin{aligned} \ln x &= \ln\left(a^{\log_a x}\right) && \text{Inverse Property for } a^x \text{ and } \log_a x \\ &= (\log_a x)(\ln a). && \text{Power Rule for logarithms, with } r = \log_a x \end{aligned}$$

Rewriting this equation as $\log_a x = (\ln x)/(\ln a)$ shows that every logarithmic function is a constant multiple of the natural logarithm $\ln x$. This allows us to extend the algebraic properties for $\ln x$ to $\log_a x$. For instance, $\log_a bx = \log_a b + \log_a x$.

> **Change of Base Formula**
> Every logarithmic function is a constant multiple of the natural logarithm.
>
> $$\log_a x = \frac{\ln x}{\ln a} \qquad (a > 0, a \neq 1)$$

EXAMPLE 2 The **half-life** of a radioactive element is the time required for half of the radioactive nuclei present in a sample to decay. It is a remarkable fact that the half-life is a constant that does not depend on the number of radioactive nuclei initially present in the sample, but only on the radioactive substance.

To see why, let y_0 be the number of radioactive nuclei initially present in the sample. Then the number y present at any later time t will be $y = y_0 e^{-rt}$. (See Example 5, Section 2.6.) We seek the value of t at which the number of radioactive nuclei present equals half the original number:

$$\begin{aligned} y_0 e^{-rt} &= \frac{1}{2} y_0 \\ e^{-rt} &= \frac{1}{2} \\ -rt &= \ln \frac{1}{2} = -\ln 2 && \text{Reciprocal Rule for logarithms} \\ t &= \frac{\ln 2}{r}, \qquad r > 0. && (1) \end{aligned}$$

This value of t is the half-life of the element. It depends only on the value of r; the number y_0 does not have any effect.

The effective radioactive lifetime of polonium-210 is so short that we measure it in days rather than years. The number of radioactive atoms remaining after t days in a sample that starts with y_0 radioactive atoms is

$$y = y_0 e^{-5\times 10^{-3} t}.$$

The element's half-life is

$$\begin{aligned} \text{Half-life} &= \frac{\ln 2}{r} && \text{Eq. (1)} \\ &= \frac{\ln 2}{5 \times 10^{-3}} && \text{The } r \text{ from polonium's decay equation} \\ &\approx 139 \text{ days.} \end{aligned}$$

■

Derivative of the Natural Logarithm Function

Since we know the exponential function $f(x) = e^x$ is differentiable everywhere, we can apply Theorem 4 to find the derivative of its inverse $f^{-1}(x) = \ln x$:

$$\begin{aligned} (f^{-1})'(x) &= \frac{1}{f'(f^{-1}(x))} && \text{Theorem 4} \\ &= \frac{1}{e^{f^{-1}(x)}} && f'(u) = e^u \\ &= \frac{1}{e^{\ln x}} \\ &= \frac{1}{x} && \text{Inverse function relationship} \end{aligned}$$

Alternate Derivation Instead of applying Theorem 4 directly, we can find the derivative of $y = \ln x$ using implicit differentiation, as follows:

$$\begin{aligned} y &= \ln x \\ e^y &= x && \text{Inverse function relationship} \\ \frac{d}{dx}(e^y) &= \frac{d}{dx}(x) && \text{Differentiate implicitly} \\ e^y \frac{dy}{dx} &= 1 && \text{Chain Rule} \\ \frac{dy}{dx} &= \frac{1}{e^y} = \frac{1}{x} && e^y = x \end{aligned}$$

No matter which derivation we use, the derivative of $y = \ln x$ with respect to x is

$$\frac{d}{dx}(\ln x) = \frac{1}{x}, \quad x > 0.$$

If u is a differentiable function of x with $u > 0$, we apply the Chain Rule to get

$$\frac{d}{dx}(\ln u) = \frac{1}{u}\frac{du}{dx}, \qquad u > 0 \tag{2}$$

EXAMPLE 3

(a) $\frac{d}{dx}\ln 2x = \frac{1}{2x}\frac{d}{dx}(2x) = \frac{1}{2x}(2) = \frac{1}{x}$

(b) Equation (2) with $u = x^2 + 3$ gives

$$\frac{d}{dx}\ln(x^2+3) = \frac{1}{x^2+3}\cdot\frac{d}{dx}(x^2+3) = \frac{1}{x^2+3}\cdot 2x = \frac{2x}{x^2+3}.$$ ■

Notice the remarkable occurrence in Example 3(a). The function $y = \ln 2x$ has the same derivative as the function $y = \ln x$. This is true of $y = \ln bx$ for any constant b, provided that $bx > 0$:

$$\frac{d}{dx}\ln bx = \frac{1}{bx}\cdot\frac{d}{dx}(bx) = \frac{1}{bx}(b) = \frac{1}{x}. \tag{3}$$

If $x < 0$ and $b < 0$, then $bx > 0$ and Equation (3) still applies. In particular, if $x < 0$ and $b = -1$ we get

$$\frac{d}{dx}\ln(-x) = \frac{1}{x} \qquad \text{for } x < 0.$$

Since $|x| = x$ when $x > 0$ and $|x| = -x$ when $x < 0$, we have the important result

$$\frac{d}{dx}\ln|x| = \frac{1}{x}, \quad x \neq 0 \tag{4}$$

The Derivatives of a^u and $\log_a u$

We start with the equation $a^x = e^{\ln(a^x)} = e^{x\ln a}$:

$$\frac{d}{dx}a^x = \frac{d}{dx}e^{x\ln a} = e^{x\ln a}\cdot\frac{d}{dx}(x\ln a) \qquad \frac{d}{dx}e^u = e^u\frac{du}{dx}$$

$$= a^x \ln a.$$

In summary, if $a > 0$, then

$$\frac{d}{dx}a^x = a^x \ln a.$$

This equation shows why e^x is the exponential function preferred in calculus. If $a = e$, then $\ln a = 1$ and the derivative of a^x simplifies to

$$\frac{d}{dx}e^x = e^x \ln e = e^x.$$

With the Chain Rule, we get a more general form for the derivative of an exponential function with base a.

If $a > 0$ and u is a differentiable function of x, then a^u is a differentiable function of x and

$$\frac{d}{dx}a^u = a^u \ln a\,\frac{du}{dx}. \tag{5}$$

In Section 2.6 we looked at the derivative $f'(0)$ for the exponential functions $f(x) = a^x$ at various values of the base a. The number $f'(0)$ is the limit, $\lim_{h\to 0}(a^h - 1)/h$, and gives the slope of the graph of a^x when it crosses the y-axis at the point (0, 1). We now see that the value of this slope is

$$\lim_{h\to 0}\frac{a^h - 1}{h} = \ln a. \tag{6}$$

In particular, when $a = e$ we obtain

$$\lim_{h\to 0}\frac{e^h - 1}{h} = \ln e = 1.$$

However, we have not fully justified that these limits actually exist. While all of the arguments given in deriving the derivatives of the exponential and logarithmic functions are correct, they do assume the existence of these limits. We establish their existence in Appendix 8.

To find the derivative of $\log_a u$ for an arbitrary base ($a > 0$, $a \neq 1$), we start with the change-of-base formula

$$\log_a x = \frac{\ln x}{\ln a}.$$

Taking derivatives, we have

$$\begin{aligned}
\frac{d}{dx}\log_a x &= \frac{d}{dx}\left(\frac{\ln x}{\ln a}\right)\\
&= \frac{1}{\ln a}\cdot\frac{d}{dx}\ln x \qquad \text{$\ln a$ is a constant}\\
&= \frac{1}{\ln a}\cdot\frac{1}{x}\\
&= \frac{1}{x\ln a}.
\end{aligned}$$

If u is a differentiable function of x and $u > 0$, the Chain Rule gives the following formula.

For $a > 0$ and $a \neq 1$,

$$\frac{d}{dx}\log_a u = \frac{1}{u\ln a}\frac{du}{dx}. \tag{7}$$

Logarithmic Differentiation

The derivatives of positive functions given by formulas that involve products, quotients, and powers can often be found more quickly if we take the natural logarithm of both sides before differentiating. This enables us to use the laws of logarithms to simplify the formulas before differentiating. The process, called **logarithmic differentiation**, is illustrated in the next example.

EXAMPLE 4 Find dy/dx if

$$y = \frac{(x^2 + 1)(x + 3)^{1/2}}{x - 1}, \qquad x > 1.$$

Solution We take the natural logarithm of both sides and simplify the result with the algebraic properties of logarithms:

$$\begin{aligned}
\ln y &= \ln \frac{(x^2+1)(x+3)^{1/2}}{x-1} \\
&= \ln\left((x^2+1)(x+3)^{1/2}\right) - \ln(x-1) && \text{Rule 2} \\
&= \ln(x^2+1) + \ln(x+3)^{1/2} - \ln(x-1) && \text{Rule 1} \\
&= \ln(x^2+1) + \frac{1}{2}\ln(x+3) - \ln(x-1). && \text{Rule 3}
\end{aligned}$$

We then take derivatives of both sides with respect to x, using Equation (2) on the left:

$$\frac{1}{y}\frac{dy}{dx} = \frac{1}{x^2+1}\cdot 2x + \frac{1}{2}\cdot\frac{1}{x+3} - \frac{1}{x-1}. \qquad \text{Eq. (2) with } u = y$$

Next we solve for dy/dx:

$$\frac{dy}{dx} = y\left(\frac{2x}{x^2+1} + \frac{1}{2x+6} - \frac{1}{x-1}\right).$$

Finally, we substitute for y:

$$\frac{dy}{dx} = \frac{(x^2+1)(x+3)^{1/2}}{x-1}\left(\frac{2x}{x^2+1} + \frac{1}{2x+6} - \frac{1}{x-1}\right).$$

■

Proof of the Power Rule (General Version)

We can now define x^n for any $x > 0$ and any real number n as $x^n = e^{n \ln x}$. Therefore, the n in the equation $\ln x^n = n \ln x$ no longer needs to be rational—it can be any fixed number as long as $x > 0$:

$$\ln x^n = \ln(e^{n \ln x}) = n \ln x \qquad \ln e^u = u, \text{ any } u$$

Together, the law $a^x/a^y = a^{x-y}$ and the definition $x^n = e^{n \ln x}$ enable us to establish the Power Rule for differentiation for any real number n. Differentiating x^n with respect to x gives

$$\begin{aligned}
\frac{d}{dx}x^n &= \frac{d}{dx}e^{n \ln x} && \text{Definition of } x^n, x > 0 \\
&= e^{n \ln x}\cdot\frac{d}{dx}(n \ln x) && \text{Chain Rule for } e^u \\
&= x^n \cdot \frac{n}{x} && \text{The definition again} \\
&= nx^{n-1}.
\end{aligned}$$

In short, as long as $x > 0$,

$$\frac{d}{dx}x^n = nx^{n-1}.$$

If $y = x^n$, y', and x^{n-1} all exist for $x < 0$, then

$$\ln|y| = \ln|x|^n = n \ln|x|.$$

Using logarithmic differentiation and Equation (4) we have

$$\frac{y'}{y} = \frac{n}{x}.$$

Solving for the derivative,

$$y' = n\frac{y}{x} = n\frac{x^n}{x} = nx^{n-1}.$$

(Notice in this case the need to *assume* the existence of the derivative y'.)

It can also be shown directly from the definition of the derivative that the derivative equals 0 when $x = 0$ and $n \neq 0$. This completes the proof of the general version of the Power Rule for all values of x. ■

EXAMPLE 5 Differentiate $f(x) = x^x, x > 0$.

Solution We note that $f(x) = x^x = e^{x \ln x}$, so differentiation gives

$$\begin{aligned} f'(x) &= \frac{d}{dx}(e^{x \ln x}) \\ &= e^{x \ln x}\frac{d}{dx}(x \ln x) && \tfrac{d}{dx}e^u,\ u = x \ln x \\ &= e^{x \ln x}\left(\ln x + x \cdot \frac{1}{x}\right) \\ &= x^x(\ln x + 1) && x > 0 \end{aligned}$$

■

The Number e Expressed as a Limit

In Section 2.6 we defined the number e as the base value for which the exponential function $y = a^x$ has slope 1 when it crosses the y-axis at (0, 1). Thus e is the constant that satisfies the equation

$$\lim_{h \to 0} \frac{e^h - 1}{h} = \ln e = 1. \qquad \text{Slope equals } \ln e \text{ from Eq. (6)}$$

We also stated that e could be calculated as $\lim_{y \to \infty}(1 + 1/y)^y$, or by substituting $y = 1/x$, as $\lim_{x \to 0}(1 + x)^{1/x}$. We now prove this result.

> **THEOREM 5—The Number e as a Limit** The number e can be calculated as the limit
>
> $$e = \lim_{x \to 0}(1 + x)^{1/x}.$$

Proof If $f(x) = \ln x$, then $f'(x) = 1/x$, so $f'(1) = 1$. But, by the definition of derivative,

$$\begin{aligned} f'(1) &= \lim_{h \to 0} \frac{f(1 + h) - f(1)}{h} = \lim_{x \to 0} \frac{f(1 + x) - f(1)}{x} \\ &= \lim_{x \to 0} \frac{\ln(1 + x) - \ln 1}{x} = \lim_{x \to 0} \frac{1}{x}\ln(1 + x) && \ln 1 = 0 \\ &= \lim_{x \to 0} \ln(1 + x)^{1/x} = \ln\left[\lim_{x \to 0}(1 + x)^{1/x}\right] && \text{Theorem 10 in Chapter 1} \end{aligned}$$

Because $f'(1) = 1$, we have

$$\ln\left[\lim_{x\to 0}(1+x)^{1/x}\right] = 1$$

Therefore, exponentiating both sides we get

$$\lim_{x\to 0}(1+x)^{1/x} = e.$$ ■

Approximating the limit in Theorem 5 by taking x very small gives approximations to e. Its value is $e \approx 2.718281828459045$ to 15 decimal places.

EXERCISES 2.10

1. Express the following logarithms in terms of ln 2 and ln 3.
   a. $\ln 0.75$ b. $\ln(4/9)$
   c. $\ln(1/2)$ d. $\ln\sqrt[3]{9}$
   e. $\ln 3\sqrt{2}$ f. $\ln\sqrt{13.5}$
2. Express the following logarithms in terms of ln 5 and ln 7.
   a. $\ln(1/125)$ b. $\ln 9.8$
   c. $\ln 7\sqrt{7}$ d. $\ln 1225$
   e. $\ln 0.056$ f. $(\ln 35 + \ln(1/7))/(\ln 25)$

Use the properties of logarithms to simplify the expressions in Exercises 3 and 4.

3. a. $\ln\sin\theta - \ln\left(\frac{\sin\theta}{5}\right)$ b. $\ln(3x^2 - 9x) + \ln\left(\frac{1}{3x}\right)$
   c. $\frac{1}{2}\ln(4t^4) - \ln 2$
4. a. $\ln\sec\theta + \ln\cos\theta$ b. $\ln(8x+4) - 2\ln 2$
   c. $3\ln\sqrt[3]{t^2-1} - \ln(t+1)$

Find simpler expressions for the quantities in Exercises 5–8.

5. a. $e^{\ln 7.2}$ b. $e^{-\ln x^2}$ c. $e^{\ln x - \ln y}$
6. a. $e^{\ln(x^2+y^2)}$ b. $e^{-\ln 0.3}$ c. $e^{\ln \pi x - \ln 2}$
7. a. $2\ln\sqrt{e}$ b. $\ln(\ln e^e)$ c. $\ln(e^{-x^2-y^2})$
8. a. $\ln(e^{\sec\theta})$ b. $\ln(e^{(e^x)})$ c. $\ln(e^{2\ln x})$

In Exercises 9–14, solve for y in terms of t or x, as appropriate.

9. $\ln y = 2t + 4$
10. $\ln y = -t + 5$
11. $\ln(y - 40) = 5t$
12. $\ln(1 - 2y) = t$
13. $\ln(y-1) - \ln 2 = x + \ln x$
14. $\ln(y^2 - 1) - \ln(y+1) = \ln(\sin x)$

In Exercises 15 and 16, solve for k.

15. a. $e^{2k} = 4$ b. $100e^{10k} = 200$ c. $e^{k/1000} = a$
16. a. $e^{5k} = \frac{1}{4}$ b. $80e^k = 1$ c. $e^{(\ln 0.8)k} = 0.8$

In Exercises 17–20, solve for t.

17. a. $e^{-0.3t} = 27$ b. $e^{kt} = \frac{1}{2}$ c. $e^{(\ln 0.2)t} = 0.4$
18. a. $e^{-0.01t} = 1000$ b. $e^{kt} = \frac{1}{10}$ c. $e^{(\ln 2)t} = \frac{1}{2}$
19. $e^{\sqrt{t}} = x^2$
20. $e^{(x^2)}e^{(2x+1)} = e^t$

Simplify the expressions in Exercises 21–24.

21. a. $5^{\log_5 7}$ b. $8^{\log_8 \sqrt{2}}$ c. $1.3^{\log_{1.3} 75}$
    d. $\log_4 16$ e. $\log_3 \sqrt{3}$ f. $\log_4\left(\frac{1}{4}\right)$
22. a. $2^{\log_2 3}$ b. $10^{\log_{10}(1/2)}$ c. $\pi^{\log_\pi 7}$
    d. $\log_{11} 121$ e. $\log_{121} 11$ f. $\log_3\left(\frac{1}{9}\right)$
23. a. $2^{\log_4 x}$ b. $9^{\log_3 x}$ c. $\log_2(e^{(\ln 2)(\sin x)})$
24. a. $25^{\log_5(3x^2)}$ b. $\log_e(e^x)$ c. $\log_4(2^{e^x \sin x})$

Express the ratios in Exercises 25 and 26 as ratios of natural logarithms and simplify.

25. a. $\frac{\log_2 x}{\log_3 x}$ b. $\frac{\log_2 x}{\log_8 x}$ c. $\frac{\log_x a}{\log_{x^2} a}$
26. a. $\frac{\log_9 x}{\log_3 x}$ b. $\frac{\log_{\sqrt{10}} x}{\log_{\sqrt{2}} x}$ c. $\frac{\log_a b}{\log_b a}$

In Exercises 27–56, find the derivative of y with respect to x, t, or θ, as appropriate.

27. $y = \ln 3x$
28. $y = \ln kx$, k constant
29. $y = \ln(t^2)$
30. $y = \ln(t^{3/2})$

31. $y = \ln \frac{3}{x}$ **32.** $y = \ln \frac{10}{x}$

33. $y = \ln(\theta + 1)$ **34.** $y = \ln(2\theta + 2)$

35. $y = \ln x^3$ **36.** $y = (\ln x)^3$

37. $y = t(\ln t)^2$ **38.** $y = t\sqrt{\ln t}$

39. $y = \frac{x^4}{4}\ln x - \frac{x^4}{16}$ **40.** $y = \frac{x^3}{3}\ln x - \frac{x^3}{9}$

41. $y = \frac{\ln t}{t}$ **42.** $y = \frac{1 + \ln t}{t}$

43. $y = \frac{\ln x}{1 + \ln x}$ **44.** $y = \frac{x \ln x}{1 + \ln x}$

45. $y = \ln(\ln x)$ **46.** $y = \ln(\ln(\ln x))$

47. $y = \theta(\sin(\ln\theta) + \cos(\ln\theta))$

48. $y = \ln(\sec\theta + \tan\theta)$

49. $y = \ln \frac{1}{x\sqrt{x + 1}}$ **50.** $y = \frac{1}{2}\ln \frac{1 + x}{1 - x}$

51. $y = \frac{1 + \ln t}{1 - \ln t}$ **52.** $y = \sqrt{\ln \sqrt{t}}$

53. $y = \ln(\sec(\ln\theta))$ **54.** $y = \ln\left(\frac{\sqrt{\sin\theta\cos\theta}}{1 + 2\ln\theta}\right)$

55. $y = \ln\left(\frac{(x^2 + 1)^5}{\sqrt{1 - x}}\right)$ **56.** $y = \ln\sqrt{\frac{(x + 1)^5}{(x + 2)^{20}}}$

In Exercises 57–70, use logarithmic differentiation to find the derivative of y with respect to the given independent variable.

57. $y = \sqrt{x(x + 1)}$ **58.** $y = \sqrt{(x^2 + 1)(x - 1)^2}$

59. $y = \sqrt{\frac{t}{t + 1}}$ **60.** $y = \sqrt{\frac{1}{t(t + 1)}}$

61. $y = \sqrt{\theta + 3}\sin\theta$ **62.** $y = (\tan\theta)\sqrt{2\theta + 1}$

63. $y = t(t + 1)(t + 2)$ **64.** $y = \frac{1}{t(t + 1)(t + 2)}$

65. $y = \frac{\theta + 5}{\theta\cos\theta}$ **66.** $y = \frac{\theta\sin\theta}{\sqrt{\sec\theta}}$

67. $y = \frac{x\sqrt{x^2 + 1}}{(x + 1)^{2/3}}$ **68.** $y = \sqrt{\frac{(x + 1)^{10}}{(2x + 1)^5}}$

69. $y = \sqrt[3]{\frac{x(x - 2)}{x^2 + 1}}$ **70.** $y = \sqrt[3]{\frac{x(x + 1)(x - 2)}{(x^2 + 1)(2x + 3)}}$

In Exercises 71–78, find the derivative of y with respect to x, t, or θ, as appropriate.

71. $y = \ln(\cos^2\theta)$ **72.** $y = \ln(3\theta e^{-\theta})$

73. $y = \ln(3te^{-t})$ **74.** $y = \ln(2e^{-t}\sin t)$

75. $y = \ln\left(\frac{e^\theta}{1 + e^\theta}\right)$ **76.** $y = \ln\left(\frac{\sqrt{\theta}}{1 + \sqrt{\theta}}\right)$

77. $y = e^{(\cos t + \ln t)}$ **78.** $y = e^{\sin t}(\ln t^2 + 1)$

In Exercises 79–82, find dy/dx.

79. $\ln y = e^y \sin x$ **80.** $\ln xy = e^{x+y}$

81. $x^y = y^x$ **82.** $\tan y = e^x + \ln x$

In Exercises 83–104, find the derivative of y with respect to the given independent variable.

83. $y = 2^x$ **84.** $y = 3^{-x}$

85. $y = 5^{\sqrt{s}}$ **86.** $y = 2^{(s^2)}$

87. $y = x^\pi$ **88.** $y = t^{1-e}$

89. $y = \log_2 5\theta$ **90.** $y = \log_3(1 + \theta\ln 3)$

91. $y = \log_4 x + \log_4 x^2$ **92.** $y = \log_{25} e^x - \log_5\sqrt{x}$

93. $y = \log_2 r \cdot \log_4 r$ **94.** $y = \log_3 r \cdot \log_9 r$

95. $y = \log_3\left(\left(\frac{x + 1}{x - 1}\right)^{\ln 3}\right)$ **96.** $y = \log_5\sqrt{\left(\frac{7x}{3x + 2}\right)^{\ln 5}}$

97. $y = \theta\sin(\log_7\theta)$ **98.** $y = \log_7\left(\frac{\sin\theta\cos\theta}{e^\theta 2^\theta}\right)$

99. $y = \log_5 e^x$ **100.** $y = \log_2\left(\frac{x^2e^2}{2\sqrt{x + 1}}\right)$

101. $y = 3^{\log_2 t}$ **102.** $y = 3\log_8(\log_2 t)$

103. $y = \log_2(8t^{\ln 2})$ **104.** $y = t\log_3\left(e^{(\sin t)(\ln 3)}\right)$

In Exercises 105–112, use logarithmic differentiation to find the derivative of y with respect to the given independent variable.

105. $y = (x + 1)^x$ **106.** $y = x^{(x+1)}$

107. $y = (\sqrt{t})^t$ **108.** $y = t^{\sqrt{t}}$

109. $y = (\sin x)^x$ **110.** $y = x^{\sin x}$

111. $y = x^{\ln x}$ **112.** $y = (\ln x)^{\ln x}$

113. Start with the graph of $y = \ln x$. Find an equation of the graph that results from

a. shifting down 3 units.

b. shifting right 1 unit.

c. shifting left 1, up 3 units.

d. shifting down 4, right 2 units.

e. reflecting about the y-axis.

f. reflecting about the line $y = x$.

T **114.** Could $x^{\ln 2}$ possibly be the same as $2^{\ln x}$ for $x > 0$? Graph the two functions and explain what you see.

115. If $y = A\sin(\ln x) + B\cos(\ln x)$, where A and B are constants, show that

$$x^2y'' + xy' + y = 0.$$

116. Show that $\lim_{n\to\infty}\left(1 + \frac{x}{n}\right)^n = e^x$ for any $x > 0$.

117. Radioactive decay The half-life of a certain radioactive substance is 12 hours. There are 8 grams present initially.

a. Express the amount of substance remaining as a function of time t.

b. When will there be 1 gram remaining?

118. Radon-222 The decay equation for radon-222 gas is known to be $y = y_0 e^{-0.18t}$, with t in days. About how long will it take the radon in a sealed sample of air to fall to 90% of its original value?

2.11 Inverse Trigonometric Functions

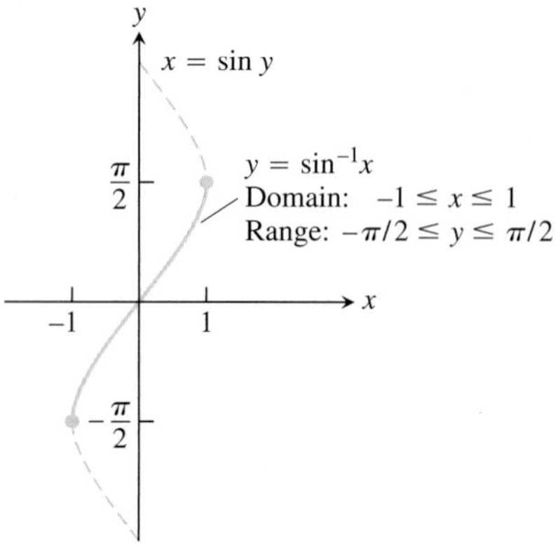

FIGURE 2.42 The graph of $y = \sin^{-1} x$.

Inverse trigonometric functions arise when we want to calculate angles from side measurements in triangles. This section shows how these functions are defined, graphed, evaluated, and how their derivatives are computed.

Defining the Inverses

The six basic trigonometric functions are not one-to-one (their values repeat periodically). However we can restrict their domains to intervals on which they are one-to-one. The sine function increases from -1 at $x = -\pi/2$ to $+1$ at $x = \pi/2$. By restricting its domain to the interval $[-\pi/2, \pi/2]$, we make it one-to-one, so that it has an inverse $\sin^{-1} x$ (Figure 2.42). Similar domain restrictions can be applied to all six trigonometric functions.

Domain restrictions that make the trigonometric functions one-to-one

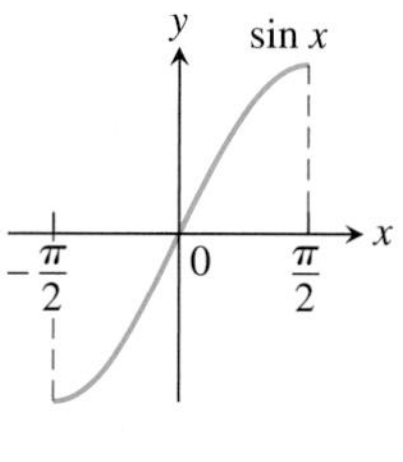

$y = \sin x$
Domain: $[-\pi/2, \pi/2]$
Range: $[-1, 1]$

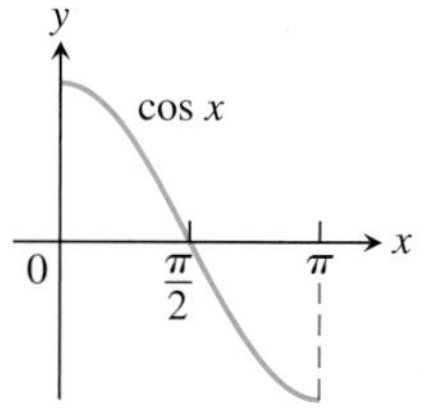

$y = \cos x$
Domain: $[0, \pi]$
Range: $[-1, 1]$

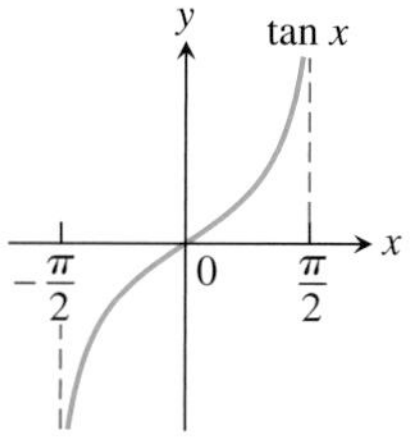

$y = \tan x$
Domain: $(-\pi/2, \pi/2)$
Range: $(-\infty, \infty)$

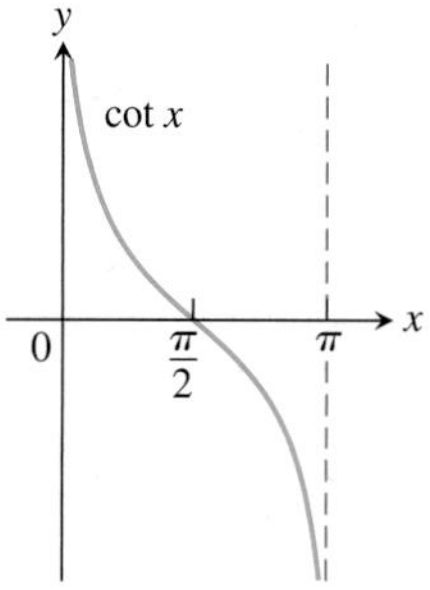

$y = \cot x$
Domain: $(0, \pi)$
Range: $(-\infty, \infty)$

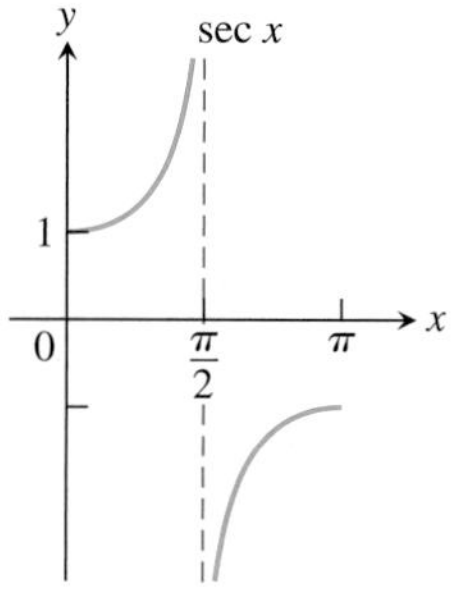

$y = \sec x$
Domain: $[0, \pi/2) \cup (\pi/2, \pi]$
Range: $(-\infty, -1] \cup [1, \infty)$

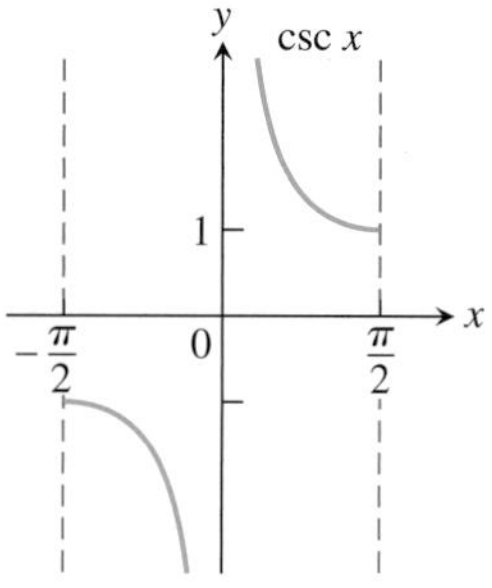

$y = \csc x$
Domain: $[-\pi/2, 0) \cup (0, \pi/2]$
Range: $(-\infty, -1] \cup [1, \infty)$

Since these restricted functions are now one-to-one, they have inverses, which we denote by

$$\begin{aligned} y &= \sin^{-1} x \quad &\text{or}\quad y &= \arcsin x \\ y &= \cos^{-1} x \quad &\text{or}\quad y &= \arccos x \\ y &= \tan^{-1} x \quad &\text{or}\quad y &= \arctan x \\ y &= \cot^{-1} x \quad &\text{or}\quad y &= \operatorname{arccot} x \\ y &= \sec^{-1} x \quad &\text{or}\quad y &= \operatorname{arcsec} x \\ y &= \csc^{-1} x \quad &\text{or}\quad y &= \operatorname{arccsc} x \end{aligned}$$

These equations are read "y equals the arcsine of x" or "y equals arcsin x" and so on.

Caution The -1 in the expressions for the inverse means "inverse." It does *not* mean reciprocal. For example, the *reciprocal* of $\sin x$ is $(\sin x)^{-1} = 1/\sin x = \csc x$.

The graphs of the six inverse trigonometric functions are shown in Figure 2.43. We can obtain these graphs by reflecting the graphs of the restricted trigonometric functions through the line $y = x$, as in Section 2.9. We now take a closer look at these functions and their derivatives.

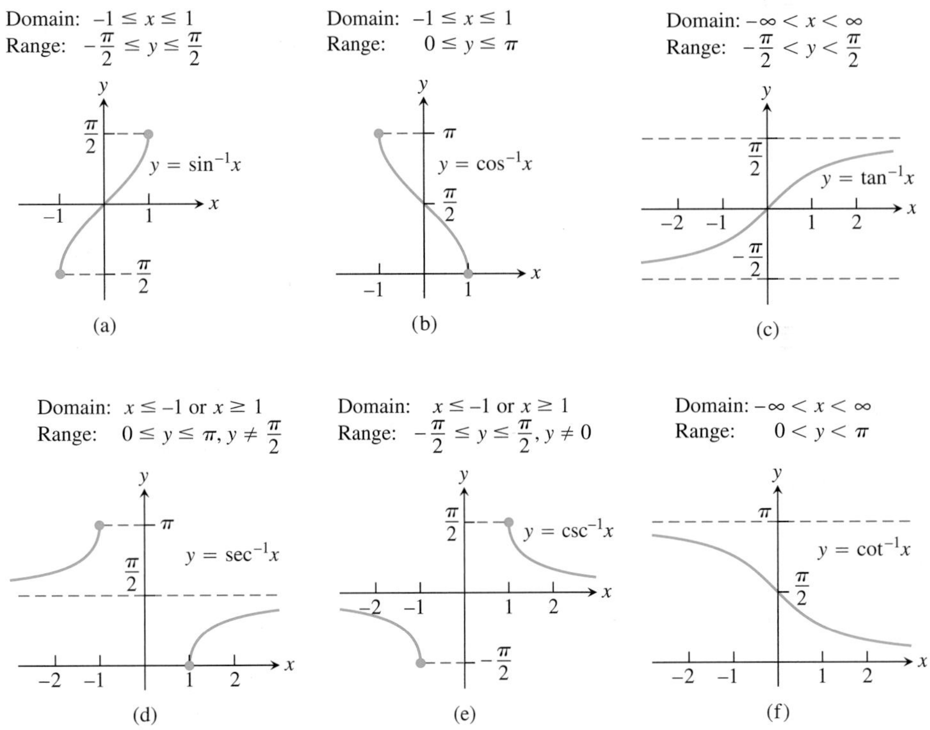

FIGURE 2.43 Graphs of the six basic inverse trigonometric functions.

The Arcsine and Arccosine Functions

The arcsine of x is the angle in $[-\pi/2, \pi/2]$ whose sine is x. The arccosine is an angle in $[0, \pi]$ whose cosine is x.

DEFINITION

$\mathbf{y = \sin^{-1} x}$ is the number in $[-\pi/2, \pi/2]$ for which $\sin y = x$.
$\mathbf{y = \cos^{-1} x}$ is the number in $[0, \pi]$ for which $\cos y = x$.

The graph of $y = \sin^{-1} x$ (Figure 2.44b) is symmetric about the origin (it lies along the graph of $x = \sin y$). The arcsine is therefore an odd function:

$$\sin^{-1}(-x) = -\sin^{-1} x. \tag{1}$$

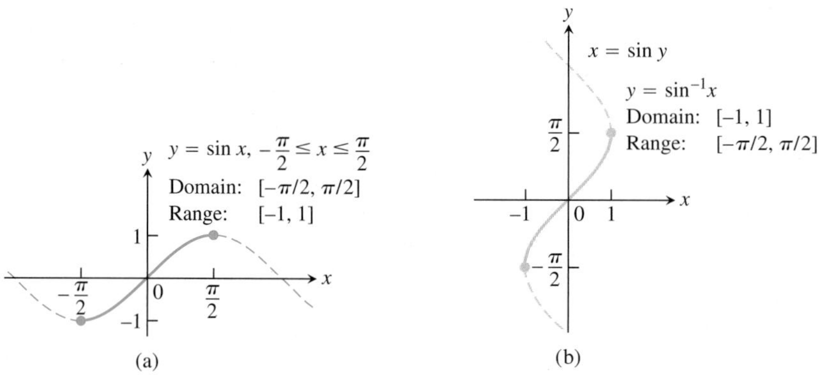

FIGURE 2.44 The graphs of (a) $y = \sin x$, $-\pi/2 \le x \le \pi/2$, and (b) its inverse, $y = \sin^{-1} x$. The graph of $\sin^{-1} x$, obtained by reflection across the line $y = x$, is a portion of the curve $x = \sin y$.

The graph of $y = \cos^{-1} x$ (Figure 2.45b) has no such symmetry.

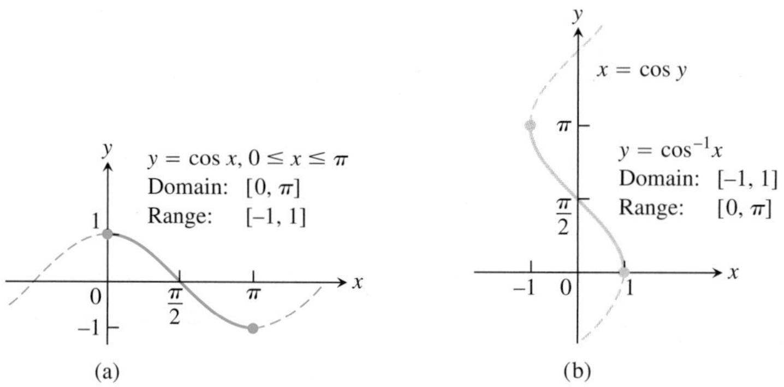

FIGURE 2.45 The graphs of (a) $y = \cos x$, $0 \le x \le \pi$, and (b) its inverse, $y = \cos^{-1} x$. The graph of $\cos^{-1} x$, obtained by reflection across the line $y = x$, is a portion of the curve $x = \cos y$.

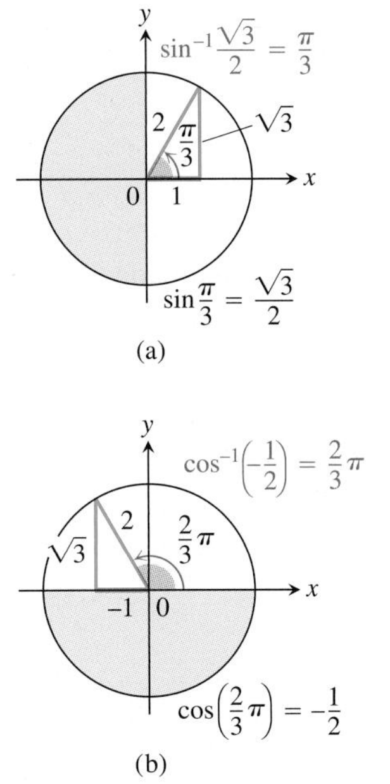

FIGURE 2.46 Values of the arcsine and arccosine functions (Example 1).

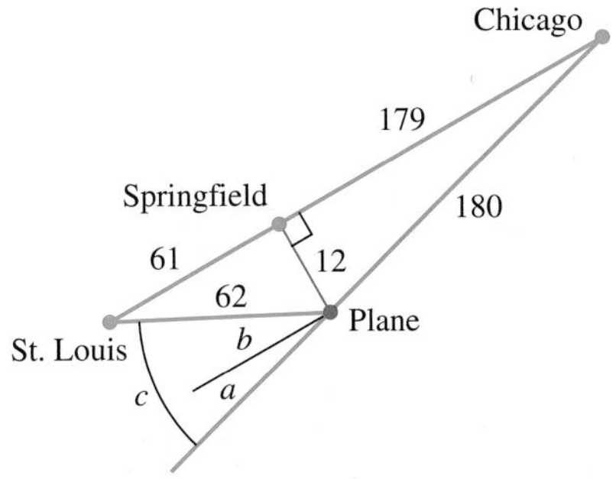

FIGURE 2.47 Diagram for drift correction (Example 2), with distances rounded to the nearest mile (drawing not to scale).

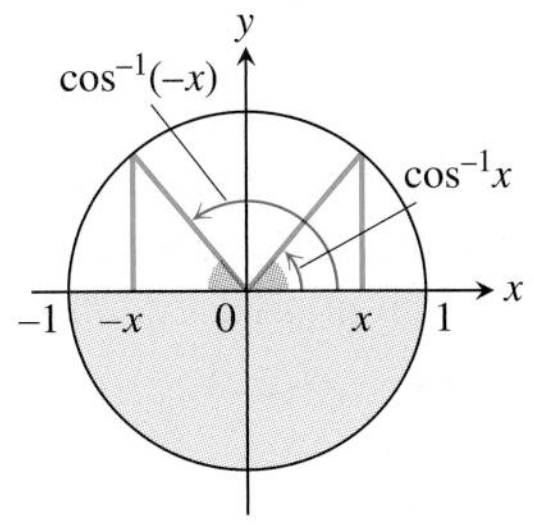

FIGURE 2.48 $\cos^{-1} x$ and $\cos^{-1}(-x)$ are supplementary angles (so their sum is π).

EXAMPLE 1 Evaluate **(a)** $\sin^{-1}\left(\frac{\sqrt{3}}{2}\right)$ and **(b)** $\cos^{-1}\left(-\frac{1}{2}\right)$.

Solution

(a) We see that

$$\sin^{-1}\left(\frac{\sqrt{3}}{2}\right) = \frac{\pi}{3}$$

because $\sin(\pi/3) = \sqrt{3}/2$ and $\pi/3$ belongs to the range $[-\pi/2, \pi/2]$ of the arcsine function. See Figure 2.46a.

(b) We have

$$\cos^{-1}\left(-\frac{1}{2}\right) = \frac{2\pi}{3}$$

because $\cos(2\pi/3) = -1/2$ and $2\pi/3$ belongs to the range $[0, \pi]$ of the arccosine function. See Figure 2.46b. ■

Using the same procedure illustrated in Example 1, we can create the following table of common values for the arcsine and arccosine functions.

x	$\sin^{-1} x$	$\cos^{-1} x$
$\sqrt{3}/2$	$\pi/3$	$\pi/6$
$\sqrt{2}/2$	$\pi/4$	$\pi/4$
$1/2$	$\pi/6$	$\pi/3$
$-1/2$	$-\pi/6$	$2\pi/3$
$-\sqrt{2}/2$	$-\pi/4$	$3\pi/4$
$-\sqrt{3}/2$	$-\pi/3$	$5\pi/6$

EXAMPLE 2 During an airplane flight from Chicago to St. Louis, the navigator determines that the plane is 12 mi off course, as shown in Figure 2.47. Find the angle a for a course parallel to the original correct course, the angle b, and the drift correction angle $c = a + b$.

Solution

$$a = \sin^{-1}\frac{12}{180} \approx 0.067 \text{ radian} \approx 3.8°$$

$$b = \sin^{-1}\frac{12}{62} \approx 0.195 \text{ radian} \approx 11.2°$$

$$c = a + b \approx 15°.$$

■

Identities Involving Arcsine and Arccosine

As we can see from Figure 2.48, the arccosine of x satisfies the identity

$$\cos^{-1} x + \cos^{-1}(-x) = \pi, \tag{2}$$

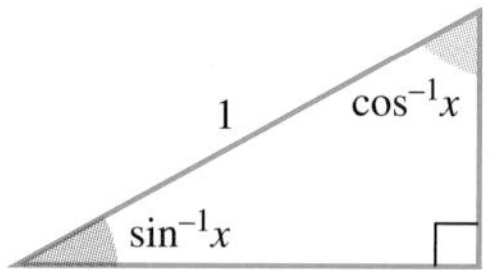

FIGURE 2.49 $\sin^{-1} x$ and $\cos^{-1} x$ are complementary angles (so their sum is $\pi/2$).

or

$$\cos^{-1}(-x) = \pi - \cos^{-1} x. \tag{3}$$

Also, we can see from the triangle in Figure 2.49 that for $x > 0$,

$$\sin^{-1} x + \cos^{-1} x = \pi/2. \tag{4}$$

Equation (4) holds for the other values of x in $[-1, 1]$ as well, but we cannot conclude this from the triangle in Figure 2.49. It is, however, a consequence of Equations (1) and (3) (Exercise 45).

Inverses of tan x, cot x, sec x, and csc x

The arctangent of x is an angle whose tangent is x. The arccotangent of x is an angle whose cotangent is x.

DEFINITION

$y = \tan^{-1} x$ is the number in $(-\pi/2, \pi/2)$ for which $\tan y = x$.

$y = \cot^{-1} x$ is the number in $(0, \pi)$ for which $\cot y = x$.

We use open intervals to avoid values where the tangent and cotangent are undefined.

The graph of $y = \tan^{-1} x$ is symmetric about the origin because it is a branch of the graph $x = \tan y$ that is symmetric about the origin (Figure 2.43c). Algebraically this means that

$$\tan^{-1}(-x) = -\tan^{-1} x;$$

the arctangent is an odd function. The graph of $y = \cot^{-1} x$ has no such symmetry (Figure 2.43f). Notice from Figure 2.43c that the graph of the arctangent function has two horizontal asymptotes; one at $y = \pi/2$ and the other at $y = -\pi/2$.

The inverses of the restricted forms of sec x and csc x are chosen to be the functions graphed in Figures 2.43d and 2.43e.

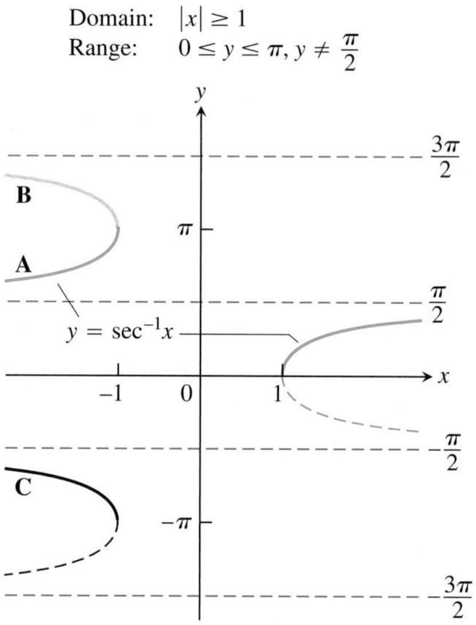

FIGURE 2.50 There are several logical choices for the left-hand branch of $y = \sec^{-1} x$. With choice **A**, $\sec^{-1} x = \cos^{-1}(1/x)$, a useful identity employed by many calculators.

Caution There is no general agreement about how to define $\sec^{-1} x$ for negative values of x. We chose angles in the second quadrant between $\pi/2$ and π. This choice makes $\sec^{-1} x = \cos^{-1}(1/x)$. It also makes $\sec^{-1} x$ an increasing function on each interval of its domain. Some tables choose $\sec^{-1} x$ to lie in $[-\pi, -\pi/2)$ for $x < 0$ and some texts choose it to lie in $[\pi, 3\pi/2)$ (Figure 2.50). These choices simplify the formula for the derivative (our formula needs absolute value signs) but fail to satisfy the computational equation $\sec^{-1} x = \cos^{-1}(1/x)$. From this, we can derive the identity

$$\sec^{-1} x = \cos^{-1}\left(\frac{1}{x}\right) = \frac{\pi}{2} - \sin^{-1}\left(\frac{1}{x}\right) \tag{5}$$

by applying Equation (4).

EXAMPLE 3 The accompanying figures show two values of $\tan^{-1} x$.

x	$\tan^{-1} x$
$\sqrt{3}$	$\pi/3$
1	$\pi/4$
$\sqrt{3}/3$	$\pi/6$
$-\sqrt{3}/3$	$-\pi/6$
-1	$-\pi/4$
$-\sqrt{3}$	$-\pi/3$

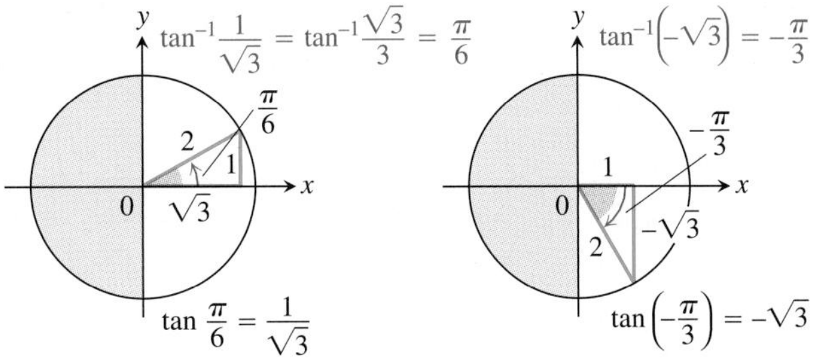

The angles come from the first and fourth quadrants because the range of $\tan^{-1} x$ is $(-\pi/2, \pi/2)$. ■

The Derivative of $y = \sin^{-1} u$

We know that the function $x = \sin y$ is differentiable in the interval $-\pi/2 < y < \pi/2$ and that its derivative, the cosine, is positive there. Theorem 4 therefore assures us that the inverse function $y = \sin^{-1} x$ is differentiable throughout the interval $-1 < x < 1$. We cannot expect it to be differentiable at $x = 1$ or $x = -1$ because the tangents to the graph are vertical at these points (see Figure 2.51).

We find the derivative of $y = \sin^{-1} x$ by applying Theorem 4 with $f(x) = \sin x$ and $f^{-1}(x) = \sin^{-1} x$:

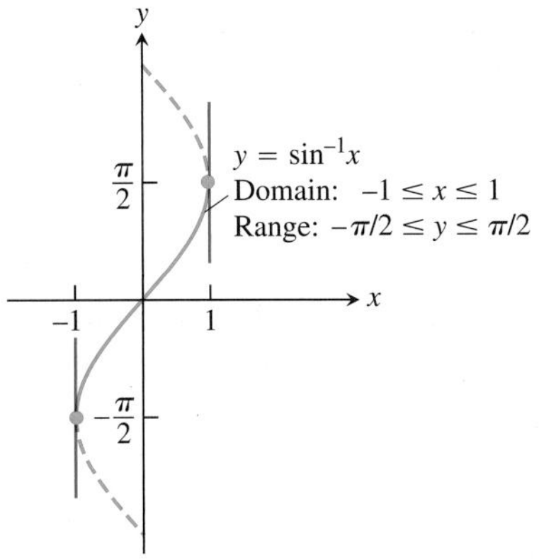

FIGURE 2.51 The graph of $y = \sin^{-1} x$ has vertical tangents at $x = -1$ and $x = 1$.

$$
\begin{aligned}
(f^{-1})'(x) &= \frac{1}{f'(f^{-1}(x))} && \text{Theorem 4} \\
&= \frac{1}{\cos(\sin^{-1} x)} && f'(u) = \cos u \\
&= \frac{1}{\sqrt{1 - \sin^2(\sin^{-1} x)}} && \cos u = \sqrt{1 - \sin^2 u} \\
&= \frac{1}{\sqrt{1 - x^2}} && \sin(\sin^{-1} x) = x
\end{aligned}
$$

If u is a differentiable function of x with $|u| < 1$, we apply the Chain Rule to get

$$\frac{d}{dx}(\sin^{-1} u) = \frac{1}{\sqrt{1 - u^2}}\frac{du}{dx}, \qquad |u| < 1.$$

EXAMPLE 4

$$\frac{d}{dx}(\sin^{-1} x^2) = \frac{1}{\sqrt{1 - (x^2)^2}} \cdot \frac{d}{dx}(x^2) = \frac{2x}{\sqrt{1 - x^4}}$$ ■

The Derivative of $y = \tan^{-1} u$

We find the derivative of $y = \tan^{-1} x$ by applying Theorem 4 with $f(x) = \tan x$ and $f^{-1}(x) = \tan^{-1} x$. Theorem 4 can be applied because the derivative of $\tan x$ is positive for $-\pi/2 < x < \pi/2$:

$$\begin{aligned}
(f^{-1})'(x) &= \frac{1}{f'(f^{-1}(x))} && \text{Theorem 4} \\
&= \frac{1}{\sec^2(\tan^{-1} x)} && f'(u) = \sec^2 u \\
&= \frac{1}{1 + \tan^2(\tan^{-1} x)} && \sec^2 u = 1 + \tan^2 u \\
&= \frac{1}{1 + x^2} && \tan(\tan^{-1} x) = x
\end{aligned}$$

The derivative is defined for all real numbers. If u is a differentiable function of x, we get the Chain Rule form:

$$\frac{d}{dx}(\tan^{-1} u) = \frac{1}{1 + u^2}\frac{du}{dx}.$$

The Derivative of $y = \sec^{-1} u$

Since the derivative of $\sec x$ is positive for $0 < x < \pi/2$ and $\pi/2 < x < \pi$, Theorem 4 says that the inverse function $y = \sec^{-1} x$ is differentiable. Instead of applying the formula in Theorem 4 directly, we find the derivative of $y = \sec^{-1} x$, $|x| > 1$, using implicit differentiation and the Chain Rule as follows:

$$\begin{aligned}
y &= \sec^{-1} x \\
\sec y &= x && \text{Inverse function relationship} \\
\frac{d}{dx}(\sec y) &= \frac{d}{dx} x && \text{Differentiate both sides.} \\
\sec y \tan y \frac{dy}{dx} &= 1 && \text{Chain Rule} \\
\frac{dy}{dx} &= \frac{1}{\sec y \tan y} && \text{Since } |x| > 1, y \text{ lies in } (0, \pi/2) \cup (\pi/2, \pi) \text{ and } \sec y \tan y \neq 0.
\end{aligned}$$

To express the result in terms of x, we use the relationships

$$\sec y = x \quad \text{and} \quad \tan y = \pm\sqrt{\sec^2 y - 1} = \pm\sqrt{x^2 - 1}$$

to get

$$\frac{dy}{dx} = \pm\frac{1}{x\sqrt{x^2 - 1}}.$$

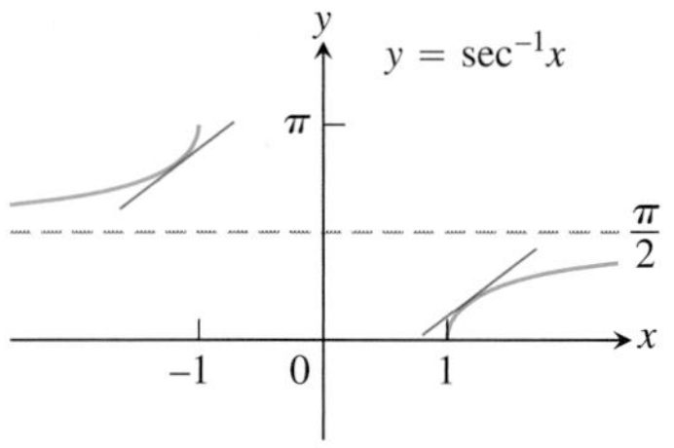

FIGURE 2.52 The slope of the curve $y = \sec^{-1} x$ is positive for both $x < -1$ and $x > 1$.

Can we do anything about the $\pm$ sign? A glance at Figure 2.52 shows that the slope of the graph $y = \sec^{-1} x$ is always positive. Thus,

$$\frac{d}{dx}\sec^{-1} x = \begin{cases} +\dfrac{1}{x\sqrt{x^2-1}} & \text{if } x > 1 \\ -\dfrac{1}{x\sqrt{x^2-1}} & \text{if } x < -1. \end{cases}$$

With the absolute value symbol, we can write a single expression that eliminates the "$\pm$" ambiguity:

$$\frac{d}{dx}\sec^{-1} x = \frac{1}{|x|\sqrt{x^2-1}}.$$

If u is a differentiable function of x with $|u| > 1$, we have the formula

$$\frac{d}{dx}(\sec^{-1} u) = \frac{1}{|u|\sqrt{u^2-1}}\frac{du}{dx}, \qquad |u| > 1.$$

EXAMPLE 5

$$\begin{aligned}\frac{d}{dx}\sec^{-1}(5x^4) &= \frac{1}{|5x^4|\sqrt{(5x^4)^2-1}}\frac{d}{dx}(5x^4) \\ &= \frac{1}{5x^4\sqrt{25x^8-1}}(20x^3) \qquad 5x^4 > 1 > 0 \\ &= \frac{4}{x\sqrt{25x^8-1}}\end{aligned}$$

■

Derivatives of the Other Three

We could use the same techniques to find the derivatives of the other three inverse trigonometric functions—arccosine, arccotangent, and arccosecant—but there is a much easier way, thanks to the following identities.

Inverse Function–Inverse Cofunction Identities

$$\cos^{-1} x = \pi/2 - \sin^{-1} x$$
$$\cot^{-1} x = \pi/2 - \tan^{-1} x$$
$$\csc^{-1} x = \pi/2 - \sec^{-1} x$$

We saw the first of these identities in Equation (4). The others are derived in a similar way. It follows easily that the derivatives of the inverse cofunctions are the negatives of the derivatives of the corresponding inverse functions. For example, the derivative of $\cos^{-1} x$ is calculated as follows:

$$\begin{aligned}\frac{d}{dx}(\cos^{-1} x) &= \frac{d}{dx}\left(\frac{\pi}{2} - \sin^{-1} x\right) && \text{Identity}\\ &= -\frac{d}{dx}(\sin^{-1} x)\\ &= -\frac{1}{\sqrt{1-x^2}} && \text{Derivative of arcsine}\end{aligned}$$

TABLE 2.1 Derivatives of the inverse trigonometric functions

1. $\dfrac{d(\sin^{-1} u)}{dx} = \dfrac{du/dx}{\sqrt{1-u^2}}, \quad |u| < 1$

2. $\dfrac{d(\cos^{-1} u)}{dx} = -\dfrac{du/dx}{\sqrt{1-u^2}}, \quad |u| < 1$

3. $\dfrac{d(\tan^{-1} u)}{dx} = \dfrac{du/dx}{1+u^2}$

4. $\dfrac{d(\cot^{-1} u)}{dx} = -\dfrac{du/dx}{1+u^2}$

5. $\dfrac{d(\sec^{-1} u)}{dx} = \dfrac{du/dx}{|u|\sqrt{u^2-1}}, \quad |u| > 1$

6. $\dfrac{d(\csc^{-1} u)}{dx} = -\dfrac{du/dx}{|u|\sqrt{u^2-1}}, \quad |u| > 1$

EXERCISES 2.11

Use reference triangles like that in Example 1 to find the angles in Exercises 1–12.

1. a. $\tan^{-1} 1$ **b.** $\tan^{-1}(-\sqrt{3})$ **c.** $\tan^{-1}\left(\dfrac{1}{\sqrt{3}}\right)$

2. a. $\tan^{-1}(-1)$ **b.** $\tan^{-1}\sqrt{3}$ **c.** $\tan^{-1}\left(\dfrac{-1}{\sqrt{3}}\right)$

3. a. $\sin^{-1}\left(\dfrac{-1}{2}\right)$ **b.** $\sin^{-1}\left(\dfrac{1}{\sqrt{2}}\right)$ **c.** $\sin^{-1}\left(\dfrac{-\sqrt{3}}{2}\right)$

4. a. $\sin^{-1}\left(\dfrac{1}{2}\right)$ **b.** $\sin^{-1}\left(\dfrac{-1}{\sqrt{2}}\right)$ **c.** $\sin^{-1}\left(\dfrac{\sqrt{3}}{2}\right)$

5. a. $\cos^{-1}\left(\dfrac{1}{2}\right)$ **b.** $\cos^{-1}\left(\dfrac{-1}{\sqrt{2}}\right)$ **c.** $\cos^{-1}\left(\dfrac{\sqrt{3}}{2}\right)$

6. a. $\cos^{-1}\left(\dfrac{-1}{2}\right)$ **b.** $\cos^{-1}\left(\dfrac{1}{\sqrt{2}}\right)$ **c.** $\cos^{-1}\left(\dfrac{-\sqrt{3}}{2}\right)$

7. **a.** $\sec^{-1}(-\sqrt{2})$ **b.** $\sec^{-1}\left(\frac{2}{\sqrt{3}}\right)$ **c.** $\sec^{-1}(-2)$

8. **a.** $\sec^{-1}\sqrt{2}$ **b.** $\sec^{-1}\left(\frac{-2}{\sqrt{3}}\right)$ **c.** $\sec^{-1} 2$

9. **a.** $\csc^{-1}\sqrt{2}$ **b.** $\csc^{-1}\left(\frac{-2}{\sqrt{3}}\right)$ **c.** $\csc^{-1} 2$

10. **a.** $\csc^{-1}(-\sqrt{2})$ **b.** $\csc^{-1}\left(\frac{2}{\sqrt{3}}\right)$ **c.** $\csc^{-1}(-2)$

11. **a.** $\cot^{-1}(-1)$ **b.** $\cot^{-1}(\sqrt{3})$ **c.** $\cot^{-1}\left(\frac{-1}{\sqrt{3}}\right)$

12. **a.** $\cot^{-1}(1)$ **b.** $\cot^{-1}(-\sqrt{3})$ **c.** $\cot^{-1}\left(\frac{1}{\sqrt{3}}\right)$

Find the limits in Exercises 13–20. (If in doubt, look at the function's graph.)

13. $\lim_{x\to 1^-} \sin^{-1} x$ **14.** $\lim_{x\to -1^+} \cos^{-1} x$

15. $\lim_{x\to\infty} \tan^{-1} x$ **16.** $\lim_{x\to -\infty} \tan^{-1} x$

17. $\lim_{x\to\infty} \sec^{-1} x$ **18.** $\lim_{x\to -\infty} \sec^{-1} x$

19. $\lim_{x\to\infty} \csc^{-1} x$ **20.** $\lim_{x\to -\infty} \csc^{-1} x$

In Exercises 21–42, find the derivative of y with respect to the appropriate variable.

21. $y = \cos^{-1}(x^2)$ **22.** $y = \cos^{-1}(1/x)$

23. $y = \sin^{-1}\sqrt{2}\,t$ **24.** $y = \sin^{-1}(1 - t)$

25. $y = \sec^{-1}(2s + 1)$ **26.** $y = \sec^{-1} 5s$

27. $y = \csc^{-1}(x^2 + 1), \quad x > 0$

28. $y = \csc^{-1}\frac{x}{2}$

29. $y = \sec^{-1}\frac{1}{t}, \quad 0 < t < 1$ **30.** $y = \sin^{-1}\frac{3}{t^2}$

31. $y = \cot^{-1}\sqrt{t}$ **32.** $y = \cot^{-1}\sqrt{t - 1}$

33. $y = \ln(\tan^{-1} x)$ **34.** $y = \tan^{-1}(\ln x)$

35. $y = \csc^{-1}(e^t)$ **36.** $y = \cos^{-1}(e^{-t})$

37. $y = s\sqrt{1 - s^2} + \cos^{-1} s$ **38.** $y = \sqrt{s^2 - 1} - \sec^{-1} s$

39. $y = \tan^{-1}\sqrt{x^2 - 1} + \csc^{-1} x, \quad x > 1$

40. $y = \cot^{-1}\frac{1}{x} - \tan^{-1} x$ **41.** $y = x\sin^{-1} x + \sqrt{1 - x^2}$

42. $y = \ln(x^2 + 4) - x\tan^{-1}\left(\frac{x}{2}\right)$

43. You are sitting in a classroom next to the wall looking at the blackboard at the front of the room. The blackboard is 12 ft long and starts 3 ft from the wall you are sitting next to. Show that your viewing angle is

$$\alpha = \cot^{-1}\frac{x}{15} - \cot^{-1}\frac{x}{3}$$

if you are x ft from the front wall.

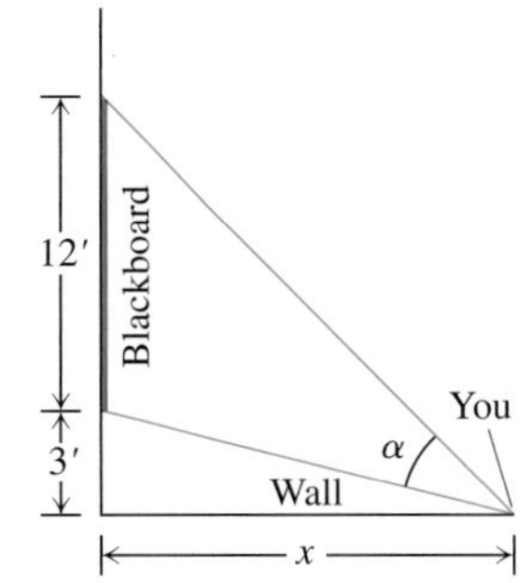

44. Two derivations of the identity $\sec^{-1}(-x) = \pi - \sec^{-1} x$

a. (*Geometric*) Here is a pictorial proof that $\sec^{-1}(-x) = \pi - \sec^{-1} x$. See if you can tell what is going on.

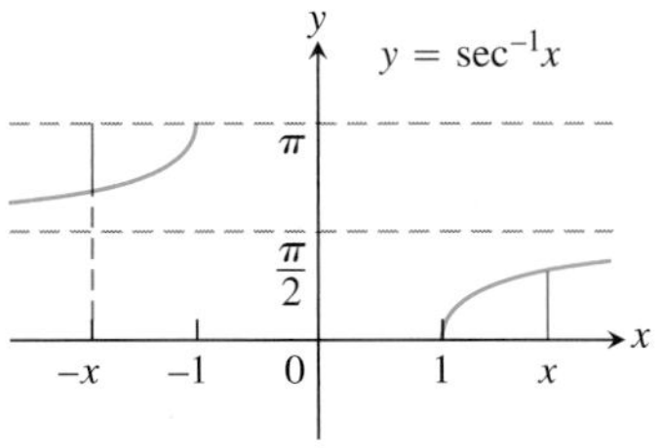

b. (*Algebraic*) Derive the identity $\sec^{-1}(-x) = \pi - \sec^{-1} x$ by combining the following two equations from the text:

$$\cos^{-1}(-x) = \pi - \cos^{-1} x \qquad \text{Eq. (3)}$$

$$\sec^{-1} x = \cos^{-1}(1/x) \qquad \text{Eq. (5)}$$

45. The identity $\sin^{-1} x + \cos^{-1} x = \pi/2$ Figure 2.49 establishes the identity for $0 < x < 1$. To establish it for the rest of $[-1, 1]$, verify by direct calculation that it holds for $x = 1$, 0, and -1. Then, for values of x in $(-1, 0)$, let $x = -a$, $a > 0$, and apply Eqs. (1) and (3) to the sum $\sin^{-1}(-a) + \cos^{-1}(-a)$.

46. Show that the sum $\tan^{-1} x + \tan^{-1}(1/x)$ is constant.

Which of the expressions in Exercises 47 and 48 are defined, and which are not? Give reasons for your answers.

47. **a.** $\tan^{-1} 2$ **b.** $\cos^{-1} 2$

c. $\sin^{-1}\sqrt{2}$

48. **a.** $\sec^{-1} 0$ **b.** $\cot^{-1}(-1/2)$

c. $\csc^{-1}(1/2)$

49. Use the identity

$$\csc^{-1} u = \frac{\pi}{2} - \sec^{-1} u$$

to derive the formula for the derivative of $\csc^{-1} u$ in Table 2.1 from the formula for the derivative of $\sec^{-1} u$.

50. Derive the formula

$$\frac{dy}{dx} = \frac{1}{1 + x^2}$$

for the derivative of $y = \tan^{-1} x$ by differentiating both sides of the equivalent equation $\tan y = x$.

51. Use the Derivative Rule in Section 2.9, Theorem 4, to derive

$$\frac{d}{dx}\sec^{-1} x = \frac{1}{|x|\sqrt{x^2 - 1}}, \quad |x| > 1.$$

52. Use the identity

$$\cot^{-1} u = \frac{\pi}{2} - \tan^{-1} u$$

to derive the formula for the derivative of $\cot^{-1} u$ in Table 2.1 from the formula for the derivative of $\tan^{-1} u$.

T In Exercises 53–55, find the domain and range of each composite function. Then graph the composites on separate screens. Do the graphs make sense in each case? Give reasons for your answers. Comment on any differences you see.

53. **a.** $y = \tan^{-1}(\tan x)$ **b.** $y = \tan(\tan^{-1} x)$

54. **a.** $y = \sin^{-1}(\sin x)$ **b.** $y = \sin(\sin^{-1} x)$

55. **a.** $y = \cos^{-1}(\cos x)$ **b.** $y = \cos(\cos^{-1} x)$

T Use your graphing utility for Exercises 56–58.

56. Graph $y = \sec(\sec^{-1} x) = \sec(\cos^{-1}(1/x))$. Explain what you see.

57. **Newton's serpentine** Graph Newton's serpentine, $y = 4x/(x^2 + 1)$. Then graph $y = 2\sin(2\tan^{-1} x)$ in the same graphing window. What do you see? Explain.

58. Graph $f(x) = \sin^{-1} x$ together with its first two derivatives. Comment on the behavior of f and the shape of its graph in relation to the signs and values of f' and f''.

2.12 Related Rates

In this section we look at problems that ask for the rate at which some variable changes when it is known how the rate of some other related variable (or perhaps several variables) changes. The problem of finding a rate of change from other known rates of change is called a *related rates problem*.

Related Rates Equations

Suppose we are pumping air into a spherical balloon. Both the volume and radius of the balloon are increasing over time. If V is the volume and r is the radius of the balloon at an instant of time, then

$$V = \frac{4}{3}\pi r^3.$$

Using the Chain Rule, we differentiate both sides with respect to t to find an equation relating the rates of change of V and r,

$$\frac{dV}{dt} = \frac{dV}{dr}\frac{dr}{dt} = 4\pi r^2 \frac{dr}{dt}.$$

So if we know the radius r of the balloon and the rate dV/dt at which the volume is increasing at a given instant of time, then we can solve this last equation for dr/dt to find how fast the radius is increasing at that instant. Note that it is easier to directly measure the rate of increase of the volume (the rate at which air is being pumped into the balloon) than it is to measure the increase in the radius. The related rates equation allows us to calculate dr/dt from dV/dt.

Very often the key to relating the variables in a related rates problem is drawing a picture that shows the geometric relations between them, as illustrated in the following example.

EXAMPLE 1 How rapidly will the fluid level inside a vertical cylindrical tank drop if we pump the fluid out at the rate of 3000 L/min?

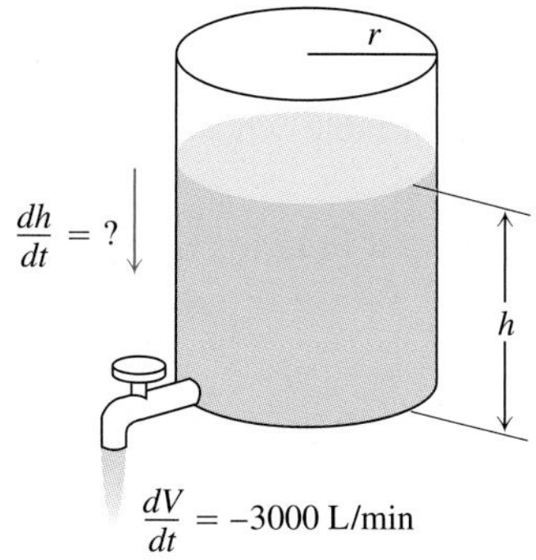

FIGURE 2.53 The rate of change of fluid volume in a cylindrical tank is related to the rate of change of fluid level in the tank (Example 1).

Solution We draw a picture of a partially filled vertical cylindrical tank, calling its radius r and the height of the fluid h (Figure 2.53). Call the volume of the fluid V.

As time passes, the radius remains constant, but V and h change. We think of V and h as differentiable functions of time and use t to represent time. We are told that

$$\frac{dV}{dt} = -3000.$$

We pump out at the rate of 3000 L/min. The rate is negative because the volume is decreasing.

We are asked to find

$$\frac{dh}{dt}.$$

How fast will the fluid level drop?

To find dh/dt, we first write an equation that relates h to V. The equation depends on the units chosen for V, r, and h. With V in liters and r and h in meters, the appropriate equation for the cylinder's volume is

$$V = 1000\pi r^2 h$$

because a cubic meter contains 1000 L.

Since V and h are differentiable functions of t, we can differentiate both sides of the equation $V = 1000\pi r^2 h$ with respect to t to get an equation that relates dh/dt to dV/dt:

$$\frac{dV}{dt} = 1000\pi r^2 \frac{dh}{dt}.$$

r is a constant.

We substitute the known value $dV/dt = -3000$ and solve for dh/dt:

$$\frac{dh}{dt} = \frac{-3000}{1000\pi r^2} = -\frac{3}{\pi r^2}.$$

The fluid level will drop at the rate of $3/(\pi r^2)$ m/min.

The equation $dh/dt = -3/\pi r^2$ shows how the rate at which the fluid level drops depends on the tank's radius. If r is small, dh/dt will be large; if r is large, dh/dt will be small.

$$\text{If } r = 1 \text{ m:} \quad \frac{dh}{dt} = -\frac{3}{\pi} \approx -0.95 \text{ m/min} = -95 \text{ cm/min}.$$

$$\text{If } r = 10 \text{ m:} \quad \frac{dh}{dt} = -\frac{3}{100\pi} \approx -0.0095 \text{ m/min} = -0.95 \text{ cm/min}.$$

■

Related Rates Problem Strategy

1. *Draw a picture and name the variables and constants.* Use t for time. Assume that all variables are differentiable functions of t.
2. *Write down the numerical information* (in terms of the symbols you have chosen).
3. *Write down what you are asked to find* (usually a rate, expressed as a derivative).
4. *Write an equation that relates the variables.* You may have to combine two or more equations to get a single equation that relates the variable whose rate you want to the variables whose rates you know.
5. *Differentiate with respect to t.* Then express the rate you want in terms of the rate and variables whose values you know.
6. *Evaluate.* Use known values to find the unknown rate.

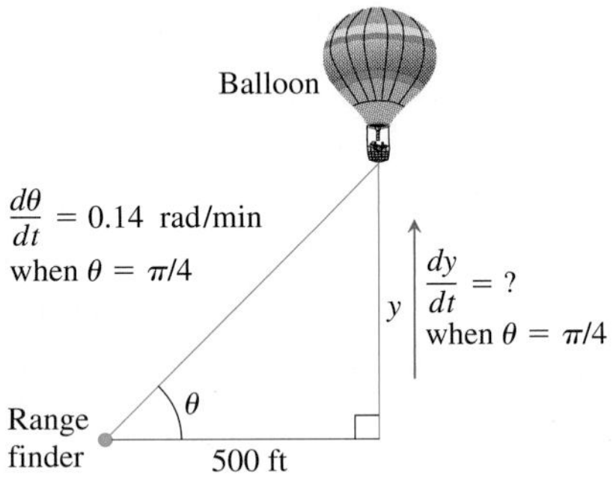

FIGURE 2.54 The rate of change of the balloon's height is related to the rate of change of the angle the range finder makes with the ground (Example 2).

EXAMPLE 2 A hot air balloon rising straight up from a level field is tracked by a range finder 500 ft from the liftoff point. At the moment the range finder's elevation angle is $\pi/4$, the angle is increasing at the rate of 0.14 rad/min. How fast is the balloon rising at that moment?

Solution We answer the question in six steps.

1. *Draw a picture and name the variables and constants* (Figure 2.54). The variables in the picture are

   $\theta =$ the angle in radians the range finder makes with the ground.

   $y =$ the height in feet of the balloon.

We let t represent time in minutes and assume that θ and y are differentiable functions of t.

The one constant in the picture is the distance from the range finder to the liftoff point (500 ft). There is no need to give it a special symbol.

2. *Write down the additional numerical information.*

$$\frac{d\theta}{dt} = 0.14 \text{ rad/min} \qquad \text{when} \qquad \theta = \frac{\pi}{4}$$

3. *Write down what we are to find.* We want dy/dt when $\theta = \pi/4$.
4. *Write an equation that relates the variables y and θ.*

$$\frac{y}{500} = \tan\theta \qquad \text{or} \qquad y = 500\tan\theta$$

5. *Differentiate with respect to t using the Chain Rule.* The result tells how dy/dt (which we want) is related to $d\theta/dt$ (which we know).

$$\frac{dy}{dt} = 500\,(\sec^2\theta)\,\frac{d\theta}{dt}$$

6. *Evaluate with $\theta = \pi/4$ and $d\theta/dt = 0.14$ to find dy/dt.*

$$\frac{dy}{dt} = 500\left(\sqrt{2}\right)^2(0.14) = 140 \qquad \sec\frac{\pi}{4} = \sqrt{2}$$

At the moment in question, the balloon is rising at the rate of 140 ft/min. ■

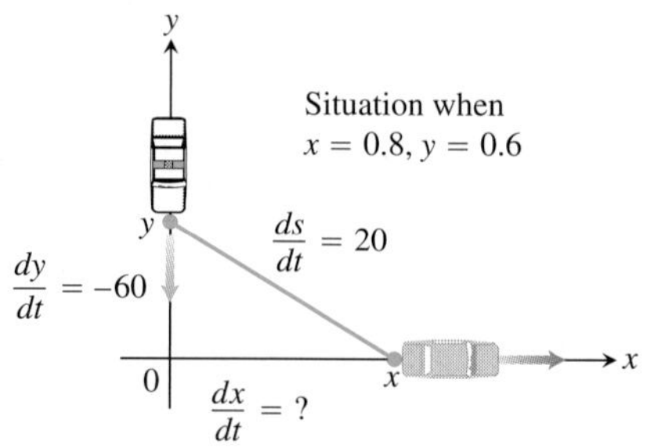

FIGURE 2.55 The speed of the car is related to the speed of the police cruiser and the rate of change of the distance between them (Example 3).

EXAMPLE 3 A police cruiser, approaching a right-angled intersection from the north, is chasing a speeding car that has turned the corner and is now moving straight east. When the cruiser is 0.6 mi north of the intersection and the car is 0.8 mi to the east, the police determine with radar that the distance between them and the car is increasing at 20 mph. If the cruiser is moving at 60 mph at the instant of measurement, what is the speed of the car?

Solution We picture the car and cruiser in the coordinate plane, using the positive x-axis as the eastbound highway and the positive y-axis as the southbound highway (Figure 2.55). We let t represent time and set

$$x = \text{position of car at time } t$$

$$y = \text{position of cruiser at time } t$$

$$s = \text{distance between car and cruiser at time } t.$$

We assume that x, y, and s are differentiable functions of t.

We want to find dx/dt when

$$x = 0.8 \text{ mi}, \qquad y = 0.6 \text{ mi}, \qquad \frac{dy}{dt} = -60 \text{ mph}, \qquad \frac{ds}{dt} = 20 \text{ mph}.$$

Note that dy/dt is negative because y is decreasing.

We differentiate the distance equation

$$s^2 = x^2 + y^2$$

(we could also use $s = \sqrt{x^2 + y^2}$), and obtain

$$2s\frac{ds}{dt} = 2x\frac{dx}{dt} + 2y\frac{dy}{dt}$$

$$\frac{ds}{dt} = \frac{1}{s}\left(x\frac{dx}{dt} + y\frac{dy}{dt}\right)$$

$$= \frac{1}{\sqrt{x^2 + y^2}}\left(x\frac{dx}{dt} + y\frac{dy}{dt}\right).$$

Finally, use $x = 0.8$, $y = 0.6$, $dy/dt = -60$, $ds/dt = 20$, and solve for dx/dt.

$$20 = \frac{1}{\sqrt{(0.8)^2 + (0.6)^2}}\left(0.8\frac{dx}{dt} + (0.6)(-60)\right)$$

$$\frac{dx}{dt} = \frac{20\sqrt{(0.8)^2 + (0.6)^2} + (0.6)(60)}{0.8} = 70$$

At the moment in question, the car's speed is 70 mph. ■

EXAMPLE 4 Water runs into a conical tank at the rate of 9 ft³/min. The tank stands point down and has a height of 10 ft and a base radius of 5 ft. How fast is the water level rising when the water is 6 ft deep?

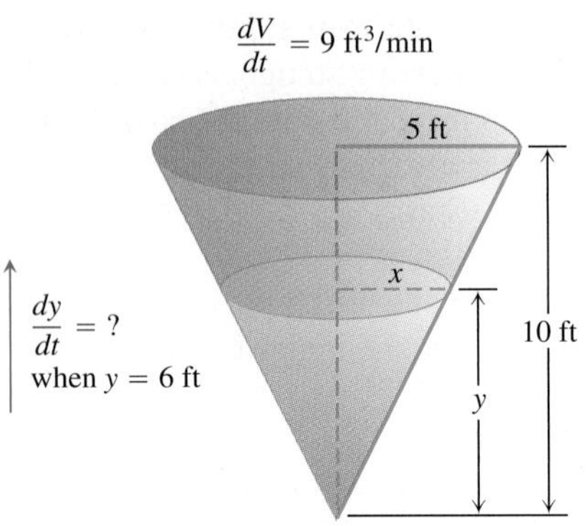

FIGURE 2.56 The geometry of the conical tank and the rate at which water fills the tank determine how fast the water level rises (Example 4).

Solution Figure 2.56 shows a partially filled conical tank. The variables in the problem are

$V =$ volume (ft^3) of the water in the tank at time t (min)

$x =$ radius (ft) of the surface of the water at time t

$y =$ depth (ft) of water in tank at time t.

We assume that V, x, and y are differentiable functions of t. The constants are the dimensions of the tank. We are asked for dy/dt when

$$y = 6 \text{ ft} \quad \text{and} \quad \frac{dV}{dt} = 9 \text{ ft}^3/\text{min}.$$

The water forms a cone with volume

$$V = \frac{1}{3}\pi x^2 y.$$

This equation involves x as well as V and y. Because no information is given about x and dx/dt at the time in question, we need to eliminate x. The similar triangles in Figure 2.56 give us a way to express x in terms of y:

$$\frac{x}{y} = \frac{5}{10} \quad \text{or} \quad x = \frac{y}{2}.$$

Therefore,

$$V = \frac{1}{3}\pi\left(\frac{y}{2}\right)^2 y = \frac{\pi}{12}y^3$$

to give the derivative

$$\frac{dV}{dt} = \frac{\pi}{12} \cdot 3y^2 \frac{dy}{dt} = \frac{\pi}{4}y^2 \frac{dy}{dt}.$$

Finally, use $y = 6$ and $dV/dt = 9$ to solve for dy/dt.

$$9 = \frac{\pi}{4}(6)^2 \frac{dy}{dt}$$

$$\frac{dy}{dt} = \frac{1}{\pi} \approx 0.32$$

At the moment in question, the water level is rising at about 0.32 ft/min. ■

EXERCISES 2.12

1. **Area** Suppose that the radius r and area $A = \pi r^2$ of a circle are differentiable functions of t. Write an equation that relates dA/dt to dr/dt.

2. **Surface area** Suppose that the radius r and surface area $S = 4\pi r^2$ of a sphere are differentiable functions of t. Write an equation that relates dS/dt to dr/dt.

3. **Volume** The radius r and height h of a right circular cylinder are related to the cylinder's volume V by the formula $V = \pi r^2 h$.
   a. How is dV/dt related to dh/dt if r is constant?
   b. How is dV/dt related to dr/dt if h is constant?
   c. How is dV/dt related to dr/dt and dh/dt if neither r nor h is constant?

4. **Volume** The radius r and height h of a right circular cone are related to the cone's volume V by the equation $V = (1/3)\pi r^2 h$.
   a. How is dV/dt related to dh/dt if r is constant?
   b. How is dV/dt related to dr/dt if h is constant?
   c. How is dV/dt related to dr/dt and dh/dt if neither r nor h is constant?

5. **Changing voltage** The voltage V (volts), current I (amperes), and resistance R (ohms) of an electric circuit like the one shown

here are related by the equation $V = IR$. Suppose that V is increasing at the rate of 1 volt/sec while I is decreasing at the rate of 1/3 amp/sec. Let t denote time in seconds.

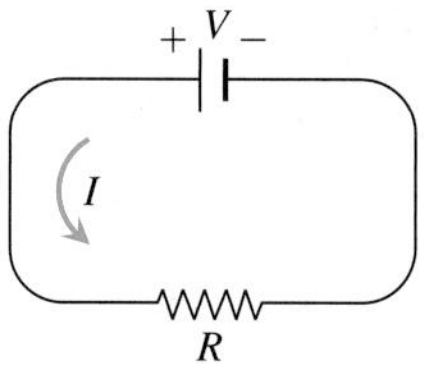

a. What is the value of dV/dt?

b. What is the value of dI/dt?

c. What equation relates dR/dt to dV/dt and dI/dt?

d. Find the rate at which R is changing when $V = 12$ volts and $I = 2$ amp. Is R increasing, or decreasing?

6. Electrical power The power P (watts) of an electric circuit is related to the circuit's resistance R (ohms) and current I (amperes) by the equation $P = RI^2$.

a. How are dP/dt, dR/dt, and dI/dt related if none of P, R, and I are constant?

b. How is dR/dt related to dI/dt if P is constant?

7. Distance Let x and y be differentiable functions of t and let $s = \sqrt{x^2 + y^2}$ be the distance between the points $(x, 0)$ and $(0, y)$ in the xy-plane.

a. How is ds/dt related to dx/dt if y is constant?

b. How is ds/dt related to dx/dt and dy/dt if neither x nor y is constant?

c. How is dx/dt related to dy/dt if s is constant?

8. Diagonals If x, y, and z are lengths of the edges of a rectangular box, the common length of the box's diagonals is $s = \sqrt{x^2 + y^2 + z^2}$.

a. Assuming that x, y, and z are differentiable functions of t, how is ds/dt related to dx/dt, dy/dt, and dz/dt?

b. How is ds/dt related to dy/dt and dz/dt if x is constant?

c. How are dx/dt, dy/dt, and dz/dt related if s is constant?

9. Area The area A of a triangle with sides of lengths a and b enclosing an angle of measure θ is

$$A = \frac{1}{2}ab\sin\theta.$$

a. How is dA/dt related to $d\theta/dt$ if a and b are constant?

b. How is dA/dt related to $d\theta/dt$ and da/dt if only b is constant?

c. How is dA/dt related to $d\theta/dt$, da/dt, and db/dt if none of a, b, and θ are constant?

10. Heating a plate When a circular plate of metal is heated in an oven, its radius increases at the rate of 0.01 cm/min. At what rate is the plate's area increasing when the radius is 50 cm?

11. Changing dimensions in a rectangle The length l of a rectangle is decreasing at the rate of 2 cm/sec while the width w is increasing at the rate of 2 cm/sec. When $l = 12$ cm and $w = 5$ cm, find the rates of change of **(a)** the area, **(b)** the perimeter, and **(c)** the lengths of the diagonals of the rectangle. Which of these quantities are decreasing, and which are increasing?

12. Changing dimensions in a rectangular box Suppose that the edge lengths x, y, and z of a closed rectangular box are changing at the following rates:

$$\frac{dx}{dt} = 1 \text{ m/sec}, \quad \frac{dy}{dt} = -2 \text{ m/sec}, \quad \frac{dz}{dt} = 1 \text{ m/sec}.$$

Find the rates at which the box's **(a)** volume, **(b)** surface area, and **(c)** diagonal length $s = \sqrt{x^2 + y^2 + z^2}$ are changing at the instant when $x = 4$, $y = 3$, and $z = 2$.

13. A sliding ladder A 13-ft ladder is leaning against a house when its base starts to slide away. By the time the base is 12 ft from the house, the base is moving at the rate of 5 ft/sec.

a. How fast is the top of the ladder sliding down the wall then?

b. At what rate is the area of the triangle formed by the ladder, wall, and ground changing then?

c. At what rate is the angle θ between the ladder and the ground changing then?

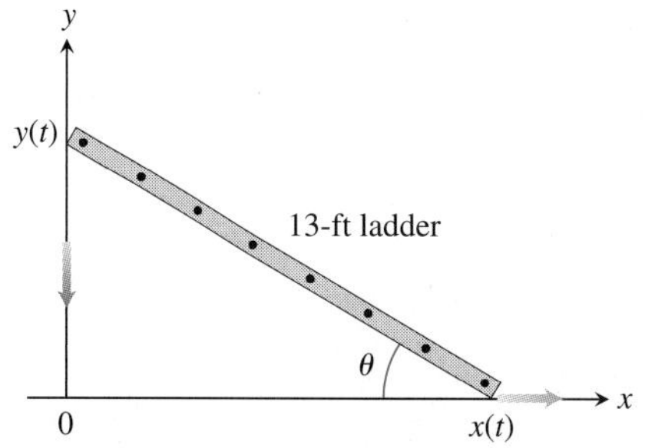

14. Commercial air traffic Two commercial airplanes are flying at 40,000 ft along straight-line courses that intersect at right angles. Plane A is approaching the intersection point at a speed of 442 knots (nautical miles per hour; a nautical mile is 2000 yd). Plane B is approaching the intersection at 481 knots. At what rate is the distance between the planes changing when A is 5 nautical miles from the intersection point and B is 12 nautical miles from the intersection point?

15. Flying a kite A girl flies a kite at a height of 300 ft, the wind carrying the kite horizontally away from her at a rate of 25 ft/sec. How fast must she let out the string when the kite is 500 ft away from her?

16. Boring a cylinder The mechanics at Lincoln Automotive are reboring a 6-in.-deep cylinder to fit a new piston. The machine they are using increases the cylinder's radius one-thousandth of an inch every 3 min. How rapidly is the cylinder volume increasing when the bore (diameter) is 3.800 in.?

17. A growing sand pile Sand falls from a conveyor belt at the rate of 10 m^3/min onto the top of a conical pile. The height of the pile is always three-eighths of the base diameter. How fast are the **(a)** height and **(b)** radius changing when the pile is 4 m high? Answer in centimeters per minute.

18. A draining conical reservoir Water is flowing at the rate of 50 m^3/min from a shallow concrete conical reservoir (vertex down) of base radius 45 m and height 6 m.

a. How fast (centimeters per minute) is the water level falling when the water is 5 m deep?

b. How fast is the radius of the water's surface changing then? Answer in centimeters per minute.

19. A draining hemispherical reservoir Water is flowing at the rate of 6 m^3/min from a reservoir shaped like a hemispherical bowl of radius 13 m, shown here in profile. Answer the following questions, given that the volume of water in a hemispherical bowl of radius R is $V = (\pi/3)y^2(3R - y)$ when the water is y meters deep.

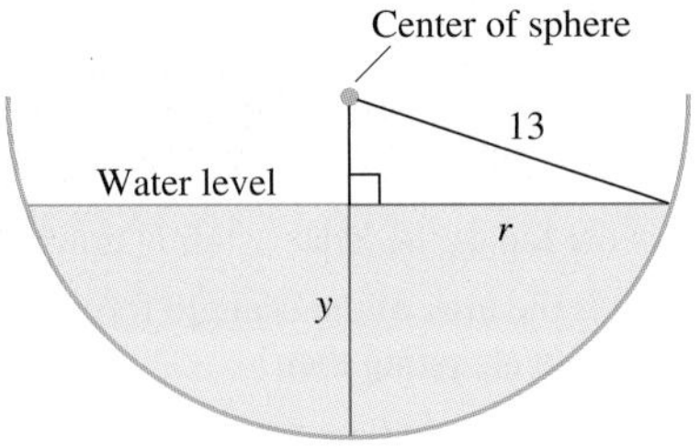

a. At what rate is the water level changing when the water is 8 m deep?

b. What is the radius r of the water's surface when the water is y m deep?

c. At what rate is the radius r changing when the water is 8 m deep?

20. A growing raindrop Suppose that a drop of mist is a perfect sphere and that, through condensation, the drop picks up moisture at a rate proportional to its surface area. Show that under these circumstances the drop's radius increases at a constant rate.

21. The radius of an inflating balloon A spherical balloon is inflated with helium at the rate of $100\pi \text{ ft}^3/\text{min}$. How fast is the balloon's radius increasing at the instant the radius is 5 ft? How fast is the surface area increasing?

22. Hauling in a dinghy A dinghy is pulled toward a dock by a rope from the bow through a ring on the dock 6 ft above the bow. The rope is hauled in at the rate of 2 ft/sec.

a. How fast is the boat approaching the dock when 10 ft of rope are out?

b. At what rate is the angle θ changing then (see the figure)?

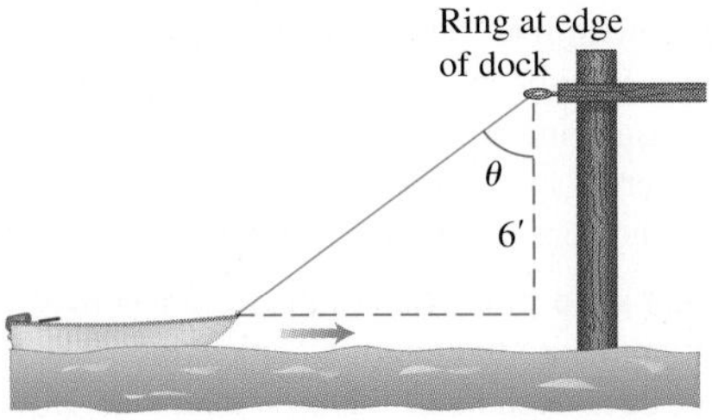

23. A balloon and a bicycle A balloon is rising vertically above a level, straight road at a constant rate of 1 ft/sec. Just when the balloon is 65 ft above the ground, a bicycle moving at a constant rate of 17 ft/sec passes under it. How fast is the distance $s(t)$ between the bicycle and balloon increasing 3 sec later?

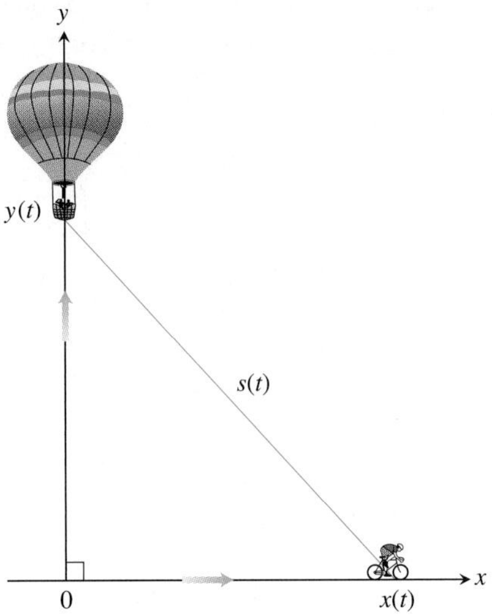

24. Making coffee Coffee is draining from a conical filter into a cylindrical coffeepot at the rate of $10 \text{ in}^3/\text{min}$.

a. How fast is the level in the pot rising when the coffee in the cone is 5 in. deep?

b. How fast is the level in the cone falling then?

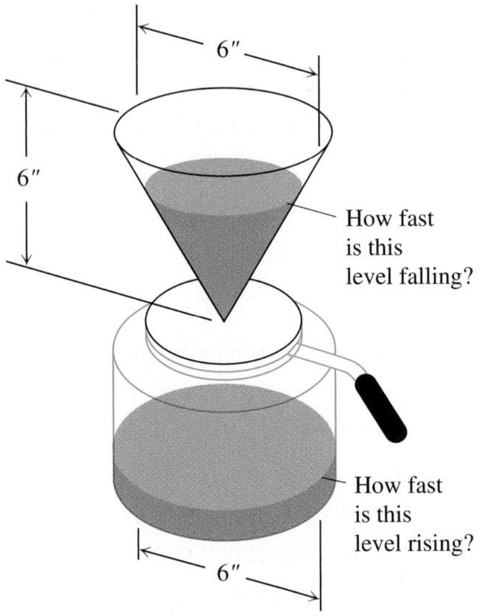

25. Moving along a parabola A particle moves along the parabola $y = x^2$ in the first quadrant in such a way that its x-coordinate (measured in meters) increases at a steady 10 m/sec. How fast is the angle of inclination θ of the line joining the particle to the origin changing when $x = 3$ m?

26. Moving along another parabola A particle moves from right to left along the parabolic curve $y = \sqrt{-x}$ in such a way that its x-coordinate (measured in meters) decreases at the rate of 8 m/sec. How fast is the angle of inclination θ of the line joining the particle to the origin changing when $x = -4$?

27. Motion in the plane The coordinates of a particle in the metric xy-plane are differentiable functions of time t with

$dx/dt = -1$ m/sec and $dy/dt = -5$ m/sec. How fast is the particle's distance from the origin changing as it passes through the point (5, 12)?

28. A moving shadow A man 6 ft tall walks at the rate of 5 ft/sec toward a streetlight that is 16 ft above the ground. At what rate is the tip of his shadow moving? At what rate is the length of his shadow changing when he is 10 ft from the base of the light?

29. Another moving shadow A light shines from the top of a pole 50 ft high. A ball is dropped from the same height from a point 30 ft away from the light. (See accompanying figure.) How fast is the shadow of the ball moving along the ground 1/2 sec later? (Assume the ball falls a distance $s = 16t^2$ ft in t sec.)

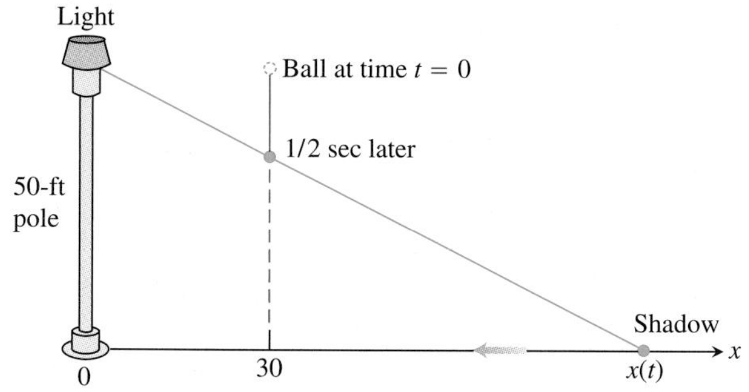

30. Videotaping a moving car You are videotaping a race from a stand 132 ft from the track, following a car that is moving at 180 mi/h (264 ft/sec). How fast will your camera angle θ be changing when the car is right in front of you? A half second later?

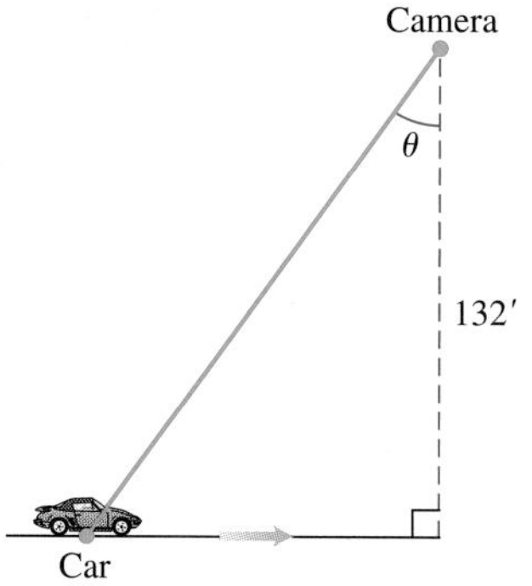

31. A melting ice layer A spherical iron ball 8 in. in diameter is coated with a layer of ice of uniform thickness. If the ice melts at the rate of 10 in^3/min, how fast is the thickness of the ice decreasing when it is 2 in. thick? How fast is the outer surface area of ice decreasing?

32. Highway patrol A highway patrol plane flies 3 mi above a level, straight road at a steady 120 mi/h. The pilot sees an oncoming car and with radar determines that at the instant the line-of-sight distance from plane to car is 5 mi, the line-of-sight distance is decreasing at the rate of 160 mi/h. Find the car's speed along the highway.

33. A building's shadow On a morning of a day when the sun will pass directly overhead, the shadow of an 80-ft building on level ground is 60 ft long. At the moment in question, the angle θ the sun makes with the ground is increasing at the rate of 0.27°/min. At what rate is the shadow decreasing? (Remember to use radians. Express your answer in inches per minute, to the nearest tenth.)

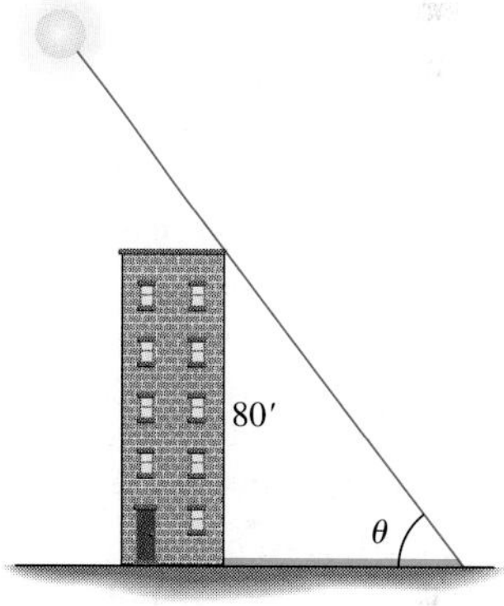

34. Ships Two ships are steaming straight away from a point O along routes that make a 120° angle. Ship A moves at 14 knots (nautical miles per hour; a nautical mile is 2000 yd). Ship B moves at 21 knots. How fast are the ships moving apart when $OA = 5$ and $OB = 3$ nautical miles?

2.13 Linearization and Differentials

Sometimes we can approximate complicated functions with simpler ones that give the accuracy we want for specific applications and are easier to work with. The approximating functions discussed in this section are called *linearizations*, and they are based on tangent lines. Other approximating functions, such as polynomials, are discussed in Chapter 7.

We introduce new variables dx and dy, called *differentials*, and define them in a way that makes Leibniz's notation for the derivative dy/dx a true ratio. We use dy to estimate error in measurement, which then provides for a precise proof of the Chain Rule (Section 2.7).

Linearization

As you can see in Figure 2.57, the tangent to the curve $y = x^2$ lies close to the curve near the point of tangency. For a brief interval to either side, the y-values along the tangent line give good approximations to the y-values on the curve. We observe this phenomenon by zooming in on the two graphs at the point of tangency or by looking at tables of values for the difference between $f(x)$ and its tangent line near the x-coordinate of the point of tangency. The phenomenon is true not just for parabolas; every differentiable curve behaves locally like its tangent line.

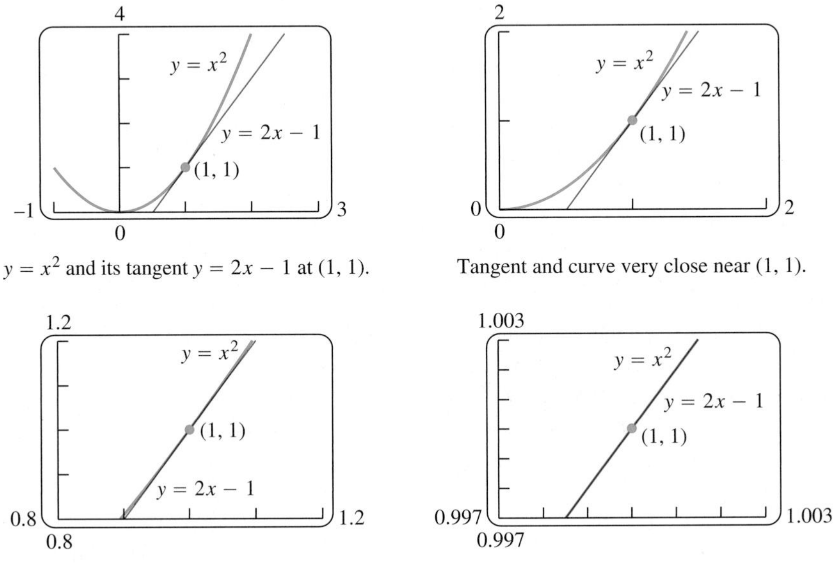

$y = x^2$ and its tangent $y = 2x - 1$ at (1, 1).

Tangent and curve very close near (1, 1).

Tangent and curve very close throughout entire x-interval shown.

Tangent and curve closer still. Computer screen cannot distinguish tangent from curve on this x-interval.

FIGURE 2.57 The more we magnify the graph of a function near a point where the function is differentiable, the flatter the graph becomes and the more it resembles its tangent.

In general, the tangent to $y = f(x)$ at a point $x = a$, where f is differentiable (Figure 2.58), passes through the point $(a, f(a))$, so its point-slope equation is

$$y = f(a) + f'(a)(x - a).$$

Thus, this tangent line is the graph of the linear function

$$L(x) = f(a) + f'(a)(x - a).$$

For as long as this line remains close to the graph of f, $L(x)$ gives a good approximation to $f(x)$.

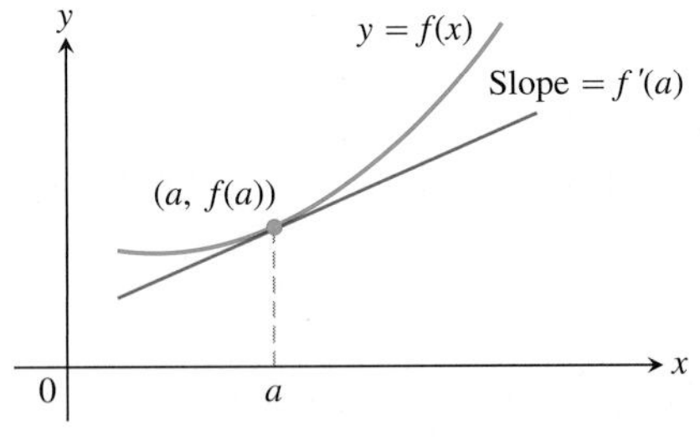

FIGURE 2.58 The tangent to the curve $y = f(x)$ at $x = a$ is the line $L(x) = f(a) + f'(a)(x - a)$.

DEFINITIONS If f is differentiable at $x = a$, then the approximating function

$$L(x) = f(a) + f'(a)(x - a)$$

is the **linearization** of f at a. The approximation

$$f(x) \approx L(x)$$

of f by L is the **standard linear approximation** of f at a. The point $x = a$ is the **center** of the approximation.

EXAMPLE 1 Find the linearization of $f(x) = \sqrt{1 + x}$ at $x = 0$ (Figure 2.59).

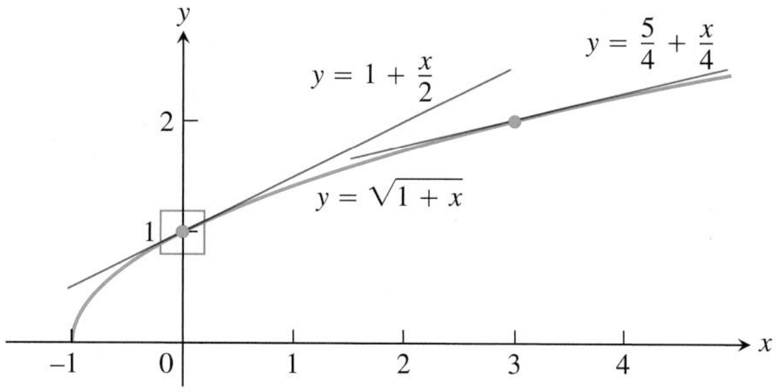

FIGURE 2.59 The graph of $y = \sqrt{1 + x}$ and its linearizations at $x = 0$ and $x = 3$. Figure 2.60 shows a magnified view of the small window about 1 on the y-axis.

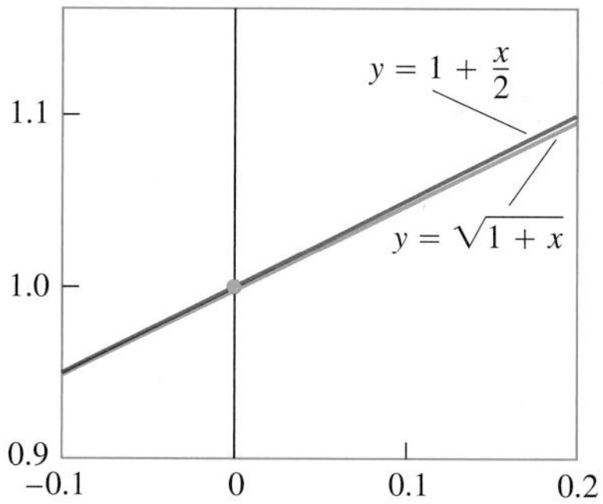

FIGURE 2.60 Magnified view of the window in Figure 2.59.

Solution Since

$$f'(x) = \frac{1}{2}\left(1 + x\right)^{-1/2},$$

we have $f(0) = 1$ and $f'(0) = 1/2$, giving the linearization

$$L(x) = f(a) + f'(a)(x - a) = 1 + \frac{1}{2}\left(x - 0\right) = 1 + \frac{x}{2}.$$

See Figure 2.60. ■

A linear approximation normally loses accuracy away from its center. As Figure 2.59 suggests, the approximation $\sqrt{1 + x} \approx 1 + (x/2)$ will probably be too crude to be useful near $x = 3$. There, we need the linearization at $x = 3$.

EXAMPLE 2 Find the linearization of $f(x) = \sqrt{1 + x}$ at $x = 3$.

Solution We evaluate the equation defining $L(x)$ at $a = 3$. With

$$f(3) = 2, \qquad f'(3) = \frac{1}{2}\left(1 + x\right)^{-1/2}\bigg|_{x=3} = \frac{1}{4},$$

we have

$$L(x) = 2 + \frac{1}{4}(x - 3) = \frac{5}{4} + \frac{x}{4}.$$ ■

At $x = 3.2$, the linearization in Example 2 gives

$$\sqrt{1 + x} = \sqrt{1 + 3.2} \approx \frac{5}{4} + \frac{3.2}{4} = 1.250 + 0.800 = 2.050,$$

which differs from the true value $\sqrt{4.2} \approx 2.04939$ by less than one one-thousandth. The linearization in Example 1 gives the value $\sqrt{4.2} \approx 2.6$.

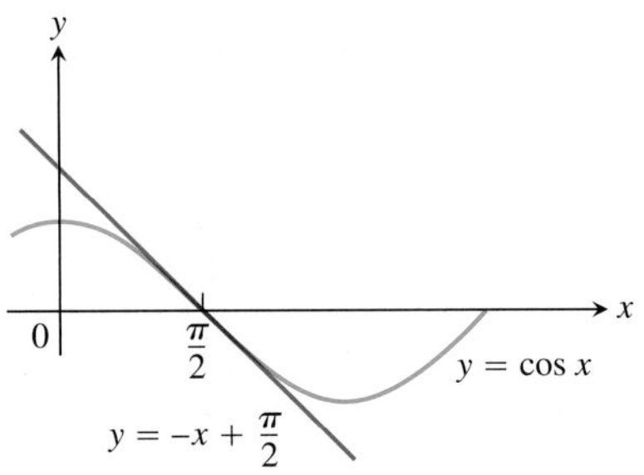

FIGURE 2.61 The graph of $f(x) = \cos x$ and its linearization at $x = \pi/2$. Near $x = \pi/2$, $\cos x \approx -x + (\pi/2)$ (Example 3).

EXAMPLE 3 Find the linearization of $f(x) = \cos x$ at $x = \pi/2$ (Figure 2.61).

Solution Since $f(\pi/2) = \cos(\pi/2) = 0$, $f'(x) = -\sin x$, and $f'(\pi/2) = -\sin(\pi/2) = -1$, we have

$$\begin{aligned} L(x) &= f(a) + f'(a)(x - a) \\ &= 0 + (-1)\left(x - \frac{\pi}{2}\right) \\ &= -x + \frac{\pi}{2}. \end{aligned}$$ ■

An important linear approximation for roots and powers is

$$(1 + x)^k \approx 1 + kx \qquad (x \text{ near } 0;\ \text{any number } k)$$

(Exercise 15). This approximation, good for values of x sufficiently close to zero, has broad application. For example, when x is small,

$$\frac{1}{1 - x} = (1 - x)^{-1} \approx 1 + (-1)(-x) = 1 + x \qquad k = -1;\ \text{replace } x \text{ by } -x.$$

$$\frac{1}{\sqrt{1 - x^2}} = (1 - x^2)^{-1/2} \approx 1 + \left(-\frac{1}{2}\right)(-x^2) = 1 + \frac{1}{2}x^2 \qquad k = -1/2;\ \text{replace } x \text{ by } -x^2.$$

Differentials

We sometimes use the Leibniz notation dy/dx to represent the derivative of y with respect to x. Contrary to its appearance, it is not a ratio. We now introduce two new variables dx and dy with the property that if their ratio exists, it will be equal to the derivative.

DEFINITION Let $y = f(x)$ be a differentiable function. The **differential dx** is an independent variable. The **differential dy** is

$$dy = f'(x)\,dx.$$

Unlike the independent variable dx, the variable dy is always a dependent variable. It depends on both x and dx. If dx is given a specific value and x is a particular number in the domain of the function f, then these values determine the numerical value of dy.

The geometric meaning of differentials is shown in Figure 2.62. Let $x = a$ and set $dx = \Delta x$. The corresponding change in $y = f(x)$ is

$$\Delta y = f(a + dx) - f(a).$$

The corresponding change in the tangent line L is

$$\begin{aligned} \Delta L &= L(a + dx) - L(a) \\ &= \underbrace{f(a) + f'(a)[(a + dx) - a]}_{L(a+dx)} - \underbrace{f(a)}_{L(a)} \\ &= f'(a)\,dx. \end{aligned}$$

That is, the change in the linearization of f is precisely the value of the differential dy when $x = a$ and $dx = \Delta x$. Therefore, dy represents the amount the tangent line rises or falls when x changes by an amount $dx = \Delta x$.

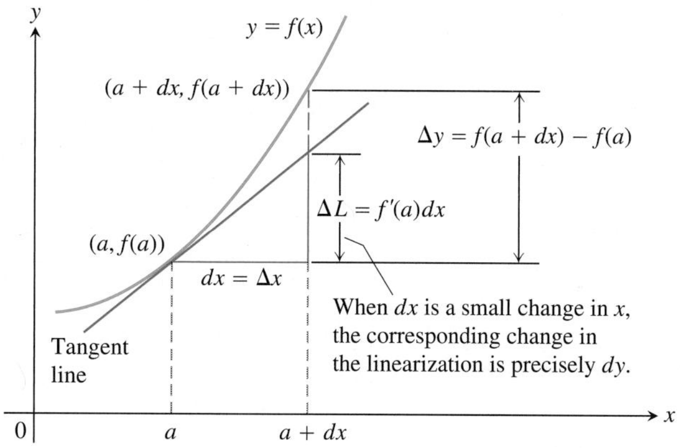

FIGURE 2.62 Geometrically, the differential dy is the change ΔL in the linearization of f when $x = a$ changes by an amount $dx = \Delta x$.

EXAMPLE 4

(a) Find dy if $y = x^5 + 37x$.

(b) Find the value of dy when $x = 1$ and $dx = 0.2$.

Solution

(a) $dy = (5x^4 + 37)\,dx$

(b) Substituting $x = 1$ and $dx = 0.2$ in the expression for dy, we have

$$dy = (5 \cdot 1^4 + 37)0.2 = 8.4.$$

■

If $dx \neq 0$, then the quotient of the differential dy by the differential dx is equal to the derivative $f'(x)$ because

$$dy \div dx = \frac{f'(x)\,dx}{dx} = f'(x) = \frac{dy}{dx}.$$

We sometimes write

$$df = f'(x)\,dx$$

in place of $dy = f'(x)\,dx$, calling df the **differential of f**.

Estimating with Differentials

Suppose we know the value of a differentiable function $f(x)$ at a point a and want to predict how much this value will change if we move to a nearby point $a + dx$. If $dx = \Delta x$ is small, then we can see from Figure 2.62 that Δy is approximately equal to the differential dy. Since

$$f(a + dx) = f(a) + \Delta y,$$

the differential approximation gives

$$f(a + dx) \approx f(a) + dy$$

where $dx = \Delta x$. Thus the approximation $\Delta y \approx dy$ can be used to calculate $f(a + dx)$ when $f(a)$ is known and dx is small.

EXAMPLE 5 The radius r of a circle increases from $a = 10$ m to 10.1 m (Figure 2.63). Use dA to estimate the increase in the circle's area A. Estimate the area of the enlarged circle and compare your estimate to the true area.

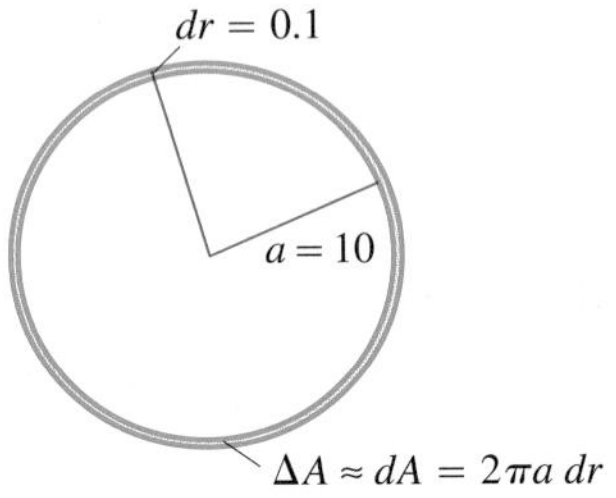

FIGURE 2.63 When dr is small compared with a, as it is when $dr = 0.1$ and $a = 10$, the differential $dA = 2\pi a\, dr$ gives a way to estimate the area of the circle with radius $r = a + dr$ (Example 5).

Solution Since $A = \pi r^2$, the estimated increase is

$$dA = A'(a)\,dr = 2\pi a\,dr = 2\pi(10)(0.1) = 2\pi \text{ m}^2.$$

Thus,

$$\begin{aligned} A(10 + 0.1) &\approx A(10) + 2\pi \\ &= \pi(10)^2 + 2\pi = 102\pi. \end{aligned}$$

The area of a circle of radius 10.1 m is approximately 102π m^2.

The true area is

$$\begin{aligned} A(10.1) &= \pi(10.1)^2 \\ &= 102.01\pi \text{ m}^2. \end{aligned}$$

The error in our estimate is 0.01π m^2, which is the difference $\Delta A - dA$. ■

The following result describes how the differential approximation estimates the true change.

> **Change in $y = f(x)$ near $x = a$**
>
> If $y = f(x)$ is differentiable at $x = a$ and x changes from a to $a + \Delta x$, the change Δy in f is given by an equation of the form
>
> $$\Delta y = f'(a)\,\Delta x + \epsilon\,\Delta x \tag{1}$$
>
> in which $\epsilon \to 0$ as $\Delta x \to 0$.

The result in Equation (1) is obtained in the following way:

$$\begin{aligned}\text{Approximation error} &= \Delta f - df \\ &= \Delta f - f'(a)\Delta x \\ &= \underbrace{f(a+\Delta x) - f(a)}_{\Delta f} - f'(a)\Delta x \\ &= \underbrace{\left(\frac{f(a+\Delta x) - f(a)}{\Delta x} - f'(a)\right)}_{\text{Call this part } \epsilon} \cdot \Delta x \\ &= \epsilon \cdot \Delta x.\end{aligned}$$

As $\Delta x \to 0$, the difference quotient

$$\frac{f(a+\Delta x) - f(a)}{\Delta x}$$

approaches $f'(a)$ (remember the definition of $f'(a)$), so the quantity in parentheses becomes a very small number (which is why we called it ϵ). In fact, $\epsilon \to 0$ as $\Delta x \to 0$. When Δx is small, the approximation error $\epsilon\ \Delta x$ is smaller still.

$$\underbrace{\Delta f}_{\substack{\text{true}\\ \text{change}}} = \underbrace{f'(a)\Delta x}_{\substack{\text{estimated}\\ \text{change}}} + \underbrace{\epsilon\ \Delta x}_{\text{error}}$$

In Example 5 we found that

$$\Delta A = \pi(10.1)^2 - \pi(10)^2 = (102.01 - 100)\pi = (\underbrace{2\pi}_{dA} + \underbrace{0.01\pi}_{\text{error}})\ \text{m}^2$$

so the approximation error is $\Delta A - dA = \epsilon\ \Delta r = 0.01\pi$ and $\epsilon = 0.01\pi/\Delta r = 0.01\pi/0.1 = 0.1\pi$ m.

The equation $df = f'(x)\,dx$ tells how *sensitive* the output of f is to a change in input at different values of x. The larger the value of f' at $x = a$, the greater the effect of a given change dx.

	True	**Estimated**
Absolute change	$\Delta f = f(a+dx) - f(a)$	$df = f'(a)\,dx$
Relative change	$\dfrac{\Delta f}{f(a)}$	$\dfrac{df}{f(a)}$
Percentage change	$\dfrac{\Delta f}{f(a)} \times 100$	$\dfrac{df}{f(a)} \times 100$

The estimated relative change in Example 5 is $dA/A = 2\pi/100\pi = 0.02$, or about 2%.

Proof of the Chain Rule

Equation (1) enables us to prove the Chain Rule correctly. Our goal is to show that if $f(u)$ is a differentiable function of u and $u = g(x)$ is a differentiable function of x, then the composite $y = f(g(x))$ is a differentiable function of x. More precisely, if g is differentiable at x_0 and f is differentiable at $g(x_0)$, then the composite is differentiable at x_0 and

$$\left.\frac{dy}{dx}\right|_{x=x_0} = f'(g(x_0)) \cdot g'(x_0).$$

Let Δx be an increment in x and let Δu and Δy be the corresponding increments in u and y. Applying Equation (1) we have

$$\Delta u = g'(x_0)\Delta x + \epsilon_1 \Delta x = (g'(x_0) + \epsilon_1)\Delta x,$$

where $\epsilon_1 \to 0$ as $\Delta x \to 0$. Similarly,

$$\Delta y = f'(u_0)\Delta u + \epsilon_2 \Delta u = (f'(u_0) + \epsilon_2)\Delta u,$$

where $\epsilon_2 \to 0$ as $\Delta u \to 0$. Notice also that $\Delta u \to 0$ as $\Delta x \to 0$. Combining the equations for Δu and Δy gives

$$\Delta y = (f'(u_0) + \epsilon_2)(g'(x_0) + \epsilon_1)\Delta x,$$

so

$$\frac{\Delta y}{\Delta x} = f'(u_0)g'(x_0) + \epsilon_2 g'(x_0) + f'(u_0)\epsilon_1 + \epsilon_2\epsilon_1.$$

Since ϵ_1 and ϵ_2 go to zero as Δx goes to zero, three of the four terms on the right vanish in the limit, leaving

$$\left.\frac{dy}{dx}\right|_{x=x_0} = \lim_{\Delta x \to 0} \frac{\Delta y}{\Delta x} = f'(u_0)g'(x_0) = f'(g(x_0)) \cdot g'(x_0).$$ ■

EXERCISES 2.13

In Exercises 1–5, find the linearization $L(x)$ of $f(x)$ at $x = a$.

1. $f(x) = x^3 - 2x + 3, \quad a = 2$

2. $f(x) = \sqrt{x^2 + 9}, \quad a = -4$

3. $f(x) = x + \frac{1}{x}, \quad a = 1$

4. $f(x) = \sqrt[3]{x}, \quad a = -8$

5. $f(x) = \tan x, \quad a = \pi$

6. Common linear approximations at $x = 0$ Find the linearizations of the following functions at $x = 0$.

(a) $\sin x$ **(b)** $\cos x$ **(c)** $\tan x$ **(d)** e^x **(e)** $\ln(1 + x)$

You want linearizations that will replace the functions in Exercises 7–14 over intervals that include the given points x_0. To make your subsequent work as simple as possible, you want to center each linearization not at x_0 but at a nearby integer $x = a$ at which the given function and its derivative are easy to evaluate. What linearization do you use in each case?

7. $f(x) = x^2 + 2x, \quad x_0 = 0.1$

8. $f(x) = x^{-1}, \quad x_0 = 0.9$

9. $f(x) = 2x^2 + 4x - 3, \quad x_0 = -0.9$

10. $f(x) = 1 + x, \quad x_0 = 8.1$

11. $f(x) = \sqrt[3]{x}, \quad x_0 = 8.5$

12. $f(x) = \frac{x}{x + 1}, \quad x_0 = 1.3$

13. $f(x) = e^{-x}, \quad x_0 = -0.1$

14. $f(x) = \sin^{-1} x, \quad x_0 = \pi/12$

15. Show that the linearization of $f(x) = (1 + x)^k$ at $x = 0$ is $L(x) = 1 + kx$.

16. Use the linear approximation $(1 + x)^k \approx 1 + kx$ to find an approximation for the function $f(x)$ for values of x near zero.

a. $f(x) = (1 - x)^6$

b. $f(x) = \frac{2}{1 - x}$

c. $f(x) = \frac{1}{\sqrt{1 + x}}$

d. $f(x) = \sqrt{2 + x^2}$

e. $f(x) = (4 + 3x)^{1/3}$

f. $f(x) = \sqrt[3]{\left(1 - \frac{1}{2 + x}\right)^2}$

17. Faster than a calculator Use the approximation $(1 + x)^k \approx 1 + kx$ to estimate the following.

a. $(1.0002)^{50}$

b. $\sqrt[3]{1.009}$

18. Find the linearization of $f(x) = \sqrt{x + 1} + \sin x$ at $x = 0$. How is it related to the individual linearizations of $\sqrt{x + 1}$ and $\sin x$ at $x = 0$?

In Exercises 19–28, find dy.

19. $y = x^3 - 3\sqrt{x}$

20. $y = \frac{2x}{1 + x^2}$

21. $2y^{3/2} + xy - x = 0$

22. $xy^2 - 4x^{3/2} - y = 0$

23. $y = \sin(5\sqrt{x})$

24. $y = 4\tan(x^3/3)$

25. $y = e^{\sqrt{x}}$

26. $y = xe^{-x}$

27. $y = \ln(1 + x^2)$

28. $y = \sec^{-1}(e^{-x})$

In Exercises 29–32, each function $f(x)$ changes value when x changes from x_0 to $x_0 + dx$. Find

a. the change $\Delta f = f(x_0 + dx) - f(x_0)$;

b. the value of the estimate $df = f'(x_0)\,dx$; and

c. the approximation error $|\Delta f - df|$.

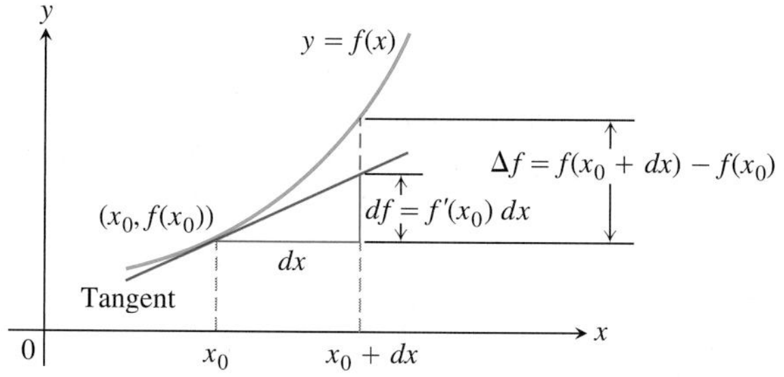

29. $f(x) = x^2 + 2x,\quad x_0 = 1,\quad dx = 0.1$

30. $f(x) = x^3 - x,\quad x_0 = 1,\quad dx = 0.1$

31. $f(x) = x^{-1},\quad x_0 = 0.5,\quad dx = 0.1$

32. $f(x) = x^3 - 2x + 3,\quad x_0 = 2,\quad dx = 0.1$

In Exercises 33–36, write a differential formula that estimates the given change in volume or surface area.

33. The change in the volume $V = (4/3)\pi r^3$ of a sphere when the radius changes from r_0 to $r_0 + dr$

34. The change in the surface area $S = 6x^2$ of a cube when the edge lengths change from x_0 to $x_0 + dx$

35. The change in the volume $V = \pi r^2 h$ of a right circular cylinder when the radius changes from r_0 to $r_0 + dr$ and the height does not change

36. The change in the lateral surface area $S = \pi r\sqrt{r^2 + h^2}$ of a right circular cone when the radius changes from r_0 to $r_0 + dr$ and the height does not change

37. The radius of a circle is increased from 2.00 to 2.02 m.

a. Estimate the resulting change in area.

b. Express the estimate as a percentage of the circle's original area.

38. The diameter of a tree was 10 in. During the following year, the circumference increased 2 in. About how much did the tree's diameter increase? The tree's cross-section area?

39. Estimating volume Estimate the volume of material in a cylindrical shell with height 30 in., radius 6 in., and shell thickness 0.5 in.

40. Estimating height of a building A surveyor, standing 30 ft from the base of a building, measures the angle of elevation to the top of the building to be 75°. How accurately must the angle be measured for the percentage error in estimating the height of the building to be less than 4%?

41. Tolerance The height and radius of a right circular cylinder are equal, so the cylinder's volume is $V = \pi h^3$. The volume is to be calculated with an error of no more than 1% of the true value. Find approximately the greatest error that can be tolerated in the measurement of h, expressed as a percentage of h.

42. Tolerance

a. About how accurately must the interior diameter of a 10-m-high cylindrical storage tank be measured to calculate the tank's volume to within 1% of its true value?

b. About how accurately must the tank's exterior diameter be measured to calculate the amount of paint it will take to paint the side of the tank to within 5% of the true amount?

43. Minting coins A manufacturer contracts to mint coins for the federal government. How much variation dr in the radius of the coins can be tolerated if the coins are to weigh within $1/1000$ of their ideal weight? Assume that the thickness does not vary.

44. Measuring acceleration of gravity When the length L of a clock pendulum is held constant by controlling its temperature, the pendulum's period T depends on the acceleration of gravity g. The period will therefore vary slightly as the clock is moved from place to place on the earth's surface, depending on the change in g. By keeping track of ΔT, we can estimate the variation in g from the equation $T = 2\pi(L/g)^{1/2}$ that relates T, g, and L.

a. With L held constant and g as the independent variable, calculate dT and use it to answer parts (b) and (c).

b. If g increases, will T increase or decrease? Will a pendulum clock speed up or slow down? Explain.

c. A clock with a 100-cm pendulum is moved from a location where $g = 980\ \text{cm/sec}^2$ to a new location. This increases the period by $dT = 0.001$ sec. Find dg and estimate the value of g at the new location.

45. The edge of a cube is measured as 10 cm with an error of 1%. The cube's volume is to be calculated from this measurement. Estimate the percentage error in the volume calculation.

46. About how accurately should you measure the side of a square to be sure of calculating the area within 2% of its true value?

47. The diameter of a sphere is measured as 100 ± 1 cm and the volume is calculated from this measurement. Estimate the percentage error in the volume calculation.

48. Estimate the allowable percentage error in measuring the diameter D of a sphere if the volume is to be calculated correctly to within 3%.

49. The linearization is the best linear approximation (This is why we use the linearization.) Suppose that $y = f(x)$ is differentiable at $x = a$ and that $g(x) = m(x - a) + c$ is a linear function in which m and c are constants. If the error $E(x) = f(x) - g(x)$ were small enough near $x = a$, we might think of using g as a linear approximation of f instead of the linearization $L(x) = f(a) + f'(a)(x - a)$. Show that if we impose on g the conditions

1. $E(a) = 0$ — The approximation error is zero at $x = a$.

2. $\lim_{x \to a} \dfrac{E(x)}{x - a} = 0$ — The error is negligible when compared with $x - a$.

then $g(x) = f(a) + f'(a)(x - a)$. Thus, the linearization $L(x)$ gives the only linear approximation whose error is both zero at $x = a$ and negligible in comparison with $x - a$.

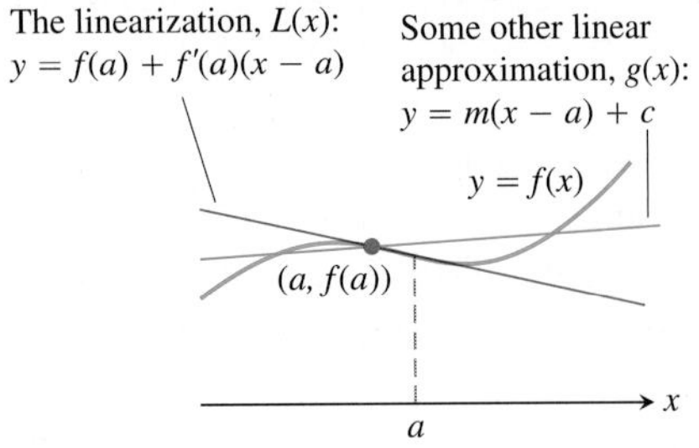

50. Quadratic approximations

a. Let $Q(x) = b_0 + b_1(x - a) + b_2(x - a)^2$ be a quadratic approximation to $f(x)$ at $x = a$ with the properties:

i. $Q(a) = f(a)$

ii. $Q'(a) = f'(a)$

iii. $Q''(a) = f''(a)$

Determine the coefficients b_0, b_1, and b_2.

b. Find the quadratic approximation to $f(x) = 1/(1 - x)$ at $x = 0$.

T c. Find the quadratic approximation to $g(x) = 1/x$ at $x = 1$. Graph g and its quadratic approximation together. Comment on what you see.

T d. Find the quadratic approximation to $h(x) = \sqrt{1 + x}$ at $x = 0$. Graph h and its quadratic approximation together. Comment on what you see.

51. The linearization of 2^x

a. Find the linearization of $f(x) = 2^x$ at $x = 0$. Then round its coefficients to two decimal places.

T b. Graph the linearization and function together for $-3 \le x \le 3$ and $-1 \le x \le 1$.

52. The linearization of $\log_3 x$

a. Find the linearization of $f(x) = \log_3 x$ at $x = 3$. Then round its coefficients to two decimal places.

T b. Graph the linearization and function together in the window $0 \le x \le 8$ and $2 \le x \le 4$.

COMPUTER EXPLORATIONS

In Exercises 53–58, use a CAS to estimate the magnitude of the error in using the linearization in place of the function over a specified interval I. Perform the following steps:

a. Plot the function f over I.

b. Find the linearization L of the function at the point a.

c. Plot f and L together on a single graph.

d. Plot the absolute error $|f(x) - L(x)|$ over I and find its maximum value.

e. From your graph in part (d), estimate as large a $\delta > 0$ as you can, satisfying

$$|x - a| < \delta \quad \Rightarrow \quad |f(x) - L(x)| < \epsilon$$

for $\epsilon = 0.5, 0.1,$ and 0.01. Then check graphically to see if your δ-estimate holds true.

53. $f(x) = x^3 + x^2 - 2x, \quad [-1, 2], \quad a = 1$

54. $f(x) = \dfrac{x - 1}{4x^2 + 1}, \quad \left[-\dfrac{3}{4}, 1\right], \quad a = \dfrac{1}{2}$

55. $f(x) = x^{2/3}(x - 2), \quad [-2, 3], \quad a = 2$

56. $f(x) = \sqrt{x} - \sin x, \quad [0, 2\pi], \quad a = 2$

57. $f(x) = x2^x, \quad [0, 2], \quad a = 1$

58. $f(x) = \sqrt{x}\sin^{-1} x, \ [0, 1], \quad a = \dfrac{1}{2}$

Chapter 2 Questions to Guide Your Review

1. What is the derivative of a function f? How is its domain related to the domain of f? Give examples.
2. What role does the derivative play in defining slopes, tangents, and rates of change?
3. How can you sometimes graph the derivative of a function when all you have is a table of the function's values?
4. What does it mean for a function to be differentiable on an open interval? On a closed interval?
5. How are derivatives and one-sided derivatives related?
6. Describe geometrically when a function typically does *not* have a derivative at a point.
7. How is a function's differentiability at a point related to its continuity there, if at all?
8. Could the unit step function

$$U(x) = \begin{cases} 0, & x < 0 \\ 1, & x \ge 0 \end{cases}$$

possibly be the derivative of some other function on $[-1, 1]$? Explain.

9. What rules do you know for calculating derivatives? Give some examples.

10. Explain how the three formulas

a. $\frac{d}{dx}(x^n) = nx^{n-1}$

b. $\frac{d}{dx}(cu) = c\frac{du}{dx}$

c. $\frac{d}{dx}(u_1 + u_2 + \cdots + u_n) = \frac{du_1}{dx} + \frac{du_2}{dx} + \cdots + \frac{du_n}{dx}$

enable us to differentiate any polynomial.

11. What formula do we need, in addition to the three listed in Question 10, to differentiate rational functions?

12. What is a second derivative? A third derivative? How many derivatives do the functions you know have? Give examples.

13. What is the relationship between a function's average and instantaneous rates of change? Give an example.

14. How do derivatives arise in the study of motion? What can you learn about a body's motion along a line by examining the derivatives of the body's position function? Give examples.

15. How can derivatives arise in economics?

16. What do the limits $\lim_{h\to 0}((\sin h)/h)$ and $\lim_{h\to 0}((\cos h - 1)/h)$ have to do with the derivatives of the sine and cosine functions? What *are* the derivatives of these functions?

17. Once you know the derivatives of $\sin x$ and $\cos x$, how can you find the derivatives of $\tan x$, $\cot x$, $\sec x$, and $\csc x$? What *are* the derivatives of these functions?

18. At what points are the six basic trigonometric functions continuous? How do you know?

19. What is the rule for calculating the derivative of a composite of two differentiable functions? How is such a derivative evaluated? Give examples.

20. If u is a differentiable function of x, how do you find $(d/dx)(u^n)$ if n is an integer? If n is a real number? Give examples.

21. What is the derivative of the exponential function e^x? How does the domain of the derivative compare with the domain of the function?

22. What is implicit differentiation? When do you need it? Give examples.

23. What is the derivative of the exponential function a^x, $a > 0$ and $a \neq 1$? What is the geometric significance of the limit of $(a^h - 1)/h$ as $h \to 0$? What is the limit when a is the number e?

24. What is the derivative of the natural logarithm function $\ln x$? How does the domain of the derivative compare with the domain of the function?

25. What is the derivative of $\log_a x$? Are there any restrictions on a?

26. What is logarithmic differentiation? Give an example.

27. How can you write any real power of x as a power of e? Are there any restrictions on x? How does this lead to the Power Rule for differentiating arbitrary real powers?

28. What is one way of expressing the special number e as a limit? What is an approximate numerical value of e correct to 7 decimal places?

29. What are the derivatives of the inverse trigonometric functions? How do the domains of the derivatives compare with the domains of the functions?

30. How do related rates problems arise? Give examples.

31. Outline a strategy for solving related rates problems. Illustrate with an example.

32. What is the linearization $L(x)$ of a function $f(x)$ at a point $x = a$? What is required of f at a for the linearization to exist? How are linearizations used? Give examples.

33. If x moves from a to a nearby value $a + dx$, how do you estimate the corresponding change in the value of a differentiable function $f(x)$? How do you estimate the relative change? The percentage change? Give an example.

Chapter 2 Practice and Additional Exercises

Find the derivatives of the functions in Exercises 1–64.

1. $y = x^5 - 0.125x^2 + 0.25x$

2. $y = 3 - 0.7x^3 + 0.3x^7$

3. $y = x^3 - 3(x^2 + \pi^2)$

4. $y = x^7 + \sqrt{7x} - \frac{1}{\pi + 1}$

5. $y = (x + 1)^2(x^2 + 2x)$

6. $y = (2x - 5)(4 - x)^{-1}$

7. $y = (\theta^2 + \sec\theta + 1)^3$

8. $y = \left(-1 - \frac{\csc\theta}{2} - \frac{\theta^2}{4}\right)^2$

9. $s = \frac{\sqrt{t}}{1 + \sqrt{t}}$

10. $s = \frac{1}{\sqrt{t} - 1}$

11. $y = 2\tan^2 x - \sec^2 x$

12. $y = \frac{1}{\sin^2 x} - \frac{2}{\sin x}$

13. $s = \cos^4(1 - 2t)$

14. $s = \cot^3\left(\frac{2}{t}\right)$

15. $s = (\sec t + \tan t)^5$

16. $s = \csc^5(1 - t + 3t^2)$

17. $r = \sqrt{2\theta \sin\theta}$

18. $r = 2\theta\sqrt{\cos\theta}$

19. $r = \sin\sqrt{2\theta}$

20. $r = \sin\left(\theta + \sqrt{\theta + 1}\right)$

21. $y = \frac{1}{2}x^2 \csc\frac{2}{x}$

22. $y = 2\sqrt{x}\sin\sqrt{x}$

23. $y = x^{-1/2}\sec(2x)^2$

24. $y = \sqrt{x}\csc(x + 1)^3$

25. $y = 5\cot x^2$

26. $y = x^2\cot 5x$

27. $y = x^2\sin^2(2x^2)$

28. $y = x^{-2}\sin^2(x^3)$

29. $s = \left(\frac{4t}{t+1}\right)^{-2}$

30. $s = \frac{-1}{15(15t-1)^3}$

31. $y = \left(\frac{\sqrt{x}}{1+x}\right)^2$

32. $y = \left(\frac{2\sqrt{x}}{2\sqrt{x}+1}\right)^2$

33. $y = \sqrt{\frac{x^2+x}{x^2}}$

34. $y = 4x\sqrt{x+\sqrt{x}}$

35. $r = \left(\frac{\sin\theta}{\cos\theta - 1}\right)^2$

36. $r = \left(\frac{1+\sin\theta}{1-\cos\theta}\right)^2$

37. $y = (2x+1)\sqrt{2x+1}$

38. $y = 20(3x-4)^{1/4}(3x-4)^{-1/5}$

39. $y = \frac{3}{(5x^2+\sin 2x)^{3/2}}$

40. $y = (3+\cos^3 3x)^{-1/3}$

41. $y = 10e^{-x/5}$

42. $y = \sqrt{2}e^{\sqrt{2}x}$

43. $y = \frac{1}{4}xe^{4x} - \frac{1}{16}e^{4x}$

44. $y = x^2e^{-2/x}$

45. $y = \ln(\sin^2\theta)$

46. $y = \ln(\sec^2\theta)$

47. $y = \log_2(x^2/2)$

48. $y = \log_5(3x-7)$

49. $y = 8^{-t}$

50. $y = 9^{2t}$

51. $y = 5x^{3.6}$

52. $y = \sqrt{2}x^{-\sqrt{2}}$

53. $y = (x+2)^{x+2}$

54. $y = 2(\ln x)^{x/2}$

55. $y = \sin^{-1}\sqrt{1-u^2},\quad 0 < u < 1$

56. $y = \sin^{-1}\left(\frac{1}{\sqrt{v}}\right),\quad v > 1$

57. $y = \ln\cos^{-1}x$

58. $y = z\cos^{-1}z - \sqrt{1-z^2}$

59. $y = t\tan^{-1}t - \frac{1}{2}\ln t$

60. $y = (1+t^2)\cot^{-1}2t$

61. $y = z\sec^{-1}z - \sqrt{z^2-1},\quad z > 1$

62. $y = 2\sqrt{x-1}\sec^{-1}\sqrt{x}$

63. $y = \csc^{-1}(\sec\theta),\quad 0 < \theta < \pi/2$

64. $y = (1+x^2)e^{\tan^{-1}x}$

In Exercises 65–78, find dy/dx by implicit differentiation.

65. $xy + 2x + 3y = 1$

66. $x^2 + xy + y^2 - 5x = 2$

67. $x^3 + 4xy - 3y^{4/3} = 2x$

68. $5x^{4/5} + 10y^{6/5} = 15$

69. $\sqrt{xy} = 1$

70. $x^2y^2 = 1$

71. $y^2 = \frac{x}{x+1}$

72. $y^2 = \sqrt{\frac{1+x}{1-x}}$

73. $e^{x+2y} = 1$

74. $y^2 = 2e^{-1/x}$

75. $\ln(x/y) = 1$

76. $x\sin^{-1}y = 1 + x^2$

77. $ye^{\tan^{-1}x} = 2$

78. $x^y = \sqrt{2}$

In Exercises 79 and 80, find dp/dq.

79. $p^3 + 4pq - 3q^2 = 2$

80. $q = (5p^2 + 2p)^{-3/2}$

In Exercises 81 and 82, find dr/ds.

81. $r\cos 2s + \sin^2 s = \pi$

82. $2rs - r - s + s^2 = -3$

83. Find d^2y/dx^2 by implicit differentiation:

a. $x^3 + y^3 = 1$ **b.** $y^2 = 1 - \frac{2}{x}$

84. **a.** By differentiating $x^2 - y^2 = 1$ implicitly, show that $dy/dx = x/y$.

b. Then show that $d^2y/dx^2 = -1/y^3$.

85. Suppose that functions $f(x)$ and $g(x)$ and their first derivatives have the following values at $x = 0$ and $x = 1$.

x	$f(x)$	$g(x)$	$f'(x)$	$g'(x)$
0	1	1	−3	1/2
1	3	5	1/2	−4

Find the first derivatives of the following combinations at the given value of x.

a. $6f(x) - g(x),\quad x = 1$ **b.** $f(x)g^2(x),\quad x = 0$

c. $\frac{f(x)}{g(x)+1},\quad x = 1$ **d.** $f(g(x)),\quad x = 0$

e. $g(f(x)),\quad x = 0$ **f.** $(x + f(x))^{3/2},\quad x = 1$

g. $f(x + g(x)),\quad x = 0$

86. Suppose that the function $f(x)$ and its first derivative have the following values at $x = 0$ and $x = 1$.

x	$f(x)$	$f'(x)$
0	9	−2
1	−3	1/5

Find the first derivatives of the following combinations at the given value of x.

a. $\sqrt{x}f(x),\quad x = 1$ **b.** $\sqrt{f(x)},\quad x = 0$

c. $f(\sqrt{x}),\quad x = 1$ **d.** $f(1 - 5\tan x),\quad x = 0$

e. $\frac{f(x)}{2+\cos x},\quad x = 0$ **f.** $10\sin\left(\frac{\pi x}{2}\right)f^2(x),\quad x = 1$

87. Find the value of dy/dt at $t = 0$ if $y = 3\sin 2x$ and $x = t^2 + \pi$.

88. Find the value of ds/du at $u = 2$ if $s = t^2 + 5t$ and $t = (u^2 + 2u)^{1/3}$.

89. Find the value of dw/ds at $s = 0$ if $w = \sin\left(e^{\sqrt{r}}\right)$ and $r = 3\sin(s + \pi/6)$.

90. Find the value of dr/dt at $t = 0$ if $r = (\theta^2 + 7)^{1/3}$ and $\theta^2 e^t + \theta = 1$.

91. If $y^3 + y = 2\cos x$, find the value of d^2y/dx^2 at the point (0, 1).

92. If $x^{1/3} + y^{1/3} = 4$, find d^2y/dx^2 at the point (8, 8).

In Exercises 93 and 94, find the derivative using the definition.

93. $f(t) = \dfrac{1}{2t + 1}$ **94.** $g(x) = 2x^2 + 1$

95. a. Graph the function

$$f(x) = \begin{cases} x^2, & -1 \le x < 0 \\ -x^2, & 0 \le x \le 1. \end{cases}$$

b. Is f continuous at $x = 0$?

c. Is f differentiable at $x = 0$?

Give reasons for your answers.

96. a. Graph the function

$$f(x) = \begin{cases} x, & -1 \le x < 0 \\ \tan x, & 0 \le x \le \pi/4. \end{cases}$$

b. Is f continuous at $x = 0$?

c. Is f differentiable at $x = 0$?

Give reasons for your answers.

97. Tangents with specified slope Are there any points on the curve $y = (x/2) + 1/(2x - 4)$ where the slope is $-3/2$? If so, find them.

98. Tangents with specified slope Are there any points on the curve $y = x - e^{-x}$ where the slope is 2? If so, find them.

99. Tangents perpendicular or parallel to lines Find the points on the curve $y = 2x^3 - 3x^2 - 12x + 20$ where the tangent is

a. perpendicular to the line $y = 1 - (x/24)$.

b. parallel to the line $y = \sqrt{2} - 12x$.

100. Tangent and normal lines Find equations for the tangent and normal to the curve $y = 1 + \cos x$ at the point $(\pi/2, 1)$. Sketch the curve, tangent, and normal together, labeling each with its equation.

101. Tangent parabola The parabola $y = x^2 + C$ is to be tangent to the line $y = x$. Find C.

102. Slope of tangent Show that the tangent to the curve $y = x^3$ at any point (a, a^3) meets the curve again at a point where the slope is four times the slope at (a, a^3).

In Exercises 103–106, find equations for the lines that are tangent and normal to the curve at the given point.

103. $x^2 + 2y^2 = 9, \quad (1, 2)$

104. $e^x + y^2 = 2, \quad (0, 1)$

105. $xy + 2x - 5y = 2, \quad (3, 2)$

106. $x + \sqrt{xy} = 6, \quad (4, 1)$

Find the limits in Exercises 107–112.

107. $\displaystyle\lim_{x\to 0} \frac{\sin x}{2x^2 - x}$ **108.** $\displaystyle\lim_{x\to 0} \frac{3x - \tan 7x}{2x}$

109. $\displaystyle\lim_{r\to 0} \frac{\sin r}{\tan 2r}$ **110.** $\displaystyle\lim_{\theta\to 0} \frac{\sin(\sin\theta)}{\theta}$

111. $\displaystyle\lim_{\theta\to(\pi/2)^-} \frac{4\tan^2\theta + \tan\theta + 1}{\tan^2\theta + 5}$

112. $\displaystyle\lim_{x\to 0} \frac{x\sin x}{2 - 2\cos x}$

In Exercises 113–118, use logarithmic differentiation to find the derivative of y with respect to the appropriate variable.

113. $y = \dfrac{2(x^2 + 1)}{\sqrt{\cos 2x}}$ **114.** $y = \sqrt[10]{\dfrac{3x + 4}{2x - 4}}$

115. $y = \left(\dfrac{(t + 1)(t - 1)}{(t - 2)(t + 3)}\right)^5, \quad t > 2$

116. $y = \dfrac{2u2^u}{\sqrt{u^2 + 1}}$

117. $y = (\sin\theta)^{\sqrt{\theta}}$ **118.** $y = (\ln x)^{1/(\ln x)}$

119. Right circular cylinder The total surface area S of a right circular cylinder is related to the base radius r and height h by the equation $S = 2\pi r^2 + 2\pi rh$.

a. How is dS/dt related to dr/dt if h is constant?

b. How is dS/dt related to dh/dt if r is constant?

c. How is dS/dt related to dr/dt and dh/dt if neither r nor h is constant?

d. How is dr/dt related to dh/dt if S is constant?

120. Right circular cone The lateral surface area S of a right circular cone is related to the base radius r and height h by the equation $S = \pi r\sqrt{r^2 + h^2}$.

a. How is dS/dt related to dr/dt if h is constant?

b. How is dS/dt related to dh/dt if r is constant?

c. How is dS/dt related to dr/dt and dh/dt if neither r nor h is constant?

121. Resistors connected in parallel If two resistors of R_1 and R_2 ohms are connected in parallel in an electric circuit to make an R-ohm resistor, the value of R can be found from the equation

$$\frac{1}{R} = \frac{1}{R_1} + \frac{1}{R_2}.$$

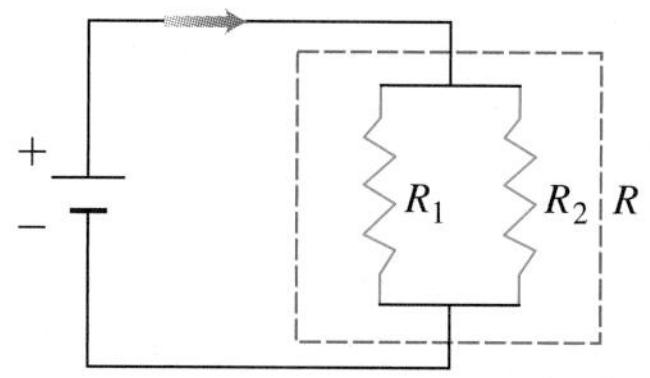

If R_1 is decreasing at the rate of 1 ohm/sec and R_2 is increasing at the rate of 0.5 ohm/sec, at what rate is R changing when $R_1 = 75$ ohms and $R_2 = 50$ ohms?

122. Impedance in a series circuit The impedance Z (ohms) in a series circuit is related to the resistance R (ohms) and reactance

X (ohms) by the equation $Z = \sqrt{R^2 + X^2}$. If R is increasing at 3 ohms/sec and X is decreasing at 2 ohms/sec, at what rate is Z changing when $R = 10$ ohms and $X = 20$ ohms?

123. Speed of moving particle The coordinates of a particle moving in the metric xy-plane are differentiable functions of time t with $dx/dt = 10$ m/sec and $dy/dt = 5$ m/sec. How fast is the particle moving away from the origin as it passes through the point $(3, -4)$?

124. Motion of a particle A particle moves along the curve $y = x^{3/2}$ in the first quadrant in such a way that its distance from the origin increases at the rate of 11 units per second. Find dx/dt when $x = 3$.

125. Draining a tank Water drains from the conical tank shown in the accompanying figure at the rate of 5 ft^3/min.

a. What is the relation between the variables h and r in the figure?

b. How fast is the water level dropping when $h = 6$ ft?

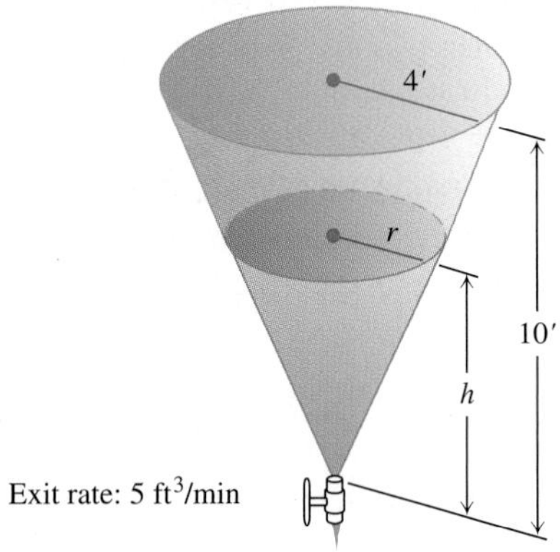

126. Rotating spool As television cable is pulled from a large spool to be strung from the telephone poles along a street, it unwinds from the spool in layers of constant radius (see accompanying figure). If the truck pulling the cable moves at a steady 6 ft/sec (a touch over 4 mph), use the equation $s = r\theta$ to find how fast (radians per second) the spool is turning when the layer of radius 1.2 ft is being unwound.

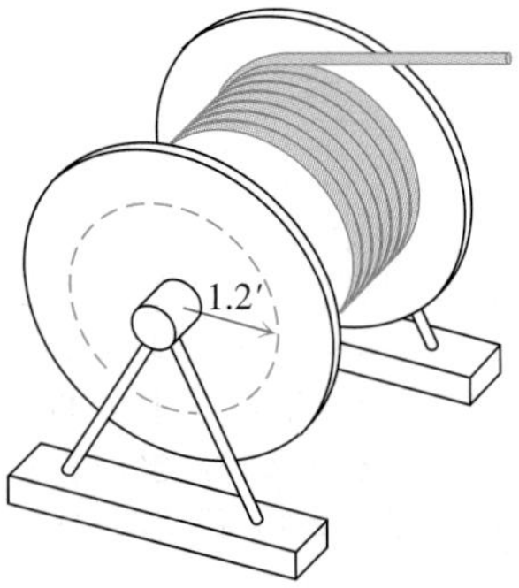

127. Moving searchlight beam The figure shows a boat 1 km offshore, sweeping the shore with a searchlight. The light turns at a constant rate, $d\theta/dt = -0.6$ rad/sec.

a. How fast is the light moving along the shore when it reaches point A?

b. How many revolutions per minute is 0.6 rad/sec?

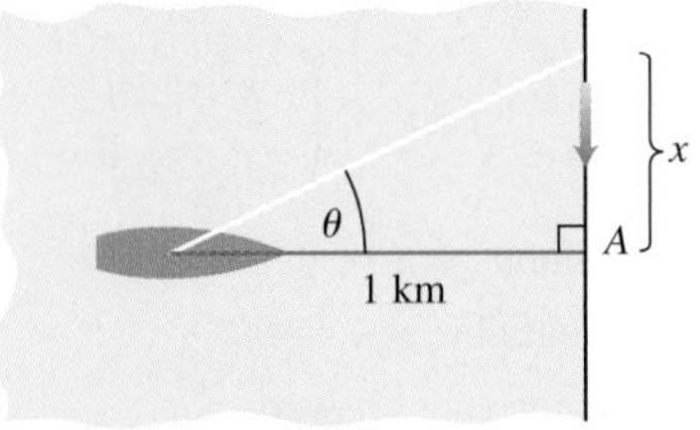

128. Points moving on coordinate axes Points A and B move along the x- and y-axes, respectively, in such a way that the distance r (meters) along the perpendicular from the origin to the line AB remains constant. How fast is OA changing, and is it increasing, or decreasing, when $OB = 2r$ and B is moving toward O at the rate of $0.3r$ m/sec?

129. Find the linearizations of

a. $\tan x$ at $x = -\pi/4$ **b.** $\sec x$ at $x = -\pi/4$.

Graph the curves and linearizations together.

130. Find the linearization of $f(x) = \sqrt{1 + x} + \sin x - 0.5$ at $x = 0$.

131. Compounding error The circumference of the equator of a sphere is measured as 10 cm with a possible error of 0.4 cm. This measurement is then used to calculate the radius. The radius is then used to calculate the surface area and volume of the sphere. Estimate the percentage errors in the calculated values of

a. the radius.

b. the surface area.

c. the volume.

132. Finding height To find the height of a lamppost (see accompanying figure), you stand a 6 ft pole 20 ft from the lamp and measure the length a of its shadow, finding it to be 15 ft, give or take an inch. Calculate the height of the lamppost using the value $a = 15$ and estimate the possible error in the result.

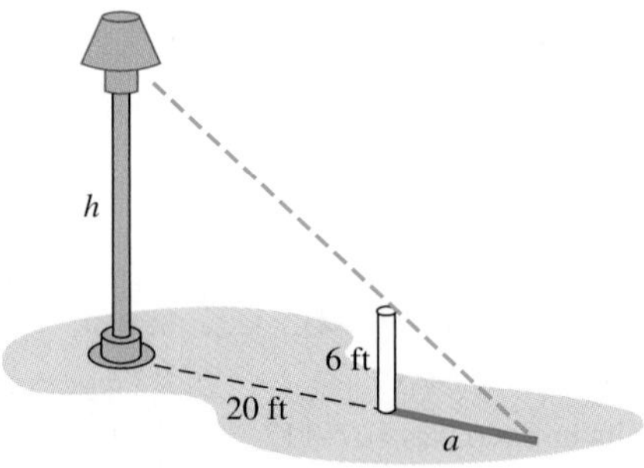

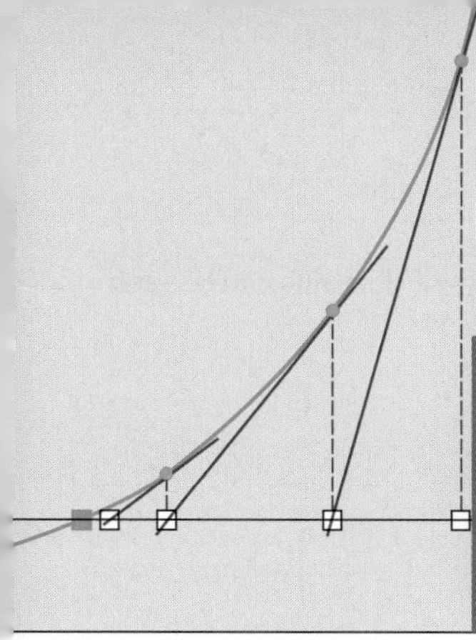

Chapter 3

APPLICATIONS OF DERIVATIVES

OVERVIEW In this chapter we use derivatives to find extreme values of functions, to determine and analyze the shapes of graphs, to calculate limits of fractions whose numerators and denominators both approach zero or infinity, and to find numerically where a function equals zero. The key to many of these applications is the Mean Value Theorem, whose corollaries pave the way to integral calculus in Chapter 4.

3.1 Extreme Values of Functions

This section shows how to locate and identify extreme (maximum or minimum) values of a continuous function from its derivative. Once we can do this, we can solve a variety of problems in which we find the optimal (best) way to do something in a given situation (see Section 3.6).

DEFINITIONS Let f be a function with domain D. Then f has an **absolute maximum** value on D at a point c if

$$f(x) \leq f(c) \qquad \text{for all } x \text{ in } D$$

and an **absolute minimum** value on D at c if

$$f(x) \geq f(c) \qquad \text{for all } x \text{ in } D.$$

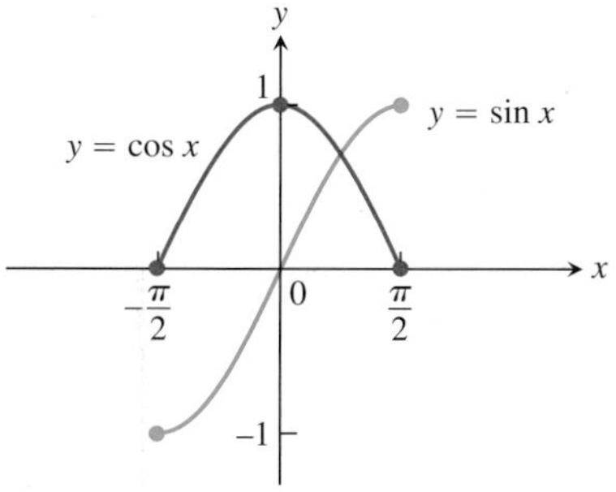

FIGURE 3.1 Absolute extrema for the sine and cosine functions on $[-\pi/2, \pi/2]$. These values can depend on the domain of a function.

Absolute maxima or minima are also referred to as **global** maxima or minima.

For example, on the closed interval $[-\pi/2, \pi/2]$ the function $f(x) = \cos x$ takes on an absolute maximum value of 1 (once) and an absolute minimum value of 0 (twice). On the same interval, the function $g(x) = \sin x$ takes on a maximum value of 1 and a minimum value of -1 (Figure 3.1).

Functions with the same defining rule can have different extrema (maximum or minimum values), depending on the domain. We see this in the following example.

EXAMPLE 1 The absolute extrema of the following functions on their domains can be seen in Figure 3.2.

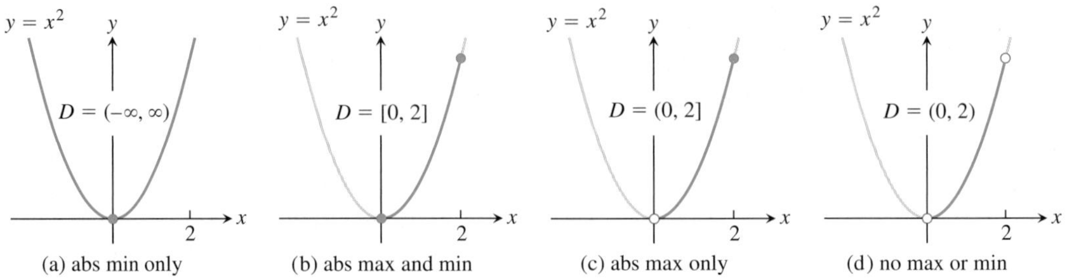

FIGURE 3.2 Graphs for Example 1.

Function rule	Domain D	Absolute extrema on D
(a) $y = x^2$	$(-\infty, \infty)$	No absolute maximum. Absolute minimum of 0 at $x = 0$.
(b) $y = x^2$	$[0, 2]$	Absolute maximum of 4 at $x = 2$. Absolute minimum of 0 at $x = 0$.
(c) $y = x^2$	$(0, 2]$	Absolute maximum of 4 at $x = 2$. No absolute minimum.
(d) $y = x^2$	$(0, 2)$	No absolute extrema.

■

HISTORICAL BIOGRAPHY

Daniel Bernoulli
(1700–1789)

The following theorem asserts that a function which is continuous at every point of a closed interval $[a, b]$ has an absolute maximum and an absolute minimum value on the interval. As stated earlier, the maximum and minimum values are called **extreme values** and we look for these values when we graph a function.

THEOREM 1—The Extreme Value Theorem If f is continuous on a closed interval $[a, b]$, then f attains both an absolute maximum value M and an absolute minimum value m in $[a, b]$. That is, there are numbers x_1 and x_2 in $[a, b]$ with $f(x_1) = m$, $f(x_2) = M$, and $m \leq f(x) \leq M$ for every other x in $[a, b]$.

The proof of the Extreme Value Theorem requires a detailed knowledge of the real number system (see Appendix 7) and we will not give it here. Figure 3.3 illustrates possible locations for the absolute extrema of a continuous function on a closed interval $[a, b]$.

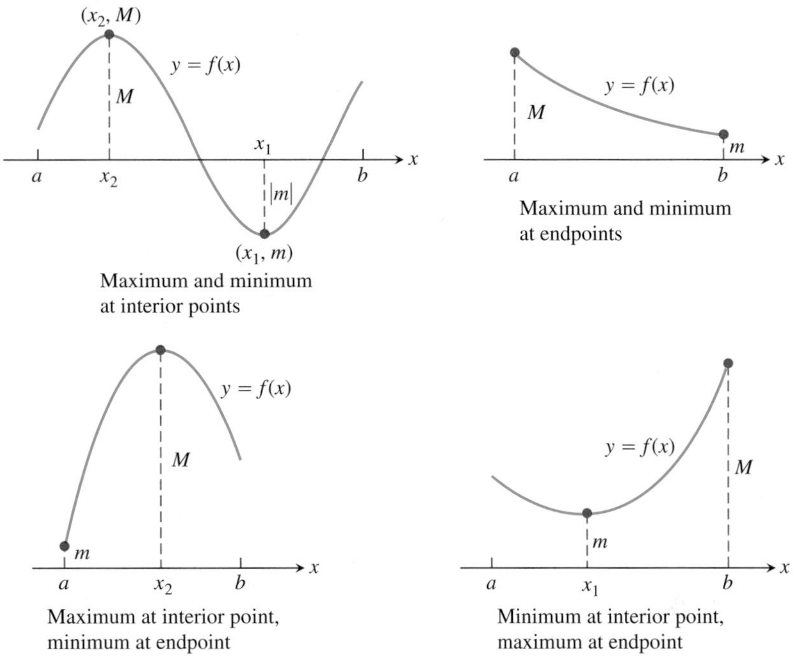

FIGURE 3.3 Some possibilities for a continuous function's maximum and minimum on a closed interval $[a, b]$.

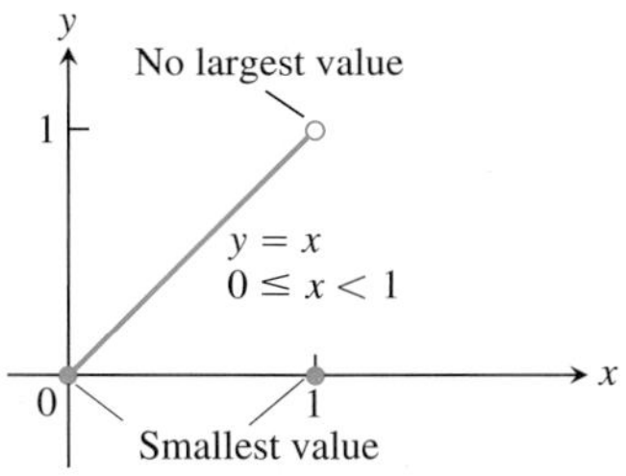

FIGURE 3.4 Even a single point of discontinuity can keep a function from having either a maximum or minimum value on a closed interval. The function

$$y = \begin{cases} x, & 0 \le x < 1 \\ 0, & x = 1 \end{cases}$$

is continuous at every point of $[0, 1]$ except $x = 1$, yet its graph over $[0, 1]$ does not have a highest point.

As we observed for the function $y = \cos x$, it is possible that an absolute minimum (or absolute maximum) may occur at two or more different points of the interval.

The requirements in Theorem 1 that the interval be closed and finite, and that the function be continuous, are key ingredients. Without them, the conclusion of the theorem need not hold. Example 1 shows that an absolute extreme value may not exist if the interval fails to be both closed and finite. Figure 3.4 shows that the continuity requirement cannot be omitted.

Local (Relative) Extreme Values

Figure 3.5 shows a graph with five points where a function has extreme values on its domain $[a, b]$. The function's absolute minimum occurs at a even though at e the function's value is smaller than at any other point *nearby*. The curve rises to the left and falls to the right around c, making $f(c)$ a maximum locally. The function attains its absolute maximum at d.

DEFINITIONS A function f has a **local maximum** value at an interior point c of its domain if

$$f(x) \le f(c) \qquad \text{for all } x \text{ in some open interval containing } c.$$

A function f has a **local minimum** value at an interior point c of its domain if

$$f(x) \ge f(c) \qquad \text{for all } x \text{ in some open interval containing } c.$$

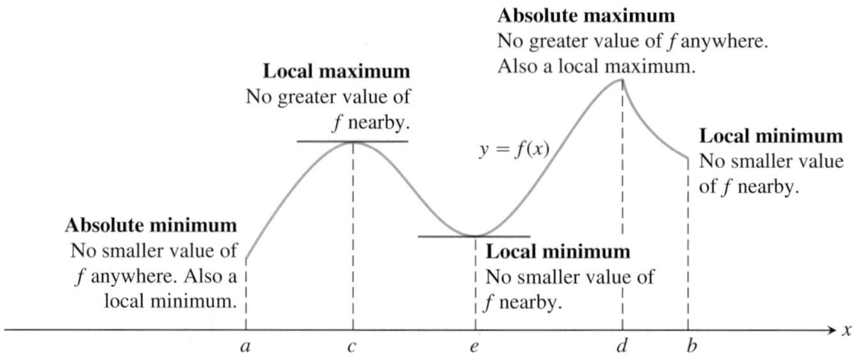

FIGURE 3.5 How to classify maxima and minima.

We can extend the definitions of local extrema to the endpoints of intervals by defining f to have a **local maximum** or **local minimum** value *at an endpoint* c if the appropriate inequality holds for all x in some half-open interval in its domain containing c. In Figure 3.5, the function f has local maxima at c and d and local minima at a, e, and b. Local extrema are also called **relative extrema**.

An absolute maximum is also a local maximum. Being the largest value overall, it is also the largest value in its immediate neighborhood. Hence, *a list of all local maxima will automatically include the absolute maximum if there is one*. Similarly, *a list of all local minima will include the absolute minimum if there is one*.

Finding Extrema

The next theorem explains why we usually need to investigate only a few values to find a function's extrema.

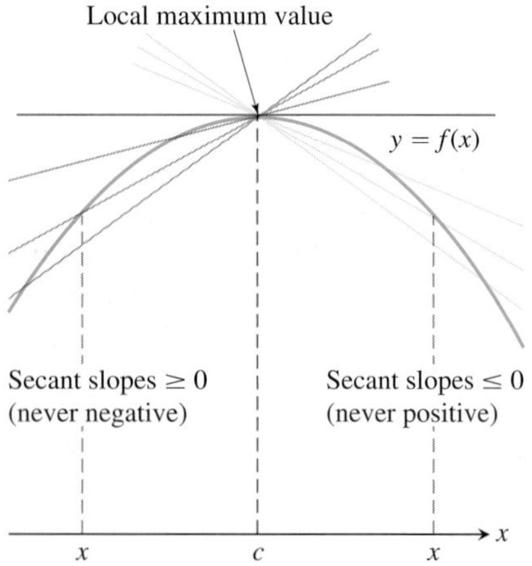

FIGURE 3.6 A curve with a local maximum value. The slope at c, simultaneously the limit of nonpositive numbers and nonnegative numbers, is zero.

THEOREM 2—The First Derivative Theorem for Local Extreme Values If f has a local maximum or minimum value at an interior point c of its domain, and if f' is defined at c, then

$$f'(c) = 0.$$

Proof To prove that $f'(c)$ is zero at a local extremum, we show first that $f'(c)$ cannot be positive and second that $f'(c)$ cannot be negative. The only number that is neither positive nor negative is zero, so that is what $f'(c)$ must be.

To begin, suppose that f has a local maximum value at $x = c$ (Figure 3.6) so that $f(x) - f(c) \le 0$ for all values of x near enough to c. Since c is an interior point of f's domain, $f'(c)$ is defined by the two-sided limit

$$\lim_{x \to c} \frac{f(x) - f(c)}{x - c}.$$

This means that the right-hand and left-hand limits both exist at $x = c$ and equal $f'(c)$. When we examine these limits separately, we find that

$$f'(c) = \lim_{x \to c^+} \frac{f(x) - f(c)}{x - c} \leq 0. \qquad \text{Because } (x - c) > 0 \text{ and } f(x) \leq f(c) \qquad (1)$$

Similarly,

$$f'(c) = \lim_{x \to c^-} \frac{f(x) - f(c)}{x - c} \geq 0. \qquad \text{Because } (x - c) < 0 \text{ and } f(x) \leq f(c) \qquad (2)$$

Together, Equations (1) and (2) imply $f'(c) = 0$.

This proves the theorem for local maximum values. To prove it for local minimum values, we simply use $f(x) \geq f(c)$, which reverses the inequalities in Equations (1) and (2). ■

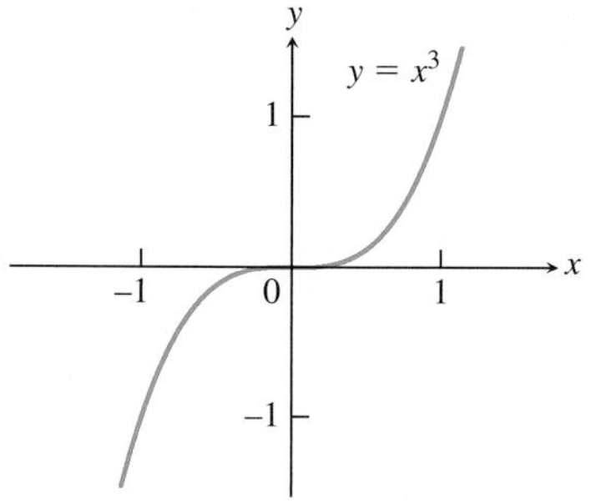

(a)

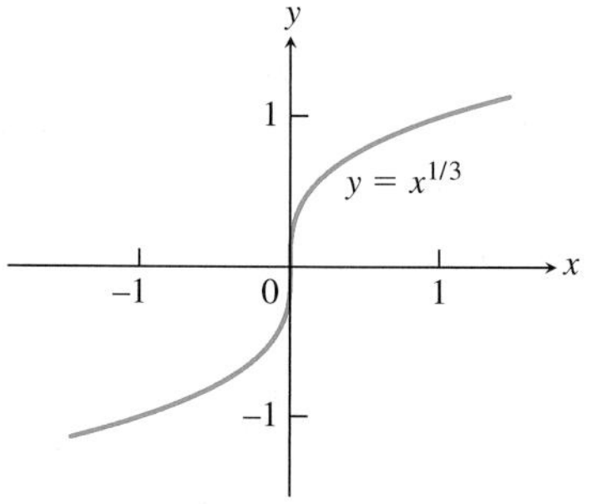

(b)

FIGURE 3.7 Critical points without extreme values. (a) $y' = 3x^2$ is 0 at $x = 0$, but $y = x^3$ has no extremum there. (b) $y' = (1/3)x^{-2/3}$ is undefined at $x = 0$, but $y = x^{1/3}$ has no extremum there.

Theorem 2 says that a function's first derivative is always zero at an interior point where the function has a local extreme value and the derivative is defined. Hence the only places where a function f can possibly have an extreme value (local or global) are

1. interior points where $f' = 0$,
2. interior points where f' is undefined,
3. endpoints of the domain of f.

The following definition helps us to summarize.

> **DEFINITION** An interior point of the domain of a function f where f' is zero or undefined is a **critical point** of f.

Thus the only domain points where a function can assume extreme values are critical points and endpoints. However, be careful not to misinterpret what is being said here. A function may have a critical point at $x = c$ without having a local extreme value there. For instance, both of the functions $y = x^3$ and $y = x^{1/3}$ have critical points at the origin and a zero value there, but each function is positive to the right of the origin and negative to the left. So neither function has a local extreme value at the origin. Instead, each function has a *point of inflection* there (see Figure 3.7). Inflection points are defined and explored further in Section 3.4.

Most problems that ask for extreme values call for finding the absolute extrema of a continuous function on a closed and finite interval. Theorem 1 assures us that such values exist; Theorem 2 tells us that they are taken on only at critical points and endpoints. Often we can simply list these points and calculate the corresponding function values to find what the largest and smallest values are, and where they are located.

> **How to Find the Absolute Extrema of a Continuous Function f on a Finite Closed Interval**
>
> 1. Evaluate f at all critical points and endpoints.
> 2. Take the largest and smallest of these values.

EXAMPLE 2 Find the absolute maximum and minimum values of $f(x) = 10x(2 - \ln x)$ on the interval $[1, e^2]$.

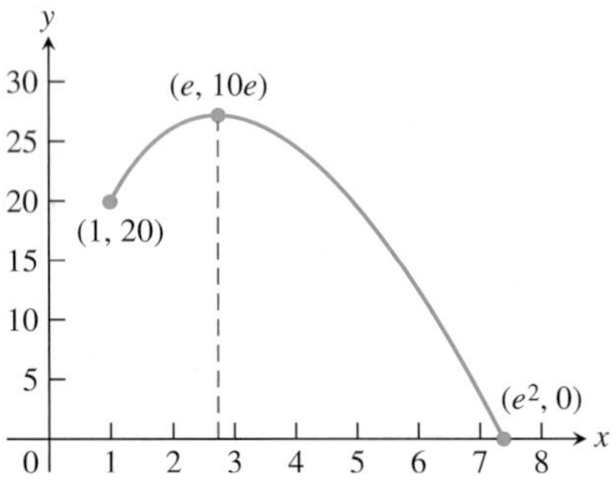

FIGURE 3.8 The extreme values of $f(x) = 10x(2 - \ln x)$ on $[1, e^2]$ occur at $x = e$ and $x = e^2$ (Example 2).

Solution Figure 3.8 suggests that f has its absolute maximum value near $x = 3$ and its absolute minimum value of 0 at $x = e^2$. Let's check this observation.

We evaluate the function at the critical points and endpoints and take the largest and smallest of the resulting values.

The first derivative is

$$f'(x) = 10(2 - \ln x) - 10x\left(\frac{1}{x}\right) = 10(1 - \ln x).$$

The only critical point in the domain $[1, e^2]$ is the point $x = e$, where $\ln x = 1$. The values of f at this one critical point and at the endpoints are

$$\begin{aligned} &\text{Critical point value:} && f(e) = 10e \\ &\text{Endpoint values:} && f(1) = 10(2 - \ln 1) = 20 \\ & && f(e^2) = 10e^2(2 - 2\ln e) = 0. \end{aligned}$$

We can see from this list that the function's absolute maximum value is $10e \approx 27.2$; it occurs at the critical interior point $x = e$. The absolute minimum value is 0 and occurs at the right endpoint $x = e^2$. ■

EXAMPLE 3 Find the absolute maximum and minimum values of $f(x) = x^{2/3}$ on the interval $[-2, 3]$.

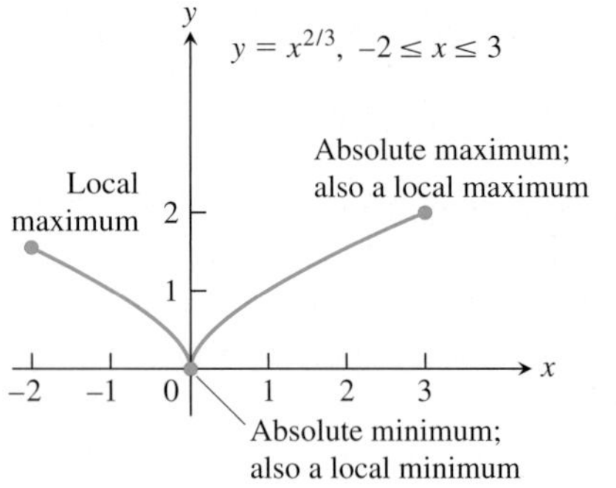

FIGURE 3.9 The extreme values of $f(x) = x^{2/3}$ on $[-2, 3]$ occur at $x = 0$ and $x = 3$ (Example 3).

Solution We evaluate the function at the critical points and endpoints and take the largest and smallest of the resulting values.

The first derivative

$$f'(x) = \frac{2}{3}x^{-1/3} = \frac{2}{3\sqrt[3]{x}}$$

has no zeros but is undefined at the interior point $x = 0$. The values of f at this one critical point and at the endpoints are

$$\begin{aligned} &\text{Critical point value:} && f(0) = 0 \\ &\text{Endpoint values:} && f(-2) = (-2)^{2/3} = \sqrt[3]{4} \\ & && f(3) = (3)^{2/3} = \sqrt[3]{9}. \end{aligned}$$

We can see from this list that the function's absolute maximum value is $\sqrt[3]{9} \approx 2.08$, and it occurs at the right endpoint $x = 3$. The absolute minimum value is 0, and it occurs at the interior point $x = 0$ where the graph has a cusp (Figure 3.9). ■

EXERCISES 3.1

In Exercises 1–6, determine from the graph whether the function has any absolute extreme values on $[a, b]$. Then explain how your answer is consistent with Theorem 1.

1.

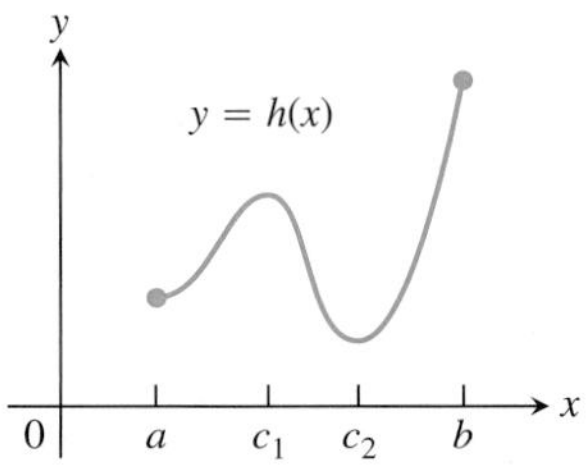

2.

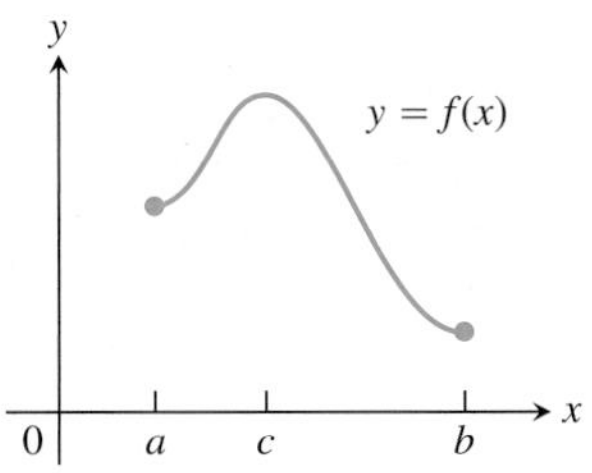

3.

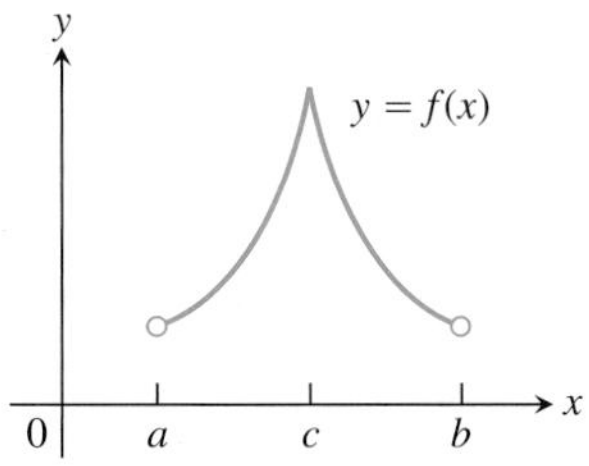

4.

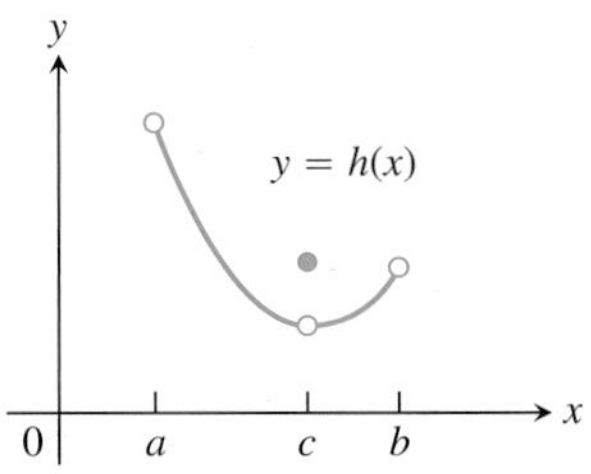

5.

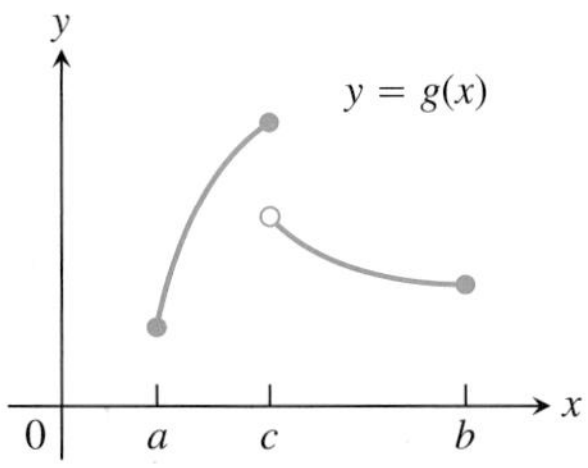

6.

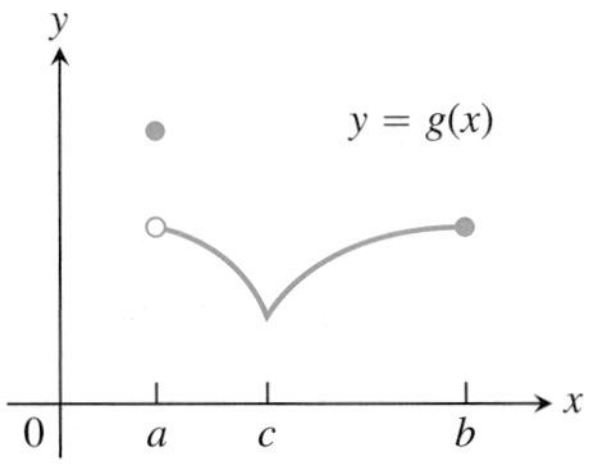

In Exercises 7–10, find the extreme values and where they occur.

7.

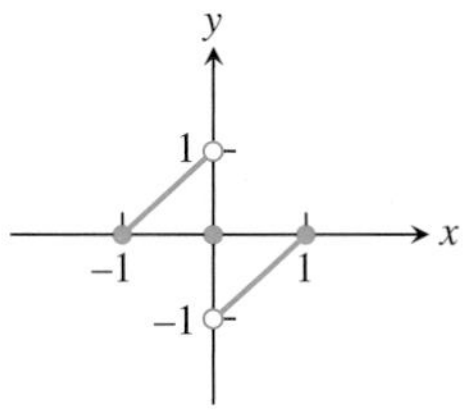

8.

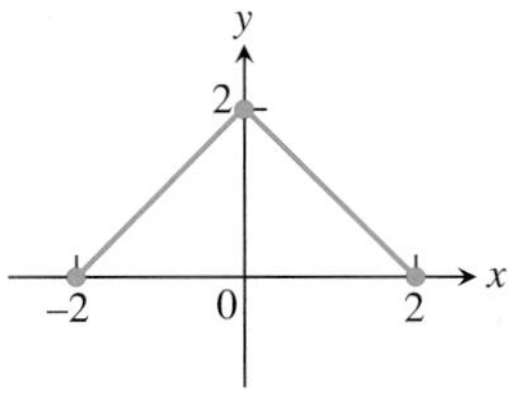

9.

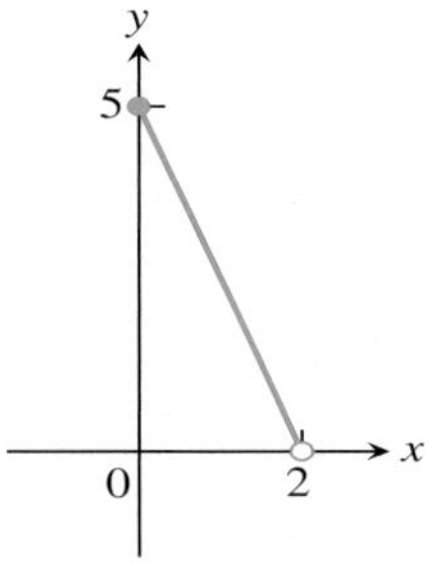

10.

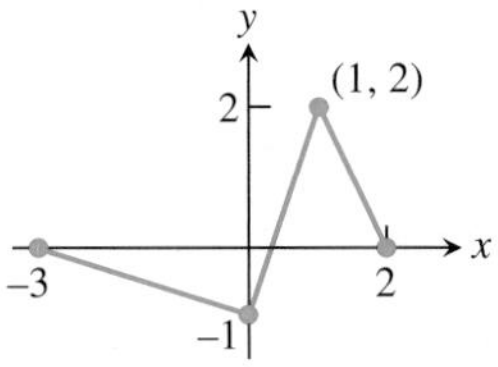

In Exercises 11–14, match the table with a graph.

11.

x	$f'(x)$
a	0
b	0
c	5

12.

x	$f'(x)$
a	0
b	0
c	-5

13.

x	$f'(x)$
a	does not exist
b	0
c	-2

14.

x	$f'(x)$
a	does not exist
b	does not exist
c	-1.7

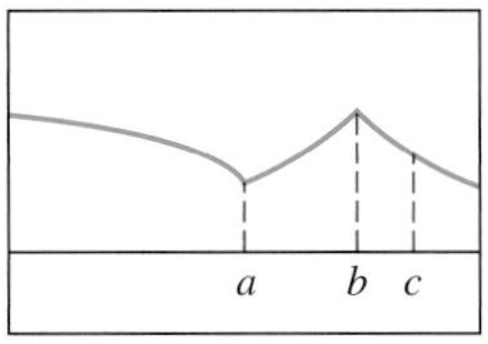

(a)

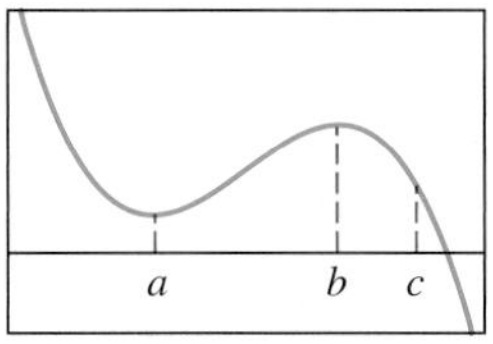

(b)

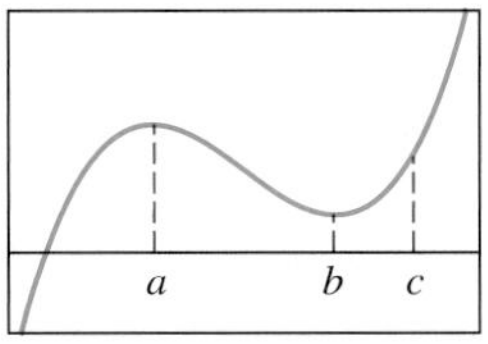

(c)

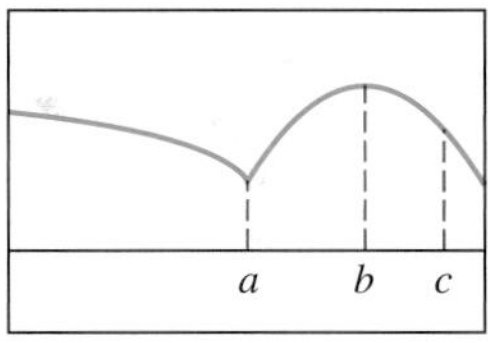

(d)

In Exercises 15–30, find the absolute maximum and minimum values of each function on the given interval. Then graph the function. Identify the points on the graph where the absolute extrema occur, and include their coordinates.

15. $f(x) = \frac{2}{3}x - 5, \quad -2 \le x \le 3$

16. $f(x) = x^2 - 1, \quad -1 \le x \le 2$

17. $F(x) = -\frac{1}{x^2}, \quad 0.5 \le x \le 2$

18. $g(x) = \sqrt{4 - x^2}, \quad -2 \le x \le 1$

19. $h(x) = \sqrt[3]{x}, \quad -1 \le x \le 8$

20. $h(x) = -3x^{2/3}, \quad -1 \le x \le 1$

21. $f(\theta) = \sin\theta, \quad -\frac{\pi}{2} \le \theta \le \frac{5\pi}{6}$

22. $f(\theta) = \tan\theta, \quad -\frac{\pi}{3} \le \theta \le \frac{\pi}{4}$

23. $g(x) = \csc x, \quad \frac{\pi}{3} \le x \le \frac{2\pi}{3}$

24. $f(t) = 2 - |t|, \quad -1 \le t \le 3$

25. $g(x) = xe^{-x}, \quad -1 \le x \le 1$

26. $h(x) = \ln(x + 1), \quad 0 \le x \le 3$

27. $f(x) = \frac{1}{x} + \ln x, \quad 0.5 \le x \le 4$

28. $g(x) = e^{-x^2}, \quad -2 \le x \le 1$

29. $f(x) = x^{4/3}, \quad -1 \le x \le 8$

30. $g(\theta) = \theta^{3/5}, \quad -32 \le \theta \le 1$

In Exercises 31–44, find the extreme values of the function and where they occur.

31. $y = 2x^2 - 8x + 9$

32. $y = x^3 - 2x + 4$

33. $y = x^3 + x^2 - 8x + 5$

34. $y = x^3 - 3x^2 + 3x - 2$

35. $y = \sqrt{x^2 - 1}$

36. $y = \frac{1}{\sqrt[3]{1 - x^2}}$

37. $y = \frac{x}{x^2 + 1}$

38. $y = \frac{x + 1}{x^2 + 2x + 2}$

39. $y = e^x + e^{-x}$

40. $y = e^x - e^{-x}$

41. $y = x \ln x$

42. $y = x^2 \ln x$

43. $y = \cos^{-1}(x^2)$

44. $y = \sin^{-1}(e^x)$

In Exercises 45–50, find the derivative at each critical point and determine the local extreme values.

45. $y = x^{2/3}(x + 2)$

46. $y = x^{2/3}(x^2 - 4)$

47. $y = x\sqrt{4 - x^2}$

48. $y = x^2\sqrt{3 - x}$

49. $y = \begin{cases} 4 - 2x, & x \le 1 \\ x + 1, & x > 1 \end{cases}$

50. $y = \begin{cases} -x^2 - 2x + 4, & x \le 1 \\ -x^2 + 6x - 4, & x > 1 \end{cases}$

51. A minimum with no derivative The function $f(x) = |x|$ has an absolute minimum value at $x = 0$ even though f is not differentiable at $x = 0$. Is this consistent with Theorem 2? Give reasons for your answer.

52. a. Even functions If an even function $f(x)$ has a local maximum value at $x = c$, can anything be said about the value of f at $x = -c$? Give reasons for your answer.

b. Odd functions If an odd function $g(x)$ has a local minimum value at $x = c$, can anything be said about the value of g at $x = -c$? Give reasons for your answer.

53. The function

$$V(x) = x(10 - 2x)(16 - 2x), \qquad 0 < x < 5,$$

models the volume of a box.

a. Find the extreme values of V.

b. Interpret any values found in part (a) in terms of volume of the box.

54. Cubic functions Consider the cubic function

$$f(x) = ax^3 + bx^2 + cx + d.$$

a. Show that f can have 0, 1, or 2 critical points. Give examples and graphs to support your argument.

b. How many local extreme values can f have?

55. Maximum height of a vertically moving body The height of a body moving vertically is given by

$$s = -\frac{1}{2}gt^2 + v_0 t + s_0, \qquad g > 0,$$

with s in meters and t in seconds. Find the body's maximum height.

56. Peak alternating current Suppose that at any given time t (in seconds) the current i (in amperes) in an alternating current circuit is $i = 2\cos t + 2\sin t$. What is the peak current for this circuit (largest magnitude)?

T Graph the functions in Exercises 57 and 58. Then find the extreme values of the function on the interval and say where they occur.

57. $f(x) = |x - 2| + |x + 3|, \quad -5 \le x \le 5$

58. $h(x) = |x + 2| - |x - 3|, \quad -\infty < x < \infty$

COMPUTER EXPLORATIONS

In Exercises 59–66, you will use a CAS to help find the absolute extrema of the given function over the specified closed interval. Perform the following steps.

a. Plot the function over the interval to see its general behavior there.

b. Find the interior points where $f' = 0$. (In some exercises, you may have to use the numerical equation solver to approximate a solution.) You may want to plot f' as well.

c. Find the interior points where f' does not exist.

d. Evaluate the function at all points found in parts (b) and (c) and at the endpoints of the interval.

e. Find the function's absolute extreme values on the interval and identify where they occur.

59. $f(x) = x^4 - 8x^2 + 4x + 2, \quad [-20/25, 64/25]$

60. $f(x) = -x^4 + 4x^3 - 4x + 1, \quad [-3/4, 3]$

61. $f(x) = x^{2/3}(3 - x), \quad [-2, 2]$

62. $f(x) = 2 + 2x - 3x^{2/3}, \quad [-1, 10/3]$

63. $f(x) = \sqrt{x} + \cos x, \quad [0, 2\pi]$

64. $f(x) = x^{3/4} - \sin x + \frac{1}{2}, \quad [0, 2\pi]$

65. $f(x) = \pi x^2 e^{-3x/2}, \quad [0, 5]$

66. $f(x) = \ln(2x + x\sin x), \quad [1, 15]$

3.2 The Mean Value Theorem

We know that constant functions have zero derivatives, but could there be a more complicated function whose derivative is always zero? If two functions have identical derivatives over an interval, how are the functions related? We answer these and other questions in this chapter by applying the Mean Value Theorem. First we introduce a special case, known as Rolle's Theorem, which is used to prove the Mean Value Theorem.

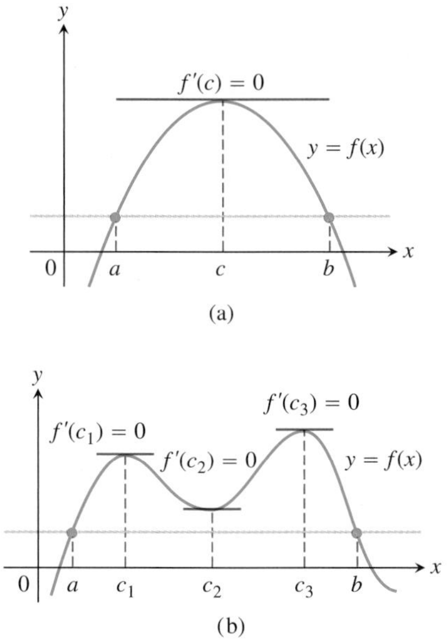

FIGURE 3.10 Rolle's Theorem says that a differentiable curve has at least one horizontal tangent between any two points where it crosses a horizontal line. It may have just one (a), or it may have more (b).

Rolle's Theorem

As suggested by its graph, if a differentiable function crosses a horizontal line at two different points, there is at least one point between them where the tangent to the graph is horizontal and the derivative is zero (Figure 3.10). We now state and prove this result.

> **THEOREM 3—Rolle's Theorem** Suppose that $y = f(x)$ is continuous at every point of the closed interval $[a, b]$ and differentiable at every point of its interior (a, b). If $f(a) = f(b)$, then there is at least one number c in (a, b) at which $f'(c) = 0$.

Proof Being continuous, f assumes absolute maximum and minimum values on $[a, b]$ by Theorem 1. These can occur only

1. at interior points where f' is zero,
2. at interior points where f' does not exist,
3. at the endpoints of the function's domain, in this case a and b.

By hypothesis, f has a derivative at every interior point. That rules out possibility (2), leaving us with interior points where $f' = 0$ and with the two endpoints a and b.

If either the maximum or the minimum occurs at a point c between a and b, then $f'(c) = 0$ by Theorem 2 in Section 3.1, and we have found a point for Rolle's Theorem.

If both the absolute maximum and the absolute minimum occur at the endpoints, then because $f(a) = f(b)$ it must be the case that f is a constant function with $f(x) = f(a) = f(b)$ for every $x \in [a, b]$. Therefore $f'(x) = 0$ and the point c can be taken anywhere in the interior (a, b). ■

HISTORICAL BIOGRAPHY

Michel Rolle
(1652–1719)

The hypotheses of Theorem 3 are essential. If they fail at even one point, the graph may not have a horizontal tangent (Figure 3.11).

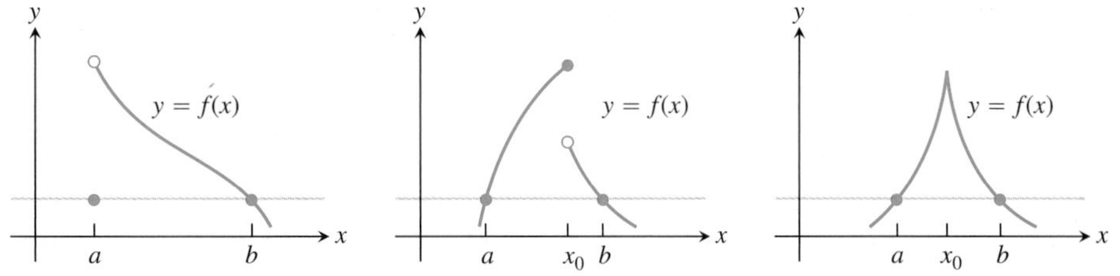

(a) Discontinuous at an endpoint of $[a, b]$

(b) Discontinuous at an interior point of $[a, b]$

(c) Continuous on $[a, b]$ but not differentiable at an interior point

FIGURE 3.11 There may be no horizontal tangent if the hypotheses of Rolle's Theorem do not hold.

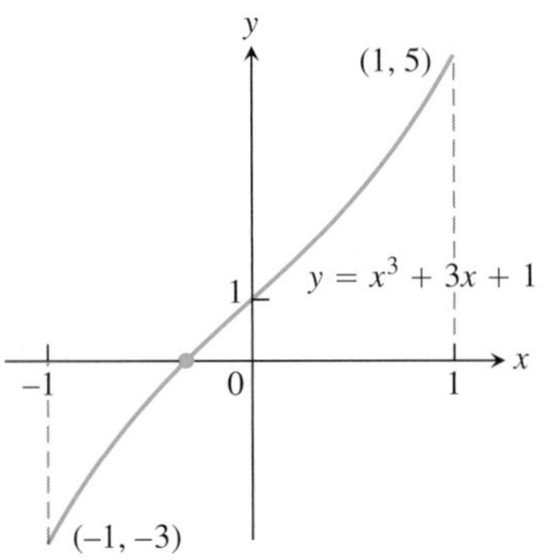

FIGURE 3.12 The only real zero of the polynomial $y = x^3 + 3x + 1$ is the one shown here where the curve crosses the x-axis between -1 and 0 (Example 1).

Rolle's Theorem may be combined with the Intermediate Value Theorem to show when there is only one real solution of an equation $f(x) = 0$ between two points where the values of a differentiable function f have opposite signs.

EXAMPLE 1 Show that the equation

$$x^3 + 3x + 1 = 0$$

has exactly one real solution.

Solution Let

$$y = f(x) = x^3 + 3x + 1.$$

Then the derivative

$$f'(x) = 3x^2 + 3$$

is never zero (because it is always positive). Now, if there were even two points $x = a$ and $x = b$ where $f(x)$ was zero, Rolle's Theorem would guarantee the existence of a point $x = c$ in between them where f' was zero. Therefore, f has no more than one zero. It does in fact have one zero, because the Intermediate Value Theorem tells us that the graph of $y = f(x)$ crosses the x-axis somewhere between $x = -1$ (where $y = -3$) and $x = 0$ (where $y = 1$). (See Figure 3.12.) ■

Our main use of Rolle's Theorem is in proving the Mean Value Theorem.

The Mean Value Theorem

The Mean Value Theorem, which was first stated by Joseph-Louis Lagrange, is a slanted version of Rolle's Theorem (Figure 3.13). The Mean Value Theorem guarantees that there is a point where the tangent line is parallel to the chord AB.

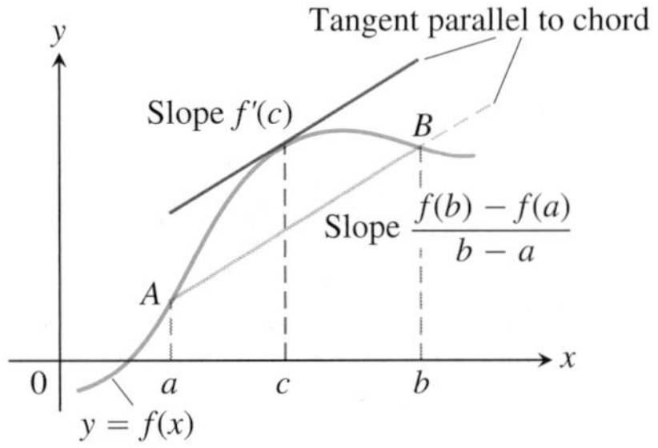

FIGURE 3.13 Geometrically, the Mean Value Theorem says that somewhere between A and B the curve has at least one tangent parallel to chord AB.

THEOREM 4—The Mean Value Theorem Suppose $y = f(x)$ is continuous on a closed interval $[a, b]$ and differentiable on the interval's interior (a, b). Then there is at least one point c in (a, b) at which

$$\frac{f(b) - f(a)}{b - a} = f'(c). \tag{1}$$

Proof We picture the graph of f and draw a line through the points $A(a, f(a))$ and $B(b, f(b))$ (see Figure 3.14). The line is the graph of the function

$$g(x) = f(a) + \frac{f(b) - f(a)}{b - a}(x - a) \tag{2}$$

(point-slope equation). The vertical difference between the graphs of f and g at x is

$$\begin{aligned} h(x) &= f(x) - g(x) \\ &= f(x) - f(a) - \frac{f(b) - f(a)}{b - a}(x - a). \end{aligned} \tag{3}$$

Figure 3.15 shows the graphs of f, g, and h together.

Historical Biography

Joseph-Louis Lagrange
(1736–1813)

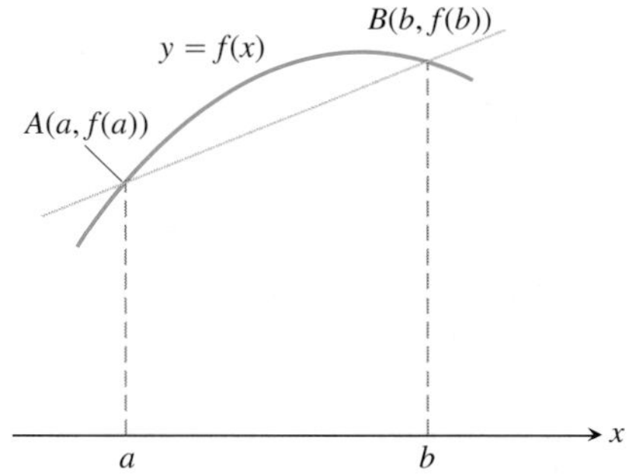

FIGURE 3.14 The graph of f and the chord AB over the interval $[a, b]$.

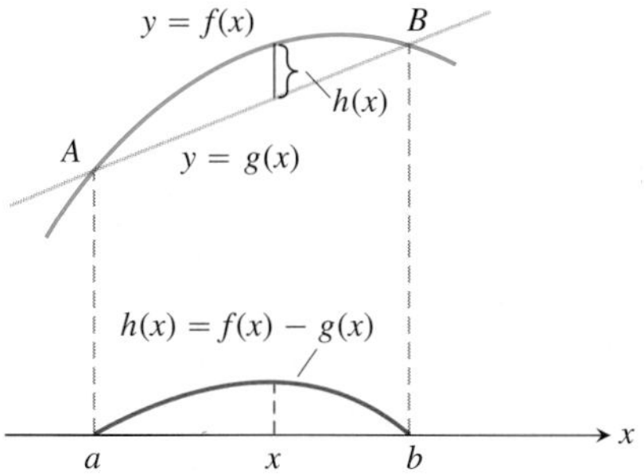

FIGURE 3.15 The chord AB is the graph of the function $g(x)$. The function $h(x) = f(x) - g(x)$ gives the vertical distance between the graphs of f and g at x.

The function h satisfies the hypotheses of Rolle's Theorem on $[a, b]$. It is continuous on $[a, b]$ and differentiable on (a, b) because both f and g are. Also, $h(a) = h(b) = 0$ because the graphs of f and g both pass through A and B. Therefore $h'(c) = 0$ at some point $c \in (a, b)$. This is the point we want for Equation (1).

To verify Equation (1), we differentiate both sides of Equation (3) with respect to x and then set $x = c$:

$$h'(x) = f'(x) - \frac{f(b) - f(a)}{b - a} \qquad \text{Derivative of Eq. (3) . . .}$$

$$h'(c) = f'(c) - \frac{f(b) - f(a)}{b - a} \qquad \text{. . . with } x = c$$

$$0 = f'(c) - \frac{f(b) - f(a)}{b - a} \qquad h'(c) = 0$$

$$f'(c) = \frac{f(b) - f(a)}{b - a}, \qquad \text{Rearranged}$$

which is what we set out to prove. ■

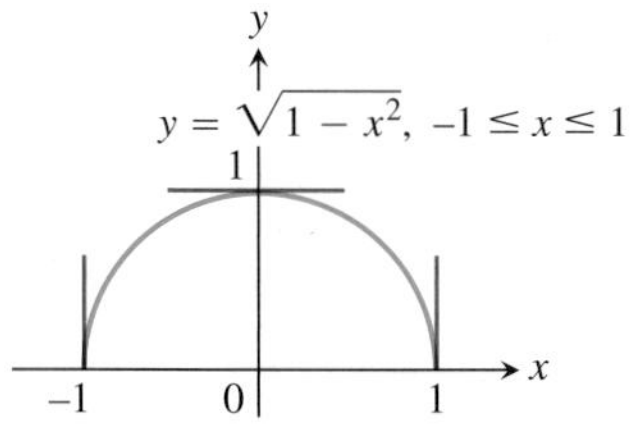

FIGURE 3.16 The function $f(x) = \sqrt{1 - x^2}$ satisfies the hypotheses (and conclusion) of the Mean Value Theorem on $[-1, 1]$ even though f is not differentiable at -1 and 1.

The hypotheses of the Mean Value Theorem do not require f to be differentiable at either a or b. Continuity at a and b is enough (Figure 3.16).

EXAMPLE 2 The function $f(x) = x^2$ (Figure 3.17) is continuous for $0 \le x \le 2$ and differentiable for $0 < x < 2$. Since $f(0) = 0$ and $f(2) = 4$, the Mean Value Theorem says that at some point c in the interval, the derivative $f'(x) = 2x$ must have the value $(4 - 0)/(2 - 0) = 2$. In this (exceptional) case we can identify c by solving the equation $2c = 2$ to get $c = 1$. ■

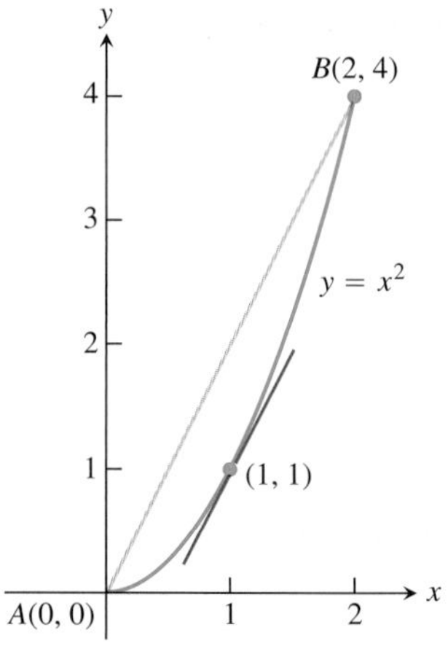

FIGURE 3.17 As we find in Example 2, $c = 1$ is where the tangent is parallel to the chord.

A Physical Interpretation

Think of the number $(f(b) - f(a))/(b - a)$ as the average change in f over $[a, b]$ and $f'(c)$ as an instantaneous change. Then the Mean Value Theorem says that at some interior point the instantaneous change must equal the average change over the entire interval.

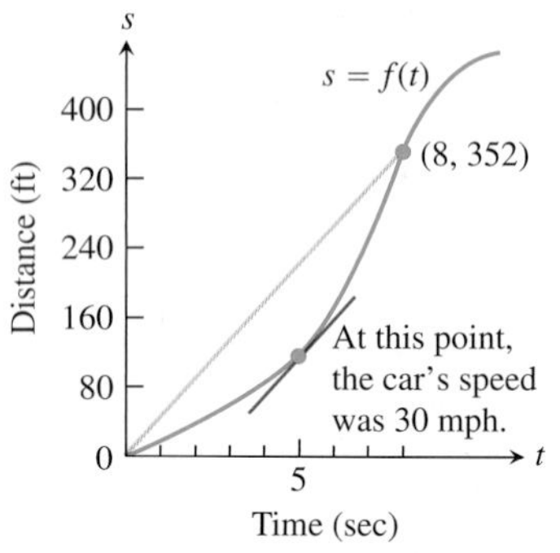

FIGURE 3.18 Distance versus elapsed time for the car in Example 3.

EXAMPLE 3 If a car accelerating from zero takes 8 sec to go 352 ft, its average velocity for the 8-sec interval is $352/8 = 44$ ft/sec. At some point during the acceleration, the Mean Value Theorem says, the speedometer must read exactly 30 mph (44 ft/sec) (Figure 3.18). ■

Mathematical Consequences

At the beginning of the section, we asked what kind of function has a zero derivative over an interval. The first corollary of the Mean Value Theorem provides the answer.

> **COROLLARY 1** If $f'(x) = 0$ at each point x of an open interval (a, b), then $f(x) = C$ for all $x \in (a, b)$, where C is a constant.

Proof We want to show that f has a constant value on the interval (a, b). We do so by showing that if x_1 and x_2 are any two points in (a, b), then $f(x_1) = f(x_2)$. Numbering x_1 and x_2 from left to right, we have $x_1 < x_2$. Then f satisfies the hypotheses of the Mean Value Theorem on $[x_1, x_2]$: It is differentiable at every point of $[x_1, x_2]$ and hence continuous at every point as well. Therefore,

$$\frac{f(x_2) - f(x_1)}{x_2 - x_1} = f'(c)$$

at some point c between x_1 and x_2. Since $f' = 0$ throughout (a, b), this equation translates successively into

$$\frac{f(x_2) - f(x_1)}{x_2 - x_1} = 0, \qquad f(x_2) - f(x_1) = 0, \qquad \text{and} \qquad f(x_1) = f(x_2). \qquad \blacksquare$$

At the beginning of this section, we also asked about the relationship between two functions that have identical derivatives over an interval. The next corollary tells us that their values on the interval have a constant difference. (See Figure 3.19)

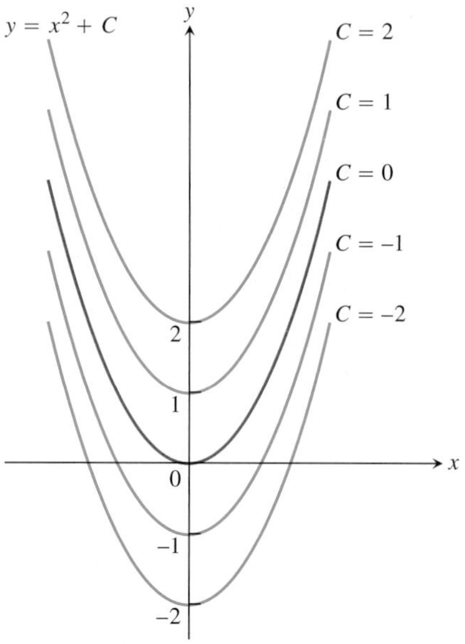

FIGURE 3.19 From a geometric point of view, Corollary 2 of the Mean Value Theorem says that the graphs of functions with identical derivatives on an interval can differ only by a vertical shift there. The graphs of the functions with derivative $2x$ are the parabolas $y = x^2 + C$, shown here for selected values of C.

> **COROLLARY 2** If $f'(x) = g'(x)$ at each point x in an open interval (a, b), then there exists a constant C such that $f(x) = g(x) + C$ for all $x \in (a, b)$. That is, $f - g$ is a constant on (a, b).

Proof At each point $x \in (a, b)$ the derivative of the difference function $h = f - g$ is

$$h'(x) = f'(x) - g'(x) = 0.$$

Thus, $h(x) = C$ on (a, b) by Corollary 1. That is, $f(x) - g(x) = C$ on (a, b), so $f(x) = g(x) + C$. ■

Corollaries 1 and 2 are also true if the open interval (a, b) fails to be finite. That is, they remain true if the interval is (a, ∞), $(-\infty, b)$, or $(-\infty, \infty)$. Corollary 2 plays an important role when we discuss antiderivatives in Section 4.1.

EXAMPLE 4 Find the function $f(x)$ whose derivative is $\sin x$ and whose graph passes through the point $(0, 2)$.

Solution Since $f(x)$ has the same derivative as $g(x) = -\cos x$, we know that $f(x) = -\cos x + C$ for some constant C. The value of C can be determined from the condition that $f(0) = 2$ (the graph of f passes through $(0, 2)$):

$$f(0) = -\cos(0) + C = 2, \qquad \text{so} \qquad C = 3.$$

The function is $f(x) = -\cos x + 3$. ∎

EXERCISES 3.2

Find the value or values of c that satisfy the equation

$$\frac{f(b) - f(a)}{b - a} = f'(c)$$

in the conclusion of the Mean Value Theorem for the functions and intervals in Exercises 1–4.

1. $f(x) = x^2 + 2x - 1, \quad [0, 1]$
2. $f(x) = x^{2/3}, \quad [0, 1]$
3. $f(x) = \sin^{-1} x, \quad [-1, 1]$
4. $f(x) = \ln(x - 1), \quad [2, 4]$

Which of the functions in Exercises 5–8 satisfy the hypotheses of the Mean Value Theorem on the given interval, and which do not? Give reasons for your answers.

5. $f(x) = x^{2/3}, \quad [-1, 8]$
6. $f(x) = x^{4/5}, \quad [0, 1]$
7. $f(x) = \sqrt{x(1 - x)}, \quad [0, 1]$
8. $f(x) = \begin{cases} \dfrac{\sin x}{x}, & -\pi \le x < 0 \\ 0, & x = 0 \end{cases}$

9. **a.** Plot the zeros of each polynomial on a line together with the zeros of its first derivative.
   - **i)** $y = x^2 - 4$
   - **ii)** $y = x^2 + 8x + 15$
   - **iii)** $y = x^3 - 3x^2 + 4 = (x + 1)(x - 2)^2$
   - **iv)** $y = x^3 - 33x^2 + 216x = x(x - 9)(x - 24)$

   b. Use Rolle's Theorem to prove that between every two zeros of $x^n + a_{n-1}x^{n-1} + \cdots + a_1 x + a_0$ there lies a zero of

   $$nx^{n-1} + (n - 1)a_{n-1}x^{n-2} + \cdots + a_1.$$

10. Suppose that f'' is continuous on $[a, b]$ and that f has three zeros in the interval. Show that f'' has at least one zero in (a, b). Generalize this result.

11. Show that if $f'' > 0$ throughout an interval $[a, b]$, then f' has at most one zero in $[a, b]$. What if $f'' < 0$ throughout $[a, b]$ instead?

12. Show that a cubic polynomial can have at most three real zeros.

13. Suppose that $f(-1) = 3$ and that $f'(x) = 0$ for all x. Must $f(x) = 3$ for all x? Give reasons for your answer.

14. Suppose that $f'(x) = 2x$ for all x. Find $f(2)$ if
    a. $f(0) = 0$ **b.** $f(1) = 0$ **c.** $f(-2) = 3$.

In Exercises 15–18, find all possible functions with the given derivative.

15. **a.** $y' = x$ **b.** $y' = x^2$ **c.** $y' = x^3$
16. **a.** $y' = 2x$ **b.** $y' = 2x - 1$ **c.** $y' = 3x^2 + 2x - 1$
17. **a.** $y' = -\dfrac{1}{x}$ **b.** $y' = 1 - \dfrac{1}{x}$ **c.** $y' = 5 + \dfrac{1}{x}$
18. **a.** $y' = \sin 2t$ **b.** $y' = \cos\dfrac{t}{2}$ **c.** $y' = \sin 2t + \cos\dfrac{t}{2}$

In Exercises 19–22, find the function with the given derivative whose graph passes through the point P.

19. $f'(x) = 2x - 1, \quad P(0, 0)$
20. $g'(x) = \dfrac{1}{x} + 2x, \quad P(1, -1)$
21. $f'(x) = e^{2x}, \quad P\left(0, \dfrac{3}{2}\right)$
22. $r'(t) = \sec t \tan t - 1, \quad P(0, 0)$

Exercises 23–26 give the velocity $v = ds/dt$ or acceleration a, with appropriate initial conditions, of a body moving along a coordinate line. Find the body's position at time t.

23. $v = 9.8t + 5, \quad s(0) = 10$
24. $v = \sin \pi t, \quad s(0) = 0$
25. $a = e^t, \quad v(0) = 20, \quad s(0) = 5$
26. $a = -4 \sin 2t, \quad v(0) = 2, \quad s(0) = -3$
27. **Temperature change** It took 14 sec for a mercury thermometer to rise from -19°C to 100°C when it was taken from a freezer and placed in boiling water. Show that somewhere along the way the mercury was rising at the rate of 8.5°C/sec.

28. A trucker handed in a ticket at a toll booth showing that in 2 hours she had covered 159 mi on a toll road with speed limit 65 mph. The trucker was cited for speeding. Why?

29. The geometric mean of a and b The *geometric mean* of two positive numbers a and b is the number $\sqrt{ab}$. Show that the value of c in the conclusion of the Mean Value Theorem for $f(x) = 1/x$ on an interval of positive numbers $[a, b]$ is $c = \sqrt{ab}$.

30. The arithmetic mean of a and b The *arithmetic mean* of two numbers a and b is the number $(a + b)/2$. Show that the value of c in the conclusion of the Mean Value Theorem for $f(x) = x^2$ on any interval $[a, b]$ is $c = (a + b)/2$.

31. Unique solution Assume that f is continuous on $[a, b]$ and differentiable on (a, b). Also assume that $f(a)$ and $f(b)$ have opposite signs and that $f' \neq 0$ between a and b. Show that $f(x) = 0$ exactly once between a and b.

32. Parallel tangents Assume that f and g are differentiable on $[a, b]$ and that $f(a) = g(a)$ and $f(b) = g(b)$. Show that there is at least one point between a and b where the tangents to the graphs of f and g are parallel or the same line. Illustrate with a sketch.

33. Assume that f is differentiable on $a \leq x \leq b$ and that $f(b) < f(a)$. Show that f' is negative at some point between a and b.

34. Show that for any numbers a and b, the sine inequality $|\sin b - \sin a| \leq |b - a|$ is true.

35. Let f be differentiable at every value of x and suppose that $f(1) = 1$, that $f' < 0$ on $(-\infty, 1)$, and that $f' > 0$ on $(1, \infty)$.

a. Show that $f(x) \geq 1$ for all x.

b. Must $f'(1) = 0$? Explain.

36. Let $f(x) = px^2 + qx + r$ be a quadratic function defined on a closed interval $[a, b]$. Show that there is exactly one point c in (a, b) at which f satisfies the conclusion of the Mean Value Theorem.

3.3 Monotonic Functions and the First Derivative Test

In sketching the graph of a differentiable function it is useful to know where it increases (rises from left to right) and where it decreases (falls from left to right) over an interval. This section gives a test to determine where it increases and where it decreases. We also show how to test the critical points of a function to identify whether local extreme values are present.

Increasing Functions and Decreasing Functions

We show that the only functions with positive derivatives are increasing functions; the only functions with negative derivatives are decreasing functions. A function that is increasing or decreasing on an interval is said to be **monotonic** on the interval.

COROLLARY 3 Suppose that f is continuous on $[a, b]$ and differentiable on (a, b).

If $f'(x) > 0$ at each point $x \in (a, b)$, then f is increasing on $[a, b]$.
If $f'(x) < 0$ at each point $x \in (a, b)$, then f is decreasing on $[a, b]$.

Proof Let x_1 and x_2 be any two points in $[a, b]$ with $x_1 < x_2$. The Mean Value Theorem applied to f on $[x_1, x_2]$ says that

$$f(x_2) - f(x_1) = f'(c)(x_2 - x_1)$$

for some c between x_1 and x_2. The sign of the right-hand side of this equation is the same as the sign of $f'(c)$ because $x_2 - x_1$ is positive. Therefore, $f(x_2) > f(x_1)$ if f' is positive on (a, b) and $f(x_2) < f(x_1)$ if f' is negative on (a, b). ∎

Here is how to apply the First Derivative Test to find where a function is increasing and decreasing. If $a < b$ are two critical points for a function f, and if f' exists but is not zero on the interval (a, b), then f' must be positive on (a, b) or negative there (Theorem 2, Section 2.2). One way we can determine the sign of f' on the interval is simply by evaluating f' for some point x in (a, b). Then we apply Corollary 3. Corollary 3 is valid for infinite as well as finite intervals, and we use this fact in our analysis in the next example.

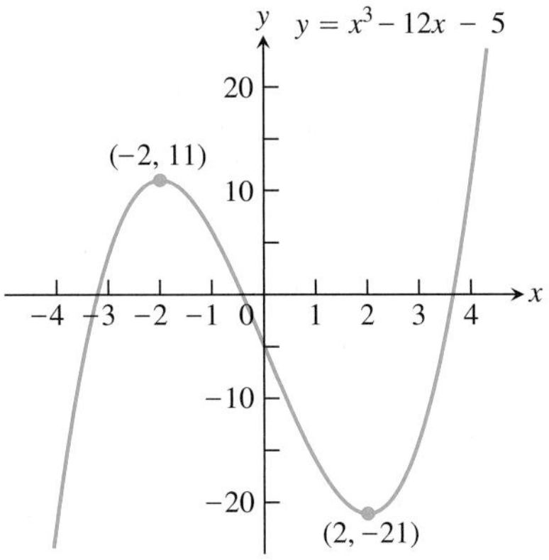

FIGURE 3.20 The function $f(x) = x^3 - 12x - 5$ is monotonic on three separate intervals (Example 1).

EXAMPLE 1 Find the critical points of $f(x) = x^3 - 12x - 5$ and identify the intervals on which f is increasing and on which f is decreasing.

Solution The function f is everywhere continuous and differentiable. The first derivative

$$\begin{aligned} f'(x) &= 3x^2 - 12 = 3(x^2 - 4) \\ &= 3(x + 2)(x - 2) \end{aligned}$$

is zero at $x = -2$ and $x = 2$. These critical points subdivide the domain of f into intervals $(-\infty, -2)$, $(-2, 2)$, and $(2, \infty)$ on which f' is either positive or negative. We determine the sign of f' by evaluating f at a convenient point in each subinterval. The behavior of f is determined by then applying Corollary 3 to each subinterval. The results are summarized in the following table, and the graph of f is given in Figure 3.20.

Interval	$-\infty < x < -2$	$-2 < x < 2$	$2 < x < \infty$
f' Evaluated	$f'(-3) = 15$	$f'(0) = -12$	$f'(3) = 15$
Sign of f'	$+$	$-$	$+$
Behavior of f	increasing	decreasing	increasing

∎

We used strict less than inequalities specifying the intervals in the summary table for Example 1. Corollary 3 says that we could use $\leq$ inequalities as well. That is, the function f in the example is increasing on $-\infty < x \leq -2$, decreasing on $-2 \leq x \leq 2$, and increasing on $2 \leq x < \infty$. We do not talk about whether a function is increasing or decreasing at a single point.

HISTORICAL BIOGRAPHY

Edmund Halley
(1656–1742)

First Derivative Test for Local Extrema

In Figure 3.21, at the points where f has a minimum value, $f' < 0$ immediately to the left and $f' > 0$ immediately to the right. (If the point is an endpoint, there is only one side to consider.) Thus, the function is decreasing on the left of the minimum value and it is increasing on its right. Similarly, at the points where f has a maximum value, $f' > 0$ immediately to the left and $f' < 0$ immediately to the right. Thus, the function is increasing on the left of the maximum value and decreasing on its right. In summary, at a local extreme point, the sign of $f'(x)$ changes.

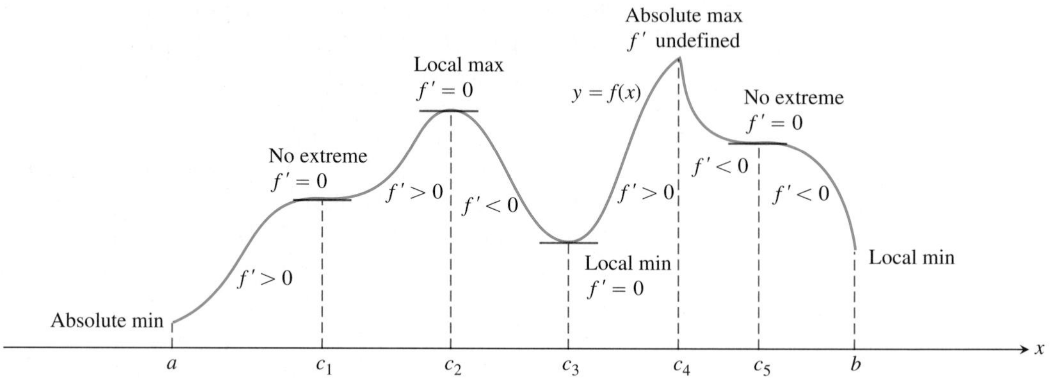

FIGURE 3.21 A function's first derivative tells how the graph rises and falls.

These observations lead to a test for the presence and nature of local extreme values of differentiable functions.

> **First Derivative Test for Local Extrema**
>
> Suppose that c is a critical point of a continuous function f, and that f is differentiable at every point in some interval containing c except possibly at c itself. Moving across c from left to right,
>
> 1. if f' changes from negative to positive at c, then f has a local minimum at c;
> 2. if f' changes from positive to negative at c, then f has a local maximum at c;
> 3. if f' does not change sign at c (that is, f' is positive on both sides of c or negative on both sides), then f has no local extremum at c.

The test for local extrema at endpoints is similar, but there is only one side to consider.

Proof Part (1). Since the sign of f' changes from negative to positive at c, there are numbers a and b such that $f' < 0$ on (a, c) and $f' > 0$ on (c, b). If $x \in (a, c)$, then $f(c) < f(x)$ because $f' < 0$ implies that f is decreasing on $[a, c]$. If $x \in (c, b)$, then $f(c) < f(x)$ because $f' > 0$ implies that f is increasing on $[c, b]$. Therefore, $f(x) \geq f(c)$ for every $x \in (a, b)$. By definition, f has a local minimum at c.

Parts (2) and (3) are proved similarly. ∎

EXAMPLE 2 Find the critical points of

$$f(x) = (x^2 - 3)e^x.$$

Identify the intervals on which f is increasing and decreasing. Find the function's local and absolute extreme values.

Solution The function f is continuous and differentiable for all real numbers, so the critical points occur only at the zeros of f'. Using the Product Rule, we find the derivative

$$\begin{aligned} f'(x) &= (x^2 - 3) \cdot \frac{d}{dx} e^x + \frac{d}{dx}(x^2 - 3) \cdot e^x \\ &= (x^2 - 3) \cdot e^x + (2x) \cdot e^x \\ &= (x^2 + 2x - 3)e^x. \end{aligned}$$

Since e^x is never zero, the first derivative is zero if and only if

$$x^2 + 2x - 3 = 0$$
$$(x + 3)(x - 1) = 0.$$

The zeros $x = -3$ and $x = 1$ partition the x-axis into intervals as follows.

Interval	$x < -3$	$-3 < x < 1$	$1 < x$
Sign of f'	$+$	$-$	$+$
Behavior of f	increasing	decreasing	increasing

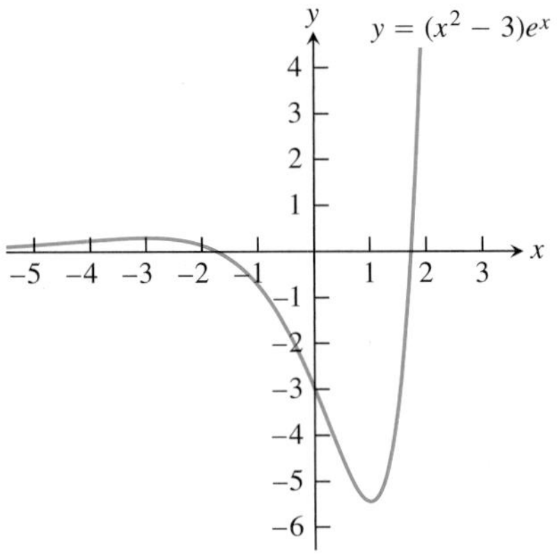

FIGURE 3.22 The graph of $f(x) = (x^2 - 3)e^x$. (Example 2)

We can see from the table that there is a local maximum (about 0.299) at $x = -3$ and a local minimum (about -5.437) at $x = 1$. The local minimum value is also an absolute minimum because $f(x) > 0$ for $|x| > \sqrt{3}$. There is no absolute maximum. The function increases on $(-\infty, -3)$ and $(1, \infty)$ and decreases on $(-3, 1)$. Figure 3.22 shows the graph. ■

EXAMPLE 3 Find the critical points of

$$f(x) = x^{1/3}(x - 4) = x^{4/3} - 4x^{1/3}.$$

Identify the intervals on which f is increasing and decreasing. Find the function's local and absolute extreme values.

Solution The function f is continuous at all x since it is the product of two continuous functions, $x^{1/3}$ and $(x - 4)$. The first derivative

$$\begin{aligned} f'(x) &= \frac{d}{dx}\left(x^{4/3} - 4x^{1/3}\right) = \frac{4}{3}x^{1/3} - \frac{4}{3}x^{-2/3} \\ &= \frac{4}{3}x^{-2/3}\left(x - 1\right) = \frac{4(x - 1)}{3x^{2/3}} \end{aligned}$$

is zero at $x = 1$ and undefined at $x = 0$. There are no endpoints in the domain, so the critical points $x = 0$ and $x = 1$ are the only places where f might have an extreme value.

The critical points partition the x-axis into intervals on which f' is either positive or negative. The sign pattern of f' reveals the behavior of f between and at the critical points. We can display the information in a table like the following:

Interval	$x < 0$	$0 < x < 1$	$x > 1$
Sign of f'	$-$	$-$	$+$
Behavior of f	decreasing	decreasing	increasing

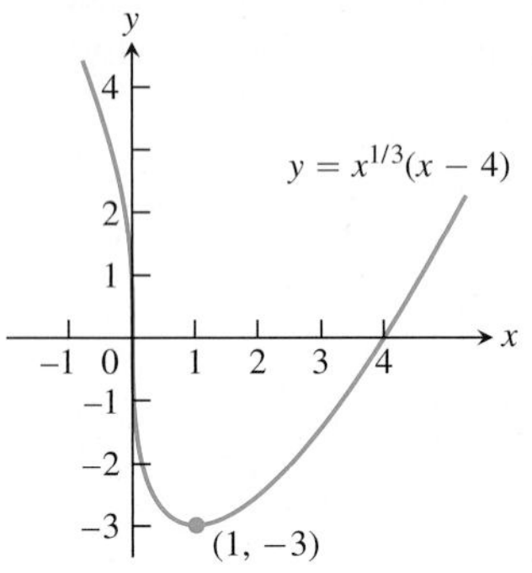

FIGURE 3.23 The function $f(x) = x^{1/3}(x - 4)$ decreases when $x < 1$ and increases when $x > 1$ (Example 3).

Corollary 3 to the Mean Value Theorem tells us that f decreases on $(-\infty, 0)$, decreases on $(0, 1)$, and increases on $(1, \infty)$. The First Derivative Test for Local Extrema tells us that f does not have an extreme value at $x = 0$ (f' does not change sign) and that f has a local minimum at $x = 1$ (f' changes from negative to positive).

The value of the local minimum is $f(1) = 1^{1/3}(1 - 4) = -3$. This is also an absolute minimum. Figure 3.23 shows this value in relation to the function's graph.

Note that $\lim_{x\to 0} f'(x) = -\infty$, so the graph of f has a vertical tangent at the origin. ■

EXERCISES 3.3

Answer the following questions about the functions whose derivatives are given in Exercises 1–8:

a. What are the critical points of f?

b. On what intervals is f increasing or decreasing?

c. At what points, if any, does f assume local maximum and minimum values?

1. $f'(x) = x(x - 1)$

2. $f'(x) = (x - 1)(x + 2)$

3. $f'(x) = (x - 1)^2(x + 2)$

4. $f'(x) = (x - 1)^2(x + 2)^2$

5. $f'(x) = (x - 1)e^{-x}$

6. $f'(x) = (x - 7)(x + 1)(x + 5)$

7. $f'(x) = x^{-1/3}(x + 2)$

8. $f'(x) = x^{-1/2}(x - 3)$

In Exercises 9–32:

a. Find the intervals on which the function is increasing and decreasing.

b. Then identify the function's local extreme values, if any, saying where they are taken on.

c. Which, if any, of the extreme values are absolute?

T **d.** Support your findings with a graphing calculator or computer grapher.

9. $g(t) = -t^2 - 3t + 3$

10. $g(t) = -3t^2 + 9t + 5$

11. $h(x) = -x^3 + 2x^2$

12. $h(x) = 2x^3 - 18x$

13. $f(\theta) = 3\theta^2 - 4\theta^3$

14. $f(\theta) = 6\theta - \theta^3$

15. $f(r) = 3r^3 + 16r$

16. $h(r) = (r + 7)^3$

17. $f(x) = x^4 - 8x^2 + 16$

18. $g(x) = x^4 - 4x^3 + 4x^2$

19. $H(t) = \frac{3}{2}t^4 - t^6$

20. $K(t) = 15t^3 - t^5$

21. $g(x) = x\sqrt{8 - x^2}$

22. $g(x) = x^2\sqrt{5 - x}$

23. $f(x) = \frac{x^2 - 3}{x - 2}, \quad x \neq 2$

24. $f(x) = \frac{x^3}{3x^2 + 1}$

25. $f(x) = x^{1/3}(x + 8)$

26. $g(x) = x^{2/3}(x + 5)$

27. $h(x) = x^{1/3}(x^2 - 4)$

28. $k(x) = x^{2/3}(x^2 - 4)$

29. $f(x) = e^{2x} + e^{-x}$

30. $f(x) = e^{\sqrt{x}}$

31. $f(x) = x \ln x$

32. $f(x) = x^2 \ln x$

In Exercises 33–40:

a. Identify the function's local extreme values in the given domain, and say where they are assumed.

b. Which of the extreme values, if any, are absolute?

T **c.** Support your findings with a graphing calculator or computer grapher.

33. $f(x) = 2x - x^2, \quad -\infty < x \leq 2$

34. $f(x) = (x + 1)^2, \quad -\infty < x \leq 0$

35. $g(x) = x^2 - 4x + 4, \quad 1 \leq x < \infty$

36. $g(x) = -x^2 - 6x - 9, \quad -4 \leq x < \infty$

37. $f(t) = 12t - t^3, \quad -3 \leq t < \infty$

38. $f(t) = t^3 - 3t^2, \quad -\infty < t \leq 3$

39. $h(x) = \frac{x^3}{3} - 2x^2 + 4x, \quad 0 \leq x < \infty$

40. $k(x) = x^3 + 3x^2 + 3x + 1, \quad -\infty < x \leq 0$

In Exercises 41–44:

a. Find the local extrema of each function on the given interval, and say where they occur.

T **b.** Graph the function and its derivative together. Comment on the behavior of f in relation to the signs and values of f'.

41. $f(x) = \dfrac{x}{2} - 2\sin\dfrac{x}{2}, \quad 0 \le x \le 2\pi$

42. $f(x) = -2\cos x - \cos^2 x, \quad -\pi \le x \le \pi$

43. $f(x) = \csc^2 x - 2\cot x, \quad 0 < x < \pi$

44. $f(x) = \sec^2 x - 2\tan x, \quad \dfrac{-\pi}{2} < x < \dfrac{\pi}{2}$

Show that the functions in Exercises 45 and 46 have local extreme values at the given values of θ, and say which kind of local extreme the function has.

45. $h(\theta) = 3\cos\dfrac{\theta}{2}, \quad 0 \le \theta \le 2\pi, \quad$ at $\theta = 0$ and $\theta = 2\pi$

46. $h(\theta) = 5\sin\dfrac{\theta}{2}, \quad 0 \le \theta \le \pi, \quad$ at $\theta = 0$ and $\theta = \pi$

47. Sketch the graph of a differentiable function $y = f(x)$ through the point $(1, 1)$ if $f'(1) = 0$ and

a. $f'(x) > 0$ for $x < 1$ and $f'(x) < 0$ for $x > 1$;

b. $f'(x) < 0$ for $x < 1$ and $f'(x) > 0$ for $x > 1$;

c. $f'(x) > 0$ for $x \ne 1$;

d. $f'(x) < 0$ for $x \ne 1$.

48. Sketch the graph of a differentiable function $y = f(x)$ that has

a. a local minimum at $(1, 1)$ and a local maximum at $(3, 3)$;

b. a local maximum at $(1, 1)$ and a local minimum at $(3, 3)$;

c. local maxima at $(1, 1)$ and $(3, 3)$;

d. local minima at $(1, 1)$ and $(3, 3)$.

49. Sketch the graph of a continuous function $y = g(x)$ such that

a. $g(2) = 2$, $0 < g' < 1$ for $x < 2$, $g'(x) \to 1^-$ as $x \to 2^-$, $-1 < g' < 0$ for $x > 2$, and $g'(x) \to -1^+$ as $x \to 2^+$;

b. $g(2) = 2$, $g' < 0$ for $x < 2$, $g'(x) \to -\infty$ as $x \to 2^-$, $g' > 0$ for $x > 2$, and $g'(x) \to \infty$ as $x \to 2^+$.

50. Sketch the graph of a continuous function $y = h(x)$ such that

a. $h(0) = 0$, $-2 \le h(x) \le 2$ for all x, $h'(x) \to \infty$ as $x \to 0^-$, and $h'(x) \to \infty$ as $x \to 0^+$;

b. $h(0) = 0$, $-2 \le h(x) \le 0$ for all x, $h'(x) \to \infty$ as $x \to 0^-$, and $h'(x) \to -\infty$ as $x \to 0^+$.

51. Locate and identify the absolute extreme values of

a. $\ln(\cos x)$ on $[-\pi/4, \pi/3]$,

b. $\cos(\ln x)$ on $[1/2, 2]$.

52. a. Prove that $f(x) = x - \ln x$ is increasing for $x > 1$.

b. Using part (a), show that $\ln x < x$ if $x > 1$.

53. Find the absolute maximum and minimum values of $f(x) = e^x - 2x$ on $[0, 1]$.

54. Where does the periodic function $f(x) = 2e^{\sin(x/2)}$ take on its extreme values and what are these values?

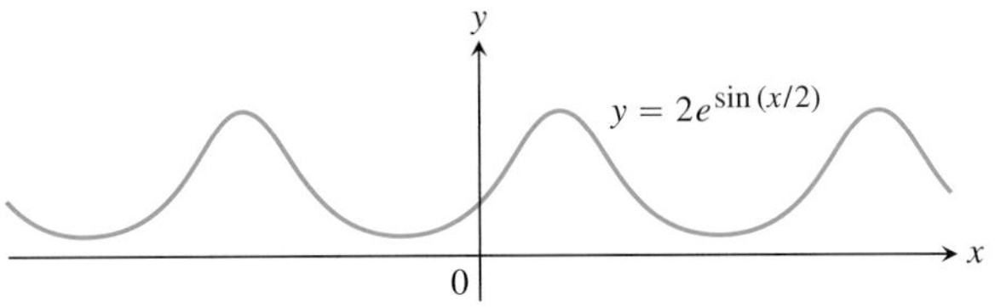

55. Find the absolute maximum value of $f(x) = x^2 \ln(1/x)$ and say where it occurs.

56. Find the intervals on which the function $f(x) = ax^2 + bx + c$, $a \ne 0$, is increasing and decreasing. Describe the reasoning behind your answer.

57. As x moves from left to right through the point $c = 2$, is the graph of $f(x) = x^3 - 3x + 2$ rising, or is it falling? Give reasons for your answer.

58. a. Prove that $e^x \ge 1 + x$ if $x \ge 0$.

b. Use the result in part (a) to show that

$$e^x \ge 1 + x + \frac{1}{2}x^2.$$

59. Show that increasing functions and decreasing functions are one-to-one. That is, show that for any x_1 and x_2 in I, $x_2 \ne x_1$ implies $f(x_2) \ne f(x_1)$.

Use the results of Exercise 59 to show that the functions in Exercises 60–64 have inverses over their domains. Find a formula for df^{-1}/dx using Theorem 4, Section 2.9.

60. $f(x) = (1/3)x + (5/6)$

61. $f(x) = 27x^3$

62. $f(x) = 1 - 8x^3$

63. $f(x) = (1 - x)^3$

64. $f(x) = x^{5/3}$

3.4 Concavity and Curve Sketching

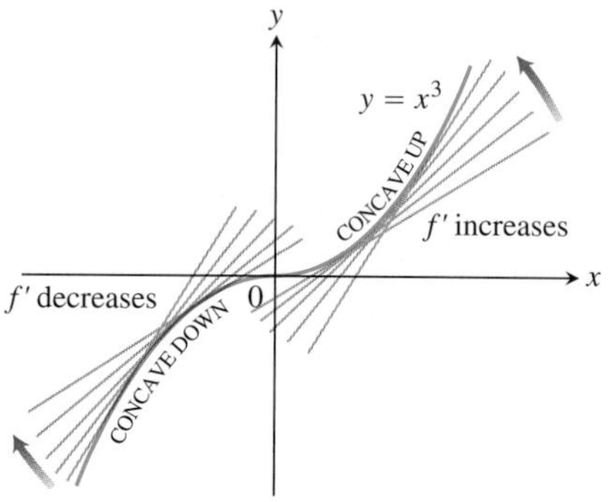

FIGURE 3.24 The graph of $f(x) = x^3$ is concave down on $(-\infty, 0)$ and concave up on $(0, \infty)$ (Example 1a).

In Section 3.3 we saw how the first derivative tells us where a function is increasing and where it is decreasing. At a critical point of a differentiable function, the First Derivative Test tells us whether there is a local maximum or a local minimum, or whether the graph just continues to rise or fall there.

In this section we see how the second derivative gives information about the way the graph of a differentiable function bends or turns. This additional information reveals key aspects of the behavior of a function, so we can present these features in a sketch of the graph.

Concavity

As you can see in Figure 3.24, the curve $y = x^3$ rises as x increases, but the portions defined on the intervals $(-\infty, 0)$ and $(0, \infty)$ turn in different ways. As we approach the origin from the left along the curve, the curve turns to our right and falls below its tangents. The slopes of the tangents are decreasing on the interval $(-\infty, 0)$. As we move away from the origin along the curve to the right, the curve turns to our left and rises above its tangents. The slopes of the tangents are increasing on the interval $(0, \infty)$. This turning or bending behavior defines the *concavity* of the curve.

DEFINITION The graph of a differentiable function $y = f(x)$ is

(a) **concave up** on an open interval I if f' is increasing on I

(b) **concave down** on an open interval I if f' is decreasing on I.

If $y = f(x)$ has a second derivative, we can apply Corollary 3 of the Mean Value Theorem to conclude that f' increases if $f'' > 0$ on I, and decreases if $f'' < 0$.

The Second Derivative Test for Concavity

Let $y = f(x)$ be twice-differentiable on an interval I.

1. If $f'' > 0$ on I, the graph of f over I is concave up.
2. If $f'' < 0$ on I, the graph of f over I is concave down.

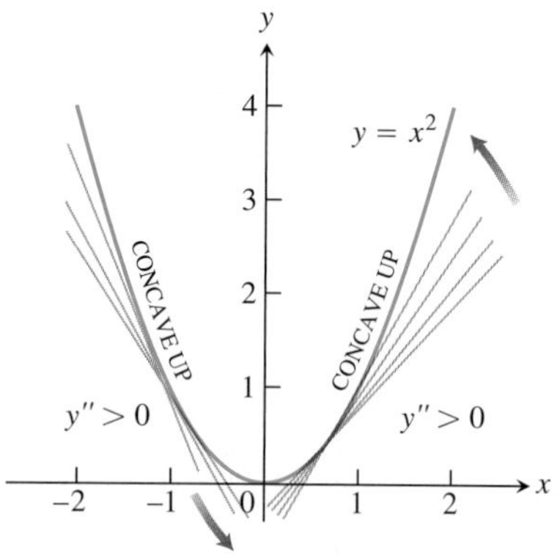

FIGURE 3.25 The graph of $f(x) = x^2$ is concave up on every interval (Example 1b).

If $y = f(x)$ is twice-differentiable, we will use the notations f'' and y'' interchangeably when denoting the second derivative.

EXAMPLE 1

(a) The curve $y = x^3$ (Figure 3.24) is concave down on $(-\infty, 0)$ where $y'' = 6x < 0$ and concave up on $(0, \infty)$ where $y'' = 6x > 0$.

(b) The curve $y = x^2$ (Figure 3.25) is concave up on $(-\infty, \infty)$ because its second derivative $y'' = 2$ is always positive. ■

EXAMPLE 2 Determine the concavity of $y = 3 + \sin x$ on $[0, 2\pi]$.

Solution The graph of $y = 3 + \sin x$ is concave down on $(0, \pi)$, where $y'' = -\sin x$ is negative. It is concave up on $(\pi, 2\pi)$, where $y'' = -\sin x$ is positive (Figure 3.26). ■

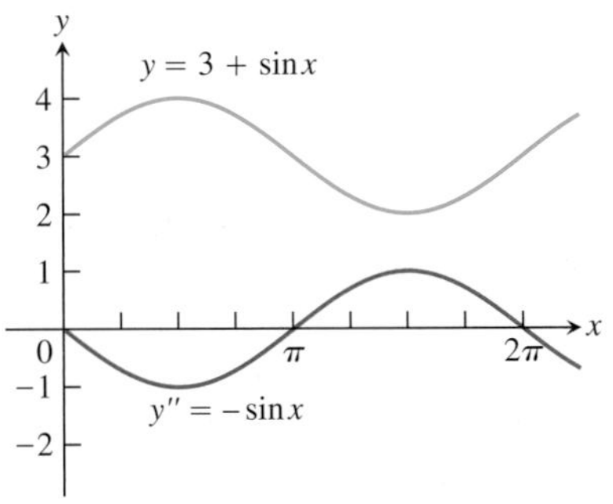

FIGURE 3.26 Using the graph of y'' to determine the concavity of y (Example 2).

Points of Inflection

The curve $y = 3 + \sin x$ in Example 2 changes concavity at the point $(\pi, 3)$. We call $(\pi, 3)$ a *point of inflection* of the curve. We assume the curve has a tangent line (including the possibility of a vertical tangent) at such a point.

> **DEFINITION** A point where the graph of a function has a tangent line and where the concavity changes is a **point of inflection**.

A point on a curve where y'' is positive on one side and negative on the other is a point of inflection. At such a point, y'' is either zero (because derivatives have the Intermediate Value Property) or undefined. If y is a twice-differentiable function, then $y'' = 0$ at a point of inflection. However, the next example shows an inflection point may not exist where $y'' = 0$.

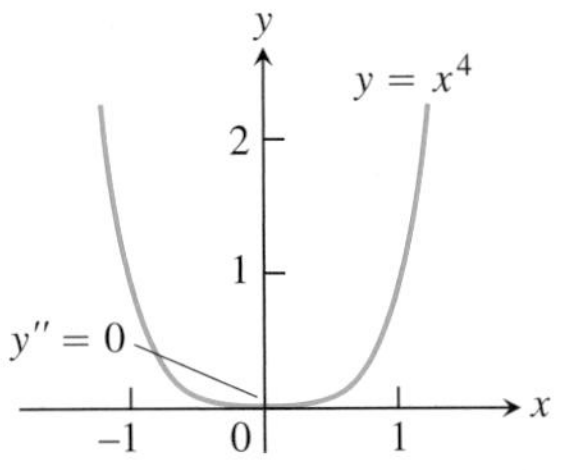

FIGURE 3.27 The graph of $y = x^4$ has no inflection point at the origin, even though $y'' = 0$ there (Example 3).

EXAMPLE 3 The curve $y = x^4$ has no inflection point at $x = 0$ (Figure 3.27). Even though $y'' = 12x^2$ is zero there, it does not change sign. ■

From the following example we see that inflection points can also occur where the second derivative is not defined.

EXAMPLE 4 The curve $y = x^{1/3}$ has a vertical tangent and a point of inflection at $x = 0$ (Figure 3.28), but y'' does not exist there as can be seen from the computation

$$y'' = \frac{d^2}{dx^2}\left(x^{1/3}\right) = \frac{d}{dx}\left(\frac{1}{3}x^{-2/3}\right) = -\frac{2}{9}x^{-5/3}.$$ ■

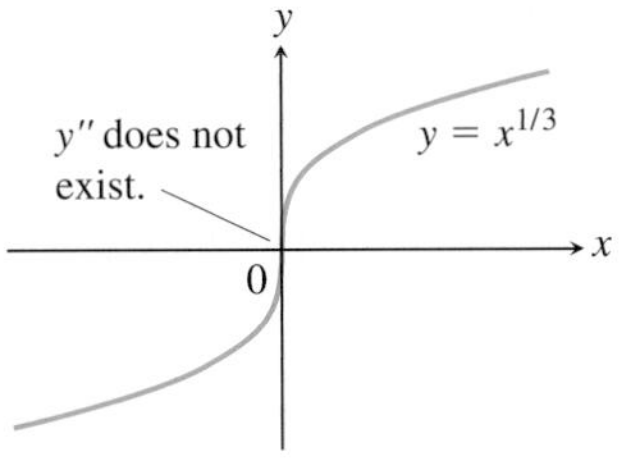

FIGURE 3.28 A point where y'' fails to exist can be a point of inflection (Example 4).

To study the motion of a body moving along a line as a function of time, we often are interested in knowing when the body's acceleration, given by the second derivative, is positive or negative. The points of inflection on the graph of the body's position function reveal where the acceleration changes sign.

EXAMPLE 5 A particle is moving along a horizontal line with position function

$$s(t) = 2t^3 - 14t^2 + 22t - 5, \qquad t \geq 0.$$

Find the velocity and acceleration, and describe the motion of the particle.

Solution The velocity is

$$v(t) = s'(t) = 6t^2 - 28t + 22 = 2(t - 1)(3t - 11),$$

and the acceleration is

$$a(t) = v'(t) = s''(t) = 12t - 28 = 4(3t - 7).$$

When the function $s(t)$ is increasing, the particle is moving to the right; when $s(t)$ is decreasing, the particle is moving to the left.

Notice that the first derivative $(v = s')$ is zero when $t = 1$ and $t = 11/3$.

Interval	$0 < t < 1$	$1 < t < 11/3$	$11/3 < t$
Sign of $v = s'$	$+$	$-$	$+$
Behavior of s	increasing	decreasing	increasing
Particle motion	right	left	right

The particle is moving to the right in the time intervals $[0, 1)$ and $(11/3, \infty)$, and moving to the left in $(1, 11/3)$. It is momentarily stationary (at rest), at $t = 1$ and $t = 11/3$.

The acceleration $a(t) = s''(t) = 4(3t - 7)$ is zero when $t = 7/3$.

Interval	$0 < t < 7/3$	$7/3 < t$
Sign of $a = s''$	$-$	$+$
Graph of s	concave down	concave up

The body starts out moving to the right while slowing down and then begins moving to the left at $t = 1$ under the influence of the leftward acceleration over the time interval $[0, 7/3]$. The acceleration then changes direction at $t = 7/3$ but the body continues moving leftward, slowing down under the rightward acceleration. At $t = 11/3$ the body reverses direction again: moving to the right in the same direction as the acceleration. ■

Second Derivative Test for Local Extrema

Instead of looking for sign changes in f' at critical points, we can sometimes use the following test to determine the presence and character of local extrema.

THEOREM 5—Second Derivative Test for Local Extrema Suppose f'' is continuous on an open interval that contains $x = c$.

1. If $f'(c) = 0$ and $f''(c) < 0$, then f has a local maximum at $x = c$.
2. If $f'(c) = 0$ and $f''(c) > 0$, then f has a local minimum at $x = c$.
3. If $f'(c) = 0$ and $f''(c) = 0$, then the test fails. The function f may have a local maximum, a local minimum, or neither.

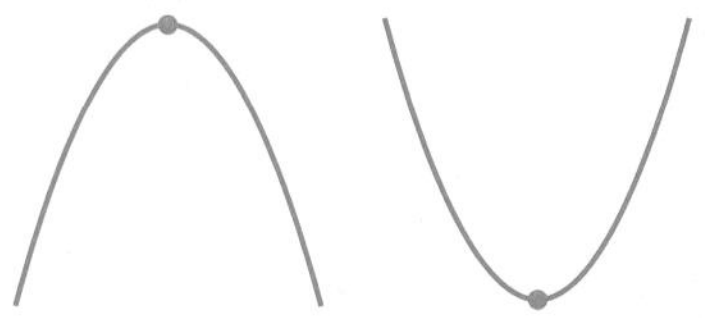

$f' = 0, f'' < 0 \Rightarrow$ local max $\quad$ $f' = 0, f'' > 0 \Rightarrow$ local min

Proof Part (1). If $f''(c) < 0$, then $f''(x) < 0$ on some open interval I containing the point c, since f'' is continuous. Therefore, f' is decreasing on I. Since $f'(c) = 0$, the sign of f' changes from positive to negative at c so f has a local maximum at c by the First Derivative Test.

The proof of Part (2) is similar.

For Part (3), consider the three functions $y = x^4$, $y = -x^4$, and $y = x^3$. For each function, the first and second derivatives are zero at $x = 0$. Yet the function $y = x^4$ has a local minimum there, $y = -x^4$ has a local maximum, and $y = x^3$ is increasing in any open interval containing $x = 0$ (having neither a maximum nor a minimum there). Thus the test fails. ■

This test requires us to know f'' *only at c itself* and not in an interval about c. This makes the test easy to apply. However, the test is inconclusive if $f'' = 0$ or if f'' does not exist at $x = c$. When this happens, use the First Derivative Test for local extreme values.

Together f' and f'' tell us the shape of the function's graph, that is, where the critical points are located and what happens at a critical point, where the function is increasing and where it is decreasing, and how the curve is turning or bending as defined by its concavity. We use this information to sketch a graph of the function that captures its key features.

EXAMPLE 6 Sketch a graph of the function

$$f(x) = x^4 - 4x^3 + 10$$

using the following steps.

(a) Identify where the extrema of f occur.

(b) Find the intervals on which f is increasing and the intervals on which f is decreasing.

(c) Find where the graph of f is concave up and where it is concave down.

(d) Sketch the general shape of the graph for f.

(e) Plot some specific points, such as local maximum and minimum points, points of inflection, and intercepts. Then sketch the curve.

Solution f is continuous since $f'(x) = 4x^3 - 12x^2$ exists. The domain of f is $(-\infty, \infty)$, and the domain of f' is also $(-\infty, \infty)$. Thus, the critical points of f occur only at the zeros of f'. Since

$$f'(x) = 4x^3 - 12x^2 = 4x^2(x - 3)$$

the first derivative is zero at $x = 0$ and $x = 3$.

Interval	$x < 0$	$0 < x < 3$	$3 < x$
Sign of f'	$-$	$-$	$+$
Behavior of f	decreasing	decreasing	increasing

(a) Using the First Derivative Test for local extrema and the table above, we see that there is no extremum at $x = 0$ and a local minimum at $x = 3$.

(b) Using the table above, we see that f is decreasing on $(-\infty, 0]$ and $[0, 3]$, and increasing on $[3, \infty)$.

(c) $f''(x) = 12x^2 - 24x = 12x(x - 2)$ is zero at $x = 0$ and $x = 2$.

Interval	$x < 0$	$0 < x < 2$	$2 < x$
Sign of f''	$+$	$-$	$+$
Behavior of f	concave up	concave down	concave up

We see that f is concave up on the intervals $(-\infty, 0)$ and $(2, \infty)$, and concave down on $(0, 2)$.

(d) Summarizing the information in the two preceding tables, we obtain

$x < 0$	$0 < x < 2$	$2 < x < 3$	$3 < x$
decreasing concave up	decreasing concave down	decreasing concave up	increasing concave up

The general shape of the curve is

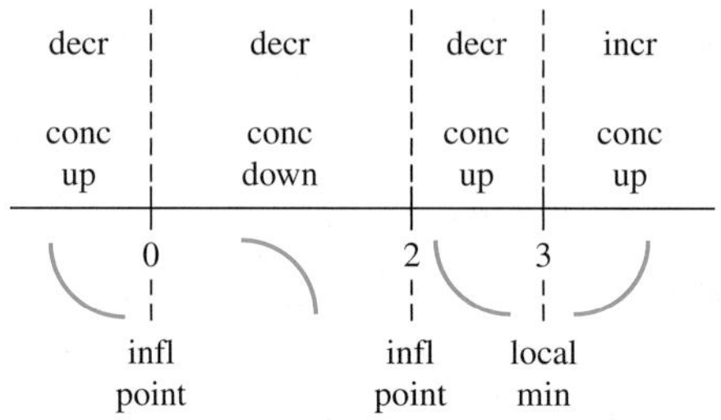

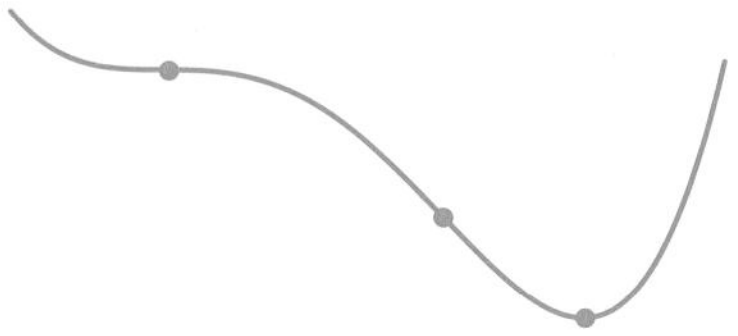

General shape.

(e) Plot the curve's intercepts (if possible) and the points where y' and y'' are zero. Indicate any local extreme values and inflection points. Use the general shape as a guide to sketch the curve. (Plot additional points as needed.) Figure 3.29 shows the graph of f. ■

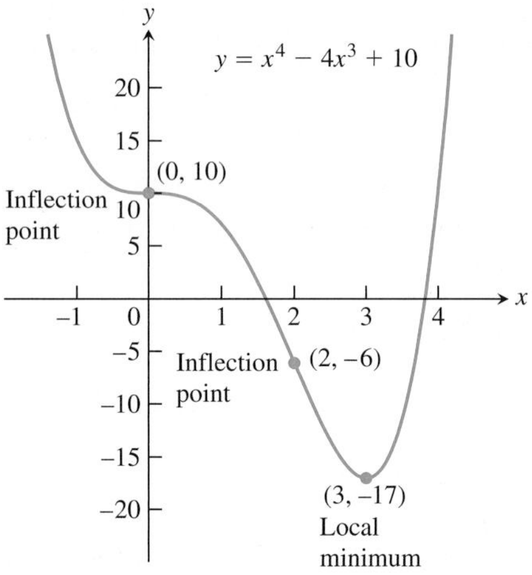

FIGURE 3.29 The graph of $f(x) = x^4 - 4x^3 + 10$ (Example 6).

The steps in Example 6 give a procedure for graphing the key features of a function.

> **Strategy for Graphing $y = f(x)$**
>
> 1. Identify the domain of f and any symmetries the curve may have.
> 2. Find y' and y''.
> 3. Find any critical points of f, and identify the function's behavior at each one.
> 4. Find where the curve is increasing and where it is decreasing.
> 5. Find the points of inflection, if any occur, and determine the concavity of the curve.
> 6. Identify any asymptotes.
> 7. Plot key points, such as the intercepts and the points found in Steps 3–5, and sketch the curve.

EXAMPLE 7 Sketch the graph of $f(x) = \dfrac{(x + 1)^2}{1 + x^2}$.

Solution

1. The domain of f is $(-\infty, \infty)$ and there are no symmetries about either axis or the origin (Section 1.1).

2. *Find f' and f''.*

$$f(x) = \frac{(x+1)^2}{1+x^2}$$

x-intercept at $x = -1$, y-intercept ($y = 1$) at $x = 0$

$$f'(x) = \frac{(1+x^2)\cdot 2(x+1) - (x+1)^2\cdot 2x}{(1+x^2)^2}$$

$$= \frac{2(1-x^2)}{(1+x^2)^2}$$

Critical points: $x = -1, x = 1$

$$f''(x) = \frac{(1+x^2)^2\cdot 2(-2x) - 2(1-x^2)[2(1+x^2)\cdot 2x]}{(1+x^2)^4}$$

$$= \frac{4x(x^2-3)}{(1+x^2)^3}$$

After some algebra

3. *Behavior at critical points.* The critical points occur only at $x = \pm 1$ where $f'(x) = 0$ (Step 2) since f' exists everywhere over the domain of f. At $x = -1$, $f''(-1) = 1 > 0$ yielding a relative minimum by the Second Derivative Test. At $x = 1$, $f''(1) = -1 < 0$ yielding a relative maximum by the Second Derivative test.

4. *Increasing and decreasing.* We see that on the interval $(-\infty, -1)$ the derivative $f'(x) < 0$, and the curve is decreasing. On the interval $(-1, 1)$, $f'(x) > 0$ and the curve is increasing; it is decreasing on $(1, \infty)$ where $f'(x) < 0$ again.

5. *Inflection points.* Notice that the denominator of the second derivative (Step 2) is always positive. The second derivative f'' is zero when $x = -\sqrt{3}, 0$, and $\sqrt{3}$. The second derivative changes sign at each of these points: negative on $\left(-\infty, -\sqrt{3}\right)$, positive on $\left(-\sqrt{3}, 0\right)$, negative on $\left(0, \sqrt{3}\right)$, and positive again on $\left(\sqrt{3}, \infty\right)$. Thus each point is a point of inflection. The curve is concave down on the interval $\left(-\infty, -\sqrt{3}\right)$, concave up on $\left(-\sqrt{3}, 0\right)$, concave down on $\left(0, \sqrt{3}\right)$, and concave up again on $\left(\sqrt{3}, \infty\right)$.

6. *Asymptotes.* Expanding the numerator of $f(x)$ and then dividing both numerator and denominator by x^2 gives

$$f(x) = \frac{(x+1)^2}{1+x^2} = \frac{x^2+2x+1}{1+x^2}$$

Expanding numerator

$$= \frac{1 + (2/x) + (1/x^2)}{(1/x^2) + 1}.$$

Dividing by x^2

We see that $f(x) \to 1^+$ as $x \to \infty$ and that $f(x) \to 1^-$ as $x \to -\infty$. Thus, the line $y = 1$ is a horizontal asymptote.

Since f decreases on $(-\infty, -1)$ and then increases on $(-1, 1)$, we know that $f(-1) = 0$ is a local minimum. Although f decreases on $(1, \infty)$, it never crosses the horizontal asymptote $y = 1$ on that interval (it approaches the asymptote

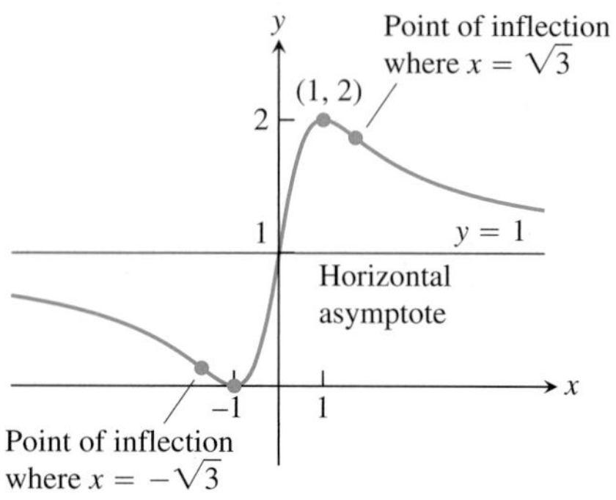

FIGURE 3.30 The graph of $y = \dfrac{(x+1)^2}{1+x^2}$ (Example 7).

from above). So the graph never becomes negative, and $f(-1) = 0$ is an absolute minimum as well. Likewise, $f(1) = 2$ is an absolute maximum because the graph never crosses the asymptote $y = 1$ on the interval $(-\infty, -1)$, approaching it from below. Therefore, there are no vertical asymptotes (the range of f is $0 \le y \le 2$).

7. The graph of f is sketched in Figure 3.30. Notice how the graph is concave down as it approaches the horizontal asymptote $y = 1$ as $x \to -\infty$, and concave up in its approach to $y = 1$ as $x \to \infty$. ■

EXAMPLE 8 Sketch the graph of $f(x) = e^{2/x}$.

Solution The domain of f is $(-\infty, 0) \cup (0, \infty)$ and there are no symmetries about either axis or the origin. The derivatives of f are

$$f'(x) = e^{2/x}\left(-\frac{2}{x^2}\right) = -\frac{2e^{2/x}}{x^2}$$

and

$$f''(x) = \frac{x^2(2e^{2/x})(-2/x^2) - 2e^{2/x}(2x)}{x^4} = \frac{4e^{2/x}(1+x)}{x^4}.$$

Both derivatives exist everywhere over the domain of f. Moreover, since $e^{2/x}$ and x^2 are both positive for all $x \neq 0$, we see that $f' < 0$ everywhere over the domain and the graph is everywhere decreasing. Examining the second derivative, we see that $f''(x) = 0$ at $x = -1$. Since $e^{2/x} > 0$ and $x^4 > 0$, we have $f'' < 0$ for $x < -1$ and $f'' > 0$ for $x > -1, x \neq 0$. Therefore, the point $(-1, e^{-2})$ is a point of inflection. The curve is concave down on the interval $(-\infty, -1)$ and concave up over $(-1, 0) \cup (0, \infty)$.

As in Example 4, Section 2.6, we see that $\lim_{x \to 0^-} f(x) = 0$. As $x \to 0^+$, we see that $2/x \to \infty$, so $\lim_{x \to 0^+} f(x) = \infty$ and the y-axis is a vertical asymptote. Also, as $x \to -\infty$, $2/x \to 0^-$ and so $\lim_{x \to -\infty} f(x) = e^0 = 1$. Therefore, $y = 1$ is a horizontal asymptote. There are no absolute extrema since f never takes on the value 0. The graph of f is sketched in Figure 3.31. ■

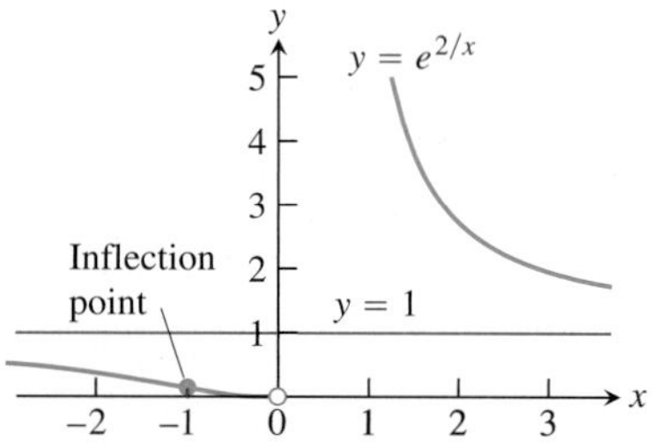

FIGURE 3.31 The graph of $y = e^{2/x}$ has a point of inflection at $(-1, e^{-2})$. The line $y = 1$ is a horizontal asymptote and $x = 0$ is a vertical asymptote (Example 8).

Graphical Behavior of Functions from Derivatives

As we saw in Examples 6 through 8, we can learn much about a twice-differentiable function $y = f(x)$ by examining its first derivative. We can find where the function's graph rises and falls and where any local extrema are located. We can differentiate y' to learn how the graph bends as it passes over the intervals of rise and fall. We can determine the shape of the function's graph. Information we cannot get from the derivative is how to place the graph in the xy-plane. But, as we discovered in Section 3.2, the only additional information we need to position the graph is the value of f at one point. Information about the asymptotes is found using limits (Section 1.8).

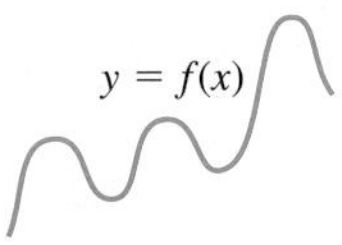 Differentiable $\Rightarrow$ smooth, connected; graph may rise and fall	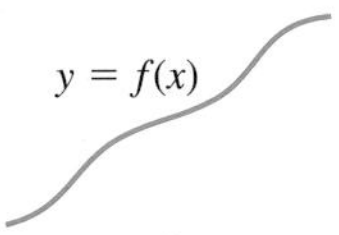 $y' > 0 \Rightarrow$ rises from left to right; may be wavy	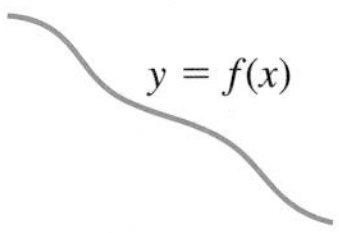 $y' < 0 \Rightarrow$ falls from left to right; may be wavy
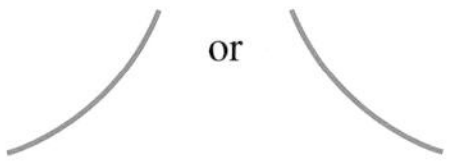 $y'' > 0 \Rightarrow$ concave up throughout; no waves; graph may rise or fall	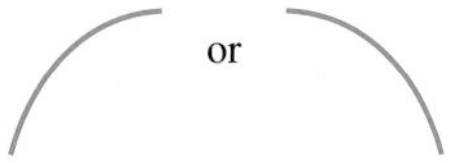 $y'' < 0 \Rightarrow$ concave down throughout; no waves; graph may rise or fall	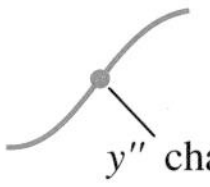 y'' changes sign Inflection point
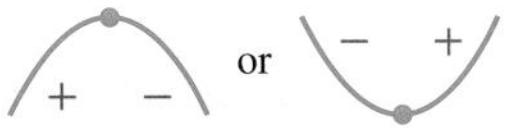 y' changes sign $\Rightarrow$ graph has local maximum or local minimum	 $y' = 0$ and $y'' < 0$ at a point; graph has local maximum	 $y' = 0$ and $y'' > 0$ at a point; graph has local minimum

EXERCISES 3.4

Identify the inflection points and local maxima and minima of the functions graphed in Exercises 1–8. Identify the intervals on which the functions are concave up and concave down.

1. $y = \frac{x^3}{3} - \frac{x^2}{2} - 2x + \frac{1}{3}$

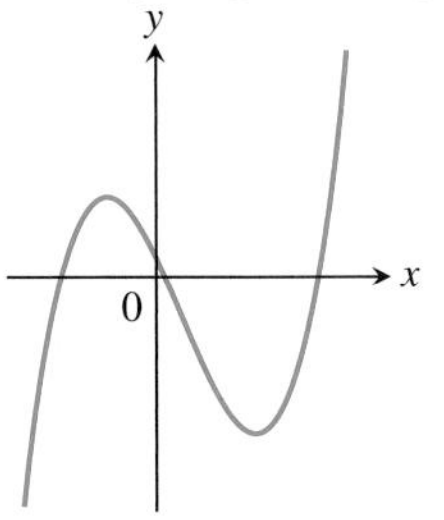

2. $y = \frac{x^4}{4} - 2x^2 + 4$

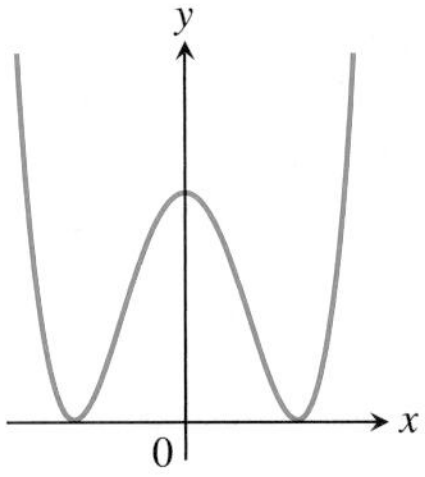

3. $y = \frac{3}{4}(x^2 - 1)^{2/3}$

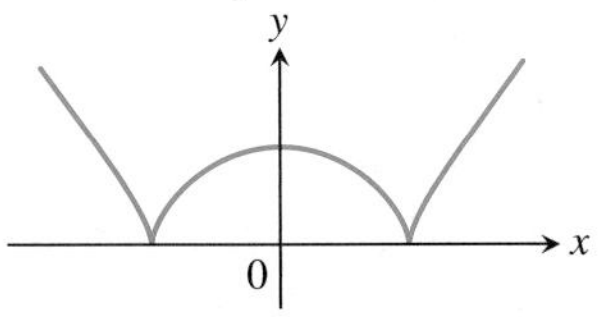

4. $y = \frac{9}{14}x^{1/3}(x^2 - 7)$

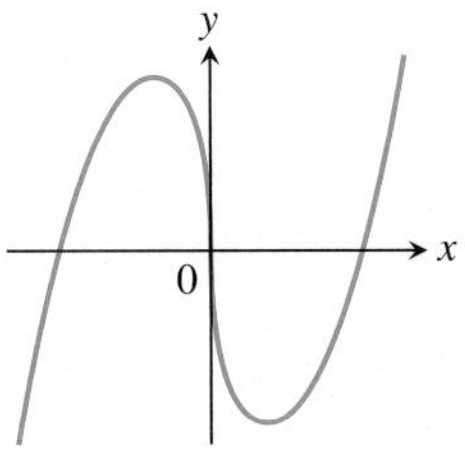

5. $y = x + \sin 2x, -\frac{2\pi}{3} \le x \le \frac{2\pi}{3}$

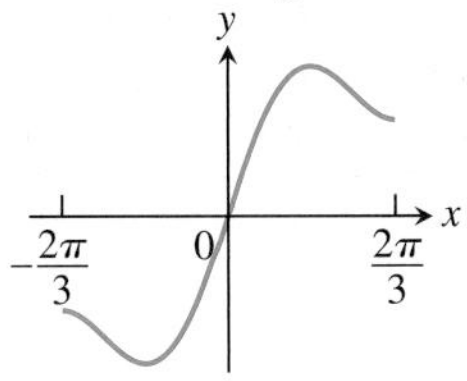

6. $y = \tan x - 4x, -\frac{\pi}{2} < x < \frac{\pi}{2}$

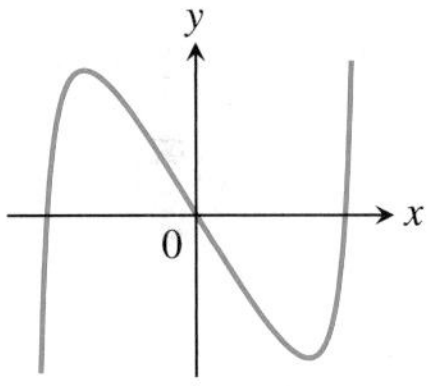

7. $y = \sin |x|, -2\pi \le x \le 2\pi$

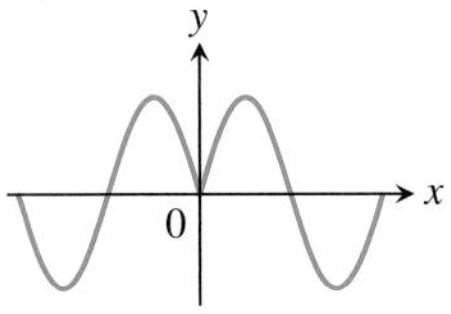

NOT TO SCALE

8. $y = 2\cos x - \sqrt{2}x, -\pi \le x \le \frac{3\pi}{2}$

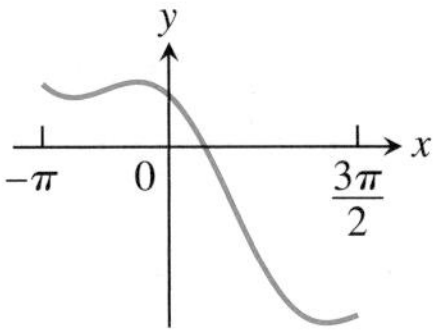

Use the steps of the graphing procedure on page 198 to graph the equations in Exercises 9–42. Include the coordinates of any local extreme points and inflection points.

9. $y = x^2 - 4x + 3$

10. $y = 6 - 2x - x^2$

11. $y = x^3 - 3x + 3$

12. $y = x(6 - 2x)^2$

13. $y = -2x^3 + 6x^2 - 3$

14. $y = (x - 2)^3 + 1$

15. $y = x^4 - 2x^2 = x^2(x^2 - 2)$

16. $y = -x^4 + 6x^2 - 4 = x^2(6 - x^2) - 4$

17. $y = 4x^3 - x^4 = x^3(4 - x)$

18. $y = x^4 + 2x^3 = x^3(x + 2)$

19. $y = x^5 - 5x^4 = x^4(x - 5)$

20. $y = x\left(\dfrac{x}{2} - 5\right)^4$

21. $y = x + \sin x, \quad 0 \le x \le 2\pi$

22. $y = x - \sin x, \quad 0 \le x \le 2\pi$

23. $y = x^{1/5}$

24. $y = x^{2/5}$

25. $y = 2x - 3x^{2/3}$

26. $y = x^{2/3}\left(\dfrac{5}{2} - x\right)$

27. $y = x\sqrt{8 - x^2}$

28. $y = (2 - x^2)^{3/2}$

29. $y = \dfrac{x^2 - 3}{x - 2}, \quad x \ne 2$

30. $y = \dfrac{x^3}{3x^2 + 1}$

31. $y = |x^2 - 1|$

32. $y = \sqrt{|x|} = \begin{cases} \sqrt{-x}, & x < 0 \\ \sqrt{x}, & x \ge 0 \end{cases}$

33. $y = xe^{1/x}$

34. $y = \dfrac{e^x}{x}$

35. $y = \ln(3 - x^2)$

36. $y = x(\ln x)^2$

37. $y = e^x - 2e^{-x} - 3x$

38. $y = xe^{-x}$

39. $y = \ln(\cos x)$

40. $y = \dfrac{\ln x}{\sqrt{x}}$

41. $y = \dfrac{1}{1 + e^{-x}}$

42. $y = \dfrac{e^x}{1 + e^x}$

Each of Exercises 43–62 gives the first derivative of a continuous function $y = f(x)$. Find y'' and then use steps 2–4 of the graphing procedure on page 198 to sketch the general shape of the graph of f.

43. $y' = 2 + x - x^2$

44. $y' = x^2 - x - 6$

45. $y' = x(x - 3)^2$

46. $y' = x^2(2 - x)$

47. $y' = x(x^2 - 12)$

48. $y' = (x^2 - 2x)(x - 5)^2$

49. $y' = \sec^2 x, \quad -\dfrac{\pi}{2} < x < \dfrac{\pi}{2}$

50. $y' = \tan x, \quad -\dfrac{\pi}{2} < x < \dfrac{\pi}{2}$

51. $y' = \cot\dfrac{\theta}{2}, \quad 0 < \theta < 2\pi$

52. $y' = \csc^2\dfrac{\theta}{2}, \quad 0 < \theta < 2\pi$

53. $y' = \tan^2\theta - 1, \quad -\dfrac{\pi}{2} < \theta < \dfrac{\pi}{2}$

54. $y' = 1 - \cot^2\theta, \quad 0 < \theta < \pi$

55. $y' = \cos t, \quad 0 \le t \le 2\pi$

56. $y' = \sin t, \quad 0 \le t \le 2\pi$

57. $y' = (x + 1)^{-2/3}$

58. $y' = (x - 2)^{-1/3}$

59. $y' = x^{-2/3}(x - 1)$

60. $y' = x^{-4/5}(x + 1)$

61. $y' = 2|x| = \begin{cases} -2x, & x < 0 \\ 2x, & x \ge 0 \end{cases}$

62. $y' = \begin{cases} -x^2, & x \le 0 \\ x^2, & x > 0 \end{cases}$

63. Sketch the graph of a twice-differentiable function $y = f(x)$ with the following properties. Label coordinates where possible.

x	y	Derivatives
$x < 2$		$y' < 0, \quad y'' > 0$
2	1	$y' = 0, \quad y'' > 0$
$2 < x < 4$		$y' > 0, \quad y'' > 0$
4	4	$y' > 0, \quad y'' = 0$
$4 < x < 6$		$y' > 0, \quad y'' < 0$
6	7	$y' = 0, \quad y'' < 0$
$x > 6$		$y' < 0, \quad y'' < 0$

64. Sketch the graph of a twice-differentiable function $y = f(x)$ that passes through the points $(-2, 2), (-1, 1), (0, 0), (1, 1)$ and $(2, 2)$ and whose first two derivatives have the following sign patterns:

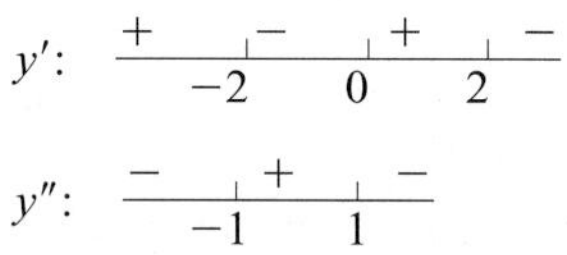

Motion Along a Line The graphs in Exercises 65 and 66 show the position $s = f(t)$ of a body moving back and forth on a coordinate line. **(a)** When is the body moving away from the origin? Toward the origin? At approximately what times is the **(b)** velocity equal to zero? **(c)** Acceleration equal to zero? **(d)** When is the acceleration positive? Negative?

65.

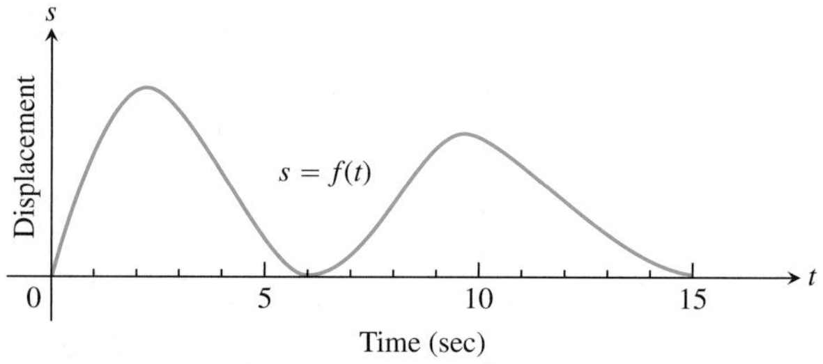

66.

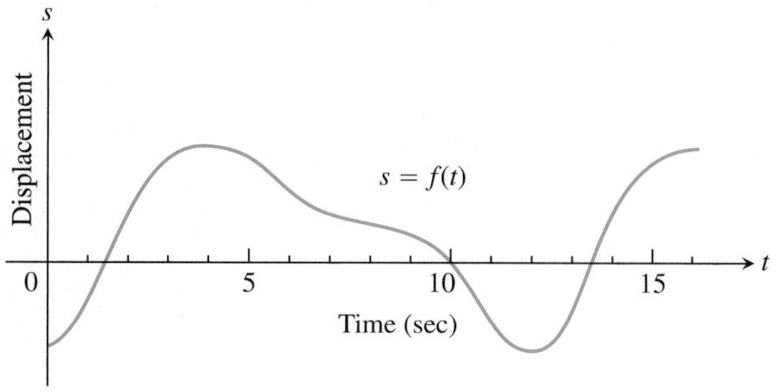

67. Suppose the derivative of the function $y = f(x)$ is

$$y' = (x - 1)^2(x - 2).$$

At what points, if any, does the graph of f have a local minimum, local maximum, or point of inflection? (*Hint:* Draw the sign pattern for y'.)

68. Suppose the derivative of the function $y = f(x)$ is

$$y' = (x - 1)^2(x - 2)(x - 4).$$

At what points, if any, does the graph of f have a local minimum, local maximum, or point of inflection?

69. Parabolas

a. Find the coordinates of the vertex of the parabola $y = ax^2 + bx + c, a \neq 0$.

b. When is the parabola concave up? Concave down? Give reasons for your answers.

70. Is it true that the concavity of the graph of a twice-differentiable function $y = f(x)$ changes every time $f''(x) = 0$? Give reasons for your answer.

71. Quadratic curves What can you say about the inflection points of a quadratic curve $y = ax^2 + bx + c, a \neq 0$? Give reasons for your answer.

72. Cubic curves What can you say about the inflection points of a cubic curve $y = ax^3 + bx^2 + cx + d, a \neq 0$? Give reasons for your answer.

COMPUTER EXPLORATIONS

In Exercises 73–76, find the inflection points (if any) on the graph of the function and the coordinates of the points on the graph where the function has a local maximum or local minimum value. Then graph the function in a region large enough to show all these points simultaneously. Add to your picture the graphs of the function's first and second derivatives. How are the values at which these graphs intersect the x-axis related to the graph of the function? In what other ways are the graphs of the derivatives related to the graph of the function?

73. $y = x^5 - 5x^4 - 240$

74. $y = x^3 - 12x^2$

75. $y = \frac{4}{5}x^5 + 16x^2 - 25$

76. $y = \frac{x^4}{4} - \frac{x^3}{3} - 4x^2 + 12x + 20$

77. a. Graph $y = x^{2/3}(x^2 - 2)$ for $-3 \leq x \leq 3$. Then use calculus to confirm what the screen shows about concavity, rise, and fall. (Depending on your grapher, you may have to enter $x^{2/3}$ as $(x^2)^{1/3}$ to obtain a plot for negative values of x.)

b. Does the curve have a cusp at $x = 0$, or does it just have a corner with different right-hand and left-hand derivatives?

78. a. Graph $y = 9x^{2/3}(x - 1)$ for $-0.5 \leq x \leq 1.5$. Then use calculus to confirm what the screen shows about concavity, rise, and fall. What concavity does the curve have to the left of the origin? (Depending on your grapher, you may have to enter $x^{2/3}$ as $(x^2)^{1/3}$ to obtain a plot for negative values of x.)

b. Does the curve have a cusp at $x = 0$, or does it just have a corner with different right-hand and left-hand derivatives?

3.5 Parametrizations of Plane Curves

Instead of describing a curve by expressing the y-coordinate of a point $P(x, y)$ on the curve as a function of x, it is sometimes more convenient to describe the curve by expressing *both* coordinates as functions of a third variable t. In this section we study curves described in this way and apply the chain rule to find their derivatives.

Parametric Equations

Figure 3.32 shows the path of a moving particle. Notice that the path fails the vertical line test, so it cannot be described as the graph of a function of the variable x. However, we can sometimes describe the path by a pair of equations, $x = f(t)$ and $y = g(t)$, where f and g

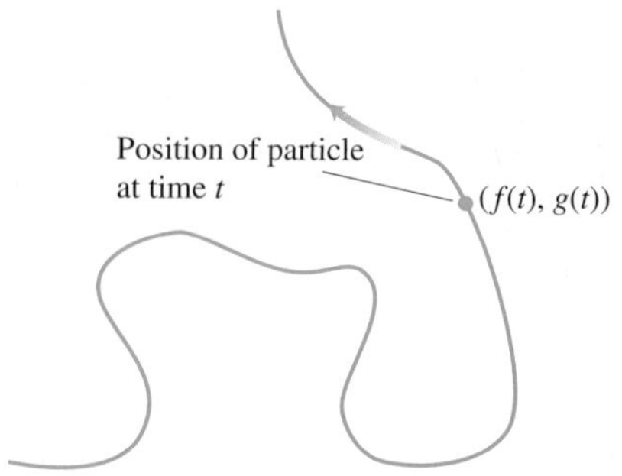

FIGURE 3.32 The path traced by a particle moving in the xy-plane is not always the graph of a function of x or a function of y.

are continuous functions. When studying motion, t usually denotes time. Equations like these describe more general curves than those like $y = f(x)$ and provide not only the graph of the path traced out, but also the location of the particle $(x, y) = (f(t), g(t))$ at any time t.

DEFINITION If x and y are given as functions

$$x = f(t), \qquad y = g(t)$$

over an interval I of t-values, then the set of points $(x, y) = (f(t), g(t))$ defined by these equations is a **parametric curve**. The equations are **parametric equations** for the curve.

The variable t is a **parameter** for the curve, and its domain I is the **parameter interval**. If I is a closed interval, $a \le t \le b$, the point $(f(a), g(a))$ is the **initial point** of the curve. The point $(f(b), g(b))$ is the **terminal point**. When we give parametric equations and a parameter interval for a curve, we say that we have **parametrized** the curve. The equations and interval together constitute a **parametrization** of the curve. A given curve can be represented by different sets of parametric equations. (See Exercises 13 and 14.)

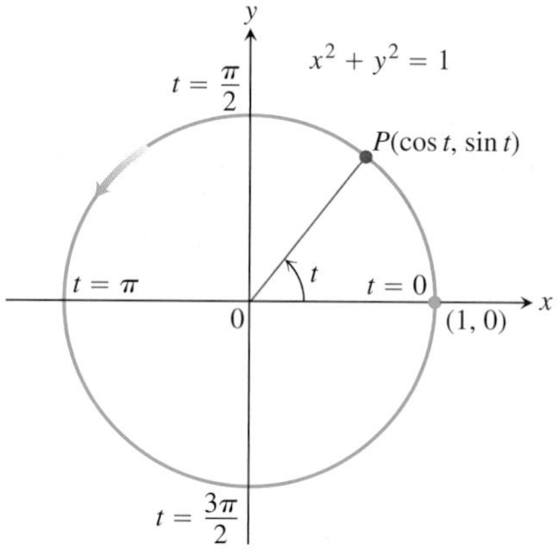

FIGURE 3.33 The equations $x = \cos t$ and $y = \sin t$ describe motion on the circle $x^2 + y^2 = 1$. The arrow shows the direction of increasing t (Example 1).

EXAMPLE 1 Graph the parametric curves

(a) $x = \cos t, \qquad y = \sin t, \qquad 0 \le t \le 2\pi.$

(b) $x = a\cos t, \qquad y = a\sin t, \qquad 0 \le t \le 2\pi.$

Solution

(a) Since $x^2 + y^2 = \cos^2 t + \sin^2 t = 1$, the parametric curve lies along the unit circle $x^2 + y^2 = 1$. As t increases from 0 to 2π, the point $(x, y) = (\cos t, \sin t)$ starts at $(1, 0)$ and traces the entire circle once counterclockwise (Figure 3.33).

(b) For $x = a\cos t$, $y = a\sin t$, $0 \le t \le 2\pi$, we have $x^2 + y^2 = a^2\cos^2 t + a^2\sin^2 t = a^2$. The parametrization describes a motion that begins at the point $(a, 0)$ and traverses the circle $x^2 + y^2 = a^2$ once counterclockwise, returning to $(a, 0)$ at $t = 2\pi$. The graph is a circle centered at the origin with radius $r = a$ and coordinate points $(a\cos t, a\sin t)$. ■

EXAMPLE 2 The position $P(x, y)$ of a particle moving in the xy-plane is given by the equations and parameter interval

$$x = \sqrt{t}, \qquad y = t, \qquad t \ge 0.$$

Identify the path traced by the particle and describe the motion.

Solution We try to identify the path by eliminating t between the equations $x = \sqrt{t}$ and $y = t$. With any luck, this will produce a recognizable algebraic relation between x and y. We find that

$$y = t = \left(\sqrt{t}\right)^2 = x^2.$$

Thus, the particle's position coordinates satisfy the equation $y = x^2$, so the particle moves along the parabola $y = x^2$.

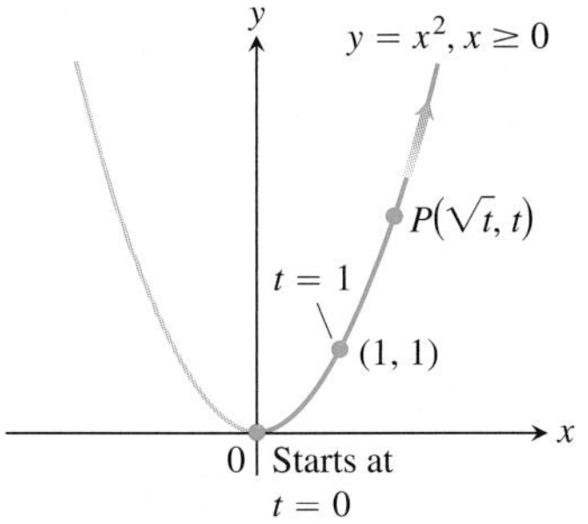

FIGURE 3.34 The equations $x = \sqrt{t}$ and $y = t$ and the interval $t \geq 0$ describe the motion of a particle that traces the right-hand half of the parabola $y = x^2$ (Example 2).

It would be a mistake, however, to conclude that the particle's path is the entire parabola $y = x^2$; it is only half the parabola. The particle's x-coordinate is never negative. The particle starts at (0, 0) when $t = 0$ and rises into the first quadrant as t increases (Figure 3.34). The parameter interval is $[0, \infty)$ and there is no terminal point. ■

EXAMPLE 3 Sketch the path traced by the point $P(x, y)$ if

$$x = t + \frac{1}{t}, \qquad y = t - \frac{1}{t}, \qquad t > 0.$$

Solution We make a brief table of values, plot the points, and draw a smooth curve through them. (See Figure 3.35.) From Table 3.1, or from the defining parametric equations, if t is replaced by $1/t$, the value of $x = t + (1/t)$ is unchanged, but the sign of $y = t - (1/t)$ changes. So we can see that the graph is symmetric with respect to the x-axis; for every (positive) value of x, both points, (x, y) and $(x, -y)$, lie on the graph.

Taking the difference between x and y as given by the parametric equations, we find that

$$x - y = \left(t + \frac{1}{t}\right) - \left(t - \frac{1}{t}\right) = \frac{2}{t}.$$

If we add the two parametric equations, we get

$$x + y = \left(t + \frac{1}{t}\right) + \left(t - \frac{1}{t}\right) = 2t.$$

We can then eliminate the parameter t by multiplying these last equations together:

$$(x - y)(x + y) = \left(\frac{2}{t}\right)(2t) = 4,$$

or

$$x^2 - y^2 = 4. \tag{1}$$

Thus the coordinates of all the points $P(x, y)$ described by the parametric equations satisfy Equation (1). However, since Equation (1) does not require that the x-coordinate be positive, there are points (x, y) satisfying Equation (1) that do not satisfy the parametric equation $x = t + (1/t)$, $t > 0$, since x is always positive in this equation. For small positive values of t, the point lies in the fourth quadrant and rises into the first quadrant as t increases, crossing the x-axis when $t = 1$ (Figure 3.35). The parameter domain is $(0, \infty)$ and there is no starting point and no terminal point. ■

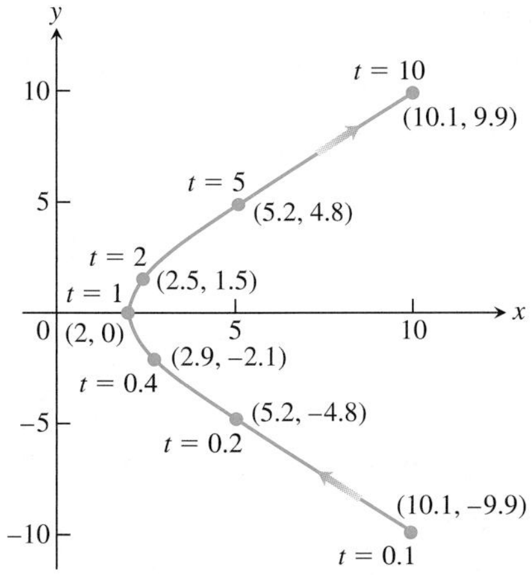

FIGURE 3.35 The graph of $x = t + (1/t)$, $y = t - (1/t)$, $t > 0$ in Example 3. (The part shown is for $0.1 \leq t \leq 10$.)

TABLE 3.1 Values of $x = t + (1/t)$ and $y = t - (1/t)$ for selected values of t.

t	$1/t$	x	y
0.1	10.0	10.1	−9.9
0.2	5.0	5.2	−4.8
0.4	2.5	2.9	−2.1
1.0	1.0	2.0	0.0
2.0	0.5	2.5	1.5
5.0	0.2	5.2	4.8
10.0	0.1	10.1	9.9

EXAMPLE 4 Find a parametrization for the line segment with endpoints $(-2, 1)$ and $(3, 5)$.

Solution Using $(-2, 1)$ we create the parametric equations

$$x = -2 + at, \qquad y = 1 + bt.$$

These represent a line, as we can see by solving each equation for t and equating to obtain

$$\frac{x + 2}{a} = \frac{y - 1}{b}.$$

This line goes through the point $(-2, 1)$ when $t = 0$. We determine a and b so that the line goes through (3, 5) when $t = 1$.

$$3 = -2 + a \quad \Rightarrow \quad a = 5 \qquad x = 3 \text{ when } t = 1.$$
$$5 = 1 + b \quad \Rightarrow \quad b = 4 \qquad y = 5 \text{ when } t = 1.$$

Therefore,

$$x = -2 + 5t, \qquad y = 1 + 4t, \qquad 0 \le t \le 1$$

is a parametrization of the line segment with initial point $(-2, 1)$ and terminal point (3, 5). ■

Slopes of Parametrized Curves

A parametrized curve $x = f(t)$ and $y = g(t)$ is **differentiable** at t if f and g are differentiable at t. At a point on a differentiable parametrized curve where y is also a differentiable function of x, the derivatives dy/dt, dx/dt, and dy/dx are related by the Chain Rule:

$$\frac{dy}{dt} = \frac{dy}{dx} \cdot \frac{dx}{dt}.$$

If $dx/dt \neq 0$, we may divide both sides of this equation by dx/dt to solve for dy/dx.

Parametric Formula for dy/dx

If all three derivatives exist and $dx/dt \neq 0$,

$$\frac{dy}{dx} = \frac{dy/dt}{dx/dt}. \tag{2}$$

EXAMPLE 5 Describe the motion of a particle whose position $P(x, y)$ at time t is given by

$$x = a \cos t, \qquad y = b \sin t, \qquad 0 \le t \le 2\pi.$$

Find the line tangent to the curve at the point $(a/\sqrt{2}, b/\sqrt{2})$, where $t = \pi/4$. (The constants a and b are both positive.)

Solution We find a Cartesian equation for the particle's coordinates by eliminating t between the equations

$$\cos t = \frac{x}{a}, \qquad \sin t = \frac{y}{b}.$$

The identity $\cos^2 t + \sin^2 t = 1$ yields

$$\left(\frac{x}{a}\right)^2 + \left(\frac{y}{b}\right)^2 = 1, \qquad \text{or} \qquad \frac{x^2}{a^2} + \frac{y^2}{b^2} = 1.$$

The particle's coordinates (x, y) satisfy the equation $(x^2/a^2) + (y^2/b^2) = 1$, so the particle moves along this ellipse. When $t = 0$, the particle's coordinates are

$$x = a\cos(0) = a, \qquad y = b\sin(0) = 0,$$

so the motion starts at $(a, 0)$. As t increases, the particle rises and moves toward the left, moving counterclockwise. It traverses the ellipse once, returning to its starting position $(a, 0)$ at $t = 2\pi$, thereby repeating the starting point.

The slope of the tangent line to the ellipse when $t = \pi/4$ is

$$\left.\frac{dy}{dx}\right|_{t=\pi/4} = \left.\frac{dy/dt}{dx/dt}\right|_{t=\pi/4} \qquad \text{Eq. (2)}$$
$$= \left.\frac{b\cos t}{-a\sin t}\right|_{t=\pi/4}$$
$$= \frac{b/\sqrt{2}}{-a/\sqrt{2}} = -\frac{b}{a}.$$

The tangent line is

$$y - \frac{b}{\sqrt{2}} = -\frac{b}{a}\left(x - \frac{a}{\sqrt{2}}\right)$$
$$y = \frac{b}{\sqrt{2}} - \frac{b}{a}\left(x - \frac{a}{\sqrt{2}}\right)$$

or

$$y = -\frac{b}{a}x + \sqrt{2}b.$$ ■

If parametric equations define y as a twice-differentiable function of x, we can apply Equation (2) to the function $dy/dx = y'$ to calculate d^2y/dx^2 as a function of t:

$$\frac{d^2y}{dx^2} = \frac{d}{dx}(y') = \frac{dy'/dt}{dx/dt}. \qquad \text{Eq. (2) with } y' \text{ in place of } y$$

Parametric Formula for d^2y/dx^2

If the equations $x = f(t)$, $y = g(t)$ define y as a twice-differentiable function of x, then at any point where $dx/dt \neq 0$ and $y' = dy/dx$,

$$\frac{d^2y}{dx^2} = \frac{dy'/dt}{dx/dt}. \qquad (3)$$

Finding d^2y/dx^2 in Terms of t

1. Express $y' = dy/dx$ in terms of t.
2. Find dy'/dt.
3. Divide dy'/dt by dx/dt.

EXAMPLE 6 Find d^2y/dx^2 as a function of t if $x = t - t^2$, $y = t - t^3$.

Solution

1. Express $y' = dy/dx$ in terms of t.

$$y' = \frac{dy}{dx} = \frac{dy/dt}{dx/dt} = \frac{1 - 3t^2}{1 - 2t}$$

2. Differentiate y' with respect to t.

$$\frac{dy'}{dt} = \frac{d}{dt}\left(\frac{1 - 3t^2}{1 - 2t}\right) = \frac{2 - 6t + 6t^2}{(1 - 2t)^2} \qquad \text{Quotient Rule}$$

3. Divide dy'/dt by dx/dt.

$$\frac{d^2y}{dx^2} = \frac{dy'/dt}{dx/dt} = \frac{(2 - 6t + 6t^2)/(1 - 2t)^2}{1 - 2t} = \frac{2 - 6t + 6t^2}{(1 - 2t)^3} \qquad \text{Eq. (3)}$$ ■

Parametrizing Inverse Functions

We can graph or represent any function $y = f(x)$ parametrically as

$$x = t \quad \text{and} \quad y = f(t).$$

Interchanging t and $f(t)$ produces parametric equations for the inverse:

$$x = f(t) \quad \text{and} \quad y = t.$$

For example, to graph the one-to-one function $f(x) = x^2, x \geq 0$, on a graphing device together with its inverse and the line $y = x, x \geq 0$, use the parametric graphing option with

$$\text{Graph of } f\text{:} \quad x_1 = t, \quad y_1 = t^2, \quad t \geq 0$$

$$\text{Graph of } f^{-1}\text{:} \quad x_2 = t^2, \quad y_2 = t$$

$$\text{Graph of } y = x\text{:} \quad x_3 = t, \quad y_3 = t$$

EXERCISES 3.5

Exercises 1–12 give parametric equations and parameter intervals for the motion of a particle in the xy-plane. Identify the particle's path by finding a Cartesian equation for it. Graph the Cartesian equation. (The graphs will vary with the equation used.) Indicate the portion of the graph traced by the particle and the direction of motion.

1. $x = \cos 2t, \quad y = \sin 2t, \quad 0 \leq t \leq \pi$
2. $x = \cos(\pi - t), \quad y = \sin(\pi - t), \quad 0 \leq t \leq \pi$
3. $x = 4\cos t, \quad y = 2\sin t, \quad 0 \leq t \leq 2\pi$
4. $x = 4\sin t, \quad y = 5\cos t, \quad 0 \leq t \leq 2\pi$
5. $x = 3t, \quad y = 9t^2, \quad -\infty < t < \infty$
6. $x = -\sqrt{t}, \quad y = t, \quad t \geq 0$
7. $x = 2t - 5, \quad y = 4t - 7, \quad -\infty < t < \infty$
8. $x = 3 - 3t, \quad y = 2t, \quad 0 \leq t \leq 1$
9. $x = t, \quad y = \sqrt{1 - t^2}, \quad -1 \leq t \leq 0$
10. $x = \sqrt{t + 1}, \quad y = \sqrt{t}, \quad t \geq 0$
11. $x = \sec^2 t - 1, \quad y = \tan t, \quad -\pi/2 < t < \pi/2$
12. $x = -\sec t, \quad y = \tan t, \quad -\pi/2 < t < \pi/2$
13. Find parametric equations and a parameter interval for the motion of a particle that starts at $(a, 0)$ and traces the circle $x^2 + y^2 = a^2$
    - **a.** once clockwise.
    - **b.** once counterclockwise.
    - **c.** twice clockwise.
    - **d.** twice counterclockwise.

    (There are many ways to do these, so your answers may not be the same as the ones in the back of the book.)
14. Find parametric equations and a parameter interval for the motion of a particle that starts at $(a, 0)$ and traces the ellipse $(x^2/a^2) + (y^2/b^2) = 1$
    - **a.** once clockwise.
    - **b.** once counterclockwise.
    - **c.** twice clockwise.
    - **d.** twice counterclockwise.

    (As in Exercise 13, there are many correct answers.)

In Exercises 15–20, find a parametrization for the curve.

15. the line segment with endpoints $(-1, -3)$ and $(4, 1)$
16. the line segment with endpoints $(-1, 3)$ and $(3, -2)$
17. the lower half of the parabola $x - 1 = y^2$
18. the left half of the parabola $y = x^2 + 2x$
19. the ray (half line) with initial point $(2, 3)$ that passes through the point $(-1, -1)$
20. the ray (half line) with initial point $(-1, 2)$ that passes through the point $(0, 0)$

In Exercises 21–28, find an equation for the line tangent to the curve at the point defined by the given value of t. Also, find the value of d^2y/dx^2 at this point.

21. $x = 2\cos t, \quad y = 2\sin t, \quad t = \pi/4$
22. $x = \cos t, \quad y = \sqrt{3}\cos t, \quad t = 2\pi/3$
23. $x = t, \quad y = \sqrt{t}, \quad t = 1/4$
24. $x = -\sqrt{t + 1}, \quad y = \sqrt{3t}, \quad t = 3$
25. $x = 2t^2 + 3, \quad y = t^4, \quad t = -1$
26. $x = t - \sin t, \quad y = 1 - \cos t, \quad t = \pi/3$
27. $x = \cos t, \quad y = 1 + \sin t, \quad t = \pi/2$
28. $x = \sec^2 t - 1, \quad y = \tan t, \quad t = -\pi/4$

Assuming that the equations in Exercises 29–32 define x and y implicitly as differentiable functions $x = f(t), y = g(t)$, find the slope of the curve $x = f(t), y = g(t)$ at the given value of t.

29. $x^3 + 2t^2 = 9, \quad 2y^3 - 3t^2 = 4, \quad t = 2$

30. $x = \sqrt{5 - \sqrt{t}}, \quad y(t - 1) = \sqrt{t}, \quad t = 4$

31. $x + 2x^{3/2} = t^2 + t, \quad y\sqrt{t + 1} + 2t\sqrt{y} = 4, \quad t = 0$

32. $x \sin t + 2x = t, \quad t \sin t - 2t = y, \quad t = \pi$

T The curves in Exercises 33 and 34 are called *Bowditch curves* or *Lissajous figures*. In each case, find the point in the interior of the first quadrant where the tangent to the curve is horizontal, and find the equations of the two tangents at the origin.

33.

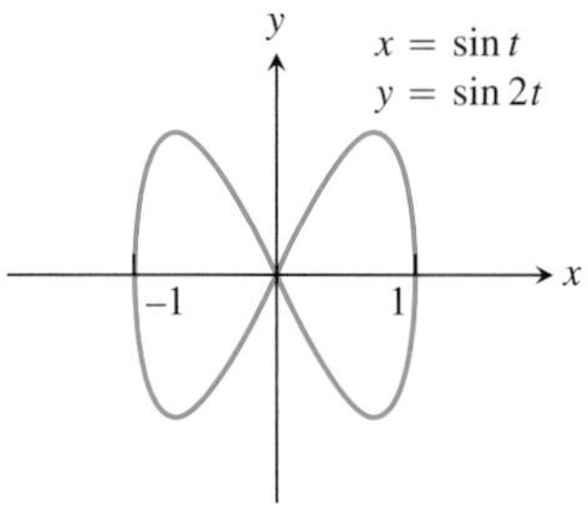

34.

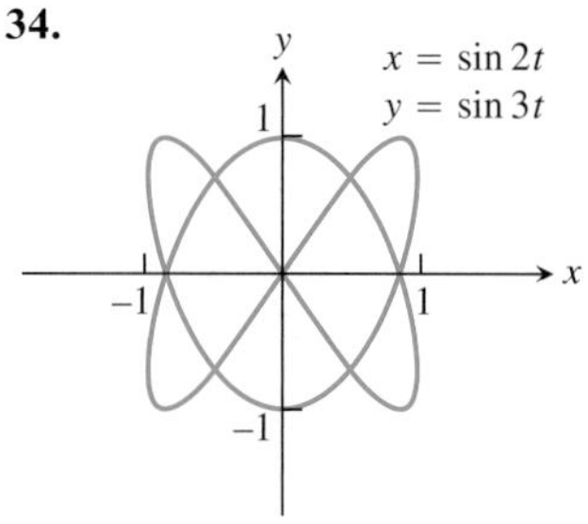

3.6 Applied Optimization

To optimize something often means to maximize or minimize some aspect of it. For example, what are the dimensions of a rectangle with fixed perimeter having *maximum area*? What are the dimensions for the *least expensive* cylindrical can of a given volume? How many items should be produced for the *most profitable* production run? In this section we use derivatives to solve a variety of optimization problems in business, mathematics, physics, and economics.

Examples from Business and Industry

EXAMPLE 1 An open-top box is to be made by cutting small congruent squares from the corners of a 12-in.-by-12-in. sheet of tin and bending up the sides. How large should the squares cut from the corners be to make the box hold as much as possible?

Solution We start with a picture (Figure 3.36). In the figure, the corner squares are x in. on a side. The volume of the box is a function of this variable:

$$V(x) = x(12 - 2x)^2 = 144x - 48x^2 + 4x^3. \qquad V = hlw$$

Since the sides of the sheet of tin are only 12 in. long, $x \le 6$ and the domain of V is the interval $0 \le x \le 6$.

A graph of V (Figure 3.37) suggests a minimum value of 0 at $x = 0$ and $x = 6$ and a maximum near $x = 2$. To learn more, we examine the first derivative of V with respect to x:

$$\frac{dV}{dx} = 144 - 96x + 12x^2 = 12(12 - 8x + x^2) = 12(2 - x)(6 - x).$$

Of the two zeros, $x = 2$ and $x = 6$, only $x = 2$ lies in the interior of the function's domain and makes the critical-point list. The values of V at this one critical point and two endpoints are

$$\text{Critical-point value: } V(2) = 128$$
$$\text{Endpoint values: } V(0) = 0, \qquad V(6) = 0.$$

The maximum volume is 128 in.^{3}. The cutout squares should be 2 in. on a side. ■

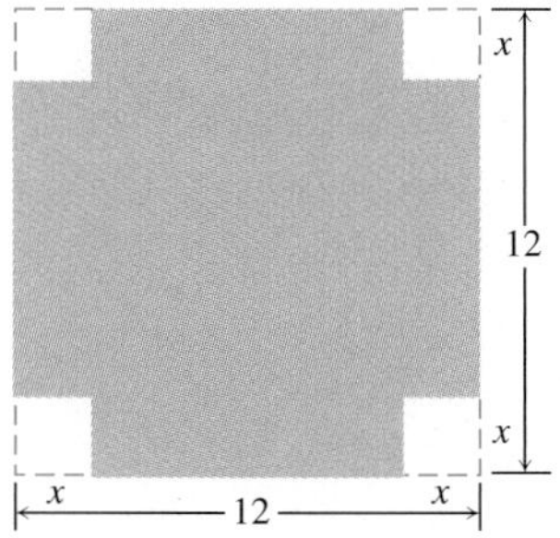

(a)

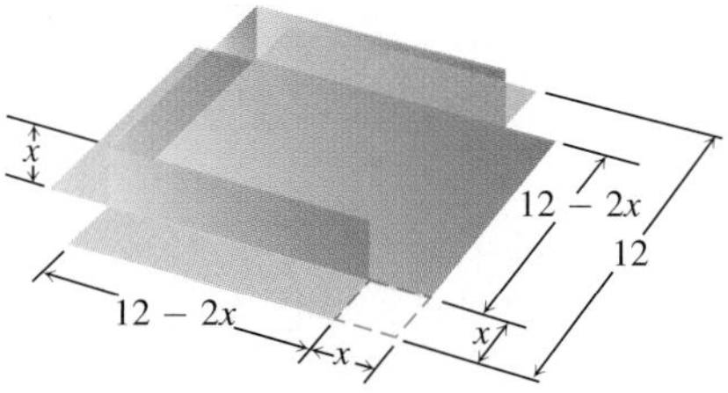

(b)

FIGURE 3.36 An open box made by cutting the corners from a square sheet of tin. What size corners maximize the box's volume (Example 1)?

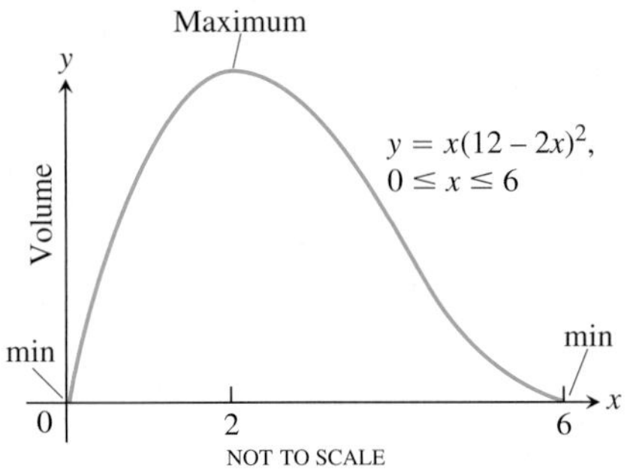

FIGURE 3.37 The volume of the box in Figure 3.36 graphed as a function of x.

EXAMPLE 2 You have been asked to design a 1-liter can shaped like a right circular cylinder (Figure 3.38). What dimensions will use the least material?

Solution *Volume of can:* If r and h are measured in centimeters, then the volume of the can in cubic centimeters is

$$\pi r^2 h = 1000. \qquad \text{1 liter} = 1000 \text{ cm}^3$$

Surface area of can:
$$A = \underbrace{2\pi r^2}_{\text{circular ends}} + \underbrace{2\pi r h}_{\text{cylindrical wall}}$$

How can we interpret the phrase "least material"? First, it is customary to ignore the thickness of the material and the waste in manufacturing. Then we ask for dimensions r and h that make the total surface area as small as possible while satisfying the constraint $\pi r^2 h = 1000$.

To express the surface area as a function of one variable, we solve for one of the variables in $\pi r^2 h = 1000$ and substitute that expression into the surface area formula. Solving for h is easier:

$$h = \frac{1000}{\pi r^2}.$$

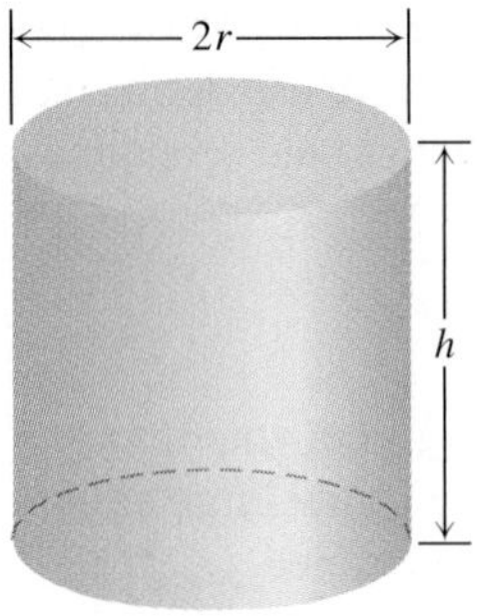

FIGURE 3.38 This 1-L can uses the least material when $h = 2r$ (Example 2).

Thus,

$$\begin{aligned} A &= 2\pi r^2 + 2\pi r h \\ &= 2\pi r^2 + 2\pi r\left(\frac{1000}{\pi r^2}\right) \\ &= 2\pi r^2 + \frac{2000}{r}. \end{aligned}$$

Our goal is to find a value of $r > 0$ that minimizes the value of A. Figure 3.39 suggests that such a value exists.

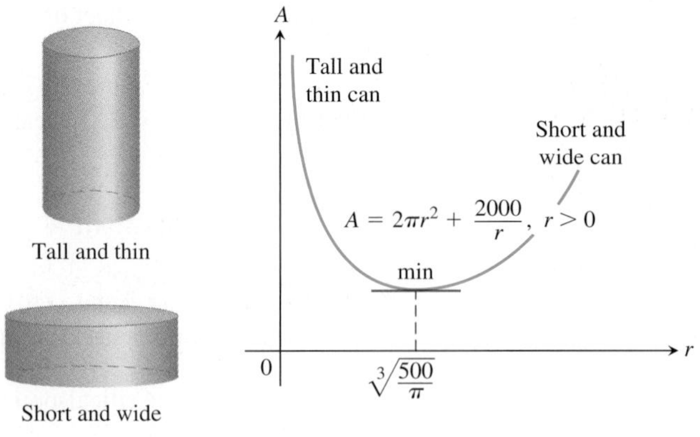

FIGURE 3.39 The graph of $A = 2\pi r^2 + 2000/r$ is concave up.

Notice from the graph that for small r (a tall thin container, like a piece of pipe), the term $2000/r$ dominates and A is large. For large r (a short wide container, like a pizza pan), the term $2\pi r^2$ dominates and A again is large.

Since A is differentiable on $r > 0$, an interval with no endpoints, it can have a minimum value only where its first derivative is zero.

$$\begin{aligned} \frac{dA}{dr} &= 4\pi r - \frac{2000}{r^2} \\ 0 &= 4\pi r - \frac{2000}{r^2} && \text{Set } dA/dr = 0. \\ 4\pi r^3 &= 2000 && \text{Multiply by } r^2. \\ r &= \sqrt[3]{\frac{500}{\pi}} \approx 5.42 && \text{Solve for } r. \end{aligned}$$

What happens at $r = \sqrt[3]{500/\pi}$?

The second derivative

$$\frac{d^2A}{dr^2} = 4\pi + \frac{4000}{r^3}$$

is positive throughout the domain of A. The graph is therefore everywhere concave up and the value of A at $r = \sqrt[3]{500/\pi}$ an absolute minimum.

The corresponding value of h (after a little algebra) is

$$h = \frac{1000}{\pi r^2} = 2\sqrt[3]{\frac{500}{\pi}} = 2r.$$

The 1-L can that uses the least material has height equal to the diameter, here with $r \approx 5.42$ cm and $h \approx 10.84$ cm. ■

Examples from Mathematics and Physics

EXAMPLE 3 A rectangle is to be inscribed in a semicircle of radius 2. What is the largest area the rectangle can have, and what are its dimensions?

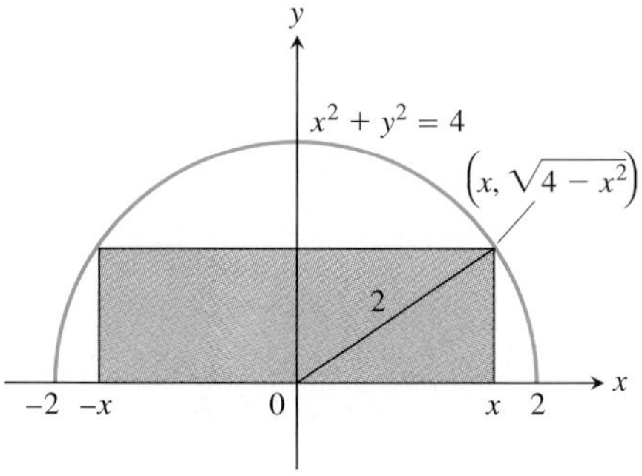

FIGURE 3.40 The rectangle inscribed in the semicircle in Example 3.

Solution Let $(x, \sqrt{4 - x^2})$ be the coordinates of the corner of the rectangle obtained by placing the circle and rectangle in the coordinate plane (Figure 3.40). The length, height, and area of the rectangle can then be expressed in terms of the position x of the lower right-hand corner:

$$\text{Length: } 2x, \qquad \text{Height: } \sqrt{4 - x^2}, \qquad \text{Area: } 2x \cdot \sqrt{4 - x^2}.$$

Notice that the values of x are to be found in the interval $0 \le x \le 2$, where the selected corner of the rectangle lies.

Our goal is to find the absolute maximum value of the function

$$A(x) = 2x\sqrt{4 - x^2}$$

on the domain [0, 2].

The derivative

$$\frac{dA}{dx} = \frac{-2x^2}{\sqrt{4 - x^2}} + 2\sqrt{4 - x^2}$$

is not defined when $x = 2$ and is equal to zero when

$$\begin{aligned} \frac{-2x^2}{\sqrt{4 - x^2}} + 2\sqrt{4 - x^2} &= 0 \\ -2x^2 + 2(4 - x^2) &= 0 \\ 8 - 4x^2 &= 0 \\ x^2 = 2 \text{ or } x &= \pm\sqrt{2}. \end{aligned}$$

Of the two zeros, $x = \sqrt{2}$ and $x = -\sqrt{2}$, only $x = \sqrt{2}$ lies in the interior of A's domain and makes the critical-point list. The values of A at the endpoints and at this one critical point are

$$\begin{aligned} &\text{Critical-point value:} \quad A\left(\sqrt{2}\right) = 2\sqrt{2}\sqrt{4 - 2} = 4 \\ &\text{Endpoint values:} \quad A(0) = 0, \qquad A(2) = 0. \end{aligned}$$

The area has a maximum value of 4 when the rectangle is $\sqrt{4 - x^2} = \sqrt{2}$ units high and $2x = 2\sqrt{2}$ units long. ■

HISTORICAL BIOGRAPHY

Willebrord Snell van Royen (1580–1626)

EXAMPLE 4 The speed of light depends on the medium through which it travels, and is generally slower in denser media.

Fermat's principle in optics states that light travels from one point to another along a path for which the time of travel is a minimum. Find the path that a ray of light will follow in going from a point A in a medium where the speed of light is c_1 to a point B in a second medium where its speed is c_2.

Solution Since light traveling from A to B follows the quickest route, we look for a path that will minimize the travel time. We assume that A and B lie in the xy-plane and that the line separating the two media is the x-axis (Figure 3.41).

In a uniform medium, where the speed of light remains constant, "shortest time" means "shortest path," and the ray of light will follow a straight line. Thus the path from A to B will consist of a line segment from A to a boundary point P, followed by another line segment from P to B. Distance equals rate times time, so

$$\text{Time} = \frac{\text{distance}}{\text{rate}}.$$

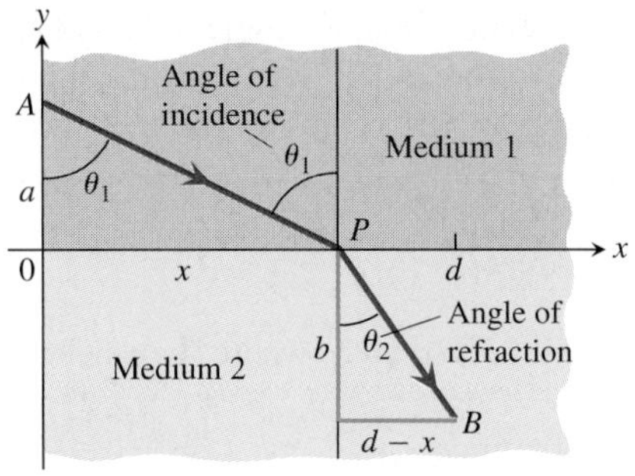

FIGURE 3.41 A light ray refracted (deflected from its path) as it passes from one medium to a denser medium (Example 4).

The time required for light to travel from A to P is

$$t_1 = \frac{AP}{c_1} = \frac{\sqrt{a^2 + x^2}}{c_1}.$$

From P to B, the time is

$$t_2 = \frac{PB}{c_2} = \frac{\sqrt{b^2 + (d - x)^2}}{c_2}.$$

The time from A to B is the sum of these:

$$t = t_1 + t_2 = \frac{\sqrt{a^2 + x^2}}{c_1} + \frac{\sqrt{b^2 + (d - x)^2}}{c_2}.$$

This equation expresses t as a differentiable function of x whose domain is $[0, d]$. We want to find the absolute minimum value of t on this closed interval. We find the derivative

$$\frac{dt}{dx} = \frac{x}{c_1\sqrt{a^2 + x^2}} - \frac{d - x}{c_2\sqrt{b^2 + (d - x)^2}}$$

and observe it is continuous. In terms of the angles θ_1 and θ_2 in Figure 3.41,

$$\frac{dt}{dx} = \frac{\sin\theta_1}{c_1} - \frac{\sin\theta_2}{c_2}.$$

If we restrict x to the interval $0 \le x \le d$, then t has a negative derivative at $x = 0$ and a positive derivative at $x = d$. By the Intermediate Value Theorem (Section 1.7), there is a point $x_0 \in [0, d]$ where $dt/dx = 0$ (Figure 3.42). There is only one such point because dt/dx is an increasing function of x (Exercise 38). At this point

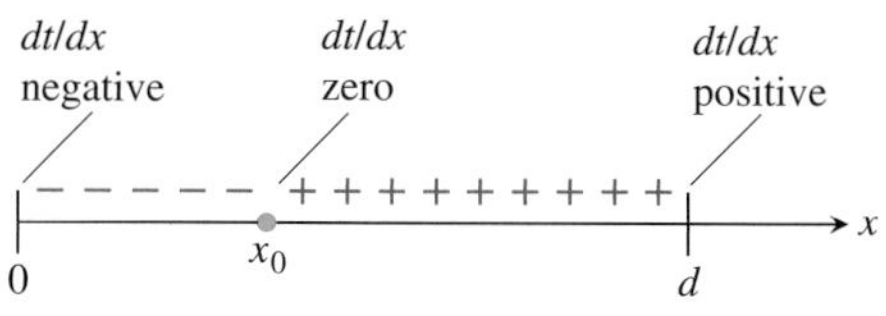

FIGURE 3.42 The sign pattern of dt/dx in Example 4.

$$\frac{\sin\theta_1}{c_1} = \frac{\sin\theta_2}{c_2}.$$

This equation is **Snell's Law** or the **Law of Refraction**, and is an important principle in the theory of optics. It describes the path the ray of light follows. ■

Examples from Economics

Suppose that

$$r(x) = \text{the revenue from selling } x \text{ items}$$
$$c(x) = \text{the cost of producing the } x \text{ items}$$
$$p(x) = r(x) - c(x) = \text{the profit from producing and selling } x \text{ items}.$$

The **marginal revenue**, **marginal cost**, and **marginal profit** when producing and selling x items are the derivatives $r'(x)$, $c'(x)$, and $p'(x)$ respectively. Let's consider the relationship of p to these derivatives.

If $r(x)$ and $c(x)$ are differentiable for all $x > 0$, and if $p(x) = r(x) - c(x)$ has a maximum value, it occurs at a production level at which $p'(x) = 0$. Since $p'(x) = r'(x) - c'(x)$, $p'(x) = 0$ implies that

$$r'(x) - c'(x) = 0 \quad \text{or} \quad r'(x) = c'(x).$$

Therefore

> At a production level yielding maximum profit, marginal revenue equals marginal cost (Figure 3.43).

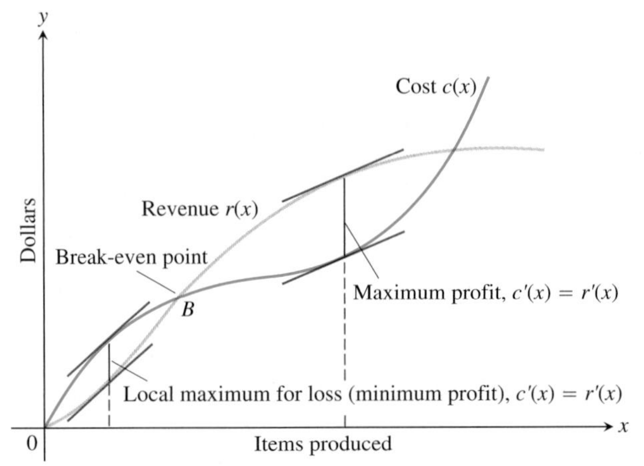

FIGURE 3.43 The graph of a typical cost function starts concave down and later turns concave up. It crosses the revenue curve at the break-even point B. To the left of B, the company operates at a loss. To the right, the company operates at a profit, with the maximum profit occurring where $c'(x) = r'(x)$. Farther to the right, cost exceeds revenue (perhaps because of a combination of rising labor and material costs and market saturation) and production levels become unprofitable again.

EXAMPLE 5 Suppose that $r(x) = 9x$ and $c(x) = x^3 - 6x^2 + 15x$, where x represents thousands of units. Is there a production level that maximizes profit? If so, what is it?

Solution Notice that $r'(x) = 9$ and $c'(x) = 3x^2 - 12x + 15$.

$$3x^2 - 12x + 15 = 9 \qquad \text{Set } c'(x) = r'(x).$$
$$3x^2 - 12x + 6 = 0$$

The two solutions of the quadratic equation are

$$x_1 = \frac{12 - \sqrt{72}}{6} = 2 - \sqrt{2} \approx 0.586 \qquad \text{and}$$

$$x_2 = \frac{12 + \sqrt{72}}{6} = 2 + \sqrt{2} \approx 3.414.$$

The possible production levels for maximum profit are $x \approx 0.586$ thousand units or $x \approx 3.414$ thousand units. The second derivative of $p(x) = r(x) - c(x)$ is $p''(x) = -c''(x)$ since $r''(x)$ is everywhere zero. Thus, $p''(x) = 6(2 - x)$, which is negative at $x = 2 + \sqrt{2}$ and positive at $x = 2 - \sqrt{2}$. By the Second Derivative Test, a maximum profit occurs at about $x = 3.414$ (where revenue exceeds costs) and maximum loss occurs at about $x = 0.586$. The graphs of $r(x)$ and $c(x)$ are shown in Figure 3.44. ■

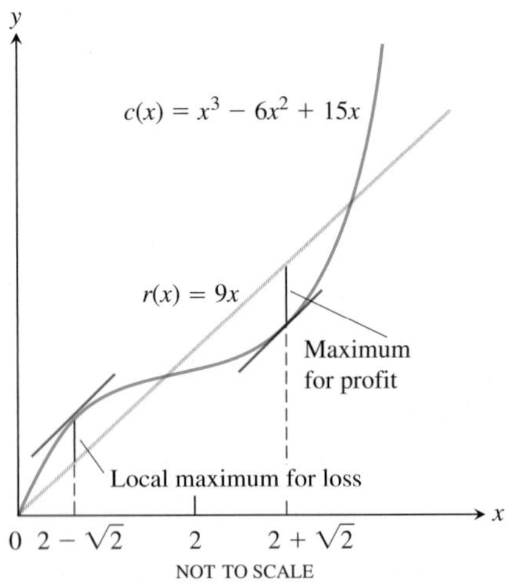

FIGURE 3.44 The cost and revenue curves for Example 5.

EXERCISES 3.6

Whenever you are maximizing or minimizing a function of a single variable, we urge you to graph it over the domain that is appropriate to the problem you are solving. The graph will provide insight before you calculate and will furnish a visual context for understanding your answer.

1. **Minimizing perimeter** What is the smallest perimeter possible for a rectangle whose area is 16 in.^{2}, and what are its dimensions?
2. Show that among all rectangles with an 8-m perimeter, the one with largest area is a square.
3. The figure shows a rectangle inscribed in an isosceles right triangle whose hypotenuse is 2 units long.
   a. Express the y-coordinate of P in terms of x. (*Hint:* Write an equation for the line AB.)
   b. Express the area of the rectangle in terms of x.
   c. What is the largest area the rectangle can have, and what are its dimensions?

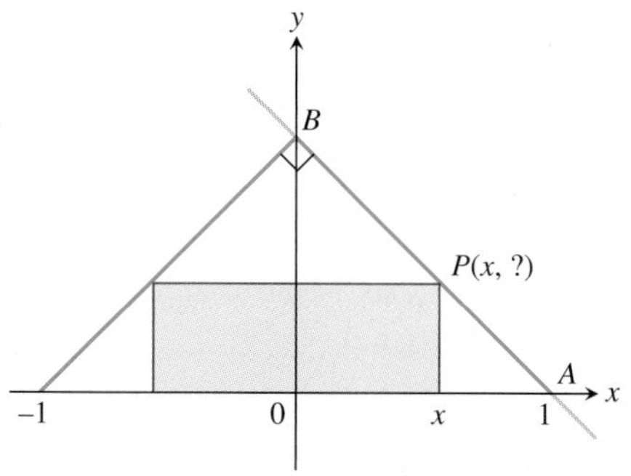

4. A rectangle has its base on the x-axis and its upper two vertices on the parabola $y = 12 - x^2$. What is the largest area the rectangle can have, and what are its dimensions?
5. You are planning to make an open rectangular box from an 8-in.-by-15-in. piece of cardboard by cutting congruent squares from the corners and folding up the sides. What are the dimensions of the box of largest volume you can make this way, and what is its volume?
6. You are planning to close off a corner of the first quadrant with a line segment 20 units long running from $(a, 0)$ to $(0, b)$. Show that the area of the triangle enclosed by the segment is largest when $a = b$.
7. **The best fencing plan** A rectangular plot of farmland will be bounded on one side by a river and on the other three sides by a single-strand electric fence. With 800 m of wire at your disposal, what is the largest area you can enclose, and what are its dimensions?
8. **The shortest fence** A 216 m^2 rectangular pea patch is to be enclosed by a fence and divided into two equal parts by another fence parallel to one of the sides. What dimensions for the outer rectangle will require the smallest total length of fence? How much fence will be needed?
9. **Designing a tank** Your iron works has contracted to design and build a 500 ft^3, square-based, open-top, rectangular steel holding tank for a paper company. The tank is to be made by welding thin stainless steel plates together along their edges. As the production engineer, your job is to find dimensions for the base and height that will make the tank weigh as little as possible.
   a. What dimensions do you tell the shop to use?
   b. Briefly describe how you took weight into account.
10. **Catching rainwater** A 1125 ft^3 open-top rectangular tank with a square base x ft on a side and y ft deep is to be built with its top flush with the ground to catch runoff water. The costs associated with the tank involve not only the material from which the tank is made but also an excavation charge proportional to the product xy.
    a. If the total cost is
    $$c = 5(x^2 + 4xy) + 10xy,$$
    what values of x and y will minimize it?
    b. Give a possible scenario for the cost function in part (a).
11. **Designing a poster** You are designing a rectangular poster to contain 50 in.^{2} of printing with a 4-in. margin at the top and bottom and a 2-in. margin at each side. What overall dimensions will minimize the amount of paper used?
12. Find the volume of the largest right circular cone that can be inscribed in a sphere of radius 3.

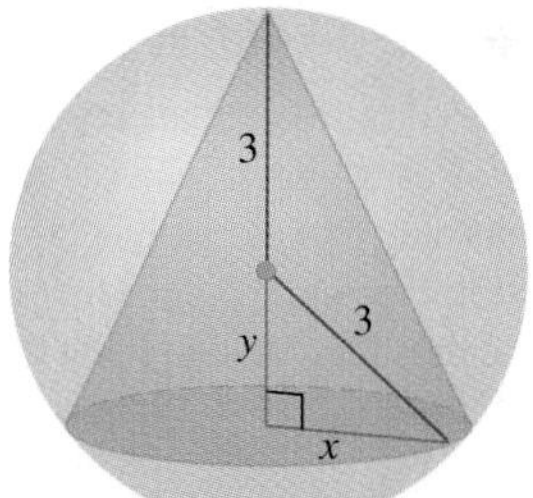

13. Two sides of a triangle have lengths a and b, and the angle between them is θ. What value of θ will maximize the triangle's area? (*Hint:* $A = (1/2)ab \sin \theta$.)
14. **Designing a can** What are the dimensions of the lightest open-top right circular cylindrical can that will hold a volume of 1000 cm^3? Compare the result here with the result in Example 2.
15. **Designing a can** You are designing a 1000 cm^3 right circular cylindrical can whose manufacture will take waste into account. There is no waste in cutting the aluminum for the side, but the top and bottom of radius r will be cut from squares that measure $2r$

units on a side. The total amount of aluminum used up by the can will therefore be

$$A = 8r^2 + 2\pi rh$$

rather than the $A = 2\pi r^2 + 2\pi rh$ in Example 2. In Example 2, the ratio of h to r for the most economical can was 2 to 1. What is the ratio now?

T **16. Designing a box with a lid** A piece of cardboard measures 10 in. by 15 in. Two equal squares are removed from the corners of a 10-in. side as shown in the figure. Two equal rectangles are removed from the other corners so that the tabs can be folded to form a rectangular box with lid.

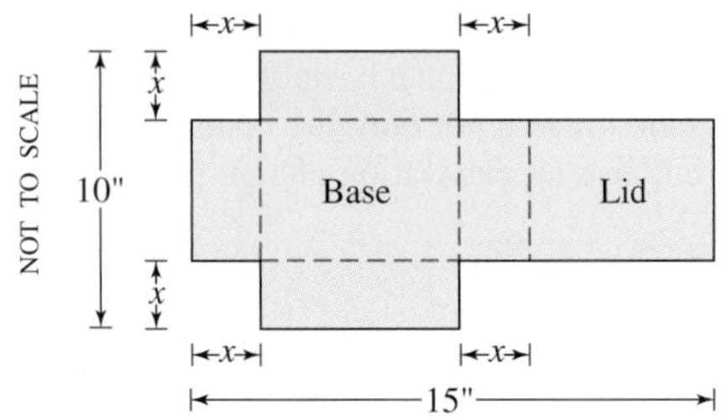

a. Write a formula $V(x)$ for the volume of the box.

b. Find the domain of V for the problem situation and graph V over this domain.

c. Use a graphical method to find the maximum volume and the value of x that gives it.

d. Confirm your result in part (c) analytically.

17. Find the dimensions of a right circular cylinder of maximum volume that can be inscribed in a sphere of radius 10 cm. What is the maximum volume?

18. A window is in the form of a rectangle surmounted by a semicircle. The rectangle is of clear glass, whereas the semicircle is of tinted glass that transmits only half as much light per unit area as clear glass does. The total perimeter is fixed. Find the proportions of the window that will admit the most light. Neglect the thickness of the frame.

19. A silo (base not included) is to be constructed in the form of a cylinder surmounted by a hemisphere. The cost of construction per square unit of surface area is twice as great for the hemisphere as it is for the cylindrical sidewall. Determine the dimensions to be used if the volume is fixed and the cost of construction is to be kept to a minimum. Neglect the thickness of the silo and waste in construction.

20. The trough in the figure is to be made to the dimensions shown. Only the angle θ can be varied. What value of θ will maximize the trough's volume?

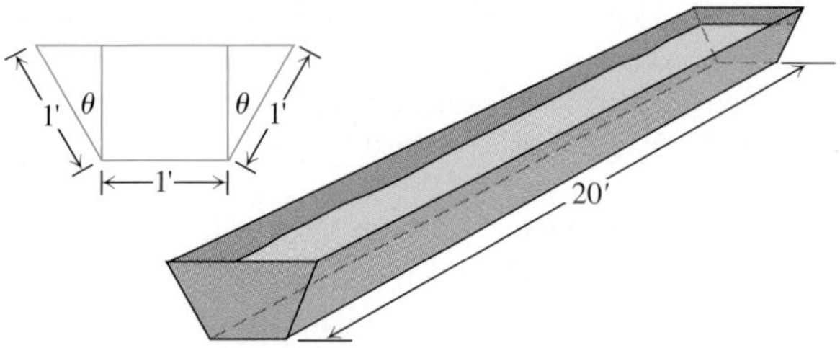

21. Constructing cones A right triangle whose hypotenuse is $\sqrt{3}$ m long is revolved about one of its legs to generate a right circular cone. Find the radius, height, and volume of the cone of greatest volume that can be made this way.

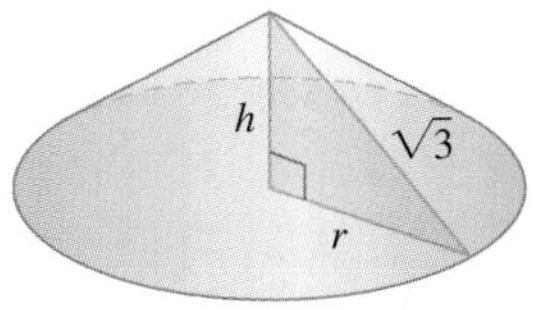

22. What value of a makes $f(x) = x^2 + (a/x)$ have

a. local minimum at $x = 2$?

b. a point of inflection at $x = 1$?

23. Vertical motion The height of an object moving vertically is given by

$$s = -16t^2 + 96t + 112,$$

with s in feet and t in seconds. Find

a. the object's velocity when $t = 0$

b. its maximum height and when it occurs

c. its velocity when $s = 0$.

24. Quickest route Jane is 2 mi offshore in a boat and wishes to reach a coastal village 6 mi down a straight shoreline from the point nearest the boat. She can row 2 mph and can walk 5 mph. Where should she land her boat to reach the village in the least amount of time?

25. Shortest beam The 8-ft wall shown here stands 27 ft from the building. Find the length of the shortest straight beam that will reach to the side of the building from the ground outside the wall.

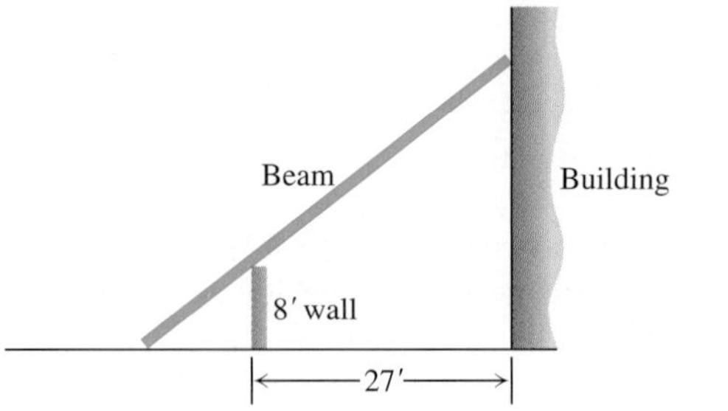

26. Motion on a line The positions of two particles on the s-axis are $s_1 = \sin t$ and $s_2 = \sin(t + \pi/3)$, with s_1 and s_2 in meters and t in seconds.

a. At what time(s) in the interval $0 \le t \le 2\pi$ do the particles meet?

b. What is the farthest apart that the particles ever get?

c. When in the interval $0 \le t \le 2\pi$ is the distance between the particles changing the fastest?

27. Frictionless cart A small frictionless cart, attached to the wall by a spring, is pulled 10 cm from its rest position and released at time $t = 0$ to roll back and forth for 4 sec. Its position at time t is $s = 10 \cos \pi t$.

a. What is the cart's maximum speed? When is the cart moving that fast? Where is it then? What is the magnitude of the acceleration then?

b. Where is the cart when the magnitude of the acceleration is greatest? What is the cart's speed then?

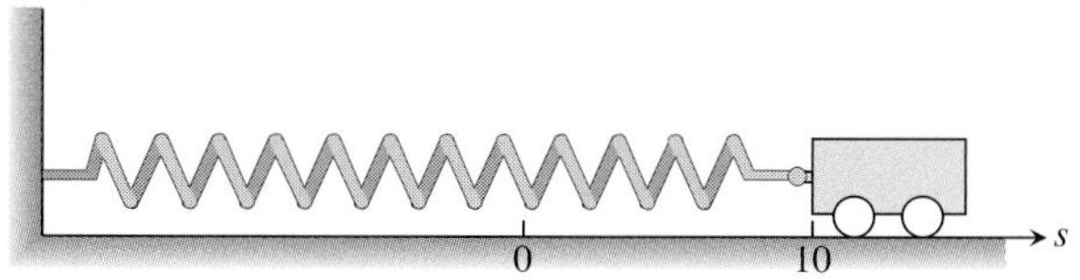

28. Two masses hanging side by side from springs have positions $s_1 = 2 \sin t$ and $s_2 = \sin 2t$, respectively.

a. At what times in the interval $0 < t$ do the masses pass each other? (*Hint:* $\sin 2t = 2 \sin t \cos t$.)

b. When in the interval $0 \le t \le 2\pi$ is the vertical distance between the masses the greatest? What is this distance? (*Hint:* $\cos 2t = 2\cos^2 t - 1$.)

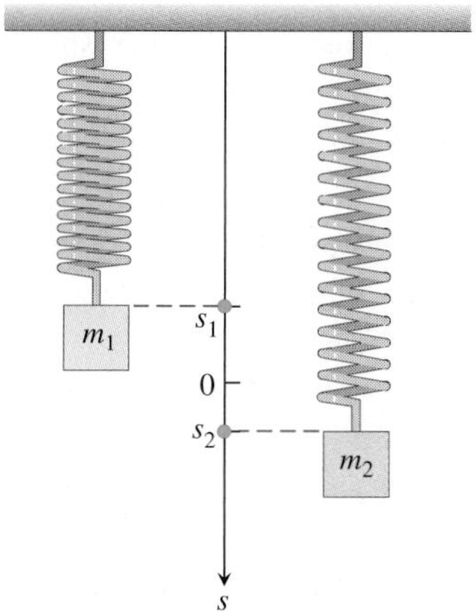

29. Locating a solar station You are under contract to build a solar station at ground level on the east–west line between the two buildings shown here. How far from the taller building should you place the station to maximize the number of hours it will be in the sun on a day when the sun passes directly overhead? Begin by observing that

$$\theta = \pi - \cot^{-1}\frac{x}{60} - \cot^{-1}\frac{50 - x}{30}.$$

Then find the value of x that maximizes θ.

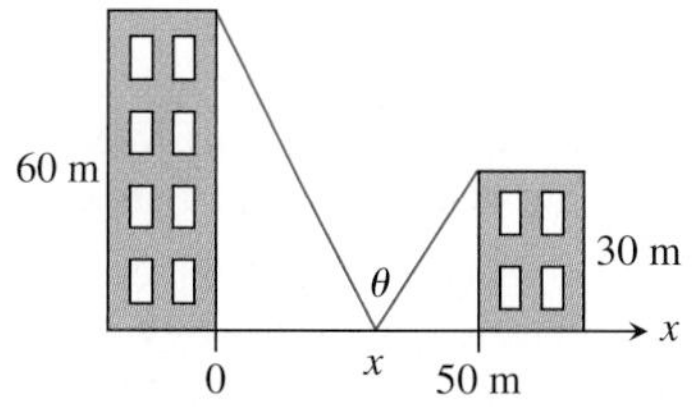

30. Fermat's principle in optics Fermat's principle in optics states that light always travels from one point to another along a path that minimizes the travel time. Light from a source A is reflected by a plane mirror to a receiver at point B, as shown in the figure. Show that for the light to obey Fermat's principle, the angle of incidence must equal the angle of reflection, both measured from the line normal to the reflecting surface. (This result can also be derived without calculus. There is a purely geometric argument, which you may prefer.)

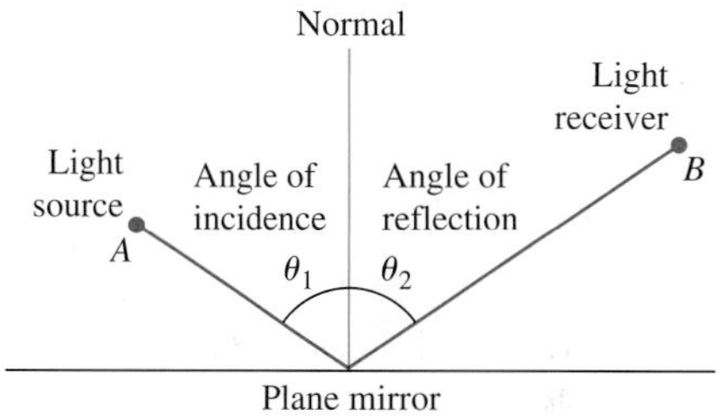

31. It costs you c dollars each to manufacture and distribute backpacks. If the backpacks sell at x dollars each, the number sold is given by

$$n = \frac{a}{x - c} + b(100 - x),$$

where a and b are positive constants. What selling price will bring a maximum profit?

32. You operate a tour service that offers the following rates:

$200 per person if 50 people (the minimum number to book the tour) go on the tour.

For each additional person, up to a maximum of 80 people total, the rate per person is reduced by $2.

It costs $6000 (a fixed cost) plus $32 per person to conduct the tour. How many people does it take to maximize your profit?

33. Wilson lot size formula One of the formulas for inventory management says that the average weekly cost of ordering, paying for, and holding merchandise is

$$A(q) = \frac{km}{q} + cm + \frac{hq}{2},$$

where q is the quantity you order when things run low (shoes, radios, brooms, or whatever the item might be), k is the cost of placing an order (the same, no matter how often you order), c is the cost of one item (a constant), m is the number of items sold each week (a constant), and h is the weekly holding cost per item (a constant that takes into account things such as space, utilities, insurance, and security).

a. Your job, as the inventory manager for your store, is to find the quantity that will minimize $A(q)$. What is it? (The formula you get for the answer is called the *Wilson lot size formula.*)

b. Shipping costs sometimes depend on order size. When they do, it is more realistic to replace k by $k + bq$, the sum of k and a constant multiple of q. What is the most economical quantity to order now?

34. Production level Prove that the production level (if any) at which average cost is smallest is a level at which the average cost equals marginal cost.

35. Sensitivity to medicine (*Continuation of Exercise 50, Section 2.3.*) Find the amount of medicine to which the body is most sensitive by finding the value of M that maximizes the derivative dR/dM, where

$$R = M^2\left(\frac{C}{2} - \frac{M}{3}\right)$$

and C is a constant.

36. How we cough

a. When we cough, the trachea (windpipe) contracts to increase the velocity of the air going out. This raises the questions of how much it should contract to maximize the velocity and whether it really contracts that much when we cough.

Under reasonable assumptions about the elasticity of the tracheal wall and about how the air near the wall is slowed by friction, the average flow velocity υ can be modeled by the equation

$$\upsilon = c(r_0 - r)r^2 \text{ cm/sec}, \qquad \frac{r_0}{2} \le r \le r_0,$$

where r_0 is the rest radius of the trachea in centimeters and c is a positive constant whose value depends in part on the length of the trachea.

Show that υ is greatest when $r = (2/3)r_0$, that is, when the trachea is about 33% contracted. The remarkable fact is that X-ray photographs confirm that the trachea contracts about this much during a cough.

T **b.** Take r_0 to be 0.5 and c to be 1 and graph υ over the interval $0 \le r \le 0.5$. Compare what you see with the claim that υ is at a maximum when $r = (2/3)r_0$.

37. An inequality for positive integers Show that if a, b, c, and d are positive integers, then

$$\frac{(a^2+1)(b^2+1)(c^2+1)(d^2+1)}{abcd} \ge 16.$$

38. The derivative dt/dx in Example 4

a. Show that

$$f(x) = \frac{x}{\sqrt{a^2 + x^2}}$$

is an increasing function of x.

b. Show that

$$g(x) = \frac{d - x}{\sqrt{b^2 + (d - x)^2}}$$

is a decreasing function of x.

c. Show that

$$\frac{dt}{dx} = \frac{x}{c_1\sqrt{a^2 + x^2}} - \frac{d - x}{c_2\sqrt{b^2 + (d - x)^2}}$$

is an increasing function of x.

39. Let $f(x)$ and $g(x)$ be the differentiable functions graphed here. Point c is the point where the vertical distance between the curves is the greatest. Is there anything special about the tangents to the two curves at c? Give reasons for your answer.

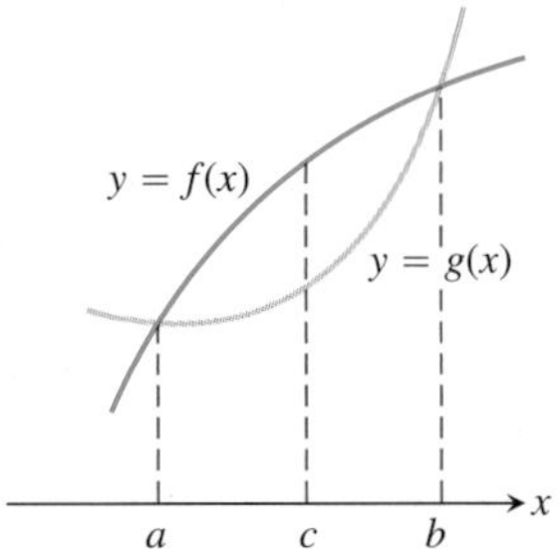

40. You have been asked to determine whether the function $f(x) = 3 + 4\cos x + \cos 2x$ is ever negative.

a. Explain why you need to consider values of x only in the interval $[0, 2\pi]$.

b. Is f ever negative? Explain.

41. a. The function $y = \cot x - \sqrt{2}\csc x$ has an absolute maximum value on the interval $0 < x < \pi$. Find it.

T **b.** Graph the function and compare what you see with your answer in part (a).

42. a. The function $y = \tan x + 3\cot x$ has an absolute minimum value on the interval $0 < x < \pi/2$. Find it.

T **b.** Graph the function and compare what you see with your answer in part (a).

43. a. How close does the curve $y = \sqrt{x}$ come to the point $(3/2, 0)$? (*Hint*: If you minimize the *square* of the distance, you can avoid square roots.)

T **b.** Graph the distance function and $y = \sqrt{x}$ together and reconcile what you see with your answer in part (a).

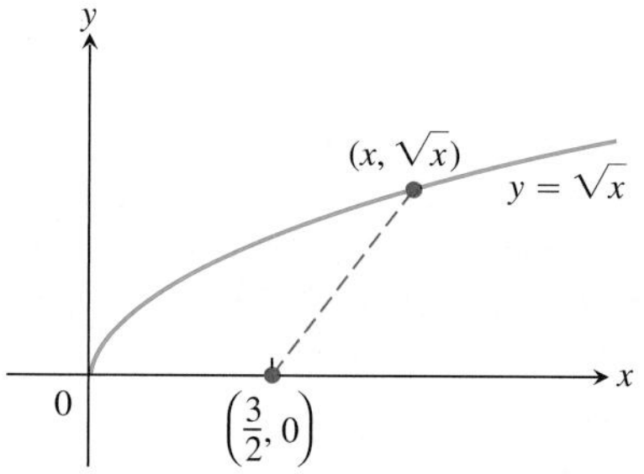

44. **a.** How close does the semicircle $y = \sqrt{16 - x^2}$ come to the point $\left(1, \sqrt{3}\right)$?

T **b.** Graph the distance function and $y = \sqrt{16 - x^2}$ together and reconcile what you see with your answer in part (a).

3.7 Indeterminate Forms and L'Hôpital's Rule

HISTORICAL BIOGRAPHY

Guillaume François Antoine de l'Hôpital (1661–1704)

Johann Bernoulli (1667–1748)

John (Johann) Bernoulli discovered a rule using derivatives to calculate limits of fractions whose numerators and denominators both approach zero or $+\infty$. The rule is known today as **l'Hôpital's Rule**, after Guillaume de l'Hôpital. He was a French nobleman who wrote the first introductory differential calculus text, where the rule first appeared in print.

Indeterminate Form 0/0

If we want to know how the function

$$F(x) = \frac{x - \sin x}{x^3}$$

behaves *near* $x = 0$ (where it is undefined), we can examine the limit of $F(x)$ as $x \to 0$. We cannot apply the Quotient Rule for limits (Theorem 1 of Chapter 1) because the limit of the denominator is 0. Moreover, in this case, *both* the numerator and denominator approach 0, and 0/0 is undefined. Such limits may or may not exist in general, but the limit does exist for the function $F(x)$ under discussion by applying l'Hôpital's Rule (see Example 1d on the next page).

If the continuous functions $f(x)$ and $g(x)$ are both zero at $x = a$, then

$$\lim_{x \to a} \frac{f(x)}{g(x)}$$

cannot be found by substituting $x = a$. The substitution produces $0/0$, a meaningless expression, which we cannot evaluate. We use $0/0$ as a notation for an expression known as an **indeterminate form**. Sometimes, but not always, limits that lead to indeterminate forms may be found by cancellation, rearrangement of terms, or other algebraic manipulations. This was our experience in Chapter 1. It took considerable analysis in Section 1.6 to find $\lim_{x \to 0} (\sin x)/x$. But we have had success with the limit

$$f'(a) = \lim_{x \to a} \frac{f(x) - f(a)}{x - a},$$

from which we calculate derivatives and which always produces the equivalent of $0/0$ when we substitute $x = a$. L'Hôpital's Rule enables us to draw on our success with derivatives to evaluate limits that otherwise lead to indeterminate forms.

THEOREM 6—L'Hôpital's Rule Suppose that $f(a) = g(a) = 0$, that f and g are differentiable on an open interval I containing a, and that $g'(x) \neq 0$ on I if $x \neq a$. Then

$$\lim_{x \to a} \frac{f(x)}{g(x)} = \lim_{x \to a} \frac{f'(x)}{g'(x)},$$

assuming that the limit on the right side of this equation exists.

A proof of Theorem 6 is given in Appendix 5.

Caution

To apply l'Hôpital's Rule to f/g, divide the derivative of f by the derivative of g. Do not fall into the trap of taking the derivative of f/g. The quotient to use is f'/g', not $(f/g)'$.

EXAMPLE 1

(a) $$\lim_{x \to 0} \frac{3x - \sin x}{x} = \lim_{x \to 0} \frac{3 - \cos x}{1} = \left. \frac{3 - \cos x}{1} \right|_{x=0} = 2$$

(b) $$\lim_{x \to 0} \frac{\sqrt{1 + x} - 1}{x} = \lim_{x \to 0} \frac{\dfrac{1}{2\sqrt{1 + x}}}{1} = \frac{1}{2}$$

(c) $$\lim_{x \to 0} \frac{\sqrt{1 + x} - 1 - x/2}{x^2} \qquad \frac{0}{0}$$

$$= \lim_{x \to 0} \frac{(1/2)(1 + x)^{-1/2} - 1/2}{2x} \qquad \text{Still } \frac{0}{0}\text{; differentiate again.}$$

$$= \lim_{x \to 0} \frac{-(1/4)(1 + x)^{-3/2}}{2} = -\frac{1}{8} \qquad \text{Not } \frac{0}{0}\text{; limit is found.}$$

(d) $$\lim_{x \to 0} \frac{x - \sin x}{x^3} \qquad \frac{0}{0}$$

$$= \lim_{x \to 0} \frac{1 - \cos x}{3x^2} \qquad \text{Still } \frac{0}{0}$$

$$= \lim_{x \to 0} \frac{\sin x}{6x} \qquad \text{Still } \frac{0}{0}$$

$$= \lim_{x \to 0} \frac{\cos x}{6} = \frac{1}{6} \qquad \text{Not } \frac{0}{0}\text{; limit is found.}$$ ■

Here is a summary of the procedure we followed in Example 1.

Using L'Hôpital's Rule

To find

$$\lim_{x \to a} \frac{f(x)}{g(x)}$$

by l'Hôpital's Rule, continue to differentiate f and g, so long as we still get the form $0/0$ at $x = a$. But as soon as one or the other of these derivatives is different from zero at $x = a$ we stop differentiating. L'Hôpital's Rule does not apply when either the numerator or denominator has a finite nonzero limit.

EXAMPLE 2 Be careful to apply l'Hôpital's Rule correctly:

$$\lim_{x \to 0} \frac{1 - \cos x}{x + x^2} \qquad \frac{0}{0}$$

$$= \lim_{x \to 0} \frac{\sin x}{1 + 2x} = \frac{0}{1} = 0. \qquad \text{Not } \frac{0}{0}\text{; limit is found.}$$

Up to now the calculation is correct, but if we continue to differentiate in an attempt to apply l'Hôpital's Rule once more, we get

$$\lim_{x \to 0} \frac{1 - \cos x}{x + x^2} = \lim_{x \to 0} \frac{\sin x}{1 + 2x} = \lim_{x \to 0} \frac{\cos x}{2} = \frac{1}{2},$$

which is wrong. L'Hôpital's Rule can only be applied to limits which give indeterminate forms, and $0/1$ is not an indeterminate form. ■

L'Hôpital's Rule applies to one-sided limits as well.

EXAMPLE 3

Recall that ∞ and $+\infty$ mean the same thing.

(a) $$\lim_{x \to 0^+} \frac{\sin x}{x^2} \qquad \frac{0}{0}$$

$$= \lim_{x \to 0^+} \frac{\cos x}{2x} = \infty \qquad \text{Positive for } x > 0.$$

(b) $$\lim_{x \to 0^-} \frac{\sin x}{x^2} \qquad \frac{0}{0}$$

$$= \lim_{x \to 0^-} \frac{\cos x}{2x} = -\infty \qquad \text{Negative for } x < 0.$$ ■

Indeterminate Forms ∞/∞, $\infty \cdot 0$, $\infty - \infty$

Sometimes when we try to evaluate a limit as $x \to a$ by substituting $x = a$ we get an ambiguous expression like ∞/∞, $\infty \cdot 0$, or $\infty - \infty$, instead of $0/0$. We first consider the form ∞/∞.

In more advanced books it is proved that l'Hôpital's Rule applies to the indeterminate form ∞/∞ as well as to $0/0$. If $f(x) \to \pm\infty$ and $g(x) \to \pm\infty$ as $x \to a$, then

$$\lim_{x\to a}\frac{f(x)}{g(x)} = \lim_{x\to a}\frac{f'(x)}{g'(x)}$$

provided the limit on the right exists. In the notation $x \to a$, a may be either finite or infinite. Moreover $x \to a$ may be replaced by the one-sided limits $x \to a^+$ or $x \to a^-$.

EXAMPLE 4 Find

(a) $\displaystyle\lim_{x\to\pi/2}\frac{\sec x}{1+\tan x}$ **(b)** $\displaystyle\lim_{x\to\infty}\frac{\ln x}{2\sqrt{x}}$ **(c)** $\displaystyle\lim_{x\to\infty}\frac{e^x}{x^2}$.

Solution

(a) The numerator and denominator are discontinuous at $x = \pi/2$, so we investigate the one-sided limits there. To apply l'Hôpital's Rule, we can choose I to be any open interval with $x = \pi/2$ as an endpoint.

$$\lim_{x\to(\pi/2)^-}\frac{\sec x}{1+\tan x} \qquad \frac{\infty}{\infty} \text{ from the left}$$

$$= \lim_{x\to(\pi/2)^-}\frac{\sec x\tan x}{\sec^2 x} = \lim_{x\to(\pi/2)^-}\sin x = 1$$

The right-hand limit is 1 also, with $(-\infty)/(-\infty)$ as the indeterminate form. Therefore, the two-sided limit is equal to 1.

(b) $$\lim_{x\to\infty}\frac{\ln x}{2\sqrt{x}} = \lim_{x\to\infty}\frac{1/x}{1/\sqrt{x}} = \lim_{x\to\infty}\frac{1}{\sqrt{x}} = 0$$

(c) $$\lim_{x\to\infty}\frac{e^x}{x^2} = \lim_{x\to\infty}\frac{e^x}{2x} = \lim_{x\to\infty}\frac{e^x}{2} = \infty$$ ■

Next we turn our attention to the indeterminate forms $\infty \cdot 0$ and $\infty - \infty$. Sometimes these forms can be handled by using algebra to convert them to a $0/0$ or ∞/∞ form. Here again we do not mean to suggest that $\infty \cdot 0$ or $\infty - \infty$ is a number. They are only notations for functional behaviors when considering limits. Here are examples of how we might work with these indeterminate forms.

EXAMPLE 5 Find

(a) $\displaystyle\lim_{x\to\infty}\left(x\sin\frac{1}{x}\right)$ **(b)** $\displaystyle\lim_{x\to 0^+}\sqrt{x}\ln x$

Solution

(a) $$\lim_{x\to\infty}\left(x\sin\frac{1}{x}\right) \qquad \infty\cdot 0$$

$$= \lim_{h\to 0^+}\left(\frac{1}{h}\sin h\right) = 1 \qquad \text{Let } h = 1/x.$$

(b) $$\lim_{x\to 0^+} \sqrt{x}\ln x = \lim_{x\to 0^+} \frac{\ln x}{1/\sqrt{x}} \qquad \infty/\infty$$

$$= \lim_{x\to 0^+} \frac{1/x}{-1/2x^{3/2}} = \lim_{x\to 0^+}\left(-2\sqrt{x}\right) = 0$$ ■

EXAMPLE 6 Find

$$\lim_{x\to 0}\left(\frac{1}{\sin x} - \frac{1}{x}\right).$$

Solution If $x \to 0^+$, then $\sin x \to 0^+$ and

$$\frac{1}{\sin x} - \frac{1}{x} \to \infty - \infty.$$

Similarly, if $x \to 0^-$, then $\sin x \to 0^-$ and

$$\frac{1}{\sin x} - \frac{1}{x} \to -\infty - (-\infty) = -\infty + \infty.$$

Neither form reveals what happens in the limit. To find out, we first combine the fractions:

$$\frac{1}{\sin x} - \frac{1}{x} = \frac{x - \sin x}{x\sin x} \qquad \text{Common denominator is } x\sin x.$$

Then apply l'Hôpital's Rule to the result:

$$\lim_{x\to 0}\left(\frac{1}{\sin x} - \frac{1}{x}\right) = \lim_{x\to 0} \frac{x - \sin x}{x\sin x} \qquad \frac{0}{0}$$

$$= \lim_{x\to 0} \frac{1 - \cos x}{\sin x + x\cos x} \qquad \text{Still } \frac{0}{0}$$

$$= \lim_{x\to 0} \frac{\sin x}{2\cos x - x\sin x} = \frac{0}{2} = 0.$$ ■

Indeterminate Powers

Limits that lead to the indeterminate forms 1^∞, 0^0, and ∞^0 can sometimes be handled by first taking the logarithm of the function. We use l'Hôpital's Rule to find the limit of the logarithm expression and then exponentiate the result to find the original function limit. This procedure is justified by the continuity of the exponential function and Theorem 10 in Section 1.7, and it is formulated as follows.

> If $\lim_{x\to a} \ln f(x) = L$, then
>
> $$\lim_{x\to a} f(x) = \lim_{x\to a} e^{\ln f(x)} = e^L.$$
>
> Here a may be either finite or infinite.

EXAMPLE 7 Apply l'Hôpital's Rule to show that $\lim_{x\to 0^+}(1+x)^{1/x} = e$.

Solution The limit leads to the indeterminate form 1^∞. We let $f(x) = (1+x)^{1/x}$ and find $\lim_{x\to 0^+} \ln f(x)$. Since

$$\ln f(x) = \ln(1+x)^{1/x} = \frac{1}{x}\ln(1+x),$$

l' Hôpital's Rule now applies to give

$$\begin{aligned}
\lim_{x\to 0^+} \ln f(x) &= \lim_{x\to 0^+} \frac{\ln(1+x)}{x} \qquad \frac{0}{0}\\
&= \lim_{x\to 0^+} \frac{\frac{1}{1+x}}{1}\\
&= \frac{1}{1} = 1.
\end{aligned}$$

Therefore, $\lim_{x\to 0^+}(1+x)^{1/x} = \lim_{x\to 0^+} f(x) = \lim_{x\to 0^+} e^{\ln f(x)} = e^1 = e.$ ■

EXAMPLE 8 Find $\lim_{x\to\infty} x^{1/x}$.

Solution The limit leads to the indeterminate form ∞^0. We let $f(x) = x^{1/x}$ and find $\lim_{x\to\infty} \ln f(x)$. Since

$$\ln f(x) = \ln x^{1/x} = \frac{\ln x}{x},$$

l'Hôpital's Rule gives

$$\begin{aligned}
\lim_{x\to\infty} \ln f(x) &= \lim_{x\to\infty} \frac{\ln x}{x} \qquad \frac{\infty}{\infty}\\
&= \lim_{x\to\infty} \frac{1/x}{1}\\
&= \frac{0}{1} = 0.
\end{aligned}$$

Therefore, $\lim_{x\to\infty} x^{1/x} = \lim_{x\to\infty} f(x) = \lim_{x\to\infty} e^{\ln f(x)} = e^0 = 1.$ ■

EXERCISES 3.7

In Exercises 1–6, use l'Hôpital's Rule to evaluate the limit. Then evaluate the limit using a method studied in Chapter 1.

1. $\displaystyle\lim_{x\to 2} \frac{x-2}{x^2-4}$
2. $\displaystyle\lim_{x\to 0} \frac{\sin 5x}{x}$
3. $\displaystyle\lim_{x\to\infty} \frac{5x^2-3x}{7x^2+1}$
4. $\displaystyle\lim_{x\to 1} \frac{x^3-1}{4x^3-x-3}$
5. $\displaystyle\lim_{x\to 0} \frac{1-\cos x}{x^2}$
6. $\displaystyle\lim_{x\to\infty} \frac{2x^2+3x}{x^3+x+1}$

Use l'Hôpital's rule to find the limits in Exercises 7–46.

7. $\displaystyle\lim_{x\to 2} \frac{x-2}{x^2-4}$
8. $\displaystyle\lim_{x\to -5} \frac{x^2-25}{x+5}$
9. $\displaystyle\lim_{t\to -3} \frac{t^3-4t+15}{t^2-t-12}$
10. $\displaystyle\lim_{t\to 1} \frac{t^3-1}{4t^3-t-3}$
11. $\displaystyle\lim_{x\to\infty} \frac{5x^3-2x}{7x^3+3}$
12. $\displaystyle\lim_{x\to\infty} \frac{x-8x^2}{12x^2+5x}$

13. $\lim_{t\to 0} \dfrac{\sin t^2}{t}$

14. $\lim_{t\to 0} \dfrac{\sin 5t}{2t}$

15. $\lim_{x\to 0} \dfrac{8x^2}{\cos x - 1}$

16. $\lim_{x\to 0} \dfrac{\sin x - x}{x^3}$

17. $\lim_{\theta\to\pi/2} \dfrac{2\theta - \pi}{\cos(2\pi - \theta)}$

18. $\lim_{\theta\to -\pi/3} \dfrac{3\theta + \pi}{\sin(\theta + (\pi/3))}$

19. $\lim_{\theta\to\pi/2} \dfrac{1 - \sin\theta}{1 + \cos 2\theta}$

20. $\lim_{x\to 1} \dfrac{x - 1}{\ln x - \sin \pi x}$

21. $\lim_{x\to 0} \dfrac{x^2}{\ln(\sec x)}$

22. $\lim_{x\to\pi/2} \dfrac{\ln(\csc x)}{(x - (\pi/2))^2}$

23. $\lim_{t\to 0} \dfrac{t(1 - \cos t)}{t - \sin t}$

24. $\lim_{t\to 0} \dfrac{t \sin t}{1 - \cos t}$

25. $\lim_{x\to(\pi/2)^-} \left(x - \dfrac{\pi}{2}\right) \sec x$

26. $\lim_{x\to(\pi/2)^-} \left(\dfrac{\pi}{2} - x\right) \tan x$

27. $\lim_{\theta\to 0} \dfrac{3^{\sin\theta} - 1}{\theta}$

28. $\lim_{\theta\to 0} \dfrac{(1/2)^\theta - 1}{\theta}$

29. $\lim_{x\to 0} \dfrac{x2^x}{2^x - 1}$

30. $\lim_{x\to 0} \dfrac{3^x - 1}{2^x - 1}$

31. $\lim_{x\to\infty} \dfrac{\ln(x + 1)}{\log_2 x}$

32. $\lim_{x\to\infty} \dfrac{\log_2 x}{\log_3(x + 3)}$

33. $\lim_{x\to 0^+} \dfrac{\ln(x^2 + 2x)}{\ln x}$

34. $\lim_{x\to 0^+} \dfrac{\ln(e^x - 1)}{\ln x}$

35. $\lim_{y\to 0} \dfrac{\sqrt{5y + 25} - 5}{y}$

36. $\lim_{y\to 0} \dfrac{\sqrt{ay + a^2} - a}{y}, \quad a > 0$

37. $\lim_{x\to\infty} (\ln 2x - \ln(x + 1))$

38. $\lim_{x\to 0^+} (\ln x - \ln \sin x)$

39. $\lim_{h\to 0} \dfrac{\sin(a + h) - \sin a}{h}$

40. $\lim_{x\to 0^+} \left(\dfrac{3x + 1}{x} - \dfrac{1}{\sin x}\right)$

41. $\lim_{x\to 1^+} \left(\dfrac{1}{x - 1} - \dfrac{1}{\ln x}\right)$

42. $\lim_{x\to 0^+} (\csc x - \cot x + \cos x)$

43. $\lim_{\theta\to 0} \dfrac{\cos\theta - 1}{e^\theta - \theta - 1}$

44. $\lim_{h\to 0} \dfrac{e^h - (1 + h)}{h^2}$

45. $\lim_{t\to\infty} \dfrac{e^t + t^2}{e^t - t}$

46. $\lim_{x\to\infty} x^2 e^{-x}$

Find the limits in Exercise 47–56.

47. $\lim_{x\to 1^+} x^{1/(1-x)}$

48. $\lim_{x\to 1^+} x^{1/(x-1)}$

49. $\lim_{x\to\infty} (\ln x)^{1/x}$

50. $\lim_{x\to e^+} (\ln x)^{1/(x-e)}$

51. $\lim_{x\to 0^+} x^{-1/\ln x}$

52. $\lim_{x\to\infty} x^{1/\ln x}$

53. $\lim_{x\to\infty} (1 + 2x)^{1/(2\ln x)}$

54. $\lim_{x\to 0} (e^x + x)^{1/x}$

55. $\lim_{x\to 0^+} x^x$

56. $\lim_{x\to 0^+} \left(1 + \dfrac{1}{x}\right)^x$

L'Hôpital's Rule does not help with the limits in Exercises 57–60. Try it—you just keep on cycling. Find the limits some other way.

57. $\lim_{x\to\infty} \dfrac{\sqrt{9x + 1}}{\sqrt{x + 1}}$

58. $\lim_{x\to 0^+} \dfrac{\sqrt{x}}{\sqrt{\sin x}}$

59. $\lim_{x\to(\pi/2)^-} \dfrac{\sec x}{\tan x}$

60. $\lim_{x\to 0^+} \dfrac{\cot x}{\csc x}$

61. Which one is correct, and which one is wrong? Give reasons for your answers.

a. $\lim_{x\to 3} \dfrac{x - 3}{x^2 - 3} = \lim_{x\to 3} \dfrac{1}{2x} = \dfrac{1}{6}$

b. $\lim_{x\to 3} \dfrac{x - 3}{x^2 - 3} = \dfrac{0}{6} = 0$

62. Which one is correct, and which one is wrong? Give reasons for your answers.

a. $\lim_{x\to 0} \dfrac{x^2 - 2x}{x^2 - \sin x} = \lim_{x\to 0} \dfrac{2x - 2}{2x - \cos x}$

$= \lim_{x\to 0} \dfrac{2}{2 + \sin x} = \dfrac{2}{2 + 0} = 1$

b. $\lim_{x\to 0} \dfrac{x^2 - 2x}{x^2 - \sin x} = \lim_{x\to 0} \dfrac{2x - 2}{2x - \cos x} = \dfrac{-2}{0 - 1} = 2$

63. Only one of these calculations is correct. Which one? Why are the others wrong? Give reasons for your answers.

a. $\lim_{x\to 0^+} x \ln x = 0 \cdot (-\infty) = 0$

b. $\lim_{x\to 0^+} x \ln x = 0 \cdot (-\infty) = -\infty$

c. $\lim_{x\to 0^+} x \ln x = \lim_{x\to 0^+} \dfrac{\ln x}{(1/x)} = \dfrac{-\infty}{\infty} = -1$

d. $\lim_{x\to 0^+} x \ln x = \lim_{x\to 0^+} \dfrac{\ln x}{(1/x)}$

$= \lim_{x\to 0^+} \dfrac{(1/x)}{(-1/x^2)} = \lim_{x\to 0^+} (-x) = 0$

T **64. $\infty - \infty$ Form**

a. Estimate the value of

$$\lim_{x\to\infty} \left(x - \sqrt{x^2 + x}\right)$$

by graphing $f(x) = x - \sqrt{x^2 + x}$ over a suitably large interval of x-values.

b. Now confirm your estimate by finding the limit with l'Hôpital's Rule. As the first step, multiply $f(x)$ by the fraction $(x + \sqrt{x^2 + x})/(x + \sqrt{x^2 + x})$ and simplify the new numerator.

T **65. 0/0 Form** Estimate the value of

$$\lim_{x\to 1} \frac{2x^2 - (3x + 1)\sqrt{x} + 2}{x - 1}$$

by graphing. Then confirm your estimate with l'Hôpital's Rule.

66. This exercise explores the difference between the limit

$$\lim_{x\to\infty}\left(1+\frac{1}{x^2}\right)^x$$

and the limit

$$\lim_{x\to\infty}\left(1+\frac{1}{x}\right)^x = e.$$

a. Use l'Hôpital's Rule to show that

$$\lim_{x\to\infty}\left(1+\frac{1}{x}\right)^x = e.$$

T **b.** Graph

$$f(x) = \left(1+\frac{1}{x^2}\right)^x \quad\text{and}\quad g(x) = \left(1+\frac{1}{x}\right)^x$$

together for $x \geq 0$. How does the behavior of f compare with that of g? Estimate the value of $\lim_{x\to\infty} f(x)$.

c. Confirm your estimate of $\lim_{x\to\infty} f(x)$ by calculating it with l'Hôpital's Rule.

67. Show that

$$\lim_{k\to\infty}\left(1+\frac{r}{k}\right)^k = e^r.$$

68. Given that $x > 0$, find the maximum value, if any, of

a. $x^{1/x}$

b. x^{1/x^2}

c. x^{1/x^n} (n a positive integer)

d. Show that $\lim_{x\to\infty} x^{1/x^n} = 1$ for every positive integer n.

3.8 Newton's Method

In this section we study a numerical method, called *Newton's method* or the *Newton–Raphson method*, which is a technique to approximate the solution to an equation $f(x) = 0$. Essentially it uses tangent lines in place of the graph of $y = f(x)$ near the points where f is zero. (A value of x where f is zero is a *root* of the function f and a *solution* of the equation $f(x) = 0$.)

Procedure for Newton's Method

The goal of Newton's method for estimating a solution of an equation $f(x) = 0$ is to produce a sequence of approximations that approach the solution. We pick the first number x_0 of the sequence. Then, under favorable circumstances, the method does the rest by moving step by step toward a point where the graph of f crosses the x-axis (Figure 3.45). At each step the method approximates a zero of f with a zero of one of its linearizations. Here is how it works.

The initial estimate, x_0, may be found by graphing or just plain guessing. The method then uses the tangent to the curve $y = f(x)$ at $(x_0, f(x_0))$ to approximate the curve, calling the point x_1 where the tangent meets the x-axis (Figure 3.45). The number x_1 is usually a better approximation to the solution than is x_0. The point x_2 where the tangent to the curve at $(x_1, f(x_1))$ crosses the x-axis is the next approximation in the sequence. We continue on, using each approximation to generate the next, until we are close enough to the root to stop.

We can derive a formula for generating the successive approximations in the following way. Given the approximation x_n, the point-slope equation for the tangent to the curve at $(x_n, f(x_n))$ is

$$y = f(x_n) + f'(x_n)(x - x_n).$$

We can find where it crosses the x-axis by setting $y = 0$ (Figure 3.46).

$$0 = f(x_n) + f'(x_n)(x - x_n)$$

$$-\frac{f(x_n)}{f'(x_n)} = x - x_n$$

$$x = x_n - \frac{f(x_n)}{f'(x_n)} \qquad \text{If } f'(x_n) \neq 0$$

This value of x is the next approximation x_{n+1}. Here is a summary of Newton's method.

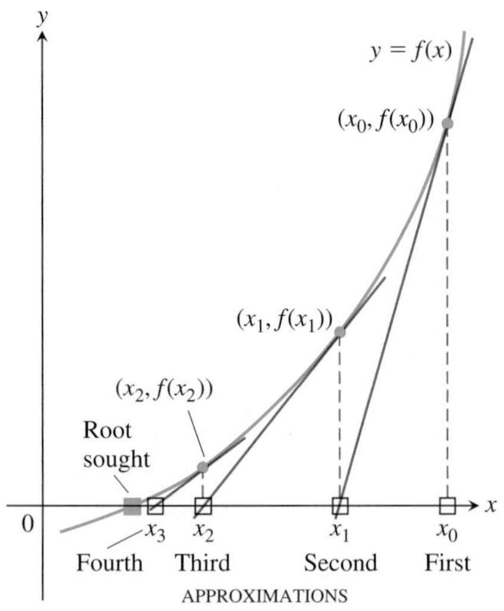

FIGURE 3.45 Newton's method starts with an initial guess x_0 and (under favorable circumstances) improves the guess one step at a time.

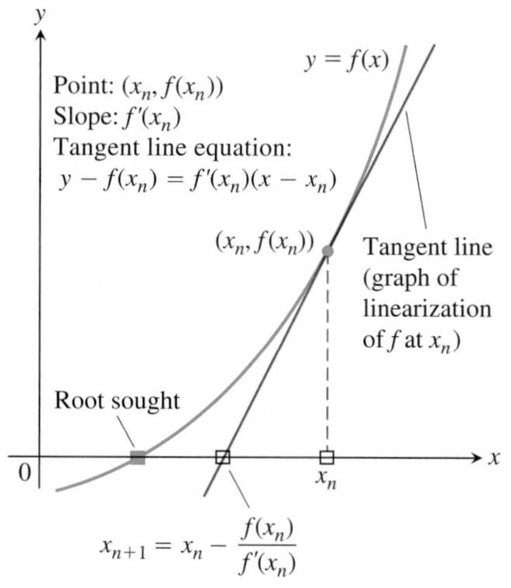

FIGURE 3.46 The geometry of the successive steps of Newton's method. From x_n we go up to the curve and follow the tangent line down to find x_{n+1}.

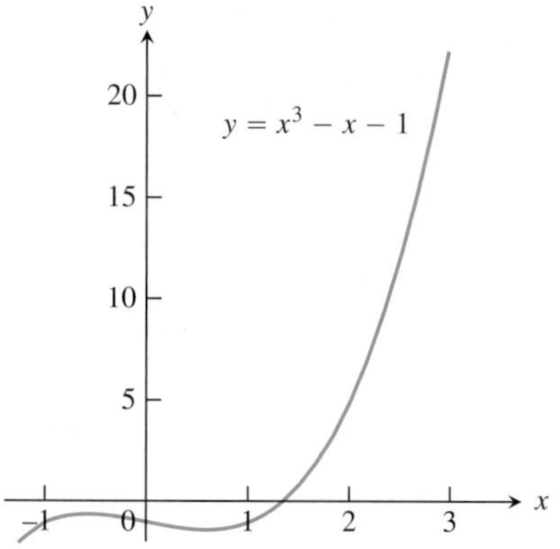

FIGURE 3.47 The graph of $f(x) = x^3 - x - 1$ crosses the x-axis once; this is the root we want to find (Example 1).

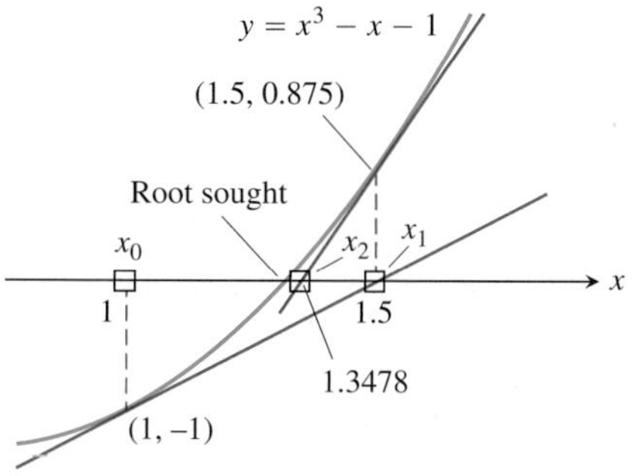

FIGURE 3.48 The first three x-values in Table 3.2 (four decimal places).

Newton's Method

1. Guess a first approximation to a solution of the equation $f(x) = 0$. A graph of $y = f(x)$ may help.
2. Use the first approximation to get a second, the second to get a third, and so on, using the formula

$$x_{n+1} = x_n - \frac{f(x_n)}{f'(x_n)}, \qquad \text{if } f'(x_n) \neq 0 \tag{1}$$

Applying Newton's Method

Applications of Newton's method generally involve many numerical computations, making them well suited for computers or calculators. Nevertheless, even when the calculations are done by hand (which may be very tedious), they give a powerful way to find solutions of equations.

EXAMPLE 1 Find the x-coordinate of the point where the curve $y = x^3 - x$ crosses the horizontal line $y = 1$.

Solution The curve crosses the line when $x^3 - x = 1$ or $x^3 - x - 1 = 0$. When does $f(x) = x^3 - x - 1$ equal zero? Since $f(1) = -1$ and $f(2) = 5$, we know by the Intermediate Value Theorem (Section 1.7) there is a root in the interval (1, 2) (Figure 3.47).

We apply Newton's method to f with the starting value $x_0 = 1$. The results are displayed in Table 3.2 and Figure 3.48.

At $n = 5$, we come to the result $x_6 = x_5 = 1.3247\ 17957$. When $x_{n+1} = x_n$, Equation (1) shows that $f(x_n) = 0$. We have found a solution of $f(x) = 0$ to nine decimals. ■

TABLE 3.2 The result of applying Newton's method to $f(x) = x^3 - x - 1$ with $x_0 = 1$

n	x_n	$f(x_n)$	$f'(x_n)$	$x_{n+1} = x_n - \frac{f(x_n)}{f'(x_n)}$
0	1	−1	2	1.5
1	1.5	0.875	5.75	1.3478 26087
2	1.3478 26087	0.1006 82173	4.4499 05482	1.3252 00399
3	1.3252 00399	0.0020 58362	4.2684 68292	1.3247 18174
4	1.3247 18174	0.0000 00924	4.2646 34722	1.3247 17957
5	1.3247 17957	−1.8672E-13	4.2646 32999	1.3247 17957

In Figure 3.49 we have indicated that the process in Example 1 might have started at the point $B_0(3, 23)$ on the curve, with $x_0 = 3$. Point B_0 is quite far from the x-axis, but the tangent at B_0 crosses the x-axis at about (2.12, 0), so x_1 is still an improvement over x_0. If

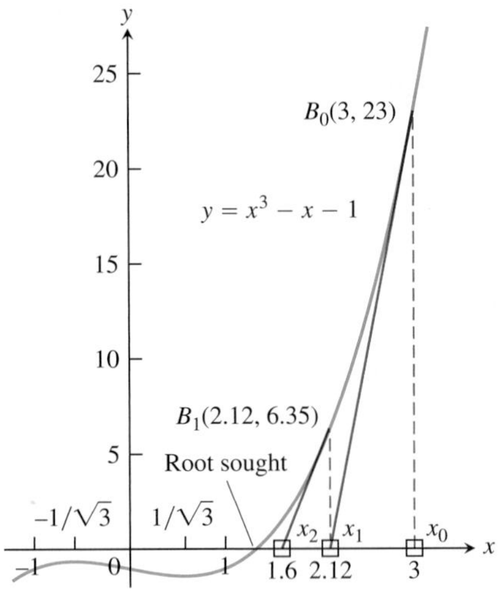

FIGURE 3.49 Any starting value x_0 to the right of $x = 1/\sqrt{3}$ will lead to the root.

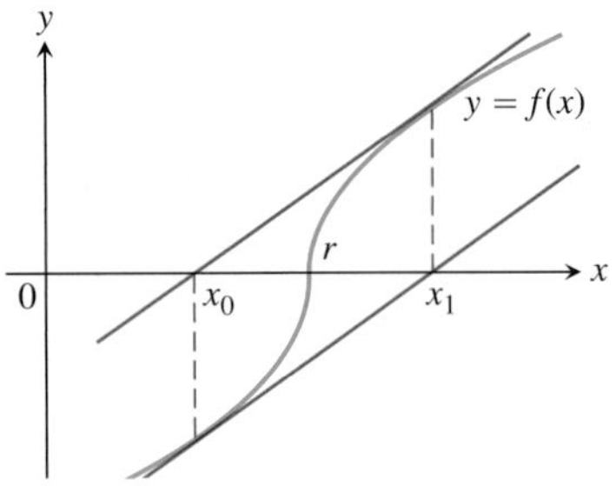

FIGURE 3.50 Newton's method fails to converge. You go from x_0 to x_1 and back to x_0, never getting any closer to r.

we use Equation (1) repeatedly as before, with $f(x) = x^3 - x - 1$ and $f'(x) = 3x^2 - 1$, we obtain the nine-place solution $x_7 = x_6 = 1.3247\ 17957$ in seven steps.

Convergence of the Approximations

In Chapter 7 we define precisely the idea of *convergence* for the approximations x_n in Newton's method. Intuitively, we mean that as the number n of approximations increases without bound, the values x_n get arbitrarily close to the desired root r. (This notion is similar to the idea of the limit of a function $g(t)$ as t approaches infinity, as defined in Section 1.8.)

In practice, Newton's method usually gives convergence with impressive speed, but this is not guaranteed. One way to test convergence is to begin by graphing the function to estimate a good starting value for x_0. You can test that you are getting closer to a zero of the function by evaluating $|f(x_n)|$ and check that the approximations are converging by evaluating $|x_n - x_{n+1}|$.

Newton's method does not always converge. For instance, if

$$f(x) = \begin{cases} -\sqrt{r - x}, & x < r \\ \sqrt{x - r}, & x \geq r, \end{cases}$$

the graph will be like the one in Figure 3.50. If we begin with $x_0 = r - h$, we get $x_1 = r + h$, and successive approximations go back and forth between these two values. No amount of iteration brings us closer to the root than our first guess.

If Newton's method does converge, it converges to a root. Be careful, however. There are situations in which the method appears to converge but there is no root there. Fortunately, such situations are rare.

When Newton's method converges to a root, it may not be the root you have in mind. Figure 3.51 shows two ways this can happen.

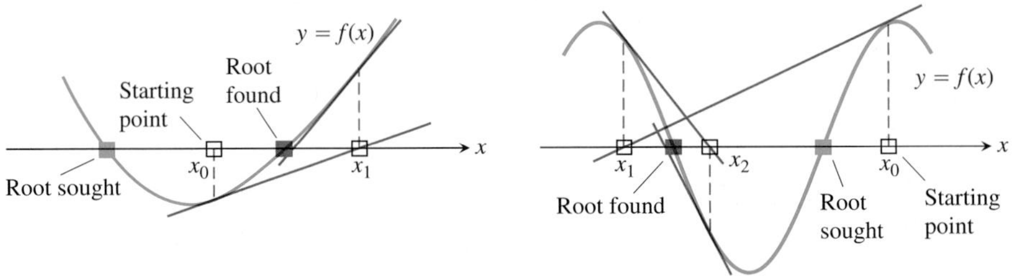

FIGURE 3.51 If you start too far away, Newton's method may miss the root you want.

EXERCISES 3.8

1. Use Newton's method to estimate the solutions of the equation $x^2 + x - 1 = 0$. Start with $x_0 = -1$ for the left-hand solution and with $x_0 = 1$ for the solution on the right. Then, in each case, find x_2.

2. Use Newton's method to estimate the one real solution of $x^3 + 3x + 1 = 0$. Start with $x_0 = 0$ and then find x_2.

3. Use Newton's method to estimate the two zeros of the function $f(x) = x^4 + x - 3$. Start with $x_0 = -1$ for the left-hand zero and with $x_0 = 1$ for the zero on the right. Then, in each case, find x_2.

4. Use Newton's method to estimate the two zeros of the function $f(x) = 2x - x^2 + 1$. Start with $x_0 = 0$ for the left-hand zero and with $x_0 = 2$ for the zero on the right. Then, in each case, find x_2.

In Exercises 5 and 6, use Newton's method to find all roots of the equation correct to six decimal places.

5. $e^{-x} = 2x + 1$

6. $\tan^{-1} x = 1 - 2x$

7. **Guessing a root** Suppose that your first guess is lucky, in the sense that x_0 is a root of $f(x) = 0$. Assuming that $f'(x_0)$ is defined and not 0, what happens to x_1 and later approximations?

8. **Estimating pi** You plan to estimate $\pi/2$ to five decimal places by using Newton's method to solve the equation $\cos x = 0$. Does it matter what your starting value is? Give reasons for your answer.

9. **Oscillation** Show that if $h > 0$, applying Newton's method to

$$f(x) = \begin{cases} \sqrt{x}, & x \geq 0 \\ \sqrt{-x}, & x < 0 \end{cases}$$

leads to $x_1 = -h$ if $x_0 = h$ and to $x_1 = h$ if $x_0 = -h$. Draw a picture that shows what is going on.

10. **Approximations that get worse and worse** Apply Newton's method to $f(x) = x^{1/3}$ with $x_0 = 1$ and calculate x_1, x_2, x_3, and x_4. Find a formula for $|x_n|$. What happens to $|x_n|$ as $n \to \infty$? Draw a picture that shows what is going on.

11. Explain why the following four statements ask for the same information:

    i) Find the roots of $f(x) = x^3 - 3x - 1$.

    ii) Find the x-coordinates of the intersections of the curve $y = x^3$ with the line $y = 3x + 1$.

    iii) Find the x-coordinates of the points where the curve $y = x^3 - 3x$ crosses the horizontal line $y = 1$.

    iv) Find the values of x where the derivative of $g(x) = (1/4)x^4 - (3/2)x^2 - x + 5$ equals zero.

12. **Locating a planet** To calculate a planet's space coordinates, we have to solve equations like $x = 1 + 0.5 \sin x$. Graphing the function $f(x) = x - 1 - 0.5 \sin x$ suggests that the function has a root near $x = 1.5$. Use one application of Newton's method to improve this estimate. That is, start with $x_0 = 1.5$ and find x_1. (The value of the root is 1.49870 to five decimal places.) Remember to use radians.

T 13. **Intersecting curves** The curve $y = \tan x$ crosses the line $y = 2x$ between $x = 0$ and $x = \pi/2$. Use Newton's method to find where.

14. **Intersection of curves**

    a. Does $\cos 3x$ ever equal x? Give reasons for your answer.

    b. Use Newton's method to find where.

T 15. a. How many solutions does the equation $\sin 3x = 0.99 - x^2$ have?

    b. Use Newton's method to find them.

T 16. **Real solutions of a quartic** Use Newton's method to find the two real solutions of the equation $x^4 - 2x^3 - x^2 - 2x + 2 = 0$.

17. **Intersection of curves** At what value(s) of x does $e^{-x^2} = x^2 - x + 1$?

18. **Intersection of curves** At what value(s) of x does $\ln(1 - x^2) = x - 1$?

T 19. **Converging to different zeros** Use Newton's method to find the zeros of $f(x) = 4x^4 - 4x^2$ using the given starting values.

    a. $x_0 = -2$ and $x_0 = -0.8$, lying in $\left(-\infty, -\sqrt{2}/2\right)$

    b. $x_0 = -0.5$ and $x_0 = 0.25$, lying in $\left(-\sqrt{21}/7, \sqrt{21}/7\right)$

    c. $x_0 = 0.8$ and $x_0 = 2$, lying in $\left(\sqrt{2}/2, \infty\right)$

    d. $x_0 = -\sqrt{21}/7$ and $x_0 = \sqrt{21}/7$

20. **The sonobuoy problem** In submarine location problems, it is often necessary to find a submarine's closest point of approach (CPA) to a sonobuoy (sound detector) in the water. Suppose that the submarine travels on the parabolic path $y = x^2$ and that the buoy is located at the point $(2, -1/2)$.

    a. Show that the value of x that minimizes the distance between the submarine and the buoy is a solution of the equation $x = 1/(x^2 + 1)$.

    b. Solve the equation $x = 1/(x^2 + 1)$ with Newton's method.

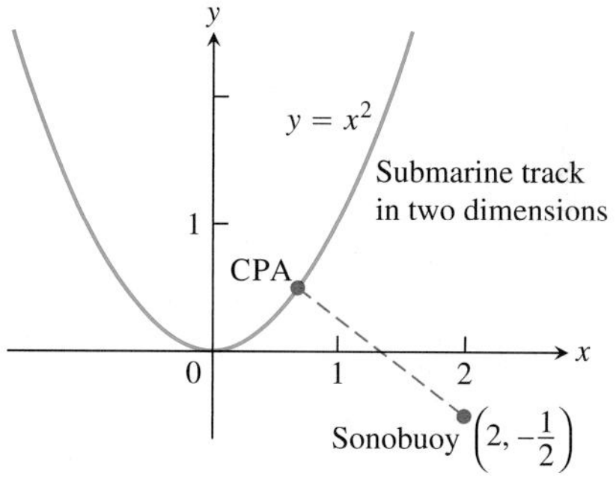

3.9 Hyperbolic Functions

The hyperbolic functions are formed by taking combinations of the two exponential functions e^x and e^{-x}. The hyperbolic functions simplify many mathematical expressions and occur frequently in mathematical applications. In this section we give a brief introduction to hyperbolic functions, their graphs, and their derivatives.

Definitions and Identities

The hyperbolic sine and hyperbolic cosine functions are defined by the equations

$$\sinh x = \frac{e^x - e^{-x}}{2} \quad \text{and} \quad \cosh x = \frac{e^x + e^{-x}}{2}.$$

We often read the notation $\sinh x$ as "cinch x," rhyming with "pinch x," and $\cosh x$ as "kosh x," rhyming with "gosh x." From this basic pair, we can then define the hyperbolic tangent, cotangent, secant, and cosecant functions. The defining equations and graphs of these functions are shown in Table 3.3. We will see that the hyperbolic functions bear many similarities to the trigonometric functions after which they are named.

TABLE 3.3 The six basic hyperbolic functions

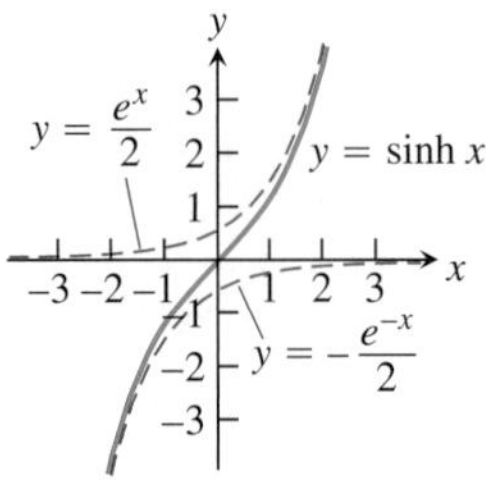

(a)

Hyperbolic sine:

$$\sinh x = \frac{e^x - e^{-x}}{2}$$

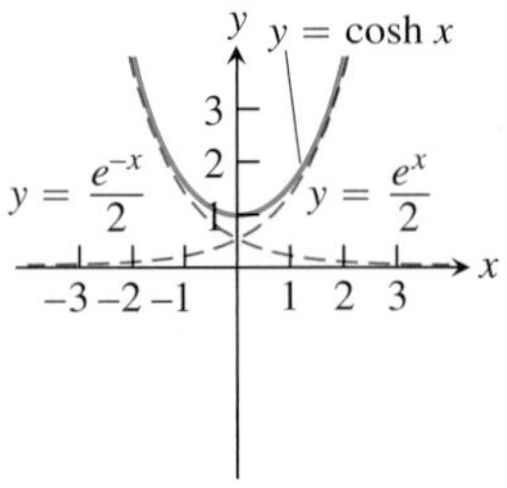

(b)

Hyperbolic cosine:

$$\cosh x = \frac{e^x + e^{-x}}{2}$$

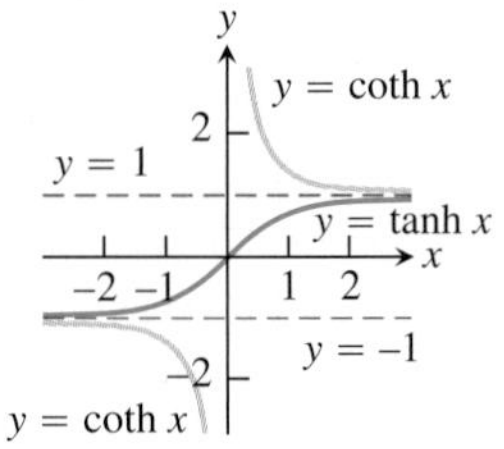

(c)

Hyperbolic tangent:

$$\tanh x = \frac{\sinh x}{\cosh x} = \frac{e^x - e^{-x}}{e^x + e^{-x}}$$

Hyperbolic cotangent:

$$\coth x = \frac{\cosh x}{\sinh x} = \frac{e^x + e^{-x}}{e^x - e^{-x}}$$

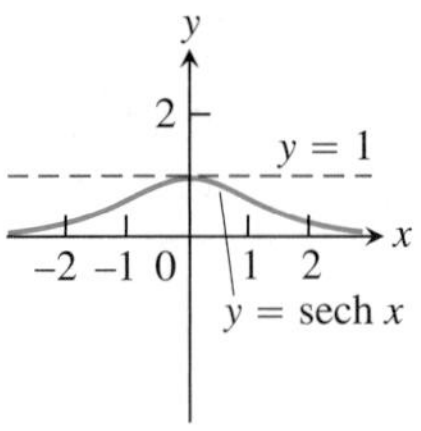

(d)

Hyperbolic secant:

$$\operatorname{sech} x = \frac{1}{\cosh x} = \frac{2}{e^x + e^{-x}}$$

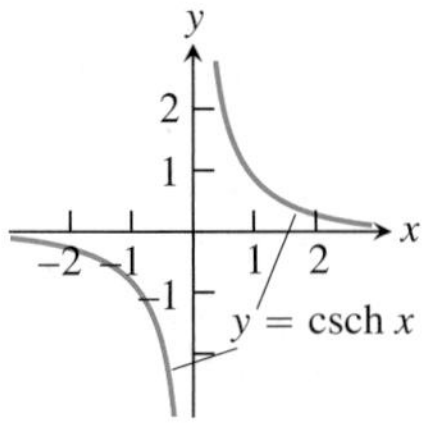

(e)

Hyperbolic cosecant:

$$\operatorname{csch} x = \frac{1}{\sinh x} = \frac{2}{e^x - e^{-x}}$$

TABLE 3.4 Identities for hyperbolic functions

$$\cosh^2 x - \sinh^2 x = 1$$
$$\sinh 2x = 2 \sinh x \cosh x$$
$$\cosh 2x = \cosh^2 x + \sinh^2 x$$
$$\cosh^2 x = \frac{\cosh 2x + 1}{2}$$
$$\sinh^2 x = \frac{\cosh 2x - 1}{2}$$
$$\tanh^2 x = 1 - \operatorname{sech}^2 x$$
$$\coth^2 x = 1 + \operatorname{csch}^2 x$$

Hyperbolic functions satisfy the identities in Table 3.4. Except for differences in sign, these resemble identities we know for the trigonometric functions. The identities are proved directly from the definitions, as we show here for the second one:

$$\begin{aligned} 2 \sinh x \cosh x &= 2\left(\frac{e^x - e^{-x}}{2}\right)\left(\frac{e^x + e^{-x}}{2}\right) \\ &= \frac{e^{2x} - e^{-2x}}{2} \\ &= \sinh 2x. \end{aligned}$$

The other identities are obtained similarly, by substituting in the definitions of the hyperbolic functions and using algebra. Like many standard functions, hyperbolic functions and their inverses are easily evaluated with calculators, which have special keys or keystroke sequences for that purpose.

For any real number u, we know the point with coordinates $(\cos u, \sin u)$ lies on the unit circle $x^2 + y^2 = 1$. So the trigonometric functions are sometimes called the *circular* functions. Because of the first identity

$$\cosh^2 u - \sinh^2 u = 1,$$

with u substituted for x in Table 3.4, the point having coordinates $(\cosh u, \sinh u)$ lies on the right-hand branch of the hyperbola $x^2 - y^2 = 1$. This is where the *hyperbolic* functions get their name.

Derivatives of Hyperbolic Functions

The six hyperbolic functions, being rational combinations of the differentiable functions e^x and e^{-x}, have derivatives at every point at which they are defined (Table 3.5). Again, there are similarities with trigonometric functions.

The derivative formulas are derived from the derivative of e^u:

$$\begin{aligned} \frac{d}{dx}(\sinh u) &= \frac{d}{dx}\left(\frac{e^u - e^{-u}}{2}\right) && \text{Definition of } \sinh u \\ &= \frac{e^u\, du/dx + e^{-u}\, du/dx}{2} && \text{Derivative of } e^u \\ &= \cosh u \frac{du}{dx} && \text{Definition of } \cosh u \end{aligned}$$

TABLE 3.5 Derivatives of hyperbolic functions

$$\frac{d}{dx}(\sinh u) = \cosh u \frac{du}{dx}$$

$$\frac{d}{dx}(\cosh u) = \sinh u \frac{du}{dx}$$

$$\frac{d}{dx}(\tanh u) = \operatorname{sech}^2 u \frac{du}{dx}$$

$$\frac{d}{dx}(\coth u) = -\operatorname{csch}^2 u \frac{du}{dx}$$

$$\frac{d}{dx}(\operatorname{sech} u) = -\operatorname{sech} u \tanh u \frac{du}{dx}$$

$$\frac{d}{dx}(\operatorname{csch} u) = -\operatorname{csch} u \coth u \frac{du}{dx}$$

This gives the first derivative formula. The calculation

$$\begin{aligned} \frac{d}{dx}(\operatorname{csch} u) &= \frac{d}{dx}\left(\frac{1}{\sinh u}\right) && \text{Definition of } \operatorname{csch} u \\ &= -\frac{\cosh u}{\sinh^2 u}\frac{du}{dx} && \text{Quotient Rule} \\ &= -\frac{1}{\sinh u}\frac{\cosh u}{\sinh u}\frac{du}{dx} && \text{Rearrange terms.} \\ &= -\operatorname{csch} u \coth u \frac{du}{dx} && \text{Definitions of } \operatorname{csch} u \text{ and } \coth u \end{aligned}$$

gives the last formula. The others are obtained similarly.

EXAMPLE 1 From Table 3.5, we find the derivative

$$\begin{aligned} \frac{d}{dt}\left(\tanh \sqrt{1 + t^2}\right) &= \operatorname{sech}^2 \sqrt{1 + t^2} \cdot \frac{d}{dt}\left(\sqrt{1 + t^2}\right) \\ &= \frac{t}{\sqrt{1 + t^2}} \operatorname{sech}^2 \sqrt{1 + t^2} \end{aligned}$$

■

Inverse Hyperbolic Functions

The inverses of the six basic hyperbolic functions are very useful in integration as we will see in Chapter 5. Since $d(\sinh x)/dx = \cosh x > 0$, the hyperbolic sine is an increasing function of x. We denote its inverse by

$$y = \sinh^{-1} x.$$

For every value of x in the interval $-\infty < x < \infty$, the value of $y = \sinh^{-1} x$ is the number whose hyperbolic sine is x. The graphs of $y = \sinh x$ and $y = \sinh^{-1} x$ are shown in Figure 3.52a.

The function $y = \cosh x$ is not one-to-one, as we can see from its graph in Table 3.3. The restricted function $y = \cosh x, x \geq 0$, however, is one-to-one and therefore has an inverse, denoted by

$$y = \cosh^{-1} x.$$

For every value of $x \geq 1, y = \cosh^{-1} x$ is the number in the interval $0 \leq y < \infty$ whose hyperbolic cosine is x. The graphs of $y = \cosh x, x \geq 0$, and $y = \cosh^{-1} x$ are shown in Figure 3.52b.

Like $y = \cosh x$, the function $y = \operatorname{sech} x = 1/\cosh x$ fails to be one-to-one, but its restriction to nonnegative values of x does have an inverse, denoted by

$$y = \operatorname{sech}^{-1} x.$$

For every value of x in the interval $(0, 1]$, $y = \operatorname{sech}^{-1} x$ is the nonnegative number whose hyperbolic secant is x. The graphs of $y = \operatorname{sech} x, x \geq 0$, and $y = \operatorname{sech}^{-1} x$ are shown in Figure 3.52c.

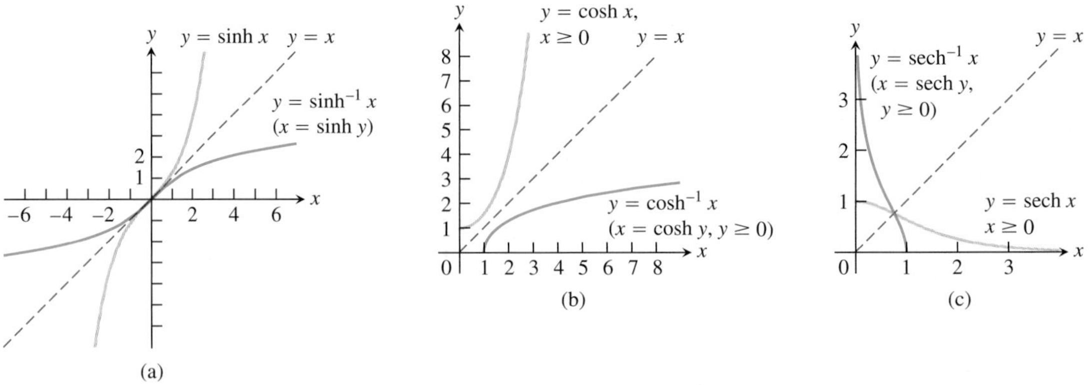

FIGURE 3.52 The graphs of the inverse hyperbolic sine, cosine, and secant of x. Notice the symmetries about the line $y = x$.

The hyperbolic tangent, cotangent, and cosecant are one-to-one on their domains and therefore have inverses, denoted by

$$y = \tanh^{-1} x, \qquad y = \coth^{-1} x, \qquad y = \operatorname{csch}^{-1} x.$$

These functions are graphed in Figure 3.53.

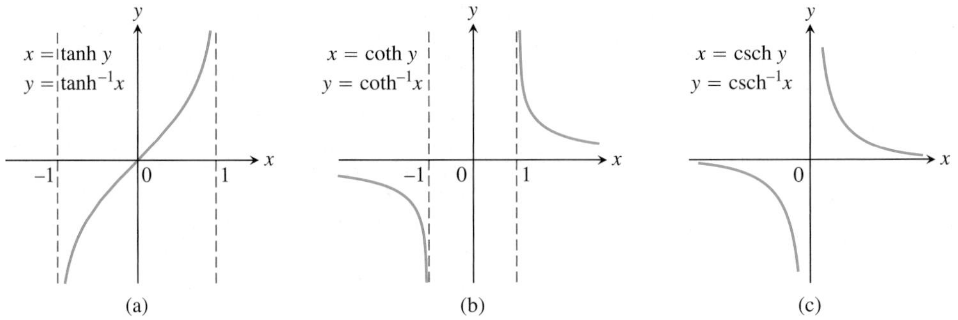

FIGURE 3.53 The graphs of the inverse hyperbolic tangent, cotangent, and cosecant of x.

TABLE 3.6 Identities for inverse hyperbolic functions

$\operatorname{sech}^{-1} x = \cosh^{-1}\frac{1}{x}$
$\operatorname{csch}^{-1} x = \sinh^{-1}\frac{1}{x}$
$\coth^{-1} x = \tanh^{-1}\frac{1}{x}$

Useful Identities

We use the identities in Table 3.6 to calculate the values of $\operatorname{sech}^{-1} x$, $\operatorname{csch}^{-1} x$, and $\coth^{-1} x$ on calculators that give only $\cosh^{-1} x$, $\sinh^{-1} x$, and $\tanh^{-1} x$. These identities are direct consequences of the definitions. For example, if $0 < x \leq 1$, then

$$\operatorname{sech}\left(\cosh^{-1}\left(\frac{1}{x}\right)\right) = \frac{1}{\cosh\left(\cosh^{-1}\left(\frac{1}{x}\right)\right)} = \frac{1}{\left(\frac{1}{x}\right)} = x$$

so

$$\cosh^{-1}\left(\frac{1}{x}\right) = \operatorname{sech}^{-1} x$$

since the hyperbolic secant is one-to-one on $(0, 1]$.

Derivatives of Inverse Hyperbolic Functions

We will see that the chief use of inverse hyperbolic functions lies in integrations that reverse the derivative formulas in Table 3.7.

TABLE 3.7 Derivatives of inverse hyperbolic functions

$\frac{d(\sinh^{-1} u)}{dx} = \frac{1}{\sqrt{1+u^2}}\frac{du}{dx}$	
$\frac{d(\cosh^{-1} u)}{dx} = \frac{1}{\sqrt{u^2-1}}\frac{du}{dx},$	$u > 1$
$\frac{d(\tanh^{-1} u)}{dx} = \frac{1}{1-u^2}\frac{du}{dx},$	$\lvert u\rvert < 1$
$\frac{d(\coth^{-1} u)}{dx} = \frac{1}{1-u^2}\frac{du}{dx},$	$\lvert u\rvert > 1$
$\frac{d(\operatorname{sech}^{-1} u)}{dx} = \frac{-du/dx}{u\sqrt{1-u^2}},$	$0 < u < 1$
$\frac{d(\operatorname{csch}^{-1} u)}{dx} = \frac{-du/dx}{\lvert u\rvert\sqrt{1+u^2}},$	$u \neq 0$

The restrictions $|u| < 1$ and $|u| > 1$ on the derivative formulas for $\tanh^{-1} u$ and $\coth^{-1} u$ come from the natural restrictions on the values of these functions. (See Figure 3.53a and b.) The distinction between $|u| < 1$ and $|u| > 1$ becomes important when we convert the derivative formulas into integral formulas in Chapter 4, Table 4.7.

We illustrate how the derivatives of the inverse hyperbolic functions are found in Example 2, where we calculate $d(\cosh^{-1} u)/dx$. The other derivatives are obtained by similar calculations.

EXAMPLE 2 Show that if u is a differentiable function of x whose values are greater than 1, then

$$\frac{d}{dx}(\cosh^{-1} u) = \frac{1}{\sqrt{u^2-1}}\frac{du}{dx}.$$

Solution First we find the derivative of $y = \cosh^{-1} x$ for $x > 1$ by applying Theorem 4, Section 2.9, with $f(x) = \cosh x$ and $f^{-1}(x) = \cosh^{-1} x$. Theorem 4 can be applied because the derivative of $\cosh x$ is positive for $x > 0$.

$$\begin{aligned}(f^{-1})'(x) &= \frac{1}{f'(f^{-1}(x))} && \text{Theorem 4, Section 2.9}\\ &= \frac{1}{\sinh(\cosh^{-1} x)} && f'(u) = \sinh u\\ &= \frac{1}{\sqrt{\cosh^2(\cosh^{-1} x) - 1}} && \cosh^2 u - \sinh^2 u = 1,\ \sinh u = \sqrt{\cosh^2 u - 1}\\ &= \frac{1}{\sqrt{x^2 - 1}} && \cosh(\cosh^{-1} x) = x\end{aligned}$$

HISTORICAL BIOGRAPHY

Sonya Kovalevsky
(1850–1891)

The Chain Rule gives the final result:

$$\frac{d}{dx}(\cosh^{-1} u) = \frac{1}{\sqrt{u^2 - 1}}\frac{du}{dx}.$$

■

EXERCISES 3.9

Each of Exercises 1–4 gives a value of $\sinh x$ or $\cosh x$. Use the definitions and the identity $\cosh^2 x - \sinh^2 x = 1$ to find the values of the remaining five hyperbolic functions.

1. $\sinh x = -\dfrac{3}{4}$
2. $\sinh x = \dfrac{4}{3}$

3. $\cosh x = \dfrac{17}{15},\quad x > 0$
4. $\cosh x = \dfrac{13}{5},\quad x > 0$

Rewrite the expressions in Exercises 5–10 in terms of exponentials and simplify the results as much as you can.

5. $2\cosh(\ln x)$
6. $\sinh(2\ln x)$
7. $\cosh 5x + \sinh 5x$
8. $\cosh 3x - \sinh 3x$
9. $(\sinh x + \cosh x)^4$
10. $\ln(\cosh x + \sinh x) + \ln(\cosh x - \sinh x)$

11. Use the identities

$$\sinh(x + y) = \sinh x \cosh y + \cosh x \sinh y$$
$$\cosh(x + y) = \cosh x \cosh y + \sinh x \sinh y$$

to show that

a. $\sinh 2x = 2\sinh x \cosh x$

b. $\cosh 2x = \cosh^2 x + \sinh^2 x$.

12. Use the definitions of $\cosh x$ and $\sinh x$ to show that

$$\cosh^2 x - \sinh^2 x = 1.$$

In Exercises 13–24, find the derivative of y with respect to the appropriate variable.

13. $y = 6\sinh\dfrac{x}{3}$
14. $y = \dfrac{1}{2}\sinh(2x + 1)$
15. $y = 2\sqrt{t}\tanh\sqrt{t}$
16. $y = t^2\tanh\dfrac{1}{t}$
17. $y = \ln(\sinh z)$
18. $y = \ln(\cosh z)$
19. $y = \operatorname{sech}\theta(1 - \ln\operatorname{sech}\theta)$
20. $y = \operatorname{csch}\theta(1 - \ln\operatorname{csch}\theta)$
21. $y = \ln\cosh v - \dfrac{1}{2}\tanh^2 v$
22. $y = \ln\sinh v - \dfrac{1}{2}\coth^2 v$

23. $y = (x^2 + 1)\operatorname{sech}(\ln x)$
(*Hint:* Before differentiating, express in terms of exponentials and simplify.)

24. $y = (4x^2 - 1)\operatorname{csch}(\ln 2x)$

In Exercises 25–36, find the derivative of y with respect to the appropriate variable.

25. $y = \sinh^{-1}\sqrt{x}$
26. $y = \cosh^{-1} 2\sqrt{x + 1}$

27. $y = (1 - \theta)\tanh^{-1}\theta$

28. $y = (\theta^2 + 2\theta)\tanh^{-1}(\theta + 1)$

29. $y = (1 - t)\coth^{-1}\sqrt{t}$

30. $y = (1 - t^2)\coth^{-1} t$

31. $y = \cos^{-1} x - x\,\mathrm{sech}^{-1} x$

32. $y = \ln x + \sqrt{1 - x^2}\,\mathrm{sech}^{-1} x$

33. $y = \mathrm{csch}^{-1}\left(\frac{1}{2}\right)^{\theta}$

34. $y = \mathrm{csch}^{-1} 2^{\theta}$

35. $y = \sinh^{-1}(\tan x)$

36. $y = \cosh^{-1}(\sec x), \quad 0 < x < \pi/2$

When hyperbolic function keys are not available on a calculator, it is still possible to evaluate the inverse hyperbolic functions by expressing them as logarithms, as shown here.

$$\sinh^{-1} x = \ln\left(x + \sqrt{x^2 + 1}\right), \qquad -\infty < x < \infty$$
$$\cosh^{-1} x = \ln\left(x + \sqrt{x^2 - 1}\right), \qquad x \geq 1$$
$$\tanh^{-1} x = \frac{1}{2}\ln\frac{1 + x}{1 - x}, \qquad |x| < 1$$
$$\mathrm{sech}^{-1} x = \ln\left(\frac{1 + \sqrt{1 - x^2}}{x}\right), \qquad 0 < x \leq 1$$
$$\mathrm{csch}^{-1} x = \ln\left(\frac{1}{x} + \frac{\sqrt{1 + x^2}}{|x|}\right), \qquad x \neq 0$$
$$\coth^{-1} x = \frac{1}{2}\ln\frac{x + 1}{x - 1}, \qquad |x| > 1$$

Use the formulas in the box here to express the numbers in Exercises 37–42 in terms of natural logarithms.

37. $\sinh^{-1}(-5/12)$

38. $\cosh^{-1}(5/3)$

39. $\tanh^{-1}(-1/2)$

40. $\coth^{-1}(5/4)$

41. $\mathrm{sech}^{-1}(3/5)$

42. $\mathrm{csch}^{-1}(-1/\sqrt{3})$

43. Use the definitions of the hyperbolic functions to find each of the following limits

(a) $\lim_{x\to\infty} \tanh x$ **(b)** $\lim_{x\to-\infty} \tanh x$

(c) $\lim_{x\to\infty} \sinh x$ **(d)** $\lim_{x\to-\infty} \sinh x$

(e) $\lim_{x\to\infty} \mathrm{sech}\, x$ **(f)** $\lim_{x\to\infty} \coth x$

(g) $\lim_{x\to0^+} \coth x$ **(h)** $\lim_{x\to0^-} \coth x$

(i) $\lim_{x\to-\infty} \mathrm{csch}\, x$

44. Derive the formula $\sinh^{-1} x = \ln\left(x + \sqrt{x^2 + 1}\right)$, $-\infty < x < \infty$. Explain in your derivation why the plus sign is used with the square root instead of the minus sign.

45. **Skydiving** If a body of mass m falling from rest under the action of gravity encounters an air resistance proportional to the square of the velocity, then the body's velocity t sec into the fall satisfies the *differential equation*

$$m\frac{dv}{dt} = mg - kv^2,$$

where k is a constant that depends on the body's aerodynamic properties and the density of the air. (We assume that the fall is short enough so that the variation in the air's density will not affect the outcome significantly.)

a. Show that

$$v = \sqrt{\frac{mg}{k}}\tanh\left(\sqrt{\frac{gk}{m}}\,t\right)$$

satisfies the differential equation and the initial condition that $v = 0$ when $t = 0$.

b. Find the body's *limiting velocity*, $\lim_{t\to\infty} v$.

c. For a 160-lb skydiver ($mg = 160$), with time in seconds and distance in feet, a typical value for k is 0.005. What is the diver's limiting velocity?

46. **Accelerations whose magnitudes are proportional to displacement** Suppose that the position of a body moving along a coordinate line at time t is

a. $s = a\cos kt + b\sin kt$

b. $s = a\cosh kt + b\sinh kt$.

Show in both cases that the acceleration d^2s/dt^2 is proportional to s but that in the first case it is directed toward the origin, whereas in the second case it is directed away from the origin.

47. **Hanging cables** Imagine a cable, like a telephone line or TV cable, strung from one support to another and hanging freely. The cable's weight per unit length is a constant w and the horizontal tension at its lowest point is a *vector* of length H. If we choose a coordinate system for the plane of the cable in which the x-axis is horizontal, the force of gravity is straight down, the positive y-axis points straight up, and the lowest point of the cable lies at the point $y = H/w$ on the y-axis (see accompanying figure), then it can be shown that the cable lies along the graph of the hyperbolic cosine

$$y = \frac{H}{w}\cosh\frac{w}{H}x.$$

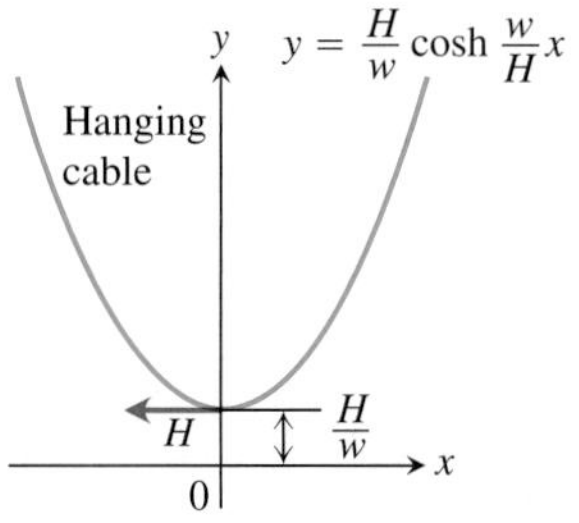

Such a curve is sometimes called a **chain curve** or a **catenary**, the latter deriving from the Latin *catena*, meaning "chain."

a. Let $P(x, y)$ denote an arbitrary point on the cable. The next accompanying figure displays the tension at P as a vector of length (magnitude) T, as well as the tension H at the lowest point A. Show that the cable's slope at P is

$$\tan\phi = \frac{dy}{dx} = \sinh\frac{w}{H}x.$$

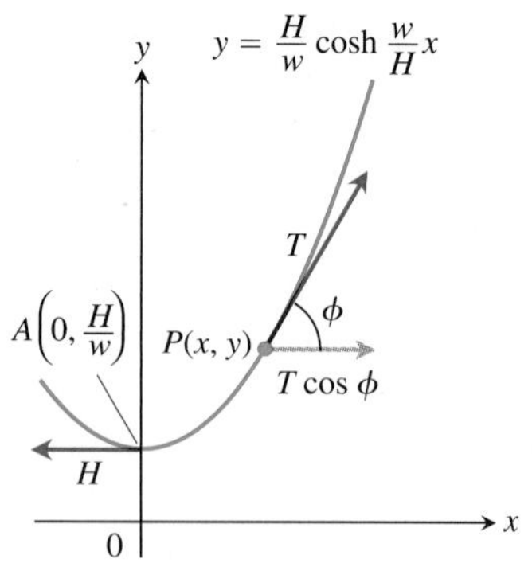

b. Using the result from part (a) and the fact that the horizontal tension at P must equal H (the cable is not moving), show that $T = wy$. Hence, the magnitude of the tension at $P(x, y)$ is exactly equal to the weight of y units of cable.

48. (*Continuation of Exercise 47.*) The length of arc AP in the Exercise 47 figure is $s = (1/a)\sinh ax$, where $a = w/H$. Show that the coordinates of P may be expressed in terms of s as

$$x = \frac{1}{a}\sinh^{-1} as, \qquad y = \sqrt{s^2 + \frac{1}{a^2}}.$$

Chapter 3 Questions to Guide Your Review

1. What can be said about the extreme values of a function that is continuous on a closed interval?
2. What does it mean for a function to have a local extreme value on its domain? An absolute extreme value? How are local and absolute extreme values related, if at all? Give examples.
3. How do you find the absolute extrema of a continuous function on a closed interval? Give examples.
4. What are the hypotheses and conclusion of Rolle's Theorem? Are the hypotheses really necessary? Explain.
5. What are the hypotheses and conclusion of the Mean Value Theorem? What physical interpretations might the theorem have?
6. State the Mean Value Theorem's three corollaries.
7. How can you sometimes identify a function $f(x)$ by knowing f' and knowing the value of f at a point $x = x_0$? Give an example.
8. What is the First Derivative Test for Local Extreme Values? Give examples of how it is applied.
9. How do you test a twice-differentiable function to determine where its graph is concave up or concave down? Give examples.
10. What is an inflection point? Give an example. What physical significance do inflection points sometimes have?
11. What is the Second Derivative Test for Local Extreme Values? Give examples of how it is applied.
12. What do the derivatives of a function tell you about the shape of its graph?
13. List the steps you would take to graph a polynomial function. Illustrate with an example.
14. What is a cusp? Give examples.
15. List the steps you would take to graph a rational function. Illustrate with an example.
16. Outline a general strategy for solving max-min problems. Give examples.
17. What is the formula for the slope dy/dx of a parametrized curve $x = f(t), y = g(t)$? When does the formula apply? When can you expect to be able to find d^2y/dx^2 as well? Give examples.
18. Describe l'Hôpital's Rule. How do you know when to use the rule and when to stop? Give an example.
19. How can you sometimes handle limits that lead to indeterminate forms ∞/∞, $\infty \cdot 0$, and $\infty - \infty$? Give examples.
20. How can you sometimes handle limits that lead to indeterminate forms 1^∞, 0^0, and ∞^∞? Give examples.
21. Describe Newton's method for solving equations. Give an example. What is the theory behind the method? What are some of the things to watch out for when you use the method?

Chapter 3 Practice and Additional Exercises

1. Does $f(x) = x^3 + 2x + \tan x$ have any local maximum or minimum values? Give reasons for your answer.
2. Does $g(x) = \csc x + 2 \cot x$ have any local maximum values? Give reasons for your answer.
3. Does $f(x) = (7 + x)(11 - 3x)^{1/3}$ have an absolute minimum value? An absolute maximum? If so, find them or give reasons why they fail to exist. List all critical points of f.
4. Find values of a and b such that the function
$$f(x) = \frac{ax + b}{x^2 - 1}$$
has a local extreme value of 1 at $x = 3$. Is this extreme value a local maximum, or a local minimum? Give reasons for your answer.
5. Does $g(x) = e^x - x$ have an absolute minimum value? An absolute maximum? If so, find them or give reasons why they fail to exist. List all critical points of g.
6. Does $f(x) = 2e^x/(1 + x^2)$ have an absolute minimum value? An absolute maximum? If so, find them or give reasons why they fail to exist. List all critical points of f.

In Exercises 7 and 8, find the absolute maximum and absolute minimum values of f over the interval.

7. $f(x) = x - 2 \ln x, \quad 1 \le x \le 3$
8. $f(x) = (4/x) + \ln x^2, \quad 1 \le x \le 4$
9. The greatest integer function $f(x) = \lfloor x \rfloor$, defined for all values of x, assumes a local maximum value of 0 at each point of $[0, 1)$. Could any of these local maximum values also be local minimum values of f? Give reasons for your answer.
10. **a.** Give an example of a differentiable function f whose first derivative is zero at some point c even though f has neither a local maximum nor a local minimum at c.

    b. How is this consistent with Theorem 2 in Section 3.1? Give reasons for your answer.
11. The function $y = 1/x$ does not take on either a maximum or a minimum on the interval $0 < x < 1$ even though the function is continuous on this interval. Does this contradict the Extreme Value Theorem for continuous functions? Why?
12. What are the maximum and minimum values of the function $y = |x|$ on the interval $-1 \le x < 1$? Notice that the interval is not closed. Is this consistent with the Extreme Value Theorem for continuous functions? Why?
13. **a.** Show that $g(t) = \sin^2 t - 3t$ decreases on every interval in its domain.

    b. How many solutions does the equation $\sin^2 t - 3t = 5$ have? Give reasons for your answer.
14. **a.** Show that $y = \tan \theta$ increases on every interval in its domain.

    b. If the conclusion in part (a) is really correct, how do you explain the fact that $\tan \pi = 0$ is less than $\tan(\pi/4) = 1$?
15. **a.** Show that the equation $x^4 + 2x^2 - 2 = 0$ has exactly one solution on $[0, 1]$.

    T **b.** Find the solution to as many decimal places as you can.
16. **a.** Show that $f(x) = x/(x + 1)$ increases on every interval in its domain.

    b. Show that $f(x) = x^3 + 2x$ has no local maximum or minimum values.
17. **Water in a reservoir** As a result of a heavy rain, the volume of water in a reservoir increased by 1400 acre-ft in 24 hours. Show that at some instant during that period the reservoir's volume was increasing at a rate in excess of 225,000 gal/min. (An acre-foot is 43,560 ft^3, the volume that would cover 1 acre to the depth of 1 ft. A cubic foot holds 7.48 gal.)
18. The formula $F(x) = 3x + C$ gives a different function for each value of C. All of these functions, however, have the same derivative with respect to x, namely $F'(x) = 3$. Are these the only differentiable functions whose derivative is 3? Could there be any others? Give reasons for your answers.
19. Show that
$$\frac{d}{dx}\left(\frac{x}{x+1}\right) = \frac{d}{dx}\left(-\frac{1}{x+1}\right)$$
even though
$$\frac{x}{x+1} \ne -\frac{1}{x+1}.$$
Doesn't this contradict Corollary 2 of the Mean Value Theorem? Give reasons for your answer.
20. Calculate the first derivatives of $f(x) = x^2/(x^2 + 1)$ and $g(x) = -1/(x^2 + 1)$. What can you conclude about the graphs of these functions?

In Exercises 21 and 22, use the graph to answer the questions.

21. Identify any global extreme values of f and the values of x at which they occur.

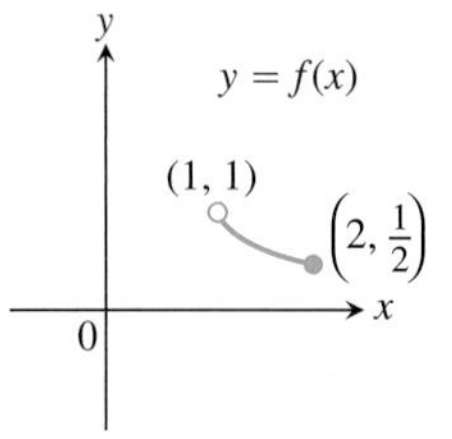

22. Estimate the intervals on which the function $y = f(x)$ is

a. increasing.

b. decreasing.

c. Use the given graph of f' to indicate where any local extreme values of the function occur, and whether each extreme is a relative maximum or minimum.

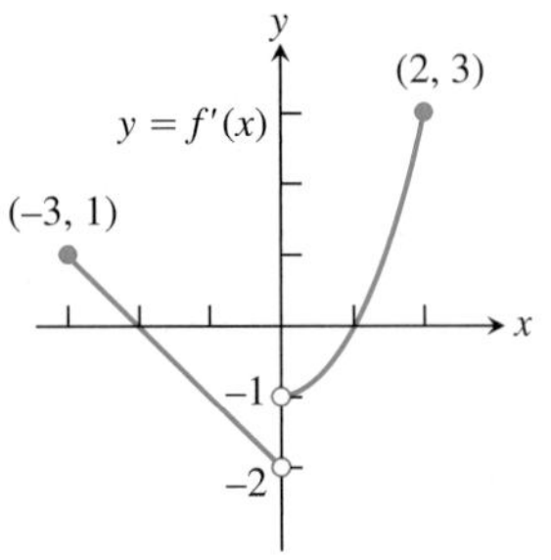

Graph the curves in Exercises 23–38.

23. $y = x^2 - (x^3/6)$

24. $y = x^3 - 3x^2 + 3$

25. $y = -x^3 + 6x^2 - 9x + 3$

26. $y = (1/8)(x^3 + 3x^2 - 9x - 27)$

27. $y = x^3(8 - x)$

28. $y = x^2(2x^2 - 9)$

29. $y = x - 3x^{2/3}$

30. $y = x^{1/3}(x - 4)$

31. $y = x\sqrt{3 - x}$

32. $y = x\sqrt{4 - x^2}$

33. $y = (x - 3)^2 e^x$

34. $y = xe^{-x^2}$

35. $y = \ln(x^2 - 4x + 3)$

36. $y = \ln(\sin x)$

37. $y = \sin^{-1}\left(\frac{1}{x}\right)$

38. $y = \tan^{-1}\left(\frac{1}{x}\right)$

Each of Exercises 39–44 gives the first derivative of a function $y = f(x)$. **(a)** At what points, if any, does the graph of f have a local maximum, local minimum, or inflection point? **(b)** Sketch the general shape of the graph.

39. $y' = 16 - x^2$

40. $y' = x^2 - x - 6$

41. $y' = 6x(x + 1)(x - 2)$

42. $y' = x^2(6 - 4x)$

43. $y' = x^4 - 2x^2$

44. $y' = 4x^2 - x^4$

Sketch the graphs of the functions in Exercises 45–50.

45. $y = \frac{x + 1}{x - 3}$

46. $y = \frac{2x}{x + 5}$

47. $y = \frac{x^2 + 1}{x}$

48. $y = \frac{x^2 - x + 1}{x}$

49. $y = \frac{x^3 + 2}{2x}$

50. $y = \frac{x^2}{x^2 - 4}$

Use l'Hôpital's Rule to find the limits in Exercises 51–60.

51. $\lim_{x\to 1} \frac{x^2 + 3x - 4}{x - 1}$

52. $\lim_{x\to 1} \frac{x^a - 1}{x^b - 1}$

53. $\lim_{x\to \pi} \frac{\tan x}{x}$

54. $\lim_{x\to 0} \frac{\tan x}{x + \sin x}$

55. $\lim_{x\to 0} \frac{\sin^2 x}{\tan(x^2)}$

56. $\lim_{x\to 0} \frac{\sin mx}{\sin nx}$

57. $\lim_{x\to \pi/2^-} \sec 7x \cos 3x$

58. $\lim_{x\to 0^+} \sqrt{x} \sec x$

59. $\lim_{x\to 0} (\csc x - \cot x)$

60. $\lim_{x\to \infty} \left(\sqrt{x^2 + x + 1} - \sqrt{x^2 - x}\right)$

Find the limits in Exercises 61–68.

61. $\lim_{x\to 0} \frac{10^x - 1}{x}$

62. $\lim_{x\to 0} \frac{2^{\sin x} - 1}{e^x - 1}$

63. $\lim_{x\to 0} \frac{5 - 5\cos x}{e^x - x - 1}$

64. $\lim_{x\to 0} \frac{4 - 4e^x}{xe^x}$

65. $\lim_{t\to 0^+} \frac{t - \ln(1 + 2t)}{t^2}$

66. $\lim_{t\to 0^+} \left(\frac{e^t}{t} - \frac{1}{t}\right)$

67. $\lim_{x\to \infty} \left(1 + \frac{b}{x}\right)^{kx}$

68. $\lim_{x\to \infty} \left(1 + \frac{2}{x} + \frac{7}{x^2}\right)$

69. The sum of two nonnegative numbers is 36. Find the numbers if

a. the difference of their square roots is to be as large as possible.

b. the sum of their square roots is to be as large as possible.

70. The sum of two nonnegative numbers is 20. Find the numbers

a. if the product of one number and the square root of the other is to be as large as possible.

b. if one number plus the square root of the other is to be as large as possible.

71. An isosceles triangle has its vertex at the origin and its base parallel to the x-axis with the vertices above the axis on the curve $y = 27 - x^2$. Find the largest area the triangle can have.

72. Find the height and radius of the largest right circular cylinder that can be put in a sphere of radius $\sqrt{3}$.

T **73. Open-top box** An open-top rectangular box is constructed from a 10-in.-by-16-in. piece of cardboard by cutting squares of equal side length from the corners and folding up the sides. Find analytically the dimensions of the box of largest volume and the maximum volume. Support your answers graphically.

74. The ladder problem What is the approximate length (in feet) of the longest ladder you can carry horizontally around the corner of the corridor shown here? Round your answer down to the nearest foot.

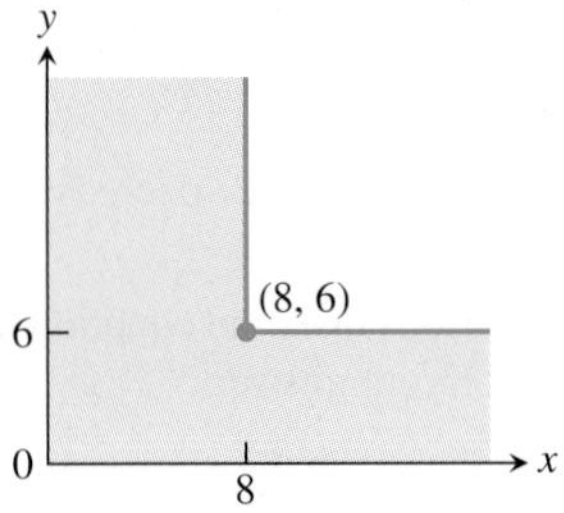

75. Let $f(x) = 3x - x^3$. Show that the equation $f(x) = -4$ has a solution in the interval [2, 3] and use Newton's method to find it.

76. Let $f(x) = x^4 - x^3$. Show that the equation $f(x) = 75$ has a solution in the interval [3, 4] and use Newton's method to find it.

77. The rectangle shown here has one side on the positive y-axis, one side on the positive x-axis, and its upper right-hand vertex on the curve $y = e^{-x^2}$. What dimensions give the rectangle its largest area, and what is that area?

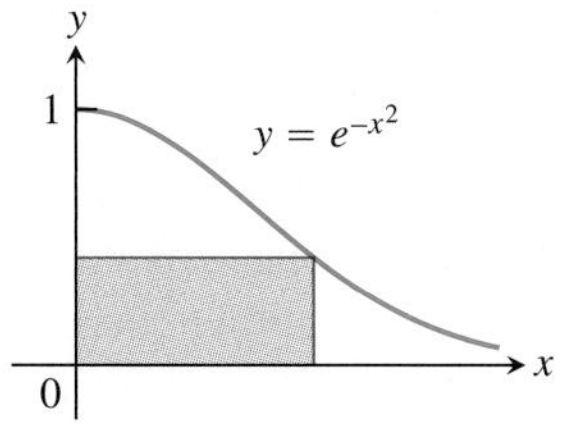

78. The rectangle shown here has one side on the positive y-axis, one side on the positive x-axis, and its upper right-hand vertex on the curve $y = (\ln x)/x^2$. What dimensions give the rectangle its largest area, and what is that area?

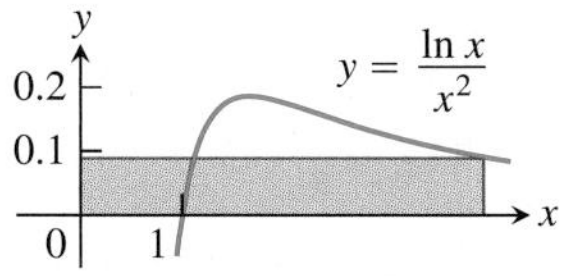

In Exercises 79 and 80, find the absolute maximum and minimum values of each function on the given interval.

79. $y = x \ln 2x - x, \quad \left[\frac{1}{2e}, \frac{e}{2}\right]$

80. $y = 10x(2 - \ln x), \quad (0, e^2]$

In Exercises 81 and 82, find the absolute maxima and minima of the functions and say where they occur.

81. $f(x) = e^{x/\sqrt{x^4+1}}$

82. $g(x) = e^{\sqrt{3-2x-x^2}}$

T **83.** Graph the following functions and use what you see to locate and estimate the extreme values, identify the coordinates of the inflection points, and identify the intervals on which the graphs are concave up and concave down. Then confirm your estimates by working with the functions' derivatives.

a. $y = (\ln x)/\sqrt{x}$ **b.** $y = e^{-x^2}$

c. $y = (1 + x)e^{-x}$

T **84.** Graph $f(x) = x \ln x$. Does the function appear to have an absolute minimum value? Confirm your answer with calculus.

In Exercises 85 and 86, find an equation for the line in the xy-plane that is tangent to the curve at the point corresponding to the given value of t. Also, find the value of d^2y/dx^2 at this point.

85. $x = (1/2)\tan t, \quad y = (1/2)\sec t, \quad t = \pi/3$

86. $x = 1 + 1/t^2, \quad y = 1 - 3/t, \quad t = 2$

In Exercises 87–96, find dy/dx.

87. $y = \sinh^2 3x$

88. $\tan x = \tanh^2 y$

89. $\sin^{-1} x = \operatorname{sech} y$

90. $\sinh y = \sec x$

91. $\tan^{-1} y = \tanh^{-1} x$

92. $y = \tanh(\ln x)$

93. $x = \cosh(\ln y)$

94. $y = \sinh(\tan^{-1} e^{3x})$

95. $y = \sinh^{-1}(\tan x)$

96. $y^2 + x\cosh y + \sinh^2 x = 50$

Chapter 4 INTEGRATION

OVERVIEW One of the great achievements of classical geometry was to obtain formulas for the areas and volumes of triangles, spheres, and cones. In this chapter we study a method to calculate the areas and volumes of these and other more general shapes. The method we develop, called *integration*, is a tool for calculating much more than areas and volumes. The *integral* is of fundamental importance in statistics, the sciences, and engineering. It allows us to calculate quantities ranging from probabilities and averages to energy consumption and the forces against a dam's floodgates.

4.1 Antiderivatives

We have studied how to find the derivative of a function. However, many problems require that we recover a function from its known derivative (from its known rate of change). For instance, we may know the velocity function of an object falling from an initial height and need to know its height at any time. More generally, we want to find a function F from its derivative f. If such a function F exists, it is called an *antiderivative* of f.

Finding Antiderivatives

DEFINITION A function F is an **antiderivative** of f on an interval I if $F'(x) = f(x)$ for all x in I.

The process of recovering a function $F(x)$ from its derivative $f(x)$ is called *antidifferentiation*. We use capital letters such as F to represent an antiderivative of a function f, G to represent an antiderivative of g, and so forth. Antiderivatives play a key role in computing limits of sums, an unexpected and wonderfully useful role that is described in a central result of this chapter, called the Fundamental Theorem of Calculus.

EXAMPLE 1 Find an antiderivative for each of the following functions.

(a) $f(x) = 2x$ **(b)** $g(x) = \cos x$ **(c)** $h(x) = \dfrac{1}{x} + 2e^{2x}$

Solution

(a) $F(x) = x^2$ **(b)** $G(x) = \sin x$ **(c)** $H(x) = \ln|x| + e^{2x}$

Each answer can be checked by differentiating. The derivative of $F(x) = x^2$ is $2x$. The derivative of $G(x) = \sin x$ is $\cos x$ and the derivative of $H(x) = \ln|x| + e^{2x}$ is $(1/x) + 2e^{2x}$. ■

The function $F(x) = x^2$ is not the only function whose derivative is $2x$. The function $x^2 + 1$ has the same derivative. So does $x^2 + C$ for any constant C. Are there others?

Corollary 2 of the Mean Value Theorem in Section 3.2 gives the answer: Any two antiderivatives of a function differ by a constant. So the functions $x^2 + C$, where C is an **arbitrary constant**, form *all* the antiderivatives of $f(x) = 2x$. More generally, we have the following result.

THEOREM 1 If F is an antiderivative of f on an interval I, then the most general antiderivative of f on I is

$$F(x) + C$$

where C is an arbitrary constant.

Thus the most general antiderivative of f on I is a *family* of functions $F(x) + C$ whose graphs are vertical translates of one another. We can select a particular antiderivative from this family by assigning a specific value to C. Here is an example showing how such an assignment might be made.

EXAMPLE 2 Find an antiderivative of $f(x) = 3x^2$ that satisfies $F(1) = -1$.

Solution Since the derivative of x^3 is $3x^2$, the general antiderivative

$$F(x) = x^3 + C$$

gives all the antiderivatives of $f(x)$. The condition $F(1) = -1$ determines a specific value for C. Substituting $x = 1$ into $F(x) = x^3 + C$ gives

$$F(1) = (1)^3 + C = 1 + C.$$

Since $F(1) = -1$, solving for C gives $C = -2$. So

$$F(x) = x^3 - 2$$

is the antiderivative satisfying $F(1) = -1$. Notice that this assignment for C selects the particular curve from the family of curves $y = x^3 + C$ that passes through the point $(1, -1)$ in the plane (Figure 4.1). ■

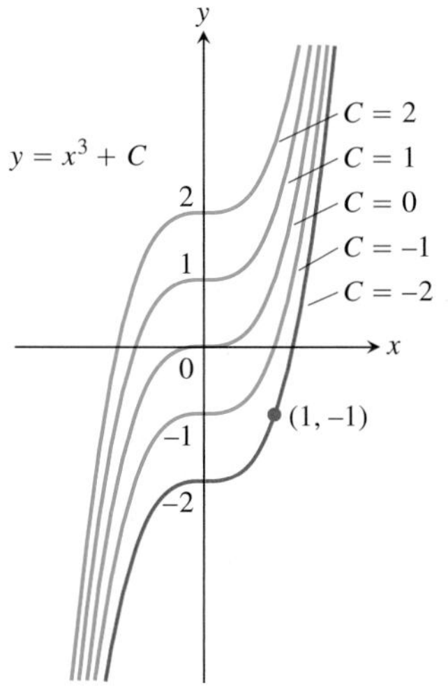

FIGURE 4.1 The curves $y = x^3 + C$ fill the coordinate plane without overlapping. In Example 2, we identify the curve $y = x^3 - 2$ as the one that passes through the given point $(1, -1)$.

By working backward from assorted differentiation rules, we can derive formulas and rules for antiderivatives. In each case there is an arbitrary constant C in the general expression representing all antiderivatives of a given function. Table 4.1 gives antiderivative formulas for a number of important functions.

The rules in Table 4.1 are easily verified by differentiating the general antiderivative formula to obtain the function to its left. For example, the derivative of $(\tan kx)/k + C$ is $\sec^2 kx$, whatever the value of the constants C or $k \neq 0$, and this establishes the formula for the most general antiderivative of $\sec^2 kx$.

TABLE 4.1 Antiderivative formulas, k a nonzero constant

	Function	General antiderivative		Function	General antiderivative
1.	x^n	$\frac{1}{n+1}x^{n+1} + C, \quad n \neq -1$	**8.**	e^{kx}	$\frac{1}{k}e^{kx} + C$
2.	$\sin kx$	$-\frac{1}{k}\cos kx + C$	**9.**	$\frac{1}{x}$	$\ln\lvert x\rvert + C, \quad x \neq 0$
3.	$\cos kx$	$\frac{1}{k}\sin kx + C$	**10.**	$\frac{1}{\sqrt{1-k^2x^2}}$	$\frac{1}{k}\sin^{-1} kx + C$
4.	$\sec^2 kx$	$\frac{1}{k}\tan kx + C$	**11.**	$\frac{1}{1+k^2x^2}$	$\frac{1}{k}\tan^{-1} kx + C$
5.	$\csc^2 kx$	$-\frac{1}{k}\cot kx + C$	**12.**	$\frac{1}{x\sqrt{k^2x^2-1}}$	$\sec^{-1} kx + C, \quad kx > 1$
6.	$\sec kx \tan kx$	$\frac{1}{k}\sec kx + C$	**13.**	a^{kx}	$\left(\frac{1}{k\ln a}\right)a^{kx} + C, \quad a > 0,\ a \neq 1$
7.	$\csc kx \cot kx$	$-\frac{1}{k}\csc kx + C$			

EXAMPLE 3 Find the general antiderivative of each of the following functions.

(a) $f(x) = x^5$ **(b)** $g(x) = \frac{1}{\sqrt{x}}$ **(c)** $h(x) = \sin 2x$

(d) $i(x) = \cos \frac{x}{2}$ **(e)** $j(x) = e^{-3x}$ **(f)** $k(x) = 2^x$

Solution

(a) $F(x) = \frac{x^6}{6} + C$ Formula 1 with $n = 5$

(b) $g(x) = x^{-1/2}$, so

$$G(x) = \frac{x^{1/2}}{1/2} + C = 2\sqrt{x} + C$$ Formula 1 with $n = -1/2$

(c) $H(x) = \frac{-\cos 2x}{2} + C$ Formula 2 with $k = 2$

(d) $I(x) = \frac{\sin(x/2)}{1/2} + C = 2\sin\frac{x}{2} + C$ Formula 3 with $k = 1/2$

(e) $J(x) = -\frac{1}{3}e^{-3x} + C$ Formula 8 with $k = -3$

(f) $K(x) = \left(\frac{1}{\ln 2}\right)2^x + C$ Formula 13 with $a = 2$, $k = 1$ ■

Other derivative rules also lead to corresponding antiderivative rules. We can add and subtract antiderivatives, and multiply them by constants.

The formulas in Table 4.2 are easily proved by differentiating the antiderivatives and verifying that the result agrees with the original function. Formula 2 is the special case $k = -1$ in Formula 1.

TABLE 4.2 Antiderivative linearity rules

		Function	General antiderivative
1.	*Constant Multiple Rule:*	$kf(x)$	$kF(x) + C$, k a constant
2.	*Negative Rule:*	$-f(x)$	$-F(x) + C$
3.	*Sum or Difference Rule:*	$f(x) \pm g(x)$	$F(x) \pm G(x) + C$

EXAMPLE 4 Find the general antiderivative of

$$f(x) = \frac{3}{\sqrt{x}} + \sin 2x.$$

Solution We have that $f(x) = 3g(x) + h(x)$ for the functions g and h in Example 3. Since $G(x) = 2\sqrt{x}$ is an antiderivative of $g(x)$ from Example 3b, it follows from the Constant Multiple Rule for antiderivatives that $3G(x) = 3 \cdot 2\sqrt{x} = 6\sqrt{x}$ is an antiderivative of $3g(x) = 3/\sqrt{x}$. Likewise, from Example 3c we know that $H(x) = (-1/2)\cos 2x$ is an antiderivative of $h(x) = \sin 2x$. From the Sum Rule for antiderivatives, we then get that

$$\begin{aligned} F(x) &= 3G(x) + H(x) + C \\ &= 6\sqrt{x} - \frac{1}{2}\cos 2x + C \end{aligned}$$

is the general antiderivative formula for $f(x)$, where C is an arbitrary constant. ■

Antiderivatives play several important roles, and methods and techniques for finding them are a major part of calculus. (This is the subject of Chapter 5.)

Initial Value Problems and Differential Equations

Finding an antiderivative for a function $f(x)$ is the same problem as finding a function $y(x)$ that satisfies the equation

$$\frac{dy}{dx} = f(x).$$

This is called a **differential equation**, since it is an equation involving an unknown function y that is being differentiated. To solve it, we need a function $y(x)$ that satisfies the equation. This function is found by taking the antiderivative of $f(x)$. We fix the arbitrary constant arising in the antidifferentiation process by specifying an initial condition

$$y(x_0) = y_0.$$

This condition means the function $y(x)$ has the value y_0 when $x = x_0$. The combination of a differential equation and an initial condition is called an **initial value problem**. Such problems play important roles in all branches of science.

The most general antiderivative $F(x) + C$ (which is $x^3 + C$ in Example 2) of the function $f(x)$ gives the **general solution** $y = F(x) + C$ of the differential equation $dy/dx = f(x)$. The general solution gives *all* the solutions of the equation (there are infinitely many, one for each value of C). We **solve** the differential equation by finding its general solution. We then solve the initial value problem by finding the **particular solution** that satisfies the initial condition $y(x_0) = y_0$. In Example 2, the function $y = x^3 - 2$ is the particular solution of the differential equation $dy/dx = 3x^2$ satisfying the initial condition $y(1) = -1$.

Antiderivatives and Motion

We have seen that the derivative of the position of an object gives its velocity, and the derivative of its velocity gives its acceleration. If we know an object's acceleration, then by finding an antiderivative we can recover the velocity, and from an antiderivative of the velocity we can recover its position function. This procedure was used as an application of Corollary 2 in Section 3.2. Now that we have a terminology and conceptual framework in terms of antiderivatives, we revisit the problem from the point of view of differential equations.

EXAMPLE 5 A balloon ascending at the rate of 12 ft/sec is at a height 80 ft above the ground when a package is dropped. How long does it take the package to reach the ground?

Solution Let $v(t)$ denote the velocity of the package at time t, and let $s(t)$ denote its height above the ground. The acceleration of gravity near the surface of the earth is 32 ft/sec^2. Assuming no other forces act on the dropped package, we have

$$\frac{dv}{dt} = -32.$$

Negative because gravity acts in the direction of decreasing s.

This leads to the following initial value problem.

Differential equation: $\dfrac{dv}{dt} = -32$

Initial condition: $v(0) = 12$ Balloon rising initially

This is our mathematical model for the package's motion. We solve the initial value problem to obtain the velocity of the package.

1. *Solve the differential equation*: The general formula for an antiderivative of -32 is

$$v = -32t + C.$$

Having found the general solution of the differential equation, we use the initial condition to find the particular solution that solves our problem.

2. *Evaluate C*:

$$12 = -32(0) + C \qquad \text{Initial condition } v(0) = 12$$
$$C = 12.$$

The solution of the initial value problem is

$$v = -32t + 12.$$

Since velocity is the derivative of height and the height of the package is 80 ft at the time $t = 0$ when it is dropped, we now have a second initial value problem.

Differential equation: $\dfrac{ds}{dt} = -32t + 12$ Set $v = ds/dt$ in the last equation.

Initial condition: $s(0) = 80$

We solve this initial value problem to find the height as a function of t.

1. *Solve the differential equation*: Finding the general antiderivative of $-32t + 12$ gives

$$s = -16t^2 + 12t + C.$$

2. *Evaluate C*:

$$80 = -16(0)^2 + 12(0) + C \qquad \text{Initial condition } s(0) = 80$$
$$C = 80.$$

The package's height above ground at time t is

$$s = -16t^2 + 12t + 80.$$

Use the solution: To find how long it takes the package to reach the ground, we set s equal to 0 and solve for t:

$$-16t^2 + 12t + 80 = 0$$
$$-4t^2 + 3t + 20 = 0$$
$$t = \frac{-3 \pm \sqrt{329}}{-8} \qquad \text{Quadratic formula}$$
$$t \approx -1.89, \qquad t \approx 2.64.$$

The package hits the ground about 2.64 sec after it is dropped from the balloon. (The negative root has no physical meaning.) ■

Indefinite Integrals

A special symbol is used to denote the collection of all antiderivatives of a function f.

> **DEFINITION** The collection of all antiderivatives of f represents the **indefinite integral** of f with respect to x, and is denoted by
>
> $$\int f(x)\,dx.$$
>
> The symbol $\int$ is an **integral sign**. The function f is the **integrand** of the integral, and x is the **variable of integration**.

Using this notation, we restate the solutions of Example 1, as follows:

$$\int 2x\,dx = x^2 + C,$$

$$\int \cos x\,dx = \sin x + C,$$

$$\int \left(\frac{1}{x} + 2e^{2x}\right) dx = \ln|x| + e^{2x} + C.$$

This notation is related to the main application of antiderivatives, which will be explored in Section 4.5.

EXAMPLE 6 Evaluate

$$\int (x^2 - 2x + 5)\, dx.$$

Solution If we recognize that $(x^3/3) - x^2 + 5x$ is an antiderivative of $x^2 - 2x + 5$, we can evaluate the integral as

$$\int (x^2 - 2x + 5)\, dx = \overbrace{\frac{x^3}{3} - x^2 + 5x}^{\text{antiderivative}} + \underbrace{C.}_{\text{arbitrary constant}}$$

If we do not recognize the antiderivative right away, we can generate it term-by-term with the Sum, Difference, and Constant Multiple Rules:

$$\begin{aligned}\int (x^2 - 2x + 5)\, dx &= \int x^2\, dx - \int 2x\, dx + \int 5\, dx \\ &= \int x^2\, dx - 2\int x\, dx + 5\int 1\, dx \\ &= \left(\frac{x^3}{3} + C_1\right) - 2\left(\frac{x^2}{2} + C_2\right) + 5(x + C_3) \\ &= \frac{x^3}{3} + C_1 - x^2 - 2C_2 + 5x + 5C_3.\end{aligned}$$

This formula is more complicated than it needs to be. If we combine C_1, $-2C_2$, and $5C_3$ into a single arbitrary constant $C = C_1 - 2C_2 + 5C_3$, the formula simplifies to

$$\frac{x^3}{3} - x^2 + 5x + C$$

and *still* gives all the antiderivatives there are. For this reason, we recommend that you go right to the final form even if you elect to integrate term-by-term. Write

$$\begin{aligned}\int (x^2 - 2x + 5)\, dx &= \int x^2\, dx - \int 2x\, dx + \int 5\, dx \\ &= \frac{x^3}{3} - x^2 + 5x + C.\end{aligned}$$

Find the simplest antiderivative you can for each part and add the arbitrary constant of integration at the end. ■

EXERCISES 4.1

In Exercises 1–24, find an antiderivative for each function. Do as many as you can mentally. Check your answers by differentiation.

1. **a.** $2x$ **b.** x^2 **c.** $x^2 - 2x + 1$
2. **a.** $6x$ **b.** x^7 **c.** $x^7 - 6x + 8$
3. **a.** $-3x^{-4}$ **b.** x^{-4} **c.** $x^{-4} + 2x + 3$
4. **a.** $2x^{-3}$ **b.** $\dfrac{x^{-3}}{2} + x^2$ **c.** $-x^{-3} + x - 1$
5. **a.** $\dfrac{1}{x^2}$ **b.** $\dfrac{5}{x^2}$ **c.** $2 - \dfrac{5}{x^2}$
6. **a.** $-\dfrac{2}{x^3}$ **b.** $\dfrac{1}{2x^3}$ **c.** $x^3 - \dfrac{1}{x^3}$

7. a. $\frac{3}{2}\sqrt{x}$ b. $\frac{1}{2\sqrt{x}}$ c. $\sqrt{x}+\frac{1}{\sqrt{x}}$

8. a. $\frac{4}{3}\sqrt[3]{x}$ b. $\frac{1}{3\sqrt[3]{x}}$ c. $\sqrt[3]{x}+\frac{1}{\sqrt[3]{x}}$

9. a. $\frac{2}{3}x^{-1/3}$ b. $\frac{1}{3}x^{-2/3}$ c. $-\frac{1}{3}x^{-4/3}$

10. a. $\frac{1}{2}x^{-1/2}$ b. $-\frac{1}{2}x^{-3/2}$ c. $-\frac{3}{2}x^{-5/2}$

11. a. $\frac{1}{x}$ b. $\frac{7}{x}$ c. $1-\frac{5}{x}$

12. a. $\frac{1}{3x}$ b. $\frac{2}{5x}$ c. $1+\frac{4}{3x}-\frac{1}{x^2}$

13. a. $-\pi\sin\pi x$ b. $3\sin x$ c. $\sin\pi x-3\sin 3x$

14. a. $\pi\cos\pi x$ b. $\frac{\pi}{2}\cos\frac{\pi x}{2}$ c. $\cos\frac{\pi x}{2}+\pi\cos x$

15. a. $\sec^2 x$ b. $\frac{2}{3}\sec^2\frac{x}{3}$ c. $-\sec^2\frac{3x}{2}$

16. a. $\csc^2 x$ b. $-\frac{3}{2}\csc^2\frac{3x}{2}$ c. $1-8\csc^2 2x$

17. a. $\csc x\cot x$ b. $-\csc 5x\cot 5x$ c. $-\pi\csc\frac{\pi x}{2}\cot\frac{\pi x}{2}$

18. a. $\sec x\tan x$ b. $4\sec 3x\tan 3x$ c. $\sec\frac{\pi x}{2}\tan\frac{\pi x}{2}$

19. a. e^{3x} b. e^{-x} c. $e^{x/2}$

20. a. e^{-2x} b. $e^{4x/3}$ c. $e^{-x/5}$

21. a. 3^x b. 2^{-x} c. $\left(\frac{5}{3}\right)^x$

22. a. $x^{\sqrt{3}}$ b. x^{π} c. $x^{\sqrt{2}-1}$

23. a. $\frac{2}{\sqrt{1-x^2}}$ b. $\frac{1}{2(x^2+1)}$ c. $\frac{1}{1+4x^2}$

24. a. $x-\left(\frac{1}{2}\right)^x$ b. x^2+2^x c. π^x-x^{-1}

In Exercises 25–70, find the most general antiderivative or indefinite integral. Check your answers by differentiation.

25. $\int(x+1)\,dx$ 26. $\int(5-6x)\,dx$

27. $\int\left(3t^2+\frac{t}{2}\right)dt$ 28. $\int\left(\frac{t^2}{2}+4t^3\right)dt$

29. $\int(2x^3-5x+7)\,dx$ 30. $\int(1-x^2-3x^5)\,dx$

31. $\int\left(\frac{1}{x^2}-x^2-\frac{1}{3}\right)dx$ 32. $\int\left(\frac{1}{5}-\frac{2}{x^3}+2x\right)dx$

33. $\int x^{-1/3}\,dx$ 34. $\int x^{-5/4}\,dx$

35. $\int\left(\sqrt{x}+\sqrt[3]{x}\right)dx$ 36. $\int\left(\frac{\sqrt{x}}{2}+\frac{2}{\sqrt{x}}\right)dx$

37. $\int\left(8y-\frac{2}{y^{1/4}}\right)dy$ 38. $\int\left(\frac{1}{7}-\frac{1}{y^{5/4}}\right)dy$

39. $\int 2x(1-x^{-3})\,dx$ 40. $\int x^{-3}(x+1)\,dx$

41. $\int\frac{t\sqrt{t}+\sqrt{t}}{t^2}\,dt$ 42. $\int\frac{4+\sqrt{t}}{t^3}\,dt$

43. $\int(-2\cos t)\,dt$ 44. $\int(-5\sin t)\,dt$

45. $\int 7\sin\frac{\theta}{3}\,d\theta$ 46. $\int 3\cos 5\theta\,d\theta$

47. $\int(-3\csc^2 x)\,dx$ 48. $\int\left(-\frac{\sec^2 x}{3}\right)dx$

49. $\int\frac{\csc\theta\cot\theta}{2}\,d\theta$ 50. $\int\frac{2}{5}\sec\theta\tan\theta\,d\theta$

51. $\int(e^{3x}+5e^{-x})\,dx$ 52. $\int(2e^x-3e^{-2x})\,dx$

53. $\int(e^{-x}+4^x)\,dx$ 54. $\int(1.3)^x\,dx$

55. $\int(4\sec x\tan x-2\sec^2 x)\,dx$

56. $\int\frac{1}{2}(\csc^2 x-\csc x\cot x)\,dx$

57. $\int(\sin 2x-\csc^2 x)\,dx$ 58. $\int(2\cos 2x-3\sin 3x)\,dx$

59. $\int\frac{1+\cos 4t}{2}\,dt$ 60. $\int\frac{1-\cos 6t}{2}\,dt$

61. $\int\left(\frac{1}{x}-\frac{5}{x^2+1}\right)dx$ 62. $\int\left(\frac{2}{\sqrt{1-y^2}}-\frac{1}{y^{1/4}}\right)dy$

63. $\int 3x^{\sqrt{3}}\,dx$ 64. $\int x^{\sqrt{2}-1}\,dx$

65. $\int(1+\tan^2\theta)\,d\theta$ 66. $\int(2+\tan^2\theta)\,d\theta$

(*Hint:* $1+\tan^2\theta=\sec^2\theta$)

67. $\int\cot^2 x\,dx$ 68. $\int(1-\cot^2 x)\,dx$

(*Hint:* $1+\cot^2 x=\csc^2 x$)

69. $\int\cos\theta\,(\tan\theta+\sec\theta)\,d\theta$ 70. $\int\frac{\csc\theta}{\csc\theta-\sin\theta}\,d\theta$

Verify the formulas in Exercises 71–82 by differentiation.

71. $\int(7x-2)^3\,dx=\frac{(7x-2)^4}{28}+C$

72. $\int(3x+5)^{-2}\,dx=-\frac{(3x+5)^{-1}}{3}+C$

73. $\int\sec^2(5x-1)\,dx=\frac{1}{5}\tan(5x-1)+C$

74. $\int \csc^2\left(\frac{x-1}{3}\right) dx = -3\cot\left(\frac{x-1}{3}\right) + C$

75. $\int \frac{1}{(x+1)^2}\,dx = -\frac{1}{x+1} + C$

76. $\int \frac{1}{(x+1)^2}\,dx = \frac{x}{x+1} + C$

77. $\int \frac{1}{x+1}\,dx = \ln(x+1) + C, \quad x > -1$

78. $\int xe^x\,dx = xe^x - e^x + C$

79. $\int \frac{dx}{a^2+x^2} = \frac{1}{a}\tan^{-1}\left(\frac{x}{a}\right) + C$

80. $\int \frac{dx}{\sqrt{a^2-x^2}} = \sin^{-1}\left(\frac{x}{a}\right) + C$

81. $\int \frac{\tan^{-1}x}{x^2}\,dx = \ln x - \frac{1}{2}\ln(1+x^2) - \frac{\tan^{-1}x}{x} + C$

82. $\int (\sin^{-1}x)^2\,dx = x(\sin^{-1}x)^2 - 2x + 2\sqrt{1-x^2}\sin^{-1}x + C$

83. Right, or wrong? Say which for each formula and give a brief reason for each answer.

a. $\int x\sin x\,dx = \frac{x^2}{2}\sin x + C$

b. $\int x\sin x\,dx = -x\cos x + C$

c. $\int x\sin x\,dx = -x\cos x + \sin x + C$

84. Right, or wrong? Say which for each formula and give a brief reason for each answer.

a. $\int \sqrt{2x+1}\,dx = \sqrt{x^2+x+C}$

b. $\int \sqrt{2x+1}\,dx = \sqrt{x^2+x} + C$

c. $\int \sqrt{2x+1}\,dx = \frac{1}{3}\left(\sqrt{2x+1}\right)^3 + C$

Solve the initial value problems in Exercises 85–100.

85. $\frac{dy}{dx} = 2x - 7, \quad y(2) = 0$

86. $\frac{dy}{dx} = 9x^2 - 4x + 5, \quad y(-1) = 0$

87. $\frac{dy}{dx} = \frac{1}{x^2} + x, \quad x > 0; \quad y(2) = 1$

88. $\frac{dy}{dx} = \frac{1}{2\sqrt{x}}, \quad y(4) = 0$

89. $\frac{dy}{dx} = 3x^{-2/3}, \quad y(-1) = -5$

90. $\frac{ds}{dt} = 1 + \cos t, \quad s(0) = 4$

91. $\frac{dr}{d\theta} = -\pi\sin\pi\theta, \quad r(0) = 0$

92. $\frac{dr}{d\theta} = \cos\pi\theta, \quad r(0) = 1$

93. $\frac{dv}{dt} = \frac{1}{2}\sec t\tan t, \quad v(0) = 1$

94. $\frac{dv}{dt} = 8t + \csc^2 t, \quad v\left(\frac{\pi}{2}\right) = -7$

95. $\frac{dv}{dt} = \frac{3}{t\sqrt{t^2-1}}, \quad t > 1, v(2) = 0$

96. $\frac{dv}{dt} = \frac{8}{1+t^2} + \sec^2 t, \quad v(0) = 1$

97. $\frac{d^2y}{dx^2} = 2 - 6x; \quad y'(0) = 4, \quad y(0) = 1$

98. $\frac{d^2r}{dt^2} = \frac{2}{t^3}; \quad \left.\frac{dr}{dt}\right|_{t=1} = 1, \quad r(1) = 1$

99. $\frac{d^3y}{dx^3} = 6; \quad y''(0) = -8, \quad y'(0) = 0, \quad y(0) = 5$

100. $y^{(4)} = -\sin t + \cos t;$

$y'''(0) = 7, \quad y''(0) = y'(0) = -1, \quad y(0) = 0$

101. Find the curve $y = f(x)$ in the xy-plane that passes through the point (9, 4) and whose slope at each point is $3\sqrt{x}$.

102. a. Find a curve $y = f(x)$ with the following properties:

i) $\frac{d^2y}{dx^2} = 6x$

ii) Its graph passes through the point (0, 1), and has a horizontal tangent there.

b. How many curves like this are there? How do you know?

103. Finding displacement from an antiderivative of velocity

a. Suppose that the velocity of a body moving along the s-axis is

$$\frac{ds}{dt} = v = 9.8t - 3.$$

i) Find the body's displacement over the time interval from $t = 1$ to $t = 3$ given that $s = 5$ when $t = 0$.

ii) Find the body's displacement from $t = 1$ to $t = 3$ given that $s = -2$ when $t = 0$.

iii) Now find the body's displacement from $t = 1$ to $t = 3$ given that $s = s_0$ when $t = 0$.

b. Suppose that the position s of a body moving along a coordinate line is a differentiable function of time t. Is it true that once you know an antiderivative of the velocity function ds/dt you can find the body's displacement from $t = a$ to $t = b$ even if you do not know the body's exact position at either of those times? Give reasons for your answer.

104. Liftoff from Earth A rocket lifts off the surface of Earth with a constant acceleration of 20 m/sec^2. How fast will the rocket be going 1 min later?

105. Motion along a coordinate line A particle moves on a coordinate line with acceleration $a = d^2s/dt^2 = 15\sqrt{t} - (3/\sqrt{t})$, subject to the conditions that $ds/dt = 4$ and $s = 0$ when $t = 1$. Find

a. the velocity $v = ds/dt$ in terms of t

b. the position s in terms of t.

106. Motion with constant acceleration The standard equation for the position s of a body moving with a constant acceleration a along a coordinate line is

$$s = \frac{a}{2}t^2 + v_0 t + s_0,$$

where v_0 and s_0 are the body's velocity and position at time $t = 0$. Derive this equation by solving the initial value problem

Differential equation: $\dfrac{d^2s}{dt^2} = a$

Initial conditions: $\dfrac{ds}{dt} = v_0$ and $s = s_0$ when $t = 0$.

107. Suppose that

$$f(x) = \frac{d}{dx}\left(1 - \sqrt{x}\right) \quad \text{and} \quad g(x) = \frac{d}{dx}(x + 2).$$

Find:

a. $\int f(x)\,dx$ **b.** $\int g(x)\,dx$

c. $\int [-f(x)]\,dx$ **d.** $\int [-g(x)]\,dx$

e. $\int [f(x) + g(x)]\,dx$ **f.** $\int [f(x) - g(x)]\,dx$

108. Uniqueness of solutions If differentiable functions $y = F(x)$ and $y = G(x)$ both solve the initial value problem

$$\frac{dy}{dx} = f(x), \qquad y(x_0) = y_0,$$

on an interval I, must $F(x) = G(x)$ for every x in I? Give reasons for your answer.

COMPUTER EXPLORATIONS

Use a CAS to solve the initial value problems in Exercises 109–112. Plot the solution curves.

109. $y' = \cos^2 x + \sin x, \quad y(\pi) = 1$

110. $y' = \dfrac{1}{x} + x, \quad y(1) = -1$

111. $y' = \dfrac{1}{\sqrt{4 - x^2}}, \quad y(0) = 2$

112. $y'' = \dfrac{2}{x} + \sqrt{x}, \quad y(1) = 0, \quad y'(1) = 0$

4.2 Area and Estimating with Finite Sums

This section shows how area, average values, and the distance traveled by an object over time can all be approximated by finite sums. In Section 4.4 we will see that finite sums are the basis for defining the *definite integral*.

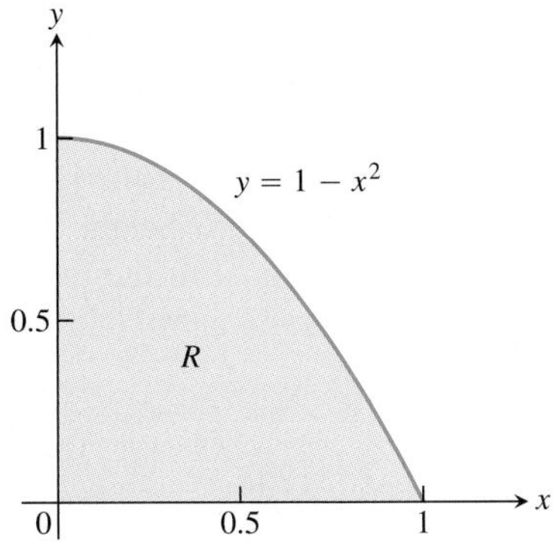

FIGURE 4.2 The area of the region R cannot be found by a simple formula.

Area

The area of a region with a curved boundary can be approximated by summing the areas of a collection of rectangles. Using more rectangles increases the accuracy of the approximation.

Suppose, for instance, we want to find the area of the shaded region R that lies above the x-axis, below the graph of $y = 1 - x^2$, and between the vertical lines $x = 0$ and $x = 1$ (Figure 4.2). Unfortunately, there is no simple geometric formula for calculating the areas of general shapes having curved boundaries like the region R.

While we do not yet have a method for determining the exact area of R, we can approximate it in a simple way. Figure 4.3a shows two rectangles that together contain the region R. Each rectangle has width $1/2$ and they have heights 1 and $3/4$, moving from

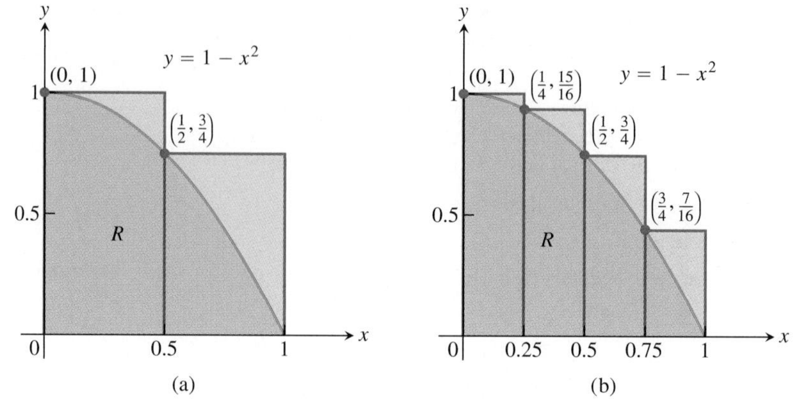

FIGURE 4.3 (a) We get an upper estimate of the area of R by using two rectangles containing R. (b) Four rectangles give a better upper estimate. Both estimates overshoot the true value for the area.

left to right. The height of each rectangle is the maximum value of the function f, obtained by evaluating f at the left endpoint of the subinterval of $[0, 1]$ forming the base of the rectangle. The total area of the two rectangles approximates the area A of the region R,

$$A \approx 1 \cdot \frac{1}{2} + \frac{3}{4} \cdot \frac{1}{2} = \frac{7}{8} = 0.875.$$

This estimate is larger than the true area A, since the two rectangles contain R. We say that 0.875 is an **upper sum** because it is obtained by taking the height of each rectangle as the maximum (uppermost) value of $f(x)$ for x a point in the base interval of the rectangle. In Figure 4.3b, we improve our estimate by using four thinner rectangles, each of width $1/4$, which taken together contain the region R. These four rectangles give the approximation

$$A \approx 1 \cdot \frac{1}{4} + \frac{15}{16} \cdot \frac{1}{4} + \frac{3}{4} \cdot \frac{1}{4} + \frac{7}{16} \cdot \frac{1}{4} = \frac{25}{32} = 0.78125,$$

which is still greater than A since the four rectangles contain R.

Suppose instead we use four rectangles contained *inside* the region R to estimate the area, as in Figure 4.4a. Each rectangle has width $1/4$ as before, but the rectangles are shorter and lie entirely beneath the graph of f. The function $f(x) = 1 - x^2$ is decreasing on $[0, 1]$, so the height of each of these rectangles is given by the value of f at the right endpoint of the subinterval forming its base. The fourth rectangle has zero height and therefore contributes no area. Summing these rectangles with heights equal to the minimum value of $f(x)$ for x a point in each base subinterval, gives a **lower sum** approximation to the area,

$$A \approx \frac{15}{16} \cdot \frac{1}{4} + \frac{3}{4} \cdot \frac{1}{4} + \frac{7}{16} \cdot \frac{1}{4} + 0 \cdot \frac{1}{4} = \frac{17}{32} = 0.53125.$$

This estimate is smaller than the area A since the rectangles all lie inside of the region R. The true value of A lies somewhere between these lower and upper sums:

$$0.53125 < A < 0.78125.$$

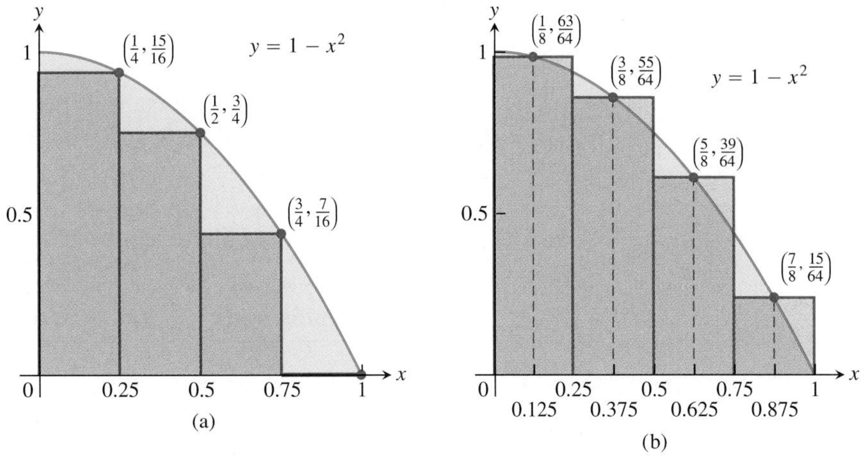

FIGURE 4.4 (a) Rectangles contained in R give an estimate for the area that undershoots the true value. (b) The midpoint rule uses rectangles whose height is the value of $y = f(x)$ at the midpoints of their bases.

By considering both lower and upper sum approximations we get not only estimates for the area, but also a bound on the size of the possible error in these estimates since the true value of the area lies somewhere between them. Here the error cannot be greater than the difference $0.78125 - 0.53125 = 0.25$.

Yet another estimate can be obtained by using rectangles whose heights are the values of f at the midpoints of their bases (Figure 4.4b). This method of estimation is called the **midpoint rule** for approximating the area. The midpoint rule gives an estimate that is between a lower sum and an upper sum, but it is not clear whether it overestimates or underestimates the true area. With four rectangles of width $1/4$ as before, the midpoint rule estimates the area of R to be

$$A \approx \frac{63}{64} \cdot \frac{1}{4} + \frac{55}{64} \cdot \frac{1}{4} + \frac{39}{64} \cdot \frac{1}{4} + \frac{15}{64} \cdot \frac{1}{4} = \frac{172}{64} \cdot \frac{1}{4} = 0.671875.$$

In each of our computed sums, the interval $[a, b]$ over which the function f is defined was subdivided into n subintervals of equal width (also called length) $\Delta x = (b - a)/n$, and f was evaluated at a point in each subinterval: c_1 in the first subinterval, c_2 in the second subinterval, and so on. The finite sums then all take the form

$$f(c_1)\,\Delta x + f(c_2)\,\Delta x + f(c_3)\,\Delta x + \cdots + f(c_n)\,\Delta x.$$

By taking more and more rectangles, with each rectangle thinner than before, it appears that these finite sums give better and better approximations to the true area of the region R.

Figure 4.5a shows a lower sum approximation for the area of R using 16 rectangles of equal width. The sum of their areas is 0.634765625, which appears close to the true area, but is still smaller since the rectangles lie inside R.

Figure 4.5b shows an upper sum approximation using 16 rectangles of equal width. The sum of their areas is 0.697265625, which is somewhat larger than the true area because the rectangles taken together contain R. The midpoint rule for 16 rectangles gives a total area approximation of 0.6669921875, but it is not immediately clear whether this estimate is larger or smaller than the true area.

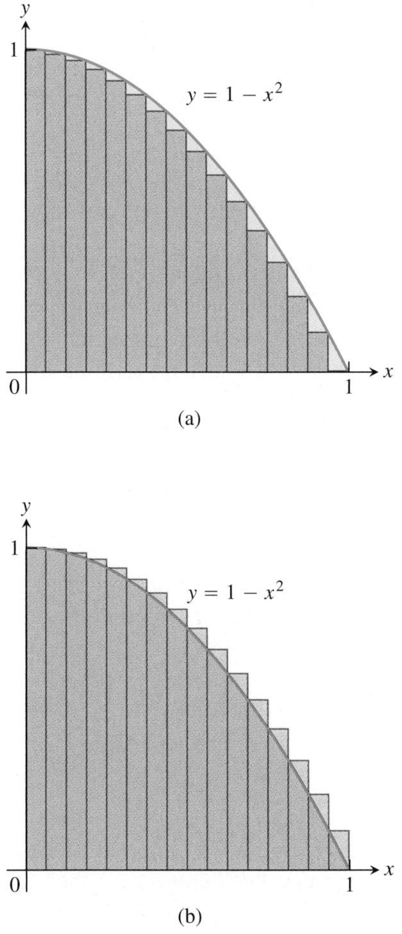

FIGURE 4.5 (a) A lower sum using 16 rectangles of equal width $\Delta x = 1/16$. (b) An upper sum using 16 rectangles.

EXAMPLE 1 Table 4.3 shows the values of upper and lower sum approximations to the area of R using up to 1000 rectangles. In Section 4.3 we will see how to get an exact value of the areas of regions such as R by taking a limit as the base width of each rectangle goes to zero and the number of rectangles goes to infinity. With the techniques developed there, we will be able to show that the area of R is exactly $2/3$. ■

TABLE 4.3 Finite approximations for the area of R

Number of subintervals	Lower sum	Midpoint rule	Upper sum
2	.375	.6875	.875
4	.53125	.671875	.78125
16	.634765625	.6669921875	.697265625
50	.6566	.6667	.6766
100	.66165	.666675	.67165
1000	.6661665	.66666675	.6671665

Distance Traveled

Suppose we know the velocity function $v(t)$ of a car moving down a highway, without changing direction, and want to know how far it traveled between times $t = a$ and $t = b$. If we already know an antiderivative $F(t)$ of $v(t)$ we can find the car's position function $s(t)$ by setting $s(t) = F(t) + C$. The distance traveled can then be found by calculating the change in position, $s(b) - s(a)$ (see Exercise 103, Section 4.1). If the velocity function is determined by recording a speedometer reading at various times on the car, then we have no formula from which to obtain an antiderivative function for velocity. So what do we do in this situation?

When we don't know an antiderivative for the velocity function $v(t)$, we can apply the same principal of approximating the distance traveled with finite sums as before. We subdivide the interval $[a, b]$ into short time intervals on each of which the velocity is considered to be fairly constant. Then we approximate the distance traveled on each time subinterval with the usual distance formula

$$\text{distance} = \text{velocity} \times \text{time}$$

and add the results across $[a, b]$.

Suppose the subdivided interval looks like

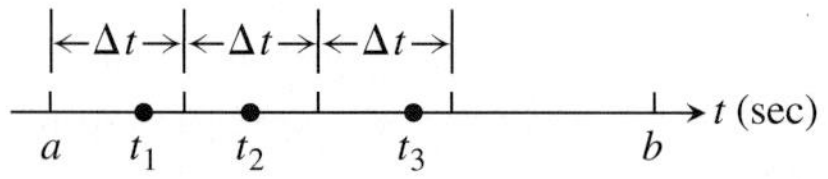

with the subintervals all of equal length Δt. Pick a number t_1 in the first interval. If Δt is so small that the velocity barely changes over a short time interval of duration Δt, then the distance traveled in the first time interval is about $v(t_1)\,\Delta t$. If t_2 is a number in the second interval, the distance traveled in the second time interval is about $v(t_2)\,\Delta t$. The sum of the distances traveled over all the time intervals is

$$D \approx v(t_1)\,\Delta t + v(t_2)\,\Delta t + \cdots + v(t_n)\,\Delta t,$$

where n is the total number of subintervals.

EXAMPLE 2 The velocity function of a projectile fired straight into the air is $f(t) = 160 - 9.8t$ m/sec. Use the summation technique just described to estimate how far the projectile rises during the first 3 sec. How close do the sums come to the exact value of 435.9 m?

Solution We explore the results for different numbers of intervals and different choices of evaluation points. Notice that $f(t)$ is decreasing, so choosing left endpoints gives an upper sum estimate; choosing right endpoints gives a lower sum estimate.

(a) *Three subintervals of length* 1, *with* f *evaluated at left endpoints giving an upper sum*:

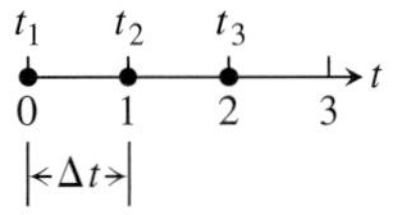

With f evaluated at $t = 0, 1$, and 2, we have

$$\begin{aligned} D &\approx f(t_1)\,\Delta t + f(t_2)\,\Delta t + f(t_3)\,\Delta t \\ &= [160 - 9.8(0)](1) + [160 - 9.8(1)](1) + [160 - 9.8(2)](1) \\ &= 450.6. \end{aligned}$$

(b) *Three subintervals of length* 1, *with* f *evaluated at right endpoints giving a lower sum*:

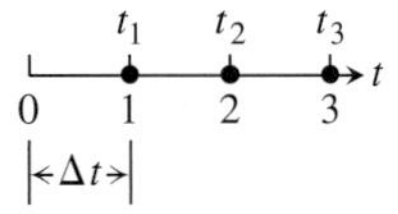

With f evaluated at $t = 1, 2$, and 3, we have

$$\begin{aligned} D &\approx f(t_1)\,\Delta t + f(t_2)\,\Delta t + f(t_3)\,\Delta t \\ &= [160 - 9.8(1)](1) + [160 - 9.8(2)](1) + [160 - 9.8(3)](1) \\ &= 421.2. \end{aligned}$$

(c) *With six subintervals of length* $1/2$, *we get*

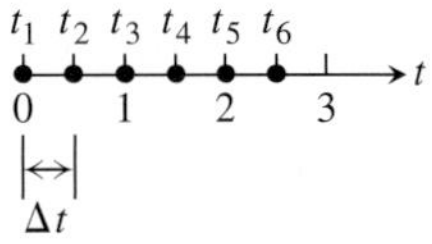

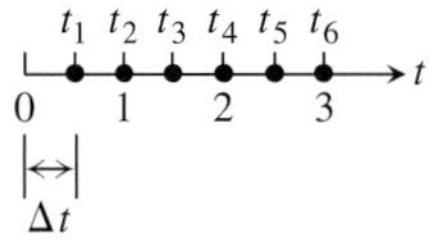

An upper sum using left endpoints: $D \approx 443.25$; a lower sum using right endpoints: $D \approx 428.55$.

These six-interval estimates are somewhat closer than the three-interval estimates. The results improve as the subintervals get shorter.

As we can see in Table 4.4, the left-endpoint upper sums approach the true value 435.9 from above, whereas the right-endpoint lower sums approach it from below. The true value lies between these upper and lower sums. The magnitude of the error in the closest entries is 0.23, a small percentage of the true value.

$$\begin{aligned} \text{Error magnitude} &= |\text{true value} - \text{calculated value}| \\ &= |435.9 - 435.67| = 0.23. \end{aligned}$$

$$\text{Error percentage} = \frac{0.23}{435.9} \approx 0.05\%.$$

TABLE 4.4 Travel-distance estimates

Number of subintervals	Length of each subinterval	Upper sum	Lower sum
3	1	450.6	421.2
6	1/2	443.25	428.55
12	1/4	439.57	432.22
24	1/8	437.74	434.06
48	1/16	436.82	434.98
96	1/32	436.36	435.44
192	1/64	436.13	435.67

It would be reasonable to conclude from the table's last entries that the projectile rose about 436 m during its first 3 sec of flight. ■

Displacement Versus Distance Traveled

If a body with position function $s(t)$ moves along a coordinate line without changing direction, we can calculate the total distance it travels from $t = a$ to $t = b$ by summing the distance traveled over small intervals, as in Example 2. If the body reverses direction one or more times during the trip, then we need to use the body's *speed* $|v(t)|$, which is the absolute value of its velocity function, $v(t)$, to find the total distance traveled. Using the velocity itself, as in Example 2, only gives an estimate to the body's **displacement**, $s(b) - s(a)$, the difference between its initial and final positions.

To see why, partition the time interval $[a, b]$ into small enough equal subintervals Δt so that the body's velocity does not change very much from time t_{k-1} to t_k. Then $v(t_k)$ gives a good approximation of the velocity throughout the interval. Accordingly, the change in the body's position coordinate during the time interval is about

$$v(t_k)\,\Delta t.$$

The change is positive if $v(t_k)$ is positive and negative if $v(t_k)$ is negative.

In either case, the distance traveled during the subinterval is about

$$|v(t_k)|\,\Delta t.$$

The **total distance traveled** is approximately the sum

$$|v(t_1)|\,\Delta t + |v(t_2)|\,\Delta t + \cdots + |v(t_n)|\,\Delta t.$$

Average Value of a Nonnegative Function

The average value of a collection of n numbers $x_1, x_2, \ldots, x_n$ is obtained by adding them together and dividing by n. But what is the average value of a continuous function f on an interval $[a, b]$? Such a function can assume infinitely many values. For example, the temperature at a certain location in a town is a continuous function that goes up and down each day. What does it mean to say that the average temperature in the town over the course of a day is 73 degrees?

When a function is constant, this question is easy to answer. A function with constant value c on an interval $[a, b]$ has average value c. When c is positive, its graph over $[a, b]$ gives a rectangle of height c. The average value of the function can then be interpreted geometrically as the area of this rectangle divided by its width $b - a$ (Figure 4.6a).

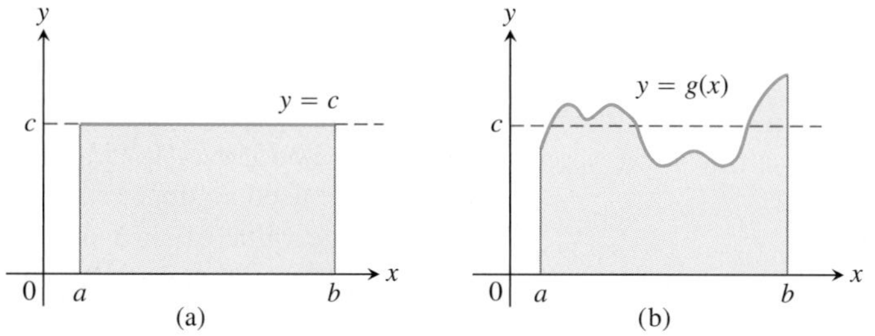

FIGURE 4.6 (a) The average value of $f(x) = c$ on $[a, b]$ is the area of the rectangle divided by $b - a$. (b) The average value of $g(x)$ on $[a, b]$ is the area beneath its graph divided by $b - a$.

What if we want to find the average value of a nonconstant function, such as the function g in Figure 4.6b? We can think of this graph as a snapshot of the height of some water that is sloshing around in a tank, between enclosing walls at $x = a$ and $x = b$. As the water moves, its height over each point changes, but its average height remains the same. To get the average height of the water, we let it settle down until it is level and its height is constant. The resulting height c equals the area under the graph of g divided by $b - a$. We are led to *define* the average value of a nonnegative function on an interval $[a, b]$ to be the area under its graph divided by $b - a$. For this definition to be valid, we need a precise understanding of what is meant by the area under a graph. This will be obtained in Section 4.4, but for now we look at two simple examples.

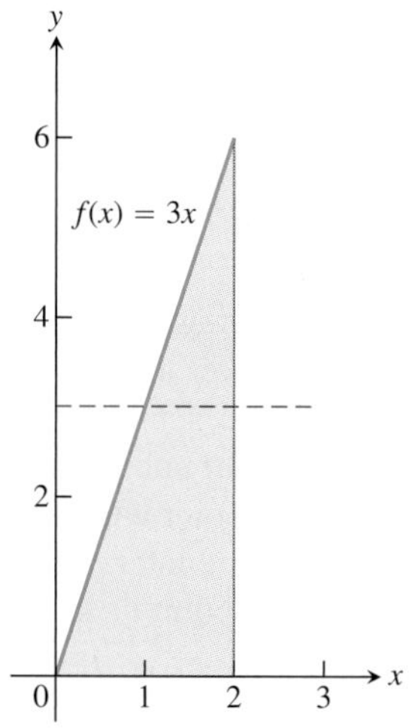

FIGURE 4.7 The average value of $f(x) = 3x$ over $[0, 2]$ is 3 (Example 3).

EXAMPLE 3 What is the average value of the function $f(x) = 3x$ on the interval $[0, 2]$?

Solution The average equals the area under the graph divided by the width of the interval. In this case we do not need finite approximation to estimate the area of the region under the graph: a triangle of height 6 and base 2 has area 6 (Figure 4.7). The width of the interval is $b - a = 2 - 0 = 2$. The average value of the function is $6/2 = 3$. ■

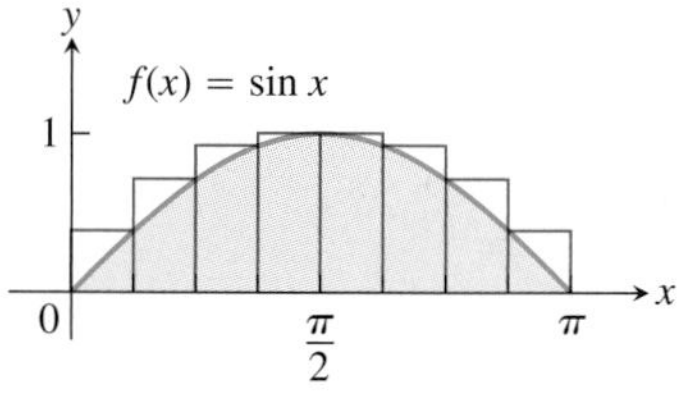

(b)

FIGURE 4.8 Approximating the area under $f(x) = \sin x$ between 0 and π to compute the average value of $\sin x$ over $[0, \pi]$, using eight rectangles (Example 4).

EXAMPLE 4 Estimate the average value of the function $f(x) = \sin x$ on the interval $[0, \pi]$.

Solution Looking at the graph of $\sin x$ between 0 and π in Figure 4.8, we can see that its average height is somewhere between 0 and 1. To find the average we need to calculate the area A under the graph and then divide this area by the length of the interval, $\pi - 0 = \pi$.

We do not have a simple way to determine the area, so we approximate it with finite sums. To get an upper sum approximation, we add the areas of eight rectangles of equal width $\pi/8$ that together contain the region beneath the graph of $y = \sin x$ and above the x-axis on $[0, \pi]$. We choose the heights of the rectangles to be the largest value of $\sin x$ on each subinterval. Over a particular subinterval, this largest value may occur at the left endpoint, the right endpoint, or somewhere between them. We evaluate $\sin x$ at this point to get the height of the rectangle for an upper sum. The sum of the rectangle areas then estimates the total area (Figure 4.8):

$$A \approx \left(\sin\frac{\pi}{8} + \sin\frac{\pi}{4} + \sin\frac{3\pi}{8} + \sin\frac{\pi}{2} + \sin\frac{\pi}{2} + \sin\frac{5\pi}{8} + \sin\frac{3\pi}{4} + \sin\frac{7\pi}{8}\right)\cdot\frac{\pi}{8}$$

$$\approx (.38 + .71 + .92 + 1 + 1 + .92 + .71 + .38)\cdot\frac{\pi}{8} = (6.02)\cdot\frac{\pi}{8} \approx 2.365.$$

To estimate the average value of $\sin x$ we divide the estimated area by π and obtain the approximation $2.365/\pi \approx 0.753$.

Since we used an upper sum to approximate the area, this estimate is greater than the actual average value of $\sin x$ over $[0, \pi]$. If we use more and more rectangles, with each rectangle getting thinner and thinner, we get closer and closer to the true average value. Using the techniques covered in Section 4.4, we will show that the true average value is $2/\pi \approx 0.64$.

As before, we could just as well have used rectangles lying under the graph of $y = \sin x$ and calculated a lower sum approximation, or we could have used the midpoint rule. In Section 4.4 we will see that in each case, the approximations are close to the true area if all the rectangles are sufficiently thin. ■

Summary

The area under the graph of a positive function, the distance traveled by a moving object that doesn't change direction, and the average value of a nonnegative function over an interval can all be approximated by finite sums. First we subdivide the interval into subintervals, treating the appropriate function f as if it were constant over each particular subinterval. Then we multiply the width of each subinterval by the value of f at some point within it, and add these products together. If the interval $[a, b]$ is subdivided into n subintervals of equal widths $\Delta x = (b - a)/n$, and if $f(c_k)$ is the value of f at the chosen point c_k in the kth subinterval, this process gives a finite sum of the form

$$f(c_1)\,\Delta x + f(c_2)\,\Delta x + f(c_3)\,\Delta x + \cdots + f(c_n)\,\Delta x.$$

The choices for the c_k could maximize or minimize the value of f in the kth subinterval, or give some value in between. The true value lies somewhere between the approximations given by upper sums and lower sums. The finite sum approximations we looked at improved as we took more subintervals of thinner width.

EXERCISES 4.2

In Exercises 1–4 use finite approximations to estimate the area under the graph of the function using

a. a lower sum with two rectangles of equal width.

b. a lower sum with four rectangles of equal width.

c. an upper sum with two rectangles of equal width.

d. an upper sum with four rectangles of equal width.

1. $f(x) = x^2$ between $x = 0$ and $x = 1$.

2. $f(x) = x^3$ between $x = 0$ and $x = 1$.

3. $f(x) = 1/x$ between $x = 1$ and $x = 5$.

4. $f(x) = 4 - x^2$ between $x = -2$ and $x = 2$.

Using rectangles whose height is given by the value of the function at the midpoint of the rectangle's base (*the midpoint rule*) estimate the area under the graphs of the following functions, using first two and then four rectangles.

5. $f(x) = x^2$ between $x = 0$ and $x = 1$.
6. $f(x) = x^3$ between $x = 0$ and $x = 1$.
7. $f(x) = 1/x$ between $x = 1$ and $x = 5$.
8. $f(x) = 4 - x^2$ between $x = -2$ and $x = 2$.
9. **Distance traveled** The accompanying table shows the velocity of a model train engine moving along a track for 10 sec. Estimate the distance traveled by the engine using 10 subintervals of length 1 with
   a. left-endpoint values.
   b. right-endpoint values.

Time (sec)	Velocity (in./sec)	Time (sec)	Velocity (in./sec)
0	0	6	11
1	12	7	6
2	22	8	2
3	10	9	6
4	5	10	0
5	13		

10. **Length of a road** You and a companion are about to drive a twisty stretch of dirt road in a car whose speedometer works but whose odometer (mileage counter) is broken. To find out how long this particular stretch of road is, you record the car's velocity at 10-sec intervals, with the results shown in the accompanying table. Estimate the length of the road using
    a. left-endpoint values.
    b. right-endpoint values.

Time (sec)	Velocity (converted to ft/sec) (30 mi/h = 44 ft/sec)	Time (sec)	Velocity (converted to ft/sec) (30 mi/h = 44 ft/sec)
0	0	70	15
10	44	80	22
20	15	90	35
30	35	100	44
40	30	110	30
50	44	120	35
60	35		

11. **Free fall with air resistance** An object is dropped straight down from a helicopter. The object falls faster and faster but its acceleration (rate of change of its velocity) decreases over time because of air resistance. The acceleration is measured in ft/sec^2 and recorded every second after the drop for 5 sec, as shown:

t	0	1	2	3	4	5
a	32.00	19.41	11.77	7.14	4.33	2.63

   a. Find an upper estimate for the speed when $t = 5$.
   b. Find a lower estimate for the speed when $t = 5$.
   c. Find an upper estimate for the distance fallen when $t = 3$.

12. **Distance traveled by a projectile** An object is shot straight upward from sea level with an initial velocity of 400 ft/sec.
    a. Assuming that gravity is the only force acting on the object, give an upper estimate for its velocity after 5 sec have elapsed. Use $g = 32$ ft/sec^2 for the gravitational acceleration.
    b. Find a lower estimate for the height attained after 5 sec.

In Exercises 13 and 14, use a finite sum to estimate the average value of f on the given interval by partitioning the interval into four subintervals of equal length and evaluating f at the subinterval midpoints.

13. $f(x) = x^3$ on $[0, 2]$
14. $f(t) = (1/2) + \sin^2 \pi t$ on $[0, 2]$

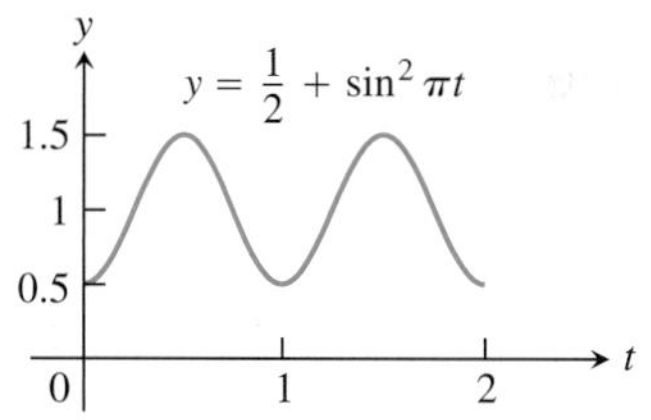

15. **Water pollution** Oil is leaking out of a tanker damaged at sea. The damage to the tanker is worsening as evidenced by the increased leakage each hour, recorded in the following table.

Time (h)	0	1	2	3	4
Leakage (gal/h)	50	70	97	136	190

Time (h)	5	6	7	8
Leakage (gal/h)	265	369	516	720

   a. Give an upper and a lower estimate of the total quantity of oil that has escaped after 5 hours.
   b. Repeat part (a) for the quantity of oil that has escaped after 8 hours.
   c. The tanker continues to leak 720 gal/h after the first 8 hours. If the tanker originally contained 25,000 gal of oil, approximately how many more hours will elapse in the worst case before all the oil has spilled? In the best case?

16. a. Inscribe a regular n-sided polygon inside a circle of radius 1 and compute the area of one of the n congruent triangles formed by drawing radii to the vertices of the polygon.
    b. Compute the limit of the area of the inscribed polygon as $n \to \infty$.
    c. Repeat the computations in parts (a) and (b) for a circle of radius r.

COMPUTER EXPLORATIONS

In Exercises 17–20, use a CAS to perform the following steps.

a. Plot the functions over the given interval.

b. Subdivide the interval into $n = 100, 200$, and 1000 subintervals of equal length and evaluate the function at the midpoint of each subinterval.

c. Compute the average value of the function values generated in part (b).

d. Solve the equation $f(x) =$ (average value) for x using the average value calculated in part (c) for the $n = 1000$ partitioning.

17. $f(x) = \sin x$ on $[0, \pi]$

18. $f(x) = \sin^2 x$ on $[0, \pi]$

19. $f(x) = x \sin \frac{1}{x}$ on $\left[\frac{\pi}{4}, \pi\right]$

20. $f(x) = x \sin^2 \frac{1}{x}$ on $\left[\frac{\pi}{4}, \pi\right]$

4.3 Sigma Notation and Limits of Finite Sums

In estimating with finite sums in Section 4.2, we encountered sums with many terms (up to 1000 in Table 4.3, for instance). In this section we introduce a more convenient notation for sums with a large number of terms. After describing the notation and stating several of its properties, we look at what happens to a finite sum approximation as the number of terms approaches infinity.

Finite Sums and Sigma Notation

Sigma notation enables us to write a sum with many terms in the compact form

$$\sum_{k=1}^{n} a_k = a_1 + a_2 + a_3 + \cdots + a_{n-1} + a_n.$$

The Greek letter Σ (capital sigma, corresponding to our letter S), stands for "sum." The **index of summation** k tells us where the sum begins (at the number below the Σ symbol) and where it ends (at the number above Σ). Any letter can be used to denote the index, but the letters i, j, and k are customary.

$$\sum_{k=1}^{n} a_k$$

The index k ends at $k = n$.

The summation symbol (Greek letter sigma)

a_k is a formula for the kth term.

The index k starts at $k = 1$.

Thus we can write

$$1^2 + 2^2 + 3^2 + 4^2 + 5^2 + 6^2 + 7^2 + 8^2 + 9^2 + 10^2 + 11^2 = \sum_{k=1}^{11} k^2,$$

and

$$f(1) + f(2) + f(3) + \cdots + f(100) = \sum_{i=1}^{100} f(i).$$

EXAMPLE 1

The sum in sigma notation	The sum written out, one term for each value of k	The value of the sum
$\sum_{k=1}^{5} k$	$1 + 2 + 3 + 4 + 5$	15
$\sum_{k=1}^{3} (-1)^k k$	$(-1)^1(1) + (-1)^2(2) + (-1)^3(3)$	$-1 + 2 - 3 = -2$
$\sum_{k=1}^{2} \frac{k}{k+1}$	$\frac{1}{1+1} + \frac{2}{2+1}$	$\frac{1}{2} + \frac{2}{3} = \frac{7}{6}$
$\sum_{k=4}^{5} \frac{k^2}{k-1}$	$\frac{4^2}{4-1} + \frac{5^2}{5-1}$	$\frac{16}{3} + \frac{25}{4} = \frac{139}{12}$

■

The lower limit of summation does not have to be 1; it can be any integer.

EXAMPLE 2 Express the sum $1 + 3 + 5 + 7 + 9$ in sigma notation.

Solution The formula generating the terms changes with the lower limit of summation, but the terms generated remain the same. It is often simplest to start with $k = 0$ or $k = 1$.

Starting with k = 0: $$1 + 3 + 5 + 7 + 9 = \sum_{k=0}^{4} (2k + 1)$$

Starting with k = 1: $$1 + 3 + 5 + 7 + 9 = \sum_{k=1}^{5} (2k - 1)$$

Starting with k = 2: $$1 + 3 + 5 + 7 + 9 = \sum_{k=2}^{6} (2k - 3)$$

Starting with k = −3: $$1 + 3 + 5 + 7 + 9 = \sum_{k=-3}^{1} (2k + 7)$$ ■

When we have a sum such as

$$\sum_{k=1}^{3} (k + k^2)$$

we can rearrange its terms,

$$\begin{aligned}\sum_{k=1}^{3} (k + k^2) &= (1 + 1^2) + (2 + 2^2) + (3 + 3^2) \\ &= (1 + 2 + 3) + (1^2 + 2^2 + 3^2) \qquad \text{Regroup terms.} \\ &= \sum_{k=1}^{3} k + \sum_{k=1}^{3} k^2\end{aligned}$$

This illustrates a general rule for finite sums:

$$\sum_{k=1}^{n} (a_k + b_k) = \sum_{k=1}^{n} a_k + \sum_{k=1}^{n} b_k$$

Four such rules are given below. A proof that they are valid can be obtained using mathematical induction (see Appendix 2).

Algebra Rules for Finite Sums

1. *Sum Rule*: $\sum_{k=1}^{n}(a_k + b_k) = \sum_{k=1}^{n} a_k + \sum_{k=1}^{n} b_k$

2. *Difference Rule*: $\sum_{k=1}^{n}(a_k - b_k) = \sum_{k=1}^{n} a_k - \sum_{k=1}^{n} b_k$

3. *Constant Multiple Rule*: $\sum_{k=1}^{n} ca_k = c \cdot \sum_{k=1}^{n} a_k$ (Any number c)

4. *Constant Value Rule*: $\sum_{k=1}^{n} c = n \cdot c$ (c is any constant value.)

EXAMPLE 3

(a) $\sum_{k=1}^{n}(3k - k^2) = 3\sum_{k=1}^{n} k - \sum_{k=1}^{n} k^2$ — Difference Rule and Constant Multiple Rule

(b) $\sum_{k=1}^{n}(-a_k) = \sum_{k=1}^{n}(-1)\cdot a_k = -1\cdot\sum_{k=1}^{n} a_k = -\sum_{k=1}^{n} a_k$ — Constant Multiple Rule

(c) $\sum_{k=1}^{3}(k + 4) = \sum_{k=1}^{3} k + \sum_{k=1}^{3} 4$ — Sum Rule

$= (1 + 2 + 3) + (3\cdot 4)$ — Constant Value Rule

$= 6 + 12 = 18$

(d) $\sum_{k=1}^{n}\frac{1}{n} = n\cdot\frac{1}{n} = 1$ — Constant Value Rule ($1/n$ is constant) ■

HISTORICAL BIOGRAPHY

Carl Friedrich Gauss
(1777–1855)

Over the years people have discovered a variety of formulas for the values of finite sums. The most famous of these are the formula for the sum of the first n integers (Gauss is said to have discovered it at age 8) and the formulas for the sums of the squares and cubes of the first n integers.

EXAMPLE 4 Show that the sum of the first n integers is

$$\sum_{k=1}^{n} k = \frac{n(n+1)}{2}.$$

Solution: The formula tells us that the sum of the first 4 integers is

$$\frac{(4)(5)}{2} = 10.$$

Addition verifies this prediction:

$$1 + 2 + 3 + 4 = 10.$$

To prove the formula in general, we write out the terms in the sum twice, once forward and once backward.

$$\begin{array}{ccccccccc} 1 & + & 2 & + & 3 & + & \cdots & + & n \\ n & + & (n-1) & + & (n-2) & + & \cdots & + & 1 \end{array}$$

If we add the two terms in the first column we get $1 + n = n + 1$. Similarly, if we add the two terms in the second column we get $2 + (n - 1) = n + 1$. The two terms in any column sum to $n + 1$. When we add the n columns together we get n terms, each equal to $n + 1$, for a total of $n(n + 1)$. Since this is twice the desired quantity, the sum of the first n integers is $(n)(n + 1)/2$. ■

Formulas for the sums of the squares and cubes of the first n integers are proved using mathematical induction (see Appendix 2). We state them here.

The first n squares: $\displaystyle\sum_{k=1}^{n} k^2 = \frac{n(n+1)(2n+1)}{6}$

The first n cubes: $\displaystyle\sum_{k=1}^{n} k^3 = \left(\frac{n(n+1)}{2}\right)^2$

Limits of Finite Sums

The finite sum approximations we considered in Section 4.2 got more accurate as the number of terms increased and the subinterval widths (lengths) became thinner. The next example shows how to calculate a limiting value as the widths of the subintervals go to zero and their number grows to infinity.

EXAMPLE 5 Find the limiting value of lower sum approximations to the area of the region R below the graph of $y = 1 - x^2$ and above the interval [0, 1] on the x-axis using equal width rectangles whose widths approach zero and whose number approaches infinity. (See Figure 4.5a.)

Solution We compute a lower sum approximation using n rectangles of equal width $\Delta x = (1 - 0)/n$, and then we see what happens as $n \to \infty$. We start by subdividing [0, 1] into n equal width subintervals

$$\left[0, \frac{1}{n}\right], \left[\frac{1}{n}, \frac{2}{n}\right], \ldots, \left[\frac{n-1}{n}, \frac{n}{n}\right].$$

Each subinterval has width $1/n$. The function $1 - x^2$ is decreasing on [0, 1], and its smallest value in a subinterval occurs at the subinterval's right endpoint. So a lower sum is constructed with rectangles whose height over the subinterval $[(k - 1)/n, k/n]$ is $f(k/n) = 1 - (k/n)^2$, giving the sum

$$f\left(\frac{1}{n}\right)\left(\frac{1}{n}\right) + f\left(\frac{2}{n}\right)\left(\frac{1}{n}\right) + \cdots + f\left(\frac{k}{n}\right)\left(\frac{1}{n}\right) + \cdots + f\left(\frac{n}{n}\right)\left(\frac{1}{n}\right).$$

We write this in sigma notation and simplify,

$$\begin{aligned}
\sum_{k=1}^{n} f\left(\frac{k}{n}\right)\left(\frac{1}{n}\right) &= \sum_{k=1}^{n}\left(1-\left(\frac{k}{n}\right)^2\right)\left(\frac{1}{n}\right) \\
&= \sum_{k=1}^{n}\left(\frac{1}{n}-\frac{k^2}{n^3}\right) \\
&= \sum_{k=1}^{n}\frac{1}{n} - \sum_{k=1}^{n}\frac{k^2}{n^3} && \text{Difference Rule} \\
&= n\cdot\frac{1}{n} - \frac{1}{n^3}\sum_{k=1}^{n}k^2 && \text{Constant Value and Constant Multiple Rules} \\
&= 1 - \left(\frac{1}{n^3}\right)\frac{(n)(n+1)(2n+1)}{6} && \text{Sum of the First } n \text{ Squares} \\
&= 1 - \frac{2n^3+3n^2+n}{6n^3}. && \text{Numerator expanded}
\end{aligned}$$

We have obtained an expression for the lower sum that holds for any n. Taking the limit of this expression as $n \to \infty$, we see that the lower sums converge as the number of subintervals increases and the subinterval widths approach zero:

$$\lim_{n\to\infty}\left(1-\frac{2n^3+3n^2+n}{6n^3}\right) = 1-\frac{2}{6} = \frac{2}{3}.$$

The lower sum approximations converge to $2/3$. A similar calculation shows that the upper sum approximations also converge to $2/3$ (Exercise 35). Any finite sum approximation, in the sense of our summary at the end of Section 4.2, also converges to the same value, $2/3$. This is because it is possible to show that any finite sum approximation is trapped between the lower and upper sum approximations. For this reason we are led to *define* the area of the region R as this limiting value. In Section 4.4 we study the limits of such finite approximations in a general setting. ■

Riemann Sums

HISTORICAL BIOGRAPHY

Georg Friedrich Bernhard Riemann (1826–1866)

The theory of limits of finite approximations was made precise by the German mathematician Bernhard Riemann. We now introduce the notion of a *Riemann sum*, which underlies the theory of the definite integral studied in the next section.

We begin with an arbitrary function f defined on a closed interval $[a, b]$. Like the function pictured in Figure 4.9, f may have negative as well as positive values. We subdivide the interval $[a, b]$ into subintervals, not necessarily of equal widths (or lengths), and form sums in the same way as for the finite approximations in Section 4.2. To do so, we choose $n - 1$ points $\{x_1, x_2, x_3, \ldots, x_{n-1}\}$ between a and b and satisfying

$$a < x_1 < x_2 < \cdots < x_{n-1} < b.$$

To make the notation consistent, we denote a by x_0 and b by x_n, so that

$$a = x_0 < x_1 < x_2 < \cdots < x_{n-1} < x_n = b.$$

The set

$$P = \{x_0, x_1, x_2, \ldots, x_{n-1}, x_n\}$$

is called a **partition** of $[a, b]$.

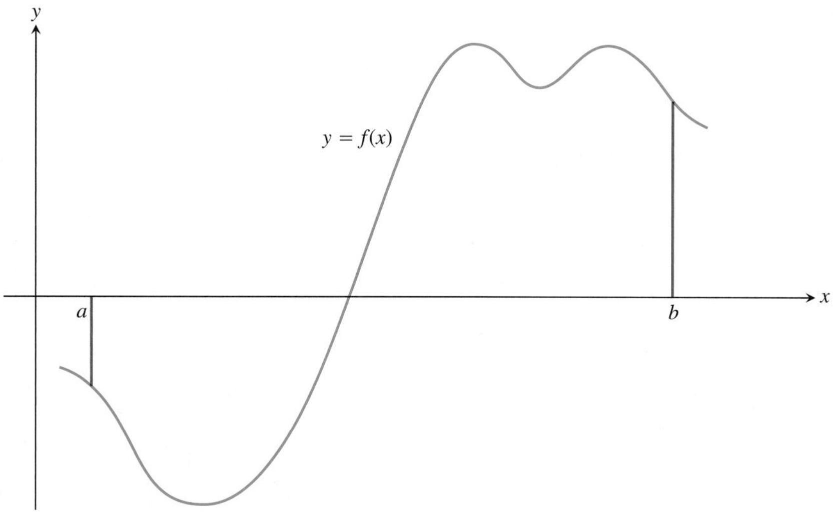

FIGURE 4.9 A typical continuous function $y = f(x)$ over a closed interval $[a, b]$.

The partition P divides $[a, b]$ into n closed subintervals

$$[x_0, x_1], [x_1, x_2], \ldots, [x_{n-1}, x_n].$$

The first of these subintervals is $[x_0, x_1]$, the second is $[x_1, x_2]$, and the ***k*th subinterval of** P is $[x_{k-1}, x_k]$, for k an integer between 1 and n.

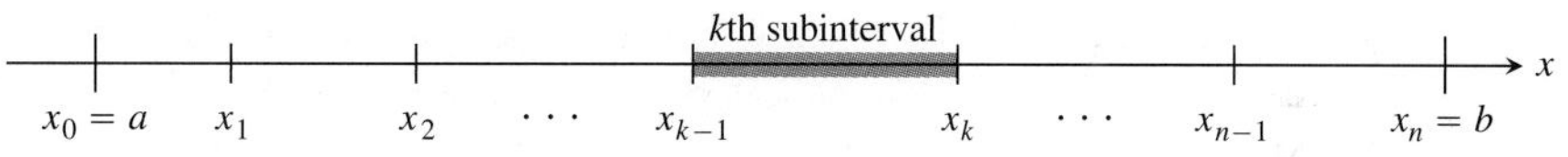

The width of the first subinterval $[x_0, x_1]$ is denoted Δx_1, the width of the second $[x_1, x_2]$ is denoted Δx_2, and the width of the kth subinterval is $\Delta x_k = x_k - x_{k-1}$. If all n subintervals have equal width, then the common width Δx is equal to $(b - a)/n$.

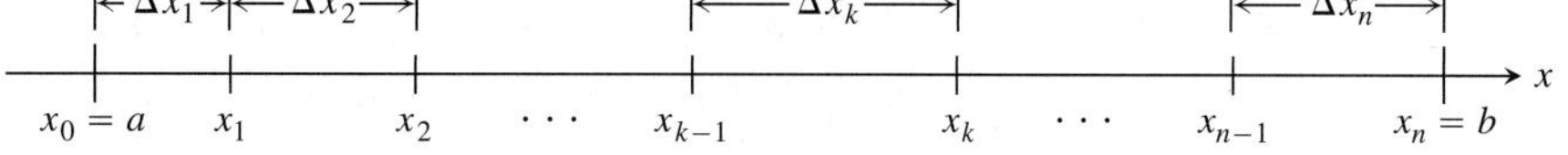

In each subinterval we select some point. The point chosen in the kth subinterval $[x_{k-1}, x_k]$ is called c_k. Then on each subinterval we stand a vertical rectangle that stretches from the x-axis to touch the curve at $(c_k, f(c_k))$. These rectangles can be above or below the x-axis, depending on whether $f(c_k)$ is positive or negative, or on it if $f(c_k) = 0$ (Figure 4.10).

On each subinterval we form the product $f(c_k) \cdot \Delta x_k$. This product is positive, negative, or zero, depending on the sign of $f(c_k)$. When $f(c_k) > 0$, the product $f(c_k) \cdot \Delta x_k$ is the area of a rectangle with height $f(c_k)$ and width Δx_k. When $f(c_k) < 0$, the product $f(c_k) \cdot \Delta x_k$ is a negative number, the negative of the area of a rectangle of width Δx_k that drops from the x-axis to the negative number $f(c_k)$.

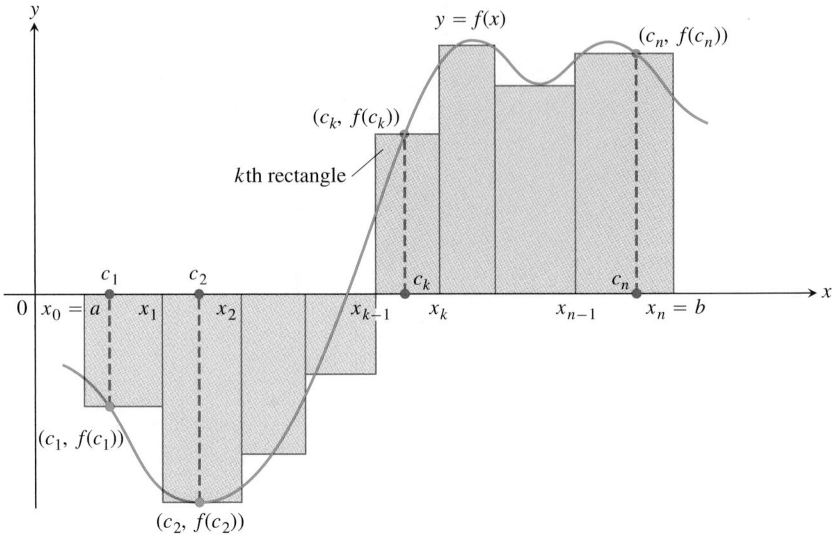

FIGURE 4.10 The rectangles approximate the region between the graph of the function $y = f(x)$ and the x-axis.

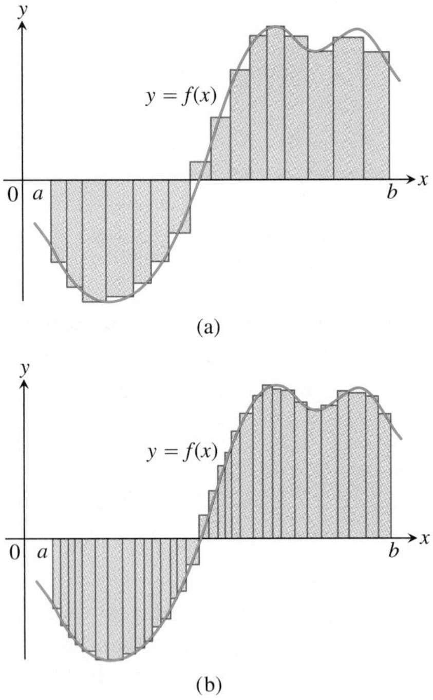

FIGURE 4.11 The curve of Figure 4.10 with rectangles from finer partitions of $[a, b]$. Finer partitions create collections of rectangles with thinner bases that approximate the region between the graph of f and the x-axis with increasing accuracy.

Finally we sum all these products to get

$$S_P = \sum_{k=1}^{n} f(c_k)\,\Delta x_k .$$

The sum S_P is called a **Riemann sum for f on the interval $[a, b]$**. There are many such sums, depending on the partition P we choose, and the choices of the points c_k in the subintervals.

In Example 5, where the subintervals all had equal widths $\Delta x = 1/n$, we could make them thinner by simply increasing their number n. When a partition has subintervals of varying widths, we can ensure they are all thin by controlling the width of a widest (longest) subinterval. We define the **norm** of a partition P, written $\|P\|$, to be the largest of all the subinterval widths. If $\|P\|$ is a small number, then all of the subintervals in the partition P have a small width. Let's look at an example of these ideas.

EXAMPLE 6 The set $P = \{0, 0.2, 0.6, 1, 1.5, 2\}$ is a partition of $[0, 2]$. There are five subintervals of P: $[0, 0.2]$, $[0.2, 0.6]$, $[0.6, 1]$, $[1, 1.5]$, and $[1.5, 2]$:

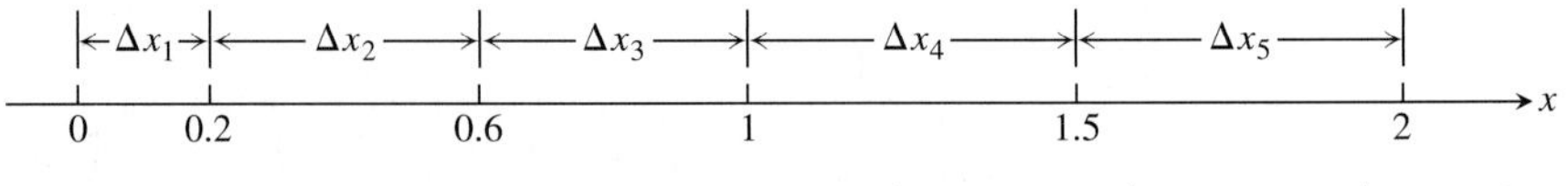

The lengths of the subintervals are $\Delta x_1 = 0.2$, $\Delta x_2 = 0.4$, $\Delta x_3 = 0.4$, $\Delta x_4 = 0.5$, and $\Delta x_5 = 0.5$. The longest subinterval length is 0.5, so the norm of the partition is $\|P\| = 0.5$. In this example, there are two subintervals of this length. ■

Any Riemann sum associated with a partition of a closed interval $[a, b]$ defines rectangles that approximate the region between the graph of a continuous function f and the x-axis. Partitions with norm approaching zero lead to collections of rectangles that approximate this region with increasing accuracy, as suggested by Figure 4.11. We will see in the next section that if the function f is continuous over the closed interval $[a, b]$, then no matter how we choose the partition P and the points c_k in its subintervals to construct a Riemann sum, a single limiting value is approached as the subinterval widths, controlled by the norm of the partition, approach zero.

EXERCISES 4.3

Write the sums in Exercises 1–6 without sigma notation. Then evaluate them.

1. $\sum_{k=1}^{2} \frac{6k}{k+1}$ **2.** $\sum_{k=1}^{3} \frac{k-1}{k}$

3. $\sum_{k=1}^{4} \cos k\pi$ **4.** $\sum_{k=1}^{5} \sin k\pi$

5. $\sum_{k=1}^{3} (-1)^{k+1} \sin \frac{\pi}{k}$ **6.** $\sum_{k=1}^{4} (-1)^{k} \cos k\pi$

7. Which of the following express $1 + 2 + 4 + 8 + 16 + 32$ in sigma notation?

a. $\sum_{k=1}^{6} 2^{k-1}$ b. $\sum_{k=0}^{5} 2^{k}$ c. $\sum_{k=-1}^{4} 2^{k+1}$

8. Which of the following express $1 - 2 + 4 - 8 + 16 - 32$ in sigma notation?

a. $\sum_{k=1}^{6} (-2)^{k-1}$ b. $\sum_{k=0}^{5} (-1)^{k} 2^{k}$ c. $\sum_{k=-2}^{3} (-1)^{k+1} 2^{k+2}$

9. Which formula is not equivalent to the other two?

a. $\sum_{k=2}^{4} \frac{(-1)^{k-1}}{k-1}$ b. $\sum_{k=0}^{2} \frac{(-1)^{k}}{k+1}$ c. $\sum_{k=-1}^{1} \frac{(-1)^{k}}{k+2}$

10. Which formula is not equivalent to the other two?

a. $\sum_{k=1}^{4} (k-1)^2$ b. $\sum_{k=-1}^{3} (k+1)^2$ c. $\sum_{k=-3}^{-1} k^2$

Express the sums in Exercises 11–16 in sigma notation. The form of your answer will depend on your choice of the lower limit of summation.

11. $1 + 2 + 3 + 4 + 5 + 6$ **12.** $1 + 4 + 9 + 16$

13. $\frac{1}{2} + \frac{1}{4} + \frac{1}{8} + \frac{1}{16}$ **14.** $2 + 4 + 6 + 8 + 10$

15. $1 - \frac{1}{2} + \frac{1}{3} - \frac{1}{4} + \frac{1}{5}$ **16.** $-\frac{1}{5} + \frac{2}{5} - \frac{3}{5} + \frac{4}{5} - \frac{5}{5}$

17. Suppose that $\sum_{k=1}^{n} a_k = -5$ and $\sum_{k=1}^{n} b_k = 6$. Find the values of

a. $\sum_{k=1}^{n} 3a_k$ b. $\sum_{k=1}^{n} \frac{b_k}{6}$ c. $\sum_{k=1}^{n} (a_k + b_k)$

d. $\sum_{k=1}^{n} (a_k - b_k)$ e. $\sum_{k=1}^{n} (b_k - 2a_k)$

18. Suppose that $\sum_{k=1}^{n} a_k = 0$ and $\sum_{k=1}^{n} b_k = 1$. Find the values of

a. $\sum_{k=1}^{n} 8a_k$ b. $\sum_{k=1}^{n} 250b_k$

c. $\sum_{k=1}^{n} (a_k + 1)$ d. $\sum_{k=1}^{n} (b_k - 1)$

Evaluate the sums in Exercises 19–28.

19. a. $\sum_{k=1}^{10} k$ b. $\sum_{k=1}^{10} k^2$ c. $\sum_{k=1}^{10} k^3$

20. a. $\sum_{k=1}^{13} k$ b. $\sum_{k=1}^{13} k^2$ c. $\sum_{k=1}^{13} k^3$

21. $\sum_{k=1}^{7} (-2k)$ **22.** $\sum_{k=1}^{5} \frac{\pi k}{15}$

23. $\sum_{k=1}^{6} (3 - k^2)$ **24.** $\sum_{k=1}^{6} (k^2 - 5)$

25. $\sum_{k=1}^{5} k(3k + 5)$ **26.** $\sum_{k=1}^{7} k(2k + 1)$

27. $\sum_{k=1}^{5} \frac{k^3}{225} + \left(\sum_{k=1}^{5} k\right)^3$ **28.** $\left(\sum_{k=1}^{7} k\right)^2 - \sum_{k=1}^{7} \frac{k^3}{4}$

In Exercises 29–32, graph each function $f(x)$ over the given interval. Partition the interval into four subintervals of equal length. Then add to your sketch the rectangles associated with the Riemann sum $\sum_{k=1}^{4} f(c_k)\,\Delta x_k$, given that c_k is the (a) left-hand endpoint, (b) right-hand endpoint, (c) midpoint of the kth subinterval. (Make a separate sketch for each set of rectangles.)

29. $f(x) = x^2 - 1$, $[0, 2]$ **30.** $f(x) = -x^2$, $[0, 1]$

31. $f(x) = \sin x$, $[-\pi, \pi]$ **32.** $f(x) = \sin x + 1$, $[-\pi, \pi]$

33. Find the norm of the partition $P = \{0, 1.2, 1.5, 2.3, 2.6, 3\}$.

34. Find the norm of the partition $P = \{-2, -1.6, -0.5, 0, 0.8, 1\}$.

For the functions in Exercises 35–40 find a formula for the upper sum obtained by dividing the interval $[a, b]$ into n equal subintervals. Then take a limit of these sums as $n \to \infty$ to calculate the area under the curve over $[a, b]$.

35. $f(x) = 1 - x^2$ over the interval $[0, 1]$.

36. $f(x) = 2x$ over the interval $[0, 3]$.

37. $f(x) = x^2 + 1$ over the interval $[0, 3]$.

38. $f(x) = 3x^2$ over the interval $[0, 1]$.

39. $f(x) = x + x^2$ over the interval $[0, 1]$.

40. $f(x) = 3x + 2x^2$ over the interval $[0, 1]$.

4.4 The Definite Integral

In Section 4.3 we investigated the limit of a finite sum for a function defined over a closed interval $[a, b]$ using n subintervals of equal width (or length), $(b - a)/n$. In this section we consider the limit of more general Riemann sums as the norm of the partitions of $[a, b]$ approaches zero. For general Riemann sums the subintervals of the partitions need not have equal widths. The limiting process then leads to the definition of the *definite integral* of a function over a closed interval $[a, b]$.

Limits of Riemann Sums

The definition of the definite integral is based on the idea that for certain functions, as the norm of the partitions of $[a, b]$ approaches zero, the values of the corresponding Riemann sums approach a limiting value J. What we mean by this converging idea is that a Riemann sum will be close to the number J provided that the norm of its partition is sufficiently small (so that all of its subintervals have thin enough widths). We introduce the symbol ϵ as a small positive number that specifies how close to J the Riemann sum must be, and the symbol δ as a second small positive number that specifies how small the norm of a partition must be in order for convergence to happen. Here is a precise formulation.

DEFINITION Let $f(x)$ be a function defined on a closed interval $[a, b]$. We say that a number J is the **definite integral of f over $[a, b]$** and that J is the limit of the Riemann sums $\sum_{k=1}^{n} f(c_k)\,\Delta x_k$ if the following condition is satisfied:

Given any number $\epsilon > 0$ there is a corresponding number $\delta > 0$ such that for every partition $P = \{x_0, x_1, \ldots, x_n\}$ of $[a, b]$ with $\|P\| < \delta$ and any choice of c_k in $[x_{k-1}, x_k]$, we have

$$\left| \sum_{k=1}^{n} f(c_k)\,\Delta x_k - J \right| < \epsilon.$$

Leibniz introduced a notation for the definite integral that captures its construction as a limit of Riemann sums. He envisioned the finite sums $\sum_{k=1}^{n} f(c_k)\,\Delta x_k$ becoming an infinite sum of function values $f(x)$ multiplied by "infinitesimal" subinterval widths dx. The sum symbol $\sum$ is replaced in the limit by the integral symbol $\int$, whose origin is in the letter "S." The function values $f(c_k)$ are replaced by a continuous selection of function values $f(x)$. The subinterval widths Δx_k become the differential dx. It is as if we are summing all products of the form $f(x) \cdot dx$ as x goes from a to b. While this notation captures the process of constructing an integral, it is Riemann's definition that gives a precise meaning to the definite integral.

Notation and Existence of the Definite Integral

The symbol for the number J in the definition of the definite integral is

$$\int_a^b f(x)\,dx$$

which is read as "the integral from a to b of f of x dee x" or sometimes as "the integral from a to b of f of x with respect to x." The component parts in the integral symbol also have names:

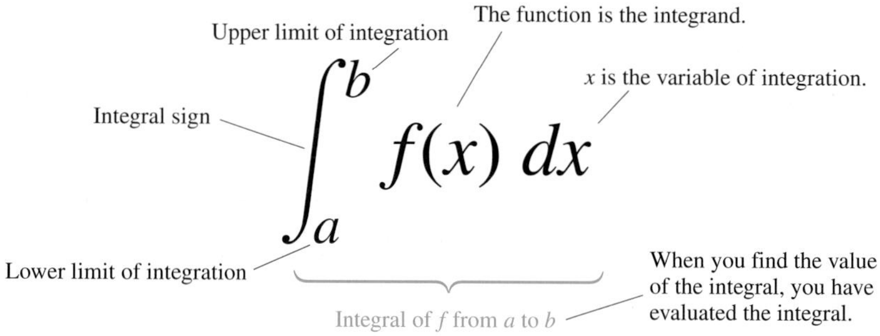

When the condition in the definition is satisfied, we say the Riemann sums of f on $[a, b]$ **converge** to the definite integral $J = \int_a^b f(x)\,dx$ and that f is **integrable** over $[a, b]$. We have many choices for a partition P with norm going to zero, and many choices of points c_k for each partition. The definite integral exists when we always get the same limit J, no matter what choices are made. When the limit exists we write it as the definite integral

$$\lim_{\|P\|\to 0} \sum_{k=1}^{n} f(c_k)\,\Delta x_k = J = \int_a^b f(x)\,dx.$$

When each partition has n equal subintervals, each of width $\Delta x = (b - a)/n$, we will also write

$$\lim_{n\to\infty} \sum_{k=1}^{n} f(c_k)\,\Delta x = J = \int_a^b f(x)\,dx.$$

The limit of any Riemann sum is always taken as the norm of the partitions approaches zero and the number of subintervals goes to infinity.

The value of the definite integral of a function over any particular interval depends on the function, not on the letter we choose to represent its independent variable. If we decide to use t or u instead of x, we simply write the integral as

$$\int_a^b f(t)\,dt \qquad \text{or} \qquad \int_a^b f(u)\,du \qquad \text{instead of} \qquad \int_a^b f(x)\,dx.$$

No matter how we write the integral, it is still the same number, defined as a limit of Riemann sums. Since it does not matter what letter we use, the variable of integration is called a **dummy variable**.

Since there are so many choices to be made in taking a limit of Riemann sums, it might seem difficult to show that such a limit exists. It turns out, however, that no matter what choices are made, the Riemann sums associated with a *continuous* function converge to the same limit.

THEOREM 2—The Existence of Definite Integrals A continuous function is integrable. That is, if a function f is continuous on an interval $[a, b]$, then its definite integral over $[a, b]$ exists.

By the Extreme Value Theorem (Theorem 1, Section 3.1), when f is continuous we can choose c_k so that $f(c_k)$ gives the maximum value of f on $[x_{k-1}, x_k]$, giving an upper sum. We can choose c_k to give the minimum value of f on $[x_{k-1}, x_k]$, giving a lower sum. We can pick

c_k to be the midpoint of $[x_{k-1}, x_k]$, the rightmost point x_k, or a random point. We can take the partitions of equal or varying widths. In each case we get the same limit for $\sum_{k=1}^{n} f(c_k)\,\Delta x_k$ as $\|P\| \to 0$. The idea behind Theorem 2 is that a Riemann sum associated with a partition is no more than the upper sum of that partition and no less than the lower sum. The upper and lower sums converge to the same value when $\|P\| \to 0$. All other Riemann sums lie between the upper and lower sums and have the same limit. A proof of Theorem 2 involves a careful analysis of functions, partitions, and limits along this line of thinking and is left to a more advanced text. An indication of this proof is given in Exercises 60 and 61.

Theorem 2 says nothing about how to *calculate* definite integrals. A method of calculation will be developed in Section 4.5, through a connection to the process of taking antiderivatives.

Integrable and Nonintegrable Functions

Theorem 2 tells us that functions continuous over the interval $[a, b]$ are integrable there. Functions that are not continuous may or may not be integrable. Discontinuous functions that are integrable include those that are increasing on $[a, b]$, and the *piecewise-continuous functions* defined in Practice and Additional Exercises 137–142 at the end of this chapter. (The latter are continuous except at a finite number of points in $[a, b]$.) For integrability to fail, a function needs to be sufficiently discontinuous so that the region between its graph and the x-axis cannot be approximated well by increasingly thin rectangles. Here is an example of a function that is not integrable.

EXAMPLE 1 The function

$$f(x) = \begin{cases} 1, & \text{if } x \text{ is rational} \\ 0, & \text{if } x \text{ is irrational} \end{cases}$$

has no Riemann integral over $[0, 1]$. Underlying this is the fact that between any two numbers there is both a rational number and an irrational number. Thus the function jumps up and down too erratically over $[0, 1]$ to allow the region beneath its graph and above the x-axis to be approximated by rectangles, no matter how thin they are. We show, in fact, that upper sum approximations and lower sum approximations converge to different limiting values.

If we pick a partition P of $[0, 1]$ and choose c_k to be the point giving the maximum value for f on $[x_{k-1}, x_k]$ then the corresponding Riemann sum is

$$U = \sum_{k=1}^{n} f(c_k)\,\Delta x_k = \sum_{k=1}^{n} (1)\,\Delta x_k = 1,$$

since each subinterval $[x_{k-1}, x_k]$ contains a rational number where $f(c_k) = 1$. Note that the lengths of the intervals in the partition sum to 1, $\sum_{k=1}^{n} \Delta x_k = 1$. So each such Riemann sum equals 1, and a limit of Riemann sums using these choices equals 1.

On the other hand, if we pick c_k to be the point giving the minimum value for f on $[x_{k-1}, x_k]$, then the Riemann sum is

$$L = \sum_{k=1}^{n} f(c_k)\,\Delta x_k = \sum_{k=1}^{n} (0)\,\Delta x_k = 0,$$

since each subinterval $[x_{k-1}, x_k]$ contains an irrational number c_k where $f(c_k) = 0$. The limit of Riemann sums using these choices equals zero. Since the limit depends on the choices of c_k, the function f is not integrable. ■

Properties of Definite Integrals

In defining $\int_a^b f(x)\,dx$ as a limit of sums $\sum_{k=1}^{n} f(c_k)\,\Delta x_k$, we moved from left to right across the interval $[a, b]$. What would happen if we instead move right to left, starting with $x_0 = b$

and ending at $x_n = a$? Each Δx_k in the Riemann sum would change its sign, with $x_k - x_{k-1}$ now negative instead of positive. With the same choices of c_k in each subinterval, the sign of any Riemann sum would change, as would the sign of the limit, the integral $\int_b^a f(x)\,dx$. Since we have not previously given a meaning to integrating backward, we are led to define

$$\int_b^a f(x)\,dx = -\int_a^b f(x)\,dx.$$

Another extension of the integral is to an interval of zero width, when $a = b$. Since $f(c_k)\,\Delta x_k$ is zero when the interval width $\Delta x_k = 0$, whenever $f(a)$ exists we define

$$\int_a^a f(x)\,dx = 0.$$

Theorem 3 states seven properties of integrals, given as rules that they satisfy, including the two above. These rules become very useful in the process of computing integrals. We will refer to them repeatedly to simplify our calculations.

Rules 2 through 7 have geometric interpretations, shown in Figure 4.12. The graphs in these figures are of positive functions, but the rules apply to general integrable functions.

THEOREM 3 When f and g are integrable, the definite integral satisfies Rules 1 to 7 in Table 4.5.

TABLE 4.5 Rules satisfied by definite integrals

1.	*Order of Integration:*	$\int_b^a f(x)\,dx = -\int_a^b f(x)\,dx$	A Definition
2.	*Zero Width Interval:*	$\int_a^a f(x)\,dx = 0$	Definition when $f(a)$ exists
3.	*Constant Multiple:*	$\int_a^b kf(x)\,dx = k\int_a^b f(x)\,dx$	Any Number k
		$\int_a^b -f(x)\,dx = -\int_a^b f(x)\,dx$	$k = -1$
4.	*Sum and Difference:*	$\int_a^b (f(x) \pm g(x))\,dx = \int_a^b f(x)\,dx \pm \int_a^b g(x)\,dx$	
5.	*Additivity:*	$\int_a^b f(x)\,dx + \int_b^c f(x)\,dx = \int_a^c f(x)\,dx$	
6.	*Max-Min Inequality:*	If f has maximum value $\max f$ and minimum value $\min f$ on $[a, b]$, then $\min f \cdot (b-a) \le \int_a^b f(x)\,dx \le \max f \cdot (b-a)$.	
7.	*Domination:*	$f(x) \ge g(x)$ on $[a, b] \Rightarrow \int_a^b f(x)\,dx \ge \int_a^b g(x)\,dx$	
		$f(x) \ge 0$ on $[a, b] \Rightarrow \int_a^b f(x)\,dx \ge 0$	(Special Case)

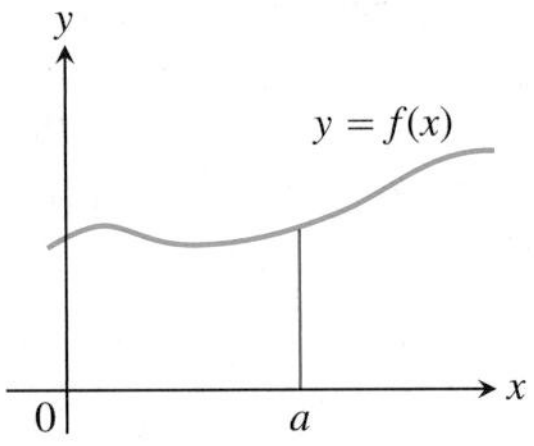

(a) *Zero Width Interval:*

$$\int_a^a f(x)\,dx = 0.$$

(The area under a point is 0.)

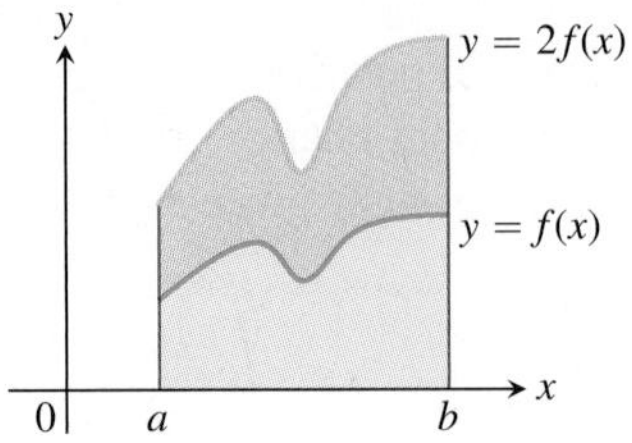

(b) *Constant Multiple:*

$$\int_a^b kf(x)\,dx = k\int_a^b f(x)\,dx.$$

(Shown for $k = 2$.)

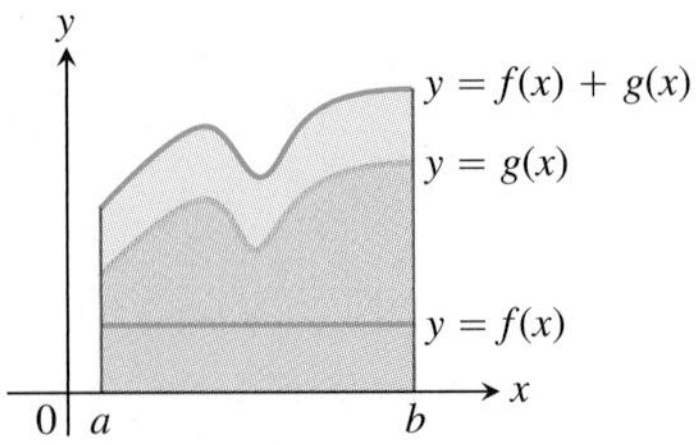

(c) *Sum:*

$$\int_a^b (f(x) + g(x))\,dx = \int_a^b f(x)\,dx + \int_a^b g(x)\,dx$$

(Areas add)

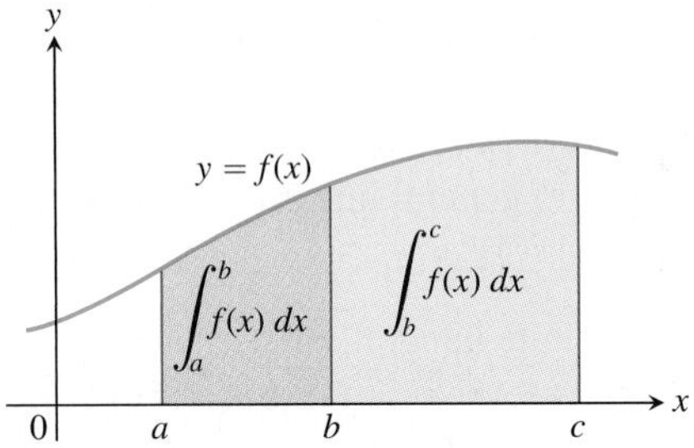

(d) *Additivity for definite integrals:*

$$\int_a^b f(x)\,dx + \int_b^c f(x)\,dx = \int_a^c f(x)\,dx$$

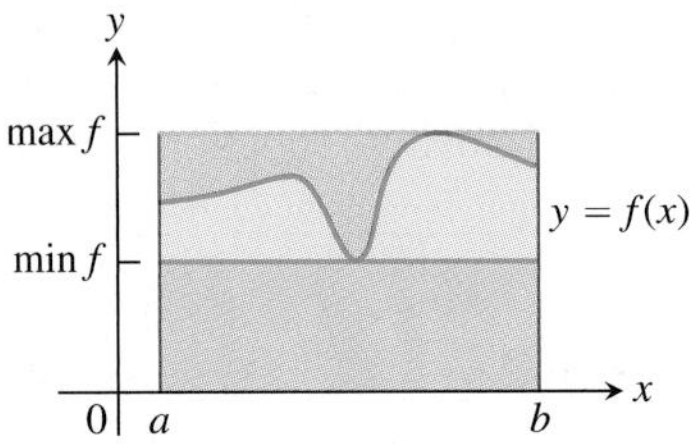

(e) *Max-Min Inequality:*

$$\min f \cdot (b - a) \le \int_a^b f(x)\,dx$$
$$\le \max f \cdot (b - a)$$

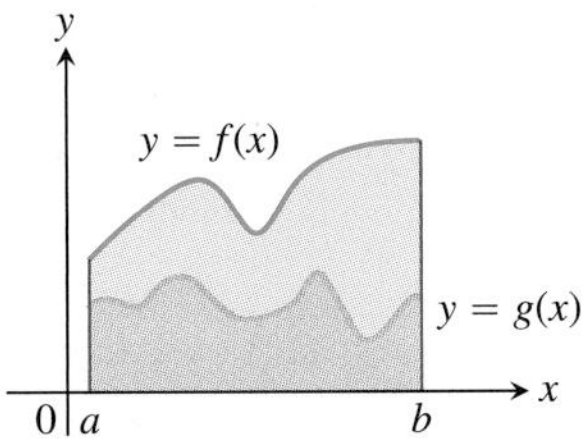

(f) *Domination:*

$f(x) \ge g(x)$ on $[a, b]$

$$\Rightarrow \int_a^b f(x)\,dx \ge \int_a^b g(x)\,dx$$

FIGURE 4.12 Geometric interpretations of Rules 2–7 in Table 4.5.

While Rules 1 and 2 are definitions, Rules 3 to 7 of Table 4.5 must be proved. The following is a proof of one of these rules. Similar proofs can be given to verify the other properties in Table 4.5.

Proof of Rule 6 Rule 6 says that the integral of f over $[a, b]$ is never smaller than the minimum value of f times the length of the interval and never larger than the maximum value of f times the length of the interval. The reason is that for every partition of $[a, b]$ and for every choice of the points c_k,

$$\begin{aligned}
\min f \cdot (b - a) &= \min f \cdot \sum_{k=1}^{n} \Delta x_k && \sum_{k=1}^{n} \Delta x_k = b - a \\
&= \sum_{k=1}^{n} \min f \cdot \Delta x_k && \text{Constant Multiple Rule} \\
&\le \sum_{k=1}^{n} f(c_k)\,\Delta x_k && \min f \le f(c_k) \\
&\le \sum_{k=1}^{n} \max f \cdot \Delta x_k && f(c_k) \le \max f \\
&= \max f \cdot \sum_{k=1}^{n} \Delta x_k && \text{Constant Multiple Rule} \\
&= \max f \cdot (b - a).
\end{aligned}$$

In short, all Riemann sums for f on $[a, b]$ satisfy the inequality

$$\min f \cdot (b - a) \leq \sum_{k=1}^{n} f(c_k)\,\Delta x_k \leq \max f \cdot (b - a).$$

Hence their limit, the integral, does too. ■

EXAMPLE 2 Suppose that

$$\int_{-1}^{1} f(x)\,dx = 5, \qquad \int_{1}^{4} f(x)\,dx = -2, \qquad \int_{-1}^{1} h(x)\,dx = 7.$$

Then

1. $\displaystyle\int_{4}^{1} f(x)\,dx = -\int_{1}^{4} f(x)\,dx = -(-2) = 2$ — Rule 1

2. $\displaystyle\int_{-1}^{1} [2f(x) + 3h(x)]\,dx = 2\int_{-1}^{1} f(x)\,dx + 3\int_{-1}^{1} h(x)\,dx$ — Rules 3 and 4

   $\displaystyle= 2(5) + 3(7) = 31$

3. $\displaystyle\int_{-1}^{4} f(x)\,dx = \int_{-1}^{1} f(x)\,dx + \int_{1}^{4} f(x)\,dx = 5 + (-2) = 3$ — Rule 5 ■

EXAMPLE 3 Show that the value of $\int_0^1 \sqrt{1 + \cos x}\,dx$ is less than or equal to $\sqrt{2}$.

Solution The Max-Min Inequality for definite integrals (Rule 6) says that $\min f \cdot (b - a)$ is a *lower bound* for the value of $\int_a^b f(x)\,dx$ and that $\max f \cdot (b - a)$ is an *upper bound*. The maximum value of $\sqrt{1 + \cos x}$ on $[0, 1]$ is $\sqrt{1 + 1} = \sqrt{2}$, so

$$\int_0^1 \sqrt{1 + \cos x}\,dx \leq \sqrt{2} \cdot (1 - 0) = \sqrt{2}.$$ ■

Area Under the Graph of a Nonnegative Function

We now make precise the notion of the area of a region with curved boundary, capturing the idea of approximating a region by increasingly many rectangles. The area under the graph of a nonnegative continuous function is defined to be a definite integral.

> **DEFINITION** If $y = f(x)$ is nonnegative and integrable over a closed interval $[a, b]$, then the **area under the curve $y = f(x)$ over $[a, b]$** is the integral of f from a to b,
>
> $$A = \int_a^b f(x)\,dx.$$

For the first time we have a rigorous definition for the area of a region whose boundary is the graph of any continuous function. We now apply this to a simple example, the

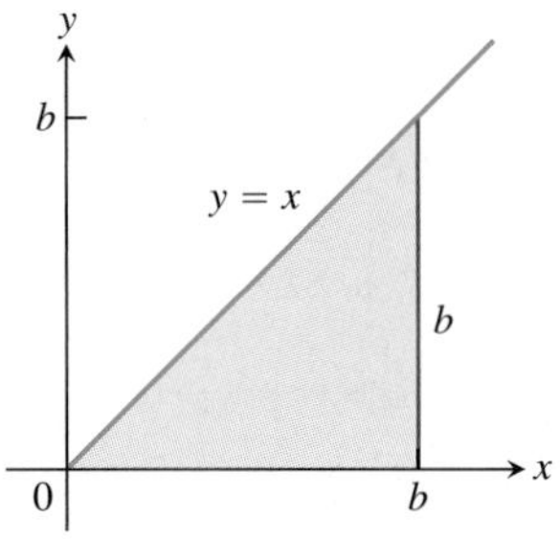

FIGURE 4.13 The region in Example 4 is a triangle.

area under a straight line, where we can verify that our new definition agrees with our previous notion of area.

EXAMPLE 4 Compute $\int_0^b x\,dx$ and find the area A under $y = x$ over the interval $[0, b]$, $b > 0$.

Solution The region of interest is a triangle (Figure 4.13). We compute the area in two ways.

(a) To compute the definite integral as the limit of Riemann sums, we calculate $\lim_{\|P\|\to 0} \sum_{k=1}^{n} f(c_k)\,\Delta x_k$ for partitions whose norms go to zero. Theorem 1 tells us that it does not matter how we choose the partitions or the points c_k as long as the norms approach zero. All choices give the exact same limit. So we consider the partition P that subdivides the interval $[0, b]$ into n subintervals of equal width $\Delta x = (b - 0)/n = b/n$, and we choose c_k to be the right endpoint in each subinterval. The partition is $P = \left\{0, \frac{b}{n}, \frac{2b}{n}, \frac{3b}{n}, \ldots, \frac{nb}{n}\right\}$ and $c_k = \frac{kb}{n}$. So

$$\begin{aligned}
\sum_{k=1}^{n} f(c_k)\,\Delta x &= \sum_{k=1}^{n} \frac{kb}{n} \cdot \frac{b}{n} && f(c_k) = c_k \\
&= \sum_{k=1}^{n} \frac{kb^2}{n^2} \\
&= \frac{b^2}{n^2} \sum_{k=1}^{n} k && \text{Constant Multiple Rule} \\
&= \frac{b^2}{n^2} \cdot \frac{n(n+1)}{2} && \text{Sum of First } n \text{ Integers} \\
&= \frac{b^2}{2}\left(1 + \frac{1}{n}\right)
\end{aligned}$$

As $n \to \infty$ and $\|P\| \to 0$, this last expression on the right has the limit $b^2/2$. Therefore,

$$\int_0^b x\,dx = \frac{b^2}{2}.$$

(b) Since the area equals the definite integral for a nonnegative function, we can quickly derive the definite integral by using the formula for the area of a triangle having base length b and height $y = b$. The area is $A = (1/2)\,b \cdot b = b^2/2$. Again we conclude that $\int_0^b x\,dx = b^2/2$. ■

Example 4 can be generalized to integrate $f(x) = x$ over any closed interval $[a, b]$, $0 < a < b$.

$$\begin{aligned}
\int_a^b x\,dx &= \int_a^0 x\,dx + \int_0^b x\,dx && \text{Rule 5} \\
&= -\int_0^a x\,dx + \int_0^b x\,dx && \text{Rule 1} \\
&= -\frac{a^2}{2} + \frac{b^2}{2}. && \text{Example 4}
\end{aligned}$$

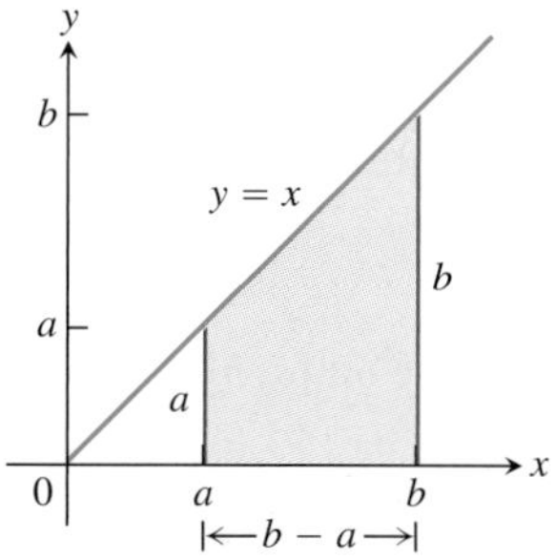

FIGURE 4.14 The area of this trapezoidal region is $A = (b^2 - a^2)/2$.

In conclusion, we have the following rule for integrating $f(x) = x$:

$$\int_a^b x\,dx = \frac{b^2}{2} - \frac{a^2}{2}, \qquad a < b \tag{1}$$

This computation gives the area of a trapezoid (Figure 4.14). Equation (1) remains valid when a and b are negative. When $a < b < 0$, the definite integral value $(b^2 - a^2)/2$ is a negative number, the negative of the area of a trapezoid dropping down to the line $y = x$ below the x-axis. When $a < 0$ and $b > 0$, Equation (1) is still valid and the definite integral gives the difference between two areas, the area under the graph and above $[0, b]$ minus the area below $[a, 0]$ and over the graph.

The following results can also be established using a Riemann sum calculation similar to that in Example 4 (Exercises 55 and 56).

$$\int_a^b c\,dx = c(b - a), \qquad c \text{ any constant} \tag{2}$$

$$\int_a^b x^2\,dx = \frac{b^3}{3} - \frac{a^3}{3}, \qquad a < b \tag{3}$$

Average Value of a Continuous Function Revisited

In Section 4.2 we introduced informally the average value of a nonnegative continuous function f over an interval $[a, b]$, leading us to define this average as the area under the graph of $y = f(x)$ divided by $b - a$. In integral notation we write this as

$$\text{Average} = \frac{1}{b - a}\int_a^b f(x)\,dx.$$

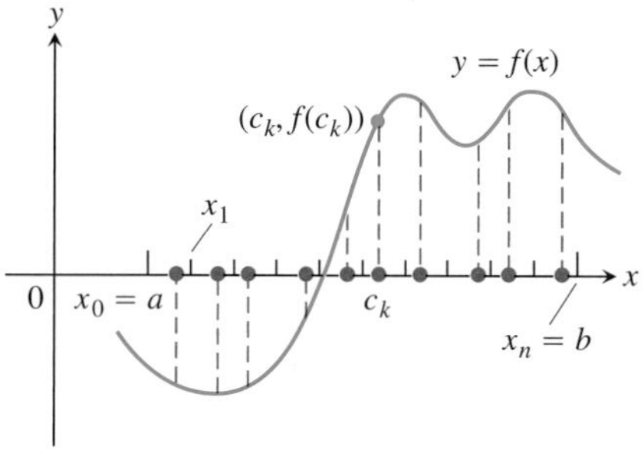

FIGURE 4.15 A sample of values of a function on an interval $[a, b]$.

We can use this formula to give a precise definition of the average value of any continuous (or integrable) function, whether positive, negative, or both.

Alternately, we can use the following reasoning. We start with the idea from arithmetic that the average of n numbers is their sum divided by n. A continuous function f on $[a, b]$ may have infinitely many values, but we can still sample them in an orderly way. We divide $[a, b]$ into n subintervals of equal width $\Delta x = (b - a)/n$ and evaluate f at a point c_k in each (Figure 4.15). The average of the n sampled values is

$$\begin{aligned}\frac{f(c_1) + f(c_2) + \cdots + f(c_n)}{n} &= \frac{1}{n}\sum_{k=1}^{n} f(c_k)\\ &= \frac{\Delta x}{b - a}\sum_{k=1}^{n} f(c_k) && \Delta x = \frac{b-a}{n},\ \text{so } \frac{1}{n} = \frac{\Delta x}{b-a}\\ &= \frac{1}{b - a}\sum_{k=1}^{n} f(c_k)\,\Delta x\end{aligned}$$

The average is obtained by dividing a Riemann sum for f on $[a, b]$ by $(b - a)$. As we increase the size of the sample and let the norm of the partition approach zero, the average approaches $(1/(b - a))\int_a^b f(x)\,dx$. Both points of view lead us to the following definition.

DEFINITION If f is integrable on $[a, b]$, then its **average value on $[a, b]$**, also called its **mean value**, is

$$\text{av}(f) = \frac{1}{b - a}\int_a^b f(x)\,dx.$$

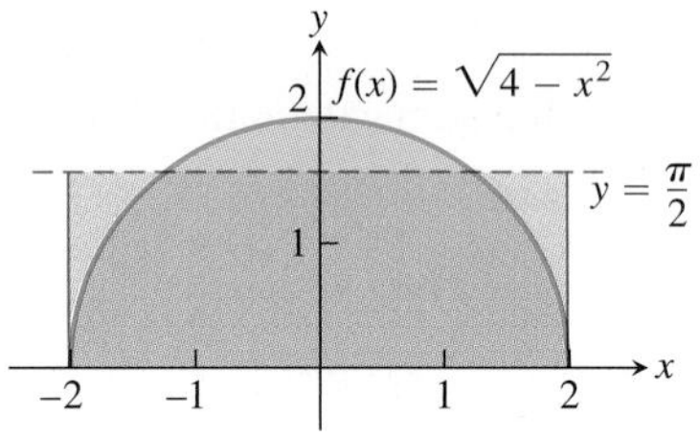

FIGURE 4.16 The average value of $f(x) = \sqrt{4 - x^2}$ on $[-2, 2]$ is $\pi/2$ (Example 5).

EXAMPLE 5 Find the average value of $f(x) = \sqrt{4 - x^2}$ on $[-2, 2]$.

Solution We recognize $f(x) = \sqrt{4 - x^2}$ as a function whose graph is the upper semicircle of radius 2 centered at the origin (Figure 4.16).

The area between the semicircle and the x-axis from -2 to 2 can be computed using the geometry formula

$$\text{Area} = \frac{1}{2}\cdot\pi r^2 = \frac{1}{2}\cdot\pi(2)^2 = 2\pi.$$

Because f is nonnegative, the area is also the value of the integral of f from -2 to 2,

$$\int_{-2}^{2}\sqrt{4 - x^2}\,dx = 2\pi.$$

Therefore, the average value of f is

$$\text{av}(f) = \frac{1}{2 - (-2)}\int_{-2}^{2}\sqrt{4 - x^2}\,dx = \frac{1}{4}(2\pi) = \frac{\pi}{2}.$$ ■

EXERCISES 4.4

Express the limits in Exercises 1–8 as definite integrals.

1. $\lim_{\|P\|\to 0}\sum_{k=1}^{n} c_k^2\,\Delta x_k$, where P is a partition of $[0, 2]$

2. $\lim_{\|P\|\to 0}\sum_{k=1}^{n} 2c_k^3\,\Delta x_k$, where P is a partition of $[-1, 0]$

3. $\lim_{\|P\|\to 0}\sum_{k=1}^{n} (c_k^2 - 3c_k)\,\Delta x_k$, where P is a partition of $[-7, 5]$

4. $\lim_{\|P\|\to 0}\sum_{k=1}^{n} \left(\frac{1}{c_k}\right)\Delta x_k$, where P is a partition of $[1, 4]$

5. $\lim_{\|P\|\to 0}\sum_{k=1}^{n} \frac{1}{1 - c_k}\,\Delta x_k$, where P is a partition of $[2, 3]$

6. $\lim_{\|P\|\to 0}\sum_{k=1}^{n} \sqrt{4 - c_k^2}\,\Delta x_k$, where P is a partition of $[0, 1]$

7. $\lim_{\|P\|\to 0}\sum_{k=1}^{n} (\sec c_k)\,\Delta x_k$, where P is a partition of $[-\pi/4, 0]$

8. $\lim_{\|P\|\to 0}\sum_{k=1}^{n} (\tan c_k)\,\Delta x_k$, where P is a partition of $[0, \pi/4]$

9. Suppose that f and g are integrable and that

$$\int_1^2 f(x)\,dx = -4,\quad \int_1^5 f(x)\,dx = 6,\quad \int_1^5 g(x)\,dx = 8.$$

Use the rules in Table 4.5 to find

a. $\int_2^2 g(x)\,dx$ b. $\int_5^1 g(x)\,dx$

c. $\int_1^2 3f(x)\,dx$ d. $\int_2^5 f(x)\,dx$

e. $\int_1^5 [f(x) - g(x)]\,dx$ f. $\int_1^5 [4f(x) - g(x)]\,dx$

10. Suppose that f and h are integrable and that

$$\int_1^9 f(x)\,dx = -1, \quad \int_7^9 f(x)\,dx = 5, \quad \int_7^9 h(x)\,dx = 4.$$

Use the rules in Table 4.5 to find

a. $\int_1^9 -2f(x)\,dx$ b. $\int_7^9 [f(x) + h(x)]\,dx$

c. $\int_7^9 [2f(x) - 3h(x)]\,dx$ d. $\int_9^1 f(x)\,dx$

e. $\int_1^7 f(x)\,dx$ f. $\int_9^7 [h(x) - f(x)]\,dx$

11. Suppose that $\int_1^2 f(x)\,dx = 5$. Find

a. $\int_1^2 f(u)\,du$ b. $\int_1^2 \sqrt{3}f(z)\,dz$

c. $\int_2^1 f(t)\,dt$ d. $\int_1^2 [-f(x)]\,dx$

12. Suppose that $\int_{-3}^0 g(t)\,dt = \sqrt{2}$. Find

a. $\int_0^{-3} g(t)\,dt$ b. $\int_{-3}^0 g(u)\,du$

c. $\int_{-3}^0 [-g(x)]\,dx$ d. $\int_{-3}^0 \frac{g(r)}{\sqrt{2}}\,dr$

13. Suppose that f is integrable and that $\int_0^3 f(z)\,dz = 3$ and $\int_0^4 f(z)\,dz = 7$. Find

a. $\int_3^4 f(z)\,dz$ b. $\int_4^3 f(t)\,dt$

14. Suppose that h is integrable and that $\int_{-1}^1 h(r)\,dr = 0$ and $\int_{-1}^3 h(r)\,dr = 6$. Find

a. $\int_1^3 h(r)\,dr$ b. $-\int_3^1 h(u)\,du$

In Exercises 15–22, graph the integrands and use areas to evaluate the integrals.

15. $\int_{-2}^4 \left(\frac{x}{2} + 3\right) dx$ **16.** $\int_{1/2}^{3/2} (-2x + 4)\,dx$

17. $\int_{-3}^3 \sqrt{9 - x^2}\,dx$ **18.** $\int_{-4}^0 \sqrt{16 - x^2}\,dx$

19. $\int_{-2}^1 |x|\,dx$ **20.** $\int_{-1}^1 (1 - |x|)\,dx$

21. $\int_{-1}^1 (2 - |x|)\,dx$ **22.** $\int_{-1}^1 \left(1 + \sqrt{1 - x^2}\right) dx$

Use areas to evaluate the integrals in Exercises 23 and 24.

23. $\int_0^b \frac{x}{2}\,dx, \quad b > 0$ **24.** $\int_a^b 2s\,ds, \quad 0 < a < b$

Use the results of Equations (1) and (3) to evaluate the integrals in Exercises 25–30.

25. $\int_1^{\sqrt{2}} x\,dx$ **26.** $\int_\pi^{2\pi} \theta\,d\theta$ **27.** $\int_0^{\sqrt[3]{7}} x^2\,dx$

28. $\int_0^{1/2} t^2\,dt$ **29.** $\int_a^{2a} x\,dx$ **30.** $\int_0^{\sqrt[3]{b}} x^2\,dx$

Use the rules in Table 4.5 and Equations (1)–(3) to evaluate the integrals in Exercises 31–36.

31. $\int_3^1 7\,dx$ **32.** $\int_0^2 5x\,dx$

33. $\int_0^2 (2t - 3)\,dt$ **34.** $\int_2^1 \left(1 + \frac{z}{2}\right) dz$

35. $\int_1^2 3u^2\,du$ **36.** $\int_1^0 (3x^2 + x - 5)\,dx$

In Exercises 37 and 38 use a definite integral to find the area of the region between the given curve and the x-axis on the interval $[0, b]$.

37. $y = 3x^2$ **38.** $y = 2x$

In Exercises 39–44, graph the function and find its average value over the given interval.

39. $f(x) = x^2 - 1$ on $\left[0, \sqrt{3}\right]$

40. $f(x) = -3x^2 - 1$ on $[0, 1]$

41. $f(t) = (t - 1)^2$ on $[0, 3]$

42. $f(t) = t^2 - t$ on $[-2, 1]$

43. $g(x) = |x| - 1$ on **a.** $[-1, 1]$, **b.** $[1, 3]$, and **c.** $[-1, 3]$

44. $h(x) = -|x|$ on **a.** $[-1, 0]$, **b.** $[0, 1]$, and **c.** $[-1, 1]$

45. What values of a and b maximize the value of

$$\int_a^b (x - x^2)\,dx?$$

(*Hint:* Where is the integrand positive?)

46. What values of a and b minimize the value of

$$\int_a^b (x^4 - 2x^2)\,dx?$$

47. Use the Max-Min Inequality to find upper and lower bounds for the value of

$$\int_0^1 \frac{1}{1 + x^2}\,dx.$$

48. (*Continuation of Exercise 47.*) Use the Max-Min Inequality to find upper and lower bounds for

$$\int_0^{0.5} \frac{1}{1+x^2}\,dx \quad \text{and} \quad \int_{0.5}^{1} \frac{1}{1+x^2}\,dx.$$

Add these to arrive at an improved estimate of

$$\int_0^1 \frac{1}{1+x^2}\,dx.$$

49. Show that the value of $\int_0^1 \sin(x^2)\,dx$ cannot possibly be 2.

50. Show that the value of $\int_0^1 \sqrt{x+8}\,dx$ lies between $2\sqrt{2} \approx 2.8$ and 3.

51. Integrals of nonnegative functions Use the Max-Min Inequality to show that if f is integrable then

$$f(x) \geq 0 \quad \text{on} \quad [a, b] \quad \Rightarrow \quad \int_a^b f(x)\,dx \geq 0.$$

52. Use the inequality $\sin x \leq x$, which holds for $x \geq 0$, to find an upper bound for the value of $\int_0^1 \sin x\,dx$.

53. If av(f) really is a typical value of the integrable function $f(x)$ on $[a, b]$, then the constant function av(f) should have the same integral over $[a, b]$ as f. Does it? That is, does

$$\int_a^b \operatorname{av}(f)\,dx = \int_a^b f(x)\,dx?$$

Give reasons for your answer.

54. It would be nice if average values of integrable functions obeyed the following rules on an interval $[a, b]$.

a. $\operatorname{av}(f+g) = \operatorname{av}(f) + \operatorname{av}(g)$

b. $\operatorname{av}(kf) = k\operatorname{av}(f)$ (any number k)

c. $\operatorname{av}(f) \leq \operatorname{av}(g)$ if $f(x) \leq g(x)$ on $[a, b]$.

Do these rules ever hold? Give reasons for your answers.

55. Use limits of Riemann sums as in Example 4a to establish Equation (2).

56. Use limits of Riemann sums as in Example 4a to establish Equation (3).

57. Upper and lower sums for increasing functions

a. Suppose the graph of a continuous function $f(x)$ rises steadily as x moves from left to right across an interval $[a, b]$. Let P be a partition of $[a, b]$ into n subintervals of length $\Delta x = (b-a)/n$. Show by referring to the accompanying figure that the difference between the upper and lower sums for f on this partition can be represented graphically as the area of a rectangle R whose dimensions are $[f(b) - f(a)]$ by Δx. (*Hint*: The difference $U - L$ is the sum of areas of rectangles whose diagonals $Q_0Q_1, Q_1Q_2, \ldots, Q_{n-1}Q_n$ lie along the curve. There is no overlapping when these rectangles are shifted horizontally onto R.)

b. Suppose that instead of being equal, the lengths Δx_k of the subintervals of the partition of $[a, b]$ vary in size. Show that

$$U - L \leq |f(b) - f(a)|\,\Delta x_{\max},$$

where $\Delta x_{\max}$ is the norm of P, and hence that $\lim_{\|P\|\to 0}(U - L) = 0$.

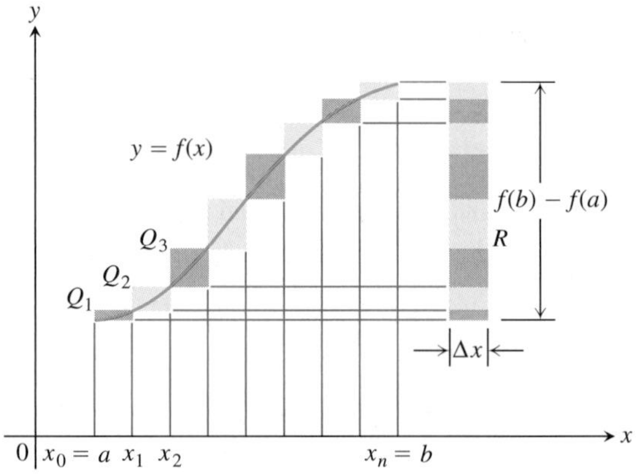

58. Upper and lower sums for decreasing functions (*Continuation of Exercise 57.*)

a. Draw a figure like the one in Exercise 57 for a continuous function $f(x)$ whose values decrease steadily as x moves from left to right across the interval $[a, b]$. Let P be a partition of $[a, b]$ into subintervals of equal length. Find an expression for $U - L$ that is analogous to the one you found for $U - L$ in Exercise 57a.

b. Suppose that instead of being equal, the lengths Δx_k of the subintervals of P vary in size. Show that the inequality

$$U - L \leq |f(b) - f(a)|\,\Delta x_{\max}$$

of Exercise 57b still holds and hence that $\lim_{\|P\|\to 0}(U - L) = 0$.

59. Use the formula

$$\sin h + \sin 2h + \sin 3h + \cdots + \sin mh = \frac{\cos(h/2) - \cos((m + (1/2))h)}{2\sin(h/2)}$$

to find the area under the curve $y = \sin x$ from $x = 0$ to $x = \pi/2$ in two steps:

a. Partition the interval $[0, \pi/2]$ into n subintervals of equal length and calculate the corresponding upper sum U; then

b. Find the limit of U as $n \to \infty$ and $\Delta x = (b-a)/n \to 0$.

60. Suppose that f is continuous and nonnegative over $[a, b]$, as in the figure on the next page. By inserting points

$$x_1, x_2, \ldots, x_{k-1}, x_k, \ldots, x_{n-1}$$

as shown, divide $[a, b]$ into n subintervals of lengths $\Delta x_1 = x_1 - a$, $\Delta x_2 = x_2 - x_1, \ldots, \Delta x_n = b - x_{n-1}$, which need not be equal.

a. If $m_k = \min\{f(x) \text{ for } x \text{ in the } k\text{th subinterval}\}$, explain the connection between the *lower sum*

$$L = m_1\,\Delta x_1 + m_2\,\Delta x_2 + \cdots + m_n\,\Delta x_n$$

and the shaded region in the first part of the figure.

b. If $M_k = \max\{f(x) \text{ for } x \text{ in the } k\text{th subinterval}\}$, explain the connection between the *upper sum*

$$U = M_1\,\Delta x_1 + M_2\,\Delta x_2 + \cdots + M_n\,\Delta x_n$$

and the shaded region in the second part of the figure.

c. Explain the connection between $U - L$ and the shaded regions along the curve in the third part of the figure.

61. We say f is **uniformly continuous** on $[a, b]$ if given any $\epsilon > 0$ there is a $\delta > 0$ such that if x_1, x_2 are in $[a, b]$ and $|x_1 - x_2| < \delta$ then $|f(x_1) - f(x_2)| < \epsilon$. It can be shown that a continuous function on $[a, b]$ is uniformly continuous. Use this and the figure below to show that if f is continuous and $\epsilon > 0$ is given, it is possible to make $U - L \le \epsilon \cdot (b - a)$ by making the largest of the Δx_k's sufficiently small.

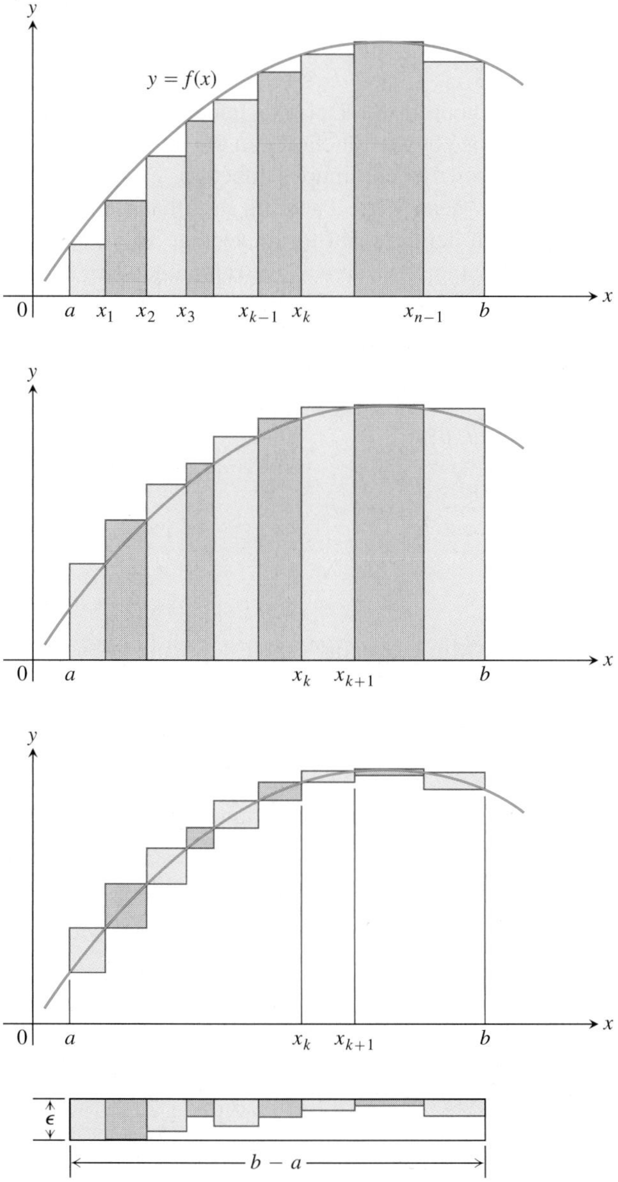

62. If you average 30 mi/h on a 150-mi trip and then return over the same 150 mi at the rate of 50 mi/h, what is your average speed for the trip? Give reasons for your answer. (*Source:* David H. Pleacher, *The Mathematics Teacher*, Vol. 85, No. 6, pp. 445–446, September 1992.)

COMPUTER EXPLORATIONS

If your CAS can draw rectangles associated with Riemann sums, use it to draw rectangles associated with Riemann sums that converge to the integrals in Exercises 63–68. Use $n = 4, 10, 20,$ and 50 subintervals of equal length in each case.

63. $\displaystyle\int_0^1 (1 - x)\,dx = \frac{1}{2}$

64. $\displaystyle\int_0^1 (x^2 + 1)\,dx = \frac{4}{3}$

65. $\displaystyle\int_{-\pi}^{\pi} \cos x\,dx = 0$

66. $\displaystyle\int_0^{\pi/4} \sec^2 x\,dx = 1$

67. $\displaystyle\int_{-1}^{1} |x|\,dx = 1$

68. $\displaystyle\int_1^2 \frac{1}{x}\,dx$ (The integral's value is about 0.693.)

In Exercises 69–76, use a CAS to perform the following steps:

a. Plot the functions over the given interval.

b. Partition the interval into $n = 100, 200,$ and 1000 subintervals of equal length, and evaluate the function at the midpoint of each subinterval.

c. Compute the average value of the function values generated in part (b).

d. Solve the equation $f(x) = (\text{average value})$ for x using the average value calculated in part (c) for the $n = 1000$ partitioning.

69. $f(x) = \sin x$ on $[0, \pi]$

70. $f(x) = \sin^2 x$ on $[0, \pi]$

71. $f(x) = x \sin\dfrac{1}{x}$ on $\left[\dfrac{\pi}{4}, \pi\right]$

72. $f(x) = x \sin^2\dfrac{1}{x}$ on $\left[\dfrac{\pi}{4}, \pi\right]$

73. $f(x) = xe^{-x}$ on $[0, 1]$

74. $f(x) = e^{-x^2}$ on $[0, 1]$

75. $f(x) = \dfrac{\ln x}{x}$ on $[2, 5]$

76. $f(x) = \dfrac{1}{\sqrt{1 - x^2}}$ on $\left[0, \dfrac{1}{2}\right]$

4.5 The Fundamental Theorem of Calculus

HISTORICAL BIOGRAPHY

Sir Isaac Newton
(1642–1727)

In this section we present the Fundamental Theorem of Calculus, which is the central theorem of integral calculus. It connects integration and differentiation, enabling us to compute integrals using an antiderivative of the integrand function rather than by taking limits of Riemann sums as we did in Section 4.4. Leibniz and Newton exploited this relationship and started mathematical developments that fueled the scientific revolution for the next 200 years.

Along the way, we present an integral version of the Mean Value Theorem, which is another important theorem of integral calculus and used to prove the Fundamental Theorem.

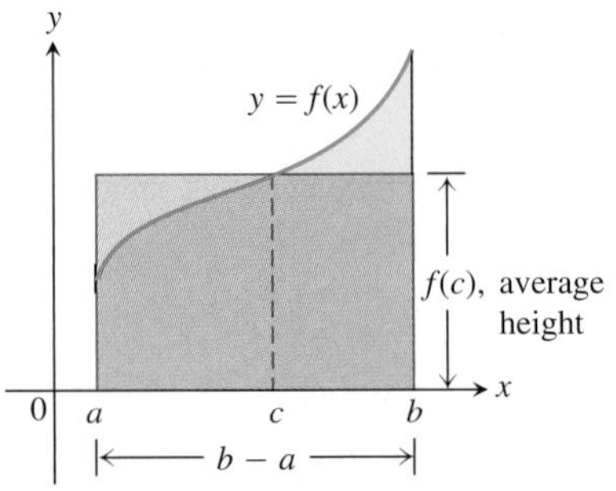

FIGURE 4.17 The value $f(c)$ in the Mean Value Theorem is, in a sense, the average (or *mean*) height of f on $[a, b]$. When $f \geq 0$, the area of the rectangle is the area under the graph of f from a to b,

$$f(c)(b - a) = \int_a^b f(x)\,dx.$$

Mean Value Theorem for Definite Integrals

In the previous section, we defined the average value of a continuous function over a closed interval $[a, b]$ as the definite integral $\int_a^b f(x)\,dx$ divided by the length or width $b - a$ of the interval. The Mean Value Theorem for Definite Integrals asserts that this average value is *always* taken on at least once by the function f in the interval.

The graph in Figure 4.17 shows a *positive* continuous function $y = f(x)$ defined over the interval $[a, b]$. Geometrically, the Mean Value Theorem says that there is a number c in $[a, b]$ such that the rectangle with height equal to the average value $f(c)$ of the function and base width $b - a$ has exactly the same area as the region beneath the graph of f from a to b.

THEOREM 4—The Mean Value Theorem for Definite Integrals If f is continuous on $[a, b]$, then at some point c in $[a, b]$,

$$f(c) = \frac{1}{b - a}\int_a^b f(x)\,dx.$$

Proof If we divide both sides of the Max-Min Inequality (Table 4.5, Rule 6) by $(b - a)$, we obtain

$$\min f \leq \frac{1}{b - a}\int_a^b f(x)\,dx \leq \max f.$$

Since f is continuous, the Intermediate Value Theorem for Continuous Functions (Section 1.7) says that f must assume every value between $\min f$ and $\max f$. It must therefore assume the value $(1/(b - a))\int_a^b f(x)\,dx$ at some point c in $[a, b]$. ■

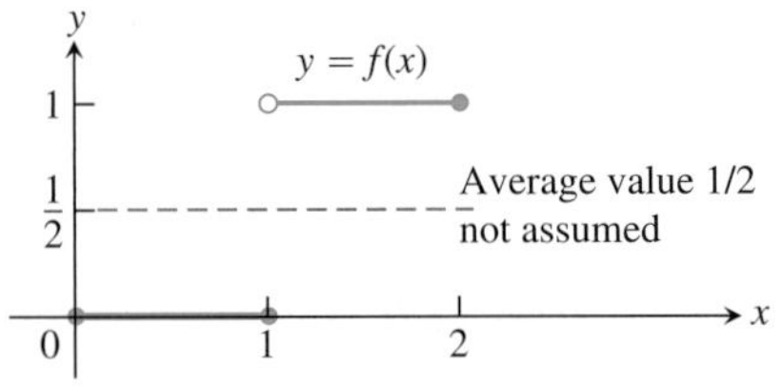

FIGURE 4.18 A discontinuous function need not assume its average value.

The continuity of f is important here. It is possible that a discontinuous function never equals its average value (Figure 4.18).

EXAMPLE 1 Show that if f is continuous on $[a, b]$, $a \neq b$, and if

$$\int_a^b f(x)\,dx = 0,$$

then $f(x) = 0$ at least once in $[a, b]$.

Solution The average value of f on $[a, b]$ is

$$\text{av}(f) = \frac{1}{b-a}\int_a^b f(x)\,dx = \frac{1}{b-a}\cdot 0 = 0.$$

By the Mean Value Theorem, f assumes this value at some point $c \in [a, b]$. ■

Fundamental Theorem, Part 1

If $f(t)$ is an integrable function over a finite interval I, then the integral from any fixed number $a \in I$ to another number $x \in I$ defines a new function F whose value at x is

$$F(x) = \int_a^x f(t)\,dt. \tag{1}$$

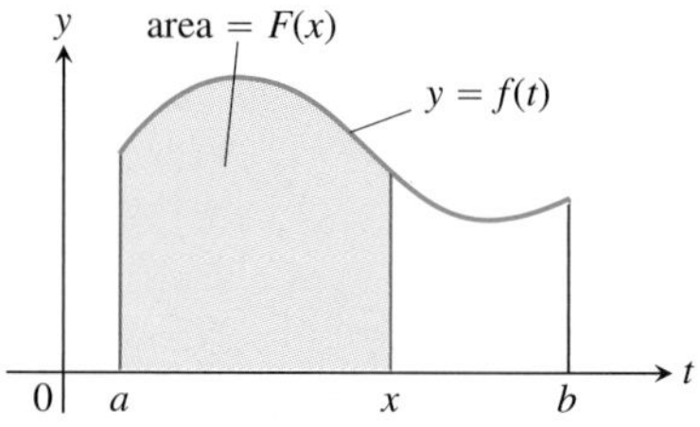

FIGURE 4.19 The function $F(x)$ defined by Equation (1) gives the area under the graph of f from a to x when f is nonnegative and $x > a$.

For example, if f is nonnegative and x lies to the right of a, then $F(x)$ is the area under the graph from a to x (Figure 4.19). The variable x is the upper limit of integration of an integral, but F is just like any other real-valued function of a real variable. For each value of the input x, there is a well-defined numerical output, in this case the definite integral of f from a to x.

Equation (1) gives a way to define new functions, (see Appendix 8), but its importance now is the connection it makes between integrals and derivatives. If f is any continuous function, then the Fundamental Theorem asserts that F is a differentiable function of x whose derivative is f itself. At every value of x, it asserts that

$$\frac{d}{dx}F(x) = f(x).$$

To gain some insight into why this result holds, we look at the geometry behind it.

If $f \ge 0$ on $[a, b]$, then the computation of $F'(x)$ from the definition of the derivative means taking the limit as $h \to 0$ of the difference quotient

$$\frac{F(x+h) - F(x)}{h}.$$

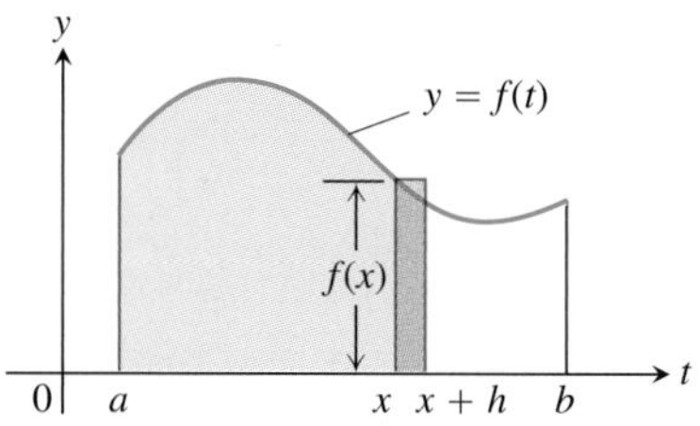

FIGURE 4.20 In Equation (1), $F(x)$ is the area to the left of x. Also, $F(x + h)$ is the area to the left of $x + h$. The difference quotient $[F(x + h) - F(x)]/h$ is then approximately equal to $f(x)$, the height of the rectangle shown here.

For $h > 0$, the numerator is obtained by subtracting two areas, so it is the area under the graph of f from x to $x + h$ (Figure 4.20). If h is small, this area is approximately equal to the area of the rectangle of height $f(x)$ and width h, which can be seen from Figure 4.20. That is,

$$F(x+h) - F(x) \approx hf(x).$$

Dividing both sides of this approximation by h and letting $h \to 0$, it is reasonable to expect that

$$F'(x) = \lim_{h\to 0}\frac{F(x+h) - F(x)}{h} = f(x).$$

This result is true even if the function f is not positive, and it forms the first part of the Fundamental Theorem of Calculus.

> **THEOREM 5—The Fundamental Theorem of Calculus, Part 1** If f is continuous on $[a, b]$ then $F(x) = \int_a^x f(t)\,dt$ is continuous on $[a, b]$ and differentiable on (a, b) and its derivative is $f(x)$;
>
> $$F'(x) = \frac{d}{dx}\int_a^x f(t)\,dt = f(x). \tag{2}$$

Before proving Theorem 5, we look at several examples to gain a better understanding of what it says.

EXAMPLE 2 Use the Fundamental Theorem to find dy/dx if

(a) $y = \int_a^x \cos t\, dt$ **(b)** $y = \int_x^5 3t \sin t\, dt$ **(c)** $y = \int_1^{x^2} \cos t\, dt$

Solution

(a) $\dfrac{dy}{dx} = \dfrac{d}{dx}\displaystyle\int_a^x \cos t\, dt = \cos x$ Eq. (2) with $f(t) = \cos t$

(b)
$$\begin{aligned} \frac{dy}{dx} = \frac{d}{dx}\int_x^5 3t \sin t\, dt &= \frac{d}{dx}\left(-\int_5^x 3t\sin t\, dt\right) && \text{Table 4.5, Rule 1} \\ &= -\frac{d}{dx}\int_5^x 3t \sin t\, dt \\ &= -3x \sin x && \text{Eq. (2) with } f(t) = 3t \sin t \end{aligned}$$

(c) The upper limit of integration is not x but x^2. This makes y a composite of the two functions,

$$y = \int_1^u \cos t\, dt \quad \text{and} \quad u = x^2.$$

We must therefore apply the Chain Rule when finding dy/dx.

$$\begin{aligned} \frac{dy}{dx} &= \frac{dy}{du} \cdot \frac{du}{dx} \\ &= \left(\frac{d}{du}\int_1^u \cos t\, dt\right) \cdot \frac{du}{dx} \\ &= \cos u \cdot \frac{du}{dx} \\ &= \cos(x^2) \cdot 2x \\ &= 2x \cos x^2 \end{aligned}$$

■

Proof of Theorem 5 We prove the Fundamental Theorem by applying the definition of the derivative directly to the function $F(x)$, when x and $x + h$ are in (a, b). This means writing out the difference quotient

$$\frac{F(x + h) - F(x)}{h} \tag{3}$$

and showing that its limit as $h \to 0$ is the number $f(x)$ for each x in (a, b). Thus,

$$\begin{aligned} F'(x) &= \lim_{h \to 0} \frac{F(x + h) - F(x)}{h} \\ &= \lim_{h \to 0} \frac{1}{h}\left[\int_a^{x+h} f(t)\, dt - \int_a^x f(t)\, dt\right] \\ &= \lim_{h \to 0} \frac{1}{h}\int_x^{x+h} f(t)\, dt && \text{Table 4.5, Rule 5} \end{aligned}$$

According to the Mean Value Theorem for Definite Integrals, the value before taking the limit in the last expression is one of the values taken on by f in the interval between x and $x + h$. That is, for some number c in this interval,

$$\frac{1}{h}\int_{x}^{x+h} f(t)\,dt = f(c). \tag{4}$$

As $h \to 0$, $x + h$ approaches x, forcing c to approach x also (because c is trapped between x and $x + h$). Since f is continuous at x, $f(c)$ approaches $f(x)$:

$$\lim_{h \to 0} f(c) = f(x). \tag{5}$$

In conclusion, we have

$$\begin{aligned} F'(x) &= \lim_{h \to 0} \frac{1}{h}\int_{x}^{x+h} f(t)\,dt \\ &= \lim_{h \to 0} f(c) && \text{Eq. (4)} \\ &= f(x). && \text{Eq. (5)} \end{aligned}$$

If $x = a$ or b, then the limit of Equation (3) is interpreted as a one-sided limit with $h \to 0^+$ or $h \to 0^-$, respectively. Then Theorem 1 in Section 2.2 shows that F is continuous for every point of $[a, b]$. This concludes the proof. ∎

Fundamental Theorem, Part 2 (The Evaluation Theorem)

We now come to the second part of the Fundamental Theorem of Calculus. This part describes how to evaluate definite integrals without having to calculate limits of Riemann sums. Instead we find and evaluate an antiderivative at the upper and lower limits of integration.

> **THEOREM 5 (Continued)—The Fundamental Theorem of Calculus, Part 2** If f is continuous at every point of $[a, b]$ and F is any antiderivative of f on $[a, b]$, then
>
> $$\int_{a}^{b} f(x)\,dx = F(b) - F(a).$$

Proof Part 1 of the Fundamental Theorem tells us that an antiderivative of f exists, namely

$$G(x) = \int_{a}^{x} f(t)\,dt.$$

Thus, if F is *any* antiderivative of f, then $F(x) = G(x) + C$ for some constant C for $a < x < b$ (by Corollary 2 of the Mean Value Theorem for Derivatives, Section 3.2). Since both F and G are continuous on $[a, b]$, we see that $F(x) = G(x) + C$ also holds when $x = a$ and $x = b$ by taking one-sided limits (as $x \to a^+$ and $x \to b^-$).

Evaluating $F(b) - F(a)$, we have

$$\begin{aligned} F(b) - F(a) &= [G(b) + C] - [G(a) + C] \\ &= G(b) - G(a) \\ &= \int_a^b f(t)\,dt - \int_a^a f(t)\,dt \\ &= \int_a^b f(t)\,dt - 0 \\ &= \int_a^b f(t)\,dt. \end{aligned}$$ ■

The theorem says that to calculate the definite integral of f over $[a, b]$ all we need to do is:

1. Find an antiderivative F of f, and
2. Calculate the number $\int_a^b f(x)\,dx = F(b) - F(a)$.

The usual notation for $F(b) - F(a)$ is

$$F(x)\Big]_a^b \quad \text{or} \quad \Big[F(x)\Big]_a^b,$$

depending on whether F has one or more terms.

EXAMPLE 3

(a) $$\int_0^{\pi} \cos x\,dx = \sin x\Big]_0^{\pi} = \sin \pi - \sin 0 = 0 - 0 = 0$$

(b) $$\int_0^{1/2} \frac{dx}{\sqrt{1 - x^2}} = \sin^{-1} x\Big]_0^{1/2} = \sin^{-1}\frac{1}{2} - \sin^{-1} 0 = \frac{\pi}{6} - 0 = \frac{\pi}{6}$$

(c) $$\begin{aligned} \int_1^4 \left(\frac{3}{2}\sqrt{x} - \frac{2}{x}\right) dx &= \Big[x^{3/2} - 2\ln x\Big]_1^4 = [4^{3/2} - 2\ln 4] - [1^{3/2} - 2\ln 1] \\ &= [8 - \ln 16] - [1 - 0] = 7 - \ln 16. \end{aligned}$$ ■

The process used in Example 3 was much easier than a Riemann sum computation.

The conclusions of the Fundamental Theorem tell us several things. Equation (2) can be rewritten as

$$\frac{d}{dx}\int_a^x f(t)\,dt = \frac{dF}{dx} = f(x),$$

which says that if you first integrate the function f and then differentiate the result, you get the function f back again. Likewise, the equation

$$\int_a^x \frac{dF}{dt}\,dt = \int_a^x f(t)\,dt = F(x) - F(a)$$

says that if you first differentiate the function F and then integrate the result, you get the function F back (adjusted by an integration constant). In a sense, the processes of integration and differentiation are "inverses" of each other. The Fundamental Theorem also says that every continuous function f has an antiderivative F, and we now see the importance of finding antiderivatives in order to evaluate definite integrals easily. Furthermore, the theorem says that the differential equation $dy/dx = f(x)$ has a solution (namely, any of the functions $y = F(x) + C$) for every continuous function f.

Total Area

The Riemann sum contains terms such as $f(c_k)\,\Delta_k$ which give the area of a rectangle when $f(c_k)$ is positive. When $f(c_k)$ is negative, then the product $f(c_k)\,\Delta_k$ is the negative of the rectangle's area. When we add up such terms for a negative function we get the negative of the area between the curve and the x-axis. If we then take the absolute value, we obtain the correct positive area.

To compute the area of the region bounded by the graph of a function $y = f(x)$ and the x-axis when the function takes on both positive and negative values, we must be careful to break up the interval $[a, b]$ into subintervals on which the function doesn't change sign. Otherwise we might get cancellation between positive and negative signed areas, leading to an incorrect total. The correct total area is obtained by adding the absolute value of the definite integral over each subinterval where $f(x)$ does not change sign. The term "area" will be taken to mean *total area*.

EXAMPLE 4 Figure 4.21 shows the graph of the function $f(x) = \sin x$ between $x = 0$ and $x = 2\pi$. Compute

(a) the definite integral of $f(x)$ over $[0, 2\pi]$.

(b) the area between the graph of $f(x)$ and the x-axis over $[0, 2\pi]$.

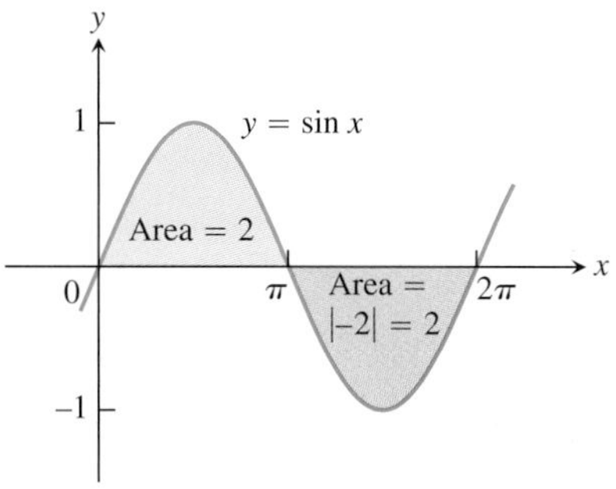

FIGURE 4.21 The total area between $y = \sin x$ and the x-axis for $0 \le x \le 2\pi$ is the sum of the absolute values of two integrals (Example 4).

Solution The definite integral for $f(x) = \sin x$ is given by

$$\int_0^{2\pi} \sin x\, dx = -\cos x\Big]_0^{2\pi} = -[\cos 2\pi - \cos 0] = -[1 - 1] = 0.$$

The definite integral is zero because the portions of the graph above and below the x-axis make canceling contributions.

The area between the graph of $f(x)$ and the x-axis over $[0, 2\pi]$ is calculated by breaking up the domain of $\sin x$ into two pieces: the interval $[0, \pi]$ over which it is nonnegative and the interval $[\pi, 2\pi]$ over which it is nonpositive.

$$\int_0^{\pi} \sin x\, dx = -\cos x\Big]_0^{\pi} = -[\cos \pi - \cos 0] = -[-1 - 1] = 2.$$

$$\int_{\pi}^{2\pi} \sin x\, dx = -\cos x\Big]_{\pi}^{2\pi} = -[\cos 2\pi - \cos \pi] = -[1 - (-1)] = -2.$$

The second integral gives a negative value. The area between the graph and the axis is obtained by adding the absolute values

$$\text{Area} = |2| + |-2| = 4.$$

■

Summary:

To find the area between the graph of $y = f(x)$ and the x-axis over the interval $[a, b]$:

1. Subdivide $[a, b]$ at the zeros of f.
2. Integrate f over each subinterval.
3. Add the absolute values of the integrals.

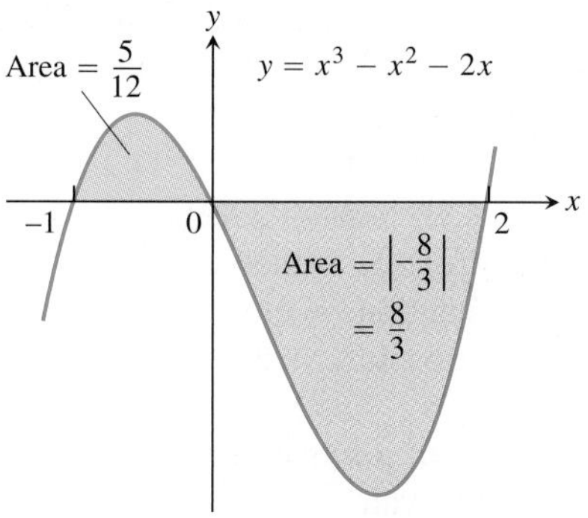

FIGURE 4.22 The region between the curve $y = x^3 - x^2 - 2x$ and the x-axis (Example 5).

EXAMPLE 5 Find the area of the region between the x-axis and the graph of $f(x) = x^3 - x^2 - 2x, -1 \le x \le 2$.

Solution First find the zeros of f. Since

$$f(x) = x^3 - x^2 - 2x = x(x^2 - x - 2) = x(x + 1)(x - 2),$$

the zeros are $x = 0, -1$, and 2 (Figure 4.22). The zeros subdivide $[-1, 2]$ into two subintervals: $[-1, 0]$, on which $f \ge 0$, and $[0, 2]$, on which $f \le 0$. We integrate f over each subinterval and add the absolute values of the calculated integrals.

$$\int_{-1}^{0} (x^3 - x^2 - 2x)\,dx = \left[\frac{x^4}{4} - \frac{x^3}{3} - x^2\right]_{-1}^{0} = 0 - \left[\frac{1}{4} + \frac{1}{3} - 1\right] = \frac{5}{12}$$

$$\int_{0}^{2} (x^3 - x^2 - 2x)\,dx = \left[\frac{x^4}{4} - \frac{x^3}{3} - x^2\right]_{0}^{2} = \left[4 - \frac{8}{3} - 4\right] - 0 = -\frac{8}{3}$$

The total enclosed area is obtained by adding the absolute values of the calculated integrals,

$$\text{Total enclosed area} = \frac{5}{12} + \left|-\frac{8}{3}\right| = \frac{37}{12}.$$

■

EXERCISES 4.5

Evaluate the integrals in Exercises 1–32.

1. $\int_{-2}^{0} (2x + 5)\,dx$
2. $\int_{-3}^{4} \left(5 - \frac{x}{2}\right) dx$
3. $\int_{0}^{4} \left(3x - \frac{x^3}{4}\right) dx$
4. $\int_{-2}^{2} (x^3 - 2x + 3)\,dx$
5. $\int_{0}^{1} \left(x^2 + \sqrt{x}\right) dx$
6. $\int_{0}^{5} x^{3/2}\,dx$
7. $\int_{1}^{32} x^{-6/5}\,dx$
8. $\int_{-2}^{-1} \frac{2}{x^2}\,dx$
9. $\int_{0}^{\pi} \sin x\,dx$
10. $\int_{0}^{\pi} (1 + \cos x)\,dx$
11. $\int_{0}^{\pi/3} 2\sec^2 x\,dx$
12. $\int_{\pi/6}^{5\pi/6} \csc^2 x\,dx$
13. $\int_{\pi/4}^{3\pi/4} \csc\theta\cot\theta\,d\theta$
14. $\int_{0}^{\pi/3} 4\sec u\tan u\,du$
15. $\int_{\pi/2}^{0} \frac{1 + \cos 2t}{2}\,dt$
16. $\int_{-\pi/3}^{\pi/3} \frac{1 - \cos 2t}{2}\,dt$
17. $\int_{-\pi/2}^{\pi/2} (8y^2 + \sin y)\,dy$
18. $\int_{-\pi/3}^{-\pi/4} \left(4\sec^2 t + \frac{\pi}{t^2}\right) dt$
19. $\int_{1}^{-1} (r + 1)^2\,dr$
20. $\int_{-\sqrt{3}}^{\sqrt{3}} (t + 1)(t^2 + 4)\,dt$
21. $\int_{\sqrt{2}}^{1} \left(\frac{u^7}{2} - \frac{1}{u^5}\right) du$
22. $\int_{1/2}^{1} \left(\frac{1}{v^3} - \frac{1}{v^4}\right) dv$
23. $\int_{1}^{\sqrt{2}} \frac{s^2 + \sqrt{s}}{s^2}\,ds$
24. $\int_{9}^{4} \frac{1 - \sqrt{u}}{\sqrt{u}}\,du$
25. $\int_{-4}^{4} |x|\,dx$
26. $\int_{0}^{\pi} \frac{1}{2}(\cos x + |\cos x|)\,dx$
27. $\int_{0}^{\ln 2} e^{3x}\,dx$
28. $\int_{1}^{2} \left(\frac{1}{x} - e^{-x}\right) dx$
29. $\int_{0}^{1} \frac{4}{1 + x^2}\,dx$
30. $\int_{2}^{5} \frac{x\,dx}{\sqrt{1 + x^2}}$
31. $\int_{2}^{4} x^{\pi - 1}\,dx$
32. $\int_{-1}^{0} \pi^{x-1}\,dx$

In Exercises 33 and 34, guess an antiderivative for the integrand function. Validate your guess by differentiation and then evaluate the given definite integral. (*Hint:* Keep in mind the Chain Rule in guessing an antiderivative. You will learn how to find such antiderivatives in the next section.)

33. $\int_{0}^{1} xe^{x^2}\,dx$
34. $\int_{1}^{2} \frac{\ln x}{x}\,dx$

Find the derivatives in Exercises 35–40.

a. by evaluating the integral and differentiating the result.

b. by differentiating the integral directly.

35. $\dfrac{d}{dx}\displaystyle\int_0^{\sqrt{x}} \cos t\, dt$

36. $\dfrac{d}{dx}\displaystyle\int_1^{\sin x} 3t^2\, dt$

37. $\dfrac{d}{dt}\displaystyle\int_0^{t^4} \sqrt{u}\, du$

38. $\dfrac{d}{d\theta}\displaystyle\int_0^{\tan\theta} \sec^2 y\, dy$

39. $\dfrac{d}{dx}\displaystyle\int_0^{x^3} e^{-t}\, dt$

40. $\dfrac{d}{dt}\displaystyle\int_0^{\sqrt{t}} \left(x^4 + \dfrac{3}{\sqrt{1-x^2}}\right) dx$

Find dy/dx in Exercises 41–50.

41. $y = \displaystyle\int_0^x \sqrt{1+t^2}\, dt$

42. $y = \displaystyle\int_1^x \frac{1}{t}\, dt, \quad x > 0$

43. $y = \displaystyle\int_{\sqrt{x}}^0 \sin(t^2)\, dt$

44. $y = \displaystyle\int_0^{x^2} \cos\sqrt{t}\, dt$

45. $y = \displaystyle\int_0^{\sin x} \frac{dt}{\sqrt{1-t^2}}, \quad |x| < \frac{\pi}{2}$

46. $y = \displaystyle\int_{\tan x}^0 \frac{dt}{1+t^2}$

47. $y = \displaystyle\int_0^{e^{x^2}} \frac{1}{\sqrt{t}}\, dt$

48. $y = \displaystyle\int_{2^x}^1 \sqrt[3]{t}\, dt$

49. $y = \displaystyle\int_0^{\sin^{-1} x} \cos t\, dt$

50. $y = \displaystyle\int_{-1}^{x^{1/\pi}} \sin^{-1} t\, dt$

In Exercises 51–56, find the total area between the region and the x-axis.

51. $y = -x^2 - 2x, \quad -3 \le x \le 2$

52. $y = 3x^2 - 3, \quad -2 \le x \le 2$

53. $y = x^3 - 3x^2 + 2x, \quad 0 \le x \le 2$

54. $y = x^3 - 4x, \quad -2 \le x \le 2$

55. $y = x^{1/3}, \quad -1 \le x \le 8$

56. $y = x^{1/3} - x, \quad -1 \le x \le 8$

Find the areas of the shaded regions in Exercises 57–60.

57.

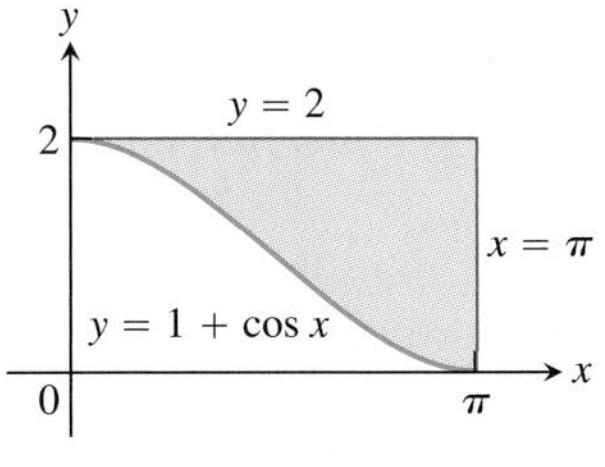

58.

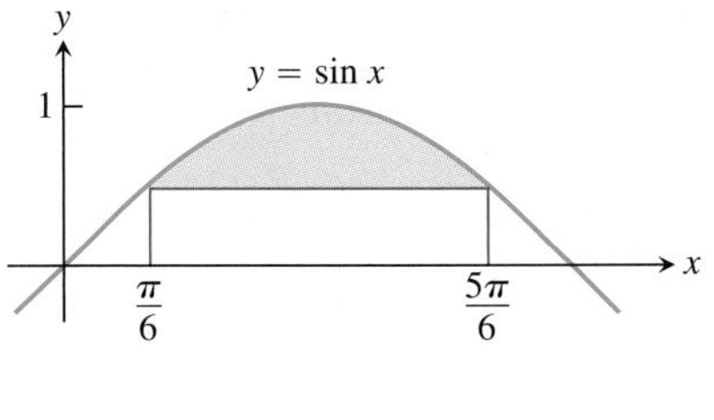

59.

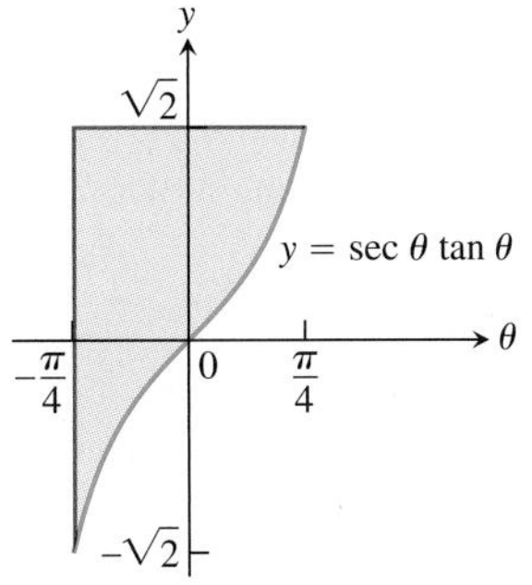

60.

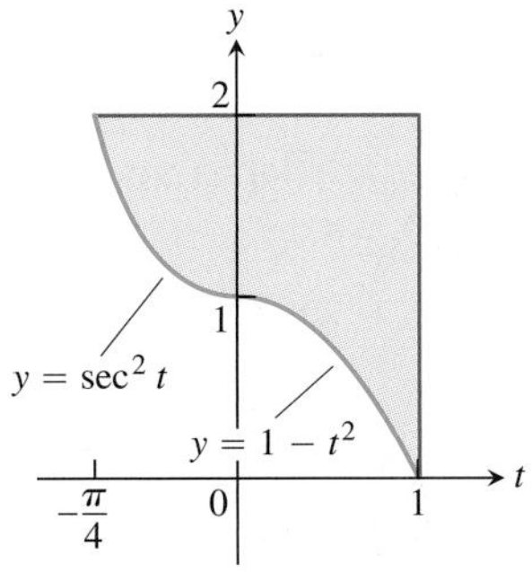

Each of the following functions solves one of the initial value problems in Exercises 61–64. Which function solves which problem? Give brief reasons for your answers.

a. $y = \displaystyle\int_1^x \frac{1}{t}\, dt - 3$

b. $y = \displaystyle\int_0^x \sec t\, dt + 4$

c. $y = \displaystyle\int_{-1}^x \sec t\, dt + 4$

d. $y = \displaystyle\int_\pi^x \frac{1}{t}\, dt - 3$

61. $\dfrac{dy}{dx} = \dfrac{1}{x}, \quad y(\pi) = -3$

62. $y' = \sec x, \quad y(-1) = 4$

63. $y' = \sec x, \quad y(0) = 4$

64. $y' = \dfrac{1}{x}, \quad y(1) = -3$

Express the solutions of the initial value problems in Exercises 65–68 in terms of integrals.

65. $\dfrac{dy}{dx} = \sec x, \quad y(2) = 3$

66. $\dfrac{dy}{dx} = \sqrt{1+x^2}, \quad y(1) = -2$

67. $\dfrac{ds}{dt} = f(t), \quad s(t_0) = s_0$

68. $\dfrac{dv}{dt} = g(t), \quad v(t_0) = v_0$

69. Archimedes' area formula for parabolas Archimedes (287–212 B.C.), inventor, military engineer, physicist, and the greatest mathematician of classical times in the Western world, discovered that the area under a parabolic arch is two-thirds the base times the height. Sketch the parabolic arch $y = h - (4h/b^2)x^2$, $-b/2 \le x \le b/2$, assuming that h and b are positive. Then use calculus to find the area of the region enclosed between the arch and the x-axis.

70. Revenue from marginal revenue Suppose that a company's marginal revenue from the manufacture and sale of egg beaters is

$$\frac{dr}{dx} = 2 - 2/(x+1)^2,$$

where r is measured in thousands of dollars and x in thousands of units. How much money should the company expect from a production run of $x = 3$ thousand egg beaters? To find out, integrate the marginal revenue from $x = 0$ to $x = 3$.

71. Cost from marginal cost The marginal cost of printing a poster when x posters have been printed is

$$\frac{dc}{dx} = \frac{1}{2\sqrt{x}}$$

dollars. Find $c(100) - c(1)$, the cost of printing posters 2–100.

72. Find

$$\lim_{x \to 0} \frac{1}{x^3} \int_0^x \frac{t^2}{t^4 + 1}\, dt.$$

73. Suppose $\int_1^x f(t)\, dt = x^2 - 2x + 1$. Find $f(x)$.

74. Find the linearization of

$$f(x) = 2 - \int_2^{x+1} \frac{9}{1 + t}\, dt$$

at $x = 1$.

COMPUTER EXPLORATIONS

In Exercises 75–78, let $F(x) = \int_a^x f(t)\, dt$ for the specified function f and interval $[a, b]$. Use a CAS to perform the following steps and answer the questions posed.

a. Plot the functions f and F together over $[a, b]$.

b. Solve the equation $F'(x) = 0$. What can you see to be true about the graphs of f and F at points where $F'(x) = 0$? Is your observation borne out by Part 1 of the Fundamental Theorem coupled with information provided by the first derivative? Explain your answer.

c. Over what intervals (approximately) is the function F increasing and decreasing? What is true about f over those intervals?

d. Calculate the derivative f' and plot it together with F. What can you see to be true about the graph of F at points where $f'(x) = 0$? Is your observation borne out by Part 1 of the Fundamental Theorem? Explain your answer.

75. $f(x) = x^3 - 4x^2 + 3x, \quad [0, 4]$

76. $f(x) = 2x^4 - 17x^3 + 46x^2 - 43x + 12, \quad \left[0, \frac{9}{2}\right]$

77. $f(x) = \sin 2x \cos \frac{x}{3}, \quad [0, 2\pi]$

78. $f(x) = x \cos \pi x, \quad [0, 2\pi]$

4.6 Indefinite Integrals and the Substitution Rule

The Fundamental Theorem of Calculus says that a definite integral of a continuous function can be computed directly if we can find an antiderivative of the function. In Section 4.1 we defined the **indefinite integral** of the function f with respect to x as the set of *all* antiderivatives of f, symbolized by

$$\int f(x)\, dx.$$

The connection between antiderivatives and the definite integral stated in the Fundamental Theorem now explains this notation. When finding the indefinite integral of a function f, remember that it always includes an arbitrary constant C.

We must distinguish carefully between definite and indefinite integrals. A definite integral $\int_a^b f(x)\, dx$ is a *number*. An indefinite integral $\int f(x)\, dx$ is a *function* plus an arbitrary constant C.

So far, we have only been able to find antiderivatives of functions that are clearly recognizable as derivatives. In this section we begin to develop more general techniques for finding antiderivatives.

Substitution: Running the Chain Rule Backwards

If u is a differentiable function of x and n is any number different from -1, the Chain Rule tells us that

$$\frac{d}{dx}\left(\frac{u^{n+1}}{n+1}\right) = u^n \frac{du}{dx}.$$

From another point of view, this same equation says that $u^{n+1}/(n+1)$ is one of the antiderivatives of the function $u^n(du/dx)$. Therefore,

$$\int\left(u^n \frac{du}{dx}\right) dx = \frac{u^{n+1}}{n+1} + C.$$

The integral on the left-hand side of this equation is usually written in the simpler "differential" form, giving

$$\int u^n\, du = \frac{u^{n+1}}{n+1} + C \qquad (n \neq -1, n \text{ any number}).$$

In deriving this equation, we assumed u to be a differentiable function of the variable x, but the name of the variable does not matter and does not appear in the final formula. We could have represented the variable with θ, t, y, or any other letter.

EXAMPLE 1 Find

$$\int \sqrt{2x+1}\, dx$$

Solution The integral does not fit the formula

$$\int u^n\, du$$

with $u = 2x + 1$ and $n = 1/2$, because

$$du = \frac{du}{dx} \cdot dx = 2\, dx$$

is not precisely dx. The constant factor 2 is missing from the integral. However, we can introduce this factor after the integral sign if we compensate for it by a factor of $1/2$ in front of the integral sign. So we write

$$\int \sqrt{2x+1}\, dx = \frac{1}{2}\int \sqrt{2x+1} \cdot 2\, dx$$

$$= \frac{1}{2}\int u^{1/2}\, du \qquad \text{Let } u = 2x+1, \quad du = 2\, dx$$

$$= \frac{1}{2}\frac{u^{3/2}}{3/2} + C$$

$$= \frac{1}{3}(2x+1)^{3/2} + C \qquad \text{Substitute } 2x+1 \text{ for } u.$$

■

The substitution in Example 1 is an instance of the following general rule.

THEOREM 6—The Substitution Rule If $u = g(x)$ is a differentiable function whose range is an interval I and f is continuous on I, then

$$\int f(g(x))g'(x)\, dx = \int f(u)\, du.$$

Proof The rule is true because, by the Chain Rule, $F(g(x))$ is an antiderivative of $f(g(x)) \cdot g'(x)$ whenever F is an antiderivative of f:

$$\begin{aligned} \frac{d}{dx} F(g(x)) &= F'(g(x)) \cdot g'(x) && \text{Chain Rule} \\ &= f(g(x)) \cdot g'(x). && \text{Because } F' = f \end{aligned}$$

If we make the substitution $u = g(x)$ then

$$\begin{aligned} \int f(g(x))g'(x)\,dx &= \int \frac{d}{dx} F(g(x))\,dx \\ &= F(g(x)) + C && \text{Fundamental Theorem} \\ &= F(u) + C && u = g(x) \\ &= \int F'(u)\,du && \text{Fundamental Theorem} \\ &= \int f(u)\,du && F' = f \end{aligned}$$

■

The Substitution Rule provides the following method to evaluate the integral

$$\int f(g(x))g'(x)\,dx,$$

when f and g' are continuous functions:

1. Substitute $u = g(x)$ and $du = g'(x)\,dx$ to obtain the integral
$$\int f(u)\,du.$$
2. Integrate with respect to u.
3. Replace u by $g(x)$ in the result.

EXAMPLE 2

$$\begin{aligned} \int \cos(7\theta + 5)\,d\theta &= \int \cos u \cdot \frac{1}{7}\,du && \text{Let } u = 7\theta + 5,\ du = 7\,d\theta,\ (1/7)\,du = d\theta. \\ &= \frac{1}{7}\int \cos u\,du && \text{With the (1/7) out front, the integral is now in standard form.} \\ &= \frac{1}{7}\sin u + C && \text{Integrate with respect to } u, \text{ Table 4.1.} \\ &= \frac{1}{7}\sin(7\theta + 5) + C && \text{Replace } u \text{ by } 7\theta + 5. \end{aligned}$$

We can verify this solution by differentiating and checking that we obtain the original function $\cos(7\theta + 5)$. ■

EXAMPLE 3

$$\begin{aligned} \int x^2 e^{x^3}\,dx &= \int e^{x^3} \cdot x^2\,dx \\ &= \int e^u \cdot \frac{1}{3}\,du && \text{Let } u = x^3,\ du = 3x^2\,dx,\ (1/3)\,du = x^2\,dx. \end{aligned}$$

$$= \frac{1}{3}\int e^u \, du$$

$$= \frac{1}{3}e^u + C \qquad \text{Integrate with respect to } u.$$

$$= \frac{1}{3}e^{x^3} + C \qquad \text{Replace } u \text{ by } x^3 + 1. \qquad \blacksquare$$

An integrand may require some algebraic manipulation before the substitution method can be applied. The next example gives an important integral obtained by multiplying the integrand by an algebraic form equal to 1, leading to an appropriate substitution.

HISTORICAL BIOGRAPHY

George David Birkhoff (1884–1944)

EXAMPLE 4

$$\int \sec x \, dx = \int (\sec x)(1) \, dx = \int \sec x \cdot \frac{\sec x + \tan x}{\sec x + \tan x} \, dx \qquad \frac{\sec x + \tan x}{\sec x + \tan x} \text{ is a form of 1}$$

$$= \int \frac{\sec^2 x + \sec x \tan x}{\sec x + \tan x} \, dx$$

$$= \int \frac{du}{u} \qquad u = \tan x + \sec x,\ du = (\sec^2 x + \sec x \tan x)\, dx$$

$$= \ln|u| + C = \ln|\sec x + \tan x| + C. \qquad \blacksquare$$

EXAMPLE 5

$$\int \frac{\ln x^2}{x} \, dx = \int \frac{2 \ln x}{x} \, dx \qquad \text{Power Rule for logarithms}$$

$$= \int 2 \ln x \cdot \frac{1}{x} \, dx$$

$$= \int 2u \, du \qquad \text{Let } u = \ln x,\ du = (1/x)\, dx.$$

$$= u^2 + C \qquad \text{Integrate with respect to } u.$$

$$= (\ln x)^2 + C \qquad \text{Replace } u \text{ by } \ln x. \qquad \blacksquare$$

EXAMPLE 6

$$\int \frac{1}{\cos^2 2x} \, dx = \int \sec^2 2x \, dx \qquad \frac{1}{\cos 2x} = \sec 2x$$

$$= \int \sec^2 u \cdot \frac{1}{2} \, du \qquad u = 2x,\ du = 2\, dx,\ dx = (1/2)\, du$$

$$= \frac{1}{2}\int \sec^2 u \, du$$

$$= \frac{1}{2}\tan u + C \qquad \frac{d}{du}\tan u = \sec^2 u$$

$$= \frac{1}{2}\tan 2x + C \qquad u = 2x \qquad \blacksquare$$

The success of the substitution method depends on finding a substitution that changes an integral we cannot evaluate directly into one that we can. If the first substitution fails, try to simplify the integrand further with additional substitutions.

EXAMPLE 7 Evaluate

$$\int \frac{2z\,dz}{\sqrt[3]{z^2+1}}.$$

Solution We can use the substitution method of integration as an exploratory tool: Substitute for the most troublesome part of the integrand and see how things work out. For the integral here, we might try $u = z^2 + 1$ or we might even press our luck and take u to be the entire cube root. Here is what happens in each case.

Solution 1: Substitute $u = z^2 + 1$.

$$\begin{aligned}
\int \frac{2z\,dz}{\sqrt[3]{z^2+1}} &= \int \frac{du}{u^{1/3}} && \text{Let } u = z^2+1,\ du = 2z\,dz. \\
&= \int u^{-1/3}\,du && \text{In the form } \textstyle\int u^n\,du \\
&= \frac{u^{2/3}}{2/3} + C && \text{Integrate with respect to } u. \\
&= \frac{3}{2}u^{2/3} + C \\
&= \frac{3}{2}(z^2+1)^{2/3} + C && \text{Replace } u \text{ by } z^2+1.
\end{aligned}$$

Solution 2: Substitute $u = \sqrt[3]{z^2+1}$ instead.

$$\begin{aligned}
\int \frac{2z\,dz}{\sqrt[3]{z^2+1}} &= \int \frac{3u^2\,du}{u} && \text{Let } u = \sqrt[3]{z^2+1},\ u^3 = z^2+1,\ 3u^2\,du = 2z\,dz. \\
&= 3\int u\,du \\
&= 3\cdot\frac{u^2}{2} + C && \text{Integrate with respect to } u. \\
&= \frac{3}{2}(z^2+1)^{2/3} + C && \text{Replace } u \text{ by } (z^2+1)^{1/3}.
\end{aligned}$$

■

The Integrals of $\sin^2 x$ and $\cos^2 x$

Sometimes we can use trigonometric identities to transform integrals we do not know how to evaluate into ones we can evaluate using the substitution rule.

EXAMPLE 8

(a)
$$\begin{aligned}
\int \sin^2 x\,dx &= \int \frac{1-\cos 2x}{2}\,dx && \sin^2 x = \frac{1-\cos 2x}{2} \\
&= \frac{1}{2}\int (1-\cos 2x)\,dx = \frac{1}{2}\int dx - \frac{1}{2}\int \cos 2x\,dx \\
&= \frac{1}{2}x - \frac{1}{2}\frac{\sin 2x}{2} + C = \frac{x}{2} - \frac{\sin 2x}{4} + C
\end{aligned}$$

(b)
$$\begin{aligned}
\int \cos^2 x\,dx &= \int \frac{1+\cos 2x}{2}\,dx && \cos^2 x = \frac{1+\cos 2x}{2} \\
&= \frac{x}{2} + \frac{\sin 2x}{4} + C
\end{aligned}$$

■

EXERCISES 4.6

Evaluate the indefinite integrals in Exercises 1–12 by using the given substitutions to reduce the integrals to standard form.

1. $\int \sin 3x\, dx, \quad u = 3x$

2. $\int x \sin(2x^2)\, dx, \quad u = 2x^2$

3. $\int \sec 2t \tan 2t\, dt, \quad u = 2t$

4. $\int \left(1 - \cos\frac{t}{2}\right)^2 \sin\frac{t}{2}\, dt, \quad u = 1 - \cos\frac{t}{2}$

5. $\int 28(7x - 2)^{-5}\, dx, \quad u = 7x - 2$

6. $\int x^3(x^4 - 1)^2\, dx, \quad u = x^4 - 1$

7. $\int \frac{9r^2\, dr}{\sqrt{1 - r^3}}, \quad u = 1 - r^3$

8. $\int 12(y^4 + 4y^2 + 1)^2(y^3 + 2y)\, dy, \quad u = y^4 + 4y^2 + 1$

9. $\int \sqrt{x} \sin^2(x^{3/2} - 1)\, dx, \quad u = x^{3/2} - 1$

10. $\int \frac{1}{x^2}\cos^2\left(\frac{1}{x}\right) dx, \quad u = -\frac{1}{x}$

11. $\int \csc^2 2\theta \cot 2\theta\, d\theta$

   a. Using $u = \cot 2\theta$

   b. Using $u = \csc 2\theta$

12. $\int \frac{dx}{\sqrt{5x + 8}}$

   a. Using $u = 5x + 8$

   b. Using $u = \sqrt{5x + 8}$

Evaluate the integrals in Exercises 13–54.

13. $\int \sqrt{3 - 2s}\, ds$

14. $\int (2x + 1)^3\, dx$

15. $\int \frac{1}{\sqrt{5s + 4}}\, ds$

16. $\int \frac{3\, dx}{(2 - x)^2}$

17. $\int \theta \sqrt[4]{1 - \theta^2}\, d\theta$

18. $\int \frac{4y\, dy}{\sqrt{2y^2 + 1}}$

19. $\int \frac{1}{\sqrt{x}(1 + \sqrt{x})^2}\, dx$

20. $\int \frac{(1 + \sqrt{x})^3}{\sqrt{x}}\, dx$

21. $\int \cos(3z + 4)\, dz$

22. $\int \tan^2 x \sec^2 x\, dx$

23. $\int \tan x\, dx$

24. $\int \tan^7 \frac{x}{2} \sec^2 \frac{x}{2}\, dx$

25. $\int r^2 \left(\frac{r^3}{18} - 1\right)^5 dr$

26. $\int r^4 \left(7 - \frac{r^5}{10}\right)^3 dr$

27. $\int x^{1/2} \sin(x^{3/2} + 1)\, dx$

28. $\int x^{1/3} \sin(x^{4/3} - 8)\, dx$

29. $\int \frac{\sin(2t + 1)}{\cos^2(2t + 1)}\, dt$

30. $\int \frac{6 \cos t}{(2 + \sin t)^3}\, dt$

31. $\int \frac{1}{\theta^2} \sin\frac{1}{\theta} \cos\frac{1}{\theta}\, d\theta$

32. $\int \frac{\sec z \tan z}{\sqrt{\sec z}}\, dz$

33. $\int \frac{1}{t^2} \cos\left(\frac{1}{t} - 1\right) dt$

34. $\int \frac{\cos\sqrt{\theta}}{\sqrt{\theta}\sin^2\sqrt{\theta}}\, d\theta$

35. $\int (s^3 + 2s^2 - 5s + 5)(3s^2 + 4s - 5)\, ds$

36. $\int t^3(1 + t^4)^3\, dt$

37. $\int \sqrt{\frac{x - 1}{x^5}}\, dx$

38. $\int x^3 \sqrt{x^2 + 1}\, dx$

39. $\int (\cos x)\, e^{\sin x}\, dx$

40. $\int (\sin 2\theta)\, e^{\sin^2\theta}\, d\theta$

41. $\int \frac{1}{\sqrt{x e^{-\sqrt{x}}}} \sec^2(e^{\sqrt{x}} + 1)\, dx$

42. $\int \frac{1}{x^2} e^{1/x} \sec(1 + e^{1/x}) \tan(1 + e^{1/x})\, dx$

43. $\int \frac{dx}{x \ln x}$

44. $\int \frac{\ln\sqrt{t}}{t}\, dt$

45. $\int \frac{dz}{1 + e^z}$

46. $\int \frac{dx}{x\sqrt{x^4 - 1}}$

47. $\int \frac{5}{9 + 4r^2}\, dr$

48. $\int \frac{1}{\sqrt{e^{2\theta} - 1}}\, d\theta$

49. $\int \frac{e^{\sin^{-1} x}\, dx}{\sqrt{1 - x^2}}$

50. $\int \frac{e^{\cos^{-1} x}\, dx}{\sqrt{1 - x^2}}$

51. $\int \frac{(\sin^{-1} x)^2\, dx}{\sqrt{1 - x^2}}$

52. $\int \frac{\sqrt{\tan^{-1} x}\, dx}{1 + x^2}$

53. $\int \frac{dy}{(\tan^{-1} y)(1 + y^2)}$

54. $\int \frac{dy}{(\sin^{-1} y)\sqrt{1 - y^2}}$

Solve the initial value problems in Exercises 55–58.

55. $\frac{ds}{dt} = 12t(3t^2 - 1)^3, \quad s(1) = 3$

56. $\dfrac{ds}{dt} = 8\sin^2\left(t + \dfrac{\pi}{12}\right), \quad s(0) = 8$

57. $\dfrac{d^2s}{dt^2} = -4\sin\left(2t - \dfrac{\pi}{2}\right), \quad s'(0) = 100, \quad s(0) = 0$

58. $\dfrac{d^2y}{dx^2} = 4\sec^2 2x \tan 2x, \quad y'(0) = 4, \quad y(0) = -1$

59. The velocity of a particle moving back and forth on a line is $v = ds/dt = 6\sin 2t$ m/sec for all t. If $s = 0$ when $t = 0$, find the value of s when $t = \pi/2$ sec.

60. The acceleration of a particle moving back and forth on a line is $a = d^2s/dt^2 = \pi^2\cos\pi t$ m/sec^2 for all t. If $s = 0$ and $v = 8$ m/sec when $t = 0$, find s when $t = 1$ sec.

4.7 Substitution and Area Between Curves

There are two methods for evaluating a definite integral by substitution. The first method is to find an antiderivative using substitution, and then to evaluate the definite integral by applying the Fundamental Theorem. The second method extends the process of substitution directly to *definite* integrals by changing the limits of integration. We apply the new formula introduced here to the problem of computing the area between two curves.

Substitution Formula

The following formula shows how the limits of integration change when the variable of integration is changed by substitution.

THEOREM 7—Substitution in Definite Integrals If g' is continuous on the interval $[a, b]$ and f is continuous on the range of g, then

$$\int_a^b f(g(x)) \cdot g'(x)\, dx = \int_{g(a)}^{g(b)} f(u)\, du$$

Proof Let F denote any antiderivative of f. Then,

$$\begin{aligned}\int_a^b f(g(x)) \cdot g'(x)\, dx &= F(g(x))\Big]_{x=a}^{x=b} && \frac{d}{dx}F(g(x)) = F'(g(x))g'(x) = f(g(x))g'(x)\\ &= F(g(b)) - F(g(a)) \\ &= F(u)\Big]_{u=g(a)}^{u=g(b)} \\ &= \int_{g(a)}^{g(b)} f(u)\, du. && \text{Fundamental Theorem, Part 2}\end{aligned}$$

■

To use the formula, make the same u-substitution $u = g(x)$ and $du = g'(x)\,dx$ you would use to evaluate the corresponding indefinite integral. Then integrate the transformed integral with respect to u from the value $g(a)$ (the value of u at $x = a$) to the value $g(b)$ (the value of u at $x = b$).

EXAMPLE 1 Evaluate $\displaystyle\int_{-1}^{1} 3x^2\sqrt{x^3 + 1}\, dx$.

Solution We have two choices.

Method 1: Transform the integral and evaluate the transformed integral with the transformed limits given in Theorem 7.

$$\int_{-1}^{1} 3x^2\sqrt{x^3+1}\,dx \qquad \text{Let } u = x^3+1,\ du = 3x^2\,dx. \text{ When } x=-1,\ u=(-1)^3+1=0. \text{ When } x=1,\ u=(1)^3+1=2.$$

$$= \int_0^2 \sqrt{u}\,du$$

$$= \frac{2}{3}u^{3/2}\Big]_0^2 \qquad \text{Evaluate the new definite integral.}$$

$$= \frac{2}{3}\left[2^{3/2} - 0^{3/2}\right] = \frac{2}{3}\left[2\sqrt{2}\right] = \frac{4\sqrt{2}}{3}$$

Method 2: Transform the integral as an indefinite integral, integrate, change back to x, and use the original x-limits.

$$\int 3x^2\sqrt{x^3+1}\,dx = \int \sqrt{u}\,du \qquad \text{Let } u = x^3+1,\ du = 3x^2\,dx.$$

$$= \frac{2}{3}u^{3/2} + C \qquad \text{Integrate with respect to } u.$$

$$= \frac{2}{3}(x^3+1)^{3/2} + C \qquad \text{Replace } u \text{ by } x^3+1.$$

$$\int_{-1}^{1} 3x^2\sqrt{x^3+1}\,dx = \frac{2}{3}(x^3+1)^{3/2}\Big]_{-1}^{1} \qquad \text{Use the integral just found, with limits of integration for } x.$$

$$= \frac{2}{3}\left[((1)^3+1)^{3/2} - ((-1)^3+1)^{3/2}\right]$$

$$= \frac{2}{3}\left[2^{3/2} - 0^{3/2}\right] = \frac{2}{3}\left[2\sqrt{2}\right] = \frac{4\sqrt{2}}{3}$$

■

Which method is better—evaluating the transformed definite integral with transformed limits using Theorem 7, or transforming the integral, integrating, and transforming back to use the original limits of integration? In Example 1, the first method seems easier, but that is not always the case. Generally, it is best to know both methods and to use whichever one seems better at the time.

EXAMPLE 2

(a)

$$\int_0^{\ln 2} e^{3x}\,dx = \int_0^{\ln 8} e^u \cdot \frac{1}{3}\,du \qquad u = 3x,\ \frac{1}{3}du = dx,\ u(0)=0,\ u(\ln 2) = 3\ln 2 = \ln 2^3 = \ln 8$$

$$= \frac{1}{3}\int_0^{\ln 8} e^u\,du$$

$$= \frac{1}{3}e^u\Big]_0^{\ln 8}$$

$$= \frac{1}{3}[8-1] = \frac{7}{3}$$

$$\textbf{(b)} \int_{-\pi/4}^{\pi/4} \tan x\, dx = \int_{-\pi/4}^{\pi/4} \frac{\sin x}{\cos x}\, dx$$

$$= -\int_{\sqrt{2}/2}^{\sqrt{2}/2} \frac{du}{u}$$ $u = \cos x$, $du = -\sin x\, dx$. When $x = -\pi/4$, $u = \sqrt{2}/2$. When $x = \pi/4$, $u = \sqrt{2}/2$.

$$= -\ln|u| \Big]_{\sqrt{2}/2}^{\sqrt{2}/2} = 0$$ Integrate, zero width interval ■

Definite Integrals of Symmetric Functions

The Substitution Formula in Theorem 7 simplifies the calculation of definite integrals of even and odd functions (Section 1.1) over a symmetric interval $[-a, a]$ (Figure 4.23). The following theorem explains the zero integral in Example 2b.

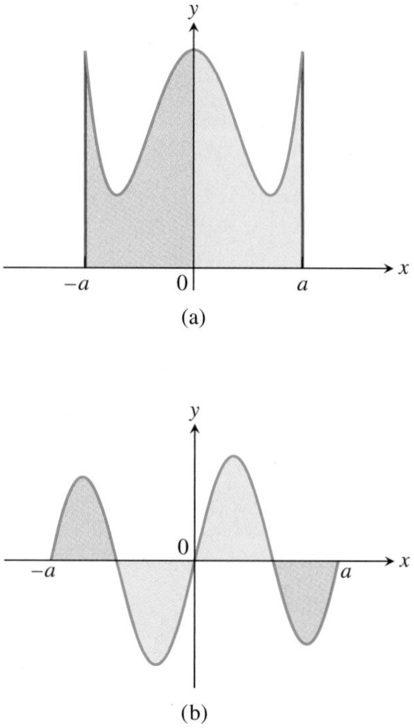

FIGURE 4.23 (a) f even, $\int_{-a}^{a} f(x)\, dx = 2\int_0^a f(x)\, dx$ (b) f odd, $\int_{-a}^{a} f(x)\, dx = 0$

THEOREM 8 Let f be continuous on the symmetric interval $[-a, a]$.

(a) If f is even, then $\int_{-a}^{a} f(x)\, dx = 2\int_0^a f(x)\, dx$.

(b) If f is odd, then $\int_{-a}^{a} f(x)\, dx = 0$.

Proof of Part (a)

$$\int_{-a}^{a} f(x)\, dx = \int_{-a}^{0} f(x)\, dx + \int_0^a f(x)\, dx$$ Additivity Rule for Definite Integrals

$$= -\int_0^{-a} f(x)\, dx + \int_0^a f(x)\, dx$$ Order of Integration Rule

$$= -\int_0^a f(-u)(-du) + \int_0^a f(x)\, dx$$ Let $u = -x$, $du = -dx$. When $x = 0$, $u = 0$. When $x = -a$, $u = a$.

$$= \int_0^a f(-u)\, du + \int_0^a f(x)\, dx$$

$$= \int_0^a f(u)\, du + \int_0^a f(x)\, dx$$ f is even, so $f(-u) = f(u)$.

$$= 2\int_0^a f(x)\, dx$$

The proof of part (b) is entirely similar and you are asked to give it in Exercise 88. ■

The assertions of Theorem 8 remain true when f is an integrable function (rather than having the stronger property of being continuous), but the proof is more difficult and best left to a more advanced course.

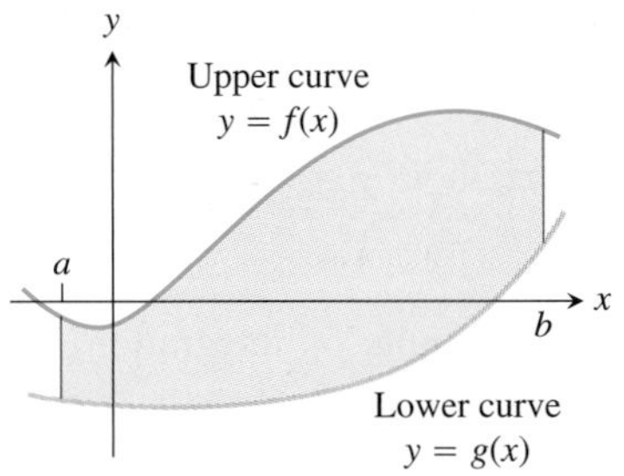

FIGURE 4.24 The region between the curves $y = f(x)$ and $y = g(x)$ and the lines $x = a$ and $x = b$.

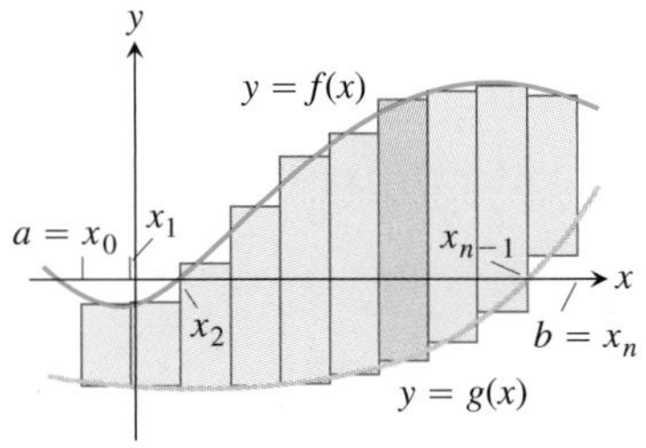

FIGURE 4.25 We approximate the region with rectangles perpendicular to the x-axis.

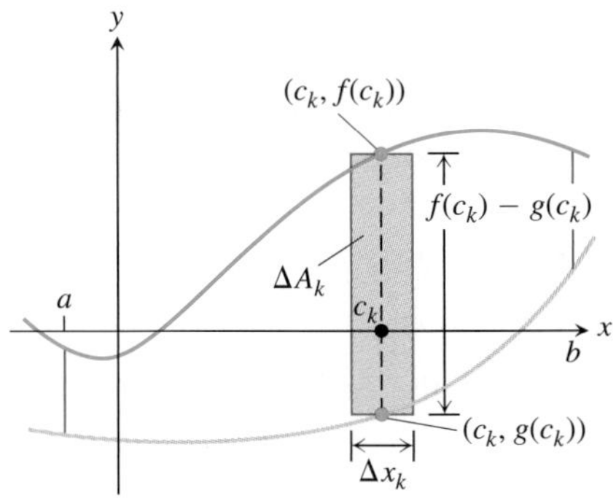

FIGURE 4.26 The area ΔA_k of the kth rectangle is the product of its height, $f(c_k) - g(c_k)$, and its width, Δx_k.

EXAMPLE 3 Evaluate $\displaystyle\int_{-2}^{2} (x^4 - 4x^2 + 6)\, dx$.

Solution Since $f(x) = x^4 - 4x^2 + 6$ satisfies $f(-x) = f(x)$, it is even on the symmetric interval $[-2, 2]$, so

$$\begin{aligned}\int_{-2}^{2} (x^4 - 4x^2 + 6)\, dx &= 2\int_{0}^{2} (x^4 - 4x^2 + 6)\, dx \\ &= 2\left[\frac{x^5}{5} - \frac{4}{3}x^3 + 6x\right]_0^2 \\ &= 2\left(\frac{32}{5} - \frac{32}{3} + 12\right) = \frac{232}{15}.\end{aligned}$$ ■

Areas Between Curves

Suppose we want to find the area of a region that is bounded above by the curve $y = f(x)$, below by the curve $y = g(x)$, and on the left and right by the lines $x = a$ and $x = b$ (Figure 4.24). The region might accidentally have a shape whose area we could find with geometry, but if f and g are arbitrary continuous functions, we usually have to find the area with an integral.

To see what the integral should be, we first approximate the region with n vertical rectangles based on a partition $P = \{x_0, x_1, \ldots, x_n\}$ of $[a, b]$ (Figure 4.25). The area of the kth rectangle (Figure 4.26) is

$$\Delta A_k = \text{height} \times \text{width} = [f(c_k) - g(c_k)]\, \Delta x_k.$$

We then approximate the area of the region by adding the areas of the n rectangles:

$$A \approx \sum_{k=1}^{n} \Delta A_k = \sum_{k=1}^{n} [f(c_k) - g(c_k)]\, \Delta x_k. \qquad \text{Riemann sum}$$

As $\|P\| \to 0$, the sums on the right approach the limit $\int_a^b [f(x) - g(x)]\, dx$ because f and g are continuous. We take the area of the region to be the value of this integral. That is,

$$A = \lim_{\|P\| \to 0} \sum_{k=1}^{n} [f(c_k) - g(c_k)]\, \Delta x_k = \int_a^b [f(x) - g(x)]\, dx.$$

DEFINITION If f and g are continuous with $f(x) \geq g(x)$ throughout $[a, b]$, then the **area of the region between the curves $y = f(x)$ and $y = g(x)$ from a to b** is the integral of $(f - g)$ from a to b:

$$A = \int_a^b [f(x) - g(x)]\, dx.$$

When applying this definition it is helpful to graph the curves. The graph reveals which curve is the upper curve f and which is the lower curve g. It also helps you find the limits of integration if they are not given. You may need to find where the curves intersect to determine the limits of integration, and this may involve solving the equation $f(x) = g(x)$

for values of x. Then you can integrate the function $f - g$ for the area between the intersections.

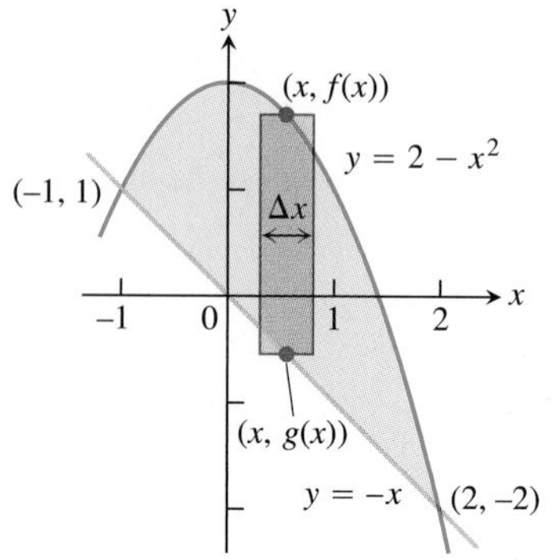

FIGURE 4.27 The region in Example 4 with a typical approximating rectangle.

EXAMPLE 4 Find the area of the region enclosed by the parabola $y = 2 - x^2$ and the line $y = -x$.

Solution First we sketch the two curves (Figure 4.27). The limits of integration are found by solving $y = 2 - x^2$ and $y = -x$ simultaneously for x.

$$
\begin{aligned}
2 - x^2 &= -x && \text{Equate } f(x) \text{ and } g(x). \\
x^2 - x - 2 &= 0 && \text{Rewrite.} \\
(x + 1)(x - 2) &= 0 && \text{Factor.} \\
x = -1, \qquad x &= 2. && \text{Solve.}
\end{aligned}
$$

The region runs from $x = -1$ to $x = 2$. The limits of integration are $a = -1, b = 2$.

The area between the curves is

$$
\begin{aligned}
A &= \int_a^b [f(x) - g(x)]\, dx = \int_{-1}^{2} [(2 - x^2) - (-x)]\, dx \\
&= \int_{-1}^{2} (2 + x - x^2)\, dx = \left[2x + \frac{x^2}{2} - \frac{x^3}{3}\right]_{-1}^{2} \\
&= \left(4 + \frac{4}{2} - \frac{8}{3}\right) - \left(-2 + \frac{1}{2} + \frac{1}{3}\right) = \frac{9}{2}
\end{aligned}
$$

■

If the formula for a bounding curve changes at one or more points, we subdivide the region into subregions that correspond to the formula changes and apply the formula for the area between curves to each subregion.

HISTORICAL BIOGRAPHY

Richard Dedekind (1831–1916)

EXAMPLE 5 Find the area of the region in the first quadrant that is bounded above by $y = \sqrt{x}$ and below by the x-axis and the line $y = x - 2$.

Solution The sketch (Figure 4.28) shows that the region's upper boundary is the graph of $f(x) = \sqrt{x}$. The lower boundary changes from $g(x) = 0$ for $0 \le x \le 2$ to $g(x) = x - 2$ for $2 \le x \le 4$ (both formulas agree at $x = 2$). We subdivide the region at $x = 2$ into subregions A and B, shown in Figure 4.28.

The limits of integration for region A are $a = 0$ and $b = 2$. The left-hand limit for region B is $a = 2$. To find the right-hand limit, we solve the equations $y = \sqrt{x}$ and $y = x - 2$ simultaneously for x:

$$
\begin{aligned}
\sqrt{x} &= x - 2 && \text{Equate } f(x) \text{ and } g(x). \\
x &= (x - 2)^2 = x^2 - 4x + 4 && \text{Square both sides.} \\
x^2 - 5x + 4 &= 0 && \text{Rewrite.} \\
(x - 1)(x - 4) &= 0 && \text{Factor.} \\
x = 1, \qquad x &= 4. && \text{Solve.}
\end{aligned}
$$

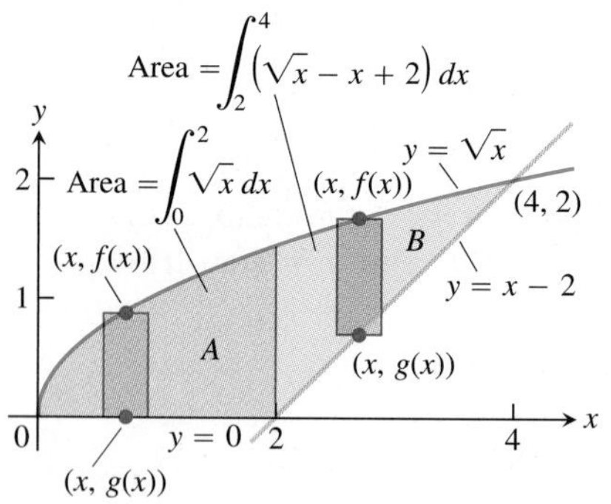

FIGURE 4.28 When the formula for a bounding curve changes, the area integral changes to become the sum of integrals to match, one integral for each of the shaded regions shown here for Example 5.

Only the value $x = 4$ satisfies the equation $\sqrt{x} = x - 2$. The value $x = 1$ is an extraneous root introduced by squaring. The right-hand limit is $b = 4$.

For $0 \le x \le 2$: $\quad f(x) - g(x) = \sqrt{x} - 0 = \sqrt{x}$

For $2 \le x \le 4$: $\quad f(x) - g(x) = \sqrt{x} - (x - 2) = \sqrt{x} - x + 2$

We add the area of subregions A and B to find the total area:

$$\text{Total area} = \underbrace{\int_0^2 \sqrt{x}\,dx}_{\text{area of } A} + \underbrace{\int_2^4 (\sqrt{x} - x + 2)\,dx}_{\text{area of } B}$$

$$= \left[\frac{2}{3}x^{3/2}\right]_0^2 + \left[\frac{2}{3}x^{3/2} - \frac{x^2}{2} + 2x\right]_2^4$$

$$= \frac{2}{3}(2)^{3/2} - 0 + \left(\frac{2}{3}(4)^{3/2} - 8 + 8\right) - \left(\frac{2}{3}(2)^{3/2} - 2 + 4\right)$$

$$= \frac{2}{3}(8) - 2 = \frac{10}{3}.$$ ■

Integration with Respect to *y*

If a region's bounding curves are described by functions of y, the approximating rectangles are horizontal instead of vertical and the basic formula has y in place of x.

For regions like these

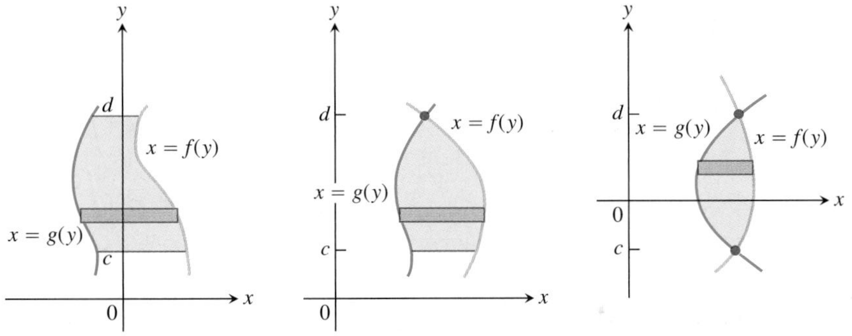

use the formula

$$A = \int_c^d [f(y) - g(y)]\,dy.$$

In this equation f always denotes the right-hand curve and g the left-hand curve, so $f(y) - g(y)$ is nonnegative.

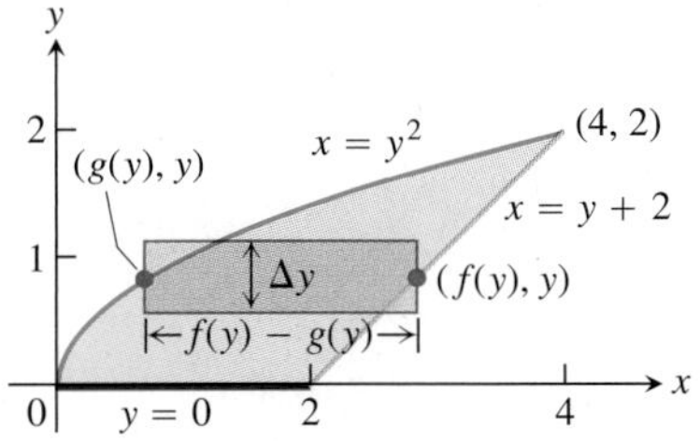

FIGURE 4.29 It takes two integrations to find the area of this region if we integrate with respect to x. It takes only one if we integrate with respect to y (Example 6).

EXAMPLE 6 Find the area of the region in Example 5 by integrating with respect to y.

Solution We first sketch the region and a typical *horizontal* rectangle based on a partition of an interval of y-values (Figure 4.29). The region's right-hand boundary is the line $x = y + 2$, so $f(y) = y + 2$. The left-hand boundary is the curve $x = y^2$, so $g(y) = y^2$. The lower limit of integration is $y = 0$. We find the upper limit by solving $x = y + 2$ and $x = y^2$ simultaneously for y:

$$y + 2 = y^2 \quad \text{Equate } f(y) = y + 2 \text{ and } g(y) = y^2.$$

$$y^2 - y - 2 = 0 \quad \text{Rewrite.}$$

$$(y + 1)(y - 2) = 0 \quad \text{Factor.}$$

$$y = -1, \quad y = 2 \quad \text{Solve.}$$

The upper limit of integration is $b = 2$. (The value $y = -1$ gives a point of intersection *below* the x-axis.)

The area of the region is

$$\begin{aligned} A = \int_a^b [f(y) - g(y)]\,dy &= \int_0^2 [y + 2 - y^2]\,dy \\ &= \int_0^2 [2 + y - y^2]\,dy \\ &= \left[2y + \frac{y^2}{2} - \frac{y^3}{3}\right]_0^2 \\ &= 4 + \frac{4}{2} - \frac{8}{3} = \frac{10}{3}. \end{aligned}$$

This is the result of Example 5, found with less work. ■

EXERCISES 4.7

Use the Substitution Formula in Theorem 7 to evaluate the integrals in Exercises 1–46.

1. a. $\int_0^3 \sqrt{y+1}\,dy$ b. $\int_{-1}^0 \sqrt{y+1}\,dy$
2. a. $\int_0^1 r\sqrt{1-r^2}\,dr$ b. $\int_{-1}^1 r\sqrt{1-r^2}\,dr$
3. a. $\int_0^{\pi/4} \tan x \sec^2 x\,dx$ b. $\int_{-\pi/4}^0 \tan x \sec^2 x\,dx$
4. a. $\int_0^{\pi} 3\cos^2 x \sin x\,dx$ b. $\int_{2\pi}^{3\pi} 3\cos^2 x \sin x\,dx$
5. a. $\int_0^1 t^3(1+t^4)^3\,dt$ b. $\int_{-1}^1 t^3(1+t^4)^3\,dt$
6. a. $\int_0^{\sqrt{7}} t(t^2+1)^{1/3}\,dt$ b. $\int_{-\sqrt{7}}^0 t(t^2+1)^{1/3}\,dt$
7. a. $\int_{-1}^1 \frac{5r}{(4+r^2)^2}\,dr$ b. $\int_0^1 \frac{5r}{(4+r^2)^2}\,dr$
8. a. $\int_0^1 \frac{10\sqrt{v}}{(1+v^{3/2})^2}\,dv$ b. $\int_1^4 \frac{10\sqrt{v}}{(1+v^{3/2})^2}\,dv$
9. a. $\int_0^{\sqrt{3}} \frac{4x}{\sqrt{x^2+1}}\,dx$ b. $\int_{-\sqrt{3}}^{\sqrt{3}} \frac{4x}{\sqrt{x^2+1}}\,dx$
10. a. $\int_0^1 \frac{x^3}{\sqrt{x^4+9}}\,dx$ b. $\int_{-1}^0 \frac{x^3}{\sqrt{x^4+9}}\,dx$
11. a. $\int_0^{\pi/6} (1-\cos 3t)\sin 3t\,dt$ b. $\int_{\pi/6}^{\pi/3} (1-\cos 3t)\sin 3t\,dt$
12. a. $\int_{-\pi/2}^0 \left(2+\tan\frac{t}{2}\right)\sec^2\frac{t}{2}\,dt$ b. $\int_{-\pi/2}^{\pi/2} \left(2+\tan\frac{t}{2}\right)\sec^2\frac{t}{2}\,dt$
13. a. $\int_0^{2\pi} \frac{\cos z}{\sqrt{4+3\sin z}}\,dz$ b. $\int_{-\pi}^{\pi} \frac{\cos z}{\sqrt{4+3\sin z}}\,dz$
14. a. $\int_{-\pi/2}^0 \frac{\sin w}{(3+2\cos w)^2}\,dw$ b. $\int_0^{\pi/2} \frac{\sin w}{(3+2\cos w)^2}\,dw$
15. $\int_0^1 \sqrt{t^5+2t}\,(5t^4+2)\,dt$
16. $\int_1^4 \frac{dy}{2\sqrt{y}\,(1+\sqrt{y})^2}$
17. $\int_0^{\pi/6} \cos^{-3} 2\theta \sin 2\theta\,d\theta$
18. $\int_{\pi}^{3\pi/2} \cot^5\left(\frac{\theta}{6}\right)\sec^2\left(\frac{\theta}{6}\right)d\theta$
19. $\int_0^{\pi} 5(5-4\cos t)^{1/4}\sin t\,dt$
20. $\int_0^{\pi/4} (1-\sin 2t)^{3/2}\cos 2t\,dt$
21. $\int_0^1 (4y-y^2+4y^3+1)^{-2/3}(12y^2-2y+4)\,dy$
22. $\int_0^1 (y^3+6y^2-12y+9)^{-1/2}(y^2+4y-4)\,dy$
23. $\int_0^{\sqrt[3]{\pi^2}} \sqrt{\theta}\cos^2(\theta^{3/2})\,d\theta$
24. $\int_{-1}^{-1/2} t^{-2}\sin^2\left(1+\frac{1}{t}\right)dt$
25. $\int_0^{\pi/4} (1+e^{\tan\theta})\sec^2\theta\,d\theta$
26. $\int_{\pi/4}^{\pi/2} (1+e^{\cot\theta})\csc^2\theta\,d\theta$
27. $\int_0^{\pi} \frac{\sin t}{2-\cos t}\,dt$
28. $\int_0^{\pi/3} \frac{4\sin\theta}{1-4\cos\theta}\,d\theta$
29. $\int_1^2 \frac{2\ln x}{x}\,dx$
30. $\int_2^4 \frac{dx}{x\ln x}$
31. $\int_2^4 \frac{dx}{x(\ln x)^2}$
32. $\int_2^{16} \frac{dx}{2x\sqrt{\ln x}}$
33. $\int_0^{\pi/2} \tan\frac{x}{2}\,dx$
34. $\int_{\pi/4}^{\pi/2} \cot t\,dt$

35. $\int_{\pi/2}^{\pi} 2\cot\frac{\theta}{3}\,d\theta$

36. $\int_{0}^{\pi/12} 6\tan 3x\,dx$

37. $\int_{-\pi/2}^{\pi/2} \frac{2\cos\theta\,d\theta}{1+(\sin\theta)^2}$

38. $\int_{\pi/6}^{\pi/4} \frac{\csc^2 x\,dx}{1+(\cot x)^2}$

39. $\int_{0}^{\ln\sqrt{3}} \frac{e^x\,dx}{1+e^{2x}}$

40. $\int_{1}^{e^{\pi/4}} \frac{4\,dt}{t(1+\ln^2 t)}$

41. $\int_{0}^{1} \frac{4\,ds}{\sqrt{4-s^2}}$

42. $\int_{0}^{3\sqrt{2}/4} \frac{ds}{\sqrt{9-4s^2}}$

43. $\int_{\sqrt{2}}^{2} \frac{\sec^2(\sec^{-1}x)\,dx}{x\sqrt{x^2-1}}$

44. $\int_{2/\sqrt{3}}^{2} \frac{\cos(\sec^{-1}x)\,dx}{x\sqrt{x^2-1}}$

45. $\int_{-1}^{-\sqrt{2}/2} \frac{dy}{y\sqrt{4y^2-1}}$

46. $\int_{-2/3}^{-\sqrt{2}/3} \frac{dy}{y\sqrt{9y^2-1}}$

Find the total areas of the shaded regions in Exercises 47–54.

47.

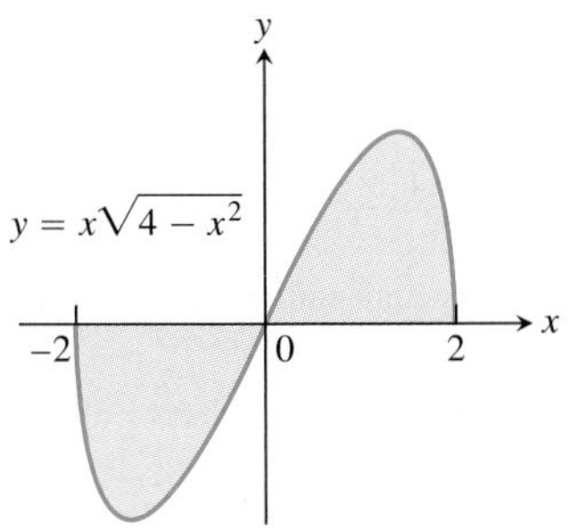

48.

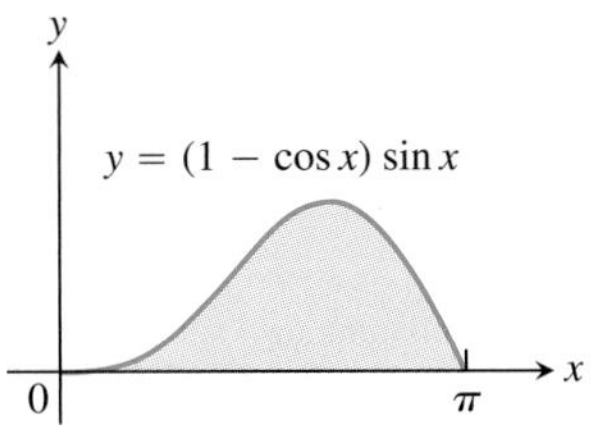

49.

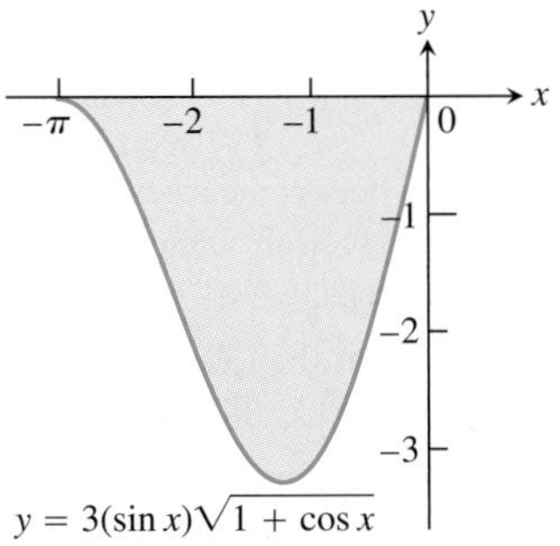

50.

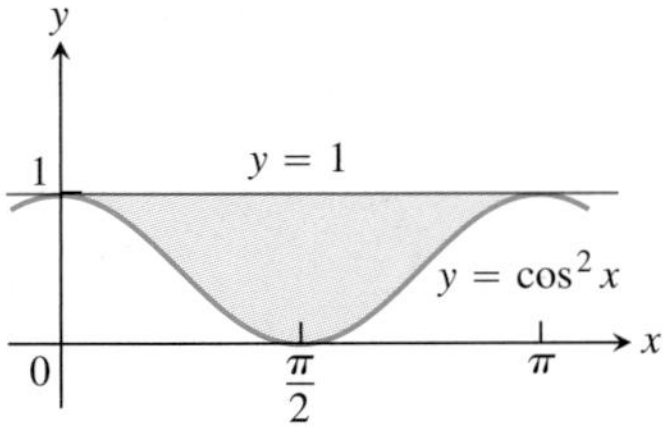

51.

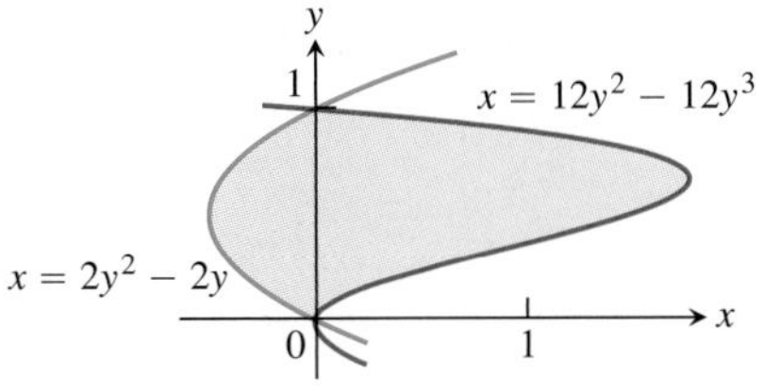

52.

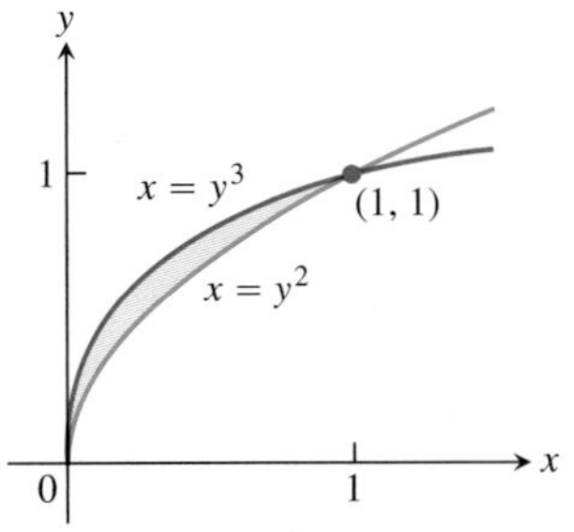

53.

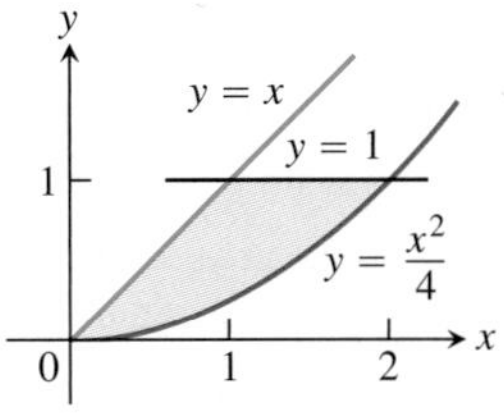

54.

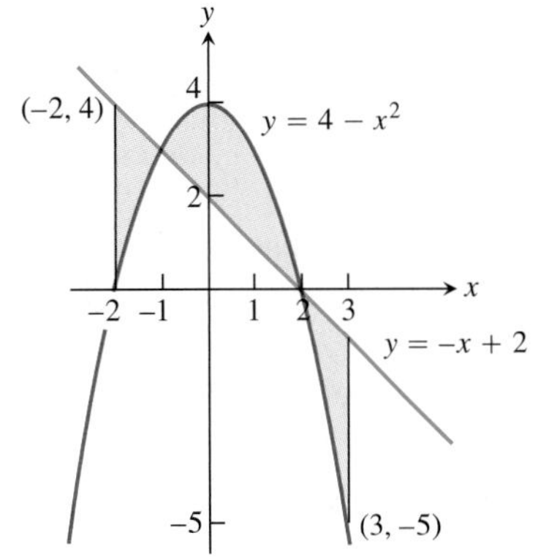

Find the areas of the regions enclosed by the lines and curves in Exercises 55–62.

55. $y = x^2 - 2$ and $y = 2$

56. $y = x^4$ and $y = 8x$

57. $y = x^2$ and $y = -x^2 + 4x$

58. $y = x^2 - 2x$ and $y = x$

59. $y = x^4 - 4x^2 + 4$ and $y = x^2$

60. $y = x\sqrt{a^2 - x^2}$, $a > 0$, and $y = 0$

61. $y = \sqrt{|x|}$ and $5y = x + 6$ (How many intersection points are there?)

62. $y = |x^2 - 4|$ and $y = (x^2/2) + 4$

Find the areas of the regions enclosed by the lines and/or curves in Exercises 63–74.

63. $x = 2y^2$, $x = 0$, and $y = 3$

64. $y^2 - 4x = 4$ and $4x - y = 16$

65. $x + y^2 = 0$ and $x + 3y^2 = 2$

66. $x - y^{2/3} = 0$ and $x + y^4 = 2$

67. $x = y^2 - 1$ and $x = |y|\sqrt{1 - y^2}$

68. $x = y^3 - y^2$ and $x = 2y$

69. $4x^2 + y = 4$ and $x^4 - y = 1$

70. $x^3 - y = 0$ and $3x^2 - y = 4$

71. $y = 2\sin x$ and $y = \sin 2x$, $0 \le x \le \pi$

72. $y = 8\cos x$ and $y = \sec^2 x, \quad -\pi/3 \le x \le \pi/3$

73. $y = \sec^2 x, \quad y = \tan^2 x, \quad x = -\pi/4, \quad$ and $\quad x = \pi/4$

74. $x = 3\sin y\sqrt{\cos y}$ and $x = 0, \quad 0 \le y \le \pi/2$

75. Find the area of the propeller-shaped region enclosed by the curve $x - y^3 = 0$ and the line $x - y = 0$.

76. Find the area of the region in the first quadrant bounded by the line $y = x$, the line $x = 2$, the curve $y = 1/x^2$, and the x-axis.

77. Find the area between the curves $y = \ln x$ and $y = \ln 2x$ from $x = 1$ to $x = 5$.

78. Find the area of the "triangular" region in the first quadrant bounded on the left by the y-axis and on the right by the curves $y = \sin x$ and $y = \cos x$.

79. Find the area of the "triangular" region in the first quadrant that is bounded above by the curve $y = e^{2x}$, below by the curve $y = e^x$, and on the right by the line $x = \ln 3$.

80. Find the area of the "triangular" region in the first quadrant that is bounded above by the curve $y = e^{x/2}$, below by the curve $y = e^{-x/2}$, and on the right by the line $x = 2\ln 2$.

81. Find the area of the region between the curve $y = 2x/(1 + x^2)$ and the interval $-2 \le x \le 2$ of the x-axis.

82. Find the area of the region between the curve $y = 2^{1-x}$ and the interval $-1 \le x \le 1$ of the x-axis.

83. Find the area of the region in the first quadrant bounded on the left by the y-axis, below by the line $y = x/4$, above left by the curve $y = 1 + \sqrt{x}$, and above right by the curve $y = 2/\sqrt{x}$.

84. The figure here shows triangle AOC inscribed in the region cut from the parabola $y = x^2$ by the line $y = a^2$. Find the limit of the ratio of the area of the triangle to the area of the parabolic region as a approaches zero.

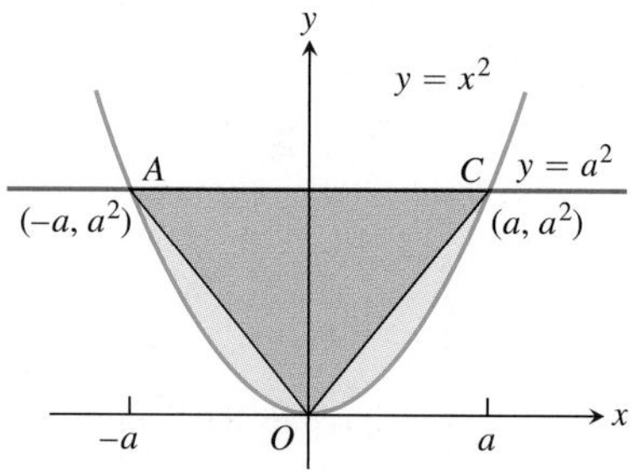

85. Suppose that $F(x)$ is an antiderivative of $f(x) = (\sin x)/x, x > 0$. Express

$$\int_1^3 \frac{\sin 2x}{x}\,dx$$

in terms of F.

86. Show that if f is continuous, then

$$\int_0^1 f(x)\,dx = \int_0^1 f(1 - x)\,dx.$$

87. Suppose that

$$\int_0^1 f(x)\,dx = 3.$$

Find

$$\int_{-1}^0 f(x)\,dx$$

if **a.** f is odd, **b.** f is even.

88. **a.** Show that if f is odd on $[-a, a]$, then

$$\int_{-a}^a f(x)\,dx = 0.$$

b. Test the result in part (a) with $f(x) = \sin x$ and $a = \pi/2$.

89. If f is a continuous function, find the value of the integral

$$I = \int_0^a \frac{f(x)\,dx}{f(x) + f(a - x)}$$

by making the substitution $u = a - x$ and adding the resulting integral to I.

90. By using a substitution, prove that for all positive numbers x and y,

$$\int_x^{xy} \frac{1}{t}\,dt = \int_1^y \frac{1}{t}\,dt.$$

COMPUTER EXPLORATIONS

In Exercises 91–94, you will find the area between curves in the plane when you cannot find their points of intersection using simple algebra. Use a CAS to perform the following steps:

a. Plot the curves together to see what they look like and how many points of intersection they have.

b. Use the numerical equation solver in your CAS to find all the points of intersection.

c. Integrate $|f(x) - g(x)|$ over consecutive pairs of intersection values.

d. Sum together the integrals found in part (c).

91. $f(x) = \dfrac{x^3}{3} - \dfrac{x^2}{2} - 2x + \dfrac{1}{3}, \quad g(x) = x - 1$

92. $f(x) = \dfrac{x^4}{2} - 3x^3 + 10, \quad g(x) = 8 - 12x$

93. $f(x) = x + \sin(2x), \quad g(x) = x^3$

94. $f(x) = x^2\cos x, \quad g(x) = x^3 - x$

Chapter 4 Questions to Guide Your Review

1. What is an indefinite integral? How do you evaluate one? What general formulas do you know for finding indefinite integrals?
2. What is an initial value problem? How do you solve one? Give an example.
3. How can you sometimes estimate quantities like distance traveled, area, and average value with finite sums? Why might you want to do so?
4. What is sigma notation? What advantage does it offer? Give examples.
5. What is a Riemann sum? Why might you want to consider such a sum?
6. What is the norm of a partition of a closed interval?
7. What is the definite integral of a function f over a closed interval $[a, b]$? When can you be sure it exists?
8. What is the relation between definite integrals and area? Describe some other interpretations of definite integrals.
9. What is the average value of an integrable function over a closed interval? Must the function assume its average value? Explain.
10. Describe the rules for working with definite integrals (Table 4.5). Give examples.
11. What is the Fundamental Theorem of Calculus? Why is it so important? Illustrate each part of the theorem with an example.
12. How does the Fundamental Theorem provide a solution to the initial value problem $dy/dx = f(x), y(x_0) = y_0$, when f is continuous?
13. How is integration by substitution related to the Chain Rule?
14. How can you sometimes evaluate indefinite integrals by substitution? Give examples.
15. How does the method of substitution work for definite integrals? Give examples.
16. How do you define and calculate the area of the region between the graphs of two continuous functions? Give an example.

Chapter 4 Practice and Additional Exercises

Find the indefinite integrals (most general antiderivatives) in Exercises 1–16. Check your answers by differentiation.

1. $\int (x^3 + 5x - 7)\, dx$

2. $\int \left(3\sqrt{t} + \frac{4}{t^2}\right) dt$

3. $\int \frac{dr}{(r+5)^2}$

4. $\int \frac{6\, dr}{\left(r - \sqrt{2}\right)^3}$

5. $\int 3\theta\sqrt{\theta^2 + 1}\, d\theta$

6. $\int \frac{\theta}{\sqrt{7 + \theta^2}}\, d\theta$

7. $\int x^3(1 + x^4)^{-1/4}\, dx$

8. $\int (2 - x)^{3/5}\, dx$

9. $\int \sec^2 \frac{s}{10}\, ds$

10. $\int \sec \frac{\theta}{3} \tan \frac{\theta}{3}\, d\theta$

11. $\int \sin^2 \frac{x}{4}\, dx$ $\left(\textit{Hint: } \sin^2\theta = \frac{1 - \cos 2\theta}{2}\right)$

12. $\int \left(\frac{5}{x^2} + \frac{2}{x^2 + 1}\right) dx$

13. $\int \left(\frac{3}{x} - x\right) dx$

14. $\int \left(\frac{1}{2}e^t - e^{-t}\right) dt$

15. $\int \theta^{1-\pi}\, d\theta$

16. $\int (5^s + s^5)\, ds$

Solve the initial value problems in Exercises 17–20.

17. $\frac{dy}{dx} = \frac{x^2 + 1}{x^2}, \quad y(1) = -1$

18. $\frac{dy}{dx} = \left(x + \frac{1}{x}\right)^2, \quad y(1) = 1$

19. $\frac{d^2r}{dt^2} = 15\sqrt{t} + \frac{3}{\sqrt{t}}; \quad r'(1) = 8, \quad r(1) = 0$

20. $\frac{d^3r}{dt^3} = -\cos t; \quad r''(0) = r'(0) = 0, \quad r(0) = -1$

21. The accompanying figure shows the graph of the velocity (ft/sec) of a model rocket for the first 8 sec after launch. The rocket accelerated straight up for the first 2 sec and then coasted to reach its maximum height at $t = 8$ sec.

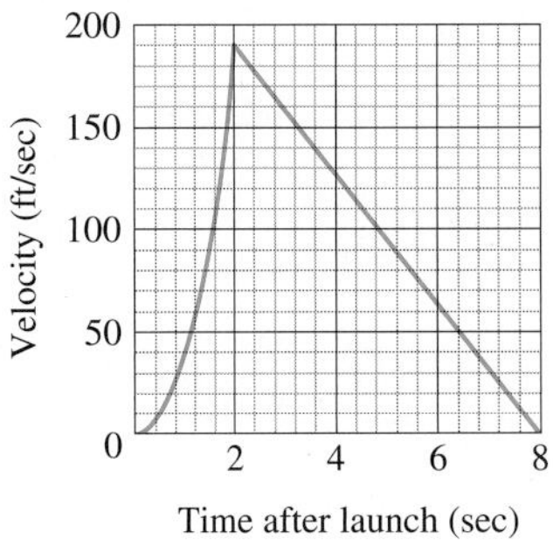

a. Assuming that the rocket was launched from ground level, about how high did it go?

b. Sketch a graph of the rocket's height above ground as a function of time for $0 \le t \le 8$.

22. a. The accompanying figure shows the velocity (m/sec) of a body moving along the s-axis during the time interval from $t = 0$ to $t = 10$ sec. About how far did the body travel during those 10 sec?

b. Sketch a graph of s as a function of t for $0 \le t \le 10$ assuming $s(0) = 0$.

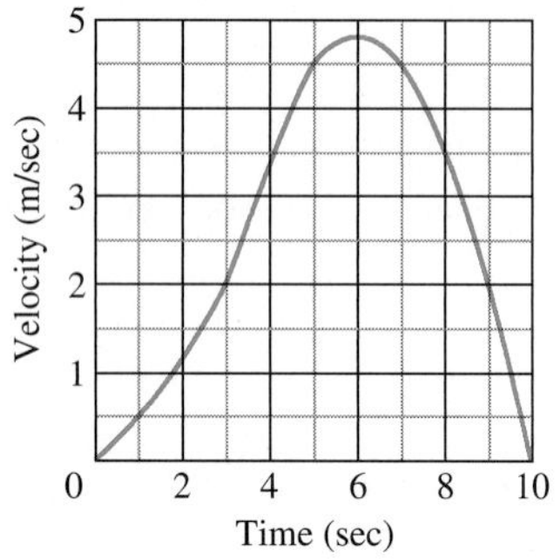

23. Suppose that $\sum_{k=1}^{10} a_k = -2$ and $\sum_{k=1}^{10} b_k = 25$. Find the value of

a. $\sum_{k=1}^{10} \frac{a_k}{4}$ **b.** $\sum_{k=1}^{10} (b_k - 3a_k)$

c. $\sum_{k=1}^{10} (a_k + b_k - 1)$ **d.** $\sum_{k=1}^{10} \left(\frac{5}{2} - b_k\right)$

24. Suppose that $\sum_{k=1}^{20} a_k = 0$ and $\sum_{k=1}^{20} b_k = 7$. Find the values of

a. $\sum_{k=1}^{20} 3a_k$ **b.** $\sum_{k=1}^{20} (a_k + b_k)$

c. $\sum_{k=1}^{20} \left(\frac{1}{2} - \frac{2b_k}{7}\right)$ **d.** $\sum_{k=1}^{20} (a_k - 2)$

In Exercises 25–28, express each limit as a definite integral. Then evaluate the integral to find the value of the limit. In each case, P is a partition of the given interval and the numbers c_k are chosen from the subintervals of P.

25. $\lim_{\|P\| \to 0} \sum_{k=1}^{n} (2c_k - 1)^{-1/2} \Delta x_k$, where P is a partition of $[1, 5]$

26. $\lim_{\|P\| \to 0} \sum_{k=1}^{n} c_k(c_k^2 - 1)^{1/3} \Delta x_k$, where P is a partition of $[1, 3]$

27. $\lim_{\|P\| \to 0} \sum_{k=1}^{n} \left(\cos\left(\frac{c_k}{2}\right)\right) \Delta x_k$, where P is a partition of $[-\pi, 0]$

28. $\lim_{\|P\| \to 0} \sum_{k=1}^{n} (\sin c_k)(\cos c_k) \Delta x_k$, where P is a partition of $[0, \pi/2]$

29. If $\int_{-2}^{2} 3f(x)\,dx = 12$, $\int_{-2}^{5} f(x)\,dx = 6$, and $\int_{-2}^{5} g(x)\,dx = 2$, find the values of the following.

a. $\int_{-2}^{2} f(x)\,dx$ **b.** $\int_{2}^{5} f(x)\,dx$

c. $\int_{5}^{-2} g(x)\,dx$ **d.** $\int_{-2}^{5} (-\pi g(x))\,dx$

e. $\int_{-2}^{5} \left(\frac{f(x) + g(x)}{5}\right) dx$

30. If $\int_0^2 f(x)\,dx = \pi$, $\int_0^2 7g(x)\,dx = 7$, and $\int_0^1 g(x)\,dx = 2$, find the values of the following.

a. $\int_0^2 g(x)\,dx$ **b.** $\int_1^2 g(x)\,dx$

c. $\int_2^0 f(x)\,dx$ **d.** $\int_0^2 \sqrt{2}\, f(x)\,dx$

e. $\int_0^2 (g(x) - 3f(x))\,dx$

In Exercises 31–34, find the total area of the region between the graph of f and the x-axis.

31. $f(x) = x^2 - 4x + 3, \quad 0 \le x \le 3$

32. $f(x) = 1 - (x^2/4), \quad -2 \le x \le 3$

33. $f(x) = 5 - 5x^{2/3}, \quad -1 \le x \le 8$

34. $f(x) = 1 - \sqrt{x}, \quad 0 \le x \le 4$

Find the areas of the regions enclosed by the curves and lines in Exercises 35–42.

35. $y = x, \quad y = 1/x^2, \quad x = 2$

36. $y = x, \quad y = 1/\sqrt{x}, \quad x = 2$

37. $x = 2y^2, \quad x = 0, \quad y = 3$ **38.** $x = 4 - y^2, \quad x = 0$

39. $y^2 = 4x, \quad y = 4x - 2$

40. $y^2 = 4x + 4, \quad y = 4x - 16$

41. $y = \sin x, \quad y = x, \quad 0 \le x \le \pi/4$

42. $y = 8\cos x, \quad y = \sec^2 x, \quad -\pi/3 \le x \le \pi/3$

43. Find the area of the "triangular" region bounded on the left by $x + y = 2$, on the right by $y = x^2$, and above by $y = 2$.

44. Find the area of the "triangular" region bounded on the left by $y = \sqrt{x}$, on the right by $y = 6 - x$, and below by $y = 1$.

45. Find the extreme values of $f(x) = x^3 - 3x^2$ and find the area of the region enclosed by the graph of f and the x-axis.

46. Find the area of the region cut from the first quadrant by the curve $x^{1/2} + y^{1/2} = a^{1/2}$.

47. Find the total area of the region enclosed by the curve $x = y^{2/3}$ and the lines $x = y$ and $y = -1$.

48. Find the total area of the region between the curves $y = \sin x$ and $y = \cos x$ for $0 \le x \le 3\pi/2$.

49. Area Find the area between the curve $y = 2(\ln x)/x$ and the x-axis from $x = 1$ to $x = e$.

50. a. Show that the area between the curve $y = 1/x$ and the x-axis from $x = 10$ to $x = 20$ is the same as the area between the curve and the x-axis from $x = 1$ to $x = 2$.

b. Show that the area between the curve $y = 1/x$ and the x-axis from ka to kb is the same as the area between the curve and the x-axis from $x = a$ to $x = b$ $(0 < a < b, k > 0)$.

Evaluate the integrals in Exercises 51–76.

51. $\int 2(\cos x)^{-1/2} \sin x\, dx$

52. $\int (\tan x)^{-3/2} \sec^2 x\, dx$

53. $\int (2\theta + 1 + 2\cos(2\theta + 1))\, d\theta$

54. $\int \left(\frac{1}{\sqrt{2\theta - \pi}} + 2\sec^2(2\theta - \pi)\right) d\theta$

55. $\int \left(t - \frac{2}{t}\right)\left(t + \frac{2}{t}\right) dt$

56. $\int \frac{(t+1)^2 - 1}{t^4}\, dt$

57. $\int \sqrt{t} \sin(2t^{3/2})\, dt$

58. $\int \sec\theta \tan\theta \sqrt{1 + \sec\theta}\, d\theta$

59. $\int e^x \sec^2(e^x - 7)\, dx$

60. $\int e^y \csc(e^y + 1) \cot(e^y + 1)\, dy$

61. $\int \sec^2(x) e^{\tan x}\, dx$

62. $\int \csc^2 x\, e^{\cot x}\, dx$

63. $\int_{-1}^{1} \frac{dx}{3x - 4}$

64. $\int_{1}^{e} \frac{\sqrt{\ln x}}{x}\, dx$

65. $\int_{0}^{4} \frac{2t}{t^2 - 25}\, dt$

66. $\int \frac{\tan(\ln v)}{v}\, dv$

67. $\int \frac{(\ln x)^{-3}}{x}\, dx$

68. $\int \frac{1}{r} \csc^2(1 + \ln r)\, dr$

69. $\int x 3^{x^2}\, dx$

70. $\int 2^{\tan x} \sec^2 x\, dx$

71. $\int \frac{3\, dr}{\sqrt{1 - 4(r-1)^2}}$

72. $\int \frac{dx}{2 + (x-1)^2}$

73. $\int \frac{dx}{(2x - 1)\sqrt{(2x-1)^2 - 4}}$

74. $\int \frac{e^{\sin^{-1}\sqrt{x}}\, dx}{2\sqrt{x - x^2}}$

75. $\int \frac{dy}{\sqrt{\tan^{-1} y}\,(1 + y^2)}$

76. $\int \frac{(\tan^{-1} x)^2\, dx}{1 + x^2}$

Evaluate the integrals in Exercises 77–112.

77. $\int_{-1}^{1} (3x^2 - 4x + 7)\, dx$

78. $\int_{0}^{1} (8s^3 - 12s^2 + 5)\, ds$

79. $\int_{1}^{2} \frac{4}{v^2}\, dv$

80. $\int_{1}^{27} x^{-4/3}\, dx$

81. $\int_{1}^{4} \frac{dt}{t\sqrt{t}}$

82. $\int_{0}^{1} \frac{36\, dx}{(2x+1)^3}$

83. $\int_{1/8}^{1} x^{-1/3}(1 - x^{2/3})^{3/2}\, dx$

84. $\int_{0}^{1/2} x^3(1 + 9x^4)^{-3/2}\, dx$

85. $\int_{0}^{\pi} \sin^2 5r\, dr$

86. $\int_{\pi/4}^{3\pi/4} \csc^2 x\, dx$

87. $\int_{\pi}^{3\pi} \cot^2 \frac{x}{6}\, dx$

88. $\int_{0}^{\pi} \tan^2 \frac{\theta}{3}\, d\theta$

89. $\int_{-\pi/3}^{0} \sec x \tan x\, dx$

90. $\int_{\pi/4}^{3\pi/4} \csc z \cot z\, dz$

91. $\int_{0}^{\pi/2} 5(\sin x)^{3/2} \cos x\, dx$

92. $\int_{-\pi/2}^{\pi/2} 15 \sin^4 3x \cos 3x\, dx$

93. $\int_{0}^{\pi/2} \frac{3 \sin x \cos x}{\sqrt{1 + 3\sin^2 x}}\, dx$

94. $\int_{0}^{\pi/4} \frac{\sec^2 x}{(1 + 7\tan x)^{2/3}}\, dx$

95. $\int_{1}^{4} \left(\frac{x}{8} + \frac{1}{2x}\right) dx$

96. $\int_{1}^{8} \left(\frac{2}{3x} - \frac{8}{x^2}\right) dx$

97. $\int_{-2}^{-1} e^{-(x+1)}\, dx$

98. $\int_{-\ln 2}^{0} e^{2w}\, dw$

99. $\int_{0}^{\ln 5} e^r(3e^r + 1)^{-3/2}\, dr$

100. $\int_{0}^{\ln 9} e^{\theta}(e^{\theta} - 1)^{1/2}\, d\theta$

101. $\int_{1}^{e} \frac{1}{x}(1 + 7\ln x)^{-1/3}\, dx$

102. $\int_{1}^{3} \frac{(\ln(v+1))^2}{v + 1}\, dv$

103. $\int_{1}^{8} \frac{\log_4 \theta}{\theta}\, d\theta$

104. $\int_{1}^{e} \frac{8 \ln 3 \log_3 \theta}{\theta}\, d\theta$

105. $\int_{-3/4}^{3/4} \frac{6\, dx}{\sqrt{9 - 4x^2}}$

106. $\int_{-1/5}^{1/5} \frac{6\, dx}{\sqrt{4 - 25x^2}}$

107. $\int_{-2}^{2} \frac{3\, dt}{4 + 3t^2}$

108. $\int_{\sqrt{3}}^{3} \frac{dt}{3 + t^2}$

109. $\int \frac{dy}{y\sqrt{4y^2 - 1}}$

110. $\int \frac{24\, dy}{y\sqrt{y^2 - 16}}$

111. $\int_{\sqrt{2}/3}^{2/3} \frac{dy}{|y|\sqrt{9y^2 - 1}}$

112. $\int_{-2/\sqrt{5}}^{-\sqrt{6}/\sqrt{5}} \frac{dy}{|y|\sqrt{5y^2 - 3}}$

113. Find the average value of $f(x) = mx + b$

a. over $[-1, 1]$

b. over $[-k, k]$

114. Find the average value of

a. $y = \sqrt{3x}$ over $[0, 3]$

b. $y = \sqrt{ax}$ over $[0, a]$

115. a. Show that $\int \ln x\, dx = x \ln x - x + C$.

b. Find the average value of $\ln x$ over $[1, e]$.

116. Find the average value of $f(x) = 1/x$ on $[1, 2]$.

In Exercises 117–122, find dy/dx.

117. $y = \int_{2}^{x} \sqrt{2 + \cos^3 t}\, dt$

118. $y = \int_{x}^{1} \frac{6}{3 + t^4}\, dt$

119. $y = \int_{\ln x^2}^{0} e^{\cos t}\, dt$

120. $y = \int_{1}^{e^{\sqrt{x}}} \ln (t^2 + 1)\, dt$

121. $y = \int_{0}^{\sin^{-1} x} \frac{dt}{\sqrt{1 - 2t^2}}$

122. $y = \int_{\tan^{-1} x}^{\pi/4} e^{\sqrt{t}}\, dt$

123. Is it true that every function $y = f(x)$ that is differentiable on $[a, b]$ is itself the derivative of some function on $[a, b]$? Give reasons for your answer.

124. Find dy/dx if $y = \int_x^1 \sqrt{1 + t^2}\, dt$. Explain the main steps in your calculation.

125. A particle is traveling upward and to the right along the curve $y = \ln x$. Its x-coordinate is increasing at the rate $(dx/dt) = \sqrt{x}$ m/sec. At what rate is the y-coordinate changing at the point $(e^2, 2)$?

126. Skydivers A and B are in a helicopter hovering at 6400 ft. Skydiver A jumps and descends for 4 sec before opening her parachute. The helicopter then climbs to 7000 ft and hovers there. Forty-five seconds after A leaves the aircraft, B jumps and descends for 13 sec before opening his parachute. Both skydivers descend at 16 ft/sec with parachutes open. Assume that the skydivers fall freely (no effective air resistance) before their parachutes open.

a. At what altitude does A's parachute open?

b. At what altitude does B's parachute open?

c. Which skydiver lands first?

127. Verify the integral formulas in Table 4.6.

128. Verify the integral formulas in Table 4.7.

Evaluate the integrals in Exercises 129–136 in terms of

a. inverse hyperbolic functions.

b. natural logarithms.

129. $\int_0^{2\sqrt{3}} \frac{dx}{\sqrt{4 + x^2}}$

130. $\int_0^{1/3} \frac{6\, dx}{\sqrt{1 + 9x^2}}$

131. $\int_{5/4}^{2} \frac{dx}{1 - x^2}$

132. $\int_0^{1/2} \frac{dx}{1 - x^2}$

133. $\int_{1/5}^{3/13} \frac{dx}{x\sqrt{1 - 16x^2}}$

134. $\int_1^{2} \frac{dx}{x\sqrt{4 + x^2}}$

135. $\int_0^{\pi} \frac{\cos x\, dx}{\sqrt{1 + \sin^2 x}}$

136. $\int_1^{e} \frac{dx}{x\sqrt{1 + (\ln x)^2}}$

Piecewise Continuous Functions Although we are mainly interested in continuous functions, many functions in applications are piecewise continuous. A function $f(x)$ is **piecewise continuous on a closed interval *I*** if f has only finitely many discontinuities in I, the limits

$$\lim_{x \to c^-} f(x) \qquad \text{and} \qquad \lim_{x \to c^+} f(x)$$

exist and are finite at every interior point of I, and the appropriate one-sided limits exist and are finite at the endpoints of I. All piecewise continuous functions are integrable. The points of discontinuity subdivide I into open and half-open subintervals on which f is continuous, and the limit criteria above guarantee that f has a continuous extension to the closure of each subinterval. To integrate a piecewise continuous function, we integrate the individual extensions and add the results. The integral of

$$f(x) = \begin{cases} 1 - x, & -1 \le x < 0 \\ x^2, & 0 \le x < 2 \\ -1, & 2 \le x \le 3 \end{cases}$$

TABLE 4.6 Integral formulas for hyperbolic functions

$$\int \sinh u\, du = \cosh u + C$$

$$\int \cosh u\, du = \sinh u + C$$

$$\int \operatorname{sech}^2 u\, du = \tanh u + C$$

$$\int \operatorname{csch}^2 u\, du = -\coth u + C$$

$$\int \operatorname{sech} u \tanh u\, du = -\operatorname{sech} u + C$$

$$\int \operatorname{csch} u \coth u\, du = -\operatorname{csch} u + C$$

TABLE 4.7 Integral formulas for inverse hyperbolic functions

$$\int \frac{du}{\sqrt{a^2 + u^2}} = \sinh^{-1}\left(\frac{u}{a}\right) + C, \qquad a > 0$$

$$\int \frac{du}{\sqrt{u^2 - a^2}} = \cosh^{-1}\left(\frac{u}{a}\right) + C, \qquad u > a > 0$$

$$\int \frac{du}{a^2 - u^2} = \begin{cases} \frac{1}{a}\tanh^{-1}\left(\frac{u}{a}\right) + C & \text{if } u^2 < a^2 \\ \frac{1}{a}\coth^{-1}\left(\frac{u}{a}\right) + C, & \text{if } u^2 > a^2 \end{cases}$$

$$\int \frac{du}{u\sqrt{a^2 - u^2}} = -\frac{1}{a}\operatorname{sech}^{-1}\left(\frac{u}{a}\right) + C, \qquad 0 < u < a$$

$$\int \frac{du}{u\sqrt{a^2 + u^2}} = -\frac{1}{a}\operatorname{csch}^{-1}\left|\frac{u}{a}\right| + C, \qquad u \ne 0 \text{ and } a > 0$$

over $[-1, 3]$ is

$$\begin{aligned}\int_{-1}^{3} f(x)\,dx &= \int_{-1}^{0}(1-x)\,dx + \int_{0}^{2} x^2\,dx + \int_{2}^{3}(-1)\,dx \\ &= \left[x - \frac{x^2}{2}\right]_{-1}^{0} + \left[\frac{x^3}{3}\right]_{0}^{2} + \left[-x\right]_{2}^{3} \\ &= \frac{3}{2} + \frac{8}{3} - 1 = \frac{19}{6}.\end{aligned}$$

The Fundamental Theorem applies to piecewise continuous functions with the restriction that $(d/dx)\int_a^x f(t)\,dt$ is expected to equal $f(x)$ only at values of x at which f is continuous.

Integrate the functions in Exercises 137–142 over their domains.

137. $f(x) = \begin{cases} x^{2/3}, & -8 \le x < 0 \\ -4, & 0 \le x \le 3 \end{cases}$

138. $f(x) = \begin{cases} \sqrt{-x}, & -4 \le x < 0 \\ x^2 - 4, & 0 \le x \le 3 \end{cases}$

139. $g(t) = \begin{cases} t, & 0 \le t < 1 \\ \sin \pi t, & 1 \le t \le 2 \end{cases}$

140. $h(z) = \begin{cases} \sqrt{1 - z}, & 0 \le z < 1 \\ (7z - 6)^{-1/3}, & 1 \le z \le 2 \end{cases}$

141. $f(x) = \begin{cases} 1, & -2 \le x < -1 \\ 1 - x^2, & -1 \le x < 1 \\ 2, & 1 \le x \le 2 \end{cases}$

142. $h(r) = \begin{cases} r, & -1 \le r < 0 \\ 1 - r^2, & 0 \le r < 1 \\ 1, & 1 \le r \le 2 \end{cases}$

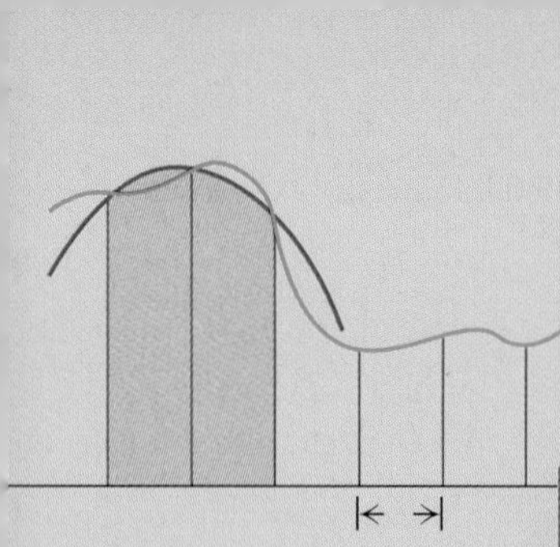

Chapter 5

TECHNIQUES OF INTEGRATION

OVERVIEW The Fundamental Theorem tells us how to evaluate a definite integral once we have an antiderivative for the integrand function. Table 5.1 summarizes the forms of antiderivatives for many of the functions we have studied so far, and the substitution method helps us use the table to evaluate more complicated functions involving these basic ones. In this chapter we study a number of other important techniques for finding antiderivatives (or indefinite integrals) for even more functions.

TABLE 5.1 Basic integration formulas

1. $\int k\,dx = kx + C \qquad$ (any number k)	11. $\int \csc x \cot x\,dx = -\csc x + C$
2. $\int x^n\,dx = \dfrac{x^{n+1}}{n+1} + C \qquad (n \neq -1)$	12. $\int \tan x\,dx = \ln\lvert\sec x\rvert + C$
3. $\int \dfrac{dx}{x} = \ln\lvert x\rvert + C$	13. $\int \cot x\,dx = \ln\lvert\sin x\rvert + C$
4. $\int e^x\,dx = e^x + C$	14. $\int \sinh x\,dx = \cosh x + C$
5. $\int a^x\,dx = \dfrac{a^x}{\ln a} + C \qquad (a > 0, a \neq 1)$	15. $\int \cosh x\,dx = \sinh x + C$
6. $\int \sin x\,dx = -\cos x + C$	16. $\int \dfrac{dx}{\sqrt{a^2 - x^2}} = \sin^{-1}\left(\dfrac{x}{a}\right) + C$
7. $\int \cos x\,dx = \sin x + C$	17. $\int \dfrac{dx}{a^2 + x^2} = \dfrac{1}{a}\tan^{-1}\left(\dfrac{x}{a}\right) + C$
8. $\int \sec^2 x\,dx = \tan x + C$	18. $\int \dfrac{dx}{x\sqrt{x^2 - a^2}} = \dfrac{1}{a}\sec^{-1}\left\lvert\dfrac{x}{a}\right\rvert + C$
9. $\int \csc^2 x\,dx = -\cot x + C$	19. $\int \dfrac{dx}{\sqrt{a^2 + x^2}} = \sinh^{-1}\left(\dfrac{x}{a}\right) + C \qquad (a > 0)$
10. $\int \sec x \tan x\,dx = \sec x + C$	20. $\int \dfrac{dx}{\sqrt{x^2 - a^2}} = \cosh^{-1}\left(\dfrac{x}{a}\right) + C \qquad (x > a > 0)$

5.1 Integration by Parts

Integration by parts is a technique for simplifying integrals of the form

$$\int f(x)g(x)\,dx.$$

It is useful when f can be differentiated repeatedly and g can be integrated repeatedly without difficulty. The integral

$$\int xe^x\,dx$$

is such an integral because $f(x) = x$ can be differentiated twice to become zero and $g(x) = e^x$ can be integrated repeatedly without difficulty. Integration by parts also applies to integrals like

$$\int e^x \sin x\,dx$$

in which each part of the integrand appears again after repeated differentiation or integration.

Product Rule in Integral Form

If f and g are differentiable functions of x, the Product Rule says

$$\frac{d}{dx}[f(x)g(x)] = f'(x)g(x) + f(x)g'(x).$$

In terms of indefinite integrals, this equation becomes

$$\int \frac{d}{dx}[f(x)g(x)]\,dx = \int [f'(x)g(x) + f(x)g'(x)]\,dx$$

or

$$\int \frac{d}{dx}[f(x)g(x)]\,dx = \int f'(x)g(x)\,dx + \int f(x)g'(x)\,dx.$$

Rearranging the terms of this last equation, we get

$$\int f(x)g'(x)\,dx = \int \frac{d}{dx}[f(x)g(x)]\,dx - \int f'(x)g(x)\,dx$$

leading to the **integration by parts** formula

$$\int f(x)g'(x)\,dx = f(x)g(x) - \int f'(x)g(x)\,dx \qquad (1)$$

Sometimes it is easier to remember the formula if we write it in differential form. Let $u = f(x)$ and $v = g(x)$. Then $du = f'(x)\,dx$ and $dv = g'(x)\,dx$. Using the Substitution Rule, the integration by parts formula becomes

Integration by Parts Formula

$$\int u\,dv = uv - \int v\,du \tag{2}$$

This formula expresses one integral, $\int u\,dv$, in terms of a second integral, $\int v\,du$. With a proper choice of u and v, the second integral may be easier to evaluate than the first. In using the formula, various choices may be available for u and dv. The next examples illustrate the technique.

EXAMPLE 1 Find

$$\int x \cos x\,dx.$$

Solution We use the formula $\displaystyle\int u\,dv = uv - \int v\,du$ with

$$u = x, \qquad dv = \cos x\,dx,$$
$$du = dx, \qquad v = \sin x. \qquad \text{Simplest antiderivative of } \cos x$$

Then

$$\int x \cos x\,dx = x \sin x - \int \sin x\,dx = x \sin x + \cos x + C. \quad \blacksquare$$

There are four choices available for u and dv in Example 1:

1. Let $u = 1$ and $dv = x \cos x\,dx$.
2. Let $u = x$ and $dv = \cos x\,dx$.
3. Let $u = x \cos x$ and $dv = dx$.
4. Let $u = \cos x$ and $dv = x\,dx$.

Choice 2 was used in Example 1. The other three choices lead to integrals we don't know how to integrate. For instance, Choice 3 leads to the integral

$$\int (x \cos x - x^2 \sin x)\,dx.$$

The goal of integration by parts is to go from an integral $\int u\,dv$ that we don't see how to evaluate to an integral $\int v\,du$ that we can evaluate. Generally, you choose dv first to be as much of the integrand, including dx, as you can readily integrate; u is the leftover part. Keep in mind that integration by parts does not always work.

EXAMPLE 2 Find

$$\int \ln x\,dx.$$

Solution Since $\int \ln x\,dx$ can be written as $\int \ln x \cdot 1\,dx$, we use the formula $\int u\,dv = uv - \int v\,du$ with

$$u = \ln x \qquad \text{Simplifies when differentiated} \qquad dv = dx \qquad \text{Easy to integrate}$$

$$du = \frac{1}{x}\,dx, \qquad v = x. \qquad \text{Simplest antiderivative}$$

Then

$$\int \ln x \, dx = x \ln x - \int x \cdot \frac{1}{x} \, dx = x \ln x - \int dx = x \ln x - x + C. \qquad \blacksquare$$

Sometimes we have to use integration by parts more than once.

EXAMPLE 3 Evaluate

$$\int x^2 e^x \, dx.$$

Solution With $u = x^2$, $dv = e^x \, dx$, $du = 2x \, dx$, and $v = e^x$, we have

$$\int x^2 e^x \, dx = x^2 e^x - 2 \int x e^x \, dx.$$

The new integral is less complicated than the original because the exponent on x is reduced by one. To evaluate the integral on the right, we integrate by parts again with $u = x$, $dv = e^x \, dx$. Then $du = dx$, $v = e^x$, and

$$\int x e^x \, dx = x e^x - \int e^x \, dx = x e^x - e^x + C.$$

Using this last evaluation, we then obtain

$$\begin{aligned} \int x^2 e^x \, dx &= x^2 e^x - 2 \int x e^x \, dx \\ &= x^2 e^x - 2x e^x + 2e^x + C. \end{aligned} \qquad \blacksquare$$

The technique of Example 3 works for any integral $\int x^n e^x \, dx$ in which n is a positive integer, because differentiating x^n will eventually lead to zero and integrating e^x is easy.

Integrals like the one in the next example occur in electrical engineering. Their evaluation requires two integrations by parts, followed by solving for the unknown integral.

EXAMPLE 4 Evaluate

$$\int e^x \cos x \, dx.$$

Solution Let $u = e^x$ and $dv = \cos x \, dx$. Then $du = e^x \, dx$, $v = \sin x$, and

$$\int e^x \cos x \, dx = e^x \sin x - \int e^x \sin x \, dx.$$

The second integral is like the first except that it has $\sin x$ in place of $\cos x$. To evaluate it, we use integration by parts with

$$u = e^x, \qquad dv = \sin x \, dx, \qquad v = -\cos x, \qquad du = e^x \, dx.$$

Then

$$\begin{aligned} \int e^x \cos x \, dx &= e^x \sin x - \left(-e^x \cos x - \int (-\cos x)(e^x \, dx) \right) \\ &= e^x \sin x + e^x \cos x - \int e^x \cos x \, dx. \end{aligned}$$

The unknown integral now appears on both sides of the equation. Adding the integral to both sides and adding the constant of integration give

$$2\int e^x \cos x\, dx = e^x \sin x + e^x \cos x + C_1.$$

Dividing by 2 and renaming the constant of integration give

$$\int e^x \cos x\, dx = \frac{e^x \sin x + e^x \cos x}{2} + C.$$

■

EXAMPLE 5 Obtain a formula that expresses the integral

$$\int \cos^n x\, dx$$

in terms of an integral of a lower power of $\cos x$.

Solution We may think of $\cos^n x$ as $\cos^{n-1} x \cdot \cos x$. Then we let

$$u = \cos^{n-1} x \qquad \text{and} \qquad dv = \cos x\, dx,$$

so that

$$du = (n-1)\cos^{n-2} x\,(-\sin x\, dx) \qquad \text{and} \qquad v = \sin x.$$

Integration by parts then gives

$$\begin{aligned}\int \cos^n x\, dx &= \cos^{n-1} x \sin x + (n-1)\int \sin^2 x \cos^{n-2} x\, dx\\ &= \cos^{n-1} x \sin x + (n-1)\int (1-\cos^2 x)\cos^{n-2} x\, dx,\\ &= \cos^{n-1} x \sin x + (n-1)\int \cos^{n-2} x\, dx - (n-1)\int \cos^n x\, dx.\end{aligned}$$

If we add

$$(n-1)\int \cos^n x\, dx$$

to both sides of this equation, we obtain

$$n\int \cos^n x\, dx = \cos^{n-1} x \sin x + (n-1)\int \cos^{n-2} x\, dx.$$

We then divide through by n, and the final result is

$$\int \cos^n x\, dx = \frac{\cos^{n-1} x \sin x}{n} + \frac{n-1}{n}\int \cos^{n-2} x\, dx.$$

This allows us to reduce the exponent on $\cos x$ by 2 and is a very useful formula. When n is a positive integer, we may apply the formula repeatedly until the remaining integral is either

$$\int \cos x\, dx = \sin x + C \qquad \text{or} \qquad \int \cos^0 x\, dx = \int dx = x + C.$$

■

For example, the result in Example 5 tells us,

$$\int \cos^3 x\, dx = \frac{\cos^2 x \sin x}{3} + \frac{2}{3}\int \cos x\, dx$$
$$= \frac{1}{3}\cos^2 x \sin x + \frac{2}{3}\sin x + C.$$

Evaluating Definite Integrals by Parts

The integration by parts formula in Equation (1) can be combined with Part 2 of the Fundamental Theorem in order to evaluate definite integrals by parts. Assuming that both f' and g' are continuous over the interval $[a, b]$, Part 2 of the Fundamental Theorem gives

Integration by Parts Formula for Definite Integrals

$$\int_a^b f(x)g'(x)\, dx = f(x)g(x)\Big]_a^b - \int_a^b f'(x)g(x)\, dx \qquad (3)$$

In applying Equation (3), we normally use the u and v notation from Equation (2) because it is easier to remember. Here is an example.

EXAMPLE 6 Find the area of the region bounded by the curve $y = xe^{-x}$ and the x-axis from $x = 0$ to $x = 4$.

Solution The region is shaded in Figure 5.1. Its area is

$$\int_0^4 xe^{-x}\, dx.$$

Let $u = x$, $dv = e^{-x}\, dx$, $v = -e^{-x}$, and $du = dx$. Then,

$$\int_0^4 xe^{-x}\, dx = -xe^{-x}\Big]_0^4 - \int_0^4 (-e^{-x})\, dx$$
$$= [-4e^{-4} - (0)] + \int_0^4 e^{-x}\, dx$$
$$= -4e^{-4} - e^{-x}\Big]_0^4$$
$$= -4e^{-4} - e^{-4} - (-e^0) = 1 - 5e^{-4} \approx 0.91.$$

■

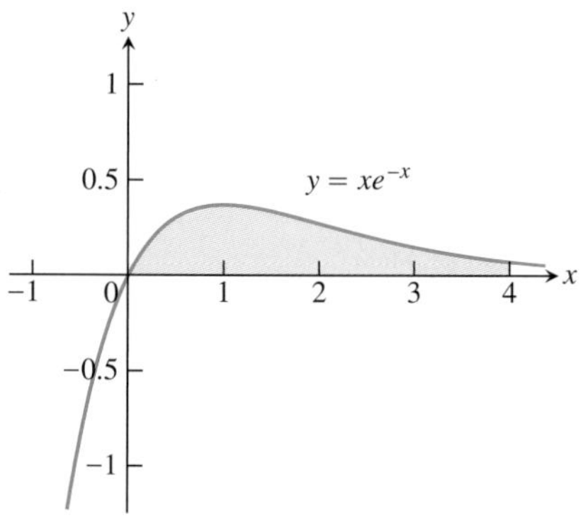

FIGURE 5.1 The region in Example 6.

EXERCISES 5.1

Evaluate the integrals in Exercises 1–24.

1. $\int x \sin \frac{x}{2}\, dx$
2. $\int \theta \cos \pi\theta\, d\theta$
3. $\int t^2 \cos t\, dt$
4. $\int x^2 \sin x\, dx$
5. $\int_1^2 x \ln x\, dx$
6. $\int_1^e x^3 \ln x\, dx$
7. $\int \tan^{-1} y\, dy$
8. $\int \sin^{-1} y\, dy$
9. $\int x \sec^2 x\, dx$
10. $\int 4x \sec^2 2x\, dx$

11. $\int x^3 e^x\,dx$

12. $\int p^4 e^{-p}\,dp$

13. $\int (x^2 - 5x)e^x\,dx$

14. $\int (r^2 + r + 1)e^r\,dr$

15. $\int x^5 e^x\,dx$

16. $\int t^2 e^{4t}\,dt$

17. $\int_0^{\pi/2} \theta^2 \sin 2\theta\,d\theta$

18. $\int_0^{\pi/2} x^3 \cos 2x\,dx$

19. $\int_{2/\sqrt{3}}^{2} t\sec^{-1} t\,dt$

20. $\int_0^{1/\sqrt{2}} 2x\sin^{-1}(x^2)\,dx$

21. $\int e^\theta \sin\theta\,d\theta$

22. $\int e^{-y}\cos y\,dy$

23. $\int e^{2x}\cos 3x\,dx$

24. $\int e^{-2x}\sin 2x\,dx$

Evaluate the integrals in Exercises 25–30 by using a substitution prior to integration by parts.

25. $\int e^{\sqrt{3s+9}}\,ds$

26. $\int_0^1 x\sqrt{1-x}\,dx$

27. $\int_0^{\pi/3} x\tan^2 x\,dx$

28. $\int \ln(x + x^2)\,dx$

29. $\int \sin(\ln x)\,dx$

30. $\int z(\ln z)^2\,dz$

31. Finding area Find the area of the region enclosed by the curve $y = x\sin x$ and the x-axis (see the accompanying figure) for

a. $0 \le x \le \pi$

b. $\pi \le x \le 2\pi$

c. $2\pi \le x \le 3\pi$.

d. What pattern do you see here? What is the area between the curve and the x-axis for $n\pi \le x \le (n+1)\pi$, n an arbitrary nonnegative integer? Give reasons for your answer.

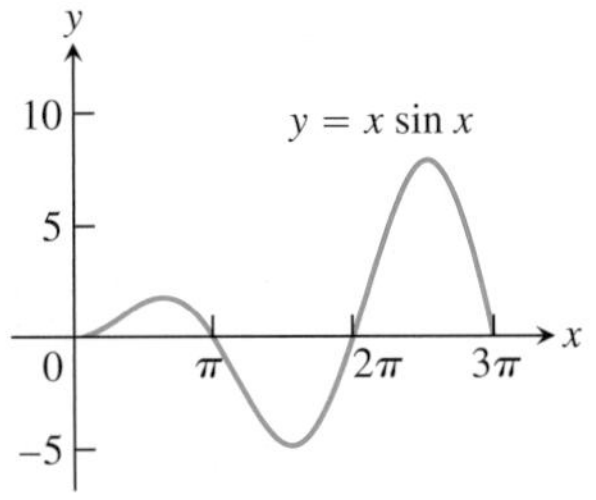

32. Finding area Find the area of the region enclosed by the curve $y = x\cos x$ and the x-axis (see the accompanying figure) for

a. $\pi/2 \le x \le 3\pi/2$

b. $3\pi/2 \le x \le 5\pi/2$

c. $5\pi/2 \le x \le 7\pi/2$.

d. What pattern do you see? What is the area between the curve and the x-axis for

$$\left(\frac{2n-1}{2}\right)\pi \le x \le \left(\frac{2n+1}{2}\right)\pi,$$

n an arbitrary positive integer? Give reasons for your answer.

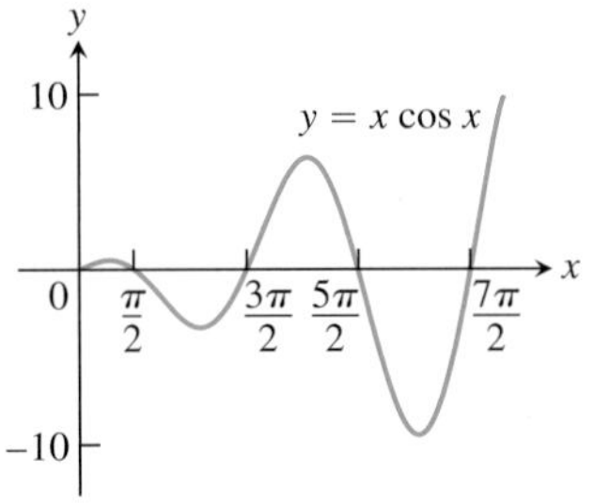

33. Average value A retarding force, symbolized by the dashpot in the accompanying figure, slows the motion of the weighted spring so that the mass's position at time t is

$$y = 2e^{-t}\cos t, \qquad t \ge 0.$$

Find the average value of y over the interval $0 \le t \le 2\pi$.

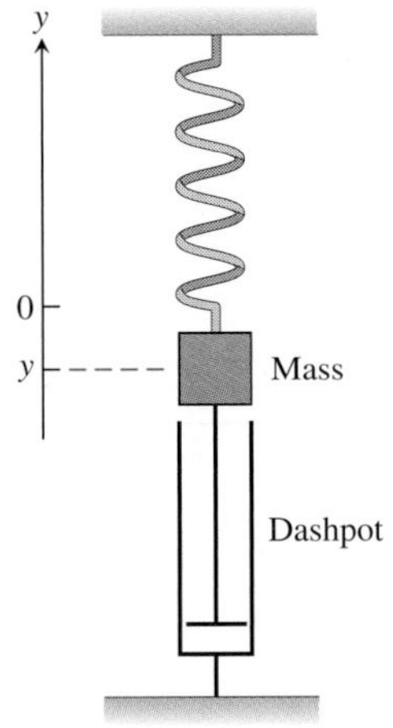

34. Average value In a mass-spring-dashpot system like the one in Exercise 33, the mass's position at time t is

$$y = 4e^{-t}(\sin t - \cos t), \qquad t \ge 0.$$

Find the average value of y over the interval $0 \le t \le 2\pi$.

Reduction Formulas In Exercises 35–38, use integration by parts to establish the *reduction formula*.

35. $\int x^n \cos x\,dx = x^n \sin x - n\int x^{n-1}\sin x\,dx$

36. $\int x^n \sin x\,dx = -x^n \cos x + n\int x^{n-1}\cos x\,dx$

37. $\int x^n e^{ax}\,dx = \dfrac{x^n e^{ax}}{a} - \dfrac{n}{a}\int x^{n-1}e^{ax}\,dx, \quad a \neq 0$

38. $\int (\ln x)^n\,dx = x(\ln x)^n - n\int (\ln x)^{n-1}\,dx$

Integrating Inverses of Functions Integration by parts leads to a rule for integrating inverses that usually gives good results:

$$\int f^{-1}(x)\,dx = \int y f'(y)\,dy \qquad \begin{aligned} y &= f^{-1}(x), \quad x = f(y) \\ dx &= f'(y)\,dy \end{aligned}$$

$$= yf(y) - \int f(y)\,dy \qquad \text{Integration by parts with } u = y,\ dv = f'(y)\,dy$$

$$= xf^{-1}(x) - \int f(y)\,dy$$

The idea is to take the most complicated part of the integral, in this case $f^{-1}(x)$, and simplify it first. For the integral of $\ln x$, we get

$$\int \ln x\,dx = \int ye^y\,dy \qquad \begin{aligned} y &= \ln x, \quad x = e^y \\ dx &= e^y\,dy \end{aligned}$$

$$= ye^y - e^y + C$$

$$= x\ln x - x + C.$$

For the integral of $\cos^{-1} x$ we get

$$\int \cos^{-1} x\,dx = x\cos^{-1} x - \int \cos y\,dy \qquad y = \cos^{-1} x$$

$$= x\cos^{-1} x - \sin y + C$$

$$= x\cos^{-1} x - \sin(\cos^{-1} x) + C.$$

Use the formula

$$\int f^{-1}(x)\,dx = xf^{-1}(x) - \int f(y)\,dy \qquad y = f^{-1}(x) \tag{4}$$

to evaluate the integrals in Exercises 39–42. Express your answers in terms of x.

39. $\int \sin^{-1} x\,dx$

40. $\int \tan^{-1} x\,dx$

41. $\int \sec^{-1} x\,dx$

42. $\int \log_2 x\,dx$

Another way to integrate $f^{-1}(x)$ (when f^{-1} is integrable, of course) is to use integration by parts with $u = f^{-1}(x)$ and $dv = dx$ to rewrite the integral of f^{-1} as

$$\int f^{-1}(x)\,dx = xf^{-1}(x) - \int x\left(\frac{d}{dx}f^{-1}(x)\right)dx. \tag{5}$$

Exercises 43 and 44 compare the results of using Equations (4) and (5).

43. Equations (4) and (5) give different formulas for the integral of $\cos^{-1} x$:

a. $\int \cos^{-1} x\,dx = x\cos^{-1} x - \sin(\cos^{-1} x) + C$ Eq. (4)

b. $\int \cos^{-1} x\,dx = x\cos^{-1} x - \sqrt{1 - x^2} + C$ Eq. (5)

Can both integrations be correct? Explain.

44. Equations (4) and (5) lead to different formulas for the integral of $\tan^{-1} x$:

a. $\int \tan^{-1} x\,dx = x\tan^{-1} x - \ln\sec(\tan^{-1} x) + C$ Eq. (4)

b. $\int \tan^{-1} x\,dx = x\tan^{-1} x - \ln\sqrt{1 + x^2} + C$ Eq. (5)

Can both integrations be correct? Explain.

Evaluate the integrals in Exercises 45 and 46 with (**a**) Eq. (4) and (**b**) Eq. (5). In each case, check your work by differentiating your answer with respect to x.

45. $\int \sinh^{-1} x\,dx$

46. $\int \tanh^{-1} x\,dx$

5.2 Trigonometric Integrals

Trigonometric integrals involve algebraic combinations of the six basic trigonometric functions. In principle, we can always express such integrals in terms of sines and cosines, but it is often simpler to work with other functions, as in the integral

$$\int \sec^2 x\,dx = \tan x + C.$$

The general idea is to use identities to transform the integrals we have to find into integrals that are easier to work with.

Products of Powers of Sines and Cosines

We begin with integrals of the form:

$$\int \sin^m x \cos^n x \, dx,$$

where m and n are nonnegative integers (positive or zero). We can divide the appropriate substitution into three cases.

Case 1 If m is odd, we write m as $2k + 1$ and use the identity $\sin^2 x = 1 - \cos^2 x$ to obtain

$$\sin^m x = \sin^{2k+1} x = (\sin^2 x)^k \sin x = (1 - \cos^2 x)^k \sin x. \qquad (1)$$

Then we combine the single $\sin x$ with dx in the integral and set $\sin x \, dx$ equal to $-d(\cos x)$.

Case 2 If m is even and n is odd in $\int \sin^m x \cos^n x \, dx$, we write n as $2k + 1$ and use the identity $\cos^2 x = 1 - \sin^2 x$ to obtain

$$\cos^n x = \cos^{2k+1} x = (\cos^2 x)^k \cos x = (1 - \sin^2 x)^k \cos x.$$

We then combine the single $\cos x$ with dx and set $\cos x \, dx$ equal to $d(\sin x)$.

Case 3 If both m and n are even in $\int \sin^m x \cos^n x \, dx$, we substitute

$$\sin^2 x = \frac{1 - \cos 2x}{2}, \qquad \cos^2 x = \frac{1 + \cos 2x}{2} \qquad (2)$$

to reduce the integrand to one in lower powers of $\cos 2x$.

Here are some examples illustrating each case.

EXAMPLE 1 Evaluate

$$\int \sin^3 x \cos^2 x \, dx.$$

Solution

$$\begin{aligned}
\int \sin^3 x \cos^2 x \, dx &= \int \sin^2 x \cos^2 x \sin x \, dx && m \text{ is odd} \\
&= \int (1 - \cos^2 x) \cos^2 x \, (-d\,(\cos x)) \\
&= \int (1 - u^2)(u^2)(-du) && u = \cos x \\
&= \int (u^4 - u^2) \, du \\
&= \frac{u^5}{5} - \frac{u^3}{3} + C \\
&= \frac{\cos^5 x}{5} - \frac{\cos^3 x}{3} + C.
\end{aligned}$$

■

EXAMPLE 2 Evaluate

$$\int \cos^5 x\,dx.$$

Solution

$$\begin{aligned}\int \cos^5 x\,dx &= \int \cos^4 x \cos x\,dx = \int (1 - \sin^2 x)^2\,d(\sin x) && m = 0, n = 5\\ &= \int (1 - u^2)^2\,du && u = \sin x\\ &= \int (1 - 2u^2 + u^4)\,du\\ &= u - \frac{2}{3}u^3 + \frac{1}{5}u^5 + C = \sin x - \frac{2}{3}\sin^3 x + \frac{1}{5}\sin^5 x + C.\end{aligned}$$

■

EXAMPLE 3 Evaluate

$$\int \sin^2 x \cos^4 x\,dx.$$

Solution

$$\begin{aligned}\int \sin^2 x \cos^4 x\,dx &= \int \left(\frac{1 - \cos 2x}{2}\right)\left(\frac{1 + \cos 2x}{2}\right)^2 dx && m \text{ and } n \text{ both even}\\ &= \frac{1}{8}\int (1 - \cos 2x)(1 + 2\cos 2x + \cos^2 2x)\,dx\\ &= \frac{1}{8}\int (1 + \cos 2x - \cos^2 2x - \cos^3 2x)\,dx\\ &= \frac{1}{8}\left[x + \frac{1}{2}\sin 2x - \int (\cos^2 2x + \cos^3 2x)\,dx\right].\end{aligned}$$

For the term involving $\cos^2 2x$, we use

$$\begin{aligned}\int \cos^2 2x\,dx &= \frac{1}{2}\int (1 + \cos 4x)\,dx\\ &= \frac{1}{2}\left(x + \frac{1}{4}\sin 4x\right). && \text{Omitting the constant of integration until the final result}\end{aligned}$$

For the $\cos^3 2x$ term, we have

$$\begin{aligned}\int \cos^3 2x\,dx &= \int (1 - \sin^2 2x)\cos 2x\,dx && u = \sin 2x,\ du = 2\cos 2x\,dx\\ &= \frac{1}{2}\int (1 - u^2)\,du = \frac{1}{2}\left(\sin 2x - \frac{1}{3}\sin^3 2x\right). && \text{Again omitting } C\end{aligned}$$

Combining everything and simplifying, we get

$$\int \sin^2 x \cos^4 x\, dx = \frac{1}{16}\left(x - \frac{1}{4}\sin 4x + \frac{1}{3}\sin^3 2x\right) + C. \qquad \blacksquare$$

Eliminating Square Roots

In the next example, we use the identity $\cos^2\theta = (1 + \cos 2\theta)/2$ to eliminate a square root.

EXAMPLE 4 Evaluate

$$\int_0^{\pi/4} \sqrt{1 + \cos 4x}\, dx.$$

Solution To eliminate the square root, we use the identity

$$\cos^2\theta = \frac{1 + \cos 2\theta}{2} \qquad \text{or} \qquad 1 + \cos 2\theta = 2\cos^2\theta.$$

With $\theta = 2x$, this becomes

$$1 + \cos 4x = 2\cos^2 2x.$$

Therefore,

$$\begin{aligned}\int_0^{\pi/4} \sqrt{1 + \cos 4x}\, dx &= \int_0^{\pi/4} \sqrt{2\cos^2 2x}\, dx = \int_0^{\pi/4} \sqrt{2}\sqrt{\cos^2 2x}\, dx \\ &= \sqrt{2}\int_0^{\pi/4} |\cos 2x|\, dx = \sqrt{2}\int_0^{\pi/4} \cos 2x\, dx \qquad \begin{matrix}\cos 2x \ge 0 \\ \text{on } [0, \pi/4]\end{matrix} \\ &= \sqrt{2}\left[\frac{\sin 2x}{2}\right]_0^{\pi/4} = \frac{\sqrt{2}}{2}[1 - 0] = \frac{\sqrt{2}}{2}. \qquad \blacksquare\end{aligned}$$

Integrals of Powers of tan *x* and sec *x*

We know how to integrate the tangent and secant and their squares. To integrate higher powers, we use the identities $\tan^2 x = \sec^2 x - 1$ and $\sec^2 x = \tan^2 x + 1$, and integrate by parts when necessary to reduce the higher powers to lower powers.

EXAMPLE 5 Evaluate

$$\int \tan^4 x\, dx.$$

Solution

$$\begin{aligned}\int \tan^4 x\, dx &= \int \tan^2 x \cdot \tan^2 x\, dx = \int \tan^2 x \cdot (\sec^2 x - 1)\, dx \\ &= \int \tan^2 x \sec^2 x\, dx - \int \tan^2 x\, dx\end{aligned}$$

$$= \int \tan^2 x \sec^2 x\, dx - \int (\sec^2 x - 1)\, dx$$

$$= \int \tan^2 x \sec^2 x\, dx - \int \sec^2 x\, dx + \int dx.$$

In the first integral, we let

$$u = \tan x, \qquad du = \sec^2 x\, dx$$

and have

$$\int u^2\, du = \frac{1}{3}u^3 + C_1.$$

The remaining integrals are standard forms, so

$$\int \tan^4 x\, dx = \frac{1}{3}\tan^3 x - \tan x + x + C. \qquad \blacksquare$$

EXAMPLE 6 Evaluate

$$\int \sec^3 x\, dx.$$

Solution We integrate by parts using

$$u = \sec x, \qquad dv = \sec^2 x\, dx, \qquad v = \tan x, \qquad du = \sec x \tan x\, dx.$$

Then

$$\int \sec^3 x\, dx = \sec x \tan x - \int (\tan x)(\sec x \tan x\, dx)$$

$$= \sec x \tan x - \int (\sec^2 x - 1) \sec x\, dx \qquad \tan^2 x = \sec^2 x - 1$$

$$= \sec x \tan x + \int \sec x\, dx - \int \sec^3 x\, dx.$$

Combining the two secant-cubed integrals gives

$$2\int \sec^3 x\, dx = \sec x \tan x + \int \sec x\, dx$$

and

$$\int \sec^3 x\, dx = \frac{1}{2}\sec x \tan x + \frac{1}{2}\ln|\sec x + \tan x| + C. \qquad \blacksquare$$

Products of Sines and Cosines

The integrals

$$\int \sin mx \sin nx\, dx, \qquad \int \sin mx \cos nx\, dx, \qquad \text{and} \qquad \int \cos mx \cos nx\, dx$$

arise in many places where trigonometric functions are applied to problems in mathematics and science. We can evaluate these integrals through integration by parts, but two such integrations are required in each case. It is simpler to use the identities

$$\sin mx \sin nx = \frac{1}{2}[\cos(m-n)x - \cos(m+n)x], \quad (3)$$

$$\sin mx \cos nx = \frac{1}{2}[\sin(m-n)x + \sin(m+n)x], \quad (4)$$

$$\cos mx \cos nx = \frac{1}{2}[\cos(m-n)x + \cos(m+n)x]. \quad (5)$$

These identities come from the angle sum formulas for the sine and cosine functions (Appendix 4). They give functions whose antiderivatives are easily found.

EXAMPLE 7 Evaluate

$$\int \sin 3x \cos 5x \, dx.$$

Solution From Equation (4) with $m = 3$ and $n = 5$, we get

$$\begin{aligned}\int \sin 3x \cos 5x \, dx &= \frac{1}{2}\int [\sin(-2x) + \sin 8x] \, dx \\ &= \frac{1}{2}\int (\sin 8x - \sin 2x) \, dx \\ &= -\frac{\cos 8x}{16} + \frac{\cos 2x}{4} + C.\end{aligned}$$

■

EXERCISES 5.2

Evaluate the integrals in Exercises 1–14.

1. $\int \sin^5 x \, dx$
2. $\int_0^{\pi} \sin^5 \frac{x}{2} \, dx$
3. $\int \cos^3 x \, dx$
4. $\int_0^{\pi/6} 3\cos^5 3x \, dx$
5. $\int_0^{\pi/2} \sin^7 y \, dy$
6. $\int 7\cos^7 t \, dt$
7. $\int_0^{\pi} 8\sin^4 x \, dx$
8. $\int 8\cos^4 2\pi x \, dx$
9. $\int 16\sin^2 x \cos^2 x \, dx$
10. $\int_0^{\pi} 8\sin^4 y \cos^2 y \, dy$
11. $\int 35\sin^4 x \cos^3 x \, dx$
12. $\int \sin 2x \cos^2 2x \, dx$
13. $\int 8\cos^3 2\theta \sin 2\theta \, d\theta$
14. $\int_0^{\pi/2} \sin^2 2\theta \cos^3 2\theta \, d\theta$

Evaluate the integrals in Exercises 15–22.

15. $\int_0^{2\pi} \sqrt{\frac{1-\cos x}{2}} \, dx$
16. $\int_0^{\pi} \sqrt{1-\cos 2x} \, dx$
17. $\int_0^{\pi} \sqrt{1-\sin^2 t} \, dt$
18. $\int_0^{\pi} \sqrt{1-\cos^2 \theta} \, d\theta$
19. $\int_{-\pi/4}^{\pi/4} \sqrt{1+\tan^2 x} \, dx$
20. $\int_{-\pi/4}^{\pi/4} \sqrt{\sec^2 x - 1} \, dx$

21. $\int_0^{\pi/2} \theta\sqrt{1 - \cos 2\theta}\, d\theta$

22. $\int_{-\pi}^{\pi} (1 - \cos^2 t)^{3/2}\, dt$

Evaluate the integrals in Exercises 23–32.

23. $\int_{-\pi/3}^{0} 2\sec^3 x\, dx$

24. $\int e^x \sec^3 e^x\, dx$

25. $\int \sec^4 \theta\, d\theta$

26. $\int 3\sec^4 3x\, dx$

27. $\int_{\pi/4}^{\pi/2} \csc^4 \theta\, d\theta$

28. $\int_{\pi/2}^{\pi} 3\csc^4 \frac{\theta}{2}\, d\theta$

29. $\int 4\tan^3 x\, dx$

30. $\int_{-\pi/4}^{\pi/4} 6\tan^4 x\, dx$

31. $\int_{\pi/6}^{\pi/3} \cot^3 x\, dx$

32. $\int 8\cot^4 t\, dt$

Evaluate the integrals in Exercises 33–38.

33. $\int \sin 3x \cos 2x\, dx$

34. $\int \sin 2x \cos 3x\, dx$

35. $\int_{-\pi}^{\pi} \sin 3x \sin 3x\, dx$

36. $\int_0^{\pi/2} \sin x \cos x\, dx$

37. $\int \cos 3x \cos 4x\, dx$

38. $\int_{-\pi/2}^{\pi/2} \cos x \cos 7x\, dx$

39. **Area** Find the area between the x-axis and the curve $y = \sqrt{1 + \cos 4x}$, $0 \le x \le \pi$.

40. **Area** Find the area between the x-axis and the curve $y = \tan^4 x$, $0 \le x \le \pi/4$.

5.3 Trigonometric Substitutions

Trigonometric substitutions can be effective in transforming integrals involving $\sqrt{a^2 - x^2}$, $\sqrt{a^2 + x^2}$, and $\sqrt{x^2 - a^2}$ into integrals we can evaluate directly.

The most common substitutions are $x = a\tan\theta$, $x = a\sin\theta$, and $x = a\sec\theta$. They come from the reference right triangles in Figure 5.2.

With $x = a\tan\theta$,

$$a^2 + x^2 = a^2 + a^2\tan^2\theta = a^2(1 + \tan^2\theta) = a^2\sec^2\theta.$$

With $x = a\sin\theta$,

$$a^2 - x^2 = a^2 - a^2\sin^2\theta = a^2(1 - \sin^2\theta) = a^2\cos^2\theta.$$

With $x = a\sec\theta$,

$$x^2 - a^2 = a^2\sec^2\theta - a^2 = a^2(\sec^2\theta - 1) = a^2\tan^2\theta.$$

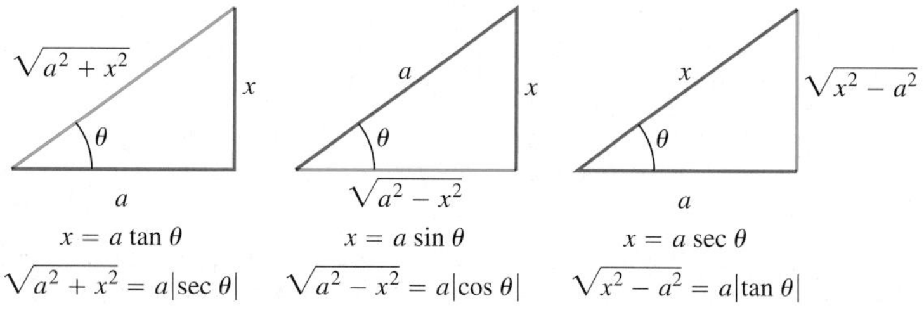

FIGURE 5.2 Reference triangles for the three basic substitutions identifying the sides labeled x and a for each substitution.

We want any substitution we use in an integration to be reversible so that we can change back to the original variable afterward. For example, if $x = a\tan\theta$, we want to be able to set

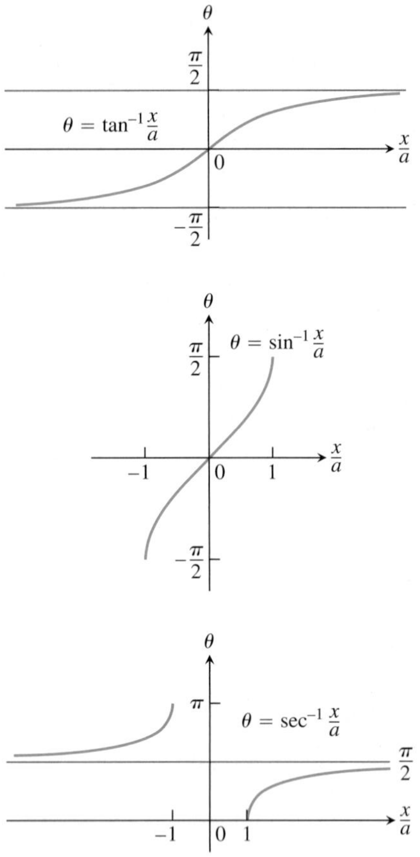

FIGURE 5.3 The arctangent, arcsine, and arcsecant of x/a, graphed as functions of x/a.

$\theta = \tan^{-1}(x/a)$ after the integration takes place. If $x = a \sin\theta$, we want to be able to set $\theta = \sin^{-1}(x/a)$ when we're done, and similarly for $x = a \sec\theta$.

As we know from Section 2.11, the functions in these substitutions have inverses only for selected values of θ (Figure 5.3). For reversibility,

$$x = a\tan\theta \quad \text{requires} \quad \theta = \tan^{-1}\left(\frac{x}{a}\right) \quad \text{with} \quad -\frac{\pi}{2} < \theta < \frac{\pi}{2},$$

$$x = a\sin\theta \quad \text{requires} \quad \theta = \sin^{-1}\left(\frac{x}{a}\right) \quad \text{with} \quad -\frac{\pi}{2} \le \theta \le \frac{\pi}{2},$$

$$x = a\sec\theta \quad \text{requires} \quad \theta = \sec^{-1}\left(\frac{x}{a}\right) \quad \text{with} \quad \begin{cases} 0 \le \theta < \frac{\pi}{2} & \text{if } \frac{x}{a} \ge 1, \\ \frac{\pi}{2} < \theta \le \pi & \text{if } \frac{x}{a} \le -1. \end{cases}$$

To simplify calculations with the substitution $x = a\sec\theta$, we will restrict its use to integrals in which $x/a \ge 1$. This will place θ in $[0, \pi/2)$ and make $\tan\theta \ge 0$. We will then have $\sqrt{x^2 - a^2} = \sqrt{a^2\tan^2\theta} = |a\tan\theta| = a\tan\theta$, free of absolute values, provided $a > 0$.

EXAMPLE 1 Evaluate

$$\int \frac{dx}{\sqrt{4 + x^2}}.$$

Solution We set

$$x = 2\tan\theta, \qquad dx = 2\sec^2\theta\, d\theta, \qquad -\frac{\pi}{2} < \theta < \frac{\pi}{2},$$

$$4 + x^2 = 4 + 4\tan^2\theta = 4(1 + \tan^2\theta) = 4\sec^2\theta.$$

Then

$$\int \frac{dx}{\sqrt{4 + x^2}} = \int \frac{2\sec^2\theta\, d\theta}{\sqrt{4\sec^2\theta}} = \int \frac{\sec^2\theta\, d\theta}{|\sec\theta|} \qquad \sqrt{\sec^2\theta} = |\sec\theta|$$

$$= \int \sec\theta\, d\theta \qquad \sec\theta > 0 \text{ for } -\frac{\pi}{2} < \theta < \frac{\pi}{2}$$

$$= \ln|\sec\theta + \tan\theta| + C$$

$$= \ln\left|\frac{\sqrt{4 + x^2}}{2} + \frac{x}{2}\right| + C \qquad \text{From Fig. 5.4}$$

$$= \ln\left|\sqrt{4 + x^2} + x\right| + C'. \qquad \text{Taking } C' = C - \ln 2$$

Notice how we expressed $\ln|\sec\theta + \tan\theta|$ in terms of x: We drew a reference triangle for the original substitution $x = 2\tan\theta$ (Figure 5.4) and read the ratios from the triangle. ■

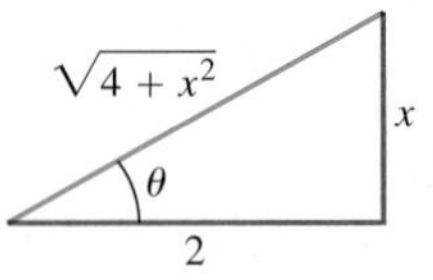

FIGURE 5.4 Reference triangle for $x = 2\tan\theta$ (Example 1):

$$\tan\theta = \frac{x}{2}$$

and

$$\sec\theta = \frac{\sqrt{4 + x^2}}{2}.$$

EXAMPLE 2 Evaluate

$$\int \frac{x^2\, dx}{\sqrt{9 - x^2}}.$$

Solution We set

$$x = 3\sin\theta, \qquad dx = 3\cos\theta\, d\theta, \qquad -\frac{\pi}{2} < \theta < \frac{\pi}{2}$$
$$9 - x^2 = 9 - 9\sin^2\theta = 9(1 - \sin^2\theta) = 9\cos^2\theta.$$

Then

$$\begin{aligned}\int \frac{x^2\,dx}{\sqrt{9 - x^2}} &= \int \frac{9\sin^2\theta \cdot 3\cos\theta\, d\theta}{|3\cos\theta|} \\ &= 9\int \sin^2\theta\, d\theta && \cos\theta > 0 \text{ for } -\frac{\pi}{2} < \theta < \frac{\pi}{2} \\ &= 9\int \frac{1 - \cos 2\theta}{2}\, d\theta \\ &= \frac{9}{2}\left(\theta - \frac{\sin 2\theta}{2}\right) + C \\ &= \frac{9}{2}(\theta - \sin\theta\cos\theta) + C && \sin 2\theta = 2\sin\theta\cos\theta \\ &= \frac{9}{2}\left(\sin^{-1}\frac{x}{3} - \frac{x}{3}\cdot\frac{\sqrt{9 - x^2}}{3}\right) + C && \text{Fig. 5.5} \\ &= \frac{9}{2}\sin^{-1}\frac{x}{3} - \frac{x}{2}\sqrt{9 - x^2} + C.\end{aligned}$$

■

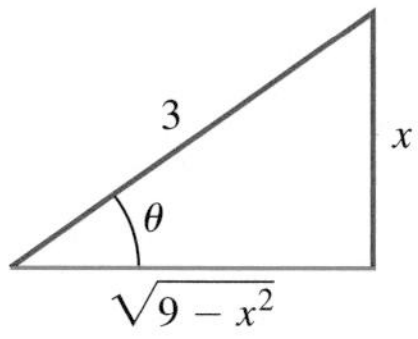

FIGURE 5.5 Reference triangle for $x = 3\sin\theta$ (Example 2):

$$\sin\theta = \frac{x}{3}$$

and

$$\cos\theta = \frac{\sqrt{9 - x^2}}{3}.$$

EXAMPLE 3 Evaluate

$$\int \frac{dx}{\sqrt{25x^2 - 4}}, \qquad x > \frac{2}{5}.$$

Solution We first rewrite the radical as

$$\begin{aligned}\sqrt{25x^2 - 4} &= \sqrt{25\left(x^2 - \frac{4}{25}\right)} \\ &= 5\sqrt{x^2 - \left(\frac{2}{5}\right)^2}\end{aligned}$$

to put the radicand in the form $x^2 - a^2$. We then substitute

$$x = \frac{2}{5}\sec\theta, \qquad dx = \frac{2}{5}\sec\theta\tan\theta\, d\theta, \qquad 0 < \theta < \frac{\pi}{2}$$

$$\begin{aligned}x^2 - \left(\frac{2}{5}\right)^2 &= \frac{4}{25}\sec^2\theta - \frac{4}{25} \\ &= \frac{4}{25}(\sec^2\theta - 1) = \frac{4}{25}\tan^2\theta\end{aligned}$$

$$\sqrt{x^2 - \left(\frac{2}{5}\right)^2} = \frac{2}{5}|\tan\theta| = \frac{2}{5}\tan\theta. \qquad \begin{array}{l}\tan\theta > 0 \text{ for} \\ 0 < \theta < \pi/2\end{array}$$

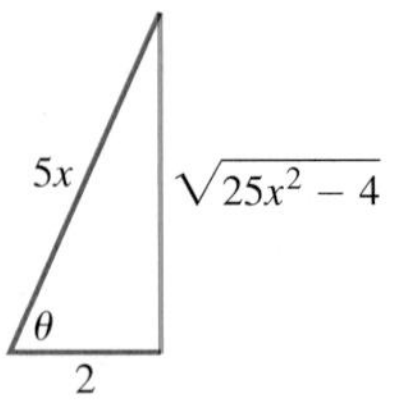

FIGURE 5.6 If $x = (2/5)\sec\theta$, $0 < \theta < \pi/2$, then $\theta = \sec^{-1}(5x/2)$, and we can read the values of the other trigonometric functions of θ from this right triangle (Example 3).

With these substitutions, we have

$$\int \frac{dx}{\sqrt{25x^2-4}} = \int \frac{dx}{5\sqrt{x^2-(4/25)}} = \int \frac{(2/5)\sec\theta\tan\theta\,d\theta}{5\cdot(2/5)\tan\theta}$$

$$= \frac{1}{5}\int \sec\theta\,d\theta = \frac{1}{5}\ln|\sec\theta+\tan\theta| + C$$

$$= \frac{1}{5}\ln\left|\frac{5x}{2} + \frac{\sqrt{25x^2-4}}{2}\right| + C. \quad \text{Fig. 5.6} \quad \blacksquare$$

EXERCISES 5.3

Evaluate the integrals in Exercises 1–28.

1. $\int \frac{dy}{\sqrt{9+y^2}}$
2. $\int \frac{3\,dy}{\sqrt{1+9y^2}}$
3. $\int_{-2}^{2} \frac{dx}{4+x^2}$
4. $\int_{0}^{2} \frac{dx}{8+2x^2}$
5. $\int_{0}^{3/2} \frac{dx}{\sqrt{9-x^2}}$
6. $\int_{0}^{1/2\sqrt{2}} \frac{2\,dx}{\sqrt{1-4x^2}}$
7. $\int \sqrt{25-t^2}\,dt$
8. $\int \sqrt{1-9t^2}\,dt$
9. $\int \frac{dx}{\sqrt{4x^2-49}}, \quad x > \frac{7}{2}$
10. $\int \frac{5\,dx}{\sqrt{25x^2-9}}, \quad x > \frac{3}{5}$
11. $\int \frac{\sqrt{y^2-49}}{y}\,dy, \quad y > 7$
12. $\int \frac{\sqrt{y^2-25}}{y^3}\,dy, \quad y > 5$
13. $\int \frac{dx}{x^2\sqrt{x^2-1}}, \quad x > 1$
14. $\int \frac{2\,dx}{x^3\sqrt{x^2-1}}, \quad x > 1$
15. $\int \frac{x^3\,dx}{\sqrt{x^2+4}}$
16. $\int \frac{dx}{x^2\sqrt{x^2+1}}$
17. $\int \frac{8\,dw}{w^2\sqrt{4-w^2}}$
18. $\int \frac{\sqrt{9-w^2}}{w^2}\,dw$
19. $\int_{0}^{\sqrt{3}/2} \frac{4x^2\,dx}{(1-x^2)^{3/2}}$
20. $\int_{0}^{1} \frac{dx}{(4-x^2)^{3/2}}$
21. $\int \frac{dx}{(x^2-1)^{3/2}}, \quad x > 1$
22. $\int \frac{x^2\,dx}{(x^2-1)^{5/2}}, \quad x > 1$
23. $\int \frac{(1-x^2)^{3/2}}{x^6}\,dx$
24. $\int \frac{(1-x^2)^{1/2}}{x^4}\,dx$
25. $\int \frac{8\,dx}{(4x^2+1)^2}$
26. $\int \frac{6\,dt}{(9t^2+1)^2}$
27. $\int \frac{v^2\,dv}{(1-v^2)^{5/2}}$
28. $\int \frac{(1-r^2)^{5/2}}{r^8}\,dr$

In Exercises 29–36, use an appropriate substitution and then a trigonometric substitution to evaluate the integrals.

29. $\int_{0}^{\ln 4} \frac{e^t\,dt}{\sqrt{e^{2t}+9}}$
30. $\int_{\ln(3/4)}^{\ln(4/3)} \frac{e^t\,dt}{(1+e^{2t})^{3/2}}$
31. $\int_{1/12}^{1/4} \frac{2\,dt}{\sqrt{t}+4t\sqrt{t}}$
32. $\int_{1}^{e} \frac{dy}{y\sqrt{1+(\ln y)^2}}$
33. $\int \frac{dx}{x\sqrt{x^2-1}}$
34. $\int \frac{dx}{1+x^2}$
35. $\int \frac{x\,dx}{\sqrt{x^2-1}}$
36. $\int \frac{dx}{\sqrt{1-x^2}}$

Solve the initial value problems in Exercises 37–40 for y as a function of x.

37. $x\frac{dy}{dx} = \sqrt{x^2-4}, \quad x \geq 2, \quad y(2) = 0$
38. $\sqrt{x^2-9}\,\frac{dy}{dx} = 1, \quad x > 3, \quad y(5) = \ln 3$
39. $(x^2+4)\frac{dy}{dx} = 3, \quad y(2) = 0$
40. $(x^2+1)^2\frac{dy}{dx} = \sqrt{x^2+1}, \quad y(0) = 1$
41. **Area** Find the area of the region in the first quadrant that is enclosed by the coordinate axes and the curve $y = \sqrt{9-x^2}/3$.
42. **Area** Find the area enclosed by the ellipse

$$\frac{x^2}{a^2} + \frac{y^2}{b^2} = 1.$$

5.4 Integration of Rational Functions by Partial Fractions

This section shows how to express a rational function (a quotient of polynomials) as a sum of simpler fractions, called *partial fractions*, which are easily integrated. For instance, the rational function $(5x - 3)/(x^2 - 2x - 3)$ can be rewritten as

$$\frac{5x - 3}{x^2 - 2x - 3} = \frac{2}{x + 1} + \frac{3}{x - 3},$$

which can be verified algebraically by placing the fractions on the right side over a common denominator $(x + 1)(x - 3)$. The skill acquired in writing rational functions as such a sum is useful in other settings as well (for instance, when using certain transform methods to solve differential equations). To integrate the rational function $(5x - 3)/(x^2 - 2x - 3)$ on the left side of our previous expression, we simply sum the integrals of the fractions on the right side:

$$\begin{aligned}\int \frac{5x - 3}{(x + 1)(x - 3)}\,dx &= \int \frac{2}{x + 1}\,dx + \int \frac{3}{x - 3}\,dx \\ &= 2 \ln |x + 1| + 3 \ln |x - 3| + C.\end{aligned}$$

The method for rewriting rational functions as a sum of simpler fractions is called **the method of partial fractions**. In the case of the above example, it consists of finding constants A and B such that

$$\frac{5x - 3}{x^2 - 2x - 3} = \frac{A}{x + 1} + \frac{B}{x - 3}. \tag{1}$$

(Pretend for a moment that we do not know that $A = 2$ and $B = 3$ will work.) We call the fractions $A/(x + 1)$ and $B/(x - 3)$ **partial fractions** because their denominators are only part of the original denominator $x^2 - 2x - 3$. We call A and B **undetermined coefficients** until proper values for them have been found.

To find A and B, we first clear Equation (1) of fractions and regroup in powers of x, obtaining

$$5x - 3 = A(x - 3) + B(x + 1) = (A + B)x - 3A + B.$$

This will be an identity in x if and only if the coefficients of like powers of x on the two sides are equal:

$$A + B = 5, \qquad -3A + B = -3.$$

Solving these equations simultaneously gives $A = 2$ and $B = 3$.

General Description of the Method

Success in writing a rational function $f(x)/g(x)$ as a sum of partial fractions depends on two things:

- *The degree of $f(x)$ must be less than the degree of $g(x)$.* That is, the fraction must be proper. If it isn't, divide $f(x)$ by $g(x)$ and work with the remainder term. See Example 3 of this section.

- *We must know the factors of* $g(x)$. In theory, any polynomial with real coefficients can be written as a product of real linear factors and real quadratic factors. In practice, the factors may be hard to find.

Here is how we find the partial fractions of a proper fraction $f(x)/g(x)$ when the factors of g are known.

Method of Partial Fractions ($f(x)/g(x)$ Proper)

1. Let $x - r$ be a linear factor of $g(x)$. Suppose that $(x - r)^m$ is the highest power of $x - r$ that divides $g(x)$. Then, to this factor, assign the sum of the m partial fractions:

$$\frac{A_1}{x - r} + \frac{A_2}{(x - r)^2} + \cdots + \frac{A_m}{(x - r)^m}.$$

Do this for each distinct linear factor of $g(x)$.

2. Let $x^2 + px + q$ be a quadratic factor of $g(x)$. Suppose that $(x^2 + px + q)^n$ is the highest power of this factor that divides $g(x)$ and that $x^2 + px + q$ has no real roots. Then, to this factor, assign the sum of the n partial fractions:

$$\frac{B_1x + C_1}{x^2 + px + q} + \frac{B_2x + C_2}{(x^2 + px + q)^2} + \cdots + \frac{B_nx + C_n}{(x^2 + px + q)^n}.$$

Do this for each distinct quadratic factor of $g(x)$.

3. Set the original fraction $f(x)/g(x)$ equal to the sum of all these partial fractions. Clear the resulting equation of fractions and arrange the terms in decreasing powers of x.
4. Equate the coefficients of corresponding powers of x and solve the resulting equations for the undetermined coefficients.

EXAMPLE 1 Evaluate

$$\int \frac{x^2 + 4x + 1}{(x - 1)(x + 1)(x + 3)}\,dx$$

using partial fractions.

Solution The partial fraction decomposition has the form

$$\frac{x^2 + 4x + 1}{(x - 1)(x + 1)(x + 3)} = \frac{A}{x - 1} + \frac{B}{x + 1} + \frac{C}{x + 3}.$$

To find the values of the undetermined coefficients A, B, and C we clear fractions and get

$$\begin{aligned} x^2 + 4x + 1 &= A(x + 1)(x + 3) + B(x - 1)(x + 3) + C(x - 1)(x + 1) \\ &= (A + B + C)x^2 + (4A + 2B)x + (3A - 3B - C). \end{aligned}$$

The polynomials on both sides of the above equation are identical, so we equate coefficients of like powers of x obtaining

$$\begin{aligned}
&\text{Coefficient of } x^2: && A + B + C = 1 \\
&\text{Coefficient of } x^1: && 4A + 2B = 4 \\
&\text{Coefficient of } x^0: && 3A - 3B - C = 1
\end{aligned}$$

There are several ways for solving such a system of linear equations for the unknowns A, B, and C, including elimination of variables, or the use of a calculator or computer. Whatever method is used, the solution is $A = 3/4$, $B = 1/2$, and $C = -1/4$. Hence we have

$$\begin{aligned}
\int \frac{x^2 + 4x + 1}{(x - 1)(x + 1)(x + 3)}\,dx &= \int \left[\frac{3}{4}\frac{1}{x - 1} + \frac{1}{2}\frac{1}{x + 1} - \frac{1}{4}\frac{1}{x + 3}\right] dx \\
&= \frac{3}{4}\ln|x - 1| + \frac{1}{2}\ln|x + 1| - \frac{1}{4}\ln|x + 3| + K,
\end{aligned}$$

where K is the arbitrary constant of integration (to avoid confusion with the undetermined coefficient we labeled as C). ■

EXAMPLE 2 Evaluate

$$\int \frac{6x + 7}{(x + 2)^2}\,dx.$$

Solution First we express the integrand as a sum of partial fractions with undetermined coefficients.

$$\begin{aligned}
\frac{6x + 7}{(x + 2)^2} &= \frac{A}{x + 2} + \frac{B}{(x + 2)^2} \\
6x + 7 &= A(x + 2) + B && \text{Multiply both sides by } (x + 2)^2. \\
&= Ax + (2A + B)
\end{aligned}$$

Equating coefficients of corresponding powers of x gives

$$A = 6 \quad \text{and} \quad 2A + B = 12 + B = 7, \quad \text{or} \quad A = 6 \quad \text{and} \quad B = -5.$$

Therefore,

$$\begin{aligned}
\int \frac{6x + 7}{(x + 2)^2}\,dx &= \int \left(\frac{6}{x + 2} - \frac{5}{(x + 2)^2}\right) dx \\
&= 6\int \frac{dx}{x + 2} - 5\int (x + 2)^{-2}\,dx \\
&= 6\ln|x + 2| + 5(x + 2)^{-1} + C
\end{aligned}$$

■

EXAMPLE 3 Evaluate

$$\int \frac{2x^3 - 4x^2 - x - 3}{x^2 - 2x - 3}\,dx.$$

Solution First we divide the denominator into the numerator to get a polynomial plus a proper fraction.

$$\begin{array}{r} 2x \qquad\qquad\qquad \\ x^2 - 2x - 3\overline{)2x^3 - 4x^2 - x - 3} \\ \underline{2x^3 - 4x^2 - 6x} \\ 5x - 3 \end{array}$$

Then we write the improper fraction as a polynomial plus a proper fraction.

$$\frac{2x^3 - 4x^2 - x - 3}{x^2 - 2x - 3} = 2x + \frac{5x - 3}{x^2 - 2x - 3}$$

We found the partial fraction decomposition of the fraction on the right in the opening example, so

$$\begin{aligned} \int \frac{2x^3 - 4x^2 - x - 3}{x^2 - 2x - 3}\,dx &= \int 2x\,dx + \int \frac{5x - 3}{x^2 - 2x - 3}\,dx \\ &= \int 2x\,dx + \int \frac{2}{x + 1}\,dx + \int \frac{3}{x - 3}\,dx \\ &= x^2 + 2\ln|x + 1| + 3\ln|x - 3| + C. \end{aligned}$$

■

A quadratic polynomial is **irreducible** if it cannot be written as the product of two linear factors with real coefficients. That is, the polynomial has no real roots.

EXAMPLE 4 Evaluate

$$\int \frac{-2x + 4}{(x^2 + 1)(x - 1)^2}\,dx$$

using partial fractions.

Solution The denominator has an irreducible quadratic factor as well as a repeated linear factor, so we write

$$\frac{-2x + 4}{(x^2 + 1)(x - 1)^2} = \frac{Ax + B}{x^2 + 1} + \frac{C}{x - 1} + \frac{D}{(x - 1)^2}. \tag{2}$$

Clearing the equation of fractions gives

$$\begin{aligned} -2x + 4 &= (Ax + B)(x - 1)^2 + C(x - 1)(x^2 + 1) + D(x^2 + 1) \\ &= (A + C)x^3 + (-2A + B - C + D)x^2 \\ &\quad + (A - 2B + C)x + (B - C + D). \end{aligned}$$

Equating coefficients of like terms gives

Coefficients of x^3: $\quad 0 = A + C$

Coefficients of x^2: $\quad 0 = -2A + B - C + D$

Coefficients of x^1: $\quad -2 = A - 2B + C$

Coefficients of x^0: $\quad 4 = B - C + D$

We solve these equations simultaneously to find the values of A, B, C, and D:

$$\begin{aligned} -4 &= -2A, \quad A = 2 && \text{Subtract fourth equation from second.} \\ C &= -A = -2 && \text{From the first equation} \\ B &= 1 && A = 2 \text{ and } C = -2 \text{ in third equation.} \\ D &= 4 - B + C = 1. && \text{From the fourth equation} \end{aligned}$$

We substitute these values into Equation (2), obtaining

$$\frac{-2x + 4}{(x^2 + 1)(x - 1)^2} = \frac{2x + 1}{x^2 + 1} - \frac{2}{x - 1} + \frac{1}{(x - 1)^2}.$$

Finally, using the expansion above we can integrate:

$$\begin{aligned} \int \frac{-2x + 4}{(x^2 + 1)(x - 1)^2}\, dx &= \int \left(\frac{2x + 1}{x^2 + 1} - \frac{2}{x - 1} + \frac{1}{(x - 1)^2} \right) dx \\ &= \int \left(\frac{2x}{x^2 + 1} + \frac{1}{x^2 + 1} - \frac{2}{x - 1} + \frac{1}{(x - 1)^2} \right) dx \\ &= \ln(x^2 + 1) + \tan^{-1} x - 2 \ln |x - 1| - \frac{1}{x - 1} + C. \quad \blacksquare \end{aligned}$$

EXAMPLE 5 Evaluate

$$\int \frac{dx}{x(x^2 + 1)^2}.$$

Solution The form of the partial fraction decomposition is

$$\frac{1}{x(x^2 + 1)^2} = \frac{A}{x} + \frac{Bx + C}{x^2 + 1} + \frac{Dx + E}{(x^2 + 1)^2}.$$

Multiplying by $x(x^2 + 1)^2$, we have

$$\begin{aligned} 1 &= A(x^2 + 1)^2 + (Bx + C)x(x^2 + 1) + (Dx + E)x \\ &= A(x^4 + 2x^2 + 1) + B(x^4 + x^2) + C(x^3 + x) + Dx^2 + Ex \\ &= (A + B)x^4 + Cx^3 + (2A + B + D)x^2 + (C + E)x + A. \end{aligned}$$

If we equate coefficients, we get the system

$$A + B = 0, \qquad C = 0, \qquad 2A + B + D = 0, \qquad C + E = 0, \qquad A = 1.$$

Solving this system gives $A = 1$, $B = -1$, $C = 0$, $D = -1$, and $E = 0$. Thus,

$$\begin{aligned} \int \frac{dx}{x(x^2 + 1)^2} &= \int \left[\frac{1}{x} + \frac{-x}{x^2 + 1} + \frac{-x}{(x^2 + 1)^2} \right] dx \\ &= \int \frac{dx}{x} - \int \frac{x\, dx}{x^2 + 1} - \int \frac{x\, dx}{(x^2 + 1)^2} \end{aligned}$$

$$= \int \frac{dx}{x} - \frac{1}{2}\int \frac{du}{u} - \frac{1}{2}\int \frac{du}{u^2} \qquad \begin{array}{l} u = x^2 + 1, \\ du = 2x\,dx \end{array}$$

$$= \ln|x| - \frac{1}{2}\ln|u| + \frac{1}{2u} + K$$

$$= \ln|x| - \frac{1}{2}\ln(x^2+1) + \frac{1}{2(x^2+1)} + K$$

$$= \ln\frac{|x|}{\sqrt{x^2+1}} + \frac{1}{2(x^2+1)} + K.$$ ■

EXERCISES 5.4

Expand the quotients in Exercises 1–8 by partial fractions.

1. $\dfrac{5x-13}{(x-3)(x-2)}$
2. $\dfrac{5x-7}{x^2-3x+2}$
3. $\dfrac{x+4}{(x+1)^2}$
4. $\dfrac{2x+2}{x^2-2x+1}$
5. $\dfrac{z+1}{z^2(z-1)}$
6. $\dfrac{z}{z^3-z^2-6z}$
7. $\dfrac{t^2+8}{t^2-5t+6}$
8. $\dfrac{t^4+9}{t^4+9t^2}$

In Exercises 9–16, express the integrands as a sum of partial fractions and evaluate the integrals.

9. $\displaystyle\int \frac{dx}{1-x^2}$
10. $\displaystyle\int \frac{dx}{x^2+2x}$
11. $\displaystyle\int \frac{x+4}{x^2+5x-6}\,dx$
12. $\displaystyle\int \frac{2x+1}{x^2-7x+12}\,dx$
13. $\displaystyle\int_4^8 \frac{y\,dy}{y^2-2y-3}$
14. $\displaystyle\int_{1/2}^1 \frac{y+4}{y^2+y}\,dy$
15. $\displaystyle\int \frac{dt}{t^3+t^2-2t}$
16. $\displaystyle\int \frac{x+3}{2x^3-8x}\,dx$

In Exercises 17–20, express the integrands as a sum of partial fractions and evaluate the integrals.

17. $\displaystyle\int_0^1 \frac{x^3\,dx}{x^2+2x+1}$
18. $\displaystyle\int_{-1}^0 \frac{x^3\,dx}{x^2-2x+1}$
19. $\displaystyle\int \frac{dx}{(x^2-1)^2}$
20. $\displaystyle\int \frac{x^2\,dx}{(x-1)(x^2+2x+1)}$

In Exercises 21–28, express the integrands as a sum of partial fractions and evaluate the integrals.

21. $\displaystyle\int_0^1 \frac{dx}{(x+1)(x^2+1)}$
22. $\displaystyle\int_1^{\sqrt{3}} \frac{3t^2+t+4}{t^3+t}\,dt$
23. $\displaystyle\int \frac{y^2+2y+1}{(y^2+1)^2}\,dy$
24. $\displaystyle\int \frac{8x^2+8x+2}{(4x^2+1)^2}\,dx$
25. $\displaystyle\int \frac{2s+2}{(s^2+1)(s-1)^3}\,ds$
26. $\displaystyle\int \frac{s^4+81}{s(s^2+9)^2}\,ds$
27. $\displaystyle\int \frac{2\theta^3+5\theta^2+8\theta+4}{(\theta^2+2\theta+2)^2}\,d\theta$
28. $\displaystyle\int \frac{\theta^4-4\theta^3+2\theta^2-3\theta+1}{(\theta^2+1)^3}\,d\theta$

In Exercises 29–34, perform long division on the integrand, write the proper fraction as a sum of partial fractions, and then evaluate the integral.

29. $\displaystyle\int \frac{2x^3-2x^2+1}{x^2-x}\,dx$
30. $\displaystyle\int \frac{x^4}{x^2-1}\,dx$
31. $\displaystyle\int \frac{9x^3-3x+1}{x^3-x^2}\,dx$
32. $\displaystyle\int \frac{16x^3}{4x^2-4x+1}\,dx$
33. $\displaystyle\int \frac{y^4+y^2-1}{y^3+y}\,dy$
34. $\displaystyle\int \frac{2y^4}{y^3-y^2+y-1}\,dy$

Evaluate the integrals in Exercises 35–40.

35. $\displaystyle\int \frac{e^t\,dt}{e^{2t}+3e^t+2}$
36. $\displaystyle\int \frac{e^{4t}+2e^{2t}-e^t}{e^{2t}+1}\,dt$
37. $\displaystyle\int \frac{\cos y\,dy}{\sin^2 y+\sin y-6}$
38. $\displaystyle\int \frac{\sin\theta\,d\theta}{\cos^2\theta+\cos\theta-2}$
39. $\displaystyle\int \frac{(x-2)^2\tan^{-1}(2x)-12x^3-3x}{(4x^2+1)(x-2)^2}\,dx$
40. $\displaystyle\int \frac{(x+1)^2\tan^{-1}(3x)+9x^3+x}{(9x^2+1)(x+1)^2}\,dx$

Solve the initial value problems in Exercises 41–44 for x as a function of t.

41. $(t^2-3t+2)\dfrac{dx}{dt} = 1 \quad (t>2), \quad x(3)=0$
42. $(3t^4+4t^2+1)\dfrac{dx}{dt} = 2\sqrt{3}, \quad x(1) = -\pi\sqrt{3}/4$

43. $(t^2 + 2t)\dfrac{dx}{dt} = 2x + 2 \quad (t, x > 0), \quad x(1) = 1$

44. $(t + 1)\dfrac{dx}{dt} = x^2 + 1 \quad (t > -1), \quad x(0) = 0$

45. Social diffusion Sociologists sometimes use the phrase "social diffusion" to describe the way information spreads through a population. The information might be a rumor, a cultural fad, or news about a technical innovation. In a sufficiently large population, the number of people x who have the information is treated as a differentiable function of time t, and the rate of diffusion, dx/dt, is assumed to be proportional to the number of people who have the information times the number of people who do not. This leads to the equation

$$\frac{dx}{dt} = kx(N - x),$$

where N is the number of people in the population.

Suppose t is in days, $k = 1/250$, and two people start a rumor at time $t = 0$ in a population of $N = 1000$ people.

a. Find x as a function of t.

b. When will half the population have heard the rumor? (This is when the rumor will be spreading the fastest.)

46. Second-order chemical reactions Many chemical reactions are the result of the interaction of two molecules that undergo a change to produce a new product. The rate of the reaction typically depends on the concentrations of the two kinds of molecules. If a is the amount of substance A and b is the amount of substance B at time $t = 0$, and if x is the amount of product at time t, then the rate of formation of x may be given by the differential equation

$$\frac{dx}{dt} = k(a - x)(b - x),$$

or

$$\frac{1}{(a - x)(b - x)}\frac{dx}{dt} = k,$$

where k is a constant for the reaction. Integrate both sides of this equation to obtain a relation between x and t (**a**) if $a = b$, and (**b**) if $a \neq b$. Assume in each case that $x = 0$ when $t = 0$.

5.5 Integral Tables and Computer Algebra Systems

In this section we discuss how to use tables and computer algebra systems to evaluate integrals.

Integral Tables

A Brief Table of Integrals is provided at the back of the book, after the index. (More extensive tables appear in compilations such as *CRC Mathematical Tables*, which contain thousands of integrals.) The integration formulas are stated in terms of constants a, b, c, m, n, and so on. These constants can usually assume any real value and need not be integers. Occasional limitations on their values are stated with the formulas. Formula 5 requires $n \neq -1$, for example, and Formula 11 requires $n \neq 2$.

The formulas also assume that the constants do not take on values that require dividing by zero or taking even roots of negative numbers. For example, Formula 8 assumes that $a \neq 0$, and Formulas 13(a) and (b) cannot be used unless b is positive.

EXAMPLE 1 Find

$$\int x(2x + 5)^{-1}\,dx.$$

Solution We use Formula 8 (not 7, which requires $n \neq -1$):

$$\int x(ax + b)^{-1}\,dx = \frac{x}{a} - \frac{b}{a^2}\ln|ax + b| + C.$$

With $a = 2$ and $b = 5$, we have

$$\int x(2x + 5)^{-1}\, dx = \frac{x}{2} - \frac{5}{4}\ln|2x + 5| + C. \qquad \blacksquare$$

EXAMPLE 2 Find

$$\int \frac{dx}{x\sqrt{2x - 4}}.$$

Solution We use Formula 13(a):

$$\int \frac{dx}{x\sqrt{ax - b}} = \frac{2}{\sqrt{b}}\tan^{-1}\sqrt{\frac{ax - b}{b}} + C.$$

With $a = 2$ and $b = 4$, we have

$$\int \frac{dx}{x\sqrt{2x - 4}} = \frac{2}{\sqrt{4}}\tan^{-1}\sqrt{\frac{2x - 4}{4}} + C = \tan^{-1}\sqrt{\frac{x - 2}{2}} + C. \qquad \blacksquare$$

EXAMPLE 3 Find

$$\int x \sin^{-1} x\, dx.$$

Solution We use Formula 99:

$$\int x^n \sin^{-1} ax\, dx = \frac{x^{n+1}}{n + 1}\sin^{-1} ax - \frac{a}{n + 1}\int \frac{x^{n+1}\, dx}{\sqrt{1 - a^2x^2}}, \qquad n \neq -1.$$

With $n = 1$ and $a = 1$, we have

$$\int x \sin^{-1} x\, dx = \frac{x^2}{2}\sin^{-1} x - \frac{1}{2}\int \frac{x^2\, dx}{\sqrt{1 - x^2}}.$$

The integral on the right is found in the table as Formula 33:

$$\int \frac{x^2}{\sqrt{a^2 - x^2}}\, dx = \frac{a^2}{2}\sin^{-1}\left(\frac{x}{a}\right) - \frac{1}{2}x\sqrt{a^2 - x^2} + C.$$

With $a = 1$,

$$\int \frac{x^2\, dx}{\sqrt{1 - x^2}} = \frac{1}{2}\sin^{-1} x - \frac{1}{2}x\sqrt{1 - x^2} + C.$$

The combined result is

$$\begin{aligned} \int x \sin^{-1} x\, dx &= \frac{x^2}{2}\sin^{-1} x - \frac{1}{2}\left(\frac{1}{2}\sin^{-1} x - \frac{1}{2}x\sqrt{1 - x^2} + C\right) \\ &= \left(\frac{x^2}{2} - \frac{1}{4}\right)\sin^{-1} x + \frac{1}{4}x\sqrt{1 - x^2} + C'. \qquad \blacksquare \end{aligned}$$

Reduction Formulas

The time required for repeated integrations by parts can sometimes be shortened by applying formulas like

$$\int \tan^n x\,dx = \frac{1}{n-1}\tan^{n-1} x - \int \tan^{n-2} x\,dx \tag{1}$$

$$\int (\ln x)^n\,dx = x(\ln x)^n - n\int (\ln x)^{n-1}\,dx \tag{2}$$

$$\int \sin^n x\cos^m x\,dx = -\frac{\sin^{n-1} x\cos^{m+1} x}{m+n} + \frac{n-1}{m+n}\int \sin^{n-2} x\cos^m x\,dx \qquad (n \neq -m). \tag{3}$$

Formulas like these are called **reduction formulas** because they replace an integral containing some power of a function with an integral of the same form with the power reduced. By applying such a formula repeatedly, we can eventually express the original integral in terms of a power low enough to be evaluated directly.

EXAMPLE 4 Find

$$\int \tan^5 x\,dx.$$

Solution We apply Equation (1) with $n = 5$ to get

$$\int \tan^5 x\,dx = \frac{1}{4}\tan^4 x - \int \tan^3 x\,dx.$$

We then apply Equation (1) again, with $n = 3$, to evaluate the remaining integral:

$$\int \tan^3 x\,dx = \frac{1}{2}\tan^2 x - \int \tan x\,dx = \frac{1}{2}\tan^2 x + \ln|\cos x| + C.$$

The combined result is

$$\int \tan^5 x\,dx = \frac{1}{4}\tan^4 x - \frac{1}{2}\tan^2 x - \ln|\cos x| + C'.$$ ■

As their form suggests, reduction formulas are derived by integration by parts. (See Example 5 in Section 5.1.)

Integration with a CAS

A powerful capability of computer algebra systems is their ability to integrate symbolically. This is performed with the **integrate command** specified by the particular system (for example, **int** in Maple, **Integrate** in Mathematica).

EXAMPLE 5 Suppose that you want to evaluate the indefinite integral of the function

$$f(x) = x^2\sqrt{a^2 + x^2}.$$

Using Maple, you first define or name the function:

$$> f{:=}\ x\hat{\ }2 * \operatorname{sqrt}(a\hat{\ }2 + x\hat{\ }2);$$

Then you use the integrate command on f, identifying the variable of integration:

$$> \operatorname{int}(f, x);$$

Maple returns the answer

$$\frac{1}{4}x(a^2 + x^2)^{3/2} - \frac{1}{8}a^2x\sqrt{a^2 + x^2} - \frac{1}{8}a^4 \ln\left(x + \sqrt{a^2 + x^2}\right).$$

If you want to see if the answer can be simplified, enter

$$> \operatorname{simplify}(\%);$$

Maple returns

$$\frac{1}{8}a^2x\sqrt{a^2 + x^2} + \frac{1}{4}x^3\sqrt{a^2 + x^2} - \frac{1}{8}a^4 \ln\left(x + \sqrt{a^2 + x^2}\right).$$

If you want the definite integral for $0 \le x \le \pi/2$, you can use the format

$$> \operatorname{int}(f, x = 0..\text{Pi}/2);$$

Maple will return the expression

$$\frac{1}{64}\pi(4a^2 + \pi^2)^{(3/2)} - \frac{1}{32}a^2\pi\sqrt{4a^2 + \pi^2} + \frac{1}{8}a^4 \ln(2) - \frac{1}{8}a^4 \ln\left(\pi + \sqrt{4a^2 + \pi^2}\right) + \frac{1}{16}a^4 \ln(a^2).$$

You can also find the definite integral for a particular value of the constant a:

$$> a{:=}\ 1;$$
$$> \operatorname{int}(f, x = 0..1);$$

Maple returns the numerical answer

$$\frac{3}{8}\sqrt{2} + \frac{1}{8}\ln\left(\sqrt{2} - 1\right).$$ ■

EXAMPLE 6 Use a CAS to find

$$\int \sin^2 x \cos^3 x\, dx.$$

Solution With Maple, we have the entry

$$> \operatorname{int}((\sin\hat{\ }2)(x) * (\cos\hat{\ }3)(x), x);$$

with the immediate return

$$-\frac{1}{5}\sin(x)\cos(x)^4 + \frac{1}{15}\cos(x)^2\sin(x) + \frac{2}{15}\sin(x).$$ ■

Computer algebra systems vary in how they process integrations. We used Maple in Examples 5 and 6. Mathematica would have returned somewhat different results:

1. In Example 5, given

In [1]:= Integrate [$x\hat{\ }2$ * Sqrt [$a\hat{\ }2 + x\hat{\ }2$], x]

Mathematica returns

$$Out\ [1]= \sqrt{a^2 + x^2}\left(\frac{a^2 x}{8} + \frac{x^3}{4}\right) - \frac{1}{8}a^4 \,\text{Log}\left[x + \sqrt{a^2 + x^2}\right]$$

without having to simplify an intermediate result. The answer is close to Formula 22 in the integral tables.

2. The Mathematica answer to the integral

$$In\ [2]:= \text{Integrate [Sin }[x]\text{^}2 * \text{Cos }[x]\text{^}3, x]$$

in Example 6 is

$$Out\ [2]= \frac{\text{Sin }[x]}{8} - \frac{1}{48}\text{Sin }[3\,x] - \frac{1}{80}\text{Sin }[5\,x]$$

differing from the Maple answer.

Although a CAS is very powerful and can aid us in solving difficult problems, each CAS has its own limitations. There are even situations where a CAS may further complicate a problem (in the sense of producing an answer that is extremely difficult to use or interpret). Note, too, that neither Maple nor Mathematica return an arbitrary constant $+C$. On the other hand, a little mathematical thinking on your part may reduce the problem to one that is quite easy to handle. We provide an example in Exercise 57.

Nonelementary Integrals

The development of computers and calculators that find antiderivatives by symbolic manipulation has led to a renewed interest in determining which antiderivatives can be expressed as finite combinations of elementary functions (the functions we have been studying) and which cannot. Integrals of functions that do not have elementary antiderivatives are called **nonelementary** integrals. They require infinite series (Chapter 7) or numerical methods for their evaluation, which give only an approximation. Examples of nonelementary integrals include the error function (which measures the probability of random errors)

$$\text{erf}\,(x) = \frac{2}{\sqrt{\pi}}\int_0^x e^{-t^2}\,dt$$

and integrals such as

$$\int \sin x^2\,dx \qquad \text{and} \qquad \int \sqrt{1 + x^4}\,dx$$

that arise in engineering and physics. These and a number of others, such as

$$\int \frac{e^x}{x}\,dx, \quad \int e^{(e^x)}\,dx, \quad \int \frac{1}{\ln x}\,dx, \quad \int \ln\,(\ln x)\,dx, \quad \int \frac{\sin x}{x}\,dx,$$

$$\int \sqrt{1 - k^2 \sin^2 x}\,dx, \qquad 0 < k < 1,$$

look so easy they tempt us to try them just to see how they turn out. It can be proved, however, that there is no way to express these integrals as finite combinations of elementary functions.

The same applies to integrals that can be changed into these by substitution. The integrands all have antiderivatives, as a consequence of the Fundamental Theorem of Calculus, Part 1, because they are continuous. However, none of the antiderivatives is elementary.

None of the integrals you are asked to evaluate in the present chapter falls into this category, but you may encounter nonelementary integrals in your other work.

EXERCISES 5.5

Use the table of integrals at the back of the book to evaluate the integrals in Exercises 1–26.

1. $\int \frac{dx}{x\sqrt{x-3}}$
2. $\int \frac{dx}{x\sqrt{x+4}}$
3. $\int \frac{x\,dx}{\sqrt{x-2}}$
4. $\int \frac{x\,dx}{(2x+3)^{3/2}}$
5. $\int x\sqrt{2x-3}\,dx$
6. $\int x(7x+5)^{3/2}\,dx$
7. $\int \frac{\sqrt{9-4x}}{x^2}\,dx$
8. $\int \frac{dx}{x^2\sqrt{4x-9}}$
9. $\int x\sqrt{4x-x^2}\,dx$
10. $\int \frac{\sqrt{x-x^2}}{x}\,dx$
11. $\int \frac{dx}{x\sqrt{7+x^2}}$
12. $\int \frac{dx}{x\sqrt{7-x^2}}$
13. $\int \frac{\sqrt{4-x^2}}{x}\,dx$
14. $\int \frac{\sqrt{x^2-4}}{x}\,dx$
15. $\int e^{2t}\cos 3t\,dt$
16. $\int e^{-3t}\sin 4t\,dt$
17. $\int x\cos^{-1} x\,dx$
18. $\int x\tan^{-1} x\,dx$
19. $\int x^2\tan^{-1} x\,dx$
20. $\int \frac{\tan^{-1} x}{x^2}\,dx$
21. $\int \sin 3x\cos 2x\,dx$
22. $\int \sin 2x\cos 3x\,dx$
23. $\int 8\sin 4t\sin\frac{t}{2}\,dt$
24. $\int \sin\frac{t}{3}\sin\frac{t}{6}\,dt$
25. $\int \cos\frac{\theta}{3}\cos\frac{\theta}{4}\,d\theta$
26. $\int \cos\frac{\theta}{2}\cos 7\theta\,d\theta$

In Exercises 27–36, use a substitution to change the integral into one you can find in the table. Then evaluate the integral.

27. $\int \frac{x^3+x+1}{(x^2+1)^2}\,dx$
28. $\int \frac{x^2+6x}{(x^2+3)^2}\,dx$
29. $\int \sin^{-1}\sqrt{x}\,dx$
30. $\int \frac{\cos^{-1}\sqrt{x}}{\sqrt{x}}\,dx$
31. $\int \frac{\sqrt{x}}{\sqrt{1-x}}\,dx$
32. $\int \frac{\sqrt{2-x}}{\sqrt{x}}\,dx$
33. $\int \cot t\sqrt{1-\sin^2 t}\,dt,\quad 0 < t < \pi/2$
34. $\int \frac{dt}{\tan t\sqrt{4-\sin^2 t}}$
35. $\int \frac{dy}{y\sqrt{3+(\ln y)^2}}$
36. $\int \tan^{-1}\sqrt{y}\,dy$

Use reduction formulas to evaluate the integrals in Exercises 37–46.

37. $\int \sin^5 2x\,dx$
38. $\int 8\cos^4 2\pi t\,dt$
39. $\int \sin^2 2\theta\cos^3 2\theta\,d\theta$
40. $\int 2\sin^2 t\sec^4 t\,dt$
41. $\int 4\tan^3 2x\,dx$
42. $\int 8\cot^4 t\,dt$
43. $\int 2\sec^3 \pi x\,dx$
44. $\int 3\sec^4 3x\,dx$
45. $\int \csc^5 x\,dx$
46. $\int 16x^3(\ln x)^2\,dx$

Evaluate the integrals in Exercises 47–52 by making a substitution (possibly trigonometric) and then applying a reduction formula.

47. $\int e^t\sec^3(e^t-1)\,dt$
48. $\int \frac{\csc^3\sqrt{\theta}}{\sqrt{\theta}}\,d\theta$
49. $\int_0^1 2\sqrt{x^2+1}\,dx$
50. $\int_0^{\sqrt{3}/2} \frac{dy}{(1-y^2)^{5/2}}$
51. $\int_1^2 \frac{(r^2-1)^{3/2}}{r}\,dr$
52. $\int_0^{1/\sqrt{3}} \frac{dt}{(t^2+1)^{7/2}}$
53. What is the largest value

$$\int_a^b \sqrt{x-x^2}\,dx$$

can have for any a and b? Give reasons for your answer.

54. What is the largest value

$$\int_a^b x\sqrt{2x - x^2}\, dx$$

can have for any a and b? Give reasons for your answer.

COMPUTER EXPLORATIONS

In Exercises 55 and 56, use a CAS to perform the integrations.

55. Evaluate the integrals

a. $\int x \ln x \, dx$ **b.** $\int x^2 \ln x \, dx$ **c.** $\int x^3 \ln x \, dx.$

d. What pattern do you see? Predict the formula for $\int x^4 \ln x \, dx$ and then see if you are correct by evaluating it with a CAS.

e. What is the formula for $\int x^n \ln x \, dx$, $n \geq 1$? Check your answer using a CAS.

56. Evaluate the integrals

a. $\int \frac{\ln x}{x^2}\, dx$ **b.** $\int \frac{\ln x}{x^3}\, dx$ **c.** $\int \frac{\ln x}{x^4}\, dx.$

d. What pattern do you see? Predict the formula for

$$\int \frac{\ln x}{x^5}\, dx$$

and then see if you are correct by evaluating it with a CAS.

e. What is the formula for

$$\int \frac{\ln x}{x^n}\, dx, \quad n \geq 2?$$

Check your answer using a CAS.

57. a. Use a CAS to evaluate

$$\int_0^{\pi/2} \frac{\sin^n x}{\sin^n x + \cos^n x}\, dx$$

where n is an arbitrary positive integer. Does your CAS find the result?

b. In succession, find the integral when $n = 1, 2, 3, 5, 7$. Comment on the complexity of the results.

c. Now substitute $x = (\pi/2) - u$ and add the new and old integrals. What is the value of

$$\int_0^{\pi/2} \frac{\sin^n x}{\sin^n x + \cos^n x}\, dx?$$

This exercise illustrates how a little mathematical ingenuity solves a problem not immediately amenable to solution by a CAS.

5.6 Numerical Integration

The antiderivatives of some functions, like $\sin(x^2)$, $1/\ln x$, and $\sqrt{1 + x^4}$, have no elementary formulas. When we cannot find a workable antiderivative for a function f that we have to integrate, we can partition the interval of integration, replace f by a closely fitting polynomial on each subinterval, integrate the polynomials, and add the results to approximate the integral of f. This procedure is an example of numerical integration. In this section we study two such methods, the *Trapezoidal Rule* and *Simpson's Rule*. In our presentation we assume that f is positive, but the only requirement is for it to be continuous over the interval of integration $[a, b]$.

Trapezoidal Approximations

The Trapezoidal Rule for the value of a definite integral is based on approximating the region between a curve and the x-axis with trapezoids instead of rectangles, as in Figure 5.7. It is not necessary for the subdivision points $x_0, x_1, x_2, \ldots, x_n$ in the figure to be evenly spaced, but the resulting formula is simpler if they are. We therefore assume that the length of each subinterval is

$$\Delta x = \frac{b - a}{n}.$$

The length $\Delta x = (b - a)/n$ is called the **step size** or **mesh size**. The area of the trapezoid that lies above the ith subinterval is

$$\Delta x\left(\frac{y_{i-1} + y_i}{2}\right) = \frac{\Delta x}{2}(y_{i-1} + y_i),$$

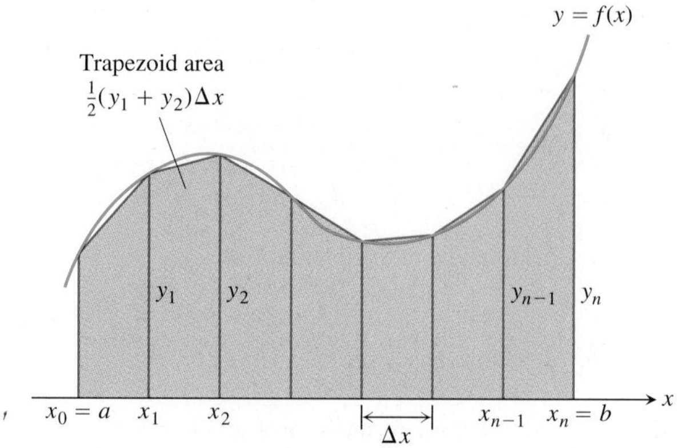

FIGURE 5.7 The Trapezoidal Rule approximates short stretches of the curve $y = f(x)$ with line segments. To approximate the integral of f from a to b, we add the areas of the trapezoids made by joining the ends of the segments to the x-axis.

where $y_{i-1} = f(x_{i-1})$ and $y_i = f(x_i)$. This area is the length Δx of the trapezoid's horizontal "altitude" times the average of its two vertical "bases." (See Figure 5.7.) The area below the curve $y = f(x)$ and above the x-axis is then approximated by adding the areas of all the trapezoids:

$$\begin{aligned} T &= \frac{1}{2}(y_0 + y_1)\Delta x + \frac{1}{2}(y_1 + y_2)\Delta x + \cdots \\ &\quad + \frac{1}{2}(y_{n-2} + y_{n-1})\Delta x + \frac{1}{2}(y_{n-1} + y_n)\Delta x \\ &= \Delta x\left(\frac{1}{2}y_0 + y_1 + y_2 + \cdots + y_{n-1} + \frac{1}{2}y_n\right) \\ &= \frac{\Delta x}{2}(y_0 + 2y_1 + 2y_2 + \cdots + 2y_{n-1} + y_n), \end{aligned}$$

where

$$y_0 = f(a), \qquad y_1 = f(x_1), \qquad \ldots, \qquad y_{n-1} = f(x_{n-1}), \qquad y_n = f(b).$$

The Trapezoidal Rule says: Use T to estimate the integral of f from a to b.

The Trapezoidal Rule

To approximate $\int_a^b f(x)\,dx$, use

$$T = \frac{\Delta x}{2}\left(y_0 + 2y_1 + 2y_2 + \cdots + 2y_{n-1} + y_n\right).$$

The y's are the values of f at the partition points

$$x_0 = a, x_1 = a + \Delta x, x_2 = a + 2\Delta x, \ldots, x_{n-1} = a + (n-1)\Delta x, x_n = b,$$

where $\Delta x = (b - a)/n$.

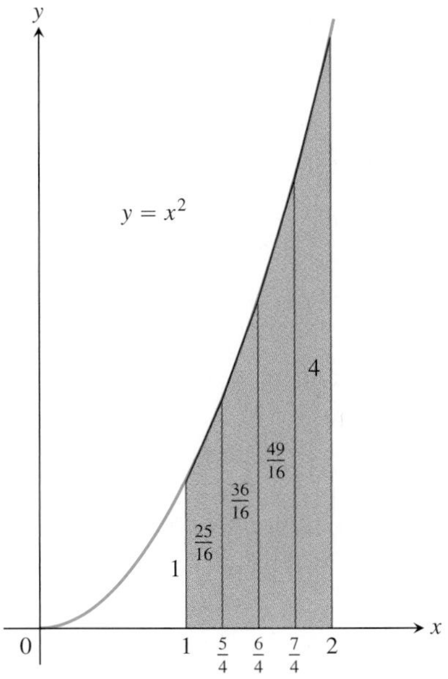

FIGURE 5.8 The trapezoidal approximation of the area under the graph of $y = x^2$ from $x = 1$ to $x = 2$ is a slight overestimate (Example 1).

TABLE 5.2

x	$y = x^2$
1	1
$\frac{5}{4}$	$\frac{25}{16}$
$\frac{6}{4}$	$\frac{36}{16}$
$\frac{7}{4}$	$\frac{49}{16}$
2	4

EXAMPLE 1 Use the Trapezoidal Rule with $n = 4$ to estimate $\int_1^2 x^2\,dx$. Compare the estimate with the exact value.

Solution Partition [1, 2] into four subintervals of equal length (Figure 5.8). Then evaluate $y = x^2$ at each partition point (Table 5.2).

Using these y values, $n = 4$, and $\Delta x = (2 - 1)/4 = 1/4$ in the Trapezoidal Rule, we have

$$\begin{aligned} T &= \frac{\Delta x}{2}\left(y_0 + 2y_1 + 2y_2 + 2y_3 + y_4\right) \\ &= \frac{1}{8}\left(1 + 2\left(\frac{25}{16}\right) + 2\left(\frac{36}{16}\right) + 2\left(\frac{49}{16}\right) + 4\right) \\ &= \frac{75}{32} = 2.34375. \end{aligned}$$

Since the parabola is concave *up*, the approximating segments lie above the curve, giving each trapezoid slightly more area than the corresponding strip under the curve. The exact value of the integral is

$$\int_1^2 x^2\,dx = \frac{x^3}{3}\bigg]_1^2 = \frac{8}{3} - \frac{1}{3} = \frac{7}{3}.$$

The T approximation overestimates the integral by about half a percent of its true value of $7/3$. The percentage error is $(2.34375 - 7/3)/(7/3) \approx 0.00446$, or 0.446%. ■

Simpson's Rule: Approximations Using Parabolas

Another rule for approximating the definite integral of a continuous function results from using parabolas instead of the straight line segments that produced trapezoids. As before, we partition the interval $[a, b]$ into n subintervals of equal length $h = \Delta x = (b - a)/n$, but this time we require that n be an *even* number. On each consecutive pair of intervals we approximate the curve $y = f(x) \geq 0$ by a parabola, as shown in Figure 5.9. A typical parabola passes through three consecutive points (x_{i-1}, y_{i-1}), (x_i, y_i), and (x_{i+1}, y_{i+1}) on the curve.

Let's calculate the shaded area beneath a parabola passing through three consecutive points. To simplify our calculations, we first take the case where $x_0 = -h$, $x_1 = 0$, and $x_2 = h$ (Figure 5.10), where $h = \Delta x = (b - a)/n$. The area under the parabola will be the same if we shift the y-axis to the left or right. The parabola has an equation of the form

$$y = Ax^2 + Bx + C,$$

so the area under it from $x = -h$ to $x = h$ is

$$\begin{aligned} A_p &= \int_{-h}^{h} (Ax^2 + Bx + C)\,dx \\ &= \frac{Ax^3}{3} + \frac{Bx^2}{2} + Cx\bigg]_{-h}^{h} \\ &= \frac{2Ah^3}{3} + 2Ch = \frac{h}{3}(2Ah^2 + 6C). \end{aligned}$$

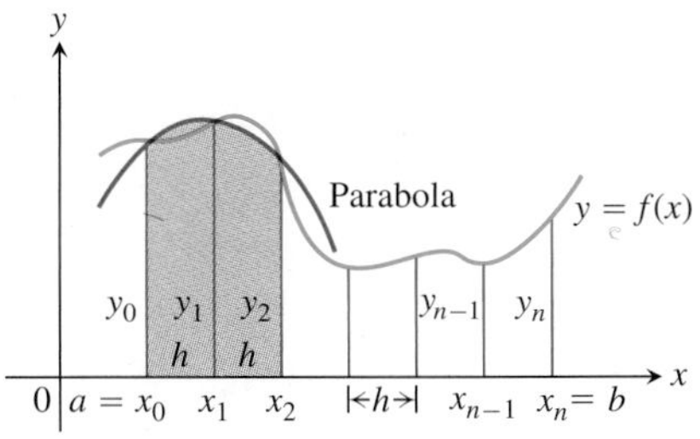

FIGURE 5.9 Simpson's Rule approximates short stretches of the curve with parabolas.

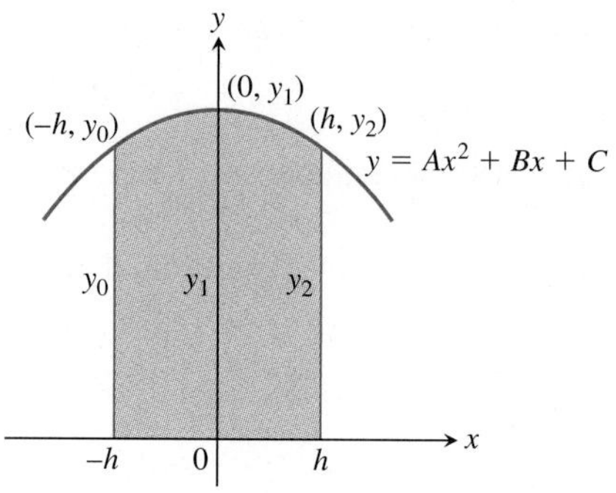

FIGURE 5.10 By integrating from $-h$ to h, we find the shaded area to be

$$\frac{h}{3}(y_0 + 4y_1 + y_2).$$

Since the curve passes through the three points $(-h, y_0)$, $(0, y_1)$, and (h, y_2), we also have

$$y_0 = Ah^2 - Bh + C, \qquad y_1 = C, \qquad y_2 = Ah^2 + Bh + C,$$

from which we obtain

$$\begin{aligned} C &= y_1, \\ Ah^2 - Bh &= y_0 - y_1, \\ Ah^2 + Bh &= y_2 - y_1, \\ 2Ah^2 &= y_0 + y_2 - 2y_1. \end{aligned}$$

Hence, expressing the area A_p in terms of the ordinates y_0, y_1, and y_2, we have

$$A_p = \frac{h}{3}(2Ah^2 + 6C) = \frac{h}{3}((y_0 + y_2 - 2y_1) + 6y_1) = \frac{h}{3}(y_0 + 4y_1 + y_2).$$

Now shifting the parabola horizontally to its shaded position in Figure 5.9 does not change the area under it. Thus the area under the parabola through (x_0, y_0), (x_1, y_1), and (x_2, y_2) in Figure 5.9 is still

$$\frac{h}{3}(y_0 + 4y_1 + y_2).$$

Similarly, the area under the parabola through the points (x_2, y_2), (x_3, y_3), and (x_4, y_4) is

$$\frac{h}{3}(y_2 + 4y_3 + y_4).$$

Computing the areas under all the parabolas and adding the results gives the approximation

$$\begin{aligned} \int_a^b f(x)\,dx &\approx \frac{h}{3}(y_0 + 4y_1 + y_2) + \frac{h}{3}(y_2 + 4y_3 + y_4) + \cdots \\ &\quad + \frac{h}{3}(y_{n-2} + 4y_{n-1} + y_n) \\ &= \frac{h}{3}(y_0 + 4y_1 + 2y_2 + 4y_3 + 2y_4 + \cdots + 2y_{n-2} + 4y_{n-1} + y_n). \end{aligned}$$

HISTORICAL BIOGRAPHY

Thomas Simpson
(1720–1761)

The result is known as Simpson's Rule. The function need not be positive, as in our derivation, but the number n of subintervals must be even to apply the rule because each parabolic arc uses two subintervals.

Simpson's Rule

To approximate $\int_a^b f(x)\,dx$, use

$$S = \frac{\Delta x}{3}(y_0 + 4y_1 + 2y_2 + 4y_3 + \cdots + 2y_{n-2} + 4y_{n-1} + y_n).$$

The y's are the values of f at the partition points

$$x_0 = a, x_1 = a + \Delta x, x_2 = a + 2\Delta x, \ldots, x_{n-1} = a + (n-1)\Delta x, x_n = b.$$

The number n is even, and $\Delta x = (b - a)/n$.

Note the pattern of the coefficients in the above rule: 1, 4, 2, 4, 2, 4, 2, . . . , 4, 2, 1.

TABLE 5.3

x	$y = 5x^4$
0	0
$\frac{1}{2}$	$\frac{5}{16}$
1	5
$\frac{3}{2}$	$\frac{405}{16}$
2	80

EXAMPLE 2 Use Simpson's Rule with $n = 4$ to approximate $\int_0^2 5x^4\,dx$.

Solution Partition [0, 2] into four subintervals and evaluate $y = 5x^4$ at the partition points (Table 5.3). Then apply Simpson's Rule with $n = 4$ and $\Delta x = 1/2$:

$$\begin{aligned} S &= \frac{\Delta x}{3}\left(y_0 + 4y_1 + 2y_2 + 4y_3 + y_4\right) \\ &= \frac{1}{6}\left(0 + 4\left(\frac{5}{16}\right) + 2(5) + 4\left(\frac{405}{16}\right) + 80\right) \\ &= 32\frac{1}{12}. \end{aligned}$$

This estimate differs from the exact value (32) by only $1/12$, a percentage error of less than three-tenths of one percent, and this was with just four subintervals. ■

Error Analysis

Whenever we use an approximation technique, the issue arises as to how accurate the approximation might be. The following theorem gives formulas for estimating the errors when using the Trapezoidal Rule and Simpson's Rule. The **error** is the difference between the approximation obtained by the rule and the actual value of the definite integral $\int_a^b f(x)\,dx$.

THEOREM 1—Error Estimates in the Trapezoidal and Simpson's Rules If f'' is continuous and M is any upper bound for the values of $|f''|$ on $[a, b]$, then the error E_T in the trapezoidal approximation of the integral of f from a to b for n steps satisfies the inequality

$$|E_T| \le \frac{M(b-a)^3}{12n^2}. \qquad \text{Trapezoidal Rule}$$

If $f^{(4)}$ is continuous and M is any upper bound for the values of $|f^{(4)}|$ on $[a, b]$, then the error E_S in the Simpson's Rule approximation of the integral of f from a to b for n steps satisfies the inequality

$$|E_S| \le \frac{M(b-a)^5}{180n^4}. \qquad \text{Simpson's Rule}$$

Theorem 1 tells us how large the step size $\Delta x = (b - a)/n$ should be for a given tolerance of error. To see why the theorem is true in the case of the Trapezoidal Rule, we begin with a result from advanced calculus, which says that if f'' is continuous on the interval $[a, b]$, then

$$\int_a^b f(x)\,dx = T - \frac{b-a}{12}\cdot f''(c)(\Delta x)^2$$

for some number c between a and b. Thus, as Δx approaches zero, the error defined by

$$E_T = -\frac{b-a}{12}\cdot f''(c)(\Delta x)^2$$

approaches zero as the *square* of Δx.

The inequality

$$|E_T| \leq \frac{b-a}{12}\max|f''(x)|(\Delta x)^2,$$

where max refers to the interval $[a, b]$, gives an upper bound for the magnitude of the error. In practice, we usually cannot find the exact value of $\max|f''(x)|$ and have to estimate an upper bound or "worst case" value for it instead. If M is any upper bound for the values of $|f''(x)|$ on $[a, b]$, so that $|f''(x)| \leq M$ on $[a, b]$, then

$$|E_T| \leq \frac{b-a}{12}M(\Delta x)^2.$$

If we substitute $(b-a)/n$ for Δx, we get

$$|E_T| \leq \frac{M(b-a)^3}{12n^2}.$$

To estimate the error in Simpson's rule, we start with a result from advanced calculus that says that if the fourth derivative $f^{(4)}$ is continuous, then

$$\int_a^b f(x)\,dx = S - \frac{b-a}{180}\cdot f^{(4)}(c)(\Delta x)^4$$

for some point c between a and b. Thus, as Δx approaches zero, the error,

$$E_S = -\frac{b-a}{180}\cdot f^{(4)}(c)(\Delta x)^4,$$

approaches zero as the *fourth power* of Δx. (This helps to explain why Simpson's Rule is likely to give better results than the Trapezoidal Rule.)

The inequality

$$|E_S| \leq \frac{b-a}{180}\max|f^{(4)}(x)|\,(\Delta x)^4$$

where max refers to the interval $[a, b]$, gives an upper bound for the magnitude of the error. As with $\max|f''|$ in the error formula for the Trapezoidal Rule, we usually cannot find the exact value of $\max|f^{(4)}(x)|$ and have to replace it with an upper bound. If M is any upper bound for the values of $|f^{(4)}|$ on $[a, b]$, then

$$|E_S| \leq \frac{b-a}{180}M(\Delta x)^4.$$

Substituting $(b-a)/n$ for Δx in this last expression gives

$$|E_S| \leq \frac{M(b-a)^5}{180n^4}.$$

Here is an example using the error estimate for Simpson's Rule.

EXAMPLE 3 Find an upper bound for the error in estimating $\int_0^2 5x^4\,dx$ using Simpson's Rule with $n = 4$ (Example 2).

Solution To estimate the error, we first find an upper bound M for the magnitude of the fourth derivative of $f(x) = 5x^4$ on the interval $0 \leq x \leq 2$. Since the fourth derivative has the constant value $f^{(4)}(x) = 120$, we take $M = 120$. With $b - a = 2$ and $n = 4$, the error estimate for Simpson's Rule gives

$$|E_S| \le \frac{M(b-a)^5}{180n^4} = \frac{120(2)^5}{180 \cdot 4^4} = \frac{1}{12}.$$

■

EXAMPLE 4 Since $\ln x$ is an antiderivative of $1/x$ and $\ln 1 = 0$, the value of $\ln 2$ can be calculated from the integral

$$\ln 2 = \int_1^2 \frac{1}{x}\,dx.$$

Table 5.4 shows T and S values for approximations of $\int_1^2 (1/x)\,dx$ using various values of n. Notice how Simpson's Rule dramatically improves over the Trapezoidal Rule. In particular, notice that when we double the value of n (thereby halving the value of $h = \Delta x$), the T error is divided by 2 *squared*, whereas the S error is divided by 2 *to the fourth*.

This has a dramatic effect as $\Delta x = (2 - 1)/n$ gets very small. The Simpson approximation for $n = 50$ rounds accurately to seven places and for $n = 100$ agrees to nine decimal places (billionths)! ■

If $f(x)$ is a polynomial of degree less than four, then its fourth derivative is zero, and

$$E_S = -\frac{b-a}{180} f^{(4)}(c)(\Delta x)^4 = -\frac{b-a}{180}(0)(\Delta x)^4 = 0.$$

Thus, there will be no error in the Simpson approximation of any integral of f. In other words, if f is a constant, a linear function, or a quadratic or cubic polynomial, Simpson's Rule will give the value of any integral of f exactly, whatever the number of subdivisions. Similarly, if f is a constant or a linear function, then its second derivative is zero and

$$E_T = -\frac{b-a}{12} f''(c)(\Delta x)^2 = -\frac{b-a}{12}(0)(\Delta x)^2 = 0.$$

The Trapezoidal Rule will therefore give the exact value of any integral of f. This is no surprise, for the trapezoids fit the graph perfectly. Although decreasing the step size Δx reduces the error in the Simpson and Trapezoidal approximations in theory, it may fail to do so in practice.

TABLE 5.4 Trapezoidal Rule approximations (T_n) and Simpson's Rule approximations (S_n) of $\ln 2 = \int_1^2 (1/x)\,dx$

n	T_n	|Error| less than . . .	S_n	|Error| less than . . .
10	0.6937714032	0.0006242227	0.6931502307	0.0000030502
20	0.6933033818	0.0001562013	0.6931473747	0.0000001942
30	0.6932166154	0.0000694349	0.6931472190	0.0000000385
40	0.6931862400	0.0000390595	0.6931471927	0.0000000122
50	0.6931721793	0.0000249988	0.6931471856	0.0000000050
100	0.6931534305	0.0000062500	0.6931471809	0.0000000004

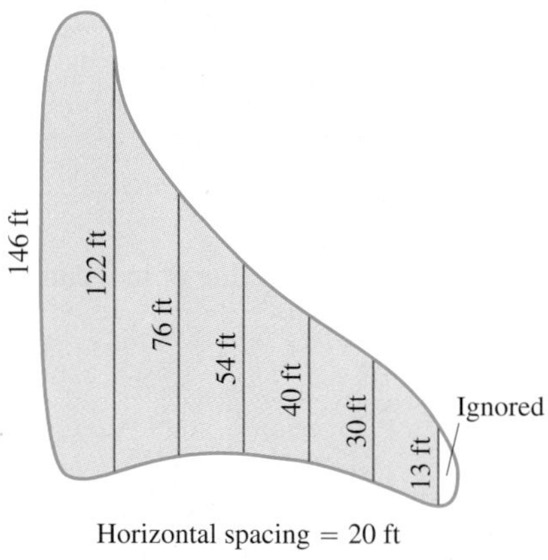

FIGURE 5.11 The dimensions of the swamp in Example 5.

When Δx is very small, say $\Delta x = 10^{-5}$, computer or calculator round-off errors in the arithmetic required to evaluate S and T may accumulate to such an extent that the error formulas no longer describe what is going on. Shrinking Δx below a certain size can actually make things worse. Although this is not an issue in this book, you should consult a text on numerical analysis for alternative methods if you are having problems with round-off.

EXAMPLE 5 A town wants to drain and fill a small polluted swamp (Figure 5.11). The swamp averages 5 ft deep. About how many cubic yards of dirt will it take to fill the area after the swamp is drained?

Solution To calculate the volume of the swamp, we estimate the surface area and multiply by 5. To estimate the area, we use Simpson's Rule with $\Delta x = 20$ ft and the y's equal to the distances measured across the swamp, as shown in Figure 5.11.

$$S = \frac{\Delta x}{3}(y_0 + 4y_1 + 2y_2 + 4y_3 + 2y_4 + 4y_5 + y_6)$$
$$= \frac{20}{3}(146 + 488 + 152 + 216 + 80 + 120 + 13) = 8100$$

The volume is about $(8100)(5) = 40{,}500$ ft^3 or 1500 yd^3. ■

EXERCISES 5.6

The instructions for the integrals in Exercises 1–10 have two parts, one for the Trapezoidal Rule and one for Simpson's Rule.

I. Using the Trapezoidal Rule

a. Estimate the integral with $n = 4$ steps and find an upper bound for $|E_T|$.

b. Evaluate the integral directly and find $|E_T|$.

c. Use the formula $(|E_T|/(\text{true value})) \times 100$ to express $|E_T|$ as a percentage of the integral's true value.

II. Using Simpson's Rule

a. Estimate the integral with $n = 4$ steps and find an upper bound for $|E_S|$.

b. Evaluate the integral directly and find $|E_S|$.

c. Use the formula $(|E_S|/(\text{true value})) \times 100$ to express $|E_S|$ as a percentage of the integral's true value.

1. $\int_1^2 x\,dx$

2. $\int_1^3 (2x - 1)\,dx$

3. $\int_{-1}^1 (x^2 + 1)\,dx$

4. $\int_{-2}^0 (x^2 - 1)\,dx$

5. $\int_0^2 (t^3 + t)\,dt$

6. $\int_{-1}^1 (t^3 + 1)\,dt$

7. $\int_1^2 \frac{1}{s^2}\,ds$

8. $\int_2^4 \frac{1}{(s-1)^2}\,ds$

9. $\int_0^\pi \sin t\,dt$

10. $\int_0^1 \sin \pi t\,dt$

In Exercises 11–22, estimate the minimum number of subintervals needed to approximate the integrals with an error of magnitude less than 10^{-4} by **(a)** the Trapezoidal Rule and **(b)** Simpson's Rule. (The integrals in Exercises 11–18 are the integrals from Exercises 1–8.)

11. $\int_1^2 x\,dx$

12. $\int_1^3 (2x - 1)\,dx$

13. $\int_{-1}^1 (x^2 + 1)\,dx$

14. $\int_{-2}^0 (x^2 - 1)\,dx$

15. $\int_0^2 (t^3 + t)\,dt$

16. $\int_{-1}^1 (t^3 + 1)\,dt$

17. $\int_1^2 \frac{1}{s^2}\,ds$

18. $\int_2^4 \frac{1}{(s-1)^2}\,ds$

19. $\int_0^3 \sqrt{x + 1}\,dx$

20. $\int_0^3 \frac{1}{\sqrt{x + 1}}\,dx$

21. $\int_0^2 \sin(x+1)\,dx$ **22.** $\int_{-1}^1 \cos(x+\pi)\,dx$

23. Volume of water in a swimming pool A rectangular swimming pool is 30 ft wide and 50 ft long. The table shows the depth $h(x)$ of the water at 5-ft intervals from one end of the pool to the other. Estimate the volume of water in the pool using the Trapezoidal Rule with $n = 10$, applied to the integral

$$V = \int_0^{50} 30 \cdot h(x)\,dx.$$

Position (ft) x	Depth (ft) $h(x)$	Position (ft) x	Depth (ft) $h(x)$
0	6.0	30	11.5
5	8.2	35	11.9
10	9.1	40	12.3
15	9.9	45	12.7
20	10.5	50	13.0
25	11.0		

24. Wing design The design of a new airplane requires a gasoline tank of constant cross-sectional area in each wing. A scale drawing of a cross-section is shown here. The tank must hold 5000 lb of gasoline, which has a density of 42 lb/ft^3. Estimate the length of the tank.

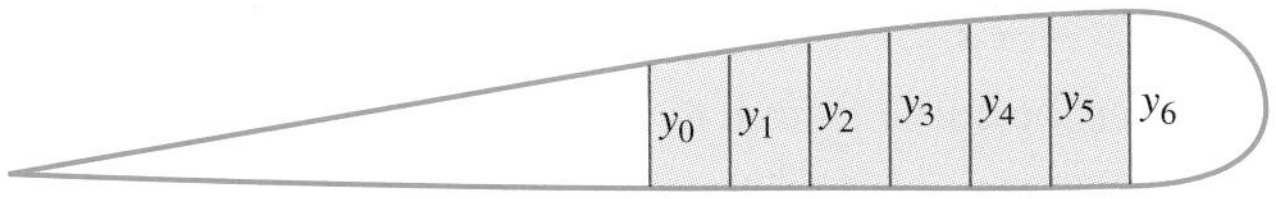

$y_0 = 1.5$ ft, $y_1 = 1.6$ ft, $y_2 = 1.8$ ft, $y_3 = 1.9$ ft,
$y_4 = 2.0$ ft, $y_5 = y_6 = 2.1$ ft Horizontal spacing = 1 ft

25. Usable values of the sine-integral function *The sine-integral function,*

$$\text{Si}(x) = \int_0^x \frac{\sin t}{t}\,dt, \qquad \text{“Sine integral of } x\text{”}$$

is one of the many functions in engineering whose formulas cannot be simplified. There is no elementary formula for the antiderivative of $(\sin t)/t$. The values of Si(x), however, are readily estimated by numerical integration.

Although the notation does not show it explicitly, the function being integrated is

$$f(t) = \begin{cases} \dfrac{\sin t}{t}, & t \neq 0 \\ 1, & t = 0, \end{cases}$$

the continuous extension of $(\sin t)/t$ to the interval $[0, x]$. The function has derivatives of all orders at every point of its domain. Its graph is smooth, and you can expect good results from Simpson's Rule.

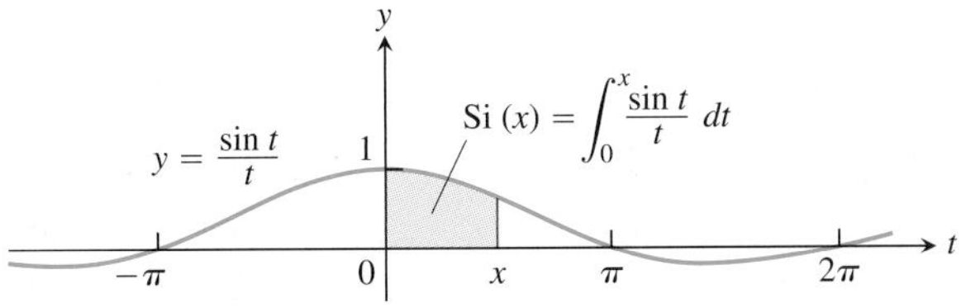

a. Use the fact that $|f^{(4)}| \le 1$ on $[0, \pi/2]$ to give an upper bound for the error that will occur if

$$\text{Si}\left(\frac{\pi}{2}\right) = \int_0^{\pi/2} \frac{\sin t}{t}\,dt$$

is estimated by Simpson's Rule with $n = 4$.

b. Estimate Si$(\pi/2)$ by Simpson's Rule with $n = 4$.

c. Express the error bound you found in part (a) as a percentage of the value you found in part (b).

26. The error function *The error function,*

$$\text{erf}(x) = \frac{2}{\sqrt{\pi}} \int_0^x e^{-t^2}\,dt,$$

important in probability and in the theories of heat flow and signal transmission, must be evaluated numerically because there is no elementary expression for the antiderivative of e^{-t^2}.

a. Use Simpson's Rule with $n = 10$ to estimate erf(1).

b. In [0, 1],

$$\left| \frac{d^4}{dt^4}\left(e^{-t^2}\right) \right| \le 12.$$

Give an upper bound for the magnitude of the error of the estimate in part (a).

27. Prove that the sum T in the Trapezoidal Rule for $\int_a^b f(x)\,dx$ is a Riemann sum for f continuous on $[a, b]$. (*Hint:* Use the Intermediate Value Theorem to show the existence of c_k in the subinterval $[x_{k-1}, x_k]$ satisfying $f(c_k) = (f(x_{k-1}) + f(x_k))/2$.)

28. Prove that the sum S in Simpson's Rule for $\int_a^b f(x)\,dx$ is a Riemann sum for f continuous on $[a, b]$. (See Exercise 27.)

29. Use numerical integration to estimate the value of

$$\sin^{-1} 0.6 = \int_0^{0.6} \frac{dx}{\sqrt{1 - x^2}}.$$

For reference, $\sin^{-1} 0.6 = 0.64350$ to five decimal places.

30. Use numerical integration to estimate the value of

$$\pi = 4\int_0^1 \frac{1}{1 + x^2}\,dx.$$

5.7 Improper Integrals

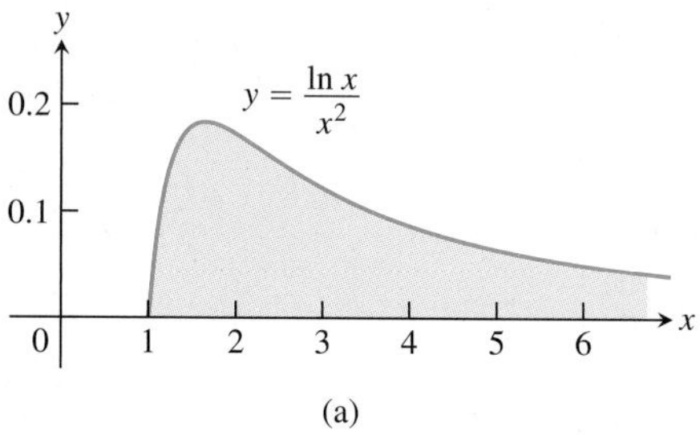

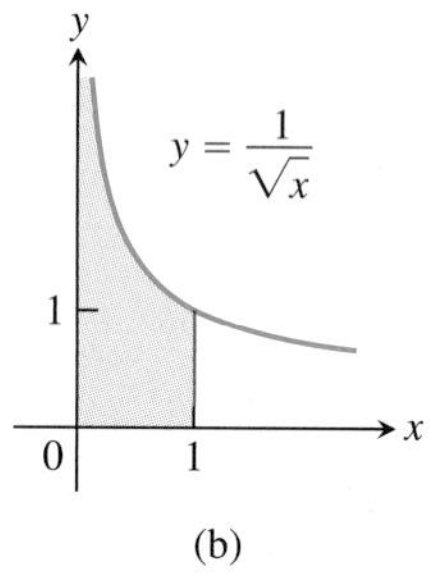

FIGURE 5.12 Are the areas under these infinite curves finite?

Up to now, we have required definite integrals to have two properties. First, that the domain of integration $[a, b]$ be finite. Second, that the range of the integrand be finite on this domain. In practice, we may encounter problems that fail to meet one or both of these conditions. The integral for the area under the curve $y = (\ln x)/x^2$ from $x = 1$ to $x = \infty$ is an example for which the domain is infinite (Figure 5.12a). The integral for the area under the curve of $y = 1/\sqrt{x}$ between $x = 0$ and $x = 1$ is an example for which the range of the integrand is infinite (Figure 5.12b). In either case, the integrals are said to be *improper* and are calculated as limits. We will see in Chapter 7 that improper integrals play an important role when investigating the convergence of certain infinite series.

Infinite Limits of Integration

Consider the infinite region that lies under the curve $y = e^{-x/2}$ in the first quadrant (Figure 5.13a). You might think this region has infinite area, but we will see that the value is finite. Here is how to assign a value to the area. First find the area $A(b)$ of the portion of the region that is bounded on the right by $x = b$ (Figure 5.13b).

$$A(b) = \int_0^b e^{-x/2}\,dx = -2e^{-x/2}\Big]_0^b = -2e^{-b/2} + 2$$

Then find the limit of $A(b)$ as $b \to \infty$

$$\lim_{b\to\infty} A(b) = \lim_{b\to\infty} (-2e^{-b/2} + 2) = 2.$$

The value we assign to the area under the curve from 0 to ∞ is

$$\int_0^\infty e^{-x/2}\,dx = \lim_{b\to\infty} \int_0^b e^{-x/2}\,dx = 2.$$

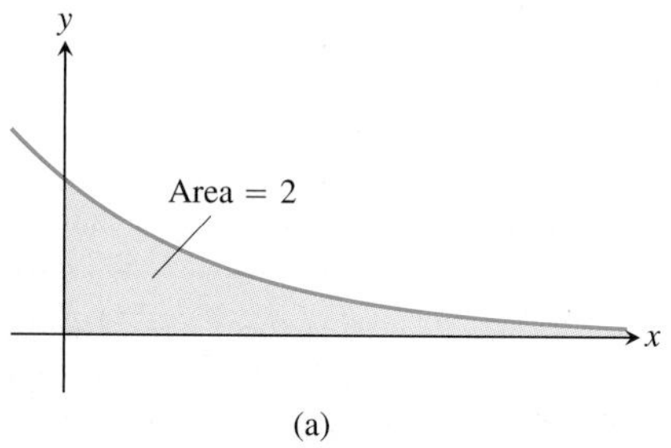

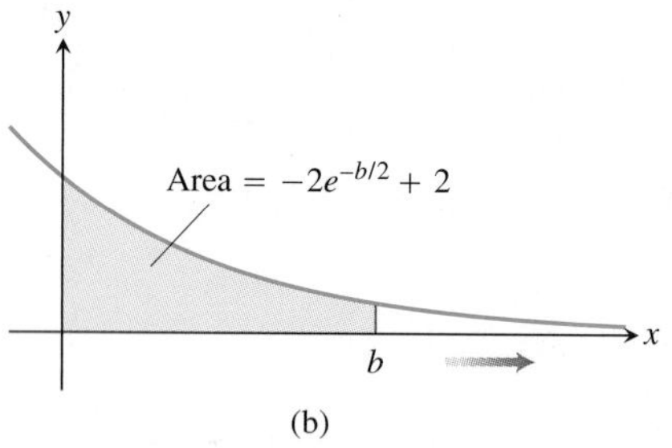

FIGURE 5.13 (a) The area in the first quadrant under the curve $y = e^{-x/2}$ (b) The area is an improper integral of the first type.

DEFINITION Integrals with infinite limits of integration are **improper integrals of Type I**.

1. If $f(x)$ is continuous on $[a, \infty)$, then

$$\int_a^\infty f(x)\,dx = \lim_{b\to\infty} \int_a^b f(x)\,dx.$$

2. If $f(x)$ is continuous on $(-\infty, b]$, then

$$\int_{-\infty}^b f(x)\,dx = \lim_{a\to-\infty} \int_a^b f(x)\,dx.$$

3. If $f(x)$ is continuous on $(-\infty, \infty)$, then

$$\int_{-\infty}^\infty f(x)\,dx = \int_{-\infty}^c f(x)\,dx + \int_c^\infty f(x)\,dx,$$

where c is any real number.

In each case, if the limit is finite we say that the improper integral **converges** and that the limit is the **value** of the improper integral. If the limit fails to exist, the improper integral **diverges**.

It can be shown that the choice of c in Part 3 of the definition is unimportant. We can evaluate or determine the convergence or divergence of $\int_{-\infty}^{\infty} f(x)\,dx$ with any convenient choice.

Any of the integrals in the above definition can be interpreted as an area if $f \geq 0$ on the interval of integration. For instance, we interpreted the improper integral in Figure 5.13 as an area. In that case, the area has the finite value 2. If $f \geq 0$ and the improper integral diverges, we say the area under the curve is **infinite**.

EXAMPLE 1 Is the area under the curve $y = (\ln x)/x^2$ from $x = 1$ to $x = \infty$ finite? If so, what is its value?

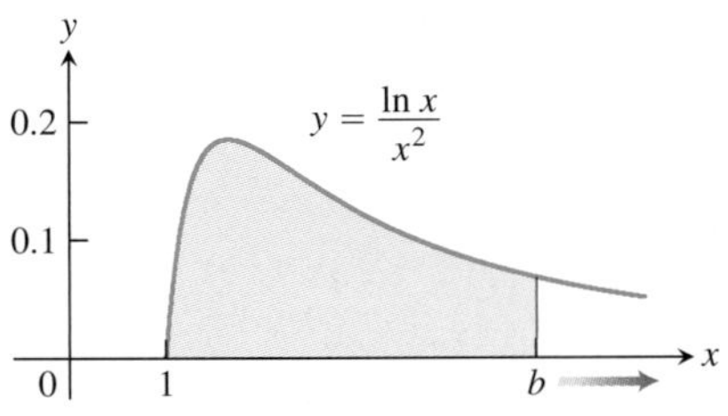

FIGURE 5.14 The area under this curve is an improper integral (Example 1).

Solution We find the area under the curve from $x = 1$ to $x = b$ and examine the limit as $b \to \infty$. If the limit is finite, we take it to be the area under the curve (Figure 5.14). The area from 1 to b is

$$\begin{aligned}\int_1^b \frac{\ln x}{x^2}\,dx &= \left[(\ln x)\left(-\frac{1}{x}\right)\right]_1^b - \int_1^b \left(-\frac{1}{x}\right)\left(\frac{1}{x}\right)dx \\ &= -\frac{\ln b}{b} - \left[\frac{1}{x}\right]_1^b \\ &= -\frac{\ln b}{b} - \frac{1}{b} + 1.\end{aligned}$$

Integration by parts with $u = \ln x$, $dv = dx/x^2$, $du = dx/x$, $v = -1/x$

The limit of the area as $b \to \infty$ is

$$\begin{aligned}\int_1^\infty \frac{\ln x}{x^2}\,dx &= \lim_{b\to\infty} \int_1^b \frac{\ln x}{x^2}\,dx \\ &= \lim_{b\to\infty}\left[-\frac{\ln b}{b} - \frac{1}{b} + 1\right] \\ &= -\left[\lim_{b\to\infty} \frac{\ln b}{b}\right] - 0 + 1 \\ &= -\left[\lim_{b\to\infty} \frac{1/b}{1}\right] + 1 = 0 + 1 = 1.\end{aligned}$$

l'Hôpital's Rule

Thus, the improper integral converges and the area has finite value 1. ■

EXAMPLE 2 Evaluate

$$\int_{-\infty}^{\infty} \frac{dx}{1 + x^2}.$$

Solution According to the definition (Part 3), we can choose $c = 0$ and write

$$\int_{-\infty}^{\infty} \frac{dx}{1 + x^2} = \int_{-\infty}^{0} \frac{dx}{1 + x^2} + \int_{0}^{\infty} \frac{dx}{1 + x^2}.$$

Next we evaluate each improper integral on the right side of the equation above.

$$\int_{-\infty}^{0} \frac{dx}{1 + x^2} = \lim_{a\to-\infty} \int_a^0 \frac{dx}{1 + x^2}$$

HISTORICAL BIOGRAPHY

Lejeune Dirichlet
(1805–1859)

$$= \lim_{a\to-\infty} \tan^{-1} x \Big]_a^0$$

$$= \lim_{a\to-\infty} (\tan^{-1} 0 - \tan^{-1} a) = 0 - \left(-\frac{\pi}{2}\right) = \frac{\pi}{2}$$

$$\int_0^\infty \frac{dx}{1+x^2} = \lim_{b\to\infty} \int_0^b \frac{dx}{1+x^2}$$

$$= \lim_{b\to\infty} \tan^{-1} x \Big]_0^b$$

$$= \lim_{b\to\infty} (\tan^{-1} b - \tan^{-1} 0) = \frac{\pi}{2} - 0 = \frac{\pi}{2}$$

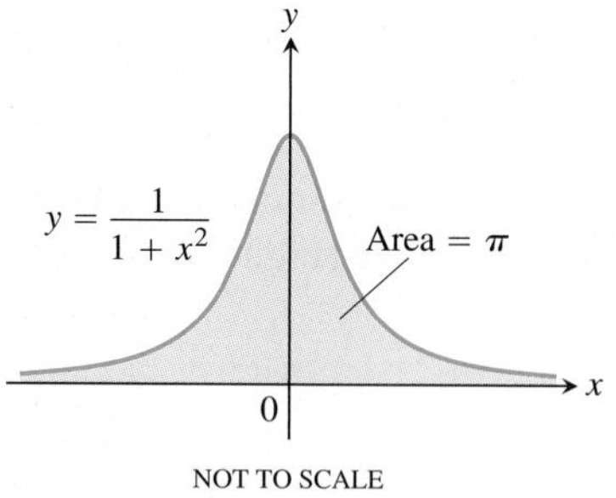

FIGURE 5.15 The area under this curve is finite (Example 2).

Thus,

$$\int_{-\infty}^\infty \frac{dx}{1+x^2} = \frac{\pi}{2} + \frac{\pi}{2} = \pi.$$

Since $1/(1+x^2) > 0$, the improper integral can be interpreted as the (finite) area beneath the curve and above the x-axis (Figure 5.15). ■

The Integral $\int_1^\infty \frac{dx}{x^p}$

The function $y = 1/x$ is the boundary between the convergent and divergent improper integrals with integrands of the form $y = 1/x^p$. As the next example shows, the improper integral converges if $p > 1$ and diverges if $p \le 1$.

EXAMPLE 3 For what values of p does the integral $\int_1^\infty dx/x^p$ converge? When the integral does converge, what is its value?

Solution If $p \neq 1$,

$$\int_1^b \frac{dx}{x^p} = \frac{x^{-p+1}}{-p+1}\Big]_1^b = \frac{1}{1-p}(b^{-p+1} - 1) = \frac{1}{1-p}\left(\frac{1}{b^{p-1}} - 1\right).$$

Thus,

$$\int_1^\infty \frac{dx}{x^p} = \lim_{b\to\infty} \int_1^b \frac{dx}{x^p}$$

$$= \lim_{b\to\infty}\left[\frac{1}{1-p}\left(\frac{1}{b^{p-1}} - 1\right)\right] = \begin{cases} \dfrac{1}{p-1}, & p > 1 \\ \infty, & p < 1 \end{cases}$$

because

$$\lim_{b\to\infty} \frac{1}{b^{p-1}} = \begin{cases} 0, & p > 1 \\ \infty, & p < 1. \end{cases}$$

Therefore, the integral converges to the value $1/(p - 1)$ if $p > 1$ and it diverges if $p < 1$.

If $p = 1$, the integral also diverges:

$$\begin{aligned}\int_1^\infty \frac{dx}{x^p} &= \int_1^\infty \frac{dx}{x} \\ &= \lim_{b\to\infty} \int_1^b \frac{dx}{x} \\ &= \lim_{b\to\infty} \ln x \Big]_1^b \\ &= \lim_{b\to\infty} (\ln b - \ln 1) = \infty.\end{aligned}$$

■

Integrands with Vertical Asymptotes

Another type of improper integral arises when the integrand has a vertical asymptote—an infinite discontinuity—at a limit of integration or at some point between the limits of integration. If the integrand f is positive over the interval of integration, we can again interpret the improper integral as the area under the graph of f and above the x-axis between the limits of integration.

Consider the region in the first quadrant that lies under the curve $y = 1/\sqrt{x}$ from $x = 0$ to $x = 1$ (Figure 5.12b). First we find the area of the portion from a to 1 (Figure 5.16).

$$\int_a^1 \frac{dx}{\sqrt{x}} = 2\sqrt{x}\Big]_a^1 = 2 - 2\sqrt{a}$$

Then we find the limit of this area as $a \to 0^+$:

$$\lim_{a\to 0^+} \int_a^1 \frac{dx}{\sqrt{x}} = \lim_{a\to 0^+} \left(2 - 2\sqrt{a}\right) = 2.$$

The area under the curve from 0 to 1 is finite and equals

$$\int_0^1 \frac{dx}{\sqrt{x}} = \lim_{a\to 0^+} \int_a^1 \frac{dx}{\sqrt{x}} = 2.$$

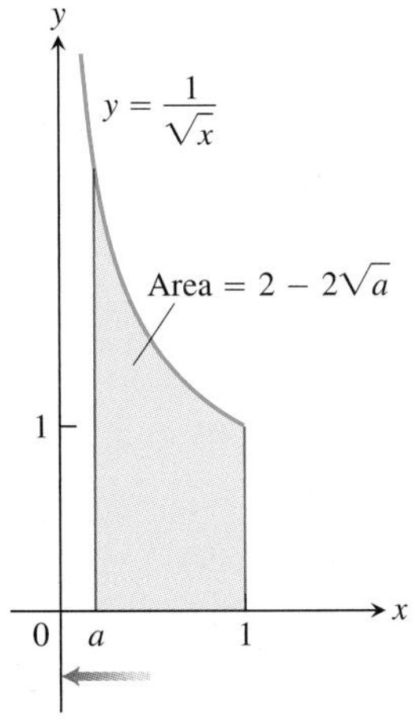

FIGURE 5.16 The area under this curve is an example of an improper integral of the second kind.

DEFINITION Integrals of functions that become infinite at a point within the interval of integration are **improper integrals of Type II**.

1. If $f(x)$ is continuous on $(a, b]$ and is discontinuous at a then
$$\int_a^b f(x)\,dx = \lim_{c\to a^+} \int_c^b f(x)\,dx.$$

2. If $f(x)$ is continuous on $[a, b)$ and is discontinuous at b, then
$$\int_a^b f(x)\,dx = \lim_{c\to b^-} \int_a^c f(x)\,dx.$$

continued

3. If $f(x)$ is discontinuous at c, where $a < c < b$, and continuous on $[a, c) \cup (c, b]$, then

$$\int_a^b f(x)\,dx = \int_a^c f(x)\,dx + \int_c^b f(x)\,dx.$$

In each case, if the limit is finite we say the improper integral **converges** and that the limit is the **value** of the improper integral. If the limit does not exist, the integral **diverges**.

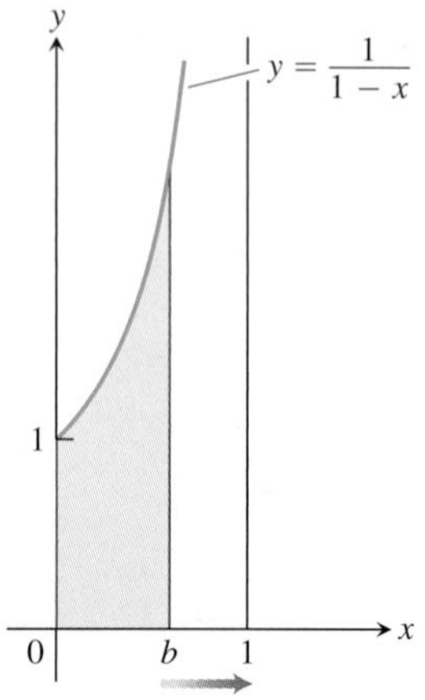

FIGURE 5.17 The area beneath the curve and above the x-axis for [0, 1) is not a real number (Example 4).

In Part 3 of the definition, the integral on the left side of the equation converges if *both* integrals on the right side converge; otherwise it diverges.

EXAMPLE 4 Investigate the convergence of

$$\int_0^1 \frac{1}{1-x}\,dx.$$

Solution The integrand $f(x) = 1/(1-x)$ is continuous on [0, 1) but is discontinuous at $x = 1$ and becomes infinite as $x \rightarrow 1^-$ (Figure 5.17). We evaluate the integral as

$$\begin{aligned}\lim_{b\to 1^-}\int_0^b \frac{1}{1-x}\,dx &= \lim_{b\to 1^-}\Big[-\ln|1-x|\Big]_0^b \\ &= \lim_{b\to 1^-}[-\ln(1-b)+0] = \infty.\end{aligned}$$

The limit is infinite, so the integral diverges. ■

EXAMPLE 5 Evaluate

$$\int_0^3 \frac{dx}{(x-1)^{2/3}}.$$

Solution The integrand has a vertical asymptote at $x = 1$ and is continuous on [0, 1) and (1, 3] (Figure 5.18). Thus, by Part 3 of the definition above,

$$\int_0^3 \frac{dx}{(x-1)^{2/3}} = \int_0^1 \frac{dx}{(x-1)^{2/3}} + \int_1^3 \frac{dx}{(x-1)^{2/3}}.$$

Next, we evaluate each improper integral on the right-hand side of this equation.

$$\begin{aligned}\int_0^1 \frac{dx}{(x-1)^{2/3}} &= \lim_{b\to 1^-}\int_0^b \frac{dx}{(x-1)^{2/3}} \\ &= \lim_{b\to 1^-} 3(x-1)^{1/3}\Big]_0^b \\ &= \lim_{b\to 1^-}[3(b-1)^{1/3}+3] = 3\end{aligned}$$

$$\begin{aligned}\int_1^3 \frac{dx}{(x-1)^{2/3}} &= \lim_{c\to 1^+}\int_c^3 \frac{dx}{(x-1)^{2/3}} \\ &= \lim_{c\to 1^+} 3(x-1)^{1/3}\Big]_c^3 \\ &= \lim_{c\to 1^+}\Big[3(3-1)^{1/3} - 3(c-1)^{1/3}\Big] = 3\sqrt[3]{2}\end{aligned}$$

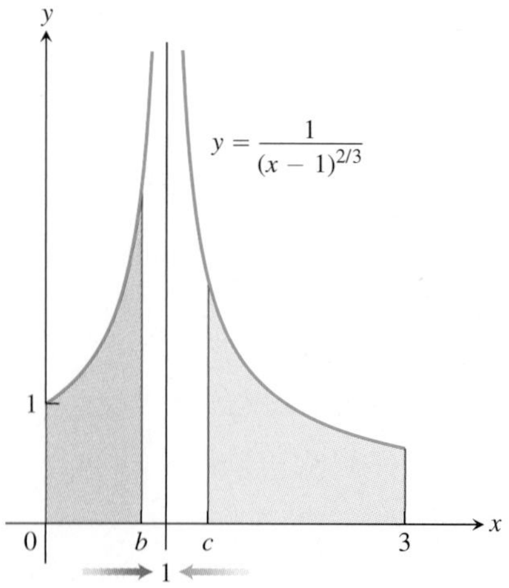

FIGURE 5.18 Example 5 shows that the area under the curve exists (so it is a real number).

We conclude that

$$\int_0^3 \frac{dx}{(x-1)^{2/3}} = 3 + 3\sqrt[3]{2}. \qquad \blacksquare$$

Computer algebra systems can evaluate many convergent improper integrals. To evaluate the integral

$$\int_2^\infty \frac{x+3}{(x-1)(x^2+1)}\,dx$$

(which converges) using Maple, enter

$$> f := (x+3)/((x-1)*(x\hat{}2+1));$$

Then use the integration command

$$> \text{int}(f, x = 2..\text{infinity});$$

Maple returns the answer

$$-\frac{1}{2}\pi + \ln(5) + \arctan(2).$$

To obtain a numerical result, use the evaluation command **evalf** and specify the number of digits, as follows:

$$> \text{evalf}(\%, 6);$$

The symbol % instructs the computer to evaluate the last expression on the screen, in this case $(-1/2)\pi + \ln(5) + \arctan(2)$. Maple returns 1.14579.

Using Mathematica, entering

$$\textit{In [1]:=}\ \text{Integrate}\,[(x+3)/((x-1)(x\hat{}2+1)),\ \{x, 2, \text{Infinity}\}]$$

returns

$$\textit{Out [1]=}\ \frac{-\pi}{2} + \text{ArcTan}\,[2] + \text{Log}\,[5].$$

To obtain a numerical result with six digits, use the command "N[%, 6]"; it also yields 1.14579.

Tests for Convergence and Divergence

When we cannot evaluate an improper integral directly, we try to determine whether it converges or diverges. If the integral diverges, that's the end of the story. If it converges, we can use numerical methods to approximate its value. The principal tests for convergence or divergence are the Direct Comparison Test and the Limit Comparison Test.

EXAMPLE 6 Does the integral $\int_1^\infty e^{-x^2}\,dx$ converge?

Solution By definition,

$$\int_1^\infty e^{-x^2}\,dx = \lim_{b\to\infty}\int_1^b e^{-x^2}\,dx.$$

We cannot evaluate this integral directly because it is nonelementary. But we *can* show that its limit as $b \to \infty$ is finite. We know that $\int_1^b e^{-x^2}\,dx$ is an increasing function

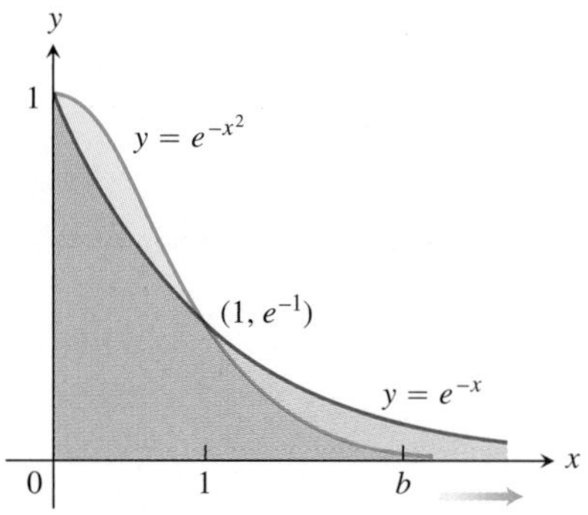

FIGURE 5.19 The graph of e^{-x^2} lies below the graph of e^{-x} for $x > 1$ (Example 6).

HISTORICAL BIOGRAPHY

Karl Weierstrass
(1815–1897)

of b. Therefore either it becomes infinite as $b \to \infty$ or it has a finite limit as $b \to \infty$. It does not become infinite: For every value of $x \geq 1$ we have $e^{-x^2} \leq e^{-x}$ (Figure 5.19), so that

$$\int_1^b e^{-x^2}\,dx \leq \int_1^b e^{-x}\,dx = -e^{-b} + e^{-1} < e^{-1} \approx 0.36788.$$

Hence

$$\int_1^\infty e^{-x^2}\,dx = \lim_{b\to\infty}\int_1^b e^{-x^2}\,dx$$

converges to some definite finite value. We do not know exactly what the value is except that it is something positive and less than 0.37. Here we are relying on the completeness property of the real numbers, discussed in Appendix 7. ■

The comparison of e^{-x^2} and e^{-x} in Example 6 is a special case of the following test.

THEOREM 2—Direct Comparison Test Let f and g be continuous on $[a, \infty)$ with $0 \leq f(x) \leq g(x)$ for all $x \geq a$. Then

1. $\int_a^\infty f(x)\,dx$ converges if $\int_a^\infty g(x)\,dx$ converges.

2. $\int_a^\infty g(x)\,dx$ diverges if $\int_a^\infty f(x)\,dx$ diverges.

Proof The reasoning behind the argument establishing Theorem 2 is similar to that in Example 6. If $0 \leq f(x) \leq g(x)$ for $x \geq a$, then

$$\int_a^b f(x)\,dx \leq \int_a^b g(x)\,dx, \qquad b > a.$$

From this it can be argued, as in Example 6, that

$$\int_a^\infty f(x)\,dx \quad \text{converges if} \quad \int_a^\infty g(x)\,dx \quad \text{converges.}$$

Turning this around says that

$$\int_a^\infty g(x)\,dx \quad \text{diverges if} \quad \int_a^\infty f(x)\,dx \quad \text{diverges.}$$

■

EXAMPLE 7

(a) $\displaystyle\int_1^\infty \frac{\sin^2 x}{x^2}\,dx$ converges because

$$0 \leq \frac{\sin^2 x}{x^2} \leq \frac{1}{x^2} \text{ on } [1, \infty) \quad \text{and} \quad \int_1^\infty \frac{1}{x^2}\,dx \quad \text{converges.}$$

Example 3

(b) $\displaystyle\int_1^{\infty} \frac{1}{\sqrt{x^2 - 0.1}}\,dx$ diverges because

$$\frac{1}{\sqrt{x^2 - 0.1}} \geq \frac{1}{x} \text{ on } [1, \infty) \quad \text{and} \quad \int_1^{\infty} \frac{1}{x}\,dx \quad \text{diverges.} \qquad \text{Example 3}$$ ■

THEOREM 3—Limit Comparison Test If the positive functions f and g are continuous on $[a, \infty)$ and if

$$\lim_{x\to\infty} \frac{f(x)}{g(x)} = L, \qquad 0 < L < \infty,$$

then

$$\int_a^{\infty} f(x)\,dx \quad \text{and} \quad \int_a^{\infty} g(x)\,dx$$

both converge or both diverge.

We omit the more advanced proof of Theorem 3.

Although the improper integrals of two functions from a to ∞ may both converge, this does not mean that their integrals necessarily have the same value, as the next example shows.

EXAMPLE 8 Show that

$$\int_1^{\infty} \frac{dx}{1 + x^2}$$

converges by comparison with $\int_1^{\infty} (1/x^2)\,dx$. Find and compare the two integral values.

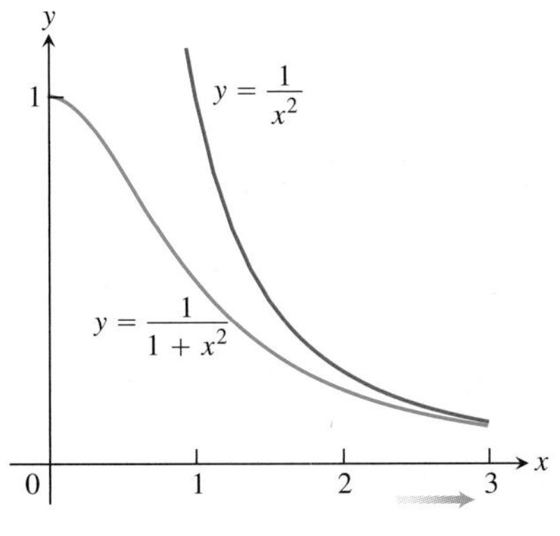

FIGURE 5.20 The functions in Example 8.

Solution The functions $f(x) = 1/x^2$ and $g(x) = 1/(1 + x^2)$ are positive and continuous on $[1, \infty)$. Also,

$$\lim_{x\to\infty} \frac{f(x)}{g(x)} = \lim_{x\to\infty} \frac{1/x^2}{1/(1 + x^2)} = \lim_{x\to\infty} \frac{1 + x^2}{x^2}$$
$$= \lim_{x\to\infty} \left(\frac{1}{x^2} + 1\right) = 0 + 1 = 1,$$

a positive finite limit (Figure 5.20). Therefore, $\displaystyle\int_1^{\infty} \frac{dx}{1 + x^2}$ converges because $\displaystyle\int_1^{\infty} \frac{dx}{x^2}$ converges.

The integrals converge to different values, however.

$$\int_1^{\infty} \frac{dx}{x^2} = \frac{1}{2 - 1} = 1 \qquad \text{Example 3}$$

and

$$\int_1^{\infty} \frac{dx}{1 + x^2} = \lim_{b\to\infty} \int_1^{b} \frac{dx}{1 + x^2}$$
$$= \lim_{b\to\infty} [\tan^{-1} b - \tan^{-1} 1] = \frac{\pi}{2} - \frac{\pi}{4} = \frac{\pi}{4}$$ ■

EXERCISES 5.7

Evaluate the integrals in Exercises 1–34 without using tables.

1. $\int_0^\infty \frac{dx}{x^2+1}$

2. $\int_1^\infty \frac{dx}{x^{1.001}}$

3. $\int_0^1 \frac{dx}{\sqrt{x}}$

4. $\int_0^4 \frac{dx}{\sqrt{4-x}}$

5. $\int_{-1}^1 \frac{dx}{x^{2/3}}$

6. $\int_{-8}^1 \frac{dx}{x^{1/3}}$

7. $\int_0^1 \frac{dx}{\sqrt{1-x^2}}$

8. $\int_0^1 \frac{dr}{r^{0.999}}$

9. $\int_{-\infty}^{-2} \frac{2\,dx}{x^2-1}$

10. $\int_{-\infty}^2 \frac{2\,dx}{x^2+4}$

11. $\int_2^\infty \frac{2}{v^2-v}\,dv$

12. $\int_2^\infty \frac{2\,dt}{t^2-1}$

13. $\int_{-\infty}^\infty \frac{2x\,dx}{(x^2+1)^2}$

14. $\int_{-\infty}^\infty \frac{x\,dx}{(x^2+4)^{3/2}}$

15. $\int_0^1 \frac{\theta+1}{\sqrt{\theta^2+2\theta}}\,d\theta$

16. $\int_0^2 \frac{s+1}{\sqrt{4-s^2}}\,ds$

17. $\int_0^\infty \frac{dx}{(1+x)\sqrt{x}}$

18. $\int_1^\infty \frac{1}{x\sqrt{x^2-1}}\,dx$

19. $\int_0^\infty \frac{dv}{(1+v^2)(1+\tan^{-1} v)}$

20. $\int_0^\infty \frac{16\tan^{-1} x}{1+x^2}\,dx$

21. $\int_{-\infty}^0 \theta e^\theta\,d\theta$

22. $\int_0^\infty 2e^{-\theta}\sin\theta\,d\theta$

23. $\int_{-\infty}^0 e^{-|x|}\,dx$

24. $\int_{-\infty}^\infty 2xe^{-x^2}\,dx$

25. $\int_0^1 x\ln x\,dx$

26. $\int_0^1 (-\ln x)\,dx$

27. $\int_0^2 \frac{ds}{\sqrt{4-s^2}}$

28. $\int_0^1 \frac{4r\,dr}{\sqrt{1-r^4}}$

29. $\int_1^2 \frac{ds}{s\sqrt{s^2-1}}$

30. $\int_2^4 \frac{dt}{t\sqrt{t^2-4}}$

31. $\int_{-1}^4 \frac{dx}{\sqrt{|x|}}$

32. $\int_0^2 \frac{dx}{\sqrt{|x-1|}}$

33. $\int_{-1}^\infty \frac{d\theta}{\theta^2+5\theta+6}$

34. $\int_0^\infty \frac{dx}{(x+1)(x^2+1)}$

In Exercises 35–64, use integration, the Direct Comparison Test, or the Limit Comparison Test to test the integrals for convergence. If more than one method applies, use whatever method you prefer.

35. $\int_0^{\pi/2} \tan\theta\,d\theta$

36. $\int_0^{\pi/2} \cot\theta\,d\theta$

37. $\int_0^\pi \frac{\sin\theta\,d\theta}{\sqrt{\pi-\theta}}$

38. $\int_{-\pi/2}^{\pi/2} \frac{\cos\theta\,d\theta}{(\pi-2\theta)^{1/3}}$

39. $\int_0^{\ln 2} x^{-2}e^{-1/x}\,dx$

40. $\int_0^1 \frac{e^{-\sqrt{x}}}{\sqrt{x}}\,dx$

41. $\int_0^\pi \frac{dt}{\sqrt{t}+\sin t}$

42. $\int_0^1 \frac{dt}{t-\sin t}$ (*Hint:* $t \ge \sin t$ for $t \ge 0$)

43. $\int_0^2 \frac{dx}{1-x^2}$

44. $\int_0^2 \frac{dx}{1-x}$

45. $\int_{-1}^1 \ln|x|\,dx$

46. $\int_{-1}^1 -x\ln|x|\,dx$

47. $\int_1^\infty \frac{dx}{x^3+1}$

48. $\int_4^\infty \frac{dx}{\sqrt{x}-1}$

49. $\int_2^\infty \frac{dv}{\sqrt{v-1}}$

50. $\int_0^\infty \frac{d\theta}{1+e^\theta}$

51. $\int_0^\infty \frac{dx}{\sqrt{x^6+1}}$

52. $\int_2^\infty \frac{dx}{\sqrt{x^2-1}}$

53. $\int_1^\infty \frac{\sqrt{x+1}}{x^2}\,dx$

54. $\int_2^\infty \frac{x\,dx}{\sqrt{x^4-1}}$

55. $\int_\pi^\infty \frac{2+\cos x}{x}\,dx$

56. $\int_\pi^\infty \frac{1+\sin x}{x^2}\,dx$

57. $\int_4^\infty \frac{2\,dt}{t^{3/2}-1}$

58. $\int_2^\infty \frac{1}{\ln x}\,dx$

59. $\int_1^\infty \frac{e^x}{x}\,dx$

60. $\int_{e^e}^\infty \ln(\ln x)\,dx$

61. $\int_1^\infty \frac{1}{\sqrt{e^x-x}}\,dx$

62. $\int_1^\infty \frac{1}{e^x-2^x}\,dx$

63. $\int_{-\infty}^\infty \frac{dx}{\sqrt{x^4+1}}$

64. $\int_{-\infty}^\infty \frac{dx}{e^x+e^{-x}}$

65. Find the values of p for which each integral converges.

a. $\int_1^2 \frac{dx}{x(\ln x)^p}$ b. $\int_2^\infty \frac{dx}{x(\ln x)^p}$

66. **$\int_{-\infty}^\infty f(x)\,dx$ may not equal $\lim_{b\to\infty}\int_{-b}^b f(x)\,dx$** Show that

$$\int_0^\infty \frac{2x\,dx}{x^2+1}$$

diverges and hence that

$$\int_{-\infty}^\infty \frac{2x\,dx}{x^2+1}$$

diverges. Then show that

$$\lim_{b\to\infty}\int_{-b}^{b}\frac{2x\,dx}{x^2+1}=0.$$

67. Find the area between the curve $y = e^{-x}$ and the x-axis.

68. Find the area of the region that lies between the curves $y = \sec x$ and $y = \tan x$ from $x = 0$ to $x = \pi/2$.

69. Estimating the value of a convergent improper integral whose domain is infinite Show that

$$\int_3^{\infty} e^{-3x}\,dx = \frac{1}{3}e^{-9} < 0.000042,$$

and hence that $\int_3^{\infty} e^{-x^2}\,dx < 0.000042$. Explain why this means that $\int_0^{\infty} e^{-x^2}\,dx$ can be replaced by $\int_0^{3} e^{-x^2}\,dx$ without introducing an error of magnitude greater than 0.000042.

T **70.** (*Continuation of Exercise 69.*) Evaluate $\int_0^3 e^{-x^2}\,dx$ numerically.

71. Sine-integral function The integral

$$\mathrm{Si}(x) = \int_0^x \frac{\sin t}{t}\,dt,$$

called the *sine-integral function*, has important applications in optics.

T **a.** Plot the integrand $(\sin t)/t$ for $t > 0$. Is the Si function everywhere increasing or decreasing? Do you think $\mathrm{Si}(x) = 0$ for $x > 0$? Check your answers by graphing the function $\mathrm{Si}(x)$ for $0 \le x \le 25$.

b. Explore the convergence of

$$\int_0^{\infty}\frac{\sin t}{t}\,dt.$$

If it converges, what is its value?

72. Error function The function

$$\mathrm{erf}(x) = \int_0^x \frac{2e^{-t^2}}{\sqrt{\pi}}\,dt,$$

called the *error function*, has important applications in probability and statistics.

T **a.** Plot the error function for $0 \le x \le 25$.

b. Explore the convergence of

$$\int_0^{\infty}\frac{2e^{-t^2}}{\sqrt{\pi}}\,dt.$$

If it converges, what appears to be its value? You will see how to confirm your estimate in Section 12.4, Exercise 31.

73. Normal probability distribution function The function

$$f(x) = \frac{1}{\sigma\sqrt{2\pi}}e^{-\frac{1}{2}\left(\frac{x-\mu}{\sigma}\right)^2}$$

is called the *normal probability density function* with mean μ and standard deviation σ. The number μ tells where the distribution is centered, and σ measures the "scatter" around the mean.

From the theory of probability, it is known that

$$\int_{-\infty}^{\infty} f(x)\,dx = 1.$$

In what follows, let $\mu = 0$ and $\sigma = 1$.

T **a.** Draw the graph of f. Find the intervals on which f is increasing, the intervals on which f is decreasing, and any local extreme values and where they occur.

b. Evaluate

$$\int_{-n}^{n} f(x)\,dx$$

for $n = 1, 2, 3$.

c. Give a convincing argument that

$$\int_{-\infty}^{\infty} f(x)\,dx = 1.$$

(*Hint:* Show that $0 < f(x) < e^{-x/2}$ for $x > 1$, and for $b > 1$,

$$\int_b^{\infty} e^{-x/2}\,dx \to 0 \quad \text{as} \quad b\to\infty.)$$

74. Show that if $f(x)$ is integrable on every interval of real numbers and a and b are real numbers with $a < b$, then

a. $\int_{-\infty}^{a} f(x)\,dx$ and $\int_a^{\infty} f(x)\,dx$ both converge if and only if $\int_{-\infty}^{b} f(x)\,dx$ and $\int_b^{\infty} f(x)\,dx$ both converge.

b. $\int_{-\infty}^{a} f(x)\,dx + \int_a^{\infty} f(x)\,dx = \int_{-\infty}^{b} f(x)\,dx + \int_b^{\infty} f(x)\,dx$ when the integrals involved converge.

COMPUTER EXPLORATIONS

In Exercises 75–78, use a CAS to explore the integrals for various values of p (include noninteger values). For what values of p does the integral converge? What is the value of the integral when it does converge? Plot the integrand for various values of p.

75. $\displaystyle\int_0^e x^p \ln x\,dx$

76. $\displaystyle\int_e^{\infty} x^p \ln x\,dx$

77. $\displaystyle\int_0^{\infty} x^p \ln x\,dx$

78. $\displaystyle\int_{-\infty}^{\infty} x^p \ln|x|\,dx$

Chapter 5 Questions to Guide Your Review

1. What is the formula for integration by parts? Where does it come from? Why might you want to use it?
2. When applying the formula for integration by parts, how do you choose the u and dv? How can you apply integration by parts to an integral of the form $\int f(x)\,dx$?
3. What is the goal of the method of partial fractions?
4. When the degree of a polynomial $f(x)$ is less than the degree of a polynomial $g(x)$, how do you write $f(x)/g(x)$ as a sum of partial fractions if $g(x)$
   a. is a product of distinct linear factors?
   b. consists of a repeated linear factor?
   c. contains an irreducible quadratic factor?

   What do you do if the degree of f is *not* less than the degree of g?
5. If an integrand is a product of the form $\sin^n x \cos^m x$, where m and n are nonnegative integers, how do you evaluate the integral? Give a specific example of each case.
6. What substitutions are made to evaluate integrals of $\sin mx \sin nx$, $\sin mx \cos nx$, and $\cos mx \cos nx$? Give an example of each case.
7. What substitutions are sometimes used to transform integrals involving $\sqrt{a^2 - x^2}$, $\sqrt{a^2 + x^2}$, and $\sqrt{x^2 - a^2}$ into integrals that can be evaluated directly? Give an example of each case.
8. What restrictions can you place on the variables involved in the three basic trigonometric substitutions to make sure the substitutions are reversible (have inverses)?
9. How are integral tables typically used? What do you do if a particular integral you want to evaluate is not listed in the table?
10. What is a reduction formula? How are reduction formulas used? Give an example.
11. You are collaborating to produce a short "how-to" manual for numerical integration, and you are writing about the Trapezoidal Rule. **(a)** What would you say about the rule itself and how to use it? How to achieve accuracy? **(b)** What would you say if you were writing about Simpson's Rule instead?
12. How would you compare the relative merits of Simpson's Rule and the Trapezoidal Rule?
13. What is an improper integral of Type I? Type II? How are the values of various types of improper integrals defined? Give examples.
14. What tests are available for determining the convergence and divergence of improper integrals that cannot be evaluated directly? Give examples of their use.

Chapter 5 Practice and Additional Exercises

Evaluate the integrals in Exercises 1–8 using integration by parts.

1. $\int \ln(x+1)\,dx$
2. $\int x^2 \ln x\,dx$
3. $\int \tan^{-1} 3x\,dx$
4. $\int \cos^{-1}\left(\frac{x}{2}\right) dx$
5. $\int (x+1)^2 e^x\,dx$
6. $\int x^2 \sin(1-x)\,dx$
7. $\int e^x \cos 2x\,dx$
8. $\int e^{-2x} \sin 3x\,dx$

Evaluate the integrals in Exercises 9–28. It may be necessary to use a substitution first.

9. $\int \frac{x\,dx}{x^2 - 3x + 2}$
10. $\int \frac{x\,dx}{x^2 + 4x + 3}$
11. $\int \frac{dx}{x(x+1)^2}$
12. $\int \frac{x+1}{x^2(x-1)}\,dx$
13. $\int \frac{\sin\theta\,d\theta}{\cos^2\theta + \cos\theta - 2}$
14. $\int \frac{\cos\theta\,d\theta}{\sin^2\theta + \sin\theta - 6}$
15. $\int \frac{3x^2 + 4x + 4}{x^3 + x}\,dx$
16. $\int \frac{4x\,dx}{x^3 + 4x}$
17. $\int \frac{v+3}{2v^3 - 8v}\,dv$
18. $\int \frac{(3v-7)\,dv}{(v-1)(v-2)(v-3)}$
19. $\int \frac{dt}{t^4 + 4t^2 + 3}$
20. $\int \frac{t\,dt}{t^4 - t^2 - 2}$
21. $\int \frac{x^3 + x^2}{x^2 + x - 2}\,dx$
22. $\int \frac{x^3 + 1}{x^3 - x}\,dx$
23. $\int \frac{x^3 + 4x^2}{x^2 + 4x + 3}\,dx$
24. $\int \frac{2x^3 + x^2 - 21x + 24}{x^2 + 2x - 8}\,dx$
25. $\int \frac{dx}{x(3\sqrt{x+1})}$
26. $\int \frac{dx}{x(1 + \sqrt[3]{x})}$

27. $\displaystyle\int \frac{ds}{e^s - 1}$ **28.** $\displaystyle\int \frac{ds}{\sqrt{e^s + 1}}$

Evaluate the integrals in Exercises 29–32 **(a)** without using a trigonometric substitution, **(b)** using a trigonometric substitution.

29. $\displaystyle\int \frac{y\,dy}{\sqrt{16 - y^2}}$ **30.** $\displaystyle\int \frac{x\,dx}{\sqrt{4 + x^2}}$

31. $\displaystyle\int \frac{x\,dx}{4 - x^2}$ **32.** $\displaystyle\int \frac{t\,dt}{\sqrt{4t^2 - 1}}$

Evaluate the integrals in Exercises 33–44.

33. $\displaystyle\int \frac{x\,dx}{9 - x^2}$ **34.** $\displaystyle\int \frac{dx}{x(9 - x^2)}$

35. $\displaystyle\int \frac{dx}{9 - x^2}$ **36.** $\displaystyle\int \frac{dx}{\sqrt{9 - x^2}}$

37. $\displaystyle\int \sin^3 x \cos^4 x\,dx$ **38.** $\displaystyle\int \cos^5 x \sin^5 x\,dx$

39. $\displaystyle\int \tan^4 x \sec^2 x\,dx$ **40.** $\displaystyle\int \tan^3 x \sec^3 x\,dx$

41. $\displaystyle\int \sin 5\theta \cos 6\theta\,d\theta$ **42.** $\displaystyle\int \cos 3\theta \cos 3\theta\,d\theta$

43. $\displaystyle\int \sqrt{1 + \cos(t/2)}\,dt$ **44.** $\displaystyle\int e^t\sqrt{\tan^2 e^t + 1}\,dt$

45. A direct calculation shows that

$$\int_0^{\pi} 2\sin^2 x\,dx = \pi.$$

How close do you come to this value by using the Trapezoidal Rule with $n = 6$? Simpson's Rule with $n = 6$? Try them and find out.

46. A brief calculation shows that if $0 \le x \le 1$, then the second derivative of $f(x) = \sqrt{1 + x^4}$ lies between 0 and 8. Based on this, about how many subdivisions would you need to estimate the integral of f from 0 to 1 with an error no greater than 10^{-3} in absolute value using the Trapezoidal Rule?

47. Mean temperature Compute the average value of the temperature function

$$f(x) = 37\sin\left(\frac{2\pi}{365}(x - 101)\right) + 25$$

for a 365-day year. This is one way to estimate the annual mean air temperature in Fairbanks, Alaska. The National Weather Service's official figure, a numerical average of the daily normal mean air temperatures for the year, is 25.7°F, which is slightly higher than the average value of $f(x)$.

48. Heat capacity of a gas Heat capacity C_v is the amount of heat required to raise the temperature of a given mass of gas with constant volume by 1°C, measured in units of cal/deg-mol (calories per degree gram molecular weight). The heat capacity of oxygen depends on its temperature T and satisfies the formula

$$C_v = 8.27 + 10^{-5}(26T - 1.87T^2).$$

Find the average value of C_v for $20° \le T \le 675°$C and the temperature at which it is attained.

Evaluate the improper integrals in Exercises 49–58.

49. $\displaystyle\int_0^3 \frac{dx}{\sqrt{9 - x^2}}$ **50.** $\displaystyle\int_0^1 \ln x\,dx$

51. $\displaystyle\int_{-1}^1 \frac{dy}{y^{2/3}}$ **52.** $\displaystyle\int_{-2}^0 \frac{d\theta}{(\theta + 1)^{3/5}}$

53. $\displaystyle\int_3^{\infty} \frac{2\,du}{u^2 - 2u}$ **54.** $\displaystyle\int_1^{\infty} \frac{3v - 1}{4v^3 - v^2}\,dv$

55. $\displaystyle\int_0^{\infty} x^2 e^{-x}\,dx$ **56.** $\displaystyle\int_{-\infty}^0 xe^{3x}\,dx$

57. $\displaystyle\int_{-\infty}^{\infty} \frac{dx}{4x^2 + 9}$ **58.** $\displaystyle\int_{-\infty}^{\infty} \frac{4\,dx}{x^2 + 16}$

Which of the improper integrals in Exercises 59–64 converge and which diverge?

59. $\displaystyle\int_6^{\infty} \frac{d\theta}{\sqrt{\theta^2 + 1}}$ **60.** $\displaystyle\int_0^{\infty} e^{-u}\cos u\,du$

61. $\displaystyle\int_1^{\infty} \frac{\ln z}{z}\,dz$ **62.** $\displaystyle\int_1^{\infty} \frac{e^{-t}}{\sqrt{t}}\,dt$

63. $\displaystyle\int_{-\infty}^{\infty} \frac{2\,dx}{e^x + e^{-x}}$ **64.** $\displaystyle\int_{-\infty}^{\infty} \frac{dx}{x^2(1 + e^x)}$

Evaluate the integrals in Exercises 65–94. The integrals are listed in random order.

65. $\displaystyle\int \frac{x\,dx}{1 + \sqrt{x}}$ **66.** $\displaystyle\int \frac{x^3 + 2}{4 - x^2}\,dx$

67. $\displaystyle\int \frac{dx}{x(x^2 + 1)^2}$ **68.** $\displaystyle\int \frac{dx}{\sqrt{-2x - x^2}}$

69. $\displaystyle\int \frac{2 - \cos x + \sin x}{\sin^2 x}\,dx$ **70.** $\displaystyle\int \frac{\sin^2\theta}{\cos^2\theta}\,d\theta$

71. $\displaystyle\int \frac{9\,dv}{81 - v^4}$ **72.** $\displaystyle\int_2^{\infty} \frac{dx}{(x - 1)^2}$

73. $\displaystyle\int \theta\cos(2\theta + 1)\,d\theta$ **74.** $\displaystyle\int \frac{x^3\,dx}{x^2 - 2x + 1}$

75. $\displaystyle\int \frac{\sin 2\theta\,d\theta}{(1 + \cos 2\theta)^2}$ **76.** $\displaystyle\int_{\pi/4}^{\pi/2} \sqrt{1 + \cos 4x}\,dx$

77. $\displaystyle\int \frac{x\,dx}{\sqrt{2 - x}}$ **78.** $\displaystyle\int \frac{\sqrt{1 - v^2}}{v^2}\,dv$

79. $\displaystyle\int \frac{dy}{y^2 - 2y + 2}$ **80.** $\displaystyle\int \frac{x\,dx}{\sqrt{8 - 2x^2 - x^4}}$

81. $\displaystyle\int \frac{z+1}{z^2(z^2+4)}\,dz$

82. $\displaystyle\int x^3 e^{(x^2)}\,dx$

83. $\displaystyle\int \frac{t\,dt}{\sqrt{9-4t^2}}$

84. $\displaystyle\int \frac{\tan^{-1} x}{x^2}\,dx$

85. $\displaystyle\int \frac{e^t\,dt}{e^{2t}+3e^t+2}$

86. $\displaystyle\int \tan^3 t\,dt$

87. $\displaystyle\int_1^{\infty} \frac{\ln y}{y^3}\,dy$

88. $\displaystyle\int \frac{\cot v\,dv}{\ln \sin v}$

89. $\displaystyle\int e^{\ln\sqrt{x}}\,dx$

90. $\displaystyle\int e^{\theta}\sqrt{3+4e^{\theta}}\,d\theta$

91. $\displaystyle\int \frac{\sin 5t\,dt}{1+(\cos 5t)^2}$

92. $\displaystyle\int \frac{dv}{\sqrt{e^{2v}-1}}$

93. $\displaystyle\int \frac{dr}{1+\sqrt{r}}$

94. $\displaystyle\int \frac{4x^3-20x}{x^4-10x^2+9}\,dx$

Chapter 6

APPLICATIONS OF DEFINITE INTEGRALS

OVERVIEW In Chapter 4 we saw that a continuous function over a closed interval has a definite integral, which is the limit of any Riemann sum for the function. We proved that we could evaluate definite integrals using the Fundamental Theorem of Calculus. We also found that the area under a curve or the area between two curves could be computed as definite integrals.

In this chapter we extend the applications of definite integrals to finding volumes, lengths of plane curves, and areas of surfaces of revolution. We also use integrals to solve problems involving the work done by a force, the location of an object's center of mass, and the fluid force against a planar wall.

6.1 Volumes by Slicing and Rotation About an Axis

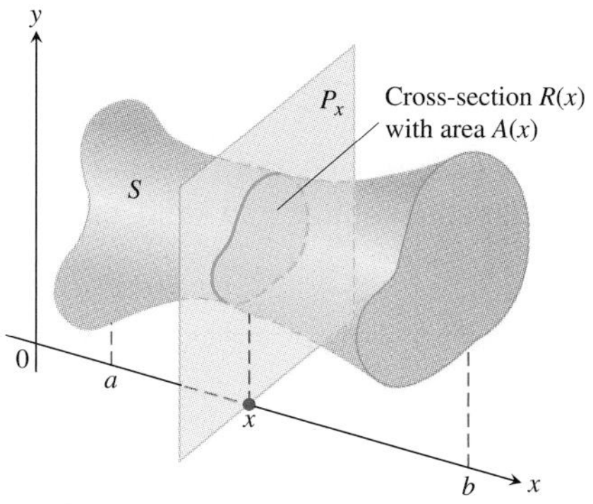

FIGURE 6.1 A cross-section of the solid S formed by intersecting S with a plane P_x perpendicular to the x-axis through the point x in the interval $[a, b]$.

In this section we define volumes of solids whose cross-sections are plane regions. A **cross-section** of a solid S is the plane region formed by intersecting S with a plane (Figure 6.1).

Suppose we want to find the volume of a solid S like the one in Figure 6.1. We begin by extending the definition of a cylinder from classical geometry to cylindrical solids with arbitrary bases (Figure 6.2). If the cylindrical solid has a known base area A and height h, then the volume of the cylindrical solid is

$$\text{Volume} = \text{area} \times \text{height} = A \cdot h.$$

This equation forms the basis for defining the volumes of many solids that are not cylinders, like the one in Figure 6.1, by the *method of slicing*.

If the cross-section of the solid S at each point x in the interval $[a, b]$ is a region $R(x)$ of area $A(x)$, and A is a continuous function of x, we can define and calculate the volume of the solid S as a definite integral in the following way.

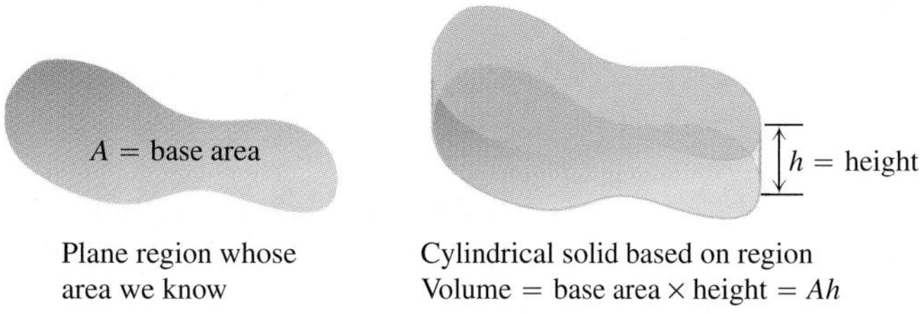

FIGURE 6.2 The volume of a cylindrical solid is always defined to be its base area times its height.

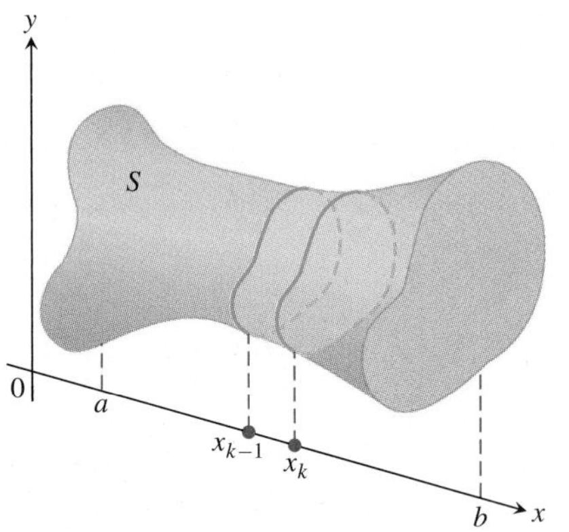

FIGURE 6.3 A typical thin slab in the solid S.

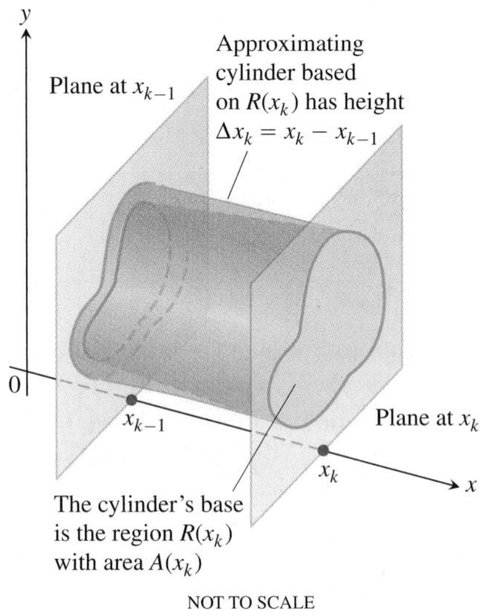

FIGURE 6.4 The solid thin slab in Figure 6.3 is approximated by the cylindrical solid with base $R(x_k)$ having area $A(x_k)$ and height $\Delta x_k = x_k - x_{k-1}$.

We partition $[a, b]$ into subintervals of width (length) Δx_k and slice the solid, as we would a loaf of bread, by planes perpendicular to the x-axis at the partition points $a = x_0 < x_1 < \cdots < x_n = b$. The planes P_{x_k}, perpendicular to the x-axis at the partition points, slice S into thin "slabs" (like thin slices of a loaf of bread). A typical slab is shown in Figure 6.3. We approximate the slab between the plane at x_{k-1} and the plane at x_k by a cylindrical solid with base area $A(x_k)$ and height $\Delta x_k = x_k - x_{k-1}$ (Figure 6.4). The volume V_k of this cylindrical solid is $A(x_k) \cdot \Delta x_k$, which is approximately the same volume as that of the slab:

$$\text{Volume of the } k\text{th slab} \approx V_k = A(x_k)\,\Delta x_k.$$

The volume V of the entire solid S is therefore approximated by the sum of these cylindrical volumes,

$$V \approx \sum_{k=1}^{n} V_k = \sum_{k=1}^{n} A(x_k)\,\Delta x_k.$$

This is a Riemann sum for the function $A(x)$ on $[a, b]$. We expect the approximations from these sums to improve as the norm of the partition of $[a, b]$ goes to zero. Taking a partition of $[a, b]$ into n subintervals with $\|P\| \to 0$ gives

$$\lim_{n\to\infty} \sum_{k=1}^{n} A(x_k)\,\Delta x_k = \int_a^b A(x)\,dx.$$

So we define their limiting definite integral to be the volume of the solid S.

DEFINITION The **volume** of a solid of known integrable cross-sectional area $A(x)$ from $x = a$ to $x = b$ is the integral of A from a to b,

$$V = \int_a^b A(x)\,dx.$$

This definition applies whenever $A(x)$ is continuous, or more generally, when it is integrable. To apply the formula in the definition to calculate the volume of a solid, take the following steps:

Calculating the Volume of a Solid

1. *Sketch the solid and a typical cross-section.*
2. *Find a formula for* $A(x)$, the area of a typical cross-section.
3. *Find the limits of integration.*
4. *Integrate* $A(x)$ using the Fundamental Theorem.

EXAMPLE 1 A pyramid 3 m high has a square base that is 3 m on a side. The cross-section of the pyramid perpendicular to the altitude x m down from the vertex is a square x m on a side. Find the volume of the pyramid.

Solution

1. *A sketch.* We draw the pyramid with its altitude along the x-axis and its vertex at the origin and include a typical cross-section (Figure 6.5).

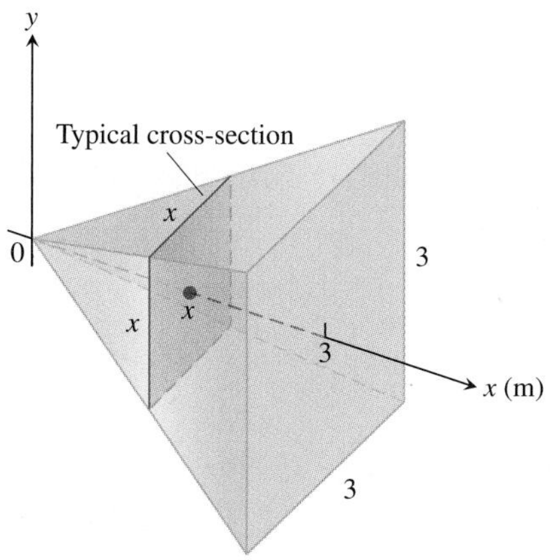

FIGURE 6.5 The cross-sections of the pyramid in Example 1 are squares.

2. *A formula for A(x).* The cross-section at x is a square x meters on a side, so its area is

$$A(x) = x^2.$$

3. *The limits of integration.* The squares lie on the planes from $x = 0$ to $x = 3$.
4. *Integrate to find the volume.*

$$V = \int_0^3 A(x)\,dx = \int_0^3 x^2\,dx = \frac{x^3}{3}\bigg]_0^3 = 9\text{ m}^3$$

■

EXAMPLE 2 Cavalieri's principle says that solids with equal altitudes and identical cross-sectional areas at each height have the same volume (Figure 6.6). This follows immediately from the definition of volume, because the cross-sectional area function $A(x)$ and the interval $[a, b]$ are the same for both solids.

HISTORICAL BIOGRAPHY

Bonaventura Cavalieri
(1598–1647)

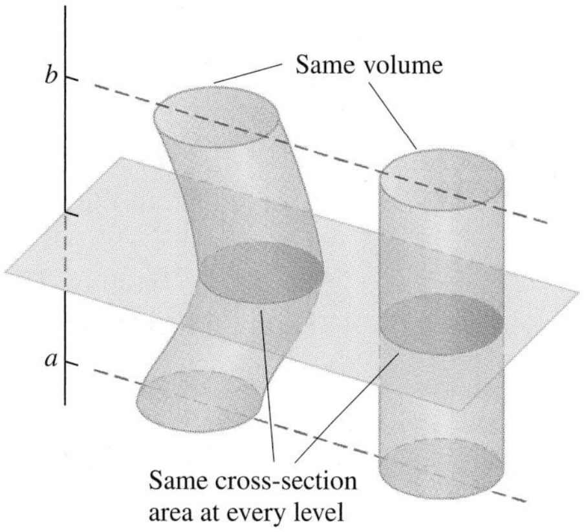

FIGURE 6.6 *Cavalieri's principle:* These solids have the same volume, which can be illustrated with stacks of coins. ■

EXAMPLE 3 A curved wedge is cut from a circular cylinder of radius 3 by two planes. One plane is perpendicular to the axis of the cylinder. The second plane crosses the first plane at a 45° angle at the center of the cylinder. Find the volume of the wedge.

Solution We draw the wedge and sketch a typical cross-section perpendicular to the x-axis (Figure 6.7). The cross-section at x is a rectangle of area

$$A(x) = (\text{height})(\text{width}) = (x)\left(2\sqrt{9 - x^2}\right) = 2x\sqrt{9 - x^2}.$$

The rectangles run from $x = 0$ to $x = 3$, so we have

$$V = \int_a^b A(x)\,dx = \int_0^3 2x\sqrt{9 - x^2}\,dx$$

$$= -\frac{2}{3}(9 - x^2)^{3/2}\bigg]_0^3$$

Let $u = 9 - x^2$, $du = -2x\,dx$, integrate, and substitute back.

$$= 0 + \frac{2}{3}(9)^{3/2}$$

$$= 18.$$

■

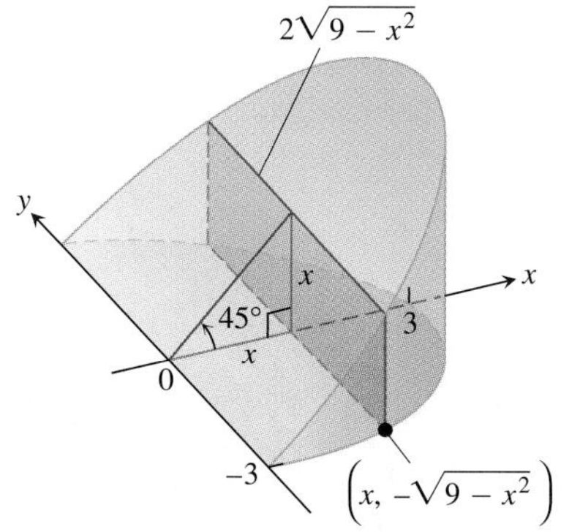

FIGURE 6.7 The wedge of Example 3, sliced perpendicular to the x-axis. The cross-sections are rectangles.

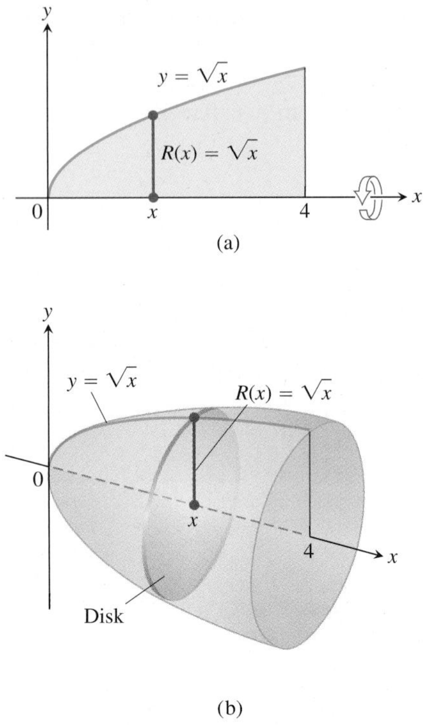

FIGURE 6.8 The region (a) and solid of revolution (b) in Example 4.

Solids of Revolution: The Disk Method

The solid generated by rotating (or revolving) a plane region about an axis in its plane is called a **solid of revolution**. To find the volume of a solid like the one shown in Figure 6.8, we need only observe that the cross-sectional area $A(x)$ is the area of a disk of radius $R(x)$, the distance of the planar region's boundary from the axis of revolution. The area is then

$$A(x) = \pi(\text{radius})^2 = \pi[R(x)]^2.$$

So the definition of volume gives

$$V = \int_a^b A(x)\,dx = \int_a^b \pi[R(x)]^2\,dx.$$

This method for calculating the volume of a solid of revolution is often called the **disk method** because a cross-section is a circular disk of radius $R(x)$.

EXAMPLE 4 The region between the curve $y = \sqrt{x}$, $0 \le x \le 4$, and the x-axis is revolved about the x-axis to generate a solid. Find its volume.

Solution We draw figures showing the region, a typical radius, and the generated solid (Figure 6.8). The volume is

$$\begin{aligned} V &= \int_a^b \pi[R(x)]^2\,dx \\ &= \int_0^4 \pi\left[\sqrt{x}\right]^2 dx \qquad R(x) = \sqrt{x} \\ &= \pi\int_0^4 x\,dx = \pi\frac{x^2}{2}\bigg]_0^4 = \pi\frac{(4)^2}{2} = 8\pi. \end{aligned}$$

■

EXAMPLE 5 The circle

$$x^2 + y^2 = a^2$$

is rotated about the x-axis to generate a sphere. Find its volume.

Solution We imagine the sphere cut into thin slices by planes perpendicular to the x-axis (Figure 6.9). The cross-sectional area at a typical point x between $-a$ and a is

$$A(x) = \pi y^2 = \pi(a^2 - x^2).$$

Therefore, the volume is

$$V = \int_{-a}^{a} A(x)\,dx = \int_{-a}^{a} \pi(a^2 - x^2)\,dx = \pi\left[a^2x - \frac{x^3}{3}\right]_{-a}^{a} = \frac{4}{3}\pi a^3.$$

■

The axis of revolution in the next example is not the x-axis, but the rule for calculating the volume is the same: Integrate $\pi(\text{radius})^2$ between appropriate limits.

EXAMPLE 6 Find the volume of the solid generated by revolving the region bounded by $y = \sqrt{x}$ and the lines $y = 1$, $x = 4$ about the line $y = 1$.

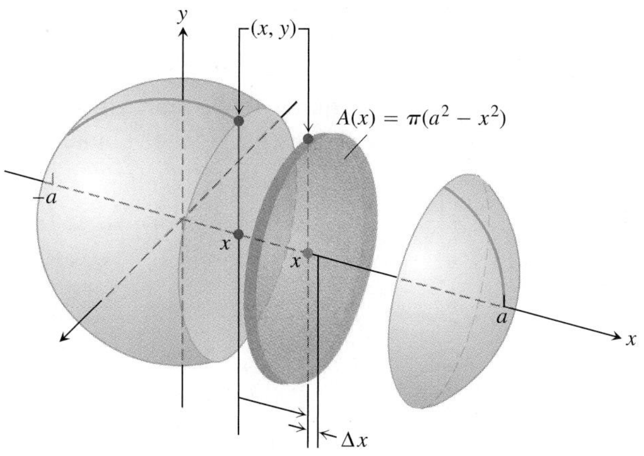

FIGURE 6.9 The sphere generated by rotating the circle $x^2 + y^2 = a^2$ about the x-axis. The radius is $R(x) = y = \sqrt{a^2 - x^2}$ (Example 5).

Solution We draw figures showing the region, a typical radius, and the generated solid (Figure 6.10). The volume is

$$\begin{aligned} V &= \int_1^4 \pi[R(x)]^2\, dx \\ &= \int_1^4 \pi\left[\sqrt{x} - 1\right]^2 dx \\ &= \pi \int_1^4 \left[x - 2\sqrt{x} + 1\right] dx \\ &= \pi\left[\frac{x^2}{2} - 2\cdot\frac{2}{3}x^{3/2} + x\right]_1^4 = \frac{7\pi}{6}. \end{aligned}$$

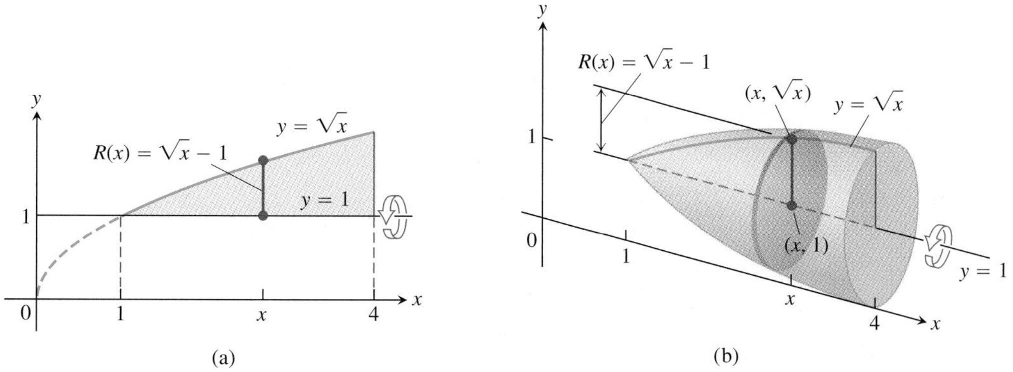

FIGURE 6.10 The region (a) and solid of revolution (b) in Example 6. ■

To find the volume of a solid generated by revolving a region between the y-axis and a curve $x = R(y)$, $c \le y \le d$, about the y-axis, we use the same method with x replaced by y. In this case, the circular cross-section is

$$A(y) = \pi[\text{radius}]^2 = \pi[R(y)]^2.$$

EXAMPLE 7 Find the volume of the solid generated by revolving the region between the y-axis and the curve $x = 2/y$, $1 \le y \le 4$, about the y-axis.

Solution We draw figures showing the region, a typical radius, and the generated solid (Figure 6.11). The volume is

$$\begin{aligned} V &= \int_1^4 \pi[R(y)]^2\, dy \\ &= \int_1^4 \pi\left(\frac{2}{y}\right)^2 dy \\ &= \pi \int_1^4 \frac{4}{y^2}\, dy = 4\pi\left[-\frac{1}{y}\right]_1^4 = 4\pi\left[\frac{3}{4}\right] = 3\pi. \end{aligned}$$

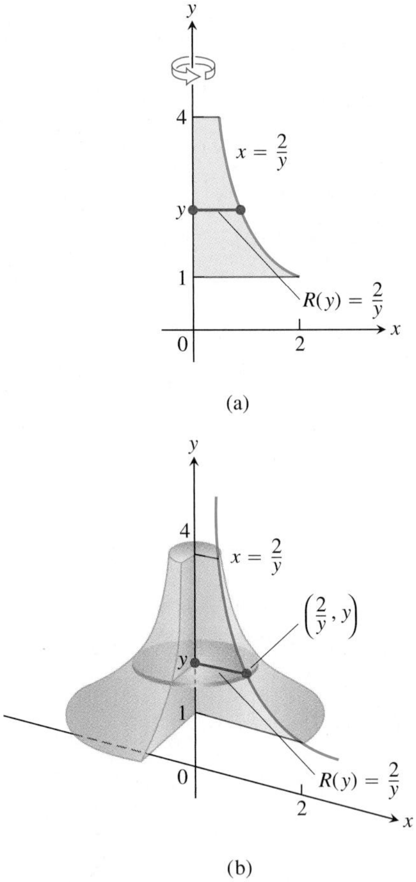

FIGURE 6.11 The region (a) and part of the solid of revolution (b) in Example 7.

EXAMPLE 8 Find the volume of the solid generated by revolving the region between the parabola $x = y^2 + 1$ and the line $x = 3$ about the line $x = 3$.

Solution We draw figures showing the region, a typical radius, and the generated solid (Figure 6.12). Note that the cross-sections are perpendicular to the line $x = 3$ and have y coordinates from $-\sqrt{2}$ to $\sqrt{2}$. The volume is

$$\begin{aligned} V &= \int_{-\sqrt{2}}^{\sqrt{2}} \pi[R(y)]^2\, dy && y = \pm\sqrt{2} \text{ when } x = 3 \\ &= \int_{-\sqrt{2}}^{\sqrt{2}} \pi[2 - y^2]^2\, dy && R(y) = 3 - (y^2 + 1) = 2 - y^2 \\ &= \pi \int_{-\sqrt{2}}^{\sqrt{2}} [4 - 4y^2 + y^4]\, dy \\ &= \pi\left[4y - \frac{4}{3}y^3 + \frac{y^5}{5}\right]_{-\sqrt{2}}^{\sqrt{2}} \\ &= \frac{64\pi\sqrt{2}}{15}. \end{aligned}$$

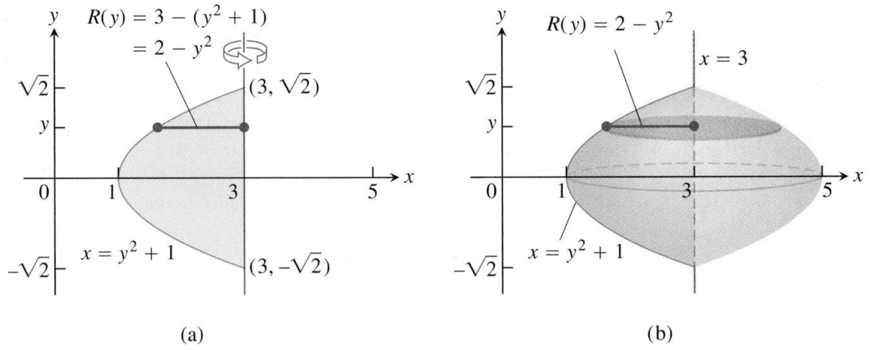

FIGURE 6.12 The region (a) and solid of revolution (b) in Example 8.

Solids of Revolution: The Washer Method

If the region we revolve to generate a solid does not border on or cross the axis of revolution, the solid has a hole in it (Figure 6.13). The cross-sections perpendicular to the axis of

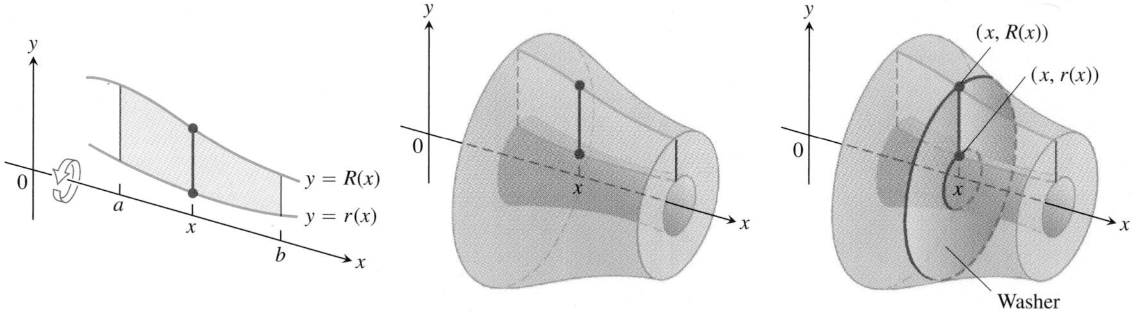

FIGURE 6.13 The cross-sections of the solid of revolution generated here are washers, not disks, so the integral $\int_a^b A(x)\,dx$ leads to a slightly different formula.

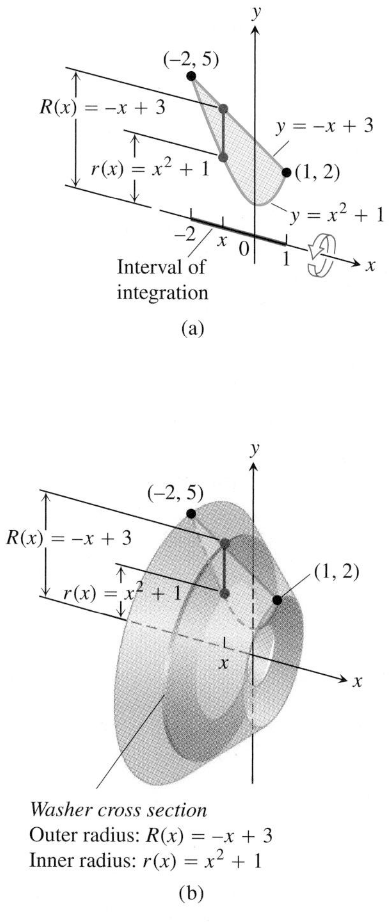

FIGURE 6.14 (a) The region in Example 9 spanned by a line segment perpendicular to the axis of revolution. (b) When the region is revolved about the x-axis, the line segment generates a washer.

revolution are *washers* (the purplish circular surface in Figure 6.13) instead of disks. The dimensions of a typical washer are

Outer radius: $R(x)$

Inner radius: $r(x)$

The washer's area is

$$A(x) = \pi[R(x)]^2 - \pi[r(x)]^2 = \pi([R(x)]^2 - [r(x)]^2).$$

Consequently, the definition of volume gives

$$V = \int_a^b A(x)\,dx = \int_a^b \pi([R(x)]^2 - [r(x)]^2)\,dx.$$

This method for calculating the volume of a solid of revolution is called the **washer method** because a slab is a circular washer of outer radius $R(x)$ and inner radius $r(x)$.

EXAMPLE 9 The region bounded by the curve $y = x^2 + 1$ and the line $y = -x + 3$ is revolved about the x-axis to generate a solid. Find the volume of the solid.

Solution

1. Draw the region and sketch a line segment across it perpendicular to the axis of revolution (the red segment in Figure 6.14).
2. Find the outer and inner radii of the washer that would be swept out by the line segment if it were revolved about the x-axis along with the region.

   These radii are the distances of the ends of the line segment from the axis of revolution (Figure 6.14).

Outer radius: $R(x) = -x + 3$

Inner radius: $r(x) = x^2 + 1$

3. Find the limits of integration by finding the x-coordinates of the intersection points of the curve and line in Figure 6.14a.

$$\begin{aligned} x^2 + 1 &= -x + 3 \\ x^2 + x - 2 &= 0 \\ (x + 2)(x - 1) &= 0 \\ x = -2, &\quad x = 1 \end{aligned}$$

4. Evaluate the volume integral.

$$\begin{aligned} V &= \int_a^b \pi([R(x)]^2 - [r(x)]^2)\, dx \\ &= \int_{-2}^{1} \pi((-x + 3)^2 - (x^2 + 1)^2)\, dx \qquad \text{Values from Steps 2 and 3} \\ &= \pi \int_{-2}^{1} (8 - 6x - x^2 - x^4)\, dx \\ &= \pi \left[8x - 3x^2 - \frac{x^3}{3} - \frac{x^5}{5} \right]_{-2}^{1} = \frac{117\pi}{5} \end{aligned}$$

■

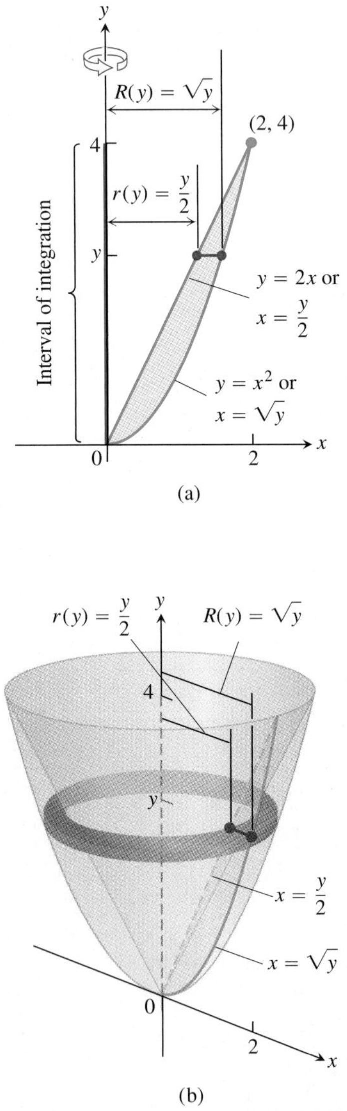

FIGURE 6.15 (a) The region being rotated about the y-axis, the washer radii, and limits of integration in Example 10. (b) The washer swept out by the line segment in part (a).

To find the volume of a solid formed by revolving a region about the y-axis, we use the same procedure as in Example 9, but integrate with respect to y instead of x. In this situation the line segment sweeping out a typical washer is perpendicular to the y-axis (the axis of revolution), and the outer and inner radii of the washer are functions of y.

EXAMPLE 10 The region bounded by the parabola $y = x^2$ and the line $y = 2x$ in the first quadrant is revolved about the y-axis to generate a solid. Find the volume of the solid.

Solution First we sketch the region and draw a line segment across it perpendicular to the axis of revolution (the y-axis). See Figure 6.15a.

The radii of the washer swept out by the line segment are $R(y) = \sqrt{y}$, $r(y) = y/2$ (Figure 6.15).

The line and parabola intersect at $y = 0$ and $y = 4$, so the limits of integration are $c = 0$ and $d = 4$. We integrate to find the volume:

$$\begin{aligned} V &= \int_c^d \pi([R(y)]^2 - [r(y)]^2)\, dy \\ &= \int_0^4 \pi\left(\left[\sqrt{y}\right]^2 - \left[\frac{y}{2}\right]^2\right) dy \\ &= \pi \int_0^4 \left(y - \frac{y^2}{4}\right) dy = \pi\left[\frac{y^2}{2} - \frac{y^3}{12}\right]_0^4 = \frac{8}{3}\pi. \end{aligned}$$

■

EXERCISES 6.1

Find the volumes of the solids in Exercises 1–10.

1. The solid lies between planes perpendicular to the x-axis at $x = 0$ and $x = 4$. The cross-sections perpendicular to the axis on the interval $0 \le x \le 4$ are squares whose diagonals run from the parabola $y = -\sqrt{x}$ to the parabola $y = \sqrt{x}$.

2. The solid lies between planes perpendicular to the x-axis at $x = -1$ and $x = 1$. The cross-sections perpendicular to the x-axis are circular disks whose diameters run from the parabola $y = x^2$ to the parabola $y = 2 - x^2$.

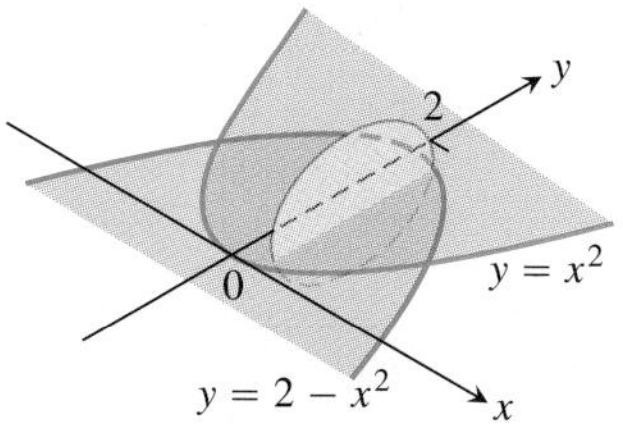

3. The solid lies between planes perpendicular to the x-axis at $x = -1$ and $x = 1$. The cross-sections perpendicular to the x-axis between these planes are squares whose bases run from the semicircle $y = -\sqrt{1 - x^2}$ to the semicircle $y = \sqrt{1 - x^2}$.

4. The solid lies between planes perpendicular to the x-axis at $x = -1$ and $x = 1$. The cross-sections perpendicular to the x-axis between these planes are squares whose diagonals run from the semicircle $y = -\sqrt{1 - x^2}$ to the semicircle $y = \sqrt{1 - x^2}$.

5. The base of a solid is the region between the curve $y = 2\sqrt{\sin x}$ and the interval $[0, \pi]$ on the x-axis. The cross-sections perpendicular to the x-axis are

   a. equilateral triangles with bases running from the x-axis to the curve as shown in the figure.

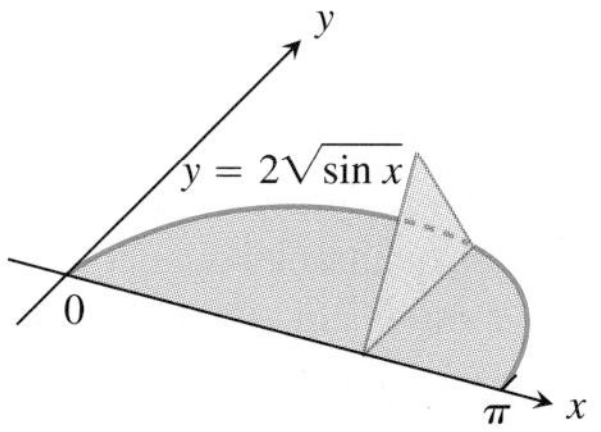

   b. squares with bases running from the x-axis to the curve.

6. The solid lies between planes perpendicular to the x-axis at $x = -\pi/3$ and $x = \pi/3$. The cross-sections perpendicular to the x-axis are

   a. circular disks with diameters running from the curve $y = \tan x$ to the curve $y = \sec x$.

   b. squares whose bases run from the curve $y = \tan x$ to the curve $y = \sec x$.

7. The solid lies between planes perpendicular to the y-axis at $y = 0$ and $y = 2$. The cross-sections perpendicular to the y-axis are circular disks with diameters running from the y-axis to the parabola $x = \sqrt{5}y^2$.

8. The base of the solid is the disk $x^2 + y^2 \le 1$. The cross-sections by planes perpendicular to the y-axis between $y = -1$ and $y = 1$ are isosceles right triangles with one leg in the disk.

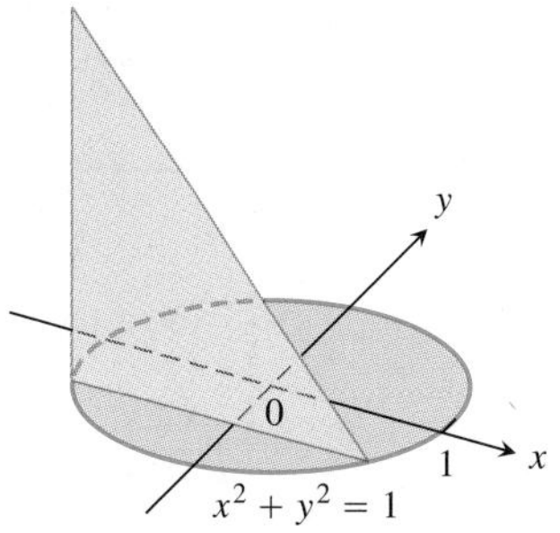

9. The solid lies between planes perpendicular to the x-axis at $x = -1$ and $x = 1$. The cross-sections perpendicular to the x-axis are

   a. circles whose diameters stretch from the curve $y = -1/\sqrt{1 + x^2}$ to the curve $y = 1/\sqrt{1 + x^2}$.

   b. vertical squares whose base edges run from the curve $y = -1/\sqrt{1 + x^2}$ to the curve $y = 1/\sqrt{1 + x^2}$.

10. The solid lies between planes perpendicular to the x-axis at $x = -\sqrt{2}/2$ and $x = \sqrt{2}/2$. The cross-sections perpendicular to the x-axis are

    a. circles whose diameters stretch from the x-axis to the curve $y = 2/\sqrt[4]{1 - x^2}$.

    b. squares whose diagonals stretch from the x-axis to the curve $y = 2/\sqrt[4]{1 - x^2}$.

11. **A twisted solid** A square of side length s lies in a plane perpendicular to a line L. One vertex of the square lies on L. As this square moves a distance h along L, the square turns one revolution about L to generate a corkscrew-like column with square cross-sections.

    a. Find the volume of the column.

    b. What will the volume be if the square turns twice instead of once? Give reasons for your answer.

12. **Cavalieri's principle** A solid lies between planes perpendicular to the x-axis at $x = 0$ and $x = 12$. The cross-sections by planes perpendicular to the x-axis are circular disks whose diameters run from the line $y = x/2$ to the line $y = x$ as shown in the accompanying figure. Explain why the solid has the same volume as a right circular cone with base radius 3 and height 12.

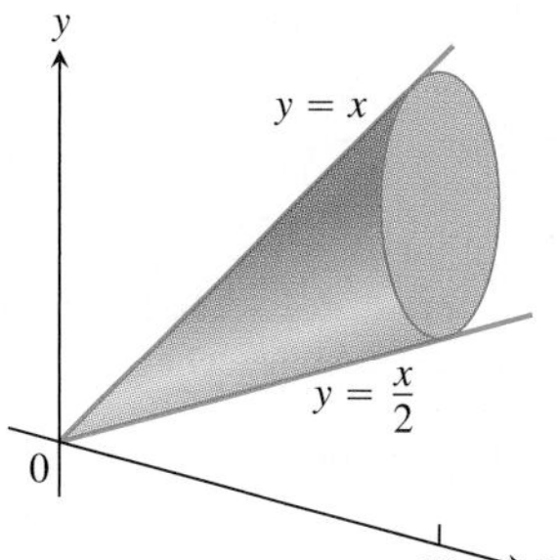

In Exercises 13–16, find the volume of the solid generated by revolving the shaded region about the given axis.

13. About the x-axis

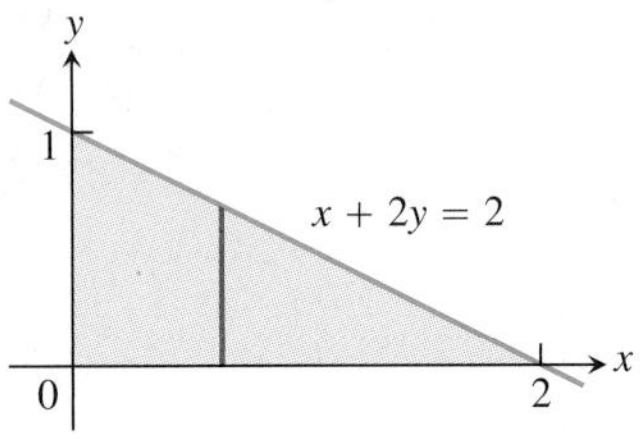

14. About the y-axis

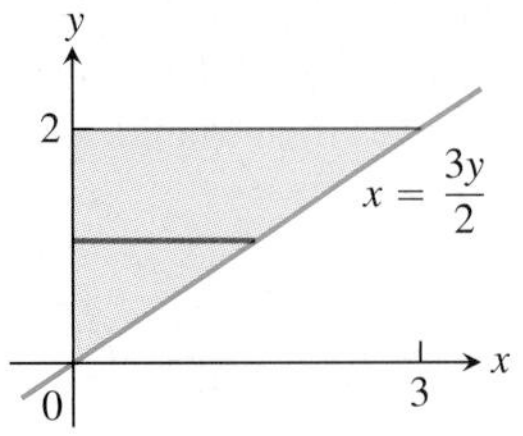

15. About the y-axis

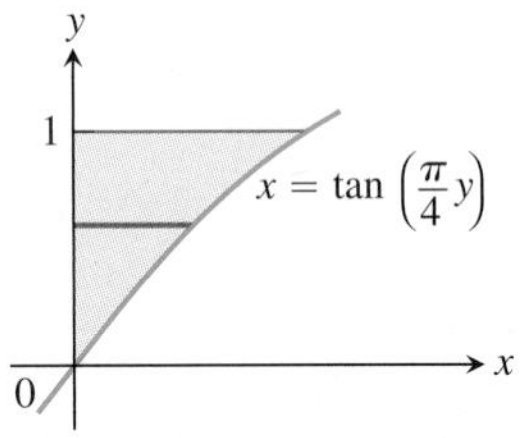

16. About the x-axis

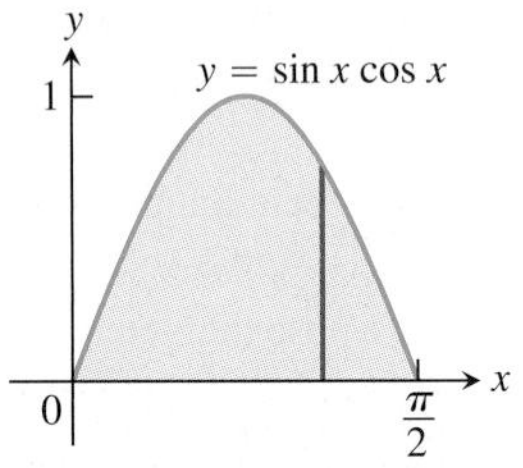

Find the volumes of the solids generated by revolving the regions bounded by the lines and curves in Exercises 17–26 about the x-axis.

17. $y = x^2, \quad y = 0, \quad x = 2$

18. $y = x^3, \quad y = 0, \quad x = 2$

19. $y = \sqrt{9 - x^2}, \quad y = 0$

20. $y = x - x^2, \quad y = 0$

21. $y = \sqrt{\cos x}, \quad 0 \le x \le \pi/2, \quad y = 0, \quad x = 0$

22. $y = \sec x, \quad y = 0, \quad x = -\pi/4, \quad x = \pi/4$

23. $y = e^{-x}, \quad y = 0, \quad x = 0, \quad x = 1$

24. The region between the curve $y = \sqrt{\cot x}$ and the x-axis from $x = \pi/6$ to $x = \pi/2$.

25. The region between the curve $y = 1/(2\sqrt{x})$ and the x-axis from $x = 1/4$ to $x = 4$.

26. The region bounded by the x-axis and one arch of the cycloid $x = \theta - \sin\theta, y = 1 - \cos\theta$.
(*Hint:* $dV = \pi y^2\, dx = \pi y^2\, (dx/d\theta)\, d\theta$.)

In Exercises 27 and 28, find the volume of the solid generated by revolving the region about the given line.

27. The region in the first quadrant bounded above by the line $y = \sqrt{2}$, below by the curve $y = \sec x \tan x$, and on the left by the y-axis, about the line $y = \sqrt{2}$

28. The region in the first quadrant bounded above by the line $y = 2$, below by the curve $y = 2 \sin x, 0 \le x \le \pi/2$, and on the left by the y-axis, about the line $y = 2$

Find the volumes of the solids generated by revolving the regions bounded by the lines and curves in Exercises 29–34 about the y-axis.

29. The region enclosed by $x = \sqrt{5}\, y^2, \quad x = 0, \quad y = -1, \quad y = 1$

30. The region enclosed by $x = y^{3/2}, \quad x = 0, \quad y = 2$

31. The region enclosed by $x = \sqrt{2 \sin 2y}, \quad 0 \le y \le \pi/2, \quad x = 0$

32. The region enclosed by $x = \sqrt{\cos(\pi y/4)}, \quad -2 \le y \le 0, \quad x = 0$

33. The region in the first quadrant bounded by the coordinate axes, the line $y = 3$, and the curve $x = 2/\sqrt{y + 1}$

34. $x = \sqrt{2y}/(y^2 + 1), \quad x = 0, \quad y = 1$

Find the volumes of the solids generated by revolving the shaded regions in Exercises 35 and 36 about the indicated axes.

35. The x-axis

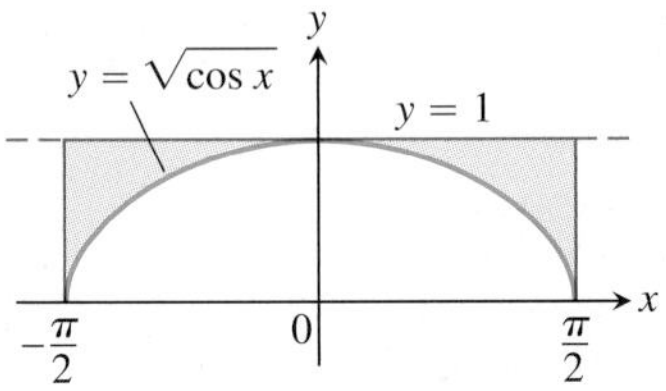

36. The y-axis

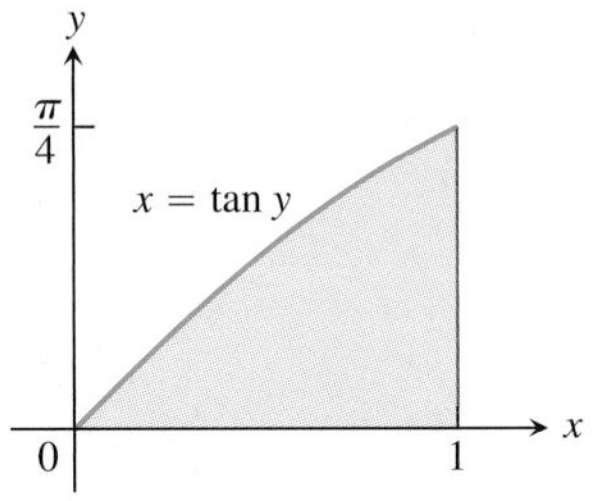

Find the volumes of the solids generated by revolving the regions bounded by the lines and curves in Exercises 37–42 about the x-axis.

37. $y = x, \quad y = 1, \quad x = 0$

38. $y = 2\sqrt{x}, \quad y = 2, \quad x = 0$

39. $y = x^2 + 1, \quad y = x + 3$

40. $y = 4 - x^2, \quad y = 2 - x$

41. $y = \sec x, \quad y = \sqrt{2}, \quad -\pi/4 \le x \le \pi/4$

42. $y = \sec x, \quad y = \tan x, \quad x = 0, \quad x = 1$

In Exercises 43–46, find the volume of the solid generated by revolving each region about the y-axis.

43. The region enclosed by the triangle with vertices (1, 0), (2, 1), and (1, 1)

44. The region enclosed by the triangle with vertices (0, 1), (1, 0), and (1, 1)

45. The region in the first quadrant bounded above by the parabola $y = x^2$, below by the x-axis, and on the right by the line $x = 2$

46. The region in the first quadrant bounded on the left by the circle $x^2 + y^2 = 3$, on the right by the line $x = \sqrt{3}$, and above by the line $y = \sqrt{3}$

In Exercises 47 and 48, find the volume of the solid generated by revolving each region about the given axis.

47. The region in the first quadrant bounded above by the curve $y = x^2$, below by the x-axis, and on the right by the line $x = 1$, about the line $x = -1$

48. The region in the second quadrant bounded above by the curve $y = -x^3$, below by the x-axis, and on the left by the line $x = -1$, about the line $x = -2$

49. Find the volume of the solid generated by revolving the region bounded by $y = \sqrt{x}$ and the lines $y = 2$ and $x = 0$ about

a. the x-axis. **b.** the y-axis.

c. the line $y = 2$. **d.** the line $x = 4$.

50. Find the volume of the solid generated by revolving about the x-axis the region in the first quadrant enclosed by the coordinate axes, the curve $y = 2/(1 + x^2)$, and the line $x = 1$.

51. Find the volume of the solid generated by revolving the region bounded by the parabola $y = x^2$ and the line $y = 1$ about

a. the line $y = 1$. **b.** the line $y = 2$.

c. the line $y = -1$.

52. Find the volume of the solid generated by revolving the region between the curve $y = e^{-x}$ and the x-axis, about the x-axis.

6.2 Volumes by Cylindrical Shells

In Section 6.1 we defined the volume of a solid as the definite integral $V = \int_a^b A(x)\,dx$, where $A(x)$ is an integrable cross-sectional area of the solid from $x = a$ to $x = b$. The area $A(x)$ was obtained by slicing through the solid with a plane perpendicular to the x-axis. However, this method of slicing is sometimes awkward to apply, as we will illustrate in our first example. To overcome this difficulty, we use the same integral definition for volume, but obtain the area by slicing through the solid in a different way.

In this section we slice through the solid using circular cylinders of increasing radii, like cookie cutters. We slice straight down through the solid so that the axis of each cylinder is parallel to the y-axis. The vertical axis of each cylinder is the same line, but the radii of the cylinders increase with each slice. In this way the solid is sliced up into thin cylindrical shells of constant thickness that grow outward from their common axis, like circular tree rings. Unrolling a cylindrical shell shows that its volume is approximately that of a rectangular slab with area $A(x)$ and thickness Δx. This allows us to apply the same integral definition for volume as before. The following example provides some insight before we derive the method in general.

EXAMPLE 1 The region enclosed by the x-axis and the parabola $y = f(x) = 3x - x^2$ is revolved about the vertical line $x = -1$ to generate the shape of a solid (Figure 6.16). Find the volume of the solid.

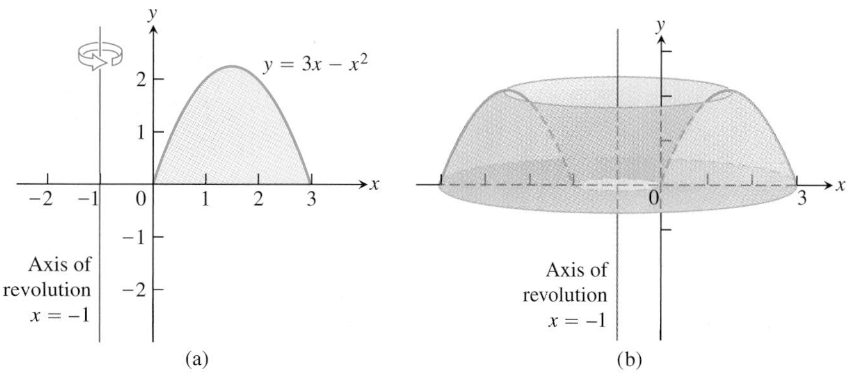

FIGURE 6.16 (a) The graph of the region in Example 1, before revolution. (b) The solid formed when the region in part (a) is revolved about the axis of revolution $x = -1$.

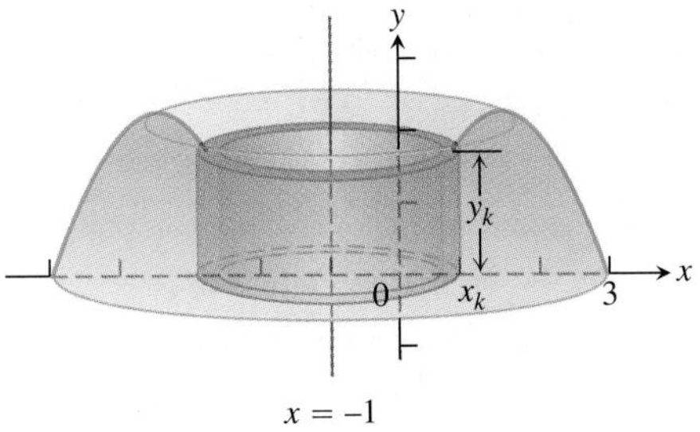

FIGURE 6.17 A cylindrical shell of height y_k obtained by rotating a vertical strip of thickness Δx about the line $x = -1$. The outer radius of the cylinder occurs at x_k, where the height of the parabola is $y_k = 3x_k - x_k^2$ (Example 1).

Solution Using the washer method from Section 6.1 would be awkward here because we would need to express the x-values of the left and right branches of the parabola in terms of y. (These x-values are the inner and outer radii for a typical washer, leading to complicated formulas.) Instead of rotating a horizontal strip of thickness Δy, we rotate a *vertical strip* of thickness Δx. This rotation produces a *cylindrical shell* of height y_k above a point x_k within the base of the vertical strip, and of thickness Δx. An example of a cylindrical shell is shown as the orange-shaded region in Figure 6.17. We can think of the cylindrical shell shown in the figure as approximating a slice of the solid obtained by cutting straight down through it, parallel to the axis of revolution, all the way around close to the inside hole. We then cut another cylindrical slice around the enlarged hole, then another, and so on, obtaining n cylinders. The radii of the cylinders gradually increase, and the heights of the cylinders follow the contour of the parabola: shorter to taller, then back to shorter (Figure 6.16a).

Each slice is sitting over a subinterval of the x-axis of length (width) Δx. Its radius is approximately $(1 + x_k)$, and its height is approximately $3x_k - x_k^2$. If we unroll the cylinder at x_k and flatten it out, it becomes (approximately) a rectangular slab with thickness Δx (Figure 6.18). The outer circumference of the kth cylinder is $2\pi \cdot \text{radius} = 2\pi(1 + x_k)$, and this is the length of the rolled-out rectangular slab. Its volume is approximated by that of a rectangular solid,

$$\begin{aligned}\Delta V_k &= \text{circumference} \times \text{height} \times \text{thickness} \\ &= 2\pi(1 + x_k) \cdot \left(3x_k - x_k^2\right) \cdot \Delta x.\end{aligned}$$

Summing together the volumes ΔV_k of the individual cylindrical shells over the interval $[0, 3]$ gives the Riemann sum

$$\sum_{k=1}^{n} \Delta V_k = \sum_{k=1}^{n} 2\pi(x_k + 1)\left(3x_k - x_k^2\right) \Delta x.$$

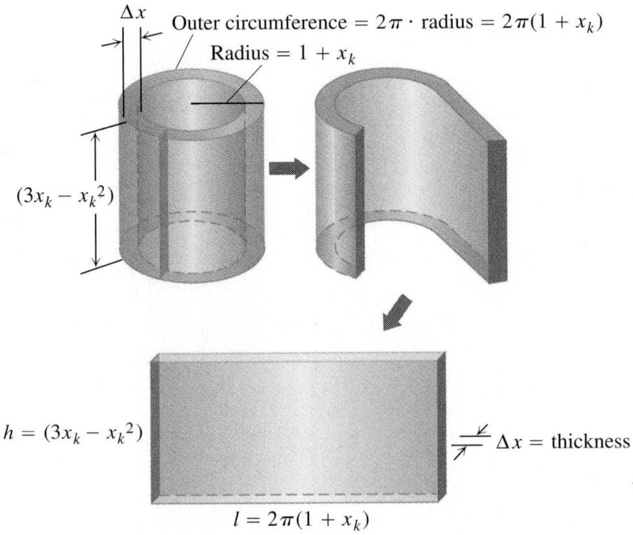

FIGURE 6.18 Imagine cutting and unrolling a cylindrical shell to get a flat (nearly) rectangular solid (Example 1).

Taking the limit as the thickness $\Delta x \to 0$ gives the volume integral

$$\begin{aligned} V &= \lim_{n\to\infty} \sum_{k=1}^{n} 2\pi(x_k + 1)(3x_k - x_k^2)\,\Delta x \\ &= \int_0^3 2\pi(x + 1)(3x - x^2)\,dx \\ &= \int_0^3 2\pi(3x^2 + 3x - x^3 - x^2)\,dx \\ &= 2\pi\int_0^3 (2x^2 + 3x - x^3)\,dx \\ &= 2\pi\left[\frac{2}{3}x^3 + \frac{3}{2}x^2 - \frac{1}{4}x^4\right]_0^3 = \frac{45\pi}{2}. \end{aligned}$$

■

We now generalize the procedure used in Example 1.

The Shell Method

Suppose the region bounded by the graph of a nonnegative continuous function $y = f(x)$ and the x-axis over the finite closed interval $[a, b]$ lies to the right of the vertical line $x = L$ (Figure 6.19a). We assume $a \geq L$, so the vertical line may touch the region, but not pass through it. We generate a solid S by rotating this region about the vertical line L.

Let P be a partition of the interval $[a, b]$ by the points $a = x_0 < x_1 < \cdots < x_n = b$, and let c_k be the midpoint of the kth subinterval $[x_{k-1}, x_k]$. We approximate the region in Figure 6.19a with rectangles based on this partition of $[a, b]$. A typical approximating rectangle has height $f(c_k)$ and width $\Delta x_k = x_k - x_{k-1}$. If this rectangle is rotated about the vertical line $x = L$, then a shell is swept out, as in Figure 6.19b. A formula from geometry tells us that the volume of the shell swept out by the rectangle is

$$\begin{aligned} \Delta V_k &= 2\pi \times \text{average shell radius} \times \text{shell height} \times \text{thickness} \\ &= 2\pi \cdot (c_k - L) \cdot f(c_k) \cdot \Delta x_k. \end{aligned}$$

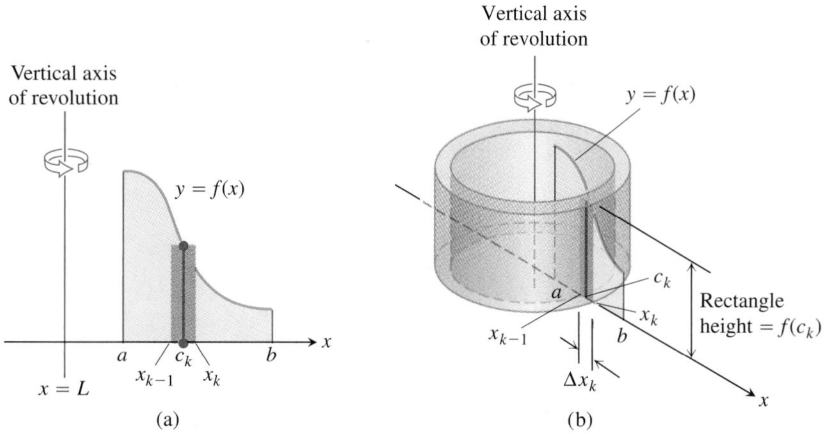

FIGURE 6.19 When the region shown in (a) is revolved about the vertical line $x = L$, a solid is produced which can be sliced into cylindrical shells. A typical shell is shown in (b).

We approximate the volume of the solid S by summing the volumes of the shells swept out by the n rectangles based on P:

$$V \approx \sum_{k=1}^{n} \Delta V_k.$$

The limit of this Riemann sum as $\|P\| \to 0$ gives the volume of the solid as a definite integral:

$$V = \lim_{\|P\|\to 0} \sum_{k=1}^{n} \Delta V_k = \int_a^b 2\pi(\text{shell radius})(\text{shell height})\, dx.$$

$$= \int_a^b 2\pi(x - L)f(x)\, dx.$$

We refer to the variable of integration, here x, as the **thickness variable**. We use the first integral, rather than the second containing a formula for the integrand, to emphasize the *process* of the shell method. This will allow for rotations about a horizontal line L as well.

Shell Formula for Revolution About a Vertical Line

The volume of the solid generated by revolving the region between the x-axis and the graph of a continuous function $y = f(x) \geq 0$, $L \leq a \leq x \leq b$, about a vertical line $x = L$ is

$$V = \int_a^b 2\pi \begin{pmatrix}\text{shell}\\ \text{radius}\end{pmatrix} \begin{pmatrix}\text{shell}\\ \text{height}\end{pmatrix} dx.$$

EXAMPLE 2 The region bounded by the curve $y = \sqrt{x}$, the x-axis, and the line $x = 4$ is revolved about the y-axis to generate a solid. Find the volume of the solid.

Solution Sketch the region and draw a line segment across it *parallel* to the axis of revolution (Figure 6.20a). Label the segment's height (shell height) and distance from the axis of revolution (shell radius). (We drew the shell in Figure 6.20b, but you need not do that.)

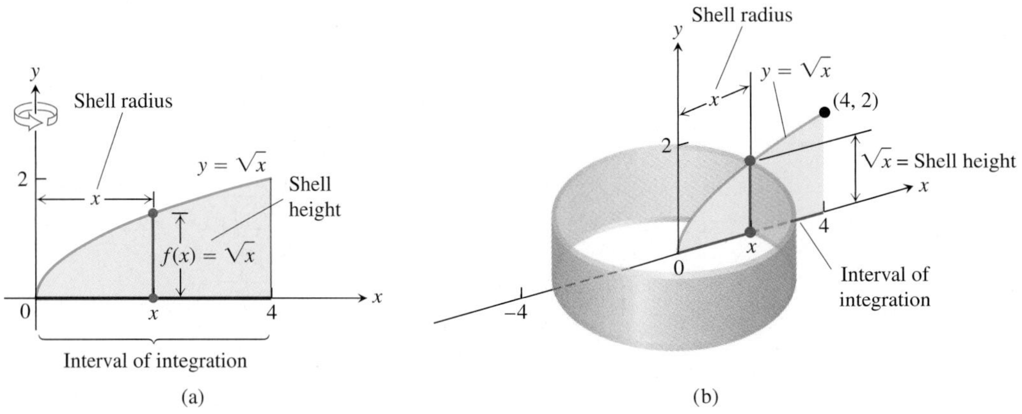

FIGURE 6.20 (a) The region, shell dimensions, and interval of integration in Example 2. (b) The shell swept out by the vertical segment in part (a) with a width Δx.

The shell thickness variable is x, so the limits of integration for the shell formula are $a = 0$ and $b = 4$ (Figure 6.20). The volume is then

$$\begin{aligned} V &= \int_a^b 2\pi \begin{pmatrix}\text{shell}\\ \text{radius}\end{pmatrix}\begin{pmatrix}\text{shell}\\ \text{height}\end{pmatrix} dx \\ &= \int_0^4 2\pi(x)\left(\sqrt{x}\right) dx \\ &= 2\pi \int_0^4 x^{3/2}\, dx = 2\pi\left[\frac{2}{5}x^{5/2}\right]_0^4 = \frac{128\pi}{5}. \end{aligned}$$

■

So far, we have used vertical axes of revolution. For horizontal axes, we replace the x's with y's.

EXAMPLE 3 The region bounded by the curve $y = \sqrt{x}$, the x-axis, and the line $x = 4$ is revolved about the x-axis to generate a solid. Find the volume of the solid by the shell method.

Solution This is the solid whose volume was found by the disk method in Example 4, Section 6.1. Now we find its volume by the shell method. First, sketch the region and draw a line segment across it *parallel* to the axis of revolution (Figure 6.21a). Label the segment's length (shell height) and distance from the axis of revolution (shell radius). (We drew the shell in Figure 6.21b, but you need not do that.)

In this case, the shell thickness variable is y, so the limits of integration for the shell formula method are $a = 0$ and $b = 2$ (along the y-axis in Figure 6.21). The volume of the solid is

$$\begin{aligned} V &= \int_a^b 2\pi \begin{pmatrix}\text{shell}\\ \text{radius}\end{pmatrix}\begin{pmatrix}\text{shell}\\ \text{height}\end{pmatrix} dy \\ &= \int_0^2 2\pi(y)(4 - y^2)\, dy \\ &= \int_0^2 2\pi(4y - y^3)\, dy = 2\pi\left[2y^2 - \frac{y^4}{4}\right]_0^2 = 8\pi. \end{aligned}$$

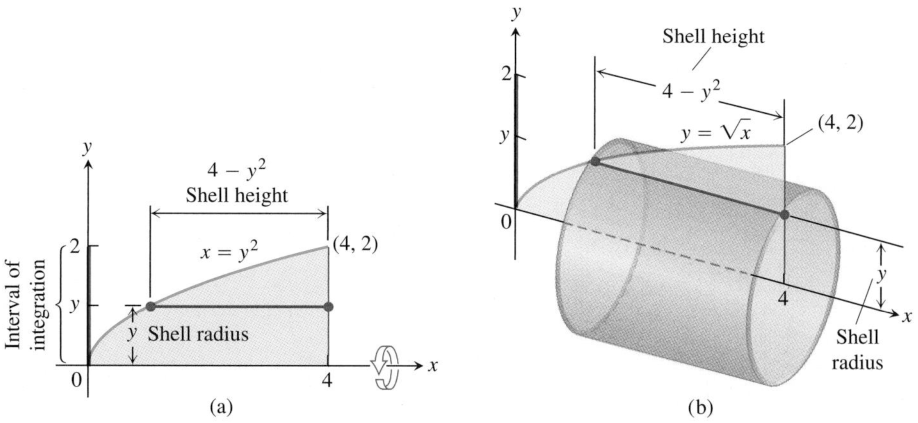

FIGURE 6.21 (a) The region, shell dimensions, and interval of integration in Example 3. (b) The shell swept out by the horizontal segment in part (a) with a width Δy.

■

Summary of the Shell Method

Regardless of the position of the axis of revolution (horizontal or vertical), the steps for implementing the shell method are these.

1. *Draw the region and sketch a line segment* across it *parallel* to the axis of revolution. *Label* the segment's height or length (shell height) and distance from the axis of revolution (shell radius).
2. *Find* the limits of integration for the thickness variable.
3. *Integrate* the product 2π (shell radius) (shell height) with respect to the thickness variable (x or y) to find the volume.

The shell method gives the same answer as the washer method when both are used to calculate the volume of a region. We do not prove that result here, but the proof is outlined in Exercise 37. Both volume formulas are actually special cases of a general volume formula we look at in studying double and triple integrals in Chapter 12. That general formula also allows for computing volumes of solids other than those swept out by regions of revolution.

EXERCISES 6.2

In Exercises 1–6, use the shell method to find the volumes of the solids generated by revolving the shaded region about the indicated axis.

1.

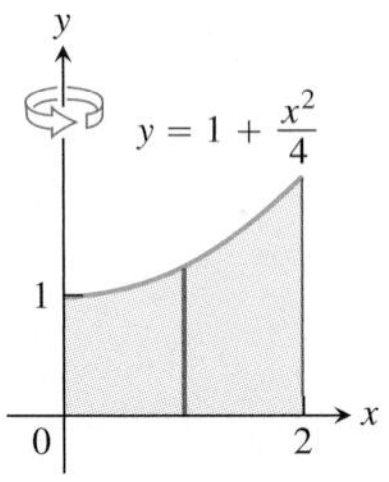

2.

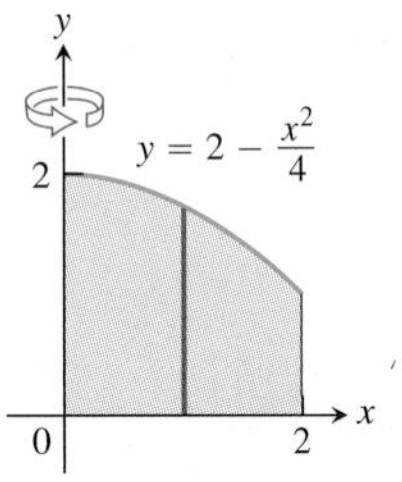

3.

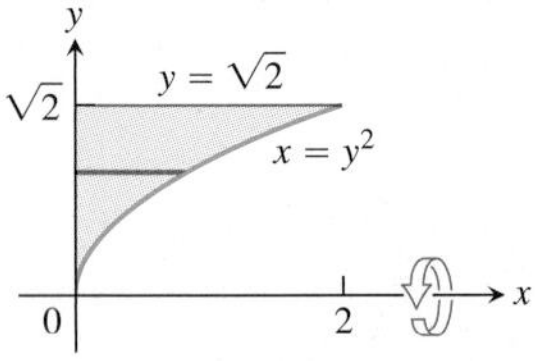

4.

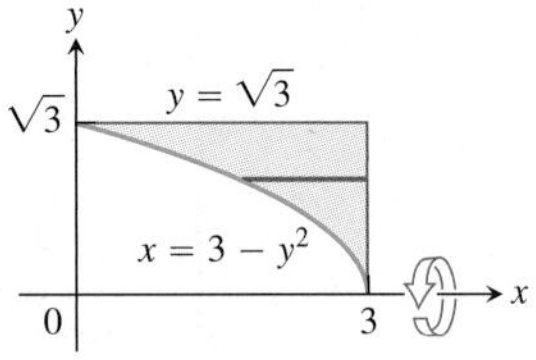

5. The y-axis

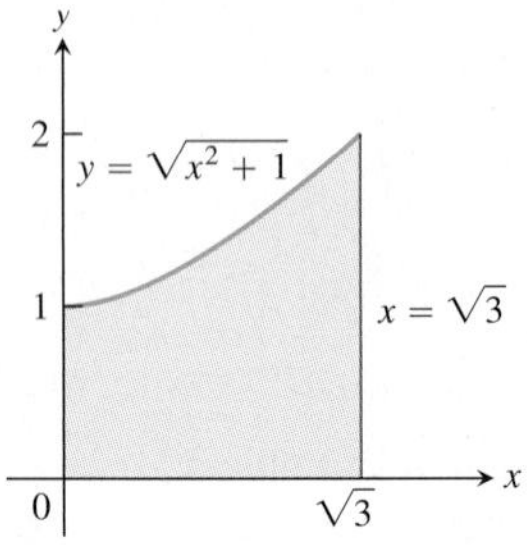

6. The y-axis

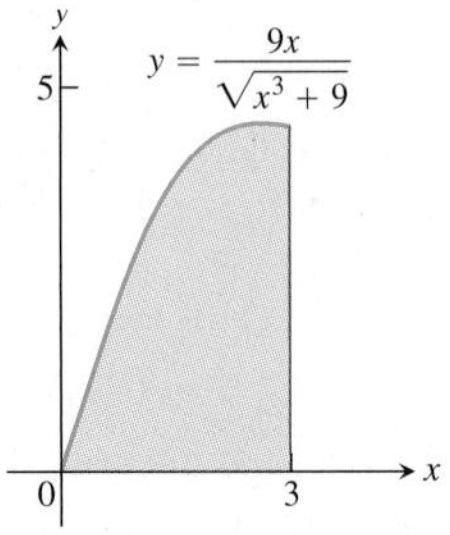

Use the shell method to find the volumes of the solids generated by revolving the regions bounded by the curves and lines in Exercises 7–14 about the y-axis.

7. $y = x, \quad y = -x/2, \quad x = 2$

8. $y = 2x, \quad y = x/2, \quad x = 1$

9. $y = x^2, \quad y = 2 - x, \quad x = 0, \quad \text{for } x \geq 0$

10. $y = 2 - x^2, \quad y = x^2, \quad x = 0$

11. $y = 2x - 1, \quad y = \sqrt{x}, \quad x = 0$

12. $y = 3/(2\sqrt{x}), \quad y = 0, \quad x = 1, \quad x = 4$

13. Let $f(x) = \begin{cases} (\sin x)/x, & 0 < x \leq \pi \\ 1, & x = 0 \end{cases}$

a. Show that $x f(x) = \sin x, 0 \leq x \leq \pi$.

b. Find the volume of the solid generated by revolving the accompanying shaded region about the y-axis.

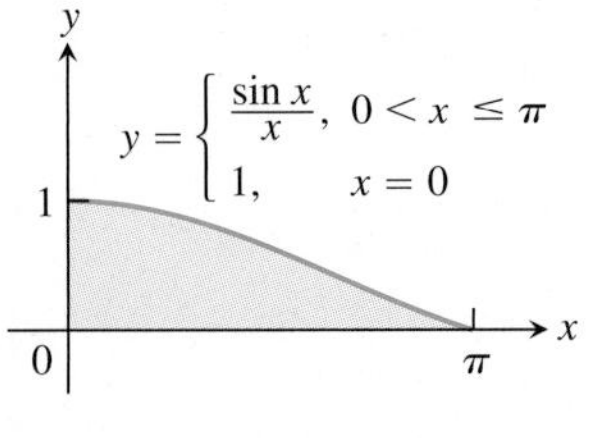

14. Let $g(x) = \begin{cases} (\tan x)^2/x, & 0 < x \leq \pi/4 \\ 0, & x = 0 \end{cases}$

a. Show that $x g(x) = (\tan x)^2, 0 \leq x \leq \pi/4$.

b. Find the volume of the solid generated by revolving the accompanying shaded region about the y-axis.

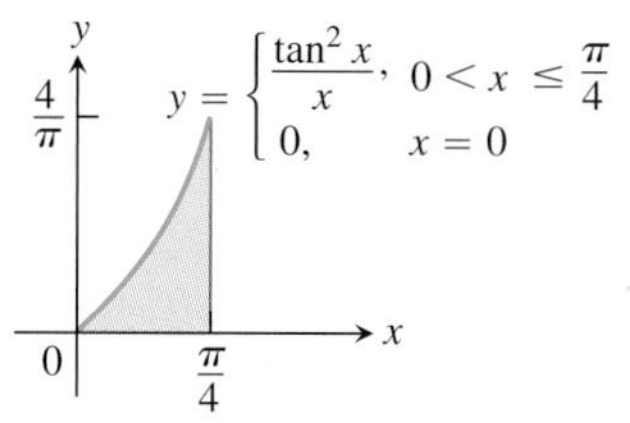

Use the shell method to find the volumes of the solids generated by revolving the regions bounded by the curves and lines in Exercises 15–22 about the x-axis.

15. $x = \sqrt{y}, \quad x = -y, \quad y = 2$

16. $x = y^2, \quad x = -y, \quad y = 2, \quad y \ge 0$

17. $x = 2y - y^2, \quad x = 0$

18. $x = 2y - y^2, \quad x = y$

19. $y = |x|, \quad y = 1$

20. $y = x, \quad y = 2x, \quad y = 2$

21. $y = \sqrt{x}, \quad y = 0, \quad y = x - 2$

22. $y = \sqrt{x}, \quad y = 0, \quad y = 2 - x$

In Exercises 23 and 24, use the shell method to find the volumes of the solids generated by revolving the shaded regions about the indicated axes.

23. **a.** The x-axis **b.** The line $y = 1$

c. The line $y = 8/5$ **d.** The line $y = -2/5$

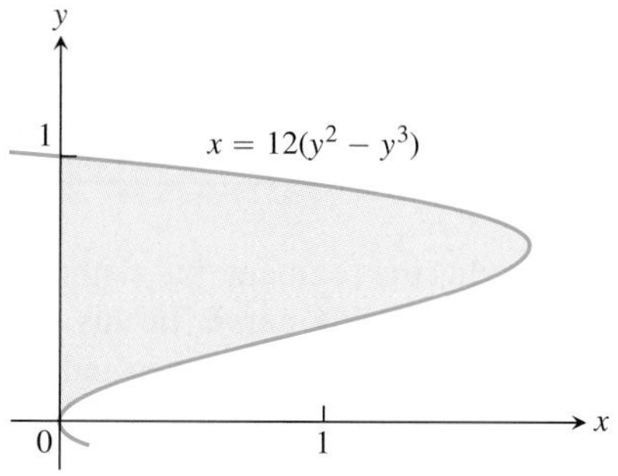

24. **a.** The x-axis **b.** The line $y = 2$

c. The line $y = 5$ **d.** The line $y = -5/8$

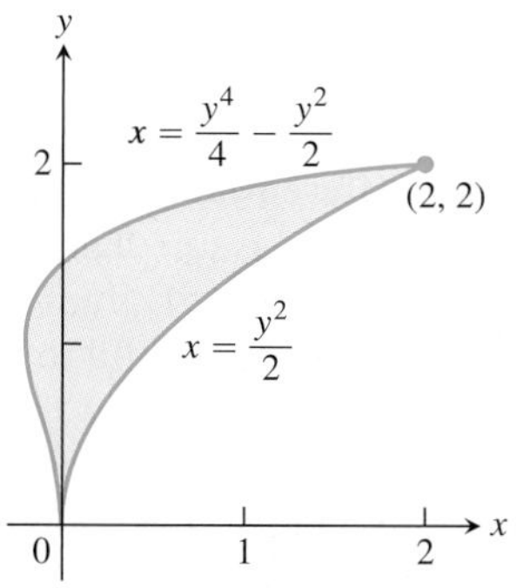

For some regions, both the washer and shell methods work well for the solid generated by revolving the region about the coordinate axes, but this is not always the case. When a region is revolved about the y-axis, for example, and washers are used, we must integrate with respect to y. It may not be possible, however, to express the integrand in terms of y. In such a case, the shell method allows us to integrate with respect to x instead. Exercises 25 and 26 provide some insight.

25. Compute the volume of the solid generated by revolving the region bounded by $y = x$ and $y = x^2$ about each coordinate axis using

a. the shell method. **b.** the washer method.

26. Compute the volume of the solid generated by revolving the triangular region bounded by the lines $2y = x + 4$, $y = x$, and $x = 0$ about

a. the x-axis using the washer method.

b. the y-axis using the shell method.

c. the line $x = 4$ using the shell method.

d. the line $y = 8$ using the washer method.

In Exercises 27–34, find the volumes of the solids generated by revolving the regions about the given axes. If you think it would be better to use washers in any given instance, feel free to do so.

27. The triangle with vertices (1, 1), (1, 2), and (2, 2) about

a. the x-axis **b.** the y-axis

c. the line $x = 10/3$ **d.** the line $y = 1$

28. The region bounded by $y = \sqrt{x}$, $y = 2$, $x = 0$ about

a. the x-axis **b.** the y-axis

c. the line $x = 4$ **d.** the line $y = 2$

29. The region in the first quadrant bounded by the curve $x = y - y^3$ and the y-axis about

a. the x-axis **b.** the line $y = 1$

30. The region in the first quadrant bounded by $x = y - y^3$, $x = 1$, and $y = 1$ about

a. the x-axis **b.** the y-axis

c. the line $x = 1$ **d.** the line $y = 1$

31. The region bounded by $y = \sqrt{x}$ and $y = x^2/8$ about

a. the x-axis **b.** the y-axis

32. The region bounded by $y = 2x - x^2$ and $y = x$ about

a. the y-axis **b.** the line $x = 1$

33. The region in the first quadrant bounded by the coordinate axes and the curve $y = \cos x$, $0 \le x \le \pi/2$, about

a. the y-axis **b.** the line $x = \pi/2$

34. The region in the first quadrant bounded by the coordinate axes, the curve $y = e^{-x}$, and the line $x = 1$ about

a. the y-axis **b.** the line $x = 1$

35. The region shown here is to be revolved about the x-axis to generate a solid. Which of the methods (disk, washer, shell) could you use to find the volume of the solid? How many integrals would be required in each case? Explain.

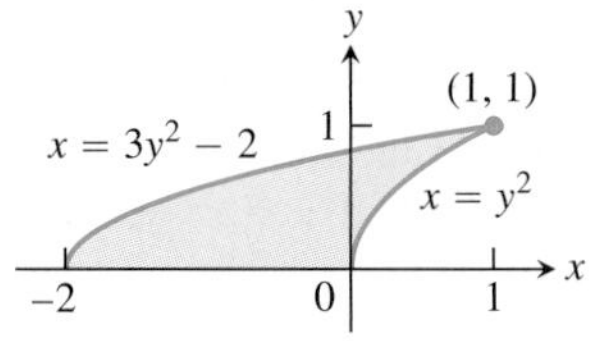

36. Find the volume of the solid generated by revolving the region between the curve $y = e^{-x}$ and the x-axis, about the y-axis.

37. Equivalence of the washer and shell methods for finding volume Let f be differentiable and increasing on the interval $a \le x \le b$, with $a > 0$, and suppose that f has a differentiable inverse, f^{-1}. Revolve about the y-axis the region bounded by the graph of f and the lines $x = a$ and $y = f(b)$ to generate a solid. Then the values of the integrals given by the washer and shell methods for the volume have identical values:

$$\int_{f(a)}^{f(b)} \pi((f^{-1}(y))^2 - a^2)\,dy = \int_a^b 2\pi x(f(b) - f(x))\,dx.$$

To prove this equality, define

$$W(t) = \int_{f(a)}^{f(t)} \pi((f^{-1}(y))^2 - a^2)\,dy$$

$$S(t) = \int_a^t 2\pi x(f(t) - f(x))\,dx.$$

Then show that the functions W and S agree at a point of $[a, b]$ and have identical derivatives on $[a, b]$. As you saw in Section 4.1, Exercise 108, this will guarantee $W(t) = S(t)$ for all t in $[a, b]$. In particular, $W(b) = S(b)$. (*Source:* "Disks and Shells Revisited," by Walter Carlip, *American Mathematical Monthly*, Vol. 98, No. 2, Feb. 1991, pp. 154–156.)

38. The region between the curve $y = \sec^{-1} x$ and the x-axis from $x = 1$ to $x = 2$ is revolved about the y-axis to generate a solid. Find the volume of the solid.

39. Find the volume of the solid generated by revolving the region enclosed by the graphs of $y = e^{-x^2}$, $y = 0$, $x = 0$, and $x = 1$ about the y-axis.

40. Find the volume of the solid generated by revolving the region enclosed by the graphs of $y = e^{x/2}$, $y = 1$, and $x = \ln 3$ about the x-axis.

6.3 Lengths of Plane Curves

We know what is meant by the length of a straight line segment, but without calculus, we have no precise notion of the length of a general winding curve. In this section we first subdivide the curve running from point A to point B into many pieces and join successive points of division by straight line segments. We then sum the lengths of all these line segments, go to the limit as their number increases, and take the limiting value to be the length of the curve.

Length of a Parametrically Defined Curve

Let C be a curve given parametrically (Section 3.5) by the equations

$$x = f(t) \quad \text{and} \quad y = g(t), \qquad a \le t \le b.$$

We assume the functions f and g are **continuously differentiable** (meaning they have continuous first derivatives) on the interval $[a, b]$. We also assume that the derivatives $f'(t)$ and $g'(t)$ are not simultaneously zero, which prevents the curve C from having any corners or cusps. Such a curve is called a **smooth curve**. It may be helpful to imagine the curve as the path of a particle moving from point $A = (f(a), g(a))$ at time $t = a$ to point $B = (f(b), g(b))$ at time $t = b$ in Figure 6.22. We subdivide the path (or arc) AB into n pieces at points $A = P_0, P_1, P_2, \dots, P_n = B$. These points correspond to a partition of the interval $[a, b]$ by $a = t_0 < t_1 < t_2 < \cdots < t_n = b$, where $P_k = (f(t_k), g(t_k))$. Join successive points of this subdivision by straight line segments (Figure 6.22). A representative line segment has length

$$\begin{aligned} L_k &= \sqrt{(\Delta x_k)^2 + (\Delta y_k)^2} \\ &= \sqrt{[f(t_k) - f(t_{k-1})]^2 + [g(t_k) - g(t_{k-1})]^2} \end{aligned}$$

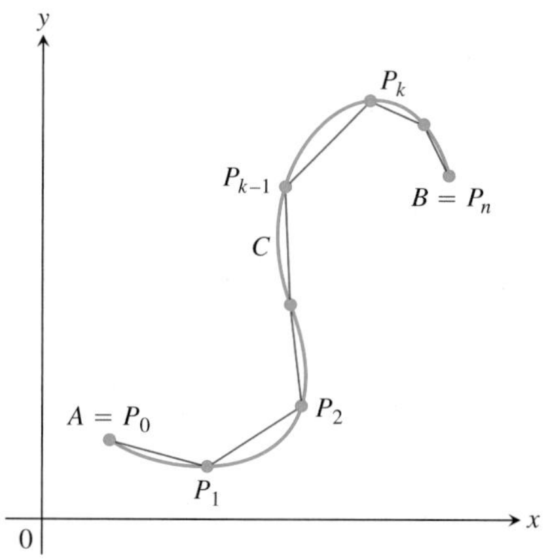

FIGURE 6.22 The curve C defined parametrically by the equations $x = f(t)$ and $y = g(t)$, $a \le t \le b$. The length of the curve from A to B is approximated by the sum of the lengths of the polygonal path (straight line segments) starting at $A = P_0$, then to P_1, and so on, ending at $B = P_n$.

(see Figure 6.23). If Δt_k is small, the length L_k is approximately the length of arc $P_{k-1}P_k$. By the Mean Value Theorem there are numbers t_k^* and t_k^{**} in $[t_{k-1}, t_k]$ such that

$$\begin{aligned} \Delta x_k &= f(t_k) - f(t_{k-1}) = f'(t_k^*)\,\Delta t_k, \\ \Delta y_k &= g(t_k) - g(t_{k-1}) = g'(t_k^{**})\,\Delta t_k. \end{aligned}$$

Assuming the path from A to B is traversed exactly once as t increases from $t = a$ to $t = b$, with no doubling back or retracing, an approximation to the "length" of the curve AB is the sum of all the lengths L_k:

$$\begin{aligned} \sum_{k=1}^{n} L_k &= \sum_{k=1}^{n} \sqrt{(\Delta x_k)^2 + (\Delta y_k)^2} \\ &= \sum_{k=1}^{n} \sqrt{[f'(t_k^*)]^2 + [g'(t_k^{**})]^2}\,\Delta t_k. \end{aligned}$$

Although this last sum on the right is not exactly a Riemann sum (because f' and g' are evaluated at different points), it can be shown that its limit, as the norm of the partition tends to zero, is the definite integral

$$\lim_{\|P\| \to 0} \sum_{k=1}^{n} \sqrt{[f'(t_k^*)]^2 + [g'(t_k^{**})]^2}\,\Delta t_k = \int_a^b \sqrt{[f'(t)]^2 + [g'(t)]^2}\,dt.$$

Therefore, it is reasonable to define the length of the curve from A to B as this integral.

DEFINITION If a curve C is defined parametrically by $x = f(t)$ and $y = g(t)$, $a \le t \le b$, where f' and g' are continuous and not simultaneously zero on $[a, b]$, and C is traversed exactly once as t increases from $t = a$ to $t = b$, then **the length of C** is the definite integral

$$L = \int_a^b \sqrt{[f'(t)]^2 + [g'(t)]^2}\,dt.$$

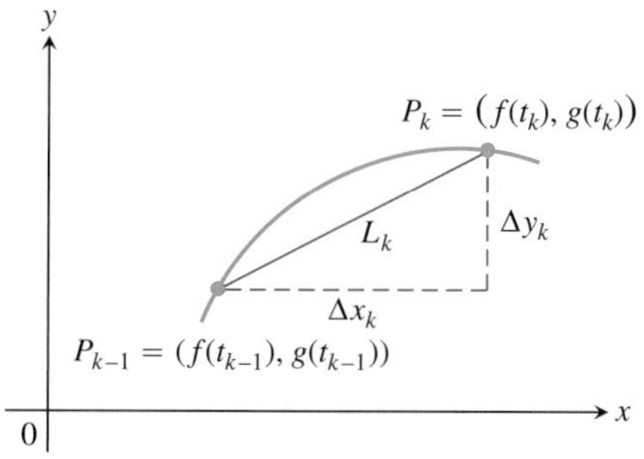

FIGURE 6.23 The arc $P_{k-1}P_k$ is approximated by the straight line segment shown here, which has length $L_k = \sqrt{(\Delta x_k)^2 + (\Delta y_k)^2}$.

A smooth curve C does not double back or reverse the direction of motion over the time interval $[a, b]$ since $(f')^2 + (g')^2 > 0$ throughout the interval.

If $x = f(t)$ and $y = g(t)$, then using the Leibniz notation we have the following result for arc length:

$$L = \int_a^b \sqrt{\left(\frac{dx}{dt}\right)^2 + \left(\frac{dy}{dt}\right)^2}\,dt. \qquad (1)$$

What if there are two different parametrizations for a curve C whose length we want to find; does it matter which one we use? The answer is no, as long as the parametrization we choose meets the conditions stated in the definition of the length of C (see Exercise 36 for an example).

EXAMPLE 1 We use the definition of length to find the length of the circle of radius r defined parametrically by

$$x = r\cos t \quad \text{and} \quad y = r\sin t, \quad 0 \le t \le 2\pi.$$

Solution As t varies from 0 to 2π, the circle is traversed exactly once, so the circumference is

$$L = \int_0^{2\pi} \sqrt{\left(\frac{dx}{dt}\right)^2 + \left(\frac{dy}{dt}\right)^2}\, dt.$$

We find

$$\frac{dx}{dt} = -r\sin t, \quad \frac{dy}{dt} = r\cos t$$

and

$$\left(\frac{dx}{dt}\right)^2 + \left(\frac{dy}{dt}\right)^2 = r^2(\sin^2 t + \cos^2 t) = r^2.$$

So

$$L = \int_0^{2\pi} \sqrt{r^2}\, dt = r\Big[t\Big]_0^{2\pi} = 2\pi r.$$

■

EXAMPLE 2 Find the length of the astroid (Figure 6.24)

$$x = \cos^3 t, \quad y = \sin^3 t, \quad 0 \le t \le 2\pi.$$

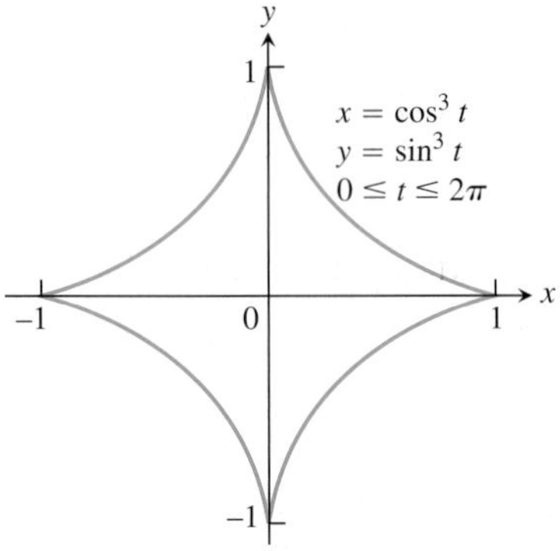

FIGURE 6.24 The astroid in Example 2.

Solution Because of the curve's symmetry with respect to the coordinate axes, its length is four times the length of the first-quadrant portion. We have

$$x = \cos^3 t, \quad y = \sin^3 t$$

$$\left(\frac{dx}{dt}\right)^2 = [3\cos^2 t(-\sin t)]^2 = 9\cos^4 t\sin^2 t$$

$$\left(\frac{dy}{dt}\right)^2 = [3\sin^2 t(\cos t)]^2 = 9\sin^4 t\cos^2 t$$

$$\begin{aligned}\sqrt{\left(\frac{dx}{dt}\right)^2 + \left(\frac{dy}{dt}\right)^2} &= \sqrt{9\cos^2 t\sin^2 t\underbrace{(\cos^2 t + \sin^2 t)}_{1}} \\ &= \sqrt{9\cos^2 t\sin^2 t} \\ &= 3|\cos t\sin t| \\ &= 3\cos t\sin t. \qquad \begin{array}{l}\cos t\sin t \ge 0 \text{ for} \\ 0 \le t \le \pi/2\end{array}\end{aligned}$$

Therefore,

$$\begin{aligned}\text{Length of first-quadrant portion} &= \int_0^{\pi/2} 3\cos t\sin t\, dt \\ &= \frac{3}{2}\int_0^{\pi/2} \sin 2t\, dt \qquad \begin{array}{l}\cos t\sin t = \\ (1/2)\sin 2t\end{array} \\ &= -\frac{3}{4}\cos 2t\Big]_0^{\pi/2} = \frac{3}{2}.\end{aligned}$$

The length of the astroid is four times this: $4(3/2) = 6$. ■

HISTORICAL BIOGRAPHY

Gregory St. Vincent
(1584–1667)

Length of a Curve $y = f(x)$

Given a continuously differentiable function $y = f(x)$, $a \le x \le b$, we can assign $x = t$ as a parameter. The graph of the function f is then the curve C defined parametrically by

$$x = t \qquad \text{and} \qquad y = f(t), \qquad a \le t \le b,$$

a special case of what we considered before. Then,

$$\frac{dx}{dt} = 1 \qquad \text{and} \qquad \frac{dy}{dt} = f'(t).$$

From our calculations in Section 3.5, we have

$$\frac{dy}{dx} = \frac{dy/dt}{dx/dt} = f'(t)$$

giving

$$\begin{aligned} \left(\frac{dx}{dt}\right)^2 + \left(\frac{dy}{dt}\right)^2 &= 1 + [f'(t)]^2 \\ &= 1 + \left(\frac{dy}{dx}\right)^2 \\ &= 1 + [f'(x)]^2. \end{aligned}$$

Substitution into Equation (1) gives the arc length formula for the graph of $y = f(x)$.

Formula for the Length of $y = f(x)$, $\quad a \le x \le b$
If f is continuously differentiable on the closed interval $[a, b]$, the length of the curve (graph) $y = f(x)$ from $x = a$ to $x = b$ is

$$L = \int_a^b \sqrt{1 + \left(\frac{dy}{dx}\right)^2}\, dx = \int_a^b \sqrt{1 + [f'(x)]^2}\, dx. \qquad (2)$$

EXAMPLE 3 Find the length of the curve

$$y = \ln(\cos x), \qquad 0 \le x \le \pi/3.$$

Solution We use Equation (2) with $a = 0$, $b = \pi/3$, and

$$\begin{aligned} y &= \ln(\cos x) \\ \frac{dy}{dx} &= -\frac{\sin x}{\cos x} = -\tan x \\ \left(\frac{dy}{dx}\right)^2 &= \tan^2 x. \end{aligned}$$

The length of the curve from $x = 0$ to $x = \pi/3$ is

$$L = \int_0^{\pi/3} \sqrt{1 + \left(\frac{dy}{dx}\right)^2}\, dx = \int_0^{\pi/3} \sqrt{1 + \tan^2 x}\, dx \qquad \text{Eq. (2) with } a = 0,\ b = \pi/3$$

$$= \int_0^{\pi/3} |\sec x|\, dx = \left[\ln|\sec x + \tan x|\right]_0^{\pi/3}$$

$$= \ln\left(2 + \sqrt{3}\right) - \ln(1 + 0) = \ln\left(2 + \sqrt{3}\right) \approx 1.32.$$

■

Dealing with Discontinuities in *dy/dx*

At a point on a curve where dy/dx fails to exist, dx/dy may exist and we may be able to find the curve's length by expressing x as a function of y and applying the following analogue of Equation (2):

Formula for the Length of $x = g(y)$, $c \le y \le d$

If g is continuously differentiable on $[c, d]$, the length of the curve $x = g(y)$ from $y = c$ to $y = d$ is

$$L = \int_c^d \sqrt{1 + \left(\frac{dx}{dy}\right)^2}\, dy = \int_c^d \sqrt{1 + [g'(y)]^2}\, dy. \qquad (3)$$

EXAMPLE 4 Find the length of the curve $y = (x/2)^{2/3}$ from $x = 0$ to $x = 2$.

Solution The derivative

$$\frac{dy}{dx} = \frac{2}{3}\left(\frac{x}{2}\right)^{-1/3}\left(\frac{1}{2}\right) = \frac{1}{3}\left(\frac{2}{x}\right)^{1/3}$$

is not defined at $x = 0$, so we cannot find the curve's length with Equation (2).

We therefore rewrite the equation to express x in terms of y:

$$y = \left(\frac{x}{2}\right)^{2/3}$$

$$y^{3/2} = \frac{x}{2} \qquad \text{Raise both sides to the power 3/2.}$$

$$x = 2y^{3/2}. \qquad \text{Solve for } x.$$

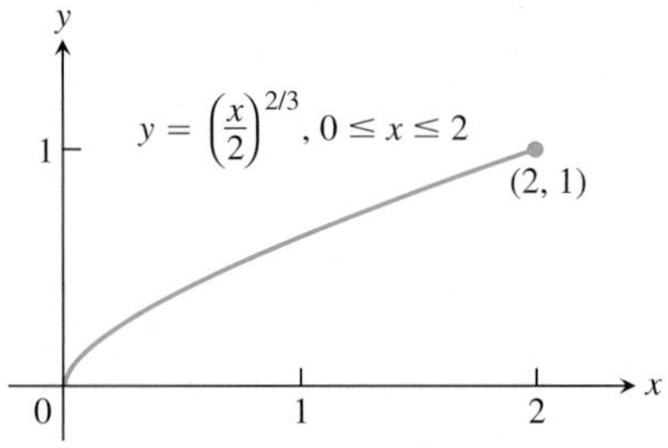

FIGURE 6.25 The graph of $y = (x/2)^{2/3}$ from $x = 0$ to $x = 2$ is also the graph of $x = 2y^{3/2}$ from $y = 0$ to $y = 1$ (Example 4).

From this we see that the curve whose length we want is also the graph of $x = 2y^{3/2}$ from $y = 0$ to $y = 1$ (Figure 6.25).

The derivative

$$\frac{dx}{dy} = 2\left(\frac{3}{2}\right)y^{1/2} = 3y^{1/2}$$

is continuous on [0, 1]. We may therefore use Equation (3) to find the curve's length:

$$L = \int_c^d \sqrt{1 + \left(\frac{dx}{dy}\right)^2}\, dy = \int_0^1 \sqrt{1 + 9y}\, dy$$

$$= \frac{1}{9} \cdot \frac{2}{3}(1 + 9y)^{3/2}\Big]_0^1$$

$$= \frac{2}{27}\left(10\sqrt{10} - 1\right) \approx 2.27.$$

Eq. (3) with $c = 0, d = 1$. Let $u = 1 + 9y$, $du/9 = dy$, integrate, and substitute back.

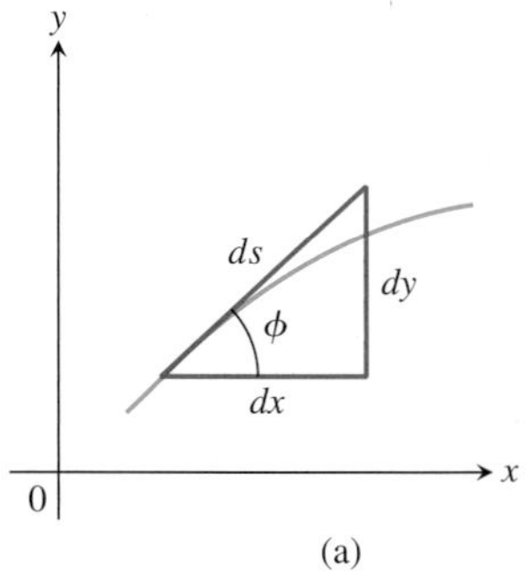

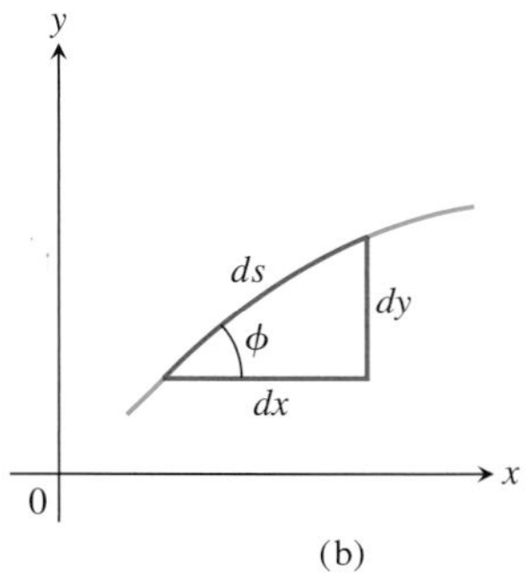

FIGURE 6.26 Diagrams for remembering the equation $ds = \sqrt{dx^2 + dy^2}$.

The Short Differential Formula

Equation (1) is frequently written in terms of differentials in place of derivatives. This is done formally by writing $(dt)^2$ under the radical in place of the dt outside the radical, and then writing

$$\left(\frac{dx}{dt}\right)^2 (dt)^2 = \left(\frac{dx}{dt}\, dt\right)^2 = (dx)^2$$

and

$$\left(\frac{dy}{dt}\right)^2 (dt)^2 = \left(\frac{dy}{dt}\, dt\right)^2 = (dy)^2.$$

It is also customary to eliminate the parentheses in $(dx)^2$ and write dx^2 instead, so that Equation (1) is written

$$L = \int \sqrt{dx^2 + dy^2}. \tag{4}$$

To compute an integral, dx and dy must both be expressed in terms of one and the same variable, and appropriate limits must be supplied in Equation (4).

A useful way to remember Equation (4) is to write

$$ds = \sqrt{dx^2 + dy^2} \tag{5}$$

and treat ds as the differential of arc length, which can be integrated between appropriate limits to give the total length of a curve. From this point of view, all the arc length formulas are simply different expressions for the equation $L = \int ds$. Figure 6.26a gives the exact interpretation of ds corresponding to Equation (5). Figure 6.26b is not strictly accurate, but is to be thought of as a simplified approximation of Figure 6.26a.

EXERCISES 6.3

Find the lengths of the curves in Exercises 1–8.

1. $x = 1 - t, \quad y = 2 + 3t, \quad -2/3 \le t \le 1$
2. $x = \cos t, \quad y = t + \sin t, \quad 0 \le t \le \pi$
3. $x = t^3, \quad y = 3t^2/2, \quad 0 \le t \le \sqrt{3}$
4. $x = t^2/2, \quad y = (2t + 1)^{3/2}/3, \quad 0 \le t \le 4$
5. $x = (2t + 3)^{3/2}/3, \quad y = t + t^2/2, \quad 0 \le t \le 3$
6. $x = 8\cos t + 8t\sin t, \quad y = 8\sin t - 8t\cos t, \quad 0 \le t \le \pi/2$
7. $x = e^t - t, \quad y = 4e^{t/2}, \quad 0 \le t \le 3$
8. $x = e^t \cos t, \quad y = e^t \sin t, \quad 0 \le t \le \pi$

Find the lengths of the curves in Exercises 9–18. If you have a grapher, you may want to graph these curves to see what they look like.

9. $y = (1/3)(x^2 + 2)^{3/2}$ from $x = 0$ to $x = 3$

10. $y = x^{3/2}$ from $x = 0$ to $x = 4$

11. $x = (y^3/3) + 1/(4y)$ from $y = 1$ to $y = 3$
(*Hint:* $1 + (dx/dy)^2$ is a perfect square.)

12. $x = (y^{3/2}/3) - y^{1/2}$ from $y = 1$ to $y = 9$
(*Hint:* $1 + (dx/dy)^2$ is a perfect square.)

13. $x = (y^4/4) + 1/(8y^2)$ from $y = 1$ to $y = 2$
(*Hint:* $1 + (dx/dy)^2$ is a perfect square.)

14. $x = (y^3/6) + 1/(2y)$ from $y = 2$ to $y = 3$
(*Hint:* $1 + (dx/dy)^2$ is a perfect square.)

15. $y = (3/4)x^{4/3} - (3/8)x^{2/3} + 5, \quad 1 \le x \le 8$

16. $y = (x^3/3) + x^2 + x + 1/(4x + 4), \quad 0 \le x \le 2$

17. $y = \sqrt{1 - x^2}, \quad -1/2 \le x \le 1/2$

18. $x = \int_0^y \sqrt{\sec^4 t - 1}\, dt, \quad -\pi/4 \le y \le \pi/4$

T In Exercises 19–26, do the following.

a. Set up an integral for the length of the curve.

b. Graph the curve to see what it looks like.

c. Use your grapher's or computer's integral evaluator to find the curve's length numerically.

19. $y = x^2, \quad -1 \le x \le 2$

20. $y = \tan x, \quad -\pi/3 \le x \le 0$

21. $x = \sin y, \quad 0 \le y \le \pi$

22. $x = \sqrt{1 - y^2}, \quad -1/2 \le y \le 1/2$

23. $y^2 + 2y = 2x + 1$ from $(-1, -1)$ to $(7, 3)$

24. $y = \sin x - x \cos x, \quad 0 \le x \le \pi$

25. $y = \int_0^x \tan t\, dt, \quad 0 \le x \le \pi/6$

26. $x = \int_0^y \sqrt{\sec^2 t - 1}\, dt, \quad -\pi/3 \le y \le \pi/4$

27. Find the length of the curve

$$y = \ln(\sec x), \quad 0 \le x \le \pi/4.$$

28. Using tangent fins to derive the length formula for curves Assume that f is smooth on $[a, b]$ and partition the interval $[a, b]$ in the usual way. In each subinterval $[x_{k-1}, x_k]$, construct the *tangent fin* at the point $(x_{k-1}, f(x_{k-1}))$, as shown in the accompanying figure.

a. Show that the length of the kth tangent fin over the interval $[x_{k-1}, x_k]$ equals $\sqrt{(\Delta x_k)^2 + (f'(x_{k-1})\,\Delta x_k)^2}$.

b. Show that

$$\lim_{n\to\infty} \sum_{k=1}^{n} (\text{length of } k\text{th tangent fin}) = \int_a^b \sqrt{1 + (f'(x))^2}\, dx,$$

which is the length L of the curve $y = f(x)$ from a to b.

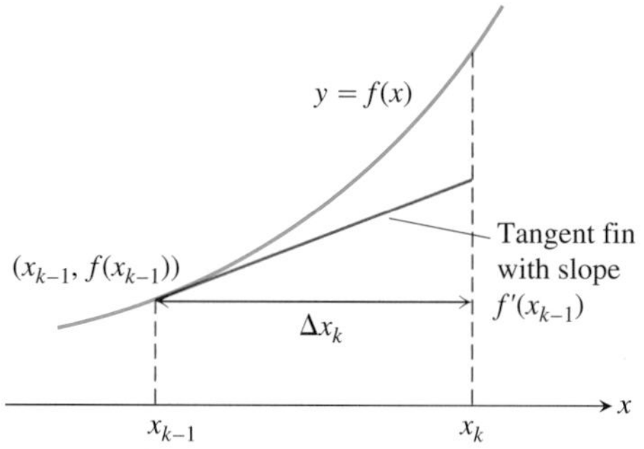

29. a. Find a curve through the point $(1, 1)$ whose length integral is

$$L = \int_1^4 \sqrt{1 + \frac{1}{4x}}\, dx.$$

b. How many such curves are there? Give reasons for your answer.

30. a. Find a curve through the point $(0, 1)$ whose length integral is

$$L = \int_1^2 \sqrt{1 + \frac{1}{y^4}}\, dy.$$

b. How many such curves are there? Give reasons for your answer.

31. Find a curve through the origin in the xy-plane whose length from $x = 0$ to $x = 1$ is

$$L = \int_0^1 \sqrt{1 + \frac{1}{4}e^x}\, dx.$$

32. Find a curve through the point $(1, 0)$ whose length from $x = 1$ to $x = 2$ is

$$L = \int_1^2 \sqrt{1 + \frac{1}{x^2}}\, dx.$$

33. Find the length of the curve

$$x = \ln(\sec t + \tan t) - \sin t$$
$$y = \cos t, \quad 0 \le t \le \pi/3$$

34. Find the length of one arch of the cycloid $x = a(\theta - \sin\theta)$, $y = a(1 - \cos\theta)$, $0 \le \theta \le 2\pi$, shown in the accompanying figure. A **cycloid** is the curve traced out by a point P on the circumference of a circle rolling along a straight line, such as the x-axis.

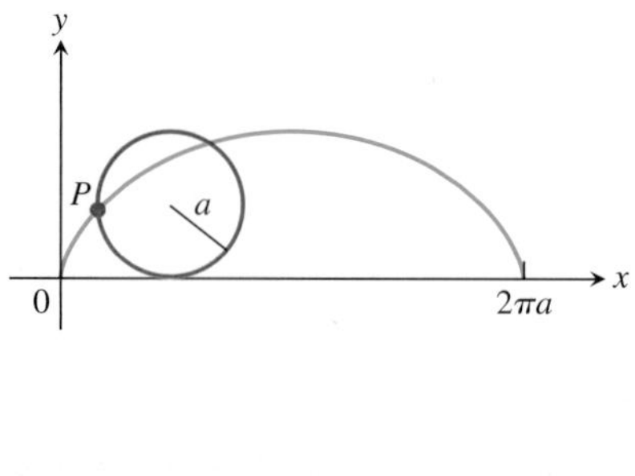

35. Find the length of the curve

$$x = e^t + e^{-t}, \quad y = 3 - 2t, \quad 0 \le t \le 3.$$

36. Length is independent of parametrization To illustrate the fact that the numbers we get for length do not depend on the way we parametrize our curves (except for the mild restrictions preventing doubling back mentioned earlier), calculate the length of the semicircle $y = \sqrt{1 - x^2}$ with these two different parametrizations:

a. $x = \cos 2t, \quad y = \sin 2t, \quad 0 \le t \le \pi/2$

b. $x = \sin \pi t, \quad y = \cos \pi t, \quad -1/2 \le t \le 1/2$

COMPUTER EXPLORATIONS

In Exercises 37–46, use a CAS to perform the following steps for the given curve over the closed interval.

a. Plot the curve together with the polygonal path approximations for $n = 2, 4, 8$ partition points over the interval. (See Figure 6.22.)

b. Find the corresponding approximation to the length of the curve by summing the lengths of the line segments.

c. Evaluate the length of the curve using an integral. Compare your approximations for $n = 2, 4, 8$ with the actual length given by the integral. How does the actual length compare with the approximations as n increases? Explain your answer.

37. $f(x) = \sqrt{1 - x^2}, \quad -1 \le x \le 1$

38. $f(x) = x^{1/3} + x^{2/3}, \quad 0 \le x \le 2$

39. $f(x) = \sin(\pi x^2), \quad 0 \le x \le \sqrt{2}$

40. $f(x) = x^2 \cos x, \quad 0 \le x \le \pi$

41. $f(x) = \dfrac{x - 1}{4x^2 + 1}, \quad -\dfrac{1}{2} \le x \le 1$

42. $f(x) = x^3 - x^2, \quad -1 \le x \le 1$

43. $x = \dfrac{1}{3}t^3, \quad y = \dfrac{1}{2}t^2, \quad 0 \le t \le 1$

44. $x = 2t^3 - 16t^2 + 25t + 5, \quad y = t^2 + t - 3, \quad 0 \le t \le 6$

45. $x = t - \cos t, \quad y = 1 + \sin t, \quad -\pi \le t \le \pi$

46. $x = \ln t, \quad y = \sqrt{t + 2}, \quad 1 \le t \le 4$

6.4 Exponential Change and Separable Differential Equations

In this section we derive the law of exponential change and describe some of the applications that account for the importance of exponential and logarithmic functions.

Exponential Change

In modeling many real-world situations, a quantity y increases or decreases at a rate proportional to its size at a given time t. Examples of such quantities include the amount of a decaying radioactive material, the size of a population, and the temperature difference between a hot object and its surrounding medium. Such quantities are said to undergo *exponential change.*

If the amount present at time $t = 0$ is called y_0, then we can find y as a function of t by solving the following initial value problem:

$$\text{Differential equation:} \quad \frac{dy}{dt} = ky$$

$$\text{Initial condition:} \quad y = y_0 \quad \text{when} \quad t = 0. \tag{1}$$

If y is positive and increasing, then k is positive, and we use Equation (1) to say that the rate of growth is proportional to what has already been accumulated. If y is positive and decreasing, then k is negative, and we use Equation (1) to say that the rate of decay is proportional to the amount still left.

We see right away that the constant function $y = 0$ is a solution of Equation (1) if $y_0 = 0$. To find the nonzero solutions, we divide Equation (1) by y:

$$\begin{aligned}
\frac{1}{y} \cdot \frac{dy}{dt} &= k && y \neq 0 \\
\int \frac{1}{y} \frac{dy}{dt}\, dt &= \int k\, dt && \text{Integrate with respect to } t; \\
\ln |y| &= kt + C && \textstyle\int (1/u)\, du = \ln |u| + C. \\
|y| &= e^{kt+C} && \text{Exponentiate.} \\
|y| &= e^{C} \cdot e^{kt} && e^{a+b} = e^{a} \cdot e^{b} \\
y &= \pm e^{C} e^{kt} && \text{If } |y| = r, \text{ then } y = \pm r. \\
y &= Ae^{kt}. && A \text{ is a shorter name for } \pm e^{C}.
\end{aligned}$$

By allowing A to take on the value 0 in addition to all possible values $\pm e^{C}$, we can include the solution $y = 0$ in the formula.

We find the value of A for the initial value problem by solving for A when $y = y_0$ and $t = 0$:

$$y_0 = Ae^{k \cdot 0} = A.$$

The solution of the initial value problem is therefore

$$y = y_0 e^{kt}. \tag{2}$$

Quantities changing in this way are said to undergo **exponential growth** if $k > 0$, and **exponential decay** if $k < 0$. The number k is called the **rate constant** of the change.

The derivation of Equation (2) shows also that the only functions that are their own derivatives are constant multiples of the exponential function.

Before presenting several examples of exponential change, let's consider the process we used to derive it.

Separable Differential Equations

Exponential change is modeled by a differential equation of the form $dy/dx = ky$, for some nonzero constant k. More generally, suppose we have a differential equation of the form

$$\frac{dy}{dx} = f(x, y), \tag{3}$$

where f is a function of *both* the independent and dependent variables. A **solution** of the equation is a differentiable function $y = y(x)$ defined on an interval of x-values (perhaps infinite) such that

$$\frac{d}{dx} y(x) = f(x, y(x))$$

on that interval. That is, when $y(x)$ and its derivative $y'(x)$ are substituted into the differential equation, the resulting equation is true for all x in the solution interval. The **general solution** is a solution $y(x)$ that contains all possible solutions and it always contains an arbitrary constant.

Equation (3) is **separable** if f can be expressed as a product of a function of x and a function of y. The differential equation then has the form

$$\frac{dy}{dx} = g(x)H(y). \qquad \text{g is a function of x, H is a function of y}$$

When we rewrite this equation in the form

$$\frac{dy}{dx} = \frac{g(x)}{h(y)}, \qquad H(y) = \frac{1}{h(y)}$$

its differential form allows us to collect all y terms with dy and all x terms with dx:

$$h(y)\,dy = g(x)\,dx.$$

Now we simply integrate both sides of this equation:

$$\int h(y)\,dy = \int g(x)\,dx. \tag{4}$$

After completing the integrations we obtain the solution y defined implicitly as a function of x.

The justification that we can simply integrate both sides in Equation (4) is based on the Substitution Rule (Section 4.6):

$$\begin{aligned}
\int h(y)\,dy &= \int h(y(x))\frac{dy}{dx}\,dx \\
&= \int h(y(x))\frac{g(x)}{h(y(x))}\,dx \qquad \frac{dy}{dx} = \frac{g(x)}{h(y)} \\
&= \int g(x)\,dx.
\end{aligned}$$

EXAMPLE 1 Solve the differential equation

$$\frac{dy}{dx} = (1 + y^2)e^x.$$

Solution Since $1 + y^2$ is never zero, we can solve the equation by separating the variables.

$$\begin{aligned}
\frac{dy}{dx} &= (1 + y^2)e^x \\
dy &= (1 + y^2)e^x\,dx && \text{Treat dy/dx as a quotient of differentials and multiply both sides by dx.} \\
\frac{dy}{1 + y^2} &= e^x\,dx && \text{Divide by $(1 + y^2)$.} \\
\int \frac{dy}{1 + y^2} &= \int e^x\,dx && \text{Integrate both sides.} \\
\tan^{-1} y &= e^x + C && \text{C represents the combined constants of integration.}
\end{aligned}$$

The equation $\tan^{-1} y = e^x + C$ gives y as an implicit function of x. When $-\pi/2 < e^x + C < \pi/2$, we can solve for y as an explicit function of x by taking the tangent of both sides:

$$\begin{aligned}
\tan(\tan^{-1} y) &= \tan(e^x + C) \\
y &= \tan(e^x + C).
\end{aligned}$$

■

EXAMPLE 2 Solve the equation

$$(x + 1)\frac{dy}{dx} = x(y^2 + 1).$$

Solution We change to differential form, separate the variables, and integrate:

$$\begin{aligned}(x + 1)\,dy &= x(y^2 + 1)\,dx \\ \frac{dy}{y^2 + 1} &= \frac{x\,dx}{x + 1} \qquad x \neq -1 \\ \int \frac{dy}{1 + y^2} &= \int \left(1 - \frac{1}{x + 1}\right) dx \\ \tan^{-1} y &= x - \ln|x + 1| + C.\end{aligned}$$

■

The initial value problem

$$\frac{dy}{dt} = ky, \qquad y(0) = y_0$$

involves a separable differential equation, and the solution $y = y_0 e^{kt}$ expresses exponential change. We now present several examples of such change.

Unlimited Population Growth

Strictly speaking, the number of individuals in a population (of people, plants, animals, or bacteria, for example) is a discontinuous function of time because it takes on discrete values. However, when the number of individuals becomes large enough, the population can be approximated by a continuous function. Differentiability of the approximating function is another reasonable hypothesis in many settings, allowing for the use of calculus to model and predict population sizes.

If we assume that the proportion of reproducing individuals remains constant and assume a constant fertility, then at any instant t the birth rate is proportional to the number $y(t)$ of individuals present. Let's assume, too, that the death rate of the population is stable and proportional to $y(t)$. If, further, we neglect departures and arrivals, the growth rate dy/dt is the birth rate minus the death rate, which is the difference of the two proportionalities under our assumptions. In other words, $dy/dt = ky$, so that $y = y_0 e^{kt}$, where y_0 is the size of the population at time $t = 0$. As with all kinds of growth, there may be limitations imposed by the surrounding environment, but we will not go into these here. The proportionality $dy/dt = ky$ models *unlimited population growth.*

In the following example we assume this population model to look at how the number of individuals infected by a disease within a given population decreases as the disease is appropriately treated.

EXAMPLE 3 One model for the way diseases die out when properly treated assumes that the rate dy/dt at which the number of infected people changes is proportional to the number y. The number of people cured is proportional to the number y that are infected with the disease. Suppose that in the course of any given year the number of cases of a disease is reduced by 20%. If there are 10,000 cases today, how many years will it take to reduce the number to 1000?

Solution We use the equation $y = y_0 e^{kt}$. There are three things to find: the value of y_0, the value of k, and the time t when $y = 1000$.

The value of y_0. We are free to count time beginning anywhere we want. If we count from today, then $y = 10{,}000$ when $t = 0$, so $y_0 = 10{,}000$. Our equation is now

$$y = 10{,}000 e^{kt}. \tag{5}$$

The value of k. When $t = 1$ year, the number of cases will be 80% of its present value, or 8000. Hence,

$$\begin{aligned} 8000 &= 10{,}000 e^{k(1)} && \text{Eq. (5) with } t = 1 \text{ and } y = 8000 \\ e^k &= 0.8 \\ \ln(e^k) &= \ln 0.8 && \text{Logs of both sides} \\ k &= \ln 0.8 < 0. \end{aligned}$$

At any given time t,

$$y = 10{,}000 e^{(\ln 0.8)t}. \tag{6}$$

The value of t that makes $y = 1000$. We set y equal to 1000 in Equation (6) and solve for t:

$$\begin{aligned} 1000 &= 10{,}000 e^{(\ln 0.8)t} \\ e^{(\ln 0.8)t} &= 0.1 \\ (\ln 0.8)t &= \ln 0.1 && \text{Logs of both sides} \\ t &= \frac{\ln 0.1}{\ln 0.8} \approx 10.32 \text{ years}. \end{aligned}$$

It will take a little more than 10 years to reduce the number of cases to 1000. ■

Radioactivity

Some atoms are unstable and can spontaneously emit mass or radiation. This process is called **radioactive decay**, and an element whose atoms go spontaneously through this process is called **radioactive**. Sometimes when an atom emits some of its mass through this process of radioactivity, the remainder of the atom re-forms to make an atom of some new element. For example, radioactive carbon-14 decays into nitrogen; radium, through a number of intermediate radioactive steps, decays into lead.

Experiments have shown that at any given time the rate at which a radioactive element decays (as measured by the number of nuclei that change per unit time) is approximately proportional to the number of radioactive nuclei present. Thus, the decay of a radioactive element is described by the equation $dy/dt = -ky$, $k > 0$. It is conventional to use $-k$, with $k > 0$, to emphasize that y is decreasing. If y_0 is the number of radioactive nuclei present at time zero, the number still present at any later time t will be

$$y = y_0 e^{-kt}, \qquad k > 0.$$

For radon-222 gas, t is measured in days and $k = 0.18$. For radium-226, which used to be painted on watch dials to make them glow at night (a dangerous practice), t is measured in years and $k = 4.3 \times 10^{-4}$.

The **half-life** of a radioactive element is the time required for half of the radioactive nuclei present in a sample to decay. It is an interesting fact that the half-life is a constant that does not depend on the number of radioactive nuclei initially present in the sample, but only on the radioactive substance.

To see why, let y_0 be the number of radioactive nuclei initially present in the sample. Then the number y present at any later time t will be $y = y_0 e^{-kt}$. We seek the value of t at which the number of radioactive nuclei present equals half the original number:

$$\begin{aligned} y_0 e^{-kt} &= \frac{1}{2} y_0 \\ e^{-kt} &= \frac{1}{2} \\ -kt &= \ln \frac{1}{2} = -\ln 2 && \text{Reciprocal Rule for logarithms} \\ t &= \frac{\ln 2}{k} \end{aligned}$$

This time value for t is the half-life of the element. It depends only on the value of k; the number y_0 does not enter in.

$$\text{Half-life} = \frac{\ln 2}{k} \qquad (7)$$

For example, the half-life for radon-222 is

$$\text{Half-life} = \frac{\ln 2}{0.18} \approx 3.9 \text{ days.}$$

EXAMPLE 4 The decay of radioactive elements can sometimes be used to date events from the Earth's past. In a living organism, the ratio of radioactive carbon, carbon-14, to ordinary carbon stays fairly constant during the lifetime of the organism, being approximately equal to the ratio in the organism's surroundings at the time. After the organism's death, however, no new carbon is ingested, and the proportion of carbon-14 in the organism's remains decreases as the carbon-14 decays.

Scientists who do carbon-14 dating use a figure of 5700 years for its half-life. Find the age of a sample in which 10% of the radioactive nuclei originally present have decayed.

Solution We use the decay equation $y = y_0 e^{-kt}$. There are two things to find: the value of k and the value of t when y is $0.9y_0$ (90% of the radioactive nuclei are still present). That is, find t when $y_0 e^{-kt} = 0.9y_0$, or $e^{-kt} = 0.9$.

The value of k. We use the half-life Equation (7):

$$k = \frac{\ln 2}{\text{half-life}} = \frac{\ln 2}{5700} \qquad (\text{about } 1.2 \times 10^{-4})$$

The value of t that makes $e^{-kt} = 0.9$:

$$\begin{aligned} e^{-kt} &= 0.9 \\ e^{-(\ln 2/5700)t} &= 0.9 \\ -\frac{\ln 2}{5700} t &= \ln 0.9 && \text{Logs of both sides} \\ t &= -\frac{5700 \ln 0.9}{\ln 2} \approx 866 \text{ years.} \end{aligned}$$

The sample is about 866 years old. ■

Heat Transfer: Newton's Law of Cooling

Hot soup left in a tin cup cools to the temperature of the surrounding air. A hot silver ingot immersed in a large tub of water cools to the temperature of the surrounding water. In situations like these, the rate at which an object's temperature is changing at any given time is roughly proportional to the difference between its temperature and the temperature of the surrounding medium. This observation is called *Newton's law of cooling*, although it applies to warming as well.

If H is the temperature of the object at time t and H_S is the constant surrounding temperature, then the differential equation is

$$\frac{dH}{dt} = -k(H - H_S). \tag{8}$$

If we substitute y for $(H - H_S)$, then

$$\begin{aligned}
\frac{dy}{dt} &= \frac{d}{dt}(H - H_S) = \frac{dH}{dt} - \frac{d}{dt}(H_S) \\
&= \frac{dH}{dt} - 0 && H_S \text{ is a constant.} \\
&= \frac{dH}{dt} \\
&= -k(H - H_S) && \text{Eq. (8)} \\
&= -ky. && H - H_S = y.
\end{aligned}$$

Now we know that the solution of $dy/dt = -ky$ is $y = y_0 e^{-kt}$, where $y(0) = y_0$. Substituting $(H - H_S)$ for y, this says that

$$H - H_S = (H_0 - H_S)e^{-kt}, \tag{9}$$

where H_0 is the temperature at $t = 0$. This equation is the solution to Newton's Law of Cooling.

EXAMPLE 5 A hard-boiled egg at 98°C is put in a sink of 18°C water. After 5 min, the egg's temperature is 38°C. Assuming that the water has not warmed appreciably, how much longer will it take the egg to reach 20°C?

Solution We find how long it would take the egg to cool from 98°C to 20°C and subtract the 5 min that have already elapsed. Using Equation (9) with $H_S = 18$ and $H_0 = 98$, the egg's temperature t min after it is put in the sink is

$$H = 18 + (98 - 18)e^{-kt} = 18 + 80e^{-kt}.$$

To find k, we use the information that $H = 38$ when $t = 5$:

$$\begin{aligned}
38 &= 18 + 80e^{-5k} \\
e^{-5k} &= \frac{1}{4} \\
-5k &= \ln\frac{1}{4} = -\ln 4 \\
k &= \frac{1}{5}\ln 4 = 0.2\ln 4 \qquad \text{(about 0.28)}.
\end{aligned}$$

The egg's temperature at time t is $H = 18 + 80e^{-(0.2 \ln 4)t}$. Now find the time t when $H = 20$:

$$\begin{aligned} 20 &= 18 + 80e^{-(0.2 \ln 4)t} \\ 80e^{-(0.2 \ln 4)t} &= 2 \\ e^{-(0.2 \ln 4)t} &= \frac{1}{40} \\ -(0.2 \ln 4)t &= \ln \frac{1}{40} = -\ln 40 \\ t &= \frac{\ln 40}{0.2 \ln 4} \approx 13 \text{ min}. \end{aligned}$$

The egg's temperature will reach 20°C about 13 min after it is put in the water to cool. Since it took 5 min to reach 38°C, it will take about 8 min more to reach 20°C. ■

EXERCISES 6.4

In Exercises 1 and 2, show that each function $y = f(x)$ is a solution of the accompanying differential equation.

1. $2y' + 3y = e^{-x}$

a. $y = e^{-x}$ **b.** $y = e^{-x} + e^{-(3/2)x}$

c. $y = e^{-x} + Ce^{-(3/2)x}$

2. $y' = y^2$

a. $y = -\dfrac{1}{x}$ **b.** $y = -\dfrac{1}{x+3}$ **c.** $y = -\dfrac{1}{x+C}$

In Exercises 3 and 4, show that the function $y = f(x)$ is a solution of the given differential equation.

3. $y = \dfrac{1}{x}\displaystyle\int_1^x \frac{e^t}{t}\,dt, \quad x^2y' + xy = e^x$

4. $y = \dfrac{1}{\sqrt{1+x^4}}\displaystyle\int_1^x \sqrt{1+t^4}\,dt, \quad y' + \dfrac{2x^3}{1+x^4}y = 1$

In Exercises 5–8, show that each function is a solution of the given initial value problem.

Differential equation	Initial condition	Solution candidate
5. $y' + y = \dfrac{2}{1+4e^{2x}}$	$y(-\ln 2) = \dfrac{\pi}{2}$	$y = e^{-x}\tan^{-1}(2e^x)$
6. $y' = e^{-x^2} - 2xy$	$y(2) = 0$	$y = (x-2)e^{-x^2}$
7. $xy' + y = -\sin x,$ $x > 0$	$y\left(\dfrac{\pi}{2}\right) = 0$	$y = \dfrac{\cos x}{x}$
8. $x^2y' = xy - y^2,$ $x > 1$	$y(e) = e$	$y = \dfrac{x}{\ln x}$

Solve the differential equation in Exercises 9–18.

9. $2\sqrt{xy}\dfrac{dy}{dx} = 1, \quad x, y > 0$

10. $\dfrac{dy}{dx} = x^2\sqrt{y}, \quad y > 0$

11. $\dfrac{dy}{dx} = e^{x-y}$

12. $\dfrac{dy}{dx} = 3x^2 e^{-y}$

13. $\dfrac{dy}{dx} = \sqrt{y}\cos^2\sqrt{y}$

14. $\sqrt{2xy}\dfrac{dy}{dx} = 1$

15. $\sqrt{x}\dfrac{dy}{dx} = e^{y+\sqrt{x}}, \quad x > 0$

16. $(\sec x)\dfrac{dy}{dx} = e^{y+\sin x}$

17. $\dfrac{dy}{dx} = 2x\sqrt{1-y^2}, \quad -1 < y < 1$

18. $\dfrac{dy}{dx} = \dfrac{e^{2x-y}}{e^{x+y}}$

T The answers to most of the following exercises are in terms of logarithms and exponentials. A calculator can be helpful, enabling you to express the answers in decimal form.

19. First-order chemical reactions In some chemical reactions, the rate at which the amount of a substance changes with time is proportional to the amount present. For the change of δ-glucono lactone into gluconic acid, for example,

$$\frac{dy}{dt} = -0.6y$$

when t is measured in hours. If there are 100 grams of δ-glucono lactone present when $t = 0$, how many grams will be left after the first hour?

20. The inversion of sugar The processing of raw sugar has a step called "inversion" that changes the sugar's molecular structure.

Once the process has begun, the rate of change of the amount of raw sugar is proportional to the amount of raw sugar remaining. If 1000 kg of raw sugar reduces to 800 kg of raw sugar during the first 10 hours, how much raw sugar will remain after another 14 hours?

21. **Working underwater** The intensity $L(x)$ of light x feet beneath the surface of the ocean satisfies the differential equation

$$\frac{dL}{dx} = -kL.$$

As a diver, you know from experience that diving to 18 ft in the Caribbean Sea cuts the intensity in half. You cannot work without artificial light when the intensity falls below one-tenth of the surface value. About how deep can you expect to work without artificial light?

22. **Voltage in a discharging capacitor** Suppose that electricity is draining from a capacitor at a rate that is proportional to the voltage V across its terminals and that, if t is measured in seconds,

$$\frac{dV}{dt} = -\frac{1}{40}V.$$

Solve this equation for V, using V_0 to denote the value of V when $t = 0$. How long will it take the voltage to drop to 10% of its original value?

23. **Cholera bacteria** Suppose that the bacteria in a colony can grow unchecked, by the law of exponential change. The colony starts with 1 bacterium and doubles every half-hour. How many bacteria will the colony contain at the end of 24 hours? (Under favorable laboratory conditions, the number of cholera bacteria can double every 30 min. In an infected person, many bacteria are destroyed, but this example helps explain why a person who feels well in the morning may be dangerously ill by evening.)

24. **Growth of bacteria** A colony of bacteria is grown under ideal conditions in a laboratory so that the population increases exponentially with time. At the end of 3 hours there are 10,000 bacteria. At the end of 5 hours there are 40,000. How many bacteria were present initially?

25. **The incidence of a disease** (*Continuation of Example 3.*) Suppose that in any given year the number of cases can be reduced by 25% instead of 20%.

    a. How long will it take to reduce the number of cases to 1000?

    b. How long will it take to eradicate the disease, that is, reduce the number of cases to less than 1?

26. **The U.S. population** The Museum of Science in Boston displays a running total of the U.S. population. On May 11, 1993, the total was increasing at the rate of 1 person every 14 sec. The displayed population figure for 3:45 P.M. that day was 257,313,431.

    a. Assuming exponential growth at a constant rate, find the rate constant for the population's growth (people per 365-day year).

    b. At this rate, what will the U.S. population be at 3:45 P.M. Boston time on May 11, 2012?

27. **Oil depletion** Suppose the amount of oil pumped from one of the canyon wells in Whittier, California, decreases at the continuous rate of 10% per year. When will the well's output fall to one-fifth of its present value?

28. **Population growth** The population of Glenbrook is 375,000 and is increasing at the rate of 2.25% per year. Predict when the population will be 1 million.

29. **Plutonium-239** The half-life of the plutonium isotope is 24,360 years. If 10 g of plutonium is released into the atmosphere by a nuclear accident, how many years will it take for 80% of the isotope to decay?

30. **Polonium-210** The half-life of polonium is 139 days, but your sample will not be useful to you after 95% of the radioactive nuclei present on the day the sample arrives has disintegrated. For about how many days after the sample arrives will you be able to use the polonium?

31. **The mean life of a radioactive nucleus** Physicists using the radioactivity equation $y = y_0 e^{-kt}$ call the number $1/k$ the *mean life* of a radioactive nucleus. The mean life of a radon nucleus is about $1/0.18 = 5.6$ days. The mean life of a carbon-14 nucleus is more than 8000 years. Show that 95% of the radioactive nuclei originally present in a sample will disintegrate within three mean lifetimes, i.e., by time $t = 3/k$. Thus, the mean life of a nucleus gives a quick way to estimate how long the radioactivity of a sample will last.

32. **Californium-252** What costs \$27 million per gram and can be used to treat brain cancer, analyze coal for its sulfur content, and detect explosives in luggage? The answer is californium-252, a radioactive isotope so rare that only 8 g of it have been made in the western world since its discovery by Glenn Seaborg in 1950. The half-life of the isotope is 2.645 years—long enough for a useful service life and short enough to have a high radioactivity per unit mass. One microgram of the isotope releases 170 million neutrons per second.

    a. What is the value of k in the decay equation for this isotope?

    b. What is the isotope's mean life? (See Exercise 31.)

    c. How long will it take 95% of a sample's radioactive nuclei to disintegrate?

33. **Cooling soup** Suppose that a cup of soup cooled from 90°C to 60°C after 10 min in a room whose temperature was 20°C. Use Newton's law of cooling to answer the following questions.

    a. How much longer would it take the soup to cool to 35°C?

    b. Instead of being left to stand in the room, the cup of 90°C soup is put in a freezer whose temperature is -15°C. How long will it take the soup to cool from 90°C to 35°C?

34. **A beam of unknown temperature** An aluminum beam was brought from the outside cold into a machine shop where the temperature was held at 65°F. After 10 min, the beam warmed to 35°F and after another 10 min it was 50°F. Use Newton's law of cooling to estimate the beam's initial temperature.

35. **Surrounding medium of unknown temperature** A pan of warm water (46°C) was put in a refrigerator. Ten minutes later, the

water's temperature was 39°C; 10 min after that, it was 33°C. Use Newton's law of cooling to estimate how cold the refrigerator was.

36. Silver cooling in air The temperature of an ingot of silver is 60°C above room temperature right now. Twenty minutes ago, it was 70°C above room temperature. How far above room temperature will the silver be

a. 15 min from now? **b.** 2 hours from now?

c. When will the silver be 10°C above room temperature?

37. The age of Crater Lake The charcoal from a tree killed in the volcanic eruption that formed Crater Lake in Oregon contained 44.5% of the carbon-14 found in living matter. About how old is Crater Lake?

38. The sensitivity of carbon-14 dating to measurement To see the effect of a relatively small error in the estimate of the amount of carbon-14 in a sample being dated, consider this hypothetical situation:

a. A fossilized bone found in central Illinois in the year A.D. 2000 contains 17% of its original carbon-14 content. Estimate the year the animal died.

b. Repeat part (a) assuming 18% instead of 17%.

c. Repeat part (a) assuming 16% instead of 17%.

39. Art forgery A painting attributed to Vermeer (1632–1675), which should contain no more than 96.2% of its original carbon-14, contains 99.5% instead. About how old is the forgery?

6.5 Work and Fluid Forces

In everyday life, *work* means an activity that requires muscular or mental effort. In science, the term refers specifically to a force acting on a body (or object) and the body's subsequent displacement. This section shows how to calculate work. It also shows how to calculate the fluid force against a vertical flat plate.

Work Done by a Constant Force

When a body moves a distance d along a straight line as a result of being acted on by a force of constant magnitude F in the direction of motion, we define the **work** W done by the force on the body with the formula

$$W = Fd \qquad \text{(Constant-force formula for work).} \qquad (1)$$

From Equation (1) we see that the unit of work in any system is the unit of force multiplied by the unit of distance. In SI units (SI stands for *Système International*, or International System), the unit of force is a newton, the unit of distance is a meter, and the unit of work is a newton-meter $(\text{N} \cdot \text{m})$. This combination appears so often it has a special name, the **joule**. In the British system, the unit of work is the foot-pound, a unit frequently used by engineers.

Joules

The joule, abbreviated J and pronounced "jewel," is named after the English physicist James Prescott Joule (1818–1889). The defining equation is

$$1 \text{ joule} = (1 \text{ newton})(1 \text{ meter}).$$

In symbols, $1 \text{ J} = 1 \text{ N} \cdot \text{m}$.

EXAMPLE 1 If you jack up the side of a 2000-lb car 1.25 ft to change a tire (you have to apply a constant vertical force of about 1000 lb) you will perform $1000 \times 1.25 = 1250$ ft-lb of work on the car. In SI units, you have applied a force of 4448 N through a distance of 0.381 m to do $4448 \times 0.381 \approx 1695$ J of work. ■

Work Done by a Variable Force Along a Line

If the force you apply varies along the way, as it will if you are compressing a spring, the formula $W = Fd$ has to be replaced by an integral formula that takes the variation in F into account.

Suppose that the force performing the work acts on an object moving along a straight line, which we take to be the x-axis. We assume that the magnitude of the force is a continuous function F of the object's position x. We want to find the work done over the interval from $x = a$ to $x = b$. We partition $[a, b]$ in the usual way and choose an arbitrary point c_k in each subinterval $[x_{k-1}, x_k]$. If the subinterval is short enough, F, being continuous, will not vary much from x_{k-1} to x_k. The amount of work done across the interval will be about $F(c_k)$ times the distance Δx_k, the same as it would be if F were constant and we could apply Equation (1). The total work done from a to b is therefore approximated by the Riemann sum

$$\text{Work} \approx \sum_{k=1}^{n} F(c_k)\,\Delta x_k.$$

We expect the approximation to improve as the norm of the partition goes to zero, so we define the work done by the force from a to b to be the integral of F from a to b:

$$\lim_{n\to\infty} \sum_{k=1}^{n} F(c_k)\,\Delta x_k = \int_a^b F(x)\,dx.$$

DEFINITION The **work** done by a variable force $F(x)$ in the direction of motion along the x-axis from $x = a$ to $x = b$ is

$$W = \int_a^b F(x)\,dx. \qquad (2)$$

The units of the integral are joules if F is in newtons and x is in meters, and foot-pounds if F is in pounds and x in feet. So, the work done by a force of $F(x) = 1/x^2$ newtons in moving an object along the x-axis from $x = 1$ m to $x = 10$ m is

$$W = \int_1^{10} \frac{1}{x^2}\,dx = -\frac{1}{x}\Big]_1^{10} = -\frac{1}{10} + 1 = 0.9 \text{ J}.$$

Hooke's Law for Springs: $F = kx$

Hooke's Law says that the force it takes to stretch or compress a spring x length units from its natural (unstressed) length is proportional to x. In symbols,

$$F = kx. \qquad (3)$$

The constant k, measured in force units per unit length, is a characteristic of the spring, called the **force constant** (or **spring constant**) of the spring. Hooke's Law, Equation (3), gives good results as long as the force doesn't distort the metal in the spring. We assume that the forces in this section are too small to do that.

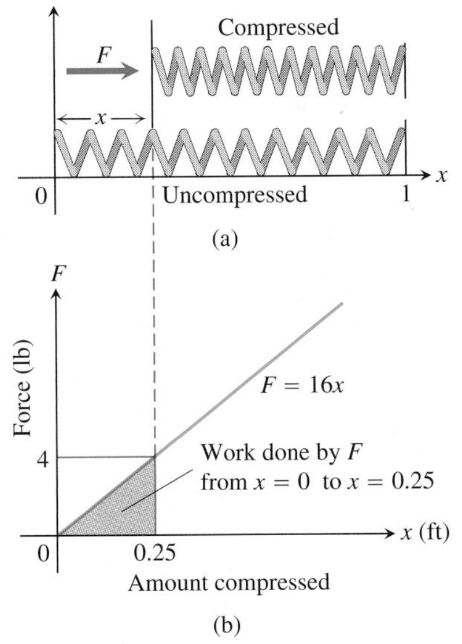

FIGURE 6.27 The force F needed to hold a spring under compression increases linearly as the spring is compressed (Example 2).

EXAMPLE 2 Find the work required to compress a spring from its natural length of 1 ft to a length of 0.75 ft if the force constant is $k = 16$ lb/ft.

Solution We picture the uncompressed spring laid out along the x-axis with its movable end at the origin and its fixed end at $x = 1$ ft (Figure 6.27). This enables us to describe the force required to compress the spring from 0 to x with the formula $F = 16x$. To compress the spring from 0 to 0.25 ft, the force must increase from

$$F(0) = 16 \cdot 0 = 0 \text{ lb} \quad \text{to} \quad F(0.25) = 16 \cdot 0.25 = 4 \text{ lb}.$$

The work done by F over this interval is

$$W = \int_0^{0.25} 16x\,dx = 8x^2\Big]_0^{0.25} = 0.5 \text{ ft-lb}. \qquad \text{Eq. (2) with } a = 0,\ b = 0.25,\ F(x) = 16x$$ ■

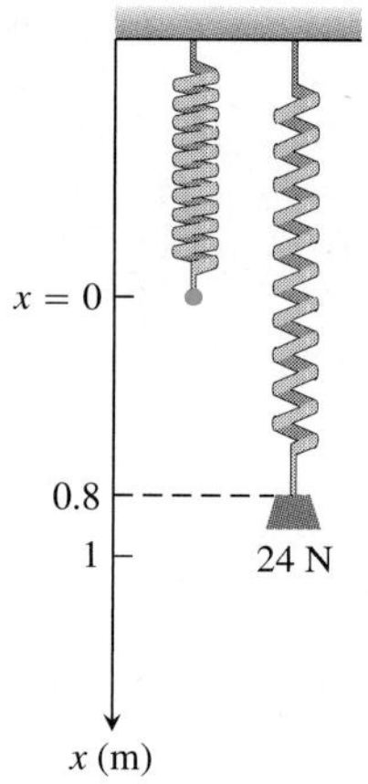

FIGURE 6.28 A 24-N weight stretches this spring 0.8 m beyond its unstressed length (Example 3).

EXAMPLE 3 A spring has a natural length of 1 m. A force of 24 N stretches the spring to a length of 1.8 m.

(a) Find the force constant k.

(b) How much work will it take to stretch the spring 2 m beyond its natural length?

(c) How far will a 45-N force stretch the spring?

Solution

(a) *The force constant.* We find the force constant from Equation (3). A force of 24 N stretches the spring 0.8 m, so

$$24 = k(0.8) \qquad \text{Eq. (3) with } F = 24, x = 0.8$$
$$k = 24/0.8 = 30 \text{ N/m}.$$

(b) *The work to stretch the spring* 2 m. We imagine the unstressed spring hanging along the x-axis with its free end at $x = 0$ (Figure 6.28). The force required to stretch the spring x m beyond its natural length is the force required to pull the free end of the spring x units from the origin. Hooke's Law with $k = 30$ says that this force is

$$F(x) = 30x.$$

The work done by F on the spring from $x = 0$ m to $x = 2$ m is

$$W = \int_0^2 30x\,dx = 15x^2\Big]_0^2 = 60 \text{ J}.$$

(c) *How far will a* 45-N *force stretch the spring?* We substitute $F = 45$ in the equation $F = 30x$ to find

$$45 = 30x, \qquad \text{or} \qquad x = 1.5 \text{ m}.$$

A 45-N force will stretch the spring 1.5 m. No calculus is required to find this. ■

The work integral is useful to calculate the work done in lifting objects whose weights vary with their elevation.

EXAMPLE 4 A 5-lb bucket is lifted from the ground into the air by pulling in 20 ft of rope at a constant speed (Figure 6.29). The rope weighs 0.08 lb/ft. How much work was spent lifting the bucket and rope?

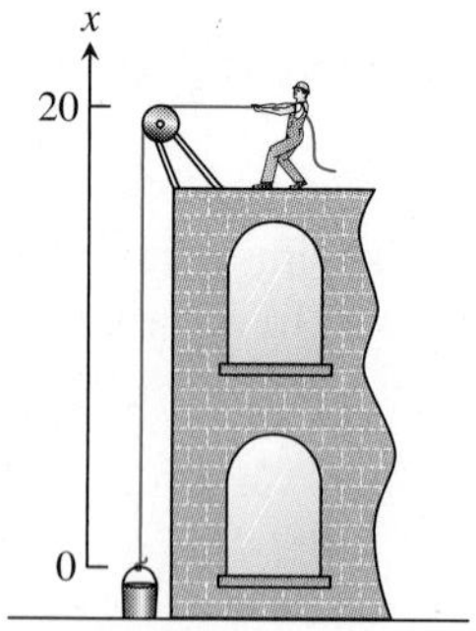

FIGURE 6.29 Lifting the bucket in Example 4.

Solution The bucket has constant weight so the work done lifting it alone is weight × distance $= 5 \cdot 20 = 100$ ft-lb.

The weight of the rope varies with the bucket's elevation, because less of it is freely hanging. When the bucket is x ft off the ground, the remaining proportion of the rope still being lifted weighs $(0.08) \cdot (20 - x)$ lb. So the work in lifting the rope is

$$\text{Work on rope} = \int_0^{20} (0.08)(20 - x)\,dx = \int_0^{20} (1.6 - 0.08x)\,dx$$
$$= \left[1.6x - 0.04x^2\right]_0^{20} = 32 - 16 = 16 \text{ ft-lb}.$$

The total work for the bucket and rope combined is

$$100 + 16 = 116 \text{ ft-lb}.$$

■

Pumping Liquids from Containers

How much work does it take to pump all or part of the liquid from a container? To find out, we imagine lifting the liquid out one thin horizontal slab at a time and applying the equation $W = Fd$ to each slab. We then evaluate the integral this leads to as the slabs become thinner and more numerous. The integral we get each time depends on the weight of the liquid and the dimensions of the container, but the way we find the integral is always the same. The next example shows what to do.

EXAMPLE 5 The conical tank in Figure 6.30 is filled to within 2 ft of the top with olive oil weighing 57 lb/ft^3. How much work does it take to pump the oil to the rim of the tank?

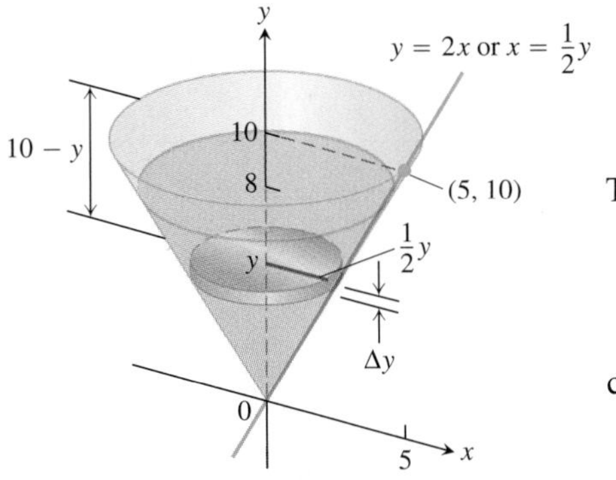

FIGURE 6.30 The olive oil and tank in Example 5.

Solution We imagine the oil divided into thin slabs by planes perpendicular to the y-axis at the points of a partition of the interval [0, 8].

The typical slab between the planes at y and $y + \Delta y$ has a volume of about

$$\Delta V = \pi(\text{radius})^2(\text{thickness}) = \pi\left(\frac{1}{2}y\right)^2 \Delta y = \frac{\pi}{4}y^2\,\Delta y \text{ ft}^3.$$

The force $F(y)$ required to lift this slab is equal to its weight,

$$F(y) = 57\,\Delta V = \frac{57\pi}{4}y^2\,\Delta y \text{ lb.}$$

Weight = weight per unit volume × volume

The distance through which $F(y)$ must act to lift this slab to the level of the rim of the cone is about $(10 - y)$ ft, so the work done lifting the slab is about

$$\Delta W = \frac{57\pi}{4}(10 - y)y^2\,\Delta y \text{ ft-lb.}$$

Assuming there are n slabs associated with the partition of [0, 8], and that $y = y_k$ denotes the plane associated with the kth slab of thickness Δy_k, we can approximate the work done lifting all of the slabs with the Riemann sum

$$W \approx \sum_{k=1}^{n} \frac{57\pi}{4}(10 - y_k)y_k{}^2\,\Delta y_k \text{ ft-lb.}$$

The work of pumping the oil to the rim is the limit of these sums as the norm of the partition goes to zero.

$$\begin{aligned} W = \lim_{n\to\infty}\sum_{k=1}^{n}\frac{57\pi}{4}(10 - y_k)y_k^2\,\Delta y_k &= \int_0^8 \frac{57\pi}{4}(10 - y)y^2\,dy \\ &= \frac{57\pi}{4}\int_0^8 (10y^2 - y^3)\,dy \\ &= \frac{57\pi}{4}\left[\frac{10y^3}{3} - \frac{y^4}{4}\right]_0^8 \approx 30{,}561 \text{ ft-lb.} \end{aligned}$$ ■

Fluid Pressures and Forces

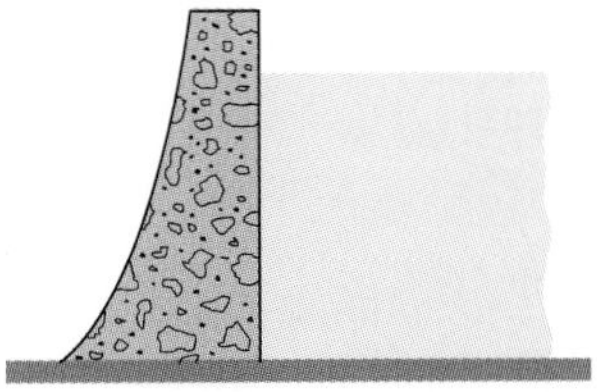

FIGURE 6.31 To withstand the increasing pressure, dams are built thicker as they go down.

We make dams thicker at the bottom than at the top (Figure 6.31) because the pressure against them increases with depth. The pressure at any point on a dam depends only on how far below the surface the point is and not on how much the surface of the dam happens to be tilted at that point. The pressure, in pounds per square foot at a point h feet below the surface, is always $62.4h$. The number 62.4 is the weight-density of water in pounds per cubic foot. The pressure h feet below the surface of any fluid is the fluid's *weight-density* times h.

Weight-density

A fluid's weight-density w is its weight per unit volume. Typical values (lb/ft^3) are

Gasoline	42
Mercury	849
Milk	64.5
Molasses	100
Olive oil	57
Seawater	64
Water	62.4

The Pressure-Depth Equation

In a fluid that is standing still, the pressure p at depth h is the fluid's weight-density w times h:

$$p = wh. \tag{4}$$

In a container of fluid with a flat horizontal base, the total force exerted by the fluid against the base can be calculated by multiplying the area of the base by the pressure at the base. We can do this because total force equals force per unit area (pressure) times area. (See Figure 6.32.) If F, p, and A are the total force, pressure, and area, then

$$\begin{aligned} F &= \text{total force} = \text{force per unit area} \times \text{area} \\ &= \text{pressure} \times \text{area} = pA \\ &= whA. && p = wh \text{ from Eq. (4)} \end{aligned}$$

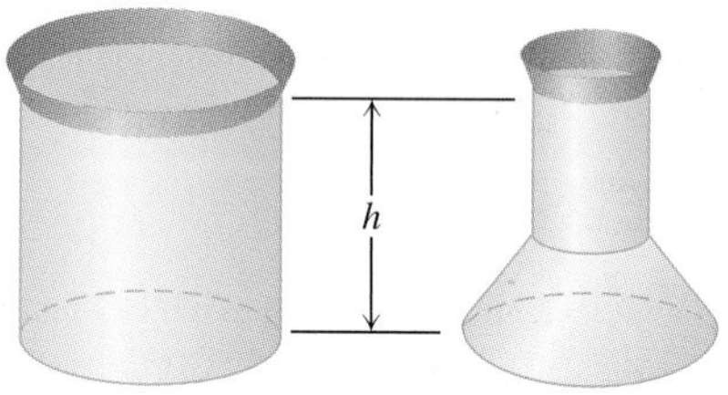

FIGURE 6.32 These containers are filled with water to the same depth and have the same base area. The total force is therefore the same on the bottom of each container. The containers' shapes do not matter here.

Fluid Force on a Constant-Depth Surface

$$F = pA = whA \tag{5}$$

For example, the weight-density of water is 62.4 lb/ft^3, so the fluid force at the bottom of a 10 ft $\times$ 20 ft rectangular swimming pool 3 ft deep is

$$\begin{aligned} F = whA &= (62.4 \text{ lb/ft}^3)(3 \text{ ft})(10 \cdot 20 \text{ ft}^2) \\ &= 37{,}440 \text{ lb}. \end{aligned}$$

For a flat plate submerged *horizontally*, like the bottom of the swimming pool just discussed, the downward force acting on its upper face due to liquid pressure is given by Equation (5). If the plate is submerged *vertically*, however, then the pressure against it will be different at different depths and Equation (5) no longer is usable in that form (because h varies).

Suppose we want to know the force exerted by a fluid against one side of a vertical plate submerged in a fluid of weight-density w. To find it, we model the plate as a region extending from $y = a$ to $y = b$ in the xy-plane (Figure 6.33). We partition $[a, b]$ in the usual way and imagine the region to be cut into thin horizontal strips by planes perpendicular to the y-axis at the partition points. The typical strip from y to $y + \Delta y$ is Δy units wide by $L(y)$ units long. We assume $L(y)$ to be a continuous function of y.

The pressure varies across the strip from top to bottom. If the strip is narrow enough, however, the pressure will remain close to its bottom-edge value of $w \times$ (strip depth). The force exerted by the fluid against one side of the strip will be about

$$\begin{aligned} \Delta F &= (\text{pressure along bottom edge}) \times (\text{area}) \\ &= w \cdot (\text{strip depth}) \cdot L(y)\, \Delta y. \end{aligned}$$

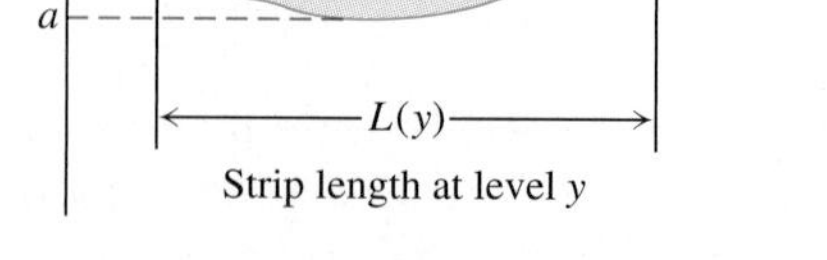

FIGURE 6.33 The force exerted by a fluid against one side of a thin, flat horizontal strip is about ΔF = pressure $\times$ area = $w \times$ (strip depth) $\times L(y)\,\Delta y$.

Assume there are n strips associated with the partition of $a \le y \le b$ and that y_k is the bottom edge of the kth strip having length $L(y_k)$ and width Δy_k. The force against the entire plate is approximated by summing the forces against each strip, giving the Riemann sum

$$F \approx \sum_{k=1}^{n} (w \cdot (\text{strip depth})_k \cdot L(y_k))\, \Delta y_k. \tag{6}$$

The sum in Equation (6) is a Riemann sum for a continuous function on $[a, b]$, and we expect the approximations to improve as the norm of the partition goes to zero. The force against the plate is the limit of these sums:

$$\lim_{n\to\infty}\sum_{k=1}^{n}(w\cdot(\text{strip depth})_k\cdot L(y_k))\,\Delta y_k = \int_a^b w\cdot(\text{strip depth})\cdot L(y)\,dy.$$

The Integral for Fluid Force Against a Vertical Flat Plate

Suppose that a plate submerged vertically in fluid of weight-density w runs from $y = a$ to $y = b$ on the y-axis. Let $L(y)$ be the length of the horizontal strip measured from left to right along the surface of the plate at level y. Then the force exerted by the fluid against one side of the plate is

$$F = \int_a^b w\cdot(\text{strip depth})\cdot L(y)\,dy. \tag{7}$$

EXAMPLE 6 A flat isosceles right triangular plate with base 6 ft and height 3 ft is submerged vertically, base up, 2 ft below the surface of a swimming pool. Find the force exerted by the water against one side of the plate.

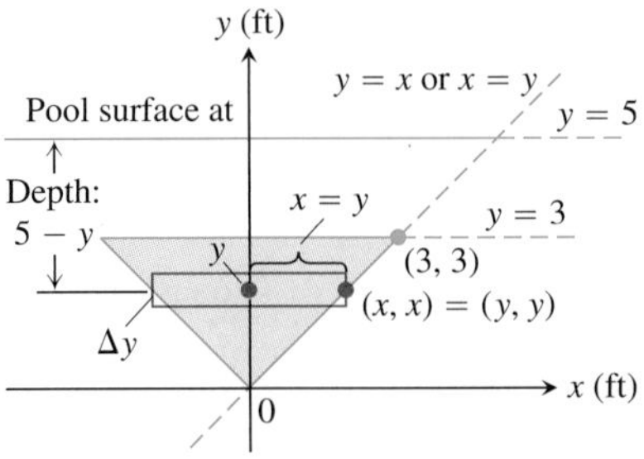

FIGURE 6.34 To find the force on one side of the submerged plate in Example 6, we can use a coordinate system like the one here.

Solution We establish a coordinate system to work in by placing the origin at the plate's bottom vertex and running the y-axis upward along the plate's axis of symmetry (Figure 6.34). The surface of the pool lies along the line $y = 5$ and the plate's top edge along the line $y = 3$. The plate's right-hand edge lies along the line $y = x$, with the upper right vertex at $(3, 3)$. The length of a thin strip at level y is

$$L(y) = 2x = 2y.$$

The depth of the strip beneath the surface is $(5 - y)$. The force exerted by the water against one side of the plate is therefore

$$\begin{aligned} F &= \int_a^b w\cdot\begin{pmatrix}\text{strip}\\ \text{depth}\end{pmatrix}\cdot L(y)\,dy \qquad \text{Eq. (7)} \\ &= \int_0^3 62.4(5 - y)2y\,dy \\ &= 124.8\int_0^3 (5y - y^2)\,dy \\ &= 124.8\left[\frac{5}{2}y^2 - \frac{y^3}{3}\right]_0^3 = 1684.8\text{ lb}. \end{aligned}$$

■

EXERCISES 6.5

1. **Spring constant** It took 1800 J of work to stretch a spring from its natural length of 2 m to a length of 5 m. Find the spring's force constant.

2. **Stretching a spring** A spring has a natural length of 10 in. An 800-lb force stretches the spring to 14 in.
   a. Find the force constant.
   b. How much work is done in stretching the spring from 10 in. to 12 in.?
   c. How far beyond its natural length will a 1600-lb force stretch the spring?

3. **Stretching a rubber band** A force of 2 N will stretch a rubber band 2 cm (0.02 m). Assuming that Hooke's Law applies, how far will a 4-N force stretch the rubber band? How much work does it take to stretch the rubber band this far?

4. **Stretching a spring** If a force of 90 N stretches a spring 1 m beyond its natural length, how much work does it take to stretch the spring 5 m beyond its natural length?

5. **Subway car springs** It takes a force of 21,714 lb to compress a coil spring assembly on a New York City Transit Authority subway car from its free height of 8 in. to its fully compressed height of 5 in.

   a. What is the assembly's force constant?

   b. How much work does it take to compress the assembly the first half inch? the second half inch? Answer to the nearest in.-lb.

6. **Bathroom scale** A bathroom scale is compressed $1/16$ in. when a 150-lb person stands on it. Assuming that the scale behaves like a spring that obeys Hooke's Law, how much does someone who compresses the scale $1/8$ in. weigh? How much work is done compressing the scale $1/8$ in.?

7. **Lifting a rope** A mountain climber is about to haul up a 50 m length of hanging rope. How much work will it take if the rope weighs 0.624 N/m?

8. **Leaky sandbag** A bag of sand originally weighing 144 lb was lifted at a constant rate. As it rose, sand also leaked out at a constant rate. The sand was half gone by the time the bag had been lifted to 18 ft. How much work was done lifting the sand this far? (Neglect the weight of the bag and lifting equipment.)

9. **Lifting an elevator cable** An electric elevator with a motor at the top has a multistrand cable weighing 4.5 lb/ft. When the car is at the first floor, 180 ft of cable are paid out, and effectively 0 ft are out when the car is at the top floor. How much work does the motor do just lifting the cable when it takes the car from the first floor to the top?

10. **Force of attraction** When a particle of mass m is at $(x, 0)$, it is attracted toward the origin with a force whose magnitude is k/x^2. If the particle starts from rest at $x = b$ and is acted on by no other forces, find the work done on it by the time it reaches $x = a$, $0 < a < b$.

11. **Leaky bucket** Assume the bucket in Example 4 is leaking. It starts with 2 gal of water (16 lb) and leaks at a constant rate. It finishes draining just as it reaches the top. How much work was spent lifting the water alone? (*Hint:* Do not include the rope and bucket, and find the proportion of water left at elevation x ft.)

12. (*Continuation of Exercise 11.*) The workers in Example 4 and Exercise 11 changed to a larger bucket that held 5 gal (40 lb) of water, but the new bucket had an even larger leak so that it, too, was empty by the time it reached the top. Assuming that the water leaked out at a steady rate, how much work was done lifting the water alone? (Do not include the rope and bucket.)

13. **Pumping water** The rectangular tank shown here, with its top at ground level, is used to catch runoff water. Assume that the water weighs 62.4 lb/ft^3.

    a. How much work does it take to empty the tank by pumping the water back to ground level once the tank is full?

    b. If the water is pumped to ground level with a $(5/11)$-horsepower (hp) motor (work output 250 ft-lb/sec), how long will it take to empty the full tank (to the nearest minute)?

    c. Show that the pump in part (b) will lower the water level 10 ft (halfway) during the first 25 min of pumping.

    d. The weight of water What are the answers to parts (a) and (b) in a location where water weighs 62.26 lb/ft^3? 62.59 lb/ft^3?

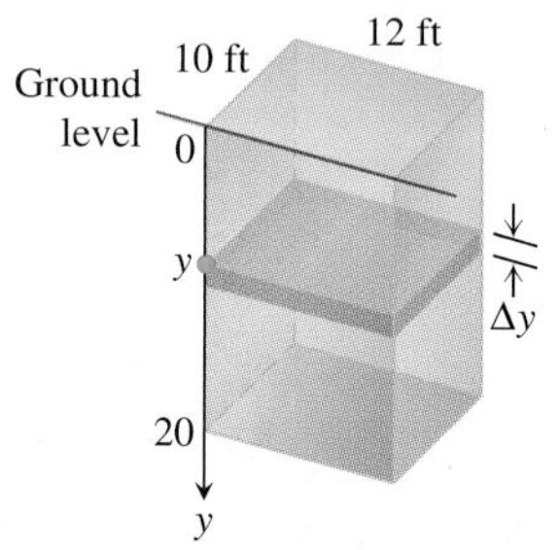

14. **Emptying a cistern** The rectangular cistern (storage tank for rainwater) shown has its top 10 ft below ground level. The cistern, currently full, is to be emptied for inspection by pumping its contents to ground level.

    a. How much work will it take to empty the cistern?

    b. How long will it take a $1/2$ hp pump, rated at 275 ft-lb/sec, to pump the tank dry?

    c. How long will it take the pump in part (b) to empty the tank halfway? (It will be less than half the time required to empty the tank completely.)

    d. The weight of water What are the answers to parts (a) through (c) in a location where water weighs 62.26 lb/ft^3? 62.59 lb/ft^3?

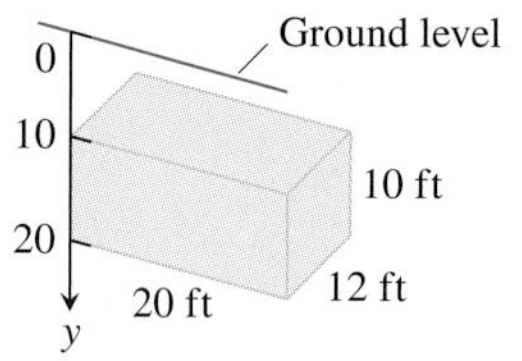

15. **Pumping oil** How much work would it take to pump oil from the tank in Example 5 to the level of the top of the tank if the tank were completely full?

16. **Pumping a half-full tank** Suppose that, instead of being full, the tank in Example 5 is only half full. How much work does it take to pump the remaining oil to a level 4 ft above the top of the tank?

17. **Emptying a tank** A vertical right circular cylindrical tank measures 30 ft high and 20 ft in diameter. It is full of kerosene weighing 51.2 lb/ft^3. How much work does it take to pump the kerosene to the level of the top of the tank?

18. **a. Pumping milk** Suppose that the conical container in Example 5 contains milk (weighing 64.5 lb/ft^3) instead of olive oil. How much work will it take to pump the contents to the rim?

    b. Pumping oil How much work will it take to pump the oil in Example 5 to a level 3 ft above the cone's rim?

19. **Emptying a water reservoir** We model pumping from spherical containers the way we do from other containers, with the axis of integration along the vertical axis of the sphere. Use the figure here to find how much work it takes to empty a full hemispherical water reservoir of radius 5 m by pumping the water to a height of 4 m above the top of the reservoir. Water weighs 9800 N/m^3.

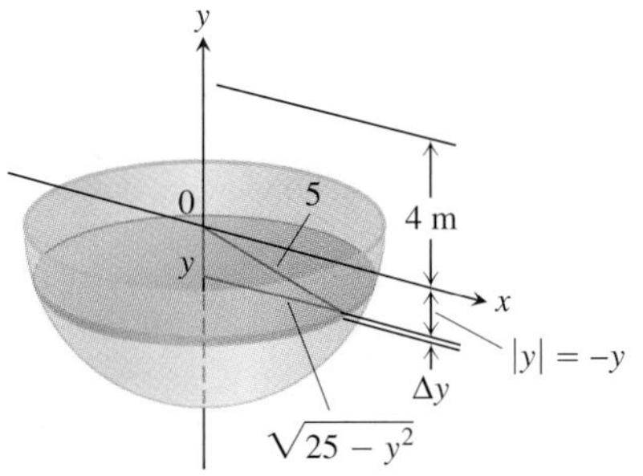

20. You are in charge of the evacuation and repair of the storage tank shown here. The tank is a hemisphere of radius 10 ft and is full of benzene weighing 56 lb/ft^3. A firm you contacted says it can empty the tank for 1/2¢ per foot-pound of work. Find the work required to empty the tank by pumping the benzene to an outlet 2 ft above the top of the tank. If you have $5000 budgeted for the job, can you afford to hire the firm?

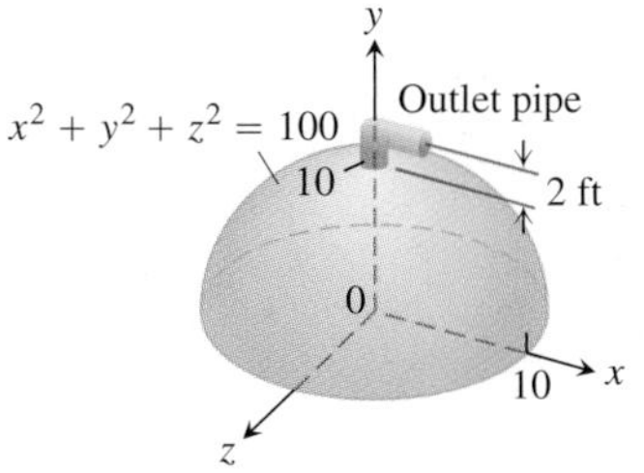

21. **Kinetic energy** If a variable force of magnitude $F(x)$ moves a body of mass m along the x-axis from x_1 to x_2, the body's velocity v can be written as dx/dt (where t represents time). Use Newton's second law of motion $F = m(dv/dt)$ and the Chain Rule

$$\frac{dv}{dt} = \frac{dv}{dx}\frac{dx}{dt} = v\frac{dv}{dx}$$

to show that the net work done by the force in moving the body from x_1 to x_2 is

$$W = \int_{x_1}^{x_2} F(x)\,dx = \frac{1}{2}m{v_2}^2 - \frac{1}{2}m{v_1}^2,$$

where v_1 and v_2 are the body's velocities at x_1 and x_2. In physics, the expression $(1/2)mv^2$ is called the *kinetic energy* of a body of mass m moving with velocity v. Therefore, *the work done by the force equals the change in the body's kinetic energy*, and we can find the work by calculating this change.

In Exercises 22–26, use the result of Exercise 21.

22. **Tennis** A 2-oz tennis ball was served at 160 ft/sec (about 109 mph). How much work was done on the ball to make it go this fast? (To find the ball's mass from its weight, express the weight in pounds and divide by 32 ft/sec^2, the acceleration of gravity.)

23. **Baseball** How many foot-pounds of work does it take to throw a baseball 90 mph? A baseball weighs 5 oz, or 0.3125 lb.

24. **Golf** A 1.6-oz golf ball is driven off the tee at a speed of 280 ft/sec (about 191 mph). How many foot-pounds of work are done on the ball getting it into the air?

25. **Tennis** During the match in which Pete Sampras won the 1990 U.S. Open men's tennis championship, Sampras hit a serve that was clocked at a phenomenal 124 mph. How much work did Sampras have to do on the 2-oz ball to get it to that speed?

26. **Softball** How much work has to be performed on a 6.5-oz softball to pitch it 132 ft/sec (90 mph)?

27. **Drinking a milkshake** The truncated conical container shown here is full of strawberry milkshake that weighs 4/9 oz/in.^{3} As you can see, the container is 7 in. deep, 2.5 in. across at the base, and 3.5 in. across at the top (a standard size at Brigham's in Boston). The straw sticks up an inch above the top. About how much work does it take to suck up the milkshake through the straw (neglecting friction)? Answer in inch-ounces.

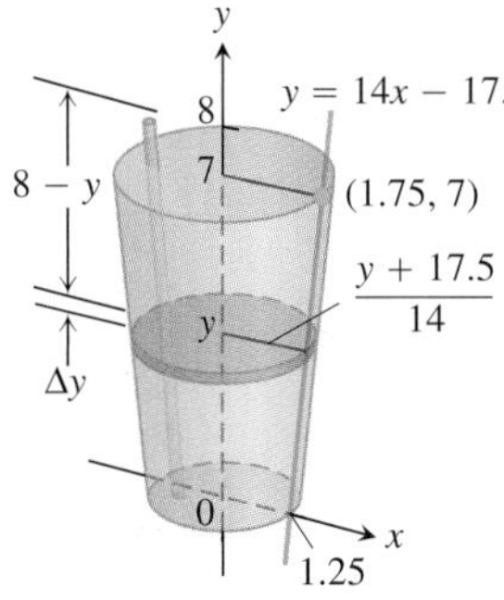

Dimensions in inches

28. **Putting a satellite in orbit** The strength of Earth's gravitational field varies with the distance r from Earth's center, and the

magnitude of the gravitational force experienced by a satellite of mass m during and after launch is

$$F(r) = \frac{mMG}{r^2}.$$

Here, $M = 5.975 \times 10^{24}$ kg is Earth's mass, $G = 6.6720 \times 10^{-11}\ \text{N} \cdot \text{m}^2\ \text{kg}^{-2}$ is the universal gravitational constant, and r is measured in meters. The work it takes to lift a 1000-kg satellite from Earth's surface to a circular orbit 35,780 km above Earth's center is therefore given by the integral

$$\text{Work} = \int_{6{,}370{,}000}^{35{,}780{,}000} \frac{1000MG}{r^2}\, dr \text{ joules.}$$

Evaluate the integral. The lower limit of integration is Earth's radius in meters at the launch site. (This calculation does not take into account energy spent lifting the launch vehicle or energy spent bringing the satellite to orbit velocity.)

29. Triangular plate Calculate the fluid force on one side of the plate in Example 6 using the coordinate system shown here.

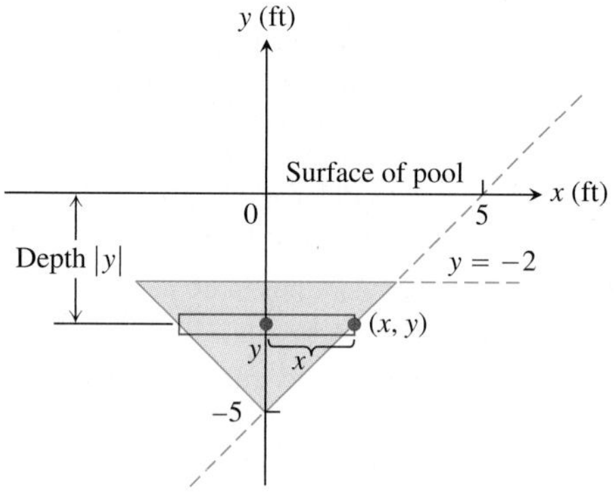

30. Triangular plate Calculate the fluid force on one side of the plate in Example 6 using the coordinate system shown here.

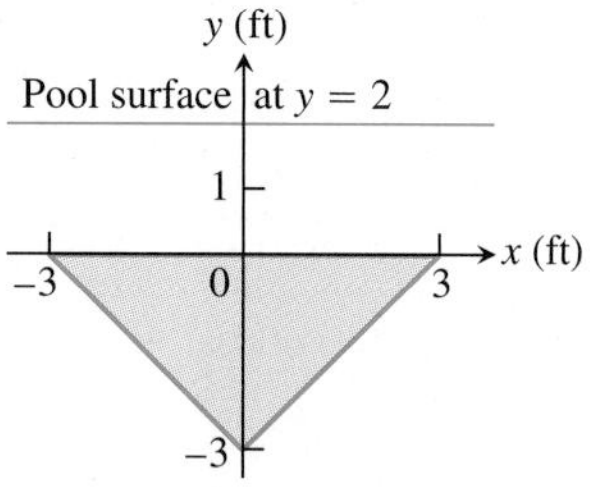

31. Triangular plate The isosceles triangular plate shown here is submerged vertically 1 ft below the surface of a freshwater lake.

a. Find the fluid force against one face of the plate.

b. What would be the fluid force on one side of the plate if the water were seawater instead of freshwater?

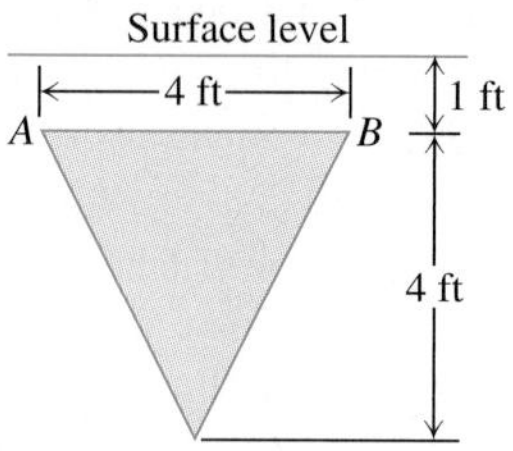

32. Rotated triangular plate The plate in Exercise 31 is revolved 180° about line AB so that part of the plate sticks out of the lake, as shown here. What force does the water exert on one face of the plate now?

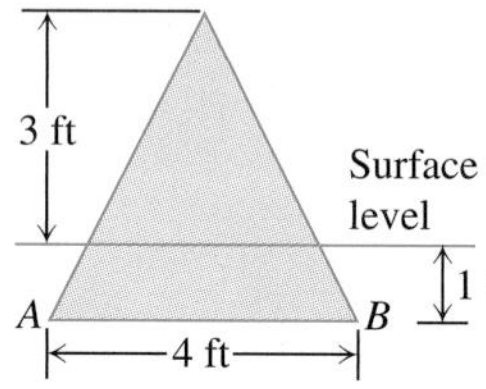

33. New England Aquarium The viewing portion of the rectangular glass window in a typical fish tank at the New England Aquarium in Boston is 63 in. wide and runs from 0.5 in. below the water's surface to 33.5 in. below the surface. Find the fluid force against this portion of the window. The weight-density of seawater is $64\ \text{lb/ft}^3$. (In case you were wondering, the glass is 3/4 in. thick and the tank walls extend 4 in. above the water to keep the fish from jumping out.)

34. Semicircular plate A semicircular plate 2 ft in diameter sticks straight down into freshwater with the diameter along the surface. Find the force exerted by the water on one side of the plate.

35. The cubical metal tank shown here has a parabolic gate held in place by bolts and designed to withstand a fluid force of 160 lb without rupturing. The liquid you plan to store has a weight-density of $50\ \text{lb/ft}^3$.

a. What is the fluid force on the gate when the liquid is 2 ft deep?

b. What is the maximum height to which the container can be filled without exceeding the gate's design limitation?

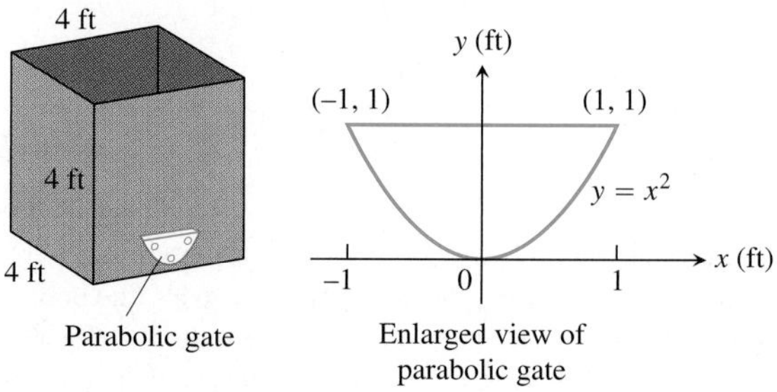

36. The end plates of the trough shown here were designed to withstand a fluid force of 6667 lb. How many cubic feet of water can the tank hold without exceeding this limitation? Round down to the nearest cubic foot.

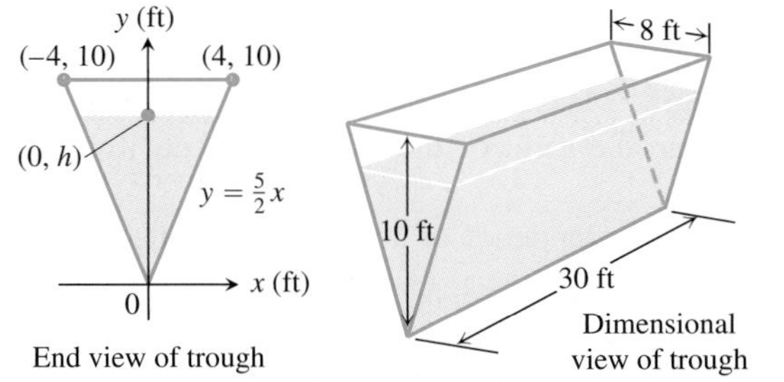

37. A vertical rectangular plate a units long by b units wide is submerged in a fluid of weight-density w with its long edges parallel to the fluid's surface. Find the average value of the pressure along the vertical dimension of the plate. Explain your answer.

38. *(Continuation of Exercise 37.)* Show that the force exerted by the fluid on one side of the plate is the average value of the pressure (found in Exercise 37) times the area of the plate.

39. Water pours into the tank here at the rate of $4\ \text{ft}^3/\text{min}$. The tank's cross-sections are 4-ft-diameter semicircles. One end of the tank is movable, but moving it to increase the volume compresses a spring. The spring constant is $k = 100\ \text{lb/ft}$. If the end of the tank moves 5 ft against the spring, the water will drain out of a safety hole in the bottom at the rate of $5\ \text{ft}^3/\text{min}$. Will the movable end reach the hole before the tank overflows?

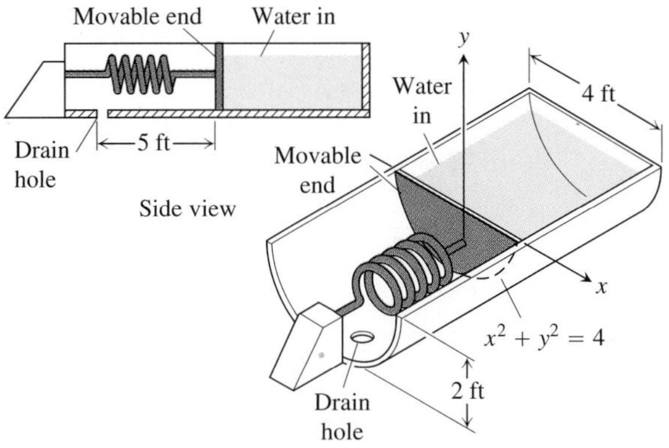

40. **Watering trough** The vertical ends of a watering trough are squares 3 ft on a side.
   a. Find the fluid force against the ends when the trough is full.
   b. How many inches do you have to lower the water level in the trough to reduce the fluid force by 25%?

6.6 Moments and Centers of Mass

Many structures and mechanical systems behave as if their masses were concentrated at a single point, called the *center of mass* (Figure 6.35). It is important to know how to locate this point, and doing so is basically a mathematical enterprise. For the moment, we deal with one- and two-dimensional objects. Three-dimensional objects are best done with the multiple integrals of Chapter 12.

Masses Along a Line

We develop our mathematical model in stages. The first stage is to imagine masses m_1, m_2, and m_3 on a rigid x-axis supported by a fulcrum at the origin.

The resulting system might balance, or it might not, depending on how large the masses are and how they are arranged.

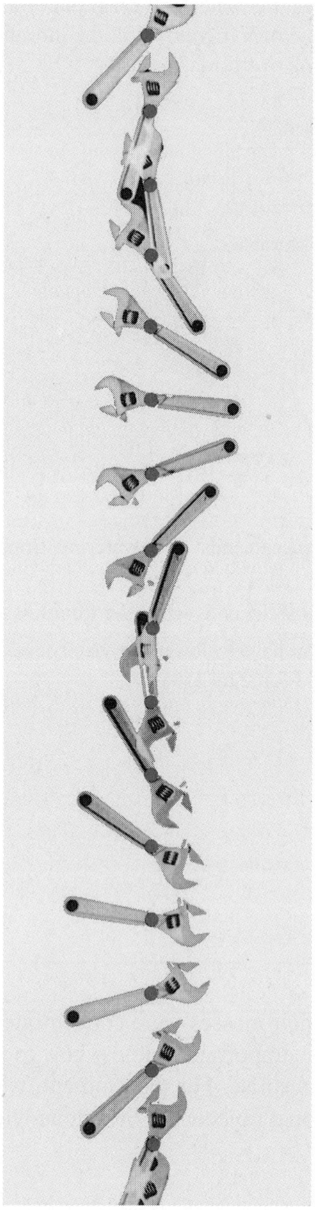

FIGURE 6.35 A wrench gliding on ice turning about its center of mass as the center glides in a straight line.

Each mass m_k exerts a downward force $m_k g$ (the weight of m_k) equal to the magnitude of the mass times the acceleration due to gravity. Each of these forces has a tendency to turn the axis about the origin, the way you turn a seesaw. This turning effect, called a **torque**, is measured by multiplying the force $m_k g$ by the signed distance x_k from the point of application to the origin. Masses to the left of the origin exert negative (counterclockwise) torque. Masses to the right of the origin exert positive (clockwise) torque.

The sum of the torques measures the tendency of a system to rotate about the origin. This sum is called the **system torque**.

$$\text{System torque} = m_1 g x_1 + m_2 g x_2 + m_3 g x_3 \tag{1}$$

The system will balance if and only if its torque is zero.

If we factor out the g in Equation (1), we see that the system torque is

$$\underbrace{g}_{\text{a feature of the environment}} \cdot \underbrace{(m_1 x_1 + m_2 x_2 + m_3 x_3)}_{\text{a feature of the system}}$$

Thus, the torque is the product of the gravitational acceleration g, which is a feature of the environment in which the system happens to reside, and the number $(m_1 x_1 + m_2 x_2 + m_3 x_3)$, which is a feature of the system itself, a constant that stays the same no matter where the system is placed.

The number $(m_1 x_1 + m_2 x_2 + m_3 x_3)$ is called the **moment of the system about the origin**. It is the sum of the **moments** $m_1 x_1, m_2 x_2, m_3 x_3$ of the individual masses.

$$M_0 = \text{Moment of system about origin} = \sum m_k x_k.$$

(We shift to sigma notation here to allow for sums with more terms.)

We usually want to know where to place the fulcrum to make the system balance, that is, at what point $\bar{x}$ to place it to make the torques add to zero.

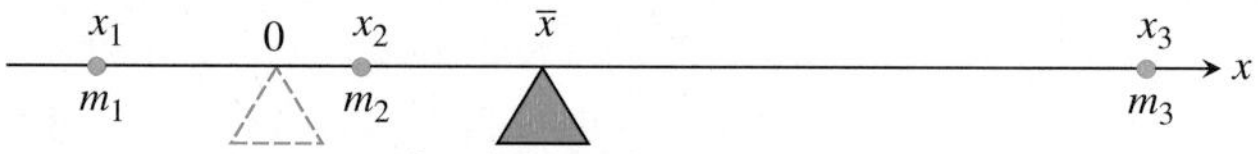

The torque of each mass about the fulcrum in this special location is

$$\begin{aligned} \text{Torque of } m_k \text{ about } \bar{x} &= \begin{pmatrix}\text{signed distance} \\ \text{of } m_k \text{ from } \bar{x}\end{pmatrix}\begin{pmatrix}\text{downward} \\ \text{force}\end{pmatrix} \\ &= (x_k - \bar{x}) m_k g. \end{aligned}$$

When we write the equation that says that the sum of these torques is zero, we get an equation we can solve for $\bar{x}$:

$$\sum (x_k - \bar{x}) m_k g = 0 \qquad \text{Sum of the torques equals zero}$$

$$\bar{x} = \frac{\sum m_k x_k}{\sum m_k}. \qquad \text{Solved for } \bar{x}$$

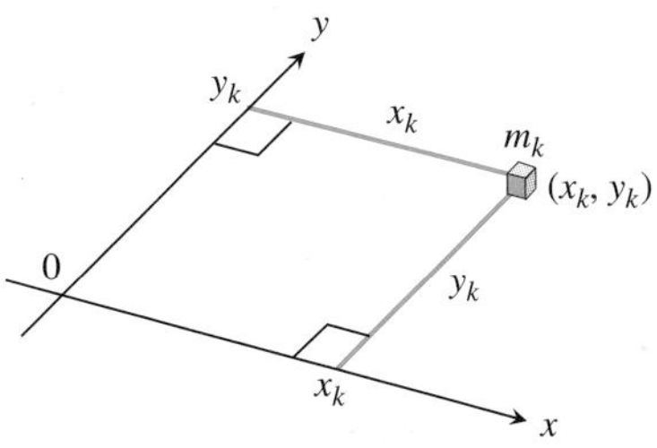

FIGURE 6.36 Each mass m_k has a moment about each axis.

This last equation tells us to find $\bar{x}$ by dividing the system's moment about the origin by the system's total mass:

$$\bar{x} = \frac{\sum m_k x_k}{\sum m_k} = \frac{\text{system moment about origin}}{\text{system mass}}. \tag{2}$$

The point $\bar{x}$ is called the system's **center of mass**.

Masses Distributed over a Plane Region

Suppose that we have a finite collection of masses located in the plane, with mass m_k at the point (x_k, y_k) (see Figure 6.36). The mass of the system is

$$\text{System mass:} \qquad M = \sum m_k.$$

Each mass m_k has a moment about each axis. Its moment about the x-axis is $m_k y_k$, and its moment about the y-axis is $m_k x_k$. The moments of the entire system about the two axes are

$$\text{Moment about } x\text{-axis:} \qquad M_x = \sum m_k y_k,$$

$$\text{Moment about } y\text{-axis:} \qquad M_y = \sum m_k x_k.$$

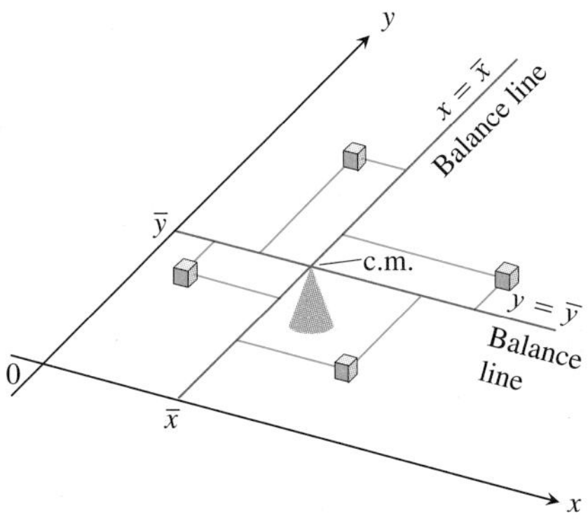

FIGURE 6.37 A two-dimensional array of masses balances on its center of mass.

The x-coordinate of the system's center of mass is defined to be

$$\bar{x} = \frac{M_y}{M} = \frac{\sum m_k x_k}{\sum m_k}. \tag{3}$$

With this choice of $\bar{x}$, as in the one-dimensional case, the system balances about the line $x = \bar{x}$ (Figure 6.37).

The y-coordinate of the system's center of mass is defined to be

$$\bar{y} = \frac{M_x}{M} = \frac{\sum m_k y_k}{\sum m_k}. \tag{4}$$

With this choice of $\bar{y}$, the system balances about the line $y = \bar{y}$ as well. The torques exerted by the masses about the line $y = \bar{y}$ cancel out. Thus, as far as balance is concerned, the system behaves as if all its mass were at the single point $(\bar{x}, \bar{y})$. We call this point the system's **center of mass**.

Thin, Flat Plates

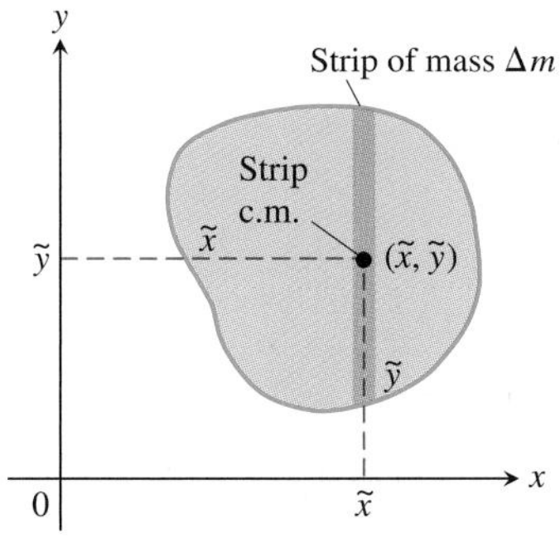

FIGURE 6.38 A plate cut into thin strips parallel to the y-axis. The moment exerted by a typical strip about each axis is the moment its mass Δm would exert if concentrated at the strip's center of mass $(\tilde{x}, \tilde{y})$.

In many applications, we need to find the center of mass of a thin, flat plate: a disk of aluminum, say, or a triangular sheet of steel. In such cases, we assume the distribution of mass to be continuous, and the formulas we use to calculate $\bar{x}$ and $\bar{y}$ contain integrals instead of finite sums. The integrals arise in the following way.

Imagine the plate occupying a region in the xy-plane, cut into thin strips parallel to one of the axes (in Figure 6.38, the y-axis). The center of mass of a typical strip is $(\tilde{x}, \tilde{y})$. We treat the strip's mass Δm as if it were concentrated at $(\tilde{x}, \tilde{y})$. The moment of the strip about the y-axis is then $\tilde{x}\,\Delta m$. The moment of the strip about the x-axis is $\tilde{y}\,\Delta m$. Equations (3) and (4) then become

$$\bar{x} = \frac{M_y}{M} = \frac{\sum \tilde{x}\,\Delta m}{\sum \Delta m}, \qquad \bar{y} = \frac{M_x}{M} = \frac{\sum \tilde{y}\,\Delta m}{\sum \Delta m}.$$

The sums are Riemann sums for integrals and approach these integrals as limiting values as the strips into which the plate is cut become narrower and narrower. We write these integrals symbolically as

$$\bar{x} = \frac{\int \tilde{x}\, dm}{\int dm} \quad \text{and} \quad \bar{y} = \frac{\int \tilde{y}\, dm}{\int dm}.$$

Moments, Mass, and Center of Mass of a Thin Plate Covering a Region in the *xy*-Plane

$$\begin{aligned} &\text{Moment about the } x\text{-axis:} && M_x = \int \tilde{y}\, dm \\ &\text{Moment about the } y\text{-axis:} && M_y = \int \tilde{x}\, dm \\ &\text{Mass:} && M = \int dm \\ &\text{Center of mass:} && \bar{x} = \frac{M_y}{M}, \quad \bar{y} = \frac{M_x}{M} \end{aligned} \tag{5}$$

Density
A material's density is its mass per unit area. For wires, rods, and narrow strips, we use mass per unit length.

The differential dm is the mass of the strip. Assuming the density δ of the plate to be a continuous function, the mass differential dm equals the product $\delta\, dA$ (mass per unit area times area). Here dA represents the area of the strip.

To evaluate the integrals in Equations (5), we picture the plate in the coordinate plane and sketch a strip of mass parallel to one of the coordinate axes. We then express the strip's mass dm and the coordinates $(\tilde{x}, \tilde{y})$ of the strip's center of mass in terms of x or y. Finally, we integrate $\tilde{y}\, dm$, $\tilde{x}\, dm$, and dm between limits of integration determined by the plate's location in the plane.

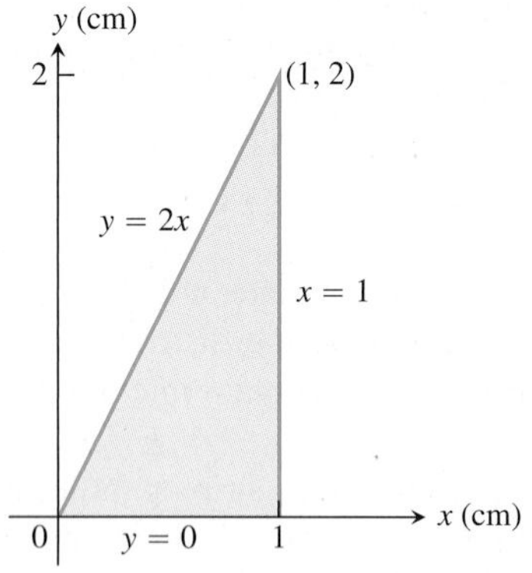

FIGURE 6.39 The plate in Example 1.

EXAMPLE 1 The triangular plate shown in Figure 6.39 has a constant density of $\delta = 3\ \text{g/cm}^2$. Find

(a) the plate's moment M_y about the y-axis. **(b)** the plate's mass M.

(c) the x-coordinate of the plate's center of mass (c.m.).

Solution **Method 1: Vertical Strips** (Figure 6.40)

(a) The moment M_y: The typical vertical strip has

center of mass (c.m.): $(\tilde{x}, \tilde{y}) = (x, x)$
length: $2x$
width: dx
area: $dA = 2x\, dx$
mass: $dm = \delta\, dA = 3 \cdot 2x\, dx = 6x\, dx$
distance of c.m. from y-axis: $\tilde{x} = x$.

The moment of the strip about the y-axis is

$$\tilde{x}\, dm = x \cdot 6x\, dx = 6x^2\, dx.$$

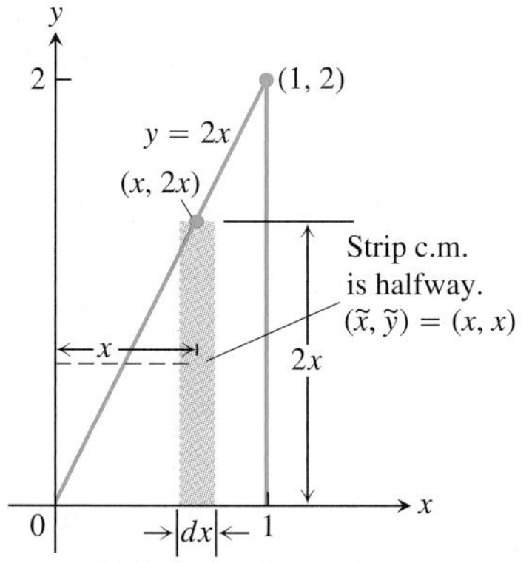

FIGURE 6.40 Modeling the plate in Example 1 with vertical strips.

The moment of the plate about the y-axis is therefore

$$M_y = \int \widetilde{x}\, dm = \int_0^1 6x^2\, dx = 2x^3\Big]_0^1 = 2 \text{ g} \cdot \text{cm}.$$

(b) The plate's mass:

$$M = \int dm = \int_0^1 6x\, dx = 3x^2\Big]_0^1 = 3 \text{ g}.$$

(c) The x-coordinate of the plate's center of mass:

$$\bar{x} = \frac{M_y}{M} = \frac{2 \text{ g} \cdot \text{cm}}{3 \text{ g}} = \frac{2}{3} \text{ cm}.$$

By a similar computation, we could find M_x and $\bar{y} = M_x/M$.

Method 2: Horizontal Strips (Figure 6.41)

(a) The moment M_y: The y-coordinate of the center of mass of a typical horizontal strip is y (see the figure), so

$$\widetilde{y} = y.$$

The x-coordinate is the x-coordinate of the point halfway across the triangle. This makes it the average of $y/2$ (the strip's left-hand x-value) and 1 (the strip's right-hand x-value):

$$\widetilde{x} = \frac{(y/2) + 1}{2} = \frac{y}{4} + \frac{1}{2} = \frac{y + 2}{4}.$$

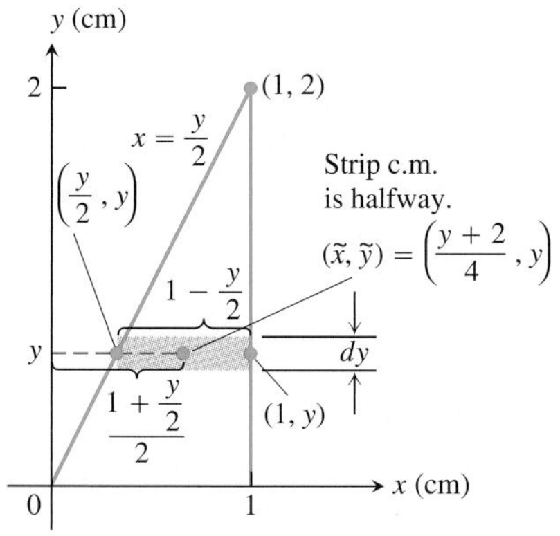

FIGURE 6.41 Modeling the plate in Example 1 with horizontal strips.

We also have

length: $1 - \frac{y}{2} = \frac{2 - y}{2}$

width: dy

area: $dA = \frac{2 - y}{2}\, dy$

mass: $dm = \delta\, dA = 3 \cdot \frac{2 - y}{2}\, dy$

distance of c.m. to y-axis: $\widetilde{x} = \frac{y + 2}{4}.$

The moment of the strip about the y-axis is

$$\widetilde{x}\, dm = \frac{y + 2}{4} \cdot 3 \cdot \frac{2 - y}{2}\, dy = \frac{3}{8}(4 - y^2)\, dy.$$

The moment of the plate about the y-axis is

$$M_y = \int \widetilde{x}\, dm = \int_0^2 \frac{3}{8}(4 - y^2)\, dy = \frac{3}{8}\left[4y - \frac{y^3}{3}\right]_0^2 = \frac{3}{8}\left(\frac{16}{3}\right) = 2 \text{ g} \cdot \text{cm}.$$

(b) The plate's mass:

$$M = \int dm = \int_0^2 \frac{3}{2}(2 - y)\, dy = \frac{3}{2}\left[2y - \frac{y^2}{2}\right]_0^2 = \frac{3}{2}(4 - 2) = 3 \text{ g}.$$

(c) The x-coordinate of the plate's center of mass:

$$\bar{x} = \frac{M_y}{M} = \frac{2\text{ g}\cdot\text{cm}}{3\text{ g}} = \frac{2}{3}\text{ cm}.$$

By a similar computation, we could find M_x and $\bar{y}$. ■

If the distribution of mass in a thin, flat plate has an axis of symmetry, the center of mass will lie on this axis. If there are two axes of symmetry, the center of mass will lie at their intersection. These facts often help to simplify our work.

EXAMPLE 2 Find the center of mass of a thin plate covering the region bounded above by the parabola $y = 4 - x^2$ and below by the x-axis (Figure 6.42). Assume the density of the plate at the point (x, y) is $\delta = 2x^2$, which is twice the square of the distance from the point to the y-axis.

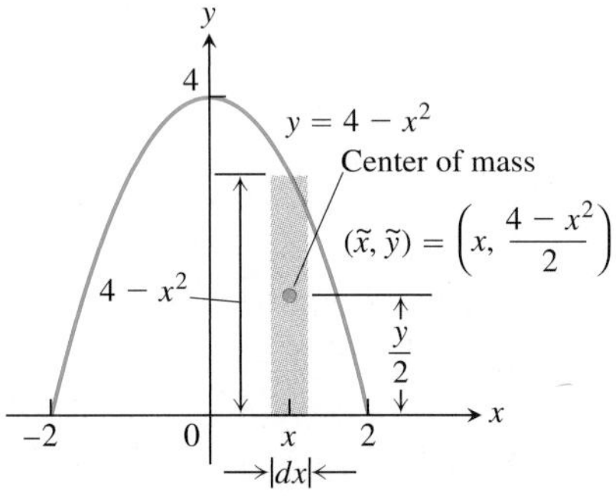

FIGURE 6.42 Modeling the plate in Example 2 with vertical strips.

Solution The mass distribution is symmetric about the y-axis, so $\bar{x} = 0$. We model the distribution of mass with vertical strips since the density is given as a function of the variable x. The typical vertical strip (see Figure 6.42) has

center of mass (c.m.): $(\tilde{x}, \tilde{y}) = \left(x, \dfrac{4 - x^2}{2}\right)$

length: $4 - x^2$

width: dx

area: $dA = (4 - x^2)\, dx$

mass: $dm = \delta\, dA = \delta(4 - x^2)\, dx$

distance from c.m. to x-axis: $\tilde{y} = \dfrac{4 - x^2}{2}.$

The moment of the strip about the x-axis is

$$\tilde{y}\, dm = \frac{4 - x^2}{2} \cdot \delta(4 - x^2)\, dx = \frac{\delta}{2}(4 - x^2)^2\, dx.$$

The moment of the plate about the x-axis is

$$\begin{aligned} M_x &= \int \tilde{y}\, dm = \int_{-2}^{2} \frac{\delta}{2}(4 - x^2)^2\, dx = \int_{-2}^{2} x^2(4 - x^2)^2\, dx \\ &= \int_{-2}^{2} (16x^2 - 8x^4 + x^6)\, dx = \frac{2048}{105} \end{aligned}$$

$$\begin{aligned} M &= \int dm = \int_{-2}^{2} \delta(4 - x^2)\, dx = \int_{-2}^{2} 2x^2(4 - x^2)\, dx \\ &= \int_{-2}^{2} (8x^2 - 2x^4)\, dx = \frac{256}{15}. \end{aligned}$$

Therefore,

$$\bar{y} = \frac{M_x}{M} = \frac{2048}{105} \cdot \frac{15}{256} = \frac{8}{7}.$$

The plate's center of mass is

$$(\bar{x}, \bar{y}) = \left(0, \frac{8}{7}\right).$$ ■

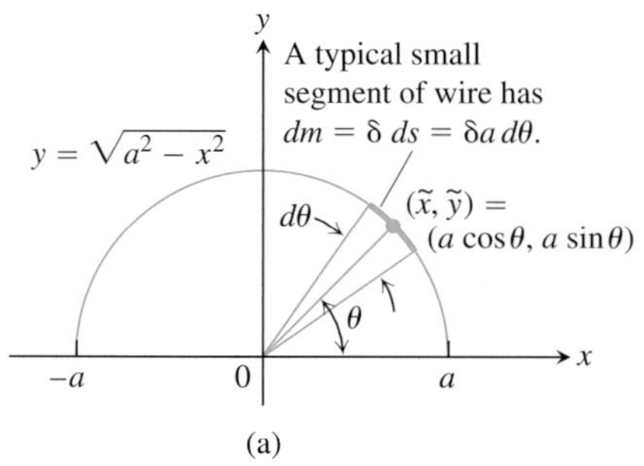

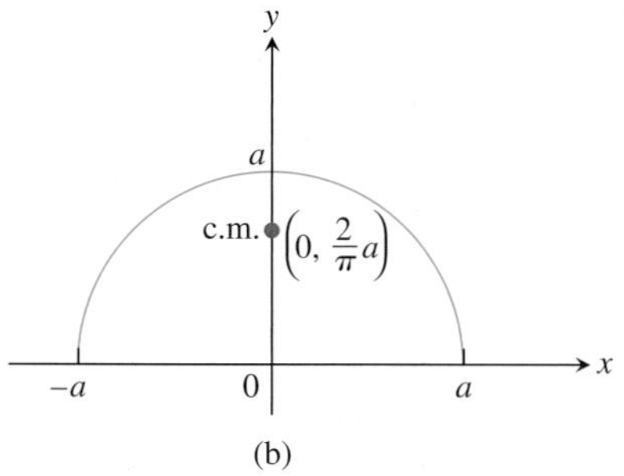

FIGURE 6.43 The semicircular wire in Example 3. (a) The dimensions and variables used in finding the center of mass. (b) The center of mass does not lie on the wire.

Centroids

When the density function is constant, it cancels out of the numerator and denominator of the formulas for $\bar{x}$ and $\bar{y}$. Thus, when the density is constant, the location of the center of mass is a feature of the geometry of the object and not of the material from which it is made. In such cases, engineers may call the center of mass the **centroid** of the shape, as in "Find the centroid of a triangle or a solid cone." To do so, just set δ equal to 1 and proceed to find $\bar{x}$ and $\bar{y}$ as before, by dividing moments by masses.

EXAMPLE 3 Find the center of mass (centroid) of a wire of constant density δ shaped like a semicircle of radius a.

Solution We model the wire with the semicircle $y = \sqrt{a^2 - x^2}$ (Figure 6.43). The distribution of mass is symmetric about the y-axis, so $\bar{x} = 0$. To find $\bar{y}$, we imagine the wire divided into short segments. The typical segment (Figure 6.43a) has

length: $ds = a\, d\theta$

mass: $dm = \delta\, ds = \delta a\, d\theta$ — Mass per unit length times length

distance of c.m. to x-axis: $\tilde{y} = a \sin\theta$.

Hence,

$$\bar{y} = \frac{\int \tilde{y}\, dm}{\int dm} = \frac{\int_0^{\pi} a \sin\theta \cdot \delta a\, d\theta}{\int_0^{\pi} \delta a\, d\theta} = \frac{\delta a^2 \left[-\cos\theta\right]_0^{\pi}}{\delta a \pi} = \frac{2}{\pi} a.$$

The center of mass lies on the axis of symmetry at the point $(0, 2a/\pi)$, about two-thirds of the way up from the origin (Figure 6.43b). Notice how δ cancels in the equation for $\bar{y}$, so we could have set $\delta = 1$ everywhere and obtained the same value for $\bar{y}$. ■

EXERCISES 6.6

In Exercises 1–14, find the center of mass of a thin plate of constant density δ covering the given region.

1. The region bounded by the parabola $y = x^2$ and the line $y = 4$
2. The region bounded by the parabola $y = 25 - x^2$ and the x-axis
3. The region bounded by the parabola $y = x - x^2$ and the line $y = -x$
4. The region enclosed by the parabolas $y = x^2 - 3$ and $y = -2x^2$
5. The region bounded by the y-axis and the curve $x = y - y^3$, $0 \le y \le 1$
6. The region bounded by the parabola $x = y^2 - y$ and the line $y = x$
7. The region bounded by the x-axis and the curve $y = \cos x$, $-\pi/2 \le x \le \pi/2$
8. The region between the x-axis and the curve $y = \sec^2 x$, $-\pi/4 \le x \le \pi/4$
9. T The region between the curve $y = 1/x$ and the x-axis from $x = 1$ to $x = 2$. Give the coordinates to two decimal places.
10. **a.** The region cut from the first quadrant by the circle $x^2 + y^2 = 9$
    b. The region bounded by the x-axis and the semicircle $y = \sqrt{9 - x^2}$
    Compare your answer in part (b) with the answer in part (a).
11. The region in the first and fourth quadrants enclosed by the curves $y = 1/(1 + x^2)$ and $y = -1/(1 + x^2)$ and by the lines $x = 0$ and $x = 1$
12. The region bounded by the parabolas $y = 2x^2 - 4x$ and $y = 2x - x^2$
13. The region between the curve $y = 1/\sqrt{x}$ and the x-axis from $x = 1$ to $x = 16$
14. The region bounded above by the curve $y = 1/x^3$, below by the curve $y = -1/x^3$, and on the left and right by the lines $x = 1$ and $x = a > 1$. Also, find $\lim_{a\to\infty} \bar{x}$.
15. Find the center of mass of a thin plate covering the region between the x-axis and the curve $y = 2/x^2$, $1 \le x \le 2$, if the plate's density at the point (x, y) is $\delta(x) = x^2$.

16. Find the center of mass of a thin plate covering the region bounded below by the parabola $y = x^2$ and above by the line $y = x$ if the plate's density at the point (x, y) is $\delta(x) = 12x$.

17. Find the center of mass of the thin plate in Exercise 13 if, instead of being constant, the density function is $\delta(x) = 4/\sqrt{x}$.

18. The region between the curve $y = 2/x$ and the x-axis from $x = 1$ to $x = 4$ is revolved about the x-axis to generate a solid.

a. Find the volume of the solid.

b. Find the center of mass of a thin plate covering the region if the plate's density at the point (x, y) is $\delta(x) = \sqrt{x}$.

c. Sketch the plate and show the center of mass in your sketch.

19. The region bounded by the curves $y = \pm 4/\sqrt{x}$ and the lines $x = 1$ and $x = 4$ is revolved about the y-axis to generate a solid.

a. Find the volume of the solid.

b. Find the center of mass of a thin plate covering the region if the plate's density at the point (x, y) is $\delta(x) = 1/x$.

c. Sketch the plate and show the center of mass in your sketch.

20. The region between the curve $y = 1/(2\sqrt{x})$ and the x-axis from $x = 1/4$ to $x = 4$ is revolved about the x-axis to generate a solid.

a. Find the volume of the solid.

b. Find the centroid of the region.

21. Centroid of a triangle Show that the centroid of any triangle lies at the intersection of the medians. To do so, denote the vertices by $(a, 0)$, $(0, b)$, and $(c, 0)$, so that one side of the triangle lies along the x-axis. A median is a line segment from a vertex to the midpoint of the opposite side, and the medians intersect at the point inside the triangle that lies one-third of the way from each side (along the median).

Use the result in Exercise 21 to find the centroids of the triangles whose vertices appear in Exercises 22–26. Assume $a, b > 0$.

22. $(-1, 0), (1, 0), (0, 3)$

23. $(0, 0), (1, 0), (0, 1)$

24. $(0, 0), (a, 0), (0, a)$

25. $(0, 0), (a, 0), (0, b)$

26. $(0, 0), (a, 0), (a/2, b)$

27. Constant density Find the moment about the x-axis of a wire of constant density that lies along the curve $y = \sqrt{x}$ from $x = 0$ to $x = 2$.

28. Constant density Find the moment about the x-axis of a wire of constant density that lies along the curve $y = x^3$ from $x = 0$ to $x = 1$.

29. Variable density Suppose that the density of the wire in Example 3 is $\delta = k \sin \theta$ (k constant). Find the center of mass.

30. Variable density Suppose that the density of the wire in Example 3 is $\delta = 1 + k|\cos \theta|$ (k constant). Find the center of mass.

Verify the statements and formulas in Exercises 31 and 32.

31. The coordinates of the centroid of a differentiable plane curve are

$$\bar{x} = \frac{\int x\, ds}{\text{length}}, \qquad \bar{y} = \frac{\int y\, ds}{\text{length}}.$$

32. Whatever the value of $p > 0$ in the equation $y = x^2/(4p)$, the y-coordinate of the centroid of the parabolic segment shown here is $\bar{y} = (3/5)a$.

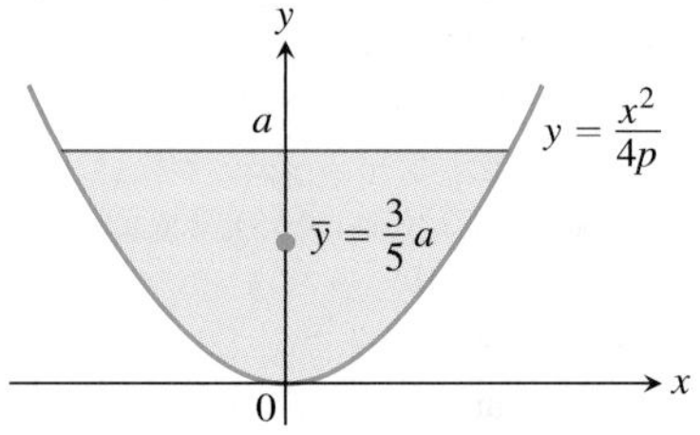

Chapter 6 Questions to Guide Your Review

1. How do you define and calculate the volumes of solids by the method of slicing? Give an example.

2. How are the disk and washer methods for calculating volumes derived from the method of slicing? Give examples of volume calculations by these methods.

3. Describe the method of cylindrical shells. Give an example.

4. How do you define the length of a smooth parametrized curve $x = f(t), y = g(t), a \le t \le b$? What does smoothness have to do with length? What else do you need to know about the parametrization to find the curve's length? Give examples.

5. How do you find the length of the graph of a smooth function over a closed interval? Give an example. What about functions that do not have continuous first derivatives?

6. How do you define and calculate the area of the surface swept out by revolving the graph of a smooth function $y = f(x)$, $a \le x \le b$, about the x-axis? Give an example.

7. Under what conditions can you find the area of the surface generated by revolving a curve $x = f(t), y = g(t), a \leq t \leq b$, about the x-axis? The y-axis? Give examples.
8. How do you solve separable first-order differential equations?
9. What is the law of exponential change? How can it be derived from an initial value problem? What are some of the applications of the law?
10. How do you define and calculate the work done by a variable force directed along a portion of the x-axis? How do you calculate the work it takes to pump a liquid from a tank? Give examples.
11. How do you calculate the force exerted by a liquid against a portion of a vertical wall? Give an example.
12. What is a center of mass?
13. How do you locate the center of mass of a thin flat plate of material? Give an example.

Chapter 6 Practice and Additional Exercises

Find the volumes of the solids in Exercises 1–16.

1. The solid lies between planes perpendicular to the x-axis at $x = 0$ and $x = 1$. The cross-sections perpendicular to the x-axis between these planes are circular disks whose diameters run from the parabola $y = x^2$ to the parabola $y = \sqrt{x}$.
2. The base of the solid is the region in the first quadrant between the line $y = x$ and the parabola $y = 2\sqrt{x}$. The cross-sections of the solid perpendicular to the x-axis are equilateral triangles whose bases stretch from the line to the curve.
3. The solid lies between planes perpendicular to the x-axis at $x = \pi/4$ and $x = 5\pi/4$. The cross-sections between these planes are circular disks whose diameters run from the curve $y = 2\cos x$ to the curve $y = 2\sin x$.
4. The solid lies between planes perpendicular to the x-axis at $x = 0$ and $x = 6$. The cross-sections between these planes are squares whose bases run from the x-axis up to the curve $x^{1/2} + y^{1/2} = \sqrt{6}$.

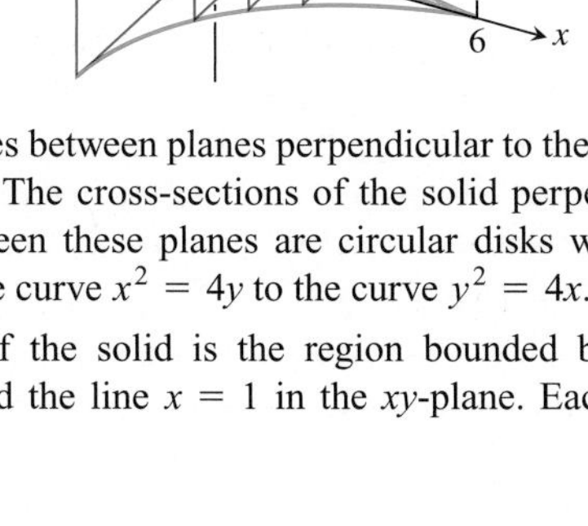

5. The solid lies between planes perpendicular to the x-axis at $x = 0$ and $x = 4$. The cross-sections of the solid perpendicular to the x-axis between these planes are circular disks whose diameters run from the curve $x^2 = 4y$ to the curve $y^2 = 4x$.
6. The base of the solid is the region bounded by the parabola $y^2 = 4x$ and the line $x = 1$ in the xy-plane. Each cross-section perpendicular to the x-axis is an equilateral triangle with one edge in the plane. (The triangles all lie on the same side of the plane.)
7. Find the volume of the solid generated by revolving the region bounded by the x-axis, the curve $y = 3x^4$, and the lines $x = 1$ and $x = -1$ about **(a)** the x-axis; **(b)** the y-axis; **(c)** the line $x = 1$; **(d)** the line $y = 3$.
8. Find the volume of the solid generated by revolving the "triangular" region bounded by the curve $y = 4/x^3$ and the lines $x = 1$ and $y = 1/2$ about **(a)** the x-axis; **(b)** the y-axis; **(c)** the line $x = 2$; **(d)** the line $y = 4$.
9. Find the volume of the solid generated by revolving the region bounded on the left by the parabola $x = y^2 + 1$ and on the right by the line $x = 5$ about **(a)** the x-axis; **(b)** the y-axis; **(c)** the line $x = 5$.
10. Find the volume of the solid generated by revolving the region bounded by the parabola $y^2 = 4x$ and the line $y = x$ about **(a)** the x-axis; **(b)** the y-axis; **(c)** the line $x = 4$; **(d)** the line $y = 4$.
11. Find the volume of the solid generated by revolving the "triangular" region bounded by the x-axis, the line $x = \pi/3$, and the curve $y = \tan x$ in the first quadrant about the x-axis.
12. Find the volume of the solid generated by revolving the region bounded by the curve $y = \sin x$ and the lines $x = 0, x = \pi$, and $y = 2$ about the line $y = 2$.
13. Find the volume of the solid generated by revolving the region bounded by the curve $x = e^{y^2}$ and the lines $y = 0$, $x = 0$, and $y = 1$ about the x-axis.
14. Find the volume of the solid generated by revolving about the x-axis the region bounded by $y = 2\tan x, y = 0, x = -\pi/4$, and $x = \pi/4$. (The region lies in the first and third quadrants and resembles a skewed bowtie.)
15. **Volume of a solid sphere hole** A round hole of radius $\sqrt{3}$ ft is bored through the center of a solid sphere of a radius 2 ft. Find the volume of material removed from the sphere.

16. Volume of a football The profile of a football resembles the ellipse shown here. Find the football's volume to the nearest cubic inch.

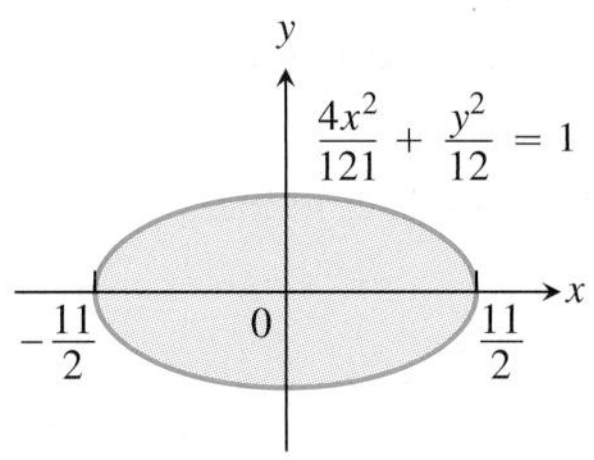

Find the lengths of the curves in Exercises 17–23.

17. $y = x^{1/2} - (1/3)x^{3/2}, \quad 1 \le x \le 4$

18. $x = y^{2/3}, \quad 1 \le y \le 8$

19. $y = x^2 - (\ln x)/8, \quad 1 \le x \le 2$

20. $x = (y^3/12) + (1/y), \quad 1 \le y \le 2$

21. $x = 5\cos t - \cos 5t, \quad y = 5\sin t - \sin 5t, \quad 0 \le t \le \pi/2$

22. $x = t^3 - 6t^2, \quad y = t^3 + 6t^2, \quad 0 \le t \le 1$

23. $x = 3\cos\theta, \quad y = 3\sin\theta, \quad 0 \le \theta \le \frac{3\pi}{2}$

24. Find the length of the enclosed loop $x = t^2, y = (t^3/3) - t$ shown here. The loop starts at $t = -\sqrt{3}$ and ends at $t = \sqrt{3}$.

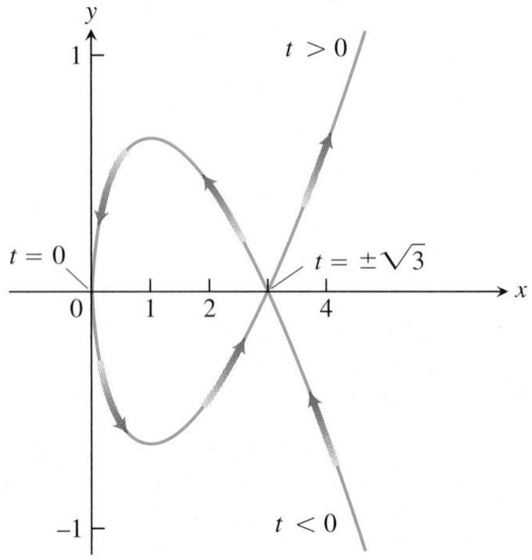

In Exercises 25–30, find the areas of the surfaces generated by revolving the curves about the given axes.

25. $y = \sqrt{2x + 1}, \quad 0 \le x \le 3; \quad x\text{-axis}$

26. $y = x^3/3, \quad 0 \le x \le 1; \quad x\text{-axis}$

27. $x = \sqrt{4y - y^2}, \quad 1 \le y \le 2; \quad y\text{-axis}$

28. $x = \sqrt{y}, \quad 2 \le y \le 6; \quad y\text{-axis}$

29. $x = t^2/2, \quad y = 2t, \quad 0 \le t \le \sqrt{5}; \quad x\text{-axis}$

30. $x = t^2 + 1/(2t), \quad y = 4\sqrt{t}, \quad 1/\sqrt{2} \le t \le 1; \quad y\text{-axis}$

In Exercises 31–34 solve the differential equation.

31. $\frac{dy}{dx} = \sqrt{y}\cos^2\sqrt{y}$ **32.** $y' = \frac{3y(x+1)^2}{y-1}$

33. $yy' = \sec y^2 \sec^2 x$ **34.** $y\cos^2 x\, dy + \sin x\, dx = 0$

In Exercises 35–38 solve the initial value problem.

35. $\frac{dy}{dx} = e^{-x-y-2}, \quad y(0) = -2$ **36.** $\frac{dy}{dx} = \frac{y \ln y}{1 + x^2}, \quad y(0) = e^2$

37. $x\,dy - \left(y + \sqrt{y}\right)dx = 0, \quad y(1) = 1$

38. $y^{-2}\frac{dx}{dy} = \frac{e^x}{e^{2x} + 1}, \quad y(0) = 1$

39. What is the age of a sample of charcoal in which 90% of the carbon-14 originally present has decayed?

40. Cooling a pie A deep-dish apple pie, whose internal temperature was 220°F when removed from the oven, was set out on a breezy 40°F porch to cool. Fifteen minutes later, the pie's internal temperature was 180°F. How long did it take the pie to cool from there to 70°F?

41. Lifting equipment A rock climber is about to haul up 100 N (about 22.5 lb) of equipment that has been hanging beneath her on 40 m of rope that weighs 0.8 newton per meter. How much work will it take? (*Hint:* Solve for the rope and equipment separately, then add.)

42. Leaky tank truck You drove an 800-gal tank truck of water from the base of Mt. Washington to the summit and discovered on arrival that the tank was only half full. You started with a full tank, climbed at a steady rate, and accomplished the 4750-ft elevation change in 50 min. Assuming that the water leaked out at a steady rate, how much work was spent in carrying water to the top? Do not count the work done in getting yourself and the truck there. Water weighs 8 lb/U.S. gal.

43. Stretching a spring If a force of 20 lb is required to hold a spring 1 ft beyond its unstressed length, how much work does it take to stretch the spring this far? An additional foot?

44. Garage door spring A force of 200 N will stretch a garage door spring 0.8 m beyond its unstressed length. How far will a 300-N force stretch the spring? How much work does it take to stretch the spring this far from its unstressed length?

45. Pumping a reservoir A reservoir shaped like a right circular cone, point down, 20 ft across the top and 8 ft deep, is full of water. How much work does it take to pump the water to a level 6 ft above the top?

46. Pumping a reservoir (*Continuation of Exercise 45.*) The reservoir is filled to a depth of 5 ft, and the water is to be pumped to the same level as the top. How much work does it take?

47. Pumping a conical tank A right circular conical tank, point down, with top radius 5 ft and height 10 ft is filled with a liquid whose weight-density is 60 lb/ft^3. How much work does it take to pump the liquid to a point 2 ft above the tank? If the pump is driven by a motor rated at 275 ft-lb/sec (1/2 hp), how long will it take to empty the tank?

48. Pumping a cylindrical tank A storage tank is a right circular cylinder 20 ft long and 8 ft in diameter with its axis horizontal.

If the tank is half full of olive oil weighing 57 lb/ft^3, find the work done in emptying it through a pipe that runs from the bottom of the tank to an outlet that is 6 ft above the top of the tank.

49. Find the centroid of a thin, flat plate covering the region enclosed by the parabolas $y = 2x^2$ and $y = 3 - x^2$.

50. Find the centroid of a thin, flat plate covering the region enclosed by the x-axis, the lines $x = 2$ and $x = -2$, and the parabola $y = x^2$.

51. Find the centroid of a thin, flat plate covering the "triangular" region in the first quadrant bounded by the y-axis, the parabola $y = x^2/4$, and the line $y = 4$.

52. Find the centroid of a thin, flat plate covering the region enclosed by the parabola $y^2 = x$ and the line $x = 2y$.

53. Find the center of mass of a thin, flat plate covering the region enclosed by the parabola $y^2 = x$ and the line $x = 2y$ if the density function is $\delta(y) = 1 + y$. (Use horizontal strips.)

54. a. Find the center of mass of a thin plate of constant density covering the region between the curve $y = 3/x^{3/2}$ and the x-axis from $x = 1$ to $x = 9$.

b. Find the plate's center of mass if, instead of being constant, the density is $\delta(x) = x$. (Use vertical strips.)

55. Trough of water The vertical triangular plate shown here is the end plate of a trough full of water ($w = 62.4$). What is the fluid force against the plate?

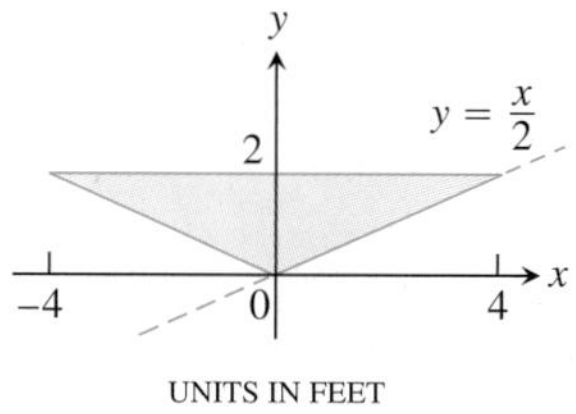

56. Trough of maple syrup The vertical trapezoid plate shown here is the end plate of a trough full of maple syrup weighing 75 lb/ft^3. What is the force exerted by the syrup against the end plate of the trough when the syrup is 10 in. deep?

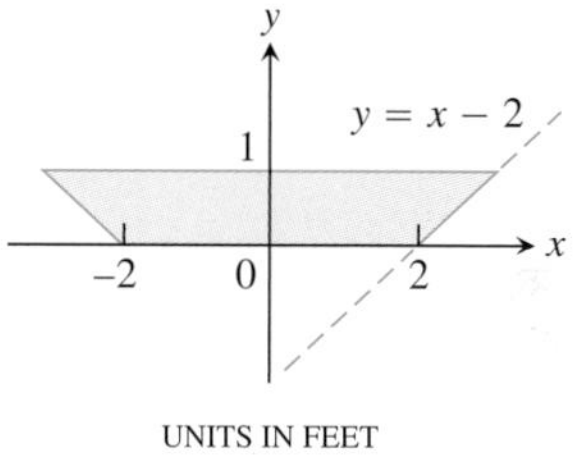

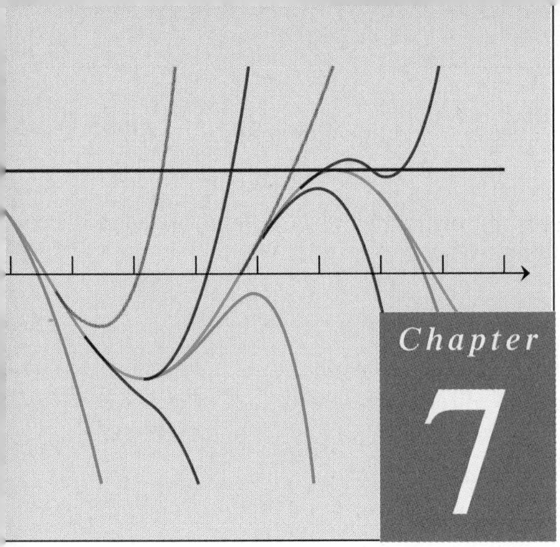

Chapter 7

INFINITE SEQUENCES AND SERIES

OVERVIEW Everyone knows how to add two numbers together, or even several. But how do you add infinitely many numbers together? In this chapter we answer this question, which is part of the theory of infinite sequences and series.

An important application of this theory is a method for representing a known differentiable function $f(x)$ as an infinite sum of powers of x, so it looks like a "polynomial with infinitely many terms." Moreover, the method extends our knowledge of how to evaluate, differentiate, and integrate polynomials, so we can work with even more general functions than those encountered so far. These new functions often solve important problems in science and engineering.

7.1 Sequences

HISTORICAL ESSAY

Sequences and Series

A sequence is a list of numbers

$$a_1, a_2, a_3, \ldots, a_n, \ldots$$

in a given order. Each of a_1, a_2, a_3 and so on represents a number. These are the **terms** of the sequence. For example the sequence

$$2, 4, 6, 8, 10, 12, \ldots, 2n, \ldots$$

has first term $a_1 = 2$, second term $a_2 = 4$, and nth term $a_n = 2n$. The integer n is called the **index** of a_n, and indicates where a_n occurs in the list. Order is important. The sequence $2, 4, 6, 8 \ldots$ is not the same as the sequence $4, 2, 6, 8 \ldots$.

We can think of the sequence

$$a_1, a_2, a_3, \ldots, a_n, \ldots$$

as a function that sends 1 to a_1, 2 to a_2, 3 to a_3, and in general sends the positive integer n to the nth term a_n. This leads to the formal definition of a sequence.

> **DEFINITION** An **infinite sequence** of numbers is a function whose domain is the set of positive integers.

The function associated with the sequence

$$2, 4, 6, 8, 10, 12, \ldots, 2n, \ldots$$

sends 1 to $a_1 = 2$, 2 to $a_2 = 4$, and so on. The general behavior of this sequence is described by the formula $a_n = 2n$.

We can equally well make the domain the integers larger than a given number n_0, and we allow sequences of this type also. For example, the sequence

$$12, 14, 16, 18, 20, 22 \ldots$$

is described by the formula $a_n = 10 + 2n$. It can also be described by the simpler formula $b_n = 2n$, where the index n starts at 6 and increases. To allow such simpler formulas, we let the first index of the sequence be any integer. In the sequence above, $\{a_n\}$ starts with a_1 while $\{b_n\}$ starts with b_6.

Sequences can be described by writing rules that specify their terms, such as

$$a_n = \sqrt{n}, \qquad b_n = (-1)^{n+1}\frac{1}{n}, \qquad c_n = \frac{n-1}{n}, \qquad d_n = (-1)^{n+1}$$

or by listing terms,

$$\{a_n\} = \left\{\sqrt{1}, \sqrt{2}, \sqrt{3}, \ldots, \sqrt{n}, \ldots\right\}$$

$$\{b_n\} = \left\{1, -\frac{1}{2}, \frac{1}{3}, -\frac{1}{4}, \ldots, (-1)^{n+1}\frac{1}{n}, \ldots\right\}$$

$$\{c_n\} = \left\{0, \frac{1}{2}, \frac{2}{3}, \frac{3}{4}, \frac{4}{5}, \ldots, \frac{n-1}{n}, \ldots\right\}$$

$$\{d_n\} = \{1, -1, 1, -1, 1, -1, \ldots, (-1)^{n+1}, \ldots\}.$$

We also sometimes write

$$\{a_n\} = \left\{\sqrt{n}\right\}_{n=1}^{\infty}$$

Figure 7.1 shows two ways to represent sequences graphically. The first marks the first few points from $a_1, a_2, a_3, \ldots, a_n, \ldots$ on the real axis. The second method shows the graph of the function defining the sequence. The function is defined only on integer inputs, and the graph consists of some points in the xy-plane, located at $(1, a_1), (2, a_2), \ldots, (n, a_n), \ldots$.

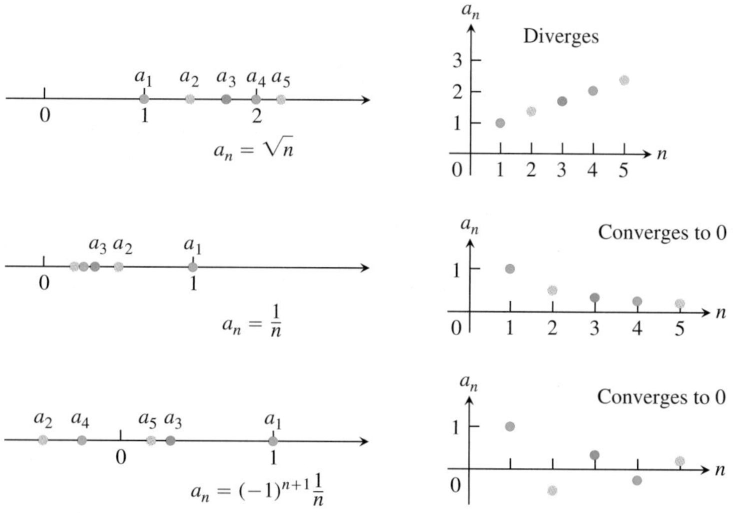

FIGURE 7.1 Sequences can be represented as points on the real line or as points in the plane where the horizontal axis n is the index number of the term and the vertical axis a_n is its value.

Convergence and Divergence

Sometimes the numbers in a sequence approach a single value as the index n increases. This happens in the sequence

$$\left\{1, \frac{1}{2}, \frac{1}{3}, \frac{1}{4}, \ldots, \frac{1}{n}, \ldots\right\}$$

whose terms approach 0 as n gets large, and in the sequence

$$\left\{0, \frac{1}{2}, \frac{2}{3}, \frac{3}{4}, \frac{4}{5}, \ldots, 1 - \frac{1}{n}, \ldots\right\}$$

whose terms approach 1. On the other hand, sequences like

$$\{\sqrt{1}, \sqrt{2}, \sqrt{3}, \ldots, \sqrt{n}, \ldots\}$$

have terms that get larger than any number as n increases, and sequences like

$$\{1, -1, 1, -1, 1, -1, \ldots, (-1)^{n+1}, \ldots\}$$

bounce back and forth between 1 and -1, never converging to a single value. The following definition captures the meaning of having a sequence converge to a limiting value. It says that if we go far enough out in the sequence, by taking the index n to be larger than some value N, the difference between a_n and the limit of the sequence becomes less than any preselected number $\epsilon > 0$.

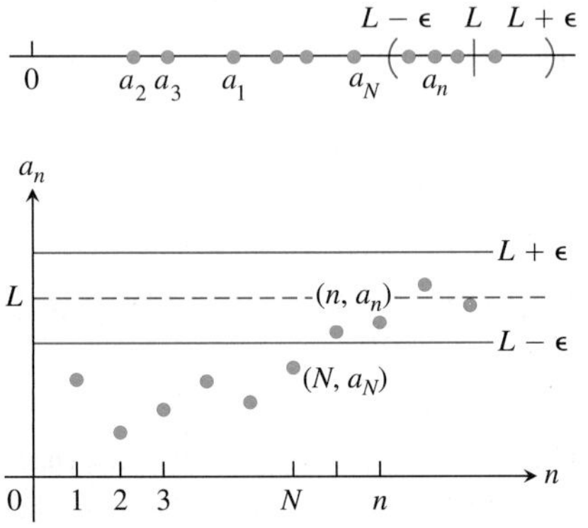

FIGURE 7.2 $a_n \to L$ if $y = L$ is a horizontal asymptote of the sequence of points $\{(n, a_n)\}$. In this figure, all the a_n's after a_N lie within ϵ of L.

DEFINITIONS The sequence $\{a_n\}$ **converges** to the number L if for every positive number ϵ there corresponds an integer N such that for all n,

$$n > N \quad \Rightarrow \quad |a_n - L| < \epsilon.$$

If no such number L exists, we say that $\{a_n\}$ **diverges.**

If $\{a_n\}$ converges to L, we write $\lim_{n\to\infty} a_n = L$, or simply $a_n \to L$, and call L the **limit** of the sequence (Figure 7.2).

The definition is very similar to the definition of the limit of a function $f(x)$ as x tends to ∞ ($\lim_{x\to\infty} f(x)$ in Section 1.8). We will exploit this connection to calculate limits of sequences.

HISTORICAL BIOGRAPHY

Nicole Oresme
(ca. 1320–1382)

EXAMPLE 1 Show that

(a) $\displaystyle\lim_{n\to\infty} \frac{1}{n} = 0$ **(b)** $\displaystyle\lim_{n\to\infty} k = k$ (any constant k)

Solution

(a) Let $\epsilon > 0$ be given. We must show that there exists an integer N such that for all n,

$$n > N \quad \Rightarrow \quad \left|\frac{1}{n} - 0\right| < \epsilon.$$

This implication will hold if $(1/n) < \epsilon$ or $n > 1/\epsilon$. If N is any integer greater than $1/\epsilon$, the implication will hold for all $n > N$. This proves that $\lim_{n\to\infty}(1/n) = 0$.

(b) Let $\epsilon > 0$ be given. We must show that there exists an integer N such that for all n,

$$n > N \quad \Rightarrow \quad |k - k| < \epsilon.$$

Since $k - k = 0$, we can use any positive integer for N and the implication will hold. This proves that $\lim_{n\to\infty} k = k$ for any constant k. ■

EXAMPLE 2 Show that the sequence $\{1, -1, 1, -1, 1, -1, \ldots, (-1)^{n+1}, \ldots\}$ diverges.

Solution Suppose the sequence converges to some number L. By choosing $\epsilon = 1/2$ in the definition of the limit, all terms a_n of the sequence with index n larger than some N must lie within $\epsilon = 1/2$ of L. Since the number 1 appears repeatedly as every other term of the sequence, we must have that the number 1 lies within the distance $\epsilon = 1/2$ of L. It follows that $|L - 1| < 1/2$, or equivalently, $1/2 < L < 3/2$. Likewise, the number -1 appears repeatedly in the sequence with arbitrarily high index. So we must also have that $|L - (-1)| < 1/2$, or equivalently, $-3/2 < L < -1/2$. But the number L cannot lie in both of the intervals $(1/2, 3/2)$ and $(-3/2, -1/2)$ because they have no overlap. Therefore, no such limit L exists and so the sequence diverges.

Note that the same argument works for any positive number ϵ smaller than 1, not just $1/2$. ■

The sequence $\{\sqrt{n}\}$ also diverges, but for a different reason. As n increases, its terms become larger than any fixed number. We describe the behavior of this sequence by writing

$$\lim_{n\to\infty} \sqrt{n} = \infty.$$

In writing infinity as the limit of a sequence, we are not saying that the differences between the terms a_n and ∞ become small as n increases. Nor are we asserting that there is some number infinity that the sequence approaches. We are merely using a notation that captures the idea that a_n eventually gets and stays larger than any fixed number as n gets large.

DEFINITION The sequence $\{a_n\}$ **diverges to infinity** if for every number M there is an integer N such that for all n larger than N, $a_n > M$. If this condition holds we write

$$\lim_{n\to\infty} a_n = \infty \qquad \text{or} \qquad a_n \to \infty.$$

Similarly if for every number m there is an integer N such that for all $n > N$ we have $a_n < m$, then we say $\{a_n\}$ **diverges to negative infinity** and write

$$\lim_{n\to\infty} a_n = -\infty \qquad \text{or} \qquad a_n \to -\infty.$$

A sequence may diverge without diverging to infinity or negative infinity, as we saw in Example 2. The sequences $\{1, -2, 3, -4, 5, -6, 7, -8, \ldots\}$ and $\{1, 0, 2, 0, 3, 0, \ldots\}$ are also examples of such divergence.

Calculating Limits of Sequences

Since sequences are functions with domain restricted to the positive integers, it is not surprising that the theorems on limits of functions given in Chapter 1 have versions for sequences.

THEOREM 1 Let $\{a_n\}$ and $\{b_n\}$ be sequences of real numbers and let A and B be real numbers. The following rules hold if $\lim_{n\to\infty} a_n = A$ and $\lim_{n\to\infty} b_n = B$.

1. *Sum Rule:* $\lim_{n\to\infty}(a_n + b_n) = A + B$
2. *Difference Rule:* $\lim_{n\to\infty}(a_n - b_n) = A - B$
3. *Product Rule:* $\lim_{n\to\infty}(a_n \cdot b_n) = A \cdot B$
4. *Constant Multiple Rule:* $\lim_{n\to\infty}(k \cdot b_n) = k \cdot B$ (Any number k)
5. *Quotient Rule:* $\lim_{n\to\infty} \frac{a_n}{b_n} = \frac{A}{B}$ if $B \neq 0$

The proof is similar to that of Theorem 1 of Section 1.4, and is omitted.

EXAMPLE 3 By combining Theorem 1 with the limits of Example 1, we have:

(a) $\lim_{n\to\infty}\left(-\frac{1}{n}\right) = -1 \cdot \lim_{n\to\infty}\frac{1}{n} = -1 \cdot 0 = 0$ Constant Multiple Rule and Example 1a

(b) $\lim_{n\to\infty}\left(\frac{n-1}{n}\right) = \lim_{n\to\infty}\left(1 - \frac{1}{n}\right) = \lim_{n\to\infty} 1 - \lim_{n\to\infty}\frac{1}{n} = 1 - 0 = 1$ Difference Rule and Example 1a

(c) $\lim_{n\to\infty}\frac{5}{n^2} = 5 \cdot \lim_{n\to\infty}\frac{1}{n} \cdot \lim_{n\to\infty}\frac{1}{n} = 5 \cdot 0 \cdot 0 = 0$ Product Rule

(d) $\lim_{n\to\infty}\frac{4 - 7n^6}{n^6 + 3} = \lim_{n\to\infty}\frac{(4/n^6) - 7}{1 + (3/n^6)} = \frac{0 - 7}{1 + 0} = -7.$ Sum and Quotient rules ■

Be cautious in applying Theorem 1. It does not say, for example, that each of the sequences $\{a_n\}$ and $\{b_n\}$ have limits if their sum $\{a_n + b_n\}$ has a limit. For instance, $\{a_n\} = \{1, 2, 3, \dots\}$ and $\{b_n\} = \{-1, -2, -3, \dots\}$ both diverge, but their sum $\{a_n + b_n\} = \{0, 0, 0, \dots\}$ clearly converges to 0.

One consequence of Theorem 1 is that every nonzero multiple of a divergent sequence $\{a_n\}$ diverges. For suppose, to the contrary, that $\{ca_n\}$ converges for some number $c \neq 0$. Then, by taking $k = 1/c$ in the Constant Multiple Rule in Theorem 1, we see that the sequence

$$\left\{\frac{1}{c} \cdot ca_n\right\} = \{a_n\}$$

converges. Thus, $\{ca_n\}$ cannot converge unless $\{a_n\}$ also converges. If $\{a_n\}$ does not converge, then $\{ca_n\}$ does not converge.

The next theorem is the sequence version of the Sandwich Theorem in Section 1.4. You are asked to prove the theorem in Exercise 93.

THEOREM 2—The Sandwich Theorem for Sequences Let $\{a_n\}$, $\{b_n\}$, and $\{c_n\}$ be sequences of real numbers. If $a_n \le b_n \le c_n$ holds for all n beyond some index N, and if $\lim_{n\to\infty} a_n = \lim_{n\to\infty} c_n = L$, then $\lim_{n\to\infty} b_n = L$ also.

An immediate consequence of Theorem 2 is that, if $|b_n| \le c_n$ and $c_n \to 0$, then $b_n \to 0$ because $-c_n \le b_n \le c_n$. We use this fact in the next example.

EXAMPLE 4 Since $1/n \to 0$, we know that

(a) $\dfrac{\cos n}{n} \to 0$ because $-\dfrac{1}{n} \le \dfrac{\cos n}{n} \le \dfrac{1}{n}$;

(b) $\dfrac{1}{2^n} \to 0$ because $0 \le \dfrac{1}{2^n} \le \dfrac{1}{n}$;

(c) $(-1)^n \dfrac{1}{n} \to 0$ because $-\dfrac{1}{n} \le (-1)^n \dfrac{1}{n} \le \dfrac{1}{n}$. ■

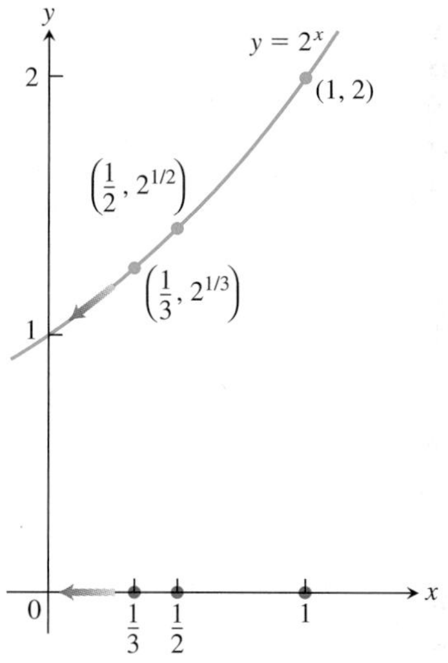

FIGURE 7.3 As $n \to \infty$, $1/n \to 0$ and $2^{1/n} \to 2^0$ (Example 6).

The application of Theorems 1 and 2 is broadened by a theorem stating that applying a continuous function to a convergent sequence produces a convergent sequence. We state the theorem without proof (Exercise 94).

THEOREM 3—The Continuous Function Theorem for Sequences Let $\{a_n\}$ be a sequence of real numbers. If $a_n \to L$ and if f is a function that is continuous at L and defined at all a_n, then $f(a_n) \to f(L)$.

EXAMPLE 5 Show that $\sqrt{(n+1)/n} \to 1$.

Solution We know that $(n+1)/n \to 1$. Taking $f(x) = \sqrt{x}$ and $L = 1$ in Theorem 3 gives $\sqrt{(n+1)/n} \to \sqrt{1} = 1$. ■

EXAMPLE 6 The sequence $\{1/n\}$ converges to 0. By taking $a_n = 1/n$, $f(x) = 2^x$, and $L = 0$ in Theorem 3, we see that $2^{1/n} = f(1/n) \to f(L) = 2^0 = 1$. The sequence $\{2^{1/n}\}$ converges to 1 (Figure 7.3). ■

Using l'Hôpital's Rule

The next theorem enables us to use l'Hôpital's Rule to find the limits of some sequences. It formalizes the connection between $\lim_{n\to\infty} a_n$ and $\lim_{x\to\infty} f(x)$.

THEOREM 4 Suppose that $f(x)$ is a function defined for all $x \ge n_0$ and that $\{a_n\}$ is a sequence of real numbers such that $a_n = f(n)$ for $n \ge n_0$. Then

$$\lim_{x\to\infty} f(x) = L \quad \Rightarrow \quad \lim_{n\to\infty} a_n = L.$$

Proof Suppose that $\lim_{x\to\infty} f(x) = L$. Then for each positive number ϵ there is a number M such that for all x,

$$x > M \quad \Rightarrow \quad |f(x) - L| < \epsilon.$$

Let N be an integer greater than M and greater than or equal to n_0. Then

$$n > N \quad \Rightarrow \quad a_n = f(n) \quad \text{and} \quad |a_n - L| = |f(n) - L| < \epsilon. \quad ■$$

EXAMPLE 7 Show that

$$\lim_{n\to\infty} \frac{\ln n}{n} = 0.$$

Solution The function $(\ln x)/x$ is defined for all $x \geq 1$ and agrees with the given sequence at positive integers. Therefore, by Theorem 4, $\lim_{n\to\infty}(\ln n)/n$ will equal $\lim_{x\to\infty}(\ln x)/x$ if the latter exists. A single application of l'Hôpital's Rule shows that

$$\lim_{x\to\infty} \frac{\ln x}{x} = \lim_{x\to\infty} \frac{1/x}{1} = \frac{0}{1} = 0.$$

We conclude that $\lim_{n\to\infty}(\ln n)/n = 0$. ■

When we use l'Hôpital's Rule to find the limit of a sequence, we often treat n as a continuous real variable and differentiate directly with respect to n. This saves us from having to rewrite the formula for a_n as we did in Example 7.

EXAMPLE 8 Does the sequence whose nth term is

$$a_n = \left(\frac{n+1}{n-1}\right)^n$$

converge? If so, find $\lim_{n\to\infty} a_n$.

Solution The limit leads to the indeterminate form 1^∞. We can apply l'Hôpital's Rule if we first change the form to $\infty \cdot 0$ by taking the natural logarithm of a_n:

$$\begin{aligned}\ln a_n &= \ln\left(\frac{n+1}{n-1}\right)^n \\ &= n\ln\left(\frac{n+1}{n-1}\right).\end{aligned}$$

Then,

$$\begin{aligned}\lim_{n\to\infty} \ln a_n &= \lim_{n\to\infty} n\ln\left(\frac{n+1}{n-1}\right) && \infty \cdot 0 \\ &= \lim_{n\to\infty} \frac{\ln\left(\frac{n+1}{n-1}\right)}{1/n} && \frac{0}{0} \\ &= \lim_{n\to\infty} \frac{-2/(n^2-1)}{-1/n^2} && \text{l'Hôpital's Rule} \\ &= \lim_{n\to\infty} \frac{2n^2}{n^2-1} = 2.\end{aligned}$$

Since $\ln a_n \to 2$ and $f(x) = e^x$ is continuous, Theorem 4 tells us that

$$a_n = e^{\ln a_n} \to e^2.$$

The sequence $\{a_n\}$ converges to e^2. ■

Commonly Occurring Limits

The next theorem gives some limits that arise frequently.

THEOREM 5 The following six sequences converge to the limits listed below:

1. $\lim_{n\to\infty} \frac{\ln n}{n} = 0$

2. $\lim_{n\to\infty} \sqrt[n]{n} = 1$

3. $\lim_{n\to\infty} x^{1/n} = 1 \qquad (x > 0)$

4. $\lim_{n\to\infty} x^n = 0 \qquad (|x| < 1)$

5. $\lim_{n\to\infty} \left(1 + \frac{x}{n}\right)^n = e^x \qquad (\text{any } x)$

6. $\lim_{n\to\infty} \frac{x^n}{n!} = 0 \qquad (\text{any } x)$

In Formulas (3) through (6), x remains fixed as $n \to \infty$.

Factorial Notation

The notation $n!$ ("n factorial") means the product $1 \cdot 2 \cdot 3 \cdots n$ of the integers from 1 to n. Notice that $(n + 1)! = (n + 1) \cdot n!$. Thus, $4! = 1 \cdot 2 \cdot 3 \cdot 4 = 24$ and $5! = 1 \cdot 2 \cdot 3 \cdot 4 \cdot 5 = 5 \cdot 4! = 120$. We define $0!$ to be 1. Factorials grow even faster than exponentials, as the table suggests.

n	e^n (rounded)	$n!$
1	3	1
5	148	120
10	22,026	3,628,800
20	4.9×10^8	2.4×10^{18}

Proof The first limit was computed in Example 7. The next two can be proved by taking logarithms and applying Theorem 4 (Exercises 91 and 92). The remaining proofs are given in Appendix 6. ■

EXAMPLE 9

(a) $\frac{\ln (n^2)}{n} = \frac{2 \ln n}{n} \to 2 \cdot 0 = 0$ Formula 1

(b) $\sqrt[n]{n^2} = n^{2/n} = (n^{1/n})^2 \to (1)^2 = 1$ Formula 2

(c) $\sqrt[n]{3n} = 3^{1/n}(n^{1/n}) \to 1 \cdot 1 = 1$ Formula 3 with $x = 3$ and Formula 2

(d) $\left(-\frac{1}{2}\right)^n \to 0$ Formula 4 with $x = -\frac{1}{2}$

(e) $\left(\frac{n-2}{n}\right)^n = \left(1 + \frac{-2}{n}\right)^n \to e^{-2}$ Formula 5 with $x = -2$

(f) $\frac{100^n}{n!} \to 0$ Formula 6 with $x = 100$ ■

Recursive Definitions

So far, we have calculated each a_n directly from the value of n. But sequences are often defined **recursively** by giving

1. The value(s) of the initial term or terms, and
2. A rule, called a **recursion formula**, for calculating any later term from terms that precede it.

EXAMPLE 10

(a) The statements $a_1 = 1$ and $a_n = a_{n-1} + 1$ define the sequence $1, 2, 3, \ldots, n, \ldots$ of positive integers. With $a_1 = 1$, we have $a_2 = a_1 + 1 = 2$, $a_3 = a_2 + 1 = 3$, and so on.

(b) The statements $a_1 = 1$ and $a_n = n \cdot a_{n-1}$ define the sequence $1, 2, 6, 24, \ldots, n!, \ldots$ of factorials. With $a_1 = 1$, we have $a_2 = 2 \cdot a_1 = 2$, $a_3 = 3 \cdot a_2 = 6$, $a_4 = 4 \cdot a_3 = 24$, and so on.

(c) The statements $a_1 = 1$, $a_2 = 1$, and $a_{n+1} = a_n + a_{n-1}$ define the sequence $1, 1, 2, 3, 5, \ldots$ of **Fibonacci numbers**. With $a_1 = 1$ and $a_2 = 1$, we have $a_3 = 1 + 1 = 2$, $a_4 = 2 + 1 = 3$, $a_5 = 3 + 2 = 5$, and so on.

(d) As we can see by applying Newton's method, the statements $x_0 = 1$ and $x_{n+1} = x_n - [(\sin x_n - x_n^2)/(\cos x_n - 2x_n)]$ define a sequence that, when it converges, gives a solution to the equation $\sin x - x^2 = 0$. ■

Bounded Nondecreasing Sequences

The terms of a general sequence can bounce around, sometimes getting larger, sometimes smaller. An important special kind of sequence is one for which each term is at least as large as its predecessor.

> **DEFINITION** A sequence $\{a_n\}$ with the property that $a_n \leq a_{n+1}$ for all n is called a **nondecreasing sequence**.

EXAMPLE 11 The following sequences are nondecreasing.

(a) The sequence $1, 2, 3, \ldots, n, \ldots$ of natural numbers

(b) The sequence $\frac{1}{2}, \frac{2}{3}, \frac{3}{4}, \ldots, \frac{n}{n+1}, \ldots$

(c) The constant sequence $\{3\}$ ■

There are two kinds of nondecreasing sequences—those whose terms increase beyond any finite bound and those whose terms do not.

> **DEFINITIONS** A sequence $\{a_n\}$ is **bounded from above** if there exists a number M such that $a_n \leq M$ for all n. The number M is an **upper bound** for $\{a_n\}$. If M is an upper bound for $\{a_n\}$ but no number less than M is an upper bound for $\{a_n\}$, then M is the **least upper bound** for $\{a_n\}$.

EXAMPLE 12

(a) The sequence $1, 2, 3, \ldots, n, \ldots$ has no upper bound.

(b) The sequence $\frac{1}{2}, \frac{2}{3}, \frac{3}{4}, \ldots, \frac{n}{n+1}, \ldots$ is bounded above by $M = 1$.

No number less than 1 is an upper bound for the sequence, so 1 is the least upper bound (Exercise 111). ■

A nondecreasing sequence that is bounded from above always has a least upper bound. This is the completeness property of the real numbers, discussed in Appendix 7. We will prove that if L is the least upper bound then the sequence converges to L.

Suppose we plot the points $(1, a_1), (2, a_2), \ldots, (n, a_n), \ldots$ in the xy-plane. If M is an upper bound of the sequence, all these points will lie on or below the line $y = M$

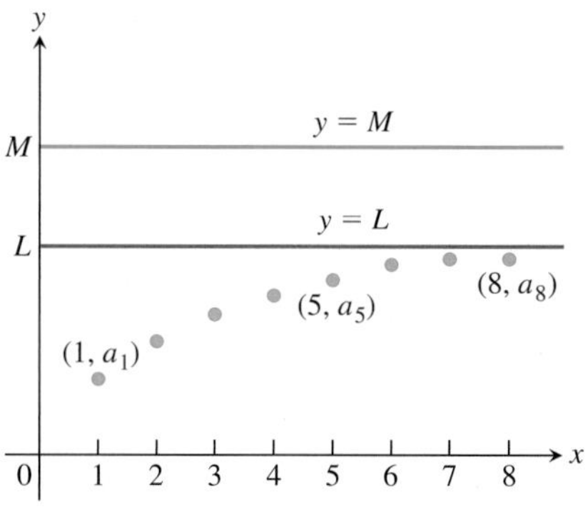

FIGURE 7.4 If the terms of a nondecreasing sequence have an upper bound M, they have a limit $L \le M$.

(Figure 7.4). The line $y = L$ is the lowest such line. None of the points (n, a_n) lies above $y = L$, but some do lie above any lower line $y = L - \epsilon$, if ϵ is a positive number. The sequence converges to L because

(a) $a_n \le L$ for *all* values of n and

(b) given any $\epsilon > 0$, there exists at least one integer N for which $a_N > L - \epsilon$.

The fact that $\{a_n\}$ is nondecreasing tells us further that

$$a_n \ge a_N > L - \epsilon \qquad \text{for all } n \ge N.$$

Thus, *all* the numbers a_n beyond the Nth number lie within ϵ of L. This is precisely the condition for L to be the limit of the sequence $\{a_n\}$.

The facts for nondecreasing sequences are summarized in the following theorem. A similar result holds for nonincreasing sequences (Exercise 105).

THEOREM 6—The Nondecreasing Sequence Theorem A nondecreasing sequence of real numbers converges if and only if it is bounded from above. If a nondecreasing sequence converges, it converges to its least upper bound.

Theorem 6 implies that a nondecreasing sequence converges when it is bounded from above. It diverges to infinity if it is not bounded from above. Exercise 10.5 shows that analogous results hold for nonincreasing sequences.

EXERCISES 7.1

Each of Exercises 1–6 gives a formula for the nth term a_n of a sequence $\{a_n\}$. Find the values of a_1, a_2, a_3, and a_4.

1. $a_n = \dfrac{1-n}{n^2}$

2. $a_n = \dfrac{1}{n!}$

3. $a_n = \dfrac{(-1)^{n+1}}{2n-1}$

4. $a_n = 2 + (-1)^n$

5. $a_n = \dfrac{2^n}{2^{n+1}}$

6. $a_n = \dfrac{2^n - 1}{2^n}$

Each of Exercises 7–12 gives the first term or two of a sequence along with a recursion formula for the remaining terms. Write out the first ten terms of the sequence.

7. $a_1 = 1, \quad a_{n+1} = a_n + (1/2^n)$

8. $a_1 = 1, \quad a_{n+1} = a_n/(n+1)$

9. $a_1 = 2, \quad a_{n+1} = (-1)^{n+1}a_n/2$

10. $a_1 = -2, \quad a_{n+1} = na_n/(n+1)$

11. $a_1 = a_2 = 1, \quad a_{n+2} = a_{n+1} + a_n$

12. $a_1 = 2, \quad a_2 = -1, \quad a_{n+2} = a_{n+1}/a_n$

In Exercises 13–22, find a formula for the nth term of the sequence.

13. The sequence $1, -1, 1, -1, 1, \dots$ 1's with alternating signs

14. The sequence $-1, 1, -1, 1, -1, \dots$ 1's with alternating signs

15. The sequence $1, -4, 9, -16, 25, \dots$ Squares of the positive integers; with alternating signs

16. The sequence $1, -\frac{1}{4}, \frac{1}{9}, -\frac{1}{16}, \frac{1}{25}, \dots$ Reciprocals of squares of the positive integers, with alternating signs

17. The sequence $0, 3, 8, 15, 24, \dots$ Squares of the positive integers diminished by 1

18. The sequence $-3, -2, -1, 0, 1, \dots$ Integers beginning with -3

19. The sequence $1, 5, 9, 13, 17, \dots$ Every other odd positive integer

20. The sequence $2, 6, 10, 14, 18, \dots$ Every other even positive integer

21. The sequence $1, 0, 1, 0, 1, \dots$ Alternating 1's and 0's

22. The sequence $0, 1, 1, 2, 2, 3, 3, 4, \dots$ Each positive integer repeated

Which of the sequences $\{a_n\}$ in Exercises 23–84 converge, and which diverge? Find the limit of each convergent sequence.

23. $a_n = 2 + (0.1)^n$

24. $a_n = \dfrac{n + (-1)^n}{n}$

25. $a_n = \dfrac{1-2n}{1+2n}$

26. $a_n = \dfrac{2n+1}{1-3\sqrt{n}}$

27. $a_n = \dfrac{1-5n^4}{n^4+8n^3}$

28. $a_n = \dfrac{n+3}{n^2+5n+6}$

29. $a_n = \dfrac{n^2 - 2n + 1}{n - 1}$

30. $a_n = \dfrac{1 - n^3}{70 - 4n^2}$

31. $a_n = 1 + (-1)^n$

32. $a_n = (-1)^n\left(1 - \dfrac{1}{n}\right)$

33. $a_n = \left(\dfrac{n+1}{2n}\right)\left(1 - \dfrac{1}{n}\right)$

34. $a_n = \left(2 - \dfrac{1}{2^n}\right)\left(3 + \dfrac{1}{2^n}\right)$

35. $a_n = \dfrac{(-1)^{n+1}}{2n - 1}$

36. $a_n = \left(-\dfrac{1}{2}\right)^n$

37. $a_n = \sqrt{\dfrac{2n}{n + 1}}$

38. $a_n = \dfrac{1}{(0.9)^n}$

39. $a_n = \sin\left(\dfrac{\pi}{2} + \dfrac{1}{n}\right)$

40. $a_n = n\pi \cos(n\pi)$

41. $a_n = \dfrac{\sin n}{n}$

42. $a_n = \dfrac{\sin^2 n}{2^n}$

43. $a_n = \dfrac{n}{2^n}$

44. $a_n = \dfrac{3^n}{n^3}$

45. $a_n = \dfrac{\ln(n + 1)}{\sqrt{n}}$

46. $a_n = \dfrac{\ln n}{\ln 2n}$

47. $a_n = 8^{1/n}$

48. $a_n = (0.03)^{1/n}$

49. $a_n = \left(1 + \dfrac{7}{n}\right)^n$

50. $a_n = \left(1 - \dfrac{1}{n}\right)^n$

51. $a_n = \sqrt[n]{10n}$

52. $a_n = \sqrt[n]{n^2}$

53. $a_n = \left(\dfrac{3}{n}\right)^{1/n}$

54. $a_n = (n + 4)^{1/(n+4)}$

55. $a_n = \dfrac{\ln n}{n^{1/n}}$

56. $a_n = \ln n - \ln(n + 1)$

57. $a_n = \sqrt[n]{4^n n}$

58. $a_n = \sqrt[n]{3^{2n+1}}$

59. $a_n = \dfrac{n!}{n^n}$ (*Hint:* Compare with $1/n$.)

60. $a_n = \dfrac{(-4)^n}{n!}$

61. $a_n = \dfrac{n!}{10^{6n}}$

62. $a_n = \dfrac{n!}{2^n \cdot 3^n}$

63. $a_n = \left(\dfrac{1}{n}\right)^{1/(\ln n)}$

64. $a_n = \ln\left(1 + \dfrac{1}{n}\right)^n$

65. $a_n = \left(\dfrac{3n + 1}{3n - 1}\right)^n$

66. $a_n = \left(\dfrac{n}{n + 1}\right)^n$

67. $a_n = \left(\dfrac{x^n}{2n + 1}\right)^{1/n}, \quad x > 0$

68. $a_n = \left(1 - \dfrac{1}{n^2}\right)^n$

69. $a_n = \dfrac{3^n \cdot 6^n}{2^{-n} \cdot n!}$

70. $a_n = \dfrac{(10/11)^n}{(9/10)^n + (11/12)^n}$

71. $a_n = \tanh n$

72. $a_n = \sinh(\ln n)$

73. $a_n = \dfrac{n^2}{2n - 1}\sin\dfrac{1}{n}$

74. $a_n = n\left(1 - \cos\dfrac{1}{n}\right)$

75. $a_n = \tan^{-1} n$

76. $a_n = \dfrac{1}{\sqrt{n}}\tan^{-1} n$

77. $a_n = \left(\dfrac{1}{3}\right)^n + \dfrac{1}{\sqrt{2^n}}$

78. $a_n = \sqrt[n]{n^2 + n}$

79. $a_n = \dfrac{(\ln n)^{200}}{n}$

80. $a_n = \dfrac{(\ln n)^5}{\sqrt{n}}$

81. $a_n = n - \sqrt{n^2 - n}$

82. $a_n = \dfrac{1}{\sqrt{n^2 - 1} - \sqrt{n^2 + n}}$

83. $a_n = \dfrac{1}{n}\displaystyle\int_1^n \dfrac{1}{x}\,dx$

84. $a_n = \displaystyle\int_1^n \dfrac{1}{x^p}\,dx, \quad p > 1$

85. Newton's method The following sequences come from the recursion formula for Newton's method,

$$x_{n+1} = x_n - \frac{f(x_n)}{f'(x_n)}.$$

Do the sequences converge? If so, to what value? In each case, begin by identifying the function f that generates the sequence.

a. $x_0 = 1, \quad x_{n+1} = x_n - \dfrac{x_n^2 - 2}{2x_n} = \dfrac{x_n}{2} + \dfrac{1}{x_n}$

b. $x_0 = 1, \quad x_{n+1} = x_n - \dfrac{\tan x_n - 1}{\sec^2 x_n}$

c. $x_0 = 1, \quad x_{n+1} = x_n - 1$

86. a. Suppose that $f(x)$ is differentiable for all x in $[0, 1]$ and that $f(0) = 0$. Define the sequence $\{a_n\}$ by the rule $a_n = nf(1/n)$. Show that $\lim_{n\to\infty} a_n = f'(0)$.

Use the result in part (a) to find the limits of the following sequences $\{a_n\}$.

b. $a_n = n\tan^{-1}\dfrac{1}{n}$

c. $a_n = n(e^{1/n} - 1)$

d. $a_n = n\ln\left(1 + \dfrac{2}{n}\right)$

87. Pythagorean triples A triple of positive integers a, b, and c is called a **Pythagorean triple** if $a^2 + b^2 = c^2$. Let a be an odd positive integer and let

$$b = \left\lfloor \frac{a^2}{2} \right\rfloor \quad \text{and} \quad c = \left\lceil \frac{a^2}{2} \right\rceil$$

be, respectively, the integer floor and ceiling for $a^2/2$. Show that $a^2 + b^2 = c^2$. (*Hint:* Let $a = 2n + 1$ and express b and c in terms of n.)

88. By direct calculation using Exercise 87, find

$$\lim_{a\to\infty} \frac{\left\lfloor \dfrac{a^2}{2} \right\rfloor}{\left\lceil \dfrac{a^2}{2} \right\rceil}.$$

89. a. Assuming that $\lim_{n\to\infty}(1/n^c) = 0$ if c is any positive constant, show that

$$\lim_{n\to\infty}\frac{\ln n}{n^c} = 0$$

if c is any positive constant.

b. Prove that $\lim_{n\to\infty}(1/n^c) = 0$ if c is any positive constant. (*Hint:* If $\epsilon = 0.001$ and $c = 0.04$, how large should N be to ensure that $|1/n^c - 0| < \epsilon$ if $n > N$?)

90. The zipper theorem Prove the "zipper theorem" for sequences: If $\{a_n\}$ and $\{b_n\}$ both converge to L, then the sequence

$$a_1, b_1, a_2, b_2, \ldots, a_n, b_n, \ldots$$

converges to L.

91. Prove that $\lim_{n\to\infty}\sqrt[n]{n} = 1$.

92. Prove that $\lim_{n\to\infty}x^{1/n} = 1, (x > 0)$.

93. Prove Theorem 2.

94. Prove Theorem 3.

In Exercises 95–98, determine if the sequence is nondecreasing and if it is bounded from above.

95. $a_n = \dfrac{3n+1}{n+1}$

96. $a_n = \dfrac{(2n+3)!}{(n+1)!}$

97. $a_n = \dfrac{2^n 3^n}{n!}$

98. $a_n = 2 - \dfrac{2}{n} - \dfrac{1}{2^n}$

Which of the sequences in Exercises 99–104 converge, and which diverge? Give reasons for your answers.

99. $a_n = 1 - \dfrac{1}{n}$

100. $a_n = n - \dfrac{1}{n}$

101. $a_n = \dfrac{2^n - 1}{2^n}$

102. $a_n = \dfrac{2^n - 1}{3^n}$

103. $a_n = ((-1)^n + 1)\left(\dfrac{n+1}{n}\right)$

104. The first term of a sequence is $x_1 = \cos(1)$. The next terms are $x_2 = x_1$ or $\cos(2)$, whichever is larger; and $x_3 = x_2$ or $\cos(3)$, whichever is larger (farther to the right). In general,

$$x_{n+1} = \max\{x_n, \cos(n+1)\}.$$

105. Nonincreasing sequences A sequence of numbers $\{a_n\}$ in which $a_n \ge a_{n+1}$ for every n is called a **nonincreasing sequence**. A sequence $\{a_n\}$ is **bounded from below** if there is a number M with $M \le a_n$ for every n. Such a number M is called a **lower bound** for the sequence. Deduce from Theorem 6 that a nonincreasing sequence that is bounded from below converges and that a nonincreasing sequence that is not bounded from below diverges.

(*Continuation of Exercise 105.*) Using the conclusion of Exercise 105, determine which of the sequences in Exercises 106–110 converge and which diverge.

106. $a_n = \dfrac{n+1}{n}$

107. $a_n = \dfrac{1+\sqrt{2n}}{\sqrt{n}}$

108. $a_n = \dfrac{1-4^n}{2^n}$

109. $a_n = \dfrac{4^{n+1}+3^n}{4^n}$

110. $a_1 = 1, \quad a_{n+1} = 2a_n - 3$

111. The sequence $\{n/(n+1)\}$ has a least upper bound of 1 Show that if M is a number less than 1, then the terms of $\{n/(n+1)\}$ eventually exceed M. That is, if $M < 1$ there is an integer N such that $n/(n+1) > M$ whenever $n > N$. Since $n/(n+1) < 1$ for every n, this proves that 1 is a least upper bound for $\{n/(n+1)\}$.

112. Uniqueness of least upper bounds Show that if M_1 and M_2 are least upper bounds for the sequence $\{a_n\}$, then $M_1 = M_2$. That is, a sequence cannot have two different least upper bounds.

113. Is it true that a sequence $\{a_n\}$ of positive numbers must converge if it is bounded from above? Give reasons for your answer.

114. Prove that if $\{a_n\}$ is a convergent sequence, then to every positive number ϵ there corresponds an integer N such that for all m and n,

$$m > N \quad \text{and} \quad n > N \quad \Rightarrow \quad |a_m - a_n| < \epsilon.$$

115. Uniqueness of limits Prove that limits of sequences are unique. That is, show that if L_1 and L_2 are numbers such that $a_n \to L_1$ and $a_n \to L_2$, then $L_1 = L_2$.

116. Limits and subsequences If the terms of one sequence appear in another sequence in their given order, we call the first sequence a **subsequence** of the second. Prove that if two subsequences of a sequence $\{a_n\}$ have different limits $L_1 \ne L_2$, then $\{a_n\}$ diverges.

117. For a sequence $\{a_n\}$ the terms of even index are denoted by a_{2k} and the terms of odd index by a_{2k+1}. Prove that if $a_{2k} \to L$ and $a_{2k+1} \to L$, then $a_n \to L$.

118. Prove that a sequence $\{a_n\}$ converges to 0 if and only if the sequence of absolute values $\{|a_n|\}$ converges to 0.

119. Sequences generated by Newton's method Newton's method, applied to a differentiable function $f(x)$, begins with a starting value x_0 and constructs from it a sequence of numbers $\{x_n\}$ that under favorable circumstances converges to a zero of f. The recursion formula for the sequence is

$$x_{n+1} = x_n - \frac{f(x_n)}{f'(x_n)}.$$

a. Show that the recursion formula for $f(x) = x^2 - a, a > 0$, can be written as $x_{n+1} = (x_n + a/x_n)/2$.

b. Starting with $x_0 = 1$ and $a = 3$, calculate successive terms of the sequence until the display begins to repeat. What number is being approximated? Explain.

120. (*Continuation of Exercise 119.*) Repeat part (b) of Exercise 119 with $a = 2$ in place of $a = 3$.

COMPUTER EXPLORATIONS

Use a CAS to perform the following steps for the sequences in Exercises 121–132.

a. Calculate and then plot the first 25 terms of the sequence. Does the sequence appear to be bounded from above or below? Does it appear to converge or diverge? If it does converge, what is the limit L?

b. If the sequence converges, find an integer N such that $|a_n - L| \le 0.01$ for $n \ge N$. How far in the sequence do you have to get for the terms to lie within 0.0001 of L?

121. $a_n = \sqrt[n]{n}$

122. $a_n = \left(1 + \frac{0.5}{n}\right)^n$

123. $a_1 = 1, \quad a_{n+1} = a_n + \frac{1}{5^n}$

124. $a_1 = 1, \quad a_{n+1} = a_n + (-2)^n$

125. $a_n = \sin n$

126. $a_n = n \sin \frac{1}{n}$

127. $a_n = \frac{\sin n}{n}$

128. $a_n = \frac{\ln n}{n}$

129. $a_n = (0.9999)^n$

130. $a_n = 123456^{1/n}$

131. $a_n = \frac{8^n}{n!}$

132. $a_n = \frac{n^{41}}{19^n}$

7.2 Infinite Series

An *infinite series* is the sum of an infinite sequence of numbers

$$a_1 + a_2 + a_3 + \cdots + a_n + \cdots$$

The goal of this section is to understand the meaning of such an infinite sum and to develop methods to calculate it. Since there are infinitely many terms to add in an infinite series, we cannot just keep adding to see what comes out. Instead we look at the result of summing the first n terms of the sequence and stopping. The sum of the first n terms

$$s_n = a_1 + a_2 + a_3 + \cdots + a_n$$

is an ordinary finite sum and can be calculated by normal addition. It is called the *nth partial sum*. As n gets larger, we expect the partial sums to get closer and closer to a limiting value in the same sense that the terms of a sequence approach a limit, as discussed in Section 7.1.

For example, to assign meaning to an expression like

$$1 + \frac{1}{2} + \frac{1}{4} + \frac{1}{8} + \frac{1}{16} + \cdots$$

we add the terms one at a time from the beginning and look for a pattern in how these partial sums grow.

Partial sum		**Suggestive expression for partial sum**	**Value**
First:	$s_1 = 1$	$2 - 1$	1
Second:	$s_2 = 1 + \frac{1}{2}$	$2 - \frac{1}{2}$	$\frac{3}{2}$
Third:	$s_3 = 1 + \frac{1}{2} + \frac{1}{4}$	$2 - \frac{1}{4}$	$\frac{7}{4}$
$\vdots$	$\vdots$	$\vdots$	$\vdots$
nth:	$s_n = 1 + \frac{1}{2} + \frac{1}{4} + \cdots + \frac{1}{2^{n-1}}$	$2 - \frac{1}{2^{n-1}}$	$\frac{2^n - 1}{2^{n-1}}$

Indeed there is a pattern. The partial sums form a sequence whose nth term is

$$s_n = 2 - \frac{1}{2^{n-1}}.$$

This sequence of partial sums converges to 2 because $\lim_{n\to\infty}(1/2^{n-1}) = 0$. We say

$$\text{“the sum of the infinite series } 1 + \frac{1}{2} + \frac{1}{4} + \cdots + \frac{1}{2^{n-1}} + \cdots \text{ is 2.”}$$

Is the sum of any finite number of terms in this series equal to 2? No. Can we actually add an infinite number of terms one by one? No. But we can still define their sum by defining it to be the limit of the sequence of partial sums as $n \to \infty$, in this case 2 (Figure 7.5). Our knowledge of sequences and limits enables us to break away from the confines of finite sums.

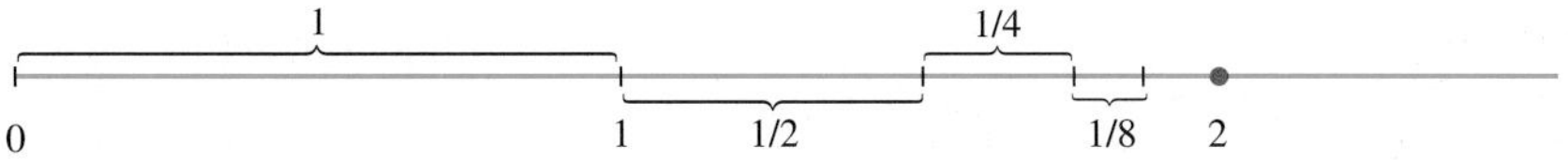

FIGURE 7.5 As the lengths $1, \frac{1}{2}, \frac{1}{4}, \frac{1}{8}, \ldots$ are added one by one, the sum approaches 2.

HISTORICAL BIOGRAPHY

Blaise Pascal
(1623–1662)

DEFINITIONS Given a sequence of numbers $\{a_n\}$, an expression of the form

$$a_1 + a_2 + a_3 + \cdots + a_n + \cdots$$

is an **infinite series**. The number a_n is the ***n*th term** of the series. The sequence $\{s_n\}$ defined by

$$\begin{aligned} s_1 &= a_1 \\ s_2 &= a_1 + a_2 \\ &\vdots \\ s_n &= a_1 + a_2 + \cdots + a_n = \sum_{k=1}^{n} a_k \\ &\vdots \end{aligned}$$

is the **sequence of partial sums** of the series, the number s_n being the ***n*th partial sum**. If the sequence of partial sums converges to a limit L, we say that the series **converges** and that its **sum** is L. In this case, we also write

$$a_1 + a_2 + \cdots + a_n + \cdots = \sum_{n=1}^{\infty} a_n = L.$$

If the sequence of partial sums of the series does not converge, we say that the series **diverges**.

When we begin to study a given series $a_1 + a_2 + \cdots + a_n + \cdots$, we might not know whether it converges or diverges. In either case, it is convenient to use sigma notation to write the series as

$$\sum_{n=1}^{\infty} a_n, \qquad \sum_{k=1}^{\infty} a_k, \qquad \text{or} \qquad \sum a_n$$

A useful shorthand when summation from 1 to ∞ is understood

Geometric Series

Geometric series are series of the form

$$a + ar + ar^2 + \cdots + ar^{n-1} + \cdots = \sum_{n=1}^{\infty} ar^{n-1}$$

in which a and r are fixed real numbers and $a \neq 0$. The series can also be written as $\sum_{n=0}^{\infty} ar^n$. The **ratio** r can be positive, as in

$$1 + \frac{1}{2} + \frac{1}{4} + \cdots + \left(\frac{1}{2}\right)^{n-1} + \cdots, \qquad r = 1/2, a = 1$$

or negative, as in

$$1 - \frac{1}{3} + \frac{1}{9} - \cdots + \left(-\frac{1}{3}\right)^{n-1} + \cdots. \qquad r = -1/3, a = 1$$

If $r = 1$, the nth partial sum of the geometric series is

$$s_n = a + a(1) + a(1)^2 + \cdots + a(1)^{n-1} = na,$$

and the series diverges because $\lim_{n\to\infty} s_n = \pm\infty$, depending on the sign of a. If $r = -1$, the series diverges because the nth partial sums alternate between a and 0. If $|r| \neq 1$, we can determine the convergence or divergence of the series in the following way:

$$\begin{aligned}
s_n &= a + ar + ar^2 + \cdots + ar^{n-1} \\
rs_n &= ar + ar^2 + \cdots + ar^{n-1} + ar^n && \text{Multiply } s_n \text{ by } r. \\
s_n - rs_n &= a - ar^n && \text{Subtract } rs_n \text{ from } s_n. \text{ Most of the terms on the right cancel.} \\
s_n(1 - r) &= a(1 - r^n) && \text{Factor.} \\
s_n &= \frac{a(1 - r^n)}{1 - r}, \quad (r \neq 1). && \text{We can solve for } s_n \text{ if } r \neq 1.
\end{aligned}$$

If $|r| < 1$, then $r^n \to 0$ as $n \to \infty$ (as in Section 7.1) and $s_n \to a/(1 - r)$. If $|r| > 1$, then $|r^n| \to \infty$ and the series diverges.

> If $|r| < 1$, the geometric series $a + ar + ar^2 + \cdots + ar^{n-1} + \cdots$ converges to $a/(1 - r)$:
>
> $$\sum_{n=1}^{\infty} ar^{n-1} = \frac{a}{1 - r}, \qquad |r| < 1.$$
>
> If $|r| \geq 1$, the series diverges.

We have determined when a geometric series converges or diverges, and to what value. Often we can determine that a series converges without knowing the value to which it converges, as we will see in the next several sections. The formula $a/(1 - r)$ for the sum of a geometric series applies *only* when the summation index begins with $n = 1$ in the expression $\sum_{n=1}^{\infty} ar^{n-1}$ (or with the index $n = 0$ if we write the series as $\sum_{n=0}^{\infty} ar^n$).

EXAMPLE 1 The geometric series with $a = 1/9$ and $r = 1/3$ is

$$\frac{1}{9} + \frac{1}{27} + \frac{1}{81} + \cdots = \sum_{n=1}^{\infty} \frac{1}{9}\left(\frac{1}{3}\right)^{n-1} = \frac{1/9}{1 - (1/3)} = \frac{1}{6}.$$

■

EXAMPLE 2 The series

$$\sum_{n=0}^{\infty} \frac{(-1)^n 5}{4^n} = 5 - \frac{5}{4} + \frac{5}{16} - \frac{5}{64} + \cdots$$

is a geometric series with $a = 5$ and $r = -1/4$. It converges to

$$\frac{a}{1-r} = \frac{5}{1 + (1/4)} = 4. \quad ■$$

EXAMPLE 3 You drop a ball from a meters above a flat surface. Each time the ball hits the surface after falling a distance h, it rebounds a distance rh, where r is positive but less than 1. Find the total distance the ball travels up and down (Figure 7.6).

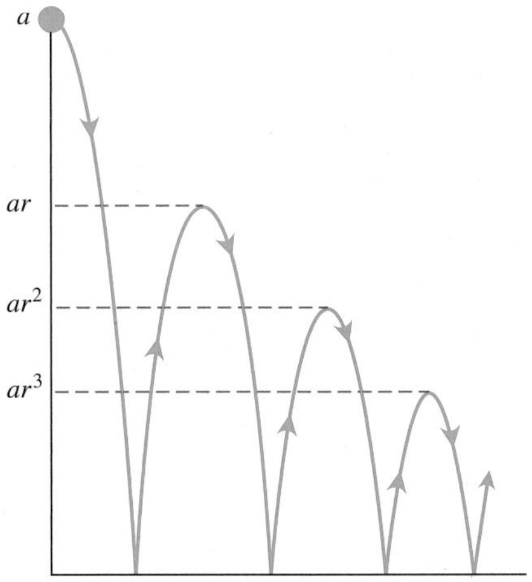

(a)

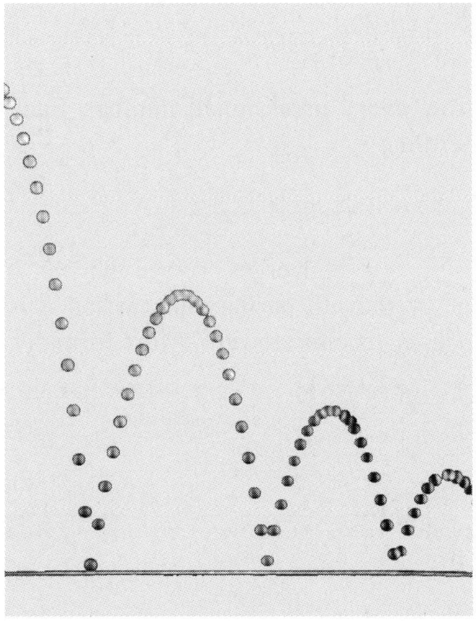

(b)

FIGURE 7.6 (a) Example 3 shows how to use a geometric series to calculate the total vertical distance traveled by a bouncing ball if the height of each rebound is reduced by the factor r. (b) A stroboscopic photo of a bouncing ball.

Solution The total distance is

$$s = a + \underbrace{2ar + 2ar^2 + 2ar^3 + \cdots}_{\text{This sum is } 2ar/(1-r).} = a + \frac{2ar}{1-r} = a\frac{1+r}{1-r}.$$

If $a = 6$ m and $r = 2/3$, for instance, the distance is

$$s = 6\frac{1 + (2/3)}{1 - (2/3)} = 6\left(\frac{5/3}{1/3}\right) = 30 \text{ m}. \quad ■$$

EXAMPLE 4 Express the repeating decimal 5.232323 . . . as the ratio of two integers.

Solution

$$5.232323\ldots = 5 + \frac{23}{100} + \frac{23}{(100)^2} + \frac{23}{(100)^3} + \cdots$$

$$= 5 + \frac{23}{100}\underbrace{\left(1 + \frac{1}{100} + \left(\frac{1}{100}\right)^2 + \cdots\right)}_{1/(1-0.01)} \qquad a = 1,\ r = 1/100$$

$$= 5 + \frac{23}{100}\left(\frac{1}{0.99}\right) = 5 + \frac{23}{99} = \frac{518}{99} \quad ■$$

Unfortunately, formulas like the one for the sum of a convergent geometric series are rare and we usually have to settle for an estimate of a series' sum (more about this later). The next example, however, is another case in which we can find the sum exactly.

EXAMPLE 5 Find the sum of the series $\displaystyle\sum_{n=1}^{\infty} \frac{1}{n(n+1)}$.

Solution We look for a pattern in the sequence of partial sums that might lead to a formula for s_k. The key observation is the partial fraction decomposition

$$\frac{1}{n(n+1)} = \frac{1}{n} - \frac{1}{n+1},$$

so

$$\sum_{n=1}^{k} \frac{1}{n(n+1)} = \sum_{n=1}^{k} \left(\frac{1}{n} - \frac{1}{n+1}\right)$$

and the right-hand sum "telescopes" with

$$s_k = \left(\frac{1}{1} - \frac{1}{2}\right) + \left(\frac{1}{2} - \frac{1}{3}\right) + \left(\frac{1}{3} - \frac{1}{4}\right) + \cdots + \left(\frac{1}{k} - \frac{1}{k+1}\right).$$

Removing parentheses and canceling adjacent terms of opposite sign collapses the sum to

$$s_k = 1 - \frac{1}{k+1}.$$

We now see that $s_k \to 1$ as $k \to \infty$. The series converges, and its sum is 1:

$$\sum_{n=1}^{\infty} \frac{1}{n(n+1)} = 1.$$

■

Divergent Series

One reason that a series may fail to converge is that its terms don't become small.

EXAMPLE 6 The series

$$\sum_{n=1}^{\infty} \frac{n+1}{n} = \frac{2}{1} + \frac{3}{2} + \frac{4}{3} + \cdots + \frac{n+1}{n} + \cdots$$

diverges because the partial sums eventually outgrow every preassigned number. Each term is greater than 1, so the sum of n terms is greater than n. ■

The *n*th-Term Test for Divergence

Notice that $\lim_{n\to\infty} a_n$ must equal zero if the series $\sum_{n=1}^{\infty} a_n$ converges. To see why, let S represent the series' sum and $s_n = a_1 + a_2 + \cdots + a_n$ the nth partial sum. When n is large, both s_n and s_{n-1} are close to S, so their difference, a_n, is close to zero. More formally,

$$a_n = s_n - s_{n-1} \quad \to \quad S - S = 0.$$

Difference Rule for sequences

This establishes the following theorem.

Caution

Theorem 7 *does not say* that $\sum_{n=1}^{\infty} a_n$ converges if $a_n \to 0$. It is possible for a series to diverge when $a_n \to 0$.

THEOREM 7 If $\sum_{n=1}^{\infty} a_n$ converges, then $a_n \to 0$.

Theorem 7 leads to a test for detecting the kind of divergence that occurred in Example 6.

The *n*th-Term Test for Divergence

$\sum_{n=1}^{\infty} a_n$ diverges if $\lim_{n\to\infty} a_n$ fails to exist or is different from zero.

EXAMPLE 7

(a) $\sum_{n=1}^{\infty} n^2$ diverges because $n^2 \to \infty$

(b) $\sum_{n=1}^{\infty} \frac{n+1}{n}$ diverges because $\frac{n+1}{n} \to 1$ $\quad \lim_{n\to\infty} a_n \neq 0$

(c) $\sum_{n=1}^{\infty} (-1)^{n+1}$ diverges because $\lim_{n\to\infty}(-1)^{n+1}$ does not exist

(d) $\sum_{n=1}^{\infty} \frac{-n}{2n+5}$ diverges because $\lim_{n\to\infty} \frac{-n}{2n+5} = -\frac{1}{2} \neq 0.$ ■

EXAMPLE 8 The series

$$1 + \underbrace{\frac{1}{2} + \frac{1}{2}}_{2\text{ terms}} + \underbrace{\frac{1}{4} + \frac{1}{4} + \frac{1}{4} + \frac{1}{4}}_{4\text{ terms}} + \cdots + \underbrace{\frac{1}{2^n} + \frac{1}{2^n} + \cdots + \frac{1}{2^n}}_{2^n\text{ terms}} + \cdots$$

diverges because the terms can be grouped into clusters that add to 1, so the partial sums increase without bound. However, the terms of the series form a sequence that converges to 0. Example 1 of Section 7.3 shows that the harmonic series also behaves in this manner. ■

Combining Series

Whenever we have two convergent series, we can add them term by term, subtract them term by term, or multiply them by constants to make new convergent series.

THEOREM 8 If $\sum a_n = A$ and $\sum b_n = B$ are convergent series, then

1. *Sum Rule:* $\sum(a_n + b_n) = \sum a_n + \sum b_n = A + B$
2. *Difference Rule:* $\sum(a_n - b_n) = \sum a_n - \sum b_n = A - B$
3. *Constant Multiple Rule:* $\sum ka_n = k\sum a_n = kA$ (Any number k).

Proof The three rules for series follow from the analogous rules for sequences in Theorem 1, Section 7.1. To prove the Sum Rule for series, let

$$A_n = a_1 + a_2 + \cdots + a_n, \qquad B_n = b_1 + b_2 + \cdots + b_n.$$

Then the partial sums of $\sum(a_n + b_n)$ are

$$\begin{aligned} s_n &= (a_1 + b_1) + (a_2 + b_2) + \cdots + (a_n + b_n) \\ &= (a_1 + \cdots + a_n) + (b_1 + \cdots + b_n) \\ &= A_n + B_n. \end{aligned}$$

Since $A_n \to A$ and $B_n \to B$, we have $s_n \to A + B$ by the Sum Rule for sequences. The proof of the Difference Rule is similar.

To prove the Constant Multiple Rule for series, observe that the partial sums of $\sum ka_n$ form the sequence

$$s_n = ka_1 + ka_2 + \cdots + ka_n = k(a_1 + a_2 + \cdots + a_n) = kA_n,$$

which converges to kA by the Constant Multiple Rule for sequences. ■

As corollaries of Theorem 8, we have the following results. We omit the proofs.

> **1.** Every nonzero constant multiple of a divergent series diverges.
>
> **2.** If $\sum a_n$ converges and $\sum b_n$ diverges, then $\sum(a_n + b_n)$ and $\sum(a_n - b_n)$ both diverge.

Caution Remember that $\sum(a_n + b_n)$ can converge when $\sum a_n$ and $\sum b_n$ both diverge. For example, $\sum a_n = 1 + 1 + 1 + \cdots$ and $\sum b_n = (-1) + (-1) + (-1) + \cdots$ diverge, whereas $\sum(a_n + b_n) = 0 + 0 + 0 + \cdots$ converges to 0.

EXAMPLE 9 Find the sums of the following series.

(a)
$$\begin{aligned}\sum_{n=1}^{\infty} \frac{3^{n-1} - 1}{6^{n-1}} &= \sum_{n=1}^{\infty} \left(\frac{1}{2^{n-1}} - \frac{1}{6^{n-1}}\right) \\ &= \sum_{n=1}^{\infty} \frac{1}{2^{n-1}} - \sum_{n=1}^{\infty} \frac{1}{6^{n-1}} && \text{Difference Rule} \\ &= \frac{1}{1 - (1/2)} - \frac{1}{1 - (1/6)} && \text{Geometric series with } a = 1 \text{ and } r = 1/2,\ 1/6 \\ &= 2 - \frac{6}{5} = \frac{4}{5}\end{aligned}$$

(b)
$$\begin{aligned}\sum_{n=0}^{\infty} \frac{4}{2^n} &= 4\sum_{n=0}^{\infty} \frac{1}{2^n} && \text{Constant Multiple Rule} \\ &= 4\left(\frac{1}{1 - (1/2)}\right) && \text{Geometric series with } a = 1,\ r = 1/2 \\ &= 8\end{aligned}$$

■

Adding or Deleting Terms

We can add a finite number of terms to a series or delete a finite number of terms without altering the series' convergence or divergence, although in the case of convergence this will usually change the sum. If $\sum_{n=1}^{\infty} a_n$ converges, then $\sum_{n=k}^{\infty} a_n$ converges for any $k > 1$ and

$$\sum_{n=1}^{\infty} a_n = a_1 + a_2 + \cdots + a_{k-1} + \sum_{n=k}^{\infty} a_n .$$

Conversely, if $\sum_{n=k}^{\infty} a_n$ converges for any $k > 1$, then $\sum_{n=1}^{\infty} a_n$ converges. Thus,

$$\sum_{n=1}^{\infty} \frac{1}{5^n} = \frac{1}{5} + \frac{1}{25} + \frac{1}{125} + \sum_{n=4}^{\infty} \frac{1}{5^n}$$

and

$$\sum_{n=4}^{\infty} \frac{1}{5^n} = \left(\sum_{n=1}^{\infty} \frac{1}{5^n}\right) - \frac{1}{5} - \frac{1}{25} - \frac{1}{125} .$$

HISTORICAL BIOGRAPHY

Richard Dedekind
(1831–1916)

Reindexing

As long as we preserve the order of its terms, we can reindex any series without altering its convergence. To raise the starting value of the index h units, replace the n in the formula for a_n by $n - h$:

$$\sum_{n=1}^{\infty} a_n = \sum_{n=1+h}^{\infty} a_{n-h} = a_1 + a_2 + a_3 + \cdots.$$

To lower the starting value of the index h units, replace the n in the formula for a_n by $n + h$:

$$\sum_{n=1}^{\infty} a_n = \sum_{n=1-h}^{\infty} a_{n+h} = a_1 + a_2 + a_3 + \cdots.$$

It works like a horizontal shift. We saw this in starting a geometric series with the index $n = 0$ instead of the index $n = 1$, but we can use any other starting index value as well. We usually give preference to indexings that lead to simple expressions.

EXAMPLE 10 We can write the geometric series

$$\sum_{n=1}^{\infty} \frac{1}{2^{n-1}} = 1 + \frac{1}{2} + \frac{1}{4} + \cdots$$

as

$$\sum_{n=0}^{\infty} \frac{1}{2^n}, \qquad \sum_{n=5}^{\infty} \frac{1}{2^{n-5}}, \qquad \text{or even} \qquad \sum_{n=-4}^{\infty} \frac{1}{2^{n+4}}.$$

The partial sums remain the same no matter what indexing we choose. ■

EXERCISES 7.2

In Exercises 1–6, find a formula for the nth partial sum of each series and use it to find the series' sum if the series converges.

1. $2 + \frac{2}{3} + \frac{2}{9} + \frac{2}{27} + \cdots + \frac{2}{3^{n-1}} + \cdots$

2. $\frac{9}{100} + \frac{9}{100^2} + \frac{9}{100^3} + \cdots + \frac{9}{100^n} + \cdots$

3. $1 - \frac{1}{2} + \frac{1}{4} - \frac{1}{8} + \cdots + (-1)^{n-1}\frac{1}{2^{n-1}} + \cdots$

4. $1 - 2 + 4 - 8 + \cdots + (-1)^{n-1}2^{n-1} + \cdots$

5. $\frac{1}{2\cdot 3} + \frac{1}{3\cdot 4} + \frac{1}{4\cdot 5} + \cdots + \frac{1}{(n+1)(n+2)} + \cdots$

6. $\frac{5}{1\cdot 2} + \frac{5}{2\cdot 3} + \frac{5}{3\cdot 4} + \cdots + \frac{5}{n(n+1)} + \cdots$

In Exercises 7–14, write out the first few terms of each series to show how the series starts. Then find the sum of the series.

7. $\sum_{n=0}^{\infty} \frac{(-1)^n}{4^n}$

8. $\sum_{n=2}^{\infty} \frac{1}{4^n}$

9. $\sum_{n=1}^{\infty} \frac{7}{4^n}$

10. $\sum_{n=0}^{\infty} (-1)^n \frac{5}{4^n}$

11. $\sum_{n=0}^{\infty} \left(\frac{5}{2^n} + \frac{1}{3^n}\right)$

12. $\sum_{n=0}^{\infty} \left(\frac{5}{2^n} - \frac{1}{3^n}\right)$

13. $\sum_{n=0}^{\infty} \left(\frac{1}{2^n} + \frac{(-1)^n}{5^n}\right)$

14. $\sum_{n=0}^{\infty} \left(\frac{2^{n+1}}{5^n}\right)$

Use partial fractions to find the sum of each series in Exercises 15–22.

15. $\sum_{n=1}^{\infty} \frac{4}{(4n-3)(4n+1)}$

16. $\sum_{n=1}^{\infty} \frac{6}{(2n-1)(2n+1)}$

17. $\sum_{n=1}^{\infty} \frac{40n}{(2n-1)^2(2n+1)^2}$

18. $\sum_{n=1}^{\infty} \frac{2n+1}{n^2(n+1)^2}$

19. $\sum_{n=1}^{\infty} \left(\frac{1}{\sqrt{n}} - \frac{1}{\sqrt{n+1}}\right)$

20. $\sum_{n=1}^{\infty} \left(\frac{1}{2^{1/n}} - \frac{1}{2^{1/(n+1)}}\right)$

21. $\sum_{n=1}^{\infty} \left(\frac{1}{\ln(n+2)} - \frac{1}{\ln(n+1)}\right)$

22. $\sum_{n=1}^{\infty} (\tan^{-1}(n) - \tan^{-1}(n+1))$

Which series in Exercises 23–40 converge, and which diverge? Give reasons for your answers. If a series converges, find its sum.

23. $\sum_{n=0}^{\infty}\left(\frac{1}{\sqrt{2}}\right)^n$

24. $\sum_{n=0}^{\infty}\left(\sqrt{2}\right)^n$

25. $\sum_{n=1}^{\infty}(-1)^{n+1}\frac{3}{2^n}$

26. $\sum_{n=1}^{\infty}(-1)^{n+1}n$

27. $\sum_{n=0}^{\infty}\cos n\pi$

28. $\sum_{n=0}^{\infty}\frac{\cos n\pi}{5^n}$

29. $\sum_{n=0}^{\infty}e^{-2n}$

30. $\sum_{n=1}^{\infty}\ln\frac{1}{n}$

31. $\sum_{n=1}^{\infty}\frac{2}{10^n}$

32. $\sum_{n=0}^{\infty}\frac{1}{x^n},\quad |x|>1$

33. $\sum_{n=0}^{\infty}\frac{2^n-1}{3^n}$

34. $\sum_{n=1}^{\infty}\left(1-\frac{1}{n}\right)^n$

35. $\sum_{n=0}^{\infty}\frac{n!}{1000^n}$

36. $\sum_{n=1}^{\infty}\frac{n^n}{n!}$

37. $\sum_{n=1}^{\infty}\ln\left(\frac{n}{n+1}\right)$

38. $\sum_{n=1}^{\infty}\ln\left(\frac{n}{2n+1}\right)$

39. $\sum_{n=0}^{\infty}\left(\frac{e}{\pi}\right)^n$

40. $\sum_{n=0}^{\infty}\frac{e^{n\pi}}{\pi^{ne}}$

In each of the geometric series in Exercises 41–44, write out the first few terms of the series to find a and r, and find the sum of the series. Then express the inequality $|r|<1$ in terms of x and find the values of x for which the inequality holds and the series converges.

41. $\sum_{n=0}^{\infty}(-1)^n x^n$

42. $\sum_{n=0}^{\infty}(-1)^n x^{2n}$

43. $\sum_{n=0}^{\infty}3\left(\frac{x-1}{2}\right)^n$

44. $\sum_{n=0}^{\infty}\frac{(-1)^n}{2}\left(\frac{1}{3+\sin x}\right)^n$

In Exercises 45–50, find the values of x for which the given geometric series converges. Also, find the sum of the series (as a function of x) for those values of x.

45. $\sum_{n=0}^{\infty}2^n x^n$

46. $\sum_{n=0}^{\infty}(-1)^n x^{-2n}$

47. $\sum_{n=0}^{\infty}(-1)^n(x+1)^n$

48. $\sum_{n=0}^{\infty}\left(-\frac{1}{2}\right)^n(x-3)^n$

49. $\sum_{n=0}^{\infty}\sin^n x$

50. $\sum_{n=0}^{\infty}(\ln x)^n$

Express each of the numbers in Exercises 51–58 as the ratio of two integers.

51. $0.\overline{23} = 0.23\ 23\ 23\ldots$

52. $0.\overline{234} = 0.234\ 234\ 234\ldots$

53. $0.\overline{7} = 0.7777\ldots$

54. $0.\overline{d} = 0.dddd\ldots,$ where d is a digit

55. $0.0\overline{6} = 0.06666\ldots$

56. $1.\overline{414} = 1.414\ 414\ 414\ldots$

57. $1.24\overline{123} = 1.24\ 123\ 123\ 123\ldots$

58. $3.\overline{142857} = 3.142857\ 142857\ldots$

59. Show by example that $\Sigma(a_n/b_n)$ may diverge even though Σa_n and Σb_n converge and no b_n equals 0.

60. Find convergent geometric series $A=\Sigma a_n$ and $B=\Sigma b_n$ that illustrate the fact that $\Sigma a_n b_n$ may converge without being equal to AB.

61. Show by example that $\Sigma(a_n/b_n)$ may converge to something other than A/B even when $A=\Sigma a_n$, $B=\Sigma b_n\neq 0$, and no b_n equals 0.

62. If Σa_n converges and $a_n>0$ for all n, can anything be said about $\Sigma(1/a_n)$? Give reasons for your answer.

63. What happens if you add a finite number of terms to a divergent series or delete a finite number of terms from a divergent series? Give reasons for your answer.

64. Find the value of b for which

$$1+e^b+e^{2b}+e^{3b}+\cdots=9.$$

65. For what values of r does the infinite series

$$1+2r+r^2+2r^3+r^4+2r^5+r^6+\cdots$$

converge? Find the sum of the series when it converges.

66. Show that the error $(L-s_n)$ obtained by replacing a convergent geometric series with one of its partial sums s_n is $ar^n/(1-r)$.

7.3 The Integral Test

Given a series, we want to know whether it converges or not. In this section and the next two, we study series with nonnegative terms. Such a series converges if its sequence of partial sums is bounded. If we establish that a given series does converge, we generally do not have a formula available for its sum, so we investigate methods to approximate the sum instead.

Nondecreasing Partial Sums

Suppose that $\sum_{n=1}^{\infty} a_n$ is an infinite series with $a_n \geq 0$ for all n. Then each partial sum is greater than or equal to its predecessor because $s_{n+1} = s_n + a_n$:

$$s_1 \leq s_2 \leq s_3 \leq \cdots \leq s_n \leq s_{n+1} \leq \cdots.$$

Since the partial sums form a nondecreasing sequence, the Nondecreasing Sequence Theorem (Theorem 6, Section 7.1) tells us that the series will converge if and only if the partial sums are bounded from above.

> **Corollary of Theorem 6** A series $\sum_{n=1}^{\infty} a_n$ of nonnegative terms converges if and only if its partial sums are bounded from above.

EXAMPLE 1 The series

$$\sum_{n=1}^{\infty} \frac{1}{n} = 1 + \frac{1}{2} + \frac{1}{3} + \cdots + \frac{1}{n} + \cdots$$

is called the **harmonic series**. The harmonic series is divergent, but this doesn't follow from the nth-Term Test. The nth term $1/n$ does go to zero, but the series still diverges. The reason it diverges is because there is no upper bound for its partial sums. To see why, group the terms of the series in the following way:

$$1 + \frac{1}{2} + \underbrace{\left(\frac{1}{3} + \frac{1}{4}\right)}_{> \frac{2}{4} = \frac{1}{2}} + \underbrace{\left(\frac{1}{5} + \frac{1}{6} + \frac{1}{7} + \frac{1}{8}\right)}_{> \frac{4}{8} = \frac{1}{2}} + \underbrace{\left(\frac{1}{9} + \frac{1}{10} + \cdots + \frac{1}{16}\right)}_{> \frac{8}{16} = \frac{1}{2}} + \cdots.$$

The sum of the first two terms is 1.5. The sum of the next two terms is $1/3 + 1/4$, which is greater than $1/4 + 1/4 = 1/2$. The sum of the next four terms is $1/5 + 1/6 + 1/7 + 1/8$, which is greater than $1/8 + 1/8 + 1/8 + 1/8 = 1/2$. The sum of the next eight terms is $1/9 + 1/10 + 1/11 + 1/12 + 1/13 + 1/14 + 1/15 + 1/16$, which is greater than $8/16 = 1/2$. The sum of the next 16 terms is greater than $16/32 = 1/2$, and so on. In general, the sum of 2^n terms ending with $1/2^{n+1}$ is greater than $2^n/2^{n+1} = 1/2$. The sequence of partial sums is not bounded from above: If $n = 2^k$, the partial sum s_n is greater than $k/2$. The harmonic series diverges. ■

The Integral Test

We introduce the Integral Test with a series that is related to the harmonic series, but whose nth term is $1/n^2$ instead of $1/n$.

EXAMPLE 2 Does the following series converge?

$$\sum_{n=1}^{\infty} \frac{1}{n^2} = 1 + \frac{1}{4} + \frac{1}{9} + \frac{1}{16} + \cdots + \frac{1}{n^2} + \cdots$$

Solution We determine the convergence of $\sum_{n=1}^{\infty}(1/n^2)$ by comparing it with $\int_1^{\infty}(1/x^2)\,dx$. To carry out the comparison, we think of the terms of the series as values of

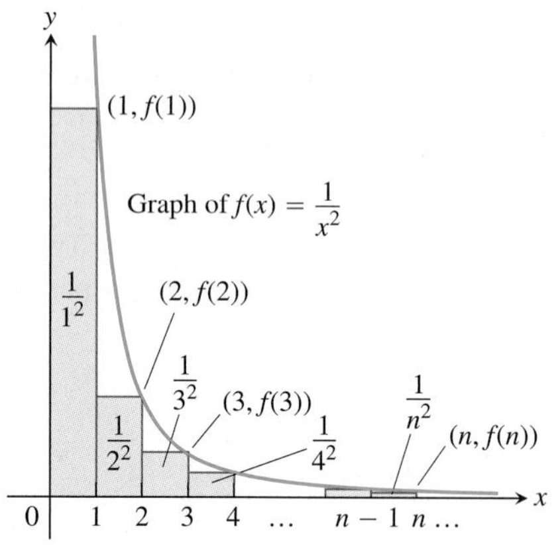

FIGURE 7.7 The sum of the areas of the rectangles under the graph of $f(x) = 1/x^2$ is less than the area under the graph (Example 2).

Caution

The series and integral need not have the same value in the convergent case. As we noted in Example 2, $\sum_{n=1}^{\infty}(1/n^2) = \pi^2/6$ while $\int_1^{\infty}(1/x^2)\,dx = 1$.

the function $f(x) = 1/x^2$ and interpret these values as the areas of rectangles under the curve $y = 1/x^2$.

As Figure 7.7 shows,

$$\begin{aligned} s_n &= \frac{1}{1^2} + \frac{1}{2^2} + \frac{1}{3^2} + \cdots + \frac{1}{n^2} \\ &= f(1) + f(2) + f(3) + \cdots + f(n) \\ &< f(1) + \int_1^n \frac{1}{x^2}\,dx && \text{Rectangle areas sum to less than area under graph.} \\ &< 1 + \int_1^{\infty} \frac{1}{x^2}\,dx && \int_1^n (1/x^2)\,dx < \int_1^{\infty}(1/x^2)\,dx \\ &< 1 + 1 = 2. && \text{As in Section 5.7, Example 3, } \int_1^{\infty}(1/x^2)\,dx = 1. \end{aligned}$$

Thus the partial sums of $\sum_{n=1}^{\infty} 1/n^2$ are bounded from above (by 2) and the series converges. The sum of the series is known to be $\pi^2/6 \approx 1.64493$. ■

THEOREM 9—The Integral Test Let $\{a_n\}$ be a sequence of positive terms. Suppose that $a_n = f(n)$, where f is a continuous, positive, decreasing function of x for all $x \geq N$ (N a positive integer). Then the series $\sum_{n=N}^{\infty} a_n$ and the integral $\int_N^{\infty} f(x)\,dx$ both converge or both diverge.

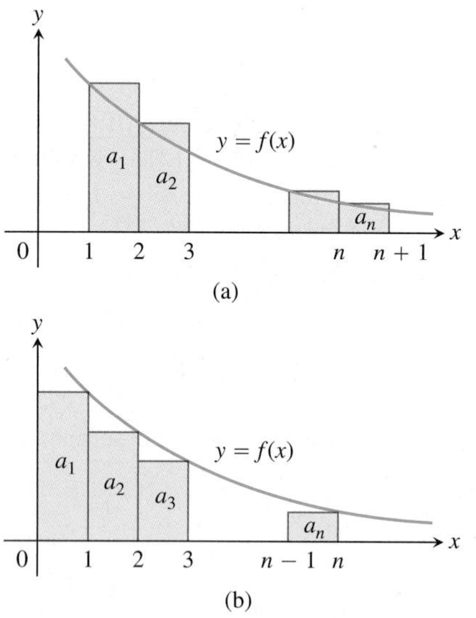

FIGURE 7.8 Subject to the conditions of the Integral Test, the series $\sum_{n=1}^{\infty} a_n$ and the integral $\int_1^{\infty}(x)\,dx$ both converge or both diverge.

Proof We establish the test for the case $N = 1$. The proof for general N is similar.

We start with the assumption that f is a decreasing function with $f(n) = a_n$ for every n. This leads us to observe that the rectangles in Figure 7.8a, which have areas $a_1, a_2, \ldots, a_n$, collectively enclose more area than that under the curve $y = f(x)$ from $x = 1$ to $x = n + 1$. That is,

$$\int_1^{n+1} f(x)\,dx \leq a_1 + a_2 + \cdots + a_n.$$

In Figure 7.8b the rectangles have been faced to the left instead of to the right. If we momentarily disregard the first rectangle, of area a_1, we see that

$$a_2 + a_3 + \cdots + a_n \leq \int_1^n f(x)\,dx.$$

If we include a_1, we have

$$a_1 + a_2 + \cdots + a_n \leq a_1 + \int_1^n f(x)\,dx.$$

Combining these results gives

$$\int_1^{n+1} f(x)\,dx \leq a_1 + a_2 + \cdots + a_n \leq a_1 + \int_1^n f(x)\,dx.$$

These inequalities hold for each n, and continue to hold as $n \to \infty$.

If $\int_1^\infty f(x)\,dx$ is finite, the right-hand inequality shows that Σa_n is finite. If $\int_1^\infty f(x)\,dx$ is infinite, the left-hand inequality shows that Σa_n is infinite. Hence the series and the integral are both finite or both infinite. ■

EXAMPLE 3 Show that the ***p*-series**

$$\sum_{n=1}^{\infty} \frac{1}{n^p} = \frac{1}{1^p} + \frac{1}{2^p} + \frac{1}{3^p} + \cdots + \frac{1}{n^p} + \cdots$$

(p a real constant) converges if $p > 1$, and diverges if $p \le 1$.

Solution If $p > 1$, then $f(x) = 1/x^p$ is a positive decreasing function of x. Since

$$\int_1^\infty \frac{1}{x^p}\,dx = \int_1^\infty x^{-p}\,dx = \lim_{b\to\infty}\left[\frac{x^{-p+1}}{-p+1}\right]_1^b$$

$$= \frac{1}{1-p}\lim_{b\to\infty}\left(\frac{1}{b^{p-1}} - 1\right)$$

$$= \frac{1}{1-p}(0-1) = \frac{1}{p-1},$$

$b^{p-1} \to \infty$ as $b \to \infty$ because $p - 1 > 0$.

the series converges by the Integral Test. We emphasize that the sum of the p-series is *not* $1/(p-1)$. The series converges, but we don't know the value it converges to.

If $p < 1$, then $1 - p > 0$ and

$$\int_1^\infty \frac{1}{x^p}\,dx = \frac{1}{1-p}\lim_{b\to\infty}(b^{1-p} - 1) = \infty.$$

The series diverges by the Integral Test.

If $p = 1$, we have the (divergent) harmonic series

$$1 + \frac{1}{2} + \frac{1}{3} + \cdots + \frac{1}{n} + \cdots.$$

We have convergence for $p > 1$ but divergence for every other value of p. ■

The p-series with $p = 1$ is the **harmonic series** (Example 1). The p-Series Test shows that the harmonic series is just *barely* divergent; if we increase p to 1.000000001, for instance, the series converges!

Error Estimation

If a series Σa_n is shown to be convergent by the integral test, we may want to estimate the size of the **remainder** R_n between the total sum S of the series and its nth partial sum s_n. That is, we wish to estimate

$$R_n = S - s_n = a_{n+1} + a_{n+2} + a_{n+3} + \cdots.$$

To get a lower bound for the remainder, we compare the sum of the areas of the rectangles with the area under the curve $y = f(x)$ for $x \geq n$ (see Figure 7.8a). We see that

$$R_n = a_{n+1} + a_{n+2} + a_{n+3} + \cdots \geq \int_{n+1}^{\infty} f(x)\,dx.$$

Similarly, from Figure 7.8b, we find an upper bound with

$$R_n = a_{n+1} + a_{n+2} + a_{n+3} + \cdots \leq \int_{n}^{\infty} f(x)\,dx.$$

These comparisons prove the following result giving bounds on the size of the remainder.

Bounds for the Remainder in the Integral Test

Suppose $\{a_k\}$ is a sequence of positive terms with $a_k = f(k)$, where f is a continuous positive decreasing function of x for all $x \geq n$, and that Σa_n converges to S. Then the remainder $R_n = S - s_n$ satisfies the inequalities

$$\int_{n+1}^{\infty} f(x)\,dx \leq R_n \leq \int_{n}^{\infty} f(x)\,dx. \qquad (1)$$

If we add the partial sum s_n to each side of the inequalities in (1), we get

$$s_n + \int_{n+1}^{\infty} f(x)\,dx \leq S \leq s_n + \int_{n}^{\infty} f(x)\,dx \qquad (2)$$

since $s_n + R_n = S$. The inequalities in (2) are useful for estimating the error in approximating the sum of a convergent series. The error can be no larger than the length of the interval containing S, as given by (2).

EXAMPLE 4 Estimate the sum of the series $\Sigma(1/n^2)$ using the inequalities in (2) and $n = 10$.

Solution We have that

$$\int_{n}^{\infty} \frac{1}{x^2}\,dx = \lim_{b\to\infty}\left[-\frac{1}{x}\right]_n^b = \lim_{b\to\infty}\left(-\frac{1}{b} + \frac{1}{n}\right) = \frac{1}{n}.$$

Using this result with the inequalities in (2), we get

$$s_{10} + \frac{1}{11} \leq S \leq s_{10} + \frac{1}{10}.$$

Taking $s_{10} = 1 + (1/4) + (1/9) + (1/16) + \cdots + (1/100) \approx 1.54977$, these last inequalities give

$$1.64068 \leq S \leq 1.64997.$$

If we approximate the sum S by the midpoint of this interval, we find that

$$\sum_{n=1}^{\infty} \frac{1}{n^2} \approx 1.6453.$$

The error in this approximation is less than half the length of the interval, so the error is less than 0.005. ■

EXERCISES 7.3

Which of the series in Exercises 1–30 converge, and which diverge? Give reasons for your answers. (When you check an answer, remember that there may be more than one way to determine the series' convergence or divergence.)

1. $\sum_{n=1}^{\infty} \frac{1}{10^n}$ **2.** $\sum_{n=1}^{\infty} e^{-n}$ **3.** $\sum_{n=1}^{\infty} \frac{n}{n+1}$

4. $\sum_{n=1}^{\infty} \frac{5}{n+1}$ **5.** $\sum_{n=1}^{\infty} \frac{3}{\sqrt{n}}$ **6.** $\sum_{n=1}^{\infty} \frac{-2}{n\sqrt{n}}$

7. $\sum_{n=1}^{\infty} -\frac{1}{8^n}$ **8.** $\sum_{n=1}^{\infty} \frac{-8}{n}$ **9.** $\sum_{n=2}^{\infty} \frac{\ln n}{n}$

10. $\sum_{n=2}^{\infty} \frac{\ln n}{\sqrt{n}}$ **11.** $\sum_{n=1}^{\infty} \frac{2^n}{3^n}$ **12.** $\sum_{n=1}^{\infty} \frac{5^n}{4^n+3}$

13. $\sum_{n=0}^{\infty} \frac{-2}{n+1}$ **14.** $\sum_{n=1}^{\infty} \frac{1}{2n-1}$ **15.** $\sum_{n=1}^{\infty} \frac{2^n}{n+1}$

16. $\sum_{n=1}^{\infty} \frac{1}{\sqrt{n}(\sqrt{n}+1)}$ **17.** $\sum_{n=2}^{\infty} \frac{\sqrt{n}}{\ln n}$ **18.** $\sum_{n=1}^{\infty} \left(1+\frac{1}{n}\right)^n$

19. $\sum_{n=1}^{\infty} \frac{1}{(\ln 2)^n}$ **20.** $\sum_{n=1}^{\infty} \frac{1}{(\ln 3)^n}$

21. $\sum_{n=3}^{\infty} \frac{(1/n)}{(\ln n)\sqrt{\ln^2 n - 1}}$ **22.** $\sum_{n=1}^{\infty} \frac{1}{n(1+\ln^2 n)}$

23. $\sum_{n=1}^{\infty} n \sin \frac{1}{n}$ **24.** $\sum_{n=1}^{\infty} n \tan \frac{1}{n}$

25. $\sum_{n=1}^{\infty} \frac{e^n}{1+e^{2n}}$ **26.** $\sum_{n=1}^{\infty} \frac{2}{1+e^n}$

27. $\sum_{n=1}^{\infty} \frac{8 \tan^{-1} n}{1+n^2}$ **28.** $\sum_{n=1}^{\infty} \frac{n}{n^2+1}$

29. $\sum_{n=1}^{\infty} \operatorname{sech} n$ **30.** $\sum_{n=1}^{\infty} \operatorname{sech}^2 n$

For what values of a, if any, do the series in Exercises 31 and 32 converge?

31. $\sum_{n=1}^{\infty} \left(\frac{a}{n+2} - \frac{1}{n+4}\right)$ **32.** $\sum_{n=3}^{\infty} \left(\frac{1}{n-1} - \frac{2a}{n+1}\right)$

33. The Cauchy condensation test The Cauchy condensation test says: Let $\{a_n\}$ be a nonincreasing sequence ($a_n \geq a_{n+1}$ for all n) of positive terms that converges to 0. Then $\sum a_n$ converges if and only if $\sum 2^n a_{2^n}$ converges. For example, $\sum(1/n)$ diverges because $\sum 2^n \cdot (1/2^n) = \sum 1$ diverges. Show why the test works.

34. Use the Cauchy condensation test from Exercise 33 to show that

a. $\sum_{n=2}^{\infty} \frac{1}{n \ln n}$ diverges;

b. $\sum_{n=1}^{\infty} \frac{1}{n^p}$ converges if $p > 1$ and diverges if $p \leq 1$.

35. Logarithmic *p*-series

a. Show that

$$\int_2^{\infty} \frac{dx}{x(\ln x)^p} \quad (p \text{ a positive constant})$$

converges if and only if $p > 1$.

b. What implications does the fact in part (a) have for the convergence of the series

$$\sum_{n=2}^{\infty} \frac{1}{n(\ln n)^p}\ ?$$

Give reasons for your answer.

36. (*Continuation of Exercise 35.*) Use the result in Exercise 35 to determine which of the following series converge and which diverge. Support your answer in each case.

a. $\sum_{n=2}^{\infty} \frac{1}{n(\ln n)}$ **b.** $\sum_{n=2}^{\infty} \frac{1}{n(\ln n)^{1.01}}$

c. $\sum_{n=2}^{\infty} \frac{1}{n \ln(n^3)}$ **d.** $\sum_{n=2}^{\infty} \frac{1}{n(\ln n)^3}$

37. Euler's constant

a. By taking $f(x) = 1/x$ in the proof of Theorem 9, show that

$$\ln(n+1) \leq 1 + \frac{1}{2} + \cdots + \frac{1}{n} \leq 1 + \ln n$$

or

$$0 < \ln(n+1) - \ln n \leq 1 + \frac{1}{2} + \cdots + \frac{1}{n} - \ln n \leq 1.$$

Thus, the sequence

$$a_n = 1 + \frac{1}{2} + \cdots + \frac{1}{n} - \ln n$$

is bounded from below and from above.

b. Show that

$$\frac{1}{n+1} < \int_n^{n+1} \frac{1}{x}\, dx = \ln(n+1) - \ln n,$$

and use this result to show that the sequence $\{a_n\}$ in part (a) is decreasing.

Since a decreasing sequence that is bounded from below converges (Exercise 105 in Section 7.1), the numbers a_n defined in part (a) converge:

$$1 + \frac{1}{2} + \cdots + \frac{1}{n} - \ln n \rightarrow \gamma.$$

The number γ, whose value is $0.5772\ldots$, is called *Euler's constant*.

38. Use the integral test to show that

$$\sum_{n=0}^{\infty} e^{-n^2}$$

converges.

39. a. For the series $\Sigma(1/n^3)$, use the inequalities in (2) with $n = 10$ to find an interval containing the sum S.

b. As in Example 4, use the midpoint of the interval found in part (a) to approximate the sum of the series. What is the maximum error for your approximation?

40. Repeat Exercise 39 using the series $\Sigma(1/n^4)$.

7.4 Comparison Tests

We have seen how to determine the convergence of geometric series, p-series, and a few others. We can test the convergence of many more series by comparing their terms to those of a series whose convergence is known.

THEOREM 10—The Comparison Test Let Σa_n be a series with nonnegative terms.

(a) Σa_n converges if there is a convergent series Σc_n with $a_n \leq c_n$ for all $n > N$, for some integer N.

(b) Σa_n diverges if there is a divergent series of nonnegative terms Σd_n with $a_n \geq d_n$ for all $n > N$, for some integer N.

HISTORICAL BIOGRAPHY

Albert of Saxony
(ca. 1316–1390)

Proof In Part (a), the partial sums of Σa_n are bounded above by

$$M = a_1 + a_2 + \cdots + a_N + \sum_{n=N+1}^{\infty} c_n.$$

They therefore form a nondecreasing sequence with a limit $L \leq M$.

In Part (b), the partial sums of Σa_n are not bounded from above. If they were, the partial sums for Σd_n would be bounded by

$$M^* = d_1 + d_2 + \cdots + d_N + \sum_{n=N+1}^{\infty} a_n$$

and Σd_n would have to converge instead of diverge. ■

EXAMPLE 1

(a) The series

$$\sum_{n=1}^{\infty} \frac{5}{5n - 1}$$

diverges because its nth term

$$\frac{5}{5n-1} = \frac{1}{n - \frac{1}{5}} > \frac{1}{n}$$

is greater than the nth term of the divergent harmonic series.

(b) The series

$$\sum_{n=0}^{\infty} \frac{1}{n!} = 1 + \frac{1}{1!} + \frac{1}{2!} + \frac{1}{3!} + \cdots$$

converges because its terms are all positive and less than or equal to the corresponding terms of

$$1 + \sum_{n=0}^{\infty} \frac{1}{2^n} = 1 + 1 + \frac{1}{2} + \frac{1}{2^2} + \cdots.$$

The geometric series on the left converges and we have

$$1 + \sum_{n=0}^{\infty} \frac{1}{2^n} = 1 + \frac{1}{1 - (1/2)} = 3.$$

The fact that 3 is an upper bound for the partial sums of $\sum_{n=0}^{\infty} (1/n!)$ does not mean that the series converges to 3. As we will see in Section 7.9, the series converges to e. ■

The Limit Comparison Test

We now introduce a comparison test that is particularly useful for series in which a_n is a rational function of n.

> **THEOREM 11—Limit Comparison Test** Suppose that $a_n > 0$ and $b_n > 0$ for all $n \geq N$ (N an integer).
>
> **1.** If $\lim_{n\to\infty} \frac{a_n}{b_n} = c > 0$, then $\sum a_n$ and $\sum b_n$ both converge or both diverge.
>
> **2.** If $\lim_{n\to\infty} \frac{a_n}{b_n} = 0$ and $\sum b_n$ converges, then $\sum a_n$ converges.
>
> **3.** If $\lim_{n\to\infty} \frac{a_n}{b_n} = \infty$ and $\sum b_n$ diverges, then $\sum a_n$ diverges.

Proof We will prove Part 1. Parts 2 and 3 are left as Exercises 37(a) and (b).

Since $c/2 > 0$, there exists an integer N such that for all n

$$n > N \Rightarrow \left| \frac{a_n}{b_n} - c \right| < \frac{c}{2}.$$

Limit definition with $\epsilon = c/2$, $L = c$, and a_n replaced by a_n/b_n

Thus, for $n > N$,

$$-\frac{c}{2} < \frac{a_n}{b_n} - c < \frac{c}{2},$$

$$\frac{c}{2} < \frac{a_n}{b_n} < \frac{3c}{2},$$

$$\left(\frac{c}{2}\right) b_n < a_n < \left(\frac{3c}{2}\right) b_n.$$

If $\sum b_n$ converges, then $\sum (3c/2) b_n$ converges and $\sum a_n$ converges by the Direct Comparison Test. If $\sum b_n$ diverges, then $\sum (c/2) b_n$ diverges and $\sum a_n$ diverges by the Direct Comparison Test. ■

EXAMPLE 2 Which of the following series converge, and which diverge?

(a) $\frac{3}{4} + \frac{5}{9} + \frac{7}{16} + \frac{9}{25} + \cdots = \sum_{n=1}^{\infty} \frac{2n+1}{(n+1)^2} = \sum_{n=1}^{\infty} \frac{2n+1}{n^2 + 2n + 1}$

(b) $\frac{1}{1} + \frac{1}{3} + \frac{1}{7} + \frac{1}{15} + \cdots = \sum_{n=1}^{\infty} \frac{1}{2^n - 1}$

Solution

(a) Let $a_n = (2n + 1)/(n^2 + 2n + 1)$. For large n, we expect a_n to behave like $2n/n^2 = 2/n$ since the leading terms dominate for large n, so we let $b_n = 1/n$. Since

$$\sum_{n=1}^{\infty} b_n = \sum_{n=1}^{\infty} \frac{1}{n} \text{ diverges}$$

and

$$\lim_{n\to\infty} \frac{a_n}{b_n} = \lim_{n\to\infty} \frac{2n^2 + n}{n^2 + 2n + 1} = 2,$$

Σa_n diverges by Part 1 of the Limit Comparison Test. We could just as well have taken $b_n = 2/n$, but $1/n$ is simpler.

(b) Let $a_n = 1/(2^n - 1)$. For large n, we expect a_n to behave like $1/2^n$, so we let $b_n = 1/2^n$. Since

$$\sum_{n=1}^{\infty} b_n = \sum_{n=1}^{\infty} \frac{1}{2^n} \text{ converges}$$

and

$$\begin{aligned}\lim_{n\to\infty} \frac{a_n}{b_n} &= \lim_{n\to\infty} \frac{2^n}{2^n - 1} \\ &= \lim_{n\to\infty} \frac{1}{1 - (1/2^n)} \\ &= 1,\end{aligned}$$

Σa_n converges by Part 1 of the Limit Comparison Test. ■

EXAMPLE 3 Does $\sum_{n=1}^{\infty} \frac{\ln n}{n^{3/2}}$ converge?

Solution Because $\ln n$ grows more slowly than n^c for any positive constant c (Section 7.1, Exercise 89), we would expect to have

$$\frac{\ln n}{n^{3/2}} < \frac{n^{1/4}}{n^{3/2}} = \frac{1}{n^{5/4}}$$

for n sufficiently large. Indeed, taking $a_n = (\ln n)/n^{3/2}$ and $b_n = 1/n^{5/4}$, we have

$$\begin{aligned}\lim_{n\to\infty} \frac{a_n}{b_n} &= \lim_{n\to\infty} \frac{\ln n}{n^{1/4}} \\ &= \lim_{n\to\infty} \frac{1/n}{(1/4)n^{-3/4}} && \text{l'Hôpital's Rule} \\ &= \lim_{n\to\infty} \frac{4}{n^{1/4}} = 0.\end{aligned}$$

Since $\Sigma b_n = \Sigma(1/n^{5/4})$ (a p-series with $p > 1$) converges, Σa_n converges by Part 2 of the Limit Comparison Test. ■

EXERCISES 7.4

Which of the series in Exercises 1–36 converge, and which diverge? Give reasons for your answers.

1. $\sum_{n=1}^{\infty} \frac{1}{2\sqrt{n} + \sqrt[3]{n}}$
2. $\sum_{n=1}^{\infty} \frac{3}{n + \sqrt{n}}$
3. $\sum_{n=1}^{\infty} \frac{\sin^2 n}{2^n}$
4. $\sum_{n=1}^{\infty} \frac{1 + \cos n}{n^2}$
5. $\sum_{n=1}^{\infty} \frac{2n}{3n - 1}$
6. $\sum_{n=1}^{\infty} \frac{n + 1}{n^2\sqrt{n}}$
7. $\sum_{n=1}^{\infty} \left(\frac{n}{3n + 1}\right)^n$
8. $\sum_{n=1}^{\infty} \frac{1}{\sqrt{n^3 + 2}}$
9. $\sum_{n=3}^{\infty} \frac{1}{\ln(\ln n)}$
10. $\sum_{n=2}^{\infty} \frac{1}{(\ln n)^2}$
11. $\sum_{n=1}^{\infty} \frac{(\ln n)^2}{n^3}$
12. $\sum_{n=1}^{\infty} \frac{(\ln n)^3}{n^3}$
13. $\sum_{n=2}^{\infty} \frac{1}{\sqrt{n} \ln n}$
14. $\sum_{n=1}^{\infty} \frac{(\ln n)^2}{n^{3/2}}$
15. $\sum_{n=1}^{\infty} \frac{1}{1 + \ln n}$
16. $\sum_{n=1}^{\infty} \frac{1}{(1 + \ln n)^2}$
17. $\sum_{n=2}^{\infty} \frac{\ln(n + 1)}{n + 1}$
18. $\sum_{n=1}^{\infty} \frac{1}{(1 + \ln^2 n)}$
19. $\sum_{n=2}^{\infty} \frac{1}{n\sqrt{n^2 - 1}}$
20. $\sum_{n=1}^{\infty} \frac{\sqrt{n}}{n^2 + 1}$
21. $\sum_{n=1}^{\infty} \frac{1 - n}{n2^n}$
22. $\sum_{n=1}^{\infty} \frac{n + 2^n}{n^2 2^n}$
23. $\sum_{n=1}^{\infty} \frac{1}{3^{n-1} + 1}$
24. $\sum_{n=1}^{\infty} \frac{3^{n-1} + 1}{3^n}$
25. $\sum_{n=1}^{\infty} \sin \frac{1}{n}$
26. $\sum_{n=1}^{\infty} \tan \frac{1}{n}$
27. $\sum_{n=1}^{\infty} \frac{10n + 1}{n(n + 1)(n + 2)}$
28. $\sum_{n=3}^{\infty} \frac{5n^3 - 3n}{n^2(n - 2)(n^2 + 5)}$
29. $\sum_{n=1}^{\infty} \frac{\tan^{-1} n}{n^{1.1}}$
30. $\sum_{n=1}^{\infty} \frac{\sec^{-1} n}{n^{1.3}}$
31. $\sum_{n=1}^{\infty} \frac{\coth n}{n^2}$
32. $\sum_{n=1}^{\infty} \frac{\tanh n}{n^2}$
33. $\sum_{n=1}^{\infty} \frac{1}{n\sqrt[n]{n}}$
34. $\sum_{n=1}^{\infty} \frac{\sqrt[n]{n}}{n^2}$
35. $\sum_{n=1}^{\infty} \frac{1}{1 + 2 + 3 + \cdots + n}$
36. $\sum_{n=1}^{\infty} \frac{1}{1 + 2^2 + 3^2 + \cdots + n^2}$
37. Prove **(a)** Part 2 and **(b)** Part 3 of the Limit Comparison Test.
38. If $\sum_{n=1}^{\infty} a_n$ is a convergent series of nonnegative numbers, can anything be said about $\sum_{n=1}^{\infty}(a_n/n)$? Explain.
39. Suppose that $a_n > 0$ and $b_n > 0$ for $n \geq N$ (N an integer). If $\lim_{n\to\infty}(a_n/b_n) = \infty$ and $\sum a_n$ converges, can anything be said about $\sum b_n$? Give reasons for your answer.
40. Prove that if $\sum a_n$ is a convergent series of nonnegative terms, then $\sum a_n^2$ converges.

7.5 The Ratio and Root Tests

The Ratio Test measures the rate of growth (or decline) of a series by examining the ratio a_{n+1}/a_n. For a geometric series $\sum ar^n$, this rate is a constant $((ar^{n+1})/(ar^n) = r)$, and the series converges if and only if its ratio is less than 1 in absolute value. The Ratio Test is a powerful rule extending that result.

THEOREM 12—The Ratio Test Let $\sum a_n$ be a series with positive terms and suppose that

$$\lim_{n\to\infty} \frac{a_{n+1}}{a_n} = \rho.$$

Then **(a)** the series *converges* if $\rho < 1$, **(b)** the series *diverges* if $\rho > 1$ or ρ is infinite, **(c)** the test is *inconclusive* if $\rho = 1$.

Proof

(a) $\boldsymbol{\rho < 1}$**.** Let r be a number between ρ and 1. Then the number $\epsilon = r - \rho$ is positive. Since

$$\frac{a_{n+1}}{a_n} \to \rho,$$

a_{n+1}/a_n must lie within ϵ of ρ when n is large enough, say for all $n \geq N$. In particular

$$\frac{a_{n+1}}{a_n} < \rho + \epsilon = r, \qquad \text{when } n \geq N.$$

That is,

$$\begin{aligned} a_{N+1} &< ra_N, \\ a_{N+2} &< ra_{N+1} < r^2 a_N, \\ a_{N+3} &< ra_{N+2} < r^3 a_N, \\ &\vdots \\ a_{N+m} &< ra_{N+m-1} < r^m a_N. \end{aligned}$$

These inequalities show that the terms of our series, after the Nth term, approach zero more rapidly than the terms in a geometric series with ratio $r < 1$. More precisely, consider the series $\sum c_n$, where $c_n = a_n$ for $n = 1, 2, \ldots, N$ and $c_{N+1} = ra_N$, $c_{N+2} = r^2 a_N, \ldots, c_{N+m} = r^m a_N, \ldots$. Now $a_n \le c_n$ for all n, and

$$\begin{aligned} \sum_{n=1}^{\infty} c_n &= a_1 + a_2 + \cdots + a_{N-1} + a_N + ra_N + r^2 a_N + \cdots \\ &= a_1 + a_2 + \cdots + a_{N-1} + a_N(1 + r + r^2 + \cdots). \end{aligned}$$

The geometric series $1 + r + r^2 + \cdots$ converges because $|r| < 1$, so $\sum c_n$ converges. Since $a_n \le c_n$, $\sum a_n$ also converges.

(b) $1 < \rho \le \infty$. From some index M on,

$$\frac{a_{n+1}}{a_n} > 1 \qquad \text{and} \qquad a_M < a_{M+1} < a_{M+2} < \cdots.$$

The terms of the series do not approach zero as n becomes infinite, and the series diverges by the nth-Term Test.

(c) $\rho = 1$. The two series

$$\sum_{n=1}^{\infty} \frac{1}{n} \qquad \text{and} \qquad \sum_{n=1}^{\infty} \frac{1}{n^2}$$

show that some other test for convergence must be used when $\rho = 1$.

$$\text{For } \sum_{n=1}^{\infty} \frac{1}{n}: \qquad \frac{a_{n+1}}{a_n} = \frac{1/(n+1)}{1/n} = \frac{n}{n+1} \to 1.$$

$$\text{For } \sum_{n=1}^{\infty} \frac{1}{n^2}: \qquad \frac{a_{n+1}}{a_n} = \frac{1/(n+1)^2}{1/n^2} = \left(\frac{n}{n+1}\right)^2 \to 1^2 = 1.$$

In both cases, $\rho = 1$, yet the first series diverges, whereas the second converges. ■

The Ratio Test is often effective when the terms of a series contain factorials of expressions involving n or expressions raised to a power involving n.

EXAMPLE 1 Investigate the convergence of the following series.

(a) $\displaystyle\sum_{n=0}^{\infty} \frac{2^n + 5}{3^n}$ **(b)** $\displaystyle\sum_{n=1}^{\infty} \frac{(2n)!}{n!n!}$

Solution

(a) For the series $\sum_{n=0}^{\infty} (2^n + 5)/3^n$,

$$\frac{a_{n+1}}{a_n} = \frac{(2^{n+1} + 5)/3^{n+1}}{(2^n + 5)/3^n} = \frac{1}{3} \cdot \frac{2^{n+1} + 5}{2^n + 5} = \frac{1}{3} \cdot \left(\frac{2 + 5 \cdot 2^{-n}}{1 + 5 \cdot 2^{-n}}\right) \rightarrow \frac{1}{3} \cdot \frac{2}{1} = \frac{2}{3}.$$

The series converges because $\rho = 2/3$ is less than 1. This does *not* mean that $2/3$ is the sum of the series. In fact,

$$\sum_{n=0}^{\infty} \frac{2^n + 5}{3^n} = \sum_{n=0}^{\infty} \left(\frac{2}{3}\right)^n + \sum_{n=0}^{\infty} \frac{5}{3^n} = \frac{1}{1 - (2/3)} + \frac{5}{1 - (1/3)} = \frac{21}{2}.$$

(b) If $a_n = \dfrac{(2n)!}{n!n!}$, then $a_{n+1} = \dfrac{(2n + 2)!}{(n + 1)!(n + 1)!}$ and

$$\begin{aligned}\frac{a_{n+1}}{a_n} &= \frac{n!n!(2n + 2)(2n + 1)(2n)!}{(n + 1)!(n + 1)!(2n)!}\\ &= \frac{(2n + 2)(2n + 1)}{(n + 1)(n + 1)} = \frac{4n + 2}{n + 1} \rightarrow 4.\end{aligned}$$

The series diverges because $\rho = 4$ is greater than 1. ■

The Root Test

The following test is often convenient to use when the nth term of a series involves the nth roots $\sqrt[n]{a_n}$.

THEOREM 13—The Root Test Let $\sum a_n$ be a series with $a_n \geq 0$ for $n \geq N$, and suppose that

$$\lim_{n \to \infty} \sqrt[n]{a_n} = \rho.$$

Then **(a)** the series *converges* if $\rho < 1$, **(b)** the series *diverges* if $\rho > 1$ or ρ is infinite, **(c)** the test is *inconclusive* if $\rho = 1$.

The proof of Theorem 13 is similar to the proof of the Ratio Test and is omitted.

EXAMPLE 2 Which of the following series converge, and which diverge?

(a) $\displaystyle\sum_{n=1}^{\infty} \frac{n^2}{2^n}$ **(b)** $\displaystyle\sum_{n=1}^{\infty} \frac{2^n}{n^2}$ **(c)** $\displaystyle\sum_{n=1}^{\infty} \left(\frac{1}{1 + n}\right)^n$

Solution

(a) $\displaystyle\sum_{n=1}^{\infty} \frac{n^2}{2^n}$ converges because $\displaystyle\sqrt[n]{\frac{n^2}{2^n}} = \frac{\sqrt[n]{n^2}}{\sqrt[n]{2^n}} = \frac{\left(\sqrt[n]{n}\right)^2}{2} \rightarrow \frac{1}{2} < 1.$

(b) $\sum_{n=1}^{\infty} \frac{2^n}{n^2}$ diverges because $\sqrt[n]{\frac{2^n}{n^2}} = \frac{2}{(\sqrt[n]{n})^2} \to \frac{2}{1} > 1.$

(c) $\sum_{n=1}^{\infty} \left(\frac{1}{1+n}\right)^n$ converges because $\sqrt[n]{\left(\frac{1}{1+n}\right)^n} = \frac{1}{1+n} \to 0 < 1.$ ■

EXERCISES 7.5

Which of the series in Exercises 1–26 converge, and which diverge? Give reasons for your answers. (When checking your answers, remember there may be more than one way to determine a series' convergence or divergence.)

1. $\sum_{n=1}^{\infty} \frac{n^{\sqrt{2}}}{2^n}$

2. $\sum_{n=1}^{\infty} n^2 e^{-n}$

3. $\sum_{n=1}^{\infty} n! e^{-n}$

4. $\sum_{n=1}^{\infty} \frac{n!}{10^n}$

5. $\sum_{n=1}^{\infty} \frac{n^{10}}{10^n}$

6. $\sum_{n=1}^{\infty} \left(\frac{n-2}{n}\right)^n$

7. $\sum_{n=1}^{\infty} \frac{2+(-1)^n}{1.25^n}$

8. $\sum_{n=1}^{\infty} \frac{(-2)^n}{3^n}$

9. $\sum_{n=1}^{\infty} \left(1 - \frac{3}{n}\right)^n$

10. $\sum_{n=1}^{\infty} \left(1 - \frac{1}{3n}\right)^n$

11. $\sum_{n=1}^{\infty} \frac{\ln n}{n^3}$

12. $\sum_{n=1}^{\infty} \frac{(\ln n)^n}{n^n}$

13. $\sum_{n=1}^{\infty} \left(\frac{1}{n} - \frac{1}{n^2}\right)$

14. $\sum_{n=1}^{\infty} \left(\frac{1}{n} - \frac{1}{n^2}\right)^n$

15. $\sum_{n=1}^{\infty} \frac{\ln n}{n}$

16. $\sum_{n=1}^{\infty} \frac{n \ln n}{2^n}$

17. $\sum_{n=1}^{\infty} \frac{(n+1)(n+2)}{n!}$

18. $\sum_{n=1}^{\infty} e^{-n}(n^3)$

19. $\sum_{n=1}^{\infty} \frac{(n+3)!}{3!n!3^n}$

20. $\sum_{n=1}^{\infty} \frac{n2^n(n+1)!}{3^n n!}$

21. $\sum_{n=1}^{\infty} \frac{n!}{(2n+1)!}$

22. $\sum_{n=1}^{\infty} \frac{n!}{n^n}$

23. $\sum_{n=2}^{\infty} \frac{n}{(\ln n)^n}$

24. $\sum_{n=2}^{\infty} \frac{n}{(\ln n)^{(n/2)}}$

25. $\sum_{n=1}^{\infty} \frac{n! \ln n}{n(n+2)!}$

26. $\sum_{n=1}^{\infty} \frac{3^n}{n^3 2^n}$

Which of the series $\sum_{n=1}^{\infty} a_n$ defined by the formulas in Exercises 27–38 converge, and which diverge? Give reasons for your answers.

27. $a_1 = 2, \quad a_{n+1} = \frac{1 + \sin n}{n} a_n$

28. $a_1 = 1, \quad a_{n+1} = \frac{1 + \tan^{-1} n}{n} a_n$

29. $a_1 = \frac{1}{3}, \quad a_{n+1} = \frac{3n-1}{2n+5} a_n$

30. $a_1 = 3, \quad a_{n+1} = \frac{n}{n+1} a_n$

31. $a_1 = 2, \quad a_{n+1} = \frac{2}{n} a_n$

32. $a_1 = 5, \quad a_{n+1} = \frac{\sqrt[n]{n}}{2} a_n$

33. $a_1 = 1, \quad a_{n+1} = \frac{1 + \ln n}{n} a_n$

34. $a_1 = \frac{1}{2}, \quad a_{n+1} = \frac{n + \ln n}{n + 10} a_n$

35. $a_1 = \frac{1}{3}, \quad a_{n+1} = \sqrt[n]{a_n}$

36. $a_1 = \frac{1}{2}, \quad a_{n+1} = (a_n)^{n+1}$

37. $a_n = \frac{2^n n! n!}{(2n)!}$

38. $a_n = \frac{(3n)!}{n!(n+1)!(n+2)!}$

Which of the series in Exercises 39–44 converge, and which diverge? Give reasons for your answers.

39. $\sum_{n=1}^{\infty} \frac{(n!)^n}{(n^n)^2}$

40. $\sum_{n=1}^{\infty} \frac{(n!)^n}{n^{(n^2)}}$

41. $\sum_{n=1}^{\infty} \frac{n^n}{2^{(n^2)}}$

42. $\sum_{n=1}^{\infty} \frac{n^n}{(2^n)^2}$

43. $\sum_{n=1}^{\infty} \frac{1 \cdot 3 \cdot \cdots \cdot (2n-1)}{4^n 2^n n!}$

44. $\sum_{n=1}^{\infty} \frac{1 \cdot 3 \cdot \cdots \cdot (2n-1)}{[2 \cdot 4 \cdot \cdots \cdot (2n)](3^n + 1)}$

7.6 Alternating Series, Absolute and Conditional Convergence

A series in which the terms are alternately positive and negative is an **alternating series**. Here are three examples:

$$1 - \frac{1}{2} + \frac{1}{3} - \frac{1}{4} + \frac{1}{5} - \cdots + \frac{(-1)^{n+1}}{n} + \cdots \tag{1}$$

$$-2 + 1 - \frac{1}{2} + \frac{1}{4} - \frac{1}{8} + \cdots + \frac{(-1)^n 4}{2^n} + \cdots \tag{2}$$

$$1 - 2 + 3 - 4 + 5 - 6 + \cdots + (-1)^{n+1} n + \cdots \tag{3}$$

Series (1), called the **alternating harmonic series**, converges, as we will see in a moment. Series (2), a geometric series with ratio $r = -1/2$, converges to $-2/[1 + (1/2)] = -4/3$. Series (3) diverges because the nth term does not approach zero.

We prove the convergence of the alternating harmonic series by applying the Alternating Series Test.

THEOREM 14—The Alternating Series Test (Leibniz's Theorem) The series

$$\sum_{n=1}^{\infty} (-1)^{n+1} u_n = u_1 - u_2 + u_3 - u_4 + \cdots$$

converges if all three of the following conditions are satisfied:

1. The u_n's are all positive.
2. $u_n \geq u_{n+1}$ for all $n \geq N$, for some integer N.
3. $u_n \rightarrow 0$.

Proof If n is an even integer, say $n = 2m$, then the sum of the first n terms is

$$\begin{aligned} s_{2m} &= (u_1 - u_2) + (u_3 - u_4) + \cdots + (u_{2m-1} - u_{2m}) \\ &= u_1 - (u_2 - u_3) - (u_4 - u_5) - \cdots - (u_{2m-2} - u_{2m-1}) - u_{2m}. \end{aligned}$$

The first equality shows that s_{2m} is the sum of m nonnegative terms, since each term in parentheses is positive or zero. Hence $s_{2m+2} \geq s_{2m}$, and the sequence $\{s_{2m}\}$ is nondecreasing. The second equality shows that $s_{2m} \leq u_1$. Since $\{s_{2m}\}$ is nondecreasing and bounded from above, it has a limit, say

$$\lim_{m \rightarrow \infty} s_{2m} = L. \tag{4}$$

If n is an odd integer, say $n = 2m + 1$, then the sum of the first n terms is $s_{2m+1} = s_{2m} + u_{2m+1}$. Since $u_n \rightarrow 0$,

$$\lim_{m \rightarrow \infty} u_{2m+1} = 0$$

and, as $m \rightarrow \infty$,

$$s_{2m+1} = s_{2m} + u_{2m+1} \rightarrow L + 0 = L. \tag{5}$$

Combining the results of Equations (4) and (5) gives $\lim_{n \rightarrow \infty} s_n = L$ (Section 7.1, Exercise 117). ■

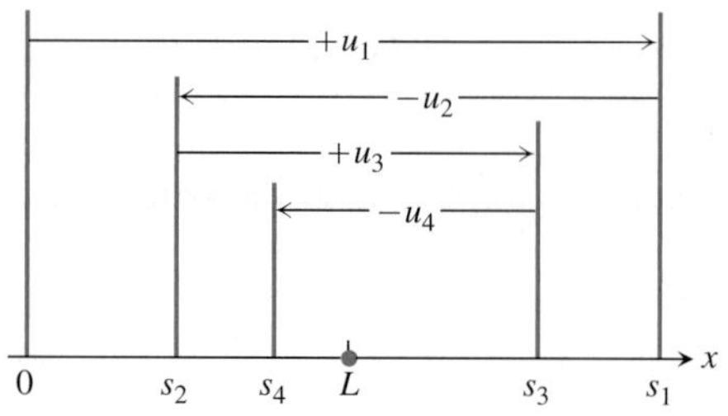

FIGURE 7.9 The partial sums of an alternating series that satisfies the hypotheses of Theorem 14 for $N = 1$ straddle the limit from the beginning.

EXAMPLE 1 The alternating harmonic series

$$\sum_{n=1}^{\infty}(-1)^{n+1}\frac{1}{n} = 1 - \frac{1}{2} + \frac{1}{3} - \frac{1}{4} + \cdots$$

satisfies the three requirements of Theorem 14 with $N = 1$; it therefore converges. ■

A graphical interpretation of the partial sums (Figure 7.9) shows how an alternating series converges to its limit L when the three conditions of Theorem 14 are satisfied with $N = 1$. Starting from the origin of the x-axis, we lay off the positive distance $s_1 = u_1$. To find the point corresponding to $s_2 = u_1 - u_2$, we back up a distance equal to u_2. Since $u_2 \le u_1$, we do not back up any farther than the origin. We continue in this seesaw fashion, backing up or going forward as the signs in the series demand. But for $n \ge N$, each forward or backward step is shorter than (or at most the same size as) the preceding step, because $u_{n+1} \le u_n$. And since the nth term approaches zero as n increases, the size of step we take forward or backward gets smaller and smaller. We oscillate across the limit L, and the amplitude of oscillation approaches zero. The limit L lies between any two successive sums s_n and s_{n+1} and hence differs from s_n by an amount less than u_{n+1}.

Because

$$|L - s_n| < u_{n+1} \qquad \text{for } n \ge N,$$

we can make useful estimates of the sums of convergent alternating series.

THEOREM 15—The Alternating Series Estimation Theorem If the alternating series $\sum_{n=1}^{\infty}(-1)^{n+1}u_n$ satisfies the three conditions of Theorem 14, then for $n \ge N$,

$$s_n = u_1 - u_2 + \cdots + (-1)^{n+1}u_n$$

approximates the sum L of the series with an error whose absolute value is less than u_{n+1}, the absolute value of the first unused term. Furthermore, the remainder, $L - s_n$, has the same sign as the first unused term.

We leave the verification of the sign of the remainder for Exercise 53.

EXAMPLE 2 We try Theorem 15 on a series whose sum we know:

$$\sum_{n=0}^{\infty}(-1)^n\frac{1}{2^n} = 1 - \frac{1}{2} + \frac{1}{4} - \frac{1}{8} + \frac{1}{16} - \frac{1}{32} + \frac{1}{64} - \frac{1}{128} \,\vdots + \frac{1}{256} - \cdots.$$

The theorem says that if we truncate the series after the eighth term, we throw away a total that is positive and less than $1/256$. The sum of the first eight terms is 0.6640625. The sum of the series is

$$\frac{1}{1 - (-1/2)} = \frac{1}{3/2} = \frac{2}{3}.$$

The difference, $(2/3) - 0.6640625 = 0.0026041666\ldots$, is positive and less than $(1/256) = 0.00390625$. ■

Absolute and Conditional Convergence

DEFINITION A series $\sum a_n$ **converges absolutely** (is **absolutely convergent**) if the corresponding series of absolute values, $\sum |a_n|$, converges.

The geometric series in Example 2 converges absolutely because the corresponding series of absolute values

$$1 + \frac{1}{2} + \frac{1}{4} + \frac{1}{8} + \cdots$$

converges. The alternating harmonic series does not converge absolutely. The corresponding series of absolute values is the (divergent) harmonic series.

DEFINITION A series that converges but does not converge absolutely **converges conditionally**.

The alternating harmonic series converges conditionally.

Absolute convergence is important for two reasons. First, we have good tests for convergence of series of positive terms. Second, if a series converges absolutely, then it converges, as we now prove.

THEOREM 16—The Absolute Convergence Test If $\sum_{n=1}^{\infty} |a_n|$ converges, then $\sum_{n=1}^{\infty} a_n$ converges.

Proof For each n,

$$-|a_n| \le a_n \le |a_n|, \qquad \text{so} \qquad 0 \le a_n + |a_n| \le 2|a_n|.$$

If $\sum_{n=1}^{\infty} |a_n|$ converges, then $\sum_{n=1}^{\infty} 2|a_n|$ converges and, by the Direct Comparison Test, the nonnegative series $\sum_{n=1}^{\infty} (a_n + |a_n|)$ converges. The equality $a_n = (a_n + |a_n|) - |a_n|$ now lets us express $\sum_{n=1}^{\infty} a_n$ as the difference of two convergent series:

$$\sum_{n=1}^{\infty} a_n = \sum_{n=1}^{\infty} (a_n + |a_n| - |a_n|) = \sum_{n=1}^{\infty} (a_n + |a_n|) - \sum_{n=1}^{\infty} |a_n|.$$

Therefore, $\sum_{n=1}^{\infty} a_n$ converges. ■

Caution We can rephrase Theorem 16 to say that every absolutely convergent series converges. However, the converse statement is false: Many convergent series do not converge absolutely (such as the alternating harmonic series in Example 1).

EXAMPLE 3

(a) For $\sum_{n=1}^{\infty}(-1)^{n+1}\frac{1}{n^2} = 1 - \frac{1}{4} + \frac{1}{9} - \frac{1}{16} + \cdots$, the corresponding series of absolute values is the convergent series

$$\sum_{n=1}^{\infty}\frac{1}{n^2} = 1 + \frac{1}{4} + \frac{1}{9} + \frac{1}{16} + \cdots.$$

The original series converges because it converges absolutely.

(b) For $\sum_{n=1}^{\infty}\frac{\sin n}{n^2} = \frac{\sin 1}{1} + \frac{\sin 2}{4} + \frac{\sin 3}{9} + \cdots$, the corresponding series of absolute values is

$$\sum_{n=1}^{\infty}\left|\frac{\sin n}{n^2}\right| = \frac{|\sin 1|}{1} + \frac{|\sin 2|}{4} + \cdots,$$

which converges by comparison with $\sum_{n=1}^{\infty}(1/n^2)$ because $|\sin n| \le 1$ for every n. The original series converges absolutely; therefore it converges. ■

EXAMPLE 4 If p is a positive constant, the sequence $\{1/n^p\}$ is a decreasing sequence with limit zero. Therefore the alternating p-series

$$\sum_{n=1}^{\infty}\frac{(-1)^{n-1}}{n^p} = 1 - \frac{1}{2^p} + \frac{1}{3^p} - \frac{1}{4^p} + \cdots, \qquad p > 0$$

converges.

If $p > 1$, the series converges absolutely. If $0 < p \le 1$, the series converges conditionally.

$$\text{Conditional convergence:} \qquad 1 - \frac{1}{\sqrt{2}} + \frac{1}{\sqrt{3}} - \frac{1}{\sqrt{4}} + \cdots$$

$$\text{Absolute convergence:} \qquad 1 - \frac{1}{2^{3/2}} + \frac{1}{3^{3/2}} - \frac{1}{4^{3/2}} + \cdots$$ ■

Rearranging Series

We can always rearrange the terms of a *finite* sum. The same result is true for an infinite series that is absolutely convergent (we omit the proof).

THEOREM 17—The Rearrangement Theorem for Absolutely Convergent Series If $\sum_{n=1}^{\infty} a_n$ converges absolutely, and $b_1, b_2, \ldots, b_n, \ldots$ is any arrangement of the sequence $\{a_n\}$, then $\sum b_n$ converges absolutely and

$$\sum_{n=1}^{\infty} b_n = \sum_{n=1}^{\infty} a_n.$$

If we rearrange the terms of a conditionally convergent series, we get different results. In fact, it can be proved that for any real number r, a given conditionally convergent series can be rearranged so its sum is equal to r. (We also omit the proof of this fact.) Therefore we must always add the terms of a conditionally convergent series in the order given.

We have now developed several tests for convergence and divergence of series. In summary:

1. **The *n*th-Term Test:** Unless $a_n \to 0$, the series diverges.
2. **Geometric series:** $\sum ar^n$ converges if $|r| < 1$; otherwise it diverges.
3. ***p*-series:** $\sum 1/n^p$ converges if $p > 1$; otherwise it diverges.
4. **Series with nonnegative terms:** Try the Integral Test, Ratio Test, or Root Test. Try comparing to a known series with the Comparison Test.
5. **Series with some negative terms:** Does $\sum |a_n|$ converge? If yes, so does $\sum a_n$, since absolute convergence implies convergence.
6. **Alternating series:** $\sum a_n$ converges if the series satisfies the conditions of the Alternating Series Test.

EXERCISES 7.6

Which of the alternating series in Exercises 1–10 converge, and which diverge? Give reasons for your answers.

1. $\sum_{n=1}^{\infty} (-1)^{n+1} \frac{1}{n^2}$

2. $\sum_{n=1}^{\infty} (-1)^{n+1} \frac{1}{n^{3/2}}$

3. $\sum_{n=1}^{\infty} (-1)^{n+1} \left(\frac{n}{10}\right)^n$

4. $\sum_{n=1}^{\infty} (-1)^{n+1} \frac{10^n}{n^{10}}$

5. $\sum_{n=2}^{\infty} (-1)^{n+1} \frac{1}{\ln n}$

6. $\sum_{n=1}^{\infty} (-1)^{n+1} \frac{\ln n}{n}$

7. $\sum_{n=2}^{\infty} (-1)^{n+1} \frac{\ln n}{\ln n^2}$

8. $\sum_{n=1}^{\infty} (-1)^{n} \ln\left(1 + \frac{1}{n}\right)$

9. $\sum_{n=1}^{\infty} (-1)^{n+1} \frac{\sqrt{n} + 1}{n + 1}$

10. $\sum_{n=1}^{\infty} (-1)^{n+1} \frac{3\sqrt{n + 1}}{\sqrt{n} + 1}$

Which of the series in Exercises 11–44 converge absolutely, which converge, and which diverge? Give reasons for your answers.

11. $\sum_{n=1}^{\infty} (-1)^{n+1} (0.1)^n$

12. $\sum_{n=1}^{\infty} (-1)^{n+1} \frac{(0.1)^n}{n}$

13. $\sum_{n=1}^{\infty} (-1)^{n} \frac{1}{\sqrt{n}}$

14. $\sum_{n=1}^{\infty} \frac{(-1)^n}{1 + \sqrt{n}}$

15. $\sum_{n=1}^{\infty} (-1)^{n+1} \frac{n}{n^3 + 1}$

16. $\sum_{n=1}^{\infty} (-1)^{n+1} \frac{n!}{2^n}$

17. $\sum_{n=1}^{\infty} (-1)^{n} \frac{1}{n + 3}$

18. $\sum_{n=1}^{\infty} (-1)^{n} \frac{\sin n}{n^2}$

19. $\sum_{n=1}^{\infty} (-1)^{n+1} \frac{3 + n}{5 + n}$

20. $\sum_{n=2}^{\infty} (-1)^{n} \frac{1}{\ln (n^3)}$

21. $\sum_{n=1}^{\infty} (-1)^{n+1} \frac{1 + n}{n^2}$

22. $\sum_{n=1}^{\infty} \frac{(-2)^{n+1}}{n + 5^n}$

23. $\sum_{n=1}^{\infty} (-1)^n n^2 (2/3)^n$

24. $\sum_{n=1}^{\infty} (-1)^{n+1} \left(\sqrt[n]{10}\right)$

25. $\sum_{n=1}^{\infty} (-1)^n \frac{\tan^{-1} n}{n^2 + 1}$

26. $\sum_{n=2}^{\infty} (-1)^{n+1} \frac{1}{n \ln n}$

27. $\sum_{n=1}^{\infty} (-1)^n \frac{n}{n + 1}$

28. $\sum_{n=1}^{\infty} (-1)^n \frac{\ln n}{n - \ln n}$

29. $\sum_{n=1}^{\infty} \frac{(-100)^n}{n!}$

30. $\sum_{n=1}^{\infty} (-5)^{-n}$

31. $\sum_{n=1}^{\infty} \frac{(-1)^{n-1}}{n^2 + 2n + 1}$

32. $\sum_{n=2}^{\infty} (-1)^n \left(\frac{\ln n}{\ln n^2}\right)^n$

33. $\sum_{n=1}^{\infty} \frac{\cos n\pi}{n\sqrt{n}}$

34. $\sum_{n=1}^{\infty} \frac{\cos n\pi}{n}$

35. $\sum_{n=1}^{\infty} \frac{(-1)^n (n + 1)^n}{(2n)^n}$

36. $\sum_{n=1}^{\infty} \frac{(-1)^{n+1} (n!)^2}{(2n)!}$

37. $\sum_{n=1}^{\infty} (-1)^n \frac{(2n)!}{2^n n! n}$

38. $\sum_{n=1}^{\infty} (-1)^n \frac{(n!)^2 3^n}{(2n + 1)!}$

39. $\sum_{n=1}^{\infty} (-1)^n \left(\sqrt{n + 1} - \sqrt{n}\right)$

40. $\sum_{n=1}^{\infty} (-1)^n \left(\sqrt{n^2 + n} - n\right)$

41. $\sum_{n=1}^{\infty} (-1)^n \left(\sqrt{n + \sqrt{n}} - \sqrt{n}\right)$

42. $\sum_{n=1}^{\infty} \frac{(-1)^n}{\sqrt{n} + \sqrt{n + 1}}$

43. $\sum_{n=1}^{\infty} (-1)^n \operatorname{sech} n$

44. $\sum_{n=1}^{\infty} (-1)^n \operatorname{csch} n$

In Exercises 45–48, estimate the magnitude of the error involved in using the sum of the first four terms to approximate the sum of the entire series.

45. $\sum_{n=1}^{\infty}(-1)^{n+1}\frac{1}{n}$ It can be shown that the sum is ln 2.

46. $\sum_{n=1}^{\infty}(-1)^{n+1}\frac{1}{10^n}$

47. $\sum_{n=1}^{\infty}(-1)^{n+1}\frac{(0.01)^n}{n}$ As you will see in Section 7.7, the sum is ln(1.01).

48. $\frac{1}{1+t}=\sum_{n=0}^{\infty}(-1)^n t^n, \quad 0<t<1$

T Approximate the sums in Exercises 49 and 50 with an error of magnitude less than 5×10^{-6}.

49. $\sum_{n=0}^{\infty}(-1)^n\frac{1}{(2n)!}$ As you will see in Section 7.9, the sum is cos 1, the cosine of 1 radian.

50. $\sum_{n=0}^{\infty}(-1)^n\frac{1}{n!}$ As you will see in Section 7.9, the sum is e^{-1}.

51. a. The series

$$\frac{1}{3}-\frac{1}{2}+\frac{1}{9}-\frac{1}{4}+\frac{1}{27}-\frac{1}{8}+\cdots+\frac{1}{3^n}-\frac{1}{2^n}+\cdots$$

does not meet one of the conditions of Theorem 14. Which one?

b. Find the sum of the series in part (a).

T **52.** The limit L of an alternating series that satisfies the conditions of Theorem 14 lies between the values of any two consecutive partial sums. This suggests using the average

$$\frac{s_n+s_{n+1}}{2}=s_n+\frac{1}{2}(-1)^{n+2}a_{n+1}$$

to estimate L. Compute

$$s_{20}+\frac{1}{2}\cdot\frac{1}{21}$$

as an approximation to the sum of the alternating harmonic series. The exact sum is $\ln 2 = 0.6931\ldots$.

53. The sign of the remainder of an alternating series that satisfies the conditions of Theorem 14 Prove the assertion in Theorem 15 that whenever an alternating series satisfying the conditions of Theorem 14 is approximated with one of its partial sums, then the remainder (sum of the unused terms) has the same sign as the first unused term. (*Hint:* Group the remainder's terms in consecutive pairs.)

54. Show that the sum of the first $2n$ terms of the series

$$1-\frac{1}{2}+\frac{1}{2}-\frac{1}{3}+\frac{1}{3}-\frac{1}{4}+\frac{1}{4}-\frac{1}{5}+\frac{1}{5}-\frac{1}{6}+\cdots$$

is the same as the sum of the first n terms of the series

$$\frac{1}{1\cdot 2}+\frac{1}{2\cdot 3}+\frac{1}{3\cdot 4}+\frac{1}{4\cdot 5}+\frac{1}{5\cdot 6}+\cdots.$$

Do these series converge? What is the sum of the first $2n+1$ terms of the first series? If the series converge, what is their sum?

55. Show that if $\sum_{n=1}^{\infty} a_n$ diverges, then $\sum_{n=1}^{\infty} |a_n|$ diverges.

56. Show that if $\sum_{n=1}^{\infty} a_n$ converges absolutely, then

$$\left|\sum_{n=1}^{\infty} a_n\right| \le \sum_{n=1}^{\infty} |a_n|.$$

57. Show that if $\sum_{n=1}^{\infty} a_n$ and $\sum_{n=1}^{\infty} b_n$ both converge absolutely, then so does

a. $\sum_{n=1}^{\infty}(a_n+b_n)$ **b.** $\sum_{n=1}^{\infty}(a_n-b_n)$

c. $\sum_{n=1}^{\infty} ka_n$ (k any number)

58. Show by example that $\sum_{n=1}^{\infty} a_n b_n$ may diverge even if $\sum_{n=1}^{\infty} a_n$ and $\sum_{n=1}^{\infty} b_n$ both converge.

7.7 Power Series

Now that we can test infinite series for convergence, we can study sums that look like "infinite polynomials." We call these sums *power series* because they are defined as infinite series of powers of some variable, in our case x. Like polynomials, power series can be added, subtracted, multiplied, differentiated, and integrated to give new power series.

Power Series and Convergence

We begin with the formal definition.

DEFINITIONS A **power series about $x = 0$** is a series of the form

$$\sum_{n=0}^{\infty} c_n x^n = c_0 + c_1 x + c_2 x^2 + \cdots + c_n x^n + \cdots. \tag{1}$$

A **power series about $x = a$** is a series of the form

$$\sum_{n=0}^{\infty} c_n (x-a)^n = c_0 + c_1(x-a) + c_2(x-a)^2 + \cdots + c_n(x-a)^n + \cdots \tag{2}$$

in which the **center** a and the **coefficients** $c_0, c_1, c_2, \ldots, c_n, \ldots$ are constants.

Equation (1) is the special case obtained by taking $a = 0$ in Equation (2).

EXAMPLE 1 Taking all the coefficients to be 1 in Equation (1) gives the geometric power series

$$\sum_{n=0}^{\infty} x^n = 1 + x + x^2 + \cdots + x^n + \cdots.$$

This is the geometric series with first term 1 and ratio x. It converges to $1/(1-x)$ for $|x| < 1$. We express this fact by writing

$$\frac{1}{1-x} = 1 + x + x^2 + \cdots + x^n + \cdots, \qquad -1 < x < 1. \tag{3}$$

■

Up to now, we have used Equation (3) as a formula for the sum of the series on the right. We now change the focus: We think of the partial sums of the series on the right as polynomials $P_n(x)$ that approximate the function on the left. For values of x near zero, we need take only a few terms of the series to get a good approximation. As we move toward $x = 1$, or -1, we must take more terms. Figure 7.10 shows the graphs of

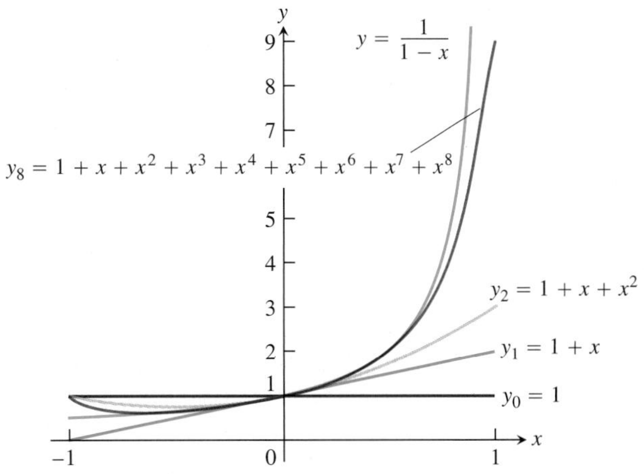

FIGURE 7.10 The graphs of $f(x) = 1/(1-x)$ and four of its polynomial approximations (Example 1).

$f(x) = 1/(1-x)$, and the approximating polynomials $y_n = P_n(x)$ for $n = 0, 1, 2$, and 8. The function $f(x) = 1/(1-x)$ is not continuous on intervals containing $x = 1$, where it has a vertical asymptote. The approximations do not apply when $x \geq 1$.

EXAMPLE 2 The power series

$$1 - \frac{1}{2}(x-2) + \frac{1}{4}(x-2)^2 - \cdots + \left(-\frac{1}{2}\right)^n (x-2)^n + \cdots \tag{4}$$

matches Equation (2) with $a = 2, c_0 = 1, c_1 = -1/2, c_2 = 1/4, \ldots, c_n = (-1/2)^n$. This is a geometric series with first term 1 and ratio $r = -\frac{x-2}{2}$. The series converges for $\left|\frac{x-2}{2}\right| < 1$ or $0 < x < 4$. The sum is

$$\frac{1}{1-r} = \frac{1}{1 + \frac{x-2}{2}} = \frac{2}{x},$$

so

$$\frac{2}{x} = 1 - \frac{(x-2)}{2} + \frac{(x-2)^2}{4} - \cdots + \left(-\frac{1}{2}\right)^n (x-2)^n + \cdots, \qquad 0 < x < 4.$$

Series (4) generates useful polynomial approximations of $f(x) = 2/x$ for values of x near 2:

$$P_0(x) = 1$$

$$P_1(x) = 1 - \frac{1}{2}(x-2) = 2 - \frac{x}{2}$$

$$P_2(x) = 1 - \frac{1}{2}(x-2) + \frac{1}{4}(x-2)^2 = 3 - \frac{3x}{2} + \frac{x^2}{4},$$

and so on (Figure 7.11). ■

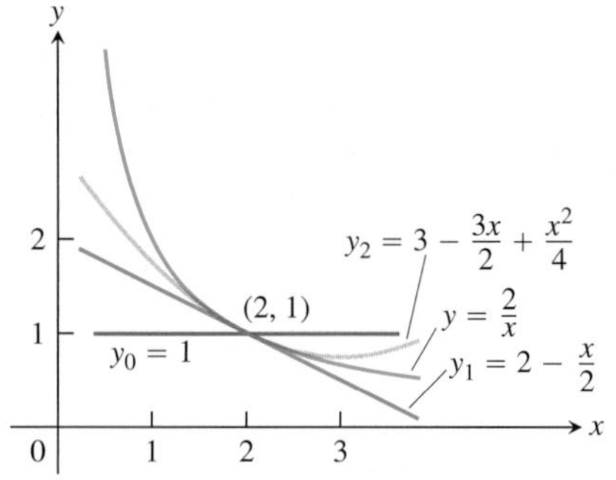

FIGURE 7.11 The graphs of $f(x) = 2/x$ and its first three polynomial approximations (Example 2).

The following example illustrates how we usually test a power series for convergence and the possible results.

EXAMPLE 3 For what values of x do the following power series converge?

(a) $\displaystyle\sum_{n=1}^{\infty} (-1)^{n-1} \frac{x^n}{n} = x - \frac{x^2}{2} + \frac{x^3}{3} - \cdots$

(b) $\displaystyle\sum_{n=1}^{\infty} (-1)^{n-1} \frac{x^{2n-1}}{2n-1} = x - \frac{x^3}{3} + \frac{x^5}{5} - \cdots$

(c) $\displaystyle\sum_{n=0}^{\infty} \frac{x^n}{n!} = 1 + x + \frac{x^2}{2!} + \frac{x^3}{3!} + \cdots$

(d) $\displaystyle\sum_{n=0}^{\infty} n!x^n = 1 + x + 2!x^2 + 3!x^3 + \cdots$

Solution Apply the Ratio Test to the series $\sum |u_n|$, where u_n is the nth term of the series in question.

(a) $$\left|\frac{u_{n+1}}{u_n}\right| = \frac{n}{n+1}|x| \to |x|.$$

The series converges absolutely for $|x| < 1$. It diverges if $|x| > 1$ because the nth term does not converge to zero. At $x = 1$, we get the alternating harmonic series $1 - 1/2 + 1/3 - 1/4 + \cdots$, which converges. At $x = -1$ we get $-1 - 1/2 - 1/3 - 1/4 - \cdots$, the negative of the harmonic series; it diverges. Series (a) converges for $-1 < x \le 1$ and diverges elsewhere.

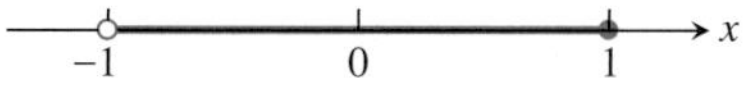

(b) $$\left|\frac{u_{n+1}}{u_n}\right| = \frac{2n-1}{2n+1}x^2 \to x^2.$$

The series converges absolutely for $x^2 < 1$. It diverges for $x^2 > 1$ because the nth term does not converge to zero. At $x = 1$ the series becomes $1 - 1/3 + 1/5 - 1/7 + \cdots$, which converges by the Alternating Series Theorem. It also converges at $x = -1$ because it is again an alternating series that satisfies the conditions for convergence. The value at $x = -1$ is the negative of the value at $x = 1$. Series (b) converges for $-1 \le x \le 1$ and diverges elsewhere.

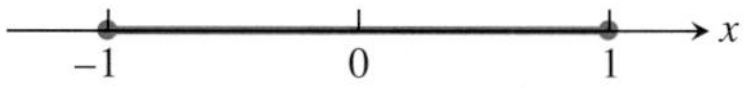

(c) $$\left|\frac{u_{n+1}}{u_n}\right| = \left|\frac{x^{n+1}}{(n+1)!} \cdot \frac{n!}{x^n}\right| = \frac{|x|}{n+1} \to 0 \text{ for every } x.$$

The series converges absolutely for all x.

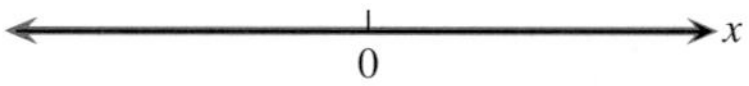

(d) $$\left|\frac{u_{n+1}}{u_n}\right| = \left|\frac{(n+1)!x^{n+1}}{n!x^n}\right| = (n+1)|x| \to \infty \text{ unless } x = 0.$$

The series diverges for all values of x except $x = 0$.

■

The previous example illustrated how a power series might converge. The next result shows that if a power series converges at more than one value, then it converges over an entire interval of values. A proof of the result is given in Appendix 9.

THEOREM 18—The Convergence Theorem for Power Series If the power series $\sum_{n=0}^{\infty} a_n x^n = a_0 + a_1 x + a_2 x^2 + \cdots$ converges for $x = c \ne 0$, then it converges absolutely for all x with $|x| < |c|$. If the series diverges for $x = d$, then it diverges for all x with $|x| > |d|$.

To simplify the notation, Theorem 18 deals with the convergence of series of the form $\sum a_n x^n$. For series of the form $\sum a_n(x - a)^n$ we can replace $x - a$ by x' and apply the results to the series $\sum a_n(x')^n$.

The Radius of Convergence of a Power Series

The theorem we have just stated and the examples we have studied lead to the conclusion that a power series $\sum c_n(x - a)^n$ behaves in one of three possible ways. It might converge only at $x = a$, or converge everywhere, or converge on some interval of radius R centered at $x = a$. We state this as a Corollary to Theorem 18. (The proof is in Appendix 9.)

COROLLARY TO THEOREM 18 The convergence of the series $\sum c_n(x - a)^n$ is described by one of the following three possibilities:

1. There is a positive number R such that the series diverges for x with $|x - a| > R$ but converges absolutely for x with $|x - a| < R$. The series may or may not converge at either of the endpoints $x = a - R$ and $x = a + R$.
2. The series converges absolutely for every x ($R = \infty$).
3. The series converges at $x = a$ and diverges elsewhere ($R = 0$).

R is called the **radius of convergence** of the power series and the interval of radius R centered at $x = a$ is called the **interval of convergence**. The interval of convergence may be open, closed, or half-open, depending on the particular series. At points x with $|x - a| < R$, the series converges absolutely. If the series converges for all values of x, we say its radius of convergence is infinite. If it converges only at $x = a$, we say its radius of convergence is zero.

How to Test a Power Series for Convergence

1. *Use the Ratio Test (or nth-Root Test) to find the interval where the series converges absolutely.* Ordinarily, this is an open interval
$$|x - a| < R \qquad \text{or} \qquad a - R < x < a + R.$$
2. *If the interval of absolute convergence is finite, test for convergence or divergence at each endpoint*, as in Examples 3a and b. Use a Comparison Test, the Integral Test, or the Alternating Series Test.
3. *If the interval of absolute convergence is $a - R < x < a + R$, the series diverges for $|x - a| > R$* (it does not even converge conditionally), because the nth term does not approach zero for those values of x.

Term-by-Term Differentiation

A theorem from advanced calculus says that a power series can be differentiated term by term at each interior point of its interval of convergence.

THEOREM 19—The Term-by-Term Differentiation Theorem If $\sum c_n(x-a)^n$ converges for $a - R < x < a + R$ for some $R > 0$, it defines a function f:

$$f(x) = \sum_{n=0}^{\infty} c_n(x-a)^n, \qquad a - R < x < a + R.$$

Such a function f has derivatives of all orders inside the interval of convergence. We can obtain the derivatives by differentiating the original series term by term:

$$f'(x) = \sum_{n=1}^{\infty} nc_n(x-a)^{n-1}$$

$$f''(x) = \sum_{n=2}^{\infty} n(n-1)c_n(x-a)^{n-2},$$

and so on. Each of these derived series converges at every interior point of the interval of convergence of the original series.

EXAMPLE 4 Find series for $f'(x)$ and $f''(x)$ if

$$f(x) = \frac{1}{1-x} = 1 + x + x^2 + x^3 + x^4 + \cdots + x^n + \cdots$$

$$= \sum_{n=0}^{\infty} x^n, \qquad -1 < x < 1$$

Solution

$$f'(x) = \frac{1}{(1-x)^2} = 1 + 2x + 3x^2 + 4x^3 + \cdots + nx^{n-1} + \cdots$$

$$= \sum_{n=1}^{\infty} nx^{n-1}, \qquad -1 < x < 1$$

$$f''(x) = \frac{2}{(1-x)^3} = 2 + 6x + 12x^2 + \cdots + n(n-1)x^{n-2} + \cdots$$

$$= \sum_{n=2}^{\infty} n(n-1)x^{n-2}, \qquad -1 < x < 1$$

■

Caution Term-by-term differentiation might not work for other kinds of series. For example, the trigonometric series

$$\sum_{n=1}^{\infty} \frac{\sin(n!x)}{n^2}$$

converges for all x. But if we differentiate term by term we get the series

$$\sum_{n=1}^{\infty} \frac{n!\cos(n!x)}{n^2},$$

which diverges for all x. This is not a power series, since it is not a sum of positive integer powers of x.

Term-by-Term Integration

It is also true that a power series can be integrated term by term throughout its interval of convergence. This result is proved in a more advanced course.

THEOREM 20—The Term-by-Term Integration Theorem Suppose that

$$f(x) = \sum_{n=0}^{\infty} c_n (x - a)^n$$

converges for $a - R < x < a + R \ (R > 0)$. Then

$$\sum_{n=0}^{\infty} c_n \frac{(x - a)^{n+1}}{n + 1}$$

converges for $a - R < x < a + R$ and

$$\int f(x)\, dx = \sum_{n=0}^{\infty} c_n \frac{(x - a)^{n+1}}{n + 1} + C$$

for $a - R < x < a + R$.

EXAMPLE 5 Identify the function

$$f(x) = x - \frac{x^3}{3} + \frac{x^5}{5} - \cdots, \qquad -1 \le x \le 1.$$

Solution We differentiate the original series term by term and get

$$f'(x) = 1 - x^2 + x^4 - x^6 + \cdots, \qquad -1 < x < 1.$$

This is a geometric series with first term 1 and ratio $-x^2$, so

$$f'(x) = \frac{1}{1 - (-x^2)} = \frac{1}{1 + x^2}.$$

We can now integrate $f'(x) = 1/(1 + x^2)$ to get

$$\int f'(x)\, dx = \int \frac{dx}{1 + x^2} = \tan^{-1} x + C.$$

The series for $f(x)$ is zero when $x = 0$, so $C = 0$. Hence

$$f(x) = x - \frac{x^3}{3} + \frac{x^5}{5} - \frac{x^7}{7} + \cdots = \tan^{-1} x, \qquad -1 < x < 1. \tag{5}$$

It can be shown that the series also converges to $\tan^{-1} x$ at $x = \pm 1$, but we omit the proof. ■

Notice that the original series in Example 5 converges at both endpoints of the original interval of convergence, but Theorem 20 can guarantee the convergence of the differentiated series only inside the interval.

EXAMPLE 6 The series

$$\frac{1}{1+t} = 1 - t + t^2 - t^3 + \cdots$$

converges on the open interval $-1 < t < 1$. Therefore,

$$\ln(1+x) = \int_0^x \frac{1}{1+t}\,dt = t - \frac{t^2}{2} + \frac{t^3}{3} - \frac{t^4}{4} + \cdots \Big]_0^x \qquad \text{Theorem 20}$$

$$= x - \frac{x^2}{2} + \frac{x^3}{3} - \frac{x^4}{4} + \cdots, \qquad -1 < x < 1.$$

It can also be shown that the series converges at $x = 1$ to the number ln 2, but that was not guaranteed by the theorem. ■

Multiplication of Power Series

Another theorem states that absolutely converging power series can be multiplied the way we multiply polynomials. We again omit the proof.

THEOREM 21—The Series Multiplication Theorem for Power Series If $A(x) = \sum_{n=0}^{\infty} a_n x^n$ and $B(x) = \sum_{n=0}^{\infty} b_n x^n$ converge absolutely for $|x| < R$, and

$$c_n = a_0 b_n + a_1 b_{n-1} + a_2 b_{n-2} + \cdots + a_{n-1} b_1 + a_n b_0 = \sum_{k=0}^{n} a_k b_{n-k},$$

then $\sum_{n=0}^{\infty} c_n x^n$ converges absolutely to $A(x)B(x)$ for $|x| < R$, and

$$\left(\sum_{n=0}^{\infty} a_n x^n\right) \cdot \left(\sum_{n=0}^{\infty} b_n x^n\right) = \sum_{n=0}^{\infty} c_n x^n.$$

Finding the general coefficient c_n in the product of two given power series can be very tedious. In practice, we usually find just the first few terms instead. For instance, from Examples 1 and 6, we have

$$\frac{1}{1-x}\ln(1+x) = \left(1 + x + x^2 + \cdots\right)\left(x - \frac{x^2}{2} + \frac{x^3}{3} - \cdots\right)$$

$$= x + \frac{x^2}{2} + \frac{5x^3}{6} - \frac{x^4}{6} \cdots.$$

EXERCISES 7.7

In Exercises 1–32, **(a)** find the series' radius and interval of convergence. For what values of x does the series converge **(b)** absolutely, **(c)** conditionally?

1. $\sum_{n=0}^{\infty} x^n$

2. $\sum_{n=0}^{\infty} (x+5)^n$

3. $\sum_{n=0}^{\infty} (-1)^n (4x+1)^n$

4. $\sum_{n=1}^{\infty} \frac{(3x-2)^n}{n}$

5. $\sum_{n=0}^{\infty} \frac{(x-2)^n}{10^n}$

6. $\sum_{n=0}^{\infty} (2x)^n$

7. $\sum_{n=0}^{\infty} \frac{nx^n}{n+2}$

8. $\sum_{n=1}^{\infty} \frac{(-1)^n (x+2)^n}{n}$

9. $\sum_{n=1}^{\infty} \frac{x^n}{n\sqrt{n}\,3^n}$

10. $\sum_{n=1}^{\infty} \frac{(x-1)^n}{\sqrt{n}}$

11. $\sum_{n=0}^{\infty} \frac{(-1)^n x^n}{n!}$

12. $\sum_{n=0}^{\infty} \frac{3^n x^n}{n!}$

13. $\sum_{n=0}^{\infty} \frac{x^{2n+1}}{n!}$

14. $\sum_{n=0}^{\infty} \frac{(2x+3)^{2n+1}}{n!}$

15. $\sum_{n=0}^{\infty} \frac{x^n}{\sqrt{n^2+3}}$

16. $\sum_{n=0}^{\infty} \frac{(-1)^n x^n}{\sqrt{n^2+3}}$

17. $\sum_{n=0}^{\infty} \frac{n(x+3)^n}{5^n}$

18. $\sum_{n=0}^{\infty} \frac{nx^n}{4^n(n^2+1)}$

19. $\sum_{n=0}^{\infty} \frac{\sqrt{n}x^n}{3^n}$

20. $\sum_{n=1}^{\infty} \sqrt[n]{n}(2x+5)^n$

21. $\sum_{n=1}^{\infty} \left(1+\frac{1}{n}\right)^n x^n$

22. $\sum_{n=1}^{\infty} (\ln n)x^n$

23. $\sum_{n=1}^{\infty} n^n x^n$

24. $\sum_{n=0}^{\infty} n!(x-4)^n$

25. $\sum_{n=1}^{\infty} \frac{(-1)^{n+1}(x+2)^n}{n2^n}$

26. $\sum_{n=0}^{\infty} (-2)^n(n+1)(x-1)^n$

27. $\sum_{n=2}^{\infty} \frac{x^n}{n(\ln n)^2}$ Get the information you need about $\sum 1/(n(\ln n)^2)$ from Section 7.3, Exercise 35.

28. $\sum_{n=2}^{\infty} \frac{x^n}{n \ln n}$ Get the information you need about $\sum 1/(n \ln n)$ from Section 7.3, Exercise 34.

29. $\sum_{n=1}^{\infty} \frac{(4x-5)^{2n+1}}{n^{3/2}}$

30. $\sum_{n=1}^{\infty} \frac{(3x+1)^{n+1}}{2n+2}$

31. $\sum_{n=1}^{\infty} \frac{(x+\pi)^n}{\sqrt{n}}$

32. $\sum_{n=0}^{\infty} \frac{\left(x-\sqrt{2}\right)^{2n+1}}{2^n}$

In Exercises 33–38, find the series' interval of convergence and, within this interval, the sum of the series as a function of x.

33. $\sum_{n=0}^{\infty} \frac{(x-1)^{2n}}{4^n}$

34. $\sum_{n=0}^{\infty} \frac{(x+1)^{2n}}{9^n}$

35. $\sum_{n=0}^{\infty} \left(\frac{\sqrt{x}}{2}-1\right)^n$

36. $\sum_{n=0}^{\infty} (\ln x)^n$

37. $\sum_{n=0}^{\infty} \left(\frac{x^2+1}{3}\right)^n$

38. $\sum_{n=0}^{\infty} \left(\frac{x^2-1}{2}\right)^n$

39. For what values of x does the series

$$1-\frac{1}{2}(x-3)+\frac{1}{4}(x-3)^2+\cdots+\left(-\frac{1}{2}\right)^n(x-3)^n+\cdots$$

converge? What is its sum? What series do you get if you differentiate the given series term by term? For what values of x does the new series converge? What is its sum?

40. If you integrate the series in Exercise 39 term by term, what new series do you get? For what values of x does the new series converge, and what is another name for its sum?

41. The series

$$\sin x = x - \frac{x^3}{3!} + \frac{x^5}{5!} - \frac{x^7}{7!} + \frac{x^9}{9!} - \frac{x^{11}}{11!} + \cdots$$

converges to $\sin x$ for all x.

a. Find the first six terms of a series for $\cos x$. For what values of x should the series converge?

b. By replacing x by $2x$ in the series for $\sin x$, find a series that converges to $\sin 2x$ for all x.

c. Using the result in part (a) and series multiplication, calculate the first six terms of a series for $2 \sin x \cos x$. Compare your answer with the answer in part (b).

42. The series

$$e^x = 1 + x + \frac{x^2}{2!} + \frac{x^3}{3!} + \frac{x^4}{4!} + \frac{x^5}{5!} + \cdots$$

converges to e^x for all x.

a. Find a series for $(d/dx)e^x$. Do you get the series for e^x? Explain your answer.

b. Find a series for $\int e^x \, dx$. Do you get the series for e^x? Explain your answer.

c. Replace x by $-x$ in the series for e^x to find a series that converges to e^{-x} for all x. Then multiply the series for e^x and e^{-x} to find the first six terms of a series for $e^{-x} \cdot e^x$.

43. The series

$$\tan x = x + \frac{x^3}{3} + \frac{2x^5}{15} + \frac{17x^7}{315} + \frac{62x^9}{2835} + \cdots$$

converges to $\tan x$ for $-\pi/2 < x < \pi/2$.

a. Find the first five terms of the series for $\ln|\sec x|$. For what values of x should the series converge?

b. Find the first five terms of the series for $\sec^2 x$. For what values of x should this series converge?

c. Check your result in part (b) by squaring the series given for $\sec x$ in Exercise 44.

44. The series

$$\sec x = 1 + \frac{x^2}{2} + \frac{5}{24}x^4 + \frac{61}{720}x^6 + \frac{277}{8064}x^8 + \cdots$$

converges to $\sec x$ for $-\pi/2 < x < \pi/2$.

a. Find the first five terms of a power series for the function $\ln|\sec x + \tan x|$. For what values of x should the series converge?

b. Find the first four terms of a series for $\sec x \tan x$. For what values of x should the series converge?

c. Check your result in part (b) by multiplying the series for $\sec x$ by the series given for $\tan x$ in Exercise 43.

45. Uniqueness of convergent power series

a. Show that if two power series $\sum_{n=0}^{\infty} a_n x^n$ and $\sum_{n=0}^{\infty} b_n x^n$ are convergent and equal for all values of x in an open interval $(-c, c)$, then $a_n = b_n$ for every n. (*Hint:* Let $f(x) = \sum_{n=0}^{\infty} a_n x^n = \sum_{n=0}^{\infty} b_n x^n$. Differentiate term by term to show that a_n and b_n both equal $f^{(n)}(0)/(n!)$.

b. Show that if $\sum_{n=0}^{\infty} a_n x^n = 0$ for all x in an open interval $(-c, c)$, then $a_n = 0$ for every n.

46. The sum of the series $\sum_{n=0}^{\infty} (n^2/2^n)$ To find the sum of this series, express $1/(1 - x)$ as a geometric series, differentiate both sides of the resulting equation with respect to x, multiply both sides of the result by x, differentiate again, multiply by x again, and set x equal to $1/2$. What do you get?

47. Convergence at endpoints Show by examples that the convergence of a power series at an endpoint of its interval of convergence may be either conditional or absolute.

48. Use the power series for $\tan^{-1} x$ in Equation (5) to prove the following expression for π as the sum of an infinite series:

$$\pi = 2\sqrt{3} \sum_{n=0}^{\infty} \frac{(-1)^n}{(2n + 1)3^n}$$

7.8 Taylor and Maclaurin Series

This section shows how functions that are infinitely differentiable generate power series called Taylor series. In many cases, these series can provide useful polynomial approximations of the generating functions.

Series Representations

We know from Theorem 19 that within its interval of convergence the sum of a power series is a continuous function with derivatives of all orders. But what about the other way around? If a function $f(x)$ has derivatives of all orders on an interval I, can it be expressed as a power series on I? And if it can, what will its coefficients be?

We can answer the last question readily if we assume that $f(x)$ is the sum of a power series

$$\begin{aligned} f(x) &= \sum_{n=0}^{\infty} a_n (x - a)^n \\ &= a_0 + a_1(x - a) + a_2(x - a)^2 + \cdots + a_n(x - a)^n + \cdots \end{aligned}$$

with a positive radius of convergence. By repeated term-by-term differentiation within the interval of convergence I we obtain

$$\begin{aligned} f'(x) &= a_1 + 2a_2(x - a) + 3a_3(x - a)^2 + \cdots + na_n(x - a)^{n-1} + \cdots \\ f''(x) &= 1 \cdot 2a_2 + 2 \cdot 3a_3(x - a) + 3 \cdot 4a_4(x - a)^2 + \cdots \\ f'''(x) &= 1 \cdot 2 \cdot 3a_3 + 2 \cdot 3 \cdot 4a_4(x - a) + 3 \cdot 4 \cdot 5a_5(x - a)^2 + \cdots, \end{aligned}$$

with the nth derivative, for all n, being

$$f^{(n)}(x) = n!a_n + \text{a sum of terms with } (x - a) \text{ as a factor}.$$

Since these equations all hold at $x = a$, we have

$$f'(a) = a_1, \qquad f''(a) = 1 \cdot 2a_2, \qquad f'''(a) = 1 \cdot 2 \cdot 3a_3,$$

and, in general,

$$f^{(n)}(a) = n!a_n.$$

These formulas reveal a pattern in the coefficients of any power series $\sum_{n=0}^{\infty} a_n(x - a)^n$ that converges to the values of f on I ("represents f on I"). If there *is* such a series (still an open question), then there is only one such series and its nth coefficient is

$$a_n = \frac{f^{(n)}(a)}{n!}.$$

If f has a series representation, then the series must be

$$f(x) = f(a) + f'(a)(x - a) + \frac{f''(a)}{2!}(x - a)^2 + \cdots + \frac{f^{(n)}(a)}{n!}(x - a)^n + \cdots. \quad (1)$$

But if we start with an arbitrary function f that is infinitely differentiable on an interval I centered at $x = a$ and use it to generate the series in Equation (1), will the series then converge to $f(x)$ at each x in the interior of I? The answer is maybe—for some functions it will but for other functions it will not, as we will see.

Taylor and Maclaurin Series

HISTORICAL BIOGRAPHIES

Brook Taylor (1685–1731)

Colin Maclaurin (1698–1746)

DEFINITIONS Let f be a function with derivatives of all orders throughout some interval containing a as an interior point. Then the **Taylor series generated by f at $x = a$ is**

$$\sum_{k=0}^{\infty} \frac{f^{(k)}(a)}{k!}(x - a)^k = f(a) + f'(a)(x - a) + \frac{f''(a)}{2!}(x - a)^2 + \cdots + \frac{f^{(n)}(a)}{n!}(x - a)^n + \cdots.$$

The **Maclaurin series generated by f is**

$$\sum_{k=0}^{\infty} \frac{f^{(k)}(0)}{k!}x^k = f(0) + f'(0)x + \frac{f''(0)}{2!}x^2 + \cdots + \frac{f^{(n)}(0)}{n!}x^n + \cdots,$$

the Taylor series generated by f at $x = 0$.

The Maclaurin series generated by f is often just called the Taylor series of f.

EXAMPLE 1 Find the Taylor series generated by $f(x) = 1/x$ at $a = 2$. Where, if anywhere, does the series converge to $1/x$?

Solution We need to find $f(2), f'(2), f''(2), \ldots$. Taking derivatives we get

$$f(x) = x^{-1}, \quad f'(x) = -x^{-2}, \quad f''(x) = 2!x^{-3}, \quad \ldots, \quad f^{(n)}(x) = (-1)^n n! x^{-(n+1)},$$

so that

$$f(2) = 2^{-1} = \frac{1}{2}, \qquad f'(2) = -\frac{1}{2^2}, \qquad \frac{f''(2)}{2!} = 2^{-3} = \frac{1}{2^3}, \qquad \ldots \qquad \frac{f^{(n)}(2)}{n!} = \frac{(-1)^n}{2^{n+1}}.$$

The Taylor series is

$$\begin{aligned} f(2) &+ f'(2)(x-2) + \frac{f''(2)}{2!}(x-2)^2 + \cdots + \frac{f^{(n)}(2)}{n!}(x-2)^n + \cdots \\ &= \frac{1}{2} - \frac{(x-2)}{2^2} + \frac{(x-2)^2}{2^3} - \cdots + (-1)^n \frac{(x-2)^n}{2^{n+1}} + \cdots. \end{aligned}$$

This is a geometric series with first term $1/2$ and ratio $r = -(x-2)/2$. It converges absolutely for $|x-2| < 2$ and its sum is

$$\frac{1/2}{1 + (x-2)/2} = \frac{1}{2 + (x-2)} = \frac{1}{x}.$$

In this example the Taylor series generated by $f(x) = 1/x$ at $a = 2$ converges to $1/x$ for $|x-2| < 2$ or $0 < x < 4$. ■

Taylor Polynomials

The linearization of a differentiable function f at a point a is the polynomial of degree one given by

$$P_1(x) = f(a) + f'(a)(x-a).$$

In Section 2.13 we used this linearization to approximate $f(x)$ at values of x near a. If f has derivatives of higher order at a, then it has higher-order polynomial approximations as well, one for each available derivative. These polynomials are called the Taylor polynomials of f.

DEFINITION Let f be a function with derivatives of order k for $k = 1, 2, \ldots, N$ in some interval containing a as an interior point. Then for any integer n from 0 through N, the **Taylor polynomial of order *n*** generated by f at $x = a$ is the polynomial

$$\begin{aligned} P_n(x) = f(a) + f'(a)(x-a) &+ \frac{f''(a)}{2!}(x-a)^2 + \cdots \\ &+ \frac{f^{(k)}(a)}{k!}(x-a)^k + \cdots + \frac{f^{(n)}(a)}{n!}(x-a)^n. \end{aligned}$$

We speak of a Taylor polynomial of *order n* rather than *degree n* because $f^{(n)}(a)$ may be zero. The first two Taylor polynomials of $f(x) = \cos x$ at $x = 0$, for example, are $P_0(x) = 1$ and $P_1(x) = 1$. The first-order Taylor polynomial has degree zero, not one.

Just as the linearization of f at $x = a$ provides the best linear approximation of f in the neighborhood of a, the higher-order Taylor polynomials provide the “best” polynomial approximations of their respective degrees. (See Exercise 32.)

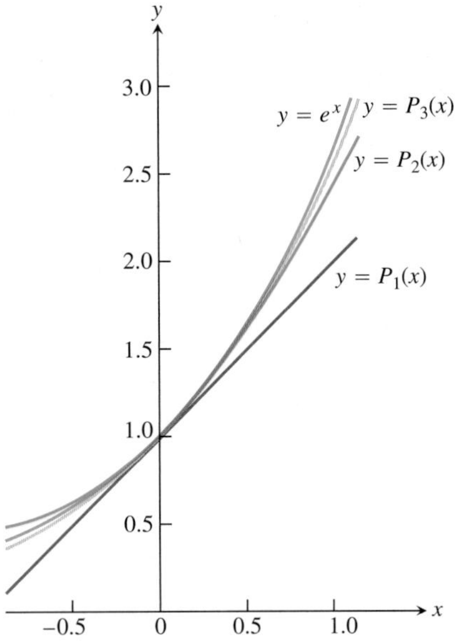

FIGURE 7.12 The graph of $f(x) = e^x$ and its Taylor polynomials

$P_1(x) = 1 + x$
$P_2(x) = 1 + x + (x^2/2!)$
$P_3(x) = 1 + x + (x^2/2!) + (x^3/3!)$.

Notice the very close agreement near the center $x = 0$ (Example 2).

EXAMPLE 2 Find the Taylor series and the Taylor polynomials generated by $f(x) = e^x$ at $x = 0$.

Solution Since

$$f(x) = e^x, \qquad f'(x) = e^x, \qquad \ldots, \qquad f^{(n)}(x) = e^x, \qquad \ldots,$$

we have

$$f(0) = e^0 = 1, \qquad f'(0) = 1, \qquad \ldots, \qquad f^{(n)}(0) = 1, \qquad \ldots.$$

The Taylor series generated by f at $x = 0$ is

$$\begin{aligned} f(0) + f'(0)x + \frac{f''(0)}{2!}x^2 + \cdots + \frac{f^{(n)}(0)}{n!}x^n + \cdots \\ &= 1 + x + \frac{x^2}{2} + \cdots + \frac{x^n}{n!} + \cdots \\ &= \sum_{k=0}^{\infty} \frac{x^k}{k!}. \end{aligned}$$

This is also the Maclaurin series for e^x. In Section 7.9 we will see that the series converges to e^x at every x.

The Taylor polynomial of order n at $x = 0$ is

$$P_n(x) = 1 + x + \frac{x^2}{2} + \cdots + \frac{x^n}{n!}.$$

See Figure 7.12. ■

EXAMPLE 3 Find the Taylor series and Taylor polynomials generated by $f(x) = \cos x$ at $x = 0$.

Solution The cosine and its derivatives are

$$\begin{aligned} f(x) &= \cos x, & f'(x) &= -\sin x, \\ f''(x) &= -\cos x, & f^{(3)}(x) &= \sin x, \\ &\vdots & &\vdots \\ f^{(2n)}(x) &= (-1)^n \cos x, & f^{(2n+1)}(x) &= (-1)^{n+1} \sin x. \end{aligned}$$

At $x = 0$, the cosines are 1 and the sines are 0, so

$$f^{(2n)}(0) = (-1)^n, \qquad f^{(2n+1)}(0) = 0.$$

The Taylor series generated by f at 0 is

$$\begin{aligned} f(0) + f'(0)x + \frac{f''(0)}{2!}x^2 + \frac{f'''(0)}{3!}x^3 + \cdots + \frac{f^{(n)}(0)}{n!}x^n + \cdots \\ &= 1 + 0 \cdot x - \frac{x^2}{2!} + 0 \cdot x^3 + \frac{x^4}{4!} + \cdots + (-1)^n \frac{x^{2n}}{(2n)!} + \cdots \\ &= \sum_{k=0}^{\infty} \frac{(-1)^k x^{2k}}{(2k)!}. \end{aligned}$$

This is also the Maclaurin series for $\cos x$. In Section 7.9, we will see that the series converges to $\cos x$ at every x.

Because $f^{(2n+1)}(0) = 0$, the Taylor polynomials of orders $2n$ and $2n+1$ are identical:

$$P_{2n}(x) = P_{2n+1}(x) = 1 - \frac{x^2}{2!} + \frac{x^4}{4!} - \cdots + (-1)^n \frac{x^{2n}}{(2n)!}.$$

Figure 7.13 shows how well these polynomials approximate $f(x) = \cos x$ near $x = 0$. Only the right-hand portions of the graphs are given because the graphs are symmetric about the y-axis. ■

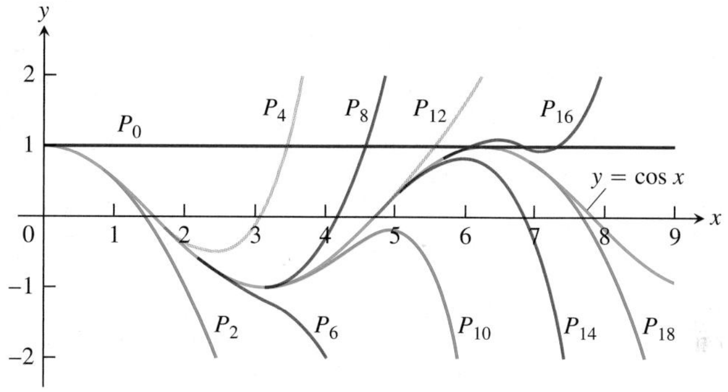

FIGURE 7.13 The polynomials

$$P_{2n}(x) = \sum_{k=0}^{n} \frac{(-1)^k x^{2k}}{(2k)!}$$

converge to $\cos x$ as $n \to \infty$. We can deduce the behavior of $\cos x$ arbitrarily far away solely from knowing the values of the cosine and its derivatives at $x = 0$ (Example 3).

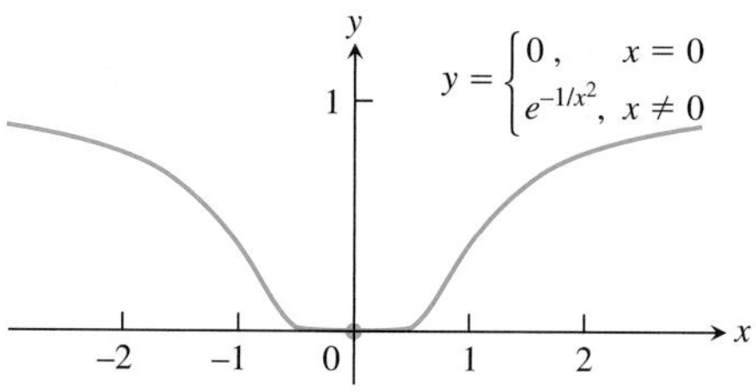

FIGURE 7.14 The graph of the function in Example 4 is so flat at the origin that all of its derivatives there are zero.

EXAMPLE 4 It can be shown (though not easily) that

$$f(x) = \begin{cases} 0, & x = 0 \\ e^{-1/x^2}, & x \neq 0 \end{cases}$$

(Figure 7.14) has derivatives of all orders at $x = 0$ and that $f^{(n)}(0) = 0$ for all n. This means that the Taylor series generated by f at $x = 0$ is

$$f(0) + f'(0)x + \frac{f''(0)}{2!}x^2 + \cdots + \frac{f^{(n)}(0)}{n!}x^n + \cdots = 0 + 0 + \cdots + 0 + \cdots.$$

The series converges for every x (its sum is 0) but converges to $f(x)$ only at $x = 0$. ■

Two questions still remain.

1. For what values of x can we normally expect a Taylor series to converge to its generating function?
2. How accurately do a function's Taylor polynomials approximate the function on a given interval?

The answers are provided by a theorem of Taylor in the next section.

EXERCISES 7.8

In Exercises 1–8, find the Taylor polynomials of orders 0, 1, 2, and 3 generated by f at a.

1. $f(x) = \ln x, \quad a = 1$
2. $f(x) = \ln(1 + x), \quad a = 0$
3. $f(x) = 1/x, \quad a = 2$
4. $f(x) = 1/(x + 2), \quad a = 0$
5. $f(x) = \sin x, \quad a = \pi/4$
6. $f(x) = \cos x, \quad a = \pi/4$
7. $f(x) = \sqrt{x}, \quad a = 4$
8. $f(x) = \sqrt{x + 4}, \quad a = 0$

Find the Maclaurin series for the functions in Exercises 9–20.

9. e^{-x}
10. $e^{x/2}$
11. $\dfrac{1}{1 + x}$
12. $\dfrac{1}{1 - x}$
13. $\sin 3x$
14. $\sin \dfrac{x}{2}$
15. $7 \cos(-x)$
16. $5 \cos \pi x$
17. $\cosh x = \dfrac{e^x + e^{-x}}{2}$
18. $\sinh x = \dfrac{e^x - e^{-x}}{2}$
19. $x^4 - 2x^3 - 5x + 4$
20. $(x + 1)^2$

In Exercises 21–28, find the Taylor series generated by f at $x = a$.

21. $f(x) = x^3 - 2x + 4, \quad a = 2$
22. $f(x) = 2x^3 + x^2 + 3x - 8, \quad a = 1$
23. $f(x) = x^4 + x^2 + 1, \quad a = -2$
24. $f(x) = 3x^5 - x^4 + 2x^3 + x^2 - 2, \quad a = -1$
25. $f(x) = 1/x^2, \quad a = 1$
26. $f(x) = x/(1 - x), \quad a = 0$
27. $f(x) = e^x, \quad a = 2$
28. $f(x) = 2^x, \quad a = 1$
29. Use the Taylor series generated by e^x at $x = a$ to show that
$$e^x = e^a\left[1 + (x - a) + \frac{(x - a)^2}{2!} + \cdots\right].$$
30. (*Continuation of Exercise 29.*) Find the Taylor series generated by e^x at $x = 1$. Compare your answer with the formula in Exercise 29.
31. Let $f(x)$ have derivatives through order n at $x = a$. Show that the Taylor polynomial of order n and its first n derivatives have the same values that f and its first n derivatives have at $x = a$.
32. **Approximation properties of Taylor polynomials** Suppose that $f(x)$ is differentiable on an interval centered at $x = a$ and that $g(x) = b_0 + b_1(x - a) + \cdots + b_n(x - a)^n$ is a polynomial of degree n with constant coefficients $b_0, \ldots, b_n$. Let $E(x) = f(x) - g(x)$. Show that if we impose on g the conditions

    a. $E(a) = 0$ — The approximation error is zero at $x = a$.

    b. $\displaystyle\lim_{x \to a} \frac{E(x)}{(x - a)^n} = 0,$ — The error is negligible when compared to $(x - a)^n$.

    then
$$g(x) = f(a) + f'(a)(x - a) + \frac{f''(a)}{2!}(x - a)^2 + \cdots + \frac{f^{(n)}(a)}{n!}(x - a)^n.$$
    Thus, the Taylor polynomial $P_n(x)$ is the only polynomial of degree less than or equal to n whose error is both zero at $x = a$ and negligible when compared with $(x - a)^n$.

Quadratic Approximations The Taylor polynomial of order 2 generated by a twice-differentiable function $f(x)$ at $x = a$ is called the *quadratic approximation* of f at $x = a$. In Exercises 33–38, find the **(a)** linearization (Taylor polynomial of order 1) and **(b)** quadratic approximation of f at $x = 0$.

33. $f(x) = \ln(\cos x)$
34. $f(x) = e^{\sin x}$
35. $f(x) = 1/\sqrt{1 - x^2}$
36. $f(x) = \cosh x$
37. $f(x) = \sin x$
38. $f(x) = \tan x$

7.9 Convergence of Taylor Series

In the last section, we asked when a Taylor series for a function can be expected to converge to that (generating) function. We answer the question in this section with the following theorem.

THEOREM 22—Taylor's Theorem If f and its first n derivatives $f', f'', \ldots, f^{(n)}$ are continuous on the closed interval between a and b, and $f^{(n)}$ is differentiable on the open interval between a and b, then there exists a number c between a and b such that

$$f(b) = f(a) + f'(a)(b-a) + \frac{f''(a)}{2!}(b-a)^2 + \cdots + \frac{f^{(n)}(a)}{n!}(b-a)^n + \frac{f^{(n+1)}(c)}{(n+1)!}(b-a)^{n+1}.$$

Taylor's Theorem is a generalization of the Mean Value Theorem (Exercise 31). There is a proof of Taylor's Theorem in Appendix 9.

When we apply Taylor's Theorem, we usually want to hold a fixed and treat b as an independent variable. Taylor's formula is easier to use in circumstances like these if we change b to x. Here is a version of the theorem with this change.

Taylor's Formula

If f has derivatives of all orders in an open interval I containing a, then for each positive integer n and for each x in I,

$$f(x) = f(a) + f'(a)(x-a) + \frac{f''(a)}{2!}(x-a)^2 + \cdots + \frac{f^{(n)}(a)}{n!}(x-a)^n + R_n(x), \qquad (1)$$

where

$$R_n(x) = \frac{f^{(n+1)}(c)}{(n+1)!}(x-a)^{n+1} \quad \text{for some } c \text{ between } a \text{ and } x. \qquad (2)$$

When we state Taylor's theorem this way, it says that for each $x \in I$,

$$f(x) = P_n(x) + R_n(x).$$

The function $R_n(x)$ is determined by the value of the $(n+1)$st derivative $f^{(n+1)}$ at a point c that depends on both a and x, and which lies somewhere between them. For any value of n we want, the equation gives both a polynomial approximation of f of that order and a formula for the error involved in using that approximation over the interval I.

Equation (1) is called **Taylor's formula**. The function $R_n(x)$ is called the **remainder of order n** or the **error term** for the approximation of f by $P_n(x)$ over I. If $R_n(x) \to 0$ as $n \to \infty$ for all $x \in I$, we say that the Taylor series generated by f at $x = a$ **converges** to f on I, and we write

$$f(x) = \sum_{k=0}^{\infty} \frac{f^{(k)}(a)}{k!}(x-a)^k.$$

Often we can estimate R_n without knowing the value of c, as the following example illustrates.

EXAMPLE 1 Show that the Taylor series generated by $f(x) = e^x$ at $x = 0$ converges to $f(x)$ for every real value of x.

Solution The function has derivatives of all orders throughout the interval $I = (-\infty, \infty)$. Equations (1) and (2) with $f(x) = e^x$ and $a = 0$ give

$$e^x = 1 + x + \frac{x^2}{2!} + \cdots + \frac{x^n}{n!} + R_n(x)$$

Polynomial from Section 7.8, Example 2

and

$$R_n(x) = \frac{e^c}{(n+1)!}x^{n+1} \quad \text{for some } c \text{ between 0 and } x.$$

Since e^x is an increasing function of x, e^c lies between $e^0 = 1$ and e^x. When x is negative, so is c, and $e^c < 1$. When x is zero, $e^x = 1$ and $R_n(x) = 0$. When x is positive, so is c, and $e^c < e^x$. Thus, for $R_n(x)$ given as above,

$$|R_n(x)| \le \frac{|x|^{n+1}}{(n+1)!} \quad \text{when } x \le 0, \qquad e^c < 1$$

and

$$|R_n(x)| < e^x \frac{x^{n+1}}{(n+1)!} \quad \text{when } x > 0. \qquad e^c < e^x$$

Finally, because

$$\lim_{n\to\infty} \frac{x^{n+1}}{(n+1)!} = 0 \quad \text{for every } x, \qquad \text{Section 7.1}$$

$\lim_{n\to\infty} R_n(x) = 0$ and the series converges to e^x for every x. Thus,

$$e^x = \sum_{k=0}^{\infty} \frac{x^k}{k!} = 1 + x + \frac{x^2}{2!} + \cdots + \frac{x^k}{k!} + \cdots. \tag{3}$$

■

We can use the result of Example 1 with $x = 1$ to write

$$e = 1 + 1 + \frac{1}{2!} + \cdots + \frac{1}{n!} + R_n(1),$$

where for some c between 0 and 1,

$$R_n(1) = e^c \frac{1}{(n+1)!} < \frac{e}{(n+1)!}. \qquad e^c < e^1$$

Estimating the Remainder

It is often possible to estimate $R_n(x)$ as we did in Example 1. This method of estimation is so convenient that we state it as a theorem for future reference.

THEOREM 23—The Remainder Estimation Theorem If there is a positive constant M such that $|f^{(n+1)}(t)| \le M$ for all t between x and a, inclusive, then the remainder term $R_n(x)$ in Taylor's Theorem satisfies the inequality

$$|R_n(x)| \le M \frac{|x-a|^{n+1}}{(n+1)!}.$$

If this condition holds for every n and the other conditions of Taylor's Theorem are satisfied by f, then the series converges to $f(x)$.

Here's an application of Theorem 23.

EXAMPLE 2 Show that the Taylor series for $\sin x$ at $x = 0$ converges for all x.

Solution The function and its derivatives are

$$\begin{aligned} f(x) &= \sin x, & f'(x) &= \cos x, \\ f''(x) &= -\sin x, & f'''(x) &= -\cos x, \\ &\vdots & &\vdots \\ f^{(2k)}(x) &= (-1)^k \sin x, & f^{(2k+1)}(x) &= (-1)^k \cos x, \end{aligned}$$

so

$$f^{(2k)}(0) = 0 \quad \text{and} \quad f^{(2k+1)}(0) = (-1)^k.$$

The series has only odd-powered terms and, for $n = 2k + 1$, Taylor's Theorem gives

$$\sin x = x - \frac{x^3}{3!} + \frac{x^5}{5!} - \cdots + \frac{(-1)^k x^{2k+1}}{(2k+1)!} + R_{2k+1}(x).$$

All the derivatives of $\sin x$ have absolute values less than or equal to 1, so we can apply the Remainder Estimation Theorem with $M = 1$ to obtain

$$|R_{2k+1}(x)| \le 1 \cdot \frac{|x|^{2k+2}}{(2k+2)!}.$$

Since $(|x|^{2k+2}/(2k+2)!) \to 0$ as $k \to \infty$, whatever the value of x, $R_{2k+1}(x) \to 0$, and the Maclaurin series for $\sin x$ converges to $\sin x$ for every x. Thus,

$$\sin x = \sum_{k=0}^{\infty} \frac{(-1)^k x^{2k+1}}{(2k+1)!} = x - \frac{x^3}{3!} + \frac{x^5}{5!} - \frac{x^7}{7!} + \cdots. \tag{4}$$

■

Similarly, we can show that the Taylor series for $\cos x$ at $x = 0$ converges to $\cos x$ for every value of x. (See Section 7.8, Example 3, for the cosine series.)

EXAMPLE 3 For what values of x can we replace $\sin x$ by $x - (x^3/3!)$ with an error of magnitude no greater than 3×10^{-4}?

Solution Here we can take advantage of the fact that the Taylor series for $\sin x$ is an alternating series for every nonzero value of x. According to the Alternating Series Estimation Theorem (Section 7.6), the error in truncating

$$\sin x = x - \frac{x^3}{3!} \;\Big|\; + \frac{x^5}{5!} - \cdots$$

after $(x^3/3!)$ is no greater than

$$\left|\frac{x^5}{5!}\right| = \frac{|x|^5}{120}.$$

Therefore the error will be less than or equal to 3×10^{-4} if

$$\frac{|x|^5}{120} < 3 \times 10^{-4} \quad \text{or} \quad |x| < \sqrt[5]{360 \times 10^{-4}} \approx 0.514.$$

Rounded down, to be safe

The Alternating Series Estimation Theorem tells us something that the Remainder Estimation Theorem does not: namely, that the estimate $x - (x^3/3!)$ for $\sin x$ is an underestimate when x is positive because then $x^5/120$ is positive.

Figure 7.15 shows the graph of $\sin x$, along with the graphs of a number of its approximating Taylor polynomials. The graph of $P_3(x) = x - (x^3/3!)$ is almost indistinguishable from the sine curve when $0 \le x \le 1$. ■

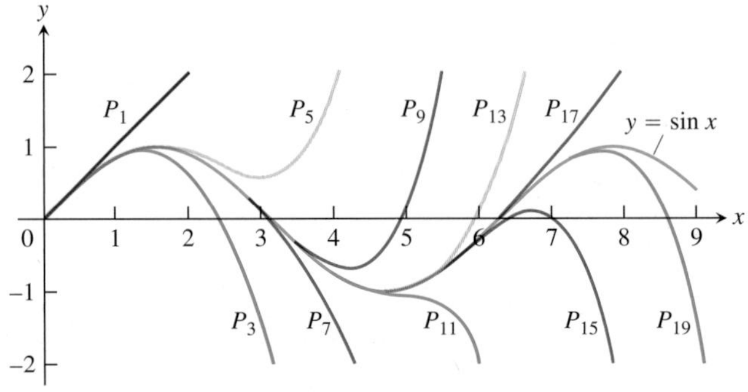

FIGURE 7.15 The polynomials

$$P_{2n+1}(x) = \sum_{k=0}^{n} \frac{(-1)^k x^{2k+1}}{(2k+1)!}$$

converge to $\sin x$ as $n \to \infty$. Notice how closely $P_3(x)$ approximates the sine curve for $x \le 1$ (Example 3).

Applying Taylor Series

On the intersection of their intervals of convergence, Taylor series can be added, subtracted, and multiplied by constants, and the results are once again Taylor series. The Taylor series for $f(x) + g(x)$ is the sum of the Taylor series for $f(x)$ and $g(x)$ because the nth derivative of $f + g$ is $f^{(n)} + g^{(n)}$, and so on. Thus the Taylor series for $\sin x + \cos x$ is the term-by-term sum of the Taylor series for $\sin x$ and $\cos x$.

Taylor series can be used to express nonelementary integrals in terms of series. For example, integrals like $\int \sin x^2\, dx$ arise in the study of the diffraction of light. However, there is no simple formula for an antiderivative of $\sin x^2$.

EXAMPLE 4 Express $\int \sin x^2\, dx$ as a power series.

Solution From the series for $\sin x$ we substitute x^2 for x and obtain

$$\sin x^2 = x^2 - \frac{x^6}{3!} + \frac{x^{10}}{5!} - \frac{x^{14}}{7!} + \frac{x^{18}}{9!} - \cdots.$$

Therefore,

$$\int \sin x^2\, dx = C + \frac{x^3}{3} - \frac{x^7}{7 \cdot 3!} + \frac{x^{11}}{11 \cdot 5!} - \frac{x^{15}}{15 \cdot 7!} + \frac{x^{10}}{19 \cdot 9!} - \cdots.$$ ■

EXAMPLE 5 Estimate $\int_0^1 \sin x^2\, dx$ with an error of less than 0.001.

Solution From the indefinite integral in Example 4,

$$\int_0^1 \sin x^2\, dx = \frac{1}{3} - \frac{1}{7 \cdot 3!} + \frac{1}{11 \cdot 5!} - \frac{1}{15 \cdot 7!} + \frac{1}{19 \cdot 9!} - \cdots.$$

The series alternates, and we find by experiment that

$$\frac{1}{11 \cdot 5!} \approx 0.00076$$

is the first term to be numerically less than 0.001. The sum of the preceding two terms gives

$$\int_0^1 \sin x^2\, dx \approx \frac{1}{3} - \frac{1}{42} \approx 0.310.$$

■

Euler's Identity

As you may recall, a complex number is a number of the form $a + bi$, where a and b are real numbers and $i = \sqrt{-1}$. If we substitute $x = i\theta$ (θ real) in the Taylor series for e^x and use the relations

$$i^2 = -1, \qquad i^3 = i^2 i = -i, \qquad i^4 = i^2 i^2 = 1, \qquad i^5 = i^4 i = i,$$

and so on, to simplify the result, we obtain

$$\begin{aligned} e^{i\theta} &= 1 + \frac{i\theta}{1!} + \frac{i^2\theta^2}{2!} + \frac{i^3\theta^3}{3!} + \frac{i^4\theta^4}{4!} + \frac{i^5\theta^5}{5!} + \frac{i^6\theta^6}{6!} + \cdots \\ &= \left(1 - \frac{\theta^2}{2!} + \frac{\theta^4}{4!} - \frac{\theta^6}{6!} + \cdots\right) + i\left(\theta - \frac{\theta^3}{3!} + \frac{\theta^5}{5!} - \cdots\right) = \cos\theta + i\sin\theta. \end{aligned}$$

This does not *prove* that $e^{i\theta} = \cos\theta + i\sin\theta$ because we have not yet defined what it means to raise e to an imaginary power. Rather, it says how to define $e^{i\theta}$ to be consistent with other things we know.

DEFINITION

For any real number θ, $e^{i\theta} = \cos\theta + i\sin\theta$. (5)

Equation (5), called **Euler's identity**, enables us to define e^{a+bi} to be $e^a \cdot e^{bi}$ for any complex number $a + bi$. One consequence of the identity is the equation

$$e^{i\pi} = -1.$$

When written in the form $e^{i\pi} + 1 = 0$, this equation combines five of the most important constants in mathematics.

EXERCISES 7.9

Use substitution (as in Example 3) to find the Taylor series at $x = 0$ of the functions in Exercises 1–6.

1. e^{-5x} **2.** $e^{-x/2}$ **3.** $5 \sin(-x)$

4. $\sin\left(\frac{\pi x}{2}\right)$ **5.** $\cos\sqrt{x+1}$ **6.** $\cos\left(x^{3/2}/\sqrt{2}\right)$

Find Taylor series at $x = 0$ for the functions in Exercises 7–18.

7. xe^x **8.** $x^2 \sin x$ **9.** $\frac{x^2}{2} - 1 + \cos x$

10. $\sin x - x + \frac{x^3}{3!}$ **11.** $x \cos \pi x$ **12.** $x^2 \cos(x^2)$

13. $\cos^2 x$ (*Hint:* $\cos^2 x = (1 + \cos 2x)/2$.)

14. $\sin^2 x$ **15.** $\frac{x^2}{1-2x}$ **16.** $x \ln(1+2x)$

17. $\frac{1}{(1-x)^2}$ **18.** $\frac{2}{(1-x)^3}$

19. For approximately what values of x can you replace $\sin x$ by $x - (x^3/6)$ with an error of magnitude no greater than 5×10^{-4}? Give reasons for your answer.

20. If $\cos x$ is replaced by $1 - (x^2/2)$ and $|x| < 0.5$, what estimate can be made of the error? Does $1 - (x^2/2)$ tend to be too large, or too small? Give reasons for your answer.

21. How close is the approximation $\sin x = x$ when $|x| < 10^{-3}$? For which of these values of x is $x < \sin x$?

22. The estimate $\sqrt{1+x} = 1 + (x/2)$ is used when x is small. Estimate the error when $|x| < 0.01$.

23. The approximation $e^x = 1 + x + (x^2/2)$ is used when x is small. Use the Remainder Estimation Theorem to estimate the error when $|x| < 0.1$.

24. (*Continuation of Exercise 23.*) When $x < 0$, the series for e^x is an alternating series. Use the Alternating Series Estimation Theorem to estimate the error that results from replacing e^x by $1 + x + (x^2/2)$ when $-0.1 < x < 0$. Compare your estimate with the one you obtained in Exercise 23.

In Exercises 25–28, find a polynomial that will approximate $F(x)$ throughout the given interval with an error of magnitude less than 10^{-3}.

25. $F(x) = \int_0^x \sin t^2\, dt, \quad [0, 1]$

26. $F(x) = \int_0^x t^2 e^{-t^2}\, dt, \quad [0, 1]$

27. $F(x) = \int_0^x \tan^{-1} t\, dt,$ **(a)** $[0, 0.5]$ **(b)** $[0, 1]$

28. $F(x) = \int_0^x \frac{\ln(1+t)}{t}\, dt,$ **(a)** $[0, 0.5]$ **(b)** $[0, 1]$

29. Multiply the Maclaurin series for e^x and $\sin x$ together to find the first five nonzero terms of the Maclaurin series for $e^x \sin x$.

30. Use the identity $\sin^2 x = (1 - \cos 2x)/2$ to obtain the Maclaurin series for $\sin^2 x$. Then differentiate this series to obtain the Maclaurin series for $2 \sin x \cos x$. Check that this is the series for $\sin 2x$.

31. Taylor's Theorem and the Mean Value Theorem Explain how the Mean Value Theorem (Section 3.2, Theorem 4) is a special case of Taylor's Theorem.

32. Linearizations at inflection points Show that if the graph of a twice-differentiable function $f(x)$ has an inflection point at $x = a$, then the linearization of f at $x = a$ is also the quadratic approximation of f at $x = a$. This explains why tangent lines fit so well at inflection points.

33. The (second) second derivative test Use the equation

$$f(x) = f(a) + f'(a)(x-a) + \frac{f''(c_2)}{2}(x-a)^2$$

to establish the following test.

Let f have continuous first and second derivatives and suppose that $f'(a) = 0$. Then

a. f has a local maximum at a if $f'' \le 0$ throughout an interval whose interior contains a;

b. f has a local minimum at a if $f'' \ge 0$ throughout an interval whose interior contains a.

34. A cubic approximation Use Taylor's formula with $a = 0$ and $n = 3$ to find the standard cubic approximation of $f(x) = 1/(1-x)$ at $x = 0$. Give an upper bound for the magnitude of the error in the approximation when $|x| \le 0.1$.

35. a. Use Taylor's formula with $n = 2$ to find the quadratic approximation of $f(x) = (1+x)^k$ at $x = 0$ (k a constant).

b. If $k = 3$, for approximately what values of x in the interval $[0, 1]$ will the error in the quadratic approximation be less than $1/100$?

36. Improving approximations to π

a. Let P be an approximation of π accurate to n decimals. Show that $P + \sin P$ gives an approximation correct to $3n$ decimals. (*Hint:* Let $P = \pi + x$.)

T **b.** Try it with a calculator.

37. The Taylor series generated by $f(x) = \sum_{n=0}^{\infty} a_n x^n$ is $\sum_{n=0}^{\infty} a_n x^n$ A function defined by a power series $\sum_{n=0}^{\infty} a_n x^n$ with a radius of convergence $c > 0$ has a Taylor series that converges to the function at every point of $(-c, c)$. Show this by showing that the Taylor series generated by $f(x) = \sum_{n=0}^{\infty} a_n x^n$ is the series $\sum_{n=0}^{\infty} a_n x^n$ itself.

38. Taylor series for even functions and odd functions Suppose that $f(x) = \sum_{n=0}^{\infty} a_n x^n$ converges for all x in an open interval $(-c, c)$. Show that

a. If f is even, then $a_1 = a_3 = a_5 = \cdots = 0$, i.e., the Taylor series for f at $x = 0$ contains only even powers of x.

b. If f is odd, then $a_0 = a_2 = a_4 = \cdots = 0$, i.e., the Taylor series for f at $x = 0$ contains only odd powers of x.

39. Use Equation (5) to write the following powers of e in the form $a + bi$.

a. $e^{-i\pi}$ **b.** $e^{i\pi/4}$ **c.** $e^{-i\pi/2}$

40. Use Equation (5) to show that

$$\cos\theta = \frac{e^{i\theta} + e^{-i\theta}}{2} \quad \text{and} \quad \sin\theta = \frac{e^{i\theta} - e^{-i\theta}}{2i}.$$

41. Establish the equations in Exercise 40 by combining the formal Taylor series for $e^{i\theta}$ and $e^{-i\theta}$.

42. Show that

a. $\cosh i\theta = \cos\theta$, **b.** $\sinh i\theta = i\sin\theta$.

43. Use the definition of $e^{i\theta}$ to show that for any real numbers θ, θ_1, and θ_2,

a. $e^{i\theta_1}e^{i\theta_2} = e^{i(\theta_1+\theta_2)}$, **b.** $e^{-i\theta} = 1/e^{i\theta}$.

44. When a and b are real, we define $e^{(a+ib)x}$ with the equation

$$e^{(a+ib)x} = e^{ax} \cdot e^{ibx} = e^{ax}(\cos bx + i\sin bx).$$

Differentiate the right-hand side of this equation to show that

$$\frac{d}{dx}e^{(a+ib)x} = (a + ib)e^{(a+ib)x}.$$

Thus the familiar rule $(d/dx)e^{kx} = ke^{kx}$ holds for k complex as well as real.

COMPUTER EXPLORATIONS

Taylor's formula with $n = 1$ and $a = 0$ gives the linearization of a function at $x = 0$. With $n = 2$ and $n = 3$ we obtain the standard quadratic and cubic approximations. In these exercises we explore the errors associated with these approximations. We seek answers to two questions:

a. For what values of x can the function be replaced by each approximation with an error less than 10^{-2}?

b. What is the maximum error we could expect if we replace the function by each approximation over the specified interval?

Using a CAS, perform the following steps to aid in answering questions (a) and (b) for the functions and intervals in Exercises 45–50.

Step 1: Plot the function over the specified interval.

Step 2: Find the Taylor polynomials $P_1(x)$, $P_2(x)$, and $P_3(x)$ at $x = 0$.

Step 3: Calculate the $(n + 1)$st derivative $f^{(n+1)}(c)$ associated with the remainder term for each Taylor polynomial. Plot the derivative as a function of c over the specified interval and estimate its maximum absolute value, M.

Step 4: Calculate the remainder $R_n(x)$ for each polynomial. Using the estimate M from Step 3 in place of $f^{(n+1)}(c)$, plot $R_n(x)$ over the specified interval. Then estimate the values of x that answer question (a).

Step 5: Compare your estimated error with the actual error $E_n(x) = |f(x) - P_n(x)|$ by plotting $E_n(x)$ over the specified interval. This will help answer question (b).

Step 6: Graph the function and its three Taylor approximations together. Discuss the graphs in relation to the information discovered in Steps 4 and 5.

45. $f(x) = \dfrac{1}{\sqrt{1 + x}}, \quad |x| \le \dfrac{3}{4}$

46. $f(x) = (1 + x)^{3/2}, \quad -\dfrac{1}{2} \le x \le 2$

47. $f(x) = \dfrac{x}{x^2 + 1}, \quad |x| \le 2$

48. $f(x) = (\cos x)(\sin 2x), \quad |x| \le 2$

49. $f(x) = e^{-x}\cos 2x, \quad |x| \le 1$

50. $f(x) = e^{x/3}\sin 2x, \quad |x| \le 2$

7.10 The Binomial Series

This section introduces the binomial series for estimating powers and roots of binomial expressions $(1 + x)^m$. A reference table of frequently used series is provided in the back pages of this book.

The Binomial Series for Powers and Roots

The Taylor series generated by $f(x) = (1 + x)^m$, when m is constant, is

$$1 + mx + \frac{m(m-1)}{2!}x^2 + \frac{m(m-1)(m-2)}{3!}x^3 + \cdots$$
$$+ \frac{m(m-1)(m-2)\cdots(m-k+1)}{k!}x^k + \cdots. \quad (1)$$

This series, called the **binomial series**, converges absolutely for $|x| < 1$. To derive the series, we first list the function and its derivatives:

$$
\begin{aligned}
f(x) &= (1+x)^m \\
f'(x) &= m(1+x)^{m-1} \\
f''(x) &= m(m-1)(1+x)^{m-2} \\
f'''(x) &= m(m-1)(m-2)(1+x)^{m-3} \\
&\vdots \\
f^{(k)}(x) &= m(m-1)(m-2)\cdots(m-k+1)(1+x)^{m-k}.
\end{aligned}
$$

We then evaluate these at $x = 0$ and substitute into the Taylor series formula to obtain Series (1).

If m is an integer greater than or equal to zero, the series stops after $(m+1)$ terms because the coefficients from $k = m + 1$ on are zero.

If m is not a positive integer or zero, the series is infinite and converges for $|x| < 1$. To see why, let u_k be the term involving x^k. Then apply the Ratio Test for absolute convergence to see that

$$\left|\frac{u_{k+1}}{u_k}\right| = \left|\frac{m-k}{k+1}x\right| \to |x| \qquad \text{as } k \to \infty.$$

Our derivation of the binomial series shows only that it is generated by $(1+x)^m$ and converges for $|x| < 1$. The derivation does not show that the series converges to $(1+x)^m$. It does, but we omit the proof. (See Exercise 20.)

The Binomial Series

For $-1 < x < 1$,

$$(1+x)^m = 1 + \sum_{k=1}^{\infty} \binom{m}{k} x^k,$$

where we define

$$\binom{m}{1} = m, \qquad \binom{m}{2} = \frac{m(m-1)}{2!},$$

and

$$\binom{m}{k} = \frac{m(m-1)(m-2)\cdots(m-k+1)}{k!} \qquad \text{for } k \geq 3.$$

EXAMPLE 1 If $m = -1$,

$$\binom{-1}{1} = -1, \qquad \binom{-1}{2} = \frac{-1(-2)}{2!} = 1,$$

and

$$\binom{-1}{k} = \frac{-1(-2)(-3)\cdots(-1-k+1)}{k!} = (-1)^k\left(\frac{k!}{k!}\right) = (-1)^k.$$

With these coefficient values and with x replaced by $-x$, the binomial series formula gives the familiar geometric series

$$(1+x)^{-1} = 1 + \sum_{k=1}^{\infty}(-1)^k x^k = 1 - x + x^2 - x^3 + \cdots + (-1)^k x^k + \cdots. \qquad \blacksquare$$

EXAMPLE 2 We know from Section 2.13, Example 1, that $\sqrt{1+x} \approx 1 + (x/2)$ for $|x|$ small. With $m = 1/2$, the binomial series gives quadratic and higher-order approximations as well, along with error estimates that come from the Alternating Series Estimation Theorem:

$$\begin{aligned}(1+x)^{1/2} &= 1 + \frac{x}{2} + \frac{\left(\frac{1}{2}\right)\left(-\frac{1}{2}\right)}{2!}x^2 + \frac{\left(\frac{1}{2}\right)\left(-\frac{1}{2}\right)\left(-\frac{3}{2}\right)}{3!}x^3 \\ &\quad + \frac{\left(\frac{1}{2}\right)\left(-\frac{1}{2}\right)\left(-\frac{3}{2}\right)\left(-\frac{5}{2}\right)}{4!}x^4 + \cdots \\ &= 1 + \frac{x}{2} - \frac{x^2}{8} + \frac{x^3}{16} - \frac{5x^4}{128} + \cdots.\end{aligned}$$

Substitution for x gives still other approximations. For example,

$$\sqrt{1-x^2} \approx 1 - \frac{x^2}{2} - \frac{x^4}{8} \qquad \text{for } |x^2| \text{ small}$$

$$\sqrt{1-\frac{1}{x}} \approx 1 - \frac{1}{2x} - \frac{1}{8x^2} \qquad \text{for } \left|\frac{1}{x}\right| \text{ small, that is, } |x| \text{ large}. \qquad \blacksquare$$

EXERCISES 7.10

Find the first four terms of the binomial series for the functions in Exercises 1–10.

1. $(1+x)^{1/2}$ **2.** $(1+x)^{1/3}$ **3.** $(1-x)^{-1/2}$

4. $(1-2x)^{1/2}$ **5.** $\left(1+\frac{x}{2}\right)^{-2}$ **6.** $\left(1-\frac{x}{2}\right)^{-2}$

7. $(1+x^3)^{-1/2}$ **8.** $(1+x^2)^{-1/3}$

9. $\left(1+\frac{1}{x}\right)^{1/2}$ **10.** $\left(1-\frac{2}{x}\right)^{1/3}$

Find the binomial series for the functions in Exercises 11–14.

11. $(1+x)^4$ **12.** $(1+x^2)^3$

13. $(1-2x)^3$ **14.** $\left(1-\frac{x}{2}\right)^4$

15. a. Use the binomial series and the fact that

$$\frac{d}{dx}\sin^{-1}x = (1-x^2)^{-1/2}$$

to generate the first four nonzero terms of the Taylor series for $\sin^{-1}x$. What is the radius of convergence?

b. Series for $\cos^{-1}x$ Use your result in part (a) to find the first five nonzero terms of the Taylor series for $\cos^{-1}x$.

16. a. Series for $\sinh^{-1}x$ Find the first four nonzero terms of the Taylor series for

$$\sinh^{-1}x = \int_0^x \frac{dt}{\sqrt{1+t^2}}.$$

T **b.** Use the first *three* terms of the series in part (a) to estimate $\sinh^{-1} 0.25$. Give an upper bound for the magnitude of the estimation error.

17. Obtain the Taylor series for $1/(1+x)^2$ from the binomial series for $-1/(1+x)$.

18. Use the binomial series for $1/(1-x^2)$ to obtain a power series for $2x/(1-x^2)^2$.

19. Series for $\sin^{-1} x$ Integrate the binomial series for $(1-x^2)^{-1/2}$ to show that for $|x| < 1$,

$$\sin^{-1} x = x + \sum_{n=1}^{\infty} \frac{1 \cdot 3 \cdot 5 \cdot \cdots \cdot (2n-1)}{2 \cdot 4 \cdot 6 \cdot \cdots \cdot (2n)} \frac{x^{2n+1}}{2n+1}.$$

20. Use the following steps to prove Equation (1).

a. Differentiate the series

$$f(x) = 1 + \sum_{k=1}^{\infty} \binom{m}{k} x^k$$

to show that

$$f'(x) = \frac{mf(x)}{1+x}, \quad -1 < x < 1.$$

b. Define $g(x) = (1+x)^{-m} f(x)$ and show that $g'(x) = 0$.

c. From part (b), show that

$$f(x) = (1+x)^m.$$

Chapter 7 Questions to Guide Your Review

1. What is an infinite sequence? What does it mean for such a sequence to converge? To diverge? Give examples.

2. What is a nondecreasing sequence? Under what circumstances does such a sequence have a limit? Give examples.

3. What theorems are available for calculating limits of sequences? Give examples.

4. What theorem sometimes enables us to use l'Hôpital's Rule to calculate the limit of a sequence? Give an example.

5. What six sequence limits are likely to arise when you work with sequences and series?

6. What is an infinite series? What does it mean for such a series to converge? To diverge? Give examples.

7. What is a geometric series? When does such a series converge? Diverge? When it does converge, what is its sum? Give examples.

8. Besides geometric series, what other convergent and divergent series do you know?

9. What is the nth-Term Test for Divergence? What is the idea behind the test?

10. What can be said about term-by-term sums and differences of convergent series? About constant multiples of convergent and divergent series?

11. What happens if you add a finite number of terms to a convergent series? A divergent series? What happens if you delete a finite number of terms from a convergent series? A divergent series?

12. How do you reindex a series? Why might you want to do this?

13. Under what circumstances will an infinite series of nonnegative terms converge? Diverge? Why study series of nonnegative terms?

14. What is the Integral Test? What is the reasoning behind it? Give an example of its use.

15. When do p-series converge? Diverge? How do you know? Give examples of convergent and divergent p-series.

16. What are the Direct Comparison Test and the Limit Comparison Test? What is the reasoning behind these tests? Give examples of their use.

17. What are the Ratio and Root Tests? Do they always give you the information you need to determine convergence or divergence? Give examples.

18. What is an alternating series? What theorem is available for determining the convergence of such a series?

19. How can you estimate the error involved in approximating the sum of an alternating series with one of the series' partial sums? What is the reasoning behind the estimate?

20. What is absolute convergence? Conditional convergence? How are the two related?

21. What do you know about rearranging the terms of an absolutely convergent series? Of a conditionally convergent series?

22. What is a power series? How do you test a power series for convergence? What are the possible outcomes?

23. What are the basic facts about

a. term-by-term differentiation of power series?

b. term-by-term integration of power series?

c. multiplication of power series?

Give examples.

24. What is the Taylor series generated by a function $f(x)$ at a point $x = a$? What information do you need about f to construct the series? Give an example.

25. What is a Maclaurin series?

26. Does a Taylor series always converge to its generating function? Explain.

27. What are Taylor polynomials? Of what use are they?

28. What is Taylor's formula? What does it say about the errors involved in using Taylor polynomials to approximate functions? In particular, what does Taylor's formula say about the error in a linearization? A quadratic approximation?

29. What is the binomial series? On what interval does it converge? How is it used?

30. What are the Taylor series for $1/(1-x)$, $1/(1+x)$, e^x, $\sin x$, $\cos x$, $\ln(1+x)$, $\ln[(1+x)/(1-x)]$, and $\tan^{-1} x$? How do you estimate the errors involved in replacing these series with their partial sums?

Chapter 7 Practice and Additional Exercises

Which of the sequences whose nth terms appear in Exercises 1–18 converge, and which diverge? Find the limit of each convergent sequence.

1. $a_n = 1 + \dfrac{(-1)^n}{n}$

2. $a_n = \dfrac{1-(-1)^n}{\sqrt{n}}$

3. $a_n = \dfrac{1-2^n}{2^n}$

4. $a_n = 1 + (0.9)^n$

5. $a_n = \sin \dfrac{n\pi}{2}$

6. $a_n = \sin n\pi$

7. $a_n = \dfrac{\ln(n^2)}{n}$

8. $a_n = \dfrac{\ln(2n+1)}{n}$

9. $a_n = \dfrac{n + \ln n}{n}$

10. $a_n = \dfrac{\ln(2n^3+1)}{n}$

11. $a_n = \left(\dfrac{n-5}{n}\right)^n$

12. $a_n = \left(1 + \dfrac{1}{n}\right)^{-n}$

13. $a_n = \sqrt[n]{\dfrac{3^n}{n}}$

14. $a_n = \left(\dfrac{3}{n}\right)^{1/n}$

15. $a_n = n(2^{1/n} - 1)$

16. $a_n = \sqrt[n]{2n+1}$

17. $a_n = \dfrac{(n+1)!}{n!}$

18. $a_n = \dfrac{(-4)^n}{n!}$

Find the sums of the series in Exercises 19–24.

19. $\displaystyle\sum_{n=3}^{\infty} \frac{1}{(2n-3)(2n-1)}$

20. $\displaystyle\sum_{n=2}^{\infty} \frac{-2}{n(n+1)}$

21. $\displaystyle\sum_{n=1}^{\infty} \frac{9}{(3n-1)(3n+2)}$

22. $\displaystyle\sum_{n=3}^{\infty} \frac{-8}{(4n-3)(4n+1)}$

23. $\displaystyle\sum_{n=0}^{\infty} e^{-n}$

24. $\displaystyle\sum_{n=1}^{\infty} (-1)^n \frac{3}{4^n}$

Which of the series in Exercises 25–40 converge absolutely, which converge conditionally, and which diverge? Give reasons for your answers.

25. $\displaystyle\sum_{n=1}^{\infty} \frac{1}{\sqrt{n}}$

26. $\displaystyle\sum_{n=1}^{\infty} \frac{-5}{n}$

27. $\displaystyle\sum_{n=1}^{\infty} \frac{(-1)^n}{\sqrt{n}}$

28. $\displaystyle\sum_{n=1}^{\infty} \frac{1}{2n^3}$

29. $\displaystyle\sum_{n=1}^{\infty} \frac{(-1)^n}{\ln(n+1)}$

30. $\displaystyle\sum_{n=2}^{\infty} \frac{1}{n(\ln n)^2}$

31. $\displaystyle\sum_{n=1}^{\infty} \frac{\ln n}{n^3}$

32. $\displaystyle\sum_{n=3}^{\infty} \frac{\ln n}{\ln(\ln n)}$

33. $\displaystyle\sum_{n=1}^{\infty} \frac{(-1)^n}{n\sqrt{n^2+1}}$

34. $\displaystyle\sum_{n=1}^{\infty} \frac{(-1)^n 3n^2}{n^3+1}$

35. $\displaystyle\sum_{n=1}^{\infty} \frac{n+1}{n!}$

36. $\displaystyle\sum_{n=1}^{\infty} \frac{(-1)^n(n^2+1)}{2n^2+n-1}$

37. $\displaystyle\sum_{n=1}^{\infty} \frac{(-3)^n}{n!}$

38. $\displaystyle\sum_{n=1}^{\infty} \frac{2^n 3^n}{n^n}$

39. $\displaystyle\sum_{n=1}^{\infty} \frac{1}{\sqrt{n(n+1)(n+2)}}$

40. $\displaystyle\sum_{n=2}^{\infty} \frac{1}{n\sqrt{n^2-1}}$

In Exercises 41–50, **(a)** find the series' radius and interval of convergence. Then identify the values of x for which the series converges **(b)** absolutely and **(c)** conditionally.

41. $\displaystyle\sum_{n=1}^{\infty} \frac{(x+4)^n}{n3^n}$

42. $\displaystyle\sum_{n=1}^{\infty} \frac{(x-1)^{2n-2}}{(2n-1)!}$

43. $\displaystyle\sum_{n=1}^{\infty} \frac{(-1)^{n-1}(3x-1)^n}{n^2}$

44. $\displaystyle\sum_{n=0}^{\infty} \frac{(n+1)(2x+1)^n}{(2n+1)2^n}$

45. $\displaystyle\sum_{n=1}^{\infty} \frac{x^n}{n^n}$

46. $\displaystyle\sum_{n=1}^{\infty} \frac{x^n}{\sqrt{n}}$

47. $\displaystyle\sum_{n=0}^{\infty} \frac{(n+1)x^{2n-1}}{3^n}$

48. $\displaystyle\sum_{n=0}^{\infty} \frac{(-1)^n(x-1)^{2n+1}}{2n+1}$

49. $\displaystyle\sum_{n=1}^{\infty} (\operatorname{csch} n)x^n$

50. $\displaystyle\sum_{n=1}^{\infty} (\coth n)x^n$

Each of the series in Exercises 51–56 is the value of the Taylor series at $x = 0$ of a function $f(x)$ at a particular point. What function and what point? What is the sum of the series?

51. $1 - \dfrac{1}{4} + \dfrac{1}{16} - \cdots + (-1)^n \dfrac{1}{4^n} + \cdots$

52. $\dfrac{2}{3} - \dfrac{4}{18} + \dfrac{8}{81} - \cdots + (-1)^{n-1} \dfrac{2^n}{n3^n} + \cdots$

53. $\pi - \dfrac{\pi^3}{3!} + \dfrac{\pi^5}{5!} - \cdots + (-1)^n \dfrac{\pi^{2n+1}}{(2n+1)!} + \cdots$

54. $1 - \dfrac{\pi^2}{9 \cdot 2!} + \dfrac{\pi^4}{81 \cdot 4!} - \cdots + (-1)^n \dfrac{\pi^{2n}}{3^{2n}(2n)!} + \cdots$

55. $1 + \ln 2 + \dfrac{(\ln 2)^2}{2!} + \cdots + \dfrac{(\ln 2)^n}{n!} + \cdots$

56. $\dfrac{1}{\sqrt{3}} - \dfrac{1}{9\sqrt{3}} + \dfrac{1}{45\sqrt{3}} - \cdots + (-1)^{n-1}\dfrac{1}{(2n-1)(\sqrt{3})^{2n-1}} + \cdots$

Find Taylor series at $x = 0$ for the functions in Exercises 57–64.

57. $\dfrac{1}{1-2x}$

58. $\dfrac{1}{1+x^3}$

59. $\sin \pi x$

60. $\sin \dfrac{2x}{3}$

61. $\cos (x^{5/2})$

62. $\cos \sqrt{5x}$

63. $e^{(\pi x/2)}$

64. e^{-x^2}

In Exercises 65–68, find the first four nonzero terms of the Taylor series generated by f at $x = a$.

65. $f(x) = \sqrt{3 + x^2}$ at $x = -1$

66. $f(x) = 1/(1 - x)$ at $x = 2$

67. $f(x) = 1/(x + 1)$ at $x = 3$

68. $f(x) = 1/x$ at $x = a > 0$

Use series to approximate the values of the integrals in Exercises 69 and 70 with an error of magnitude less than 10^{-8}. (The answer section gives the integrals' values rounded to 10 decimal places.)

69. $\displaystyle\int_0^{1/2} e^{-x^3}\, dx$

70. $\displaystyle\int_0^{1/2} \frac{\tan^{-1} x}{x}\, dx$

71. Use a series representation of $\sin 3x$ to find values of r and s for which

$$\lim_{x\to 0}\left(\frac{\sin 3x}{x^3} + \frac{r}{x^2} + s\right) = 0.$$

T **72.** Compare the accuracies of the approximations $\sin x \approx x$ and $\sin x \approx 6x/(6 + x^2)$ by comparing the graphs of $f(x) = \sin x - x$ and $g(x) = \sin x - (6x/(6 + x^2))$. Describe what you find.

73. Find the radius of convergence of the series

$$\sum_{n=1}^{\infty} \frac{2\cdot 5\cdot 8\cdot \cdots \cdot(3n-1)}{2\cdot 4\cdot 6\cdot \cdots \cdot(2n)} x^n.$$

74. Find the radius of convergence of the series

$$\sum_{n=1}^{\infty} \frac{3\cdot 5\cdot 7\cdot \cdots \cdot(2n+1)}{4\cdot 9\cdot 14\cdot \cdots \cdot(5n-1)}(x-1)^n.$$

75. **a.** Find the interval of convergence of the series

$$y = 1 + \frac{1}{6}x^3 + \frac{1}{180}x^6 + \cdots + \frac{1\cdot 4\cdot 7\cdot \cdots \cdot(3n-2)}{(3n)!}x^{3n} + \cdots.$$

b. Show that the function defined by the series satisfies a differential equation of the form

$$\frac{d^2y}{dx^2} = x^a y + b$$

and find the values of the constants a and b.

76. **a.** Find the Maclaurin series for the function $x^2/(1 + x)$.

b. Does the series converge at $x = 1$? Explain.

77. Prove that the sequence $\{x_n\}$ and the series $\sum_{k=1}^{\infty}(x_{k+1} - x_k)$ both converge or both diverge.

78. Prove that $\sum_{n=1}^{\infty}(a_n/(1 + a_n))$ converges if $a_n > 0$ for all n and $\sum_{n=1}^{\infty} a_n$ converges.

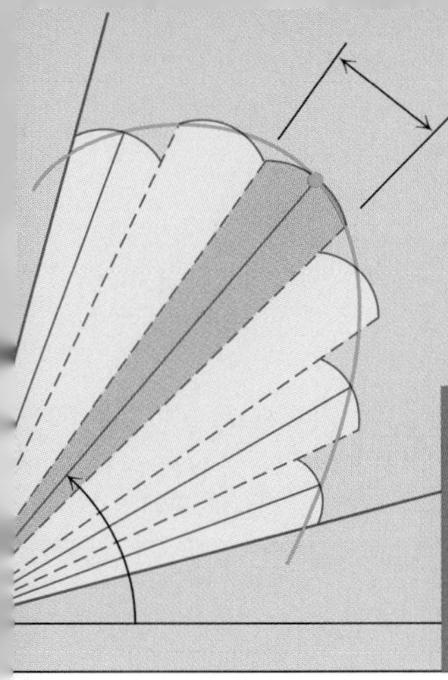

Chapter

8

Polar Coordinates and Conics

OVERVIEW In this chapter we study curves, derivatives, and integrals in polar coordinates. Polar coordinates are useful for describing rotational motion, as in planetary and satellite motion, and they can also simplify the calculations of multiple integrals (Chapter 12). Additionally, we review the geometric definitions of parabolas, ellipses, and hyperbolas and derive their standard equations. These curves are called *conic sections*, or *conics*, and model the paths traveled by projectiles, planets, and any other object moving under the sole influence of a gravitational or electromagnetic force.

8.1 Polar Coordinates

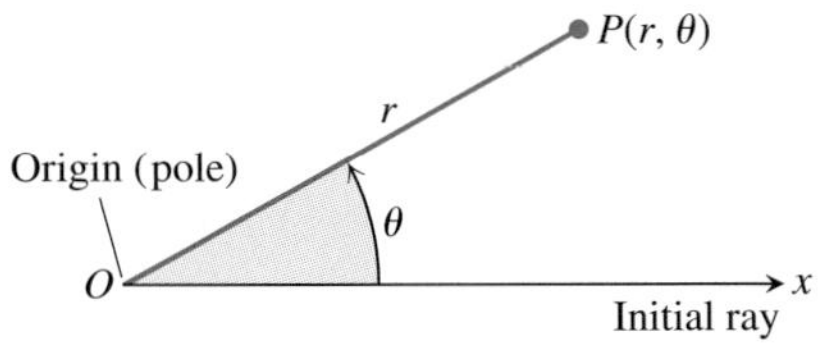

FIGURE 8.1 To define polar coordinates for the plane, we start with an origin, called the pole, and an initial ray.

In this section, we study polar coordinates and their relation to Cartesian coordinates. While a point in the plane has just one pair of Cartesian coordinates, it has infinitely many pairs of polar coordinates.

Definition of Polar Coordinates

To define polar coordinates, we first fix an **origin** O (called the **pole**) and an **initial ray** from O (Figure 8.1). Then each point P can be located by assigning to it a **polar coordinate pair** (r, θ) in which r gives the directed distance from O to P and θ gives the directed angle from the initial ray to ray OP.

> **Polar Coordinates**
>
> $P(r, \theta)$
>
> Directed distance from O to P — Directed angle from initial ray to OP

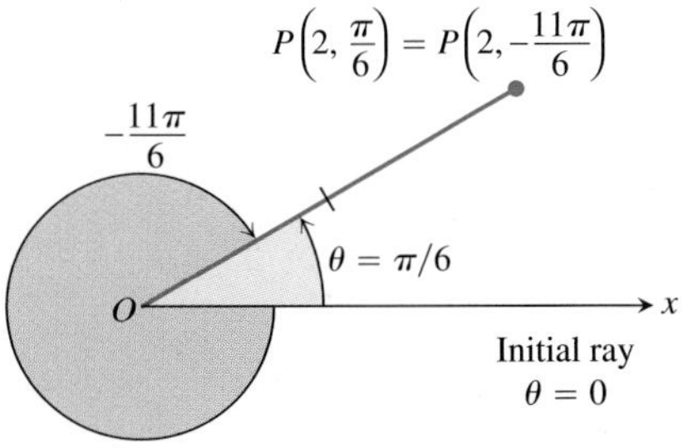

FIGURE 8.2 Polar coordinates are not unique.

As in trigonometry, θ is positive when measured counterclockwise and negative when measured clockwise. The angle associated with a given point is not unique. For instance, the point 2 units from the origin along the ray $\theta = \pi/6$ has polar coordinates $r = 2$, $\theta = \pi/6$. It also has coordinates $r = 2, \theta = -11\pi/6$ (Figure 8.2). In some situations we allow r to be negative. That is why we use directed distance in defining $P(r, \theta)$. The point $P(2, 7\pi/6)$ can be reached by turning $7\pi/6$ radians counterclockwise from the initial ray and going forward 2 units (Figure 8.3). It can also be reached by turning $\pi/6$ radians

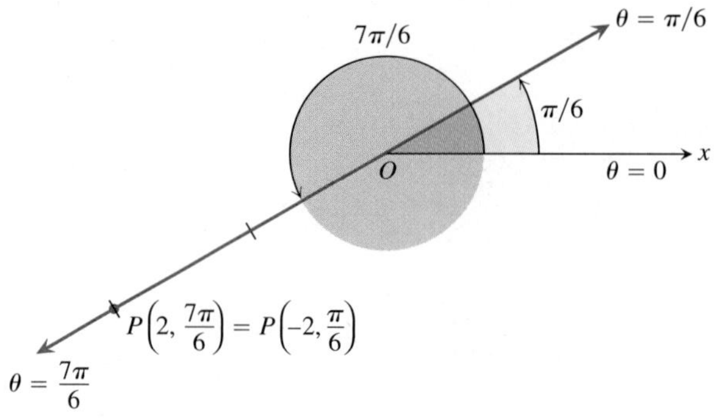

FIGURE 8.3 Polar coordinates can have negative r-values.

counterclockwise from the initial ray and going *backward* 2 units. So the point also has polar coordinates $r = -2, \theta = \pi/6$.

EXAMPLE 1 Find all the polar coordinates of the point $P(2, \pi/6)$.

Solution We sketch the initial ray of the coordinate system, draw the ray from the origin that makes an angle of $\pi/6$ radians with the initial ray, and mark the point $(2, \pi/6)$ (Figure 8.4). We then find the angles for the other coordinate pairs of P in which $r = 2$ and $r = -2$.

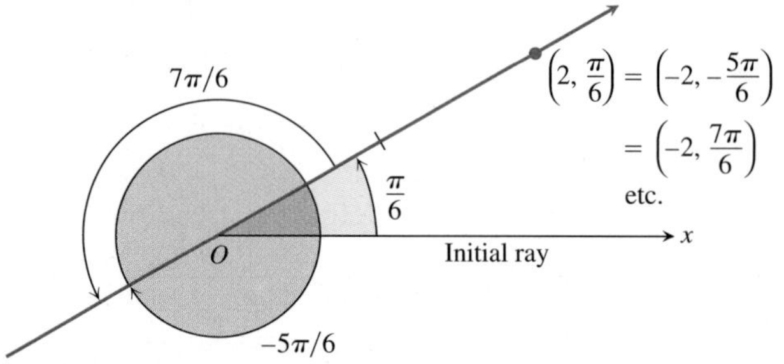

FIGURE 8.4 The point $P(2, \pi/6)$ has infinitely many polar coordinate pairs (Example 1).

For $r = 2$, the complete list of angles is

$$\frac{\pi}{6}, \quad \frac{\pi}{6} \pm 2\pi, \quad \frac{\pi}{6} \pm 4\pi, \quad \frac{\pi}{6} \pm 6\pi, \quad \ldots.$$

For $r = -2$, the angles are

$$-\frac{5\pi}{6}, \quad -\frac{5\pi}{6} \pm 2\pi, \quad -\frac{5\pi}{6} \pm 4\pi, \quad -\frac{5\pi}{6} \pm 6\pi, \quad \ldots.$$

The corresponding coordinate pairs of P are

$$\left(2, \frac{\pi}{6} + 2n\pi\right), \qquad n = 0, \pm 1, \pm 2, \ldots$$

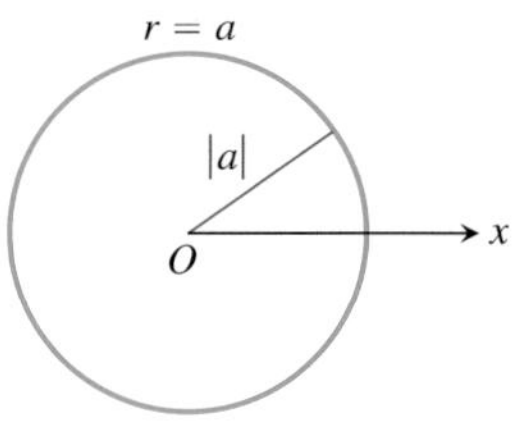

FIGURE 8.5 The polar equation for a circle is $r = a$.

and

$$\left(-2, -\frac{5\pi}{6} + 2n\pi\right), \qquad n = 0, \pm 1, \pm 2, \ldots .$$

When $n = 0$, the formulas give $(2, \pi/6)$ and $(-2, -5\pi/6)$. When $n = 1$, they give $(2, 13\pi/6)$ and $(-2, 7\pi/6)$, and so on. ■

Polar Equations and Graphs

If we hold r fixed at a constant value $r = a \neq 0$, the point $P(r, \theta)$ will lie $|a|$ units from the origin O. As θ varies over any interval of length 2π, P then traces a circle of radius $|a|$ centered at O (Figure 8.5).

If we hold θ fixed at a constant value $\theta = \theta_0$ and let r vary between $-\infty$ and ∞, the point $P(r, \theta)$ traces the line through O that makes an angle of measure θ_0 with the initial ray.

Equation	Graph
$r = a$	Circle of radius $\lvert a \rvert$ centered at O
$\theta = \theta_0$	Line through O making an angle θ_0 with the initial ray

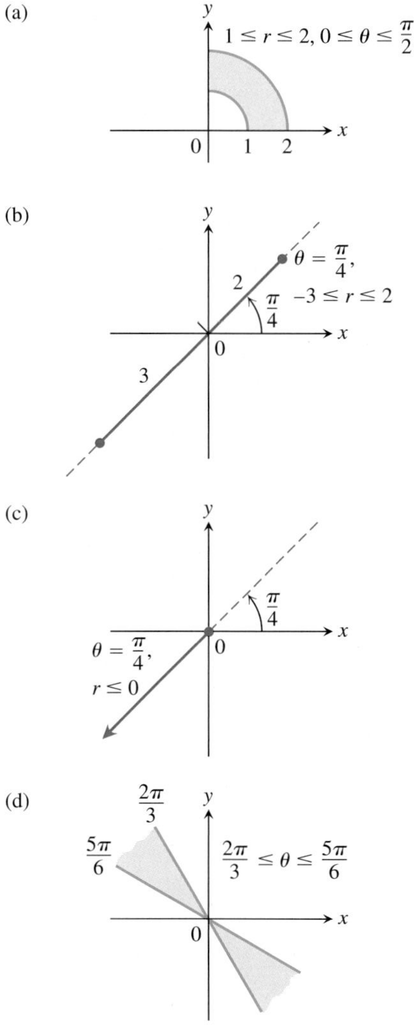

FIGURE 8.6 The graphs of typical inequalities in r and θ (Example 3).

EXAMPLE 2

(a) $r = 1$ and $r = -1$ are equations for the circle of radius 1 centered at O.

(b) $\theta = \pi/6, \theta = 7\pi/6$, and $\theta = -5\pi/6$ are equations for the line in Figure 8.4. ■

Equations of the form $r = a$ and $\theta = \theta_0$ can be combined to define regions, segments, and rays.

EXAMPLE 3 Graph the sets of points whose polar coordinates satisfy the following conditions.

(a) $1 \le r \le 2 \quad$ and $\quad 0 \le \theta \le \dfrac{\pi}{2}$

(b) $-3 \le r \le 2 \quad$ and $\quad \theta = \dfrac{\pi}{4}$

(c) $r \le 0 \quad$ and $\quad \theta = \dfrac{\pi}{4}$

(d) $\dfrac{2\pi}{3} \le \theta \le \dfrac{5\pi}{6} \quad$ (no restriction on r)

Solution The graphs are shown in Figure 8.6. ■

Relating Polar and Cartesian Coordinates

When we use both polar and Cartesian coordinates in a plane, we place the two origins together and take the initial polar ray as the positive x-axis. The ray $\theta = \pi/2, r > 0$,

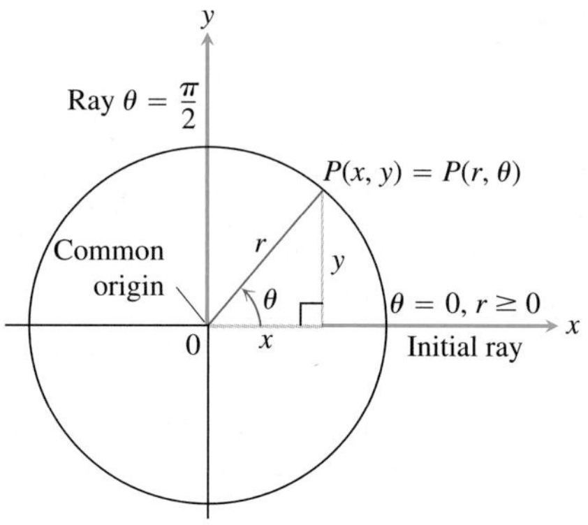

FIGURE 8.7 The usual way to relate polar and Cartesian coordinates.

becomes the positive y-axis (Figure 8.7). The two coordinate systems are then related by the following equations.

> **Equations Relating Polar and Cartesian Coordinates**
>
> $$x = r\cos\theta, \qquad y = r\sin\theta, \qquad x^2 + y^2 = r^2$$

The first two of these equations uniquely determine the Cartesian coordinates x and y given the polar coordinates r and θ. On the other hand, if x and y are given, the third equation gives two possible choices for r (a positive and a negative value). For each selection, there is a unique $\theta \in [0, 2\pi)$ satisfying the first two equations, each then giving a polar coordinate representation of the Cartesian point (x, y). The other polar coordinate representations for the point can be determined from these two, as in Example 1.

EXAMPLE 4 Here are some equivalent equations.

Polar equation	Cartesian equivalent
$r\cos\theta = 2$	$x = 2$
$r^2\cos\theta\sin\theta = 4$	$xy = 4$
$r^2\cos^2\theta - r^2\sin^2\theta = 1$	$x^2 - y^2 = 1$
$r = 1 + 2r\cos\theta$	$y^2 - 3x^2 - 4x - 1 = 0$
$r = 1 - \cos\theta$	$x^4 + y^4 + 2x^2y^2 + 2x^3 + 2xy^2 - y^2 = 0$

With some curves, we are better off with polar coordinates; with others, we aren't. ■

EXAMPLE 5 Find a polar equation for the circle $x^2 + (y - 3)^2 = 9$ (Figure 8.8).

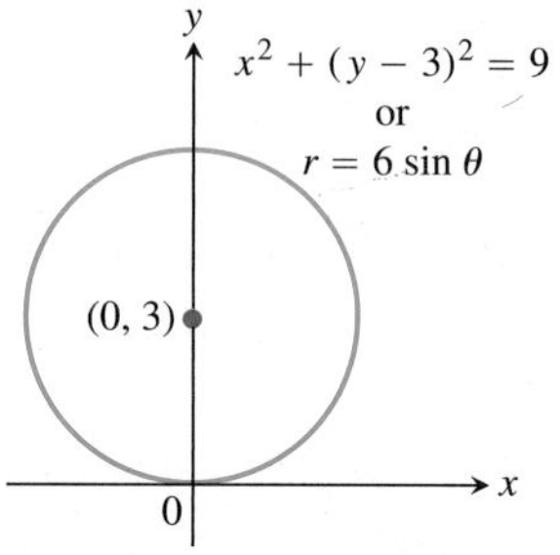

FIGURE 8.8 The circle in Example 5.

Solution

$$\begin{aligned} x^2 + y^2 - 6y + 9 &= 9 && \text{Expand } (y-3)^2. \\ x^2 + y^2 - 6y &= 0 && \text{The 9's cancel.} \\ r^2 - 6r\sin\theta &= 0 && x^2 + y^2 = r^2 \\ r = 0 \quad \text{or} \quad r - 6\sin\theta &= 0 && \\ r &= 6\sin\theta && \text{Includes both possibilities} \end{aligned}$$

■

EXAMPLE 6 Replace the following polar equations by equivalent Cartesian equations, and identify their graphs.

(a) $r\cos\theta = -4$

(b) $r^2 = 4r\cos\theta$

(c) $r = \dfrac{4}{2\cos\theta - \sin\theta}$

Solution We use the substitutions $r\cos\theta = x$, $r\sin\theta = y$, $r^2 = x^2 + y^2$.

(a) $r\cos\theta = -4$

The Cartesian equation: $\quad r\cos\theta = -4$

$$x = -4$$

The graph: Vertical line through $x = -4$ on the x-axis

(b) $r^2 = 4r\cos\theta$

The Cartesian equation:
$$\begin{aligned} r^2 &= 4r\cos\theta \\ x^2 + y^2 &= 4x \\ x^2 - 4x + y^2 &= 0 \\ x^2 - 4x + 4 + y^2 &= 4 \quad \text{Completing the square} \\ (x-2)^2 + y^2 &= 4 \end{aligned}$$

The graph: Circle, radius 2, center $(h, k) = (2, 0)$

(c) $r = \dfrac{4}{2\cos\theta - \sin\theta}$

The Cartesian equation:
$$\begin{aligned} r(2\cos\theta - \sin\theta) &= 4 \\ 2r\cos\theta - r\sin\theta &= 4 \\ 2x - y &= 4 \\ y &= 2x - 4 \end{aligned}$$

The graph: Line, slope $m = 2$, y-intercept $b = -4$ ■

EXERCISES 8.1

1. Which polar coordinate pairs label the same point?

a. $(3, 0)$ **b.** $(-3, 0)$ **c.** $(2, 2\pi/3)$
d. $(2, 7\pi/3)$ **e.** $(-3, \pi)$ **f.** $(2, \pi/3)$
g. $(-3, 2\pi)$ **h.** $(-2, -\pi/3)$

2. Which polar coordinate pairs label the same point?

a. $(-2, \pi/3)$ **b.** $(2, -\pi/3)$ **c.** (r, θ)
d. $(r, \theta + \pi)$ **e.** $(-r, \theta)$ **f.** $(2, -2\pi/3)$
g. $(-r, \theta + \pi)$ **h.** $(-2, 2\pi/3)$

3. Plot the following points (given in polar coordinates). Then find all the polar coordinates of each point.

a. $(2, \pi/2)$ **b.** $(2, 0)$
c. $(-2, \pi/2)$ **d.** $(-2, 0)$

4. Plot the following points (given in polar coordinates). Then find all the polar coordinates of each point.

a. $(3, \pi/4)$ **b.** $(-3, \pi/4)$
c. $(3, -\pi/4)$ **d.** $(-3, -\pi/4)$

5. Find the Cartesian coordinates of the points in Exercise 1.

6. Find the Cartesian coordinates of the following points (given in polar coordinates).

a. $(\sqrt{2}, \pi/4)$ **b.** $(1, 0)$
c. $(0, \pi/2)$ **d.** $(-\sqrt{2}, \pi/4)$
e. $(-3, 5\pi/6)$ **f.** $(5, \tan^{-1}(4/3))$
g. $(-1, 7\pi)$ **h.** $(2\sqrt{3}, 2\pi/3)$

Graph the sets of points whose polar coordinates satisfy the equations and inequalities in Exercises 7–22.

7. $r = 2$
8. $0 \le r \le 2$
9. $r \ge 1$
10. $1 \le r \le 2$
11. $0 \le \theta \le \pi/6, \quad r \ge 0$
12. $\theta = 2\pi/3, \quad r \le -2$
13. $\theta = \pi/3, \quad -1 \le r \le 3$
14. $\theta = 11\pi/4, \quad r \ge -1$
15. $\theta = \pi/2, \quad r \ge 0$
16. $\theta = \pi/2, \quad r \le 0$
17. $0 \le \theta \le \pi, \quad r = 1$
18. $0 \le \theta \le \pi, \quad r = -1$
19. $\pi/4 \le \theta \le 3\pi/4, \quad 0 \le r \le 1$
20. $-\pi/4 \le \theta \le \pi/4, \quad -1 \le r \le 1$
21. $-\pi/2 \le \theta \le \pi/2, \quad 1 \le r \le 2$
22. $0 \le \theta \le \pi/2, \quad 1 \le |r| \le 2$

Replace the polar equations in Exercises 23–48 by equivalent Cartesian equations. Then describe or identify the graph.

23. $r\cos\theta = 2$
24. $r\sin\theta = -1$
25. $r\sin\theta = 0$
26. $r\cos\theta = 0$

27. $r = 4 \csc \theta$

28. $r = -3 \sec \theta$

29. $r \cos \theta + r \sin \theta = 1$

30. $r \sin \theta = r \cos \theta$

31. $r^2 = 1$

32. $r^2 = 4r \sin \theta$

33. $r = \dfrac{5}{\sin \theta - 2 \cos \theta}$

34. $r^2 \sin 2\theta = 2$

35. $r = \cot \theta \csc \theta$

36. $r = 4 \tan \theta \sec \theta$

37. $r = \csc \theta\, e^{r \cos \theta}$

38. $r \sin \theta = \ln r + \ln \cos \theta$

39. $r^2 + 2r^2 \cos \theta \sin \theta = 1$

40. $\cos^2 \theta = \sin^2 \theta$

41. $r^2 = -4r \cos \theta$

42. $r^2 = -6r \sin \theta$

43. $r = 8 \sin \theta$

44. $r = 3 \cos \theta$

45. $r = 2 \cos \theta + 2 \sin \theta$

46. $r = 2 \cos \theta - \sin \theta$

47. $r \sin \left(\theta + \dfrac{\pi}{6}\right) = 2$

48. $r \sin \left(\dfrac{2\pi}{3} - \theta\right) = 5$

Replace the Cartesian equations in Exercises 49–62 by equivalent polar equations.

49. $x = 7$

50. $y = 1$

51. $x = y$

52. $x - y = 3$

53. $x^2 + y^2 = 4$

54. $x^2 - y^2 = 1$

55. $\dfrac{x^2}{9} + \dfrac{y^2}{4} = 1$

56. $xy = 2$

57. $y^2 = 4x$

58. $x^2 + xy + y^2 = 1$

59. $x^2 + (y - 2)^2 = 4$

60. $(x - 5)^2 + y^2 = 25$

61. $(x - 3)^2 + (y + 1)^2 = 4$

62. $(x + 2)^2 + (y - 5)^2 = 16$

63. Find all polar coordinates of the origin.

64. Vertical and horizontal lines

a. Show that every vertical line in the xy-plane has a polar equation of the form $r = a \sec \theta$.

b. Find the analogous polar equation for horizontal lines in the xy-plane.

8.2 Graphing in Polar Coordinates

This section describes techniques for graphing equations in polar coordinates.

Symmetry

Figure 8.9 illustrates the standard polar coordinate tests for symmetry.

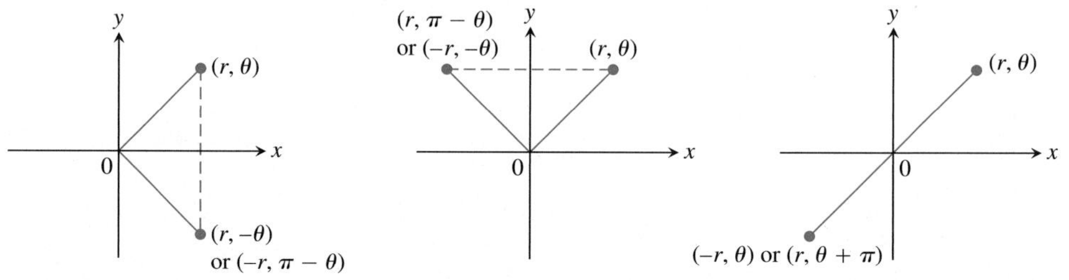

(a) About the x-axis (b) About the y-axis (c) About the origin

FIGURE 8.9 Three tests for symmetry in polar coordinates.

Symmetry Tests for Polar Graphs

1. *Symmetry about the x-axis:* If the point (r, θ) lies on the graph, the point $(r, -\theta)$ or $(-r, \pi - \theta)$ lies on the graph (Figure 8.9a).
2. *Symmetry about the y-axis:* If the point (r, θ) lies on the graph, the point $(r, \pi - \theta)$ or $(-r, -\theta)$ lies on the graph (Figure 8.9b).
3. *Symmetry about the origin:* If the point (r, θ) lies on the graph, the point $(-r, \theta)$ or $(r, \theta + \pi)$ lies on the graph (Figure 8.9c).

Slope

The slope of a polar curve $r = f(\theta)$ is given by dy/dx, not by $r' = df/d\theta$. To see why, think of the graph of f as the graph of the parametric equations

$$x = r\cos\theta = f(\theta)\cos\theta, \qquad y = r\sin\theta = f(\theta)\sin\theta.$$

If f is a differentiable function of θ, then so are x and y and, when $dx/d\theta \neq 0$, we can calculate dy/dx from the parametric formula

$$\frac{dy}{dx} = \frac{dy/d\theta}{dx/d\theta} \qquad \text{Section 3.5, Eq. (2) with } t = \theta$$

$$= \frac{\dfrac{d}{d\theta}(f(\theta)\cdot\sin\theta)}{\dfrac{d}{d\theta}(f(\theta)\cdot\cos\theta)}$$

$$= \frac{\dfrac{df}{d\theta}\sin\theta + f(\theta)\cos\theta}{\dfrac{df}{d\theta}\cos\theta - f(\theta)\sin\theta} \qquad \text{Product Rule for derivatives}$$

Slope of the Curve $r = f(\theta)$

$$\left.\frac{dy}{dx}\right|_{(r,\,\theta)} = \frac{f'(\theta)\sin\theta + f(\theta)\cos\theta}{f'(\theta)\cos\theta - f(\theta)\sin\theta},$$

provided $dx/d\theta \neq 0$ at (r, θ).

If the curve $r = f(\theta)$ passes through the origin at $\theta = \theta_0$, then $f(\theta_0) = 0$, and the slope equation gives

$$\left.\frac{dy}{dx}\right|_{(0,\,\theta_0)} = \frac{f'(\theta_0)\sin\theta_0}{f'(\theta_0)\cos\theta_0} = \tan\theta_0.$$

If the graph of $r = f(\theta)$ passes through the origin at the value $\theta = \theta_0$, the slope of the curve there is $\tan\theta_0$. The reason we say "slope at $(0, \theta_0)$" and not just "slope at the origin" is that a polar curve may pass through the origin (or any point) more than once, with different slopes at different θ-values. This is not the case in our first example, however.

EXAMPLE 1 Graph the curve $r = 1 - \cos\theta$.

Solution The curve is symmetric about the x-axis because

$$(r, \theta) \text{ on the graph} \Rightarrow r = 1 - \cos\theta$$

$$\Rightarrow r = 1 - \cos(-\theta) \qquad \cos\theta = \cos(-\theta)$$

$$\Rightarrow (r, -\theta) \text{ on the graph}.$$

θ	$r = 1 - \cos\theta$
0	0
$\frac{\pi}{3}$	$\frac{1}{2}$
$\frac{\pi}{2}$	1
$\frac{2\pi}{3}$	$\frac{3}{2}$
π	2

(a)

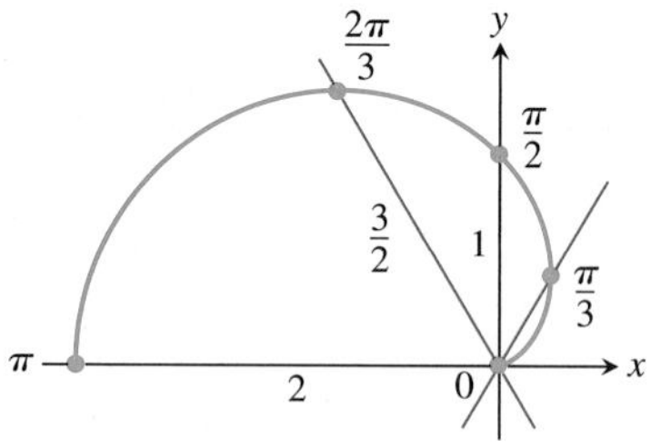

(b)

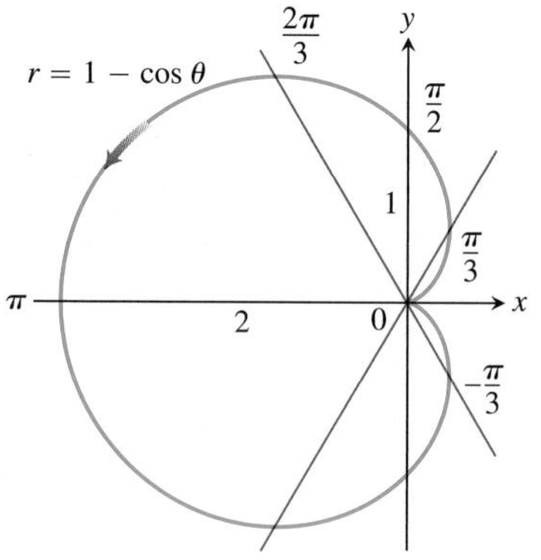

(c)

FIGURE 8.10 The steps in graphing the cardioid $r = 1 - \cos\theta$ (Example 1). The arrow shows the direction of increasing θ.

As θ increases from 0 to π, $\cos\theta$ decreases from 1 to -1, and $r = 1 - \cos\theta$ increases from a minimum value of 0 to a maximum value of 2. As θ continues on from π to 2π, $\cos\theta$ increases from -1 back to 1 and r decreases from 2 back to 0. The curve starts to repeat when $\theta = 2\pi$ because the cosine has period 2π.

The curve leaves the origin with slope $\tan(0) = 0$ and returns to the origin with slope $\tan(2\pi) = 0$.

We make a table of values from $\theta = 0$ to $\theta = \pi$, plot the points, draw a smooth curve through them with a horizontal tangent at the origin, and reflect the curve across the x-axis to complete the graph (Figure 8.10). The curve is called a *cardioid* because of its heart shape. ■

EXAMPLE 2 Graph the curve $r^2 = 4\cos\theta$.

Solution The equation $r^2 = 4\cos\theta$ requires $\cos\theta \geq 0$, so we get the entire graph by running θ from $-\pi/2$ to $\pi/2$. The curve is symmetric about the x-axis because

$$
\begin{aligned}
(r, \theta) \text{ on the graph} &\Rightarrow r^2 = 4\cos\theta \\
&\Rightarrow r^2 = 4\cos(-\theta) \qquad \cos\theta = \cos(-\theta) \\
&\Rightarrow (r, -\theta) \text{ on the graph.}
\end{aligned}
$$

The curve is also symmetric about the origin because

$$
\begin{aligned}
(r, \theta) \text{ on the graph} &\Rightarrow r^2 = 4\cos\theta \\
&\Rightarrow (-r)^2 = 4\cos\theta \\
&\Rightarrow (-r, \theta) \text{ on the graph.}
\end{aligned}
$$

Together, these two symmetries imply symmetry about the y-axis.

The curve passes through the origin when $\theta = -\pi/2$ and $\theta = \pi/2$. It has a vertical tangent both times because $\tan\theta$ is infinite.

For each value of θ in the interval between $-\pi/2$ and $\pi/2$, the formula $r^2 = 4\cos\theta$ gives two values of r:

$$r = \pm 2\sqrt{\cos\theta}.$$

θ	$\cos\theta$	$r = \pm 2\sqrt{\cos\theta}$
0	1	± 2
$\pm\frac{\pi}{6}$	$\frac{\sqrt{3}}{2}$	$\approx \pm 1.9$
$\pm\frac{\pi}{4}$	$\frac{1}{\sqrt{2}}$	$\approx \pm 1.7$
$\pm\frac{\pi}{3}$	$\frac{1}{2}$	$\approx \pm 1.4$
$\pm\frac{\pi}{2}$	0	0

(a)

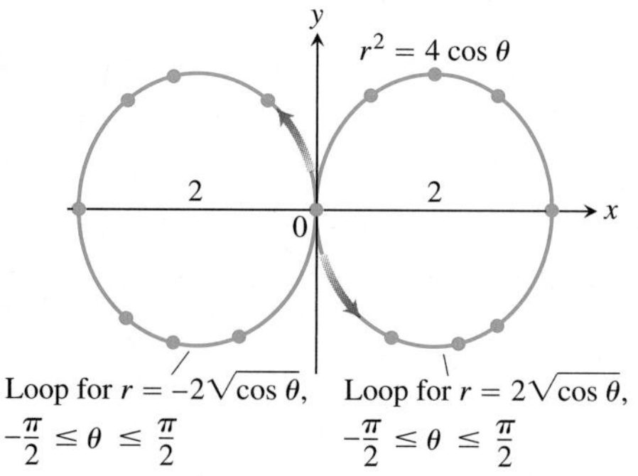

(b)

FIGURE 8.11 The graph of $r^2 = 4\cos\theta$. The arrows show the direction of increasing θ. The values of r in the table are rounded (Example 2).

We make a short table of values, plot the corresponding points, and use information about symmetry and tangents to guide us in connecting the points with a smooth curve (Figure 8.11). ■

USING TECHNOLOGY Graphing Polar Curves Parametrically

For complicated polar curves we may need to use a graphing calculator or computer to graph the curve. If the device does not plot polar graphs directly, we can convert $r = f(\theta)$ into parametric form using the equations

$$x = r\cos\theta = f(\theta)\cos\theta, \qquad y = r\sin\theta = f(\theta)\sin\theta.$$

Then we use the device to draw a parametrized curve in the Cartesian xy-plane. It may be necessary to use the parameter t rather than θ for the graphing device.

EXERCISES 8.2

Identify the symmetries of the curves in Exercises 1–12. Then sketch the curves.

1. $r = 1 + \cos\theta$
2. $r = 2 - 2\cos\theta$
3. $r = 1 - \sin\theta$
4. $r = 1 + \sin\theta$
5. $r = 2 + \sin\theta$
6. $r = 1 + 2\sin\theta$
7. $r = \sin(\theta/2)$
8. $r = \cos(\theta/2)$
9. $r^2 = \cos\theta$
10. $r^2 = \sin\theta$
11. $r^2 = -\sin\theta$
12. $r^2 = -\cos\theta$

Graph the lemniscates in Exercises 13–16. What symmetries do these curves have?

13. $r^2 = 4\cos 2\theta$
14. $r^2 = 4\sin 2\theta$
15. $r^2 = -\sin 2\theta$
16. $r^2 = -\cos 2\theta$

Find the slopes of the curves in Exercises 17–20 at the given points. Sketch the curves along with their tangents at these points.

17. Cardioid $r = -1 + \cos\theta;\quad \theta = \pm\pi/2$

18. Cardioid $r = -1 + \sin\theta;\quad \theta = 0, \pi$

19. Four-leaved rose $r = \sin 2\theta;\quad \theta = \pm\pi/4, \pm 3\pi/4$

20. Four-leaved rose $r = \cos 2\theta;\quad \theta = 0, \pm\pi/2, \pi$

Graph the limaçons in Exercises 21–24. Limaçon ("*lee*-ma-sahn") is Old French for "snail." You will understand the name when you graph the limaçons in Exercise 21. Equations for limaçons have the form $r = a \pm b\cos\theta$ or $r = a \pm b\sin\theta$. There are four basic shapes.

21. Limaçons with an inner loop

a. $r = \dfrac{1}{2} + \cos\theta$
b. $r = \dfrac{1}{2} + \sin\theta$

22. Cardioids

a. $r = 1 - \cos\theta$
b. $r = -1 + \sin\theta$

23. Dimpled limaçons

a. $r = \dfrac{3}{2} + \cos\theta$
b. $r = \dfrac{3}{2} - \sin\theta$

24. Oval limaçons

a. $r = 2 + \cos\theta$
b. $r = -2 + \sin\theta$

25. Sketch the region defined by the inequalities $-1 \le r \le 2$ and $-\pi/2 \le \theta \le \pi/2$.

26. Sketch the region defined by the inequalities $0 \le r \le 2\sec\theta$ and $-\pi/4 \le \theta \le \pi/4$.

In Exercises 27 and 28, sketch the region defined by the inequality.

27. $0 \le r \le 2 - 2\cos\theta$
28. $0 \le r^2 \le \cos\theta$

T **29.** Which of the following has the same graph as $r = 1 - \cos\theta$?

a. $r = -1 - \cos\theta$

b. $r = 1 + \cos\theta$

Confirm your answer with algebra.

T **30.** Which of the following has the same graph as $r = \cos 2\theta$?

a. $r = -\sin(2\theta + \pi/2)$

b. $r = -\cos(\theta/2)$

Confirm your answer with algebra.

T **31. A rose within a rose** Graph the equation $r = 1 - 2\sin 3\theta$.

T **32. The nephroid of Freeth** Graph the nephroid of Freeth:

$$r = 1 + 2\sin\frac{\theta}{2}.$$

T **33. Roses** Graph the roses $r = \cos m\theta$ for $m = 1/3, 2, 3$, and 7.

T **34. Spirals** Polar coordinates are just the thing for defining spirals. Graph the following spirals.

a. $r = \theta$

b. $r = -\theta$

c. *A logarithmic spiral:* $r = e^{\theta/10}$

d. *A hyperbolic spiral:* $r = 8/\theta$

e. *An equilateral hyperbola:* $r = \pm 10/\sqrt{\theta}$

(Use different colors for the two branches.)

8.3 Areas and Lengths in Polar Coordinates

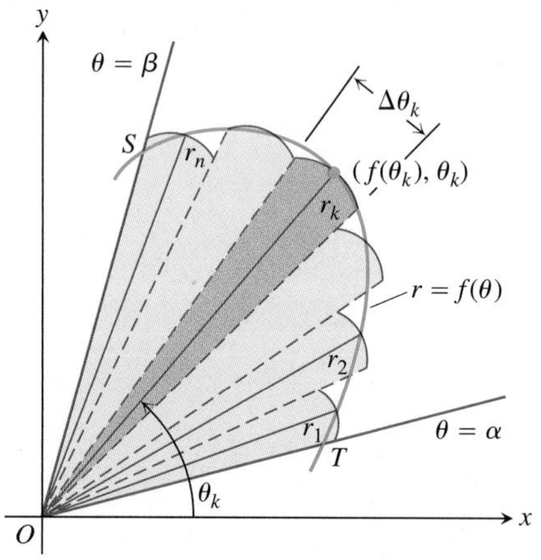

FIGURE 8.12 To derive a formula for the area of region *OTS*, we approximate the region with fan-shaped circular sectors.

This section shows how to calculate areas of plane regions and lengths of curves in polar coordinates.

Area in the Plane

The region *OTS* in Figure 8.12 is bounded by the rays $\theta = \alpha$ and $\theta = \beta$ and the curve $r = f(\theta)$. We approximate the region with n nonoverlapping fan-shaped circular sectors based on a partition P of angle *TOS*. The typical sector has radius $r_k = f(\theta_k)$ and central angle of radian measure $\Delta\theta_k$. Its area is $\Delta\theta_k/2\pi$ times the area of a circle of radius r_k, or

$$A_k = \frac{1}{2} r_k^2 \, \Delta\theta_k = \frac{1}{2}\left(f(\theta_k)\right)^2 \Delta\theta_k .$$

The area of region *OTS* is approximately

$$\sum_{k=1}^{n} A_k = \sum_{k=1}^{n} \frac{1}{2}\left(f(\theta_k)\right)^2 \Delta\theta_k .$$

If f is continuous, we expect the approximations to improve as the norm of the partition P goes to zero, where the norm of P is the largest value of $\Delta\theta_k$. We are then led to the following formula for the region's area:

$$\begin{aligned} A &= \lim_{\|P\| \to 0} \sum_{k=1}^{n} \frac{1}{2}\left(f(\theta_k)\right)^2 \Delta\theta_k \\ &= \int_{\alpha}^{\beta} \frac{1}{2}\left(f(\theta)\right)^2 d\theta . \end{aligned}$$

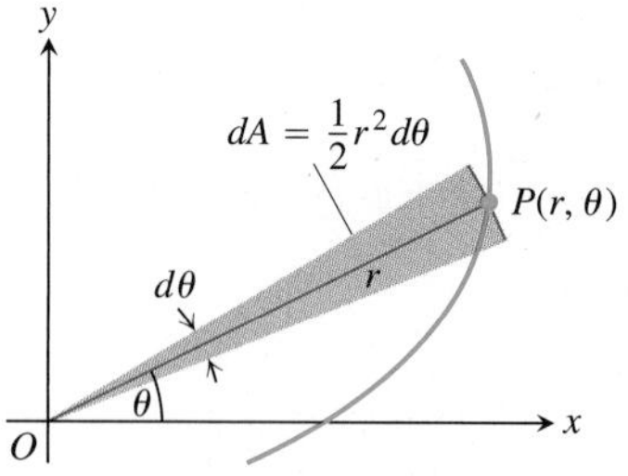

FIGURE 8.13 The area differential dA for the curve $r = f(\theta)$.

Area of the Fan-Shaped Region Between the Origin and the Curve $r = f(\theta), \alpha \le \theta \le \beta$

$$A = \int_{\alpha}^{\beta} \frac{1}{2} r^2 \, d\theta .$$

This is the integral of the **area differential** (Figure 8.13)

$$dA = \frac{1}{2} r^2 \, d\theta = \frac{1}{2}\left(f(\theta)\right)^2 d\theta .$$

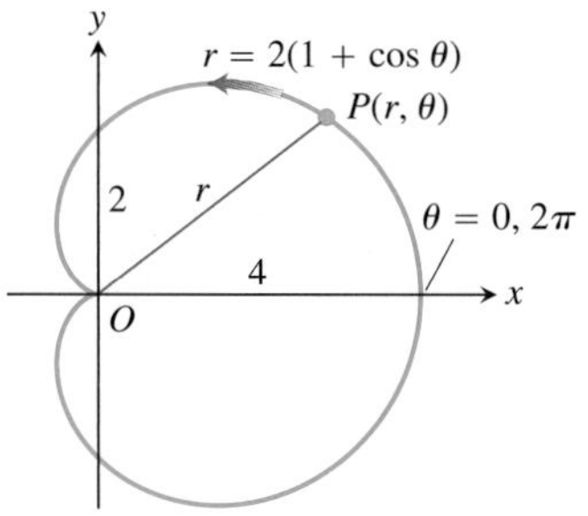

FIGURE 8.14 The cardioid in Example 1.

EXAMPLE 1 Find the area of the region in the plane enclosed by the cardioid $r = 2(1 + \cos\theta)$.

Solution We graph the cardioid (Figure 8.14) and determine that the radius OP sweeps out the region exactly once as θ runs from 0 to 2π. The area is therefore

$$\begin{aligned}\int_{\theta=0}^{\theta=2\pi} \frac{1}{2} r^2\, d\theta &= \int_0^{2\pi} \frac{1}{2}\cdot 4(1+\cos\theta)^2\, d\theta \\ &= \int_0^{2\pi} 2(1 + 2\cos\theta + \cos^2\theta)\, d\theta \\ &= \int_0^{2\pi}\left(2 + 4\cos\theta + 2\,\frac{1+\cos 2\theta}{2}\right) d\theta \\ &= \int_0^{2\pi} (3 + 4\cos\theta + \cos 2\theta)\, d\theta \\ &= \left[3\theta + 4\sin\theta + \frac{\sin 2\theta}{2}\right]_0^{2\pi} = 6\pi - 0 = 6\pi.\end{aligned}$$

■

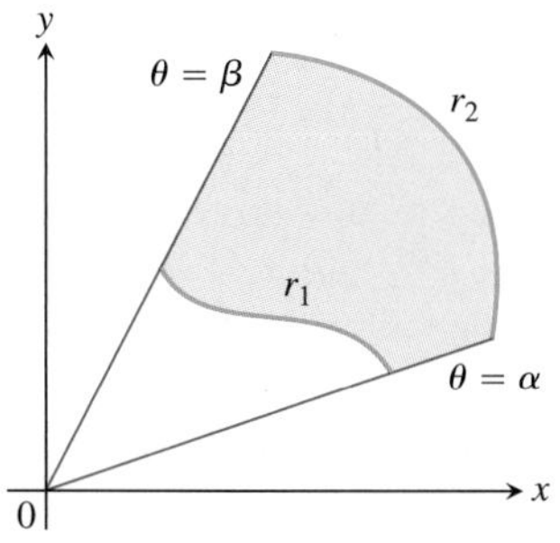

FIGURE 8.15 The area of the shaded region is calculated by subtracting the area of the region between r_1 and the origin from the area of the region between r_2 and the origin.

To find the area of a region like the one in Figure 8.15, which lies between two polar curves $r_1 = r_1(\theta)$ and $r_2 = r_2(\theta)$ from $\theta = \alpha$ to $\theta = \beta$, we subtract the integral of $(1/2)r_1^2\, d\theta$ from the integral of $(1/2)r_2^2\, d\theta$. This leads to the following formula.

Area of the Region $0 \le r_1(\theta) \le r \le r_2(\theta), \quad \alpha \le \theta \le \beta$

$$A = \int_\alpha^\beta \frac{1}{2} r_2^2\, d\theta - \int_\alpha^\beta \frac{1}{2} r_1^2\, d\theta = \int_\alpha^\beta \frac{1}{2}\left(r_2^2 - r_1^2\right) d\theta \qquad (1)$$

EXAMPLE 2 Find the area of the region that lies inside the circle $r = 1$ and outside the cardioid $r = 1 - \cos\theta$.

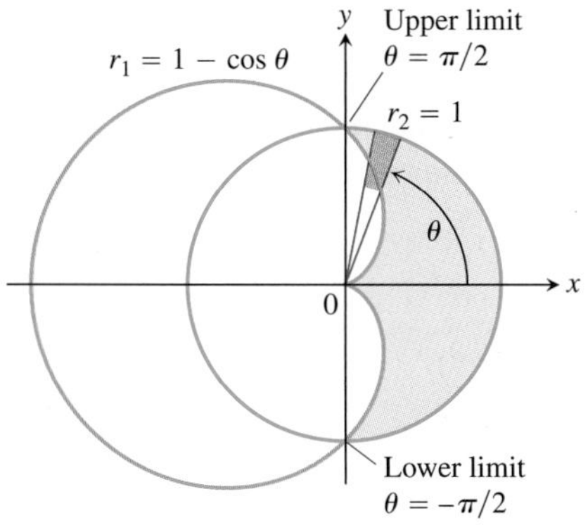

FIGURE 8.16 The region and limits of integration in Example 2.

Solution We sketch the region to determine its boundaries and find the limits of integration (Figure 8.16). The outer curve is $r_2 = 1$, the inner curve is $r_1 = 1 - \cos\theta$, and θ runs from $-\pi/2$ to $\pi/2$. The area, from Equation (1), is

$$\begin{aligned}A &= \int_{-\pi/2}^{\pi/2} \frac{1}{2}\left(r_2^2 - r_1^2\right) d\theta \\ &= 2\int_0^{\pi/2} \frac{1}{2}\left(r_2^2 - r_1^2\right) d\theta \qquad \text{Symmetry} \\ &= \int_0^{\pi/2} (1 - (1 - 2\cos\theta + \cos^2\theta))\, d\theta \\ &= \int_0^{\pi/2} (2\cos\theta - \cos^2\theta)\, d\theta = \int_0^{\pi/2}\left(2\cos\theta - \frac{1+\cos 2\theta}{2}\right) d\theta \\ &= \left[2\sin\theta - \frac{\theta}{2} - \frac{\sin 2\theta}{4}\right]_0^{\pi/2} = 2 - \frac{\pi}{4}.\end{aligned}$$

■

The fact that we can represent a point in different ways in polar coordinates requires extra care in deciding when a point lies on the graph of a polar equation and in determining the points in which polar graphs intersect. (We needed intersection points in Example 2.) In Cartesian coordinates, we can always find the points where two curves cross by solving their equations simultaneously. In polar coordinates, the story is different. Simultaneous solution may reveal some intersection points without revealing others. One way to identify all the points of intersection is to graph the equations.

Length of a Polar Curve

We can obtain a polar coordinate formula for the length of a curve $r = f(\theta), \alpha \le \theta \le \beta$, by parametrizing the curve as

$$x = r\cos\theta = f(\theta)\cos\theta, \qquad y = r\sin\theta = f(\theta)\sin\theta, \qquad \alpha \le \theta \le \beta. \tag{2}$$

The parametric length formula, Equation (1) from Section 6.3, then gives the length as

$$L = \int_\alpha^\beta \sqrt{\left(\frac{dx}{d\theta}\right)^2 + \left(\frac{dy}{d\theta}\right)^2}\, d\theta.$$

This equation becomes

$$L = \int_\alpha^\beta \sqrt{r^2 + \left(\frac{dr}{d\theta}\right)^2}\, d\theta$$

when Equations (2) are substituted for x and y (Exercise 27).

Length of a Polar Curve

If $r = f(\theta)$ has a continuous first derivative for $\alpha \le \theta \le \beta$ and if the point $P(r, \theta)$ traces the curve $r = f(\theta)$ exactly once as θ runs from α to β, then the length of the curve is

$$L = \int_\alpha^\beta \sqrt{r^2 + \left(\frac{dr}{d\theta}\right)^2}\, d\theta. \tag{3}$$

EXAMPLE 3 Find the length of the cardioid $r = 1 - \cos\theta$.

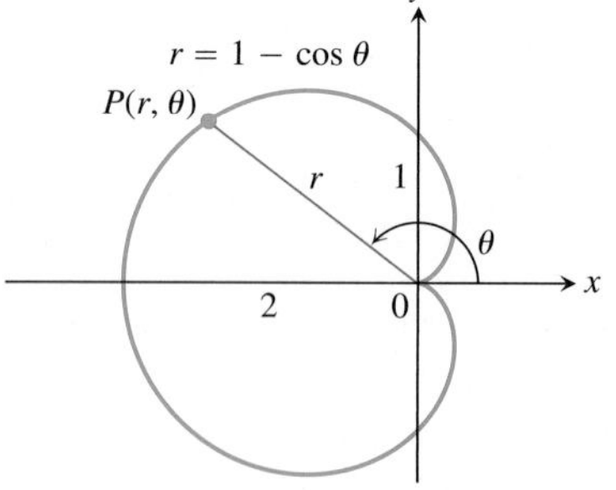

FIGURE 8.17 Calculating the length of a cardioid (Example 3).

Solution We sketch the cardioid to determine the limits of integration (Figure 8.17). The point $P(r, \theta)$ traces the curve once, counterclockwise as θ runs from 0 to 2π, so these are the values we take for α and β.

With

$$r = 1 - \cos\theta, \qquad \frac{dr}{d\theta} = \sin\theta,$$

we have

$$\begin{aligned} r^2 + \left(\frac{dr}{d\theta}\right)^2 &= (1 - \cos\theta)^2 + (\sin\theta)^2 \\ &= 1 - 2\cos\theta + \underbrace{\cos^2\theta + \sin^2\theta}_{1} = 2 - 2\cos\theta \end{aligned}$$

and

$$\begin{aligned}
L &= \int_{\alpha}^{\beta} \sqrt{r^2 + \left(\frac{dr}{d\theta}\right)^2}\, d\theta = \int_0^{2\pi} \sqrt{2 - 2\cos\theta}\, d\theta \\
&= \int_0^{2\pi} \sqrt{4\sin^2\frac{\theta}{2}}\, d\theta \qquad 1 - \cos\theta = 2\sin^2\frac{\theta}{2} \\
&= \int_0^{2\pi} 2\left|\sin\frac{\theta}{2}\right| d\theta \\
&= \int_0^{2\pi} 2\sin\frac{\theta}{2}\, d\theta \qquad \sin\frac{\theta}{2} \geq 0 \quad \text{for} \quad 0 \leq \theta \leq 2\pi \\
&= \left[-4\cos\frac{\theta}{2}\right]_0^{2\pi} = 4 + 4 = 8.
\end{aligned}$$

■

EXERCISES 8.3

Find the areas of the regions in Exercises 1–6.

1. Inside the oval limaçon $r = 4 + 2\cos\theta$
2. Inside the cardioid $r = a(1 + \cos\theta), \quad a > 0$
3. Inside one leaf of the four-leaved rose $r = \cos 2\theta$
4. Inside the lemniscate $r^2 = 2a^2\cos 2\theta, \quad a > 0$
5. Inside one loop of the lemniscate $r^2 = 4\sin 2\theta$
6. Inside the six-leaved rose $r^2 = 2\sin 3\theta$

Find the areas of the regions in Exercises 7–16.

7. Shared by the circles $r = 2\cos\theta$ and $r = 2\sin\theta$
8. Shared by the circles $r = 1$ and $r = 2\sin\theta$
9. Shared by the circle $r = 2$ and the cardioid $r = 2(1 - \cos\theta)$
10. Shared by the cardioids $r = 2(1 + \cos\theta)$ and $r = 2(1 - \cos\theta)$
11. Inside the lemniscate $r^2 = 6\cos 2\theta$ and outside the circle $r = \sqrt{3}$
12. Inside the circle $r = 3a\cos\theta$ and outside the cardioid $r = a(1 + \cos\theta), a > 0$
13. Inside the circle $r = -2\cos\theta$ and outside the circle $r = 1$
14. Inside the circle $r = 6$ above the line $r = 3\csc\theta$
15. **a.** Find the area of the shaded region in the accompanying figure.

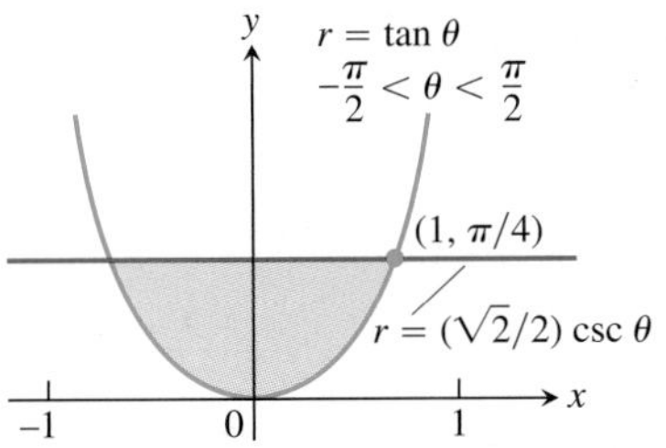

b. It looks as if the graph of $r = \tan\theta, -\pi/2 < \theta < \pi/2$, could be asymptotic to the lines $x = 1$ and $x = -1$. Is it? Give reasons for your answer.

16. The area of the region that lies inside the cardioid curve $r = \cos\theta + 1$ and outside the circle $r = \cos\theta$ is not

$$\frac{1}{2}\int_0^{2\pi} [(\cos\theta + 1)^2 - \cos^2\theta]\, d\theta = \pi.$$

Why not? What *is* the area? Give reasons for your answers.

Find the lengths of the curves in Exercises 17–25.

17. The spiral $r = \theta^2, \quad 0 \leq \theta \leq \sqrt{5}$
18. The spiral $r = e^{\theta}/\sqrt{2}, \quad 0 \leq \theta \leq \pi$
19. The cardioid $r = 1 + \cos\theta$
20. The curve $r = a\sin^2(\theta/2), \quad 0 \leq \theta \leq \pi, \quad a > 0$
21. The parabolic segment $r = 6/(1 + \cos\theta), \quad 0 \leq \theta \leq \pi/2$
22. The parabolic segment $r = 2/(1 - \cos\theta), \quad \pi/2 \leq \theta \leq \pi$
23. The curve $r = \cos^3(\theta/3), \quad 0 \leq \theta \leq \pi/4$
24. The curve $r = \sqrt{1 + \sin 2\theta}, \quad 0 \leq \theta \leq \pi\sqrt{2}$
25. The curve $r = \sqrt{1 + \cos 2\theta}, \quad 0 \leq \theta \leq \pi\sqrt{2}$
26. **Circumferences of circles** As usual, when faced with a new formula, it is a good idea to try it on familiar objects to be sure it gives results consistent with past experience. Use the length formula in Equation (3) to calculate the circumferences of the following circles $(a > 0)$:

    a. $r = a$ **b.** $r = a\cos\theta$ **c.** $r = a\sin\theta$

27. **The length of the curve $r = f(\theta), \alpha \leq \theta \leq \beta$** Assuming that the necessary derivatives are continuous, show how the substitutions

$$x = f(\theta)\cos\theta, \quad y = f(\theta)\sin\theta$$

(Equations 2 in the text) transform

$$L = \int_{\alpha}^{\beta} \sqrt{\left(\frac{dx}{d\theta}\right)^2 + \left(\frac{dy}{d\theta}\right)^2}\, d\theta$$

into

$$L = \int_{\alpha}^{\beta} \sqrt{r^2 + \left(\frac{dr}{d\theta}\right)^2}\, d\theta.$$

28. Average value If f is continuous, the average value of the polar coordinate r over the curve $r = f(\theta)$, $\alpha \le \theta \le \beta$, with respect to θ is given by the formula

$$r_{\text{av}} = \frac{1}{\beta - \alpha} \int_{\alpha}^{\beta} f(\theta)\, d\theta.$$

Use this formula to find the average value of r with respect to θ over the following curves ($a > 0$).

a. The cardioid $r = a(1 - \cos\theta)$

b. The circle $r = a$

c. The circle $r = a\cos\theta$, $-\pi/2 \le \theta \le \pi/2$

29. $r = f(\theta)$ *vs.* $r = 2f(\theta)$ Can anything be said about the relative lengths of the curves $r = f(\theta)$, $\alpha \le \theta \le \beta$, and $r = 2f(\theta)$, $\alpha \le \theta \le \beta$? Give reasons for your answer.

8.4 Conics in Polar Coordinates

Polar coordinates are important in astronomy and astronautical engineering because satellites, moons, planets, and comets all move approximately along ellipses, parabolas, and hyperbolas that can be described with a single relatively simple polar coordinate equation. We develop that equation here after first introducing the idea of a conic section's *eccentricity*. The eccentricity reveals the conic section's type (circle, ellipse, parabola, or hyperbola), and the degree to which it is "squashed" or flattened. (Conic sections are reviewed on the website.)

Eccentricity

Although the center-to-focus distance c (Figure 8.18) does not appear in the equation

$$\frac{x^2}{a^2} + \frac{y^2}{b^2} = 1, \qquad (a > b)$$

for an ellipse, we can still determine c from the equation $c = \sqrt{a^2 - b^2}$. If we fix a and vary c over the interval $0 \le c \le a$, the resulting ellipses will vary in shape. They are circles if $c = 0$ (so that $a = b$) and flatten as c increases. If $c = a$, the foci and vertices overlap and the ellipse degenerates into a line segment. Thus we are led to consider the ratio $e = c/a$. We use this ratio for hyperbolas as well, only in this case c equals $\sqrt{a^2 + b^2}$ instead of $\sqrt{a^2 - b^2}$, and define these ratios with the somewhat familiar term *eccentricity*.

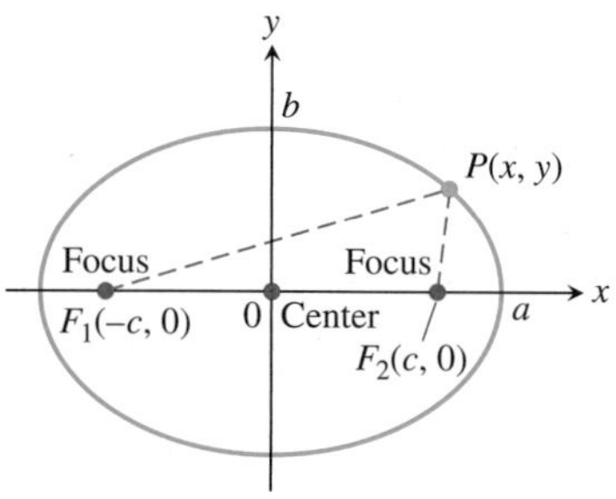

FIGURE 8.18 The ellipse defined by the equation $PF_1 + PF_2 = 2a$ is the graph of the equation $(x^2/a^2) + (y^2/b^2) = 1$, where $b^2 = a^2 - c^2$.

DEFINITION

The **eccentricity** of the ellipse $(x^2/a^2) + (y^2/b^2) = 1$ $(a > b)$ is

$$e = \frac{c}{a} = \frac{\sqrt{a^2 - b^2}}{a}.$$

The **eccentricity** of the hyperbola $(x^2/a^2) - (y^2/b^2) = 1$ is

$$e = \frac{c}{a} = \frac{\sqrt{a^2 + b^2}}{a}.$$

The **eccentricity** of a parabola is $e = 1$.

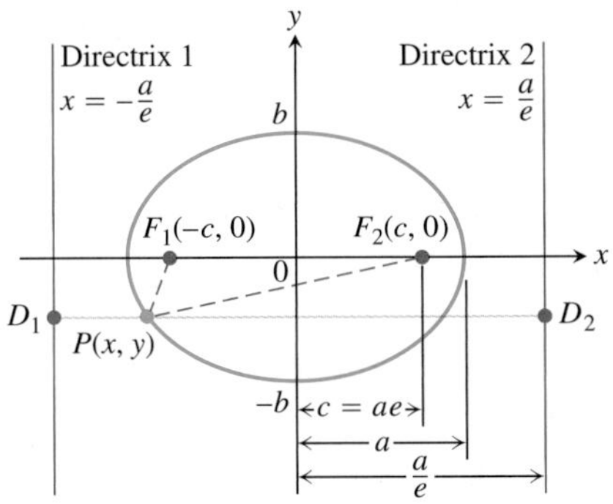

FIGURE 8.19 The foci and directrices of the ellipse $(x^2/a^2) + (y^2/b^2) = 1$. Directrix 1 corresponds to focus F_1, and directrix 2 to focus F_2.

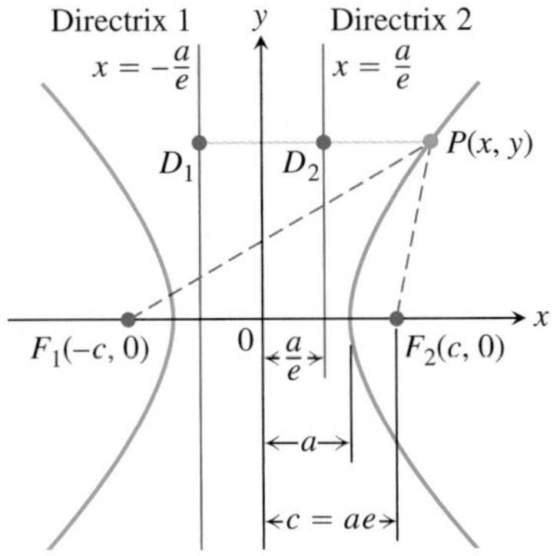

FIGURE 8.20 The foci and directrices of the hyperbola $(x^2/a^2) - (y^2/b^2) = 1$. No matter where P lies on the hyperbola, $PF_1 = e \cdot PD_1$ and $PF_2 = e \cdot PD_2$.

Whereas a parabola has one focus and one directrix, each **ellipse** has two foci and two **directrices**. These are the lines perpendicular to the major axis at distances $\pm a/e$ from the center. The parabola has the property that

$$PF = 1 \cdot PD \tag{1}$$

for any point P on it, where F is the focus and D is the point nearest P on the directrix. For an ellipse, it can be shown that the equations that replace Equation (1) are

$$PF_1 = e \cdot PD_1, \qquad PF_2 = e \cdot PD_2. \tag{2}$$

Here, e is the eccentricity, P is any point on the ellipse, F_1 and F_2 are the foci, and D_1 and D_2 are the points on the directrices nearest P (Figure 8.19).

In both Equations (2) the directrix and focus must correspond; that is, if we use the distance from P to F_1, we must also use the distance from P to the directrix at the same end of the ellipse. The directrix $x = -a/e$ corresponds to $F_1(-c, 0)$, and the directrix $x = a/e$ corresponds to $F_2(c, 0)$.

As with the ellipse, it can be shown that the lines $x = \pm a/e$ act as **directrices** for the **hyperbola** and that

$$PF_1 = e \cdot PD_1 \quad \text{and} \quad PF_2 = e \cdot PD_2. \tag{3}$$

Here P is any point on the hyperbola, F_1 and F_2 are the foci, and D_1 and D_2 are the points nearest P on the directrices (Figure 8.20).

In both the ellipse and the hyperbola, the eccentricity is the ratio of the distance between the foci to the distance between the vertices (because $c/a = 2c/2a$).

$$\text{Eccentricity} = \frac{\text{distance between foci}}{\text{distance between vertices}}$$

In an ellipse, the foci are closer together than the vertices and the ratio is less than 1. In a hyperbola, the foci are farther apart than the vertices and the ratio is greater than 1.

The "focus–directrix" equation $PF = e \cdot PD$ unites the parabola, ellipse, and hyperbola in the following way. Suppose that the distance PF of a point P from a fixed point F (the focus) is a constant multiple of its distance from a fixed line (the directrix). That is, suppose

$$PF = e \cdot PD, \tag{4}$$

where e is the constant of proportionality. Then the path traced by P is

(a) a *parabola* if $e = 1$,

(b) an *ellipse* of eccentricity e if $e < 1$, and

(c) a *hyperbola* of eccentricity e if $e > 1$.

There are no coordinates in Equation (4) and when we try to translate it into coordinate form it translates in different ways, depending on the size of e. At least, that is what happens in Cartesian coordinates. However, as we will see, in polar coordinates the equation $PF = e \cdot PD$ translates into a single equation regardless of the value of e.

Given the focus and corresponding directrix of a hyperbola centered at the origin and with foci on the x-axis, we can use the dimensions shown in Figure 8.20 to find e. Knowing

e, we can derive a Cartesian equation for the hyperbola from the equation $PF = e \cdot PD$, as in the next example. We can find equations for ellipses centered at the origin and with foci on the x-axis in a similar way, using the dimensions shown in Figure 8.19.

EXAMPLE 1 Find a Cartesian equation for the hyperbola centered at the origin that has a focus at $(3, 0)$ and the line $x = 1$ as the corresponding directrix.

Solution We first use the dimensions shown in Figure 8.20 to find the hyperbola's eccentricity. The focus is

$$(c, 0) = (3, 0) \quad \text{so} \quad c = 3.$$

The directrix is the line

$$x = \frac{a}{e} = 1, \quad \text{so} \quad a = e.$$

When combined with the equation $e = c/a$ that defines eccentricity, these results give

$$e = \frac{c}{a} = \frac{3}{e}, \quad \text{so} \quad e^2 = 3 \quad \text{and} \quad e = \sqrt{3}.$$

Knowing e, we can now derive the equation we want from the equation $PF = e \cdot PD$. In the notation of Figure 8.21, we have

$$\begin{aligned} PF &= e \cdot PD && \text{Eq. (4)} \\ \sqrt{(x-3)^2 + (y-0)^2} &= \sqrt{3}\,|x-1| && e = \sqrt{3} \\ x^2 - 6x + 9 + y^2 &= 3(x^2 - 2x + 1) \\ 2x^2 - y^2 &= 6 \\ \frac{x^2}{3} - \frac{y^2}{6} &= 1. \end{aligned}$$

■

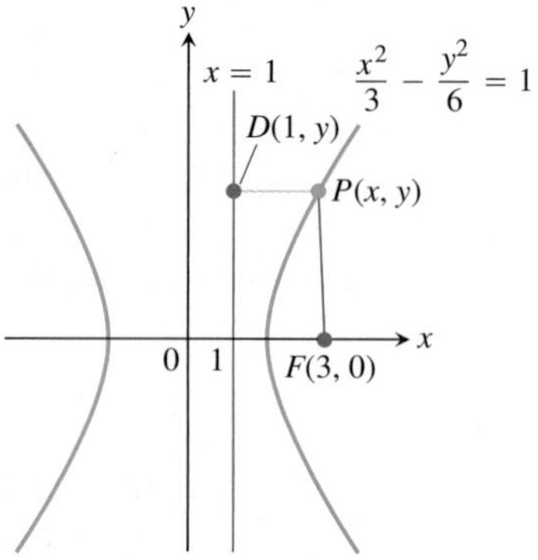

FIGURE 8.21 The hyperbola and directrix in Example 1.

Polar Equations

To find polar equations for ellipses, parabolas, and hyperbolas, we place one focus at the origin and the corresponding directrix to the right of the origin along the vertical line $x = k$ (Figure 8.22). This makes

$$PF = r$$

and

$$PD = k - FB = k - r\cos\theta.$$

The conic's focus–directrix equation $PF = e \cdot PD$ then becomes

$$r = e(k - r\cos\theta),$$

which can be solved for r to obtain the following expression.

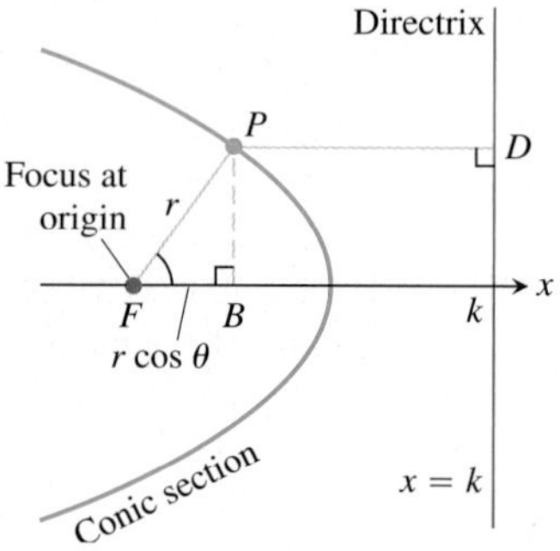

FIGURE 8.22 If a conic section is put in the position with its focus placed at the origin and a directrix perpendicular to the initial ray and right of the origin, we can find its polar equation from the conic's focus–directrix equation.

Polar Equation for a Conic with Eccentricity e

$$r = \frac{ke}{1 + e\cos\theta}, \qquad (5)$$

where $x = k > 0$ is the vertical directrix.

EXAMPLE 2 Here are polar equations for three conics.

$$e = \frac{1}{2}: \quad \text{ellipse} \quad r = \frac{k}{2 + \cos\theta}$$

$$e = 1: \quad \text{parabola} \quad r = \frac{k}{1 + \cos\theta}$$

$$e = 2: \quad \text{hyperbola} \quad r = \frac{2k}{1 + 2\cos\theta}$$

■

You may see variations of Equation (5), depending on the location of the directrix. If the directrix is the line $x = -k$ to the left of the origin (the origin is still a focus), we replace Equation (5) by

$$r = \frac{ke}{1 - e\cos\theta}.$$

The denominator now has a $(-)$ instead of a $(+)$. If the directrix is either of the lines $y = k$ or $y = -k$, the equations have sines in them instead of cosines, as shown in Figure 8.23.

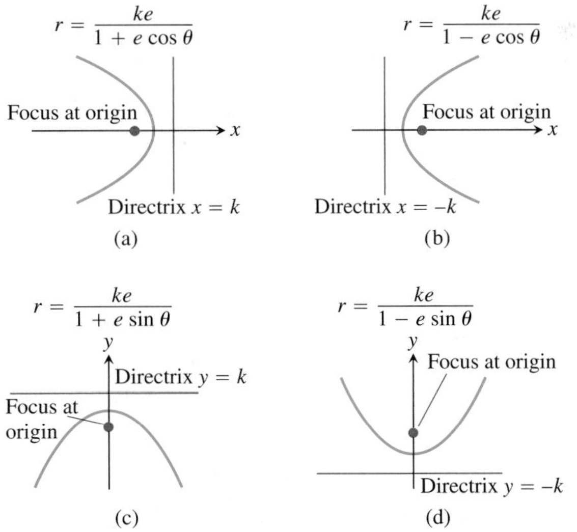

FIGURE 8.23 Equations for conic sections with eccentricity $e > 0$, but different locations of the directrix. The graphs here show a parabola, so $e = 1$.

EXAMPLE 3 Find an equation for the hyperbola with eccentricity $3/2$ and directrix $x = 2$.

Solution We use Equation (5) with $k = 2$ and $e = 3/2$:

$$r = \frac{2(3/2)}{1 + (3/2)\cos\theta} \quad \text{or} \quad r = \frac{6}{2 + 3\cos\theta}.$$

■

EXAMPLE 4 Find the directrix of the parabola

$$r = \frac{25}{10 + 10\cos\theta}.$$

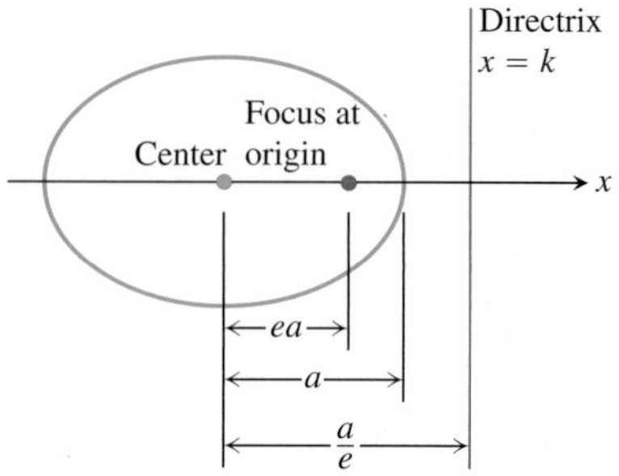

FIGURE 8.24 In an ellipse with semimajor axis a, the focus–directrix distance is $k = (a/e) - ea$, so $ke = a(1 - e^2)$.

Solution We divide the numerator and denominator by 10 to put the equation in standard polar form:

$$r = \frac{5/2}{1 + \cos\theta}.$$

This is the equation

$$r = \frac{ke}{1 + e\cos\theta}$$

with $k = 5/2$ and $e = 1$. The equation of the directrix is $x = 5/2$. ■

From the ellipse diagram in Figure 8.24, we see that k is related to the eccentricity e and the semimajor axis a by the equation

$$k = \frac{a}{e} - ea.$$

From this, we find that $ke = a(1 - e^2)$. Replacing ke in Equation (5) by $a(1 - e^2)$ gives the standard polar equation for an ellipse.

Polar Equation for the Ellipse with Eccentricity e and Semimajor Axis a

$$r = \frac{a(1 - e^2)}{1 + e\cos\theta} \tag{6}$$

Notice that when $e = 0$, Equation (6) becomes $r = a$, which represents a circle.

Lines

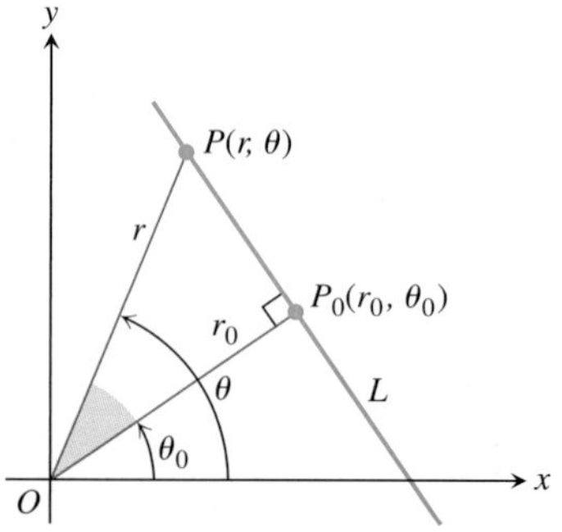

FIGURE 8.25 We can obtain a polar equation for line L by reading the relation $r_0 = r\cos(\theta - \theta_0)$ from the right triangle OP_0P.

Suppose the perpendicular from the origin to line L meets L at the point $P_0(r_0, \theta_0)$, with $r_0 \geq 0$ (Figure 8.25). Then, if $P(r, \theta)$ is any other point on L, the points P, P_0, and O are the vertices of a right triangle, from which we can read the relation

$$r_0 = r\cos(\theta - \theta_0).$$

The Standard Polar Equation for Lines

If the point $P_0(r_0, \theta_0)$ is the foot of the perpendicular from the origin to the line L, and $r_0 \geq 0$, then an equation for L is

$$r\cos(\theta - \theta_0) = r_0. \tag{7}$$

For example, if $\theta_0 = \pi/3$, we find that

$$r\cos\left(\theta - \frac{\pi}{3}\right) = 2$$

$$r\left(\cos\theta\cos\frac{\pi}{3} + \sin\theta\sin\frac{\pi}{3}\right) = 2$$

$$\frac{1}{2}r\cos\theta + \frac{\sqrt{3}}{2}r\sin\theta = 2, \qquad \text{or} \qquad x + \sqrt{3}\,y = 4.$$

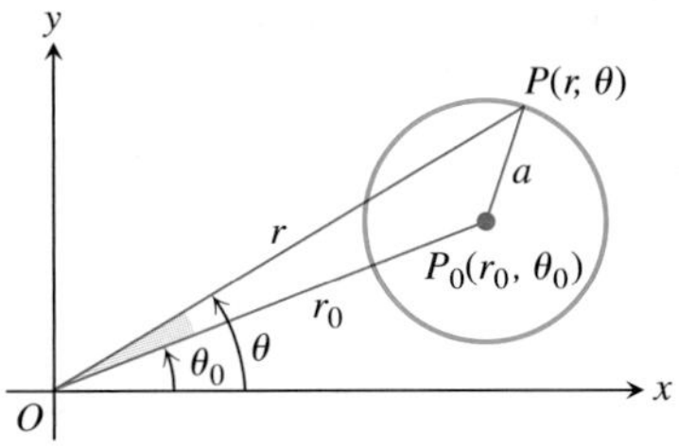

FIGURE 8.26 We can get a polar equation for this circle by applying the Law of Cosines to triangle OP_0P.

Circles

To find a polar equation for the circle of radius a centered at $P_0(r_0, \theta_0)$, we let $P(r, \theta)$ be a point on the circle and apply the Law of Cosines to triangle OP_0P (Figure 8.26). This gives

$$a^2 = r_0^2 + r^2 - 2r_0 r \cos(\theta - \theta_0).$$

If the circle passes through the origin, then $r_0 = a$ and this equation simplifies to

$$\begin{aligned} a^2 &= a^2 + r^2 - 2ar\cos(\theta - \theta_0) \\ r^2 &= 2ar\cos(\theta - \theta_0) \\ r &= 2a\cos(\theta - \theta_0). \end{aligned}$$

If the circle's center lies on the positive x-axis, $\theta_0 = 0$ and we get the further simplification

$$r = 2a\cos\theta. \tag{8}$$

If the center lies on the positive y-axis, $\theta = \pi/2$, $\cos(\theta - \pi/2) = \sin\theta$, and the equation $r = 2a\cos(\theta - \theta_0)$ becomes

$$r = 2a\sin\theta. \tag{9}$$

Equations for circles through the origin centered on the negative x- and y-axes can be obtained by replacing r with $-r$ in the above equations.

EXAMPLE 5 Here are several polar equations for circles through the origin.

Radius	Center (polar coordinates)	Polar equation
3	$(3, 0)$	$r = 6\cos\theta$
2	$(2, \pi/2)$	$r = 4\sin\theta$
1/2	$(-1/2, 0)$	$r = -\cos\theta$
1	$(-1, \pi/2)$	$r = -2\sin\theta$

■

EXERCISES 8.4

In Exercises 1–8, find the eccentricity of the ellipse. Then find and graph the ellipse's foci and directrices.

1. $16x^2 + 25y^2 = 400$
2. $7x^2 + 16y^2 = 112$
3. $2x^2 + y^2 = 2$
4. $2x^2 + y^2 = 4$
5. $3x^2 + 2y^2 = 6$
6. $9x^2 + 10y^2 = 90$
7. $6x^2 + 9y^2 = 54$
8. $169x^2 + 25y^2 = 4225$

Exercises 9–12 give the foci or vertices and the eccentricities of ellipses centered at the origin of the xy-plane. In each case, find the ellipse's standard-form equation in Cartesian coordinates.

9. Foci: $(0, \pm 3)$
   Eccentricity: 0.5
10. Foci: $(\pm 8, 0)$
    Eccentricity: 0.2
11. Vertices: $(0, \pm 70)$
    Eccentricity: 0.1
12. Vertices: $(\pm 10, 0)$
    Eccentricity: 0.24

Exercises 13–16 give foci and corresponding directrices of ellipses centered at the origin of the xy-plane. In each case, use the dimensions in Figure 8.19 to find the eccentricity of the ellipse. Then find the ellipse's standard-form equation in Cartesian coordinates.

13. Focus: $(\sqrt{5}, 0)$
    Directrix: $x = \dfrac{9}{\sqrt{5}}$
14. Focus: $(4, 0)$
    Directrix: $x = \dfrac{16}{3}$
15. Focus: $(-4, 0)$
    Directrix: $x = -16$
16. Focus: $(-\sqrt{2}, 0)$
    Directrix: $x = -2\sqrt{2}$

In Exercises 17–24, find the eccentricity of the hyperbola. Then find and graph the hyperbola's foci and directrices.

17. $x^2 - y^2 = 1$

18. $9x^2 - 16y^2 = 144$

19. $y^2 - x^2 = 8$

20. $y^2 - x^2 = 4$

21. $8x^2 - 2y^2 = 16$

22. $y^2 - 3x^2 = 3$

23. $8y^2 - 2x^2 = 16$

24. $64x^2 - 36y^2 = 2304$

Exercises 25–28 give the eccentricities and the vertices or foci of hyperbolas centered at the origin of the xy-plane. In each case, find the hyperbola's standard-form equation in Cartesian coordinates.

25. Eccentricity: 3
Vertices: $(0, \pm 1)$

26. Eccentricity: 2
Vertices: $(\pm 2, 0)$

27. Eccentricity: 3
Foci: $(\pm 3, 0)$

28. Eccentricity: 1.25
Foci: $(0, \pm 5)$

Exercises 29–36 give the eccentricities of conic sections with one focus at the origin, along with the directrix corresponding to that focus. Find a polar equation for each conic section.

29. $e = 1, \quad x = 2$

30. $e = 1, \quad y = 2$

31. $e = 5, \quad y = -6$

32. $e = 2, \quad x = 4$

33. $e = 1/2, \quad x = 1$

34. $e = 1/4, \quad x = -2$

35. $e = 1/5, \quad y = -10$

36. $e = 1/3, \quad y = 6$

Sketch the parabolas and ellipses in Exercises 37–44. Include the directrix that corresponds to the focus at the origin. Label the vertices with appropriate polar coordinates. Label the centers of the ellipses as well.

37. $r = \dfrac{1}{1 + \cos\theta}$

38. $r = \dfrac{6}{2 + \cos\theta}$

39. $r = \dfrac{25}{10 - 5\cos\theta}$

40. $r = \dfrac{4}{2 - 2\cos\theta}$

41. $r = \dfrac{400}{16 + 8\sin\theta}$

42. $r = \dfrac{12}{3 + 3\sin\theta}$

43. $r = \dfrac{8}{2 - 2\sin\theta}$

44. $r = \dfrac{4}{2 - \sin\theta}$

Sketch the lines in Exercises 45–48 and find Cartesian equations for them.

45. $r\cos\left(\theta - \dfrac{\pi}{4}\right) = \sqrt{2}$

46. $r\cos\left(\theta + \dfrac{3\pi}{4}\right) = 1$

47. $r\cos\left(\theta - \dfrac{2\pi}{3}\right) = 3$

48. $r\cos\left(\theta + \dfrac{\pi}{3}\right) = 2$

Find a polar equation in the form $r\cos(\theta - \theta_0) = r_0$ for each of the lines in Exercises 49–52.

49. $\sqrt{2}x + \sqrt{2}y = 6$

50. $\sqrt{3}x - y = 1$

51. $y = -5$

52. $x = -4$

Sketch the circles in Exercises 53–56. Give polar coordinates for their centers and identify their radii.

53. $r = 4\cos\theta$

54. $r = 6\sin\theta$

55. $r = -2\cos\theta$

56. $r = -8\sin\theta$

Find polar equations for the circles in Exercises 57–64. Sketch each circle in the coordinate plane and label it with both its Cartesian and polar equations.

57. $(x - 6)^2 + y^2 = 36$

58. $(x + 2)^2 + y^2 = 4$

59. $x^2 + (y - 5)^2 = 25$

60. $x^2 + (y + 7)^2 = 49$

61. $x^2 + 2x + y^2 = 0$

62. $x^2 - 16x + y^2 = 0$

63. $x^2 + y^2 + y = 0$

64. $x^2 + y^2 - \frac{4}{3}y = 0$

T Graph the lines and conic sections in Exercises 65–74.

65. $r = 3\sec(\theta - \pi/3)$

66. $r = 4\sec(\theta + \pi/6)$

67. $r = 4\sin\theta$

68. $r = -2\cos\theta$

69. $r = 8/(4 + \cos\theta)$

70. $r = 8/(4 + \sin\theta)$

71. $r = 1/(1 - \sin\theta)$

72. $r = 1/(1 + \cos\theta)$

73. $r = 1/(1 + 2\sin\theta)$

74. $r = 1/(1 + 2\cos\theta)$

75. Perihelion and aphelion A planet travels about its sun in an ellipse whose semimajor axis has length a. (See accompanying figure.)

a. Show that $r = a(1 - e)$ when the planet is closest to the sun and that $r = a(1 + e)$ when the planet is farthest from the sun.

b. Use the data in the table in Exercise 76 to find how close each planet in our solar system comes to the sun and how far away each planet gets from the sun.

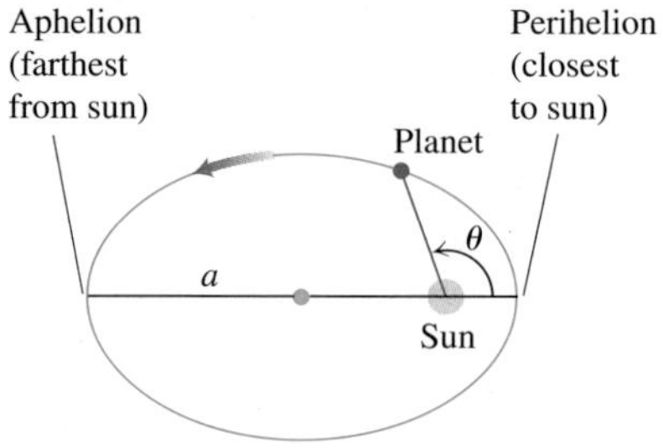

76. Planetary orbits Use the data in the table below and Equation (6) to find polar equations for the orbits of the planets.

Planet	Semimajor axis (astronomical units)	Eccentricity
Mercury	0.3871	0.2056
Venus	0.7233	0.0068
Earth	1.000	0.0167
Mars	1.524	0.0934
Jupiter	5.203	0.0484
Saturn	9.539	0.0543
Uranus	19.18	0.0460
Neptune	30.06	0.0082
Pluto	39.44	0.2481

For example, the orbit of Pluto is

$$r = \frac{39.44(1 - (0.25)^2)}{1 + 0.25\cos\theta} = \frac{147.9}{4 + \cos\theta}$$

In Exercises 77–80, find a polar equation for the given curve. In each case, sketch a typical curve.

77. $x^2 + y^2 - 2ay = 0$

78. $y^2 = 4ax + 4a^2$

79. $x \cos \alpha + y \sin \alpha = p$ (α, p constant)

80. $(x^2 + y^2)^2 + 2ax(x^2 + y^2) - a^2y^2 = 0$

COMPUTER EXPLORATIONS

81. Use a CAS to plot the polar equation

$$r = \frac{ke}{1 + e \cos \theta}$$

for various values of k and e, $-\pi \leq \theta \leq \pi$. Answer the following questions.

a. Take $k = -2$. Describe what happens to the plots as you take e to be $3/4$, 1, and $5/4$. Repeat for $k = 2$.

b. Take $k = -1$. Describe what happens to the plots as you take e to be $7/6$, $5/4$, $4/3$, $3/2$, 2, 3, 5, 10, and 20. Repeat for $e = 1/2, 1/3, 1/4, 1/10$, and $1/20$.

c. Now keep $e > 0$ fixed and describe what happens as you take k to be $-1, -2, -3, -4$, and -5. Be sure to look at graphs for parabolas, ellipses, and hyperbolas.

82. Use a CAS to plot the polar ellipse

$$r = \frac{a(1 - e^2)}{1 + e \cos \theta}$$

for various values of $a > 0$ and $0 < e < 1$, $-\pi \leq \theta \leq \pi$.

a. Take $e = 9/10$. Describe what happens to the plots as you let a equal 1, $3/2$, 2, 3, 5, and 10. Repeat with $e = 1/4$.

b. Take $a = 2$. Describe what happens as you take e to be $9/10$, $8/10$, $7/10$, $\ldots$, $1/10$, $1/20$, and $1/50$.

8.5 Conics and Parametric Equations; The Cycloid

Curves in the Cartesian plane defined by parametric equations, and the calculation of their derivatives, were introduced in Section 3.5. There we studied parametrizations of lines, circles, and ellipses. In this section we discuss parametrization of other plane curves, including parabolas, hyperbolas, cycloids, brachistocrones, and tautocrones.

Parabolas and Hyperbolas

In Section 3.5 we used the parametrization

$$x = \sqrt{t}, \qquad y = t, \qquad t \geq 0$$

to describe the motion of a particle moving along the right branch of the parabola $y = x^2$. In the following example we obtain a parametrization of the entire parabola, not just its right branch.

EXAMPLE 1 The position $P(x, y)$ of a particle moving in the xy-plane is given by the equations and parameter interval

$$x = t, \qquad y = t^2, \qquad -\infty < t < \infty.$$

Identify the particle's path and describe the motion.

Solution We identify the path by eliminating t between the equations $x = t$ and $y = t^2$, obtaining

$$y = (t)^2 = x^2.$$

The particle's position coordinates satisfy the equation $y = x^2$, so the particle moves along this curve.

In contrast to Example 2 in Section 3.5, the particle now traverses the entire parabola. As t increases from $-\infty$ to ∞, the particle comes down the left-hand side, passes through the origin, and moves up the right-hand side (Figure 8.27). ■

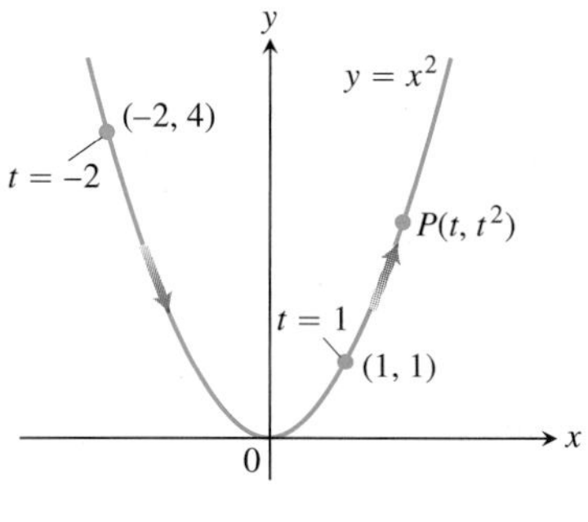

FIGURE 8.27 The path defined by $x = t, y = t^2, -\infty < t < \infty$ is the entire parabola $y = x^2$ (Example 1).

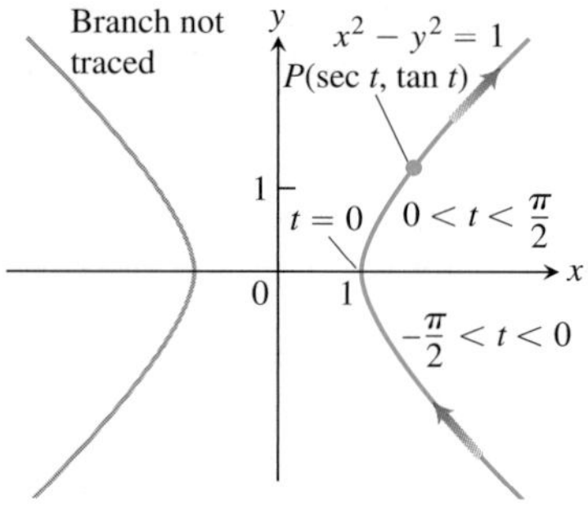

FIGURE 8.28 The equations $x = \sec t, y = \tan t$ and interval $-\pi/2 < t < \pi/2$ describe the right-hand branch of the hyperbola $x^2 - y^2 = 1$ (Example 2).

As Example 1 illustrates, any curve $y = f(x)$ has the parametrization $x = t, y = f(t)$. This is so simple we usually do not use it, but the point of view is occasionally helpful.

EXAMPLE 2 Describe the motion of the particle whose position $P(x, y)$ at time t is given by

$$x = \sec t, \qquad y = \tan t, \qquad -\frac{\pi}{2} < t < \frac{\pi}{2}.$$

Solution We find a Cartesian equation for the coordinates of P by eliminating t between the equations

$$\sec t = x, \qquad \tan t = y.$$

We accomplish this with the identity $\sec^2 t - \tan^2 t = 1$, which yields

$$x^2 - y^2 = 1.$$

Since the particle's coordinates (x, y) satisfy the equation $x^2 - y^2 = 1$, the motion takes place somewhere on this hyperbola. As t runs between $-\pi/2$ and $\pi/2$, $x = \sec t$ remains positive and $y = \tan t$ runs between $-\infty$ and ∞, so P traverses the hyperbola's right-hand branch. It comes in along the branch's lower half as $t \to 0^-$, reaches $(1, 0)$ at $t = 0$, and moves out into the first quadrant as t increases toward $\pi/2$ (Figure 8.28). ■

HISTORICAL BIOGRAPHY

Christiaan Huygens (1629–1695)

Cycloids

The problem with a pendulum clock whose bob swings in a circular arc is that the frequency of the swing depends on the amplitude of the swing. The wider the swing, the longer it takes the bob to return to center (its lowest position).

This does not happen if the bob can be made to swing in a *cycloid*. In 1673, Christiaan Huygens designed a pendulum clock whose bob would swing in a cycloid, a curve we define in Example 3. He hung the bob from a fine wire constrained by guards that caused it to draw up as it swung away from center (Figure 8.29).

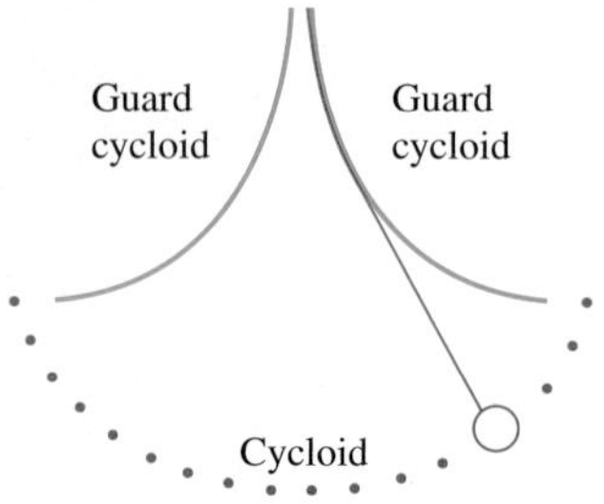

FIGURE 8.29 In Huygens' pendulum clock, the bob swings in a cycloid, so the frequency is independent of the amplitude.

EXAMPLE 3 A wheel of radius a rolls along a horizontal straight line. Find parametric equations for the path traced by a point P on the wheel's circumference. The path is called a **cycloid**.

Solution We take the line to be the x-axis, mark a point P on the wheel, start the wheel with P at the origin, and roll the wheel to the right. As parameter, we use the angle t through which the wheel turns, measured in radians. Figure 8.30 shows the wheel a short while later, when its base lies at units from the origin. The wheel's center C lies at (at, a) and the coordinates of P are

$$x = at + a\cos\theta, \qquad y = a + a\sin\theta.$$

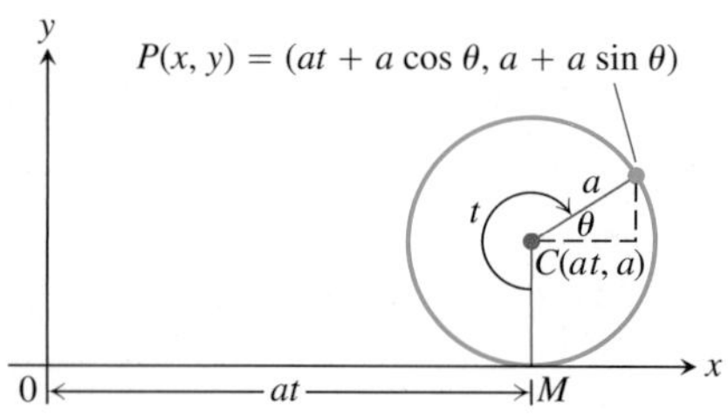

FIGURE 8.30 The position of $P(x, y)$ on the rolling wheel at angle t (Example 3).

To express θ in terms of t, we observe that $t + \theta = 3\pi/2$ in the figure, so that

$$\theta = \frac{3\pi}{2} - t.$$

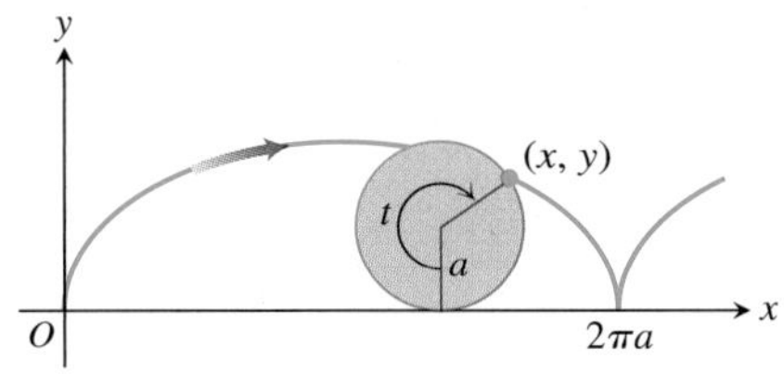

FIGURE 8.31 The cycloid $x = a(t - \sin t), y = a(1 - \cos t)$, for $t \geq 0$.

This makes

$$\cos\theta = \cos\left(\frac{3\pi}{2} - t\right) = -\sin t, \qquad \sin\theta = \sin\left(\frac{3\pi}{2} - t\right) = -\cos t.$$

The equations we seek are

$$x = at - a\sin t, \qquad y = a - a\cos t.$$

These are usually written with the a factored out:

$$x = a(t - \sin t), \qquad y = a(1 - \cos t). \tag{1}$$

Figure 8.31 shows the first arch of the cycloid and part of the next. ■

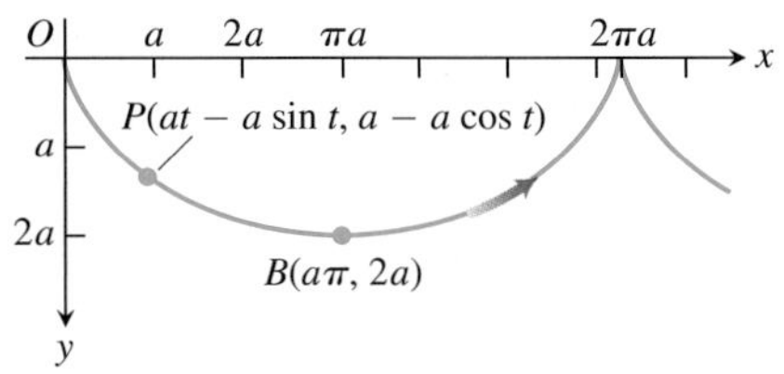

FIGURE 8.32 To study motion along an upside-down cycloid under the influence of gravity, we turn Figure 8.31 upside down.

Brachistochrones and Tautochrones

If we turn Figure 8.31 upside down, Equations (1) still apply and the resulting curve (Figure 8.32) has two interesting physical properties. The first relates to the origin O and the point B at the bottom of the first arch. Among all smooth curves joining these points, the cycloid is the curve along which a frictionless bead, subject only to the force of gravity, will slide from O to B the fastest. This makes the cycloid a **brachistochrone** ("brah-*kiss*-toe-krone"), or shortest time curve for these points. The second property is that even if you start the bead partway down the curve toward B, it will still take the bead the same amount of time to reach B. This makes the cycloid a **tautochrone** ("*taw*-toe-krone"), or same-time curve for O and B. It can be shown that the cycloid from O to B is the one and only brachistochrone for O and B. We omit the argument here.

EXERCISE 8.5

Exercises 1–12 give parametric equations and parameter intervals for the motion of a particle in the xy-plane. Identify the particle's path by finding a Cartesian equation for it. Graph the Cartesian equation. (The graphs will vary with the equation used.) Indicate the portion of the graph traced by the particle and the direction of motion.

1. $x = \cos t, \quad y = \sin t, \quad 0 \leq t \leq \pi$
2. $x = \sin(2\pi(1 - t)), \quad y = \cos(2\pi(1 - t)); \quad 0 \leq t \leq 1$
3. $x = 4\cos t, \quad y = 5\sin t; \quad 0 \leq t \leq \pi$
4. $x = 4\sin t, \quad y = 5\cos t; \quad 0 \leq t \leq 2\pi$
5. $x = t, \quad y = \sqrt{t}; \quad t \geq 0$
6. $x = \sec^2 t - 1, \quad y = \tan t; \quad -\pi/2 < t < \pi/2$
7. $x = -\sec t, \quad y = \tan t; \quad -\pi/2 < t < \pi/2$
8. $x = \csc t, \quad y = \cot t; \quad 0 < t < \pi$
9. $x = t, \quad y = \sqrt{4 - t^2}; \quad 0 \leq t \leq 2$
10. $x = t^2, \quad y = \sqrt{t^4 + 1}; \quad t \geq 0$
11. $x = -\cosh t, \quad y = \sinh t; \quad -\infty < t < \infty$
12. $x = 2\sinh t, \quad y = 2\cosh t; \quad -\infty < t < \infty$
13. **Hypocycloids** When a circle rolls on the inside of a fixed circle, any point P on the circumference of the rolling circle describes a *hypocycloid*. Let the fixed circle be $x^2 + y^2 = a^2$, let the radius of the rolling circle be b, and let the initial position of the tracing point P be $A(a, 0)$. Find parametric equations for the hypocycloid, using as the parameter the angle θ from the positive x-axis to the line joining the circles' centers. In particular, if

$b = a/4$, as in the accompanying figure, show that the hypocycloid is the astroid

$$x = a\cos^3\theta, \quad y = a\sin^3\theta.$$

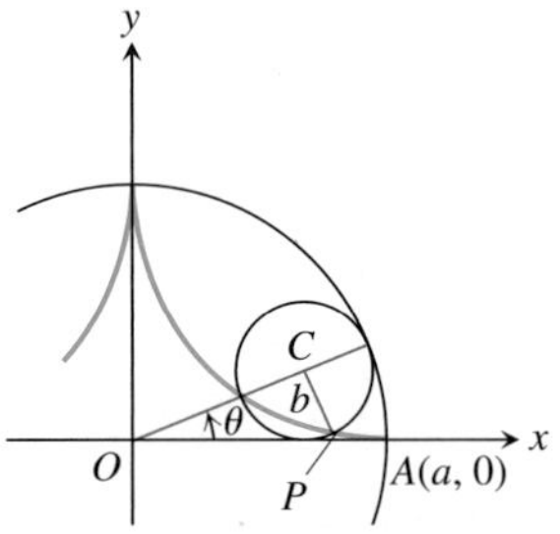

14. More about hypocycloids The accompanying figure shows a circle of radius a tangent to the inside of a circle of radius $2a$. The point P, shown as the point of tangency in the figure, is attached to the smaller circle. What path does P trace as the smaller circle rolls around the inside of the larger circle?

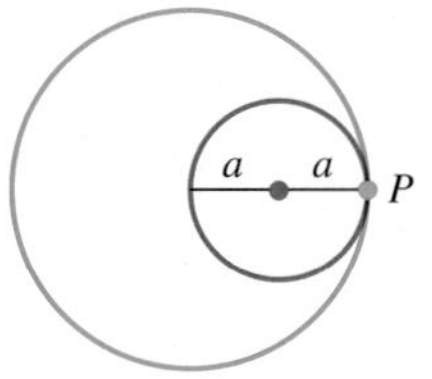

15. As the point N moves along the line $y = a$ in the accompanying figure, P moves in such a way that $OP = MN$. Find parametric equations for the coordinates of P as functions of the angle t that the line ON makes with the positive y-axis.

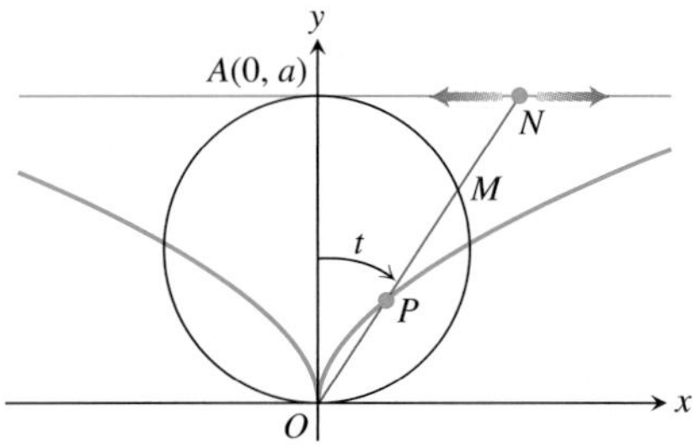

16. Trochoids A wheel of radius a rolls along a horizontal straight line without slipping. Find parametric equations for the curve traced out by a point P on a spoke of the wheel b units from its center. As parameter, use the angle θ through which the wheel turns. The curve is called a *trochoid*, which is a cycloid when $b = a$.

17. Find the point on the parabola $x = t, y = t^2, -\infty < t < \infty$, closest to the point $(2, 1/2)$. (*Hint:* Minimize the square of the distance as a function of t.)

18. Find the point on the ellipse $x = 2\cos t, y = \sin t, 0 \le t \le 2\pi$ closest to the point $(3/4, 0)$. (*Hint:* Minimize the square of the distance as a function of t.)

T If you have a parametric equation grapher, graph the equations in Exercises 19–24 over the given intervals.

19. A nice curve (a deltoid)

$$x = 2\cos t + \cos 2t, \quad y = 2\sin t - \sin 2t; \quad 0 \le t \le 2\pi$$

What happens if you replace 2 with -2 in the equations for x and y? Graph the new equations and find out.

20. An even nicer curve

$$x = 3\cos t + \cos 3t, \quad y = 3\sin t - \sin 3t; \quad 0 \le t \le 2\pi$$

What happens if you replace 3 with -3 in the equations for x and y? Graph the new equations and find out.

21. Epicycloid

$$x = 9\cos t - \cos 9t, \quad y = 9\sin t - \sin 9t; \quad 0 \le t \le 2\pi$$

22. Hypocycloid

$$x = 8\cos t + 2\cos 4t, \quad y = 8\sin t - 2\sin 4t; \quad 0 \le t \le 2\pi$$

23. Hypotrochoid

$$x = \cos t + 5\cos 3t, \quad y = 6\cos t - 5\sin 3t; \quad 0 \le t \le 2\pi$$

24. $x = 6\cos t + 5\cos 3t, \quad y = 6\sin t - 5\sin 3t; \quad 0 \le t \le 2\pi$

Chapter 8 Questions to Guide Your Review

1. What are polar coordinates? What equations relate polar coordinates to Cartesian coordinates? Why might you want to change from one coordinate system to the other?
2. What consequence does the lack of uniqueness of polar coordinates have for graphing? Give an example.
3. How do you graph equations in polar coordinates? Include in your discussion symmetry, slope, behavior at the origin, and the use of Cartesian graphs. Give examples.
4. How do you find the area of a region $0 \le r_1(\theta) \le r \le r_2(\theta)$, $\alpha \le \theta \le \beta$, in the polar coordinate plane? Give examples.
5. Under what conditions can you find the length of a curve $r = f(\theta)$, $\alpha \le \theta \le \beta$, in the polar coordinate plane? Give an example of a typical calculation.
6. What is the eccentricity of a conic section? How can you classify conic sections by eccentricity? How are an ellipse's shape and eccentricity related?

7. Explain the equation $PF = e \cdot PD$.

8. What are the standard equations for lines and conic sections in polar coordinates? Give examples.

9. What are some typical parametrizations for conic sections?

10. What is a cycloid? What are typical parametric equations for cycloids? What physical properties account for the importance of cycloids?

Chapter 8 Practice and Additional Exercises

Sketch the lines in Exercises 1–6. Also, find a Cartesian equation for each line.

1. $r \cos\left(\theta + \dfrac{\pi}{3}\right) = 2\sqrt{3}$ **2.** $r \cos\left(\theta - \dfrac{3\pi}{4}\right) = \dfrac{\sqrt{2}}{2}$

3. $r = 2 \sec \theta$ **4.** $r = -\sqrt{2} \sec \theta$

5. $r = -(3/2) \csc \theta$ **6.** $r = \left(3\sqrt{3}\right) \csc \theta$

Find Cartesian equations for the circles in Exercises 7–10. Sketch each circle in the coordinate plane and label it with both its Cartesian and polar equations.

7. $r = -4 \sin \theta$ **8.** $r = 3\sqrt{3} \sin \theta$

9. $r = 2\sqrt{2} \cos \theta$ **10.** $r = -6 \cos \theta$

Find polar equations for the circles in Exercises 11–14. Sketch each circle in the coordinate plane and label it with both its Cartesian and polar equations.

11. $x^2 + y^2 + 5y = 0$ **12.** $x^2 + y^2 - 2y = 0$

13. $x^2 + y^2 - 3x = 0$ **14.** $x^2 + y^2 + 4x = 0$

Sketch the regions defined by the polar coordinate inequalities in Exercises 15 and 16.

15. $0 \le r \le 6 \cos \theta$ **16.** $-4 \sin \theta \le r \le 0$

Match each graph in Exercises 17–24 with the appropriate equation (a)–(l). There are more equations than graphs, so some equations will not be matched.

a. $r = \cos 2\theta$ **b.** $r \cos \theta = 1$

c. $r = \dfrac{6}{1 - 2\cos\theta}$ **d.** $r = \sin 2\theta$

e. $r = \theta$ **f.** $r^2 = \cos 2\theta$

g. $r = 1 + \cos \theta$ **h.** $r = 1 - \sin \theta$

i. $r = \dfrac{2}{1 - \cos\theta}$ **j.** $r^2 = \sin 2\theta$

k. $r = -\sin \theta$ **l.** $r = 2\cos\theta + 1$

17. Four-leaved rose

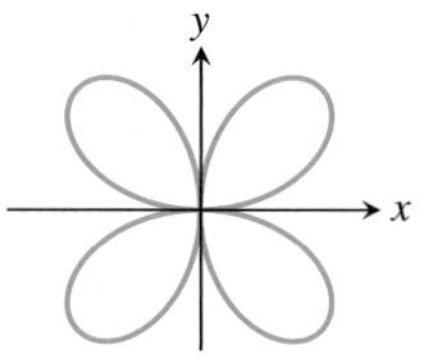

18. Spiral

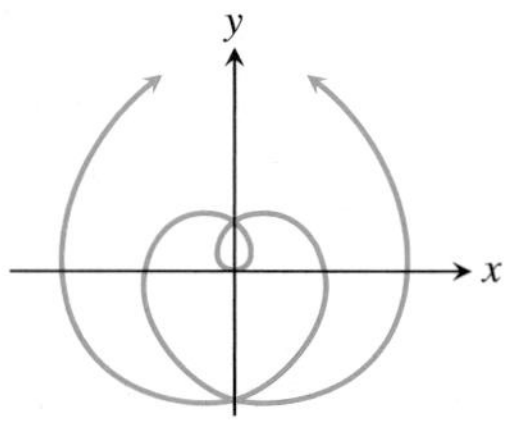

19. Limaçon

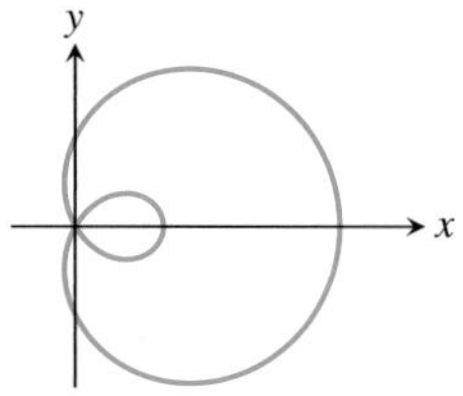

20. Lemniscate

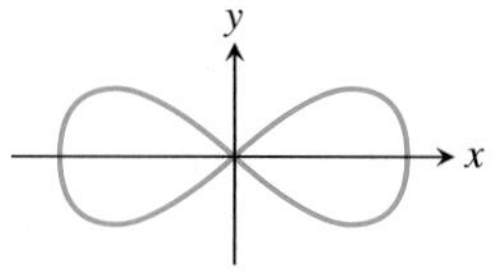

21. Circle

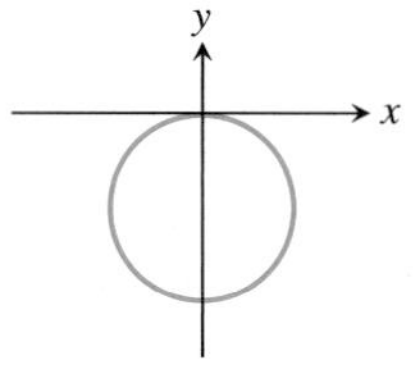

22. Cardioid

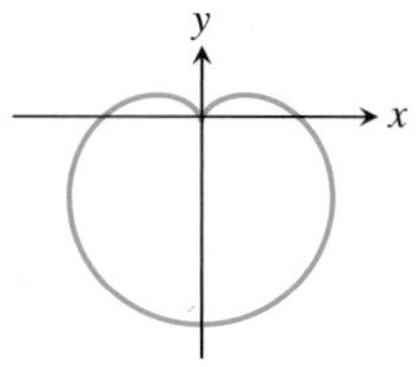

23. Parabola

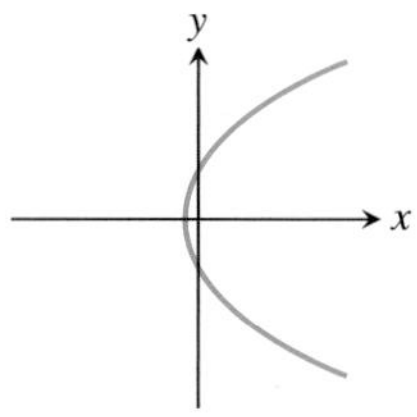

24. Lemniscate

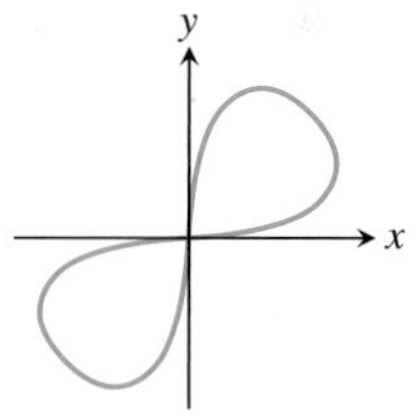

Find the areas of the regions in the polar coordinate plane described in Exercises 25–28.

25. Enclosed by the limaçon $r = 2 - \cos\theta$

26. Enclosed by one leaf of the three-leaved rose $r = \sin 3\theta$

27. Inside the "figure eight" $r = 1 + \cos 2\theta$ and outside the circle $r = 1$

28. Inside the cardioid $r = 2(1 + \sin\theta)$ and outside the circle $r = 2\sin\theta$

Find the lengths of the curves given by the polar coordinate equations in Exercises 29–32.

29. $r = -1 + \cos\theta$

30. $r = 2\sin\theta + 2\cos\theta, \quad 0 \le \theta \le \pi/2$

31. $r = 8 \sin^3 (\theta/3), \quad 0 \le \theta \le \pi/4$

32. $r = \sqrt{1 + \cos 2\theta}, \quad -\pi/2 \le \theta \le \pi/2$

Sketch the conic sections whose polar coordinate equations are given in Exercises 33–36. Give polar coordinates for the vertices and, in the case of ellipses, for the centers as well.

33. $r = \dfrac{2}{1 + \cos \theta}$

34. $r = \dfrac{8}{2 + \cos \theta}$

35. $r = \dfrac{6}{1 - 2 \cos \theta}$

36. $r = \dfrac{12}{3 + \sin \theta}$

Exercises 37–40 give the eccentricities of conic sections with one focus at the origin of the polar coordinate plane, along with the directrix for that focus. Find a polar equation for each conic section.

37. $e = 2, \quad r \cos \theta = 2$

38. $e = 1, \quad r \cos \theta = -4$

39. $e = 1/2, \quad r \sin \theta = 2$

40. $e = 1/3, \quad r \sin \theta = -6$

41. Find the volume of the solid generated by revolving the region enclosed by the ellipse $9x^2 + 4y^2 = 36$ about **(a)** the x-axis, **(b)** the y-axis.

42. The "triangular" region in the first quadrant bounded by the x-axis, the line $x = 4$, and the hyperbola $9x^2 - 4y^2 = 36$ is revolved about the x-axis to generate a solid. Find the volume of the solid.

43. Show that the equations $x = r \cos \theta, y = r \sin \theta$ transform the polar equation

$$r = \frac{k}{1 + e \cos \theta}$$

into the Cartesian equation

$$(1 - e^2)x^2 + y^2 + 2kex - k^2 = 0.$$

44. **Archimedes spirals** The graph of an equation of the form $r = a\theta$, where a is a nonzero constant, is called an *Archimedes spiral*. Is there anything special about the widths between the successive turns of such a spiral?

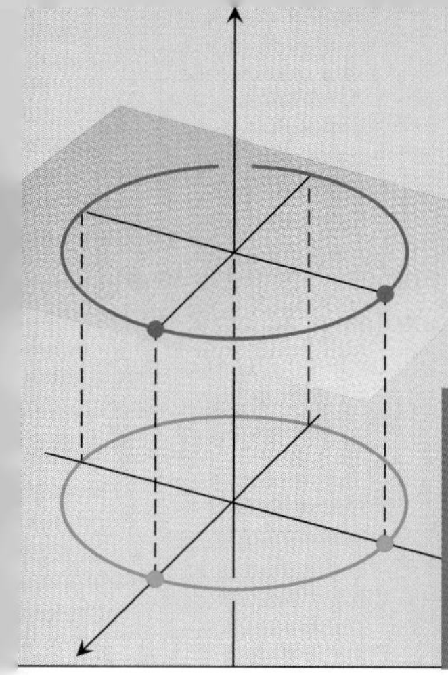

Chapter 9

Vectors and the Geometry of Space

OVERVIEW To apply calculus in many real-world situations and in higher mathematics, we need a mathematical description of three-dimensional space. In this chapter we introduce three-dimensional coordinate systems and vectors. Building on what we already know about coordinates in the xy-plane, we establish coordinates in space by adding a third axis that measures distance above and below the xy-plane. Vectors are used to study the analytic geometry of space, where they give simple ways to describe lines, planes, surfaces, and curves in space. We use these geometric ideas later in the book to study motion in space and the calculus of functions of several variables, with their many important applications in science, engineering, economics, and higher mathematics.

9.1 Three-Dimensional Coordinate Systems

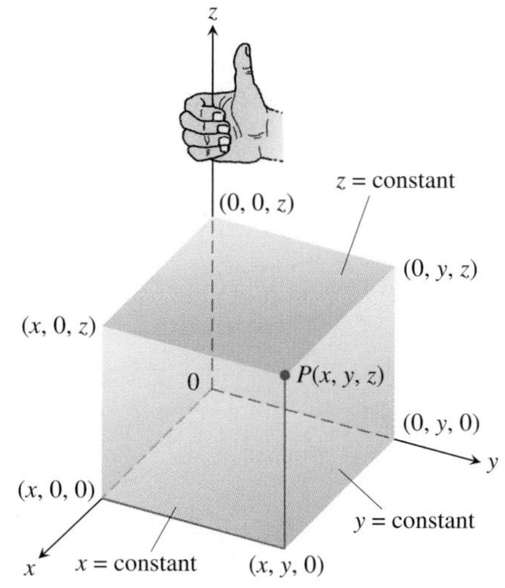

FIGURE 9.1 The Cartesian coordinate system is right-handed.

To locate a point in space, we use three mutually perpendicular coordinate axes, arranged as in Figure 9.1. The axes shown there make a *right-handed* coordinate frame. When you hold your right hand so that the fingers curl from the positive x-axis toward the positive y-axis, your thumb points along the positive z-axis. So when you look down on the xy-plane from the positive direction of the z-axis, positive angles in the plane are measured counterclockwise from the positive x-axis and around the positive z-axis. (In a *left-handed* coordinate frame, the z-axis would point downward in Figure 9.1 and angles in the plane would be positive when measured clockwise from the positive x-axis. Right-handed and left-handed coordinate frames are not equivalent.)

The Cartesian coordinates (x, y, z) of a point P in space are the values at which the planes through P perpendicular to the axes cut the axes. Cartesian coordinates for space are also called **rectangular coordinates** because the axes that define them meet at right angles. Points on the x-axis have y- and z-coordinates equal to zero. That is, they have coordinates of the form $(x, 0, 0)$. Similarly, points on the y-axis have coordinates of the form $(0, y, 0)$, and points on the z-axis have coordinates of the form $(0, 0, z)$.

The planes determined by the coordinates axes are the ***xy*-plane**, whose standard equation is $z = 0$; the ***yz*-plane**, whose standard equation is $x = 0$; and the ***xz*-plane**, whose standard equation is $y = 0$. They meet at the **origin** $(0, 0, 0)$ (Figure 9.2). The origin is also identified by simply 0 or sometimes the letter O.

The three **coordinate planes** $x = 0, y = 0$, and $z = 0$ divide space into eight cells called **octants**. The octant in which the point coordinates are all positive is called the **first octant**; there is no convention for numbering the other seven octants.

The points in a plane perpendicular to the x-axis all have the same x-coordinate, this being the number at which that plane cuts the x-axis. The y- and z-coordinates can be any numbers.

Similarly, the points in a plane perpendicular to the y-axis have a common y-coordinate and the points in a plane perpendicular to the z-axis have a common z-coordinate. To write equations for these planes, we name the common coordinate's value. The plane $x = 2$ is the plane perpendicular to the x-axis at $x = 2$. The plane $y = 3$ is the plane perpendicular to the y-axis at $y = 3$. The plane $z = 5$ is the plane perpendicular to the z-axis at $z = 5$. Figure 9.3 shows the planes $x = 2$, $y = 3$, and $z = 5$, together with their intersection point (2, 3, 5).

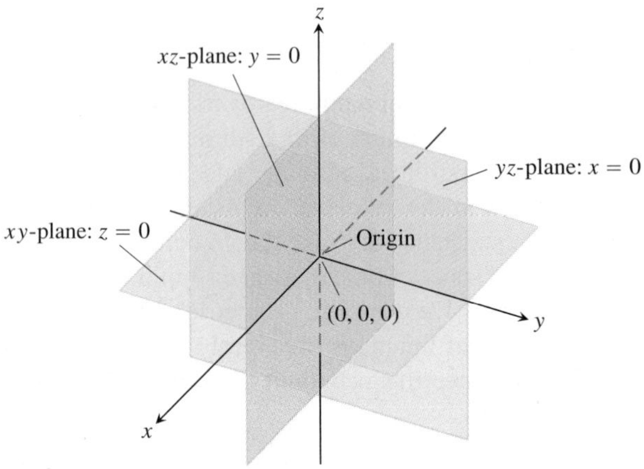

FIGURE 9.2 The planes $x = 0$, $y = 0$, and $z = 0$ divide space into eight octants.

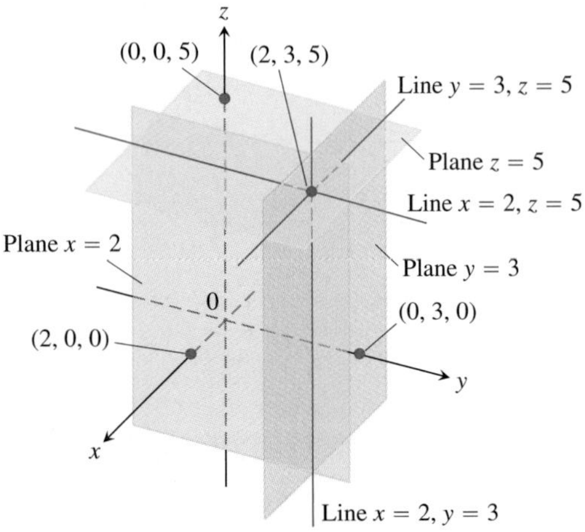

FIGURE 9.3 The planes $x = 2$, $y = 3$, and $z = 5$ determine three lines through the point (2, 3, 5).

The planes $x = 2$ and $y = 3$ in Figure 9.3 intersect in a line parallel to the z-axis. This line is described by the *pair* of equations $x = 2$, $y = 3$. A point (x, y, z) lies on the line if and only if $x = 2$ and $y = 3$. Similarly, the line of intersection of the planes $y = 3$ and $z = 5$ is described by the equation pair $y = 3$, $z = 5$. This line runs parallel to the x-axis. The line of intersection of the planes $x = 2$ and $z = 5$, parallel to the y-axis, is described by the equation pair $x = 2$, $z = 5$.

In the following examples, we match coordinate equations and inequalities with the sets of points they define in space.

EXAMPLE 1 We interpret these equations and inequalities geometrically.

(a) $z \geq 0$ — The half-space consisting of the points on and above the xy-plane.

(b) $x = -3$ — The plane perpendicular to the x-axis at $x = -3$. This plane lies parallel to the yz-plane and 3 units behind it.

(c) $z = 0, x \leq 0, y \geq 0$ — The second quadrant of the xy-plane.

(d) $x \geq 0, y \geq 0, z \geq 0$ — The first octant.

(e) $-1 \leq y \leq 1$ — The slab between the planes $y = -1$ and $y = 1$ (planes included).

(f) $y = -2, z = 2$ — The line in which the planes $y = -2$ and $z = 2$ intersect. Alternatively, the line through the point $(0, -2, 2)$ parallel to the x-axis. ■

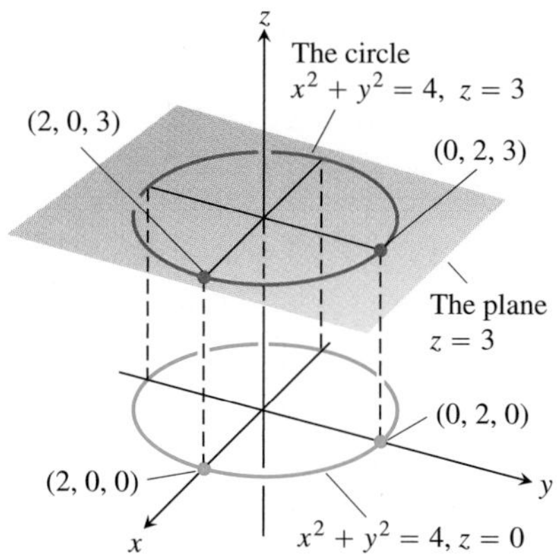

FIGURE 9.4 The circle $x^2 + y^2 = 4$ in the plane $z = 3$ (Example 2).

EXAMPLE 2 What points $P(x, y, z)$ satisfy the equations

$$x^2 + y^2 = 4 \qquad \text{and} \qquad z = 3?$$

Solution The points lie in the horizontal plane $z = 3$ and, in this plane, make up the circle $x^2 + y^2 = 4$. We call this set of points "the circle $x^2 + y^2 = 4$ in the plane $z = 3$" or, more simply, "the circle $x^2 + y^2 = 4, z = 3$" (Figure 9.4). ■

Distance and Spheres in Space

The formula for the distance between two points in the xy-plane extends to points in space.

The Distance Between $P_1(x_1, y_1, z_1)$ and $P_2(x_2, y_2, z_2)$ is

$$|P_1P_2| = \sqrt{(x_2 - x_1)^2 + (y_2 - y_1)^2 + (z_2 - z_1)^2}$$

Proof We construct a rectangular box with faces parallel to the coordinate planes and the points P_1 and P_2 at opposite corners of the box (Figure 9.5). If $A(x_2, y_1, z_1)$ and $B(x_2, y_2, z_1)$ are the vertices of the box indicated in the figure, then the three box edges P_1A, AB, and BP_2 have lengths

$$|P_1A| = |x_2 - x_1|, \qquad |AB| = |y_2 - y_1|, \qquad |BP_2| = |z_2 - z_1|.$$

Because triangles P_1BP_2 and P_1AB are both right-angled, two applications of the Pythagorean theorem give

$$|P_1P_2|^2 = |P_1B|^2 + |BP_2|^2 \qquad \text{and} \qquad |P_1B|^2 = |P_1A|^2 + |AB|^2$$

(see Figure 9.5). So

$$\begin{aligned} |P_1P_2|^2 &= |P_1B|^2 + |BP_2|^2 \\ &= |P_1A|^2 + |AB|^2 + |BP_2|^2 \qquad \text{Substitute } |P_1B|^2 = |P_1A|^2 + |AB|^2. \\ &= |x_2 - x_1|^2 + |y_2 - y_1|^2 + |z_2 - z_1|^2 \\ &= (x_2 - x_1)^2 + (y_2 - y_1)^2 + (z_2 - z_1)^2 \end{aligned}$$

Therefore

$$|P_1P_2| = \sqrt{(x_2 - x_1)^2 + (y_2 - y_1)^2 + (z_2 - z_1)^2}$$ ■

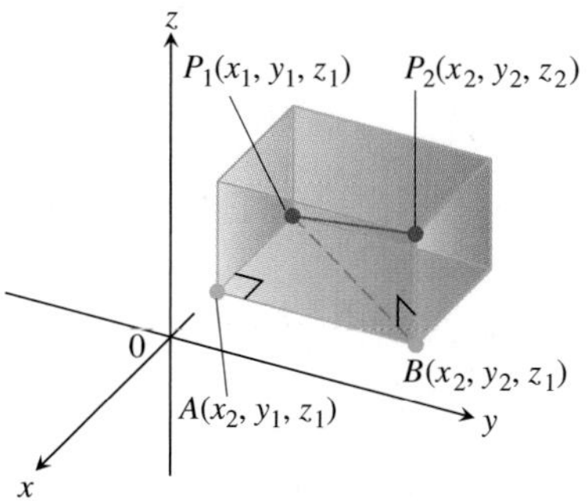

FIGURE 9.5 We find the distance between P_1 and P_2 by applying the Pythagorean theorem to the right triangles P_1AB and P_1BP_2.

EXAMPLE 3 The distance between $P_1(2, 1, 5)$ and $P_2(-2, 3, 0)$ is

$$\begin{aligned} |P_1P_2| &= \sqrt{(-2 - 2)^2 + (3 - 1)^2 + (0 - 5)^2} \\ &= \sqrt{16 + 4 + 25} \\ &= \sqrt{45} \approx 6.708. \end{aligned}$$ ■

We can use the distance formula to write equations for spheres in space (Figure 9.6). A point $P(x, y, z)$ lies on the sphere of radius a centered at $P_0(x_0, y_0, z_0)$ precisely when $|P_0P| = a$ or

$$(x - x_0)^2 + (y - y_0)^2 + (z - z_0)^2 = a^2.$$

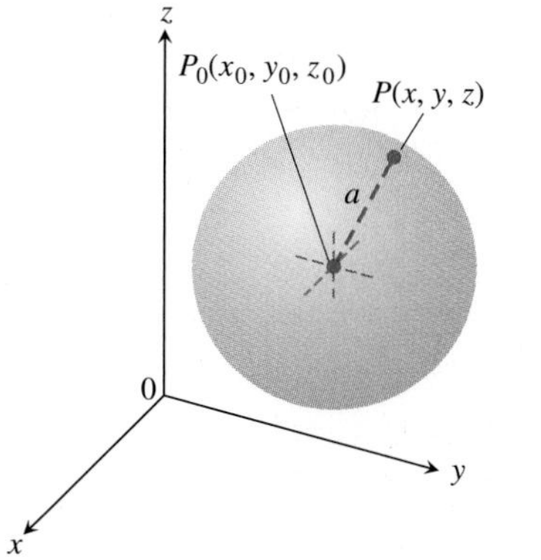

FIGURE 9.6 The sphere of radius a centered at the point (x_0, y_0, z_0).

The Standard Equation for the Sphere of Radius a and Center (x_0, y_0, z_0)

$$(x - x_0)^2 + (y - y_0)^2 + (z - z_0)^2 = a^2$$

EXAMPLE 4 Find the center and radius of the sphere

$$x^2 + y^2 + z^2 + 3x - 4z + 1 = 0.$$

Solution We find the center and radius of a sphere the way we find the center and radius of a circle: Complete the squares on the x-, y-, and z-terms as necessary and write each quadratic as a squared linear expression. Then, from the equation in standard form, read off the center and radius. For the sphere here, we have

$$x^2 + y^2 + z^2 + 3x - 4z + 1 = 0$$
$$(x^2 + 3x) + y^2 + (z^2 - 4z) = -1$$
$$\left(x^2 + 3x + \left(\frac{3}{2}\right)^2\right) + y^2 + \left(z^2 - 4z + \left(\frac{-4}{2}\right)^2\right) = -1 + \left(\frac{3}{2}\right)^2 + \left(\frac{-4}{2}\right)^2$$
$$\left(x + \frac{3}{2}\right)^2 + y^2 + (z - 2)^2 = -1 + \frac{9}{4} + 4 = \frac{21}{4}.$$

From this standard form, we read that $x_0 = -3/2$, $y_0 = 0$, $z_0 = 2$, and $a = \sqrt{21}/2$. The center is $(-3/2, 0, 2)$. The radius is $\sqrt{21}/2$. ■

EXAMPLE 5 Here are some geometric interpretations.

(a) $x^2 + y^2 + z^2 < 4$ — The interior of the sphere $x^2 + y^2 + z^2 = 4$.

(b) $x^2 + y^2 + z^2 \le 4$ — The solid ball bounded by the sphere $x^2 + y^2 + z^2 = 4$. Alternatively, the sphere $x^2 + y^2 + z^2 = 4$ together with its interior.

(c) $x^2 + y^2 + z^2 > 4$ — The exterior of the sphere $x^2 + y^2 + z^2 = 4$.

(d) $x^2 + y^2 + z^2 = 4, z \le 0$ — The lower hemisphere cut from the sphere $x^2 + y^2 + z^2 = 4$ by the xy-plane (the plane $z = 0$). ■

Just as polar coordinates give another way to locate points in the xy-plane (Section 8.1), alternative coordinate systems, different from the Cartesian coordinate system developed here, exist for three-dimensional space. We examine two of these coordinate systems in Section 12.7.

EXERCISES 9.1

In Exercises 1–12, give a geometric description of the set of points in space whose coordinates satisfy the given pairs of equations.

1. $x = 2, \quad y = 3$
2. $x = -1, \quad z = 0$
3. $y = 0, \quad z = 0$
4. $x = 1, \quad y = 0$
5. $x^2 + y^2 = 4, \quad z = 0$
6. $x^2 + y^2 = 4, \quad z = -2$
7. $x^2 + z^2 = 4, \quad y = 0$
8. $y^2 + z^2 = 1, \quad x = 0$
9. $x^2 + y^2 + z^2 = 1, \quad x = 0$
10. $x^2 + y^2 + z^2 = 25, \quad y = -4$
11. $x^2 + y^2 + (z + 3)^2 = 25, \quad z = 0$
12. $x^2 + (y - 1)^2 + z^2 = 4, \quad y = 0$

In Exercises 13–18, describe the sets of points in space whose coordinates satisfy the given inequalities or combinations of equations and inequalities.

13. **a.** $x \geq 0, \quad y \geq 0, \quad z = 0$ **b.** $x \geq 0, \quad y \leq 0, \quad z = 0$

14. **a.** $0 \leq x \leq 1$ **b.** $0 \leq x \leq 1, \quad 0 \leq y \leq 1$

c. $0 \leq x \leq 1, \quad 0 \leq y \leq 1, \quad 0 \leq z \leq 1$

15. **a.** $x^2 + y^2 + z^2 \leq 1$ **b.** $x^2 + y^2 + z^2 > 1$

16. **a.** $x^2 + y^2 \leq 1, \quad z = 0$ **b.** $x^2 + y^2 \leq 1, \quad z = 3$

c. $x^2 + y^2 \leq 1$, no restriction on z

17. **a.** $x^2 + y^2 + z^2 = 1, \quad z \geq 0$

b. $x^2 + y^2 + z^2 \leq 1, \quad z \geq 0$

18. **a.** $x = y, \quad z = 0$ **b.** $x = y$, no restriction on z

In Exercises 19–28, describe the given set with a single equation or with a pair of equations.

19. The plane perpendicular to the

a. x-axis at $(3, 0, 0)$ **b.** y-axis at $(0, -1, 0)$

c. z-axis at $(0, 0, -2)$

20. The plane through the point $(3, -1, 2)$ perpendicular to the

a. x-axis **b.** y-axis **c.** z-axis

21. The plane through the point $(3, -1, 1)$ parallel to the

a. xy-plane **b.** yz-plane **c.** xz-plane

22. The circle of radius 2 centered at $(0, 0, 0)$ and lying in the

a. xy-plane **b.** yz-plane **c.** xz-plane

23. The circle of radius 2 centered at $(0, 2, 0)$ and lying in the

a. xy-plane **b.** yz-plane **c.** plane $y = 2$

24. The circle of radius 1 centered at $(-3, 4, 1)$ and lying in a plane parallel to the

a. xy-plane **b.** yz-plane **c.** xz-plane

25. The line through the point $(1, 3, -1)$ parallel to the

a. x-axis **b.** y-axis **c.** z-axis

26. The set of points in space equidistant from the origin and the point $(0, 2, 0)$

27. The circle in which the plane through the point $(1, 1, 3)$ perpendicular to the z-axis meets the sphere of radius 5 centered at the origin

28. The set of points in space that lie 2 units from the point $(0, 0, 1)$ and, at the same time, 2 units from the point $(0, 0, -1)$

Write inequalities to describe the sets in Exercises 29–34.

29. The slab bounded by the planes $z = 0$ and $z = 1$ (planes included)

30. The solid cube in the first octant bounded by the coordinate planes and the planes $x = 2$, $y = 2$, and $z = 2$

31. The half-space consisting of the points on and below the xy-plane

32. The upper hemisphere of the sphere of radius 1 centered at the origin

33. The **(a)** interior and **(b)** exterior of the sphere of radius 1 centered at the point $(1, 1, 1)$

34. The closed region bounded by the spheres of radius 1 and radius 2 centered at the origin. (*Closed* means the spheres are to be included. Had we wanted the spheres left out, we would have asked for the *open* region bounded by the spheres. This is analogous to the way we use *closed* and *open* to describe intervals: *closed* means endpoints included, *open* means endpoints left out. Closed sets include boundaries; open sets leave them out.)

In Exercises 35–40, find the distance between points P_1 and P_2.

35. $P_1(1, 1, 1), \quad P_2(3, 3, 0)$

36. $P_1(-1, 1, 5), \quad P_2(2, 5, 0)$

37. $P_1(1, 4, 5), \quad P_2(4, -2, 7)$

38. $P_1(3, 4, 5), \quad P_2(2, 3, 4)$

39. $P_1(0, 0, 0), \quad P_2(2, -2, -2)$

40. $P_1(5, 3, -2), \quad P_2(0, 0, 0)$

Find the centers and radii of the spheres in Exercises 41–44.

41. $(x + 2)^2 + y^2 + (z - 2)^2 = 8$

42. $\left(x + \frac{1}{2}\right)^2 + \left(y + \frac{1}{2}\right)^2 + \left(z + \frac{1}{2}\right)^2 = \frac{21}{4}$

43. $\left(x - \sqrt{2}\right)^2 + \left(y - \sqrt{2}\right)^2 + \left(z + \sqrt{2}\right)^2 = 2$

44. $x^2 + \left(y + \frac{1}{3}\right)^2 + \left(z - \frac{1}{3}\right)^2 = \frac{29}{9}$

Find equations for the spheres whose centers and radii are given in Exercises 45–48.

	Center	**Radius**
45.	$(1, 2, 3)$	$\sqrt{14}$
46.	$(0, -1, 5)$	2
47.	$(-2, 0, 0)$	$\sqrt{3}$
48.	$(0, -7, 0)$	7

Find the centers and radii of the spheres in Exercises 49–52.

49. $x^2 + y^2 + z^2 + 4x - 4z = 0$

50. $x^2 + y^2 + z^2 - 6y + 8z = 0$

51. $2x^2 + 2y^2 + 2z^2 + x + y + z = 9$

52. $3x^2 + 3y^2 + 3z^2 + 2y - 2z = 9$

53. Find a formula for the distance from the point $P(x, y, z)$ to the

a. x-axis **b.** y-axis **c.** z-axis

54. Find a formula for the distance from the point $P(x, y, z)$ to the

a. xy-plane **b.** yz-plane **c.** xz-plane

55. Find the perimeter of the triangle with vertices $A(-1, 2, 1)$, $B(1, -1, 3)$, and $C(3, 4, 5)$.

56. Show that the point $P(3, 1, 2)$ is equidistant from the points $A(2, -1, 3)$ and $B(4, 3, 1)$.

9.2 Vectors

Some of the things we measure are determined simply by their magnitudes. To record mass, length, or time, for example, we need only write down a number and name an appropriate unit of measure. We need more information to describe a force, displacement, or velocity. To describe a force, we need to record the direction in which it acts as well as how large it is. To describe a body's displacement, we have to say in what direction it moved as well as how far. To describe a body's velocity, we have to know where the body is headed as well as how fast it is going. In this section we show how to represent things that have both magnitude and direction in the plane or in space.

Component Form

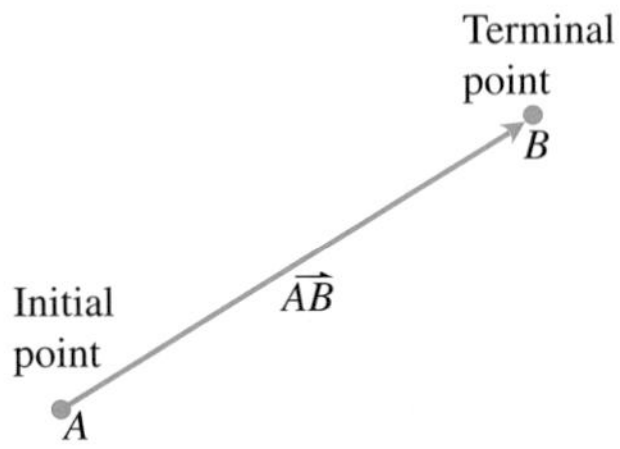

FIGURE 9.7 The directed line segment $\overrightarrow{AB}$.

A quantity such as force, displacement, or velocity is called a *vector* and is represented by a **directed line segment** (Figure 9.7). The arrow points in the direction of the action and its length gives the magnitude of the action in terms of a suitably chosen unit. For example, a force vector points in the direction in which the force acts; its length is a measure of the force's strength; a velocity vector points in the direction of motion and its length is the speed of the moving object. Figure 9.8 displays the velocity vector **v** at a specific location for a particle moving along a path in the plane or in space. (This application of vectors is studied in Chapter 10.)

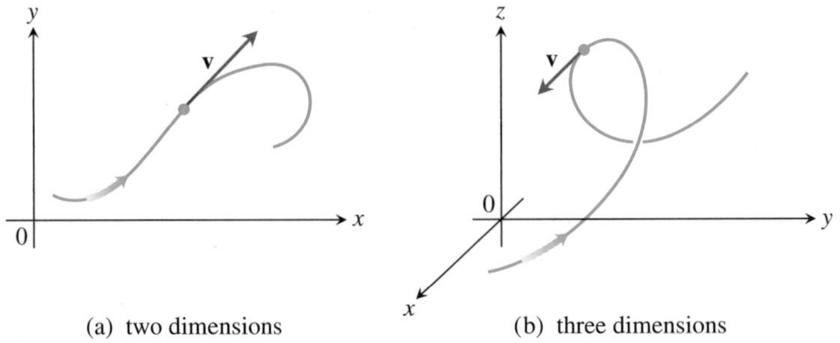

FIGURE 9.8 The velocity vector of a particle moving along a path (a) in the plane (b) in space. The arrowhead on the path indicates the direction of motion of the particle.

> **DEFINITIONS** A **vector** is a directed line segment. The directed line segment $\overrightarrow{AB}$ has **initial point** A and **terminal point** B; its **length** is denoted by $|\overrightarrow{AB}|$. Two vectors are **equal** if they have the same length and direction.

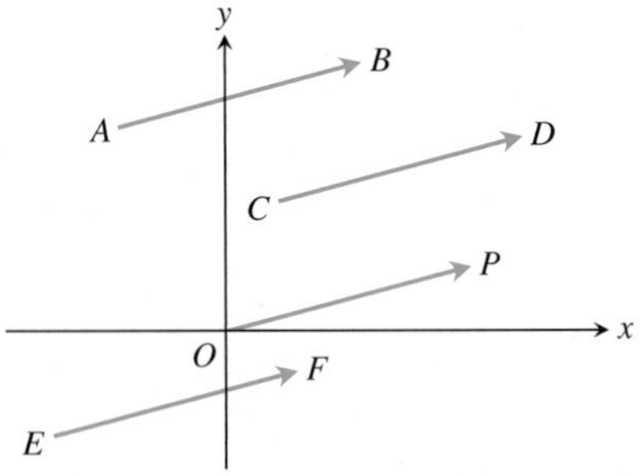

FIGURE 9.9 The four arrows in the plane (directed line segments) shown here have the same length and direction. They therefore represent the same vector, and we write $\overrightarrow{AB} = \overrightarrow{CD} = \overrightarrow{OP} = \overrightarrow{EF}$.

The arrows we use when we draw vectors are understood to represent the same vector if they have the same length, are parallel, and point in the same direction (Figure 9.9) regardless of the initial point.

In textbooks, vectors are usually written in lowercase, boldface letters, for example **u**, **v**, and **w**. Sometimes we use uppercase boldface letters, such as **F**, to denote a force vector. In handwritten form, it is customary to draw small arrows above the letters, for example $\vec{u}$, $\vec{v}$, $\vec{w}$, and $\vec{F}$.

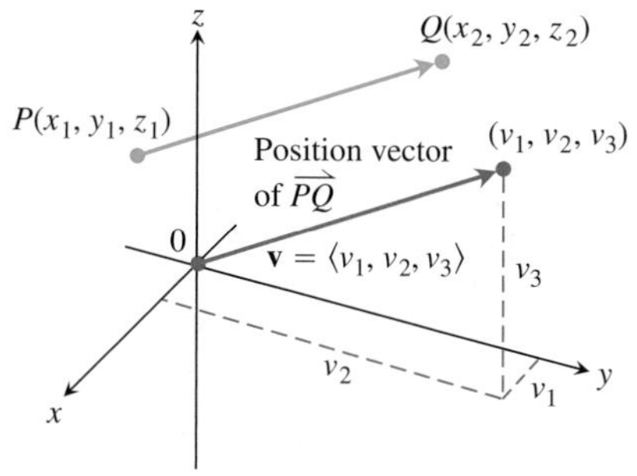

FIGURE 9.10 A vector $\overrightarrow{PQ}$ in standard position has its initial point at the origin. The directed line segments $\overrightarrow{PQ}$ and $\mathbf{v}$ are parallel and have the same length.

We need a way to represent vectors algebraically so that we can be more precise about the direction of a vector. Let $\mathbf{v} = \overrightarrow{PQ}$. There is one directed line segment equal to $\overrightarrow{PQ}$ whose initial point is the origin (Figure 9.10). It is the representative of $\mathbf{v}$ in **standard position** and is the vector we normally use to represent $\mathbf{v}$. We can specify $\mathbf{v}$ by writing the coordinates of its terminal point (v_1, v_2, v_3) when $\mathbf{v}$ is in standard position. If $\mathbf{v}$ is a vector in the plane its terminal point (v_1, v_2) has two coordinates.

DEFINITION

If $\mathbf{v}$ is a **two-dimensional** vector in the plane equal to the vector with initial point at the origin and terminal point (v_1, v_2), then the **component form** of $\mathbf{v}$ is

$$\mathbf{v} = \langle v_1, v_2 \rangle.$$

If $\mathbf{v}$ is a **three-dimensional** vector equal to the vector with initial point at the origin and terminal point (v_1, v_2, v_3), then the **component form** of $\mathbf{v}$ is

$$\mathbf{v} = \langle v_1, v_2, v_3 \rangle.$$

So a two-dimensional vector is an ordered pair $\mathbf{v} = \langle v_1, v_2 \rangle$ of real numbers, and a three-dimensional vector is an ordered triple $\mathbf{v} = \langle v_1, v_2, v_3 \rangle$ of real numbers. The numbers v_1, v_2, and v_3 are the **components** of $\mathbf{v}$.

If $\mathbf{v} = \langle v_1, v_2, v_3 \rangle$ is represented by the directed line segment $\overrightarrow{PQ}$, where the initial point is $P(x_1, y_1, z_1)$ and the terminal point is $Q(x_2, y_2, z_2)$, then $x_1 + v_1 = x_2$, $y_1 + v_2 = y_2$, and $z_1 + v_3 = z_2$ (see Figure 9.10). Thus, $v_1 = x_2 - x_1$, $v_2 = y_2 - y_1$, and $v_3 = z_2 - z_1$ are the components of $\overrightarrow{PQ}$.

In summary, given the points $P(x_1, y_1, z_1)$ and $Q(x_2, y_2, z_2)$, the standard position vector $\mathbf{v} = \langle v_1, v_2, v_3 \rangle$ equal to $\overrightarrow{PQ}$ is

$$\mathbf{v} = \langle x_2 - x_1, y_2 - y_1, z_2 - z_1 \rangle.$$

If $\mathbf{v}$ is two-dimensional with $P(x_1, y_1)$ and $Q(x_2, y_2)$ as points in the plane, then $\mathbf{v} = \langle x_2 - x_1, y_2 - y_1 \rangle$. There is no third component for planar vectors. With this understanding, we will develop the algebra of three-dimensional vectors and simply drop the third component when the vector is two-dimensional (a planar vector).

Two vectors are equal if and only if their standard position vectors are identical. Thus $\langle u_1, u_2, u_3 \rangle$ and $\langle v_1, v_2, v_3 \rangle$ are equal if and only if $u_1 = v_1$, $u_2 = v_2$, and $u_3 = v_3$.

The **magnitude** or **length** of the vector $\overrightarrow{PQ}$ is the length of any of its equivalent directed line segment representations. In particular, if $\mathbf{v} = \langle x_2 - x_1, y_2 - y_1, z_2 - z_1 \rangle$ is the standard position vector for $\overrightarrow{PQ}$, then the distance formula gives the magnitude or length of $\mathbf{v}$, denoted by the symbol $|\mathbf{v}|$ or $\|\mathbf{v}\|$.

The **magnitude** or **length** of the vector $\mathbf{v} = \overrightarrow{PQ}$ is the nonnegative number

$$|\mathbf{v}| = \sqrt{v_1^2 + v_2^2 + v_3^2} = \sqrt{(x_2 - x_1)^2 + (y_2 - y_1)^2 + (z_2 - z_1)^2}$$

(See Figure 9.10.)

The only vector with length 0 is the **zero vector** $\mathbf{0} = \langle 0, 0 \rangle$ or $\mathbf{0} = \langle 0, 0, 0 \rangle$. This vector is also the only vector with no specific direction.

EXAMPLE 1 Find the **(a)** component form and **(b)** length of the vector with initial point $P(-3, 4, 1)$ and terminal point $Q(-5, 2, 2)$.

Solution

(a) The standard position vector **v** representing $\overrightarrow{PQ}$ has components

$$v_1 = x_2 - x_1 = -5 - (-3) = -2, \qquad v_2 = y_2 - y_1 = 2 - 4 = -2,$$

and

$$v_3 = z_2 - z_1 = 2 - 1 = 1.$$

The component form of $\overrightarrow{PQ}$ is

$$\mathbf{v} = \langle -2, -2, 1 \rangle.$$

(b) The length or magnitude of $\mathbf{v} = \overrightarrow{PQ}$ is

$$|\mathbf{v}| = \sqrt{(-2)^2 + (-2)^2 + (1)^2} = \sqrt{9} = 3.$$ ■

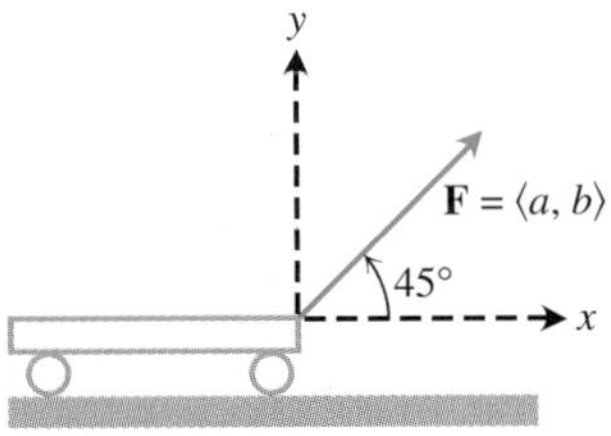

FIGURE 9.11 The force pulling the cart forward is represented by the vector **F** of magnitude 20 (pounds) making an angle of 45° with the horizontal ground (positive x-axis) (Example 2).

EXAMPLE 2 A small cart is being pulled along a smooth horizontal floor with a 20-lb force **F** making a 45° angle to the floor (Figure 9.11). What is the *effective* force moving the cart forward?

Solution The effective force is the horizontal component of $\mathbf{F} = \langle a, b \rangle$, given by

$$a = |\mathbf{F}| \cos 45° = (20)\left(\frac{\sqrt{2}}{2}\right) \approx 14.14 \text{ lb}.$$

Notice that **F** is a two-dimensional vector. ■

Vector Algebra Operations

Two principal operations involving vectors are *vector addition* and *scalar multiplication.* A **scalar** is simply a real number, and is called such when we want to draw attention to its differences from vectors. Scalars can be positive, negative, or zero and are used to “scale” a vector by multiplication.

DEFINITIONS Let $\mathbf{u} = \langle u_1, u_2, u_3 \rangle$ and $\mathbf{v} = \langle v_1, v_2, v_3 \rangle$ be vectors with k a scalar.

Addition: $\mathbf{u} + \mathbf{v} = \langle u_1 + v_1, u_2 + v_2, u_3 + v_3 \rangle$

Scalar multiplication: $k\mathbf{u} = \langle ku_1, ku_2, ku_3 \rangle$

We add vectors by adding the corresponding components of the vectors. We multiply a vector by a scalar by multiplying each component by the scalar. The definitions apply to planar vectors except there are only two components, $\langle u_1, u_2 \rangle$ and $\langle v_1, v_2 \rangle$.

The definition of vector addition is illustrated geometrically for planar vectors in Figure 9.12a, where the initial point of one vector is placed at the terminal point of the other. Another interpretation is shown in Figure 9.12b (called the **parallelogram law** of addition), where the sum, called the **resultant vector**, is the diagonal of the parallelogram. In physics, forces add vectorially as do velocities, accelerations, and so on. So the force acting on a particle subject to two gravitational forces, for example, is obtained by adding the two force vectors.

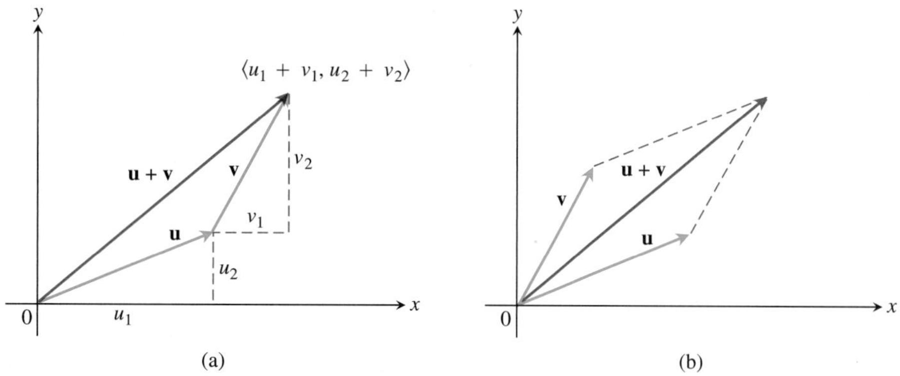

FIGURE 9.12 (a) Geometric interpretation of the vector sum. (b) The parallelogram law of vector addition.

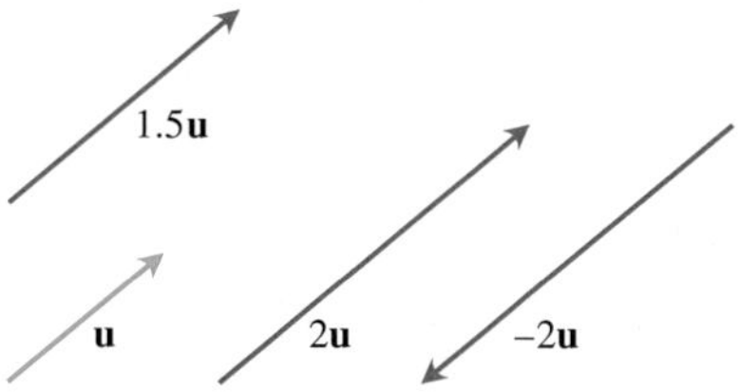

FIGURE 9.13 Scalar multiples of **u**.

Figure 9.13 displays a geometric interpretation of the product $k\mathbf{u}$ of the scalar k and vector $\mathbf{u}$. If $k > 0$, then $k\mathbf{u}$ has the same direction as $\mathbf{u}$; if $k < 0$, then the direction of $k\mathbf{u}$ is opposite to that of $\mathbf{u}$. Comparing the lengths of $\mathbf{u}$ and $k\mathbf{u}$, we see that

$$\begin{aligned}|k\mathbf{u}| &= \sqrt{(ku_1)^2 + (ku_2)^2 + (ku_3)^2} = \sqrt{k^2(u_1^2 + u_2^2 + u_3^2)} \\ &= \sqrt{k^2}\sqrt{u_1^2 + u_2^2 + u_3^2} = |k|\,|\mathbf{u}|.\end{aligned}$$

The length of $k\mathbf{u}$ is the absolute value of the scalar k times the length of $\mathbf{u}$. The vector $(-1)\mathbf{u} = -\mathbf{u}$ has the same length as $\mathbf{u}$ but points in the opposite direction.

The **difference** $\mathbf{u} - \mathbf{v}$ of two vectors is defined by

$$\mathbf{u} - \mathbf{v} = \mathbf{u} + (-\mathbf{v}).$$

If $\mathbf{u} = \langle u_1, u_2, u_3 \rangle$ and $\mathbf{v} = \langle v_1, v_2, v_3 \rangle$, then

$$\mathbf{u} - \mathbf{v} = \langle u_1 - v_1, u_2 - v_2, u_3 - v_3 \rangle.$$

Note that $(\mathbf{u} - \mathbf{v}) + \mathbf{v} = \mathbf{u}$, so adding the vector $(\mathbf{u} - \mathbf{v})$ to $\mathbf{v}$ gives $\mathbf{u}$ (Figure 9.14a). Figure 9.14b shows the difference $\mathbf{u} - \mathbf{v}$ as the sum $\mathbf{u} + (-\mathbf{v})$.

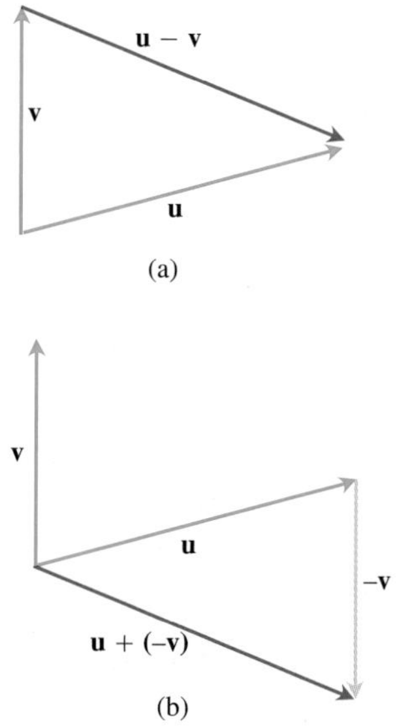

FIGURE 9.14 (a) The vector $\mathbf{u} - \mathbf{v}$, when added to $\mathbf{v}$, gives $\mathbf{u}$. (b) $\mathbf{u} - \mathbf{v} = \mathbf{u} + (-\mathbf{v})$.

EXAMPLE 3 Let $\mathbf{u} = \langle -1, 3, 1 \rangle$ and $\mathbf{v} = \langle 4, 7, 0 \rangle$. Find

(a) $2\mathbf{u} + 3\mathbf{v}$ **(b)** $\mathbf{u} - \mathbf{v}$ **(c)** $\left|\frac{1}{2}\mathbf{u}\right|$.

Solution

(a) $2\mathbf{u} + 3\mathbf{v} = 2\langle -1, 3, 1 \rangle + 3\langle 4, 7, 0 \rangle = \langle -2, 6, 2 \rangle + \langle 12, 21, 0 \rangle = \langle 10, 27, 2 \rangle$

(b) $\mathbf{u} - \mathbf{v} = \langle -1, 3, 1 \rangle - \langle 4, 7, 0 \rangle = \langle -1 - 4, 3 - 7, 1 - 0 \rangle = \langle -5, -4, 1 \rangle$

(c) $\left|\frac{1}{2}\mathbf{u}\right| = \left|\left\langle -\frac{1}{2}, \frac{3}{2}, \frac{1}{2} \right\rangle\right| = \sqrt{\left(-\frac{1}{2}\right)^2 + \left(\frac{3}{2}\right)^2 + \left(\frac{1}{2}\right)^2} = \frac{1}{2}\sqrt{11}.$ ■

Vector operations have many of the properties of ordinary arithmetic. These properties are readily verified using the definitions of vector addition and multiplication by a scalar.

Properties of Vector Operations

Let $\mathbf{u}$, $\mathbf{v}$, $\mathbf{w}$ be vectors and a, b be scalars.

1. $\mathbf{u} + \mathbf{v} = \mathbf{v} + \mathbf{u}$
2. $(\mathbf{u} + \mathbf{v}) + \mathbf{w} = \mathbf{u} + (\mathbf{v} + \mathbf{w})$
3. $\mathbf{u} + \mathbf{0} = \mathbf{u}$
4. $\mathbf{u} + (-\mathbf{u}) = \mathbf{0}$
5. $0\mathbf{u} = \mathbf{0}$
6. $1\mathbf{u} = \mathbf{u}$
7. $a(b\mathbf{u}) = (ab)\mathbf{u}$
8. $a(\mathbf{u} + \mathbf{v}) = a\mathbf{u} + a\mathbf{v}$
9. $(a + b)\mathbf{u} = a\mathbf{u} + b\mathbf{u}$

An important application of vectors occurs in navigation.

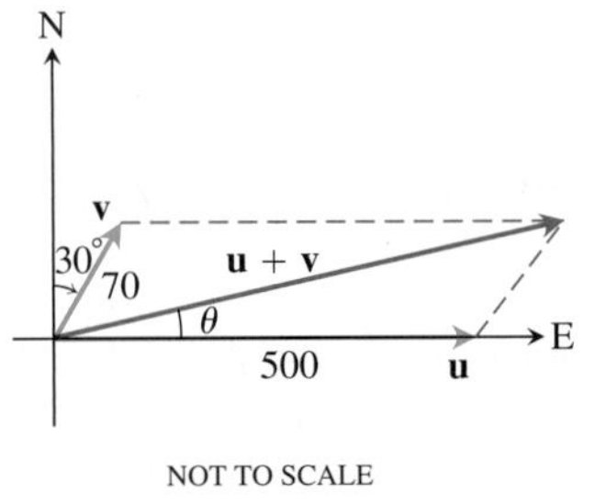

FIGURE 9.15 Vectors representing the velocities of the airplane $\mathbf{u}$ and tailwind $\mathbf{v}$ in Example 4.

EXAMPLE 4 A jet airliner, flying due east at 500 mph in still air, encounters a 70-mph tailwind blowing in the direction 60° north of east. The airplane holds its compass heading due east but, because of the wind, acquires a new ground speed and direction. What are they?

Solution If $\mathbf{u}$ = the velocity of the airplane alone and $\mathbf{v}$ = the velocity of the tailwind, then $|\mathbf{u}| = 500$ and $|\mathbf{v}| = 70$ (Figure 9.15). The velocity of the airplane with respect to the ground is given by the magnitude and direction of the *resultant vector* $\mathbf{u} + \mathbf{v}$. If we let the positive x-axis represent east and the positive y-axis represent north, then the component forms of $\mathbf{u}$ and $\mathbf{v}$ are

$$\mathbf{u} = \langle 500, 0 \rangle \quad \text{and} \quad \mathbf{v} = \langle 70 \cos 60°, 70 \sin 60° \rangle = \langle 35, 35\sqrt{3} \rangle.$$

Therefore,

$$\mathbf{u} + \mathbf{v} = \langle 535, 35\sqrt{3} \rangle$$

$$|\mathbf{u} + \mathbf{v}| = \sqrt{535^2 + (35\sqrt{3})^2} \approx 538.4$$

and

$$\theta = \tan^{-1} \frac{35\sqrt{3}}{535} \approx 6.5°. \qquad \text{Figure 9.15}$$

The new ground speed of the airplane is about 538.4 mph, and its new direction is about 6.5° north of east. ■

Unit Vectors

A vector $\mathbf{v}$ of length 1 is called a **unit vector**. The **standard unit vectors** are

$$\mathbf{i} = \langle 1, 0, 0 \rangle, \qquad \mathbf{j} = \langle 0, 1, 0 \rangle, \qquad \text{and} \quad \mathbf{k} = \langle 0, 0, 1 \rangle.$$

Any vector $\mathbf{v} = \langle v_1, v_2, v_3 \rangle$ can be written as a *linear combination* of the standard unit vectors as follows:

$$\begin{aligned} \mathbf{v} = \langle v_1, v_2, v_3 \rangle &= \langle v_1, 0, 0 \rangle + \langle 0, v_2, 0 \rangle + \langle 0, 0, v_3 \rangle \\ &= v_1\langle 1, 0, 0 \rangle + v_2\langle 0, 1, 0 \rangle + v_3\langle 0, 0, 1 \rangle \\ &= v_1\mathbf{i} + v_2\mathbf{j} + v_3\mathbf{k}. \end{aligned}$$

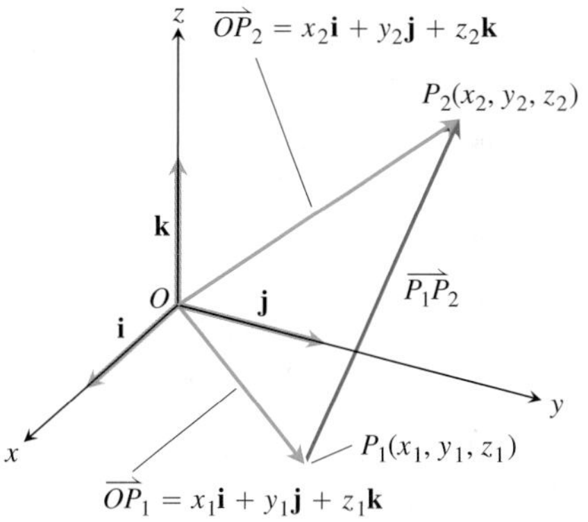

FIGURE 9.16 The vector from P_1 to P_2 is $\overrightarrow{P_1P_2} = (x_2 - x_1)\mathbf{i} + (y_2 - y_1)\mathbf{j} + (z_2 - z_1)\mathbf{k}$.

We call the scalar (or number) v_1 the **i-component** of the vector $\mathbf{v}$, v_2 the **j-component**, and v_3 the **k-component**. In component form, the vector from $P_1(x_1, y_1, z_1)$ to $P_2(x_2, y_2, z_2)$ is

$$\overrightarrow{P_1P_2} = (x_2 - x_1)\mathbf{i} + (y_2 - y_1)\mathbf{j} + (z_2 - z_1)\mathbf{k}$$

(Figure 9.16).

Whenever $\mathbf{v} \neq \mathbf{0}$, its length $|\mathbf{v}|$ is not zero and

$$\left|\frac{1}{|\mathbf{v}|}\mathbf{v}\right| = \frac{1}{|\mathbf{v}|}|\mathbf{v}| = 1.$$

That is, $\mathbf{v}/|\mathbf{v}|$ is a unit vector in the direction of $\mathbf{v}$, called **the direction** of the nonzero vector $\mathbf{v}$.

EXAMPLE 5 Find a unit vector $\mathbf{u}$ in the direction of the vector from $P_1(1, 0, 1)$ to $P_2(3, 2, 0)$.

Solution We divide $\overrightarrow{P_1P_2}$ by its length:

$$\overrightarrow{P_1P_2} = (3 - 1)\mathbf{i} + (2 - 0)\mathbf{j} + (0 - 1)\mathbf{k} = 2\mathbf{i} + 2\mathbf{j} - \mathbf{k}$$

$$|\overrightarrow{P_1P_2}| = \sqrt{(2)^2 + (2)^2 + (-1)^2} = \sqrt{4 + 4 + 1} = \sqrt{9} = 3$$

$$\mathbf{u} = \frac{\overrightarrow{P_1P_2}}{|\overrightarrow{P_1P_2}|} = \frac{2\mathbf{i} + 2\mathbf{j} - \mathbf{k}}{3} = \frac{2}{3}\mathbf{i} + \frac{2}{3}\mathbf{j} - \frac{1}{3}\mathbf{k}.$$

The unit vector $\mathbf{u}$ is the direction of $\overrightarrow{P_1P_2}$. ■

EXAMPLE 6 If $\mathbf{v} = 3\mathbf{i} - 4\mathbf{j}$ is a velocity vector, express $\mathbf{v}$ as a product of its speed times a unit vector in the direction of motion.

Solution Speed is the magnitude (length) of $\mathbf{v}$:

$$|\mathbf{v}| = \sqrt{(3)^2 + (-4)^2} = \sqrt{9 + 16} = 5.$$

The unit vector $\mathbf{v}/|\mathbf{v}|$ has the same direction as $\mathbf{v}$:

$$\frac{\mathbf{v}}{|\mathbf{v}|} = \frac{3\mathbf{i} - 4\mathbf{j}}{5} = \frac{3}{5}\mathbf{i} - \frac{4}{5}\mathbf{j}.$$

So

$$\mathbf{v} = 3\mathbf{i} - 4\mathbf{j} = \underbrace{5}_{\text{Length (speed)}}\underbrace{\left(\frac{3}{5}\mathbf{i} - \frac{4}{5}\mathbf{j}\right)}_{\text{Direction of motion}}.$$ ■

HISTORICAL BIOGRAPHY

Hermann Grassmann
(1809–1877)

In summary, we can express any nonzero vector $\mathbf{v}$ in terms of its two important features, length and direction, by writing $\mathbf{v} = |\mathbf{v}|\dfrac{\mathbf{v}}{|\mathbf{v}|}$.

If $\mathbf{v} \neq \mathbf{0}$, then

1. $\dfrac{\mathbf{v}}{|\mathbf{v}|}$ is a unit vector in the direction of $\mathbf{v}$;
2. the equation $\mathbf{v} = |\mathbf{v}|\dfrac{\mathbf{v}}{|\mathbf{v}|}$ expresses $\mathbf{v}$ in terms of its length and direction.

EXAMPLE 7 A force of 6 newtons is applied in the direction of the vector $\mathbf{v} = 2\mathbf{i} + 2\mathbf{j} - \mathbf{k}$. Express the force $\mathbf{F}$ as a product of its magnitude and direction.

Solution The force vector has magnitude 6 and direction $\dfrac{\mathbf{v}}{|\mathbf{v}|}$, so

$$\mathbf{F} = 6\frac{\mathbf{v}}{|\mathbf{v}|} = 6\frac{2\mathbf{i} + 2\mathbf{j} - \mathbf{k}}{\sqrt{2^2 + 2^2 + (-1)^2}} = 6\frac{2\mathbf{i} + 2\mathbf{j} - \mathbf{k}}{3}$$
$$= 6\left(\frac{2}{3}\mathbf{i} + \frac{2}{3}\mathbf{j} - \frac{1}{3}\mathbf{k}\right).$$

■

Midpoint of a Line Segment

Vectors are often useful in geometry. For example, the coordinates of the midpoint of a line segment are found by averaging.

The **midpoint** M of the line segment joining points $P_1(x_1, y_1, z_1)$ and $P_2(x_2, y_2, z_2)$ is the point

$$\left(\frac{x_1 + x_2}{2}, \frac{y_1 + y_2}{2}, \frac{z_1 + z_2}{2}\right).$$

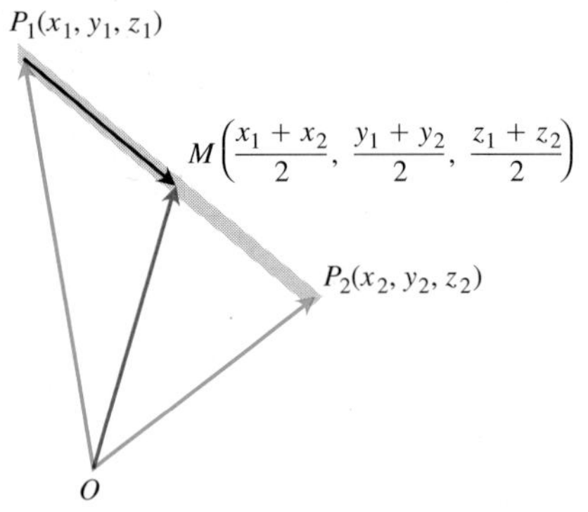

FIGURE 9.17 The coordinates of the midpoint are the averages of the coordinates of P_1 and P_2.

To see why, observe (Figure 9.17) that

$$\overrightarrow{OM} = \overrightarrow{OP_1} + \frac{1}{2}(\overrightarrow{P_1P_2}) = \overrightarrow{OP_1} + \frac{1}{2}(\overrightarrow{OP_2} - \overrightarrow{OP_1})$$
$$= \frac{1}{2}(\overrightarrow{OP_1} + \overrightarrow{OP_2})$$
$$= \frac{x_1 + x_2}{2}\mathbf{i} + \frac{y_1 + y_2}{2}\mathbf{j} + \frac{z_1 + z_2}{2}\mathbf{k}.$$

EXAMPLE 8 The midpoint of the segment joining $P_1(3, -2, 0)$ and $P_2(7, 4, 4)$ is

$$\left(\frac{3 + 7}{2}, \frac{-2 + 4}{2}, \frac{0 + 4}{2}\right) = (5, 1, 2).$$

■

EXERCISES 9.2

In Exercises 1–8, let $\mathbf{u} = \langle 3, -2 \rangle$ and $\mathbf{v} = \langle -2, 5 \rangle$. Find the **(a)** component form and **(b)** magnitude (length) of the vector.

1. $3\mathbf{u}$
2. $-2\mathbf{v}$
3. $\mathbf{u} + \mathbf{v}$
4. $\mathbf{u} - \mathbf{v}$
5. $2\mathbf{u} - 3\mathbf{v}$
6. $-2\mathbf{u} + 5\mathbf{v}$
7. $\frac{3}{5}\mathbf{u} + \frac{4}{5}\mathbf{v}$
8. $-\frac{5}{13}\mathbf{u} + \frac{12}{13}\mathbf{v}$

In Exercises 9–16, find the component form of the vector.

9. The vector $\overrightarrow{PQ}$, where $P = (1, 3)$ and $Q = (2, -1)$
10. The vector $\overrightarrow{OP}$ where O is the origin and P is the midpoint of segment RS, where $R = (2, -1)$ and $S = (-4, 3)$
11. The vector from the point $A = (2, 3)$ to the origin
12. The sum of $\overrightarrow{AB}$ and $\overrightarrow{CD}$, where $A = (1, -1)$, $B = (2, 0)$, $C = (-1, 3)$, and $D = (-2, 2)$
13. The unit vector that makes an angle $\theta = 2\pi/3$ with the positive x-axis
14. The unit vector that makes an angle $\theta = -3\pi/4$ with the positive x-axis
15. The unit vector obtained by rotating the vector $\langle 0, 1 \rangle$ 120° counterclockwise about the origin
16. The unit vector obtained by rotating the vector $\langle 1, 0 \rangle$ 135° counterclockwise about the origin

In Exercises 17–22, express each vector in the form $\mathbf{v} = v_1\mathbf{i} + v_2\mathbf{j} + v_3\mathbf{k}$.

17. $\overrightarrow{P_1P_2}$ if P_1 is the point $(5, 7, -1)$ and P_2 is the point $(2, 9, -2)$
18. $\overrightarrow{P_1P_2}$ if P_1 is the point $(1, 2, 0)$ and P_2 is the point $(-3, 0, 5)$
19. $\overrightarrow{AB}$ if A is the point $(-7, -8, 1)$ and B is the point $(-10, 8, 1)$
20. $\overrightarrow{AB}$ if A is the point $(1, 0, 3)$ and B is the point $(-1, 4, 5)$
21. $5\mathbf{u} - \mathbf{v}$ if $\mathbf{u} = \langle 1, 1, -1 \rangle$ and $\mathbf{v} = \langle 2, 0, 3 \rangle$
22. $-2\mathbf{u} + 3\mathbf{v}$ if $\mathbf{u} = \langle -1, 0, 2 \rangle$ and $\mathbf{v} = \langle 1, 1, 1 \rangle$

In Exercises 23 and 24, copy vectors **u**, **v**, and **w** head to tail as needed to sketch the indicated vector.

23.

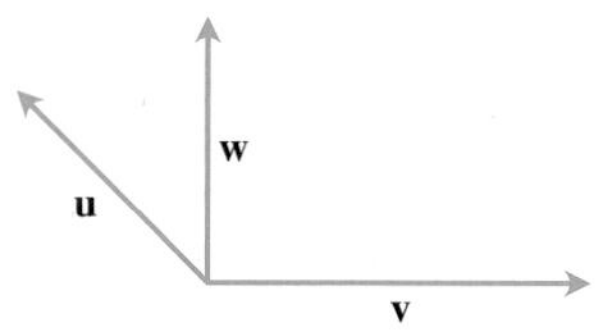

a. $\mathbf{u} + \mathbf{v}$ b. $\mathbf{u} + \mathbf{v} + \mathbf{w}$
c. $\mathbf{u} - \mathbf{v}$ d. $\mathbf{u} - \mathbf{w}$

24.

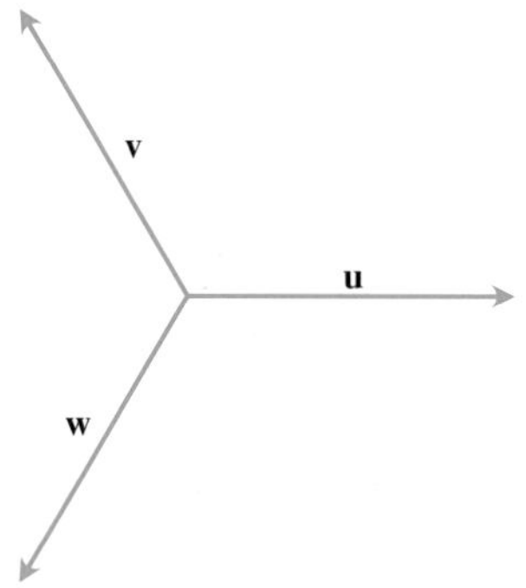

a. $\mathbf{u} - \mathbf{v}$ b. $\mathbf{u} - \mathbf{v} + \mathbf{w}$
c. $2\mathbf{u} - \mathbf{v}$ d. $\mathbf{u} + \mathbf{v} + \mathbf{w}$

In Exercises 25–30, express each vector as a product of its length and direction.

25. $2\mathbf{i} + \mathbf{j} - 2\mathbf{k}$
26. $9\mathbf{i} - 2\mathbf{j} + 6\mathbf{k}$
27. $5\mathbf{k}$
28. $\frac{3}{5}\mathbf{i} + \frac{4}{5}\mathbf{k}$
29. $\frac{1}{\sqrt{6}}\mathbf{i} - \frac{1}{\sqrt{6}}\mathbf{j} - \frac{1}{\sqrt{6}}\mathbf{k}$
30. $\frac{\mathbf{i}}{\sqrt{3}} + \frac{\mathbf{j}}{\sqrt{3}} + \frac{\mathbf{k}}{\sqrt{3}}$
31. Find the vectors whose lengths and directions are given. Try to do the calculations without writing.

	Length	Direction
a.	2	$\mathbf{i}$
b.	$\sqrt{3}$	$-\mathbf{k}$
c.	$\frac{1}{2}$	$\frac{3}{5}\mathbf{j} + \frac{4}{5}\mathbf{k}$
d.	7	$\frac{6}{7}\mathbf{i} - \frac{2}{7}\mathbf{j} + \frac{3}{7}\mathbf{k}$

32. Find the vectors whose lengths and directions are given. Try to do the calculations without writing.

	Length	Direction
a.	7	$-\mathbf{j}$
b.	$\sqrt{2}$	$-\frac{3}{5}\mathbf{i} - \frac{4}{5}\mathbf{k}$
c.	$\frac{13}{12}$	$\frac{3}{13}\mathbf{i} - \frac{4}{13}\mathbf{j} - \frac{12}{13}\mathbf{k}$
d.	$a > 0$	$\frac{1}{\sqrt{2}}\mathbf{i} + \frac{1}{\sqrt{3}}\mathbf{j} - \frac{1}{\sqrt{6}}\mathbf{k}$

33. Find a vector of magnitude 7 in the direction of $\mathbf{v} = 12\mathbf{i} - 5\mathbf{k}$.
34. Find a vector of magnitude 3 in the direction opposite to the direction of $\mathbf{v} = (1/2)\mathbf{i} - (1/2)\mathbf{j} - (1/2)\mathbf{k}$.

In Exercises 35–38, find

a. the direction of $\overrightarrow{P_1P_2}$ and

b. the midpoint of line segment P_1P_2.

35. $P_1(-1, 1, 5)$ $P_2(2, 5, 0)$

36. $P_1(1, 4, 5)$ $P_2(4, -2, 7)$

37. $P_1(3, 4, 5)$ $P_2(2, 3, 4)$

38. $P_1(0, 0, 0)$ $P_2(2, -2, -2)$

39. If $\overrightarrow{AB} = \mathbf{i} + 4\mathbf{j} - 2\mathbf{k}$ and B is the point $(5, 1, 3)$, find A.

40. If $\overrightarrow{AB} = -7\mathbf{i} + 3\mathbf{j} + 8\mathbf{k}$ and A is the point $(-2, -3, 6)$, find B.

41. **Linear combination** Let $\mathbf{u} = 2\mathbf{i} + \mathbf{j}$, $\mathbf{v} = \mathbf{i} + \mathbf{j}$, and $\mathbf{w} = \mathbf{i} - \mathbf{j}$. Find scalars a and b such that $\mathbf{u} = a\mathbf{v} + b\mathbf{w}$.

42. **Linear combination** Let $\mathbf{u} = \mathbf{i} - 2\mathbf{j}$, $\mathbf{v} = 2\mathbf{i} + 3\mathbf{j}$, and $\mathbf{w} = \mathbf{i} + \mathbf{j}$. Write $\mathbf{u} = \mathbf{u}_1 + \mathbf{u}_2$, where $\mathbf{u}_1$ is parallel to $\mathbf{v}$ and $\mathbf{u}_2$ is parallel to $\mathbf{w}$. (See Exercise 41.)

43. **Velocity** An airplane is flying in the direction 25° west of north at 800 km/h. Find the component form of the velocity of the airplane, assuming that the positive x-axis represents due east and the positive y-axis represents due north.

44. **Velocity** An airplane is flying in the direction 10° east of south at 600 km/h. Find the component form of the velocity of the airplane, assuming that the positive x-axis represents due east and the positive y-axis represents due north.

45. **Medians of a triangle** Suppose that A, B, and C are the corner points of the thin triangular plate of constant density shown here.

a. Find the vector from C to the midpoint M of side AB.

b. Find the vector from C to the point that lies two-thirds of the way from C to M on the median CM.

c. Find the coordinates of the point in which the medians of ΔABC intersect. This point is the plate's center of mass.

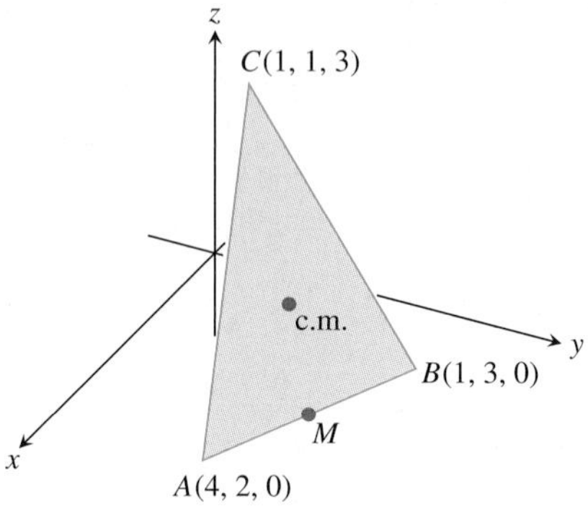

46. Use similar triangles to find the coordinates of the point Q that divides the segment from $P_1(x_1, y_1, z_1)$ to $P_2(x_2, y_2, z_2)$ into two lengths whose ratio is $p/q = r$.

47. Let $ABCD$ be a general, not necessarily planar, quadrilateral in space. Show that the two segments joining the midpoints of opposite sides of $ABCD$ bisect each other. (*Hint:* Show that the segments have the same midpoint.)

48. Find the vector from the origin to the point of intersection of the medians of the triangle whose vertices are

$$A(1, -1, 2), \quad B(2, 1, 3), \quad \text{and} \quad C(-1, 2, -1).$$

49. Suppose that A, B, and C are vertices of a triangle and that a, b, and c are, respectively, the midpoints of the opposite sides. Show that $\overrightarrow{Aa} + \overrightarrow{Bb} + \overrightarrow{Cc} = \mathbf{0}$.

50. **Unit vectors in the plane** Show that a unit vector in the plane can be expressed as $\mathbf{u} = (\cos\theta)\mathbf{i} + (\sin\theta)\mathbf{j}$, obtained by rotating $\mathbf{i}$ through an angle θ in the counterclockwise direction. Explain why this form gives *every* unit vector in the plane.

9.3 The Dot Product

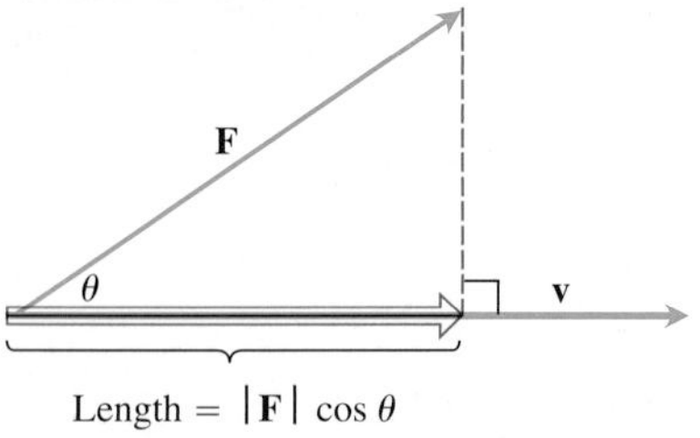

FIGURE 9.18 The magnitude of the force $\mathbf{F}$ in the direction of vector $\mathbf{v}$ is the length $|\mathbf{F}|\cos\theta$ of the projection of $\mathbf{F}$ onto $\mathbf{v}$.

If a force $\mathbf{F}$ is applied to a particle moving along a path, we often need to know the magnitude of the force in the direction of motion. If $\mathbf{v}$ is parallel to the tangent line to the path at the point where $\mathbf{F}$ is applied, then we want the magnitude of $\mathbf{F}$ in the direction of $\mathbf{v}$. Figure 9.18 shows that the scalar quantity we seek is the length $|\mathbf{F}|\cos\theta$, where θ is the angle between the two vectors $\mathbf{F}$ and $\mathbf{v}$.

In this section we show how to calculate easily the angle between two vectors directly from their components. A key part of the calculation is an expression called the *dot product*. Dot products are also called *inner* or *scalar* products because the product results in a scalar, not a vector. After investigating the dot product, we apply it to finding the projection of one vector onto another (as displayed in Figure 9.18) and to finding the work done by a constant force acting through a displacement.

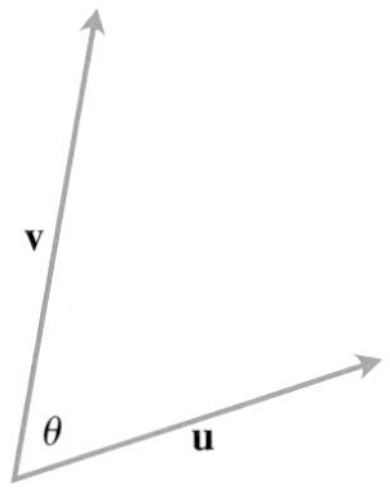

FIGURE 9.19 The angle between **u** and **v**.

Angle Between Vectors

When two nonzero vectors **u** and **v** are placed so their initial points coincide, they form an angle θ of measure $0 \le \theta \le \pi$ (Figure 9.19). If the vectors do not lie along the same line, the angle θ is measured in the plane containing both of them. If they do lie along the same line, the angle between them is 0 if they point in the same direction, and π if they point in opposite directions. The angle θ is the **angle between u** and **v**. Theorem 1 gives a formula to determine this angle.

THEOREM 1—Angle Between Two Vectors The angle θ between two nonzero vectors $\mathbf{u} = \langle u_1, u_2, u_3 \rangle$ and $\mathbf{v} = \langle v_1, v_2, v_3 \rangle$ is given by

$$\theta = \cos^{-1}\left(\frac{u_1 v_1 + u_2 v_2 + u_3 v_3}{|\mathbf{u}||\mathbf{v}|}\right).$$

Before proving Theorem 1, we focus attention on the expression $u_1 v_1 + u_2 v_2 + u_3 v_3$ in the calculation for θ.

DEFINITION The **dot product u · v** ("**u** dot **v**") of vectors $\mathbf{u} = \langle u_1, u_2, u_3 \rangle$ and $\mathbf{v} = \langle v_1, v_2, v_3 \rangle$ is

$$\mathbf{u} \cdot \mathbf{v} = u_1 v_1 + u_2 v_2 + u_3 v_3.$$

EXAMPLE 1

(a) $$\langle 1, -2, -1 \rangle \cdot \langle -6, 2, -3 \rangle = (1)(-6) + (-2)(2) + (-1)(-3) = -6 - 4 + 3 = -7$$

(b) $$\left(\frac{1}{2}\mathbf{i} + 3\mathbf{j} + \mathbf{k}\right) \cdot (4\mathbf{i} - \mathbf{j} + 2\mathbf{k}) = \left(\frac{1}{2}\right)(4) + (3)(-1) + (1)(2) = 1$$ ■

The dot product of a pair of two-dimensional vectors is defined in a similar fashion:

$$\langle u_1, u_2 \rangle \cdot \langle v_1, v_2 \rangle = u_1 v_1 + u_2 v_2.$$

Proof of Theorem 1 Applying the law of cosines (Equation (8), Appendix 4) to the triangle in Figure 9.20, we find that

$$|\mathbf{w}|^2 = |\mathbf{u}|^2 + |\mathbf{v}|^2 - 2|\mathbf{u}||\mathbf{v}|\cos\theta \qquad \text{Law of cosines}$$

$$2|\mathbf{u}||\mathbf{v}|\cos\theta = |\mathbf{u}|^2 + |\mathbf{v}|^2 - |\mathbf{w}|^2.$$

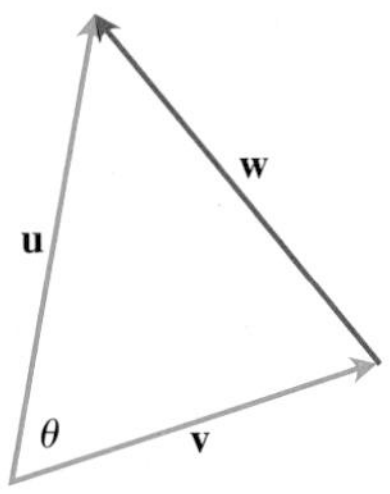

FIGURE 9.20 The parallelogram law of addition of vectors gives **w** = **u** − **v**.

Because $\mathbf{w} = \mathbf{u} - \mathbf{v}$, the component form of **w** is $\langle u_1 - v_1, u_2 - v_2, u_3 - v_3 \rangle$. So

$$|\mathbf{u}|^2 = \left(\sqrt{u_1^2 + u_2^2 + u_3^2}\right)^2 = u_1^2 + u_2^2 + u_3^2$$

$$|\mathbf{v}|^2 = \left(\sqrt{v_1^2 + v_2^2 + v_3^2}\right)^2 = v_1^2 + v_2^2 + v_3^2$$

$$\begin{aligned}|\mathbf{w}|^2 &= \left(\sqrt{(u_1 - v_1)^2 + (u_2 - v_2)^2 + (u_3 - v_3)^2}\right)^2 \\ &= (u_1 - v_1)^2 + (u_2 - v_2)^2 + (u_3 - v_3)^2 \\ &= u_1^2 - 2u_1v_1 + v_1^2 + u_2^2 - 2u_2v_2 + v_2^2 + u_3^2 - 2u_3v_3 + v_3^2\end{aligned}$$

and

$$|\mathbf{u}|^2 + |\mathbf{v}|^2 - |\mathbf{w}|^2 = 2(u_1v_1 + u_2v_2 + u_3v_3).$$

Therefore,

$$\begin{aligned} 2|\mathbf{u}||\mathbf{v}|\cos\theta &= |\mathbf{u}|^2 + |\mathbf{v}|^2 - |\mathbf{w}|^2 = 2(u_1v_1 + u_2v_2 + u_3v_3) \\ |\mathbf{u}||\mathbf{v}|\cos\theta &= u_1v_1 + u_2v_2 + u_3v_3 \\ \cos\theta &= \frac{u_1v_1 + u_2v_2 + u_3v_3}{|\mathbf{u}||\mathbf{v}|}. \end{aligned}$$

Since $0 \le \theta < \pi$, we have

$$\theta = \cos^{-1}\left(\frac{u_1v_1 + u_2v_2 + u_3v_3}{|\mathbf{u}||\mathbf{v}|}\right). \qquad \blacksquare$$

With the notation of the dot product, the angle between two vectors **u** and **v** can be written as

$$\theta = \cos^{-1}\left(\frac{\mathbf{u}\cdot\mathbf{v}}{|\mathbf{u}||\mathbf{v}|}\right).$$

EXAMPLE 2 Find the angle between $\mathbf{u} = \mathbf{i} - 2\mathbf{j} - 2\mathbf{k}$ and $\mathbf{v} = 6\mathbf{i} + 3\mathbf{j} + 2\mathbf{k}$.

Solution We use the formula above:

$$\begin{aligned} \mathbf{u}\cdot\mathbf{v} &= (1)(6) + (-2)(3) + (-2)(2) = 6 - 6 - 4 = -4 \\ |\mathbf{u}| &= \sqrt{(1)^2 + (-2)^2 + (-2)^2} = \sqrt{9} = 3 \\ |\mathbf{v}| &= \sqrt{(6)^2 + (3)^2 + (2)^2} = \sqrt{49} = 7 \\ \theta &= \cos^{-1}\left(\frac{\mathbf{u}\cdot\mathbf{v}}{|\mathbf{u}||\mathbf{v}|}\right) \\ &= \cos^{-1}\left(\frac{-4}{(3)(7)}\right) \approx 1.76 \text{ radians}. \end{aligned} \qquad \blacksquare$$

The angle formula applies to two-dimensional vectors as well.

EXAMPLE 3 Find the angle θ in the triangle ABC determined by the vertices $A = (0, 0)$, $B = (3, 5)$, and $C = (5, 2)$ (Figure 9.21).

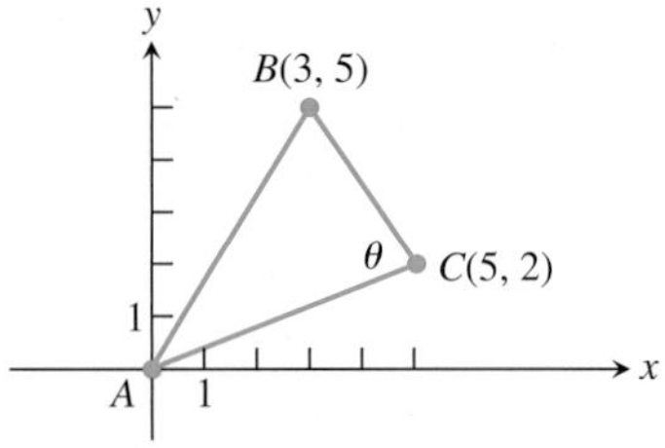

FIGURE 9.21 The triangle in Example 3.

Solution The angle θ is the angle between the vectors $\overrightarrow{CA}$ and $\overrightarrow{CB}$. The component forms of these two vectors are

$$\overrightarrow{CA} = \langle -5, -2\rangle \qquad \text{and} \qquad \overrightarrow{CB} = \langle -2, 3\rangle.$$

First we calculate the dot product and magnitudes of these two vectors.

$$\begin{aligned} \overrightarrow{CA}\cdot\overrightarrow{CB} &= (-5)(-2) + (-2)(3) = 4 \\ |\overrightarrow{CA}| &= \sqrt{(-5)^2 + (-2)^2} = \sqrt{29} \\ |\overrightarrow{CB}| &= \sqrt{(-2)^2 + (3)^2} = \sqrt{13} \end{aligned}$$

Then applying the angle formula, we have

$$\begin{aligned}\theta &= \cos^{-1}\left(\frac{\overrightarrow{CA}\cdot\overrightarrow{CB}}{|\overrightarrow{CA}||\overrightarrow{CB}|}\right)\\ &= \cos^{-1}\left(\frac{4}{\left(\sqrt{29}\right)\left(\sqrt{13}\right)}\right)\\ &\approx 78.1^\circ \quad \text{or} \quad 1.36 \text{ radians}.\end{aligned}$$

■

Perpendicular (Orthogonal) Vectors

Two nonzero vectors **u** and **v** are perpendicular or **orthogonal** if the angle between them is $\pi/2$. For such vectors, we have $\mathbf{u}\cdot\mathbf{v} = 0$ because $\cos(\pi/2) = 0$. The converse is also true. If **u** and **v** are nonzero vectors with $\mathbf{u}\cdot\mathbf{v} = |\mathbf{u}||\mathbf{v}|\cos\theta = 0$, then $\cos\theta = 0$ and $\theta = \cos^{-1} 0 = \pi/2$.

> **DEFINITION** Vectors **u** and **v** are **orthogonal** (or **perpendicular**) if and only if $\mathbf{u}\cdot\mathbf{v} = 0$.

EXAMPLE 4

(a) $\mathbf{u} = \langle 3, -2\rangle$ and $\mathbf{v} = \langle 4, 6\rangle$ are orthogonal because $\mathbf{u}\cdot\mathbf{v} = (3)(4) + (-2)(6) = 0$.

(b) $\mathbf{u} = 3\mathbf{i} - 2\mathbf{j} + \mathbf{k}$ and $\mathbf{v} = 2\mathbf{j} + 4\mathbf{k}$ are orthogonal because $\mathbf{u}\cdot\mathbf{v} = (3)(0) + (-2)(2) + (1)(4) = 0$.

(c) **0** is orthogonal to every vector **u** since

$$\begin{aligned}\mathbf{0}\cdot\mathbf{u} &= \langle 0, 0, 0\rangle\cdot\langle u_1, u_2, u_3\rangle\\ &= (0)(u_1) + (0)(u_2) + (0)(u_3)\\ &= 0.\end{aligned}$$

■

Dot Product Properties and Vector Projections

The dot product obeys many of the laws that hold for ordinary products of real numbers (scalars).

> **Properties of the Dot Product**
> If **u**, **v**, and **w** are any vectors and c is a scalar, then
> 1. $\mathbf{u}\cdot\mathbf{v} = \mathbf{v}\cdot\mathbf{u}$
> 2. $(c\mathbf{u})\cdot\mathbf{v} = \mathbf{u}\cdot(c\mathbf{v}) = c(\mathbf{u}\cdot\mathbf{v})$
> 3. $\mathbf{u}\cdot(\mathbf{v} + \mathbf{w}) = \mathbf{u}\cdot\mathbf{v} + \mathbf{u}\cdot\mathbf{w}$
> 4. $\mathbf{u}\cdot\mathbf{u} = |\mathbf{u}|^2$
> 5. $\mathbf{0}\cdot\mathbf{u} = 0$.

HISTORICAL BIOGRAPHY

Carl Friedrich Gauss
(1777–1855)

Proofs of Properties 1 and 3 The properties are easy to prove using the definition. For instance, here are the proofs of Properties 1 and 3.

1. $\mathbf{u}\cdot\mathbf{v} = u_1v_1 + u_2v_2 + u_3v_3 = v_1u_1 + v_2u_2 + v_3u_3 = \mathbf{v}\cdot\mathbf{u}$

3.
$$\begin{aligned}\mathbf{u}\cdot(\mathbf{v}+\mathbf{w}) &= \langle u_1, u_2, u_3\rangle\cdot\langle v_1 + w_1, v_2 + w_2, v_3 + w_3\rangle \\ &= u_1(v_1 + w_1) + u_2(v_2 + w_2) + u_3(v_3 + w_3) \\ &= u_1v_1 + u_1w_1 + u_2v_2 + u_2w_2 + u_3v_3 + u_3w_3 \\ &= (u_1v_1 + u_2v_2 + u_3v_3) + (u_1w_1 + u_2w_2 + u_3w_3) \\ &= \mathbf{u}\cdot\mathbf{v} + \mathbf{u}\cdot\mathbf{w}\end{aligned}$$
■

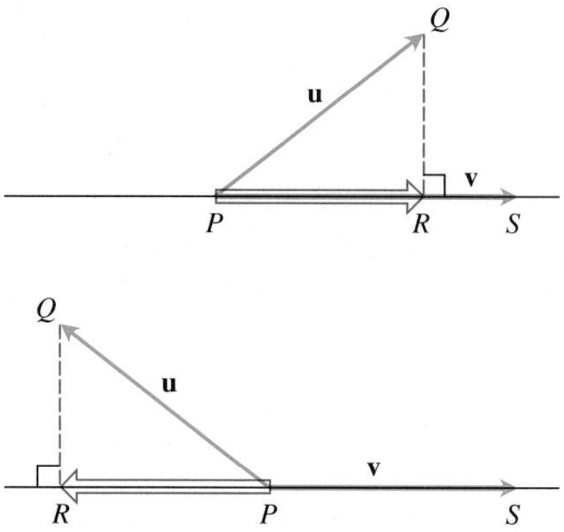

FIGURE 9.22 The vector projection of **u** onto **v**.

We now return to the problem of projecting one vector onto another, posed in the opening to this section. The **vector projection** of $\mathbf{u} = \overrightarrow{PQ}$ onto a nonzero vector $\mathbf{v} = \overrightarrow{PS}$ (Figure 9.22) is the vector $\overrightarrow{PR}$ determined by dropping a perpendicular from Q to the line PS. The notation for this vector is

$$\text{proj}_{\mathbf{v}}\,\mathbf{u} \qquad (\text{"the vector projection of } \mathbf{u} \text{ onto } \mathbf{v}\text{"}).$$

If **u** represents a force, then $\text{proj}_{\mathbf{v}}\,\mathbf{u}$ represents the effective force in the direction of **v** (Figure 9.23).

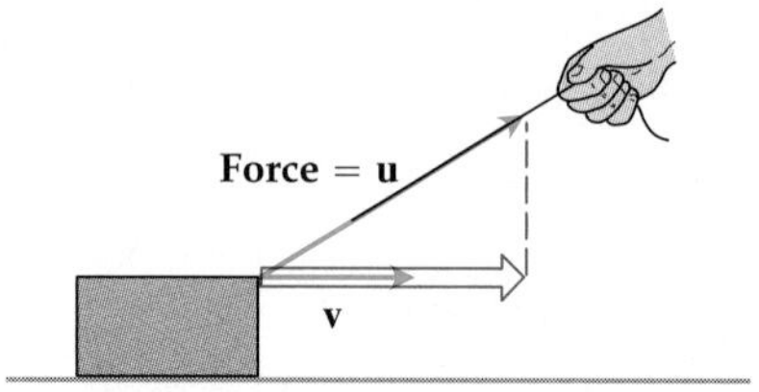

FIGURE 9.23 If we pull on the box with force **u**, the effective force moving the box forward in the direction **v** is the projection of **u** onto **v**.

If the angle θ between **u** and **v** is acute, $\text{proj}_{\mathbf{v}}\,\mathbf{u}$ has length $|\mathbf{u}|\cos\theta$ and direction $\mathbf{v}/|\mathbf{v}|$ (Figure 9.24). If θ is obtuse, $\cos\theta < 0$ and $\text{proj}_{\mathbf{v}}\,\mathbf{u}$ has length $-|\mathbf{u}|\cos\theta$ and direction $-\mathbf{v}/|\mathbf{v}|$. In both cases,

$$\begin{aligned}\text{proj}_{\mathbf{v}}\,\mathbf{u} &= (|\mathbf{u}|\cos\theta)\frac{\mathbf{v}}{|\mathbf{v}|} \\ &= \left(\frac{\mathbf{u}\cdot\mathbf{v}}{|\mathbf{v}|}\right)\frac{\mathbf{v}}{|\mathbf{v}|} \qquad |\mathbf{u}|\cos\theta = \frac{|\mathbf{u}||\mathbf{v}|\cos\theta}{|\mathbf{v}|} = \frac{\mathbf{u}\cdot\mathbf{v}}{|\mathbf{v}|} \\ &= \left(\frac{\mathbf{u}\cdot\mathbf{v}}{|\mathbf{v}|^2}\right)\mathbf{v}.\end{aligned}$$

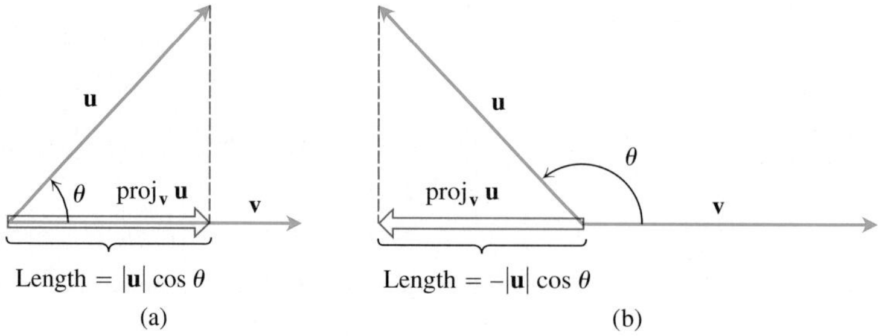

FIGURE 9.24 The length of $\text{proj}_{\mathbf{v}}\,\mathbf{u}$ is (a) $|\mathbf{u}|\cos\theta$ if $\cos\theta \geq 0$ and (b) $-|\mathbf{u}|\cos\theta$ if $\cos\theta < 0$.

The number $|\mathbf{u}| \cos \theta$ is called the **scalar component of u in the direction of v**. To summarize,

Vector projection of **u** onto **v**:

$$\operatorname{proj}_{\mathbf{v}} \mathbf{u} = \left(\frac{\mathbf{u} \cdot \mathbf{v}}{|\mathbf{v}|^2}\right)\mathbf{v} \tag{1}$$

Scalar component of **u** in the direction of **v**:

$$|\mathbf{u}| \cos \theta = \frac{\mathbf{u} \cdot \mathbf{v}}{|\mathbf{v}|} = \mathbf{u} \cdot \frac{\mathbf{v}}{|\mathbf{v}|} \tag{2}$$

Note that both the vector projection of **u** onto **v** and the scalar component of **u** onto **v** depend only on the direction of the vector **v** and not its length (because we dot **u** with $\mathbf{v}/|\mathbf{v}|$, which is the direction of **v**).

EXAMPLE 5 Find the vector projection of $\mathbf{u} = 6\mathbf{i} + 3\mathbf{j} + 2\mathbf{k}$ onto $\mathbf{v} = \mathbf{i} - 2\mathbf{j} - 2\mathbf{k}$ and the scalar component of **u** in the direction of **v**.

Solution We find $\operatorname{proj}_{\mathbf{v}} \mathbf{u}$ from Equation (1):

$$\begin{aligned} \operatorname{proj}_{\mathbf{v}} \mathbf{u} &= \frac{\mathbf{u} \cdot \mathbf{v}}{\mathbf{v} \cdot \mathbf{v}}\mathbf{v} = \frac{6 - 6 - 4}{1 + 4 + 4}(\mathbf{i} - 2\mathbf{j} - 2\mathbf{k}) \\ &= -\frac{4}{9}(\mathbf{i} - 2\mathbf{j} - 2\mathbf{k}) = -\frac{4}{9}\mathbf{i} + \frac{8}{9}\mathbf{j} + \frac{8}{9}\mathbf{k}. \end{aligned}$$

We find the scalar component of **u** in the direction of **v** from Equation (2):

$$\begin{aligned} |\mathbf{u}| \cos \theta &= \mathbf{u} \cdot \frac{\mathbf{v}}{|\mathbf{v}|} = (6\mathbf{i} + 3\mathbf{j} + 2\mathbf{k}) \cdot \left(\frac{1}{3}\mathbf{i} - \frac{2}{3}\mathbf{j} - \frac{2}{3}\mathbf{k}\right) \\ &= 2 - 2 - \frac{4}{3} = -\frac{4}{3}. \end{aligned}$$

■

Equations (1) and (2) also apply to two-dimensional vectors.

EXAMPLE 6 Find the vector projection of a force $\mathbf{F} = 5\mathbf{i} + 2\mathbf{j}$ onto $\mathbf{v} = \mathbf{i} - 3\mathbf{j}$ and the scalar component of **F** in the direction of **v**.

Solution The vector projection is

$$\begin{aligned} \operatorname{proj}_{\mathbf{v}} \mathbf{F} &= \left(\frac{\mathbf{F} \cdot \mathbf{v}}{|\mathbf{v}|^2}\right)\mathbf{v} \\ &= \frac{5 - 6}{1 + 9}(\mathbf{i} - 3\mathbf{j}) = -\frac{1}{10}(\mathbf{i} - 3\mathbf{j}) \\ &= -\frac{1}{10}\mathbf{i} + \frac{3}{10}\mathbf{j}. \end{aligned}$$

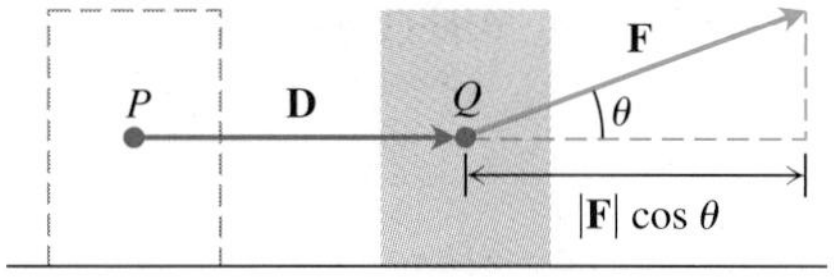

FIGURE 9.25 The work done by a constant force **F** during a displacement **D** is $(|\mathbf{F}| \cos\theta)|\mathbf{D}|$.

The scalar component of **F** in the direction of **v** is

$$|\mathbf{F}|\cos\theta = \frac{\mathbf{F}\cdot\mathbf{v}}{|\mathbf{v}|} = \frac{5-6}{\sqrt{1+9}} = -\frac{1}{\sqrt{10}}.$$ ■

Work

In Chapter 6, we calculated the work done by a constant force of magnitude F in moving an object through a distance d as $W = Fd$. That formula holds only if the force is directed along the line of motion. If a force **F** moving an object through a displacement $\mathbf{D} = \overrightarrow{PQ}$ has some other direction, the work is performed by the component of **F** in the direction of **D**. If θ is the angle between **F** and **D** (Figure 9.25), then

$$\begin{aligned}\text{Work} &= \begin{pmatrix}\text{scalar component of } \mathbf{F}\\ \text{in the direction of } \mathbf{D}\end{pmatrix}(\text{length of } \mathbf{D})\\ &= (|\mathbf{F}|\cos\theta)|\mathbf{D}|\\ &= \mathbf{F}\cdot\mathbf{D}.\end{aligned}$$

> **DEFINITION** The **work** done by a constant force **F** acting through a displacement $\mathbf{D} = \overrightarrow{PQ}$ is
>
> $$W = \mathbf{F}\cdot\mathbf{D} = |\mathbf{F}||\mathbf{D}|\cos\theta,$$
>
> where θ is the angle between **F** and **D**.

EXAMPLE 7 If $|\mathbf{F}| = 40$ N (newtons), $|\mathbf{D}| = 3$ m, and $\theta = 60°$, the work done by **F** in acting from P to Q is

$$\begin{aligned}\text{Work} &= |\mathbf{F}||\mathbf{D}|\cos\theta && \text{Definition}\\ &= (40)(3)\cos 60° && \text{Given values}\\ &= (120)(1/2)\\ &= 60 \text{ J (joules)}.\end{aligned}$$ ■

We encounter more challenging work problems in Chapter 13 when we learn to find the work done by a variable force along a *path* in space.

EXERCISES 9.3

In Exercises 1–8, find

a. $\mathbf{v}\cdot\mathbf{u}$, $|\mathbf{v}|$, $|\mathbf{u}|$

b. the cosine of the angle between **v** and **u**

c. the scalar component of **u** in the direction of **v**

d. the vector $\text{proj}_{\mathbf{v}}\,\mathbf{u}$.

1. $\mathbf{v} = 2\mathbf{i} - 4\mathbf{j} + \sqrt{5}\mathbf{k}$, $\mathbf{u} = -2\mathbf{i} + 4\mathbf{j} - \sqrt{5}\mathbf{k}$

2. $\mathbf{v} = (3/5)\mathbf{i} + (4/5)\mathbf{k}$, $\mathbf{u} = 5\mathbf{i} + 12\mathbf{j}$

3. $\mathbf{v} = 10\mathbf{i} + 11\mathbf{j} - 2\mathbf{k}$, $\mathbf{u} = 3\mathbf{j} + 4\mathbf{k}$

4. $\mathbf{v} = 2\mathbf{i} + 10\mathbf{j} - 11\mathbf{k}$, $\mathbf{u} = 2\mathbf{i} + 2\mathbf{j} + \mathbf{k}$

5. $\mathbf{v} = 5\mathbf{j} - 3\mathbf{k}$, $\mathbf{u} = \mathbf{i} + \mathbf{j} + \mathbf{k}$

6. $\mathbf{v} = -\mathbf{i} + \mathbf{j}$, $\mathbf{u} = \sqrt{2}\mathbf{i} + \sqrt{3}\mathbf{j} + 2\mathbf{k}$

7. $\mathbf{v} = 5\mathbf{i} + \mathbf{j}$, $\mathbf{u} = 2\mathbf{i} + \sqrt{17}\mathbf{j}$

8. $\mathbf{v} = \left\langle \frac{1}{\sqrt{2}}, \frac{1}{\sqrt{3}} \right\rangle$, $\mathbf{u} = \left\langle \frac{1}{\sqrt{2}}, -\frac{1}{\sqrt{3}} \right\rangle$

T Find the angles between the vectors in Exercises 9–12 to the nearest hundredth of a radian.

9. $\mathbf{u} = 2\mathbf{i} + \mathbf{j}, \quad \mathbf{v} = \mathbf{i} + 2\mathbf{j} - \mathbf{k}$

10. $\mathbf{u} = 2\mathbf{i} - 2\mathbf{j} + \mathbf{k}, \quad \mathbf{v} = 3\mathbf{i} + 4\mathbf{k}$

11. $\mathbf{u} = \sqrt{3}\mathbf{i} - 7\mathbf{j}, \quad \mathbf{v} = \sqrt{3}\mathbf{i} + \mathbf{j} - 2\mathbf{k}$

12. $\mathbf{u} = \mathbf{i} + \sqrt{2}\mathbf{j} - \sqrt{2}\mathbf{k}, \quad \mathbf{v} = -\mathbf{i} + \mathbf{j} + \mathbf{k}$

13. Triangle Find the measures of the angles of the triangle whose vertices are $A = (-1, 0)$, $B = (2, 1)$, and $C = (1, -2)$.

14. Rectangle Find the measures of the angles between the diagonals of the rectangle whose vertices are $A = (1, 0)$, $B = (0, 3)$, $C = (3, 4)$, and $D = (4, 1)$.

15. Direction angles and direction cosines The *direction angles* α, β, and γ of a vector $\mathbf{v} = a\mathbf{i} + b\mathbf{j} + c\mathbf{k}$ are defined as follows:

α is the angle between $\mathbf{v}$ and the positive x-axis $(0 \le \alpha \le \pi)$

β is the angle between $\mathbf{v}$ and the positive y-axis $(0 \le \beta \le \pi)$

γ is the angle between $\mathbf{v}$ and the positive z-axis $(0 \le \gamma \le \pi)$.

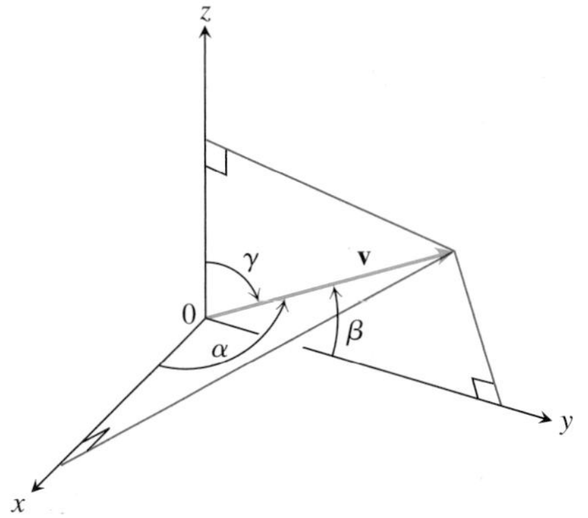

a. Show that

$$\cos\alpha = \frac{a}{|\mathbf{v}|}, \qquad \cos\beta = \frac{b}{|\mathbf{v}|}, \qquad \cos\gamma = \frac{c}{|\mathbf{v}|},$$

and $\cos^2\alpha + \cos^2\beta + \cos^2\gamma = 1$. These cosines are called the *direction cosines* of $\mathbf{v}$.

b. Unit vectors are built from direction cosines Show that if $\mathbf{v} = a\mathbf{i} + b\mathbf{j} + c\mathbf{k}$ is a unit vector, then a, b, and c are the direction cosines of $\mathbf{v}$.

16. Water main construction A water main is to be constructed with a 20% grade in the north direction and a 10% grade in the east direction. Determine the angle θ required in the water main for the turn from north to east.

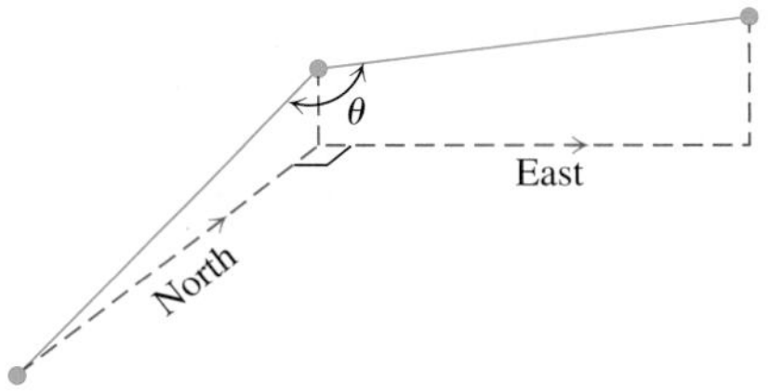

17. Diagonals of a rhombus Show that the diagonals of a rhombus (parallelogram with sides of equal length) are perpendicular.

18. Perpendicular diagonals Show that squares are the only rectangles with perpendicular diagonals.

19. When parallelograms are rectangles Prove that a parallelogram is a rectangle if and only if its diagonals are equal in length. (This fact is often exploited by carpenters.)

20. Diagonal of parallelogram Show that the indicated diagonal of the parallelogram determined by vectors $\mathbf{u}$ and $\mathbf{v}$ bisects the angle between $\mathbf{u}$ and $\mathbf{v}$ if $|\mathbf{u}| = |\mathbf{v}|$.

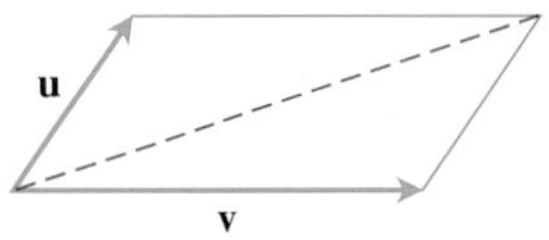

21. Projectile motion A gun with muzzle velocity of 1200 ft/sec is fired at an angle of 8° above the horizontal. Find the horizontal and vertical components of the velocity.

22. Inclined plane Suppose that a box is being towed up an inclined plane as shown in the figure. Find the force $\mathbf{w}$ needed to make the component of the force parallel to the inclined plane equal to 2.5 lb.

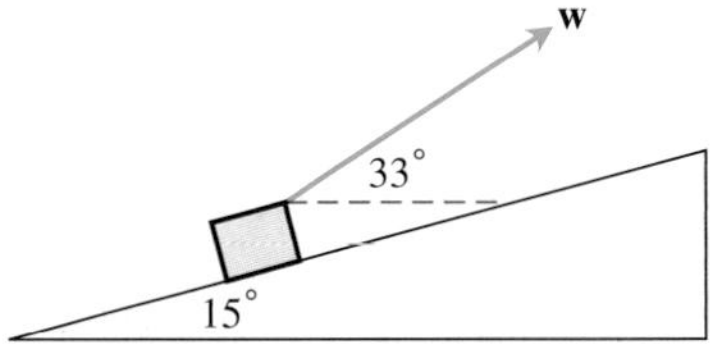

23. a. Cauchy-Schwartz inequality Use the fact that $\mathbf{u}\cdot\mathbf{v} = |\mathbf{u}||\mathbf{v}|\cos\theta$ to show that the inequality $|\mathbf{u}\cdot\mathbf{v}| \le |\mathbf{u}||\mathbf{v}|$ holds for any vectors $\mathbf{u}$ and $\mathbf{v}$.

b. Under what circumstances, if any, does $|\mathbf{u}\cdot\mathbf{v}|$ equal $|\mathbf{u}||\mathbf{v}|$? Give reasons for your answer.

24. Cancellation in dot products In real-number multiplication, if $uv_1 = uv_2$ and $u \ne 0$, we can cancel the u and conclude that $v_1 = v_2$. Does the same rule hold for the dot product: If $\mathbf{u}\cdot\mathbf{v}_1 = \mathbf{u}\cdot\mathbf{v}_2$ and $\mathbf{u} \ne \mathbf{0}$, can you conclude that $\mathbf{v}_1 = \mathbf{v}_2$? Give reasons for your answer.

25. Line perpendicular to a vector Show that the vector $\mathbf{v} = a\mathbf{i} + b\mathbf{j}$ is perpendicular to the line $ax + by = c$ by establishing that the slope of $\mathbf{v}$ is the negative reciprocal of the slope of the given line.

26. Line parallel to a vector Show that the vector $\mathbf{v} = a\mathbf{i} + b\mathbf{j}$ is parallel to the line $bx - ay = c$ by establishing that the slope of the line segment representing $\mathbf{v}$ is the same as the slope of the given line.

In Exercises 27–30, use the result of Exercise 25 to find an equation for the line through P perpendicular to $\mathbf{v}$. Then sketch the line. Include $\mathbf{v}$ in your sketch *as a vector starting at the origin.*

27. $P(2, 1), \quad \mathbf{v} = \mathbf{i} + 2\mathbf{j}$

28. $P(-1, 2), \quad \mathbf{v} = -2\mathbf{i} - \mathbf{j}$

29. $P(-2, -7), \quad \mathbf{v} = -2\mathbf{i} + \mathbf{j}$

30. $P(11, 10), \quad \mathbf{v} = 2\mathbf{i} - 3\mathbf{j}$

In Exercises 31–34, use the result of Exercise 26 to find an equation for the line through P parallel to $\mathbf{v}$. Then sketch the line. Include $\mathbf{v}$ in your sketch *as a vector starting at the origin*.

31. $P(-2, 1)$, $\mathbf{v} = \mathbf{i} - \mathbf{j}$

32. $P(0, -2)$, $\mathbf{v} = 2\mathbf{i} + 3\mathbf{j}$

33. $P(1, 2)$, $\mathbf{v} = -\mathbf{i} - 2\mathbf{j}$

34. $P(1, 3)$, $\mathbf{v} = 3\mathbf{i} - 2\mathbf{j}$

35. Work along a line Find the work done by a force $\mathbf{F} = 5\mathbf{i}$ (magnitude 5 N) in moving an object along the line from the origin to the point (1, 1) (distance in meters).

36. Inclined plane How much work does it take to slide a crate 20 m along a loading dock by pulling on it with a 200 N force at an angle of 30° from the horizontal?

The **acute angle between intersecting lines** that do not cross at right angles is the same as the angle determined by vectors normal to the lines or by the vectors parallel to the lines.

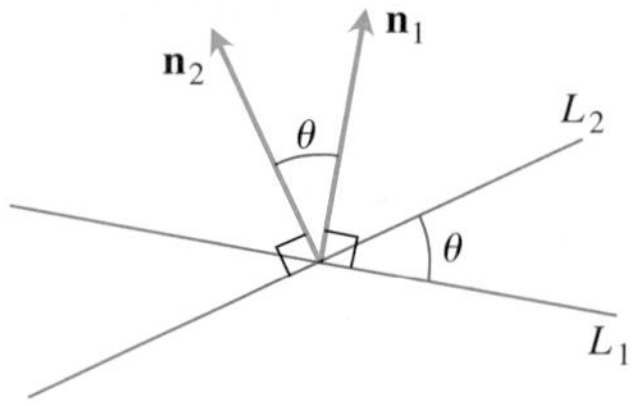

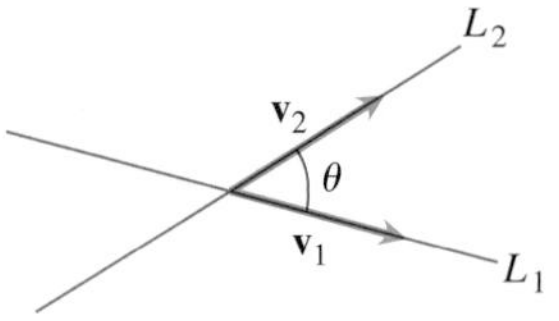

Use this fact and the results of Exercise 25 or 26 to find the acute angles between the lines in Exercises 37–40.

37. $3x + y = 5$, $2x - y = 4$

38. $y = \sqrt{3}x - 1$, $y = -\sqrt{3}x + 2$

39. $\sqrt{3}x - y = -2$, $x - \sqrt{3}y = 1$

40. $3x - 4y = 3$, $x - y = 7$

9.4 The Cross Product

In studying lines in the plane, when we needed to describe how a line was tilting, we used the notions of slope and angle of inclination. In space, we want a way to describe how a *plane* is tilting. We accomplish this by multiplying two vectors in the plane together to get a third vector perpendicular to the plane. The direction of this third vector tells us the "inclination" of the plane. The product we use to multiply the vectors together is the *vector* or *cross product*, the second of the two vector multiplication methods. We study the cross product in this section.

The Cross Product of Two Vectors in Space

We start with two nonzero vectors $\mathbf{u}$ and $\mathbf{v}$ in space. If $\mathbf{u}$ and $\mathbf{v}$ are not parallel, they determine a plane. We select a unit vector $\mathbf{n}$ perpendicular to the plane by the **right-hand rule**. This means that we choose $\mathbf{n}$ to be the unit (normal) vector that points the way your right thumb points when your fingers curl through the angle θ from $\mathbf{u}$ to $\mathbf{v}$ (Figure 9.26). Then the **cross product $\mathbf{u} \times \mathbf{v}$** ("**u** cross **v**") is the *vector* defined as follows.

DEFINITION

$$\mathbf{u} \times \mathbf{v} = (|\mathbf{u}||\mathbf{v}| \sin \theta)\, \mathbf{n}$$

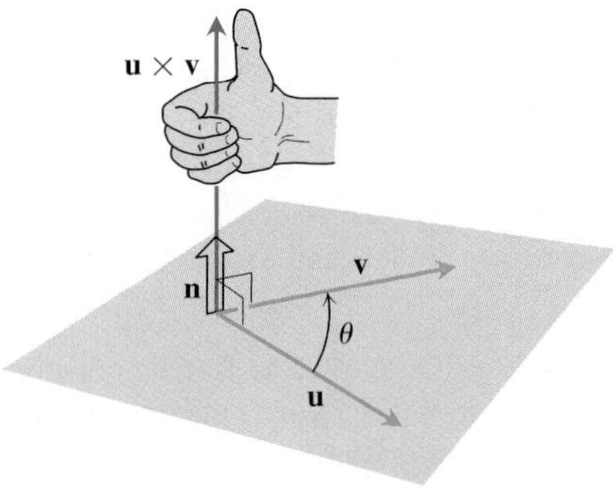

FIGURE 9.26 The construction of $\mathbf{u} \times \mathbf{v}$.

Unlike the dot product, the cross product is a vector. For this reason it's also called the **vector product** of $\mathbf{u}$ and $\mathbf{v}$, and applies *only* to vectors in space. The vector $\mathbf{u} \times \mathbf{v}$ is orthogonal to both $\mathbf{u}$ and $\mathbf{v}$ because it is a scalar multiple of $\mathbf{n}$.

There is a straightforward way to calculate the cross product of two vectors from their components. The method does not require that we know the angle between them (as suggested by the definition), but we postpone that calculation momentarily so we can focus first on the properties of the cross product.

Since the sines of 0 and π are both zero, it makes sense to define the cross product of two parallel nonzero vectors to be $\mathbf{0}$. If one or both of $\mathbf{u}$ and $\mathbf{v}$ are zero, we also define $\mathbf{u} \times \mathbf{v}$ to be zero. This way, the cross product of two vectors $\mathbf{u}$ and $\mathbf{v}$ is zero if and only if $\mathbf{u}$ and $\mathbf{v}$ are parallel or one or both of them are zero.

Parallel Vectors

Nonzero vectors $\mathbf{u}$ and $\mathbf{v}$ are parallel if and only if $\mathbf{u} \times \mathbf{v} = \mathbf{0}$.

The cross product obeys the following laws.

Properties of the Cross Product

If $\mathbf{u}$, $\mathbf{v}$, and $\mathbf{w}$ are any vectors and r, s are scalars, then

1. $(r\mathbf{u}) \times (s\mathbf{v}) = (rs)(\mathbf{u} \times \mathbf{v})$
2. $\mathbf{u} \times (\mathbf{v} + \mathbf{w}) = \mathbf{u} \times \mathbf{v} + \mathbf{u} \times \mathbf{w}$
3. $(\mathbf{v} + \mathbf{w}) \times \mathbf{u} = \mathbf{v} \times \mathbf{u} + \mathbf{w} \times \mathbf{u}$
4. $\mathbf{v} \times \mathbf{u} = -(\mathbf{u} \times \mathbf{v})$
5. $\mathbf{0} \times \mathbf{u} = \mathbf{0}$
6. $\mathbf{u} \times (\mathbf{v} \times \mathbf{w}) = (\mathbf{u} \cdot \mathbf{w})\mathbf{v} - (\mathbf{u} \cdot \mathbf{v})\mathbf{w}$

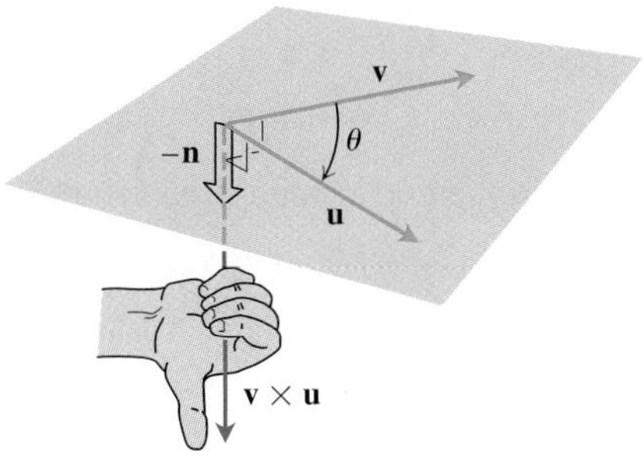

FIGURE 9.27 The construction of $\mathbf{v} \times \mathbf{u}$.

To visualize Property 4, for example, notice that when the fingers of a right hand curl through the angle θ from $\mathbf{v}$ to $\mathbf{u}$, the thumb points the opposite way and the unit vector we choose in forming $\mathbf{v} \times \mathbf{u}$ is the negative of the one we choose in forming $\mathbf{u} \times \mathbf{v}$ (Figure 9.27).

Property 1 can be verified by applying the definition of cross product to both sides of the equation and comparing the results. Property 2 is proved in Appendix 10. Property 3 follows by multiplying both sides of the equation in Property 2 by -1 and reversing the order of the products using Property 4. Property 5 is a definition. The cross product $\mathbf{u} \times (\mathbf{v} \times \mathbf{w})$ that appears in Property 6 is called the **vector triple product.** Its proof is left as Exercise 32.

When we apply the definition to calculate the pairwise cross products of $\mathbf{i}$, $\mathbf{j}$, and $\mathbf{k}$, we find (Figure 9.28)

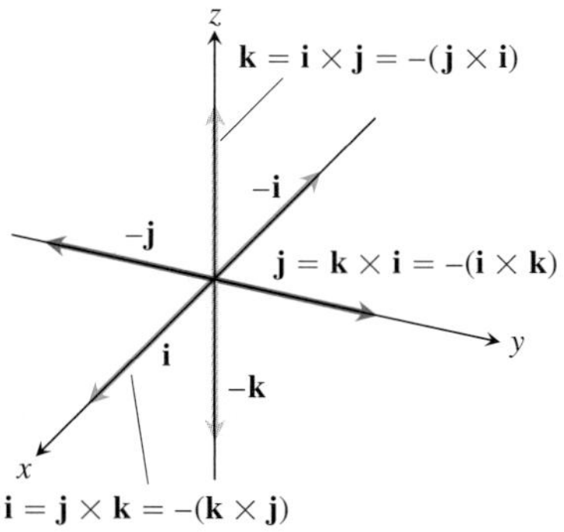

FIGURE 9.28 The pairwise cross products of $\mathbf{i}$, $\mathbf{j}$, and $\mathbf{k}$.

$$\mathbf{i} \times \mathbf{j} = -(\mathbf{j} \times \mathbf{i}) = \mathbf{k}$$
$$\mathbf{j} \times \mathbf{k} = -(\mathbf{k} \times \mathbf{j}) = \mathbf{i}$$
$$\mathbf{k} \times \mathbf{i} = -(\mathbf{i} \times \mathbf{k}) = \mathbf{j}$$

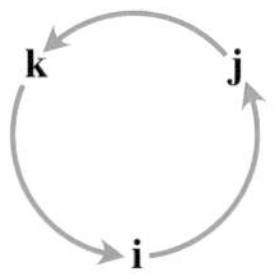

Diagram for recalling these products

and

$$\mathbf{i} \times \mathbf{i} = \mathbf{j} \times \mathbf{j} = \mathbf{k} \times \mathbf{k} = \mathbf{0}.$$

$|\mathbf{u} \times \mathbf{v}|$ Is the Area of a Parallelogram

Because $\mathbf{n}$ is a unit vector, the magnitude of $\mathbf{u} \times \mathbf{v}$ is

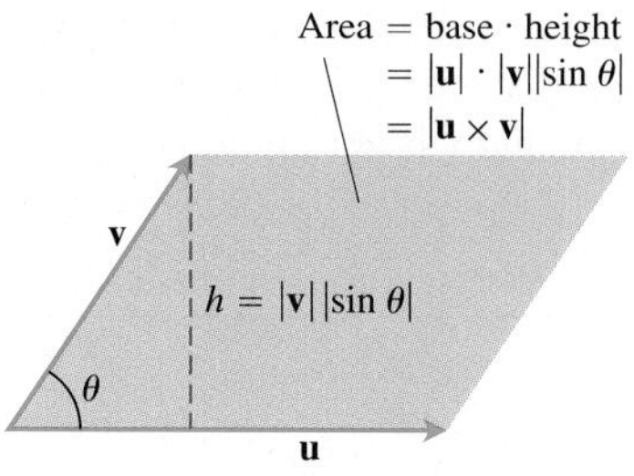

FIGURE 9.29 The parallelogram determined by **u** and **v**.

$$|\mathbf{u} \times \mathbf{v}| = |\mathbf{u}||\mathbf{v}|\,|\sin\theta|\,|\mathbf{n}| = |\mathbf{u}||\mathbf{v}|\sin\theta.$$

This is the area of the parallelogram determined by **u** and **v** (Figure 9.29), $|\mathbf{u}|$ being the base of the parallelogram and $|\mathbf{v}||\sin\theta|$ the height.

Determinant Formula for u × v

Our next objective is to calculate $\mathbf{u} \times \mathbf{v}$ from the components of **u** and **v** relative to a Cartesian coordinate system.

Suppose that

$$\mathbf{u} = u_1\mathbf{i} + u_2\mathbf{j} + u_3\mathbf{k}, \qquad \mathbf{v} = v_1\mathbf{i} + v_2\mathbf{j} + v_3\mathbf{k}.$$

Then the distributive laws and the rules for multiplying **i**, **j**, and **k** tell us that

$$\begin{aligned}
\mathbf{u} \times \mathbf{v} &= (u_1\mathbf{i} + u_2\mathbf{j} + u_3\mathbf{k}) \times (v_1\mathbf{i} + v_2\mathbf{j} + v_3\mathbf{k}) \\
&= u_1v_1\mathbf{i}\times\mathbf{i} + u_1v_2\mathbf{i}\times\mathbf{j} + u_1v_3\mathbf{i}\times\mathbf{k} \\
&\quad + u_2v_1\mathbf{j}\times\mathbf{i} + u_2v_2\mathbf{j}\times\mathbf{j} + u_2v_3\mathbf{j}\times\mathbf{k} \\
&\quad + u_3v_1\mathbf{k}\times\mathbf{i} + u_3v_2\mathbf{k}\times\mathbf{j} + u_3v_3\mathbf{k}\times\mathbf{k} \\
&= (u_2v_3 - u_3v_2)\mathbf{i} - (u_1v_3 - u_3v_1)\mathbf{j} + (u_1v_2 - u_2v_1)\mathbf{k}.
\end{aligned}$$

The terms in the last line are the same as the terms in the expansion of the symbolic determinant

$$\begin{vmatrix} \mathbf{i} & \mathbf{j} & \mathbf{k} \\ u_1 & u_2 & u_3 \\ v_1 & v_2 & v_3 \end{vmatrix}.$$

We therefore have the following rule.

> **Calculating Cross Products Using Determinants**
>
> If $\mathbf{u} = u_1\mathbf{i} + u_2\mathbf{j} + u_3\mathbf{k}$ and $\mathbf{v} = v_1\mathbf{i} + v_2\mathbf{j} + v_3\mathbf{k}$, then
>
> $$\mathbf{u} \times \mathbf{v} = \begin{vmatrix} \mathbf{i} & \mathbf{j} & \mathbf{k} \\ u_1 & u_2 & u_3 \\ v_1 & v_2 & v_3 \end{vmatrix}.$$

EXAMPLE 1 Find $\mathbf{u} \times \mathbf{v}$ and $\mathbf{v} \times \mathbf{u}$ if $\mathbf{u} = 2\mathbf{i} + \mathbf{j} + \mathbf{k}$ and $\mathbf{v} = -4\mathbf{i} + 3\mathbf{j} + \mathbf{k}$.

Solution

$$\mathbf{u} \times \mathbf{v} = \begin{vmatrix} \mathbf{i} & \mathbf{j} & \mathbf{k} \\ 2 & 1 & 1 \\ -4 & 3 & 1 \end{vmatrix} = \begin{vmatrix} 1 & 1 \\ 3 & 1 \end{vmatrix}\mathbf{i} - \begin{vmatrix} 2 & 1 \\ -4 & 1 \end{vmatrix}\mathbf{j} + \begin{vmatrix} 2 & 1 \\ -4 & 3 \end{vmatrix}\mathbf{k}$$

$$= -2\mathbf{i} - 6\mathbf{j} + 10\mathbf{k}$$

$$\mathbf{v} \times \mathbf{u} = -(\mathbf{u} \times \mathbf{v}) = 2\mathbf{i} + 6\mathbf{j} - 10\mathbf{k}$$ ■

Determinants

2×2 and 3×3 determinants are evaluated as follows:

$$\begin{vmatrix} a & b \\ c & d \end{vmatrix} = ad - bc$$

EXAMPLE

$$\begin{vmatrix} 2 & 1 \\ -4 & 3 \end{vmatrix} = (2)(3) - (1)(-4) = 6 + 4 = 10$$

$$\begin{vmatrix} a_1 & a_2 & a_3 \\ b_1 & b_2 & b_3 \\ c_1 & c_2 & c_3 \end{vmatrix} = a_1\begin{vmatrix} b_2 & b_3 \\ c_2 & c_3 \end{vmatrix} - a_2\begin{vmatrix} b_1 & b_3 \\ c_1 & c_3 \end{vmatrix} + a_3\begin{vmatrix} b_1 & b_2 \\ c_1 & c_2 \end{vmatrix}$$

EXAMPLE

$$\begin{vmatrix} -5 & 3 & 1 \\ 2 & 1 & 1 \\ -4 & 3 & 1 \end{vmatrix} = (-5)\begin{vmatrix} 1 & 1 \\ 3 & 1 \end{vmatrix} - (3)\begin{vmatrix} 2 & 1 \\ -4 & 1 \end{vmatrix} + (1)\begin{vmatrix} 2 & 1 \\ -4 & 3 \end{vmatrix}$$

$$= -5(1 - 3) - 3(2 + 4) + 1(6 + 4)$$

$$= 10 - 18 + 10 = 2$$

(For more information, see the Web site at **www.aw-bc.com/thomas**.)

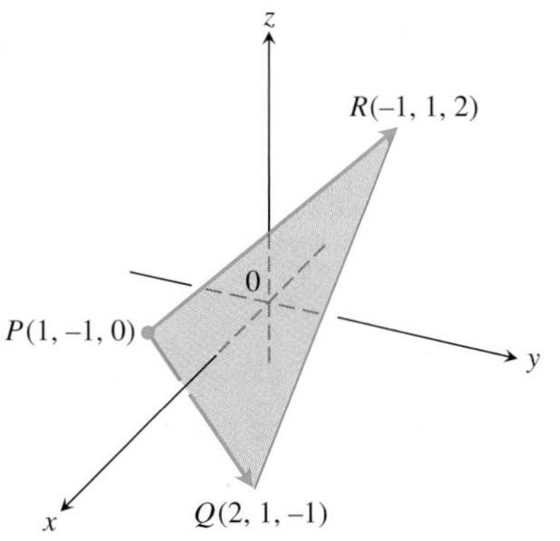

FIGURE 9.30 The area of triangle PQR is half of $|\overrightarrow{PQ} \times \overrightarrow{PR}|$ (Example 2).

EXAMPLE 2 Find a vector perpendicular to the plane of $P(1, -1, 0)$, $Q(2, 1, -1)$, and $R(-1, 1, 2)$ (Figure 9.30).

Solution The vector $\overrightarrow{PQ} \times \overrightarrow{PR}$ is perpendicular to the plane because it is perpendicular to both vectors. In terms of components,

$$\overrightarrow{PQ} = (2-1)\mathbf{i} + (1+1)\mathbf{j} + (-1-0)\mathbf{k} = \mathbf{i} + 2\mathbf{j} - \mathbf{k}$$

$$\overrightarrow{PR} = (-1-1)\mathbf{i} + (1+1)\mathbf{j} + (2-0)\mathbf{k} = -2\mathbf{i} + 2\mathbf{j} + 2\mathbf{k}$$

$$\overrightarrow{PQ} \times \overrightarrow{PR} = \begin{vmatrix} \mathbf{i} & \mathbf{j} & \mathbf{k} \\ 1 & 2 & -1 \\ -2 & 2 & 2 \end{vmatrix} = \begin{vmatrix} 2 & -1 \\ 2 & 2 \end{vmatrix}\mathbf{i} - \begin{vmatrix} 1 & -1 \\ -2 & 2 \end{vmatrix}\mathbf{j} + \begin{vmatrix} 1 & 2 \\ -2 & 2 \end{vmatrix}\mathbf{k}$$

$$= 6\mathbf{i} + 6\mathbf{k}.$$ ■

EXAMPLE 3 Find the area of the triangle with vertices $P(1, -1, 0)$, $Q(2, 1, -1)$, and $R(-1, 1, 2)$ (Figure 9.30).

Solution The area of the parallelogram determined by P, Q, and R is

$$|\overrightarrow{PQ} \times \overrightarrow{PR}| = |6\mathbf{i} + 6\mathbf{k}| \qquad \text{Values from Example 2.}$$

$$= \sqrt{(6)^2 + (6)^2} = \sqrt{2 \cdot 36} = 6\sqrt{2}.$$

The triangle's area is half of this, or $3\sqrt{2}$. ■

EXAMPLE 4 Find a unit vector perpendicular to the plane of $P(1, -1, 0)$, $Q(2, 1, -1)$, and $R(-1, 1, 2)$.

Solution Since $\overrightarrow{PQ} \times \overrightarrow{PR}$ is perpendicular to the plane, its direction $\mathbf{n}$ is a unit vector perpendicular to the plane. Taking values from Examples 2 and 3, we have

$$\mathbf{n} = \frac{\overrightarrow{PQ} \times \overrightarrow{PR}}{|\overrightarrow{PQ} \times \overrightarrow{PR}|} = \frac{6\mathbf{i} + 6\mathbf{k}}{6\sqrt{2}} = \frac{1}{\sqrt{2}}\mathbf{i} + \frac{1}{\sqrt{2}}\mathbf{k}.$$ ■

For ease in calculating the cross product using determinants, we usually write vectors in the form $\mathbf{v} = v_1\mathbf{i} + v_2\mathbf{j} + v_3\mathbf{k}$ rather than as ordered triples $\mathbf{v} = \langle v_1, v_2, v_3 \rangle$.

Torque

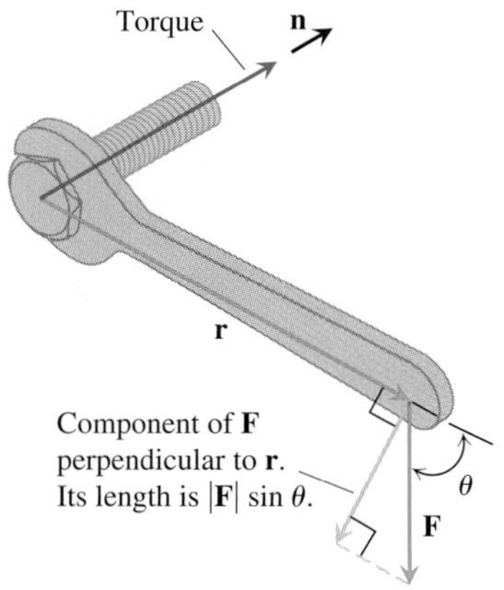

FIGURE 9.31 The torque vector describes the tendency of the force $\mathbf{F}$ to drive the bolt forward.

When we turn a bolt by applying a force $\mathbf{F}$ to a wrench (Figure 9.31), the torque we produce acts along the axis of the bolt to drive the bolt forward. The magnitude of the torque depends on how far out on the wrench the force is applied and on how much of the force is perpendicular to the wrench at the point of application. The number we use to measure the torque's magnitude is the product of the length of the lever arm $\mathbf{r}$ and the scalar component of $\mathbf{F}$ perpendicular to $\mathbf{r}$. In the notation of Figure 9.31,

$$\text{Magnitude of torque vector} = |\mathbf{r}|\,|\mathbf{F}|\sin\theta,$$

or $|\mathbf{r} \times \mathbf{F}|$. If we let $\mathbf{n}$ be a unit vector along the axis of the bolt in the direction of the torque, then a complete description of the torque vector is $\mathbf{r} \times \mathbf{F}$, or

$$\text{Torque vector} = (|\mathbf{r}||\mathbf{F}| \sin\theta)\,\mathbf{n}.$$

Recall that we defined $\mathbf{u} \times \mathbf{v}$ to be $\mathbf{0}$ when $\mathbf{u}$ and $\mathbf{v}$ are parallel. This is consistent with the torque interpretation as well. If the force $\mathbf{F}$ in Figure 9.31 is parallel to the wrench, meaning that we are trying to turn the bolt by pushing or pulling along the line of the wrench's handle, the torque produced is zero.

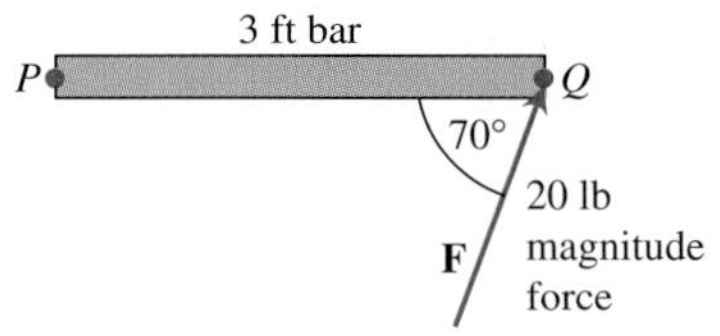

FIGURE 9.32 The magnitude of the torque exerted by **F** at P is about 56.4 ft-lb (Example 5).

EXAMPLE 5 The magnitude of the torque generated by force $\mathbf{F}$ at the pivot point P in Figure 9.32 is

$$\begin{aligned} |\overrightarrow{PQ} \times \mathbf{F}| &= |\overrightarrow{PQ}||\mathbf{F}| \sin 70^\circ \\ &\approx (3)(20)(0.94) \\ &\approx 56.4 \text{ ft-lb}. \end{aligned}$$

Triple Scalar or Box Product

The product $(\mathbf{u} \times \mathbf{v}) \cdot \mathbf{w}$ is called the **triple scalar product** of $\mathbf{u}$, $\mathbf{v}$, and $\mathbf{w}$ (in that order). As you can see from the formula

$$|(\mathbf{u} \times \mathbf{v}) \cdot \mathbf{w}| = |\mathbf{u} \times \mathbf{v}||\mathbf{w}||\cos\theta|,$$

the absolute value of the product is the volume of the parallelepiped (parallelogram-sided box) determined by $\mathbf{u}$, $\mathbf{v}$, and $\mathbf{w}$ (Figure 9.33). The number $|\mathbf{u} \times \mathbf{v}|$ is the area of the base parallelogram. The number $|\mathbf{w}||\cos\theta|$ is the parallelepiped's height. Because of this geometry, $(\mathbf{u} \times \mathbf{v}) \cdot \mathbf{w}$ is also called the **box product** of $\mathbf{u}$, $\mathbf{v}$, and $\mathbf{w}$.

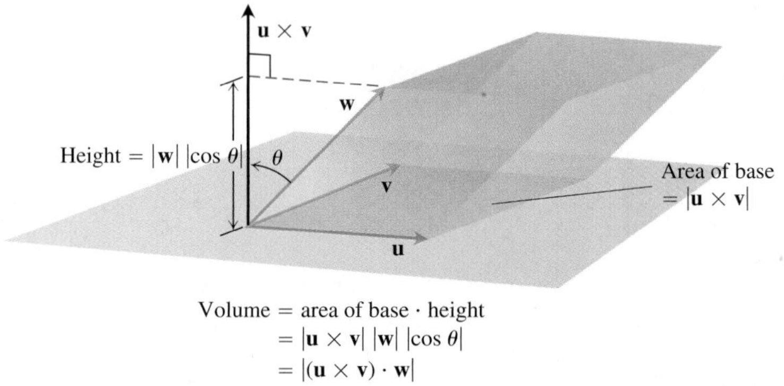

FIGURE 9.33 The number $|(\mathbf{u} \times \mathbf{v}) \cdot \mathbf{w}|$ is the volume of a parallelepiped.

By treating the planes of $\mathbf{v}$ and $\mathbf{w}$ and of $\mathbf{w}$ and $\mathbf{u}$ as the base planes of the parallelepiped determined by $\mathbf{u}$, $\mathbf{v}$, and $\mathbf{w}$, we see that

$$(\mathbf{u} \times \mathbf{v}) \cdot \mathbf{w} = (\mathbf{v} \times \mathbf{w}) \cdot \mathbf{u} = (\mathbf{w} \times \mathbf{u}) \cdot \mathbf{v}.$$

The dot and cross may be interchanged in a triple scalar product without altering its value.

Since the dot product is commutative, we also have

$$(\mathbf{u} \times \mathbf{v}) \cdot \mathbf{w} = \mathbf{u} \cdot (\mathbf{v} \times \mathbf{w}).$$

The triple scalar product can be evaluated as a determinant:

$$(\mathbf{u}\times\mathbf{v})\cdot\mathbf{w} = \left[\begin{vmatrix} u_2 & u_3 \\ v_2 & v_3 \end{vmatrix}\mathbf{i} - \begin{vmatrix} u_1 & u_3 \\ v_1 & v_3 \end{vmatrix}\mathbf{j} + \begin{vmatrix} u_1 & u_2 \\ v_1 & v_2 \end{vmatrix}\mathbf{k}\right]\cdot\mathbf{w}$$

$$= w_1\begin{vmatrix} u_2 & u_3 \\ v_2 & v_3 \end{vmatrix} - w_2\begin{vmatrix} u_1 & u_3 \\ v_1 & v_3 \end{vmatrix} + w_3\begin{vmatrix} u_1 & u_2 \\ v_1 & v_2 \end{vmatrix}$$

$$= \begin{vmatrix} u_1 & u_2 & u_3 \\ v_1 & v_2 & v_3 \\ w_1 & w_2 & w_3 \end{vmatrix}.$$

Calculating the Triple Scalar Product

$$(\mathbf{u}\times\mathbf{v})\cdot\mathbf{w} = \begin{vmatrix} u_1 & u_2 & u_3 \\ v_1 & v_2 & v_3 \\ w_1 & w_2 & w_3 \end{vmatrix}$$

EXAMPLE 6 Find the volume of the box (parallelepiped) determined by $\mathbf{u} = \mathbf{i} + 2\mathbf{j} - \mathbf{k}$, $\mathbf{v} = -2\mathbf{i} + 3\mathbf{k}$, and $\mathbf{w} = 7\mathbf{j} - 4\mathbf{k}$.

Solution Using the rule for calculating determinants, we find

$$(\mathbf{u}\times\mathbf{v})\cdot\mathbf{w} = \begin{vmatrix} 1 & 2 & -1 \\ -2 & 0 & 3 \\ 0 & 7 & -4 \end{vmatrix} = -23.$$

The volume is $|(\mathbf{u}\times\mathbf{v})\cdot\mathbf{w}| = 23$ units cubed. ■

EXERCISES 9.4

In Exercises 1–8, find the length and direction (when defined) of $\mathbf{u}\times\mathbf{v}$ and $\mathbf{v}\times\mathbf{u}$.

1. $\mathbf{u} = 2\mathbf{i} - 2\mathbf{j} - \mathbf{k}, \quad \mathbf{v} = \mathbf{i} - \mathbf{k}$
2. $\mathbf{u} = 2\mathbf{i} + 3\mathbf{j}, \quad \mathbf{v} = -\mathbf{i} + \mathbf{j}$
3. $\mathbf{u} = 2\mathbf{i} - 2\mathbf{j} + 4\mathbf{k}, \quad \mathbf{v} = -\mathbf{i} + \mathbf{j} - 2\mathbf{k}$
4. $\mathbf{u} = \mathbf{i} + \mathbf{j} - \mathbf{k}, \quad \mathbf{v} = \mathbf{0}$
5. $\mathbf{u} = 2\mathbf{i}, \quad \mathbf{v} = -3\mathbf{j}$
6. $\mathbf{u} = \mathbf{i}\times\mathbf{j}, \quad \mathbf{v} = \mathbf{j}\times\mathbf{k}$
7. $\mathbf{u} = -8\mathbf{i} - 2\mathbf{j} - 4\mathbf{k}, \quad \mathbf{v} = 2\mathbf{i} + 2\mathbf{j} + \mathbf{k}$
8. $\mathbf{u} = \frac{3}{2}\mathbf{i} - \frac{1}{2}\mathbf{j} + \mathbf{k}, \quad \mathbf{v} = \mathbf{i} + \mathbf{j} + 2\mathbf{k}$

In Exercises 9–14, sketch the coordinate axes and then include the vectors $\mathbf{u}$, $\mathbf{v}$ and $\mathbf{u}\times\mathbf{v}$ as vectors starting at the origin.

9. $\mathbf{u} = \mathbf{i}, \quad \mathbf{v} = \mathbf{j}$
10. $\mathbf{u} = \mathbf{i} - \mathbf{k}, \quad \mathbf{v} = \mathbf{j}$
11. $\mathbf{u} = \mathbf{i} - \mathbf{k}, \quad \mathbf{v} = \mathbf{j} + \mathbf{k}$
12. $\mathbf{u} = 2\mathbf{i} - \mathbf{j}, \quad \mathbf{v} = \mathbf{i} + 2\mathbf{j}$
13. $\mathbf{u} = \mathbf{i} + \mathbf{j}, \quad \mathbf{v} = \mathbf{i} - \mathbf{j}$
14. $\mathbf{u} = \mathbf{j} + 2\mathbf{k}, \quad \mathbf{v} = \mathbf{i}$

In Exercises 15–18,

a. Find the area of the triangle determined by the points P, Q, and R.
b. Find a unit vector perpendicular to plane PQR.

15. $P(1, -1, 2), \quad Q(2, 0, -1), \quad R(0, 2, 1)$
16. $P(1, 1, 1), \quad Q(2, 1, 3), \quad R(3, -1, 1)$

17. $P(2, -2, 1)$, $Q(3, -1, 2)$, $R(3, -1, 1)$

18. $P(-2, 2, 0)$, $Q(0, 1, -1)$, $R(-1, 2, -2)$

In Exercises 19–22, verify that $(\mathbf{u} \times \mathbf{v}) \cdot \mathbf{w} = (\mathbf{v} \times \mathbf{w}) \cdot \mathbf{u} = (\mathbf{w} \times \mathbf{u}) \cdot \mathbf{v}$ and find the volume of the parallelepiped (box) determined by **u**, **v**, and **w**.

	u	**v**	**w**
19.	$2\mathbf{i}$	$2\mathbf{j}$	$2\mathbf{k}$
20.	$\mathbf{i} - \mathbf{j} + \mathbf{k}$	$2\mathbf{i} + \mathbf{j} - 2\mathbf{k}$	$-\mathbf{i} + 2\mathbf{j} - \mathbf{k}$
21.	$2\mathbf{i} + \mathbf{j}$	$2\mathbf{i} - \mathbf{j} + \mathbf{k}$	$\mathbf{i} + 2\mathbf{k}$
22.	$\mathbf{i} + \mathbf{j} - 2\mathbf{k}$	$-\mathbf{i} - \mathbf{k}$	$2\mathbf{i} + 4\mathbf{j} - 2\mathbf{k}$

23. Parallel and perpendicular vectors Let $\mathbf{u} = 5\mathbf{i} - \mathbf{j} + \mathbf{k}$, $\mathbf{v} = \mathbf{j} - 5\mathbf{k}$, $\mathbf{w} = -15\mathbf{i} + 3\mathbf{j} - 3\mathbf{k}$. Which vectors, if any, are **(a)** perpendicular? **(b)** Parallel? Give reasons for your answers.

24. Parallel and perpendicular vectors Let $\mathbf{u} = \mathbf{i} + 2\mathbf{j} - \mathbf{k}$, $\mathbf{v} = -\mathbf{i} + \mathbf{j} + \mathbf{k}$, $\mathbf{w} = \mathbf{i} + \mathbf{k}$, $\mathbf{r} = -(\pi/2)\mathbf{i} - \pi\mathbf{j} + (\pi/2)\mathbf{k}$. Which vectors, if any, are **(a)** perpendicular? **(b)** Parallel? Give reasons for your answers.

In Exercises 25 and 26, find the magnitude of the torque exerted by **F** on the bolt at P if $|\overrightarrow{PQ}| = 8$ in. and $|\mathbf{F}| = 30$ lb. Answer in foot-pounds.

25.

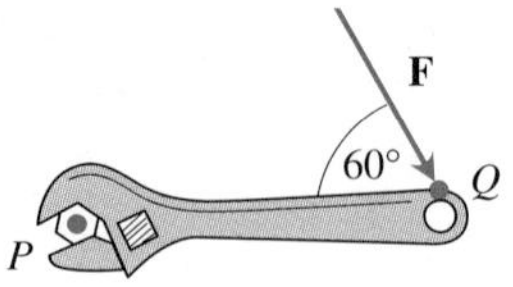

26.

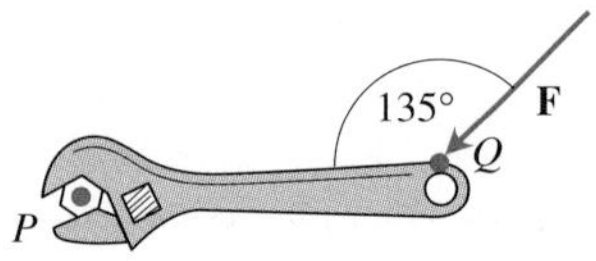

27. Which of the following are *always true*, and which are *not always true*? Give reasons for your answers.

a. $|\mathbf{u}| = \sqrt{\mathbf{u} \cdot \mathbf{u}}$

b. $\mathbf{u} \cdot \mathbf{u} = |\mathbf{u}|$

c. $\mathbf{u} \times \mathbf{0} = \mathbf{0} \times \mathbf{u} = \mathbf{0}$

d. $\mathbf{u} \times (-\mathbf{u}) = \mathbf{0}$

e. $\mathbf{u} \times \mathbf{v} = \mathbf{v} \times \mathbf{u}$

f. $\mathbf{u} \times (\mathbf{v} + \mathbf{w}) = \mathbf{u} \times \mathbf{v} + \mathbf{u} \times \mathbf{w}$

g. $(\mathbf{u} \times \mathbf{v}) \cdot \mathbf{v} = 0$

h. $(\mathbf{u} \times \mathbf{v}) \cdot \mathbf{w} = \mathbf{u} \cdot (\mathbf{v} \times \mathbf{w})$

28. Which of the following are *always true*, and which are *not always true*? Give reasons for your answers.

a. $\mathbf{u} \cdot \mathbf{v} = \mathbf{v} \cdot \mathbf{u}$ **b.** $\mathbf{u} \times \mathbf{v} = -(\mathbf{v} \times \mathbf{u})$

c. $(-\mathbf{u}) \times \mathbf{v} = -(\mathbf{u} \times \mathbf{v})$

d. $(c\mathbf{u}) \cdot \mathbf{v} = \mathbf{u} \cdot (c\mathbf{v}) = c(\mathbf{u} \cdot \mathbf{v})$ (any number c)

e. $c(\mathbf{u} \times \mathbf{v}) = (c\mathbf{u}) \times \mathbf{v} = \mathbf{u} \times (c\mathbf{v})$ (any number c)

f. $\mathbf{u} \cdot \mathbf{u} = |\mathbf{u}|^2$ **g.** $(\mathbf{u} \times \mathbf{u}) \cdot \mathbf{u} = 0$

h. $(\mathbf{u} \times \mathbf{v}) \cdot \mathbf{u} = \mathbf{v} \cdot (\mathbf{u} \times \mathbf{v})$

29. Given nonzero vectors **u**, **v**, and **w**, use dot product and cross product notation, as appropriate, to describe the following.

a. The vector projection of **u** onto **v**

b. A vector orthogonal to **u** and **v**

c. A vector orthogonal to $\mathbf{u} \times \mathbf{v}$ and **w**

d. The volume of the parallelepiped determined by **u**, **v**, and **w**

e. A vector orthogonal to $\mathbf{u} \times \mathbf{v}$ and $\mathbf{u} \times \mathbf{w}$

f. A vector of length $|\mathbf{u}|$ in the direction of **v**

30. Compute $(\mathbf{i} \times \mathbf{j}) \times \mathbf{j}$ and $\mathbf{i} \times (\mathbf{j} \times \mathbf{j})$. What can you conclude about the associativity of the cross product?

31. Let **u**, **v**, and **w** be vectors. Which of the following make sense, and which do not? Give reasons for your answers.

a. $(\mathbf{u} \times \mathbf{v}) \cdot \mathbf{w}$

b. $\mathbf{u} \times (\mathbf{v} \cdot \mathbf{w})$

c. $\mathbf{u} \times (\mathbf{v} \times \mathbf{w})$

d. $\mathbf{u} \cdot (\mathbf{v} \cdot \mathbf{w})$

32. Vector triple product Prove Property 6 of the cross product.

33. Cancellation in cross products If $\mathbf{u} \times \mathbf{v} = \mathbf{u} \times \mathbf{w}$ and $\mathbf{u} \neq \mathbf{0}$, then does $\mathbf{v} = \mathbf{w}$? Give reasons for your answer.

34. Double cancellation If $\mathbf{u} \neq \mathbf{0}$ and if $\mathbf{u} \times \mathbf{v} = \mathbf{u} \times \mathbf{w}$ and $\mathbf{u} \cdot \mathbf{v} = \mathbf{u} \cdot \mathbf{w}$, then does $\mathbf{v} = \mathbf{w}$? Give reasons for your answer.

Find the areas of the parallelograms whose vertices are given in Exercises 35–38.

35. $A(1, 0)$, $B(0, 1)$, $C(-1, 0)$, $D(0, -1)$

36. $A(0, 0)$, $B(7, 3)$, $C(9, 8)$, $D(2, 5)$

37. $A(-1, 2)$, $B(2, 0)$, $C(7, 1)$, $D(4, 3)$

38. $A(-6, 0)$, $B(1, -4)$, $C(3, 1)$, $D(-4, 5)$

Find the areas of the triangles whose vertices are given in Exercises 39–42.

39. $A(0, 0)$, $B(-2, 3)$, $C(3, 1)$

40. $A(-1, -1)$, $B(3, 3)$, $C(2, 1)$

41. $A(-5, 3)$, $B(1, -2)$, $C(6, -2)$

42. $A(-6, 0)$, $B(10, -5)$, $C(-2, 4)$

9.5 Lines and Planes in Space

This section shows how to use scalar and vector products to write equations for lines, line segments, and planes in space. We will use these representations throughout the rest of the book.

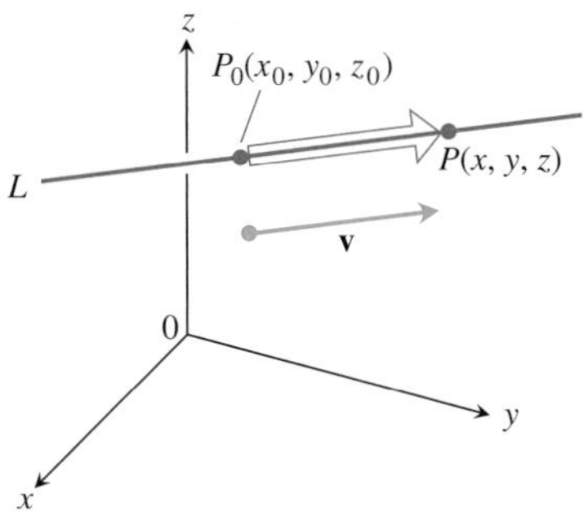

FIGURE 9.34 A point P lies on L through P_0 parallel to $\mathbf{v}$ if and only if $\overrightarrow{P_0P}$ is a scalar multiple of $\mathbf{v}$.

Lines and Line Segments in Space

In the plane, a line is determined by a point and a number giving the slope of the line. In space a line is determined by a point and a *vector* giving the direction of the line.

Suppose that L is a line in space passing through a point $P_0(x_0, y_0, z_0)$ parallel to a vector $\mathbf{v} = v_1\mathbf{i} + v_2\mathbf{j} + v_3\mathbf{k}$. Then L is the set of all points $P(x, y, z)$ for which $\overrightarrow{P_0P}$ is parallel to $\mathbf{v}$ (Figure 9.34). Thus, $\overrightarrow{P_0P} = t\mathbf{v}$ for some scalar parameter t. The value of t depends on the location of the point P along the line, and the domain of t is $(-\infty, \infty)$. The expanded form of the equation $\overrightarrow{P_0P} = t\mathbf{v}$ is

$$(x - x_0)\mathbf{i} + (y - y_0)\mathbf{j} + (z - z_0)\mathbf{k} = t(v_1\mathbf{i} + v_2\mathbf{j} + v_3\mathbf{k}),$$

which can be rewritten as

$$x\mathbf{i} + y\mathbf{j} + z\mathbf{k} = x_0\mathbf{i} + y_0\mathbf{j} + z_0\mathbf{k} + t(v_1\mathbf{i} + v_2\mathbf{j} + v_3\mathbf{k}). \tag{1}$$

If $\mathbf{r}(t)$ is the position vector of a point $P(x, y, z)$ on the line and $\mathbf{r}_0$ is the position vector of the point $P_0(x_0, y_0, z_0)$, then Equation (1) gives the following vector form for the equation of a line in space.

Vector Equation for a Line

A vector equation for the line L through $P_0(x_0, y_0, z_0)$ parallel to $\mathbf{v}$ is

$$\mathbf{r}(t) = \mathbf{r}_0 + t\mathbf{v}, \qquad -\infty < t < \infty, \tag{2}$$

where $\mathbf{r}$ is the position vector of a point $P(x, y, z)$ on L and $\mathbf{r}_0$ is the position vector of $P_0(x_0, y_0, z_0)$.

Equating the corresponding components of the two sides of Equation (1) gives three scalar equations involving the parameter t:

$$x = x_0 + tv_1, \qquad y = y_0 + tv_2, \qquad z = z_0 + tv_3.$$

These equations give us the standard parametrization of the line for the parameter interval $-\infty < t < \infty$.

Parametric Equations for a Line

The standard parametrization of the line through $P_0(x_0, y_0, z_0)$ parallel to $\mathbf{v} = v_1\mathbf{i} + v_2\mathbf{j} + v_3\mathbf{k}$ is

$$x = x_0 + tv_1, \quad y = y_0 + tv_2, \quad z = z_0 + tv_3, \quad -\infty < t < \infty \tag{3}$$

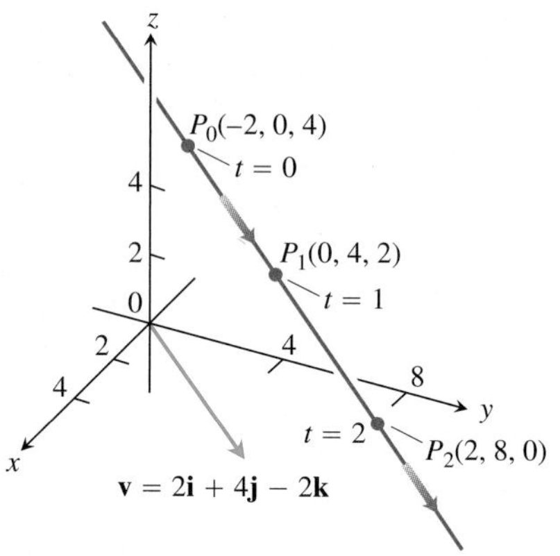

FIGURE 9.35 Selected points and parameter values on the line in Example 1. The arrows show the direction of increasing t.

EXAMPLE 1 Find parametric equations for the line through $(-2, 0, 4)$ parallel to $\mathbf{v} = 2\mathbf{i} + 4\mathbf{j} - 2\mathbf{k}$ (Figure 9.35).

Solution With $P_0(x_0, y_0, z_0)$ equal to $(-2, 0, 4)$ and $v_1\mathbf{i} + v_2\mathbf{j} + v_3\mathbf{k}$ equal to $2\mathbf{i} + 4\mathbf{j} - 2\mathbf{k}$, Equations (3) become

$$x = -2 + 2t, \qquad y = 4t, \qquad z = 4 - 2t.$$

■

EXAMPLE 2 Find parametric equations for the line through $P(-3, 2, -3)$ and $Q(1, -1, 4)$.

Solution The vector

$$\begin{aligned}\overrightarrow{PQ} &= (1 - (-3))\mathbf{i} + (-1 - 2)\mathbf{j} + (4 - (-3))\mathbf{k}\\ &= 4\mathbf{i} - 3\mathbf{j} + 7\mathbf{k}\end{aligned}$$

is parallel to the line, and Equations (3) with $(x_0, y_0, z_0) = (-3, 2, -3)$ give

$$x = -3 + 4t, \qquad y = 2 - 3t, \qquad z = -3 + 7t.$$

We could have chosen $Q(1, -1, 4)$ as the "base point" and written

$$x = 1 + 4t, \qquad y = -1 - 3t, \qquad z = 4 + 7t.$$

These equations serve as well as the first; they simply place you at a different point on the line for a given value of t. ■

Notice that parametrizations are not unique. Not only can the "base point" change, but so can the parameter. The equations $x = -3 + 4t^3$, $y = 2 - 3t^3$, and $z = -3 + 7t^3$ also parametrize the line in Example 2.

To parametrize a line segment joining two points, we first parametrize the line through the points. We then find the t-values for the endpoints and restrict t to lie in the closed interval bounded by these values. The line equations together with this added restriction parametrize the segment.

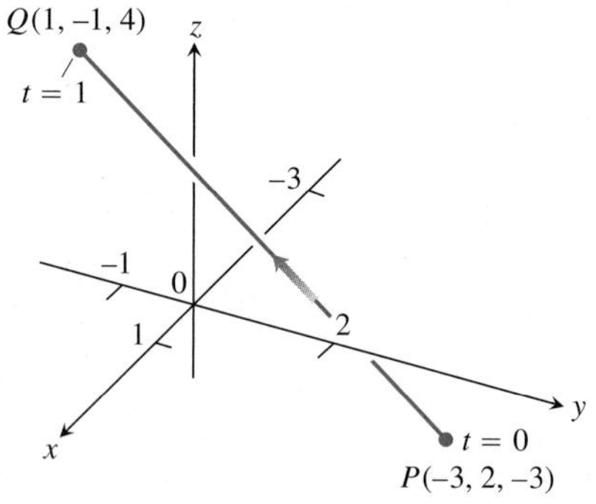

FIGURE 9.36 Example 3 derives a parametrization of line segment PQ. The arrow shows the direction of increasing t.

EXAMPLE 3 Parametrize the line segment joining the points $P(-3, 2, -3)$ and $Q(1, -1, 4)$ (Figure 9.36).

Solution We begin with equations for the line through P and Q, taking them, in this case, from Example 2:

$$x = -3 + 4t, \qquad y = 2 - 3t, \qquad z = -3 + 7t.$$

We observe that the point

$$(x, y, z) = (-3 + 4t, 2 - 3t, -3 + 7t)$$

on the line passes through $P(-3, 2, -3)$ at $t = 0$ and $Q(1, -1, 4)$ at $t = 1$. We add the restriction $0 \le t \le 1$ to parametrize the segment:

$$x = -3 + 4t, \qquad y = 2 - 3t, \qquad z = -3 + 7t, \qquad 0 \le t \le 1.$$

■

The vector form (Equation (2)) for a line in space is more revealing if we think of a line as the path of a particle starting at position $P_0(x_0, y_0, z_0)$ and moving in the direction of vector $\mathbf{v}$. Rewriting Equation (2), we have

$$\mathbf{r}(t) = \mathbf{r}_0 + t\mathbf{v}$$

$$= \underbrace{\mathbf{r}_0}_{\text{Initial position}} + \underbrace{t}_{\text{Time}}\underbrace{|\mathbf{v}|}_{\text{Speed}}\underbrace{\frac{\mathbf{v}}{|\mathbf{v}|}}_{\text{Direction}}. \tag{4}$$

In other words, the position of the particle at time t is its initial position plus its distance moved (speed $\times$ time) in the direction $\mathbf{v}/|\mathbf{v}|$ of its straight-line motion.

The Distance from a Point to a Line in Space

To find the distance from a point S to a line that passes through a point P parallel to a vector $\mathbf{v}$, we find the absolute value of the scalar component of $\overrightarrow{PS}$ in the direction of a vector normal to the line (Figure 9.37). In the notation of the figure, the absolute value of the scalar component is $|\overrightarrow{PS}|\sin\theta$, which is $\dfrac{|\overrightarrow{PS}\times\mathbf{v}|}{|\mathbf{v}|}$.

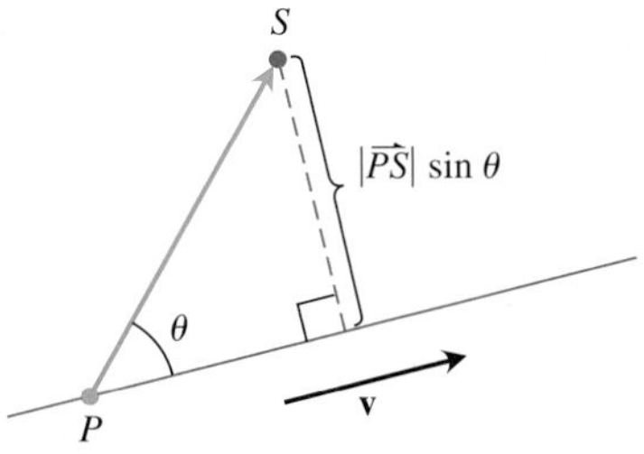

FIGURE 9.37 The distance from S to the line through P parallel to $\mathbf{v}$ is $|\overrightarrow{PS}|\sin\theta$, where θ is the angle between $\overrightarrow{PS}$ and $\mathbf{v}$.

Distance from a Point S to a Line Through P Parallel to v

$$d = \frac{|\overrightarrow{PS}\times\mathbf{v}|}{|\mathbf{v}|} \tag{5}$$

EXAMPLE 4 Find the distance from the point $S(1, 1, 5)$ to the line

$$L: \quad x = 1 + t, \quad y = 3 - t, \quad z = 2t.$$

Solution We see from the equations for L that L passes through $P(1, 3, 0)$ parallel to $\mathbf{v} = \mathbf{i} - \mathbf{j} + 2\mathbf{k}$. With

$$\overrightarrow{PS} = (1-1)\mathbf{i} + (1-3)\mathbf{j} + (5-0)\mathbf{k} = -2\mathbf{j} + 5\mathbf{k}$$

and

$$\overrightarrow{PS}\times\mathbf{v} = \begin{vmatrix} \mathbf{i} & \mathbf{j} & \mathbf{k} \\ 0 & -2 & 5 \\ 1 & -1 & 2 \end{vmatrix} = \mathbf{i} + 5\mathbf{j} + 2\mathbf{k},$$

Equation (5) gives

$$d = \frac{|\overrightarrow{PS}\times\mathbf{v}|}{|\mathbf{v}|} = \frac{\sqrt{1+25+4}}{\sqrt{1+1+4}} = \frac{\sqrt{30}}{\sqrt{6}} = \sqrt{5}.$$ ■

An Equation for a Plane in Space

A plane in space is determined by knowing a point on the plane and its “tilt” or orientation. This “tilt” is defined by specifying a vector that is perpendicular or normal to the plane.

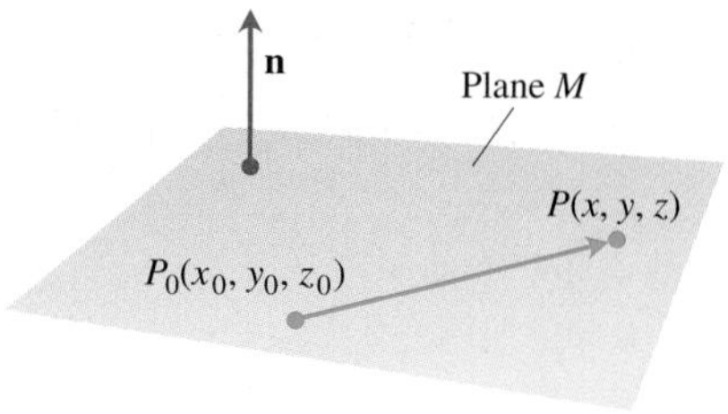

FIGURE 9.38 The standard equation for a plane in space is defined in terms of a vector normal to the plane: A point P lies in the plane through P_0 normal to $\mathbf{n}$ if and only if $\mathbf{n} \cdot \overrightarrow{P_0P} = 0$.

Suppose that plane M passes through a point $P_0(x_0, y_0, z_0)$ and is normal to the nonzero vector $\mathbf{n} = A\mathbf{i} + B\mathbf{j} + C\mathbf{k}$. Then M is the set of all points $P(x, y, z)$ for which $\overrightarrow{P_0P}$ is orthogonal to $\mathbf{n}$ (Figure 9.38). Thus, the dot product $\mathbf{n} \cdot \overrightarrow{P_0P} = 0$. This equation is equivalent to

$$(A\mathbf{i} + B\mathbf{j} + C\mathbf{k}) \cdot [(x - x_0)\mathbf{i} + (y - y_0)\mathbf{j} + (z - z_0)\mathbf{k}] = 0$$

or

$$A(x - x_0) + B(y - y_0) + C(z - z_0) = 0.$$

Equation for a Plane

The plane through $P_0(x_0, y_0, z_0)$ normal to $\mathbf{n} = A\mathbf{i} + B\mathbf{j} + C\mathbf{k}$ has

Vector equation:	$\mathbf{n} \cdot \overrightarrow{P_0P} = 0$
Component equation:	$A(x - x_0) + B(y - y_0) + C(z - z_0) = 0$
Component equation simplified:	$Ax + By + Cz = D$, where $D = Ax_0 + By_0 + Cz_0$

EXAMPLE 5 Find an equation for the plane through $P_0(-3, 0, 7)$ perpendicular to $\mathbf{n} = 5\mathbf{i} + 2\mathbf{j} - \mathbf{k}$.

Solution The component equation is

$$5(x - (-3)) + 2(y - 0) + (-1)(z - 7) = 0.$$

Simplifying, we obtain

$$5x + 15 + 2y - z + 7 = 0$$
$$5x + 2y - z = -22.$$ ■

Notice in Example 5 how the components of $\mathbf{n} = 5\mathbf{i} + 2\mathbf{j} - \mathbf{k}$ became the coefficients of x, y, and z in the equation $5x + 2y - z = -22$. The vector $\mathbf{n} = A\mathbf{i} + B\mathbf{j} + C\mathbf{k}$ is normal to the plane $Ax + By + Cz = D$.

EXAMPLE 6 Find an equation for the plane through $A(0, 0, 1)$, $B(2, 0, 0)$, and $C(0, 3, 0)$.

Solution We find a vector normal to the plane and use it with one of the points (it does not matter which) to write an equation for the plane.

The cross product

$$\overrightarrow{AB} \times \overrightarrow{AC} = \begin{vmatrix} \mathbf{i} & \mathbf{j} & \mathbf{k} \\ 2 & 0 & -1 \\ 0 & 3 & -1 \end{vmatrix} = 3\mathbf{i} + 2\mathbf{j} + 6\mathbf{k}$$

is normal to the plane. We substitute the components of this vector and the coordinates of $A(0, 0, 1)$ into the component form of the equation to obtain

$$3(x - 0) + 2(y - 0) + 6(z - 1) = 0$$
$$3x + 2y + 6z = 6.$$ ■

Lines of Intersection

Just as lines are parallel if and only if they have the same direction, two planes are **parallel** if and only if their normals are parallel, or $\mathbf{n}_1 = k\mathbf{n}_2$ for some scalar k. Two planes that are not parallel intersect in a line.

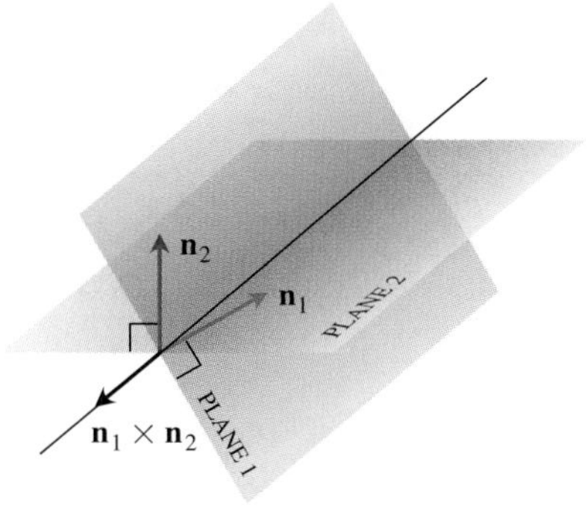

FIGURE 9.39 How the line of intersection of two planes is related to the planes' normal vectors (Example 7).

EXAMPLE 7 Find a vector parallel to the line of intersection of the planes $3x - 6y - 2z = 15$ and $2x + y - 2z = 5$.

Solution The line of intersection of two planes is perpendicular to both planes' normal vectors $\mathbf{n}_1$ and $\mathbf{n}_2$ (Figure 9.39) and therefore parallel to $\mathbf{n}_1 \times \mathbf{n}_2$. Turning this around, $\mathbf{n}_1 \times \mathbf{n}_2$ is a vector parallel to the planes' line of intersection. In our case,

$$\mathbf{n}_1 \times \mathbf{n}_2 = \begin{vmatrix} \mathbf{i} & \mathbf{j} & \mathbf{k} \\ 3 & -6 & -2 \\ 2 & 1 & -2 \end{vmatrix} = 14\mathbf{i} + 2\mathbf{j} + 15\mathbf{k}.$$

Any nonzero scalar multiple of $\mathbf{n}_1 \times \mathbf{n}_2$ will do as well. ■

EXAMPLE 8 Find parametric equations for the line in which the planes $3x - 6y - 2z = 15$ and $2x + y - 2z = 5$ intersect.

Solution We find a vector parallel to the line and a point on the line and use Equations (3).

Example 7 identifies $\mathbf{v} = 14\mathbf{i} + 2\mathbf{j} + 15\mathbf{k}$ as a vector parallel to the line. To find a point on the line, we can take any point common to the two planes. Substituting $z = 0$ in the plane equations and solving for x and y simultaneously identifies one of these points as $(3, -1, 0)$. The line is

$$x = 3 + 14t, \qquad y = -1 + 2t, \qquad z = 15t.$$

The choice $z = 0$ is arbitrary and we could have chosen $z = 1$ or $z = -1$ just as well. Or we could have let $x = 0$ and solved for y and z. The different choices would simply give different parametrizations of the same line. ■

Sometimes we want to know where a line and a plane intersect. For example, if we are looking at a flat plate and a line segment passes through it, we may be interested in knowing what portion of the line segment is hidden from our view by the plate. This application is used in computer graphics (Exercise 74).

EXAMPLE 9 Find the point where the line

$$x = \frac{8}{3} + 2t, \qquad y = -2t, \qquad z = 1 + t$$

intersects the plane $3x + 2y + 6z = 6$.

Solution The point

$$\left(\frac{8}{3} + 2t, -2t, 1 + t\right)$$

lies in the plane if its coordinates satisfy the equation of the plane, that is, if

$$\begin{aligned} 3\left(\frac{8}{3}+2t\right)+2(-2t)+6(1+t) &= 6 \\ 8+6t-4t+6+6t &= 6 \\ 8t &= -8 \\ t &= -1. \end{aligned}$$

The point of intersection is

$$(x, y, z)\big|_{t=-1} = \left(\frac{8}{3}-2, 2, 1-1\right) = \left(\frac{2}{3}, 2, 0\right).$$

■

The Distance from a Point to a Plane

If P is a point on a plane with normal $\mathbf{n}$, then the distance from any point S to the plane is the length of the vector projection of $\overrightarrow{PS}$ onto $\mathbf{n}$. That is, the distance from S to the plane is

$$d = \left|\overrightarrow{PS}\cdot\frac{\mathbf{n}}{|\mathbf{n}|}\right| \tag{6}$$

where $\mathbf{n} = A\mathbf{i} + B\mathbf{j} + C\mathbf{k}$ is normal to the plane.

EXAMPLE 10 Find the distance from $S(1, 1, 3)$ to the plane $3x + 2y + 6z = 6$.

Solution We find a point P in the plane and calculate the length of the vector projection of $\overrightarrow{PS}$ onto a vector $\mathbf{n}$ normal to the plane (Figure 9.40). The coefficients in the equation $3x + 2y + 6z = 6$ give

$$\mathbf{n} = 3\mathbf{i} + 2\mathbf{j} + 6\mathbf{k}.$$

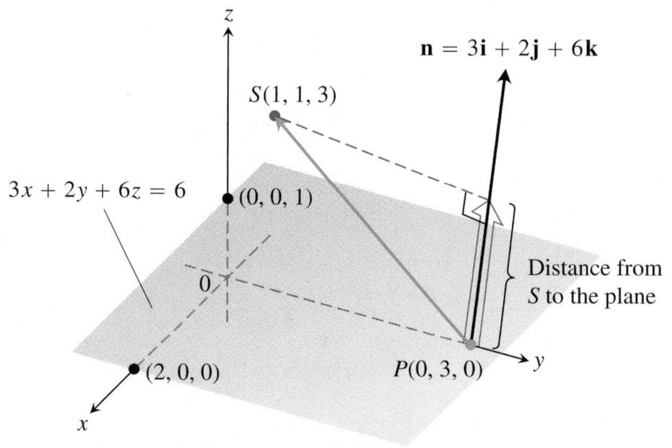

FIGURE 9.40 The distance from S to the plane is the length of the vector projection of $\overrightarrow{PS}$ onto $\mathbf{n}$ (Example 10).

The points on the plane easiest to find from the plane's equation are the intercepts. If we take P to be the y-intercept (0, 3, 0), then

$$\begin{aligned}\overrightarrow{PS} &= (1-0)\mathbf{i} + (1-3)\mathbf{j} + (3-0)\mathbf{k} \\ &= \mathbf{i} - 2\mathbf{j} + 3\mathbf{k}, \\ |\mathbf{n}| &= \sqrt{(3)^2 + (2)^2 + (6)^2} = \sqrt{49} = 7.\end{aligned}$$

The distance from S to the plane is

$$\begin{aligned}d &= \left|\overrightarrow{PS} \cdot \frac{\mathbf{n}}{|\mathbf{n}|}\right| \qquad \text{length of } \mathrm{proj}_{\mathbf{n}}\overrightarrow{PS} \\ &= \left|(\mathbf{i} - 2\mathbf{j} + 3\mathbf{k}) \cdot \left(\frac{3}{7}\mathbf{i} + \frac{2}{7}\mathbf{j} + \frac{6}{7}\mathbf{k}\right)\right| \\ &= \left|\frac{3}{7} - \frac{4}{7} + \frac{18}{7}\right| = \frac{17}{7}.\end{aligned}$$

■

Angles Between Planes

The angle between two intersecting planes is defined to be the (acute) angle between their normal vectors (Figure 9.41).

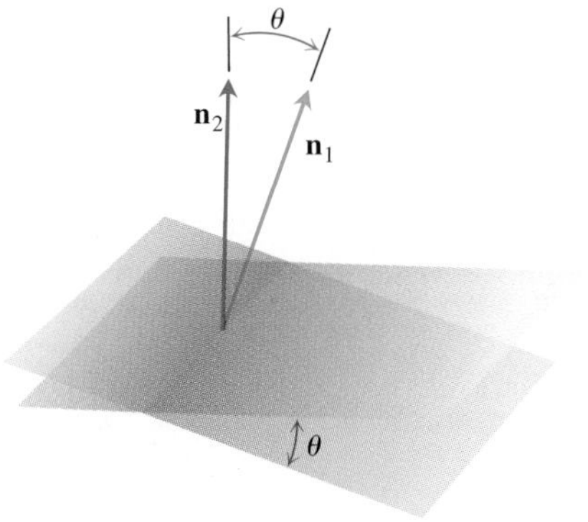

FIGURE 9.41 The angle between two planes is obtained from the angle between their normals.

EXAMPLE 11 Find the angle between the planes $3x - 6y - 2z = 15$ and $2x + y - 2z = 5$.

Solution The vectors

$$\mathbf{n}_1 = 3\mathbf{i} - 6\mathbf{j} - 2\mathbf{k}, \qquad \mathbf{n}_2 = 2\mathbf{i} + \mathbf{j} - 2\mathbf{k}$$

are normals to the planes. The angle between them is

$$\begin{aligned}\theta &= \cos^{-1}\left(\frac{\mathbf{n}_1 \cdot \mathbf{n}_2}{|\mathbf{n}_1||\mathbf{n}_2|}\right) \\ &= \cos^{-1}\left(\frac{4}{21}\right) \\ &\approx 1.38 \text{ radians.} \qquad \text{About 79 deg}\end{aligned}$$

■

EXERCISES 9.5

Find parametric equations for the lines in Exercises 1–12.

1. The line through the point $P(3, -4, -1)$ parallel to the vector $\mathbf{i} + \mathbf{j} + \mathbf{k}$
2. The line through $P(1, 2, -1)$ and $Q(-1, 0, 1)$
3. The line through $P(-2, 0, 3)$ and $Q(3, 5, -2)$
4. The line through $P(1, 2, 0)$ and $Q(1, 1, -1)$
5. The line through the origin parallel to the vector $2\mathbf{j} + \mathbf{k}$
6. The line through the point $(3, -2, 1)$ parallel to the line $x = 1 + 2t, y = 2 - t, z = 3t$
7. The line through (1, 1, 1) parallel to the z-axis
8. The line through (2, 4, 5) perpendicular to the plane $3x + 7y - 5z = 21$
9. The line through $(0, -7, 0)$ perpendicular to the plane $x + 2y + 2z = 13$
10. The line through (2, 3, 0) perpendicular to the vectors $\mathbf{u} = \mathbf{i} + 2\mathbf{j} + 3\mathbf{k}$ and $\mathbf{v} = 3\mathbf{i} + 4\mathbf{j} + 5\mathbf{k}$
11. The x-axis
12. The z-axis

Find parametrizations for the line segments joining the points in Exercises 13–20. Draw coordinate axes and sketch each segment, indicating the direction of increasing t for your parametrization.

13. $(0, 0, 0)$, $(1, 1, 3/2)$ **14.** $(0, 0, 0)$, $(1, 0, 0)$

15. $(1, 0, 0)$, $(1, 1, 0)$ **16.** $(1, 1, 0)$, $(1, 1, 1)$

17. $(0, 1, 1)$, $(0, -1, 1)$ **18.** $(0, 2, 0)$, $(3, 0, 0)$

19. $(2, 0, 2)$, $(0, 2, 0)$ **20.** $(1, 0, -1)$, $(0, 3, 0)$

Find equations for the planes in Exercises 21–26.

21. The plane through $P_0(0, 2, -1)$ normal to $\mathbf{n} = 3\mathbf{i} - 2\mathbf{j} - \mathbf{k}$

22. The plane through $(1, -1, 3)$ parallel to the plane

$$3x + y + z = 7$$

23. The plane through $(1, 1, -1)$, $(2, 0, 2)$, and $(0, -2, 1)$

24. The plane through $(2, 4, 5)$, $(1, 5, 7)$, and $(-1, 6, 8)$

25. The plane through $P_0(2, 4, 5)$ perpendicular to the line

$$x = 5 + t, \quad y = 1 + 3t, \quad z = 4t$$

26. The plane through $A(1, -2, 1)$ perpendicular to the vector from the origin to A.

27. Find the point of intersection of the lines $x = 2t + 1$, $y = 3t + 2$, $z = 4t + 3$, and $x = s + 2$, $y = 2s + 4$, $z = -4s - 1$, and then find the plane determined by these lines.

28. Find the point of intersection of the lines $x = t$, $y = -t + 2$, $z = t + 1$, and $x = 2s + 2$, $y = s + 3$, $z = 5s + 6$, and then find the plane determined by these lines.

In Exercises 29 and 30, find the plane determined by the intersecting lines.

29. $L1$: $x = -1 + t$, $y = 2 + t$, $z = 1 - t$; $-\infty < t < \infty$

$L2$: $x = 1 - 4s$, $y = 1 + 2s$, $z = 2 - 2s$; $-\infty < s < \infty$

30. $L1$: $x = t$, $y = 3 - 3t$, $z = -2 - t$; $-\infty < t < \infty$

$L2$: $x = 1 + s$, $y = 4 + s$, $z = -1 + s$; $-\infty < s < \infty$

31. Find a plane through $P_0(2, 1, -1)$ and perpendicular to the line of intersection of the planes $2x + y - z = 3$, $x + 2y + z = 2$.

32. Find a plane through the points $P_1(1, 2, 3)$, $P_2(3, 2, 1)$ and perpendicular to the plane $4x - y + 2z = 7$.

In Exercises 33–38, find the distance from the point to the line.

33. $(0, 0, 12)$; $x = 4t$, $y = -2t$, $z = 2t$

34. $(0, 0, 0)$; $x = 5 + 3t$, $y = 5 + 4t$, $z = -3 - 5t$

35. $(2, 1, 3)$; $x = 2 + 2t$, $y = 1 + 6t$, $z = 3$

36. $(2, 1, -1)$; $x = 2t$, $y = 1 + 2t$, $z = 2t$

37. $(3, -1, 4)$; $x = 4 - t$, $y = 3 + 2t$, $z = -5 + 3t$

38. $(-1, 4, 3)$; $x = 10 + 4t$, $y = -3$, $z = 4t$

In Exercises 39–44, find the distance from the point to the plane.

39. $(2, -3, 4)$, $x + 2y + 2z = 13$

40. $(0, 0, 0)$, $3x + 2y + 6z = 6$

41. $(0, 1, 1)$, $4y + 3z = -12$

42. $(2, 2, 3)$, $2x + y + 2z = 4$

43. $(0, -1, 0)$, $2x + y + 2z = 4$

44. $(1, 0, -1)$, $-4x + y + z = 4$

45. Find the distance from the plane $x + 2y + 6z = 1$ to the plane $x + 2y + 6z = 10$.

46. Find the distance from the line $x = 2 + t$, $y = 1 + t$, $z = -(1/2) - (1/2)t$ to the plane $x + 2y + 6z = 10$.

Find the angles between the planes in Exercises 47 and 48.

47. $x + y = 1$, $2x + y - 2z = 2$

48. $5x + y - z = 10$, $x - 2y + 3z = -1$

T Use a calculator to find the acute angles between the planes in Exercises 49–52 to the nearest hundredth of a radian.

49. $2x + 2y + 2z = 3$, $2x - 2y - z = 5$

50. $x + y + z = 1$, $z = 0$ (the xy-plane)

51. $2x + 2y - z = 3$, $x + 2y + z = 2$

52. $4y + 3z = -12$, $3x + 2y + 6z = 6$

In Exercises 53–56, find the point in which the line meets the plane.

53. $x = 1 - t$, $y = 3t$, $z = 1 + t$; $2x - y + 3z = 6$

54. $x = 2$, $y = 3 + 2t$, $z = -2 - 2t$; $6x + 3y - 4z = -12$

55. $x = 1 + 2t$, $y = 1 + 5t$, $z = 3t$; $x + y + z = 2$

56. $x = -1 + 3t$, $y = -2$, $z = 5t$; $2x - 3z = 7$

Find parametrizations for the lines in which the planes in Exercises 57–60 intersect.

57. $x + y + z = 1$, $x + y = 2$

58. $3x - 6y - 2z = 3$, $2x + y - 2z = 2$

59. $x - 2y + 4z = 2$, $x + y - 2z = 5$

60. $5x - 2y = 11$, $4y - 5z = -17$

Given two lines in space, either they are parallel, or they intersect, or they are skew (imagine, for example, the flight paths of two planes in the sky). Exercises 61 and 62 each give three lines. In each exercise, determine whether the lines, taken two at a time, are parallel, intersect, or are skew. If they intersect, find the point of intersection.

61. $L1$: $x = 3 + 2t$, $y = -1 + 4t$, $z = 2 - t$; $-\infty < t < \infty$

$L2$: $x = 1 + 4s$, $y = 1 + 2s$, $z = -3 + 4s$; $-\infty < s < \infty$

$L3$: $x = 3 + 2r$, $y = 2 + r$, $z = -2 + 2r$; $-\infty < r < \infty$

62. $L1$: $x = 1 + 2t$, $y = -1 - t$, $z = 3t$; $-\infty < t < \infty$

$L2$: $x = 2 - s$, $y = 3s$, $z = 1 + s$; $-\infty < s < \infty$

$L3$: $x = 5 + 2r$, $y = 1 - r$, $z = 8 + 3r$; $-\infty < r < \infty$

63. Use Equations (3) to generate a parametrization of the line through $P(2, -4, 7)$ parallel to $\mathbf{v}_1 = 2\mathbf{i} - \mathbf{j} + 3\mathbf{k}$. Then generate another parametrization of the line using the point $P_2(-2, -2, 1)$ and the vector $\mathbf{v}_2 = -\mathbf{i} + (1/2)\mathbf{j} - (3/2)\mathbf{k}$.

64. Use the component form to generate an equation for the plane through $P_1(4, 1, 5)$ normal to $\mathbf{n}_1 = \mathbf{i} - 2\mathbf{j} + \mathbf{k}$. Then generate another equation for the same plane using the point $P_2(3, -2, 0)$ and the normal vector $\mathbf{n}_2 = -\sqrt{2}\mathbf{i} + 2\sqrt{2}\mathbf{j} - \sqrt{2}\mathbf{k}$.

65. Find the points in which the line $x = 1 + 2t$, $y = -1 - t$, $z = 3t$ meets the coordinate planes. Describe the reasoning behind your answer.

66. Find equations for the line in the plane $z = 3$ that makes an angle of $\pi/6$ rad with $\mathbf{i}$ and an angle of $\pi/3$ rad with $\mathbf{j}$. Describe the reasoning behind your answer.

67. Is the line $x = 1 - 2t$, $y = 2 + 5t$, $z = -3t$ parallel to the plane $2x + y - z = 8$? Give reasons for your answer.

68. How can you tell when two planes $A_1x + B_1y + C_1z = D_1$ and $A_2x + B_2y + C_2z = D_2$ are parallel? Perpendicular? Give reasons for your answer.

69. Find two different planes whose intersection is the line $x = 1 + t$, $y = 2 - t$, $z = 3 + 2t$. Write equations for each plane in the form $Ax + By + Cz = D$.

70. Find a plane through the origin that meets the plane M: $2x + 3y + z = 12$ in a right angle. How do you know that your plane is perpendicular to M?

71. For any nonzero numbers a, b, and c, the graph of $(x/a) + (y/b) + (z/c) = 1$ is a plane. Which planes have an equation of this form?

72. Suppose L_1 and L_2 are disjoint (nonintersecting) nonparallel lines. Is it possible for a nonzero vector to be perpendicular to both L_1 and L_2? Give reasons for your answer.

73. **Perspective in computer graphics** In computer graphics and perspective drawing, we need to represent objects seen by the eye in space as images on a two-dimensional plane. Suppose that the eye is at $E(x_0, 0, 0)$ as shown here and that we want to represent a point $P_1(x_1, y_1, z_1)$ as a point on the yz-plane. We do this by projecting P_1 onto the plane with a ray from E. The point P_1 will be portrayed as the point $P(0, y, z)$. The problem for us as graphics designers is to find y and z given E and P_1.

a. Write a vector equation that holds between $\overrightarrow{EP}$ and $\overrightarrow{EP_1}$. Use the equation to express y and z in terms of x_0, x_1, y_1, and z_1.

b. Test the formulas obtained for y and z in part (a) by investigating their behavior at $x_1 = 0$ and $x_1 = x_0$ and by seeing what happens as $x_0 \to \infty$. What do you find?

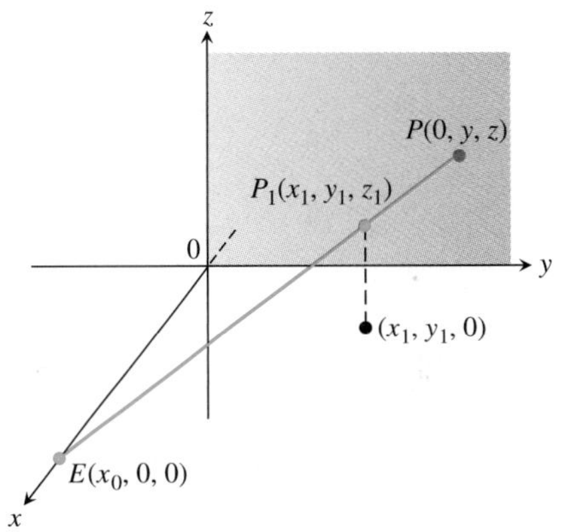

74. **Hidden lines** Here is another typical problem in computer graphics. Your eye is at (4, 0, 0). You are looking at a triangular plate whose vertices are at (1, 0, 1), (1, 1, 0), and $(-2, 2, 2)$. The line segment from (1, 0, 0) to (0, 2, 2) passes through the plate. What portion of the line segment is hidden from your view by the plate? (This is an exercise in finding intersections of lines and planes.)

9.6 Cylinders and Quadric Surfaces

Up to now, we have studied two special types of surfaces: spheres and planes. In this section, we extend our inventory to include a variety of cylinders and quadric surfaces. Quadric surfaces are surfaces defined by second-degree equations in x, y, and z. Spheres are quadric surfaces, but there are others of equal interest.

Cylinders

A **cylinder** is a surface that is generated by moving a straight line along a given planar curve while holding the line parallel to a given fixed line. The curve is called a **generating curve** for the cylinder (Figure 9.42). In solid geometry, where *cylinder* means *circular cylinder*, the generating curves are circles, but now we allow generating curves of any kind. The cylinder in our first example is generated by a parabola.

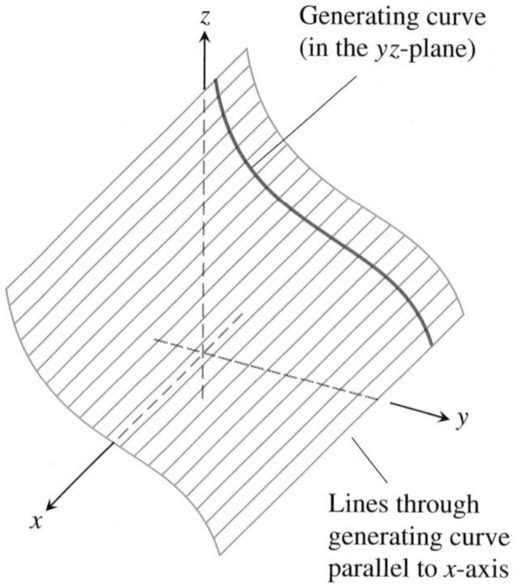

FIGURE 9.42 A cylinder and generating curve.

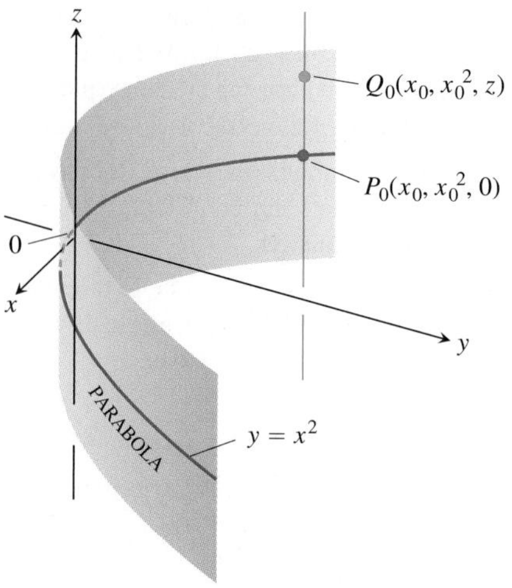

FIGURE 9.43 Every point of the cylinder in Example 1 has coordinates of the form (x_0, x_0^2, z). We call it "the cylinder $y = x^2$."

EXAMPLE 1 Find an equation for the cylinder made by the lines parallel to the z-axis that pass through the parabola $y = x^2, z = 0$ (Figure 9.43).

Solution The point $P_0(x_0, x_0^2, 0)$ lies on the parabola $y = x^2$ in the xy-plane. Then, for any value of z, the point $Q(x_0, x_0^2, z)$ lies on the cylinder because it lies on the line $x = x_0, y = x_0^2$ through P_0 parallel to the z-axis. Conversely, any point $Q(x_0, x_0^2, z)$ whose y-coordinate is the square of its x-coordinate lies on the cylinder because it lies on the line $x = x_0, y = x_0^2$ through P_0 parallel to the z-axis (Figure 9.43).

Regardless of the value of z, therefore, the points on the surface are the points whose coordinates satisfy the equation $y = x^2$. This makes $y = x^2$ an equation for the cylinder. Because of this, we call the cylinder "the cylinder $y = x^2$." ■

As Example 1 suggests, any curve $f(x, y) = c$ in the xy-plane defines a cylinder parallel to the z-axis whose equation is also $f(x, y) = c$. For instance, the equation $x^2 + y^2 = 1$ defines the circular cylinder made by the lines parallel to the z-axis that pass through the circle $x^2 + y^2 = 1$ in the xy-plane.

In a similar way, any curve $g(x, z) = c$ in the xz-plane defines a cylinder parallel to the y-axis whose space equation is also $g(x, z) = c$. Any curve $h(y, z) = c$ defines a cylinder parallel to the x-axis whose space equation is also $h(y, z) = c$. The axis of a cylinder need not be parallel to a coordinate axis, however.

Quadric Surfaces

A **quadric surface** is the graph in space of a second-degree equation in x, y, and z. We focus on the special equation

$$Ax^2 + By^2 + Cz^2 + Dz = E,$$

where A, B, C, D, and E are constants. The basic quadric surfaces are **ellipsoids**, **paraboloids**, **elliptical cones**, and **hyperboloids**. Spheres are special cases of ellipsoids. We present a few examples illustrating how to sketch a quadric surface, and then give a summary table of graphs of the basic types.

EXAMPLE 2 The **ellipsoid**

$$\frac{x^2}{a^2} + \frac{y^2}{b^2} + \frac{z^2}{c^2} = 1$$

(Figure 9.44) cuts the coordinate axes at $(\pm a, 0, 0)$, $(0, \pm b, 0)$, and $(0, 0, \pm c)$. It lies within the rectangular box defined by the inequalities $|x| \le a$, $|y| \le b$, and $|z| \le c$. The surface is symmetric with respect to each of the coordinate planes because each variable in the defining equation is squared.

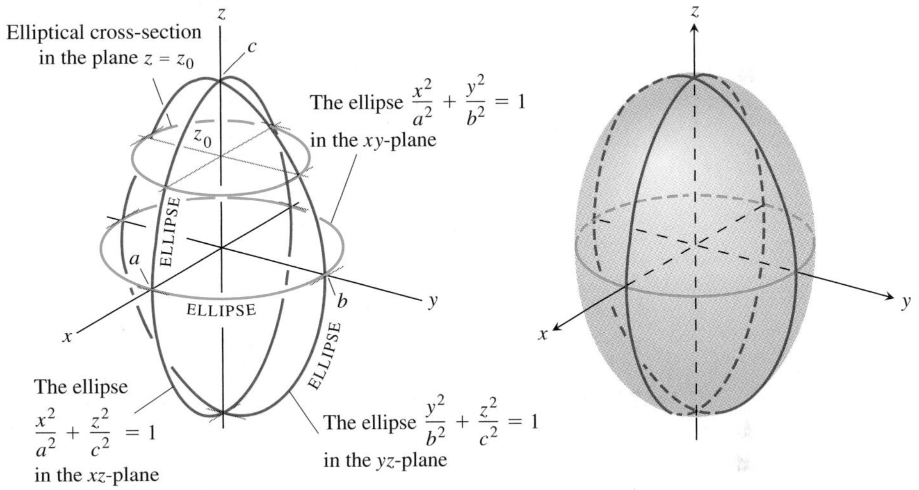

FIGURE 9.44 The ellipsoid

$$\frac{x^2}{a^2} + \frac{y^2}{b^2} + \frac{z^2}{c^2} = 1$$

in Example 2 has elliptical cross-sections in each of the three coordinate planes.

The curves in which the three coordinate planes cut the surface are ellipses. For example,

$$\frac{x^2}{a^2} + \frac{y^2}{b^2} = 1 \qquad \text{when} \qquad z = 0.$$

The curve cut from the surface by the plane $z = z_0$, $|z_0| < c$, is the ellipse

$$\frac{x^2}{a^2(1 - (z_0/c)^2)} + \frac{y^2}{b^2(1 - (z_0/c)^2)} = 1.$$

If any two of the semiaxes a, b, and c are equal, the surface is an **ellipsoid of revolution**. If all three are equal, the surface is a sphere. ■

EXAMPLE 3 The **hyperbolic paraboloid**

$$\frac{y^2}{b^2} - \frac{x^2}{a^2} = \frac{z}{c}, \qquad c > 0$$

has symmetry with respect to the planes $x = 0$ and $y = 0$ (Figure 9.45). The cross-sections in these planes are

$$x = 0: \quad \text{the parabola } z = \frac{c}{b^2}y^2. \tag{1}$$

$$y = 0: \quad \text{the parabola } z = -\frac{c}{a^2}x^2. \tag{2}$$

In the plane $x = 0$, the parabola opens upward from the origin. The parabola in the plane $y = 0$ opens downward.

If we cut the surface by a plane $z = z_0 > 0$, the cross-section is a hyperbola,

$$\frac{y^2}{b^2} - \frac{x^2}{a^2} = \frac{z_0}{c},$$

with its focal axis parallel to the y-axis and its vertices on the parabola in Equation (1). If z_0 is negative, the focal axis is parallel to the x-axis and the vertices lie on the parabola in Equation (2).

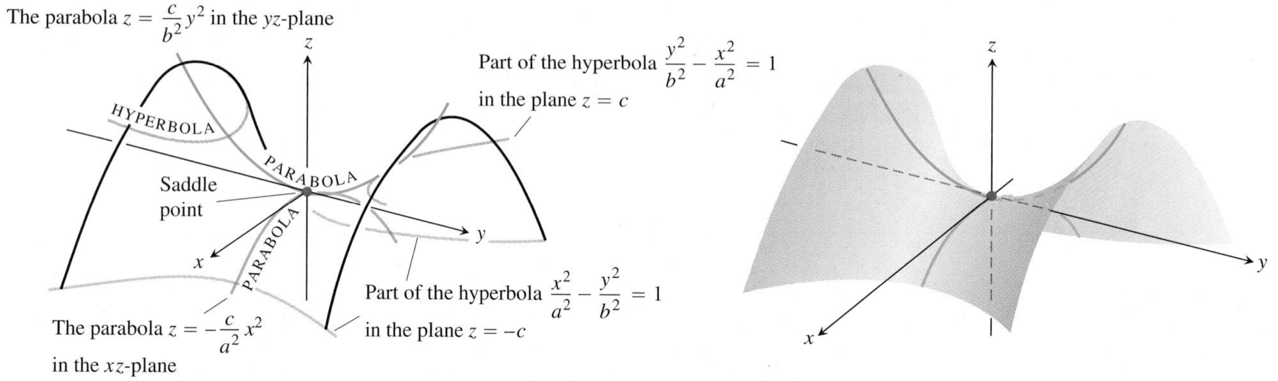

FIGURE 9.45 The hyperbolic paraboloid $(y^2/b^2) - (x^2/a^2) = z/c, c > 0$. The cross-sections in planes perpendicular to the z-axis above and below the xy-plane are hyperbolas. The cross-sections in planes perpendicular to the other axes are parabolas.

Near the origin, the surface is shaped like a saddle or mountain pass. To a person traveling along the surface in the yz-plane the origin looks like a minimum. To person traveling the xz-plane the origin looks like a maximum. Such a point is called a **saddle point** of a surface. We will say more about saddle points in Section 11.7. ■

Table 9.1 shows graphs of the six basic types of quadric surfaces. Each surface shown is symmetric with respect to the z-axis, but other coordinate axes can serve as well (with appropriate changes to the equation).

TABLE 9.1 Graphs of Quadric Surfaces

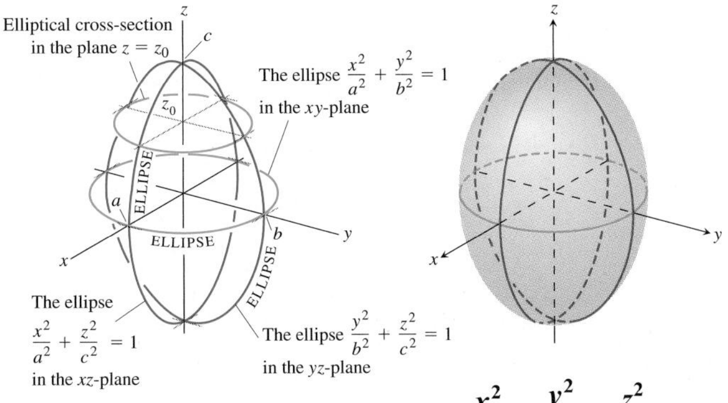

ELLIPSOID $$\frac{x^2}{a^2} + \frac{y^2}{b^2} + \frac{z^2}{c^2} = 1$$

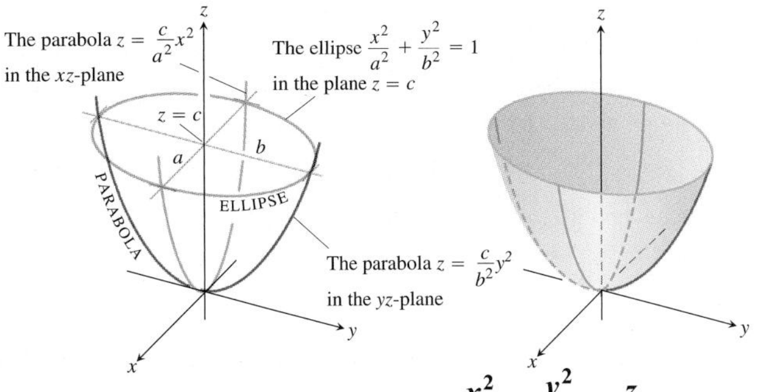

ELLIPTICAL PARABOLOID $$\frac{x^2}{a^2} + \frac{y^2}{b^2} = \frac{z}{c}$$

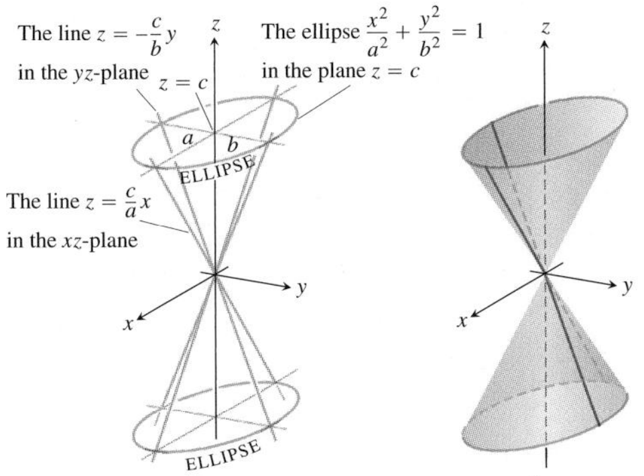

ELLIPTICAL CONE $$\frac{x^2}{a^2} + \frac{y^2}{b^2} = \frac{z^2}{c^2}$$

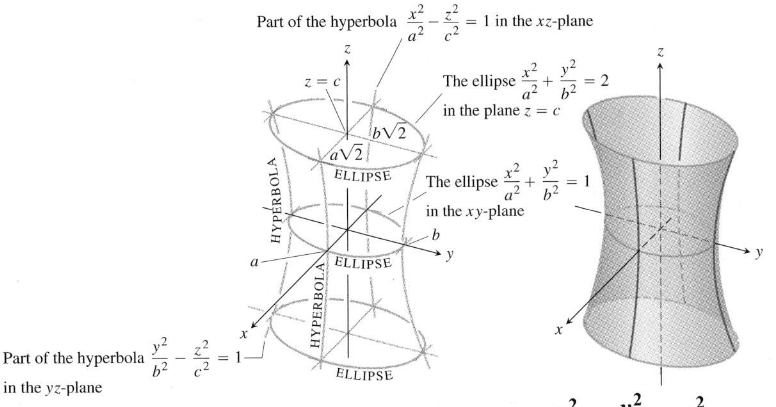

HYPERBOLOID OF ONE SHEET $$\frac{x^2}{a^2} + \frac{y^2}{b^2} - \frac{z^2}{c^2} = 1$$

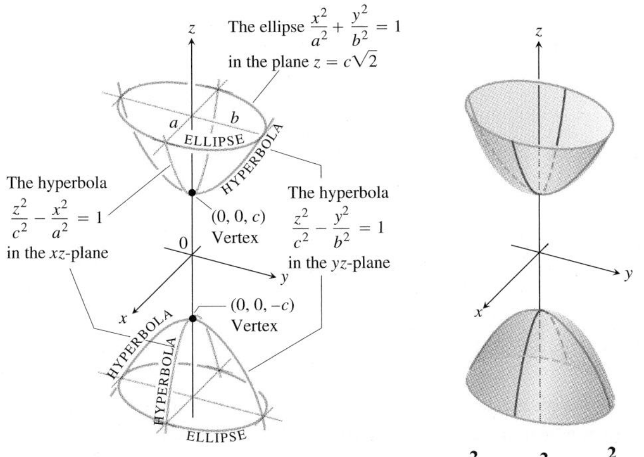

HYPERBOLOID OF TWO SHEETS $$\frac{z^2}{c^2} - \frac{x^2}{a^2} - \frac{y^2}{b^2} = 1$$

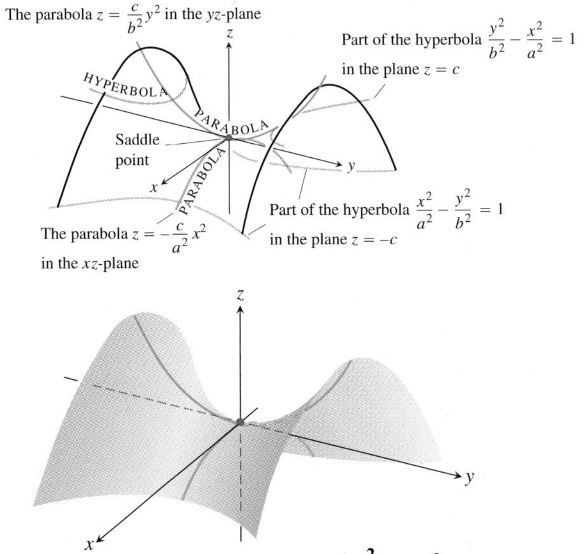

HYPERBOLIC PARABOLOID $$\frac{y^2}{b^2} - \frac{x^2}{a^2} = \frac{z}{c}, \quad c > 0$$

EXERCISES 9.6

In Exercises 1–12, match the equation with the surface it defines. Also, identify each surface by type (paraboloid, ellipsoid, etc.) The surfaces are labeled (a)–(l).

1. $x^2 + y^2 + 4z^2 = 10$ **2.** $z^2 + 4y^2 - 4x^2 = 4$

3. $9y^2 + z^2 = 16$ **4.** $y^2 + z^2 = x^2$

5. $x = y^2 - z^2$ **6.** $x = -y^2 - z^2$

7. $x^2 + 2z^2 = 8$ **8.** $z^2 + x^2 - y^2 = 1$

9. $x = z^2 - y^2$ **10.** $z = -4x^2 - y^2$

11. $x^2 + 4z^2 = y^2$ **12.** $9x^2 + 4y^2 + 2z^2 = 36$

a.

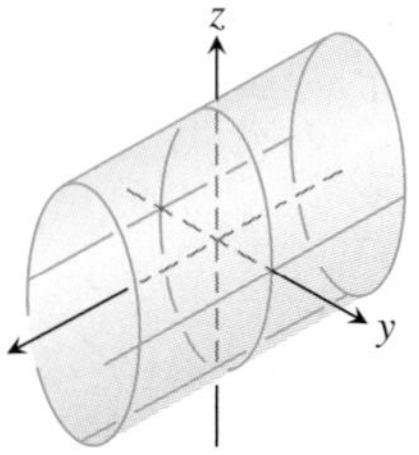

b.

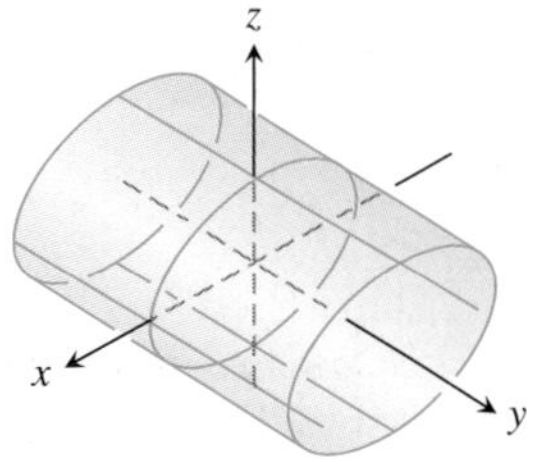

c.

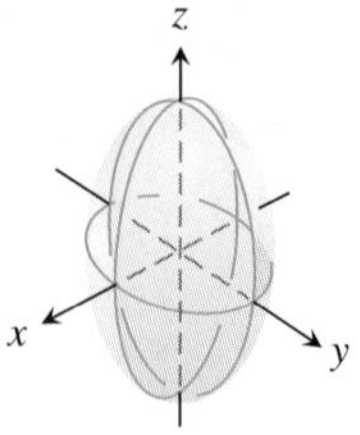

d.

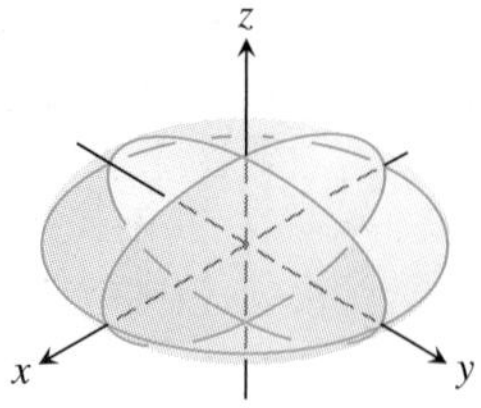

e.

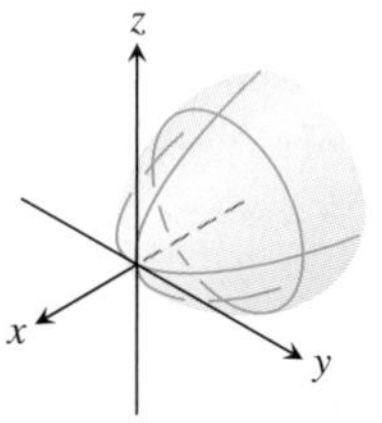

f.

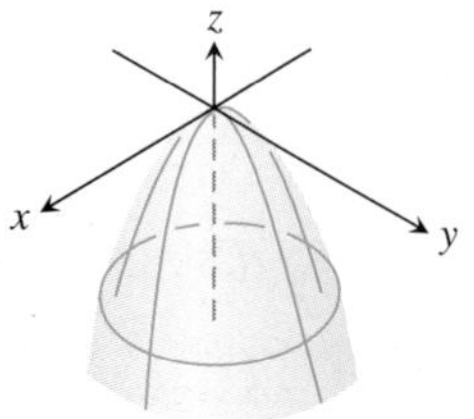

g.

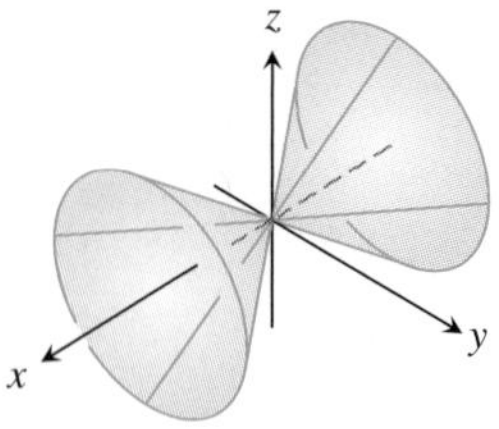

h.

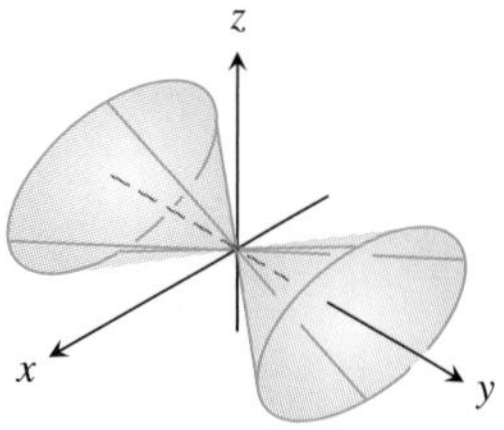

i.

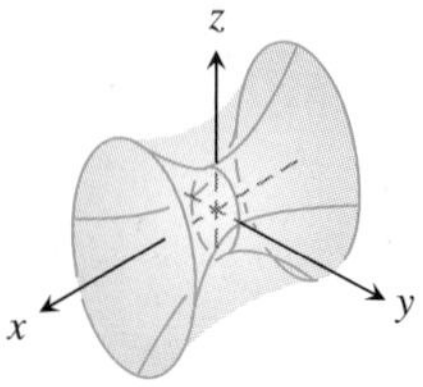

j.

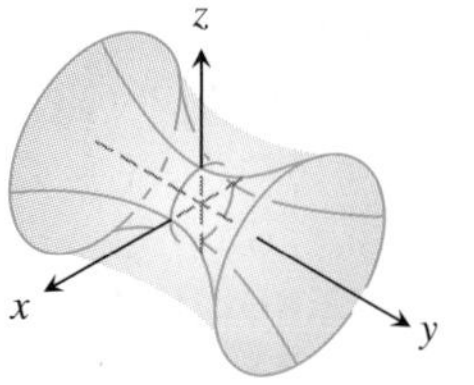

k.

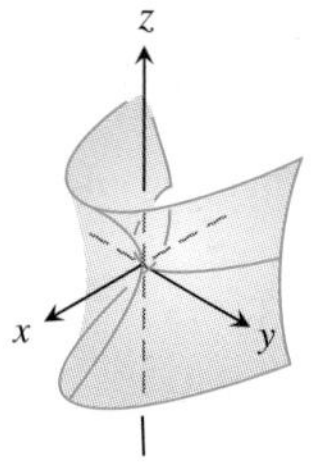

l.

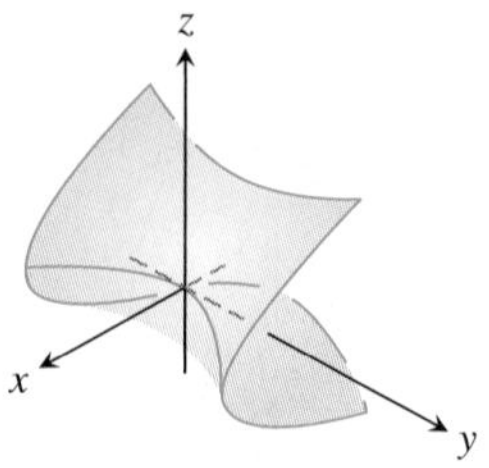

Sketch the surfaces in Exercises 13–44.

CYLINDERS

13. $x^2 + y^2 = 4$ **14.** $z = y^2 - 1$

15. $x^2 + 4z^2 = 16$ **16.** $4x^2 + y^2 = 36$

ELLIPSOIDS

17. $9x^2 + y^2 + z^2 = 9$ **18.** $4x^2 + 4y^2 + z^2 = 16$

19. $4x^2 + 9y^2 + 4z^2 = 36$ **20.** $9x^2 + 4y^2 + 36z^2 = 36$

PARABOLOIDS AND CONES

21. $z = x^2 + 4y^2$ **22.** $z = 8 - x^2 - y^2$

23. $x = 4 - 4y^2 - z^2$ **24.** $y = 1 - x^2 - z^2$

25. $x^2 + y^2 = z^2$ **26.** $4x^2 + 9z^2 = 9y^2$

HYPERBOLOIDS

27. $x^2 + y^2 - z^2 = 1$ **28.** $y^2 + z^2 - x^2 = 1$

29. $z^2 - x^2 - y^2 = 1$ **30.** $(y^2/4) - (x^2/4) - z^2 = 1$

HYPERBOLIC PARABOLOIDS

31. $y^2 - x^2 = z$ **32.** $x^2 - y^2 = z$

ASSORTED

33. $z = 1 + y^2 - x^2$ **34.** $4x^2 + 4y^2 = z^2$

35. $y = -(x^2 + z^2)$ **36.** $16x^2 + 4y^2 = 1$

37. $x^2 + y^2 - z^2 = 4$ **38.** $x^2 + z^2 = y$

39. $x^2 + z^2 = 1$ **40.** $16y^2 + 9z^2 = 4x^2$

41. $z = -(x^2 + y^2)$ **42.** $y^2 - x^2 - z^2 = 1$

43. $4y^2 + z^2 - 4x^2 = 4$ **44.** $x^2 + y^2 = z$

45. a. Express the area A of the cross-section cut from the ellipsoid

$$x^2 + \frac{y^2}{4} + \frac{z^2}{9} = 1$$

by the plane $z = c$ as a function of c. (The area of an ellipse with semiaxes a and b is πab.)

b. Use slices perpendicular to the z-axis to find the volume of the ellipsoid in part (a).

c. Now find the volume of the ellipsoid

$$\frac{x^2}{a^2} + \frac{y^2}{b^2} + \frac{z^2}{c^2} = 1.$$

Does your formula give the volume of a sphere of radius a if $a = b = c$?

46. Show that the volume of the segment cut from the paraboloid

$$\frac{x^2}{a^2} + \frac{y^2}{b^2} = \frac{z}{c}$$

by the plane $z = h$ equals half the segment's base times its altitude.

COMPUTER EXPLORATIONS

Use a CAS to plot the surfaces in Exercises 47–52. Identify the type of quadric surface from your graph.

47. $\frac{x^2}{9} + \frac{y^2}{36} = 1 - \frac{z^2}{25}$

48. $\frac{x^2}{9} - \frac{z^2}{9} = 1 - \frac{y^2}{16}$

49. $5x^2 = z^2 - 3y^2$

50. $\frac{y^2}{16} = 1 - \frac{x^2}{9} + z$

51. $\frac{x^2}{9} - 1 = \frac{y^2}{16} + \frac{z^2}{2}$

52. $y - \sqrt{4 - z^2} = 0$

Chapter 9 Questions to Guide Your Review

1. When do directed line segments in the plane represent the same vector?
2. How are vectors added and subtracted geometrically? Algebraically?
3. How do you find a vector's magnitude and direction?
4. If a vector is multiplied by a positive scalar, how is the result related to the original vector? What if the scalar is zero? Negative?
5. Define the *dot product* (*scalar product*) of two vectors. Which algebraic laws are satisfied by dot products? Give examples. When is the dot product of two vectors equal to zero?
6. What geometric interpretation does the dot product have? Give examples.
7. What is the vector projection of a vector **u** onto a vector **v**? Give an example of a useful application of a vector projection.
8. Define the *cross product* (*vector product*) of two vectors. Which algebraic laws are satisfied by cross products, and which are not? Give examples. When is the cross product of two vectors equal to zero?
9. What geometric or physical interpretations do cross products have? Give examples.
10. What is the determinant formula for calculating the cross product of two vectors relative to the Cartesian **i**, **j**, **k**-coordinate system? Use it in an example.
11. How do you find equations for lines, line segments, and planes in space? Give examples. Can you express a line in space by a single equation? A plane?
12. How do you find the distance from a point to a line in space? From a point to a plane? Give examples.
13. What are box products? What significance do they have? How are they evaluated? Give an example.
14. How do you find equations for spheres in space? Give examples.
15. How do you find the intersection of two lines in space? A line and a plane? Two planes? Give examples.
16. What is a cylinder? Give examples of equations that define cylinders in Cartesian coordinates.
17. What are quadric surfaces? Give examples of different kinds of ellipsoids, paraboloids, cones, and hyperboloids (equations and sketches).

Chapter 9 Practice and Additional Exercises

In Exercises 1–4, let $\mathbf{u} = \langle -3, 4\rangle$ and $\mathbf{v} = \langle 2, -5\rangle$. Find **(a)** the component form of the vector and **(b)** its magnitude.

1. $3\mathbf{u} - 4\mathbf{v}$

2. $\mathbf{u} + \mathbf{v}$

3. $-2\mathbf{u}$

4. $5\mathbf{v}$

In Exercises 5–8, find the component form of the vector.

5. The vector obtained by rotating $\langle 0, 1\rangle$ through an angle of $2\pi/3$ radians

6. The unit vector that makes an angle of $\pi/6$ radian with the positive x-axis

7. The vector 2 units long in the direction $4\mathbf{i} - \mathbf{j}$

8. The vector 5 units long in the direction opposite to the direction of $(3/5)\mathbf{i} + (4/5)\mathbf{j}$

Express the vectors in Exercises 9–12 in terms of their lengths and directions.

9. $\sqrt{2}\mathbf{i} + \sqrt{2}\mathbf{j}$

10. $-\mathbf{i} - \mathbf{j}$

11. Velocity vector $\mathbf{v} = (-2 \sin t)\mathbf{i} + (2 \cos t)\mathbf{j}$ when $t = \pi/2$.

12. Velocity vector $\mathbf{v} = (e^t \cos t - e^t \sin t)\mathbf{i} + (e^t \sin t + e^t \cos t)\mathbf{j}$ when $t = \ln 2$.

Express the vectors in Exercises 13 and 14 in terms of their lengths and directions.

13. $2\mathbf{i} - 3\mathbf{j} + 6\mathbf{k}$

14. $\mathbf{i} + 2\mathbf{j} - \mathbf{k}$

15. Find a vector 2 units long in the direction of $\mathbf{v} = 4\mathbf{i} - \mathbf{j} + 4\mathbf{k}$.

16. Find a vector 5 units long in the direction opposite to the direction of $\mathbf{v} = (3/5)\,\mathbf{i} + (4/5)\mathbf{k}$.

In Exercises 17 and 18, find $|\mathbf{v}|$, $|\mathbf{u}|$, $\mathbf{v}\cdot\mathbf{u}$, $\mathbf{u}\cdot\mathbf{v}$, $\mathbf{v}\times\mathbf{u}$, $\mathbf{u}\times\mathbf{v}$, $|\mathbf{v}\times\mathbf{u}|$, the angle between $\mathbf{v}$ and $\mathbf{u}$, the scalar component of $\mathbf{u}$ in the direction of $\mathbf{v}$, and the vector projection of $\mathbf{u}$ onto $\mathbf{v}$.

17. $\mathbf{v} = \mathbf{i} + \mathbf{j}$
$\mathbf{u} = 2\mathbf{i} + \mathbf{j} - 2\mathbf{k}$

18. $\mathbf{v} = \mathbf{i} + \mathbf{j} + 2\mathbf{k}$
$\mathbf{u} = -\mathbf{i} - \mathbf{k}$

In Exercises 19 and 20, find $\text{proj}_{\mathbf{v}}\,\mathbf{u}$.

19. $\mathbf{v} = 2\mathbf{i} + \mathbf{j} - \mathbf{k}$
$\mathbf{u} = \mathbf{i} + \mathbf{j} - 5\mathbf{k}$

20. $\mathbf{u} = \mathbf{i} - 2\mathbf{j}$
$\mathbf{v} = \mathbf{i} + \mathbf{j} + \mathbf{k}$

In Exercises 21 and 22, draw coordinate axes and then sketch $\mathbf{u}$, $\mathbf{v}$, and $\mathbf{u}\times\mathbf{v}$ as vectors at the origin.

21. $\mathbf{u} = \mathbf{i}$, $\mathbf{v} = \mathbf{i} + \mathbf{j}$

22. $\mathbf{u} = \mathbf{i} - \mathbf{j}$, $\mathbf{v} = \mathbf{i} + \mathbf{j}$

23. If $|\mathbf{v}| = 2$, $|\mathbf{w}| = 3$, and the angle between $\mathbf{v}$ and $\mathbf{w}$ is $\pi/3$, find $|\mathbf{v} - 2\mathbf{w}|$.

24. For what value or values of a will the vectors $\mathbf{u} = 2\mathbf{i} + 4\mathbf{j} - 5\mathbf{k}$ and $\mathbf{v} = -4\mathbf{i} - 8\mathbf{j} + a\mathbf{k}$ be parallel?

In Exercises 25 and 26, find **(a)** the area of the parallelogram determined by vectors $\mathbf{u}$ and $\mathbf{v}$ and **(b)** the volume of the parallelepiped determined by the vectors $\mathbf{u}$, $\mathbf{v}$, and $\mathbf{w}$.

25. $\mathbf{u} = \mathbf{i} + \mathbf{j} - \mathbf{k}$, $\mathbf{v} = 2\mathbf{i} + \mathbf{j} + \mathbf{k}$, $\mathbf{w} = -\mathbf{i} - 2\mathbf{j} + 3\mathbf{k}$

26. $\mathbf{u} = \mathbf{i} + \mathbf{j}$, $\mathbf{v} = \mathbf{j}$, $\mathbf{w} = \mathbf{i} + \mathbf{j} + \mathbf{k}$

27. Suppose that $\mathbf{n}$ is normal to a plane and that $\mathbf{v}$ is parallel to the plane. Describe how you would find a vector $\mathbf{n}$ that is both perpendicular to $\mathbf{v}$ and parallel to the plane.

28. Find a vector in the plane parallel to the line $ax + by = c$.

In Exercises 29 and 30, find the distance from the point to the line.

29. $(2, 2, 0)$; $x = -t$, $y = t$, $z = -1 + t$

30. $(0, 4, 1)$; $x = 2 + t$, $y = 2 + t$, $z = t$

31. Parametrize the line that passes through the point $(1, 2, 3)$ parallel to the vector $\mathbf{v} = -3\mathbf{i} + 7\mathbf{k}$.

32. Parametrize the line segment joining the points $P(1, 2, 0)$ and $Q(1, 3, -1)$.

In Exercises 33 and 34, find the distance from the point to the plane.

33. $(6, 0, -6)$, $x - y = 4$

34. $(3, 0, 10)$, $2x + 3y + z = 2$

35. Find an equation for the plane that passes through the point $(3, -2, 1)$ normal to the vector $\mathbf{n} = 2\mathbf{i} + \mathbf{j} + \mathbf{k}$.

36. Find an equation for the plane that passes through the point $(-1, 6, 0)$ perpendicular to the line $x = -1 + t$, $y = 6 - 2t$, $z = 3t$.

In Exercises 37 and 38, find an equation for the plane through points P, Q, and R.

37. $P(1, -1, 2)$, $Q(2, 1, 3)$, $R(-1, 2, -1)$

38. $P(1, 0, 0)$, $Q(0, 1, 0)$, $R(0, 0, 1)$

39. Find the points in which the line $x = 1 + 2t$, $y = -1 - t$, $z = 3t$ meets the three coordinate planes.

40. Find the point in which the line through the origin perpendicular to the plane $2x - y - z = 4$ meets the plane $3x - 5y + 2z = 6$.

41. Find the acute angle between the planes $x = 7$ and $x + y + \sqrt{2}z = -3$.

42. Find the acute angle between the planes $x + y = 1$ and $y + z = 1$.

43. Find parametric equations for the line in which the planes $x + 2y + z = 1$ and $x - y + 2z = -8$ intersect.

44. Show that the line in which the planes

$$x + 2y - 2z = 5 \quad \text{and} \quad 5x - 2y - z = 0$$

intersect is parallel to the line

$$x = -3 + 2t, \quad y = 3t, \quad z = 1 + 4t.$$

45. The planes $3x + 6z = 1$ and $2x + 2y - z = 3$ intersect in a line.

a. Show that the planes are orthogonal.

b. Find equations for the line of intersection.

46. Find an equation for the plane that passes through the point $(1, 2, 3)$ parallel to $\mathbf{u} = 2\mathbf{i} + 3\mathbf{j} + \mathbf{k}$ and $\mathbf{v} = \mathbf{i} - \mathbf{j} + 2\mathbf{k}$.

47. Is $\mathbf{v} = 2\mathbf{i} - 4\mathbf{j} + \mathbf{k}$ related in any special way to the plane $2x + y = 5$? Give reasons for your answer.

48. The equation $\mathbf{n}\cdot\overrightarrow{P_0P} = 0$ represents the plane through P_0 normal to $\mathbf{n}$. What set does the inequality $\mathbf{n}\cdot\overrightarrow{P_0P} > 0$ represent?

49. Find the distance from the point $P(1, 4, 0)$ to the plane through $A(0, 0, 0)$, $B(2, 0, -1)$, and $C(2, -1, 0)$.

50. Find the distance from the point $(2, 2, 3)$ to the plane $2x + 3y + 5z = 0$.

51. Find a vector parallel to the plane $2x - y - z = 4$ and orthogonal to $\mathbf{i} + \mathbf{j} + \mathbf{k}$.

52. Find a unit vector orthogonal to $\mathbf{A}$ in the plane of $\mathbf{B}$ and $\mathbf{C}$ if $\mathbf{A} = 2\mathbf{i} - \mathbf{j} + \mathbf{k}$, $\mathbf{B} = \mathbf{i} + 2\mathbf{j} + \mathbf{k}$, and $\mathbf{C} = \mathbf{i} + \mathbf{j} - 2\mathbf{k}$.

53. Find a vector of magnitude 2 parallel to the line of intersection of the planes $x + 2y + z - 1 = 0$ and $x - y + 2z + 7 = 0$.

54. Find the point in which the line through the origin perpendicular to the plane $2x - y - z = 4$ meets the plane $3x - 5y + 2z = 6$.

55. Find the point in which the line through $P(3, 2, 1)$ normal to the plane $2x - y + 2z = -2$ meets the plane.

56. What angle does the line of intersection of the planes $2x + y - z = 0$ and $x + y + 2z = 0$ make with the positive x-axis?

57. The line

$$L: \quad x = 3 + 2t, \quad y = 2t, \quad z = t$$

intersects the plane $x + 3y - z = -4$ in a point P. Find the coordinates of P and find equations for the line in the plane through P perpendicular to L.

58. Show that for every real number k the plane

$$x - 2y + z + 3 + k(2x - y - z + 1) = 0$$

contains the line of intersection of the planes

$$x - 2y + z + 3 = 0 \quad \text{and} \quad 2x - y - z + 1 = 0.$$

59. Find an equation for the plane through $A(-2, 0, -3)$ and $B(1, -2, 1)$ that lies parallel to the line through $C(-2, -13/5, 26/5)$ and $D(16/5, -13/5, 0)$.

60. Is the line $x = 1 + 2t$, $y = -2 + 3t$, $z = -5t$ related in any way to the plane $-4x - 6y + 10z = 9$? Give reasons for your answer.

61. Distance between lines Find the distance between the line L_1 through the points $A(1, 0, -1)$ and $B(-1, 1, 0)$ and the line L_2 through the points $C(3, 1, -1)$ and $D(4, 5, -2)$. The distance is to be measured along the line perpendicular to the two lines. First find a vector $\mathbf{n}$ perpendicular to both lines. Then project $\overrightarrow{AC}$ onto $\mathbf{n}$.

62. (*Continuation of Exercise 61.*) Find the distance between the line through $A(4, 0, 2)$ and $B(2, 4, 1)$ and the line through $C(1, 3, 2)$ and $D(2, 2, 4)$.

Identify and sketch the surfaces in Exercises 63–74.

63. $x^2 + y^2 + z^2 = 4$

64. $x^2 + (y - 1)^2 + z^2 = 1$

65. $4x^2 + 4y^2 + z^2 = 4$

66. $36x^2 + 9y^2 + 4z^2 = 36$

67. $z = -(x^2 + y^2)$

68. $y = -(x^2 + z^2)$

69. $x^2 + y^2 = z^2$

70. $x^2 + z^2 = y^2$

71. $x^2 + y^2 - z^2 = 4$

72. $4y^2 + z^2 - 4x^2 = 4$

73. $y^2 - x^2 - z^2 = 1$

74. $z^2 - x^2 - y^2 = 1$

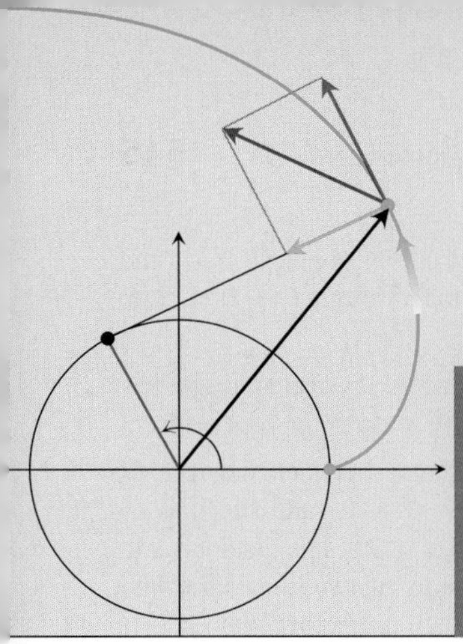

Chapter 10

VECTOR-VALUED FUNCTIONS AND MOTION IN SPACE

OVERVIEW In this chapter we introduce the calculus of vector-valued functions, including their derivatives and integrals. Here we use vector calculus to study the paths, velocities, and accelerations of moving bodies. As we proceed, we will see how questions are answered about the paths and motions of projectiles and other objects moving in a plane or in space.

10.1 Vector Functions and Their Derivatives

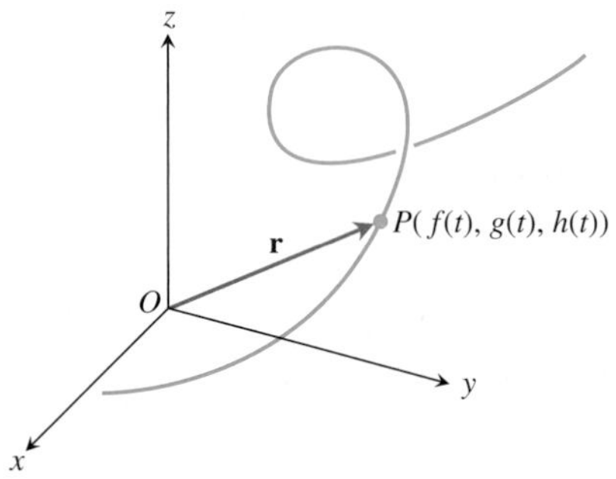

FIGURE 10.1 The position vector $\mathbf{r} = \overrightarrow{OP}$ of a particle moving through space is a function of time.

When a particle moves through space during a time interval I, we think of the particle's coordinates as functions defined on I:

$$x = f(t), \qquad y = g(t), \qquad z = h(t), \qquad t \in I. \tag{1}$$

The points $(x, y, z) = (f(t), g(t), h(t))$, $t \in I$, make up the **curve** in space that we call the particle's **path**. The equations and interval in Equation (1) **parametrize** the curve.

A curve in space can also be represented in vector form. The vector

$$\mathbf{r}(t) = \overrightarrow{OP} = f(t)\mathbf{i} + g(t)\mathbf{j} + h(t)\mathbf{k} \tag{2}$$

from the origin to the particle's **position** $P(f(t), g(t), h(t))$ at time t is the particle's **position vector** (Figure 10.1). The functions f, g, and h are the **component functions (components)** of the position vector. We think of the particle's path as the **curve traced by r** during the time interval I. Figure 10.2 displays several space curves generated by a computer graphing program. It would not be easy to plot these curves by hand.

Equation (2) defines $\mathbf{r}$ as a vector function of the real variable t on the interval I. More generally, a **vector-valued function** or **vector function** on a domain set D is a rule that assigns a vector in space to each element in D. For now, the domains will be intervals of real numbers resulting in a space curve. Later, in Chapter 13, the domains will be regions in the plane. Vector functions will then represent surfaces in space. Vector functions on a domain in the plane or space also give rise to "vector fields," which are important to the study of the flow of a fluid, gravitational fields, and electromagnetic phenomena. We investigate vector fields and their applications in Chapter 13.

Real-valued functions are called **scalar functions** to distinguish them from vector functions. The components of $\mathbf{r}$ are scalar functions of t. The domain of a vector-valued function is the common domain of its components.

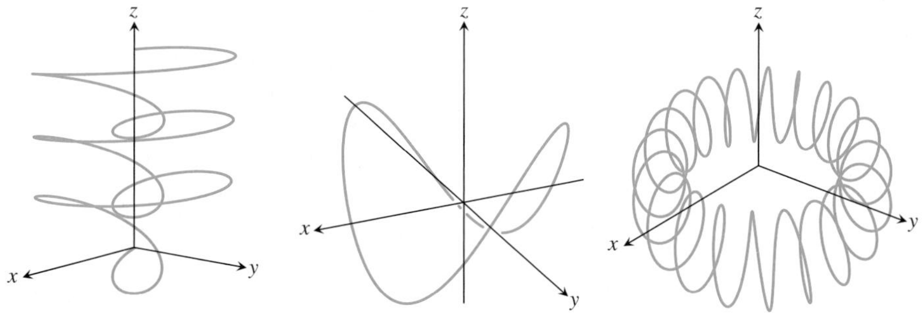

$\mathbf{r}(t) = (\sin 3t)(\cos t)\mathbf{i} + (\sin 3t)(\sin t)\mathbf{j} + t\mathbf{k}$

(a)

$\mathbf{r}(t) = (\cos t)\mathbf{i} + (\sin t)\mathbf{j} + (\sin 2t)\mathbf{k}$

(b)

$\mathbf{r}(t) = (4 + \sin 20t)(\cos t)\mathbf{i} + (4 + \sin 20t)(\sin t)\mathbf{j} + (\cos 20t)\mathbf{k}$

(c)

FIGURE 10.2 Computer-generated space curves are defined by the position vectors $\mathbf{r}(t)$.

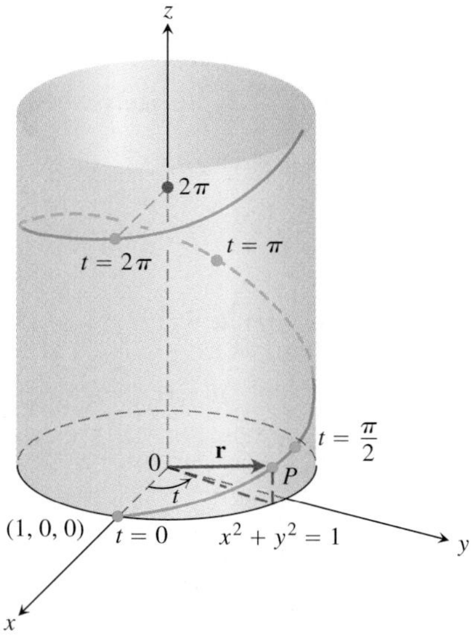

FIGURE 10.3 The upper half of the helix $\mathbf{r}(t) = (\cos t)\mathbf{i} + (\sin t)\mathbf{j} + t\mathbf{k}$ (Example 1).

EXAMPLE 1 Graph the vector function

$$\mathbf{r}(t) = (\cos t)\mathbf{i} + (\sin t)\mathbf{j} + t\mathbf{k}.$$

Solution The vector function

$$\mathbf{r}(t) = (\cos t)\mathbf{i} + (\sin t)\mathbf{j} + t\mathbf{k}$$

is defined for all real values of t. The curve traced by $\mathbf{r}$ winds around the circular cylinder $x^2 + y^2 = 1$ (Figure 10.3). The curve lies on the cylinder because the **i**- and **j**-components of $\mathbf{r}$, being the x- and y-coordinates of the tip of $\mathbf{r}$, satisfy the cylinder's equation:

$$x^2 + y^2 = (\cos t)^2 + (\sin t)^2 = 1.$$

The curve rises as the **k**-component $z = t$ increases. Each time t increases by 2π, the curve completes one turn around the cylinder. The curve is called a **helix** (from an old Greek word for "spiral"). The equations

$$x = \cos t, \qquad y = \sin t, \qquad z = t$$

parametrize the helix, the interval $-\infty < t < \infty$ being understood. You will find more helices in Figure 10.4. ■

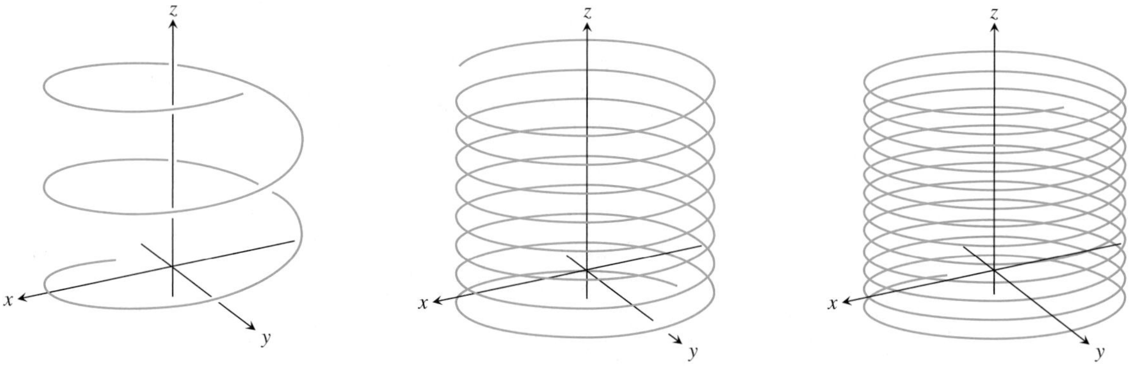

$\mathbf{r}(t) = (\cos t)\mathbf{i} + (\sin t)\mathbf{j} + t\mathbf{k}$ $\qquad$ $\mathbf{r}(t) = (\cos t)\mathbf{i} + (\sin t)\mathbf{j} + 0.3t\mathbf{k}$ $\qquad$ $\mathbf{r}(t) = (\cos 5t)\mathbf{i} + (\sin 5t)\mathbf{j} + t\mathbf{k}$

FIGURE 10.4 Helices drawn by computer.

Limits and Continuity

The way we define limits of vector-valued functions is similar to the way we define limits of real-valued functions.

DEFINITION Let $\mathbf{r}(t) = f(t)\mathbf{i} + g(t)\mathbf{j} + h(t)\mathbf{k}$ be a vector function and $\mathbf{L}$ a vector. We say that $\mathbf{r}$ has **limit L** as t approaches t_0 and write

$$\lim_{t \to t_0} \mathbf{r}(t) = \mathbf{L}$$

if, for every number $\epsilon > 0$, there exists a corresponding number $\delta > 0$ such that for all t

$$0 < |t - t_0| < \delta \quad \Rightarrow \quad |\mathbf{r}(t) - \mathbf{L}| < \epsilon.$$

If $\mathbf{L} = L_1\mathbf{i} + L_2\mathbf{j} + L_3\mathbf{k}$, then it can be shown that $\lim_{t \to t_0} \mathbf{r}(t) = \mathbf{L}$ precisely when

$$\lim_{t \to t_0} f(t) = L_1, \qquad \lim_{t \to t_0} g(t) = L_2, \qquad \text{and} \qquad \lim_{t \to t_0} h(t) = L_3.$$

We omit the proof. The equation

$$\lim_{t \to t_0} \mathbf{r}(t) = \left(\lim_{t \to t_0} f(t)\right)\mathbf{i} + \left(\lim_{t \to t_0} g(t)\right)\mathbf{j} + \left(\lim_{t \to t_0} h(t)\right)\mathbf{k} \tag{3}$$

provides a practical way to calculate limits of vector functions.

EXAMPLE 2 If $\mathbf{r}(t) = (\cos t)\mathbf{i} + (\sin t)\mathbf{j} + t\mathbf{k}$, then

$$\begin{aligned}\lim_{t \to \pi/4} \mathbf{r}(t) &= \left(\lim_{t \to \pi/4} \cos t\right)\mathbf{i} + \left(\lim_{t \to \pi/4} \sin t\right)\mathbf{j} + \left(\lim_{t \to \pi/4} t\right)\mathbf{k} \\ &= \frac{\sqrt{2}}{2}\mathbf{i} + \frac{\sqrt{2}}{2}\mathbf{j} + \frac{\pi}{4}\mathbf{k}.\end{aligned}$$

■

We define continuity for vector functions the same way we define continuity for scalar functions.

DEFINITION A vector function $\mathbf{r}(t)$ is **continuous at a point** $t = t_0$ in its domain if $\lim_{t \to t_0} \mathbf{r}(t) = \mathbf{r}(t_0)$. The function is **continuous** if it is continuous at every point in its domain.

From Equation (3), we see that $\mathbf{r}(t)$ is continuous at $t = t_0$ if and only if each component function is continuous there (Exercise 31).

EXAMPLE 3

(a) All the space curves shown in Figures 10.2 and 10.4 are continuous because their component functions are continuous at every value of t in $(-\infty, \infty)$.

(b) The function

$$\mathbf{g}(t) = (\cos t)\mathbf{i} + (\sin t)\mathbf{j} + \lfloor t \rfloor \mathbf{k}$$

is discontinuous at every integer, where the greatest integer function $\lfloor t \rfloor$ is discontinuous. ■

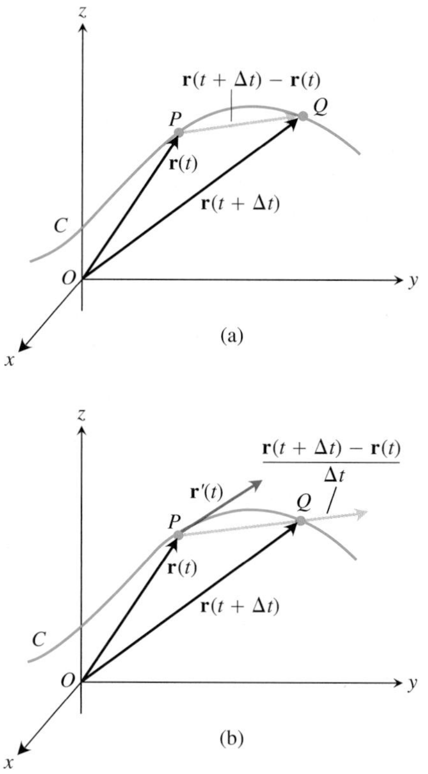

FIGURE 10.5 As $\Delta t \to 0$, the point Q approaches the point P along the curve C. In the limit, the vector $\overrightarrow{PQ}/\Delta t$ becomes the tangent vector $\mathbf{r}'(t)$.

Derivatives and Motion

Suppose that $\mathbf{r}(t) = f(t)\mathbf{i} + g(t)\mathbf{j} + h(t)\mathbf{k}$ is the position vector of a particle moving along a curve in space and that f, g, and h are differentiable functions of t. Then the difference between the particle's positions at time t and time $t + \Delta t$ is

$$\Delta\mathbf{r} = \mathbf{r}(t + \Delta t) - \mathbf{r}(t)$$

(Figure 10.5a). In terms of components,

$$\begin{aligned}\Delta\mathbf{r} &= \mathbf{r}(t + \Delta t) - \mathbf{r}(t)\\ &= [f(t + \Delta t)\mathbf{i} + g(t + \Delta t)\mathbf{j} + h(t + \Delta t)\mathbf{k}]\\ &\quad - [f(t)\mathbf{i} + g(t)\mathbf{j} + h(t)\mathbf{k}]\\ &= [f(t + \Delta t) - f(t)]\mathbf{i} + [g(t + \Delta t) - g(t)]\mathbf{j} + [h(t + \Delta t) - h(t)]\mathbf{k}.\end{aligned}$$

As Δt approaches zero, three things seem to happen simultaneously. First, Q approaches P along the curve. Second, the secant line PQ seems to approach a limiting position tangent to the curve at P. Third, the quotient $\Delta\mathbf{r}/\Delta t$ (Figure 10.5b) approaches the limit

$$\begin{aligned}\lim_{\Delta t \to 0} \frac{\Delta\mathbf{r}}{\Delta t} &= \left[\lim_{\Delta t \to 0} \frac{f(t + \Delta t) - f(t)}{\Delta t}\right]\mathbf{i} + \left[\lim_{\Delta t \to 0} \frac{g(t + \Delta t) - g(t)}{\Delta t}\right]\mathbf{j}\\ &\quad + \left[\lim_{\Delta t \to 0} \frac{h(t + \Delta t) - h(t)}{\Delta t}\right]\mathbf{k}\\ &= \left[\frac{df}{dt}\right]\mathbf{i} + \left[\frac{dg}{dt}\right]\mathbf{j} + \left[\frac{dh}{dt}\right]\mathbf{k}.\end{aligned}$$

We are therefore led by past experience to the following definition.

DEFINITION The vector function $\mathbf{r}(t) = f(t)\mathbf{i} + g(t)\mathbf{j} + h(t)\mathbf{k}$ has a **derivative (is differentiable) at *t*** if f, g, and h have derivatives at t. The derivative is the vector function

$$\mathbf{r}'(t) = \frac{d\mathbf{r}}{dt} = \lim_{\Delta t \to 0} \frac{\mathbf{r}(t + \Delta t) - \mathbf{r}(t)}{\Delta t} = \frac{df}{dt}\mathbf{i} + \frac{dg}{dt}\mathbf{j} + \frac{dh}{dt}\mathbf{k}.$$

A vector function $\mathbf{r}$ is **differentiable** if it is differentiable at every point of its domain. The curve traced by $\mathbf{r}$ is **smooth** if $d\mathbf{r}/dt$ is continuous and never $\mathbf{0}$, that is, if f, g, and h have continuous first derivatives that are not simultaneously 0.

The geometric significance of the definition of derivative is shown in Figure 10.5. The points P and Q have position vectors $\mathbf{r}(t)$ and $\mathbf{r}(t + \Delta t)$, and the vector $\overrightarrow{PQ}$ is represented by $\mathbf{r}(t + \Delta t) - \mathbf{r}(t)$. For $\Delta t > 0$, the scalar multiple $(1/\Delta t)(\mathbf{r}(t + \Delta t) - \mathbf{r}(t))$

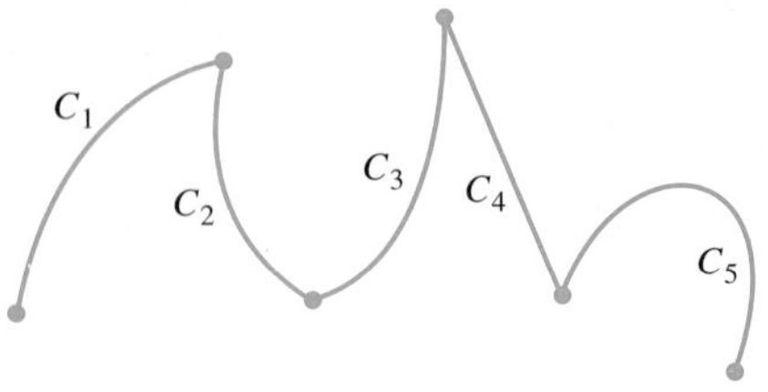

FIGURE 10.6 A piecewise smooth curve made up of five smooth curves connected end to end in continuous fashion. The curve here is not smooth at the points joining the five smooth curves.

points in the same direction as the vector $\overrightarrow{PQ}$. As $\Delta t \to 0$, this vector approaches a vector that is tangent to the curve at P (Figure 10.5b). The vector $\mathbf{r}'(t)$, when different from $\mathbf{0}$, is defined to be the vector **tangent** to the curve at P. The **tangent line** to the curve at a point $(f(t_0), g(t_0), h(t_0))$ is defined to be the line through the point parallel to $\mathbf{r}'(t_0)$. We require $d\mathbf{r}/dt \neq \mathbf{0}$ for a smooth curve to make sure the curve has a continuously turning tangent at each point. On a smooth curve, there are no sharp corners or cusps.

A curve that is made up of a finite number of smooth curves pieced together in a continuous fashion is called **piecewise smooth** (Figure 10.6).

Look once again at Figure 10.5. We drew the figure for Δt positive, so $\Delta\mathbf{r}$ points forward, in the direction of the motion. The vector $\Delta\mathbf{r}/\Delta t$, having the same direction as $\Delta\mathbf{r}$, points forward too. Had Δt been negative, $\Delta\mathbf{r}$ would have pointed backward, against the direction of motion. The quotient $\Delta\mathbf{r}/\Delta t$, however, being a negative scalar multiple of $\Delta\mathbf{r}$, would once again have pointed forward. No matter how $\Delta\mathbf{r}$ points, $\Delta\mathbf{r}/\Delta t$ points forward and we expect the vector $d\mathbf{r}/dt = \lim_{\Delta t \to 0} \Delta\mathbf{r}/\Delta t$, when different from $\mathbf{0}$, to do the same. This means that the derivative $d\mathbf{r}/dt$, which is the rate of change of position with respect to time, always points in the direction of motion. For a smooth curve, $d\mathbf{r}/dt$ is never zero; the particle does not stop or reverse direction.

DEFINITIONS If $\mathbf{r}$ is the position vector of a particle moving along a smooth curve in space, then

$$\mathbf{v}(t) = \frac{d\mathbf{r}}{dt}$$

is the particle's **velocity vector**, tangent to the curve. At any time t, the direction of $\mathbf{v}$ is the **direction of motion**, the magnitude of $\mathbf{v}$ is the particle's **speed**, and the derivative $\mathbf{a} = d\mathbf{v}/dt$, when it exists, is the particle's **acceleration vector**. In summary,

1. Velocity is the derivative of position: $\mathbf{v} = \dfrac{d\mathbf{r}}{dt}$.
2. Speed is the magnitude of velocity: Speed $= |\mathbf{v}|$.
3. Acceleration is the derivative of velocity: $\mathbf{a} = \dfrac{d\mathbf{v}}{dt} = \dfrac{d^2\mathbf{r}}{dt^2}$.
4. The unit vector $\mathbf{v}/|\mathbf{v}|$ is the direction of motion at time t.

We can express the velocity of a moving particle as the product of its speed and direction:

$$\text{Velocity} = |\mathbf{v}|\left(\frac{\mathbf{v}}{|\mathbf{v}|}\right) = (\text{speed})(\text{direction}).$$

Differentiation Rules

Because the derivatives of vector functions may be computed component by component, the rules for differentiating vector functions have the same form as the rules for differentiating scalar functions.

Differentiation Rules for Vector Functions

Let **u** and **v** be differentiable vector functions of t, **C** a constant vector, c any scalar, and f any differentiable scalar function.

1. *Constant Function Rule:* $\dfrac{d}{dt}\mathbf{C} = \mathbf{0}$
2. *Scalar Multiple Rules:* $\dfrac{d}{dt}[c\mathbf{u}(t)] = c\mathbf{u}'(t)$

   $\dfrac{d}{dt}[f(t)\mathbf{u}(t)] = f'(t)\mathbf{u}(t) + f(t)\mathbf{u}'(t)$
3. *Sum Rule:* $\dfrac{d}{dt}[\mathbf{u}(t) + \mathbf{v}(t)] = \mathbf{u}'(t) + \mathbf{v}'(t)$
4. *Difference Rule:* $\dfrac{d}{dt}[\mathbf{u}(t) - \mathbf{v}(t)] = \mathbf{u}'(t) - \mathbf{v}'(t)$
5. *Dot Product Rule:* $\dfrac{d}{dt}[\mathbf{u}(t)\cdot\mathbf{v}(t)] = \mathbf{u}'(t)\cdot\mathbf{v}(t) + \mathbf{u}(t)\cdot\mathbf{v}'(t)$
6. *Cross Product Rule:* $\dfrac{d}{dt}[\mathbf{u}(t)\times\mathbf{v}(t)] = \mathbf{u}'(t)\times\mathbf{v}(t) + \mathbf{u}(t)\times\mathbf{v}'(t)$
7. *Chain Rule:* $\dfrac{d}{dt}[\mathbf{u}(f(t))] = f'(t)\mathbf{u}'(f(t))$

When you use the Cross Product Rule, remember to preserve the order of the factors. If **u** comes first on the left side of the equation, it must also come first on the right or the signs will be wrong.

We will prove the Dot Product Rule and Chain Rule, but leave the other rules as exercises.

Proof of the Dot Product Rule Suppose that

$$\mathbf{u} = u_1(t)\mathbf{i} + u_2(t)\mathbf{j} + u_3(t)\mathbf{k}$$

and

$$\mathbf{v} = v_1(t)\mathbf{i} + v_2(t)\mathbf{j} + v_3(t)\mathbf{k}.$$

Then

$$\begin{aligned}\frac{d}{dt}(\mathbf{u}\cdot\mathbf{v}) &= \frac{d}{dt}(u_1v_1 + u_2v_2 + u_3v_3)\\ &= \underbrace{u_1'v_1 + u_2'v_2 + u_3'v_3}_{\mathbf{u}'\cdot\mathbf{v}} + \underbrace{u_1v_1' + u_2v_2' + u_3v_3'}_{\mathbf{u}\cdot\mathbf{v}'}.\end{aligned}$$

■

Proof of the Chain Rule Suppose that $\mathbf{u}(s) = a(s)\mathbf{i} + b(s)\mathbf{j} + c(s)\mathbf{k}$ is a differentiable vector function of s and that $s = f(t)$ is a differentiable scalar function of t. Then a, b, and c are differentiable functions of t, and the Chain Rule for differentiable real-valued functions gives

$$\begin{aligned}\frac{d}{dt}[\mathbf{u}(s)] &= \frac{da}{dt}\mathbf{i} + \frac{db}{dt}\mathbf{j} + \frac{dc}{dt}\mathbf{k}\\ &= \frac{da}{ds}\frac{ds}{dt}\mathbf{i} + \frac{db}{ds}\frac{ds}{dt}\mathbf{j} + \frac{dc}{ds}\frac{ds}{dt}\mathbf{k}\end{aligned}$$

As an algebraic convenience, we sometimes write the product of a scalar c and a vector **v** as **v**c instead of c**v**. This permits us, for instance, to write the Chain Rule in a familiar form:

$$\frac{d\mathbf{u}}{dt} = \frac{d\mathbf{u}}{ds}\frac{ds}{dt},$$

where $s = f(t)$.

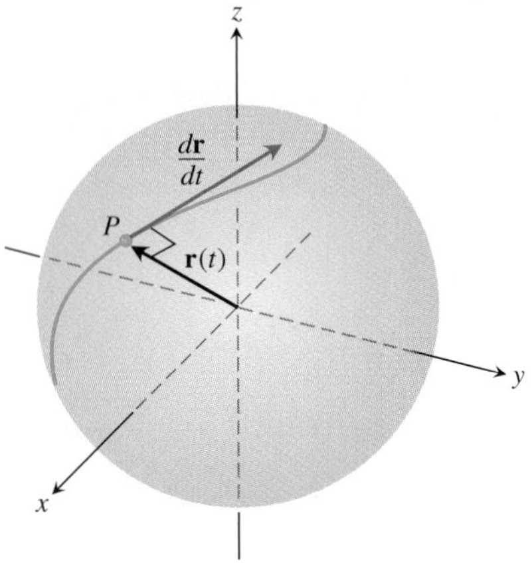

FIGURE 10.7 If a particle moves on a sphere in such a way that its position $\mathbf{r}$ is a differentiable function of time, then $\mathbf{r} \cdot (d\mathbf{r}/dt) = 0$.

$$= \frac{ds}{dt}\left(\frac{da}{ds}\mathbf{i} + \frac{db}{ds}\mathbf{j} + \frac{dc}{ds}\mathbf{k}\right)$$

$$= \frac{ds}{dt}\frac{d\mathbf{u}}{ds}$$

$$= f'(t)\mathbf{u}'(f(t)). \qquad s = f(t)$$

■

Vector Functions of Constant Length

When we track a particle moving on a sphere centered at the origin (Figure 10.7), the position vector has a constant length equal to the radius of the sphere. The velocity vector $d\mathbf{r}/dt$, tangent to the path of motion, is tangent to the sphere and hence perpendicular to $\mathbf{r}$. This is always the case for a differentiable vector function of constant length: The vector and its first derivative are orthogonal. By direct calculation,

$$\mathbf{r}(t) \cdot \mathbf{r}(t) = c^2 \qquad |\mathbf{r}(t)| = c \text{ is constant.}$$

$$\frac{d}{dt}[\mathbf{r}(t) \cdot \mathbf{r}(t)] = 0 \qquad \text{Differentiate both sides.}$$

$$\mathbf{r}'(t) \cdot \mathbf{r}(t) + \mathbf{r}(t) \cdot \mathbf{r}'(t) = 0 \qquad \text{Rule 5 with } \mathbf{r}(t) = \mathbf{u}(t) = \mathbf{v}(t)$$

$$2\mathbf{r}'(t) \cdot \mathbf{r}(t) = 0.$$

The vectors $\mathbf{r}'(t)$ and $\mathbf{r}(t)$ are orthogonal because their dot product is 0. In summary,

> If $\mathbf{r}$ is a differentiable vector function of t of constant length, then
>
> $$\mathbf{r} \cdot \frac{d\mathbf{r}}{dt} = 0. \qquad (4)$$

We will use this observation repeatedly in Section 10.4.

EXERCISES 10.1

In Exercises 1–4, $\mathbf{r}(t)$ is the position of a particle in the xy-plane at time t. Find an equation in x and y whose graph is the path of the particle. Then find the particle's velocity and acceleration vectors at the given value of t.

1. $\mathbf{r}(t) = (t + 1)\mathbf{i} + (t^2 - 1)\mathbf{j}, \quad t = 1$
2. $\mathbf{r}(t) = (t^2 + 1)\mathbf{i} + (2t - 1)\mathbf{j}, \quad t = 1/2$
3. $\mathbf{r}(t) = e^t\mathbf{i} + \frac{2}{9}e^{2t}\mathbf{j}, \quad t = \ln 3$
4. $\mathbf{r}(t) = (\cos 2t)\mathbf{i} + (3 \sin 2t)\mathbf{j}, \quad t = 0$

Exercises 5–8 give the position vectors of particles moving along various curves in the xy-plane. In each case, find the particle's velocity and acceleration vectors at the stated times and sketch them as vectors on the curve.

5. **Motion on the circle $x^2 + y^2 = 1$**

$$\mathbf{r}(t) = (\sin t)\mathbf{i} + (\cos t)\mathbf{j}; \quad t = \pi/4 \text{ and } \pi/2$$

6. **Motion on the circle $x^2 + y^2 = 16$**

$$\mathbf{r}(t) = \left(4 \cos \frac{t}{2}\right)\mathbf{i} + \left(4 \sin \frac{t}{2}\right)\mathbf{j}; \quad t = \pi \text{ and } 3\pi/2$$

7. **Motion on the cycloid $x = t - \sin t, y = 1 - \cos t$**

$$\mathbf{r}(t) = (t - \sin t)\mathbf{i} + (1 - \cos t)\mathbf{j}; \quad t = \pi \text{ and } 3\pi/2$$

8. **Motion on the parabola $y = x^2 + 1$**

$$\mathbf{r}(t) = t\mathbf{i} + (t^2 + 1)\mathbf{j}; \quad t = -1, 0, \text{ and } 1$$

In Exercises 9–14, $\mathbf{r}(t)$ is the position of a particle in space at time t. Find the particle's velocity and acceleration vectors. Then find the particle's speed and direction of motion at the given value of t. Write the particle's velocity at that time as the product of its speed and direction.

9. $\mathbf{r}(t) = (t+1)\mathbf{i} + (t^2-1)\mathbf{j} + 2t\mathbf{k}, \quad t = 1$

10. $\mathbf{r}(t) = (1+t)\mathbf{i} + \dfrac{t^2}{\sqrt{2}}\mathbf{j} + \dfrac{t^3}{3}\mathbf{k}, \quad t = 1$

11. $\mathbf{r}(t) = (2\cos t)\mathbf{i} + (3\sin t)\mathbf{j} + 4t\mathbf{k}, \quad t = \pi/2$

12. $\mathbf{r}(t) = (\sec t)\mathbf{i} + (\tan t)\mathbf{j} + \dfrac{4}{3}t\mathbf{k}, \quad t = \pi/6$

13. $\mathbf{r}(t) = (2\ln(t+1))\mathbf{i} + t^2\mathbf{j} + \dfrac{t^2}{2}\mathbf{k}, \quad t = 1$

14. $\mathbf{r}(t) = (e^{-t})\mathbf{i} + (2\cos 3t)\mathbf{j} + (2\sin 3t)\mathbf{k}, \quad t = 0$

In Exercises 15–18, $\mathbf{r}(t)$ is the position of a particle in space at time t. Find the angle between the velocity and acceleration vectors at time $t = 0$.

15. $\mathbf{r}(t) = (3t+1)\mathbf{i} + \sqrt{3}t\mathbf{j} + t^2\mathbf{k}$

16. $\mathbf{r}(t) = \left(\dfrac{\sqrt{2}}{2}t\right)\mathbf{i} + \left(\dfrac{\sqrt{2}}{2}t - 16t^2\right)\mathbf{j}$

17. $\mathbf{r}(t) = (\ln(t^2+1))\mathbf{i} + (\tan^{-1} t)\mathbf{j} + \sqrt{t^2+1}\,\mathbf{k}$

18. $\mathbf{r}(t) = \dfrac{4}{9}(1+t)^{3/2}\mathbf{i} + \dfrac{4}{9}(1-t)^{3/2}\mathbf{j} + \dfrac{1}{3}t\mathbf{k}$

As mentioned in the text, the **tangent line** to a smooth curve $\mathbf{r}(t) = f(t)\mathbf{i} + g(t)\mathbf{j} + h(t)\mathbf{k}$ at $t = t_0$ is the line that passes through the point $(f(t_0), g(t_0), h(t_0))$ parallel to $\mathbf{v}(t_0)$, the curve's velocity vector at t_0. In Exercises 19–22, find parametric equations for the line that is tangent to the given curve at the given parameter value $t = t_0$.

19. $\mathbf{r}(t) = (\sin t)\mathbf{i} + (t^2 - \cos t)\mathbf{j} + e^t\mathbf{k}, \quad t_0 = 0$

20. $\mathbf{r}(t) = (2\sin t)\mathbf{i} + (2\cos t)\mathbf{j} + 5t\mathbf{k}, \quad t_0 = 4\pi$

21. $\mathbf{r}(t) = (a\sin t)\mathbf{i} + (a\cos t)\mathbf{j} + bt\mathbf{k}, \quad t_0 = 2\pi$

22. $\mathbf{r}(t) = (\cos t)\mathbf{i} + (\sin t)\mathbf{j} + (\sin 2t)\mathbf{k}, \quad t_0 = \dfrac{\pi}{2}$

23. Each of the following equations in parts (a)–(e) describes the motion of a particle having the same path, namely the unit circle $x^2 + y^2 = 1$. Although the path of each particle in parts (a)–(e) is the same, the behavior, or "dynamics," of each particle is different. For each particle, answer the following questions.

i. Does the particle have constant speed? If so, what is its constant speed?

ii. Is the particle's acceleration vector always orthogonal to its velocity vector?

iii. Does the particle move clockwise or counterclockwise around the circle?

iv. Does the particle begin at the point $(1, 0)$?

a. $\mathbf{r}(t) = (\cos t)\mathbf{i} + (\sin t)\mathbf{j}, \quad t \ge 0$

b. $\mathbf{r}(t) = \cos(2t)\mathbf{i} + \sin(2t)\mathbf{j}, \quad t \ge 0$

c. $\mathbf{r}(t) = \cos(t - \pi/2)\mathbf{i} + \sin(t - \pi/2)\mathbf{j}, \quad t \ge 0$

d. $\mathbf{r}(t) = (\cos t)\mathbf{i} - (\sin t)\mathbf{j}, \quad t \ge 0$

e. $\mathbf{r}(t) = \cos(t^2)\mathbf{i} + \sin(t^2)\mathbf{j}, \quad t \ge 0$

24. Show that the vector-valued function

$$\mathbf{r}(t) = (2\mathbf{i} + 2\mathbf{j} + \mathbf{k}) + \cos t\left(\frac{1}{\sqrt{2}}\mathbf{i} - \frac{1}{\sqrt{2}}\mathbf{j}\right) + \sin t\left(\frac{1}{\sqrt{3}}\mathbf{i} + \frac{1}{\sqrt{3}}\mathbf{j} + \frac{1}{\sqrt{3}}\mathbf{k}\right)$$

describes the motion of a particle moving in the circle of radius 1 centered at the point $(2, 2, 1)$ and lying in the plane $x + y - 2z = 2$.

25. Motion along a parabola A particle moves along the top of the parabola $y^2 = 2x$ from left to riikht at a constant speed of 5 units per second. Find the velocity of the particle as it moves through the point (2, 2).

26. Motion along a cycloid A particle moves in the xy-plane in such a way that its position at time t is

$$\mathbf{r}(t) = (t - \sin t)\mathbf{i} + (1 - \cos t)\mathbf{j}.$$

T **a.** Graph $\mathbf{r}(t)$. The resulting curve is a cycloid.

b. Find the maximum and minimum values of $|\mathbf{v}|$ and $|\mathbf{a}|$. (*Hint:* Find the extreme values of $|\mathbf{v}|^2$ and $|\mathbf{a}|^2$ first and take square roots later.)

27. Let $\mathbf{v}$ be a differentiable vector function of t. Show that if $\mathbf{v} \cdot (d\mathbf{v}/dt) = 0$ for all t, then $|\mathbf{v}|$ is constant.

28. Derivatives of triple scalar products

a. Show that if $\mathbf{u}$, $\mathbf{v}$, and $\mathbf{w}$ are differentiable vector functions of t, then

$$\frac{d}{dt}(\mathbf{u}\cdot\mathbf{v}\times\mathbf{w}) = \frac{d\mathbf{u}}{dt}\cdot\mathbf{v}\times\mathbf{w} + \mathbf{u}\cdot\frac{d\mathbf{v}}{dt}\times\mathbf{w} + \mathbf{u}\cdot\mathbf{v}\times\frac{d\mathbf{w}}{dt}.$$

b. Show that

$$\frac{d}{dt}\left(\mathbf{r}\cdot\frac{d\mathbf{r}}{dt}\times\frac{d^2\mathbf{r}}{dt^2}\right) = \mathbf{r}\cdot\left(\frac{d\mathbf{r}}{dt}\times\frac{d^3\mathbf{r}}{dt^3}\right).$$

(*Hint:* Differentiate on the left and look for vectors whose products are zero.)

29. Prove the two Scalar Multiple Rules for vector functions.

30. Prove the Sum and Difference Rules for vector functions.

31. Component Test for Continuity at a Point Show that the vector function $\mathbf{r}$ defined by $\mathbf{r}(t) = f(t)\mathbf{i} + g(t)\mathbf{j} + h(t)\mathbf{k}$ is continuous at $t = t_0$ if and only if f, g, and h are continuous at t_0.

32. Limits of cross products of vector functions Suppose that $\mathbf{r}_1(t) = f_1(t)\mathbf{i} + f_2(t)\mathbf{j} + f_3(t)\mathbf{k}$, $\mathbf{r}_2(t) = g_1(t)\mathbf{i} + g_2(t)\mathbf{j} + g_3(t)\mathbf{k}$, $\lim_{t\to t_0}\mathbf{r}_1(t) = \mathbf{A}$, and $\lim_{t\to t_0}\mathbf{r}_2(t) = \mathbf{B}$. Use the determinant formula for cross products and the Limit Product Rule for scalar functions to show that

$$\lim_{t \to t_0} (\mathbf{r}_1(t) \times \mathbf{r}_2(t)) = \mathbf{A} \times \mathbf{B}.$$

33. Differentiable vector functions are continuous Show that if $\mathbf{r}(t) = f(t)\mathbf{i} + g(t)\mathbf{j} + h(t)\mathbf{k}$ is differentiable at $t = t_0$, then it is continuous at t_0 as well.

34. Prove the Cross Product Rule for vector functions. You will need to use the results of Exercises 32 and 33.

10.2 Integrals of Vector Functions

In this section we investigate integrals of vector functions and their application to motion along a path in space or in the plane.

Integrals of Vector Functions

A differentiable vector function $\mathbf{R}(t)$ is an **antiderivative** of a vector function $\mathbf{r}(t)$ on an interval I if $d\mathbf{R}/dt = \mathbf{r}$ at each point of I. If $\mathbf{R}$ is an antiderivative of $\mathbf{r}$ on I, it can be shown, working one component at a time, that every antiderivative of $\mathbf{r}$ on I has the form $\mathbf{R} + \mathbf{C}$ for some constant vector $\mathbf{C}$ (Exercise 27). The set of all antiderivatives of $\mathbf{r}$ on I is the **indefinite integral** of $\mathbf{r}$ on I.

DEFINITION The **indefinite integral** of $\mathbf{r}$ with respect to t is the set of all antiderivatives of $\mathbf{r}$, denoted by $\int \mathbf{r}(t)\, dt$. If $\mathbf{R}$ is any antiderivative of $\mathbf{r}$, then

$$\int \mathbf{r}(t)\, dt = \mathbf{R}(t) + \mathbf{C}.$$

The usual arithmetic rules for indefinite integrals apply.

EXAMPLE 1

$$\int ((\cos t)\mathbf{i} + \mathbf{j} - 2t\mathbf{k})\, dt = \left(\int \cos t\, dt\right)\mathbf{i} + \left(\int dt\right)\mathbf{j} - \left(\int 2t\, dt\right)\mathbf{k} \quad (1)$$

$$= (\sin t + C_1)\mathbf{i} + (t + C_2)\mathbf{j} - (t^2 + C_3)\mathbf{k} \quad (2)$$

$$= (\sin t)\mathbf{i} + t\mathbf{j} - t^2\mathbf{k} + \mathbf{C} \qquad \mathbf{C} = C_1\mathbf{i} + C_2\mathbf{j} - C_3\mathbf{k}$$

As in the integration of scalar functions, we recommend that you skip the steps in Equations (1) and (2) and go directly to the final form. Find an antiderivative for each component and add a constant vector at the end. ■

Definite integrals of vector functions are best defined in terms of components.

DEFINITION If the components of $\mathbf{r}(t) = f(t)\mathbf{i} + g(t)\mathbf{j} + h(t)\mathbf{k}$ are integrable over $[a, b]$, then so is $\mathbf{r}$, and the **definite integral** of $\mathbf{r}$ from a to b is

$$\int_a^b \mathbf{r}(t)\,dt = \left(\int_a^b f(t)\,dt\right)\mathbf{i} + \left(\int_a^b g(t)\,dt\right)\mathbf{j} + \left(\int_a^b h(t)\,dt\right)\mathbf{k}.$$

EXAMPLE 2

$$\begin{aligned}\int_0^{\pi} ((\cos t)\mathbf{i} + \mathbf{j} - 2t\mathbf{k})\,dt &= \left(\int_0^{\pi} \cos t\,dt\right)\mathbf{i} + \left(\int_0^{\pi} dt\right)\mathbf{j} - \left(\int_0^{\pi} 2t\,dt\right)\mathbf{k} \\ &= \Big[\sin t\Big]_0^{\pi}\mathbf{i} + \Big[t\Big]_0^{\pi}\mathbf{j} - \Big[t^2\Big]_0^{\pi}\mathbf{k} \\ &= [0 - 0]\mathbf{i} + [\pi - 0]\mathbf{j} - [\pi^2 - 0^2]\mathbf{k} \\ &= \pi\mathbf{j} - \pi^2\mathbf{k}\end{aligned}$$

■

The Fundamental Theorem of Calculus for continuous vector functions says that

$$\int_a^b \mathbf{r}(t)\,dt = \mathbf{R}(t)\Big]_a^b = \mathbf{R}(b) - \mathbf{R}(a)$$

where $\mathbf{R}$ is any antiderivative of $\mathbf{r}$, so that $\mathbf{R}'(t) = \mathbf{r}(t)$ (Exercise 28).

EXAMPLE 3 Suppose we do not know the path of a hang glider, but only its acceleration vector $\mathbf{a}(t) = -(3\cos t)\mathbf{i} - (3\sin t)\mathbf{j} + 2\mathbf{k}$. We also know that initially (at time $t = 0$), the glider departed from the point $(3, 0, 0)$ with velocity $\mathbf{v}(0) = 3\mathbf{j}$. Find the glider's position as a function of t.

Solution Our goal is to find $\mathbf{r}(t)$ knowing

The differential equation: $\mathbf{a} = \dfrac{d^2\mathbf{r}}{dt^2} = -(3\cos t)\mathbf{i} - (3\sin t)\mathbf{j} + 2\mathbf{k}$

The initial conditions: $\mathbf{v}(0) = 3\mathbf{j}$ and $\mathbf{r}(0) = 3\mathbf{i} + 0\mathbf{j} + 0\mathbf{k}$.

Integrating both sides of the differential equation with respect to t gives

$$\mathbf{v}(t) = -(3\sin t)\mathbf{i} + (3\cos t)\mathbf{j} + 2t\mathbf{k} + \mathbf{C}_1.$$

We use $\mathbf{v}(0) = 3\mathbf{j}$ to find $\mathbf{C}_1$:

$$\begin{aligned}3\mathbf{j} &= -(3\sin 0)\mathbf{i} + (3\cos 0)\mathbf{j} + (0)\mathbf{k} + \mathbf{C}_1 \\ 3\mathbf{j} &= 3\mathbf{j} + \mathbf{C}_1 \\ \mathbf{C}_1 &= \mathbf{0}.\end{aligned}$$

The glider's velocity as a function of time is

$$\frac{d\mathbf{r}}{dt} = \mathbf{v}(t) = -(3\sin t)\mathbf{i} + (3\cos t)\mathbf{j} + 2t\mathbf{k}.$$

Integrating both sides of this last differential equation gives

$$\mathbf{r}(t) = (3\cos t)\mathbf{i} + (3\sin t)\mathbf{j} + t^2\mathbf{k} + \mathbf{C}_2.$$

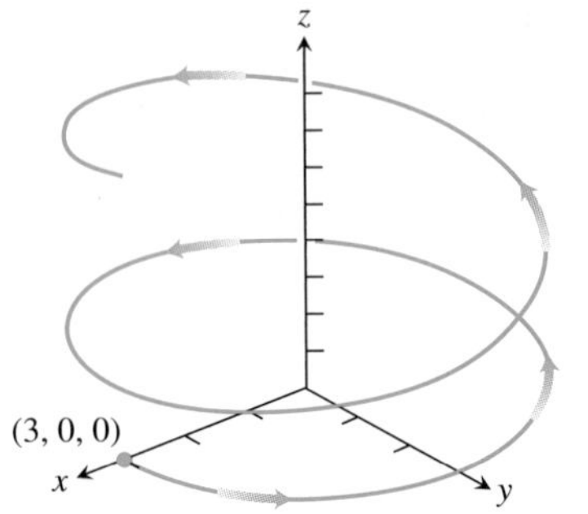

FIGURE 10.8 The path of a hang glider with position vector $\mathbf{r}(t) = (3\cos t)\mathbf{i} + (3\sin t)\mathbf{j} + t^2\mathbf{k}$ (Example 3).

We then use the initial condition $\mathbf{r}(0) = 3\mathbf{i}$ to find $\mathbf{C}_2$:

$$3\mathbf{i} = (3\cos 0)\mathbf{i} + (3\sin 0)\mathbf{j} + (0^2)\mathbf{k} + \mathbf{C}_2$$
$$3\mathbf{i} = 3\mathbf{i} + (0)\mathbf{j} + (0)\mathbf{k} + \mathbf{C}_2$$
$$\mathbf{C}_2 = \mathbf{0}.$$

The glider's position as a function of t is

$$\mathbf{r}(t) = (3\cos t)\mathbf{i} + (3\sin t)\mathbf{j} + t^2\mathbf{k}.$$

This is the path of the glider shown in Figure 10.8.

Note: It turned out in this example that both of the constant vectors of integration, $\mathbf{C}_1$ and $\mathbf{C}_2$, are $\mathbf{0}$. Exercises 11 and 12 give examples for which the constant vectors of integration are not $\mathbf{0}$. ■

The Vector and Parametric Equations for Ideal Projectile Motion

To derive equations for projectile motion, we assume that the projectile behaves like a particle moving in a vertical coordinate plane and that the only force acting on the projectile during its flight is the constant force of gravity, which always points straight down. (In practice, none of these assumptions really holds. The ground moves beneath the projectile as the earth turns, the air creates a frictional force that varies with the projectile's speed and altitude, and the force of gravity changes as the projectile moves along.)

We assume that the projectile is launched from the origin at time $t = 0$ into the first quadrant with an initial velocity $\mathbf{v}_0$ (Figure 10.9). If $\mathbf{v}_0$ makes an angle α with the horizontal, then

$$\mathbf{v}_0 = (|\mathbf{v}_0|\cos\alpha)\mathbf{i} + (|\mathbf{v}_0|\sin\alpha)\mathbf{j}.$$

If we use the simpler notation v_0 for the initial speed $|\mathbf{v}_0|$, then

$$\mathbf{v}_0 = (v_0\cos\alpha)\mathbf{i} + (v_0\sin\alpha)\mathbf{j}. \tag{3}$$

The projectile's initial position is

$$\mathbf{r}_0 = 0\mathbf{i} + 0\mathbf{j} = \mathbf{0}. \tag{4}$$

Newton's second law of motion says that the force acting on the projectile is equal to the projectile's mass m times its acceleration, or $m(d^2\mathbf{r}/dt^2)$ if $\mathbf{r}$ is the projectile's position vector and t is time. If the force is solely the gravitational force $-mg\mathbf{j}$, then

$$m\frac{d^2\mathbf{r}}{dt^2} = -mg\mathbf{j} \quad \text{and} \quad \frac{d^2\mathbf{r}}{dt^2} = -g\mathbf{j}.$$

We find $\mathbf{r}$ as a function of t by solving the following initial value problem.

Differential equation: $\dfrac{d^2\mathbf{r}}{dt^2} = -g\mathbf{j}$

Initial conditions: $\mathbf{r} = \mathbf{r}_0$ and $\dfrac{d\mathbf{r}}{dt} = \mathbf{v}_0$ when $t = 0$

The first integration gives

$$\frac{d\mathbf{r}}{dt} = -(gt)\mathbf{j} + \mathbf{v}_0.$$

A second integration gives

$$\mathbf{r} = -\frac{1}{2}gt^2\mathbf{j} + \mathbf{v}_0 t + \mathbf{r}_0.$$

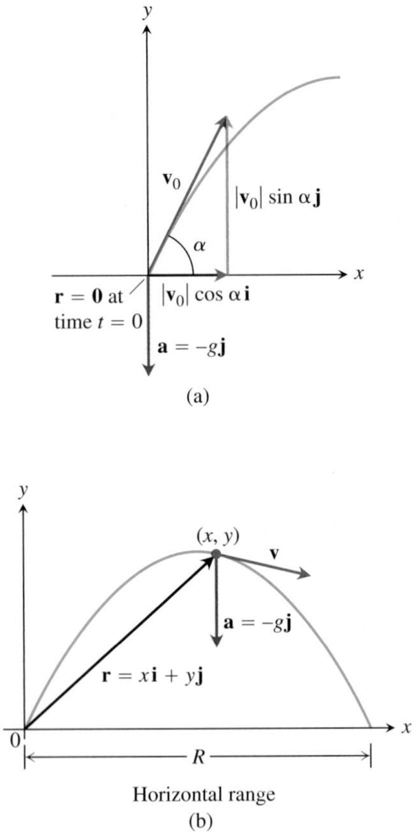

FIGURE 10.9 (a) Position, velocity, acceleration, and launch angle at $t = 0$. (b) Position, velocity, and acceleration at a later time t.

Substituting the values of $\mathbf{v}_0$ and $\mathbf{r}_0$ from Equations (3) and (4) gives

$$\mathbf{r} = -\frac{1}{2}gt^2\mathbf{j} + \underbrace{(v_0 \cos\alpha)t\mathbf{i} + (v_0 \sin\alpha)t\mathbf{j}}_{\mathbf{v}_0 t} + \mathbf{0}.$$

Collecting terms, we have

Ideal Projectile Motion Equation

$$\mathbf{r} = (v_0 \cos\alpha)t\mathbf{i} + \left((v_0 \sin\alpha)t - \frac{1}{2}gt^2\right)\mathbf{j}. \tag{5}$$

Equation (5) is the *vector equation* for ideal projectile motion. The angle α is the projectile's **launch angle (firing angle, angle of elevation)**, and v_0, as we said before, is the projectile's **initial speed**. The components of $\mathbf{r}$ give the parametric equations

$$x = (v_0 \cos\alpha)t \qquad \text{and} \qquad y = (v_0 \sin\alpha)t - \frac{1}{2}gt^2, \tag{6}$$

where x is the distance downrange and y is the height of the projectile at time $t \geq 0$.

EXAMPLE 4 A projectile is fired from the origin over horizontal ground at an initial speed of 500 m/sec and a launch angle of 60°. Where will the projectile be 10 sec later?

Solution We use Equation (5) with $v_0 = 500$, $\alpha = 60°$, $g = 9.8$, and $t = 10$ to find the projectile's components 10 sec after firing.

$$\begin{aligned}
\mathbf{r} &= (v_0 \cos\alpha)t\mathbf{i} + \left((v_0 \sin\alpha)t - \frac{1}{2}gt^2\right)\mathbf{j} \\
&= (500)\left(\frac{1}{2}\right)(10)\mathbf{i} + \left((500)\left(\frac{\sqrt{3}}{2}\right)10 - \left(\frac{1}{2}\right)(9.8)(100)\right)\mathbf{j} \\
&\approx 2500\mathbf{i} + 3840\mathbf{j}.
\end{aligned}$$

Ten seconds after firing, the projectile is about 3840 m in the air and 2500 m downrange. ■

Ideal projectiles move along parabolas, as we can see from Equations (6). If we substitute $t = x/(v_0 \cos\alpha)$ from the first equation into the second, we obtain the Cartesian-coordinate equation

$$y = -\left(\frac{g}{2v_0^2 \cos^2\alpha}\right)x^2 + (\tan\alpha)x.$$

This equation has the form $y = ax^2 + bx$, so its graph is a parabola.

A projectile reaches its highest point when its vertical velocity component is zero. When fired over horizontal ground, the projectile lands when its vertical component equals zero in Equation (5), and the **range R** is the distance from the origin to the point of impact. We summarize the results here, which you are asked to verify in Exercise 22.

Height, Flight Time, and Range for Ideal Projectile Motion

For ideal projectile motion when an object is launched from the origin over a horizontal surface with initial speed v_0 and launch angle α:

Maximum height: $y_{max} = \dfrac{(v_0 \sin\alpha)^2}{2g}$

Flight time: $t = \dfrac{2v_0 \sin\alpha}{g}$

Range: $R = \dfrac{v_0^2}{g}\sin 2\alpha .$

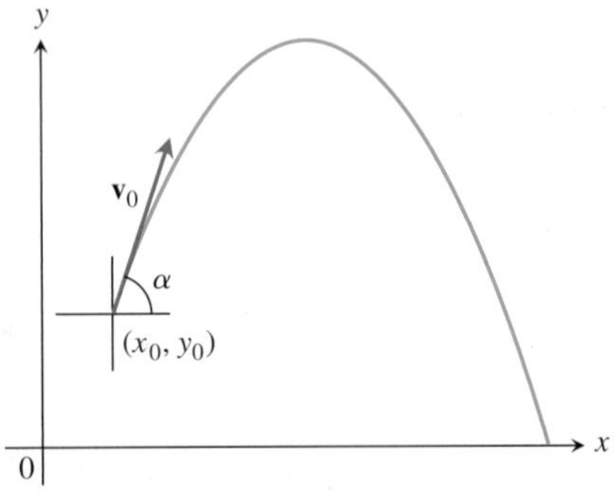

FIGURE 10.10 The path of a projectile fired from (x_0, y_0) with an initial velocity $\mathbf{v}_0$ at an angle of α degrees with the horizontal.

If we fire our ideal projectile from the point (x_0, y_0) instead of the origin (Figure 10.10), the position vector for the path of motion is

$$\mathbf{r} = (x_0 + (v_0 \cos\alpha)t)\mathbf{i} + \left(y_0 + (v_0 \sin\alpha)t - \frac{1}{2}gt^2\right)\mathbf{j}, \tag{7}$$

as you are asked to show in Exercise 23.

EXERCISES 10.2

Evaluate the integrals in Exercises 1–6.

1. $\displaystyle\int_0^1 [t^3\mathbf{i} + 7\mathbf{j} + (t+1)\mathbf{k}]\,dt$

2. $\displaystyle\int_1^2 \left[(6-6t)\mathbf{i} + 3\sqrt{t}\mathbf{j} + \left(\frac{4}{t^2}\right)\mathbf{k}\right] dt$

3. $\displaystyle\int_{-\pi/4}^{\pi/4} [(\sin t)\mathbf{i} + (1+\cos t)\mathbf{j} + (\sec^2 t)\mathbf{k}]\,dt$

4. $\displaystyle\int_0^{\pi/3} [(\sec t \tan t)\mathbf{i} + (\tan t)\mathbf{j} + (2\sin t \cos t)\mathbf{k}]\,dt$

5. $\displaystyle\int_1^4 \left[\frac{1}{t}\mathbf{i} + \frac{1}{5-t}\mathbf{j} + \frac{1}{2t}\mathbf{k}\right] dt$

6. $\displaystyle\int_0^1 \left[\frac{2}{\sqrt{1-t^2}}\mathbf{i} + \frac{\sqrt{3}}{1+t^2}\mathbf{k}\right] dt$

Solve the initial value problems in Exercises 7–12 for $\mathbf{r}$ as a vector function of t.

7. Differential equation: $\dfrac{d\mathbf{r}}{dt} = -t\mathbf{i} - t\mathbf{j} - t\mathbf{k}$

   Initial condition: $\mathbf{r}(0) = \mathbf{i} + 2\mathbf{j} + 3\mathbf{k}$

8. Differential equation: $\dfrac{d\mathbf{r}}{dt} = (180t)\mathbf{i} + (180t - 16t^2)\mathbf{j}$

   Initial condition: $\mathbf{r}(0) = 100\mathbf{j}$

9. Differential equation: $\dfrac{d\mathbf{r}}{dt} = \dfrac{3}{2}(t+1)^{1/2}\mathbf{i} + e^{-t}\mathbf{j} + \dfrac{1}{t+1}\mathbf{k}$

   Initial condition: $\mathbf{r}(0) = \mathbf{k}$

10. Differential equation: $\dfrac{d\mathbf{r}}{dt} = (t^3 + 4t)\mathbf{i} + t\mathbf{j} + 2t^2\mathbf{k}$

    Initial condition: $\mathbf{r}(0) = \mathbf{i} + \mathbf{j}$

11. Differential equation: $\dfrac{d^2\mathbf{r}}{dt^2} = -32\mathbf{k}$

    Initial conditions: $\mathbf{r}(0) = 100\mathbf{k}$ and $\left.\dfrac{d\mathbf{r}}{dt}\right|_{t=0} = 8\mathbf{i} + 8\mathbf{j}$

12. Differential equation: $\dfrac{d^2\mathbf{r}}{dt^2} = -(\mathbf{i} + \mathbf{j} + \mathbf{k})$

    Initial conditions: $\mathbf{r}(0) = 10\mathbf{i} + 10\mathbf{j} + 10\mathbf{k}$ and $\left.\dfrac{d\mathbf{r}}{dt}\right|_{t=0} = \mathbf{0}$

13. At time $t = 0$, a particle is located at the point (1, 2, 3). It travels in a straight line to the point (4, 1, 4), has speed 2 at (1, 2, 3) and constant acceleration $3\mathbf{i} - \mathbf{j} + \mathbf{k}$. Find an equation for the position vector $\mathbf{r}(t)$ of the particle at time t.

14. A particle traveling in a straight line is located at the point $(1, -1, 2)$ and has speed 2 at time $t = 0$. The particle moves toward the point (3, 0, 3) with constant acceleration $2\mathbf{i} + \mathbf{j} + \mathbf{k}$. Find its position vector $\mathbf{r}(t)$ at time t.

Projectile flights in the following exercises are to be treated as ideal unless stated otherwise. All launch angles are assumed to be measured from the horizontal. All projectiles are assumed to be launched from the origin over a horizontal surface unless stated otherwise.

15. **Travel time** A projectile is fired at a speed of 840 m/sec at an angle of 60°. How long will it take to get 21 km downrange?

16. **Finding muzzle speed** Find the muzzle speed of a gun whose maximum range is 24.5 km.

17. **Flight time and height** A projectile is fired with an initial speed of 500 m/sec at an angle of elevation of 45°.
    a. When and how far away will the projectile strike?
    b. How high overhead will the projectile be when it is 5 km downrange?
    c. What is the greatest height reached by the projectile?

18. **Throwing a baseball** A baseball is thrown from the stands 32 ft above the field at an angle of 30° up from the horizontal. When and how far away will the ball strike the ground if its initial speed is 32 ft/sec?

19. **Firing golf balls** A spring gun at ground level fires a golf ball at an angle of 45°. The ball lands 10 m away.
    a. What was the ball's initial speed?
    b. For the same initial speed, find the two firing angles that make the range 6 m.

20. **Beaming electrons** An electron in a TV tube is beamed horizontally at a speed of 5×10^6 m/sec toward the face of the tube 40 cm away. About how far will the electron drop before it hits?

21. **Equal-range firing angles** What two angles of elevation will enable a projectile to reach a target 16 km downrange on the same level as the gun if the projectile's initial speed is 400 m/sec?

22. Verify the results given in the text (following Example 4) for the maximum height, flight time, and range for ideal projectile motion.

23. **Firing from (x_0, y_0)** Derive the equations

$$x = x_0 + (v_0 \cos \alpha)t,$$
$$y = y_0 + (v_0 \sin \alpha)t - \frac{1}{2}gt^2,$$

(see Equation (7) in the text) by solving the following initial value problem for a vector **r** in the plane.

Differential equation: $\dfrac{d^2\mathbf{r}}{dt^2} = -g\mathbf{j}$

Initial conditions: $\mathbf{r}(0) = x_0\mathbf{i} + y_0\mathbf{j}$

$$\frac{d\mathbf{r}}{dt}(0) = (v_0 \cos \alpha)\mathbf{i} + (v_0 \sin \alpha)\mathbf{j}$$

24. **Where trajectories crest** For a projectile fired from the ground at launch angle α with initial speed v_0, consider α as a variable and v_0 as a fixed constant. For each α, $0 < \alpha < \pi/2$, we obtain a parabolic trajectory as shown in the accompanying figure. Show that the points in the plane that give the maximum heights of these parabolic trajectories all lie on the ellipse

$$x^2 + 4\left(y - \frac{v_0^2}{4g}\right)^2 = \frac{v_0^4}{4g^2},$$

where $x \geq 0$.

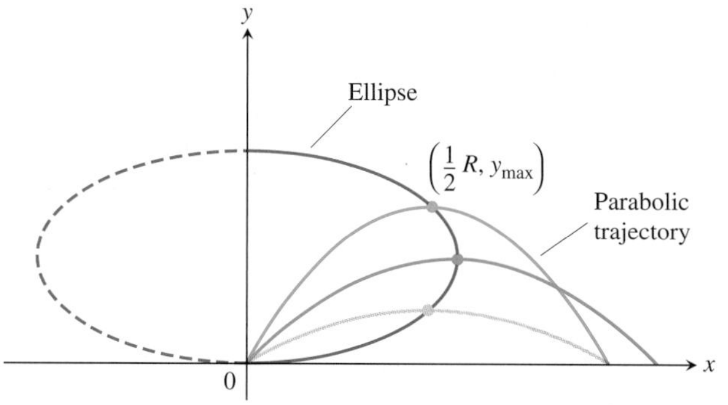

25. Establish the following properties of integrable vector functions.
    a. The *Constant Scalar Multiple Rule:*

$$\int_a^b k\mathbf{r}(t)\,dt = k\int_a^b \mathbf{r}(t)\,dt \quad \text{(any scalar } k\text{)}$$

The *Rule for Negatives,*

$$\int_a^b (-\mathbf{r}(t))\,dt = -\int_a^b \mathbf{r}(t)\,dt,$$

is obtained by taking $k = -1$.

    b. The *Sum and Difference Rules:*

$$\int_a^b (\mathbf{r}_1(t) \pm \mathbf{r}_2(t))\,dt = \int_a^b \mathbf{r}_1(t)\,dt \pm \int_a^b \mathbf{r}_2(t)\,dt$$

    c. The *Constant Vector Multiple Rules:*

$$\int_a^b \mathbf{C}\cdot\mathbf{r}(t)\,dt = \mathbf{C}\cdot\int_a^b \mathbf{r}(t)\,dt \quad \text{(any constant vector } \mathbf{C}\text{)}$$

and

$$\int_a^b \mathbf{C}\times\mathbf{r}(t)\,dt = \mathbf{C}\times\int_a^b \mathbf{r}(t)\,dt \quad \text{(any constant vector } \mathbf{C}\text{)}$$

26. **Products of scalar and vector functions** Suppose that the scalar function $u(t)$ and the vector function $\mathbf{r}(t)$ are both defined for $a \leq t \leq b$.
    a. Show that $u\mathbf{r}$ is continuous on $[a, b]$ if u and $\mathbf{r}$ are continuous on $[a, b]$.
    b. If u and $\mathbf{r}$ are both differentiable on $[a, b]$, show that $u\mathbf{r}$ is differentiable on $[a, b]$ and that

$$\frac{d}{dt}(u\mathbf{r}) = u\frac{d\mathbf{r}}{dt} + \mathbf{r}\frac{du}{dt}.$$

27. Antiderivatives of vector functions

a. Use Corollary 2 of the Mean Value Theorem for scalar functions to show that if two vector functions $\mathbf{R}_1(t)$ and $\mathbf{R}_2(t)$ have identical derivatives on an interval I, then the functions differ by a constant vector value throughout I.

b. Use the result in part (a) to show that if $\mathbf{R}(t)$ is any antiderivative of $\mathbf{r}(t)$ on I, then any other antiderivative of $\mathbf{r}$ on I equals $\mathbf{R}(t) + \mathbf{C}$ for some constant vector $\mathbf{C}$.

28. The Fundamental Theorem of Calculus The Fundamental Theorem of Calculus for scalar functions of a real variable holds for vector functions of a real variable as well. Prove this by using the theorem for scalar functions to show first that if a vector function $\mathbf{r}(t)$ is continuous for $a \leq t \leq b$, then

$$\frac{d}{dt}\int_a^t \mathbf{r}(\tau)\, d\tau = \mathbf{r}(t)$$

at every point t of (a, b). Then use the conclusion in part (b) of Exercise 27 to show that if $\mathbf{R}$ is any antiderivative of $\mathbf{r}$ on $[a, b]$ then

$$\int_a^b \mathbf{r}(t)\, dt = \mathbf{R}(b) - \mathbf{R}(a).$$

29. Volleyball A volleyball is hit when it is 4 ft above the ground and 12 ft from a 6-ft-high net. It leaves the point of impact with an initial velocity of 35 ft/sec at an angle of 27° and slips by the opposing team untouched.

a. Find a vector equation for the path of the volleyball.

b. How high does the volleyball go, and when does it reach maximum height?

c. Find its range and flight time.

d. When is the volleyball 7 ft above the ground? How far (ground distance) is the volleyball from where it will land?

e. Suppose that the net is raised to 8 ft. Does this change things? Explain.

30. Shot put In Moscow in 1987, Natalya Lisouskaya set a women's world record by putting an 8 lb 13 oz shot 73 ft 10 in. Assuming that she launched the shot at a 40° angle to the horizontal from 6.5 ft above the ground, what was the shot's initial speed?

10.3 Arc Length in Space

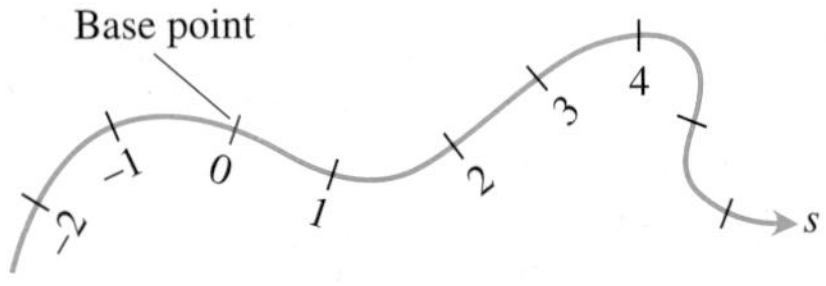

FIGURE 10.11 Smooth curves can be scaled like number lines, the coordinate of each point being its directed distance along the curve from a preselected base point.

In this and the next two sections, we study the mathematical features of a curve's shape that describe the sharpness of its turning and its twisting.

Arc Length Along a Space Curve

One of the features of smooth space curves is that they have a measurable length. This enables us to locate points along these curves by giving their directed distance s along the curve from some base point, the way we locate points on coordinate axes by giving their directed distance from the origin (Figure 10.11).

To measure distance along a smooth curve in space, we add a z-term to the formula we use for curves in the plane.

DEFINITION The **length** of a smooth curve $\mathbf{r}(t) = x(t)\mathbf{i} + y(t)\mathbf{j} + z(t)\mathbf{k}$, $a \leq t \leq b$, that is traced exactly once as t increases from $t = a$ to $t = b$, is

$$L = \int_a^b \sqrt{\left(\frac{dx}{dt}\right)^2 + \left(\frac{dy}{dt}\right)^2 + \left(\frac{dz}{dt}\right)^2}\, dt. \qquad (1)$$

Just as for plane curves, we can calculate the length of a curve in space from any convenient parametrization that meets the stated conditions. We omit the proof.

The square root in Equation (1) is $|\mathbf{v}|$, the length of a velocity vector $d\mathbf{r}/dt$. This enables us to write the formula for length a shorter way.

Arc Length Formula

$$L = \int_a^b |\mathbf{v}|\,dt \tag{2}$$

EXAMPLE 1 A glider is soaring upward along the helix $\mathbf{r}(t) = (\cos t)\mathbf{i} + (\sin t)\mathbf{j} + t\mathbf{k}$. How long is the glider's path from $t = 0$ to $t = 2\pi$?

Solution The path segment during this time corresponds to one full turn of the helix (Figure 10.12). The length of this portion of the curve is

$$L = \int_a^b |\mathbf{v}|\,dt = \int_0^{2\pi} \sqrt{(-\sin t)^2 + (\cos t)^2 + (1)^2}\,dt$$

$$= \int_0^{2\pi} \sqrt{2}\,dt = 2\pi\sqrt{2} \text{ units of length.}$$

This is $\sqrt{2}$ times the circumference of the circle in the xy-plane over which the helix stands. ■

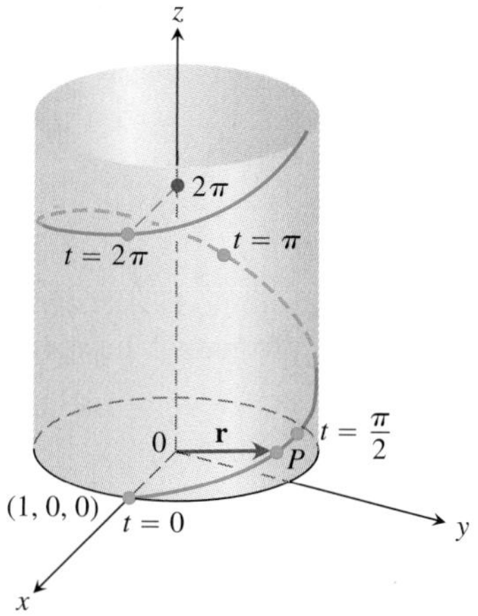

FIGURE 10.12 The helix $\mathbf{r}(t) = (\cos t)\mathbf{i} + (\sin t)\mathbf{j} + t\mathbf{k}$ in Example 1.

If we choose a base point $P(t_0)$ on a smooth curve C parametrized by t, each value of t determines a point $P(t) = (x(t), y(t), z(t))$ on C and a "directed distance"

$$s(t) = \int_{t_0}^{t} |\mathbf{v}(\tau)|\,d\tau,$$

measured along C from the base point (Figure 10.13). If $t > t_0$, $s(t)$ is the distance along the curve from $P(t_0)$ to $P(t)$. If $t < t_0$, $s(t)$ is the negative of the distance. Each value of s determines a point on C and this parametrizes C with respect to s. We call s an **arc length parameter** for the curve. The parameter's value increases in the direction of increasing t. The arc length parameter is particularly effective for investigating the turning and twisting nature of a space curve.

Arc Length Parameter with Base Point $P(t_0)$

$$s(t) = \int_{t_0}^{t} \sqrt{[x'(\tau)]^2 + [y'(\tau)]^2 + [z'(\tau)]^2}\,d\tau = \int_{t_0}^{t} |\mathbf{v}(\tau)|\,d\tau \tag{3}$$

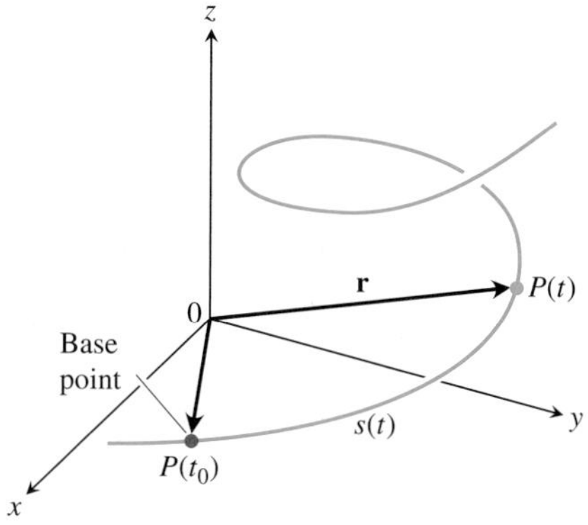

FIGURE 10.13 The directed distance along the curve from $P(t_0)$ to any point $P(t)$ is

$$s(t) = \int_{t_0}^{t} |\mathbf{v}(\tau)|\,d\tau.$$

We use the Greek letter τ ("tau") as the variable of integration in Equation (3) because the letter t is already in use as the upper limit.

If a curve $\mathbf{r}(t)$ is already given in terms of some parameter t and $s(t)$ is the arc length function given by Equation (3), then we may be able to solve for t as a function of s: $t = t(s)$. Then the curve can be reparametrized in terms of s by substituting for t: $\mathbf{r} = \mathbf{r}(t(s))$. The new parametrization identifies a point on the curve with its directed distance along the curve from the base point.

EXAMPLE 2 If $t_0 = 0$, the arc length parameter along the helix

$$\mathbf{r}(t) = (\cos t)\mathbf{i} + (\sin t)\mathbf{j} + t\mathbf{k}$$

from t_0 to t is

$$\begin{aligned} s(t) &= \int_{t_0}^{t} |\mathbf{v}(\tau)|\, d\tau && \text{Eq. (3)} \\ &= \int_{0}^{t} \sqrt{2}\, d\tau && \text{Value from Example 1} \\ &= \sqrt{2}\, t. \end{aligned}$$

Solving this equation for t gives $t = s/\sqrt{2}$. Substituting into the position vector $\mathbf{r}$ gives the following arc length parametrization for the helix:

$$\mathbf{r}(t(s)) = \left(\cos \frac{s}{\sqrt{2}}\right)\mathbf{i} + \left(\sin \frac{s}{\sqrt{2}}\right)\mathbf{j} + \frac{s}{\sqrt{2}}\mathbf{k}.$$ ■

Unlike Example 2, the arc length parametrization is generally difficult to find analytically for a curve already given in terms of some other parameter t. Fortunately, however, we rarely need an exact formula for $s(t)$ or its inverse $t(s)$.

HISTORICAL BIOGRAPHY

Josiah Willard Gibbs (1839–1903)

Speed on a Smooth Curve

Since the derivatives beneath the radical in Equation (3) are continuous (the curve is smooth), the Fundamental Theorem of Calculus tells us that s is a differentiable function of t with derivative

$$\frac{ds}{dt} = |\mathbf{v}(t)|. \tag{4}$$

As we already knew, the speed with which a particle moves along its path is the magnitude of $\mathbf{v}$.

Although the base point $P(t_0)$ plays a role in defining s in Equation (3), it plays no role in Equation (4). The rate at which a moving particle covers distance along its path is independent of how far away it is from the base point.

Notice that $ds/dt > 0$ since, by definition, $|\mathbf{v}|$ is never zero for a smooth curve. We see once again that s is an increasing function of t.

Unit Tangent Vector T

We already know the velocity vector $\mathbf{v} = d\mathbf{r}/dt$ is tangent to the curve and that the vector

$$\mathbf{T} = \frac{\mathbf{v}}{|\mathbf{v}|}$$

is therefore a unit vector tangent to the (smooth) curve. Since $ds/dt > 0$ for the curves we are considering, s is one-to-one and has an inverse that gives t as a differentiable function of s (Section 2.9). The derivative of the inverse is

$$\frac{dt}{ds} = \frac{1}{ds/dt} = \frac{1}{|\mathbf{v}|}.$$

This makes $\mathbf{r}$ a differentiable function of s whose derivative can be calculated with the Chain Rule to be

$$\frac{d\mathbf{r}}{ds} = \frac{d\mathbf{r}}{dt}\frac{dt}{ds} = \mathbf{v}\frac{1}{|\mathbf{v}|} = \frac{\mathbf{v}}{|\mathbf{v}|} = \mathbf{T}.$$

This equation says that $d\mathbf{r}/ds$ is the unit tangent vector in the direction of the velocity vector $\mathbf{v}$ (Figure 10.14).

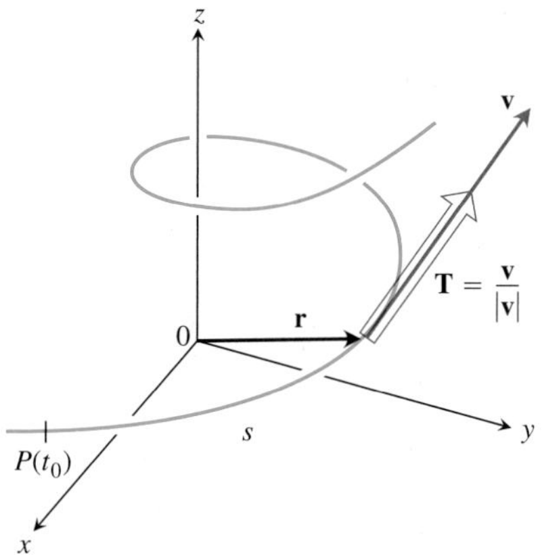

FIGURE 10.14 We find the unit tangent vector $\mathbf{T}$ by dividing $\mathbf{v}$ by $|\mathbf{v}|$.

DEFINITION The **unit tangent vector** of a smooth curve $\mathbf{r}(t)$ is

$$\mathbf{T} = \frac{d\mathbf{r}}{ds} = \frac{d\mathbf{r}/dt}{ds/dt} = \frac{\mathbf{v}}{|\mathbf{v}|}. \qquad (5)$$

The unit tangent vector $\mathbf{T}$ is a differentiable function of t whenever $\mathbf{v}$ is a differentiable function of t. As we will see in Section 10.5, $\mathbf{T}$ is one of three unit vectors in a traveling reference frame that is used to describe the motion of objects traveling in three dimensions.

EXAMPLE 3 Find the unit tangent vector of the curve

$$\mathbf{r}(t) = (3\cos t)\mathbf{i} + (3\sin t)\mathbf{j} + t^2\mathbf{k}$$

representing the path of the glider in Example 3, Section 10.2.

Solution In that example, we found

$$\mathbf{v} = \frac{d\mathbf{r}}{dt} = -(3\sin t)\mathbf{i} + (3\cos t)\mathbf{j} + 2t\mathbf{k}$$

and

$$|\mathbf{v}| = \sqrt{9 + 4t^2}.$$

Thus,

$$\mathbf{T} = \frac{\mathbf{v}}{|\mathbf{v}|} = -\frac{3\sin t}{\sqrt{9+4t^2}}\mathbf{i} + \frac{3\cos t}{\sqrt{9+4t^2}}\mathbf{j} + \frac{2t}{\sqrt{9+4t^2}}\mathbf{k}.$$ ■

For the counterclockwise motion

$$\mathbf{r}(t) = (\cos t)\mathbf{i} + (\sin t)\mathbf{j}$$

around the unit circle, we see that

$$\mathbf{v} = (-\sin t)\mathbf{i} + (\cos t)\mathbf{j}$$

is already a unit vector, so $\mathbf{T} = \mathbf{v}$ (Figure 10.15).

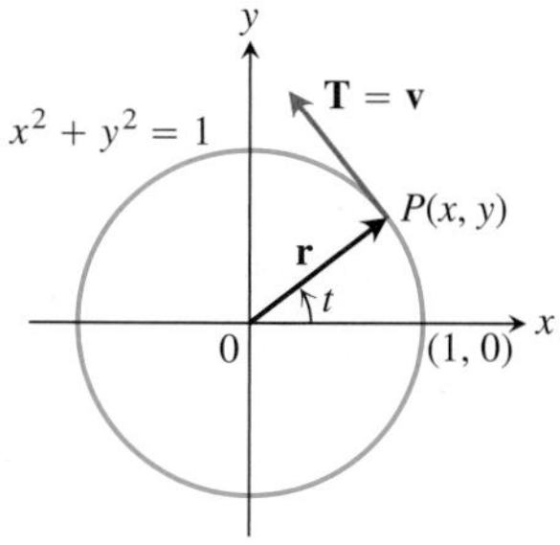

FIGURE 10.15 Counterclockwise motion around the unit circle.

EXERCISES 10.3

In Exercises 1–8, find the curve's unit tangent vector. Also, find the length of the indicated portion of the curve.

1. $\mathbf{r}(t) = (2\cos t)\mathbf{i} + (2\sin t)\mathbf{j} + \sqrt{5}t\mathbf{k}, \quad 0 \le t \le \pi$

2. $\mathbf{r}(t) = (6\sin 2t)\mathbf{i} + (6\cos 2t)\mathbf{j} + 5t\mathbf{k}, \quad 0 \le t \le \pi$

3. $\mathbf{r}(t) = t\mathbf{i} + (2/3)t^{3/2}\mathbf{k}, \quad 0 \le t \le 8$

4. $\mathbf{r}(t) = (2 + t)\mathbf{i} - (t + 1)\mathbf{j} + t\mathbf{k}, \quad 0 \le t \le 3$

5. $\mathbf{r}(t) = (\cos^3 t)\mathbf{j} + (\sin^3 t)\mathbf{k}, \quad 0 \le t \le \pi/2$

6. $\mathbf{r}(t) = 6t^3\mathbf{i} - 2t^3\mathbf{j} - 3t^3\mathbf{k}, \quad 1 \le t \le 2$

7. $\mathbf{r}(t) = (t\cos t)\mathbf{i} + (t\sin t)\mathbf{j} + \left(2\sqrt{2}/3\right)t^{3/2}\mathbf{k}, \quad 0 \le t \le \pi$

8. $\mathbf{r}(t) = (t\sin t + \cos t)\mathbf{i} + (t\cos t - \sin t)\mathbf{j}, \quad \sqrt{2} \le t \le 2$

9. Find the point on the curve

$$\mathbf{r}(t) = (5\sin t)\mathbf{i} + (5\cos t)\mathbf{j} + 12t\mathbf{k}$$

at a distance 26π units along the curve from the origin in the direction of increasing arc length.

10. Find the point on the curve

$$\mathbf{r}(t) = (12\sin t)\mathbf{i} - (12\cos t)\mathbf{j} + 5t\mathbf{k}$$

at a distance 13π units along the curve from the origin in the direction opposite to the direction of increasing arc length.

In Exercises 11–14, find the arc length parameter along the curve from the point where $t = 0$ by evaluating the integral

$$s = \int_0^t |\mathbf{v}(\tau)|\,d\tau$$

from Equation (3). Then find the length of the indicated portion of the curve.

11. $\mathbf{r}(t) = (4\cos t)\mathbf{i} + (4\sin t)\mathbf{j} + 3t\mathbf{k}, \quad 0 \le t \le \pi/2$

12. $\mathbf{r}(t) = (\cos t + t\sin t)\mathbf{i} + (\sin t - t\cos t)\mathbf{j}, \quad \pi/2 \le t \le \pi$

13. $\mathbf{r}(t) = (e^t\cos t)\mathbf{i} + (e^t\sin t)\mathbf{j} + e^t\mathbf{k}, \quad -\ln 4 \le t \le 0$

14. $\mathbf{r}(t) = (1 + 2t)\mathbf{i} + (1 + 3t)\mathbf{j} + (6 - 6t)\mathbf{k}, \quad -1 \le t \le 0$

15. Arc length Find the length of the curve

$$\mathbf{r}(t) = \left(\sqrt{2}t\right)\mathbf{i} + \left(\sqrt{2}t\right)\mathbf{j} + (1 - t^2)\mathbf{k}$$

from $(0, 0, 1)$ to $\left(\sqrt{2}, \sqrt{2}, 0\right)$.

16. Length is independent of parametrization To illustrate that the length of a smooth space curve does not depend on the parametrization you use to compute it, calculate the length of one turn of the helix in Example 1 with the following parametrizations.

a. $\mathbf{r}(t) = (\cos 4t)\mathbf{i} + (\sin 4t)\mathbf{j} + 4t\mathbf{k}, \quad 0 \le t \le \pi/2$

b. $\mathbf{r}(t) = [\cos(t/2)]\mathbf{i} + [\sin(t/2)]\mathbf{j} + (t/2)\mathbf{k}, \quad 0 \le t \le 4\pi$

c. $\mathbf{r}(t) = (\cos t)\mathbf{i} - (\sin t)\mathbf{j} - t\mathbf{k}, \quad -2\pi \le t \le 0$

17. The involute of a circle If a string wound around a fixed circle is unwound while held taut in the plane of the circle, its end P traces an *involute* of the circle. In the accompanying figure, the circle in question is the circle $x^2 + y^2 = 1$ and the tracing point starts at $(1, 0)$. The unwound portion of the string is tangent to the circle at Q, and t is the radian measure of the angle from the positive x-axis to segment OQ. Derive the parametric equations

$$x = \cos t + t\sin t, \quad y = \sin t - t\cos t, \quad t > 0$$

of the point $P(x, y)$ for the involute.

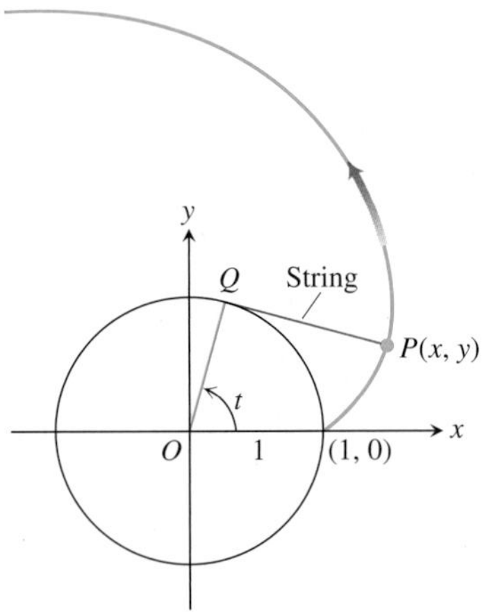

18. (*Continuation of Exercise 17.*) Find the unit tangent vector to the involute of the circle at the point $P(x, y)$.

19. Distance along a line Show that if $\mathbf{u}$ is a unit vector, then the arc length parameter along the line $\mathbf{r}(t) = P_0 + t\mathbf{u}$ from the point $P_0(x_0, y_0, z_0)$ where $t = 0$, is t itself.

20. Use Simpson's Rule with $n = 10$ to approximate the length of arc of $\mathbf{r}(t) = t\mathbf{i} + t^2\mathbf{j} + t^3\mathbf{k}$ from the origin to the point $(2, 4, 8)$.

10.4 Curvature of a Curve

In this section we study how a curve turns or bends. We look first at curves in the coordinate plane, and then at curves in space.

Curvature of a Plane Curve

As a particle moves along a smooth curve in the plane, $\mathbf{T} = d\mathbf{r}/ds$ turns as the curve bends. Since $\mathbf{T}$ is a unit vector, its length remains constant and only its direction changes as the particle moves along the curve. The rate at which $\mathbf{T}$ turns per unit of length along the curve is called the *curvature* (Figure 10.16). The traditional symbol for the curvature function is the Greek letter κ ("kappa").

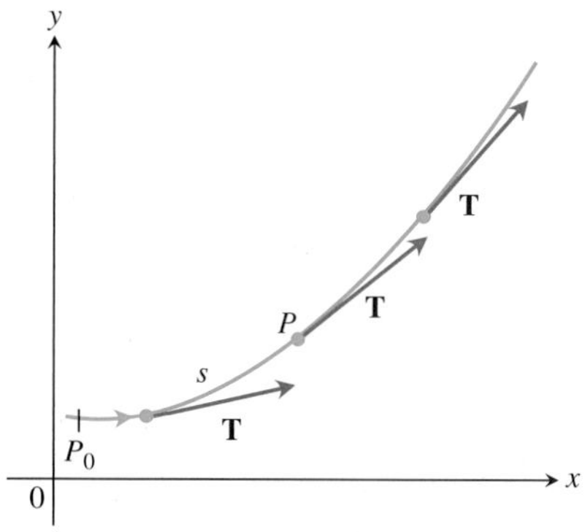

FIGURE 10.16 As P moves along the curve in the direction of increasing arc length, the unit tangent vector turns. The value of $|d\mathbf{T}/ds|$ at P is called the *curvature* of the curve at P.

DEFINITION If **T** is the unit vector of a smooth curve, the **curvature** function of the curve is

$$\kappa = \left|\frac{d\mathbf{T}}{ds}\right|.$$

If $|d\mathbf{T}/ds|$ is large, **T** turns sharply as the particle passes through P, and the curvature at P is large. If $|d\mathbf{T}/ds|$ is close to zero, **T** turns more slowly and the curvature at P is smaller.

If a smooth curve r(t) is already given in terms of some parameter t other than the arc length parameter s, we can calculate the curvature as

$$\begin{aligned}\kappa &= \left|\frac{d\mathbf{T}}{ds}\right| = \left|\frac{d\mathbf{T}}{dt}\frac{dt}{ds}\right| && \text{Chain Rule}\\ &= \frac{1}{|ds/dt|}\left|\frac{d\mathbf{T}}{dt}\right|\\ &= \frac{1}{|\mathbf{v}|}\left|\frac{d\mathbf{T}}{dt}\right|. && \frac{ds}{dt} = |\mathbf{v}|\end{aligned}$$

Formula for Calculating Curvature

If $\mathbf{r}(t)$ is a smooth curve, then the curvature is

$$\kappa = \frac{1}{|\mathbf{v}|}\left|\frac{d\mathbf{T}}{dt}\right|, \tag{1}$$

where $\mathbf{T} = \mathbf{v}/|\mathbf{v}|$ is the unit tangent vector.

Testing the definition, we see in Examples 1 and 2 below that the curvature is constant for straight lines and circles.

EXAMPLE 1 A straight line is parametrized by $\mathbf{r}(t) = \mathbf{C} + t\mathbf{v}$ for constant vectors **C** and **v**. Thus, $\mathbf{r}'(t) = \mathbf{v}$, and the unit tangent vector $\mathbf{T} = \mathbf{v}/|\mathbf{v}|$ is a constant vector which always points in the same direction and has derivative **0** (Figure 10.17). It follows that, for any value of the parameter t, the curvature of the straight line is

$$\kappa = \frac{1}{|\mathbf{v}|}\left|\frac{d\mathbf{T}}{dt}\right| = \frac{1}{|\mathbf{v}|}|\mathbf{0}| = 0. \qquad \blacksquare$$

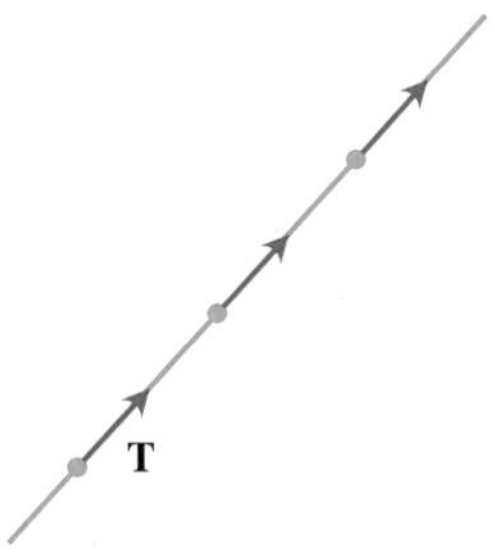

FIGURE 10.17 Along a straight line, **T** always points in the same direction. The curvature, $|d\mathbf{T}/ds|$, is zero (Example 1).

EXAMPLE 2 Here we find the curvature of a circle. We begin with the parametrization

$$\mathbf{r}(t) = (a\cos t)\mathbf{i} + (a\sin t)\mathbf{j}$$

of a circle of radius a. Then,

$$\begin{aligned}\mathbf{v} &= \frac{d\mathbf{r}}{dt} = -(a\sin t)\mathbf{i} + (a\cos t)\mathbf{j}\\ |\mathbf{v}| &= \sqrt{(-a\sin t)^2 + (a\cos t)^2} = \sqrt{a^2} = |a| = a. && \text{Since } a > 0,\ |a| = a.\end{aligned}$$

From this we find

$$\begin{aligned}
\mathbf{T} &= \frac{\mathbf{v}}{|\mathbf{v}|} = -(\sin t)\mathbf{i} + (\cos t)\mathbf{j} \\
\frac{d\mathbf{T}}{dt} &= -(\cos t)\mathbf{i} - (\sin t)\mathbf{j} \\
\left|\frac{d\mathbf{T}}{dt}\right| &= \sqrt{\cos^2 t + \sin^2 t} = 1.
\end{aligned}$$

Hence, for any value of the parameter t, the curvature of the circle is

$$\kappa = \frac{1}{|\mathbf{v}|}\left|\frac{d\mathbf{T}}{dt}\right| = \frac{1}{a}(1) = \frac{1}{a} = \frac{1}{\text{radius}}.$$ ■

Although the formula for calculating κ in Equation (1) is also valid for space curves, in the next section we find a computational formula that is usually more convenient to apply.

Among the vectors orthogonal to the unit tangent vector $\mathbf{T}$ is one of particular significance because it points in the direction in which the curve is turning. Since $\mathbf{T}$ has constant length (namely, 1), the derivative $d\mathbf{T}/ds$ is orthogonal to $\mathbf{T}$ (Equation 4, Section 10.1). Therefore, if we divide $d\mathbf{T}/ds$ by its length κ, we obtain a *unit* vector $\mathbf{N}$ orthogonal to $\mathbf{T}$ (Figure 10.18).

DEFINITION At a point where $\kappa \neq 0$, the **principal unit normal** vector for a smooth curve in the plane is

$$\mathbf{N} = \frac{1}{\kappa}\frac{d\mathbf{T}}{ds}.$$

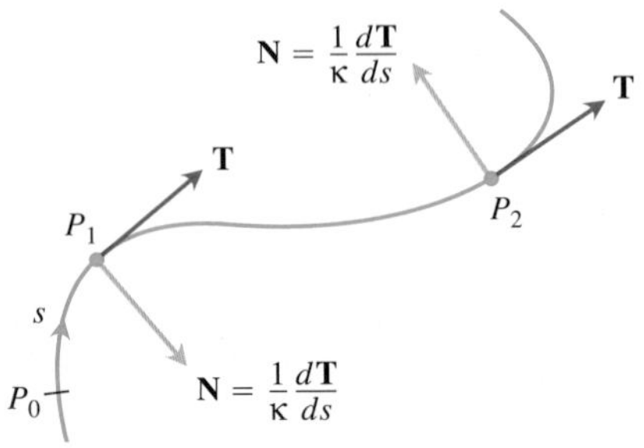

FIGURE 10.18 The vector $d\mathbf{T}/ds$, normal to the curve, always points in the direction in which $\mathbf{T}$ is turning. The unit normal vector $\mathbf{N}$ is the direction of $d\mathbf{T}/ds$.

The vector $d\mathbf{T}/ds$ points in the direction in which $\mathbf{T}$ turns as the curve bends. Therefore, if we face in the direction of increasing arc length, the vector $d\mathbf{T}/ds$ points toward the right if $\mathbf{T}$ turns clockwise and toward the left if $\mathbf{T}$ turns counterclockwise. In other words, the principal normal vector $\mathbf{N}$ will point toward the concave side of the curve (Figure 10.18).

If a smooth curve $\mathbf{r}(t)$ is already given in terms of some parameter t other than the arc length parameter s, we can use the Chain Rule to calculate $\mathbf{N}$ directly:

$$\begin{aligned}
\mathbf{N} &= \frac{d\mathbf{T}/ds}{|d\mathbf{T}/ds|} \\
&= \frac{(d\mathbf{T}/dt)(dt/ds)}{|d\mathbf{T}/dt||dt/ds|} \\
&= \frac{d\mathbf{T}/dt}{|d\mathbf{T}/dt|}. \qquad \frac{dt}{ds} = \frac{1}{ds/dt} > 0 \text{ cancels}
\end{aligned}$$

This formula enables us to find $\mathbf{N}$ without having to find κ and s first.

Formula for Calculating N

If $\mathbf{r}(t)$ is a smooth curve, then the principal unit normal is

$$\mathbf{N} = \frac{d\mathbf{T}/dt}{|d\mathbf{T}/dt|}, \tag{2}$$

where $\mathbf{T} = \mathbf{v}/|\mathbf{v}|$ is the unit tangent vector.

EXAMPLE 3 Find **T** and **N** for the circular motion

$$\mathbf{r}(t) = (\cos 2t)\mathbf{i} + (\sin 2t)\mathbf{j}.$$

Solution We first find **T**:

$$\begin{aligned}\mathbf{v} &= -(2\sin 2t)\mathbf{i} + (2\cos 2t)\mathbf{j}\\ |\mathbf{v}| &= \sqrt{4\sin^2 2t + 4\cos^2 2t} = 2\\ \mathbf{T} &= \frac{\mathbf{v}}{|\mathbf{v}|} = -(\sin 2t)\mathbf{i} + (\cos 2t)\mathbf{j}.\end{aligned}$$

From this we find

$$\begin{aligned}\frac{d\mathbf{T}}{dt} &= -(2\cos 2t)\mathbf{i} - (2\sin 2t)\mathbf{j}\\ \left|\frac{d\mathbf{T}}{dt}\right| &= \sqrt{4\cos^2 2t + 4\sin^2 2t} = 2\end{aligned}$$

and

$$\begin{aligned}\mathbf{N} &= \frac{d\mathbf{T}/dt}{|d\mathbf{T}/dt|}\\ &= -(\cos 2t)\mathbf{i} - (\sin 2t)\mathbf{j}. \qquad \text{Eq. (2)}\end{aligned}$$

Notice that $\mathbf{T}\cdot\mathbf{N} = 0$, verifying that **N** is orthogonal to **T**. Notice too, that for the circular motion here, **N** points from $\mathbf{r}(t)$ towards the circle's center at the origin. ■

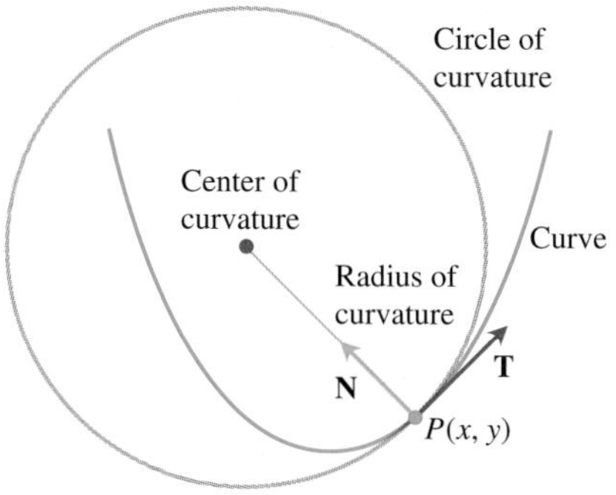

FIGURE 10.19 The osculating circle at $P(x, y)$ lies toward the inner side of the curve.

Circle of Curvature for Plane Curves

The **circle of curvature** or **osculating circle** at a point P on a plane curve where $\kappa \neq 0$ is the circle in the plane of the curve that

1. is tangent to the curve at P (has the same tangent line the curve has)
2. has the same curvature the curve has at P
3. lies toward the concave or inner side of the curve (as in Figure 10.19).

The **radius of curvature** of the curve at P is the radius of the circle of curvature, which, according to Example 2, is

$$\text{Radius of curvature} = \rho = \frac{1}{\kappa}.$$

To find ρ, we find κ and take the reciprocal. The **center of curvature** of the curve at P is the center of the circle of curvature.

EXAMPLE 4 Find and graph the osculating circle of the parabola $y = x^2$ at the origin.

Solution We parametrize the parabola using the parameter $t = x$ (Section 8.5, Example 1)

$$\mathbf{r}(t) = t\mathbf{i} + t^2\mathbf{j}.$$

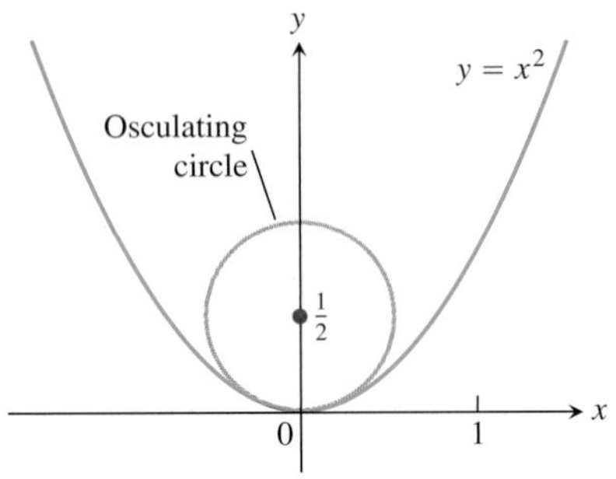

FIGURE 10.20 The osculating circle for the parabola $y = x^2$ at the origin (Example 4).

First we find the curvature of the parabola at the origin, using Equation (1):

$$\mathbf{v} = \frac{d\mathbf{r}}{dt} = \mathbf{i} + 2t\mathbf{j}$$
$$|\mathbf{v}| = \sqrt{1 + 4t^2}$$

so that

$$\mathbf{T} = \frac{\mathbf{v}}{|\mathbf{v}|} = (1 + 4t^2)^{-1/2}\mathbf{i} + 2t(1 + 4t^2)^{-1/2}\mathbf{j}.$$

From this we find

$$\frac{d\mathbf{T}}{dt} = -4t(1 + 4t^2)^{-3/2}\mathbf{i} + [2(1 + 4t^2)^{-1/2} - 8t^2(1 + 4t^2)^{-3/2}]\,\mathbf{j}.$$

At the origin, $t = 0$, so the curvature is

$$\begin{aligned}\kappa(0) &= \frac{1}{|\mathbf{v}(0)|}\left|\frac{d\mathbf{T}}{dt}(0)\right| &&\text{Eq. (1)}\\ &= \frac{1}{\sqrt{1}}|0\mathbf{i} + 2\mathbf{j}|\\ &= (1)\sqrt{0^2 + 2^2} = 2.\end{aligned}$$

Therefore, the radius of curvature is $1/\kappa = 1/2$. We see in Figure 10.20 that at the origin, $\mathbf{T} = \mathbf{i}$ so $\mathbf{N} = \mathbf{j}$ and the center of the circle is (0, 1/2). The equation of the osculating circle is

$$(x - 0)^2 + \left(y - \frac{1}{2}\right)^2 = \left(\frac{1}{2}\right)^2.$$

You can see from Figure 10.20 that the osculating circle is a better approximation to the parabola at the origin than is the tangent line approximation $y = 0$. ■

Curvature and Normal Vectors for Space Curves

If a smooth curve in space is specified by the position vector $\mathbf{r}(t)$ as a function of some parameter t, and if s is the arc length parameter of the curve, then the unit tangent vector $\mathbf{T}$ is $d\mathbf{r}/ds = \mathbf{v}/|\mathbf{v}|$. The **curvature** in space is then defined to be

$$\kappa = \left|\frac{d\mathbf{T}}{ds}\right| = \frac{1}{|\mathbf{v}|}\left|\frac{d\mathbf{T}}{dt}\right| \tag{3}$$

just as for plane curves. The vector $d\mathbf{T}/ds$ is orthogonal to $\mathbf{T}$, and we define the **principal unit normal** to be

$$\mathbf{N} = \frac{1}{\kappa}\frac{d\mathbf{T}}{ds} = \frac{d\mathbf{T}/dt}{|d\mathbf{T}/dt|}. \tag{4}$$

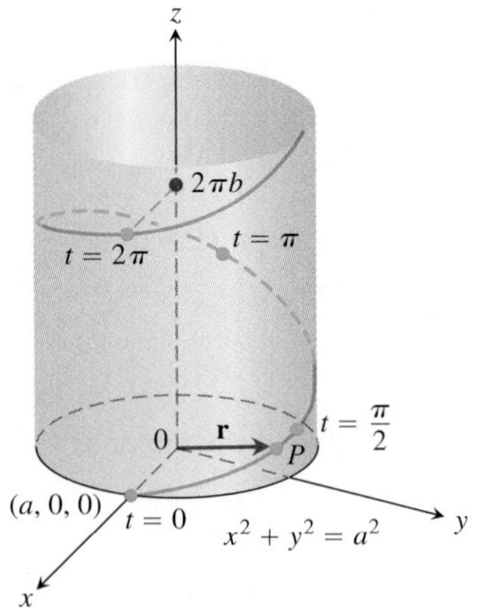

FIGURE 10.21 The helix

$\mathbf{r}(t) = (a\cos t)\mathbf{i} + (a\sin t)\mathbf{j} + bt\mathbf{k}$,

drawn with a and b positive and $t \geq 0$ (Example 5).

EXAMPLE 5 Find the curvature for the helix (Figure 10.21)

$$\mathbf{r}(t) = (a\cos t)\mathbf{i} + (a\sin t)\mathbf{j} + bt\mathbf{k}, \qquad a, b \geq 0, \qquad a^2 + b^2 \neq 0.$$

Solution We calculate $\mathbf{T}$ from the velocity vector $\mathbf{v}$:

$$\mathbf{v} = -(a\sin t)\mathbf{i} + (a\cos t)\mathbf{j} + b\mathbf{k}$$
$$|\mathbf{v}| = \sqrt{a^2\sin^2 t + a^2\cos^2 t + b^2} = \sqrt{a^2 + b^2}$$
$$\mathbf{T} = \frac{\mathbf{v}}{|\mathbf{v}|} = \frac{1}{\sqrt{a^2 + b^2}}[-(a\sin t)\mathbf{i} + (a\cos t)\mathbf{j} + b\mathbf{k}].$$

Then using Equation (3),

$$\begin{aligned}\kappa &= \frac{1}{|\mathbf{v}|}\left|\frac{d\mathbf{T}}{dt}\right| \\ &= \frac{1}{\sqrt{a^2+b^2}}\left|\frac{1}{\sqrt{a^2+b^2}}[-(a\cos t)\mathbf{i} - (a\sin t)\mathbf{j}]\right| \\ &= \frac{a}{a^2+b^2}|-(\cos t)\mathbf{i} - (\sin t)\mathbf{j}| \\ &= \frac{a}{a^2+b^2}\sqrt{(\cos t)^2 + (\sin t)^2} = \frac{a}{a^2+b^2}.\end{aligned}$$

From this equation, we see that increasing b for a fixed a decreases the curvature. Decreasing a for a fixed b eventually decreases the curvature as well.

If $b = 0$, the helix reduces to a circle of radius a and its curvature reduces to $1/a$, as it should. If $a = 0$, the helix becomes the z-axis, and its curvature reduces to 0, again as it should. ■

EXAMPLE 6 Find **N** for the helix in Example 5 and describe how the vector is pointing.

Solution We have

$$\frac{d\mathbf{T}}{dt} = -\frac{1}{\sqrt{a^2+b^2}}[(a\cos t)\mathbf{i} + (a\sin t)\mathbf{j}] \qquad \text{Example 5}$$

$$\left|\frac{d\mathbf{T}}{dt}\right| = \frac{1}{\sqrt{a^2+b^2}}\sqrt{a^2\cos^2 t + a^2\sin^2 t} = \frac{a}{\sqrt{a^2+b^2}}$$

$$\begin{aligned}\mathbf{N} &= \frac{d\mathbf{T}/dt}{|d\mathbf{T}/dt|} \qquad \text{Eq. (4)} \\ &= -\frac{\sqrt{a^2+b^2}}{a}\cdot\frac{1}{\sqrt{a^2+b^2}}[(a\cos t)\mathbf{i} + (a\sin t)\mathbf{j}] \\ &= -(\cos t)\mathbf{i} - (\sin t)\mathbf{j}.\end{aligned}$$

Thus, **N** is parallel to the xy-plane and always points toward the z-axis. ■

EXERCISES 10.4

Find **T**, **N**, and κ for the plane curves in Exercises 1–4.

1. $\mathbf{r}(t) = t\mathbf{i} + (\ln\cos t)\mathbf{j}, \quad -\pi/2 < t < \pi/2$
2. $\mathbf{r}(t) = (\ln\sec t)\mathbf{i} + t\mathbf{j}, \quad -\pi/2 < t < \pi/2$
3. $\mathbf{r}(t) = (2t+3)\mathbf{i} + (5-t^2)\mathbf{j}$
4. $\mathbf{r}(t) = (\cos t + t\sin t)\mathbf{i} + (\sin t - t\cos t)\mathbf{j}, \quad t > 0$
5. **A formula for the curvature of the graph of a function in the xy-plane**
   a. The graph $y = f(x)$ in the xy-plane automatically has the parametrization $x = x$, $y = f(x)$, and the vector formula $\mathbf{r}(x) = x\mathbf{i} + f(x)\mathbf{j}$. Use this formula to show that if f is a twice-differentiable function of x, then

$$\kappa(x) = \frac{|f''(x)|}{\left[1 + (f'(x))^2\right]^{3/2}}.$$

   b. Use the formula for κ in part (a) to find the curvature of $y = \ln(\cos x)$, $-\pi/2 < x < \pi/2$. Compare your answer with the answer in Exercise 1.
   c. Show that the curvature is zero at a point of inflection.

6. A formula for the curvature of a parametrized plane curve

a. Show that the curvature of a smooth curve $\mathbf{r}(t) = f(t)\mathbf{i} + g(t)\mathbf{j}$ defined by twice-differentiable functions $x = f(t)$ and $y = g(t)$ is given by the formula

$$\kappa = \frac{|\dot{x}\ddot{y} - \dot{y}\ddot{x}|}{(\dot{x}^2 + \dot{y}^2)^{3/2}}.$$

The dots in the formula denote differentiation with respect to t, one derivative for each dot. Apply the formula to find the curvatures of the following curves.

b. $\mathbf{r}(t) = t\mathbf{i} + (\ln \sin t)\mathbf{j}, \quad 0 < t < \pi$

c. $\mathbf{r}(t) = [\tan^{-1}(\sinh t)]\mathbf{i} + (\ln \cosh t)\mathbf{j}$

7. Normals to plane curves

a. Show that $\mathbf{n}(t) = -g'(t)\mathbf{i} + f'(t)\mathbf{j}$ and $-\mathbf{n}(t) = g'(t)\mathbf{i} - f'(t)\mathbf{j}$ are both normal to the curve $\mathbf{r}(t) = f(t)\mathbf{i} + g(t)\mathbf{j}$ at the point $(f(t), g(t))$.

To obtain **N** for a particular plane curve, we can choose the one of **n** or $-\mathbf{n}$ from part (a) that points toward the concave side of the curve, and make it into a unit vector. (See Figure 10.18.) Apply this method to find **N** for the following curves.

b. $\mathbf{r}(t) = t\mathbf{i} + e^{2t}\mathbf{j}$

c. $\mathbf{r}(t) = \sqrt{4 - t^2}\,\mathbf{i} + t\mathbf{j}, \quad -2 \le t \le 2$

8. (*Continuation of Exercise 7.*)

a. Use the method of Exercise 7 to find **N** for the curve $\mathbf{r}(t) = t\mathbf{i} + (1/3)t^3\mathbf{j}$ when $t < 0$; when $t > 0$.

b. Calculate **N** for $t \neq 0$ directly from **T** using Eq. (4) for the curve in part (a). Does **N** exist at $t = 0$? Graph the curve and explain what is happening to **N** as t passes from negative to positive values.

Find **T**, **N**, and κ for the space curves in Exercises 9–16.

9. $\mathbf{r}(t) = (3 \sin t)\mathbf{i} + (3 \cos t)\mathbf{j} + 4t\mathbf{k}$

10. $\mathbf{r}(t) = (\cos t + t \sin t)\mathbf{i} + (\sin t - t \cos t)\mathbf{j} + 3\mathbf{k}$

11. $\mathbf{r}(t) = (e^t \cos t)\mathbf{i} + (e^t \sin t)\mathbf{j} + 2\mathbf{k}$

12. $\mathbf{r}(t) = (6 \sin 2t)\mathbf{i} + (6 \cos 2t)\mathbf{j} + 5t\mathbf{k}$

13. $\mathbf{r}(t) = (t^3/3)\mathbf{i} + (t^2/2)\mathbf{j}, \quad t > 0$

14. $\mathbf{r}(t) = (\cos^3 t)\mathbf{i} + (\sin^3 t)\mathbf{j}, \quad 0 < t < \pi/2$

15. $\mathbf{r}(t) = t\mathbf{i} + (a \cosh (t/a))\mathbf{j}, \quad a > 0$

16. $\mathbf{r}(t) = (\cosh t)\mathbf{i} - (\sinh t)\mathbf{j} + t\mathbf{k}$

17. Show that the parabola $y = ax^2, a \neq 0$, has its largest curvature at its vertex and has no minimum curvature. (*Note:* Since the curvature of a curve remains the same if the curve is translated or rotated, this result is true for any parabola.)

18. Show that the ellipse $x = a \cos t, y = b \sin t, a > b > 0$, has its largest curvature on its major axis and its smallest curvature on its minor axis. (As in Exercise 17, the same is true for any ellipse.)

19. Maximizing the curvature of a helix In Example 5, we found the curvature of the helix $\mathbf{r}(t) = (a \cos t)\mathbf{i} + (a \sin t)\mathbf{j} + bt\mathbf{k}$ $(a, b \ge 0)$ to be $\kappa = a/(a^2 + b^2)$. What is the largest value κ can have for a given value of b? Give reasons for your answer.

20. Total curvature We find the *total curvature* of the portion of a smooth curve that runs from $s = s_0$ to $s = s_1 > s_0$ by integrating κ from s_0 to s_1. If the curve has some other parameter, say t, then the total curvature is

$$K = \int_{s_0}^{s_1} \kappa\, ds = \int_{t_0}^{t_1} \kappa \frac{ds}{dt}\, dt = \int_{t_0}^{t_1} \kappa |\mathbf{v}|\, dt,$$

where t_0 and t_1 correspond to s_0 and s_1. Find the total curvatures of

a. The portion of the helix $\mathbf{r}(t) = (3 \cos t)\mathbf{i} + (3 \sin t)\mathbf{j} + t\mathbf{k}$, $0 \le t \le 4\pi$.

b. The parabola $y = x^2, -\infty < x < \infty$.

21. Find an equation for the circle of curvature of the curve $\mathbf{r}(t) = t\mathbf{i} + (\sin t)\mathbf{j}$ at the point $(\pi/2, 1)$. (The curve parametrizes the graph of $y = \sin x$ in the xy-plane.)

22. Find an equation for the circle of curvature of the curve $\mathbf{r}(t) = (2 \ln t)\mathbf{i} - [t + (1/t)]\mathbf{j}, e^{-2} \le t \le e^2$, at the point $(0, -2)$, where $t = 1$.

T The formula

$$\kappa(x) = \frac{|f''(x)|}{\left[1 + (f'(x))^2\right]^{3/2}},$$

derived in Exercise 5, expresses the curvature $\kappa(x)$ of a twice-differentiable plane curve $y = f(x)$ as a function of x. Find the curvature function of each of the curves in Exercises 23–26. Then graph $f(x)$ together with $\kappa(x)$ over the given interval. You will find some surprises.

23. $y = x^2, \quad -2 \le x \le 2$

24. $y = x^4/4, \quad -2 \le x \le 2$

25. $y = \sin x, \quad 0 \le x \le 2\pi$

26. $y = e^x, \quad -1 \le x \le 2$

COMPUTER EXPLORATIONS

In Exercises 27–34 you will use a CAS to explore the osculating circle at a point P on a plane curve where $\kappa \neq 0$. Use a CAS to perform the following steps:

a. Plot the plane curve given in parametric or function form over the specified interval to see what it looks like.

b. Calculate the curvature κ of the curve at the given value t_0 using the appropriate formula from Exercise 5 or 6. Use the parametrization $x = t$ and $y = f(t)$ if the curve is given as a function $y = f(x)$.

c. Find the unit normal vector **N** at t_0. Notice that the signs of the components of **N** depend on whether the unit tangent vector **T** is turning clockwise or counterclockwise at $t = t_0$. (See Exercise 7.)

d. If $\mathbf{C} = a\mathbf{i} + b\mathbf{j}$ is the vector from the origin to the center (a, b) of the osculating circle, find the center **C** from the vector equation

$$\mathbf{C} = \mathbf{r}(t_0) + \frac{1}{\kappa(t_0)}\mathbf{N}(t_0).$$

The point $P(x_0, y_0)$ on the curve is given by the position vector $\mathbf{r}(t_0)$.

e. Plot implicitly the equation $(x - a)^2 + (y - b)^2 = 1/\kappa^2$ of the osculating circle. Then plot the curve and osculating circle together. You may need to experiment with the size of the viewing window, but be sure it is square.

27. $\mathbf{r}(t) = (3\cos t)\mathbf{i} + (5\sin t)\mathbf{j}, \quad 0 \le t \le 2\pi, \quad t_0 = \pi/4$

28. $\mathbf{r}(t) = (\cos^3 t)\mathbf{i} + (\sin^3 t)\mathbf{j}, \quad 0 \le t \le 2\pi, \quad t_0 = \pi/4$

29. $\mathbf{r}(t) = t^2\mathbf{i} + (t^3 - 3t)\mathbf{j}, \quad -4 \le t \le 4, \quad t_0 = 3/5$

30. $\mathbf{r}(t) = (t^3 - 2t^2 - t)\mathbf{i} + \dfrac{3t}{\sqrt{1 + t^2}}\mathbf{j}, \quad -2 \le t \le 5, \quad t_0 = 1$

31. $\mathbf{r}(t) = (2t - \sin t)\mathbf{i} + (2 - 2\cos t)\mathbf{j}, \quad 0 \le t \le 3\pi,$ $t_0 = 3\pi/2$

32. $\mathbf{r}(t) = (e^{-t}\cos t)\mathbf{i} + (e^{-t}\sin t)\mathbf{j}, \quad 0 \le t \le 6\pi, \quad t_0 = \pi/4$

33. $y = x^2 - x, \quad -2 \le x \le 5, \quad x_0 = 1$

34. $y = x(1 - x)^{2/5}, \quad -1 \le x \le 2, \quad x_0 = 1/2$

10.5 Tangential and Normal Components of Acceleration

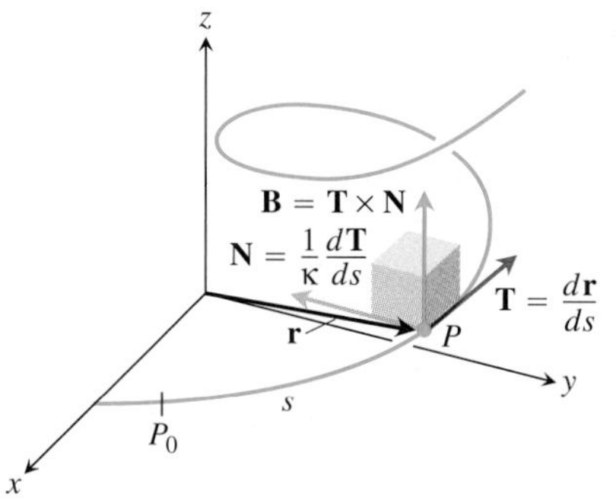

FIGURE 10.22 The **TNB** frame of mutually orthogonal unit vectors traveling along a curve in space.

If you are traveling along a space curve, the Cartesian **i**, **j**, and **k** coordinate system for representing the vectors describing your motion is not truly relevant to you. What is meaningful instead are the vectors representative of your forward direction (the unit tangent vector **T**), the direction in which your path is turning (the unit normal vector **N**), and the tendency of your motion to "twist" out of the plane created by these vectors in the direction perpendicular to this plane (defined by the *unit binormal vector* $\mathbf{B} = \mathbf{T} \times \mathbf{N}$). Expressing the acceleration vector along the curve as a linear combination of this **TNB** frame of mutually orthogonal unit vectors traveling with the motion (Figure 10.22) is particularly revealing of the nature of the path and motion along it.

The TNB Frame

The **binormal vector** of a curve in space is $\mathbf{B} = \mathbf{T} \times \mathbf{N}$, a unit vector orthogonal to both **T** and **N** (Figure 10.23). Together **T**, **N**, and **B** define a moving right-handed vector frame that plays a significant role in calculating the paths of particles moving through space. It is called the **Frenet** ("fre-*nay*") **frame** (after Jean-Frédéric Frenet, 1816–1900), or the **TNB frame**.

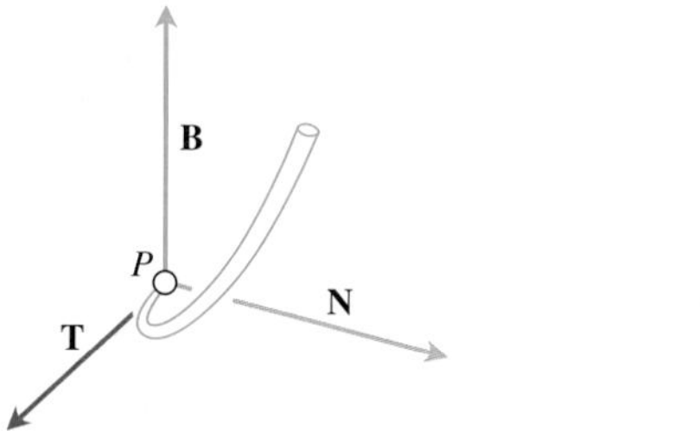

FIGURE 10.23 The vectors **T**, **N**, and **B** (in that order) make a right-handed frame of mutually orthogonal unit vectors in space.

Tangential and Normal Components of Acceleration

When a body is accelerated by gravity, brakes, or a combination of rocket motors, we usually want to know how much of the acceleration acts in the direction of motion, in the tangential direction **T**. We can calculate this using the Chain Rule to rewrite **v** as

$$\mathbf{v} = \frac{d\mathbf{r}}{dt} = \frac{d\mathbf{r}}{ds}\frac{ds}{dt} = \mathbf{T}\frac{ds}{dt}$$

and differentiating both ends of this string of equalities to get

$$\mathbf{a} = \frac{d\mathbf{v}}{dt} = \frac{d}{dt}\left(\mathbf{T}\frac{ds}{dt}\right) = \frac{d^2s}{dt^2}\mathbf{T} + \frac{ds}{dt}\frac{d\mathbf{T}}{dt}$$

$$= \frac{d^2s}{dt^2}\mathbf{T} + \frac{ds}{dt}\left(\frac{d\mathbf{T}}{ds}\frac{ds}{dt}\right) = \frac{d^2s}{dt^2}\mathbf{T} + \frac{ds}{dt}\left(\kappa\mathbf{N}\frac{ds}{dt}\right) \qquad \frac{d\mathbf{T}}{ds} = \kappa\mathbf{N}$$

$$= \frac{d^2s}{dt^2}\mathbf{T} + \kappa\left(\frac{ds}{dt}\right)^2\mathbf{N}.$$

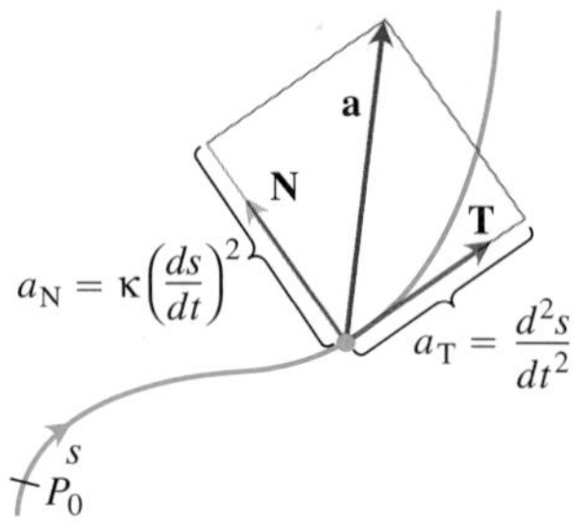

FIGURE 10.24 The tangential and normal components of acceleration. The acceleration **a** always lies in the plane of **T** and **N**, orthogonal to **B**.

DEFINITION If the acceleration vector is written as

$$\mathbf{a} = a_T\mathbf{T} + a_N\mathbf{N}, \tag{1}$$

then

$$a_T = \frac{d^2s}{dt^2} = \frac{d}{dt}|\mathbf{v}| \quad \text{and} \quad a_N = \kappa\left(\frac{ds}{dt}\right)^2 = \kappa|\mathbf{v}|^2 \tag{2}$$

are the **tangential** and **normal** scalar components of acceleration.

Notice that the binormal vector **B** does not appear in Equation (1). No matter how the path of the moving body we are watching may appear to twist and turn in space, the acceleration **a** *always lies in the plane of* **T** and **N** orthogonal to **B**. The equation also tells us exactly how much of the acceleration takes place tangent to the motion (d^2s/dt^2) and how much takes place normal to the motion [$\kappa(ds/dt)^2$] (Figure 10.24).

What information can we discover from Equations (2)? By definition, acceleration **a** is the rate of change of velocity **v**, and in general, both the length and direction of **v** change as a body moves along its path. The tangential component of acceleration a_T measures the rate of change of the *length* of **v** (that is, the change in the speed). The normal component of acceleration a_N measures the rate of change of the *direction* of **v**.

Notice that the normal scalar component of the acceleration is the curvature times the *square* of the speed. This explains why you have to hold on when your car makes a sharp (large κ), high-speed (large $|\mathbf{v}|$) turn. If you double the speed of your car, you will experience four times the normal component of acceleration for the same curvature.

If a body moves in a circle at a constant speed, d^2s/dt^2 is zero and all the acceleration points along **N** toward the circle's center. If the body is speeding up or slowing down, **a** has a nonzero tangential component (Figure 10.25).

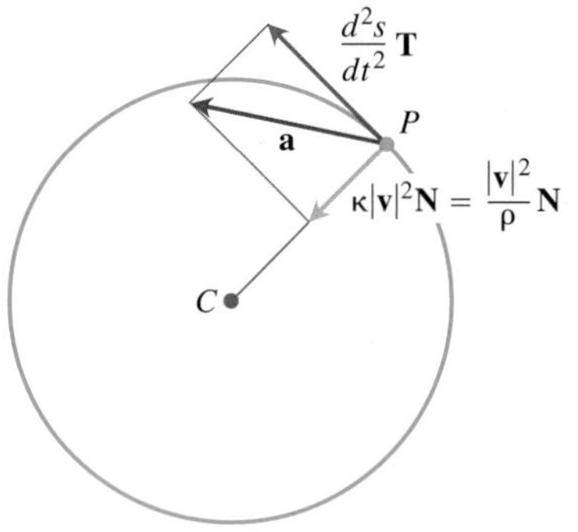

FIGURE 10.25 The tangential and normal components of the acceleration of a body that is speeding up as it moves counterclockwise around a circle of radius ρ.

To calculate a_N, we usually use the formula $a_N = \sqrt{|\mathbf{a}|^2 - a_T^2}$, which comes from solving the equation $|\mathbf{a}|^2 = \mathbf{a}\cdot\mathbf{a} = a_T^2 + a_N^2$ for a_N. With this formula, we can find a_N without having to calculate κ first.

Formula for Calculating the Normal Component of Acceleration

$$a_N = \sqrt{|\mathbf{a}|^2 - a_T^2} \tag{3}$$

EXAMPLE 1 Without finding **T** and **N**, write the acceleration of the motion

$$\mathbf{r}(t) = (\cos t + t\sin t)\mathbf{i} + (\sin t - t\cos t)\mathbf{j}, \qquad t > 0$$

in the form $\mathbf{a} = a_T\mathbf{T} + a_N\mathbf{N}$. (The path of the motion is the involute of the circle in Figure 10.26. See also Section 10.3, Exercise 17.)

Solution We use the first of Equations (2) to find a_T:

$$\begin{aligned}\mathbf{v} &= \frac{d\mathbf{r}}{dt} = (-\sin t + \sin t + t\cos t)\mathbf{i} + (\cos t - \cos t + t\sin t)\mathbf{j}\\ &= (t\cos t)\mathbf{i} + (t\sin t)\mathbf{j}\end{aligned}$$

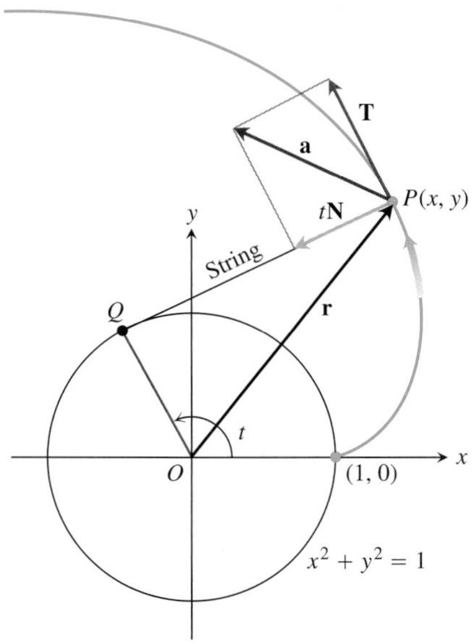

FIGURE 10.26 The tangential and normal components of the acceleration of the motion $\mathbf{r}(t) = (\cos t + t\sin t)\mathbf{i} + (\sin t - t\cos t)\mathbf{j}$, for $t > 0$. If a string wound around a fixed circle is unwound while held taut in the plane of the circle, its end P traces an involute of the circle (Example 1).

$$|\mathbf{v}| = \sqrt{t^2\cos^2 t + t^2\sin^2 t} = \sqrt{t^2} = |t| = t \qquad t > 0$$

$$a_{\mathrm{T}} = \frac{d}{dt}|\mathbf{v}| = \frac{d}{dt}(t) = 1. \qquad \text{Eq. (2)}$$

Knowing a_{T}, we use Equation (3) to find a_{N}:

$$\mathbf{a} = (\cos t - t\sin t)\mathbf{i} + (\sin t + t\cos t)\mathbf{j} \qquad \text{After some algebra}$$

$$|\mathbf{a}|^2 = t^2 + 1$$

$$a_{\mathrm{N}} = \sqrt{|\mathbf{a}|^2 - a_{\mathrm{T}}^2}$$

$$= \sqrt{(t^2 + 1) - (1)} = \sqrt{t^2} = t.$$

We then use Equation (1) to find $\mathbf{a}$:

$$\mathbf{a} = a_{\mathrm{T}}\mathbf{T} + a_{\mathrm{N}}\mathbf{N} = (1)\mathbf{T} + (t)\mathbf{N} = \mathbf{T} + t\mathbf{N}.$$

■

Torsion

How does $d\mathbf{B}/ds$ behave in relation to $\mathbf{T}$, $\mathbf{N}$, and $\mathbf{B}$? From the rule for differentiating a cross product, we have

$$\frac{d\mathbf{B}}{ds} = \frac{d(\mathbf{T}\times\mathbf{N})}{ds} = \frac{d\mathbf{T}}{ds}\times\mathbf{N} + \mathbf{T}\times\frac{d\mathbf{N}}{ds}.$$

Since $\mathbf{N}$ is the direction of $d\mathbf{T}/ds$, $(d\mathbf{T}/ds)\times\mathbf{N} = \mathbf{0}$ and

$$\frac{d\mathbf{B}}{ds} = \mathbf{0} + \mathbf{T}\times\frac{d\mathbf{N}}{ds} = \mathbf{T}\times\frac{d\mathbf{N}}{ds}.$$

From this we see that $d\mathbf{B}/ds$ is orthogonal to $\mathbf{T}$ since a cross product is orthogonal to its factors.

Since $d\mathbf{B}/ds$ is also orthogonal to $\mathbf{B}$ (the latter has constant length), it follows that $d\mathbf{B}/ds$ is orthogonal to the plane of $\mathbf{B}$ and $\mathbf{T}$. In other words, $d\mathbf{B}/ds$ is parallel to $\mathbf{N}$, so $d\mathbf{B}/ds$ is a scalar multiple of $\mathbf{N}$. In symbols,

$$\frac{d\mathbf{B}}{ds} = -\tau\mathbf{N}.$$

The negative sign in this equation is traditional. The scalar τ is called the *torsion* along the curve. Notice that

$$\frac{d\mathbf{B}}{ds}\cdot\mathbf{N} = -\tau\mathbf{N}\cdot\mathbf{N} = -\tau(1) = -\tau.$$

We use this equation for our next definition.

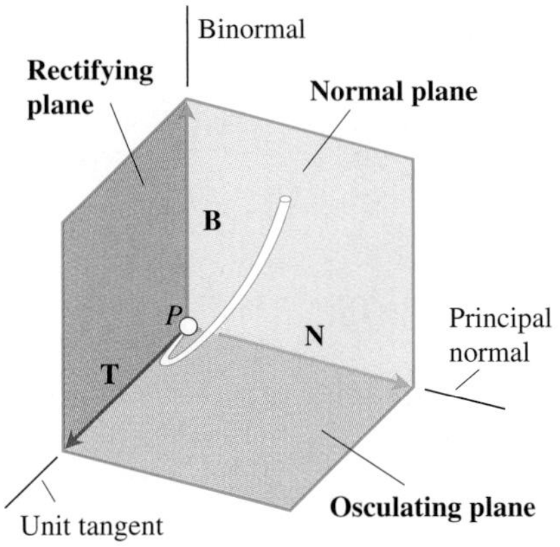

FIGURE 10.27 The names of the three planes determined by **T**, **N**, and **B**.

> **DEFINITION** Let $\mathbf{B} = \mathbf{T} \times \mathbf{N}$. The **torsion** function of a smooth curve is
>
> $$\tau = -\frac{d\mathbf{B}}{ds} \cdot \mathbf{N}. \qquad (4)$$

Unlike the curvature κ, which is never negative, the torsion τ may be positive, negative, or zero.

The three planes determined by **T**, **N**, and **B** are named and shown in Figure 10.27. The curvature $\kappa = |d\mathbf{T}/ds|$ can be thought of as the rate at which the normal plane turns as the point P moves along its path. Similarly, the torsion $\tau = -(d\mathbf{B}/ds) \cdot \mathbf{N}$ is the rate at which the osculating plane turns about **T** as P moves along the curve. Torsion measures how the curve twists.

Look at Figure 10.28. If P is a train climbing up a curved track, the rate at which the headlight turns from side to side per unit distance is the curvature of the track. The rate at which the engine tends to twist out of the plane formed by **T** and **N** is the torsion.

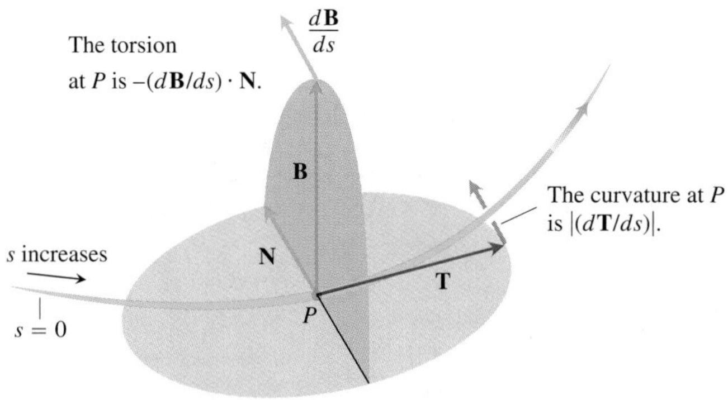

FIGURE 10.28 Every moving body travels with a **TNB** frame that characterizes the geometry of its path of motion.

Computational Formula for Torsion

The most widely used formula for torsion, derived in more advanced texts, is

$$\tau = \frac{\begin{vmatrix} \dot{x} & \dot{y} & \dot{z} \\ \ddot{x} & \ddot{y} & \ddot{z} \\ \dddot{x} & \dddot{y} & \dddot{z} \end{vmatrix}}{|\mathbf{v} \times \mathbf{a}|^2} \qquad (\text{if } \mathbf{v} \times \mathbf{a} \neq \mathbf{0}). \qquad (5)$$

The dots in Equation (5) denote differentiation with respect to t, one derivative for each dot. Thus, $\dot{x}$ ("x dot") means dx/dt, $\ddot{x}$ ("x double dot") means d^2x/dt^2, and $\dddot{x}$ ("x triple dot") means d^3x/dt^3. Similarly, $\dot{y} = dy/dt$, and so on.

EXERCISES 10.5

In Exercises 1 and 2, write **a** in the form $\mathbf{a} = a_{\mathrm{T}}\mathbf{T} + a_{\mathrm{N}}\mathbf{N}$ without finding **T** and **N**.

1. $\mathbf{r}(t) = (a \cos t)\mathbf{i} + (a \sin t)\mathbf{j} + bt\mathbf{k}$

2. $\mathbf{r}(t) = (1 + 3t)\mathbf{i} + (t - 2)\mathbf{j} - 3t\mathbf{k}$

In Exercises 3–6, write **a** in the form $\mathbf{a} = a_{\mathrm{T}}\mathbf{T} + a_{\mathrm{N}}\mathbf{N}$ at the given value of t without finding **T** and **N**.

3. $\mathbf{r}(t) = (t + 1)\mathbf{i} + 2t\mathbf{j} + t^2\mathbf{k}, \quad t = 1$

4. $\mathbf{r}(t) = (t \cos t)\mathbf{i} + (t \sin t)\mathbf{j} + t^2\mathbf{k}, \quad t = 0$

5. $\mathbf{r}(t) = t^2\mathbf{i} + (t + (1/3)t^3)\mathbf{j} + (t - (1/3)t^3)\mathbf{k}, \quad t = 0$

6. $\mathbf{r}(t) = (e^t \cos t)\mathbf{i} + (e^t \sin t)\mathbf{j} + \sqrt{2}e^t\mathbf{k}, \quad t = 0$

In Exercises 7 and 8, find $\mathbf{r}$, $\mathbf{T}$, $\mathbf{N}$, and $\mathbf{B}$ at the given value of t. Then find equations for the osculating, normal, and rectifying planes at that value of t.

7. $\mathbf{r}(t) = (\cos t)\mathbf{i} + (\sin t)\mathbf{j} - \mathbf{k}, \quad t = \pi/4$

8. $\mathbf{r}(t) = (\cos t)\mathbf{i} + (\sin t)\mathbf{j} + t\mathbf{k}, \quad t = 0$

In Exercises 9–16 of Section 10.4, you found $\mathbf{T}$, $\mathbf{N}$, and κ. Now, in Exercises 9–16, find $\mathbf{B}$ and τ for these space curves.

9. $\mathbf{r}(t) = (3 \sin t)\mathbf{i} + (3 \cos t)\mathbf{j} + 4t\mathbf{k}$

10. $\mathbf{r}(t) = (\cos t + t \sin t)\mathbf{i} + (\sin t - t \cos t)\mathbf{j} + 3\mathbf{k}$

11. $\mathbf{r}(t) = (e^t \cos t)\mathbf{i} + (e^t \sin t)\mathbf{j} + 2\mathbf{k}$

12. $\mathbf{r}(t) = (6 \sin 2t)\mathbf{i} + (6 \cos 2t)\mathbf{j} + 5t\mathbf{k}$

13. $\mathbf{r}(t) = (t^3/3)\mathbf{i} + (t^2/2)\mathbf{j}, \quad t > 0$

14. $\mathbf{r}(t) = (\cos^3 t)\mathbf{i} + (\sin^3 t)\mathbf{j}, \quad 0 < t < \pi/2$

15. $\mathbf{r}(t) = t\mathbf{i} + (a \cosh(t/a))\mathbf{j}, \quad a > 0$

16. $\mathbf{r}(t) = (\cosh t)\mathbf{i} - (\sinh t)\mathbf{j} + t\mathbf{k}$

17. The speedometer on your car reads a steady 35 mph. Could you be accelerating? Explain.

18. Can anything be said about the acceleration of a particle that is moving at a constant speed? Give reasons for your answer.

19. Can anything be said about the speed of a particle whose acceleration is always orthogonal to its velocity? Give reasons for your answer.

20. An object of mass m travels along the parabola $y = x^2$ with a constant speed of 10 units/sec. What is the force on the object due to its acceleration at (0, 0)? at $(2^{1/2}, 2)$? Write your answers in terms of $\mathbf{i}$ and $\mathbf{j}$. (Remember Newton's law, $\mathbf{F} = m\mathbf{a}$.)

21. **Vector formula for curvature** For a smooth curve, derive the curvature formula

$$\kappa = \frac{|\mathbf{v} \times \mathbf{a}|}{|\mathbf{v}|^3}.$$

22. Show that a moving particle will move in a straight line if the normal component of its acceleration is zero.

23. **A sometime shortcut to curvature** If you already know $|a_N|$ and $|\mathbf{v}|$, then the formula $a_N = \kappa|\mathbf{v}|^2$ gives a convenient way to find the curvature. Use it to find the curvature and radius of curvature of the curve

$$\mathbf{r}(t) = (\cos t + t \sin t)\mathbf{i} + (\sin t - t \cos t)\mathbf{j}, \quad t > 0.$$

(Take a_N and $|\mathbf{v}|$ from Example 1.)

24. Show that κ and τ are both zero for the line

$$\mathbf{r}(t) = (x_0 + At)\mathbf{i} + (y_0 + Bt)\mathbf{j} + (z_0 + Ct)\mathbf{k}.$$

25. What can be said about the torsion of a smooth plane curve $\mathbf{r}(t) = f(t)\mathbf{i} + g(t)\mathbf{j}$? Give reasons for your answer.

26. **The torsion of a helix** Show that the torsion of the helix

$$\mathbf{r}(t) = (a \cos t)\mathbf{i} + (a \sin t)\mathbf{j} + bt\mathbf{k}, \quad a, b \geq 0$$

is $\tau = b/(a^2 + b^2)$. What is the largest value τ can have for a given value of a? Give reasons for your answer.

27. **Differentiable curves with zero torsion lie in planes** That a sufficiently differentiable curve with zero torsion lies in a plane is a special case of the fact that a particle whose velocity remains perpendicular to a fixed vector $\mathbf{C}$ moves in a plane perpendicular to $\mathbf{C}$. This, in turn, can be viewed as the solution of the following problem in calculus.

Suppose $\mathbf{r}(t) = f(t)\mathbf{i} + g(t)\mathbf{j} + h(t)\mathbf{k}$ is twice differentiable for all t in an interval $[a, b]$, that $\mathbf{r} = 0$ when $t = a$, and that $\mathbf{v} \cdot \mathbf{k} = 0$ for all t in $[a, b]$. Then $h(t) = 0$ for all t in $[a, b]$.

Solve this problem. (*Hint:* Start with $\mathbf{a} = d^2\mathbf{r}/dt^2$ and apply the initial conditions in reverse order.)

28. **A formula that calculates τ from B and v** If we start with the definition $\tau = -(d\mathbf{B}/ds) \cdot \mathbf{N}$ and apply the Chain Rule to rewrite $d\mathbf{B}/ds$ as

$$\frac{d\mathbf{B}}{ds} = \frac{d\mathbf{B}}{dt}\frac{dt}{ds} = \frac{d\mathbf{B}}{dt}\frac{1}{|\mathbf{v}|},$$

we arrive at the formula

$$\tau = -\frac{1}{|\mathbf{v}|}\left(\frac{d\mathbf{B}}{dt} \cdot \mathbf{N}\right).$$

The advantage of this formula over Equation (5) is that it is easier to derive and state. The disadvantage is that it can take a lot of work to evaluate without a computer. Use the new formula to find the torsion of the helix in Exercise 26.

COMPUTER EXPLORATIONS

Rounding the answers to four decimal places, use a CAS to find $\mathbf{v}$, $\mathbf{a}$, speed, $\mathbf{T}$, $\mathbf{N}$, $\mathbf{B}$, κ, τ, and the tangential and normal components of acceleration for the curves in Exercises 29–32 at the given values of t.

29. $\mathbf{r}(t) = (t \cos t)\mathbf{i} + (t \sin t)\mathbf{j} + t\mathbf{k}, \quad t = \sqrt{3}$

30. $\mathbf{r}(t) = (e^t \cos t)\mathbf{i} + (e^t \sin t)\mathbf{j} + e^t\mathbf{k}, \quad t = \ln 2$

31. $\mathbf{r}(t) = (t - \sin t)\mathbf{i} + (1 - \cos t)\mathbf{j} + \sqrt{-t}\mathbf{k}, \quad t = -3\pi$

32. $\mathbf{r}(t) = (3t - t^2)\mathbf{i} + (3t^2)\mathbf{j} + (3t + t^3)\mathbf{k}, \quad t = 1$

10.6 Velocity and Acceleration in Polar Coordinates

In this section we derive equations for velocity and acceleration in polar coordinates. We then apply them to establish Kepler's second law of planetary motion as a consequence of Newton's law of gravitation.

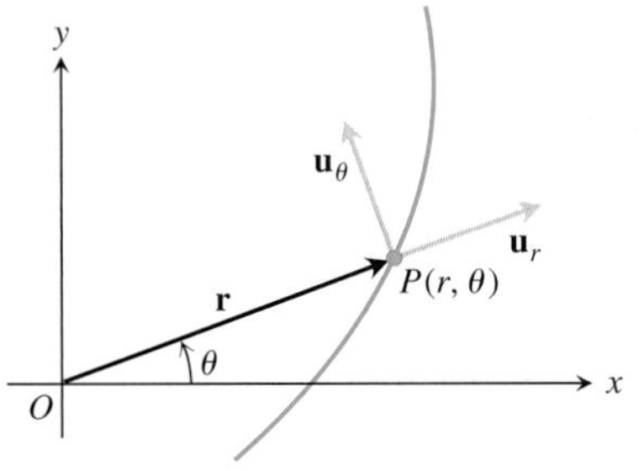

FIGURE 10.29 The length of $\mathbf{r}$ is the positive polar coordinate r of the point P. Thus, $\mathbf{u}_r$, which is $\mathbf{r}/|\mathbf{r}|$, is also $\mathbf{r}/r$. Equations (1) express $\mathbf{u}_r$ and $\mathbf{u}_\theta$ in terms of $\mathbf{i}$ and $\mathbf{j}$.

Motion in Polar and Cylindrical Coordinates

When a particle at $P(r, \theta)$ moves along a curve in the polar coordinate plane, we express its position, velocity, and acceleration in terms of the moving unit vectors

$$\mathbf{u}_r = (\cos\theta)\mathbf{i} + (\sin\theta)\mathbf{j}, \qquad \mathbf{u}_\theta = -(\sin\theta)\mathbf{i} + (\cos\theta)\mathbf{j}, \tag{1}$$

shown in Figure 10.29. The vector $\mathbf{u}_r$ points along the position vector $\overrightarrow{OP}$, so $\mathbf{r} = r\mathbf{u}_r$. The vector $\mathbf{u}_\theta$, orthogonal to $\mathbf{u}_r$, points in the direction of increasing θ.

We find from Equations (1) that

$$\frac{d\mathbf{u}_r}{d\theta} = -(\sin\theta)\mathbf{i} + (\cos\theta)\mathbf{j} = \mathbf{u}_\theta$$
$$\frac{d\mathbf{u}_\theta}{d\theta} = -(\cos\theta)\mathbf{i} - (\sin\theta)\mathbf{j} = -\mathbf{u}_r.$$

When we differentiate $\mathbf{u}_r$ and $\mathbf{u}_\theta$ with respect to t to find how they change with time, the Chain Rule gives

$$\dot{\mathbf{u}}_r = \frac{d\mathbf{u}_r}{d\theta}\dot{\theta} = \dot{\theta}\mathbf{u}_\theta, \qquad \dot{\mathbf{u}}_\theta = \frac{d\mathbf{u}_\theta}{d\theta}\dot{\theta} = -\dot{\theta}\mathbf{u}_r. \tag{2}$$

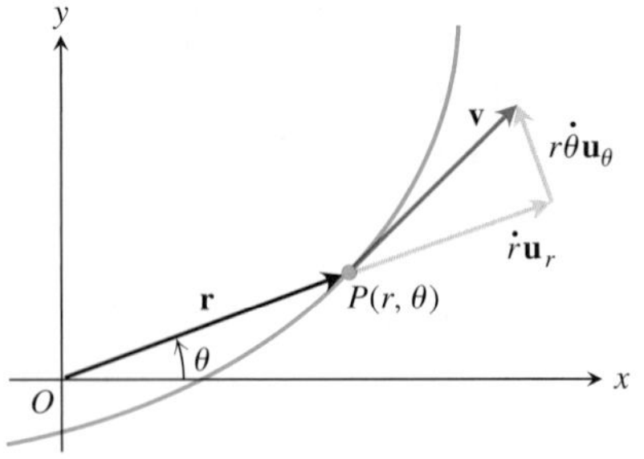

FIGURE 10.30 In polar coordinates, the velocity vector is

$$\mathbf{v} = \dot{r}\mathbf{u}_r + r\dot{\theta}\mathbf{u}_\theta$$

Hence,

$$\mathbf{v} = \dot{\mathbf{r}} = \frac{d}{dt}\left(r\mathbf{u}_r\right) = \dot{r}\mathbf{u}_r + r\dot{\mathbf{u}}_r = \dot{r}\mathbf{u}_r + r\dot{\theta}\,\mathbf{u}_\theta.$$

See Figure 10.30. As in the previous section, we use Newton's dot notation for time derivatives to keep the formulas as simple as we can: $\dot{\mathbf{u}}_r$ means $d\mathbf{u}_r/dt$, $\dot{\theta}$ means $d\theta/dt$, and so on.

The acceleration is

$$\mathbf{a} = \dot{\mathbf{v}} = (\ddot{r}\mathbf{u}_r + \dot{r}\dot{\mathbf{u}}_r) + (\dot{r}\dot{\theta}\mathbf{u}_\theta + r\ddot{\theta}\mathbf{u}_\theta + r\dot{\theta}\dot{\mathbf{u}}_\theta).$$

When Equations (2) are used to evaluate $\dot{\mathbf{u}}_r$ and $\dot{\mathbf{u}}_\theta$ and the components are separated, the equation for acceleration becomes

$$\mathbf{a} = (\ddot{r} - r\dot{\theta}^2)\mathbf{u}_r + (r\ddot{\theta} + 2\dot{r}\dot{\theta})\mathbf{u}_\theta.$$

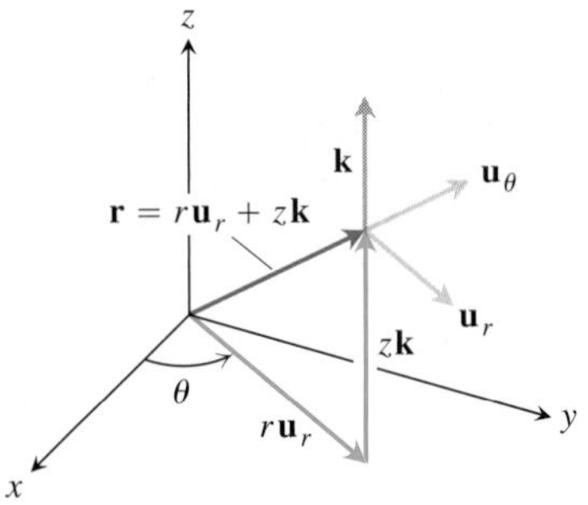

FIGURE 10.31 Position vector and basic unit vectors in cylindrical coordinates. Notice that $|\mathbf{r}| \neq r$ if $z \neq 0$.

To extend these equations of motion to space, we add $z\mathbf{k}$ to the right-hand side of the equation $\mathbf{r} = r\mathbf{u}_r$. Then, in these *cylindrical coordinates*, we have

$$\begin{aligned} \mathbf{r} &= r\mathbf{u}_r + z\mathbf{k} \\ \mathbf{v} &= \dot{r}\mathbf{u}_r + r\dot{\theta}\mathbf{u}_\theta + \dot{z}\mathbf{k} \\ \mathbf{a} &= (\ddot{r} - r\dot{\theta}^2)\mathbf{u}_r + (r\ddot{\theta} + 2\dot{r}\dot{\theta})\mathbf{u}_\theta + \ddot{z}\mathbf{k}. \end{aligned} \tag{3}$$

The vectors $\mathbf{u}_r$, $\mathbf{u}_\theta$, and $\mathbf{k}$ make a right-handed frame (Figure 10.31) in which

$$\mathbf{u}_r \times \mathbf{u}_\theta = \mathbf{k}, \qquad \mathbf{u}_\theta \times \mathbf{k} = \mathbf{u}_r, \qquad \mathbf{k} \times \mathbf{u}_r = \mathbf{u}_\theta.$$

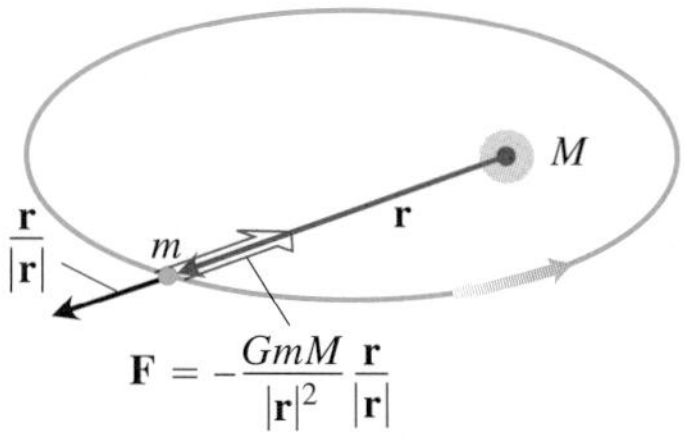

FIGURE 10.32 The force of gravity is directed along the line joining the centers of mass.

Planets Move in Planes

Newton's law of gravitation says that if $\mathbf{r}$ is the radius vector from the center of a sun of mass M to the center of a planet of mass m, then the force $\mathbf{F}$ of the gravitational attraction between the planet and sun is

$$\mathbf{F} = -\frac{GmM}{|\mathbf{r}|^2}\frac{\mathbf{r}}{|\mathbf{r}|}$$

(Figure 10.32). The number G is the **universal gravitational constant**. If we measure mass in kilograms, force in newtons, and distance in meters, G is about $6.6726 \times 10^{-11}\ \mathrm{Nm^2\ kg^{-2}}$.

Combining the gravitation law with Newton's second law, $\mathbf{F} = m\ddot{\mathbf{r}}$, for the force acting on the planet gives

$$m\ddot{\mathbf{r}} = -\frac{GmM}{|\mathbf{r}|^2}\frac{\mathbf{r}}{|\mathbf{r}|},$$

$$\ddot{\mathbf{r}} = -\frac{GM}{|\mathbf{r}|^2}\frac{\mathbf{r}}{|\mathbf{r}|}.$$

The planet is accelerated toward the sun's center of mass at all times.

Since $\ddot{\mathbf{r}}$ is a scalar multiple of $\mathbf{r}$, we have

$$\mathbf{r} \times \ddot{\mathbf{r}} = \mathbf{0}.$$

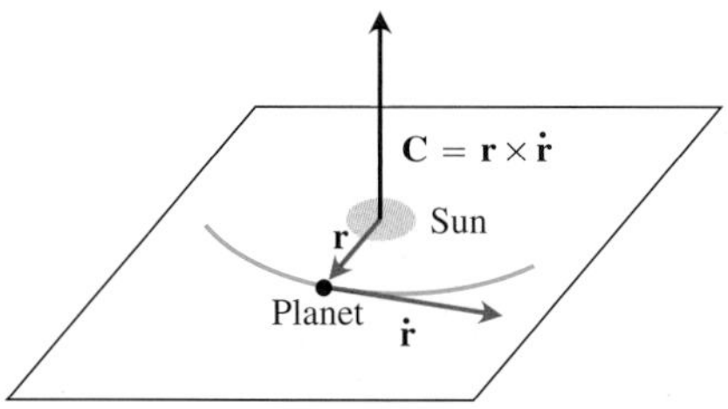

FIGURE 10.33 A planet that obeys Newton's laws of gravitation and motion travels in the plane through the sun's center of mass perpendicular to $\mathbf{C} = \mathbf{r} \times \dot{\mathbf{r}}$.

From this last equation,

$$\frac{d}{dt}(\mathbf{r} \times \dot{\mathbf{r}}) = \underbrace{\dot{\mathbf{r}} \times \dot{\mathbf{r}}}_{0} + \mathbf{r} \times \ddot{\mathbf{r}} = \mathbf{r} \times \ddot{\mathbf{r}} = \mathbf{0}.$$

It follows that

$$\mathbf{r} \times \dot{\mathbf{r}} = \mathbf{C} \tag{4}$$

for some constant vector C.

Equation (4) tells us that $\mathbf{r}$ and $\dot{\mathbf{r}}$ always lie in a plane perpendicular to $\mathbf{C}$. Hence, the planet moves in a fixed plane through the center of its sun (Figure 10.33).

HISTORICAL BIOGRAPHY

Johannes Kepler
(1571–1630)

Kepler's First Law (Ellipse Law)

Kepler's first law says that a planet's path is an ellipse with the sun at one focus. The eccentricity of the ellipse is

$$e = \frac{r_0 v_0^2}{GM} - 1 \tag{5}$$

and the polar equation is

$$r = \frac{(1 + e)r_0}{1 + e\cos\theta}. \tag{6}$$

Here v_0 is the speed when the planet is positioned at its minimum distance r_0 from the sun. We omit the lengthy proof. The sun's mass M is 1.99×10^{30} kg.

Kepler's Second Law (Equal Area Law)

Kepler's second law says that the radius vector from the sun to a planet (the vector **r** in our model) sweeps out equal areas in equal times (Figure 10.34). To derive the law, we use Equation (3) to evaluate the cross product $\mathbf{C} = \mathbf{r} \times \dot{\mathbf{r}}$ from Equation (4):

$$\begin{aligned} \mathbf{C} &= \mathbf{r} \times \dot{\mathbf{r}} = \mathbf{r} \times \mathbf{v} \\ &= r\mathbf{u}_r \times (\dot{r}\mathbf{u}_r + r\dot{\theta}\mathbf{u}_\theta) && \text{Eq. (3), } \dot{z} = 0 \\ &= r\dot{r}\underbrace{(\mathbf{u}_r \times \mathbf{u}_r)}_{\mathbf{0}} + r(r\dot{\theta})\underbrace{(\mathbf{u}_r \times \mathbf{u}_\theta)}_{\mathbf{k}} \\ &= r(r\dot{\theta})\mathbf{k}. \end{aligned} \quad (7)$$

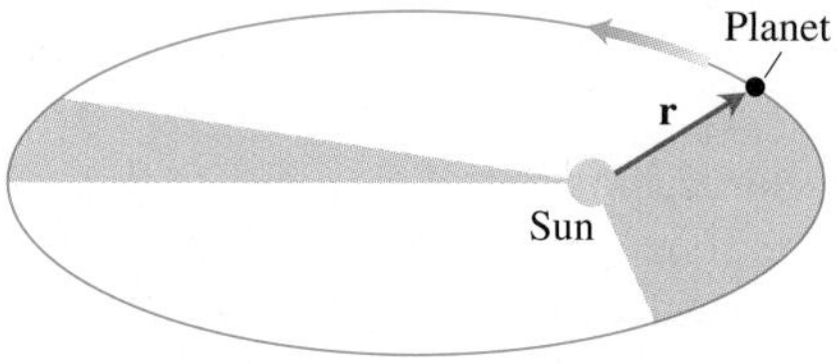

FIGURE 10.34 The line joining a planet to its sun sweeps over equal areas in equal times.

Setting t equal to zero shows that

$$\mathbf{C} = [r(r\dot{\theta})]_{t=0}\,\mathbf{k} = r_0 v_0 \mathbf{k}.$$

Substituting this value for **C** in Equation (7) gives

$$r_0 v_0 \mathbf{k} = r^2\dot{\theta}\mathbf{k}, \qquad \text{or} \qquad r^2\dot{\theta} = r_0 v_0.$$

This is where the area comes in. The area differential in polar coordinates is

$$dA = \frac{1}{2} r^2\, d\theta$$

(Section 8.3). Accordingly, dA/dt has the constant value

$$\frac{dA}{dt} = \frac{1}{2} r^2\dot{\theta} = \frac{1}{2} r_0 v_0. \quad (8)$$

So dA/dt is constant, giving Kepler's second law.

Kepler's Third Law (Time–Distance Law)

The time T it takes a planet to go around its sun once is the planet's **orbital period.** *Kepler's third law* says that T and the orbit's semimajor axis a are related by the equation

$$\frac{T^2}{a^3} = \frac{4\pi^2}{GM}.$$

Since the right-hand side of this equation is constant within a given solar system, the ratio of T^2 to a^3 *is the same for every planet in the system.*

Here is a partial derivation of Kepler's third law. The area enclosed by the planet's elliptical orbit is calculated as follows:

$$\begin{aligned} \text{Area} &= \int_0^T dA \\ &= \int_0^T \frac{1}{2} r_0 v_0\, dt && \text{Eq. (8)} \\ &= \frac{1}{2} T r_0 v_0. \end{aligned}$$

If b is the semiminor axis, the area of the ellipse is πab, so

$$T = \frac{2\pi ab}{r_0 v_0} = \frac{2\pi a^2}{r_0 v_0}\sqrt{1 - e^2}. \qquad \text{For any ellipse, } b = a\sqrt{1 - e^2} \quad (9)$$

It remains only to express a and e in terms of r_0, v_0, G, and M. Equation (5) does this for e. For a, we observe that setting θ equal to π in Equation (6) gives

$$r_{\max} = r_0 \frac{1 + e}{1 - e}.$$

Hence, from Figure 10.34,

$$2a = r_0 + r_{\max} = \frac{2r_0}{1 - e} = \frac{2r_0 GM}{2GM - r_0 v_0^2}. \tag{10}$$

Squaring both sides of Equation (9) and substituting the results of Equations (5) and (10) produces Kepler's third law (Exercise 9).

EXERCISES 10.6

In Problems 1–5, find the velocity and acceleration vectors in terms of $\mathbf{u}_r$ and $\mathbf{u}_\theta$.

1. $r = a(1 - \cos\theta)$ and $\dfrac{d\theta}{dt} = 3$
2. $r = a \sin 2\theta$ and $\dfrac{d\theta}{dt} = 2t$
3. $r = e^{a\theta}$ and $\dfrac{d\theta}{dt} = 2$
4. $r = a(1 + \sin t)$ and $\theta = 1 - e^{-t}$
5. $r = 2\cos 4t$ and $\theta = 2t$
6. **Type of orbit** For what values of v_0 in Equation (5) is the orbit in Equation (6) a circle? An ellipse? A parabola? A hyperbola?
7. **Circular orbits** Show that a planet in a circular orbit moves with a constant speed. (*Hint:* This is a consequence of one of Kepler's laws.)
8. Suppose that $\mathbf{r}$ is the position vector of a particle moving along a plane curve and dA/dt is the rate at which the vector sweeps out area. Without introducing coordinates, and assuming the necessary derivatives exist, give a geometric argument based on increments and limits for the validity of the equation

$$\frac{dA}{dt} = \frac{1}{2}|\mathbf{r} \times \dot{\mathbf{r}}|.$$

9. **Kepler's third law** Complete the derivation of Kepler's third law (the part following Equation (10)).
10. Find the length of the major axis of the Earth's orbit using Kepler's third law and the fact that the Earth's orbital period is 365.256 days.

Chapter 10 Questions to Guide Your Review

1. State the rules for differentiating and integrating vector functions. Give examples.
2. How do you define and calculate the velocity, speed, direction of motion, and acceleration of a body moving along a sufficiently differentiable space curve? Give an example.
3. What is special about the derivatives of vector functions of constant length? Give an example.
4. What are the vector and parametric equations for ideal projectile motion? How do you find a projectile's maximum height, flight time, and range? Give examples.
5. How do you define and calculate the length of a segment of a smooth space curve? Give an example. What mathematical assumptions are involved in the definition?
6. How do you measure distance along a smooth curve in space from a preselected base point? Give an example.
7. What is a differentiable curve's unit tangent vector? Give an example.
8. Define curvature, circle of curvature (osculating circle), center of curvature, and radius of curvature for twice-differentiable curves in the plane. Give examples. What curves have zero curvature? Constant curvature?
9. What is a plane curve's principal normal vector? When is it defined? Which way does it point? Give an example.
10. How do you define $\mathbf{N}$ and κ for curves in space? How are these quantities related? Give examples.
11. What is a curve's binormal vector? Give an example. How is this vector related to the curve's torsion? Give an example.
12. What formulas are available for writing a moving body's acceleration as a sum of its tangential and normal components? Give an example. Why might one want to write the acceleration this way? What if the body moves at a constant speed? At a constant speed around a circle?
13. State Kepler's laws.

Chapter 10 Practice and Additional Exercises

In Exercises 1 and 2, graph the curves and sketch their velocity and acceleration vectors at the given values of t. Then write $\mathbf{a}$ in the form $\mathbf{a} = a_T\mathbf{T} + a_N\mathbf{N}$ without finding $\mathbf{T}$ and $\mathbf{N}$, and find the value of κ at the given values of t.

1. $\mathbf{r}(t) = (4\cos t)\mathbf{i} + \left(\sqrt{2}\sin t\right)\mathbf{j}, \quad t = 0$ and $\pi/4$
2. $\mathbf{r}(t) = \left(\sqrt{3}\sec t\right)\mathbf{i} + \left(\sqrt{3}\tan t\right)\mathbf{j}, \quad t = 0$
3. The position of a particle in the plane at time t is

$$\mathbf{r} = \frac{1}{\sqrt{1+t^2}}\mathbf{i} + \frac{t}{\sqrt{1+t^2}}\mathbf{j}.$$

Find the particle's highest speed.

4. Suppose $\mathbf{r}(t) = (e^t\cos t)\mathbf{i} + (e^t\sin t)\mathbf{j}$. Show that the angle between $\mathbf{r}$ and $\mathbf{a}$ never changes. What *is* the angle?
5. **Finding curvature** At point P, the velocity and acceleration of a particle moving in the plane are $\mathbf{v} = 3\mathbf{i} + 4\mathbf{j}$ and $\mathbf{a} = 5\mathbf{i} + 15\mathbf{j}$. Find the curvature of the particle's path at P.
6. Find the point on the curve $y = e^x$ where the curvature is greatest.
7. A particle moves around the unit circle in the xy-plane. Its position at time t is $\mathbf{r} = x\mathbf{i} + y\mathbf{j}$, where x and y are differentiable functions of t. Find dy/dt if $\mathbf{v}\cdot\mathbf{i} = y$. Is the motion clockwise, or counterclockwise?
8. You send a message through a pneumatic tube that follows the curve $9y = x^3$ (distance in meters). At the point (3, 3), $\mathbf{v}\cdot\mathbf{i} = 4$ and $\mathbf{a}\cdot\mathbf{i} = -2$. Find the values of $\mathbf{v}\cdot\mathbf{j}$ and $\mathbf{a}\cdot\mathbf{j}$ at (3, 3).
9. **Characterizing circular motion** A particle moves in the plane so that its velocity and position vectors are always orthogonal. Show that the particle moves in a circle centered at the origin.
10. **Speed along a cycloid** A circular wheel with radius 1 ft and center C rolls to the right along the x-axis at a half-turn per second. (See the accompanying figure.) At time t seconds, the position vector of the point P on the wheel's circumference is

$$\mathbf{r} = (\pi t - \sin\pi t)\mathbf{i} + (1 - \cos\pi t)\mathbf{j}.$$

   a. Sketch the curve traced by P during the interval $0 \le t \le 3$.
   b. Find $\mathbf{v}$ and $\mathbf{a}$ at $t = 0, 1, 2,$ and 3 and add these vectors to your sketch.
   c. At any given time, what is the forward speed of the topmost point of the wheel? Of C?

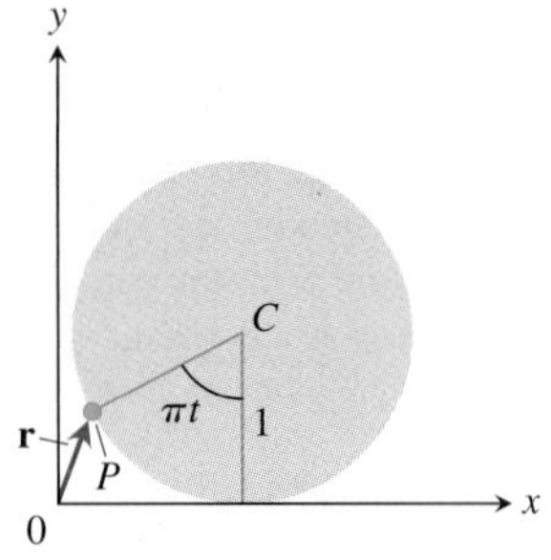

11. **Shot put** A shot leaves the thrower's hand 6.5 ft above the ground at a 45° angle at 44 ft/sec. Where is it 3 sec later?
12. **Javelin** A javelin leaves the thrower's hand 7 ft above the ground at a 45° angle at 80 ft/sec. How high does it go?

Find the lengths of the curves in Exercises 13 and 14.

13. $\mathbf{r}(t) = (2\cos t)\mathbf{i} + (2\sin t)\mathbf{j} + t^2\mathbf{k}, \quad 0 \le t \le \pi/4$
14. $\mathbf{r}(t) = (3\cos t)\mathbf{i} + (3\sin t)\mathbf{j} + 2t^{3/2}\mathbf{k}, \quad 0 \le t \le 3$

In Exercises 15–18, find $\mathbf{T}$, $\mathbf{N}$, $\mathbf{B}$, κ, and τ at the given value of t.

15. $\mathbf{r}(t) = \frac{4}{9}(1+t)^{3/2}\mathbf{i} + \frac{4}{9}(1-t)^{3/2}\mathbf{j} + \frac{1}{3}t\mathbf{k}, \quad t = 0$
16. $\mathbf{r}(t) = (e^t\sin 2t)\mathbf{i} + (e^t\cos 2t)\mathbf{j} + 2e^t\mathbf{k}, \quad t = 0$
17. $\mathbf{r}(t) = t\mathbf{i} + \frac{1}{2}e^{2t}\mathbf{j}, \quad t = \ln 2$
18. $\mathbf{r}(t) = (3\cosh 2t)\mathbf{i} + (3\sinh 2t)\mathbf{j} + 6t\mathbf{k}, \quad t = \ln 2$

In Exercises 19 and 20, write $\mathbf{a}$ in the form $\mathbf{a} = a_T\mathbf{T} + a_N\mathbf{N}$ at $t = 0$ without finding $\mathbf{T}$ and $\mathbf{N}$.

19. $\mathbf{r}(t) = (2 + 3t + 3t^2)\mathbf{i} + (4t + 4t^2)\mathbf{j} - (6\cos t)\mathbf{k}$
20. $\mathbf{r}(t) = (2 + t)\mathbf{i} + (t + 2t^2)\mathbf{j} + (1 + t^2)\mathbf{k}$
21. Find $\mathbf{T}$, $\mathbf{N}$, $\mathbf{B}$, κ, and τ as functions of t if $\mathbf{r}(t) = (\sin t)\mathbf{i} + \left(\sqrt{2}\cos t\right)\mathbf{j} + (\sin t)\mathbf{k}$.
22. At what times in the interval $0 \le t \le \pi$ are the velocity and acceleration vectors of the motion $\mathbf{r}(t) = \mathbf{i} + (5\cos t)\mathbf{j} + (3\sin t)\mathbf{k}$ orthogonal?
23. Find parametric equations for the line that is tangent to the curve $\mathbf{r}(t) = e^t\mathbf{i} + (\sin t)\mathbf{j} + \ln(1 - t)\mathbf{k}$ at $t = 0$.
24. **Synchronous curves** By eliminating α from the ideal projectile equations

$$x = (v_0\cos\alpha)t, \quad y = (v_0\sin\alpha)t - \frac{1}{2}gt^2,$$

show that $x^2 + (y + gt^2/2)^2 = v_0^2t^2$. This shows that projectiles launched simultaneously from the origin at the same initial speed will, at any given instant, all lie on the circle of radius v_0t centered at $(0, -gt^2/2)$, regardless of their launch angle. These circles are the *synchronous curves* of the launching.

25. **An alternative definition of curvature in the plane** An alternative definition gives the curvature of a sufficiently differentiable plane curve to be $|d\phi/ds|$, where ϕ is the angle between $\mathbf{T}$ and $\mathbf{i}$ (Part (a) of the accompanying figure). Part (b) shows the distance s measured counterclockwise around the circle $x^2 + y^2 = a^2$ from the point $(a, 0)$ to a point P, along with the angle ϕ at P. Calculate the circle's curvature using the alternative definition. (*Hint:* $\phi = \theta + \pi/2$.)

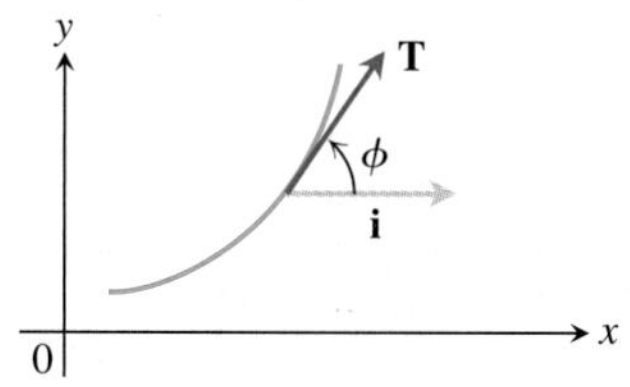

(a)

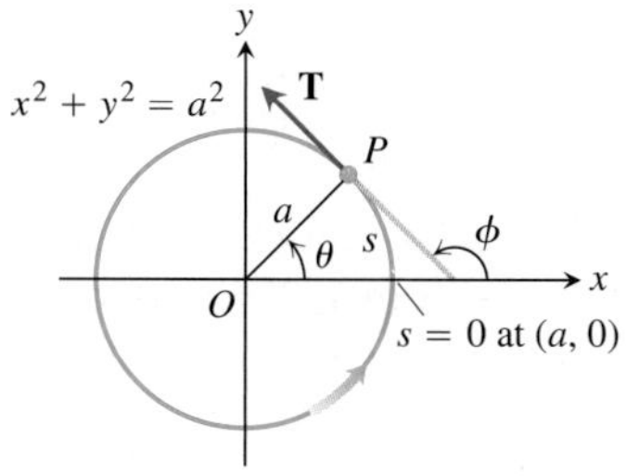

(b)

26. Radius of curvature Show that the radius of curvature of a twice-differentiable plane curve $\mathbf{r}(t) = f(t)\mathbf{i} + g(t)\mathbf{j}$ is given by the formula

$$\rho = \frac{\dot{x}^2 + \dot{y}^2}{\sqrt{\ddot{x}^2 + \ddot{y}^2 - \ddot{s}^2}}, \quad \text{where} \quad \ddot{s} = \frac{d}{dt}\sqrt{\dot{x}^2 + \dot{y}^2}.$$

Chapter 11 PARTIAL DERIVATIVES

OVERVIEW Many functions depend on more than one independent variable. For instance, the volume of a right circular cylinder is a function $V = \pi r^2 h$ of its radius and its height, so it is a function $V(r, h)$ of two variables. In this chapter we extend the basic ideas of single variable calculus to functions of several variables. Their derivatives are more varied and interesting because of the different ways the variables can interact. Their integrals (studied in the next chapter) lead to a great variety of applications.

11.1 Functions of Several Variables

In this section we define functions of more than one independent variable and discuss ways to graph them.

Real-valued functions of several independent real variables are defined similarly to functions in the single-variable case. The domains are sets of ordered pairs (triples, quadruples, n-tuples) of real numbers, and the ranges are sets of real numbers of the kind we have worked with all along.

DEFINITIONS Suppose D is a set of n-tuples of real numbers $(x_1, x_2, \ldots, x_n)$. A **real-valued function** f on D is a rule that assigns a unique (single) real number

$$w = f(x_1, x_2, \ldots, x_n)$$

to each element in D. The set D is the function's **domain**. The set of w-values taken on by f is the function's **range**. The symbol w is the **dependent variable** of f, and f is said to be a function of the n **independent variables** x_1 to x_n. We also call the x_j's the function's **input variables** and call w the function's **output variable**.

If f is a function of two independent variables, we usually call the independent variables x and y and picture the domain of f as a region in the xy-plane. If f is a function of three independent variables, we call the variables x, y, and z and picture the domain as a region in space.

In applications, we tend to use letters that remind us of what the variables stand for. To say that the volume of a right circular cylinder is a function of its radius and height, we might write $V = f(r, h)$. To be more specific, we might replace the notation $f(r, h)$ by the formula that calculates the value of V from the values of r and h, and write $V = \pi r^2 h$. In either case, r and h would be the independent variables and V the dependent variable of the function.

As usual, we evaluate functions defined by formulas by substituting the values of the independent variables in the formula and calculating the corresponding value of the dependent variable. For example, the value of $f(x, y, z) = \sqrt{x^2 + y^2 + z^2}$ at the point $(3, 0, 4)$ is

$$f(3, 0, 4) = \sqrt{(3)^2 + (0)^2 + (4)^2} = \sqrt{25} = 5.$$

Domains and Ranges

In defining a function of more than one variable, we follow the usual practice of excluding inputs that lead to complex numbers or division by zero. If $f(x, y) = \sqrt{y - x^2}$, y cannot be less than x^2. If $f(x, y) = 1/(xy)$, xy cannot be zero. The domain of a function is assumed to be the largest set for which the defining rule generates real numbers, unless the domain is otherwise specified explicitly. The range consists of the set of output values for the dependent variable.

EXAMPLE 1 (a) These are functions of two variables.

Function	Domain	Range
$w = \sqrt{y - x^2}$	$y \geq x^2$	$[0, \infty)$
$w = \dfrac{1}{xy}$	$xy \neq 0$	$(-\infty, 0) \cup (0, \infty)$
$w = \sin xy$	Entire plane	$[-1, 1]$

(b) These are functions of three variables.

Function	Domain	Range
$w = \sqrt{x^2 + y^2 + z^2}$	Entire space	$[0, \infty)$
$w = \dfrac{1}{x^2 + y^2 + z^2}$	$(x, y, z) \neq (0, 0, 0)$	$(0, \infty)$
$w = xy \ln z$	Half-space $z > 0$	$(-\infty, \infty)$

■

Functions of Two Variables

Regions in the plane can have interior points and boundary points just like intervals on the real line. Closed intervals $[a, b]$ include their boundary points, open intervals (a, b) don't include their boundary points, and intervals such as $[a, b)$ are neither open nor closed.

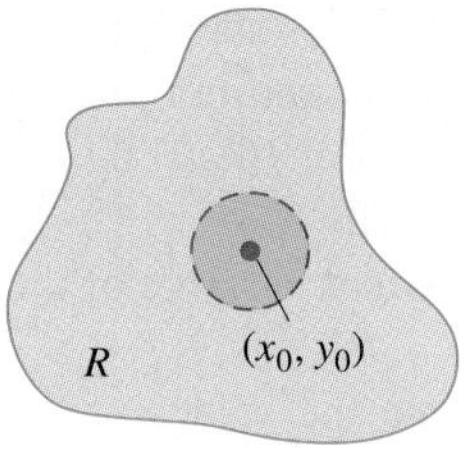

(a) Interior point

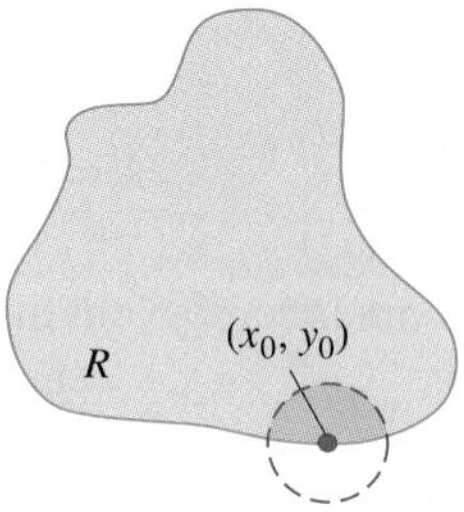

(b) Boundary point

FIGURE 11.1 Interior points and boundary points of a plane region R. An interior point is necessarily a point of R. A boundary point of R need not belong to R.

DEFINITIONS A point (x_0, y_0) in a region (set) R in the xy-plane is an **interior point** of R if it is the center of a disk of positive radius that lies entirely in R (Figure 11.1). A point (x_0, y_0) is a **boundary point** of R if every disk centered at (x_0, y_0) contains points that lie outside of R as well as points that lie in R. (The boundary point itself need not belong to R.)

The interior points of a region, as a set, make up the **interior** of the region. The region's boundary points make up its **boundary**. A region is **open** if it consists entirely of interior points. A region is **closed** if it contains all its boundary points (Figure 11.2).

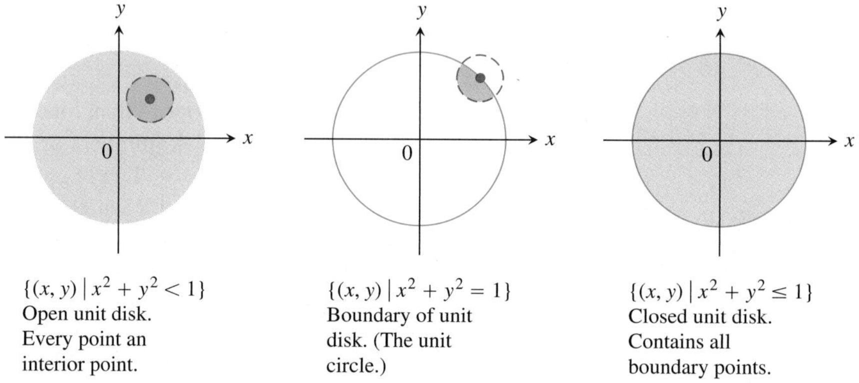

FIGURE 11.2 Interior points and boundary points of the unit disk in the plane.

As with a half-open interval of real numbers $[a, b)$, some regions in the plane are neither open nor closed. If you start with the open disk in Figure 11.2 and add to it some of but not all its boundary points, the resulting set is neither open nor closed. The boundary points that *are* there keep the set from being open. The absence of the remaining boundary points keeps the set from being closed.

DEFINITIONS A region in the plane is **bounded** if it lies inside a disk of fixed radius. A region is **unbounded** if it is not bounded.

Examples of *bounded* sets in the plane include line segments, triangles, interiors of triangles, rectangles, circles, and disks. Examples of *unbounded* sets in the plane include lines, coordinate axes, the graphs of functions defined on infinite intervals, quadrants, half-planes, and the plane itself.

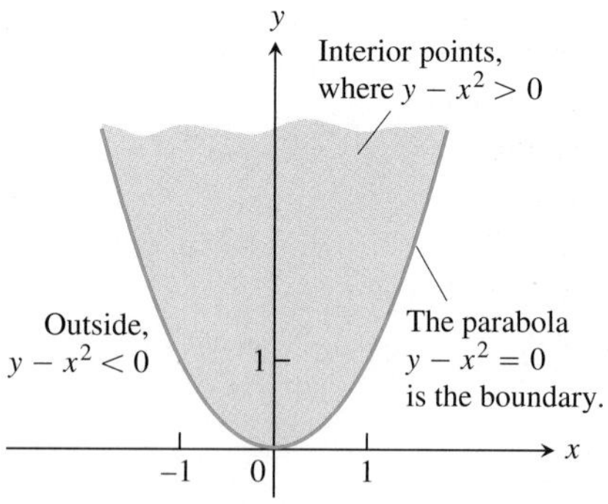

FIGURE 11.3 The domain of $f(x, y)$ in Example 2 consists of the shaded region and its bounding parabola.

EXAMPLE 2 Describe the domain of the function $f(x, y) = \sqrt{y - x^2}$.

Solution Since f is defined only where $y - x^2 \ge 0$, the domain is the closed, unbounded region shown in Figure 11.3. The parabola $y = x^2$ is the boundary of the domain. The points above the parabola make up the domain's interior. ■

Graphs, Level Curves, and Contours of Functions of Two Variables

There are two standard ways to picture the values of a function $f(x, y)$. One is to draw and label curves in the domain on which f has a constant value. The other is to sketch the surface $z = f(x, y)$ in space.

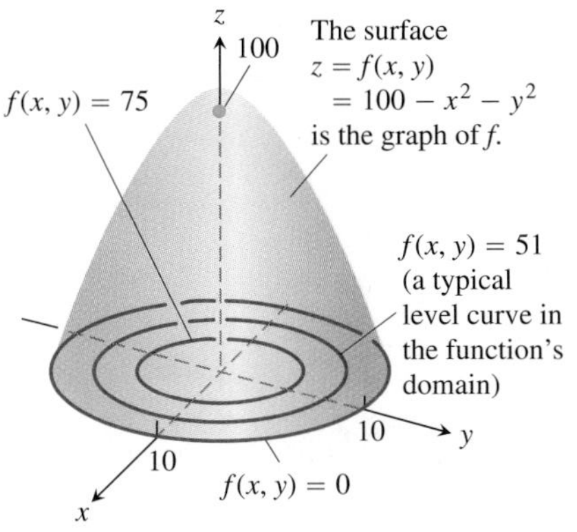

FIGURE 11.4 The graph and selected level curves of the function $f(x, y)$ in Example 3.

> **DEFINITIONS** The set of points in the plane where a function $f(x, y)$ has a constant value $f(x, y) = c$ is called a **level curve** of f. The set of all points $(x, y, f(x, y))$ in space, for (x, y) in the domain of f, is called the **graph** of f. The graph of f is also called the **surface** $z = f(x, y)$.

EXAMPLE 3 Graph $f(x, y) = 100 - x^2 - y^2$ and plot the level curves $f(x, y) = 0$, $f(x, y) = 51$, and $f(x, y) = 75$ in the domain of f in the plane.

Solution The domain of f is the entire xy-plane, and the range of f is the set of real numbers less than or equal to 100. The graph is the paraboloid $z = 100 - x^2 - y^2$, a portion of which is shown in Figure 11.4.

The level curve $f(x, y) = 0$ is the set of points in the xy-plane at which

$$f(x, y) = 100 - x^2 - y^2 = 0, \qquad \text{or} \qquad x^2 + y^2 = 100,$$

which is the circle of radius 10 centered at the origin. Similarly, the level curves $f(x, y) = 51$ and $f(x, y) = 75$ (Figure 11.4) are the circles

$$f(x, y) = 100 - x^2 - y^2 = 51, \qquad \text{or} \qquad x^2 + y^2 = 49$$

$$f(x, y) = 100 - x^2 - y^2 = 75, \qquad \text{or} \qquad x^2 + y^2 = 25.$$

The level curve $f(x, y) = 100$ consists of the origin alone. (It is still a level curve.) ■

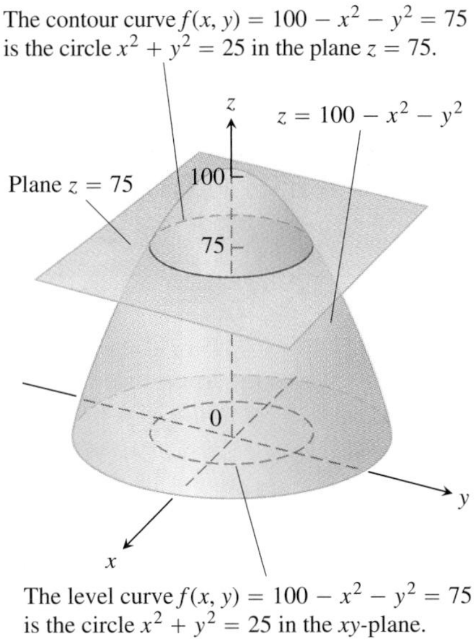

FIGURE 11.5 A plane $z = c$ parallel to the xy-plane intersecting a surface $z = f(x, y)$ produces a contour curve.

The curve in space in which the plane $z = c$ cuts a surface $z = f(x, y)$ is made up of the points that represent the function value $f(x, y) = c$. It is called the **contour curve** $f(x, y) = c$ to distinguish it from the level curve $f(x, y) = c$ in the domain of f. Figure 11.5 shows the contour curve $f(x, y) = 75$ on the surface $z = 100 - x^2 - y^2$ defined by the function $f(x, y) = 100 - x^2 - y^2$. The contour curve lies directly above the circle $x^2 + y^2 = 25$, which is the level curve $f(x, y) = 75$ in the function's domain.

Not everyone makes this distinction, however, and you may wish to call both kinds of curves by a single name and rely on context to convey which one you have in mind. On most maps, for example, the curves that represent constant elevation (height above sea level) are called contours, not level curves (Figure 11.6).

Functions of Three Variables

In the plane, the points where a function of two independent variables has a constant value $f(x, y) = c$ make a curve in the function's domain. In space, the points where a function of three independent variables has a constant value $f(x, y, z) = c$ make a surface in the function's domain.

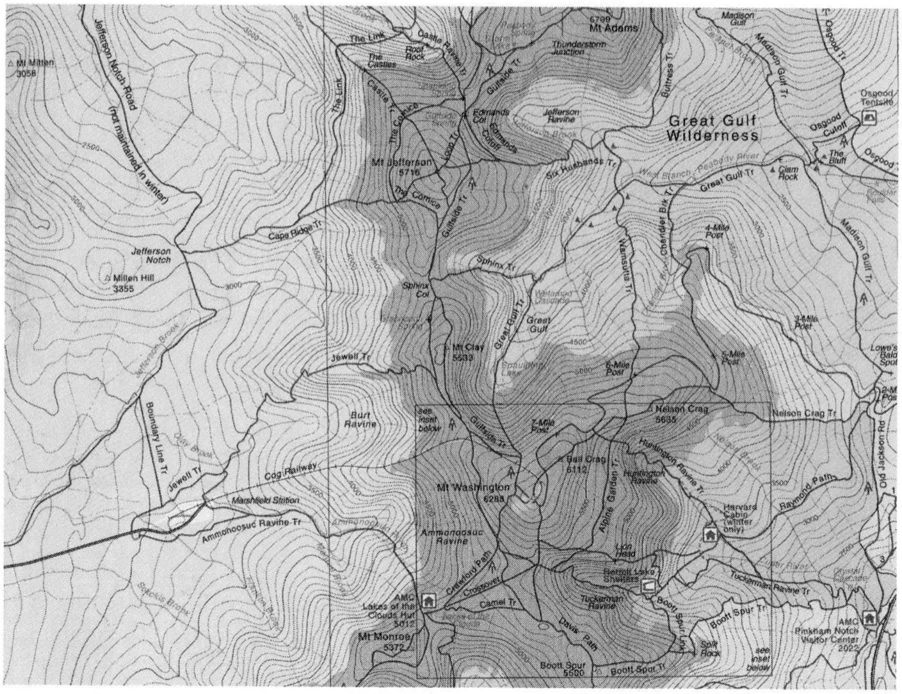

FIGURE 11.6 Contours on Mt. Washington in New Hampshire. (Reproduced by permission from the Appalachian Mountain Club.)

> **DEFINITION** The set of points (x, y, z) in space where a function of three independent variables has a constant value $f(x, y, z) = c$ is called a **level surface** of f.

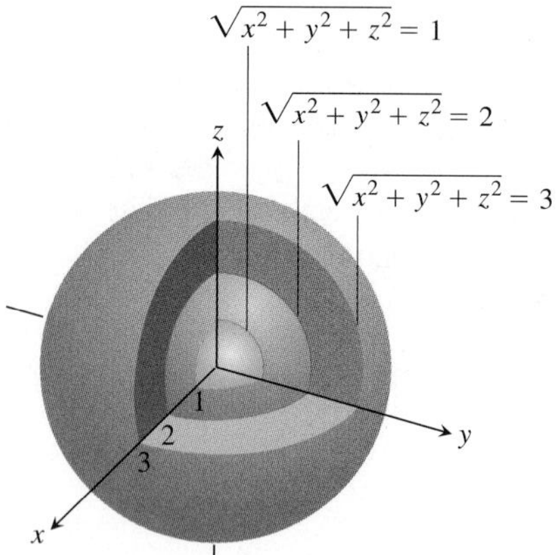

FIGURE 11.7 The level surfaces of $f(x, y, z) = \sqrt{x^2 + y^2 + z^2}$ are concentric spheres (Example 4).

Since the graphs of functions of three variables consist of points $(x, y, z, f(x, y, z))$ lying in a four-dimensional space, we cannot sketch them effectively in our three-dimensional frame of reference. We can see how the function behaves, however, by looking at its three-dimensional level surfaces.

EXAMPLE 4 Describe the level surfaces of the function

$$f(x, y, z) = \sqrt{x^2 + y^2 + z^2}.$$

Solution The value of f is the distance from the origin to the point (x, y, z). Each level surface $\sqrt{x^2 + y^2 + z^2} = c, c > 0$, is a sphere of radius c centered at the origin. Figure 11.7 shows a cutaway view of three of these spheres. The level surface $\sqrt{x^2 + y^2 + z^2} = 0$ consists of the origin alone.

We are not graphing the function here; we are looking at level surfaces in the function's domain. The level surfaces show how the function's values change as we move through its domain. If we remain on a sphere of radius c centered at the origin, the function maintains a constant value, namely c. If we move from a point on one sphere to a point on another, the function's value changes. It increases if we move away from the origin and decreases if we move toward the origin. The way the values change depends on the direction we take. The dependence of change on direction is important. We return to it in Section 11.5. ■

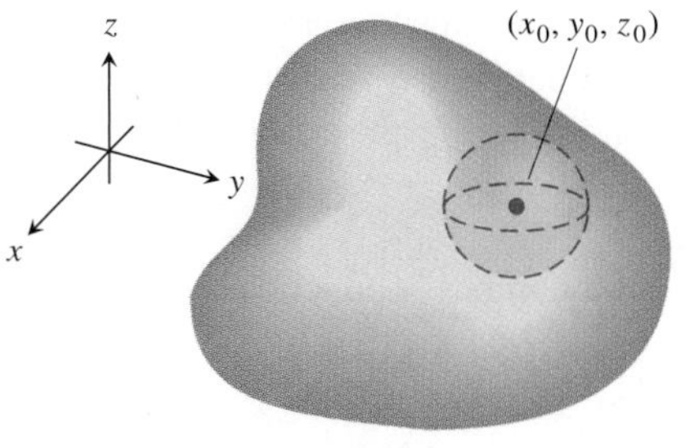

(a) Interior point

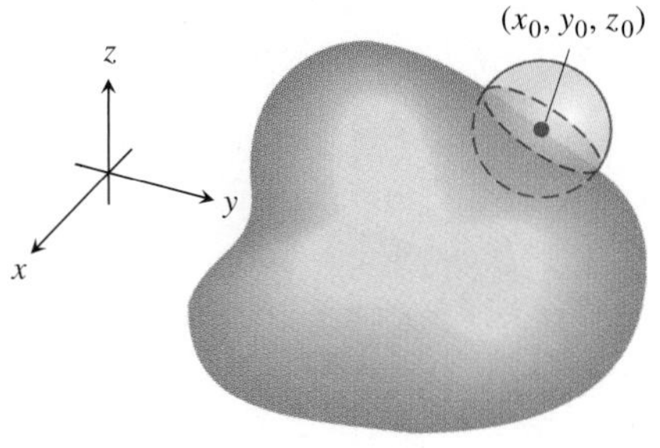

(b) Boundary point

FIGURE 11.8 Interior points and boundary points of a region in space.

The definitions of interior, boundary, open, closed, bounded, and unbounded for regions in space are similar to those for regions in the plane. To accommodate the extra dimension, we use solid balls of positive radius instead of disks.

DEFINITIONS A point (x_0, y_0, z_0) in a region R in space is an **interior point** of R if it is the center of a solid ball that lies entirely in R (Figure 11.8a). A point (x_0, y_0, z_0) is a **boundary point** of R if every solid ball centered at (x_0, y_0, z_0) contains points that lie outside of R as well as points that lie inside R (Figure 11.8b). The **interior** of R is the set of interior points of R. The **boundary** of R is the set of boundary points of R.

A region is **open** if it consists entirely of interior points. A region is **closed** if it contains its entire boundary.

Examples of *open* sets in space include the interior of a sphere, the open half-space $z > 0$, the first octant (where x, y, and z are all positive), and space itself.

Examples of *closed* sets in space include lines, planes, and the closed half-space $z \geq 0$.

A solid sphere with part of its boundary removed or a solid cube with a missing face, edge, or corner point is *neither open nor closed.*

Functions of more than three independent variables are also important. For example, the temperature on a surface in space may depend not only on the location of the point $P(x, y, z)$ on the surface, but also on the time t when it is visited, so we would write $T = f(x, y, z, t)$.

Computer Graphing

Three-dimensional graphing programs for computers and calculators make it possible to graph functions of two variables with only a few keystrokes. We can often get information more quickly from a graph than from a formula.

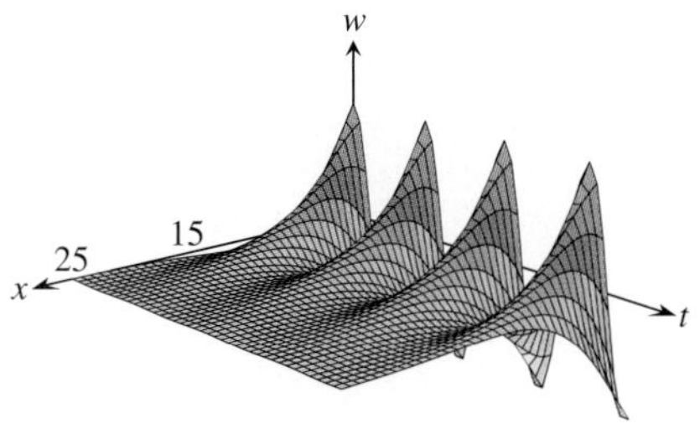

FIGURE 11.9 This computer-generated graph shows the seasonal variation of the temperature below ground as a fraction of surface temperature. (Example 5). (Adapted from art provided by Norton Starr.)

EXAMPLE 5 The temperature w beneath the Earth's surface is a function of the depth x beneath the surface and the time t of the year. If we measure x in feet and t as the number of days elapsed from the expected date of the yearly highest surface temperature, we can model the variation in temperature with the function

$$w = \cos(1.7 \times 10^{-2}t - 0.2x)e^{-0.2x}.$$

(The temperature at 0 ft is scaled to vary from $+1$ to -1, so that the variation at x feet can be interpreted as a fraction of the variation at the surface.)

Figure 11.9 shows a computer-generated graph of the function. At a depth of 15 ft, the variation (change in vertical amplitude in the figure) is about 5% of the surface variation. At 25 ft, there is almost no variation during the year.

The graph also shows that the temperature 15 ft below the surface is about half a year out of phase with the surface temperature. When the temperature is lowest on the surface (late January, say), it is at its highest 15 ft below. Fifteen feet below the ground, the seasons are reversed. ■

Figure 11.10 shows computer-generated graphs of a number of functions of two variables together with their level curves.

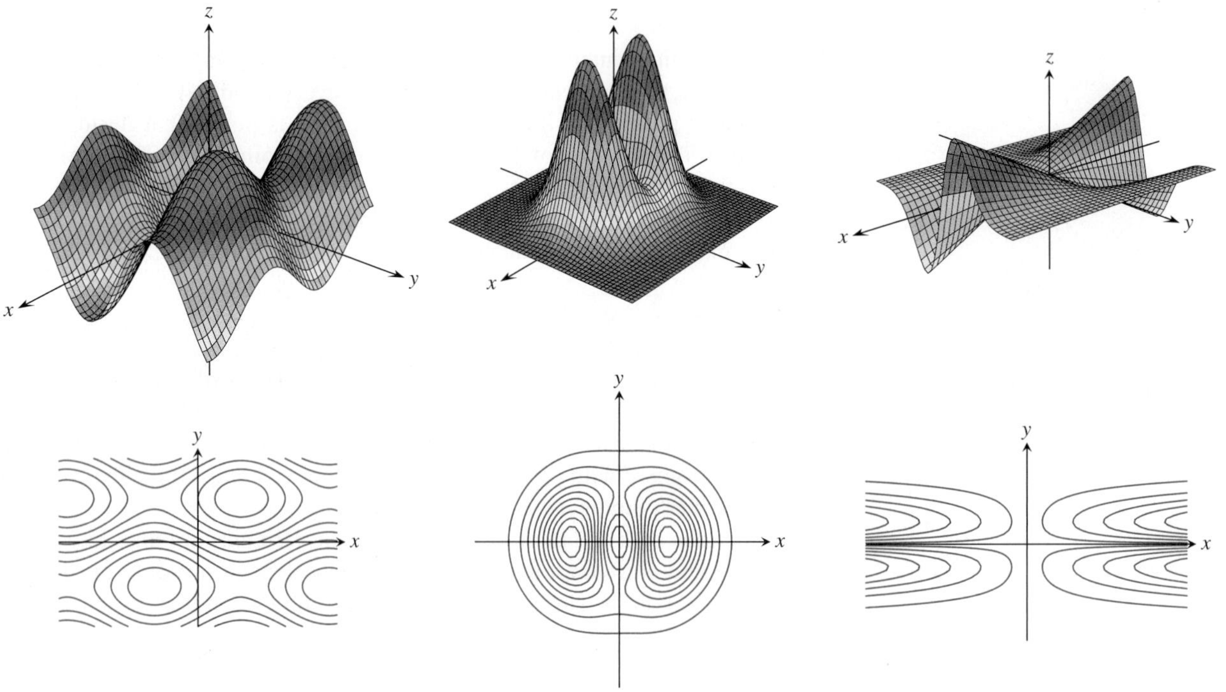

(a) $z = \sin x + 2 \sin y$ (b) $z = (4x^2 + y^2)e^{-x^2-y^2}$ (c) $z = xye^{-y^2}$

FIGURE 11.10 Computer-generated graphs and level curves of typical functions of two variables.

EXERCISES 11.1

In Exercises 1–12, **(a)** find the function's domain, **(b)** find the function's range, **(c)** describe the function's level curves, **(d)** find the boundary of the function's domain, **(e)** determine if the domain is an open region, a closed region, or neither, and **(f)** decide if the domain is bounded or unbounded.

1. $f(x, y) = y - x$

2. $f(x, y) = \sqrt{y - x}$

3. $f(x, y) = 4x^2 + 9y^2$

4. $f(x, y) = x^2 - y^2$

5. $f(x, y) = xy$

6. $f(x, y) = y/x^2$

7. $f(x, y) = \dfrac{1}{\sqrt{16 - x^2 - y^2}}$

8. $f(x, y) = \sqrt{9 - x^2 - y^2}$

9. $f(x, y) = \ln(x^2 + y^2)$

10. $f(x, y) = e^{-(x^2+y^2)}$

11. $f(x, y) = \sin^{-1}(y - x)$

12. $f(x, y) = \tan^{-1}\left(\dfrac{y}{x}\right)$

Exercises 13–18 show level curves for the functions graphed in (a)–(f). Match each set of curves with the appropriate function.

13.

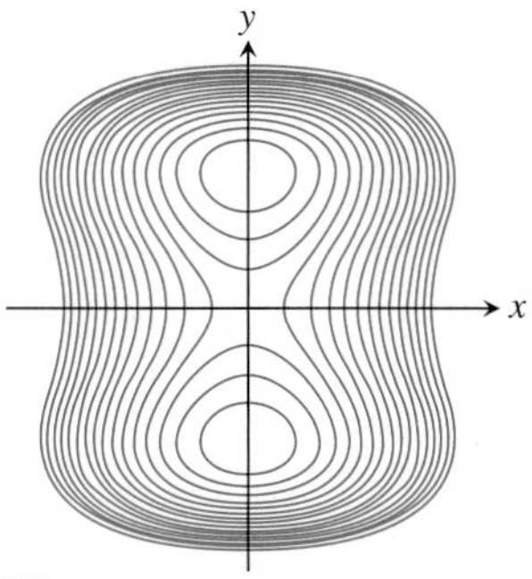

14.

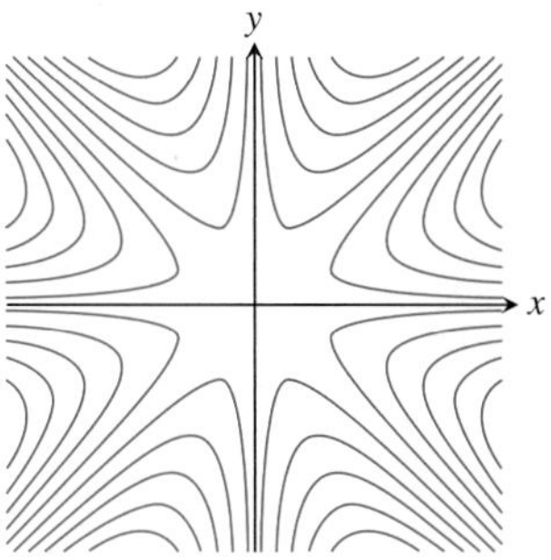

15.

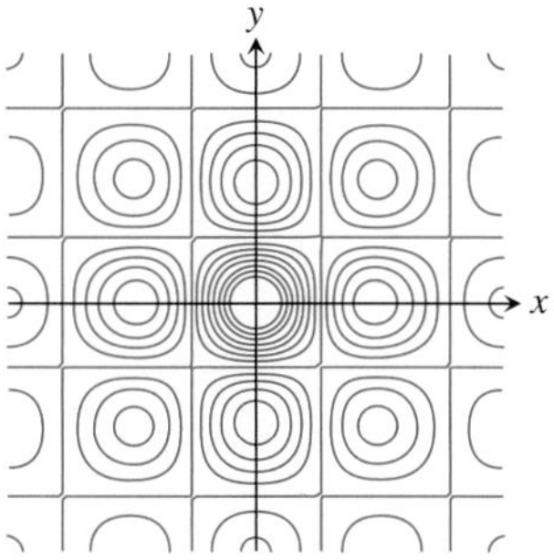

16.

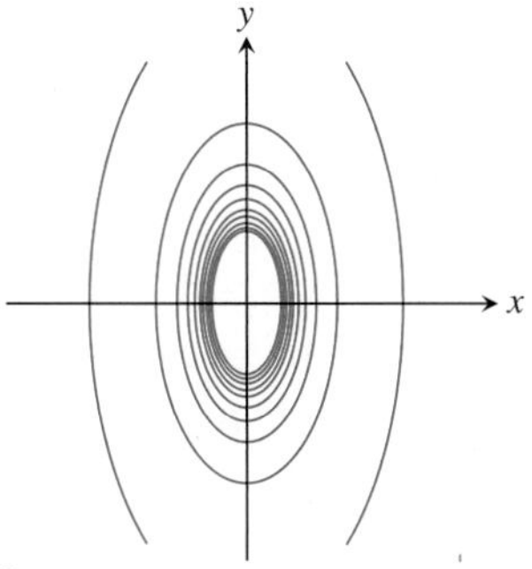

17.

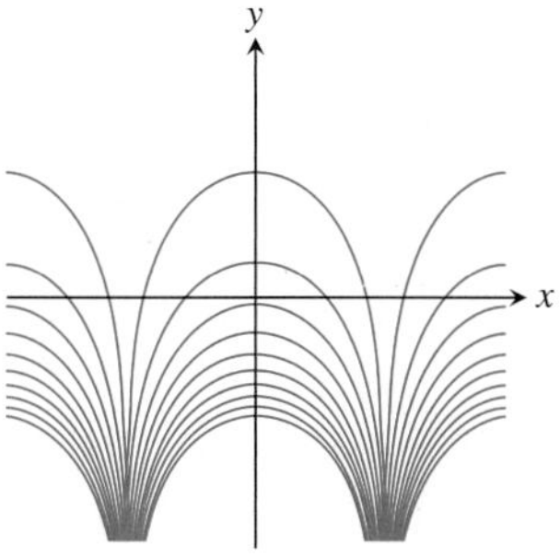

18.

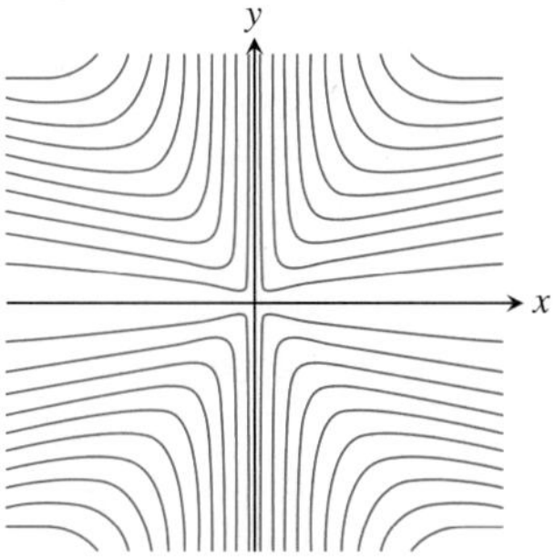

a.

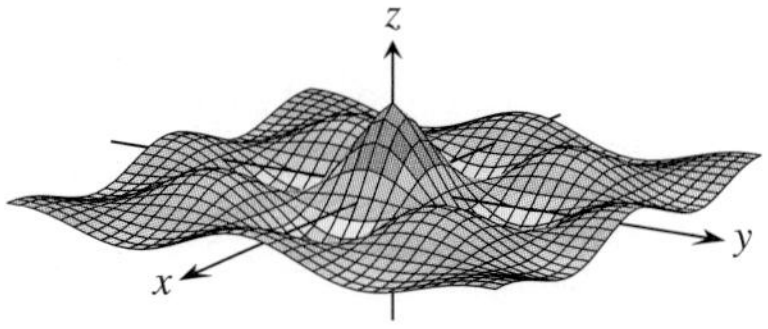

$z = (\cos x)(\cos y)\, e^{-\sqrt{x^2 + y^2}/4}$

b.

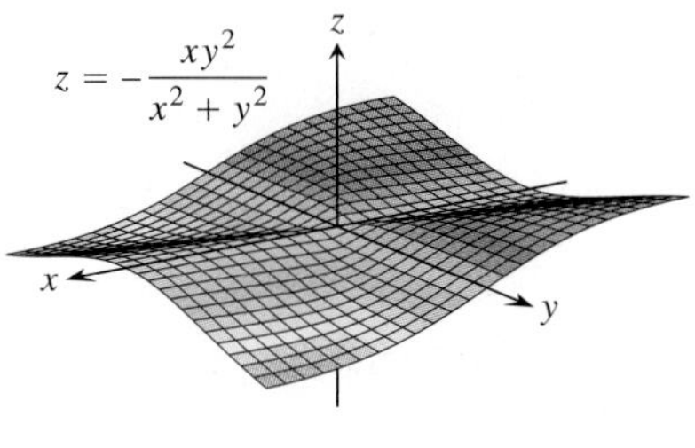

$z = -\dfrac{xy^2}{x^2 + y^2}$

c.

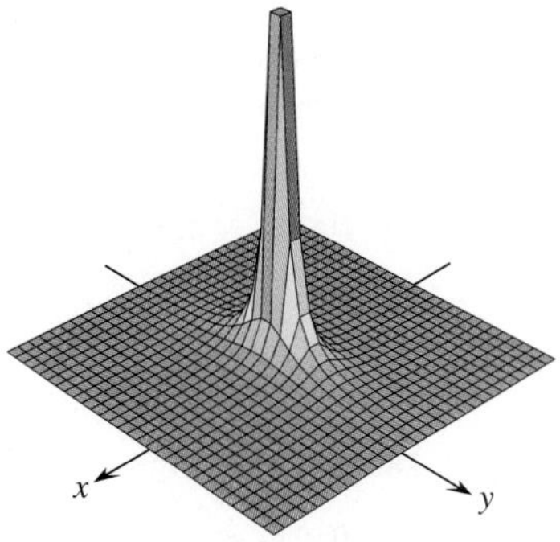

$z = \dfrac{1}{4x^2 + y^2}$

d.

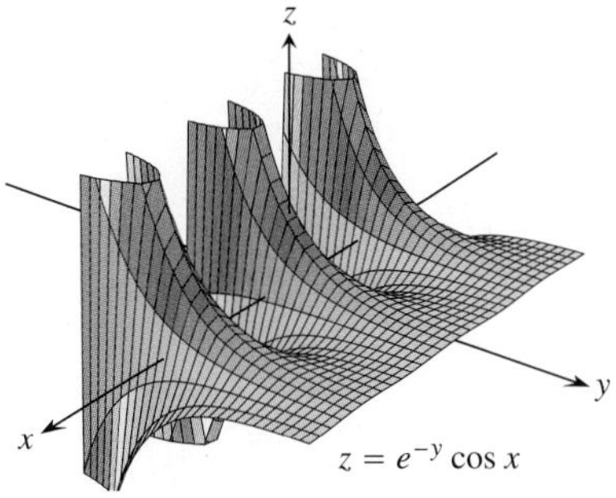

$z = e^{-y} \cos x$

e.

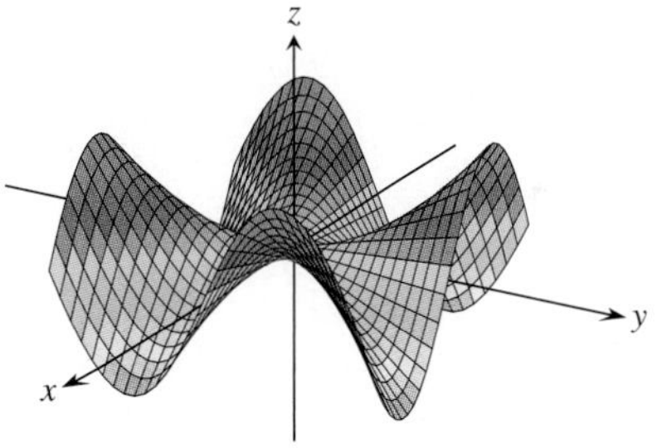

$z = \dfrac{xy(x^2 - y^2)}{x^2 + y^2}$

f.

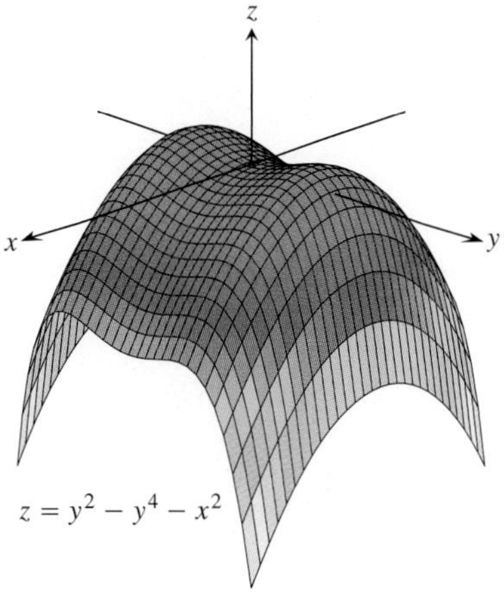

$z = y^2 - y^4 - x^2$

Display the values of the functions in Exercises 19–28 in two ways: **(a)** by sketching the surface $z = f(x, y)$ and **(b)** by drawing an assortment of level curves in the function's domain. Label each level curve with its function value.

19. $f(x, y) = y^2$

20. $f(x, y) = 4 - y^2$

21. $f(x, y) = x^2 + y^2$

22. $f(x, y) = \sqrt{x^2 + y^2}$

23. $f(x, y) = -(x^2 + y^2)$

24. $f(x, y) = 4 - x^2 - y^2$

25. $f(x, y) = 4x^2 + y^2$

26. $f(x, y) = 4x^2 + y^2 + 1$

27. $f(x, y) = 1 - |y|$

28. $f(x, y) = 1 - |x| - |y|$

In Exercises 29–32, find an equation for the level curve of the function $f(x, y)$ that passes through the given point.

29. $f(x, y) = 16 - x^2 - y^2, \quad \left(2\sqrt{2}, \sqrt{2}\right)$

30. $f(x, y) = \sqrt{x^2 - 1}, \quad (1, 0)$

31. $f(x, y) = \displaystyle\int_x^y \frac{dt}{1 + t^2}, \quad \left(-\sqrt{2}, \sqrt{2}\right)$

32. $f(x, y) = \displaystyle\sum_{n=0}^{\infty} \left(\frac{x}{y}\right)^n, \quad (1, 2)$

In Exercises 33–40, sketch a typical level surface for the function.

33. $f(x, y, z) = x^2 + y^2 + z^2$

34. $f(x, y, z) = \ln(x^2 + y^2 + z^2)$

35. $f(x, y, z) = x + z$

36. $f(x, y, z) = z$

37. $f(x, y, z) = x^2 + y^2$

38. $f(x, y, z) = y^2 + z^2$

39. $f(x, y, z) = z - x^2 - y^2$

40. $f(x, y, z) = (x^2/25) + (y^2/16) + (z^2/9)$

In Exercises 41–44, find an equation for the level surface of the function through the given point.

41. $f(x, y, z) = \sqrt{x - y} - \ln z, \quad (3, -1, 1)$

42. $f(x, y, z) = \ln(x^2 + y + z^2), \quad (-1, 2, 1)$

43. $g(x, y, z) = \displaystyle\sum_{n=0}^{\infty} \frac{(x + y)^n}{n!z^n}, \quad (\ln 2, \ln 4, 3)$

44. $g(x, y, z) = \displaystyle\int_x^y \frac{d\theta}{\sqrt{1 - \theta^2}} + \int_{\sqrt{2}}^z \frac{dt}{t\sqrt{t^2 - 1}}, \quad (0, 1/2, 2)$

COMPUTER EXPLORATIONS

Use a CAS to perform the following steps for each of the functions in Exercises 45–48.

a. Plot the surface over the given rectangle.

b. Plot several level curves in the rectangle.

c. Plot the level curve of f through the given point.

45. $f(x, y) = x \sin\dfrac{y}{2} + y \sin 2x, \quad 0 \le x \le 5\pi \quad 0 \le y \le 5\pi,$
$P(3\pi, 3\pi)$

46. $f(x, y) = (\sin x)(\cos y)e^{\sqrt{x^2+y^2}/8}, \quad 0 \le x \le 5\pi,$
$0 \le y \le 5\pi, \quad P(4\pi, 4\pi)$

47. $f(x, y) = \sin(x + 2\cos y), \quad -2\pi \le x \le 2\pi,$
$-2\pi \le y \le 2\pi, \quad P(\pi, \pi)$

48. $f(x, y) = e^{(x^{0.1} - y)} \sin(x^2 + y^2), \quad 0 \le x \le 2\pi,$
$-2\pi \le y \le \pi, \quad P(\pi, -\pi)$

Use a CAS to plot the implicitly defined level surfaces in Exercises 49–52.

49. $4 \ln(x^2 + y^2 + z^2) = 1$

50. $x^2 + z^2 = 1$

51. $x + y^2 - 3z^2 = 1$

52. $\sin\left(\dfrac{x}{2}\right) - (\cos y)\sqrt{x^2 + z^2} = 2$

Parametrized Surfaces Just as you describe curves in the plane parametrically with a pair of equations $x = f(t), y = g(t)$ defined on some parameter interval I, you can sometimes describe surfaces in space with a triple of equations $x = f(u, v), y = g(u, v), z = h(u, v)$ defined on some parameter rectangle $a \le u \le b, c \le v \le d$. Many computer algebra systems permit you to plot such surfaces in *parametric mode*. (Parametrized surfaces are discussed in detail in Section 13.5.) Use a CAS to plot the surfaces in Exercises 53–56. Also plot several level curves in the xy-plane.

53. $x = u\cos v, \quad y = u\sin v, \quad z = u, \quad 0 \le u \le 2,$
$0 \le v \le 2\pi$

54. $x = u\cos v, \quad y = u\sin v, \quad z = v, \quad 0 \le u \le 2,$
$0 \le v \le 2\pi$

55. $x = (2 + \cos u)\cos v, \quad y = (2 + \cos u)\sin v, \quad z = \sin u,$
$0 \le u \le 2\pi, \quad 0 \le v \le 2\pi$

56. $x = 2\cos u\cos v, \quad y = 2\cos u\sin v, \quad z = 2\sin u,$
$0 \le u \le 2\pi, \quad 0 \le v \le \pi$

11.2 Limits and Continuity in Higher Dimensions

This section treats limits and continuity for multivariable functions.

Limits

If the values of $f(x, y)$ lie arbitrarily close to a fixed real number L for all points (x, y) sufficiently close to a point (x_0, y_0), we say that f approaches the limit L as (x, y) approaches (x_0, y_0). This is similar to the informal definition for the limit of a function of a single variable. Notice, however, that if (x_0, y_0) lies in the interior of f's domain, (x, y) can approach (x_0, y_0) from any direction. The direction of approach can be an issue, as in some of the examples that follow.

DEFINITION We say that a function $f(x, y)$ approaches the **limit** L as (x, y) approaches (x_0, y_0), and write

$$\lim_{(x, y)\to(x_0, y_0)} f(x, y) = L$$

if, for every number $\epsilon > 0$, there exists a corresponding number $\delta > 0$ such that for all (x, y) in the domain of f,

$$|f(x, y) - L| < \epsilon \qquad \text{whenever} \qquad 0 < \sqrt{(x - x_0)^2 + (y - y_0)^2} < \delta.$$

The definition of limit says that the distance between $f(x, y)$ and L becomes arbitrarily small whenever the distance from (x, y) to (x_0, y_0) is made sufficiently small (but not 0). The definition applies to interior points (x_0, y_0) as well as boundary points of the domain of f, although a boundary point need not lie within the domain. However, the points (x, y) that approach (x_0, y_0) are always taken to be in the domain of f.

As for functions of a single variable, it can be shown that

$$\lim_{(x, y)\to(x_0, y_0)} x = x_0$$

$$\lim_{(x, y)\to(x_0, y_0)} y = y_0$$

$$\lim_{(x, y)\to(x_0, y_0)} k = k \qquad \text{(any number } k\text{)}.$$

For example, in the first limit statement above, $f(x, y) = x$ and $L = x_0$. Using the definition of limit, suppose that $\epsilon > 0$ is chosen. If we let δ equal this ϵ, we see that

$$0 < \sqrt{(x - x_0)^2 + (y - y_0)^2} < \delta = \epsilon$$

implies

$$\sqrt{(x - x_0)^2} < \epsilon \qquad (x - x_0)^2 \le (x - x_0)^2 + (y - y_0)^2$$

$$|x - x_0| < \epsilon \qquad \sqrt{a^2} = |a|$$

$$|f(x, y) - x_0| < \epsilon \qquad x = f(x, y)$$

That is,

$$|f(x, y) - x_0| < \epsilon \qquad \text{whenever} \qquad 0 < \sqrt{(x - x_0)^2 + (y - y_0)^2} < \delta.$$

So

$$\lim_{(x, y) \to (x_0, y_0)} f(x, y) = \lim_{(x, y) \to (x_0, y_0)} x = x_0.$$

As with single-variable functions, the limit of the sum of two functions is the sum of their limits (when they both exist), with similar results for the limits of the differences, products, constant multiples, quotients, powers, and roots.

THEOREM 1—Properties of Limits of Functions of Two Variables The following rules hold if L, M, and k are real numbers and

$$\lim_{(x, y) \to (x_0, y_0)} f(x, y) = L \qquad \text{and} \qquad \lim_{(x, y) \to (x_0, y_0)} g(x, y) = M.$$

1. *Sum Rule:* $\displaystyle\lim_{(x, y) \to (x_0, y_0)} (f(x, y) + g(x, y)) = L + M$
2. *Difference Rule:* $\displaystyle\lim_{(x, y) \to (x_0, y_0)} (f(x, y) - g(x, y)) = L - M$
3. *Product Rule:* $\displaystyle\lim_{(x, y) \to (x_0, y_0)} (f(x, y) \cdot g(x, y)) = L \cdot M$
4. *Constant Multiple Rule:* $\displaystyle\lim_{(x, y) \to (x_0, y_0)} (kf(x, y)) = kL$ (any number k)
5. *Quotient Rule:* $\displaystyle\lim_{(x, y) \to (x_0, y_0)} \frac{f(x, y)}{g(x, y)} = \frac{L}{M}$ $\quad M \neq 0$
6. *Power Rule:* $\displaystyle\lim_{(x, y) \to (x_0, y_0)} [f(x, y)]^n = L^n$, n a positive integer
7. *Root Rule:* $\displaystyle\lim_{(x, y) \to (x_0, y_0)} \sqrt[n]{f(x, y)} = \sqrt[n]{L} = L^{1/n}$, n a positive integer

   (If n is even, we assume that $L > 0$.)

While we won't prove Theorem 1 here, we give an informal discussion of why it's true. If (x, y) is sufficiently close to (x_0, y_0), then $f(x, y)$ is close to L and $g(x, y)$ is close to M (from the informal interpretation of limits). It is then reasonable that $f(x, y) + g(x, y)$ is close to $L + M$; $f(x, y) - g(x, y)$ is close to $L - M$; $f(x, y)g(x, y)$ is close to LM; $kf(x, y)$ is close to kL; and that $f(x, y)/g(x, y)$ is close to L/M if $M \neq 0$.

When we apply Theorem 1 to polynomials and rational functions, we obtain the useful result that the limits of these functions as $(x, y) \to (x_0, y_0)$ can be calculated by evaluating the functions at (x_0, y_0). The only requirement is that the rational functions be defined at (x_0, y_0).

EXAMPLE 1

(a) $$\lim_{(x, y) \to (0,1)} \frac{x - xy + 3}{x^2y + 5xy - y^3} = \frac{0 - (0)(1) + 3}{(0)^2(1) + 5(0)(1) - (1)^3} = -3$$

(b) $$\lim_{(x, y) \to (3, -4)} \sqrt{x^2 + y^2} = \sqrt{(3)^2 + (-4)^2} = \sqrt{25} = 5$$ ■

EXAMPLE 2 Find

$$\lim_{(x,y)\to(0,0)} \frac{x^2 - xy}{\sqrt{x} - \sqrt{y}}.$$

Solution Since the denominator $\sqrt{x} - \sqrt{y}$ approaches 0 as $(x, y) \to (0, 0)$, we cannot use the Quotient Rule from Theorem 1. If we multiply numerator and denominator by $\sqrt{x} + \sqrt{y}$, however, we produce an equivalent fraction whose limit we *can* find:

$$\begin{aligned}
\lim_{(x,y)\to(0,0)} \frac{x^2 - xy}{\sqrt{x} - \sqrt{y}} &= \lim_{(x,y)\to(0,0)} \frac{\left(x^2 - xy\right)\left(\sqrt{x} + \sqrt{y}\right)}{\left(\sqrt{x} - \sqrt{y}\right)\left(\sqrt{x} + \sqrt{y}\right)} \\
&= \lim_{(x,y)\to(0,0)} \frac{x(x - y)\left(\sqrt{x} + \sqrt{y}\right)}{x - y} && \text{Algebra} \\
&= \lim_{(x,y)\to(0,0)} x\left(\sqrt{x} + \sqrt{y}\right) && \text{Cancel the nonzero factor } (x - y). \\
&= 0\left(\sqrt{0} + \sqrt{0}\right) = 0
\end{aligned}$$

We can cancel the factor $(x - y)$ because the path $y = x$ (along which $x - y = 0$) is *not* in the domain of the function

$$\frac{x^2 - xy}{\sqrt{x} - \sqrt{y}}.$$

■

EXAMPLE 3 Find $\displaystyle\lim_{(x,y)\to(0,0)} \frac{4xy^2}{x^2 + y^2}$ if it exists.

Solution We first observe that along the line $x = 0$, the function always has value 0 when $y \neq 0$. Likewise, along the line $y = 0$, the function has value 0 provided $x \neq 0$. So if the limit does exist as (x, y) approaches $(0, 0)$, the value of the limit must be 0. To see if this is true, we apply the definition of limit.

Let $\epsilon > 0$ be given, but arbitrary. We want to find a $\delta > 0$ such that

$$\left|\frac{4xy^2}{x^2 + y^2} - 0\right| < \epsilon \quad \text{whenever} \quad 0 < \sqrt{x^2 + y^2} < \delta$$

or

$$\frac{4|x|y^2}{x^2 + y^2} < \epsilon \quad \text{whenever} \quad 0 < \sqrt{x^2 + y^2} < \delta.$$

Since $y^2 \leq x^2 + y^2$ we have that

$$\frac{4|x|y^2}{x^2 + y^2} \leq 4|x| = 4\sqrt{x^2} \leq 4\sqrt{x^2 + y^2}. \qquad \frac{y^2}{x^2 + y^2} \leq 1$$

So if we choose $\delta = \epsilon/4$ and let $0 < \sqrt{x^2 + y^2} < \delta$, we get

$$\left| \frac{4xy^2}{x^2 + y^2} - 0 \right| \leq 4\sqrt{x^2 + y^2} < 4\delta = 4\left(\frac{\epsilon}{4}\right) = \epsilon.$$

It follows from the definition that

$$\lim_{(x, y) \to (0,0)} \frac{4xy^2}{x^2 + y^2} = 0. \qquad \blacksquare$$

Continuity

As with functions of a single variable, continuity is defined in terms of limits.

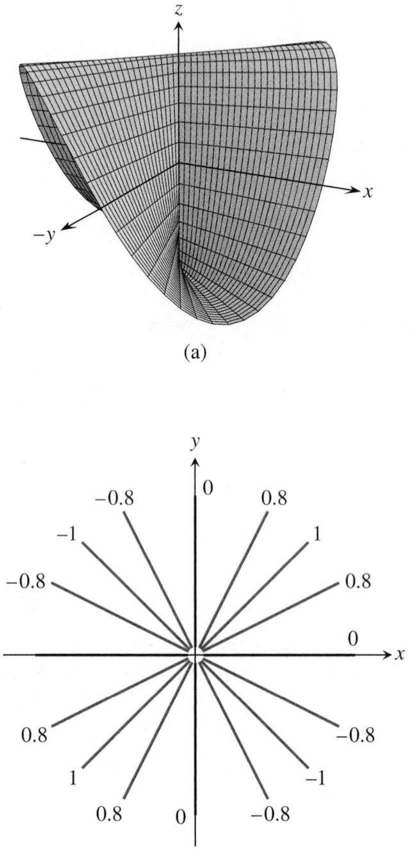

FIGURE 11.11 (a) The graph of

$$f(x, y) = \begin{cases} \dfrac{2xy}{x^2 + y^2}, & (x, y) \neq (0, 0) \\ 0, & (x, y) = (0, 0). \end{cases}$$

The function is continuous at every point except the origin. (b) The level curves of f (Example 4).

> **DEFINITION** A function $f(x, y)$ is **continuous at the point (x_0, y_0)** if
>
> 1. f is defined at (x_0, y_0),
> 2. $\lim_{(x, y) \to (x_0, y_0)} f(x, y)$ exists,
> 3. $\lim_{(x, y) \to (x_0, y_0)} f(x, y) = f(x_0, y_0)$.
>
> A function is **continuous** if it is continuous at every point of its domain.

As with the definition of limit, the definition of continuity applies at boundary points as well as interior points of the domain of f. The only requirement is that the point (x, y) be in the domain of f.

A consequence of Theorem 1 is that algebraic combinations of continuous functions are continuous at every point at which all the functions involved are defined. This means that sums, differences, products, constant multiples, quotients, and powers of continuous functions are continuous where defined. In particular, polynomials and rational functions of two variables are continuous at every point at which they are defined.

EXAMPLE 4 Show that

$$f(x, y) = \begin{cases} \dfrac{2xy}{x^2 + y^2}, & (x, y) \neq (0, 0) \\ 0, & (x, y) = (0, 0) \end{cases}$$

is continuous at every point except the origin (Figure 11.11).

Solution The function f is continuous at any point $(x, y) \neq (0, 0)$ because its values are then given by a rational function of x and y.

At $(0, 0)$, the value of f is defined, but f, we claim, has no limit as $(x, y) \to (0, 0)$. The reason is that different paths of approach to the origin can lead to different results, as we now see.

For every value of m, the function f has a constant value on the "punctured" line $y = mx, x \neq 0$, because

$$f(x, y)\Big|_{y=mx} = \frac{2xy}{x^2 + y^2}\Big|_{y=mx} = \frac{2x(mx)}{x^2 + (mx)^2} = \frac{2mx^2}{x^2 + m^2x^2} = \frac{2m}{1 + m^2}.$$

Therefore, f has this number as its limit as (x, y) approaches $(0, 0)$ along the line:

$$\lim_{\substack{(x, y)\to(0,0) \\ \text{along } y=mx}} f(x, y) = \lim_{(x, y)\to(0,0)} \left[f(x, y)\Big|_{y=mx} \right] = \frac{2m}{1 + m^2}.$$

This limit changes with m. There is therefore no single number we may call the limit of f as (x, y) approaches the origin. The limit fails to exist, and the function is not continuous. ■

Example 4 illustrates an important point about limits of functions of two variables (or even more variables, for that matter). For a limit to exist at a point, the limit must be the same along every approach path. This result is analogous to the single-variable case where both the left- and right-sided limits had to have the same value. For functions of two or more variables, if we ever find paths with different limits, we know the function has no limit at the point they approach.

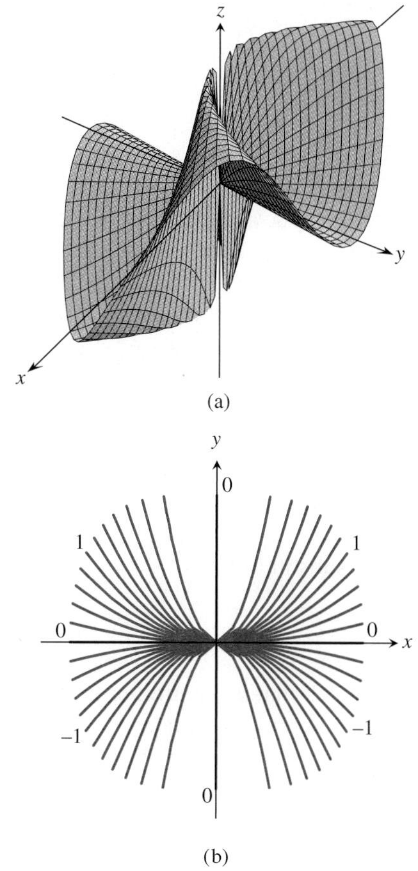

FIGURE 11.12 (a) The graph of $f(x, y) = 2x^2y/(x^4 + y^2)$. As the graph suggests and the level-curve values in part (b) confirm, $\lim_{(x, y)\to(0,0)} f(x, y)$ does not exist (Example 5).

Two-Path Test for Nonexistence of a Limit
If a function $f(x, y)$ has different limits along two different paths in the domain of f as (x, y) approaches (x_0, y_0), then $\lim_{(x, y)\to(x_0, y_0)} f(x, y)$ does not exist.

EXAMPLE 5 Show that the function

$$f(x, y) = \frac{2x^2y}{x^4 + y^2}$$

(Figure 11.12) has no limit as (x, y) approaches $(0, 0)$.

Solution The limit cannot be found by direct substitution, which gives the indeterminate form $0/0$. We examine the values of f along curves that end at $(0, 0)$. Along the curve $y = kx^2, x \neq 0$, the function has the constant value

$$f(x, y)\Big|_{y=kx^2} = \frac{2x^2y}{x^4 + y^2}\Big|_{y=kx^2} = \frac{2x^2(kx^2)}{x^4 + (kx^2)^2} = \frac{2kx^4}{x^4 + k^2x^4} = \frac{2k}{1 + k^2}.$$

Therefore,

$$\lim_{\substack{(x, y)\to(0,0) \\ \text{along } y=kx^2}} f(x, y) = \lim_{(x, y)\to(0,0)} \left[f(x, y)\Big|_{y=kx^2} \right] = \frac{2k}{1 + k^2}.$$

This limit varies with the path of approach. If (x, y) approaches $(0, 0)$ along the parabola $y = x^2$, for instance, $k = 1$ and the limit is 1. If (x, y) approaches $(0, 0)$ along the x-axis, $k = 0$ and the limit is 0. By the two-path test, f has no limit as (x, y) approaches $(0, 0)$. ■

It can be shown that the function in Example 5 has limit 0 along every path $y = mx$ (Exercise 43). Having the same limit along all straight lines approaching (x_0, y_0) does not imply a limit exists at (x_0, y_0).

Compositions of continuous functions are also continuous. The proof, omitted here, is similar to that for functions of a single variable (Theorem 9 in Section 1.7).

Continuity of Composites
If f is continuous at (x_0, y_0) and g is a single-variable function continuous at $f(x_0, y_0)$, then the composite function $h = g \circ f$ defined by $h(x, y) = g(f(x, y))$ is continuous at (x_0, y_0).

For example, the composite functions

$$e^{x-y}, \qquad \cos\frac{xy}{x^2+1}, \qquad \ln(1 + x^2y^2)$$

are continuous at every point (x, y).

As with functions of a single variable, the general rule is that composites of continuous functions are continuous. The only requirement is that each function be continuous where it is applied.

Functions of More Than Two Variables

The definitions of limit and continuity for functions of two variables and the conclusions about limits and continuity for sums, products, quotients, powers, and composites all extend to functions of three or more variables. Functions like

$$\ln(x + y + z) \qquad \text{and} \qquad \frac{y \sin z}{x - 1}$$

are continuous throughout their domains, and limits like

$$\lim_{P \to (1,0,-1)} \frac{e^{x+z}}{z^2 + \cos\sqrt{xy}} = \frac{e^{1-1}}{(-1)^2 + \cos 0} = \frac{1}{2},$$

where P denotes the point (x, y, z), may be found by direct substitution.

Extreme Values of Continuous Functions on Closed, Bounded Sets

The Extreme Value Theorem (Theorem 1, Section 3.1) states that a function of a single variable that is continuous throughout a closed, bounded interval $[a, b]$ takes on an absolute maximum value and an absolute minimum value at least once in $[a, b]$. The same is true of a function $z = f(x, y)$ that is continuous on a closed, bounded set R in the plane (like a line segment, a disk, or a filled-in triangle). The function takes on an absolute maximum value at some point in R and an absolute minimum value at some point in R.

Similar results hold for functions of three or more variables. A continuous function $w = f(x, y, z)$, for example, must take on absolute maximum and minimum values on any closed, bounded set (solid ball or cube, spherical shell, rectangular solid) on which it is defined. We will learn how to find these extreme values in Section 11.7.

EXERCISES 11.2

Find the limits in Exercises 1–12.

1. $\lim_{(x,y)\to(0,0)} \dfrac{3x^2 - y^2 + 5}{x^2 + y^2 + 2}$

2. $\lim_{(x,y)\to(0,4)} \dfrac{x}{\sqrt{y}}$

3. $\lim_{(x,y)\to(3,4)} \sqrt{x^2 + y^2 - 1}$

4. $\lim_{(x,y)\to(2,-3)} \left(\dfrac{1}{x} + \dfrac{1}{y}\right)^2$

5. $\lim_{(x,y)\to(0,\pi/4)} \sec x \tan y$

6. $\lim_{(x,y)\to(0,0)} \cos \dfrac{x^2 + y^3}{x + y + 1}$

7. $\lim_{(x,y)\to(0,\ln 2)} e^{x-y}$

8. $\lim_{(x,y)\to(1,1)} \ln |1 + x^2 y^2|$

9. $\lim_{(x,y)\to(0,0)} \dfrac{e^y \sin x}{x}$

10. $\lim_{(x,y)\to(1,1)} \cos \sqrt[3]{|xy| - 1}$

11. $\lim_{(x,y)\to(1,0)} \dfrac{x \sin y}{x^2 + 1}$

12. $\lim_{(x,y)\to(\pi/2,0)} \dfrac{\cos y + 1}{y - \sin x}$

Find the limits in Exercises 13–20 by rewriting the fractions first.

13. $\lim_{\substack{(x,y)\to(1,1)\\ x\neq y}} \dfrac{x^2 - 2xy + y^2}{x - y}$

14. $\lim_{\substack{(x,y)\to(1,1)\\ x\neq y}} \dfrac{x^2 - y^2}{x - y}$

15. $\lim_{\substack{(x,y)\to(1,1)\\ x\neq 1}} \dfrac{xy - y - 2x + 2}{x - 1}$

16. $\lim_{\substack{(x,y)\to(2,-4)\\ y\neq -4,\, x\neq x^2}} \dfrac{y + 4}{x^2 y - xy + 4x^2 - 4x}$

17. $\lim_{\substack{(x,y)\to(0,0)\\ x\neq y}} \dfrac{x - y + 2\sqrt{x} - 2\sqrt{y}}{\sqrt{x} - \sqrt{y}}$

18. $\lim_{\substack{(x,y)\to(2,2)\\ x+y\neq 4}} \dfrac{x + y - 4}{\sqrt{x + y} - 2}$

19. $\lim_{\substack{(x,y)\to(2,0)\\ 2x-y\neq 4}} \dfrac{\sqrt{2x - y} - 2}{2x - y - 4}$

20. $\lim_{\substack{(x,y)\to(4,3)\\ x\neq y+1}} \dfrac{\sqrt{x} - \sqrt{y + 1}}{x - y - 1}$

Find the limits in Exercises 21–26.

21. $\lim_{P\to(1,3,4)} \left(\dfrac{1}{x} + \dfrac{1}{y} + \dfrac{1}{z}\right)$

22. $\lim_{P\to(1,-1,-1)} \dfrac{2xy + yz}{x^2 + z^2}$

23. $\lim_{P\to(3,3,0)} (\sin^2 x + \cos^2 y + \sec^2 z)$

24. $\lim_{P\to(-1/4,\pi/2,2)} \tan^{-1} xyz$

25. $\lim_{P\to(\pi,0,3)} ze^{-2y} \cos 2x$

26. $\lim_{P\to(0,-2,0)} \ln \sqrt{x^2 + y^2 + z^2}$

At what points (x, y) in the plane are the functions in Exercises 27–30 continuous?

27. **a.** $f(x, y) = \sin (x + y)$ **b.** $f(x, y) = \ln (x^2 + y^2)$

28. **a.** $f(x, y) = \dfrac{x + y}{x - y}$ **b.** $f(x, y) = \dfrac{y}{x^2 + 1}$

29. **a.** $g(x, y) = \sin \dfrac{1}{xy}$ **b.** $g(x, y) = \dfrac{x + y}{2 + \cos x}$

30. **a.** $g(x, y) = \dfrac{x^2 + y^2}{x^2 - 3x + 2}$ **b.** $g(x, y) = \dfrac{1}{x^2 - y}$

At what points (x, y, z) in space are the functions in Exercises 31–34 continuous?

31. **a.** $f(x, y, z) = x^2 + y^2 - 2z^2$
b. $f(x, y, z) = \sqrt{x^2 + y^2 - 1}$

32. **a.** $f(x, y, z) = \ln xyz$ **b.** $f(x, y, z) = e^{x+y} \cos z$

33. **a.** $h(x, y, z) = xy \sin \dfrac{1}{z}$ **b.** $h(x, y, z) = \dfrac{1}{x^2 + z^2 - 1}$

34. **a.** $h(x, y, z) = \dfrac{1}{|y| + |z|}$ **b.** $h(x, y, z) = \dfrac{1}{|xy| + |z|}$

By considering different paths of approach, show that the functions in Exercises 35–42 have no limit as $(x, y) \to (0, 0)$.

35. $f(x, y) = -\dfrac{x}{\sqrt{x^2 + y^2}}$

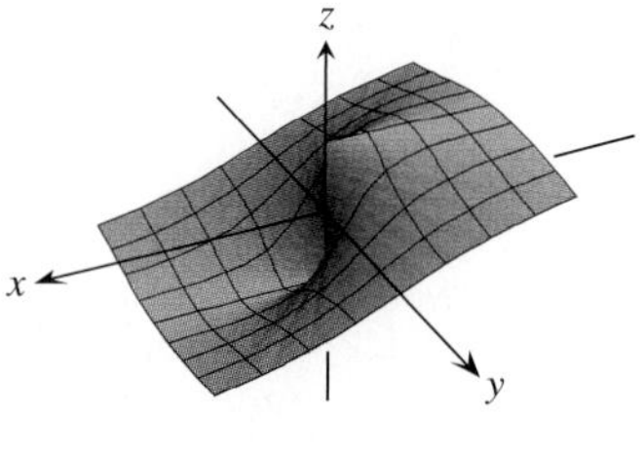

36. $f(x, y) = \dfrac{x^4}{x^4 + y^2}$

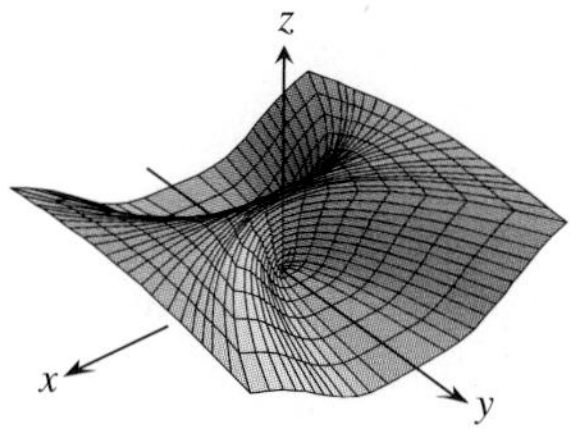

37. $f(x, y) = \dfrac{x^4 - y^2}{x^4 + y^2}$

38. $f(x, y) = \dfrac{xy}{|xy|}$

39. $g(x, y) = \dfrac{x - y}{x + y}$

40. $g(x, y) = \dfrac{x + y}{x - y}$

41. $h(x, y) = \dfrac{x^2 + y}{y}$

42. $h(x, y) = \dfrac{x^2}{x^2 - y}$

43. Show that the function in Example 5 has limit 0 along every straight line approaching (0, 0).

44. If $f(x_0, y_0) = 3$, what can you say about

$$\lim_{(x,y)\to(x_0,y_0)} f(x, y)$$

if f is continuous at (x_0, y_0)? If f is not continuous at (x_0, y_0)? Give reasons for your answers.

The Sandwich Theorem For functions of two variables states that if $g(x, y) \le f(x, y) \le h(x, y)$ for all $(x, y) \ne (x_0, y_0)$ in a disk centered at (x_0, y_0) and if g and h have the same finite limit L as $(x, y) \to (x_0, y_0)$, then

$$\lim_{(x,y)\to(x_0,y_0)} f(x, y) = L.$$

Use this result to support your answers to the questions in Exercises 45–48.

45. Does knowing that

$$1 - \frac{x^2y^2}{3} < \frac{\tan^{-1} xy}{xy} < 1$$

tell you anything about

$$\lim_{(x,y)\to(0,0)} \frac{\tan^{-1} xy}{xy}?$$

Give reasons for your answer.

46. Does knowing that

$$2|xy| - \frac{x^2y^2}{6} < 4 - 4\cos\sqrt{|xy|} < 2|xy|$$

tell you anything about

$$\lim_{(x,y)\to(0,0)} \frac{4 - 4\cos\sqrt{|xy|}}{|xy|}?$$

Give reasons for your answer.

47. Does knowing that $|\sin(1/x)| \le 1$ tell you anything about

$$\lim_{(x,y)\to(0,0)} y \sin\frac{1}{x}?$$

Give reasons for your answer.

48. Does knowing that $|\cos(1/y)| \le 1$ tell you anything about

$$\lim_{(x,y)\to(0,0)} x \cos\frac{1}{y}?$$

Give reasons for your answer.

49. (*Continuation of Example 4.*)

a. Reread Example 4. Then substitute $m = \tan\theta$ into the formula

$$f(x, y)\bigg|_{y=mx} = \frac{2m}{1 + m^2}$$

and simplify the result to show how the value of f varies with the line's angle of inclination.

b. Use the formula you obtained in part (a) to show that the limit of f as $(x, y) \to (0, 0)$ along the line $y = mx$ varies from -1 to 1 depending on the angle of approach.

50. Continuous extension Define $f(0, 0)$ in a way that *extends*

$$f(x, y) = xy\frac{x^2 - y^2}{x^2 + y^2}$$

to be continuous at the origin.

Changing to Polar Coordinates If you cannot make any headway with $\lim_{(x,y)\to(0,0)} f(x, y)$ in rectangular coordinates, try changing to polar coordinates. Substitute $x = r\cos\theta$, $y = r\sin\theta$, and investigate the limit of the resulting expression as $r \to 0$.

In Exercises 51–56, use polar coordinates to find the limit of f as $(x, y) \to (0, 0)$ or show that the limit does not exist.

51. $f(x, y) = \dfrac{x^3 - xy^2}{x^2 + y^2}$

52. $f(x, y) = \cos\left(\dfrac{x^3 - y^3}{x^2 + y^2}\right)$

53. $f(x, y) = \dfrac{y^2}{x^2 + y^2}$

54. $f(x, y) = \dfrac{2x}{x^2 + x + y^2}$

55. $f(x, y) = \tan^{-1}\left(\dfrac{|x| + |y|}{x^2 + y^2}\right)$

56. $f(x, y) = \dfrac{x^2 - y^2}{x^2 + y^2}$

Each of Exercises 57–60 gives a function $f(x, y)$ and a positive number ϵ. In each exercise, show that there exists a $\delta > 0$ such that for all (x, y),

$$\sqrt{x^2 + y^2} < \delta \quad \Rightarrow \quad |f(x, y) - f(0, 0)| < \epsilon.$$

57. $f(x, y) = x^2 + y^2$, $\epsilon = 0.01$

58. $f(x, y) = y/(x^2 + 1)$, $\epsilon = 0.05$

59. $f(x, y) = (x + y)/(x^2 + 1)$, $\epsilon = 0.01$

60. $f(x, y) = (x + y)/(2 + \cos x)$, $\epsilon = 0.02$

Each of Exercises 61–64 gives a function $f(x, y, z)$ and a positive number ϵ. In each exercise, show that there exists a $\delta > 0$ such that for all (x, y, z),

$$\sqrt{x^2 + y^2 + z^2} < \delta \quad \Rightarrow \quad |f(x, y, z) - f(0, 0, 0)| < \epsilon.$$

61. $f(x, y, z) = x^2 + y^2 + z^2$, $\epsilon = 0.015$

62. $f(x, y, z) = xyz$, $\epsilon = 0.008$

63. $f(x, y, z) = \dfrac{x + y + z}{x^2 + y^2 + z^2 + 1}$, $\epsilon = 0.015$

64. $f(x, y, z) = \tan^2 x + \tan^2 y + \tan^2 z$, $\epsilon = 0.03$

65. Show that $f(x, y, z) = x + y - z$ is continuous at every point (x_0, y_0, z_0).

66. Show that $f(x, y, z) = x^2 + y^2 + z^2$ is continuous at the origin.

11.3 Partial Derivatives

The calculus of several variables is basically single-variable calculus applied to several variables one at a time. When we hold all but one of the independent variables of a function constant and differentiate with respect to that one variable, we get a "partial" derivative. This section shows how partial derivatives are defined and interpreted geometrically, and how to calculate them by applying the rules for differentiating functions of a single variable.

Partial Derivatives of a Function of Two Variables

If (x_0, y_0) is a point in the domain of a function $f(x, y)$, the vertical plane $y = y_0$ will cut the surface $z = f(x, y)$ in the curve $z = f(x, y_0)$ (Figure 11.13). This curve is the graph of the function $z = f(x, y_0)$ in the plane $y = y_0$. The horizontal coordinate in this plane is x; the vertical coordinate is z. The y-value is held constant at y_0, so y is not a variable.

We define the partial derivative of f with respect to x at the point (x_0, y_0) as the ordinary derivative of $f(x, y_0)$ with respect to x at the point $x = x_0$. To distinguish partial derivatives from ordinary derivatives we use the symbol ∂ rather than the d previously used.

DEFINITION The **partial derivative of $f(x, y)$ with respect to x** at the point (x_0, y_0) is

$$\left.\frac{\partial f}{\partial x}\right|_{(x_0, y_0)} = \lim_{h \to 0} \frac{f(x_0 + h, y_0) - f(x_0, y_0)}{h},$$

provided the limit exists.

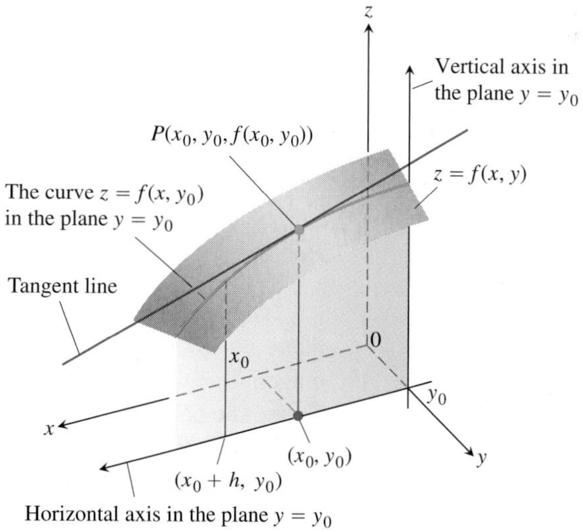

FIGURE 11.13 The intersection of the plane $y = y_0$ with the surface $z = f(x, y)$, viewed from above the first quadrant of the xy-plane.

An equivalent expression for the partial derivative is

$$\frac{d}{dx} f(x, y_0)\bigg|_{x=x_0}.$$

The slope of the curve $z = f(x, y_0)$ at the point $P(x_0, y_0, f(x_0, y_0))$ in the plane $y = y_0$ is the value of the partial derivative of f with respect to x at (x_0, y_0). (In Figure 11.13 this slope is negative.) The tangent line to the curve at P is the line in the plane $y = y_0$ that passes through P with this slope. The partial derivative $\partial f/\partial x$ at (x_0, y_0) gives the rate of change of f with respect to x when y is held fixed at the value y_0.

The notation for a partial derivative depends on what we want to emphasize:

$\dfrac{\partial f}{\partial x}(x_0, y_0)$ or $f_x(x_0, y_0)$ — "Partial derivative of f with respect to x at (x_0, y_0)" or "f sub x at (x_0, y_0)." Convenient for stressing the point (x_0, y_0).

$\dfrac{\partial z}{\partial x}\bigg|_{(x_0, y_0)}$ — "Partial derivative of z with respect to x at (x_0, y_0)." Common in science and engineering when you are dealing with variables and do not mention the function explicitly.

$f_x, \dfrac{\partial f}{\partial x}, z_x$, or $\dfrac{\partial z}{\partial x}$ — "Partial derivative of f (or z) with respect to x." Convenient when you regard the partial derivative as a function in its own right.

The definition of the partial derivative of $f(x, y)$ with respect to y at a point (x_0, y_0) is similar to the definition of the partial derivative of f with respect to x. We hold x fixed at the value x_0 and take the ordinary derivative of $f(x_0, y)$ with respect to y at y_0.

DEFINITION The **partial derivative of $f(x, y)$ with respect to y** at the point (x_0, y_0) is

$$\frac{\partial f}{\partial y}\bigg|_{(x_0, y_0)} = \frac{d}{dy} f(x_0, y)\bigg|_{y=y_0} = \lim_{h \to 0} \frac{f(x_0, y_0 + h) - f(x_0, y_0)}{h},$$

provided the limit exists.

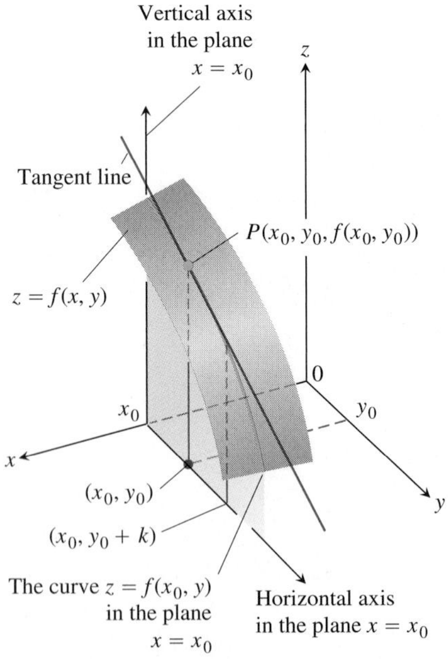

FIGURE 11.14 The intersection of the plane $x = x_0$ with the surface $z = f(x, y)$, viewed from above the first quadrant of the xy-plane.

The slope of the curve $z = f(x_0, y)$ at the point $P(x_0, y_0, f(x_0, y_0))$ in the vertical plane $x = x_0$ (Figure 11.14) is the partial derivative of f with respect to y at (x_0, y_0). The tangent line to the curve at P is the line in the plane $x = x_0$ that passes through P with this slope. The partial derivative gives the rate of change of f with respect to y at (x_0, y_0) when x is held fixed at the value x_0.

The partial derivative with respect to y is denoted the same way as the partial derivative with respect to x:

$$\frac{\partial f}{\partial y}(x_0, y_0), \qquad f_y(x_0, y_0), \qquad \frac{\partial f}{\partial y}, \qquad f_y.$$

Notice that we now have two tangent lines associated with the surface $z = f(x, y)$ at the point $P(x_0, y_0, f(x_0, y_0))$ (Figure 11.15). Is the plane they determine tangent to the

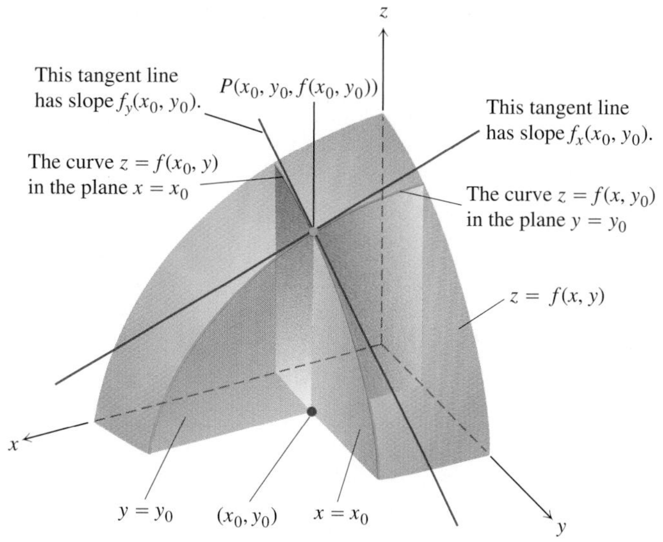

FIGURE 11.15 Figures 11.13 and 11.14 combined. The tangent lines at the point $(x_0, y_0, f(x_0, y_0))$ determine a plane that, in this picture at least, appears to be tangent to the surface.

surface at P? We will see that it is, but we have to learn more about partial derivatives before we can find out why.

Calculations

The definitions of $\partial f/\partial x$ and $\partial f/\partial y$ give us two different ways of differentiating f at a point: with respect to x in the usual way while treating y as a constant and with respect to y in the usual way while treating x as constant. As the following examples show, the values of these partial derivatives are usually different at a given point (x_0, y_0).

EXAMPLE 1 Find the values of $\partial f/\partial x$ and $\partial f/\partial y$ at the point $(4, -5)$ if

$$f(x, y) = x^2 + 3xy + y - 1.$$

Solution To find $\partial f/\partial x$, we treat y as a constant and differentiate with respect to x:

$$\frac{\partial f}{\partial x} = \frac{\partial}{\partial x}(x^2 + 3xy + y - 1) = 2x + 3 \cdot 1 \cdot y + 0 - 0 = 2x + 3y.$$

The value of $\partial f/\partial x$ at $(4, -5)$ is $2(4) + 3(-5) = -7$.

To find $\partial f/\partial y$, we treat x as a constant and differentiate with respect to y:

$$\frac{\partial f}{\partial y} = \frac{\partial}{\partial y}(x^2 + 3xy + y - 1) = 0 + 3 \cdot x \cdot 1 + 1 - 0 = 3x + 1.$$

The value of $\partial f/\partial y$ at $(4, -5)$ is $3(4) + 1 = 13$. ■

EXAMPLE 2 Find $\partial f/\partial y$ as a function if $f(x, y) = y \sin xy$.

Solution We treat x as a constant and f as a product of y and $\sin xy$:

$$\frac{\partial f}{\partial y} = \frac{\partial}{\partial y}(y \sin xy) = y\frac{\partial}{\partial y}\sin xy + (\sin xy)\frac{\partial}{\partial y}(y)$$

$$= (y \cos xy)\frac{\partial}{\partial y}(xy) + \sin xy = xy \cos xy + \sin xy.$$

■

EXAMPLE 3 Find f_x and f_y as functions if

$$f(x, y) = \frac{2y}{y + \cos x}.$$

Solution We treat f as a quotient. With y held constant, we get

$$f_x = \frac{\partial}{\partial x}\left(\frac{2y}{y + \cos x}\right) = \frac{(y + \cos x)\frac{\partial}{\partial x}(2y) - 2y\frac{\partial}{\partial x}(y + \cos x)}{(y + \cos x)^2}$$

$$= \frac{(y + \cos x)(0) - 2y(-\sin x)}{(y + \cos x)^2} = \frac{2y \sin x}{(y + \cos x)^2}.$$

With x held constant, we get

$$f_y = \frac{\partial}{\partial y}\left(\frac{2y}{y + \cos x}\right) = \frac{(y + \cos x)\frac{\partial}{\partial y}(2y) - 2y\frac{\partial}{dy}(y + \cos x)}{(y + \cos x)^2}$$

$$= \frac{(y + \cos x)(2) - 2y(1)}{(y + \cos x)^2} = \frac{2 \cos x}{(y + \cos x)^2}.$$

■

Implicit differentiation works for partial derivatives the way it works for ordinary derivatives, as the next example illustrates.

EXAMPLE 4 Find $\partial z/\partial x$ if the equation

$$yz - \ln z = x + y$$

defines z as a function of the two independent variables x and y and the partial derivative exists.

Solution We differentiate both sides of the equation with respect to x, holding y constant and treating z as a differentiable function of x:

$$\frac{\partial}{\partial x}(yz) - \frac{\partial}{\partial x}\ln z = \frac{\partial x}{\partial x} + \frac{\partial y}{\partial x}$$

$$y\frac{\partial z}{\partial x} - \frac{1}{z}\frac{\partial z}{\partial x} = 1 + 0$$

With y constant, $\frac{\partial}{\partial x}(yz) = y\frac{\partial z}{\partial x}$.

$$\left(y - \frac{1}{z}\right)\frac{\partial z}{\partial x} = 1$$

$$\frac{\partial z}{\partial x} = \frac{z}{yz - 1}.$$

■

EXAMPLE 5 The plane $x = 1$ intersects the paraboloid $z = x^2 + y^2$ in a parabola. Find the slope of the tangent to the parabola at (1, 2, 5) (Figure 11.16).

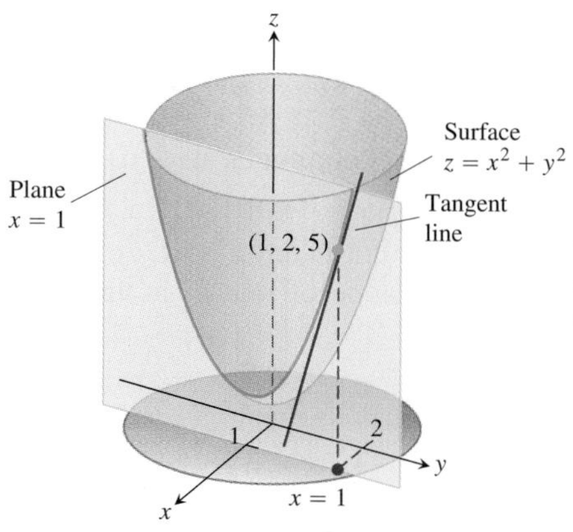

FIGURE 11.16 The tangent to the curve of intersection of the plane $x = 1$ and surface $z = x^2 + y^2$ at the point (1, 2, 5) (Example 5).

Solution The slope is the value of the partial derivative $\partial z/\partial y$ at (1, 2):

$$\left.\frac{\partial z}{\partial y}\right|_{(1,2)} = \left.\frac{\partial}{\partial y}(x^2 + y^2)\right|_{(1,2)} = \left.2y\right|_{(1,2)} = 2(2) = 4.$$

As a check, we can treat the parabola as the graph of the single-variable function $z = (1)^2 + y^2 = 1 + y^2$ in the plane $x = 1$ and ask for the slope at $y = 2$. The slope, calculated now as an ordinary derivative, is

$$\left.\frac{dz}{dy}\right|_{y=2} = \left.\frac{d}{dy}(1 + y^2)\right|_{y=2} = \left.2y\right|_{y=2} = 4.$$ ■

Functions of More Than Two Variables

The definitions of the partial derivatives of functions of more than two independent variables are like the definitions for functions of two variables. They are ordinary derivatives with respect to one variable, taken while the other independent variables are held constant.

EXAMPLE 6 If x, y, and z are independent variables and

$$f(x, y, z) = x \sin(y + 3z),$$

then

$$\frac{\partial f}{\partial z} = \frac{\partial}{\partial z}[x \sin(y + 3z)] = x\frac{\partial}{\partial z}\sin(y + 3z)$$
$$= x\cos(y + 3z)\frac{\partial}{\partial z}(y + 3z) = 3x\cos(y + 3z).$$ ■

Partial Derivatives and Continuity

A function $f(x, y)$ can have partial derivatives with respect to both x and y at a point without the function being continuous there. This is different from functions of a single variable, where the existence of a derivative implies continuity. If the partial derivatives of $f(x, y)$ exist and are continuous throughout a disk centered at (x_0, y_0), however, then f *is* continuous at (x_0, y_0), as we see at the end of this section.

EXAMPLE 7 Let

$$f(x, y) = \begin{cases} 0, & xy \neq 0 \\ 1, & xy = 0 \end{cases}$$

(Figure 11.17).

(a) Find the limit of f as (x, y) approaches (0, 0) along the line $y = x$.

(b) Prove that f is not continuous at the origin.

(c) Show that both partial derivatives $\partial f/\partial x$ and $\partial f/\partial y$ exist at the origin.

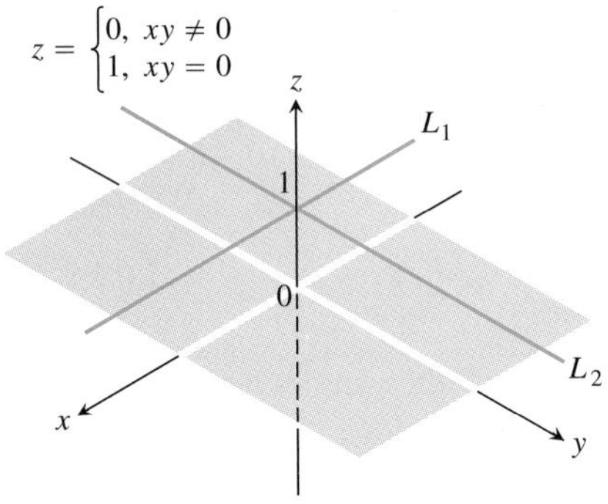

FIGURE 11.17 The graph of

$$f(x, y) = \begin{cases} 0, & xy \neq 0 \\ 1, & xy = 0 \end{cases}$$

consists of the lines L_1 and L_2 and the four open quadrants of the xy-plane. The function has partial derivatives at the origin but is not continuous there (Example 7).

Solution

(a) Since $f(x, y)$ is constantly zero along the line $y = x$ (except at the origin), we have

$$\lim_{(x, y) \to (0,0)} f(x, y) \Big|_{y=x} = \lim_{(x, y) \to (0,0)} 0 = 0.$$

(b) Since $f(0, 0) = 1$, the limit in part (a) proves that f is not continuous at $(0, 0)$.

(c) To find $\partial f/\partial x$ at $(0, 0)$, we hold y fixed at $y = 0$. Then $f(x, y) = 1$ for all x, and the graph of f is the line L_1 in Figure 11.17. The slope of this line at any x is $\partial f/\partial x = 0$. In particular, $\partial f/\partial x = 0$ at $(0, 0)$. Similarly, $\partial f/\partial y$ is the slope of line L_2 at any y, so $\partial f/\partial y = 0$ at $(0, 0)$. ■

Example 7 notwithstanding, it is still true in higher dimensions that *differentiability* at a point implies continuity. What Example 7 suggests is that we need a stronger requirement for differentiability in higher dimensions than the mere existence of the partial derivatives. We define differentiability for functions of two variables at the end of this section and revisit the connection to continuity.

Second-Order Partial Derivatives

When we differentiate a function $f(x, y)$ twice, we produce its second-order derivatives. These derivatives are usually denoted by

$\dfrac{\partial^2 f}{\partial x^2}$	"d squared $f dx$ squared"	or	f_{xx}	"f sub xx"
$\dfrac{\partial^2 f}{\partial y^2}$	"d squared $f dy$ squared"	or	f_{yy}	"f sub yy"
$\dfrac{\partial^2 f}{\partial x \partial y}$	"d squared $f dx\, dy$"	or	f_{yx}	"f sub yx"
$\dfrac{\partial^2 f}{\partial y \partial x}$	"d squared $f dy\, dx$"	or	f_{xy}	"f sub xy"

The defining equations are

$$\frac{\partial^2 f}{\partial x^2} = \frac{\partial}{\partial x}\left(\frac{\partial f}{\partial x}\right), \qquad \frac{\partial^2 f}{\partial x \partial y} = \frac{\partial}{\partial x}\left(\frac{\partial f}{\partial y}\right),$$

and so on. Notice the order in which the derivatives are taken:

$\dfrac{\partial^2 f}{\partial x \partial y}$ Differentiate first with respect to y, then with respect to x.

$f_{yx} = (f_y)_x$ Means the same thing.

HISTORICAL BIOGRAPHY

Pierre-Simon Laplace
(1749–1827)

EXAMPLE 8 If $f(x, y) = x \cos y + ye^x$, find the second-order derivatives

$$\frac{\partial^2 f}{\partial x^2}, \quad \frac{\partial^2 f}{\partial y \partial x}, \quad \frac{\partial^2 f}{\partial y^2}, \quad \text{and} \quad \frac{\partial^2 f}{\partial x \partial y}.$$

Solution

$$\frac{\partial f}{\partial x} = \frac{\partial}{\partial x}(x\cos y + ye^x) = \cos y + ye^x$$

$$\frac{\partial f}{\partial y} = \frac{\partial}{\partial y}(x\cos y + ye^x) = -x\sin y + e^x$$

So

$$\frac{\partial^2 f}{\partial y \partial x} = \frac{\partial}{\partial y}\left(\frac{\partial f}{\partial x}\right) = -\sin y + e^x$$

$$\frac{\partial^2 f}{\partial x^2} = \frac{\partial}{\partial x}\left(\frac{\partial f}{\partial x}\right) = ye^x.$$

So

$$\frac{\partial^2 f}{\partial x \partial y} = \frac{\partial}{\partial x}\left(\frac{\partial f}{\partial y}\right) = -\sin y + e^x$$

$$\frac{\partial^2 f}{\partial y^2} = \frac{\partial}{\partial y}\left(\frac{\partial f}{\partial y}\right) = -x\cos y.$$ ■

The Mixed Derivative Theorem

You may have noticed that the "mixed" second-order partial derivatives

$$\frac{\partial^2 f}{\partial y \partial x} \quad \text{and} \quad \frac{\partial^2 f}{\partial x \partial y}$$

in Example 8 were equal. This was not a coincidence. They must be equal whenever f, f_x, f_y, f_{xy}, and f_{yx} are continuous, as stated in the following theorem.

THEOREM 2—The Mixed Derivative Theorem If $f(x, y)$ and its partial derivatives f_x, f_y, f_{xy}, and f_{yx} are defined throughout an open region containing a point (a, b) and are all continuous at (a, b), then

$$f_{xy}(a, b) = f_{yx}(a, b).$$

HISTORICAL BIOGRAPHY

Alexis Clairaut
(1713–1765)

Theorem 2 is also known as Clairaut's Theorem, named after the French mathematician Alexis Clairaut who discovered it. A proof is given in Appendix 11. Theorem 2 says that to calculate a mixed second-order derivative, we may differentiate in either order, provided the continuity conditions are satisfied. This can work to our advantage.

EXAMPLE 9 Find $\partial^2 w/\partial x \partial y$ if

$$w = xy + \frac{e^y}{y^2 + 1}.$$

Solution The symbol $\partial^2 w/\partial x \partial y$ tells us to differentiate first with respect to y and then with respect to x. If we postpone the differentiation with respect to y and differentiate first with respect to x, however, we get the answer more quickly. In two steps,

$$\frac{\partial w}{\partial x} = y \quad \text{and} \quad \frac{\partial^2 w}{\partial y \partial x} = 1.$$

If we differentiate first with respect to y, we obtain $\partial^2 w/\partial x \partial y = 1$ as well. We can differentiate in either order because the conditions of Theorem 2 hold for w at all points (x_0, y_0). ■

Partial Derivatives of Still Higher Order

Although we will deal mostly with first- and second-order partial derivatives, because these appear the most frequently in applications, there is no theoretical limit to how many times we can differentiate a function as long as the derivatives involved exist. Thus, we get third- and fourth-order derivatives denoted by symbols like

$$\frac{\partial^3 f}{\partial x \partial y^2} = f_{yyx}$$

$$\frac{\partial^4 f}{\partial x^2 \partial y^2} = f_{yyxx},$$

and so on. As with second-order derivatives, the order of differentiation is immaterial as long as all the derivatives through the order in question are continuous.

EXAMPLE 10 Find f_{yxyz} if $f(x, y, z) = 1 - 2xy^2z + x^2y$.

Solution We first differentiate with respect to the variable y, then x, then y again, and finally with respect to z:

$$\begin{aligned} f_y &= -4xyz + x^2 \\ f_{yx} &= -4yz + 2x \\ f_{yxy} &= -4z \\ f_{yxyz} &= -4 \end{aligned}$$

■

Differentiability

The starting point for differentiability is not the difference quotient we saw in studying single-variable functions, but rather the idea of increment. Recall from our work with functions of a single variable in Section 2.13 that if $y = f(x)$ is differentiable at $x = x_0$, then the change in the value of f that results from changing x from x_0 to $x_0 + \Delta x$ is given by an equation of the form

$$\Delta y = f'(x_0)\Delta x + \epsilon \Delta x$$

in which $\epsilon \to 0$ as $\Delta x \to 0$. For functions of two variables, the analogous property becomes the definition of differentiability. The Increment Theorem (from advanced calculus) tells us when to expect the property to hold.

THEOREM 3—The Increment Theorem for Functions of Two Variables Suppose that the first partial derivatives of $f(x, y)$ are defined throughout an open region R containing the point (x_0, y_0) and that f_x and f_y are continuous at (x_0, y_0). Then the change

$$\Delta z = f(x_0 + \Delta x, y_0 + \Delta y) - f(x_0, y_0)$$

in the value of f that results from moving from (x_0, y_0) to another point $(x_0 + \Delta x, y_0 + \Delta y)$ in R satisfies an equation of the form

$$\Delta z = f_x(x_0, y_0)\Delta x + f_y(x_0, y_0)\Delta y + \epsilon_1 \Delta x + \epsilon_2 \Delta y,$$

in which each of $\epsilon_1, \epsilon_2 \to 0$ as both $\Delta x, \Delta y \to 0$.

You can see where the epsilons come from in the proof in Appendix 11. You will also see that similar results hold for functions of more than two independent variables.

DEFINITION A function $z = f(x, y)$ is **differentiable at** (x_0, y_0) if $f_x(x_0, y_0)$ and $f_y(x_0, y_0)$ exist and Δz satisfies an equation of the form

$$\Delta z = f_x(x_0, y_0)\Delta x + f_y(x_0, y_0)\Delta y + \epsilon_1\Delta x + \epsilon_2\Delta y,$$

in which each of $\epsilon_1, \epsilon_2 \to 0$ as both $\Delta x, \Delta y \to 0$. We call f **differentiable** if it is differentiable at every point in its domain.

In light of this definition, we have the immediate corollary of Theorem 3 that a function is differentiable if its first partial derivatives are *continuous*.

COROLLARY OF THEOREM 3 If the partial derivatives f_x and f_y of a function $f(x, y)$ are continuous throughout an open region R, then f is differentiable at every point of R.

If $z = f(x, y)$ is differentiable, then the definition of differentiability assures that $\Delta z = f(x_0 + \Delta x, y_0 + \Delta y) - f(x_0, y_0)$ approaches 0 as Δx and Δy approach 0. This tells us that a function of two variables is continuous at every point where it is differentiable.

THEOREM 4—Differentiability Implies Continuity If a function $f(x, y)$ is differentiable at (x_0, y_0), then f is continuous at (x_0, y_0).

As we can see from Theorems 3 and 4, a function $f(x, y)$ must be continuous at a point (x_0, y_0) if f_x and f_y are continuous throughout an open region containing (x_0, y_0). Remember, however, that it is still possible for a function of two variables to be discontinuous at a point where its first partial derivatives exist, as we saw in Example 7. Existence alone of the partial derivatives at that point is not enough.

EXERCISES 11.3

In Exercises 1–22, find $\partial f/\partial x$ and $\partial f/\partial y$.

1. $f(x, y) = 2x^2 - 3y - 4$

2. $f(x, y) = x^2 - xy + y^2$

3. $f(x, y) = (x^2 - 1)(y + 2)$

4. $f(x, y) = 5xy - 7x^2 - y^2 + 3x - 6y + 2$

5. $f(x, y) = (xy - 1)^2$

6. $f(x, y) = (2x - 3y)^3$

7. $f(x, y) = \sqrt{x^2 + y^2}$

8. $f(x, y) = (x^3 + (y/2))^{2/3}$

9. $f(x, y) = 1/(x + y)$

10. $f(x, y) = x/(x^2 + y^2)$

11. $f(x, y) = (x + y)/(xy - 1)$

12. $f(x, y) = \tan^{-1}(y/x)$

13. $f(x, y) = e^{(x+y+1)}$

14. $f(x, y) = e^{-x}\sin(x + y)$

15. $f(x, y) = \ln(x + y)$

16. $f(x, y) = e^{xy}\ln y$

17. $f(x, y) = \sin^2(x - 3y)$

18. $f(x, y) = \cos^2(3x - y^2)$

19. $f(x, y) = x^y$

20. $f(x, y) = \log_y x$

21. $f(x, y) = \int_x^y g(t)\,dt$ (g continuous for all t)

22. $f(x, y) = \sum_{n=0}^{\infty}(xy)^n$ $(|xy| < 1)$

In Exercises 23–34, find f_x, f_y, and f_z.

23. $f(x, y, z) = 1 + xy^2 - 2z^2$ **24.** $f(x, y, z) = xy + yz + xz$

25. $f(x, y, z) = x - \sqrt{y^2 + z^2}$

26. $f(x, y, z) = (x^2 + y^2 + z^2)^{-1/2}$

27. $f(x, y, z) = \sin^{-1}(xyz)$ **28.** $f(x, y, z) = \sec^{-1}(x + yz)$

29. $f(x, y, z) = \ln(x + 2y + 3z)$

30. $f(x, y, z) = yz \ln(xy)$ **31.** $f(x, y, z) = e^{-(x^2+y^2+z^2)}$

32. $f(x, y, z) = e^{-xyz}$

33. $f(x, y, z) = \tanh(x + 2y + 3z)$

34. $f(x, y, z) = \sinh(xy - z^2)$

In Exercises 35–40, find the partial derivative of the function with respect to each variable.

35. $f(t, \alpha) = \cos(2\pi t - \alpha)$ **36.** $g(u, v) = v^2 e^{(2u/v)}$

37. $h(\rho, \phi, \theta) = \rho \sin\phi \cos\theta$ **38.** $g(r, \theta, z) = r(1 - \cos\theta) - z$

39. $W(P, V, \delta, v, g) = PV + \dfrac{V\delta v^2}{2g}$

40. Wilson lot size formula (Section 3.6, Exercise 33)

$$A(c, h, k, m, q) = \frac{km}{q} + cm + \frac{hq}{2}$$

Find all the second-order partial derivatives of the functions in Exercises 41–46.

41. $f(x, y) = x + y + xy$ **42.** $f(x, y) = \sin xy$

43. $g(x, y) = x^2 y + \cos y + y \sin x$

44. $h(x, y) = xe^y + y + 1$ **45.** $r(x, y) = \ln(x + y)$

46. $s(x, y) = \tan^{-1}(y/x)$

In Exercises 47–50, verify that $w_{xy} = w_{yx}$.

47. $w = \ln(2x + 3y)$ **48.** $w = e^x + x \ln y + y \ln x$

49. $w = xy^2 + x^2y^3 + x^3y^4$ **50.** $w = x \sin y + y \sin x + xy$

51. Which order of differentiation will calculate f_{xy} faster: x first or y first? Try to answer without writing anything down.

a. $f(x, y) = x \sin y + e^y$

b. $f(x, y) = 1/x$

c. $f(x, y) = y + (x/y)$

d. $f(x, y) = y + x^2y + 4y^3 - \ln(y^2 + 1)$

e. $f(x, y) = x^2 + 5xy + \sin x + 7e^x$

f. $f(x, y) = x \ln xy$

52. The fifth-order partial derivative $\partial^5 f/\partial x^2 \partial y^3$ is zero for each of the following functions. To show this as quickly as possible, which variable would you differentiate with respect to first: x or y? Try to answer without writing anything down.

a. $f(x, y) = y^2x^4e^x + 2$

b. $f(x, y) = y^2 + y(\sin x - x^4)$

c. $f(x, y) = x^2 + 5xy + \sin x + 7e^x$

d. $f(x, y) = xe^{y^2/2}$

In Exercises 53 and 54, use the limit definition of partial derivative to compute the partial derivatives of the functions at the specified points.

53. $f(x, y) = 1 - x + y - 3x^2y$, $\dfrac{\partial f}{\partial x}$ and $\dfrac{\partial f}{\partial y}$ at $(1, 2)$

54. $f(x, y) = 4 + 2x - 3y - xy^2$, $\dfrac{\partial f}{\partial x}$ and $\dfrac{\partial f}{\partial y}$ at $(-2, 1)$

55. Three variables Let $w = f(x, y, z)$ be a function of three independent variables and write the formal definition of the partial derivative $\partial f/\partial z$ at (x_0, y_0, z_0). Use this definition to find $\partial f/\partial z$ at $(1, 2, 3)$ for $f(x, y, z) = x^2yz^2$.

56. Three variables Let $w = f(x, y, z)$ be a function of three independent variables and write the formal definition of the partial derivative $\partial f/\partial y$ at (x_0, y_0, z_0). Use this definition to find $\partial f/\partial y$ at $(-1, 0, 3)$ for $f(x, y, z) = -2xy^2 + yz^2$.

57. Find the value of $\partial z/\partial x$ at the point $(1, 1, 1)$ if the equation

$$xy + z^3x - 2yz = 0$$

defines z as a function of the two independent variables x and y and the partial derivative exists.

58. Find the value of $\partial x/\partial z$ at the point $(1, -1, -3)$ if the equation

$$xz + y \ln x - x^2 + 4 = 0$$

defines x as a function of the two independent variables y and z and the partial derivative exists.

Exercises 59 and 60 are about the triangle shown here.

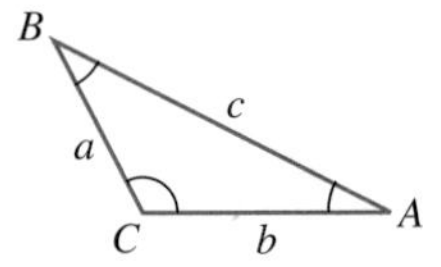

59. Express A implicitly as a function of a, b, and c and calculate $\partial A/\partial a$ and $\partial A/\partial b$.

60. Express a implicitly as a function of A, b, and B and calculate $\partial a/\partial A$ and $\partial a/\partial B$.

61. Two dependent variables Express v_x in terms of u and y if the equations $x = v \ln u$ and $y = u \ln v$ define u and v as functions of the independent variables x and y, and if v_x exists. (*Hint:* Differentiate both equations with respect to x and solve for v_x by eliminating u_x.)

62. Two dependent variables Find $\partial x/\partial u$ and $\partial y/\partial u$ if the equations $u = x^2 - y^2$ and $v = x^2 - y$ define x and y as functions of the independent variables u and v, and the partial derivatives exist. (See the hint in Exercise 61.) Then let $s = x^2 + y^2$ and find $\partial s/\partial u$.

The **three-dimensional Laplace equation**

$$\frac{\partial^2 f}{\partial x^2} + \frac{\partial^2 f}{\partial y^2} + \frac{\partial^2 f}{\partial z^2} = 0$$

is satisfied by steady-state temperature distributions $T = f(x, y, z)$ in space, by gravitational potentials, and by electrostatic potentials. The **two-dimensional Laplace equation**, obtained by dropping the $\partial^2 f/\partial z^2$ term from the previous equation, applies to a plane.

Show that each function in Exercises 63–68 satisfies a Laplace equation.

63. $f(x, y, z) = x^2 + y^2 - 2z^2$

64. $f(x, y, z) = 2z^3 - 3(x^2 + y^2)z$

65. $f(x, y) = e^{-2y} \cos 2x$

66. $f(x, y) = \ln\sqrt{x^2 + y^2}$

67. $f(x, y, z) = (x^2 + y^2 + z^2)^{-1/2}$

68. $f(x, y, z) = e^{3x+4y} \cos 5z$

The Wave Equation The **one-dimensional wave equation** is given by

$$\frac{\partial^2 w}{\partial t^2} = c^2 \frac{\partial^2 w}{\partial x^2},$$

where w is the wave height, x is the distance variable, t is the time variable, and c is the velocity with which the waves are propagated. It describes the motion of a waveform such as an ocean wave or a sound wave. The number c varies with the medium and type of wave.

Show that the functions in Exercises 69–75 are all solutions of the wave equation.

69. $w = \sin(x + ct)$

70. $w = \cos(2x + 2ct)$

71. $w = \sin(x + ct) + \cos(2x + 2ct)$

72. $w = \ln(2x + 2ct)$

73. $w = \tan(2x - 2ct)$

74. $w = 5\cos(3x + 3ct) + e^{x+ct}$

75. $w = f(u)$, where f is a differentiable function of u, and $u = a(x + ct)$, where a is a constant

76. Does a function $f(x, y)$ with continuous first partial derivatives throughout an open region R have to be continuous on R? Give reasons for your answer.

77. If a function $f(x, y)$ has continuous second partial derivatives throughout an open region R, must the first-order partial derivatives of f be continuous on R? Give reasons for your answer.

11.4 The Chain Rule

The Chain Rule for functions of a single variable studied in Section 2.7 said that when $w = f(x)$ was a differentiable function of x and $x = g(t)$ was a differentiable function of t, w became a differentiable function of t and dw/dt could be calculated with the formula

$$\frac{dw}{dt} = \frac{dw}{dx}\frac{dx}{dt}.$$

For functions of two or more variables the Chain Rule has several forms. The form depends on how many variables are involved but works like the Chain Rule in Section 2.7 once we account for the presence of additional variables.

Functions of Two Variables

The Chain Rule formula for a function $w = f(x, y)$ when $x = x(t)$ and $y = y(t)$ are both differentiable functions of t is given in the following theorem.

THEOREM 5—Chain Rule for Functions of Two Independent Variables If $w = f(x, y)$ has continuous partial derivatives f_x and f_y and if $x = x(t), y = y(t)$ are differentiable functions of t, then the composite $w = f(x(t), y(t))$ is a differentiable function of t and

$$\frac{dw}{dt} = f_x(x(t), y(t)) \cdot x'(t) + f_y(x(t), y(t)) \cdot y'(t),$$

or

$$\frac{dw}{dt} = \frac{\partial w}{\partial x}\frac{dx}{dt} + \frac{\partial w}{\partial y}\frac{dy}{dt}.$$

Each of $\frac{\partial f}{\partial x}, \frac{\partial w}{\partial x}, f_x$ indicates the partial derivative of f with respect to x.

Proof The proof consists of showing that if x and y are differentiable at $t = t_0$, then w is differentiable at t_0 and

$$\left(\frac{dw}{dt}\right)_{t_0} = \left(\frac{\partial w}{\partial x}\right)_{P_0}\left(\frac{dx}{dt}\right)_{t_0} + \left(\frac{\partial w}{\partial y}\right)_{P_0}\left(\frac{dy}{dt}\right)_{t_0},$$

where $P_0 = (x(t_0), y(t_0))$. The subscripts indicate where each of the derivatives are to be evaluated.

Let Δx, Δy, and Δw be the increments that result from changing t from t_0 to $t_0 + \Delta t$. Since f is differentiable (see the definition in Section 11.3),

$$\Delta w = \left(\frac{\partial w}{\partial x}\right)_{P_0}\Delta x + \left(\frac{\partial w}{\partial y}\right)_{P_0}\Delta y + \epsilon_1\Delta x + \epsilon_2\Delta y,$$

where $\epsilon_1, \epsilon_2 \to 0$ as $\Delta x, \Delta y \to 0$. To find dw/dt, we divide this equation through by Δt and let Δt approach zero. The division gives

$$\frac{\Delta w}{\Delta t} = \left(\frac{\partial w}{\partial x}\right)_{P_0}\frac{\Delta x}{\Delta t} + \left(\frac{\partial w}{\partial y}\right)_{P_0}\frac{\Delta y}{\Delta t} + \epsilon_1\frac{\Delta x}{\Delta t} + \epsilon_2\frac{\Delta y}{\Delta t}.$$

Letting Δt approach zero gives

$$\left(\frac{dw}{dt}\right)_{t_0} = \lim_{\Delta t \to 0}\frac{\Delta w}{\Delta t}$$

$$= \left(\frac{\partial w}{\partial x}\right)_{P_0}\left(\frac{dx}{dt}\right)_{t_0} + \left(\frac{\partial w}{\partial y}\right)_{P_0}\left(\frac{dy}{dt}\right)_{t_0} + 0 \cdot \left(\frac{dx}{dt}\right)_{t_0} + 0 \cdot \left(\frac{dy}{dt}\right)_{t_0}.$$ ■

To remember the Chain Rule picture the diagram below. To find dw/dt, start at w and read down each route to t, multiplying derivatives along the way. Then add the products.

Chain Rule

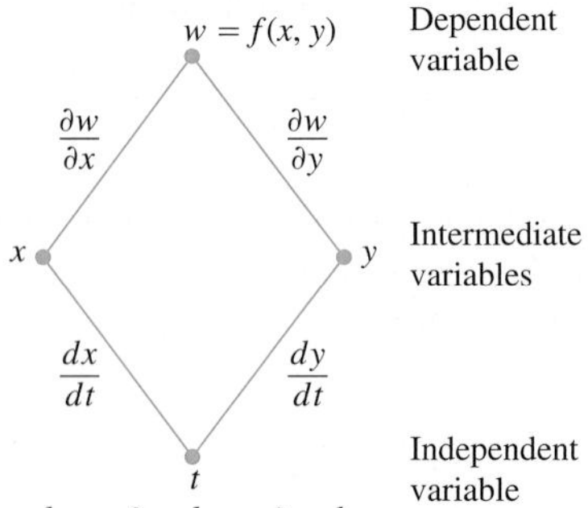

$$\frac{dw}{dt} = \frac{\partial w}{\partial x}\frac{dx}{dt} + \frac{\partial w}{\partial y}\frac{dy}{dt}$$

The **tree diagram** in the margin provides a convenient way to remember the Chain Rule. Notice that w is both a function of x and y and a function of t. In the diagram, when $t = t_0$, the derivatives dx/dt and dy/dt are evaluated at t_0. The value of t_0 then determines the value x_0 for the differentiable function x and the value y_0 for the differentiable function y. The partial derivatives $\partial w/\partial x$ and $\partial w/\partial y$ (which are themselves functions of x and y) are evaluated at the point $P_0(x_0, y_0)$ corresponding to t_0. The "true" independent variable is t, whereas x and y are *intermediate variables* (controlled by t) and w is the dependent variable.

A more precise notation for the Chain Rule shows where the various derivatives in Theorem 5 are evaluated:

$$\frac{dw}{dt}(t_0) = \frac{\partial f}{\partial x}(x_0, y_0) \cdot \frac{dx}{dt}(t_0) + \frac{\partial f}{\partial y}(x_0, y_0) \cdot \frac{dy}{dt}(t_0).$$

EXAMPLE 1 Use the Chain Rule to find the derivative of

$$w = xy$$

with respect to t along the path $x = \cos t, y = \sin t$. What is the derivative's value at $t = \pi/2$?

Solution We apply the Chain Rule to find dw/dt as follows:

$$\begin{aligned}
\frac{dw}{dt} &= \frac{\partial w}{\partial x}\frac{dx}{dt} + \frac{\partial w}{\partial y}\frac{dy}{dt} \\
&= \frac{\partial(xy)}{\partial x} \cdot \frac{d}{dt}(\cos t) + \frac{\partial(xy)}{\partial y} \cdot \frac{d}{dt}(\sin t) \\
&= (y)(-\sin t) + (x)(\cos t) \\
&= (\sin t)(-\sin t) + (\cos t)(\cos t) \\
&= -\sin^2 t + \cos^2 t \\
&= \cos 2t.
\end{aligned}$$

In this example, we can check the result with a more direct calculation. As a function of t,

$$w = xy = \cos t \sin t = \frac{1}{2}\sin 2t,$$

so

$$\frac{dw}{dt} = \frac{d}{dt}\left(\frac{1}{2}\sin 2t\right) = \frac{1}{2} \cdot 2\cos 2t = \cos 2t.$$

In either case, at the given value of t,

$$\left(\frac{dw}{dt}\right)_{t=\pi/2} = \cos\left(2 \cdot \frac{\pi}{2}\right) = \cos \pi = -1.$$

■

Functions of Three Variables

You can probably predict the Chain Rule for functions of three variables, as it only involves adding the expected third term to the two-variable formula.

THEOREM 6—Chain Rule for Functions of Three Independent Variables If $w = f(x, y, z)$ is differentiable and x, y, and z are differentiable functions of t, then w is a differentiable function of t and

$$\frac{dw}{dt} = \frac{\partial w}{\partial x}\frac{dx}{dt} + \frac{\partial w}{\partial y}\frac{dy}{dt} + \frac{\partial w}{\partial z}\frac{dz}{dt}.$$

Here we have three routes from w to t instead of two, but finding dw/dt is still the same. Read down each route, multiplying derivatives along the way; then add.

Chain Rule

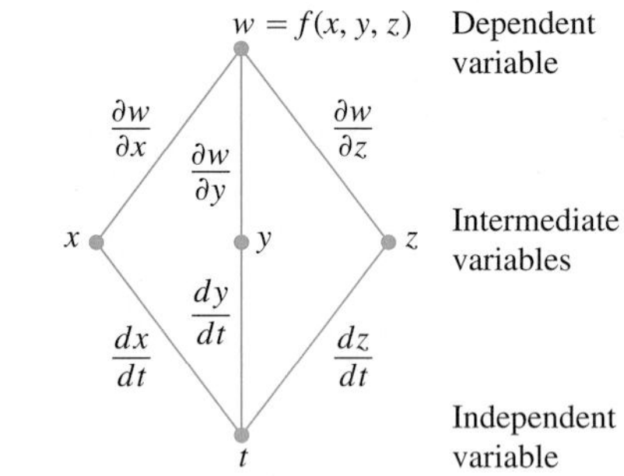

$$\frac{dw}{dt} = \frac{\partial w}{\partial x}\frac{dx}{dt} + \frac{\partial w}{\partial y}\frac{dy}{dt} + \frac{\partial w}{\partial z}\frac{dz}{dt}$$

The proof is identical with the proof of Theorem 5 except that there are now three intermediate variables instead of two. The diagram we use for remembering the new equation is similar as well, with three routes from w to t.

EXAMPLE 2 Find dw/dt if

$$w = xy + z, \qquad x = \cos t, \qquad y = \sin t, \qquad z = t.$$

In this example the values of $w(t)$ are changing along the path of a helix (Section 10.1) as t changes. What is the derivative's value at $t = 0$?

Solution

$$\begin{aligned} \frac{dw}{dt} &= \frac{\partial w}{\partial x}\frac{dx}{dt} + \frac{\partial w}{\partial y}\frac{dy}{dt} + \frac{\partial w}{\partial z}\frac{dz}{dt} \\ &= (y)(-\sin t) + (x)(\cos t) + (1)(1) \\ &= (\sin t)(-\sin t) + (\cos t)(\cos t) + 1 \\ &= -\sin^2 t + \cos^2 t + 1 = 1 + \cos 2t. \end{aligned}$$

Substitute for the intermediate variables.

$$\left(\frac{dw}{dt}\right)_{t=0} = 1 + \cos(0) = 2.$$

■

For a physical interpretation of change along a curve think of an object whose position is changing with time t. If $w = T(x, y, z)$ is the temperature at each point (x, y, z) along a curve C with parametric equations $x = x(t)$, $y = y(t)$, and $z = z(t)$, then the composite function $w = T(x(t), y(t), z(t))$ represents the temperature relative to t along the curve. The derivative dw/dt is then the instantaneous rate of change of temperature due to the motion along the curve, as calculated in Theorem 6.

Functions Defined on Surfaces

If we are interested in the temperature $w = f(x, y, z)$ at points (x, y, z) on a globe in space, we might prefer to think of x, y, and z as functions of the variables r and s that give the points' longitudes and latitudes. If $x = g(r, s)$, $y = h(r, s)$, and $z = k(r, s)$, we could then express the temperature as a function of r and s with the composite function

$$w = f(g(r, s), h(r, s), k(r, s)).$$

Under the conditions stated below, w has partial derivatives with respect to both r and s that can be calculated in the following way.

THEOREM 7—Chain Rule for Two Independent Variables and Three Intermediate Variables Suppose that $w = f(x, y, z)$, $x = g(r, s)$, $y = h(r, s)$, and $z = k(r, s)$. If all four functions are differentiable, then w has partial derivatives with respect to r and s, given by the formulas

$$\frac{\partial w}{\partial r} = \frac{\partial w}{\partial x}\frac{\partial x}{\partial r} + \frac{\partial w}{\partial y}\frac{\partial y}{\partial r} + \frac{\partial w}{\partial z}\frac{\partial z}{\partial r}$$

$$\frac{\partial w}{\partial s} = \frac{\partial w}{\partial x}\frac{\partial x}{\partial s} + \frac{\partial w}{\partial y}\frac{\partial y}{\partial s} + \frac{\partial w}{\partial z}\frac{\partial z}{\partial s}.$$

The first of these equations can be derived from the Chain Rule in Theorem 6 by holding s fixed and treating r as t. The second can be derived in the same way, holding r fixed and treating s as t. The tree diagrams for both equations are shown in Figure 11.18.

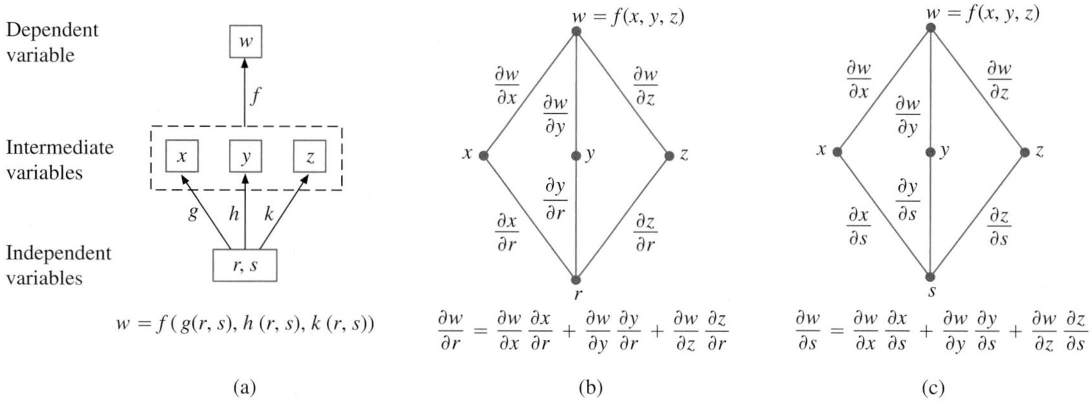

FIGURE 11.18 Composite function and tree diagrams for Theorem 7.

EXAMPLE 3 Express $\partial w/\partial r$ and $\partial w/\partial s$ in terms of r and s if

$$w = x + 2y + z^2, \qquad x = \frac{r}{s}, \qquad y = r^2 + \ln s, \qquad z = 2r.$$

Solution

$$\begin{aligned}\frac{\partial w}{\partial r} &= \frac{\partial w}{\partial x}\frac{\partial x}{\partial r} + \frac{\partial w}{\partial y}\frac{\partial y}{\partial r} + \frac{\partial w}{\partial z}\frac{\partial z}{\partial r}\\ &= (1)\left(\frac{1}{s}\right) + (2)(2r) + (2z)(2)\\ &= \frac{1}{s} + 4r + (4r)(2) = \frac{1}{s} + 12r\end{aligned}$$

Substitute for intermediate variable z.

$$\begin{aligned}\frac{\partial w}{\partial s} &= \frac{\partial w}{\partial x}\frac{\partial x}{\partial s} + \frac{\partial w}{\partial y}\frac{\partial y}{\partial s} + \frac{\partial w}{\partial z}\frac{\partial z}{\partial s}\\ &= (1)\left(-\frac{r}{s^2}\right) + (2)\left(\frac{1}{s}\right) + (2z)(0) = \frac{2}{s} - \frac{r}{s^2}\end{aligned}$$

■

If f is a function of two variables instead of three, each equation in Theorem 7 becomes correspondingly one term shorter.

Chain Rule

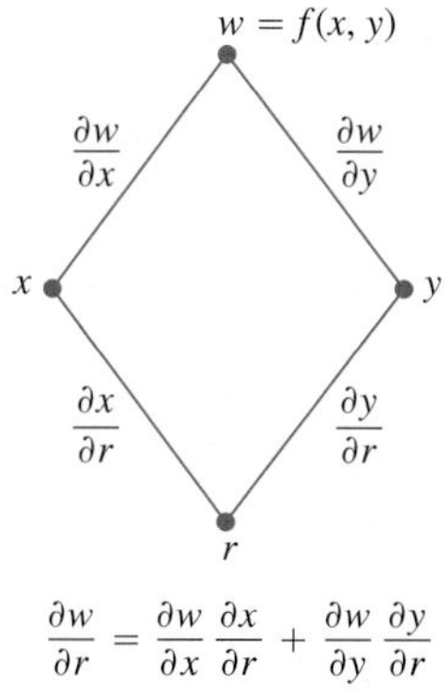

FIGURE 11.19 Tree diagram for the equation
$$\frac{\partial w}{\partial r} = \frac{\partial w}{\partial x}\frac{\partial x}{\partial r} + \frac{\partial w}{\partial y}\frac{\partial y}{\partial r}.$$

> If $w = f(x, y)$, $x = g(r, s)$, and $y = h(r, s)$, then
>
> $$\frac{\partial w}{\partial r} = \frac{\partial w}{\partial x}\frac{\partial x}{\partial r} + \frac{\partial w}{\partial y}\frac{\partial y}{\partial r} \quad\text{and}\quad \frac{\partial w}{\partial s} = \frac{\partial w}{\partial x}\frac{\partial x}{\partial s} + \frac{\partial w}{\partial y}\frac{\partial y}{\partial s}.$$

Figure 11.19 shows the tree diagram for the first of these equations. The diagram for the second equation is similar; just replace r with s.

EXAMPLE 4 Express $\partial w/\partial r$ and $\partial w/\partial s$ in terms of r and s if

$$w = x^2 + y^2, \qquad x = r - s, \qquad y = r + s.$$

Solution

$$\begin{aligned} \frac{\partial w}{\partial r} &= \frac{\partial w}{\partial x}\frac{\partial x}{\partial r} + \frac{\partial w}{\partial y}\frac{\partial y}{\partial r} & \frac{\partial w}{\partial s} &= \frac{\partial w}{\partial x}\frac{\partial x}{\partial s} + \frac{\partial w}{\partial y}\frac{\partial y}{\partial s} \\ &= (2x)(1) + (2y)(1) & &= (2x)(-1) + (2y)(1) \\ &= 2(r - s) + 2(r + s) & &= -2(r - s) + 2(r + s) \\ &= 4r & &= 4s \end{aligned}$$

Substitute for the intermediate variables.

■

If f is a function of x alone, our equations are even simpler.

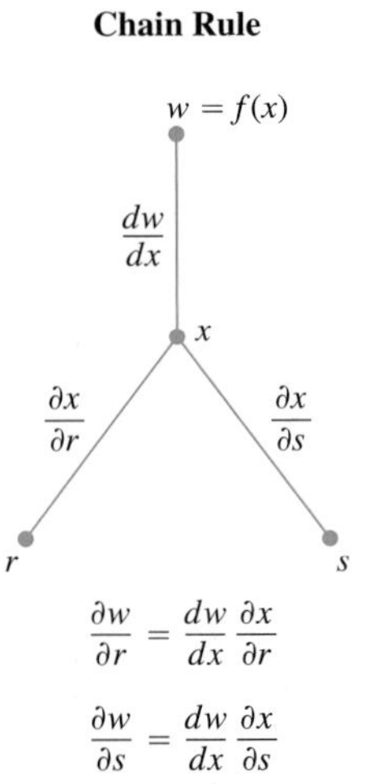

FIGURE 11.20 Tree diagram for differentiating f as a composite function of r and s with one intermediate variable.

> If $w = f(x)$ and $x = g(r, s)$, then
>
> $$\frac{\partial w}{\partial r} = \frac{dw}{dx}\frac{\partial x}{\partial r} \quad \text{and} \quad \frac{\partial w}{\partial s} = \frac{dw}{dx}\frac{\partial x}{\partial s}.$$

In this case, we use the ordinary (single-variable) derivative, dw/dx. The tree diagram is shown in Figure 11.20.

Implicit Differentiation Revisited

The two-variable Chain Rule in Theorem 5 leads to a formula that takes some of the algebra out of implicit differentiation. Suppose that

1. The function $F(x, y)$ is differentiable and
2. The equation $F(x, y) = 0$ defines y implicitly as a differentiable function of x, say $y = h(x)$.

Since $w = F(x, y) = 0$, the derivative dw/dx must be zero. Computing the derivative from the Chain Rule (tree diagram in Figure 11.21), we find

$$\begin{aligned} 0 = \frac{dw}{dx} &= F_x\frac{dx}{dx} + F_y\frac{dy}{dx} \\ &= F_x \cdot 1 + F_y \cdot \frac{dy}{dx}. \end{aligned}$$

Theorem 5 with $t = x$ and $f = F$

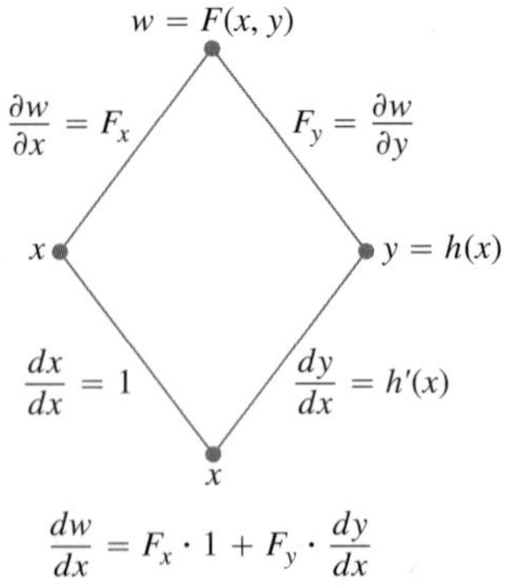

FIGURE 11.21 Tree diagram for differentiating $w = F(x, y)$ with respect to x. Setting $dw/dx = 0$ leads to a simple computational formula for implicit differentiation (Theorem 8).

If $F_y = \partial w/\partial y \neq 0$, we can solve this equation for dy/dx to get

$$\frac{dy}{dx} = -\frac{F_x}{F_y}.$$

We state this result formally.

THEOREM 8—A Formula for Implicit Differentiation Suppose that $F(x, y)$ is differentiable and that the equation $F(x, y) = 0$ defines y as a differentiable function of x. Then at any point where $F_y \neq 0$,

$$\frac{dy}{dx} = -\frac{F_x}{F_y}.$$

EXAMPLE 5 Use Theorem 8 to find dy/dx if $y^2 - x^2 - \sin xy = 0$.

Solution Take $F(x, y) = y^2 - x^2 - \sin xy$. Then

$$\begin{aligned}\frac{dy}{dx} &= -\frac{F_x}{F_y} = -\frac{-2x - y\cos xy}{2y - x\cos xy}\\ &= \frac{2x + y\cos xy}{2y - x\cos xy}.\end{aligned}$$

This calculation is significantly shorter than a single-variable calculation using implicit differentiation. ■

Functions of Many Variables

We have seen several different forms of the Chain Rule in this section, but you do not have to memorize them all if you can see them as special cases of the same general formula. When solving particular problems, it may help to draw the appropriate tree diagram by placing the dependent variable on top, the intermediate variables in the middle, and the selected independent variable at the bottom. To find the derivative of the dependent variable with respect to the selected independent variable, start at the dependent variable and read down each route of the tree to the independent variable, calculating and multiplying the derivatives along each route. Then add the products you found for the different routes.

In general, suppose that $w = f(x, y, \dots, v)$ is a differentiable function of the variables $x, y, \dots, v$ (a finite set) and the $x, y, \dots, v$ are differentiable functions of $p, q, \dots, t$ (another finite set). Then w is a differentiable function of the variables p through t and the partial derivatives of w with respect to these variables are given by equations of the form

$$\frac{\partial w}{\partial p} = \frac{\partial w}{\partial x}\frac{\partial x}{\partial p} + \frac{\partial w}{\partial y}\frac{\partial y}{\partial p} + \cdots + \frac{\partial w}{\partial v}\frac{\partial v}{\partial p}.$$

The other equations are obtained by replacing p by $q, \dots, t$, one at a time.

One way to remember this equation is to think of the right-hand side as the dot product of two vectors with components

$$\underbrace{\left(\frac{\partial w}{\partial x}, \frac{\partial w}{\partial y}, \dots, \frac{\partial w}{\partial v}\right)}_{\substack{\text{Derivatives of } w \text{ with}\\ \text{respect to the}\\ \text{intermediate variables}}} \quad \text{and} \quad \underbrace{\left(\frac{\partial x}{\partial p}, \frac{\partial y}{\partial p}, \dots, \frac{\partial v}{\partial p}\right)}_{\substack{\text{Derivatives of the intermediate}\\ \text{variables with respect to the}\\ \text{selected independent variable}}}.$$

EXERCISES 11.4

In Exercises 1–6, **(a)** express dw/dt as a function of t, both by using the Chain Rule and by expressing w in terms of t and differentiating directly with respect to t. Then **(b)** evaluate dw/dt at the given value of t.

1. $w = x^2 + y^2, \quad x = \cos t, \quad y = \sin t; \quad t = \pi$

2. $w = x^2 + y^2, \quad x = \cos t + \sin t, \quad y = \cos t - \sin t; \quad t = 0$

3. $w = \frac{x}{z} + \frac{y}{z}, \quad x = \cos^2 t, \quad y = \sin^2 t, \quad z = 1/t; \quad t = 3$

4. $w = \ln(x^2 + y^2 + z^2), \quad x = \cos t, \quad y = \sin t, \quad z = 4\sqrt{t};$
$t = 3$

5. $w = 2ye^x - \ln z, \quad x = \ln(t^2 + 1), \quad y = \tan^{-1} t, \quad z = e^t;$
$t = 1$

6. $w = z - \sin xy, \quad x = t, \quad y = \ln t, \quad z = e^{t-1}; \quad t = 1$

In Exercises 7 and 8, **(a)** express $\partial z/\partial u$ and $\partial z/\partial v$ as functions of u and v both by using the Chain Rule and by expressing z directly in terms of u and v before differentiating. Then **(b)** evaluate $\partial z/\partial u$ and $\partial z/\partial v$ at the given point (u, v).

7. $z = 4e^x \ln y, \quad x = \ln(u \cos v), \quad y = u \sin v;$
$(u, v) = (2, \pi/4)$

8. $z = \tan^{-1}(x/y), \quad x = u \cos v, \quad y = u \sin v;$
$(u, v) = (1.3, \pi/6)$

In Exercises 9 and 10, **(a)** express $\partial w/\partial u$ and $\partial w/\partial v$ as functions of u and v both by using the Chain Rule and by expressing w directly in terms of u and v before differentiating. Then **(b)** evaluate $\partial w/\partial u$ and $\partial w/\partial v$ at the given point (u, v).

9. $w = xy + yz + xz, \quad x = u + v, \quad y = u - v, \quad z = uv;$
$(u, v) = (1/2, 1)$

10. $w = \ln(x^2 + y^2 + z^2), \quad x = ue^v \sin u, \quad y = ue^v \cos u,$
$z = ue^v; \quad (u, v) = (-2, 0)$

In Exercises 11 and 12, **(a)** express $\partial u/\partial x$, $\partial u/\partial y$, and $\partial u/\partial z$ as functions of x, y, and z both by using the Chain Rule and by expressing u directly in terms of x, y, and z before differentiating. Then **(b)** evaluate $\partial u/\partial x$, $\partial u/\partial y$, and $\partial u/\partial z$ at the given point (x, y, z).

11. $u = \dfrac{p - q}{q - r}, \quad p = x + y + z, \quad q = x - y + z,$
$r = x + y - z; \quad (x, y, z) = \left(\sqrt{3}, 2, 1\right)$

12. $u = e^{qr} \sin^{-1} p, \quad p = \sin x, \quad q = z^2 \ln y, \quad r = 1/z;$
$(x, y, z) = (\pi/4, 1/2, -1/2)$

In Exercises 13–24, draw a tree diagram and write a Chain Rule formula for each derivative.

13. $\dfrac{dz}{dt}$ for $z = f(x, y), \quad x = g(t), \quad y = h(t)$

14. $\dfrac{dz}{dt}$ for $z = f(u, v, w), \quad u = g(t), \quad v = h(t), \quad w = k(t)$

15. $\dfrac{\partial w}{\partial u}$ and $\dfrac{\partial w}{\partial v}$ for $w = h(x, y, z), \quad x = f(u, v), \quad y = g(u, v),$
$z = k(u, v)$

16. $\dfrac{\partial w}{\partial x}$ and $\dfrac{\partial w}{\partial y}$ for $w = f(r, s, t), \quad r = g(x, y), \quad s = h(x, y),$
$t = k(x, y)$

17. $\dfrac{\partial w}{\partial u}$ and $\dfrac{\partial w}{\partial v}$ for $w = g(x, y), \quad x = h(u, v), \quad y = k(u, v)$

18. $\dfrac{\partial w}{\partial x}$ and $\dfrac{\partial w}{\partial y}$ for $w = g(u, v), \quad u = h(x, y), \quad v = k(x, y)$

19. $\dfrac{\partial z}{\partial t}$ and $\dfrac{\partial z}{\partial s}$ for $z = f(x, y), \quad x = g(t, s), \quad y = h(t, s)$

20. $\dfrac{\partial y}{\partial r}$ for $y = f(u), \quad u = g(r, s)$

21. $\dfrac{\partial w}{\partial s}$ and $\dfrac{\partial w}{\partial t}$ for $w = g(u), \quad u = h(s, t)$

22. $\dfrac{\partial w}{\partial p}$ for $w = f(x, y, z, v), \quad x = g(p, q), \quad y = h(p, q),$
$z = j(p, q), \quad v = k(p, q)$

23. $\dfrac{\partial w}{\partial r}$ and $\dfrac{\partial w}{\partial s}$ for $w = f(x, y), \quad x = g(r), \quad y = h(s)$

24. $\dfrac{\partial w}{\partial s}$ for $w = g(x, y), \quad x = h(r, s, t), \quad y = k(r, s, t)$

Assuming that the equations in Exercises 25–28 define y as a differentiable function of x, use Theorem 8 to find the value of dy/dx at the given point.

25. $x^3 - 2y^2 + xy = 0, \quad (1, 1)$

26. $xy + y^2 - 3x - 3 = 0, \quad (-1, 1)$

27. $x^2 + xy + y^2 - 7 = 0, \quad (1, 2)$

28. $xe^y + \sin xy + y - \ln 2 = 0, \quad (0, \ln 2)$

Three-Variable Implicit Differentiation Theorem 8 can be generalized to functions of three variables and even more. The three-variable version goes like this: If the equation $F(x, y, z) = 0$ determines z as a differentiable function of x and y, then, at points where $F_z \neq 0$,

$$\frac{\partial z}{\partial x} = -\frac{F_x}{F_z} \quad \text{and} \quad \frac{\partial z}{\partial y} = -\frac{F_y}{F_z}.$$

Use these equations to find the values of $\partial z/\partial x$ and $\partial z/\partial y$ at the points in Exercises 29–32.

29. $z^3 - xy + yz + y^3 - 2 = 0, \quad (1, 1, 1)$

30. $\frac{1}{x} + \frac{1}{y} + \frac{1}{z} - 1 = 0, \quad (2, 3, 6)$

31. $\sin(x + y) + \sin(y + z) + \sin(x + z) = 0, \quad (\pi, \pi, \pi)$

32. $xe^y + ye^z + 2\ln x - 2 - 3\ln 2 = 0, \quad (1, \ln 2, \ln 3)$

33. Find $\partial w/\partial r$ when $r = 1$, $s = -1$ if $w = (x + y + z)^2$, $x = r - s$, $y = \cos(r + s)$, $z = \sin(r + s)$.

34. Find $\partial w/\partial v$ when $u = -1$, $v = 2$ if $w = xy + \ln z$, $x = v^2/u$, $y = u + v$, $z = \cos u$.

35. Find $\partial w/\partial v$ when $u = 0$, $v = 0$ if $w = x^2 + (y/x)$, $x = u - 2v + 1$, $y = 2u + v - 2$.

36. Find $\partial z/\partial u$ when $u = 0$, $v = 1$ if $z = \sin xy + x\sin y$, $x = u^2 + v^2$, $y = uv$.

37. Find $\partial z/\partial u$ and $\partial z/\partial v$ when $u = \ln 2$, $v = 1$ if $z = 5\tan^{-1} x$ and $x = e^u + \ln v$.

38. Find $\partial z/\partial u$ and $\partial z/\partial v$ when $u = 1$ and $v = -2$ if $z = \ln q$ and $q = \sqrt{v + 3}\tan^{-1} u$.

39. Changing voltage in a circuit The voltage V in a circuit that satisfies the law $V = IR$ is slowly dropping as the battery wears out. At the same time, the resistance R is increasing as the resistor heats up. Use the equation

$$\frac{dV}{dt} = \frac{\partial V}{\partial I}\frac{dI}{dt} + \frac{\partial V}{\partial R}\frac{dR}{dt}$$

to find how the current is changing at the instant when $R = 600$ ohms, $I = 0.04$ amp, $dR/dt = 0.5$ ohm/sec, and $dV/dt = -0.01$ volt/sec.

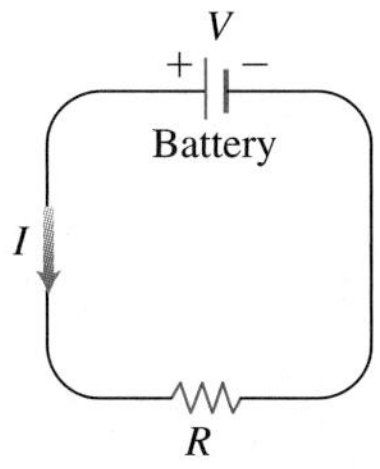

40. Changing dimensions in a box The lengths a, b, and c of the edges of a rectangular box are changing with time. At the instant in question, $a = 1$ m, $b = 2$ m, $c = 3$ m, $da/dt = db/dt = 1$ m/sec, and $dc/dt = -3$ m/sec. At what rates are the box's volume V and surface area S changing at that instant? Are the box's interior diagonals increasing in length or decreasing?

41. If $f(u, v, w)$ is differentiable and $u = x - y$, $v = y - z$, and $w = z - x$, show that

$$\frac{\partial f}{\partial x} + \frac{\partial f}{\partial y} + \frac{\partial f}{\partial z} = 0.$$

42. Polar coordinates Suppose that we substitute polar coordinates $x = r\cos\theta$ and $y = r\sin\theta$ in a differentiable function $w = f(x, y)$.

a. Show that

$$\frac{\partial w}{\partial r} = f_x\cos\theta + f_y\sin\theta$$

and

$$\frac{1}{r}\frac{\partial w}{\partial \theta} = -f_x\sin\theta + f_y\cos\theta.$$

b. Solve the equations in part (a) to express f_x and f_y in terms of $\partial w/\partial r$ and $\partial w/\partial\theta$.

c. Show that

$$(f_x)^2 + (f_y)^2 = \left(\frac{\partial w}{\partial r}\right)^2 + \frac{1}{r^2}\left(\frac{\partial w}{\partial \theta}\right)^2.$$

43. Laplace equations Show that if $w = f(u, v)$ satisfies the Laplace equation $f_{uu} + f_{vv} = 0$ and if $u = (x^2 - y^2)/2$ and $v = xy$, then w satisfies the Laplace equation $w_{xx} + w_{yy} = 0$.

44. Laplace equations Let $w = f(u) + g(v)$, where $u = x + iy$ and $v = x - iy$ and $i = \sqrt{-1}$. Show that w satisfies the Laplace equation $w_{xx} + w_{yy} = 0$ if all the necessary functions are differentiable.

45. Extreme values on a helix Suppose that the partial derivatives of a function $f(x, y, z)$ at points on the helix $x = \cos t$, $y = \sin t$, $z = t$ are

$$f_x = \cos t, \qquad f_y = \sin t, \qquad f_z = t^2 + t - 2.$$

At what points on the curve, if any, can f take on extreme values?

46. A space curve Let $w = x^2e^{2y}\cos 3z$. Find the value of dw/dt at the point $(1, \ln 2, 0)$ on the curve $x = \cos t$, $y = \ln(t + 2)$, $z = t$.

47. Temperature on a circle Let $T = f(x, y)$ be the temperature at the point (x, y) on the circle $x = \cos t$, $y = \sin t$, $0 \le t \le 2\pi$ and suppose that

$$\frac{\partial T}{\partial x} = 8x - 4y, \qquad \frac{\partial T}{\partial y} = 8y - 4x.$$

a. Find where the maximum and minimum temperatures on the circle occur by examining the derivatives dT/dt and d^2T/dt^2.

b. Suppose that $T = 4x^2 - 4xy + 4y^2$. Find the maximum and minimum values of T on the circle.

48. Temperature on an ellipse Let $T = g(x, y)$ be the temperature at the point (x, y) on the ellipse

$$x = 2\sqrt{2}\cos t, \qquad y = \sqrt{2}\sin t, \qquad 0 \le t \le 2\pi,$$

and suppose that

$$\frac{\partial T}{\partial x} = y, \qquad \frac{\partial T}{\partial y} = x.$$

a. Locate the maximum and minimum temperatures on the ellipse by examining dT/dt and d^2T/dt^2.

b. Suppose that $T = xy - 2$. Find the maximum and minimum values of T on the ellipse.

Differentiating Integrals Under mild continuity restrictions, it is true that if

$$F(x) = \int_a^b g(t, x)\,dt,$$

then $F'(x) = \int_a^b g_x(t, x)\,dt$. Using this fact and the Chain Rule, we can find the derivative of

$$F(x) = \int_a^{f(x)} g(t, x)\,dt$$

by letting

$$G(u, x) = \int_a^{u} g(t, x)\,dt,$$

where $u = f(x)$. Find the derivatives of the functions in Exercises 49 and 50.

49. $F(x) = \int_0^{x^2} \sqrt{t^4 + x^3}\,dt$

50. $F(x) = \int_{x^2}^{1} \sqrt{t^3 + x^2}\,dt$

11.5 Directional Derivatives and Gradient Vectors

If you look at the map (Figure 11.22) showing contours on the West Point Area along the Hudson River in New York, you will notice that the tributary streams flow perpendicular to the contours. The streams are following paths of steepest descent so the waters reach the Hudson as quickly as possible. Therefore, the instantaneous rate of change in a stream's altitude above sea level has a particular direction. In this section, you see why this direction, called the "downhill" direction, is perpendicular to the contours.

Directional Derivatives in the Plane

We know from Section 11.4 that if $f(x, y)$ is differentiable, then the rate at which f changes with respect to t along a differentiable curve $x = g(t)$, $y = h(t)$ is

$$\frac{df}{dt} = \frac{\partial f}{\partial x}\frac{dx}{dt} + \frac{\partial f}{\partial y}\frac{dy}{dt}.$$

At any point $P_0(x_0, y_0) = P_0(g(t_0), h(t_0))$, this equation gives the rate of change of f with respect to increasing t and therefore depends, among other things, on the direction of

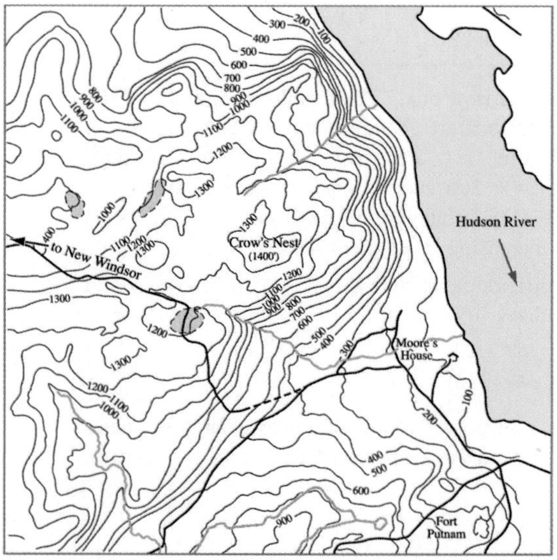

FIGURE 11.22 Contours of the West Point Area in New York show streams, which follow paths of steepest descent, running perpendicular to the contours.

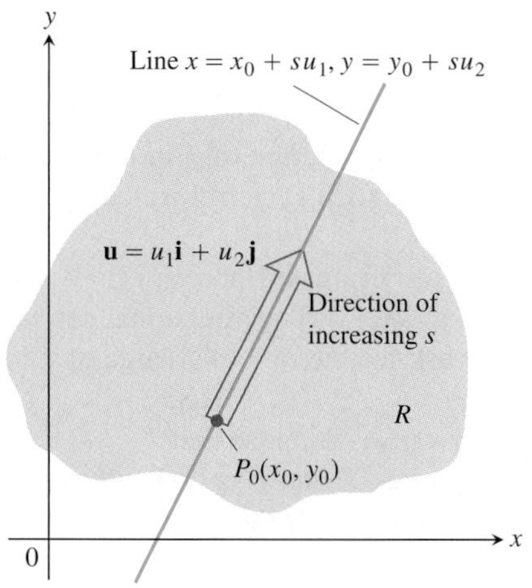

FIGURE 11.23 The rate of change of f in the direction of $\mathbf{u}$ at a point P_0 is the rate at which f changes along this line at P_0.

motion along the curve. If the curve is a straight line and t is the arc length parameter along the line measured from P_0 in the direction of a given unit vector $\mathbf{u}$, then df/dt is the rate of change of f with respect to distance in its domain in the direction of $\mathbf{u}$. By varying $\mathbf{u}$, we find the rates at which f changes with respect to distance as we move through P_0 in different directions. We now define this idea more precisely.

Suppose that the function $f(x, y)$ is defined throughout a region R in the xy-plane, that $P_0(x_0, y_0)$ is a point in R, and that $\mathbf{u} = u_1\mathbf{i} + u_2\mathbf{j}$ is a unit vector. Then the equations

$$x = x_0 + su_1, \qquad y = y_0 + su_2$$

parametrize the line through P_0 parallel to $\mathbf{u}$. If the parameter s measures arc length from P_0 in the direction of $\mathbf{u}$, we find the rate of change of f at P_0 in the direction of $\mathbf{u}$ by calculating df/ds at P_0 (Figure 11.23).

DEFINITION The **derivative of f at $P_0(x_0, y_0)$ in the direction of the unit vector $\mathbf{u} = u_1\mathbf{i} + u_2\mathbf{j}$** is the number

$$\left(\frac{df}{ds}\right)_{\mathbf{u},P_0} = \lim_{s\to 0} \frac{f(x_0 + su_1, y_0 + su_2) - f(x_0, y_0)}{s}, \tag{1}$$

provided the limit exists.

The directional derivative is also denoted by

$$(D_{\mathbf{u}}f)_{P_0}.$$

"The derivative of f at P_0 in the direction of $\mathbf{u}$"

EXAMPLE 1 Using the definition, find the derivative of

$$f(x, y) = x^2 + xy$$

at $P_0(1, 2)$ in the direction of the unit vector $\mathbf{u} = \left(1/\sqrt{2}\right)\mathbf{i} + \left(1/\sqrt{2}\right)\mathbf{j}$.

Solution

$$\begin{aligned}
\left(\frac{df}{ds}\right)_{\mathbf{u},P_0} &= \lim_{s\to 0} \frac{f(x_0 + su_1, y_0 + su_2) - f(x_0, y_0)}{s} \qquad \text{Eq. (1)}\\
&= \lim_{s\to 0} \frac{f\left(1 + s\cdot\frac{1}{\sqrt{2}}, 2 + s\cdot\frac{1}{\sqrt{2}}\right) - f(1, 2)}{s}\\
&= \lim_{s\to 0} \frac{\left(1 + \frac{s}{\sqrt{2}}\right)^2 + \left(1 + \frac{s}{\sqrt{2}}\right)\left(2 + \frac{s}{\sqrt{2}}\right) - (1^2 + 1\cdot 2)}{s}\\
&= \lim_{s\to 0} \frac{\left(1 + \frac{2s}{\sqrt{2}} + \frac{s^2}{2}\right) + \left(2 + \frac{3s}{\sqrt{2}} + \frac{s^2}{2}\right) - 3}{s}\\
&= \lim_{s\to 0} \frac{\frac{5s}{\sqrt{2}} + s^2}{s} = \lim_{s\to 0}\left(\frac{5}{\sqrt{2}} + s\right) = \frac{5}{\sqrt{2}}.
\end{aligned}$$

The rate of change of $f(x, y) = x^2 + xy$ at $P_0(1, 2)$ in the direction $\mathbf{u}$ is $5/\sqrt{2}$. ■

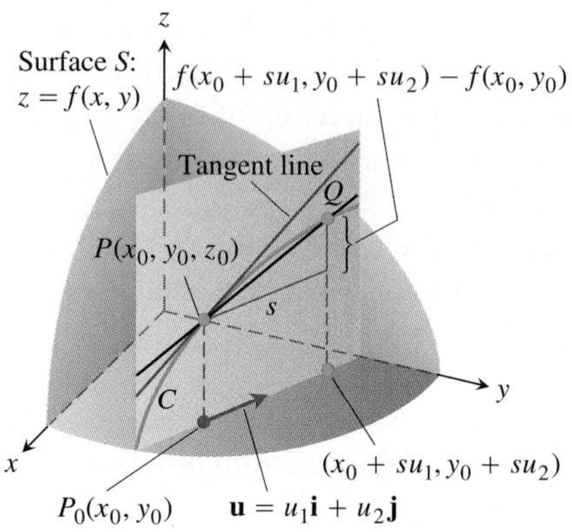

FIGURE 11.24 The slope of curve C at P_0 is $\lim_{Q\to P}$ slope (PQ); this is the directional derivative

$$\left(\frac{df}{ds}\right)_{\mathbf{u},P_0} = (D_{\mathbf{u}}f)_{P_0}.$$

Interpretation of the Directional Derivative

The equation $z = f(x, y)$ represents a surface S in space. If $z_0 = f(x_0, y_0)$, then the point $P(x_0, y_0, z_0)$ lies on S. The vertical plane that passes through P and $P_0(x_0, y_0)$ parallel to $\mathbf{u}$ intersects S in a curve C (Figure 11.24). The rate of change of f in the direction of $\mathbf{u}$ is the slope of the tangent to C at P in the right-handed system formed by the vectors $\mathbf{u}$ and $\mathbf{k}$.

When $\mathbf{u} = \mathbf{i}$, the directional derivative at P_0 is $\partial f/\partial x$ evaluated at (x_0, y_0). When $\mathbf{u} = \mathbf{j}$, the directional derivative at P_0 is $\partial f/\partial y$ evaluated at (x_0, y_0). The directional derivative generalizes the two partial derivatives. We can now ask for the rate of change of f in any direction $\mathbf{u}$, not just the directions $\mathbf{i}$ and $\mathbf{j}$.

Here's a physical interpretation of the directional derivative. Suppose that $T = f(x, y)$ is the temperature at each point (x, y) over a region in the plane. Then $f(x_0, y_0)$ is the temperature at the point $P_0(x_0, y_0)$ and $(D_{\mathbf{u}}f)_{P_0}$ is the instantaneous rate of change of the temperature at P_0 stepping off in the direction $\mathbf{u}$.

Calculation and Gradients

We now develop an efficient formula to calculate the directional derivative for a differentiable function f. We begin with the line

$$x = x_0 + su_1, \qquad y = y_0 + su_2, \tag{2}$$

through $P_0(x_0, y_0)$, parametrized with the arc length parameter s increasing in the direction of the unit vector $\mathbf{u} = u_1\mathbf{i} + u_2\mathbf{j}$. Then

$$\begin{aligned}
\left(\frac{df}{ds}\right)_{\mathbf{u},P_0} &= \left(\frac{\partial f}{\partial x}\right)_{P_0}\frac{dx}{ds} + \left(\frac{\partial f}{\partial y}\right)_{P_0}\frac{dy}{ds} && \text{Chain Rule for differentiable } f\\
&= \left(\frac{\partial f}{\partial x}\right)_{P_0}\cdot u_1 + \left(\frac{\partial f}{\partial y}\right)_{P_0}\cdot u_2 && \text{From Eqs. (2), } dx/ds = u_1 \text{ and } dy/ds = u_2\\
&= \underbrace{\left[\left(\frac{\partial f}{\partial x}\right)_{P_0}\mathbf{i} + \left(\frac{\partial f}{\partial y}\right)_{P_0}\mathbf{j}\right]}_{\text{Gradient of } f \text{ at } P_0}\cdot\underbrace{\left[u_1\mathbf{i} + u_2\mathbf{j}\right]}_{\text{Direction } \mathbf{u}}. && (3)
\end{aligned}$$

DEFINITION The **gradient vector (gradient)** of $f(x, y)$ at a point $P_0(x_0, y_0)$ is the vector

$$\nabla f = \frac{\partial f}{\partial x}\mathbf{i} + \frac{\partial f}{\partial y}\mathbf{j}$$

obtained by evaluating the partial derivatives of f at P_0.

The notation ∇f is read "grad f" as well as "gradient of f" and "del f." The symbol ∇ by itself is read "del." Another notation for the gradient is grad f, read the way it is written.

Equation (3) says that the derivative of a differentiable function f in the direction of $\mathbf{u}$ at P_0 is the dot product of $\mathbf{u}$ with the gradient of f at P_0.

THEOREM 9—The Directional Derivative Is a Dot Product If $f(x, y)$ is differentiable in an open region containing $P_0(x_0, y_0)$, then

$$\left(\frac{df}{ds}\right)_{\mathbf{u},P_0} = (\nabla f)_{P_0} \cdot \mathbf{u}, \tag{4}$$

the dot product of the gradient f at P_0 and $\mathbf{u}$.

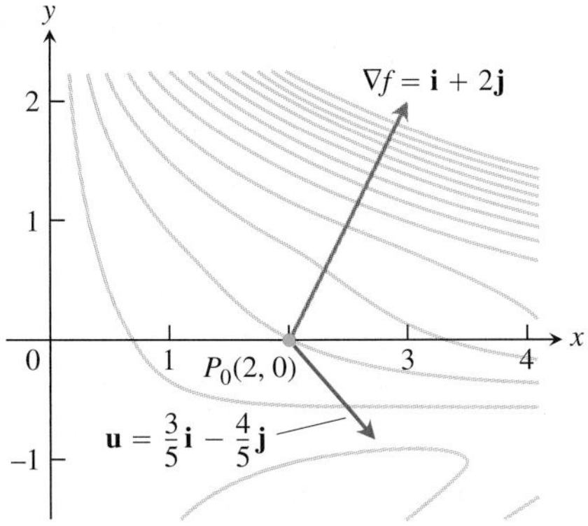

FIGURE 11.25 Picture ∇f as a vector in the domain of f. In the case of $f(x, y) = xe^y + \cos(xy)$, the domain is the entire plane. The rate at which f changes at $(2, 0)$ in the direction $\mathbf{u} = (3/5)\mathbf{i} - (4/5)\mathbf{j}$ is $\nabla f \cdot \mathbf{u} = -1$ (Example 2).

EXAMPLE 2 Find the derivative of $f(x, y) = xe^y + \cos(xy)$ at the point $(2, 0)$ in the direction of $\mathbf{v} = 3\mathbf{i} - 4\mathbf{j}$.

Solution The direction of $\mathbf{v}$ is the unit vector obtained by dividing $\mathbf{v}$ by its length:

$$\mathbf{u} = \frac{\mathbf{v}}{|\mathbf{v}|} = \frac{\mathbf{v}}{5} = \frac{3}{5}\mathbf{i} - \frac{4}{5}\mathbf{j}.$$

The partial derivatives of f are everywhere continuous and at $(2, 0)$ are given by

$$f_x(2, 0) = (e^y - y\sin(xy))_{(2,0)} = e^0 - 0 = 1$$

$$f_y(2, 0) = (xe^y - x\sin(xy))_{(2,0)} = 2e^0 - 2 \cdot 0 = 2.$$

The gradient of f at $(2, 0)$ is

$$\nabla f|_{(2,0)} = f_x(2, 0)\mathbf{i} + f_y(2, 0)\mathbf{j} = \mathbf{i} + 2\mathbf{j}$$

(Figure 11.25). The derivative of f at $(2, 0)$ in the direction of $\mathbf{v}$ is therefore

$$(D_{\mathbf{u}}f)|_{(2,0)} = \nabla f|_{(2,0)} \cdot \mathbf{u} \qquad \text{Eq. (4)}$$

$$= (\mathbf{i} + 2\mathbf{j}) \cdot \left(\frac{3}{5}\mathbf{i} - \frac{4}{5}\mathbf{j}\right) = \frac{3}{5} - \frac{8}{5} = -1. \qquad \blacksquare$$

Evaluating the dot product in the formula

$$D_{\mathbf{u}}f = \nabla f \cdot \mathbf{u} = |\nabla f||\mathbf{u}|\cos\theta = |\nabla f|\cos\theta,$$

where θ is the angle between the vectors $\mathbf{u}$ and ∇f, reveals the following properties.

Properties of the Directional Derivative $D_{\mathbf{u}}f = \nabla f \cdot \mathbf{u} = |\nabla f|\cos\theta$

1. The function f increases most rapidly when $\cos\theta = 1$ or when $\mathbf{u}$ is the direction of ∇f. That is, at each point P in its domain, f increases most rapidly in the direction of the gradient vector ∇f at P. The derivative in this direction is
$$D_{\mathbf{u}}f = |\nabla f|\cos(0) = |\nabla f|.$$
2. Similarly, f decreases most rapidly in the direction of $-\nabla f$. The derivative in this direction is $D_{\mathbf{u}}f = |\nabla f|\cos(\pi) = -|\nabla f|$.
3. Any direction $\mathbf{u}$ orthogonal to a gradient $\nabla f \neq 0$ is a direction of zero change in f because θ then equals $\pi/2$ and
$$D_{\mathbf{u}}f = |\nabla f|\cos(\pi/2) = |\nabla f| \cdot 0 = 0.$$

As we discuss later, these properties hold in three dimensions as well as two.

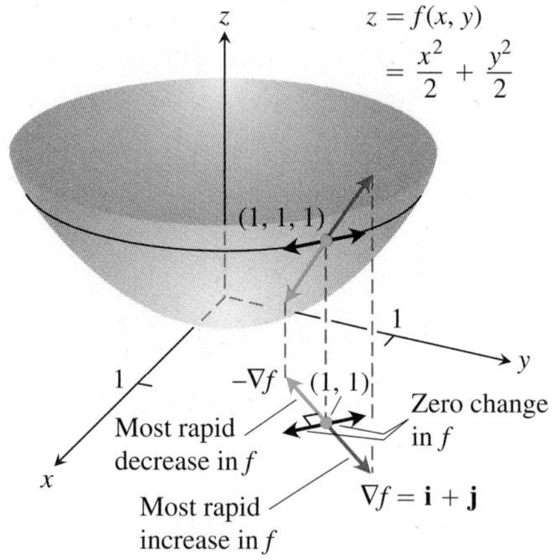

FIGURE 11.26 The direction in which $f(x, y)$ increases most rapidly at (1, 1) is the direction of $\nabla f|_{(1,1)} = \mathbf{i} + \mathbf{j}$. It corresponds to the direction of steepest ascent on the surface at (1, 1, 1) (Example 3).

EXAMPLE 3 Find the directions in which $f(x, y) = (x^2/2) + (y^2/2)$

(a) Increases most rapidly at the point (1, 1).

(b) Decreases most rapidly at (1, 1).

(c) What are the directions of zero change in f at (1, 1)?

Solution

(a) The function increases most rapidly in the direction of ∇f at (1, 1). The gradient there is

$$(\nabla f)_{(1,1)} = (x\mathbf{i} + y\mathbf{j})_{(1,1)} = \mathbf{i} + \mathbf{j}.$$

Its direction is

$$\mathbf{u} = \frac{\mathbf{i} + \mathbf{j}}{|\mathbf{i} + \mathbf{j}|} = \frac{\mathbf{i} + \mathbf{j}}{\sqrt{(1)^2 + (1)^2}} = \frac{1}{\sqrt{2}}\mathbf{i} + \frac{1}{\sqrt{2}}\mathbf{j}.$$

(b) The function decreases most rapidly in the direction of $-\nabla f$ at (1, 1), which is

$$-\mathbf{u} = -\frac{1}{\sqrt{2}}\mathbf{i} - \frac{1}{\sqrt{2}}\mathbf{j}.$$

(c) The directions of zero change at (1, 1) are the directions orthogonal to ∇f:

$$\mathbf{n} = -\frac{1}{\sqrt{2}}\mathbf{i} + \frac{1}{\sqrt{2}}\mathbf{j} \quad \text{and} \quad -\mathbf{n} = \frac{1}{\sqrt{2}}\mathbf{i} - \frac{1}{\sqrt{2}}\mathbf{j}.$$

See Figure 11.26.

Gradients and Tangents to Level Curves

If a differentiable function $f(x, y)$ has a constant value c along a smooth curve $\mathbf{r} = g(t)\mathbf{i} + h(t)\mathbf{j}$ (making the curve a level curve of f), then $f(g(t), h(t)) = c$. Differentiating both sides of this equation with respect to t leads to the equations

$$\frac{d}{dt} f(g(t), h(t)) = \frac{d}{dt}(c)$$

$$\frac{\partial f}{\partial x}\frac{dg}{dt} + \frac{\partial f}{\partial y}\frac{dh}{dt} = 0 \qquad \text{Chain Rule}$$

$$\underbrace{\left(\frac{\partial f}{\partial x}\mathbf{i} + \frac{\partial f}{\partial y}\mathbf{j}\right)}_{\nabla f} \cdot \underbrace{\left(\frac{dg}{dt}\mathbf{i} + \frac{dh}{dt}\mathbf{j}\right)}_{\frac{d\mathbf{r}}{dt}} = 0. \qquad (5)$$

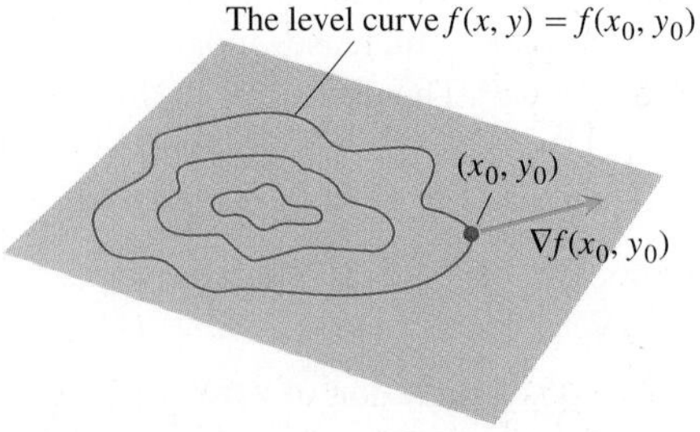

FIGURE 11.27 The gradient of a differentiable function of two variables at a point is always normal to the function's level curve through that point.

Equation (5) says that ∇f is normal to the tangent vector $d\mathbf{r}/dt$, so it is normal to the curve.

> At every point (x_0, y_0) in the domain of a differentiable function $f(x, y)$, the gradient of f is normal to the level curve through (x_0, y_0) (Figure 11.27).

Equation (5) validates our observation that streams flow perpendicular to the contours in topographical maps (see Figure 11.22). Since the downflowing stream will reach its

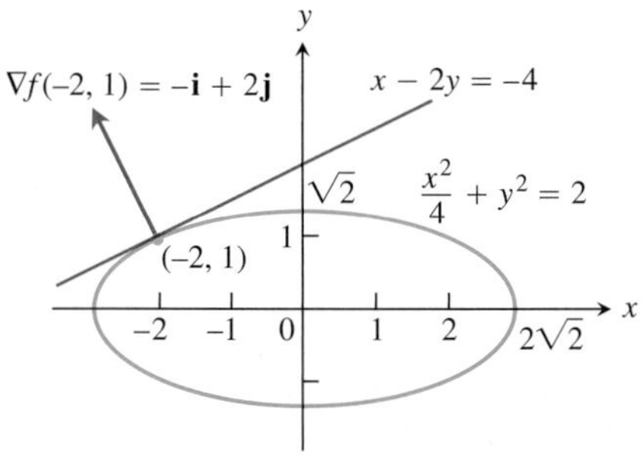

FIGURE 11.28 We can find the tangent to the ellipse $(x^2/4) + y^2 = 2$ by treating the ellipse as a level curve of the function $f(x, y) = (x^2/4) + y^2$ (Example 4).

destination in the fastest way, it must flow in the direction of the negative gradient vectors from Property 2 for the directional derivative. Equation (5) tells us these directions are perpendicular to the level curves.

This observation also enables us to find equations for tangent lines to level curves. They are the lines normal to the gradients. The line through a point $P_0(x_0, y_0)$ normal to a vector $\mathbf{N} = A\mathbf{i} + B\mathbf{j}$ has the equation

$$A(x - x_0) + B(y - y_0) = 0$$

(Exercise 35). If $\mathbf{N}$ is the gradient $(\nabla f)_{(x_0, y_0)} = f_x(x_0, y_0)\mathbf{i} + f_y(x_0, y_0)\mathbf{j}$, the equation is the tangent line given by

$$f_x(x_0, y_0)(x - x_0) + f_y(x_0, y_0)(y - y_0) = 0. \tag{6}$$

EXAMPLE 4 Find an equation for the tangent to the ellipse

$$\frac{x^2}{4} + y^2 = 2$$

(Figure 11.28) at the point $(-2, 1)$.

Solution The ellipse is a level curve of the function

$$f(x, y) = \frac{x^2}{4} + y^2.$$

The gradient of f at $(-2, 1)$ is

$$\nabla f\big|_{(-2,1)} = \left(\frac{x}{2}\mathbf{i} + 2y\mathbf{j}\right)_{(-2,1)} = -\mathbf{i} + 2\mathbf{j}.$$

The tangent is the line

$$(-1)(x + 2) + (2)(y - 1) = 0 \qquad \text{Eq. (6)}$$
$$x - 2y = -4.$$

■

If we know the gradients of two functions f and g, we automatically know the gradients of their constant multiples, sum, difference, product, and quotient. You are asked to establish the following rules in Exercise 36. Notice that these rules have the same form as the corresponding rules for derivatives of single-variable functions.

Algebra Rules for Gradients

1. *Constant Multiple Rule:* $\nabla(kf) = k\nabla f$ (any number k)
2. *Sum Rule:* $\nabla(f + g) = \nabla f + \nabla g$
3. *Difference Rule:* $\nabla(f - g) = \nabla f - \nabla g$
4. *Product Rule:* $\nabla(fg) = f\nabla g + g\nabla f$
5. *Quotient Rule:* $\nabla\left(\frac{f}{g}\right) = \frac{g\nabla f - f\nabla g}{g^2}$

EXAMPLE 5 We illustrate two of the rules with

$$f(x, y) = x - y \qquad g(x, y) = 3y$$
$$\nabla f = \mathbf{i} - \mathbf{j} \qquad \nabla g = 3\mathbf{j}.$$

We have

1. $\nabla(f - g) = \nabla(x - 4y) = \mathbf{i} - 4\mathbf{j} = \nabla f - \nabla g$ Rule 3

2. $$\begin{aligned}\nabla(fg) &= \nabla(3xy - 3y^2) = 3y\mathbf{i} + (3x - 6y)\mathbf{j} \\ &= 3y(\mathbf{i} - \mathbf{j}) + 3y\mathbf{j} + (3x - 6y)\mathbf{j} \\ &= 3y(\mathbf{i} - \mathbf{j}) + (3x - 3y)\mathbf{j} \\ &= 3y(\mathbf{i} - \mathbf{j}) + (x - y)3\mathbf{j} = g\nabla f + f\nabla g \quad \text{Rule 4}\end{aligned}$$

■

Functions of Three Variables

For a differentiable function $f(x, y, z)$ and a unit vector $\mathbf{u} = u_1\mathbf{i} + u_2\mathbf{j} + u_3\mathbf{k}$ in space, we have

$$\nabla f = \frac{\partial f}{\partial x}\mathbf{i} + \frac{\partial f}{\partial y}\mathbf{j} + \frac{\partial f}{\partial z}\mathbf{k}$$

and

$$D_{\mathbf{u}}f = \nabla f \cdot \mathbf{u} = \frac{\partial f}{\partial x}u_1 + \frac{\partial f}{\partial y}u_2 + \frac{\partial f}{\partial z}u_3.$$

The directional derivative can once again be written in the form

$$D_{\mathbf{u}}f = \nabla f \cdot \mathbf{u} = |\nabla f||u|\cos\theta = |\nabla f|\cos\theta,$$

so the properties listed earlier for functions of two variables continue to hold. At any given point, f increases most rapidly in the direction of ∇f and decreases most rapidly in the direction of $-\nabla f$. In any direction orthogonal to ∇f, the derivative is zero.

EXAMPLE 6

(a) Find the derivative of $f(x, y, z) = x^3 - xy^2 - z$ at $P_0(1, 1, 0)$ in the direction of $\mathbf{v} = 2\mathbf{i} - 3\mathbf{j} + 6\mathbf{k}$.

(b) In what directions does f change most rapidly at P_0, and what are the rates of change in these directions?

Solution

(a) The direction of $\mathbf{v}$ is obtained by dividing $\mathbf{v}$ by its length:

$$|\mathbf{v}| = \sqrt{(2)^2 + (-3)^2 + (6)^2} = \sqrt{49} = 7$$

$$\mathbf{u} = \frac{\mathbf{v}}{|\mathbf{v}|} = \frac{2}{7}\mathbf{i} - \frac{3}{7}\mathbf{j} + \frac{6}{7}\mathbf{k}.$$

The partial derivatives of f at P_0 are

$f_x = (3x^2 - y^2)_{(1,1,0)} = 2, \quad f_y = -2xy|_{(1,1,0)} = -2, \quad f_z = -1|_{(1,1,0)} = -1.$

The gradient of f at P_0 is

$$\nabla f|_{(1,1,0)} = 2\mathbf{i} - 2\mathbf{j} - \mathbf{k}.$$

The derivative of f at P_0 in the direction of $\mathbf{v}$ is therefore

$$(D_{\mathbf{u}}f)_{(1,1,0)} = \nabla f|_{(1,1,0)} \cdot \mathbf{u} = (2\mathbf{i} - 2\mathbf{j} - \mathbf{k}) \cdot \left(\frac{2}{7}\mathbf{i} - \frac{3}{7}\mathbf{j} + \frac{6}{7}\mathbf{k}\right)$$
$$= \frac{4}{7} + \frac{6}{7} - \frac{6}{7} = \frac{4}{7}.$$

(b) The function increases most rapidly in the direction of $\nabla f = 2\mathbf{i} - 2\mathbf{j} - \mathbf{k}$ and decreases most rapidly in the direction of $-\nabla f$. The rates of change in the directions are, respectively,

$$|\nabla f| = \sqrt{(2)^2 + (-2)^2 + (-1)^2} = \sqrt{9} = 3 \quad \text{and} \quad -|\nabla f| = -3. \qquad \blacksquare$$

EXERCISES 11.5

In Exercises 1–4, find the gradient of the function at the given point. Then sketch the gradient together with the level curve that passes through the point.

1. $f(x, y) = y - x, \quad (2, 1)$ **2.** $f(x, y) = \ln(x^2 + y^2), \quad (1, 1)$

3. $g(x, y) = y - x^2, \quad (-1, 0)$ **4.** $g(x, y) = \frac{x^2}{2} - \frac{y^2}{2}, \quad (\sqrt{2}, 1)$

In Exercises 5–8, find ∇f at the given point.

5. $f(x, y, z) = x^2 + y^2 - 2z^2 + z \ln x, \quad (1, 1, 1)$

6. $f(x, y, z) = 2z^3 - 3(x^2 + y^2)z + \tan^{-1} xz, \quad (1, 1, 1)$

7. $f(x, y, z) = (x^2 + y^2 + z^2)^{-1/2} + \ln(xyz), \quad (-1, 2, -2)$

8. $f(x, y, z) = e^{x+y} \cos z + (y + 1) \sin^{-1} x, \quad (0, 0, \pi/6)$

In Exercises 9–16, find the derivative of the function at P_0 in the direction of $\mathbf{A}$.

9. $f(x, y) = 2xy - 3y^2, \quad P_0(5, 5), \quad \mathbf{A} = 4\mathbf{i} + 3\mathbf{j}$

10. $f(x, y) = 2x^2 + y^2, \quad P_0(-1, 1), \quad \mathbf{A} = 3\mathbf{i} - 4\mathbf{j}$

11. $g(x, y) = x - (y^2/x) + \sqrt{3} \sec^{-1}(2xy), \quad P_0(1, 1),$
$\mathbf{A} = 12\mathbf{i} + 5\mathbf{j}$

12. $h(x, y) = \tan^{-1}(y/x) + \sqrt{3} \sin^{-1}(xy/2), \quad P_0(1, 1),$
$\mathbf{A} = 3\mathbf{i} - 2\mathbf{j}$

13. $f(x, y, z) = xy + yz + zx, \quad P_0(1, -1, 2), \quad \mathbf{A} = 3\mathbf{i} + 6\mathbf{j} - 2\mathbf{k}$

14. $f(x, y, z) = x^2 + 2y^2 - 3z^2, \quad P_0(1, 1, 1), \quad \mathbf{A} = \mathbf{i} + \mathbf{j} + \mathbf{k}$

15. $g(x, y, z) = 3e^x \cos yz, \quad P_0(0, 0, 0), \quad \mathbf{A} = 2\mathbf{i} + \mathbf{j} - 2\mathbf{k}$

16. $h(x, y, z) = \cos xy + e^{yz} + \ln zx, \quad P_0(1, 0, 1/2),$
$\mathbf{A} = \mathbf{i} + 2\mathbf{j} + 2\mathbf{k}$

In Exercises 17–22, find the directions in which the functions increase and decrease most rapidly at P_0. Then find the derivatives of the functions in these directions.

17. $f(x, y) = x^2 + xy + y^2, \quad P_0(-1, 1)$

18. $f(x, y) = x^2y + e^{xy} \sin y, \quad P_0(1, 0)$

19. $f(x, y, z) = (x/y) - yz, \quad P_0(4, 1, 1)$

20. $g(x, y, z) = xe^y + z^2, \quad P_0(1, \ln 2, 1/2)$

21. $f(x, y, z) = \ln xy + \ln yz + \ln xz, \quad P_0(1, 1, 1)$

22. $h(x, y, z) = \ln(x^2 + y^2 - 1) + y + 6z, \quad P_0(1, 1, 0)$

In Exercises 23–26, sketch the curve $f(x, y) = c$ together with ∇f and the tangent line at the given point. Then write an equation for the tangent line.

23. $x^2 + y^2 = 4, \quad (\sqrt{2}, \sqrt{2})$

24. $x^2 - y = 1, \quad (\sqrt{2}, 1)$

25. $xy = -4, \quad (2, -2)$

26. $x^2 - xy + y^2 = 7, \quad (-1, 2)$

27. Zero directional derivative In what direction is the derivative of $f(x, y) = xy + y^2$ at $P(3, 2)$ equal to zero?

28. Zero directional derivative In what directions is the derivative of $f(x, y) = (x^2 - y^2)/(x^2 + y^2)$ at $P(1, 1)$ equal to zero?

29. Is there a direction $\mathbf{u}$ in which the rate of change of $f(x, y) = x^2 - 3xy + 4y^2$ at $P(1, 2)$ equals 14? Give reasons for your answer.

30. Changing temperature along a circle Is there a direction $\mathbf{u}$ in which the rate of change of the temperature function $T(x, y, z) = 2xy - yz$ (temperature in degrees Celsius, distance in feet) at $P(1, -1, 1)$ is $-3°\text{C/ft}$? Give reasons for your answer.

31. The derivative of $f(x, y)$ at $P_0(1, 2)$ in the direction of $\mathbf{i} + \mathbf{j}$ is $2\sqrt{2}$ and in the direction of $-2\mathbf{j}$ is -3. What is the derivative of f in the direction of $-\mathbf{i} - 2\mathbf{j}$? Give reasons for your answer.

32. The derivative of $f(x, y, z)$ at a point P is greatest in the direction of $\mathbf{v} = \mathbf{i} + \mathbf{j} - \mathbf{k}$. In this direction, the value of the derivative is $2\sqrt{3}$.

a. What is ∇f at P? Give reasons for your answer.

b. What is the derivative of f at P in the direction of $\mathbf{i} + \mathbf{j}$?

33. Directional derivatives and scalar components How is the derivative of a differentiable function $f(x, y, z)$ at a point P_0 in the direction of a unit vector $\mathbf{u}$ related to the scalar component of $(\nabla f)_{P_0}$ in the direction of $\mathbf{u}$? Give reasons for your answer.

34. Directional derivatives and partial derivatives Assuming that the necessary derivatives of $f(x, y, z)$ are defined, how are $D_{\mathbf{i}}f$, $D_{\mathbf{j}}f$, and $D_{\mathbf{k}}f$ related to f_x, f_y, and f_z? Give reasons for your answer.

35. Lines in the *xy*-plane Show that $A(x - x_0) + B(y - y_0) = 0$ is an equation for the line in the xy-plane through the point (x_0, y_0) normal to the vector $\mathbf{N} = A\mathbf{i} + B\mathbf{j}$.

36. The algebra rules for gradients Given a constant k and the gradients

$$\nabla f = \frac{\partial f}{\partial x}\mathbf{i} + \frac{\partial f}{\partial y}\mathbf{j} + \frac{\partial f}{\partial z}\mathbf{k}, \qquad \nabla g = \frac{\partial g}{\partial x}\mathbf{i} + \frac{\partial g}{\partial y}\mathbf{j} + \frac{\partial g}{\partial z}\mathbf{k},$$

establish the algebra rules for gradients.

11.6 Tangent Planes and Differentials

In this section we define the tangent plane at a point on a smooth surface in space. Then we show how to calculate an equation of the tangent plane from the partial derivatives of the function defining the surface. This idea is similar to the definition of the tangent line at a point on a curve in the coordinate plane for single-variable functions (Section 2.1). We then study the total differential and linearization of functions of several variables.

Tangent Planes and Normal Lines

If $\mathbf{r} = g(t)\mathbf{i} + h(t)\mathbf{j} + k(t)\mathbf{k}$ is a smooth curve on the level surface $f(x, y, z) = c$ of a differentiable function f, then $f(g(t), h(t), k(t)) = c$. Differentiating both sides of this equation with respect to t leads to

$$\frac{d}{dt}f(g(t), h(t), k(t)) = \frac{d}{dt}(c)$$

$$\frac{\partial f}{\partial x}\frac{dg}{dt} + \frac{\partial f}{\partial y}\frac{dh}{dt} + \frac{\partial f}{\partial z}\frac{dk}{dt} = 0 \qquad \text{Chain Rule}$$

$$\underbrace{\left(\frac{\partial f}{\partial x}\mathbf{i} + \frac{\partial f}{\partial y}\mathbf{j} + \frac{\partial f}{\partial z}\mathbf{k}\right)}_{\nabla f} \cdot \underbrace{\left(\frac{dg}{dt}\mathbf{i} + \frac{dh}{dt}\mathbf{j} + \frac{dk}{dt}\mathbf{k}\right)}_{d\mathbf{r}/dt} = 0. \qquad (1)$$

At every point along the curve, ∇f is orthogonal to the curve's velocity vector.

Now let us restrict our attention to the curves that pass through P_0 (Figure 11.29). All the velocity vectors at P_0 are orthogonal to ∇f at P_0, so the curves' tangent lines all lie in the plane through P_0 normal to ∇f. We call this plane the tangent plane of the surface at P_0. The line through P_0 perpendicular to the plane is the surface's normal line at P_0.

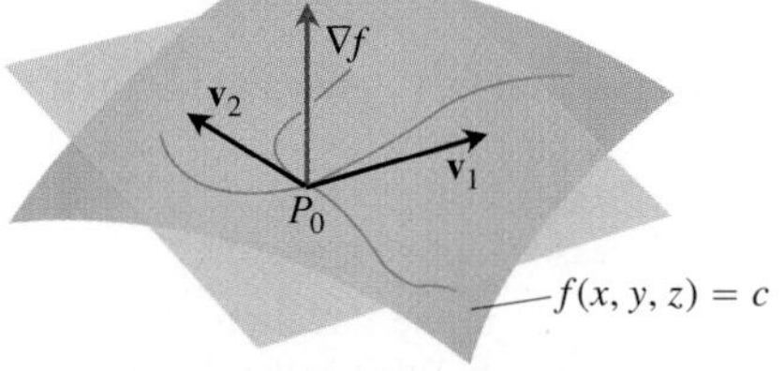

FIGURE 11.29 The gradient ∇f is orthogonal to the velocity vector of every smooth curve in the surface through P_0. The velocity vectors at P_0 therefore lie in a common plane, which we call the tangent plane at P_0.

DEFINITIONS The **tangent plane** at the point $P_0(x_0, y_0, z_0)$ on the level surface $f(x, y, z) = c$ of a differentiable function f is the plane through P_0 normal to $\nabla f|_{P_0}$.

The **normal line** of the surface at P_0 is the line through P_0 parallel to $\nabla f|_{P_0}$.

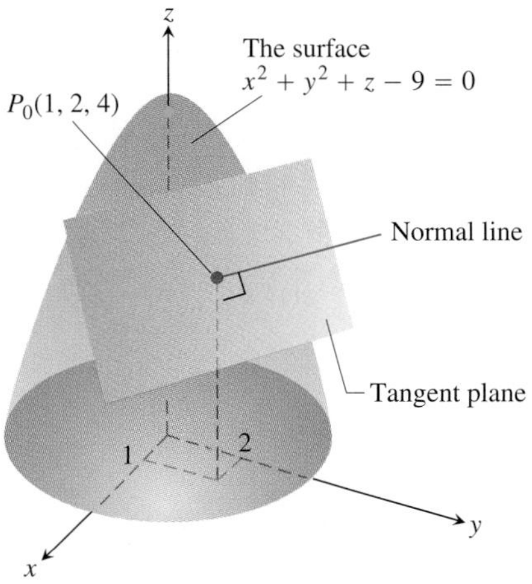

FIGURE 11.30 The tangent plane and normal line to this surface at P_0 (Example 1).

Thus, from Section 9.5, the tangent plane and normal line have the following equations:

Tangent Plane to $f(x, y, z) = c$ at $P_0(x_0, y_0, z_0)$

$$f_x(P_0)(x - x_0) + f_y(P_0)(y - y_0) + f_z(P_0)(z - z_0) = 0 \quad (2)$$

Normal Line to $f(x, y, z) = c$ at $P_0(x_0, y_0, z_0)$

$$x = x_0 + f_x(P_0)t, \qquad y = y_0 + f_y(P_0)t, \qquad z = z_0 + f_z(P_0)t \quad (3)$$

EXAMPLE 1 Find the tangent plane and normal line of the surface

$$f(x, y, z) = x^2 + y^2 + z - 9 = 0 \qquad \text{A circular paraboloid}$$

at the point $P_0(1, 2, 4)$.

Solution The surface is shown in Figure 11.30.

The tangent plane is the plane through P_0 perpendicular to the gradient of f at P_0. The gradient is

$$\nabla f|_{P_0} = (2x\mathbf{i} + 2y\mathbf{j} + \mathbf{k})_{(1,2,4)} = 2\mathbf{i} + 4\mathbf{j} + \mathbf{k}.$$

The tangent plane is therefore the plane

$$2(x - 1) + 4(y - 2) + (z - 4) = 0, \qquad \text{or} \qquad 2x + 4y + z = 14.$$

The line normal to the surface at P_0 is

$$x = 1 + 2t, \qquad y = 2 + 4t, \qquad z = 4 + t.$$ ■

To find an equation for the plane tangent to a smooth surface $z = f(x, y)$ at a point $P_0(x_0, y_0, z_0)$ where $z_0 = f(x_0, y_0)$, we first observe that the equation $z = f(x, y)$ is equivalent to $f(x, y) - z = 0$. The surface $z = f(x, y)$ is therefore the zero level surface of the function $F(x, y, z) = f(x, y) - z$. The partial derivatives of F are

$$F_x = \frac{\partial}{\partial x}(f(x, y) - z) = f_x - 0 = f_x$$

$$F_y = \frac{\partial}{\partial y}(f(x, y) - z) = f_y - 0 = f_y$$

$$F_z = \frac{\partial}{\partial z}(f(x, y) - z) = 0 - 1 = -1.$$

The formula

$$F_x(P_0)(x - x_0) + F_y(P_0)(y - y_0) + F_z(P_0)(z - z_0) = 0$$

for the plane tangent to the level surface at P_0 therefore reduces to

$$f_x(x_0, y_0)(x - x_0) + f_y(x_0, y_0)(y - y_0) - (z - z_0) = 0.$$

Plane Tangent to a Surface $z = f(x, y)$ at $(x_0, y_0, f(x_0, y_0))$

The plane tangent to the surface $z = f(x, y)$ of a differentiable function f at the point $P_0(x_0, y_0, z_0) = (x_0, y_0, f(x_0, y_0))$ is

$$f_x(x_0, y_0)(x - x_0) + f_y(x_0, y_0)(y - y_0) - (z - z_0) = 0. \quad (4)$$

EXAMPLE 2 Find the plane tangent to the surface $z = x\cos y - ye^x$ at $(0, 0, 0)$.

Solution We calculate the partial derivatives of $f(x, y) = x\cos y - ye^x$ and use Equation (4):

$$f_x(0, 0) = (\cos y - ye^x)_{(0,0)} = 1 - 0 \cdot 1 = 1$$

$$f_y(0, 0) = (-x\sin y - e^x)_{(0,0)} = 0 - 1 = -1.$$

The tangent plane is therefore

$$1 \cdot (x - 0) - 1 \cdot (y - 0) - (z - 0) = 0, \qquad \text{Eq. (4)}$$

or

$$x - y - z = 0.$$

■

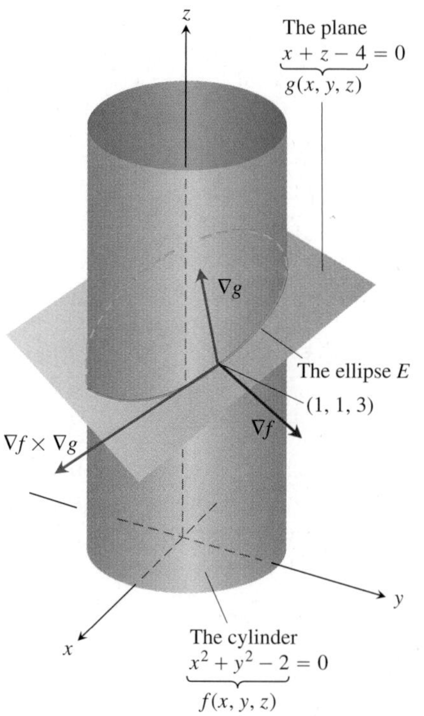

FIGURE 11.31 This cylinder and plane intersect in an ellipse E (Example 3).

EXAMPLE 3 The surfaces

$$f(x, y, z) = x^2 + y^2 - 2 = 0 \qquad \text{A cylinder}$$

and

$$g(x, y, z) = x + z - 4 = 0 \qquad \text{A plane}$$

meet in an ellipse E (Figure 11.31). Find parametric equations for the line tangent to E at the point $P_0(1, 1, 3)$.

Solution The tangent line is orthogonal to both ∇f and ∇g at P_0, and therefore parallel to $\mathbf{v} = \nabla f \times \nabla g$. The components of $\mathbf{v}$ and the coordinates of P_0 give us equations for the line. We have

$$\nabla f|_{(1,1,3)} = (2x\mathbf{i} + 2y\mathbf{j})_{(1,1,3)} = 2\mathbf{i} + 2\mathbf{j}$$

$$\nabla g|_{(1,1,3)} = (\mathbf{i} + \mathbf{k})_{(1,1,3)} = \mathbf{i} + \mathbf{k}$$

$$\mathbf{v} = (2\mathbf{i} + 2\mathbf{j}) \times (\mathbf{i} + \mathbf{k}) = \begin{vmatrix} \mathbf{i} & \mathbf{j} & \mathbf{k} \\ 2 & 2 & 0 \\ 1 & 0 & 1 \end{vmatrix} = 2\mathbf{i} - 2\mathbf{j} - 2\mathbf{k}.$$

The tangent line is

$$x = 1 + 2t, \qquad y = 1 - 2t, \qquad z = 3 - 2t.$$

■

Estimating Change in a Specific Direction

The directional derivative plays the role of an ordinary derivative when we want to estimate how much the value of a function f changes if we move a small distance ds from a point P_0 to another point nearby. If f were a function of a single variable, we would have

$$df = f'(P_0)\,ds. \qquad \text{Ordinary derivative} \times \text{increment}$$

For a function of two or more variables, we use the formula

$$df = (\nabla f|_{P_0} \cdot \mathbf{u})\,ds, \qquad \text{Directional derivative} \times \text{increment}$$

where $\mathbf{u}$ is the direction of the motion away from P_0.

Estimating the Change in f in a Direction u

To estimate the change in the value of a differentiable function f when we move a small distance ds from a point P_0 in a particular direction $\mathbf{u}$, use the formula

$$df = \underbrace{(\nabla f|_{P_0} \cdot \mathbf{u})}_{\text{Directional derivative}} \cdot \underbrace{ds}_{\text{Distance increment}}$$

EXAMPLE 4 Estimate how much the value of

$$f(x, y, z) = y \sin x + 2yz$$

will change if the point $P(x, y, z)$ moves 0.1 unit from $P_0(0, 1, 0)$ straight toward $P_1(2, 2, -2)$.

Solution We first find the derivative of f at P_0 in the direction of the vector $\overrightarrow{P_0P_1} = 2\mathbf{i} + \mathbf{j} - 2\mathbf{k}$. The direction of this vector is

$$\mathbf{u} = \frac{\overrightarrow{P_0P_1}}{|\overrightarrow{P_0P_1}|} = \frac{\overrightarrow{P_0P_1}}{3} = \frac{2}{3}\mathbf{i} + \frac{1}{3}\mathbf{j} - \frac{2}{3}\mathbf{k}.$$

The gradient of f at P_0 is

$$\nabla f|_{(0,1,0)} = ((y \cos x)\mathbf{i} + (\sin x + 2z)\mathbf{j} + 2y\mathbf{k})_{(0,1,0)} = \mathbf{i} + 2\mathbf{k}.$$

Therefore,

$$\nabla f|_{P_0} \cdot \mathbf{u} = (\mathbf{i} + 2\mathbf{k}) \cdot \left(\frac{2}{3}\mathbf{i} + \frac{1}{3}\mathbf{j} - \frac{2}{3}\mathbf{k}\right) = \frac{2}{3} - \frac{4}{3} = -\frac{2}{3}.$$

The change df in f that results from moving $ds = 0.1$ unit away from P_0 in the direction of $\mathbf{u}$ is approximately

$$df = (\nabla f|_{P_0} \cdot \mathbf{u})(ds) = \left(-\frac{2}{3}\right)(0.1) \approx -0.067 \text{ unit.}$$ ■

How to Linearize a Function of Two Variables

Functions of two variables can be complicated, and we sometimes need to replace them with simpler ones that give the accuracy required for specific applications without being so difficult to work with. We do this in a way that is similar to the way we find linear replacements for functions of a single variable (Section 2.13).

Suppose the function we wish to replace is $z = f(x, y)$ and that we want the replacement to be effective near a point (x_0, y_0) at which we know the values of f, f_x, and f_y and at which f is differentiable. If we move from (x_0, y_0) to any point (x, y) by increments $\Delta x = x - x_0$ and $\Delta y = y - y_0$, then the definition of differentiability from Section 11.3 gives the change

$$f(x, y) - f(x_0, y_0) = f_x(x_0, y_0)\Delta x + f_y(x_0, y_0)\Delta y + \epsilon_1 \Delta x + \epsilon_2 \Delta y,$$

where $\epsilon_1, \epsilon_2 \to 0$ as $\Delta x, \Delta y \to 0$. If the increments Δx and Δy are small, the products $\epsilon_1 \Delta x$ and $\epsilon_2 \Delta y$ will eventually be smaller still and we will have

$$f(x, y) \approx \underbrace{f(x_0, y_0) + f_x(x_0, y_0)(x - x_0) + f_y(x_0, y_0)(y - y_0)}_{L(x, y)}.$$

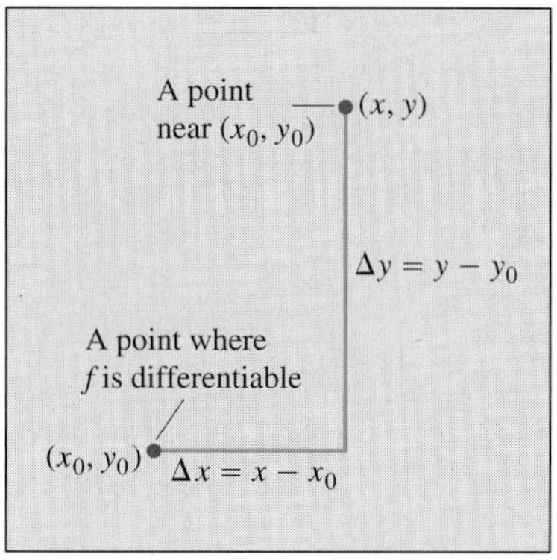

FIGURE 11.32 If f is differentiable at (x_0, y_0), then the value of f at any point (x, y) nearby is approximately $f(x_0, y_0) + f_x(x_0, y_0)\Delta x + f_y(x_0, y_0)\Delta y$.

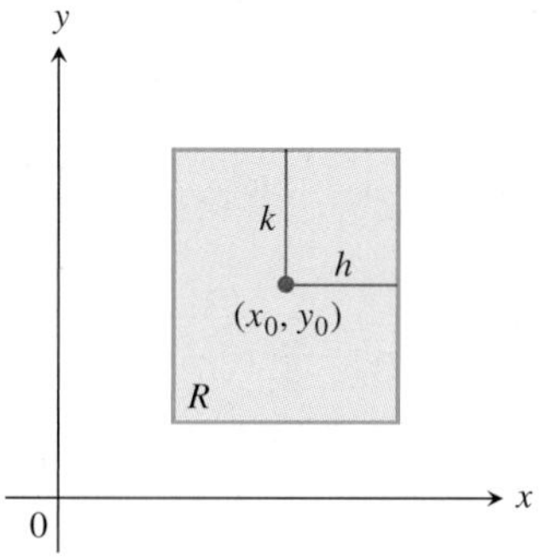

FIGURE 11.33 The rectangular region R: $|x - x_0| \le h$, $|y - y_0| \le k$ in the xy-plane.

In other words, as long as Δx and Δy are small, f will have approximately the same value as the linear function L. If f is hard to use, and our work can tolerate the error involved, we may approximate f by L (Figure 11.32).

DEFINITIONS The **linearization** of a function $f(x, y)$ at a point (x_0, y_0) where f is differentiable is the function

$$L(x, y) = f(x_0, y_0) + f_x(x_0, y_0)(x - x_0) + f_y(x_0, y_0)(y - y_0). \quad (5)$$

The approximation

$$f(x, y) \approx L(x, y)$$

is the **standard linear approximation** of f at (x_0, y_0).

From Equation (4), we see that the plane $z = L(x, y)$ is tangent to the surface $z = f(x, y)$ at the point (x_0, y_0). Thus, the linearization of a function of two variables is a tangent-*plane* approximation in the same way that the linearization of a function of a single variable is a tangent-*line* approximation.

When approximating a differentiable function $f(x, y)$ by its linearization $L(x, y)$ at (x_0, y_0), an important question is how accurate the approximation might be.

If we can find a common upper bound M for $|f_{xx}|$, $|f_{yy}|$, and $|f_{xy}|$ on a rectangle R centered at (x_0, y_0) (Figure 11.33), then we can bound the error E throughout R by using a simple formula (derived in Appendix 12). The **error** is defined by $E(x, y) = f(x, y) - L(x, y)$.

The Error in the Standard Linear Approximation

If f has continuous first and second partial derivatives throughout an open set containing a rectangle R centered at (x_0, y_0) and if M is any upper bound for the values of $|f_{xx}|$, $|f_{yy}|$, and $|f_{xy}|$ on R, then the error $E(x, y)$ incurred in replacing $f(x, y)$ on R by its linearization

$$L(x, y) = f(x_0, y_0) + f_x(x_0, y_0)(x - x_0) + f_y(x_0, y_0)(y - y_0)$$

satisfies the inequality

$$|E(x, y)| \le \frac{1}{2}M(|x - x_0| + |y - y_0|)^2.$$

To make $|E(x, y)|$ small for a given M, we just make $|x - x_0|$ and $|y - y_0|$ small.

Differentials

Recall from Section 2.13 that for a function of a single variable, $y = f(x)$, we defined the change in f as x changes from a to $a + \Delta x$ by

$$\Delta f = f(a + \Delta x) - f(a)$$

and the differential of f as

$$df = f'(a)\Delta x.$$

We now consider a function of two variables.

Suppose a differentiable function $f(x, y)$ and its partial derivatives exist at a point (x_0, y_0). If we move to a nearby point $(x_0 + \Delta x, y_0 + \Delta y)$, the change in f is

$$\Delta f = f(x_0 + \Delta x, y_0 + \Delta y) - f(x_0, y_0).$$

A straightforward calculation from the definition of $L(x, y)$, using the notation $x - x_0 = \Delta x$ and $y - y_0 = \Delta y$, shows that the corresponding change in L is

$$\begin{aligned}\Delta L &= L(x_0 + \Delta x, y_0 + \Delta y) - L(x_0, y_0)\\ &= f_x(x_0, y_0)\Delta x + f_y(x_0, y_0)\Delta y.\end{aligned}$$

The **differentials** dx and dy are independent variables, so they can be assigned any values. Often we take $dx = \Delta x = x - x_0$, and $dy = \Delta y = y - y_0$. We then have the following definition of the differential or *total* differential of f.

DEFINITION If we move from (x_0, y_0) to a point $(x_0 + dx, y_0 + dy)$ nearby, the resulting change

$$df = f_x(x_0, y_0)\,dx + f_y(x_0, y_0)\,dy$$

in the linearization of f is called the **total differential of f**.

EXAMPLE 5 Your company manufactures right circular cylindrical molasses storage tanks that are 25 ft high with a radius of 5 ft. How sensitive are the tanks' volumes to small variations in height and radius?

Solution With $V = \pi r^2 h$, the total differential gives the approximation for the change in volume as

$$\begin{aligned}dV &= V_r(5, 25)\,dr + V_h(5, 25)\,dh\\ &= (2\pi rh)_{(5,25)}\,dr + (\pi r^2)_{(5,25)}\,dh\\ &= 250\pi\,dr + 25\pi\,dh.\end{aligned}$$

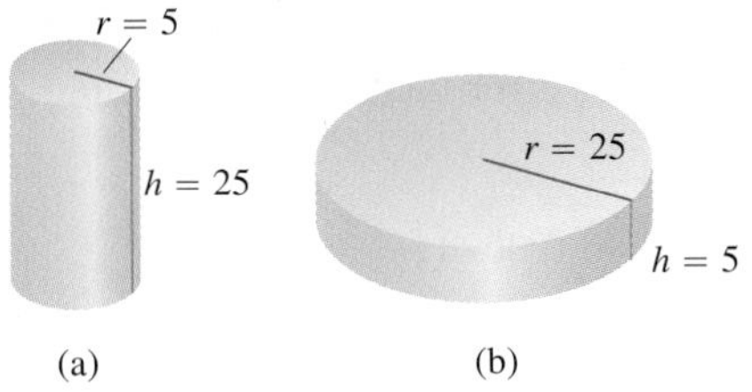

FIGURE 11.34 The volume of cylinder (a) is more sensitive to a small change in r than it is to an equally small change in h. The volume of cylinder (b) is more sensitive to small changes in h than it is to small changes in r (Example 5).

Thus, a 1-unit change in r will change V by about 250π units. A 1-unit change in h will change V by about 25π units. The tank's volume is 10 times more sensitive to a small change in r than it is to a small change of equal size in h. As a quality control engineer concerned with being sure the tanks have the correct volume, you would want to pay special attention to their radii.

In contrast, if the values of r and h are reversed to make $r = 25$ and $h = 5$, then the total differential in V becomes

$$dV = (2\pi rh)_{(25,5)}\,dr + (\pi r^2)_{(25,5)}\,dh = 250\pi\,dr + 625\pi\,dh.$$

Now the volume is more sensitive to changes in h than to changes in r (Figure 11.34).

The general rule is that functions are most sensitive to small changes in the variables that generate the largest partial derivatives. ■

Functions of More Than Two Variables

Analogous results hold for differentiable functions of more than two variables.

1. The **linearization** of $f(x, y, z)$ at a point $P_0(x_0, y_0, z_0)$ is

$$L(x, y, z) = f(P_0) + f_x(P_0)(x - x_0) + f_y(P_0)(y - y_0) + f_z(P_0)(z - z_0).$$

2. Suppose that R is a closed rectangular solid centered at P_0 and lying in an open region on which the second partial derivatives of f are continuous. Suppose also that $|f_{xx}|, |f_{yy}|, |f_{zz}|, |f_{xy}|, |f_{xz}|$, and $|f_{yz}|$ are all less than or equal to M throughout R. Then the **error** $E(x, y, z) = f(x, y, z) - L(x, y, z)$ in the approximation of f by L is bounded throughout R by the inequality

$$|E| \leq \frac{1}{2} M(|x - x_0| + |y - y_0| + |z - z_0|)^2.$$

3. If the second partial derivatives of f are continuous and if x, y, and z change from x_0, y_0, and z_0 by small amounts dx, dy, and dz, the **total differential**

$$df = f_x(P_0)\,dx + f_y(P_0)\,dy + f_z(P_0)\,dz$$

gives a good approximation of the resulting change in f.

EXAMPLE 6 Find the linearization $L(x, y, z)$ of

$$f(x, y, z) = x^2 - xy + 3\sin z$$

at the point $(x_0, y_0, z_0) = (2, 1, 0)$. Find an upper bound for the error incurred in replacing f by L on the rectangle

$$R: \quad |x - 2| \leq 0.01, \qquad |y - 1| \leq 0.02, \qquad |z| \leq 0.01.$$

Solution Routine calculations give

$$f(2, 1, 0) = 2, \qquad f_x(2, 1, 0) = 3, \qquad f_y(2, 1, 0) = -2, \qquad f_z(2, 1, 0) = 3.$$

Thus,

$$L(x, y, z) = 2 + 3(x - 2) + (-2)(y - 1) + 3(z - 0) = 3x - 2y + 3z - 2.$$

Since

$$f_{xx} = 2, \qquad f_{yy} = 0, \qquad f_{zz} = -3\sin z, \qquad f_{xy} = -1, \qquad f_{xz} = 0, \qquad f_{yz} = 0,$$

we may safely take M to be $\max|-3\sin z| = 3$. Hence, the error incurred by replacing f by L on R satisfies

$$|E| \leq \frac{1}{2}(3)(0.01 + 0.02 + 0.01)^2 = 0.0024.$$

■

EXERCISES 11.6

In Exercises 1–8, find equations for the

(a) tangent plane and **(b)** normal line at the point P_0 on the given surface.

1. $x^2 + y^2 + z^2 = 3, \quad P_0(1, 1, 1)$
2. $x^2 + y^2 - z^2 = 18, \quad P_0(3, 5, -4)$
3. $2z - x^2 = 0, \quad P_0(2, 0, 2)$
4. $x^2 + 2xy - y^2 + z^2 = 7, \quad P_0(1, -1, 3)$
5. $\cos \pi x - x^2y + e^{xz} + yz = 4, \quad P_0(0, 1, 2)$
6. $x^2 - xy - y^2 - z = 0, \quad P_0(1, 1, -1)$
7. $x + y + z = 1, \quad P_0(0, 1, 0)$
8. $x^2 + y^2 - 2xy - x + 3y - z = -4, \quad P_0(2, -3, 18)$

In Exercises 9–12, find an equation for the plane that is tangent to the given surface at the given point.

9. $z = \ln(x^2 + y^2), \quad (1, 0, 0)$
10. $z = e^{-(x^2+y^2)}, \quad (0, 0, 1)$
11. $z = \sqrt{y - x}, \quad (1, 2, 1)$
12. $z = 4x^2 + y^2, \quad (1, 1, 5)$

In Exercises 13–18, find parametric equations for the line tangent to the curve of intersection of the surfaces at the given point.

13. Surfaces: $x + y^2 + 2z = 4, \quad x = 1$

Point: $(1, 1, 1)$

14. Surfaces: $xyz = 1, \quad x^2 + 2y^2 + 3z^2 = 6$

Point: $(1, 1, 1)$

15. Surfaces: $x^2 + 2y + 2z = 4, \quad y = 1$

Point: $(1, 1, 1/2)$

16. Surfaces: $x + y^2 + z = 2, \quad y = 1$

Point: $(1/2, 1, 1/2)$

17. Surfaces: $x^3 + 3x^2y^2 + y^3 + 4xy - z^2 = 0,$
$x^2 + y^2 + z^2 = 11$

Point: $(1, 1, 3)$

18. Surfaces: $x^2 + y^2 = 4, \quad x^2 + y^2 - z = 0$

Point: $\left(\sqrt{2}, \sqrt{2}, 4\right)$

19. By about how much will

$$f(x, y, z) = \ln\sqrt{x^2 + y^2 + z^2}$$

change if the point $P(x, y, z)$ moves from $P_0(3, 4, 12)$ a distance of $ds = 0.1$ unit in the direction of $3\mathbf{i} + 6\mathbf{j} - 2\mathbf{k}$?

20. By about how much will

$$f(x, y, z) = e^x \cos yz$$

change as the point $P(x, y, z)$ moves from the origin a distance of $ds = 0.1$ unit in the direction of $2\mathbf{i} + 2\mathbf{j} - 2\mathbf{k}$?

21. By about how much will

$$g(x, y, z) = x + x\cos z - y\sin z + y$$

change if the point $P(x, y, z)$ moves from $P_0(2, -1, 0)$ a distance of $ds = 0.2$ unit toward the point $P_1(0, 1, 2)$?

22. By about how much will

$$h(x, y, z) = \cos(\pi xy) + xz^2$$

change if the point $P(x, y, z)$ moves from $P_0(-1, -1, -1)$ a distance of $ds = 0.1$ unit toward the origin?

23. Temperature change along a circle Suppose that the Celsius temperature at the point (x, y) in the xy-plane is $T(x, y) = x\sin 2y$ and that distance in the xy-plane is measured in meters. A particle is moving *clockwise* around the circle of radius 1 m centered at the origin at the constant rate of 2 m/sec.

a. How fast is the temperature experienced by the particle changing in degrees Celsius per meter at the point $P\left(1/2, \sqrt{3}/2\right)$?

b. How fast is the temperature experienced by the particle changing in degrees Celsius per second at P?

24. Changing temperature along a space curve The Celsius temperature in a region in space is given by $T(x, y, z) = 2x^2 - xyz$. A particle is moving in this region and its position at time t is given by $x = 2t^2, y = 3t, z = -t^2$, where time is measured in seconds and distance in meters.

a. How fast is the temperature experienced by the particle changing in degrees Celsius per meter when the particle is at the point $P(8, 6, -4)$?

b. How fast is the temperature experienced by the particle changing in degrees Celsius per second at P?

In Exercises 25–30, find the linearization $L(x, y)$ of the function at each point.

25. $f(x, y) = x^2 + y^2 + 1$ at **a.** $(0, 0)$, **b.** $(1, 1)$

26. $f(x, y) = (x + y + 2)^2$ at **a.** $(0, 0)$, **b.** $(1, 2)$

27. $f(x, y) = 3x - 4y + 5$ at **a.** $(0, 0)$, **b.** $(1, 1)$

28. $f(x, y) = x^3y^4$ at **a.** $(1, 1)$, **b.** $(0, 0)$

29. $f(x, y) = e^x \cos y$ at **a.** $(0, 0)$, **b.** $(0, \pi/2)$

30. $f(x, y) = e^{2y-x}$ at **a.** $(0, 0)$, **b.** $(1, 2)$

In Exercises 31–36, find the linearization $L(x, y)$ of the function $f(x, y)$ at P_0. Then find an upper bound for the magnitude $|E|$ of the error in the approximation $f(x, y) \approx L(x, y)$ over the rectangle R.

31. $f(x, y) = x^2 - 3xy + 5$ at $P_0(2, 1)$,
$R: \quad |x - 2| \le 0.1, \quad |y - 1| \le 0.1$

32. $f(x, y) = (1/2)x^2 + xy + (1/4)y^2 + 3x - 3y + 4$ at $P_0(2, 2)$,
$R: \quad |x - 2| \le 0.1, \quad |y - 2| \le 0.1$

33. $f(x, y) = 1 + y + x\cos y$ at $P_0(0, 0)$,
$R: \quad |x| \le 0.2, \quad |y| \le 0.2$
(Use $|\cos y| \le 1$ and $|\sin y| \le 1$ in estimating E.)

34. $f(x, y) = xy^2 + y\cos(x - 1)$ at $P_0(1, 2)$,
$R: \quad |x - 1| \le 0.1, \quad |y - 2| \le 0.1$

35. $f(x, y) = e^x \cos y$ at $P_0(0, 0)$,
$R: \quad |x| \le 0.1, \quad |y| \le 0.1$
(Use $e^x \le 1.11$ and $|\cos y| \le 1$ in estimating E.)

36. $f(x, y) = \ln x + \ln y$ at $P_0(1, 1)$,
$R: \quad |x - 1| \le 0.2, \quad |y - 1| \le 0.2$

Find the linearizations $L(x, y, z)$ of the functions in Exercises 37–42 at the given points.

37. $f(x, y, z) = xy + yz + xz$ at
a. $(1, 1, 1)$ **b.** $(1, 0, 0)$ **c.** $(0, 0, 0)$

38. $f(x, y, z) = x^2 + y^2 + z^2$ at
a. $(1, 1, 1)$ **b.** $(0, 1, 0)$ **c.** $(1, 0, 0)$

39. $f(x, y, z) = \sqrt{x^2 + y^2 + z^2}$ at
a. $(1, 0, 0)$ **b.** $(1, 1, 0)$ **c.** $(1, 2, 2)$

40. $f(x, y, z) = (\sin xy)/z$ at
a. $(\pi/2, 1, 1)$ **b.** $(2, 0, 1)$

41. $f(x, y, z) = e^x + \cos(y + z)$ at
a. $(0, 0, 0)$ **b.** $\left(0, \frac{\pi}{2}, 0\right)$ **c.** $\left(0, \frac{\pi}{4}, \frac{\pi}{4}\right)$

42. $f(x, y, z) = \tan^{-1}(xyz)$ at
a. $(1, 0, 0)$ **b.** $(1, 1, 0)$ **c.** $(1, 1, 1)$

In Exercises 43–46, find the linearization $L(x, y, z)$ of the function $f(x, y, z)$ at P_0. Then find an upper bound for the magnitude of the error E in the approximation $f(x, y, z) \approx L(x, y, z)$ over the region R.

43. $f(x, y, z) = xz - 3yz + 2$ at $P_0(1, 1, 2)$

R: $|x - 1| \leq 0.01$, $|y - 1| \leq 0.01$, $|z - 2| \leq 0.02$

44. $f(x, y, z) = x^2 + xy + yz + (1/4)z^2$ at $P_0(1, 1, 2)$

R: $|x - 1| \leq 0.01$, $|y - 1| \leq 0.01$, $|z - 2| \leq 0.08$

45. $f(x, y, z) = xy + 2yz - 3xz$ at $P_0(1, 1, 0)$

R: $|x - 1| \leq 0.01$, $|y - 1| \leq 0.01$, $|z| \leq 0.01$

46. $f(x, y, z) = \sqrt{2} \cos x \sin (y + z)$ at $P_0(0, 0, \pi/4)$

R: $|x| \leq 0.01$, $|y| \leq 0.01$, $|z - \pi/4| \leq 0.01$

47. Estimating maximum error Suppose that T is to be found from the formula $T = x(e^y + e^{-y})$, where x and y are found to be 2 and ln 2 with maximum possible errors of $|dx| = 0.1$ and $|dy| = 0.02$. Estimate the maximum possible error in the computed value of T.

48. a. Around the point (1, 0), is $f(x, y) = x^2(y + 1)$ more sensitive to changes in x or to changes in y? Give reasons for your answer.

b. What ratio of dx to dy will make df equal zero at (1, 0)?

49. Value of a 2 × 2 determinant If $|a|$ is much greater than $|b|, |c|$, and $|d|$, to which of a, b, c, and d is the value of the determinant

$$f(a, b, c, d) = \begin{vmatrix} a & b \\ c & d \end{vmatrix}$$

most sensitive? Give reasons for your answer.

50. The Wilson lot size formula The Wilson lot size formula in economics says that the most economical quantity Q of goods (radios, shoes, brooms, whatever) for a store to order is given by the formula $Q = \sqrt{2KM/h}$, where K is the cost of placing the order, M is the number of items sold per week, and h is the weekly holding cost for each item (cost of space, utilities, security, and so on). To which of the variables K, M, and h is Q most sensitive near the point $(K_0, M_0, h_0) = (2, 20, 0.05)$? Give reasons for your answer.

51. The linearization of $f(x, y)$ is a tangent-plane approximation Show that the tangent plane at the point $P_0(x_0, y_0, f(x_0, y_0))$ on the surface $z = f(x, y)$ defined by a differentiable function f is the plane

$$f_x(x_0, y_0)(x - x_0) + f_y(x_0, y_0)(y - y_0) - (z - f(x_0, y_0)) = 0$$

or

$$z = f(x_0, y_0) + f_x(x_0, y_0)(x - x_0) + f_y(x_0, y_0)(y - y_0).$$

Thus, the tangent plane at P_0 is the graph of the linearization of f at P_0 (see accompanying figure).

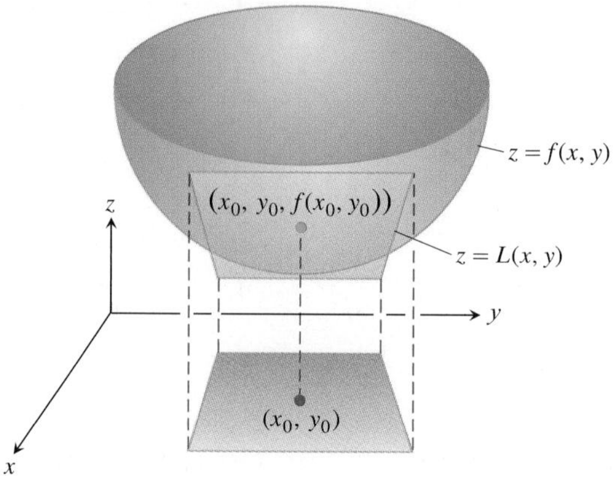

52. Change along the involute of a circle Find the derivative of $f(x, y) = x^2 + y^2$ in the direction of the unit tangent vector of the curve

$$\mathbf{r}(t) = (\cos t + t \sin t)\mathbf{i} + (\sin t - t \cos t)\mathbf{j}, \quad t > 0.$$

53. Change along a helix Find the derivative of $f(x, y, z) = x^2 + y^2 + z^2$ in the direction of the unit tangent vector of the helix

$$\mathbf{r}(t) = (\cos t)\mathbf{i} + (\sin t)\mathbf{j} + t\mathbf{k}$$

at the points where $t = -\pi/4$, 0, and $\pi/4$. The function f gives the square of the distance from a point $P(x, y, z)$ on the helix to the origin. The derivatives calculated here give the rates at which the square of the distance is changing with respect to t as P moves through the points where $t = -\pi/4$, 0, and $\pi/4$.

54. Normal curves A smooth curve is *normal* to a surface $f(x, y, z) = c$ at a point of intersection if the curve's velocity vector is a nonzero scalar multiple of ∇f at the point.

Show that the curve

$$\mathbf{r}(t) = \sqrt{t}\mathbf{i} + \sqrt{t}\mathbf{j} - \frac{1}{4}(t + 3)\mathbf{k}$$

is normal to the surface $x^2 + y^2 - z = 3$ when $t = 1$.

55. Tangent curves A smooth curve is *tangent* to the surface at a point of intersection if its velocity vector is orthogonal to ∇f there.

Show that the curve

$$\mathbf{r}(t) = \sqrt{t}\mathbf{i} + \sqrt{t}\mathbf{j} + (2t - 1)\mathbf{k}$$

is tangent to the surface $x^2 + y^2 - z = 1$ when $t = 1$.

11.7 Extreme Values and Saddle Points

Continuous functions of two variables assume extreme values on closed, bounded domains (see Figures 11.35 and 11.36). We see in this section that we can narrow the search for these extreme values by examining the functions' first partial derivatives. A function of two variables can assume extreme values only at domain boundary points or at interior domain points where both first partial derivatives are zero or where one or both of the first partial derivatives fails to exist. However, the vanishing of derivatives at an interior point (a, b) does not always signal the presence of an extreme value. The surface that is the graph of the function might be shaped like a saddle right above (a, b) and cross its tangent plane there.

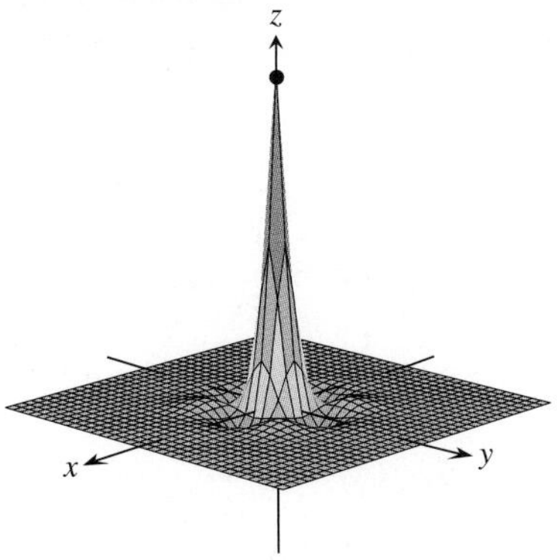

FIGURE 11.35 The function
$$z = (\cos x)(\cos y)e^{-\sqrt{x^2+y^2}}$$
has a maximum value of 1 and a minimum value of about -0.067 on the square region $|x| \le 3\pi/2, |y| \le 3\pi/2$.

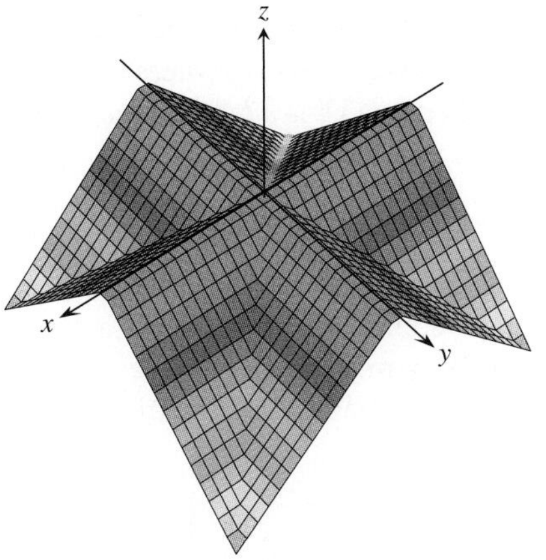

FIGURE 11.36 The "roof surface"
$$z = \frac{1}{2}\left(\big||x| - |y|\big| - |x| - |y|\right)$$
viewed from the point (10, 15, 20). The defining function has a maximum value of 0 and a minimum value of $-a$ on the square region $|x| \le a, |y| \le a$.

HISTORICAL BIOGRAPHY

Siméon-Denis Poisson
(1781–1840)

Derivative Tests for Local Extreme Values

To find the local extreme values of a function of a single variable, we look for points where the graph has a horizontal tangent line. At such points, we then look for local maxima, local minima, and points of inflection. For a function $f(x, y)$ of two variables, we look for points where the surface $z = f(x, y)$ has a horizontal tangent *plane*. At such points, we then look for local maxima, local minima, and saddle points. We begin by defining maxima and minima.

DEFINITIONS Let $f(x, y)$ be defined on a region R containing the point (a, b). Then

1. $f(a, b)$ is a **local maximum** value of f if $f(a, b) \ge f(x, y)$ for all domain points (x, y) in an open disk centered at (a, b).
2. $f(a, b)$ is a **local minimum** value of f if $f(a, b) \le f(x, y)$ for all domain points (x, y) in an open disk centered at (a, b).

Local maxima correspond to mountain peaks on the surface $z = f(x, y)$ and local minima correspond to valley bottoms (Figure 11.37). At such points the tangent planes, when they exist, are horizontal. Local extrema are also called **relative extrema**.

As with functions of a single variable, the key to identifying the local extrema is a first derivative test.

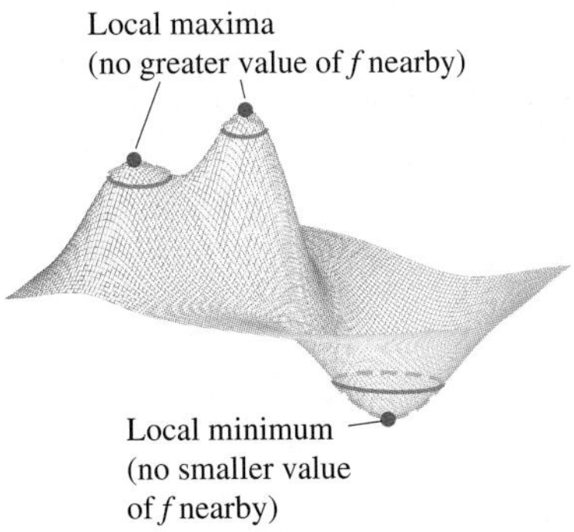

FIGURE 11.37 A local maximum is a mountain peak and a local minimum is a valley low.

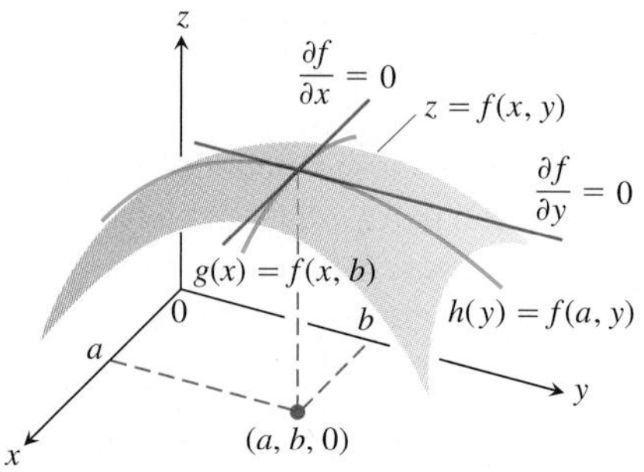

FIGURE 11.38 If a local maximum of f occurs at $x = a, y = b$, then the first partial derivatives $f_x(a, b)$ and $f_y(a, b)$ are both zero.

THEOREM 10—First Derivative Test for Local Extreme Values If $f(x, y)$ has a local maximum or minimum value at an interior point (a, b) of its domain and if the first partial derivatives exist there, then $f_x(a, b) = 0$ and $f_y(a, b) = 0$.

Proof If f has a local extremum at (a, b), then the function $g(x) = f(x, b)$ has a local extremum at $x = a$ (Figure 11.38). Therefore, $g'(a) = 0$ (Chapter 3, Theorem 2). Now $g'(a) = f_x(a, b)$, so $f_x(a, b) = 0$. A similar argument with the function $h(y) = f(a, y)$ shows that $f_y(a, b) = 0$. ■

If we substitute the values $f_x(a, b) = 0$ and $f_y(a, b) = 0$ into the equation

$$f_x(a, b)(x - a) + f_y(a, b)(y - b) - (z - f(a, b)) = 0$$

for the tangent plane to the surface $z = f(x, y)$ at (a, b), the equation reduces to

$$0 \cdot (x - a) + 0 \cdot (y - b) - z + f(a, b) = 0$$

or

$$z = f(a, b).$$

Thus, Theorem 10 says that the surface does indeed have a horizontal tangent plane at a local extremum, provided there is a tangent plane there.

DEFINITION An interior point of the domain of a function $f(x, y)$ where both f_x and f_y are zero or where one or both of f_x and f_y do not exist is a **critical point** of f.

Theorem 10 says that the only points where a function $f(x, y)$ can assume extreme values are critical points and boundary points. As with differentiable functions of a single variable, not every critical point gives rise to a local extremum. A differentiable function of a single variable might have a point of inflection. A differentiable function of two variables might have a *saddle point*.

DEFINITION A differentiable function $f(x, y)$ has a **saddle point** at a critical point (a, b) if in every open disk centered at (a, b) there are domain points (x, y) where $f(x, y) > f(a, b)$ and domain points (x, y) where $f(x, y) < f(a, b)$. The corresponding point $(a, b, f(a, b))$ on the surface $z = f(x, y)$ is called a saddle point of the surface (Figure 11.39).

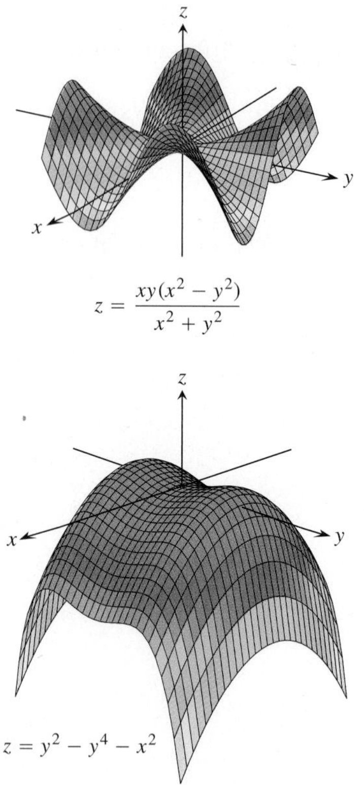

FIGURE 11.39 Saddle points at the origin.

EXAMPLE 1 Find the local extreme values of $f(x, y) = x^2 + y^2$.

Solution The domain of f is the entire plane (so there are no boundary points) and the partial derivatives $f_x = 2x$ and $f_y = 2y$ exist everywhere. Therefore, local extreme values can occur only where

$$f_x = 2x = 0 \qquad \text{and} \qquad f_y = 2y = 0.$$

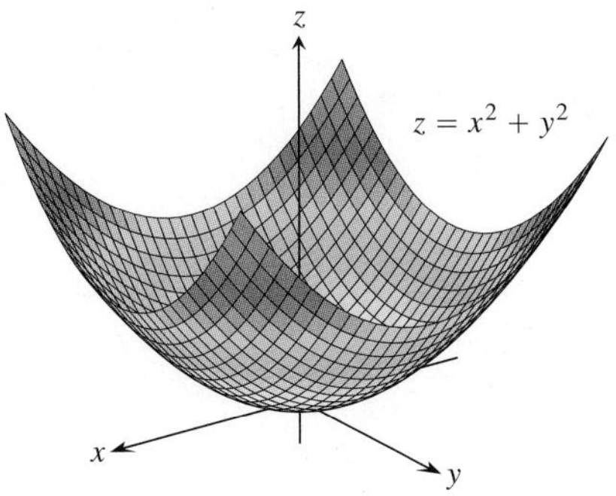

FIGURE 11.40 The graph of the function $f(x, y) = x^2 + y^2$ is the paraboloid $z = x^2 + y^2$. The function has a local minimum value of 0 at the origin (Example 1).

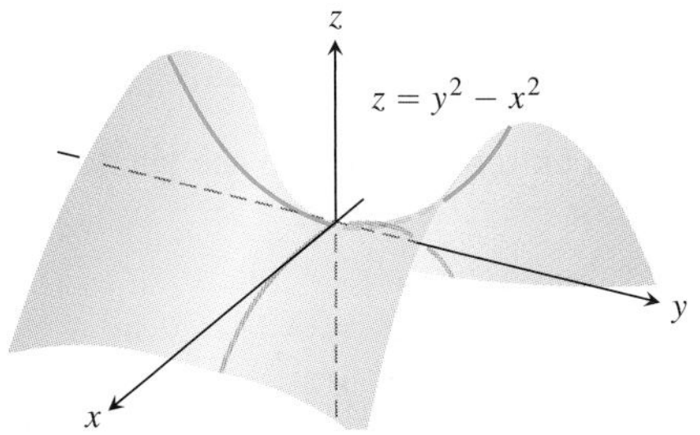

FIGURE 11.41 The origin is a saddle point of the function $f(x, y) = y^2 - x^2$. There are no local extreme values (Example 2).

The only possibility is the origin, where the value of f is zero. Since f is never negative, we see that the origin gives a local minimum (Figure 11.40). ■

EXAMPLE 2 Find the local extreme values (if any) of $f(x, y) = y^2 - x^2$.

Solution The domain of f is the entire plane (so there are no boundary points) and the partial derivatives $f_x = -2x$ and $f_y = 2y$ exist everywhere. Therefore, local extrema can occur only at the origin $(0, 0)$. Along the positive x-axis, however, f has the value $f(x, 0) = -x^2 < 0$; along the positive y-axis, f has the value $f(0, y) = y^2 > 0$. Therefore, every open disk in the xy-plane centered at $(0, 0)$ contains points where the function is positive and points where it is negative. The function has a saddle point at the origin (Figure 11.41) instead of a local extreme value. We conclude that the function has no local extreme values. ■

That $f_x = f_y = 0$ at an interior point (a, b) of R does not guarantee f has a local extreme value there. If f and its first and second partial derivatives are continuous on R, however, we may be able to learn more from the following theorem, proved in Appendix 12.

THEOREM 11—Second Derivative Test for Local Extreme Values Suppose that $f(x, y)$ and its first and second partial derivatives are continuous throughout a disk centered at (a, b) and that $f_x(a, b) = f_y(a, b) = 0$. Then

i. f has a **local maximum** at (a, b) if $f_{xx} < 0$ and $f_{xx}f_{yy} - f_{xy}^2 > 0$ at (a, b).
ii. f has a **local minimum** at (a, b) if $f_{xx} > 0$ and $f_{xx}f_{yy} - f_{xy}^2 > 0$ at (a, b).
iii. f has a **saddle point** at (a, b) if $f_{xx}f_{yy} - f_{xy}^2 < 0$ at (a, b).
iv. **The test is inconclusive** at (a, b) if $f_{xx}f_{yy} - f_{xy}^2 = 0$ at (a, b). In this case, we must find some other way to determine the behavior of f at (a, b).

The expression $f_{xx}f_{yy} - f_{xy}^2$ is called the **discriminant** or **Hessian** of f. It is sometimes easier to remember it in determinant form,

$$f_{xx}f_{yy} - f_{xy}^2 = \begin{vmatrix} f_{xx} & f_{xy} \\ f_{xy} & f_{yy} \end{vmatrix}.$$

Theorem 11 says that if the discriminant is positive at the point (a, b), then the surface curves the same way in all directions: downward if $f_{xx} < 0$, giving rise to a local maximum, and upward if $f_{xx} > 0$, giving a local minimum. On the other hand, if the discriminant is negative at (a, b), then the surface curves up in some directions and down in others, so we have a saddle point.

EXAMPLE 3 Find the local extreme values of the function

$$f(x, y) = xy - x^2 - y^2 - 2x - 2y + 4.$$

Solution The function is defined and differentiable for all x and y and its domain has no boundary points. The function therefore has extreme values only at the points where f_x and f_y are simultaneously zero. This leads to

$$f_x = y - 2x - 2 = 0, \qquad f_y = x - 2y - 2 = 0,$$

or

$$x = y = -2.$$

Therefore, the point $(-2, -2)$ is the only point where f may take on an extreme value. To see if it does so, we calculate

$$f_{xx} = -2, \qquad f_{yy} = -2, \qquad f_{xy} = 1.$$

The discriminant of f at $(a, b) = (-2, -2)$ is

$$f_{xx}f_{yy} - f_{xy}^2 = (-2)(-2) - (1)^2 = 4 - 1 = 3.$$

The combination

$$f_{xx} < 0 \quad \text{and} \quad f_{xx}f_{yy} - f_{xy}^2 > 0$$

tells us that f has a local maximum at $(-2, -2)$. The value of f at this point is $f(-2, -2) = 8$. ■

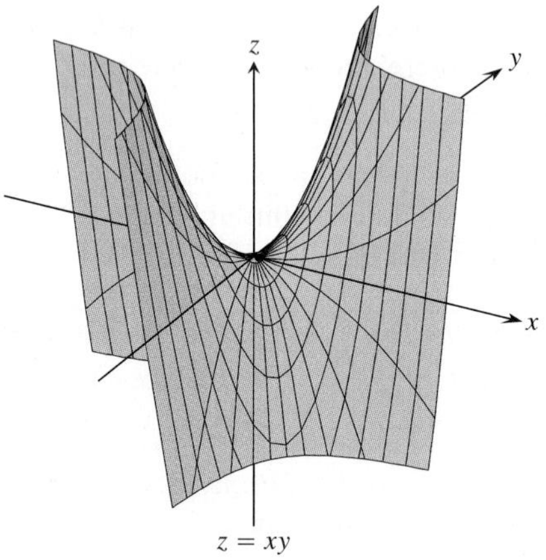

FIGURE 11.42 The surface $z = xy$ has a saddle point at the origin (Example 4).

EXAMPLE 4 Find the local extreme values of $f(x, y) = xy$.

Solution Since f is differentiable everywhere (Figure 11.42), it can assume extreme values only where

$$f_x = y = 0 \quad \text{and} \quad f_y = x = 0.$$

Thus, the origin is the only point where f might have an extreme value. To see what happens there, we calculate

$$f_{xx} = 0, \qquad f_{yy} = 0, \qquad f_{xy} = 1.$$

The discriminant,

$$f_{xx}f_{yy} - f_{xy}^2 = -1,$$

is negative. Therefore, the function has a saddle point at $(0, 0)$. We conclude that $f(x, y) = xy$ has no local extreme values. ■

Absolute Maxima and Minima on Closed Bounded Regions

We organize the search for the absolute extrema of a continuous function $f(x, y)$ on a closed and bounded region R into three steps.

1. *List the interior points of R* where f may have local maxima and minima and evaluate f at these points. These are the critical points of f.
2. *List the boundary points of R* where f has local maxima and minima and evaluate f at these points. We show how to do this shortly.
3. *Look through the lists* for the maximum and minimum values of f. These will be the absolute maximum and minimum values of f on R. Since absolute maxima and minima are also local maxima and minima, the absolute maximum and minimum values of f appear somewhere in the lists made in Steps 1 and 2.

EXAMPLE 5 Find the absolute maximum and minimum values of

$$f(x, y) = 2 + 2x + 2y - x^2 - y^2$$

on the triangular region in the first quadrant bounded by the lines $x = 0, y = 0, y = 9 - x$.

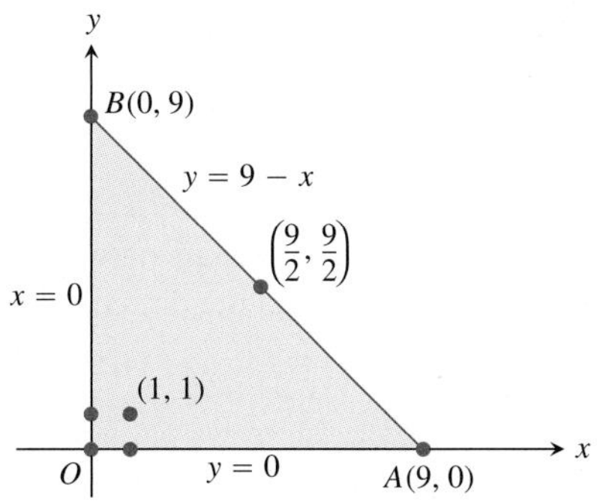

FIGURE 11.43 This triangular region is the domain of the function in Example 5.

Solution Since f is differentiable, the only places where f can assume these values are points inside the triangle (Figure 11.43) where $f_x = f_y = 0$ and points on the boundary.

(a) Interior points. For these we have

$$f_x = 2 - 2x = 0, \qquad f_y = 2 - 2y = 0,$$

yielding the single point $(x, y) = (1, 1)$. The value of f there is

$$f(1, 1) = 4.$$

(b) Boundary points. We take the triangle one side at a time:

(i) On the segment OA, $y = 0$. The function

$$f(x, y) = f(x, 0) = 2 + 2x - x^2$$

may now be regarded as a function of x defined on the closed interval $0 \le x \le 9$. Its extreme values (we know from Chapter 3) may occur at the endpoints

$$x = 0 \quad \text{where} \quad f(0, 0) = 2$$
$$x = 9 \quad \text{where} \quad f(9, 0) = 2 + 18 - 81 = -61$$

and at the interior points where $f'(x, 0) = 2 - 2x = 0$. The only interior point where $f'(x, 0) = 0$ is $x = 1$, where

$$f(x, 0) = f(1, 0) = 3.$$

(ii) On the segment OB, $x = 0$ and

$$f(x, y) = f(0, y) = 2 + 2y - y^2.$$

We know from the symmetry of f in x and y and from the analysis we just carried out that the candidates on this segment are

$$f(0, 0) = 2, \qquad f(0, 9) = -61, \qquad f(0, 1) = 3.$$

(iii) We have already accounted for the values of f at the endpoints of AB, so we need only look at the interior points of AB. With $y = 9 - x$, we have

$$f(x, y) = 2 + 2x + 2(9 - x) - x^2 - (9 - x)^2 = -61 + 18x - 2x^2.$$

Setting $f'(x, 9 - x) = 18 - 4x = 0$ gives

$$x = \frac{18}{4} = \frac{9}{2}.$$

At this value of x,

$$y = 9 - \frac{9}{2} = \frac{9}{2} \quad \text{and} \quad f(x, y) = f\left(\frac{9}{2}, \frac{9}{2}\right) = -\frac{41}{2}.$$

Summary We list all the candidates: $4, 2, -61, 3, -(41/2)$. The maximum is 4, which f assumes at $(1, 1)$. The minimum is -61, which f assumes at $(0, 9)$ and $(9, 0)$. ■

Solving extreme value problems with algebraic constraints on the variables usually requires the method of Lagrange multipliers introduced in the next section. But sometimes we can solve such problems directly, as in the next example.

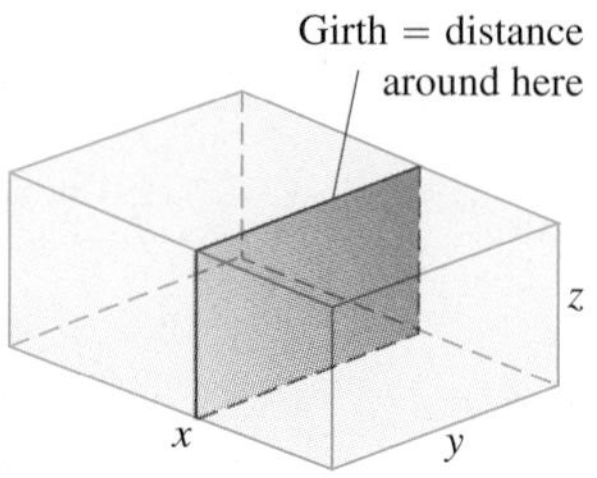

FIGURE 11.44 The box in Example 6.

EXAMPLE 6 A delivery company accepts only rectangular boxes the sum of whose length and girth (perimeter of a cross-section) does not exceed 108 in. Find the dimensions of an acceptable box of largest volume.

Solution Let x, y, and z represent the length, width, and height of the rectangular box, respectively. Then the girth is $2y + 2z$. We want to maximize the volume $V = xyz$ of the box (Figure 11.44) satisfying $x + 2y + 2z = 108$ (the largest box accepted by the delivery company). Thus, we can write the volume of the box as a function of two variables:

$$\begin{aligned} V(y, z) &= (108 - 2y - 2z)yz \qquad && V = xyz \text{ and } x = 108 - 2y - 2z \\ &= 108yz - 2y^2z - 2yz^2. \end{aligned}$$

Setting the first partial derivatives equal to zero,

$$\begin{aligned} V_y(y, z) &= 108z - 4yz - 2z^2 = (108 - 4y - 2z)z = 0 \\ V_z(y, z) &= 108y - 2y^2 - 4yz = (108 - 2y - 4z)y = 0, \end{aligned}$$

gives the critical points (0, 0), (0, 54),(54, 0), and (18, 18). The volume is zero at (0, 0), (0, 54), (54, 0), which are not maximum values. At the point (18, 18), we apply the Second Derivative Test (Theorem 11):

$$V_{yy} = -4z, \qquad V_{zz} = -4y, \qquad V_{yz} = 108 - 4y - 4z.$$

Then

$$V_{yy}V_{zz} - V_{yz}^2 = 16yz - 16(27 - y - z)^2.$$

Thus,

$$V_{yy}(18, 18) = -4(18) < 0$$

and

$$\left[V_{yy}V_{zz} - V_{yz}^2\right]_{(18,18)} = 16(18)(18) - 16(-9)^2 > 0$$

imply that (18, 18), gives a local maximum volume. The volume $V = xyz$ is defined on the closed and bounded triangular plane $x + 2y + 2z = 108$ where x, y, and z are nonnegative. On the boundary of this plane, one of the variables x, y, z is zero, so the local maximum is the absolute maximum volume. The dimensions of the package are $x = 108 - 2(18) - 2(18) = 36$ in., $y = 18$ in., and $z = 18$ in. The maximum volume is $V = (36)(18)(18) = 11{,}664 \text{ in.}^3$, or 6.75 ft^3. ■

Despite the power of Theorem 10, we urge you to remember its limitations. It does not apply to boundary points of a function's domain, where it is possible for a function to have extreme values along with nonzero derivatives. Also, it does not apply to points where either f_x or f_y fails to exist.

EXERCISES 11.7

Find all the local maxima, local minima, and saddle points of the functions in Exercises 1–30.

1. $f(x, y) = x^2 + xy + y^2 + 3x - 3y + 4$

2. $f(x, y) = x^2 + 3xy + 3y^2 - 6x + 3y - 6$

3. $f(x, y) = 2xy - 5x^2 - 2y^2 + 4x + 4y - 4$

4. $f(x, y) = 2xy - 5x^2 - 2y^2 + 4x - 4$

5. $f(x, y) = x^2 + xy + 3x + 2y + 5$

6. $f(x, y) = y^2 + xy - 2x - 2y + 2$

7. $f(x, y) = 5xy - 7x^2 + 3x - 6y + 2$

8. $f(x, y) = 2xy - x^2 - 2y^2 + 3x + 4$

9. $f(x, y) = x^2 - 4xy + y^2 + 6y + 2$

10. $f(x, y) = 3x^2 + 6xy + 7y^2 - 2x + 4y$

11. $f(x, y) = 2x^2 + 3xy + 4y^2 - 5x + 2y$

12. $f(x, y) = 4x^2 - 6xy + 5y^2 - 20x + 26y$

13. $f(x, y) = x^2 - y^2 - 2x + 4y + 6$

14. $f(x, y) = x^2 - 2xy + 2y^2 - 2x + 2y + 1$

15. $f(x, y) = x^2 + 2xy$

16. $f(x, y) = 3 + 2x + 2y - 2x^2 - 2xy - y^2$

17. $f(x, y) = x^3 - y^3 - 2xy + 6$

18. $f(x, y) = x^3 + 3xy + y^3$

19. $f(x, y) = 6x^2 - 2x^3 + 3y^2 + 6xy$

20. $f(x, y) = 3y^2 - 2y^3 - 3x^2 + 6xy$

21. $f(x, y) = 9x^3 + y^3/3 - 4xy$

22. $f(x, y) = 8x^3 + y^3 + 6xy$

23. $f(x, y) = x^3 + y^3 + 3x^2 - 3y^2 - 8$

24. $f(x, y) = 2x^3 + 2y^3 - 9x^2 + 3y^2 - 12y$

25. $f(x, y) = 4xy - x^4 - y^4$

26. $f(x, y) = x^4 + y^4 + 4xy$

27. $f(x, y) = \dfrac{1}{x^2 + y^2 - 1}$

28. $f(x, y) = \dfrac{1}{x} + xy + \dfrac{1}{y}$

29. $f(x, y) = y \sin x$

30. $f(x, y) = e^{2x} \cos y$

In Exercises 31–38, find the absolute maxima and minima of the functions on the given domains.

31. $f(x, y) = 2x^2 - 4x + y^2 - 4y + 1$ on the closed triangular plate bounded by the lines $x = 0$, $y = 2$, $y = 2x$ in the first quadrant

32. $D(x, y) = x^2 - xy + y^2 + 1$ on the closed triangular plate in the first quadrant bounded by the lines $x = 0$, $y = 4$, $y = x$

33. $f(x, y) = x^2 + y^2$ on the closed triangular plate bounded by the lines $x = 0$, $y = 0$, $y + 2x = 2$ in the first quadrant

34. $T(x, y) = x^2 + xy + y^2 - 6x$ on the rectangular plate $0 \le x \le 5$, $-3 \le y \le 3$

35. $T(x, y) = x^2 + xy + y^2 - 6x + 2$ on the rectangular plate $0 \le x \le 5$, $-3 \le y \le 0$

36. $f(x, y) = 48xy - 32x^3 - 24y^2$ on the rectangular plate $0 \le x \le 1$, $0 \le y \le 1$

37. $f(x, y) = (4x - x^2) \cos y$ on the rectangular plate $1 \le x \le 3$, $-\pi/4 \le y \le \pi/4$ (see accompanying figure).

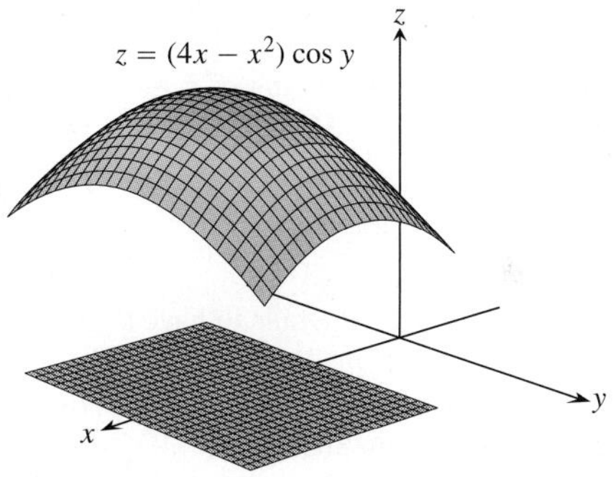

38. $f(x, y) = 4x - 8xy + 2y + 1$ on the triangular plate bounded by the lines $x = 0$, $y = 0$, $x + y = 1$ in the first quadrant

39. Find two numbers a and b with $a \le b$ such that

$$\int_a^b (6 - x - x^2)\, dx$$

has its largest value.

40. Find two numbers a and b with $a \le b$ such that

$$\int_a^b (24 - 2x - x^2)^{1/3}\, dx$$

has its largest value.

41. Find the maxima, minima, and saddle points of $f(x, y)$, if any, given that

a. $f_x = 2x - 4y$ and $f_y = 2y - 4x$

b. $f_x = 2x - 2$ and $f_y = 2y - 4$

c. $f_x = 9x^2 - 9$ and $f_y = 2y + 4$

Describe your reasoning in each case.

42. The discriminant $f_{xx} f_{yy} - f_{xy}^2$ is zero at the origin for each of the following functions, so the Second Derivative Test fails there. Determine whether the function has a maximum, a minimum, or neither at the origin by imagining what the surface $z = f(x, y)$ looks like. Describe your reasoning in each case.

a. $f(x, y) = x^2 y^2$ **b.** $f(x, y) = 1 - x^2 y^2$

c. $f(x, y) = xy^2$ **d.** $f(x, y) = x^3 y^2$

e. $f(x, y) = x^3 y^3$ **f.** $f(x, y) = x^4 y^4$

43. Show that (0, 0) is a critical point of $f(x, y) = x^2 + kxy + y^2$ no matter what value the constant k has. (*Hint:* Consider two cases: $k = 0$ and $k \neq 0$.)

44. For what values of the constant k does the Second Derivative Test guarantee that $f(x, y) = x^2 + kxy + y^2$ will have a saddle point at (0, 0)? A local minimum at (0, 0)? For what values of k is the Second Derivative Test inconclusive? Give reasons for your answers.

45. If $f_x(a, b) = f_y(a, b) = 0$, must f have a local maximum or minimum value at (a, b)? Give reasons for your answer.

46. Can you conclude anything about $f(a, b)$ if f and its first and second partial derivatives are continuous throughout a disk centered at the critical point (a, b) and $f_{xx}(a, b)$ and $f_{yy}(a, b)$ differ in sign? Give reasons for your answer.

47. Among all the points on the graph of $z = 10 - x^2 - y^2$ that lie above the plane $x + 2y + 3z = 0$, find the point farthest from the plane.

48. Find the point on the graph of $z = x^2 + y^2 + 10$ nearest the plane $x + 2y - z = 0$.

49. The function $f(x, y) = x + y$ fails to have an absolute maximum value in the closed first quadrant $x \geq 0$ and $y \geq 0$. Does this contradict the discussion on finding absolute extrema given in the text? Give reasons for your answer.

50. Consider the function $f(x, y) = x^2 + y^2 + 2xy - x - y + 1$ over the square $0 \leq x \leq 1$ and $0 \leq y \leq 1$.

a. Show that f has an absolute minimum along the line segment $2x + 2y = 1$ in this square. What *is* the absolute minimum value?

b. Find the absolute maximum value of f over the square.

Extreme Values on Parametrized Curves To find the extreme values of a function $f(x, y)$ on a curve $x = x(t), y = y(t)$, we treat f as a function of the single variable t and use the Chain Rule to find where df/dt is zero. As in any other single-variable case, the extreme values of f are then found among the values at the

a. critical points (points where df/dt is zero or fails to exist), and

b. endpoints of the parameter domain.

Find the absolute maximum and minimum values of the following functions on the given curves.

51. Functions:

a. $f(x, y) = x + y$ **b.** $g(x, y) = xy$

c. $h(x, y) = 2x^2 + y^2$

Curves:

i. The semicircle $x^2 + y^2 = 4, \quad y \geq 0$

ii. The quarter circle $x^2 + y^2 = 4, \quad x \geq 0, \quad y \geq 0$

Use the parametric equations $x = 2\cos t, y = 2\sin t$.

52. Functions:

a. $f(x, y) = 2x + 3y$ **b.** $g(x, y) = xy$

c. $h(x, y) = x^2 + 3y^2$

Curves:

i. The semi-ellipse $(x^2/9) + (y^2/4) = 1, \quad y \geq 0$

ii. The quarter ellipse $(x^2/9) + (y^2/4) = 1, \quad x \geq 0, \quad y \geq 0$

Use the parametric equations $x = 3\cos t, y = 2\sin t$.

53. Function: $f(x, y) = xy$

Curves:

i. The line $x = 2t, \quad y = t + 1$

ii. The line segment $x = 2t, \quad y = t + 1, \quad -1 \leq t \leq 0$

iii. The line segment $x = 2t, \quad y = t + 1, \quad 0 \leq t \leq 1$

54. Functions:

a. $f(x, y) = x^2 + y^2$

b. $g(x, y) = 1/(x^2 + y^2)$

Curves:

i. The line $x = t, \quad y = 2 - 2t$

ii. The line segment $x = t, \quad y = 2 - 2t, \quad 0 \leq t \leq 1$

55. Least Squares and Regression Lines When we try to fit a line $y = mx + b$ to a set of numerical data points $(x_1, y_1), (x_2, y_2), \ldots, (x_n, y_n)$ (Figure 11.45 on the next page), we usually choose the line that minimizes the sum of the squares of the vertical distances from the points to the line. In theory, this means finding the values of m and b that minimize the value of the function

$$w = (mx_1 + b - y_1)^2 + \cdots + (mx_n + b - y_n)^2. \quad (1)$$

Show that the values of m and b that do this are

$$m = \frac{\left(\sum x_k\right)\left(\sum y_k\right) - n\sum x_k y_k}{\left(\sum x_k\right)^2 - n\sum x_k^2}, \quad (2)$$

$$b = \frac{1}{n}\left(\sum y_k - m\sum x_k\right), \quad (3)$$

with all sums running from $k = 1$ to $k = n$. Many scientific calculators have these formulas built in, enabling you to find m and b with only a few key strokes after you have entered the data.

The line $y = mx + b$ determined by these values of m and b is called the **least squares line**, **regression line**, or **trend line** for the data under study. Finding a least squares line lets you

1. summarize data with a simple expression,
2. predict values of y for other, experimentally untried values of x,
3. handle data analytically.

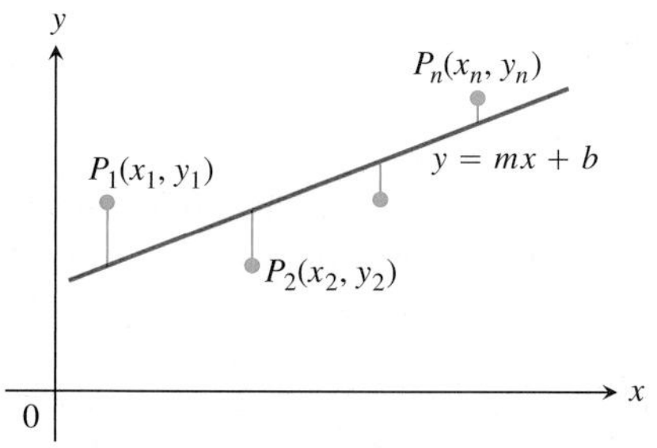

FIGURE 11.45 To fit a line to noncollinear points, we choose the line that minimizes the sum of the squares of the deviations.

In Exercises 56–58, use Equations (2) and (3) to find the least squares line for each set of data points. Then use the linear equation you obtain to predict the value of y that would correspond to $x = 4$.

56. $(-2, 0),\quad (0, 2),\quad (2, 3)$ **57.** $(-1, 2),\quad (0, 1),\quad (3, -4)$

58. $(0, 0),\quad (1, 2),\quad (2, 3)$

COMPUTER EXPLORATIONS

In Exercises 59–64, you will explore functions to identify their local extrema. Use a CAS to perform the following steps:

a. Plot the function over the given rectangle.

b. Plot some level curves in the rectangle.

c. Calculate the function's first partial derivatives and use the CAS equation solver to find the critical points. How do the critical points relate to the level curves plotted in part (b)? Which critical points, if any, appear to give a saddle point? Give reasons for your answer.

d. Calculate the function's second partial derivatives and find the discriminant $f_{xx}f_{yy} - f_{xy}^{\;2}$.

e. Using the max-min tests, classify the critical points found in part (c). Are your findings consistent with your discussion in part (c)?

59. $f(x, y) = x^2 + y^3 - 3xy,\quad -5 \le x \le 5,\quad -5 \le y \le 5$

60. $f(x, y) = x^3 - 3xy^2 + y^2,\quad -2 \le x \le 2,\quad -2 \le y \le 2$

61. $f(x, y) = x^4 + y^2 - 8x^2 - 6y + 16,\quad -3 \le x \le 3,\quad -6 \le y \le 6$

62. $f(x, y) = 2x^4 + y^4 - 2x^2 - 2y^2 + 3,\quad -3/2 \le x \le 3/2,\quad -3/2 \le y \le 3/2$

63. $f(x, y) = 5x^6 + 18x^5 - 30x^4 + 30xy^2 - 120x^3,\quad -4 \le x \le 3,\quad -2 \le y \le 2$

64. $f(x, y) = \begin{cases} x^5 \ln(x^2 + y^2), & (x, y) \ne (0, 0) \\ 0, & (x, y) = (0, 0) \end{cases}$, $\quad -2 \le x \le 2,\quad -2 \le y \le 2$

11.8 Lagrange Multipliers

HISTORICAL BIOGRAPHY

Joseph Louis Lagrange (1736–1813)

Sometimes we need to find the extreme values of a function whose domain is constrained to lie within some particular subset of the plane—a disk, for example, a closed triangular region, or along a curve. In this section, we explore a powerful method for finding extreme values of constrained functions: the method of *Lagrange multipliers*.

Constrained Maxima and Minima

We first consider a problem where a constrained minimum can be found by eliminating a variable.

EXAMPLE 1 Find the point $P(x, y, z)$ closest to the origin on the plane $2x + y - z - 5 = 0$.

Solution The problem asks us to find the minimum value of the function

$$|\overrightarrow{OP}| = \sqrt{(x - 0)^2 + (y - 0)^2 + (z - 0)^2} = \sqrt{x^2 + y^2 + z^2}$$

subject to the constraint that

$$2x + y - z - 5 = 0.$$

Since $|\overrightarrow{OP}|$ has a minimum value wherever the function

$$f(x, y, z) = x^2 + y^2 + z^2$$

has a minimum value, we may solve the problem by finding the minimum value of $f(x, y, z)$ subject to the constraint $2x + y - z - 5 = 0$ (thus avoiding square roots). If we regard x and y as the independent variables in this equation and write z as

$$z = 2x + y - 5,$$

our problem reduces to one of finding the points (x, y) at which the function

$$h(x, y) = f(x, y, 2x + y - 5) = x^2 + y^2 + (2x + y - 5)^2$$

has its minimum value or values. Since the domain of h is the entire xy-plane, the First Derivative Test of Section 11.7 tells us that any minima that h might have must occur at points where

$$h_x = 2x + 2(2x + y - 5)(2) = 0, \qquad h_y = 2y + 2(2x + y - 5) = 0.$$

This leads to

$$10x + 4y = 20, \qquad 4x + 4y = 10,$$

and the solution

$$x = \frac{5}{3}, \qquad y = \frac{5}{6}.$$

We may apply a geometric argument together with the Second Derivative Test to show that these values minimize h. The z-coordinate of the corresponding point on the plane $z = 2x + y - 5$ is

$$z = 2\left(\frac{5}{3}\right) + \frac{5}{6} - 5 = -\frac{5}{6}.$$

Therefore, the point we seek is

$$\text{Closest point:} \qquad P\left(\frac{5}{3}, \frac{5}{6}, -\frac{5}{6}\right).$$

The distance from P to the origin is $5/\sqrt{6} \approx 2.04$. ■

Attempts to solve a constrained maximum or minimum problem by substitution, as we might call the method of Example 1, do not always go smoothly. This is one of the reasons for learning the new method of this section.

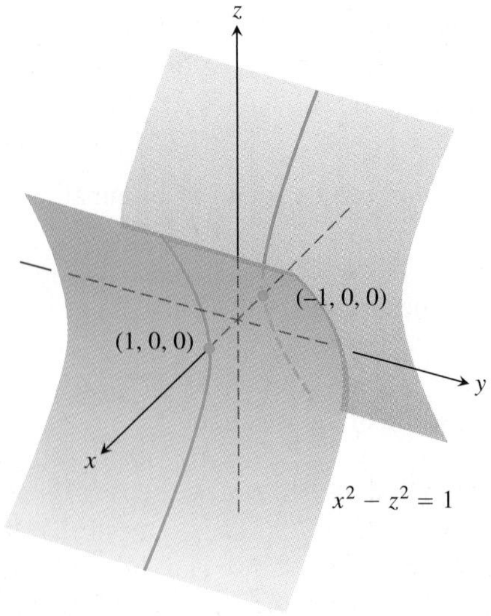

FIGURE 11.46 The hyperbolic cylinder $x^2 - z^2 - 1 = 0$ in Example 2.

EXAMPLE 2 Find the points closest to the origin on the hyperbolic cylinder $x^2 - z^2 - 1 = 0$.

Solution 1 The cylinder is shown in Figure 11.46. We seek the points on the cylinder closest to the origin. These are the points whose coordinates minimize the value of the function

$$f(x, y, z) = x^2 + y^2 + z^2 \qquad \text{Square of the distance}$$

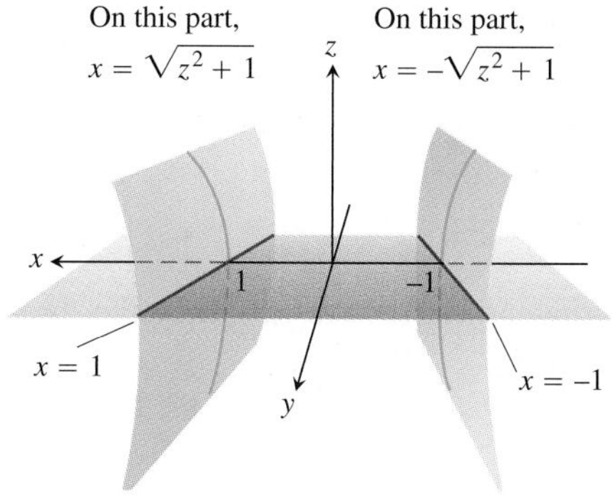

FIGURE 11.47 The region in the xy-plane from which the first two coordinates of the points (x, y, z) on the hyperbolic cylinder $x^2 - z^2 = 1$ are selected excludes the band $-1 < x < 1$ in the xy-plane (Example 2).

subject to the constraint that $x^2 - z^2 - 1 = 0$. If we regard x and y as independent variables in the constraint equation, then

$$z^2 = x^2 - 1$$

and the values of $f(x, y, z) = x^2 + y^2 + z^2$ on the cylinder are given by the function

$$h(x, y) = x^2 + y^2 + (x^2 - 1) = 2x^2 + y^2 - 1.$$

To find the points on the cylinder whose coordinates minimize f, we look for the points in the xy-plane whose coordinates minimize h. The only extreme value of h occurs where

$$h_x = 4x = 0 \qquad \text{and} \qquad h_y = 2y = 0,$$

that is, at the point $(0, 0)$. But there are no points on the cylinder where both x and y are zero. What went wrong?

What happened was that the First Derivative Test found (as it should have) the point *in the domain of h* where h has a minimum value. We, on the other hand, want the points *on the cylinder* where h has a minimum value. Although the domain of h is the entire xy-plane, the domain from which we can select the first two coordinates of the points (x, y, z) on the cylinder is restricted to the "shadow" of the cylinder on the xy-plane; it does not include the band between the lines $x = -1$ and $x = 1$ (Figure 11.47).

We can avoid this problem if we treat y and z as independent variables (instead of x and y) and express x in terms of y and z as

$$x^2 = z^2 + 1.$$

With this substitution, $f(x, y, z) = x^2 + y^2 + z^2$ becomes

$$k(y, z) = (z^2 + 1) + y^2 + z^2 = 1 + y^2 + 2z^2$$

and we look for the points where k takes on its smallest value. The domain of k in the yz-plane now matches the domain from which we select the y- and z-coordinates of the points (x, y, z) on the cylinder. Hence, the points that minimize k in the plane will have corresponding points on the cylinder. The smallest values of k occur where

$$k_y = 2y = 0 \qquad \text{and} \qquad k_z = 4z = 0,$$

or where $y = z = 0$. This leads to

$$x^2 = z^2 + 1 = 1, \qquad x = \pm 1.$$

The corresponding points on the cylinder are $(\pm 1, 0, 0)$. We can see from the inequality

$$k(y, z) = 1 + y^2 + 2z^2 \geq 1$$

that the points $(\pm 1, 0, 0)$ give a minimum value for k. We can also see that the minimum distance from the origin to a point on the cylinder is 1 unit.

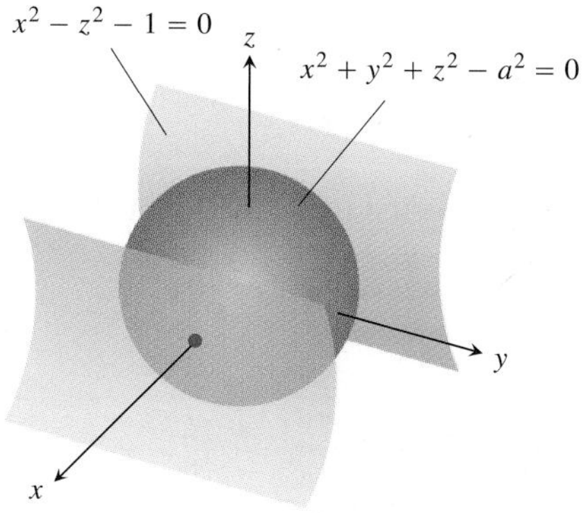

FIGURE 11.48 A sphere expanding like a soap bubble centered at the origin until it just touches the hyperbolic cylinder $x^2 - z^2 - 1 = 0$ (Example 2).

Solution 2 Another way to find the points on the cylinder closest to the origin is to imagine a small sphere centered at the origin expanding like a soap bubble until it just touches the cylinder (Figure 11.48). At each point of contact, the cylinder and sphere have the same tangent plane and normal line. Therefore, if the sphere and cylinder are represented as the level surfaces obtained by setting

$$f(x, y, z) = x^2 + y^2 + z^2 - a^2 \qquad \text{and} \qquad g(x, y, z) = x^2 - z^2 - 1$$

equal to 0, then the gradients ∇f and ∇g will be parallel where the surfaces touch. At any point of contact, we should therefore be able to find a scalar λ ("lambda") such that

$$\nabla f = \lambda \nabla g,$$

or

$$2x\mathbf{i} + 2y\mathbf{j} + 2z\mathbf{k} = \lambda(2x\mathbf{i} - 2z\mathbf{k}).$$

Thus, the coordinates x, y, and z of any point of tangency will have to satisfy the three scalar equations

$$2x = 2\lambda x, \qquad 2y = 0, \qquad 2z = -2\lambda z.$$

For what values of λ will a point (x, y, z) whose coordinates satisfy these scalar equations also lie on the surface $x^2 - z^2 - 1 = 0$? To answer this question, we use our knowledge that no point on the surface has a zero x-coordinate to conclude that $x \neq 0$. Hence, $2x = 2\lambda x$ only if

$$2 = 2\lambda, \qquad \text{or} \qquad \lambda = 1.$$

For $\lambda = 1$, the equation $2z = -2\lambda z$ becomes $2z = -2z$. If this equation is to be satisfied as well, z must be zero. Since $y = 0$ also (from the equation $2y = 0$), we conclude that the points we seek all have coordinates of the form

$$(x, 0, 0).$$

What points on the surface $x^2 - z^2 = 1$ have coordinates of this form? The answer is the points $(x, 0, 0)$ for which

$$x^2 - (0)^2 = 1, \qquad x^2 = 1, \qquad \text{or} \qquad x = \pm 1.$$

The points on the cylinder closest to the origin are the points $(\pm 1, 0, 0)$. ■

The Method of Lagrange Multipliers

In Solution 2 of Example 2, we used the **method of Lagrange multipliers**. The method says that the extreme values of a function $f(x, y, z)$ whose variables are subject to a constraint $g(x, y, z) = 0$ are to be found on the surface $g = 0$ among the points where

$$\nabla f = \lambda \nabla g$$

for some scalar λ (called a **Lagrange multiplier**).

To explore the method further and see why it works, we first make the following observation, which we state as a theorem.

THEOREM 12—The Orthogonal Gradient Theorem Suppose that $f(x, y, z)$ is differentiable in a region whose interior contains a smooth curve

$$C: \quad \mathbf{r}(t) = g(t)\mathbf{i} + h(t)\mathbf{j} + k(t)\mathbf{k}.$$

If P_0 is a point on C where f has a local maximum or minimum relative to its values on C, then ∇f is orthogonal to C at P_0.

Proof We show that ∇f is orthogonal to the curve's velocity vector at P_0. The values of f on C are given by the composite $f(g(t), h(t), k(t))$, whose derivative with respect to t is

$$\frac{df}{dt} = \frac{\partial f}{\partial x}\frac{dg}{dt} + \frac{\partial f}{\partial y}\frac{dh}{dt} + \frac{\partial f}{\partial z}\frac{dk}{dt} = \nabla f \cdot \mathbf{v}.$$

At any point P_0 where f has a local maximum or minimum relative to its values on the curve, $df/dt = 0$, so

$$\nabla f \cdot \mathbf{v} = 0.$$

■

By dropping the z-terms in Theorem 12, we obtain a similar result for functions of two variables.

COROLLARY OF THEOREM 12 At the points on a smooth curve $\mathbf{r}(t) = g(t)\mathbf{i} + h(t)\mathbf{j}$ where a differentiable function $f(x, y)$ takes on its local maxima and minima relative to its values on the curve, $\nabla f \cdot \mathbf{v} = 0$, where $\mathbf{v} = d\mathbf{r}/dt$.

Theorem 12 is the key to the method of Lagrange multipliers. Suppose that $f(x, y, z)$ and $g(x, y, z)$ are differentiable and that P_0 is a point on the surface $g(x, y, z) = 0$ where f has a local maximum or minimum value relative to its other values on the surface. We assume also that $\nabla g \neq \mathbf{0}$ at points on the surface $g(x, y, z) = 0$. Then f takes on a local maximum or minimum at P_0 relative to its values on every differentiable curve through P_0 on the surface $g(x, y, z) = 0$. Therefore, ∇f is orthogonal to the velocity vector of every such differentiable curve through P_0. So is ∇g, moreover (because ∇g is orthogonal to the level surface $g = 0$, as we saw in Section 11.5). Therefore, at P_0, ∇f is some scalar multiple λ of ∇g.

The Method of Lagrange Multipliers

Suppose that $f(x, y, z)$ and $g(x, y, z)$ are differentiable and $\nabla g \neq \mathbf{0}$ when $g(x, y, z) = 0$. To find the local maximum and minimum values of f subject to the constraint $g(x, y, z) = 0$ (if these exist), find the values of x, y, z, and λ that simultaneously satisfy the equations

$$\nabla f = \lambda \nabla g \qquad \text{and} \qquad g(x, y, z) = 0. \tag{1}$$

For functions of two independent variables, the condition is similar, but without the variable z.

Some care must be used in applying this method. An extreme value may not actually exist (Exercise 39).

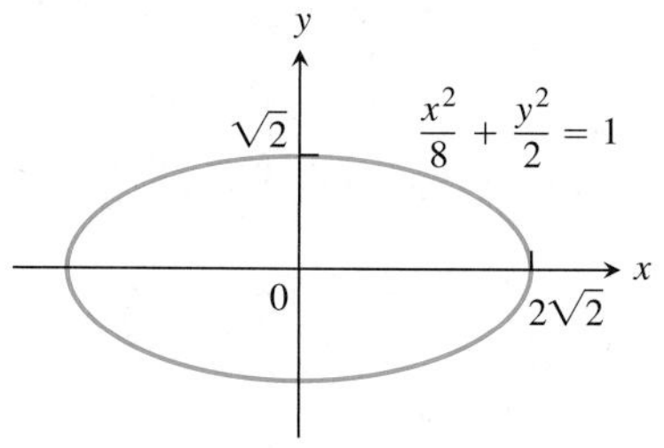

FIGURE 11.49 Example 3 shows how to find the largest and smallest values of the product xy on this ellipse.

EXAMPLE 3 Find the greatest and smallest values that the function

$$f(x, y) = xy$$

takes on the ellipse (Figure 11.49)

$$\frac{x^2}{8} + \frac{y^2}{2} = 1.$$

Solution We want the extreme values of $f(x, y) = xy$ subject to the constraint

$$g(x, y) = \frac{x^2}{8} + \frac{y^2}{2} - 1 = 0.$$

To do so, we first find the values of x, y, and λ for which

$$\nabla f = \lambda \nabla g \quad \text{and} \quad g(x, y) = 0.$$

The gradient equation in Equations (1) gives

$$y\mathbf{i} + x\mathbf{j} = \frac{\lambda}{4}x\mathbf{i} + \lambda y\mathbf{j},$$

from which we find

$$y = \frac{\lambda}{4}x, \qquad x = \lambda y, \qquad \text{and} \qquad y = \frac{\lambda}{4}(\lambda y) = \frac{\lambda^2}{4}y,$$

so that $y = 0$ or $\lambda = \pm 2$. We now consider these two cases.

Case 1: If $y = 0$, then $x = y = 0$. But $(0, 0)$ is not on the ellipse. Hence, $y \neq 0$.
Case 2: If $y \neq 0$, then $\lambda = \pm 2$ and $x = \pm 2y$. Substituting this in the equation $g(x, y) = 0$ gives

$$\frac{(\pm 2y)^2}{8} + \frac{y^2}{2} = 1, \qquad 4y^2 + 4y^2 = 8 \qquad \text{and} \qquad y = \pm 1.$$

The function $f(x, y) = xy$ therefore takes on its extreme values on the ellipse at the four points $(\pm 2, 1)$, $(\pm 2, -1)$. The extreme values are $xy = 2$ and $xy = -2$.

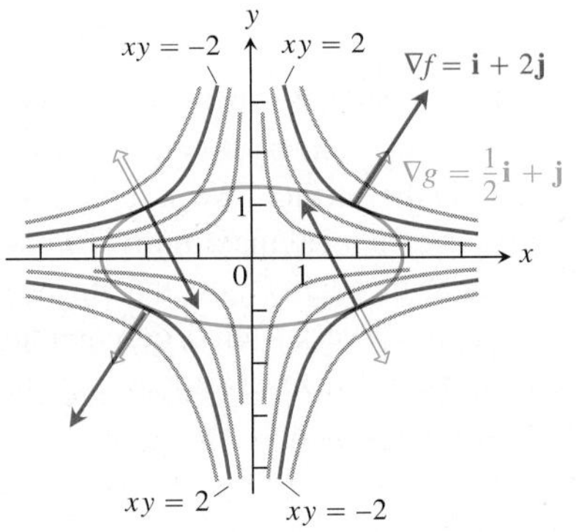

FIGURE 11.50 When subjected to the constraint $g(x, y) = x^2/8 + y^2/2 - 1 = 0$, the function $f(x, y) = xy$ takes on extreme values at the four points $(\pm 2, \pm 1)$. These are the points on the ellipse when ∇f (red) is a scalar multiple of ∇g (blue) (Example 3).

The Geometry of the Solution The level curves of the function $f(x, y) = xy$ are the hyperbolas $xy = c$ (Figure 11.50). The farther the hyperbolas lie from the origin, the larger the absolute value of f. We want to find the extreme values of $f(x, y)$, given that the point (x, y) also lies on the ellipse $x^2 + 4y^2 = 8$. Which hyperbolas intersecting the ellipse lie farthest from the origin? The hyperbolas that just graze the ellipse, the ones that are tangent to it, are farthest. At these points, any vector normal to the hyperbola is normal to the ellipse, so $\nabla f = y\mathbf{i} + x\mathbf{j}$ is a multiple $(\lambda = \pm 2)$ of $\nabla g = (x/4)\mathbf{i} + y\mathbf{j}$. At the point $(2, 1)$, for example,

$$\nabla f = \mathbf{i} + 2\mathbf{j}, \qquad \nabla g = \frac{1}{2}\mathbf{i} + \mathbf{j}, \qquad \text{and} \qquad \nabla f = 2\nabla g.$$

At the point $(-2, 1)$,

$$\nabla f = \mathbf{i} - 2\mathbf{j}, \qquad \nabla g = -\frac{1}{2}\mathbf{i} + \mathbf{j}, \qquad \text{and} \qquad \nabla f = -2\nabla g.$$ ■

EXAMPLE 4 Find the maximum and minimum values of the function $f(x, y) = 3x + 4y$ on the circle $x^2 + y^2 = 1$.

Solution We model this as a Lagrange multiplier problem with

$$f(x, y) = 3x + 4y, \qquad g(x, y) = x^2 + y^2 - 1$$

and look for the values of x, y, and λ that satisfy the equations

$$\nabla f = \lambda \nabla g: \quad 3\mathbf{i} + 4\mathbf{j} = 2x\lambda\mathbf{i} + 2y\lambda\mathbf{j}$$
$$g(x, y) = 0: \quad x^2 + y^2 - 1 = 0.$$

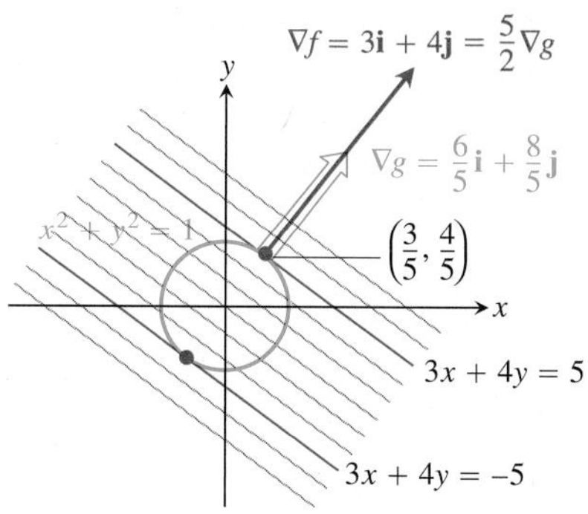

FIGURE 11.51 The function $f(x, y) = 3x + 4y$ takes on its largest value on the unit circle $g(x, y) = x^2 + y^2 - 1 = 0$ at the point $(3/5, 4/5)$ and its smallest value at the point $(-3/5, -4/5)$ (Example 4). At each of these points, ∇f is a scalar multiple of ∇g. The figure shows the gradients at the first point but not the second.

The gradient equation in Equations (1) implies that $\lambda \neq 0$ and gives

$$x = \frac{3}{2\lambda}, \qquad y = \frac{2}{\lambda}.$$

These equations tell us, among other things, that x and y have the same sign. With these values for x and y, the equation $g(x, y) = 0$ gives

$$\left(\frac{3}{2\lambda}\right)^2 + \left(\frac{2}{\lambda}\right)^2 - 1 = 0,$$

so

$$\frac{9}{4\lambda^2} + \frac{4}{\lambda^2} = 1, \qquad 9 + 16 = 4\lambda^2, \qquad 4\lambda^2 = 25, \qquad \text{and} \qquad \lambda = \pm\frac{5}{2}.$$

Thus,

$$x = \frac{3}{2\lambda} = \pm\frac{3}{5}, \qquad y = \frac{2}{\lambda} = \pm\frac{4}{5},$$

and $f(x, y) = 3x + 4y$ has extreme values at $(x, y) = \pm(3/5, 4/5)$.

By calculating the value of $3x + 4y$ at the points $\pm(3/5, 4/5)$, we see that its maximum and minimum values on the circle $x^2 + y^2 = 1$ are

$$3\left(\frac{3}{5}\right) + 4\left(\frac{4}{5}\right) = \frac{25}{5} = 5 \qquad \text{and} \qquad 3\left(-\frac{3}{5}\right) + 4\left(-\frac{4}{5}\right) = -\frac{25}{5} = -5.$$

The Geometry of the Solution The level curves of $f(x, y) = 3x + 4y$ are the lines $3x + 4y = c$ (Figure 11.51). The farther the lines lie from the origin, the larger the absolute value of f. We want to find the extreme values of $f(x, y)$ given that the point (x, y) also lies on the circle $x^2 + y^2 = 1$. Which lines intersecting the circle lie farthest from the origin? The lines tangent to the circle are farthest. At the points of tangency, any vector normal to the line is normal to the circle, so the gradient $\nabla f = 3\mathbf{i} + 4\mathbf{j}$ is a multiple $(\lambda = \pm 5/2)$ of the gradient $\nabla g = 2x\mathbf{i} + 2y\mathbf{j}$. At the point $(3/5, 4/5)$, for example,

$$\nabla f = 3\mathbf{i} + 4\mathbf{j}, \qquad \nabla g = \frac{6}{5}\mathbf{i} + \frac{8}{5}\mathbf{j}, \qquad \text{and} \qquad \nabla f = \frac{5}{2}\nabla g. \qquad \blacksquare$$

Lagrange Multipliers with Two Constraints

Many problems require us to find the extreme values of a differentiable function $f(x, y, z)$ whose variables are subject to two constraints. If the constraints are

$$g_1(x, y, z) = 0 \qquad \text{and} \qquad g_2(x, y, z) = 0$$

and g_1 and g_2 are differentiable, with ∇g_1 not parallel to ∇g_2, we find the constrained local maxima and minima of f by introducing two Lagrange multipliers λ and μ (mu, pronounced "mew"). That is, we locate the points $P(x, y, z)$ where f takes on its constrained extreme values by finding the values of x, y, z, λ, and μ that simultaneously satisfy the equations

$$\nabla f = \lambda \nabla g_1 + \mu \nabla g_2, \qquad g_1(x, y, z) = 0, \qquad g_2(x, y, z) = 0 \qquad (2)$$

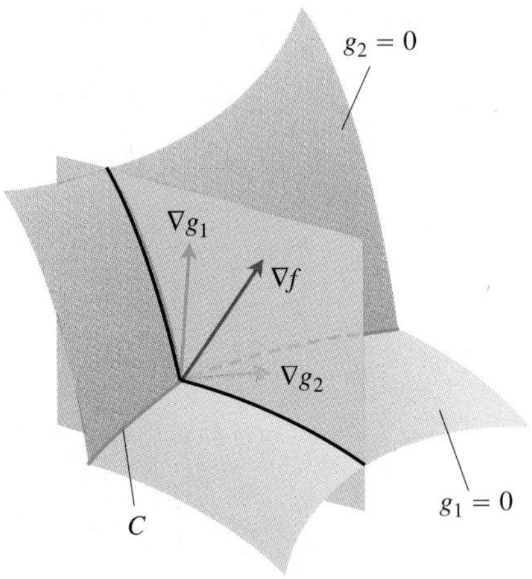

FIGURE 11.52 The vectors ∇g_1 and ∇g_2 lie in a plane perpendicular to the curve C because ∇g_1 is normal to the surface $g_1 = 0$ and ∇g_2 is normal to the surface $g_2 = 0$.

Equations (2) have a nice geometric interpretation. The surfaces $g_1 = 0$ and $g_2 = 0$ (usually) intersect in a smooth curve, say C (Figure 11.52). Along this curve we seek the points where f has local maximum and minimum values relative to its other values on the curve. These are the points where ∇f is normal to C, as we saw in Theorem 12. But ∇g_1 and ∇g_2 are also normal to C at these points because C lies in the surfaces $g_1 = 0$ and $g_2 = 0$. Therefore, ∇f lies in the plane determined by ∇g_1 and ∇g_2, which means that $\nabla f = \lambda \nabla g_1 + \mu \nabla g_2$ for some λ and μ. Since the points we seek also lie in both surfaces, their coordinates must satisfy the equations $g_1(x, y, z) = 0$ and $g_2(x, y, z) = 0$, which are the remaining requirements in Equations (2).

EXAMPLE 5 The plane $x + y + z = 1$ cuts the cylinder $x^2 + y^2 = 1$ in an ellipse (Figure 11.53). Find the points on the ellipse that lie closest to and farthest from the origin.

Solution We find the extreme values of

$$f(x, y, z) = x^2 + y^2 + z^2$$

(the square of the distance from (x, y, z) to the origin) subject to the constraints

$$g_1(x, y, z) = x^2 + y^2 - 1 = 0 \tag{3}$$

$$g_2(x, y, z) = x + y + z - 1 = 0. \tag{4}$$

The gradient equation in Equations (2) then gives

$$\nabla f = \lambda \nabla g_1 + \mu \nabla g_2$$
$$2x\mathbf{i} + 2y\mathbf{j} + 2z\mathbf{k} = \lambda(2x\mathbf{i} + 2y\mathbf{j}) + \mu(\mathbf{i} + \mathbf{j} + \mathbf{k})$$
$$2x\mathbf{i} + 2y\mathbf{j} + 2z\mathbf{k} = (2\lambda x + \mu)\mathbf{i} + (2\lambda y + \mu)\mathbf{j} + \mu\mathbf{k}$$

or

$$2x = 2\lambda x + \mu, \qquad 2y = 2\lambda y + \mu, \qquad 2z = \mu. \tag{5}$$

The scalar equations in Equations (5) yield

$$\begin{aligned} 2x &= 2\lambda x + 2z \Rightarrow (1 - \lambda)x = z, \\ 2y &= 2\lambda y + 2z \Rightarrow (1 - \lambda)y = z. \end{aligned} \tag{6}$$

Equations (6) are satisfied simultaneously if either $\lambda = 1$ and $z = 0$ or $\lambda \neq 1$ and $x = y = z/(1 - \lambda)$.

If $z = 0$, then solving Equations (3) and (4) simultaneously to find the corresponding points on the ellipse gives the two points $(1, 0, 0)$ and $(0, 1, 0)$. This makes sense when you look at Figure 11.53.

If $x = y$, then Equations (3) and (4) give

$$\begin{aligned} x^2 + x^2 - 1 &= 0 & \qquad x + x + z - 1 &= 0 \\ 2x^2 &= 1 & z &= 1 - 2x \\ x &= \pm\frac{\sqrt{2}}{2} & z &= 1 \mp \sqrt{2}. \end{aligned}$$

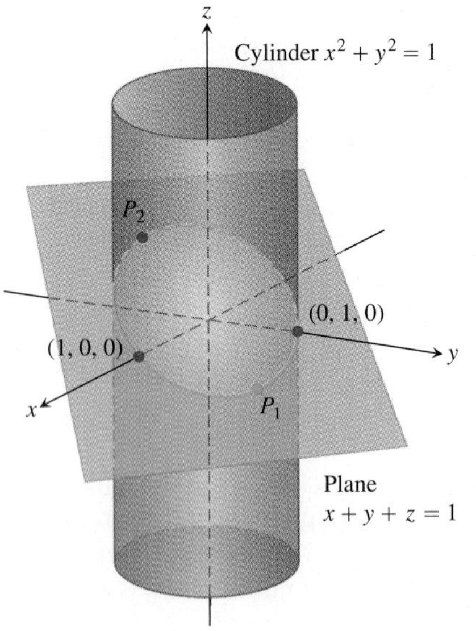

FIGURE 11.53 On the ellipse where the plane and cylinder meet, we find the points closest to and farthest from the origin. (Example 5).

The corresponding points on the ellipse are

$$P_1 = \left(\frac{\sqrt{2}}{2}, \frac{\sqrt{2}}{2}, 1 - \sqrt{2}\right) \quad \text{and} \quad P_2 = \left(-\frac{\sqrt{2}}{2}, -\frac{\sqrt{2}}{2}, 1 + \sqrt{2}\right).$$

Here we need to be careful, however. Although P_1 and P_2 both give local maxima of f on the ellipse, P_2 is farther from the origin than P_1.

The points on the ellipse closest to the origin are (1, 0, 0) and (0, 1, 0). The point on the ellipse farthest from the origin is P_2. ■

EXERCISES 11.8

1. **Extrema on an ellipse** Find the points on the ellipse $x^2 + 2y^2 = 1$ where $f(x, y) = xy$ has its extreme values.
2. **Extrema on a circle** Find the extreme values of $f(x, y) = xy$ subject to the constraint $g(x, y) = x^2 + y^2 - 10 = 0$.
3. **Maximum on a line** Find the maximum value of $f(x, y) = 49 - x^2 - y^2$ on the line $x + 3y = 10$.
4. **Extrema on a line** Find the local extreme values of $f(x, y) = x^2y$ on the line $x + y = 3$.
5. **Constrained minimum** Find the points on the curve $xy^2 = 54$ nearest the origin.
6. **Constrained minimum** Find the points on the curve $x^2y = 2$ nearest the origin.
7. Use the method of Lagrange multipliers to find
   a. **Minimum on a hyperbola** The minimum value of $x + y$, subject to the constraints $xy = 16, x > 0, y > 0$
   b. **Maximum on a line** The maximum value of xy, subject to the constraint $x + y = 16$.

   Comment on the geometry of each solution.
8. **Extrema on a curve** Find the points on the curve $x^2 + xy + y^2 = 1$ in the xy-plane that are nearest to and farthest from the origin.
9. **Minimum surface area with fixed volume** Find the dimensions of the closed right circular cylindrical can of smallest surface area whose volume is 16π cm^3.
10. **Cylinder in a sphere** Find the radius and height of the open right circular cylinder of largest surface area that can be inscribed in a sphere of radius a. What *is* the largest surface area?
11. **Rectangle of greatest area in an ellipse** Use the method of Lagrange multipliers to find the dimensions of the rectangle of greatest area that can be inscribed in the ellipse $x^2/16 + y^2/9 = 1$ with sides parallel to the coordinate axes.
12. **Rectangle of longest perimeter in an ellipse** Find the dimensions of the rectangle of largest perimeter that can be inscribed in the ellipse $x^2/a^2 + y^2/b^2 = 1$ with sides parallel to the coordinate axes. What *is* the largest perimeter?
13. **Extrema on a circle** Find the maximum and minimum values of $x^2 + y^2$ subject to the constraint $x^2 - 2x + y^2 - 4y = 0$.
14. **Extrema on a circle** Find the maximum and minimum values of $3x - y + 6$ subject to the constraint $x^2 + y^2 = 4$.
15. **Ant on a metal plate** The temperature at a point (x, y) on a metal plate is $T(x, y) = 4x^2 - 4xy + y^2$. An ant on the plate walks around the circle of radius 5 centered at the origin. What are the highest and lowest temperatures encountered by the ant?
16. **Cheapest storage tank** Your firm has been asked to design a storage tank for liquid petroleum gas. The customer's specifications call for a cylindrical tank with hemispherical ends, and the tank is to hold 8000 m^3 of gas. The customer also wants to use the smallest amount of material possible in building the tank. What radius and height do you recommend for the cylindrical portion of the tank?
17. **Minimum distance to a point** Find the point on the plane $x + 2y + 3z = 13$ closest to the point (1, 1, 1).
18. **Maximum distance to a point** Find the point on the sphere $x^2 + y^2 + z^2 = 4$ farthest from the point $(1, -1, 1)$.
19. **Minimum distance to the origin** Find the minimum distance from the surface $x^2 - y^2 - z^2 = 1$ to the origin.
20. **Minimum distance to the origin** Find the point on the surface $z = xy + 1$ nearest the origin.
21. **Minimum distance to the origin** Find the points on the surface $z^2 = xy + 4$ closest to the origin.
22. **Minimum distance to the origin** Find the point(s) on the surface $xyz = 1$ closest to the origin.
23. **Extrema on a sphere** Find the maximum and minimum values of

$$f(x, y, z) = x - 2y + 5z$$

    on the sphere $x^2 + y^2 + z^2 = 30$.
24. **Extrema on a sphere** Find the points on the sphere $x^2 + y^2 + z^2 = 25$ where $f(x, y, z) = x + 2y + 3z$ has its maximum and minimum values.
25. **Minimizing a sum of squares** Find three real numbers whose sum is 9 and the sum of whose squares is as small as possible.
26. **Maximizing a product** Find the largest product the positive numbers x, y, and z can have if $x + y + z^2 = 16$.
27. **Rectangular box of longest volume in a sphere** Find the dimensions of the closed rectangular box with maximum volume that can be inscribed in the unit sphere.
28. **Box with vertex on a plane** Find the volume of the largest closed rectangular box in the first octant having three faces in the coordinate

planes and a vertex on the plane $x/a + y/b + z/c = 1$, where $a > 0$, $b > 0$, and $c > 0$.

29. **Hottest point on a space probe** A space probe in the shape of the ellipsoid

$$4x^2 + y^2 + 4z^2 = 16$$

enters Earth's atmosphere and its surface begins to heat. After 1 hour, the temperature at the point (x, y, z) on the probe's surface is

$$T(x, y, z) = 8x^2 + 4yz - 16z + 600 .$$

Find the hottest point on the probe's surface.

30. **Extreme temperatures on a sphere** Suppose that the Celsius temperature at the point (x, y, z) on the sphere $x^2 + y^2 + z^2 = 1$ is $T = 400xyz^2$. Locate the highest and lowest temperatures on the sphere.

31. Maximize the function $f(x, y, z) = x^2 + 2y - z^2$ subject to the constraints $2x - y = 0$ and $y + z = 0$.

32. Minimize the function $f(x, y, z) = x^2 + y^2 + z^2$ subject to the constraints $x + 2y + 3z = 6$ and $x + 3y + 9z = 9$.

33. **Minimum distance to the origin** Find the point closest to the origin on the line of intersection of the planes $y + 2z = 12$ and $x + y = 6$.

34. **Maximum value on line of intersection** Find the maximum value that $f(x, y, z) = x^2 + 2y - z^2$ can have on the line of intersection of the planes $2x - y = 0$ and $y + z = 0$.

35. **Extrema on a curve of intersection** Find the extreme values of $f(x, y, z) = x^2yz + 1$ on the intersection of the plane $z = 1$ with the sphere $x^2 + y^2 + z^2 = 10$.

36. a. **Maximum on line of intersection** Find the maximum value of $w = xyz$ on the line of intersection of the two planes $x + y + z = 40$ and $x + y - z = 0$.
    b. Give a geometric argument to support your claim that you have found a maximum, and not a minimum, value of w.

37. **Extrema on a circle of intersection** Find the extreme values of the function $f(x, y, z) = xy + z^2$ on the circle in which the plane $y - x = 0$ intersects the sphere $x^2 + y^2 + z^2 = 4$.

38. **Minimum distance to the origin** Find the point closest to the origin on the curve of intersection of the plane $2y + 4z = 5$ and the cone $z^2 = 4x^2 + 4y^2$.

39. **The condition $\nabla f = \lambda \nabla g$ is not sufficient** Although $\nabla f = \lambda \nabla g$ is a necessary condition for the occurrence of an extreme value of $f(x, y)$ subject to the conditions $g(x, y) = 0$ and $\nabla g \neq \mathbf{0}$, it does not in itself guarantee that one exists. As a case in point, try using the method of Lagrange multipliers to find a maximum value of $f(x, y) = x + y$ subject to the constraint that $xy = 16$. The method will identify the two points $(4, 4)$ and $(-4, -4)$ as candidates for the location of extreme values. Yet the sum $(x + y)$ has no maximum value on the hyperbola $xy = 16$. The farther you go from the origin on this hyperbola in the first quadrant, the larger the sum $f(x, y) = x + y$ becomes.

40. **A least squares plane** The plane $z = Ax + By + C$ is to be "fitted" to the following points (x_k, y_k, z_k):

$$(0, 0, 0), \quad (0, 1, 1), \quad (1, 1, 1), \quad (1, 0, -1).$$

Find the values of A, B, and C that minimize

$$\sum_{k=1}^{4} (Ax_k + By_k + C - z_k)^2,$$

the sum of the squares of the deviations.

41. a. **Maximum on a sphere** Show that the maximum value of $a^2b^2c^2$ on a sphere of radius r centered at the origin of a Cartesian abc-coordinate system is $(r^2/3)^3$.
    b. **Geometric and arithmetic means** Using part (a), show that for nonnegative numbers a, b, and c,

$$(abc)^{1/3} \leq \frac{a + b + c}{3};$$

that is, the *geometric mean* of three nonnegative numbers is less than or equal to their *arithmetic mean.*

42. **Sum of products** Let $a_1, a_2, \ldots, a_n$ be n positive numbers. Find the maximum of $\Sigma_{i=1}^{n} a_i x_i$ subject to the constraint $\Sigma_{i=1}^{n} x_i^2 = 1$.

COMPUTER EXPLORATIONS

In Exercises 43–48, use a CAS to perform the following steps implementing the method of Lagrange multipliers for finding constrained extrema:

a. Form the function $h = f - \lambda_1 g_1 - \lambda_2 g_2$, where f is the function to optimize subject to the constraints $g_1 = 0$ and $g_2 = 0$.
b. Determine all the first partial derivatives of h, including the partials with respect to λ_1 and λ_2, and set them equal to 0.
c. Solve the system of equations found in part (b) for all the unknowns, including λ_1 and λ_2.
d. Evaluate f at each of the solution points found in part (c) and select the extreme value subject to the constraints asked for in the exercise.

43. Minimize $f(x, y, z) = xy + yz$ subject to the constraints $x^2 + y^2 - 2 = 0$ and $x^2 + z^2 - 2 = 0$.

44. Minimize $f(x, y, z) = xyz$ subject to the constraints $x^2 + y^2 - 1 = 0$ and $x - z = 0$.

45. Maximize $f(x, y, z) = x^2 + y^2 + z^2$ subject to the constraints $2y + 4z - 5 = 0$ and $4x^2 + 4y^2 - z^2 = 0$.

46. Minimize $f(x, y, z) = x^2 + y^2 + z^2$ subject to the constraints $x^2 - xy + y^2 - z^2 - 1 = 0$ and $x^2 + y^2 - 1 = 0$.

47. Minimize $f(x, y, z, w) = x^2 + y^2 + z^2 + w^2$ subject to the constraints $2x - y + z - w - 1 = 0$ and $x + y - z + w - 1 = 0$.

48. Determine the distance from the line $y = x + 1$ to the parabola $y^2 = x$. (*Hint:* Let (x, y) be a point on the line and (w, z) a point on the parabola. You want to minimize $(x - w)^2 + (y - z)^2$.)

Chapter 11 Questions to Guide Your Review

1. What is a real-valued function of two independent variables? Three independent variables? Give examples.
2. What does it mean for sets in the plane or in space to be open? Closed? Give examples. Give examples of sets that are neither open nor closed.
3. How can you display the values of a function $f(x, y)$ of two independent variables graphically? How do you do the same for a function $f(x, y, z)$ of three independent variables?
4. What does it mean for a function $f(x, y)$ to have limit L as $(x, y) \to (x_0, y_0)$? What are the basic properties of limits of functions of two independent variables?
5. When is a function of two (three) independent variables continuous at a point in its domain? Give examples of functions that are continuous at some points but not others.
6. What can be said about algebraic combinations and composites of continuous functions?
7. Explain the two-path test for nonexistence of limits.
8. How are the partial derivatives $\partial f/\partial x$ and $\partial f/\partial y$ of a function $f(x, y)$ defined? How are they interpreted and calculated?
9. How does the relation between first partial derivatives and continuity of functions of two independent variables differ from the relation between first derivatives and continuity for real-valued functions of a single independent variable? Give an example.
10. What is the Mixed Derivative Theorem for mixed second-order partial derivatives? How can it help in calculating partial derivatives of second and higher orders? Give examples.
11. What does it mean for a function $f(x, y)$ to be differentiable? What does the Increment Theorem say about differentiability?
12. How can you sometimes decide from examining f_x and f_y that a function $f(x, y)$ is differentiable? What is the relation between the differentiability of f and the continuity of f at a point?
13. What is the Chain Rule? What form does it take for functions of two independent variables? Three independent variables? Functions defined on surfaces? How do you diagram these different forms? Give examples. What pattern enables one to remember all the different forms?
14. What is the derivative of a function $f(x, y)$ at a point P_0 in the direction of a unit vector $\mathbf{u}$? What rate does it describe? What geometric interpretation does it have? Give examples.
15. What is the gradient vector of a differentiable function $f(x, y)$? How is it related to the function's directional derivatives? State the analogous results for functions of three independent variables.
16. How do you find the tangent line at a point on a level curve of a differentiable function $f(x, y)$? How do you find the tangent plane and normal line at a point on a level surface of a differentiable function $f(x, y, z)$? Give examples.
17. How can you use directional derivatives to estimate change?
18. How do you linearize a function $f(x, y)$ of two independent variables at a point (x_0, y_0) ? Why might you want to do this? How do you linearize a function of three independent variables?
19. What can you say about the accuracy of linear approximations of functions of two (three) independent variables?
20. If (x, y) moves from (x_0, y_0) to a point $(x_0 + dx, y_0 + dy)$ nearby, how can you estimate the resulting change in the value of a differentiable function $f(x, y)$? Give an example.
21. How do you define local maxima, local minima, and saddle points for a differentiable function $f(x, y)$? Give examples.
22. What derivative tests are available for determining the local extreme values of a function $f(x, y)$? How do they enable you to narrow your search for these values? Give examples.
23. How do you find the extrema of a continuous function $f(x, y)$ on a closed bounded region of the xy-plane? Give an example.
24. Describe the method of Lagrange multipliers and give examples.

Chapter 11 Practice and Additional Exercises

In Exercises 1–4, find the domain and range of the given function and identify its level curves. Sketch a typical level curve.

1. $f(x, y) = 9x^2 + y^2$
2. $f(x, y) = e^{x+y}$
3. $g(x, y) = 1/xy$
4. $g(x, y) = \sqrt{x^2 - y}$

In Exercises 5–8, find the domain and range of the given function and identify its level surfaces. Sketch a typical level surface.

5. $f(x, y, z) = x^2 + y^2 - z$
6. $g(x, y, z) = x^2 + 4y^2 + 9z^2$
7. $h(x, y, z) = \dfrac{1}{x^2 + y^2 + z^2}$
8. $k(x, y, z) = \dfrac{1}{x^2 + y^2 + z^2 + 1}$

Find the limits in Exercises 9–14.

9. $\displaystyle\lim_{(x,y)\to(\pi, \ln 2)} e^y \cos x$
10. $\displaystyle\lim_{(x,y)\to(0,0)} \frac{2 + y}{x + \cos y}$

11. $\lim_{(x,y)\to(1,1)} \dfrac{x-y}{x^2-y^2}$

12. $\lim_{(x,y)\to(1,1)} \dfrac{x^3y^3-1}{xy-1}$

13. $\lim_{P\to(1,-1,e)} \ln|x+y+z|$

14. $\lim_{P\to(1,-1,-1)} \tan^{-1}(x+y+z)$

By considering different paths of approach, show that the limits in Exercises 15 and 16 do not exist.

15. $\lim_{\substack{(x,y)\to(0,0)\\ y\neq x^2}} \dfrac{y}{x^2-y}$

16. $\lim_{\substack{(x,y)\to(0,0)\\ xy\neq 0}} \dfrac{x^2+y^2}{xy}$

Use polar coordinates to find the limits in Exercises 17 and 18.

17. $\lim_{(x,y)\to(0,0)} \dfrac{x^3-y^3}{x^2+y^2}$

18. $\lim_{(x,y)\to(0,0)} (x^2+y^2)\ln(x^2+y^2)$

In Exercises 19–24, find the partial derivative of the function with respect to each variable.

19. $g(r, \theta) = r\cos\theta + r\sin\theta$

20. $f(x, y) = \dfrac{1}{2}\ln(x^2+y^2) + \tan^{-1}\dfrac{y}{x}$

21. $f(R_1, R_2, R_3) = \dfrac{1}{R_1} + \dfrac{1}{R_2} + \dfrac{1}{R_3}$

22. $h(x, y, z) = \sin(2\pi x + y - 3z)$

23. $P(n, R, T, V) = \dfrac{nRT}{V}$ (the ideal gas law)

24. $f(r, l, T, w) = \dfrac{1}{2rl}\sqrt{\dfrac{T}{\pi w}}$

Find the second-order partial derivatives of the functions in Exercises 25–28.

25. $g(x, y) = y + \dfrac{x}{y}$

26. $g(x, y) = e^x + y\sin x$

27. $f(x, y) = x + xy - 5x^3 + \ln(x^2+1)$

28. $f(x, y) = y^2 - 3xy + \cos y + 7e^y$

29. Find dw/dt at $t = 0$ if $w = \sin(xy + \pi)$, $x = e^t$, and $y = \ln(t + 1)$.

30. Find dw/dt at $t = 1$ if $w = xe^y + y\sin z - \cos z$, $x = 2\sqrt{t}$, $y = t - 1 + \ln t$, and $z = \pi t$.

31. Find $\partial w/\partial r$ and $\partial w/\partial s$ when $r = \pi$ and $s = 0$ if $w = \sin(2x - y)$, $x = r + \sin s$, $y = rs$.

32. Find $\partial w/\partial u$ and $\partial w/\partial v$ when $u = v = 0$ if $w = \ln\sqrt{1 + x^2} - \tan^{-1}x$ and $x = 2e^u\cos v$.

33. Find the value of the derivative of $f(x, y, z) = xy + yz + xz$ with respect to t on the curve $x = \cos t$, $y = \sin t$, $z = \cos 2t$ at $t = 1$.

34. Show that if $w = f(s)$ is any differentiable function of s and if $s = y + 5x$, then

$$\frac{\partial w}{\partial x} - 5\frac{\partial w}{\partial y} = 0.$$

Assuming that the equations in Exercises 35 and 36 define y as a differentiable function of x, find the value of dy/dx at point P.

35. $1 - x - y^2 - \sin xy = 0$, $P(0, 1)$

36. $2xy + e^{x+y} - 2 = 0$, $P(0, \ln 2)$

In Exercises 37–40, find the directions in which f increases and decreases most rapidly at P_0 and find the derivative of f in each direction. Also, find the derivative of f at P_0 in the direction of the vector $\mathbf{v}$.

37. $f(x, y) = \cos x\cos y$, $P_0(\pi/4, \pi/4)$, $\mathbf{v} = 3\mathbf{i} + 4\mathbf{j}$

38. $f(x, y) = x^2e^{-2y}$, $P_0(1, 0)$, $\mathbf{v} = \mathbf{i} + \mathbf{j}$

39. $f(x, y, z) = \ln(2x + 3y + 6z)$, $P_0(-1, -1, 1)$, $\mathbf{v} = 2\mathbf{i} + 3\mathbf{j} + 6\mathbf{k}$

40. $f(x, y, z) = x^2 + 3xy - z^2 + 2y + z + 4$, $P_0(0, 0, 0)$, $\mathbf{v} = \mathbf{i} + \mathbf{j} + \mathbf{k}$

41. **Derivative in velocity direction** Find the derivative of $f(x, y, z) = xyz$ in the direction of the velocity vector of the helix

$$\mathbf{r}(t) = (\cos 3t)\mathbf{i} + (\sin 3t)\mathbf{j} + 3t\mathbf{k}$$

at $t = \pi/3$.

42. **Maximum directional derivative** What is the largest value that the directional derivative of $f(x, y, z) = xyz$ can have at the point $(1, 1, 1)$?

43. **Directional derivatives with given values** At the point $(1, 2)$, the function $f(x, y)$ has a derivative of 2 in the direction toward $(2, 2)$ and a derivative of -2 in the direction toward $(1, 1)$.

a. Find $f_x(1, 2)$ and $f_y(1, 2)$.

b. Find the derivative of f at $(1, 2)$ in the direction toward the point $(4, 6)$.

44. Which of the following statements are true if $f(x, y)$ is differentiable at (x_0, y_0)? Give reasons for your answers.

a. If $\mathbf{u}$ is a unit vector, the derivative of f at (x_0, y_0) in the direction of $\mathbf{u}$ is $(f_x(x_0, y_0)\mathbf{i} + f_y(x_0, y_0)\mathbf{j})\cdot\mathbf{u}$.

b. The derivative of f at (x_0, y_0) in the direction of $\mathbf{u}$ is a vector.

c. The directional derivative of f at (x_0, y_0) has its greatest value in the direction of ∇f.

d. At (x_0, y_0), vector ∇f is normal to the curve $f(x, y) = f(x_0, y_0)$.

In Exercises 45 and 46, sketch the surface $f(x, y, z) = c$ together with ∇f at the given points.

45. $x^2 + y + z^2 = 0$; $(0, -1, \pm 1)$, $(0, 0, 0)$

46. $y^2 + z^2 = 4$; $(2, \pm 2, 0)$, $(2, 0, \pm 2)$

In Exercises 47 and 48, find an equation for the plane tangent to the level surface $f(x, y, z) = c$ at the point P_0. Also, find parametric equations for the line that is normal to the surface at P_0.

47. $x^2 - y - 5z = 0$, $P_0(2, -1, 1)$

48. $x^2 + y^2 + z = 4$, $P_0(1, 1, 2)$

In Exercises 49 and 50, find an equation for the plane tangent to the surface $z = f(x, y)$ at the given point.

49. $z = \ln(x^2 + y^2)$, $(0, 1, 0)$

50. $z = 1/(x^2 + y^2)$, $(1, 1, 1/2)$

In Exercises 51 and 52, find equations for the lines that are tangent and normal to the level curve $f(x, y) = c$ at the point P_0. Then sketch the lines and level curve together with ∇f at P_0.

51. $y - \sin x = 1$, $P_0(\pi, 1)$ **52.** $\frac{y^2}{2} - \frac{x^2}{2} = \frac{3}{2}$, $P_0(1, 2)$

In Exercises 53 and 54, find parametric equations for the line that is tangent to the curve of intersection of the surfaces at the given point.

53. Surfaces: $x^2 + 2y + 2z = 4$, $y = 1$

Point: $(1, 1, 1/2)$

54. Surfaces: $x + y^2 + z = 2$, $y = 1$

Point: $(1/2, 1, 1/2)$

In Exercises 55 and 56, find the linearization $L(x, y)$ of the function $f(x, y)$ at the point P_0. Then find an upper bound for the magnitude of the error E in the approximation $f(x, y) \approx L(x, y)$ over the rectangle R.

55. $f(x, y) = \sin x \cos y$, $P_0(\pi/4, \pi/4)$

R: $\left|x - \frac{\pi}{4}\right| \le 0.1$, $\left|y - \frac{\pi}{4}\right| \le 0.1$

56. $f(x, y) = xy - 3y^2 + 2$, $P_0(1, 1)$

R: $|x - 1| \le 0.1$, $|y - 1| \le 0.2$

Find the linearizations of the functions in Exercises 57 and 58 at the given points.

57. $f(x, y, z) = xy + 2yz - 3xz$ at $(1, 0, 0)$ and $(1, 1, 0)$

58. $f(x, y, z) = \sqrt{2} \cos x \sin(y + z)$ at $(0, 0, \pi/4)$ and $(\pi/4, \pi/4, 0)$

Test the functions in Exercises 59–64 for local maxima and minima and saddle points. Find each function's value at these points.

59. $f(x, y) = x^2 - xy + y^2 + 2x + 2y - 4$

60. $f(x, y) = 5x^2 + 4xy - 2y^2 + 4x - 4y$

61. $f(x, y) = 2x^3 + 3xy + 2y^3$

62. $f(x, y) = x^3 + y^3 - 3xy + 15$

63. $f(x, y) = x^3 + y^3 + 3x^2 - 3y^2$

64. $f(x, y) = x^4 - 8x^2 + 3y^2 - 6y$

In Exercises 65–72, find the absolute maximum and minimum values of f on the region R.

65. $f(x, y) = x^2 + xy + y^2 - 3x + 3y$

R: The triangular region cut from the first quadrant by the line $x + y = 4$

66. $f(x, y) = x^2 - y^2 - 2x + 4y + 1$

R: The rectangular region in the first quadrant bounded by the coordinate axes and the lines $x = 4$ and $y = 2$

67. $f(x, y) = y^2 - xy - 3y + 2x$

R: The square region enclosed by the lines $x = \pm 2$ and $y = \pm 2$

68. $f(x, y) = 2x + 2y - x^2 - y^2$

R: The square region bounded by the coordinate axes and the lines $x = 2, y = 2$ in the first quadrant

69. $f(x, y) = x^2 - y^2 - 2x + 4y$

R: The triangular region bounded below by the x-axis, above by the line $y = x + 2$, and on the right by the line $x = 2$

70. $f(x, y) = 4xy - x^4 - y^4 + 16$

R: The triangular region bounded below by the line $y = -2$, above by the line $y = x$, and on the right by the line $x = 2$

71. $f(x, y) = x^3 + y^3 + 3x^2 - 3y^2$

R: The square region enclosed by the lines $x = \pm 1$ and $y = \pm 1$

72. $f(x, y) = x^3 + 3xy + y^3 + 1$

R: The square region enclosed by the lines $x = \pm 1$ and $y = \pm 1$

73. Extrema on a circle Find the extreme values of $f(x, y) = x^3 + y^2$ on the circle $x^2 + y^2 = 1$.

74. Extrema on a circle Find the extreme values of $f(x, y) = xy$ on the circle $x^2 + y^2 = 1$.

75. Extrema in a disk Find the extreme values of $f(x, y) = x^2 + 3y^2 + 2y$ on the unit disk $x^2 + y^2 \le 1$.

76. Extrema in a disk Find the extreme values of $f(x, y) = x^2 + y^2 - 3x - xy$ on the disk $x^2 + y^2 \le 9$.

77. Extrema on a sphere Find the extreme values of $f(x, y, z) = x - y + z$ on the unit sphere $x^2 + y^2 + z^2 = 1$.

78. Minimum distance to origin Find the points on the surface $x^2 - zy = 4$ closest to the origin.

79. Minimizing cost of a box A closed rectangular box is to have volume V cm^3. The cost of the material used in the box is a cents/cm^2 for top and bottom, b cents/cm^2 for front and back, and c cents/cm^2 for the remaining sides. What dimensions minimize the total cost of materials?

80. Least volume Find the plane $x/a + y/b + z/c = 1$ that passes through the point $(2, 1, 2)$ and cuts off the least volume from the first octant.

81. Extrema on curve of intersecting surfaces Find the extreme values of $f(x, y, z) = x(y + z)$ on the curve of intersection of the right circular cylinder $x^2 + y^2 = 1$ and the hyperbolic cylinder $xz = 1$.

82. Minimum distance to origin on curve of intersecting plane and cone Find the point closest to the origin on the curve of intersection of the plane $x + y + z = 1$ and the cone $z^2 = 2x^2 + 2y^2$.

83. Let $w = f(r, \theta)$, $r = \sqrt{x^2 + y^2}$, and $\theta = \tan^{-1}(y/x)$. Find $\partial w/\partial x$ and $\partial w/\partial y$ and express your answers in terms of r and θ.

84. Let $z = f(u, v)$, $u = ax + by$, and $v = ax - by$. Express z_x and z_y in terms of f_u, f_v, and the constants a and b.

85. If a and b are constants, $w = u^3 + \tanh u + \cos u$, and $u = ax + by$, show that

$$a\frac{\partial w}{\partial y} = b\frac{\partial w}{\partial x}.$$

86. Using the Chain Rule If $w = \ln(x^2 + y^2 + 2z)$, $x = r + s$, $y = r - s$, and $z = 2rs$, find w_r and w_s by the Chain Rule. Then check your answer another way.

87. Angle between vectors The equations $e^u \cos v - x = 0$ and $e^u \sin v - y = 0$ define u and v as differentiable functions of x and y. Show that the angle between the vectors

$$\frac{\partial u}{\partial x}\mathbf{i} + \frac{\partial u}{\partial y}\mathbf{j} \quad \text{and} \quad \frac{\partial v}{\partial x}\mathbf{i} + \frac{\partial v}{\partial y}\mathbf{j}$$

is constant.

88. Polar coordinates and second derivatives Introducing polar coordinates $x = r\cos\theta$ and $y = r\sin\theta$ changes $f(x, y)$ to $g(r, \theta)$. Find the value of $\partial^2 g/\partial\theta^2$ at the point $(r, \theta) = (2, \pi/2)$, given that

$$\frac{\partial f}{\partial x} = \frac{\partial f}{\partial y} = \frac{\partial^2 f}{\partial x^2} = \frac{\partial^2 f}{\partial y^2} = 1$$

at that point.

89. Normal line parallel to a plane Find the points on the surface

$$(y + z)^2 + (z - x)^2 = 16$$

where the normal line is parallel to the yz-plane.

90. Tangent plane parallel to xy-plane Find the points on the surface

$$xy + yz + zx - x - z^2 = 0$$

where the tangent plane is parallel to the xy-plane.

91. When gradient is parallel to position vector Suppose that $\nabla f(x, y, z)$ is always parallel to the position vector $x\mathbf{i} + y\mathbf{j} + z\mathbf{k}$. Show that $f(0, 0, a) = f(0, 0, -a)$ for any a.

92. One-sided directional derivative in all directions, but no gradient The one-sided directional derivative of f at $P(x_0, y_0, z_0)$ in the direction $\mathbf{u} = u_1\mathbf{i} + u_2\mathbf{j} + u_3\mathbf{k}$ is the number

$$\lim_{s\to 0^+} \frac{f(x_0 + su_1, y_0 + su_2, z_0 + su_3) - f(x_0, y_0, z_0)}{s}.$$

Show that the one-sided directional derivative of

$$f(x, y, z) = \sqrt{x^2 + y^2 + z^2}$$

at the origin equals 1 in any direction but that f has no gradient vector at the origin.

MULTIPLE INTEGRALS

OVERVIEW In this chapter we consider the integral of a function of two variables $f(x, y)$ over a region in the plane and the integral of a function of three variables $f(x, y, z)$ over a region in space. These *multiple integrals* are defined as the limit of approximating Riemann sums, much like the single-variable integrals presented in Chapter 4.

12.1 Double and Iterated Integrals over Rectangles

In Chapter 4 we defined the definite integral of a continuous function $f(x)$ over an interval $[a, b]$ as a limit of Riemann sums. In this section we extend this idea to define the *double integral* of a continuous function of two variables $f(x, y)$ over a bounded rectangle R in the plane. In both cases the integrals are limits of approximating Riemann sums. The Riemann sums for the integral of a single-variable function $f(x)$ are obtained by partitioning a finite interval into thin subintervals, multiplying the width of each subinterval by the value of f at a point c_k inside that subinterval, and then adding together all the products. A similar method of partitioning, multiplying, and summing is used to construct double integrals.

Double Integrals

We begin our investigation of double integrals by considering the simplest type of planar region, a rectangle. We consider a function $f(x, y)$ defined on a rectangular region R,

$$R: \quad a \le x \le b, \quad c \le y \le d.$$

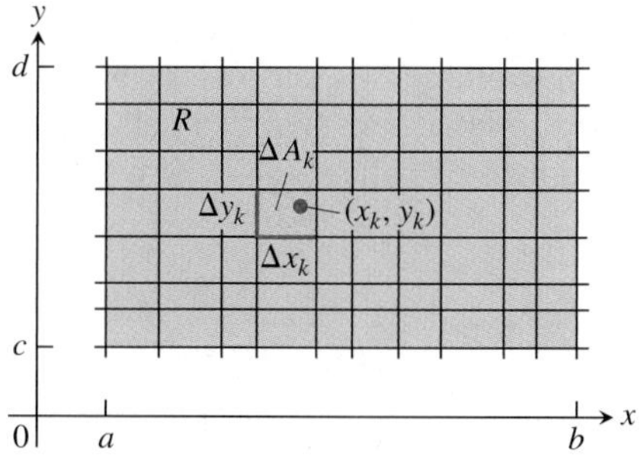

FIGURE 12.1 Rectangular grid partitioning the region R into small rectangles of area $\Delta A_k = \Delta x_k\, \Delta y_k$.

We subdivide R into small rectangles using a network of lines parallel to the x- and y-axes (Figure 12.1). The lines divide R into n rectangular pieces, where the number of such pieces n gets large as the width and height of each piece gets small. These rectangles form a **partition** of R. A small rectangular piece of width Δx and height Δy has area $\Delta A = \Delta x \Delta y$. If we number the small pieces partitioning R in some order, then their areas are given by numbers $\Delta A_1, \Delta A_2, \ldots, \Delta A_n$, where ΔA_k is the area of the kth small rectangle.

To form a Riemann sum over R, we choose a point (x_k, y_k) in the kth small rectangle, multiply the value of f at that point by the area ΔA_k, and add together the products:

$$S_n = \sum_{k=1}^{n} f(x_k, y_k)\, \Delta A_k.$$

Depending on how we pick (x_k, y_k) in the kth small rectangle, we may get different values for S_n.

We are interested in what happens to these Riemann sums as the widths and heights of all the small rectangles in the partition of R approach zero. The **norm** of a partition P, written $\|P\|$, is the largest width or height of any rectangle in the partition. If $\|P\| = 0.1$ then all the rectangles in the partition of R have width at most 0.1 and height at most 0.1. Sometimes the Riemann sums converge as the norm of P goes to zero, written $\|P\| \to 0$. The resulting limit is then written as

$$\lim_{\|P\|\to 0} \sum_{k=1}^{n} f(x_k, y_k)\,\Delta A_k.$$

As $\|P\| \to 0$ and the rectangles get narrow and short, their number n increases, so we can also write this limit as

$$\lim_{n\to\infty} \sum_{k=1}^{n} f(x_k, y_k)\,\Delta A_k.$$

with the understanding that $\Delta A_k \to 0$ as $n \to \infty$ and $\|P\| \to 0$.

There are many choices involved in a limit of this kind. The collection of small rectangles is determined by the grid of vertical and horizontal lines that determine a rectangular partition of R. In each of the resulting small rectangles there is a choice of an arbitrary point (x_k, y_k) at which f is evaluated. These choices together determine a single Riemann sum. To form a limit, we repeat the whole process again and again, choosing partitions whose rectangle widths and heights both go to zero and whose number goes to infinity.

When a limit of the sums S_n exists, giving the same limiting value no matter what choices are made, then the function f is said to be **integrable** and the limit is called the **double integral** of f over R, written as

$$\iint_R f(x, y)\,dA \qquad \text{or} \qquad \iint_R f(x, y)\,dx\,dy.$$

It can be shown that if $f(x, y)$ is a continuous function throughout R, then f is integrable, as in the single-variable case discussed in Chapter 4. Many discontinuous functions are also integrable, including functions which are discontinuous only on a finite number of points or smooth curves. We leave the proof of these facts to a more advanced text.

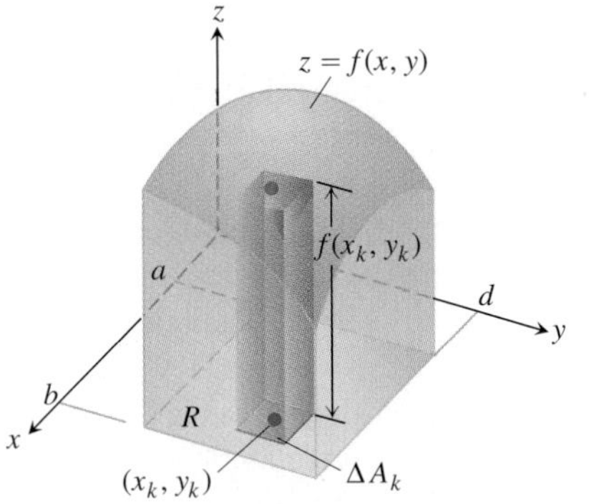

FIGURE 12.2 Approximating solids with rectangular boxes leads us to define the volumes of more general solids as double integrals. The volume of the solid shown here is the double integral of $f(x, y)$ over the base region R.

Double Integrals as Volumes

When $f(x, y)$ is a positive function over a rectangular region R in the xy-plane, we may interpret the double integral of f over R as the volume of the 3-dimensional solid region over the xy-plane bounded below by R and above by the surface $z = f(x, y)$ (Figure 12.2). Each term $f(x_k, y_k)\Delta A_k$ in the sum $S_n = \sum f(x_k, y_k)\Delta A_k$ is the volume of a vertical rectangular box that approximates the volume of the portion of the solid that stands directly above the base ΔA_k. The sum S_n thus approximates what we want to call the total volume of the solid. We *define* this volume to be

$$\text{Volume} = \lim_{n\to\infty} S_n = \iint_R f(x, y)\,dA,$$

where $\Delta A_k \to 0$ as $n \to \infty$.

As you might expect, this more general method of calculating volume agrees with the methods in Chapter 6, but we do not prove this here. Figure 12.3 shows Riemann sum approximations to the volume becoming more accurate as the number n of boxes increases.

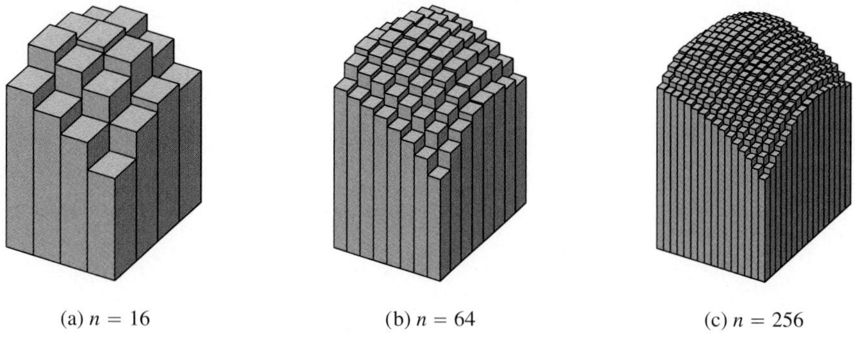

FIGURE 12.3 As n increases, the Riemann sum approximations approach the total volume of the solid shown in Figure 12.2.

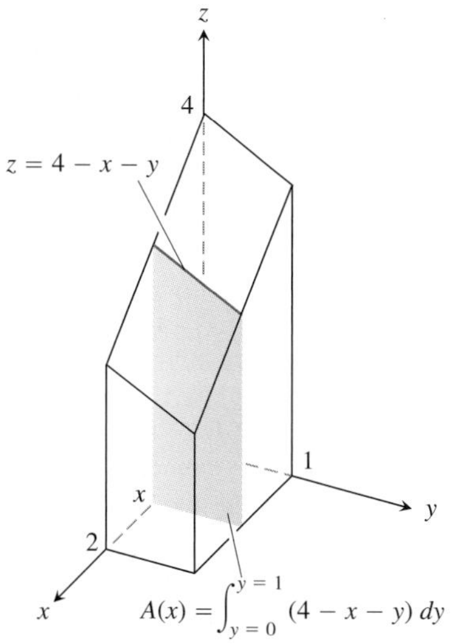

FIGURE 12.4 To obtain the cross-sectional area $A(x)$, we hold x fixed and integrate with respect to y.

Fubini's Theorem for Calculating Double Integrals

Suppose that we wish to calculate the volume under the plane $z = 4 - x - y$ over the rectangular region R: $0 \le x \le 2, 0 \le y \le 1$ in the xy-plane. If we apply the method of slicing from Section 6.1, with slices perpendicular to the x-axis (Figure 12.4), then the volume is

$$\int_{x=0}^{x=2} A(x)\, dx, \tag{1}$$

where $A(x)$ is the cross-sectional area at x. For each value of x, we may calculate $A(x)$ as the integral

$$A(x) = \int_{y=0}^{y=1} (4 - x - y)\, dy, \tag{2}$$

which is the area under the curve $z = 4 - x - y$ in the plane of the cross-section at x. In calculating $A(x)$, x is held fixed and the integration takes place with respect to y. Combining Equations (1) and (2), we see that the volume of the entire solid is

$$\begin{aligned}
\text{Volume} &= \int_{x=0}^{x=2} A(x)\, dx = \int_{x=0}^{x=2} \left(\int_{y=0}^{y=1} (4 - x - y) dy \right) dx \\
&= \int_{x=0}^{x=2} \left[4y - xy - \frac{y^2}{2} \right]_{y=0}^{y=1} dx = \int_{x=0}^{x=2} \left(\frac{7}{2} - x \right) dx \\
&= \left[\frac{7}{2}x - \frac{x^2}{2} \right]_0^2 = 5.
\end{aligned} \tag{3}$$

If we just wanted to write a formula for the volume, without carrying out any of the integrations, we could write

$$\text{Volume} = \int_0^2 \int_0^1 (4 - x - y)\, dy\, dx.$$

The expression on the right, called an **iterated** or **repeated integral**, says that the volume is obtained by integrating $4 - x - y$ with respect to y from $y = 0$ to $y = 1$, holding x fixed, and then integrating the resulting expression in x with respect to x from $x = 0$ to $x = 2$. The limits of integration 0 and 1 are associated with y, so they are placed on the integral closest to dy. The other limits of integration, 0 and 2, are associated with the variable x, so they are placed on the outside integral symbol that is paired with dx.

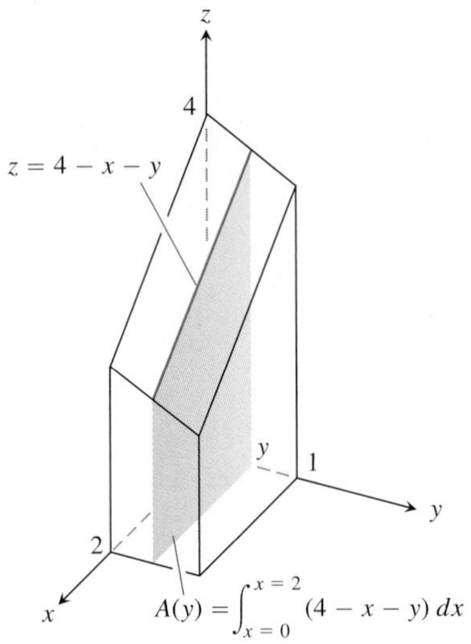

FIGURE 12.5 To obtain the cross-sectional area $A(y)$, we hold y fixed and integrate with respect to x.

What would have happened if we had calculated the volume by slicing with planes perpendicular to the y-axis (Figure 12.5)? As a function of y, the typical cross-sectional area is

$$A(y) = \int_{x=0}^{x=2} (4 - x - y)\, dx = \left[4x - \frac{x^2}{2} - xy\right]_{x=0}^{x=2} = 6 - 2y. \tag{4}$$

The volume of the entire solid is therefore

$$\text{Volume} = \int_{y=0}^{y=1} A(y)\, dy = \int_{y=0}^{y=1} (6 - 2y)\, dy = \left[6y - y^2\right]_0^1 = 5,$$

in agreement with our earlier calculation.

Again, we may give a formula for the volume as an iterated integral by writing

$$\text{Volume} = \int_0^1 \int_0^2 (4 - x - y)\, dx\, dy.$$

The expression on the right says we can find the volume by integrating $4 - x - y$ with respect to x from $x = 0$ to $x = 2$ as in Equation (4) and integrating the result with respect to y from $y = 0$ to $y = 1$. In this iterated integral, the order of integration is first x and then y, the reverse of the order in Equation (3).

What do these two volume calculations with iterated integrals have to do with the double integral

$$\iint_R (4 - x - y)\, dA$$

over the rectangle R: $0 \le x \le 2,\ 0 \le y \le 1$? The answer is that both iterated integrals give the value of the double integral. This is what we would reasonably expect, since the double integral measures the volume of the same region as the two iterated integrals. A theorem published in 1907 by Guido Fubini says that the double integral of any continuous function over a rectangle can be calculated as an iterated integral in either order of integration. (Fubini proved his theorem in greater generality, but this is what it says in our setting.)

HISTORICAL BIOGRAPHY

Guido Fubini
(1879–1943)

THEOREM 1—Fubini's Theorem (First Form) If $f(x, y)$ is continuous throughout the rectangular region R: $a \le x \le b,\ c \le y \le d$, then

$$\iint_R f(x, y)\, dA = \int_c^d \int_a^b f(x, y)\, dx\, dy = \int_a^b \int_c^d f(x, y)\, dy\, dx.$$

Fubini's Theorem says that double integrals over rectangles can be calculated as iterated integrals. Thus, we can evaluate a double integral by integrating with respect to one variable at a time.

Fubini's Theorem also says that we may calculate the double integral by integrating in *either* order, a genuine convenience. When we calculate a volume by slicing, we may use either planes perpendicular to the x-axis or planes perpendicular to the y-axis.

EXAMPLE 1 Calculate $\iint_R f(x, y)\, dA$ for

$$f(x, y) = 1 - 6x^2y \quad \text{and} \quad R:\ 0 \le x \le 2,\quad -1 \le y \le 1.$$

Solution By Fubini's Theorem,

$$\iint_R f(x, y)\, dA = \int_{-1}^{1}\int_{0}^{2}(1 - 6x^2y)\, dx\, dy = \int_{-1}^{1}\Big[x - 2x^3y\Big]_{x=0}^{x=2} dy$$

$$= \int_{-1}^{1}(2 - 16y)\, dy = \Big[2y - 8y^2\Big]_{-1}^{1} = 4.$$

Reversing the order of integration gives the same answer:

$$\int_0^2\int_{-1}^{1}(1 - 6x^2y)\, dy\, dx = \int_0^2\Big[y - 3x^2y^2\Big]_{y=-1}^{y=1} dx$$

$$= \int_0^2[(1 - 3x^2) - (-1 - 3x^2)]\, dx$$

$$= \int_0^2 2\, dx = 4.$$

■

EXAMPLE 2 Find the volume of the region bounded above by the ellipitical paraboloid $z = x^2 + 3y^2$ and below by the rectangle R: $0 \le x \le 1, 0 \le y \le 2$.

Solution The volume is given by the double integral

$$V = \iint_R (x^2 + 3y^2)\, dA = \int_0^1\int_0^2 (x^2 + 3y^2)\, dy\, dx$$

$$= \int_0^1\Big[x^2y + y^3\Big]_{y=0}^{y=2} dx$$

$$= \int_0^1 (2x^2 + 8)\, dx = \left[\frac{2}{3}x^3 + 8x\right]_0^1 = \frac{26}{3}.$$

■

EXERCISES 12.1

In Exercises 1–12, evaluate the iterated integral.

1. $\displaystyle\int_1^2\int_0^4 2xy\, dy\, dx$

2. $\displaystyle\int_0^2\int_{-1}^{1}(x - y)\, dy\, dx$

3. $\displaystyle\int_{-1}^{0}\int_{-1}^{1}(x + y + 1)\, dx\, dy$

4. $\displaystyle\int_0^1\int_0^1\left(1 - \frac{x^2 + y^2}{2}\right) dx\, dy$

5. $\displaystyle\int_0^3\int_0^2 (4 - y^2)\, dy\, dx$

6. $\displaystyle\int_0^3\int_{-2}^{0}(x^2y - 2xy)\, dy\, dx$

7. $\displaystyle\int_0^1\int_0^1 \frac{y}{1 + xy}\, dx\, dy$

8. $\displaystyle\int_1^4\int_0^4\left(\frac{x}{2} + \sqrt{y}\right) dx\, dy$

9. $\displaystyle\int_0^{\ln 2}\int_1^{\ln 5} e^{2x+y}\, dy\, dx$

10. $\displaystyle\int_0^1\int_1^2 xye^x\, dy\, dx$

11. $\displaystyle\int_{-1}^{2}\int_0^{\pi/2} y\sin x\, dx\, dy$

12. $\displaystyle\int_\pi^{2\pi}\int_0^\pi (\sin x + \cos y)\, dx\, dy$

In Exercises 13–20, evaluate the double integral over the given region R.

13. $\displaystyle\iint_R (6y^2 - 2x)\, dA,\quad R:\ 0 \le x \le 1,\ \ 0 \le y \le 2$

14. $\displaystyle\iint_R \left(\frac{\sqrt{x}}{y^2}\right) dA,\quad R:\ 0 \le x \le 4,\ \ 1 \le y \le 2$

15. $\displaystyle\iint_R xy\cos y\, dA,\quad R:\ -1 \le x \le 1,\ \ 0 \le y \le \pi$

16. $\iint_R y \sin(x + y)\, dA, \quad R: \; -\pi \le x \le 0, \; 0 \le y \le \pi$

17. $\iint_R e^{x-y}\, dA, \quad R: \; 0 \le x \le \ln 2, \; 0 \le y \le \ln 2$

18. $\iint_R xye^{xy^2}\, dA, \quad R: \; 0 \le x \le 2, \; 0 \le y \le 1$

19. $\iint_R \frac{xy^3}{x^2 + 1}\, dA, \quad R: \; 0 \le x \le 1, \; 0 \le y \le 2$

20. $\iint_R \frac{y}{x^2y^2 + 1}\, dA, \quad R: \; 0 \le x \le 1, \; 0 \le y \le 1$

In Exercises 21 and 22, integrate f over the given region.

21. **Square** $f(x, y) = 1/(xy)$ over the square $1 \le x \le 2$, $1 \le y \le 2$

22. **Rectangle** $f(x, y) = y \cos xy$ over the rectangle $0 \le x \le \pi$, $0 \le y \le 1$

23. Find the volume of the region bounded above by the paraboloid $z = x^2 + y^2$ and below by the square $R: -1 \le x \le 1, -1 \le y \le 1$.

24. Find the volume of the region bounded above by the ellipitical paraboloid $z = 16 - x^2 - y^2$ and below by the square $R: 0 \le x \le 2, 0 \le y \le 2$.

25. Find the volume of the region bounded above by the plane $z = 2 - x - y$ and below by the square $R: 0 \le x \le 1, 0 \le y \le 1$.

26. Find the volume of the region bounded above by the plane $z = y/2$ and below by the rectangle $R: 0 \le x \le 4, 0 \le y \le 2$.

27. Find the volume of the region bounded above by the surface $z = 2 \sin x \cos y$ and below by the rectangle $R: 0 \le x \le \pi/2, 0 \le y \le \pi/4$.

28. Find the volume of the region bounded above by the surface $z = 4 - y^2$ and below by the rectangle $R: 0 \le x \le 1, 0 \le y \le 2$.

12.2 Double Integrals over General Regions

In this section we define and evaluate double integrals over bounded regions in the plane which are more general than rectangles. These double integrals are also evaluated as iterated integrals, with the main practical problem being that of determining the limits of integration. Since the region of integration may have boundaries other than line segments parallel to the coordinate axes, the limits of integration often involve variables, not just constants.

Double Integrals over Bounded Nonrectangular Regions

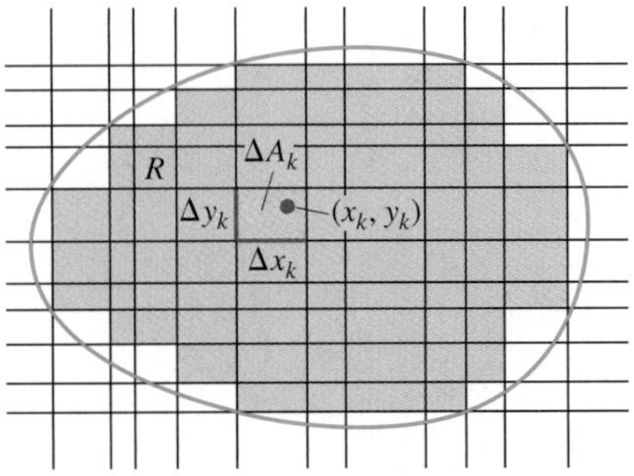

FIGURE 12.6 A rectangular grid partitioning a bounded nonrectangular region into rectangular cells.

To define the double integral of a function $f(x, y)$ over a bounded, nonrectangular region R, such as the one in Figure 12.6, we again begin by covering R with a grid of small rectangular cells whose union contains all points of R. This time, however, we cannot exactly fill R with a finite number of rectangles lying inside R, since its boundary is curved, and some of the small rectangles in the grid lie partly outside R. A partition of R is formed by taking the rectangles that lie completely inside it, not using any that are either partly or completely outside. For commonly arising regions, more and more of R is included as the norm of a partition (the largest width or height of any rectangle used) approaches zero.

Once we have a partition of R, we number the rectangles in some order from 1 to n and let ΔA_k be the area of the kth rectangle. We then choose a point (x_k, y_k) in the kth rectangle and form the Riemann sum

$$S_n = \sum_{k=1}^{n} f(x_k, y_k)\, \Delta A_k.$$

As the norm of the partition forming S_n goes to zero, $\|P\| \to 0$, the width and height of each enclosed rectangle goes to zero and their number goes to infinity. If $f(x, y)$ is a continuous function, then these Riemann sums converge to a limiting value, not dependent on any of the choices we made. This limit is called the **double integral** of $f(x, y)$ over R:

$$\lim_{\|P\| \to 0} \sum_{k=1}^{n} f(x_k, y_k)\, \Delta A_k = \iint_R f(x, y)\, dA.$$

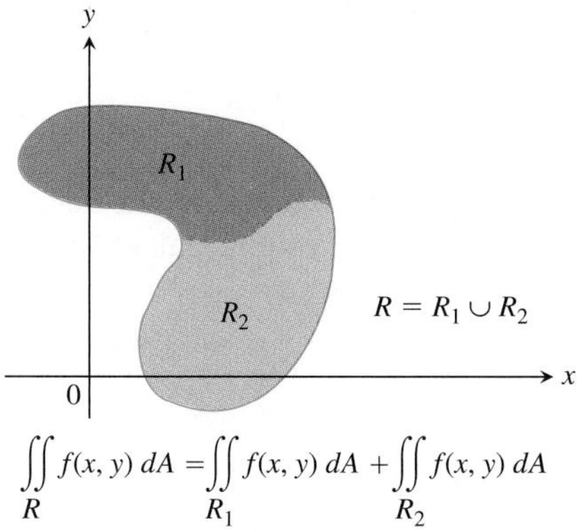

FIGURE 12.7 The Additivity Property for rectangular regions holds for regions bounded by continuous curves.

The nature of the boundary of R introduces issues not found in integrals over an interval. When R has a curved boundary, the n rectangles of a partition lie inside R but do not cover all of R. In order for a partition to approximate R well, the parts of R covered by small rectangles lying partly outside R must become negligible as the norm of the partition approaches zero. This property of being nearly filled in by a partition of small norm is satisfied by all the regions that we will encounter. There is no problem with boundaries made from polygons, circles, ellipses, and from continuous graphs over an interval, joined end to end. A curve with a "fractal" type of shape would be problematic, but such curves are not relevant for most applications. A careful discussion of which type of regions R can be used for computing double integrals is left to a more advanced text.

Double integrals of continuous functions over nonrectangular regions have algebraic properties (summarized further on) similar to those for single integrals. The Domain Additivity Property says that if R is decomposed into nonoverlapping regions R_1 and R_2 with boundaries that are made of a finite number of line segments or smooth curves (see Figure 12.7 for an example), then

$$\iint_R f(x, y)\, dA = \iint_{R_1} f(x, y)\, dA + \iint_{R_2} f(x, y)\, dA.$$

Volumes

If $f(x, y)$ is positive and continuous over R we define the volume of the solid region between R and the surface $z = f(x, y)$ to be $\iint_R f(x, y)\, dA$, as before (Figure 12.8).

If R is a region like the one shown in the xy-plane in Figure 12.9, bounded "above" and "below" by the curves $y = g_2(x)$ and $y = g_1(x)$ and on the sides by the lines $x = a, x = b$, we may again calculate the volume by the method of slicing. We first calculate the cross-sectional area

$$A(x) = \int_{y=g_1(x)}^{y=g_2(x)} f(x, y)\, dy$$

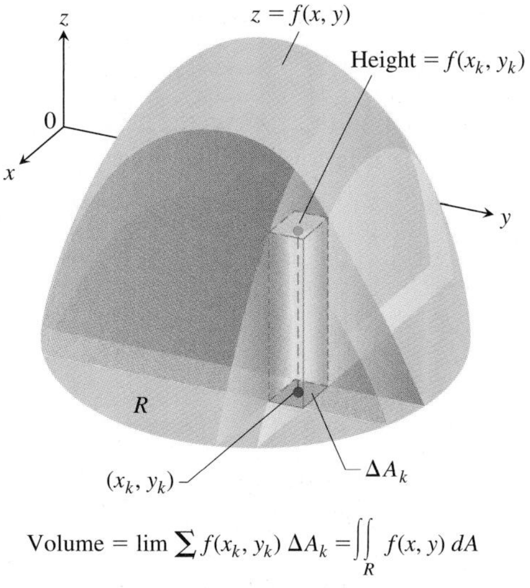

FIGURE 12.8 We define the volumes of solids with curved bases the same way we define the volumes of solids with rectangular bases.

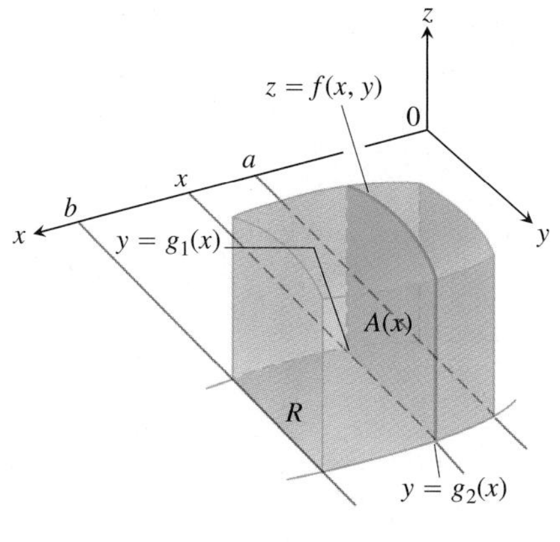

FIGURE 12.9 The area of the vertical slice shown here is $A(x)$. To calculate the volume of the solid, we integrate this area from $x = a$ to $x = b$:

$$\int_a^b A(x)\, dx = \int_a^b \int_{g_1(x)}^{g_2(x)} f(x, y)\, dy\, dx.$$

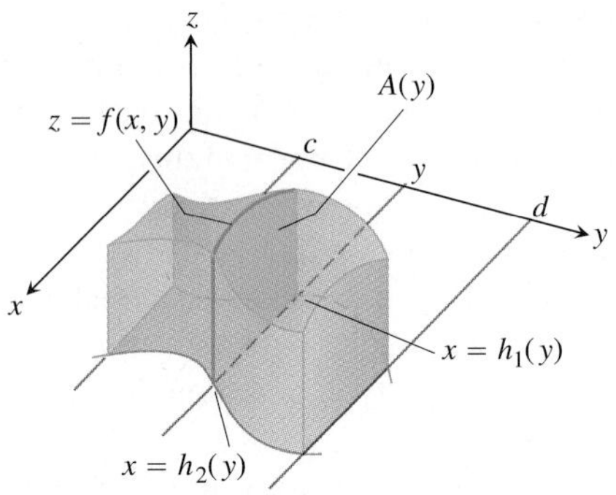

FIGURE 12.10 The volume of the solid shown here is

$$\int_c^d A(y)\,dy = \int_c^d \int_{h_1(y)}^{h_2(y)} f(x, y)\,dx\,dy.$$

It is the same volume as calculated in Figure 12.9.

and then integrate $A(x)$ from $x = a$ to $x = b$ to get the volume as an iterated integral:

$$V = \int_a^b A(x)\,dx = \int_a^b \int_{g_1(x)}^{g_2(x)} f(x, y)\,dy\,dx. \tag{1}$$

Similarly, if R is a region like the one shown in Figure 12.10, bounded by the curves $x = h_2(y)$ and $x = h_1(y)$ and the lines $y = c$ and $y = d$, then the volume calculated by slicing is given by the iterated integral

$$\text{Volume} = \int_c^d \int_{h_1(y)}^{h_2(y)} f(x, y)\,dx\,dy. \tag{2}$$

That the iterated integrals in Equations (1) and (2) both give the volume that we defined to be the double integral of f over R is a consequence of the following stronger form of Fubini's Theorem.

THEOREM 2—Fubini's Theorem (Stronger Form) Let $f(x, y)$ be continuous on a region R.

1. If R is defined by $a \le x \le b$, $g_1(x) \le y \le g_2(x)$, with g_1 and g_2 continuous on $[a, b]$, then

$$\iint_R f(x, y)\,dA = \int_a^b \int_{g_1(x)}^{g_2(x)} f(x, y)\,dy\,dx.$$

2. If R is defined by $c \le y \le d$, $h_1(y) \le x \le h_2(y)$, with h_1 and h_2 continuous on $[c, d]$, then

$$\iint_R f(x, y)\,dA = \int_c^d \int_{h_1(y)}^{h_2(y)} f(x, y)\,dx\,dy.$$

EXAMPLE 1 Find the volume of the prism whose base is the triangle in the xy-plane bounded by the x-axis and the lines $y = x$ and $x = 1$ and whose top lies in the plane

$$z = f(x, y) = 3 - x - y.$$

Solution See Figure 12.11 on the next page. For any x between 0 and 1, y may vary from $y = 0$ to $y = x$ (Figure 12.11b). Hence,

$$\begin{aligned} V &= \int_0^1 \int_0^x (3 - x - y)\,dy\,dx = \int_0^1 \left[3y - xy - \frac{y^2}{2}\right]_{y=0}^{y=x} dx \\ &= \int_0^1 \left(3x - \frac{3x^2}{2}\right) dx = \left[\frac{3x^2}{2} - \frac{x^3}{2}\right]_{x=0}^{x=1} = 1. \end{aligned}$$

When the order of integration is reversed (Figure 12.11c), the integral for the volume is

$$\begin{aligned} V &= \int_0^1 \int_y^1 (3 - x - y)\,dx\,dy = \int_0^1 \left[3x - \frac{x^2}{2} - xy\right]_{x=y}^{x=1} dy \\ &= \int_0^1 \left(3 - \frac{1}{2} - y - 3y + \frac{y^2}{2} + y^2\right) dy \\ &= \int_0^1 \left(\frac{5}{2} - 4y + \frac{3}{2}y^2\right) dy = \left[\frac{5}{2}y - 2y^2 + \frac{y^3}{2}\right]_{y=0}^{y=1} = 1. \end{aligned}$$

The two integrals are equal, as they should be. ■

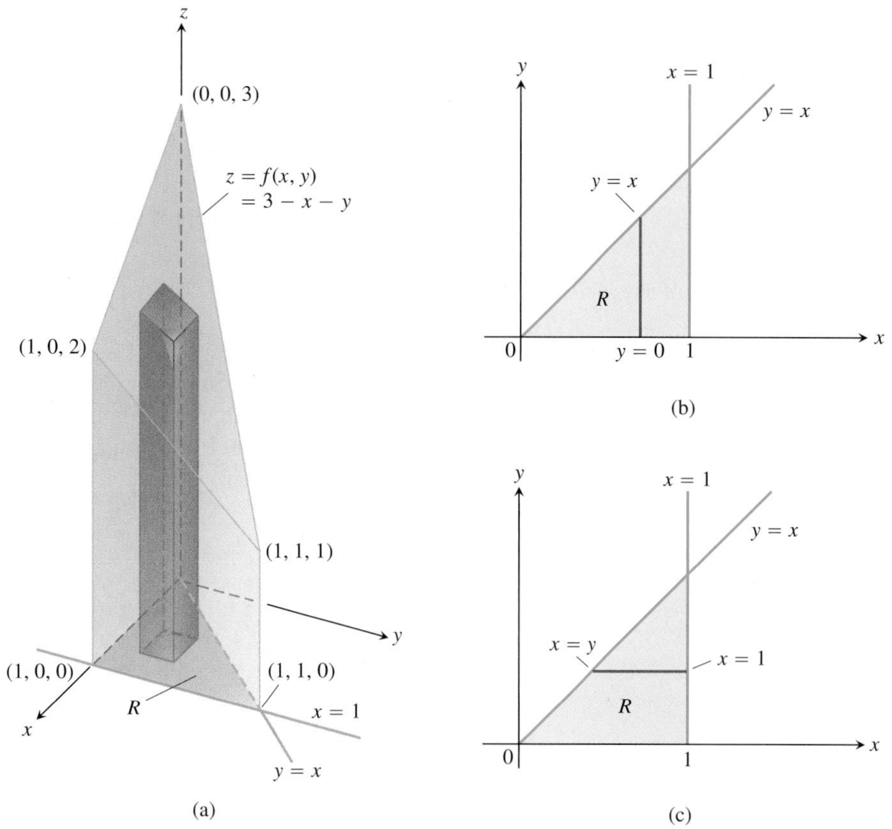

FIGURE 12.11 (a) Prism with a triangular base in the xy-plane. The volume of this prism is defined as a double integral over R. To evaluate it as an iterated integral, we may integrate first with respect to y and then with respect to x, or the other way around (Example 1). (b) Integration limits of

$$\int_{x=0}^{x=1}\int_{y=0}^{y=x} f(x, y)\, dy\, dx.$$

If we integrate first with respect to y, we integrate along a vertical line through R and then integrate from left to right to include all the vertical lines in R. (c) Integration limits of

$$\int_{y=0}^{y=1}\int_{x=y}^{x=1} f(x, y)\, dx\, dy.$$

If we integrate first with respect to x, we integrate along a horizontal line through R and then integrate from bottom to top to include all the horizontal lines in R.

Although Fubini's Theorem assures us that a double integral may be calculated as an iterated integral in either order of integration, the value of one integral may be easier to find than the value of the other. The next example shows how this can happen.

EXAMPLE 2 Calculate

$$\iint_R \frac{\sin x}{x}\, dA,$$

where R is the triangle in the xy-plane bounded by the x-axis, the line $y = x$, and the line $x = 1$.

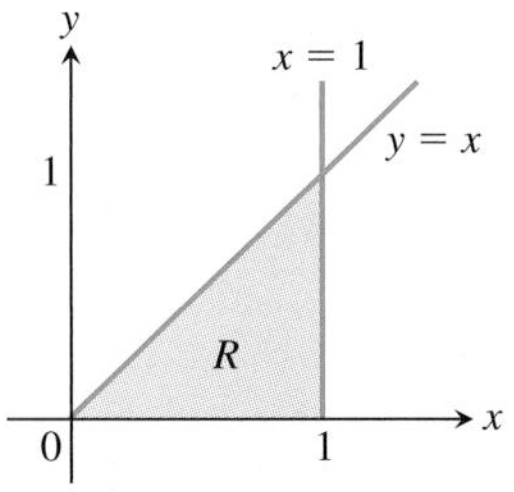

FIGURE 12.12 The region of integration in Example 2.

Solution The region of integration is shown in Figure 12.12. If we integrate first with respect to y and then with respect to x, we find

$$\int_0^1 \left(\int_0^x \frac{\sin x}{x}\, dy \right) dx = \int_0^1 \left(y \frac{\sin x}{x} \Big]_{y=0}^{y=x} \right) dx = \int_0^1 \sin x\, dx$$
$$= -\cos(1) + 1 \approx 0.46.$$

If we reverse the order of integration and attempt to calculate

$$\int_0^1 \int_y^1 \frac{\sin x}{x}\, dx\, dy,$$

we run into a problem, because $\int ((\sin x)/x)\, dx$ cannot be expressed in terms of elementary functions (there is no simple antiderivative).

There is no general rule for predicting which order of integration will be the good one in circumstances like these. If the order you first choose doesn't work, try the other. Sometimes neither order will work, and then we need to use numerical approximations. ■

Finding Limits of Integration

We now give a procedure for finding limits of integration that applies for many regions in the plane. Regions that are more complicated, and for which this procedure fails, can often be split up into pieces on which the procedure works.

When faced with evaluating $\iint_R f(x, y)\, dA$, integrating first with respect to y and then with respect to x, do the following:

1. *Sketch*. Sketch the region of integration and label the bounding curves (Figure 12.13a).
2. *Find the y-limits of integration*. Imagine a vertical line L cutting through R in the direction of increasing y. Mark the y-values where L enters and leaves. These are the y-limits of integration and are usually functions of x (instead of constants) (Figure 12.13b).
3. *Find the x-limits of integration*. Choose x-limits that include all the vertical lines through R. The integral shown here (see Figure 12.13c) is

$$\iint_R f(x, y)\, dA = \int_{x=0}^{x=1} \int_{y=1-x}^{y=\sqrt{1-x^2}} f(x, y)\, dy\, dx.$$

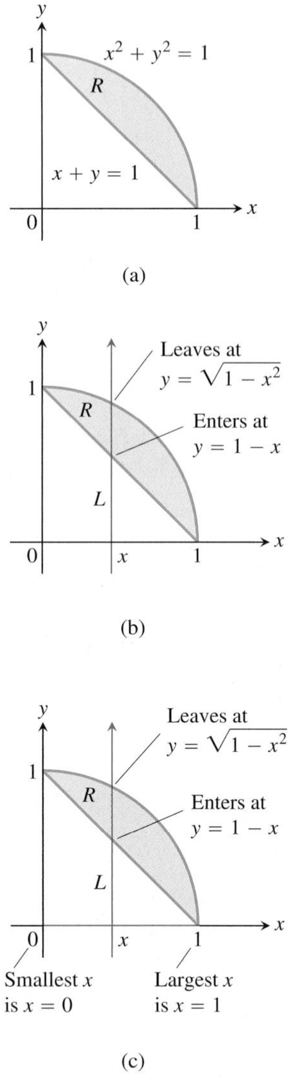

FIGURE 12.13 Finding the limits of integration when integrating first with respect to y and then with respect to x.

To evaluate the same double integral as an iterated integral with the order of integration reversed, use horizontal lines instead of vertical lines in Steps 2 and 3. The integral is

$$\iint_R f(x, y)\, dA = \int_0^1 \int_{1-y}^{\sqrt{1-y^2}} f(x, y)\, dx\, dy.$$

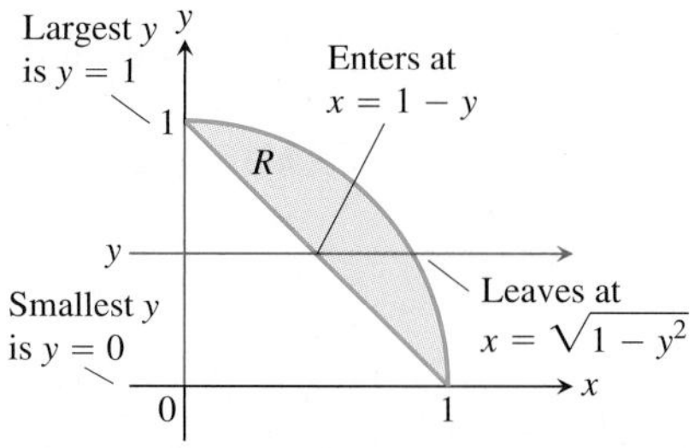

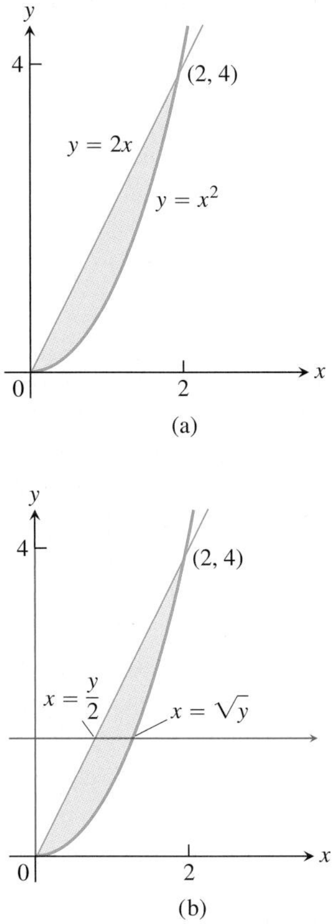

FIGURE 12.14 Region of integration for Example 3.

EXAMPLE 3 Sketch the region of integration for the integral

$$\int_0^2 \int_{x^2}^{2x} (4x + 2)\, dy\, dx$$

and write an equivalent integral with the order of integration reversed.

Solution The region of integration is given by the inequalities $x^2 \le y \le 2x$ and $0 \le x \le 2$. It is therefore the region bounded by the curves $y = x^2$ and $y = 2x$ between $x = 0$ and $x = 2$ (Figure 12.14a).

To find limits for integrating in the reverse order, we imagine a horizontal line passing from left to right through the region. It enters at $x = y/2$ and leaves at $x = \sqrt{y}$. To include all such lines, we let y run from $y = 0$ to $y = 4$ (Figure 12.14b). The integral is

$$\int_0^4 \int_{y/2}^{\sqrt{y}} (4x + 2)\, dx\, dy.$$

The common value of these integrals is 8. ■

Properties of Double Integrals

Like single integrals, double integrals of continuous functions have algebraic properties that are useful in computations and applications.

If $f(x, y)$ and $g(x, y)$ are continuous, then

1. *Constant Multiple:* $\iint_R cf(x, y)\, dA = c \iint_R f(x, y)\, dA$ (any number c)

2. *Sum and Difference:*

$$\iint_R (f(x, y) \pm g(x, y))\, dA = \iint_R f(x, y)\, dA \pm \iint_R g(x, y)\, dA$$

3. *Domination:*

(a) $\iint_R f(x, y)\, dA \ge 0$ if $f(x, y) \ge 0$ on R

(b) $\iint_R f(x, y)\, dA \ge \iint_R g(x, y)\, dA$ if $f(x, y) \ge g(x, y)$ on R

4. *Additivity:* $\iint_R f(x, y)\, dA = \iint_{R_1} f(x, y)\, dA + \iint_{R_2} f(x, y)\, dA$

if R is the union of two nonoverlapping regions R_1 and R_2 (Figure 12.7).

The idea behind these properties is that integrals behave like sums. If the function $f(x, y)$ is replaced by its constant multiple $cf(x, y)$, then a Riemann sum for f

$$S_n = \sum_{k=1}^{n} f(x_k, y_k)\,\Delta A_k$$

is replaced by a Riemann sum for cf

$$\sum_{k=1}^{n} cf(x_k, y_k)\,\Delta A_k = c\sum_{k=1}^{n} f(x_k, y_k)\,\Delta A_k = cS_n.$$

Taking limits as $n \to \infty$ shows that $c \lim_{n\to\infty} S_n = c\iint_R f\,dA$ and $\lim_{n\to\infty} cS_n = \iint_R cf\,dA$ are equal. It follows that the constant multiple property carries over from sums to double integrals.

The other properties are also easy to verify for Riemann sums, and carry over to double integrals for the same reason. While this discussion gives the idea, an actual proof that these properties hold requires a more careful analysis of how Riemann sums converge.

EXERCISES 12.2

In Exercises 1–6, sketch the region of integration and evaluate the integral.

1. $\displaystyle\int_0^{\pi}\int_0^{x} x\sin y\,dy\,dx$

2. $\displaystyle\int_0^{\pi}\int_0^{\sin x} y\,dy\,dx$

3. $\displaystyle\int_1^{\ln 8}\int_0^{\ln y} e^{x+y}\,dx\,dy$

4. $\displaystyle\int_1^{2}\int_y^{y^2} dx\,dy$

5. $\displaystyle\int_0^{1}\int_0^{y^2} 3y^3e^{xy}\,dx\,dy$

6. $\displaystyle\int_1^{4}\int_0^{\sqrt{x}} \frac{3}{2}e^{y/\sqrt{x}}\,dy\,dx$

In Exercises 7–10, integrate f over the given region.

7. Quadrilateral $f(x, y) = x/y$ over the region in the first quadrant bounded by the lines $y = x$, $y = 2x$, $x = 1$, $x = 2$

8. Triangle $f(x, y) = x^2 + y^2$ over the triangular region with vertices $(0, 0)$, $(1, 0)$, and $(0, 1)$

9. Triangle $f(u, v) = v - \sqrt{u}$ over the triangular region cut from the first quadrant of the uv-plane by the line $u + v = 1$

10. Curved region $f(s, t) = e^s \ln t$ over the region in the first quadrant of the st-plane that lies above the curve $s = \ln t$ from $t = 1$ to $t = 2$

Each of Exercises 11–14 gives an integral over a region in a Cartesian coordinate plane. Sketch the region and evaluate the integral.

11. $\displaystyle\int_{-2}^{0}\int_v^{-v} 2\,dp\,dv$ (the pv-plane)

12. $\displaystyle\int_0^{1}\int_0^{\sqrt{1-s^2}} 8t\,dt\,ds$ (the st-plane)

13. $\displaystyle\int_{-\pi/3}^{\pi/3}\int_0^{\sec t} 3\cos t\,du\,dt$ (the tu-plane)

14. $\displaystyle\int_0^{3}\int_1^{4-2u} \frac{4-2u}{v^2}\,dv\,du$ (the uv-plane)

In Exercises 15–24, sketch the region of integration and write an equivalent double integral with the order of integration reversed.

15. $\displaystyle\int_0^{1}\int_2^{4-2x} dy\,dx$

16. $\displaystyle\int_0^{2}\int_{y-2}^{0} dx\,dy$

17. $\displaystyle\int_0^{1}\int_y^{\sqrt{y}} dx\,dy$

18. $\displaystyle\int_0^{1}\int_{1-x}^{1-x^2} dy\,dx$

19. $\displaystyle\int_0^{1}\int_1^{e^x} dy\,dx$

20. $\displaystyle\int_0^{\ln 2}\int_{e^y}^{2} dx\,dy$

21. $\displaystyle\int_0^{3/2}\int_0^{9-4x^2} 16x\,dy\,dx$

22. $\displaystyle\int_0^{2}\int_0^{4-y^2} y\,dx\,dy$

23. $\displaystyle\int_0^{1}\int_{-\sqrt{1-y^2}}^{\sqrt{1-y^2}} 3y\,dx\,dy$

24. $\displaystyle\int_0^{2}\int_{-\sqrt{4-x^2}}^{\sqrt{4-x^2}} 6x\,dy\,dx$

In Exercises 25–34, sketch the region of integration, reverse the order of integration, and evaluate the integral.

25. $\displaystyle\int_0^{\pi}\int_x^{\pi} \frac{\sin y}{y}\,dy\,dx$

26. $\displaystyle\int_0^{2}\int_x^{2} 2y^2\sin xy\,dy\,dx$

27. $\displaystyle\int_0^{1}\int_y^{1} x^2e^{xy}\,dx\,dy$

28. $\displaystyle\int_0^{2}\int_0^{4-x^2} \frac{xe^{2y}}{4-y}\,dy\,dx$

29. $\int_0^{2\sqrt{\ln 3}} \int_{y/2}^{\sqrt{\ln 3}} e^{x^2}\, dx\, dy$ **30.** $\int_0^3 \int_{\sqrt{x/3}}^1 e^{y^3}\, dy\, dx$

31. $\int_0^{1/16} \int_{y^{1/4}}^{1/2} \cos(16\pi x^5)\, dx\, dy$

32. $\int_0^8 \int_{\sqrt[3]{x}}^2 \frac{dy\, dx}{y^4 + 1}$

33. Square region $\iint_R (y - 2x^2)\, dA$ where R is the region bounded by the square $|x| + |y| = 1$

34. Triangular region $\iint_R xy\, dA$ where R is the region bounded by the lines $y = x$, $y = 2x$, and $x + y = 2$

35. Find the volume of the region bounded by the paraboloid $z = x^2 + y^2$ and below by the triangle enclosed by the lines $y = x$, $x = 0$, and $x + y = 2$ in the xy-plane.

36. Find the volume of the solid that is bounded above by the cylinder $z = x^2$ and below by the region enclosed by the parabola $y = 2 - x^2$ and the line $y = x$ in the xy-plane.

37. Find the volume of the solid whose base is the region in the xy-plane that is bounded by the parabola $y = 4 - x^2$ and the line $y = 3x$, while the top of the solid is bounded by the plane $z = x + 4$.

38. Find the volume of the solid in the first octant bounded by the coordinate planes, the cylinder $x^2 + y^2 = 4$, and the plane $z + y = 3$.

39. Find the volume of the solid in the first octant bounded by the coordinate planes, the plane $x = 3$, and the parabolic cylinder $z = 4 - y^2$.

40. Find the volume of the solid cut from the first octant by the surface $z = 4 - x^2 - y$.

41. Find the volume of the wedge cut from the first octant by the cylinder $z = 12 - 3y^2$ and the plane $x + y = 2$.

42. Find the volume of the solid cut from the square column $|x| + |y| \le 1$ by the planes $z = 0$ and $3x + z = 3$.

43. Find the volume of the solid that is bounded on the front and back by the planes $x = 2$ and $x = 1$, on the sides by the cylinders $y = \pm 1/x$, and above and below by the planes $z = x + 1$ and $z = 0$.

44. Find the volume of the solid bounded on the front and back by the planes $x = \pm\pi/3$, on the sides by the cylinders $y = \pm\sec x$, above by the cylinder $z = 1 + y^2$, and below by the xy-plane.

Improper double integrals can often be computed similarly to improper integrals of one variable. The first iteration of the following improper integrals is conducted just as if they were proper integrals. One then evaluates an improper integral of a single variable by taking appropriate limits, as in Section 5.7. Evaluate the improper integrals in Exercises 45–48 as iterated integrals.

45. $\int_1^\infty \int_{e^{-x}}^1 \frac{1}{x^3 y}\, dy\, dx$

46. $\int_{-1}^1 \int_{-1/\sqrt{1-x^2}}^{1/\sqrt{1-x^2}} (2y + 1)\, dy\, dx$

47. $\int_{-\infty}^\infty \int_{-\infty}^\infty \frac{1}{(x^2 + 1)(y^2 + 1)}\, dx\, dy$

48. $\int_0^\infty \int_0^\infty xe^{-(x+2y)}\, dx\, dy$

In Exercises 49 and 50, approximate the double integral of $f(x, y)$ over the region R partitioned by the given vertical lines $x = a$ and horizontal lines $y = c$. In each subrectangle, use (x_k, y_k) as indicated for your approximation.

$$\iint_R f(x, y)\, dA \approx \sum_{k=1}^n f(x_k, y_k)\, \Delta A_k$$

49. $f(x, y) = x + y$ over the region R bounded above by the semicircle $y = \sqrt{1 - x^2}$ and below by the x-axis, using the partition $x = -1, -1/2, 0, 1/4, 1/2, 1$ and $y = 0, 1/2, 1$ with (x_k, y_k) the lower left corner in the kth subrectangle (provided the subrectangle lies within R)

50. $f(x, y) = x + 2y$ over the region R inside the circle $(x - 2)^2 + (y - 3)^2 = 1$ using the partition $x = 1, 3/2, 2, 5/2, 3$ and $y = 2, 5/2, 3, 7/2, 4$ with (x_k, y_k) the center (centroid) in the kth subrectangle (provided the subrectangle lies within R)

51. Circular sector Integrate $f(x, y) = \sqrt{4 - x^2}$ over the smaller sector cut from the disk $x^2 + y^2 \le 4$ by the rays $\theta = \pi/6$ and $\theta = \pi/2$.

52. Unbounded region Integrate $f(x, y) = 1/[(x^2 - x)(y - 1)^{2/3}]$ over the infinite rectangle $2 \le x < \infty$, $0 \le y \le 2$.

53. Noncircular cylinder A solid right (noncircular) cylinder has its base R in the xy-plane and is bounded above by the paraboloid $z = x^2 + y^2$. The cylinder's volume is

$$V = \int_0^1 \int_0^y (x^2 + y^2)\, dx\, dy + \int_1^2 \int_0^{2-y} (x^2 + y^2)\, dx\, dy.$$

Sketch the base region R and express the cylinder's volume as a single iterated integral with the order of integration reversed. Then evaluate the integral to find the volume.

54. Converting to a double integral Evaluate the integral

$$\int_0^2 (\tan^{-1}\pi x - \tan^{-1} x)\, dx.$$

(*Hint:* Write the integrand as an integral.)

55. Maximizing a double integral What region R in the xy-plane maximizes the value of

$$\iint_R (4 - x^2 - 2y^2)\, dA?$$

Give reasons for your answer.

56. Minimizing a double integral What region R in the xy-plane minimizes the value of

$$\iint_R (x^2 + y^2 - 9)\, dA?$$

Give reasons for your answer.

57. Is it possible to evaluate the integral of a continuous function $f(x, y)$ over a rectangular region in the xy-plane and get different answers depending on the order of integration? Give reasons for your answer.

58. How would you evaluate the double integral of a continuous function $f(x, y)$ over the region R in the xy-plane enclosed by the triangle with vertices (0, 1), (2, 0), and (1, 2)? Give reasons for your answer.

59. Unbounded region Prove that

$$\int_{-\infty}^{\infty}\int_{-\infty}^{\infty} e^{-x^2-y^2}\,dx\,dy = \lim_{b\to\infty}\int_{-b}^{b}\int_{-b}^{b} e^{-x^2-y^2}\,dx\,dy$$
$$= 4\left(\int_0^{\infty} e^{-x^2}\,dx\right)^2.$$

60. Improper double integral Evaluate the improper integral

$$\int_0^1\int_0^3 \frac{x^2}{(y-1)^{2/3}}\,dy\,dx.$$

COMPUTER EXPLORATIONS

Use a CAS double-integral evaluator to estimate the values of the integrals in Exercises 61–64.

61. $\displaystyle\int_1^3\int_1^x \frac{1}{xy}\,dy\,dx$

62. $\displaystyle\int_0^1\int_0^1 e^{-(x^2+y^2)}\,dy\,dx$

63. $\displaystyle\int_0^1\int_0^1 \tan^{-1} xy\,dy\,dx$

64. $\displaystyle\int_{-1}^1\int_0^{\sqrt{1-x^2}} 3\sqrt{1-x^2-y^2}\,dy\,dx$

Use a CAS double-integral evaluator to find the integrals in Exercises 65–70. Then reverse the order of integration and evaluate, again with a CAS.

65. $\displaystyle\int_0^1\int_{2y}^4 e^{x^2}\,dx\,dy$

66. $\displaystyle\int_0^3\int_{x^2}^9 x\cos(y^2)\,dy\,dx$

67. $\displaystyle\int_0^2\int_{y^3}^{4\sqrt{2y}} (x^2y - xy^2)\,dx\,dy$

68. $\displaystyle\int_0^2\int_0^{4-y^2} e^{xy}\,dx\,dy$

69. $\displaystyle\int_1^2\int_0^{x^2} \frac{1}{x+y}\,dy\,dx$

70. $\displaystyle\int_1^2\int_{y^3}^8 \frac{1}{\sqrt{x^2+y^2}}\,dx\,dy$

12.3 Area by Double Integration

In this section we show how to use double integrals to calculate the areas of bounded regions in the plane, and to find the average value of a function of two variables.

Areas of Bounded Regions in the Plane

If we take $f(x, y) = 1$ in the definition of the double integral over a region R in the preceding section, the Riemann sums reduce to

$$S_n = \sum_{k=1}^{n} f(x_k, y_k)\,\Delta A_k = \sum_{k=1}^{n} \Delta A_k. \tag{1}$$

This is simply the sum of the areas of the small rectangles in the partition of R, and approximates what we would like to call the area of R. As the norm of a partition of R approaches zero, the height and width of all rectangles in the partition approach zero, and the coverage of R becomes increasingly complete (Figure 12.6). We define the area of R to be the limit

$$\lim_{\|P\|\to 0} \sum_{k=1}^{n} \Delta A_k = \iint_R dA \tag{2}$$

DEFINITION The **area** of a closed, bounded plane region R is

$$A = \iint_R dA.$$

As with the other definitions in this chapter, the definition here applies to a greater variety of regions than does the earlier single-variable definition of area, but it agrees with the earlier definition on regions to which they both apply. To evaluate the integral in the definition of area, we integrate the constant function $f(x, y) = 1$ over R.

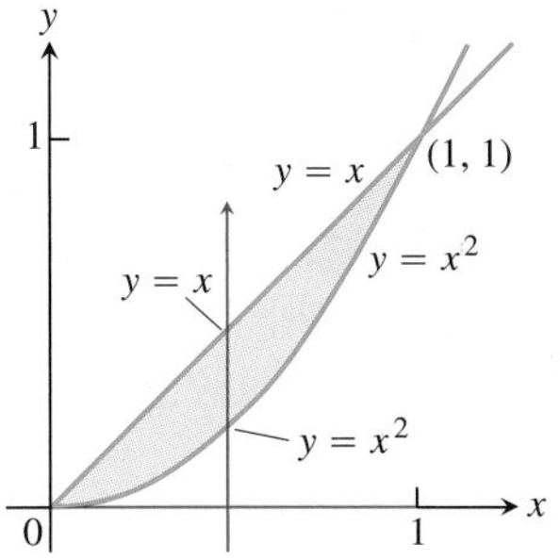

FIGURE 12.15 The region in Example 1.

EXAMPLE 1 Find the area of the region R bounded by $y = x$ and $y = x^2$ in the first quadrant.

Solution We sketch the region (Figure 12.15), noting where the two curves intersect, and calculate the area as

$$A = \int_0^1 \int_{x^2}^{x} dy\, dx = \int_0^1 \Big[y \Big]_{x^2}^{x} dx$$
$$= \int_0^1 (x - x^2)\, dx = \left[\frac{x^2}{2} - \frac{x^3}{3}\right]_0^1 = \frac{1}{6}.$$

Notice that the single integral $\int_0^1 (x - x^2)\, dx$, obtained from evaluating the inside iterated integral, is the integral for the area between these two curves using the method of Section 4.7. ■

EXAMPLE 2 Find the area of the region R enclosed by the parabola $y = x^2$ and the line $y = x + 2$.

Solution If we divide R into the regions R_1 and R_2 shown in Figure 12.16a, we may calculate the area as

$$A = \iint_{R_1} dA + \iint_{R_2} dA = \int_0^1 \int_{-\sqrt{y}}^{\sqrt{y}} dx\, dy + \int_1^4 \int_{y-2}^{\sqrt{y}} dx\, dy.$$

On the other hand, reversing the order of integration (Figure 12.16b) gives

$$A = \int_{-1}^{2} \int_{x^2}^{x+2} dy\, dx.$$

This second result, which requires only one integral, is simpler and is the only one we would bother to write down in practice. The area is

$$A = \int_{-1}^{2} \Big[y \Big]_{x^2}^{x+2} dx = \int_{-1}^{2} (x + 2 - x^2)\, dx = \left[\frac{x^2}{2} + 2x - \frac{x^3}{3}\right]_{-1}^{2} = \frac{9}{2}.$$ ■

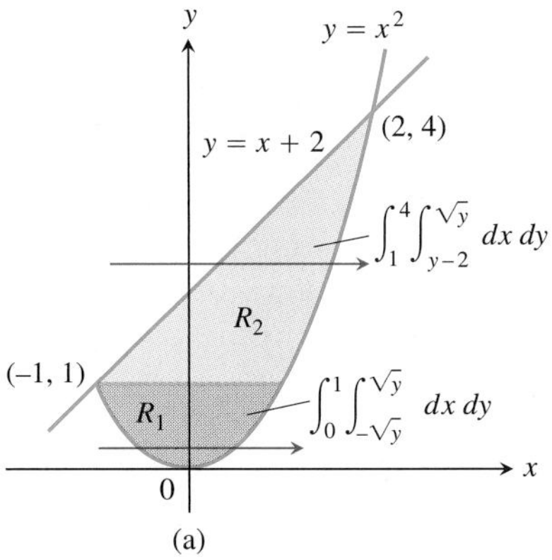

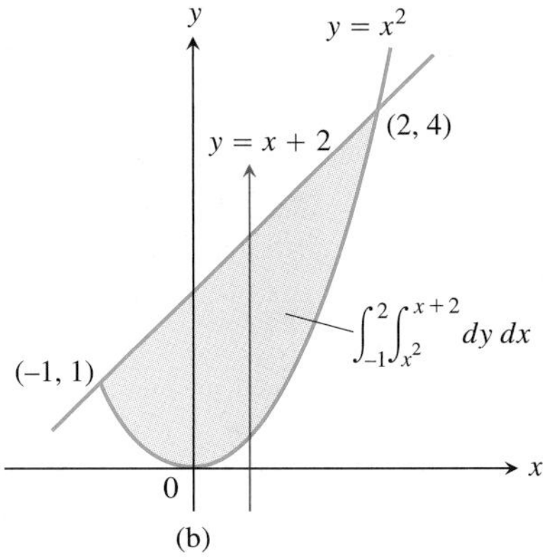

FIGURE 12.16 Calculating this area takes (a) two double integrals if the first integration is with respect to x, but (b) only one if the first integration is with respect to y (Example 2).

Average Value

The average value of an integrable function of one variable on a closed interval is the integral of the function over the interval divided by the length of the interval. For an integrable function of two variables defined on a bounded region in the plane, the average value is the integral over the region divided by the area of the region. This can be visualized by thinking of the function as giving the height at one instant of some water sloshing around in a tank whose vertical walls lie over the boundary of the region. The average height of the water in the tank can be found by letting the water settle down to a constant height. The height is then equal to the volume of water in the tank divided by the area of R. We are led to define the average value of an integrable function f over a region R as follows:

$$\textbf{Average value} \text{ of } f \text{ over } R = \frac{1}{\text{area of } R}\iint_R f\,dA. \qquad (3)$$

If f is the temperature of a thin plate covering R, then the double integral of f over R divided by the area of R is the plate's average temperature. If $f(x, y)$ is the distance from the point (x, y) to a fixed point P, then the average value of f over R is the average distance of points in R from P.

EXAMPLE 3 Find the average value of $f(x, y) = x\cos xy$ over the rectangle R: $0 \le x \le \pi, 0 \le y \le 1$.

Solution The value of the integral of f over R is

$$\int_0^{\pi}\int_0^1 x\cos xy\,dy\,dx = \int_0^{\pi}\left[\sin xy\right]_{y=0}^{y=1} dx \qquad \int x\cos xy\,dy = \sin xy + C$$

$$= \int_0^{\pi}(\sin x - 0)\,dx = -\cos x\Big]_0^{\pi} = 1 + 1 = 2.$$

The area of R is π. The average value of f over R is $2/\pi$. ■

EXERCISES 12.3

In Exercises 1–8, sketch the region bounded by the given lines and curves. Then express the region's area as an iterated double integral and evaluate the integral.

1. The coordinate axes and the line $x + y = 2$
2. The lines $x = 0$, $y = 2x$, and $y = 4$
3. The parabola $x = -y^2$ and the line $y = x + 2$
4. The parabola $x = y - y^2$ and the line $y = -x$
5. The curve $y = e^x$ and the lines $y = 0$, $x = 0$, and $x = \ln 2$
6. The curves $y = \ln x$ and $y = 2\ln x$ and the line $x = e$, in the first quadrant
7. The parabolas $x = y^2$ and $x = 2y - y^2$
8. The parabolas $x = y^2 - 1$ and $x = 2y^2 - 2$

The integrals and sums of integrals in Exercises 9–14 give the areas of regions in the xy-plane. Sketch each region, label each bounding curve with its equation, and give the coordinates of the points where the curves intersect. Then find the area of the region.

9. $\displaystyle\int_0^6\int_{y^2/3}^{2y} dx\,dy$

10. $\displaystyle\int_0^3\int_{-x}^{x(2-x)} dy\,dx$

11. $\displaystyle\int_0^{\pi/4}\int_{\sin x}^{\cos x} dy\,dx$

12. $\displaystyle\int_{-1}^2\int_{y^2}^{y+2} dx\,dy$

13. $\displaystyle\int_{-1}^0\int_{-2x}^{1-x} dy\,dx + \int_0^2\int_{-x/2}^{1-x} dy\,dx$

14. $\displaystyle\int_0^2\int_{x^2-4}^0 dy\,dx + \int_0^4\int_0^{\sqrt{x}} dy\,dx$

15. Find the average value of $f(x, y) = \sin(x + y)$ over
    a. the rectangle $0 \le x \le \pi, \quad 0 \le y \le \pi$
    b. the rectangle $0 \le x \le \pi, \quad 0 \le y \le \pi/2$

16. Which do you think will be larger, the average value of $f(x, y) = xy$ over the square $0 \le x \le 1, 0 \le y \le 1$, or the average value of f over the quarter circle $x^2 + y^2 \le 1$ in the first quadrant? Calculate them to find out.

17. Find the average height of the paraboloid $z = x^2 + y^2$ over the square $0 \leq x \leq 2, 0 \leq y \leq 2$.

18. Find the average value of $f(x, y) = 1/(xy)$ over the square $\ln 2 \leq x \leq 2 \ln 2, \ln 2 \leq y \leq 2 \ln 2$.

19. **Bacterium population** If $f(x, y) = (10{,}000e^y)/(1 + |x|/2)$ represents the "population density" of a certain bacterium on the xy-plane, where x and y are measured in centimeters, find the total population of bacteria within the rectangle $-5 \leq x \leq 5$ and $-2 \leq y \leq 0$.

20. **Regional population** If $f(x, y) = 100\,(y + 1)$ represents the population density of a planar region on Earth, where x and y are measured in miles, find the number of people in the region bounded by the curves $x = y^2$ and $x = 2y - y^2$.

21. **Average temperature in Texas** According to the *Texas Almanac*, Texas has 254 counties and a National Weather Service station in each county. Assume that at time t_0, each of the 254 weather stations recorded the local temperature. Find a formula that would give a reasonable approximation to the average temperature in Texas at time t_0. Your answer should involve information that you would expect to be readily available in the *Texas Almanac*.

22. If $y = f(x)$ is a nonnegative continuous function over the closed interval $a \leq x \leq b$, show that the double integral definition of area for the closed plane region bounded by the graph of f, the vertical lines $x = a$ and $x = b$, and the x-axis, agrees with the definition for area beneath the curve in Section 4.4.

12.4 Double Integrals in Polar Form

Integrals are sometimes easier to evaluate if we change to polar coordinates. This section shows how to accomplish the change and how to evaluate integrals over regions whose boundaries are given by polar equations.

Integrals in Polar Coordinates

When we defined the double integral of a function over a region R in the xy-plane, we began by cutting R into rectangles whose sides were parallel to the coordinate axes. These were the natural shapes to use because their sides have either constant x-values or constant y-values. In polar coordinates, the natural shape is a "polar rectangle" whose sides have constant r- and θ-values.

Suppose that a function $f(r, \theta)$ is defined over a region R that is bounded by the rays $\theta = \alpha$ and $\theta = \beta$ and by the continuous curves $r = g_1(\theta)$ and $r = g_2(\theta)$. Suppose also that $0 \leq g_1(\theta) \leq g_2(\theta) \leq a$ for every value of θ between α and β. Then R lies in a fan-shaped region Q defined by the inequalities $0 \leq r \leq a$ and $\alpha \leq \theta \leq \beta$. See Figure 12.17.

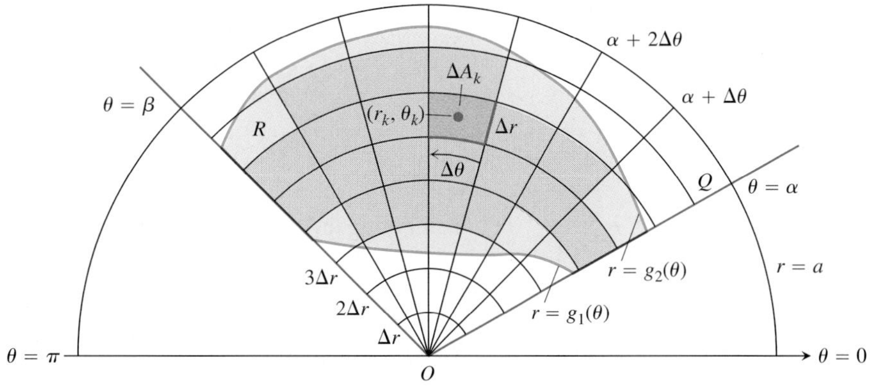

FIGURE 12.17 The region R: $g_1(\theta) \leq r \leq g_2(\theta)$, $\alpha \leq \theta \leq \beta$, is contained in the fan-shaped region Q: $0 \leq r \leq a$, $\alpha \leq \theta \leq \beta$. The partition of Q by circular arcs and rays induces a partition of R.

We cover Q by a grid of circular arcs and rays. The arcs are cut from circles centered at the origin, with radii $\Delta r, 2\Delta r, \ldots, m\Delta r$, where $\Delta r = a/m$. The rays are given by

$$\theta = \alpha, \qquad \theta = \alpha + \Delta\theta, \qquad \theta = \alpha + 2\Delta\theta, \qquad \ldots, \qquad \theta = \alpha + m'\Delta\theta = \beta,$$

where $\Delta\theta = (\beta - \alpha)/m'$. The arcs and rays partition Q into small patches called "polar rectangles."

We number the polar rectangles that lie inside R (the order does not matter), calling their areas $\Delta A_1, \Delta A_2, \ldots, \Delta A_n$. We let (r_k, θ_k) be any point in the polar rectangle whose area is ΔA_k. We then form the sum

$$S_n = \sum_{k=1}^{n} f(r_k, \theta_k)\, \Delta A_k.$$

If f is continuous throughout R, this sum will approach a limit as we refine the grid to make Δr and $\Delta\theta$ go to zero. The limit is called the double integral of f over R. In symbols,

$$\lim_{n\to\infty} S_n = \iint_R f(r, \theta)\, dA.$$

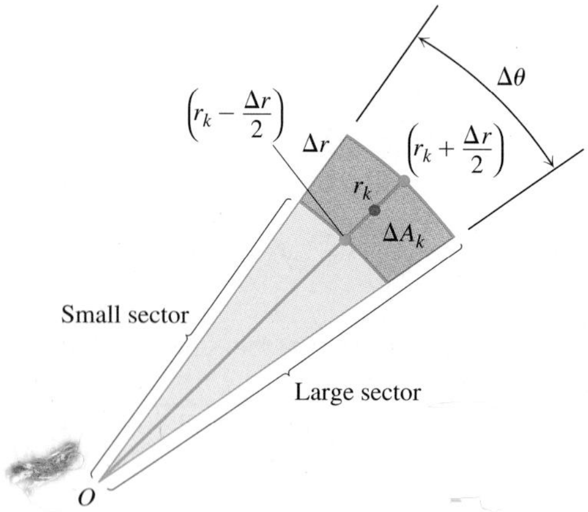

FIGURE 12.18 The observation that

$$\Delta A_k = \begin{pmatrix}\text{area of} \\ \text{large sector}\end{pmatrix} - \begin{pmatrix}\text{area of} \\ \text{small sector}\end{pmatrix}$$

leads to the formula $\Delta A_k = r_k\, \Delta r\, \Delta\theta$.

To evaluate this limit, we first have to write the sum S_n in a way that expresses ΔA_k in terms of Δr and $\Delta\theta$. For convenience we choose r_k to be the average of the radii of the inner and outer arcs bounding the kth polar rectangle ΔA_k. The radius of the inner arc bounding ΔA_k is then $r_k - (\Delta r/2)$ (Figure 12.18). The radius of the outer arc is $r_k + (\Delta r/2)$.

The area of a wedge-shaped sector of a circle having radius r and angle θ is

$$A = \frac{1}{2}\theta \cdot r^2,$$

as can be seen by multiplying πr^2, the area of the circle, by $\theta/2\pi$, the fraction of the circle's area contained in the wedge. So the areas of the circular sectors subtended by these arcs at the origin are

$$\text{Inner radius:} \qquad \frac{1}{2}\left(r_k - \frac{\Delta r}{2}\right)^2 \Delta\theta$$

$$\text{Outer radius:} \qquad \frac{1}{2}\left(r_k + \frac{\Delta r}{2}\right)^2 \Delta\theta.$$

Therefore,

$$\begin{aligned}\Delta A_k &= \text{area of large sector} - \text{area of small sector} \\ &= \frac{\Delta\theta}{2}\left[\left(r_k + \frac{\Delta r}{2}\right)^2 - \left(r_k - \frac{\Delta r}{2}\right)^2\right] = \frac{\Delta\theta}{2}(2r_k\, \Delta r) = r_k\, \Delta r\, \Delta\theta.\end{aligned}$$

Combining this result with the sum defining S_n gives

$$S_n = \sum_{k=1}^{n} f(r_k, \theta_k) r_k\, \Delta r\, \Delta\theta.$$

As $n \to \infty$ and the values of Δr and $\Delta\theta$ approach zero, these sums converge to the double integral

$$\lim_{n\to\infty} S_n = \iint_R f(r, \theta)\, r\, dr\, d\theta.$$

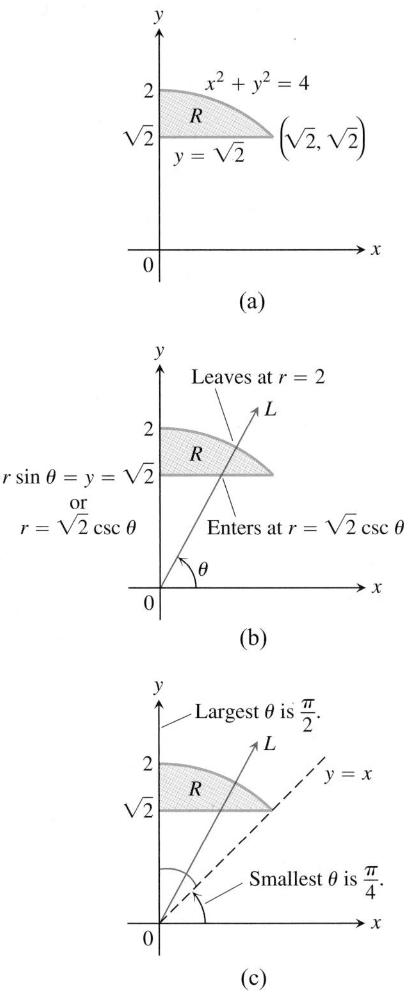

FIGURE 12.19 Finding the limits of integration in polar coordinates.

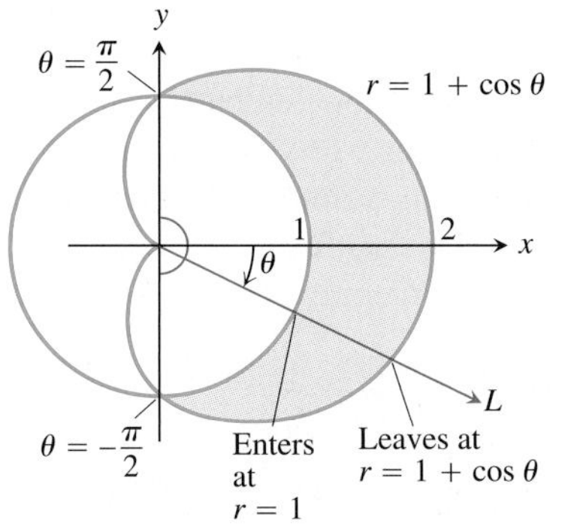

FIGURE 12.20 Finding the limits of integration in polar coordinates for the region in Example 1.

A version of Fubini's Theorem says that the limit approached by these sums can be evaluated by repeated single integrations with respect to r and θ as

$$\iint_R f(r, \theta)\, dA = \int_{\theta=\alpha}^{\theta=\beta} \int_{r=g_1(\theta)}^{r=g_2(\theta)} f(r, \theta)\, r\, dr\, d\theta.$$

Finding Limits of Integration

The procedure for finding limits of integration in rectangular coordinates also works for polar coordinates. To evaluate $\iint_R f(r, \theta)\, dA$ over a region R in polar coordinates, integrating first with respect to r and then with respect to θ, take the following steps.

1. *Sketch*: Sketch the region and label the bounding curves (Figure 12.19a).
2. *Find the r-limits of integration*: Imagine a ray L from the origin cutting through R in the direction of increasing r. Mark the r-values where L enters and leaves R. These are the r-limits of integration. They usually depend on the angle θ that L makes with the positive x-axis (Figure 12.19b).
3. *Find the θ-limits of integration*: Find the smallest and largest θ-values that bound R. These are the θ-limits of integration (Figure 12.19c). The integral is

$$\iint_R f(r, \theta)\, dA = \int_{\theta=\pi/4}^{\theta=\pi/2} \int_{r=\sqrt{2}\csc\theta}^{r=2} f(r, \theta)\, r\, dr\, d\theta.$$

EXAMPLE 1 Find the limits of integration for integrating $f(r, \theta)$ over the region R that lies inside the cardioid $r = 1 + \cos \theta$ and outside the circle $r = 1$.

Solution

1. We first sketch the region and label the bounding curves (Figure 12.20).
2. Next we find the *r-limits of integration*. A typical ray from the origin enters R where $r = 1$ and leaves where $r = 1 + \cos \theta$.
3. Finally we find the *θ-limits of integration*. The rays from the origin that intersect R run from $\theta = -\pi/2$ to $\theta = \pi/2$. The integral is

$$\int_{-\pi/2}^{\pi/2} \int_{1}^{1+\cos\theta} f(r, \theta)\, r\, dr\, d\theta.$$

■

If $f(r, \theta)$ is the constant function whose value is 1, then the integral of f over R is the area of R.

Area in Polar Coordinates

The area of a closed and bounded region R in the polar coordinate plane is

$$A = \iint_R r\, dr\, d\theta.$$

This formula for area is consistent with all earlier formulas, although we do not prove this fact.

EXAMPLE 2 Find the area enclosed by the lemniscate $r^2 = 4\cos 2\theta$.

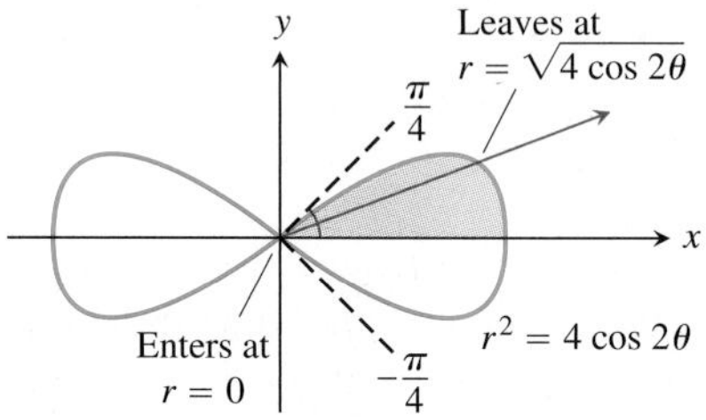

FIGURE 12.21 To integrate over the shaded region, we run r from 0 to $\sqrt{4\cos 2\theta}$ and θ from 0 to $\pi/4$ (Example 2).

Solution We graph the lemniscate to determine the limits of integration (Figure 12.21) and see from the symmetry of the region that the total area is 4 times the first-quadrant portion.

$$A = 4\int_0^{\pi/4}\int_0^{\sqrt{4\cos 2\theta}} r\,dr\,d\theta = 4\int_0^{\pi/4}\left[\frac{r^2}{2}\right]_{r=0}^{r=\sqrt{4\cos 2\theta}} d\theta$$
$$= 4\int_0^{\pi/4} 2\cos 2\theta\,d\theta = 4\sin 2\theta\Big]_0^{\pi/4} = 4.$$

Changing Cartesian Integrals into Polar Integrals

The procedure for changing a Cartesian integral $\iint_R f(x, y)\,dx\,dy$ into a polar integral has two steps. First substitute $x = r\cos\theta$ and $y = r\sin\theta$, and replace $dx\,dy$ by $r\,dr\,d\theta$ in the Cartesian integral. Then supply polar limits of integration for the boundary of R. The Cartesian integral then becomes

$$\iint_R f(x, y)\,dx\,dy = \iint_G f(r\cos\theta, r\sin\theta)\,r\,dr\,d\theta,$$

where G denotes the region of integration in polar coordinates. This is like the substitution method in Chapter 4 except that there are now two variables to substitute for instead of one. Notice that $dx\,dy$ is not replaced by $dr\,d\theta$ but by $r\,dr\,d\theta$. A more general discussion of changes of variables (substitutions) in multiple integrals is given in Section 12.8.

EXAMPLE 3 Evaluate

$$\iint_R e^{x^2+y^2}\,dy\,dx,$$

where R is the semicircular region bounded by the x-axis and the curve $y = \sqrt{1 - x^2}$ (Figure 12.22).

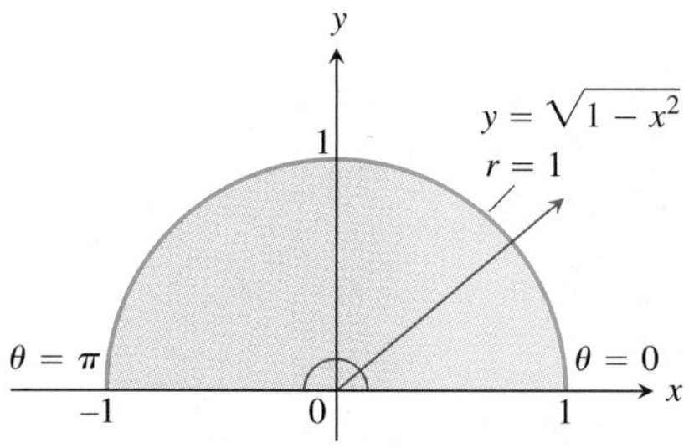

FIGURE 12.22 The semicircular region in Example 3 is the region

$$0 \le r \le 1, \qquad 0 \le \theta \le \pi.$$

Solution In Cartesian coordinates, the integral in question is a nonelementary integral and there is no direct way to integrate $e^{x^2+y^2}$ with respect to either x or y. Yet this integral and others like it are important in mathematics—in statistics, for example—and we need to find a way to evaluate it. Polar coordinates save the day. Substituting $x = r\cos\theta$, $y = r\sin\theta$ and replacing $dy\,dx$ by $r\,dr\,d\theta$ enables us to evaluate the integral as

$$\iint_R e^{x^2+y^2}\,dy\,dx = \int_0^{\pi}\int_0^1 e^{r^2}\,r\,dr\,d\theta = \int_0^{\pi}\left[\frac{1}{2}e^{r^2}\right]_0^1 d\theta$$
$$= \int_0^{\pi}\frac{1}{2}(e - 1)\,d\theta = \frac{\pi}{2}(e - 1).$$

The r in the $r\,dr\,d\theta$ was just what we needed to integrate e^{r^2}. Without it, we would have been unable to proceed.

EXERCISES 12.4

In Exercises 1–16, change the Cartesian integral into an equivalent polar integral. Then evaluate the polar integral.

1. $\int_{-1}^{1}\int_{0}^{\sqrt{1-x^2}} dy\,dx$

2. $\int_{-1}^{1}\int_{-\sqrt{1-x^2}}^{\sqrt{1-x^2}} dy\,dx$

3. $\int_{0}^{1}\int_{0}^{\sqrt{1-y^2}} (x^2+y^2)\,dx\,dy$

4. $\int_{-1}^{1}\int_{-\sqrt{1-y^2}}^{\sqrt{1-y^2}} (x^2+y^2)\,dy\,dx$

5. $\int_{-a}^{a}\int_{-\sqrt{a^2-x^2}}^{\sqrt{a^2-x^2}} dy\,dx$

6. $\int_{0}^{2}\int_{0}^{\sqrt{4-y^2}} (x^2+y^2)\,dx\,dy$

7. $\int_{0}^{6}\int_{0}^{y} x\,dx\,dy$

8. $\int_{0}^{2}\int_{0}^{x} y\,dy\,dx$

9. $\int_{-1}^{0}\int_{-\sqrt{1-x^2}}^{0} \frac{2}{1+\sqrt{x^2+y^2}}\,dy\,dx$

10. $\int_{-1}^{1}\int_{-\sqrt{1-y^2}}^{0} \frac{4\sqrt{x^2+y^2}}{1+x^2+y^2}\,dx\,dy$

11. $\int_{0}^{\ln 2}\int_{0}^{\sqrt{(\ln 2)^2-y^2}} e^{\sqrt{x^2+y^2}}\,dx\,dy$

12. $\int_{0}^{1}\int_{0}^{\sqrt{1-x^2}} e^{-(x^2+y^2)}\,dy\,dx$

13. $\int_{0}^{2}\int_{0}^{\sqrt{1-(x-1)^2}} \frac{x+y}{x^2+y^2}\,dy\,dx$

14. $\int_{0}^{2}\int_{-\sqrt{1-(y-1)^2}}^{0} xy^2\,dx\,dy$

15. $\int_{-1}^{1}\int_{-\sqrt{1-y^2}}^{\sqrt{1-y^2}} \ln(x^2+y^2+1)\,dx\,dy$

16. $\int_{-1}^{1}\int_{-\sqrt{1-x^2}}^{\sqrt{1-x^2}} \frac{2}{(1+x^2+y^2)^2}\,dy\,dx$

17. Find the area of the region cut from the first quadrant by the curve $r = 2(2 - \sin 2\theta)^{1/2}$.

18. **Cardioid overlapping a circle** Find the area of the region that lies inside the cardioid $r = 1 + \cos\theta$ and outside the circle $r = 1$.

19. **One leaf of a rose** Find the area enclosed by one leaf of the rose $r = 12\cos 3\theta$.

20. **Snail shell** Find the area of the region enclosed by the positive x-axis and spiral $r = 4\theta/3$, $0 \le \theta \le 2\pi$. The region looks like a snail shell.

21. **Cardioid in the first quadrant** Find the area of the region cut from the first quadrant by the cardioid $r = 1 + \sin\theta$.

22. **Overlapping cardioids** Find the area of the region common to the interiors of the cardioids $r = 1 + \cos\theta$ and $r = 1 - \cos\theta$.

In polar coordinates, the **average value** of a function over a region R (Section 12.3) is given by

$$\frac{1}{\text{Area}(R)}\iint_R f(r,\theta)\,r\,dr\,d\theta.$$

23. **Average height of a hemisphere** Find the average height of the hemispherical surface $z = \sqrt{a^2 - x^2 - y^2}$ above the disk $x^2 + y^2 \le a^2$ in the xy-plane.

24. **Average height of a cone** Find the average height of the (single) cone $z = \sqrt{x^2+y^2}$ above the disk $x^2 + y^2 \le a^2$ in the xy-plane.

25. **Average distance from interior of disk to center** Find the average distance from a point $P(x, y)$ in the disk $x^2 + y^2 \le a^2$ to the origin.

26. **Average distance squared from a point in a disk to a point in its boundary** Find the average value of the *square* of the distance from the point $P(x, y)$ in the disk $x^2 + y^2 \le 1$ to the boundary point $A(1, 0)$.

27. **Converting to a polar integral** Integrate $f(x, y) = [\ln(x^2 + y^2)]/\sqrt{x^2+y^2}$ over the region $1 \le x^2 + y^2 \le e$.

28. **Converting to a polar integral** Integrate $f(x, y) = [\ln(x^2 + y^2)]/(x^2 + y^2)$ over the region $1 \le x^2 + y^2 \le e^2$.

29. **Volume of noncircular right cylinder** The region that lies inside the cardioid $r = 1 + \cos\theta$ and outside the circle $r = 1$ is the base of a solid right cylinder. The top of the cylinder lies in the plane $z = x$. Find the cylinder's volume.

30. **Volume of noncircular right cylinder** The region enclosed by the lemniscate $r^2 = 2\cos 2\theta$ is the base of a solid right cylinder whose top is bounded by the sphere $z = \sqrt{2 - r^2}$. Find the cylinder's volume.

31. **Converting to polar integrals**

   a. The usual way to evaluate the improper integral $I = \int_0^\infty e^{-x^2}\,dx$ is first to calculate its square:

$$I^2 = \left(\int_0^\infty e^{-x^2}\,dx\right)\left(\int_0^\infty e^{-y^2}\,dy\right) = \int_0^\infty\int_0^\infty e^{-(x^2+y^2)}\,dx\,dy.$$

   Evaluate the last integral using polar coordinates and solve the resulting equation for I.

   b. Evaluate

$$\lim_{x\to\infty} \text{erf}(x) = \lim_{x\to\infty}\int_0^x \frac{2e^{-t^2}}{\sqrt{\pi}}\,dt.$$

32. **Converting to a polar integral** Evaluate the integral

$$\int_0^\infty\int_0^\infty \frac{1}{(1+x^2+y^2)^2}\,dx\,dy.$$

33. **Existence** Integrate the function $f(x, y) = 1/(1 - x^2 - y^2)$ over the disk $x^2 + y^2 \le 3/4$. Does the integral of $f(x, y)$ over the disk $x^2 + y^2 \le 1$ exist? Give reasons for your answer.

34. **Area formula in polar coordinates** Use the double integral in polar coordinates to derive the formula

$$A = \int_\alpha^\beta \frac{1}{2}r^2\,d\theta$$

for the area of the fan-shaped region between the origin and polar curve $r = f(\theta)$, $\alpha \le \theta \le \beta$.

35. **Average distance to a given point inside a disk** Let P_0 be a point inside a circle of radius a and let h denote the distance from

P_0 to the center of the circle. Let d denote the distance from an arbitrary point P to P_0. Find the average value of d^2 over the region enclosed by the circle. (*Hint:* Simplify your work by placing the center of the circle at the origin and P_0 on the x-axis.)

36. Area Suppose that the area of a region in the polar coordinate plane is

$$A = \int_{\pi/4}^{3\pi/4} \int_{\csc\theta}^{2\sin\theta} r\,dr\,d\theta.$$

Sketch the region and find its area.

COMPUTER EXPLORATIONS

In Exercises 37–40, use a CAS to change the Cartesian integrals into an equivalent polar integral and evaluate the polar integral. Perform the following steps in each exercise.

a. Plot the Cartesian region of integration in the xy-plane.

b. Change each boundary curve of the Cartesian region in part (a) to its polar representation by solving its Cartesian equation for r and θ.

c. Using the results in part (b), plot the polar region of integration in the $r\theta$-plane.

d. Change the integrand from Cartesian to polar coordinates. Determine the limits of integration from your plot in part (c) and evaluate the polar integral using the CAS integration utility.

37. $\displaystyle\int_0^1 \int_x^1 \frac{y}{x^2 + y^2}\,dy\,dx$ **38.** $\displaystyle\int_0^1 \int_0^{x/2} \frac{x}{x^2 + y^2}\,dy\,dx$

39. $\displaystyle\int_0^1 \int_{-y/3}^{y/3} \frac{y}{\sqrt{x^2 + y^2}}\,dx\,dy$ **40.** $\displaystyle\int_0^1 \int_y^{2-y} \sqrt{x + y}\,dx\,dy$

12.5 Triple Integrals in Rectangular Coordinates

Just as double integrals allow us to deal with more general situations than could be handled by single integrals, triple integrals enable us to solve still more general problems. We use triple integrals to calculate the volumes of three-dimensional shapes and the average value of a function over a three-dimensional region. Triple integrals also arise in the study of vector fields and fluid flow in three dimensions, as we will see in Chapter 13.

Triple Integrals

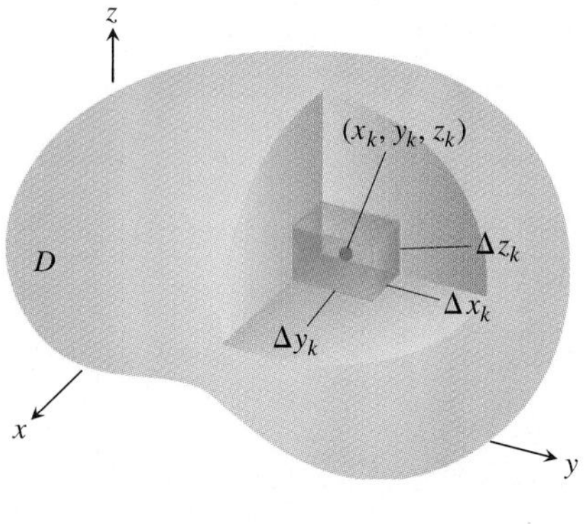

FIGURE 12.23 Partitioning a solid with rectangular cells of volume ΔV_k.

If $F(x, y, z)$ is a function defined on a closed bounded region D in space, such as the region occupied by a solid ball or a lump of clay, then the integral of F over D may be defined in the following way. We partition a rectangular boxlike region containing D into rectangular cells by planes parallel to the coordinate axes (Figure 12.23). We number the cells that lie inside D from 1 to n in some order, the kth cell having dimensions Δx_k by Δy_k by Δz_k and volume $\Delta V_k = \Delta x_k \Delta y_k \Delta z_k$. We choose a point (x_k, y_k, z_k) in each cell and form the sum

$$S_n = \sum_{k=1}^{n} F(x_k, y_k, z_k)\,\Delta V_k. \tag{1}$$

We are interested in what happens as D is partitioned by smaller and smaller cells, so that Δx_k, Δy_k, Δz_k and the norm of the partition $\|P\|$, the largest value among Δx_k, Δy_k, Δz_k, all approach zero. When a single limiting value is attained, no matter how the partitions and points (x_k, y_k, z_k) are chosen, we say that F is **integrable** over D. As before, it can be shown that when F is continuous and the bounding surface of D is formed from finitely many smooth surfaces joined together along finitely many smooth curves, then F is integrable. As $\|P\| \to 0$ and the number of cells n goes to ∞, the sums S_n approach a limit. We call this limit the **triple integral of *F* over *D*** and write

$$\lim_{n\to\infty} S_n = \iiint_D F(x, y, z)\,dV \qquad \text{or} \qquad \lim_{\|P\|\to 0} S_n = \iiint_D F(x, y, z)\,dx\,dy\,dz.$$

The regions D over which continuous functions are integrable are those having "reasonably smooth" boundaries.

Volume of a Region in Space

If F is the constant function whose value is 1, then the sums in Equation (1) reduce to

$$S_n = \sum F(x_k, y_k, z_k)\,\Delta V_k = \sum 1 \cdot \Delta V_k = \sum \Delta V_k.$$

As Δx_k, Δy_k, and Δz_k approach zero, the cells ΔV_k become smaller and more numerous and fill up more and more of D. We therefore define the volume of D to be the triple integral

$$\lim_{n\to\infty} \sum_{k=1}^{n} \Delta V_k = \iiint\limits_D dV.$$

DEFINITION The **volume** of a closed, bounded region D in space is

$$V = \iiint\limits_D dV.$$

This definition is in agreement with our previous definitions of volume, although we omit the verification of this fact. As we see in a moment, this integral enables us to calculate the volumes of solids enclosed by curved surfaces.

Finding Limits of Integration

We evaluate a triple integral by applying a three-dimensional version of Fubini's Theorem (Section 12.2) to evaluate it by three repeated single integrations. As with double integrals, there is a geometric procedure for finding the limits of integration for these single integrals.

To evaluate

$$\iiint\limits_D F(x, y, z)\,dV$$

over a region D, integrate first with respect to z, then with respect to y, finally with x.

1. *Sketch:* Sketch the region D along with its "shadow" R (vertical projection) in the xy-plane. Label the upper and lower bounding surfaces of D and the upper and lower bounding curves of R.

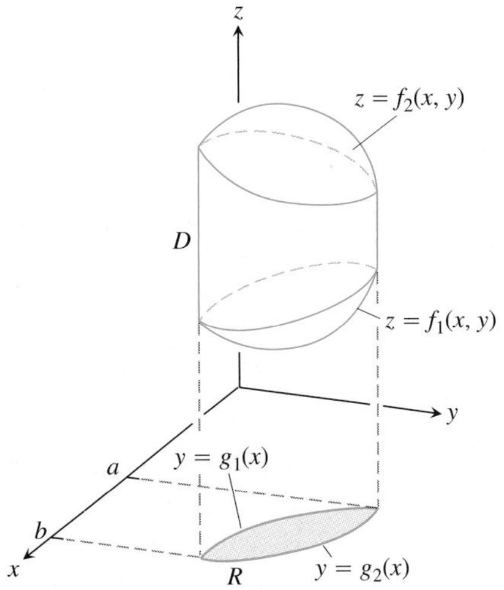

2. *Find the z-limits of integration:* Draw a line M passing through a typical point (x, y) in R parallel to the z-axis. As z increases, M enters D at $z = f_1(x, y)$ and leaves at $z = f_2(x, y)$. These are the z-limits of integration.

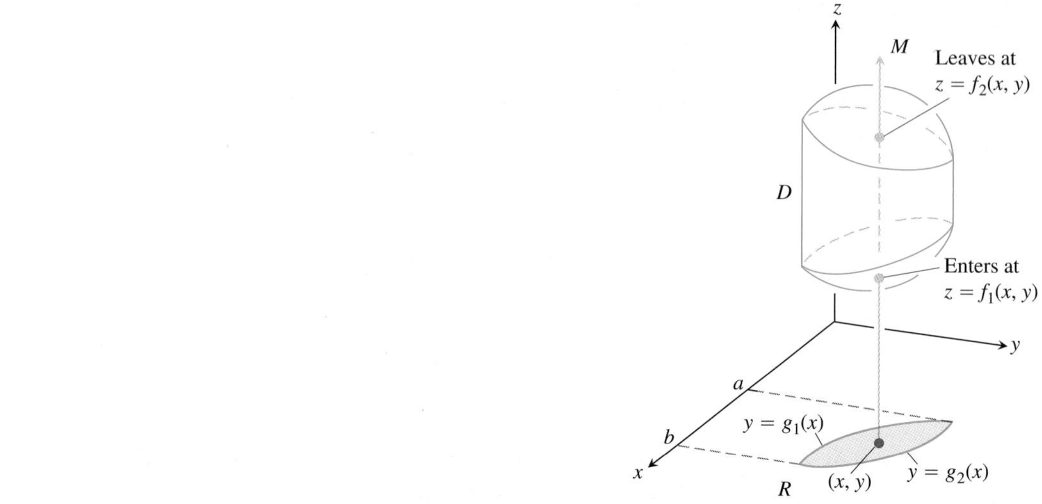

3. *Find the y-limits of integration:* Draw a line L through (x, y) parallel to the y-axis. As y increases, L enters R at $y = g_1(x)$ and leaves at $y = g_2(x)$. These are the y-limits of integration.

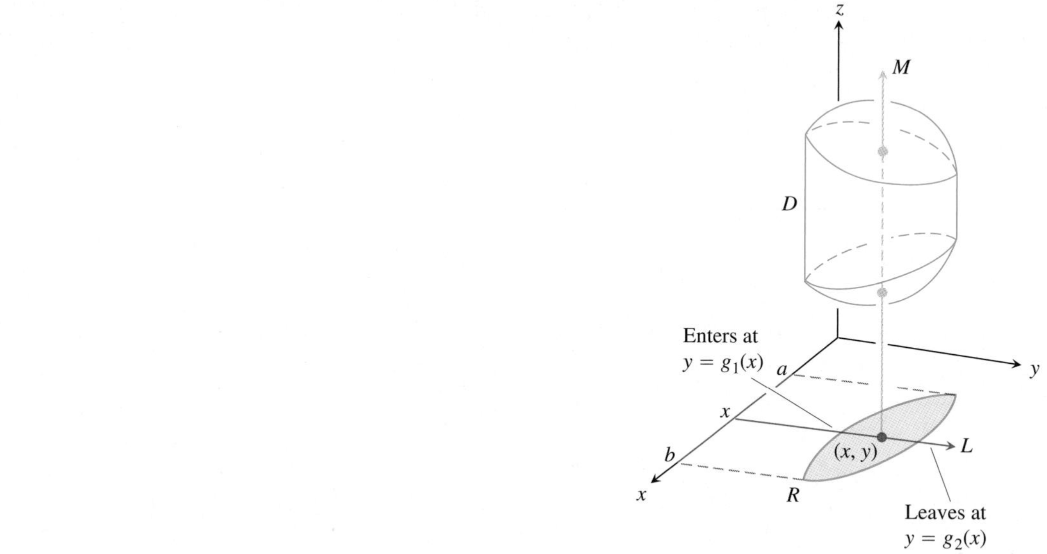

4. *Find the x-limits of integration:* Choose x-limits that include all lines through R parallel to the y-axis ($x = a$ and $x = b$ in the preceding figure). These are the x-limits of integration. The integral is

$$\int_{x=a}^{x=b} \int_{y=g_1(x)}^{y=g_2(x)} \int_{z=f_1(x, y)}^{z=f_2(x, y)} F(x, y, z)\, dz\, dy\, dx.$$

Follow similar procedures if you change the order of integration. The "shadow" of region D lies in the plane of the last two variables with respect to which the iterated integration takes place.

The preceding procedure applies whenever a solid region D is bounded above and below by a surface, and when the "shadow" region R is bounded by a lower and upper curve. It does not apply to regions with complicated holes through them, although sometimes such regions can be subdivided into simpler regions for which the procedure does apply.

EXAMPLE 1 Find the volume of the region D enclosed by the surfaces $z = x^2 + 3y^2$ and $z = 8 - x^2 - y^2$.

Solution The volume is

$$V = \iiint_D dz\, dy\, dx,$$

the integral of $F(x, y, z) = 1$ over D. To find the limits of integration for evaluating the integral, we first sketch the region. The surfaces (Figure 12.24) intersect on the elliptical cylinder $x^2 + 3y^2 = 8 - x^2 - y^2$ or $x^2 + 2y^2 = 4,\ z > 0$. The boundary of the region R, the projection of D onto the xy-plane, is an ellipse with the same equation: $x^2 + 2y^2 = 4$. The "upper" boundary of R is the curve $y = \sqrt{(4 - x^2)/2}$. The lower boundary is the curve $y = -\sqrt{(4 - x^2)/2}$.

Now we find the z-limits of integration. The line M passing through a typical point (x, y) in R parallel to the z-axis enters D at $z = x^2 + 3y^2$ and leaves at $z = 8 - x^2 - y^2$.

Next we find the y-limits of integration. The line L through (x, y) parallel to the y-axis enters R at $y = -\sqrt{(4 - x^2)/2}$ and leaves at $y = \sqrt{(4 - x^2)/2}$.

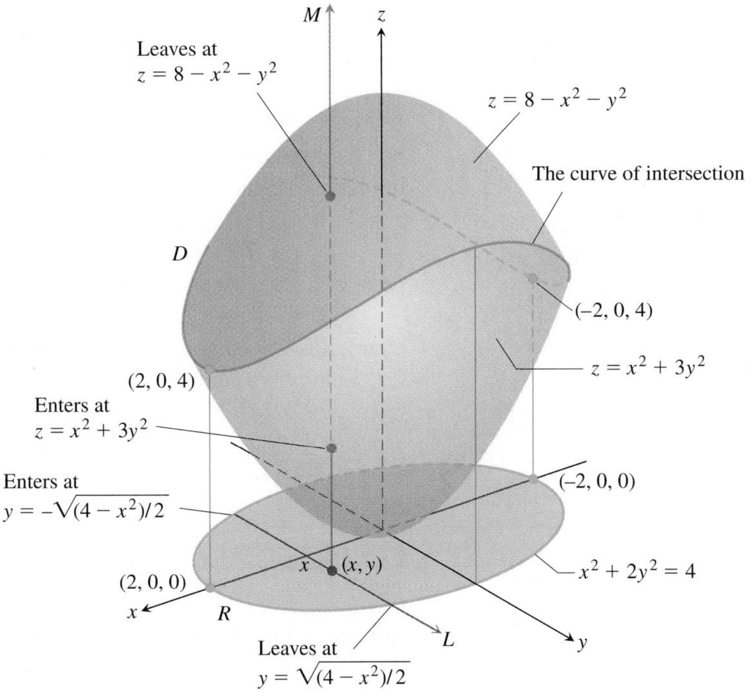

FIGURE 12.24 The volume of the region enclosed by two paraboloids, calculated in Example 1.

Finally we find the x-limits of integration. As L sweeps across R, the value of x varies from $x = -2$ at $(-2, 0, 0)$ to $x = 2$ at $(2, 0, 0)$. The volume of D is

$$\begin{aligned}
V &= \iiint_D dz\,dy\,dx \\
&= \int_{-2}^{2}\int_{-\sqrt{(4-x^2)/2}}^{\sqrt{(4-x^2)/2}}\int_{x^2+3y^2}^{8-x^2-y^2} dz\,dy\,dx \\
&= \int_{-2}^{2}\int_{-\sqrt{(4-x^2)/2}}^{\sqrt{(4-x^2)/2}} (8 - 2x^2 - 4y^2)\,dy\,dx \\
&= \int_{-2}^{2}\left[(8 - 2x^2)y - \frac{4}{3}y^3\right]_{y=-\sqrt{(4-x^2)/2}}^{y=\sqrt{(4-x^2)/2}} dx \\
&= \int_{-2}^{2}\left(2(8 - 2x^2)\sqrt{\frac{4 - x^2}{2}} - \frac{8}{3}\left(\frac{4 - x^2}{2}\right)^{3/2}\right) dx \\
&= \int_{-2}^{2}\left[8\left(\frac{4 - x^2}{2}\right)^{3/2} - \frac{8}{3}\left(\frac{4 - x^2}{2}\right)^{3/2}\right] dx = \frac{4\sqrt{2}}{3}\int_{-2}^{2}(4 - x^2)^{3/2}\,dx \\
&= 8\pi\sqrt{2}. \qquad \text{After integration with the substitution } x = 2\sin u.
\end{aligned}$$

■

In the next example, we project D onto the xz-plane instead of the xy-plane, to show how to use a different order of integration.

EXAMPLE 2 Set up the limits of integration for evaluating the triple integral of a function $F(x, y, z)$ over the tetrahedron D with vertices $(0, 0, 0)$, $(1, 1, 0)$, $(0, 1, 0)$, and $(0, 1, 1)$. Use the order of integration $dy\,dz\,dx$.

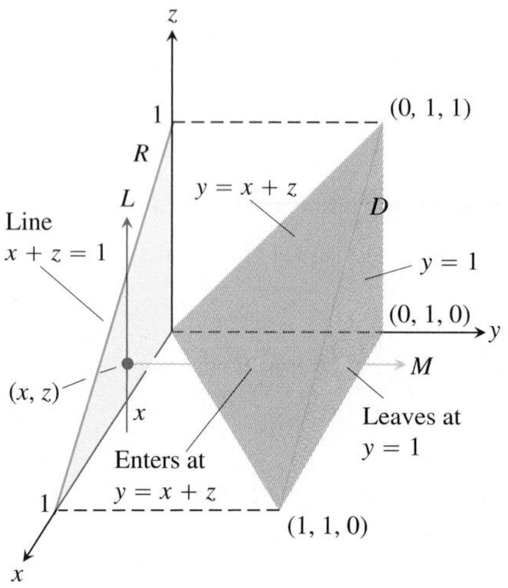

FIGURE 12.25 Finding the limits of integration for evaluating the triple integral of a function defined over the tetrahedron D (Examples 2 and 3).

Solution We sketch D along with its "shadow" R in the xz-plane (Figure 12.25). The upper (right-hand) bounding surface of D lies in the plane $y = 1$. The lower (left-hand) bounding surface lies in the plane $y = x + z$. The upper boundary of R is the line $z = 1 - x$. The lower boundary is the line $z = 0$.

First we find the y-limits of integration. The line through a typical point (x, z) in R parallel to the y-axis enters D at $y = x + z$ and leaves at $y = 1$.

Next we find the z-limits of integration. The line L through (x, z) parallel to the z-axis enters R at $z = 0$ and leaves at $z = 1 - x$.

Finally we find the x-limits of integration. As L sweeps across R, the value of x varies from $x = 0$ to $x = 1$. The integral is

$$\int_0^1\int_0^{1-x}\int_{x+z}^{1} F(x, y, z)\,dy\,dz\,dx.$$

■

EXAMPLE 3 Integrate $F(x, y, z)$ over the tetrahedron D in Example 2 in the order $dz\,dy\,dx$.

Solution First we find the z-limits of integration. A line parallel to the z-axis through a typical point (x, y) in the xy-plane "shadow" enters the tetrahedron at $z = 0$ and exits through the upper plane where $z = y - x$ (Figure 12.25).

Next we find the y-limits of integration. On the xy-plane, where $z = 0$, the sloped side of the tetrahedron crosses the plane along the line $y = x$. A line through (x, y) parallel to the y-axis enters the shadow in the xy-plane at $y = x$ and exits at $y = 1$.

Finally we find the x-limits of integration. As the line parallel to the y-axis in the previous step sweeps out the shadow, the value of x varies from $x = 0$ to $x = 1$ at the point (1, 1, 0). The integral is

$$\int_0^1 \int_x^1 \int_0^{y-x} F(x, y, z)\, dz\, dy\, dx.$$

For example, if $F(x, y, z) = 1$, we would find the volume of the tetrahedron to be

$$\begin{aligned} V &= \int_0^1 \int_x^1 \int_0^{y-x} dz\, dy\, dx \\ &= \int_0^1 \int_x^1 (y - x)\, dy\, dx \\ &= \int_0^1 \left[\frac{1}{2}y^2 - xy\right]_{y=x}^{y=1} dx \\ &= \int_0^1 \left(\frac{1}{2} - x + \frac{1}{2}x^2\right) dx \\ &= \left[\frac{1}{2}x - \frac{1}{2}x^2 + \frac{1}{6}x^3\right]_0^1 \\ &= \frac{1}{6}. \end{aligned}$$

We get the same result by integrating with the order $dy\, dz\, dx$. From Example 2,

$$V = \int_0^1 \int_0^{1-x} \int_{x+z}^1 dy\, dz\, dx = \frac{1}{6}.$$

■

Average Value of a Function in Space

The average value of a function F over a region D in space is defined by the formula

$$\textbf{Average value} \text{ of } F \text{ over } D = \frac{1}{\text{volume of } D} \iiint_D F\, dV. \tag{2}$$

For example, if $F(x, y, z) = \sqrt{x^2 + y^2 + z^2}$, then the average value of F over D is the average distance of points in D from the origin. If $F(x, y, z)$ is the temperature at (x, y, z) on a solid that occupies a region D in space, then the average value of F over D is the average temperature of the solid.

EXAMPLE 4 Find the average value of $F(x, y, z) = xyz$ over the cube bounded by the coordinate planes and the planes $x = 2$, $y = 2$, and $z = 2$ in the first octant.

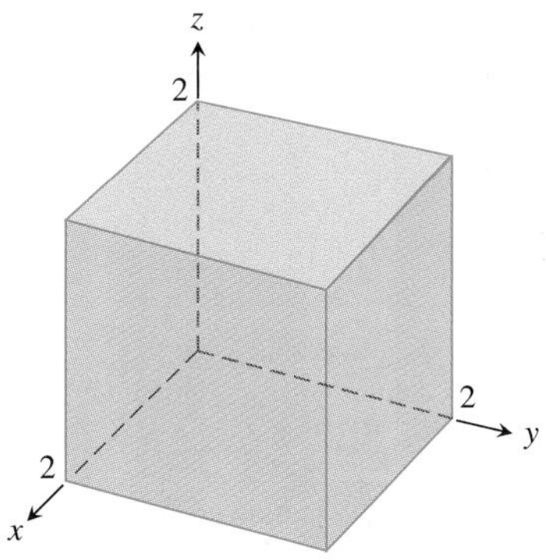

FIGURE 12.26 The region of integration in Example 4.

Solution We sketch the cube with enough detail to show the limits of integration (Figure 12.26). We then use Equation (2) to calculate the average value of F over the cube.

The volume of the cube is $(2)(2)(2) = 8$. The value of the integral of F over the cube is

$$\begin{aligned} \int_0^2 \int_0^2 \int_0^2 xyz\, dx\, dy\, dz &= \int_0^2 \int_0^2 \left[\frac{x^2}{2}yz\right]_{x=0}^{x=2} dy\, dz = \int_0^2 \int_0^2 2yz\, dy\, dz \\ &= \int_0^2 \left[y^2 z\right]_{y=0}^{y=2} dz = \int_0^2 4z\, dz = \left[2z^2\right]_0^2 = 8. \end{aligned}$$

With these values, Equation (2) gives

$$\begin{array}{c}\text{Average value of} \\ xyz \text{ over the cube}\end{array} = \frac{1}{\text{volume}} \iiint_{\text{cube}} xyz\, dV = \left(\frac{1}{8}\right)(8) = 1.$$

In evaluating the integral, we chose the order $dx\, dy\, dz$, but any of the other five possible orders would have done as well. ■

Properties of Triple Integrals

Triple integrals have the same algebraic properties as double and single integrals. Simply replace the double integrals in the four properties given in Section 12.2, page 667, with triple integrals.

EXERCISES 12.5

1. Evaluate the integral in Example 2 taking $F(x, y, z) = 1$ to find the volume of the tetrahedron.
2. **Volume of rectangular solid** Write six different iterated triple integrals for the volume of the rectangular solid in the first octant bounded by the coordinate planes and the planes $x = 1, y = 2$, and $z = 3$. Evaluate one of the integrals.
3. **Volume of tetrahedron** Write six different iterated triple integrals for the volume of the tetrahedron cut from the first octant by the plane $6x + 3y + 2z = 6$. Evaluate one of the integrals.
4. **Volume of solid** Write six different iterated triple integrals for the volume of the region in the first octant enclosed by the cylinder $x^2 + z^2 = 4$ and the plane $y = 3$. Evaluate one of the integrals.
5. **Volume enclosed by paraboloids** Let D be the region bounded by the paraboloids $z = 8 - x^2 - y^2$ and $z = x^2 + y^2$. Write six different triple iterated integrals for the volume of D. Evaluate one of the integrals.
6. **Volume inside paraboloid beneath a plane** Let D be the region bounded by the paraboloid $z = x^2 + y^2$ and the plane $z = 2y$. Write triple iterated integrals in the order $dz\, dx\, dy$ and $dz\, dy\, dx$ that give the volume of D. Do not evaluate either integral.

Evaluate the integrals in Exercises 7–20.

7. $\displaystyle\int_0^1\int_0^1\int_0^1 (x^2 + y^2 + z^2)\, dz\, dy\, dx$
8. $\displaystyle\int_0^{\sqrt{2}}\int_0^{3y}\int_{x^2+3y^2}^{8-x^2-y^2} dz\, dx\, dy$
9. $\displaystyle\int_1^e\int_1^e\int_1^e \frac{1}{xyz}\, dx\, dy\, dz$
10. $\displaystyle\int_0^1\int_0^{3-3x}\int_0^{3-3x-y} dz\, dy\, dx$
11. $\displaystyle\int_0^1\int_0^{\pi}\int_0^{\pi} y\sin z\, dx\, dy\, dz$
12. $\displaystyle\int_{-1}^1\int_{-1}^1\int_{-1}^1 (x + y + z)\, dy\, dx\, dz$
13. $\displaystyle\int_0^3\int_0^{\sqrt{9-x^2}}\int_0^{\sqrt{9-x^2}} dz\, dy\, dx$
14. $\displaystyle\int_0^2\int_{-\sqrt{4-y^2}}^{\sqrt{4-y^2}}\int_0^{2x+y} dz\, dx\, dy$
15. $\displaystyle\int_0^1\int_0^{2-x}\int_0^{2-x-y} dz\, dy\, dx$
16. $\displaystyle\int_0^1\int_0^{1-x^2}\int_3^{4-x^2-y} x\, dz\, dy\, dx$
17. $\displaystyle\int_0^{\pi}\int_0^{\pi}\int_0^{\pi} \cos(u + v + w)\, du\, dv\, dw$ (uvw-space)
18. $\displaystyle\int_1^e\int_1^e\int_1^e \ln r \ln s \ln t\, dt\, dr\, ds$ (rst-space)
19. $\displaystyle\int_0^{\pi/4}\int_0^{\ln \sec v}\int_{-\infty}^{2t} e^x\, dx\, dt\, dv$ (tvx-space)
20. $\displaystyle\int_0^7\int_0^2\int_0^{\sqrt{4-q^2}} \frac{q}{r+1}\, dp\, dq\, dr$ (pqr-space)
21. Here is the region of integration of the integral

$$\int_{-1}^1\int_{x^2}^1\int_0^{1-y} dz\, dy\, dx.$$

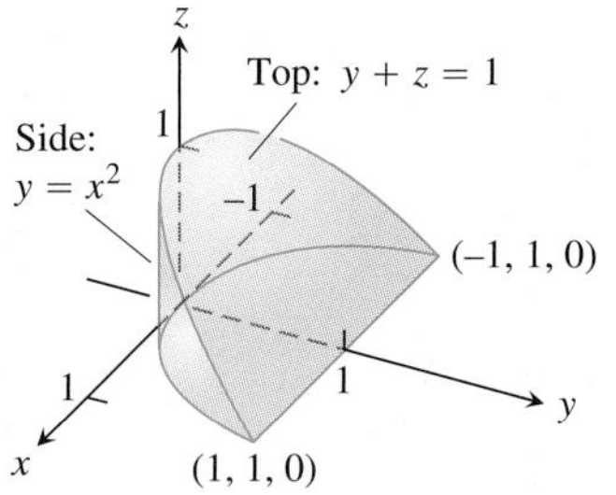

Rewrite the integral as an equivalent iterated integral in the order

a. $dy\, dz\, dx$
b. $dy\, dx\, dz$
c. $dx\, dy\, dz$
d. $dx\, dz\, dy$
e. $dz\, dx\, dy$.

22. Here is the region of integration of the integral

$$\int_0^1 \int_{-1}^0 \int_0^{y^2} dz\, dy\, dx.$$

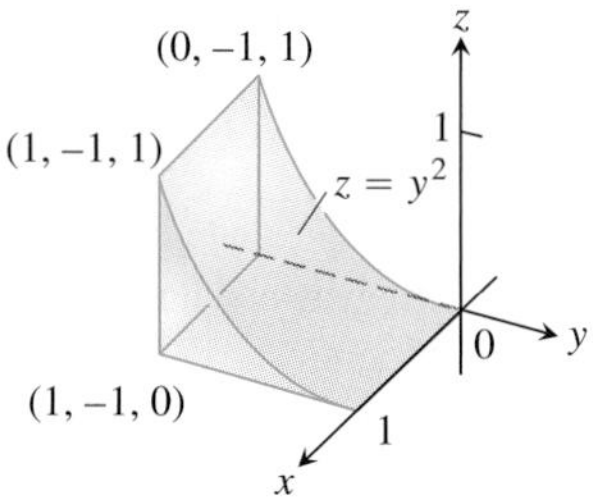

Rewrite the integral as an equivalent iterated integral in the order

a. $dy\, dz\, dx$ **b.** $dy\, dx\, dz$

c. $dx\, dy\, dz$ **d.** $dx\, dz\, dy$

e. $dz\, dx\, dy$.

Find the volumes of the regions in Exercises 23–36.

23. The region between the cylinder $z = y^2$ and the xy-plane that is bounded by the planes $x = 0, x = 1, y = -1, y = 1$

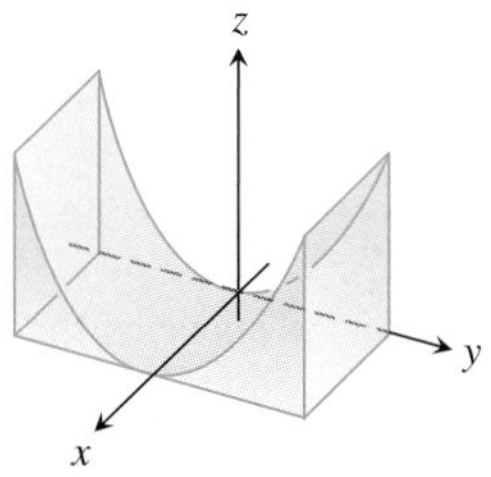

24. The region in the first octant bounded by the coordinate planes and the planes $x + z = 1, y + 2z = 2$

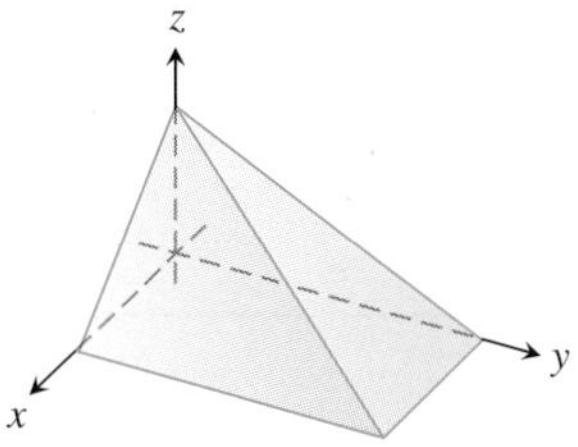

25. The region in the first octant bounded by the coordinate planes, the plane $y + z = 2$, and the cylinder $x = 4 - y^2$

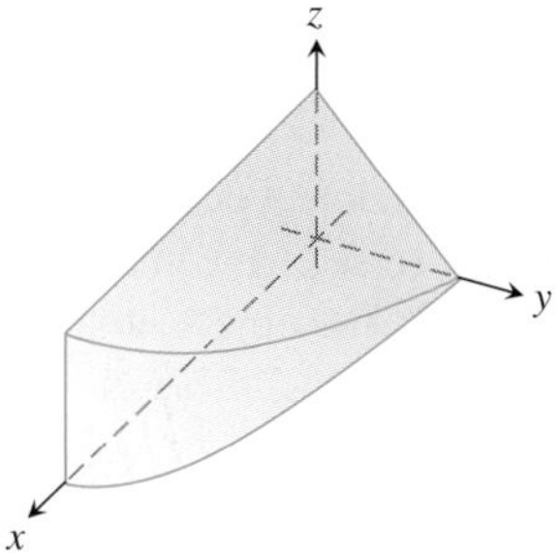

26. The wedge cut from the cylinder $x^2 + y^2 = 1$ by the planes $z = -y$ and $z = 0$

27. The tetrahedron in the first octant bounded by the coordinate planes and the plane passing through (1, 0, 0), (0, 2, 0), and (0, 0, 3)

28. The region in the first octant bounded by the coordinate planes, the plane $y = 1 - x$, and the surface $z = \cos(\pi x/2), 0 \le x \le 1$

29. The region common to the interiors of the cylinders $x^2 + y^2 = 1$ and $x^2 + z^2 = 1$, one-eighth of which is shown in the accompanying figure.

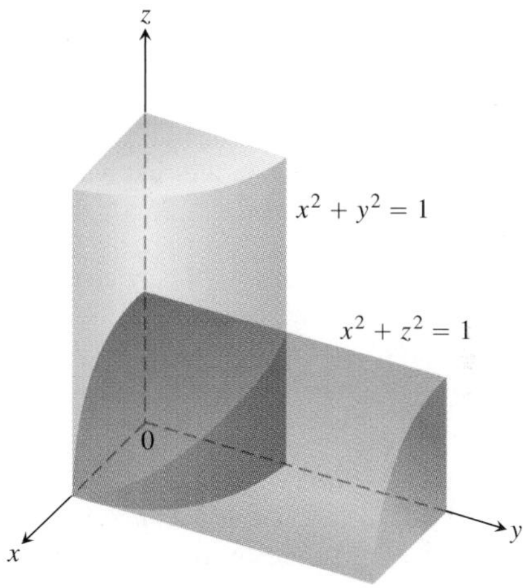

30. The region in the first octant bounded by the coordinate planes and the surface $z = 4 - x^2 - y$

31. The region in the first octant bounded by the coordinate planes, the plane $x + y = 4$, and the cylinder $y^2 + 4z^2 = 16$

32. The region cut from the cylinder $x^2 + y^2 = 4$ by the plane $z = 0$ and the plane $x + z = 3$

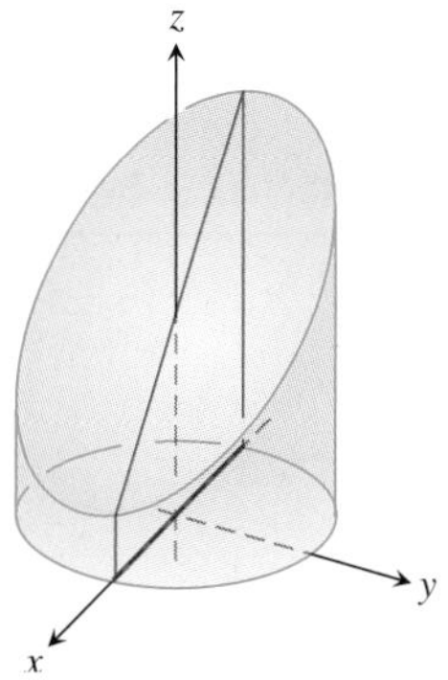

33. The region between the planes $x + y + 2z = 2$ and $2x + 2y + z = 4$ in the first octant

34. The finite region bounded by the planes $z = x, x + z = 8, z = y, y = 8$, and $z = 0$

35. The region cut from the solid elliptical cylinder $x^2 + 4y^2 \le 4$ by the xy-plane and the plane $z = x + 2$

36. The region bounded in back by the plane $x = 0$, on the front and sides by the parabolic cylinder $x = 1 - y^2$, on the top by the paraboloid $z = x^2 + y^2$, and on the bottom by the xy-plane

In Exercises 37–40, find the average value of $F(x, y, z)$ over the given region.

37. $F(x, y, z) = x^2 + 9$ over the cube in the first octant bounded by the coordinate planes and the planes $x = 2$, $y = 2$, and $z = 2$

38. $F(x, y, z) = x + y - z$ over the rectangular solid in the first octant bounded by the coordinate planes and the planes $x = 1$, $y = 1$, and $z = 2$

39. $F(x, y, z) = x^2 + y^2 + z^2$ over the cube in the first octant bounded by the coordinate planes and the planes $x = 1$, $y = 1$, and $z = 1$

40. $F(x, y, z) = xyz$ over the cube in the first octant bounded by the coordinate planes and the planes $x = 2$, $y = 2$, and $z = 2$

Evaluate the integrals in Exercises 41–44 by changing the order of integration in an appropriate way.

41. $\displaystyle\int_0^4\int_0^1\int_{2y}^2 \frac{4\cos(x^2)}{2\sqrt{z}}\,dx\,dy\,dz$

42. $\displaystyle\int_0^1\int_0^1\int_{x^2}^1 12xze^{zy^2}\,dy\,dx\,dz$

43. $\displaystyle\int_0^1\int_{\sqrt[3]{z}}^1\int_0^{\ln 3} \frac{\pi e^{2x}\sin \pi y^2}{y^2}\,dx\,dy\,dz$

44. $\displaystyle\int_0^2\int_0^{4-x^2}\int_0^x \frac{\sin 2z}{4 - z}\,dy\,dz\,dx$

45. Finding an upper limit of an iterated integral Solve for a:

$$\int_0^1\int_0^{4-a-x^2}\int_a^{4-x^2-y} dz\,dy\,dx = \frac{4}{15}.$$

46. Ellipsoid For what value of c is the volume of the ellipsoid $x^2 + (y/2)^2 + (z/c)^2 = 1$ equal to 8π?

47. Minimizing a triple integral What domain D in space minimizes the value of the integral

$$\iiint_D (4x^2 + 4y^2 + z^2 - 4)\,dV\,?$$

Give reasons for your answer.

48. Maximizing a triple integral What domain D in space maximizes the value of the integral

$$\iiint_D (1 - x^2 - y^2 - z^2)\,dV\,?$$

Give reasons for your answer.

COMPUTER EXPLORATIONS

In Exercises 49–52, use a CAS integration utility to evaluate the triple integral of the given function over the specified solid region.

49. $F(x, y, z) = x^2y^2z$ over the solid cylinder bounded by $x^2 + y^2 = 1$ and the planes $z = 0$ and $z = 1$

50. $F(x, y, z) = |xyz|$ over the solid bounded below by the paraboloid $z = x^2 + y^2$ and above by the plane $z = 1$

51. $F(x, y, z) = \dfrac{z}{(x^2 + y^2 + z^2)^{3/2}}$ over the solid bounded below by the cone $z = \sqrt{x^2 + y^2}$ and above by the plane $z = 1$

52. $F(x, y, z) = x^4 + y^2 + z^2$ over the solid sphere $x^2 + y^2 + z^2 \le 1$

12.6 Moments and Centers of Mass

This section shows how to calculate the masses and moments of two- and three-dimensional objects in Cartesian coordinates. Section 12.7 gives the calculations for cylindrical and spherical coordinates.

Masses and First Moments

If $\delta(x, y, z)$ is the density (mass per unit volume) of an object occupying a region D in space, the integral of δ over D gives the **mass** of the object. To see why, imagine partitioning the object into n mass elements like the one in Figure 12.27. The object's mass is the limit

$$M = \lim_{n\to\infty}\sum_{k=1}^{n}\Delta m_k = \lim_{n\to\infty}\sum_{k=1}^{n}\delta(x_k, y_k, z_k)\,\Delta V_k = \iiint_D \delta(x, y, z)\,dV.$$

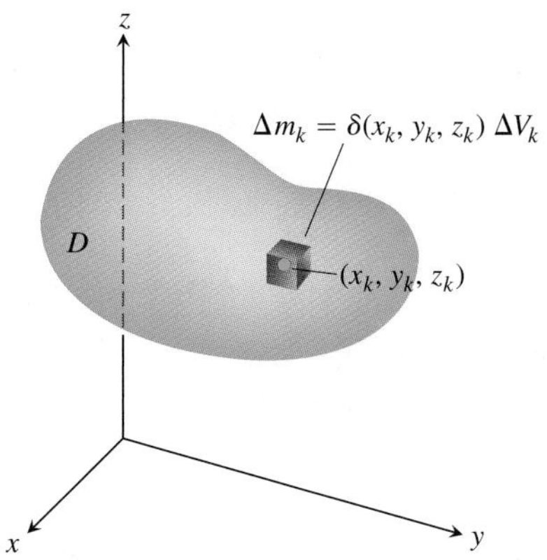

FIGURE 12.27 To define an object's mass, we first imagine it to be partitioned into a finite number of mass elements Δm_k.

The *first moment* of a solid region D about a coordinate plane is defined as the triple integral over D of the distance from a point (x, y, z) in D to the plane multiplied by the density of the solid at that point. For instance, the first moment about the yz-plane is the integral

$$M_{yz} = \iiint_D x\delta(x, y, z)\, dV.$$

The *center of mass* is found from the first moments. For instance, the x-coordinate of the center of mass is $\bar{x} = M_{yz}/M$.

For a two-dimensional object, such as a thin flat plate, we calculate first moments about the coordinate axes by simply dropping the z-coordinate. So the first moment about the y-axis is the double integral over the region R forming the plate of the distance from the axis multiplied by the density, or

$$M_y = \iint_R x\delta(x, y)\, dA.$$

Table 12.1 summarizes the formulas.

TABLE 12.1 Mass and first moment formulas

THREE-DIMENSIONAL SOLID

Mass: $M = \iiint_D \delta\, dV$ $\quad\delta = \delta(x, y, z)$ is the density at (x, y, z).

First moments about the coordinate planes:

$$M_{yz} = \iiint_D x\,\delta\, dV, \qquad M_{xz} = \iiint_D y\,\delta\, dV, \qquad M_{xy} = \iiint_D z\,\delta\, dV$$

Center of mass:

$$\bar{x} = \frac{M_{yz}}{M}, \qquad \bar{y} = \frac{M_{xz}}{M}, \qquad \bar{z} = \frac{M_{xy}}{M}$$

TWO-DIMENSIONAL PLATE

Mass: $M = \iint_R \delta(x, y)\, dA$ $\quad\delta(x, y)$ is the density at (x, y).

First moments: $M_y = \iint_R x\,\delta(x, y)\, dA, \qquad M_x = \iint_R y\,\delta(x, y)\, dA$

Center of mass: $\bar{x} = \dfrac{M_y}{M}, \qquad \bar{y} = \dfrac{M_x}{M}$

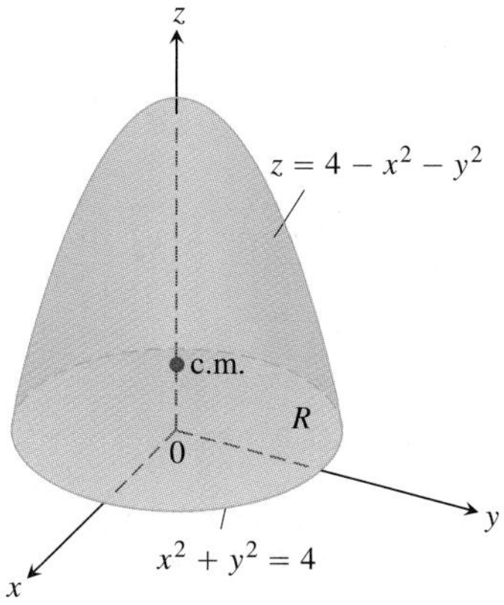

FIGURE 12.28 Finding the center of mass of a solid (Example 1).

EXAMPLE 1 Find the center of mass of a solid of constant density δ bounded below by the disk $R: x^2 + y^2 \le 4$ in the plane $z = 0$ and above by the paraboloid $z = 4 - x^2 - y^2$ (Figure 12.28).

Solution By symmetry $\bar{x} = \bar{y} = 0$. To find $\bar{z}$, we first calculate

$$M_{xy} = \iint_R \int_{z=0}^{z=4-x^2-y^2} z\,\delta\,dz\,dy\,dx = \iint_R \left[\frac{z^2}{2}\right]_{z=0}^{z=4-x^2-y^2} \delta\,dy\,dx$$

$$= \frac{\delta}{2}\iint_R (4 - x^2 - y^2)^2\,dy\,dx$$

$$= \frac{\delta}{2}\int_0^{2\pi}\int_0^2 (4 - r^2)^2 r\,dr\,d\theta \qquad \text{Polar coordinates simplify the integration.}$$

$$= \frac{\delta}{2}\int_0^{2\pi}\left[-\frac{1}{6}(4 - r^2)^3\right]_{r=0}^{r=2} d\theta = \frac{16\delta}{3}\int_0^{2\pi} d\theta = \frac{32\pi\delta}{3}.$$

A similar calculation gives the mass

$$M = \iint_R \int_0^{4-x^2-y^2} \delta\,dz\,dy\,dx = 8\pi\delta.$$

Therefore $\bar{z} = (M_{xy}/M) = 4/3$ and the center of mass is $(\bar{x}, \bar{y}, \bar{z}) = (0, 0, 4/3)$. ■

When the density of a solid object or plate is constant (as in Example 1), the center of mass is called the **centroid** of the object. To find a centroid, we set δ equal to 1 and proceed to find $\bar{x}$, $\bar{y}$, and $\bar{z}$ as before, by dividing first moments by masses.

EXAMPLE 2 Find the centroid of the region in the first quadrant that is bounded above by the line $y = x$ and below by the parabola $y = x^2$. These calculations are also valid for two-dimensional objects.

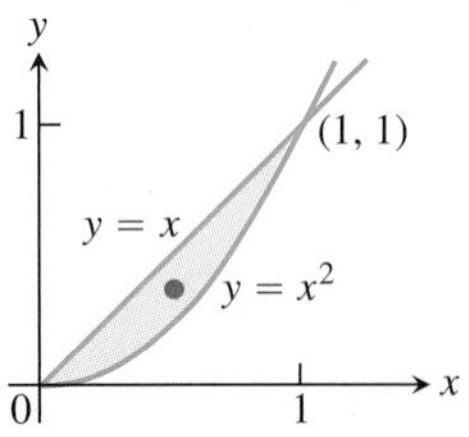

FIGURE 12.29 The centroid of this region is found in Example 2.

Solution We sketch the region and include enough detail to determine the limits of integration (Figure 12.29). We then set δ equal to 1 and evaluate the appropriate formulas from Table 12.1:

$$M = \int_0^1\int_{x^2}^{x} 1\,dy\,dx = \int_0^1 \left[y\right]_{y=x^2}^{y=x} dx = \int_0^1 (x - x^2)\,dx = \left[\frac{x^2}{2} - \frac{x^3}{3}\right]_0^1 = \frac{1}{6}$$

$$M_x = \int_0^1\int_{x^2}^{x} y\,dy\,dx = \int_0^1 \left[\frac{y^2}{2}\right]_{y=x^2}^{y=x} dx$$

$$= \int_0^1 \left(\frac{x^2}{2} - \frac{x^4}{2}\right) dx = \left[\frac{x^3}{6} - \frac{x^5}{10}\right]_0^1 = \frac{1}{15}$$

$$M_y = \int_0^1\int_{x^2}^{x} x\,dy\,dx = \int_0^1 \left[xy\right]_{y=x^2}^{y=x} dx = \int_0^1 (x^2 - x^3)\,dx = \left[\frac{x^3}{3} - \frac{x^4}{4}\right]_0^1 = \frac{1}{12}.$$

From these values of M, M_x, and M_y, we find

$$\bar{x} = \frac{M_y}{M} = \frac{1/12}{1/6} = \frac{1}{2} \quad \text{and} \quad \bar{y} = \frac{M_x}{M} = \frac{1/15}{1/6} = \frac{2}{5}.$$

The centroid is the point $(1/2, 2/5)$. ■

Moments of Inertia

A body's first moments (Table 12.1) tell us about balance and about the torque the body exerts about different axes in a gravitational field. If the body is a rotating shaft, however,

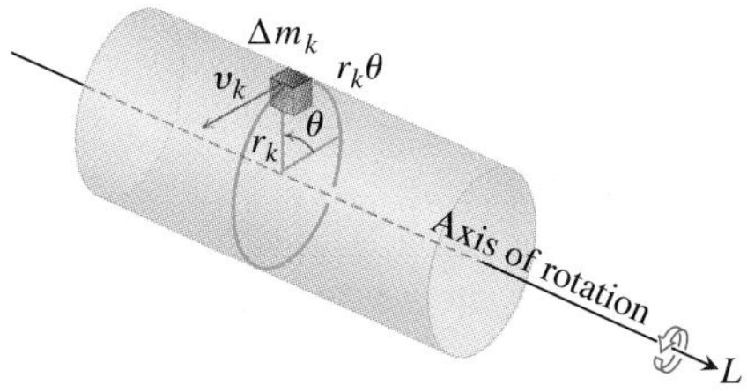

FIGURE 12.30 To find an integral for the amount of energy stored in a rotating shaft, we first imagine the shaft to be partitioned into small blocks. Each block has its own kinetic energy. We add the contributions of the individual blocks to find the kinetic energy of the shaft.

we are more likely to be interested in how much energy is stored in the shaft or about how much energy it will take to accelerate the shaft to a particular angular velocity. This is where the second moment or moment of inertia comes in.

Think of partitioning the shaft into small blocks of mass Δm_k and let r_k denote the distance from the kth block's center of mass to the axis of rotation (Figure 12.30). If the shaft rotates at a constant angular velocity of $\omega = d\theta/dt$ radians per second, the block's center of mass will trace its orbit at a linear speed of

$$v_k = \frac{d}{dt}(r_k\theta) = r_k\frac{d\theta}{dt} = r_k\omega.$$

The block's kinetic energy will be approximately

$$\frac{1}{2}\Delta m_k v_k^2 = \frac{1}{2}\Delta m_k(r_k\omega)^2 = \frac{1}{2}\omega^2 r_k^2\,\Delta m_k.$$

The kinetic energy of the shaft will be approximately

$$\sum \frac{1}{2}\omega^2 r_k^2\,\Delta m_k.$$

The integral approached by these sums as the shaft is partitioned into smaller and smaller blocks gives the shaft's kinetic energy:

$$\text{KE}_{\text{shaft}} = \int \frac{1}{2}\omega^2 r^2\,dm = \frac{1}{2}\omega^2\int r^2\,dm. \tag{1}$$

The factor

$$I = \int r^2\,dm$$

is the *moment of inertia* of the shaft about its axis of rotation, and we see from Equation (1) that the shaft's kinetic energy is

$$\text{KE}_{\text{shaft}} = \frac{1}{2}I\omega^2.$$

The moment of inertia of a shaft resembles in some ways the inertial mass of a locomotive. To start a locomotive with mass m moving at a linear velocity v, we need to provide a kinetic energy of $\text{KE} = (1/2)mv^2$. To stop the locomotive we have to remove this amount of energy. To start a shaft with moment of inertia I rotating at an angular velocity ω, we need to provide a kinetic energy of $\text{KE} = (1/2)I\omega^2$. To stop the shaft we have to take this amount of energy back out. The shaft's moment of inertia is analogous to the locomotive's mass. What makes the locomotive hard to start or stop is its mass. What makes the shaft hard to start or stop is its moment of inertia. The moment of inertia depends not only on the mass of the shaft, but also its distribution.

We now derive a formula for the moment of inertia for a solid in space. If $r(x, y, z)$ is the distance from the point (x, y, z) in D to a line L, then the moment of inertia of the mass $\Delta m_k = \delta(x_k, y_k, z_k)\Delta V_k$ about the line L (as in Figure 12.30) is approximately $\Delta I_k = r^2(x_k, y_k, z_k)\Delta m_k$. **The moment of inertia about *L*** of the entire object is

$$I_L = \lim_{n\to\infty}\sum_{k=1}^{n}\Delta I_k = \lim_{n\to\infty}\sum_{k=1}^{n} r^2(x_k, y_k, z_k)\,\delta(x_k, y_k, z_k)\,\Delta V_k = \iiint_D r^2\delta\,dV.$$

If L is the x-axis, then $r^2 = y^2 + z^2$ (Figure 12.31) and

$$I_x = \iiint_D (y^2 + z^2)\,\delta\,dV.$$

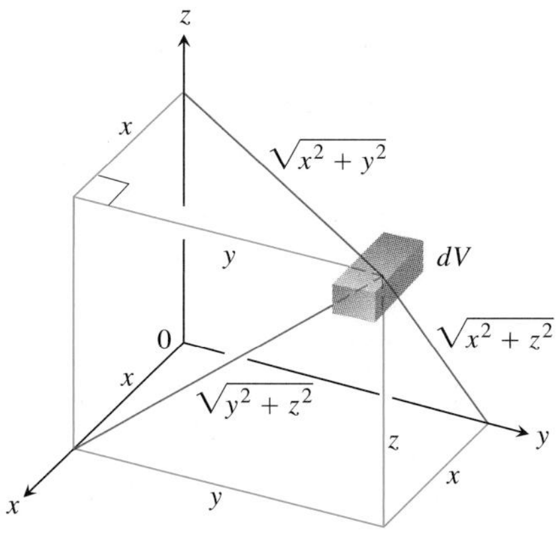

FIGURE 12.31 Distances from dV to the coordinate planes and axes.

Similarly, if L is the y-axis or z-axis we have

$$I_y = \iiint_D (x^2 + z^2)\,\delta\,dV \quad \text{and} \quad I_z = \iiint_D (x^2 + y^2)\,\delta\,dV.$$

Table 12.2 summarizes the formulas for these moments of inertia (second moments because they invoke the *squares* of the distances). It shows the definition of the *polar moment* about the origin as well.

TABLE 12.2 Moments of inertia (second moments) formulas

THREE-DIMENSIONAL SOLID		
About the x-axis:	$I_x = \iiint (y^2 + z^2)\,\delta\,dV$	$\delta = \delta(x, y, z)$
About the y-axis:	$I_y = \iiint (x^2 + z^2)\,\delta\,dV$	
About the z-axis:	$I_z = \iiint (x^2 + y^2)\,\delta\,dV$	
About a line L:	$I_L = \iiint r^2\,\delta\,dV$	$r(x, y, z)$ = distance from the point (x, y, z) to line L
TWO-DIMENSIONAL PLATE		
About the x-axis:	$I_x = \iint y^2\,\delta\,dA$	$\delta = \delta(x, y)$
About the y-axis:	$I_y = \iint x^2\,\delta\,dA$	
About a line L:	$I_L = \iint r^2(x, y)\,\delta\,dA,$	$r(x, y)$ = distance from (x, y) to L
About the origin (polar moment):	$I_0 = \iint (x^2 + y^2)\,\delta\,dA = I_x + I_y$	

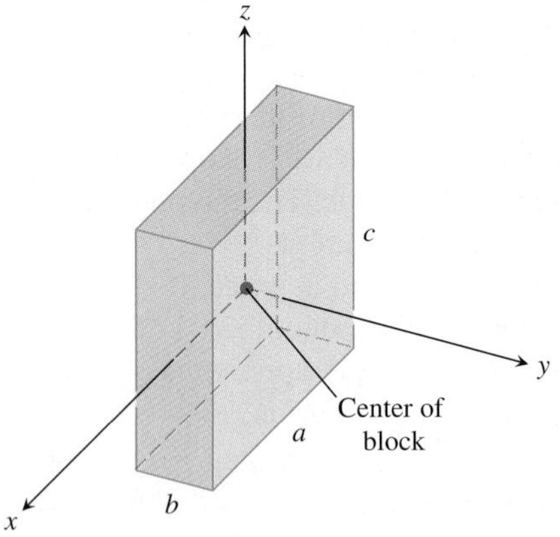

FIGURE 12.32 Finding I_x, I_y, and I_z for the block shown here. The origin lies at the center of the block (Example 3).

EXAMPLE 3 Find I_x, I_y, I_z for the rectangular solid of constant density δ shown in Figure 12.32.

Solution The formula for I_x gives

$$I_x = \int_{-c/2}^{c/2} \int_{-b/2}^{b/2} \int_{-a/2}^{a/2} (y^2 + z^2)\,\delta\,dx\,dy\,dz.$$

We can avoid some of the work of integration by observing that $(y^2 + z^2)\delta$ is an even function of x, y, and z. The rectangular solid consists of eight symmetric pieces, one in each octant. We can evaluate the integral on one of these pieces and then multiply by 8 to get the total value.

$$\begin{aligned} I_x &= 8\int_0^{c/2}\int_0^{b/2}\int_0^{a/2}(y^2+z^2)\,\delta\,dx\,dy\,dz = 4a\delta\int_0^{c/2}\int_0^{b/2}(y^2+z^2)\,dy\,dz \\ &= 4a\delta\int_0^{c/2}\left[\frac{y^3}{3}+z^2y\right]_{y=0}^{y=b/2}dz \\ &= 4a\delta\int_0^{c/2}\left(\frac{b^3}{24}+\frac{z^2b}{2}\right)dz \\ &= 4a\delta\left(\frac{b^3c}{48}+\frac{c^3b}{48}\right) = \frac{abc\delta}{12}(b^2+c^2) = \frac{M}{12}(b^2+c^2). \end{aligned}$$

Similarly,

$$I_y = \frac{M}{12}(a^2+c^2) \qquad \text{and} \qquad I_z = \frac{M}{12}(a^2+b^2).$$ ■

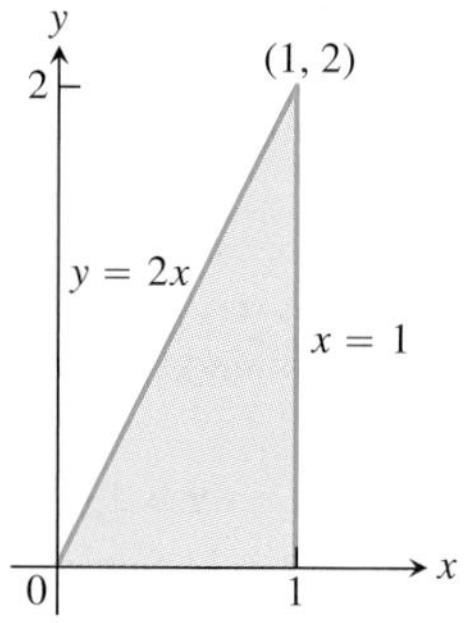

FIGURE 12.33 The triangular region covered by the plate in Example 4.

EXAMPLE 4 A thin plate covers the triangular region bounded by the x-axis and the lines $x = 1$ and $y = 2x$ in the first quadrant. The plate's density at the point (x, y) is $\delta(x, y) = 6x + 6y + 6$. Find the plate's moments of inertia about the coordinate axes and the origin.

Solution We sketch the plate and put in enough detail to determine the limits of integration for the integrals we have to evaluate (Figure 12.33). The moment of inertia about the x-axis is

$$\begin{aligned} I_x &= \int_0^1\int_0^{2x} y^2\delta(x,y)\,dy\,dx = \int_0^1\int_0^{2x}(6xy^2+6y^3+6y^2)\,dy\,dx \\ &= \int_0^1\left[2xy^3+\frac{3}{2}y^4+2y^3\right]_{y=0}^{y=2x}dx = \int_0^1(40x^4+16x^3)\,dx \\ &= \left[8x^5+4x^4\right]_0^1 = 12. \end{aligned}$$

Similarly, the moment of inertia about the y-axis is

$$I_y = \int_0^1\int_0^{2x} x^2\delta(x,y)\,dy\,dx = \frac{39}{5}.$$

Notice that we integrate y^2 times density in calculating I_x and x^2 times density to find I_y.

Since we know I_x and I_y, we do not need to evaluate an integral to find I_0; we can use the equation $I_0 = I_x + I_y$ from Table 12.2 instead:

$$I_0 = 12 + \frac{39}{5} = \frac{60+39}{5} = \frac{99}{5}.$$ ■

The moment of inertia also plays a role in determining how much a horizontal metal beam will bend under a load. The stiffness of the beam is a constant times I, the moment of inertia of a typical cross-section of the beam about the beam's longitudinal axis. The greater the value of I, the stiffer the beam and the less it will bend under a given load. That is why we use I-beams instead of beams whose cross-sections are square. The flanges at the top and bottom of the beam hold most of the beam's mass away from the longitudinal axis to maximize the value of I (Figure 12.34).

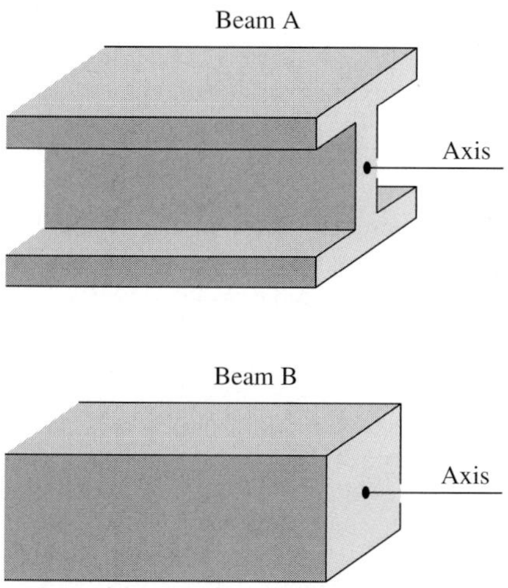

FIGURE 12.34 The greater the polar moment of inertia of the cross-section of a beam about the beam's longitudinal axis, the stiffer the beam. Beams A and B have the same cross-sectional area, but A is stiffer.

EXERCISES 12.6

1. **Finding center of mass** Find the center of mass of a thin plate of density $\delta = 3$ bounded by the lines $x = 0$, $y = x$, and the parabola $y = 2 - x^2$ in the first quadrant.

2. **Finding moments of inertia** Find the moments of inertia about the coordinate axes of a thin rectangular plate of constant density δ bounded by the lines $x = 3$ and $y = 3$ in the first quadrant.

3. **Finding a centroid** Find the centroid of the region in the first quadrant bounded by the x-axis, the parabola $y^2 = 2x$, and the line $x + y = 4$.

4. **Finding a centroid** Find the centroid of the triangular region cut from the first quadrant by the line $x + y = 3$.

5. **Finding a centroid** Find the centroid of the region cut from the first quadrant by the circle $x^2 + y^2 = a^2$.

6. **Finding a centroid** Find the centroid of the region between the x-axis and the arch $y = \sin x$, $0 \le x \le \pi$.

7. **Finding moments of inertia** Find the moment of inertia about the x-axis of a thin plate of density $\delta = 1$ bounded by the circle $x^2 + y^2 = 4$. Then use your result to find I_y and I_0 for the plate.

8. **Finding a moment of inertia** Find the moment of inertia with respect to the y-axis of a thin sheet of constant density $\delta = 1$ bounded by the curve $y = (\sin^2 x)/x^2$ and the interval $\pi \le x \le 2\pi$ of the x-axis.

9. **The centroid of an infinite region** Find the centroid of the infinite region in the second quadrant enclosed by the coordinate axes and the curve $y = e^x$. (Use improper integrals in the mass-moment formulas.)

10. **The first moment of an infinite plate** Find the first moment about the y-axis of a thin plate of density $\delta(x, y) = 1$ covering the infinite region under the curve $y = e^{-x^2/2}$ in the first quadrant.

11. **Finding a moment of inertia** Find the moment of inertia about the x-axis of a thin plate bounded by the parabola $x = y - y^2$ and the line $x + y = 0$ if $\delta(x, y) = x + y$.

12. **Finding mass** Find the mass of a thin plate occupying the smaller region cut from the ellipse $x^2 + 4y^2 = 12$ by the parabola $x = 4y^2$ if $\delta(x, y) = 5x$.

13. **Finding a center of mass** Find the center of mass of a thin triangular plate bounded by the y-axis and the lines $y = x$ and $y = 2 - x$ if $\delta(x, y) = 6x + 3y + 3$.

14. **Finding a center of mass and moment of inertia** Find the center of mass and moment of inertia about the x-axis of a thin plate bounded by the curves $x = y^2$ and $x = 2y - y^2$ if the density at the point (x, y) is $\delta(x, y) = y + 1$.

15. **Center of mass, moment of inertia** Find the center of mass and the moment of inertia about the y-axis of a thin rectangular plate cut from the first quadrant by the lines $x = 6$ and $y = 1$ if $\delta(x, y) = x + y + 1$.

16. **Center of mass, moment of inertia** Find the center of mass and the moment of inertia about the y-axis of a thin plate bounded by the line $y = 1$ and the parabola $y = x^2$ if the density is $\delta(x, y) = y + 1$.

17. **Center of mass, moment of inertia** Find the center of mass and the moment of inertia about the y-axis of a thin plate bounded by the x-axis, the lines $x = \pm 1$, and the parabola $y = x^2$ if $\delta(x, y) = 7y + 1$.

18. **Center of mass, moment of inertia** Find the center of mass and the moment of inertia about the x-axis of a thin rectangular plate bounded by the lines $x = 0$, $x = 20$, $y = -1$, and $y = 1$ if $\delta(x, y) = 1 + (x/20)$.

19. **Center of mass, moments of inertia** Find the center of mass, the moment of inertia about the coordinate axes, and the polar moment of inertia of a thin triangular plate bounded by the lines $y = x$, $y = -x$, and $y = 1$ if $\delta(x, y) = y + 1$.

20. **Center of mass, moments of inertia** Repeat Exercise 19 for $\delta(x, y) = 3x^2 + 1$.

21. **Moments of inertia** Find the moments of inertia of the rectangular solid shown here with respect to its edges by calculating I_x, I_y, and I_z.

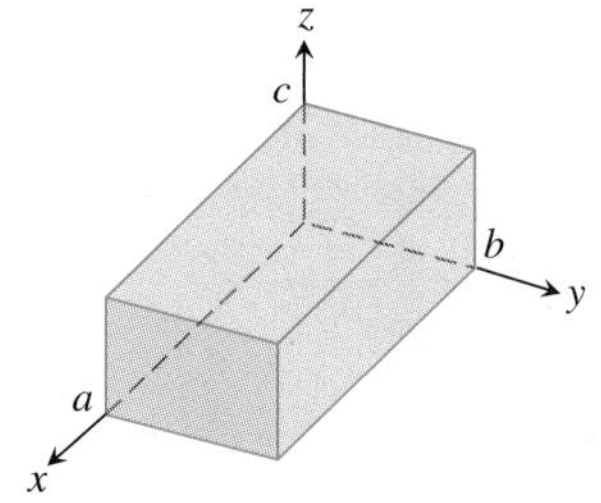

22. **Moments of inertia** The coordinate axes in the figure run through the centroid of a solid wedge parallel to the labeled edges. Find I_x, I_y, and I_z if $a = b = 6$ and $c = 4$.

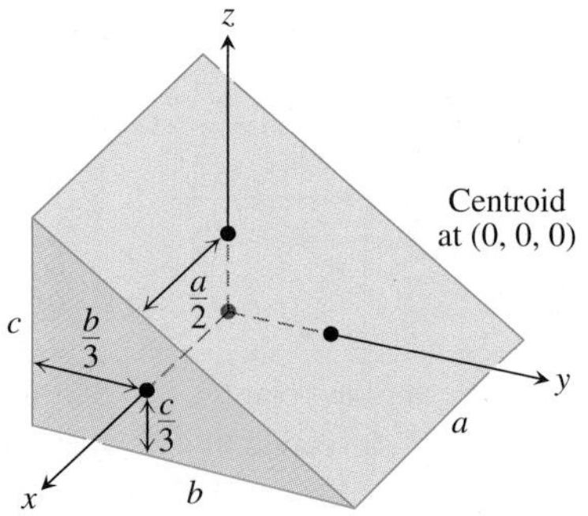

23. **Center of mass and moments of inertia** A solid "trough" of constant density is bounded below by the surface $z = 4y^2$, above

by the plane $z = 4$, and on the ends by the planes $x = 1$ and $x = -1$. Find the center of mass and the moments of inertia with respect to the three axes.

24. Center of mass A solid of constant density is bounded below by the plane $z = 0$, on the sides by the elliptical cylinder $x^2 + 4y^2 = 4$, and above by the plane $z = 2 - x$ (see the accompanying figure).

a. Find $\bar{x}$ and $\bar{y}$.

b. Evaluate the integral

$$M_{xy} = \int_{-2}^{2}\int_{-(1/2)\sqrt{4-x^2}}^{(1/2)\sqrt{4-x^2}}\int_{0}^{2-x} z\,dz\,dy\,dx$$

using integral tables to carry out the final integration with respect to x. Then divide M_{xy} by M to verify that $\bar{z} = 5/4$.

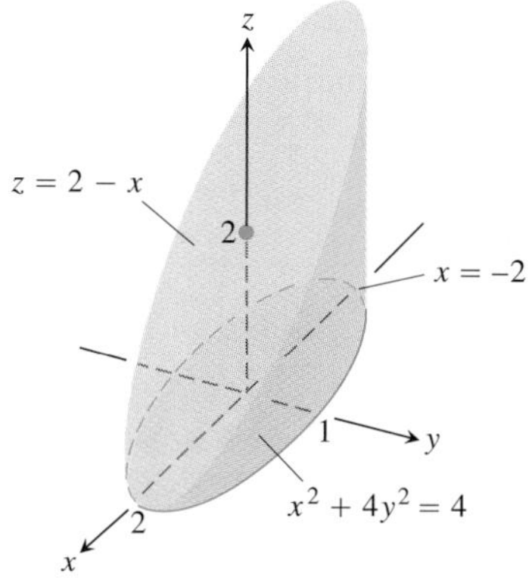

25. a. Center of mass Find the center of mass of a solid of constant density bounded below by the paraboloid $z = x^2 + y^2$ and above by the plane $z = 4$.

b. Find the plane $z = c$ that divides the solid into two parts of equal volume. This plane does not pass through the center of mass.

26. Moments A solid cube, 2 units on a side, is bounded by the planes $x = \pm 1, z = \pm 1, y = 3$, and $y = 5$. Find the center of mass and the moments of inertia about the coordinate axes.

27. Moment of inertia about a line A wedge like the one in Exercise 22 has $a = 4, b = 6$, and $c = 3$. Make a quick sketch to check for yourself that the square of the distance from a typical point (x, y, z) of the wedge to the line $L: z = 0, y = 6$ is $r^2 = (y - 6)^2 + z^2$. Then calculate the moment of inertia of the wedge about L.

28. Moment of inertia about a line A wedge like the one in Exercise 22 has $a = 4, b = 6$, and $c = 3$. Make a quick sketch to check for yourself that the square of the distance from a typical point (x, y, z) of the wedge to the line $L: x = 4, y = 0$ is $r^2 = (x - 4)^2 + y^2$. Then calculate the moment of inertia of the wedge about L.

In Exercises 29 and 30, find

a. the mass of the solid. **b.** the center of mass.

29. A solid region in the first octant is bounded by the coordinate planes and the plane $x + y + z = 2$. The density of the solid is $\delta(x, y, z) = 2x$.

30. A solid in the first octant is bounded by the planes $y = 0$ and $z = 0$ and by the surfaces $z = 4 - x^2$ and $x = y^2$. Its density function is $\delta(x, y, z) = kxy$, k a constant.

In Exercises 31 and 32, find

a. the mass of the solid.

b. the center of mass.

c. the moments of inertia about the coordinate axes.

31. A solid cube in the first octant is bounded by the coordinate planes and by the planes $x = 1, y = 1$, and $z = 1$. The density of the cube is $\delta(x, y, z) = x + y + z + 1$.

32. A wedge like the one in Exercise 22 has dimensions $a = 2, b = 6$, and $c = 3$. The density is $\delta(x, y, z) = x + 1$. Notice that if the density is constant, the center of mass will be $(0, 0, 0)$.

33. Mass Find the mass of the solid bounded by the planes $x + z = 1, x - z = -1, y = 0$ and the surface $y = \sqrt{z}$. The density of the solid is $\delta(x, y, z) = 2y + 5$.

34. Mass Find the mass of the solid region bounded by the parabolic surfaces $z = 16 - 2x^2 - 2y^2$ and $z = 2x^2 + 2y^2$ if the density of the solid is $\delta(x, y, z) = \sqrt{x^2 + y^2}$.

12.7 Triple Integrals in Cylindrical and Spherical Coordinates

When a calculation in physics, engineering, or geometry involves a cylinder, cone, or sphere, we can often simplify our work by using cylindrical or spherical coordinates, which are introduced in this section. The procedure for transforming to these coordinates and evaluating the resulting triple integrals is similar to the transformation to polar coordinates in the plane studied in Section 12.4.

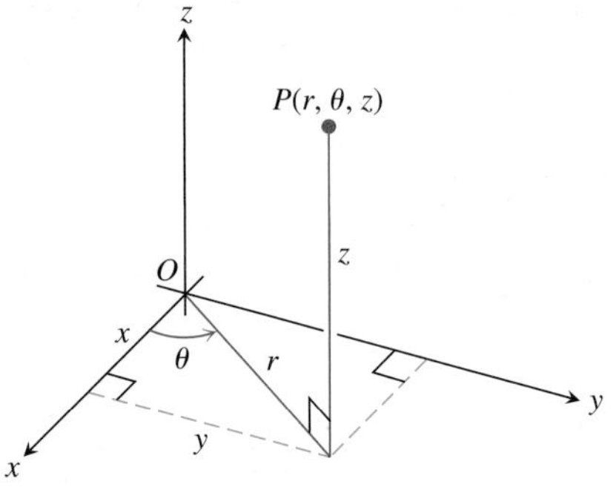

FIGURE 12.35 The cylindrical coordinates of a point in space are r, θ, and z.

Integration in Cylindrical Coordinates

We obtain cylindrical coordinates for space by combining polar coordinates in the xy-plane with the usual z-axis. This assigns to every point in space one or more coordinate triples of the form (r, θ, z), as shown in Figure 12.35.

> **DEFINITION** **Cylindrical coordinates** represent a point P in space by ordered triples (r, θ, z) in which
>
> 1. r and θ are polar coordinates for the vertical projection of P on the xy-plane
> 2. z is the rectangular vertical coordinate.

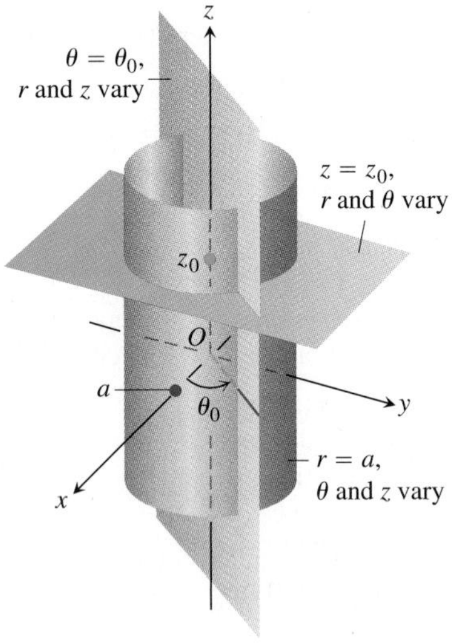

FIGURE 12.36 Constant-coordinate equations in cylindrical coordinates yield cylinders and planes.

The values of x, y, r, and θ in rectangular and cylindrical coordinates are related by the usual equations.

> **Equations Relating Rectangular (x, y, z) and Cylindrical (r, θ, z) Coordinates**
>
> $$x = r\cos\theta, \qquad y = r\sin\theta, \qquad z = z,$$
> $$r^2 = x^2 + y^2, \qquad \tan\theta = y/x$$

In cylindrical coordinates, the equation $r = a$ describes not just a circle in the xy-plane but an entire cylinder about the z-axis (Figure 12.36). The z-axis is given by $r = 0$. The equation $\theta = \theta_0$ describes the plane that contains the z-axis and makes an angle θ_0 with the positive x-axis. And, just as in rectangular coordinates, the equation $z = z_0$ describes a plane perpendicular to the z-axis.

Cylindrical coordinates are good for describing cylinders whose axes run along the z-axis and planes that either contain the z-axis or lie perpendicular to the z-axis. Surfaces like these have equations of constant coordinate value:

$r = 4$	Cylinder, radius 4, axis the z-axis
$\theta = \dfrac{\pi}{3}$	Plane containing the z-axis
$z = 2.$	Plane perpendicular to the z-axis

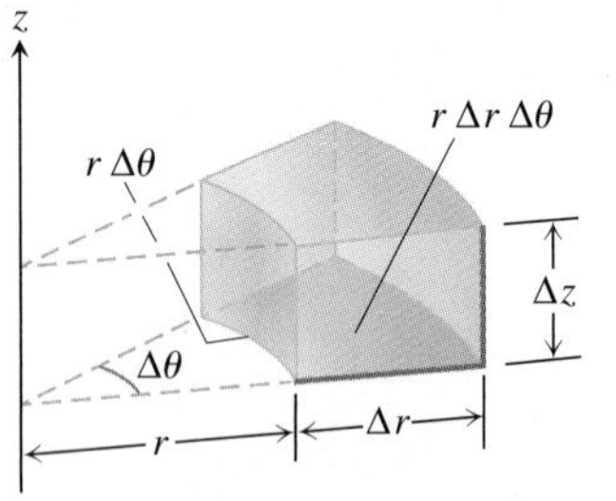

FIGURE 12.37 In cylindrical coordinates the volume of the wedge is approximated by the product $\Delta V = \Delta z\, r\, \Delta r\, \Delta\theta$.

When computing triple integrals over a region D in cylindrical coordinates, we partition the region into n small cylindrical wedges, rather than into rectangular boxes. In the kth cylindrical wedge, r, θ and z change by Δr_k, $\Delta\theta_k$, and Δz_k, and the largest of these numbers among all the cylindrical wedges is called the **norm** of the partition. We define the triple integral as a limit of Riemann sums using these wedges. The volume of such a cylindrical wedge ΔV_k is obtained by taking the area ΔA_k of its base in the $r\theta$-plane and multiplying by the height Δz (Figure 12.37).

For a point (r_k, θ_k, z_k) in the center of the kth wedge, we calculated in polar coordinates that $\Delta A_k = r_k\,\Delta r_k\,\Delta\theta_k$. So $\Delta V_k = \Delta z_k\,r_k\,\Delta r_k\,\Delta\theta_k$ and a Riemann sum for f over D has the form

$$S_n = \sum_{k=1}^{n} f(r_k, \theta_k, z_k)\,\Delta z_k\,r_k\,\Delta r_k\,\Delta\theta_k.$$

The triple integral of a function f over D is obtained by taking a limit of such Riemann sums with partitions whose norms approach zero:

$$\lim_{n\to\infty} S_n = \iiint_D f\,dV = \iiint_D f\,dz\,r\,dr\,d\theta.$$

Triple integrals in cylindrical coordinates are then evaluated as iterated integrals, as in the following example.

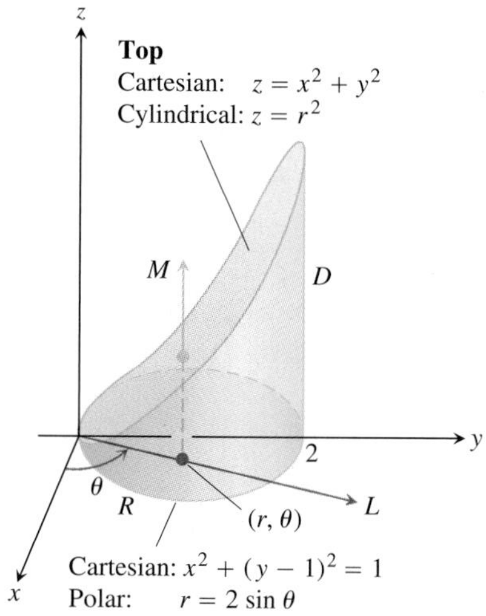

FIGURE 12.38 Finding the limits of integration for evaluating an integral in cylindrical coordinates (Example 1).

EXAMPLE 1 Find the limits of integration in cylindrical coordinates for integrating a function $f(r, \theta, z)$ over the region D bounded below by the plane $z = 0$, laterally by the circular cylinder $x^2 + (y - 1)^2 = 1$, and above by the paraboloid $z = x^2 + y^2$.

Solution The base of D is also the region's projection R on the xy-plane. The boundary of R is the circle $x^2 + (y - 1)^2 = 1$. Its polar coordinate equation is

$$\begin{aligned} x^2 + (y - 1)^2 &= 1 \\ x^2 + y^2 - 2y + 1 &= 1 \\ r^2 - 2r\sin\theta &= 0 \\ r &= 2\sin\theta. \end{aligned}$$

The region is sketched in Figure 12.38.

We find the limits of integration, starting with the z-limits. A line M through a typical point (r, θ) in R parallel to the z-axis enters D at $z = 0$ and leaves at $z = x^2 + y^2 = r^2$.

Next we find the r-limits of integration. A ray L through (r, θ) from the origin enters R at $r = 0$ and leaves at $r = 2\sin\theta$.

Finally we find the θ-limits of integration. As L sweeps across R, the angle θ it makes with the positive x-axis runs from $\theta = 0$ to $\theta = \pi$. The integral is

$$\iiint_D f(r, \theta, z)\,dV = \int_0^{\pi}\int_0^{2\sin\theta}\int_0^{r^2} f(r, \theta, z)\,dz\,r\,dr\,d\theta. \qquad \blacksquare$$

Example 1 illustrates a good procedure for finding limits of integration in cylindrical coordinates. The procedure is summarized as follows.

How to Integrate in Cylindrical Coordinates

To evaluate

$$\iiint_D f(r, \theta, z)\,dV$$

over a region D in space in cylindrical coordinates, integrating first with respect to z, then with respect to r, and finally with respect to θ, take the following steps.

1. *Sketch*. Sketch the region D along with its projection R on the xy-plane. Label the surfaces and curves that bound D and R.

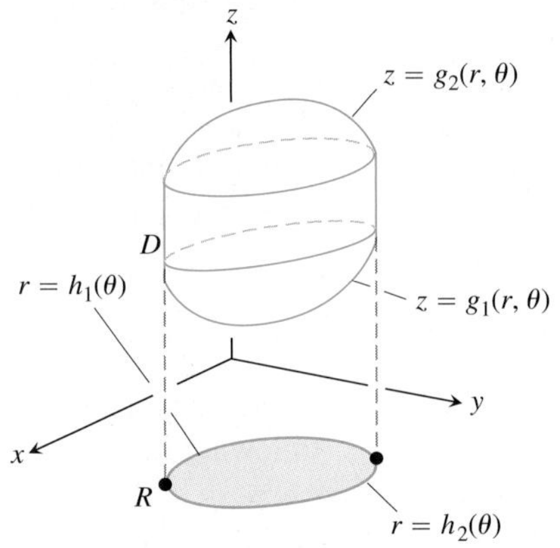

2. *Find the z-limits of integration*. Draw a line M through a typical point (r, θ) of R parallel to the z-axis. As z increases, M enters D at $z = g_1(r, \theta)$ and leaves at $z = g_2(r, \theta)$. These are the z-limits of integration.

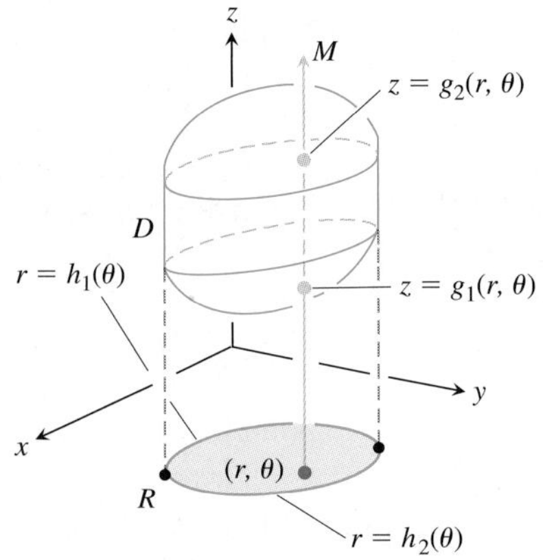

3. *Find the r-limits of integration*. Draw a ray L through (r, θ) from the origin. The ray enters R at $r = h_1(\theta)$ and leaves at $r = h_2(\theta)$. These are the r-limits of integration.

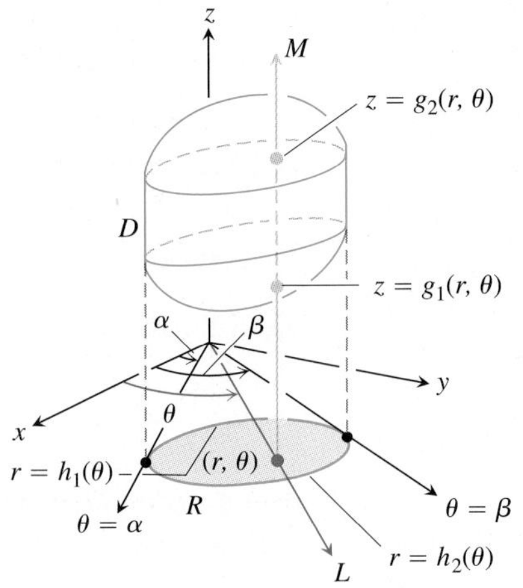

4. *Find the θ-limits of integration.* As L sweeps across R, the angle θ it makes with the positive x-axis runs from $\theta = \alpha$ to $\theta = \beta$. These are the θ-limits of integration. The integral is

$$\iiint_D f(r, \theta, z)\, dV = \int_{\theta=\alpha}^{\theta=\beta} \int_{r=h_1(\theta)}^{r=h_2(\theta)} \int_{z=g_1(r,\theta)}^{z=g_2(r,\theta)} f(r, \theta, z)\, dz\, r\, dr\, d\theta.$$

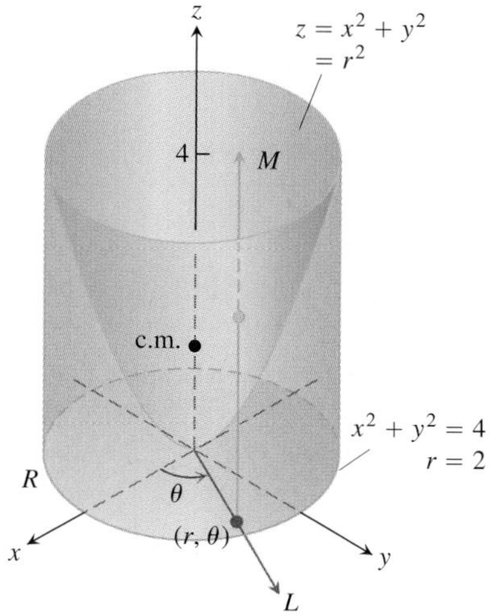

FIGURE 12.39 Example 2 shows how to find the centroid of this solid.

EXAMPLE 2 Find the centroid ($\delta = 1$) of the solid enclosed by the cylinder $x^2 + y^2 = 4$, bounded above by the paraboloid $z = x^2 + y^2$, and bounded below by the xy-plane.

Solution We sketch the solid, bounded above by the paraboloid $z = r^2$ and below by the plane $z = 0$ (Figure 12.39). Its base R is the disk $0 \le r \le 2$ in the xy-plane.

The solid's centroid $(\bar{x}, \bar{y}, \bar{z})$ lies on its axis of symmetry, here the z-axis. This makes $\bar{x} = \bar{y} = 0$. To find $\bar{z}$, we divide the first moment M_{xy} by the mass M.

To find the limits of integration for the mass and moment integrals, we continue with the four basic steps. We completed our initial sketch. The remaining steps give the limits of integration.

The z-limits. A line M through a typical point (r, θ) in the base parallel to the z-axis enters the solid at $z = 0$ and leaves at $z = r^2$.

The r-limits. A ray L through (r, θ) from the origin enters R at $r = 0$ and leaves at $r = 2$.

The θ-limits. As L sweeps over the base like a clock hand, the angle θ it makes with the positive x-axis runs from $\theta = 0$ to $\theta = 2\pi$. The value of M_{xy} is

$$M_{xy} = \int_0^{2\pi} \int_0^2 \int_0^{r^2} z\, dz\, r\, dr\, d\theta = \int_0^{2\pi} \int_0^2 \left[\frac{z^2}{2}\right]_0^{r^2} r\, dr\, d\theta$$

$$= \int_0^{2\pi} \int_0^2 \frac{r^5}{2}\, dr\, d\theta = \int_0^{2\pi} \left[\frac{r^6}{12}\right]_0^2 d\theta = \int_0^{2\pi} \frac{16}{3}\, d\theta = \frac{32\pi}{3}.$$

The value of M is

$$M = \int_0^{2\pi} \int_0^2 \int_0^{r^2} dz\, r\, dr\, d\theta = \int_0^{2\pi} \int_0^2 \left[z\right]_0^{r^2} r\, dr\, d\theta$$

$$= \int_0^{2\pi} \int_0^2 r^3\, dr\, d\theta = \int_0^{2\pi} \left[\frac{r^4}{4}\right]_0^2 d\theta = \int_0^{2\pi} 4\, d\theta = 8\pi.$$

Therefore,

$$\bar{z} = \frac{M_{xy}}{M} = \frac{32\pi}{3} \frac{1}{8\pi} = \frac{4}{3},$$

and the centroid is $(0, 0, 4/3)$. Notice that the centroid lies outside the solid. ■

Spherical Coordinates and Integration

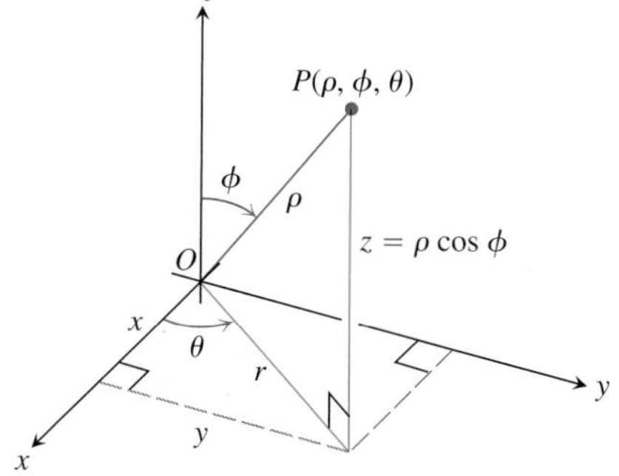

FIGURE 12.40 The spherical coordinates ρ, ϕ, and θ and their relation to x, y, z, and r.

Spherical coordinates locate points in space with two angles and one distance, as shown in Figure 12.40. The first coordinate, $\rho = |\overrightarrow{OP}|$, is the point's distance from the origin. Unlike r, *the variable ρ is never negative.* The second coordinate, ϕ, is the angle $\overrightarrow{OP}$ makes

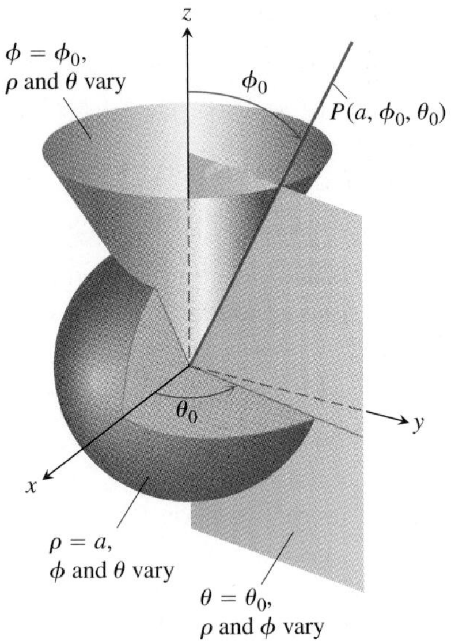

FIGURE 12.41 Constant-coordinate equations in spherical coordinates yield spheres, single cones, and half-planes.

with the positive z-axis. It is required to lie in the interval $[0, \pi]$. The third coordinate is the angle θ as measured in cylindrical coordinates.

> **DEFINITION** **Spherical coordinates** represent a point P in space by ordered triples (ρ, ϕ, θ) in which
>
> 1. ρ is the distance from P to the origin.
> 2. ϕ is the angle $\overrightarrow{OP}$ makes with the positive z-axis $(0 \le \phi \le \pi)$.
> 3. θ is the angle from cylindrical coordinates.

On maps of the Earth, θ is related to the meridian of a point on the Earth and ϕ to its latitude, while ρ is related to elevation above the Earth's surface.

The equation $\rho = a$ describes the sphere of radius a centered at the origin (Figure 12.41). The equation $\phi = \phi_0$ describes a single cone whose vertex lies at the origin and whose axis lies along the z-axis. (We broaden our interpretation to include the xy-plane as the cone $\phi = \pi/2$.) If ϕ_0 is greater than $\pi/2$, the cone $\phi = \phi_0$ opens downward. The equation $\theta = \theta_0$ describes the half-plane that contains the z-axis and makes an angle θ_0 with the positive x-axis.

> **Equations Relating Spherical Coordinates to Cartesian and Cylindrical Coordinates**
>
> $$\begin{aligned} r &= \rho \sin \phi, & x &= r \cos \theta = \rho \sin \phi \cos \theta, \\ z &= \rho \cos \phi, & y &= r \sin \theta = \rho \sin \phi \sin \theta, \end{aligned} \tag{1}$$
> $$\rho = \sqrt{x^2 + y^2 + z^2} = \sqrt{r^2 + z^2}.$$

EXAMPLE 3 Find a spherical coordinate equation for the sphere $x^2 + y^2 + (z - 1)^2 = 1$.

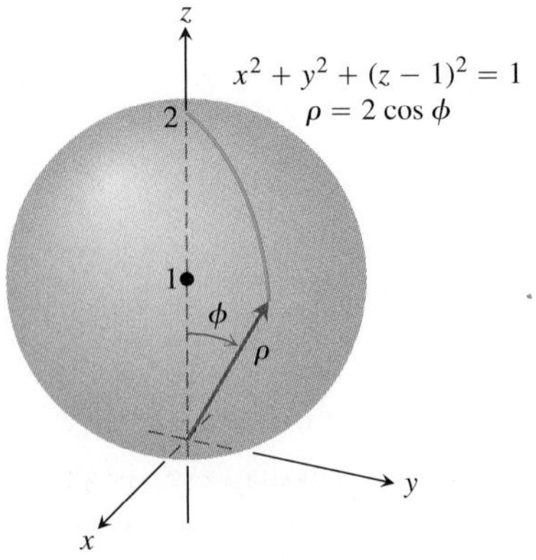

FIGURE 12.42 The sphere in Example 3.

Solution We use Equations (1) to substitute for x, y, and z:

$$x^2 + y^2 + (z - 1)^2 = 1$$
$$\rho^2 \sin^2 \phi \cos^2 \theta + \rho^2 \sin^2 \phi \sin^2 \theta + (\rho \cos \phi - 1)^2 = 1 \qquad \text{Eq. (1)}$$
$$\rho^2 \sin^2 \phi \underbrace{(\cos^2 \theta + \sin^2 \theta)}_{1} + \rho^2 \cos^2 \phi - 2\rho \cos \phi + 1 = 1$$
$$\rho^2 \underbrace{(\sin^2 \phi + \cos^2 \phi)}_{1} = 2\rho \cos \phi$$
$$\rho^2 = 2\rho \cos \phi$$
$$\rho = 2 \cos \phi .$$

(See Figure 12.42.) ■

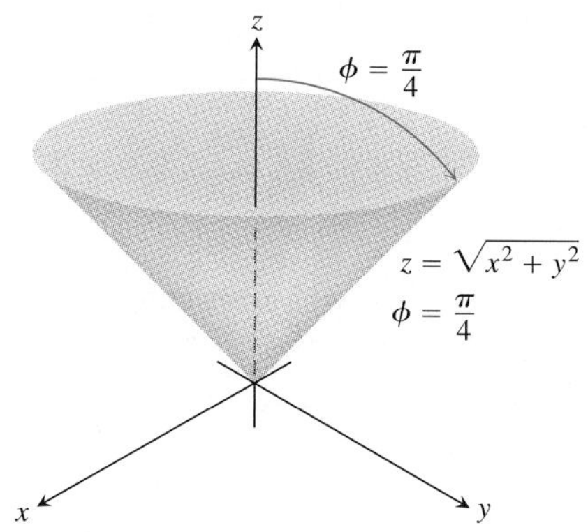

FIGURE 12.43 The cone in Example 4.

EXAMPLE 4 Find a spherical coordinate equation for the cone $z = \sqrt{x^2 + y^2}$.

Solution 1 *Use geometry.* The cone is symmetric with respect to the z-axis and cuts the first quadrant of the yz-plane along the line $z = y$. The angle between the cone and the positive z-axis is therefore $\pi/4$ radians. The cone consists of the points whose spherical coordinates have ϕ equal to $\pi/4$, so its equation is $\phi = \pi/4$. (See Figure 12.43.)

Solution 2 *Use algebra.* If we use Equations (1) to substitute for x, y, and z we obtain the same result:

$$\begin{aligned}
z &= \sqrt{x^2 + y^2} \\
\rho \cos \phi &= \sqrt{\rho^2 \sin^2 \phi} && \text{Example 3} \\
\rho \cos \phi &= \rho \sin \phi && \rho \ge 0,\ \sin \phi \ge 0 \\
\cos \phi &= \sin \phi \\
\phi &= \frac{\pi}{4}. && 0 \le \phi \le \pi
\end{aligned}$$

■

Spherical coordinates are good for describing spheres centered at the origin, half-planes hinged along the z-axis, and cones whose vertices lie at the origin and whose axes lie along the z-axis. Surfaces like these have equations of constant coordinate value:

$$\begin{aligned}
\rho &= 4 && \text{Sphere, radius 4, center at origin} \\
\phi &= \frac{\pi}{3} && \text{Cone opening up from the origin, making an angle of } \pi/3 \text{ radians with the positive } z\text{-axis} \\
\theta &= \frac{\pi}{3}. && \text{Half-plane, hinged along the } z\text{-axis, making an angle of } \pi/3 \text{ radians with the positive } x\text{-axis}
\end{aligned}$$

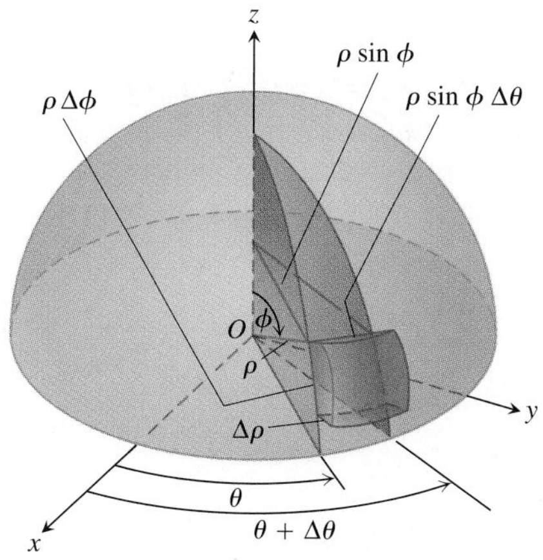

FIGURE 12.44 In spherical coordinates

$$\begin{aligned} dV &= d\rho \cdot \rho\, d\phi \cdot \rho \sin \phi\, d\theta \\ &= \rho^2 \sin \phi\, d\rho\, d\phi\, d\theta. \end{aligned}$$

When computing triple integrals over a region D in spherical coordinates, we partition the region into n spherical wedges. The size of the kth spherical wedge, which contains a point $(\rho_k, \phi_k, \theta_k)$, is given by the changes $\Delta\rho_k$, $\Delta\theta_k$, and $\Delta\phi_k$ in ρ, θ, and ϕ. Such a spherical wedge has one edge a circular arc of length $\rho_k\,\Delta\phi_k$, another edge a circular arc of length $\rho_k \sin \phi_k\,\Delta\theta_k$, and thickness $\Delta\rho_k$. The spherical wedge closely approximates a cube of these dimensions when $\Delta\rho_k$, $\Delta\theta_k$, and $\Delta\phi_k$ are all small (Figure 12.44). It can be shown that the volume of this spherical wedge ΔV_k is $\Delta V_k = \rho_k^2 \sin \phi_k\,\Delta\rho_k\,\Delta\phi_k\,\Delta\theta_k$ for $(\rho_k, \phi_k, \theta_k)$ a point chosen inside the wedge.

The corresponding Riemann sum for a function $f(\rho, \phi, \theta)$ is

$$S_n = \sum_{k=1}^{n} f(\rho_k, \phi_k, \theta_k)\,\rho_k^2 \sin \phi_k\,\Delta\rho_k\,\Delta\phi_k\,\Delta\theta_k.$$

As the norm of a partition approaches zero, and the spherical wedges get smaller, the Riemann sums have a limit when f is continuous:

$$\lim_{n\to\infty} S_n = \iiint_D f(\rho, \phi, \theta)\,dV = \iiint_D f(\rho, \phi, \theta)\,\rho^2 \sin \phi\,d\rho\,d\phi\,d\theta.$$

In spherical coordinates, we have

$$dV = \rho^2 \sin \phi\,d\rho\,d\phi\,d\theta.$$

To evaluate integrals in spherical coordinates, we usually integrate first with respect to ρ. The procedure for finding the limits of integration is shown below. We restrict our attention to integrating over domains that are solids of revolution about the z-axis (or portions thereof) and for which the limits for θ and ϕ are constant.

How to Integrate in Spherical Coordinates

To evaluate

$$\iiint_D f(\rho, \phi, \theta)\, dV$$

over a region D in space in spherical coordinates, integrating first with respect to ρ, then with respect to ϕ, and finally with respect to θ, take the following steps.

1. *Sketch.* Sketch the region D along with its projection R on the xy-plane. Label the surfaces that bound D.

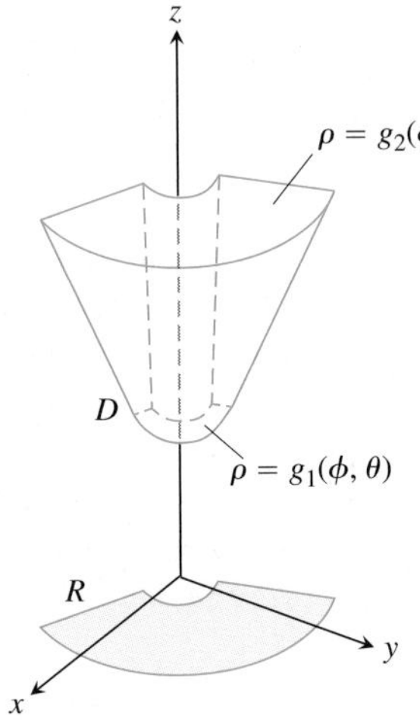

2. *Find the ρ-limits of integration.* Draw a ray M from the origin through D making an angle ϕ with the positive z-axis. Also draw the projection of M on the xy-plane (call the projection L). The ray L makes an angle θ with the positive x-axis. As ρ increases, M enters D at $\rho = g_1(\phi, \theta)$ and leaves at $\rho = g_2(\phi, \theta)$. These are the ρ-limits of integration.

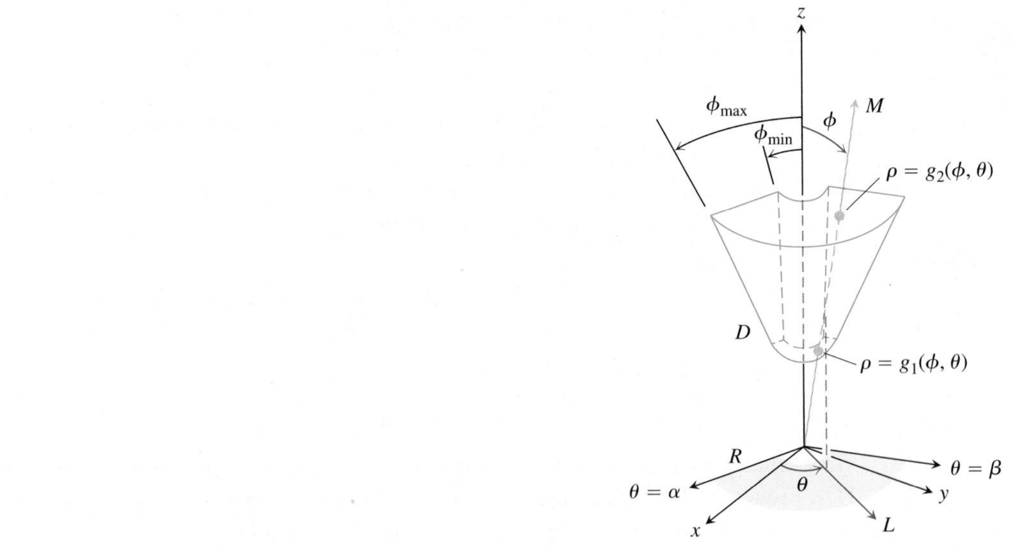

3. *Find the ϕ-limits of integration.* For any given θ, the angle ϕ that M makes with the z-axis runs from $\phi = \phi_{\min}$ to $\phi = \phi_{\max}$. These are the ϕ -limits of integration.
4. *Find the θ-limits of integration.* The ray L sweeps over R as θ runs from α to β. These are the θ-limits of integration. The integral is

$$\iiint_D f(\rho, \phi, \theta)\, dV = \int_{\theta=\alpha}^{\theta=\beta} \int_{\phi=\phi_{\min}}^{\phi=\phi_{\max}} \int_{\rho=g_1(\phi,\theta)}^{\rho=g_2(\phi,\theta)} f(\rho, \phi, \theta)\, \rho^2 \sin\phi\, d\rho\, d\phi\, d\theta.$$

EXAMPLE 5 Find the volume of the "ice cream cone" D cut from the solid sphere $\rho \le 1$ by the cone $\phi = \pi/3$.

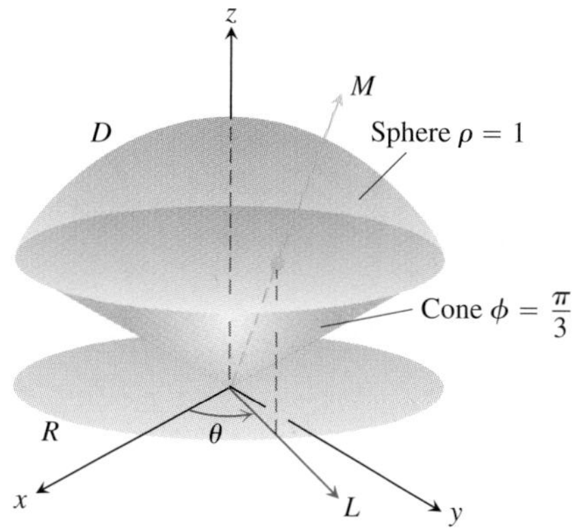

FIGURE 12.45 The ice cream cone in Example 5.

Solution The volume is $V = \iiint_D \rho^2 \sin\phi\, d\rho\, d\phi\, d\theta$, the integral of $f(\rho, \phi, \theta) = 1$ over D.

To find the limits of integration for evaluating the integral, we begin by sketching D and its projection R on the xy-plane (Figure 12.45).

The ρ-limits of integration. We draw a ray M from the origin through D making an angle ϕ with the positive z-axis. We also draw L, the projection of M on the xy-plane, along with the angle θ that L makes with the positive x-axis. Ray M enters D at $\rho = 0$ and leaves at $\rho = 1$.

The ϕ-limits of integration. The cone $\phi = \pi/3$ makes an angle of $\pi/3$ with the positive z-axis. For any given θ, the angle ϕ can run from $\phi = 0$ to $\phi = \pi/3$.

The θ-limits of integration. The ray L sweeps over R as θ runs from 0 to 2π. The volume is

$$V = \iiint_D \rho^2 \sin\phi\, d\rho\, d\phi\, d\theta = \int_0^{2\pi} \int_0^{\pi/3} \int_0^1 \rho^2 \sin\phi\, d\rho\, d\phi\, d\theta$$

$$= \int_0^{2\pi} \int_0^{\pi/3} \left[\frac{\rho^3}{3}\right]_0^1 \sin\phi\, d\phi\, d\theta = \int_0^{2\pi} \int_0^{\pi/3} \frac{1}{3} \sin\phi\, d\phi\, d\theta$$

$$= \int_0^{2\pi} \left[-\frac{1}{3}\cos\phi\right]_0^{\pi/3} d\theta = \int_0^{2\pi} \left(-\frac{1}{6} + \frac{1}{3}\right) d\theta = \frac{1}{6}(2\pi) = \frac{\pi}{3}.$$ ■

EXAMPLE 6 A solid of constant density $\delta = 1$ occupies the region D in Example 5. Find the solid's moment of inertia about the z-axis.

Solution In rectangular coordinates, the moment is

$$I_z = \iiint (x^2 + y^2)\, dV.$$

In spherical coordinates, $x^2 + y^2 = (\rho \sin\phi \cos\theta)^2 + (\rho \sin\phi \sin\theta)^2 = \rho^2 \sin^2\phi$. Hence,

$$I_z = \iiint (\rho^2 \sin^2\phi)\, \rho^2 \sin\phi\, d\rho\, d\phi\, d\theta = \iiint \rho^4 \sin^3\phi\, d\rho\, d\phi\, d\theta.$$

For the region in Example 5, this becomes

$$\begin{aligned}
I_z &= \int_0^{2\pi}\int_0^{\pi/3}\int_0^1 \rho^4 \sin^3\phi \, d\rho \, d\phi \, d\theta = \int_0^{2\pi}\int_0^{\pi/3}\left[\frac{\rho^5}{5}\right]_0^1 \sin^3\phi \, d\phi \, d\theta \\
&= \frac{1}{5}\int_0^{2\pi}\int_0^{\pi/3}(1-\cos^2\phi)\sin\phi \, d\phi \, d\theta = \frac{1}{5}\int_0^{2\pi}\left[-\cos\phi + \frac{\cos^3\phi}{3}\right]_0^{\pi/3} d\theta \\
&= \frac{1}{5}\int_0^{2\pi}\left(-\frac{1}{2} + 1 + \frac{1}{24} - \frac{1}{3}\right) d\theta = \frac{1}{5}\int_0^{2\pi}\frac{5}{24}\, d\theta = \frac{1}{24}(2\pi) = \frac{\pi}{12}.
\end{aligned}$$

■

Coordinate Conversion Formulas

CYLINDRICAL TO RECTANGULAR	SPHERICAL TO RECTANGULAR	SPHERICAL TO CYLINDRICAL
$x = r\cos\theta$	$x = \rho\sin\phi\cos\theta$	$r = \rho\sin\phi$
$y = r\sin\theta$	$y = \rho\sin\phi\sin\theta$	$z = \rho\cos\phi$
$z = z$	$z = \rho\cos\phi$	$\theta = \theta$

Corresponding formulas for dV in triple integrals:

$$\begin{aligned}
dV &= dx\,dy\,dz \\
&= dz\,r\,dr\,d\theta \\
&= \rho^2\sin\phi\,d\rho\,d\phi\,d\theta
\end{aligned}$$

In the next section we offer a more general procedure for determining dV in cylindrical and spherical coordinates. The results, of course, will be the same.

EXERCISES 12.7

Evaluate the cylindrical coordinate integrals in Exercises 1–6.

1. $\displaystyle\int_0^{2\pi}\int_0^1\int_r^{\sqrt{2-r^2}} dz\,r\,dr\,d\theta$

2. $\displaystyle\int_0^{2\pi}\int_0^3\int_{r^2/3}^{\sqrt{18-r^2}} dz\,r\,dr\,d\theta$

3. $\displaystyle\int_0^{2\pi}\int_0^{\theta/2\pi}\int_0^{3+24r^2} dz\,r\,dr\,d\theta$

4. $\displaystyle\int_0^{\pi}\int_0^{\theta/\pi}\int_{-\sqrt{4-r^2}}^{3\sqrt{4-r^2}} z\,dz\,r\,dr\,d\theta$

5. $\displaystyle\int_0^{2\pi}\int_0^1\int_r^{1/\sqrt{2-r^2}} 3\,dz\,r\,dr\,d\theta$

6. $\displaystyle\int_0^{2\pi}\int_0^1\int_{-1/2}^{1/2}(r^2\sin^2\theta + z^2)\,dz\,r\,dr\,d\theta$

The integrals we have seen so far suggest that there are preferred orders of integration for cylindrical coordinates, but other orders usually work well and are occasionally easier to evaluate. Evaluate the integrals in Exercises 7–10.

7. $\displaystyle\int_0^{2\pi}\int_0^3\int_0^{z/3} r^3\,dr\,dz\,d\theta$

8. $\displaystyle\int_{-1}^1\int_0^{2\pi}\int_0^{1+\cos\theta} 4r\,dr\,d\theta\,dz$

9. $\displaystyle\int_0^1\int_0^{\sqrt{z}}\int_0^{2\pi}(r^2\cos^2\theta + z^2)\,r\,d\theta\,dr\,dz$

10. $\displaystyle\int_0^2\int_{r-2}^{\sqrt{4-r^2}}\int_0^{2\pi}(r\sin\theta + 1)\,r\,d\theta\,dz\,dr$

11. Let D be the region bounded below by the plane $z = 0$, above by the sphere $x^2 + y^2 + z^2 = 4$, and on the sides by the cylinder $x^2 + y^2 = 1$. Set up the triple integrals in cylindrical coordinates that give the volume of D using the following orders of integration.

    a. $dz\,dr\,d\theta$ **b.** $dr\,dz\,d\theta$ **c.** $d\theta\,dz\,dr$

12. Let D be the region bounded below by the cone $z = \sqrt{x^2 + y^2}$ and above by the paraboloid $z = 2 - x^2 - y^2$. Set up the triple integrals in cylindrical coordinates that give the volume of D using the following orders of integration.

    a. $dz\,dr\,d\theta$ **b.** $dr\,dz\,d\theta$ **c.** $d\theta\,dz\,dr$

13. Give the limits of integration for evaluating the integral

$$\iiint f(r, \theta, z)\, dz\, r\, dr\, d\theta$$

as an iterated integral over the region that is bounded below by the plane $z = 0$, on the side by the cylinder $r = \cos\theta$, and on top by the paraboloid $z = 3r^2$.

14. Convert the integral

$$\int_{-1}^{1}\int_{0}^{\sqrt{1-y^2}}\int_{0}^{x} (x^2 + y^2)\, dz\, dx\, dy$$

to an equivalent integral in cylindrical coordinates and evaluate the result.

In Exercises 15–20, set up the iterated integral for evaluating $\iiint_D f(r, \theta, z)\, dz\, r\, dr\, d\theta$ over the given region D.

15. D is the right circular cylinder whose base is the circle $r = 2\sin\theta$ in the xy-plane and whose top lies in the plane $z = 4 - y$.

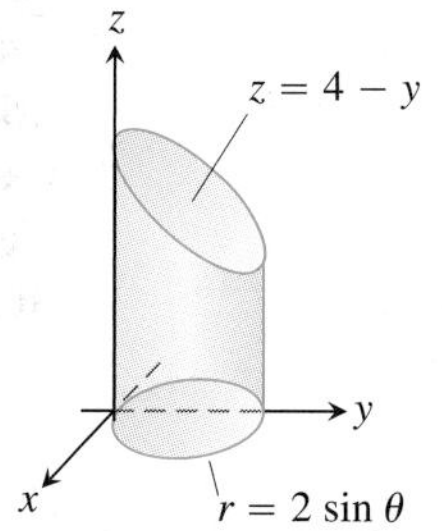

16. D is the right circular cylinder whose base is the circle $r = 3\cos\theta$ and whose top lies in the plane $z = 5 - x$.

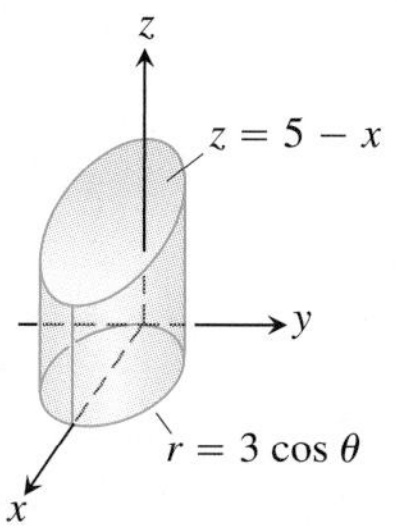

17. D is the solid right cylinder whose base is the region in the xy-plane that lies inside the cardioid $r = 1 + \cos\theta$ and outside the circle $r = 1$ and whose top lies in the plane $z = 4$.

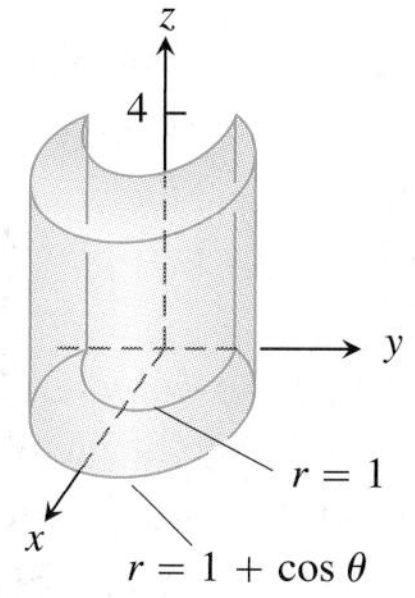

18. D is the solid right cylinder whose base is the region between the circles $r = \cos\theta$ and $r = 2\cos\theta$ and whose top lies in the plane $z = 3 - y$.

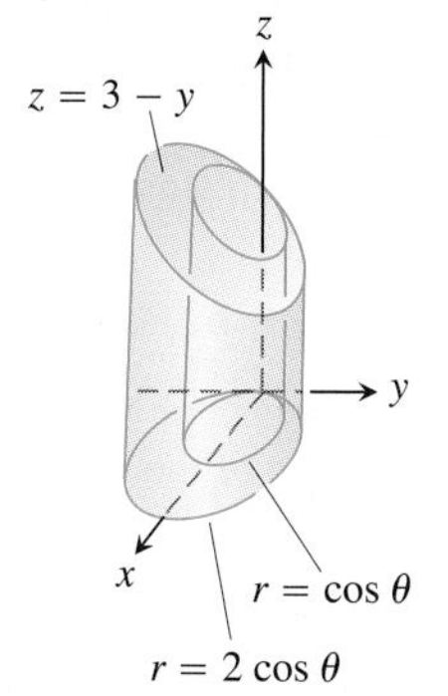

19. D is the prism whose base is the triangle in the xy-plane bounded by the x-axis and the lines $y = x$ and $x = 1$ and whose top lies in the plane $z = 2 - y$.

20. D is the prism whose base is the triangle in the xy-plane bounded by the y-axis and the lines $y = x$ and $y = 1$ and whose top lies in the plane $z = 2 - x$.

Evaluate the spherical coordinate integrals in Exercises 21–26.

21. $\displaystyle\int_{0}^{\pi}\int_{0}^{\pi}\int_{0}^{2\sin\phi} \rho^2 \sin\phi\, d\rho\, d\phi\, d\theta$

22. $\displaystyle\int_{0}^{2\pi}\int_{0}^{\pi/4}\int_{0}^{2} (\rho\cos\phi)\, \rho^2 \sin\phi\, d\rho\, d\phi\, d\theta$

23. $\displaystyle\int_{0}^{2\pi}\int_{0}^{\pi}\int_{0}^{(1-\cos\phi)/2} \rho^2 \sin\phi\, d\rho\, d\phi\, d\theta$

24. $\displaystyle\int_{0}^{3\pi/2}\int_{0}^{\pi}\int_{0}^{1} 5\rho^3 \sin^3\phi\, d\rho\, d\phi\, d\theta$

25. $\displaystyle\int_{0}^{2\pi}\int_{0}^{\pi/3}\int_{\sec\phi}^{2} 3\rho^2 \sin\phi\, d\rho\, d\phi\, d\theta$

26. $\displaystyle\int_{0}^{2\pi}\int_{0}^{\pi/4}\int_{0}^{\sec\phi} (\rho\cos\phi)\, \rho^2 \sin\phi\, d\rho\, d\phi\, d\theta$

The previous integrals suggest there are preferred orders of integration for spherical coordinates, but other orders give the same value and are occasionally easier to evaluate. Evaluate the integrals in Exercises 27–30.

27. $\displaystyle\int_{0}^{2}\int_{-\pi}^{0}\int_{\pi/4}^{\pi/2} \rho^3 \sin 2\phi\, d\phi\, d\theta\, d\rho$

28. $\displaystyle\int_{\pi/6}^{\pi/3}\int_{\csc\phi}^{2\csc\phi}\int_{0}^{2\pi} \rho^2 \sin\phi\, d\theta\, d\rho\, d\phi$

29. $\displaystyle\int_{0}^{1}\int_{0}^{\pi}\int_{0}^{\pi/4} 12\rho \sin^3\phi\, d\phi\, d\theta\, d\rho$

30. $\displaystyle\int_{\pi/6}^{\pi/2}\int_{-\pi/2}^{\pi/2}\int_{\csc\phi}^{2} 5\rho^4 \sin^3\phi\, d\rho\, d\theta\, d\phi$

31. Let D be the region in Exercise 11. Set up the triple integrals in spherical coordinates that give the volume of D using the following orders of integration.

a. $d\rho\, d\phi\, d\theta$ **b.** $d\phi\, d\rho\, d\theta$

32. Let D be the region bounded below by the cone $z = \sqrt{x^2 + y^2}$ and above by the plane $z = 1$. Set up the triple integrals in spherical coordinates that give the volume of D using the following orders of integration.

a. $d\rho\, d\phi\, d\theta$ **b.** $d\phi\, d\rho\, d\theta$

In Exercises 33–38, **(a)** find the spherical coordinate limits for the integral that calculates the volume of the given solid and then **(b)** evaluate the integral.

33. The solid between the sphere $\rho = \cos\phi$ and the hemisphere $\rho = 2, z \geq 0$

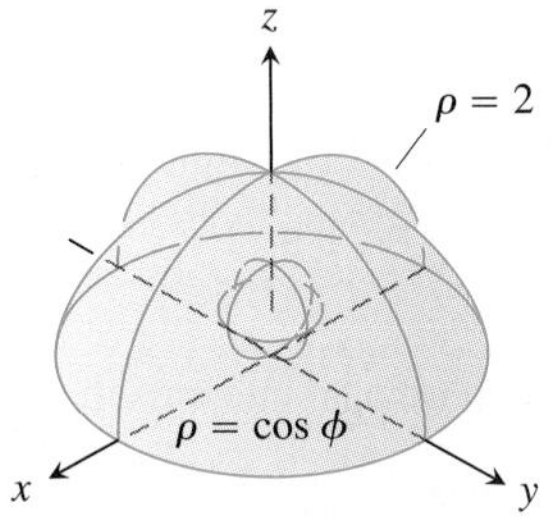

34. The solid bounded below by the hemisphere $\rho = 1, z \geq 0$, and above by the cardioid of revolution $\rho = 1 + \cos\phi$

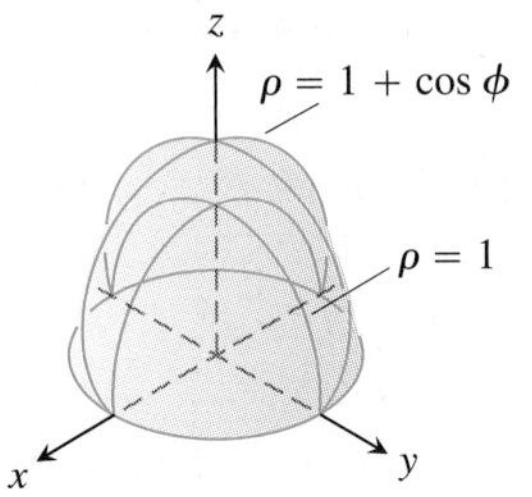

35. The solid enclosed by the cardioid of revolution $\rho = 1 - \cos\phi$

36. The upper portion cut from the solid in Exercise 35 by the xy-plane

37. The solid bounded below by the sphere $\rho = 2\cos\phi$ and above by the cone $z = \sqrt{x^2 + y^2}$

38. The solid bounded below by the xy-plane, on the sides by the sphere $\rho = 2$, and above by the cone $\phi = \pi/3$

39. Set up triple integrals for the volume of the sphere $\rho = 2$ in **(a)** spherical, **(b)** cylindrical, and **(c)** rectangular coordinates.

40. Let D be the region in the first octant that is bounded below by the cone $\phi = \pi/4$ and above by the sphere $\rho = 3$. Express the volume of D as an iterated triple integral in **(a)** cylindrical and **(b)** spherical coordinates. Then **(c)** find V.

41. Let D be the smaller cap cut from a solid ball of radius 2 units by a plane 1 unit from the center of the sphere. Express the volume of D as an iterated triple integral in **(a)** spherical, **(b)** cylindrical, and **(c)** rectangular coordinates. Then **(d)** find the volume by evaluating one of the three triple integrals.

42. Express the moment of inertia I_z of the solid hemisphere $x^2 + y^2 + z^2 \leq 1, z \geq 0$, as an iterated integral in **(a)** cylindrical and **(b)** spherical coordinates. Then **(c)** find I_z.

Find the volumes of the solids in Exercises 43–48.

43.

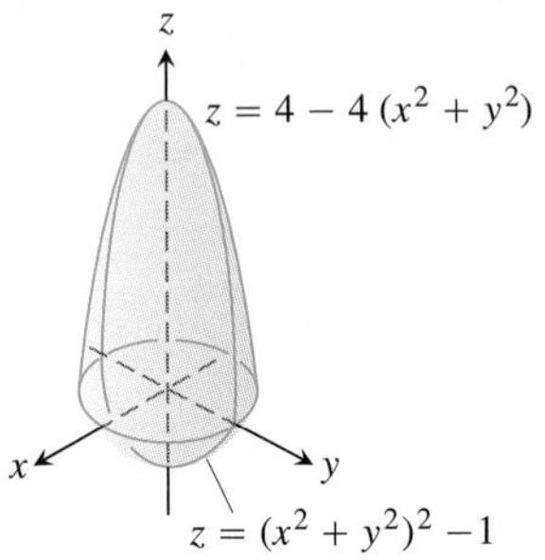

44.

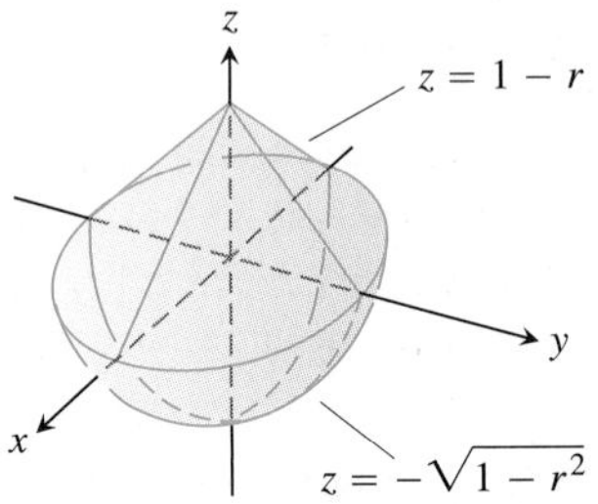

45.

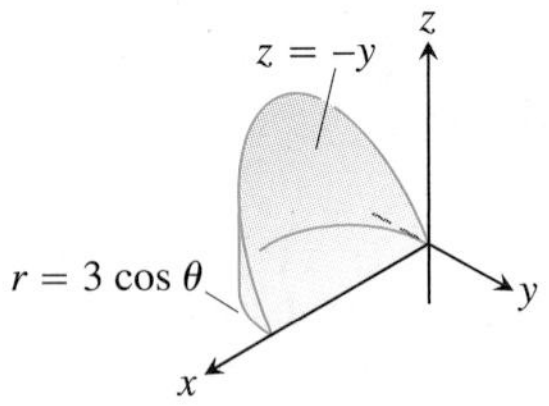

46.

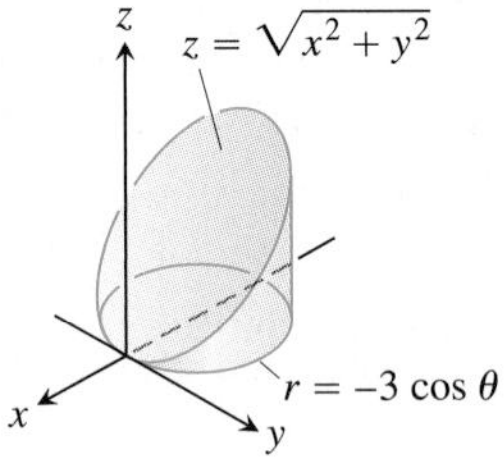

47.

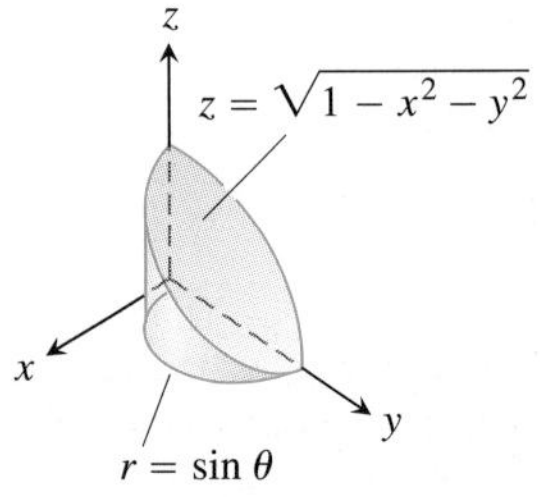

48.

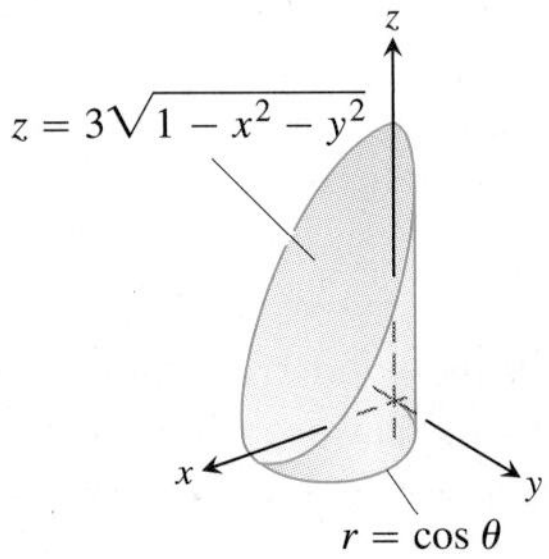

49. Sphere and cones Find the volume of the portion of the solid sphere $\rho \leq a$ that lies between the cones $\phi = \pi/3$ and $\phi = 2\pi/3$.

50. Sphere and half-planes Find the volume of the region cut from the solid sphere $\rho \leq a$ by the half-planes $\theta = 0$ and $\theta = \pi/6$ in the first octant.

51. Sphere and plane Find the volume of the smaller region cut from the solid sphere $\rho \leq 2$ by the plane $z = 1$.

52. Cone and planes Find the volume of the solid enclosed by the cone $z = \sqrt{x^2 + y^2}$ between the planes $z = 1$ and $z = 2$.

53. Cylinder and paraboloid Find the volume of the region bounded below by the plane $z = 0$, laterally by the cylinder $x^2 + y^2 = 1$, and above by the paraboloid $z = x^2 + y^2$.

54. Cylinder and paraboloids Find the volume of the region bounded below by the paraboloid $z = x^2 + y^2$, laterally by the cylinder $x^2 + y^2 = 1$, and above by the paraboloid $z = x^2 + y^2 + 1$.

55. Cylinder and cones Find the volume of the solid cut from the thick-walled cylinder $1 \le x^2 + y^2 \le 2$ by the cones $z = \pm\sqrt{x^2 + y^2}$.

56. Sphere and cylinder Find the volume of the region that lies inside the sphere $x^2 + y^2 + z^2 = 2$ and outside the cylinder $x^2 + y^2 = 1$.

57. Cylinder and planes Find the volume of the region enclosed by the cylinder $x^2 + y^2 = 4$ and the planes $z = 0$ and $y + z = 4$.

58. Cylinder and planes Find the volume of the region enclosed by the cylinder $x^2 + y^2 = 4$ and the planes $z = 0$ and $x + y + z = 4$.

59. Region trapped by paraboloids Find the volume of the region bounded above by the paraboloid $z = 5 - x^2 - y^2$ and below by the paraboloid $z = 4x^2 + 4y^2$.

60. Paraboloid and cylinder Find the volume of the region bounded above by the paraboloid $z = 9 - x^2 - y^2$, below by the xy-plane, and lying *outside* the cylinder $x^2 + y^2 = 1$.

61. Cylinder and sphere Find the volume of the region cut from the solid cylinder $x^2 + y^2 \le 1$ by the sphere $x^2 + y^2 + z^2 = 4$.

62. Sphere and paraboloid Find the volume of the region bounded above by the sphere $x^2 + y^2 + z^2 = 2$ and below by the paraboloid $z = x^2 + y^2$.

63. Find the average value of the function $f(r, \theta, z) = r$ over the region bounded by the cylinder $r = 1$ between the planes $z = -1$ and $z = 1$.

64. Find the average value of the function $f(r, \theta, z) = r$ over the solid ball bounded by the sphere $r^2 + z^2 = 1$. (This is the sphere $x^2 + y^2 + z^2 = 1$.)

65. Find the average value of the function $f(\rho, \phi, \theta) = \rho$ over the solid ball $\rho \le 1$.

66. Find the average value of the function $f(\rho, \phi, \theta) = \rho\cos\phi$ over the solid upper ball $\rho \le 1, 0 \le \phi \le \pi/2$.

67. Center of mass A solid of constant density is bounded below by the plane $z = 0$, above by the cone $z = r, r \ge 0$, and on the sides by the cylinder $r = 1$. Find the center of mass.

68. Centroid Find the centroid of the region in the first octant that is bounded above by the cone $z = \sqrt{x^2 + y^2}$, below by the plane $z = 0$, and on the sides by the cylinder $x^2 + y^2 = 4$ and the planes $x = 0$ and $y = 0$.

69. Centroid Find the centroid of the solid in Exercise 38.

70. Centroid Find the centroid of the solid bounded above by the sphere $\rho = a$ and below by the cone $\phi = \pi/4$.

71. Centroid Find the centroid of the region that is bounded above by the surface $z = \sqrt{r}$, on the sides by the cylinder $r = 4$, and below by the xy-plane.

72. Centroid Find the centroid of the region cut from the solid ball $r^2 + z^2 \le 1$ by the half-planes $\theta = -\pi/3, r \ge 0$, and $\theta = \pi/3, r \ge 0$.

73. Moment of inertia of solid cone Find the moment of inertia of a right circular cone of base radius 1 and height 1 about an axis through the vertex parallel to the base. (Take $\delta = 1$.)

74. Moment of inertia of solid sphere Find the moment of inertia of a solid sphere of radius a about a diameter. (Take $\delta = 1$.)

75. Variable density A solid is bounded below by the cone $z = \sqrt{x^2 + y^2}$ and above by the plane $z = 1$. Find the center of mass and the moment of inertia about the z-axis if the density is

a. $\delta(r, \theta, z) = z$ **b.** $\delta(r, \theta, z) = z^2$.

76. Variable density A solid ball is bounded by the sphere $\rho = a$. Find the moment of inertia about the z-axis if the density is

a. $\delta(\rho, \phi, \theta) = \rho^2$ **b.** $\delta(\rho, \phi, \theta) = r = \rho\sin\phi$.

12.8 Substitutions in Multiple Integrals

This section shows how to evaluate multiple integrals by substitution. As in single integration, the goal of substitution is to replace complicated integrals by ones that are easier to evaluate. Substitutions accomplish this by simplifying the integrand, the limits of integration, or both.

Substitutions in Double Integrals

The polar coordinate substitution of Section 12.4 is a special case of a more general substitution method for double integrals, a method that pictures changes in variables as transformations of regions.

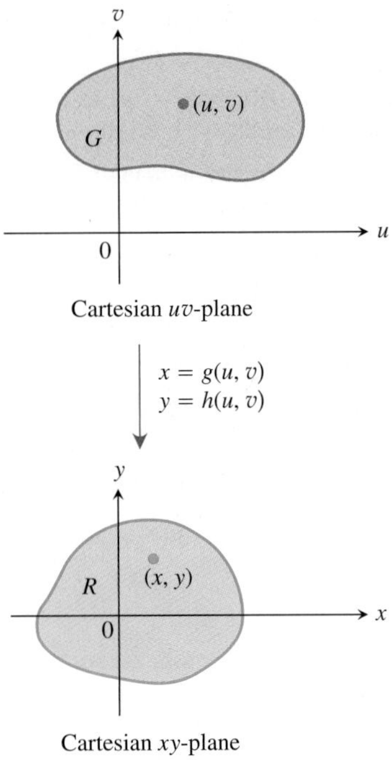

FIGURE 12.46 The equations $x = g(u, v)$ and $y = h(u, v)$ allow us to change an integral over a region R in the xy-plane into an integral over a region G in the uv-plane.

Suppose that a region G in the uv-plane is transformed one-to-one into the region R in the xy-plane by equations of the form

$$x = g(u, v), \qquad y = h(u, v),$$

as suggested in Figure 12.46. We call R the **image** of G under the transformation, and G the **preimage** of R. Any function $f(x, y)$ defined on R can be thought of as a function $f(g(u, v), h(u, v))$ defined on G as well. How is the integral of $f(x, y)$ over R related to the integral of $f(g(u, v), h(u, v))$ over G?

The answer is: If g, h, and f have continuous partial derivatives and $J(u, v)$ (to be discussed in a moment) is zero only at isolated points, if at all, then

$$\iint_R f(x, y)\,dx\,dy = \iint_G f(g(u, v), h(u, v))\,|J(u, v)|\,du\,dv. \tag{1}$$

The factor $J(u, v)$, whose absolute value appears in Equation (1), is the *Jacobian* of the coordinate transformation, named after German mathematician Carl Jacobi. It measures how much the transformation is expanding or contracting the area around a point in G as G is transformed into R.

DEFINITION The **Jacobian determinant** or **Jacobian** of the coordinate transformation $x = g(u, v)$, $y = h(u, v)$ is

$$J(u, v) = \begin{vmatrix} \dfrac{\partial x}{\partial u} & \dfrac{\partial x}{\partial v} \\ \dfrac{\partial y}{\partial u} & \dfrac{\partial y}{\partial v} \end{vmatrix} = \frac{\partial x}{\partial u}\frac{\partial y}{\partial v} - \frac{\partial y}{\partial u}\frac{\partial x}{\partial v}. \tag{2}$$

The Jacobian is also denoted by

$$J(u, v) = \frac{\partial(x, y)}{\partial(u, v)}$$

to help remember how the determinant in Equation (2) is constructed from the partial derivatives of x and y. The derivation of Equation (1) is intricate and properly belongs to a course in advanced calculus. We do not give the derivation here.

HISTORICAL BIOGRAPHY

Carl Gustav Jacob Jacobi (1804–1851)

For polar coordinates, we have r and θ in place of u and v. With $x = r\cos\theta$ and $y = r\sin\theta$, the Jacobian is

$$J(r, \theta) = \begin{vmatrix} \dfrac{\partial x}{\partial r} & \dfrac{\partial x}{\partial \theta} \\ \dfrac{\partial y}{\partial r} & \dfrac{\partial y}{\partial \theta} \end{vmatrix} = \begin{vmatrix} \cos\theta & -r\sin\theta \\ \sin\theta & r\cos\theta \end{vmatrix} = r(\cos^2\theta + \sin^2\theta) = r.$$

Hence, Equation (1) becomes

$$\iint_R f(x, y)\,dx\,dy = \iint_G f(r\cos\theta, r\sin\theta)\,|r|\,dr\,d\theta$$

$$= \iint_G f(r\cos\theta, r\sin\theta)\,r\,dr\,d\theta, \qquad \text{If } r \ge 0 \tag{3}$$

which is the equation found in Section 12.4.

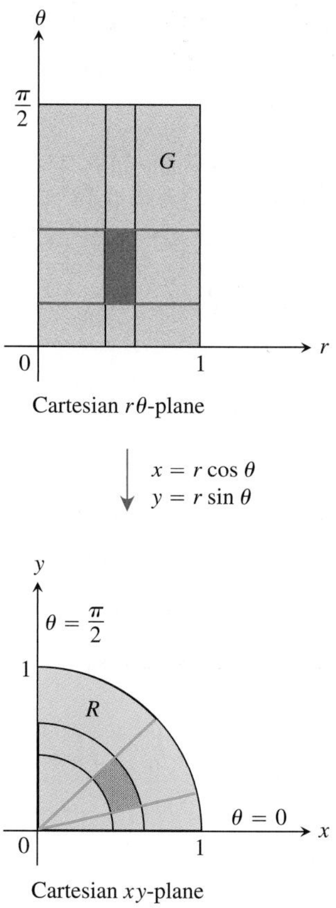

FIGURE 12.47 The equations $x = r\cos\theta$, $y = r\sin\theta$ transform G into R.

Figure 12.47 shows how the equations $x = r\cos\theta$, $y = r\sin\theta$ transform the rectangle G: $0 \le r \le 1$, $0 \le \theta \le \pi/2$ into the quarter circle R bounded by $x^2 + y^2 = 1$ in the first quadrant of the xy-plane.

Notice that the integral on the right-hand side of Equation (3) is not the integral of $f(r\cos\theta, r\sin\theta)$ over a region in the polar coordinate plane. It is the integral of the product of $f(r\cos\theta, r\sin\theta)$ and r over a region G in the *Cartesian* $r\theta$-plane.

Here is an example of another substitution.

EXAMPLE 1 Evaluate

$$\int_0^4 \int_{x=y/2}^{x=(y/2)+1} \frac{2x - y}{2}\, dx\, dy$$

by applying the transformation

$$u = \frac{2x - y}{2}, \qquad v = \frac{y}{2} \tag{4}$$

and integrating over an appropriate region in the uv-plane.

Solution We sketch the region R of integration in the xy-plane and identify its boundaries (Figure 12.48).

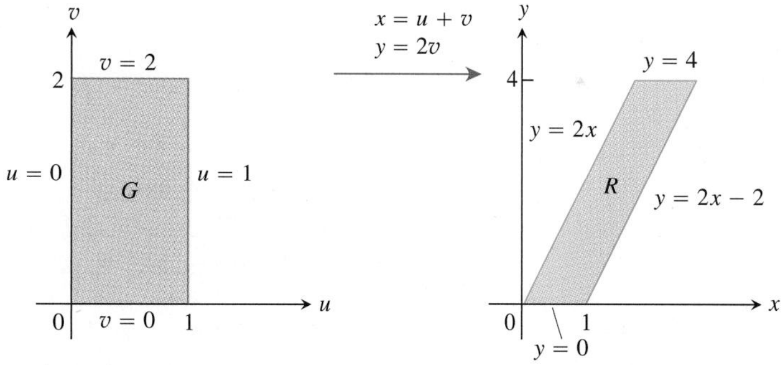

FIGURE 12.48 The equations $x = u + v$ and $y = 2v$ transform G into R. Reversing the transformation by the equations $u = (2x - y)/2$ and $v = y/2$ transforms R into G (Example 1).

To apply Equation (1), we need to find the corresponding uv-region G and the Jacobian of the transformation. To find them, we first solve Equations (4) for x and y in terms of u and v. Routine algebra gives

$$x = u + v, \qquad y = 2v. \tag{5}$$

We then find the boundaries of G by substituting these expressions into the equations for the boundaries of R (Figure 12.48).

xy-equations for the boundary of R	Corresponding uv-equations for the boundary of G	Simplified uv-equations
$x = y/2$	$u + v = 2v/2 = v$	$u = 0$
$x = (y/2) + 1$	$u + v = (2v/2) + 1 = v + 1$	$u = 1$
$y = 0$	$2v = 0$	$v = 0$
$y = 4$	$2v = 4$	$v = 2$

The Jacobian of the transformation (again from Equations (5)) is

$$J(u, v) = \begin{vmatrix} \frac{\partial x}{\partial u} & \frac{\partial x}{\partial v} \\ \frac{\partial y}{\partial u} & \frac{\partial y}{\partial v} \end{vmatrix} = \begin{vmatrix} \frac{\partial}{\partial u}(u + v) & \frac{\partial}{\partial v}(u + v) \\ \frac{\partial}{\partial u}(2v) & \frac{\partial}{\partial v}(2v) \end{vmatrix} = \begin{vmatrix} 1 & 1 \\ 0 & 2 \end{vmatrix} = 2.$$

We now have everything we need to apply Equation (1):

$$\int_0^4 \int_{x=y/2}^{x=(y/2)+1} \frac{2x - y}{2}\, dx\, dy = \int_{v=0}^{v=2} \int_{u=0}^{u=1} u\,|J(u, v)|\, du\, dv$$

$$= \int_0^2 \int_0^1 (u)(2)\, du\, dv = \int_0^2 \Big[u^2\Big]_0^1 dv = \int_0^2 dv = 2.$$

■

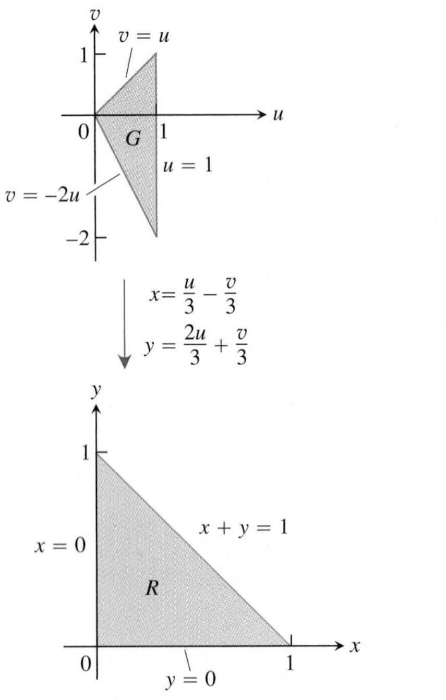

FIGURE 12.49 The equations $x = (u/3) - (v/3)$ and $y = (2u/3) + (v/3)$ transform G into R. Reversing the transformation by the equations $u = x + y$ and $v = y - 2x$ transforms R into G (Example 2).

EXAMPLE 2 Evaluate

$$\int_0^1 \int_0^{1-x} \sqrt{x + y}\,(y - 2x)^2\, dy\, dx.$$

Solution We sketch the region R of integration in the xy-plane and identify its boundaries (Figure 12.49). The integrand suggests the transformation $u = x + y$ and $v = y - 2x$. Routine algebra produces x and y as functions of u and v:

$$x = \frac{u}{3} - \frac{v}{3}, \qquad y = \frac{2u}{3} + \frac{v}{3}. \tag{6}$$

From Equations (6), we can find the boundaries of the uv-region G (Figure 12.49).

xy-equations for the boundary of R	Corresponding uv-equations for the boundary of G	Simplified uv-equations
$x + y = 1$	$\left(\frac{u}{3} - \frac{v}{3}\right) + \left(\frac{2u}{3} + \frac{v}{3}\right) = 1$	$u = 1$
$x = 0$	$\frac{u}{3} - \frac{v}{3} = 0$	$v = u$
$y = 0$	$\frac{2u}{3} + \frac{v}{3} = 0$	$v = -2u$

The Jacobian of the transformation in Equations (6) is

$$J(u, v) = \begin{vmatrix} \frac{\partial x}{\partial u} & \frac{\partial x}{\partial v} \\ \frac{\partial y}{\partial u} & \frac{\partial y}{\partial v} \end{vmatrix} = \begin{vmatrix} \frac{1}{3} & -\frac{1}{3} \\ \frac{2}{3} & \frac{1}{3} \end{vmatrix} = \frac{1}{3}.$$

Applying Equation (1), we evaluate the integral:

$$\int_0^1 \int_0^{1-x} \sqrt{x + y}\,(y - 2x)^2\, dy\, dx = \int_{u=0}^{u=1} \int_{v=-2u}^{v=u} u^{1/2} v^2 |J(u, v)|\, dv\, du$$

$$= \int_0^1 \int_{-2u}^{u} u^{1/2} v^2 \left(\frac{1}{3}\right) dv\, du = \frac{1}{3}\int_0^1 u^{1/2} \left[\frac{1}{3} v^3\right]_{v=-2u}^{v=u} du$$

$$= \frac{1}{9}\int_0^1 u^{1/2}(u^3 + 8u^3)\, du = \int_0^1 u^{7/2}\, du = \frac{2}{9} u^{9/2}\bigg]_0^1 = \frac{2}{9}.$$ ■

Substitutions in Triple Integrals

The cylindrical and spherical coordinate substitutions in Section 12.7 are special cases of a substitution method that pictures changes of variables in triple integrals as transformations of three-dimensional regions. The method is like the method for double integrals except that now we work in three dimensions instead of two.

Suppose that a region G in uvw-space is transformed one-to-one into the region D in xyz-space by differentiable equations of the form

$$x = g(u, v, w), \qquad y = h(u, v, w), \qquad z = k(u, v, w),$$

as suggested in Figure 12.50. Then any function $F(x, y, z)$ defined on D can be thought of as a function

$$F(g(u, v, w), h(u, v, w), k(u, v, w)) = H(u, v, w)$$

defined on G. If g, h, and k have continuous first partial derivatives, then the integral of $F(x, y, z)$ over D is related to the integral of $H(u, v, w)$ over G by the equation

$$\iiint_D F(x, y, z)\, dx\, dy\, dz = \iiint_G H(u, v, w)\,|J(u, v, w)|\, du\, dv\, dw. \tag{7}$$

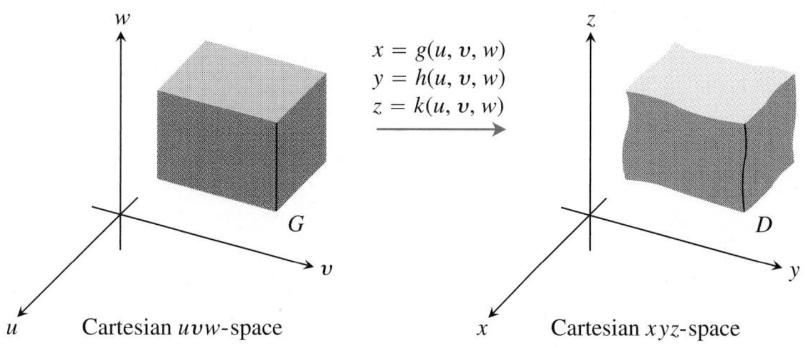

FIGURE 12.50 The equations $x = g(u, v, w)$, $y = h(u, v, w)$, and $z = k(u, v, w)$ allow us to change an integral over a region D in Cartesian xyz-space into an integral over a region G in Cartesian uvw-space.

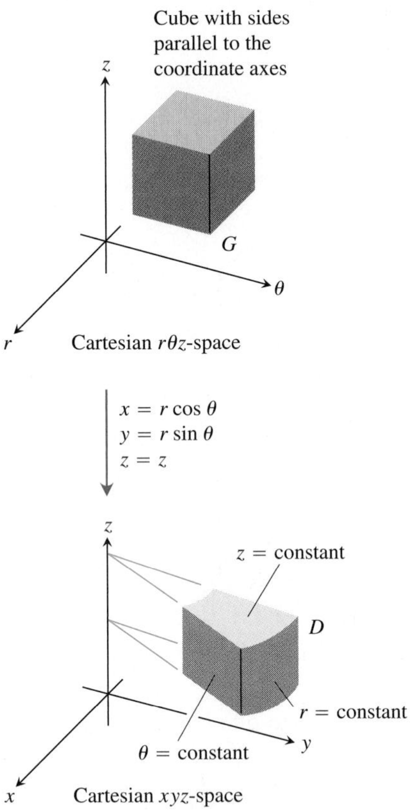

FIGURE 12.51 The equations $x = r\cos\theta$, $y = r\sin\theta$, and $z = z$ transform the cube G into a cylindrical wedge D.

The factor $J(u, v, w)$, whose absolute value appears in this equation, is the **Jacobian determinant**

$$J(u, v, w) = \begin{vmatrix} \dfrac{\partial x}{\partial u} & \dfrac{\partial x}{\partial v} & \dfrac{\partial x}{\partial w} \\[2mm] \dfrac{\partial y}{\partial u} & \dfrac{\partial y}{\partial v} & \dfrac{\partial y}{\partial w} \\[2mm] \dfrac{\partial z}{\partial u} & \dfrac{\partial z}{\partial v} & \dfrac{\partial z}{\partial w} \end{vmatrix} = \frac{\partial(x, y, z)}{\partial(u, v, w)}.$$

This determinant measures how much the volume near a point in G is being expanded or contracted by the transformation from (u, v, w) to (x, y, z) coordinates. As in the two-dimensional case, the derivation of the change-of-variable formula in Equation (7) is complicated and we do not go into it here.

For cylindrical coordinates, r, θ, and z take the place of u, v, and w. The transformation from Cartesian $r\theta z$-space to Cartesian xyz-space is given by the equations

$$x = r\cos\theta, \qquad y = r\sin\theta, \qquad z = z$$

(Figure 12.51). The Jacobian of the transformation is

$$J(r, \theta, z) = \begin{vmatrix} \dfrac{\partial x}{\partial r} & \dfrac{\partial x}{\partial \theta} & \dfrac{\partial x}{\partial z} \\[2mm] \dfrac{\partial y}{\partial r} & \dfrac{\partial y}{\partial \theta} & \dfrac{\partial y}{\partial z} \\[2mm] \dfrac{\partial z}{\partial r} & \dfrac{\partial z}{\partial \theta} & \dfrac{\partial z}{\partial z} \end{vmatrix} = \begin{vmatrix} \cos\theta & -r\sin\theta & 0 \\ \sin\theta & r\cos\theta & 0 \\ 0 & 0 & 1 \end{vmatrix}$$

$$= r\cos^2\theta + r\sin^2\theta = r.$$

The corresponding version of Equation (7) is

$$\iiint_D F(x, y, z)\, dx\, dy\, dz = \iiint_G H(r, \theta, z)\,|r|\, dr\, d\theta\, dz.$$

We can drop the absolute value signs whenever $r \geq 0$.

For spherical coordinates, ρ, ϕ, and θ take the place of u, v, and w. The transformation from Cartesian $\rho\phi\theta$-space to Cartesian xyz-space is given by

$$x = \rho\sin\phi\cos\theta, \qquad y = \rho\sin\phi\sin\theta, \qquad z = \rho\cos\phi$$

(Figure 12.52). The Jacobian of the transformation (see Exercise 17) is

$$J(\rho, \phi, \theta) = \begin{vmatrix} \dfrac{\partial x}{\partial \rho} & \dfrac{\partial x}{\partial \phi} & \dfrac{\partial x}{\partial \theta} \\[2mm] \dfrac{\partial y}{\partial \rho} & \dfrac{\partial y}{\partial \phi} & \dfrac{\partial y}{\partial \theta} \\[2mm] \dfrac{\partial z}{\partial \rho} & \dfrac{\partial z}{\partial \phi} & \dfrac{\partial z}{\partial \theta} \end{vmatrix} = \rho^2\sin\phi.$$

The corresponding version of Equation (7) is

$$\iiint_D F(x, y, z)\, dx\, dy\, dz = \iiint_G H(\rho, \phi, \theta)\,|\rho^2\sin\phi|\, d\rho\, d\phi\, d\theta.$$

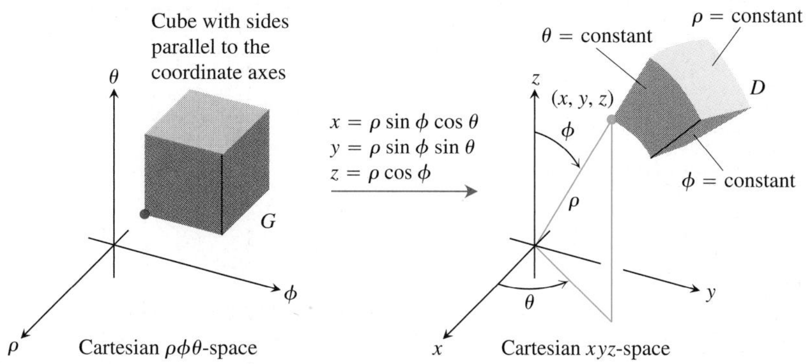

FIGURE 12.52 The equations $x = \rho \sin\phi \cos\theta$, $y = \rho \sin\phi \sin\theta$, and $z = \rho \cos\phi$ transform the cube G into the spherical wedge D.

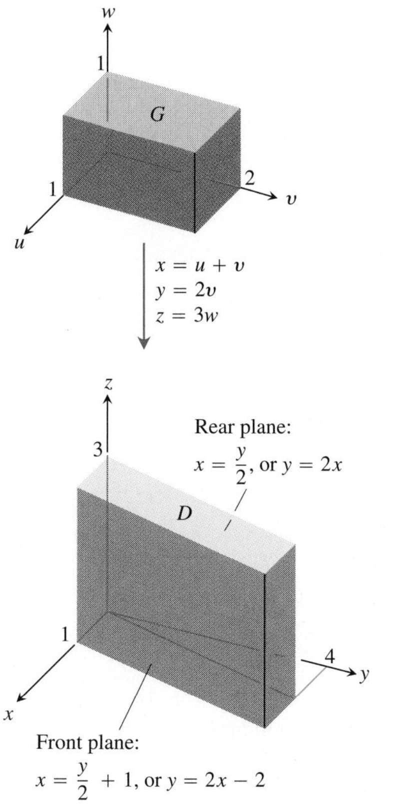

FIGURE 12.53 The equations $x = u + v$, $y = 2v$, and $z = 3w$ transform G into D. Reversing the transformation by the equations $u = (2x - y)/2$, $v = y/2$, and $w = z/3$ transforms D into G (Example 3).

We can drop the absolute value signs because $\sin\phi$ is never negative for $0 \le \phi \le \pi$. Note that this is the same result we obtained in Section 12.7.

Here is an example of another substitution. Although we could evaluate the integral in this example directly, we have chosen it to illustrate the substitution method in a simple (and fairly intuitive) setting.

EXAMPLE 3 Evaluate

$$\int_0^3 \int_0^4 \int_{x=y/2}^{x=(y/2)+1} \left(\frac{2x - y}{2} + \frac{z}{3}\right) dx\, dy\, dz$$

by applying the transformation

$$u = (2x - y)/2, \qquad v = y/2, \qquad w = z/3 \tag{8}$$

and integrating over an appropriate region in uvw-space.

Solution We sketch the region D of integration in xyz-space and identify its boundaries (Figure 12.53). In this case, the bounding surfaces are planes.

To apply Equation (7), we need to find the corresponding uvw-region G and the Jacobian of the transformation. To find them, we first solve Equations (8) for x, y, and z in terms of u, v, and w. Routine algebra gives

$$x = u + v, \qquad y = 2v, \qquad z = 3w. \tag{9}$$

We then find the boundaries of G by substituting these expressions into the equations for the boundaries of D:

xyz*-equations for the boundary of *D	**Corresponding *uvw*-equations for the boundary of *G***	**Simplified *uvw*-equations**
$x = y/2$	$u + v = 2v/2 = v$	$u = 0$
$x = (y/2) + 1$	$u + v = (2v/2) + 1 = v + 1$	$u = 1$
$y = 0$	$2v = 0$	$v = 0$
$y = 4$	$2v = 4$	$v = 2$
$z = 0$	$3w = 0$	$w = 0$
$z = 3$	$3w = 3$	$w = 1$

The Jacobian of the transformation, again from Equations (9), is

$$J(u, v, w) = \begin{vmatrix} \dfrac{\partial x}{\partial u} & \dfrac{\partial x}{\partial v} & \dfrac{\partial x}{\partial w} \\ \dfrac{\partial y}{\partial u} & \dfrac{\partial y}{\partial v} & \dfrac{\partial y}{\partial w} \\ \dfrac{\partial z}{\partial u} & \dfrac{\partial z}{\partial v} & \dfrac{\partial z}{\partial w} \end{vmatrix} = \begin{vmatrix} 1 & 1 & 0 \\ 0 & 2 & 0 \\ 0 & 0 & 3 \end{vmatrix} = 6.$$

We now have everything we need to apply Equation (7):

$$\int_0^3 \int_0^4 \int_{x=y/2}^{x=(y/2)+1} \left(\frac{2x - y}{2} + \frac{z}{3} \right) dx\, dy\, dz$$

$$= \int_0^1 \int_0^2 \int_0^1 (u + w)\,|J(u, v, w)|\, du\, dv\, dw$$

$$= \int_0^1 \int_0^2 \int_0^1 (u + w)(6)\, du\, dv\, dw = 6 \int_0^1 \int_0^2 \left[\frac{u^2}{2} + uw \right]_0^1 dv\, dw$$

$$= 6 \int_0^1 \int_0^2 \left(\frac{1}{2} + w \right) dv\, dw = 6 \int_0^1 \left[\frac{v}{2} + vw \right]_0^2 dw = 6 \int_0^1 (1 + 2w)\, dw$$

$$= 6\Big[w + w^2\Big]_0^1 = 6(2) = 12.$$

The goal of this section was to introduce you to the ideas involved in coordinate transformations. A thorough discussion of transformations, the Jacobian, and multivariable substitution is best given in an advanced calculus course after a study of linear algebra.

EXERCISES 12.8

1. a. Solve the system

$$u = x - y, \qquad v = 2x + y$$

for x and y in terms of u and y. Then find the value of the Jacobian $\partial(x, y)/\partial(u, v)$.

b. Find the image under the transformation $u = x - y$, $v = 2x + y$ of the triangular region with vertices (0, 0), (1, 1), and (1, −2) in the xy-plane. Sketch the transformed region in the uv-plane.

2. a. Solve the system

$$u = x + 2y, \qquad v = x - y$$

for x and y in terms of u and v. Then find the value of the Jacobian $\partial(x, y)/\partial(u, v)$.

b. Find the image under the transformation $u = x + 2y$, $v = x - y$ of the triangular region in the xy-plane bounded by the lines $y = 0$, $y = x$, and $x + 2y = 2$. Sketch the transformed region in the uv-plane.

3. a. Solve the system

$$u = 3x + 2y, \qquad v = x + 4y$$

for x and y in terms of u and v. Then find the value of the Jacobian $\partial(x, y)/\partial(u, v)$.

b. Find the image under the transformation $u = 3x + 2y$, $v = x + 4y$ of the triangular region in the xy-plane bounded by the x-axis, the y-axis, and the line $x + y = 1$. Sketch the transformed region in the uv-plane.

4. a. Solve the system

$$u = 2x - 3y, \qquad v = -x + y$$

for x and y in terms of u and v. Then find the value of the Jacobian $\partial(x, y)/\partial(u, v)$.

b. Find the image under the transformation $u = 2x - 3y$, $v = -x + y$ of the parallelogram R in the xy-plane with boundaries $x = -3$, $x = 0$, $y = x$, and $y = x + 1$. Sketch the transformed region in the uv-plane.

5. Evaluate the integral

$$\int_0^4 \int_{x=y/2}^{x=(y/2)+1} \frac{2x-y}{2}\, dx\, dy$$

from Example 1 directly by integration with respect to x and y to confirm that its value is 2.

6. Use the transformation in Exercise 1 to evaluate the integral

$$\iint_R (2x^2 - xy - y^2)\, dx\, dy$$

for the region R in the first quadrant bounded by the lines $y = -2x + 4$, $y = -2x + 7$, $y = x - 2$, and $y = x + 1$.

7. Use the transformation in Exercise 3 to evaluate the integral

$$\iint_R (3x^2 + 14xy + 8y^2)\, dx\, dy$$

for the region R in the first quadrant bounded by the lines $y = -(3/2)x + 1$, $y = -(3/2)x + 3$, $y = -(1/4)x$, and $y = -(1/4)x + 1$.

8. Use the transformation and parallelogram R in Exercise 4 to evaluate the integral

$$\iint_R 2(x - y)\, dx\, dy.$$

9. Let R be the region in the first quadrant of the xy-plane bounded by the hyperbolas $xy = 1$, $xy = 9$ and the lines $y = x$, $y = 4x$. Use the transformation $x = u/v$, $y = uv$ with $u > 0$ and $v > 0$ to rewrite

$$\iint_R \left(\sqrt{\frac{y}{x}} + \sqrt{xy}\right) dx\, dy$$

as an integral over an appropriate region G in the uv-plane. Then evaluate the uv-integral over G.

10. a. Find the Jacobian of the transformation $x = u$, $y = uv$, and sketch the region G: $1 \le u \le 2$, $1 \le uv \le 2$ in the uv-plane.

b. Then use Equation (1) to transform the integral

$$\int_1^2 \int_1^2 \frac{y}{x}\, dy\, dx$$

into an integral over G, and evaluate both integrals.

11. Polar moment of inertia of an elliptical plate A thin plate of constant density covers the region bounded by the ellipse $x^2/a^2 + y^2/b^2 = 1$, $a > 0$, $b > 0$, in the xy-plane. Find the first moment of the plate about the origin. (*Hint:* Use the transformation $x = ar\cos\theta$, $y = br\sin\theta$.)

12. The area of an ellipse The area πab of the ellipse $x^2/a^2 + y^2/b^2 = 1$ can be found by integrating the function $f(x, y) = 1$ over the region bounded by the ellipse in the xy-plane. Evaluating the integral directly requires a trigonometric substitution. An easier way to evaluate the integral is to use the transformation $x = au$, $y = bv$ and evaluate the transformed integral over the disk G: $u^2 + v^2 \le 1$ in the uv-plane. Find the area this way.

13. Use the transformation in Exercise 2 to evaluate the integral

$$\int_0^{2/3} \int_y^{2-2y} (x + 2y)e^{(y-x)}\, dx\, dy$$

by first writing it as an integral over a region G in the uv-plane.

14. Use the transformation $x = u + (1/2)v$, $y = v$ to evaluate the integral

$$\int_0^2 \int_{y/2}^{(y+4)/2} y^3(2x - y)e^{(2x-y)^2}\, dx\, dy$$

by first writing it as an integral over a region G in the uv-plane.

15. Find the Jacobian $\partial(x, y)/\partial(u, v)$ for the transformation

a. $x = u\cos v$, $y = u\sin v$

b. $x = u\sin v$, $y = u\cos v$.

16. Find the Jacobian $\partial(x, y, z)/\partial(u, v, w)$ of the transformation

a. $x = u\cos v$, $y = u\sin v$, $z = w$

b. $x = 2u - 1$, $y = 3v - 4$, $z = (1/2)(w - 4)$.

17. Evaluate the appropriate determinant to show that the Jacobian of the transformation from Cartesian $\rho\phi\theta$-space to Cartesian xyz-space is $\rho^2 \sin\phi$.

18. Substitutions in single integrals How can substitutions in single definite integrals be viewed as transformations of regions? What is the Jacobian in such a case? Illustrate with an example.

19. Evaluate the integral in Example 3 by integrating with respect to x, y, and z.

20. Volume of an ellipsoid Find the volume of the ellipsoid

$$\frac{x^2}{a^2} + \frac{y^2}{b^2} + \frac{z^2}{c^2} = 1.$$

(*Hint:* Let $x = au$, $y = bv$, and $z = cw$. Then find the volume of an appropriate region in uvw-space.)

21. Evaluate

$$\iiint |xyz|\, dx\, dy\, dz$$

over the solid ellipsoid

$$\frac{x^2}{a^2} + \frac{y^2}{b^2} + \frac{z^2}{c^2} \le 1.$$

(*Hint:* Let $x = au$, $y = bv$, and $z = cw$. Then integrate over an appropriate region in uvw-space.)

22. Let D be the region in xyz-space defined by the inequalities

$$1 \le x \le 2, \quad 0 \le xy \le 2, \quad 0 \le z \le 1.$$

Evaluate

$$\iiint_D (x^2 y + 3xyz)\, dx\, dy\, dz$$

by applying the transformation

$$u = x, \quad v = xy, \quad w = 3z$$

and integrating over an appropriate region G in uvw-space.

Chapter 12 Questions to Guide Your Review

1. Define the double integral of a function of two variables over a bounded region in the coordinate plane.
2. How are double integrals evaluated as iterated integrals? Does the order of integration matter? How are the limits of integration determined? Give examples.
3. How are double integrals used to calculate areas and average values. Give examples.
4. How can you change a double integral in rectangular coordinates into a double integral in polar coordinates? Why might it be worthwhile to do so? Give an example.
5. Define the triple integral of a function $f(x, y, z)$ over a bounded region in space.
6. How are triple integrals in rectangular coordinates evaluated? How are the limits of integration determined? Give an example.
7. How are double and triple integrals in rectangular coordinates used to calculate volumes, average values, masses, moments, and centers of mass. Give examples.
8. How are triple integrals defined in cylindrical and spherical coordinates? Why might one prefer working in one of these coordinate systems to working in rectangular coordinates?
9. How are triple integrals in cylindrical and spherical coordinates evaluated? How are the limits of integration found? Give examples.
10. How are substitutions in double integrals pictured as transformations of two-dimensional regions? Give a sample calculation.
11. How are substitutions in triple integrals pictured as transformations of three-dimensional regions? Give a sample calculation.

Chapter 12 Practice and Additional Exercises

In Exercises 1–4, sketch the region of integration and evaluate the double integral.

1. $\displaystyle\int_1^{10}\int_0^{1/y} ye^{xy}\,dx\,dy$

2. $\displaystyle\int_0^1\int_0^{x^3} e^{y/x}\,dy\,dx$

3. $\displaystyle\int_0^{3/2}\int_{-\sqrt{9-4t^2}}^{\sqrt{9-4t^2}} t\,ds\,dt$

4. $\displaystyle\int_0^1\int_{\sqrt{y}}^{2-\sqrt{y}} xy\,dx\,dy$

In Exercises 5–8, sketch the region of integration and write an equivalent integral with the order of integration reversed. Then evaluate both integrals.

5. $\displaystyle\int_0^4\int_{-\sqrt{4-y}}^{(y-4)/2} dx\,dy$

6. $\displaystyle\int_0^1\int_{x^2}^{x} \sqrt{x}\,dy\,dx$

7. $\displaystyle\int_0^{3/2}\int_{-\sqrt{9-4y^2}}^{\sqrt{9-4y^2}} y\,dx\,dy$

8. $\displaystyle\int_0^2\int_0^{4-x^2} 2x\,dy\,dx$

Evaluate the integrals in Exercises 9–12.

9. $\displaystyle\int_0^1\int_{2y}^{2} 4\cos(x^2)\,dx\,dy$

10. $\displaystyle\int_0^2\int_{y/2}^{1} e^{x^2}\,dx\,dy$

11. $\displaystyle\int_0^8\int_{\sqrt[3]{x}}^{2} \frac{dy\,dx}{y^4+1}$

12. $\displaystyle\int_0^1\int_{\sqrt[3]{y}}^{1} \frac{2\pi\sin \pi x^2}{x^2}\,dx\,dy$

13. **Area between line and parabola** Find the area of the region enclosed by the line $y = 2x + 4$ and the parabola $y = 4 - x^2$ in the xy-plane.

14. **Area bounded by lines and parabola** Find the area of the "triangular" region in the xy-plane that is bounded on the right by the parabola $y = x^2$, on the left by the line $x + y = 2$, and above by the line $y = 4$.

15. **Volume of the region under a paraboloid** Find the volume under the paraboloid $z = x^2 + y^2$ above the triangle enclosed by the lines $y = x$, $x = 0$, and $x + y = 2$ in the xy-plane.

16. **Volume of the region under parabolic cylinder** Find the volume under the parabolic cylinder $z = x^2$ above the region enclosed by the parabola $y = 6 - x^2$ and the line $y = x$ in the xy-plane.

Find the average value of $f(x, y) = xy$ over the regions in Exercises 17 and 18.

17. The square bounded by the lines $x = 1$, $y = 1$ in the first quadrant
18. The quarter circle $x^2 + y^2 \le 1$ in the first quadrant

Evaluate the integrals in Exercises 19 and 20 by changing to polar coordinates.

19. $\displaystyle\int_{-1}^1\int_{-\sqrt{1-x^2}}^{\sqrt{1-x^2}} \frac{2\,dy\,dx}{(1+x^2+y^2)^2}$

20. $\displaystyle\int_{-1}^1\int_{-\sqrt{1-y^2}}^{\sqrt{1-y^2}} \ln(x^2+y^2+1)\,dx\,dy$

21. **Integrating over lemniscate** Integrate the function $f(x, y) = 1/(1 + x^2 + y^2)^2$ over the region enclosed by one loop of the lemniscate $(x^2 + y^2)^2 - (x^2 - y^2) = 0$.

22. Integrate $f(x, y) = 1/(1 + x^2 + y^2)^2$ over
    a. **Triangular region** The triangle with vertices $(0, 0)$, $(1, 0)$, $(1, \sqrt{3})$.
    b. **First quadrant** The first quadrant of the xy-plane.

Evaluate the integrals in Exercises 23–26.

23. $\displaystyle\int_0^{\pi}\int_0^{\pi}\int_0^{\pi} \cos(x + y + z)\, dx\, dy\, dz$

24. $\displaystyle\int_{\ln 6}^{\ln 7}\int_0^{\ln 2}\int_{\ln 4}^{\ln 5} e^{(x+y+z)}\, dz\, dy\, dx$

25. $\displaystyle\int_0^{1}\int_0^{x^2}\int_0^{x+y} (2x - y - z)\, dz\, dy\, dx$

26. $\displaystyle\int_1^{e}\int_1^{x}\int_0^{z} \frac{2y}{z^3}\, dy\, dz\, dx$

27. Volume Find the volume of the wedge-shaped region enclosed on the side by the cylinder $x = -\cos y$, $-\pi/2 \le y \le \pi/2$, on the top by the plane $z = -2x$, and below by the xy-plane.

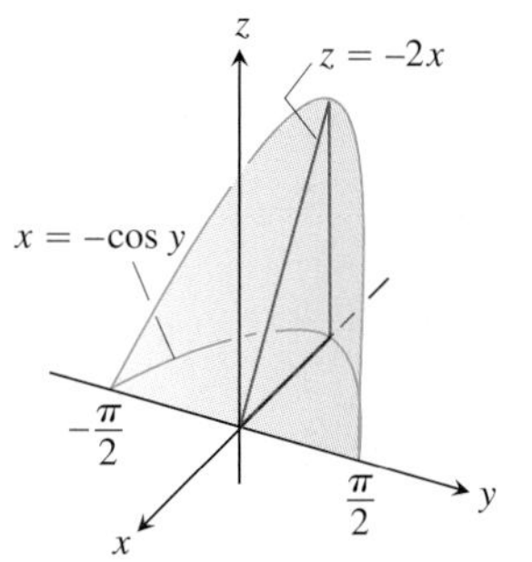

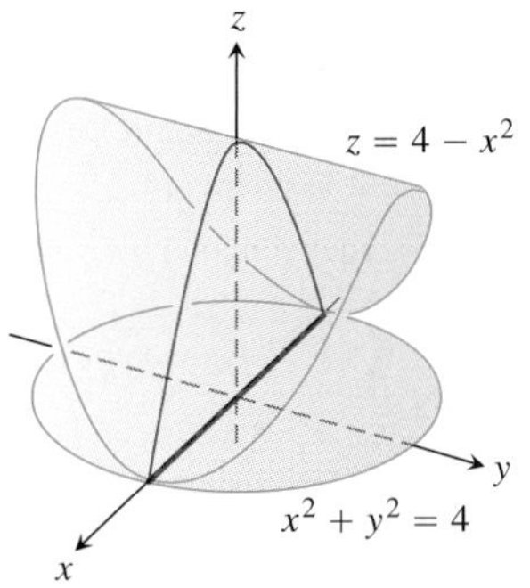

28. Volume Find the volume of the solid that is bounded above by the cylinder $z = 4 - x^2$, on the sides by the cylinder $x^2 + y^2 = 4$, and below by the xy-plane.

29. Average value Find the average value of $f(x, y, z) = 30xz\sqrt{x^2 + y}$ over the rectangular solid in the first octant bounded by the coordinate planes and the planes $x = 1$, $y = 3$, $z = 1$.

30. Average value Find the average value of ρ over the solid sphere $\rho \le a$ (spherical coordinates).

31. Cylindrical to rectangular coordinates Convert

$$\int_0^{2\pi}\int_0^{\sqrt{2}}\int_r^{\sqrt{4-r^2}} 3\, dz\, r\, dr\, d\theta, \qquad r \ge 0$$

to **(a)** rectangular coordinates with the order of integration $dz\, dx\, dy$ and **(b)** spherical coordinates. Then **(c)** evaluate one of the integrals.

32. Rectangular to cylindrical coordinates **(a)** Convert to cylindrical coordinates. Then **(b)** evaluate the new integral.

$$\int_0^{1}\int_{-\sqrt{1-x^2}}^{\sqrt{1-x^2}}\int_{-(x^2+y^2)}^{(x^2+y^2)} 21xy^2\, dz\, dy\, dx$$

33. Rectangular to spherical coordinates **(a)** Convert to spherical coordinates. Then **(b)** evaluate the new integral.

$$\int_{-1}^{1}\int_{-\sqrt{1-x^2}}^{\sqrt{1-x^2}}\int_{\sqrt{x^2+y^2}}^{1} dz\, dy\, dx$$

34. Rectangular, cylindrical, and spherical coordinates Write an iterated triple integral for the integral of $f(x, y, z) = 6 + 4y$ over the region in the first octant bounded by the cone $z = \sqrt{x^2 + y^2}$, the cylinder $x^2 + y^2 = 1$, and the coordinate planes in **(a)** rectangular coordinates, **(b)** cylindrical coordinates, and **(c)** spherical coordinates. Then **(d)** find the integral of f by evaluating one of the triple integrals.

35. Cylindrical to rectangular coordinates Set up an integral in rectangular coordinates equivalent to the integral

$$\int_0^{\pi/2}\int_1^{\sqrt{3}}\int_1^{\sqrt{4-r^2}} r^3(\sin\theta\cos\theta)z^2\, dz\, dr\, d\theta.$$

Arrange the order of integration to be z first, then y, then x.

36. Rectangular to cylindrical coordinates The volume of a solid is

$$\int_0^{2}\int_0^{\sqrt{2x-x^2}}\int_{-\sqrt{4-x^2-y^2}}^{\sqrt{4-x^2-y^2}} dz\, dy\, dx.$$

a. Describe the solid by giving equations for the surfaces that form its boundary.

b. Convert the integral to cylindrical coordinates but do not evaluate the integral.

37. Spherical versus cylindrical coordinates Triple integrals involving spherical shapes do not always require spherical coordinates for convenient evaluation. Some calculations may be accomplished more easily with cylindrical coordinates. As a case in point, find the volume of the region bounded above by the sphere $x^2 + y^2 + z^2 = 8$ and below by the plane $z = 2$ by using **(a)** cylindrical coordinates and **(b)** spherical coordinates.

38. Finding I_z in spherical coordinates Find the moment of inertia about the z-axis of a solid of constant density $\delta = 1$ that is bounded above by the sphere $\rho = 2$ and below by the cone $\phi = \pi/3$ (spherical coordinates).

39. Moment of inertia of a "thick" sphere Find the moment of inertia of a solid of constant density δ bounded by two concentric spheres of radii a and b $(a < b)$ about a diameter.

40. Moment of inertia of an apple Find the moment of inertia about the z-axis of a solid of density $\delta = 1$ enclosed by the spherical coordinate surface $\rho = 1 - \cos\phi$. The solid is the red curve rotated about the z-axis in the accompanying figure.

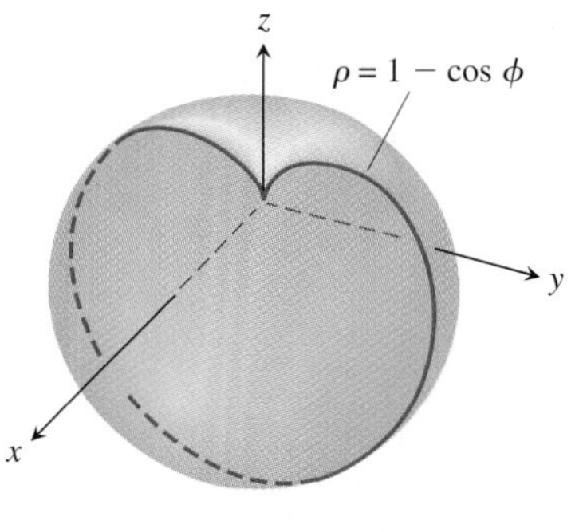

41. Centroid Find the centroid of the "triangular" region bounded by the lines $x = 2, y = 2$ and the hyperbola $xy = 2$ in the xy-plane.

42. Centroid Find the centroid of the region between the parabola $x + y^2 - 2y = 0$ and the line $x + 2y = 0$ in the xy-plane.

43. Polar moment Find the polar moment of inertia about the origin of a thin triangular plate of constant density $\delta = 3$ bounded by the y-axis and the lines $y = 2x$ and $y = 4$ in the xy-plane.

44. Polar moment Find the polar moment of inertia about the center of a thin rectangular sheet of constant density $\delta = 1$ bounded by the lines

a. $x = \pm 2, \quad y = \pm 1$ in the xy-plane

b. $x = \pm a, \quad y = \pm b$ in the xy-plane.

(*Hint:* Find I_x. Then use the formula for I_x to find I_y and add the two to find I_0).

45. Inertial moment Find the moment of inertia about the x-axis of a thin plate of constant density δ covering the triangle with vertices $(0, 0)$, $(3, 0)$, and $(3, 2)$ in the xy-plane.

46. Plate with variable density Find the center of mass and the moments of inertia about the coordinate axes of a thin plate bounded by the line $y = x$ and the parabola $y = x^2$ in the xy-plane if the density is $\delta(x, y) = x + 1$.

47. Plate with variable density Find the mass and first moments about the coordinate axes of a thin square plate bounded by the lines $x = \pm 1, y = \pm 1$ in the xy-plane if the density is $\delta(x, y) = x^2 + y^2 + 1/3$.

48. Triangles with same inertial moment Find the moment of inertia about the x-axis of a thin triangular plate of constant density δ whose base lies along the interval $[0, b]$ on the x-axis and whose vertex lies on the line $y = h$ above the x-axis. As you will see, it does not matter where on the line this vertex lies. All such triangles have the same moment of inertia about the x-axis.

49. Centroid Find the centroid of the region in the polar coordinate plane defined by the inequalities $0 \le r \le 3, -\pi/3 \le \theta \le \pi/3$.

50. Centroid Find the centroid of the region in the first quadrant bounded by the rays $\theta = 0$ and $\theta = \pi/2$ and the circles $r = 1$ and $r = 3$.

51. a. Centroid Find the centroid of the region in the polar coordinate plane that lies inside the cardioid $r = 1 + \cos\theta$ and outside the circle $r = 1$.

b. Sketch the region and show the centroid in your sketch.

52. a. Centroid Find the centroid of the plane region defined by the polar coordinate inequalities $0 \le r \le a, \ -\alpha \le \theta \le \alpha$ $(0 < \alpha \le \pi)$. How does the centroid move as $\alpha \to \pi^-$?

b. Sketch the region for $\alpha = 5\pi/6$ and show the centroid in your sketch.

53. Show that if $u = x - y$ and $v = y$, then

$$\int_0^\infty \int_0^x e^{-sx} f(x - y, y)\, dy\, dx = \int_0^\infty \int_0^\infty e^{-s(u+v)} f(u, v)\, du\, dv.$$

54. What relationship must hold between the constants a, b, and c to make

$$\int_{-\infty}^{\infty} \int_{-\infty}^{\infty} e^{-(ax^2+2bxy+cy^2)}\, dx\, dy = 1?$$

(*Hint:* Let $s = \alpha x + \beta y$ and $t = \gamma x + \delta y$, where $(\alpha\delta - \beta\gamma)^2 = ac - b^2$. Then $ax^2 + 2bxy + cy^2 = s^2 + t^2$.)

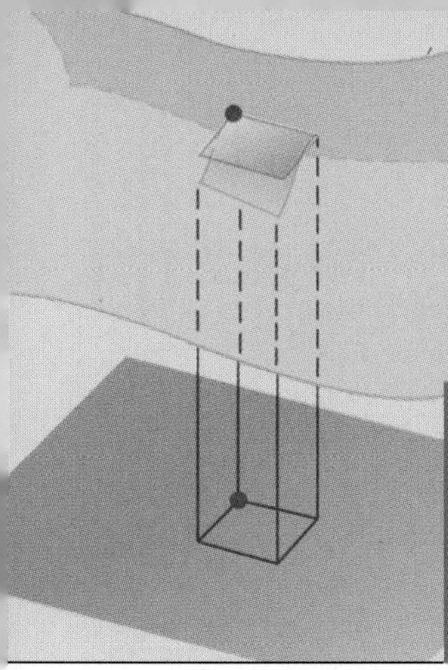

Chapter
13
INTEGRATION IN VECTOR FIELDS

OVERVIEW In this chapter we combine ideas from the preceding four chapters: space curves, vector functions, partial derivatives, and multiple integrals. The result is the development of line integrals, vector fields, and surface integrals giving powerful mathematical tools for science and engineering. Line integrals are used to find the work done by a force in moving an object along a path, and to find the mass of a wire having a varying density. Surface integrals are used to find the rate at which a fluid flows across a surface. We also present several important theorems that connect these new tools and offer insight into their mathematical calculations and physical applications.

13.1 Line Integrals

To calculate the masses of thin rods or wires lying along a *curve* in space, or to find the work done by a variable force acting along such a curve, we need a more general notion of integral than was defined in Chapter 4. We need to integrate over a curve C rather than over an interval $[a, b]$. These more general integrals are called *line integrals* (although "curve" integrals might be more descriptive). We make our definitions for space curves, remembering that curves in the xy-plane are just a special case with z-coordinate identically zero.

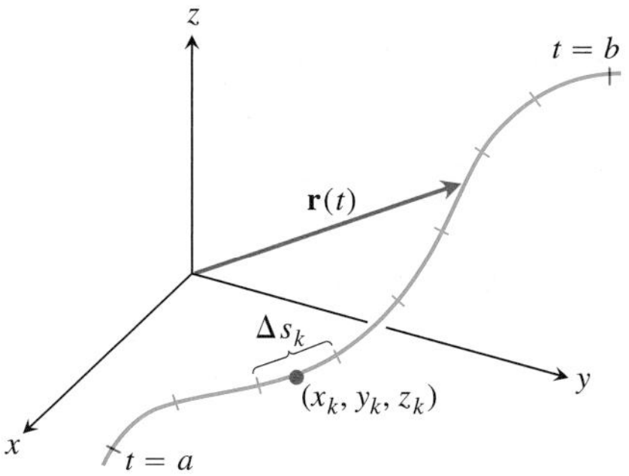

FIGURE 13.1 The curve $\mathbf{r}(t)$ partitioned into small arcs from $t = a$ to $t = b$. The length of a typical subarc is Δs_k.

Suppose that $f(x, y, z)$ is a real-valued function we wish to integrate over the curve C lying within the domain of f and parametrized by $\mathbf{r}(t) = g(t)\mathbf{i} + h(t)\mathbf{j} + k(t)\mathbf{k}$, $a \le t \le b$. The values of f along the curve are given by the composite function $f(g(t), h(t), k(t))$. We are going to integrate this composite with respect to arc length from $t = a$ to $t = b$. To begin, we first partition the curve C into a finite number n of subarcs (Figure 13.1). The typical subarc has length Δs_k. In each subarc we choose a point (x_k, y_k, z_k) and form the sum

$$S_n = \sum_{k=1}^{n} f(x_k, y_k, z_k)\, \Delta s_k,$$

which is similar to a Riemann sum. Depending on how we partition the curve C and pick (x_k, y_k, z_k) in the kth subarc, we may get different values for S_n. If f is continuous and the functions g, h, and k have continuous first derivatives, then these sums approach a limit as n increases and the lengths Δs_k approach zero. This limit gives the following definition, similar to that for a single integral.

DEFINITION If f is defined on a curve C given parametrically by $\mathbf{r}(t) = g(t)\mathbf{i} + h(t)\mathbf{j} + k(t)\mathbf{k}$, $a \le t \le b$, then the **line integral of f over C** is

$$\int_C f(x, y, z)\, ds = \lim_{\substack{n\to\infty \\ \Delta s_k \to 0}} \sum_{k=1}^{n} f(x_k, y_k, z_k)\, \Delta s_k, \tag{1}$$

provided this limit exists.

If the curve C is smooth for $a \le t \le b$ (so $\mathbf{v} = d\mathbf{r}/dt$ is continuous and never $\mathbf{0}$) and the function f is continuous on C, then the limit in Equation (1) can be shown to exist. We can then apply the Fundamental Theorem of Calculus to differentiate the arc length equation,

$$s(t) = \int_a^t |\mathbf{v}(\tau)|\, d\tau, \qquad \text{Eq. (3) of Section 10.3 with } t_0 = a$$

to express ds in Equation (1) as $ds = |\mathbf{v}(t)|\, dt$ and evaluate the integral of f over C as

$$\int_C f(x, y, z)\, ds = \int_a^b f(g(t), h(t), k(t))|\mathbf{v}(t)|\, dt. \tag{2}$$

Notice that the integral on the right side of Equation (2) is just an ordinary (single) definite integral, as defined in Chapter 4, where we are integrating with respect to the parameter t. The formula evaluates the line integral on the left side correctly no matter what parametrization is used, as long as the parametrization is smooth.

How to Evaluate a Line Integral

To integrate a continuous function $f(x, y, z)$ over a curve C:

1. Find a smooth parametrization of C,
$$\mathbf{r}(t) = g(t)\mathbf{i} + h(t)\mathbf{j} + k(t)\mathbf{k}, \qquad a \le t \le b$$
2. Evaluate the integral as
$$\int_C f(x, y, z)\, ds = \int_a^b f(g(t), h(t), k(t))|\mathbf{v}(t)|\, dt.$$

If f has the constant value 1, then the integral of f over C gives the length of C.

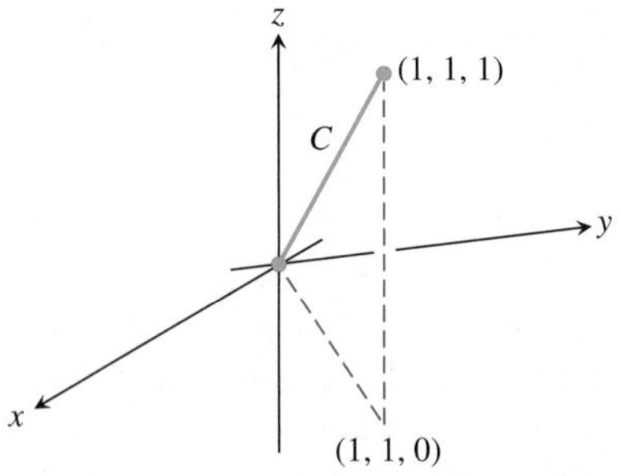

FIGURE 13.2 The integration path in Example 1.

EXAMPLE 1 Integrate $f(x, y, z) = x - 3y^2 + z$ over the line segment C joining the origin to the point (1, 1, 1) (Figure 13.2).

Solution We choose the simplest parametrization we can think of:

$$\mathbf{r}(t) = t\mathbf{i} + t\mathbf{j} + t\mathbf{k}, \qquad 0 \le t \le 1.$$

The components have continuous first derivatives and $|\mathbf{v}(t)| = |\mathbf{i} + \mathbf{j} + \mathbf{k}| = \sqrt{1^2 + 1^2 + 1^2} = \sqrt{3}$ is never 0, so the parametrization is smooth. The integral of f over C is

$$\begin{aligned}
\int_C f(x, y, z)\, ds &= \int_0^1 f(t, t, t)\left(\sqrt{3}\right) dt \qquad \text{Eq. (2)} \\
&= \int_0^1 (t - 3t^2 + t)\sqrt{3}\, dt \\
&= \sqrt{3}\int_0^1 (2t - 3t^2)\, dt = \sqrt{3}\left[t^2 - t^3\right]_0^1 = 0. \qquad \blacksquare
\end{aligned}$$

Additivity

Line integrals have the useful property that if a curve C is made by joining a finite number of curves $C_1, C_2, \ldots, C_n$ end to end, then the integral of a function over C is the sum of the integrals over the curves that make it up:

$$\int_C f\, ds = \int_{C_1} f\, ds + \int_{C_2} f\, ds + \cdots + \int_{C_n} f\, ds. \tag{3}$$

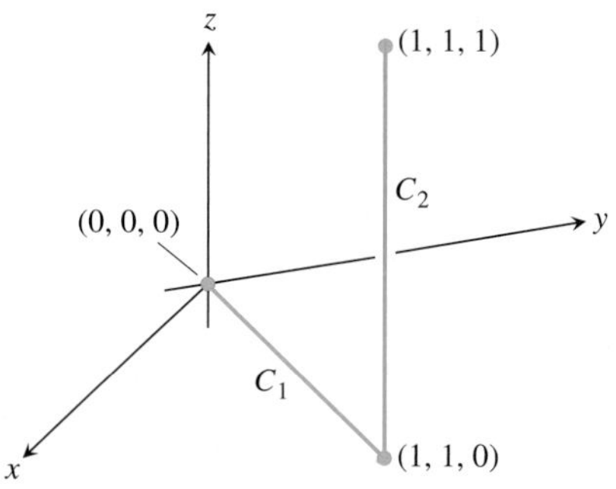

FIGURE 13.3 The path of integration in Example 2.

EXAMPLE 2 Figure 13.3 shows another path from the origin to (1, 1, 1), the union of line segments C_1 and C_2. Integrate $f(x, y, z) = x - 3y^2 + z$ over $C_1 \cup C_2$.

Solution We choose the simplest parametrizations for C_1 and C_2 we can think of, checking the lengths of the velocity vectors as we go along:

$$\begin{aligned}
C_1&: \quad \mathbf{r}(t) = t\mathbf{i} + t\mathbf{j}, \quad 0 \le t \le 1; \quad |\mathbf{v}| = \sqrt{1^2 + 1^2} = \sqrt{2} \\
C_2&: \quad \mathbf{r}(t) = \mathbf{i} + \mathbf{j} + t\mathbf{k}, \quad 0 \le t \le 1; \quad |\mathbf{v}| = \sqrt{0^2 + 0^2 + 1^2} = 1.
\end{aligned}$$

With these parametrizations we find that

$$\begin{aligned}
\int_{C_1 \cup C_2} f(x, y, z)\, ds &= \int_{C_1} f(x, y, z)\, ds + \int_{C_2} f(x, y, z)\, ds \qquad \text{Eq. (3)} \\
&= \int_0^1 f(t, t, 0)\sqrt{2}\, dt + \int_0^1 f(1, 1, t)(1)\, dt \qquad \text{Eq. (2)} \\
&= \int_0^1 (t - 3t^2 + 0)\sqrt{2}\, dt + \int_0^1 (1 - 3 + t)(1)\, dt \\
&= \sqrt{2}\left[\frac{t^2}{2} - t^3\right]_0^1 + \left[\frac{t^2}{2} - 2t\right]_0^1 = -\frac{\sqrt{2}}{2} - \frac{3}{2}. \qquad \blacksquare
\end{aligned}$$

Notice three things about the integrations in Examples 1 and 2. First, as soon as the components of the appropriate curve were substituted into the formula for f, the integration became a standard integration with respect to t. Second, the integral of f over $C_1 \cup C_2$ was obtained by integrating f over each section of the path and adding the results. Third, the integrals of f over C and $C_1 \cup C_2$ had different values. The value of the integral along a path joining two points can change if you change the path between them.

Mass and Moment Calculations

We treat coil springs and wires like masses distributed along smooth curves in space. The distribution is described by a continuous density function $\delta(x, y, z)$ (mass per unit length). The spring's or wire's mass, center of mass, and moments are then calculated with the formulas in Table 13.1. The formulas also apply to thin rods. The derivations of these formulas are similar to those in Section 6.6. Notice how alike the formulas are to those in Tables 12.1 and 12.2 for double and triple integrals. The triple integrals for solids, and the double integrals for planar regions, become line integrals for wires or rods.

TABLE 13.1 Mass and moment formulas for coil springs, thin rods, and wires lying along a smooth curve C in space

Mass: $M = \int_C \delta(x, y, z)\, ds$ $\quad \delta = \delta(x, y, z)$ is the density at (x, y, z)

First moments about the coordinate planes:

$$M_{yz} = \int_C x\,\delta\, ds, \qquad M_{xz} = \int_C y\,\delta\, ds, \qquad M_{xy} = \int_C z\,\delta\, ds$$

Coordinates of the center of mass:

$$\bar{x} = M_{yz}/M, \qquad \bar{y} = M_{xz}/M, \qquad \bar{z} = M_{xy}/M$$

Moments of inertia about axes and other lines:

$$I_x = \int_C (y^2 + z^2)\,\delta\, ds, \qquad I_y = \int_C (x^2 + z^2)\,\delta\, ds, \qquad I_z = \int_C (x^2 + y^2)\,\delta\, ds,$$

$$I_L = \int_C r^2\,\delta\, ds$$ $r(x, y, z) =$ distance from the point (x, y, z) to line L

Notice that the element of mass dm in the table is equal to $\delta\, ds$ rather than $\delta\, dV$ as in Table 12.1, and that the integrals are taken over the curve C.

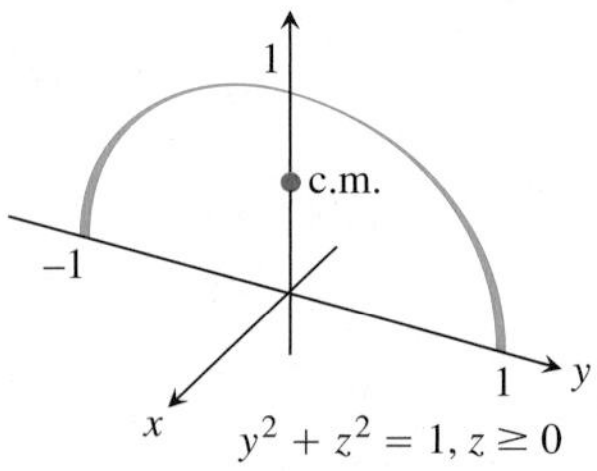

FIGURE 13.4 Example 3 shows how to find the center of mass of a circular arch of variable density.

EXAMPLE 3 A slender metal arch, denser at the bottom than top, lies along the semicircle $y^2 + z^2 = 1$, $z \geq 0$, in the yz-plane (Figure 13.4). Find the center of the arch's mass if the density at the point (x, y, z) on the arch is $\delta(x, y, z) = 2 - z$.

Solution We know that $\bar{x} = 0$ and $\bar{y} = 0$ because the arch lies in the yz-plane with its mass distributed symmetrically about the z-axis. To find $\bar{z}$, we parametrize the circle as

$$\mathbf{r}(t) = (\cos t)\mathbf{j} + (\sin t)\mathbf{k}, \qquad 0 \leq t \leq \pi.$$

For this parametrization,

$$|\mathbf{v}(t)| = \sqrt{\left(\frac{dx}{dt}\right)^2 + \left(\frac{dy}{dt}\right)^2 + \left(\frac{dz}{dt}\right)^2} = \sqrt{(0)^2 + (-\sin t)^2 + (\cos t)^2} = 1.$$

The formulas in Table 13.1 then give

$$M = \int_C \delta\, ds = \int_C (2 - z)\, ds = \int_0^{\pi} (2 - \sin t)(1)\, dt = 2\pi - 2$$

$$M_{xy} = \int_C z\delta\, ds = \int_C z(2 - z)\, ds = \int_0^{\pi} (\sin t)(2 - \sin t)\, dt$$

$$= \int_0^{\pi} (2 \sin t - \sin^2 t)\, dt = \frac{8 - \pi}{2}$$

$$\bar{z} = \frac{M_{xy}}{M} = \frac{8 - \pi}{2} \cdot \frac{1}{2\pi - 2} = \frac{8 - \pi}{4\pi - 4} \approx 0.57.$$

With $\bar{z}$ to the nearest hundredth, the center of mass is (0, 0, 0.57). ■

EXERCISES 13.1

Match the vector equations in Exercises 1–8 with the graphs (a)–(h) given here.

a.

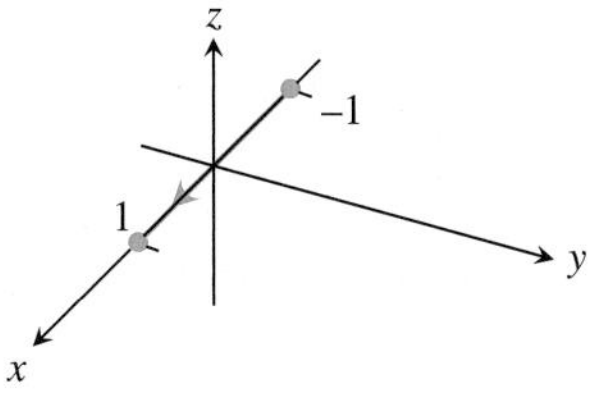

b.

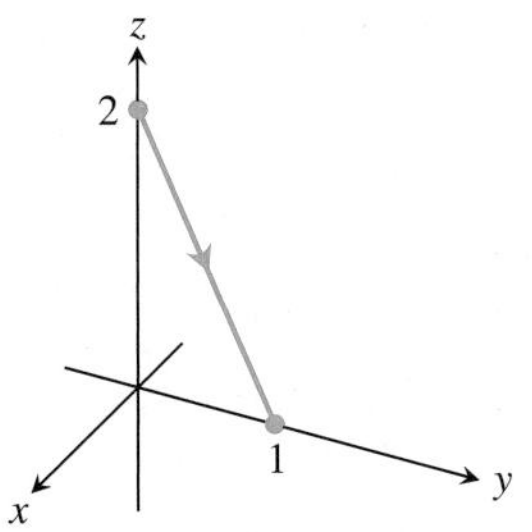

c.

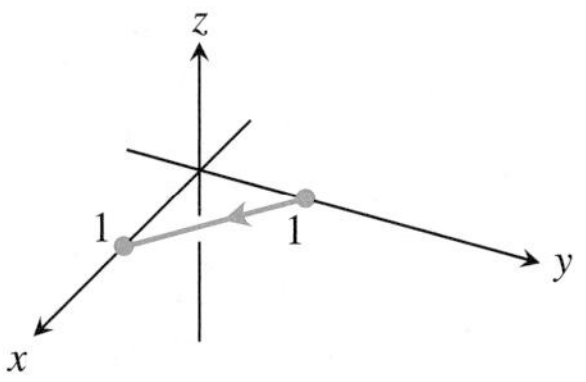

d.

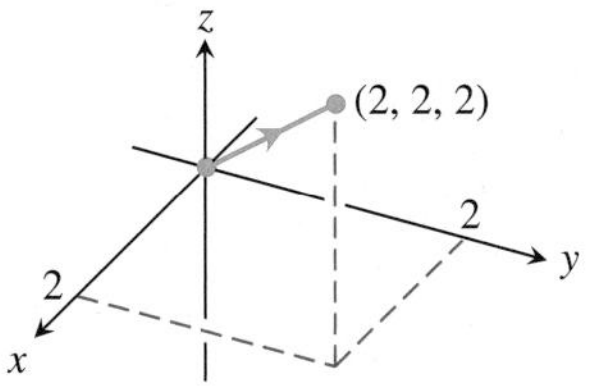

e.

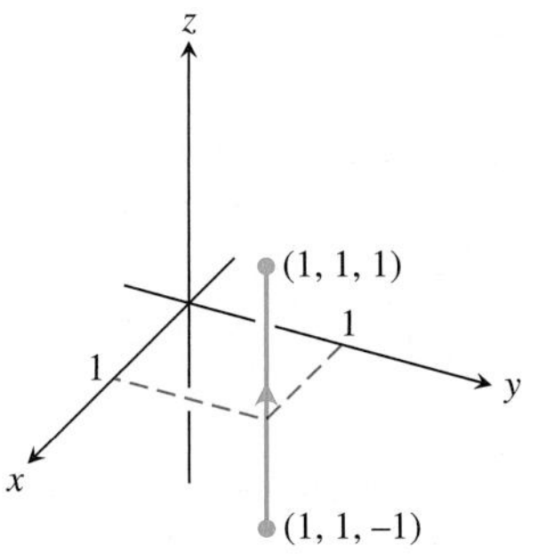

f.

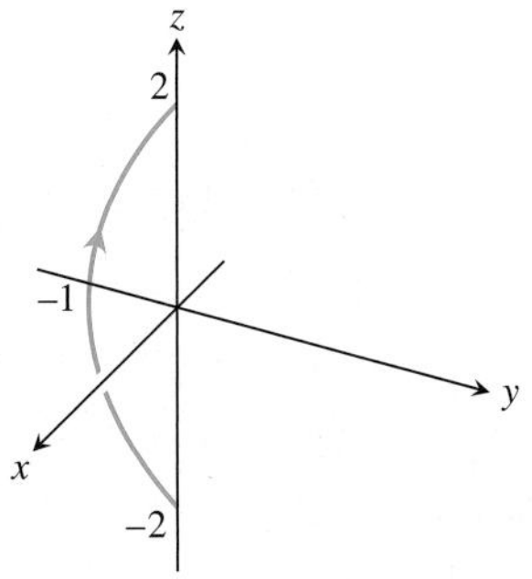

g.

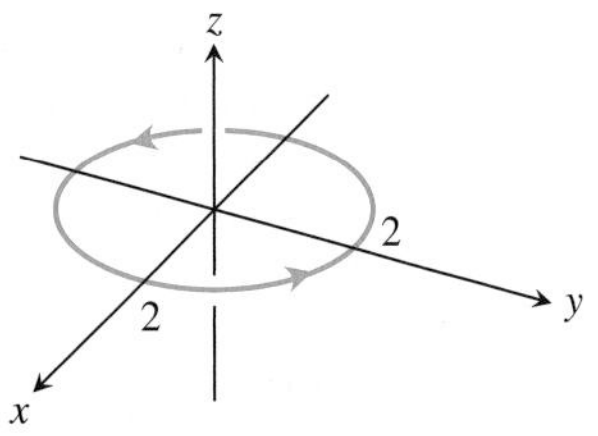

h.

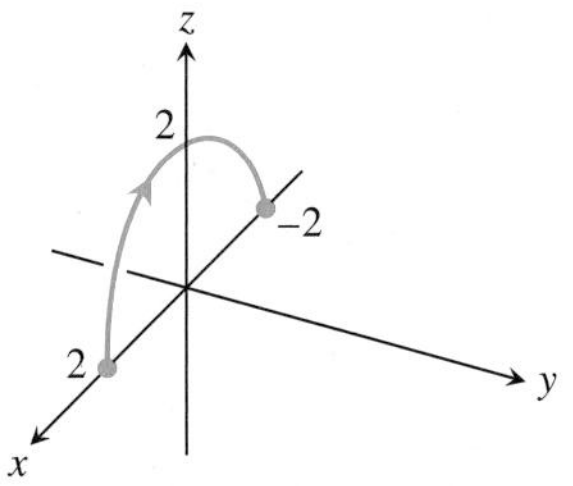

1. $\mathbf{r}(t) = t\mathbf{i} + (1 - t)\mathbf{j}, \quad 0 \le t \le 1$
2. $\mathbf{r}(t) = \mathbf{i} + \mathbf{j} + t\mathbf{k}, \quad -1 \le t \le 1$
3. $\mathbf{r}(t) = (2 \cos t)\mathbf{i} + (2 \sin t)\mathbf{j}, \quad 0 \le t \le 2\pi$
4. $\mathbf{r}(t) = t\mathbf{i}, \quad -1 \le t \le 1$
5. $\mathbf{r}(t) = t\mathbf{i} + t\mathbf{j} + t\mathbf{k}, \quad 0 \le t \le 2$
6. $\mathbf{r}(t) = t\mathbf{j} + (2 - 2t)\mathbf{k}, \quad 0 \le t \le 1$
7. $\mathbf{r}(t) = (t^2 - 1)\mathbf{j} + 2t\mathbf{k}, \quad -1 \le t \le 1$
8. $\mathbf{r}(t) = (2 \cos t)\mathbf{i} + (2 \sin t)\mathbf{k}, \quad 0 \le t \le \pi$
9. Evaluate $\int_C (x + y)\, ds$ where C is the straight-line segment $x = t$, $y = (1 - t)$, $z = 0$, from (0, 1, 0) to (1, 0, 0).
10. Evaluate $\int_C (x - y + z - 2)\, ds$ where C is the straight-line segment $x = t$, $y = (1 - t)$, $z = 1$, from (0, 1, 1) to (1, 0, 1).
11. Evaluate $\int_C (xy + y + z)\, ds$ along the curve $\mathbf{r}(t) = 2t\mathbf{i} + t\mathbf{j} + (2 - 2t)\mathbf{k}$, $0 \le t \le 1$.
12. Evaluate $\int_C \sqrt{x^2 + y^2}\, ds$ along the curve $\mathbf{r}(t) = (4 \cos t)\mathbf{i} + (4 \sin t)\mathbf{j} + 3t\mathbf{k}$, $-2\pi \le t \le 2\pi$.
13. Find the line integral of $f(x, y, z) = x + y + z$ over the straight-line segment from (1, 2, 3) to (0, −1, 1).
14. Find the line integral of $f(x, y, z) = \sqrt{3}/(x^2 + y^2 + z^2)$ over the curve $\mathbf{r}(t) = t\mathbf{i} + t\mathbf{j} + t\mathbf{k}$, $1 \le t \le \infty$.

15. Integrate $f(x, y, z) = x + \sqrt{y} - z^2$ over the path from (0, 0, 0) to (1, 1, 1) (Figure 13.5a) given by

$$C_1: \ \mathbf{r}(t) = t\mathbf{i} + t^2\mathbf{j}, \quad 0 \le t \le 1$$
$$C_2: \ \mathbf{r}(t) = \mathbf{i} + \mathbf{j} + t\mathbf{k}, \quad 0 \le t \le 1$$

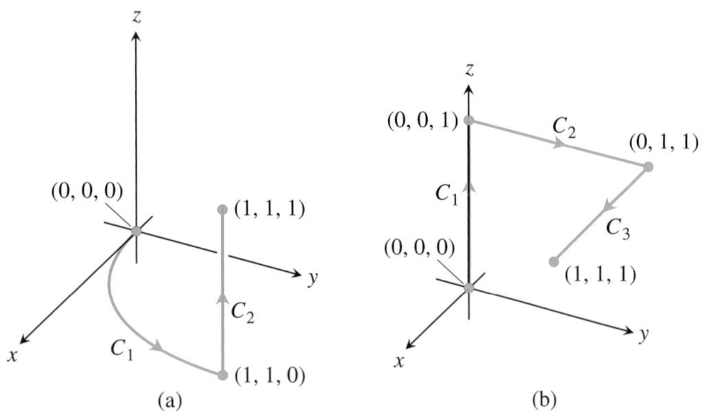

FIGURE 13.5 The paths of integration for Exercises 15 and 16.

16. Integrate $f(x, y, z) = x + \sqrt{y} - z^2$ over the path from (0, 0, 0) to (1, 1, 1) (Figure 13.5b) given by

$$C_1: \ \mathbf{r}(t) = t\mathbf{k}, \quad 0 \le t \le 1$$
$$C_2: \ \mathbf{r}(t) = t\mathbf{j} + \mathbf{k}, \quad 0 \le t \le 1$$
$$C_3: \ \mathbf{r}(t) = t\mathbf{i} + \mathbf{j} + \mathbf{k}, \quad 0 \le t \le 1$$

17. Integrate $f(x, y, z) = (x + y + z)/(x^2 + y^2 + z^2)$ over the path $\mathbf{r}(t) = t\mathbf{i} + t\mathbf{j} + t\mathbf{k}, 0 < a \le t \le b$.

18. Integrate $f(x, y, z) = -\sqrt{x^2 + z^2}$ over the circle

$$\mathbf{r}(t) = (a\cos t)\mathbf{j} + (a\sin t)\mathbf{k}, \quad 0 \le t \le 2\pi.$$

In Exercises 19–22, integrate f over the given curve.

19. $f(x, y) = x^3/y, \quad C: \ y = x^2/2, \quad 0 \le x \le 2$

20. $f(x, y) = (x + y^2)/\sqrt{1 + x^2}, \quad C: \ y = x^2/2$ from (1, 1/2) to (0, 0)

21. $f(x, y) = x + y, \quad C: \ x^2 + y^2 = 4$ in the first quadrant from (2, 0) to (0, 2)

22. $f(x, y) = x^2 - y, \quad C: \ x^2 + y^2 = 4$ in the first quadrant from (0, 2) to $(\sqrt{2}, \sqrt{2})$

23. Mass of a wire Find the mass of a wire that lies along the curve $\mathbf{r}(t) = (t^2 - 1)\mathbf{j} + 2t\mathbf{k}, 0 \le t \le 1$, if the density is $\delta = (3/2)t$.

24. Center of mass of a curved wire A wire of density $\delta(x, y, z) = 15\sqrt{y + 2}$ lies along the curve $\mathbf{r}(t) = (t^2 - 1)\mathbf{j} + 2t\mathbf{k}, -1 \le t \le 1$. Find its center of mass. Then sketch the curve and center of mass together.

25. Mass of wire with variable density Find the mass of a thin wire lying along the curve $\mathbf{r}(t) = \sqrt{2}t\mathbf{i} + \sqrt{2}t\mathbf{j} + (4 - t^2)\mathbf{k}$, $0 \le t \le 1$, if the density is **(a)** $\delta = 3t$ and **(b)** $\delta = 1$.

26. Center of mass of wire with variable density Find the center of mass of a thin wire lying along the curve $\mathbf{r}(t) = t\mathbf{i} + 2t\mathbf{j} + (2/3)t^{3/2}\mathbf{k}, 0 \le t \le 2$, if the density is $\delta = 3\sqrt{5 + t}$.

27. Moment of inertia of wire hoop A circular wire hoop of constant density δ lies along the circle $x^2 + y^2 = a^2$ in the xy-plane. Find the hoop's moment of inertia about the z-axis.

28. Inertia of a slender rod A slender rod of constant density lies along the line segment $\mathbf{r}(t) = t\mathbf{j} + (2 - 2t)\mathbf{k}, 0 \le t \le 1$, in the yz-plane. Find the moments of inertia of the rod about the three coordinate axes.

29. Two springs of constant density A spring of constant density δ lies along the helix

$$\mathbf{r}(t) = (\cos t)\mathbf{i} + (\sin t)\mathbf{j} + t\mathbf{k}, \quad 0 \le t \le 2\pi.$$

a. Find I_z.

b. Suppose that you have another spring of constant density δ that is twice as long as the spring in part (a) and lies along the helix for $0 \le t \le 4\pi$. Do you expect I_z for the longer spring to be the same as that for the shorter one, or should it be different? Check your prediction by calculating I_z for the longer spring.

30. Wire of constant density A wire of constant density $\delta = 1$ lies along the curve

$$\mathbf{r}(t) = (t\cos t)\mathbf{i} + (t\sin t)\mathbf{j} + \left(2\sqrt{2}/3\right)t^{3/2}\mathbf{k}, \quad 0 \le t \le 1.$$

Find $\bar{z}$ and I_z.

31. The arch in Example 3 Find I_x for the arch in Example 3.

32. Center of mass and moments of inertia for wire with variable density Find the center of mass and the moments of inertia about the coordinate axes of a thin wire lying along the curve

$$\mathbf{r}(t) = t\mathbf{i} + \frac{2\sqrt{2}}{3}t^{3/2}\mathbf{j} + \frac{t^2}{2}\mathbf{k}, \quad 0 \le t \le 2,$$

if the density is $\delta = 1/(t + 1)$.

COMPUTER EXPLORATIONS

In Exercises 33–36, use a CAS to perform the following steps to evaluate the line integrals.

a. Find $ds = |\mathbf{v}(t)|\,dt$ for the path $\mathbf{r}(t) = g(t)\mathbf{i} + h(t)\mathbf{j} + k(t)\mathbf{k}$.

b. Express the integrand $f(g(t), h(t), k(t))|\mathbf{v}(t)|$ as a function of the parameter t.

c. Evaluate $\int_C f\,ds$ using Equation (2) in the text.

33. $f(x, y, z) = \sqrt{1 + 30x^2 + 10y}; \quad \mathbf{r}(t) = t\mathbf{i} + t^2\mathbf{j} + 3t^2\mathbf{k}$, $0 \le t \le 2$

34. $f(x, y, z) = \sqrt{1 + x^3 + 5y^3}; \quad \mathbf{r}(t) = t\mathbf{i} + \frac{1}{3}t^2\mathbf{j} + \sqrt{t}\mathbf{k}$, $0 \le t \le 2$

35. $f(x, y, z) = x\sqrt{y} - 3z^2; \quad \mathbf{r}(t) = (\cos 2t)\mathbf{i} + (\sin 2t)\mathbf{j} + 5t\mathbf{k}$, $0 \le t \le 2\pi$

36. $f(x, y, z) = \left(1 + \frac{9}{4}z^{1/3}\right)^{1/4}; \quad \mathbf{r}(t) = (\cos 2t)\mathbf{i} + (\sin 2t)\mathbf{j} + t^{5/2}\mathbf{k}, \quad 0 \le t \le 2\pi$

13.2 Vector Fields, Work, Circulation, and Flux

Gravitational and electric forces have both direction and magnitude. They are represented by a vector at each point in their domain, producing a *vector field.* In this section we show how to compute the work done in moving an object through such a field by using a line integral over the vector field. We also discuss velocity fields, such as the vector field representing the velocity of a flowing fluid in its domain. A line integral can be used to find the rate at which the fluid flows along or across a curve within the domain.

Vector Fields

Suppose a region in the plane or in space is occupied by a moving fluid, such as air or water. The fluid is made up of a large number of particles, and at any instant of time, a particle has a velocity $\mathbf{v}$. At different points of position at a given (same) time, these velocities can vary. We can think of a velocity vector being attached to each point of the fluid representing the velocity of a particle at that point. Such a fluid flow is an example of a *vector field.* Figure 13.6 shows a velocity vector field obtained from air flowing around an airfoil in a wind tunnel. Figure 13.7 shows a vector field of velocity vectors along the streamlines of water moving through a contracting channel. Vector fields are also associated to forces such as gravitational attraction (Figure 13.8), and to magnetic force fields, electric fields, and also purely mathematical fields.

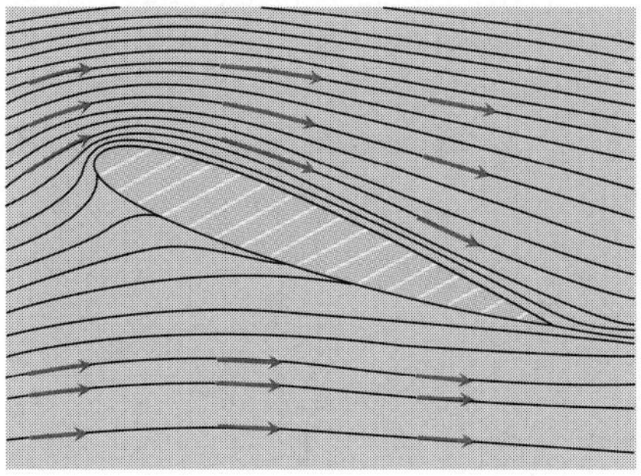

FIGURE 13.6 Velocity vectors of a flow around an airfoil in a wind tunnel.

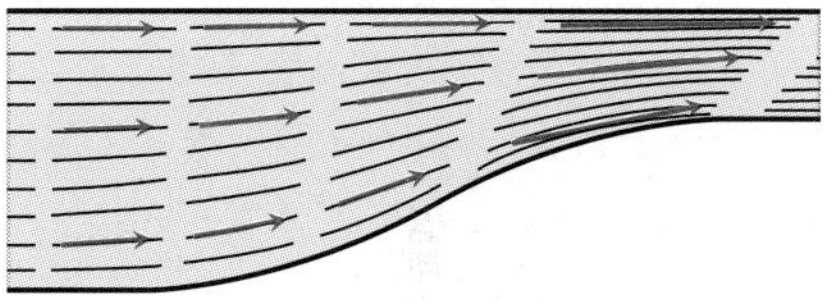

FIGURE 13.7 Streamlines in a contracting channel. The water speeds up as the channel narrows and the velocity vectors increase in length.

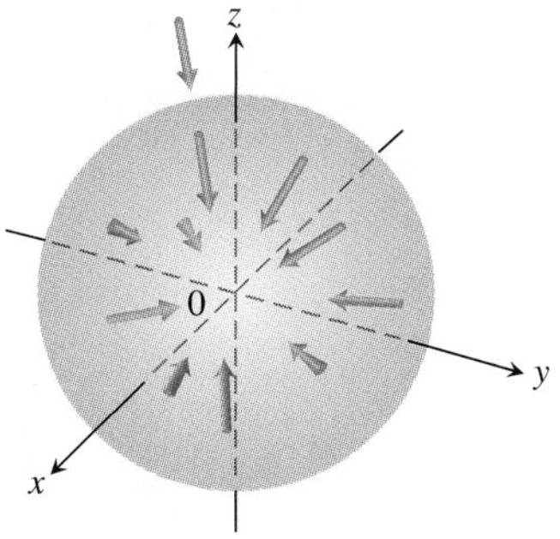

FIGURE 13.8 Vectors in a gravitational field point toward the center of mass that gives the source of the field.

Generally, a **vector field** is a function that assigns a vector to each point in its domain. A vector field on a three-dimensional domain in space might have a formula like

$$\mathbf{F}(x, y, z) = M(x, y, z)\mathbf{i} + N(x, y, z)\mathbf{j} + P(x, y, z)\mathbf{k}.$$

The field is **continuous** if the **component functions** M, N, and P are continuous; it is **differentiable** if each of the component functions is differentiable. The formula for a field of two-dimensional vectors could look like

$$\mathbf{F}(x, y) = M(x, y)\mathbf{i} + N(x, y)\mathbf{j}.$$

We encountered another type of vector field in Chapter 10. The tangent vectors $\mathbf{T}$ and normal vectors $\mathbf{N}$ for a curve in space both form vector fields along the curve. Along a curve $\mathbf{r}(t)$ they might have a formula such as

$$\mathbf{v}(t) = f(t)\mathbf{i} + g(t)\mathbf{j} + h(t)\mathbf{k}.$$

If we attach the gradient vector ∇f of a scalar function $f(x, y, z)$ to each point of a level surface of the function, we obtain a three-dimensional field on the surface. If we attach the velocity vector to each point of a flowing fluid, we have a three-dimensional field defined on a region in space. These and other fields are illustrated in Figures 13.9–13.14. To sketch the fields, we picked a representative selection of domain points and drew the vectors attached to them. The arrows are drawn with their tails, not their heads, attached to the points where the vector functions are evaluated.

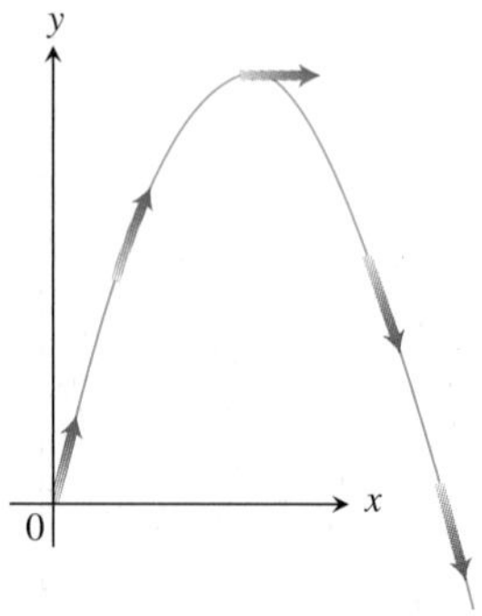

FIGURE 13.9 The velocity vectors $\mathbf{v}(t)$ of a projectile's motion make a vector field along the trajectory.

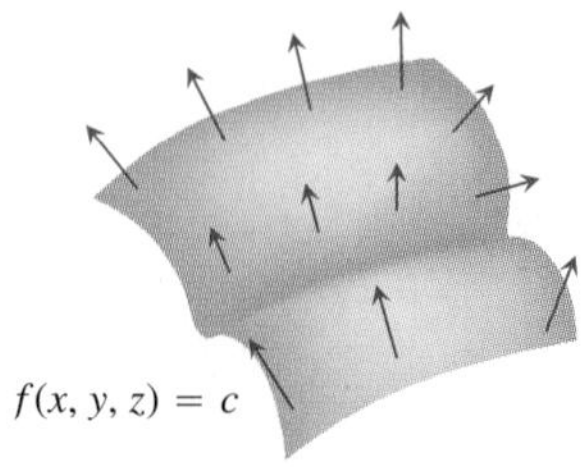

FIGURE 13.10 The field of gradient vectors ∇f on a surface $f(x, y, z) = c$.

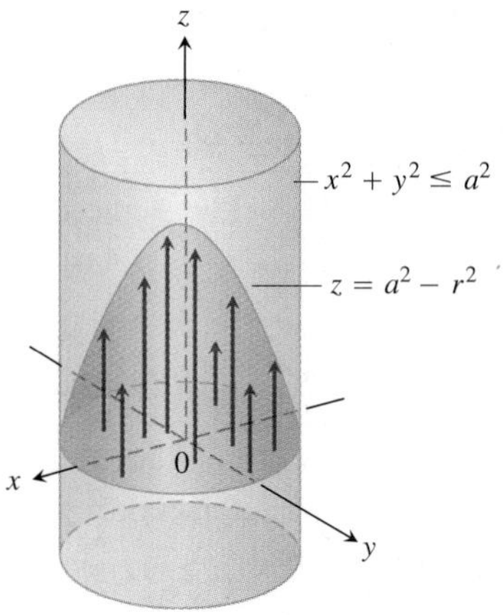

FIGURE 13.11 The flow of fluid in a long cylindrical pipe. The vectors $\mathbf{v} = (a^2 - r^2)\mathbf{k}$ inside the cylinder that have their bases in the xy-plane have their tips on the paraboloid $z = a^2 - r^2$.

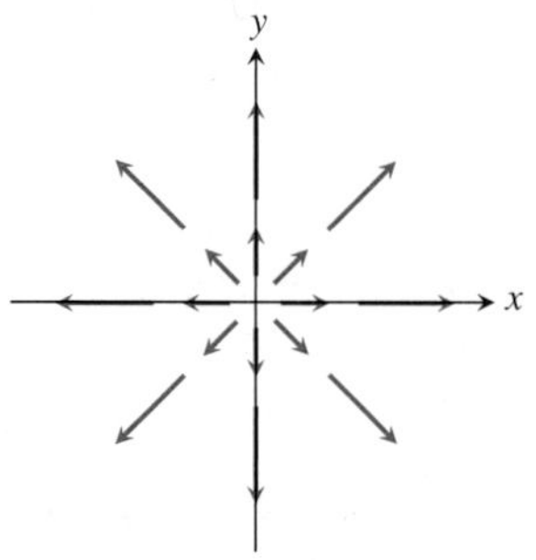

FIGURE 13.12 The radial field $\mathbf{F} = x\mathbf{i} + y\mathbf{j}$ of position vectors of points in the plane. Notice the convention that an arrow is drawn with its tail, not its head, at the point where $\mathbf{F}$ is evaluated.

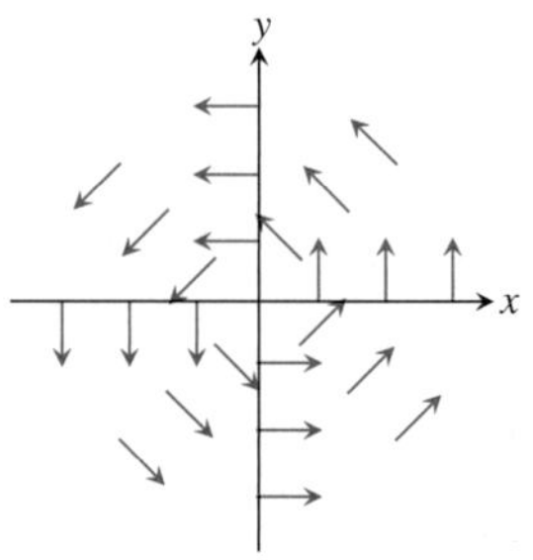

FIGURE 13.13 The circumferential or "spin" field of unit vectors

$$\mathbf{F} = (-y\mathbf{i} + x\mathbf{j})/(x^2 + y^2)^{1/2}$$

in the plane. The field is not defined at the origin.

Gradient Fields

The gradient vector of a differentiable scalar-valued function at a point gives the direction of greatest increase of the function. An important type of vector field is formed by all the gradient vectors of the function.

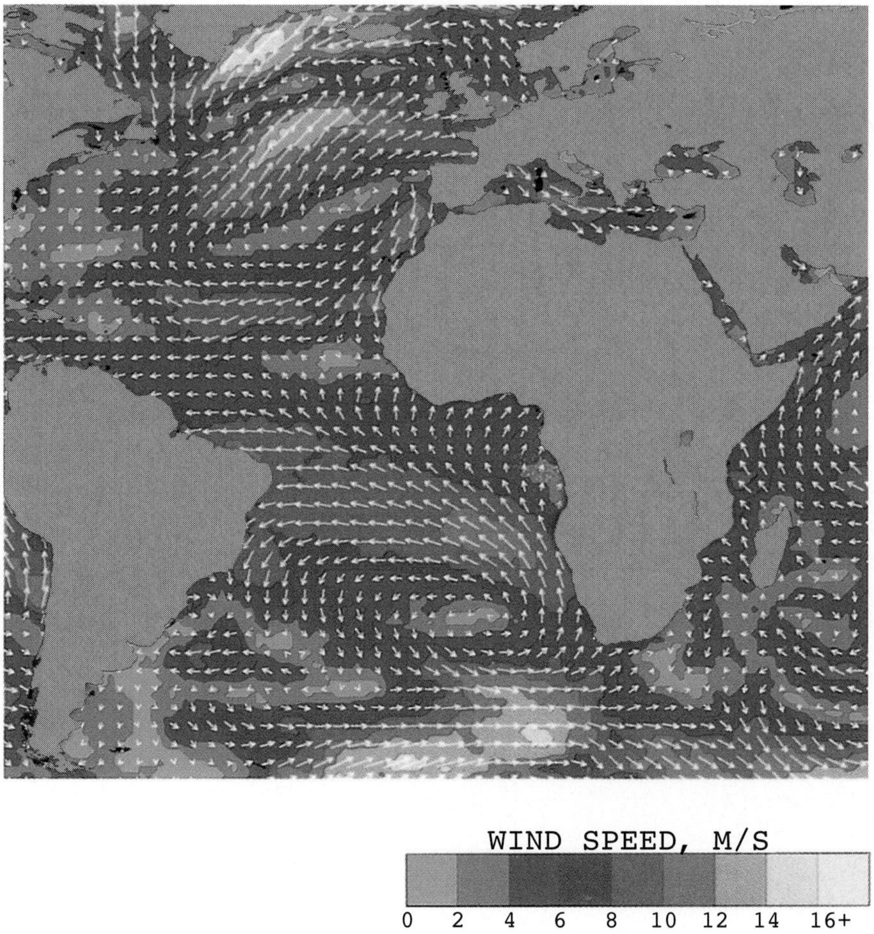

FIGURE 13.14 NASA's *Seasat* used radar to take 350,000 wind measurements over the world's oceans. The arrows show wind direction; their length and the color contouring indicate speed. Notice the heavy storm south of Greenland.

DEFINITION The **gradient field** of a differentiable function $f(x, y, z)$ is the field of gradient vectors

$$\nabla f = \frac{\partial f}{\partial x}\mathbf{i} + \frac{\partial f}{\partial y}\mathbf{j} + \frac{\partial f}{\partial z}\mathbf{k}.$$

EXAMPLE 1 Suppose that the temperature T at each point (x, y, z) in a region of space is given by

$$T = 100 - x^2 - y^2 - z^2,$$

and that $\mathbf{F}(x, y, z)$ is defined to be the gradient of T. Find the vector field $\mathbf{F}$.

Solution The gradient field $\mathbf{F}$ is the field $\mathbf{F} = \nabla T = -2x\mathbf{i} - 2y\mathbf{j} - 2z\mathbf{k}$. ■

At each point in space, the vector field $\mathbf{F}$ in Example 1 gives the direction for which the temperature is increasing most rapidly. It is not a force field and it is not a velocity field.

Work Done by a Force over a Curve in Space

Suppose that the vector field $\mathbf{F} = M(x, y, z)\mathbf{i} + N(x, y, z)\mathbf{j} + P(x, y, z)\mathbf{k}$ represents a force throughout a region in space (it might be the force of gravity or an electromagnetic force of some kind) and that

$$\mathbf{r}(t) = g(t)\mathbf{i} + h(t)\mathbf{j} + k(t)\mathbf{k}, \qquad a \le t \le b,$$

is a smooth curve in the region. The formula for the work done by the force in moving an object along the curve is motivated by the same kind of reasoning we used in Chapter 6 to derive the formula $W = \int_a^b F(x)\, dx$ for the work done by a continuous force of magnitude $F(x)$ directed along an interval of the x-axis. We divide the curve into short segments, apply the (constant-force) $\times$ (distance) formula for work to approximate the work over each curved segment, add the results to approximate the work over the entire curve, and calculate the work as the limit of the approximating sums as the segments become shorter and more numerous. To find exactly what the limiting integral should be, we partition the parameter interval $[a, b]$ in the usual way and choose a point c_k in each subinterval $[t_k, t_{k+1}]$. The partition of $[a, b]$ determines ("induces," we say) a partition of the curve, with the point P_k being the tip of the position vector $\mathbf{r}(t_k)$ and Δs_k being the length of the curve segment P_kP_{k+1} (Figure 13.15).

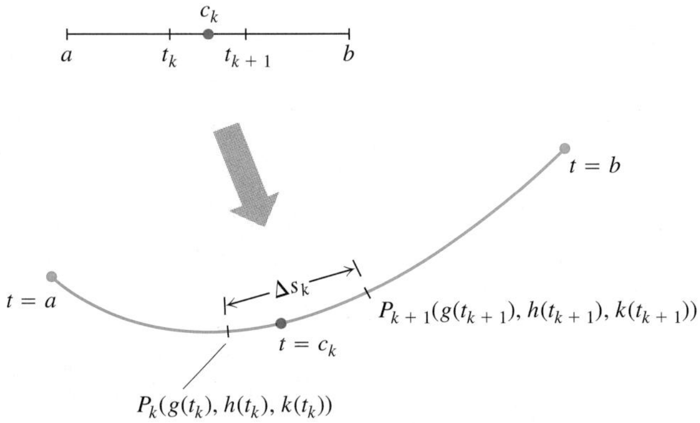

FIGURE 13.15 Each partition of $[a, b]$ induces a partition of the curve $\mathbf{r}(t) = g(t)\mathbf{i} + h(t)\mathbf{j} + k(t)\mathbf{k}$.

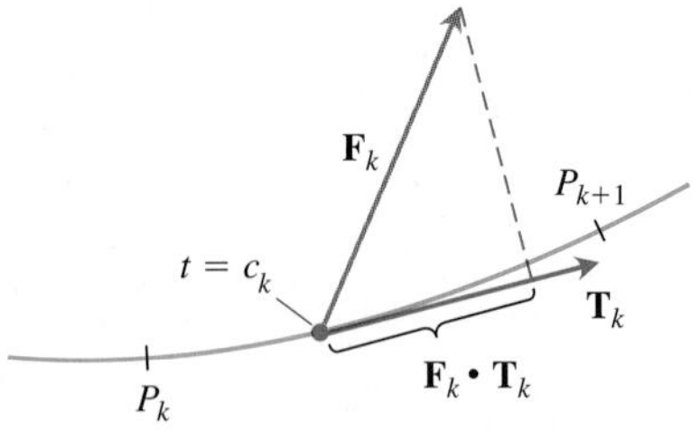

FIGURE 13.16 An enlarged view of the curve segment P_kP_{k+1} in Figure 13.15, showing the force and unit tangent vectors at the point on the curve where $t = c_k$.

If $\mathbf{F}_k$ denotes the value of the force field $\mathbf{F}$ at the point on the curve corresponding to $t = c_k$ and $\mathbf{T}_k$ denotes the curve's unit tangent vector at this point, then $\mathbf{F}_k \cdot \mathbf{T}_k$ is the scalar component of $\mathbf{F}$ in the direction of $\mathbf{T}$ at $t = c_k$ (Figure 13.16). The work done by $\mathbf{F}$ along the curve segment P_kP_{k+1} is approximately

$$\begin{pmatrix}\text{Force component in} \\ \text{direction of motion}\end{pmatrix} \times \begin{pmatrix}\text{distance} \\ \text{applied}\end{pmatrix} = \mathbf{F}_k \cdot \mathbf{T}_k \, \Delta s_k.$$

The work done by $\mathbf{F}$ in moving an object along the curve from $t = a$ to $t = b$ is approximately

$$\sum_{k=1}^{n} \mathbf{F}_k \cdot \mathbf{T}_k \, \Delta s_k.$$

As the norm of the partition of $[a, b]$ approaches zero, the norm of the induced partition of the curve approaches zero and these sums approach the line integral

$$\int_{t=a}^{t=b} \mathbf{F} \cdot \mathbf{T} \, ds.$$

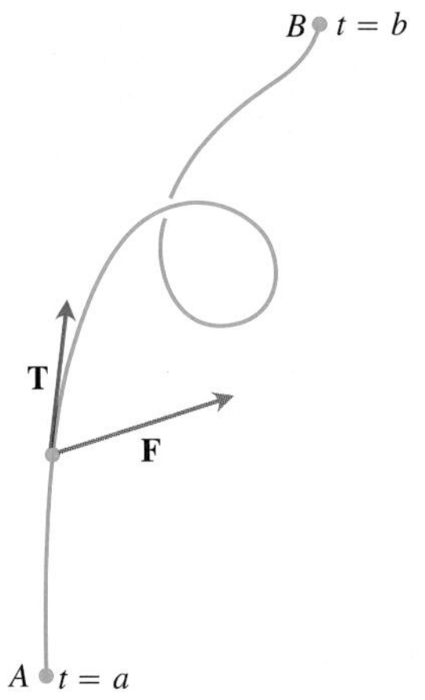

FIGURE 13.17 The work done by a force $\mathbf{F}$ is the line integral of the scalar component $\mathbf{F} \cdot \mathbf{T}$ over the smooth curve from A to B.

DEFINITION The **work** done by a force $\mathbf{F} = M\mathbf{i} + N\mathbf{j} + P\mathbf{k}$ in moving an object over a smooth curve $\mathbf{r}(t)$ from $t = a$ to $t = b$ is

$$W = \int_{t=a}^{t=b} \mathbf{F} \cdot \mathbf{T} \, ds. \tag{1}$$

The sign of the number we calculate with this integral depends on the direction in which the curve is traversed as t increases. If we reverse the direction of motion, we reverse the direction of $\mathbf{T}$ and change the sign of $\mathbf{F} \cdot \mathbf{T}$ and its integral (Figure 13.17).

Table 13.2 shows six ways to write the work integral in Equation (1). Despite their variety, the formulas in Table 13.2 are all evaluated the same way. In the table, $\mathbf{r}(t) = g(t)\mathbf{i} + h(t)\mathbf{j} + k(t)\mathbf{k} = x\mathbf{i} + y\mathbf{j} + z\mathbf{k}$ is a smooth curve, and

$$d\mathbf{r} = \frac{d\mathbf{r}}{dt}\, dt = dg\mathbf{i} + dh\mathbf{j} + dk\mathbf{k}$$

is its differential.

TABLE 13.2 Six different ways to write the work integral

$\mathbf{W} = \int_{t=a}^{t=b} \mathbf{F} \cdot \mathbf{T} \, ds$	The definition
$= \int_{t=a}^{t=b} \mathbf{F} \cdot d\mathbf{r}$	Compact differential form
$= \int_{a}^{b} \mathbf{F} \cdot \frac{d\mathbf{r}}{dt}\, dt$	Expanded to include dt; emphasizes the parameter t and velocity vector $d\mathbf{r}/dt$
$= \int_{a}^{b} \left(M\frac{dg}{dt} + N\frac{dh}{dt} + P\frac{dk}{dt} \right) dt$	Emphasizes the component functions of $\mathbf{F}$
$= \int_{a}^{b} \left(M\frac{dx}{dt} + N\frac{dy}{dt} + P\frac{dz}{dt} \right) dt$	Abbreviates the components of $\mathbf{r}$
$= \int_{a}^{b} M\,dx + N\,dy + P\,dz$	dt's canceled; the most common form

To evaluate the work integral along a smooth curve $\mathbf{r}(t)$, take these steps:

1. Evaluate $\mathbf{F}$ on the curve as a function of the parameter t.
2. Find $d\mathbf{r}/dt$.
3. Integrate $\mathbf{F} \cdot d\mathbf{r}/dt$ from $t = a$ to $t = b$.

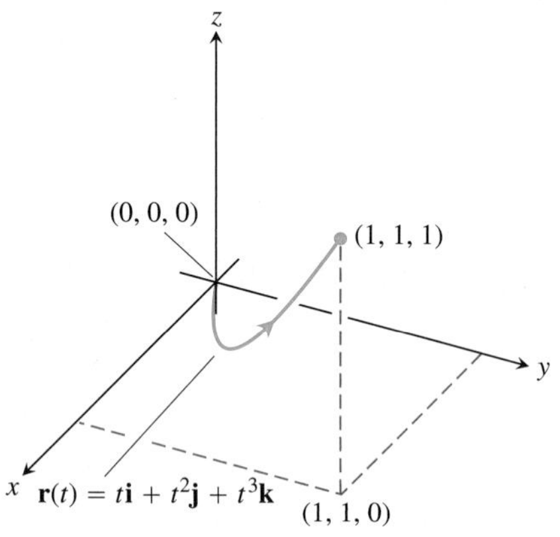

FIGURE 13.18 The curve in Example 2.

EXAMPLE 2 Find the work done by $\mathbf{F} = (y - x^2)\mathbf{i} + (z - y^2)\mathbf{j} + (x - z^2)\mathbf{k}$ over the curve $\mathbf{r}(t) = t\mathbf{i} + t^2\mathbf{j} + t^3\mathbf{k}$, $0 \le t \le 1$, from (0, 0, 0) to (1, 1, 1) (Figure 13.18).

Solution First we evaluate $\mathbf{F}$ on the curve:

$$\begin{aligned}\mathbf{F} &= (y - x^2)\mathbf{i} + (z - y^2)\mathbf{j} + (x - z^2)\mathbf{k} \\ &= \underbrace{(t^2 - t^2)}\mathbf{i} + (t^3 - t^4)\mathbf{j} + (t - t^6)\mathbf{k}\end{aligned}$$

Then we find $d\mathbf{r}/dt$,

$$\frac{d\mathbf{r}}{dt} = \frac{d}{dt}(t\mathbf{i} + t^2\mathbf{j} + t^3\mathbf{k}) = \mathbf{i} + 2t\mathbf{j} + 3t^2\mathbf{k}$$

Finally, we find $\mathbf{F} \cdot d\mathbf{r}/dt$ and integrate from $t = 0$ to $t = 1$:

$$\begin{aligned}\mathbf{F} \cdot \frac{d\mathbf{r}}{dt} &= [(t^3 - t^4)\mathbf{j} + (t - t^6)\mathbf{k}] \cdot (\mathbf{i} + 2t\mathbf{j} + 3t^2\mathbf{k}) \\ &= (t^3 - t^4)(2t) + (t - t^6)(3t^2) = 2t^4 - 2t^5 + 3t^3 - 3t^8,\end{aligned}$$

so

$$\begin{aligned}\text{Work} &= \int_0^1 (2t^4 - 2t^5 + 3t^3 - 3t^8)\, dt \\ &= \left[\frac{2}{5}t^5 - \frac{2}{6}t^6 + \frac{3}{4}t^4 - \frac{3}{9}t^9\right]_0^1 = \frac{29}{60}.\end{aligned}$$

■

Flow Integrals and Circulation for Velocity Fields

Suppose that $\mathbf{F}$ represents the velocity field of a fluid flowing through a region in space (a tidal basin or the turbine chamber of a hydroelectric generator, for example). Under these circumstances, the integral of $\mathbf{F} \cdot \mathbf{T}$ along a curve in the region gives the fluid's flow along the curve.

DEFINITIONS If $\mathbf{r}(t)$ is a smooth curve in the domain of a continuous velocity field $\mathbf{F}$, the **flow** along the curve from $t = a$ to $t = b$ is

$$\text{Flow} = \int_a^b \mathbf{F} \cdot \mathbf{T}\, ds. \qquad (2)$$

The integral in this case is called a **flow integral**. If the curve is a closed loop, the flow is called the **circulation** around the curve.

We evaluate flow integrals the same way we evaluate work integrals.

EXAMPLE 3 A fluid's velocity field is $\mathbf{F} = x\mathbf{i} + z\mathbf{j} + y\mathbf{k}$. Find the flow along the helix $\mathbf{r}(t) = (\cos t)\mathbf{i} + (\sin t)\mathbf{j} + t\mathbf{k}$, $0 \le t \le \pi/2$.

Solution We evaluate $\mathbf{F}$ on the curve,

$$\mathbf{F} = x\mathbf{i} + z\mathbf{j} + y\mathbf{k} = (\cos t)\mathbf{i} + t\mathbf{j} + (\sin t)\mathbf{k} \qquad x = \cos t,\ z = t,\ \text{and } y = \sin t \text{ on } \mathbf{r}(t)$$

and then find $d\mathbf{r}/dt$:

$$\frac{d\mathbf{r}}{dt} = (-\sin t)\mathbf{i} + (\cos t)\mathbf{j} + \mathbf{k}.$$

Then we integrate $\mathbf{F} \cdot (d\mathbf{r}/dt)$ from $t = 0$ to $t = \frac{\pi}{2}$:

$$\begin{aligned}\mathbf{F} \cdot \frac{d\mathbf{r}}{dt} &= (\cos t)(-\sin t) + (t)(\cos t) + (\sin t)(1) \\ &= -\sin t \cos t + t\cos t + \sin t\end{aligned}$$

so,

$$\begin{aligned}\text{Flow} &= \int_{t=a}^{t=b} \mathbf{F} \cdot \frac{d\mathbf{r}}{dt}\, dt = \int_0^{\pi/2} (-\sin t \cos t + t\cos t + \sin t)\, dt \\ &= \left[\frac{\cos^2 t}{2} + t\sin t\right]_0^{\pi/2} = \left(0 + \frac{\pi}{2}\right) - \left(\frac{1}{2} + 0\right) = \frac{\pi}{2} - \frac{1}{2}.\end{aligned}$$

■

EXAMPLE 4 Find the circulation of the field $\mathbf{F} = (x - y)\mathbf{i} + x\mathbf{j}$ around the circle $\mathbf{r}(t) = (\cos t)\mathbf{i} + (\sin t)\mathbf{j}$, $0 \le t \le 2\pi$.

Solution On the circle, $\mathbf{F} = (x - y)\mathbf{i} + x\mathbf{j} = (\cos t - \sin t)\mathbf{i} + (\cos t)\mathbf{j}$, and

$$\frac{d\mathbf{r}}{dt} = (-\sin t)\mathbf{i} + (\cos t)\mathbf{j}.$$

Then

$$\mathbf{F} \cdot \frac{d\mathbf{r}}{dt} = -\sin t \cos t + \underbrace{\sin^2 t + \cos^2 t}_{1}$$

gives

$$\begin{aligned}\text{Circulation} &= \int_0^{2\pi} \mathbf{F} \cdot \frac{d\mathbf{r}}{dt}\, dt = \int_0^{2\pi} (1 - \sin t \cos t)\, dt \\ &= \left[t - \frac{\sin^2 t}{2}\right]_0^{2\pi} = 2\pi.\end{aligned}$$

■

Flux Across a Plane Curve

To find the rate at which a fluid is entering or leaving a region enclosed by a smooth curve C in the xy-plane, we calculate the line integral over C of $\mathbf{F} \cdot \mathbf{n}$, the scalar component of the fluid's velocity field in the direction of the curve's outward-pointing normal vector. The value of this integral is the *flux* of $\mathbf{F}$ across C. *Flux* is Latin for *flow*, but many flux calculations involve no motion at all. If $\mathbf{F}$ were an electric field or a magnetic field, for instance, the integral of $\mathbf{F} \cdot \mathbf{n}$ would still be called the flux of the field across C.

DEFINITION If C is a smooth closed curve in the domain of a continuous vector field $\mathbf{F} = M(x, y)\mathbf{i} + N(x, y)\mathbf{j}$ in the plane and if $\mathbf{n}$ is the outward-pointing unit normal vector on C, the **flux** of $\mathbf{F}$ across C is

$$\text{Flux of } \mathbf{F} \text{ across } C = \int_C \mathbf{F} \cdot \mathbf{n}\, ds. \tag{3}$$

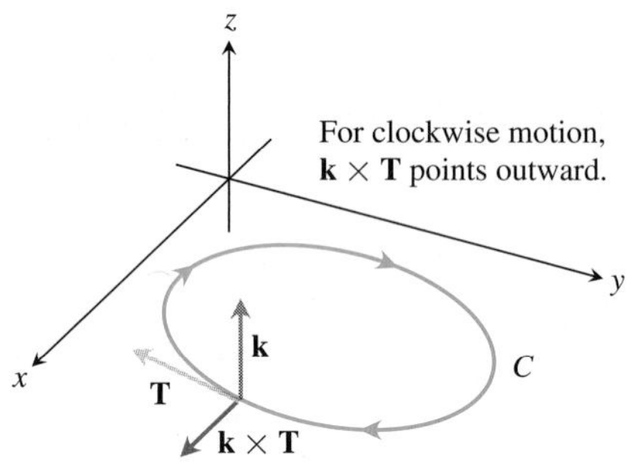

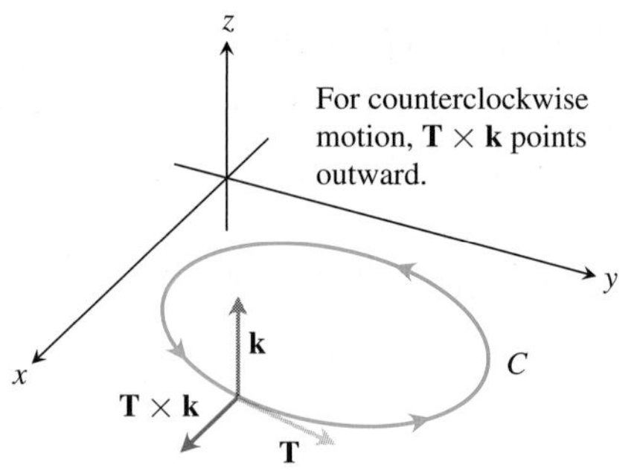

FIGURE 13.19 To find an outward unit normal vector for a smooth curve C in the xy-plane that is traversed counterclockwise as t increases, we take $\mathbf{n} = \mathbf{T} \times \mathbf{k}$. For clockwise motion, we take $\mathbf{n} = \mathbf{k} \times \mathbf{T}$.

Notice the difference between flux and circulation. The flux of $\mathbf{F}$ across C is the line integral with respect to arc length of $\mathbf{F} \cdot \mathbf{n}$, the scalar component of $\mathbf{F}$ in the direction of the outward normal. The circulation of $\mathbf{F}$ around C is the line integral with respect to arc length of $\mathbf{F} \cdot \mathbf{T}$, the scalar component of $\mathbf{F}$ in the direction of the unit tangent vector. Flux is the integral of the normal component of $\mathbf{F}$; circulation is the integral of the tangential component of $\mathbf{F}$.

To evaluate the integral in Equation (3), we begin with a smooth parametrization

$$x = g(t), \qquad y = h(t), \qquad a \le t \le b,$$

that traces the curve C exactly once as t increases from a to b. We can find the outward unit normal vector $\mathbf{n}$ by crossing the curve's unit tangent vector $\mathbf{T}$ with the vector $\mathbf{k}$. But which order do we choose, $\mathbf{T} \times \mathbf{k}$ or $\mathbf{k} \times \mathbf{T}$? Which one points outward? It depends on which way C is traversed as t increases. If the motion is clockwise, $\mathbf{k} \times \mathbf{T}$ points outward; if the motion is counterclockwise, $\mathbf{T} \times \mathbf{k}$ points outward (Figure 13.19). The usual choice is $\mathbf{n} = \mathbf{T} \times \mathbf{k}$, the choice that assumes counterclockwise motion. Thus, although the value of the arc length integral in the definition of flux in Equation (3) does not depend on which way C is traversed, the formulas we are about to derive for evaluating the integral in Equation (3) will assume counterclockwise motion.

In terms of components,

$$\mathbf{n} = \mathbf{T} \times \mathbf{k} = \left(\frac{dx}{ds}\mathbf{i} + \frac{dy}{ds}\mathbf{j}\right) \times \mathbf{k} = \frac{dy}{ds}\mathbf{i} - \frac{dx}{ds}\mathbf{j}.$$

If $\mathbf{F} = M(x, y)\mathbf{i} + N(x, y)\mathbf{j}$, then

$$\mathbf{F} \cdot \mathbf{n} = M(x, y)\frac{dy}{ds} - N(x, y)\frac{dx}{ds}.$$

Hence,

$$\int_C \mathbf{F} \cdot \mathbf{n}\, ds = \int_C \left(M\frac{dy}{ds} - N\frac{dx}{ds}\right) ds = \oint_C M\, dy - N\, dx.$$

We put a directed circle $\circlearrowleft$ on the last integral as a reminder that the integration around the closed curve C is to be in the counterclockwise direction. To evaluate this integral, we express M, dy, N, and dx in terms of t and integrate from $t = a$ to $t = b$. We do not need to know either $\mathbf{n}$ or ds to find the flux.

Calculating Flux Across a Smooth Closed Plane Curve

$$(\text{Flux of } \mathbf{F} = M\mathbf{i} + N\mathbf{j} \text{ across } C) = \oint_C M\,dy - N\,dx \tag{4}$$

The integral can be evaluated from any smooth parametrization $x = g(t), y = h(t)$, $a \le t \le b$, that traces C counterclockwise exactly once.

EXAMPLE 5 Find the flux of $\mathbf{F} = (x - y)\mathbf{i} + x\mathbf{j}$ across the circle $x^2 + y^2 = 1$ in the xy-plane.

Solution The parametrization $\mathbf{r}(t) = (\cos t)\mathbf{i} + (\sin t)\mathbf{j}$, $0 \le t \le 2\pi$, traces the circle counterclockwise exactly once. We can therefore use this parametrization in Equation (4). With

$$M = x - y = \cos t - \sin t, \qquad dy = d(\sin t) = \cos t\,dt$$

$$N = x = \cos t, \qquad dx = d(\cos t) = -\sin t\,dt,$$

we find

$$\text{Flux} = \int_C M\,dy - N\,dx = \int_0^{2\pi} (\cos^2 t - \sin t \cos t + \cos t \sin t)\,dt \qquad \text{Eq. (4)}$$

$$= \int_0^{2\pi} \cos^2 t\,dt = \int_0^{2\pi} \frac{1 + \cos 2t}{2}\,dt = \left[\frac{t}{2} + \frac{\sin 2t}{4}\right]_0^{2\pi} = \pi.$$

The flux of $\mathbf{F}$ across the circle is π. Since the answer is positive, the net flow across the curve is outward. A net inward flow would have given a negative flux. ■

EXERCISES 13.2

Find the gradient fields of the functions in Exercises 1–4.

1. $f(x, y, z) = (x^2 + y^2 + z^2)^{-1/2}$
2. $f(x, y, z) = \ln\sqrt{x^2 + y^2 + z^2}$
3. $g(x, y, z) = e^z - \ln(x^2 + y^2)$
4. $g(x, y, z) = xy + yz + xz$
5. Give a formula $\mathbf{F} = M(x, y)\mathbf{i} + N(x, y)\mathbf{j}$ for the vector field in the plane that has the property that $\mathbf{F}$ points toward the origin with magnitude inversely proportional to the square of the distance from (x, y) to the origin. (The field is not defined at $(0, 0)$.)
6. Give a formula $\mathbf{F} = M(x, y)\mathbf{i} + N(x, y)\mathbf{j}$ for the vector field in the plane that has the properties that $\mathbf{F} = \mathbf{0}$ at $(0, 0)$ and that at any other point (a, b), $\mathbf{F}$ is tangent to the circle $x^2 + y^2 = a^2 + b^2$ and points in the clockwise direction with magnitude $|\mathbf{F}| = \sqrt{a^2 + b^2}$.

In Exercises 7–12, find the work done by force $\mathbf{F}$ from $(0, 0, 0)$ to $(1, 1, 1)$ over each of the following paths (Figure 13.20):

a. The straight-line path C_1: $\mathbf{r}(t) = t\mathbf{i} + t\mathbf{j} + t\mathbf{k}$, $0 \le t \le 1$
b. The curved path C_2: $\mathbf{r}(t) = t\mathbf{i} + t^2\mathbf{j} + t^4\mathbf{k}$, $0 \le t \le 1$
c. The path $C_3 \cup C_4$ consisting of the line segment from $(0, 0, 0)$ to $(1, 1, 0)$ followed by the segment from $(1, 1, 0)$ to $(1, 1, 1)$

7. $\mathbf{F} = 3y\mathbf{i} + 2x\mathbf{j} + 4z\mathbf{k}$
8. $\mathbf{F} = [1/(x^2 + 1)]\mathbf{j}$
9. $\mathbf{F} = \sqrt{z}\mathbf{i} - 2x\mathbf{j} + \sqrt{y}\mathbf{k}$
10. $\mathbf{F} = xy\mathbf{i} + yz\mathbf{j} + xz\mathbf{k}$
11. $\mathbf{F} = (3x^2 - 3x)\mathbf{i} + 3z\mathbf{j} + \mathbf{k}$
12. $\mathbf{F} = (y + z)\mathbf{i} + (z + x)\mathbf{j} + (x + y)\mathbf{k}$

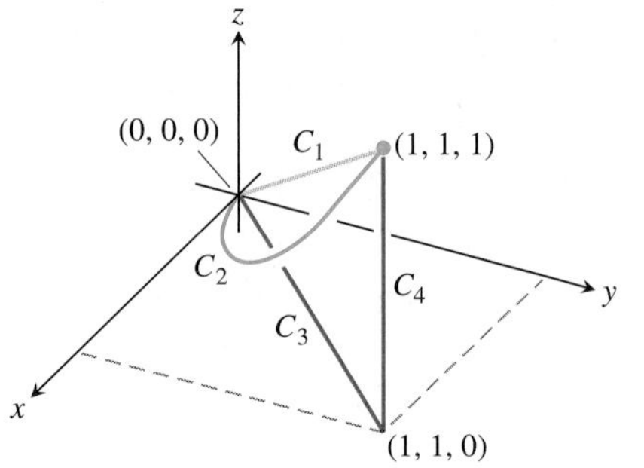

FIGURE 13.20 The paths from (0, 0, 0) to (1, 1, 1).

In Exercises 13–16, find the work done by $\mathbf{F}$ over the curve in the direction of increasing t.

13. $\mathbf{F} = xy\mathbf{i} + y\mathbf{j} - yz\mathbf{k}$
$\mathbf{r}(t) = t\mathbf{i} + t^2\mathbf{j} + t\mathbf{k}, \quad 0 \le t \le 1$

14. $\mathbf{F} = 2y\mathbf{i} + 3x\mathbf{j} + (x + y)\mathbf{k}$
$\mathbf{r}(t) = (\cos t)\mathbf{i} + (\sin t)\mathbf{j} + (t/6)\mathbf{k}, \quad 0 \le t \le 2\pi$

15. $\mathbf{F} = z\mathbf{i} + x\mathbf{j} + y\mathbf{k}$
$\mathbf{r}(t) = (\sin t)\mathbf{i} + (\cos t)\mathbf{j} + t\mathbf{k}, \quad 0 \le t \le 2\pi$

16. $\mathbf{F} = 6z\mathbf{i} + y^2\mathbf{j} + 12x\mathbf{k}$
$\mathbf{r}(t) = (\sin t)\mathbf{i} + (\cos t)\mathbf{j} + (t/6)\mathbf{k}, \quad 0 \le t \le 2\pi$

17. Evaluate $\int_C xy\,dx + (x + y)\,dy$ along the curve $y = x^2$ from $(-1, 1)$ to $(2, 4)$.

18. Evaluate $\int_C (x - y)\,dx + (x + y)\,dy$ counterclockwise around the triangle with vertices $(0, 0)$, $(1, 0)$, and $(0, 1)$.

19. Evaluate $\int_C \mathbf{F}\cdot\mathbf{T}\,ds$ for the vector field $\mathbf{F} = x^2\mathbf{i} - y\mathbf{j}$ along the curve $x = y^2$ from $(4, 2)$ to $(1, -1)$.

20. Evaluate $\int_C \mathbf{F}\cdot d\mathbf{r}$ for the vector field $\mathbf{F} = y\mathbf{i} - x\mathbf{j}$ counterclockwise along the unit circle $x^2 + y^2 = 1$ from $(1, 0)$ to $(0, 1)$.

21. **Work** Find the work done by the force $\mathbf{F} = xy\mathbf{i} + (y - x)\mathbf{j}$ over the straight line from $(1, 1)$ to $(2, 3)$.

22. **Work** Find the work done by the gradient of $f(x, y) = (x + y)^2$ counterclockwise around the circle $x^2 + y^2 = 4$ from $(2, 0)$ to itself.

23. **Circulation and flux** Find the circulation and flux of the fields

$$\mathbf{F}_1 = x\mathbf{i} + y\mathbf{j} \quad \text{and} \quad \mathbf{F}_2 = -y\mathbf{i} + x\mathbf{j}$$

around and across each of the following curves.

a. The circle $\mathbf{r}(t) = (\cos t)\mathbf{i} + (\sin t)\mathbf{j}, \quad 0 \le t \le 2\pi$

b. The ellipse $\mathbf{r}(t) = (\cos t)\mathbf{i} + (4\sin t)\mathbf{j}, \quad 0 \le t \le 2\pi$

24. **Flux across a circle** Find the flux of the fields

$$\mathbf{F}_1 = 2x\mathbf{i} - 3y\mathbf{j} \quad \text{and} \quad \mathbf{F}_2 = 2x\mathbf{i} + (x - y)\mathbf{j}$$

across the circle

$$\mathbf{r}(t) = (a\cos t)\mathbf{i} + (a\sin t)\mathbf{j}, \qquad 0 \le t \le 2\pi.$$

In Exercises 25–28, find the circulation and flux of the field $\mathbf{F}$ around and across the closed semicircular path that consists of the semicircular arch $\mathbf{r}_1(t) = (a\cos t)\mathbf{i} + (a\sin t)\mathbf{j}$, $0 \le t \le \pi$, followed by the line segment $\mathbf{r}_2(t) = t\mathbf{i}$, $-a \le t \le a$.

25. $\mathbf{F} = x\mathbf{i} + y\mathbf{j}$

26. $\mathbf{F} = x^2\mathbf{i} + y^2\mathbf{j}$

27. $\mathbf{F} = -y\mathbf{i} + x\mathbf{j}$

28. $\mathbf{F} = -y^2\mathbf{i} + x^2\mathbf{j}$

29. **Flow integrals** Find the flow of the velocity field $\mathbf{F} = (x + y)\mathbf{i} - (x^2 + y^2)\mathbf{j}$ along each of the following paths from $(1, 0)$ to $(-1, 0)$ in the xy-plane.

a. The upper half of the circle $x^2 + y^2 = 1$

b. The line segment from $(1, 0)$ to $(-1, 0)$

c. The line segment from $(1, 0)$ to $(0, -1)$ followed by the line segment from $(0, -1)$ to $(-1, 0)$.

30. **Flux across a triangle** Find the flux of the field $\mathbf{F}$ in Exercise 29 outward across the triangle with vertices $(1, 0)$, $(0, 1)$, $(-1, 0)$.

31. **Spin field** Draw the spin field

$$\mathbf{F} = -\frac{y}{\sqrt{x^2 + y^2}}\mathbf{i} + \frac{x}{\sqrt{x^2 + y^2}}\mathbf{j}$$

(see Figure 13.13) along with its horizontal and vertical components at a representative assortment of points on the circle $x^2 + y^2 = 4$.

32. **Radial field** Draw the radial field

$$\mathbf{F} = x\mathbf{i} + y\mathbf{j}$$

(see Figure 13.12) along with its horizontal and vertical components at a representative assortment of points on the circle $x^2 + y^2 = 1$.

33. **A field of tangent vectors**

a. Find a field $\mathbf{G} = P(x, y)\mathbf{i} + Q(x, y)\mathbf{j}$ in the xy-plane with the property that at any point $(a, b) \ne (0, 0)$, $\mathbf{G}$ is a vector of magnitude $\sqrt{a^2 + b^2}$ tangent to the circle $x^2 + y^2 = a^2 + b^2$ and pointing in the counterclockwise direction. (The field is undefined at $(0, 0)$.)

b. How is $\mathbf{G}$ related to the spin field $\mathbf{F}$ in Figure 13.13?

34. **A field of tangent vectors**

a. Find a field $\mathbf{G} = P(x, y)\mathbf{i} + Q(x, y)\mathbf{j}$ in the xy-plane with the property that at any point $(a, b) \ne (0, 0)$, $\mathbf{G}$ is a unit vector tangent to the circle $x^2 + y^2 = a^2 + b^2$ and pointing in the clockwise direction.

b. How is $\mathbf{G}$ related to the spin field $\mathbf{F}$ in Figure 13.13?

35. **Unit vectors pointing toward the origin** Find a field $\mathbf{F} = M(x, y)\mathbf{i} + N(x, y)\mathbf{j}$ in the xy-plane with the property that at each point $(x, y) \ne (0, 0)$, $\mathbf{F}$ is a unit vector pointing toward the origin. (The field is undefined at $(0, 0)$.)

36. Two "central" fields Find a field $\mathbf{F} = M(x, y)\mathbf{i} + N(x, y)\mathbf{j}$ in the xy-plane with the property that at each point $(x, y) \neq (0, 0)$, $\mathbf{F}$ points toward the origin and $|\mathbf{F}|$ is **(a)** the distance from (x, y) to the origin, **(b)** inversely proportional to the distance from (x, y) to the origin. (The field is undefined at $(0, 0)$.)

In Exercises 37–40, $\mathbf{F}$ is the velocity field of a fluid flowing through a region in space. Find the flow along the given curve in the direction of increasing t.

37. $\mathbf{F} = -4xy\mathbf{i} + 8y\mathbf{j} + 2\mathbf{k}$
$\mathbf{r}(t) = t\mathbf{i} + t^2\mathbf{j} + \mathbf{k}, \quad 0 \leq t \leq 2$

38. $\mathbf{F} = x^2\mathbf{i} + yz\mathbf{j} + y^2\mathbf{k}$
$r(t) = 3t\mathbf{j} + 4t\mathbf{k}, \quad 0 \leq t \leq 1$

39. $\mathbf{F} = (x - z)\mathbf{i} + x\mathbf{k}$
$\mathbf{r}(t) = (\cos t)\mathbf{i} + (\sin t)\mathbf{k}, \quad 0 \leq t \leq \pi$

40. $\mathbf{F} = -y\mathbf{i} + x\mathbf{j} + 2\mathbf{k}$
$\mathbf{r}(t) = (-2\cos t)\mathbf{i} + (2\sin t)\mathbf{j} + 2t\mathbf{k}, \quad 0 \leq t \leq 2\pi$

41. Circulation Find the circulation of $\mathbf{F} = 2x\mathbf{i} + 2z\mathbf{j} + 2y\mathbf{k}$ around the closed path consisting of the following three curves traversed in the direction of increasing t:

$$C_1: \quad \mathbf{r}(t) = (\cos t)\mathbf{i} + (\sin t)\mathbf{j} + t\mathbf{k}, \quad 0 \leq t \leq \pi/2$$

$$C_2: \quad \mathbf{r}(t) = \mathbf{j} + (\pi/2)(1 - t)\mathbf{k}, \quad 0 \leq t \leq 1$$

$$C_3: \quad \mathbf{r}(t) = t\mathbf{i} + (1 - t)\mathbf{j}, \quad 0 \leq t \leq 1$$

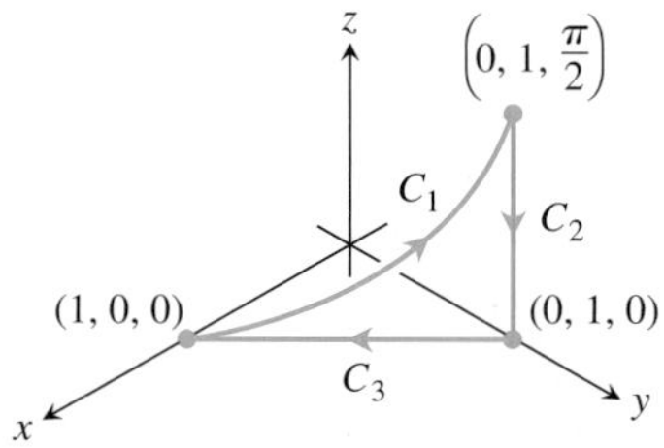

42. Zero circulation Let C be the ellipse in which the plane $2x + 3y - z = 0$ meets the cylinder $x^2 + y^2 = 12$. Show, without evaluating either line integral directly, that the circulation of the field $\mathbf{F} = x\mathbf{i} + y\mathbf{j} + z\mathbf{k}$ around C in either direction is zero.

43. Flow along a curve The field $\mathbf{F} = xy\mathbf{i} + y\mathbf{j} - yz\mathbf{k}$ is the velocity field of a flow in space. Find the flow from $(0, 0, 0)$ to $(1, 1, 1)$ along the curve of intersection of the cylinder $y = x^2$ and the plane $z = x$. (*Hint:* Use $t = x$ as the parameter.)

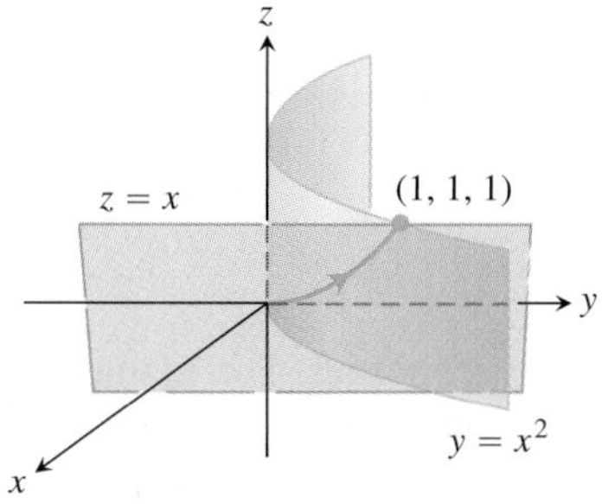

44. Flow of a gradient field Find the flow of the field $\mathbf{F} = \nabla(xy^2z^3)$:

a. Once around the curve C in Exercise 42, clockwise as viewed from above

b. Along the line segment from $(1, 1, 1)$ to $(2, 1, -1)$.

45. Work and area Suppose that $f(t)$ is differentiable and positive for $a \leq t \leq b$. Let C be the path $\mathbf{r}(t) = t\mathbf{i} + f(t)\mathbf{j}$, $a \leq t \leq b$, and $\mathbf{F} = y\mathbf{i}$. Is there any relation between the value of the work integral

$$\int_C \mathbf{F} \cdot d\mathbf{r}$$

and the area of the region bounded by the t-axis, the graph of f, and the lines $t = a$ and $t = b$? Give reasons for your answer.

46. Work done by a radial force with constant magnitude A particle moves along the smooth curve $y = f(x)$ from $(a, f(a))$ to $(b, f(b))$. The force moving the particle has constant magnitude k and always points away from the origin. Show that the work done by the force is

$$\int_C \mathbf{F} \cdot \mathbf{T}\, ds = k\left[(b^2 + (f(b))^2)^{1/2} - (a^2 + (f(a))^2)^{1/2}\right].$$

COMPUTER EXPLORATIONS

In Exercises 47–52, use a CAS to perform the following steps for finding the work done by force $\mathbf{F}$ over the given path:

a. Find $d\mathbf{r}$ for the path $\mathbf{r}(t) = g(t)\mathbf{i} + h(t)\mathbf{j} + k(t)\mathbf{k}$.

b. Evaluate the force $\mathbf{F}$ along the path.

c. Evaluate $\int_C \mathbf{F} \cdot d\mathbf{r}$.

47. $\mathbf{F} = xy^6\mathbf{i} + 3x(xy^5 + 2)\mathbf{j}; \quad \mathbf{r}(t) = (2\cos t)\mathbf{i} + (\sin t)\mathbf{j}$,
$0 \leq t \leq 2\pi$

48. $\mathbf{F} = \dfrac{3}{1 + x^2}\mathbf{i} + \dfrac{2}{1 + y^2}\mathbf{j}; \quad \mathbf{r}(t) = (\cos t)\mathbf{i} + (\sin t)\mathbf{j}$,
$0 \leq t \leq \pi$

49. $\mathbf{F} = (y + yz\cos xyz)\mathbf{i} + (x^2 + xz\cos xyz)\mathbf{j} +$
$(z + xy\cos xyz)\mathbf{k}; \quad \mathbf{r}(t) = (2\cos t)\mathbf{i} + (3\sin t)\mathbf{j} + \mathbf{k}$,
$0 \leq t \leq 2\pi$

50. $\mathbf{F} = 2xy\mathbf{i} - y^2\mathbf{j} + ze^x\mathbf{k}; \quad \mathbf{r}(t) = -t\mathbf{i} + \sqrt{t}\mathbf{j} + 3t\mathbf{k}$,
$1 \leq t \leq 4$

51. $\mathbf{F} = (2y + \sin x)\mathbf{i} + (z^2 + (1/3)\cos y)\mathbf{j} + x^4\mathbf{k}$;
$\mathbf{r}(t) = (\sin t)\mathbf{i} + (\cos t)\mathbf{j} + (\sin 2t)\mathbf{k}, \quad -\pi/2 \leq t \leq \pi/2$

52. $\mathbf{F} = (x^2y)\mathbf{i} + \frac{1}{3}x^3\mathbf{j} + xy\mathbf{k}; \quad \mathbf{r}(t) = (\cos t)\mathbf{i} + (\sin t)\mathbf{j} +$
$(2\sin^2 t - 1)\mathbf{k}, \quad 0 \leq t \leq 2\pi$

13.3 Path Independence, Potential Functions, and Conservative Fields

A **gravitational field g** is a vector field that represents the effect of gravity on a body placed within it. The gravitational force on a body of mass m placed in the field is given by $\mathbf{F} = m\mathbf{g}$. Similarly, an **electric field E** is a vector field in space that represents the effect of electric forces on a charged particle placed within it. The force on a body of charge q placed in the field is given by $\mathbf{F} = q\mathbf{E}$. In gravitational and electric fields, the amount of work it takes to move a mass or charge from one point to another depends on the initial and final positions of the object—not on which path is taken between these positions. In this section we study fields with this property and the calculation of work integrals associated with them.

Path Independence

If A and B are two points in an open region D in space, the work $\int \mathbf{F} \cdot d\mathbf{r}$ done in moving a particle from A to B by a field $\mathbf{F}$ defined on D usually depends on the path taken. For some special fields, however, the integral's value is the same for all paths from A to B.

> **DEFINITIONS** Let $\mathbf{F}$ be a field defined on an open region D in space, and suppose that for any two points A and B in D the work $\int_A^B \mathbf{F} \cdot d\mathbf{r}$ done in moving from A to B in D is the same over all paths from A to B. Then the integral $\int \mathbf{F} \cdot d\mathbf{r}$ is **path independent in *D*** and the field $\mathbf{F}$ is **conservative on *D***.

The word *conservative* comes from physics, where it refers to fields in which the principle of conservation of energy holds.

Under differentiability conditions normally met in practice, we will show that a field $\mathbf{F}$ is conservative if and only if it is the gradient field of a scalar function f; that is, if and only if $\mathbf{F} = \nabla f$ for some f. The function f then has a special name.

> **DEFINITION** If $\mathbf{F}$ is a field defined on D and $\mathbf{F} = \nabla f$ for some scalar function f on D, then f is called a **potential function for F**.

An electric potential is a scalar function whose gradient field is an electric field. A gravitational potential is a scalar function whose gradient field is a gravitational field, and so on. As we will see, once we have found a potential function f for a field $\mathbf{F}$, we can evaluate all the work integrals in the domain of $\mathbf{F}$ over any path between A and B by

$$\int_A^B \mathbf{F} \cdot d\mathbf{r} = \int_A^B \nabla f \cdot d\mathbf{r} = f(B) - f(A). \tag{1}$$

If you think of ∇f for functions of several variables as being something like the derivative f' for functions of a single variable, then you see that Equation (1) is the vector calculus analogue of the Fundamental Theorem of Calculus formula

$$\int_a^b f'(x)\, dx = f(b) - f(a).$$

Conservative fields have other remarkable properties. For example, saying that $\mathbf{F}$ is conservative on D is equivalent to saying that the integral of $\mathbf{F}$ around every closed path in D is zero. Certain conditions on the curves, fields, and domains must be satisfied for Equation (1) to be valid. We discuss these conditions next.

Assumptions on Curves, Vector Fields, and Domains

In order for the computations and results we derive below to be valid, we must assume certain properties for the curves, surfaces, domains, and vector fields we consider. We give these assumptions in the statements of theorems, and they also apply to the examples and exercises unless otherwise stated.

The curves we consider are **piecewise smooth**. Such curves are made up of finitely many smooth pieces connected end to end, as discussed in Section 10.1. We will treat vector fields **F** whose components have continuous first partial derivatives.

The solid domains D we consider are open regions in space, so every point in D is the center of an open ball that lies entirely in D (see Section 11.1). We also assume D to be **connected**. For an open region, this means that any two points in D can be joined by a smooth curve that lies in the region. Finally, we assume D is **simply connected**, which means that every loop in D can be contracted to a point in D without ever leaving D. The plane with a disk removed is a two-dimensional region that is *not* simply connected, for example. A loop in the plane that goes around the disk cannot be contracted to a point without going into the "hole" left by the removed disk. Similarly, if we remove an infinitely long solid cylinder from space, the remaining region D is *not* simply connected. A curve encircling the cylinder cannot be shrunk to a point while remaining inside D.

Connectivity and simple connectivity are not the same, and neither property implies the other. Think of connected regions as being in "one piece" and simply connected regions as not having any "loop-catching holes." All of space itself is both connected and simply connected. Figure 13.21 illustrates some of these properties.

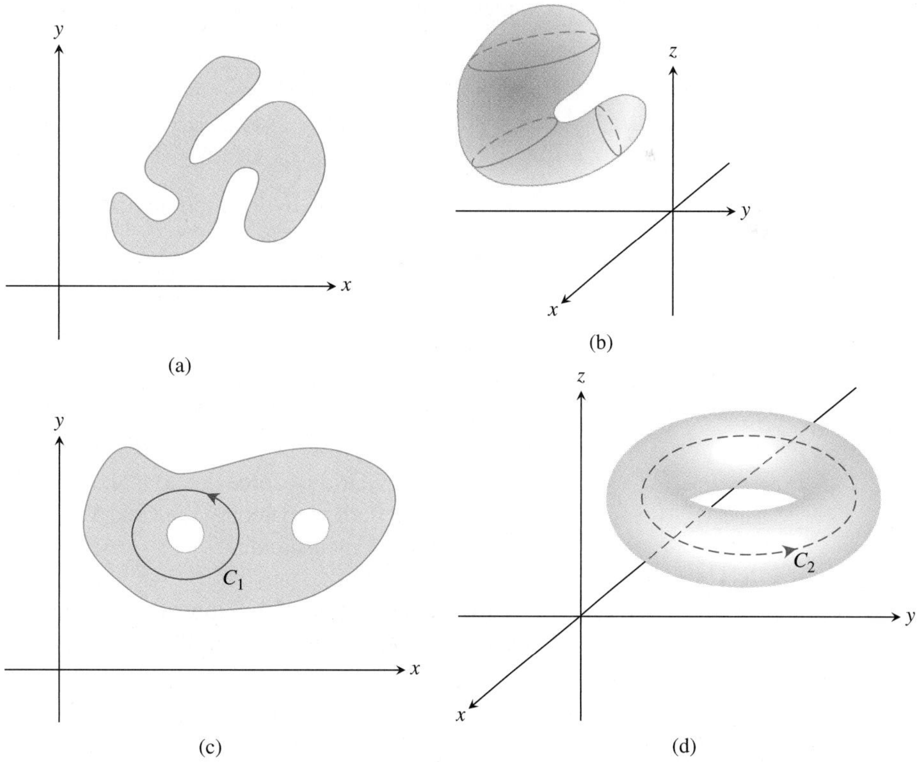

FIGURE 13.21 Four connected regions. In (a) and (b), the regions are simply connected. In (c) and (d), the regions are not simply connected because the curves C_1 and C_2 cannot be contracted to a point inside the regions containing them.

Caution Some of the results in this chapter can fail to hold if applied to situations where the conditions we've imposed do not hold. In particular, the component test for conservative fields, given later in this section, is not valid on domains that are not simply connected.

Line Integrals in Conservative Fields

Gradient fields $\mathbf{F}$ are obtained by differentiating a scalar function f. A theorem analogous to the Fundamental Theorem of Calculus gives a way to evaluate the line integrals of gradient fields.

THEOREM 1—Fundamental Theorem of Line Integrals Let C be a smooth curve joining the points A and B in the plane or in space and given by $\mathbf{r}(t)$. Let f be a differentiable function with a continuous gradient vector $\mathbf{F} = \nabla f$ on a domain containing C. Then

$$\int_C \mathbf{F} \cdot d\mathbf{r} = f(B) - f(A).$$

Like the Fundamental Theorem, Theorem 1 gives a way to evaluate line integrals without having to take limits of Riemann sums or finding the line integral directly using the method of Section 13.1. Before proving Theorem 1, we give an example.

EXAMPLE 1 Suppose the vector field $\mathbf{F} = \nabla f$ is the gradient of the function

$$f(x, y, z) = -\frac{1}{x^2 + y^2 + z^2}.$$

Find the work done by $\mathbf{F}$ in moving a particle along a smooth curve C joining $(1, 0, 0)$ to $(0, 0, 2)$.

Solution There is no need to compute a line integral directly. By applying Theorem 1, we see that the work done by $\mathbf{F}$ along any smooth curve C joining the two points is

$$\int_C \mathbf{F} \cdot d\mathbf{r} = f(0, 0, 2) - f(1, 0, 0) = -\frac{1}{4} - (-1) = \frac{3}{4}.$$ ■

The gravitational force due to a planet, and the electric force associated with a charged particle, are both modeled by the field $\mathbf{F}$ given in Example 1 up to a constant that depends on the units of measurement.

Proof of Theorem 1 Suppose that A and B are two points in region D and that C: $\mathbf{r}(t) = g(t)\mathbf{i} + h(t)\mathbf{j} + k(t)\mathbf{k}$, $a \le t \le b$, is a smooth curve in D joining A and B. Along the curve, f is a differentiable function of t and

$$\frac{df}{dt} = \frac{\partial f}{\partial x}\frac{dx}{dt} + \frac{\partial f}{\partial y}\frac{dy}{dt} + \frac{\partial f}{\partial z}\frac{dz}{dt}$$

Chain Rule with $x = g(t)$, $y = h(t)$, $z = k(t)$

$$= \nabla f \cdot \left(\frac{dx}{dt}\mathbf{i} + \frac{dy}{dt}\mathbf{j} + \frac{dz}{dt}\mathbf{k}\right) = \nabla f \cdot \frac{d\mathbf{r}}{dt} = \mathbf{F} \cdot \frac{d\mathbf{r}}{dt}.$$

Because $\mathbf{F} = \nabla f$

Therefore,

$$\int_C \mathbf{F} \cdot d\mathbf{r} = \int_{t=a}^{t=b} \mathbf{F} \cdot \frac{d\mathbf{r}}{dt}\, dt = \int_a^b \frac{df}{dt}\, dt$$

$$= f(g(t), h(t), k(t)) \Big]_a^b = f(B) - f(A). \qquad \blacksquare$$

So we see from Theorem 1 that the line integral of a gradient field $\mathbf{F} = \nabla f$ is straightforward to compute once we know the function f. Many important vector fields arising in nature are indeed gradient fields. The next result, which follows from Theorem 1, shows that any conservative field is of this type. Therefore the value of a line integral of a conservative field along a smooth curve depends only on the endpoints of the curve.

THEOREM 2—Conservative Fields are Gradient Fields Let $\mathbf{F} = M\mathbf{i} + N\mathbf{j} + P\mathbf{k}$ be a vector field whose components are continuous throughout an open connected region D in space. Then $\mathbf{F}$ is conservative if and only if $\mathbf{F}$ is a gradient field ∇f for a differentiable function f.

Theorem 2 says that $\mathbf{F} = \nabla f$ if and only if for any two points A and B in the region D, the value of line integral $\int_C \mathbf{F} \cdot d\mathbf{r}$ is independent of the path joining A to B in D.

Partial Proof of Theorem 2 If $\mathbf{F}$ is a gradient field, then $\mathbf{F} = \nabla f$ for a differentiable function f, and Theorem 1 shows that $\int_C \mathbf{F} \cdot d\mathbf{r} = f(B) - f(A)$. The value of the line integral does not depend on C, but only on its endpoints A and B. So the line integral is path independent and $\mathbf{F}$ satisfies the definition of a conservative field.

On the other hand, if $\mathbf{F}$ is conservative, we can define a scalar function f on D as follows. Pick a point A in D and set $f(A)$ equal to a constant k. Then define $f(B)$ to equal $f(A) + \int_C \mathbf{F} \cdot d\mathbf{r}$, where C is any smooth path from A to B. Notice that $f(B)$ does not depend on the choice of C since $\mathbf{F}$ is conservative. It can then be shown that $\mathbf{F} = \nabla f$, but we omit the more advanced proof here. ■

EXAMPLE 2 Find the work done by the conservative field

$$\mathbf{F} = yz\mathbf{i} + xz\mathbf{j} + xy\mathbf{k} = \nabla(xyz)$$

along any smooth curve C joining the point $A(-1, 3, 9)$ to $B(1, 6, -4)$.

Solution With $f(x, y, z) = xyz$, we have

$$\int_A^B \mathbf{F} \cdot d\mathbf{r} = \int_A^B \nabla f \cdot d\mathbf{r} \qquad \mathbf{F} = \nabla f$$

$$= f(B) - f(A) \qquad \text{Theorem 1}$$

$$= xyz\big|_{(1,6,-4)} - xyz\big|_{(-1,3,9)}$$

$$= (1)(6)(-4) - (-1)(3)(9)$$

$$= -24 + 27 = 3. \qquad \blacksquare$$

THEOREM 3—Closed-Loop Property of Conservative Fields The following statements are equivalent.

1. $\int \mathbf{F} \cdot d\mathbf{r} = 0$ around every closed loop in D.
2. The field $\mathbf{F}$ is conservative on D.

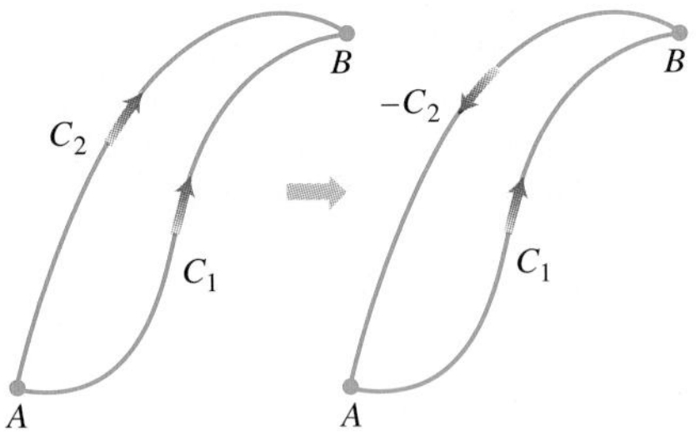

FIGURE 13.22 If we have two paths from A to B, one of them can be reversed to make a loop.

Proof that Part 1 $\Rightarrow$ Part 2 We want to show that for any two points A and B in D, the integral of $\mathbf{F} \cdot d\mathbf{r}$ has the same value over any two paths C_1 and C_2 from A to B. We reverse the direction on C_2 to make a path $-C_2$ from B to A (Figure 13.22). Together, C_1 and $-C_2$ make a closed loop C, and by assumption,

$$\int_{C_1} \mathbf{F} \cdot d\mathbf{r} - \int_{C_2} \mathbf{F} \cdot d\mathbf{r} = \int_{C_1} \mathbf{F} \cdot d\mathbf{r} + \int_{-C_2} \mathbf{F} \cdot d\mathbf{r} = \int_{C} \mathbf{F} \cdot d\mathbf{r} = 0.$$

Thus, the integrals over C_1 and C_2 give the same value. Note that the definition of line integral shows that changing the direction along a curve reverses the sign of the line integral.

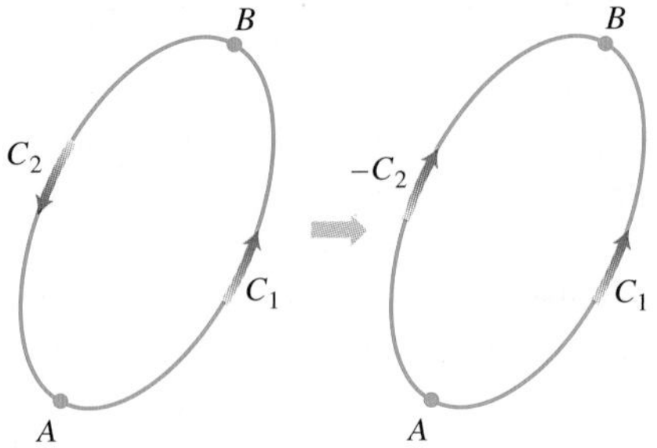

FIGURE 13.23 If A and B lie on a loop, we can reverse part of the loop to make two paths from A to B.

Proof that Part 2 $\Rightarrow$ Part 1 We want to show that the integral of $\mathbf{F} \cdot d\mathbf{r}$ is zero over any closed loop C. We pick two points A and B on C and use them to break C into two pieces: C_1 from A to B followed by C_2 from B back to A (Figure 13.23). Then

$$\oint_{C} \mathbf{F} \cdot d\mathbf{r} = \int_{C_1} \mathbf{F} \cdot d\mathbf{r} + \int_{C_2} \mathbf{F} \cdot d\mathbf{r} = \int_{A}^{B} \mathbf{F} \cdot d\mathbf{r} - \int_{A}^{B} \mathbf{F} \cdot d\mathbf{r} = 0. \qquad \blacksquare$$

The following diagram summarizes the results of Theorems 2 and 3.

	Theorem 2		Theorem 3	
$\mathbf{F} = \nabla f$ on D	$\Leftrightarrow$	$\mathbf{F}$ conservative on D	$\Leftrightarrow$	$\oint_C \mathbf{F} \cdot d\mathbf{r} = 0$ over any closed path in D

Two questions arise:

1. How do we know whether a given field $\mathbf{F}$ is conservative?
2. If $\mathbf{F}$ is in fact conservative, how do we find a potential function f (so that $\mathbf{F} = \nabla f$)?

Finding Potentials for Conservative Fields

The test for being conservative is the following.

Component Test for Conservative Fields
Let $\mathbf{F} = M(x, y, z)\mathbf{i} + N(x, y, z)\mathbf{j} + P(x, y, z)\mathbf{k}$ be a field on a connected and simply connected domain whose component functions have continuous first partial derivatives. Then, $\mathbf{F}$ is conservative if and only if

$$\frac{\partial P}{\partial y} = \frac{\partial N}{\partial z}, \quad \frac{\partial M}{\partial z} = \frac{\partial P}{\partial x}, \quad \text{and} \quad \frac{\partial N}{\partial x} = \frac{\partial M}{\partial y}. \tag{2}$$

Proof that Equations (2) hold if F is conservative There is a potential function f such that

$$\mathbf{F} = M\mathbf{i} + N\mathbf{j} + P\mathbf{k} = \frac{\partial f}{\partial x}\mathbf{i} + \frac{\partial f}{\partial y}\mathbf{j} + \frac{\partial f}{\partial z}\mathbf{k}.$$

Hence,

$$\begin{aligned}\frac{\partial P}{\partial y} &= \frac{\partial}{\partial y}\left(\frac{\partial f}{\partial z}\right) = \frac{\partial^2 f}{\partial y\,\partial z}\\ &= \frac{\partial^2 f}{\partial z\,\partial y}\\ &= \frac{\partial}{\partial z}\left(\frac{\partial f}{\partial y}\right) = \frac{\partial N}{\partial z}.\end{aligned}$$

Continuity implies that the mixed partial derivatives are equal.

The others in Equations (2) are proved similarly. ■

The second half of the proof, that Equations (2) imply that **F** is conservative, is a consequence of Stokes' Theorem, taken up in Section 13.7, and requires our assumption that the domain of **F** be simply connected.

Once we know that **F** is conservative, we usually want to find a potential function for **F**. This requires solving the equation $\nabla f = \mathbf{F}$ or

$$\frac{\partial f}{\partial x}\mathbf{i} + \frac{\partial f}{\partial y}\mathbf{j} + \frac{\partial f}{\partial z}\mathbf{k} = M\mathbf{i} + N\mathbf{j} + P\mathbf{k}$$

for f. We accomplish this by integrating the three equations

$$\frac{\partial f}{\partial x} = M, \qquad \frac{\partial f}{\partial y} = N, \qquad \frac{\partial f}{\partial z} = P,$$

as illustrated in the next example.

EXAMPLE 3 Show that $\mathbf{F} = (e^x \cos y + yz)\mathbf{i} + (xz - e^x \sin y)\mathbf{j} + (xy + z)\mathbf{k}$ is conservative and find a potential function for it.

Solution We apply the test in Equations (2) to

$$M = e^x \cos y + yz, \qquad N = xz - e^x \sin y, \qquad P = xy + z$$

and calculate

$$\frac{\partial P}{\partial y} = x = \frac{\partial N}{\partial z}, \qquad \frac{\partial M}{\partial z} = y = \frac{\partial P}{\partial x}, \qquad \frac{\partial N}{\partial x} = -e^x \sin y + z = \frac{\partial M}{\partial y}.$$

Together, these equalities tell us that there is a function f with $\nabla f = \mathbf{F}$.

We find f by integrating the equations

$$\frac{\partial f}{\partial x} = e^x \cos y + yz, \qquad \frac{\partial f}{\partial y} = xz - e^x \sin y, \qquad \frac{\partial f}{\partial z} = xy + z. \tag{3}$$

We integrate the first equation with respect to x, holding y and z fixed, to get

$$f(x, y, z) = e^x \cos y + xyz + g(y, z).$$

We write the constant of integration as a function of y and z because its value may change if y and z change. We then calculate $\partial f/\partial y$ from this equation and match it with the expression for $\partial f/\partial y$ in Equations (3). This gives

$$-e^x \sin y + xz + \frac{\partial g}{\partial y} = xz - e^x \sin y,$$

so $\partial g/\partial y = 0$. Therefore, g is a function of z alone, and

$$f(x, y, z) = e^x \cos y + xyz + h(z).$$

We now calculate $\partial f/\partial z$ from this equation and match it to the formula for $\partial f/\partial z$ in Equations (3). This gives

$$xy + \frac{dh}{dz} = xy + z, \qquad \text{or} \qquad \frac{dh}{dz} = z,$$

so

$$h(z) = \frac{z^2}{2} + C.$$

Hence,

$$f(x, y, z) = e^x \cos y + xyz + \frac{z^2}{2} + C.$$

We have infinitely many potential functions of **F**, one for each value of C. ■

EXAMPLE 4 Show that $\mathbf{F} = (2x - 3)\mathbf{i} - z\mathbf{j} + (\cos z)\mathbf{k}$ is not conservative.

Solution We apply the component test in Equations (2) and find immediately that

$$\frac{\partial P}{\partial y} = \frac{\partial}{\partial y}(\cos z) = 0, \qquad \frac{\partial N}{\partial z} = \frac{\partial}{\partial z}(-z) = -1.$$

The two are unequal, so **F** is not conservative. No further testing is required. ■

Exact Differential Forms

As we see in the next section and again later on, it is often convenient to express work and circulation integrals in the "differential" form

$$\int_A^B M\,dx + N\,dy + P\,dz$$

mentioned in Section 13.2. Such integrals are relatively easy to evaluate if $M\,dx + N\,dy + P\,dz$ is the total differential of a function f. For then

$$\begin{aligned}\int_A^B M\,dx + N\,dy + P\,dz &= \int_A^B \frac{\partial f}{\partial x}dx + \frac{\partial f}{\partial y}dy + \frac{\partial f}{\partial z}dz \\ &= \int_A^B \nabla f \cdot d\mathbf{r} \\ &= f(B) - f(A). \qquad \text{Theorem 1}\end{aligned}$$

Thus,

$$\int_A^B df = f(B) - f(A),$$

just as with differentiable functions of a single variable.

DEFINITIONS Any expression $M(x, y, z)\,dx + N(x, y, z)\,dy + P(x, y, z)\,dz$ is a **differential form**. A differential form is **exact** on a domain D in space if

$$M\,dx + N\,dy + P\,dz = \frac{\partial f}{\partial x}dx + \frac{\partial f}{\partial y}dy + \frac{\partial f}{\partial z}dz = df$$

for some scalar function f throughout D.

Notice that if $M\,dx + N\,dy + P\,dz = df$ on D, then $\mathbf{F} = M\mathbf{i} + N\mathbf{j} + P\mathbf{k}$ is the gradient field of f on D. Conversely, if $\mathbf{F} = \nabla f$, then the form $M\,dx + N\,dy + P\,dz$ is exact. The test for the form's being exact is therefore the same as the test for $\mathbf{F}$ being conservative.

Component Test for Exactness of $M\,dx + N\,dy + P\,dz$
The differential form $M\,dx + N\,dy + P\,dz$ is exact on a connected and simply connected domain if and only if

$$\frac{\partial P}{\partial y} = \frac{\partial N}{\partial z}, \quad \frac{\partial M}{\partial z} = \frac{\partial P}{\partial x}, \quad \text{and} \quad \frac{\partial N}{\partial x} = \frac{\partial M}{\partial y}.$$

This is equivalent to saying that the field $\mathbf{F} = M\mathbf{i} + N\mathbf{j} + P\mathbf{k}$ is conservative.

EXAMPLE 5 Show that $y\,dx + x\,dy + 4\,dz$ is exact and evaluate the integral

$$\int_{(1,1,1)}^{(2,3,-1)} y\,dx + x\,dy + 4\,dz$$

over the line segment from $(1, 1, 1)$ to $(2, 3, -1)$.

Solution We let $M = y, N = x, P = 4$ and apply the Test for Exactness:

$$\frac{\partial P}{\partial y} = 0 = \frac{\partial N}{\partial z}, \quad \frac{\partial M}{\partial z} = 0 = \frac{\partial P}{\partial x}, \quad \frac{\partial N}{\partial x} = 1 = \frac{\partial M}{\partial y}.$$

These equalities tell us that $y\,dx + x\,dy + 4\,dz$ is exact, so

$$y\,dx + x\,dy + 4\,dz = df$$

for some function f, and the integral's value is $f(2, 3, -1) - f(1, 1, 1)$.

We find f up to a constant by integrating the equations

$$\frac{\partial f}{\partial x} = y, \quad \frac{\partial f}{\partial y} = x, \quad \frac{\partial f}{\partial z} = 4. \tag{4}$$

From the first equation we get

$$f(x, y, z) = xy + g(y, z).$$

The second equation tells us that

$$\frac{\partial f}{\partial y} = x + \frac{\partial g}{\partial y} = x, \qquad \text{or} \qquad \frac{\partial g}{\partial y} = 0.$$

Hence, g is a function of z alone, and

$$f(x, y, z) = xy + h(z).$$

The third of Equations (4) tells us that

$$\frac{\partial f}{\partial z} = 0 + \frac{dh}{dz} = 4, \qquad \text{or} \qquad h(z) = 4z + C.$$

Therefore,

$$f(x, y, z) = xy + 4z + C.$$

The value of the integral is

$$f(2, 3, -1) - f(1, 1, 1) = 2 + C - (5 + C) = -3.$$ ■

EXERCISES 13.3

Which fields in Exercises 1–6 are conservative, and which are not?

1. $\mathbf{F} = yz\mathbf{i} + xz\mathbf{j} + xy\mathbf{k}$
2. $\mathbf{F} = (y \sin z)\mathbf{i} + (x \sin z)\mathbf{j} + (xy \cos z)\mathbf{k}$
3. $\mathbf{F} = y\mathbf{i} + (x + z)\mathbf{j} - y\mathbf{k}$
4. $\mathbf{F} = -y\mathbf{i} + x\mathbf{j}$
5. $\mathbf{F} = (z + y)\mathbf{i} + z\mathbf{j} + (y + x)\mathbf{k}$
6. $\mathbf{F} = (e^x \cos y)\mathbf{i} - (e^x \sin y)\mathbf{j} + z\mathbf{k}$

In Exercises 7–12, find a potential function f for the field $\mathbf{F}$.

7. $\mathbf{F} = 2x\mathbf{i} + 3y\mathbf{j} + 4z\mathbf{k}$
8. $\mathbf{F} = (y + z)\mathbf{i} + (x + z)\mathbf{j} + (x + y)\mathbf{k}$
9. $\mathbf{F} = e^{y+2z}(\mathbf{i} + x\mathbf{j} + 2x\mathbf{k})$
10. $\mathbf{F} = (y \sin z)\mathbf{i} + (x \sin z)\mathbf{j} + (xy \cos z)\mathbf{k}$
11. $\mathbf{F} = (\ln x + \sec^2(x + y))\mathbf{i} + \left(\sec^2(x + y) + \frac{y}{y^2 + z^2}\right)\mathbf{j} + \frac{z}{y^2 + z^2}\mathbf{k}$
12. $\mathbf{F} = \frac{y}{1 + x^2y^2}\mathbf{i} + \left(\frac{x}{1 + x^2y^2} + \frac{z}{\sqrt{1 - y^2z^2}}\right)\mathbf{j} + \left(\frac{y}{\sqrt{1 - y^2z^2}} + \frac{1}{z}\right)\mathbf{k}$

In Exercises 13–17, show that the differential forms in the integrals are exact. Then evaluate the integrals.

13. $\int_{(0,0,0)}^{(2,3,-6)} 2x\,dx + 2y\,dy + 2z\,dz$
14. $\int_{(1,1,2)}^{(3,5,0)} yz\,dx + xz\,dy + xy\,dz$
15. $\int_{(0,0,0)}^{(1,2,3)} 2xy\,dx + (x^2 - z^2)\,dy - 2yz\,dz$
16. $\int_{(0,0,0)}^{(3,3,1)} 2x\,dx - y^2\,dy - \frac{4}{1 + z^2}\,dz$
17. $\int_{(1,0,0)}^{(0,1,1)} \sin y \cos x\,dx + \cos y \sin x\,dy + dz$

Although they are not defined on all of space R^3, the fields associated with Exercises 18–22 are simply connected and the Component Test can be used to show they are conservative. Find a potential function for each field and evaluate the integrals as in Example 5.

18. $\int_{(0,2,1)}^{(1,\pi/2,2)} 2\cos y\,dx + \left(\frac{1}{y} - 2x \sin y\right)dy + \frac{1}{z}\,dz$
19. $\int_{(1,1,1)}^{(1,2,3)} 3x^2\,dx + \frac{z^2}{y}\,dy + 2z \ln y\,dz$

20. $\int_{(1,2,1)}^{(2,1,1)} (2x \ln y - yz)\, dx + \left(\frac{x^2}{y} - xz\right) dy - xy\, dz$

21. $\int_{(1,1,1)}^{(2,2,2)} \frac{1}{y}\, dx + \left(\frac{1}{z} - \frac{x}{y^2}\right) dy - \frac{y}{z^2}\, dz$

22. $\int_{(-1,-1,-1)}^{(2,2,2)} \frac{2x\, dx + 2y\, dy + 2z\, dz}{x^2 + y^2 + z^2}$

23. **Revisiting Example 5** Evaluate the integral

$$\int_{(1,1,1)}^{(2,3,-1)} y\, dx + x\, dy + 4\, dz$$

from Example 5 by finding parametric equations for the line segment from (1, 1, 1) to (2, 3, −1) and evaluating the line integral of $\mathbf{F} = y\mathbf{i} + x\mathbf{j} + 4\mathbf{k}$ along the segment. Since **F** is conservative, the integral is independent of the path.

24. Evaluate

$$\int_C x^2\, dx + yz\, dy + (y^2/2)\, dz$$

along the line segment C joining (0, 0, 0) to (0, 3, 4).

Independence of path Show that the values of the integrals in Exercises 25 and 26 do not depend on the path taken from A to B.

25. $\int_A^B z^2\, dx + 2y\, dy + 2xz\, dz$

26. $\int_A^B \frac{x\, dx + y\, dy + z\, dz}{\sqrt{x^2 + y^2 + z^2}}$

In Exercises 27 and 28, find a potential function for **F**.

27. $\mathbf{F} = \frac{2x}{y}\mathbf{i} + \left(\frac{1 - x^2}{y^2}\right)\mathbf{j}, \qquad \{(x, y): y > 0\}$

28. $\mathbf{F} = (e^x \ln y)\mathbf{i} + \left(\frac{e^x}{y} + \sin z\right)\mathbf{j} + (y \cos z)\mathbf{k}$

29. **Work along different paths** Find the work done by $\mathbf{F} = (x^2 + y)\mathbf{i} + (y^2 + x)\mathbf{j} + ze^z\mathbf{k}$ over the following paths from (1, 0, 0) to (1, 0, 1).
   a. The line segment $x = 1, y = 0, 0 \le z \le 1$
   b. The helix $\mathbf{r}(t) = (\cos t)\mathbf{i} + (\sin t)\mathbf{j} + (t/2\pi)\mathbf{k}, 0 \le t \le 2\pi$
   c. The x-axis from (1, 0, 0) to (0, 0, 0) followed by the parabola $z = x^2, y = 0$ from (0, 0, 0) to (1, 0, 1)

30. **Work along different paths** Find the work done by $\mathbf{F} = e^{yz}\mathbf{i} + (xze^{yz} + z \cos y)\mathbf{j} + (xye^{yz} + \sin y)\mathbf{k}$ over the following paths from (1, 0, 1) to $(1, \pi/2, 0)$.
   a. The line segment $x = 1, y = \pi t/2, z = 1 - t, 0 \le t \le 1$
   b. The line segment from (1, 0, 1) to the origin followed by the line segment from the origin to $(1, \pi/2, 0)$
   c. The line segment from (1, 0, 1) to (1, 0, 0), followed by the x-axis from (1, 0, 0) to the origin, followed by the parabola $y = \pi x^2/2, z = 0$ from there to $(1, \pi/2, 0)$

31. **Evaluating a work integral two ways** Let $\mathbf{F} = \nabla(x^3y^2)$ and let C be the path in the xy-plane from (−1, 1) to (1, 1) that consists of the line segment from (−1, 1) to (0, 0) followed by the line segment from (0, 0) to (1, 1). Evaluate $\int_C \mathbf{F} \cdot d\mathbf{r}$ in two ways.
   a. Find parametrizations for the segments that make up C and evaluate the integral.
   b. Use $f(x, y) = x^3y^2$ as a potential function for **F**.

32. **Integral along different paths** Evaluate the integral $\int_C 2x \cos y\, dx - x^2 \sin y\, dy$ along the following paths C in the xy-plane.
   a. The parabola $y = (x - 1)^2$ from (1, 0) to (0, 1)
   b. The line segment from $(-1, \pi)$ to (1, 0)
   c. The x-axis from (−1, 0) to (1, 0)
   d. The astroid $\mathbf{r}(t) = (\cos^3 t)\mathbf{i} + (\sin^3 t)\mathbf{j}, 0 \le t \le 2\pi$, counterclockwise from (1, 0) back to (1, 0)

33. a. **Exact differential form** How are the constants a, b, and c related if the following differential form is exact?

$$(ay^2 + 2czx)\, dx + y(bx + cz)\, dy + (ay^2 + cx^2)\, dz$$

   b. **Gradient field** For what values of b and c will

$$\mathbf{F} = (y^2 + 2czx)\mathbf{i} + y(bx + cz)\mathbf{j} + (y^2 + cx^2)\mathbf{k}$$

   be a gradient field?

34. **Gradient of a line integral** Suppose that $\mathbf{F} = \nabla f$ is a conservative vector field and

$$g(x, y, z) = \int_{(0,0,0)}^{(x,y,z)} \mathbf{F} \cdot d\mathbf{r}.$$

Show that $\nabla g = \mathbf{F}$.

35. **Path of least work** You have been asked to find the path along which a force field **F** will perform the least work in moving a particle between two locations. A quick calculation on your part shows **F** to be conservative. How should you respond? Give reasons for your answer.

36. **A revealing experiment** By experiment, you find that a force field **F** performs only half as much work in moving an object along path C_1 from A to B as it does in moving the object along path C_2 from A to B. What can you conclude about **F**? Give reasons for your answer.

37. **Work by a constant force** Show that the work done by a constant force field $\mathbf{F} = a\mathbf{i} + b\mathbf{j} + c\mathbf{k}$ in moving a particle along any path from A to B is $W = \mathbf{F} \cdot \overrightarrow{AB}$.

38. **Gravitational field**
   a. Find a potential function for the gravitational field

$$\mathbf{F} = -GmM\frac{x\mathbf{i} + y\mathbf{j} + z\mathbf{k}}{(x^2 + y^2 + z^2)^{3/2}} \qquad (G, m, \text{ and } M \text{ are constants}).$$

   b. Let P_1 and P_2 be points at distance s_1 and s_2 from the origin. Show that the work done by the gravitational field in part (a) in moving a particle from P_1 to P_2 is

$$GmM\left(\frac{1}{s_2} - \frac{1}{s_1}\right).$$

13.4 Green's Theorem in the Plane

If $\mathbf{F}$ is a conservative field, then we know $\mathbf{F} = \nabla f$ for a differentiable function f, and we can calculate the work integral over any path C joining point A to B as $\int_C \mathbf{F} \cdot d\mathbf{r} = f(B) - f(A)$. In this section we derive a method for computing a work or flux integral over a *closed* curve C in the plane when the field $\mathbf{F}$ is *not* conservative. This method, known as Green's Theorem, allows us to convert the line integral into a double integral over the region enclosed by C.

The discussion is given in terms of velocity fields of fluid flows because they are easy to visualize. However, Green's Theorem applies to any vector field, independent of any particular interpretation of the field, provided the assumptions of the theorem are satisfied.

Divergence

We need two new ideas for Green's Theorem. The first is the idea of the *divergence* of a vector field at a point, sometimes called the *flux density* of the vector field by physicists and engineers. We obtain it in the following way.

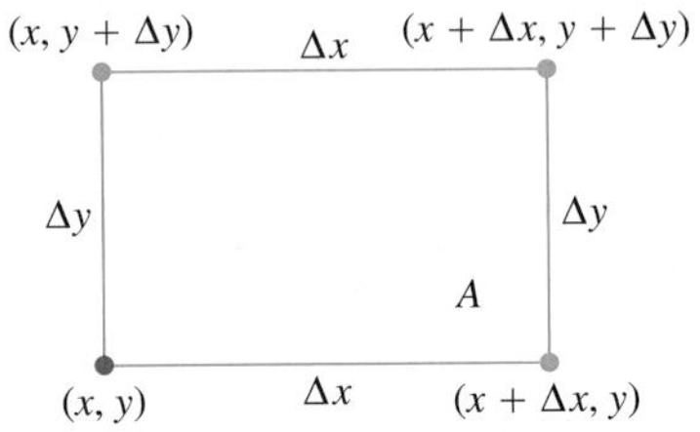

FIGURE 13.24 The rectangle for defining the divergence (flux density) of a vector field at a point (x, y).

Suppose that $\mathbf{F}(x, y) = M(x, y)\mathbf{i} + N(x, y)\mathbf{j}$ is the velocity field of a fluid flow in the plane and that the first partial derivatives of M and N are continuous at each point of a region R. Let (x, y) be a point in R and let A be a small rectangle with one corner at (x, y) that, along with its interior, lies entirely in R (Figure 13.24). The sides of the rectangle, parallel to the coordinate axes, have lengths of Δx and Δy. The rate at which fluid leaves the rectangle across the bottom edge is approximately

$$\mathbf{F}(x, y) \cdot (-\mathbf{j})\, \Delta x = -N(x, y)\Delta x.$$

This is the scalar component of the velocity at (x, y) in the direction of the outward normal times the length of the segment. If the velocity is in meters per second, for example, the exit rate will be in meters per second times meters or square meters per second. The rates at which the fluid crosses the other three sides in the directions of their outward normals can be estimated in a similar way. All told, we have

Exit Rates:	Top:	$\mathbf{F}(x, y + \Delta y) \cdot \mathbf{j}\, \Delta x = N(x, y + \Delta y)\Delta x$
	Bottom:	$\mathbf{F}(x, y) \cdot (-\mathbf{j})\, \Delta x = -N(x, y)\Delta x$
	Right:	$\mathbf{F}(x + \Delta x, y) \cdot \mathbf{i}\, \Delta y = M(x + \Delta x, y)\Delta y$
	Left:	$\mathbf{F}(x, y) \cdot (-\mathbf{i})\, \Delta y = -M(x, y)\Delta y.$

Combining opposite pairs gives

$$\text{Top and bottom:} \quad (N(x, y + \Delta y) - N(x, y))\Delta x \approx \left(\frac{\partial N}{\partial y}\Delta y\right)\Delta x$$

$$\text{Right and left:} \quad (M(x + \Delta x, y) - M(x, y))\Delta y \approx \left(\frac{\partial M}{\partial x}\Delta x\right)\Delta y.$$

Adding these last two equations gives

$$\text{Flux across rectangle boundary} \approx \left(\frac{\partial M}{\partial x} + \frac{\partial N}{\partial y}\right)\Delta x \Delta y.$$

We now divide by $\Delta x \Delta y$ to estimate the total flux per unit area or flux density for the rectangle:

$$\frac{\text{Flux across rectangle boundary}}{\text{rectangle area}} \approx \left(\frac{\partial M}{\partial x} + \frac{\partial N}{\partial y}\right).$$

Finally, we let Δx and Δy approach zero to define the *flux density* of **F** at the point (x, y). In mathematics, we call the flux density the *divergence* of **F**. The symbol for it is div **F**, pronounced "divergence of **F**" or "div **F**."

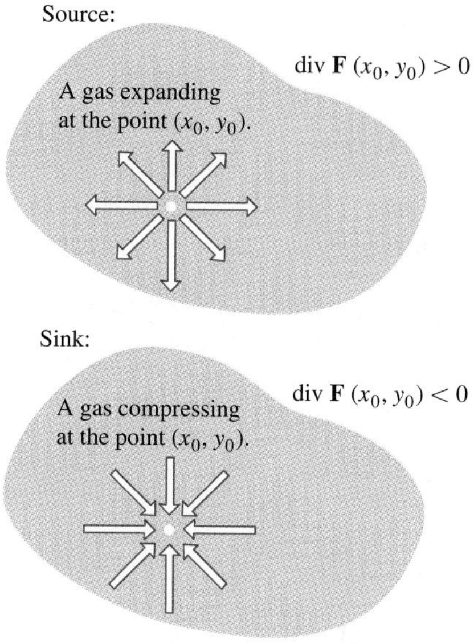

FIGURE 13.25 If a gas is expanding at a point (x_0, y_0), the lines of flow have positive divergence; if the gas is compressing, the divergence is negative.

DEFINITION The **divergence (flux density)** of a vector field $\mathbf{F} = M\mathbf{i} + N\mathbf{j}$ at the point (x, y) is

$$\text{div } \mathbf{F} = \frac{\partial M}{\partial x} + \frac{\partial N}{\partial y}. \tag{1}$$

Intuitively, if a gas is expanding at the point (x_0, y_0), the lines of flow would diverge there (hence the name) and, since the gas would be flowing out of a small rectangle about (x_0, y_0) the divergence of **F** at (x_0, y_0) would be positive. If the gas were compressing instead of expanding, the divergence would be negative (Figure 13.25).

EXAMPLE 1 Find the divergence of $\mathbf{F}(x, y) = (x^2 - y)\mathbf{i} + (xy - y^2)\mathbf{j}$.

Solution We use the formula in Equation (1):

$$\begin{aligned} \text{div } \mathbf{F} &= \frac{\partial M}{\partial x} + \frac{\partial N}{\partial y} = \frac{\partial}{\partial x}(x^2 - y) + \frac{\partial}{\partial y}(xy - y^2) \\ &= 2x + x - 2y = 3x - 2y. \end{aligned}$$

■

Spin Around an Axis: The k-Component of Curl

The second idea we need for Green's Theorem has to do with measuring how a paddle wheel spins at a point in a fluid flowing in a plane region. This idea gives some sense of how the fluid is circulating around axes located at different points and perpendicular to the region. Physicists sometimes refer to this as the *circulation density* of a vector field **F** at a point. To obtain it, we return to the velocity field

$$\mathbf{F}(x, y) = M(x, y)\mathbf{i} + N(x, y)\mathbf{j}$$

and the rectangle A. The rectangle is redrawn here as Figure 13.26.

The counterclockwise circulation of **F** around the boundary of A is the sum of flow rates along the sides. For the bottom edge, the flow rate is approximately

$$\mathbf{F}(x, y) \cdot \mathbf{i}\, \Delta x = M(x, y)\Delta x.$$

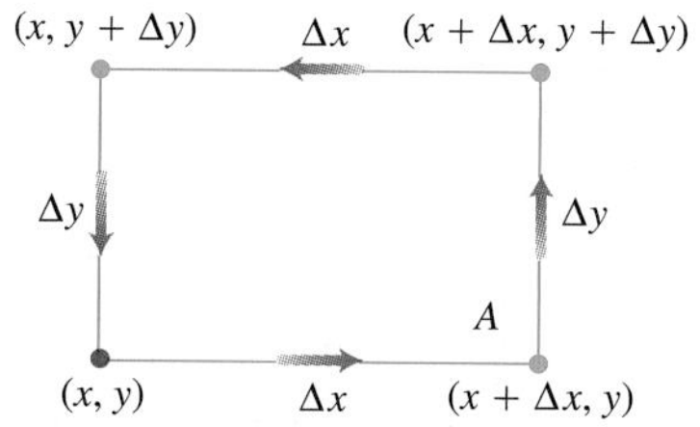

FIGURE 13.26 The rectangle for defining the curl (circulation density) of a vector field at a point (x, y).

This is the scalar component of the velocity $\mathbf{F}(x, y)$ in the direction of the tangent vector **i** times the length of the segment. The rates of flow along the other sides in the counterclockwise direction are expressed in a similar way. In all, we have

$$\begin{aligned} &\text{Top:} && \mathbf{F}(x, y + \Delta y) \cdot (-\mathbf{i})\, \Delta x = -M(x, y + \Delta y)\Delta x \\ &\text{Bottom:} && \mathbf{F}(x, y) \cdot \mathbf{i}\, \Delta x = M(x, y)\Delta x \\ &\text{Right:} && \mathbf{F}(x + \Delta x, y) \cdot \mathbf{j}\, \Delta y = N(x + \Delta x, y)\Delta y \\ &\text{Left:} && \mathbf{F}(x, y) \cdot (-\mathbf{j})\, \Delta y = -N(x, y)\Delta y. \end{aligned}$$

We add opposite pairs to get

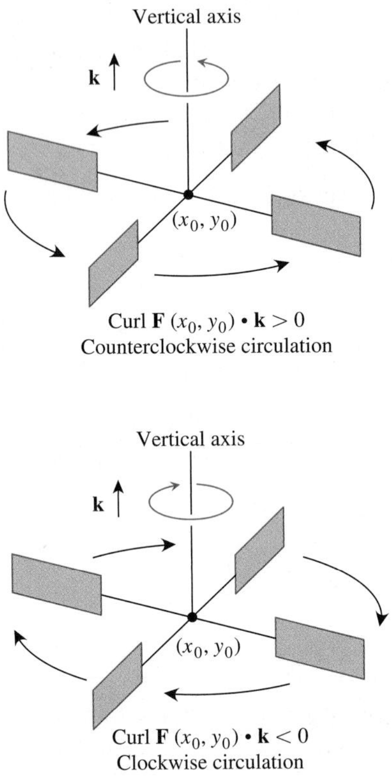

FIGURE 13.27 In the flow of an incompressible fluid over a plane region, the **k**-component of the curl measures the rate of the fluid's rotation at a point. The **k**-component of the curl is positive at points where the rotation is counterclockwise and negative where the rotation is clockwise.

Top and bottom:

$$-(M(x, y + \Delta y) - M(x, y))\Delta x \approx -\left(\frac{\partial M}{\partial y}\Delta y\right)\Delta x$$

Right and left:

$$(N(x + \Delta x, y) - N(x, y))\Delta y \approx \left(\frac{\partial N}{\partial x}\Delta x\right)\Delta y.$$

Adding these last two equations and dividing by $\Delta x \Delta y$ gives an estimate of the circulation density for the rectangle:

$$\frac{\text{Circulation around rectangle}}{\text{rectangle area}} \approx \frac{\partial N}{\partial x} - \frac{\partial M}{\partial y}.$$

We let Δx and Δy approach zero to define the *circulation density* of **F** at the point (x, y).

The positive orientation of the circulation density for the plane is the *counterclockwise* rotation around the vertical axis, looking downward on the xy-plane from the tip of the (vertical) unit vector **k** (Figure 13.27). The circulation value is actually the **k**-component of a more general circulation vector field we define in Section 13.7, called the *curl* of the vector field **F**. For Green's Theorem, we need only this **k**-component.

DEFINITION The **k-component of the curl (circulation density)** of a vector field $\mathbf{F} = M\mathbf{i} + N\mathbf{j}$ at the point (x, y) is the scalar

$$(\text{curl } \mathbf{F}) \cdot \mathbf{k} = \frac{\partial N}{\partial x} - \frac{\partial M}{\partial y}. \tag{2}$$

If water is moving about a region in the xy-plane in a thin layer, then the **k**-component of the circulation, or curl, at a point (x_0, y_0) gives a way to measure how fast and in what direction a small paddle wheel will spin if it is put into the water at (x_0, y_0) with its axis perpendicular to the plane, parallel to **k** (Figure 13.27).

EXAMPLE 2 Find the **k**-component of the curl for the vector field

$$\mathbf{F}(x, y) = (x^2 - y)\mathbf{i} + (xy - y^2)\mathbf{j}.$$

Solution We use the formula in Equation (2):

$$(\text{curl } \mathbf{F}) \cdot \mathbf{k} = \frac{\partial N}{\partial x} - \frac{\partial M}{\partial y} = \frac{\partial}{\partial x}(xy - y^2) - \frac{\partial}{\partial y}(x^2 - y) = y + 1. \quad \blacksquare$$

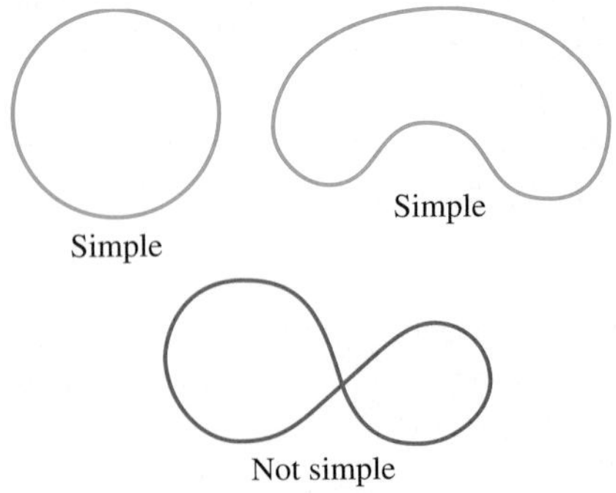

FIGURE 13.28 In proving Green's Theorem, we distinguish between two kinds of closed curves, simple and not simple. Simple curves do not cross themselves. A circle is simple but a figure 8 is not.

Two Forms for Green's Theorem

In one form, Green's Theorem says that under suitable conditions the outward flux of a vector field across a simple closed curve in the plane (Figure 13.28) equals the double integral of the divergence of the field over the region enclosed by the curve. Recall the formulas for flux in Equations (3) and (4) in Section 13.2.

THEOREM 4—Green's Theorem (Flux-Divergence or Normal Form) Let C be a piecewise smooth, simple closed curve enclosing a region R in the plane. Let $\mathbf{F} = M\mathbf{i} + N\mathbf{j}$ be a vector field with M and N having continuous first partial derivatives in an open region containing R. Then the outward flux of $\mathbf{F}$ across C equals the double integral of div $\mathbf{F}$ over the region R enclosed by C.

$$\underbrace{\oint_C \mathbf{F}\cdot\mathbf{n}\,ds}_{\text{Outward flux}} = \oint_C M\,dy - N\,dx = \underbrace{\iint_R \left(\frac{\partial M}{\partial x} + \frac{\partial N}{\partial y}\right) dx\,dy}_{\text{Divergence integral}} \tag{3}$$

In another form, Green's Theorem says that the counterclockwise circulation of a vector field around a simple closed curve is the double integral of the **k**-component of the curl of the field over the region enclosed by the curve. Recall the defining Equation (2) for circulation in Section 13.2.

THEOREM 5—Green's Theorem (Circulation-Curl or Tangential Form) Let C be a piecewise smooth, simple closed curve enclosing a region R in the plane. Let $\mathbf{F} = M\mathbf{i} + N\mathbf{j}$ be a vector field with M and N having continuous first partial derivatives in an open region containing R. Then the counterclockwise circulation of $\mathbf{F}$ around C equals the double integral of $(\text{curl }\mathbf{F})\cdot\mathbf{k}$ over R.

$$\underbrace{\oint_C \mathbf{F}\cdot\mathbf{T}\,ds}_{\text{Counterclockwise circulation}} = \oint_C M\,dx + N\,dy = \underbrace{\iint_R \left(\frac{\partial N}{\partial x} - \frac{\partial M}{\partial y}\right) dx\,dy}_{\text{Curl integral}} \tag{4}$$

The two forms of Green's Theorem are equivalent. Applying Equation (3) to the field $\mathbf{G}_1 = N\mathbf{i} - M\mathbf{j}$ gives Equation (4), and applying Equation (4) to $\mathbf{G}_2 = -N\mathbf{i} + M\mathbf{j}$ gives Equation (3).

EXAMPLE 3 Verify both forms of Green's Theorem for the field

$$\mathbf{F}(x, y) = (x - y)\mathbf{i} + x\mathbf{j}$$

and the region R bounded by the unit circle

$$C: \quad \mathbf{r}(t) = (\cos t)\mathbf{i} + (\sin t)\mathbf{j}, \qquad 0 \le t \le 2\pi.$$

Solution We have

$$M = \cos t - \sin t, \qquad dx = d(\cos t) = -\sin t\,dt,$$
$$N = \cos t, \qquad dy = d(\sin t) = \cos t\,dt,$$
$$\frac{\partial M}{\partial x} = 1, \qquad \frac{\partial M}{\partial y} = -1, \qquad \frac{\partial N}{\partial x} = 1, \qquad \frac{\partial N}{\partial y} = 0.$$

The two sides of Equation (3) are

$$\oint_C M\,dy - N\,dx = \int_{t=0}^{t=2\pi} (\cos t - \sin t)(\cos t\,dt) - (\cos t)(-\sin t\,dt)$$

$$= \int_0^{2\pi} \cos^2 t\,dt = \pi$$

$$\iint_R \left(\frac{\partial M}{\partial x} + \frac{\partial N}{\partial y}\right) dx\,dy = \iint_R (1 + 0)\,dx\,dy$$

$$= \iint_R dx\,dy = \text{area inside the unit circle} = \pi.$$

The two sides of Equation (4) are

$$\oint_C M\,dx + N\,dy = \int_{t=0}^{t=2\pi} (\cos t - \sin t)(-\sin t\,dt) + (\cos t)(\cos t\,dt)$$

$$= \int_0^{2\pi} (-\sin t \cos t + 1)\,dt = 2\pi$$

$$\iint_R \left(\frac{\partial N}{\partial x} - \frac{\partial M}{dy}\right) dx\,dy = \iint_R (1 - (-1))\,dx\,dy = 2\iint_R dx\,dy = 2\pi.$$ ■

Using Green's Theorem to Evaluate Line Integrals

If we construct a closed curve C by piecing a number of different curves end to end, the process of evaluating a line integral over C can be lengthy because there are so many different integrals to evaluate. If C bounds a region R to which Green's Theorem applies, however, we can use Green's Theorem to change the line integral around C into one double integral over R.

EXAMPLE 4 Evaluate the integral

$$\oint_C xy\,dy - y^2\,dx,$$

where C is the square cut from the first quadrant by the lines $x = 1$ and $y = 1$.

Solution We can use either form of Green's Theorem to change the line integral into a double integral over the square.

1. *With the Normal Form Equation* (3): Taking $M = xy$, $N = y^2$, and C and R as the square's boundary and interior gives

$$\oint_C xy\,dy - y^2\,dx = \iint_R (y + 2y)\,dx\,dy = \int_0^1\int_0^1 3y\,dx\,dy$$

$$= \int_0^1 \left[3xy\right]_{x=0}^{x=1} dy = \int_0^1 3y\,dy = \frac{3}{2}y^2\bigg]_0^1 = \frac{3}{2}.$$

2. *With the Tangential Form Equation* (4): Taking $M = -y^2$ and $N = xy$ gives the same result:

$$\oint_C -y^2\,dx + xy\,dy = \iint_R (y - (-2y))\,dx\,dy = \frac{3}{2}.$$ ■

EXAMPLE 5 Calculate the outward flux of the field $\mathbf{F}(x, y) = x\mathbf{i} + y^2\mathbf{j}$ across the square bounded by the lines $x = \pm 1$ and $y = \pm 1$.

Solution Calculating the flux with a line integral would take four integrations, one for each side of the square. With Green's Theorem, we can change the line integral to one double integral. With $M = x$, $N = y^2$, C the square, and R the square's interior, we have

$$\begin{aligned}\text{Flux} &= \oint_C \mathbf{F}\cdot\mathbf{n}\,ds = \oint_C M\,dy - N\,dx \\ &= \iint_R \left(\frac{\partial M}{\partial x} + \frac{\partial N}{\partial y}\right) dx\,dy \qquad \text{Green's Theorem} \\ &= \int_{-1}^{1}\int_{-1}^{1} (1 + 2y)\,dx\,dy = \int_{-1}^{1} \left[x + 2xy\right]_{x=-1}^{x=1} dy \\ &= \int_{-1}^{1} (2 + 4y)\,dy = \left[2y + 2y^2\right]_{-1}^{1} = 4.\end{aligned}$$ ■

Proof of Green's Theorem for Special Regions

Let C be a smooth simple closed curve in the xy-plane with the property that lines parallel to the axes cut it in no more than two points. Let R be the region enclosed by C and suppose that M, N, and their first partial derivatives are continuous at every point of some open region containing C and R. We want to prove the circulation-curl form of Green's Theorem,

$$\oint_C M\,dx + N\,dy = \iint_R \left(\frac{\partial N}{\partial x} - \frac{\partial M}{\partial y}\right) dx\,dy. \tag{5}$$

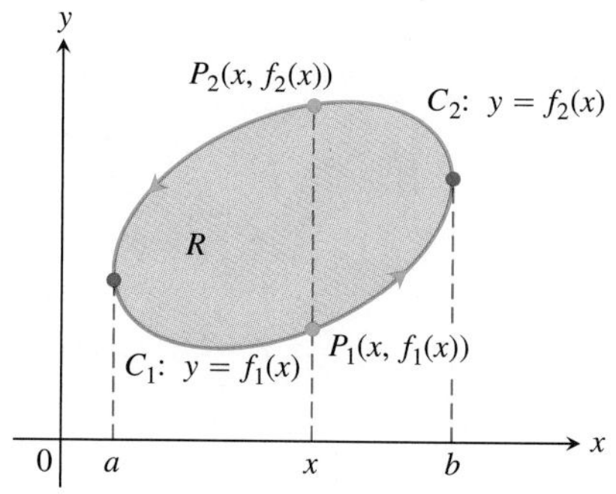

FIGURE 13.29 The boundary curve C is made up of C_1, the graph of $y = f_1(x)$, and C_2, the graph of $y = f_2(x)$.

Figure 13.29 shows C made up of two directed parts:

$$C_1:\quad y = f_1(x),\quad a \le x \le b, \qquad C_2:\quad y = f_2(x),\quad b \ge x \ge a.$$

For any x between a and b, we can integrate $\partial M/\partial y$ with respect to y from $y = f_1(x)$ to $y = f_2(x)$ and obtain

$$\int_{f_1(x)}^{f_2(x)} \frac{\partial M}{\partial y}\,dy = M(x, y)\Big]_{y=f_1(x)}^{y=f_2(x)} = M(x, f_2(x)) - M(x, f_1(x)).$$

We can then integrate this with respect to x from a to b:

$$\begin{aligned}\int_a^b\int_{f_1(x)}^{f_2(x)} \frac{\partial M}{\partial y}\,dy\,dx &= \int_a^b [M(x, f_2(x)) - M(x, f_1(x))]\,dx \\ &= -\int_b^a M(x, f_2(x))\,dx - \int_a^b M(x, f_1(x))\,dx \\ &= -\int_{C_2} M\,dx - \int_{C_1} M\,dx \\ &= -\oint_C M\,dx.\end{aligned}$$

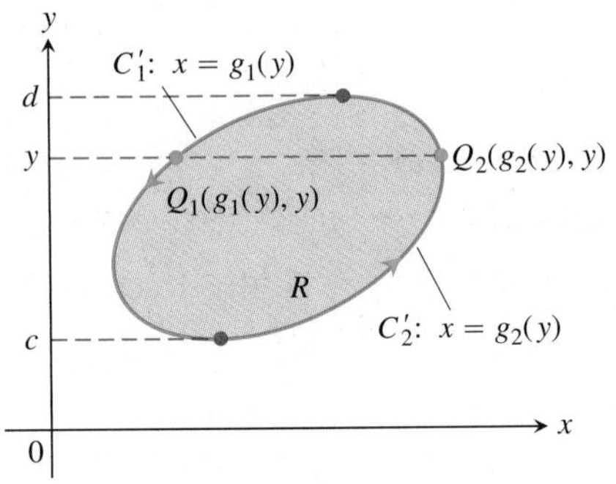

FIGURE 13.30 The boundary curve C is made up of C_1', the graph of $x = g_1(y)$, and C_2', the graph of $x = g_2(y)$.

Therefore

$$\oint_C M\,dx = \iint_R \left(-\frac{\partial M}{\partial y}\right) dx\,dy. \tag{6}$$

Equation (6) is half the result we need for Equation (5). We derive the other half by integrating $\partial N/\partial x$ first with respect to x and then with respect to y, as suggested by Figure 13.30. This shows the curve C of Figure 13.29 decomposed into the two directed parts $C_1'\colon x = g_1(y),\ d \ge y \ge c$ and $C_2'\colon x = g_2(y),\ c \le y \le d$. The result of this double integration is

$$\oint_C N\,dy = \iint_R \frac{\partial N}{\partial x}\,dx\,dy. \tag{7}$$

Summing Equations (6) and (7) gives Equation (5). This concludes the proof. ■

Green's Theorem also holds for more general regions, such as those shown in Figures 13.31 and 13.32, but we will not prove this result here. Notice that the region in Figure 13.32 is not simply connected. The curves on its boundary are oriented so that the region R is always on the left-hand side as the curves are traversed in the directions shown. With this convention, Green's Theorem is valid for regions that are not simply connected.

While we stated the theorem in the xy-plane, Green's Theorem does apply for any region R contained in a plane bounded by a curve C in space. We will see how to express the double integral over R for this more general form of Green's Theorem in Section 13.7.

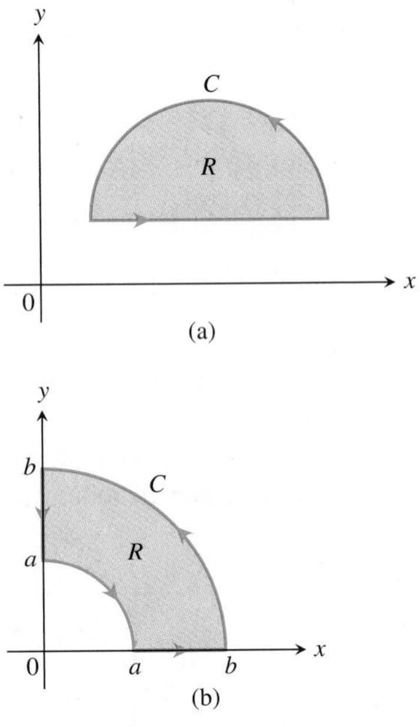

FIGURE 13.31 Other regions to which Green's Theorem applies.

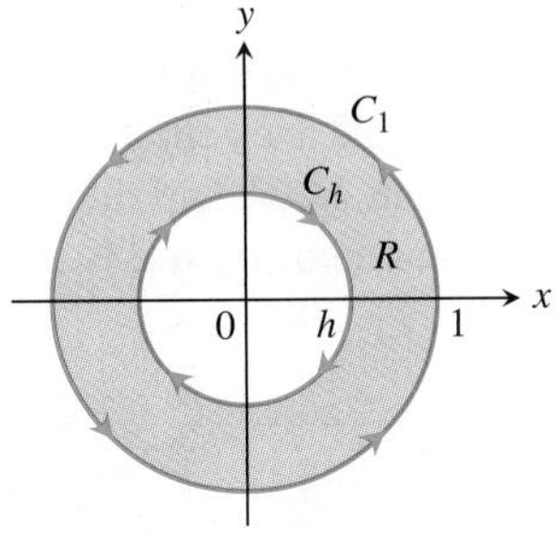

FIGURE 13.32 Green's Theorem may be applied to the annular region R by integrating along the boundaries as shown.

EXERCISES 13.4

In Exercises 1–4, verify the conclusion of Green's Theorem by evaluating both sides of Equations (3) and (4) for the field $\mathbf{F} = M\mathbf{i} + N\mathbf{j}$. Take the domains of integration in each case to be the disk $R\colon x^2 + y^2 \le a^2$ and its bounding circle $C\colon \mathbf{r} = (a\cos t)\mathbf{i} + (a\sin t)\mathbf{j},\ 0 \le t \le 2\pi$.

1. $\mathbf{F} = -y\mathbf{i} + x\mathbf{j}$

2. $\mathbf{F} = y\mathbf{i}$

3. $\mathbf{F} = 2x\mathbf{i} - 3y\mathbf{j}$

4. $\mathbf{F} = -x^2y\mathbf{i} + xy^2\mathbf{j}$

In Exercises 5–10, use Green's Theorem to find the counterclockwise circulation and outward flux for the field $\mathbf{F}$ and curve C.

5. $\mathbf{F} = (x - y)\mathbf{i} + (y - x)\mathbf{j}$

C: The square bounded by $x = 0, x = 1, y = 0, y = 1$

6. $\mathbf{F} = (x^2 + 4y)\mathbf{i} + (x + y^2)\mathbf{j}$

C: The square bounded by $x = 0, x = 1, y = 0, y = 1$

7. $\mathbf{F} = (y^2 - x^2)\mathbf{i} + (x^2 + y^2)\mathbf{j}$

C: The triangle bounded by $y = 0, x = 3$, and $y = x$

8. $\mathbf{F} = (x + y)\mathbf{i} - (x^2 + y^2)\mathbf{j}$

C: The triangle bounded by $y = 0, x = 1$, and $y = x$

9. $\mathbf{F} = (x + e^x \sin y)\mathbf{i} + (x + e^x \cos y)\mathbf{j}$

C: The right-hand loop of the lemniscate $r^2 = \cos 2\theta$

10. $\mathbf{F} = \left(\tan^{-1}\frac{y}{x}\right)\mathbf{i} + \ln(x^2 + y^2)\mathbf{j}$

C: The boundary of the region defined by the polar coordinate inequalities $1 \le r \le 2, 0 \le \theta \le \pi$

11. Find the counterclockwise circulation and outward flux of the field $\mathbf{F} = xy\mathbf{i} + y^2\mathbf{j}$ around and over the boundary of the region enclosed by the curves $y = x^2$ and $y = x$ in the first quadrant.

12. Find the counterclockwise circulation and the outward flux of the field $\mathbf{F} = (-\sin y)\mathbf{i} + (x \cos y)\mathbf{j}$ around and over the square cut from the first quadrant by the lines $x = \pi/2$ and $y = \pi/2$.

13. Find the outward flux of the field

$$\mathbf{F} = \left(3xy - \frac{x}{1 + y^2}\right)\mathbf{i} + (e^x + \tan^{-1} y)\mathbf{j}$$

across the cardioid $r = a(1 + \cos\theta)$, $a > 0$.

14. Find the counterclockwise circulation of $\mathbf{F} = (y + e^x \ln y)\mathbf{i} + (e^x/y)\mathbf{j}$ around the boundary of the region that is bounded above by the curve $y = 3 - x^2$ and below by the curve $y = x^4 + 1$.

In Exercises 15 and 16, find the work done by $\mathbf{F}$ in moving a particle once counterclockwise around the given curve.

15. $\mathbf{F} = 2xy^3\mathbf{i} + 4x^2y^2\mathbf{j}$

C: The boundary of the "triangular" region in the first quadrant enclosed by the x-axis, the line $x = 1$, and the curve $y = x^3$

16. $\mathbf{F} = (4x - 2y)\mathbf{i} + (2x - 4y)\mathbf{j}$

C: The circle $(x - 2)^2 + (y - 2)^2 = 4$

Apply Green's Theorem to evaluate the integrals in Exercises 17–20.

17. $\displaystyle\oint_C (y^2\,dx + x^2\,dy)$

C: The triangle bounded by $x = 0, x + y = 1, y = 0$

18. $\displaystyle\oint_C (3y\,dx + 2x\,dy)$

C: The boundary of $0 \le x \le \pi, 0 \le y \le \sin x$

19. $\displaystyle\oint_C (6y + x)\,dx + (y + 2x)\,dy$

C: The circle $(x - 2)^2 + (y - 3)^2 = 4$

20. $\displaystyle\oint_C (2x + y^2)\,dx + (2xy + 3y)\,dy$

C: Any simple closed curve in the plane for which Green's Theorem holds

Calculating Area with Green's Theorem If a simple closed curve C in the plane and the region R it encloses satisfy the hypotheses of Green's Theorem, the area of R is given by

Green's Theorem Area Formula

$$\text{Area of } R = \frac{1}{2}\oint_C x\,dy - y\,dx$$

The reason is that by Equation (3), run backward,

$$\text{Area of } R = \iint_R dy\,dx = \iint_R \left(\frac{1}{2} + \frac{1}{2}\right) dy\,dx$$

$$= \oint_C \frac{1}{2}x\,dy - \frac{1}{2}y\,dx.$$

Use the Green's Theorem area formula given above to find the areas of the regions enclosed by the curves in Exercises 21–24.

21. The circle $\mathbf{r}(t) = (a\cos t)\mathbf{i} + (a\sin t)\mathbf{j}$, $\quad 0 \le t \le 2\pi$

22. The ellipse $\mathbf{r}(t) = (a\cos t)\mathbf{i} + (b\sin t)\mathbf{j}$, $\quad 0 \le t \le 2\pi$

23. The astroid $\mathbf{r}(t) = (\cos^3 t)\mathbf{i} + (\sin^3 t)\mathbf{j}$, $\quad 0 \le t \le 2\pi$

24. Find the area under one arch of the cycloid $x = t - \sin t$, $\quad y = 1 - \cos t$.

25. Let C be the boundary of a region on which Green's Theorem holds. Use Green's Theorem to calculate

a. $\displaystyle\oint_C f(x)\,dx + g(y)\,dy$

b. $\displaystyle\oint_C ky\,dx + hx\,dy$ $\quad$ (k and h constants).

26. Integral dependent only on area Show that the value of

$$\oint_C xy^2\,dx + (x^2y + 2x)\,dy$$

around any square depends only on the area of the square and not on its location in the plane.

27. Area as a line integral Show that if R is a region in the plane bounded by a piecewise smooth, simple closed curve C, then

$$\text{Area of } R = \oint_C x\,dy = -\oint_C y\,dx.$$

28. Definite integral as a line integral Suppose that a nonnegative function $y = f(x)$ has a continuous first derivative on $[a, b]$. Let C be the boundary of the region in the xy-plane that is bounded below by the x-axis, above by the graph of f, and on the sides by the lines $x = a$ and $x = b$. Show that

$$\int_a^b f(x)\,dx = -\oint_C y\,dx.$$

29. Area and the centroid Let A be the area and $\bar{x}$ the x-coordinate of the centroid of a region R that is bounded by a piecewise smooth, simple closed curve C in the xy-plane. Show that

$$\frac{1}{2}\oint_C x^2\,dy = -\oint_C xy\,dx = \frac{1}{3}\oint_C x^2\,dy - xy\,dx = A\bar{x}.$$

30. Moment of inertia Let I_y be the moment of inertia about the y-axis of the region in Exercise 29. Show that

$$\frac{1}{3}\oint_C x^3\,dy = -\oint_C x^2y\,dx = \frac{1}{4}\oint_C x^3\,dy - x^2y\,dx = I_y.$$

31. Regions with many holes Green's Theorem holds for a region R with any finite number of holes as long as the bounding curves are smooth, simple, and closed and we integrate over each component of the boundary in the direction that keeps R on our immediate left as we go along (see accompanying figure).

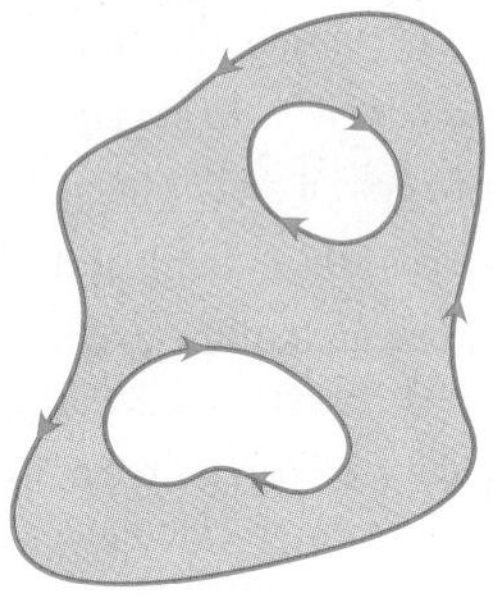

a. Let $f(x, y) = \ln(x^2 + y^2)$ and let C be the circle $x^2 + y^2 = a^2$. Evaluate the flux integral

$$\oint_C \nabla f \cdot \mathbf{n}\,ds.$$

b. Let K be an arbitrary smooth, simple closed curve in the plane that does not pass through $(0, 0)$. Use Green's Theorem to show that

$$\oint_K \nabla f \cdot \mathbf{n}\,ds$$

has two possible values, depending on whether $(0, 0)$ lies inside K or outside K.

32. Establish Equation (7) to finish the proof of the special case of Green's Theorem.

COMPUTER EXPLORATIONS

In Exercises 33–36, use a CAS and Green's Theorem to find the counterclockwise circulation of the field $\mathbf{F}$ around the simple closed curve C. Perform the following CAS steps.

a. Plot C in the xy-plane.

b. Determine the integrand $(\partial N/\partial x) - (\partial M/\partial y)$ for the curl form of Green's Theorem.

c. Determine the (double integral) limits of integration from your plot in part (a) and evaluate the curl integral for the circulation.

33. $\mathbf{F} = (2x - y)\mathbf{i} + (x + 3y)\mathbf{j}$, C: The ellipse $x^2 + 4y^2 = 4$

34. $\mathbf{F} = (2x^3 - y^3)\mathbf{i} + (x^3 + y^3)\mathbf{j}$, C: The ellipse $\dfrac{x^2}{4} + \dfrac{y^2}{9} = 1$

35. $\mathbf{F} = x^{-1}e^y\mathbf{i} + (e^y \ln x + 2x)\mathbf{j}$,

C: The boundary of the region defined by $y = 1 + x^4$ (below) and $y = 2$ (above)

36. $\mathbf{F} = xe^y\mathbf{i} + 4x^2 \ln y\,\mathbf{j}$,

C: The triangle with vertices $(0, 0)$, $(2, 0)$, and $(0, 4)$

13.5 Surfaces and Area

We have defined curves in the plane in three different ways:

Explicit form:	$y = f(x)$
Implicit form:	$F(x, y) = 0$
Parametric vector form:	$\mathbf{r}(t) = f(t)\mathbf{i} + g(t)\mathbf{j}, \quad a \le t \le b.$

We have analogous definitions of surfaces in space:

Explicit form:	$z = f(x, y)$
Implicit form:	$F(x, y, z) = 0.$

There is also a parametric form that gives the position of a point on the surface as a vector function of two variables. We discuss this new form in this section and apply the form to obtain the area of a surface as a double integral.

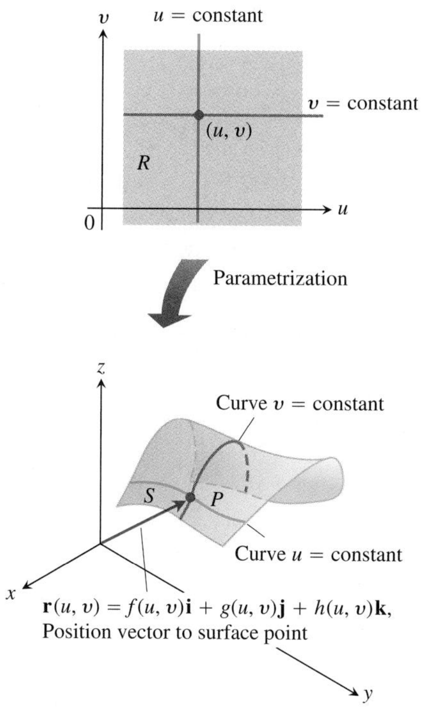

FIGURE 13.33 A parametrized surface S expressed as a vector function of two variables defined on a region R.

Parametrizations of Surfaces

Let

$$\mathbf{r}(u, v) = f(u, v)\mathbf{i} + g(u, v)\mathbf{j} + h(u, v)\mathbf{k} \tag{1}$$

be a continuous vector function that is defined on a region R in the uv-plane and one-to-one on the interior of R (Figure 13.33). We call the range of $\mathbf{r}$ the **surface** S defined or traced by $\mathbf{r}$. Equation (1) together with the domain R constitute a **parametrization** of the surface. The variables u and v are the **parameters**, and R is the **parameter domain**. To simplify our discussion, we take R to be a rectangle defined by inequalities of the form $a \le u \le b, c \le v \le d$. The requirement that $\mathbf{r}$ be one-to-one on the interior of R ensures that S does not cross itself. Notice that Equation (1) is the vector equivalent of *three* parametric equations:

$$x = f(u, v), \qquad y = g(u, v), \qquad z = h(u, v).$$

EXAMPLE 1 Find a parametrization of the cone

$$z = \sqrt{x^2 + y^2}, \qquad 0 \le z \le 1.$$

Solution Here, cylindrical coordinates provide a parametrization. A typical point (x, y, z) on the cone (Figure 13.34) has $x = r\cos\theta, y = r\sin\theta$, and $z = \sqrt{x^2 + y^2} = r$, with $0 \le r \le 1$ and $0 \le \theta \le 2\pi$. Taking $u = r$ and $v = \theta$ in Equation (1) gives the parametrization

$$\mathbf{r}(r, \theta) = (r\cos\theta)\mathbf{i} + (r\sin\theta)\mathbf{j} + r\mathbf{k}, \qquad 0 \le r \le 1, \quad 0 \le \theta \le 2\pi.$$

The parameterization is one-to-one on the interior of the domain R, though not on its boundary where $r = 0$. ■

EXAMPLE 2 Find a parametrization of the sphere $x^2 + y^2 + z^2 = a^2$.

Solution Spherical coordinates provide what we need. A typical point (x, y, z) on the sphere (Figure 13.35) has $x = a\sin\phi\cos\theta$, $y = a\sin\phi\sin\theta$, and $z = a\cos\phi$,

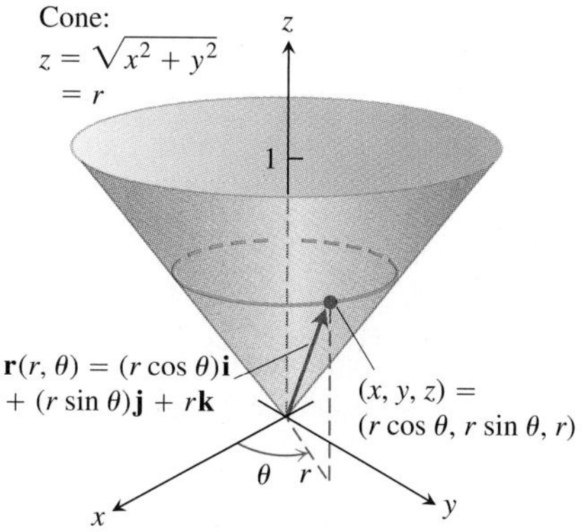

FIGURE 13.34 The cone in Example 1 can be parametrized using cylindrical coordinates.

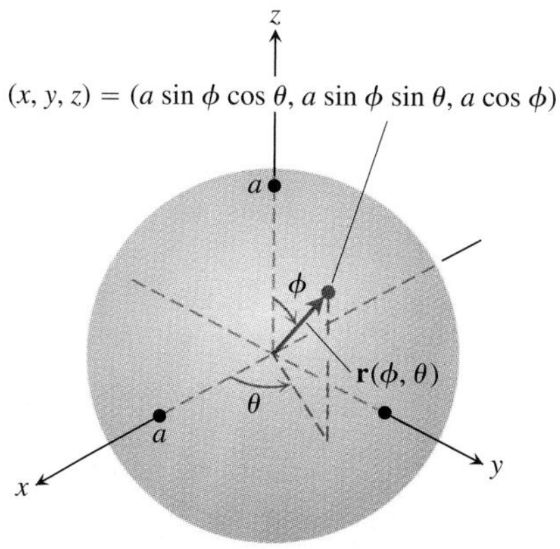

FIGURE 13.35 The sphere in Example 2 can be parametrized using spherical coordinates.

$0 \le \phi \le \pi$, $0 \le \theta \le 2\pi$. Taking $u = \phi$ and $v = \theta$ in Equation (1) gives the parametrization

$$\mathbf{r}(\phi, \theta) = (a \sin \phi \cos \theta)\mathbf{i} + (a \sin \phi \sin \theta)\mathbf{j} + (a \cos \phi)\mathbf{k},$$
$$0 \le \phi \le \pi, \quad 0 \le \theta \le 2\pi. \qquad \blacksquare$$

EXAMPLE 3 Find a parametrization of the cylinder

$$x^2 + (y - 3)^2 = 9, \qquad 0 \le z \le 5.$$

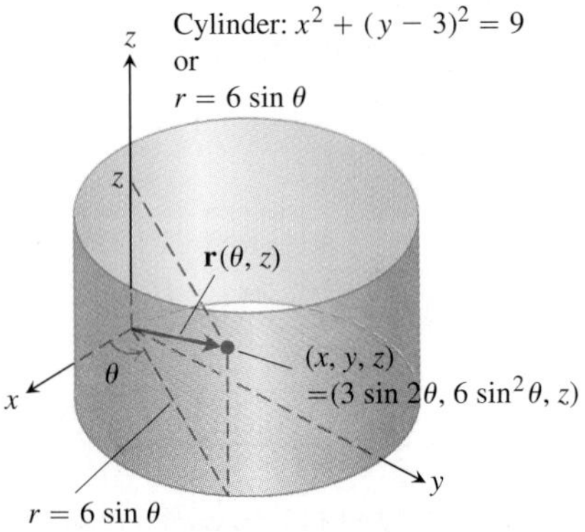

FIGURE 13.36 The cylinder in Example 3 can be parametrized using cylindrical coordinates.

Solution In cylindrical coordinates, a point (x, y, z) has $x = r \cos \theta$, $y = r \sin \theta$, and $z = z$. For points on the cylinder $x^2 + (y - 3)^2 = 9$ (Figure 13.36), the equation is the same as the polar equation for the cylinder's base in the xy-plane:

$$x^2 + (y^2 - 6y + 9) = 9$$
$$r^2 - 6r \sin \theta = 0 \qquad x^2 + y^2 = r^2,\ y = r \sin \theta$$

or

$$r = 6 \sin \theta, \qquad 0 \le \theta \le \pi.$$

A typical point on the cylinder therefore has

$$x = r \cos \theta = 6 \sin \theta \cos \theta = 3 \sin 2\theta$$
$$y = r \sin \theta = 6 \sin^2 \theta$$
$$z = z.$$

Taking $u = \theta$ and $v = z$ in Equation (1) gives the parametrization

$$\mathbf{r}(\theta, z) = (3 \sin 2\theta)\mathbf{i} + (6 \sin^2 \theta)\mathbf{j} + z\mathbf{k}, \qquad 0 \le \theta \le \pi, \qquad 0 \le z \le 5. \qquad \blacksquare$$

Surface Area

Our goal is to find a double integral for calculating the area of a curved surface S based on the parametrization

$$\mathbf{r}(u, v) = f(u, v)\mathbf{i} + g(u, v)\mathbf{j} + h(u, v)\mathbf{k}, \qquad a \le u \le b, \qquad c \le v \le d.$$

We need S to be smooth for the construction we are about to carry out. The definition of smoothness involves the partial derivatives of $\mathbf{r}$ with respect to u and v:

$$\mathbf{r}_u = \frac{\partial \mathbf{r}}{\partial u} = \frac{\partial f}{\partial u}\mathbf{i} + \frac{\partial g}{\partial u}\mathbf{j} + \frac{\partial h}{\partial u}\mathbf{k}$$
$$\mathbf{r}_v = \frac{\partial \mathbf{r}}{\partial v} = \frac{\partial f}{\partial v}\mathbf{i} + \frac{\partial g}{\partial v}\mathbf{j} + \frac{\partial h}{\partial v}\mathbf{k}.$$

> **DEFINITION** A parametrized surface $\mathbf{r}(u, v) = f(u, v)\mathbf{i} + g(u, v)\mathbf{j} + h(u, v)\mathbf{k}$ is **smooth** if $\mathbf{r}_u$ and $\mathbf{r}_v$ are continuous and $\mathbf{r}_u \times \mathbf{r}_v$ is never zero on the interior of the parameter domain.

The condition that $\mathbf{r}_u \times \mathbf{r}_v$ is never the zero vector in the definition of smoothness means that the two vectors $\mathbf{r}_u$ and $\mathbf{r}_v$ are nonzero and never lie along the same line, so they always determine a plane tangent to the surface. We relax this condition on the boundary of the domain, but this does not affect the area computations.

Now consider a small rectangle ΔA_{uv} in R with sides on the lines $u = u_0, u = u_0 + \Delta u, v = v_0$ and $v = v_0 + \Delta v$ (Figure 13.37). Each side of ΔA_{uv} maps to a curve on the surface S, and together these four curves bound a "curved area element" $\Delta\sigma_{uv}$. In the notation of the figure, the side $v = v_0$ maps to curve C_1, the side $u = u_0$ maps to C_2, and their common vertex (u_0, v_0) maps to P_0.

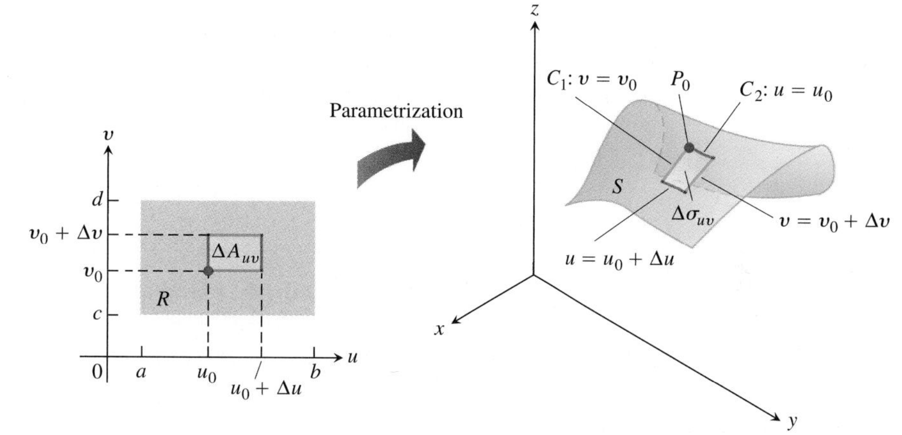

FIGURE 13.37 A rectangular area element ΔA_{uv} in the uv-plane maps onto a curved area element $\Delta\sigma_{uv}$ on S.

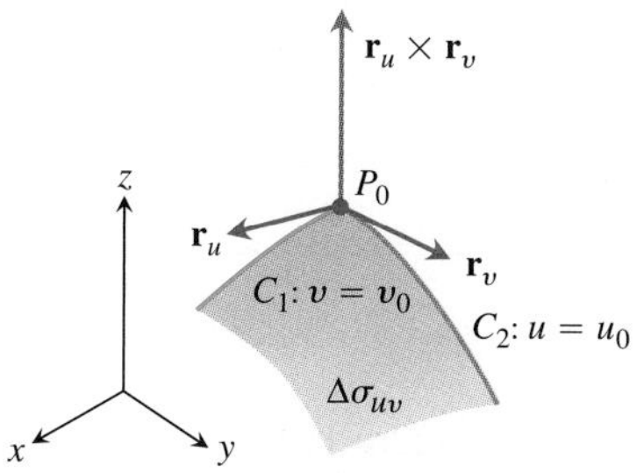

FIGURE 13.38 A magnified view of a surface area element $\Delta\sigma_{uv}$.

Figure 13.38 shows an enlarged view of $\Delta\sigma_{uv}$. The vector $\mathbf{r}_u(u_0, v_0)$ is tangent to C_1 at P_0. Likewise, $\mathbf{r}_v(u_0, v_0)$ is tangent to C_2 at P_0. The cross product $\mathbf{r}_u \times \mathbf{r}_v$ is normal to the surface at P_0. (Here is where we begin to use the assumption that S is smooth. We want to be sure that $\mathbf{r}_u \times \mathbf{r}_v \neq \mathbf{0}$.)

We next approximate the surface element $\Delta\sigma_{uv}$ by the parallelogram on the tangent plane whose sides are determined by the vectors $\Delta u\mathbf{r}_u$ and $\Delta v\mathbf{r}_v$ (Figure 13.39). The area of this parallelogram is

$$|\Delta u\mathbf{r}_u \times \Delta v\mathbf{r}_v| = |\mathbf{r}_u \times \mathbf{r}_v|\,\Delta u\,\Delta v. \tag{2}$$

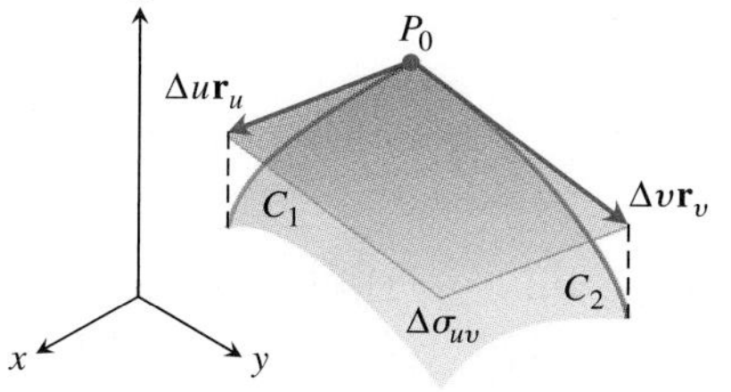

FIGURE 13.39 The parallelogram determined by the vectors $\Delta u\mathbf{r}_u$ and $\Delta v\mathbf{r}_v$ approximates the surface area element $\Delta\sigma_{uv}$.

A partition of the region R in the uv-plane by rectangular regions ΔA_{uv} induces a partition of the surface S into surface area elements $\Delta\sigma_{uv}$. We approximate the area of each surface element $\Delta\sigma_{uv}$ by the parallelogram area in Equation (2) and sum these areas together to obtain an approximation of the area of S:

$$\sum_n |\mathbf{r}_u \times \mathbf{r}_v|\,\Delta u\,\Delta v. \tag{3}$$

As Δu and Δv approach zero independently, the number of area elements n tends to ∞ and the continuity of $\mathbf{r}_u$ and $\mathbf{r}_v$ guarantees that the sum in Equation (3) approaches the double integral $\int_c^d \int_a^b |\mathbf{r}_u \times \mathbf{r}_v|\,du\,dv$. This double integral over the region R defines the area of the surface S.

DEFINITION The **area** of the smooth surface

$$\mathbf{r}(u, v) = f(u, v)\mathbf{i} + g(u, v)\mathbf{j} + h(u, v)\mathbf{k}, \qquad a \le u \le b, \quad c \le v \le d$$

is

$$A = \iint_R |\mathbf{r}_u \times \mathbf{r}_v|\, dA = \int_c^d \int_a^b |\mathbf{r}_u \times \mathbf{r}_v|\, du\, dv. \tag{4}$$

We can abbreviate the integral in Equation (4) by writing $d\sigma$ for $|\mathbf{r}_u \times \mathbf{r}_v|\, du\, dv$. The surface area differential $d\sigma$ is analogous to the arc length differential ds in Section 10.3.

Surface Area Differential for a Parametrized Surface

$$d\sigma = |\mathbf{r}_u \times \mathbf{r}_v|\, du\, dv \qquad \iint_S d\sigma \tag{5}$$

Surface area differential — Differential formula for surface area

EXAMPLE 4 Find the surface area of the cone in Example 1 (Figure 13.34).

Solution In Example 1, we found the parametrization

$$\mathbf{r}(r, \theta) = (r\cos\theta)\mathbf{i} + (r\sin\theta)\mathbf{j} + r\mathbf{k}, \qquad 0 \le r \le 1, \quad 0 \le \theta \le 2\pi.$$

To apply Equation (4), we first find $\mathbf{r}_r \times \mathbf{r}_\theta$:

$$\mathbf{r}_r \times \mathbf{r}_\theta = \begin{vmatrix} \mathbf{i} & \mathbf{j} & \mathbf{k} \\ \cos\theta & \sin\theta & 1 \\ -r\sin\theta & r\cos\theta & 0 \end{vmatrix}$$

$$= -(r\cos\theta)\mathbf{i} - (r\sin\theta)\mathbf{j} + \underbrace{(r\cos^2\theta + r\sin^2\theta)}_{r}\mathbf{k}.$$

Thus, $|\mathbf{r}_r \times \mathbf{r}_\theta| = \sqrt{r^2\cos^2\theta + r^2\sin^2\theta + r^2} = \sqrt{2r^2} = \sqrt{2}r$. The area of the cone is

$$A = \int_0^{2\pi}\int_0^1 |\mathbf{r}_r \times \mathbf{r}_\theta|\, dr\, d\theta \qquad \text{Eq. (4) with } u = r, v = \theta$$

$$= \int_0^{2\pi}\int_0^1 \sqrt{2}r\, dr\, d\theta = \int_0^{2\pi} \frac{\sqrt{2}}{2}\, d\theta = \frac{\sqrt{2}}{2}(2\pi) = \pi\sqrt{2} \text{ units squared.}$$ ■

EXAMPLE 5 Find the surface area of a sphere of radius a.

Solution We use the parametrization from Example 2:

$$\mathbf{r}(\phi, \theta) = (a \sin \phi \cos \theta)\mathbf{i} + (a \sin \phi \sin \theta)\mathbf{j} + (a \cos \phi)\mathbf{k},$$
$$0 \le \phi \le \pi, \quad 0 \le \theta \le 2\pi.$$

For $\mathbf{r}_\phi \times \mathbf{r}_\theta$, we get

$$\mathbf{r}_\phi \times \mathbf{r}_\theta = \begin{vmatrix} \mathbf{i} & \mathbf{j} & \mathbf{k} \\ a\cos\phi\cos\theta & a\cos\phi\sin\theta & -a\sin\phi \\ -a\sin\phi\sin\theta & a\sin\phi\cos\theta & 0 \end{vmatrix}$$
$$= (a^2 \sin^2 \phi \cos \theta)\mathbf{i} + (a^2 \sin^2 \phi \sin \theta)\mathbf{j} + (a^2 \sin \phi \cos \phi)\mathbf{k}.$$

Thus,

$$\begin{aligned} |\mathbf{r}_\phi \times \mathbf{r}_\theta| &= \sqrt{a^4 \sin^4 \phi \cos^2 \theta + a^4 \sin^4 \phi \sin^2 \theta + a^4 \sin^2 \phi \cos^2 \phi} \\ &= \sqrt{a^4 \sin^4 \phi + a^4 \sin^2 \phi \cos^2 \phi} = \sqrt{a^4 \sin^2 \phi\,(\sin^2 \phi + \cos^2 \phi)} \\ &= a^2 \sqrt{\sin^2 \phi} = a^2 \sin \phi, \end{aligned}$$

since $\sin \phi \ge 0$ for $0 \le \phi \le \pi$. Therefore, the area of the sphere is

$$\begin{aligned} A &= \int_0^{2\pi} \int_0^{\pi} a^2 \sin \phi \, d\phi \, d\theta \\ &= \int_0^{2\pi} \Big[-a^2 \cos \phi \Big]_0^{\pi} d\theta = \int_0^{2\pi} 2a^2 \, d\theta = 4\pi a^2 \text{ units squared.} \end{aligned}$$

This agrees with the well-known formula for the surface area of a sphere. ■

Implicit Surfaces

Surfaces are often presented as level sets of a function, described by an equation such as

$$F(x, y, z) = c,$$

for some constant c. Such a level surface does not come with an explicit parametrization, and is called an *implicitly defined surface.* Implicit surfaces arise, for example, as equipotential surfaces in electric or gravitational fields. Figure 13.40 shows a piece of such a surface. It may be difficult to find explicit formulas for the functions f, g, and h that describe the surface in the form $\mathbf{r}(u, v) = f(u, v)\mathbf{i} + g(u, v)\mathbf{j} + h(u, v)\mathbf{k}$. We now show how to compute the surface area differential $d\sigma$ for implicit surfaces.

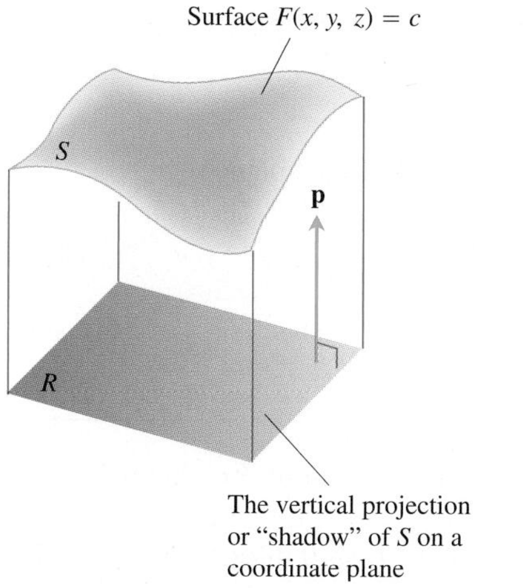

FIGURE 13.40 As we soon see, the area of a surface S in space can be calculated by evaluating a related double integral over the vertical projection or "shadow" of S on a coordinate plane. The unit vector $\mathbf{p}$ is normal to the plane.

Figure 13.40 shows a piece of an implicit surface S that lies above its "shadow" region R in the plane beneath it. The surface is defined by the equation $F(x, y, z) = c$ and $\mathbf{p}$ is a unit vector normal to the plane region R. We assume that the surface is **smooth** (F is differentiable and ∇F is nonzero and continuous on S) and that $\nabla F \cdot \mathbf{p} \neq 0$, so the surface never folds back over itself.

Assume that the normal vector $\mathbf{p}$ is the unit vector $\mathbf{k}$, so the region R in Figure 13.40 lies in the xy-plane. By assumption, we then have $\nabla F \cdot \mathbf{p} = \nabla F \cdot \mathbf{k} = F_z \neq 0$ on S. An advanced calculus theorem called the Implicit Function Theorem implies that S is then the graph of a differentiable function $z = h(x, y)$, although the function $h(x, y)$ is not explicitly known. Define the parameters u and v by $u = x$ and $v = y$. Then $z = h(u, v)$ and

$$\mathbf{r}(u, v) = u\mathbf{i} + v\mathbf{j} + h(u, v)\mathbf{k} \tag{6}$$

gives a parametrization of the surface S. We use Equation (4) to find the area of S.

Calculating the partial derivatives of $\mathbf{r}$, we find

$$\mathbf{r}_u = \mathbf{i} + \frac{\partial h}{\partial u}\mathbf{k} \qquad \text{and} \qquad \mathbf{r}_v = \mathbf{j} + \frac{\partial h}{\partial v}\mathbf{k}.$$

Applying the chain rule for implicit differentiation (Theorem 8, Section 11.4) to $F(x, y, z) = c$, where $x = u$, $y = v$, and $z = h(u, v)$, we obtain the partial derivatives

$$\frac{\partial h}{\partial u} = -\frac{F_x}{F_z} \qquad \text{and} \qquad \frac{\partial h}{\partial v} = -\frac{F_y}{F_z}.$$

Substitution of these derivatives into the derivatives of $\mathbf{r}$ gives

$$\mathbf{r}_u = \mathbf{i} - \frac{F_x}{F_z}\mathbf{k} \qquad \text{and} \qquad \mathbf{r}_v = \mathbf{j} - \frac{F_y}{F_z}\mathbf{k}.$$

From a routine calculation of the cross product we find

$$\begin{aligned}
\mathbf{r}_u \times \mathbf{r}_v &= \frac{F_x}{F_z}\mathbf{i} + \frac{F_y}{F_z}\mathbf{j} + \mathbf{k} && F_z \neq 0 \\
&= \frac{1}{F_z}(F_x\mathbf{i} + F_y\mathbf{j} + F_z\mathbf{k}) \\
&= \frac{\nabla F}{F_z} = \frac{\nabla F}{\nabla F \cdot \mathbf{k}} \\
&= \frac{\nabla F}{\nabla F \cdot \mathbf{p}}. && \mathbf{p} = \mathbf{k}
\end{aligned}$$

Therefore, the surface area differential is given by

$$d\sigma = |\mathbf{r}_u \times \mathbf{r}_v|\, du\, dv = \frac{|\nabla F|}{|\nabla F \cdot \mathbf{p}|}\, dx\, dy. \qquad u = x \text{ and } v = y$$

We obtain similar calculations if instead the vector $\mathbf{p} = \mathbf{j}$ is normal to the xz-plane when $F_y \neq 0$ on S, or if $\mathbf{p} = \mathbf{i}$ is normal to the yz-plane when $F_x \neq 0$ on S. Combining these results with Equation (4) then gives the following general formula.

Formula for Surface Area of an Implicit Surface

The area of the surface $F(x, y, z) = c$ over a closed and bounded plane region R is

$$\text{Surface area} = \iint_R \frac{|\nabla F|}{|\nabla F \cdot \mathbf{p}|}\, dA, \qquad (7)$$

where $\mathbf{p}$ is a unit vector normal to R and $\nabla F \cdot \mathbf{p} \neq 0$.

Thus, the area is the double integral over R of the magnitude of ∇F divided by the magnitude of the scalar component of ∇F normal to R.

We reached Equation (7) under the assumption that $\nabla F \cdot \mathbf{p} \neq 0$ throughout R and that ∇F is continuous. Whenever the integral exists, however, we define its value to be the area of the portion of the surface $F(x, y, z) = c$ that lies over R. (Recall that the projection is assumed to be one-to-one.)

EXAMPLE 6 Find the area of the surface cut from the bottom of the paraboloid $x^2 + y^2 - z = 0$ by the plane $z = 4$.

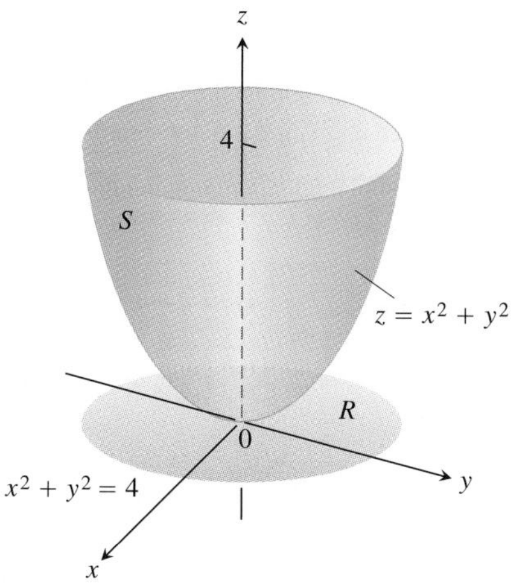

FIGURE 13.41 The area of this parabolic surface is calculated in Example 6.

Solution We sketch the surface S and the region R below it in the xy-plane (Figure 13.41). The surface S is part of the level surface $F(x, y, z) = x^2 + y^2 - z = 0$, and R is the disk $x^2 + y^2 \leq 4$ in the xy-plane. To get a unit vector normal to the plane of R, we can take $\mathbf{p} = \mathbf{k}$.

At any point (x, y, z) on the surface, we have

$$\begin{aligned} F(x, y, z) &= x^2 + y^2 - z \\ \nabla F &= 2x\mathbf{i} + 2y\mathbf{j} - \mathbf{k} \\ |\nabla F| &= \sqrt{(2x)^2 + (2y)^2 + (-1)^2} \\ &= \sqrt{4x^2 + 4y^2 + 1} \\ |\nabla F \cdot \mathbf{p}| &= |\nabla F \cdot \mathbf{k}| = |-1| = 1. \end{aligned}$$

In the region R, $dA = dx\,dy$. Therefore,

$$\begin{aligned} \text{Surface area} &= \iint_R \frac{|\nabla F|}{|\nabla F \cdot \mathbf{p}|}\,dA & \text{Eq. (7)} \\ &= \iint_{x^2+y^2\leq 4} \sqrt{4x^2 + 4y^2 + 1}\,dx\,dy \\ &= \int_0^{2\pi}\int_0^2 \sqrt{4r^2 + 1}\,r\,dr\,d\theta & \text{Polar coordinates} \\ &= \int_0^{2\pi}\left[\frac{1}{12}(4r^2 + 1)^{3/2}\right]_0^2 d\theta \\ &= \int_0^{2\pi}\frac{1}{12}(17^{3/2} - 1)\,d\theta = \frac{\pi}{6}\left(17\sqrt{17} - 1\right). \end{aligned}$$

■

For a graph $z = f(x, y)$ over a region R in the xy-plane, the surface area formula is

$$A = \iint_R \sqrt{f_x^2 + f_y^2 + 1}\,dx\,dy.$$

(See Exercises 47–53.)

EXERCISES 13.5

In Exercises 1–16, find a parametrization of the surface. (There are many correct ways to do these, so your answers may not be the same as those in the back of the book.)

1. The paraboloid $z = x^2 + y^2$, $z \leq 4$
2. The paraboloid $z = 9 - x^2 - y^2$, $z \geq 0$
3. **Cone frustum** The first-octant portion of the cone $z = \sqrt{x^2 + y^2}/2$ between the planes $z = 0$ and $z = 3$
4. **Cone frustum** The portion of the cone $z = 2\sqrt{x^2 + y^2}$ between the planes $z = 2$ and $z = 4$
5. **Spherical cap** The cap cut from the sphere $x^2 + y^2 + z^2 = 9$ by the cone $z = \sqrt{x^2 + y^2}$
6. **Spherical cap** The portion of the sphere $x^2 + y^2 + z^2 = 4$ in the first octant between the xy-plane and the cone $z = \sqrt{x^2 + y^2}$
7. **Spherical band** The portion of the sphere $x^2 + y^2 + z^2 = 3$ between the planes $z = \sqrt{3}/2$ and $z = -\sqrt{3}/2$
8. **Spherical cap** The upper portion cut from the sphere $x^2 + y^2 + z^2 = 8$ by the plane $z = -2$

9. Parabolic cylinder between planes The surface cut from the parabolic cylinder $z = 4 - y^2$ by the planes $x = 0, x = 2$, and $z = 0$

10. Parabolic cylinder between planes The surface cut from the parabolic cylinder $y = x^2$ by the planes $z = 0, z = 3$ and $y = 2$

11. Circular cylinder band The portion of the cylinder $y^2 + z^2 = 9$ between the planes $x = 0$ and $x = 3$

12. Circular cylinder band The portion of the cylinder $x^2 + z^2 = 4$ above the xy-plane between the planes $y = -2$ and $y = 2$

13. Tilted plane inside cylinder The portion of the plane $x + y + z = 1$

a. Inside the cylinder $x^2 + y^2 = 9$

b. Inside the cylinder $y^2 + z^2 = 9$

14. Tilted plane inside cylinder The portion of the plane $x - y + 2z = 2$

a. Inside the cylinder $x^2 + z^2 = 3$

b. Inside the cylinder $y^2 + z^2 = 2$

15. Circular cylinder band The portion of the cylinder $(x - 2)^2 + z^2 = 4$ between the planes $y = 0$ and $y = 3$

16. Circular cylinder band The portion of the cylinder $y^2 + (z - 5)^2 = 25$ between the planes $x = 0$ and $x = 10$

In Exercises 17–26, use a parametrization to express the area of the surface as a double integral. Then evaluate the integral. (There are many correct ways to set up the integrals, so your integrals may not be the same as those in the back of the book. They should have the same values, however.)

17. Tilted plane inside cylinder The portion of the plane $y + 2z = 2$ inside the cylinder $x^2 + y^2 = 1$

18. Plane inside cylinder The portion of the plane $z = -x$ inside the cylinder $x^2 + y^2 = 4$

19. Cone frustum The portion of the cone $z = 2\sqrt{x^2 + y^2}$ between the planes $z = 2$ and $z = 6$

20. Cone frustum The portion of the cone $z = \sqrt{x^2 + y^2}/3$ between the planes $z = 1$ and $z = 4/3$

21. Circular cylinder band The portion of the cylinder $x^2 + y^2 = 1$ between the planes $z = 1$ and $z = 4$

22. Circular cylinder band The portion of the cylinder $x^2 + z^2 = 10$ between the planes $y = -1$ and $y = 1$

23. Parabolic cap The cap cut from the paraboloid $z = 2 - x^2 - y^2$ by the cone $z = \sqrt{x^2 + y^2}$

24. Parabolic band The portion of the paraboloid $z = x^2 + y^2$ between the planes $z = 1$ and $z = 4$

25. Sawed-off sphere The lower portion cut from the sphere $x^2 + y^2 + z^2 = 2$ by the cone $z = \sqrt{x^2 + y^2}$

26. Spherical band The portion of the sphere $x^2 + y^2 + z^2 = 4$ between the planes $z = -1$ and $z = \sqrt{3}$

Planes Tangent to Parametrized Surfaces The tangent plane at a point $P_0(f(u_0, v_0), g(u_0, v_0), h(u_0, v_0))$ on a parametrized surface $\mathbf{r}(u, v) = f(u, v)\mathbf{i} + g(u, v)\mathbf{j} + h(u, v)\mathbf{k}$ is the plane through P_0 normal to the vector $\mathbf{r}_u(u_0, v_0) \times \mathbf{r}_v(u_0, v_0)$, the cross product of the tangent vectors $\mathbf{r}_u(u_0, v_0)$ and $\mathbf{r}_v(u_0, v_0)$ at P_0. In Exercises 27–30, find an equation for the plane tangent to the surface at P_0. Then find a Cartesian equation for the surface and sketch the surface and tangent plane together.

27. Cone The cone $\mathbf{r}(r, \theta) = (r\cos\theta)\mathbf{i} + (r\sin\theta)\mathbf{j} + r\mathbf{k}, r \geq 0, 0 \leq \theta \leq 2\pi$ at the point $P_0(\sqrt{2}, \sqrt{2}, 2)$ corresponding to $(r, \theta) = (2, \pi/4)$

28. Hemisphere The hemisphere surface $\mathbf{r}(\phi, \theta) = (4\sin\phi\cos\theta)\mathbf{i} + (4\sin\phi\sin\theta)\mathbf{j} + (4\cos\phi)\mathbf{k}, 0 \leq \phi \leq \pi/2, 0 \leq \theta \leq 2\pi$, at the point $P_0(\sqrt{2}, \sqrt{2}, 2\sqrt{3})$ corresponding to $(\phi, \theta) = (\pi/6, \pi/4)$

29. Circular cylinder The circular cylinder $\mathbf{r}(\theta, z) = (3\sin 2\theta)\mathbf{i} + (6\sin^2\theta)\mathbf{j} + z\mathbf{k}, 0 \leq \theta \leq \pi$, at the point $P_0(3\sqrt{3}/2, 9/2, 0)$ corresponding to $(\theta, z) = (\pi/3, 0)$ (See Example 3.)

30. Parabolic cylinder The parabolic cylinder surface $\mathbf{r}(x, y) = x\mathbf{i} + y\mathbf{j} - x^2\mathbf{k}, -\infty < x < \infty, -\infty < y < \infty$, at the point $P_0(1, 2, -1)$ corresponding to $(x, y) = (1, 2)$

31. a. A *torus of revolution* (doughnut) is obtained by rotating a circle C in the xz-plane about the z-axis in space. (See Figure 13.42.) If C has radius $r > 0$ and center $(R, 0, 0)$, show that a parametrization of the torus is

$$\mathbf{r}(u, v) = ((R + r\cos u)\cos v)\mathbf{i} + ((R + r\cos u)\sin v)\mathbf{j} + (r\sin u)\mathbf{k},$$

where $0 \leq u \leq 2\pi$ and $0 \leq v \leq 2\pi$ are the angles in the figure.

b. Show that the surface area of the torus is $A = 4\pi^2 Rr$.

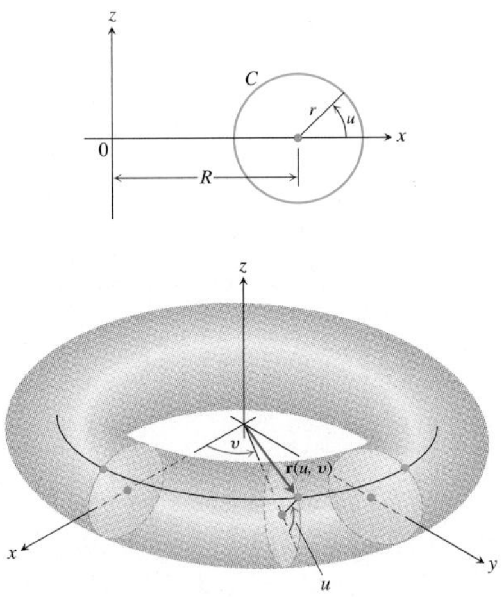

FIGURE 13.42 A torus of revolution.

32. Hyperboloid of one sheet

a. Find a parametrization for the hyperboloid of one sheet $x^2 + y^2 - z^2 = 1$ in terms of the angle θ associated with the circle $x^2 + y^2 = r^2$ and the hyperbolic parameter u associated with the hyperbolic function $r^2 - z^2 = 1$. (*Hint:* $\cosh^2 u - \sinh^2 u = 1$)

b. Generalize the result in part (a) to the hyperboloid $(x^2/a^2) + (y^2/b^2) - (z^2/c^2) = 1$.

33. (*Continuation of Exercise 32.*) Find a Cartesian equation for the plane tangent to the hyperboloid $x^2 + y^2 - z^2 = 25$ at the point $(x_0, y_0, 0)$, where $x_0^2 + y_0^2 = 25$.

34. Hyperboloid of two sheets Find a parametrization of the hyperboloid of two sheets $(z^2/c^2) - (x^2/a^2) - (y^2/b^2) = 1$.

35. Find the area of the surface cut from the paraboloid $x^2 + y^2 - z = 0$ by the plane $z = 2$.

36. Find the area of the band cut from the paraboloid $x^2 + y^2 - z = 0$ by the planes $z = 2$ and $z = 6$.

37. Find the area of the region cut from the plane $x + 2y + 2z = 5$ by the cylinder whose walls are $x = y^2$ and $x = 2 - y^2$.

38. Find the area of the portion of the surface $x^2 - 2z = 0$ that lies above the triangle bounded by the lines $x = \sqrt{3}$, $y = 0$, and $y = x$ in the xy-plane.

39. Find the area of the surface $x^2 - 2y - 2z = 0$ that lies above the triangle bounded by the lines $x = 2$, $y = 0$, and $y = 3x$ in the xy-plane.

40. Find the area of the cap cut from the sphere $x^2 + y^2 + z^2 = 2$ by the cone $z = \sqrt{x^2 + y^2}$.

41. Find the area of the ellipse cut from the plane $z = cx$ (c a constant) by the cylinder $x^2 + y^2 = 1$.

42. Find the area of the upper portion of the cylinder $x^2 + z^2 = 1$ that lies between the planes $x = \pm 1/2$ and $y = \pm 1/2$.

43. Find the area of the portion of the paraboloid $x = 4 - y^2 - z^2$ that lies above the ring $1 \le y^2 + z^2 \le 4$ in the yz-plane.

44. Find the area of the surface cut from the paraboloid $x^2 + y + z^2 = 2$ by the plane $y = 0$.

45. Find the area of the surface $x^2 - 2\ln x + \sqrt{15}y - z = 0$ above the square R: $1 \le x \le 2$, $0 \le y \le 1$, in the xy-plane.

46. Find the area of the surface $2x^{3/2} + 2y^{3/2} - 3z = 0$ above the square R: $0 \le x \le 1$, $0 \le y \le 1$, in the xy-plane.

Surface Area for Explicit Forms If S is the surface defined by a function $z = f(x, y)$ that has continuous first partial derivatives throughout a region R_{xy} in the xy-plane then S is also the level surface $F(x, y, z) = 0$ of the function $F(x, y, z) = f(x, y) - z$. Taking the unit normal to R_{xy} to be $\mathbf{p} = \mathbf{k}$ then gives

$$|\nabla F| = |f_x\mathbf{i} + f_y\mathbf{j} - \mathbf{k}| = \sqrt{f_x^2 + f_y^2 + 1}$$

$$|\nabla F \cdot \mathbf{p}| = |(f_x\mathbf{i} + f_y\mathbf{j} - \mathbf{k}) \cdot \mathbf{k}| = |-1| = 1$$

and

$$\iint_{R_{xy}} \frac{|\nabla F|}{|\nabla F \cdot \mathbf{p}|}\, dA = \iint_{R_{xy}} \sqrt{f_x^2 + f_y^2 + 1}\, dx\, dy, \tag{8}$$

Similarly, the area of a smooth surface $x = f(y, z)$ over a region R_{yz} in the yz-plane is

$$A = \iint_{R_{yz}} \sqrt{f_y^2 + f_z^2 + 1}\, dy\, dz, \tag{9}$$

and the area of a smooth $y = f(x, z)$ over a region R_{xz} in the xz-plane is

$$A = \iint_{R_{xz}} \sqrt{f_x^2 + f_z^2 + 1}\, dx\, dz. \tag{10}$$

Use Equations (8)–(10) to find the area of the surfaces in Exercises 47–52.

47. The surface cut from the bottom of the paraboloid $z = x^2 + y^2$ by the plane $z = 3$

48. The surface cut from the "nose" of the paraboloid $x = 1 - y^2 - z^2$ by the yz-plane

49. The portion of the cone $z = \sqrt{x^2 + y^2}$ that lies over the region between the circle $x^2 + y^2 = 1$ and the ellipse $9x^2 + 4y^2 = 36$ in the xy-plane. (*Hint:* Use formulas from geometry to find the area of the region.)

50. The triangle cut from the plane $2x + 6y + 3z = 6$ by the bounding planes of the first octant. Calculate the area three ways, once with each area formula.

51. The surface in the first octant cut from the cylinder $y = (2/3)z^{3/2}$ by the planes $x = 1$ and $y = 16/3$

52. The portion of the plane $y + z = 4$ that lies above the region cut from the first quadrant of the xz-plane by the parabola $x = 4 - z^2$

53. Use the parametrization

$$\mathbf{r}(x, y) = x\mathbf{i} + y\mathbf{j} + f(x, y)\,\mathbf{k}$$

and Equation (5) to derive the double integral formula for $d\sigma$ in Equation (8).

54. Let S be the surface obtained by rotating the smooth curve $y = f(x)$, $a \le x \le b$, about the x-axis, where $f(x) \ge 0$.

a. Show that the vector function

$$\mathbf{r}(x, \theta) = x\mathbf{i} + f(x)\cos\theta\,\mathbf{j} + f(x)\sin\theta\,\mathbf{k}$$

is a parametrization of S, where θ is the angle of rotation around the x-axis (see the accompanying figure).

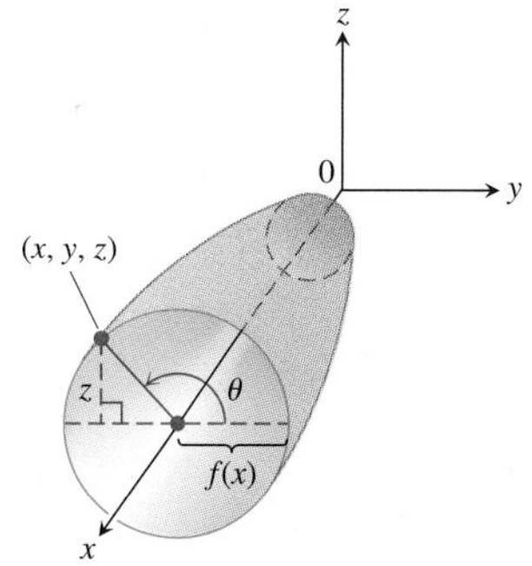

b. Use Equation (4) to show that the surface area of this surface of revolution is given by

$$A = \int_a^b 2\pi f(x)\sqrt{1 + [f'(x)]^2}\, dx.$$

13.6 Surface Integrals and Flux

To compute quantities such as the flow of liquid across a curved membrane or the upward force on a falling parachute, we need to integrate a function over a curved surface in space. This concept of a *surface integral* is an extension of the idea of a line integral when integrating over a curve.

Surface Integrals

Suppose, for example, that we have an electrical charge distributed over a surface S, and that the function $G(x, y, z)$ gives the charge per unit area (charge density) at each point on S. Then we may calculate the total charge on S as an integral in the following way.

Assume, as in Section 13.5, that the surface S is defined parametrically on a region R in the uv-plane,

$$\mathbf{r}(u, v) = f(u, v)\mathbf{i} + g(u, v)\mathbf{j} + h(u, v)\mathbf{k}, \qquad (u, v) \in R.$$

In Figure 13.37, we saw how a subdivision of R (considered as a rectangle for simplicity) divided the surface S into corresponding curved surface elements, or patches, of area

$$\Delta\sigma_{uv} \approx |\mathbf{r}_u \times \mathbf{r}_v|\, du\, dv.$$

As we did for the subdivisions when defining double integrals in Section 12.1, we number the surface element patches in some order with their areas given by $\Delta\sigma_1, \Delta\sigma_2, \ldots, \Delta\sigma_n$. To form a Riemann sum over S, we choose a point (x_k, y_k, z_k) in the kth patch, multiply the value of the function G at that point by the area $\Delta\sigma_k$, and add together the products:

$$\sum_{k=1}^{n} G(x_k, y_k, z_k)\, \Delta\sigma_k.$$

Depending on how we pick (x_k, y_k, z_k) in the kth patch, we may get different values for this Riemann sum. Then we take the limit as the number of surface patches increases and their areas shrink to zero. This limit, whenever it exists, defines the **surface integral of G over the surface S** as

$$\iint_S G(x, y, z)\, d\sigma = \lim_{n\to\infty} \sum_{k=1}^{n} G(x_k, y_k, z_k)\, \Delta\sigma_k. \tag{1}$$

Notice the analogy with the definition of the double integral (Section 12.1) and with the line integral (Section 13.1). If S is a piecewise smooth surface, and G is continuous over S, then the surface integral defined by Equation (1) can be shown to exist.

The surface integral in Equation (1) takes on different meanings in different applications. If G has the constant value 1, the integral gives the area of S. If G gives the mass density of a thin shell of material modeled by S, the integral gives the mass of the shell.

The formula for evaluating the surface integral depends on the manner in which S is described, parametrically or implicitly, as discussed in Section 13.5.

Computing a Surface Integral

1. For a smooth surface S defined parametrically as $\mathbf{r}(u, v) = f(u, v)\mathbf{i} + g(u, v)\mathbf{j} + h(u, v)\mathbf{k}$, $(u, v) \in R$, and a continuous function $G(x, y, z)$ defined on S, the surface integral of G over S is given by the double integral over R

$$\iint_S G(x, y, z)\, d\sigma = \iint_R G(f(u, v), g(u, v), h(u, v))\, |\mathbf{r}_u \times \mathbf{r}_v|\, du\, dv. \tag{2}$$

2. For a surface S given implicitly by $F(x, y, z) = c$, where F is a continuously differentiable function, with S lying above its closed and bounded shadow region R in the coordinate plane beneath it, the surface integral of the continuous function G over S is given by the double integral over R

$$\iint_S G(x, y, z)\, d\sigma = \iint_R G(x, y, z)\frac{|\nabla F|}{|\nabla F \cdot \mathbf{p}|}\, dA, \tag{3}$$

where $\mathbf{p}$ is a unit vector normal to R and $\nabla F \cdot \mathbf{p} \neq 0$.

EXAMPLE 1 Integrate $G(x, y, z) = x^2$ over the cone $z = \sqrt{x^2 + y^2}$, $0 \leq z \leq 1$.

Solution Using Equation (2) and the calculations from Example 4 in Section 13.5, we have $|\mathbf{r}_r \times \mathbf{r}_\theta| = \sqrt{2}r$ and

$$\iint_S x^2\, d\sigma = \int_0^{2\pi}\int_0^1 \left(r^2 \cos^2\theta\right)\left(\sqrt{2}r\right) dr\, d\theta \qquad x = r\cos\theta$$

$$= \sqrt{2}\int_0^{2\pi}\int_0^1 r^3 \cos^2\theta\, dr\, d\theta$$

$$= \frac{\sqrt{2}}{4}\int_0^{2\pi} \cos^2\theta\, d\theta = \frac{\sqrt{2}}{4}\left[\frac{\theta}{2} + \frac{1}{4}\sin 2\theta\right]_0^{2\pi} = \frac{\pi\sqrt{2}}{4}. \qquad \blacksquare$$

Surface integrals behave like other double integrals, the integral of the sum of two functions being the sum of their integrals and so on. The domain Additivity Property takes the form

$$\iint_S G\, d\sigma = \iint_{S_1} G\, d\sigma + \iint_{S_2} G\, d\sigma + \cdots + \iint_{S_n} G\, d\sigma.$$

The idea is that if S is partitioned by smooth curves into a finite number of nonoverlapping smooth patches (i.e., if S is **piecewise smooth**), then the integral over S is the sum of the integrals over the patches. Thus, the integral of a function over the surface of a cube is the sum of the integrals over the faces of the cube. We integrate over a turtle shell of welded plates by integrating over one plate at a time and adding the results.

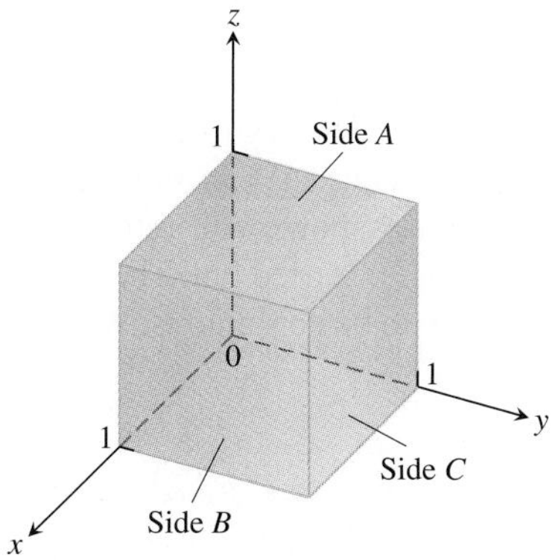

FIGURE 13.43 The cube in Example 2.

EXAMPLE 2 Integrate $G(x, y, z) = xyz$ over the surface of the cube cut from the first octant by the planes $x = 1$, $y = 1$, and $z = 1$ (Figure 13.43).

Solution We integrate xyz over each of the six sides and add the results. Since $xyz = 0$ on the sides that lie in the coordinate planes, the integral over the surface of the cube reduces to

$$\iint_{\substack{\text{Cube}\\\text{surface}}} xyz\, d\sigma = \iint_{\text{Side } A} xyz\, d\sigma + \iint_{\text{Side } B} xyz\, d\sigma + \iint_{\text{Side } C} xyz\, d\sigma.$$

Side A is the surface $f(x, y, z) = z = 1$ over the square region R_{xy}: $0 \le x \le 1,\ 0 \le y \le 1$, in the xy-plane. For this surface and region,

$$\mathbf{p} = \mathbf{k}, \qquad \nabla f = \mathbf{k}, \qquad |\nabla f| = 1, \qquad |\nabla f \cdot \mathbf{p}| = |\mathbf{k} \cdot \mathbf{k}| = 1$$

$$d\sigma = \frac{|\nabla f|}{|\nabla f \cdot \mathbf{p}|}\, dA = \frac{1}{1}\, dx\, dy = dx\, dy$$

$$xyz = xy(1) = xy$$

and

$$\iint_{\text{Side } A} xyz\, d\sigma = \iint_{R_{xy}} xy\, dx\, dy = \int_0^1 \int_0^1 xy\, dx\, dy = \int_0^1 \frac{y}{2}\, dy = \frac{1}{4}.$$

Symmetry tells us that the integrals of xyz over sides B and C are also $1/4$. Hence,

$$\iint_{\substack{\text{Cube}\\\text{surface}}} xyz\, d\sigma = \frac{1}{4} + \frac{1}{4} + \frac{1}{4} = \frac{3}{4}.$$ ■

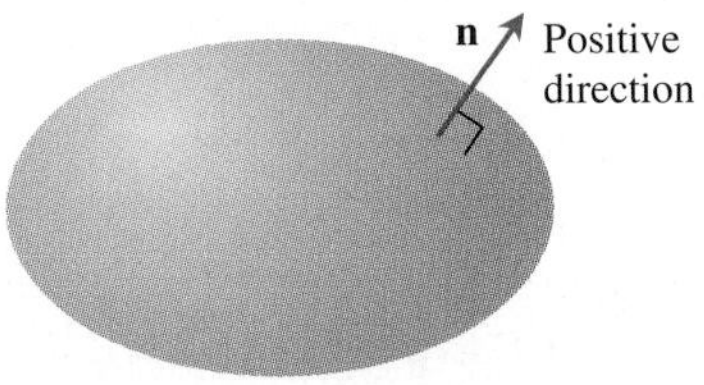

FIGURE 13.44 Smooth closed surfaces in space are orientable. The outward unit normal vector defines the positive direction at each point.

Orientation

We call a smooth surface S **orientable** or **two-sided** if it is possible to define a field $\mathbf{n}$ of unit normal vectors on S that varies continuously with position. Any patch or subportion of an orientable surface is orientable. Spheres and other smooth closed surfaces in space (smooth surfaces that enclose solids) are orientable. By convention, we choose $\mathbf{n}$ on a closed surface to point outward.

Once $\mathbf{n}$ has been chosen, we say that we have **oriented** the surface, and we call the surface together with its normal field an **oriented surface**. The vector $\mathbf{n}$ at any point is called the **positive direction** at that point (Figure 13.44).

The Möbius band in Figure 13.45 is not orientable. No matter where you start to construct a continuous unit normal field (shown as the shaft of a thumbtack in the figure), moving the vector continuously around the surface in the manner shown will return it to the starting point with a direction opposite to the one it had when it started out. The vector at that point cannot point both ways and yet it must if the field is to be continuous. We conclude that no such field exists.

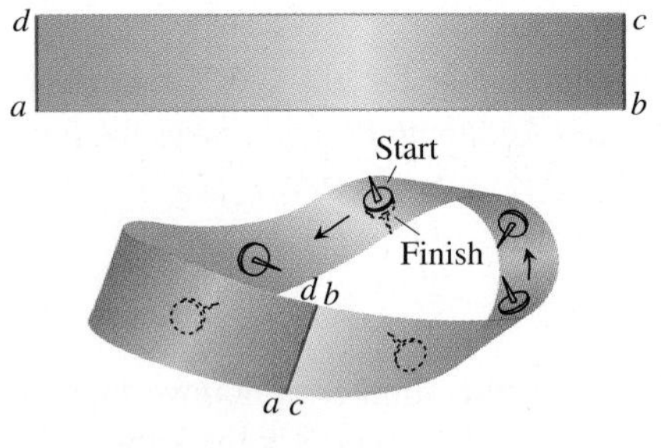

FIGURE 13.45 To make a Möbius band, take a rectangular strip of paper $abcd$, give the end bc a single twist, and paste the ends of the strip together to match a with c and b with d. The Möbius band is a nonorientable or one-sided surface.

Surface Integral for Flux

Suppose that $\mathbf{F}$ is a continuous vector field defined over an oriented surface S and that $\mathbf{n}$ is the chosen unit normal field on the surface. We call the integral of $\mathbf{F} \cdot \mathbf{n}$ over S the flux of

F across S in the positive direction. Thus, the flux is the integral over S of the scalar component of **F** in the direction of **n**.

DEFINITION The **flux** of a three-dimensional vector field **F** across an oriented surface S in the direction of **n** is

$$\text{Flux} = \iint_S \mathbf{F} \cdot \mathbf{n}\, d\sigma. \tag{4}$$

The definition is analogous to the flux of a two-dimensional field **F** across a plane curve C. In the plane (Section 13.2), the flux is

$$\int_C \mathbf{F} \cdot \mathbf{n}\, ds,$$

the integral of the scalar component of **F** normal to the curve.

If **F** is the velocity field of a three-dimensional fluid flow, the flux of **F** across S is the net rate at which fluid is crossing S in the chosen positive direction. We discuss such flows in more detail in Section 13.7.

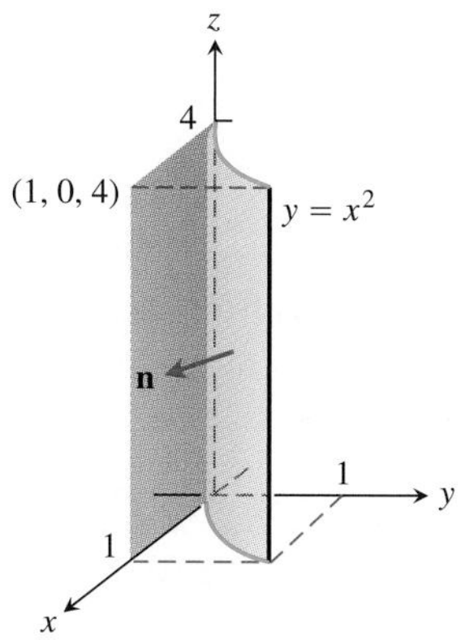

FIGURE 13.46 Finding the flux through the surface of a parabolic cylinder (Example 3).

EXAMPLE 3 Find the flux of $\mathbf{F} = yz\mathbf{i} + x\mathbf{j} - z^2\mathbf{k}$ outward through the parabolic cylinder $y = x^2$, $0 \le x \le 1$, $0 \le z \le 4$ (Figure 13.46).

Solution On the surface we have $x = x$, $y = x^2$, and $z = z$, so we automatically have the parametrization $\mathbf{r}(x, z) = x\mathbf{i} + x^2\mathbf{j} + z\mathbf{k}$, $0 \le x \le 1$, $0 \le z \le 4$. The cross product of tangent vectors is

$$\mathbf{r}_x \times \mathbf{r}_z = \begin{vmatrix} \mathbf{i} & \mathbf{j} & \mathbf{k} \\ 1 & 2x & 0 \\ 0 & 0 & 1 \end{vmatrix} = 2x\mathbf{i} - \mathbf{j}.$$

The unit normal pointing outward from the surface is

$$\mathbf{n} = \frac{\mathbf{r}_x \times \mathbf{r}_z}{|\mathbf{r}_x \times \mathbf{r}_z|} = \frac{2x\mathbf{i} - \mathbf{j}}{\sqrt{4x^2 + 1}}.$$

On the surface, $y = x^2$, so the vector field there is

$$\mathbf{F} = yz\mathbf{i} + x\mathbf{j} - z^2\mathbf{k} = x^2z\mathbf{i} + x\mathbf{j} - z^2\mathbf{k}.$$

Thus,

$$\begin{aligned} \mathbf{F} \cdot \mathbf{n} &= \frac{1}{\sqrt{4x^2 + 1}}\left((x^2z)(2x) + (x)(-1) + (-z^2)(0)\right) \\ &= \frac{2x^3z - x}{\sqrt{4x^2 + 1}}. \end{aligned}$$

The flux of $\mathbf{F}$ outward through the surface is

$$\begin{aligned}
\iint_S \mathbf{F}\cdot\mathbf{n}\,d\sigma &= \int_0^4\int_0^1 \frac{2x^3z - x}{\sqrt{4x^2+1}}\,|\mathbf{r}_x \times \mathbf{r}_z|\,dx\,dz \\
&= \int_0^4\int_0^1 \frac{2x^3z - x}{\sqrt{4x^2+1}}\sqrt{4x^2+1}\,dx\,dz \\
&= \int_0^4\int_0^1 (2x^3z - x)\,dx\,dz = \int_0^4 \left[\frac{1}{2}x^4z - \frac{1}{2}x^2\right]_{x=0}^{x=1} dz \\
&= \int_0^4 \frac{1}{2}(z-1)\,dz = \frac{1}{4}(z-1)^2\Big]_0^4 \\
&= \frac{1}{4}(9) - \frac{1}{4}(1) = 2.
\end{aligned}$$

■

If S is part of a level surface $g(x, y, z) = c$, then $\mathbf{n}$ may be taken to be one of the two fields

$$\mathbf{n} = \pm\frac{\nabla g}{|\nabla g|}, \tag{5}$$

depending on which one gives the preferred direction. The corresponding flux is

$$\begin{aligned}
\text{Flux} &= \iint_S \mathbf{F}\cdot\mathbf{n}\,d\sigma \\
&= \iint_R \left(\mathbf{F}\cdot\frac{\pm\nabla g}{|\nabla g|}\right)\frac{|\nabla g|}{|\nabla g\cdot\mathbf{p}|}\,dA \qquad \text{Eqs. (5) and (3)} \\
&= \iint_R \mathbf{F}\cdot\frac{\pm\nabla g}{|\nabla g\cdot\mathbf{p}|}\,dA.
\end{aligned} \tag{6}$$

EXAMPLE 4 Find the flux of $\mathbf{F} = yz\mathbf{j} + z^2\mathbf{k}$ outward through the surface S cut from the cylinder $y^2 + z^2 = 1$, $z \geq 0$, by the planes $x = 0$ and $x = 1$.

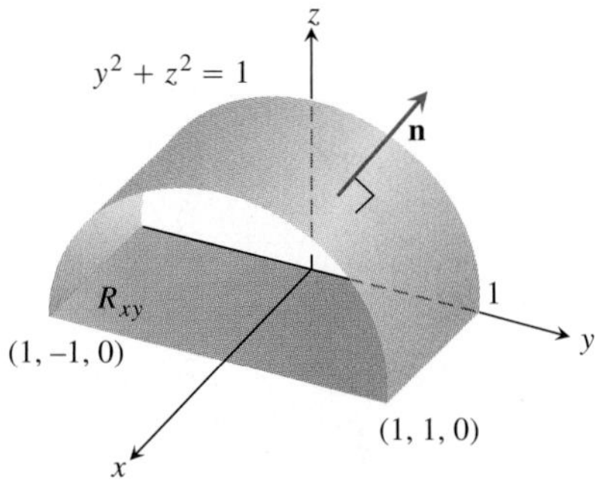

FIGURE 13.47 Calculating the flux of a vector field outward through this surface. The area of the shadow region R_{xy} is 2 (Example 4).

Solution The outward normal field on S (Figure 13.47) may be calculated from the gradient of $g(x, y, z) = y^2 + z^2$ to be

$$\mathbf{n} = +\frac{\nabla g}{|\nabla g|} = \frac{2y\mathbf{j} + 2z\mathbf{k}}{\sqrt{4y^2 + 4z^2}} = \frac{2y\mathbf{j} + 2z\mathbf{k}}{2\sqrt{1}} = y\mathbf{j} + z\mathbf{k}.$$

With $\mathbf{p} = \mathbf{k}$, we also have

$$d\sigma = \frac{|\nabla g|}{|\nabla g\cdot\mathbf{k}|}\,dA = \frac{2}{|2z|}\,dA = \frac{1}{z}\,dA.$$

We can drop the absolute value bars because $z \geq 0$ on S.

The value of $\mathbf{F}\cdot\mathbf{n}$ on the surface is

$$\begin{aligned}
\mathbf{F}\cdot\mathbf{n} &= (yz\mathbf{j} + z^2\mathbf{k})\cdot(y\mathbf{j} + z\mathbf{k}) \\
&= y^2z + z^3 = z(y^2 + z^2) \\
&= z. \qquad y^2 + z^2 = 1 \text{ on } S
\end{aligned}$$

Therefore, the flux of $\mathbf{F}$ outward through S is

$$\iint_S \mathbf{F}\cdot\mathbf{n}\,d\sigma = \iint_S (z)\left(\frac{1}{z}\,dA\right) = \iint_{R_{xy}} dA = \text{area}(R_{xy}) = 2.$$

■

Moments and Masses of Thin Shells

Thin shells of material like bowls, metal drums, and domes are modeled with surfaces. Their moments and masses are calculated with the formulas in Table 13.3. The derivations are similar to those in Section 6.6. The formulas are like those for line integrals in Table 13.1, Section 13.1.

TABLE 13.3 Mass and moment formulas for very thin shells

Mass: $M = \iint_S \delta(x, y, z)\, d\sigma$ $\qquad$ $\delta(x, y, z)$ = density at (x, y, z), mass per unit area

First moments about the coordinate planes:

$$M_{yz} = \iint_S x\,\delta\, d\sigma, \qquad M_{xz} = \iint_S y\,\delta\, d\sigma, \qquad M_{xy} = \iint_S z\,\delta\, d\sigma$$

Coordinates of center of mass:

$$\bar{x} = M_{yz}/M, \qquad \bar{y} = M_{xz}/M, \qquad \bar{z} = M_{xy}/M$$

Moments of inertia about coordinate axes:

$$I_x = \iint_S (y^2 + z^2)\,\delta\, d\sigma, \quad I_y = \iint_S (x^2 + z^2)\,\delta\, d\sigma, \quad I_z = \iint_S (x^2 + y^2)\,\delta\, d\sigma,$$

$$I_L = \iint_S r^2\delta\, d\sigma, \qquad r(x, y, z) = \text{distance from point } (x, y, z) \text{ to line } L$$

EXAMPLE 5 Find the center of mass of a thin hemispherical shell of radius a and constant density δ.

Solution We model the shell with the hemisphere

$$f(x, y, z) = x^2 + y^2 + z^2 = a^2, \qquad z \geq 0$$

(Figure 13.48). The symmetry of the surface about the z-axis tells us that $\bar{x} = \bar{y} = 0$. It remains only to find $\bar{z}$ from the formula $\bar{z} = M_{xy}/M$.

The mass of the shell is

$$M = \iint_S \delta\, d\sigma = \delta \iint_S d\sigma = (\delta)(\text{area of } S) = 2\pi a^2\delta.$$

To evaluate the integral for M_{xy}, we take $\mathbf{p} = \mathbf{k}$ and calculate

$$|\nabla f| = |2x\mathbf{i} + 2y\mathbf{j} + 2z\mathbf{k}| = 2\sqrt{x^2 + y^2 + z^2} = 2a$$

$$|\nabla f \cdot \mathbf{p}| = |\nabla f \cdot \mathbf{k}| = |2z| = 2z$$

$$d\sigma = \frac{|\nabla f|}{|\nabla f \cdot \mathbf{p}|}\, dA = \frac{a}{z}\, dA.$$

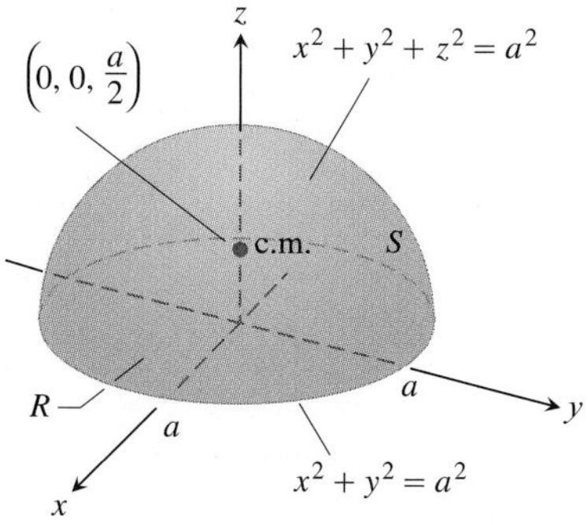

FIGURE 13.48 The center of mass of a thin hemispherical shell of constant density lies on the axis of symmetry halfway from the base to the top (Example 5).

Then

$$M_{xy} = \iint_S z\delta\, d\sigma = \delta \iint_R z\frac{a}{z}\, dA = \delta a \iint_R dA = \delta a(\pi a^2) = \delta\pi a^3$$

$$\bar{z} = \frac{M_{xy}}{M} = \frac{\pi a^3 \delta}{2\pi a^2 \delta} = \frac{a}{2}.$$

The shell's center of mass is the point $(0, 0, a/2)$. ■

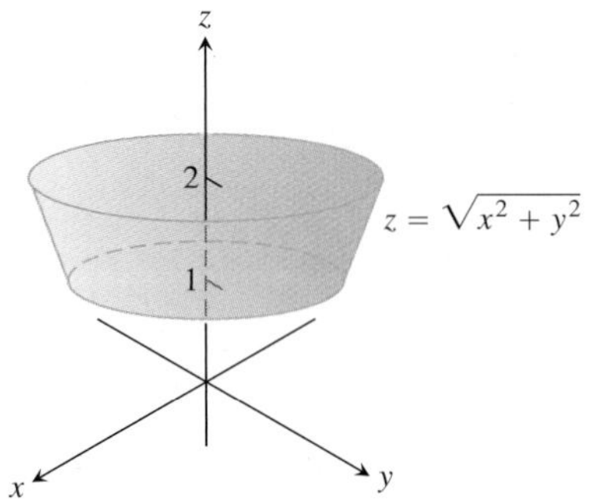

FIGURE 13.49 The cone frustum formed when the cone $z = \sqrt{x^2 + y^2}$ is cut by the planes $z = 1$ and $z = 2$ (Example 6).

EXAMPLE 6 Find the center of mass of a thin shell of constant density δ cut from the cone $z = \sqrt{x^2 + y^2}$ by the planes $z = 1$ and $z = 2$ (Figure 13.49).

Solution The symmetry of the surface about the z-axis tells us that $\bar{x} = \bar{y} = 0$. We find $\bar{z} = M_{xy}/M$. Working as in Example 4 of Section 14.5, we have

$$\mathbf{r}(r, \theta) = r\cos\theta\mathbf{i} + r\sin\theta\mathbf{j} + r\mathbf{k}, \qquad 1 \le r \le 2, \quad 0 \le \theta \le 2\pi,$$

and

$$|\mathbf{r}_r \times \mathbf{r}_\theta| = \sqrt{2}r.$$

Therefore,

$$\begin{aligned} M &= \iint_S \delta\, d\sigma = \int_0^{2\pi}\int_1^2 \delta\sqrt{2}r\, dr\, d\theta \\ &= \delta\sqrt{2}\int_0^{2\pi}\left[\frac{r^2}{2}\right]_1^2 d\theta = \delta\sqrt{2}\int_0^{2\pi}\left(2 - \frac{1}{2}\right) d\theta \\ &= \delta\sqrt{2}\left[\frac{3\theta}{2}\right]_0^{2\pi} = 3\pi\delta\sqrt{2} \\ M_{xy} &= \iint_S \delta z\, d\sigma = \int_0^{2\pi}\int_1^2 \delta r\sqrt{2}r\, dr\, d\theta \\ &= \delta\sqrt{2}\int_0^{2\pi}\int_1^2 r^2\, dr\, d\theta = \delta\sqrt{2}\int_0^{2\pi}\left[\frac{r^3}{3}\right]_1^2 d\theta \\ &= \delta\sqrt{2}\int_0^{2\pi}\frac{7}{3}\, d\theta = \frac{14}{3}\pi\delta\sqrt{2} \\ \bar{z} &= \frac{M_{xy}}{M} = \frac{14\pi\delta\sqrt{2}}{3\left(3\pi\delta\sqrt{2}\right)} = \frac{14}{9}. \end{aligned}$$

The shell's center of mass is the point $(0, 0, 14/9)$. ■

EXERCISES 13.6

In Exercises 1–8, integrate the given function over the given surface.

1. **Parabolic cylinder** $G(x, y, z) = x$, over the parabolic cylinder $y = x^2, 0 \le x \le 2, 0 \le z \le 3$
2. **Circular cylinder** $G(x, y, z) = z$, over the cylindrical surface $y^2 + z^2 = 4, z \ge 0, 1 \le x \le 4$
3. **Sphere** $G(x, y, z) = x^2$, over the unit sphere $x^2 + y^2 + z^2 = 1$
4. **Hemisphere** $G(x, y, z) = z^2$, over the hemisphere $x^2 + y^2 + z^2 = a^2, z \ge 0$
5. **Portion of plane** $F(x, y, z) = z$, over the portion of the plane $x + y + z = 4$ that lies above the square $0 \le x \le 1, 0 \le y \le 1$, in the xy-plane
6. **Cone** $F(x, y, z) = z - x$, over the cone $z = \sqrt{x^2 + y^2}, 0 \le z \le 1$

7. Parabolic dome $H(x, y, z) = x^2\sqrt{5 - 4z}$, over the parabolic dome $z = 1 - x^2 - y^2, z \geq 0$

8. Spherical cap $H(x, y, z) = yz$, over the part of the sphere $x^2 + y^2 + z^2 = 4$ that lies above the cone $z = \sqrt{x^2 + y^2}$

9. Integrate $G(x, y, z) = x + y + z$ over the surface of the cube cut from the first octant by the planes $x = a, y = a, z = a$.

10. Integrate $G(x, y, z) = y + z$ over the surface of the wedge in the first octant bounded by the coordinate planes and the planes $x = 2$ and $y + z = 1$.

11. Integrate $G(x, y, z) = xyz$ over the surface of the rectangular solid cut from the first octant by the planes $x = a, y = b$, and $z = c$.

12. Integrate $G(x, y, z) = xyz$ over the surface of the rectangular solid bounded by the planes $x = \pm a, y = \pm b$, and $z = \pm c$.

13. Integrate $G(x, y, z) = x + y + z$ over the portion of the plane $2x + 2y + z = 2$ that lies in the first octant.

14. Integrate $G(x, y, z) = x\sqrt{y^2 + 4}$ over the surface cut from the parabolic cylinder $y^2 + 4z = 16$ by the planes $x = 0, x = 1$, and $z = 0$.

In Exercises 15–24, use a parametrization to find the flux $\iint_S \mathbf{F} \cdot \mathbf{n}\, d\sigma$ across the surface in the given direction.

15. Parabolic cylinder $\mathbf{F} = z^2\mathbf{i} + x\mathbf{j} - 3z\mathbf{k}$ outward (normal away from the x-axis) through the surface cut from the parabolic cylinder $z = 4 - y^2$ by the planes $x = 0, x = 1$, and $z = 0$

16. Parabolic cylinder $\mathbf{F} = x^2\mathbf{j} - xz\mathbf{k}$ outward (normal away from the yz-plane) through the surface cut from the parabolic cylinder $y = x^2, -1 \leq x \leq 1$, by the planes $z = 0$ and $z = 2$

17. Sphere $\mathbf{F} = z\mathbf{k}$ across the portion of the sphere $x^2 + y^2 + z^2 = a^2$ in the first octant in the direction away from the origin

18. Sphere $\mathbf{F} = x\mathbf{i} + y\mathbf{j} + z\mathbf{k}$ across the sphere $x^2 + y^2 + z^2 = a^2$ in the direction away from the origin

19. Plane $\mathbf{F} = 2xy\mathbf{i} + 2yz\mathbf{j} + 2xz\mathbf{k}$ upward across the portion of the plane $x + y + z = 2a$ that lies above the square $0 \leq x \leq a, 0 \leq y \leq a$, in the xy-plane

20. Cylinder $\mathbf{F} = x\mathbf{i} + y\mathbf{j} + z\mathbf{k}$ outward through the portion of the cylinder $x^2 + y^2 = 1$ cut by the planes $z = 0$ and $z = a$

21. Cone $\mathbf{F} = xy\mathbf{i} - z\mathbf{k}$ outward (normal away from the z-axis) through the cone $z = \sqrt{x^2 + y^2}, 0 \leq z \leq 1$

22. Cone $\mathbf{F} = y^2\mathbf{i} + xz\mathbf{j} - \mathbf{k}$ outward (normal away from the z-axis) through the cone $z = 2\sqrt{x^2 + y^2}, 0 \leq z \leq 2$

23. Cone frustum $\mathbf{F} = -x\mathbf{i} - y\mathbf{j} + z^2\mathbf{k}$ outward (normal away from the z-axis) through the portion of the cone $z = \sqrt{x^2 + y^2}$ between the planes $z = 1$ and $z = 2$

24. Paraboloid $\mathbf{F} = 4x\mathbf{i} + 4y\mathbf{j} + 2\mathbf{k}$ outward (normal away from the z-axis) through the surface cut from the bottom of the paraboloid $z = x^2 + y^2$ by the plane $z = 1$

In Exercises 25 and 26, find the flux of the field $\mathbf{F}$ across the portion of the given surface in the specified direction.

25. $\mathbf{F}(x, y, z) = -\mathbf{i} + 2\mathbf{j} + 3\mathbf{k}$

S: rectangular surface $z = 0$, $0 \leq x \leq 2$, $0 \leq y \leq 3$, direction $\mathbf{k}$

26. $\mathbf{F}(x, y, z) = yx^2\mathbf{i} - 2\mathbf{j} + xz\mathbf{k}$

S: rectangular surface $y = 0$, $-1 \leq x \leq 2$, $2 \leq z \leq 7$, direction $-\mathbf{j}$

In Exercises 27–32, find the flux of the field $\mathbf{F}$ across the portion of the sphere $x^2 + y^2 + z^2 = a^2$ in the first octant in the direction away from the origin.

27. $\mathbf{F}(x, y, z) = z\mathbf{k}$

28. $\mathbf{F}(x, y, z) = -y\mathbf{i} + x\mathbf{j}$

29. $\mathbf{F}(x, y, z) = y\mathbf{i} - x\mathbf{j} + \mathbf{k}$

30. $\mathbf{F}(x, y, z) = zx\mathbf{i} + zy\mathbf{j} + z^2\mathbf{k}$

31. $\mathbf{F}(x, y, z) = x\mathbf{i} + y\mathbf{j} + z\mathbf{k}$

32. $\mathbf{F}(x, y, z) = \dfrac{x\mathbf{i} + y\mathbf{j} + z\mathbf{k}}{\sqrt{x^2 + y^2 + z^2}}$

33. Find the flux of the field $\mathbf{F}(x, y, z) = z^2\mathbf{i} + x\mathbf{j} - 3z\mathbf{k}$ outward through the surface cut from the parabolic cylinder $z = 4 - y^2$ by the planes $x = 0, x = 1$, and $z = 0$.

34. Find the flux of the field $\mathbf{F}(x, y, z) = 4x\mathbf{i} + 4y\mathbf{j} + 2\mathbf{k}$ outward (away from the z-axis) through the surface cut from the bottom of the paraboloid $z = x^2 + y^2$ by the plane $z = 1$.

35. Let S be the portion of the cylinder $y = e^x$ in the first octant that projects parallel to the x-axis onto the rectangle R_{yz}: $1 \leq y \leq 2$, $0 \leq z \leq 1$ in the yz-plane (see the accompanying figure). Let $\mathbf{n}$ be the unit vector normal to S that points away from the yz-plane. Find the flux of the field $\mathbf{F}(x, y, z) = -2\mathbf{i} + 2y\mathbf{j} + z\mathbf{k}$ across S in the direction of $\mathbf{n}$.

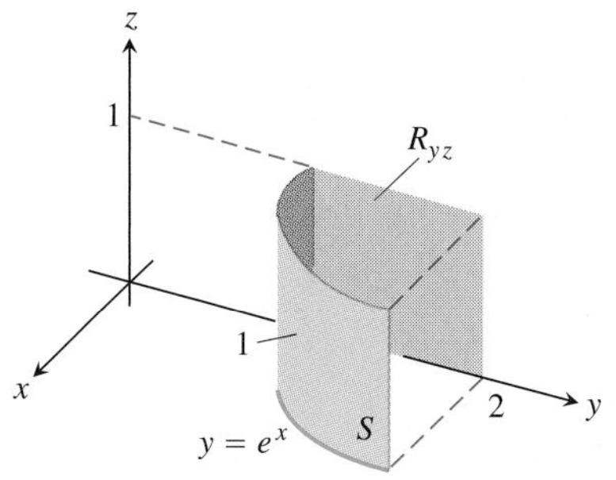

36. Let S be the portion of the cylinder $y = \ln x$ in the first octant whose projection parallel to the y-axis onto the xz-plane is the rectangle R_{xz}: $1 \leq x \leq e, 0 \leq z \leq 1$. Let $\mathbf{n}$ be the unit vector normal to S that points away from the xz-plane. Find the flux of $\mathbf{F} = 2y\mathbf{j} + z\mathbf{k}$ through S in the direction of $\mathbf{n}$.

37. Find the outward flux of the field $\mathbf{F} = 2xy\mathbf{i} + 2yz\mathbf{j} + 2xz\mathbf{k}$ across the surface of the cube cut from the first octant by the planes $x = a, y = a, z = a$.

38. Find the outward flux of the field $\mathbf{F} = xz\mathbf{i} + yz\mathbf{j} + \mathbf{k}$ across the surface of the upper cap cut from the solid sphere $x^2 + y^2 + z^2 \leq 25$ by the plane $z = 3$.

39. Centroid Find the centroid of the portion of the sphere $x^2 + y^2 + z^2 = a^2$ that lies in the first octant.

40. Thin shell of constant density Find the center of mass and the moment of inertia about the z-axis of a thin shell of constant density δ cut from the cone $x^2 + y^2 - z^2 = 0$ by the planes $z = 1$ and $z = 2$.

41. Spherical shells Find the moment of inertia about a diameter of a thin spherical shell of radius a and constant density δ. (Work with a hemispherical shell and double the result.)

42. Conical Surface Find the centroid of the lateral surface of a solid cone of base radius a and height h (cone surface minus the base).

13.7 Stokes' Theorem

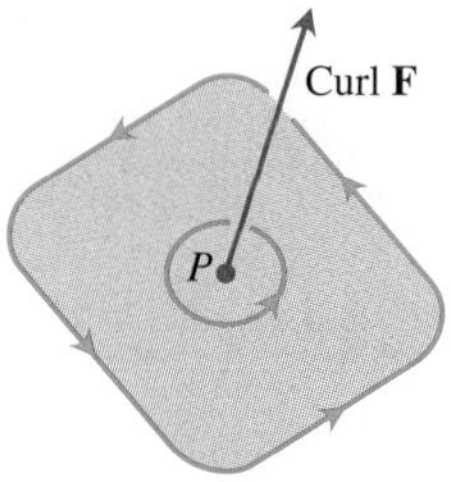

FIGURE 13.50 The circulation vector at a point P in a plane in a three-dimensional fluid flow. Notice its right-hand relation to the circulation line.

As we saw in Section 13.4, the circulation density or curl component of a two-dimensional field $\mathbf{F} = M\mathbf{i} + N\mathbf{j}$ at a point (x, y) is described by the scalar quantity $(\partial N/\partial x - \partial M/\partial y)$. In three dimensions, the circulation around a point P in a plane is described with a vector. This vector is normal to the plane of the circulation (Figure 13.50) and points in the direction that gives it a right-hand relation to the circulation line. The length of the vector gives the rate of the fluid's rotation, which usually varies as the circulation plane is tilted about P. It turns out that the vector of greatest circulation in a flow with velocity field $\mathbf{F} = M\mathbf{i} + N\mathbf{j} + P\mathbf{k}$ is the **curl vector**

$$\text{curl }\mathbf{F} = \left(\frac{\partial P}{\partial y} - \frac{\partial N}{\partial z}\right)\mathbf{i} + \left(\frac{\partial M}{\partial z} - \frac{\partial P}{\partial x}\right)\mathbf{j} + \left(\frac{\partial N}{\partial x} - \frac{\partial M}{\partial y}\right)\mathbf{k}. \tag{1}$$

We get this information from Stokes' Theorem, the generalization to space of the circulation-curl form of Green's Theorem.

Notice that $(\text{curl }\mathbf{F})\cdot\mathbf{k} = (\partial N/\partial x - \partial M/\partial y)$ is consistent with our definition in Section 13.4 when $\mathbf{F} = M(x, y)\mathbf{i} + N(x, y)\mathbf{j}$. The formula for curl $\mathbf{F}$ in Equation (1) is often written using the symbolic operator

$$\nabla = \mathbf{i}\frac{\partial}{\partial x} + \mathbf{j}\frac{\partial}{\partial y} + \mathbf{k}\frac{\partial}{\partial z}. \tag{2}$$

(The symbol ∇ is pronounced "del.") The curl of $\mathbf{F}$ is $\nabla \times \mathbf{F}$:

$$\begin{aligned}\nabla \times \mathbf{F} &= \begin{vmatrix} \mathbf{i} & \mathbf{j} & \mathbf{k} \\ \frac{\partial}{\partial x} & \frac{\partial}{\partial y} & \frac{\partial}{\partial z} \\ M & N & P \end{vmatrix} \\ &= \left(\frac{\partial P}{\partial y} - \frac{\partial N}{\partial z}\right)\mathbf{i} + \left(\frac{\partial M}{\partial z} - \frac{\partial P}{\partial x}\right)\mathbf{j} + \left(\frac{\partial N}{\partial x} - \frac{\partial M}{\partial y}\right)\mathbf{k} \\ &= \text{curl }\mathbf{F}.\end{aligned}$$

$$\text{curl }\mathbf{F} = \nabla \times \mathbf{F} \tag{3}$$

EXAMPLE 1 Find the curl of $\mathbf{F} = (x^2 - y)\mathbf{i} + 4z\mathbf{j} + x^2\mathbf{k}$.

Solution

$$\begin{aligned}
\text{curl } \mathbf{F} &= \nabla \times \mathbf{F} \qquad \text{Eq. (3)} \\
&= \begin{vmatrix} \mathbf{i} & \mathbf{j} & \mathbf{k} \\ \dfrac{\partial}{\partial x} & \dfrac{\partial}{\partial y} & \dfrac{\partial}{\partial z} \\ x^2 - y & 4z & x^2 \end{vmatrix} \\
&= \left(\frac{\partial}{\partial y}(x^2) - \frac{\partial}{\partial z}(4z)\right)\mathbf{i} - \left(\frac{\partial}{\partial x}(x^2) - \frac{\partial}{\partial z}(x^2 - y)\right)\mathbf{j} \\
&\quad + \left(\frac{\partial}{\partial x}(4z) - \frac{\partial}{\partial y}(x^2 - y)\right)\mathbf{k} \\
&= (0 - 4)\mathbf{i} - (2x - 0)\mathbf{j} + (0 + 1)\mathbf{k} \\
&= -4\mathbf{i} - 2x\mathbf{j} + \mathbf{k}
\end{aligned}$$

■

As we will see, the operator ∇ has a number of other applications. For instance, when applied to a scalar function $f(x, y, z)$, it gives the gradient of f:

$$\nabla f = \frac{\partial f}{\partial x}\mathbf{i} + \frac{\partial f}{\partial y}\mathbf{j} + \frac{\partial f}{\partial z}\mathbf{k}.$$

It is sometimes read as "del f" as well as "grad f."

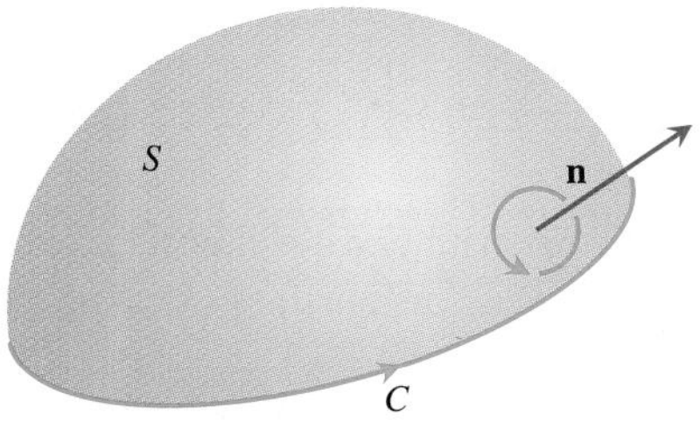

FIGURE 13.51 The orientation of the bounding curve C gives it a right-handed relation to the normal field $\mathbf{n}$. If the thumb of a right hand points along $\mathbf{n}$, the fingers curl in the direction of C.

Stokes' Theorem

Stokes' Theorem says that the circulation of a vector field around the boundary of an oriented surface in space in the direction counterclockwise with respect to the surface's unit normal vector $\mathbf{n}$ (Figure 13.51) equals the integral of the normal component of the curl of the field over the surface. We require that the surface be **piecewise smooth**, which means that it is a finite union of smooth surfaces joining along smooth curves.

THEOREM 6—Stokes' Theorem Let S be a piecewise smooth oriented surface having a piecewise smooth boundary curve C. Let $\mathbf{F} = M\mathbf{i} + N\mathbf{j} + P\mathbf{k}$ be a vector field whose components have continuous first partial derivatives on an open region containing S. Then the circulation of $\mathbf{F}$ around C in the direction counterclockwise with respect to the surface's unit normal vector $\mathbf{n}$ equals the integral of $\nabla \times \mathbf{F} \cdot \mathbf{n}$ over S.

$$\underbrace{\oint_C \mathbf{F} \cdot d\mathbf{r}}_{\text{Counterclockwise circulation}} = \underbrace{\iint_S \nabla \times \mathbf{F} \cdot \mathbf{n}\, d\sigma}_{\text{Curl integral}} \qquad (4)$$

Notice from Equation (4) that if two different oriented surfaces S_1 and S_2 have the same boundary C, their curl integrals are equal:

$$\iint_{S_1} \nabla \times \mathbf{F} \cdot \mathbf{n}_1 \, d\sigma = \iint_{S_2} \nabla \times \mathbf{F} \cdot \mathbf{n}_2 \, d\sigma.$$

Both curl integrals equal the counterclockwise circulation integral on the left side of Equation (4) as long as the unit normal vectors $\mathbf{n}_1$ and $\mathbf{n}_2$ correctly orient the surfaces.

If C is a curve in the xy-plane, oriented counterclockwise, and R is the region in the xy-plane bounded by C, then $d\sigma = dx\,dy$ and

$$(\nabla \times \mathbf{F}) \cdot \mathbf{n} = (\nabla \times \mathbf{F}) \cdot \mathbf{k} = \left(\frac{\partial N}{\partial x} - \frac{\partial M}{\partial y}\right).$$

Under these conditions, Stokes' equation becomes

$$\oint_C \mathbf{F} \cdot d\mathbf{r} = \iint_R \left(\frac{\partial N}{\partial x} - \frac{\partial M}{\partial y}\right) dx\,dy,$$

which is the circulation-curl form of the equation in Green's Theorem. Conversely, by reversing these steps we can rewrite the circulation-curl form of Green's Theorem for two-dimensional fields in del notation as

$$\oint_C \mathbf{F} \cdot d\mathbf{r} = \iint_R \nabla \times \mathbf{F} \cdot \mathbf{k}\, dA. \tag{5}$$

See Figure 13.52.

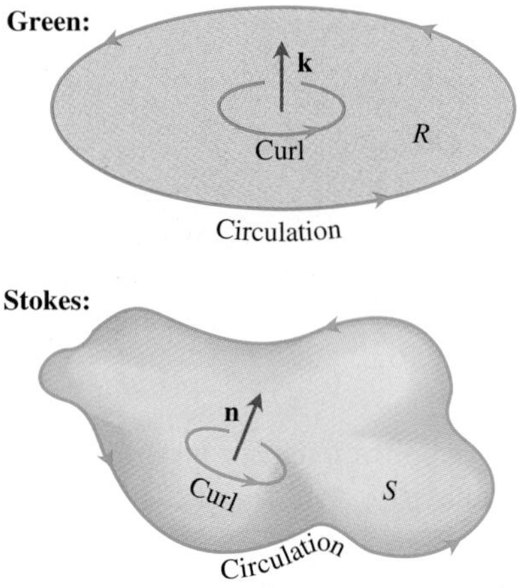

FIGURE 13.52 Comparison of Green's Theorem and Stokes' Theorem.

EXAMPLE 2 Evaluate Equation (4) for the hemisphere S: $x^2 + y^2 + z^2 = 9, z \geq 0$, its bounding circle C: $x^2 + y^2 = 9, z = 0$, and the field $\mathbf{F} = y\mathbf{i} - x\mathbf{j}$.

Solution We calculate the counterclockwise circulation around C (as viewed from above) using the parametrization $\mathbf{r}(\theta) = (3\cos\theta)\mathbf{i} + (3\sin\theta)\mathbf{j}, 0 \leq \theta \leq 2\pi$:

$$d\mathbf{r} = (-3\sin\theta\, d\theta)\mathbf{i} + (3\cos\theta\, d\theta)\mathbf{j}$$

$$\mathbf{F} = y\mathbf{i} - x\mathbf{j} = (3\sin\theta)\mathbf{i} - (3\cos\theta)\mathbf{j}$$

$$\mathbf{F} \cdot d\mathbf{r} = -9\sin^2\theta\, d\theta - 9\cos^2\theta\, d\theta = -9\, d\theta$$

$$\oint_C \mathbf{F} \cdot d\mathbf{r} = \int_0^{2\pi} -9\, d\theta = -18\pi.$$

For the curl integral of $\mathbf{F}$, we have

$$\nabla \times \mathbf{F} = \left(\frac{\partial P}{\partial y} - \frac{\partial N}{\partial z}\right)\mathbf{i} + \left(\frac{\partial M}{\partial z} - \frac{\partial P}{\partial x}\right)\mathbf{j} + \left(\frac{\partial N}{\partial x} - \frac{\partial M}{\partial y}\right)\mathbf{k}$$

$$= (0 - 0)\mathbf{i} + (0 - 0)\mathbf{j} + (-1 - 1)\mathbf{k} = -2\mathbf{k}$$

$$\mathbf{n} = \frac{x\mathbf{i} + y\mathbf{j} + z\mathbf{k}}{\sqrt{x^2 + y^2 + z^2}} = \frac{x\mathbf{i} + y\mathbf{j} + z\mathbf{k}}{3} \qquad \text{Outer unit normal}$$

$$d\sigma = \frac{3}{z}\, dA \qquad \text{Section 13.6, Example 5, with } a = 3$$

$$\nabla \times \mathbf{F} \cdot \mathbf{n}\, d\sigma = -\frac{2z}{3}\frac{3}{z}\, dA = -2\, dA$$

and

$$\iint_S \nabla \times \mathbf{F} \cdot \mathbf{n}\, d\sigma = \iint_{x^2+y^2 \le 9} -2\, dA = -18\pi.$$

The circulation around the circle equals the integral of the curl over the hemisphere, as it should. ■

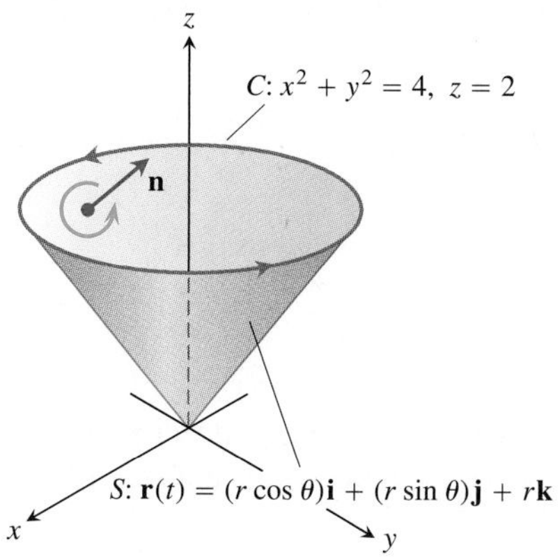

FIGURE 13.53 The curve C and cone S in Example 3.

EXAMPLE 3 Find the circulation of the field $\mathbf{F} = (x^2 - y)\mathbf{i} + 4z\mathbf{j} + x^2\mathbf{k}$ around the curve C in which the plane $z = 2$ meets the cone $z = \sqrt{x^2 + y^2}$, counterclockwise as viewed from above (Figure 13.53).

Solution Stokes' Theorem enables us to find the circulation by integrating over the surface of the cone. Traversing C in the counterclockwise direction viewed from above corresponds to taking the *inner* normal $\mathbf{n}$ to the cone, the normal with a positive z-component.

We parametrize the cone as

$$\mathbf{r}(r, \theta) = (r\cos\theta)\mathbf{i} + (r\sin\theta)\mathbf{j} + r\mathbf{k}, \qquad 0 \le r \le 2, \quad 0 \le \theta \le 2\pi.$$

We then have

$$\mathbf{n} = \frac{\mathbf{r}_r \times \mathbf{r}_\theta}{|\mathbf{r}_r \times \mathbf{r}_\theta|} = \frac{-(r\cos\theta)\mathbf{i} - (r\sin\theta)\mathbf{j} + r\mathbf{k}}{r\sqrt{2}} \qquad \text{Section 13.5, Example 4}$$

$$= \frac{1}{\sqrt{2}}\left(-(\cos\theta)\mathbf{i} - (\sin\theta)\mathbf{j} + \mathbf{k}\right)$$

$$d\sigma = r\sqrt{2}\, dr\, d\theta \qquad \text{Section 13.5, Example 4}$$

$$\nabla \times \mathbf{F} = -4\mathbf{i} - 2x\mathbf{j} + \mathbf{k} \qquad \text{Example 1}$$

$$= -4\mathbf{i} - 2r\cos\theta\mathbf{j} + \mathbf{k}. \qquad x = r\cos\theta$$

Accordingly,

$$\nabla \times \mathbf{F} \cdot \mathbf{n} = \frac{1}{\sqrt{2}}\left(4\cos\theta + 2r\cos\theta\sin\theta + 1\right)$$

$$= \frac{1}{\sqrt{2}}\left(4\cos\theta + r\sin 2\theta + 1\right)$$

and the circulation is

$$\oint_C \mathbf{F} \cdot d\mathbf{r} = \iint_S \nabla \times \mathbf{F} \cdot \mathbf{n}\, d\sigma \qquad \text{Stokes' Theorem, Eq. (4)}$$

$$= \int_0^{2\pi}\int_0^2 \frac{1}{\sqrt{2}}\left(4\cos\theta + r\sin 2\theta + 1\right)\left(r\sqrt{2}\, dr\, d\theta\right) = 4\pi.$$ ■

Paddle Wheel Interpretation of $\nabla \times \mathbf{F}$

Suppose that $\mathbf{v}(x, y, z)$ is the velocity of a moving fluid whose density at (x, y, z) is $\delta(x, y, z)$ and let $\mathbf{F} = \delta\mathbf{v}$. Then

$$\oint_C \mathbf{F} \cdot d\mathbf{r}$$

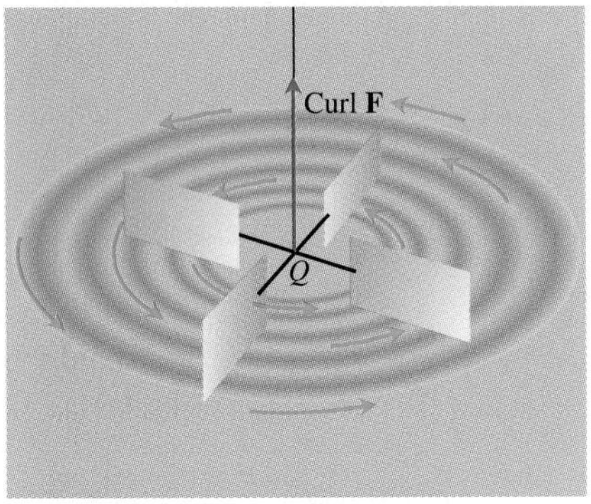

FIGURE 13.54 The paddle wheel interpretation of curl **F**.

is the circulation of the fluid around the closed curve C. By Stokes' Theorem, the circulation is equal to the flux of $\nabla \times \mathbf{F}$ through a surface S spanning C:

$$\oint_C \mathbf{F} \cdot d\mathbf{r} = \iint_S \nabla \times \mathbf{F} \cdot \mathbf{n}\, d\sigma.$$

Suppose we fix a point Q in the domain of $\mathbf{F}$ and a direction $\mathbf{u}$ at Q. Let C be a circle of radius ρ, with center at Q, whose plane is normal to $\mathbf{u}$. If $\nabla \times \mathbf{F}$ is continuous at Q, the average value of the $\mathbf{u}$-component of $\nabla \times \mathbf{F}$ over the circular disk S bounded by C approaches the $\mathbf{u}$-component of $\nabla \times \mathbf{F}$ at Q as $\rho \to 0$:

$$(\nabla \times \mathbf{F} \cdot \mathbf{u})_Q = \lim_{\rho \to 0} \frac{1}{\pi \rho^2} \iint_S \nabla \times \mathbf{F} \cdot \mathbf{u}\, d\sigma.$$

If we replace the surface integral in this last equation by the circulation, we get

$$(\nabla \times \mathbf{F} \cdot \mathbf{u})_Q = \lim_{\rho \to 0} \frac{1}{\pi \rho^2} \oint_C \mathbf{F} \cdot d\mathbf{r}. \tag{6}$$

The left-hand side of Equation (6) has its maximum value when $\mathbf{u}$ is the direction of $\nabla \times \mathbf{F}$. When ρ is small, the limit on the right-hand side of Equation (6) is approximately

$$\frac{1}{\pi \rho^2} \oint_C \mathbf{F} \cdot d\mathbf{r},$$

which is the circulation around C divided by the area of the disk (circulation density). Suppose that a small paddle wheel of radius ρ is introduced into the fluid at Q, with its axle directed along $\mathbf{u}$. The circulation of the fluid around C will affect the rate of spin of the paddle wheel. The wheel will spin fastest when the circulation integral is maximized; therefore it will spin fastest when the axle of the paddle wheel points in the direction of $\nabla \times \mathbf{F}$ (Figure 13.54).

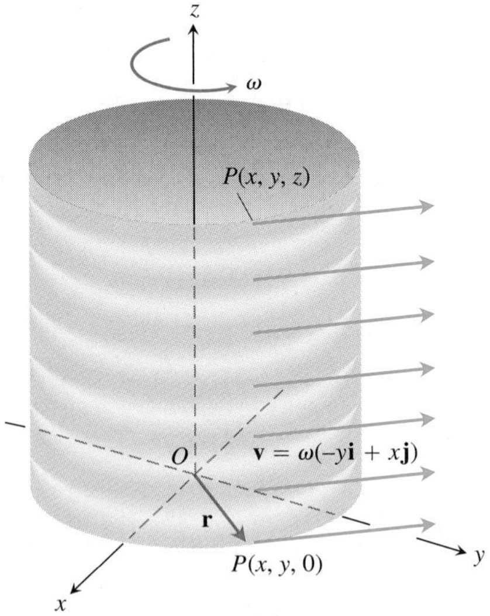

FIGURE 13.55 A steady rotational flow parallel to the xy-plane, with constant angular velocity ω in the positive (counterclockwise) direction (Example 4).

EXAMPLE 4 A fluid of constant density rotates around the z-axis with velocity $\mathbf{v} = \omega(-y\mathbf{i} + x\mathbf{j})$, where ω is a positive constant called the *angular velocity* of the rotation (Figure 13.55). If $\mathbf{F} = \mathbf{v}$, find $\nabla \times \mathbf{F}$ and relate it to the circulation density.

Solution With $\mathbf{F} = \mathbf{v} = -\omega y\mathbf{i} + \omega x\mathbf{j}$,

$$\begin{aligned}\nabla \times \mathbf{F} &= \left(\frac{\partial P}{\partial y} - \frac{\partial N}{\partial z}\right)\mathbf{i} + \left(\frac{\partial M}{\partial z} - \frac{\partial P}{\partial x}\right)\mathbf{j} + \left(\frac{\partial N}{\partial x} - \frac{\partial M}{\partial y}\right)\mathbf{k}\\ &= (0 - 0)\mathbf{i} + (0 - 0)\mathbf{j} + (\omega - (-\omega))\mathbf{k} = 2\omega\mathbf{k}.\end{aligned}$$

By Stokes' Theorem, the circulation of $\mathbf{F}$ around a circle C of radius ρ bounding a disk S in a plane normal to $\nabla \times \mathbf{F}$, say the xy-plane, is

$$\oint_C \mathbf{F} \cdot d\mathbf{r} = \iint_S \nabla \times \mathbf{F} \cdot \mathbf{n}\, d\sigma = \iint_S 2\omega\mathbf{k} \cdot \mathbf{k}\, dx\, dy = (2\omega)(\pi \rho^2).$$

Thus,

$$(\nabla \times \mathbf{F}) \cdot \mathbf{k} = 2\omega = \frac{1}{\pi \rho^2} \oint_C \mathbf{F} \cdot d\mathbf{r},$$

consistent with Equation (6) when $\mathbf{u} = \mathbf{k}$. ■

EXAMPLE 5 Use Stokes' Theorem to evaluate $\int_C \mathbf{F}\cdot d\mathbf{r}$, if $\mathbf{F} = xz\mathbf{i} + xy\mathbf{j} + 3xz\mathbf{k}$ and C is the boundary of the portion of the plane $2x + y + z = 2$ in the first octant, traversed counterclockwise as viewed from above (Figure 13.56).

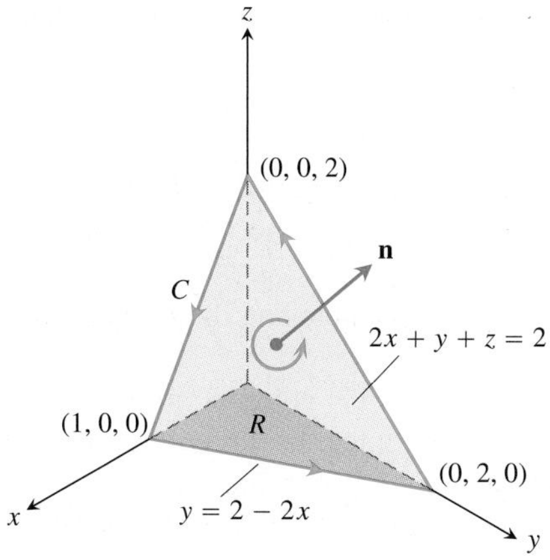

FIGURE 13.56 The planar surface in Example 5.

Solution The plane is the level surface $f(x, y, z) = 2$ of the function $f(x, y, z) = 2x + y + z$. The unit normal vector

$$\mathbf{n} = \frac{\nabla f}{|\nabla f|} = \frac{(2\mathbf{i} + \mathbf{j} + \mathbf{k})}{|2\mathbf{i} + \mathbf{j} + \mathbf{k}|} = \frac{1}{\sqrt{6}}\left(2\mathbf{i} + \mathbf{j} + \mathbf{k}\right)$$

is consistent with the counterclockwise motion around C. To apply Stokes' Theorem, we find

$$\operatorname{curl}\mathbf{F} = \nabla\times\mathbf{F} = \begin{vmatrix} \mathbf{i} & \mathbf{j} & \mathbf{k} \\ \dfrac{\partial}{\partial x} & \dfrac{\partial}{\partial y} & \dfrac{\partial}{\partial z} \\ xz & xy & 3xz \end{vmatrix} = (x - 3z)\mathbf{j} + y\mathbf{k}.$$

On the plane, z equals $2 - 2x - y$, so

$$\nabla\times\mathbf{F} = (x - 3(2 - 2x - y))\mathbf{j} + y\mathbf{k} = (7x + 3y - 6)\mathbf{j} + y\mathbf{k}$$

and

$$\nabla\times\mathbf{F}\cdot\mathbf{n} = \frac{1}{\sqrt{6}}\left(7x + 3y - 6 + y\right) = \frac{1}{\sqrt{6}}\left(7x + 4y - 6\right).$$

The surface area element is

$$d\sigma = \frac{|\nabla f|}{|\nabla f\cdot\mathbf{k}|}\,dA = \frac{\sqrt{6}}{1}\,dx\,dy.$$

The circulation is

$$\oint_C \mathbf{F}\cdot d\mathbf{r} = \iint_S \nabla\times\mathbf{F}\cdot\mathbf{n}\,d\sigma \qquad \text{Stokes' Theorem, Eq. (4)}$$

$$= \int_0^1\int_0^{2-2x} \frac{1}{\sqrt{6}}\left(7x + 4y - 6\right)\sqrt{6}\,dy\,dx$$

$$= \int_0^1\int_0^{2-2x} (7x + 4y - 6)\,dy\,dx = -1.$$ ■

Proof of Stokes' Theorem for Polyhedral Surfaces

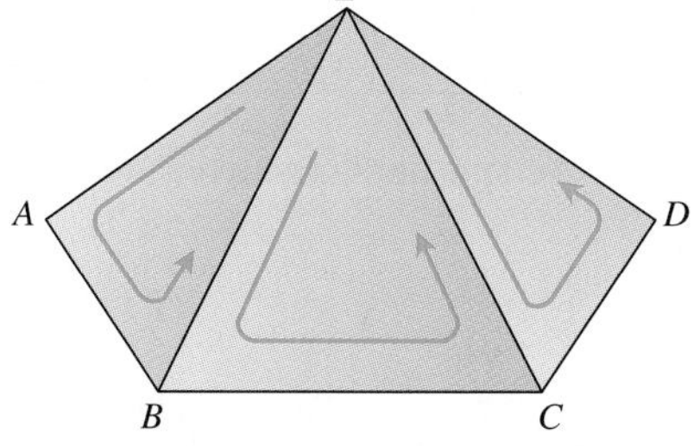

FIGURE 13.57 Part of a polyhedral surface.

Let S be a polyhedral surface consisting of a finite number of plane regions. (See Figure 13.57 for an example.) We apply Green's Theorem to each separate panel of S. There are two types of panels:

1. Those that are surrounded on all sides by other panels
2. Those that have one or more edges that are not adjacent to other panels.

The boundary Δ of S consists of those edges of the type 2 panels that are not adjacent to other panels. In Figure 13.57, the triangles EAB, BCE, and CDE represent a part of S, with

ABCD part of the boundary Δ. We apply a generalized tangential form of Green's Theorem to the three triangles in turn and add the results to get

$$\left(\oint_{EAB} + \oint_{BCE} + \oint_{CDE}\right)\mathbf{F}\cdot d\mathbf{r} = \left(\iint_{EAB} + \iint_{BCE} + \iint_{CDE}\right)\nabla\times\mathbf{F}\cdot\mathbf{n}\,d\sigma. \tag{7}$$

In the generalized form, the line integral of **F** around the curve enclosing the plane region R normal to **n** equals the double integral of (curl **F**) · **n** over R.

The three line integrals on the left-hand side of Equation (7) combine into a single line integral taken around the periphery *ABCDE* because the integrals along interior segments cancel in pairs. For example, the integral along segment *BE* in triangle *ABE* is opposite in sign to the integral along the same segment in triangle *EBC*. The same holds for segment *CE*. Hence, Equation (7) reduces to

$$\oint_{ABCDE} \mathbf{F}\cdot d\mathbf{r} = \iint_{ABCDE} \nabla\times\mathbf{F}\cdot\mathbf{n}\,d\sigma.$$

When we apply the generalized form of Green's Theorem to all the panels and add the results, we get

$$\oint_{\Delta} \mathbf{F}\cdot d\mathbf{r} = \iint_{S} \nabla\times\mathbf{F}\cdot\mathbf{n}\,d\sigma.$$

This is Stokes' Theorem for a polyhedral surface S. You can find proofs for more general surfaces in advanced calculus texts. (More general surfaces are limits of polyhedral surfaces and the proof can be extended.)

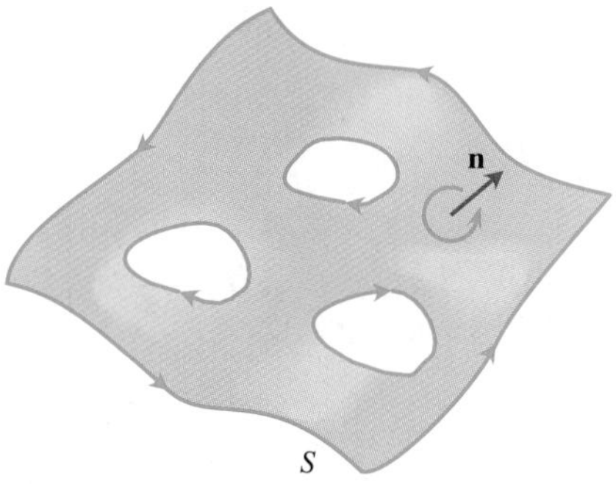

FIGURE 13.58 Stokes' Theorem also holds for oriented surfaces with holes.

Stokes' Theorem for Surfaces with Holes

Stokes' Theorem holds for an oriented surface S that has one or more holes (Figure 13.58). The surface integral over S of the normal component of $\nabla\times\mathbf{F}$ equals the sum of the line integrals around all the boundary curves of the tangential component of **F**, where the curves are to be traced in the direction induced by the orientation of S.

An Important Identity

The following identity arises frequently in mathematics and the physical sciences.

$$\text{curl grad } f = \mathbf{0} \qquad \text{or} \qquad \nabla\times\nabla f = \mathbf{0} \tag{8}$$

This identity holds for any function $f(x, y, z)$ whose second partial derivatives are continuous. The proof goes like this:

$$\nabla\times\nabla f = \begin{vmatrix} \mathbf{i} & \mathbf{j} & \mathbf{k} \\ \dfrac{\partial}{\partial x} & \dfrac{\partial}{\partial y} & \dfrac{\partial}{\partial z} \\ \dfrac{\partial f}{\partial x} & \dfrac{\partial f}{\partial y} & \dfrac{\partial f}{\partial z} \end{vmatrix} = (f_{zy} - f_{yz})\mathbf{i} - (f_{zx} - f_{xz})\mathbf{j} + (f_{yx} - f_{xy})\mathbf{k}.$$

If the second partial derivatives are continuous, the mixed second derivatives in parentheses are equal (Theorem 2, Section 11.3) and the vector is zero.

Conservative Fields and Stokes' Theorem

In Section 13.3, we found that a field $\mathbf{F}$ being conservative in an open region D in space is equivalent to the integral of $\mathbf{F}$ around every closed loop in D being zero. This, in turn, is equivalent in *simply connected* open regions to saying that $\nabla \times \mathbf{F} = \mathbf{0}$.

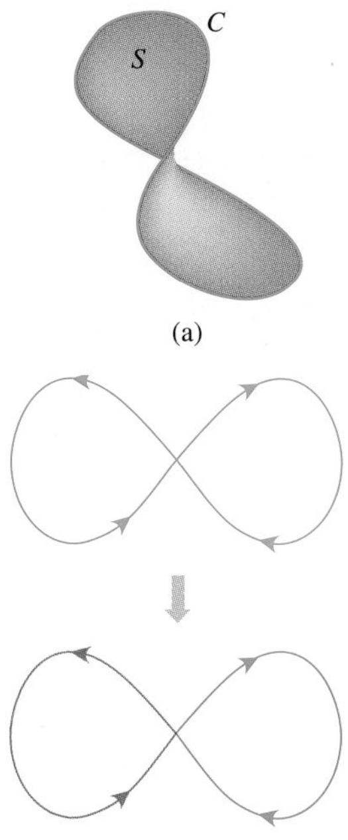

FIGURE 13.59 (a) In a simply connected open region in space, a simple closed curve C is the boundary of a smooth surface S. (b) Smooth curves that cross themselves can be divided into loops to which Stokes' Theorem applies.

THEOREM 7—Curl F = 0 Related to the Closed-Loop Property If $\nabla \times \mathbf{F} = \mathbf{0}$ at every point of a simply connected open region D in space, then on any piecewise-smooth closed path C in D,

$$\oint_C \mathbf{F} \cdot d\mathbf{r} = 0.$$

Sketch of a Proof Theorem 7 is usually proved in two steps. The first step is for simple closed curves (loops that do not cross themselves), like the one in Figure 13.59a. A theorem from topology, a branch of advanced mathematics, states that every smooth simple closed curve C in a simply connected open region D is the boundary of a smooth two-sided surface S that also lies in D. Hence, by Stokes' Theorem,

$$\oint_C \mathbf{F} \cdot d\mathbf{r} = \iint_S \nabla \times \mathbf{F} \cdot \mathbf{n}\, d\sigma = 0.$$

The second step is for curves that cross themselves, like the one in Figure 13.59b. The idea is to break these into simple loops spanned by orientable surfaces, apply Stokes' Theorem one loop at a time, and add the results. ■

The following diagram summarizes the results for conservative fields defined on connected, simply connected open regions. Therefore the four properties at the corners of the diagram are mathematically equivalent.

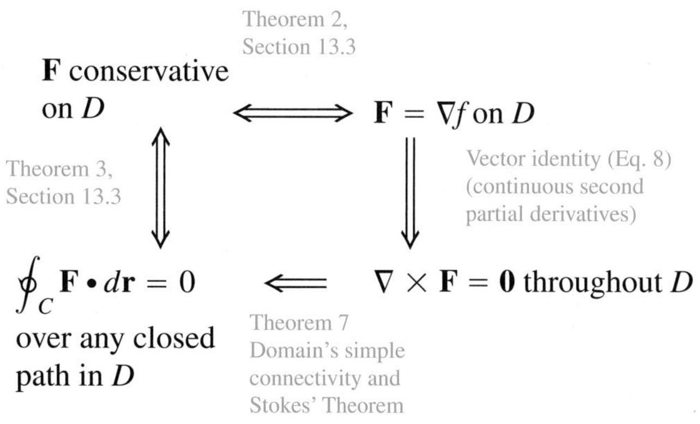

EXERCISES 13.7

In Exercises 1–6, use the surface integral in Stokes' Theorem to calculate the circulation of the field $\mathbf{F}$ around the curve C in the indicated direction.

1. $\mathbf{F} = x^2\mathbf{i} + 2x\mathbf{j} + z^2\mathbf{k}$
   C: The ellipse $4x^2 + y^2 = 4$ in the xy-plane, counterclockwise when viewed from above
2. $\mathbf{F} = 2y\mathbf{i} + 3x\mathbf{j} - z^2\mathbf{k}$
   C: The circle $x^2 + y^2 = 9$ in the xy-plane, counterclockwise when viewed from above
3. $\mathbf{F} = y\mathbf{i} + xz\mathbf{j} + x^2\mathbf{k}$
   C: The boundary of the triangle cut from the plane $x + y + z = 1$ by the first octant, counterclockwise when viewed from above
4. $\mathbf{F} = (y^2 + z^2)\mathbf{i} + (x^2 + z^2)\mathbf{j} + (x^2 + y^2)\mathbf{k}$
   C: The boundary of the triangle cut from the plane $x + y + z = 1$ by the first octant, counterclockwise when viewed from above
5. $\mathbf{F} = (y^2 + z^2)\mathbf{i} + (x^2 + y^2)\mathbf{j} + (x^2 + y^2)\mathbf{k}$
   C: The square bounded by the lines $x = \pm 1$ and $y = \pm 1$ in the xy-plane, counterclockwise when viewed from above
6. $\mathbf{F} = x^2y^3\mathbf{i} + \mathbf{j} + z\mathbf{k}$
   C: The intersection of the cylinder $x^2 + y^2 = 4$ and the hemisphere $x^2 + y^2 + z^2 = 16, z \geq 0$, counterclockwise when viewed from above.
7. Let $\mathbf{n}$ be the outer unit normal of the elliptical shell
   $$S:\quad 4x^2 + 9y^2 + 36z^2 = 36, \qquad z \geq 0,$$
   and let
   $$\mathbf{F} = y\mathbf{i} + x^2\mathbf{j} + (x^2 + y^4)^{3/2}\sin e^{\sqrt{xyz}}\,\mathbf{k}.$$
   Find the value of
   $$\iint_S \nabla \times \mathbf{F} \cdot \mathbf{n}\, d\sigma.$$
   (*Hint:* One parametrization of the ellipse at the base of the shell is $x = 3\cos t,\ y = 2\sin t,\ 0 \leq t \leq 2\pi$.)
8. Let $\mathbf{n}$ be the outer unit normal (normal away from the origin) of the parabolic shell
   $$S:\quad 4x^2 + y + z^2 = 4, \qquad y \geq 0,$$
   and let
   $$\mathbf{F} = \left(-z + \frac{1}{2+x}\right)\mathbf{i} + (\tan^{-1} y)\mathbf{j} + \left(x + \frac{1}{4+z}\right)\mathbf{k}.$$
   Find the value of
   $$\iint_S \nabla \times \mathbf{F} \cdot \mathbf{n}\, d\sigma.$$
9. Let S be the cylinder $x^2 + y^2 = a^2, 0 \leq z \leq h$, together with its top, $x^2 + y^2 \leq a^2$, $z = h$. Let $\mathbf{F} = -y\mathbf{i} + x\mathbf{j} + x^2\mathbf{k}$. Use Stokes' Theorem to find the flux of $\nabla \times \mathbf{F}$ outward through S.
10. Evaluate
    $$\iint_S \nabla \times (y\mathbf{i}) \cdot \mathbf{n}\, d\sigma,$$
    where S is the hemisphere $x^2 + y^2 + z^2 = 1, z \geq 0$.
11. **Flux of curl F** Show that
    $$\iint_S \nabla \times \mathbf{F} \cdot \mathbf{n}\, d\sigma$$
    has the same value for all oriented surfaces S that span C and that induce the same positive direction on C.
12. Let $\mathbf{F}$ be a differentiable vector field defined on a region containing a smooth closed oriented surface S and its interior. Let $\mathbf{n}$ be the unit normal vector field on S. Suppose that S is the union of two surfaces S_1 and S_2 joined along a smooth simple closed curve C. Can anything be said about
    $$\iint_S \nabla \times \mathbf{F} \cdot \mathbf{n}\, d\sigma?$$
    Give reasons for your answer.

In Exercises 13–18, use the surface integral in Stokes' Theorem to calculate the flux of the curl of the field $\mathbf{F}$ across the surface S in the direction of the outward unit normal $\mathbf{n}$.

13. $\mathbf{F} = 2z\mathbf{i} + 3x\mathbf{j} + 5y\mathbf{k}$
    S: $\mathbf{r}(r, \theta) = (r\cos\theta)\mathbf{i} + (r\sin\theta)\mathbf{j} + (4 - r^2)\mathbf{k}$,
    $0 \leq r \leq 2,\quad 0 \leq \theta \leq 2\pi$
14. $\mathbf{F} = (y - z)\mathbf{i} + (z - x)\mathbf{j} + (x + z)\mathbf{k}$
    S: $\mathbf{r}(r, \theta) = (r\cos\theta)\mathbf{i} + (r\sin\theta)\mathbf{j} + (9 - r^2)\mathbf{k}$,
    $0 \leq r \leq 3,\quad 0 \leq \theta \leq 2\pi$
15. $\mathbf{F} = x^2y\mathbf{i} + 2y^3z\mathbf{j} + 3z\mathbf{k}$
    S: $\mathbf{r}(r, \theta) = (r\cos\theta)\mathbf{i} + (r\sin\theta)\mathbf{j} + r\mathbf{k}$,
    $0 \leq r \leq 1,\quad 0 \leq \theta \leq 2\pi$
16. $\mathbf{F} = (x - y)\mathbf{i} + (y - z)\mathbf{j} + (z - x)\mathbf{k}$
    S: $\mathbf{r}(r, \theta) = (r\cos\theta)\mathbf{i} + (r\sin\theta)\mathbf{j} + (5 - r)\mathbf{k}$,
    $0 \leq r \leq 5,\quad 0 \leq \theta \leq 2\pi$
17. $\mathbf{F} = 3y\mathbf{i} + (5 - 2x)\mathbf{j} + (z^2 - 2)\mathbf{k}$
    S: $\mathbf{r}(\phi, \theta) = \left(\sqrt{3}\sin\phi\cos\theta\right)\mathbf{i} + \left(\sqrt{3}\sin\phi\sin\theta\right)\mathbf{j} + \left(\sqrt{3}\cos\phi\right)\mathbf{k}$, $\quad 0 \leq \phi \leq \pi/2,\quad 0 \leq \theta \leq 2\pi$
18. $\mathbf{F} = y^2\mathbf{i} + z^2\mathbf{j} + x\mathbf{k}$
    S: $\mathbf{r}(\phi, \theta) = (2\sin\phi\cos\theta)\mathbf{i} + (2\sin\phi\sin\theta)\mathbf{j} + (2\cos\phi)\mathbf{k}$,
    $0 \leq \phi \leq \pi/2,\quad 0 \leq \theta \leq 2\pi$
19. **Zero circulation** Use the identity $\nabla \times \nabla f = \mathbf{0}$ (Equation (8) in the text) and Stokes' Theorem to show that the circulations of the following fields around the boundary of any smooth orientable surface in space are zero.
    a. $\mathbf{F} = 2x\mathbf{i} + 2y\mathbf{j} + 2z\mathbf{k}$ **b.** $\mathbf{F} = \nabla(xy^2z^3)$
    c. $\mathbf{F} = \nabla \times (x\mathbf{i} + y\mathbf{j} + z\mathbf{k})$ **d.** $\mathbf{F} = \nabla f$

20. Zero circulation Let $f(x, y, z) = (x^2 + y^2 + z^2)^{-1/2}$. Show that the clockwise circulation of the field $\mathbf{F} = \nabla f$ around the circle $x^2 + y^2 = a^2$ in the xy-plane is zero

a. by taking $\mathbf{r} = (a \cos t)\mathbf{i} + (a \sin t)\mathbf{j}$, $0 \le t \le 2\pi$, and integrating $\mathbf{F} \cdot d\mathbf{r}$ over the circle.

b. by applying Stokes' Theorem.

21. Let C be a simple closed smooth curve in the plane $2x + 2y + z = 2$, oriented as shown here. Show that

$$\oint_C 2y\,dx + 3z\,dy - x\,dz$$

depends only on the area of the region enclosed by C and not on the position or shape of C.

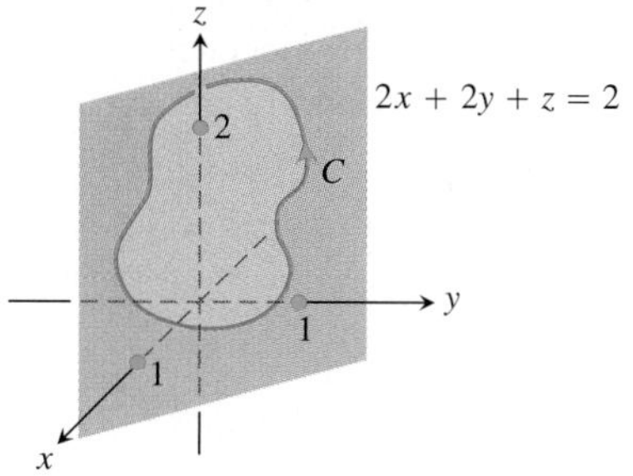

22. Show that if $\mathbf{F} = x\mathbf{i} + y\mathbf{j} + z\mathbf{k}$, then $\nabla \times \mathbf{F} = \mathbf{0}$.

23. Find a vector field with twice-differentiable components whose curl is $x\mathbf{i} + y\mathbf{j} + z\mathbf{k}$ or prove that no such field exists.

24. Does Stokes' Theorem say anything special about circulation in a field whose curl is zero? Give reasons for your answer.

25. Let R be a region in the xy-plane that is bounded by a piecewise-smooth simple closed curve C and suppose that the moments of inertia of R about the x- and y-axes are known to be I_x and I_y. Evaluate the integral

$$\oint_C \nabla(r^4) \cdot \mathbf{n}\,ds,$$

where $r = \sqrt{x^2 + y^2}$, in terms of I_x and I_y.

26. Zero curl, yet field not conservative Show that the curl of

$$\mathbf{F} = \frac{-y}{x^2 + y^2}\mathbf{i} + \frac{x}{x^2 + y^2}\mathbf{j} + z\mathbf{k}$$

is zero but that

$$\oint_C \mathbf{F} \cdot d\mathbf{r}$$

is not zero if C is the circle $x^2 + y^2 = 1$ in the xy-plane. (Theorem 7 does not apply here because the domain of $\mathbf{F}$ is not simply connected. The field $\mathbf{F}$ is not defined along the z-axis so there is no way to contract C to a point without leaving the domain of $\mathbf{F}$.)

13.8 The Divergence Theorem and a Unified Theory

The divergence form of Green's Theorem in the plane states that the net outward flux of a vector field across a simple closed curve can be calculated by integrating the divergence of the field over the region enclosed by the curve. The corresponding theorem in three dimensions, called the Divergence Theorem, states that the net outward flux of a vector field across a closed surface in space can be calculated by integrating the divergence of the field over the region enclosed by the surface. In this section, we prove the Divergence Theorem and show how it simplifies the calculation of flux. We also derive Gauss's law for flux in an electric field and the continuity equation of hydrodynamics. Finally, we unify the chapter's vector integral theorems into a single fundamental theorem.

Divergence in Three Dimensions

The **divergence** of a vector field $\mathbf{F} = M(x, y, z)\mathbf{i} + N(x, y, z)\mathbf{j} + P(x, y, z)\mathbf{k}$ is the scalar function

$$\text{div}\,\mathbf{F} = \nabla \cdot \mathbf{F} = \frac{\partial M}{\partial x} + \frac{\partial N}{\partial y} + \frac{\partial P}{\partial z}. \tag{1}$$

The symbol "div **F**" is read as "divergence of **F**" or "div **F**." The notation $\nabla \cdot \mathbf{F}$ is read "del dot **F**."

Div **F** has the same physical interpretation in three dimensions that it does in two. If **F** is the velocity field of a flowing gas, the value of div **F** at a point (x, y, z) is the rate at which the gas is compressing or expanding at (x, y, z). The divergence is the flux per unit volume or flux density at the point.

EXAMPLE 1 Find the divergence of $\mathbf{F} = 2xz\mathbf{i} - xy\mathbf{j} - z\mathbf{k}$.

Solution The divergence of $\mathbf{F}$ is

$$\nabla \cdot \mathbf{F} = \frac{\partial}{\partial x}(2xz) + \frac{\partial}{\partial y}(-xy) + \frac{\partial}{\partial z}(-z) = 2z - x - 1. \qquad \blacksquare$$

Divergence Theorem

The Divergence Theorem says that under suitable conditions, the outward flux of a vector field across a closed surface (oriented outward) equals the triple integral of the divergence of the field over the region enclosed by the surface.

> **THEOREM 8—Divergence Theorem** Let $\mathbf{F}$ be a vector field whose components have continuous first partial derivatives, and let S be a piecewise smooth oriented closed surface. The flux of $\mathbf{F}$ across S in the direction of the surface's outward unit normal field $\mathbf{n}$ equals the integral of $\nabla \cdot \mathbf{F}$ over the region D enclosed by the surface:
>
> $$\underbrace{\iint_S \mathbf{F} \cdot \mathbf{n}\, d\sigma}_{\text{Outward flux}} = \underbrace{\iiint_D \nabla \cdot \mathbf{F}\, dV}_{\text{Divergence integral}}. \qquad (2)$$

EXAMPLE 2 Evaluate both sides of Equation (2) for the field $\mathbf{F} = x\mathbf{i} + y\mathbf{j} + z\mathbf{k}$ over the sphere $x^2 + y^2 + z^2 = a^2$.

Solution The outer unit normal to S, calculated from the gradient of $f(x, y, z) = x^2 + y^2 + z^2 - a^2$, is

$$\mathbf{n} = \frac{2(x\mathbf{i} + y\mathbf{j} + z\mathbf{k})}{\sqrt{4(x^2 + y^2 + z^2)}} = \frac{x\mathbf{i} + y\mathbf{j} + z\mathbf{k}}{a}.$$

Hence,

$$\mathbf{F} \cdot \mathbf{n}\, d\sigma = \frac{x^2 + y^2 + z^2}{a}\, d\sigma = \frac{a^2}{a}\, d\sigma = a\, d\sigma$$

because $x^2 + y^2 + z^2 = a^2$ on the surface. Therefore,

$$\iint_S \mathbf{F} \cdot \mathbf{n}\, d\sigma = \iint_S a\, d\sigma = a \iint_S d\sigma = a(4\pi a^2) = 4\pi a^3.$$

The divergence of $\mathbf{F}$ is

$$\nabla \cdot \mathbf{F} = \frac{\partial}{\partial x}(x) + \frac{\partial}{\partial y}(y) + \frac{\partial}{\partial z}(z) = 3,$$

so

$$\iiint_D \nabla \cdot \mathbf{F}\, dV = \iiint_D 3\, dV = 3\left(\frac{4}{3}\pi a^3\right) = 4\pi a^3. \qquad \blacksquare$$

EXAMPLE 3 Find the flux of $\mathbf{F} = xy\mathbf{i} + yz\mathbf{j} + xz\mathbf{k}$ outward through the surface of the cube cut from the first octant by the planes $x = 1$, $y = 1$, and $z = 1$.

Solution Instead of calculating the flux as a sum of six separate integrals, one for each face of the cube, we can calculate the flux by integrating the divergence

$$\nabla \cdot \mathbf{F} = \frac{\partial}{\partial x}(xy) + \frac{\partial}{\partial y}(yz) + \frac{\partial}{\partial z}(xz) = y + z + x$$

over the cube's interior:

$$\text{Flux} = \iint_{\substack{\text{Cube} \\ \text{surface}}} \mathbf{F} \cdot \mathbf{n}\, d\sigma = \iiint_{\substack{\text{Cube} \\ \text{interior}}} \nabla \cdot \mathbf{F}\, dV \qquad \text{The Divergence Theorem}$$

$$= \int_0^1 \int_0^1 \int_0^1 (x + y + z)\, dx\, dy\, dz = \frac{3}{2}. \qquad \text{Routine integration}$$

■

Proof of the Divergence Theorem for Special Regions

To prove the Divergence Theorem, we assume that the components of $\mathbf{F}$ have continuous first partial derivatives. We also assume that D is a convex region with no holes or bubbles, such as a solid ball, cube, or ellipsoid, and that S is a piecewise smooth surface. In addition, we assume that any line perpendicular to the xy-plane at an interior point of the region R_{xy} that is the projection of D on the xy-plane intersects the surface S in exactly two points, producing surfaces

$$S_1: \quad z = f_1(x, y), \qquad (x, y) \text{ in } R_{xy}$$

$$S_2: \quad z = f_2(x, y), \qquad (x, y) \text{ in } R_{xy},$$

with $f_1 \le f_2$. We make similar assumptions about the projection of D onto the other coordinate planes. See Figure 13.60.

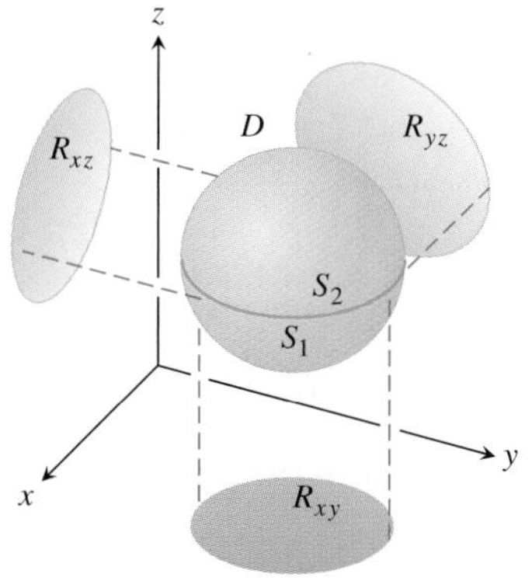

FIGURE 13.60 We prove the Divergence Theorem for the kind of three-dimensional region shown here.

The components of the unit normal vector $\mathbf{n} = n_1\mathbf{i} + n_2\mathbf{j} + n_3\mathbf{k}$ are the cosines of the angles α, β, and γ that $\mathbf{n}$ makes with $\mathbf{i}$, $\mathbf{j}$, and $\mathbf{k}$ (Figure 13.61). This is true because all the vectors involved are unit vectors. We have

$$n_1 = \mathbf{n} \cdot \mathbf{i} = |\mathbf{n}||\mathbf{i}| \cos \alpha = \cos \alpha$$

$$n_2 = \mathbf{n} \cdot \mathbf{j} = |\mathbf{n}||\mathbf{j}| \cos \beta = \cos \beta$$

$$n_3 = \mathbf{n} \cdot \mathbf{k} = |\mathbf{n}||\mathbf{k}| \cos \gamma = \cos \gamma$$

Thus,

$$\mathbf{n} = (\cos \alpha)\mathbf{i} + (\cos \beta)\mathbf{j} + (\cos \gamma)\mathbf{k}$$

and

$$\mathbf{F} \cdot \mathbf{n} = M \cos \alpha + N \cos \beta + P \cos \gamma.$$

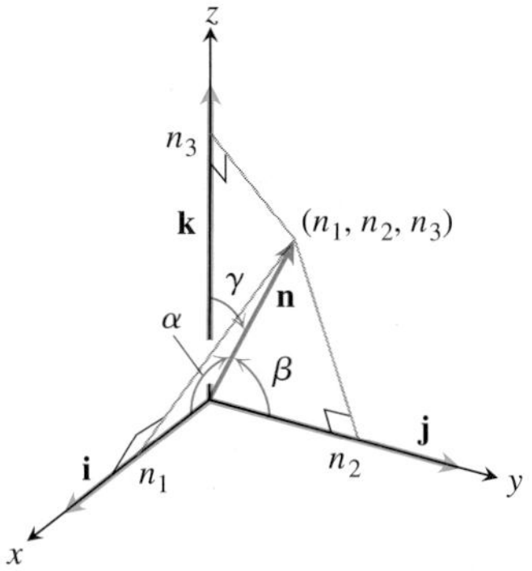

FIGURE 13.61 The components of $\mathbf{n}$ are the cosines of the angles α, β, and γ that it makes with $\mathbf{i}$, $\mathbf{j}$, and $\mathbf{k}$.

In component form, the Divergence Theorem states that

$$\iint_S (M \cos \alpha + N \cos \beta + P \cos \gamma)\, d\sigma = \iiint_D \left(\frac{\partial M}{\partial x} + \frac{\partial N}{\partial y} + \frac{\partial P}{\partial z} \right) dx\, dy\, dz.$$

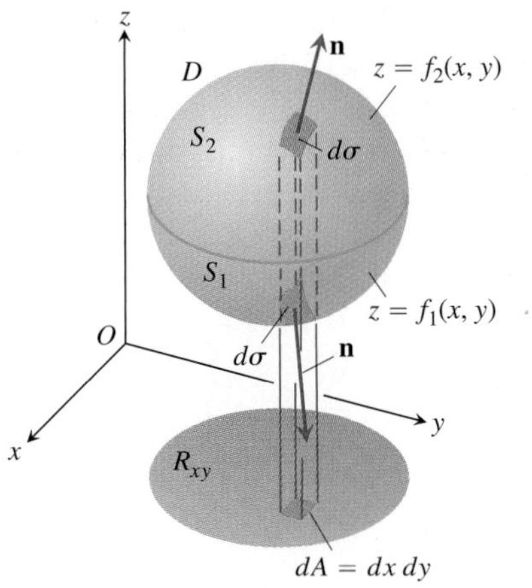

FIGURE 13.62 The region D enclosed by the surfaces S_1 and S_2 projects vertically onto R_{xy} in the xy-plane.

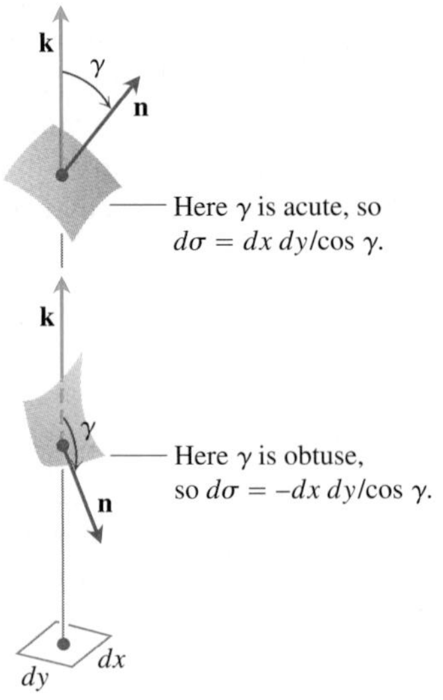

FIGURE 13.63 An enlarged view of the area patches in Figure 13.62. The relations $d\sigma = \pm dx\, dy/\cos\gamma$ come from Eq. (7) in Section 13.5.

We prove the theorem by proving the three following equalities:

$$\iint_S M\cos\alpha\, d\sigma = \iiint_D \frac{\partial M}{\partial x}\, dx\, dy\, dz \tag{3}$$

$$\iint_S N\cos\beta\, d\sigma = \iiint_D \frac{\partial N}{\partial y}\, dx\, dy\, dz \tag{4}$$

$$\iint_S P\cos\gamma\, d\sigma = \iiint_D \frac{\partial P}{\partial z}\, dx\, dy\, dz \tag{5}$$

Proof of Equation (5) We prove Equation (5) by converting the surface integral on the left to a double integral over the projection R_{xy} of D on the xy-plane (Figure 13.62). The surface S consists of an upper part S_2 whose equation is $z = f_2(x, y)$ and a lower part S_1 whose equation is $z = f_1(x, y)$. On S_2, the outer normal **n** has a positive **k**-component and

$$\cos\gamma\, d\sigma = dx\, dy \qquad \text{because} \qquad d\sigma = \frac{dA}{|\cos\gamma|} = \frac{dx\, dy}{\cos\gamma}.$$

See Figure 13.63. On S_1, the outer normal **n** has a negative **k**-component and

$$\cos\gamma\, d\sigma = -dx\, dy.$$

Therefore,

$$\begin{aligned}
\iint_S P\cos\gamma\, d\sigma &= \iint_{S_2} P\cos\gamma\, d\sigma + \iint_{S_1} P\cos\gamma\, d\sigma \\
&= \iint_{R_{xy}} P(x, y, f_2(x, y))\, dx\, dy - \iint_{R_{xy}} P(x, y, f_1(x, y))\, dx\, dy \\
&= \iint_{R_{xy}} [P(x, y, f_2(x, y)) - P(x, y, f_1(x, y))]\, dx\, dy \\
&= \iint_{R_{xy}} \left[\int_{f_1(x,y)}^{f_2(x,y)} \frac{\partial P}{\partial z}\, dz\right] dx\, dy = \iiint_D \frac{\partial P}{\partial z}\, dz\, dx\, dy.
\end{aligned}$$

This proves Equation (5). ■

The proofs for Equations (3) and (4) follow the same pattern; or just permute x, y, z; M, N, P; α, β, γ, in order, and get those results from Equation (5).

Divergence Theorem for Other Regions

The Divergence Theorem can be extended to regions that can be partitioned into a finite number of simple regions of the type just discussed and to regions that can be defined as limits of simpler regions in certain ways. For an example of one step in such a splitting process, suppose that D is the region between two concentric spheres and that **F** has

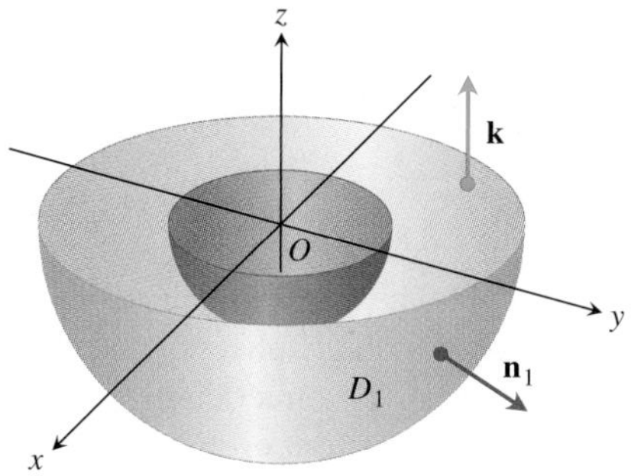

FIGURE 13.64 The lower half of the solid region between two concentric spheres.

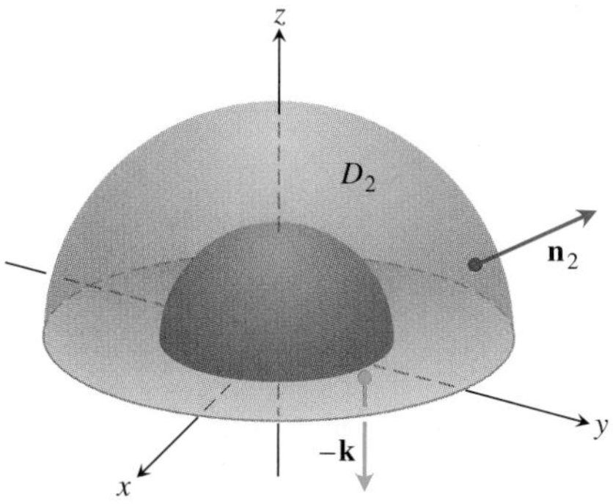

FIGURE 13.65 The upper half of the solid region between two concentric spheres.

continuously differentiable components throughout D and on the bounding surfaces. Split D by an equatorial plane and apply the Divergence Theorem to each half separately. The bottom half, D_1, is shown in Figure 13.64. The surface S_1 that bounds D_1 consists of an outer hemisphere, a plane washer-shaped base, and an inner hemisphere. The Divergence Theorem says that

$$\iint_{S_1} \mathbf{F} \cdot \mathbf{n}_1 \, d\sigma_1 = \iiint_{D_1} \nabla \cdot \mathbf{F} \, dV_1. \tag{6}$$

The unit normal $\mathbf{n}_1$ that points outward from D_1 points away from the origin along the outer surface, equals $\mathbf{k}$ along the flat base, and points toward the origin along the inner surface. Next apply the Divergence Theorem to D_2, and its surface S_2 (Figure 13.65):

$$\iint_{S_2} \mathbf{F} \cdot \mathbf{n}_2 \, d\sigma_2 = \iiint_{D_2} \nabla \cdot \mathbf{F} \, dV_2. \tag{7}$$

As we follow $\mathbf{n}_2$ over S_2, pointing outward from D_2, we see that $\mathbf{n}_2$ equals $-\mathbf{k}$ along the washer-shaped base in the xy-plane, points away from the origin on the outer sphere, and points toward the origin on the inner sphere. When we add Equations (6) and (7), the integrals over the flat base cancel because of the opposite signs of $\mathbf{n}_1$ and $\mathbf{n}_2$. We thus arrive at the result

$$\iint_{S} \mathbf{F} \cdot \mathbf{n} \, d\sigma = \iiint_{D} \nabla \cdot \mathbf{F} \, dV,$$

with D the region between the spheres, S the boundary of D consisting of two spheres, and $\mathbf{n}$ the unit normal to S directed outward from D.

EXAMPLE 4 Find the net outward flux of the field

$$\mathbf{F} = \frac{x\mathbf{i} + y\mathbf{j} + z\mathbf{k}}{\rho^3}, \qquad \rho = \sqrt{x^2 + y^2 + z^2}$$

across the boundary of the region D: $0 < a^2 \le x^2 + y^2 + z^2 \le b^2$.

Solution The flux can be calculated by integrating $\nabla \cdot \mathbf{F}$ over D. We have

$$\frac{\partial \rho}{\partial x} = \frac{1}{2}(x^2 + y^2 + z^2)^{-1/2}(2x) = \frac{x}{\rho}$$

and

$$\frac{\partial M}{\partial x} = \frac{\partial}{\partial x}(x\rho^{-3}) = \rho^{-3} - 3x\rho^{-4}\frac{\partial \rho}{\partial x} = \frac{1}{\rho^3} - \frac{3x^2}{\rho^5}.$$

Similarly,

$$\frac{\partial N}{\partial y} = \frac{1}{\rho^3} - \frac{3y^2}{\rho^5} \quad \text{and} \quad \frac{\partial P}{\partial z} = \frac{1}{\rho^3} - \frac{3z^2}{\rho^5}.$$

Hence,

$$\text{div}\,\mathbf{F} = \frac{3}{\rho^3} - \frac{3}{\rho^5}(x^2 + y^2 + z^2) = \frac{3}{\rho^3} - \frac{3\rho^2}{\rho^5} = 0$$

and

$$\iiint_D \nabla \cdot \mathbf{F}\, dV = 0. \qquad \nabla \cdot \mathbf{F} = \operatorname{div} \mathbf{F}$$

So the integral of $\nabla \cdot \mathbf{F}$ over D is zero and the net outward flux across the boundary of D is zero. There is more to learn from this example, though. The flux leaving D across the inner sphere S_a is the negative of the flux leaving D across the outer sphere S_b (because the sum of these fluxes is zero). Hence, the flux of $\mathbf{F}$ across S_a in the direction away from the origin equals the flux of $\mathbf{F}$ across S_b in the direction away from the origin. Thus, the flux of $\mathbf{F}$ across a sphere centered at the origin is independent of the radius of the sphere. What is this flux?

To find it, we evaluate the flux integral directly. The outward unit normal on the sphere of radius a is

$$\mathbf{n} = \frac{x\mathbf{i} + y\mathbf{j} + z\mathbf{k}}{\sqrt{x^2 + y^2 + z^2}} = \frac{x\mathbf{i} + y\mathbf{j} + z\mathbf{k}}{a}.$$

Hence, on the sphere,

$$\mathbf{F} \cdot \mathbf{n} = \frac{x\mathbf{i} + y\mathbf{j} + z\mathbf{k}}{a^3} \cdot \frac{x\mathbf{i} + y\mathbf{j} + z\mathbf{k}}{a} = \frac{x^2 + y^2 + z^2}{a^4} = \frac{a^2}{a^4} = \frac{1}{a^2}$$

and

$$\iint_{S_a} \mathbf{F} \cdot \mathbf{n}\, d\sigma = \frac{1}{a^2} \iint_{S_a} d\sigma = \frac{1}{a^2}(4\pi a^2) = 4\pi.$$

The outward flux of $\mathbf{F}$ across any sphere centered at the origin is 4π. ■

Gauss's Law: One of the Four Great Laws of Electromagnetic Theory

There is still more to be learned from Example 4. In electromagnetic theory, the electric field created by a point charge q located at the origin is

$$\mathbf{E}(x, y, z) = \frac{1}{4\pi\epsilon_0} \frac{q}{|\mathbf{r}|^2} \left(\frac{\mathbf{r}}{|\mathbf{r}|} \right) = \frac{q}{4\pi\epsilon_0} \frac{\mathbf{r}}{|\mathbf{r}|^3} = \frac{q}{4\pi\epsilon_0} \frac{x\mathbf{i} + y\mathbf{j} + z\mathbf{k}}{\rho^3},$$

where ϵ_0 is a physical constant, $\mathbf{r}$ is the position vector of the point (x, y, z), and $\rho = |\mathbf{r}| = \sqrt{x^2 + y^2 + z^2}$. In the notation of Example 4,

$$\mathbf{E} = \frac{q}{4\pi\epsilon_0} \mathbf{F}.$$

The calculations in Example 4 show that the outward flux of $\mathbf{E}$ across any sphere centered at the origin is q/ϵ_0, but this result is not confined to spheres. The outward flux of $\mathbf{E}$ across any closed surface S that encloses the origin (and to which the Divergence Theorem applies) is also q/ϵ_0. To see why, we have only to imagine a large sphere S_a centered at the origin and enclosing the surface S. Since

$$\nabla \cdot \mathbf{E} = \nabla \cdot \frac{q}{4\pi\epsilon_0} \mathbf{F} = \frac{q}{4\pi\epsilon_0} \nabla \cdot \mathbf{F} = 0$$

when $\rho > 0$, the integral of $\nabla \cdot \mathbf{E}$ over the region D between S and S_a is zero. Hence, by the Divergence Theorem,

$$\iint_{\substack{\text{Boundary}\\ \text{of } D}} \mathbf{E} \cdot \mathbf{n}\, d\sigma = 0,$$

and the flux of $\mathbf{E}$ across S in the direction away from the origin must be the same as the flux of $\mathbf{E}$ across S_a in the direction away from the origin, which is q/ϵ_0. This statement, called *Gauss's Law*, also applies to charge distributions that are more general than the one assumed here, as you will see in nearly any physics text.

$$\text{Gauss's law:} \qquad \iint_S \mathbf{E} \cdot \mathbf{n}\, d\sigma = \frac{q}{\epsilon_0}$$

Unifying the Integral Theorems

If we think of a two-dimensional field $\mathbf{F} = M(x, y)\mathbf{i} + N(x, y)\mathbf{j}$ as a three-dimensional field whose $\mathbf{k}$-component is zero, then $\nabla \cdot \mathbf{F} = (\partial M/\partial x) + (\partial N/\partial y)$ and the normal form of Green's Theorem can be written as

$$\oint_C \mathbf{F} \cdot \mathbf{n}\, ds = \iint_R \left(\frac{\partial M}{\partial x} + \frac{\partial N}{\partial y}\right) dx\, dy = \iint_R \nabla \cdot \mathbf{F}\, dA.$$

Similarly, $\nabla \times \mathbf{F} \cdot \mathbf{k} = (\partial N/\partial x) - (\partial M/\partial y)$, so the tangential form of Green's Theorem can be written as

$$\oint_C \mathbf{F} \cdot d\mathbf{r} = \iint_R \left(\frac{\partial N}{\partial x} - \frac{\partial M}{\partial y}\right) dx\, dy = \iint_R \nabla \times \mathbf{F} \cdot \mathbf{k}\, dA.$$

With the equations of Green's Theorem now in del notation, we can see their relationships to the equations in Stokes' Theorem and the Divergence Theorem.

Green's Theorem and Its Generalization to Three Dimensions

Normal form of Green's Theorem: $\displaystyle \oint_C \mathbf{F} \cdot \mathbf{n}\, ds = \iint_R \nabla \cdot \mathbf{F}\, dA$

Divergence Theorem: $\displaystyle \iint_S \mathbf{F} \cdot \mathbf{n}\, d\sigma = \iiint_D \nabla \cdot \mathbf{F}\, dV$

Tangential form of Green's Theorem: $\displaystyle \oint_C \mathbf{F} \cdot d\mathbf{r} = \iint_R \nabla \times \mathbf{F} \cdot \mathbf{k}\, dA$

Stokes' Theorem: $\displaystyle \oint_C \mathbf{F} \cdot d\mathbf{r} = \iint_S \nabla \times \mathbf{F} \cdot \mathbf{n}\, d\sigma$

Notice how Stokes' Theorem generalizes the tangential (curl) form of Green's Theorem from a flat surface in the plane to a surface in three-dimensional space. In each case, the integral of the normal component of curl **F** over the interior of the surface equals the circulation of **F** around the boundary.

Likewise, the Divergence Theorem generalizes the normal (flux) form of Green's Theorem from a two-dimensional region in the plane to a three-dimensional region in space. In each case, the integral of $\nabla \cdot \mathbf{F}$ over the interior of the region equals the total flux of the field across the boundary.

There is still more to be learned here. All these results can be thought of as forms of a *single fundamental theorem*. Think back to the Fundamental Theorem of Calculus in Section 4.5. It says that if $f(x)$ is differentiable on (a, b) and continuous on $[a, b]$, then

$$\int_a^b \frac{df}{dx}\,dx = f(b) - f(a).$$

FIGURE 13.66 The outward unit normals at the boundary of $[a, b]$ in one-dimensional space.

If we let $\mathbf{F} = f(x)\mathbf{i}$ throughout $[a, b]$, then $(df/dx) = \nabla \cdot \mathbf{F}$. If we define the unit vector field $\mathbf{n}$ normal to the boundary of $[a, b]$ to be $\mathbf{i}$ at b and $-\mathbf{i}$ at a (Figure 13.66), then

$$\begin{aligned} f(b) - f(a) &= f(b)\mathbf{i} \cdot (\mathbf{i}) + f(a)\mathbf{i} \cdot (-\mathbf{i}) \\ &= \mathbf{F}(b) \cdot \mathbf{n} + \mathbf{F}(a) \cdot \mathbf{n} \\ &= \text{total outward flux of } \mathbf{F} \text{ across the boundary of } [a, b]. \end{aligned}$$

The Fundamental Theorem now says that

$$\mathbf{F}(b) \cdot \mathbf{n} + \mathbf{F}(a) \cdot \mathbf{n} = \int_{[a,b]} \nabla \cdot \mathbf{F}\,dx.$$

The Fundamental Theorem of Calculus, the normal form of Green's Theorem, and the Divergence Theorem all say that the integral of the differential operator $\nabla \cdot$ operating on a field **F** over a region equals the sum of the normal field components over the boundary of the region. (Here we are interpreting the line integral in Green's Theorem and the surface integral in the Divergence Theorem as "sums" over the boundary.)

Stokes' Theorem and the tangential form of Green's Theorem say that, when things are properly oriented, the integral of the normal component of the curl operating on a field equals the sum of the tangential field components on the boundary of the surface.

The beauty of these interpretations is the observance of a single unifying principle, which we might state as follows.

> The integral of a differential operator acting on a field over a region equals the sum of the field components appropriate to the operator over the boundary of the region.

EXERCISES 13.8

In Exercises 1–4, find the divergence of the field.

1. The spin field in Figure 13.13.

2. The radial field in Figure 13.12.

3. The gravitational field in Figure 13.8 and Exercise 38a in Section 13.3.

4. The velocity field in Figure 13.11.

In Exercises 5–16, use the Divergence Theorem to find the outward flux of **F** across the boundary of the region D.

5. Cube $\mathbf{F} = (y - x)\mathbf{i} + (z - y)\mathbf{j} + (y - x)\mathbf{k}$

D: The cube bounded by the planes $x = \pm 1$, $y = \pm 1$, and $z = \pm 1$

6. $\mathbf{F} = x^2\mathbf{i} + y^2\mathbf{j} + z^2\mathbf{k}$

   a. **Cube** D: The cube cut from the first octant by the planes $x = 1, y = 1$, and $z = 1$

   b. **Cube** D: The cube bounded by the planes $x = \pm 1$, $y = \pm 1$, and $z = \pm 1$

   c. **Cylindrical can** D: The region cut from the solid cylinder $x^2 + y^2 \le 4$ by the planes $z = 0$ and $z = 1$

7. **Cylinder and paraboloid** $\mathbf{F} = y\mathbf{i} + xy\mathbf{j} - z\mathbf{k}$

   D: The region inside the solid cylinder $x^2 + y^2 \le 4$ between the plane $z = 0$ and the paraboloid $z = x^2 + y^2$

8. **Sphere** $\mathbf{F} = x^2\mathbf{i} + xz\mathbf{j} + 3z\mathbf{k}$

   D: The solid sphere $x^2 + y^2 + z^2 \le 4$

9. **Portion of sphere** $\mathbf{F} = x^2\mathbf{i} - 2xy\mathbf{j} + 3xz\mathbf{k}$

   D: The region cut from the first octant by the sphere $x^2 + y^2 + z^2 = 4$

10. **Cylindrical can** $\mathbf{F} = (6x^2 + 2xy)\mathbf{i} + (2y + x^2z)\mathbf{j} + 4x^2y^3\mathbf{k}$

    D: The region cut from the first octant by the cylinder $x^2 + y^2 = 4$ and the plane $z = 3$

11. **Wedge** $\mathbf{F} = 2xz\mathbf{i} - xy\mathbf{j} - z^2\mathbf{k}$

    D: The wedge cut from the first octant by the plane $y + z = 4$ and the elliptical cylinder $4x^2 + y^2 = 16$

12. **Sphere** $\mathbf{F} = x^3\mathbf{i} + y^3\mathbf{j} + z^3\mathbf{k}$

    D: The solid sphere $x^2 + y^2 + z^2 \le a^2$

13. **Thick sphere** $\mathbf{F} = \sqrt{x^2 + y^2 + z^2}\,(x\mathbf{i} + y\mathbf{j} + z\mathbf{k})$

    D: The region $1 \le x^2 + y^2 + z^2 \le 2$

14. **Thick sphere** $\mathbf{F} = (x\mathbf{i} + y\mathbf{j} + z\mathbf{k})/\sqrt{x^2 + y^2 + z^2}$

    D: The region $1 \le x^2 + y^2 + z^2 \le 4$

15. **Thick sphere** $\mathbf{F} = (5x^3 + 12xy^2)\mathbf{i} + (y^3 + e^y \sin z)\mathbf{j} + (5z^3 + e^y \cos z)\mathbf{k}$

    D: The solid region between the spheres $x^2 + y^2 + z^2 = 1$ and $x^2 + y^2 + z^2 = 2$

16. **Thick cylinder** $\mathbf{F} = \ln(x^2 + y^2)\mathbf{i} - \left(\frac{2z}{x}\tan^{-1}\frac{y}{x}\right)\mathbf{j} + z\sqrt{x^2 + y^2}\,\mathbf{k}$

    D: The thick-walled cylinder $1 \le x^2 + y^2 \le 2, \quad -1 \le z \le 2$

17. **div (curl G) is zero**

    a. Show that if the necessary partial derivatives of the components of the field $\mathbf{G} = M\mathbf{i} + N\mathbf{j} + P\mathbf{k}$ are continuous, then $\nabla \cdot \nabla \times \mathbf{G} = 0$.

    b. What, if anything, can you conclude about the flux of the field $\nabla \times \mathbf{G}$ across a closed surface? Give reasons for your answer.

18. Let $\mathbf{F}_1$ and $\mathbf{F}_2$ be differentiable vector fields and let a and b be arbitrary real constants. Verify the following identities.

    a. $\nabla \cdot (a\mathbf{F}_1 + b\mathbf{F}_2) = a\nabla \cdot \mathbf{F}_1 + b\nabla \cdot \mathbf{F}_2$

    b. $\nabla \times (a\mathbf{F}_1 + b\mathbf{F}_2) = a\nabla \times \mathbf{F}_1 + b\nabla \times \mathbf{F}_2$

    c. $\nabla \cdot (\mathbf{F}_1 \times \mathbf{F}_2) = \mathbf{F}_2 \cdot \nabla \times \mathbf{F}_1 - \mathbf{F}_1 \cdot \nabla \times \mathbf{F}_2$

19. Let $\mathbf{F}$ be a differentiable vector field and let $g(x, y, z)$ be a differentiable scalar function. Verify the following identities.

    a. $\nabla \cdot (g\mathbf{F}) = g\nabla \cdot \mathbf{F} + \nabla g \cdot \mathbf{F}$

    b. $\nabla \times (g\mathbf{F}) = g\nabla \times \mathbf{F} + \nabla g \times \mathbf{F}$

20. If $\mathbf{F} = M\mathbf{i} + N\mathbf{j} + P\mathbf{k}$ is a differentiable vector field, we define the notation $\mathbf{F} \cdot \nabla$ to mean

$$M\frac{\partial}{\partial x} + N\frac{\partial}{\partial y} + P\frac{\partial}{\partial z}.$$

    For differentiable vector fields $\mathbf{F}_1$ and $\mathbf{F}_2$, verify the following identities.

    a. $\nabla \times (\mathbf{F}_1 \times \mathbf{F}_2) = (\mathbf{F}_2 \cdot \nabla)\mathbf{F}_1 - (\mathbf{F}_1 \cdot \nabla)\mathbf{F}_2 + (\nabla \cdot \mathbf{F}_2)\mathbf{F}_1 - (\nabla \cdot \mathbf{F}_1)\mathbf{F}_2$

    b. $\nabla(\mathbf{F}_1 \cdot \mathbf{F}_2) = (\mathbf{F}_1 \cdot \nabla)\mathbf{F}_2 + (\mathbf{F}_2 \cdot \nabla)\mathbf{F}_1 + \mathbf{F}_1 \times (\nabla \times \mathbf{F}_2) + \mathbf{F}_2 \times (\nabla \times \mathbf{F}_1)$

21. Let $\mathbf{F}$ be a field whose components have continuous first partial derivatives throughout a portion of space containing a region D bounded by a smooth closed surface S. If $|\mathbf{F}| \le 1$, can any bound be placed on the size of

$$\iiint_D \nabla \cdot \mathbf{F}\, dV?$$

    Give reasons for your answer.

22. The base of the closed cubelike surface shown here is the unit square in the xy-plane. The four sides lie in the planes $x = 0$, $x = 1$, $y = 0$, and $y = 1$. The top is an arbitrary smooth surface whose identity is unknown. Let $\mathbf{F} = x\mathbf{i} - 2y\mathbf{j} + (z + 3)\mathbf{k}$ and suppose the outward flux of $\mathbf{F}$ through side A is 1 and through side B is -3. Can you conclude anything about the outward flux through the top? Give reasons for your answer.

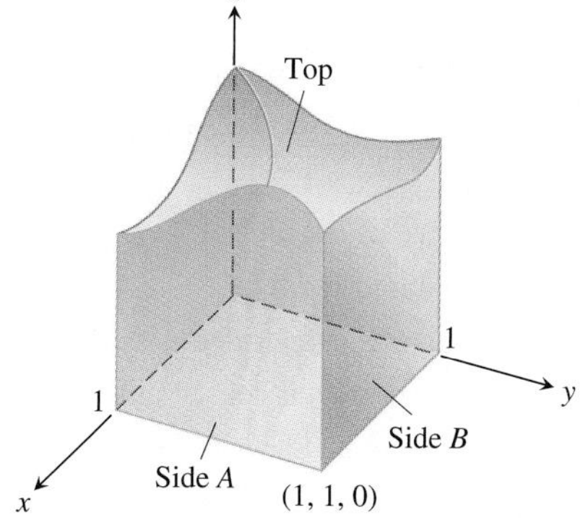

23. a. Show that the flux of the position vector field $\mathbf{F} = x\mathbf{i} + y\mathbf{j} + z\mathbf{k}$ outward through a smooth closed surface S is three times the volume of the region enclosed by the surface.

    b. Let $\mathbf{n}$ be the outward unit normal vector field on S. Show that it is not possible for $\mathbf{F}$ to be orthogonal to $\mathbf{n}$ at every point of S.

24. **Maximum flux** Among all rectangular solids defined by the inequalities $0 \le x \le a, 0 \le y \le b, 0 \le z \le 1$, find the one for

which the total flux of $\mathbf{F} = (-x^2 - 4xy)\mathbf{i} - 6yz\mathbf{j} + 12z\mathbf{k}$ outward through the six sides is greatest. What *is* the greatest flux?

25. **Volume of a solid region** Let $\mathbf{F} = x\mathbf{i} + y\mathbf{j} + z\mathbf{k}$ and suppose that the surface S and region D satisfy the hypotheses of the Divergence Theorem. Show that the volume of D is given by the formula

$$\text{Volume of } D = \frac{1}{3}\iint_S \mathbf{F}\cdot\mathbf{n}\,d\sigma.$$

26. **Flux of a constant field** Show that the outward flux of a constant vector field $\mathbf{F} = \mathbf{C}$ across any closed surface to which the Divergence Theorem applies is zero.

27. **Harmonic functions** A function $f(x, y, z)$ is said to be *harmonic* in a region D in space if it satisfies the Laplace equation

$$\nabla^2 f = \nabla\cdot\nabla f = \frac{\partial^2 f}{\partial x^2} + \frac{\partial^2 f}{\partial y^2} + \frac{\partial^2 f}{\partial z^2} = 0$$

throughout D.

a. Suppose that f is harmonic throughout a bounded region D enclosed by a smooth surface S and that $\mathbf{n}$ is the chosen unit normal vector on S. Show that the integral over S of $\nabla f\cdot\mathbf{n}$, the derivative of f in the direction of $\mathbf{n}$, is zero.

b. Show that if f is harmonic on D, then

$$\iint_S f\,\nabla f\cdot\mathbf{n}\,d\sigma = \iiint_D |\nabla f|^2\,dV.$$

28. **Flux of a gradient field** Let S be the surface of the portion of the solid sphere $x^2 + y^2 + z^2 \le a^2$ that lies in the first octant and let $f(x, y, z) = \ln\sqrt{x^2 + y^2 + z^2}$. Calculate

$$\iint_S \nabla f\cdot\mathbf{n}\,d\sigma.$$

($\nabla f\cdot\mathbf{n}$ is the derivative of f in the direction of $\mathbf{n}$.)

29. **Green's first formula** Suppose that f and g are scalar functions with continuous first- and second-order partial derivatives throughout a region D that is bounded by a closed piecewise-smooth surface S. Show that

$$\iint_S f\,\nabla g\cdot\mathbf{n}\,d\sigma = \iiint_D (f\,\nabla^2 g + \nabla f\cdot\nabla g)\,dV. \qquad (8)$$

Equation (8) is **Green's first formula**. (*Hint:* Apply the Divergence Theorem to the field $\mathbf{F} = f\,\nabla g$.)

30. **Green's second formula** (*Continuation of Exercise 29.*) Interchange f and g in Equation (8) to obtain a similar formula. Then subtract this formula from Equation (8) to show that

$$\iint_S (f\,\nabla g - g\nabla f)\cdot\mathbf{n}\,d\sigma = \iiint_D (f\,\nabla^2 g - g\nabla^2 f)\,dV. \qquad (9)$$

This equation is **Green's second formula**.

Chapter 13 Questions to Guide Your Review

1. What are line integrals? How are they evaluated? Give examples.
2. How can you use line integrals to find the centers of mass of springs? Explain.
3. What is a vector field? A gradient field? Give examples.
4. How do you calculate the work done by a force in moving a particle along a curve? Give an example.
5. What are flow, circulation, and flux?
6. What is special about path independent fields?
7. How can you tell when a field is conservative?
8. What is a potential function? Show by example how to find a potential function for a conservative field.
9. What is a differential form? What does it mean for such a form to be exact? How do you test for exactness? Give examples.
10. What is the divergence of a vector field? How can you interpret it?
11. What is the curl of a vector field? How can you interpret it?
12. What is Green's theorem? How can you interpret it?
13. How do you calculate the area of a curved surface in space? Give an example.
14. What is an oriented surface? How do you calculate the flux of a three-dimensional vector field across an oriented surface? Give an example.
15. What are surface integrals? What can you calculate with them? Give an example.
16. What is a parametrized surface? How do you find the area of such a surface? Give examples.
17. How do you integrate a function over a parametrized surface? Give an example.
18. What is Stokes' Theorem? How can you interpret it?
19. Summarize the chapter's results on conservative fields.
20. What is the Divergence Theorem? How can you interpret it?
21. How does the Divergence Theorem generalize Green's Theorem?
22. How does Stokes' Theorem generalize Green's Theorem?
23. How can Green's Theorem, Stokes' Theorem, and the Divergence Theorem be thought of as forms of a single fundamental theorem?

Chapter 13 Practice and Additional Exercises

1. The accompanying figure shows two polygonal paths in space joining the origin to the point (1, 1, 1). Integrate $f(x, y, z) = 2x - 3y^2 - 2z + 3$ over each path.

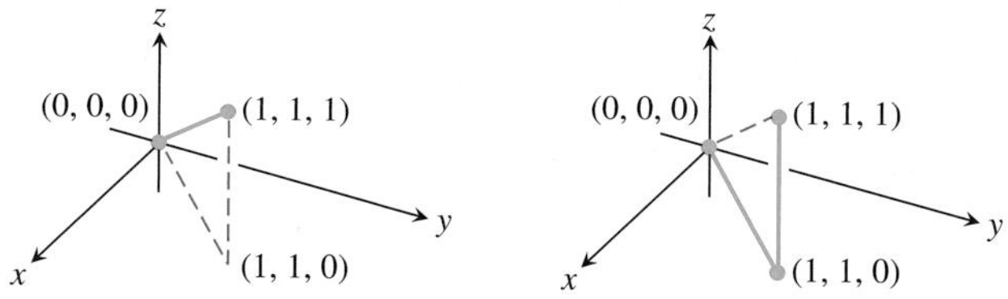

Path 1 Path 2

2. The accompanying figure shows three polygonal paths joining the origin to the point (1, 1, 1). Integrate $f(x, y, z) = x^2 + y - z$ over each path.

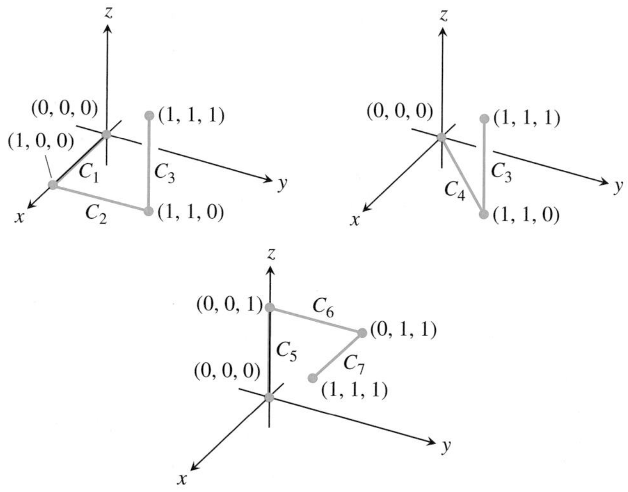

3. Integrate $f(x, y, z) = \sqrt{x^2 + z^2}$ over the circle

$$\mathbf{r}(t) = (a \cos t)\mathbf{j} + (a \sin t)\mathbf{k}, \qquad 0 \le t \le 2\pi.$$

4. Integrate $f(x, y, z) = \sqrt{x^2 + y^2}$ over the involute curve

$$\mathbf{r}(t) = (\cos t + t \sin t)\mathbf{i} + (\sin t - t \cos t)\mathbf{j}, \qquad 0 \le t \le \sqrt{3}.$$

Evaluate the integrals in Exercises 5 and 6.

5. $\displaystyle\int_{(-1,1,1)}^{(4,-3,0)} \frac{dx + dy + dz}{\sqrt{x + y + z}}$

6. $\displaystyle\int_{(1,1,1)}^{(10,3,3)} dx - \sqrt{\frac{z}{y}}\,dy - \sqrt{\frac{y}{z}}\,dz$

7. Integrate $\mathbf{F} = -(y \sin z)\mathbf{i} + (x \sin z)\mathbf{j} + (xy \cos z)\mathbf{k}$ around the circle cut from the sphere $x^2 + y^2 + z^2 = 5$ by the plane $z = -1$, clockwise as viewed from above.

8. Integrate $\mathbf{F} = 3x^2y\mathbf{i} + (x^3 + 1)\mathbf{j} + 9z^2\mathbf{k}$ around the circle cut from the sphere $x^2 + y^2 + z^2 = 9$ by the plane $x = 2$.

Evaluate the integrals in Exercises 9 and 10.

9. $\displaystyle\int_C 8x \sin y\, dx - 8y \cos x\, dy$

   C is the square cut from the first quadrant by the lines $x = \pi/2$ and $y = \pi/2$.

10. $\displaystyle\int_C y^2\, dx + x^2\, dy$

    C is the circle $x^2 + y^2 = 4$.

11. **Area of an elliptical region** Find the area of the elliptical region cut from the plane $x + y + z = 1$ by the cylinder $x^2 + y^2 = 1$.

12. **Area of a parabolic cap** Find the area of the cap cut from the paraboloid $y^2 + z^2 = 3x$ by the plane $x = 1$.

13. **Area of a spherical cap** Find the area of the cap cut from the top of the sphere $x^2 + y^2 + z^2 = 1$ by the plane $z = \sqrt{2}/2$.

14. **a. Hemisphere cut by cylinder** Find the area of the surface cut from the hemisphere $x^2 + y^2 + z^2 = 4$, $z \ge 0$, by the cylinder $x^2 + y^2 = 2x$.

    b. Find the area of the portion of the cylinder that lies inside the hemisphere. (*Hint:* Project onto the xz-plane. Or evaluate the integral $\int h\, ds$, where h is the altitude of the cylinder and ds is the element of arc length on the circle $x^2 + y^2 = 2x$ in the xy-plane.)

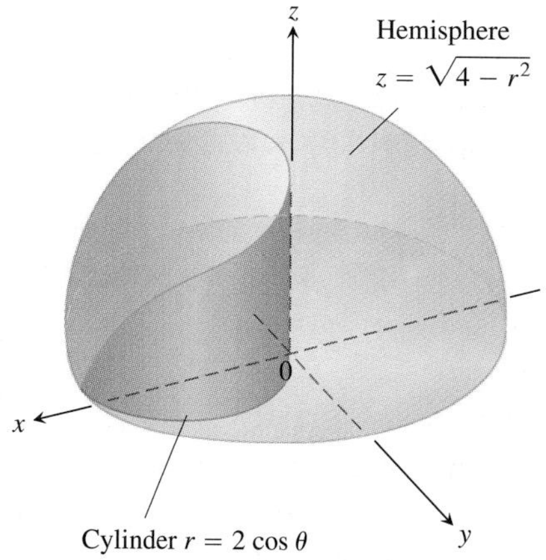

15. **Area of a triangle** Find the area of the triangle in which the plane $(x/a) + (y/b) + (z/c) = 1$ $(a, b, c > 0)$ intersects the first octant. Check your answer with an appropriate vector calculation.

16. **Parabolic cylinder cut by planes** Integrate

    a. $g(x, y, z) = \dfrac{yz}{\sqrt{4y^2 + 1}}$

b. $g(x, y, z) = \dfrac{z}{\sqrt{4y^2 + 1}}$

over the surface cut from the parabolic cylinder $y^2 - z = 1$ by the planes $x = 0, x = 3$, and $z = 0$.

17. Circular cylinder cut by planes Integrate $g(x, y, z) = x^4y(y^2 + z^2)$ over the portion of the cylinder $y^2 + z^2 = 25$ that lies in the first octant between the planes $x = 0$ and $x = 1$ and above the plane $z = 3$.

18. Area of Wyoming The state of Wyoming is bounded by the meridians 111°3′ and 104°3′ west longitude and by the circles 41° and 45° north latitude. Assuming that Earth is a sphere of radius $R = 3959$ mi, find the area of Wyoming.

Find parametrizations for the surfaces in Exercises 19–24. (There are many ways to do these, so your answers may not be the same as those in the back of the book.)

19. Spherical band The portion of the sphere $x^2 + y^2 + z^2 = 36$ between the planes $z = -3$ and $z = 3\sqrt{3}$

20. Parabolic cap The portion of the paraboloid $z = -(x^2 + y^2)/2$ above the plane $z = -2$

21. Cone The cone $z = 1 + \sqrt{x^2 + y^2}, z \le 3$

22. Plane above square The portion of the plane $4x + 2y + 4z = 12$ that lies above the square $0 \le x \le 2, 0 \le y \le 2$ in the first quadrant

23. Portion of paraboloid The portion of the paraboloid $y = 2(x^2 + z^2),\ y \le 2$, that lies above the xy-plane

24. Portion of hemisphere The portion of the hemisphere $x^2 + y^2 + z^2 = 10, y \ge 0$, in the first octant

25. Surface area Find the area of the surface

$$\mathbf{r}(u, v) = (u + v)\mathbf{i} + (u - v)\mathbf{j} + v\mathbf{k},$$
$$0 \le u \le 1,\ 0 \le v \le 1.$$

26. Surface integral Integrate $f(x, y, z) = xy - z^2$ over the surface in Exercise 25.

27. Area of a helicoid Find the surface area of the helicoid

$$\mathbf{r}(r, \theta) = (r\cos\theta)\mathbf{i} + (r\sin\theta)\mathbf{j} + \theta\mathbf{k},\quad 0 \le \theta \le 2\pi,\ 0 \le r \le 1,$$

in the accompanying figure.

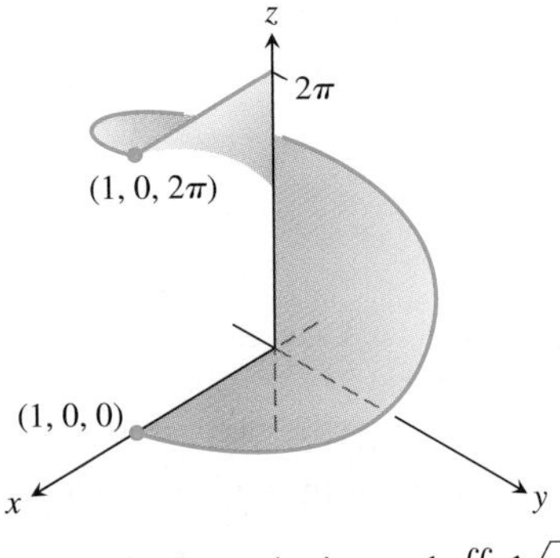

28. Surface integral Evaluate the integral $\iint_S \sqrt{x^2 + y^2 + 1}\, d\sigma$, where S is the helicoid in Exercise 27.

Which of the fields in Exercises 29–32 are conservative, and which are not?

29. $\mathbf{F} = x\mathbf{i} + y\mathbf{j} + z\mathbf{k}$

30. $\mathbf{F} = (x\mathbf{i} + y\mathbf{j} + z\mathbf{k})/(x^2 + y^2 + z^2)^{3/2}$

31. $\mathbf{F} = xe^y\mathbf{i} + ye^z\mathbf{j} + ze^x\mathbf{k}$

32. $\mathbf{F} = (\mathbf{i} + z\mathbf{j} + y\mathbf{k})/(x + yz)$

Find potential functions for the fields in Exercises 33 and 34.

33. $\mathbf{F} = 2\mathbf{i} + (2y + z)\mathbf{j} + (y + 1)\mathbf{k}$

34. $\mathbf{F} = (z\cos xz)\mathbf{i} + e^y\mathbf{j} + (x\cos xz)\mathbf{k}$

In Exercises 35 and 36, find the work done by each field along the paths from (0, 0, 0) to (1, 1, 1) in Exercise 1.

35. $\mathbf{F} = 2xy\mathbf{i} + \mathbf{j} + x^2\mathbf{k}$

36. $\mathbf{F} = 2xy\mathbf{i} + x^2\mathbf{j} + \mathbf{k}$

37. Finding work in two ways Find the work done by

$$\mathbf{F} = \frac{x\mathbf{i} + y\mathbf{j}}{(x^2 + y^2)^{3/2}}$$

over the plane curve $\mathbf{r}(t) = (e^t\cos t)\mathbf{i} + (e^t\sin t)\mathbf{j}$ from the point $(1, 0)$ to the point $(e^{2\pi}, 0)$ in two ways:

a. By using the parametrization of the curve to evaluate the work integral

b. By evaluating a potential function for **F**.

38. Flow along different paths Find the flow of the field $\mathbf{F} = \nabla(x^2ze^y)$

a. Once around the ellipse C in which the plane $x + y + z = 1$ intersects the cylinder $x^2 + z^2 = 25$, clockwise as viewed from the positive y-axis

b. Along the curved boundary of the helicoid in Exercise 27 from $(1, 0, 0)$ to $(1, 0, 2\pi)$.

39. Circulation around an ellipse Use the surface integral in Stokes' Theorem to find the circulation of the field $\mathbf{F} = y^2\mathbf{i} - y\mathbf{j} + 3z^2\mathbf{k}$ around the curve C in the indicated direction.

C: The ellipse in which the plane $2x + 6y - 3z = 6$ meets the cylinder $x^2 + y^2 = 1$, counterclockwise as viewed from above

40. Wire with different densities Find the mass of a thin wire lying along the curve $\mathbf{r}(t) = \sqrt{2}t\mathbf{i} + \sqrt{2}t\mathbf{j} + (4 - t^2)\mathbf{k}, 0 \le t \le 1$, if the density at t is **(a)** $\delta = 3t$ and **(b)** $\delta = 1$.

41. Wire with variable density Find the center of mass and the moments of inertia about the coordinate axes of a thin wire lying along the curve

$$\mathbf{r}(t) = t\mathbf{i} + \frac{2\sqrt{2}}{3}t^{3/2}\mathbf{j} + \frac{t^2}{2}\mathbf{k},\qquad 0 \le t \le 2,$$

if the density at t is $\delta = 1/(t + 1)$.

42. Center of mass of an arch A slender metal arch lies along the semicircle $y = \sqrt{a^2 - x^2}$ in the xy-plane. The density at the point (x, y) on the arch is $\delta(x, y) = 2a - y$. Find the center of mass.

43. **Wire with constant density** A wire of constant density $\delta = 1$ lies along the curve $\mathbf{r}(t) = (e^t \cos t)\mathbf{i} + (e^t \sin t)\mathbf{j} + e^t\mathbf{k}$, $0 \leq t \leq \ln 2$. Find $\bar{z}$ and I_z.

44. **Inertia and center of mass of a shell** Find I_z and the center of mass of a thin shell of density $\delta(x, y, z) = z$ cut from the upper portion of the sphere $x^2 + y^2 + z^2 = 25$ by the plane $z = 3$.

Use Green's Theorem to find the counterclockwise circulation and outward flux for the fields and curves in Exercises 45 and 46.

45. **Square** $\mathbf{F} = (2xy + x)\mathbf{i} + (xy - y)\mathbf{j}$

C: The square bounded by $x = 0, x = 1, y = 0, y = 1$

46. **Triangle** $\mathbf{F} = (y - 6x^2)\mathbf{i} + (x + y^2)\mathbf{j}$

C: The triangle made by the lines $y = 0, y = x$, and $x = 1$

In Exercises 47 and 48, find the outward flux of $\mathbf{F}$ across the boundary of D.

47. **Cube** $\mathbf{F} = 2xy\mathbf{i} + 2yz\mathbf{j} + 2xz\mathbf{k}$

D: The cube cut from the first octant by the planes $x = 1, y = 1, z = 1$

48. **Spherical cap** $\mathbf{F} = -2x\mathbf{i} - 3y\mathbf{j} + z\mathbf{k}$

D: The upper region cut from the solid sphere $x^2 + y^2 + z^2 \leq 2$ by the paraboloid $z = x^2 + y^2$

49. **Hemisphere, cylinder, and plane** Let S be the surface that is bounded on the left by the hemisphere $x^2 + y^2 + z^2 = a^2, y \leq 0$, in the middle by the cylinder $x^2 + z^2 = a^2, 0 \leq y \leq a$, and on the right by the plane $y = a$. Find the flux of $\mathbf{F} = y\mathbf{i} + z\mathbf{j} + x\mathbf{k}$ outward across S.

50. **Hemisphere** Find the flux of $\mathbf{F} = (3z + 1)\mathbf{k}$ upward across the hemisphere $x^2 + y^2 + z^2 = a^2, z \geq 0$ **(a)** with the Divergence Theorem and **(b)** by evaluating the flux integral directly.

APPENDICES

A.1-3 Real Numbers and the Real Line; Mathematical Induction; Lines, Circles, and Parabolas

These appendices can be found online at the book's Web site, **www.aw-bc.com/thomas**

A.4 Trigonometric Functions

This section reviews radian measure and the basic trigonometric functions.

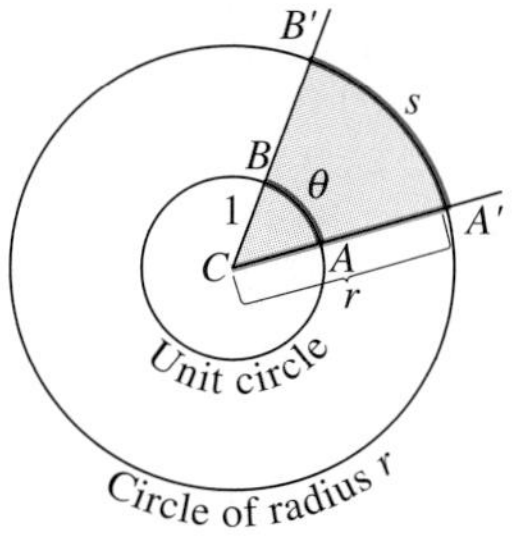

FIGURE A.21 The radian measure of the central angle $A'CB'$ is the number $\theta = s/r$. For a unit circle of radius $r = 1$, θ is the length of arc AB that central angle ACB cuts from the unit circle.

Angles

Angles are measured in degrees or radians. The number of **radians** in the central angle $A'CB'$ within a circle of radius r is defined as the number of "radius units" contained in the arc s subtended by that central angle. If we denote this central angle by θ when measured in radians, this means that $\theta = s/r$ (Figure A.21), or

$$s = r\theta \qquad (\theta \text{ in radians}). \tag{1}$$

If the circle is a unit circle having radius $r = 1$, then from Figure A.21 and Equation (1), we see that the central angle θ measured in radians is just the length of the arc that the angle cuts from the unit circle. Since one complete revolution of the unit circle is 360° or 2π radians, we have

$$\pi \text{ radians} = 180^\circ \tag{2}$$

and

$$1 \text{ radian} = \frac{180}{\pi} (\approx 57.3) \text{ degrees} \qquad \text{or} \qquad 1 \text{ degree} = \frac{\pi}{180} (\approx 0.017) \text{ radians}.$$

Table A.2 shows the equivalence between degree and radian measures for some basic angles.

TABLE A.2 Angles measured in degrees and radians

Degrees	**−180**	**−135**	**−90**	**−45**	**0**	**30**	**45**	**60**	**90**	**120**	**135**	**150**	**180**	**270**	**360**
θ (radians)	$-\pi$	$\frac{-3\pi}{4}$	$\frac{-\pi}{2}$	$\frac{-\pi}{4}$	0	$\frac{\pi}{6}$	$\frac{\pi}{4}$	$\frac{\pi}{3}$	$\frac{\pi}{2}$	$\frac{2\pi}{3}$	$\frac{3\pi}{4}$	$\frac{5\pi}{6}$	π	$\frac{3\pi}{2}$	2π

An angle in the xy plane is said to be in **standard position** if its vertex lies at the origin and its initial ray lies along the positive x-axis (Figure A.22). Angles measured counterclockwise from the positive x-axis are assigned positive measures; angles measured clockwise are assigned negative measures.

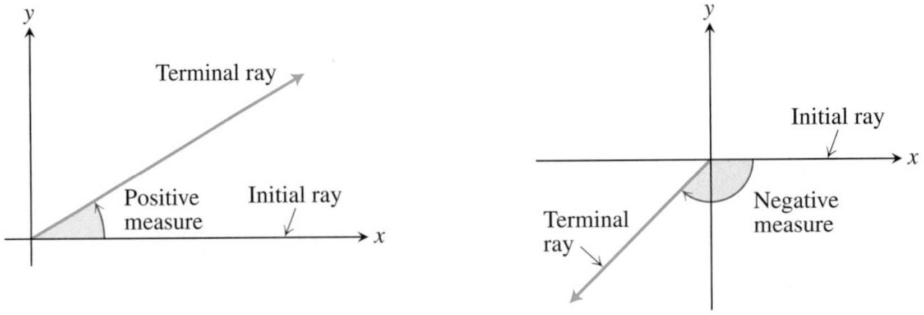

FIGURE A.22 Angles in standard position in the xy-plane.

Angles describing counterclockwise rotations can go arbitrarily far beyond 2π radians or 360°. Similarly, angles describing clockwise rotations can have negative measures of all sizes (Figure A.23).

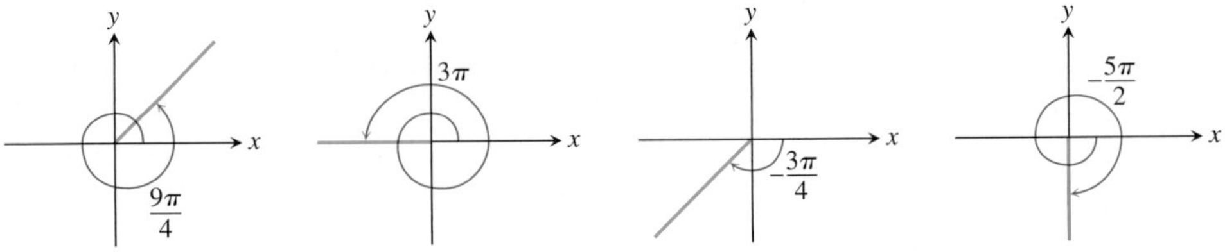

FIGURE A.23 Nonzero radian measures can be positive or negative and can go beyond 2π.

Angle Convention: Use Radians From now on, in this book it is assumed that all angles are measured in radians unless degrees or some other unit is stated explicitly. When we talk about the angle $\pi/3$, we mean $\pi/3$ radians (which is 60°), not $\pi/3$ degrees. When you do calculus, keep your calculator in radian mode.

The Six Basic Trigonometric Functions

You are probably familiar with defining the trigonometric functions of an acute angle in terms of the sides of a right triangle (Figure A.24). We extend this definition to obtuse and negative angles by first placing the angle in standard position in a circle of radius r. We

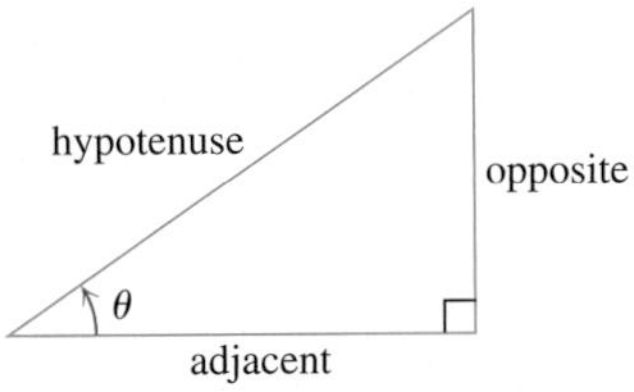

$$\sin\theta = \frac{\text{opp}}{\text{hyp}} \qquad \csc\theta = \frac{\text{hyp}}{\text{opp}}$$

$$\cos\theta = \frac{\text{adj}}{\text{hyp}} \qquad \sec\theta = \frac{\text{hyp}}{\text{adj}}$$

$$\tan\theta = \frac{\text{opp}}{\text{adj}} \qquad \cot\theta = \frac{\text{adj}}{\text{opp}}$$

FIGURE A.24 Trigonometric ratios of an acute angle.

then define the trigonometric functions in terms of the coordinates of the point $P(x, y)$ where the angle's terminal ray intersects the circle (Figure A.25).

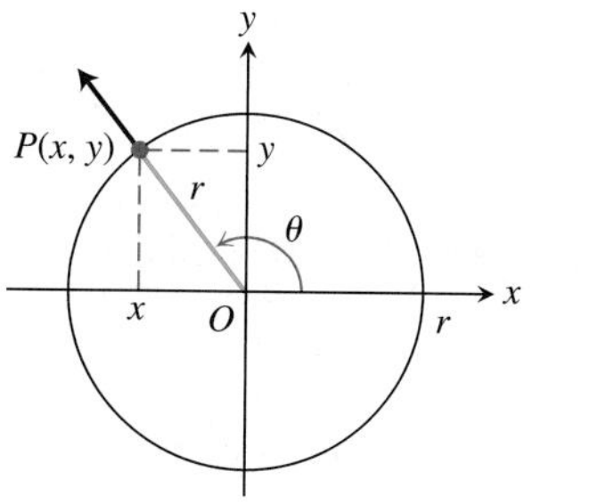

FIGURE A.25 The trigonometric functions of a general angle θ are defined in terms of x, y, and r.

$$\textbf{sine:}\quad \sin\theta = \frac{y}{r} \qquad \textbf{cosecant:}\quad \csc\theta = \frac{r}{y}$$

$$\textbf{cosine:}\quad \cos\theta = \frac{x}{r} \qquad \textbf{secant:}\quad \sec\theta = \frac{r}{x}$$

$$\textbf{tangent:}\quad \tan\theta = \frac{y}{x} \qquad \textbf{cotangent:}\quad \cot\theta = \frac{x}{y}$$

These extended definitions agree with the right-triangle definitions when the angle is acute.

Notice also the following definitions, whenever the quotients are defined.

$$\tan\theta = \frac{\sin\theta}{\cos\theta} \qquad \cot\theta = \frac{1}{\tan\theta}$$

$$\sec\theta = \frac{1}{\cos\theta} \qquad \csc\theta = \frac{1}{\sin\theta}$$

As you can see, $\tan\theta$ and $\sec\theta$ are not defined if $x = 0$. This means they are not defined if θ is $\pm\pi/2, \pm 3\pi/2, \ldots$. Similarly, $\cot\theta$ and $\csc\theta$ are not defined for values of θ for which $y = 0$, namely $\theta = 0, \pm\pi, \pm 2\pi, \ldots$.

The exact values of these trigonometric ratios for some angles can be read from the triangles in Figure A.26. For instance,

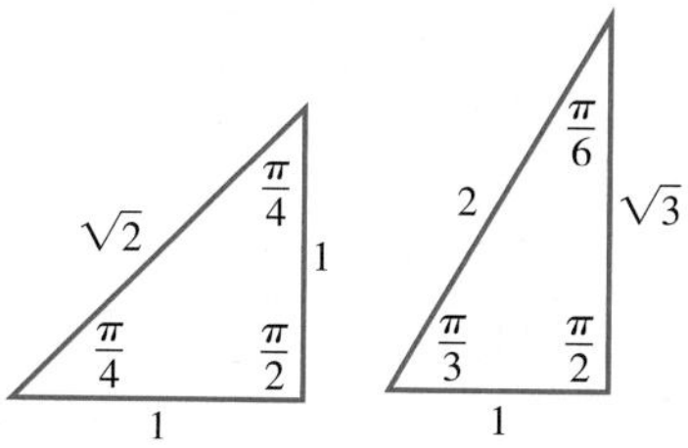

FIGURE A.26 Radian angles and side lengths of two common triangles.

$$\sin\frac{\pi}{4} = \frac{1}{\sqrt{2}} \qquad \sin\frac{\pi}{6} = \frac{1}{2} \qquad \sin\frac{\pi}{3} = \frac{\sqrt{3}}{2}$$

$$\cos\frac{\pi}{4} = \frac{1}{\sqrt{2}} \qquad \cos\frac{\pi}{6} = \frac{\sqrt{3}}{2} \qquad \cos\frac{\pi}{3} = \frac{1}{2}$$

$$\tan\frac{\pi}{4} = 1 \qquad \tan\frac{\pi}{6} = \frac{1}{\sqrt{3}} \qquad \tan\frac{\pi}{3} = \sqrt{3}$$

The CAST rule (Figure A.27) is useful for remembering when the basic trigonometric functions are positive or negative. For instance, from the triangle in Figure A.28, we see that

$$\sin\frac{2\pi}{3} = \frac{\sqrt{3}}{2}, \qquad \cos\frac{2\pi}{3} = -\frac{1}{2}, \qquad \tan\frac{2\pi}{3} = -\sqrt{3}.$$

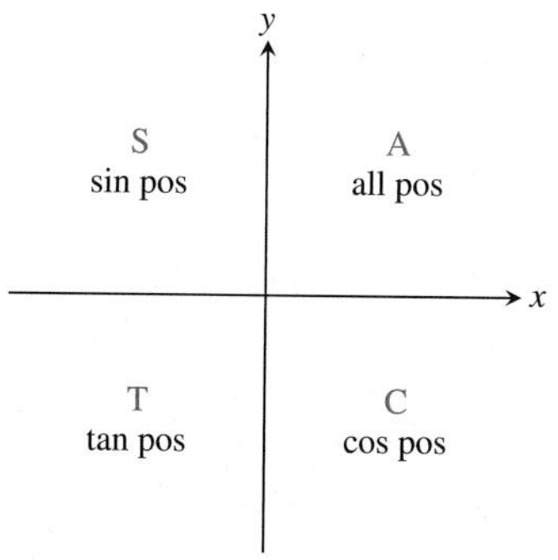

FIGURE A.27 The CAST rule, remembered by the statement "All Students Take Calculus," tells which trigonometric functions are positive in each quadrant.

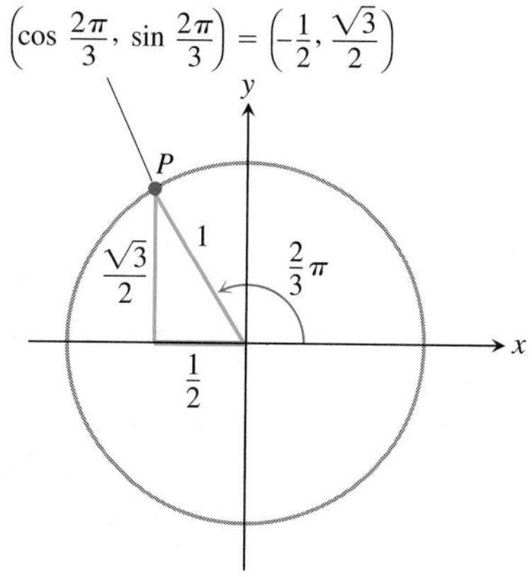

FIGURE A.28 The triangle for calculating the sine and cosine of $2\pi/3$ radians. The side lengths come from the geometry of right triangles.

Using a similar method we determined the values of $\sin\theta$, $\cos\theta$, and $\tan\theta$ shown in Table A.3.

TABLE A.3 Values of $\sin\theta$, $\cos\theta$, and $\tan\theta$ for selected values of θ

Degrees	**−180**	**−135**	**−90**	**−45**	**0**	**30**	**45**	**60**	**90**	**120**	**135**	**150**	**180**	**270**	**360**
θ (radians)	$-\pi$	$\frac{-3\pi}{4}$	$\frac{-\pi}{2}$	$\frac{-\pi}{4}$	0	$\frac{\pi}{6}$	$\frac{\pi}{4}$	$\frac{\pi}{3}$	$\frac{\pi}{2}$	$\frac{2\pi}{3}$	$\frac{3\pi}{4}$	$\frac{5\pi}{6}$	π	$\frac{3\pi}{2}$	2π
$\sin\theta$	0	$\frac{-\sqrt{2}}{2}$	-1	$\frac{-\sqrt{2}}{2}$	0	$\frac{1}{2}$	$\frac{\sqrt{2}}{2}$	$\frac{\sqrt{3}}{2}$	1	$\frac{\sqrt{3}}{2}$	$\frac{\sqrt{2}}{2}$	$\frac{1}{2}$	0	-1	0
$\cos\theta$	-1	$\frac{-\sqrt{2}}{2}$	0	$\frac{\sqrt{2}}{2}$	1	$\frac{\sqrt{3}}{2}$	$\frac{\sqrt{2}}{2}$	$\frac{1}{2}$	0	$-\frac{1}{2}$	$\frac{-\sqrt{2}}{2}$	$\frac{-\sqrt{3}}{2}$	-1	0	1
$\tan\theta$	0	1		-1	0	$\frac{\sqrt{3}}{3}$	1	$\sqrt{3}$		$-\sqrt{3}$	-1	$\frac{-\sqrt{3}}{3}$	0		0

Periodicity and Graphs of the Trigonometric Functions

When an angle of measure θ and an angle of measure $\theta + 2\pi$ are in standard position, their terminal rays coincide. The two angles therefore have the same trigonometric function values: $\sin(\theta + 2\pi) = \sin\theta$, $\tan(\theta + 2\pi) = \tan\theta$, and so on. Similarly, $\cos(\theta - 2\pi) = \cos\theta$, $\sin(\theta - 2\pi) = \sin\theta$, and so on. We describe this repeating behavior by saying that the six basic trigonometric functions are *periodic*.

Periods of Trigonometric Functions

Period π: $\tan(x + \pi) = \tan x$
$\cot(x + \pi) = \cot x$

Period 2π: $\sin(x + 2\pi) = \sin x$
$\cos(x + 2\pi) = \cos x$
$\sec(x + 2\pi) = \sec x$
$\csc(x + 2\pi) = \csc x$

> **DEFINITION** A function $f(x)$ is **periodic** if there is a positive number p such that $f(x + p) = f(x)$ for every value of x. The smallest such value of p is the **period** of f.

When we graph trigonometric functions in the coordinate plane, we usually denote the independent variable by x instead of θ. Figure A.29 shows that the tangent and cotangent

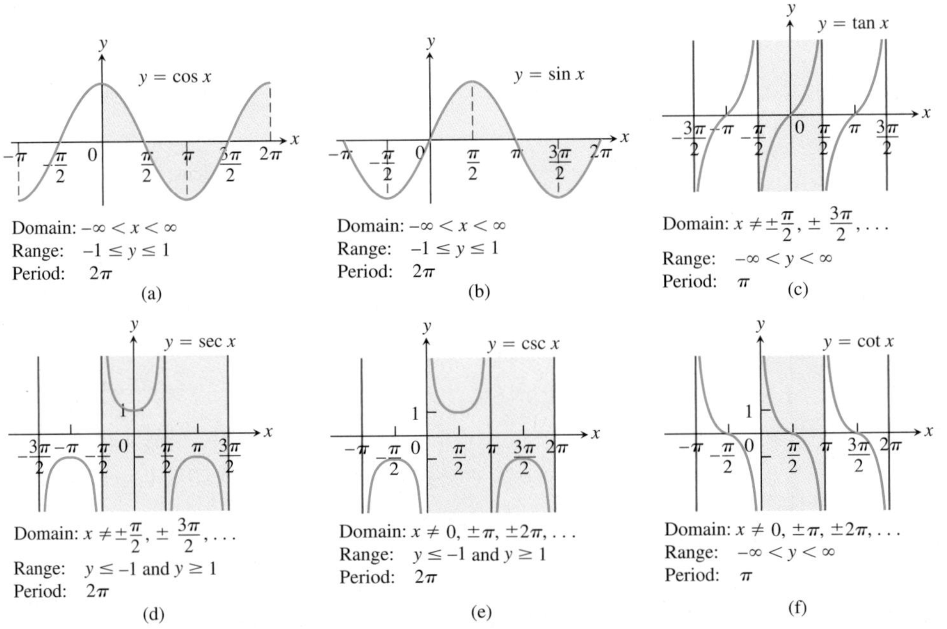

FIGURE A.29 Graphs of the six basic trigonometric functions using radian measure. The shading for each trigonometric function indicates its periodicity.

Even
$\cos(-x) = \cos x$
$\sec(-x) = \sec x$

Odd
$\sin(-x) = -\sin x$
$\tan(-x) = -\tan x$
$\csc(-x) = -\csc x$
$\cot(-x) = -\cot x$

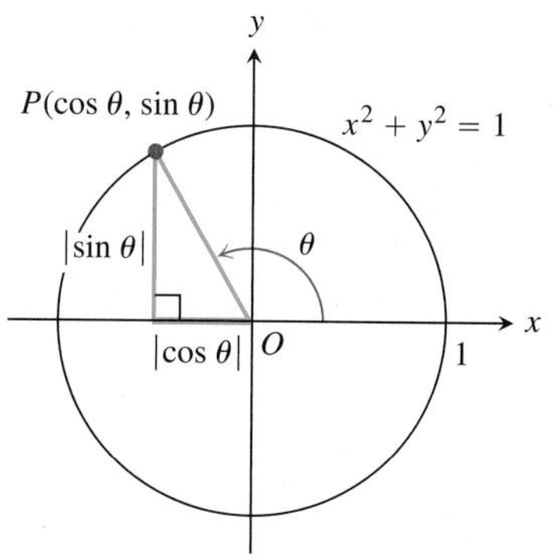

FIGURE A.30 The reference triangle for a general angle θ.

functions have period $p = \pi$, and the other four functions have period 2π. Also, the symmetries in these graphs reveal that the cosine and secant functions are even and the other four functions are odd.

Trigonometric Identities

The coordinates of any point $P(x, y)$ in the plane can be expressed in terms of the point's distance from the origin and the angle that ray OP makes with the positive x-axis (Figure A.25). Since $x/r = \cos\theta$ and $y/r = \sin\theta$, we have

$$x = r\cos\theta, \qquad y = r\sin\theta.$$

When $r = 1$ we can apply the Pythagorean theorem to the reference right triangle in Figure A.30 and obtain the equation

$$\cos^2\theta + \sin^2\theta = 1. \tag{3}$$

This equation, true for all values of θ, is the most frequently used identity in trigonometry. Dividing this identity in turn by $\cos^2\theta$ and $\sin^2\theta$ gives

$$1 + \tan^2\theta = \sec^2\theta.$$
$$1 + \cot^2\theta = \csc^2\theta.$$

The following formulas hold for all angles A and B (Exercise 32).

Addition Formulas

$$\cos(A + B) = \cos A\cos B - \sin A\sin B$$
$$\sin(A + B) = \sin A\cos B + \cos A\sin B \tag{4}$$

There are similar formulas for $\cos(A - B)$ and $\sin(A - B)$ (Exercises 21 and 22). All the trigonometric identities needed in this book derive from Equations (3) and (4). For example, substituting θ for both A and B in the addition formulas gives

Double-Angle Formulas

$$\cos 2\theta = \cos^2\theta - \sin^2\theta$$
$$\sin 2\theta = 2\sin\theta\cos\theta \tag{5}$$

Additional formulas come from combining the equations

$$\cos^2\theta + \sin^2\theta = 1, \qquad \cos^2\theta - \sin^2\theta = \cos 2\theta.$$

We add the two equations to get $2\cos^2\theta = 1 + \cos 2\theta$ and subtract the second from the first to get $2\sin^2\theta = 1 - \cos 2\theta$. This results in the following identities, which are useful in integral calculus.

Half-Angle Formulas

$$\cos^2\theta = \frac{1 + \cos 2\theta}{2} \tag{6}$$

$$\sin^2\theta = \frac{1 - \cos 2\theta}{2} \tag{7}$$

The Law of Cosines

If a, b, and c are sides of a triangle ABC and if θ is the angle opposite c, then

$$c^2 = a^2 + b^2 - 2ab\cos\theta. \tag{8}$$

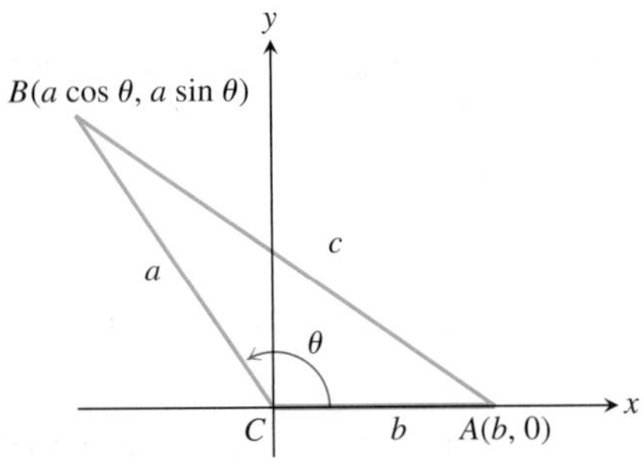

FIGURE A.31 The square of the distance between A and B gives the law of cosines.

This equation is called the **law of cosines**.

We can see why the law holds if we introduce coordinate axes with the origin at C and the positive x-axis along one side of the triangle, as in Figure A.31. The coordinates of A are $(b, 0)$; the coordinates of B are $(a\cos\theta, a\sin\theta)$. The square of the distance between A and B is therefore

$$\begin{aligned} c^2 &= (a\cos\theta - b)^2 + (a\sin\theta)^2 \\ &= a^2\underbrace{(\cos^2\theta + \sin^2\theta)}_{1} + b^2 - 2ab\cos\theta \\ &= a^2 + b^2 - 2ab\cos\theta. \end{aligned}$$

The law of cosines generalizes the Pythagorean theorem. If $\theta = \pi/2$, then $\cos\theta = 0$ and $c^2 = a^2 + b^2$.

Transformations of Trigonometric Graphs

The rules for shifting, stretching, compressing, and reflecting the graph of a function summarized in the following diagram, apply to the trigonometric functions we have discussed in this section.

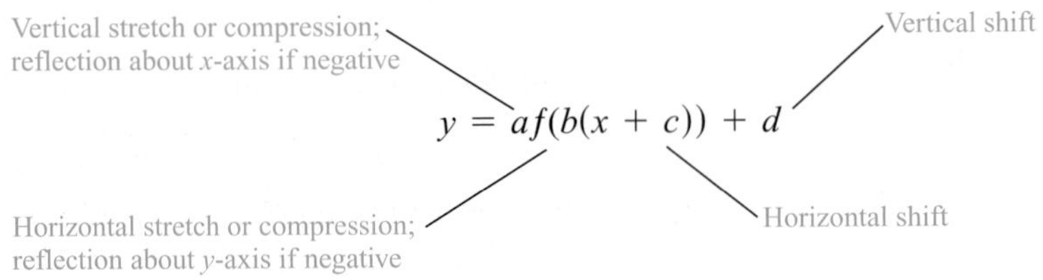

The transformation rules applied to the sine function give the **general sine function** or **sinusoid** formula

$$f(x) = A\sin\left[\frac{2\pi}{B}(x - C)\right] + D,$$

where $|A|$ is the *amplitude*, $|B|$ is the *period*, C is the *horizontal shift*, and D is the *vertical shift*. A graphical interpretation of the various terms is revealing and given below.

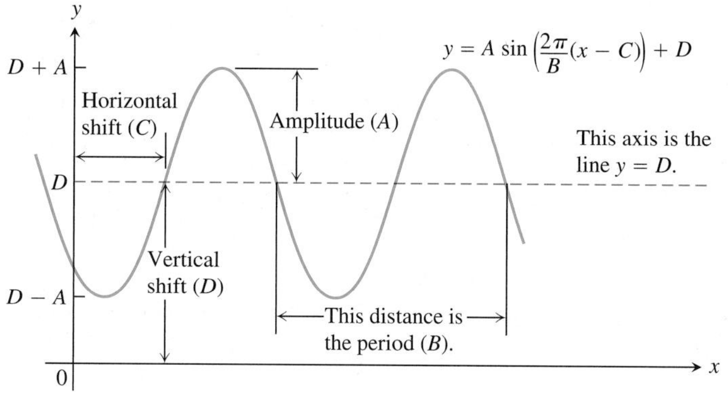

Two Special Inequalities

For any angle θ measured in radians,

$$-|\theta| \le \sin\theta \le |\theta| \quad \text{and} \quad -|\theta| \le 1 - \cos\theta \le |\theta|.$$

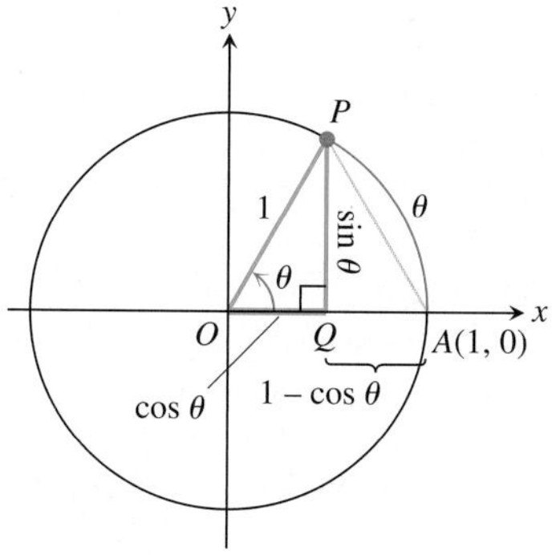

FIGURE A.32 From the geometry of this figure, drawn for $\theta > 0$, we get the inequality $\sin^2\theta + (1 - \cos\theta)^2 < \theta^2$.

To establish these inequalities, we picture θ as an angle in standard position (Figure A.32). The circle in the figure is a unit circle, so $|\theta|$ equals the length of the circular arc AP. The length of line segment AP is therefore less than $|\theta|$.

Triangle APQ is a right triangle with sides of length

$$QP = |\sin\theta|, \qquad AQ = 1 - \cos\theta.$$

From the Pythagorean theorem and the fact that $AP < |\theta|$, we get

$$\sin^2\theta + (1 - \cos\theta)^2 = (AP)^2 < \theta^2. \tag{9}$$

The terms on the left-hand side of Equation (9) are both positive, so each is smaller than their sum and hence is less than θ^2:

$$\sin^2\theta < \theta^2 \quad \text{and} \quad (1 - \cos\theta)^2 < \theta^2.$$

By taking square roots, this is equivalent to saying that

$$|\sin\theta| < |\theta| \quad \text{and} \quad |1 - \cos\theta| < |\theta|.$$

Allowing for $\theta = 0$ as well, we then have

$$-|\theta| \le \sin\theta \le |\theta| \quad \text{and} \quad -|\theta| \le 1 - \cos\theta \le |\theta|.$$

These inequalities are used in Example 10, Section 1.4.

EXERCISES A.4

1. On a circle of radius 10 m, how long is an arc that subtends a central angle of (**a**) $4\pi/5$ radians? (**b**) $110°$?
2. A central angle in a circle of radius 8 is subtended by an arc of length 10π. Find the angle's radian and degree measures.
3. You want to make an $80°$ angle by marking an arc on the perimeter of a 12-in.-diameter disk and drawing lines from the ends of the arc to the disk's center. To the nearest tenth of an inch, how long should the arc be?
4. If you roll a 1-m-diameter wheel forward 30 cm over level ground, through what angle will the wheel turn? Answer in radians (to the nearest tenth) and degrees (to the nearest degree).
5. Copy and complete the following table of function values. If the function is undefined at a given angle, enter "UND." Do not use a calculator or tables.

θ	$-\pi$	$-2\pi/3$	0	$\pi/2$	$3\pi/4$
$\sin\theta$					
$\cos\theta$					
$\tan\theta$					
$\cot\theta$					
$\sec\theta$					
$\csc\theta$					

In Exercises 6–8, one of $\sin x$, $\cos x$, and $\tan x$ is given. Find the other two if x lies in the specified interval.

6. $\sin x = \frac{3}{5}, \quad x \in \left[\frac{\pi}{2}, \pi\right]$
7. $\cos x = \frac{1}{3}, \quad x \in \left[-\frac{\pi}{2}, 0\right]$
8. $\tan x = \frac{1}{2}, \quad x \in \left[\pi, \frac{3\pi}{2}\right]$

Graph the functions in Exercises 9–13. What is the period of each function?

9. $\sin 2x$
10. $\cos \pi x$
11. $-\sin\frac{\pi x}{3}$
12. $\cos\left(x - \frac{\pi}{2}\right)$
13. $\sin\left(x - \frac{\pi}{4}\right) + 1$

Graph the functions in Exercises 14–17 in the ts-plane (t-axis horizontal, s-axis vertical). What is the period of each function? What symmetries do the graphs have?

14. $s = \cot 2t$
15. $s = -\tan \pi t$
16. $s = \sec\left(\frac{\pi t}{2}\right)$
17. $s = \csc\left(\frac{t}{2}\right)$
18. Graph $y = \sin x$ and $y = \lfloor \sin x \rfloor$ together. What are the domain and range of $\lfloor \sin x \rfloor$?

Use the addition formulas to derive the identities in Exercises 19 and 20.

19. $\cos\left(x - \frac{\pi}{2}\right) = \sin x$
20. $\sin\left(x + \frac{\pi}{2}\right) = \cos x$
21. $\cos(A - B) = \cos A\cos B + \sin A\sin B$
22. $\sin(A - B) = \sin A\cos B - \cos A\sin B$
23. What happens if you take $B = A$ in the trigonometric identity $\cos(A - B) = \cos A\cos B + \sin A\sin B$? Does the result agree with something you already know?
24. What happens if you take $B = 2\pi$ in the addition formulas? Do the results agree with something you already know?

In Exercises 25 and 26, express the given quantity in terms of $\sin x$ and $\cos x$.

25. $\cos(\pi + x)$
26. $\sin\left(\frac{3\pi}{2} - x\right)$
27. Evaluate $\sin\frac{7\pi}{12}$ as $\sin\left(\frac{\pi}{4} + \frac{\pi}{3}\right)$.
28. Evaluate $\cos\frac{\pi}{12}$.

Find the function values in Exercises 29 and 30.

29. $\cos^2\frac{\pi}{8}$
30. $\sin^2\frac{\pi}{12}$
31. **The tangent sum formula** The standard formula for the tangent of the sum of two angles is

$$\tan(A + B) = \frac{\tan A + \tan B}{1 - \tan A\tan B}.$$

Derive the formula.

32. **a.** Derive the formula for $\cos(A + B)$ by substituting $-B$ for B in the formula for $\cos(A - B)$ from Exercise 21.
    b. Apply the formula for $\cos(A - B)$ to the identity $\sin\theta = \cos\left(\frac{\pi}{2} - \theta\right)$ to obtain the addition formula for $\sin(A + B)$.
33. A triangle has sides $a = 2$ and $b = 3$ and angle $C = 60°$. Find the length of side c.
34. **The law of sines** *The law of sines* says that if a, b, and c are the sides opposite the angles A, B, and C in a triangle, then

$$\frac{\sin A}{a} = \frac{\sin B}{b} = \frac{\sin C}{c}.$$

Use the accompanying figures and the identity $\sin(\pi - \theta) = \sin\theta$, if required, to derive the law.

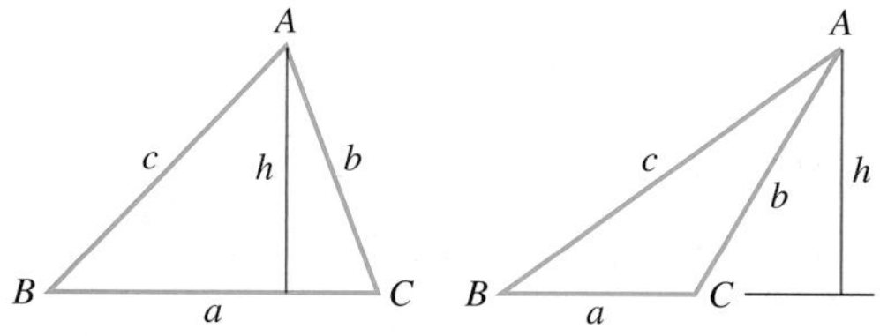

35. A triangle has sides $a = 2$ and $b = 3$ and angle $C = 60°$ (as in Exercise 33). Find the sine of angle B using the law of sines.

36. A triangle has side $c = 2$ and angles $A = \pi/4$ and $B = \pi/3$. Find the length a of the side opposite A.

T **37. The approximation $\sin x \approx x$** It is often useful to know that, when x is measured in radians, $\sin x \approx x$ for numerically small values of x. In Section 2.13, we will see why the approximation holds. The approximation error is less than 1 in 5000 if $|x| < 0.1$.

a. With your grapher in radian mode, graph $y = \sin x$ and $y = x$ together in a viewing window about the origin. What do you see happening as x nears the origin?

b. With your grapher in degree mode, graph $y = \sin x$ and $y = x$ together about the origin again. How is the picture different from the one obtained with radian mode?

c. A quick radian mode check Is your calculator in radian mode? Evaluate $\sin x$ at a value of x near the origin, say $x = 0.1$. If $\sin x \approx x$, the calculator is in radian mode; if not, it isn't. Try it.

A.5 Proofs of Limit Theorems and L'Hôpital's Rule

This appendix proves Theorem 1, Parts 2–5, and Theorem 4 from Section 1.4.

THEOREM 1—Limit Laws If L, M, c, and k are real numbers and

$$\lim_{x\to c} f(x) = L \quad \text{and} \quad \lim_{x\to c} g(x) = M, \quad \text{then}$$

1. *Sum Rule*: $\lim_{x\to c} \left(f(x) + g(x)\right) = L + M$
2. *Difference Rule*: $\lim_{x\to c} \left(f(x) - g(x)\right) = L - M$
3. *Constant Multiple Rule*: $\lim_{x\to c} \left(k \cdot f(x)\right) = k \cdot L$
4. *Product Rule*: $\lim_{x\to c} \left(f(x) \cdot g(x)\right) = L \cdot M$
5. *Quotient Rule*: $\lim_{x\to c} \dfrac{f(x)}{g(x)} = \dfrac{L}{M}, \quad M \neq 0$
6. *Power Rule*: $\lim_{x\to c} [f(x)]^n = L^n$, n a positive integer
7. *Root Rule*: $\lim_{x\to c} \sqrt[n]{f(x)} = \sqrt[n]{L} = L^{1/n}$, n a positive integer

(If n is even, we assume that $\lim_{x\to c} f(x) = L > 0$.)

We proved the Sum Rule in Section 1.5 and the Power Rule is proved in more advanced texts. We obtain the Difference Rule by replacing $g(x)$ by $-g(x)$ and M by $-M$ in the Sum Rule. The Constant Multiple Rule is the special case $g(x) = k$ of the Product Rule. This leaves only the Product and Quotient Rules.

Proof of the Limit Product Rule We show that for any $\epsilon > 0$ there exists a $\delta > 0$ such that for all x in the intersection D of the domains of f and g,

$$0 < |x - c| < \delta \quad \Rightarrow \quad |f(x)g(x) - LM| < \epsilon.$$

Suppose then that ϵ is a positive number, and write $f(x)$ and $g(x)$ as

$$f(x) = L + (f(x) - L), \qquad g(x) = M + (g(x) - M).$$

Multiply these expressions together and subtract LM:

$$\begin{aligned} f(x) \cdot g(x) - LM &= (L + (f(x) - L))(M + (g(x) - M)) - LM \\ &= LM + L(g(x) - M) + M(f(x) - L) \\ &\quad + (f(x) - L)(g(x) - M) - LM \\ &= L(g(x) - M) + M(f(x) - L) + (f(x) - L)(g(x) - M). \end{aligned} \tag{1}$$

Since f and g have limits L and M as $x \to c$, there exist positive numbers δ_1, δ_2, δ_3, and δ_4 such that for all x in D

$$\begin{aligned} 0 < |x - c| < \delta_1 &\quad \Rightarrow \quad |f(x) - L| < \sqrt{\epsilon/3} \\ 0 < |x - c| < \delta_2 &\quad \Rightarrow \quad |g(x) - M| < \sqrt{\epsilon/3} \\ 0 < |x - c| < \delta_3 &\quad \Rightarrow \quad |f(x) - L| < \epsilon/(3(1 + |M|)) \\ 0 < |x - c| < \delta_4 &\quad \Rightarrow \quad |g(x) - M| < \epsilon/(3(1 + |L|)) \end{aligned} \tag{2}$$

If we take δ to be the smallest numbers δ_1 through δ_4, the inequalities on the right-hand side of the Implications (2) will hold simultaneously for $0 < |x - c| < \delta$. Therefore, for all x in D, $0 < |x - c| < \delta$ implies

$$\begin{aligned} &|f(x) \cdot g(x) - LM| && \text{Triangle inequality applied to Eq. (1)} \\ &\quad \le |L||g(x) - M| + |M||f(x) - L| + |f(x) - L||g(x) - M| \\ &\quad \le (1 + |L|)|g(x) - M| + (1 + |M|)|f(x) - L| + |f(x) - L||g(x) - M| \\ &\quad < \frac{\epsilon}{3} + \frac{\epsilon}{3} + \sqrt{\frac{\epsilon}{3}}\sqrt{\frac{\epsilon}{3}} = \epsilon. && \text{Values from (2)} \end{aligned}$$

This completes the proof of the Limit Product Rule. ■

Proof of the Limit Quotient Rule We show that $\lim_{x\to c}(1/g(x)) = 1/M$. We can then conclude that

$$\lim_{x\to c} \frac{f(x)}{g(x)} = \lim_{x\to c}\left(f(x) \cdot \frac{1}{g(x)}\right) = \lim_{x\to c} f(x) \cdot \lim_{x\to c} \frac{1}{g(x)} = L \cdot \frac{1}{M} = \frac{L}{M}$$

by the Limit Product Rule.

Let $\epsilon > 0$ be given. To show that $\lim_{x\to c}(1/g(x)) = 1/M$, we need to show that there exists a $\delta > 0$ such that for all x

$$0 < |x - c| < \delta \quad \Rightarrow \quad \left|\frac{1}{g(x)} - \frac{1}{M}\right| < \epsilon.$$

Since $|M| > 0$, there exists a positive number δ_1 such that for all x

$$0 < |x - c| < \delta_1 \quad \Rightarrow \quad |g(x) - M| < \frac{M}{2}. \tag{3}$$

For any numbers A and B it can be shown that $|A| - |B| \leq |A - B|$ and $|B| - |A| \leq |A - B|$, from which it follows that $||A| - |B|| \leq |A - B|$. With $A = g(x)$ and $B = M$, this becomes

$$||g(x)| - |M|| \leq |g(x) - M|,$$

which can be combined with the inequality on the right in Implication (3) to get, in turn,

$$||g(x)| - |M|| < \frac{|M|}{2}$$

$$-\frac{|M|}{2} < |g(x)| - |M| < \frac{|M|}{2}$$

$$\frac{|M|}{2} < |g(x)| < \frac{3|M|}{2}$$

$$|M| < 2|g(x)| < 3|M|$$

$$\frac{1}{|g(x)|} < \frac{2}{|M|} < \frac{3}{|g(x)|} \tag{4}$$

Therefore, $0 < |x - c| < \delta_1$ implies that

$$\left|\frac{1}{g(x)} - \frac{1}{M}\right| = \left|\frac{M - g(x)}{Mg(x)}\right| \leq \frac{1}{|M|} \cdot \frac{1}{|g(x)|} \cdot |M - g(x)|$$

$$< \frac{1}{|M|} \cdot \frac{2}{|M|} \cdot |M - g(x)|. \qquad \text{Inequality (4)} \tag{5}$$

Since $(1/2)|M|^2\epsilon > 0$, there exists a number $\delta_2 > 0$ such that for all x

$$0 < |x - c| < \delta_2 \quad \Rightarrow \quad |M - g(x)| < \frac{\epsilon}{2}|M|^2. \tag{6}$$

If we take δ to be the smaller of δ_1 and δ_2, the conclusions in (5) and (6) both hold for all x such that $0 < |x - c| < \delta$. Combining these conclusions gives

$$0 < |x - c| < \delta \quad \Rightarrow \quad \left|\frac{1}{g(x)} - \frac{1}{M}\right| < \epsilon.$$

This concludes the proof of the Limit Quotient Rule. ■

THEOREM 4—The Sandwich Theorem Suppose that $g(x) \leq f(x) \leq h(x)$ for all x in some open interval I containing c, except possibly at $x = c$ itself. Suppose also that $\lim_{x\to c} g(x) = \lim_{x\to c} h(x) = L$. Then $\lim_{x\to c} f(x) = L$.

Proof for Right-Hand Limits Suppose $\lim_{x\to c^+} g(x) = \lim_{x\to c^+} h(x) = L$. Then for any $\epsilon > 0$ there exists a $\delta > 0$ such that for all x the interval $c < x < c + \delta$ is contained in I and the inequality implies

$$L - \epsilon < g(x) < L + \epsilon \quad \text{and} \quad L - \epsilon < h(x) < L + \epsilon.$$

These inequalities combine with the inequality $g(x) \leq f(x) \leq h(x)$ to give

$$\begin{aligned} L - \epsilon &< g(x) \leq f(x) \leq h(x) < L + \epsilon, \\ L - \epsilon &< f(x) < L + \epsilon, \\ -\epsilon &< f(x) - L < \epsilon. \end{aligned}$$

Therefore, for all x, the inequality $c < x < c + \delta$ implies $|f(x) - L| < \epsilon$.

Proof for Left-Hand Limits Suppose $\lim_{x \to c^-} g(x) = \lim_{x \to c^-} h(x) = L$. Then for any $\epsilon > 0$ there exists a $\delta > 0$ such that for all x the interval $c - \delta < x < c$ is contained in I and the inequality implies

$$L - \epsilon < g(x) < L + \epsilon \quad \text{and} \quad L - \epsilon < h(x) < L + \epsilon.$$

We conclude as before that for all x, $c - \delta < x < c$ implies $|f(x) - L| < \epsilon$.

Proof for Two-Sided Limits If $\lim_{x \to c} g(x) = \lim_{x \to c} h(x) = L$, then $g(x)$ and $h(x)$ both approach L as $x \to c^+$ and as $x \to c^-$; so $\lim_{x \to c^+} f(x) = L$ and $\lim_{x \to c^-} f(x) = L$. Hence $\lim_{x \to c} f(x)$ exists and equals L. ■

THEOREM 10—Limits of Continuous Functions If g is continuous at the point b and $\lim_{x \to c} f(x) = b$, then

$$\lim_{x \to c} g(f(x)) = g(b) = g(\lim_{x \to c} (f(x))).$$

Proof Let $\epsilon > 0$ be given. Since g is continuous at b, there exists a number $\delta_1 > 0$ such that

$$|g(y) - g(b)| < \epsilon \quad \text{whenever} \quad 0 < |y - b| < \delta_1.$$

Since $\lim_{x \to c} f(x) = b$, there exists a $\delta > 0$ such that

$$|f(x) - b| < \delta_1 \quad \text{whenever} \quad 0 < |x - c| < \delta.$$

If we let $y = f(x)$, we then have that

$$|y - b| < \delta_1 \quad \text{whenever} \quad 0 < |x - c| < \delta$$

which implies from the first statement that $|g(y) - g(b)| = |g(f(x)) - g(b)| < \epsilon$ whenever $0 < |x - c| < \delta$. From the definition of limit, this proves that $\lim_{x \to c} g(f(x)) = g(b)$. ■

Proof of L'Hôpital's Rule

The proof of l'Hôpital's Rule is based on Cauchy's Mean Value Theorem, a Mean Value Theorem that involves two functions instead of one. We prove Cauchy's Theorem first and then show how it leads to l'Hôpital's Rule.

HISTORICAL BIOGRAPHY

Augustin-Louis Cauchy
(1789–1857)

THEOREM—Cauchy's Mean Value Theorem Suppose functions f and g are continuous on $[a, b]$ and differentiable throughout (a, b) and also suppose $g'(x) \neq 0$ throughout (a, b). Then there exists a number c in (a, b) at which

$$\frac{f'(c)}{g'(c)} = \frac{f(b) - f(a)}{g(b) - g(a)}.$$

Proof We apply the Mean Value Theorem of Section 3.2 twice. First we use it to show that $g(a) \neq g(b)$. For if $g(b)$ did equal $g(a)$, then the Mean Value Theorem would give

$$g'(c) = \frac{g(b) - g(a)}{b - a} = 0$$

for some c between a and b, which cannot happen because $g'(x) \neq 0$ in (a, b).

We next apply the Mean Value Theorem to the function

$$F(x) = f(x) - f(a) - \frac{f(b) - f(a)}{g(b) - g(a)}[g(x) - g(a)].$$

This function is continuous and differentiable where f and g are, and $F(b) = F(a) = 0$. Therefore, there is a number c between a and b for which $F'(c) = 0$. When expressed in terms of f and g, this equation becomes

$$F'(c) = f'(c) - \frac{f(b) - f(a)}{g(b) - g(a)}[g'(c)] = 0$$

or

$$\frac{f'(c)}{g'(c)} = \frac{f(b) - f(a)}{g(b) - g(a)}. \qquad \blacksquare$$

THEOREM 6— L'Hôpital's Rule Suppose that $f(a) = g(a) = 0$, that f and g are differentiable on an open interval I containing a, and that $g'(x) \neq 0$ on I if $x \neq a$. Then

$$\lim_{x \to a} \frac{f(x)}{g(x)} = \lim_{x \to a} \frac{f'(x)}{g'(x)},$$

assuming that the limit on the right side of this equation exists.

Proof We first establish the limit equation for the case $x \to a^+$. The method needs almost no change to apply to $x \to a^-$, and the combination of these two cases establishes the result.

Suppose that x lies to the right of a. Then $g'(x) \neq 0$, and we can apply Cauchy's Mean Value Theorem to the closed interval from a to x. This step produces a number c between a and x such that

$$\frac{f'(c)}{g'(c)} = \frac{f(x) - f(a)}{g(x) - g(a)}.$$

But $f(a) = g(a) = 0$, so

$$\frac{f'(c)}{g'(c)} = \frac{f(x)}{g(x)}.$$

As x approaches a, c approaches a because it always lies between a and x. Therefore,

$$\lim_{x \to a^+} \frac{f(x)}{g(x)} = \lim_{c \to a^+} \frac{f'(c)}{g'(c)} = \lim_{x \to a^+} \frac{f'(x)}{g'(x)},$$

which establishes l'Hôpital's Rule for the case where x approaches a from above. The case where x approaches a from below is proved by applying Cauchy's Mean Value Theorem to the closed interval $[x, a]$, $x < a$. ■

EXERCISES A.5

1. Suppose that functions $f_1(x)$, $f_2(x)$, and $f_3(x)$ have limits L_1, L_2, and L_3, respectively, as $x \to c$. Show that their sum has limit $L_1 + L_2 + L_3$. Use mathematical induction (Appendix 2) to generalize this result to the sum of any finite number of functions.

2. Use mathematical induction and the Limit Product Rule in Theorem 1 to show that if functions $f_1(x), f_2(x), \ldots, f_n(x)$ have limits $L_1, L_2, \ldots, L_n$ as $x \to c$, then

$$\lim_{x \to c} f_1(x) f_2(x) \cdot \cdots \cdot f_n(x) = L_1 \cdot L_2 \cdot \cdots \cdot L_n.$$

3. Use the fact that $\lim_{x \to c} x = c$ and the result of Exercise 2 to show that $\lim_{x \to c} x^n = c^n$ for any integer $n > 1$.

4. **Limits of polynomials** Use the fact that $\lim_{x \to c}(k) = k$ for any number k together with the results of Exercises 1 and 3 to show that $\lim_{x \to c} f(x) = f(c)$ for any polynomial function

$$f(x) = a_n x^n + a_{n-1} x^{n-1} + \cdots + a_1 x + a_0.$$

5. **Limits of rational functions** Use Theorem 1 and the result of Exercise 4 to show that if $f(x)$ and $g(x)$ are polynomial functions and $g(c) \neq 0$, then

$$\lim_{x \to c} \frac{f(x)}{g(x)} = \frac{f(c)}{g(c)}.$$

6. **Composites of continuous functions** Figure A.33 gives the diagram for a proof that the composite of two continuous functions is continuous. Reconstruct the proof from the diagram. The statement to be proved is this: If f is continuous at $x = c$ and g is continuous at $f(c)$, then $g \circ f$ is continuous at c.

   Assume that c is an interior point of the domain of f and that $f(c)$ is an interior point of the domain of g. This will make the limits involved two-sided. (The arguments for the cases that involve one-sided limits are similar.)

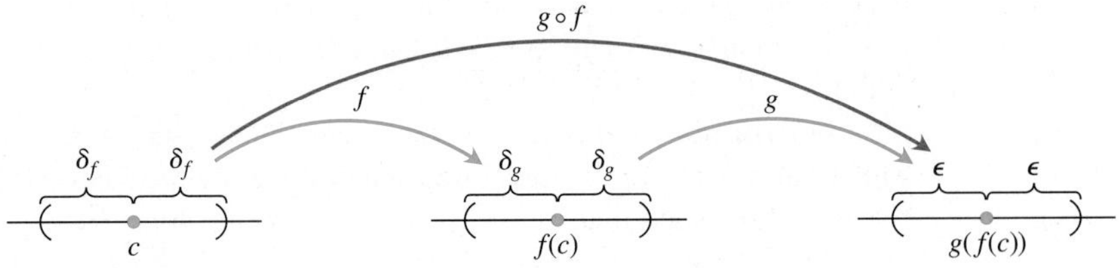

FIGURE A.33 The diagram for a proof that the composite of two continuous functions is continuous.

A.6 Commonly Occurring Limits

This appendix verifies limits (4)–(6) in Theorem 5 of Section 7.1.

Limit 4: If $|x| < 1$, $\lim_{n\to\infty} x^n = 0$ We need to show that to each $\epsilon > 0$ there corresponds an integer N so large that $|x^n| < \epsilon$ for all n greater than N. Since $\epsilon^{1/n} \to 1$, while $|x| < 1$, there exists an integer N for which $\epsilon^{1/N} > |x|$. In other words,

$$|x^N| = |x|^N < \epsilon. \tag{1}$$

This is the integer we seek because, if $|x| < 1$, then

$$|x^n| < |x^N| \quad \text{for all } n > N. \tag{2}$$

Combining (1) and (2) produces $|x^n| < \epsilon$ for all $n > N$, concluding the proof. ■

Limit 5: For any number x, $\lim_{n\to\infty}\left(1 + \frac{x}{n}\right)^n = e^x$ Let

$$a_n = \left(1 + \frac{x}{n}\right)^n.$$

Then

$$\ln a_n = \ln\left(1 + \frac{x}{n}\right)^n = n\ln\left(1 + \frac{x}{n}\right) \to x,$$

as we can see by the following application of l'Hôpital's Rule, in which we differentiate with respect to n:

$$\lim_{n\to\infty} n\ln\left(1 + \frac{x}{n}\right) = \lim_{n\to\infty}\frac{\ln(1 + x/n)}{1/n}$$

$$= \lim_{n\to\infty}\frac{\left(\dfrac{1}{1 + x/n}\right)\cdot\left(-\dfrac{x}{n^2}\right)}{-1/n^2} = \lim_{n\to\infty}\frac{x}{1 + x/n} = x.$$

Apply Theorem 3, Section 7.1, with $f(x) = e^x$ to conclude that

$$\left(1 + \frac{x}{n}\right)^n = a_n = e^{\ln a_n} \to e^x.$$ ■

Limit 6: For any number x, $\lim_{n\to\infty}\frac{x^n}{n!} = 0$ Since

$$-\frac{|x|^n}{n!} \le \frac{x^n}{n!} \le \frac{|x|^n}{n!},$$

all we need to show is that $|x|^n/n! \to 0$. We can then apply the Sandwich Theorem for Sequences (Section 7.1, Theorem 2) to conclude that $x^n/n! \to 0$.

The first step in showing that $|x|^n/n! \to 0$ is to choose an integer $M > |x|$, so that $(|x|/M) < 1$. By Limit 4, just proved, we then have $(|x|/M)^n \to 0$. We then restrict our attention to values of $n > M$. For these values of n, we can write

$$\frac{|x|^n}{n!} = \frac{|x|^n}{1 \cdot 2 \cdot \cdots \cdot M \cdot \underbrace{(M+1)(M+2) \cdot \cdots \cdot n}_{(n-M) \text{ factors}}}$$

$$\leq \frac{|x|^n}{M! M^{n-M}} = \frac{|x|^n M^M}{M! M^n} = \frac{M^M}{M!}\left(\frac{|x|}{M}\right)^n.$$

Thus,

$$0 \leq \frac{|x|^n}{n!} \leq \frac{M^M}{M!}\left(\frac{|x|}{M}\right)^n.$$

Now, the constant $M^M/M!$ does not change as n increases. Thus the Sandwich Theorem tells us that $|x|^n/n! \to 0$ because $(|x|/M)^n \to 0$. ∎

A.7 Theory of the Real Numbers

A rigorous development of calculus is based on properties of the real numbers. Many results about functions, derivatives, and integrals would be false if stated for functions defined only on the rational numbers. In this appendix we briefly examine some basic concepts of the theory of the reals that hint at what might be learned in a deeper, more theoretical study of calculus.

Three types of properties make the real numbers what they are. These are the **algebraic**, **order**, and **completeness** properties. The algebraic properties involve addition and multiplication, subtraction and division. They apply to rational or complex numbers as well as to the reals.

The structure of numbers is built around a set with addition and multiplication operations. The following properties are required of addition and multiplication.

A1 $a + (b + c) = (a + b) + c$ for all a, b, c.

A2 $a + b = b + a$ for all a, b.

A3 There is a number called "0" such that $a + 0 = a$ for all a.

A4 For each number a, there is a b such that $a + b = 0$.

M1 $a(bc) = (ab)c$ for all a, b, c.

M2 $ab = ba$ for all a, b.

M3 There is a number called "1" such that $a \cdot 1 = a$ for all a.

M4 For each nonzero a, there is a b such that $ab = 1$.

D $a(b + c) = ab + bc$ for all a, b, c.

A1 and M1 are *associative laws*, A2 and M2 are *commutativity laws*, A3 and M3 are *identity laws*, and D is the *distributive law*. Sets that have these algebraic properties are examples of **fields**, and are studied in depth in the area of theoretical mathematics called abstract algebra.

The **order** properties allow us to compare the size of any two numbers. The order properties are

O1 For any a and b, either $a \leq b$ or $b \leq a$ or both.

O2 If $a \leq b$ and $b \leq a$ then $a = b$.

O3 If $a \leq b$ and $b \leq c$ then $a \leq c$.

O4 If $a \leq b$ then $a + c \leq b + c$.

O5 If $a \leq b$ and $0 \leq c$ then $ac \leq bc$.

O3 is the *transitivity law*, and O4 and O5 relate ordering to addition and multiplication.

We can order the reals, the integers, and the rational numbers, but we cannot order the complex numbers. There is no reasonable way to decide whether a number like $i = \sqrt{-1}$ is bigger or smaller than zero. A field in which the size of any two elements can be compared as above is called an **ordered field**. Both the rational numbers and the real numbers are ordered fields, and there are many others.

We can think of real numbers geometrically, lining them up as points on a line. The **completeness property** says that the real numbers correspond to all points on the line, with no "holes" or "gaps." The rationals, in contrast, omit points such as $\sqrt{2}$ and π, and the integers even leave out fractions like $1/2$. The reals, having the completeness property, omit no points.

What exactly do we mean by this vague idea of missing holes? To answer this we must give a more precise description of completeness. A number M is an **upper bound** for a set of numbers if all numbers in the set are smaller than or equal to M. M is a **least upper bound** if it is the smallest upper bound. For example, $M = 2$ is an upper bound for the negative numbers. So is $M = 1$, showing that 2 is not a least upper bound. The least upper bound for the set of negative numbers is $M = 0$. We define a **complete** ordered field to be one in which every nonempty set bounded above has a least upper bound.

If we work with just the rational numbers, the set of numbers less than $\sqrt{2}$ is bounded, but it does not have a rational least upper bound, since any rational upper bound M can be replaced by a slightly smaller rational number that is still larger than $\sqrt{2}$. So the rationals are not complete. In the real numbers, a set that is bounded above always has a least upper bound. The reals are a complete ordered field.

The completeness property is at the heart of many results in calculus. One example occurs when searching for a maximum value for a function on a closed interval $[a, b]$, as in Section 3.1. The function $y = x - x^3$ has a maximum value on $[0, 1]$ at the point x satisfying $1 - 3x^2 = 0$, or $x = \sqrt{1/3}$. If we limited our consideration to functions defined only on rational numbers, we would have to conclude that the function has no maximum, since $\sqrt{1/3}$ is irrational (Figure A.34). The Extreme Value Theorem (Section 3.1), which implies that continuous functions on closed intervals $[a, b]$ have a maximum value, is not true for functions defined only on the rationals.

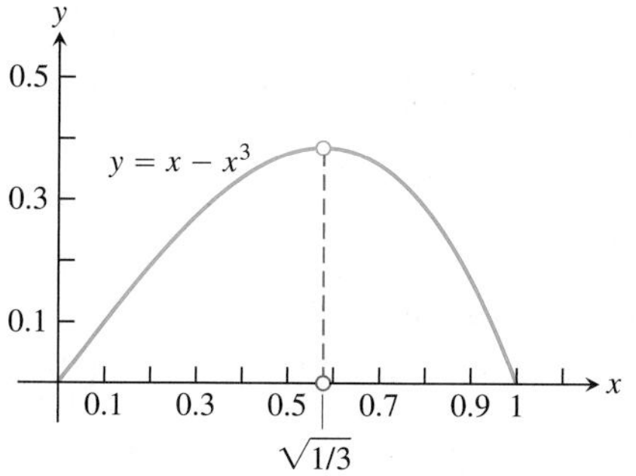

FIGURE A.34 The maximum value of $y = x - x^3$ on $[0, 1]$ occurs at the irrational number $x = \sqrt{1/3}$.

The Intermediate Value Theorem implies that a continuous function f on an interval $[a, b]$ with $f(a) < 0$ and $f(b) > 0$ must be zero somewhere in $[a, b]$. The function values cannot jump from negative to positive without there being some point x in $[a, b]$ where $f(x) = 0$. The Intermediate Value Theorem also relies on the completeness of the real numbers and is false for continuous functions defined only on the rationals. The function $f(x) = 3x^2 - 1$ has $f(0) = -1$ and $f(1) = 2$, but if we consider f only on the rational numbers, it never equals zero. The only value of x for which $f(x) = 0$ is $x = \sqrt{1/3}$, an irrational number.

We have captured the desired properties of the reals by saying that the real numbers are a complete ordered field. But we're not quite finished. Greek mathematicians in the school of Pythagoras tried to impose another property on the numbers of the real line, the condition that all numbers are ratios of integers. They learned that their effort was doomed when they discovered irrational numbers such as $\sqrt{2}$. How do we know that our efforts to specify the real numbers are not also flawed, for some unseen reason? The artist Escher drew optical illusions of spiral staircases that went up and up until they rejoined themselves at the bottom. An engineer trying to build such a staircase would find that no structure realized the plans the architect had drawn. Could it be that our design for the reals contains some subtle contradiction, and that no construction of such a number system can be made?

We resolve this issue by giving a specific description of the real numbers and verifying that the algebraic, order, and completeness properties are satisfied in this model. This is called a **construction** of the reals, and just as stairs can be built with wood, stone, or steel, there are several approaches to constructing the reals. One construction treats the reals as all the infinite decimals,

$$a.d_1d_2d_3d_4\ldots$$

In this approach a real number is an integer a followed by a sequence of decimal digits $d_1, d_2, d_3, \ldots$, each between 0 and 9. This sequence may stop, or repeat in a periodic pattern, or keep going forever with no pattern. In this form, 2.00, 0.3333333... and 3.1415926535898... represent three familiar real numbers. The real meaning of the dots "..." following these digits requires development of the theory of sequences and series, as in Chapter 7. Each real number is constructed as the limit of a sequence of rational numbers given by its finite decimal approximations. An infinite decimal is then the same as a series

$$a + \frac{d_1}{10} + \frac{d_2}{100} + \cdots.$$

This decimal construction of the real numbers is not entirely straightforward. It's easy enough to check that it gives numbers that satisfy the completeness and order properties, but verifying the algebraic properties is rather involved. Even adding or multiplying two numbers requires an infinite number of operations. Making sense of division requires a careful argument involving limits of rational approximations to infinite decimals.

A different approach was taken by Richard Dedekind (1831–1916), a German mathematician, who gave the first rigorous construction of the real numbers in 1872. Given any real number x, we can divide the rational numbers into two sets: those less than or equal to x and those greater. Dedekind cleverly reversed this reasoning and defined a real number to be a division of the rational numbers into two such sets. This seems like a strange approach, but such indirect methods of constructing new structures from old are common in theoretical mathematics.

These and other approaches can be used to construct a system of numbers having the desired algebraic, order, and completeness properties. A final issue that arises is whether all the constructions give the same thing. Is it possible that different constructions result in different number systems satisfying all the required properties? If yes, which of these is the real numbers? Fortunately, the answer turns out to be no. The reals are the only number system satisfying the algebraic, order, and completeness properties.

Confusion about the nature of the numbers and about limits caused considerable controversy in the early development of calculus. Calculus pioneers such as Newton, Leibniz, and their successors, when looking at what happens to the difference quotient

$$\frac{\Delta y}{\Delta x} = \frac{f(x + \Delta x) - f(x)}{\Delta x}$$

as each of Δy and Δx approach zero, talked about the resulting derivative being a quotient of two infinitely small quantities. These "infinitesimals," written dx and dy, were thought to be some new kind of number, smaller than any fixed number but not zero. Similarly, a definite integral was thought of as a sum of an infinite number of infinitesimals

$$f(x) \cdot dx$$

as x varied over a closed interval. While the approximating difference quotients $\Delta y / \Delta x$ were understood much as today, it was the quotient of infinitesimal quantities, rather than a limit, that was thought to encapsulate the meaning of the derivative. This way of thinking led to logical difficulties, as attempted definitions and manipulations of infinitesimals ran into contradictions and inconsistencies. The more concrete and computable difference quotients did not cause such trouble, but they were thought of merely as useful calculation tools. Difference quotients were used to work out the numerical value of the derivative and to derive general formulas for calculation, but were not considered to be at the heart of the question of what the derivative actually was. Today we realize that the logical problems associated with infinitesimals can be avoided by *defining* the derivative to be the limit of its approximating difference quotients. The ambiguities of the old approach are no longer present, and in the standard theory of calculus, infinitesimals are neither needed nor used.

A.8 The Logarithm Defined as an Integral

In Chapter 2 we introduced exponential functions informally. We defined the logarithm function as the corresponding inverse function. Our treatment was imprecise because of subtle issues arising when raising numbers to irrational exponents. (A complete treatment of these issues requires advanced mathematical analysis.)

In this appendix we give another development of the logarithmic and exponential functions solidly based on techniques of calculus. Specifically, we *define* the natural logarithm function as an integral of the function $1/x$. The exponential function e^x is then defined as the inverse of the natural logarithm.

Definition of the Natural Logarithm Function

The natural logarithm of a positive number x, written as $\ln x$, is the value of an integral.

DEFINITION

$$\ln x = \int_1^x \frac{1}{t}\,dt, \qquad x > 0 \tag{1}$$

From the Fundamental Theorem of Calculus, $\ln x$ is a continuous function. Geometrically, if $x > 1$, then $\ln x$ is the area under the curve $y = 1/t$ from $t = 1$ to $t = x$ (Figure A.35). For $0 < x < 1$, $\ln x$ gives the negative of the area under the curve from x to 1. The function is not defined for $x \le 0$. From the Zero Width Interval Rule for definite integrals, we also have

$$\ln 1 = \int_1^1 \frac{1}{t}\,dt = 0.$$

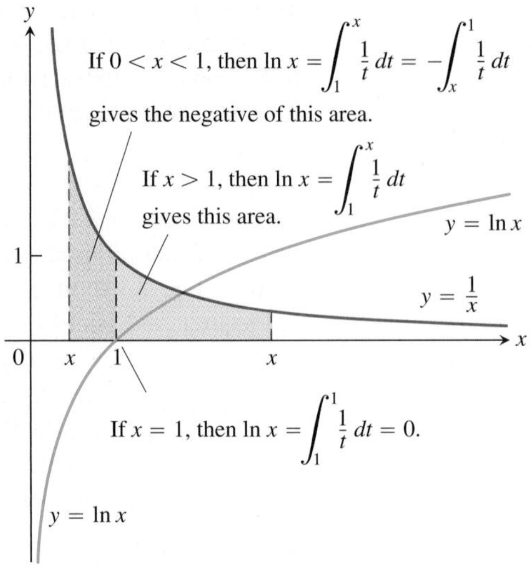

FIGURE A.35 The graph of $y = \ln x$ and its relation to the function $y = 1/x, x > 0$. The graph of the logarithm rises above the x-axis as x moves from 1 to the right, and it falls below the x-axis as x moves from 1 to the left.

Notice that we show the graph of $y = 1/x$ in Figure A.35 but use $y = 1/t$ in the integral. Using x for everything would have us writing

$$\ln x = \int_1^x \frac{1}{x}\,dx,$$

with x meaning two different things. Thus we change the variable of integration to t.

By using rectangles to obtain finite approximations of the area under the graph of $y = 1/t$ and over the interval between $t = 1$ and $t = x$, as in Section 4.2, we can approximate the values of the function $\ln x$. Several values are given in Table A.4. There is an important number between $x = 2$ and $x = 3$ whose natural logarithm equals 1. This number, which we now define, exists because $\ln x$ is a continuous function and therefore satisfies the Intermediate Value Theorem on [2, 3].

TABLE A.4 Typical 2-place values of $\ln x$

x	$\ln x$
0	undefined
0.05	−3.00
0.5	−0.69
1	0
2	0.69
3	1.10
4	1.39
10	2.30

DEFINITION The number e is that number in the domain of the natural logarithm satisfying

$$\ln(e) = 1$$

So the number e lies within the interval [2, 3] and satisfies

$$\int_1^e \frac{1}{t}\,dt = 1.$$

Interpreted geometrically, the number e corresponds to the point on the x-axis for which the area under the graph of $y = 1/t$ and above the interval $[1, e]$ equals the area of the unit square. That is, the area of the region shaded blue in Figure A.35 is 1 sq unit when $x = e$.

The Derivative of $y = \ln x$

By the first part of the Fundamental Theorem of Calculus (Section 4.5),

$$\frac{d}{dx}\ln x = \frac{d}{dx}\int_1^x \frac{1}{t}\,dt = \frac{1}{x}.$$

For every positive value of x, we have

$$\frac{d}{dx}\ln x = \frac{1}{x},$$

and the Chain Rule gives

$$\frac{d}{dx}\ln u = \frac{d}{du}\ln u \cdot \frac{du}{dx} = \frac{1}{u}\frac{du}{dx}.$$

$$\frac{d}{dx}\ln u = \frac{1}{u}\frac{du}{dx}, \qquad u > 0 \tag{2}$$

As we saw in Equation (5) of Section 2.10, the function $\ln|x|$ has derivative $1/x$ at all nonzero x.

$$\frac{d}{dx}\ln|x| = \frac{1}{x}, \qquad x \neq 0$$

Properties of Logarithms

We now show how the integral definition of the natural logarithm leads to the familiar algebraic properties that make logarithmic functions so useful. These properties were used in Section 2.10.

THEOREM—Algebraic Properties of the Natural Logarithm For any numbers $b > 0$ and $x > 0$, the natural logarithm satisfies the following rules:

1. *Product Rule:* $\ln bx = \ln b + \ln x$

2. *Quotient Rule:* $\ln \frac{b}{x} = \ln b - \ln x$

3. *Reciprocal Rule:* $\ln \frac{1}{x} = -\ln x$ Rule 2 with $b = 1$

4. *Power Rule:* $\ln x^r = r \ln x$ For r rational

For now we consider only rational exponents in Rule 4. We will see that the rule holds for irrational exponents as well, after we define exponential functions and obtain their properties.

The algebraic properties are proved by applying Corollary 2 of the Mean Value Theorem (Section 3.2) to each of them. The steps in the proof are similar to those used in solving problems involving logarithms.

Proof that $\ln bx = \ln b + \ln x$ The argument starts by observing that $\ln bx$ and $\ln x$ have the same derivative:

$$\frac{d}{dx}\ln(bx) = \frac{b}{bx} = \frac{1}{x} = \frac{d}{dx}\ln x.$$

According to Corollary 2 of the Mean Value Theorem, the functions must differ by a constant, which means that

$$\ln bx = \ln x + C$$

for some constant C.

Since this last equation holds for all positive values of x, it must hold for $x = 1$. Hence,

$$\begin{aligned} \ln(b \cdot 1) &= \ln 1 + C \\ \ln b &= 0 + C \qquad \ln 1 = 0 \\ C &= \ln b. \end{aligned}$$

By substituting we conclude,

$$\ln bx = \ln b + \ln x. \qquad \blacksquare$$

Proof that $\ln(b/x) = \ln b - \ln x$ The derivative of $\ln(b/x)$ is

$$\frac{d}{dx}\ln\left(\frac{b}{x}\right) = \frac{-b/x^2}{b/x} = -\frac{1}{x}$$

while the derivative of $\ln b - \ln x$ is also $-1/x$. We conclude that

$$\ln(b/x) = \ln b - \ln x + C.$$

The value of the constant C is zero, obtained by taking $x = b$ in the last equation. This proves Rule 2. Rule 3 is a special case of Rule 2, obtained by setting $a = 1$ and noting that $\ln 1 = 0$. $\blacksquare$

Proof that $\ln x^r = r\ln x$ (assuming r rational) We use the same-derivative argument again. For all positive values of x,

$$\begin{aligned} \frac{d}{dx}\ln x^r &= \frac{1}{x^r}\frac{d}{dx}(x^r) && \text{Eq. (2) with } u = x^r \\ &= \frac{1}{x^r} r x^{r-1} && \text{General Power Rule for derivatives, } r \text{ rational} \\ &= r \cdot \frac{1}{x} = \frac{d}{dx}(r \ln x). \end{aligned}$$

Since $\ln x^r$ and $r \ln x$ have the same derivative,

$$\ln x^r = r \ln x + C$$

for some constant C. Taking x to be 1 identifies C as zero, and we're done. Exercise 40 in Section 2.8 indicates a proof of the General Power Rule for derivatives when r is rational.

This establishes all the algebraic properties of the natural logarithm. ■

The Graph and Range of $\ln x$

The derivative $d(\ln x)/dx = 1/x$ is positive for $x > 0$, so $\ln x$ is an increasing function of x. The second derivative, $-1/x^2$, is negative, so the graph of $\ln x$ is concave down.

We can estimate the value of $\ln 2$ by considering the area under the graph of $y = 1/x$ and above the interval $[1, 2]$. In Figure A.35 a rectangle of height $1/2$ over the interval $[1, 2]$ fits under the graph. Therefore the area under the graph, which is $\ln 2$, is greater than the area, $1/2$, of the rectangle. So $\ln 2 > 1/2$. Knowing this we have,

$$\ln 2^n = n \ln 2 > n\left(\frac{1}{2}\right) = \frac{n}{2}$$

and

$$\ln 2^{-n} = -n \ln 2 < -n\left(\frac{1}{2}\right) = -\frac{n}{2}.$$

It follows that

$$\lim_{x\to\infty} \ln x = \infty \qquad \text{and} \qquad \lim_{x\to 0^+} \ln x = -\infty.$$

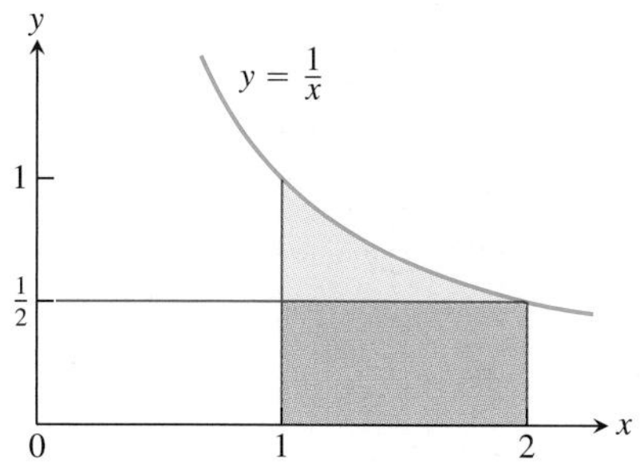

FIGURE A.36 The rectangle of height $y = 1/2$ fits beneath the graph of $y = 1/x$ for the interval $1 \le x \le 2$.

We defined $\ln x$ for $x > 0$, so the domain of $\ln x$ is the set of positive real numbers. The above discussion and the Intermediate Value Theorem show that its range is the entire real line giving the graph of $y = \ln x$ shown in Figure A.35.

The Inverse of $\ln x$ and the Number e

The function $\ln x$, being an increasing function of x with domain $(0, \infty)$ and range $(-\infty, \infty)$, has an inverse $\ln^{-1} x$ with domain $(-\infty, \infty)$ and range $(0, \infty)$. The graph of $\ln^{-1} x$ is the graph of $\ln x$ reflected across the line $y = x$. As you can see in Figure A.37,

$$\lim_{x\to\infty} \ln^{-1} x = \infty \qquad \text{and} \qquad \lim_{x\to -\infty} \ln^{-1} x = 0.$$

The function $\ln^{-1} x$ is also denoted by $\exp x$. We now show that $\ln^{-1} x = \exp x$ is an exponential function with base e.

The number e was defined to satisfy the equation $\ln(e) = 1$, so $e = \ln^{-1}(1) = \exp(1)$. We can raise the number e to a rational power r in the usual algebraic way:

$$e^2 = e \cdot e, \qquad e^{-2} = \frac{1}{e^2}, \qquad e^{1/2} = \sqrt{e},$$

and so on. Since e is positive, e^r is positive too. Thus, e^r has a logarithm. When we take the logarithm, we find that for r rational

$$\ln e^r = r \ln e = r \cdot 1 = r. \qquad \text{Rule 4}$$

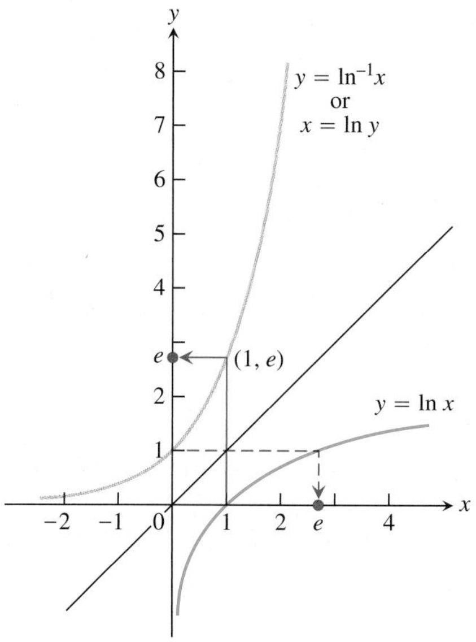

FIGURE A.37 The graphs of $y = \ln x$ and $y = \ln^{-1} x = \exp x$. The number e is $\ln^{-1} 1 = \exp(1)$.

Then applying the function $\ln^{-1}$ to both sides of the equation $\ln e^r = r$, we find that

$$e^r = \ln^{-1} r = \exp r \qquad \text{for } r \text{ rational}. \tag{3}$$

We have not yet found a way to give an obvious meaning to e^x for x irrational. But $\ln^{-1} x$ has meaning for any x, rational or irrational. So Equation (3) provides a way to extend the definition of e^x to irrational values of x. The function $\ln^{-1} x$ is defined for all x, so we use it to assign a value to e^x at every point.

DEFINITION For every real number x, $e^x = \ln^{-1} x = \exp x$.

For the first time we have a precise meaning for an irrational exponent. Usually the exponential function is denoted by e^x rather than $\exp x$. Since $\ln x$ and e^x are inverses of one another, we have the following relationships.

Inverse Equations for e^x and $\ln x$

$$e^{\ln x} = x \qquad (\text{all } x > 0)$$
$$\ln(e^x) = x \qquad (\text{all } x)$$

The Derivative and Integral of e^x

According to Theorem 4 of Section 2.9, the exponential function is differentiable because it is the inverse of a differentiable function whose derivative is never zero. We calculate its derivative using Theorem 4 and our knowledge of the derivative of $\ln x$. Let

$$f(x) = \ln x \qquad \text{and} \qquad y = e^x = \ln^{-1} x = f^{-1}(x).$$

Then,

$$\begin{aligned}
\frac{dy}{dx} = \frac{d}{dx}(e^x) &= \frac{d}{dx}\ln^{-1} x \\
&= \frac{d}{dx} f^{-1}(x) \\
&= \frac{1}{f'(f^{-1}(x))} && \text{Theorem 4, Section 2.9} \\
&= \frac{1}{f'(e^x)} && f^{-1}(x) = e^x \\
&= \frac{1}{\left(\frac{1}{e^x}\right)} && f'(z) = \frac{1}{z} \text{ with } z = e^x \\
&= e^x.
\end{aligned}$$

That is, for $y = e^x$, we find that $dy/dx = e^x$ so the natural exponential function e^x is its own derivative. Therefore if $f(x) = e^x$, then $f'(0) = e^0 = 1$. This means that the exponential function $e^x = \exp x = \ln^{-1} x$ has slope 1 as it crosses the y-axis at $x = 0$.

The Chain Rule extends the derivative result for the natural exponential function in the usual way to a more general form.

If u is any differentiable function of x, then

$$\frac{d}{dx} e^u = e^u \frac{du}{dx}.$$

Laws of Exponents

Even though e^x is defined in a seemingly roundabout way as $\ln^{-1} x$, it obeys the familiar laws of exponents from algebra. The next theorem shows that these laws are consequences of the definitions of $\ln x$ and e^x.

> **THEOREM** For all numbers x, x_1, and x_2, the natural exponential e^x obeys the following laws:
>
> **1.** $e^{x_1} \cdot e^{x_2} = e^{x_1 + x_2}$ **2.** $e^{-x} = \dfrac{1}{e^x}$
>
> **3.** $\dfrac{e^{x_1}}{e^{x_2}} = e^{x_1 - x_2}$ **4.** $(e^{x_1})^r = e^{r x_1}$, if r is rational

Proof of Law 1 Let

$$y_1 = e^{x_1} \qquad \text{and} \qquad y_2 = e^{x_2}.$$

Then

$$\begin{aligned}
x_1 &= \ln y_1 \quad \text{and} \quad x_2 = \ln y_2 && \text{Inverse equations} \\
x_1 + x_2 &= \ln y_1 + \ln y_2 \\
&= \ln y_1 y_2 && \text{Product Rule for logarithms} \\
e^{x_1 + x_2} &= e^{\ln y_1 y_2} && \text{Exponentiate.} \\
&= y_1 y_2 && e^{\ln u} = u \\
&= e^{x_1} e^{x_2}.
\end{aligned}$$

■

Proof of Law 4 Let $y = (e^{x_1})^r$. Then

$$\begin{aligned}
\ln y &= \ln (e^{x_1})^r \\
&= r \ln (e^{x_1}) && \text{Power Rule for logarithms, rational } r \\
&= r x_1 && \ln e^u = u \text{ with } u = x_1
\end{aligned}$$

Thus, exponentiating each side,

$$y = e^{r x_1}. \qquad e^{\ln y} = y$$

■

Laws 2 and 3 follow from Law 1. Like the Power Rule for logarithms, Law 4 holds for all real numbers r, as we will see shortly.

The General Exponential Function a^x

Since $a = e^{\ln a}$ for any positive number a, we can think of a^x as $(e^{\ln a})^x = e^{x \ln a}$. We therefore use the function e^x to define the other exponential functions.

> **DEFINITION** For any numbers $a > 0$ and x, the exponential function with base a is
>
> $$a^x = e^{x \ln a}.$$

When $a = e$, the definition gives $a^x = e^{x \ln a} = e^{x \ln e} = e^{x \cdot 1} = e^x$.

The algebraic laws for the natural exponential are also valid for a^x, the exponential function with base a. For example,

$$\begin{aligned} a^{x_1} \cdot a^{x_2} &= e^{x_1 \ln a} \cdot e^{x_2 \ln a} && \text{Definition of } a^x \\ &= e^{x_1 \ln a + x_2 \ln a} && \text{Law 1} \\ &= e^{(x_1 + x_2) \ln a} && \text{Factor } \ln a \\ &= a^{x_1 + x_2}. && \text{Definition of } a^x \end{aligned}$$

In Section 2.10 we showed how this definition of a^x and the rules for differentiating e^x lead to the general power rule formula:

$$\frac{d}{dx} x^r = r x^{r-1}, \quad \text{all real } r.$$

Previously when we established that $\ln x^r = r \ln x$ (Rule 4 for logarithms) we restricted our assertion to rational powers r. The proof used the power rule, which had only been established for rational exponents. Now that we have verified the power rule for all real exponents, the same argument shows that $\ln x^r = r \ln x$ for all real exponents. Therefore we have established the properties of logarithms claimed in Section 2.10.

Logarithms with Base *a*

If a is any positive number other than 1, the function a^x is one-to-one and has a nonzero derivative at every point. It therefore has a differentiable inverse. We call the inverse the **logarithm of *x* with base *a*** and denote it by $\mathbf{\log_a x}$.

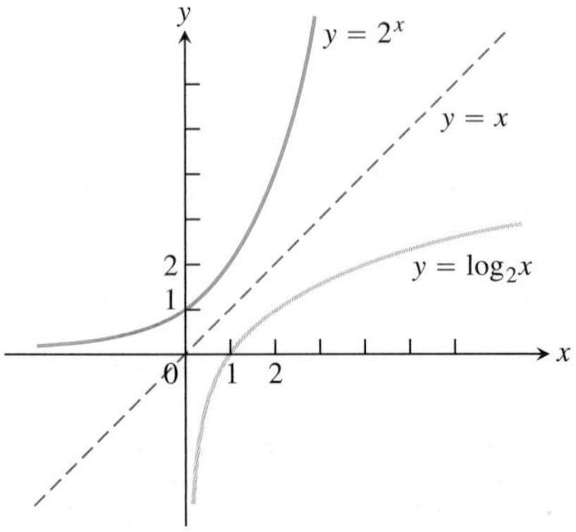

FIGURE A.38 The graph of 2^x and its inverse, $\log_2 x$.

DEFINITION For any positive number $a \neq 1$,

$\log_a x$ is the inverse function of a^x.

The graph of $y = \log_a x$ can be obtained by reflecting the graph of $y = a^x$ across the 45° line $y = x$ (Figure A.38). When $a = e$, we have $\log_e x = \text{inverse of } e^x = \ln x$. Since $\log_a x$ and a^x are inverses of one another, composing them in either order gives the identity function.

Inverse Equations for a^x and $\log_a x$

$$a^{\log_a x} = x \qquad (x > 0)$$

$$\log_a (a^x) = x \qquad (\text{all } x)$$

The function $\log_a x$ is actually just a numerical multiple of $\ln x$. To see this, we let $y = \log_a x$ and then take the natural logarithm of both sides of the equivalent equation $a^y = x$ to obtain $y \ln a = \ln x$. Solving for y gives

$$\log_a x = \frac{\ln x}{\ln a}.$$

It then follows easily that the arithmetic rules satisfied by $\log_a x$ are the same as the ones for $\ln x$. These rules, given in Table A.5, can be proved by dividing the corresponding rules for the natural logarithm function by $\ln a$. For example,

$$\ln xy = \ln x + \ln y \qquad \text{Rule 1 for natural logarithms} \ldots$$

$$\frac{\ln xy}{\ln a} = \frac{\ln x}{\ln a} + \frac{\ln y}{\ln a} \qquad \ldots \text{divided by } \ln a \ldots$$

$$\log_a xy = \log_a x + \log_a y. \qquad \ldots \text{gives Rule 1 for base } a \text{ logarithms.}$$

TABLE A.5 Rules for base a logarithms

For any numbers $x > 0$, $y > 0$, and r,

1. *Product Rule*:
   $\log_a xy = \log_a x + \log_a y$
2. *Quotient Rule*:
   $\log_a \frac{x}{y} = \log_a x - \log_a y$
3. *Reciprocal Rule*:
   $\log_a \frac{1}{y} = -\log_a y$
4. *Power Rule*:
   $\log_a x^r = r \log_a x$

Derivatives and Integrals Involving $\log_a x$

To find derivatives or integrals involving base a logarithms, we convert them to natural logarithms. For example, if u is a positive differentiable function of x, then

$$\frac{d}{dx}(\log_a u) = \frac{d}{dx}\left(\frac{\ln u}{\ln a}\right) = \frac{1}{\ln a}\frac{d}{dx}(\ln u) = \frac{1}{\ln a} \cdot \frac{1}{u}\frac{du}{dx}.$$

A.9 Convergence of Power Series and Proof of Taylor's Theorem

This appendix proves the Convergence Theorem for Power Series (Theorem 18, Section 7.7), its corollary establishing the radius of convergence (Corollary, Section 7.7), and Taylor's Theorem (Theorem 22, Section 7.9).

THEOREM 18—The Convergence Theorem for Power Series If the power series $\sum_{n=0}^{\infty} a_n x^n = a_0 + a_1 x + a_2 x^2 + \cdots$ converges for $x = c \neq 0$, then it converges absolutely for all x with $|x| < |c|$. If the series diverges for $x = d$, then it diverges for all x with $|x| > |d|$.

Proof Suppose the series $\sum_{n=0}^{\infty} a_n c^n$ converges. Then $\lim_{n\to\infty} a_n c^n = 0$. Hence, there is an integer N such that $|a_n c^n| < 1$ for all $n \geq N$. That is,

$$|a_n| < \frac{1}{|c|^n} \quad \text{for } n \geq N. \tag{1}$$

Now take any x such that $|x| < |c|$ and consider

$$|a_0| + |a_1 x| + \cdots + |a_{N-1}x^{N-1}| + |a_N x^N| + |a_{N+1}x^{N+1}| + \cdots.$$

There are only a finite number of terms prior to $|a_N x^N|$, and their sum is finite. Starting with $|a_N x^N|$ and beyond, the terms are less than

$$\left|\frac{x}{c}\right|^N + \left|\frac{x}{c}\right|^{N+1} + \left|\frac{x}{c}\right|^{N+2} + \cdots \qquad (2)$$

because of Inequality (1). But Series (2) is a geometric series with ratio $r = |x/c|$, which is less than 1, since $|x| < |c|$. Hence Series (2) converges, so the original series converges absolutely. This proves the first half of the theorem.

The second half of the theorem follows from the first. If the series diverges at $x = d$ and converges at a value x_0 with $|x_0| > |d|$, we may take $c = x_0$ in the first half of the theorem and conclude that the series converges absolutely at d. But the series cannot converge absolutely and diverge at one and the same time. Hence, if it diverges at d, it diverges for all x with $|x| > |d|$. ■

COROLLARY TO THEOREM 18 The convergence of the series $\sum c_n(x - a)^n$ is described by one of the following three possibilities:

1. There is a positive number R such that the series diverges for x with $|x - a| > R$ but converges absolutely for x with $|x - a| < R$. The series may or may not converge at either of the endpoints $x = a - R$ and $x = a + R$.
2. The series converges absolutely for every x ($R = \infty$).
3. The series converges at $x = a$ and diverges elsewhere ($R = 0$).

Proof We assume first that $a = 0$, so that the power series is centered at 0. If the series converges everywhere we are in Case 2. If it converges only at $x = 0$ we are in Case 3. Otherwise there is a nonzero number d such that $\sum c_n d^n$ diverges. The set S of values of x for which the series $\sum c_n x^n$ converges is nonempty because it contains 0 and a positive number p as well. By Theorem 18, the series diverges for all x with $|x| > |d|$, so $|x| \le |d|$ for all $x \in S$, and S is a bounded set. By the Completeness Property of the real numbers (see Appendix 7) a nonempty, bounded set has a least upper bound R. (The least upper bound is the smallest number with the property that the elements $x \in S$ satisfy $x \le R$.) If $|x| > R \ge p$, then $x \notin S$ so the series $\sum c_n x^n$ diverges. If $|x| < R$, then $|x|$ is not an upper bound for S (because it's smaller than the least upper bound) so there is a number $b \in S$ such that $b > |x|$. Since $b \in S$, the series $\sum c_n b^n$ converges and therefore the series $\sum c_n |x|^n$ converges by Theorem 18. This proves the Corollary for power series centered at $a = 0$.

For a power series centered at $a \neq 0$, we set $x' = (x - a)$ and repeat the argument with x'. Since $x' = 0$ when $x = a$, a radius R interval of convergence for $\sum c_n(x')^n$ centered at $x' = 0$ is the same as a radius R interval of convergence for $\sum c_n(x - a)^n$ centered at $x = a$. This establishes the Corollary for the general case. ■

THEOREM 22—Taylor's Theorem If f and its first n derivatives $f', f'', \ldots, f^{(n)}$ are continuous on the closed interval between a and b, and $f^{(n)}$ is differentiable on the open interval between a and b, then there exists a number c between a and b such that

$$f(b) = f(a) + f'(a)(b - a) + \frac{f''(a)}{2!}(b - a)^2 + \cdots + \frac{f^{(n)}(a)}{n!}(b - a)^n + \frac{f^{(n+1)}(c)}{(n + 1)!}(b - a)^{n+1}.$$

Proof We prove Taylor's theorem assuming $a < b$. The proof for $a > b$ is nearly the same.

The Taylor polynomial

$$P_n(x) = f(a) + f'(a)(x - a) + \frac{f''(a)}{2!}(x - a)^2 + \cdots + \frac{f^{(n)}(a)}{n!}(x - a)^n$$

and its first n derivatives match the function f and its first n derivatives at $x = a$. We do not disturb that matching if we add another term of the form $K(x - a)^{n+1}$, where K is any constant, because such a term and its first n derivatives are all equal to zero at $x = a$. The new function

$$\phi_n(x) = P_n(x) + K(x - a)^{n+1}$$

and its first n derivatives still agree with f and its first n derivatives at $x = a$.

We now choose the particular value of K that makes the curve $y = \phi_n(x)$ agree with the original curve $y = f(x)$ at $x = b$. In symbols,

$$f(b) = P_n(b) + K(b - a)^{n+1}, \qquad \text{or} \qquad K = \frac{f(b) - P_n(b)}{(b - a)^{n+1}}. \tag{3}$$

With K defined by Equation (3), the function

$$F(x) = f(x) - \phi_n(x)$$

measures the difference between the original function f and the approximating function ϕ_n for each x in $[a, b]$.

We now use Rolle's Theorem (Section 3.2). First, because $F(a) = F(b) = 0$ and both F and F' are continuous on $[a, b]$, we know that

$$F'(c_1) = 0 \qquad \text{for some } c_1 \text{ in } (a, b).$$

Next, because $F'(a) = F'(c_1) = 0$ and both F' and F'' are continuous on $[a, c_1]$, we know that

$$F''(c_2) = 0 \qquad \text{for some } c_2 \text{ in } (a, c_1).$$

Rolle's Theorem, applied successively to $F'', F''', \dots, F^{(n-1)}$ implies the existence of

$$\begin{aligned} &c_3 \quad \text{in } (a, c_2) && \text{such that } F'''(c_3) = 0, \\ &c_4 \quad \text{in } (a, c_3) && \text{such that } F^{(4)}(c_4) = 0, \\ &\ \vdots \\ &c_n \quad \text{in } (a, c_{n-1}) && \text{such that } F^{(n)}(c_n) = 0. \end{aligned}$$

Finally, because $F^{(n)}$ is continuous on $[a, c_n]$ and differentiable on (a, c_n), and $F^{(n)}(a) = F^{(n)}(c_n) = 0$, Rolle's Theorem implies that there is a number c_{n+1} in (a, c_n) such that

$$F^{(n+1)}(c_{n+1}) = 0. \tag{4}$$

If we differentiate $F(x) = f(x) - P_n(x) - K(x - a)^{n+1}$ a total of $n + 1$ times, we get

$$F^{(n+1)}(x) = f^{(n+1)}(x) - 0 - (n + 1)!K. \tag{5}$$

Equations (4) and (5) together give

$$K = \frac{f^{(n+1)}(c)}{(n + 1)!} \qquad \text{for some number } c = c_{n+1} \text{ in } (a, b). \tag{6}$$

Equations (3) and (6) give

$$f(b) = P_n(b) + \frac{f^{(n+1)}(c)}{(n + 1)!}(b - a)^{n+1}.$$

This concludes the proof. ■

A.10 The Distributive Law for Vector Cross Products

In this appendix, we prove the Distributive Law

$$\mathbf{u} \times (\mathbf{v} + \mathbf{w}) = \mathbf{u} \times \mathbf{v} + \mathbf{u} \times \mathbf{w}$$

which is Property 2 in Section 9.4.

Proof To derive the Distributive Law, we construct $\mathbf{u} \times \mathbf{v}$ a new way. We draw $\mathbf{u}$ and $\mathbf{v}$ from the common point O and construct a plane M perpendicular to $\mathbf{u}$ at O (Figure A.39). We then project $\mathbf{v}$ orthogonally onto M, yielding a vector $\mathbf{v}'$ with length $|\mathbf{v}|\sin\theta$. We rotate $\mathbf{v}'$ 90° about $\mathbf{u}$ in the positive sense to produce a vector $\mathbf{v}''$. Finally, we multiply $\mathbf{v}''$ by the length of $\mathbf{u}$. The resulting vector $|\mathbf{u}|\mathbf{v}''$ is equal to $\mathbf{u} \times \mathbf{v}$ since $\mathbf{v}''$ has the same direction as $\mathbf{u} \times \mathbf{v}$ by its construction (Figure A.39) and

$$|\mathbf{u}||\mathbf{v}''| = |\mathbf{u}||\mathbf{v}'| = |\mathbf{u}||\mathbf{v}|\sin\theta = |\mathbf{u} \times \mathbf{v}|.$$

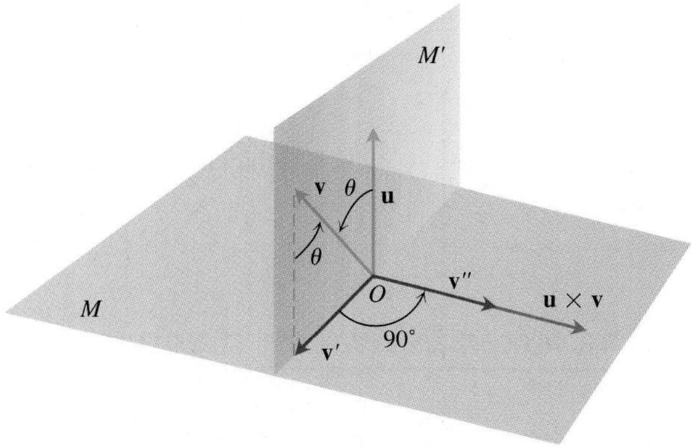

FIGURE A.39 As explained in the text, $\mathbf{u} \times \mathbf{v} = |\mathbf{u}|\mathbf{v}''$.

Now each of these three operations, namely,

1. projection onto M
2. rotation about $\mathbf{u}$ through 90°
3. multiplication by the scalar $|\mathbf{u}|$

when applied to a triangle whose plane is not parallel to $\mathbf{u}$, will produce another triangle. If we start with the triangle whose sides are $\mathbf{v}$, $\mathbf{w}$, and $\mathbf{v} + \mathbf{w}$ (Figure A.40) and apply these three steps, we successively obtain the following:

1. A triangle whose sides are $\mathbf{v}'$, $\mathbf{w}'$, and $(\mathbf{v} + \mathbf{w})'$ satisfying the vector equation

$$\mathbf{v}' + \mathbf{w}' = (\mathbf{v} + \mathbf{w})'$$

2. A triangle whose sides are $\mathbf{v}''$, $\mathbf{w}''$, and $(\mathbf{v} + \mathbf{w})''$ satisfying the vector equation

$$\mathbf{v}'' + \mathbf{w}'' = (\mathbf{v} + \mathbf{w})''$$

(the double prime on each vector has the same meaning as in Figure A.40)

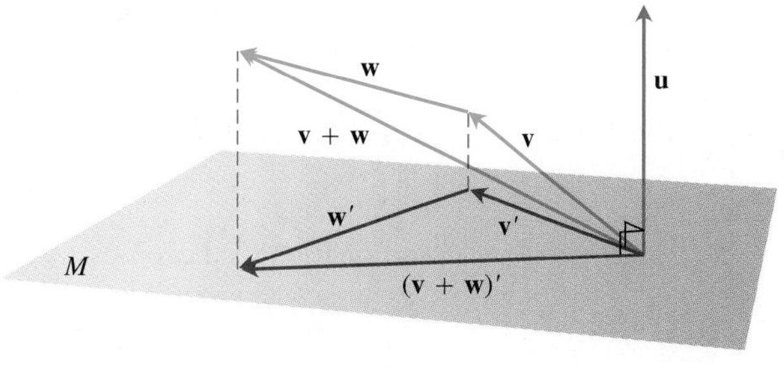

FIGURE A.40 The vectors, $\mathbf{v}$, $\mathbf{w}$, $\mathbf{v} + \mathbf{w}$, and their projections onto a plane perpendicular to $\mathbf{u}$.

3. A triangle whose sides are $|\mathbf{u}|\mathbf{v}''$, $|\mathbf{u}|\mathbf{w}''$, and $|\mathbf{u}|(\mathbf{v} + \mathbf{w})''$ satisfying the vector equation

$$|\mathbf{u}|\mathbf{v}'' + |\mathbf{u}|\mathbf{w}'' = |\mathbf{u}|(\mathbf{v} + \mathbf{w})''.$$

Substituting $|\mathbf{u}|\mathbf{v}'' = \mathbf{u} \times \mathbf{v}, |\mathbf{u}|\mathbf{w}'' = \mathbf{u} \times \mathbf{w}$, and $|\mathbf{u}|(\mathbf{v} + \mathbf{w})'' = \mathbf{u} \times (\mathbf{v} + \mathbf{w})$ from our discussion above into this last equation gives

$$\mathbf{u} \times \mathbf{v} + \mathbf{u} \times \mathbf{w} = \mathbf{u} \times (\mathbf{v} + \mathbf{w}),$$

which is the law we wanted to establish. ■

A.11 The Mixed Derivative Theorem and the Increment Theorem

This appendix derives the Mixed Derivative Theorem (Theorem 2, Section 11.3) and the Increment Theorem for Functions of Two Variables (Theorem 3, Section 11.3). Euler first published the Mixed Derivative Theorem in 1734, in a series of papers he wrote on hydrodynamics.

> **THEOREM 2—The Mixed Derivative Theorem** If $f(x, y)$ and its partial derivatives f_x, f_y, f_{xy}, and f_{yx} are defined throughout an open region containing a point (a, b) and are all continuous at (a, b), then $f_{xy}(a, b) = f_{yx}(a, b)$.

Proof The equality of $f_{xy}(a, b)$ and $f_{yx}(a, b)$ can be established by four applications of the Mean Value Theorem (Theorem 4, Section 3.2). By hypothesis, the point (a, b) lies in the interior of a rectangle R in the xy-plane on which f, f_x, f_y, f_{xy}, and f_{yx} are all defined. We let h and k be the numbers such that the point $(a + h, b + k)$ also lies in R, and we consider the difference

$$\Delta = F(a + h) - F(a), \tag{1}$$

where

$$F(x) = f(x, b + k) - f(x, b). \tag{2}$$

We apply the Mean Value Theorem to F, which is continuous because it is differentiable. Then Equation (1) becomes

$$\Delta = hF'(c_1), \tag{3}$$

where c_1 lies between a and $a + h$. From Equation (2).

$$F'(x) = f_x(x, b + k) - f_x(x, b),$$

so Equation (3) becomes

$$\Delta = h[f_x(c_1, b + k) - f_x(c_1, b)]. \tag{4}$$

Now we apply the Mean Value Theorem to the function $g(y) = f_x(c_1, y)$ and have

$$g(b + k) - g(b) = kg'(d_1),$$

or

$$f_x(c_1, b + k) - f_x(c_1, b) = kf_{xy}(c_1, d_1)$$

for some d_1 between b and $b + k$. By substituting this into Equation (4), we get

$$\Delta = hkf_{xy}(c_1, d_1) \tag{5}$$

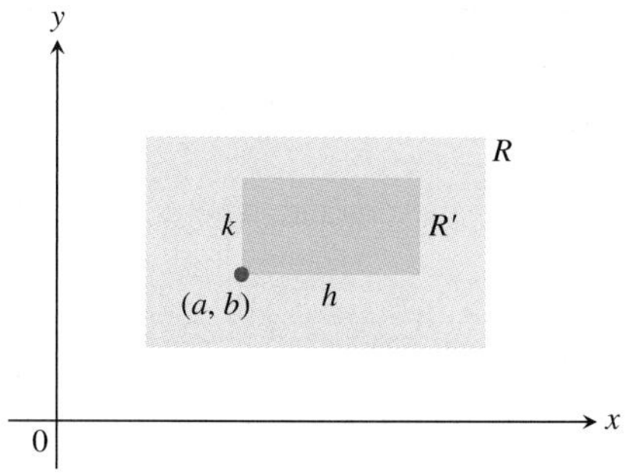

FIGURE A.41 The key to proving $f_{xy}(a, b) = f_{yx}(a, b)$ is that no matter how small R' is, f_{xy} and f_{yx} take on equal values somewhere inside R' (although not necessarily at the same point).

for some point (c_1, d_1) in the rectangle R' whose vertices are the four points (a, b), $(a + h, b)$, $(a + h, b + k)$, and $(a, b + k)$. (See Figure A.41.)

By substituting from Equation (2) into Equation (1), we may also write

$$\begin{aligned}\Delta &= f(a + h, b + k) - f(a + h, b) - f(a, b + k) + f(a, b) \\ &= [f(a + h, b + k) - f(a, b + k)] - [f(a + h, b) - f(a, b)] \\ &= \phi(b + k) - \phi(b),\end{aligned} \tag{6}$$

where

$$\phi(y) = f(a + h, y) - f(a, y). \tag{7}$$

The Mean Value Theorem applied to Equation (6) now gives

$$\Delta = k\phi'(d_2) \tag{8}$$

for some d_2 between b and $b + k$. By Equation (7),

$$\phi'(y) = f_y(a + h, y) - f_y(a, y). \tag{9}$$

Substituting from Equation (9) into Equation (8) gives

$$\Delta = k[f_y(a + h, d_2) - f_y(a, d_2)].$$

Finally, we apply the Mean Value Theorem to the expression in brackets and get

$$\Delta = khf_{yx}(c_2, d_2) \tag{10}$$

for some c_2 between a and $a + h$.

Together, Equations (5) and (10) show that

$$f_{xy}(c_1, d_1) = f_{yx}(c_2, d_2), \tag{11}$$

where (c_1, d_1) and (c_2, d_2) both lie in the rectangle R' (Figure A.41). Equation (11) is not quite the result we want, since it says only that f_{xy} has the same value at (c_1, d_1) that f_{yx} has at (c_2, d_2). The numbers h and k in our discussion, however, may be made as small as we wish. The hypothesis that f_{xy} and f_{yx} are both continuous at (a, b) means that $f_{xy}(c_1, d_1) = f_{xy}(a, b) + \epsilon_1$ and $f_{yx}(c_2, d_2) = f_{yx}(a, b) + \epsilon_2$, where each of $\epsilon_1, \epsilon_2 \to 0$ as both $h, k \to 0$. Hence, if we let h and $k \to 0$, we have $f_{xy}(a, b) = f_{yx}(a, b)$. ■

The equality of $f_{xy}(a, b)$ and $f_{yx}(a, b)$ can be proved with hypotheses weaker than the ones we assumed. For example, it is enough for f, f_x, and f_y to exist in R and for f_{xy} to be continuous at (a, b). Then f_{yx} will exist at (a, b) and equal f_{xy} at that point.

THEOREM 3—The Increment Theorem for Functions of Two Variables

Suppose that the first partial derivatives of $z = f(x, y)$ are defined throughout an open region R containing the point (x_0, y_0) and that f_x and f_y are continuous at (x_0, y_0). Then the change $\Delta z = f(x_0 + \Delta x, y_0 + \Delta y) - f(x_0, y_0)$ in the value of f that results from moving from (x_0, y_0) to another point $(x_0 + \Delta x, y_0 + \Delta y)$ in R satisfies an equation of the form

$$\Delta z = f_x(x_0, y_0)\Delta x + f_y(x_0, y_0)\Delta y + \epsilon_1\Delta x + \epsilon_2\Delta y,$$

in which each of $\epsilon_1, \epsilon_2 \to 0$ as both $\Delta x, \Delta y \to 0$.

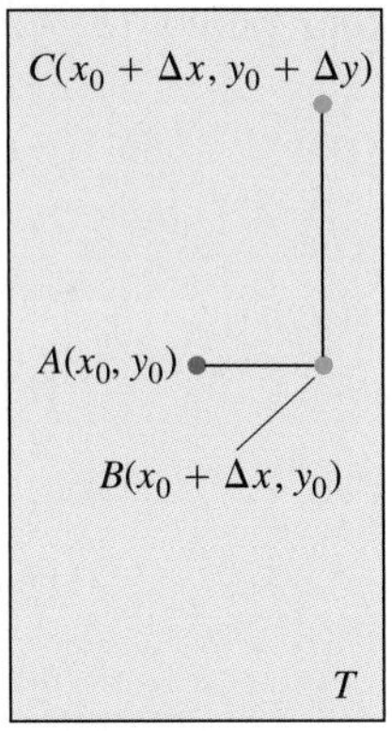

FIGURE A.42 The rectangular region T in the proof of the Increment Theorem. The figure is drawn for Δx and Δy positive, but either increment might be zero or negative.

Proof We work within a rectangle T centered at $A(x_0, y_0)$ and lying within R, and we assume that Δx and Δy are already so small that the line segment joining A to $B(x_0 + \Delta x, y_0)$ and the line segment joining B to $C(x_0 + \Delta x, y_0 + \Delta y)$ lie in the interior of T (Figure A.42).

We may think of Δz as the sum $\Delta z = \Delta z_1 + \Delta z_2$ of two increments, where

$$\Delta z_1 = f(x_0 + \Delta x, y_0) - f(x_0, y_0)$$

is the change in the value of f from A to B and

$$\Delta z_2 = f(x_0 + \Delta x, y_0 + \Delta y) - f(x_0 + \Delta x, y_0)$$

is the change in the value of f from B to C (Figure A.43).

On the closed interval of x-values joining x_0 to $x_0 + \Delta x$, the function $F(x) = f(x, y_0)$ is a differentiable (and hence continuous) function of x, with derivative

$$F'(x) = f_x(x, y_0).$$

By the Mean Value Theorem (Theorem 4, Section 3.2), there is an x-value c between x_0 and $x_0 + \Delta x$ at which

$$F(x_0 + \Delta x) - F(x_0) = F'(c)\,\Delta x$$

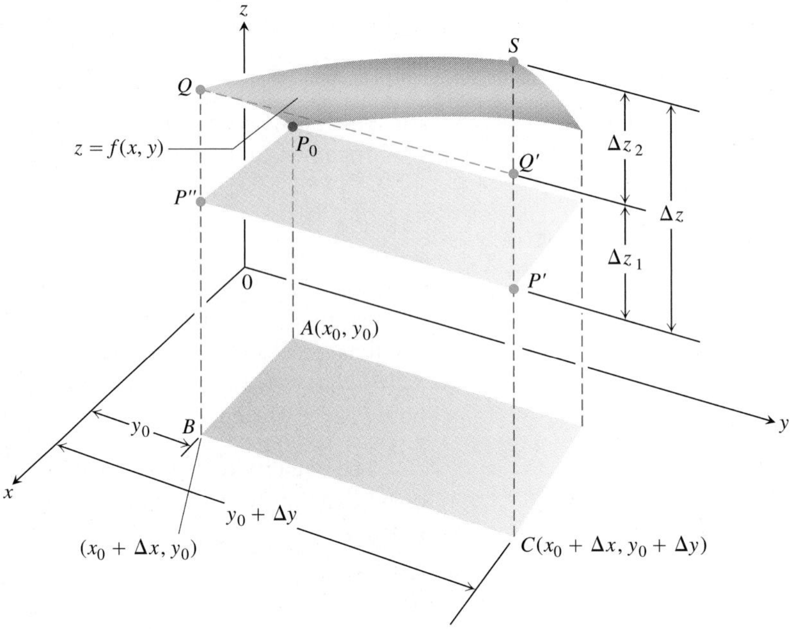

FIGURE A.43 Part of the surface $z = f(x, y)$ near $P_0(x_0, y_0, f(x_0, y_0))$. The points P_0, P', and P'' have the same height $z_0 = f(x_0, y_0)$ above the xy-plane. The change in z is $\Delta z = P'S$. The change

$$\Delta z_1 = f(x_0 + \Delta x, y_0) - f(x_0, y_0),$$

shown as $P''Q = P'Q'$, is caused by changing x from x_0 to $x_0 + \Delta x$ while holding y equal to y_0. Then, with x held equal to $x_0 + \Delta x$,

$$\Delta z_2 = f(x_0 + \Delta x, y_0 + \Delta y) - f(x_0 + \Delta x, y_0)$$

is the change in z caused by changing y_0 from $y_0 + \Delta y$, which is represented by $Q'S$. The total change in z is the sum of Δz_1 and Δz_2.

or

$$f(x_0 + \Delta x, y_0) - f(x_0, y_0) = f_x(c, y_0)\,\Delta x$$

or

$$\Delta z_1 = f_x(c, y_0)\,\Delta x. \tag{12}$$

Similarly, $G(y) = f(x_0 + \Delta x, y)$ is a differentiable (and hence continuous) function of y on the closed y-interval joining y_0 and $y_0 + \Delta y$, with derivative

$$G'(y) = f_y(x_0 + \Delta x, y).$$

Hence, there is a y-value d between y_0 and $y_0 + \Delta y$ at which

$$G(y_0 + \Delta y) - G(y_0) = G'(d)\,\Delta y$$

or

$$f(x_0 + \Delta x, y_0 + \Delta y) - f(x_0 + \Delta x, y) = f_y(x_0 + \Delta x, d)\,\Delta y$$

or

$$\Delta z_2 = f_y(x_0 + \Delta x, d)\,\Delta y. \tag{13}$$

Now, as both Δx and $\Delta y \to 0$, we know that $c \to x_0$ and $d \to y_0$. Therefore, since f_x and f_y are continuous at (x_0, y_0), the quantities

$$\begin{aligned} \epsilon_1 &= f_x(c, y_0) - f_x(x_0, y_0), \\ \epsilon_2 &= f_y(x_0 + \Delta x, d) - f_y(x_0, y_0) \end{aligned} \tag{14}$$

both approach zero as both Δx and $\Delta y \to 0$.

Finally,

$$\begin{aligned} \Delta z &= \Delta z_1 + \Delta z_2 \\ &= f_x(c, y_0)\Delta x + f_y(x_0 + \Delta x, d)\Delta y && \text{From Eqs. (12) and (13)} \\ &= [f_x(x_0, y_0) + \epsilon_1]\Delta x + [f_y(x_0, y_0) + \epsilon_2]\Delta y && \text{From Eq. (14)} \\ &= f_x(x_0, y_0)\Delta x + f_y(x_0, y_0)\Delta y + \epsilon_1\Delta x + \epsilon_2\Delta y, \end{aligned}$$

where both ϵ_1 and $\epsilon_2 \to 0$ as both Δx and $\Delta y \to 0$, which is what we set out to prove. ■

Analogous results hold for functions of any finite number of independent variables. Suppose that the first partial derivatives of $w = f(x, y, z)$ are defined throughout an open region containing the point (x_0, y_0, z_0) and that f_x, f_y, and f_z are continuous at (x_0, y_0, z_0). Then

$$\begin{aligned} \Delta w &= f(x_0 + \Delta x, y_0 + \Delta y, z_0 + \Delta z) - f(x_0, y_0, z_0) \\ &= f_x\Delta x + f_y\Delta y + f_z\Delta z + \epsilon_1\Delta x + \epsilon_2\Delta y + \epsilon_3\Delta z, \end{aligned} \tag{15}$$

where $\epsilon_1, \epsilon_2, \epsilon_3 \to 0$ as Δx, Δy, and $\Delta z \to 0$.

The partial derivatives f_x, f_y, f_z in Equation (15) are to be evaluated at the point (x_0, y_0, z_0).

Equation (15) can be proved by treating Δw as the sum of three increments,

$$\Delta w_1 = f(x_0 + \Delta x, y_0, z_0) - f(x_0, y_0, z_0) \tag{16}$$

$$\Delta w_2 = f(x_0 + \Delta x, y_0 + \Delta y, z_0) - f(x_0 + \Delta x, y_0, z_0) \tag{17}$$

$$\Delta w_3 = f(x_0 + \Delta x, y_0 + \Delta y, z_0 + \Delta z) - f(x_0 + \Delta x, y_0 + \Delta y, z_0), \tag{18}$$

and applying the Mean Value Theorem to each of these separately. Two coordinates remain constant and only one varies in each of these partial increments $\Delta w_1, \Delta w_2, \Delta w_3$. In Equation (17), for example, only y varies, since x is held equal to $x_0 + \Delta x$ and z is held equal to z_0. Since $f(x_0 + \Delta x, y, z_0)$ is a continuous function of y with a derivative f_y, it is subject to the Mean Value Theorem, and we have

$$\Delta w_2 = f_y(x_0 + \Delta x, y_1, z_0)\, \Delta y$$

for some y_1 between y_0 and $y_0 + \Delta y$.

A.12 Taylor's Formula for Two Variables

This section uses Taylor's formula to derive the Second Derivative Test for local extreme values (Section 11.7) and the error formula for linearizations of functions of two independent variables (Section 11.6). The use of Taylor's formula in these derivations leads to an extension of the formula that provides polynomial approximations of all orders for functions of two independent variables.

Derivation of the Second Derivative Test

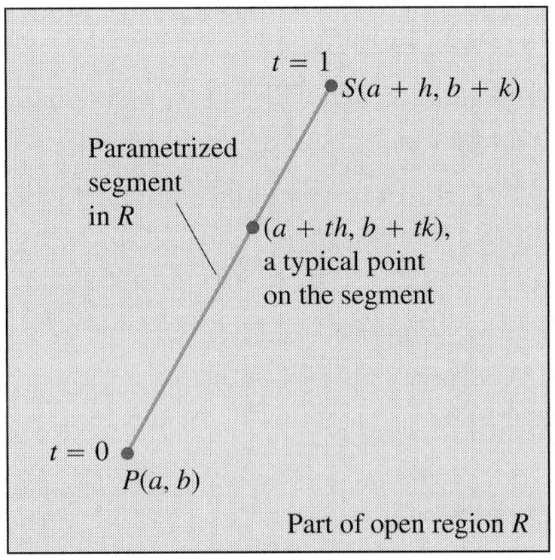

FIGURE A.44 We begin the derivation of the second derivative test at $P(a, b)$ by parametrizing a typical line segment from P to a point S nearby.

Let $f(x, y)$ have continuous partial derivatives in an open region R containing a point $P(a, b)$ where $f_x = f_y = 0$ (Figure A.44). Let h and k be increments small enough to put the point $S(a + h, b + k)$ and the line segment joining it to P inside R. We parametrize the segment PS as

$$x = a + th, \qquad y = b + tk, \qquad 0 \le t \le 1.$$

If $F(t) = f(a + th, b + tk)$, the Chain Rule gives

$$F'(t) = f_x \frac{dx}{dt} + f_y \frac{dy}{dt} = hf_x + kf_y.$$

Since f_x and f_y are differentiable (they have continuous partial derivatives), F' is a differentiable function of t and

$$\begin{aligned} F'' &= \frac{\partial F'}{\partial x}\frac{dx}{dt} + \frac{\partial F'}{\partial y}\frac{dy}{dt} = \frac{\partial}{\partial x}(hf_x + kf_y)\cdot h + \frac{\partial}{\partial y}(hf_x + kf_y)\cdot k \\ &= h^2 f_{xx} + 2hkf_{xy} + k^2 f_{yy}. \qquad f_{xy} = f_{yx} \end{aligned}$$

Since F and F' are continuous on $[0, 1]$ and F' is differentiable on $(0, 1)$, we can apply Taylor's formula with $n = 2$ and $a = 0$ to obtain

$$\begin{aligned} F(1) &= F(0) + F'(0)(1 - 0) + F''(c)\frac{(1-0)^2}{2} \\ F(1) &= F(0) + F'(0) + \frac{1}{2}F''(c) \end{aligned} \tag{1}$$

for some c between 0 and 1. Writing Equation (1) in terms of f gives

$$\begin{aligned} f(a + h, b + k) &= f(a, b) + hf_x(a, b) + kf_y(a, b) \\ &\quad + \frac{1}{2}\left(h^2 f_{xx} + 2hkf_{xy} + k^2 f_{yy}\right)\Big|_{(a+ch,\, b+ck)}. \end{aligned} \tag{2}$$

Since $f_x(a, b) = f_y(a, b) = 0$, this reduces to

$$f(a + h, b + k) - f(a, b) = \frac{1}{2}\left(h^2 f_{xx} + 2hk f_{xy} + k^2 f_{yy}\right)\Big|_{(a+ch,\, b+ck)}. \tag{3}$$

The presence of an extremum of f at (a, b) is determined by the sign of $f(a + h, b + k) - f(a, b)$. By Equation (3), this is the same as the sign of

$$Q(c) = (h^2 f_{xx} + 2hk f_{xy} + k^2 f_{yy})|_{(a+ch,b+ck)}.$$

Now, if $Q(0) \neq 0$, the sign of $Q(c)$ will be the same as the sign of $Q(0)$ for sufficiently small values of h and k. We can predict the sign of

$$Q(0) = h^2 f_{xx}(a, b) + 2hk f_{xy}(a, b) + k^2 f_{yy}(a, b) \tag{4}$$

from the signs of f_{xx} and $f_{xx} f_{yy} - f_{xy}^2$ at (a, b). Multiply both sides of Equation (4) by f_{xx} and rearrange the right-hand side to get

$$f_{xx} Q(0) = (h f_{xx} + k f_{xy})^2 + (f_{xx} f_{yy} - f_{xy}^2)k^2. \tag{5}$$

From Equation (5) we see that

1. If $f_{xx} < 0$ and $f_{xx} f_{yy} - f_{xy}^2 > 0$ at (a, b), then $Q(0) < 0$ for all sufficiently small nonzero values of h and k, and f has a *local maximum* value at (a, b).
2. If $f_{xx} > 0$ and $f_{xx} f_{yy} - f_{xy}^2 > 0$ at (a, b), then $Q(0) > 0$ for all sufficiently small nonzero values of h and k and f has a *local minimum* value at (a, b).
3. If $f_{xx} f_{yy} - f_{xy}^2 < 0$ at (a, b), there are combinations of arbitrarily small nonzero values of h and k for which $Q(0) > 0$, and other values for which $Q(0) < 0$. Arbitrarily close to the point $P_0(a, b, f(a, b))$ on the surface $z = f(x, y)$ there are points above P_0 and points below P_0, so f has a *saddle point* at (a, b).
4. If $f_{xx} f_{yy} - f_{xy}^2 = 0$, another test is needed. The possibility that $Q(0)$ equals zero prevents us from drawing conclusions about the sign of $Q(c)$.

The Error Formula for Linear Approximations

We want to show that the difference $E(x, y)$, between the values of a function $f(x, y)$, and its linearization $L(x, y)$ at (x_0, y_0) satisfies the inequality

$$|E(x, y)| \leq \frac{1}{2} M(|x - x_0| + |y - y_0|)^2.$$

The function f is assumed to have continuous second partial derivatives throughout an open set containing a closed rectangular region R centered at (x_0, y_0). The number M is an upper bound for $|f_{xx}|, |f_{yy}|$, and $|f_{xy}|$ on R.

The inequality we want comes from Equation (2). We substitute x_0 and y_0 for a and b, and $x - x_0$ and $y - y_0$ for h and k, respectively, and rearrange the result as

$$f(x, y) = \underbrace{f(x_0, y_0) + f_x(x_0, y_0)(x - x_0) + f_y(x_0, y_0)(y - y_0)}_{\text{linearization } L(x,\, y)}$$

$$\underbrace{+ \frac{1}{2}\left((x - x_0)^2 f_{xx} + 2(x - x_0)(y - y_0) f_{xy} + (y - y_0)^2 f_{yy}\right)\Big|_{(x_0 + c(x - x_0),\, y_0 + c(y - y_0))}}_{\text{error } E(x,\, y)}.$$

This equation reveals that

$$|E| \le \frac{1}{2}\left(|x - x_0|^2|f_{xx}| + 2|x - x_0||y - y_0||f_{xy}| + |y - y_0|^2|f_{yy}|\right).$$

Hence, if M is an upper bound for the values of $|f_{xx}|, |f_{xy}|$, and $|f_{yy}|$ on R,

$$\begin{aligned}|E| &\le \frac{1}{2}\left(|x - x_0|^2 M + 2|x - x_0||y - y_0|M + |y - y_0|^2 M\right)\\ &= \frac{1}{2}M(|x - x_0| + |y - y_0|)^2.\end{aligned}$$

Taylor's Formula for Functions of Two Variables

The formulas derived earlier for F' and F'' can be obtained by applying to $f(x, y)$ the operators

$$\left(h\frac{\partial}{\partial x} + k\frac{\partial}{\partial y}\right) \quad \text{and} \quad \left(h\frac{\partial}{\partial x} + k\frac{\partial}{\partial y}\right)^2 = h^2\frac{\partial^2}{\partial x^2} + 2hk\frac{\partial^2}{\partial x\,\partial y} + k^2\frac{\partial^2}{\partial y^2}.$$

These are the first two instances of a more general formula,

$$F^{(n)}(t) = \frac{d^n}{dt^n}F(t) = \left(h\frac{\partial}{\partial x} + k\frac{\partial}{\partial y}\right)^n f(x, y), \tag{6}$$

which says that applying d^n/dt^n to $F(t)$ gives the same result as applying the operator

$$\left(h\frac{\partial}{\partial x} + k\frac{\partial}{\partial y}\right)^n$$

to $f(x, y)$ after expanding it by the Binomial Theorem.

If partial derivatives of f through order $n + 1$ are continuous throughout a rectangular region centered at (a, b), we may extend the Taylor formula for $F(t)$ to

$$F(t) = F(0) + F'(0)t + \frac{F''(0)}{2!}t^2 + \cdots + \frac{F^{(n)}(0)}{n!}t^{(n)} + \text{remainder},$$

and take $t = 1$ to obtain

$$F(1) = F(0) + F'(0) + \frac{F''(0)}{2!} + \cdots + \frac{F^{(n)}(0)}{n!} + \text{remainder}.$$

When we replace the first n derivatives on the right of this last series by their equivalent expressions from Equation (6) evaluated at $t = 0$ and add the appropriate remainder term, we arrive at the following formula.

Taylor's Formula for $f(x, y)$ at the Point (a, b)

Suppose $f(x, y)$ and its partial derivatives through order $n + 1$ are continuous throughout an open rectangular region R centered at a point (a, b). Then, throughout R,

$$\begin{aligned}f(a + h, b + k) = {} & f(a, b) + (hf_x + kf_y)|_{(a,b)} + \frac{1}{2!}(h^2 f_{xx} + 2hkf_{xy} + k^2 f_{yy})|_{(a,b)}\\ &+ \frac{1}{3!}(h^3 f_{xxx} + 3h^2 kf_{xxy} + 3hk^2 f_{xyy} + k^3 f_{yyy})|_{(a,b)} + \cdots + \frac{1}{n!}\left(h\frac{\partial}{\partial x} + k\frac{\partial}{\partial y}\right)^n f\Big|_{(a,b)}\\ &+ \frac{1}{(n + 1)!}\left(h\frac{\partial}{\partial x} + k\frac{\partial}{\partial y}\right)^{n+1} f\Big|_{(a+ch,b+ck)}.\end{aligned} \tag{7}$$

The first n derivative terms are evaluated at (a, b). The last term is evaluated at some point $(a + ch, b + ck)$ on the line segment joining (a, b) and $(a + h, b + k)$.

If $(a, b) = (0, 0)$ and we treat h and k as independent variables (denoting them now by x and y), then Equation (7) assumes the following simpler form.

Taylor's Formula for $f(x, y)$ at the Origin

$$f(x, y) = f(0, 0) + xf_x + yf_y + \frac{1}{2!}(x^2 f_{xx} + 2xyf_{xy} + y^2 f_{yy})$$
$$+ \frac{1}{3!}(x^3 f_{xxx} + 3x^2 y f_{xxy} + 3xy^2 f_{xyy} + y^3 f_{yyy}) + \cdots + \frac{1}{n!}\left(x\frac{\partial}{\partial x} + y\frac{\partial}{\partial y}\right)^n f$$
$$+ \frac{1}{(n+1)!}\left(x\frac{\partial}{\partial x} + y\frac{\partial}{\partial y}\right)^{n+1} f\Bigg|_{(cx,cy)} \tag{8}$$

The first n derivative terms are evaluated at $(0, 0)$. The last term is evaluated at a point on the line segment joining the origin and (x, y).

Taylor's formula provides polynomial approximations of two-variable functions. The first n derivative terms give the polynomial; the last term gives the approximation error. The first three terms of Taylor's formula give the function's linearization. To improve on the linearization, we add higher power terms.

EXAMPLE 1 Find a quadratic approximation to $f(x, y) = \sin x \sin y$ near the origin. How accurate is the approximation if $|x| \le 0.1$ and $|y| \le 0.1$?

Solution We take $n = 2$ in Equation (8):

$$f(x, y) = f(0, 0) + (xf_x + yf_y) + \frac{1}{2}(x^2 f_{xx} + 2xyf_{xy} + y^2 f_{yy})$$
$$+ \frac{1}{6}(x^3 f_{xxx} + 3x^2 y f_{xxy} + 3xy^2 f_{xyy} + y^3 f_{yyy})_{(cx,cy)}$$

with

$$f(0,0) = \sin x \sin y|_{(0,0)} = 0, \qquad f_{xx}(0,0) = -\sin x \sin y|_{(0,0)} = 0,$$
$$f_x(0,0) = \cos x \sin y|_{(0,0)} = 0, \qquad f_{xy}(0,0) = \cos x \cos y|_{(0,0)} = 1,$$
$$f_y(0,0) = \sin x \cos y|_{(0,0)} = 0, \qquad f_{yy}(0,0) = -\sin x \sin y|_{(0,0)} = 0,$$

we have

$$\sin x \sin y \approx 0 + 0 + 0 + \frac{1}{2}(x^2(0) + 2xy(1) + y^2(0)),$$
$$\sin x \sin y \approx xy.$$

The error in the approximation is

$$E(x, y) = \frac{1}{6}(x^3 f_{xxx} + 3x^2 y f_{xxy} + 3xy^2 f_{xyy} + y^3 f_{yyy})\big|_{(cx,cy)}.$$

The third derivatives never exceed 1 in absolute value because they are products of sines and cosines. Also, $|x| \le 0.1$ and $|y| \le 0.1$. Hence

$$|E(x, y)| \le \frac{1}{6}((0.1)^3 + 3(0.1)^3 + 3(0.1)^3 + (0.1)^3) = \frac{8}{6}(0.1)^3 \le 0.00134$$

(rounded up). The error will not exceed 0.00134 if $|x| \le 0.1$ and $|y| \le 0.1$. ■

EXERCISES A.12

In Exercises 1–10, use Taylor's formula for $f(x, y)$ at the origin to find quadratic and cubic approximations of f near the origin.

1. $f(x, y) = xe^y$

2. $f(x, y) = e^x \cos y$

3. $f(x, y) = y \sin x$

4. $f(x, y) = \sin x \cos y$

5. $f(x, y) = e^x \ln(1 + y)$

6. $f(x, y) = \ln(2x + y + 1)$

7. $f(x, y) = \sin(x^2 + y^2)$

8. $f(x, y) = \cos(x^2 + y^2)$

9. $f(x, y) = \dfrac{1}{1 - x - y}$

10. $f(x, y) = \dfrac{1}{1 - x - y + xy}$

11. Use Taylor's formula to find a quadratic approximation of $f(x, y) = \cos x \cos y$ at the origin. Estimate the error in the approximation if $|x| \le 0.1$ and $|y| \le 0.1$.

12. Use Taylor's formula to find a quadratic approximation of $e^x \sin y$ at the origin. Estimate the error in the approximation if $|x| \le 0.1$ and $|y| \le 0.1$.

ANSWERS

CHAPTER 1

Section 1.1, pp. 11–13

1. $D: (-\infty, \infty)$, $R: [1, \infty)$ **3.** $D: (0, \infty)$, $R: (0, \infty)$

5. (a) Not a function of x because some values of x have two values of y

(b) A function of x because for every x there is only one possible y

7. $A = \dfrac{\sqrt{3}}{4}x^2$, $p = 3x$ **9.** $x = \dfrac{d}{\sqrt{3}}$, $A = 2d^2$, $V = \dfrac{d^3}{3\sqrt{3}}$

11. $(-\infty, \infty)$

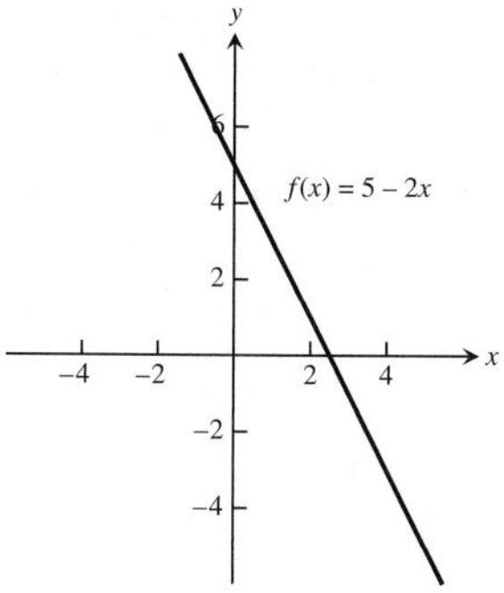

13. $(-\infty, \infty)$

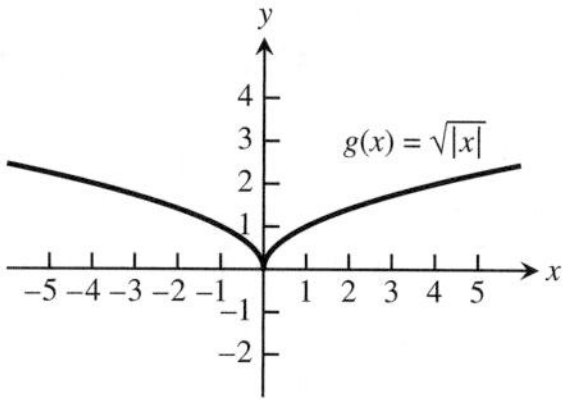

15. $(-\infty, 0) \cup (0, \infty)$

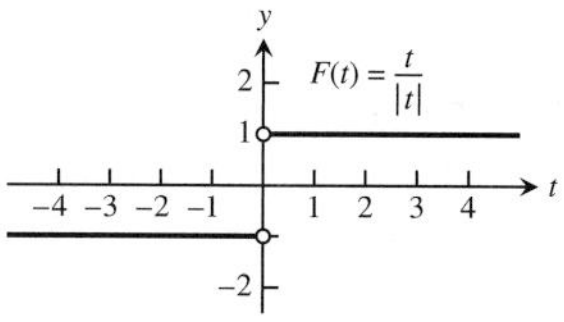

17. (a) For each positive value of x, there are two values of y.

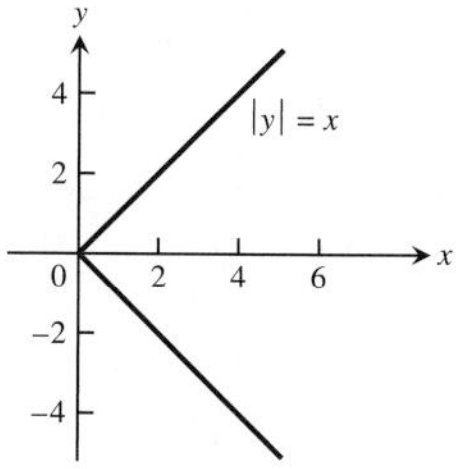

(b) For each value of $x \neq 0$, there are two values of y.

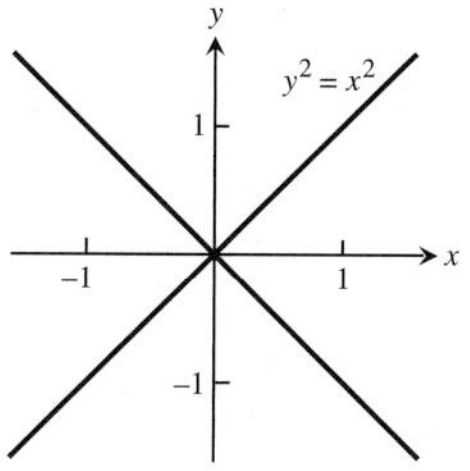

19.

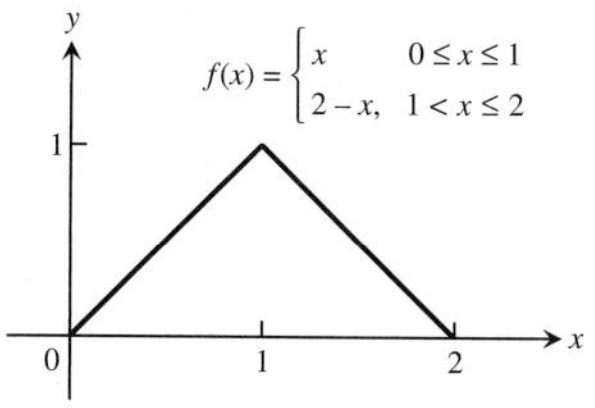

21.

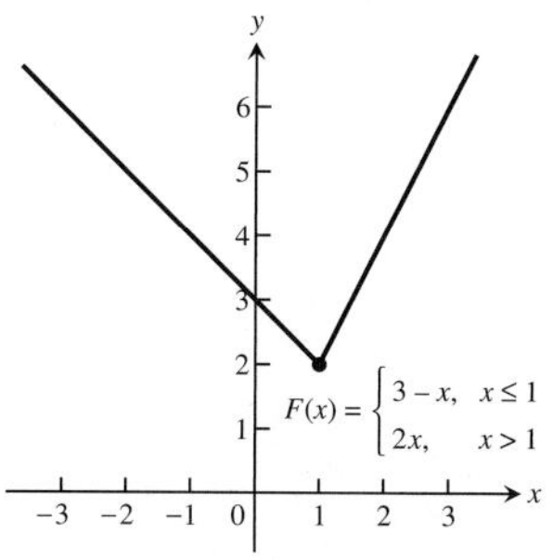

23. (a) $f(x) = \begin{cases} x, & 0 \le x \le 1 \\ -x + 2, & 1 < x \le 2 \end{cases}$

(b) $f(x) = \begin{cases} 2, & 0 \le x < 1 \\ 0, & 1 \le x < 2 \\ 2, & 2 \le x < 3 \\ 0, & 3 \le x \le 4 \end{cases}$

25. (a) $f(x) = \begin{cases} -x, & -1 \le x < 0 \\ 1, & 0 < x \le 1 \\ -\frac{1}{2}x + \frac{3}{2}, & 1 < x < 3 \end{cases}$

(b) $f(x) = \begin{cases} \frac{1}{2}x, & -2 \le x \le 0 \\ -2x + 2, & 0 < x \le 1 \\ -1, & 1 < x \le 3 \end{cases}$

27. (a) $0 \le x < 1$ **(b)** $-1 < x \le 0$ **29.** Yes

31. Symmetric about the origin

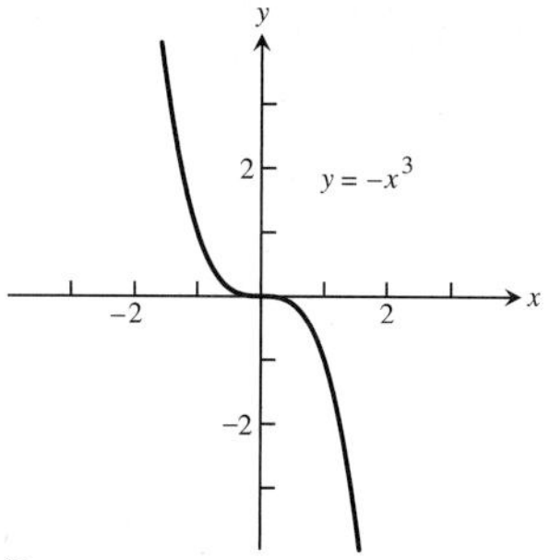

Dec. $-\infty < x < \infty$

33. Symmetric about the origin

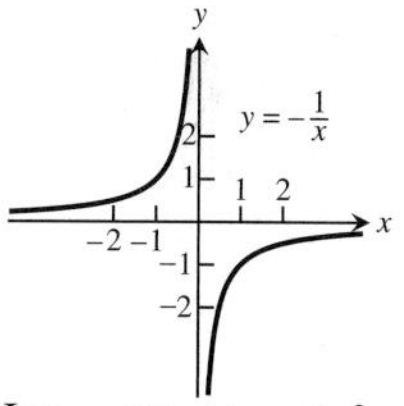

Inc. $-\infty < x < 0$ and $0 < x < \infty$

35. Symmetric about the y-axis

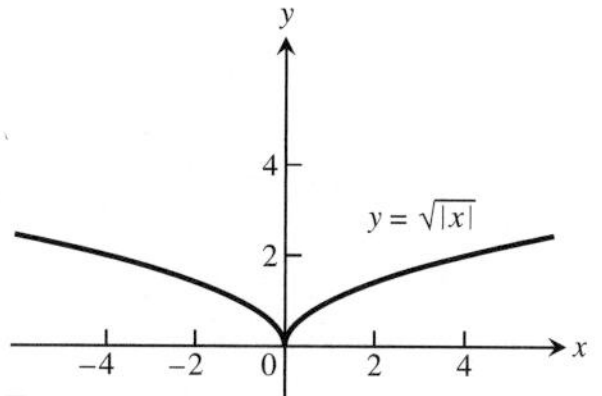

Dec. $-\infty < x \le 0$; inc. $0 \le x < \infty$

37. Symmetric about the origin

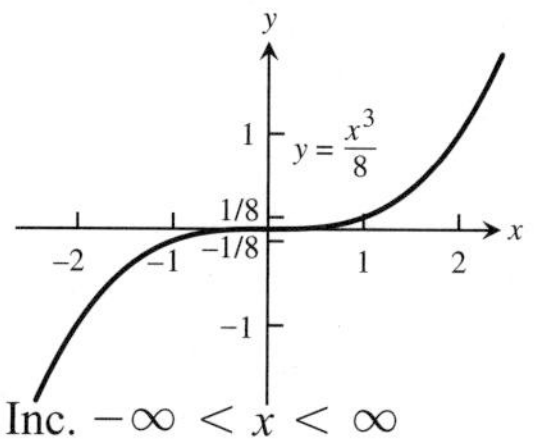

Inc. $-\infty < x < \infty$

39. No symmetry

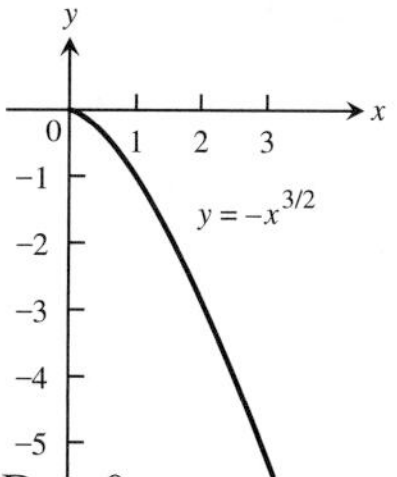

Dec. $0 \le x < \infty$

41. Symmetric about the y-axis

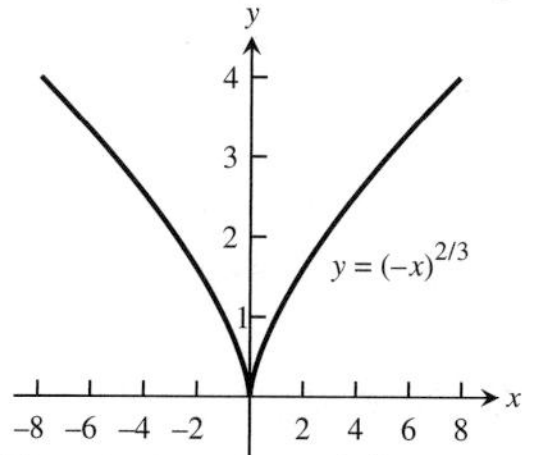

Dec. $-\infty < x \le 0$; inc. $0 \le x < \infty$

43. Even **45.** Even **47.** Odd **49.** Even **51.** Neither **53.** Neither **55.** Odd **57.** Odd **59. (a)** $(-2, 0) \cup (4, \infty)$

Section 1.2, pp. 16–18

1. $D_f: -\infty < x < \infty,\quad D_g: x \geq 1,\quad R_f: -\infty < y < \infty,$
$R_g: y \geq 0,\quad D_{f+g} = D_{f\cdot g} = D_g,\quad R_{f+g}: y \geq 1,\quad R_{f\cdot g}: y \geq 0$

3. $D_f: -\infty < x < \infty,\quad D_g: -\infty < x < \infty,\quad R_f: y = 2,$
$R_g: y \geq 1,\quad D_{f/g}: -\infty < x < \infty,\quad R_{f/g}: 0 < y \leq 2,$
$D_{g/f}: -\infty < x < \infty,\quad R_{g/f}: y \geq 1/2$

5. **(a)** 2 **(b)** 22 **(c)** $x^2 + 2$ **(d)** $x^2 + 10x + 22$ **(e)** 5 **(f)** -2 **(g)** $x + 10$ **(h)** $x^4 - 6x^2 + 6$

7. **(a)** $\dfrac{4}{x^2} - 5$ **(b)** $\dfrac{4}{x^2} - 5$ **(c)** $\left(\dfrac{4}{x} - 5\right)^2$ **(d)** $\left(\dfrac{1}{4x - 5}\right)^2$ **(e)** $\dfrac{1}{4x^2 - 5}$ **(f)** $\dfrac{1}{(4x - 5)^2}$

9. **(a)** $f(g(x))$ **(b)** $j(g(x))$ **(c)** $g(g(x))$ **(d)** $j(j(x))$ **(e)** $g(h(f(x)))$ **(f)** $h(j(f(x)))$

11.

	$g(x)$	$f(x)$	$(f \circ g)(x)$
(a)	$x - 7$	$\sqrt{x}$	$\sqrt{x - 7}$
(b)	$x + 2$	$3x$	$3x + 6$
(c)	x^2	$\sqrt{x - 5}$	$\sqrt{x^2 - 5}$
(d)	$\dfrac{x}{x - 1}$	$\dfrac{x}{x - 1}$	x
(e)	$\dfrac{1}{x - 1}$	$1 + \dfrac{1}{x}$	x
(f)	$\dfrac{1}{x}$	$\dfrac{1}{x}$	x

13. **(a)** Position 4 **(b)** Position 1 **(c)** Position 2 **(d)** Position 3

15. $(x + 2)^2 + (y + 3)^2 = 49$

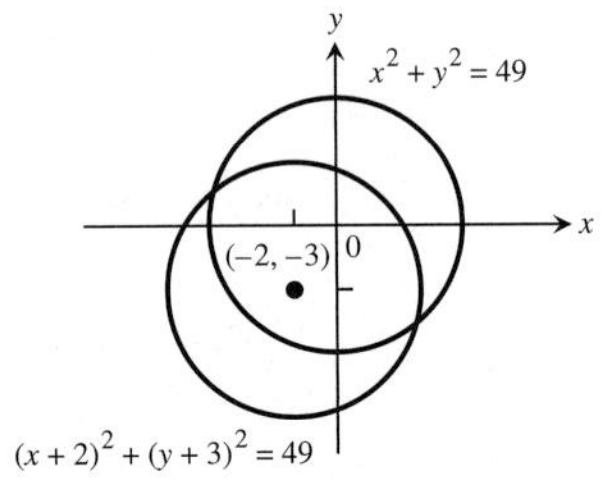

17. $y = \sqrt{x + 0.81}$

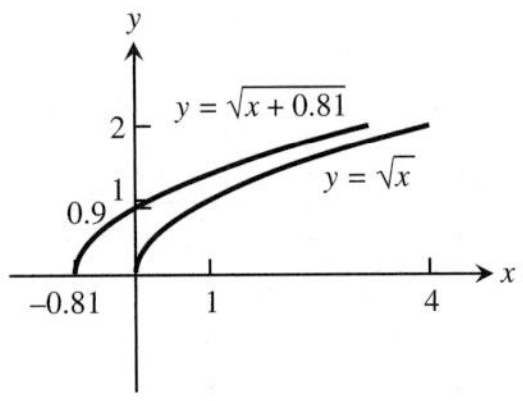

19. $y - 1 = \dfrac{1}{x - 1}$

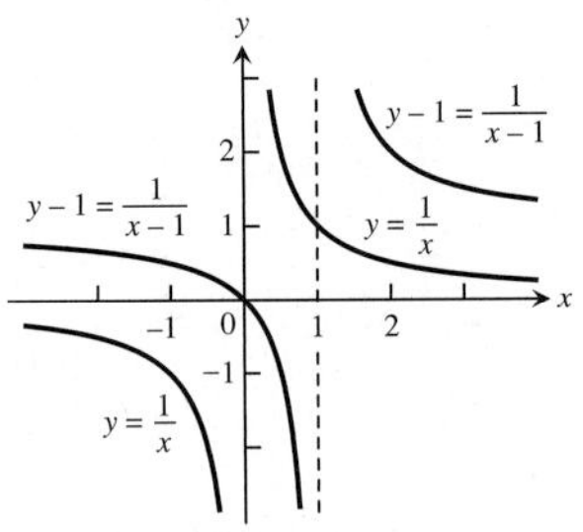

21.

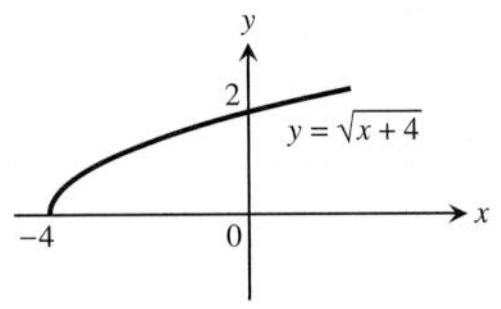

23.

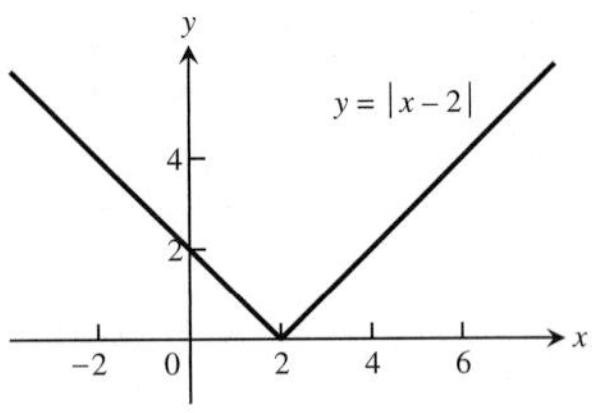

25.

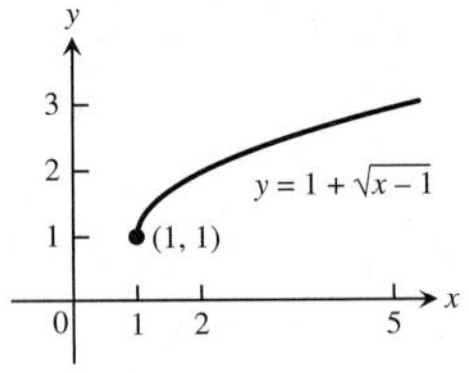

27.

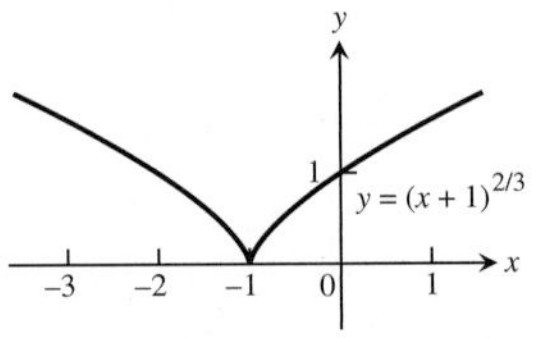

29.

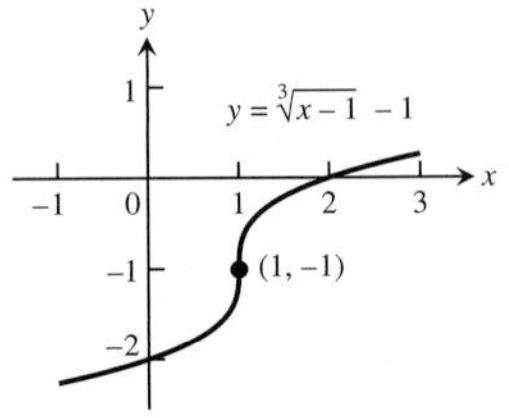

31.

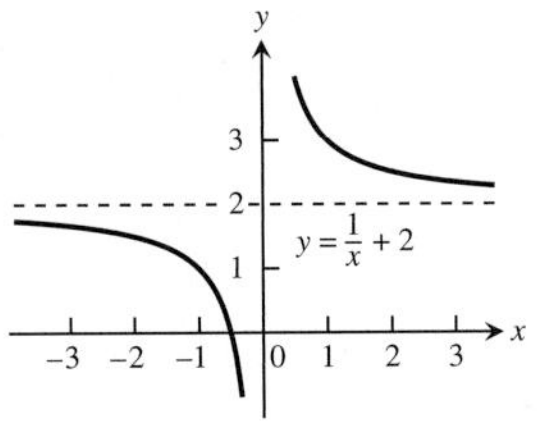

33.

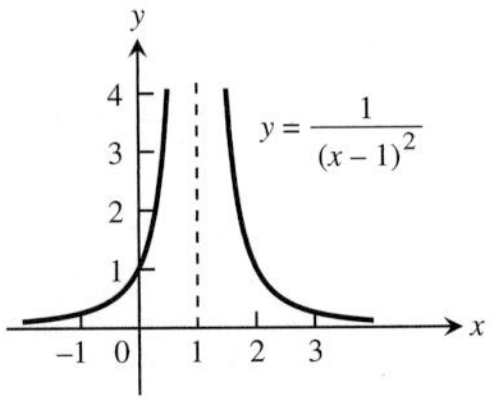

35. **(a)** $D: [0, 2],\quad R: [2, 3]$

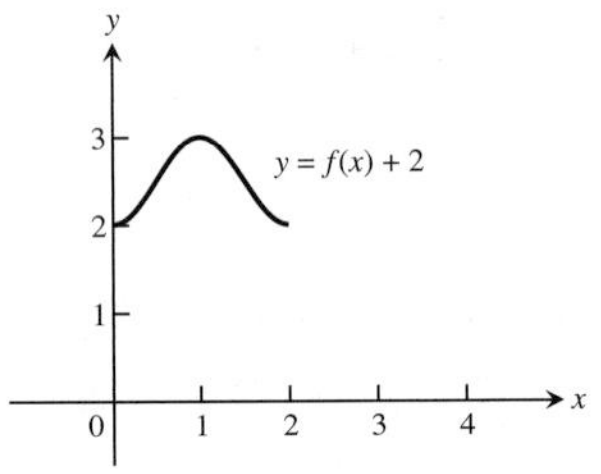

(b) $D: [0, 2],\quad R: [-1, 0]$

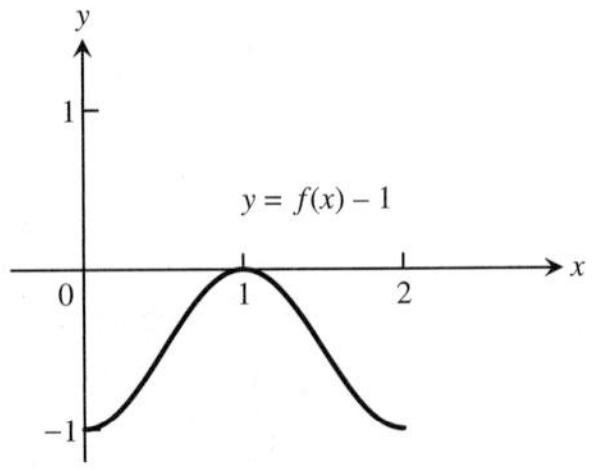

(c) $D: [0, 2],\quad R: [0, 2]$

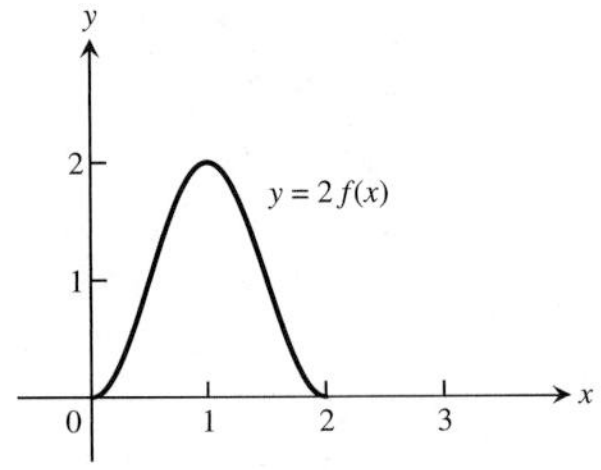

(d) $D: [0, 2],\quad R: [-1, 0]$

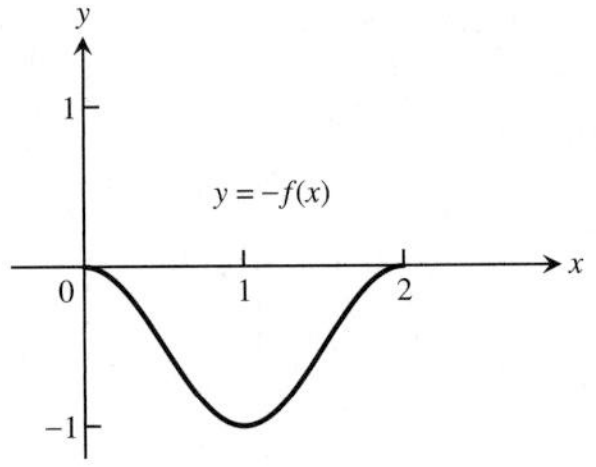

(e) $D: [-2, 0],\quad R: [0, 1]$

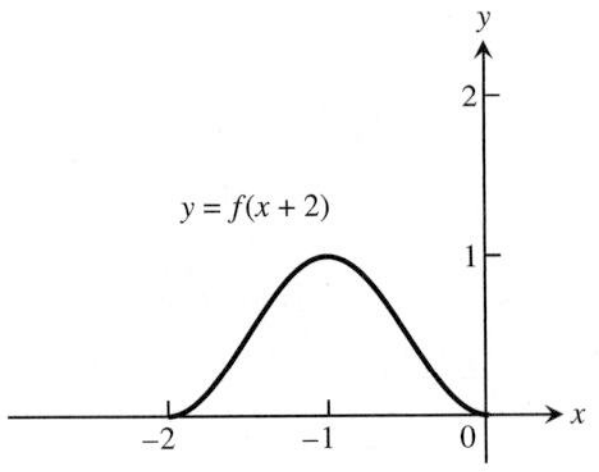

(f) $D: [1, 3],\quad R: [0, 1]$

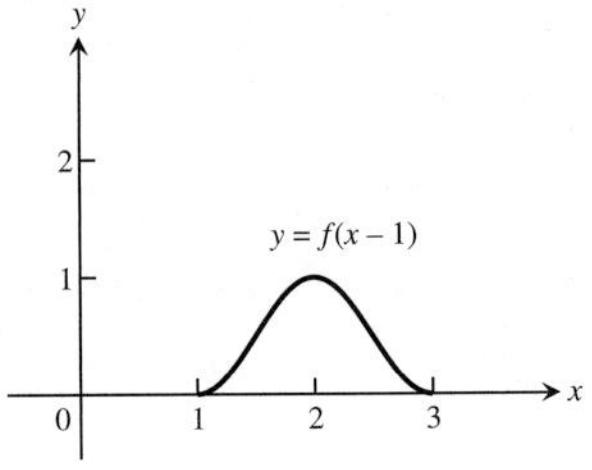

(g) $D: [-2, 0]$, $R: [0, 1]$ **(h)** $D: [-1, 1]$, $R: [0, 1]$

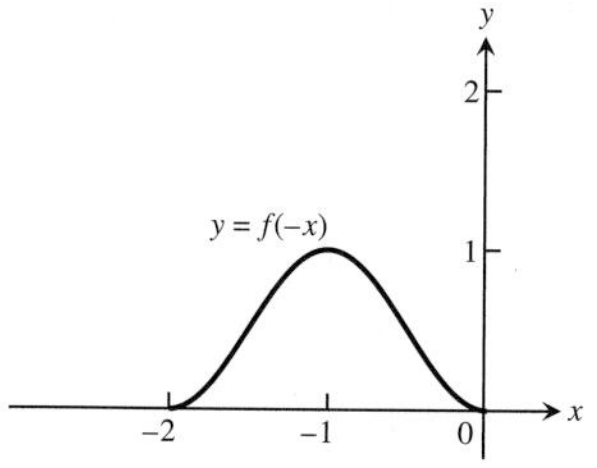

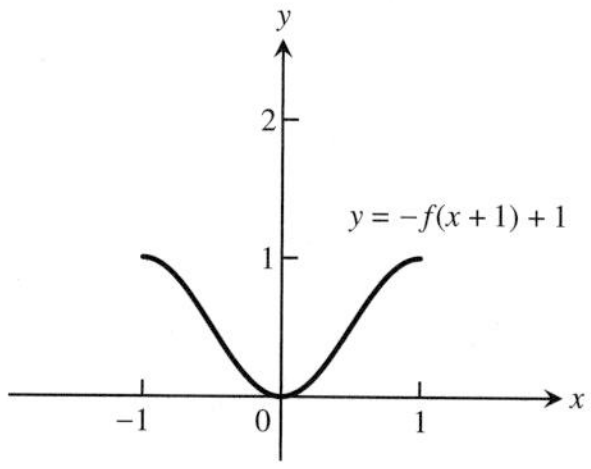

37. $y = 3x^2 - 3$ **39.** $y = \frac{1}{2} + \frac{1}{2x^2}$ **41.** $y = \sqrt{4x + 1}$

43. $y = \sqrt{4 - \frac{x^2}{4}}$ **45.** $y = 1 - 27x^3$

47.

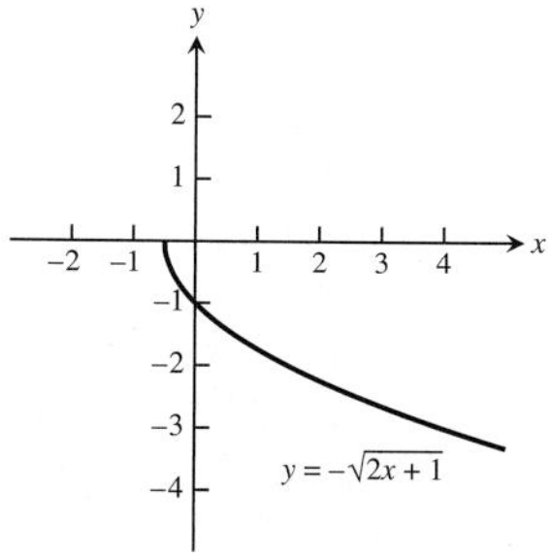

49.

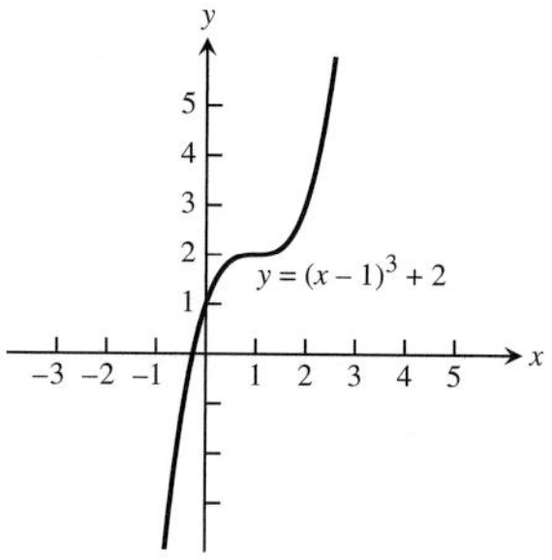

51.

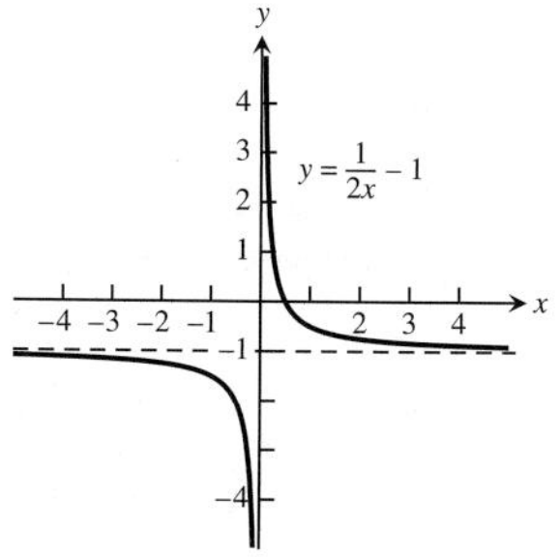

53.

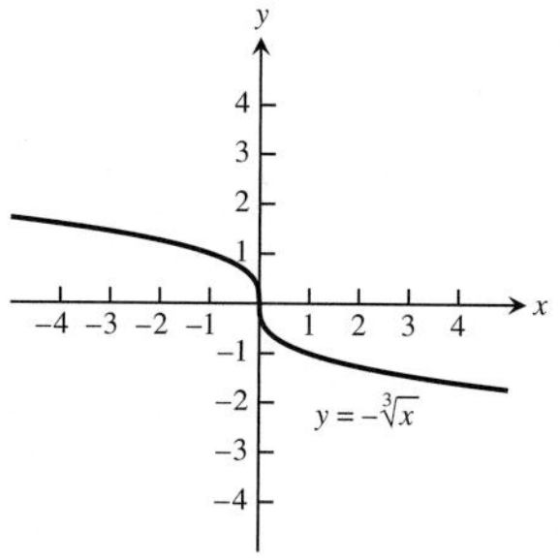

55.

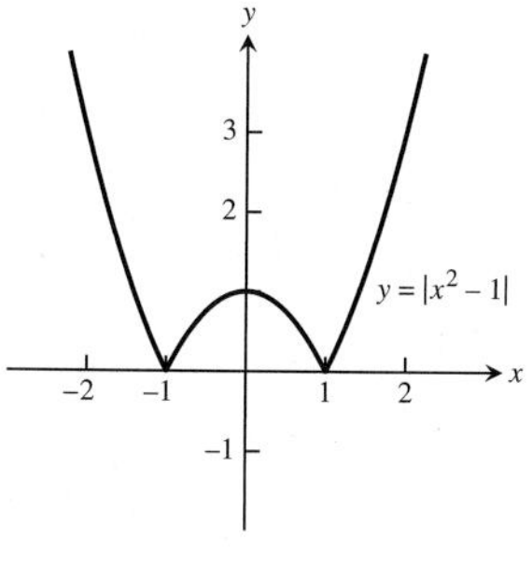

57. **(a)** Odd **(b)** Odd **(c)** Odd **(d)** Even **(e)** Even
(f) Even **(g)** Even **(h)** Even **(i)** Odd

Section 1.3, pp. 23–24

1. **(a)** 19 **(b)** 1

3. **(a)** $-\frac{4}{\pi}$ **(b)** $-\frac{3\sqrt{3}}{\pi}$ **5.** 1

7. Your estimates may not completely agree with these.

(a)

PQ_1	PQ_2	PQ_3	PQ_4
43	46	49	50

The appropriate units are m/sec.

(b) $\approx$ 50 m/sec or 180 km/h

9. **(a)** 4 **(b)** $y = 4x - 7$
11. **(a)** 2 **(b)** $y = 2x - 7$
13. **(a)** 12 **(b)** $y = 12x - 16$
15. **(a)** -9 **(b)** $y = -9x - 2$
17. **(a)**

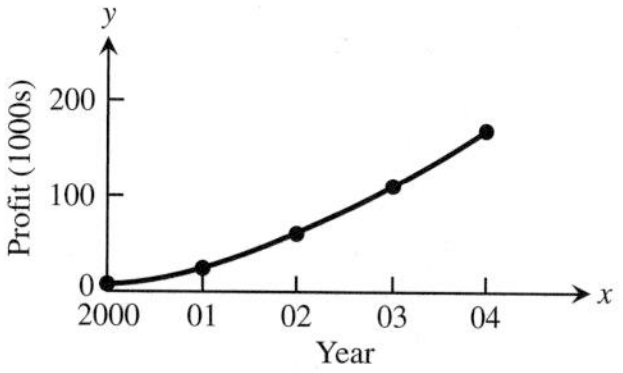

(b) $\approx$ \$56,000/year
(c) $\approx$ \$42,000/year

Section 1.4, pp. 31–34

1. **(a)** Does not exist. As x approaches 1 from the right, $g(x)$ approaches 0. As x approaches 1 from the left, $g(x)$ approaches 1. There is no single number L that all the values $g(x)$ get arbitrarily close to as $x \to 1$.
(b) 1 **(c)** 0

3. **(a)** True **(b)** True **(c)** False
(d) False **(e)** False **(f)** True

5. As x approaches 0 from the left, $x/|x|$ approaches -1. As x approaches 0 from the right, $x/|x|$ approaches 1. There is no single number L that the function values all get arbitrarily close to as $x \to 0$.

7. Nothing can be said. **9.** No; no; no

11. -9 **13.** 4 **15.** -8 **17.** 5/8 **19.** 5/2 **21.** 27 **23.** 16
25. 3/2 **27.** 3/2 **29.** 1/10 **31.** -7 **33.** 3/2 **35.** $-1/2$
37. 4/3 **39.** 1/6 **41.** 4 **43.** 1/2 **45.** 3/2 **47.** -1
49. 1 **51.** 1/3 **53.** 1
55. **(a)** -10 **(b)** -20 **(c)** -1 **(d)** 5/7
57. 2 **59.** 3 **61.** $1/(2\sqrt{7})$ **63.** $\sqrt{5}$
65. **(a)** The limit is 1.
67. **(a)** $f(x) = (x^2 - 9)/(x + 3)$

x	-3.1	-3.01	-3.001	-3.0001	-3.00001	-3.000001
$f(x)$	-6.1	-6.01	-6.001	-6.0001	-6.00001	-6.000001

x	-2.9	-2.99	-2.999	-2.9999	-2.99999	-2.999999
$f(x)$	-5.9	-5.99	-5.999	-5.9999	-5.99999	-5.999999

(c) $\lim_{x \to -3} f(x) = -6$

69. **(a)** $f(x) = (x^2 - 1)/(|x| - 1)$

x	-1.1	-1.01	-1.001	-1.0001	-1.00001	-1.000001
$f(x)$	2.1	2.01	2.001	2.0001	2.00001	2.000001

x	$-.9$	$-.99$	$-.999$	$-.9999$	$-.99999$	$-.999999$
$f(x)$	1.9	1.99	1.999	1.9999	1.99999	1.999999

(c) $\lim_{x \to -1} f(x) = 2$

71. **(a)** $g(\theta) = (\sin\theta)/\theta$

θ	.1	.01	.001	.0001	.00001	.000001
$g(\theta)$	.998334	.999983	.999999	.999999	.999999	.999999

θ	$-.1$	$-.01$	$-.001$	$-.0001$	$-.00001$	$-.000001$
$g(\theta)$	.998334	.999983	.999999	.999999	.999999	.999999

$\lim_{\theta \to 0} g(\theta) = 1$

73. $c = 0, 1, -1$; The limit is 0 at $c = 0$, and 1 at $c = 1, -1$.
75. **(a)** 5 **(b)** 5

Section 1.5, pp. 39–42

1. $\delta = 2$

3. $\delta = 1/2$

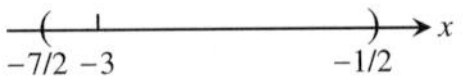

5. $\delta = 1/18$

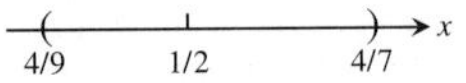

7. $\delta = 0.1$ **9.** $\delta = \sqrt{5} - 2$ **11.** $(3.99, 4.01)$, $\delta = 0.01$
13. $(-0.19, 0.21)$, $\delta = 0.19$ **15.** $(3, 15)$, $\delta = 5$
17. $(10/3, 5)$, $\delta = 2/3$
19. $(-\sqrt{4.5}, -\sqrt{3.5})$, $\delta = \sqrt{4.5} - 2 \approx 0.12$
21. $(\sqrt{15}, \sqrt{17})$, $\delta = \sqrt{17} - 4 \approx 0.12$
23. $L = -3$, $\delta = 0.01$
25. $L = 4$, $\delta = 0.05$
43. $[3.384, 3.387]$. To be safe, the left endpoint was rounded up and the right endpoint rounded down.

Section 1.6, pp. 47–49

1. **(a)** True **(b)** True **(c)** False **(d)** True **(e)** True **(f)** True **(g)** False **(h)** False **(i)** False **(j)** False **(k)** True **(l)** False
3. **(a)** 2, 1 **(b)** No, $\lim_{x \to 2^+} f(x) \neq \lim_{x \to 2^-} f(x)$ **(c)** 3, 3 **(d)** Yes, 3
5. **(a)**

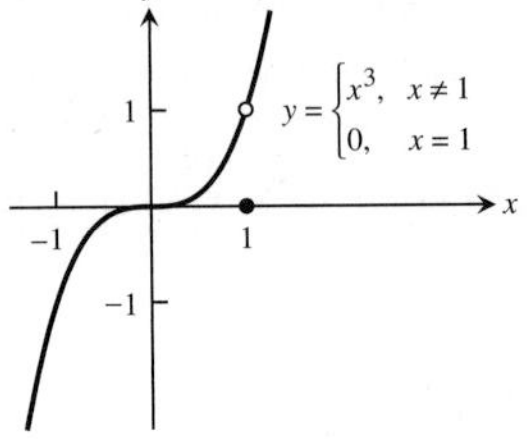

(b) 1, 1 **(c)** Yes, 1

7. **(a)** $D: 0 \le x \le 2$, $R: 0 < y \le 1$ and $y = 2$.
(b) $(0, 1) \cup (1, 2)$ **(c)** $x = 2$ **(d)** $x = 0$

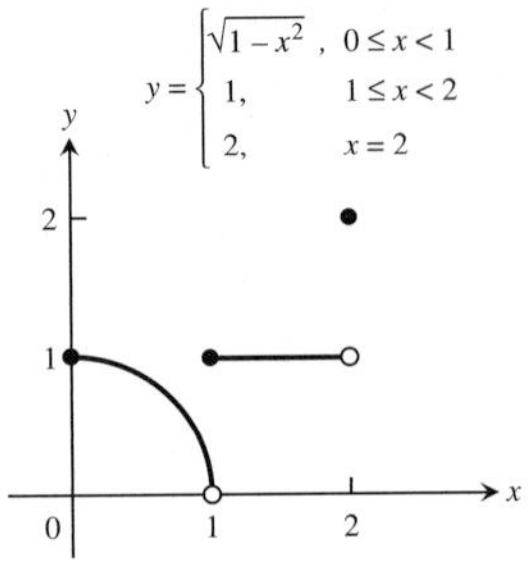

9. $\sqrt{3}$ **11.** 1 **13.** $2/\sqrt{5}$ **15.** **(a)** 1 **(b)** -1
17. **(a)** 1 **(b)** $2/3$ **19.** 1 **21.** $3/4$ **23.** 2 **25.** $1/2$
27. 2 **29.** 1 **31.** $1/2$ **33.** $3/8$
39. $\delta = \epsilon^2$, $\lim_{x \to 5^+} \sqrt{x - 5} = 0$
43. **(a)** 400 **(b)** 399 **(c)** The limit does not exist.

Section 1.7, pp. 56–57

1. No; discontinuous at $x = 2$; not defined at $x = 2$
3. Continuous **5.** **(a)** Yes **(b)** Yes **(c)** Yes **(d)** Yes
7. **(a)** No **(b)** No **9.** 1, nonremovable; 0, removable
11. All x except $x = 2$ **13.** All x except $x = 3, x = 1$
15. All x **17.** All x except $x = 0$ **19.** All x except $x = n\pi/2$, n any integer **21.** All x except $n\pi/2$, n an odd integer
23. All $x \ge -3/2$ **25.** All x **27.** 0; continuous at $x = \pi$
29. 1; continuous at $y = 1$ **31.** $a = 4/3$
49. $x \approx 1.8794, -1.5321, -0.3473$ **51.** $x \approx 1.7549$
53. $x \approx 3.5156$ **55.** $x \approx 0.7391$

Section 1.8, pp. 66–68

1. **(a)** -3 **(b)** -3 **3.** **(a)** $1/2$ **(b)** $1/2$ **5.** **(a)** $-5/3$ **(b)** $-5/3$ **7.** 0 **9.** -1 **11.** **(a)** $2/5$ **(b)** $2/5$
13. **(a)** 0 **(b)** 0 **15.** **(a)** 7 **(b)** 7 **17.** **(a)** 0 **(b)** 0
19. **(a)** $-2/3$ **(b)** $-2/3$ **21.** 0 **23.** 1 **25.** ∞
27. ∞ **29.** $-\infty$ **31.** $-\infty$ **33.** ∞ **35.** **(a)** ∞ **(b)** $-\infty$
37. ∞ **39.** ∞ **41.** $-\infty$
43. **(a)** ∞ **(b)** $-\infty$ **(c)** $-\infty$ **(d)** ∞
45. **(a)** $-\infty$ **(b)** ∞ **(c)** 0 **(d)** $3/2$
47. **(a)** $-\infty$ **(b)** $1/4$ **(c)** $1/4$ **(d)** $1/4$ **(e)** It will be $-\infty$.
49. **(a)** $-\infty$ **(b)** ∞
51. **(a)** ∞ **(b)** ∞ **(c)** ∞ **(d)** ∞
53.

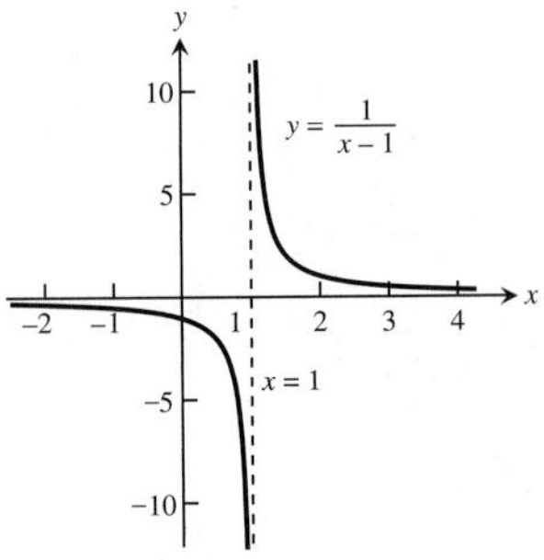

55.

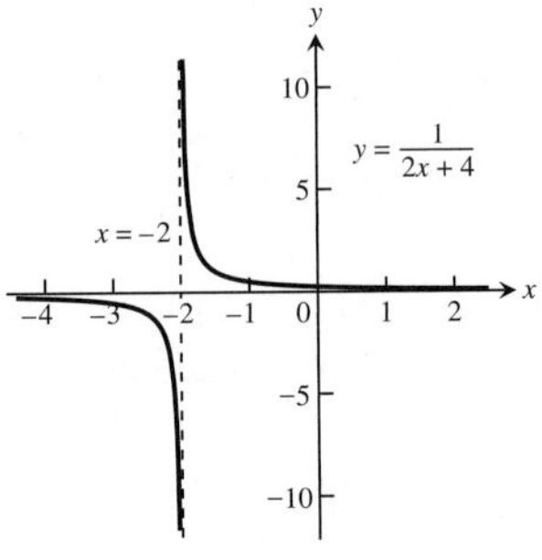

57.

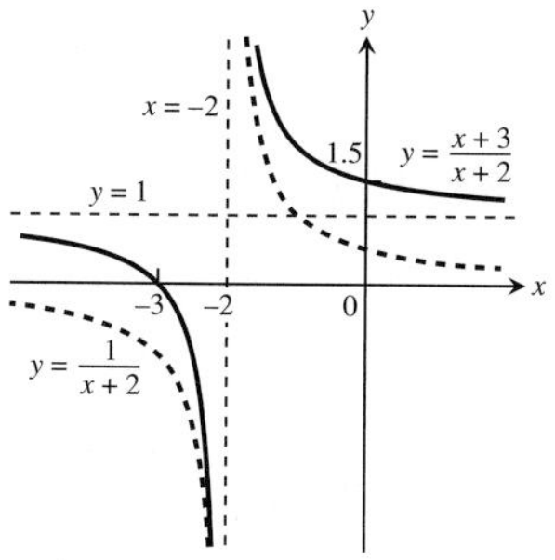

59. Here is one possibility.

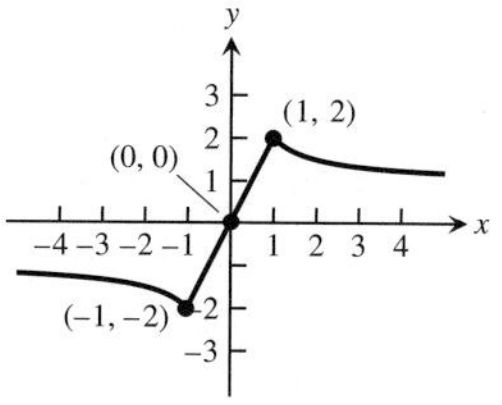

87.

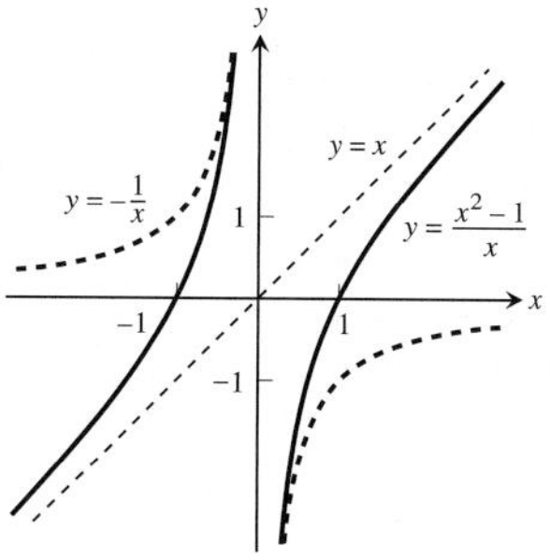

89.

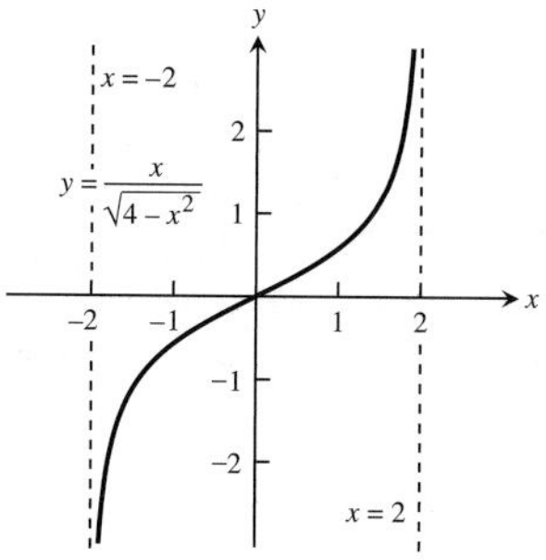

61. Here is one possibility.

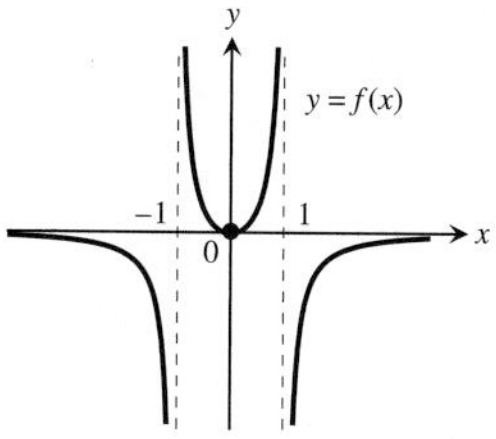

63. Here is one possibility.

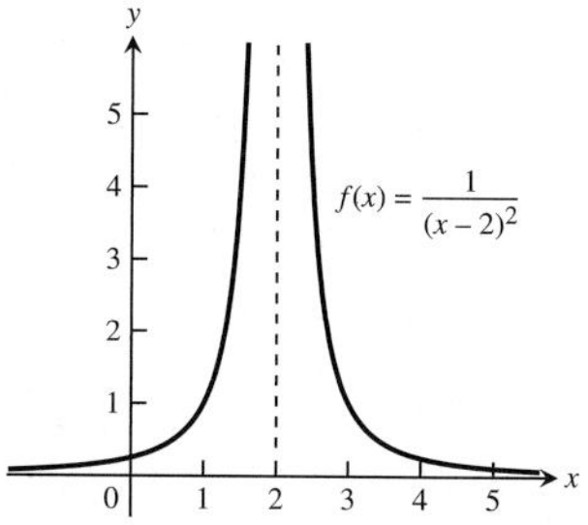

91.

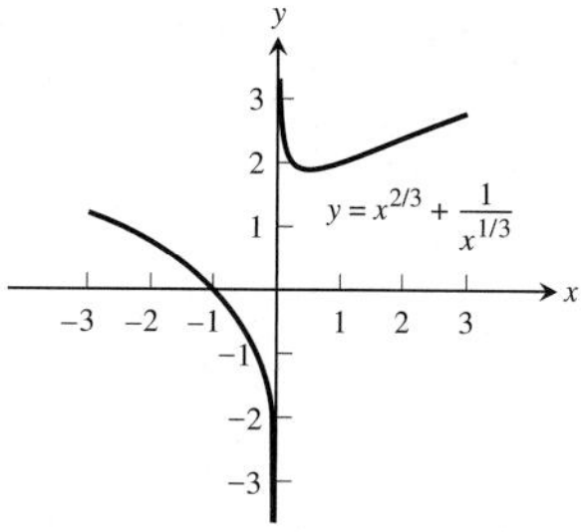

65. Here is one possibility.

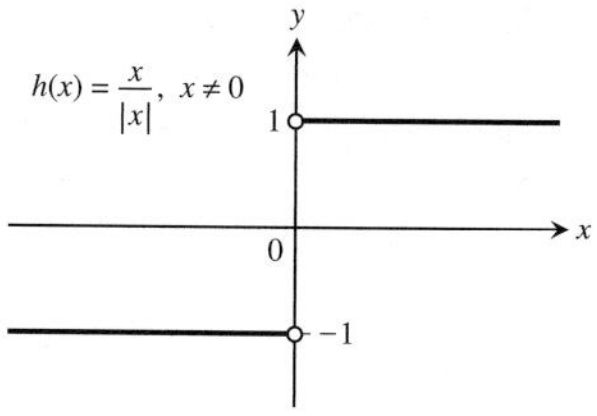

69. At most one.

93. At ∞: ∞, at $-\infty$: 0

77. (a) For every positive real number B there exists a corresponding number $\delta > 0$ such that for all x

$$x_0 - \delta < x < x_0 \quad \Rightarrow \quad f(x) > B.$$

(b) For every negative real number $-B$ there exists a corresponding number $\delta > 0$ such that for all x

$$x_0 < x < x_0 + \delta \quad \Rightarrow \quad f(x) < -B.$$

(c) For every negative real number $-B$ there exists a corresponding number $\delta > 0$ such that for all x

$$x_0 - \delta < x < x_0 \quad \Rightarrow \quad f(x) < -B.$$

83.

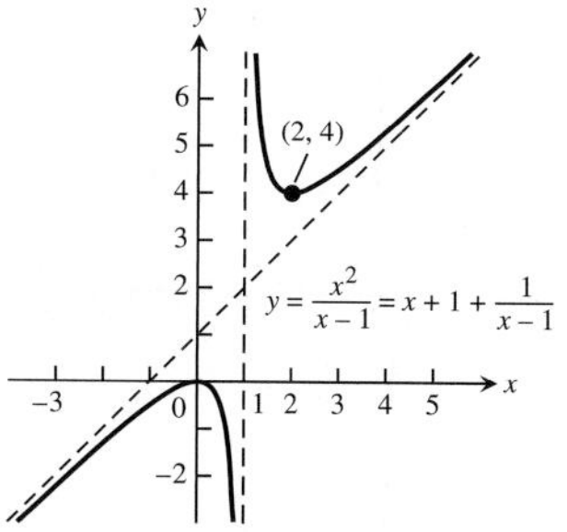

85.

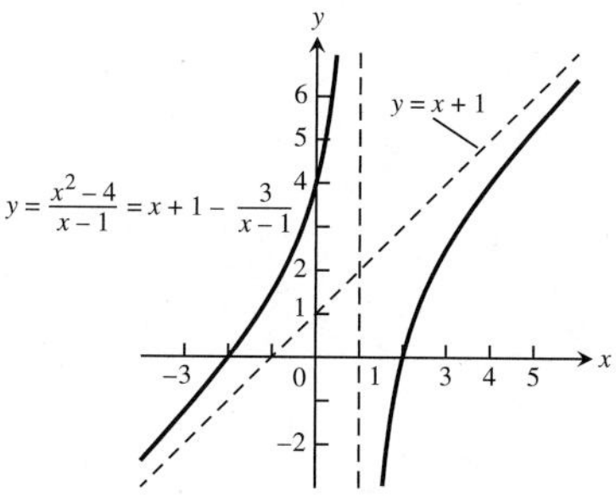

Practice and Additional Exercises, pp. 69–70

1. (a) Domain: all reals **(b)** Range: $[-2, \infty)$

3. (a) Domain: $[-4, 4]$ **(b)** Range: $[0, 4]$

5. (a) Domain: all reals **(b)** Range: $[-3, 1]$

7. (a) Domain: $[-4, 4]$ **(b)** Range: $[0, 2]$

9. $f(x) = \begin{cases} 1 - x, & 0 \le x < 1 \\ 2 - x, & 1 \le x \le 2 \end{cases}$

11. (a) 1 **(b)** $\dfrac{1}{\sqrt{2.5}} = \sqrt{\dfrac{2}{5}}$ **(c)** $x, x \neq 0$

(d) $\dfrac{1}{\sqrt{1/\sqrt{x+2}+2}}$

13. (a) $(f \circ g)(x) = -x, x \ge -2$, $(g \circ f)(x) = \sqrt{4 - x^2}$

(b) Domain $(f \circ g)$: $[-2, \infty)$, domain $(g \circ f)$: $[-2, 2]$

(c) Range $(f \circ g)$: $(-\infty, 2]$, range $(g \circ f)$: $[0, 2]$

15. Even **17.** Odd **19.** Neither

21. At $x = -1$: $\lim_{x\to-1^-} f(x) = \lim_{x\to-1^+} f(x) = 1$, so $\lim_{x\to-1} f(x) = 1 = f(-1)$; continuous at $x = -1$

At $x = 0$: $\lim_{x\to0^-} f(x) = \lim_{x\to0^+} f(x) = 0$, so $\lim_{x\to0} f(x) = 0$. However, $f(0) \neq 0$, so f is discontinuous at $x = 0$. The discontinuity can be removed by redefining $f(0)$ to be 0.

At $x = 1$: $\lim_{x\to1^-} f(x) = -1$ and $\lim_{x\to1^+} f(x) = 1$, so $\lim_{x\to1} f(x)$ does not exist. The function is discontinuous at $x = 1$, and the discontinuity is not removable.

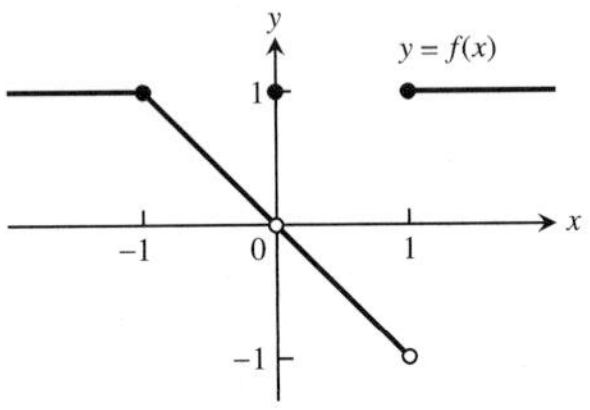

23. **(a)** -21 **(b)** 49 **(c)** 0 **(d)** 1 **(e)** 1 **(f)** 7
(g) -7 **(h)** $-\frac{1}{7}$ **25.** 4

27. **(a)** $(-\infty, +\infty)$ **(b)** $[0, \infty)$ **(c)** $(-\infty, 0)$ and $(0, \infty)$
(d) $(0, \infty)$ **29.** **(a)** Does not exist **(b)** 0

31. $\frac{1}{2}$ **33.** $2x$ **35.** $-\frac{1}{4}$ **37.** $\frac{2}{\pi}$ **39.** 1 **41.** 2

43. 0 **45.** $\frac{2}{5}$ **47.** 0 **49.** $-\infty$ **51.** 0 **53.** 1

55. **(b)** 1.324717957

CHAPTER 2

Section 2.1, pp. 74–75

1. P_1: $m_1 = 1$, P_2: $m_2 = 5$ **3.** P_1: $m_1 = 5/2$, P_2: $m_2 = -1/2$

5. $y = 2x + 4$ **7.** $y = x + 1$

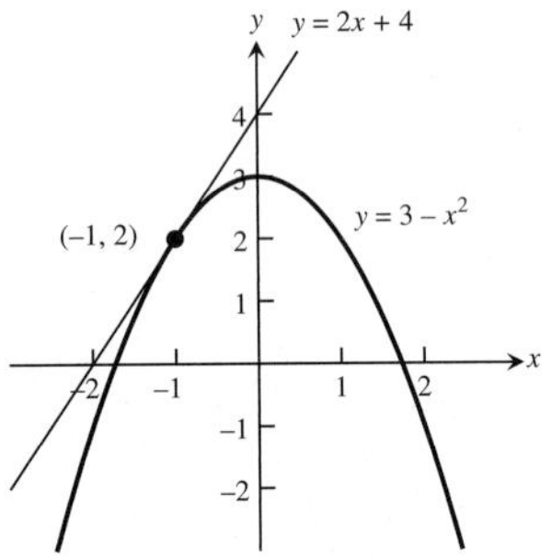

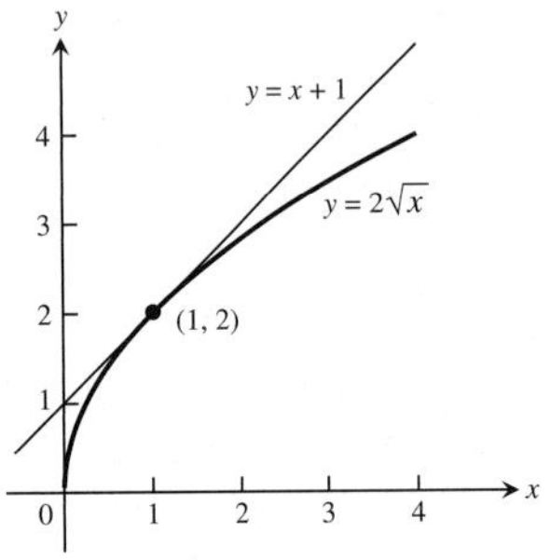

9. $y = 12x + 17$

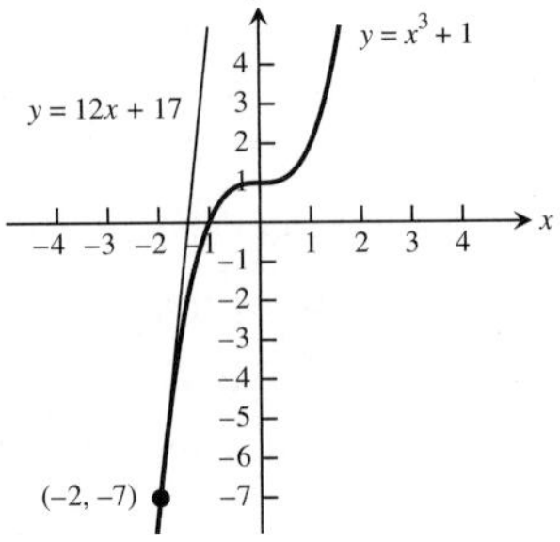

11. $m = 4$, $y - 5 = 4(x - 2)$

13. $m = -2$, $y - 3 = -2(x - 3)$

15. $m = 11$, $y - 6 = 11(t - 2)$

17. $m = \frac{1}{4}$, $y - 2 = \frac{1}{4}(x - 4)$

19. $m = -10$ **21.** $m = -1/4$ **23.** $(-2,-5)$

25. $y = -(x + 1)$, $y = -(x - 3)$ **27.** 19.6 m/sec

29. 6π **33.** Yes **35.** Yes **37.** **(a)** Nowhere

39. **(a)** At $x = 0$ **41.** **(a)** Nowhere **43.** **(a)** At $x = 1$

45. **(a)** At $x = 0$

Section 2.2, pp. 82–84

1. $-2x, 6, 0, -2$ **3.** $-\frac{2}{t^3}, 2, -\frac{1}{4}, -\frac{2}{3\sqrt{3}}$

5. $\frac{3}{2\sqrt{3\theta}}, \frac{3}{2\sqrt{3}}, \frac{1}{2}, \frac{3}{2\sqrt{2}}$ **7.** $6x^2$ **9.** $\frac{1}{(2t + 1)^2}$

11. $\frac{-1}{2(q + 1)\sqrt{q + 1}}$ **13.** $1 - \frac{9}{x^2}, 0$ **15.** $3t^2 - 2t, 5$

17. $\frac{-4}{(x - 2)\sqrt{x - 2}}$, $y - 4 = -\frac{1}{2}(x - 6)$ **19.** 6

21. $1/8$ **23.** $\frac{-1}{(x + 2)^2}$ **25.** $\frac{-1}{(x - 1)^2}$ **27.** b **29.** d

31. **(a)** $x = 0, 1, 4$ **(b)**

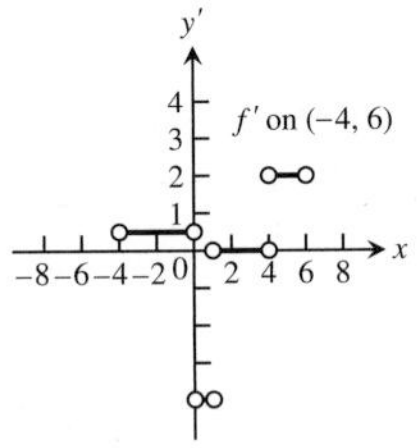

33. Since $\lim_{x\to0^+} f'(x) = 1$ while $\lim_{x\to0^-} f'(x) = 0$, $f(x)$ is not differentiable at $x = 0$.

35. Since $\lim_{x\to1^+} f'(x) = 2$ while $\lim_{x\to1^-} f'(x) = 1/2$, $f(x)$ is not differentiable at $x = 1$.

37. **(a)** $-3 \le x \le 2$ **(b)** None **(c)** None

39. **(a)** $-3 \le x < 0, 0 < x \le 3$ **(b)** None **(c)** $x = 0$

41. **(a)** $-1 \le x < 0, 0 < x \le 2$ **(b)** $x = 0$ **(c)** None

43. Yes, $y + 16 = -(x - 3)$ is tangent at $(3, -16)$

45. **(a)** $f'(0) = 0$ **(b)** $f'(0) = 0$

Section 2.3, pp. 92–93

1. $\frac{dy}{dx} = -2x, \frac{d^2y}{dx^2} = -2$

3. $\frac{ds}{dt} = 15t^2 - 15t^4, \frac{d^2s}{dt^2} = 30t - 60t^3$

5. $\frac{dy}{dx} = 4x^2 - 1, \frac{d^2y}{dx^2} = 8x0$

7. $\frac{dw}{dz} = -\frac{6}{z^3} + \frac{1}{z^2}, \frac{d^2w}{dz^2} = \frac{18}{z^4} - \frac{2}{z^3}$

9. $\frac{dy}{dx} = 12x - 10 + 10x^{-3}, \frac{d^2y}{dx^2} = 12 - 30x^{-4}$

11. $\dfrac{dr}{ds} = \dfrac{-2}{3s^3} + \dfrac{5}{2s^2}, \dfrac{d^2r}{ds^2} = \dfrac{2}{s^4} - \dfrac{5}{s^3}$

13. $y' = -5x^4 + 12x^2 - 2x - 3$

15. $y' = 3x^2 + 10x + 2 - \dfrac{1}{x^2}$ **17.** $y' = \dfrac{-19}{(3x-2)^2}$

19. $g'(x) = \dfrac{x^2 + x + 4}{(x + 0.5)^2}$ **21.** $\dfrac{dv}{dt} = \dfrac{t^2 - 2t - 1}{(1 + t^2)^2}$

23. $f'(s) = \dfrac{1}{\sqrt{s}(\sqrt{s} + 1)^2}$ **25.** $v' = -\dfrac{1}{x^2} + 2x^{-3/2}$

27. $y' = \dfrac{-4x^3 - 3x^2 + 1}{(x^2 - 1)^2(x^2 + x + 1)^2}$

29. $y' = 2x^3 - 3x - 1, y'' = 6x^2 - 3, y''' = 12x, y^{(4)} = 12,$
$y^{(n)} = 0$ for $n \geq 5$

31. $y' = 2x - 7x^{-2}, y'' = 2 + 14x^{-3}$

33. $\dfrac{dr}{d\theta} = 3\theta^{-4}, \dfrac{d^2r}{d\theta^2} = -12\theta^{-5}$

35. $\dfrac{dw}{dz} = -z^{-2} - 1, \dfrac{d^2w}{dz^2} = 2z^{-3}$

37. $\dfrac{dp}{dq} = \dfrac{1}{6}q + \dfrac{1}{6}q^{-3} + q^{-5}, \dfrac{d^2p}{dq^2} = \dfrac{1}{6} - \dfrac{1}{2}q^{-4} - 5q^{-6}$

39. **(a)** 13 **(b)** −7 **(c)** 7/25 **(d)** 20

41. **(a)** $y = -\dfrac{x}{8} + \dfrac{5}{4}$ **(b)** $m = -4$ at $(0, 1)$
(c) $y = 8x - 15, y = 8x + 17$

43. $y = 4x, y = 2$ **45.** $a = 1, b = 1, c = 0$

47. **(a)** $y = 2x + 2$ **(c)** $(2, 6)$

49. $P'(x) = na_nx^{n-1} + (n-1)a_{n-1}x^{n-2} + \cdots + 2a_2x + a_1$

51. The Product Rule is then the Constant Multiple Rule, so the latter is a special case of the Product Rule.

53. $\dfrac{dP}{dV} = -\dfrac{nRT}{(V - nb)^2} + \dfrac{2an^2}{V^3}$

Section 2.4, pp. 99–102

1. **(a)** −2 m, −1 m/sec
(b) 3 m/sec, 1 m/sec; 2 m/sec², 2 m/sec²
(c) changes direction at $t = 3/2$ sec

3. **(a)** −9 m, −3 m/sec
(b) 3 m/sec, 12 m/sec; 6 m/sec², −12 m/sec²
(c) no change in direction

5. **(a)** −20 m, −5 m/sec
(b) 45 m/sec, (1/5) m/sec; 140 m/sec², (4/25) m/sec²
(c) no change in direction

7. **(a)** $a(1) = -6$ m/sec², $a(3) = 6$ m/sec²
(b) $v(2) = 3$ m/sec **(c)** 6 m

9. Mars: ≈ 7.5 sec, Jupiter: ≈ 1.2 sec

11. **(a)** $v = -32t, |v| = 32t$ ft/sec, $a = -32$ ft/sec²
(b) $t \approx 3.3$ sec **(c)** $v \approx -107.0$ ft/sec

13. **(a)** $t = 2, t = 7$ **(b)** $3 \leq t \leq 6$
(c)

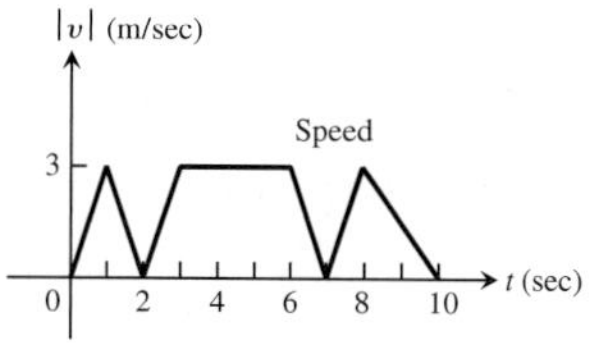

(d)

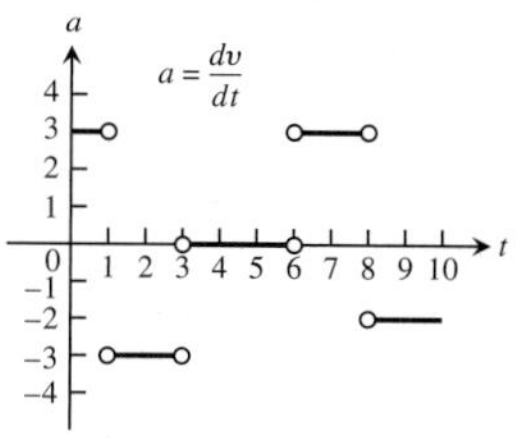

15. **(a)** 190 ft/sec **(b)** 2 sec **(c)** 8 sec, 0 ft/sec
(d) 10.8 sec, 90 ft/sec **(e)** 2.8 sec
(f) greatest acceleration happens 2 sec after launch
(g) constant acceleration between 2 and 10.8 sec, −32 ft/sec²

17. **(a)** $\dfrac{4}{7}$ sec, 280 cm/sec **(b)** 560 cm/sec, 980 cm/sec²
(c) 29.75 flashes/sec

19. **(a)** \$110/machine **(b)** \$80 **(c)** \$79.90

21. **(a)** $b'(0) = 10^4$ bacteria/h **(b)** $b'(5) = 0$ bacteria/h
(c) $b'(10) = -10^4$ bacteria/h

23. $t = 25$ sec $D = \dfrac{6250}{9}$ m

25.

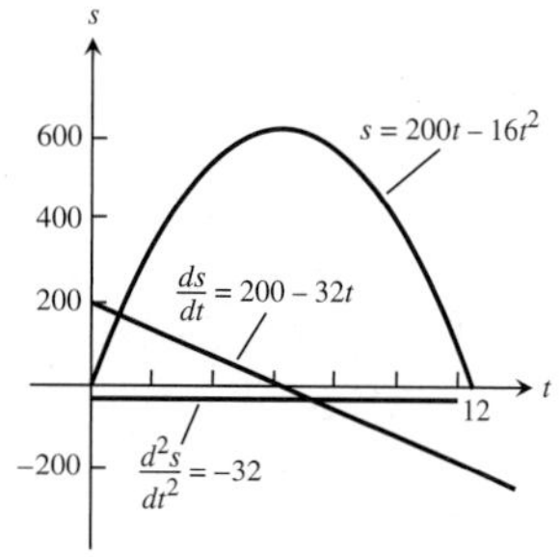

(a) $v = 0$ when $t = 6.25$ sec.
(b) $v > 0$ when $0 \leq t < 6.25 \Rightarrow$ body moves up; $v < 0$ when $6.25 < t \leq 12.5 \Rightarrow$ body moves down.
(c) Body changes direction at $t = 6.25$ sec.
(d) Body speeds up on (6.25, 12.5] and slows down on [0, 6.25).
(e) The body is moving fastest at the endpoints $t = 0$ and $t = 12.5$ when it is traveling 200 ft/sec. It's moving slowest at $t = 6.25$ when the speed is 0.
(f) When $t = 6.25$ the body is $s = 625$ m from the origin and farthest away.

27.

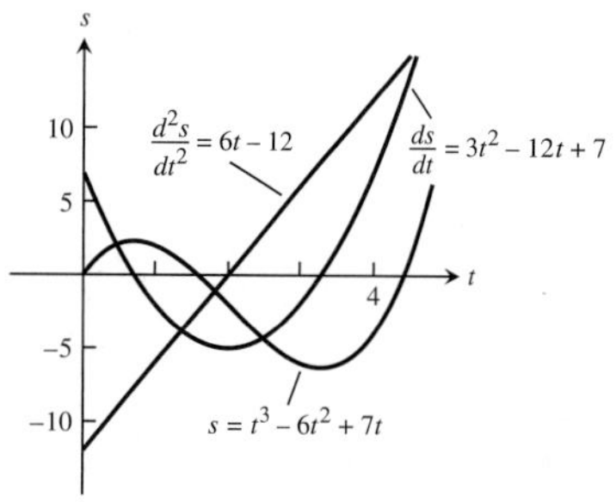

(a) $v = 0$ when $t = \dfrac{6 \pm \sqrt{15}}{3}$ sec.

(b) $v < 0$ when $\dfrac{6 - \sqrt{15}}{3} < t < \dfrac{6 + \sqrt{15}}{3} \Rightarrow$ body moves left; $v > 0$ when $0 \le t < \dfrac{6 - \sqrt{15}}{3}$ or $\dfrac{6 + \sqrt{15}}{3} < t \le 4 \Rightarrow$ body moves right.

(c) Body changes direction at $t = \dfrac{6 \pm \sqrt{15}}{3}$ sec.

(d) Body speeds up on $\left(\dfrac{6 - \sqrt{15}}{3}, 2\right) \cup \left(\dfrac{6 + \sqrt{15}}{3}, 4\right]$ and slows down on $\left[0, \dfrac{6 - \sqrt{15}}{3}\right) \cup \left(2, \dfrac{6 + \sqrt{15}}{3}\right)$.

(e) The body is moving fastest at $t = 0$ and $t = 4$ when it is moving 7 units/sec and slowest at $t = \dfrac{6 \pm \sqrt{15}}{3}$ sec.

(f) When $t = \dfrac{6 + \sqrt{15}}{3}$ the body is at position $s \approx -6.303$ units and farthest from the origin.

Section 2.5, pp. 106–107

1. $-10 - 3 \sin x$ **3.** $-\csc x \cot x - \dfrac{2}{\sqrt{x}}$ **5.** 0

7. $\dfrac{-\csc^2 x}{(1 + \cot x)^2}$ **9.** $4 \tan x \sec x - \csc^2 x$ **11.** $x^2 \cos x$

13. $\sec^2 t - 1$ **15.** $\dfrac{-2 \csc t \cot t}{(1 - \csc t)^2}$ **17.** $-\theta(\theta \cos \theta + 2 \sin \theta)$

19. $\sec \theta \csc \theta (\tan \theta - \cot \theta) = \sec^2 \theta - \csc^2 \theta$

21. $\sec^2 q$ **23.** $\sec^2 q$

25. **(a)** $2 \csc^3 x - \csc x$ **(b)** $2 \sec^3 x - \sec x$

27.

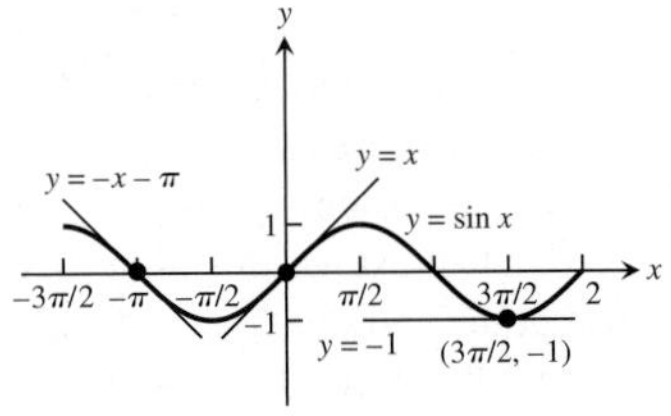

29.

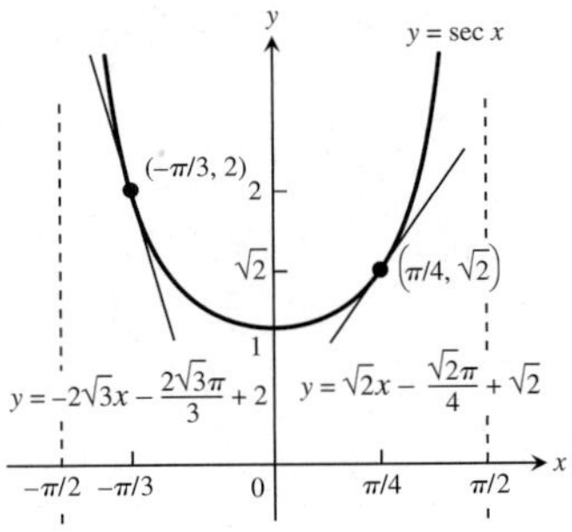

31. Yes, at $x = \pi$ **33.** No

35. $\left(-\dfrac{\pi}{4}, -1\right); \left(\dfrac{\pi}{4}, 1\right)$

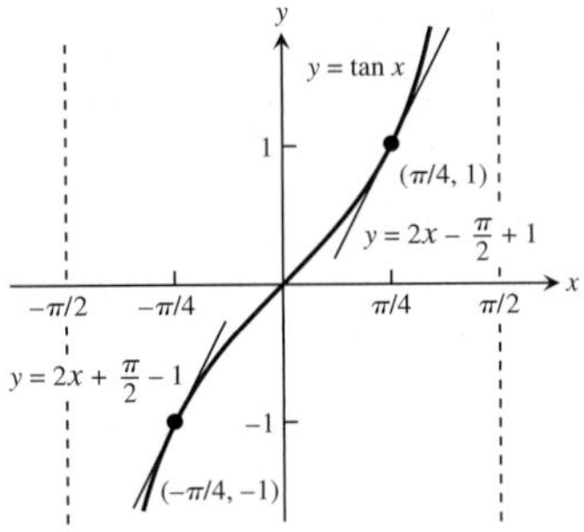

37. 0 **39.** -1 **41.** 0

43. $-\sqrt{2}$ m/sec, $\sqrt{2}$ m/sec, $\sqrt{2}$ m/sec^2, $\sqrt{2}$ m/sec^3

45. $c = 9$ **47.** $\sin x$

Section 2.6, p. 113

1.

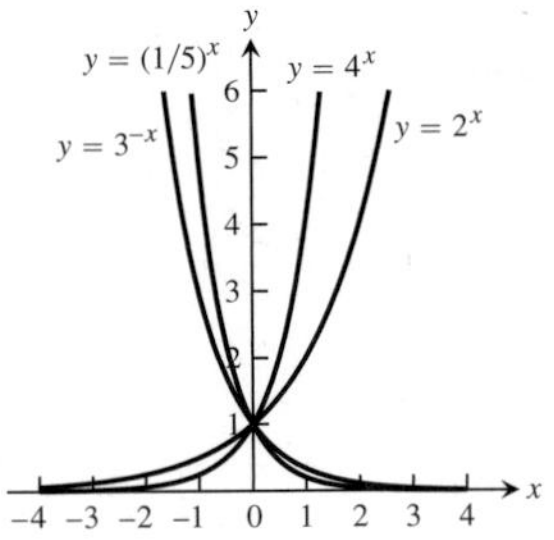

3.

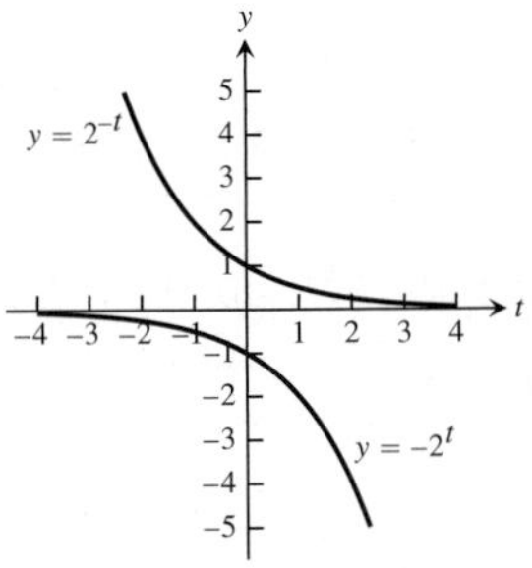

5.

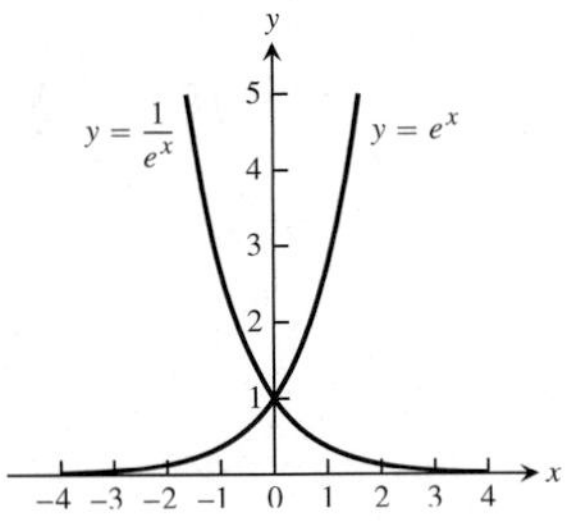

7.

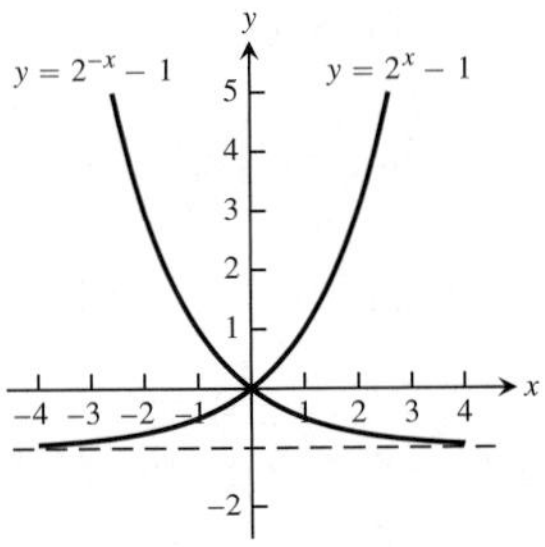

9.

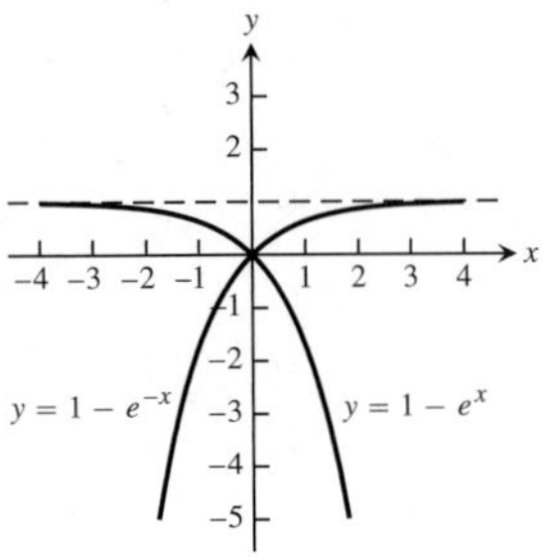

11. $16^{1/4} = 2$ **13.** $4^{1/2} = 2$ **15.** 5 **17.** $14^{\sqrt{3}}$ **19.** 4
21. Domain: $-\infty < x < \infty$ Range: $0 < y < 1/2$ **23.** Domain: $-\infty < t < \infty$ Range: $1 < y < \infty$
25. $\frac{dy}{dx} = 3x^2e^x + x^3e^x$ **27.** $y' = \frac{9}{4}x^{5/4}$ **29.** $\frac{ds}{dt} = 3t^{1/2}$
31. $y' = \frac{2}{7x^{5/7}} - ex^{e-1}$ **33.** $\frac{dr}{ds} = \frac{se^s - e^s}{s^2}$
35. $y' = 2x^3 - 3x - 1, y'' = 6x^2 - 3, y''' = 12x, y^{(4)} = 12, y^{(n)} = 0$ for $n \geq 5$
37. $y' = 2x - 7x^{-2}, y'' = 2 + 14x^{-3}$
39. $\frac{dr}{d\theta} = 3\theta^{-4}, \frac{d^2r}{d\theta^2} = -12\theta^{-5}$
41. $\frac{dw}{dz} = 6ze^z + 3z^2e^z, \frac{d^2w}{dz^2} = 6e^z + 12ze^z + 3z^2e^z$
43. $\frac{dp}{dq} = \frac{1}{6}q + \frac{1}{6}q^{-3} + q^{-5}, \frac{d^2p}{dq^2} = \frac{1}{6} - \frac{1}{2}q^{-4} - 5q^{-6}$
45. $\frac{dy}{dx} = e^{-x}(3 + 2x + x^2), \frac{d^2y}{dx^2} = e^{-x}(x^2 - 4x - 1)$
47. $x \approx 2.3219$ **49.** After 19 years
51. $\approx$ 11.433 years, or when interest is paid **53.** 0
55. -1 **57.** $x = 4$

Section 2.7, pp. 119–121

1. $12x^3$ **3.** $3\cos(3x + 1)$ **5.** $-\sin(\sin x)\cos x$
7. $10\sec^2(10x - 5)$
9. With $u = (2x + 1), y = u^5: \frac{dy}{dx} = \frac{dy}{du}\frac{du}{dx} = 5u^4 \cdot 2 = 10(2x + 1)^4$
11. With $u = (1 - (x/7)), y = u^{-7}: \frac{dy}{dx} = \frac{dy}{du}\frac{du}{dx} = -7u^{-8} \cdot \left(-\frac{1}{7}\right) = \left(1 - \frac{x}{7}\right)^{-8}$
13. With $u = 1 - 6x, y = u^{2/3}: \frac{dy}{dx} = \frac{dy}{du}\frac{du}{dx} = \frac{2}{3}u^{-1/3} \cdot (-6) = \frac{-4}{(1 - 6x)^{1/3}}$
15. With $u = \tan x, y = \sec u: \frac{dy}{dx} = \frac{dy}{du}\frac{du}{dx} = (\sec u \tan u)(\sec^2 x) = \sec(\tan x)\tan(\tan x)\sec^2 x$
17. With $u = \sin x, y = u^3: \frac{dy}{dx} = \frac{dy}{du}\frac{du}{dx} = 3u^2\cos x = 3\sin^2 x(\cos x)$
19. $y = e^u, u = -5x, \frac{dy}{dx} = -5e^{-5x}$
21. $y = e^u, u = 5 - 7x, \frac{dy}{dx} = -7e^{(5-7x)}$
23. $-\frac{1}{2\sqrt{3 - t}}$ **25.** $\frac{4}{\pi}(\cos 3t - \sin 5t)$ **27.** $\frac{\csc\theta}{\cot\theta + \csc\theta}$
29. $2x\sin^4 x + 4x^2\sin^3 x\cos x + \cos^{-2} x + 2x\cos^{-3} x\sin x$
31. $(3x - 2)^6 - \frac{1}{x^3\left(4 - \frac{1}{2x^2}\right)^2}$ **33.** $\frac{(4x + 3)^3(4x + 7)}{(x + 1)^4}$
35. $(1 - x)e^{-x} + 3e^{3x}$ **37.** $\left(\frac{5}{2}x^2 - 3x + 3\right)e^{5x/2}$
39. $\sqrt{x}\sec^2(2\sqrt{x}) + \tan(2\sqrt{x})$ **41.** $\frac{2\sin\theta}{(1 + \cos\theta)^2}$
43. $h'(\theta) = -\frac{2}{3}(\sin 2\theta)(1 + \cos 2\theta)^{-2/3}$
45. $\frac{dq}{dt} = \left(\frac{t + 2}{2(t + 1)^{3/2}}\right)\cos\left(\frac{t}{\sqrt{t + 1}}\right)$ **47.** $2\theta e^{-\theta^2}\sin(e^{-\theta^2})$
49. $2\pi\sin(\pi t - 2)\cos(\pi t - 2)$ **51.** $\frac{8\sin(2t)}{(1 + \cos 2t)^5}$
53. $\frac{dy}{dt} = -2\pi\sin(\pi t - 1) \cdot \cos(\pi t - 1) \cdot e^{\cos^2(\pi t - 1)}$
55. $-2\cos(\cos(2t - 5))(\sin(2t - 5))$
57. $\left(1 + \tan^4\left(\frac{t}{12}\right)\right)^2\left(\tan^3\left(\frac{t}{12}\right)\sec^2\left(\frac{t}{12}\right)\right)$
59. $-\frac{t\sin(t^2)}{\sqrt{1 + \cos(t^2)}}$ **61.** $\frac{6}{x^3}\left(1 + \frac{1}{x}\right)\left(1 + \frac{2}{x}\right)$
63. $2\csc^2(3x - 1)\cot(3x - 1)$ **65.** $y'' = 2(2x^2 + 1)e^{x^2}$
67. 5/2 **69.** $-\pi/4$ **71.** 0
73. **(a)** 2/3 **(b)** $2\pi + 5$ **(c)** $15 - 8\pi$ **(d)** 37/6 **(e)** -1 **(f)** $\sqrt{2}/24$ **(g)** 5/32 **(h)** $-5/(3\sqrt{17})$
75. 5 **77.** **(a)** 1 **(b)** 1
79. **(a)** $y = \pi x + 2 - \pi$ **(b)** $\pi/2$
81. $v(6) = \frac{2}{5}$ m/sec, $a(6) = -\frac{4}{125}$ m/sec^2

Section 2.8, pp. 124–125

1. $\frac{-2xy - y^2}{x^2 + 2xy}$ **3.** $\frac{1 - 2y}{2x + 2y - 1}$
5. $\frac{-2x^3 + 3x^2y - xy^2 + x}{x^2y - x^3 + y}$ **7.** $\frac{1}{y(x + 1)^2}$ **9.** $\cos^2 y$
11. $\frac{2e^{2x} - \cos(x + 3y)}{3\cos(x + 3y)}$ **13.** $\frac{-y^2}{y\sin\left(\frac{1}{y}\right) - \cos\left(\frac{1}{y}\right) + xy}$
15. $-\frac{\sqrt{r}}{\sqrt{\theta}}$ **17.** $\frac{-r}{\theta}$ **19.** $y' = -\frac{x}{y}, y'' = \frac{-y^2 - x^2}{y^3}$
21. $\frac{dy}{dx} = \frac{xe^{x^2} + 1}{y}, \frac{d^2y}{dx^2} = \frac{(2x^2y^2 + y^2 - 2x)e^{x^2} - x^2e^{2x^2} - 1}{y^3}$

23. $y' = \dfrac{\sqrt{y}}{\sqrt{y}+1},\ y'' = \dfrac{1}{2(\sqrt{y}+1)^3}$

25. -2 **27.** $(-2, 1): m = -1, (-2, -1): m = 1$

29. **(a)** $y = \frac{7}{4}x - \frac{1}{2}$ **(b)** $y = -\frac{4}{7}x + \frac{29}{7}$

31. **(a)** $y = 3x + 6$ **(b)** $y = -\frac{1}{3}x + \frac{8}{3}$

33. **(a)** $y = \frac{6}{7}x + \frac{6}{7}$ **(b)** $y = -\frac{7}{6}x - \frac{7}{6}$

35. **(a)** $y = -\frac{\pi}{2}x + \pi$ **(b)** $y = \frac{2}{\pi}x - \frac{2}{\pi} + \frac{\pi}{2}$

37. **(a)** $y = 2\pi x - 2\pi$ **(b)** $y = -\frac{x}{2\pi} + \frac{1}{2\pi}$

39. Points: $(-\sqrt{7}, 0)$ and $(\sqrt{7}, 0)$, Slope: -2

41. $m = -1$ at $\left(\frac{\sqrt{3}}{4}, \frac{\sqrt{3}}{2}\right)$, $m = \sqrt{3}$ at $\left(\frac{\sqrt{3}}{4}, \frac{1}{2}\right)$

43. $-\frac{3}{16}$

45. $\dfrac{dy}{dx} = \dfrac{y^3 + 2xy}{x^2 + 3xy^2},\ \dfrac{dx}{dy} = -\dfrac{x^2 + 3xy^2}{y^3 + 2xy},\ \dfrac{dx}{dy} = \dfrac{1}{dy/dx}$

Section 2.9, pp. 132–133

1. One-to-one **3.** Not one-to-one

5. D: $[-1, 1]$ R: $[-\pi/2, \pi/2]$

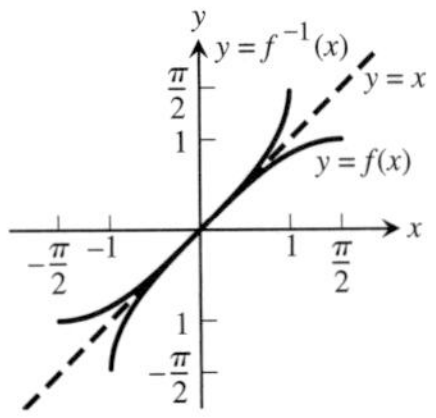

7. $f^{-1}(x) = \sqrt{x-1}$ **9.** $f^{-1}(x) = \sqrt[3]{x+1}$

11. $f^{-1}(x) = \sqrt{x} - 1$

13. $f^{-1}(x) = \sqrt[5]{x}$; domain: $-\infty < x < \infty$; range: $-\infty < y < \infty$

15. $f^{-1}(x) = \dfrac{1}{\sqrt{x}}$; domain: $x > 0$; range: $y > 0$

17. **(a)** $f^{-1}(x) = \frac{x}{2} - \frac{3}{2}$

(b)

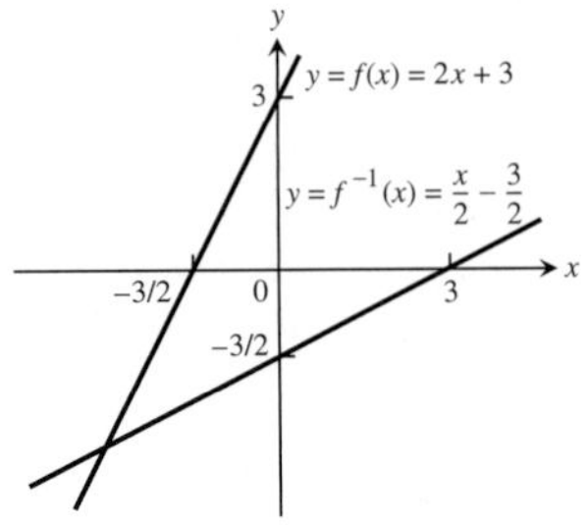

(c) 2, 1/2

19. **(a)** $f^{-1}(x) = -\frac{x}{4} + \frac{5}{4}$

(b)

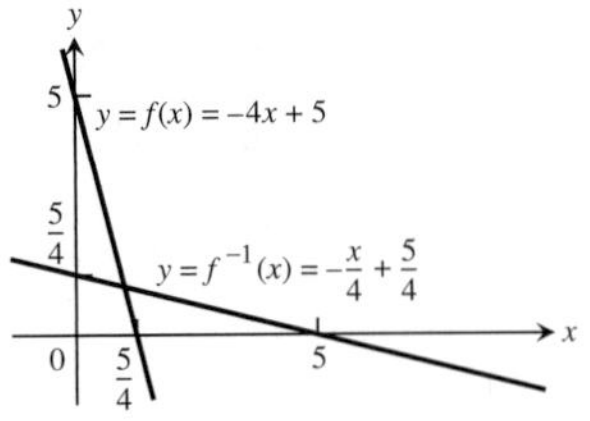

(c) $-4, -1/4$

21. **(b)**

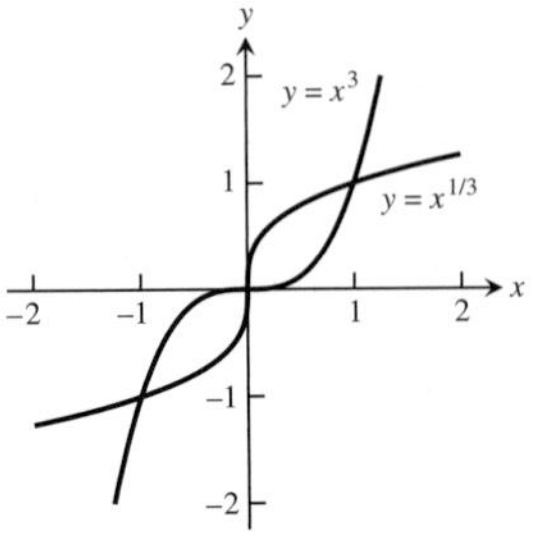

(c) Slope of f at $(1, 1)$: 3; slope of g at $(1, 1)$: 1/3; slope of f at $(-1, -1)$: 3; slope of g at $(-1, -1)$: 1/3

(d) $y = 0$ is tangent to $y = x^3$ at $x = 0$; $x = 0$ is tangent to $y = \sqrt[3]{x}$ at $x = 0$

23. 1/9 **25.** **(a)** $f^{-1}(x) = \frac{1}{m}x$

(b) The graph of f^{-1} is the line through the origin with slope $1/m$.

29. Increasing, therefore one-to-one; $df^{-1}/dx = \frac{1}{9}x^{-2/3}$

31. Decreasing, therefore one-to-one; $df^{-1}/dx = -\frac{1}{3}x^{-2/3}$

Section 2.10, pp. 142–144

1. **(a)** $\ln 3 - 2\ln 2$ **(b)** $2(\ln 2 - \ln 3)$ **(c)** $-\ln 2$

(d) $\frac{2}{3}\ln 3$ **(e)** $\ln 3 + \frac{1}{2}\ln 2$ **(f)** $\frac{1}{2}(3\ln 3 - \ln 2)$

3. **(a)** $\ln 5$ **(b)** $\ln(x - 3)$ **(c)** $\ln(t^2)$

5. **(a)** 7.2 **(b)** $\frac{1}{x^2}$ **(c)** $\frac{x}{y}$

7. **(a)** 1 **(b)** 1 **(c)** $-x^2 - y^2$

9. e^{2t+4} **11.** $e^{5t} + 40$ **13.** $y = 2xe^x + 1$

15. **(a)** $k = \ln 2$ **(b)** $k = (1/10)\ln 2$ **(c)** $k = 1000\ln a$

17. **(a)** $t = -10\ln 3$ **(b)** $t = -\frac{\ln 2}{k}$ **(c)** $t = \frac{\ln .4}{\ln .2}$

19. $4(\ln x)^2$

21. **(a)** 7 **(b)** $\sqrt{2}$ **(c)** 75 **(d)** 2 **(e)** 0.5 **(f)** -1

23. **(a)** $\sqrt{x}$ **(b)** x^2 **(c)** $\sin x$ **25.** **(a)** $\frac{\ln 3}{\ln 2}$ **(b)** 3 **(c)** 2

27. $1/x$ **29.** $2/t$ **31.** $-1/x$ **33.** $\frac{1}{\theta + 1}$ **35.** $3/x$

37. $2(\ln t) + (\ln t)^2$ **39.** $x^3 \ln x$ **41.** $\frac{1 - \ln t}{t^2}$

43. $\frac{1}{x(1 + \ln x)^2}$ **45.** $\frac{1}{x\ln x}$ **47.** $2\cos(\ln\theta)$

49. $-\dfrac{3x+2}{2x(x+1)}$ **51.** $\dfrac{2}{t(1-\ln t)^2}$ **53.** $\dfrac{\tan(\ln\theta)}{\theta}$

55. $\dfrac{10x}{x^2+1}+\dfrac{1}{2(1-x)}$

57. $\left(\dfrac{1}{2}\right)\sqrt{x(x+1)}\left(\dfrac{1}{x}+\dfrac{1}{x+1}\right)=\dfrac{2x+1}{2\sqrt{x(x+1)}}$

59. $\left(\dfrac{1}{2}\right)\sqrt{\dfrac{t}{t+1}}\left(\dfrac{1}{t}-\dfrac{1}{t+1}\right)=\dfrac{1}{2\sqrt{t}(t+1)^{3/2}}$

61. $\sqrt{\theta+3}(\sin\theta)\left(\dfrac{1}{2(\theta+3)}+\cot\theta\right)$

63. $t(t+1)(t+2)\left[\dfrac{1}{t}+\dfrac{1}{t+1}+\dfrac{1}{t+2}\right]=3t^2+6t+2$

65. $\dfrac{\theta+5}{\theta\cos\theta}\left[\dfrac{1}{\theta+5}-\dfrac{1}{\theta}+\tan\theta\right]$

67. $\dfrac{x\sqrt{x^2+1}}{(x+1)^{2/3}}\left[\dfrac{1}{x}+\dfrac{x}{x^2+1}-\dfrac{2}{3(x+1)}\right]$

69. $\dfrac{1}{3}\sqrt[3]{\dfrac{x(x-2)}{x^2+1}}\left(\dfrac{1}{x}+\dfrac{1}{x-2}-\dfrac{2x}{x^2+1}\right)$ **71.** $-2\tan\theta$

73. $\dfrac{1-t}{t}$ **75.** $1/(1+e^{\theta})$ **77.** $e^{\cos t}(1-t\sin t)$

79. $\dfrac{ye^y\cos x}{1-ye^y\sin x}$ **81.** $\dfrac{dy}{dx}=\dfrac{y^2-xy\ln y}{x^2-xy\ln x}$ **83.** $2^x\ln x$

85. $\left(\dfrac{\ln 5}{2\sqrt{s}}\right)5^{\sqrt{s}}$ **87.** $\pi x^{(\pi-1)}$ **89.** $\dfrac{1}{\theta\ln 2}$ **91.** $\dfrac{3}{x\ln 4}$

93. $\dfrac{2(\ln r)}{r(\ln 2)(\ln 4)}$ **95.** $\dfrac{-2}{(x+1)(x-1)}$

97. $\sin(\log_7\theta)+\dfrac{1}{\ln 7}\cos(\log_7\theta)$ **99.** $\dfrac{1}{\ln 5}$

101. $\dfrac{1}{t}(\log_2 3)3^{\log_2 t}$ **103.** $\dfrac{1}{t}$ **105.** $(x+1)^x\left(\dfrac{x}{x+1}+\ln(x+1)\right)$

107. $(\sqrt{t})^t\left(\dfrac{\ln t}{2}+\dfrac{1}{2}\right)$ **109.** $(\sin x)^x(\ln\sin x+x\cot x)$

111. $(x^{\ln x})\left(\dfrac{\ln x^2}{x}\right)$

113. **(a)** $y=\ln x-3$ **(b)** $y=\ln(x-1)$
(c) $y=3+\ln(x+1)$ **(d)** $y=\ln(x-2)-4$
(e) $y=\ln(-x)$ **(f)** $y=e^x$

117. **(a)** Amount $=8\left(\dfrac{1}{2}\right)^{t/12}$ **(b)** 36 hours

Section 2.11, pp. 152–154

1. **(a)** $\pi/4$ **(b)** $-\pi/3$ **(c)** $\pi/6$
3. **(a)** $-\pi/6$ **(b)** $\pi/4$ **(c)** $-\pi/3$
5. **(a)** $\pi/3$ **(b)** $3\pi/4$ **(c)** $\pi/6$
7. **(a)** $3\pi/4$ **(b)** $\pi/6$ **(c)** $2\pi/3$
9. **(a)** $\pi/4$ **(b)** $-\pi/3$ **(c)** $\pi/6$
11. **(a)** $3\pi/4$ **(b)** $\pi/6$ **(c)** $2\pi/3$

13. $\pi/2$ **15.** $\pi/2$ **17.** $\pi/2$ **19.** 0 **21.** $\dfrac{-2x}{\sqrt{1-x^4}}$

23. $\dfrac{\sqrt{2}}{\sqrt{1-2t^2}}$ **25.** $\dfrac{1}{|2s+1|\sqrt{s^2+s}}$

27. $\dfrac{-2x}{(x^2+1)\sqrt{x^4+2x^2}}$ **29.** $\dfrac{-1}{\sqrt{1-t^2}}$

31. $\dfrac{-1}{2\sqrt{t}(1+t)}$ **33.** $\dfrac{1}{(\tan^{-1}x)(1+x^2)}$

35. $\dfrac{-e^t}{|e^t|\sqrt{(e^t)^2-1}}=\dfrac{-1}{\sqrt{e^{2t}-1}}$ **37.** $\dfrac{-2s^n}{\sqrt{1-s^2}}$ **39.** 0

41. $\sin^{-1}x$

47. **(a)** Defined; there is an angle whose tangent is 2.
(b) Not defined; there is no angle whose cosine is 2.
(c) Not defined; no angle has sine $\sqrt{2}$.

53. **(a)** Domain: all real numbers except those having the form $\dfrac{\pi}{2}+k\pi$ where k is an integer; range: $-\pi/2<y<\pi/2$.
(b) Domain: $-\infty<x<\infty$; range: $-\infty<y<\infty$

55. **(a)** Domain: $-\infty<x<\infty$; range: $0\le y\le\pi$
(b) Domain: $-1\le x\le 1$; range: $-1\le y\le 1$

57. The graphs are identical.

Section 2.12, pp. 158–161

1. $\dfrac{dA}{dt}=2\pi r\dfrac{dr}{dt}$

3. **(a)** $\dfrac{dV}{dt}=\pi r^2\dfrac{dh}{dt}$ **(b)** $\dfrac{dV}{dt}=2\pi hr\dfrac{dr}{dt}$

(c) $\dfrac{dV}{dt}=\pi r^2\dfrac{dh}{dt}+2\pi hr\dfrac{dr}{dt}$

5. **(a)** 1 volt/sec **(b)** $-\dfrac{1}{3}$ amp/sec

(c) $\dfrac{dR}{dt}=\dfrac{1}{I}\left(\dfrac{dV}{dt}-\dfrac{V}{I}\dfrac{dI}{dt}\right)$

(d) 3/2 ohms/sec, R is increasing.

7. **(a)** $\dfrac{ds}{dt}=\dfrac{x}{\sqrt{x^2+y^2}}\dfrac{dx}{dt}$

(b) $\dfrac{ds}{dt}=\dfrac{x}{\sqrt{x^2+y^2}}\dfrac{dx}{dt}+\dfrac{y}{\sqrt{x^2+y^2}}\dfrac{dy}{dt}$ **(c)** $\dfrac{dx}{dt}=-\dfrac{y}{x}\dfrac{dy}{dt}$

9. **(a)** $\dfrac{dA}{dt}=\dfrac{1}{2}ab\cos\theta\dfrac{d\theta}{dt}$

(b) $\dfrac{dA}{dt}=\dfrac{1}{2}ab\cos\theta\dfrac{d\theta}{dt}+\dfrac{1}{2}b\sin\theta\dfrac{da}{dt}$

(c) $\dfrac{dA}{dt}=\dfrac{1}{2}ab\cos\theta\dfrac{d\theta}{dt}+\dfrac{1}{2}b\sin\theta\dfrac{da}{dt}+\dfrac{1}{2}a\sin\theta\dfrac{db}{dt}$

11. **(a)** 14 cm^2/sec, increasing **(b)** 0 cm/sec, constant
(c) $-14/13$ cm/sec, decreasing

13. **(a)** -12 ft/sec **(b)** -59.5 ft^2/sec **(c)** -1 rad/sec

15. 20 ft/sec

17. **(a)** $\dfrac{dh}{dt}=11.19$ cm/min **(b)** $\dfrac{dr}{dt}=14.92$ cm/min

19. **(a)** $\frac{-1}{24\pi}$ m/min **(b)** $r = \sqrt{26y - y^2}$ m, **(c)** $\frac{dr}{dt} = -\frac{5}{288\pi}$ m/min

21. 1 ft/min, 40π ft^2/min **23.** 11 ft/sec

25. 1 rad/sec **27.** -5 m/sec **29.** -1500 ft/sec

31. $\frac{5}{72\pi}$ in./min, $\frac{10}{3}$ in.^{2}/min **33.** 7.1 in./min

Section 2.13, pp. 168–170

1. $L(x) = 10x - 13$ **3.** $L(x) = 2$ **5.** $L(x) = x - \pi$

7. $2x$ **9.** -5 **11.** $\frac{1}{12}x + \frac{4}{3}$ **13.** $1 - x$

15. $f(0) = 1$. Also, $f'(x) = k(1 + x)^{k-1}$, so $f'(0) = k$. This means the linearization at $x = 0$ is $L(x) = 1 + kx$.

17. **(a)** 1.01 **(b)** 1.003 **19.** $\left(3x^2 - \frac{3}{2\sqrt{x}}\right) dx$

21. $\frac{1 - y}{3\sqrt{y} + x} dx$ **23.** $\frac{5}{2\sqrt{x}} \cos(5\sqrt{x})\, dx$

25. $\frac{1}{2\sqrt{x}} \cdot e^{\sqrt{x}}\, dx$ **27.** $\frac{2x}{1 + x^2} dx$

29. **(a)** .41 **(b)** .4 **(c)** .01

31. **(a)** $-1/3$ **(b)** $-2/5$ **(c)** $1/15$ **33.** $dV = 4\pi r_0^2\, dr$

35. $dV = 2\pi r_0 h\, dr$

37. **(a)** 0.08π m^2 **(b)** 2% **39.** $dV \approx 565.5$ in.^{3}

41. $\frac{1}{3}$% **43.** 0.05% **45.** 3% **47.** 3%

51. **(a)** $L(x) = 1 + (\ln 2)x \approx 0.69x + 1$

Practice and Additional Exercises, pp. 171–174

1. $5x^4 - 0.25x + 0.25$ **3.** $3x(x - 2)$

5. $2(x + 1)(2x^2 + 4x + 1)$

7. $3(\theta^2 + \sec\theta + 1)^2 (2\theta + \sec\theta\tan\theta)$

9. $\frac{1}{2\sqrt{t}(1 + \sqrt{t})^2}$ **11.** $2\sec^2 x \tan x$

13. $8\cos^3(1 - 2t)\sin(1 - 2t)$

15. $5(\sec t)(\sec t + \tan t)^5$

17. $\frac{\theta\cos\theta + \sin\theta}{\sqrt{2\theta\sin\theta}}$ **19.** $\frac{\cos\sqrt{2\theta}}{\sqrt{2\theta}}$

21. $x\csc\left(\frac{2}{x}\right) + \csc\left(\frac{2}{x}\right)\cot\left(\frac{2}{x}\right)$

23. $\frac{1}{2}x^{1/2}\sec(2x)^2\left[16\tan(2x)^2 - x^{-2}\right]$

25. $-10x\csc^2(x^2)$ **27.** $8x^3\sin(2x^2)\cos(2x^2) + 2x\sin^2(2x^2)$

29. $\frac{-(t + 1)}{8t^3}$ **31.** $\frac{1 - x}{(x + 1)^3}$ **33.** $\frac{-1}{2x^2\left(1 + \frac{1}{x}\right)^{1/2}}$

35. $\frac{-2\sin\theta}{(\cos\theta - 1)^2}$ **37.** $3\sqrt{2x + 1}$ **39.** $-9\left[\frac{5x + \cos 2x}{(5x^2 + \sin 2x)^{5/2}}\right]$

41. $-2e^{-x/5}$ **43.** xe^{4x} **45.** $\frac{2\sin\theta\cos\theta}{\sin^2\theta} = 2\cot\theta$

47. $\frac{2}{(\ln 2)x}$ **49.** $-8^{-t}(\ln 8)$ **51.** $18x^{2.6}$

53. $(x + 2)^{x+2}(\ln(x + 2) + 1)$ **55.** $-\frac{1}{\sqrt{1 - u^2}}$

57. $\frac{-1}{\sqrt{1 - x^2}\cos^{-1}x}$ **59.** $\tan^{-1}(t) + \frac{t}{1 + t^2} - \frac{1}{2t}$

61. $\frac{1 - z}{\sqrt{z^2 - 1}} + \sec^{-1}z$ **63.** -1 **65.** $-\frac{y + 2}{x + 3}$

67. $\frac{-3x^2 - 4y + 2}{4x - 4y^{1/3}}$ **69.** $-\frac{y}{x}$ **71.** $\frac{1}{2y(x + 1)^2}$

73. $-1/2$ **75.** y/x **77.** $\frac{2e^{-\tan^{-1}x}}{1 + x^2}$ **79.** $\frac{dp}{dq} = \frac{6q - 4p}{3p^2 + 4q}$

81. $\frac{dr}{ds} = (2r - 1)(\tan 2s)$

83. **(a)** $\frac{d^2y}{dx^2} = \frac{-2xy^3 - 2x^4}{y^5}$ **(b)** $\frac{d^2y}{dx^2} = \frac{-2xy^2 - 1}{x^4y^3}$

85. **(a)** 7 **(b)** -2 **(c)** $5/12$ **(d)** $1/4$ **(e)** 12 **(f)** $9/2$ **(g)** $3/4$

87. 0 **89.** $\frac{3\sqrt{2e^{\sqrt{3}/2}}}{4}\cos\left(e^{\sqrt{3}/2}\right)$

91. $-\frac{1}{2}$ **93.** $\frac{-2}{(2t + 1)^2}$

95. **(a)**

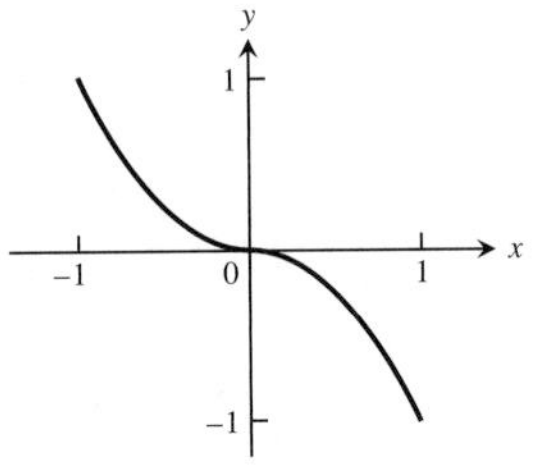

$f(x) = \begin{cases} x^2, & -1 \le x < 0 \\ -x^2, & 0 \le x < 1 \end{cases}$

(b) Yes **(c)** Yes

97. $\left(\frac{5}{2}, \frac{9}{4}\right)$ and $\left(\frac{3}{2}, -\frac{1}{4}\right)$

99. **(a)** $(-2, 16), (3, 11)$ **(b)** $(0, 20), (1, 7)$

101. $\frac{1}{4}$ **103.** Tangent: $y = -\frac{1}{4}x + \frac{9}{4}$, normal: $y = 4x - 2$

105. Tangent: $y = 2x - 4$, normal: $y = -\frac{1}{2}x + \frac{7}{2}$

107. -1 **109.** $1/2$ **111.** 4

113. $\frac{2(x^2 + 1)}{\sqrt{\cos 2x}}\left[\frac{2x}{x^2 + 1} + \tan 2x\right]$

115. $5\left[\frac{(t + 1)(t - 1)}{(t - 2)(t + 3)}\right]^5\left[\frac{1}{t + 1} + \frac{1}{t - 1} - \frac{1}{t - 2} - \frac{1}{t + 3}\right]$

117. $\frac{1}{\sqrt{\theta}}(\sin\theta)^{\sqrt{\theta}}\left(\frac{\ln\sqrt{\sin\theta}}{2}+\theta\cot\theta\right)$

119. (a) $\frac{dS}{dt}=(4\pi r+2\pi h)\frac{dr}{dt}$ **(b)** $\frac{dS}{dt}=2\pi r\frac{dh}{dt}$

(c) $\frac{dS}{dt}=(4\pi r+2\pi h)\frac{dr}{dt}+2\pi r\frac{dh}{dt}$

(d) $\frac{dr}{dt}=-\frac{r}{2r+h}\frac{dh}{dt}$

121. 0.02 ohm/sec **123.** 22 m/sec

125. (a) $r=\frac{2}{5}h$ **(b)** $-\frac{125}{144\pi}$ ft/min

127. (a) $\frac{3}{5}$ km/sec or 600 m/sec **(b)** $\frac{18}{\pi}$ rpm

129. (a) $L(x)=2x+\frac{\pi-2}{2}$

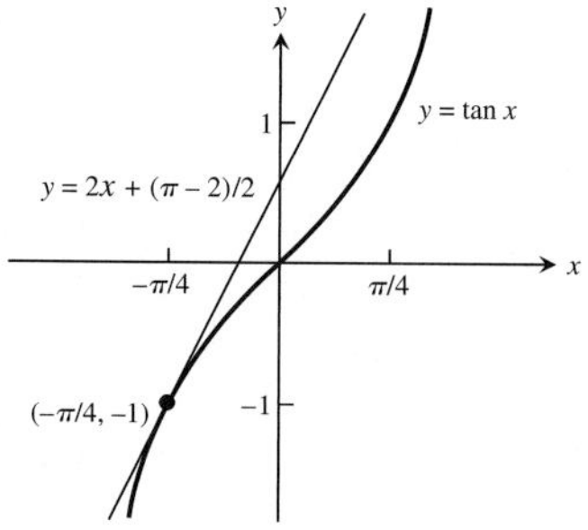

(b) $L(x)=-\sqrt{2}x+\frac{\sqrt{2}(4-\pi)}{4}$

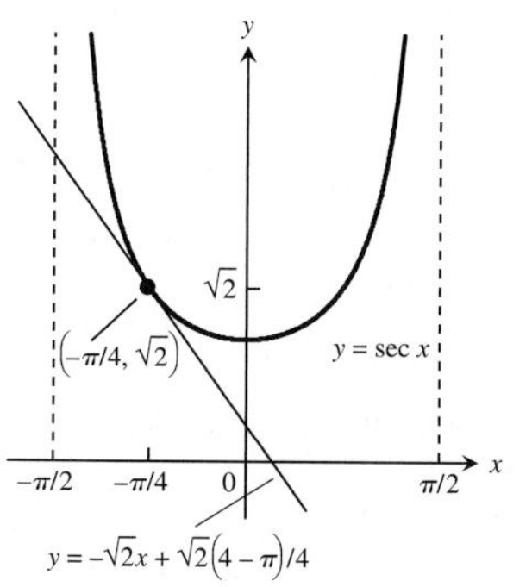

131. (a) 4% **(b)** 8% **(c)** 12%

CHAPTER 3

Section 3.1, pp. 181–182

1. Absolute minimum at $x=c_2$; absolute maximum at $x=b$
3. Absolute maximum at $x=c$; no absolute minimum
5. Absolute minimum at $x=a$; absolute maximum at $x=c$
7. Local minimum at $(-1, 0)$; local maximum at $(1, 0)$

9. Maximum at (0, 5) **11.** (c) **13.** (d)

15. Absolute maximum: -3; absolute minimum: $-19/3$

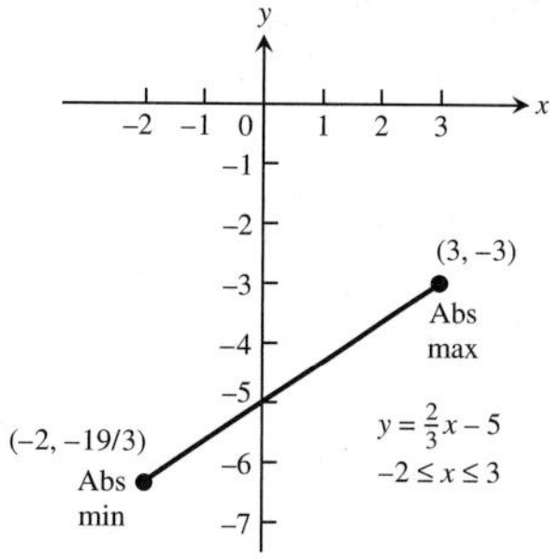

17. Absolute maximum: -0.25; absolute minimum: -4

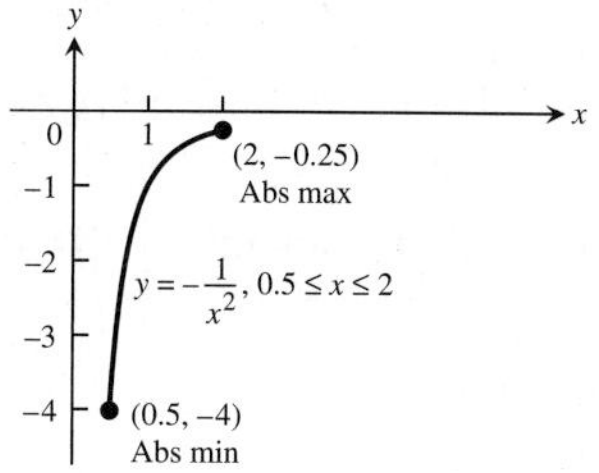

19. Absolute maximum: 2; absolute minimum: -1

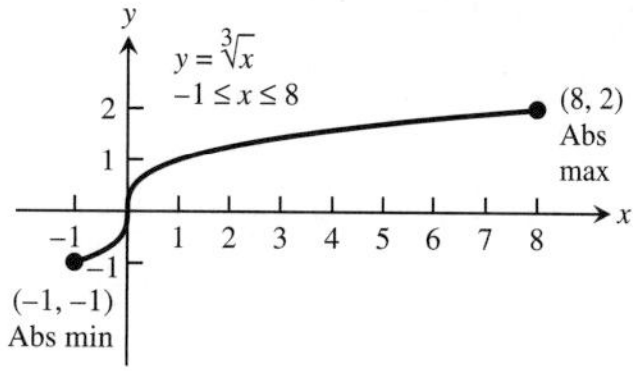

21. Absolute maximum: 1; absolute minimum: -1

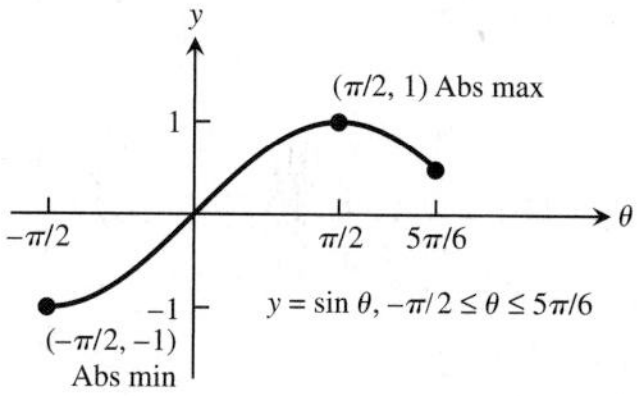

23. Absolute maximum: $2/\sqrt{3}$; absolute minimum: 1

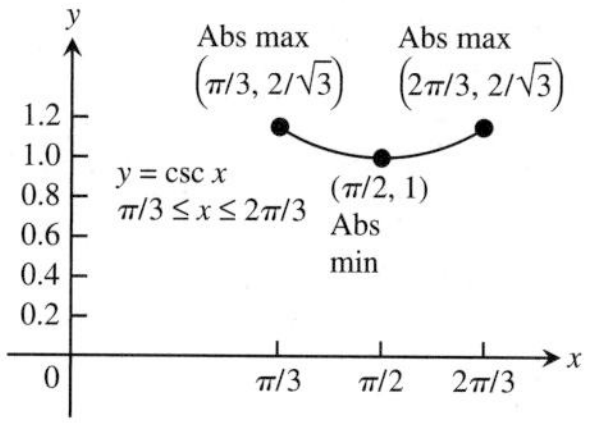

25. Absolute maximum: $1/e$; absolute minimum: $-e$

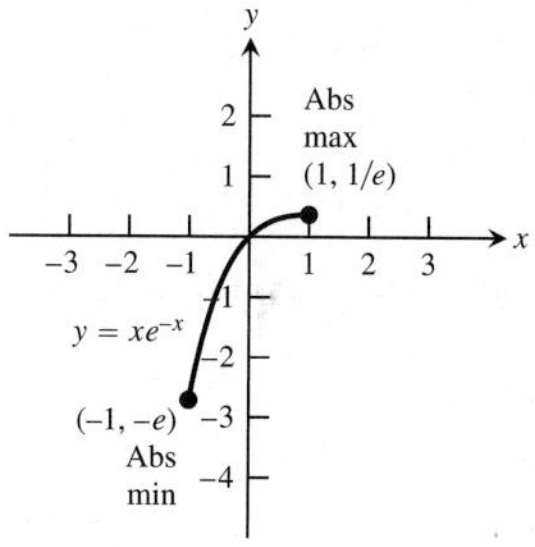

27. Absolute maximum at $(4, \frac{1}{4}+\ln 4)$; absolute minimum at (1, 1)
29. Increasing on (0, 8), decreasing on $(-1, 0)$; absolute maximum: 16 at $x=8$; absolute minimum: 0 at $x=0$
31. Minimum value is 1 at $x=2$.
33. Local maximum at $(-2, 17)$; local minimum at $\left(\frac{4}{3}, -\frac{41}{27}\right)$
35. Minimum value is 0 at $x=-1$ and $x=1$.
37. Maximum value is $\frac{1}{2}$ at $x=1$; minimum value is $-\frac{1}{2}$ at $x=-1$.
39. Minimum value is 2 at (0, 2).
41. Minimum value is $\frac{-1}{e}$ at $\left(\frac{1}{e}, \frac{-1}{e}\right)$
43. Maximum value is $\frac{\pi}{2}$ at $\left(0, \frac{\pi}{2}\right)$
Minimum value is 0 at $(\pm 1, 0)$

45.

Critical point	Derivative	Extremum	Value
$x = -\frac{4}{5}$	0	Local max	$\frac{12}{25}10^{1/3} = 1.034$
$x = 0$	Undefined	Local min	0

47.

Critical point	Derivative	Extremum	Value
$x = -2$	Undefined	Local max	0
$x = -\sqrt{2}$	0	Minimum	-2
$x = \sqrt{2}$	0	Maximum	2
$x = 2$	Undefined	Local min	0

49.

Critical point	Derivative	Extremum	Value
$x = 1$	Undefined	Minimum	2

51. Yes

53. **(a)** Maximum value is 144 at $x = 2$.
(b) The largest volume of the box is 144 cubic units, and it occurs when $x = 2$.

55. $\frac{{v_0}^2}{2g} + s_0$

57. Maximum value is 11 at $x = 5$; minimum value is 5 on the interval $[-3, 2]$; local maximum at $(-5, 9)$.

Section 3.2, pp. 187–188

1. 1/2

3. $c = \pm\sqrt{1 - \frac{4}{\pi^2}} \approx \pm 0.771$

5. Does not; f is not differentiable at the interior domain point $x = 0$.

7. Does

9. **(a)**

i)

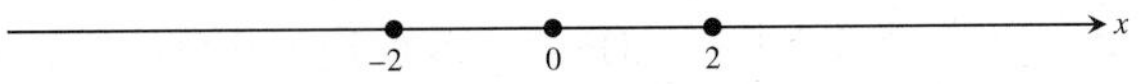

ii)

iii)

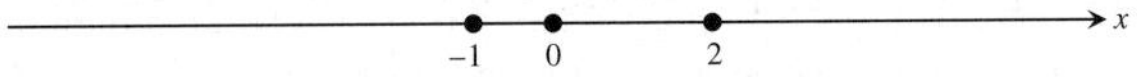

iv)

13. Yes

15. **(a)** $\frac{x^2}{2} + C$ **(b)** $\frac{x^3}{3} + C$ **(c)** $\frac{x^4}{4} + C$

17. **(a)** $y = -\ln|x| + C$ **(b)** $y = x - \ln|x| + C$
(c) $y = 5x + \ln|x| + C$

19. $f(x) = x^2 - x$

21. $f(x) = 1 + \frac{e^{2x}}{2}$

23. $s = 4.9t^2 + 5t + 10$

25. $s = e^t + 19t + 4$

27. If $T(t)$ is the temperature of the thermometer at time t, then $T(0) = -19°\text{C}$ and $T(14) = 100°\text{C}$. From the Mean Value Theorem, there exists a $0 < t_0 < 14$ such that $\frac{T(14) - T(0)}{14 - 0} = 8.5\ °\text{C/sec} = T'(t_0)$, the rate at which the temperature was changing at $t = t_0$ as measured by the rising mercury on the thermometer.

29. The conclusion of the Mean Value Theorem yields

$$\frac{\frac{1}{b} - \frac{1}{a}}{b - a} = -\frac{1}{c^2} \Rightarrow c^2\left(\frac{a - b}{ab}\right) = a - b \Rightarrow c = \sqrt{ab}.$$

31. $f(x)$ must be zero at least once between a and b by the Intermediate Value Theorem. Now suppose that $f(x)$ is zero twice between a and b. Then, by the Mean Value Theorem, $f'(x)$ would have to be zero at least once between the two zeros of $f(x)$, but this can't be true since we are given that $f'(x) \neq 0$ on this interval. Therefore, $f(x)$ is zero once and only once between a and b.

Section 3.3, pp. 192–193

1. **(a)** 0, 1
(b) increasing on $(-\infty, 0)$ and $(1, \infty)$, decreasing on $(0, 1)$
(c) local maximum at $x = 0$, local minimum at $x = 1$

3. **(a)** $-2, 1$
(b) increasing on $(-2, 1)$ and $(1, \infty)$, decreasing on $(-\infty, -2)$
(c) no local maximum, local minimum at $x = -2$

5. **(a)** Critical point at $x = 1$
(b) increasing on $[1, \infty)$, decreasing on $(-\infty, 1]$
(c) local (and absolute) minimum at $x = 1$

7. **(a)** $-2, 0$
(b) increasing on $(-\infty, -2)$ and $(0, \infty)$, decreasing on $(-2, 0)$
(c) local maximum at $x = -2$, local minimum at $x = 0$

9. **(a)** Increasing on $(-\infty, -1.5)$, decreasing on $(-1.5, \infty)$
(b) local maximum: 5.25 at $t = -1.5$
(c) absolute maximum: 5.25 at $t = -1.5$

11. **(a)** Decreasing on $(-\infty, 0)$, increasing on $(0, 4/3)$, decreasing on $(4/3, \infty)$
(b) local minimum at $x = 0\ (0, 0)$, local maximum at $x = 4/3\ (4/3, 32/27)$
(c) no absolute extrema

13. **(a)** Decreasing on $(-\infty, 0)$, increasing on $(0, 1/2)$, decreasing on $(1/2, \infty)$
(b) local minimum at $\theta = 0\ (0, 0)$, local maximum at $\theta = 1/2\ (1/2, 1/4)$
(c) no absolute extrema

15. **(a)** Increasing on $(-\infty, \infty)$, never decreasing
(b) no local extrema
(c) no absolute extrema

17. **(a)** Increasing on $(-2, 0)$ and $(2, \infty)$, decreasing on $(-\infty, -2)$ and $(0, 2)$
(b) local maximum: 16 at $x = 0$, local minimum: 0 at $x = \pm 2$
(c) no absolute maximum; absolute minimum: 0 at $x = \pm 2$

19. **(a)** Increasing on $(-\infty, -1)$, decreasing on $(-1, 0)$, increasing on $(0, 1)$, decreasing on $(1, \infty)$

(b) local maximum at $x = \pm 1$ $(1, 0.5)$, $(-1, 0.5)$, local minimum at $x = 0$ $(0, 0)$
(c) absolute maximum: $1/2$ at $x = \pm 1$; no absolute minimum

21. **(a)** Decreasing on $(-2\sqrt{2}, -2)$, increasing on $(-2, 2)$, decreasing on $(2, 2\sqrt{2})$
(b) local minima: $g(-2) = -4$, $g(2\sqrt{2}) = 0$; local maxima: $g(-2\sqrt{2}) = 0$, $g(2) = 4$
(c) absolute maximum: 4 at $x = 2$; absolute minimum: -4 at $x = -2$

23. **(a)** Increasing on $(-\infty, 1)$, decreasing when $1 < x < 2$, decreasing when $2 < x < 3$, discontinuous at $x = 2$, increasing on $(3, \infty)$
(b) local minimum at $x = 3$ $(3, 6)$, local maximum at $x = 1$ $(1, 2)$
(c) no absolute extrema

25. **(a)** Increasing on $(-2, 0)$ and $(0, \infty)$, decreasing on $(-\infty, -2)$
(b) local minimum: $-6\sqrt[3]{2}$ at $x = -2$
(c) no absolute maximum; absolute minimum: $-6\sqrt[3]{2}$ at $x = -2$

27. **(a)** Increasing on $(-\infty, -2/\sqrt{7})$ and $(2/\sqrt{7}, \infty)$, decreasing on $(-2/\sqrt{7}, 0)$ and $(0, 2/\sqrt{7})$
(b) local maximum: $24\sqrt[3]{2}/7^{7/6} \approx 3.12$ at $x = -2/\sqrt{7}$; local minimum: $-24\sqrt[3]{2}/7^{7/6} \approx -3.12$ at $x = 2/\sqrt{7}$
(c) no absolute extrema

29. **(a)** Increasing on $\left(\frac{1}{3}\ln\left(\frac{1}{2}\right)\infty\right)$, decreasing on $\left(-\infty, \frac{1}{3}\ln\left(\frac{1}{2}\right)\right)$
(b) local minimum: $\frac{3}{2^{2/3}}$ at $x = \frac{1}{3}\ln\left(\frac{1}{2}\right)$
(c) absolute minimum: $\frac{3}{2^{2/3}}$ at $x = \frac{1}{3}\ln\left(\frac{1}{2}\right)$

31. **(a)** Increasing on $\left(\frac{1}{e}, \infty\right)$, decreasing on $(0, 1/e)$
(b) local minimum: $\frac{-1}{e}$ at $x = \frac{1}{e}$
(c) absolute minimum: $\frac{-1}{e}$ at $x = \frac{1}{e}$

33. **(a)** Local maximum: 1 at $x = 1$; local minimum: 0 at $x = 2$
(b) absolute maximum: 1 at $x = 1$; no absolute minimum

35. **(a)** Local maximum: 1 at $x = 1$; local minimum: 0 at $x = 2$
(b) no absolute maximum; absolute minimum: 0 at $x = 2$

37. **(a)** Local maxima: -9 at $t = -3$ and 16 at $t = 2$; local minimum: -16 at $t = -2$
(b) absolute maximum: 16 at $t = 2$; no absolute minimum

39. **(a)** Local minimum: 0 at $x = 0$
(b) no absolute maximum; absolute minimum: 0 at $x = 0$

41. **(a)** Local minimum: $(\pi/3) - \sqrt{3}$ at $x = 2\pi/3$; local maximum: 0 at $x = 0$; local maximum: π at $x = 2\pi$

43. **(a)** Local minimum: 0 at $x = \pi/4$

45. Local maximum: 3 at $\theta = 0$; local minimum: -3 at $\theta = 2\pi$

47.

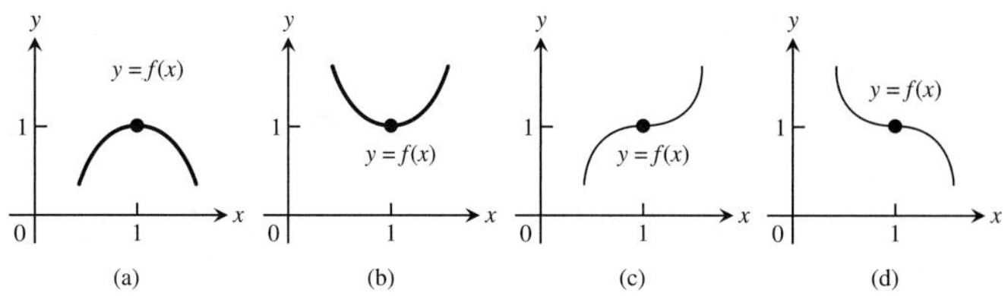

49. **(a)** **(b)**

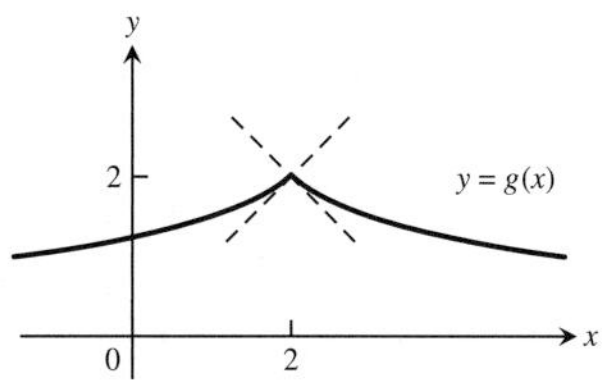

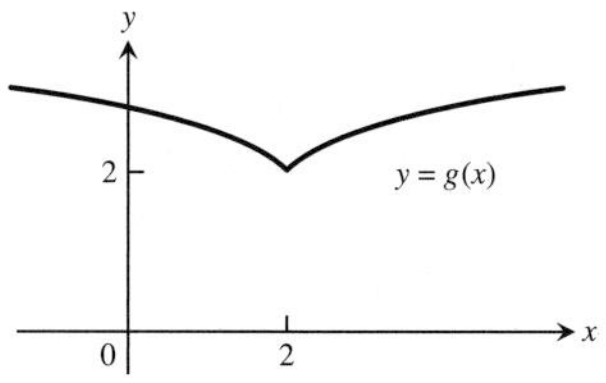

51. **(a)** Max $= 0$ at $x = 0$, min $= -\ln 2$ at $x = \pi/3$
(b) Max $= 1$ at $x = 1$, min $= \cos(\ln 2)$ at $x = \frac{1}{2}$ and $x = 2$.

53. Maximum: 1 at $x = 0$, minimum: $2 - 2\ln 2$ at $x = \ln 2$

55. Abs max of $1/(2e)$ occurs at $x = 1/\sqrt{e}$ **57.** Rising

61. Increasing, therefore one-to-one; $df^{-1}/dx = \frac{1}{9}x^{-2/3}$

63. Decreasing, therefore one-to-one; $df^{-1}/dx = -\frac{1}{3}x^{-2/3}$

Section 3.4, pp. 201–203

1. Local maximum: $3/2$ at $x = -1$, local minimum: -3 at $x = 2$, point of inflection at $(1/2, -3/4)$, rising on $(-\infty, -1)$ and $(2, \infty)$, falling on $(-1, 2)$, concave up on $(1/2, \infty)$, concave down on $(-\infty, 1/2)$

3. Local maximum: $3/4$ at $x = 0$, local minimum: 0 at $x = \pm 1$, points of inflection at $\left(-\sqrt{3}, \frac{3\sqrt[3]{4}}{4}\right)$ and $\left(\sqrt{3}, \frac{3\sqrt[3]{4}}{4}\right)$, rising on $(-1, 0)$ and $(1, \infty)$, falling on $(-\infty, -1)$ and $(0, 1)$, concave up on $(-\infty, -\sqrt{3})$ and $(\sqrt{3}, \infty)$, concave down on $(-\sqrt{3}, \sqrt{3})$

5. Local maxima: $-2\pi/3 + \sqrt{3}/2$ at $x = -2\pi/3$; $\frac{\pi}{3} + \frac{\sqrt{3}}{2}$ at $x = \frac{\pi}{3}$, local minima: $-\frac{\pi}{3} - \frac{\sqrt{3}}{2}$ at $x = -\frac{\pi}{3}$; $2\pi/3 - \sqrt{3}/2$ at $x = \frac{2\pi}{3}$, points of inflection at $(-\pi/2, -\pi/2)$, $(0, 0)$, and $(\pi/2, \pi/2)$, rising on $(-\pi/3, \pi/3)$, falling on $(-2\pi/3, -\pi/3)$ and $(\pi/3, 2\pi/3)$, concave up on $(-\pi/2, 0)$ and $(\pi/2, 2\pi/3)$, concave down on $(-2\pi/3, -\pi/2)$ and $(0, \pi/2)$

7. Local maxima: 1 at $x = -\frac{\pi}{2}$ and $x = \frac{\pi}{2}$; 0 at $x = -2\pi$ and $x = 2\pi$; local minima: -1 at $x = -\frac{3\pi}{2}$ and $x = \frac{3\pi}{2}$, 0 at $x = 0$, points of inflection at $(-\pi, 0)$ and $(\pi, 0)$, rising on $(-3\pi/2, -\pi/2)$, $(0, \pi/2)$, and $(3\pi/2, 2\pi)$, falling on

$(-2\pi, -3\pi/2)$, $(-\pi/2, 0)$, and $(\pi/2, 3\pi/2)$, concave up on $(-2\pi, -\pi)$ and $(\pi, 2\pi)$, concave down on $(-\pi, 0)$ and $(0, \pi)$

9.

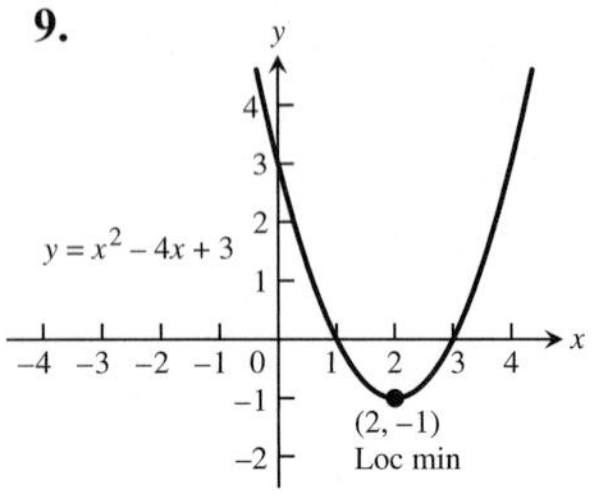

11.

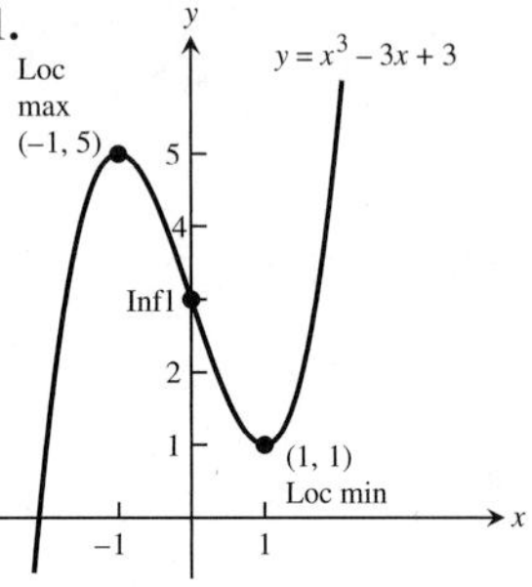

13.

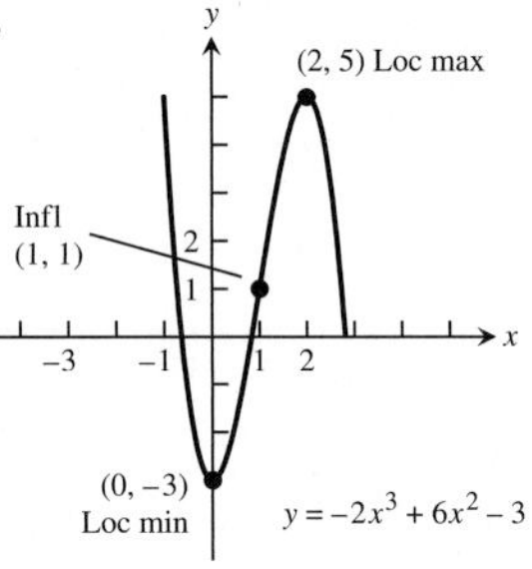

15.

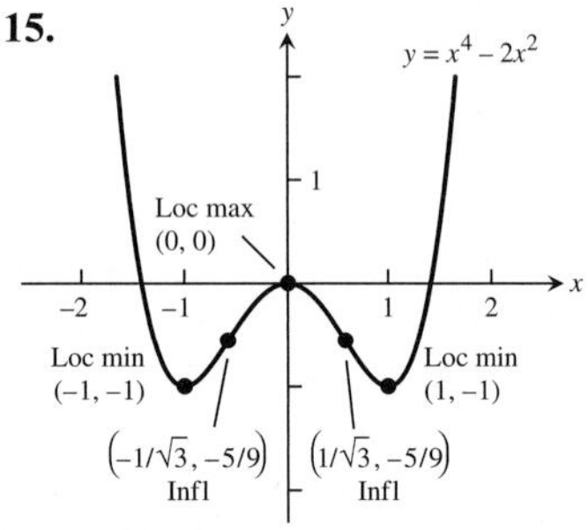

17.

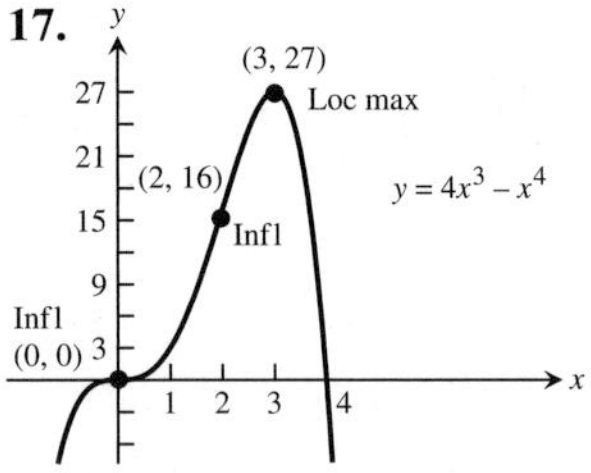

19.

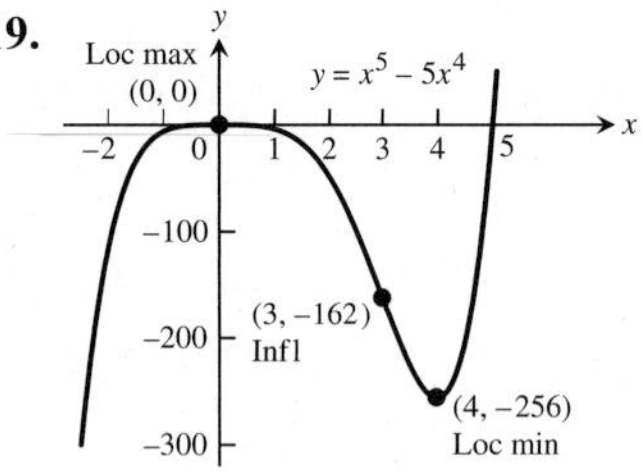

21.

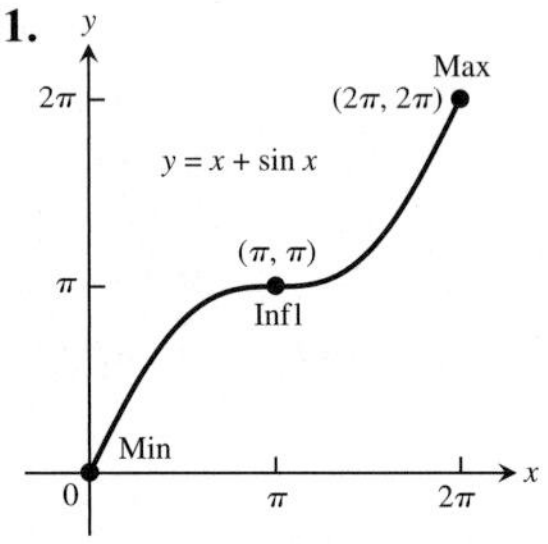

23.

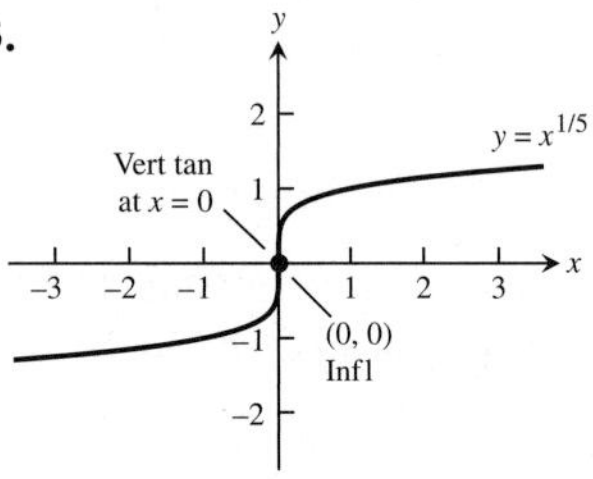

25.

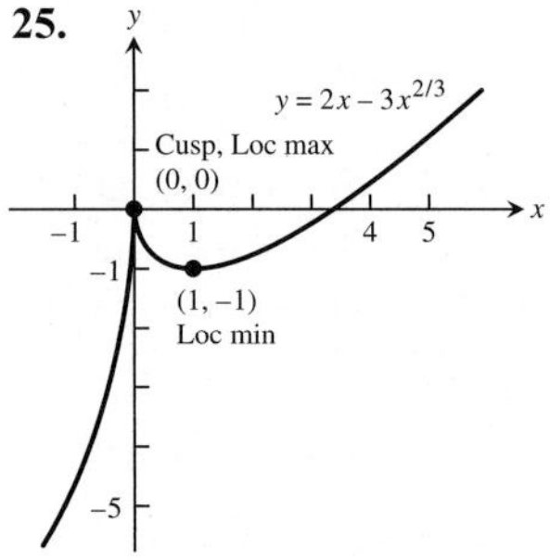

27.

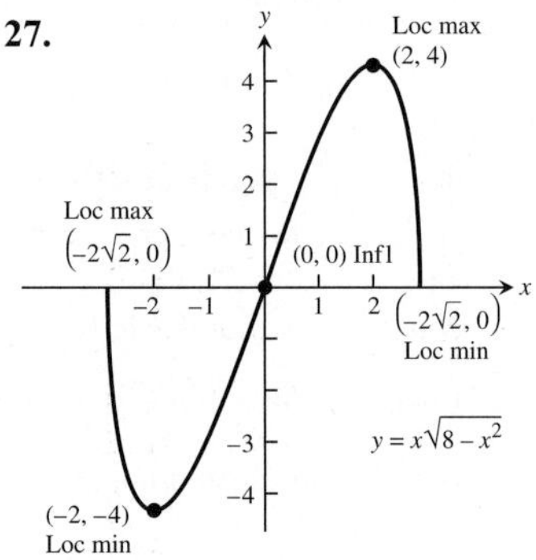

29.

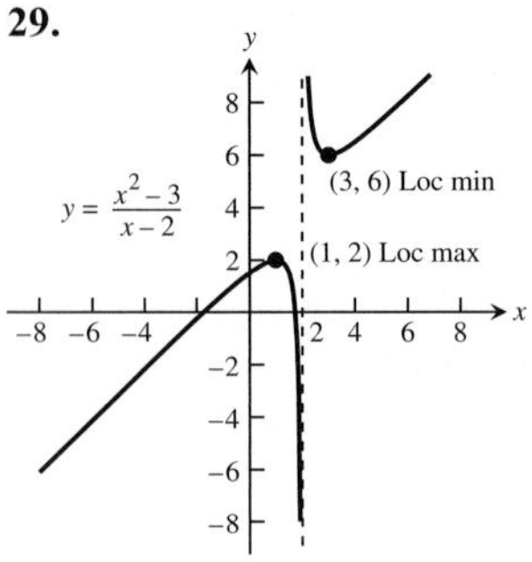

31.

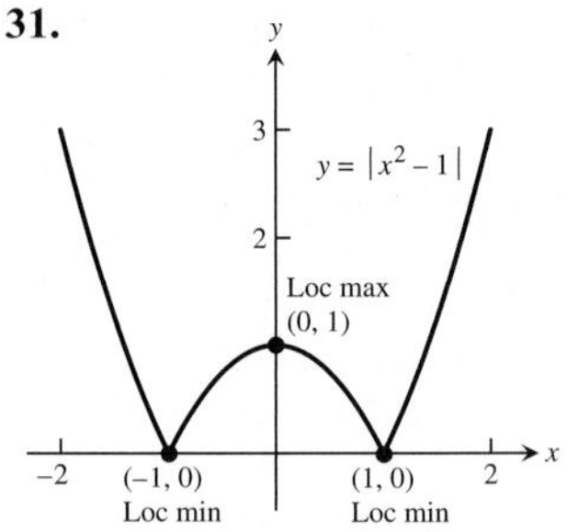

33.

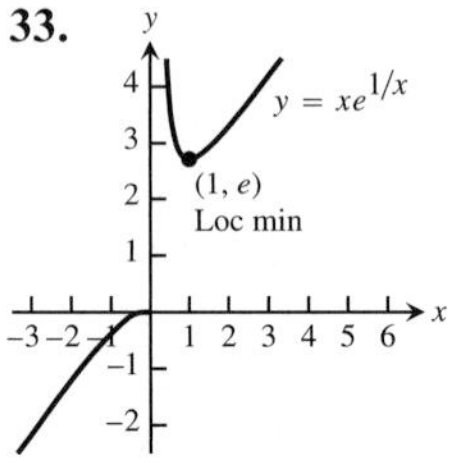

35.

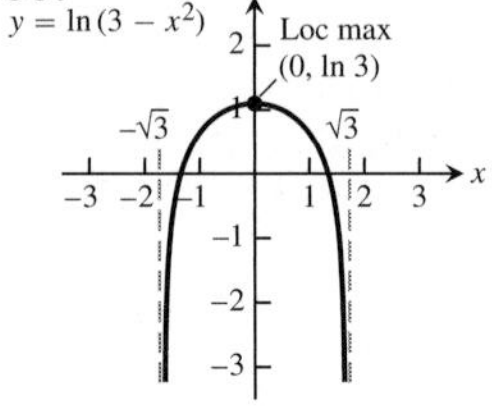

37.

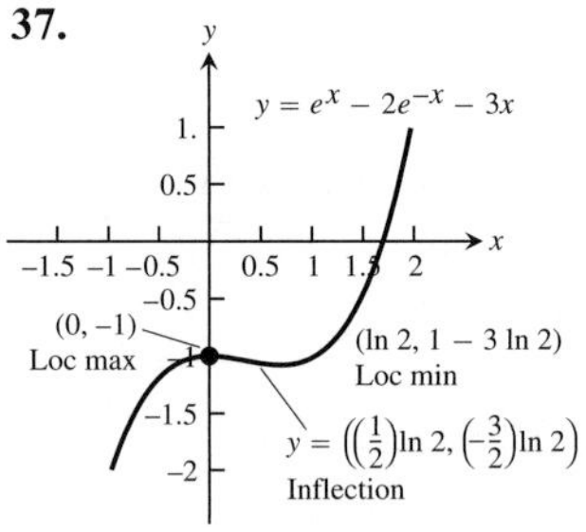

39.

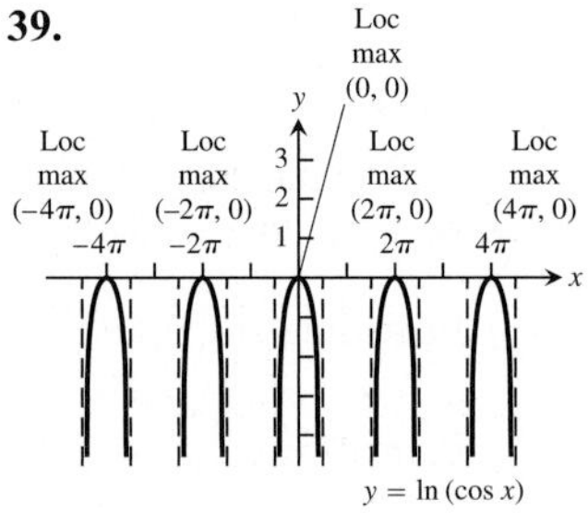

41.

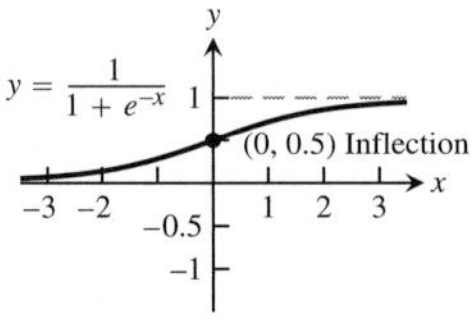

43. $y'' = 1 - 2x$

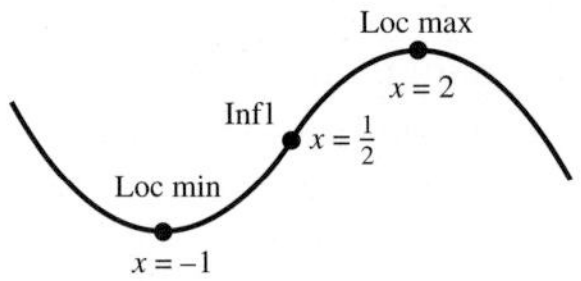

45. $y'' = 3(x - 3)(x - 1)$

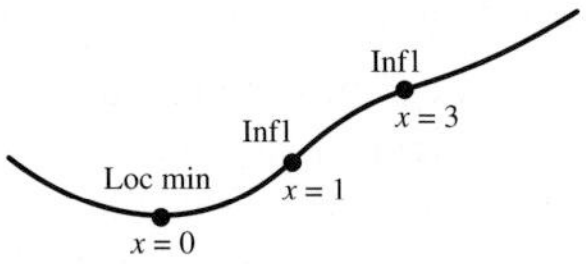

47. $y'' = 3(x - 2)(x + 2)$

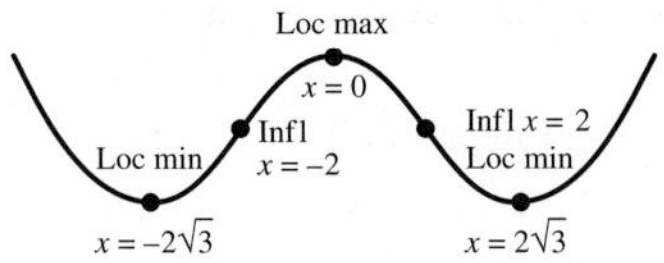

49. $y'' = 2\sec^2 x \tan x$

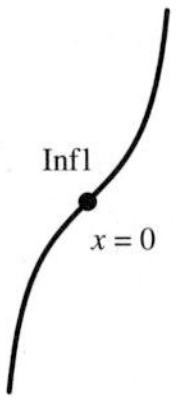

51. $y'' = -\frac{1}{2}\csc^2\frac{\theta}{2},\ 0 < \theta < 2\pi$

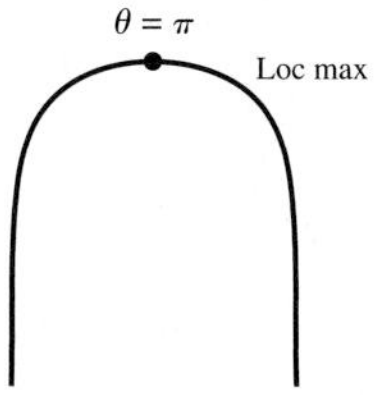

53. $y'' = 2\tan\theta\sec^2\theta,\ -\frac{\pi}{2} < \theta < \frac{\pi}{2}$

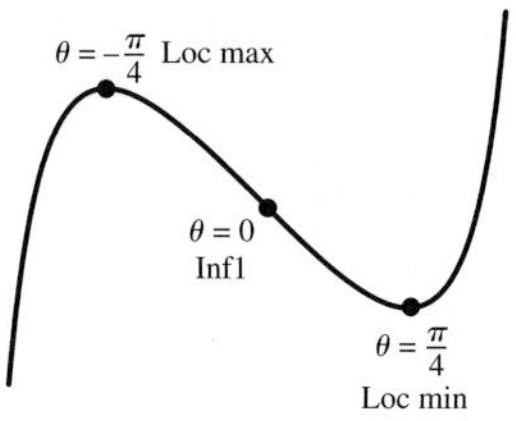

55. $y'' = -\sin t,\ 0 \le t \le 2\pi$

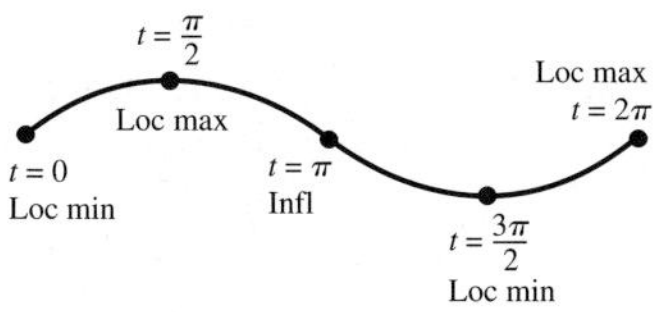

57. $y'' = -\frac{2}{3}(x+1)^{-5/3}$

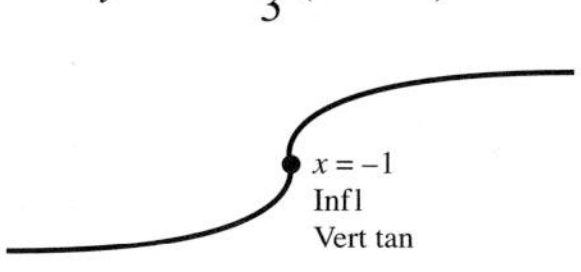

59. $y'' = \frac{1}{3}x^{-2/3} + \frac{2}{3}x^{-5/3}$

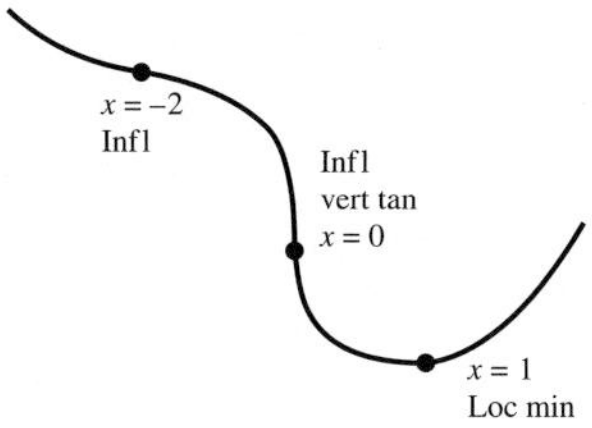

61. $y'' = \begin{cases} -2, & x < 0 \\ 2, & x > 0 \end{cases}$

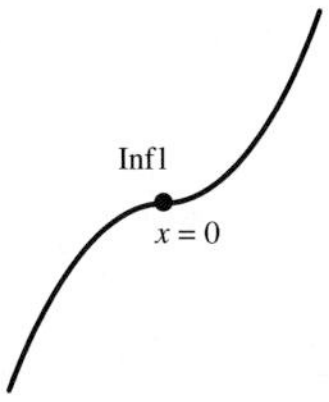

63.

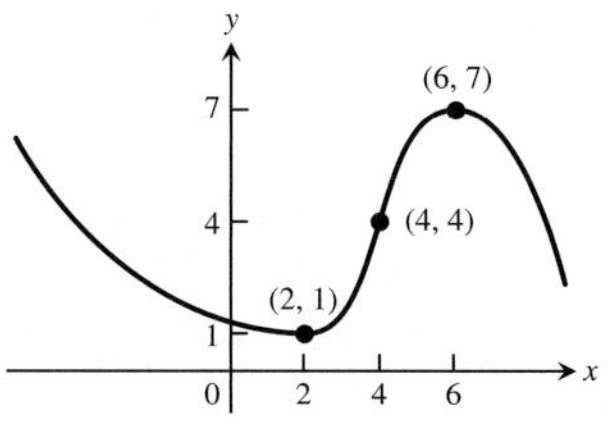

67. Local minimum at $x = 2$, inflection points at $x = 1$ and $x = 5/3$

69. **(a)** $\left(-\frac{b}{2a}, \frac{4ac - b^2}{4a}\right)$

(b) concave up if $a > 0$, concave down if $a < 0$

73. The zeros of $y' = 0$ and $y'' = 0$ include the extrema and points of inflection, respectively. Inflection at $x = 3$, local maximum at $x = 0$, local minimum at $x = 4$.

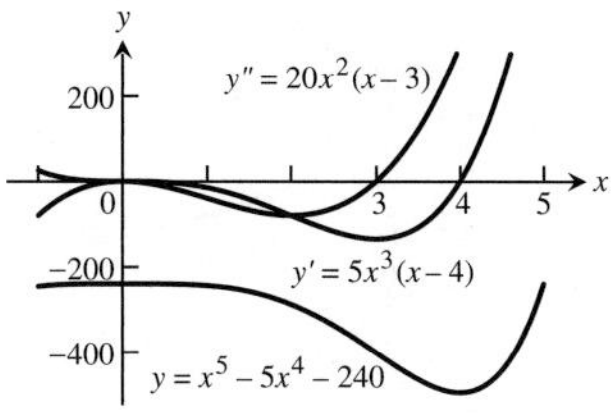

75. The zeros of $y' = 0$ and $y'' = 0$ include the extrema and points of inflection, respectively. Inflection at $x = -\sqrt[3]{2}$, local maximum at $x = -2$, local minimum at $x = 0$.

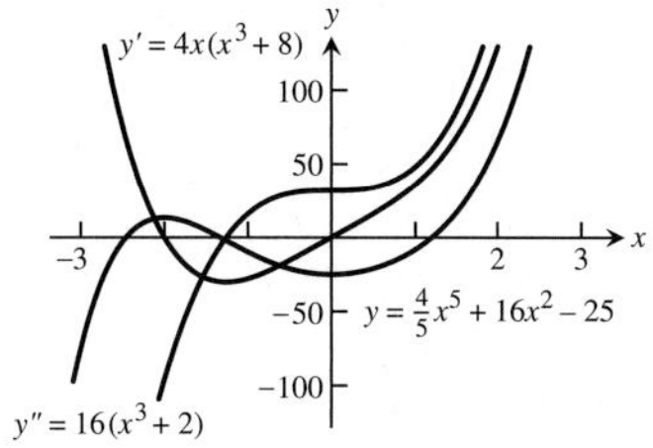

77. **(b)** A cusp since $\lim_{x\to 0^-} y' = \infty$ and $\lim_{x\to 0^+} y' = -\infty$

Section 3.5, pp. 208–209

1.

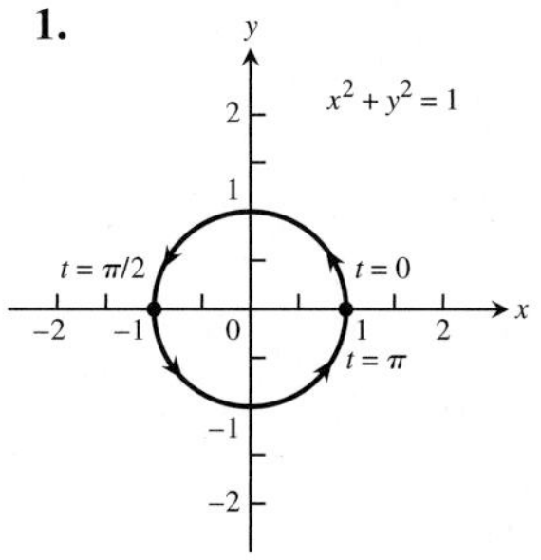

3.

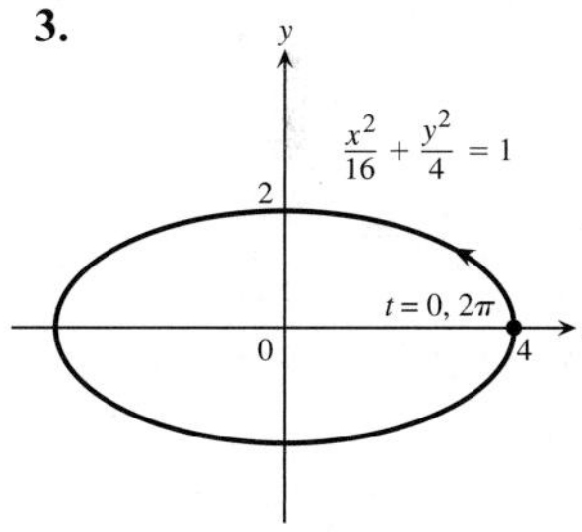

5.

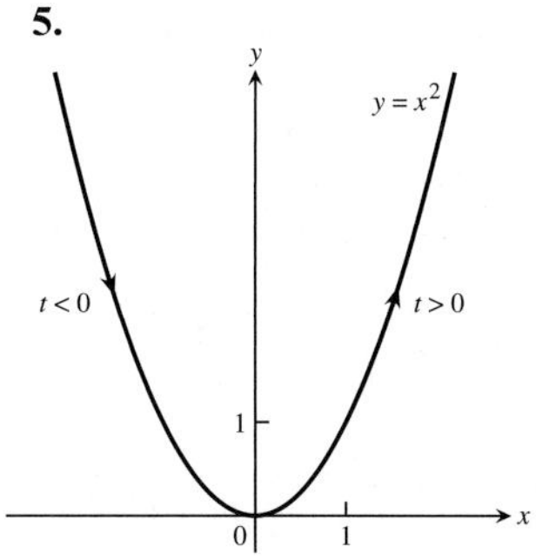

7.

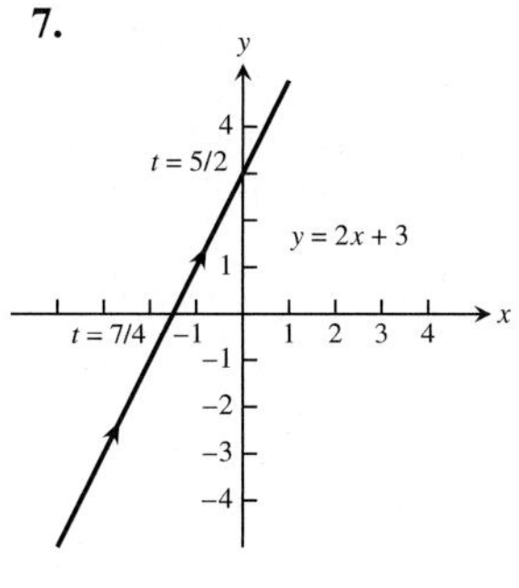

9.

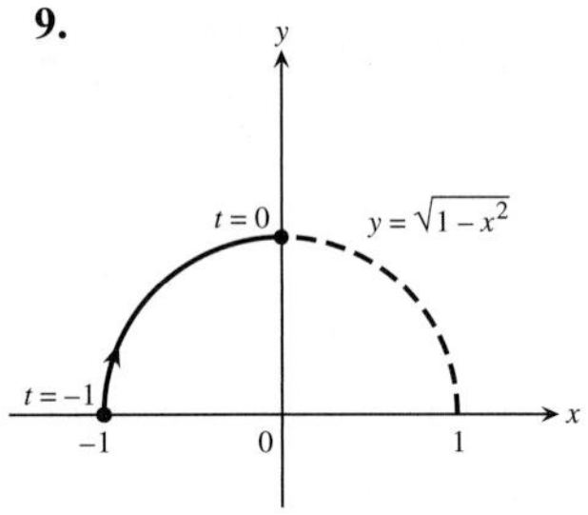

11.

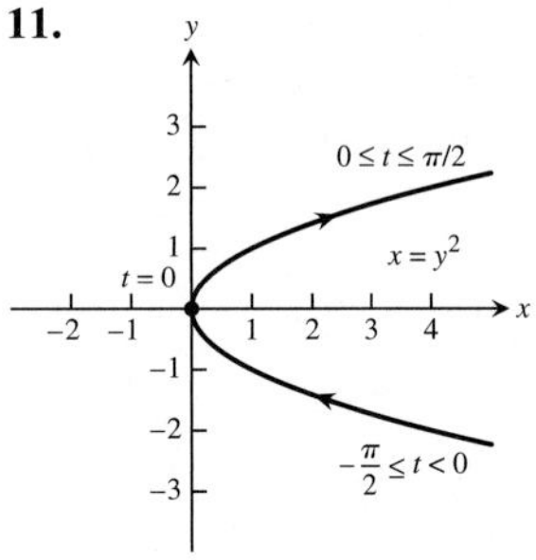

13. (a) $x = a\cos t, y = -a\sin t, 0 \le t \le 2\pi$
(b) $x = a\cos t, y = a\sin t, 0 \le t \le 2\pi$
(c) $x = a\cos t, y = -a\sin t, 0 \le t \le 4\pi$
(d) $x = a\cos t, y = a\sin t, 0 \le t \le 4\pi$
15. Possible answer: $x = -1 + 5t, y = -3 + 4t, 0 \le t \le 1$
17. Possible answer: $x = t^2 + 1, y = t, t \le 0$
19. Possible answer: $x = 2 - 3t, y = 3 - 4t, t \ge 0$
21. $y = -x + 2, \left.\frac{d^2y}{dx^2}\right|_{t=\pi/4} = -\sqrt{2}$
23. $y = x + \frac{1}{4}, \left.\frac{d^2y}{dx^2}\right|_{t=1/4} = -2$
25. $y = x - 4, \left.\frac{d^2y}{dx^2}\right|_{t=-1} = \frac{1}{2}$
27. $y = 2, \left.\frac{d^2y}{dx^2}\right|_{t=\pi/2} = -1$ **29.** $-\frac{3}{16}$ **31.** -6
33. $\left(\frac{\sqrt{2}}{2}, 1\right), y = 2x$ at $t = 0, y = -2x$ at $t = \pi$

Section 3.6, pp. 215–219

1. 16 in., 4 in. by 4 in.
3. (a) $(x, 1 - x)$ **(b)** $A(x) = 2x(1 - x)$
(c) $\frac{1}{2}$ square units, 1 by $\frac{1}{2}$
5. $\frac{14}{3} \times \frac{35}{3} \times \frac{5}{3}$ in., $\frac{2450}{27}$ in.^{3} **7.** 80,000 m^2; 400 m by 200 m
9. (a) The optimum dimensions of the tank are 10 ft on the base edges and 5 ft deep.
(b) Minimizing the surface area of the tank minimizes its weight for a given wall thickness. The thickness of the steel walls would likely be determined by other considerations such as structural requirements.
11. 9×18 in. **13.** $\frac{\pi}{2}$ **15.** $h : r = 8 : \pi$
17. ≈ 2418.40 cm^3
19. If r is the radius of the hemisphere, h the height of the cylinder, and V the volume, then $r = \left(\frac{3V}{8\pi}\right)^{1/3}$ and $h = \left(\frac{3V}{\pi}\right)^{1/3}$.
21. Radius $= \sqrt{2}$ m, height $= 1$ m, volume $\frac{2\pi}{3}$ m^3
23. (a) $v(0) = 96$ ft/sec
(b) 256 ft at $t = 3$ sec
(c) Velocity when $s = 0$ is $v(7) = -128$ ft/sec
25. ≈ 46.87 ft
27. (a) $10\pi \approx 31.42$ cm/sec; when $t = 0.5$ sec, 1.5 sec, 2.5 sec, 3.5 sec; $s = 0$, acceleration is 0
(b) 10 cm from rest position; speed is 0
29. $20\left(5 - \sqrt{17}\right)$ m
31. $\frac{c}{2} + 50$
33. (a) $\sqrt{\frac{2km}{h}}$ **(b)** $\sqrt{\frac{2km}{h}}$
35. $M = \frac{C}{2}$
41. (a) $y = -1$
43. (a) The minimum distance is $\frac{\sqrt{5}}{2}$.
(b) The minimum distance is from the point (3/2, 0) to the point (1, 1) on the graph of $y = \sqrt{x}$, and this occurs at the value $x = 1$, where $D(x)$, the distance squared, has its minimum value.

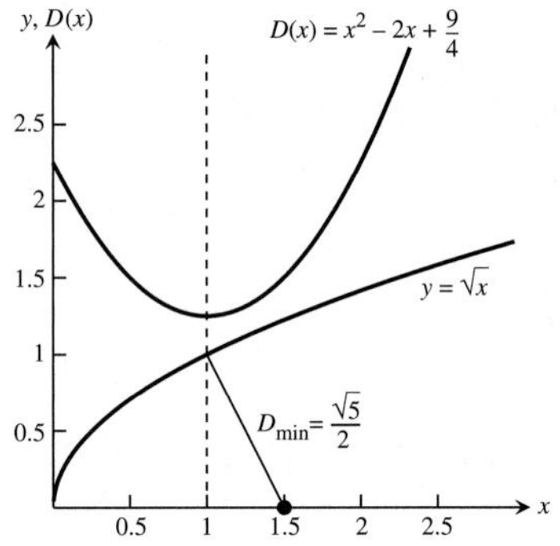

Section 3.7, pp. 224–226

1. 1/4 **3.** 5/7 **5.** 1/2 **7.** 1/4 **9.** −23/7 **11.** 5/7 **13.** 0
15. −16 **17.** −2 **19.** 1/4 **21.** 2 **23.** 3 **25.** −1
27. ln 3 **29.** $\frac{1}{\ln 2}$ **31.** ln 2 **33.** 1 **35.** 1/2 **37.** ln 2
39. cos a **41.** −1/2 **43.** −1 **45.** 1 **47.** $1/e$ **49.** 1
51. $1/e$ **53.** $e^{1/2}$ **55.** 1 **57.** 3 **59.** 1
61. (b) is correct. **63.** (d) is correct. **65.** −1

Section 3.8, pp. 228–229

1. $x_2 = -\frac{5}{3}, \frac{13}{21}$ **3.** $x_2 = -\frac{51}{31}, \frac{5763}{4945}$
7. x_1, and all later approximations will equal x_0.

9.

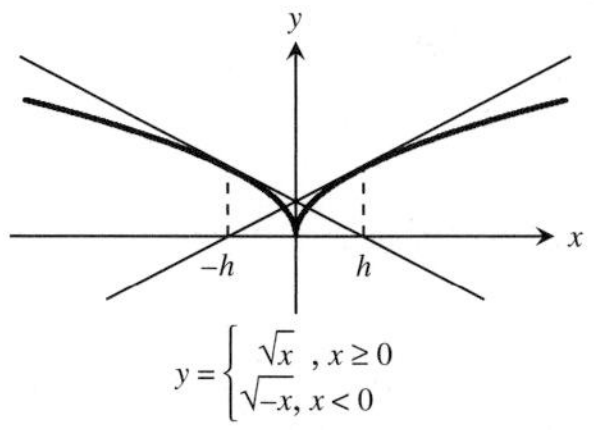

11. The points of intersection of $y = x^3$ and $y = 3x + 1$ or $y = x^3 - 3x$ and $y = 1$ have the same x-values as the roots of part (i) or the solutions of part (iv). **13.** 1.165561185
15. **(a)** Two **(b)** 0.35003501505249 and -1.0261731615301
17. 0, 0.53485
19. **(a)** For $x_0 = -2$ or $x_0 = -0.8$, $x_i \to -1$ as i gets large.
(b) For $x_0 = -0.5$ or $x_0 = 0.25$, $x_i \to 0$ as i gets large.
(c) For $x_0 = 0.8$ or $x_0 = 2$, $x_i \to 1$ as i gets large.
(d) For $x_0 = -\sqrt{21}/7$ or $x_0 = \sqrt{21}/7$, Newton's method does not converge. The values of x_i alternate between $-\sqrt{21}/7$ and $\sqrt{21}/7$ as i increases.

Section 3.9, pp. 234–236

1. $\cosh x = 5/4$, $\tanh x = -3/5$, $\coth x = -5/3$, $\operatorname{sech} x = 4/5$, $\operatorname{csch} x = -4/3$
3. $\sinh x = 8/15$, $\tanh x = 8/17$, $\coth x = 17/8$, $\operatorname{sech} x = 15/17$, $\operatorname{csch} x = 15/8$
5. $x + \dfrac{1}{x}$ **7.** e^{5x} **9.** e^{4x} **13.** $2\cosh\dfrac{x}{3}$
15. $\operatorname{sech}^2\sqrt{t} + \dfrac{\tanh\sqrt{t}}{\sqrt{t}}$ **17.** $\coth z$
19. $(\ln \operatorname{sech}\theta)(\operatorname{sech}\theta\tanh\theta)$ **21.** $\tanh^3 v$ **23.** 2
25. $\dfrac{1}{2\sqrt{x(1+x)}}$ **27.** $\dfrac{1}{1+\theta} - \tanh^{-1}\theta$
29. $\dfrac{1}{2\sqrt{t}} - \coth^{-1}\sqrt{t}$ **31.** $-\operatorname{sech}^{-1}x$ **33.** $\dfrac{\ln 2}{\sqrt{1+\left(\frac{1}{2}\right)^{2\theta}}}$
35. $|\sec x|$ **37.** $\ln(2/3)$ **39.** $\dfrac{-\ln 3}{2}$ **41.** $\ln 3$
43. **(a)** 1 **(b)** -1 **(c)** ∞ **(d)** $-\infty$
(e) 0 **(f)** 1 **(g)** ∞ **(h)** $-\infty$ **(i)** 0
45. **(b)** $\sqrt{\dfrac{mg}{k}}$ **(c)** $80\sqrt{5} \approx 178.89$ ft/sec

Practice and Additional Exercises, pp. 237–239

1. No **3.** No minimum; absolute maximum: $f(1) = 16$; critical points: $x = 1$ and $11/3$
5. Absolute minimum: $g(0) = 1$; no absolute maximum; critical point $x = 0$
7. Absolute minimum: $2 - 2\ln 2$ at $x = 2$; absolute maximum 1 at $x = 1$
9. Yes, except at $x = 0$ **11.** No **13.** **(b)** one

15. **(b)** 0.8555996772 **21.** Global minimum value of $\dfrac{1}{2}$ at $x = 2$

23.

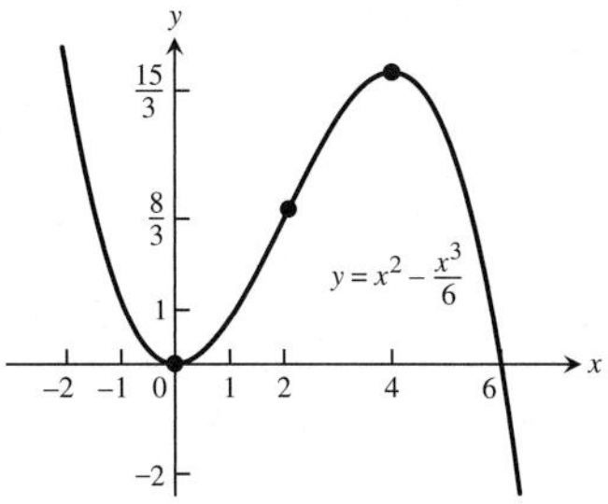

25.

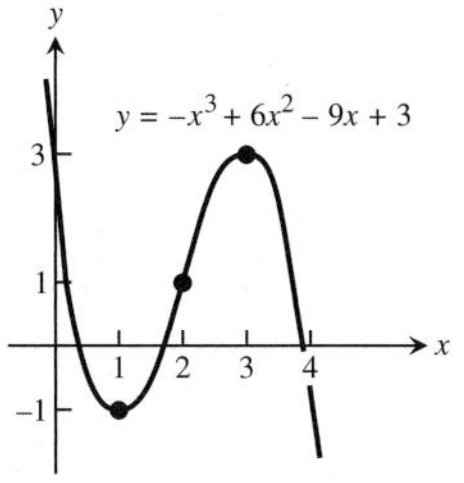

27.

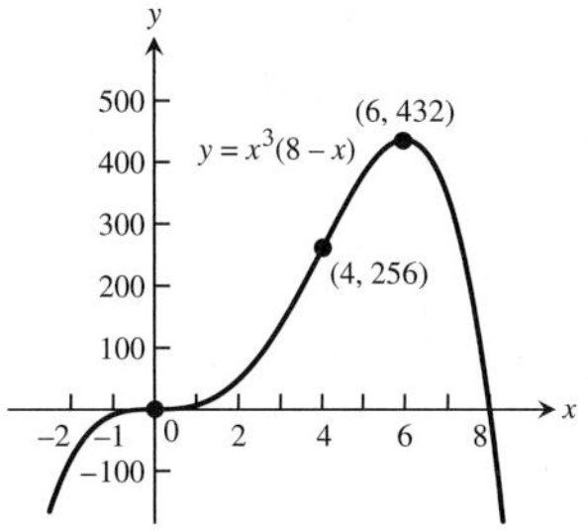

29.

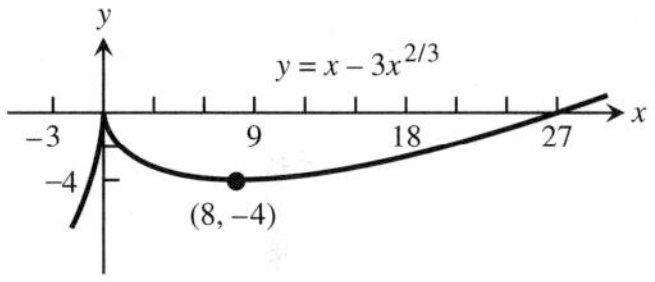

31.

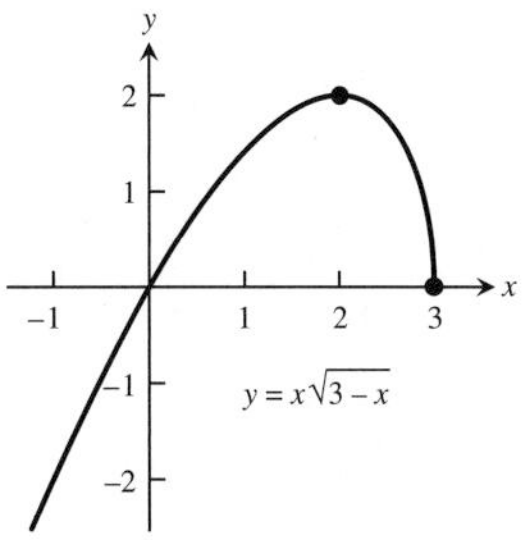

33.

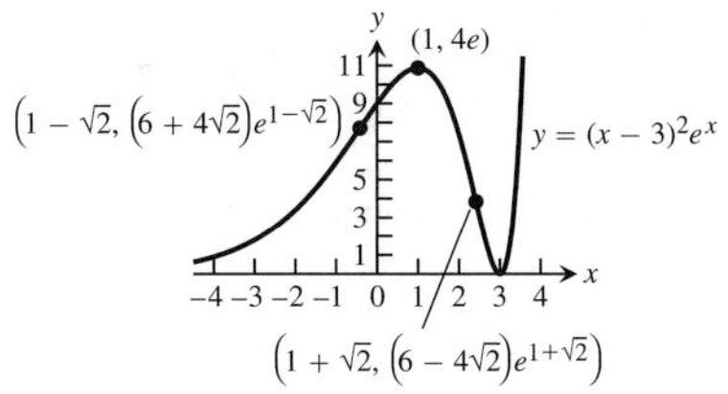

35.

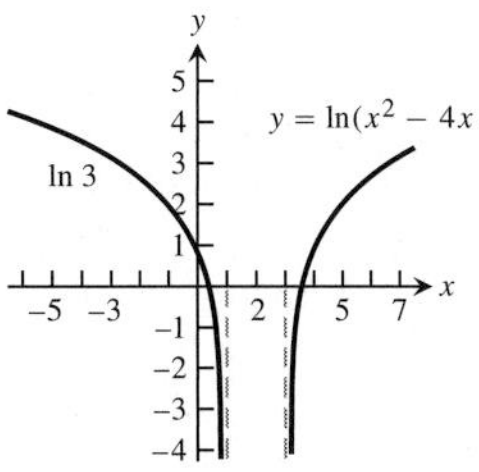

37.

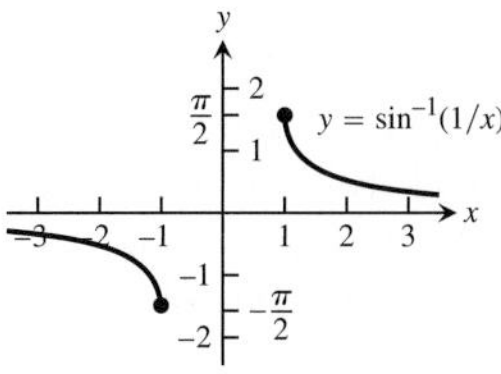

39. **(a)** Local maximum at $x = 4$, local minimum at $x = -4$, inflection point at $x = 0$

(b)

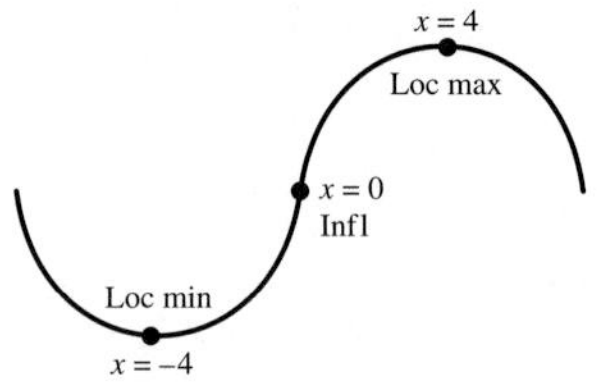

41. (a) Local maximum at $x = 0$, local minima at $x = -1$ and $x = 2$, inflection points at $x = (1 \pm \sqrt{7})/3$

(b)

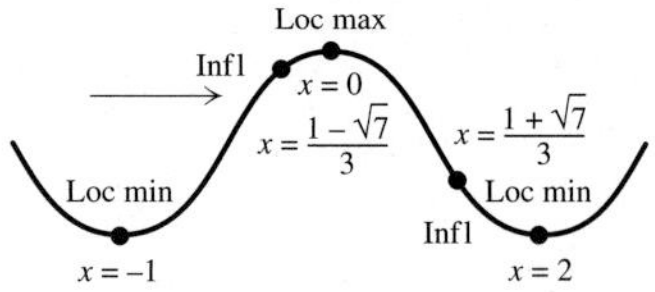

43. (a) Local maximum at $x = -\sqrt{2}$, local minimum at $x = \sqrt{2}$, inflection points at $x = \pm 1$ and 0

(b)

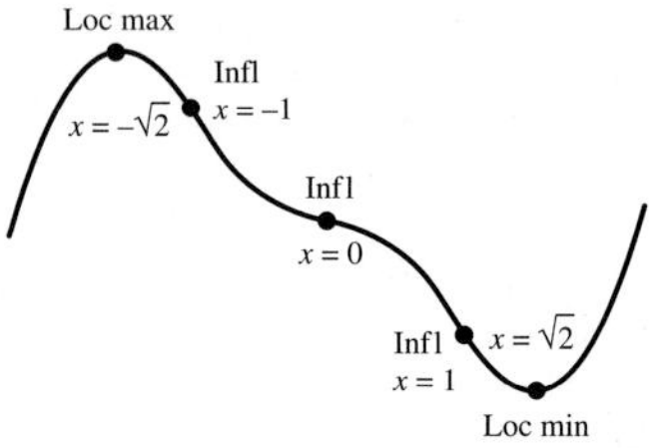

45.

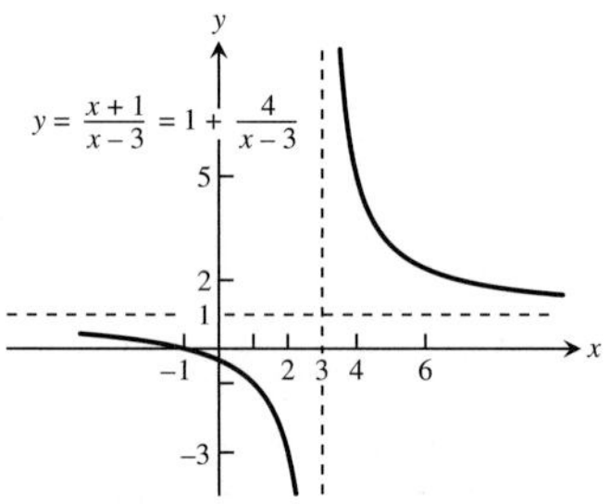

47.

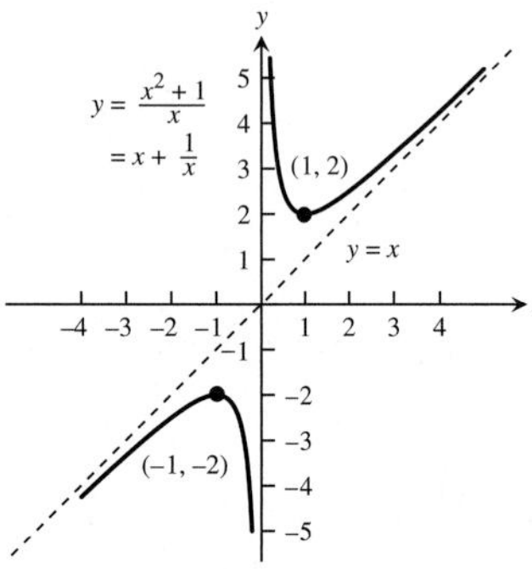

49.

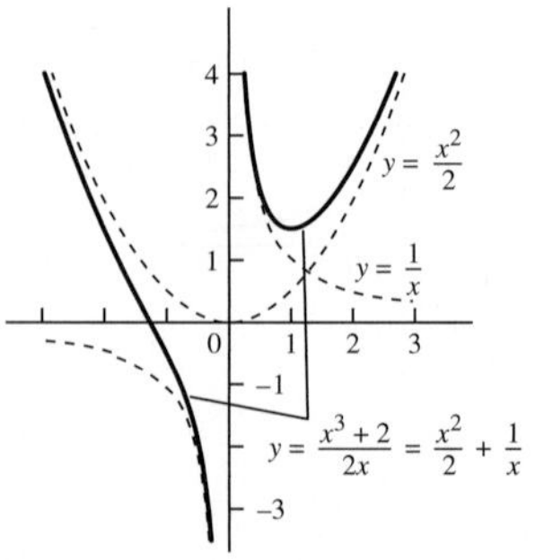

51. 5 **53.** 0 **55.** 1 **57.** $3/7$ **59.** 0 **61.** $\ln 10$
63. 5 **65.** $-\infty$ **67.** e^{bk} **69. (a)** 0, 36 **(b)** 18, 18
71. 54 square units

73. Dimensions: base is 6 in. by 12 in., height = 2 in.; maximum volume = 144 in.^{3}

75. $x_5 = 2.195823345$

77. $1/\sqrt{2}$ units long by $1/\sqrt{e}$ units high, $A = 1/\sqrt{2e} \approx 0.43$ units^2

79. Absolute maximum $= 0$ at $x = e/2$, absolute minimum $= -0.5$ at $x = 0.5$

81. $x = \pm 1$ are the critical points; $y = 1$ is a horizontal asymptote in both directions; absolute minimum value of the function is $e^{-\sqrt{2}/2}$ at $x = -1$, and absolute maximum value is $e^{\sqrt{2}/2}$ at $x = 1$.

83. (a) Absolute maximum of $2/e$ at $x = e^2$, inflection point $(e^{8/3}, (8/3)e^{-4/3})$, concave up on $(e^{8/3}, \infty)$, concave down on $(0, e^{8/3})$
(b) Absolute maximum of 1 at $x = 0$, inflection points $(\pm 1/\sqrt{2}, 1/\sqrt{e})$, concave up on $(-\infty, -1/\sqrt{2}) \cup (1/\sqrt{2}, \infty)$, concave down on $(-1/\sqrt{2}, 1/\sqrt{2})$
(c) Absolute maximum of 1 at $x = 0$, inflection point $(1, 2/e)$, concave up on $(1, \infty)$, concave down on $(-\infty, 1)$

85. $y = \left(\frac{\sqrt{3}}{2}\right)x + \frac{1}{4}, \frac{1}{4}$

87. $3 \sinh 6x$ **89.** $-(\cosh y \coth y)/\sqrt{1 - x^2}$

91. $(1 + y^2)/(1 - x^2)$ **93.** $y \operatorname{csch}(\ln y)$ **95.** $|\sec x|$

CHAPTER 4

Section 4.1, pp. 246–249

1. (a) x^2 **(b)** $\frac{x^3}{3}$ **(c)** $\frac{x^3}{3} - x^2 + x$

3. (a) x^{-3} **(b)** $-\frac{1}{3}x^{-3}$ **(c)** $-\frac{1}{3}x^{-3} + x^2 + 3x$

5. (a) $-\frac{1}{x}$ **(b)** $-\frac{5}{x}$ **(c)** $2x + \frac{5}{x}$

7. (a) $\sqrt{x^3}$ **(b)** $\sqrt{x}$ **(c)** $\frac{2\sqrt{x^3}}{3} + 2\sqrt{x}$

9. (a) $x^{2/3}$ **(b)** $x^{1/3}$ **(c)** $x^{-1/3}$

11. (a) $\ln x$ **(b)** $7 \ln x$ **(c)** $x - 5 \ln x$

13. (a) $\cos(\pi x)$ **(b)** $-3 \cos x$ **(c)** $-\frac{1}{\pi}\cos(\pi x) + \cos(3x)$

15. (a) $\tan x$ **(b)** $2 \tan\left(\frac{x}{3}\right)$ **(c)** $-\frac{2}{3}\tan\left(\frac{3x}{2}\right)$

17. (a) $-\csc x$ **(b)** $\frac{1}{5}\csc(5x)$ **(c)** $2 \csc\left(\frac{\pi x}{2}\right)$

19. (a) $\frac{1}{3}e^{3x}$ **(b)** $-e^{-x}$ **(c)** $2e^{x/2}$

21. (a) $\frac{1}{\ln 3}3^x$ **(b)** $\frac{-1}{\ln 2}2^{-x}$ **(c)** $\frac{1}{\ln(5/3)}\left(\frac{5}{3}\right)^x$

23. (a) $2 \sin^{-1} x$ **(b)** $\frac{1}{2}\tan^{-1} x$ **(c)** $\frac{1}{2}\tan^{-1} 2x$

25. $\frac{x^2}{2} + x + C$ **27.** $t^3 + \frac{t^2}{4} + C$ **29.** $\frac{x^4}{2} - \frac{5x^2}{2} + 7x + C$

31. $-\frac{1}{x} - \frac{x^3}{3} - \frac{x}{3} + C$ **33.** $\frac{3}{2}x^{2/3} + C$

35. $\frac{2}{3}x^{3/2} + \frac{3}{4}x^{4/3} + C$ **37.** $4y^2 - \frac{8}{3}y^{3/4} + C$

39. $x^2 + \frac{2}{x} + C$ **41.** $2\sqrt{t} - \frac{2}{\sqrt{t}} + C$ **43.** $-2\sin t + C$

45. $-21\cos\frac{\theta}{3} + C$ **47.** $3\cot x + C$ **49.** $-\frac{1}{2}\csc\theta + C$

51. $\frac{1}{3}e^{3x} - 5e^{-x} + C$ **53.** $-e^{-x} + \frac{4^x}{\ln 4} + C$

55. $4\sec x - 2\tan x + C$ **57.** $-\frac{1}{2}\cos 2x + \cot x + C$

59. $\frac{t}{2} + \frac{\sin 4t}{8} + C$ **61.** $\ln|x| - 5\tan^{-1}x + C$

63. $\frac{3x^{(\sqrt{3}+1)}}{\sqrt{3}+1} + C$ **65.** $\tan\theta + C$ **67.** $-\cot x - x + C$

69. $-\cos\theta + \theta + C$

83. **(a)** Wrong: $\frac{d}{dx}\left(\frac{x^2}{2}\sin x + C\right) = \frac{2x}{2}\sin x + \frac{x^2}{2}\cos x = x\sin x + \frac{x^2}{2}\cos x$

(b) Wrong: $\frac{d}{dx}(-x\cos x + C) = -\cos x + x\sin x$

(c) Right: $\frac{d}{dx}(-x\cos x + \sin x + C) = -\cos x + x\sin x + \cos x = x\sin x$

85. $y = x^2 - 7x + 10$ **87.** $y = -\frac{1}{x} + \frac{x^2}{2} - \frac{1}{2}$

89. $y = 9x^{1/3} + 4$ **91.** $r = \cos(\pi\theta) - 1$

93. $v = \frac{1}{2}\sec t + \frac{1}{2}$ **95.** $v = 3\sec^{-1}t - \pi$

97. $y = x^2 - x^3 + 4x + 1$ **99.** $y = x^3 - 4x^2 + 5$

101. $y = 2x^{3/2} - 50$

103. **(a)** **(i)** 33.2 units, **(ii)** 33.2 units, **(iii)** 33.2 units **(b)** True

105. **(a)** $v = 10t^{3/2} - 6t^{1/2}$ **(b)** $s = 4t^{5/2} - 4t^{3/2}$

107. **(a)** $-\sqrt{x} + C$ **(b)** $x + C$ **(c)** $\sqrt{x} + C$ **(d)** $-x + C$ **(e)** $x - \sqrt{x} + C$ **(f)** $-x - \sqrt{x} + C$

Section 4.2, pp. 256–258

1. **(a)** 0.125 **(b)** 0.21875 **(c)** 0.625 **(d)** 0.46875

3. **(a)** 1.066667 **(b)** 1.283333 **(c)** 2.666667 **(d)** 2.083333

5. 0.3125, 0.328125 **7.** 1.5, 1.574603

9. **(a)** 87 in. **(b)** 87 in.

11. **(a)** 74.65 ft/sec **(b)** 45.28 ft/sec **(c)** 146.59 ft

13. $\frac{31}{16}$

15. **(a)** Upper = 758 gal, lower = 543 gal
(b) Upper = 2363 gal, lower = 1693 gal
(c) ≈ 31.4 h, ≈ 32.4 h

Section 4.3, p. 265

1. $\frac{6(1)}{1+1} + \frac{6(2)}{2+1} = 7$

3. $\cos(1)\pi + \cos(2)\pi + \cos(3)\pi + \cos(4)\pi = 0$

5. $\sin\pi - \sin\frac{\pi}{2} + \sin\frac{\pi}{3} = \frac{\sqrt{3}-2}{2}$ **7.** All of them **9.** b

11. $\sum_{k=1}^{6} k$ **13.** $\sum_{k=1}^{4} \frac{1}{2^k}$ **15.** $\sum_{k=1}^{5} (-1)^{k+1}\frac{1}{k}$

17. **(a)** −15 **(b)** 1 **(c)** 1 **(d)** −11 **(e)** 16

19. **(a)** 55 **(b)** 385 **(c)** 3025

21. −56 **23.** −73 **25.** 240 **27.** 3376

29. **(a)**

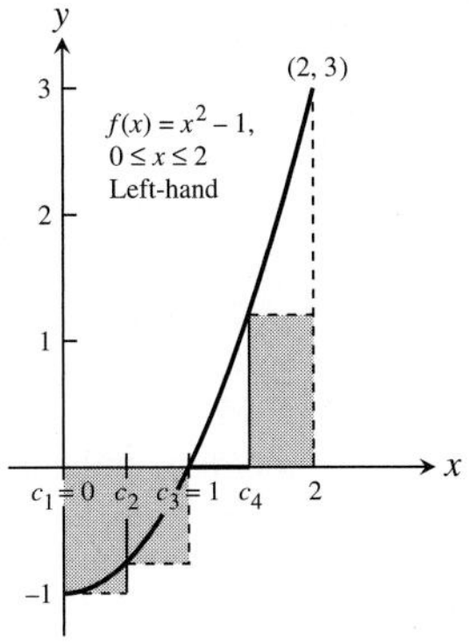

(b)

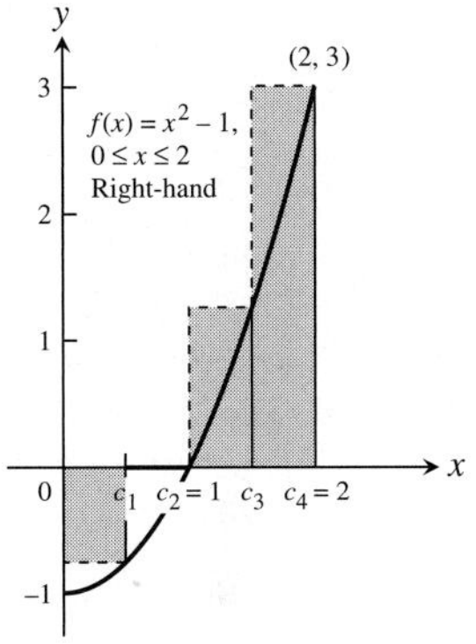

(c)

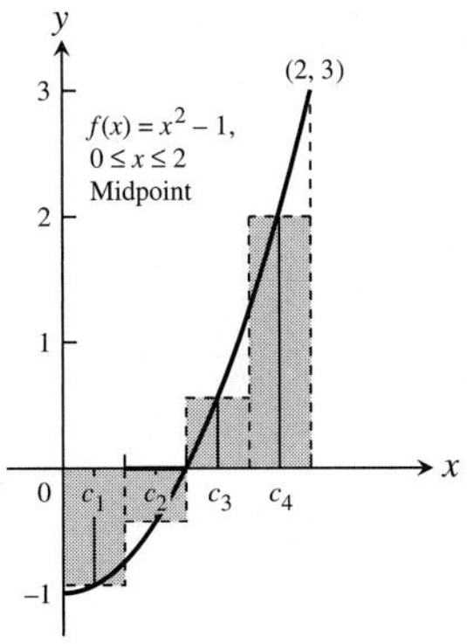

31. **(a)**

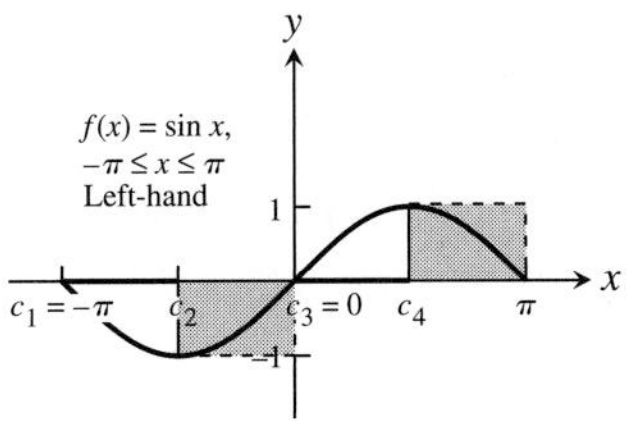

(b)

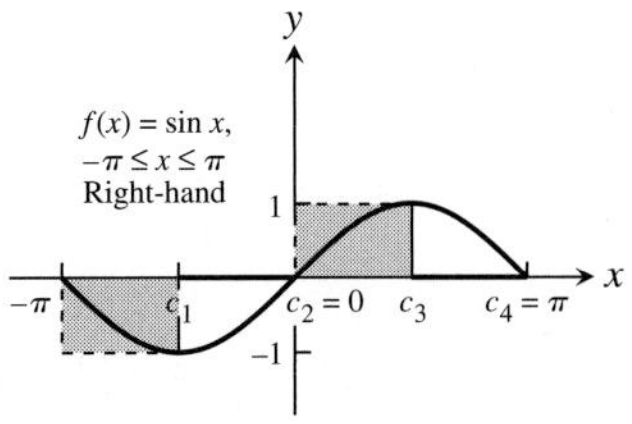

(c)

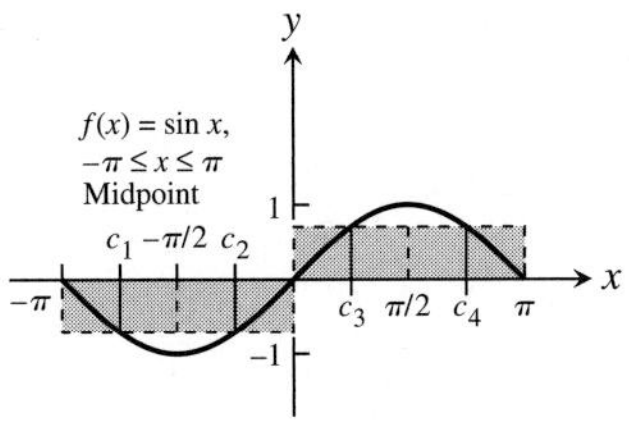

33. 1.2 **35.** $\frac{2}{3} + \frac{3n-1}{6n^2}, \frac{2}{3}$ **37.** $12 + \frac{27n+9}{2n^2}$, 12

39. $\frac{5}{6} + \frac{6n+1}{6n^2}, \frac{5}{6}$

Section 4.4, pp. 274–277

1. $\int_0^2 x^2\,dx$ **3.** $\int_{-7}^5 (x^2 - 3x)\,dx$ **5.** $\int_2^3 \frac{1}{1-x}\,dx$

7. $\int_{-\pi/4}^0 \sec x\,dx$

9. **(a)** 0 **(b)** −8 **(c)** −12 **(d)** 10 **(e)** −2 **(f)** 16

11. **(a)** 5 **(b)** $5\sqrt{3}$ **(c)** −5 **(d)** −5

13. **(a)** 4 **(b)** −4

15. Area = 21 square units

17. Area = $9\pi/2$ square units **19.** Area = 2.5 square units

21. Area = 3 square units **23.** $b^2/4$ **25.** 1/2

27. 7/3 **29.** $3a^2/2$ **31.** −14 **33.** −2 **35.** 7

37. Using n subintervals of length $\Delta x = b/n$ and right-endpoint values:

$$\text{Area} = \int_0^b 3x^2\,dx = b^3$$

39. $\text{av}(f) = 0$ **41.** $\text{av}(f) = 1$

43. **(a)** $\text{av}(g) = -1/2$ **(b)** $\text{av}(g) = 1$ **(c)** $\text{av}(g) = 1/4$

45. $a = 0$ and $b = 1$ maximize the integral.

47. Upper bound = 1, lower bound = 1/2

49. For example, $\int_0^1 \sin(x^2)\,dx \le \int_0^1 dx = 1$

51. $\int_a^b f(x)\,dx \ge \int_a^b 0\,dx = 0$

Section 4.5, pp. 284–286

1. 6 **3.** 8 **5.** 1 **7.** 5/2 **9.** 2 **11.** $2\sqrt{3}$ **13.** 0

15. $-\pi/4$ **17.** $\frac{2\pi^3}{3}$ **19.** −8/3 **21.** −3/4

23. $\sqrt{2} - \sqrt[4]{8} + 1$ **25.** 16 **27.** 7/3 **29.** π

31. $\frac{1}{\pi}(4^\pi - 2^\pi)$ **33.** $\frac{1}{2}(e-1)$ **35.** $(\cos\sqrt{x})\left(\frac{1}{2\sqrt{x}}\right)$

37. $4t^5$ **39.** $3x^2e^{-x^3}$ **41.** $\sqrt{1+x^2}$ **43.** $-\frac{1}{2}x^{-1/2}\sin x$

45. 1 **47.** $2xe^{(1/2)x^2}$ **49.** 1 **51.** 28/3 **53.** 1/2

55. 51/4 **57.** π **59.** $\frac{\sqrt{2\pi}}{2}$

61. d, since $y' = \frac{1}{x}$ and $y(\pi) = \int_\pi^\pi \frac{1}{t}\,dt - 3 = -3$

63. b, since $y' = \sec x$ and $y(0) = \int_0^0 \sec t\,dt + 4 = 4$

65. $y = \int_2^x \sec t\,dt + 3$ **67.** $s = \int_{t_0}^t f(x)\,dx + s_0$

69. $\frac{2}{3}bh$ **71.** \$9.00 **73.** $2x - 2$

Section 4.6, pp. 291–292

1. $-\frac{1}{3}\cos 3x + C$ **3.** $\frac{1}{2}\sec 2t + C$ **5.** $-(7x-2)^{-4} + C$

7. $-6(1-r^3)^{1/2} + C$ **9.** $\frac{1}{3}(x^{3/2}-1) - \frac{1}{6}\sin(2x^{3/2}-2) + C$

11. **(a)** $-\frac{1}{4}(\cot^2 2\theta) + C$ **(b)** $-\frac{1}{4}(\csc^2 2\theta) + C$

13. $-\frac{1}{3}(3-2s)^{3/2} + C$ **15.** $\frac{2}{5}(5s+4)^{1/2} + C$

17. $-\frac{2}{5}(1-\theta^2)^{5/4} + C$ **19.** $(-2/(1+\sqrt{x})) + C$

21. $\frac{1}{3}\sin(3z+4) + C$ **23.** $\ln|\sec x| + C$

25. $\left(\frac{r^3}{18} - 1\right)^6 + C$ **27.** $-\frac{2}{3}\cos(x^{3/2}+1) + C$

29. $\frac{1}{2\cos(2t+1)} + C$ **31.** $-\frac{\sin^2(1/\theta)}{2} + C$

33. $-\sin\left(\frac{1}{t} - 1\right) + C$ **35.** $\frac{(s^3+2s^2-5s+5)^2}{2} + C$

37. $\frac{2}{3}\left(1 - \frac{1}{x}\right)^{3/2} + C$ **39.** $e^{\sin x} + C$

41. $2\tan\left(e^{\sqrt{x}} + 1\right) + C$ **43.** $\ln|\ln x| + C$

45. $z - \ln(1+e^z) + C$ **47.** $\frac{5}{6}\tan^{-1}\left(\frac{2r}{3}\right) + C$

49. $e^{\sin^{-1}x} + C$ **51.** $\frac{1}{3}(\sin^{-1}x)^3 + C$ **53.** $\ln|\tan^{-1}y| + C$

55. $s = \frac{1}{2}(3t^2-1)^4 - 5$

57. $s = \sin\left(2t - \frac{\pi}{2}\right) + 100t + 1$ **59.** 6 m

Section 4.7, pp. 298–300

1. **(a)** 14/3 **(b)** 2/3 **3.** **(a)** 1/2 **(b)** −1/2

5. **(a)** 15/16 **(b)** 0 **7.** **(a)** 0 **(b)** 1/8 **9.** **(a)** 4 **(b)** 0

11. **(a)** 1/6 **(b)** 1/2 **13.** **(a)** 0 **(b)** 0 **15.** $2\sqrt{3}$

17. 3/4 **19.** $3^{5/2} - 1$ **21.** 3 **23.** $\pi/3$ **25.** e

27. ln 3 **29.** $(\ln 2)^2$ **31.** $\frac{1}{\ln 4}$ **33.** ln 2 **35.** ln 27 **37.** π

39. $\pi/12$ **41.** $2\pi/3$ **43.** $\sqrt{3} - 1$ **45.** $-\pi/12$

47. 16/3 **49.** $2^{5/2}$ **51.** 4/3 **53.** 5/6 **55.** 32/3

57. 8/3 **59.** 8 **61.** 5/3 (There are three intersection points.)

63. 18 **65.** 8/3 **67.** 2 **69.** 104/15 **71.** 4 **73.** $\pi/2$

75. 1/2 **77.** ln 16 **79.** 2 **81.** 2 ln 5 **83.** 11/3

85. $F(6) - F(2)$ **87.** **(a)** −3 **(b)** 3 **89.** $I = a/2$

Practice and Additional Exercises, pp. 301–305

1. $\frac{x^4}{4} + \frac{5}{2}x^2 - 7x + C$

3. $-\frac{1}{r+5} + C$ **5.** $(\theta^2+1)^{3/2} + C$

7. $\frac{1}{3}(1 + x^4)^{3/4} + C$ **9.** $10 \tan \frac{s}{10} + C$

11. $\frac{1}{2}x - \sin \frac{x}{2} + C$ **13.** $3 \ln x - \frac{x^2}{2} + C$

15. $\frac{\theta^{2-\pi}}{2 - \pi} + C$ **17.** $y = x - \frac{1}{x} - 1$

19. $r = 4t^{5/2} + 4t^{3/2} - 8t$

21. (a) about 680 ft **(b)**

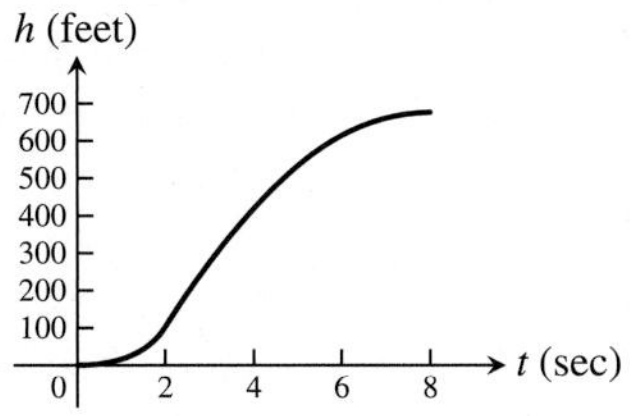

23. (a) $-1/2$ **(b)** 31 **(c)** 13 **(d)** 0

25. $\int_1^5 (2x - 1)^{-1/2}\, dx = 2$ **27.** $\int_{-\pi}^0 \cos \frac{x}{2}\, dx = 2$

29. (a) 4 **(b)** 2 **(c)** -2 **(d)** -2π **(e)** 8/5

31. 8/3 **33.** 62 **35.** 1 **37.** 18 **39.** 9/8

41. $\frac{\pi^2}{32} + \frac{\sqrt{2}}{2} - 1$ **43.** $\frac{8\sqrt{2} - 7}{6}$

45. Min: -4, max: 0, area: 27/4 **47.** 6/5 **49.** 1

51. $-4(\cos x)^{1/2} + C$ **53.** $\theta^2 + \theta + \sin(2\theta + 1) + C$

55. $\frac{t^3}{3} + \frac{4}{t} + C$ **57.** $-\frac{1}{3}\cos(2t^{3/2}) + C$

59. $\tan(e^x - 7) + C$ **61.** $e^{\tan x} + C$ **63.** $\frac{-\ln 7}{3}$

65. $\ln(9/25)$ **67.** $-\frac{1}{2}(\ln x)^{-2} + C$ **69.** $\frac{1}{2 \ln 3}\left(3^{x^2}\right) + C$

71. $\frac{3}{2}\sin^{-1} 2(r - 1) + C$

73. $\frac{1}{4}\sec^{-1}\left|\frac{2x - 1}{2}\right| + C$ **75.** $2\sqrt{\tan^{-1} y} + C$

77. 16 **79.** 2 **81.** 1 **83.** $27\sqrt{3}/160$ **85.** $\pi/2$

87. $6\sqrt{3} - 2\pi$ **89.** -1 **91.** 2 **93.** 1

95. $15/16 + \ln 2$ **97.** $e - 1$ **99.** 1/6 **101.** 9/14

103. $\frac{9 \ln 2}{4}$ **105.** π **107.** $\pi/\sqrt{3}$ **109.** $\sec^{-1}|2y| + C$

111. $\pi/12$ **113. (a)** b **(b)** b

115. (a) $\frac{d}{dx}(x \ln x - x + C) = x \cdot \frac{1}{x} + \ln x - 1 + 0 = \ln x$

(b) $\frac{1}{e - 1}$

117. $\sqrt{2 + \cos^3 x}$ **119.** $\frac{dy}{dx} = \frac{-2}{x} e^{\cos(2 \ln x)}$

121. $\frac{dy}{dx} = \frac{1}{\sqrt{1 - x^2}\sqrt{1 - 2(\sin^{-1} x)^2}}$ **123.** Yes

125. $1/e$ m/sec

129. (a) $\sinh^{-1}(\sqrt{3})$ **(b)** $\ln(\sqrt{3} + 2)$

131. (a) $\coth^{-1}(2) - \coth^{-1}(5/4)$ **(b)** $\left(\frac{1}{2}\right)\ln\left(\frac{1}{3}\right)$

133. (a) $-\text{sech}^{-1}\left(\frac{12}{13}\right) + \text{sech}^{-1}\left(\frac{4}{5}\right)$

(b) $-\ln\left(\frac{1 + \sqrt{1 - (12/13)^2}}{(12/13)}\right) + \ln\left(\frac{1 + \sqrt{1 - (4/5)^2}}{(4/5)}\right) = -\ln\left(\frac{3}{2}\right) + \ln(2) = \ln(4/3)$

135. (a) 0 **(b)** 0

137. 36/5 **139.** $\frac{1}{2} - \frac{2}{\pi}$

141. 13/3

CHAPTER 5

Section 5.1, pp. 311–313

1. $-2x\cos(x/2) + 4\sin(x/2) + C$

3. $t^2 \sin t + 2t \cos t - 2 \sin t + C$ **5.** $\ln 4 - \frac{3}{4}$

7. $y \tan^{-1}(y) - \ln\sqrt{1 + y^2} + C$

9. $x \tan x + \ln|\cos x| + C$

11. $(x^3 - 3x^2 + 6x - 6)e^x + C$

13. $(x^2 - 7x + 7)e^x + C$

15. $(x^5 - 5x^4 + 20x^3 - 60x^2 + 120x - 120)e^x + C$

17. $\frac{\pi^2 - 4}{8}$ **19.** $\frac{5\pi - 3\sqrt{3}}{9}$

21. $\frac{1}{2}(-e^\theta \cos\theta + e^\theta \sin\theta) + C$

23. $\frac{e^{2x}}{13}(3 \sin 3x + 2 \cos 3x) + C$

25. $\frac{2}{3}\left(\sqrt{3s + 9}\, e^{\sqrt{3s+9}} - e^{\sqrt{3s+9}}\right) + C$

27. $\frac{\pi\sqrt{3}}{3} - \ln(2) - \frac{\pi^2}{18}$

29. $\frac{1}{2}[-x\cos(\ln x) + x \sin(\ln x)] + C$

31. (a) π **(b)** 3π **(c)** 5π **(d)** $(2n + 1)\pi$

33. $\frac{1}{2\pi}(1 - e^{-2\pi})$ **35.** $u = x^n, dv = \cos x\, dx$

37. $u = x^n, dv = e^{ax}\, dx$ **39.** $x \sin^{-1} x + \cos(\sin^{-1} x) + C$

41. $x \sec^{-1} x - \ln\left|x + \sqrt{x^2 - 1}\right| + C$ **43.** Yes

45. (a) $x \sinh^{-1} x - \cosh(\sinh^{-1} x) + C$
(b) $x \sinh^{-1} x - (1 + x^2)^{1/2} + C$

Section 5.2, pp. 318–319

1. $-\cos x + \frac{2}{3}\cos^3 x - \frac{1}{5}\cos^5 x + C$ **3.** $\sin x - \frac{1}{3}\sin^3 x + C$

5. 16/35 **7.** 3π **9.** $2x - \frac{1}{2}\sin 4x + C$

11. $7\sin^5 x - 5\sin^7 x + C$ **13.** $-\cos^4 2\theta + C$ **15.** 4
17. 2 **19.** $2\ln\left(1+\sqrt{2}\right)$ **21.** $\sqrt{2}$
23. $2\sqrt{3} + \ln\left(2+\sqrt{3}\right)$ **25.** $\tan\theta + \frac{1}{3}\tan^3\theta + C$ **27.** 4/3
29. $2\tan^2 x - 4\ln|\sec x| + C$ **31.** $\frac{4}{3} - \ln\sqrt{3}$
33. $-\frac{1}{2}\cos x - \frac{1}{10}\cos 5x + C$ **35.** π
37. $\frac{1}{2}\sin x + \frac{1}{14}\sin 7x + C$ **39.** $2\sqrt{2}$

Section 5.3, p. 322

1. $\ln\left|\sqrt{9+y^2}+y\right| + C$ **3.** $\pi/4$ **5.** $\pi/6$
7. $\frac{25}{2}\sin^{-1}\left(\frac{t}{5}\right) + \frac{t\sqrt{25-t^2}}{2} + C$
9. $\frac{1}{2}\ln\left|\frac{2x}{7} + \frac{\sqrt{4x^2-49}}{7}\right| + C$
11. $7\left[\frac{\sqrt{y^2-49}}{7} - \sec^{-1}\left(\frac{y}{7}\right)\right] + C$ **13.** $\frac{\sqrt{x^2-1}}{x} + C$
15. $\frac{1}{3}(x^2+4)^{3/2} - 4\sqrt{x^2+4} + C$ **17.** $\frac{-2\sqrt{4-w^2}}{w} + C$
19. $4\sqrt{3} - 4\pi/3$ **21.** $-\frac{x}{\sqrt{x^2-1}} + C$
23. $-\frac{1}{5}\left(\frac{\sqrt{1-x^2}}{x}\right)^5 + C$
25. $2\tan^{-1}2x + \frac{4x}{(4x^2+1)} + C$
27. $\frac{1}{3}\left(\frac{v}{\sqrt{1-v^2}}\right)^3 + C$ **29.** $\ln 9 - \ln\left(1+\sqrt{10}\right)$
31. $\pi/6$ **33.** $\sec^{-1}|x| + C$ **35.** $\sqrt{x^2-1} + C$
37. $y = 2\left[\frac{\sqrt{x^2-4}}{2} - \sec^{-1}\left(\frac{x}{2}\right)\right]$
39. $y = \frac{3}{2}\tan^{-1}\left(\frac{x}{2}\right) - \frac{3\pi}{8}$ **41.** $3\pi/4$

Section 5.4, pp. 328–329

1. $\frac{2}{x-3} + \frac{3}{x-2}$ **3.** $\frac{1}{x+1} + \frac{3}{(x+1)^2}$
5. $\frac{-2}{z} + \frac{-1}{z^2} + \frac{2}{z-1}$ **7.** $1 + \frac{17}{t-3} + \frac{-12}{t-2}$
9. $\frac{1}{2}[\ln|1+x| - \ln|1-x|] + C$
11. $\frac{1}{7}\ln|(x+6)^2(x-1)^5| + C$ **13.** $(\ln 15)/2$
15. $-\frac{1}{2}\ln|t| + \frac{1}{6}\ln|t+2| + \frac{1}{3}\ln|t-1| + C$ **17.** $3\ln 2 - 2$
19. $\frac{1}{4}\ln\left|\frac{x+1}{x-1}\right| - \frac{x}{2(x^2-1)} + C$ **21.** $(\pi + 2\ln 2)/8$
23. $\tan^{-1}y - \frac{1}{y^2+1} + C$
25. $-(s-1)^{-2} + (s-1)^{-1} + \tan^{-1}s + C$
27. $\frac{-1}{\theta^2+2\theta+2} + \ln(\theta^2+2\theta+2) - \tan^{-1}(\theta+1) + C$
29. $x^2 + \ln\left|\frac{x-1}{x}\right| + C$
31. $9x + 2\ln|x| + \frac{1}{x} + 7\ln|x-1| + C$
33. $\frac{y^2}{2} - \ln|y| + \frac{1}{2}\ln(1+y^2) + C$ **35.** $\ln\left(\frac{e^t+1}{e^t+2}\right) + C$
37. $\frac{1}{5}\ln\left|\frac{\sin y - 2}{\sin y + 3}\right| + C$
39. $\frac{(\tan^{-1}2x)^2}{4} - 3\ln|x-2| + \frac{6}{x-2} + C$
41. $x = \ln|t-2| - \ln|t-1| + \ln 2$ **43.** $x = \frac{6t}{t+2} - 1$
45. **(a)** $x = \frac{1000e^{4t}}{499+e^{4t}}$ **(b)** 1.55 days

Section 5.5, pp. 334–335

1. $\frac{2}{\sqrt{3}}\left(\tan^{-1}\sqrt{\frac{x-3}{3}}\right) + C$
3. $\sqrt{x-2}\left(\frac{2(x-2)}{3} + 4\right) + C$
5. $\frac{(2x-3)^{3/2}(x+1)}{5} + C$
7. $\frac{-\sqrt{9-4x}}{x} - \frac{2}{3}\ln\left|\frac{\sqrt{9-4x}-3}{\sqrt{9-4x}+3}\right| + C$
9. $\frac{(x+2)(2x-6)\sqrt{4x-x^2}}{6} + 4\sin^{-1}\left(\frac{x-2}{2}\right) + C$
11. $-\frac{1}{\sqrt{7}}\ln\left|\frac{\sqrt{7}+\sqrt{7+x^2}}{x}\right| + C$
13. $\sqrt{4-x^2} - 2\ln\left|\frac{2+\sqrt{4-x^2}}{x}\right| + C$
15. $\frac{e^{2t}}{13}(2\cos 3t + 3\sin 3t) + C$
17. $\frac{x^2}{2}\cos^{-1}(x) + \frac{1}{4}\sin^{-1}(x) - \frac{1}{4}x\sqrt{1-x^2} + C$
19. $\frac{x^3}{3}\tan^{-1}x - \frac{x^2}{6} + \frac{1}{6}\ln(1+x^2) + C$
21. $-\frac{\cos 5x}{10} - \frac{\cos x}{2} + C$ **23.** $8\left[\frac{\sin(7t/2)}{7} - \frac{\sin(9t/2)}{9}\right] + C$
25. $6\sin(\theta/12) + \frac{6}{7}\sin(7\theta/12) + C$
27. $\frac{1}{2}\ln(x^2+1) + \frac{x}{2(1+x^2)} + \frac{1}{2}\tan^{-1}x + C$
29. $\left(x - \frac{1}{2}\right)\sin^{-1}\sqrt{x} + \frac{1}{2}\sqrt{x-x^2} + C$
31. $\sin^{-1}\sqrt{x} - \sqrt{x-x^2} + C$

33. $\sqrt{1-\sin^2 t} - \ln\left|\dfrac{1+\sqrt{1-\sin^2 t}}{\sin t}\right| + C$

35. $\ln\left|\ln y + \sqrt{3+(\ln y)^2}\right| + C$

37. $-\dfrac{\sin^4 2x \cos 2x}{10} - \dfrac{2\sin^2 2x \cos 2x}{15} - \dfrac{4\cos 2x}{15} + C$

39. $\dfrac{\sin^3 2\theta \cos^2 2\theta}{10} + \dfrac{\sin^3 2\theta}{15} + C$

41. $\tan^2 2x - 2\ln|\sec 2x| + C$

43. $\dfrac{(\sec \pi x)(\tan \pi x)}{\pi} + \dfrac{1}{\pi}\ln|\sec \pi x + \tan \pi x| + C$

45. $\dfrac{-\csc^3 x \cot x}{4} - \dfrac{3\csc x \cot x}{8} - \dfrac{3}{8}\ln|\csc x + \cot x| + C$

47. $\dfrac{1}{2}[\sec(e^t - 1)\tan(e^t - 1) +$
$\ln|\sec(e^t - 1) + \tan(e^t - 1)|] + C$

49. $\sqrt{2} + \ln\left(\sqrt{2}+1\right)$ **51.** $\pi/3$

53. $\pi/8$ **57.** $\pi/4$

71. **(a)**

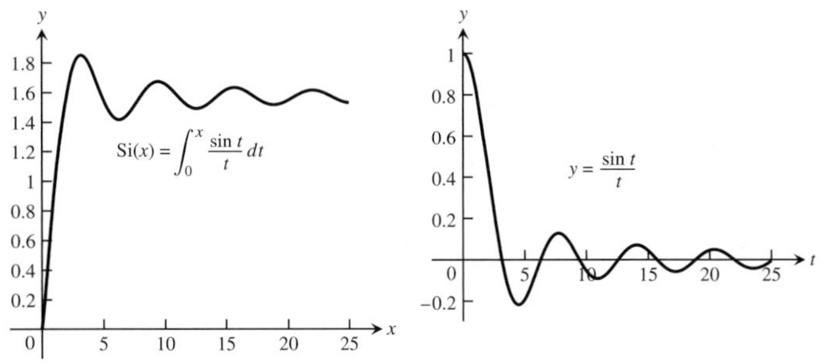

(b) $\pi/2$

73. **(a)**

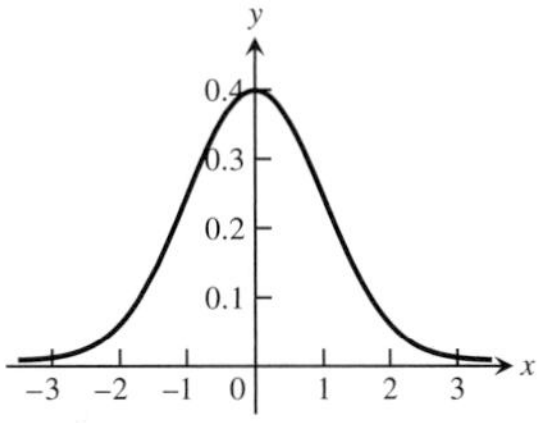

(b) ≈ 0.683, ≈ 0.954, ≈ 0.997

Section 5.6, pp. 342–343

1. **I:** **(a)** 1.5, 0 **(b)** 1.5, 0 **(c)** 0%
II: **(a)** 1.5, 0 **(b)** 1.5, 0 **(c)** 0%
3. **I:** **(a)** 2.75, 0.08 **(b)** 2.67, 0.08 **(c)** $0.0312 \approx 3\%$
II: **(a)** 2.67, 0 **(b)** 2.67, 0 **(c)** 0%
5. **I:** **(a)** 6.25, 0.5 **(b)** 6, 0.25 **(c)** $0.0417 \approx 4\%$
II: **(a)** 6, 0 **(b)** 6, 0 **(c)** 0%
7. **I:** **(a)** 0.509, 0.03125 **(b)** 0.5, 0.009 **(c)** $0.018 \approx 2\%$
II: **(a)** 0.5, 0.002604 **(b)** 0.5, 0.0004 **(c)** 0%
9. **I:** **(a)** 1.8961, 0.161 **(b)** 2, 0.1039 **(c)** $0.052 \approx 5\%$
II: **(a)** 2.0045, 0.0066 **(b)** 2, 0.00454 **(c)** 0%
11. **(a)** 1 **(b)** 2 **13.** **(a)** 116 **(b)** 2
15. **(a)** 283 **(b)** 2 **17.** **(a)** 71 **(b)** 10
19. **(a)** 76 **(b)** 12 **21.** **(a)** 82 **(b)** 8
23. 15,990 ft^3
25. **(a)** ≈ 0.00021 **(b)** ≈ 1.37079 **(c)** $\approx 0.015\%$

Section 5.7, pp. 352–353

1. $\pi/2$ **3.** 2 **5.** 6 **7.** $\pi/2$ **9.** $\ln 3$ **11.** $\ln 4$ **13.** 0
15. $\sqrt{3}$ **17.** π **19.** $\ln\left(1+\dfrac{\pi}{2}\right)$ **21.** -1 **23.** 1
25. $-1/4$ **27.** $\pi/2$ **29.** $\pi/3$ **31.** 6 **33.** $\ln 2$
35. Diverges **37.** Converges **39.** Converges **41.** Converges
43. Diverges **45.** Converges **47.** Converges **49.** Diverges
51. Converges **53.** Converges **55.** Diverges **57.** Converges
59. Diverges **61.** Converges **63.** Converges
65. **(a)** Converges when $p < 1$ **(b)** Converges when $p > 1$
67. 1

Practice and Additional Exercises, pp. 354–356

1. $(x+1)(\ln(x+1)) - (x+1) + C$

3. $x\tan^{-1}(3x) - \dfrac{1}{6}\ln(1+9x^2) + C$

5. $(x+1)^2 e^x - 2(x+1)e^x + 2e^x + C$

7. $\dfrac{2e^x \sin 2x}{5} + \dfrac{e^x \cos 2x}{5} + C$

9. $2\ln|x-2| - \ln|x-1| + C$

11. $\ln|x| - \ln|x+1| + \dfrac{1}{x+1} + C$

13. $-\dfrac{1}{3}\ln\left|\dfrac{\cos\theta - 1}{\cos\theta + 2}\right| + C$

15. $4\ln|x| - \dfrac{1}{2}\ln(x^2+1) + 4\tan^{-1}x + C$

17. $\dfrac{1}{16}\ln\left|\dfrac{(v-2)^5(v+2)}{v^6}\right| + C$

19. $\dfrac{1}{2}\tan^{-1}t - \dfrac{\sqrt{3}}{6}\tan^{-1}\dfrac{t}{\sqrt{3}} + C$

21. $\dfrac{x^2}{2} + \dfrac{4}{3}\ln|x+2| + \dfrac{2}{3}\ln|x-1| + C$

23. $\dfrac{x^2}{2} - \dfrac{9}{2}\ln|x+3| + \dfrac{3}{2}\ln|x+1| + C$

25. $\dfrac{1}{3}\ln\left|\dfrac{\sqrt{x+1}-1}{\sqrt{x+1}+1}\right| + C$ **27.** $\ln|1 - e^{-s}| + C$

29. $-\sqrt{16-y^2} + C$ **31.** $-\dfrac{1}{2}\ln|4-x^2| + C$

33. $\ln\dfrac{1}{\sqrt{9-x^2}} + C$ **35.** $\dfrac{1}{6}\ln\left|\dfrac{x+3}{x-3}\right| + C$

37. $-\dfrac{\cos^5 x}{5} + \dfrac{\cos^7 x}{7} + C$ **39.** $\dfrac{\tan^5 x}{5} + C$

41. $\dfrac{\cos\theta}{2} - \dfrac{\cos 11\theta}{22} + C$ **43.** $4\sqrt{1 - \cos(t/2)} + C$

45. $T = \pi, S = \pi$ **47.** 25°F

49. $\pi/2$ **51.** 6 **53.** ln 3 **55.** 2 **57.** $\pi/6$

59. Diverges **61.** Diverges **63.** Converges

65. $\dfrac{2x^{3/2}}{3} - x + 2\sqrt{x} - 2\ln\left(\sqrt{x} + 1\right) + C$

67. $\ln\left|\dfrac{\sqrt{x}}{\sqrt{x^2+1}}\right| - \dfrac{1}{2}\left(\dfrac{x}{\sqrt{x^2+1}}\right)^2 + C$

69. $-2\cot x - \ln|\csc x + \cot x| + \csc x + C$

71. $\dfrac{1}{12}\ln\left|\dfrac{3+v}{3-v}\right| + \dfrac{1}{6}\tan^{-1}\dfrac{v}{3} + C$

73. $\dfrac{\theta\sin(2\theta+1)}{2} + \dfrac{\cos(2\theta+1)}{4} + C$ **75.** $\dfrac{1}{4}\sec^2\theta + C$

77. $2\left(\dfrac{\left(\sqrt{2-x}\right)^3}{3} - 2\sqrt{2-x}\right) + C$ **79.** $\tan^{-1}(y-1) + C$

81. $\dfrac{1}{4}\ln|z| - \dfrac{1}{4z} - \dfrac{1}{4}\left[\dfrac{1}{2}\ln(z^2+4) + \dfrac{1}{2}\tan^{-1}\left(\dfrac{z}{2}\right)\right] + C$

83. $-\dfrac{1}{4}\sqrt{9 - 4t^2} + C$ **85.** $\ln\left(\dfrac{e^t+1}{e^t+2}\right) + C$ **87.** 1/4

89. $\dfrac{2}{3}x^{3/2} + C$ **91.** $-\dfrac{1}{5}\tan^{-1}\cos(5t) + C$

93. $2\sqrt{r} - 2\ln\left(1 + \sqrt{r}\right) + C$

CHAPTER 6

Section 6.1, pp. 365–367

1. 16 **3.** $\dfrac{16}{3}$ **5.** **(a)** $2\sqrt{3}$ **(b)** 8 **7.** 8π

9. **(a)** $\pi^2/2$ **(b)** 2π **11.** **(a)** s^2h **(b)** s^2h **13.** $\dfrac{2\pi}{3}$

15. $4 - \pi$ **17.** $\dfrac{32\pi}{5}$ **19.** 36π **21.** π **23.** $\dfrac{\pi}{2}\left(1 - \dfrac{1}{e^2}\right)$

25. $\dfrac{\pi}{2}\ln 4$ **27.** $\pi\left(\dfrac{\pi}{2} + 2\sqrt{2} - \dfrac{11}{3}\right)$ **29.** 2π **31.** 2π

33. $4\pi\ln 4$ **35.** $\pi^2 - 2\pi$ **37.** $\dfrac{2\pi}{3}$ **39.** $\dfrac{117\pi}{5}$

41. $\pi(\pi - 2)$ **43.** $\dfrac{4\pi}{3}$ **45.** 8π **47.** $\dfrac{7\pi}{6}$

49. **(a)** 8π **(b)** $\dfrac{32\pi}{5}$ **(c)** $\dfrac{8\pi}{3}$ **(d)** $\dfrac{224\pi}{15}$

51. **(a)** $\dfrac{16\pi}{15}$ **(b)** $\dfrac{56\pi}{15}$ **(c)** $\dfrac{64\pi}{15}$

Section 6.2, pp. 372–374

1. 6π **3.** 2π **5.** $\dfrac{14\pi}{3}$ **7.** 8π **9.** $\dfrac{5\pi}{6}$ **11.** $\dfrac{7\pi}{15}$

13. **(b)** 4π **15.** $\dfrac{16\pi}{15}(3\sqrt{2} + 5)$ **17.** $\dfrac{8\pi}{3}$

19. $\dfrac{4\pi}{3}$ **21.** $\dfrac{16\pi}{3}$

23. **(a)** $\dfrac{6\pi}{5}$ **(b)** $\dfrac{4\pi}{5}$ **(c)** 2π **(d)** 2π

25. **(a)** About the x-axis: $V = \dfrac{2\pi}{15}$; about the y-axis: $V = \dfrac{\pi}{6}$
(b) About the x-axis: $V = \dfrac{2\pi}{15}$; about the y-axis: $V = \dfrac{\pi}{6}$

27. **(a)** $\dfrac{5\pi}{3}$ **(b)** $\dfrac{4\pi}{3}$ **(c)** 2π **(d)** $\dfrac{2\pi}{3}$

29. **(a)** $\dfrac{4\pi}{15}$ **(b)** $\dfrac{7\pi}{30}$

31. **(a)** $\dfrac{24\pi}{5}$ **(b)** $\dfrac{48\pi}{5}$

33. **(a)** $\pi(\pi - 2)$ **(b)** 2π

35. Disk: 2 integrals; washer: 2 integrals; shell: 1 integral

39. $\pi\left(1 - \dfrac{1}{e}\right)$

Section 6.3, pp. 379–381

1. $\dfrac{5\sqrt{10}}{3}$ **3.** 7 **5.** $\dfrac{21}{2}$ **7.** $e^3 + 2$ **9.** 12 **11.** $\dfrac{53}{6}$

13. $\dfrac{123}{32}$ **15.** $\dfrac{99}{8}$ **17.** $\pi/3$

19. **(a)** $\displaystyle\int_{-1}^{2}\sqrt{1 + 4x^2}\,dx$ **(c)** ≈ 6.13

21. **(a)** $\displaystyle\int_{0}^{\pi}\sqrt{1 + \cos^2 y}\,dy$ **(c)** ≈ 3.82

23. **(a)** $\displaystyle\int_{-1}^{3}\sqrt{1 + (y+1)^2}\,dy$ **(c)** ≈ 9.29

25. **(a)** $\displaystyle\int_{0}^{\pi/6}\sec x\,dx$ **(c)** ≈ 0.55

27. $\ln(\sqrt{2} + 1)$

29. **(a)** $y = \sqrt{x}$ from (1, 1) to (4, 2)
(b) Only one. We know the derivative of the function and the value of the function at one value of x.

31. $y = e^{x/2} - 1$ **33.** $-\ln(2)$ **35.** $e^3 - \dfrac{1}{e^3}$

Section 6.4, pp. 388–390

9. $\dfrac{2}{3}y^{3/2} - x^{1/2} = C$ **11.** $e^y - e^x = C$

13. $-x + 2\tan\sqrt{y} = C$ **15.** $e^{-y} + 2e^{\sqrt{x}} = C$

17. $y = \sin(x^2 + C)$

19. 54.88 g **21.** 59.8 ft **23.** 2.8147498×10^{14}

25. **(a)** 8 years **(b)** 32.02 years
27. 15.28 years **29.** 56,562 years
33. **(a)** 17.5 min. **(b)** 13.26 min.
35. $-3°C$ **37.** About 6658 years **39.** 41 years old

Section 6.5, pp. 395–399

1. 400 N/m **3.** 4 cm, 0.08 J
5. **(a)** 7238 lb/in. **(b)** 905 in.-lb, 2714 in.-lb
7. 780 J **9.** 72,900 ft-lb **11.** 160 ft-lb
13. **(a)** 1,497,600 ft-lb **(b)** 1 hr, 40 min
(d) At 62.26 lb/ft^3: a) 1,494,240 ft-lb b) 1 hr, 40 min
At 62.59 lb/ft^3: a) 1,502,160 ft-lb b) 1 hr, 40.1 min
15. 37,306 ft-lb **17.** 7,238,229.47 ft-lb
19. 15,073,099.75 J
23. 85.1 ft-lb **25.** 64.6 ft-lb
27. 91.32 in.-oz
29. 1684.8 lb **31.** **(a)** 1164.8 lb **(b)** 1194.7 lb
33. 1309 lb **35.** **(a)** 93.33 lb **(b)** 3 ft
37. $wb/2$
39. No. The tank will overflow because the movable end will have moved only $3\frac{1}{3}$ ft by the time the tank is full.

Section 6.6, pp. 405–406

1. $\bar{x} = 0, \bar{y} = 12/5$ **3.** $\bar{x} = 1, \bar{y} = -3/5$
5. $\bar{x} = 16/105, \bar{y} = 8/15$ **7.** $\bar{x} = 0, \bar{y} = \pi/8$
9. **(a)** $\bar{x} \approx 1.44, \bar{y} \approx 0.36$
(b)

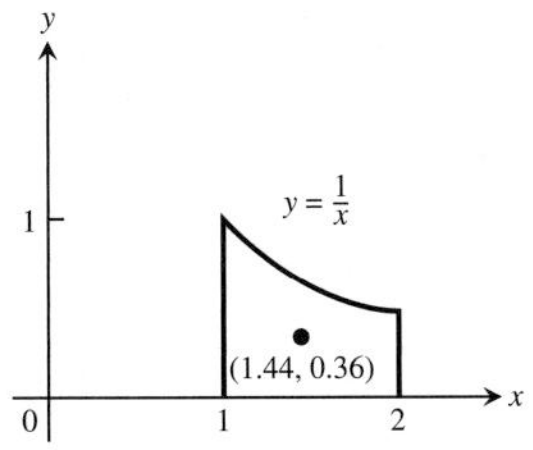

11. $\bar{x} = \dfrac{\ln 4}{\pi}, \bar{y} = 0$ **13.** $\bar{x} = 7, \bar{y} = \dfrac{\ln 16}{12}$

15. $\bar{x} = 3/2, \bar{y} = 1/2$ **17.** $\bar{x} = \dfrac{15}{\ln 16}, \bar{y} = \dfrac{3}{4 \ln 16}$

19. **(a)** $\dfrac{224\pi}{3}$ **(b)** $\bar{x} = 2, \bar{y} = 0$
(c)

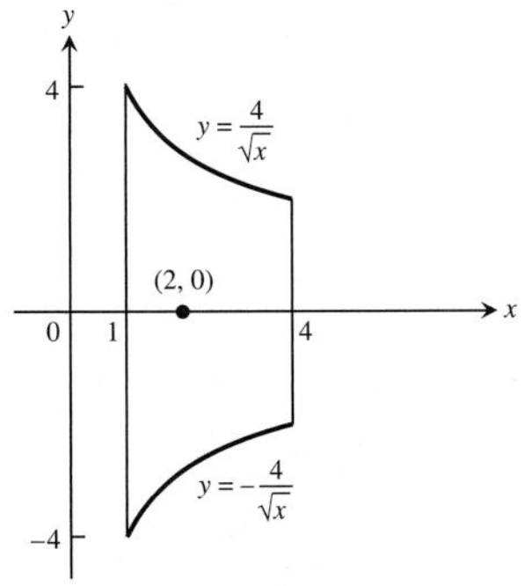

23. $\bar{x} = \bar{y} = 1/3$ **25.** $\bar{x} = a/3, \bar{y} = b/3$ **27.** $13\delta/6$

29. $\bar{x} = 0, \bar{y} = \dfrac{a\pi}{4}$

Practice and Additional Exercises, pp. 407–409

1. $\dfrac{9\pi}{280}$ **3.** π^2 **5.** $\dfrac{72\pi}{35}$
7. **(a)** 2π **(b)** π **(c)** $12\pi/5$ **(d)** $26\pi/5$
9. **(a)** 8π **(b)** $1088\pi/15$ **(c)** $512\pi/15$

11. $\pi(3\sqrt{3} - \pi)/3$ **13.** $\pi(e - 1)$ **15.** $\dfrac{28\pi}{3}$ ft^3 **17.** $\dfrac{10}{3}$

19. $3 + \dfrac{1}{8}\ln 2$ **21.** 10 **23.** $\dfrac{9\pi}{2}$

25. $28\pi\sqrt{2}/3$ **27.** 4π **29.** $76\pi/3$

31. $y = \left(\tan^{-1}\left(\dfrac{x + C}{2}\right)\right)^2$ **33.** $y^2 = \sin^{-1}(2\tan x + C)$

35. $y = -2 + \ln(2 - e^{-x})$ **37.** $4x - 4\sqrt{x} + 1$
39. 18,935 years **41.** 4640 J
43. 10 ft-lb, 30 ft-lb **45.** 418,208.81 ft-lb
47. $22{,}500\pi$ ft-lb, 257 sec **49.** $\bar{x} = 0, \bar{y} = 8/5$

51. $\bar{x} = 3/2, \bar{y} = 12/5$ **53.** $\bar{x} = 9/5, \bar{y} = 11/10$

55. 332.8 lb

CHAPTER 7

Section 7.1, pp. 419–422

1. $a_1 = 0, a_2 = -1/4, a_3 = -2/9, a_4 = -3/16$
3. $a_1 = 1, a_2 = -1/3, a_3 = 1/5, a_4 = -1/7$
5. $a_1 = 1/2, a_2 = 1/2, a_3 = 1/2, a_4 = 1/2$

7. $1, \dfrac{3}{2}, \dfrac{7}{4}, \dfrac{15}{8}, \dfrac{31}{16}, \dfrac{63}{32}, \dfrac{127}{64}, \dfrac{255}{128}, \dfrac{511}{256}, \dfrac{1023}{512}$

9. $2, 1, -\dfrac{1}{2}, -\dfrac{1}{4}, \dfrac{1}{8}, \dfrac{1}{16}, -\dfrac{1}{32}, -\dfrac{1}{64}, \dfrac{1}{128}, \dfrac{1}{256}$

11. 1, 1, 2, 3, 5, 8, 13, 21, 34, 55 **13.** $a_n = (-1)^{n+1}, n \ge 1$
15. $a_n = (-1)^{n+1}(n)^2, n \ge 1$ **17.** $a_n = n^2 - 1, n \ge 1$

19. $a_n = 4n - 3, n \ge 1$ **21.** $a_n = \dfrac{1 + (-1)^{n+1}}{2}, n \ge 1$

23. Converges, 2 **25.** Converges, -1 **27.** Converges, -5
29. Diverges **31.** Diverges **33.** Converges, 1/2
35. Converges, 0 **37.** Converges, $\sqrt{2}$ **39.** Converges, 1
41. Converges, 0 **43.** Converges, 0 **45.** Converges, 0
47. Converges, 1 **49.** Converges, e^7 **51.** Converges, 1
53. Converges, 1 **55.** Diverges **57.** Converges, 4
59. Converges, 0 **61.** Diverges **63.** Converges, e^{-1}
65. Converges, $e^{2/3}$ **67.** Converges, x $(x > 0)$
69. Converges, 0 **71.** Converges, 1 **73.** Converges, 1/2
75. Converges, $\pi/2$ **77.** Converges, 0 **79.** Converges, 0
81. Converges, 1/2 **83.** Converges, 0

85. **(a)** $f(x) = x^2 - 2$, $1.414213562 \approx \sqrt{2}$
(b) $f(x) = \tan(x) - 1$, $0.7853981635 \approx \pi/4$
(c) $f(x) = e^x$, diverges
95. Nondecreasing, bounded
97. Not nondecreasing, bounded
99. Converges, nondecreasing sequence theorem
101. Converges, nondecreasing sequence theorem
103. Diverges, definition of divergence **107.** Converges
109. Converges
119. **(b)** $\sqrt{3}$

Section 7.2, pp. 429–430

1. $s_n = \frac{2(1-(1/3)^n)}{1-(1/3)}, 3$ **3.** $s_n = \frac{1-(-1/2)^n}{1-(-1/2)}, 2/3$

5. $s_n = \frac{1}{2} - \frac{1}{n+2}, \frac{1}{2}$ **7.** $1 - \frac{1}{4} + \frac{1}{16} - \frac{1}{64} + \cdots, \frac{4}{5}$

9. $\frac{7}{4} + \frac{7}{16} + \frac{7}{64} + \cdots, \frac{7}{3}$

11. $(5+1) + \left(\frac{5}{2} + \frac{1}{3}\right) + \left(\frac{5}{4} + \frac{1}{9}\right) + \left(\frac{5}{8} + \frac{1}{27}\right) + \cdots, \frac{23}{2}$

13. $(1+1) + \left(\frac{1}{2} - \frac{1}{5}\right) + \left(\frac{1}{4} + \frac{1}{25}\right) + \left(\frac{1}{8} - \frac{1}{125}\right) + \cdots, \frac{17}{6}$

15. 1 **17.** 5 **19.** 1 **21.** $-\frac{1}{\ln 2}$ **23.** Converges, $2 + \sqrt{2}$

25. Converges, 1 **27.** Diverges **29.** Converges, $\frac{e^2}{e^2 - 1}$

31. Converges, 2/9 **33.** Converges, 3/2 **35.** Diverges

37. Diverges **39.** Converges, $\frac{\pi}{\pi - e}$

41. $a = 1, r = -x$; converges to $1/(1+x)$ for $|x| < 1$
43. $a = 3, r = (x-1)/2$; converges to $6/(3-x)$ for x in $(-1, 3)$

45. $|x| < \frac{1}{2}, \frac{1}{1-2x}$ **47.** $-2 < x < 0, \frac{1}{2+x}$

49. $x \neq (2k+1)\frac{\pi}{2}$, k an integer; $\frac{1}{1 - \sin x}$

51. 23/99 **53.** 7/9 **55.** 1/15 **57.** 41333/33300

65. $|r| < 1, \frac{1+2r}{1-r^2}$

Section 7.3, pp. 435–436

1. Converges; geometric series, $r = \frac{1}{10} < 1$

3. Diverges; $\lim_{n\to\infty} \frac{n}{n+1} = 1 \neq 0$ **5.** Diverges; p-series, $p < 1$

7. Converges; geometric series, $r = \frac{1}{8} < 1$

9. Diverges; Integral Test
11. Converges; geometric series, $r = 2/3 < 1$

13. Diverges; Integral Test **15.** Diverges; $\lim_{n\to\infty} \frac{2^n}{n+1} \neq 0$

17. Diverges; $\lim_{n\to\infty}\left(\sqrt{n}/\ln n\right) \neq 0$

19. Diverges; geometric series, $r = \frac{1}{\ln 2} > 1$

21. Converges; Integral Test **23.** Diverges; nth-Term Test
25. Converges; Integral Test **27.** Converges; Integral Test
29. Converges; Integral Test **31.** $a = 1$
39. **(a)** $1.20166 \leq S \leq 1.20253$ **(b)** $S \approx 1.2021$, error < 0.0005

Section 7.4, p. 439

1. Diverges; limit comparison with $\sum\left(1/\sqrt{n}\right)$
3. Converges; compare with $\sum\left(1/2^n\right)$
5. Diverges; nth-Term Test
7. Converges; $\left(\frac{n}{3n+1}\right)^n < \left(\frac{n}{3n}\right)^n = \left(\frac{1}{3}\right)^n$
9. Diverges; direct comparison with $\sum(1/n)$
11. Converges; limit comparison with $\sum(1/n^2)$
13. Diverges; limit comparison with $\sum(1/n)$
15. Diverges; limit comparison with $\sum(1/n)$
17. Diverges; Integral Test
19. Converges; compare with $\sum(1/n^{3/2})$
21. Converges; $\frac{1}{n2^n} \leq \frac{1}{2^n}$ **23.** Converges; $\frac{1}{3^{n-1}+1} < \frac{1}{3^{n-1}}$
25. Diverges; limit comparison with $\sum(1/n)$
27. Converges; compare with $\sum(1/n^2)$
29. Converges; $\frac{\tan^{-1} n}{n^{1.1}} < \frac{\pi/2}{n^{1.1}}$
31. Converges; compare with $\sum(1/n^2)$
33. Diverges; limit comparison with $\sum(1/n)$
35. Converges; limit comparison with $\sum(1/n^2)$

Section 7.5, p. 442

1. Converges; Ratio Test **3.** Diverges; Ratio Test
5. Converges; Ratio Test
7. Converges; compare with $\sum(3/(1.25)^n)$
9. Diverges; $\lim_{n\to\infty}\left(1 - \frac{3}{n}\right)^n = e^{-3} \neq 0$
11. Converges; compare with $\sum(1/n^2)$
13. Diverges; compare with $\sum(1/(2n))$
15. Diverges; compare with $\sum(1/n)$ **17.** Converges; Ratio Test
19. Converges; Ratio Test **21.** Converges; Ratio Test
23. Converges; Root Test
25. Converges; compare with $\sum(1/n^2)$
27. Converges; Ratio Test **29.** Diverges; Ratio Test
31. Converges; Ratio Test **33.** Converges; Ratio Test
35. Diverges; $a_n = \left(\frac{1}{3}\right)^{(1/n!)} \to 1$ **37.** Converges; Ratio Test
39. Diverges; Root Test **41.** Converges; Root Test
43. Converges; Ratio Test

Section 7.6, pp. 447–448

1. Converges by Theorem 16 **3.** Diverges; $a_n \not\to 0$
5. Converges by Theorem 16 **7.** Diverges; $a_n \to 1/2$

9. Converges by Theorem 16

11. Converges absolutely. Series of absolute values is a convergent geometric series.

13. Converges conditionally. $1/\sqrt{n} \to 0$ but $\sum_{n=1}^{\infty} \frac{1}{\sqrt{n}}$ diverges.

15. Converges absolutely. Compare with $\sum_{n=1}^{\infty}(1/n^2)$.

17. Converges conditionally. $1/(n+3) \to 0$ but $\sum_{n=1}^{\infty} \frac{1}{n+3}$ diverges (compare with $\sum_{n=1}^{\infty}(1/n)$).

19. Diverges; $\frac{3+n}{5+n} \to 1$

21. Converges conditionally; $\left(\frac{1}{n^2} + \frac{1}{n}\right) \to 0$ but $(1+n)/n^2 > 1/n$

23. Converges absolutely; Ratio Test

25. Converges absolutely by Integral Test **27.** Diverges; $a_n \not\to 0$

29. Converges absolutely by the Ratio Test

31. Converges absolutely; $\frac{1}{n^2+2n+1} < \frac{1}{n^2}$

33. Converges absolutely since $\left|\frac{\cos n\pi}{n\sqrt{n}}\right| = \left|\frac{(-1)^{n+1}}{n^{3/2}}\right| = \frac{1}{n^{3/2}}$ (convergent p-series)

35. Converges absolutely by Root Test **37.** Diverges; $a_n \to \infty$

39. Converges conditionally; $\sqrt{n+1} - \sqrt{n} = 1/\left(\sqrt{n} + \sqrt{n+1}\right) \to 0$, but series of absolute values diverges (compare with $\sum\left(1/\sqrt{n}\right)$)

41. Diverges, $a_n \to 1/2 \neq 0$

43. Converges absolutely; $\operatorname{sech} n = \frac{2}{e^n + e^{-n}} = \frac{2e^n}{e^{2n}+1} < \frac{2e^n}{e^{2n}} = \frac{2}{e^n}$, a term from a convergent geometric series.

45. $|\text{Error}| < 0.2$ **47.** $|\text{Error}| < 2 \times 10^{-11}$ **49.** 0.54030

51. **(a)** $a_n \geq a_{n+1}$ **(b)** $-1/2$

Section 7.7, pp. 455–457

1. **(a)** $1, -1 < x < 1$ **(b)** $-1 < x < 1$ **(c)** none

3. **(a)** $1/4, -1/2 < x < 0$ **(b)** $-1/2 < x < 0$ **(c)** none

5. **(a)** $10, -8 < x < 12$ **(b)** $-8 < x < 12$ **(c)** none

7. **(a)** $1, -1 < x < 1$ **(b)** $-1 < x < 1$ **(c)** none

9. **(a)** $3, -3 \leq x \leq 3$ **(b)** $-3 \leq x \leq 3$ **(c)** none

11. **(a)** ∞, for all x **(b)** for all x **(c)** none

13. **(a)** ∞, for all x **(b)** for all x **(c)** none

15. **(a)** $1, -1 \leq x < 1$ **(b)** $-1 < x < 1$ **(c)** $x = -1$

17. **(a)** $5, -8 < x < 2$ **(b)** $-8 < x < 2$ **(c)** none

19. **(a)** $3, -3 < x < 3$ **(b)** $-3 < x < 3$ **(c)** none

21. **(a)** $1, -1 < x < 1$ **(b)** $-1 < x < 1$ **(c)** none

23. **(a)** $0, x = 0$ **(b)** $x = 0$ **(c)** none

25. **(a)** $2, -4 < x \leq 0$ **(b)** $-4 < x < 0$ **(c)** $x = 0$

27. **(a)** $1, -1 \leq x \leq 1$ **(b)** $-1 \leq x \leq 1$ **(c)** none

29. **(a)** $1/4, 1 \leq x \leq 3/2$ **(b)** $1 \leq x \leq 3/2$ **(c)** none

31. **(a)** $1, (-1-\pi) \leq x < (1-\pi)$ **(b)** $(-1-\pi) < x < (1-\pi)$ **(c)** $x = -1-\pi$

33. $-1 < x < 3,\ 4/(3+2x-x^2)$

35. $0 < x < 16,\ 2/\left(4-\sqrt{x}\right)$

37. $-\sqrt{2} < x < \sqrt{2},\ 3/(2-x^2)$

39. $1 < x < 5,\ 2/(x-1),\ 1 < x < 5,\ -2/(x-1)^2$

41. **(a)** $\cos x = 1 - \frac{x^2}{2!} + \frac{x^4}{4!} - \frac{x^6}{6!} + \frac{x^8}{8!} - \frac{x^{10}}{10!} + \cdots$; converges for all x
(b) Same answer as part (c)
(c) $2x - \frac{2^3x^3}{3!} + \frac{2^5x^5}{5!} - \frac{2^7x^7}{7!} + \frac{2^9x^9}{9!} - \frac{2^{11}x^{11}}{11!} + \cdots$

43. **(a)** $\frac{x^2}{2} + \frac{x^4}{12} + \frac{x^6}{45} + \frac{17x^8}{2520} + \frac{31x^{10}}{14175},\ -\frac{\pi}{2} < x < \frac{\pi}{2}$
(b) $1 + x^2 + \frac{2x^4}{3} + \frac{17x^6}{45} + \frac{62x^8}{315} + \cdots,\ -\frac{\pi}{2} < x < \frac{\pi}{2}$

Section 7.8, p. 462

1. $P_0(x) = 0,\ P_1(x) = x-1,\ P_2(x) = (x-1) - \frac{1}{2}(x-1)^2$,
$P_3(x) = (x-1) - \frac{1}{2}(x-1)^2 + \frac{1}{3}(x-1)^3$

3. $P_0(x) = \frac{1}{2},\ P_1(x) = \frac{1}{2} - \frac{1}{4}(x-2)$,
$P_2(x) = \frac{1}{2} - \frac{1}{4}(x-2) + \frac{1}{8}(x-2)^2$,
$P_3(x) = \frac{1}{2} - \frac{1}{4}(x-2) + \frac{1}{8}(x-2)^2 - \frac{1}{16}(x-2)^3$

5. $P_0(x) = \frac{\sqrt{2}}{2},\ P_1(x) = \frac{\sqrt{2}}{2} + \frac{\sqrt{2}}{2}\left(x - \frac{\pi}{4}\right)$,
$P_2(x) = \frac{\sqrt{2}}{2} + \frac{\sqrt{2}}{2}\left(x - \frac{\pi}{4}\right) - \frac{\sqrt{2}}{4}\left(x - \frac{\pi}{4}\right)^2$,
$P_3(x) = \frac{\sqrt{2}}{2} + \frac{\sqrt{2}}{2}\left(x - \frac{\pi}{4}\right) - \frac{\sqrt{2}}{4}\left(x - \frac{\pi}{4}\right)^2 - \frac{\sqrt{2}}{12}\left(x - \frac{\pi}{4}\right)^3$

7. $P_0(x) = 2,\ P_1(x) = 2 + \frac{1}{4}(x-4)$,
$P_2(x) = 2 + \frac{1}{4}(x-4) - \frac{1}{64}(x-4)^2$,
$P_3(x) = 2 + \frac{1}{4}(x-4) - \frac{1}{64}(x-4)^2 + \frac{1}{512}(x-4)^3$

9. $\sum_{n=0}^{\infty} \frac{(-x)^n}{n!} = 1 - x + \frac{x^2}{2!} - \frac{x^3}{3!} + \frac{x^4}{4!} - \cdots$

11. $\sum_{n=0}^{\infty}(-1)^n x^n = 1 - x + x^2 - x^3 + \cdots$

13. $\sum_{n=0}^{\infty} \frac{(-1)^n 3^{2n+1} x^{2n+1}}{(2n+1)!}$ **15.** $7\sum_{n=0}^{\infty} \frac{(-1)^n x^{2n}}{(2n)!}$ **17.** $\sum_{n=0}^{\infty} \frac{x^{2n}}{(2n)!}$

19. $x^4 - 2x^3 - 5x + 4$

21. $8 + 10(x-2) + 6(x-2)^2 + (x-2)^3$

23. $21 - 36(x+2) + 25(x+2)^2 - 8(x+2)^3 + (x+2)^4$

25. $\sum_{n=0}^{\infty}(-1)^n(n+1)(x-1)^n$ **27.** $\sum_{n=0}^{\infty}\frac{e^2}{n!}(x-2)^n$

33. $L(x) = 0, Q(x) = -x^2/2$ **35.** $L(x) = 1, Q(x) = 1 + x^2/2$

37. $L(x) = x, Q(x) = x$

Section 7.9, pp. 468–469

1. $\sum_{n=0}^{\infty}\frac{(-5x)^n}{n!} = 1 - 5x + \frac{5^2x^2}{2!} - \frac{5^3x^3}{3!} + \cdots$

3. $\sum_{n=0}^{\infty}\frac{5(-1)^n(-x)^{2n+1}}{(2n+1)!} = \sum_{n=0}^{\infty}\frac{5(-1)^{n+1}x^{2n+1}}{(2n+1)!}$
$= -5x + \frac{5x^3}{3!} - \frac{5x^5}{5!} + \frac{5x^7}{7!} + \cdots$

5. $\sum_{n=0}^{\infty}\frac{(-1)^n(x+1)^n}{(2n)!}$

7. $\sum_{n=0}^{\infty}\frac{x^{n+1}}{n!} = x + x^2 + \frac{x^3}{2!} + \frac{x^4}{3!} + \frac{x^5}{4!} + \cdots$

9. $\sum_{n=2}^{\infty}\frac{(-1)^n x^{2n}}{(2n)!} = \frac{x^4}{4!} - \frac{x^6}{6!} + \frac{x^8}{8!} - \frac{x^{10}}{10!} + \cdots$

11. $x - \frac{\pi^2x^3}{2!} + \frac{\pi^4x^5}{4!} - \frac{\pi^6x^7}{6!} + \cdots = \sum_{n=0}^{\infty}\frac{(-1)^n\pi^{2n}x^{2n+1}}{(2n)!}$

13. $1 + \sum_{n=1}^{\infty}\frac{(-1)^n(2x)^{2n}}{2\cdot(2n)!} =$
$1 - \frac{(2x)^2}{2\cdot 2!} + \frac{(2x)^4}{2\cdot 4!} - \frac{(2x)^6}{2\cdot 6!} + \frac{(2x)^8}{2\cdot 8!} - \cdots$

15. $x^2\sum_{n=0}^{\infty}(2x)^n = x^2 + 2x^3 + 4x^4 + \cdots$

17. $\sum_{n=1}^{\infty}nx^{n-1} = 1 + 2x + 3x^2 + 4x^3 + \cdots$

19. $|x| < (0.06)^{1/5} < 0.56968$

21. $|\text{Error}| < (10^{-3})^3/6 < 1.67 \times 10^{-10}$, $-10^{-3} < x < 0$

23. $|\text{Error}| < (3^{0.1})(0.1)^3/6 < 1.87 \times 10^{-4}$

25. $\frac{x^3}{3} - \frac{x^7}{7\cdot 3!} + \frac{x^{11}}{11\cdot 5!}$

27. **(a)** $\frac{x^2}{2} - \frac{x^4}{12}$

(b) $\frac{x^2}{2} - \frac{x^4}{3\cdot 4} + \frac{x^6}{5\cdot 6} - \frac{x^8}{7\cdot 8} + \cdots + (-1)^{15}\frac{x^{32}}{31\cdot 32}$

29. $e^x \sin x = x + x^2 + \frac{x^3}{3} - \frac{x^5}{30} - \frac{x^6}{90}\cdots$

35. **(a)** $Q(x) = 1 + kx + \frac{k(k-1)}{2}x^2$ **(b)** for $0 \le x < 100^{-1/3}$

39. **(a)** -1 **(b)** $(1/\sqrt{2})(1+i)$ **(c)** $-i$

Section 7.10, pp. 471–472

1. $1 + \frac{x}{2} - \frac{x^2}{8} + \frac{x^3}{16}$ **3.** $1 + \frac{1}{2}x + \frac{3}{8}x^2 + \frac{5}{16}x^3 + \cdots$

5. $1 - x + \frac{3x^2}{4} - \frac{x^3}{2}$ **7.** $1 - \frac{x^3}{2} + \frac{3x^6}{8} - \frac{5x^9}{16}$

9. $1 + \frac{1}{2x} - \frac{1}{8x^2} + \frac{1}{16x^3}$

11. $(1+x)^4 = 1 + 4x + 6x^2 + 4x^3 + x^4$

13. $(1-2x)^3 = 1 - 6x + 12x^2 - 8x^3$

15. **(a)** $x + \frac{x^3}{6} + \frac{3x^5}{40} + \frac{5x^7}{112}$, radius of convergence $= 1$

(b) $\frac{\pi}{2} - x - \frac{x^3}{6} - \frac{3x^5}{40} - \frac{5x^7}{112}$

17. $1 - 2x + 3x^2 - 4x^3 + \cdots$

Practice and Additional Exercises, pp. 473–474

1. Converges to 1 **3.** Converges to -1 **5.** Diverges
7. Converges to 0 **9.** Converges to 1 **11.** Converges to e^{-5}
13. Converges to 3 **15.** Converges to ln 2 **17.** Diverges
19. $1/6$ **21.** $3/2$ **23.** $e/(e-1)$ **25.** Diverges
27. Converges conditionally **29.** Converges conditionally
31. Converges absolutely **33.** Converges absolutely
35. Converges absolutely **37.** Converges absolutely
39. Converges absolutely
41. **(a)** $3, -7 \le x < -1$ **(b)** $-7 < x < -1$ **(c)** $x = -7$
43. **(a)** $1/3, 0 \le x \le 2/3$ **(b)** $0 \le x \le 2/3$ **(c)** none
45. **(a)** ∞, for all x **(b)** for all x **(c)** none
47. **(a)** $\sqrt{3}, -\sqrt{3} < x < \sqrt{3}$ **(b)** $-\sqrt{3} < x < \sqrt{3}$
(c) none
49. **(a)** $e, -e < x < e$ **(b)** $-e < x < e$ **(c)** empty set
51. $\frac{1}{1+x}, \frac{1}{4}, \frac{4}{5}$ **53.** $\sin x, \pi, 0$ **55.** $e^x, \ln 2, 2$ **57.** $\sum_{n=0}^{\infty}2^nx^n$
59. $\sum_{n=0}^{\infty}\frac{(-1)^n\pi^{2n+1}x^{2n+1}}{(2n+1)!}$ **61.** $\sum_{n=0}^{\infty}\frac{(-1)^nx^{5n}}{(2n)!}$ **63.** $\sum_{n=0}^{\infty}\frac{((\pi x)/2)^n}{n!}$
65. $2 - \frac{(x+1)}{2\cdot 1!} + \frac{3(x+1)^2}{2^3\cdot 2!} + \frac{9(x+1)^3}{2^5\cdot 3!} + \cdots$
67. $\frac{1}{4} - \frac{1}{4^2}(x-3) + \frac{1}{4^3}(x-3)^2 - \frac{1}{4^4}(x-3)^3$
69. 0.4849171431 **71.** $r = -3, s = 9/2$ **73.** $2/3$
75. **(a)** ∞ **(b)** $a = 1$, $b = 0$

CHAPTER 8

Section 8.1, pp. 479–480

1. a, e; b, g; c, h; d, f **3.**

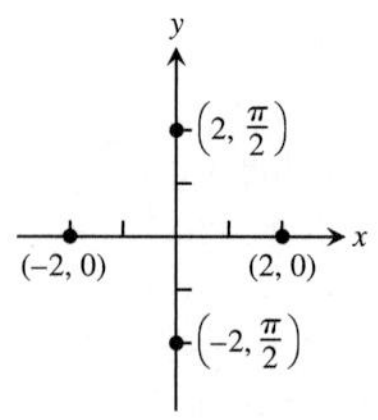

(a) $\left(2, \frac{\pi}{2} + 2n\pi\right)$ and $\left(-2, \frac{\pi}{2} + (2n+1)\pi\right)$, n an integer

(b) $(2, 2n\pi)$ and $(-2, (2n+1)\pi)$, n an integer

(c) $\left(2, \frac{3\pi}{2} + 2n\pi\right)$ and $\left(-2, \frac{3\pi}{2} + (2n+1)\pi\right)$, n an integer

(d) $(2, (2n+1)\pi)$ and $(-2, 2n\pi)$, n an integer

5. **(a)** $(3, 0)$ **(b)** $(-3, 0)$ **(c)** $\left(-1, \sqrt{3}\right)$ **(d)** $\left(1, \sqrt{3}\right)$
(e) $(3, 0)$ **(f)** $\left(1, \sqrt{3}\right)$ **(g)** $(-3, 0)$ **(h)** $\left(-1, \sqrt{3}\right)$

7.

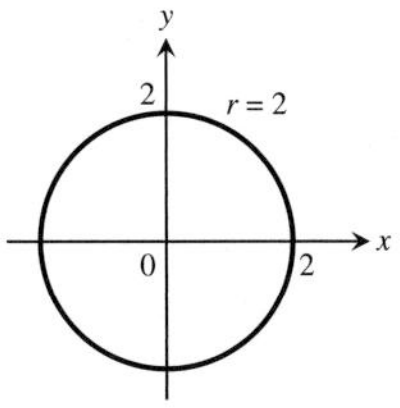

9.

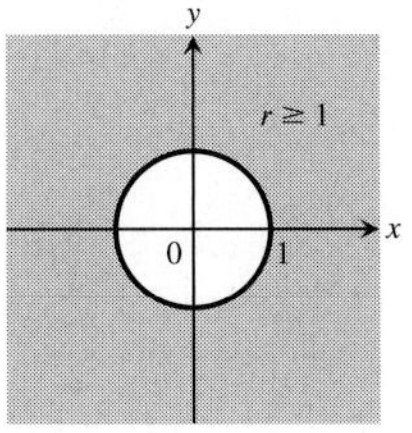

11.

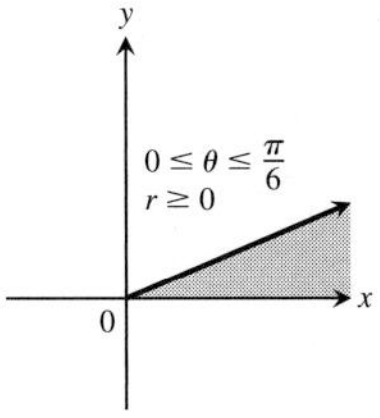

13.

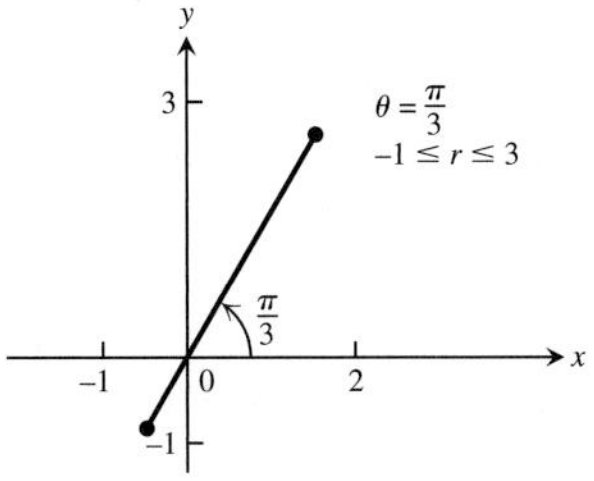

15.

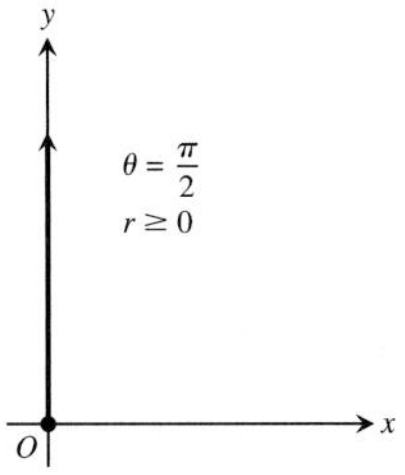

17.

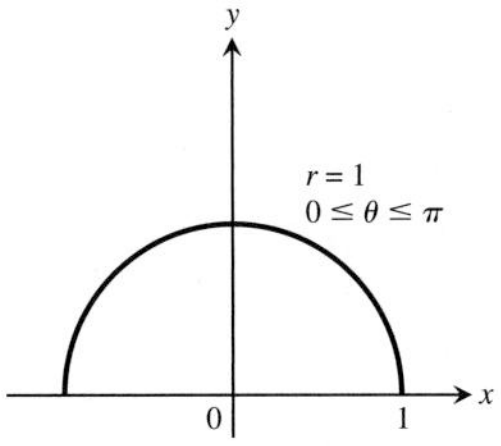

19.

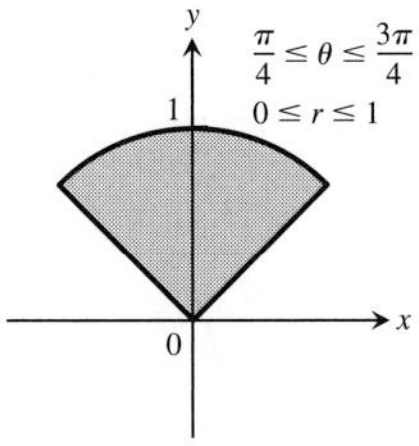

21.

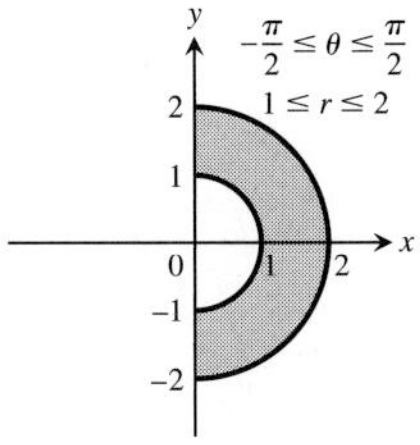

23. $x = 2$, vertical line through $(2, 0)$ **25.** $y = 0$, the x-axis
27. $y = 4$, horizontal line through $(0, 4)$
29. $x + y = 1$, line, $m = -1, b = 1$
31. $x^2 + y^2 = 1$, circle, $C(0, 0)$, radius 1
33. $y - 2x = 5$, line: $m = 2, b = 5$
35. $y^2 = x$, parabola, vertex $(0, 0)$, opens right
37. $y = e^x$, graph of natural exponential function
39. $x + y = \pm 1$, two straight lines of slope -1, y-intercepts $b = \pm 1$
41. $(x + 2)^2 + y^2 = 4$, circle, $C(-2, 0)$, radius 2
43. $x^2 + (y - 4)^2 = 16$, circle, $C(0, 4)$, radius 4
45. $(x - 1)^2 + (y - 1)^2 = 2$, circle, $C(1, 1)$, radius $\sqrt{2}$

47. $\sqrt{3}y + x = 4$ **49.** $r\cos\theta = 7$ **51.** $\theta = \pi/4$
53. $r = 2$ or $r = -2$ **55.** $4r^2\cos^2\theta + 9r^2\sin^2\theta = 36$
57. $r\sin^2\theta = 4\cos\theta$ **59.** $r = 4\sin\theta$
61. $r^2 = 6r\cos\theta - 2r\sin\theta - 6$ **63.** $(0, \theta)$, where θ is any angle

Section 8.2, pp. 483–484

1. x-axis

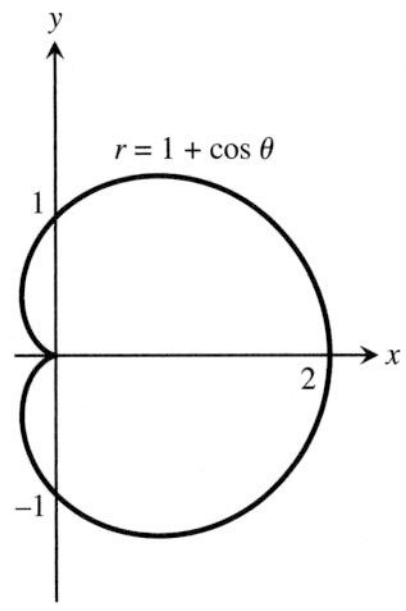

3. y-axis

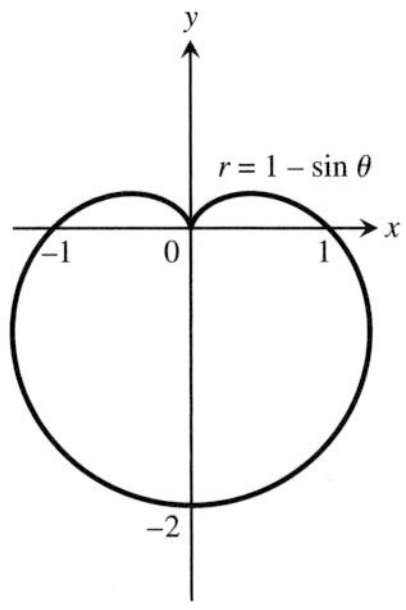

5. y-axis

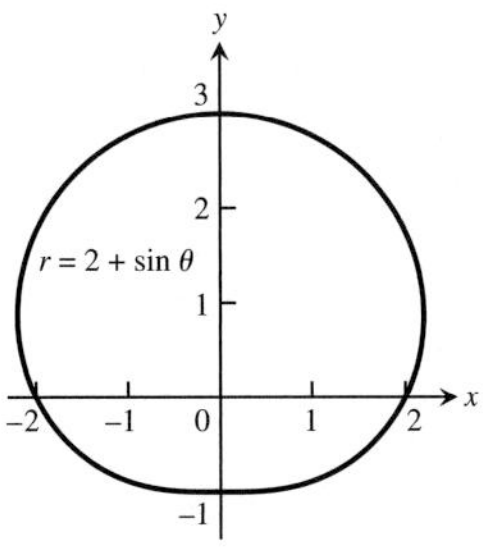

7. x-axis

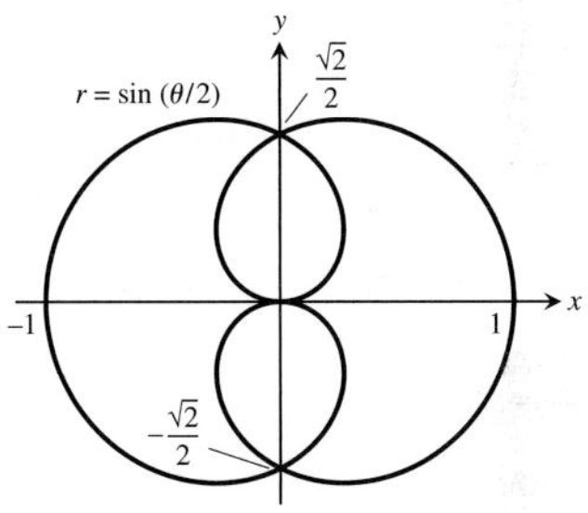

9. x-axis, y-axis, origin

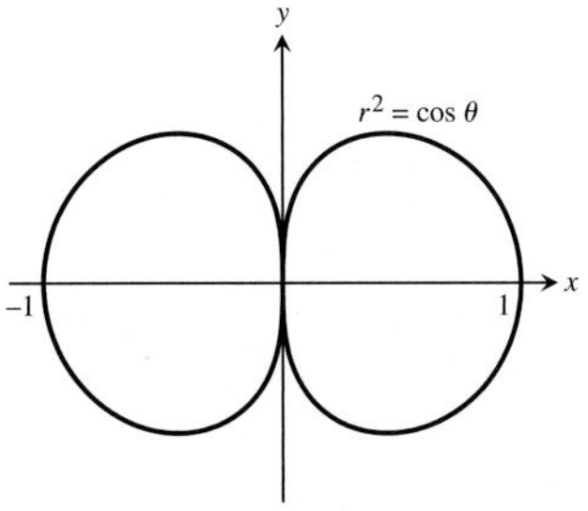

11. y-axis, x-axis, origin

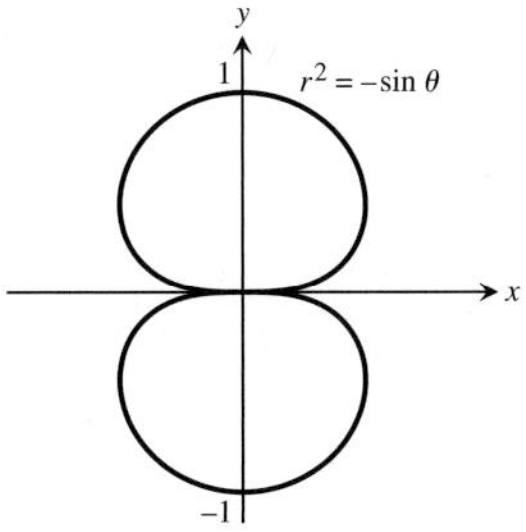

13. x-axis, y-axis, origin **15.** Origin
17. The slope at $(-1, \pi/2)$ is -1, at $(-1, -\pi/2)$ is 1.

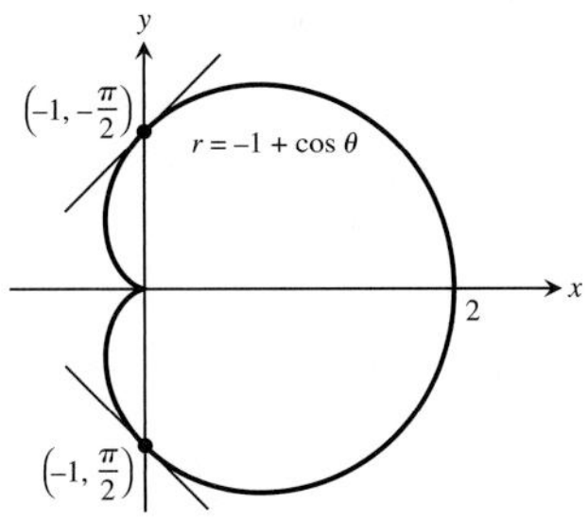

19. The slope at $(1, \pi/4)$ is -1, at $(-1, -\pi/4)$ is 1, at $(-1, 3\pi/4)$ is 1, at $(1, -3\pi/4)$ is -1.

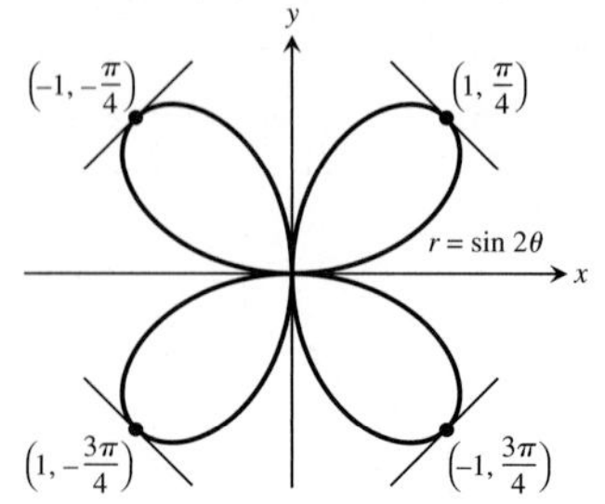

21. **(a)**

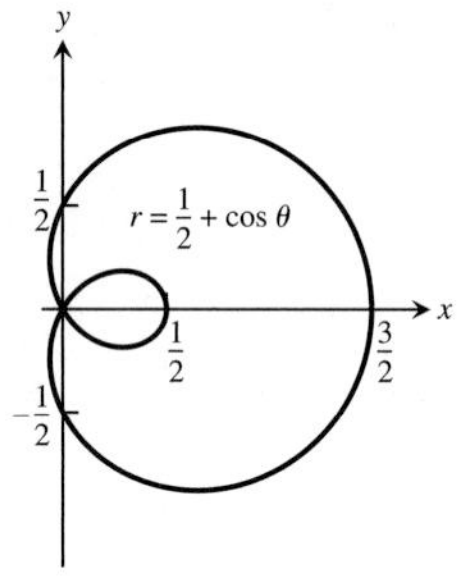

(b)

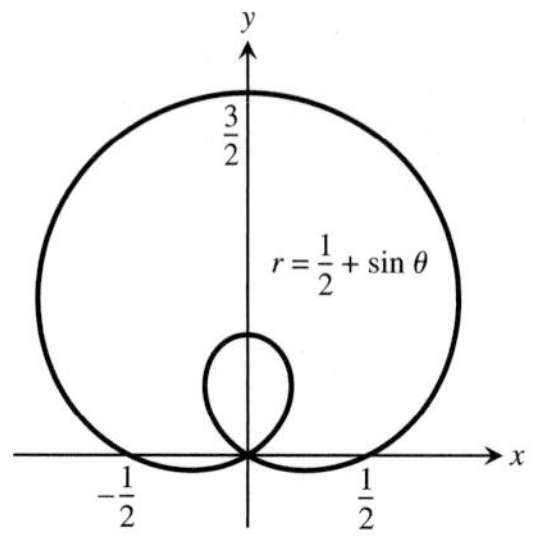

23. **(a)**

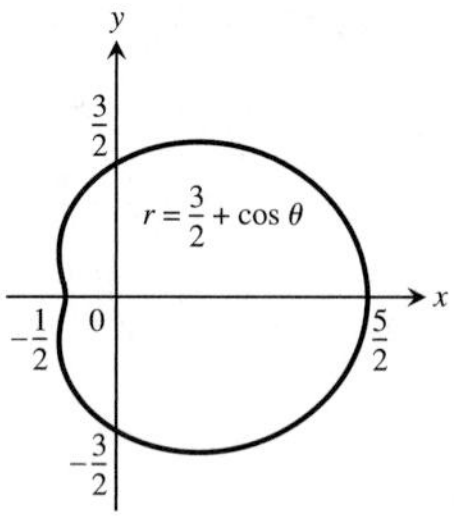

(b)

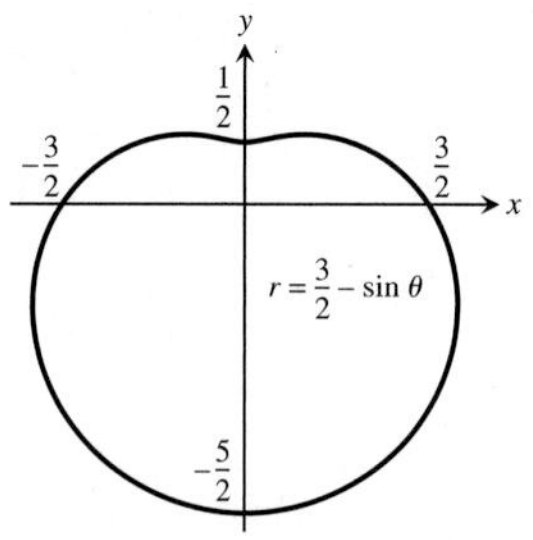

25.

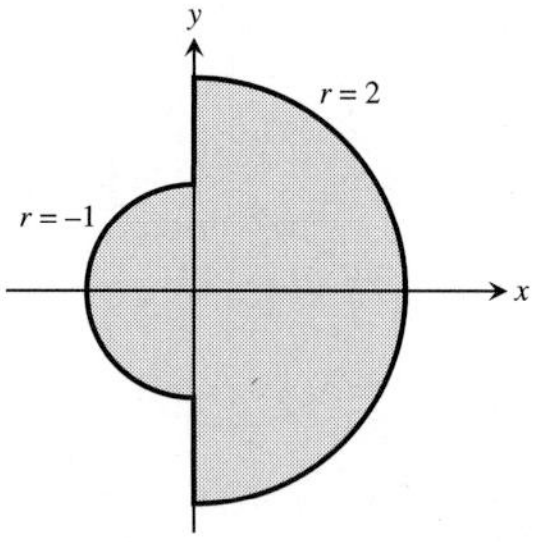

27.

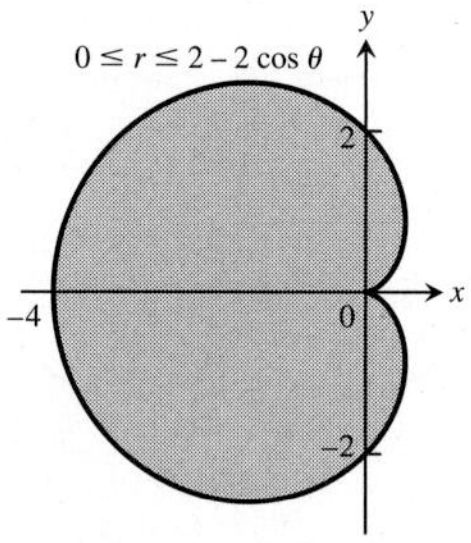

29. (a)

Section 8.3, pp. 487–488

1. 18π **3.** $\pi/8$ **5.** 2 **7.** $\frac{\pi}{2} - 1$ **9.** $5\pi - 8$

11. $3\sqrt{3} - \pi$ **13.** $\frac{\pi}{3} + \frac{\sqrt{3}}{2}$ **15.** **(a)** $\frac{3}{2} - \frac{\pi}{4}$ **17.** 19/3

19. 8 **21.** $3\left(\sqrt{2} + \ln\left(1 + \sqrt{2}\right)\right)$ **23.** $\frac{\pi}{8} + \frac{3}{8}$ **25.** 2π

Section 8.4, pp. 493–495

1. $e = 3/5, \quad F(\pm 3, 0), \quad x = \pm 25/3$
3. $e = 1/\sqrt{2}, \quad F(0, \pm 1), \quad y = \pm 2$
5. $e = 1/\sqrt{3}, \quad F(0, \pm 1), \quad y = \pm 3$
7. $e = \sqrt{3}/3, \quad F\left(\pm\sqrt{3}, 0\right), \quad x = \pm 3\sqrt{3}$
9. $\frac{x^2}{27} + \frac{y^2}{36} = 1$
11. $\frac{x^2}{4851} + \frac{y^2}{4900} = 1$
13. $e = \frac{\sqrt{5}}{3}, \quad \frac{x^2}{9} + \frac{y^2}{4} = 1$
15. $e = 1/2, \quad \frac{x^2}{64} + \frac{y^2}{48} = 1$
17. $e = \sqrt{2}, \quad F\left(\pm\sqrt{2}, 0\right), \quad x = \pm 1/\sqrt{2}$
19. $e = \sqrt{2}, \quad F(0, \pm 4), \quad y = \pm 2$
21. $e = \sqrt{5}, \quad F\left(\pm\sqrt{10}, 0\right), \quad x = \pm 2/\sqrt{10}$
23. $e = \sqrt{5}, \quad F\left(0, \pm\sqrt{10}\right), \quad y = \pm 2/\sqrt{10}$
25. $y^2 - \frac{x^2}{8} = 1$ **27.** $x^2 - \frac{y^2}{8} = 1$
29. $r = 2/(1 + \cos\theta)$
31. $r = 30/(1 - 5\sin\theta)$ **33.** $r = 1/(2 + \cos\theta)$
35. $r = 10/(5 - \sin\theta)$

37.

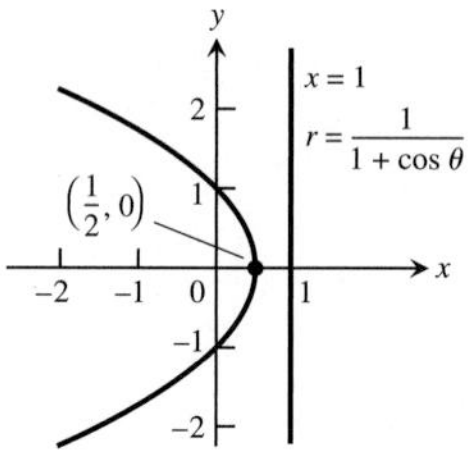

39.

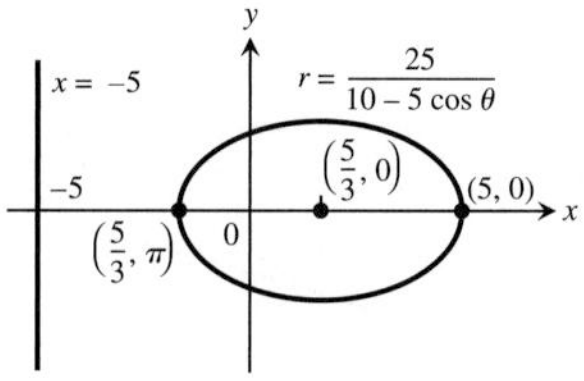

41.

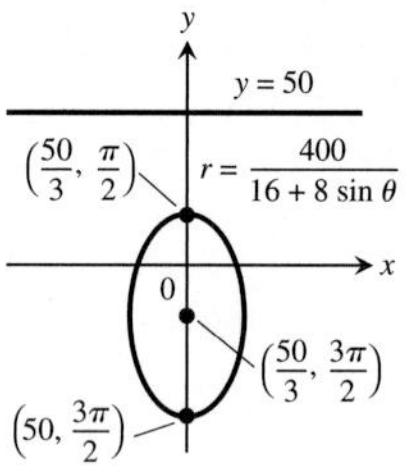

43.

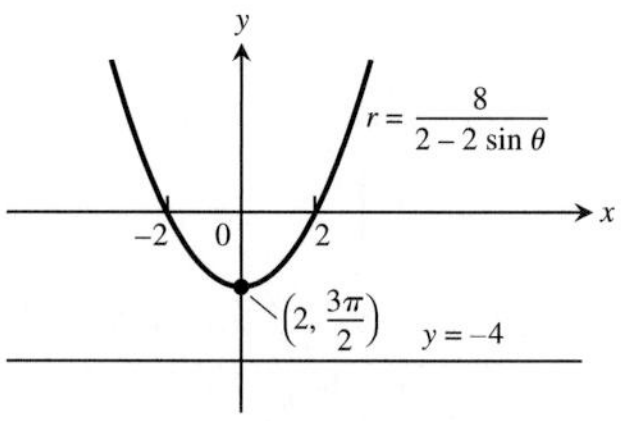

45. $y = 2 - x$ **47.** $y = \left(\sqrt{3}/3\right)x + 2\sqrt{3}$
49. $r\cos\left(\theta - \frac{\pi}{4}\right) = 3$ **51.** $r\cos\left(\theta + \frac{\pi}{2}\right) = 5$
53. $C(2, 0)$, radius $= 2$ **55.** $C(1, \pi)$, radius $= 1$
57. $(x - 6)^2 + y^2 = 36, r = 12\cos\theta$
59. $x^2 + (y - 5)^2 = 25, r = 10\sin\theta$
61. $(x + 1)^2 + y^2 = 1, r = -2\cos\theta$
63. $x^2 + (y + 1/2)^2 = 1/4, r = -\sin\theta$

75. **(b)**

Planet	Perihelion	Aphelion
Mercury	0.3075 AU	0.4667 AU
Venus	0.7184 AU	0.7282 AU
Earth	0.9833 AU	1.0167 AU
Mars	1.3817 AU	1.6663 AU
Jupiter	4.9512 AU	5.4548 AU
Saturn	9.0210 AU	10.0570 AU
Uranus	18.2977 AU	20.0623 AU
Neptune	29.8135 AU	30.3065 AU
Pluto	29.6549 AU	49.2251 AU

77. $r = 2a \sin\theta$ (a circle) **79.** $r\cos(\theta - \alpha) = p$ (a line)

Section 8.5, pp. 497–498

1.

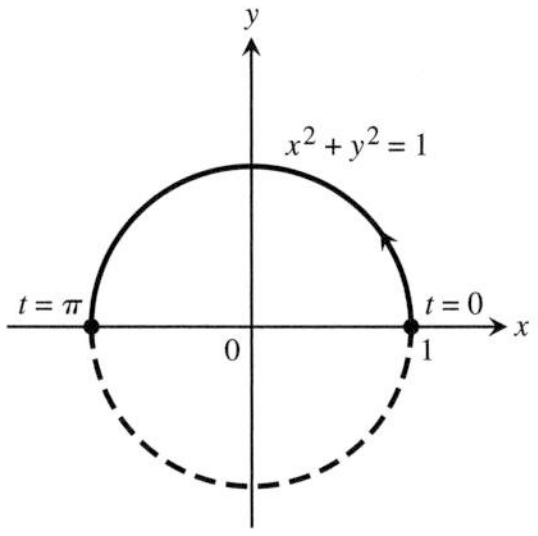

3.

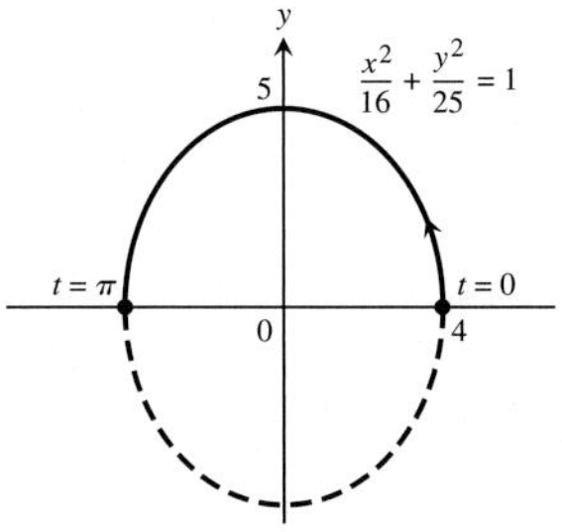

5.

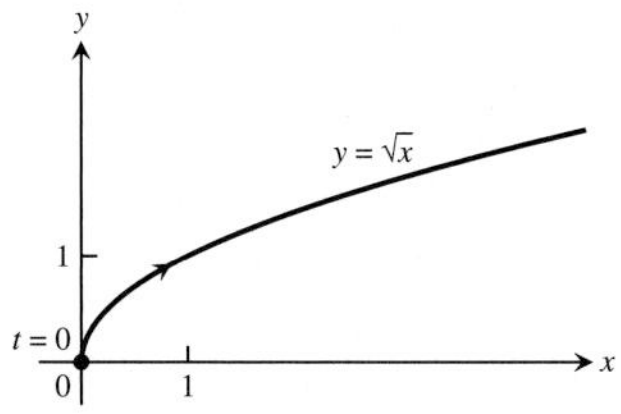

7.

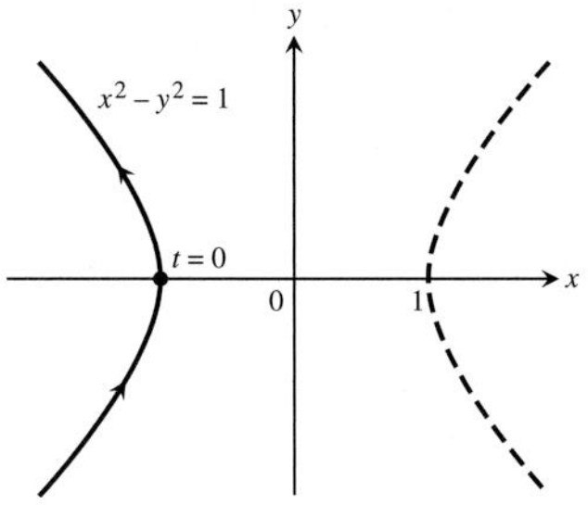

9.

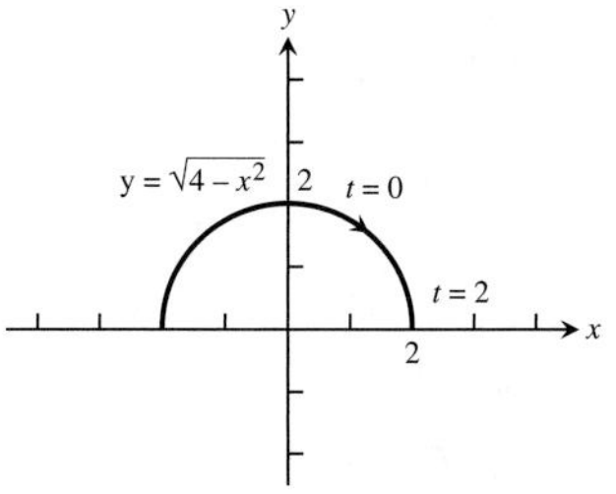

11.

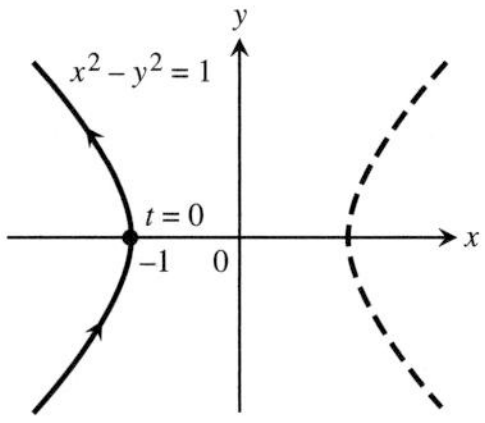

13. $x = (a - b)\cos\theta + b\cos\left(\frac{a - b}{b}\theta\right)$,

$y = (a - b)\sin\theta - b\sin\left(\frac{a - b}{b}\theta\right)$

15. $x = a\sin^2 t \tan t, y = a\sin^2 t$ **17.** $(1, 1)$

Practice and Additional Exercises, pp. 499–500

1. $y = \left(\sqrt{3}/3\right)x - 4$ **3.** $x = 2$ **5.** $y = -3/2$
7. $x^2 + (y + 2)^2 = 4$ **9.** $\left(x - \sqrt{2}\right)^2 + y^2 = 2$
11. $r = -5\sin\theta$ **13.** $r = 3\cos\theta$
15.

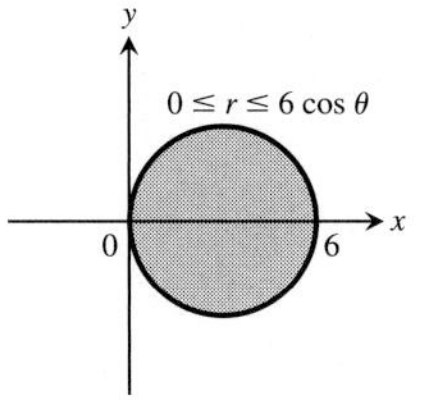

17. (d) **19.** (l) **21.** (k)

23. (i) **25.** $9\pi/2$ **27.** $2 + \pi/4$
29. 8 **31.** $\pi - 3$
33.

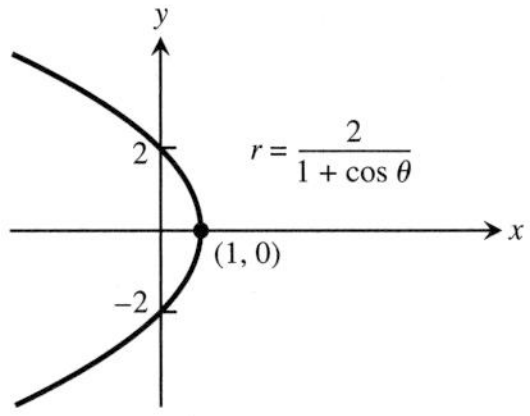

35.

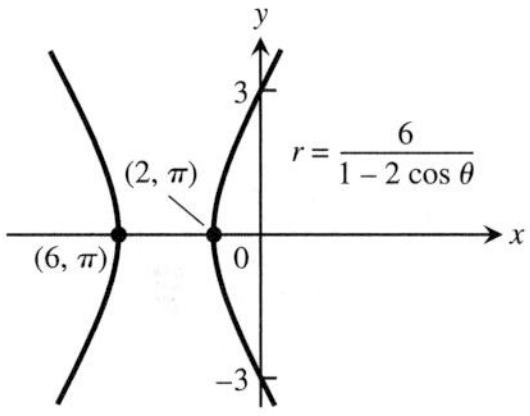

37. $r = \dfrac{4}{1 + 2\cos\theta}$ **39.** $r = \dfrac{2}{2 + \sin\theta}$

41. **(a)** 24π **(b)** 16π

CHAPTER 9

Section 9.1, pp. 504–505

1. The line through the point (2, 3, 0) parallel to the z-axis
3. The x-axis **5.** The circle $x^2 + y^2 = 4$ in the xy-plane
7. The circle $x^2 + z^2 = 4$ in the xz-plane
9. The circle $y^2 + z^2 = 1$ in the yz-plane
11. The circle $x^2 + y^2 = 16$ in the xy-plane
13. **(a)** The first quadrant of the xy-plane
(b) The fourth quadrant of the xy-plane
15. **(a)** The ball of radius 1 centered at the origin
(b) All points greater than 1 unit from the origin
17. **(a)** The upper hemisphere of radius 1 centered at the origin
(b) The solid upper hemisphere of radius 1 centered at the origin
19. **(a)** $x = 3$ **(b)** $y = -1$ **(c)** $z = -2$
21. **(a)** $z = 1$ **(b)** $x = 3$ **(c)** $y = -1$
23. **(a)** $x^2 + (y - 2)^2 = 4, z = 0$
(b) $(y - 2)^2 + z^2 = 4, x = 0$ **(c)** $x^2 + z^2 = 4, y = 2$
25. **(a)** $y = 3, z = -1$ **(b)** $x = 1, z = -1$ **(c)** $x = 1, y = 3$
27. $x^2 + y^2 + z^2 = 25, z = 3$ **29.** $0 \le z \le 1$ **31.** $z \le 0$
33. **(a)** $(x - 1)^2 + (y - 1)^2 + (z - 1)^2 < 1$
(b) $(x - 1)^2 + (y - 1)^2 + (z - 1)^2 > 1$
35. 3 **37.** 7 **39.** $2\sqrt{3}$ **41.** $C(-2, 0, 2), a = 2\sqrt{2}$

43. $C(\sqrt{2}, \sqrt{2}, -\sqrt{2}), a = \sqrt{2}$

45. $(x-1)^2 + (y-2)^2 + (z-3)^2 = 14$

47. $(x+2)^2 + y^2 + z^2 = 3$ **49.** $C(-2, 0, 2), a = \sqrt{8}$

51. $C\left(-\frac{1}{4}, -\frac{1}{4}, -\frac{1}{4}\right), a = \frac{5\sqrt{3}}{4}$

53. **(a)** $\sqrt{y^2+z^2}$ **(b)** $\sqrt{x^2+z^2}$ **(c)** $\sqrt{x^2+y^2}$

55. $\sqrt{17} + \sqrt{33} + 6$

Section 9.2, pp. 513–514

1. **(a)** $\langle 9, -6\rangle$ **(b)** $3\sqrt{13}$ **3.** **(a)** $\langle 1, 3\rangle$ **(b)** $\sqrt{10}$

5. **(a)** $\langle 12, -19\rangle$ **(b)** $\sqrt{505}$ **7.** **(a)** $\left\langle \frac{1}{5}, \frac{14}{5}\right\rangle$ **(b)** $\frac{\sqrt{197}}{5}$

9. $\langle 1, -4\rangle$ **11.** $\langle -2, -3\rangle$ **13.** $\left\langle -\frac{1}{2}, \frac{\sqrt{3}}{2}\right\rangle$

15. $\left\langle -\frac{\sqrt{3}}{2}, -\frac{1}{2}\right\rangle$ **17.** $\langle -3, 2, -1\rangle$ **19.** $\langle -3, 16, 0\rangle$

21. $\langle 3, 5, -8\rangle$

23. The vector **v** is horizontal and 1 in. long. The vectors **u** and **w** are $\frac{11}{16}$ in. long. **w** is vertical and **u** makes a 45° angle with the horizontal. All vectors must be drawn to scale.

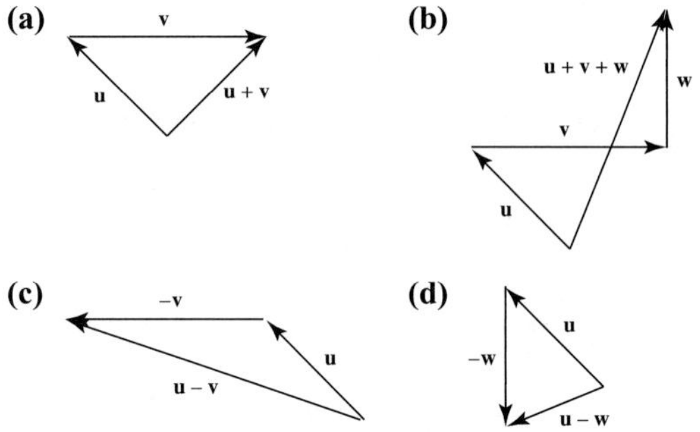

25. $3\left(\frac{2}{3}\mathbf{i} + \frac{1}{3}\mathbf{j} - \frac{2}{3}\mathbf{k}\right)$ **27.** $5(\mathbf{k})$

29. $\sqrt{\frac{1}{2}}\left(\frac{1}{\sqrt{3}}\mathbf{i} - \frac{1}{\sqrt{3}}\mathbf{j} - \frac{1}{\sqrt{3}}\mathbf{k}\right)$

31. **(a)** $2\mathbf{i}$ **(b)** $-\sqrt{3}\mathbf{k}$ **(c)** $\frac{3}{10}\mathbf{j} + \frac{2}{5}\mathbf{k}$ **(d)** $6\mathbf{i} - 2\mathbf{j} + 3\mathbf{k}$

33. $\frac{7}{13}(12\mathbf{i} - 5\mathbf{k})$

35. **(a)** $\frac{3}{5\sqrt{2}}\mathbf{i} + \frac{4}{5\sqrt{2}}\mathbf{j} - \frac{1}{\sqrt{2}}\mathbf{k}$ **(b)** $(1/2, 3, 5/2)$

37. **(a)** $-\frac{1}{\sqrt{3}}\mathbf{i} - \frac{1}{\sqrt{3}}\mathbf{j} - \frac{1}{\sqrt{3}}\mathbf{k}$ **(b)** $\left(\frac{5}{2}, \frac{7}{2}, \frac{9}{2}\right)$

39. $A(4, -3, 5)$ **41.** $a = \frac{3}{2}, b = \frac{1}{2}$

43. $\approx\langle -338.095, 725.046\rangle$

45. **(a)** $\frac{3}{2}\mathbf{i} + \frac{3}{2}\mathbf{j} - 3\mathbf{k}$ **(b)** $\mathbf{i} + \mathbf{j} - 2\mathbf{k}$ **(c)** $(2, 2, 1)$

Section 9.3, pp. 520–522

1. **(a)** $-25, 5, 5$ **(b)** -1 **(c)** -5 **(d)** $-2\mathbf{i} + 4\mathbf{j} - \sqrt{5}\mathbf{k}$

3. **(a)** $25, 15, 5$ **(b)** $\frac{1}{3}$ **(c)** $\frac{5}{3}$ **(d)** $\frac{1}{9}(10\mathbf{i} + 11\mathbf{j} - 2\mathbf{k})$

5. **(a)** $2, \sqrt{34}, \sqrt{3}$ **(b)** $\frac{2}{\sqrt{3}\sqrt{34}}$ **(c)** $\frac{2}{\sqrt{34}}$
(d) $\frac{1}{17}(5\mathbf{j} - 3\mathbf{k})$

7. **(a)** $10 + \sqrt{17}, \sqrt{26}, \sqrt{21}$ **(b)** $\frac{10 + \sqrt{17}}{\sqrt{546}}$
(c) $\frac{10 + \sqrt{17}}{\sqrt{26}}$ **(d)** $\frac{10 + \sqrt{17}}{26}(5\mathbf{i} + \mathbf{j})$

9. 0.75 rad **11.** 1.77 rad

13. Angle at $A = \cos^{-1}\left(\frac{1}{\sqrt{5}}\right) \approx 63.435$ degrees, angle at $B = \cos^{-1}\left(\frac{3}{5}\right) \approx 53.130$ degrees, angle at $C = \cos^{-1}\left(\frac{1}{\sqrt{5}}\right) \approx 63.435$ degrees.

21. Horizontal component: ≈ 1188 ft/sec, vertical component: ≈ 167 ft/sec

23. **(a)** Since $|\cos\theta| \le 1$, we have $|\mathbf{u}\cdot\mathbf{v}| = |\mathbf{u}||\mathbf{v}||\cos\theta| \le |\mathbf{u}||\mathbf{v}|(1) = |\mathbf{u}||\mathbf{v}|$.
(b) We have equality precisely when $|\cos\theta| = 1$ or when one or both of **u** and **v** are **0**. In the case of nonzero vectors, we have equality when $\theta = 0$ or π, that is, when the vectors are parallel.

27. $x + 2y = 4$

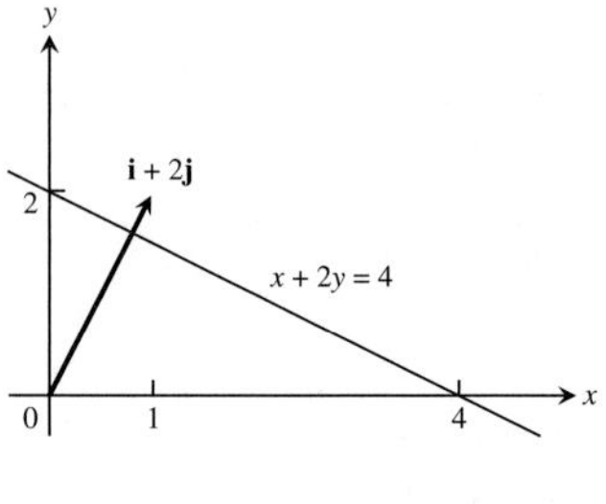

29. $-2x + y = -3$

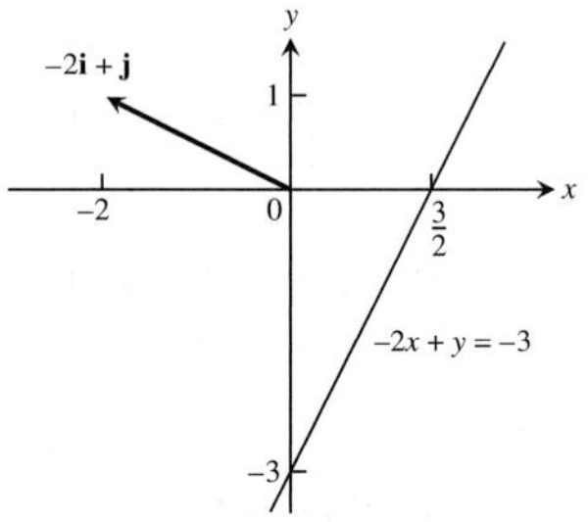

31. $x + y = -1$

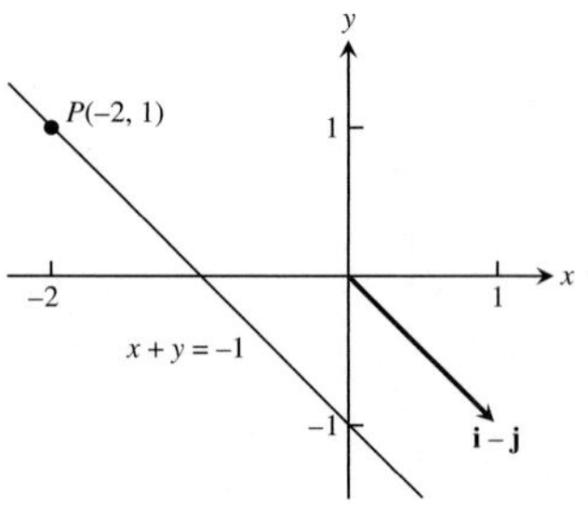

33. $2x - y = 0$

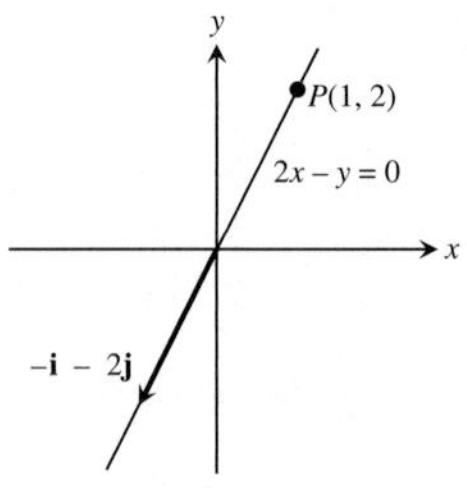

35. 5 J **37.** $\frac{\pi}{4}$ **39.** $\frac{\pi}{6}$

Section 9.4, pp. 527–528

1. $|\mathbf{u} \times \mathbf{v}| = 3$, direction is $\frac{2}{3}\mathbf{i} + \frac{1}{3}\mathbf{j} + \frac{2}{3}\mathbf{k}$; $|\mathbf{v} \times \mathbf{u}| = 3$, direction is $-\frac{2}{3}\mathbf{i} - \frac{1}{3}\mathbf{j} - \frac{2}{3}\mathbf{k}$

3. $|\mathbf{u} \times \mathbf{v}| = 0$, no direction; $|\mathbf{v} \times \mathbf{u}| = 0$, no direction

5. $|\mathbf{u} \times \mathbf{v}| = 6$, direction is $-\mathbf{k}$; $|\mathbf{v} \times \mathbf{u}| = 6$, direction is $\mathbf{k}$

7. $|\mathbf{u} \times \mathbf{v}| = 6\sqrt{5}$, direction is $\frac{1}{\sqrt{5}}\mathbf{i} - \frac{2}{\sqrt{5}}\mathbf{k}$; $|\mathbf{v} \times \mathbf{u}| = 6\sqrt{5}$, direction is $-\frac{1}{\sqrt{5}}\mathbf{i} + \frac{2}{\sqrt{5}}\mathbf{k}$

9.

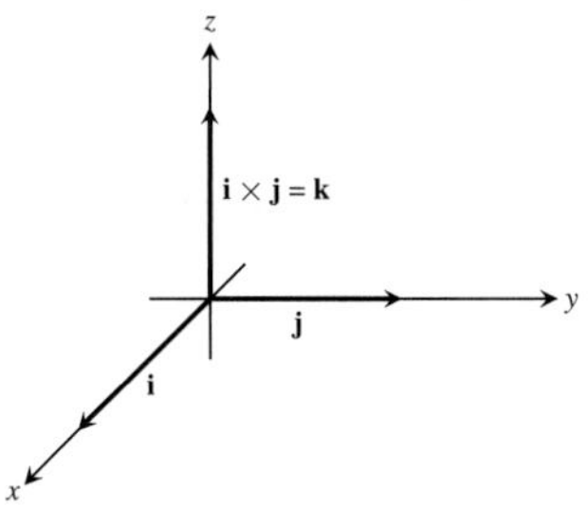

11.

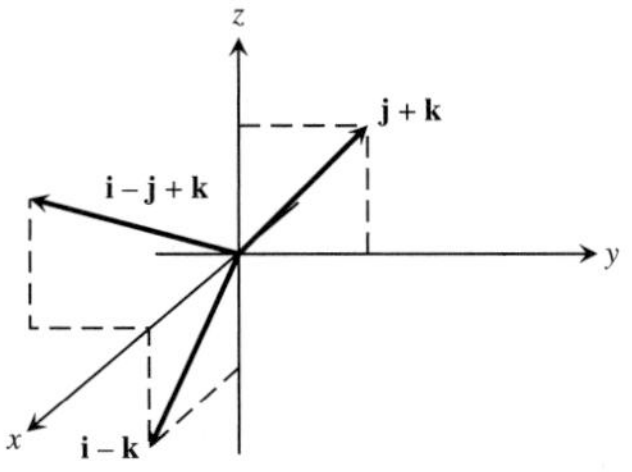

13.

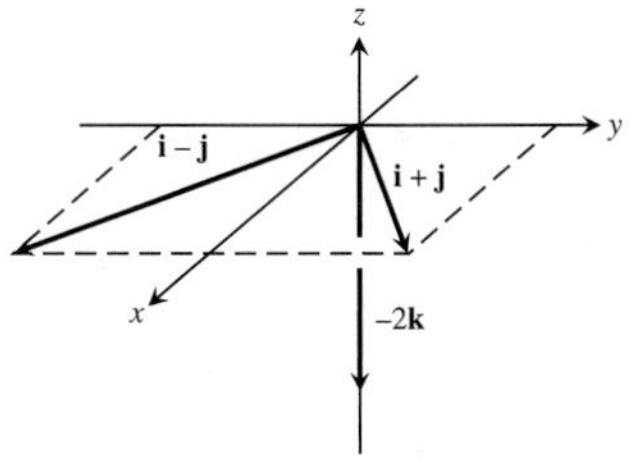

15. **(a)** $2\sqrt{6}$ **(b)** $\pm\frac{1}{\sqrt{6}}(2\mathbf{i} + \mathbf{j} + \mathbf{k})$

17. **(a)** $\frac{\sqrt{2}}{2}$ **(b)** $\pm\frac{1}{\sqrt{2}}(\mathbf{i} - \mathbf{j})$

19. 8 **21.** 7 **23.** **(a)** None **(b)** $\mathbf{u}$ and $\mathbf{w}$ **25.** $10\sqrt{3}$ ft-lb

27. **(a)** True **(b)** Not always true **(c)** True **(d)** True **(e)** Not always true **(f)** True **(g)** True **(h)** True

29. **(a)** $\text{proj}_{\mathbf{v}}\,\mathbf{u} = \frac{\mathbf{u} \cdot \mathbf{v}}{\mathbf{v} \cdot \mathbf{v}}\mathbf{v}$ **(b)** $\pm\,\mathbf{u} \times \mathbf{v}$ **(c)** $\pm\,(\mathbf{u} \times \mathbf{v}) \times \mathbf{w}$ **(d)** $|(\mathbf{u} \times \mathbf{v}) \cdot \mathbf{w}|$ **(e)** $(\mathbf{u} \times \mathbf{v}) \times (\mathbf{u} \times \mathbf{w})$ **(f)** $|\mathbf{u}|\,\frac{\mathbf{v}}{|\mathbf{v}|}$

31. **(a)** Yes **(b)** No **(c)** Yes **(d)** No

33. No, $\mathbf{v}$ need not equal $\mathbf{w}$. For example, $\mathbf{i} + \mathbf{j} \neq -\mathbf{i} + \mathbf{j}$, but $\mathbf{i} \times (\mathbf{i} + \mathbf{j}) = \mathbf{i} \times \mathbf{i} + \mathbf{i} \times \mathbf{j} = \mathbf{0} + \mathbf{k} = \mathbf{k}$ and $\mathbf{i} \times (-\mathbf{i} + \mathbf{j}) = -\mathbf{i} \times \mathbf{i} + \mathbf{i} \times \mathbf{j} = \mathbf{0} + \mathbf{k} = \mathbf{k}$.

35. 2 **37.** 13 **39.** 11/2 **41.** 25/2

Section 9.5, pp. 535–537

1. $x = 3 + t, y = -4 + t, z = -1 + t$

3. $x = -2 + 5t, y = 5t, z = 3 - 5t$ **5.** $x = 0, y = 2t, z = t$

7. $x = 1, y = 1, z = 1 + t$ **9.** $x = t, y = -7 + 2t, z = 2t$

11. $x = t, y = 0, z = 0$

13. $x = t, y = t, z = \frac{3}{2}t, 0 \le t \le 1$

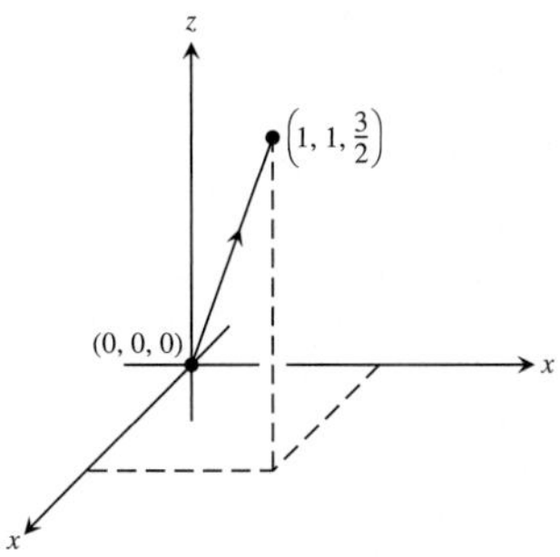

15. $x = 1, y = 1 + t, z = 0, -1 \le t \le 0$

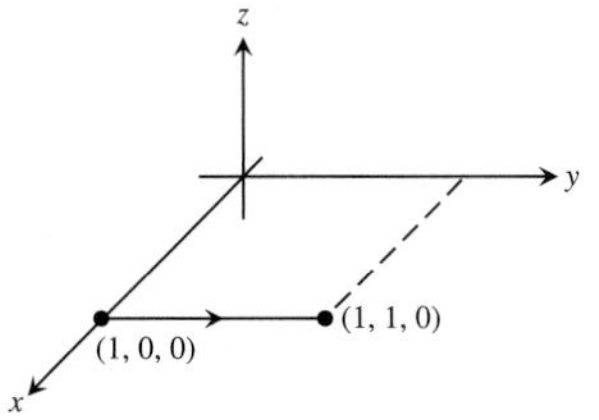

17. $x = 0, y = 1 - 2t, z = 1, 0 \le t \le 1$

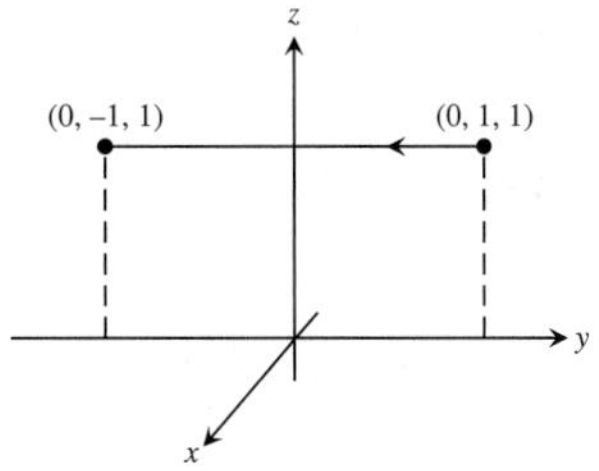

19. $x = 2 - 2t, y = 2t, z = 2 - 2t, 0 \le t \le 1$

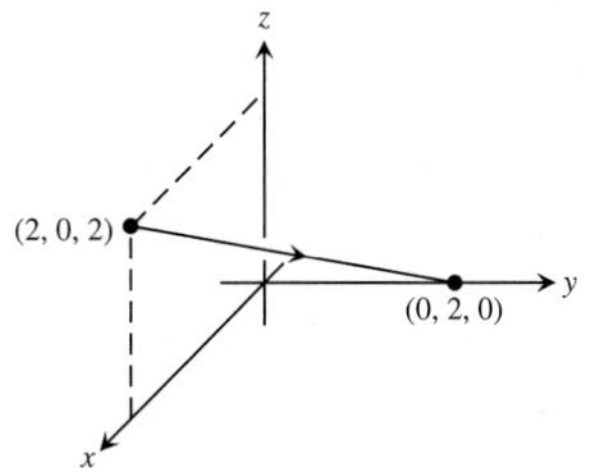

21. $3x - 2y - z = -3$ **23.** $7x - 5y - 4z = 6$

25. $x + 3y + 4z = 34$ **27.** $(1, 2, 3), -20x + 12y + z = 7$

29. $y + z = 3$ **31.** $x - y + z = 0$ **33.** $2\sqrt{30}$ **35.** 0

37. $\frac{9\sqrt{42}}{7}$ **39.** 3 **41.** 19/5 **43.** 5/3 **45.** $9/\sqrt{41}$

47. $\pi/4$ **49.** 1.76 rad **51.** 0.82 rad **53.** $\left(\frac{3}{2}, -\frac{3}{2}, \frac{1}{2}\right)$

55. $(1, 1, 0)$ **57.** $x = 1 - t, y = 1 + t, z = -1$

59. $x = 4, y = 3 + 6t, z = 1 + 3t$

61. $L1$ intersects $L2$; $L2$ is parallel to $L3$; $L1$ and $L3$ are skew.

63. $x = 2 + 2t, y = -4 - t, z = 7 + 3t$; $x = -2 - t$, $y = -2 + (1/2)t, z = 1 - (3/2)t$

65. $\left(0, -\dfrac{1}{2}, -\dfrac{3}{2}\right), (-1, 0, -3), (1, -1, 0)$

69. Many possible answers. One possibility: $x + y = 3$ and $2y + z = 7$

71. $(x/a) + (y/b) + (z/c) = 1$ describes all planes *except* those through the origin or parallel to a coordinate axis.

Section 9.6, pp. 542–543

1. **(d)**, ellipsoid **3.** **(a)**, cylinder **5.** **(l)**, hyperbolic paraboloid
7. **(b)**, cylinder **9.** **(k)**, hyperbolic paraboloid **11.** **(h)**, cone

13.

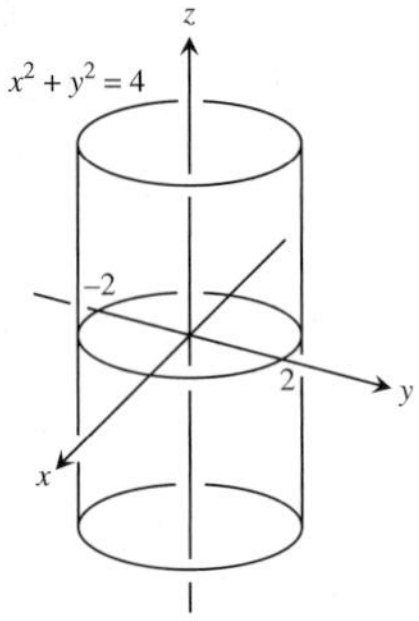

15.

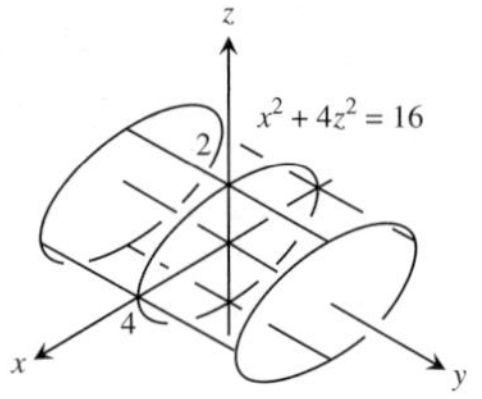

17.

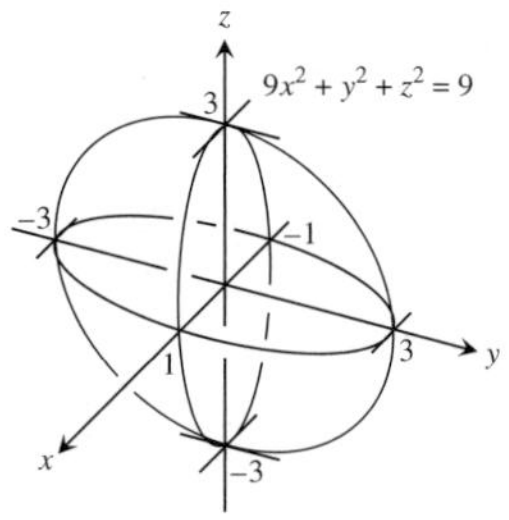

19.

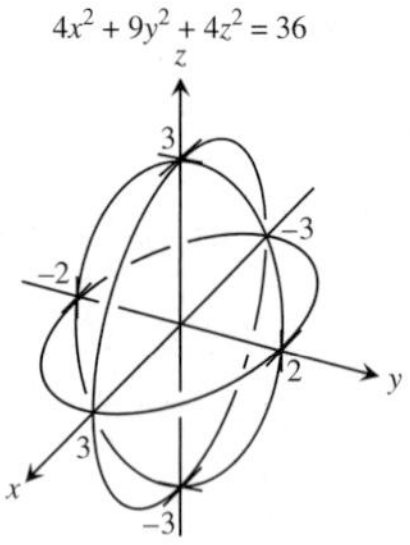

21.

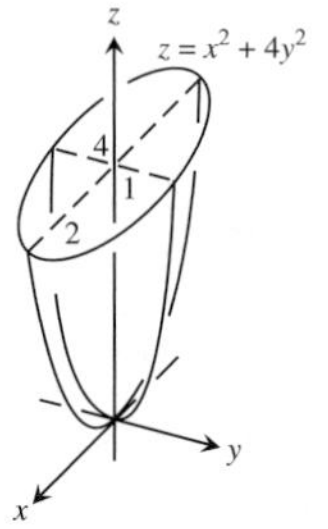

23.

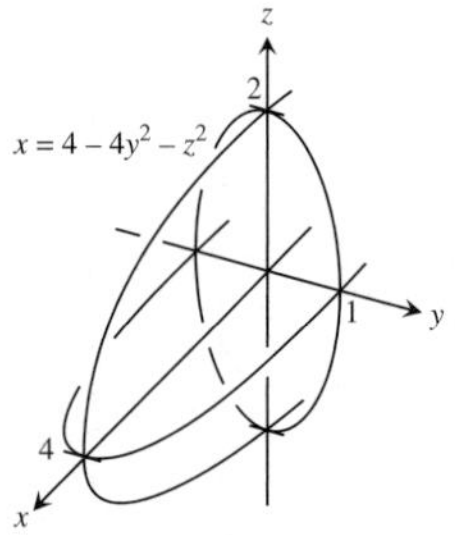

25.

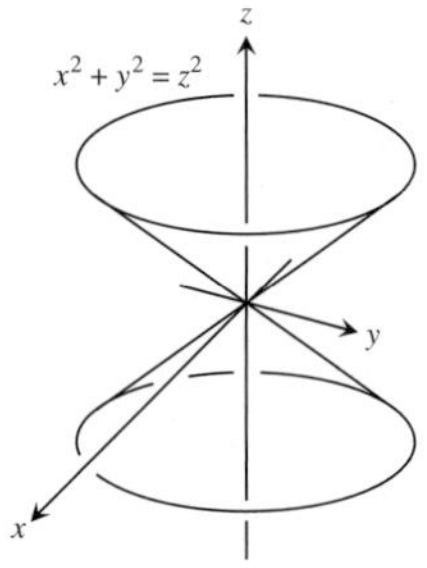

27.

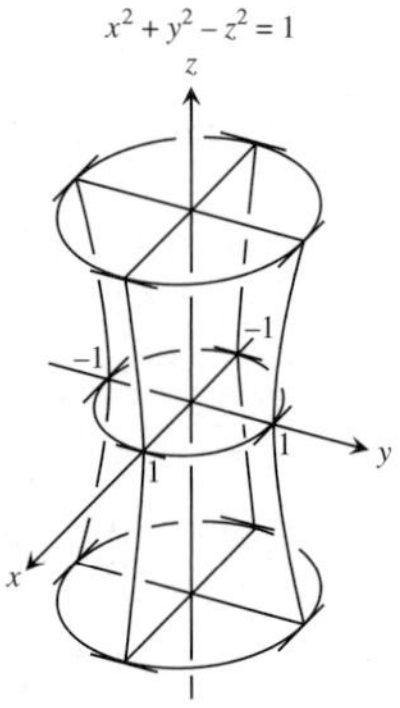

29.

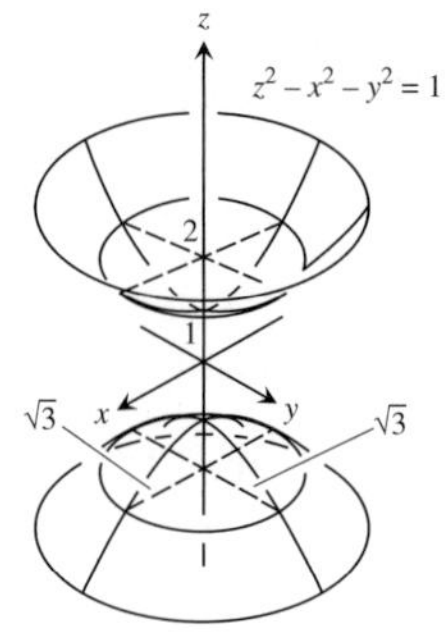

31.

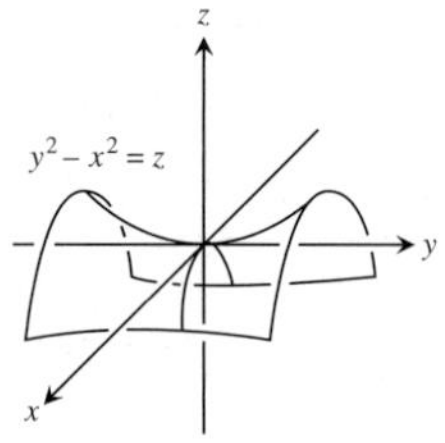

33.

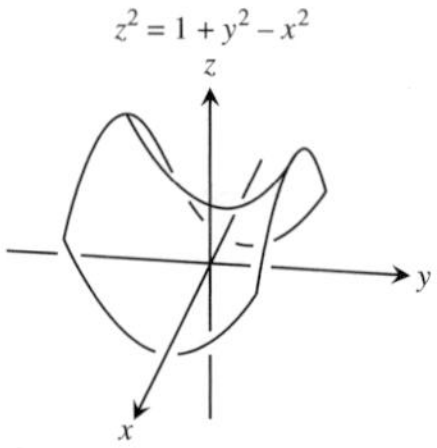

35.

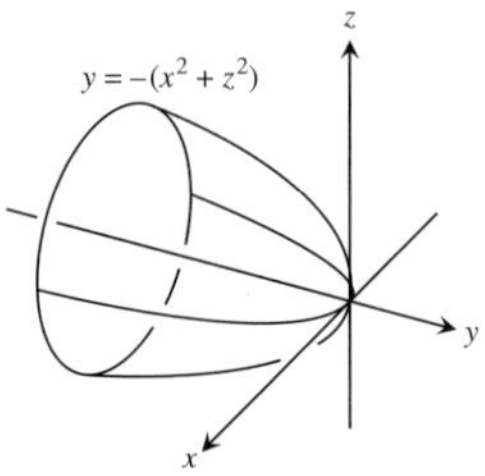

37.

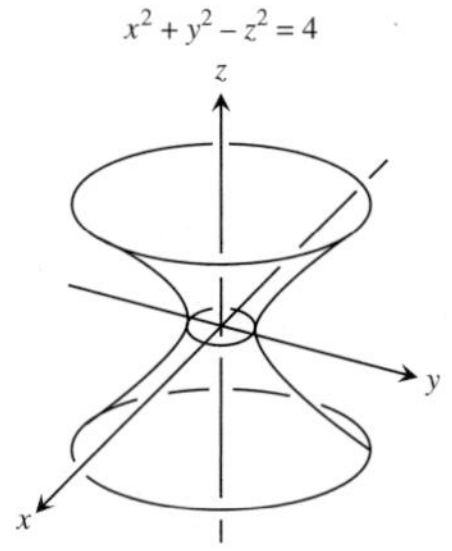

39.

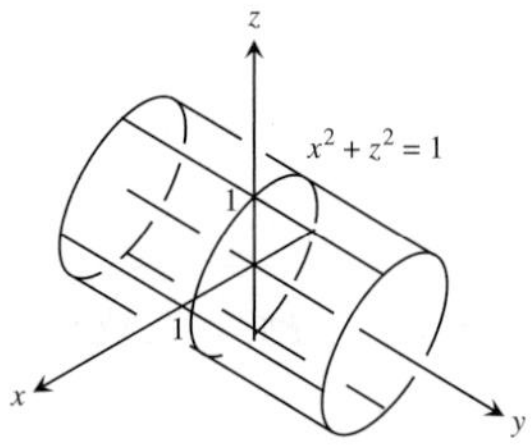

41.

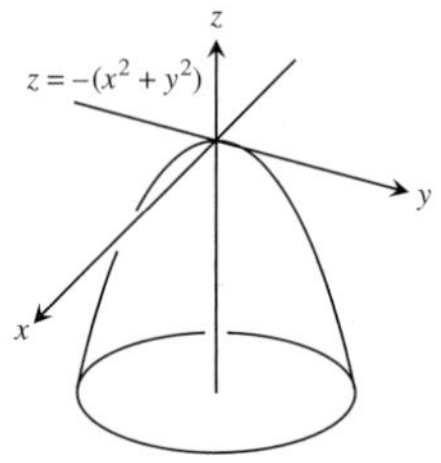

43.

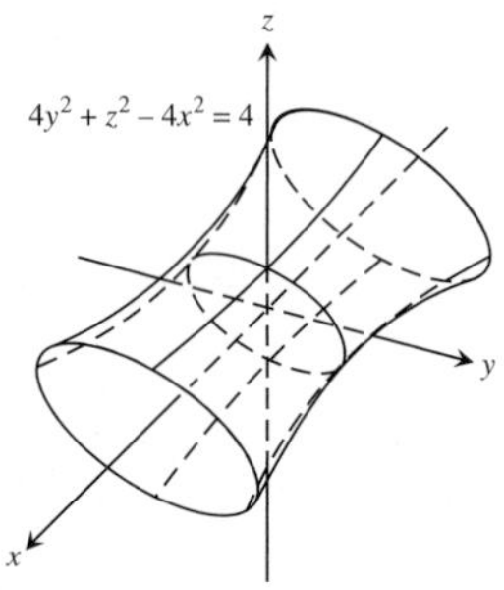

45. **(a)** $\dfrac{2\pi(9 - c^2)}{9}$ **(b)** 8π **(c)** $\dfrac{4\pi abc}{3}$

Practice and Additional Exercises, pp. 543–545

1. **(a)** $\langle -17, 32 \rangle$ **(b)** $\sqrt{1313}$

3. **(a)** $\langle 6, -8 \rangle$ **(b)** 10

5. $\left\langle -\dfrac{\sqrt{3}}{2}, -\dfrac{1}{2} \right\rangle$ [assuming counterclockwise]

7. $\left\langle \dfrac{8}{\sqrt{17}}, -\dfrac{2}{\sqrt{17}} \right\rangle$

9. Length $= 2$, direction is $\dfrac{1}{\sqrt{2}}\mathbf{i} + \dfrac{1}{\sqrt{2}}\mathbf{j}$.

11. $\mathbf{v}(\pi/2) = 2(-\mathbf{i})$

13. Length $= 7$, direction is $\frac{2}{7}\mathbf{i} - \frac{3}{7}\mathbf{j} + \frac{6}{7}\mathbf{k}$.

15. $\frac{8}{\sqrt{33}}\mathbf{i} - \frac{2}{\sqrt{33}}\mathbf{j} + \frac{8}{\sqrt{33}}\mathbf{k}$

17. $|\mathbf{v}| = \sqrt{2}, |\mathbf{u}| = 3, \mathbf{v}\cdot\mathbf{u} = \mathbf{u}\cdot\mathbf{v} = 3, \mathbf{v}\times\mathbf{u} = -2\mathbf{i} + 2\mathbf{j} - \mathbf{k}$, $\mathbf{u}\times\mathbf{v} = 2\mathbf{i} - 2\mathbf{j} + \mathbf{k}, |\mathbf{v}\times\mathbf{u}| = 3, \theta = \cos^{-1}\left(\frac{1}{\sqrt{2}}\right) = \frac{\pi}{4}$, $|\mathbf{u}|\cos\theta = \frac{3}{\sqrt{2}}$, $\text{proj}_{\mathbf{v}}\,\mathbf{u} = \frac{3}{2}(\mathbf{i} + \mathbf{j})$

19. $\frac{4}{3}(2\mathbf{i} + \mathbf{j} - \mathbf{k})$

21. $\mathbf{u}\times\mathbf{v} = \mathbf{k}$

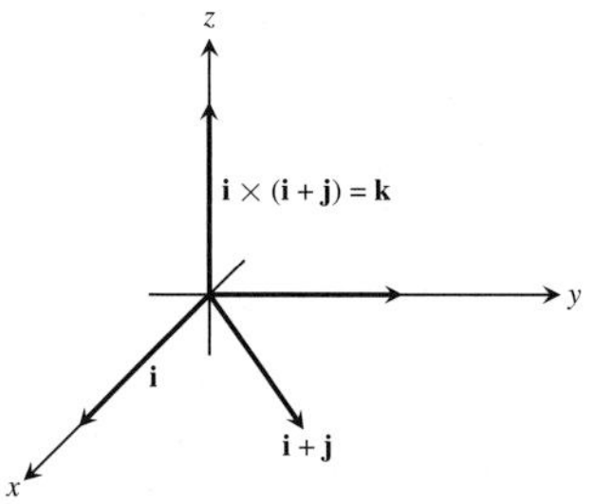

23. $2\sqrt{7}$ **25. (a)** $\sqrt{14}$ **(b)** 1 **29.** $\sqrt{78}/3$

31. $x = 1 - 3t, y = 2, z = 3 + 7t$ **33.** $\sqrt{2}$

35. $2x + y + z = 5$

37. $-9x + y + 7z = 4$

39. $\left(0, -\frac{1}{2}, -\frac{3}{2}\right), (-1, 0, -3), (1, -1, 0)$ **41.** $\pi/3$

43. $x = -5 + 5t, y = 3 - t, z = -3t$

45. (b) $x = -12t, y = 19/12 + 15t, z = 1/6 + 6t$

47. Yes; $\mathbf{v}$ is parallel to the plane.

49. 3 **51.** $-3\mathbf{j} + 3\mathbf{k}$

53. $\frac{2}{\sqrt{35}}\left(5\mathbf{i} - \mathbf{j} - 3\mathbf{k}\right)$

55. $\left(\frac{11}{9}, \frac{26}{9}, \frac{7}{9}\right)$

57. $(1, -2, -1); x = 1 - 5t, y = -2 + 3t, z = -1 + 4t$

59. $2x + 7y + 2z + 10 = 0$

61. $11/\sqrt{107}$

63.

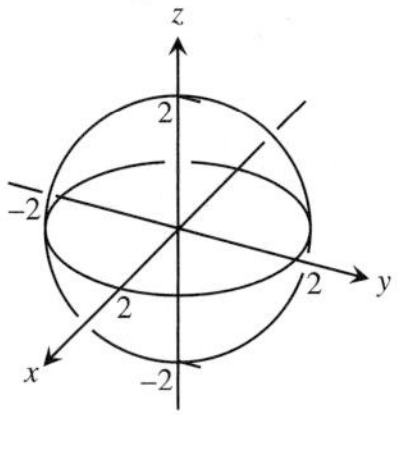

65.

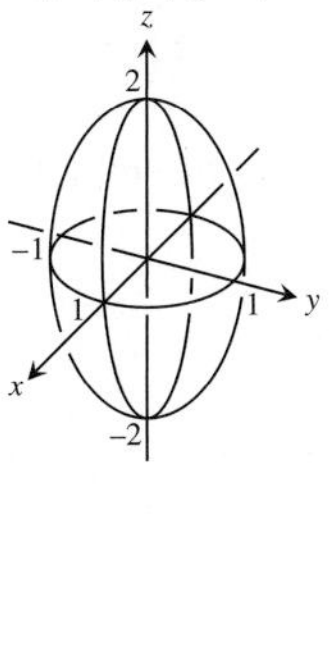

67.

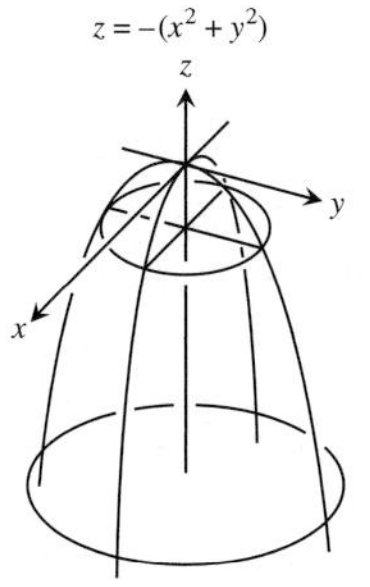

69.

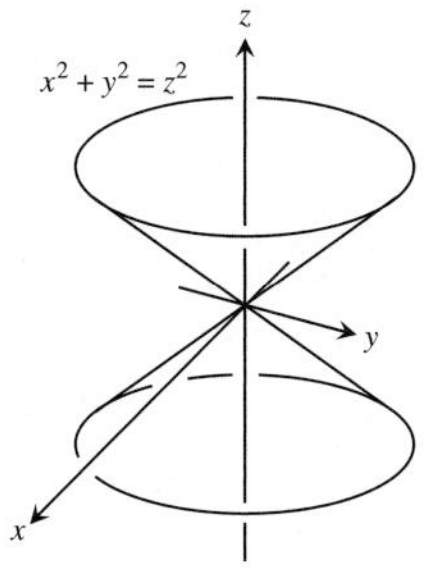

71.

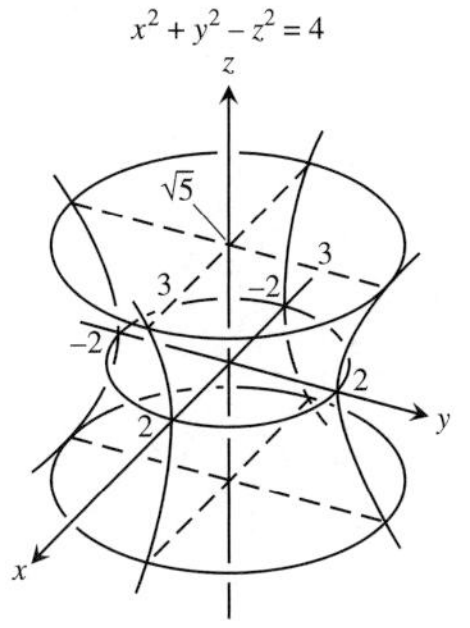

73.

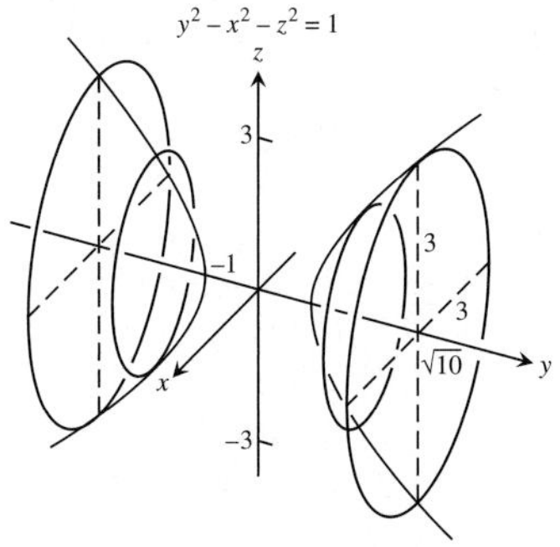

CHAPTER 10

Section 10.1, pp. 552–554

1. $y = x^2 - 2x, \mathbf{v} = \mathbf{i} + 2\mathbf{j}, \mathbf{a} = 2\mathbf{j}$

3. $y = \frac{2}{9}x^2, \mathbf{v} = 3\mathbf{i} + 4\mathbf{j}, \mathbf{a} = 3\mathbf{i} + 8\mathbf{j}$

5. $t = \frac{\pi}{4}$: $\mathbf{v} = \frac{\sqrt{2}}{2}\mathbf{i} - \frac{\sqrt{2}}{2}\mathbf{j}, \mathbf{a} = \frac{-\sqrt{2}}{2}\mathbf{i} - \frac{\sqrt{2}}{2}\mathbf{j}$; $t = \pi/2$: $\mathbf{v} = -\mathbf{j}, \mathbf{a} = -\mathbf{i}$

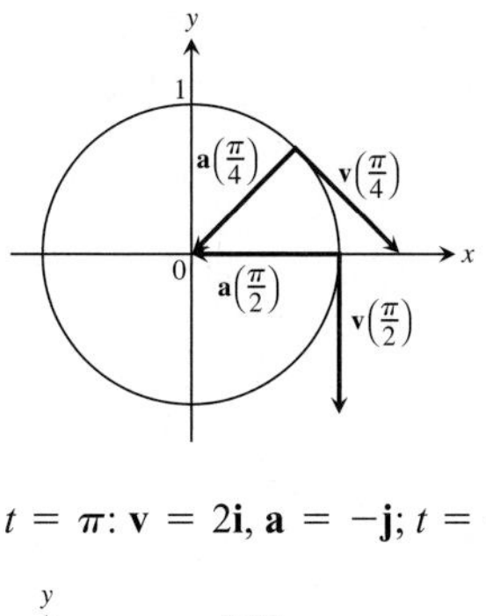

7. $t = \pi$: $\mathbf{v} = 2\mathbf{i}, \mathbf{a} = -\mathbf{j}$; $t = \frac{3\pi}{2}$: $\mathbf{v} = \mathbf{i} - \mathbf{j}, \mathbf{a} = -\mathbf{i}$

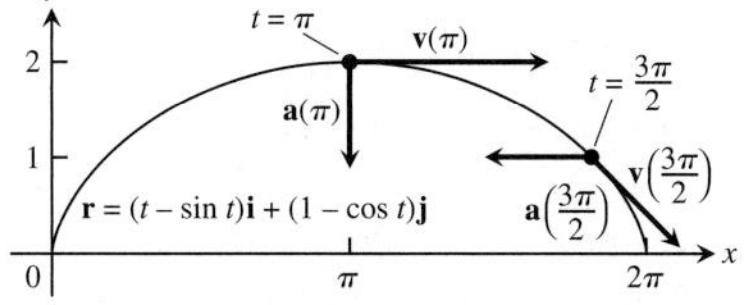

9. $\mathbf{v} = \mathbf{i} + 2t\mathbf{j} + 2\mathbf{k}$; $\mathbf{a} = 2\mathbf{j}$; speed: 3; direction: $\frac{1}{3}\mathbf{i} + \frac{2}{3}\mathbf{j} + \frac{2}{3}\mathbf{k}$; $\mathbf{v}(1) = 3\left(\frac{1}{3}\mathbf{i} + \frac{2}{3}\mathbf{j} + \frac{2}{3}\mathbf{k}\right)$

11. $\mathbf{v} = (-2\sin t)\mathbf{i} + (3\cos t)\mathbf{j} + 4\mathbf{k}$;
$\mathbf{a} = (-2\cos t)\mathbf{i} - (3\sin t)\mathbf{j}$; speed: $2\sqrt{5}$;
direction: $\left(-1/\sqrt{5}\right)\mathbf{i} + \left(2/\sqrt{5}\right)\mathbf{k}$;
$\mathbf{v}(\pi/2) = 2\sqrt{5}\left[\left(-1/\sqrt{5}\right)\mathbf{i} + \left(2/\sqrt{5}\right)\mathbf{k}\right]$

13. $\mathbf{v} = \left(\frac{2}{t+1}\right)\mathbf{i} + 2t\mathbf{j} + t\mathbf{k}$; $\mathbf{a} = \left(\frac{-2}{(t+1)^2}\right)\mathbf{i} + 2\mathbf{j} + \mathbf{k}$;
speed: $\sqrt{6}$; direction: $\frac{1}{\sqrt{6}}\mathbf{i} + \frac{2}{\sqrt{6}}\mathbf{j} + \frac{1}{\sqrt{6}}\mathbf{k}$;
$\mathbf{v}(1) = \sqrt{6}\left(\frac{1}{\sqrt{6}}\mathbf{i} + \frac{2}{\sqrt{6}}\mathbf{j} + \frac{1}{\sqrt{6}}\mathbf{k}\right)$

15. $\pi/2$ **17.** $\pi/2$

19. $x = t, y = -1, z = 1 + t$

21. $x = at, y = a, z = 2\pi b + bt$

23. **(a)** (i): It has constant speed 1 (ii): Yes
(iii): Counterclockwise (iv): Yes
(b) (i): It has constant speed 2 (ii): Yes
(iii): Counterclockwise (iv): Yes
(c) (i): It has constant speed 1 (ii): Yes
(iii): Counterclockwise
(iv): It starts at $(0, -1)$ instead of $(1, 0)$
(d) (i): It has constant speed 1 (ii): Yes
(iii): Clockwise (iv): Yes
(e) (i): It has variable speed (ii): No
(iii): Counterclockwise (iv): Yes

25. $\mathbf{v} = 2\sqrt{5}\mathbf{i} + \sqrt{5}\mathbf{j}$

Section 10.2, pp. 558–560

1. $(1/4)\mathbf{i} + 7\mathbf{j} + (3/2)\mathbf{k}$ **3.** $\left(\frac{\pi + 2\sqrt{2}}{2}\right)\mathbf{j} + 2\mathbf{k}$

5. $(\ln 4)\mathbf{i} + (\ln 4)\mathbf{j} + (\ln 2)\mathbf{k}$

7. $\mathbf{r}(t) = \left(\frac{-t^2}{2} + 1\right)\mathbf{i} + \left(\frac{-t^2}{2} + 2\right)\mathbf{j} + \left(\frac{-t^2}{2} + 3\right)\mathbf{k}$

9. $\mathbf{r}(t) = ((t+1)^{3/2} - 1)\mathbf{i} + (-e^{-t} + 1)\mathbf{j} + (\ln(t+1) + 1)\mathbf{k}$

11. $\mathbf{r}(t) = 8t\mathbf{i} + 8t\mathbf{j} + (-16t^2 + 100)\mathbf{k}$

13. $\mathbf{r}(t) = \left(\frac{3}{2}t^2 + \frac{6}{\sqrt{11}}t + 1\right)\mathbf{i} - \left(\frac{1}{2}t^2 + \frac{2}{\sqrt{11}}t - 2\right)\mathbf{j} + \left(\frac{1}{2}t^2 + \frac{2}{\sqrt{11}}t + 3\right)\mathbf{k} = \left(\frac{1}{2}t^2 + \frac{2t}{\sqrt{11}}\right)(3\mathbf{i} - \mathbf{j} + \mathbf{k}) + (\mathbf{i} + 2\mathbf{j} + 3\mathbf{k})$

15. 50 sec

17. **(a)** 72.2 sec; 25,510 m **(b)** 4020 m **(c)** 6378 m

19. **(a)** $v_0 \approx 9.9$ m/sec **(b)** $\alpha \approx 18.4°$ or $71.6°$

21. 39.3° or 50.7°

29. **(a)** (Assuming that "x" is zero at the point of impact.)
$\mathbf{r}(t) = (x(t))\mathbf{i} + (y(t))\mathbf{j}$, where $x(t) = (35\cos 27°)t$ and $y(t) = 4 + (35\sin 27°)t - 16t^2$.
(b) At $t \approx 0.497$ sec, it reaches its maximum height of about 7.945 ft.
(c) Range $\approx$ 37.45 ft; flight time $\approx$ 1.201 sec
(d) At $t \approx 0.254$ and $t \approx 0.740$ sec, when it is $\approx$ 29.532 and $\approx$ 14.376 ft from where it will land.
(e) Yes. It changes things because the ball won't clear the net.

Section 10.3, pp. 563–564

1. $\mathbf{T} = \left(-\frac{2}{3}\sin t\right)\mathbf{i} + \left(\frac{2}{3}\cos t\right)\mathbf{j} + \frac{\sqrt{5}}{3}\mathbf{k}$, 3π

3. $\mathbf{T} = \frac{1}{\sqrt{1+t}}\mathbf{i} + \frac{\sqrt{t}}{\sqrt{1+t}}\mathbf{k}$, $\frac{52}{3}$

5. $\mathbf{T} = -\cos t\mathbf{j} + \sin t\mathbf{k}$, $\frac{3}{2}$

7. $\mathbf{T} = \left(\frac{\cos t - t\sin t}{t+1}\right)\mathbf{i} + \left(\frac{\sin t + t\cos t}{t+1}\right)\mathbf{j} + \left(\frac{\sqrt{2}t^{1/2}}{t+1}\right)\mathbf{k}$, $\frac{\pi^2}{2} + \pi$

9. $(0, 5, 24\pi)$ **11.** $s(t) = 5t$, $L = \frac{5\pi}{2}$

13. $s(t) = \sqrt{3}e^t - \sqrt{3}$, $L = \frac{3\sqrt{3}}{4}$

15. $\sqrt{2} + \ln\left(1 + \sqrt{2}\right)$

Section 10.4, pp. 569–571

1. $\mathbf{T} = (\cos t)\mathbf{i} - (\sin t)\mathbf{j}$, $\mathbf{N} = (-\sin t)\mathbf{i} - (\cos t)\mathbf{j}$, $\kappa = \cos t$

3. $\mathbf{T} = \frac{1}{\sqrt{1+t^2}}\mathbf{i} - \frac{t}{\sqrt{1+t^2}}\mathbf{j}$, $\mathbf{N} = \frac{-t}{\sqrt{1+t^2}}\mathbf{i} - \frac{1}{\sqrt{1+t^2}}\mathbf{j}$,
$\kappa = \frac{1}{2\left(\sqrt{1+t^2}\right)^3}$

5. **(b)** $\cos x$

7. **(b)** $\mathbf{N} = \frac{-2e^{2t}}{\sqrt{1+4e^{4t}}}\mathbf{i} + \frac{1}{\sqrt{1+4e^{4t}}}\mathbf{j}$
(c) $\mathbf{N} = -\frac{1}{2}\left(\sqrt{4 - t^2}\,\mathbf{i} + t\mathbf{j}\right)$

9. $\mathbf{T} = \frac{3\cos t}{5}\mathbf{i} - \frac{3\sin t}{5}\mathbf{j} + \frac{4}{5}\mathbf{k}$, $\mathbf{N} = (-\sin t)\mathbf{i} - (\cos t)\mathbf{j}$,
$\kappa = \frac{3}{25}$

11. $\mathbf{T} = \left(\frac{\cos t - \sin t}{\sqrt{2}}\right)\mathbf{i} + \left(\frac{\cos t + \sin t}{\sqrt{2}}\right)\mathbf{j}$,
$\mathbf{N} = \left(\frac{-\cos t - \sin t}{\sqrt{2}}\right)\mathbf{i} + \left(\frac{-\sin t + \cos t}{\sqrt{2}}\right)\mathbf{j}$,
$\kappa = \frac{1}{e^t\sqrt{2}}$

13. $\mathbf{T} = \frac{t}{\sqrt{t^2+1}}\mathbf{i} + \frac{1}{\sqrt{t^2+1}}\mathbf{j}$, $\mathbf{N} = \frac{\mathbf{i}}{\sqrt{t^2+1}} - \frac{t\mathbf{j}}{\sqrt{t^2+1}}$,
$\kappa = \frac{1}{t(t^2+1)^{3/2}}$

15. $\mathbf{T} = \left(\operatorname{sech}\frac{t}{a}\right)\mathbf{i} + \left(\tanh\frac{t}{a}\right)\mathbf{j}$,

$\mathbf{N} = \left(-\tanh\frac{t}{a}\right)\mathbf{i} + \left(\operatorname{sech}\frac{t}{a}\right)\mathbf{j}$,

$\kappa = \frac{1}{a}\operatorname{sech}^2\frac{t}{a}$

19. $1/(2b)$ **21.** $\left(x - \frac{\pi}{2}\right)^2 + y^2 = 1$

23. $\kappa(x) = 2/(1 + 4x^2)^{3/2}$ **25.** $\kappa(x) = |\sin x|/(1 + \cos^2 x)^{3/2}$

Section 10.5, pp. 574–575

1. $\mathbf{a} = |a|\mathbf{N}$ **3.** $\mathbf{a}(1) = \frac{4}{3}\mathbf{T} + \frac{2\sqrt{5}}{3}\mathbf{N}$ **5.** $\mathbf{a}(0) = 2\mathbf{N}$

7. $\mathbf{r}\left(\frac{\pi}{4}\right) = \frac{\sqrt{2}}{2}\mathbf{i} + \frac{\sqrt{2}}{2}\mathbf{j} - \mathbf{k}$, $\mathbf{T}\left(\frac{\pi}{4}\right) = -\frac{\sqrt{2}}{2}\mathbf{i} + \frac{\sqrt{2}}{2}\mathbf{j}$,

$\mathbf{N}\left(\frac{\pi}{4}\right) = -\frac{\sqrt{2}}{2}\mathbf{i} - \frac{\sqrt{2}}{2}\mathbf{j}$, $\mathbf{B}\left(\frac{\pi}{4}\right) = \mathbf{k}$; osculating plane: $z = -1$; normal plane: $-x + y = 0$; rectifying plane: $x + y = \sqrt{2}$

9. $\mathbf{B} = \left(\frac{4}{5}\cos t\right)\mathbf{i} - \left(\frac{4}{5}\sin t\right)\mathbf{j} - \frac{3}{5}\mathbf{k}$, $\tau = -\frac{4}{25}$

11. $\mathbf{B} = \mathbf{k}, \tau = 0$ **13.** $\mathbf{B} = -\mathbf{k}, \tau = 0$ **15.** $\mathbf{B} = \mathbf{k}, \tau = 0$

17. Yes. If the car is moving on a curved path ($\kappa \neq 0$), then $a_{\mathrm{N}} = \kappa|\mathbf{v}|^2 \neq 0$ and $\mathbf{a} \neq \mathbf{0}$.

23. $\kappa = \frac{1}{t}, \rho = t$

29. Components of **v**: −1.8701, 0.7089, 1.0000
Components of **a**: −1.6960, −2.0307, 0
Speed: 2.2361; Components of **T**: −0.8364, 0.3170, 0.4472
Components of **N**: −0.4143, −0.8998, −0.1369
Components of **B**: 0.3590, −0.2998, 0.8839; Curvature: 0.5060
Torsion: 0.2813; Tangential component of acceleration: 0.7746
Normal component of acceleration: 2.5298

31. Components of **v**: 2.0000, 0, 0.1629
Components of **a**: 0, −1.0000, 0.0086; Speed: 2.0066
Components of **T**: 0.9967, 0, 0.0812
Components of **N**: −0.0007, −1.0000, 0.0086
Components of **B**: 0.0812, −0.0086, −0.9967;
Curvature: 0.2484
Torsion: −0.0411; Tangential component of acceleration: 0.0007
Normal component of acceleration: 1.0000

Section 10.6, p. 579

1. $\mathbf{v} = (3a\sin\theta)\mathbf{u}_r + 3a(1 - \cos\theta)\mathbf{u}_\theta$
$\mathbf{a} = 9a(2\cos\theta - 1)\mathbf{u}_r + (18a\sin\theta)\mathbf{u}_\theta$

3. $\mathbf{v} = 2ae^{a\theta}\mathbf{u}_r + 2e^{a\theta}\mathbf{u}_\theta$
$\mathbf{a} = 4e^{a\theta}(a^2 - 2)\mathbf{u}_r + 8ae^{a\theta}\mathbf{u}_\theta$

5. $\mathbf{v} = (-8\sin 4t)\mathbf{u}_r + (4\cos 4t)\mathbf{u}_\theta$
$\mathbf{a} = (-40\cos 4t)\mathbf{u}_r - (32\sin 4t)\mathbf{u}_\theta$

Practice and Additional Exercises, p. 580

1. $\frac{x^2}{16} + \frac{y^2}{2} = 1$

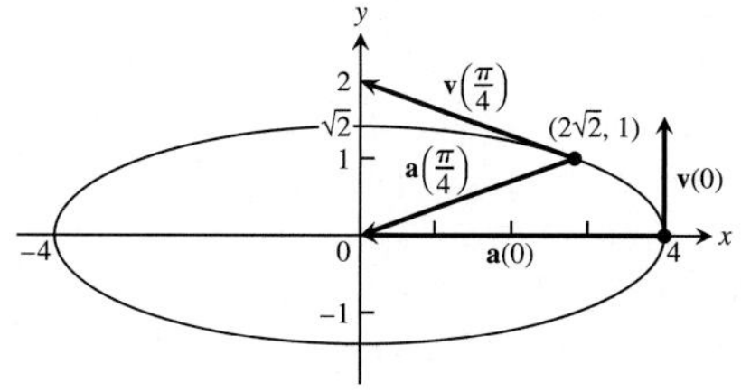

At $t = 0$: $a_{\mathrm{T}} = 0, a_{\mathrm{N}} = 4, \kappa = 2$;

At $t = \frac{\pi}{4}$: $a_{\mathrm{T}} = \frac{7}{3}, a_{\mathrm{N}} = \frac{4\sqrt{2}}{3}, \kappa = \frac{4\sqrt{2}}{27}$

3. $|\mathbf{v}|_{\max} = 1$ **5.** $\kappa = 1/5$ **7.** $dy/dt = -x$; clockwise

11. Shot put is on the ground, about 66 ft, 3 in. from the stopboard.

13. Length $= \frac{\pi}{4}\sqrt{1 + \frac{\pi^2}{16}} + \ln\left(\frac{\pi}{4} + \sqrt{1 + \frac{\pi^2}{16}}\right)$

15. $\mathbf{T}(0) = \frac{2}{3}\mathbf{i} - \frac{2}{3}\mathbf{j} + \frac{1}{3}\mathbf{k}$; $\mathbf{N}(0) = \frac{1}{\sqrt{2}}\mathbf{i} + \frac{1}{\sqrt{2}}\mathbf{j}$;

$\mathbf{B}(0) = -\frac{1}{3\sqrt{2}}\mathbf{i} + \frac{1}{3\sqrt{2}}\mathbf{j} + \frac{4}{3\sqrt{2}}\mathbf{k}$; $\kappa = \frac{\sqrt{2}}{3}$; $\tau = \frac{1}{6}$

17. $\mathbf{T}(\ln 2) = \frac{1}{\sqrt{17}}\mathbf{i} + \frac{4}{\sqrt{17}}\mathbf{j}$; $\mathbf{N}(\ln 2) = -\frac{4}{\sqrt{17}}\mathbf{i} + \frac{1}{\sqrt{17}}\mathbf{j}$;

$\mathbf{B}(\ln 2) = \mathbf{k}$; $\kappa = \frac{8}{17\sqrt{17}}$; $\tau = 0$

19. $\mathbf{a}(0) = 10\mathbf{T} + 6\mathbf{N}$

21. $\mathbf{T} = \left(\frac{1}{\sqrt{2}}\cos t\right)\mathbf{i} - (\sin t)\mathbf{j} + \left(\frac{1}{\sqrt{2}}\cos t\right)\mathbf{k}$;

$\mathbf{N} = \left(-\frac{1}{\sqrt{2}}\sin t\right)\mathbf{i} - (\cos t)\mathbf{j} - \left(\frac{1}{\sqrt{2}}\sin t\right)\mathbf{k}$;

$\mathbf{B} = \frac{1}{\sqrt{2}}\mathbf{i} - \frac{1}{\sqrt{2}}\mathbf{k}$; $\kappa = \frac{1}{\sqrt{2}}$; $\tau = 0$

23. $x = 1 + t, y = t, z = -t$ **25.** $\kappa = 1/a$

CHAPTER 11

Section 11.1, pp. 588–590

1. **(a)** All points in the xy-plane **(b)** All reals
(c) The lines $y - x = c$ **(d)** No boundary points
(e) Both open and closed **(f)** Unbounded

3. **(a)** All points in the xy-plane **(b)** $z \geq 0$
(c) For $f(x, y) = 0$, the origin; for $f(x, y) \neq 0$, ellipses with the center (0, 0), and major and minor axes, along the x- and y-axes, respectively
(d) No boundary points **(e)** Both open and closed
(f) Unbounded

5. (a) All points in the xy-plane **(b)** All reals
(c) For $f(x, y) = 0$, the x- and y-axes; for $f(x, y) \neq 0$, hyperbolas with the x- and y-axes as asymptotes
(d) No boundary points **(e)** Both open and closed
(f) Unbounded

7. (a) All (x, y) satisfying $x^2 + y^2 < 16$ **(b)** $z \geq 1/4$
(c) Circles centered at the origin with radii $r < 4$
(d) Boundary is the circle $x^2 + y^2 = 16$
(e) Open **(f)** Bounded

9. (a) $(x, y) \neq (0, 0)$ **(b)** All reals
(c) The circles with center $(0, 0)$ and radii $r > 0$
(d) Boundary is the single point $(0, 0)$
(e) Open **(f)** Unbounded

11. (a) All (x, y) satisfying $-1 \leq y - x \leq 1$
(b) $-\pi/2 \leq z \leq \pi/2$
(c) Straight lines of the form $y - x = c$ where $-1 \leq c \leq 1$
(d) Boundary is two straight lines $y = 1 + x$ and $y = -1 + x$
(e) Closed **(f)** Unbounded

13. (f) **15.** (a) **17.** (d)

19. (a)

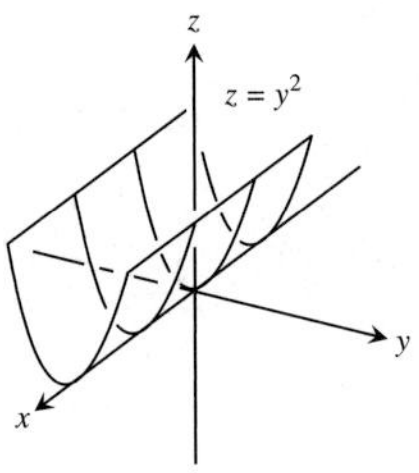

(b)

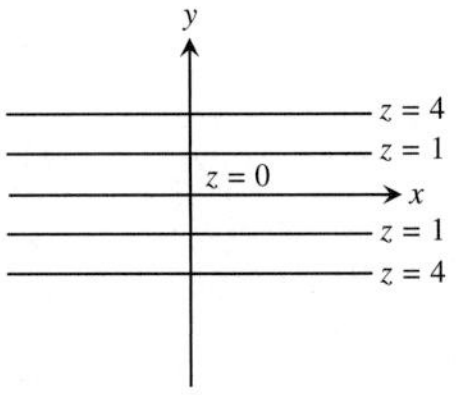

21. (a)

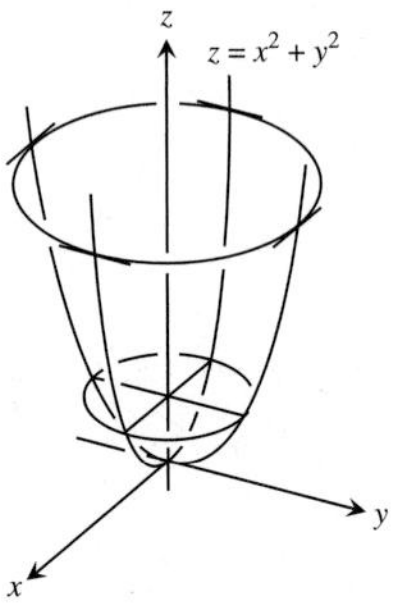

(b)

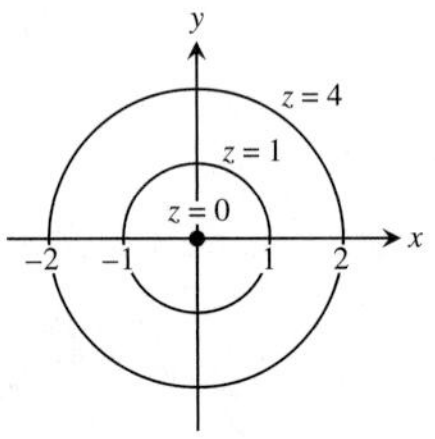

23. (a)

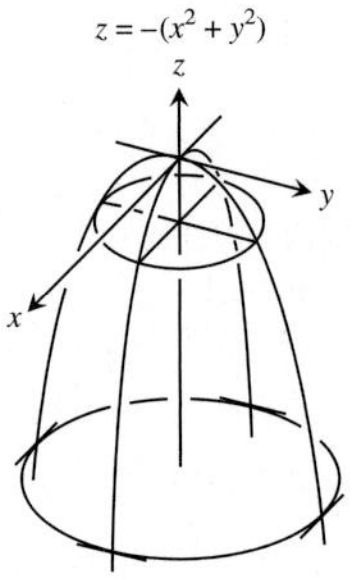

(b)

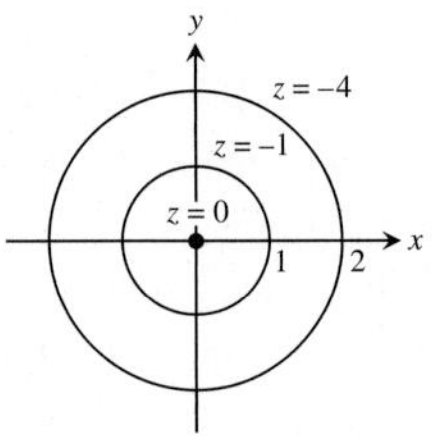

25. (a)

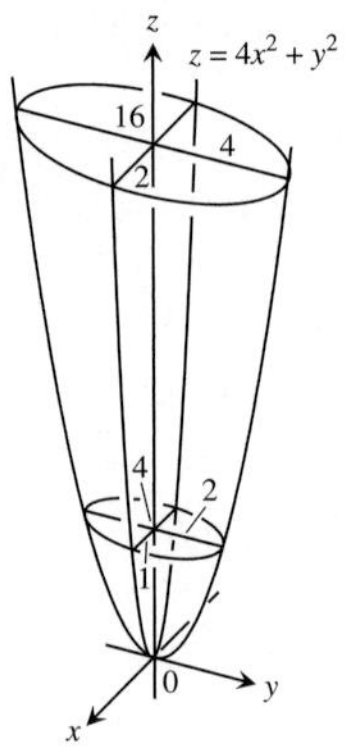

(b)

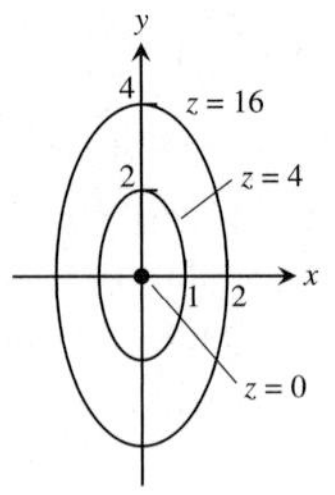

27. (a)

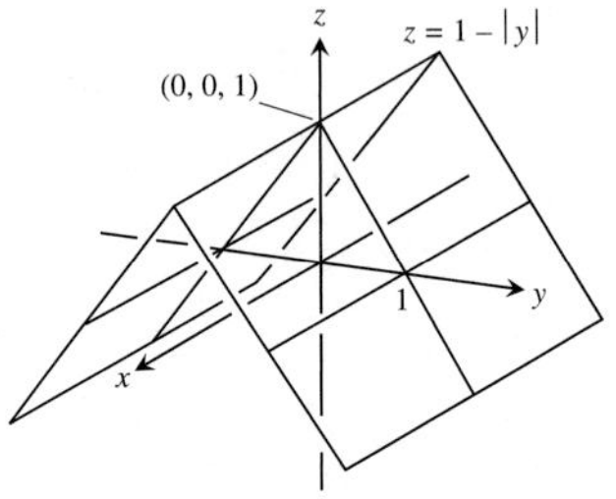

(b)

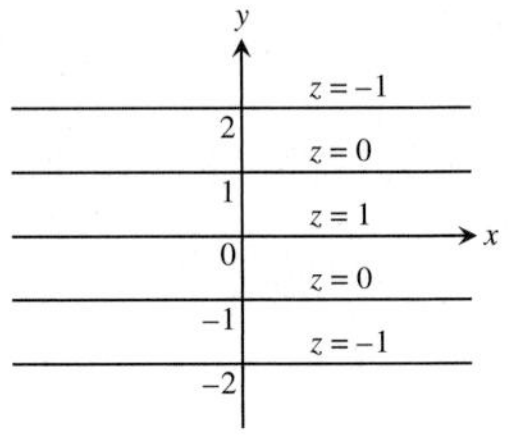

29. $x^2 + y^2 = 10$ **31.** $\tan^{-1} y - \tan^{-1} x = 2 \tan^{-1} \sqrt{2}$

33.

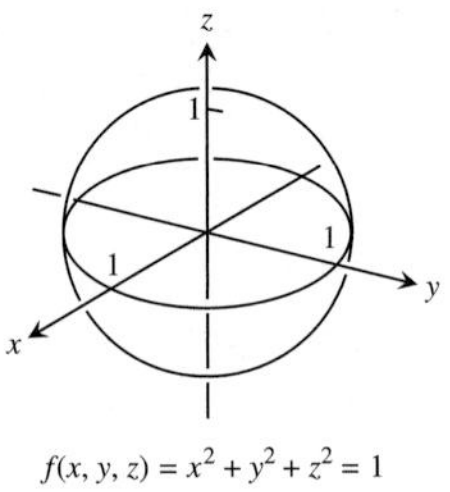

35.

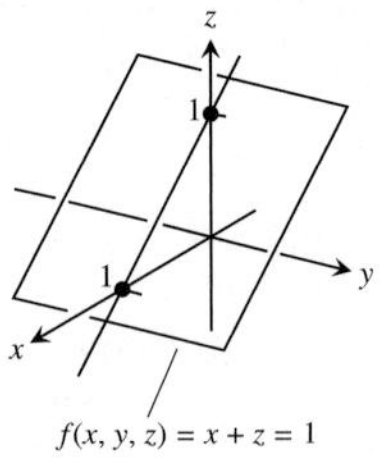

37.

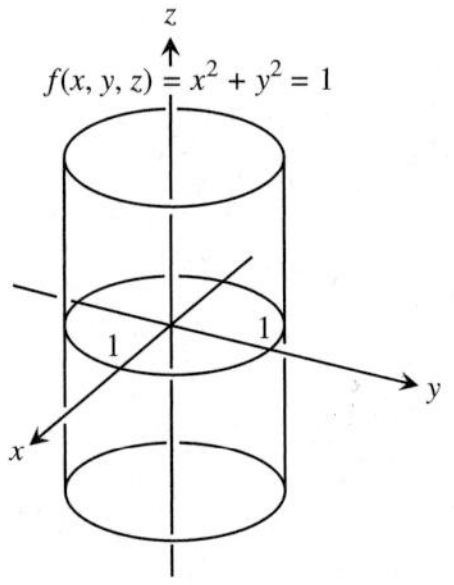

39.

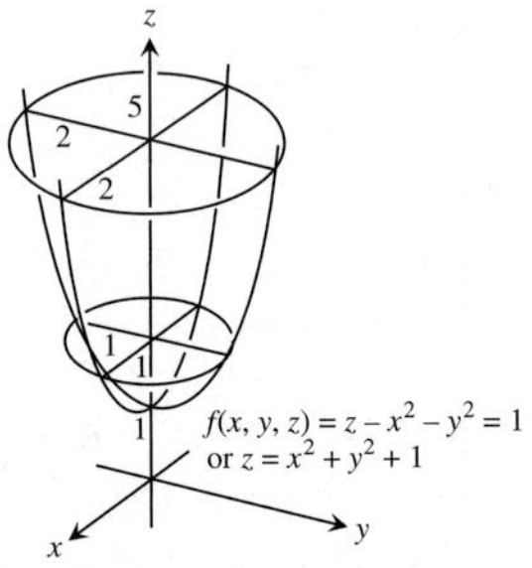

41. $\sqrt{x - y} - \ln z = 2$ **43.** $\dfrac{x + y}{z} = \ln 2$

Section 11.2, pp. 597–598

1. $5/2$ **3.** $2\sqrt{6}$ **5.** 1 **7.** $1/2$ **9.** 1 **11.** 0 **13.** 0
15. -1 **17.** 2 **19.** $1/4$ **21.** $19/12$ **23.** 2 **25.** 3
27. (a) All (x, y) **(b)** All (x, y) except $(0, 0)$
29. (a) All (x, y) except where $x = 0$ or $y = 0$ **(b)** All (x, y)

31. **(a)** All (x, y, z) **(b)** All (x, y, z) except the interior of the cylinder $x^2 + y^2 = 1$
33. **(a)** All (x, y, z) with $z \neq 0$ **(b)** All (x, y, z) with $x^2 + z^2 \neq 1$
35. Consider paths along $y = x, x > 0$, and along $y = x, x < 0$
37. Consider the paths $y = kx^2$, k a constant
39. Consider the paths $y = mx$, m a constant, $m \neq -1$
41. Consider the paths $y = kx^2$, k a constant, $k \neq 0$
45. The limit is 1 **47.** The limit is 0
49. **(a)** $f(x, y)|_{y=mx} = \sin 2\theta$ where $\tan\theta = m$ **51.** 0
53. Does not exist **55.** $\pi/2$ **57.** $\delta = 0.1$
59. $\delta = 0.005$ **61.** $\delta = \sqrt{0.015}$ **63.** $\delta = 0.005$

Section 11.3, pp. 607–609

1. $\frac{\partial f}{\partial x} = 4x, \frac{\partial f}{\partial y} = -3$ **3.** $\frac{\partial f}{\partial x} = 2x(y + 2), \frac{\partial f}{\partial y} = x^2 - 1$

5. $\frac{\partial f}{\partial x} = 2y(xy - 1), \frac{\partial f}{\partial y} = 2x(xy - 1)$

7. $\frac{\partial f}{\partial x} = \frac{x}{\sqrt{x^2 + y^2}}, \frac{\partial f}{\partial y} = \frac{y}{\sqrt{x^2 + y^2}}$

9. $\frac{\partial f}{\partial x} = \frac{-1}{(x + y)^2}, \frac{\partial f}{\partial y} = \frac{-1}{(x + y)^2}$

11. $\frac{\partial f}{\partial x} = \frac{-y^2 - 1}{(xy - 1)^2}, \frac{\partial f}{\partial y} = \frac{-x^2 - 1}{(xy - 1)^2}$

13. $\frac{\partial f}{\partial x} = e^{x+y+1}, \frac{\partial f}{\partial y} = e^{x+y+1}$ **15.** $\frac{\partial f}{\partial x} = \frac{1}{x + y}, \frac{\partial f}{\partial y} = \frac{1}{x + y}$

17. $\frac{\partial f}{\partial x} = 2\sin(x - 3y)\cos(x - 3y)$,
$\frac{\partial f}{\partial y} = -6\sin(x - 3y)\cos(x - 3y)$

19. $\frac{\partial f}{\partial x} = yx^{y-1}, \frac{\partial f}{\partial y} = x^y \ln x$ **21.** $\frac{\partial f}{\partial x} = -g(x), \frac{\partial f}{\partial y} = g(y)$

23. $f_x = y^2, f_y = 2xy, f_z = -4z$
25. $f_x = 1, f_y = -y(y^2 + z^2)^{-1/2}, f_z = -z(y^2 + z^2)^{-1/2}$

27. $f_x = \frac{yz}{\sqrt{1 - x^2y^2z^2}}, f_y = \frac{xz}{\sqrt{1 - x^2y^2z^2}}, f_z = \frac{xy}{\sqrt{1 - x^2y^2z^2}}$

29. $f_x = \frac{1}{x + 2y + 3z}, f_y = \frac{2}{x + 2y + 3z}, f_z = \frac{3}{x + 2y + 3z}$

31. $f_x = -2xe^{-(x^2+y^2+z^2)}, f_y = -2ye^{-(x^2+y^2+z^2)}, f_z = -2ze^{-(x^2+y^2+z^2)}$
33. $f_x = \operatorname{sech}^2(x + 2y + 3z), f_y = 2\operatorname{sech}^2(x + 2y + 3z)$,
$f_z = 3\operatorname{sech}^2(x + 2y + 3z)$

35. $\frac{\partial f}{\partial t} = -2\pi\sin(2\pi t - \alpha), \frac{\partial f}{\partial \alpha} = \sin(2\pi t - \alpha)$

37. $\frac{\partial h}{\partial \rho} = \sin\phi\cos\theta, \frac{\partial h}{\partial \phi} = \rho\cos\phi\cos\theta, \frac{\partial h}{\partial \theta} = -\rho\sin\phi\sin\theta$

39. $W_P(P, V, \delta, v, g) = V, W_V(P, V, \delta, v, g) = P + \frac{\delta v^2}{2g}$,
$W_\delta(P, V, \delta, v, g) = \frac{Vv^2}{2g}, W_v(P, V, \delta, v, g) = \frac{V\delta v}{g}$,
$W_g(P, V, \delta, v, g) = -\frac{V\delta v^2}{2g^2}$

41. $\frac{\partial f}{\partial x} = 1 + y, \frac{\partial f}{\partial y} = 1 + x, \frac{\partial^2 f}{\partial x^2} = 0, \frac{\partial^2 f}{\partial y^2} = 0$,
$\frac{\partial^2 f}{\partial y\,\partial x} = \frac{\partial^2 f}{\partial x\,\partial y} = 1$

43. $\frac{\partial g}{\partial x} = 2xy + y\cos x, \frac{\partial g}{\partial y} = x^2 - \sin y + \sin x$,
$\frac{\partial^2 g}{\partial x^2} = 2y - y\sin x, \frac{\partial^2 g}{\partial y^2} = -\cos y$,
$\frac{\partial^2 g}{\partial y\,\partial x} = \frac{\partial^2 g}{\partial x\,\partial y} = 2x + \cos x$

45. $\frac{\partial r}{\partial x} = \frac{1}{x + y}, \frac{\partial r}{\partial y} = \frac{1}{x + y}, \frac{\partial^2 r}{\partial x^2} = \frac{-1}{(x + y)^2}, \frac{\partial^2 r}{\partial y^2} = \frac{-1}{(x + y)^2}$,
$\frac{\partial^2 r}{\partial y\,\partial x} = \frac{\partial^2 r}{\partial x\,\partial y} = \frac{-1}{(x + y)^2}$

47. $\frac{\partial w}{\partial x} = \frac{2}{2x + 3y}, \frac{\partial w}{\partial y} = \frac{3}{2x + 3y}, \frac{\partial^2 w}{\partial y\,\partial x} = \frac{\partial^2 w}{\partial x\,\partial y} = \frac{-6}{(2x + 3y)^2}$

49. $\frac{\partial w}{\partial x} = y^2 + 2xy^3 + 3x^2y^4, \frac{\partial w}{\partial y} = 2xy + 3x^2y^2 + 4x^3y^3$,
$\frac{\partial^2 w}{\partial y\,\partial x} = \frac{\partial^2 w}{\partial x\,\partial y} = 2y + 6xy^2 + 12x^2y^3$

51. **(a)** x first **(b)** y first **(c)** x first
(d) x first **(e)** y first **(f)** y first
53. $f_x(1, 2) = -13, f_y(1, 2) = -2$ **55.** 12 **57.** -2

59. $\frac{\partial A}{\partial a} = \frac{a}{bc\sin A}, \frac{\partial A}{\partial b} = \frac{c\cos A - b}{bc\sin A}$

61. $v_x = \frac{\ln v}{(\ln u)(\ln v) - 1}$

77. Yes

Section 11.4, pp. 616–618

1. **(a)** $\frac{dw}{dt} = 0$, **(b)** $\frac{dw}{dt}(\pi) = 0$

3. **(a)** $\frac{dw}{dt} = 1$, **(b)** $\frac{dw}{dt}(3) = 1$

5. **(a)** $\frac{dw}{dt} = 4t\tan^{-1}t + 1$, **(b)** $\frac{dw}{dt}(1) = \pi + 1$

7. **(a)** $\frac{\partial z}{\partial u} = 4\cos v\ln(u\sin v) + 4\cos v$,
$\frac{\partial z}{\partial v} = -4u\sin v\ln(u\sin v) + \frac{4u\cos^2 v}{\sin v}$
(b) $\frac{\partial z}{\partial u} = \sqrt{2}(\ln 2 + 2), \frac{\partial z}{\partial v} = -2\sqrt{2}(\ln 2 - 2)$

9. **(a)** $\frac{\partial w}{\partial u} = 2u + 4uv, \frac{\partial w}{\partial v} = -2v + 2u^2$
(b) $\frac{\partial w}{\partial u} = 3, \frac{\partial w}{\partial v} = -\frac{3}{2}$

11. **(a)** $\frac{\partial u}{\partial x} = 0, \frac{\partial u}{\partial y} = \frac{z}{(z - y)^2}, \frac{\partial u}{\partial z} = \frac{-y}{(z - y)^2}$
(b) $\frac{\partial u}{\partial x} = 0, \frac{\partial u}{\partial y} = 1, \frac{\partial u}{\partial z} = -2$

13. $\dfrac{dz}{dt} = \dfrac{\partial z}{\partial x}\dfrac{dx}{dt} + \dfrac{\partial z}{\partial y}\dfrac{dy}{dt}$

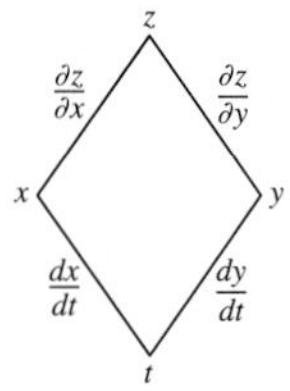

15. $\dfrac{\partial w}{\partial u} = \dfrac{\partial w}{\partial x}\dfrac{\partial x}{\partial u} + \dfrac{\partial w}{\partial y}\dfrac{\partial y}{\partial u} + \dfrac{\partial w}{\partial z}\dfrac{\partial z}{\partial u}$,
$\dfrac{\partial w}{\partial v} = \dfrac{\partial w}{\partial x}\dfrac{\partial x}{\partial v} + \dfrac{\partial w}{\partial y}\dfrac{\partial y}{\partial v} + \dfrac{\partial w}{\partial z}\dfrac{\partial z}{\partial v}$

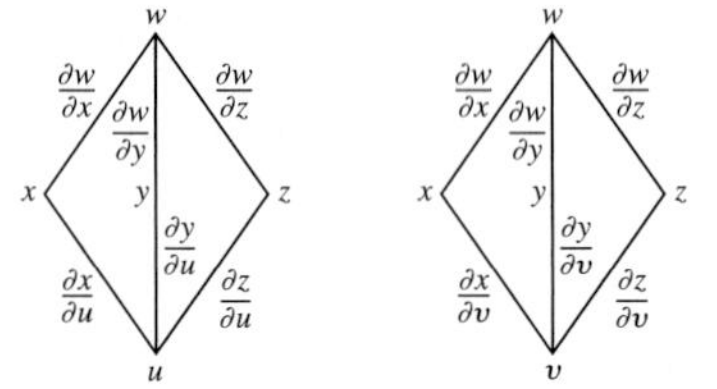

17. $\dfrac{\partial w}{\partial u} = \dfrac{\partial w}{\partial x}\dfrac{\partial x}{\partial u} + \dfrac{\partial w}{\partial y}\dfrac{\partial y}{\partial u}, \dfrac{\partial w}{\partial v} = \dfrac{\partial w}{\partial x}\dfrac{\partial x}{\partial v} + \dfrac{\partial w}{\partial y}\dfrac{\partial y}{\partial v}$

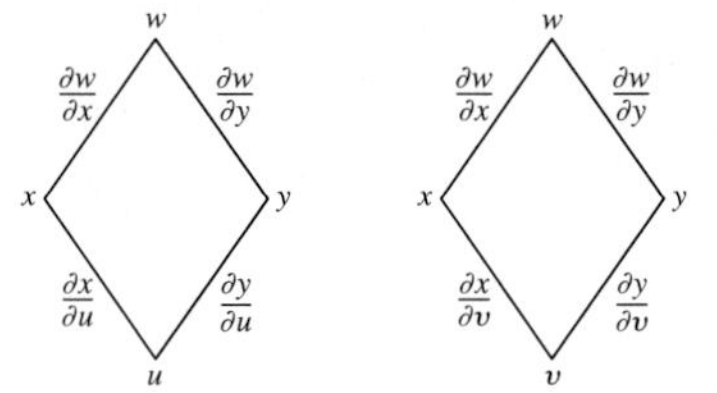

19. $\dfrac{\partial z}{\partial t} = \dfrac{\partial z}{\partial x}\dfrac{\partial x}{\partial t} + \dfrac{\partial z}{\partial y}\dfrac{\partial y}{\partial t}, \dfrac{\partial z}{\partial s} = \dfrac{\partial z}{\partial x}\dfrac{\partial x}{\partial s} + \dfrac{\partial z}{\partial y}\dfrac{\partial y}{\partial s}$

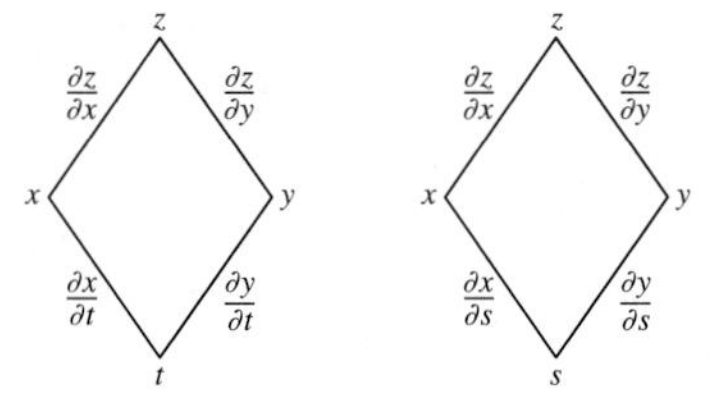

21. $\dfrac{\partial w}{\partial s} = \dfrac{dw}{du}\dfrac{\partial u}{\partial s}, \dfrac{\partial w}{\partial t} = \dfrac{dw}{du}\dfrac{\partial u}{\partial t}$

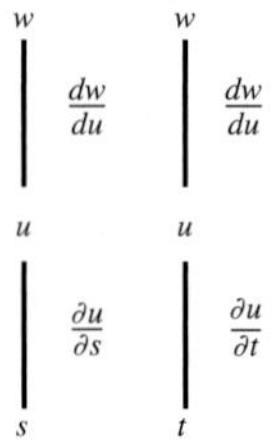

23. $\dfrac{\partial w}{\partial r} = \dfrac{\partial w}{\partial x}\dfrac{dx}{dr} + \dfrac{\partial w}{\partial y}\dfrac{dy}{dr} = \dfrac{\partial w}{\partial x}\dfrac{dx}{dr}$ since $\dfrac{dy}{dr} = 0$,
$\dfrac{\partial w}{\partial s} = \dfrac{\partial w}{\partial x}\dfrac{dx}{ds} + \dfrac{\partial w}{\partial y}\dfrac{dy}{ds} = \dfrac{\partial w}{\partial y}\dfrac{dy}{ds}$ since $\dfrac{dx}{ds} = 0$

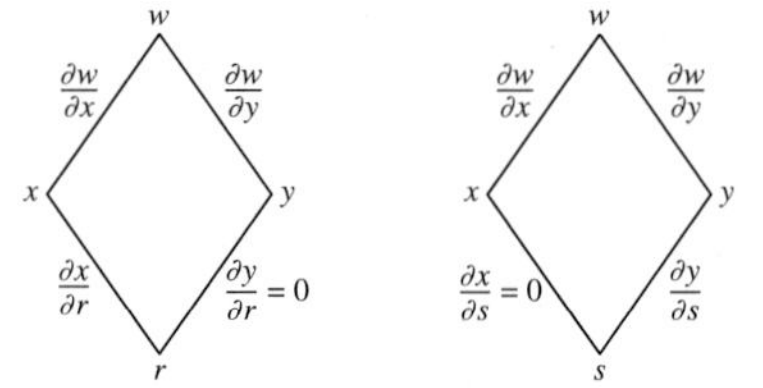

25. $4/3$ **27.** $-4/5$ **29.** $\dfrac{\partial z}{\partial x} = \dfrac{1}{4}, \dfrac{\partial z}{\partial y} = -\dfrac{3}{4}$

31. $\dfrac{\partial z}{\partial x} = -1, \dfrac{\partial z}{\partial y} = -1$ **33.** 12 **35.** -7

37. $\dfrac{\partial z}{\partial u} = 2, \dfrac{\partial z}{\partial v} = 1$ **39.** -0.00005 amps/sec

45. $(\cos 1, \sin 1, 1)$ and $(\cos(-2), \sin(-2), -2)$

47. **(a)** Maximum at $\left(-\dfrac{\sqrt{2}}{2}, \dfrac{\sqrt{2}}{2}\right)$ and $\left(\dfrac{\sqrt{2}}{2}, -\dfrac{\sqrt{2}}{2}\right)$; minimum at $\left(\dfrac{\sqrt{2}}{2}, \dfrac{\sqrt{2}}{2}\right)$ and $\left(-\dfrac{\sqrt{2}}{2}, -\dfrac{\sqrt{2}}{2}\right)$
(b) Max $= 6$, min $= 2$

49. $2x\sqrt{x^8 + x^3} + \displaystyle\int_0^{x^2} \frac{3x^2}{2\sqrt{t^4 + x^3}}\,dt$

Section 11.5, pp. 625–626

1.

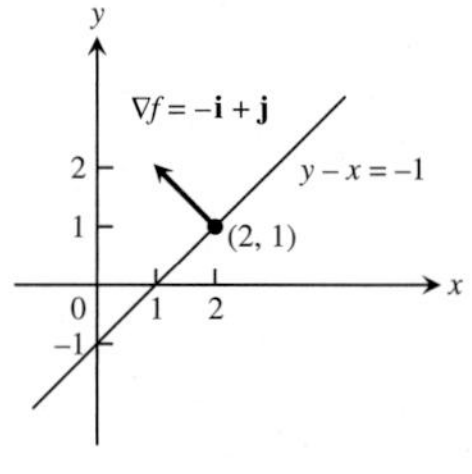

3.

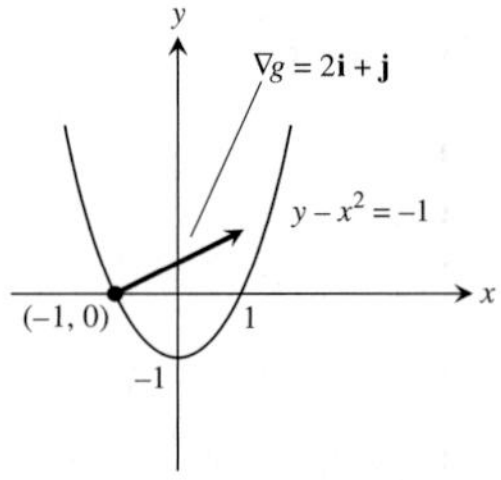

5. $\nabla f = 3\mathbf{i} + 2\mathbf{j} - 4\mathbf{k}$ **7.** $\nabla f = -\dfrac{26}{27}\mathbf{i} + \dfrac{23}{54}\mathbf{j} - \dfrac{23}{54}\mathbf{k}$

9. -4 **11.** $31/13$ **13.** 3 **15.** 2

17. $\mathbf{u} = -\dfrac{1}{\sqrt{2}}\mathbf{i} + \dfrac{1}{\sqrt{2}}\mathbf{j}, (D_{\mathbf{u}}f)_{P_0} = \sqrt{2}; -\mathbf{u} = \dfrac{1}{\sqrt{2}}\mathbf{i} - \dfrac{1}{\sqrt{2}}\mathbf{j},$
$(D_{-\mathbf{u}}f)_{P_0} = -\sqrt{2}$

19. $\mathbf{u} = \dfrac{1}{3\sqrt{3}}\mathbf{i} - \dfrac{5}{3\sqrt{3}}\mathbf{j} - \dfrac{1}{3\sqrt{3}}\mathbf{k}, (D_{\mathbf{u}}f)_{P_0} = 3\sqrt{3};$
$-\mathbf{u} = -\dfrac{1}{3\sqrt{3}}\mathbf{i} + \dfrac{5}{3\sqrt{3}}\mathbf{j} + \dfrac{1}{3\sqrt{3}}\mathbf{k}, (D_{-\mathbf{u}}f)_{P_0} = -3\sqrt{3}$

21. $\mathbf{u} = \frac{1}{\sqrt{3}}(\mathbf{i} + \mathbf{j} + \mathbf{k}),\ (D_{\mathbf{u}}f)_{P_0} = 2\sqrt{3};$

$-\mathbf{u} = -\frac{1}{\sqrt{3}}(\mathbf{i} + \mathbf{j} + \mathbf{k}),\ (D_{-\mathbf{u}}f)_{P_0} = -2\sqrt{3}$

23.

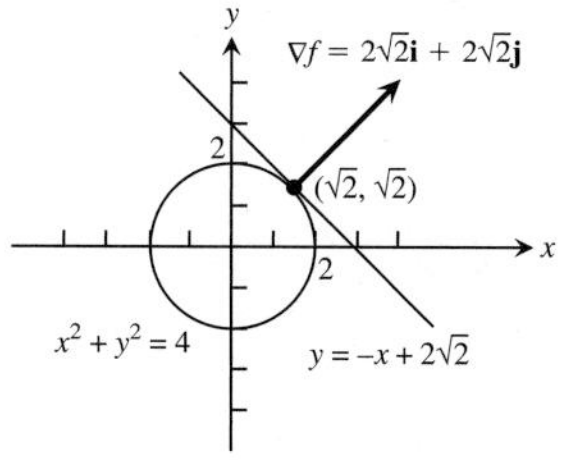

25.

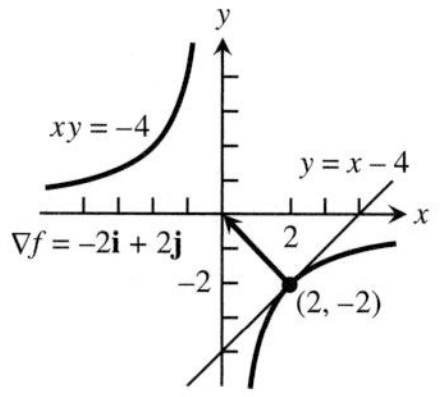

27. $\mathbf{u} = \frac{7}{\sqrt{53}}\mathbf{i} - \frac{2}{\sqrt{53}}\mathbf{j},\ -\mathbf{u} = -\frac{7}{\sqrt{53}}\mathbf{i} + \frac{2}{\sqrt{53}}\mathbf{j}$

29. No, the maximum rate of change is $\sqrt{185} < 14$. **31.** $-\frac{7}{\sqrt{5}}$

Section 11.6, pp. 632–634

1. **(a)** $x + y + z = 3$
(b) $x = 1 + 2t,\ y = 1 + 2t,\ z = 1 + 2t$

3. **(a)** $2x - z - 2 = 0$ **(b)** $x = 2 - 4t,\ y = 0,\ z = 2 + 2t$

5. **(a)** $2x + 2y + z - 4 = 0$
(b) $x = 2t,\ y = 1 + 2t,\ z = 2 + t$

7. **(a)** $x + y + z - 1 = 0$ **(b)** $x = t,\ y = 1 + t,\ z = t$

9. $2x - z - 2 = 0$ **11.** $x - y + 2z - 1 = 0$

13. $x = 1,\ y = 1 + 2t,\ z = 1 - 2t$

15. $x = 1 - 2t,\ y = 1,\ z = \frac{1}{2} + 2t$

17. $x = 1 + 90t,\ y = 1 - 90t,\ z = 3$

19. $df = \frac{9}{11{,}830} \approx 0.0008$ **21.** $dg = 0$

23. **(a)** $\frac{\sqrt{3}}{2}\sin\sqrt{3} - \frac{1}{2}\cos\sqrt{3} \approx 0.935°\text{C/ft}$

(b) $\sqrt{3}\sin\sqrt{3} - \cos\sqrt{3} \approx 1.87°\text{C/sec}$

25. **(a)** $L(x, y) = 1$ **(b)** $L(x, y) = 2x + 2y - 1$

27. **(a)** $L(x, y) = 3x - 4y + 5$ **(b)** $L(x, y) = 3x - 4y + 5$

29. **(a)** $L(x, y) = 1 + x$ **(b)** $L(x, y) = -y + \frac{\pi}{2}$

31. $L(x, y) = 7 + x - 6y;\ 0.06$ **33.** $L(x, y) = x + y + 1;\ 0.08$

35. $L(x, y) = 1 + x;\ 0.0222$

37. **(a)** $L(x, y, z) = 2x + 2y + 2z - 3$ **(b)** $L(x, y, z) = y + z$
(c) $L(x, y, z) = 0$

39. **(a)** $L(x, y, z) = x$ **(b)** $L(x, y, z) = \frac{1}{\sqrt{2}}x + \frac{1}{\sqrt{2}}y$

(c) $L(x, y, z) = \frac{1}{3}x + \frac{2}{3}y + \frac{2}{3}z$

41. **(a)** $L(x, y, z) = 2 + x$

(b) $L(x, y, z) = x - y - z + \frac{\pi}{2} + 1$

(c) $L(x, y, z) = x - y - z + \frac{\pi}{2} + 1$

43. $L(x, y, z) = 2x - 6y - 2z + 6,\ 0.0024$

45. $L(x, y, z) = x + y - z - 1,\ 0.00135$

47. Maximum error (estimate) ≤ 0.31 in magnitude

49. f is most sensitive to a change in d.

53. At $-\frac{\pi}{4}, -\frac{\pi}{2\sqrt{2}}$; at 0, 0; at $\frac{\pi}{4}, \frac{\pi}{2\sqrt{2}}$

Section 11.7, pp. 641–643

1. $f(-3, 3) = -5$, local minimum

3. $f\left(\frac{2}{3}, \frac{4}{3}\right) = 0$, local maximum

5. $f(-2, 1)$, saddle point

7. $f\left(\frac{6}{5}, \frac{69}{25}\right)$, saddle point **9.** $f(2, 1)$, saddle point

11. $f(2, -1) = -6$, local minimum **13.** $f(1, 2)$, saddle point

15. $f(0, 0)$, saddle point

17. $f(0, 0)$, saddle point; $f\left(-\frac{2}{3}, \frac{2}{3}\right) = \frac{170}{27}$, local maximum

19. $f(0, 0) = 0$, local minimum; $f(1, -1)$, saddle point

21. $f(0, 0)$, saddle point; $f\left(\frac{4}{9}, \frac{4}{3}\right) = -\frac{64}{81}$, local minimum

23. $f(0, 0)$, saddle point; $f(0, 2) = -12$, local minimum; $f(-2, 0) = -4$, local maximum; $f(-2, 2)$, saddle point

25. $f(0, 0)$, saddle point; $f(1, 1) = 2$, $f(-1, -1) = 2$, local maxima

27. $f(0, 0) = -1$, local maximum

29. $f(n\pi, 0)$, saddle point; $f(n\pi, 0) = 0$ for every n

31. Absolute maximum: 1 at (0, 0); absolute minimum: -5 at (1, 2)

33. Absolute maximum: 4 at (0, 2); absolute minimum: 0 at (0, 0)

35. Absolute maximum: 11 at $(0, -3)$; absolute minimum: -10 at $(4, -2)$

37. Absolute maximum: 4 at (2, 0); absolute minimum: $\frac{3\sqrt{2}}{2}$ at $\left(3, -\frac{\pi}{4}\right), \left(3, \frac{\pi}{4}\right), \left(1, -\frac{\pi}{4}\right)$, and $\left(1, \frac{\pi}{4}\right)$

39. $a = -3,\ b = 2$

41. **(a)** $f(0, 0)$, saddle point **(b)** $f(1, 2)$, local minimum
(c) $f(1, -2)$, local minimum; $f(-1, -2)$, saddle point

47. $\left(\frac{1}{6}, \frac{1}{3}, \frac{355}{36}\right)$

51. **(a)** On the semicircle, max $f = 2\sqrt{2}$ at $t = \pi/4$, min $f = -2$ at $t = \pi$. On the quarter circle, max $f = 2\sqrt{2}$ at $t = \pi/4$, min $f = 2$ at $t = 0, \pi/2$.
(b) On the semicircle, max $g = 2$ at $t = \pi/4$, min $g = -2$ at $t = 3\pi/4$. On the quarter circle, max $g = 2$ at $t = \pi/4$, min $g = 0$ at $t = 0, \pi/2$.
(c) On the semicircle, max $h = 8$ at $t = 0, \pi$; min $h = 4$ at $t = \pi/2$. On the quarter circle, max $h = 8$ at $t = 0$, min $h = 4$ at $t = \pi/2$.

53. i) min $f = -1/2$ at $t = -1/2$; no max ii) max $f = 0$ at $t = -1, 0$; min $f = -1/2$ at $t = -1/2$ iii) max $f = 4$ at $t = 1$; min $f = 0$ at $t = 0$

57. $y = -\frac{20}{13}x + \frac{9}{13}$, $y|_{x=4} = -\frac{71}{13}$

Section 11.8, pp. 651–652

1. $\left(\pm\frac{1}{\sqrt{2}}, \frac{1}{2}\right), \left(\pm\frac{1}{\sqrt{2}}, -\frac{1}{2}\right)$ **3.** 39

5. $(3, \pm 3\sqrt{2})$

7. **(a)** 8 **(b)** 64

9. $r = 2$ cm, $h = 4$ cm

11. Length $= 4\sqrt{2}$, width $= 3\sqrt{2}$

13. $f(0, 0) = 0$ is minimum, $f(2, 4) = 20$ is maximum.

15. Lowest $= 0°$, highest $= 125°$

17. $\left(\frac{3}{2}, 2, \frac{5}{2}\right)$ **19.** 1 **21.** $(0, 0, 2), (0, 0, -2)$

23. $f(1, -2, 5) = 30$ is maximum, $f(-1, 2, -5) = -30$ is minimum.

25. 3, 3, 3 **27.** $\frac{2}{\sqrt{3}}$ by $\frac{2}{\sqrt{3}}$ by $\frac{2}{\sqrt{3}}$ units

29. $(\pm 4/3, -4/3, -4/3)$ **31.** $f(2/3, 4/3, -4/3) = \frac{4}{3}$

33. $(2, 4, 4)$

35. Maximum is $1 + 6\sqrt{3}$ at $(\pm\sqrt{6}, \sqrt{3}, 1)$, minimum is $1 - 6\sqrt{3}$ at $(\pm\sqrt{6}, -\sqrt{3}, 1)$.

37. Maximum is 4 at $(0, 0, \pm 2)$, minimum is 2 at $(\pm\sqrt{2}, \pm\sqrt{2}, 0)$.

Practice and Additional Exercises, pp. 653–656

1. Domain: all points in the xy-plane; range: $z \geq 0$. Level curves are ellipses with major axis along the y-axis and minor axis along the x-axis.

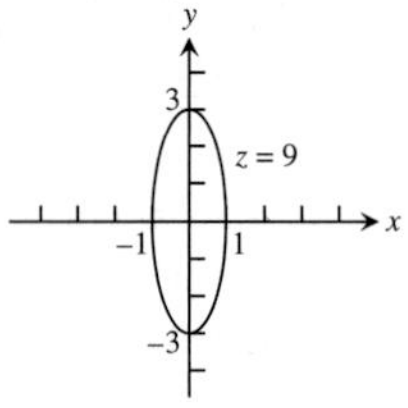

3. Domain: all (x, y) such that $x \neq 0$ and $y \neq 0$; range: $z \neq 0$. Level curves are hyperbolas with the x- and y-axes as asymptotes.

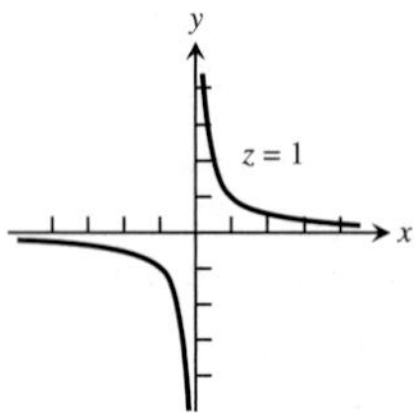

5. Domain: all points in xyz-space; range: all real numbers. Level surfaces are paraboloids of revolution with the z-axis as axis.

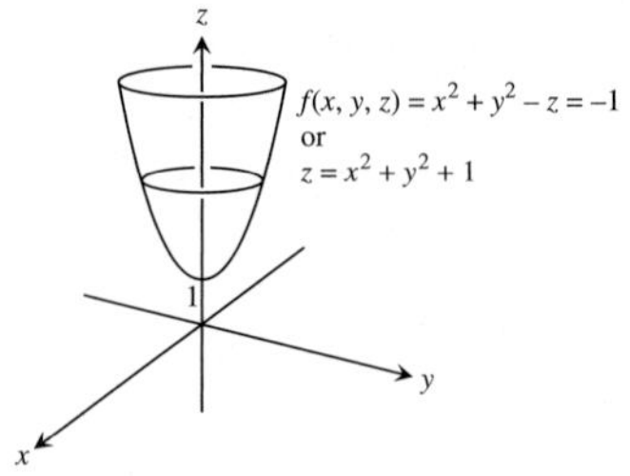

7. Domain: all (x, y, z) such that $(x, y, z) \neq (0, 0, 0)$; range: positive real numbers. Level surfaces are spheres with center $(0, 0, 0)$ and radius $r > 0$.

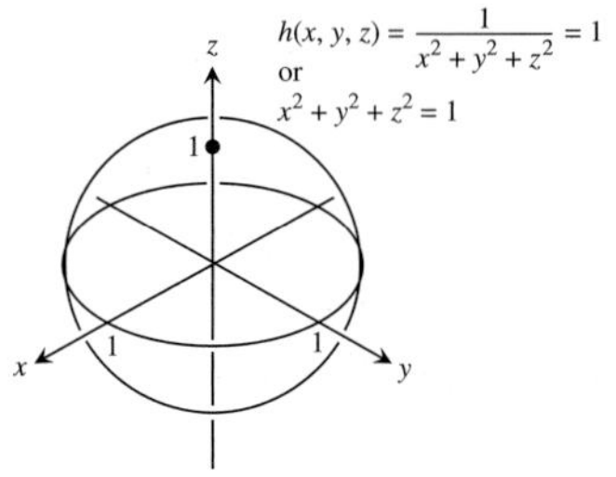

9. -2 **11.** $1/2$ **13.** 1 **15.** Let $y = kx^2, k \neq 1$ **17.** 0

19. $\frac{\partial g}{\partial r} = \cos\theta + \sin\theta, \frac{\partial g}{\partial \theta} = -r\sin\theta + r\cos\theta$

21. $\frac{\partial f}{\partial R_1} = -\frac{1}{R_1^2}, \frac{\partial f}{\partial R_2} = -\frac{1}{R_2^2}, \frac{\partial f}{\partial R_3} = -\frac{1}{R_3^2}$

23. $\frac{\partial P}{\partial n} = \frac{RT}{V}, \frac{\partial P}{\partial R} = \frac{nT}{V}, \frac{\partial P}{\partial T} = \frac{nR}{V}, \frac{\partial P}{\partial V} = -\frac{nRT}{V^2}$

25. $\frac{\partial^2 g}{\partial x^2} = 0, \frac{\partial^2 g}{\partial y^2} = \frac{2x}{y^3}, \frac{\partial^2 g}{\partial y\,\partial x} = \frac{\partial^2 g}{\partial x\,\partial y} = -\frac{1}{y^2}$

27. $\frac{\partial^2 f}{\partial x^2} = -30x + \frac{2 - 2x^2}{(x^2 + 1)^2}, \frac{\partial^2 f}{\partial y^2} = 0, \frac{\partial^2 f}{\partial y\,\partial x} = \frac{\partial^2 f}{\partial x\,\partial y} = 1$

29. $\left.\frac{dw}{dt}\right|_{t=0} = -1$

31. $\left.\frac{\partial w}{\partial r}\right|_{(r,s)=(\pi,0)} = 2, \left.\frac{\partial w}{\partial s}\right|_{(r,s)=(\pi,0)} = 2 - \pi$

33. $\left.\frac{df}{dt}\right|_{t=1} = -(\sin 1 + \cos 2)(\sin 1) + (\cos 1 + \cos 2)(\cos 1) - 2(\sin 1 + \cos 1)(\sin 2)$

35. $\left.\frac{dy}{dx}\right|_{(x,y)=(0,1)} = -1$

37. Increases most rapidly in the direction $\mathbf{u} = -\frac{\sqrt{2}}{2}\mathbf{i} - \frac{\sqrt{2}}{2}\mathbf{j}$; decreases most rapidly in the direction $-\mathbf{u} = \frac{\sqrt{2}}{2}\mathbf{i} + \frac{\sqrt{2}}{2}\mathbf{j}$; $D_{\mathbf{u}}f = \frac{\sqrt{2}}{2}$; $D_{-\mathbf{u}}f = -\frac{\sqrt{2}}{2}$; $D_{\mathbf{u}_1}f = -\frac{7}{10}$ where $\mathbf{u}_1 = \frac{\mathbf{v}}{|\mathbf{v}|}$

39. Increases most rapidly in the direction $\mathbf{u} = \frac{2}{7}\mathbf{i} + \frac{3}{7}\mathbf{j} + \frac{6}{7}\mathbf{k}$; decreases most rapidly in the direction $-\mathbf{u} = -\frac{2}{7}\mathbf{i} - \frac{3}{7}\mathbf{j} - \frac{6}{7}\mathbf{k}$; $D_{\mathbf{u}}f = 7$; $D_{-\mathbf{u}}f = -7$; $D_{\mathbf{u}_1}f = 7$ where $\mathbf{u}_1 = \frac{\mathbf{v}}{|\mathbf{v}|}$
41. $\pi/\sqrt{2}$
43. **(a)** $f_x(1, 2) = f_y(1, 2) = 2$ **(b)** $14/5$
45.

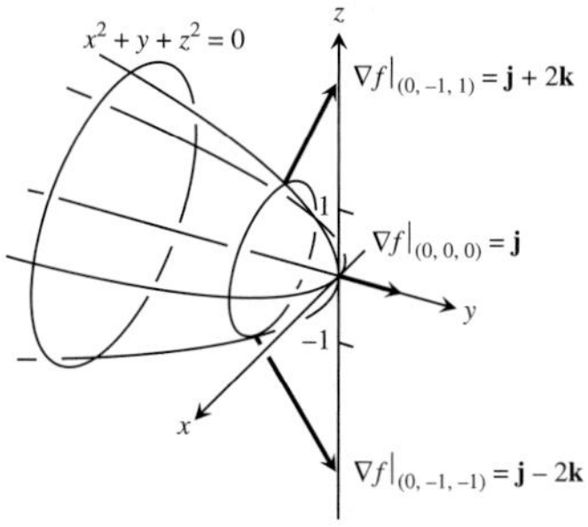

47. Tangent: $4x - y - 5z = 4$; normal line: $x = 2 + 4t, y = -1 - t, z = 1 - 5t$
49. $2y - z - 2 = 0$
51. Tangent: $x + y = \pi + 1$; normal line: $y = x - \pi + 1$

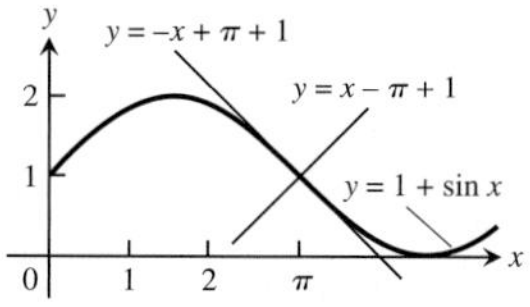

53. $x = 1 - 2t, y = 1, z = 1/2 + 2t$
55. Answers will depend on the upper bound used for $|f_{xx}|, |f_{xy}|, |f_{yy}|$. With $M = \sqrt{2}/2$, $|E| \le 0.0142$. With $M = 1$, $|E| \le 0.02$.
57. $L(x, y, z) = y - 3z$, $L(x, y, z) = x + y - z - 1$
59. Local minimum of -8 at $(-2, -2)$
61. Saddle point at $(0, 0)$, $f(0, 0) = 0$; local maximum of $1/4$ at $(-1/2, -1/2)$
63. Saddle point at $(0, 0)$, $f(0, 0) = 0$; local minimum of -4 at $(0, 2)$; local maximum of 4 at $(-2, 0)$; saddle point at $(-2, 2)$, $f(-2, 2) = 0$
65. Absolute maximum: 28 at $(0, 4)$; absolute minimum: $-9/4$ at $(3/2, 0)$
67. Absolute maximum: 18 at $(2, -2)$; absolute minimum: $-17/4$ at $(-2, 1/2)$
69. Absolute maximum: 8 at $(-2, 0)$; absolute minimum: -1 at $(1, 0)$
71. Absolute maximum: 4 at $(1, 0)$; absolute minimum: -4 at $(0, -1)$
73. Absolute maximum: 1 at $(0, \pm 1)$ and $(1, 0)$; absolute minimum: -1 at $(-1, 0)$
75. Maximum: 5 at $(0, 1)$; minimum: $-1/3$ at $(0, -1/3)$
77. Maximum: $\sqrt{3}$ at $\left(\frac{1}{\sqrt{3}}, -\frac{1}{\sqrt{3}}, \frac{1}{\sqrt{3}}\right)$; minimum: $-\sqrt{3}$ at $\left(-\frac{1}{\sqrt{3}}, \frac{1}{\sqrt{3}}, -\frac{1}{\sqrt{3}}\right)$
79. Width $= \left(\frac{c^2V}{ab}\right)^{1/3}$, depth $= \left(\frac{b^2V}{ac}\right)^{1/3}$, height $= \left(\frac{a^2V}{bc}\right)^{1/3}$
81. Maximum: $\frac{3}{2}$ at $\left(\frac{1}{\sqrt{2}}, \frac{1}{\sqrt{2}}, \sqrt{2}\right)$ and $\left(-\frac{1}{\sqrt{2}}, -\frac{1}{\sqrt{2}}, -\sqrt{2}\right)$; minimum: $\frac{1}{2}$ at $\left(-\frac{1}{\sqrt{2}}, \frac{1}{\sqrt{2}}, -\sqrt{2}\right)$ and $\left(\frac{1}{\sqrt{2}}, -\frac{1}{\sqrt{2}}, \sqrt{2}\right)$
83. $\frac{\partial w}{\partial x} = \cos\theta \frac{\partial w}{\partial r} - \frac{\sin\theta}{r}\frac{\partial w}{\partial \theta}$, $\frac{\partial w}{\partial y} = \sin\theta \frac{\partial w}{\partial r} + \frac{\cos\theta}{r}\frac{\partial w}{\partial \theta}$
89. $(t, -t \pm 4, t)$, t a real number

CHAPTER 12

Section 12.1, pp. 661–662

1. 24 **3.** 1 **5.** 16 **7.** $2\ln 2 - 1$ **9.** $(3/2)(5 - e)$ **11.** $3/2$ **13.** 14 **15.** 0 **17.** $1/2$ **19.** $2\ln 2$ **21.** $(\ln 2)^2$ **23.** $8/3$ **25.** 1 **27.** $\sqrt{2}$

Section 12.2, pp. 668–670

1. $\frac{\pi^2}{2} + 2$

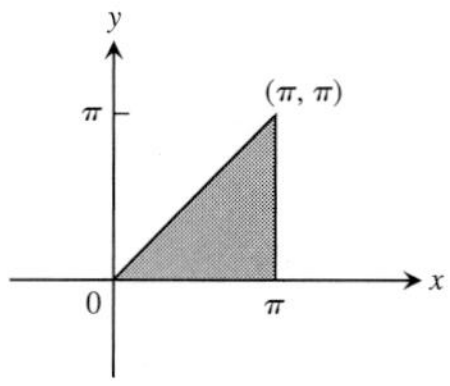

3. $8\ln 8 - 16 + e$

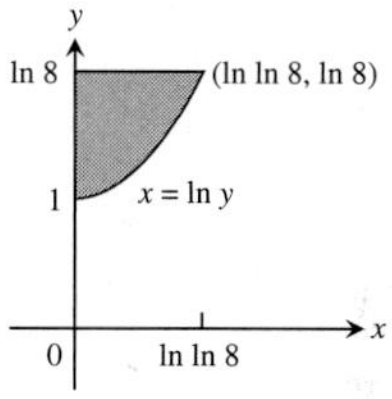

5. $e - 2$

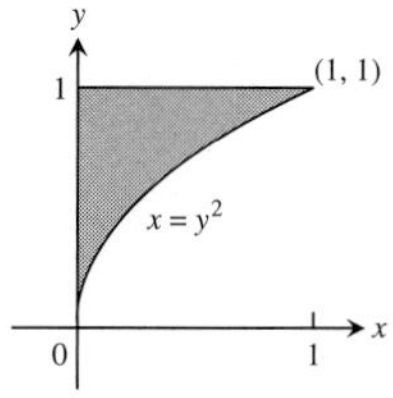

7. $\frac{3}{2}\ln 2$ **9.** $-1/10$

11. 8

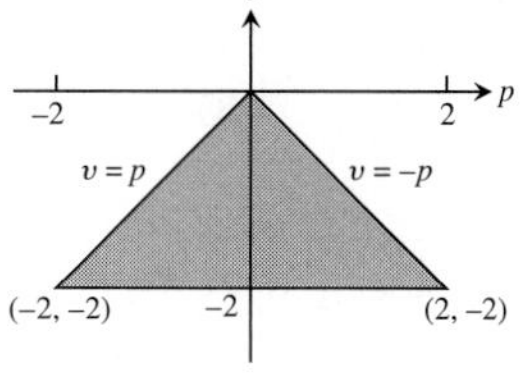

13. 2π

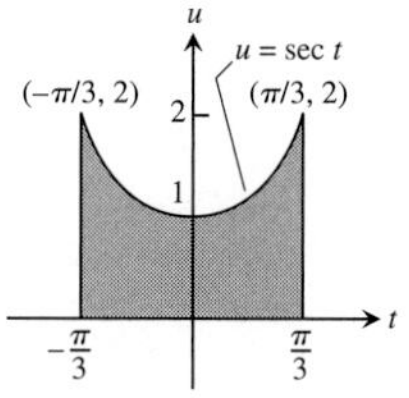

15. $\int_2^4 \int_0^{(4-y)/2} dx\, dy$

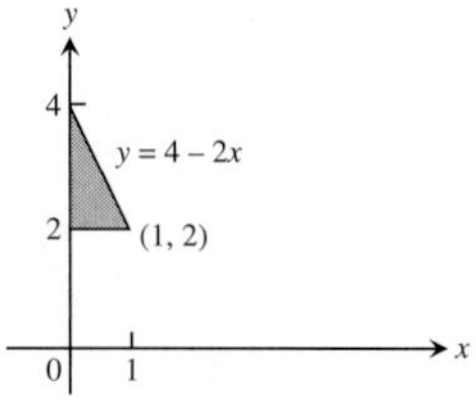

17. $\int_0^1 \int_{x^2}^{x} dy\, dx$

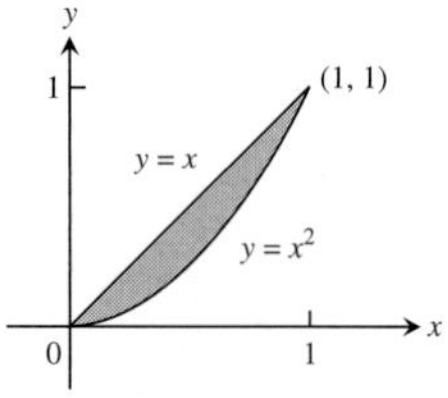

19. $\int_1^e \int_{\ln y}^{1} dx\, dy$

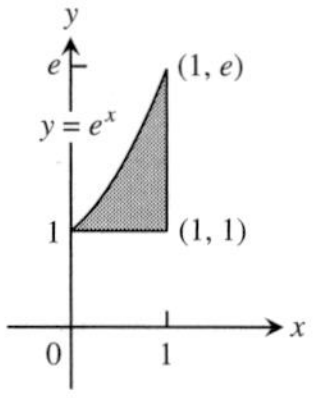

21. $\int_0^9 \int_0^{(\sqrt{9-y})/2} 16x\, dx\, dy$

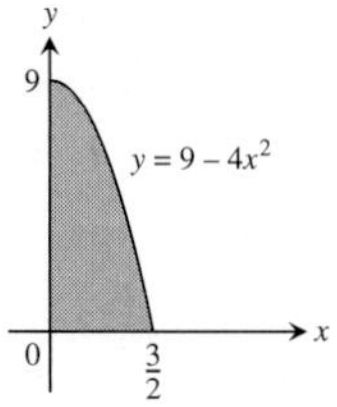

23. $\int_{-1}^1 \int_0^{\sqrt{1-x^2}} 3y\, dy\, dx$

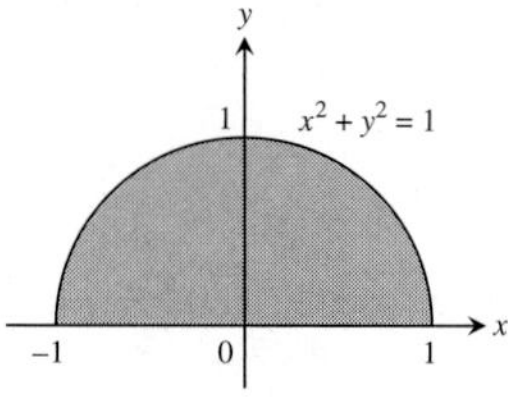

25. 2

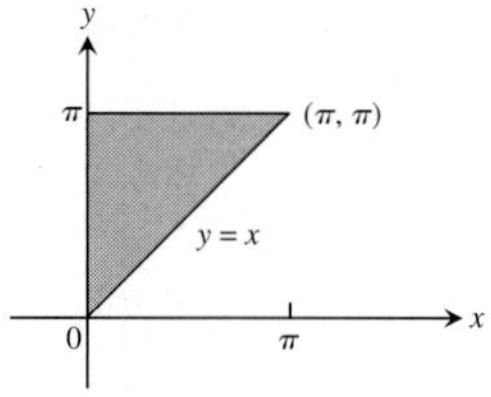

27. $\frac{e-2}{2}$

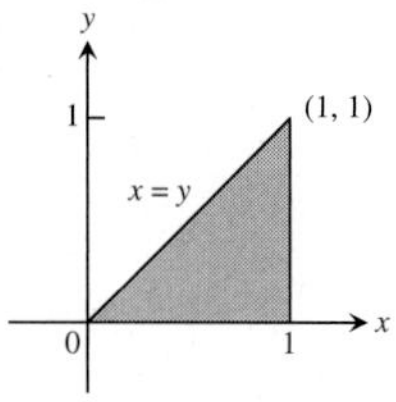

29. 2

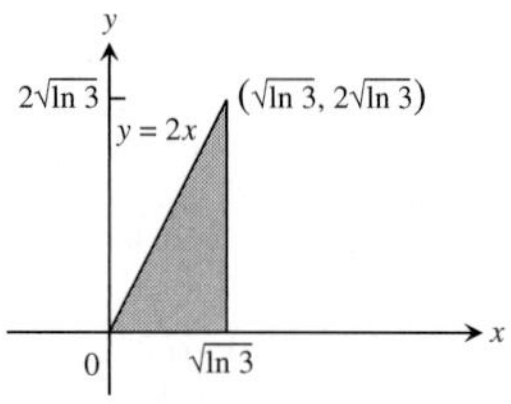

31. $1/(80\pi)$

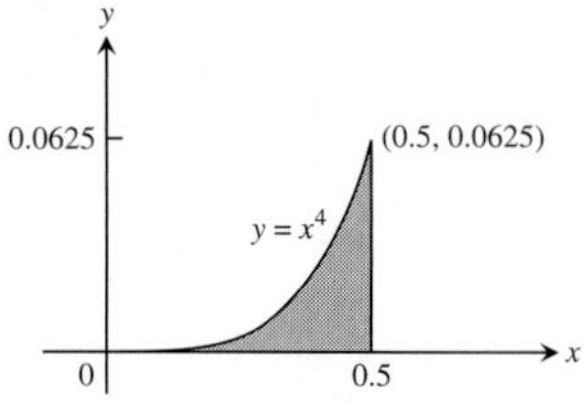

33. $-2/3$

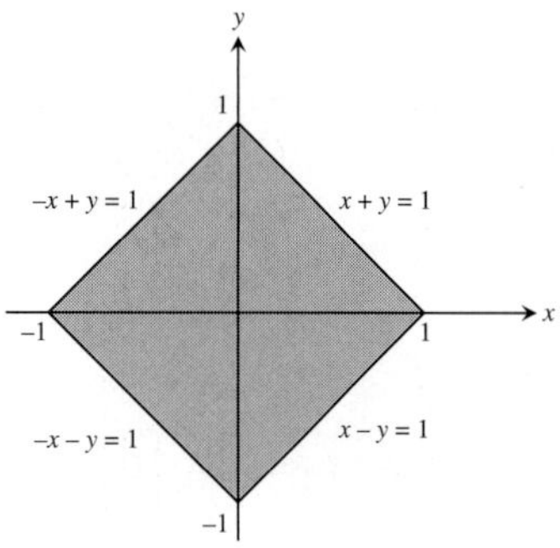

35. $4/3$ **37.** $625/12$ **39.** 16 **41.** 20 **43.** $2(1 + \ln 2)$

45. 1 **47.** π^2 **49.** $-\frac{3}{32}$ **51.** $\frac{20\sqrt{3}}{9}$

53. $\int_0^1 \int_x^{2-x} (x^2 + y^2)\, dy\, dx = \frac{4}{3}$

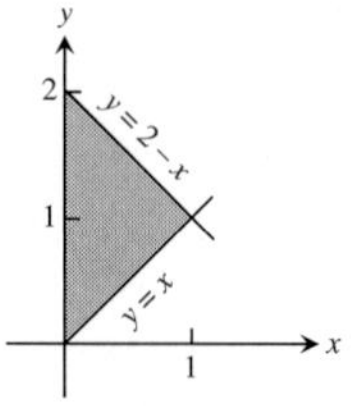

55. R is the set of points (x, y) such that $x^2 + 2y^2 < 4$

57. No, by Fubini's Theorem, the two orders of integration must give the same result.

61. 0.603 **63.** 0.233

Section 12.3, pp. 672–673

1. $\int_0^2 \int_0^{2-x} dy\, dx = 2$ or $\int_0^2 \int_0^{2-y} dx\, dy = 2$

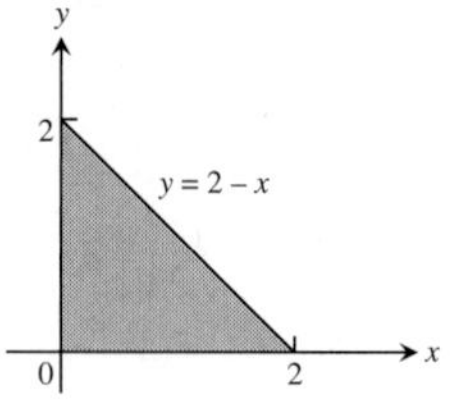

3. $\int_{-2}^1 \int_{y-2}^{-y^2} dx\, dy = \frac{9}{2}$

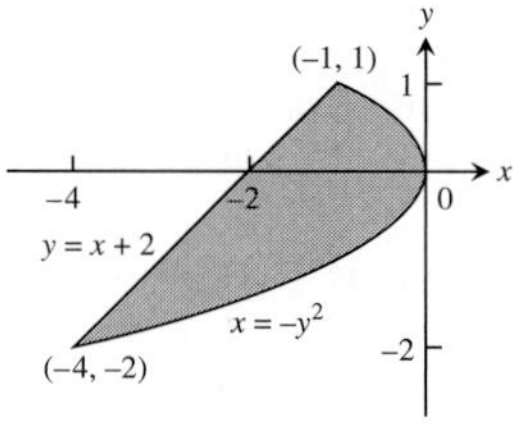

5. $\int_0^{\ln 2} \int_0^{e^x} dy\, dx = 1$

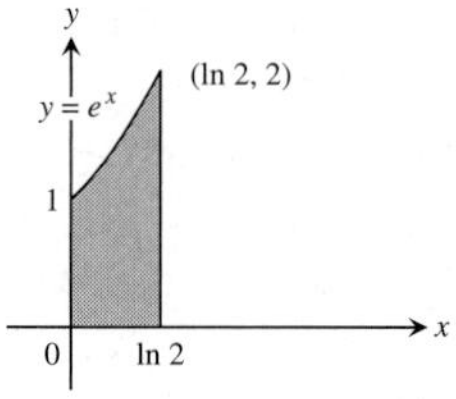

7. $\int_0^1 \int_{y^2}^{2y-y^2} dx\, dy = \frac{1}{3}$

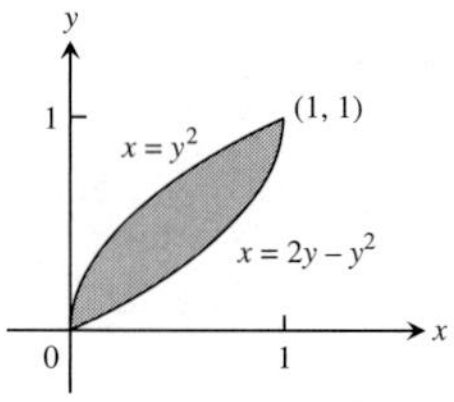

9. 12

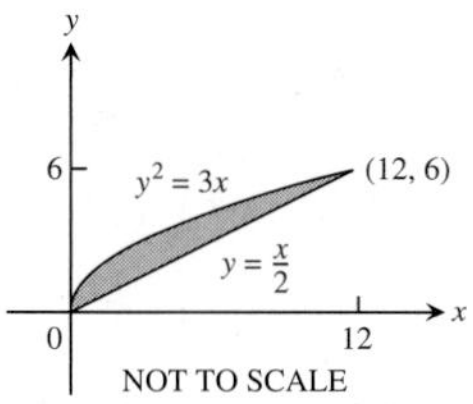

11. $\sqrt{2} - 1$

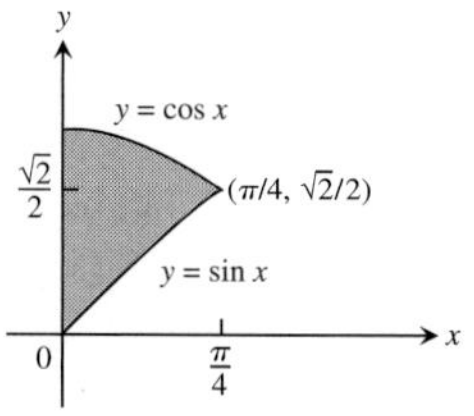

13. $\frac{3}{2}$

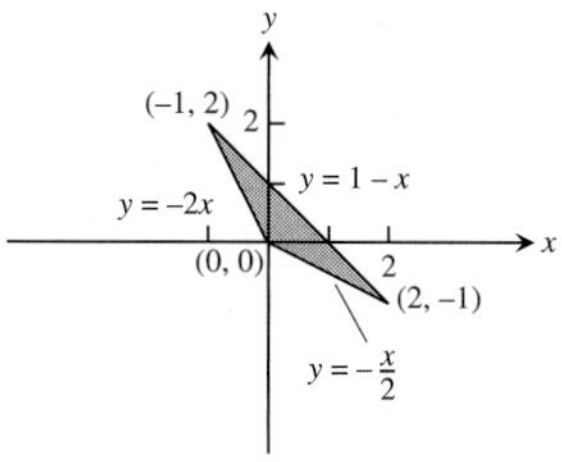

15. (a) 0 **(b)** $4/\pi^2$ **17.** $8/3$
19. $40{,}000(1 - e^{-2})\ln(7/2) \approx 43{,}329$

Section 12.4, pp. 677–678

1. $\pi/2$ **3.** $\pi/8$ **5.** πa^2 **7.** 36 **9.** $(1 - \ln 2)\pi$
11. $(2\ln 2 - 1)(\pi/2)$ **13.** $(\pi/2) + 1$ **15.** $\pi(\ln 4 - 1)$
17. $2(\pi - 1)$ **19.** 12π **21.** $(3\pi/8) + 1$ **23.** $\frac{2a}{3}$ **25.** $\frac{2a}{3}$
27. $2\pi\left(2 - \sqrt{e}\right)$ **29.** $\frac{4}{3} + \frac{5\pi}{8}$ **31. (a)** $\frac{\sqrt{\pi}}{2}$ **(b)** 1
33. $\pi \ln 4$, no **35.** $\frac{1}{2}(a^2 + 2h^2)$

Section 12.5, pp. 684–686

1. $1/6$

3. $\int_0^1\int_0^{2-2x}\int_0^{3-3x-3y/2} dz\,dy\,dx,\quad \int_0^2\int_0^{1-y/2}\int_0^{3-3x-3y/2} dz\,dx\,dy,$
$\int_0^1\int_0^{3-3x}\int_0^{2-2x-2z/3} dy\,dz\,dx,\quad \int_0^3\int_0^{1-z/3}\int_0^{2-2x-2z/3} dy\,dx\,dz,$
$\int_0^2\int_0^{3-3y/2}\int_0^{1-y/2-z/3} dx\,dz\,dy,$
$\int_0^3\int_0^{2-2z/3}\int_0^{1-y/2-z/3} dx\,dy\,dz.$
The value of all six integrals is 1.

5. $\int_{-2}^2\int_{-\sqrt{4-x^2}}^{\sqrt{4-x^2}}\int_{x^2+y^2}^{8-x^2-y^2} 1\,dz\,dx\,dy,\quad \int_{-2}^2\int_{-\sqrt{4-y^2}}^{\sqrt{4-y^2}}\int_{x^2+y^2}^{8-x^2-y^2} 1\,dz\,dx\,dy,$
$\int_{-2}^2\int_4^{8-y^2}\int_{-\sqrt{8-z-y^2}}^{\sqrt{8-z-y^2}} 1\,dx\,dz\,dy + \int_{-2}^2\int_{y^2}^4\int_{-\sqrt{z-y^2}}^{\sqrt{z-y^2}} 1\,dx\,dz\,dy,$
$\int_4^8\int_{-\sqrt{8-z}}^{\sqrt{8-z}}\int_{-\sqrt{8-z-y^2}}^{\sqrt{8-z-y^2}} 1\,dx\,dy\,dz + \int_0^4\int_{-\sqrt{z}}^{\sqrt{z}}\int_{-\sqrt{z-y^2}}^{\sqrt{z-y^2}} 1\,dx\,dy\,dz,$
$\int_{-2}^2\int_4^{8-x^2}\int_{-\sqrt{8-z-x^2}}^{\sqrt{8-z-x^2}} 1\,dy\,dz\,dx + \int_{-2}^2\int_{x^2}^4\int_{-\sqrt{z-x^2}}^{\sqrt{z-x^2}} 1\,dy\,dz\,dx,$
$\int_4^8\int_{-\sqrt{8-z}}^{\sqrt{8-z}}\int_{-\sqrt{8-z-x^2}}^{\sqrt{8-z-x^2}} 1\,dy\,dx\,dz + \int_0^4\int_{-\sqrt{z}}^{\sqrt{z}}\int_{-\sqrt{z-x^2}}^{\sqrt{z-x^2}} 1\,dy\,dx\,dz.$
The value of all six integrals is 16π.

7. 1 **9.** 1 **11.** $\frac{\pi^3}{2}(1 - \cos 1)$ **13.** 18 **15.** $7/6$
17. 0 **19.** $\frac{1}{2} - \frac{\pi}{8}$

21. (a) $\int_{-1}^1\int_0^{1-x^2}\int_{x^2}^{1-z} dy\,dz\,dx$ **(b)** $\int_0^1\int_{-\sqrt{1-z}}^{\sqrt{1-z}}\int_{x^2}^{1-z} dy\,dx\,dz$
(c) $\int_0^1\int_0^{1-z}\int_{-\sqrt{y}}^{\sqrt{y}} dx\,dy\,dz$ **(d)** $\int_0^1\int_0^{1-y}\int_{-\sqrt{y}}^{\sqrt{y}} dx\,dz\,dy$
(e) $\int_0^1\int_{-\sqrt{y}}^{\sqrt{y}}\int_0^{1-y} dz\,dx\,dy$

23. $2/3$ **25.** $20/3$ **27.** 1 **29.** $16/3$ **31.** $8\pi - \frac{32}{3}$
33. 2 **35.** 4π **37.** $31/3$ **39.** 1 **41.** $2\sin 4$ **43.** 4
45. $a = 3$ or $a = 13/3$
47. The domain is the set of all points (x, y, z) such that $4x^2 + 4y^2 + z^2 \le 4$.

Section 12.6, pp. 692–693

1. $\bar{x} = 5/14, \bar{y} = 38/35$ **3.** $\bar{x} = 64/35, \bar{y} = 5/7$
5. $\bar{x} = \bar{y} = 4a/(3\pi)$ **7.** $I_x = I_y = 4\pi, I_0 = 8\pi$
9. $\bar{x} = -1, \bar{y} = 1/4$ **11.** $I_x = 64/105$
13. $\bar{x} = 3/8, \bar{y} = 17/16$
15. $\bar{x} = 11/3, \bar{y} = 14/27, I_y = 432$
17. $\bar{x} = 0, \bar{y} = 13/31, I_y = 7/5$
19. $\bar{x} = 0, \bar{y} = 7/10; I_x = 9/10, I_y = 3/10, I_0 = 6/5$
21. $I_x = \frac{M}{3}(b^2 + c^2), I_y = \frac{M}{3}(a^2 + c^2), I_z = \frac{M}{3}(a^2 + b^2)$
23. $\bar{x} = \bar{y} = 0, \bar{z} = 12/5, I_x = 7904/105 \approx 75.28,$
$I_y = 4832/63 \approx 76.70, I_z = 256/45 \approx 5.69$
25. (a) $\bar{x} = \bar{y} = 0, \bar{z} = 8/3$ **(b)** $c = 2\sqrt{2}$
27. $I_L = 1386$
29. (a) $4/3$ **(b)** $\bar{x} = 4/5, \bar{y} = \bar{z} = 2/5$
31. (a) $5/2$ **(b)** $\bar{x} = \bar{y} = \bar{z} = 8/15$ **(c)** $I_x = I_y = I_z = 11/6$
33. 3

Section 12.7, pp. 702–705

1. $\frac{4\pi\left(\sqrt{2} - 1\right)}{3}$ **3.** $\frac{17\pi}{5}$ **5.** $\pi\left(6\sqrt{2} - 8\right)$ **7.** $\frac{3\pi}{10}$
9. $\pi/3$

11. (a) $\int_0^{2\pi}\int_0^1\int_0^{\sqrt{4-r^2}} r\,dz\,dr\,d\theta$
(b) $\int_0^{2\pi}\int_0^{\sqrt{3}}\int_0^1 r\,dr\,dz\,d\theta + \int_0^{2\pi}\int_{\sqrt{3}}^2\int_0^{\sqrt{4-z^2}} r\,dr\,dz\,d\theta$
(c) $\int_0^1\int_0^{\sqrt{4-r^2}}\int_0^{2\pi} r\,d\theta\,dz\,dr$

13. $\int_{-\pi/2}^{\pi/2}\int_0^{\cos\theta}\int_0^{3r^2} f(r,\theta,z)\,dz\,r\,dr\,d\theta$

15. $\int_0^{\pi}\int_0^{2\sin\theta}\int_0^{4-r\sin\theta} f(r,\theta,z)\,dz\,r\,dr\,d\theta$

17. $\int_{-\pi/2}^{\pi/2}\int_1^{1+\cos\theta}\int_0^{4} f(r,\theta,z)\,dz\,r\,dr\,d\theta$

19. $\int_0^{\pi/4}\int_0^{\sec\theta}\int_0^{2-r\sin\theta} f(r,\theta,z)\,dz\,r\,dr\,d\theta$ **21.** π^2 **23.** $\pi/3$

25. 5π **27.** 2π **29.** $\left(\frac{8-5\sqrt{2}}{2}\right)\pi$

31. (a) $\int_0^{2\pi}\int_0^{\pi/6}\int_0^{2}\rho^2\sin\phi\,d\rho\,d\phi\,d\theta\ +$

$\int_0^{2\pi}\int_{\pi/6}^{\pi/2}\int_0^{\csc\phi}\rho^2\sin\phi\,d\rho\,d\phi\,d\theta$

(b) $\int_0^{2\pi}\int_1^{2}\int_{\pi/6}^{\sin^{-1}(1/\rho)}\rho^2\sin\phi\,d\phi\,d\rho\,d\theta\ +$

$\int_0^{2\pi}\int_0^{2}\int_0^{\pi/6}\rho^2\sin\phi\,d\phi\,d\rho\,d\theta\ +$

$\int_0^{2\pi}\int_0^{1}\int_{\pi/6}^{\pi/2}\rho^2\sin\phi\,d\phi\,d\rho\,d\theta$

33. $\int_0^{2\pi}\int_0^{\pi/2}\int_{\cos\phi}^{2}\rho^2\sin\phi\,d\rho\,d\phi\,d\theta = \frac{31\pi}{6}$

35. $\int_0^{2\pi}\int_0^{\pi}\int_0^{1-\cos\phi}\rho^2\sin\phi\,d\rho\,d\phi\,d\theta = \frac{8\pi}{3}$

37. $\int_0^{2\pi}\int_{\pi/4}^{\pi/2}\int_0^{2\cos\phi}\rho^2\sin\phi\,d\rho\,d\phi\,d\theta = \frac{\pi}{3}$

39. (a) $8\int_0^{\pi/2}\int_0^{\pi/2}\int_0^{2}\rho^2\sin\phi\,d\rho\,d\phi\,d\theta$

(b) $8\int_0^{\pi/2}\int_0^{2}\int_0^{\sqrt{4-r^2}} r\,dz\,dr\,d\theta$

(c) $8\int_0^{2}\int_0^{\sqrt{4-x^2}}\int_0^{\sqrt{4-x^2-y^2}} dz\,dy\,dx$

41. (a) $\int_0^{2\pi}\int_0^{\pi/3}\int_{\sec\phi}^{2}\rho^2\sin\phi\,d\rho\,d\phi\,d\theta$

(b) $\int_0^{2\pi}\int_0^{\sqrt{3}}\int_1^{\sqrt{4-r^2}} r\,dz\,dr\,d\theta$

(c) $\int_{-\sqrt{3}}^{\sqrt{3}}\int_{-\sqrt{3-x^2}}^{\sqrt{3-x^2}}\int_1^{\sqrt{4-x^2-y^2}} dz\,dy\,dx$ **(d)** $5\pi/3$

43. $8\pi/3$ **45.** $9/4$ **47.** $\frac{3\pi-4}{18}$ **49.** $\frac{2\pi a^3}{3}$ **51.** $5\pi/3$

53. $\pi/2$ **55.** $\frac{4(2\sqrt{2}-1)\pi}{3}$ **57.** 16π **59.** $5\pi/2$

61. $\frac{4\pi(8-3\sqrt{3})}{3}$ **63.** $2/3$ **65.** $3/4$

67. $\bar{x}=\bar{y}=0, \bar{z}=3/8$ **69.** $(\bar{x},\bar{y},\bar{z})=(0,0,3/8)$

71. $\bar{x}=\bar{y}=0, \bar{z}=5/6$ **73.** $I_x=\pi/4$

75. (a) $(\bar{x},\bar{y},\bar{z})=\left(0,0,\frac{4}{5}\right), I_z=\frac{\pi}{12}$

(b) $(\bar{x},\bar{y},\bar{z})=\left(0,0,\frac{5}{6}\right), I_z=\frac{\pi}{14}$

Section 12.8, pp. 712–713

1. (a) $x=\frac{u+v}{3}, y=\frac{v-2u}{3}; \frac{1}{3}$

(b) Triangular region with boundaries $u=0$, $v=0$, and $u+v=3$

3. (a) $x=\frac{1}{5}(2u-v), y=\frac{1}{10}(3v-u); \frac{1}{10}$

(b) Triangular region with boundaries $3v=u$, $v=2u$, and $3u+v=10$

7. $64/5$ **9.** $\int_1^2\int_1^3(u+v)\frac{2u}{v}\,du\,dv = 8+\frac{52}{3}\ln 2$

11. $\frac{\pi ab(a^2+b^2)}{4}$ **13.** $\frac{1}{3}\left(1+\frac{3}{e^2}\right)\approx 0.4687$

15. (a) $\begin{vmatrix}\cos v & -u\sin v\\ \sin v & u\cos v\end{vmatrix} = u\cos^2 v + u\sin^2 v = u$

(b) $\begin{vmatrix}\sin v & u\cos v\\ \cos v & -u\sin v\end{vmatrix} = -u\sin^2 v - u\cos^2 v = -u$

19. 12 **21.** $\frac{a^2b^2c^2}{6}$

Practice and Additional Exercises, pp. 714–716

1. $9e-9$

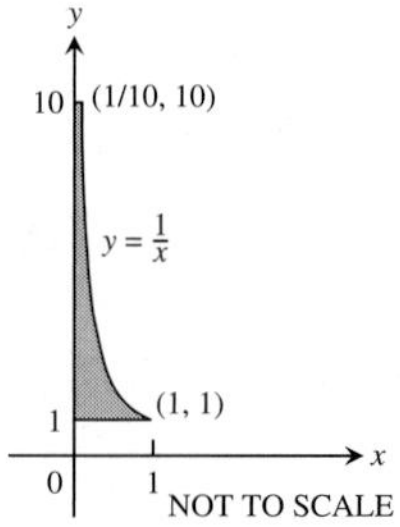

3. $9/2$

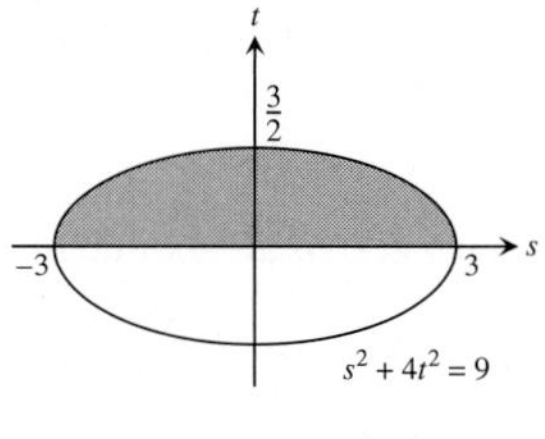

5. $\int_{-2}^{0}\int_{2x+4}^{4-x^2} dy\,dx = \frac{4}{3}$

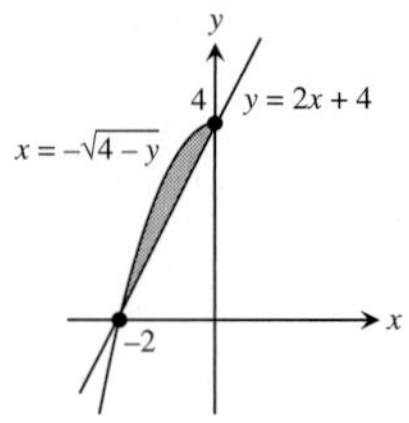

7. $\int_{-3}^{3}\int_0^{(1/2)\sqrt{9-x^2}} y\,dy\,dx = \frac{9}{2}$

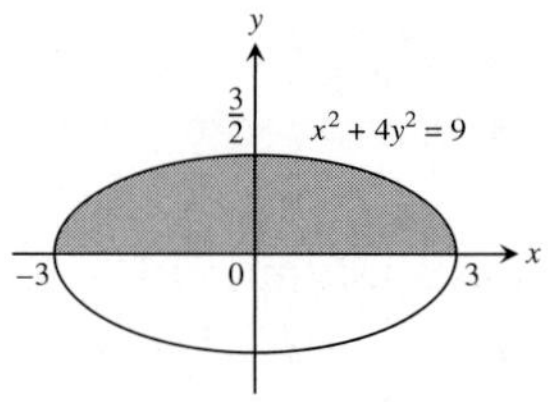

9. $\sin 4$ **11.** $\frac{\ln 17}{4}$ **13.** $4/3$ **15.** $4/3$ **17.** $1/4$ **19.** π

21. $\frac{\pi - 2}{4}$ **23.** 0 **25.** $8/35$ **27.** $\pi/2$ **29.** $\frac{2(31 - 3^{5/2})}{3}$

31. (a) $\int_{-\sqrt{2}}^{\sqrt{2}}\int_{-\sqrt{2-y^2}}^{\sqrt{2-y^2}}\int_{\sqrt{x^2+y^2}}^{\sqrt{4-x^2-y^2}} 3\,dz\,dx\,dy$

(b) $\int_0^{2\pi}\int_0^{\pi/4}\int_0^{2} 3\,\rho^2 \sin\phi\,d\rho\,d\phi\,d\theta$ **(c)** $2\pi\left(8 - 4\sqrt{2}\right)$

33. $\int_0^{2\pi}\int_0^{\pi/4}\int_0^{\sec\phi} \rho^2 \sin\phi\,d\rho\,d\phi\,d\theta = \frac{\pi}{3}$

35. $\int_0^{1}\int_{\sqrt{1-x^2}}^{\sqrt{3-x^2}}\int_1^{\sqrt{4-x^2-y^2}} z^2 xy\,dz\,dy\,dx$

$+ \int_1^{\sqrt{3}}\int_0^{\sqrt{3-x^2}}\int_1^{\sqrt{4-x^2-y^2}} z^2 xy\,dz\,dy\,dx$

37. (a) $\frac{8\pi\left(4\sqrt{2} - 5\right)}{3}$ **(b)** $\frac{8\pi\left(4\sqrt{2} - 5\right)}{3}$

39. $I_z = \frac{8\pi\delta(b^5 - a^5)}{15}$

41. $\bar{x} = \bar{y} = \frac{1}{2 - \ln 4}$ **43.** $I_0 = 104$ **45.** $I_x = 2\delta$

47. $M = 4, M_x = 0, M_y = 0$ **49.** $\bar{x} = \frac{3\sqrt{3}}{\pi}, \bar{y} = 0$

51. (a) $\bar{x} = \frac{15\pi + 32}{6\pi + 48}, \bar{y} = 0$

(b)

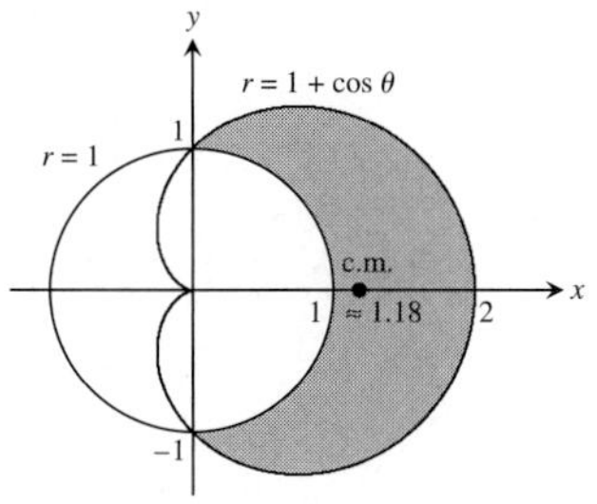

CHAPTER 13

Section 13.1, pp. 721–722

1. Graph (c) **3.** Graph (g) **5.** Graph (d) **7.** Graph (f)

9. $\sqrt{2}$ **11.** $\frac{13}{2}$ **13.** $3\sqrt{14}$ **15.** $\frac{1}{6}\left(5\sqrt{5} + 9\right)$

17. $\sqrt{3}\ln\left(\frac{b}{a}\right)$ **19.** $\frac{10\sqrt{5} - 2}{3}$ **21.** 8 **23.** $2\sqrt{2} - 1$

25. (a) $4\sqrt{2} - 2$ **(b)** $\sqrt{2} + \ln\left(1 + \sqrt{2}\right)$ **27.** $I_z = 2\pi\delta a^3$

29. (a) $I_z = 2\pi\sqrt{2}\delta$ **(b)** $I_z = 4\pi\sqrt{2}\delta$ **31.** $I_x = 2\pi - 2$

Section 13.2, pp. 731–733

1. $\nabla f = -(x\mathbf{i} + y\mathbf{j} + z\mathbf{k})(x^2 + y^2 + z^2)^{-3/2}$

3. $\nabla g = -\left(\frac{2x}{x^2 + y^2}\right)\mathbf{i} - \left(\frac{2y}{x^2 + y^2}\right)\mathbf{j} + e^z\mathbf{k}$

5. $\mathbf{F} = -\frac{kx}{(x^2 + y^2)^{3/2}}\mathbf{i} - \frac{ky}{(x^2 + y^2)^{3/2}}\mathbf{j}$, any $k > 0$

7. (a) $9/2$ **(b)** $13/3$ **(c)** $9/2$

9. (a) $1/3$ **(b)** $-1/5$ **(c)** 0

11. (a) 2 **(b)** $3/2$ **(c)** $1/2$

13. $1/2$ **15.** $-\pi$ **17.** $69/4$ **19.** $-39/2$ **21.** $25/6$

23. (a) $\text{Circ}_1 = 0$, $\text{circ}_2 = 2\pi$, $\text{flux}_1 = 2\pi$, $\text{flux}_2 = 0$

(b) $\text{Circ}_1 = 0$, $\text{circ}_2 = 8\pi$, $\text{flux}_1 = 8\pi$, $\text{flux}_2 = 0$

25. Circ $= 0$, flux $= a^2\pi$ **27.** Circ $= a^2\pi$, flux $= 0$

29. (a) $-\frac{\pi}{2}$ **(b)** 0 **(c)** 1

31.

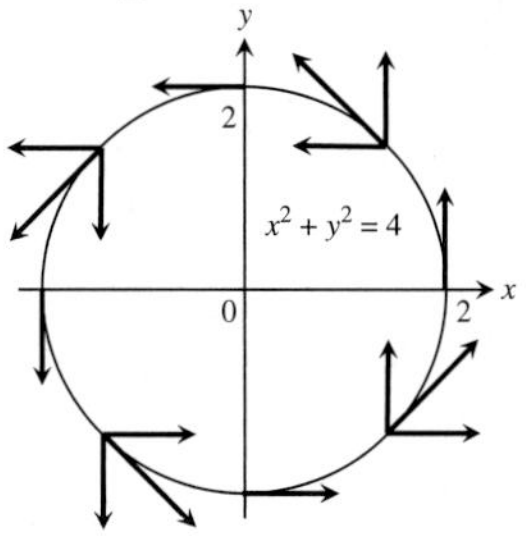

33. (a) $\mathbf{G} = -y\mathbf{i} + x\mathbf{j}$ **(b)** $\mathbf{G} = \sqrt{x^2 + y^2}\,\mathbf{F}$

35. $\mathbf{F} = -\frac{x\mathbf{i} + y\mathbf{j}}{\sqrt{x^2 + y^2}}$ **37.** 48 **39.** π **41.** 0 **43.** $\frac{1}{2}$

Section 13.3, pp. 742–743

1. Conservative **3.** Not conservative **5.** Not conservative

7. $f(x, y, z) = x^2 + \frac{3y^2}{2} + 2z^2 + C$

9. $f(x, y, z) = xe^{y+2z} + C$

11. $f(x, y, z) = x\ln x - x + \tan(x + y) + \frac{1}{2}\ln(y^2 + z^2) + C$

13. 49 **15.** -16 **17.** 1 **19.** $9\ln 2$ **21.** 0 **23.** -3

27. $\mathbf{F} = \nabla\left(\frac{x^2 - 1}{y}\right)$ **29. (a)** 1 **(b)** 1 **(c)** 1

31. (a) 2 **(b)** 2 **33. (a)** $c = b = 2a$ **(b)** $c = b = 2$

35. It does not matter what path you use. The work will be the same on any path because the field is conservative.

37. The force $\mathbf{F}$ is conservative because all partial derivatives of M, N, and P are zero. $f(x, y, z) = ax + by + cz + C$; $A = (xa, ya, za)$ and $B = (xb, yb, zb)$. Therefore, $\int \mathbf{F}\cdot d\mathbf{r} = f(B) - f(A) = a(xb - xa) + b(yb - ya) + c(zb - za) = \mathbf{F}\cdot\overrightarrow{AB}$.

Section 13.4, pp. 750–752

1. Flux $= 0$, circ $= 2\pi a^2$ **3.** Flux $= -\pi a^2$, circ $= 0$

5. Flux $= 2$, circ $= 0$ **7.** Flux $= -9$, circ $= 9$

9. Flux $= 1/2$, circ $= 1/2$ **11.** Flux $= 1/5$, circ $= -1/12$

13. 0 **15.** $2/33$ **17.** 0 **19.** -16π **21.** πa^2 **23.** $\frac{3}{8}\pi$

25. (a) 0 if C is traversed counterclockwise

(b) $(h - k)$(area of the region) **31. (a)** 0

Section 13.5, pp. 759–761

1. $\mathbf{r}(r, \theta) = (r\cos\theta)\mathbf{i} + (r\sin\theta)\mathbf{j} + r^2\mathbf{k},\ 0 \le r \le 2,\ 0 \le \theta \le 2\pi$

3. $\mathbf{r}(r, \theta) = (r\cos\theta)\mathbf{i} + (r\sin\theta)\mathbf{j} + (r/2)\mathbf{k},\ 0 \le r \le 6,\ 0 \le \theta \le \pi/2$

5. $\mathbf{r}(r, \theta) = (r\cos\theta)\mathbf{i} + (r\sin\theta)\mathbf{j} + \sqrt{9 - r^2}\,\mathbf{k},\ 0 \le r \le 3\sqrt{2}/2,\ 0 \le \theta \le 2\pi$; Also: $\mathbf{r}(\phi, \theta) = (3\sin\phi\cos\theta)\mathbf{i} + (3\sin\phi\sin\theta)\mathbf{j} + (3\cos\phi)\mathbf{k},\ 0 \le \phi \le \pi/4,\ 0 \le \theta \le 2\pi$

7. $\mathbf{r}(\phi, \theta) = \left(\sqrt{3}\sin\phi\cos\theta\right)\mathbf{i} + \left(\sqrt{3}\sin\phi\sin\theta\right)\mathbf{j} + \left(\sqrt{3}\cos\phi\right)\mathbf{k},\ \pi/3 \le \phi \le 2\pi/3,\ 0 \le \theta \le 2\pi$

9. $\mathbf{r}(x, y) = x\mathbf{i} + y\mathbf{j} + (4 - y^2)\mathbf{k},\ 0 \le x \le 2,\ -2 \le y \le 2$

11. $\mathbf{r}(u, v) = u\mathbf{i} + (3\cos v)\mathbf{j} + (3\sin v)\mathbf{k},\ 0 \le u \le 3,\ 0 \le v \le 2\pi$

13. **(a)** $\mathbf{r}(r, \theta) = (r\cos\theta)\mathbf{i} + (r\sin\theta)\mathbf{j} + (1 - r\cos\theta - r\sin\theta)\mathbf{k},\ 0 \le r \le 3,\ 0 \le \theta \le 2\pi$
(b) $\mathbf{r}(u, v) = (1 - u\cos v - u\sin v)\mathbf{i} + (u\cos v)\mathbf{j} + (u\sin v)\mathbf{k},\ 0 \le u \le 3,\ 0 \le v \le 2\pi$

15. $\mathbf{r}(u, v) = (4\cos^2 v)\mathbf{i} + u\mathbf{j} + (4\cos v\sin v)\mathbf{k},\ 0 \le u \le 3,\ -(\pi/2) \le v \le (\pi/2)$; Another way: $\mathbf{r}(u, v) = (2 + 2\cos v)\mathbf{i} + u\mathbf{j} + (2\sin v)\mathbf{k},\ 0 \le u \le 3,\ 0 \le v \le 2\pi$

17. $\displaystyle\int_0^{2\pi}\int_0^1 \frac{\sqrt{5}}{2} r\,dr\,d\theta = \frac{\pi\sqrt{5}}{2}$

19. $\displaystyle\int_0^{2\pi}\int_1^3 r\sqrt{5}\,dr\,d\theta = 8\pi\sqrt{5}$ **21.** $\displaystyle\int_0^{2\pi}\int_1^4 1\,du\,dv = 6\pi$

23. $\displaystyle\int_0^{2\pi}\int_0^1 u\sqrt{4u^2 + 1}\,du\,dv = \frac{\left(5\sqrt{5} - 1\right)}{6}\pi$

25. $\displaystyle\int_0^{2\pi}\int_{\pi/4}^{\pi} 2\sin\phi\,d\phi\,d\theta = \left(4 + 2\sqrt{2}\right)\pi$

27.

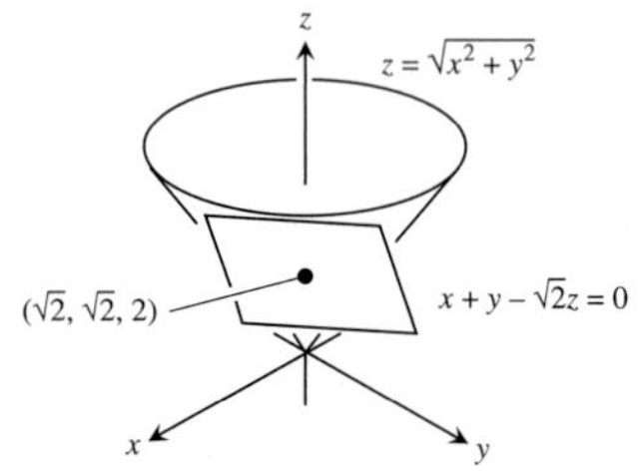

29.

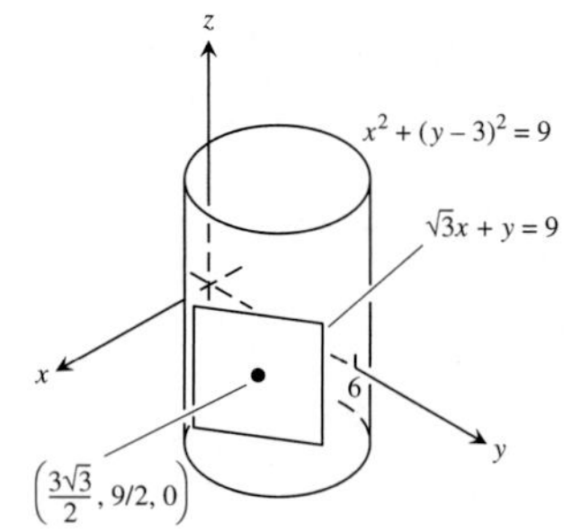

33. $x_0x + y_0y = 25$ **35.** $\frac{13}{3}\pi$ **37.** 4 **39.** $6\sqrt{6} - 2\sqrt{2}$

41. $\pi\sqrt{c^2 + 1}$ **43.** $\frac{\pi}{6}\left(17\sqrt{17} - 5\sqrt{5}\right)$ **45.** $3 + 2\ln 2$

47. $\frac{\pi}{6}\left(13\sqrt{13} - 1\right)$ **49.** $5\pi\sqrt{2}$ **51.** $\frac{2}{3}\left(5\sqrt{5} - 1\right)$

Section 13.6, pp. 768–770

1. $\displaystyle\iint_S x\,d\sigma = \int_0^3\int_0^2 u\sqrt{4u^2 + 1}\,du\,dv = \frac{17\sqrt{17} - 1}{4}$

3. $\displaystyle\iint_S x^2\,d\sigma = \int_0^{2\pi}\int_0^{\pi} \sin^3\phi\cos^2\theta\,d\phi\,d\theta = \frac{4\pi}{3}$

5. $\displaystyle\iint_S z\,d\sigma = \int_0^1\int_0^1 (4 - u - v)\sqrt{3}\,dv\,du = 3\sqrt{3}$ (for $x = u, y = v$)

7. $\displaystyle\iint_S x^2\sqrt{5 - 4z}\,d\sigma = \int_0^1\int_0^{2\pi} u^2\cos^2 v \cdot \sqrt{4u^2 + 1}\cdot u\sqrt{4u^2 + 1}\,dv\,du = \int_0^1\int_0^{2\pi} u^3(4u^2 + 1)\cos^2 v\,dv\,du = \frac{11\pi}{12}$

9. $9a^3$ **11.** $\frac{abc}{4}(ab + ac + bc)$ **13.** 2 **15.** -32

17. $\pi a^3/6$ **19.** $13a^4/6$ **21.** $2\pi/3$ **23.** $-73\pi/6$ **25.** 18

27. $\frac{\pi a^3}{6}$ **29.** $\frac{\pi a^2}{4}$ **31.** $\frac{\pi a^3}{2}$ **33.** -32 **35.** -4 **37.** $3a^4$

39. $\left(\frac{a}{2}, \frac{a}{2}, \frac{a}{2}\right)$ **41.** $\frac{8\pi}{3}a^4\delta$

Section 13.7, pp. 778–779

1. 4π **3.** $-5/6$ **5.** 0 **7.** -6π **9.** $2\pi a^2$ **13.** 12π
15. $-\pi/4$ **17.** -15π **25.** $16I_y + 16I_x$

Section 13.8, pp. 786–788

1. 0 **3.** 0 **5.** -16 **7.** -8π **9.** 3π **11.** $-40/3$
13. 12π **15.** $12\pi\left(4\sqrt{2} - 1\right)$
21. The integral's value never exceeds the surface area of *S*.

Practice and Additional Exercises, pp. 789–791

1. Path 1: $2\sqrt{3}$; path 2: $1 + 3\sqrt{2}$ **3.** $4a^2$ **5.** 0

7. $8\pi\sin(1)$ **9.** 0 **11.** $\pi\sqrt{3}$ **13.** $2\pi\left(1 - \frac{1}{\sqrt{2}}\right)$

15. $\frac{abc}{2}\sqrt{\frac{1}{a^2} + \frac{1}{b^2} + \frac{1}{c^2}}$ **17.** 50

19. $\mathbf{r}(\phi, \theta) = (6\sin\phi\cos\theta)\mathbf{i} + (6\sin\phi\sin\theta)\mathbf{j} + (6\cos\phi)\mathbf{k},\ \frac{\pi}{6} \le \phi \le \frac{2\pi}{3},\ 0 \le \theta \le 2\pi$

21. $\mathbf{r}(r, \theta) = (r\cos\theta)\mathbf{i} + (r\sin\theta)\mathbf{j} + (1 + r)\mathbf{k},\ 0 \le r \le 2,\ 0 \le \theta \le 2\pi$

23. $\mathbf{r}(u, v) = (u \cos v)\mathbf{i} + 2u^2\mathbf{j} + (u \sin v)\mathbf{k},\ 0 \le u \le 1,\ 0 \le v \le \pi$
25. $\sqrt{6}$ **27.** $\pi\left[\sqrt{2} + \ln\left(1 + \sqrt{2}\right)\right]$ **29.** Conservative
31. Not conservative **33.** $f(x, y, z) = y^2 + yz + 2x + z$
35. Path 1: 2; path 2: 8/3 **37. (a)** $1 - e^{-2\pi}$ **(b)** $1 - e^{-2\pi}$
39. 0

41. $(\bar{x}, \bar{y}, \bar{z}) = \left(1, \frac{16}{15}, \frac{2}{3}\right);\ I_x = \frac{232}{45},\ I_y = \frac{64}{15},\ I_z = \frac{56}{9}$

43. $\bar{z} = \frac{3}{2},\ I_z = \frac{7\sqrt{3}}{3}$

45. Flux: 3/2; circ: $-1/2$ **47.** 3 **49.** 0

APPENDICES

Appendix 4, pp. AP-8–AP-9

1. (a) 8π m **(b)** $\frac{55\pi}{9}$ m **3.** 8.4 in.

5.

θ	$-\pi$	$-2\pi/3$	0	$\pi/2$	$3\pi/4$
$\sin\theta$	0	$-\frac{\sqrt{3}}{2}$	0	1	$\frac{1}{\sqrt{2}}$
$\cos\theta$	-1	$-\frac{1}{2}$	1	0	$-\frac{1}{\sqrt{2}}$
$\tan\theta$	0	$\sqrt{3}$	0	UND	-1
$\cot\theta$	UND	$\frac{1}{\sqrt{3}}$	UND	0	-1
$\sec\theta$	-1	-2	1	UND	$-\sqrt{2}$
$\csc\theta$	UND	$-\frac{2}{\sqrt{3}}$	UND	1	$\sqrt{2}$

7. $\sin x = -\frac{\sqrt{8}}{3},\ \tan x = -\sqrt{8}$

9. Period π

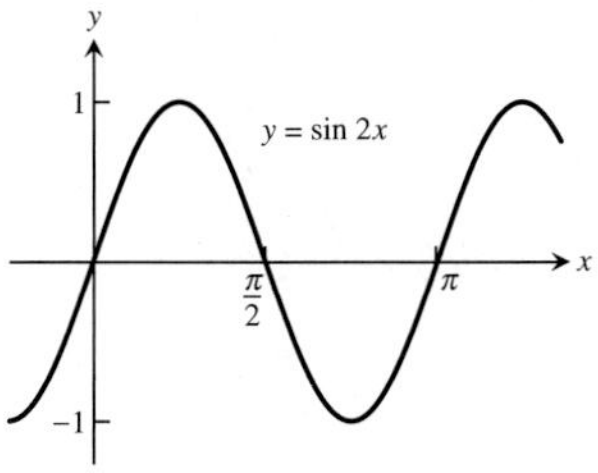

11. Period 6

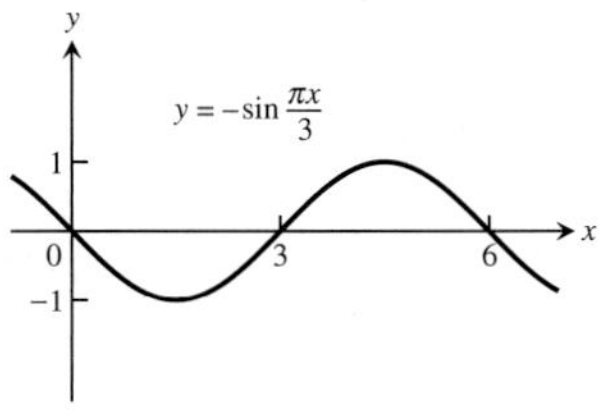

13. Period 2π

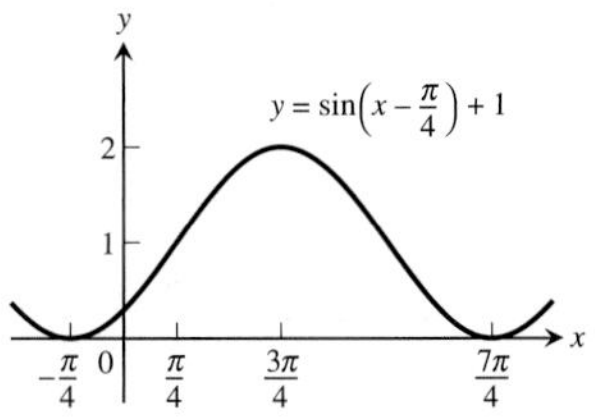

15. Period = 1, symmetric about the origin

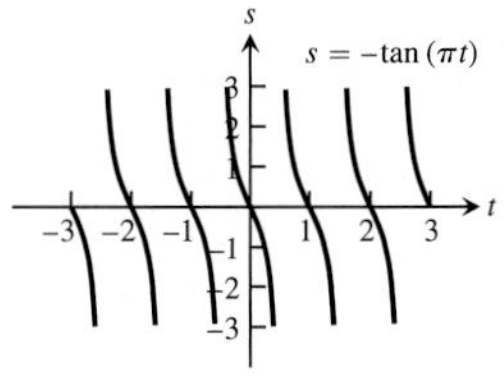

17. Period = 4π, symmetric about the origin

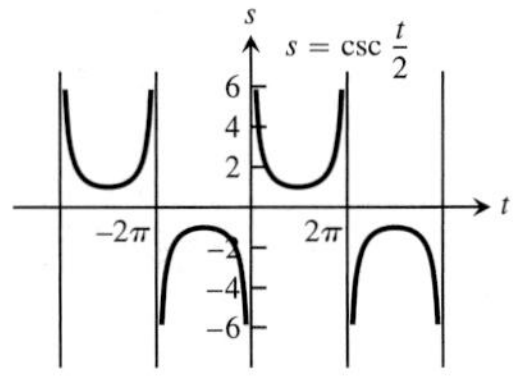

25. $-\cos x$ **27.** $\frac{\sqrt{6} + \sqrt{2}}{4}$ **29.** $\frac{2 + \sqrt{2}}{4}$

33. $\sqrt{7} \approx 2.65$ **35.** $\frac{3\sqrt{3}}{2\sqrt{7}} \approx 0.982$

Appendix 12, p. AP-40

1. Quadratic: $x + xy$; cubic: $x + xy + \frac{1}{2}xy^2$

3. Quadratic: xy; cubic: xy

5. Quadratic: $y + \frac{1}{2}(2xy - y^2)$;
cubic: $y + \frac{1}{2}(2xy - y^2) + \frac{1}{6}(3x^2y - 3xy^2 + 2y^3)$

7. Quadratic: $\frac{1}{2}(2x^2 + 2y^2) = x^2 + y^2$; cubic: $x^2 + y^2$

9. Quadratic: $1 + (x + y) + (x + y)^2$;
cubic: $1 + (x + y) + (x + y)^2 + (x + y)^3$

11. Quadratic: $1 - \frac{1}{2}x^2 - \frac{1}{2}y^2$; $E(x, y) \le 0.00134$

INDEX

A Brief Table of Integrals

1. $\int u\,dv = uv - \int v\,du$

2. $\int a^u\,du = \dfrac{a^u}{\ln a} + C, \quad a \neq 1, \quad a > 0$

3. $\int \cos u\,du = \sin u + C$

4. $\int \sin u\,du = -\cos u + C$

5. $\int (ax+b)^n\,dx = \dfrac{(ax+b)^{n+1}}{a(n+1)} + C, \quad n \neq -1$

6. $\int (ax+b)^{-1}\,dx = \dfrac{1}{a}\ln|ax+b| + C$

7. $\int x(ax+b)^n\,dx = \dfrac{(ax+b)^{n+1}}{a^2}\left[\dfrac{ax+b}{n+2} - \dfrac{b}{n+1}\right] + C, \quad n \neq -1, -2$

8. $\int x(ax+b)^{-1}\,dx = \dfrac{x}{a} - \dfrac{b}{a^2}\ln|ax+b| + C$

9. $\int x(ax+b)^{-2}\,dx = \dfrac{1}{a^2}\left[\ln|ax+b| + \dfrac{b}{ax+b}\right] + C$

10. $\int \dfrac{dx}{x(ax+b)} = \dfrac{1}{b}\ln\left|\dfrac{x}{ax+b}\right| + C$

11. $\int \left(\sqrt{ax+b}\right)^n dx = \dfrac{2}{a}\dfrac{\left(\sqrt{ax+b}\right)^{n+2}}{n+2} + C, \quad n \neq -2$

12. $\int \dfrac{\sqrt{ax+b}}{x}\,dx = 2\sqrt{ax+b} + b\int \dfrac{dx}{x\sqrt{ax+b}}$

13. (a) $\int \dfrac{dx}{x\sqrt{ax-b}} = \dfrac{2}{\sqrt{b}}\tan^{-1}\sqrt{\dfrac{ax-b}{b}} + C$ (b) $\int \dfrac{dx}{x\sqrt{ax+b}} = \dfrac{1}{\sqrt{b}}\ln\left|\dfrac{\sqrt{ax+b}-\sqrt{b}}{\sqrt{ax+b}+\sqrt{b}}\right| + C$

14. $\int \dfrac{\sqrt{ax+b}}{x^2}\,dx = -\dfrac{\sqrt{ax+b}}{x} + \dfrac{a}{2}\int \dfrac{dx}{x\sqrt{ax+b}} + C$

15. $\int \dfrac{dx}{x^2\sqrt{ax+b}} = -\dfrac{\sqrt{ax+b}}{bx} - \dfrac{a}{2b}\int \dfrac{dx}{x\sqrt{ax+b}} + C$

16. $\int \dfrac{dx}{a^2+x^2} = \dfrac{1}{a}\tan^{-1}\dfrac{x}{a} + C$

17. $\int \dfrac{dx}{(a^2+x^2)^2} = \dfrac{x}{2a^2(a^2+x^2)} + \dfrac{1}{2a^3}\tan^{-1}\dfrac{x}{a} + C$

18. $\int \dfrac{dx}{a^2-x^2} = \dfrac{1}{2a}\ln\left|\dfrac{x+a}{x-a}\right| + C$

19. $\int \dfrac{dx}{(a^2-x^2)^2} = \dfrac{x}{2a^2(a^2-x^2)} + \dfrac{1}{4a^3}\ln\left|\dfrac{x+a}{x-a}\right| + C$

20. $\int \dfrac{dx}{\sqrt{a^2+x^2}} = \sinh^{-1}\dfrac{x}{a} + C = \ln\left(x + \sqrt{a^2+x^2}\right) + C$

21. $\int \sqrt{a^2+x^2}\,dx = \dfrac{x}{2}\sqrt{a^2+x^2} + \dfrac{a^2}{2}\ln\left(x + \sqrt{a^2+x^2}\right) + C$

22. $\int x^2\sqrt{a^2+x^2}\,dx = \dfrac{x}{8}(a^2+2x^2)\sqrt{a^2+x^2} - \dfrac{a^4}{8}\ln\left(x + \sqrt{a^2+x^2}\right) + C$

23. $\displaystyle\int \frac{\sqrt{a^2+x^2}}{x}\,dx = \sqrt{a^2+x^2} - a\ln\left|\frac{a+\sqrt{a^2+x^2}}{x}\right| + C$

24. $\displaystyle\int \frac{\sqrt{a^2+x^2}}{x^2}\,dx = \ln\left(x+\sqrt{a^2+x^2}\right) - \frac{\sqrt{a^2+x^2}}{x} + C$

25. $\displaystyle\int \frac{x^2}{\sqrt{a^2+x^2}}\,dx = -\frac{a^2}{2}\ln\left(x+\sqrt{a^2+x^2}\right) + \frac{x\sqrt{a^2+x^2}}{2} + C$

26. $\displaystyle\int \frac{dx}{x\sqrt{a^2+x^2}} = -\frac{1}{a}\ln\left|\frac{a+\sqrt{a^2+x^2}}{x}\right| + C$

27. $\displaystyle\int \frac{dx}{x^2\sqrt{a^2+x^2}} = -\frac{\sqrt{a^2+x^2}}{a^2x} + C$

28. $\displaystyle\int \frac{dx}{\sqrt{a^2-x^2}} = \sin^{-1}\frac{x}{a} + C$

29. $\displaystyle\int \sqrt{a^2-x^2}\,dx = \frac{x}{2}\sqrt{a^2-x^2} + \frac{a^2}{2}\sin^{-1}\frac{x}{a} + C$

30. $\displaystyle\int x^2\sqrt{a^2-x^2}\,dx = \frac{a^4}{8}\sin^{-1}\frac{x}{a} - \frac{1}{8}x\sqrt{a^2-x^2}\,(a^2-2x^2) + C$

31. $\displaystyle\int \frac{\sqrt{a^2-x^2}}{x}\,dx = \sqrt{a^2-x^2} - a\ln\left|\frac{a+\sqrt{a^2-x^2}}{x}\right| + C$

32. $\displaystyle\int \frac{\sqrt{a^2-x^2}}{x^2}\,dx = -\sin^{-1}\frac{x}{a} - \frac{\sqrt{a^2-x^2}}{x} + C$

33. $\displaystyle\int \frac{x^2}{\sqrt{a^2-x^2}}\,dx = \frac{a^2}{2}\sin^{-1}\frac{x}{a} - \frac{1}{2}x\sqrt{a^2-x^2} + C$

34. $\displaystyle\int \frac{dx}{x\sqrt{a^2-x^2}} = -\frac{1}{a}\ln\left|\frac{a+\sqrt{a^2-x^2}}{x}\right| + C$

35. $\displaystyle\int \frac{dx}{x^2\sqrt{a^2-x^2}} = -\frac{\sqrt{a^2-x^2}}{a^2x} + C$

36. $\displaystyle\int \frac{dx}{\sqrt{x^2-a^2}} = \cosh^{-1}\frac{x}{a} + C = \ln\left|x+\sqrt{x^2-a^2}\right| + C$

37. $\displaystyle\int \sqrt{x^2-a^2}\,dx = \frac{x}{2}\sqrt{x^2-a^2} - \frac{a^2}{2}\ln\left|x+\sqrt{x^2-a^2}\right| + C$

38. $\displaystyle\int \left(\sqrt{x^2-a^2}\right)^n dx = \frac{x\left(\sqrt{x^2-a^2}\right)^n}{n+1} - \frac{na^2}{n+1}\int \left(\sqrt{x^2-a^2}\right)^{n-2} dx, \quad n \neq -1$

39. $\displaystyle\int \frac{dx}{\left(\sqrt{x^2-a^2}\right)^n} = \frac{x\left(\sqrt{x^2-a^2}\right)^{2-n}}{(2-n)a^2} - \frac{n-3}{(n-2)a^2}\int \frac{dx}{\left(\sqrt{x^2-a^2}\right)^{n-2}}, \quad n \neq 2$

40. $\displaystyle\int x\left(\sqrt{x^2-a^2}\right)^n dx = \frac{\left(\sqrt{x^2-a^2}\right)^{n+2}}{n+2} + C, \quad n \neq -2$

41. $\displaystyle\int x^2\sqrt{x^2-a^2}\,dx = \frac{x}{8}(2x^2-a^2)\sqrt{x^2-a^2} - \frac{a^4}{8}\ln\left|x+\sqrt{x^2-a^2}\right| + C$

42. $\displaystyle\int \frac{\sqrt{x^2-a^2}}{x}\,dx = \sqrt{x^2-a^2} - a\sec^{-1}\left|\frac{x}{a}\right| + C$

43. $\displaystyle\int \frac{\sqrt{x^2-a^2}}{x^2}\,dx = \ln\left|x+\sqrt{x^2-a^2}\right| - \frac{\sqrt{x^2-a^2}}{x} + C$

44. $\displaystyle\int \frac{x^2}{\sqrt{x^2-a^2}}\,dx = \frac{a^2}{2}\ln\left|x+\sqrt{x^2-a^2}\right| + \frac{x}{2}\sqrt{x^2-a^2} + C$

45. $\displaystyle\int \frac{dx}{x\sqrt{x^2-a^2}} = \frac{1}{a}\sec^{-1}\left|\frac{x}{a}\right| + C = \frac{1}{a}\cos^{-1}\left|\frac{a}{x}\right| + C$

46. $\displaystyle\int \frac{dx}{x^2\sqrt{x^2-a^2}} = \frac{\sqrt{x^2-a^2}}{a^2x} + C$

47. $\displaystyle\int \frac{dx}{\sqrt{2ax - x^2}} = \sin^{-1}\left(\frac{x-a}{a}\right) + C$

48. $\displaystyle\int \sqrt{2ax - x^2}\, dx = \frac{x-a}{2}\sqrt{2ax - x^2} + \frac{a^2}{2}\sin^{-1}\left(\frac{x-a}{a}\right) + C$

49. $\displaystyle\int \left(\sqrt{2ax - x^2}\right)^n dx = \frac{(x-a)\left(\sqrt{2ax - x^2}\right)^n}{n+1} + \frac{na^2}{n+1}\int \left(\sqrt{2ax - x^2}\right)^{n-2} dx$

50. $\displaystyle\int \frac{dx}{\left(\sqrt{2ax - x^2}\right)^n} = \frac{(x-a)\left(\sqrt{2ax - x^2}\right)^{2-n}}{(n-2)a^2} + \frac{n-3}{(n-2)a^2}\int \frac{dx}{\left(\sqrt{2ax - x^2}\right)^{n-2}}$

51. $\displaystyle\int x\sqrt{2ax - x^2}\, dx = \frac{(x+a)(2x-3a)\sqrt{2ax - x^2}}{6} + \frac{a^3}{2}\sin^{-1}\left(\frac{x-a}{a}\right) + C$

52. $\displaystyle\int \frac{\sqrt{2ax - x^2}}{x}\, dx = \sqrt{2ax - x^2} + a\sin^{-1}\left(\frac{x-a}{a}\right) + C$

53. $\displaystyle\int \frac{\sqrt{2ax - x^2}}{x^2}\, dx = -2\sqrt{\frac{2a-x}{x}} - \sin^{-1}\left(\frac{x-a}{a}\right) + C$

54. $\displaystyle\int \frac{x\, dx}{\sqrt{2ax - x^2}} = a\sin^{-1}\left(\frac{x-a}{a}\right) - \sqrt{2ax - x^2} + C$

55. $\displaystyle\int \frac{dx}{x\sqrt{2ax - x^2}} = -\frac{1}{a}\sqrt{\frac{2a-x}{x}} + C$

56. $\displaystyle\int \sin ax\, dx = -\frac{1}{a}\cos ax + C$

57. $\displaystyle\int \cos ax\, dx = \frac{1}{a}\sin ax + C$

58. $\displaystyle\int \sin^2 ax\, dx = \frac{x}{2} - \frac{\sin 2ax}{4a} + C$

59. $\displaystyle\int \cos^2 ax\, dx = \frac{x}{2} + \frac{\sin 2ax}{4a} + C$

60. $\displaystyle\int \sin^n ax\, dx = -\frac{\sin^{n-1} ax \cos ax}{na} + \frac{n-1}{n}\int \sin^{n-2} ax\, dx$

61. $\displaystyle\int \cos^n ax\, dx = \frac{\cos^{n-1} ax \sin ax}{na} + \frac{n-1}{n}\int \cos^{n-2} ax\, dx$

62. (a) $\displaystyle\int \sin ax \cos bx\, dx = -\frac{\cos(a+b)x}{2(a+b)} - \frac{\cos(a-b)x}{2(a-b)} + C, \quad a^2 \neq b^2$

(b) $\displaystyle\int \sin ax \sin bx\, dx = \frac{\sin(a-b)x}{2(a-b)} - \frac{\sin(a+b)x}{2(a+b)} + C, \quad a^2 \neq b^2$

(c) $\displaystyle\int \cos ax \cos bx\, dx = \frac{\sin(a-b)x}{2(a-b)} + \frac{\sin(a+b)x}{2(a+b)} + C, \quad a^2 \neq b^2$

63. $\displaystyle\int \sin ax \cos ax\, dx = -\frac{\cos 2ax}{4a} + C$

64. $\displaystyle\int \sin^n ax \cos ax\, dx = \frac{\sin^{n+1} ax}{(n+1)a} + C, \quad n \neq -1$

65. $\displaystyle\int \frac{\cos ax}{\sin ax}\, dx = \frac{1}{a}\ln|\sin ax| + C$

66. $\displaystyle\int \cos^n ax \sin ax\, dx = -\frac{\cos^{n+1} ax}{(n+1)a} + C, \quad n \neq -1$

67. $\displaystyle\int \frac{\sin ax}{\cos ax}\, dx = -\frac{1}{a}\ln|\cos ax| + C$

68. $\int \sin^n ax \cos^m ax\,dx = -\frac{\sin^{n-1} ax \cos^{m+1} ax}{a(m+n)} + \frac{n-1}{m+n}\int \sin^{n-2} ax \cos^m ax\,dx, \quad n \neq -m \quad (\text{reduces } \sin^n ax)$

69. $\int \sin^n ax \cos^m ax\,dx = \frac{\sin^{n+1} ax \cos^{m-1} ax}{a(m+n)} + \frac{m-1}{m+n}\int \sin^n ax \cos^{m-2} ax\,dx, \quad m \neq -n \quad (\text{reduces } \cos^m ax)$

70. $\int \frac{dx}{b + c\sin ax} = \frac{-2}{a\sqrt{b^2 - c^2}}\tan^{-1}\left[\sqrt{\frac{b-c}{b+c}}\tan\left(\frac{\pi}{4} - \frac{ax}{2}\right)\right] + C, \quad b^2 > c^2$

71. $\int \frac{dx}{b + c\sin ax} = \frac{-1}{a\sqrt{c^2 - b^2}}\ln\left|\frac{c + b\sin ax + \sqrt{c^2 - b^2}\cos ax}{b + c\sin ax}\right| + C, \quad b^2 < c^2$

72. $\int \frac{dx}{1 + \sin ax} = -\frac{1}{a}\tan\left(\frac{\pi}{4} - \frac{ax}{2}\right) + C$

73. $\int \frac{dx}{1 - \sin ax} = \frac{1}{a}\tan\left(\frac{\pi}{4} + \frac{ax}{2}\right) + C$

74. $\int \frac{dx}{b + c\cos ax} = \frac{2}{a\sqrt{b^2 - c^2}}\tan^{-1}\left[\sqrt{\frac{b-c}{b+c}}\tan\frac{ax}{2}\right] + C, \quad b^2 > c^2$

75. $\int \frac{dx}{b + c\cos ax} = \frac{1}{a\sqrt{c^2 - b^2}}\ln\left|\frac{c + b\cos ax + \sqrt{c^2 - b^2}\sin ax}{b + c\cos ax}\right| + C, \quad b^2 < c^2$

76. $\int \frac{dx}{1 + \cos ax} = \frac{1}{a}\tan\frac{ax}{2} + C$

77. $\int \frac{dx}{1 - \cos ax} = -\frac{1}{a}\cot\frac{ax}{2} + C$

78. $\int x\sin ax\,dx = \frac{1}{a^2}\sin ax - \frac{x}{a}\cos ax + C$

79. $\int x\cos ax\,dx = \frac{1}{a^2}\cos ax + \frac{x}{a}\sin ax + C$

80. $\int x^n \sin ax\,dx = -\frac{x^n}{a}\cos ax + \frac{n}{a}\int x^{n-1}\cos ax\,dx$

81. $\int x^n \cos ax\,dx = \frac{x^n}{a}\sin ax - \frac{n}{a}\int x^{n-1}\sin ax\,dx$

82. $\int \tan ax\,dx = \frac{1}{a}\ln|\sec ax| + C$

83. $\int \cot ax\,dx = \frac{1}{a}\ln|\sin ax| + C$

84. $\int \tan^2 ax\,dx = \frac{1}{a}\tan ax - x + C$

85. $\int \cot^2 ax\,dx = -\frac{1}{a}\cot ax - x + C$

86. $\int \tan^n ax\,dx = \frac{\tan^{n-1} ax}{a(n-1)} - \int \tan^{n-2} ax\,dx, \quad n \neq 1$

87. $\int \cot^n ax\,dx = -\frac{\cot^{n-1} ax}{a(n-1)} - \int \cot^{n-2} ax\,dx, \quad n \neq 1$

88. $\int \sec ax\,dx = \frac{1}{a}\ln|\sec ax + \tan ax| + C$

89. $\int \csc ax\,dx = -\frac{1}{a}\ln|\csc ax + \cot ax| + C$

90. $\int \sec^2 ax\,dx = \frac{1}{a}\tan ax + C$

91. $\int \csc^2 ax\,dx = -\frac{1}{a}\cot ax + C$

92. $\int \sec^n ax\,dx = \frac{\sec^{n-2} ax\tan ax}{a(n-1)} + \frac{n-2}{n-1}\int \sec^{n-2} ax\,dx, \quad n \neq 1$

93. $\int \csc^n ax\,dx = -\frac{\csc^{n-2} ax\cot ax}{a(n-1)} + \frac{n-2}{n-1}\int \csc^{n-2} ax\,dx, \quad n \neq 1$

94. $\int \sec^n ax\tan ax\,dx = \frac{\sec^n ax}{na} + C, \quad n \neq 0$

95. $\int \csc^n ax\cot ax\,dx = -\frac{\csc^n ax}{na} + C, \quad n \neq 0$

96. $\int \sin^{-1} ax\, dx = x \sin^{-1} ax + \frac{1}{a}\sqrt{1 - a^2x^2} + C$

97. $\int \cos^{-1} ax\, dx = x \cos^{-1} ax - \frac{1}{a}\sqrt{1 - a^2x^2} + C$

98. $\int \tan^{-1} ax\, dx = x \tan^{-1} ax - \frac{1}{2a} \ln(1 + a^2x^2) + C$

99. $\int x^n \sin^{-1} ax\, dx = \frac{x^{n+1}}{n+1} \sin^{-1} ax - \frac{a}{n+1} \int \frac{x^{n+1}\, dx}{\sqrt{1 - a^2x^2}}, \quad n \neq -1$

100. $\int x^n \cos^{-1} ax\, dx = \frac{x^{n+1}}{n+1} \cos^{-1} ax + \frac{a}{n+1} \int \frac{x^{n+1}\, dx}{\sqrt{1 - a^2x^2}}, \quad n \neq -1$

101. $\int x^n \tan^{-1} ax\, dx = \frac{x^{n+1}}{n+1} \tan^{-1} ax - \frac{a}{n+1} \int \frac{x^{n+1}\, dx}{1 + a^2x^2}, \quad n \neq -1$

102. $\int e^{ax}\, dx = \frac{1}{a} e^{ax} + C$

103. $\int b^{ax}\, dx = \frac{1}{a} \frac{b^{ax}}{\ln b} + C, \quad b > 0, b \neq 1$

104. $\int xe^{ax}\, dx = \frac{e^{ax}}{a^2}(ax - 1) + C$

105. $\int x^n e^{ax}\, dx = \frac{1}{a} x^n e^{ax} - \frac{n}{a} \int x^{n-1} e^{ax}\, dx$

106. $\int x^n b^{ax}\, dx = \frac{x^n b^{ax}}{a \ln b} - \frac{n}{a \ln b} \int x^{n-1} b^{ax}\, dx, \quad b > 0, b \neq 1$

107. $\int e^{ax} \sin bx\, dx = \frac{e^{ax}}{a^2 + b^2}(a \sin bx - b \cos bx) + C$

108. $\int e^{ax} \cos bx\, dx = \frac{e^{ax}}{a^2 + b^2}(a \cos bx + b \sin bx) + C$

109. $\int \ln ax\, dx = x \ln ax - x + C$

110. $\int x^n (\ln ax)^m\, dx = \frac{x^{n+1}(\ln ax)^m}{n+1} - \frac{m}{n+1} \int x^n (\ln ax)^{m-1}\, dx, \quad n \neq -1$

111. $\int x^{-1} (\ln ax)^m\, dx = \frac{(\ln ax)^{m+1}}{m+1} + C, \quad m \neq -1$

112. $\int \frac{dx}{x \ln ax} = \ln |\ln ax| + C$

113. $\int \sinh ax\, dx = \frac{1}{a} \cosh ax + C$

114. $\int \cosh ax\, dx = \frac{1}{a} \sinh ax + C$

115. $\int \sinh^2 ax\, dx = \frac{\sinh 2ax}{4a} - \frac{x}{2} + C$

116. $\int \cosh^2 ax\, dx = \frac{\sinh 2ax}{4a} + \frac{x}{2} + C$

117. $\int \sinh^n ax\, dx = \frac{\sinh^{n-1} ax \cosh ax}{na} - \frac{n-1}{n} \int \sinh^{n-2} ax\, dx, \quad n \neq 0$

118. $\int \cosh^n ax\, dx = \frac{\cosh^{n-1} ax \sinh ax}{na} + \frac{n-1}{n} \int \cosh^{n-2} ax\, dx, \quad n \neq 0$

119. $\int x \sinh ax\, dx = \frac{x}{a} \cosh ax - \frac{1}{a^2} \sinh ax + C$

120. $\int x \cosh ax\, dx = \frac{x}{a} \sinh ax - \frac{1}{a^2} \cosh ax + C$

121. $\int x^n \sinh ax\, dx = \frac{x^n}{a} \cosh ax - \frac{n}{a} \int x^{n-1} \cosh ax\, dx$

122. $\int x^n \cosh ax\, dx = \frac{x^n}{a} \sinh ax - \frac{n}{a} \int x^{n-1} \sinh ax\, dx$

123. $\int \tanh ax\, dx = \frac{1}{a} \ln(\cosh ax) + C$

124. $\int \coth ax\, dx = \frac{1}{a} \ln |\sinh ax| + C$

125. $\int \tanh^2 ax\,dx = x - \frac{1}{a}\tanh ax + C$

126. $\int \coth^2 ax\,dx = x - \frac{1}{a}\coth ax + C$

127. $\int \tanh^n ax\,dx = -\frac{\tanh^{n-1} ax}{(n-1)a} + \int \tanh^{n-2} ax\,dx, \quad n \neq 1$

128. $\int \coth^n ax\,dx = -\frac{\coth^{n-1} ax}{(n-1)a} + \int \coth^{n-2} ax\,dx, \quad n \neq 1$

129. $\int \operatorname{sech} ax\,dx = \frac{1}{a}\sin^{-1}(\tanh ax) + C$

130. $\int \operatorname{csch} ax\,dx = \frac{1}{a}\ln\left|\tanh\frac{ax}{2}\right| + C$

131. $\int \operatorname{sech}^2 ax\,dx = \frac{1}{a}\tanh ax + C$

132. $\int \operatorname{csch}^2 ax\,dx = -\frac{1}{a}\coth ax + C$

133. $\int \operatorname{sech}^n ax\,dx = \frac{\operatorname{sech}^{n-2} ax \tanh ax}{(n-1)a} + \frac{n-2}{n-1}\int \operatorname{sech}^{n-2} ax\,dx, \quad n \neq 1$

134. $\int \operatorname{csch}^n ax\,dx = -\frac{\operatorname{csch}^{n-2} ax \coth ax}{(n-1)a} - \frac{n-2}{n-1}\int \operatorname{csch}^{n-2} ax\,dx, \quad n \neq 1$

135. $\int \operatorname{sech}^n ax \tanh ax\,dx = -\frac{\operatorname{sech}^n ax}{na} + C, \quad n \neq 0$

136. $\int \operatorname{csch}^n ax \coth ax\,dx = -\frac{\operatorname{csch}^n ax}{na} + C, \quad n \neq 0$

137. $\int e^{ax}\sinh bx\,dx = \frac{e^{ax}}{2}\left[\frac{e^{bx}}{a+b} - \frac{e^{-bx}}{a-b}\right] + C, \quad a^2 \neq b^2$

138. $\int e^{ax}\cosh bx\,dx = \frac{e^{ax}}{2}\left[\frac{e^{bx}}{a+b} + \frac{e^{-bx}}{a-b}\right] + C, \quad a^2 \neq b^2$

139. $\int_0^\infty x^{n-1}e^{-x}\,dx = \Gamma(n) = (n-1)!, \quad n > 0$

140. $\int_0^\infty e^{-ax^2}\,dx = \frac{1}{2}\sqrt{\frac{\pi}{a}}, \quad a > 0$

141. $$\int_0^{\pi/2} \sin^n x\,dx = \int_0^{\pi/2} \cos^n x\,dx = \begin{cases} \dfrac{1\cdot 3\cdot 5\cdot \cdots \cdot (n-1)}{2\cdot 4\cdot 6\cdot \cdots \cdot n}\cdot\dfrac{\pi}{2}, & \text{if } n \text{ is an even integer} \geq 2 \\ \dfrac{2\cdot 4\cdot 6\cdot \cdots \cdot (n-1)}{3\cdot 5\cdot 7\cdot \cdots \cdot n}, & \text{if } n \text{ is an odd integer} \geq 3 \end{cases}$$

CREDITS

Page 400, **Figure 6.35**, PSC Physics 2/e, DC Heath & Co. with Education Development Center; **Page 425**, **Figure 7.6**, PSC Physics 2/e, DC Heath & Co. with Education Development Center; **Page 586**, **Figure 11.6**, Appalachian Mountain Club; **Page 618**, **Figure 11.22**, Copyrights/Permissions Department of US Military Academy, West Point; **Page 723**, **Figures 13.6** and **13.7**, Education Development Center; **Page 725**, **Figure 13.14**, InterNetwork Media, Inc.

Trigonometry Formulas

1. Definitions and Fundamental Identities

Sine: $\sin\theta = \dfrac{y}{r} = \dfrac{1}{\csc\theta}$

Cosine: $\cos\theta = \dfrac{x}{r} = \dfrac{1}{\sec\theta}$

Tangent: $\tan\theta = \dfrac{y}{x} = \dfrac{1}{\cot\theta}$

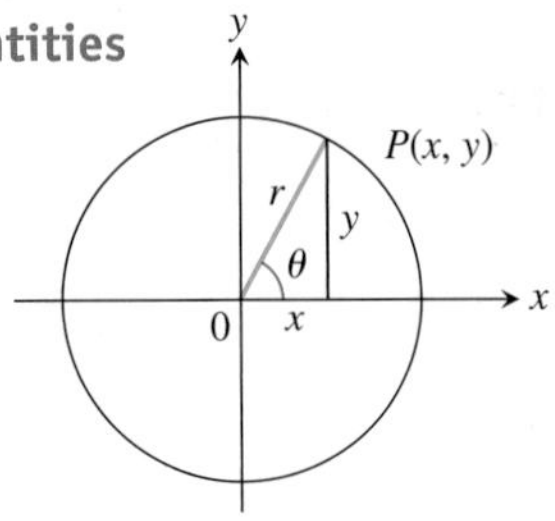

2. Identities

$$\sin(-\theta) = -\sin\theta, \quad \cos(-\theta) = \cos\theta$$

$$\sin^2\theta + \cos^2\theta = 1, \quad \sec^2\theta = 1 + \tan^2\theta, \quad \csc^2\theta = 1 + \cot^2\theta$$

$$\sin 2\theta = 2\sin\theta\cos\theta, \quad \cos 2\theta = \cos^2\theta - \sin^2\theta$$

$$\cos^2\theta = \frac{1 + \cos 2\theta}{2}, \quad \sin^2\theta = \frac{1 - \cos 2\theta}{2}$$

$$\sin(A + B) = \sin A\cos B + \cos A\sin B$$

$$\sin(A - B) = \sin A\cos B - \cos A\sin B$$

$$\cos(A + B) = \cos A\cos B - \sin A\sin B$$

$$\cos(A - B) = \cos A\cos B + \sin A\sin B$$

$$\tan(A + B) = \frac{\tan A + \tan B}{1 - \tan A\tan B}$$

$$\tan(A - B) = \frac{\tan A - \tan B}{1 + \tan A\tan B}$$

$$\sin\left(A - \frac{\pi}{2}\right) = -\cos A, \qquad \cos\left(A - \frac{\pi}{2}\right) = \sin A$$

$$\sin\left(A + \frac{\pi}{2}\right) = \cos A, \qquad \cos\left(A + \frac{\pi}{2}\right) = -\sin A$$

$$\sin A\sin B = \frac{1}{2}\cos(A - B) - \frac{1}{2}\cos(A + B)$$

$$\cos A\cos B = \frac{1}{2}\cos(A - B) + \frac{1}{2}\cos(A + B)$$

$$\sin A\cos B = \frac{1}{2}\sin(A - B) + \frac{1}{2}\sin(A + B)$$

$$\sin A + \sin B = 2\sin\frac{1}{2}(A + B)\cos\frac{1}{2}(A - B)$$

$$\sin A - \sin B = 2\cos\frac{1}{2}(A + B)\sin\frac{1}{2}(A - B)$$

$$\cos A + \cos B = 2\cos\frac{1}{2}(A + B)\cos\frac{1}{2}(A - B)$$

$$\cos A - \cos B = -2\sin\frac{1}{2}(A + B)\sin\frac{1}{2}(A - B)$$

Trigonometric Functions

Radian Measure

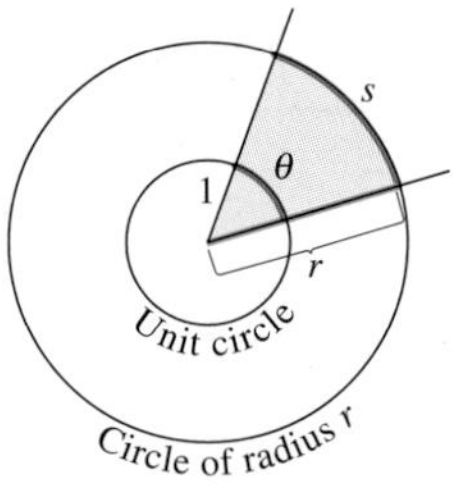

$$\frac{s}{r} = \frac{\theta}{1} = \theta \quad \text{or} \quad \theta = \frac{s}{r},$$

$180° = \pi$ radians.

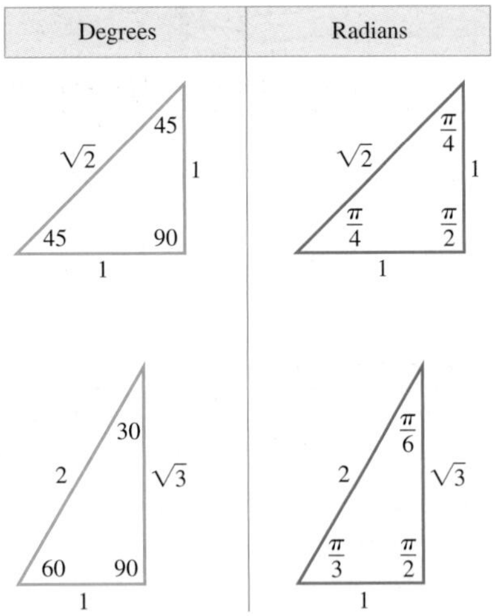

The angles of two common triangles, in degrees and radians.

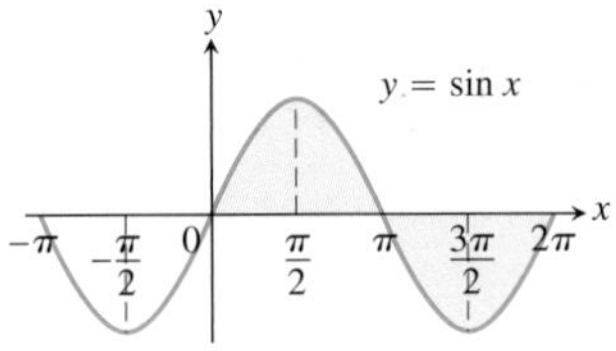

Domain: $(-\infty, \infty)$
Range: $[-1, 1]$

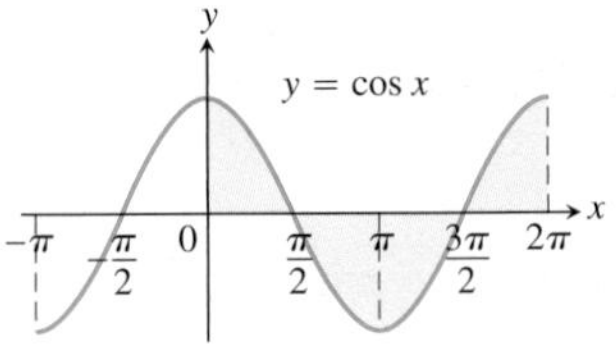

Domain: $(-\infty, \infty)$
Range: $[-1, 1]$

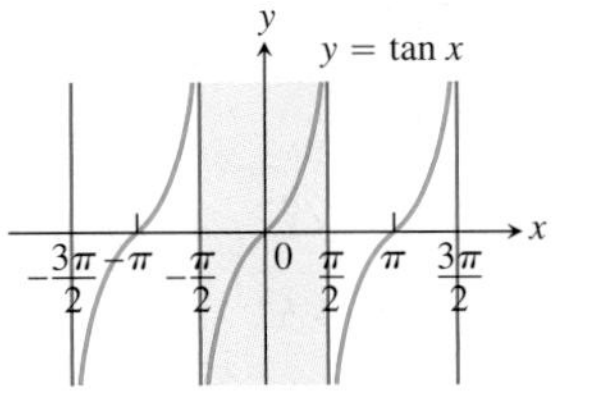

Domain: All real numbers except odd integer multiples of $\pi/2$
Range: $(-\infty, \infty)$

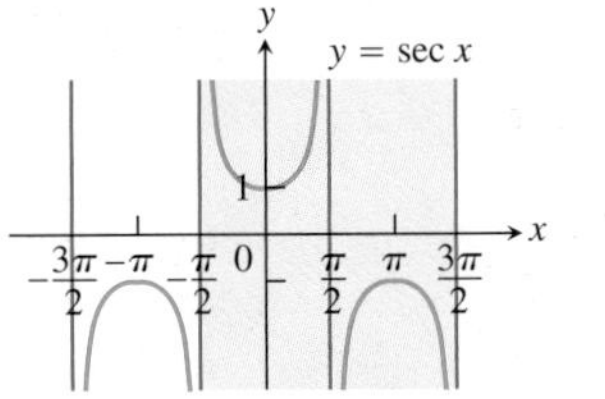

Domain: All real numbers except odd integer multiples of $\pi/2$
Range: $(-\infty, -1] \cup [1, \infty)$

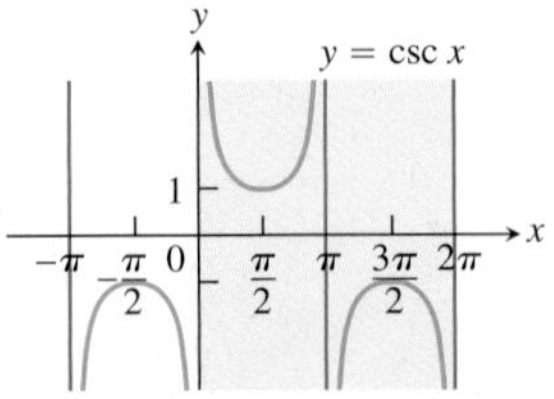

Domain: $x \neq 0, \pm\pi, \pm 2\pi, \ldots$
Range: $(-\infty, -1] \cup [1, \infty)$

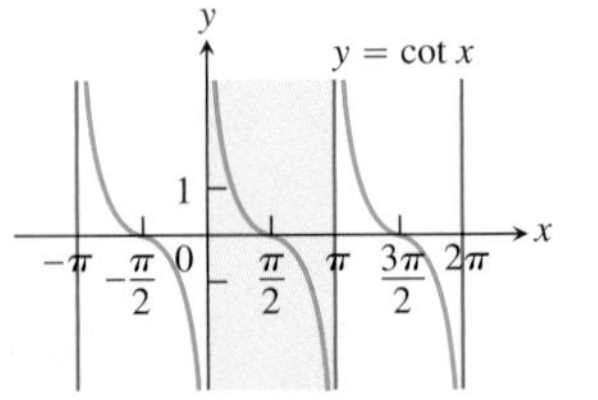

Domain: $x \neq 0, \pm\pi, \pm 2\pi, \ldots$
Range: $(-\infty, \infty)$

SERIES

Taylor Series

$$\frac{1}{1-x} = 1 + x + x^2 + \cdots + x^n + \cdots = \sum_{n=0}^{\infty} x^n, \qquad |x| < 1$$

$$\frac{1}{1+x} = 1 - x + x^2 - \cdots + (-x)^n + \cdots = \sum_{n=0}^{\infty} (-1)^n x^n, \qquad |x| < 1$$

$$e^x = 1 + x + \frac{x^2}{2!} + \cdots + \frac{x^n}{n!} + \cdots = \sum_{n=0}^{\infty} \frac{x^n}{n!}, \qquad |x| < \infty$$

$$\sin x = x - \frac{x^3}{3!} + \frac{x^5}{5!} - \cdots + (-1)^n \frac{x^{2n+1}}{(2n+1)!} + \cdots = \sum_{n=0}^{\infty} \frac{(-1)^n x^{2n+1}}{(2n+1)!}, \qquad |x| < \infty$$

$$\cos x = 1 - \frac{x^2}{2!} + \frac{x^4}{4!} - \cdots + (-1)^n \frac{x^{2n}}{(2n)!} + \cdots = \sum_{n=0}^{\infty} \frac{(-1)^n x^{2n}}{(2n)!}, \qquad |x| < \infty$$

$$\ln(1+x) = x - \frac{x^2}{2} + \frac{x^3}{3} - \cdots + (-1)^{n-1} \frac{x^n}{n} + \cdots = \sum_{n=1}^{\infty} \frac{(-1)^{n-1} x^n}{n}, \qquad -1 < x \le 1$$

$$\ln \frac{1+x}{1-x} = 2 \tanh^{-1} x = 2\left(x + \frac{x^3}{3} + \frac{x^5}{5} + \cdots + \frac{x^{2n+1}}{2n+1} + \cdots\right) = 2\sum_{n=0}^{\infty} \frac{x^{2n+1}}{2n+1}, \qquad |x| < 1$$

$$\tan^{-1} x = x - \frac{x^3}{3} + \frac{x^5}{5} - \cdots + (-1)^n \frac{x^{2n+1}}{2n+1} + \cdots = \sum_{n=0}^{\infty} \frac{(-1)^n x^{2n+1}}{2n+1}, \qquad |x| \le 1$$

Binomial Series

$$(1+x)^m = 1 + mx + \frac{m(m-1)x^2}{2!} + \frac{m(m-1)(m-2)x^3}{3!} + \cdots + \frac{m(m-1)(m-2)\cdots(m-k+1)x^k}{k!} + \cdots$$

$$= 1 + \sum_{k=1}^{\infty} \binom{m}{k} x^k, \qquad |x| < 1,$$

where

$$\binom{m}{1} = m, \qquad \binom{m}{2} = \frac{m(m-1)}{2!}, \qquad \binom{m}{k} = \frac{m(m-1)\cdots(m-k+1)}{k!} \qquad \text{for } k \ge 3.$$

VECTOR OPERATOR FORMULAS (CARTESIAN FORM)

Formulas for Grad, Div, Curl, and the Laplacian

	Cartesian (x, y, z) **i**, **j**, and **k** are unit vectors in the directions of increasing x, y, and z. M, N, and P are the scalar components of $\mathbf{F}(x, y, z)$ in these directions.
Gradient	$\nabla f = \frac{\partial f}{\partial x}\mathbf{i} + \frac{\partial f}{\partial y}\mathbf{j} + \frac{\partial f}{\partial z}\mathbf{k}$
Divergence	$\nabla \cdot \mathbf{F} = \frac{\partial M}{\partial x} + \frac{\partial N}{\partial y} + \frac{\partial P}{\partial z}$
Curl	$\nabla \times \mathbf{F} = \begin{vmatrix} \mathbf{i} & \mathbf{j} & \mathbf{k} \\ \frac{\partial}{\partial x} & \frac{\partial}{\partial y} & \frac{\partial}{\partial z} \\ M & N & P \end{vmatrix}$
Laplacian	$\nabla^2 f = \frac{\partial^2 f}{\partial x^2} + \frac{\partial^2 f}{\partial y^2} + \frac{\partial^2 f}{\partial z^2}$

Vector Triple Products

$(\mathbf{u} \times \mathbf{v}) \cdot \mathbf{w} = (\mathbf{v} \times \mathbf{w}) \cdot \mathbf{u} = (\mathbf{w} \times \mathbf{u}) \cdot \mathbf{v}$

$\mathbf{u} \times (\mathbf{v} \times \mathbf{w}) = (\mathbf{u} \cdot \mathbf{w})\mathbf{v} - (\mathbf{u} \cdot \mathbf{v})\mathbf{w}$

The Fundamental Theorem of Line Integrals

1. Let $\mathbf{F} = M\mathbf{i} + N\mathbf{j} + P\mathbf{k}$ be a vector field whose components are continuous throughout an open connected region D in space. Then there exists a differentiable function f such that

$$\mathbf{F} = \nabla f = \frac{\partial f}{\partial x}\mathbf{i} + \frac{\partial f}{\partial y}\mathbf{j} + \frac{\partial f}{\partial z}\mathbf{k}$$

if and only if for all points A and B in D the value of $\int_A^B \mathbf{F} \cdot d\mathbf{r}$ is independent of the path joining A to B in D.

2. If the integral is independent of the path from A to B, its value is

$$\int_A^B \mathbf{F} \cdot d\mathbf{r} = f(B) - f(A).$$

Green's Theorem and Its Generalization to Three Dimensions

Normal form of Green's Theorem: $\oint_C \mathbf{F} \cdot \mathbf{n}\, ds = \iint_R \nabla \cdot \mathbf{F}\, dA$

Divergence Theorem: $\iint_S \mathbf{F} \cdot \mathbf{n}\, d\sigma = \iiint_D \nabla \cdot \mathbf{F}\, dV$

Tangential form of Green's Theorem: $\oint_C \mathbf{F} \cdot d\mathbf{r} = \iint_R \nabla \times \mathbf{F} \cdot \mathbf{k}\, dA$

Stokes' Theorem: $\oint_C \mathbf{F} \cdot d\mathbf{r} = \iint_S \nabla \times \mathbf{F} \cdot \mathbf{n}\, d\sigma$

Vector Identities

In the identities here, f and g are differentiable scalar functions, $\mathbf{F}$, $\mathbf{F}_1$, and $\mathbf{F}_2$ are differentiable vector fields, and a and b are real constants.

$\nabla \times (\nabla f) = \mathbf{0}$

$\nabla(fg) = f\nabla g + g\nabla f$

$\nabla \cdot (g\mathbf{F}) = g\nabla \cdot \mathbf{F} + \nabla g \cdot \mathbf{F}$

$\nabla \times (g\mathbf{F}) = g\nabla \times \mathbf{F} + \nabla g \times \mathbf{F}$

$\nabla \cdot (a\mathbf{F}_1 + b\mathbf{F}_2) = a\nabla \cdot \mathbf{F}_1 + b\nabla \cdot \mathbf{F}_2$

$\nabla \times (a\mathbf{F}_1 + b\mathbf{F}_2) = a\nabla \times \mathbf{F}_1 + b\nabla \times \mathbf{F}_2$

$\nabla(\mathbf{F}_1 \cdot \mathbf{F}_2) = (\mathbf{F}_1 \cdot \nabla)\mathbf{F}_2 + (\mathbf{F}_2 \cdot \nabla)\mathbf{F}_1 +$
$\mathbf{F}_1 \times (\nabla \times \mathbf{F}_2) + \mathbf{F}_2 \times (\nabla \times \mathbf{F}_1)$

$\nabla \cdot (\mathbf{F}_1 \times \mathbf{F}_2) = \mathbf{F}_2 \cdot \nabla \times \mathbf{F}_1 - \mathbf{F}_1 \cdot \nabla \times \mathbf{F}_2$

$\nabla \times (\mathbf{F}_1 \times \mathbf{F}_2) = (\mathbf{F}_2 \cdot \nabla)\mathbf{F}_1 - (\mathbf{F}_1 \cdot \nabla)\mathbf{F}_2 +$
$(\nabla \cdot \mathbf{F}_2)\mathbf{F}_1 - (\nabla \cdot \mathbf{F}_1)\mathbf{F}_2$

$\nabla \times (\nabla \times \mathbf{F}) = \nabla(\nabla \cdot \mathbf{F}) - (\nabla \cdot \nabla)\mathbf{F} = \nabla(\nabla \cdot \mathbf{F}) - \nabla^2\mathbf{F}$

$(\nabla \times \mathbf{F}) \times \mathbf{F} = (\mathbf{F} \cdot \nabla)\mathbf{F} - \frac{1}{2}\nabla(\mathbf{F} \cdot \mathbf{F})$

LIMITS

General Laws

If L, M, c, and k are real numbers and

$$\lim_{x \to c} f(x) = L \quad \text{and} \quad \lim_{x \to c} g(x) = M, \quad \text{then}$$

Sum Rule: $\lim_{x \to c}(f(x) + g(x)) = L + M$

Difference Rule: $\lim_{x \to c}(f(x) - g(x)) = L - M$

Product Rule: $\lim_{x \to c}(f(x) \cdot g(x)) = L \cdot M$

Constant Multiple Rule: $\lim_{x \to c}(k \cdot f(x)) = k \cdot L$

Quotient Rule: $\lim_{x \to c} \dfrac{f(x)}{g(x)} = \dfrac{L}{M}, \quad M \neq 0$

The Sandwich Theorem

If $g(x) \leq f(x) \leq h(x)$ in an open interval containing c, except possibly at $x = c$, and if

$$\lim_{x \to c} g(x) = \lim_{x \to c} h(x) = L,$$

then $\lim_{x \to c} f(x) = L$.

Inequalities

If $f(x) \leq g(x)$ in an open interval containing c, except possibly at $x = c$, and both limits exist, then

$$\lim_{x \to c} f(x) \leq \lim_{x \to c} g(x).$$

Continuity

If g is continuous at L and $\lim_{x \to c} f(x) = L$, then

$$\lim_{x \to c} g(f(x)) = g(L).$$

Specific Formulas

If $P(x) = a_n x^n + a_{n-1} x^{n-1} + \cdots + a_0$, then

$$\lim_{x \to c} P(x) = P(c) = a_n c^n + a_{n-1} c^{n-1} + \cdots + a_0.$$

If $P(x)$ and $Q(x)$ are polynomials and $Q(c) \neq 0$, then

$$\lim_{x \to c} \frac{P(x)}{Q(x)} = \frac{P(c)}{Q(c)}.$$

If $f(x)$ is continuous at $x = c$, then

$$\lim_{x \to c} f(x) = f(c).$$

$$\lim_{x \to 0} \frac{\sin x}{x} = 1 \quad \text{and} \quad \lim_{x \to 0} \frac{1 - \cos x}{x} = 0$$

L'Hôpital's Rule

If $f(a) = g(a) = 0$, both f' and g' exist in an open interval I containing a, and $g'(x) \neq 0$ on I if $x \neq a$, then

$$\lim_{x \to a} \frac{f(x)}{g(x)} = \lim_{x \to a} \frac{f'(x)}{g'(x)},$$

assuming the limit on the right side exists.

DIFFERENTIATION RULES

General Formulas

Assume u and v are differentiable functions of x.

Constant: $\frac{d}{dx}(c) = 0$

Sum: $\frac{d}{dx}(u + v) = \frac{du}{dx} + \frac{dv}{dx}$

Difference: $\frac{d}{dx}(u - v) = \frac{du}{dx} - \frac{dv}{dx}$

Constant Multiple: $\frac{d}{dx}(cu) = c\frac{du}{dx}$

Product: $\frac{d}{dx}(uv) = u\frac{dv}{dx} + v\frac{du}{dx}$

Quotient: $\frac{d}{dx}\left(\frac{u}{v}\right) = \frac{v\frac{du}{dx} - u\frac{dv}{dx}}{v^2}$

Power: $\frac{d}{dx}x^n = nx^{n-1}$

Chain Rule: $\frac{d}{dx}(f(g(x)) = f'(g(x)) \cdot g'(x)$

Trigonometric Functions

$\frac{d}{dx}(\sin x) = \cos x$ $\quad \frac{d}{dx}(\cos x) = -\sin x$

$\frac{d}{dx}(\tan x) = \sec^2 x$ $\quad \frac{d}{dx}(\sec x) = \sec x \tan x$

$\frac{d}{dx}(\cot x) = -\csc^2 x$ $\quad \frac{d}{dx}(\csc x) = -\csc x \cot x$

Exponential and Logarithmic Functions

$\frac{d}{dx}e^x = e^x$ $\quad \frac{d}{dx}\ln x = \frac{1}{x}$

$\frac{d}{dx}a^x = a^x \ln a$ $\quad \frac{d}{dx}(\log_a x) = \frac{1}{x \ln a}$

Inverse Trigonometric Functions

$\frac{d}{dx}(\sin^{-1} x) = \frac{1}{\sqrt{1 - x^2}}$ $\quad \frac{d}{dx}(\cos^{-1} x) = -\frac{1}{\sqrt{1 - x^2}}$

$\frac{d}{dx}(\tan^{-1} x) = \frac{1}{1 + x^2}$ $\quad \frac{d}{dx}(\sec^{-1} x) = \frac{1}{|x|\sqrt{x^2 - 1}}$

$\frac{d}{dx}(\cot^{-1} x) = -\frac{1}{1 + x^2}$ $\quad \frac{d}{dx}(\csc^{-1} x) = -\frac{1}{|x|\sqrt{x^2 - 1}}$

Hyperbolic Functions

$\frac{d}{dx}(\sinh x) = \cosh x$ $\quad \frac{d}{dx}(\cosh x) = \sinh x$

$\frac{d}{dx}(\tanh x) = \operatorname{sech}^2 x$ $\quad \frac{d}{dx}(\operatorname{sech} x) = -\operatorname{sech} x \tanh x$

$\frac{d}{dx}(\coth x) = -\operatorname{csch}^2 x$ $\quad \frac{d}{dx}(\operatorname{csch} x) = -\operatorname{csch} x \coth x$

Inverse Hyperbolic Functions

$\frac{d}{dx}(\sinh^{-1} x) = \frac{1}{\sqrt{1 + x^2}}$ $\quad \frac{d}{dx}(\cosh^{-1} x) = \frac{1}{\sqrt{x^2 - 1}}$

$\frac{d}{dx}(\tanh^{-1} x) = \frac{1}{1 - x^2}$ $\quad \frac{d}{dx}(\operatorname{sech}^{-1} x) = -\frac{1}{x\sqrt{1 - x^2}}$

$\frac{d}{dx}(\coth^{-1} x) = \frac{1}{1 - x^2}$ $\quad \frac{d}{dx}(\operatorname{csch}^{-1} x) = -\frac{1}{|x|\sqrt{1 + x^2}}$

Parametric Equations

If $x = f(t)$ and $y = g(t)$ are differentiable, then

$$y' = \frac{dy}{dx} = \frac{dy/dt}{dx/dt} \quad \text{and} \quad \frac{d^2y}{dx^2} = \frac{dy'/dt}{dx/dt}.$$